multiply	by	to obtain
in/ft	1/0.012	mm/m
in^2	6.4516	cm^2
in^3	1/1728	ft^3
J	9.4778×10^{-4}	Btu
J	6.2415×10^{18}	eV
J	0.73756	ft-lbf
J	1.0	N·m
J/s	1.0	W
kg	2.2046	lbm
kg/m^3	0.06243	lbm/ft^3
kip	4.4480	kN
kip	1000.0	lbf
kip	4448.0	N
kip/ft	14.594	kN/m
kip/ft^2	47.880	kPa
kJ	0.94778	Btu
kJ	737.56	ft-lbf
kJ/kg	0.42992	Btu/lbm
kJ/kg·K	0.23885	Btu/lbm-°R
km	3280.8	ft
km	0.62138	mi
km/hr	0.62138	mi/hr
kN	0.2248	kips
kN·m	0.73757	ft-kips
kN/m	0.06852	kips/ft
kPa	9.8692×10^{-3}	atm
kPa	0.14504	lbf/in^2
kPa	1000.0	Pa
kPa	0.02089	kips/ft^2
kPa	0.01	bar
ksi	6.8948×10^6	Pa
ksi	68948.8	kPa
kW	737.56	ft-lbf/s
kW	44,250	ft-lbf/min
kW	1.3410	hp
kW	3413.0	Btu/hr
kW	0.9483	Btu/s
kW-hr	3413.0	Btu
kW-hr	3.60×10^6	J
L	1/102,790	ac-in
L	1000.0	cm^3
L	0.03531	ft^3
L	0.26417	gal
L	61.024	in^3
L	0.0010	m^3
L/s	2.1189	ft^3/min
L/s	15.850	gal/min
lbf	0.001	kips
lbf	4.4482	N
lbf/ft^2	0.01414	in Hg
lbf/ft^2	0.19234	in water
lbf/ft^2	0.00694	lbf/in^2
lbf/ft^2	47.880	Pa
lbf/ft^2	5.0×10^{-4}	tons/ft^2
lbf/in^2	0.06805	atm
lbf/in^2	144.0	lbf/ft^2
lbf/in^2	2.308	ft water
lbf/in^2	27.70	in water
lbf/in^2	2.0370	in Hg
lbf/in^2	6894.8	Pa
lbf/in^2	0.00050	tons/in^2
lbf/in^2	0.0720	tons/ft^2
lbm	7000.0	grains
lbm	453.59	g
lbm	0.45359	kg
lbm	4.5359×10^5	mg
lbm	5.0×10^{-4}	tons (mass)
lbm (of water)	0.12	gal (of water)
lbm/ac-ft-day	0.02296	lbm/1000 ft^3-day
lbm/ft^3	0.016018	g/cm^3
lbm/ft^3	16.018	kg/m^3
lbm/1000 ft^3-day	43.560	lbm/ac-ft-day
lbm/1000 ft^3-day	133.68	lbm/MG-day
lbm/MG	0.0070	grains/gal
lbm/MG	0.11983	mg/L
lbm/MG-day	0.00748	lbm/1000 ft^3-day
leagues	4428.0	m
m	1.0×10^{10}	angstroms

multiply	by	to obtain
m	3.2808	
m	39.370	in
m	2.2583×10^{-4}	leagues
m	1.0936	
m/s		n
m^2		
m^2		
m^2		e
mi^2-in		
mi^2-in,		
m^3		
m^3/m·c		gal/day-ft
m^3/m^2·	24.542	gal/day-ft^2
Meinzer unit	1.0	gal/day-ft^2
mg	2.2046×10^{-6}	lbm
mg/L	1.0	ppm
mg/L	0.05842	grains/gal
mg/L	8.3454	lbm/MG
MG	1.0×10^6	gal
MG/ac-day	22.968	gal/ft^2-day
MGD	1.5472	ft^3/sec
MGD	1×10^6	gal/day
MGD/ac (mgad)	22.957	gal/day-ft^2
mi	5280.0	ft
mi	80.0	chains
mi	1.6093	km
mi (statute)	0.86839	miles (nautical)
mi	320.0	rods
mi/hr	1.4667	ft/s
mi^2	640.0	acres
micron	1.0×10^{-6}	m
micron	0.001	mm
mil (angular)	0.05625	degrees
mil (angular)	3.375	min
min (angular)	0.29630	mils
min (angular)	2.90888×10^{-4}	radians
min (time, mean solar)	60	s
mm	1/25.4	in
mm	1000.0	microns
mm/m	0.012	in/ft
MPa	1.0×10^6	Pa
N	0.22481	lbf
N	1.0×10^5	dynes
N·m	0.73756	ft-lbf
N·m	8.8511	in-lbf
N·m	1.0	J
N/m^2	1.0	Pa
oz	28.353	g
Pa	0.001	kPa
Pa	1.4504×10^{-7}	ksi
Pa	1.4504×10^{-4}	lbf/in^2
Pa	0.02089	lbf/ft^2
Pa	1.0×10^{-6}	MPa
Pa	1.0	N/m^2
ppm	0.05842	grains/gal
radian	180/π	degrees (angular)
radian	3437.7	min (angular)
rod	0.250	chain
rod	16.50	ft
rod	1/320	mi
s (time)	1/86,400	day (mean solar)
s (time)	1.1605×10^{-5}	day (sidereal)
s (time)	1/60	min
therm	1.0×10^5	Btu
ton (force)	2000.0	lbf
ton (mass)	2000.0	lbm
ton/ft^2	2000.0	lbf/ft^2
ton/ft^2	13.889	lbf/in^2
W	3.413	Btu/hr
W	0.73756	ft-lbf/s
W	1.3410×10^{-3}	hp
W	1.0	J/s
yd	1/22	chain
yd	0.91440	m
yd^3	27	ft^3
yd^3	201.97	gal

Civil Engineering Reference Manual

for the PE Exam

Eleventh Edition

Michael R. Lindeburg, PE

The Power to Pass
www.ppi2pass.com

Professional Publications, Inc. • Belmont, California

How to Locate and Report Errata for This Book

At PPI, we do our best to bring you error-free books. But when errors do occur, we want to make sure you can view corrections and report any potential errors you find, so the errors cause as little confusion as possible.

A current list of known errata and other updates for this book is available on the PPI website at **www.ppi2pass.com/errata**. We update the errata page as often as necessary, so check in regularly. You will also find instructions for submitting suspected errata. We are grateful to every reader who takes the time to help us improve the quality of our books by pointing out an error.

CIVIL ENGINEERING REFERENCE MANUAL FOR THE PE EXAM
Eleventh Edition

Current printing of this edition: 2

Printing History

edition number	printing number	update
10	1	New edition. Major updates. Copyright update.
11	1	New edition. Major updates. Copyright update.
11	2	Minor corrections.

Printed in the United States of America.

PPI
1250 Fifth Avenue, Belmont, CA 94002
(650) 593-9119
www.ppi2pass.com

ISBN: 978-1-59126-129-2

Library of Congress Control Number: 2008923487

Topics

Topic I: Background and Support

Topic II: Water Resources

Topic III: Environmental

Topic IV: Geotechnical

Topic V: Structural

Topic VI: Transportation

Topic VII: Construction

Topic VIII: Systems, Management, and Professional

Topic IX: Support Material

Support Material

Background and Support

Water Resources

Environmental

Geotechnical

Structural

Transportation

Construction

Systems, Mgmt., and Professional

Where do I find practice problems to test what I've learned in this Reference Manual?

The *Civil Engineering Reference Manual* provides a knowledge base that will prepare you for the Civil PE exam. But there's no better way to exercise your skills than to practice solving problems. To test your knowledge, you need *Practice Problems for the Civil Engineering PE Exam: A Companion to the Civil Engineering Reference Manual*. This essential study aid will challenge you with more than 540 practice problems, each with a complete, step-by-step solution.

Practice Problems for the Civil Engineering PE Exam may be purchased from PPI at **www.ppi2pass.com** or from your favorite bookstore.

Table of Contents

Topic VII: Construction

Topic VIII: Systems, Management, and Professional

Topic IX: Support Material

Index

Appendices
Table of Contents

Preface to the Eleventh Edition

A publisher's marketing department, like the one here at PPI, usually wants the preface to quickly identify the book's purpose and positioning. The publisher's customer support department wants the preface to identify what has changed in the new edition. Booksellers prefer a preface that within the first few sentences "sells the book off the shelves." Editorial and production departments want the book to shine like a Shakespearean sonnet or da Vinci painting. Competitors would like the preface to contain some tidbit of information that will improve their abilities to produce better books and influence the market. Lawyers would like the preface to be a smoking gun that proves whatever point they care to make. Authors want to use the preface as the only place in the book where they can be humorous or add some personal anecdotes mentioning their spouses and children or tell what went on during the writing process or rant on about what inspired them in the first place. Most people who buy a book don't care about any of this or read the preface at all. This preface will attempt to honor all of these people.

First, to satisfy the examinees, I'll tell you that, "We've got you covered." I will tell you that, "Yes. The *Civil Engineering Reference Manual*, 11th edition, includes **Construction Engineering**." A lot of the construction engineering topic was already in the previous editions, but this edition adds many new subjects to that material, as well as consolidating some of the construction material into more conveniently located sections. I'll say, "Yes. CERM11 is based on the combined 13th edition of the AISC *Steel Construction Manual*. And, yes, CERM11 includes steel design by LRFD." I'll say, "Yes. CERM11 is based on the 5th edition of the AASHTO Green Book." And, so on. You're covered.

To satisfy the booksellers, I will tell them unabashedly that the *Civil Engineering Reference Manual*, 11th edition, is, hands down, the most effective tool that their customers can find on the face of the earth to help them prepare for the Professional Engineering (PE) exam in civil engineering, developed by the National Council of Examiners for Engineering and Surveying (NCEES). In the examination room, if booksellers went to that grim place, they would see that virtually every examinee has a copy of this book. Based on what 25 years of readers have told me, their customers will continue to refer to this book throughout their careers, long after they have nailed a PE license to the wall.

To satisfy the marketing department, I'll tell the marketers that this book is completely consistent with the NCEES exam content and breadth-and-depth format, and that it is equally representative of the codes and standards NCEES adopted in late 2007. I'll tell them that, due to the exam's emphasis on undergraduate engineering subjects, this book will be particularly useful in the morning (breadth) part of the PE exam. (And PPI's robust line of other products is available for even more attention to the afternoon, or depth, part of the exam.)

As an aside to you, the discriminating engineer, this book appeals to more than just PE licensing candidates. If you are a civil engineering major who needs a general reference book covering your current course of study, or if you are a practicing engineer who seeks a convenient collection of methods and data covering familiar, but perhaps faintly remembered subjects, this book is also for you. The intent of this book is to provide you with a resource for your daily work long after you have taken an exam.

For the lawyers, I will say that this book is ethical and professional. That means that actual exam content was not used (or even seen) in the development of the upgrades to this book. It means that I, as a professional engineer, understand the need to protect the public, all while helping qualified applicants to prepare for their future careers. It means that PPI and I, though not in any way associated with NCEES, share its passion for exam security. It also means that PPI went far beyond industry standards in getting content checked and reviewed, edited, and proofread. No smoking guns here.

Next, the dry material to please the PPI customer care department, which inevitably has to answer pre-purchase questions such as, "What has changed?" or, "Do I really need to purchase this book?" and, "I have the 6th edition. Can I use it?" This department also has to deal with irate customers who purchased the previous edition 17 months ago and swear they would have waited if they had known that Michael Lindeburg was writing a new edition. (Hey, People: I'm always writing a new edition.)

To answer those questions, I'll say: "Yes. You need to get this new edition. I didn't write it for nothing; I wrote it because the exam changed. This book differs from the 6th edition in several thousand ways. You wouldn't use an old building code to design your office building, would you? So, why would you take a book based on obsolete material into the most important examination of your career?"

This book continues to evolve in response to changes in building, structural, and traffic codes, as well as improvements in design methodologies, technologies, and tabulations of supporting data. Also, as with the previous ten editions, hundreds of reader comments have shaped the addition of new subjects, augmentation of appendices, clarification of explanations, and the resequencing of existing sections.

The largest replacement of content occurred in the structural and traffic/transportation chapters due to significantly revised codes and standards. The actual codes used to write this book are listed on p. xiii, and you will see that essentially all are new-millennium editions. The two notable exceptions, AASHTO's 1993 *Guide for Design of Pavement Structures* and AI's 1989 *Asphalt Handbook*, are no longer identified by NCEES as exam design standards. I have kept references to both (particularly in Ch. 75, Flexible Pavement Design, and Ch. 76, Rigid Pavement Design) as it seems unlikely to me that their content will be completely eliminated from the exam, and regardless, all civil engineers should still at least have a cursory understanding of pavement design.

Since I know that some inquisitors won't be satisfied until they hear about the structural changes in excruciating detail, I've prepared Sidebar S.1.

Sidebar S.1 Structural Code Revisions

This edition updates structural Chs. 48–69. The changes reflect the exam's reliance on ACI 318-05, ACI 530-05 and 530.1-05, AISC 13th ed., AASHTO (Green Book) 2004, ASCE7-05, NDS 2005, and IBC 2006. New material has been added featuring strength design methods for concrete, steel, and masonry. For steel, LRFD and ASD methods are presented in parallel. Now, regardless of what you choose to study—strength design or allowable stress design methods—you will be supported.

Using the massive structural revision as a vehicle for additional improvements, here is what else I've done throughout the book.

- created an entirely new construction engineering topic
- added formulas to support concepts in other chapters
- rewrote several sections to add explanatory text and to make them easier to use
- reworked examples to bring them into harmony with data, properties, and equations found elsewhere in the book
- made nomenclature more consistent among chapters
- implemented new nomenclature, terminology, and abbreviation changes consistent with new codes and standards
- added support for companion books (such as *Practice Problems for the Civil Engineering Reference Manual*)

- moved some appendix information into the respective chapters
- added approximately 300 new index entries
- incorporated errata that I became aware of since the last printing of the 10th edition, which has previously appeared only on the PPI website

For the competition, I will reveal that PPI has developed a method of keeping this book up to date on an instantaneous basis. Starting with the next edition (and I'm always writing a new edition), PPI will be able to print in one day an entire book that incorporates changes I made the previous day. This method has cut the delay between manuscript submission and delivery to the printer from more than six months down to a day or two. This ability will not change the care and attention given to all new editions that PPI produces, but it will change the timing of when the new editions appear on booksellers' shelves.

Some competitors compete on the basis of price—they rely on books sold at low prices to attract customers. Others may falsely claim to know precisely what you need and don't need to study in order to sell you a stripped-down, incomplete book. Others may use strategic alliances with legitimate associations and industries to try to establish credibility in the market. Still others may rely on a bevy of sales reps to muscle their books onto the shelves of every corner bookstore. However, a cheap, easily located, and well endorsed book will not help you if the book contains yesterday's news, and educated readers such as you know this.

That is the inescapable fact my competitors run up against in the market. In the future, everyone will expect everything to be completely and instantaneously up to date. And, for PPI, the future is here.

As the author, I'd like to write something clever, captivating, and humorous. Or even something humorous that wasn't clever or captivating. However, after ten previous editions of this book and approximately 50 editions of other books, I'm sort of out of jokes. I suppose I could go back to an earlier edition and plagiarize myself, but the publishing agreement that I signed with PPI requires everything I submit to be original. Actually, there aren't too many jokes about engineers, anyway. I started a file of engineering humor more than 20 years ago, in preparation for someday publishing the compilation. I guess the true joke is that, now, after 20 years, the file has only three or four items in it. Lucky lawyers.

To the customer who has purchased this book, well, I guess I've been writing to you the whole time. Regardless of whether you read this preface, helping you learn the covered material was always foremost in my mind. Helping you conduct an ethical review was considered in every word and in every page that I wrote. You and I are associates in the same honorable profession, and we're also members of the same species. In those regards, we exist to help each other. I'm here for you. Now you go out and do your best to serve humanity.

Michael R. Lindeburg, PE

Acknowledgments

So, who helped me with this new edition? Wow. Where do I start?

This edition shares a common developmental heritage with the ten previous editions. Looking back through the acknowledgments in those books, there are hundreds and hundreds of people to whom I owe gratitude. For many of the previous editions, I asked subject matter experts from industry and academia to help me review, update, reorganize, and write original material. The inspiration for their contributions came from code and standards changes, exam content revisions, requests from readers, suggestions from PPI's editorial staff, and outlines from me. These experts also incorporated many new topics into their respective chapters according to their own experience and understanding of what makes engineering tick.

Many of the revisions needed to bring out this edition were made by a team of structural and transportation experts. Truthfully, without their help, we'd still all be waiting for this edition. (Even if you're reading this paragraph in 2025, you'd still be waiting.) S. K. Ghosh, PhD, of S. K. Ghosh Associates, Inc., Palatine, IL, reviewed and revised the concrete chapters. William Wood, PhD, PE, director of the School of Engineering Technology at Youngstown State University in Ohio, attended to the masonry revisions. Thomas W. Schreffler, QEP, PE, a registered professional engineer and project manager for Light-Heigel and Associates, Inc., in Pennsylvania, and Peter W. Hoadley, PE, a registered professional engineer and civil engineering professor at the Virginia Military Institute, reviewed and revised the steel chapters. Similar work on the traffic and transportation chapters was completed by Jamie F. Rana, PE, PTOE, a registered professional engineer in Missouri and Texas, and a registered Professional Traffic Operations Engineer. Jim Monroe, JR, PE, a civil and forensic engineer employed by the California Department of Transportation, reviewed the new construction material. All of these contributors applied themselves to the new edition with the diligence and attention to quality that you've come to expect from PPI authors and staff.

Because the revisions to the steel chapters were extraordinarily extensive, substantial, and even onerous, and because the amount of new material written was huge, and because they, themselves, were extraordinarily responsive to the demands of PPI's quick-turnaround publishing schedule, Thomas Schreffler and Peter Hoadley deserve another mention. They were incredible.

As the book's author, I got to check accuracy, logic, and cross-references. I got to check the cancellation of units, provide SI conversions, and key the calculations several times. I got to check equations and their references against the codes, and I got to revise the contents of the index to make sure you could find everything that everyone else did. As Editor-in-Chief of PPI, I got to check, review, and revise everyone's contributions, but few changes were needed.

Within PPI, a number of staffers worked many overtime hours. Project editor Jenny Lindeburg King did the majority of the editing and proofreading. Typesetter Kate Hayes formatted the raw text into the pages that you see. (I've said before that her fingers have given life to this book over many previous editions. True to character, her fingers were incredibly quick and nimble throughout the many rounds and revisions that the CERM11 manuscript went through.) Illustrator Tom Bergstrom rendered new illustrations and redrew many existing figures to be clear and consistent with PPI's style guide. It's a pleasure to incorporate illustrations that are clear, informative, and unambiguous. Amy Schwertman created CERM's amazing new cover design. Sarah Hubbard and Cathy Schrott managed the editorial and production departments, respectively. Sometimes, it's hard to tell where one of them ends and the other begins, so closely do they work together for the common good.

It's a shame that PPI authors never mention the marketing manager (Greg Monteforte)—as if the books sell themselves. Or, the IT manager (Mitch Bakos)—as if the company's computer systems are infinitely reliable. Or, the operations manager (Patty Steinhardt)—as if the company's functions are autonomic. No, if I'm going to croon on about the common good, they need to be mentioned, also.

Elizabeth, my wife, continues to bring me tea throughout the day while I'm researching and writing that inevitably new book or new edition. That routine hasn't changed much in 25 years, although she brings me much less pie with the tea these days. And, there still isn't a day I sit down at my upstairs desk to write that a wee little one doesn't come bounding up the stairs to jump on my lap. These days, however, the wee little one is only my cat, TC (an acronym for, The Cat), not one of my children.

If you put all of the acknowledgments of previous editions end to end, you'd see how my wife and family contributed and managed to survive during the previous 25 years of authorship. The first editions of my books mentioned a wife, and then two children. Then, the children grew older and had a more significant presence in my acknowledgments (and in our lives). Within the last decade, they graduated from high school and went off to college, and I wrote in my acknowledgments they no longer had to endure a father who pecked away so continually at a keyboard, or who brought boxes of manuscripts to proofread on every family vacation they took together, even to those to far distant lands.

Both of my two daughters have selected courses of study and careers that put them squarely on the path of helping humanity. Both are attracted to careers (and, men) with technical backgrounds, both have my odd sense of humor, and both occasionally speak about the books they are going to write some day. These little girls, now grown women, are the best books that my wife, Elizabeth, and I ever wrote.

For this edition, I'm pleased to report that our youngest daughter, Katie, has finished up at the University of California, Santa Barbara. She is, apparently, now one of the most actively sought-after soils science master's candidates on the face of the earth. Recent options offered to her included studying volcanic soil in Hawaii, the methane production from swamps in Uganda (where she is, as I write this in early 2008), the high plateaus in Tibet, and the more mundane greenhouse gas-producing forests of Pennsylvania and million year old terraces in Oregon.

Jenny, my oldest daughter who finished a master's degree at The George Washington University, subsequently married and bought a house, but has yet to get a pet monkey, has proven herself to be a great project editor at PPI. It is particularly weird to have my daughter editing and having veto power over the contents of a book that I originally wrote before she was born, but I'm getting over it. Now, it's her turn to be taking boxes of manuscripts and editorial rounds with her on family vacations. The fruit doesn't fall very far from the tree.

And, as kind of an ongoing private joke between her and me, I will continue to include this secret message to Jenny in each edition of this book until she figures out how to decode it. She's getting close, but no cigar yet. [Jenny] y92 e8e 8 w5q6 wqh3 24858ht r49j 5y3 eq4i3w5 j94h8ht y974w 59 5y3 233 y974w 9r 5y3 h8ty5? 8 yqf3 j6 3oe3w5 eq7ty534 u3hh6 59 5yqhi r94 5yq5. y34 w9dd34 04qd58d3 i305 j6 go99e d84d7oq58ht 2y8o3 y34 dyq5534 qhe u9i3w 59ww3e g35233h 974 529 d9j07534 h99iw 8h j6 9rr8d3 i305 j6 w08485w w9q48ht.

So, there you go. About eleven hundred people have helped with this new edition. That's a lot of people for me to thank. Even thanking all of them would require a staff. And, then, I'd have to thank the staff. Ah, such is the life of an author. We're never finished.

Thank you, people!

Michael R. Lindeburg, PE

Codes Used to Prepare This Book[1]

The information that was used to write and update this book was the most current at the time. However, as with engineering practice itself, the PE examination is not always based on the most current codes or cutting-edge technology. Similarly, codes, standards, and regulations adopted by state and local agencies often lag issuance by several years. Thus, the codes that are current, used by you in practice, and tested on the exam can all be different.

PPI lists on its website the dates and editions of the codes, standards, and regulations on which NCEES has announced the PE exams are based. It is your responsibility to find out which codes will be tested on your exam. In the meantime, here are the codes that have been incorporated into this edition.[1]

STRUCTURAL DESIGN STANDARDS

AASHTO: *AASHTO LRFD Bridge Design Specifications*, Third ed., 2004, with 2005 and 2006 Interim Revisions, American Association of State Highway and Transportation Officials, Washington, DC

ACI 318: *Building Code Requirements for Structural Concrete*, 2005, American Concrete Institute, Farmington Hills, MI

ACI 530: *Building Code Requirements for Masonry Structures*, 2005, and ACI 530.1: *Specifications for Masonry Structures*, 2005, American Concrete Institute, Detroit, MI

AISC: *Steel Construction Manual*, Thirteenth ed., 2005, American Institute of Steel Construction, Inc., Chicago, IL

ASCE 7: *Minimum Design Standards for Buildings and Other Structures*, 2005, American Society of Civil Engineers, New York, NY

IBC: *International Building Code*, 2006 ed., International Code Council, Inc., Falls Church, VA

NDS: *National Design Specification for Wood Construction*, 2005 ASD ed., and *National Design Specification Supplement*, 2005 ASD ed., American Forest and Paper Association, Washington, DC

PCI: *PCI Design Handbook*, Sixth ed., 2004, Precast/Prestressed Concrete Institute, Chicago, IL

TRANSPORTATION DESIGN STANDARDS

AASHTO: *Guide for Design of Pavement Structures*, Fourth ed., 1993, American Association of State Highway and Transportation Officials, Washington, DC[2]

AASHTO: *A Policy on Geometric Design of Highways and Streets*, Fifth ed., 2004, American Association of State Highway and Transportation Officials, Washington, DC

AASHTO: *Roadside Design Guide*, Third ed., 2002, American Association of State Highway and Transportation Officials, Washington, DC

AI: *The Asphalt Handbook* (MS-4), 1989, Asphalt Institute, College Park, MD[2]

HCM: *Highway Capacity Manual (HCM 2000)*, Fourth ed., U.S. Customary version, 2000, Transportation Research Board, National Research Council, Washington, DC

ITE: *Traffic Engineering Handbook*, Fifth ed., 1999, Institute of Transportation Engineers, Washington, DC

MUTCD: *Manual on Uniform Traffic Control Devices*, 2003, U.S. Dept. of Transportation, Federal Highway Administration, Washington, DC

PCA: *Design and Control of Concrete Mixtures*, Fourteenth ed., 2002, Portland Cement Association, Skokie, IL

CONSTRUCTION

ACI 347-04: *Guide to Formwork for Concrete*, 2004, American Concrete Institute, Farmington Hills, MI (in ACI SP-4, Seventh ed. appendix)

ACI SP-4: *Formwork for Concrete*, Seventh ed., 2005, American Concrete Institute, Farmington Hills, MI

AISC: *Steel Construction Manual*, Thirteenth ed., 2005, American Institute of Steel Construction, Inc., Chicago, IL

ASCE 37-02: *Design Loads on Structures During Construction*, 2002, American Society of Civil Engineers, Reston, VA

---

[1]This is a list of the codes used to prepare *this* book. The Introduction contains a complete list of the references that you should consider bringing to the exam.

[2]Though these codes are not specified by NCEES, they are integral to the information presented in Chs. 75 and 76 of the *Civil Engineering Reference Manual*.

CMWB: *Standard Practice for Bracing Masonry Walls During Construction*, 2001, Council for Masonry Wall Bracing, Mason Contractors Association of America, Lombard, IL

MUTCD-Pt 6: *Manual on Uniform Traffic Control Devices*—Part 6 Temporary Traffic Control, 2003, U.S. Federal Highway Administration

NDS: *National Design Specification for Wood Construction*, 2005, American Forest & Paper Association/American Wood Council, Washington, DC

OSHA: *Occupational Safety and Health Standards for the Construction Industry*, 29 CFR Part 1926, (U.S. federal version), U.S. Department of Labor, Washington, DC

Introduction

Part 1: How You Can Use This Book

QUICKSTART

If you never read the material at the front of your books anyway, and if you're in a hurry to begin and only want to read one paragraph, here it is:

> Most chapters in this book are independent. Start with any one and look through it. Decide if you are going to work questions in that subject. Solve as many practice problems in that subject as time allows. Use the index extensively. Don't stop studying until the exam. Start right now! Quickly! Good luck.

However, if you want to begin a thorough review, you should probably try to find out everything there is to know about the PE exam. The rest of this introduction is for you.

IF YOU ARE A PRACTICING ENGINEER

If you are a practicing engineer or an engineering major and have obtained this book as a general reference handbook, it will probably sit in your bookcase until you have a specific need.

However, if you are preparing for the PE examination in civil engineering, the following suggestions may help.

- Become intimately familiar with this book. This means knowing the order of the chapters, the approximate locations of important figures and tables, what appendices are available, and so on.

- Use the subject title tabs along the side of each page. The tab names correspond to the exam organization.

- Use Tables 2 and 3 of this Introduction to learn which subjects in this book are not specific exam subjects. Some chapters in this book are supportive and do not cover exam topics. These chapters provide background and support for the other chapters.

- Skim through a chapter and its appendices. Familiarize yourself with the subjects before starting to solve practice problems.

- Identify and obtain a set of 10–30 solved practice problems for each of the exam subjects. I have written an accompanying book, *Practice Problems for the Civil Engineering PE Exam*, for this purpose. However, you may use problem sets from your old textbooks, college notes, or review course.

- Set a reasonable limit on the time you spend on each subject. It isn't necessary to solve an infinite number of practice problems. The number of practice problems you attempt will depend on how much time you have and how skilled you are in the subject.

- Use the solutions to your practice problems to check your work. If your answer isn't correct, figure out why.

- If you decide to work in customary U.S. (English) units, you will find equations in which the quantity g/g_c appears. For calculations at standard gravity, the numerical value of this fraction is 1.00. Therefore, it is necessary to incorporate this quantity only in calculations with a nonstandard gravity or when you are being meticulous with units.

- To minimize time spent in searching for often-used formulas and data, prepare a one-page summary of all the important formulas and information in each subject area. You can then use these summaries during the examination instead of searching in this book.

- Use the index extensively. Every significant term, law, theorem, and concept has been indexed in every conceivable way—backwards and forwards. If you don't recognize a term used, look for it in the index.

- Some subjects appear in more than one chapter. Use the index liberally to learn all there is to know about a particular subject.

IF YOU ARE AN INSTRUCTOR

The first two editions of this book consisted of a series of handouts prepared for the benefit of my PE review courses. These editions were intended to be compilations of all the long formulas, illustrations, and tables of data that I did not have time to put on the chalkboard. You can use this edition in the same way.

If you are teaching a review course for the PE examination without the benefit of recent, firsthand exam experience, you can use the material in this book as a guide to prepare your lectures. You should emphasize the subjects in each chapter and avoid subjects omitted. You can feel confident that subjects omitted from this book have rarely, if ever, appeared on the PE exam.

I have always tried to overprepare my students. For that reason, the homework contains practice problems that are often more difficult and more varied than actual examination questions. Also, you will appreciate the fact that it is more efficient to cover several procedural steps in one practice problem than to ask simple "one-liners" or definition questions. That is the reason that my problems are often harder and longer than actual exam problems.

To do all the homework for some chapters requires approximately 15 to 20 hours of preparation per week. "Capacity assignment" is the goal in my review courses. If you assign 20 hours of homework and a student is able to put in only 10 hours that week, that student will have worked to his or her capacity. After the PE examination, that student will honestly say that he or she could not have prepared any more than he or she did in your course.

Homework assignments in my review courses are not individually graded. Instead, students are permitted to make use of existing solutions to learn procedures and techniques to the problems in their homework set, such as those in the accompanying *Practice Problems for the Civil Engineering PE Exam*, which contains solutions to all practice problems. However, each student must turn in a completed set of problems for credit each week. Though I don't correct the homework problems, I address special needs or questions written on the assignments.

I have found that a 14-week format works well for a PE review course. It's a little rushed, but the course is over before everyone gets bored with my jokes. Each week, there is a three-hour meeting, which includes lecture and a short break. Table 1 outlines a course format that might work for you. If you can add more course time, your students will appreciate it. Another lecture covering water resources or environmental engineering would be wonderful. However, I don't think you can cover the full breadth of material in much less time or in many fewer weeks.

I have tried to order the subjects in a logical, progressive manner, keeping my eye on "playing the high-probability subjects." I cover the subjects that everyone can learn (e.g., fluids and soils) early in the course. I leave the subjects that only daily practitioners should attempt (e.g., concrete, steel, and highway capacity) to the end.

Lecture coverage of some examination subjects is necessarily brief; other subjects are not covered at all. These omissions are intentional; they are not the result of scheduling omissions. Why? First, time is not on our side in a review course. Second, some subjects rarely contribute to the examination. Third, some subjects are not well-received by the students. For example, I have

found that very few people try to become proficient in bridges, timber, and masonry if they don't already work in those areas. Most nonstructural civil engineers stick with the nonstructural basics: fluids, soils, surveying, water supply, and wastewater. Most structural engineers stick with structural subjects. Unless you have six months in which to teach your PE review, your students' time can be better spent covering other subjects.

Table 1 Typical PE Exam Review Course Format

meeting	subject covered	chapters
1	Introduction to the Exam, Engineering Economics	86, 89
2	Fluids, Conduit Flow, and Pumps	14–18
3	Open Channel Flow	19
4	Hydrology	20, 21
5	Water Supply	25, 26
6	Wastewater and the Environment	28–30, 31–34
7	Soils, Foundations, Settlement, and Retaining Walls	35–40
8	Surveying and Concrete Mixing	77
9	Highway Design and Traffic Analysis	73, 74, 78
10	Construction Engineering	79–82
11	Mechanics of Materials	41–47
12	Concrete Design	48–50
13	Steel Design	59, 61, 62

All the skipped chapters and any related practice problems are presented as floating assignments to be made up in the students' "free time."

I strongly believe in exposing my students to a realistic sample examination, but I no longer administer an in-class mock exam. Since the review course usually ends only a few days before the real PE examination, I hesitate to make students sit for several hours in the late evening to take a "final exam." Rather, I distribute and assign a take-home sample exam at the first meeting of the review course.

If the practice test is to be used as an indication of preparedness, caution your students not to even look at the sample exam prior to taking it. Looking at the sample examination or otherwise using it to direct their review will produce unwarranted specialization in subjects contained in the sample examination.

There are many ways to organize a PE review course, depending on your available time, budget, intended audience, facilities, and enthusiasm. However, all good course formats have the same result: The students struggle with the workload during the course, and then they breeze through the examination after the course.

Part 2: Everything You Ever Wanted to Know About the PE Exam

WHAT IS THE FORMAT OF THE PE EXAM?

The NCEES PE examination in civil engineering consists of two four-hour sessions separated by a one-hour lunch period. The morning "breadth" (a.m.) session is taken by all examinees. In the afternoon, you will be able to select from five "depth" (p.m.) modules: water resources and environmental, geotechnical, structural, transportation, and construction. (The depth modules may be referred to as "discipline-specific," or DS, modules, borrowing a term from the FE exam.) All five depth modules will be present in your examination booklet for review prior to making your selection. After you begin working in a depth module, you may leave it and begin working in another module. However, keep in mind that you won't be given any additional time.

Both the morning and afternoon sessions contain 40 questions in multiple-choice (i.e., "objective") format. As this is a "no-choice" exam, you must answer all questions in each session correctly to receive full credit. There are no optional questions.

WHAT SUBJECTS ARE ON THE PE EXAM?

NCEES has published a description of subjects on the examination. Irrespective of the published examination structure, the exact number of questions that will appear in each subject area cannot be predicted reliably.

There is no guarantee that any single subject will occur in any quantity. One of the reasons for this is that some of the questions span several disciplines. You might consider a pump selection question to come from the subject of fluids, while someone else might categorize it as engineering economics.

Table 2 describes the subjects in detail. Most examinees find the list to be formidable in appearance. The percentage breakdowns in Table 2 are according to NCEES, but these percentages are approximate. NCEES adds,

> The examination is developed with questions that require a variety of approaches and methodologies including design, analysis, application, and operations. Some questions may require knowledge of engineering economics. These areas are examples of the kinds of knowledge that will be tested but are not exclusive or exhaustive categories.

As you can see, the subjects in morning and afternoon sessions overlap. However, the depth of required knowledge is not consistent. The following table provides some guidance as to just "what" each of these subjects means.

WHAT IS THE TYPICAL QUESTION FORMAT?

Almost all of the questions are standalone—that is, they are completely independent. However, NCEES allows that some sets of questions may start with a statement of a "situation" that will apply to (typically) two to five following questions. Such grouped questions are expected to be increasingly rare, however.

Since the questions are multiple-choice in design, all required data should appear in the situation statement. You will not generally be required to come up with numerical data that might affect your success on the question. There will be superfluous information in the majority of questions.

Each of the questions will have four answer choices, labeled "A," "B," "C," and "D." One of the answer choices is correct (or, "most nearly correct," as described in the following section). The remaining answer choices are incorrect and may consist of one or more logical distractors.

NCEES tries hard to make sure the questions are not interrelated. Questions are independent or start with new given data. A mistake on one of the questions shouldn't cause you to get a subsequent question wrong. However, considerable time may be required to repeat previous calculations with a new set of given data.

WHAT DOES "MOST NEARLY" REALLY MEAN?

One of the more disquieting aspects of these questions is that the available answer choices are seldom exact. Answer choices generally have only two or three significant digits. Exam questions ask, "Which answer choice is most nearly the correct value?" or they instruct you to complete the sentence, "The value is approximately ..." A lot of self-confidence is required to move on to the next question when you don't find an exact match for the answer you calculated, or if you have had to split the difference because no available answer choice is close.

NCEES describes it like this:

> Many of the questions on NCEES exams require calculations to arrive at a numerical answer. Depending on the method of calculation used, it is very possible that examinees working correctly will arrive at a range of answers. The phrase "most nearly" is used to accommodate answers that have been derived correctly but that may be slightly different from the correct answer choice given on the exam. You should use good engineering judgment when selecting your choice of answer. For example, if the question asks you to calculate an electrical current or determine the load on a beam, you should literally select the answer option that is most nearly what you calculated, regardless of whether it is more or less than your calculated value. However, if the question asks you to select

Table 2 Detailed Analysis of Tested Subjects

MORNING SESSION
(40 multiple-choice questions)

Construction (20%)

Earthwork construction and layout: excavation and embankment (cut and fill), borrow pit volumes, site layout and control

Estimating quantities and costs: quantity take-off methods, cost estimating

Scheduling: construction sequencing, resource scheduling, time-cost trade-off

Material quality control and production: material testing (e.g., concrete, soil, asphalt)

Temporary structures: construction loads

Geotechnical (20%)

Subsurface exploration and sampling: soil classification, boring log interpretation (e.g., soil profile)

Engineering properties of soils and materials: permeability, pavement design criteria

Soil mechanics analysis: pressure distribution, lateral earth pressure, consolidation, compaction, effective and total stresses

Earth structures: slope stability, slabs-on-grade

Shallow foundations: bearing capacity, settlement

Earth retaining structures: gravity walls, cantilever walls, stability analysis, braced and anchored excavations

Structural (20%)

Loadings: dead and live loads, construction loads

Analysis: determinate analysis

Mechanics of materials: shear and moment diagrams, flexure, shear, tension and compression, deflection, combined stresses

Materials: reinforced and plain concrete; structural, light gage, and reinforcing steel

Member design: beams, slabs, footings

Transportation (20%)

Geometric design: horizontal curves, vertical curves, sight distance, superelevation, vertical and/or horizontal clearances, acceleration and deceleration

Water Resources and Environmental (20%)

Hydraulics—closed conduit: energy and/or continuity equation (e.g., Bernoulli), pressure conduit (e.g., single pipe, force mains), closed pipe flow equations (e.g., Hazen-Williams, Darcy-Weisbach), friction and/or minor losses, pipe network analysis (e.g., pipeline design, branch networks, loop networks), pump application and analysis

Hydraulics—open channel: open-channel flow (e.g., Manning's equation), culvert design, spillway capacity, energy dissipation (e.g., hydraulic jump, velocity control), stormwater collection (e.g., stormwater inlets, gutter flow, street flow, storm sewer pipes), flood plains/floodways, flow measurement (open channel)

Hydrology: storm characterization (e.g., rainfall measurement and distribution), storm frequency, hydrographs application, rainfall intensity, duration, and frequency (IDF) curves, time of concentration, runoff analysis including Rational and SCS methods, erosion, detention/retention ponds

Wastewater treatment: collection systems (e.g., lift stations, sewer networks, infiltration, inflow)

Water Treatment: hydraulic loading, distribution systems

AFTERNOON SESSIONS
(40 multiple-choice questions)

CIVIL/CONSTRUCTION DEPTH EXAM

Construction (90%)

Earthwork construction and layout (10%): excavation and embankment (cut and fill), borrow pit volumes, site layout and control, earthwork mass diagrams

(continued)

Table 2 *Detailed Analysis of Tested Subjects*
(continued)

CIVIL/CONSTRUCTION DEPTH EXAM (continued)

Estimating quantities and costs (17.5%): quantity take-off methods, cost estimating, engineering economics, value engineering and costing

Construction operations and methods (15%): lifting and rigging, crane selection, erection, and stability, dewatering and pumping, equipment production, productivity analysis and improvement, temporary erosion control

Scheduling (17.5%): construction sequencing, CPM network analysis, activity time analysis, resource, scheduling, time-cost trade-off

Material quality control and production (10%): material testing (e.g., concrete, soil, asphalt), welding and bolting testing, quality control process (QA/QC), concrete mix design

Temporary structures (12.5%): construction loads, formwork, falsework and scaffolding, shoring and reshoring, concrete maturity and early strength evaluation, bracing, anchorage, cofferdams (systems for temporary excavation support), codes and standards (e.g., American Society of Civil Engineers (ASCE 37), American Concrete Institute (ACI 347), American Forest and Paper Association-NDS, Masonry Wall Bracing Standard)

Worker health, safety, and environment (7.5%): OSHA regulations, safety management, safety statistics (e.g., incident rate, EMR)

Other Topics (10%)

Groundwater and well fields: groundwater control including drainage, construction dewatering

Subsurface exploration and sampling: drilling and sampling procedures

Earth retaining structures: mechanically stabilized earth wall, soil and rock anchors

Deep foundations: pile load test, pile installation

Loadings: wind loads, snow loads, load paths

Mechanics of materials: progressive collapse

Materials: concrete (prestressed, post-tensioned), timber

Traffic safety: work zone safety

CIVIL/GEOTECHNICAL DEPTH EXAM

Geotechnical (87.5%)

Subsurface exploration and sampling (7.5%): drilling and sampling procedures, in-situ testing, soil classification, boring log interpretation, soil profile development, general rock characterization (e.g., RQD, description, joints and fractures)

Engineering properties of soils and materials (12.5%): index properties, phase relationships, shear strength properties, permeability, geosynthetics, pavement design criteria, frost susceptibility

Soil mechanics analysis (12.5%): effective and total stresses, pressure distribution, lateral earth pressure, consolidation, compaction, expansive soils, seepage (e.g., exit gradient, drain fields, seepage forces, flow nets)

Earth structures (10%): slope stability, slabs-on-grade, earth dams, techniques and suitability of ground modification

Shallow foundations (15%): bearing capacity, settlement, mat and raft foundations

Deep foundations (10%): axial capacity (single pile/drilled shaft), lateral capacity and deflections (single pile/drilled shaft), settlement, behavior of pile and/or drilled shaft groups, pile dynamics (e.g., wave equation, PDA test), pile load tests, pile installation

Earth retaining structures (15%): gravity walls, cantilever walls, mechanically stabilized earth walls, braced and anchored excavations, soil and rock anchors

Earthquake engineering (5%): liquefaction, pseudo-static analysis, seismic site characterization

Other Topics (12.5%)

Groundwater and well fields: well logging and subsurface properties, aquifers (e.g., characterization), groundwater flow including Darcy's Law and seepage analysis, well analysis (steady flow only), groundwater control including drainage, construction dewatering

Loadings: earthquake loads

(continued)

Table 2 *Detailed Analysis of Tested Subjects*
  (continued)

CIVIL/GEOTECHNICAL DEPTH EXAM (continued)

Construction operations and methods: dewatering and pumping, quality control process (QA/QC) (e.g., when digging, confirming quality; writing QA processes)

Temporary structures: shoring and reshoring, concrete maturity and early strength evaluation, bracing, anchorage, cofferdams (systems for temporary excavation support)

Worker health, safety, and environment: OSHA regulations, safety management

CIVIL/STRUCTURAL DEPTH EXAM

Structural (87.5%)

Loadings (12.5%): dead and live loads, moving loads, wind loads, earthquake loads (including liquefaction, site characterization, and pseudo-static analysis), snow loads, construction loads, impact loads, load paths, load combinations

Analysis (12.5%): determinate, indeterminate

Mechanics of materials (12.5%): shear and moment diagrams, flexure, shear, torsion, tension and compression, combined stresses, deflection, progressive collapse, buckling, fatigue, thermal deformation

Materials (12.5%): reinforced and plain concrete, pre-stressed and post-tension concrete, structural steel (structural, light gage, reinforcing), timber, masonry (brick veneer, CMU), composite construction

Member design (25%): beams, slabs, columns, footings, trusses, braces and frames, connections (bolted, welded, embedded, anchored), shear and bearing walls, diaphragms (horizontal, vertical, flexible, rigid)

Design criteria (12.5%): IBC, ACI, PCI, AISC, NDS, AASHTO, ASCE7, AWS

Other Topics (12.5%)

Engineering properties of soils and materials: index properties (e.g., plasticity index; interpretation and how to use them)

Soil mechanics analysis: expansive soils

Shallow foundations: mat and raft foundations

Deep foundations: axial capacity (single pile/drilled shaft), lateral capacity and deflections (single pile/drilled shaft), settlement, behavior of pile and/or drilled shaft group

Engineering economics: value engineering and costing

Material quality control and production: welding and bolting testing

Temporary structures: formwork, falsework and scaffolding, shoring and reshoring, concrete maturity and early strength evaluation, bracing, anchorage

Worker health, safety, and environment: OSHA regulations, safety management

CIVIL/TRANSPORTATION DEPTH EXAM

Transportation (75%)

Traffic analysis (22.5%): traffic signals, speed studies, traffic capacity studies, intersection analysis, traffic volume studies, sight distance evaluation, traffic control devices, pedestrian facilities, driver behavior/performance

Transportation planning (7.5%): traffic impact studies, capacity analysis (future conditions), optimization/cost analysis (e.g., transportation route A or transportation route B)

Geometric design (30%): horizontal curves, vertical curves, sight distance, superelevation, vertical/horizontal clearances, acceleration and deceleration, intersections and/or interchanges

Traffic safety (15%): accident analysis, roadside clearance analysis, work zone safety, conflict analysis

Other Topics (25%)

Hydraulics: flow measurement—closed conduits, open channel—subcritical and supercritical flow

Hydrology: hydrograph development and synthetic hydrographs

(continued)

Table 2 *Detailed Analysis of Tested Subjects*
(continued)

CIVIL/TRANSPORTATION DEPTH EXAM (continued)

Engineering properties of soils and materials: index properties (e.g., identification of types of soils; suitable or unsuitable)

Soil mechanics analysis: expansive soils

Engineering economics: value engineering and costing

Construction operations and methods: National Pollutant Discharge Elimination System (NPDES) permitting

Temporary structures: concrete maturity and early strength evaluation

CIVIL/WATER RESOURCES AND ENVIRONMENTAL DEPTH EXAM

Water Resources and Environmental (97.5%)

Hydraulics—closed conduit (15%): energy and/or continuity equation (e.g., Bernoulli); pressure conduit (e.g., single pipe, force mains); closed pipeflow equations including Hazen-Williams, Darcy-Weisbach equations; friction, and/or minor losses; pipe network analysis (e.g., pipeline design, branch networks, loop networks); pump application and analysis; cavitation; transient analysis (e.g., water hammer); flow measurement—closed conduits, momentum equation (e.g., thrust blocks, pipeline restraints)

Hydraulics—open channel (15%): open-channel flow (e.g., Manning's equation); culvert design; spillway capacity; energy dissipation (e.g., hydraulic jump, velocity control); stormwater collection including stormwater inlets, gutter flow, street flow, storm sewer pipes; flood plain/floodway; subcritical and supercritical flow, flow measurement—open channel; gradually varied flow

Hydrology (15%): storm characterization including rainfall measurement and distribution; storm frequency; hydrographs application; hydrograph development and synthetic hydrographs; rainfall intensity, duration, and frequency (IDF) curves; time of concentration; runoff analysis including rational and SCS methods; gauging stations including runoff frequency analysis, flow calculations; depletions (e.g., transpiration, evaporation, infiltration); sedimentation; erosion; detention/retention ponds

Groundwater and well fields (7.5%): aquifers (e.g., characterization); groundwater flow including Darcy's law, seepage analysis; well analysis (steady flow only); groundwater control including drainage, construction dewatering; water quality analysis; groundwater contamination

Wastewater treatment (15%): wastewater flow rates (e.g., municipal, industrial, commercial); unit operations and processes; primary treatment (e.g., bar screens, clarification); secondary clarification; chemical treatment; collection systems (e.g., lift stations, sewer network, infiltration, inflow); National Pollutant Discharge Elimination System (NPDES) permitting; effluent limits; biological treatment; physical treatment; solids handling (e.g., thickening, drying processes); digesters; disinfection; nitrification and/or denitrification; operations (e.g., odor control, corrosion control, compliance); advanced treatment (e.g., nutrient removal, filtration, wetlands); beneficial reuse (e.g., liquids, biosolids, gas)

Water quality (15%): stream degradation (e.g., thermal, base flow, TDS, TSS, BOD, COD), oxygen dynamics (e.g., oxygenation, deoxygenation, oxygen sag curve), risk assessment and management, toxicity, biological contaminants (e.g., algae, mussels), chemical contaminants (e.g., organics, heavy metals), bioaccumulation, eutrophication, indicator organisms and testing, sampling and monitoring (e.g., QA/QC, laboratory procedures)

Water treatment (15%): demands, hydraulic loading, storage including raw and treated water, sedimentation, taste and odor control, rapid mixing, coagulation and flocculation, filtration, disinfection, softening, advanced treatment (e.g., membranes, activated carbon, desalination), distribution systems

Engineering Economics (2.5%)

Life-cycle modeling

Value engineering and costing

a fuse or circuit breaker to protect against a calculated current or to size a beam to carry a load, you should select an answer option that will safely carry the current or load. Typically, this requires selecting a value that is closest to but larger than the current or load.

The difference is significant. Suppose you were asked to calculate "most nearly" the volumetric pure water flow required to dilute a contaminated stream to an acceptable concentration. Suppose, also, that you calculated 823 gpm. If the answer choices were (A) 600 gpm, (B) 800 gpm, (C) 1000 gpm, and (D) 1200 gpm, you would go with answer choice (B), because it is most nearly what you calculated. If, however, you were asked to select a pump or pipe with the same rated capacities, you would have to go with choice (C). Got it?

HOW MUCH MATHEMATICS IS NEEDED FOR THE EXAM?

Generally, only simple algebra, trigonometry, and geometry are needed on the PE exam. You will need to use the trigonometric, logarithm, square root, and similar buttons on your calculator. There is no need to use any other method for these functions.

There are no pure mathematics questions (algebra, geometry, trigonometry, etc.) on the exam. However, you will need to apply your knowledge of these subjects to the exam questions.

Except for simple quadratic equations, you will probably not need to find the roots of higher-order equations. Occasionally, it will be convenient to use the equation-solving capability of an advanced calculator. However, other solution methods will always exist.

There is little or no use of calculus on the exam. Rarely, you may need to take a simple derivative to find a maximum or minimum of some function. Even rarer is the need to integrate to find an average or moment of inertia.

There is essentially no need to solve differential equations. Questions involving radioactive decay, seismic vibrations, control systems, chemical reactions, and fluid mixing have appeared from time to time. However, these applications are extremely rare, have usually been first-order, and could usually be handled without having to solve differential equations.

Basic statistical analysis of observed data may be necessary. Statistical calculations are generally limited to finding means, medians, standard deviations, variances, percentiles, and confidence limits. The only population distribution you need to be familiar with is the normal curve. Probability, reliability, hypothesis testing, and statistical quality control are not explicit exam subjects.

The PE exam is concerned with numerical answers, not with proofs or derivations. You will not be asked to prove or derive formulas.

Occasionally, a calculation may require an iterative solution method. Generally, there is no need to complete more than two iterations. You will not need to program your calculator to obtain an "exact" answer. Nor will you generally need to use complex numerical methods.

STRAIGHT TALK: STRUCTURAL TOPICS

There are four specialized structural subjects that require special consideration in your review: masonry, timber, bridge, and seismic design. These subjects appear regularly (i.e., on every exam). (Yes, there are simple seismic design questions on the NCEES civil PE exam!) Even if you do not intend to work in the structural afternoon depth section, you will have to know a little something about these subjects for the morning breadth section.

The masonry chapters in this book are introductions to basic concepts and the most common design applications. Exam questions often go far beyond wall and column design. If you intend to be ready for any masonry question, I recommend that you supplement your study of these chapters.

Bridge, timber, and seismic design are specialized structural fields that are covered in separate PPI publications. These are *Bridge Design for the Civil and Structural Professional Engineering Exams*, *Timber Design for the Civil and Structural Professional Engineering Exams*, and *Seismic Design of Building Structures*.

STEEL: ASD AND LRFD

Either ASD or LRFD may be used, as there will be "parallel" questions adapted to both. It is not necessary to learn both systems of design.

WHICH BUILDING CODE?

Building codes may be needed, but only to a very limited extent (e.g., for wind and seismic loadings). Since the PE exam is a national exam, you are not required to use a regional building code. NCEES has adopted the *International Building Code* as the basis of its code-related questions. The code edition (year, version, etc.) tested is a thorny issue, however. The exam can be years behind the most recently adopted code, and many years behind the latest codes being published and distributed.

HOW ABOUT ENGINEERING ECONOMICS?

For most of the early years of engineering licensing, questions on engineering economics appeared frequently on the examinations. This is no longer the case. However, in its outline of exam subjects, NCEES notes: "Some questions may require knowledge of engineering economics." What this means is that engineering economics can constitute anything from nothing to several questions on the exam. While the degree of engineering economics knowledge may have decreased somewhat, the basic economic concepts (e.g., time value of money, present worth, non-annual compounding, comparison of alternatives, etc.) are still valid test subjects.

If engineering economics is incorporated into other questions, its "disguise" may be totally transparent. For example, you might need to compare the economics of buying and operating two blowers for remediation of a hydrocarbon spill—blowers whose annual costs must be calculated from airflow rates and heads.

WHAT ABOUT PROFESSIONALISM AND ETHICS?

For many years, NCEES has discussed adding professionalism and ethics questions to the PE exam. However, these subjects are not part of the test outline, and there has yet to be a question on them in the exam.

IS THE EXAM TRICKY?

Other than providing superfluous data, the PE exam is not a "tricky exam." The exam does not overtly try to get you to fail. Examinees manage to fail on a regular basis with perfectly straightforward questions. The exam questions are difficult in their own right. NCEES does not need to provide misleading or conflicting statements. However, you will find that commonly made mistakes are represented in the available answer choices. Thus, the alternative answers (known as *distractors*) will be logical.

Questions are generally practical, dealing with common and plausible situations that you might experience in your job. You will not be asked to design a structure for reduced gravity on the moon, to design a mud-brick road, to analyze the effects of a nuclear bomb blast on a structure, or to use bamboo for tension reinforcement.

WHAT MAKES THE QUESTIONS DIFFICULT?

Some questions are difficult because the pertinent theory is not obvious. There may be only one acceptable procedure, and it may be heuristic (or defined by a code) such that nothing else will be acceptable. Many highway capacity questions are this way.

Some questions are difficult because the data needed are hard to find. Some data just aren't available unless you happen to have brought the right reference book. Many of the structural questions are of this nature. There is no way to solve most structural steel questions without the AISC *Steel Construction Manual*. Similarly, designing an eccentrically loaded concrete column without published interaction diagrams is nearly impossible in six minutes.

Some questions are difficult because they defy the imagination. Three dimensional structural questions and some surveying curve questions fit this description. If you cannot visualize the question, you probably cannot solve it.

Some questions are difficult because the computational burden is high, and they just take a long time. Pipe networking questions solved with the Hardy Cross method fall into this category.

Some questions are difficult because the terminology is obscure, and you just don't know what the terms mean. This can happen in almost any subject.

DOES THE PE EXAM USE SI UNITS?

The PE exam in civil engineering requires working in both customary U.S. units (also known as "English units," "inch-pound units," and "British units") and a variety of metric systems, including SI. Questions use the units that correspond to commonly accepted industry standards. Some questions, such as those in structural, soils, surveying, and traffic subjects, primarily use units of pounds, feet, seconds, gallons, and degrees Fahrenheit. Metric units are used in chemical-related subjects, including electrical power (watts) and water concentration (mg/L) questions. Either system can be used for fluids, although the use of metric units is still rare.

The exam does not differentiate between lbf and lbm (pounds-force and pounds-mass) as is done in this book.

WHY DOES NCEES REUSE SOME QUESTIONS?

NCEES reuses some of the more reliable questions from each exam. The percentage of repeat questions isn't high—no more than 25% of the exam. NCEES repeats questions in order to equate the performance of one group of examinees with the performance of an earlier group. The repeated questions are known as *equaters*.

Occasionally, a new question appears on the exam that very few of the examinees do well on. Usually, the reason for this is that the subject is too obscure or too difficult. Questions on water chemistry, timber, masonry, control systems, and some engineering management subjects (e.g., linear programming) fall into this category. Also, there have been cases where a low percentage of the examinees get the answer correct because the question was inadvertently stated in a poor or confusing manner. Questions that everyone gets correct are also considered defective.

NCEES tracks the usage and "success" of each of the exam questions. Such "rogue" questions are not repeated without modification. This is one of the reasons historical analysis of question types shouldn't be used as the basis of your review.

DOES NCEES USE THE EXAM TO PRE-TEST FUTURE QUESTIONS?

NCEES does not use the PE exam to "pre-test" or qualify future questions. (It does use this procedure on the FE exam, however.) All of the questions you work will contribute toward your final score.

ARE THE PRACTICE PROBLEMS REPRESENTATIVE OF THE EXAM?

The practice problems in the companion book *Practice Problems for the Civil Engineering PE Exam* cover exam subjects. However, they are generally more comprehensive and complex than actual exam problems. Many of the practice problems are marked *"Time limit: one hour."* Compared to the six-minute problems on the PE exam, such one-hour problems are considerably more time-consuming.

WHAT REFERENCE MATERIAL IS PERMITTED IN THE EXAM?

The PE examination is an open-book exam. Most states do not have any limits on the numbers and types of books you can use. Personal notes in a three-ring binder and other semipermanent covers can usually be used.

Some states use a "shake test" to eliminate loose papers from binders. Make sure that nothing escapes from your binders when they are inverted and shaken.

The references you bring into the examination room in the morning do not have to be the same as the references you use in the afternoon. However, you cannot share books with other examinees during the exam.

A few states do not permit collections of solved problems such as *Schaum's Outline Series*, sample exams, and solutions manuals. A few states maintain a formal list of banned books.

Strictly speaking, loose paper and scratch pads are not permitted in the examination. Certain types of pre-printed graphs and logarithmically scaled graph papers (which are almost never needed) should be three-hole punched and brought in a three-ring binder. An exception to this restriction may be made for laminated and oversize charts, graphs, and tables that are commonly needed for particular types of questions. However, there probably aren't any such items for the civil PE exam.

HOW MANY BOOKS SHOULD YOU BRING?

You actually won't use many books in the examination. The trouble is, you can't know in advance which ones you will need. That's the reason why many examinees show up with boxes and boxes of books. Without a doubt, there are things that you will need that are not in this book. But there are not so many that you need to bring your entire company's library. The examination is very fast-paced. You will not have time to use books

with which you are not thoroughly familiar. The exam doesn't require you to know obscure solution methods or to use difficult-to-find data.

So, it really is unnecessary to bring a large quantity of books with you. Essential books are identified in Table 3 in this front matter, and you should be able to decide which support you need for the areas in which you intend to work. This book and five to ten other references of your choice should be sufficient for most of the questions you answer.

MAY TABS BE PLACED ON PAGES?

It is common to tab pages in your books in an effort to reduce the time required to locate useful sections. Inasmuch as some states consider Post-it notes to be "loose paper," your tabs should be of the more permanent variety. Although you can purchase tabs with gummed attachment points, it is also possible simply to use transparent tape to attach the Post-its you have already placed in your books.

WHAT ABOUT CALCULATORS?

The exam requires use of a scientific calculator. However, it may not be obvious that you should bring a spare calculator with you to the examination. It is always unfortunate when an examinee is not able to finish because his or her calculator was dropped or stolen or stopped working for some unknown reason.

NCEES has banned communicating and text-editing calculators from the exam site. Only select types of calculators are permitted. Check the current list of permissible devices at the PPI website (**www.ppi2pass.com /calculators**). Contact your state board to determine if nomographs and specialty slide rules are permitted.

The exam has not been optimized for any particular brand or type of calculator. In fact, for most calculations, a \$15 scientific calculator will produce results as satisfactory as those from a \$200 calculator. There are definite benefits to having built-in statistical functions, graphing, unit-conversion, and equation-solving capabilities. However, these benefits are not so great as to give anyone an unfair advantage.

It is essential that a calculator used for the civil PE examination have the following functions.

- trigonometric and inverse trigonometric functions
- hyperbolic and inverse hyperbolic functions
- π
- $\sqrt{x}$ and x^2
- both common and natural logarithms
- y^x and e^x

For maximum speed, your calculator should also have or be programmed for the following functions.

- interpolation

- finding standard deviations and variances
- extracting roots of quadratic and higher-order equations
- calculating determinants of matrices
- linear regression
- calculating factors for economic analysis questions

You may not share calculators with other examinees.

Laptop computers are generally not permitted in the examination. Their use has been considered, and some states may actually permit them. However, considering the nature of the exam questions, it is very unlikely that laptops would provide any advantage.

You may not use a walkie-talkie, cell phone, or other communications device during the exam.

Be sure to take your calculator with you whenever you leave the examination room for any length of time.

HOW IS THE EXAM GRADED AND SCORED?

The maximum number of points you can earn on the civil engineering PE exam is 80. The minimum number of points for passing (referred to by NCEES as the cut score) varies from exam to exam. The cut score is determined through a rational procedure, without the benefit of knowing examinees' performance on the exam. That is, the exam is not graded on a curve. The cut score is selected based on what you are expected to know, not based on passing a certain percentage of engineers.

Each of the questions is worth one point. Grading is straightforward, since a computer grades your score sheet. Either you get the question right or you don't. There is no deduction for incorrect answers, so guessing is encouraged. However, if you mark two or more answers for the same problem, no credit is given for the problem.

You will receive the results of your examination from your state board (not NCEES) by mail. Allow at least four months for notification. Your score may or may not be revealed to you, depending on your state's procedure. Even when the score is reported to you, it may have been scaled or normalized to 100%.

HOW YOU SHOULD GUESS

NCEES produces defensible licensing exams. As a result, there is no pattern to the placement of correct responses. Therefore, it most likely will not help you to guess all "A," "B," "C," or "D."

The proper way to guess is as an engineer. You should use your knowledge of the subject to eliminate illogical answer choices. Illogical answer choices are those that violate good engineering principles, that are outside normal operating ranges, or that require extraordinary assumptions. Of course, this requires you to have some basic understanding of the subject in the first place. Otherwise, it's back to random guessing. That's the reason that the minimum passing score is higher than 25%.

You won't get any points using the "test-taking skills" that helped you in college—the skills that helped with tests prepared by amateurs. You won't be able to eliminate any [verb] answer choices from "Which [noun] ..." questions. You won't find problems with options of the "more than 50" and "less than 50" variety. You won't find one answer choice among the four that has a different number of significant digits, or has a verb in a different tense, or has some singular/plural discrepancy with the stem. The distractors will always match the stem, and they will be logical.

WHAT IS THE HISTORICAL PASSING RATE?

Before the civil engineering PE exam became a no-choice breadth-and-depth (B&D) exam with multiple-choice questions, the passing rate for first-timers varied considerably. It might have been 40% for one exam and 70% for the next. The passing rate for repeat examinees was even lower. The no-choice, objective, B&D format has reduced the variability in the passing rate considerably.

HOW IS THE CUT SCORE ESTABLISHED?

The raw cut score may be established by NCEES before or after the exam is administered. Final adjustments may be made following the exam date.

The NCEES uses a process known as the modified Angoff procedure to establish the cut score. This procedure starts with a small group (the cut score panel) of professional engineers selected by the NCEES. Each individual in the group reviews each problem and makes an estimate of its difficulty. Specifically, each individual estimates the number of minimally qualified engineers out of a hundred examinees who should know the correct answer to the problem. (This is equivalent to predicting the percentage of minimally qualified engineers who will answer correctly.)

Next, the panel assembles, and the estimates for each problem are openly compared and discussed. Eventually, a consensus value is obtained for each. When the panel has established a consensus value for every problem, the values are summed and divided by 100 to establish the cut score.

Various minor adjustments can be made to account for examinee population (as characterized by the average performance on equater questions) and any flawed problems.

ARE ALL OF THE DEPTH MODULES EQUAL IN DIFFICULTY?

Nothing in the modified Angoff procedure ensures that the cut score will be the same in all of the depth modules. Thus, each depth module may have a different cut score. The easier the questions, the higher the cut score will be. Accordingly, the passing rate is different for each depth module.

CHEATING AND EXAM SUBVERSION

There aren't very many ways to cheat on an open-book test. The proctors are well trained in spotting the few ways that do exist. It goes without saying that you should not talk to other examinees in the room, nor should you pass notes back and forth. The number of people who are released to use the restroom may be limited to prevent discussions.

NCEES regularly reuses good problems that have appeared on previous exams. Therefore, examination security is a serious issue with NCEES, which goes to great lengths to make sure nobody copies the questions. You may not keep your exam booklet, enter text of questions into your calculator, or copy problems into your own material.

The proctors are concerned about exam subversion, which generally means activity that might invalidate the examination or the examination process. The most common form of exam subversion involves trying to copy exam problems for future use.

Part 3: How to Prepare for and Pass the PE Exam in Civil Engineering

WHAT SHOULD YOU STUDY?

The exam covers many diverse subjects. Strictly speaking, you don't have to study every subject on the exam in order to pass. However, the more subjects you study, the more you'll improve your chances of passing. You should decide early in the preparation process which subjects you are going to study. The strategy you select will depend on your background. Following are the four most common strategies.

A broad approach is the key to success for examinees who have recently completed their academic studies. This strategy is to review the fundamentals in a broad range of undergraduate subjects (which means studying all or most of the chapters in this book). The examination includes enough fundamentals problems to make this strategy worthwhile. Overall, it's the best approach.

Engineers who have little time for preparation tend to concentrate on the subject areas in which they hope to find the most problems. By studying the list of examination subjects, some have been able to focus on those subjects that will give them the highest probability of

finding enough problems that they can answer. This strategy works as long as the examination cooperates and has enough of the types of questions they need. Too often, though, examinees who pick and choose subjects to review can't find enough problems to complete the exam.

Engineers who have been away from classroom work for a long time tend to concentrate on the subjects in which they have had extensive experience, in the hope that the exam will feature lots of problems in those subjects. This method is seldom successful.

Some engineers plan on modeling their solutions from similar problems they have found in textbooks, collections of solutions, and old exams. These engineers often spend a lot of time compiling and indexing the example and sample problem types in all of their books. This is not a legitimate preparation method, and it is almost never successful.

DO YOU NEED A CLASSROOM REVIEW COURSE?

Approximately 60% of first-time PE examinees take a review course of some form. Classroom, audio, video, correspondence, and internet courses are available for some or all of the exam topics. Courses provide several significant advantages over self-directed study, some of which may apply to you.

- A course structures and paces your review. It ensures that you keep going forward without getting bogged down in one subject.

- A course focuses you on a limited amount of material. Without a course, you might not know which subjects to study.

- A course provides you with the questions you need to solve. You won't have to spend time looking for them.

- A course spoon-feeds you the material. You may not need to read the book!

- The course instructor can answer your questions when you are stuck.

You probably already know if any of these advantages apply to you. A review course will be less valuable if you are thorough, self-motivated, and highly disciplined.

HOW LONG SHOULD YOU STUDY?

We've all heard stories of the person who didn't crack a book until the week before the exam and still passed it with flying colors. Yes, these people really exist. However, I'm not one of them, and you probably aren't either. In fact, after having taught thousands of engineers in my own classes, I'm convinced that these people are as rare as the ones who have taken the exam five times and still can't pass it.

A thorough review takes approximately 300 hours. Most of this time is spent solving problems. Some of it may be spent in class; some is spent at home. Some examinees spread this time over a year. Others cram it all into two months. Most classroom review courses last for three or four months. The best time to start studying will depend on how much time you can spend per week.

WHAT THE WELL-HEELED CIVIL ENGINEER SHOULD BEGIN ACCUMULATING

There are many references and resources that you should begin to assemble for review and for use in the examination.

It is unlikely that you could pass the PE exam without accumulating other books and resources. There certainly isn't much margin for error if you show up with only one book. True, references aren't needed to answer some fluids, hydrology, and soils questions. However, there are many depth questions that require knowledge, data, and experience that are presented and described only in books dedicated to a single subject. You would have to be truly lucky to go in "bare," find the right mix of questions, and pass.

Few examinees are able to accumulate all of the references needed to support the exam's entire body of knowledge. The accumulation process is too expensive and time-consuming, and the sources are too diverse. Like purchasing an insurance policy, what you end up with will be more a function of your budget than of your needs. In some cases, one book will satisfy several needs.

The list in Table 3 was compiled from approximately 50 administrations of the civil engineering PE exam. The books and other items listed are regularly cited by examinees as being particularly useful to them. This listing only includes the major "named" books that have become standard references in the industry. These books are in addition to any textbooks or resources that you might choose to bring.

ADDITIONAL REVIEW MATERIAL

In addition to this book and its accompanying *Practice Problems for the Civil Engineering PE Exam*, PPI can provide you with many targeted references and study aids, some of which are listed here.

- *Civil Engineering Sample Examination*
- *Quick Reference for the Civil Engineering PE Exam*
- *Engineering Unit Conversions*
- *101 Solved Civil Engineering Problems*
- *Seismic Design of Building Structures*
- *Surveying Principles for Civil Engineers*

- *Timber Design for the Civil and Structural PE Exams*
- *Bridge Design for the Civil and Structural PE Exams*

DON'T FORGET THE DOWNLOADS

Many of the tables and appendices in this book are representative abridgments with just enough data to (a) do the practice problems in the companion book and (b) give you a false sense of security. You can download or link to additional data, explanations, and references by visiting PPI's website, **www.ppi2pass.com /CEwebrefs**, where links to additional "Engineering Exam Support" sources are provided.

WHAT YOU WON'T NEED

Generally, people bring too many things to the examination. One general rule is that you shouldn't bring books that you have not looked at during your review. If you didn't need a book while doing the practice problems in this book, you won't need it during the exam.

There are some other things that you won't need.

- Books on basic and introductory subjects: You won't need books that cover trigonometry, geometry, or calculus.

- Books that cover background engineering subjects that appear on the exam, such as fluids, thermodynamics, and chemistry: The exam is more concerned with the applications of these bodies of knowledge than with the bodies of knowledge themselves.

- Books on non-exam subjects: Such subjects as materials science, statics, dynamics, mechanics of materials, drafting, history, the English language, geography, and philosophy are not part of the exam.

- Books on mathematical analysis, or extensive mathematics tabulations

- Extensive collections of properties: You will not be expected to know the properties characteristics of chemical compounds, obscure or exotic alloys, or biological organisms. Most characteristics affecting performance are provided as part of the question statement.

- Plumbing, electrical, or fire codes

- Obscure books and materials: Books that are in a different language, doctoral theses, and papers presented at technical societies won't be needed during the exam.

- Old textbooks or obsolete, rare, and ancient books: NCEES exam committees are aware of which textbooks are in use. Material that is available only in out-of-print publications and old editions won't be used.

- Handbooks in other disciplines: You probably won't need a mechanical, electrical, or industrial engineering handbook.

- The *Handbook of Chemistry and Physics*

- Computer science books: You won't need to bring books covering computer logic, programming, algorithms, program design, or subroutines for BASIC, FORTRAN, C, Pascal, HTML, Java, Active-X, or any other language.

- Crafts- and trades-oriented books: The exam does not expect to you to have detailed knowledge of trades or manufacturing operations (e.g., carpentry, plumbing, electrical wiring, roofing, sheetrocking, foundry, metal turning, sheet-metal forming, or designing jigs and fixtures).

- Manufacturer's literature and catalogs: No part of the exam requires you to be familiar with products that are proprietary to any manufacturer.

- U.S. government publications: With the exceptions of the publications mentioned in this book, no government publications are required in the PE exam.

- Your state's laws: The PE exam is a national exam. Nothing unique to your state will appear on it. (However, federal legislation affecting civil engineers, particularly in environmental areas, is fair game.)

- Local or state building codes

SHOULD YOU LOOK FOR OLD EXAMS?

The traditional approach to preparing for standardized tests includes working sample tests. However, NCEES does not release old tests or questions after they are used. Therefore, there are no official questions or tests available from legitimate sources. NCEES has published a booklet of sample questions and solutions to illustrate the format of the exam. However, these questions have been assembled from various previous exams, and the publication is not a true "old exam."

WHAT SHOULD YOU MEMORIZE?

You get lucky here, because it isn't necessary to actually memorize anything. The exam is open-book, so you can look up any procedure, formula, or piece of information that you need. You can speed up your problem-solving response time significantly if you don't have to look up the conversion from gal/min to ft^3/sec, the definition of the sine of an angle, and the chemical formula for carbon dioxide, but you don't even have to memorize these kinds of things. As you work practice problems in the companion book, you will automatically memorize the things that you come across more than a few times.

DO YOU NEED A REVIEW SCHEDULE?

It is important that you develop and adhere to a review outline and schedule. Once you have decided which subjects you are going to study, you can allocate the available time to those subjects in a manner that makes sense to you. If you are not taking a classroom review course (where the order of preparation is determined by the lectures), you should make an outline of subjects for self-study to use for scheduling your preparation. A fill-in-the-dates schedule is provided in Table 4 at the end of this Introduction. An interactive, adjustable, and personalized study schedule is also available on the PPI website.

A SIMPLE PLANNING SUGGESTION

Designate some location (a drawer, a corner, a cardboard box, or even a paper shopping bag left on the floor) as your "exam catch-all." Use your catch-all during the months before the exam when you have revelations about things you should bring with you. For example, you might realize that the plastic ruler marked off in tenths of an inch that is normally kept in the kitchen junk drawer can help you with some soil pressure questions. Or, you might decide that a certain book is particularly valuable. Or that it would be nice to have dental floss after lunch. Or that large rubber bands are useful for holding books open.

It isn't actually necessary to put these treasured items in the catch-all during your preparation. You can, of course, if it's convenient. But if these items will have other functions during the time before the exam, at least write yourself a note and put the note into the catch-all. When you go to pack your exam kit a few days before the exam, you can transfer some items immediately, and the notes will be your reminders for the other items that are back in the kitchen drawer.

HOW YOU CAN MAKE YOUR REVIEW REALISTIC

In the exam, you must be able to quickly recall solution procedures, formulas, and important data. You must remain sharp for eight hours or more. When you played a sport back in school, your coach tried to put you in game-related situations. Preparing for the PE exam isn't much different from preparing for a big game. Some part of your preparation should be realistic and representative of the examination environment.

There are several things you can do to make your review more representative. For example, if you gather most of your review resources (i.e., books) in advance and try to use them exclusively during your review, you will become more familiar with them. (Of course, you can also add to or change your references if you find inadequacies.)

Learning to use your time wisely is one of the most important lessons you can learn during your review. You

Table 3 *Required and Recommended References*[a]

Structural

Building Code Requirements for Structural Concrete (ACI-318) *with Commentary* (ACI-318R). American Concrete Institute

Notes on ACI-318: Building Code Requirements for Structural Concrete with Design Applications. Portland Cement Association

Steel Construction Manual. American Institute of Steel Construction, Inc.

LRFD Manual of Steel Construction: Vol. 1, "Structural Members, Specifications, and Codes." American Institute of Steel Construction (AISC) (Only if you intend to work in LRFD.)

Design Handbook: Beams, One-Way Slabs, Brackets, Footings, Pile Caps, Columns, Two-Way Slabs (ACI SP-017). American Concrete Institute

Building Code Requirements for Masonry Structures, ACI 530, *and Specifications for Masonry Structures*, ACI 530.1. American Concrete Institute

National Design Specification for Wood Construction. American Forest and Paper Association

AASHTO LRFD Bridge Design Specifications. American Association of State Highway and Transportation Officials (AASHTO)

International Building Code. International Code Council

PCI Design Handbook. Precast/Prestressed Concrete Institute

Minimum Design Loads for Buildings and Other Structures (ASCE 7). American Society of Civil Engineers

Transportation

A Policy on Geometric Design of Highways and Streets ("AASHTO Green Book"). AASHTO

Highway Capacity Manual. Transportation Research Board/National Research Council

Manual of Uniform Traffic Control Devices (MUTCD). U.S. Department of Transportation, Federal Highway Administration (FHA), Traffic Control Systems Divisions, Office of Traffic Operations

Roadside Design Guide. AASHTO

Thickness Design for Concrete Highway and Street Pavements. Portland Cement Association

Guide for Design of Pavement Structures, Volumes 1 and 2. AASHTO

The Asphalt Handbook, Manual MS-4. The Asphalt Institute

Design and Control of Concrete Mixtures. Portland Cement Association

Traffic Engineering Handbook. Institute of Transportation Engineers

Trip Generation. Institute of Transportation Engineers

Water Resources

Handbook of Hydraulics: For the Solution of Hydraulic Engineering Questions. Brater, King, Lindell, and Wei

Urban Hydrology for Small Watersheds (Technical Release TR-55). United States Department of Agriculture, Natural Resources Conservation Service (previously, Soil Conservation Service)

Environmental

Wastewater Engineering: Treatment, Disposal, and Reuse (Metcalf & Eddy). George Tchobanoglous and Franklin L. Burton

Recommended Standard for Wastewater Facilities ("10 States' Standards"). Health Education Services, Health Resources, Inc.

(continued)

Table 3 *Required and Recommended References[a]*
(continued)

Standard Methods for the Examination of Water and Wastewater. A joint publication of the American Public Health Association (APHA), the American Water Works Association (AWWA), and the Water Pollution Control Federation (WPCF)

Geotechnical

NAVFAC Design Manuals DM 7.1 and 7.2. Department of the Navy, Naval Facilities Engineering Command

Construction

ASCE 37-02: Design Loads on Structures During Construction. American Society of Civil Engineers

NDS: National Design Specification for Wood Construction. American Forest & Paper Association/American Wood Council

CMWB: Standard Practice for Bracing Masonry Walls During Construction. Council for Masonry Wall Bracing, Mason Contractors Association of America

AISC: Steel Construction Manual. American Institute of Steel Construction, Inc.

ACI 347-04: Guide to Formwork for Concrete. American Concrete Institute, Farmington Hills, MI (in ACI SP-4, 7th edition appendix)

ACI SP-4: Formwork for Concrete. American Concrete Institute

OSHA: Occupational Safety and Health Standards for the Construction Industry, 29 CFR Part 1926 (U.S. Federal version). U.S. Department of Labor

MUTCD-Pt 6: Manual on Uniform Traffic Control Devices—Part 6 Temporary Traffic Control. U.S. Federal Highway Administration

---

[a]Although any edition can be used to learn the subject, the exam is "edition sensitive." Since the code version, edition, or year that are tested on the exam can change without notice, this information is available at PPI's website, **www.ppi2pass.com**.

will undoubtedly encounter questions that end up taking much longer than you expected. In some instances, you will cause your own delays by spending too much time looking through books for things you need (or just by looking for the books themselves!). Other times, the questions will entail too much work. Learn to recognize these situations so that you can make an intelligent decision about skipping such questions in the exam.

WHAT TO DO A FEW DAYS BEFORE THE EXAM

There are a few things you should do a week or so before the examination. You should arrange for childcare and transportation. Since the examination does not always start or end at the designated time, make sure that your childcare and transportation arrangements are flexible.

Check PPI's website for last-minute updates and errata to any PPI books you might have and are bringing to the exam.

If you haven't already done so, read the "Advice from Examinees" section of PPI's website.

If you haven't been following along on the Engineering Exam Forum on PPI's website, use the search function to locate discussions on this bulletin board.

If it's convenient, visit the exam location in order to find the building, parking areas, examination room, and restrooms. If it's not convenient, you may find driving directions and/or site maps on the web.

Take the battery cover off your calculator and check to make sure you are bringing the correct size replacement batteries. Some calculators require a different kind of battery for their "permanent" memories. Put the cover back on and secure it with a piece of masking tape. Write your name on the tape to identify your calculator.

If your spare calculator is not the same as your primary calculator, spend a few minutes familiarizing yourself with how it works. In particular, you should verify that your spare calculator is functional.

PREPARE YOUR CAR

[ ] Gather snow chains, shovel, and tarp to lie on while installing chains.

[ ] Check tire pressures.

[ ] Check your spare tire.

[ ] Check for tire installation tools.

[ ] Verify that you have the vehicle manual.

[ ] Check fluid levels (oil, gas, water, brake fluid, transmission fluid, window-washing solution).

[ ] Fill up with gas.

[ ] Check battery and charge if necessary.

[ ] Know something about your fuse system (where they are, how to replace them, etc.).

[ ] Assemble all required maps.

[ ] Fix anything that might slow you down (missing wiper blades, etc.).

[ ] Check your taillights.

[ ] Affix the recently arrived DMV license sticker.

[ ] Fix anything that might get you pulled over on the way to the exam (burned-out taillight or headlight, broken lenses, bald tires, missing license plate, noisy muffler).

[ ] Treat the inside windows with anti-fog solution.

[ ] Put a roll of paper towels in the back seat.

[ ] Gather exact change for any bridge tolls or toll roads.

[ ] Put $20 in your glove box.

[ ] Check for current registration and proof of insurance.

[ ] Locate a spare key.

[ ] Find your AAA or other roadside-assistance cards and phone numbers.

[ ] Plan out alternate routes.

PREPARE YOUR EXAM KITS

Second in importance to your scholastic preparation is the preparation of your two examination kits. The first kit consists of a bag, box (plastic milk crates hold up better than cardboard in the rain), or wheeled travel suitcase containing items to be brought with you into the examination room.

[ ] letter admitting you to the examination

[ ] photographic identification (e.g., driver's license)

[ ] this book

[ ] other textbooks and reference books

[ ] *Merritt's Handbook for Civil Engineers* (any reasonably current edition)

[ ] regular dictionary

[ ] scientific/engineering dictionary

[ ] review course notes in a three-ring binder

[ ] cardboard boxes or plastic milk crates to use as a bookcase

[ ] primary calculator

[ ] spare calculator

[ ] instruction booklets for your calculators

[ ] extra calculator batteries

[ ] straightedge and rulers

[ ] compass

[ ] protractor

[ ] scissors

[ ] stapler

[ ] transparent tape

[ ] magnifying glass

[ ] small (jeweler's) screwdriver for fixing your glasses or for removing batteries from your calculator

[ ] unobtrusive (quiet) snacks or candies, already unwrapped

[ ] two small plastic bottles of water

[ ] travel pack of tissue (keep in your pocket)

[ ] handkerchief

[ ] headache remedy

[ ] personal medication

[ ] $3.00 in miscellaneous change

[ ] light, comfortable sweater

[ ] loose shoes or slippers

[ ] cushion for your chair

[ ] earplugs

[ ] wristwatch with alarm

[ ] several large trash bags ("raincoats" for your boxes of books)

[ ] roll of paper towels

[ ] wire coat hanger (to hang up your jacket or to get back into your car in an emergency)

[ ] extra set of car keys on a string around your neck

The second kit consists of the following items and should be left in a separate bag or box in your car in case they are needed.

[ ] copy of your application

[ ] proof of delivery

[ ] light lunch

[ ] beverage in thermos or cans

[ ] sunglasses

[ ] extra pair of prescription glasses

[ ] raincoat, boots, gloves, hat, and umbrella

[ ] street map of the examination area

[ ] parking permit

[ ] battery-powered desk lamp

[ ] your cell phone

[ ] piece of rope

The following items cannot be used during the examination and should be left at home.

[ ] personal pencils and erasers (NCEES distributes mechanical pencils at the exam.)

[ ] fountain pens

[ ] radio or tape/CD player

[ ] battery charger

[ ] extension cords

[ ] scratch paper

[ ] note pads

PREPARE FOR THE WORST

All of the occurrences listed in this section happen to examinees on a regular basis. Granted, you cannot prepare for every eventuality. But, even though each of these occurrences taken individually is a low-probability event, taken together, they are worth considering in advance.

- Imagine getting a flat tire, getting stuck in traffic, or running out of gas on the way to the exam.

- Imagine rain and snow as you are carrying your cardboard boxes of books into the exam room. Would plastic trash bags be helpful?

- Imagine arriving late. Can you get into the exam without having to make two trips from your car?

- Imagine having to park two blocks from the exam site. How are you going to get everything to the exam room? Can you actually carry everything that far? Could you use a furniture dolly, a supermarket basket, or perhaps a helpmate?

- Imagine a Star Trek convention, square-dancing contest, construction, or auction in the next room.

- Imagine a site without any heat, with poor lighting, or with sunlight streaming directly into your eyes.

- Imagine a hard folding chair and a table with one short leg.

- Imagine a site next to an airport with frequent take-offs, or next to a construction site with a pile driver, or next to the NHRA's Drag Racing Championship.

- Imagine a seat where someone nearby chews gum with an open mouth; taps his pencil or drums her fingers; or wheezes, coughs, and sneezes for eight hours.

- Imagine the distraction of someone crying or of proctors evicting yelling and screaming examinees who have been found cheating. Imagine the tragedy of another examinee's serious medical emergency.

- Imagine a delay of an hour while they find someone to unlock the building, turn on the heat, or wait for the head proctor to bring instructions.

- Imagine a power outage occurring sometime during the exam.

- Imagine a proctor who (a) tells you that one of your favorite books can't be used in the exam, (b) accuses you of cheating, or (c) calls "time up" without giving you any warning.

- Imagine not being able to get your lunch out of your car or find a restaurant.

- Imagine getting sick or nervous in the exam.

- Imagine someone stealing your calculator during lunch.

WHAT TO DO THE DAY BEFORE THE EXAM

Take the day before the examination off from work to relax. Do not cram the last night. A good night's sleep is the best way to start the examination. If you live a considerable distance from the examination site, consider getting a hotel room in which to spend the night.

Practice setting up your examination work environment. Carry your boxes to the kitchen table. Arrange your "bookcases" and supplies. Decide what stays on the floor in boxes and what gets an "honored position" on the tabletop.

Use your checklist to make sure you have everything. Make sure your exam kits are packed and ready to go. Wrap your boxes in plastic bags in case it's raining when you carry them from the car to the exam room.

Calculate your wake-up time and set the alarms on two bedroom clocks. Select and lay out your clothing items. (Dress in layers.) Select and lay out your breakfast items.

If it's going to be hot on exam day, put your (plastic) bottles of water in the freezer.

Make sure you have gas in your car and money in your wallet.

WHAT TO DO THE DAY OF THE EXAM

Turn off the quarterly and hourly alerts on your wristwatch. Leave your pager or cell phone at home. If you must bring them, change them to silent mode.

Bring or buy a morning newspaper.

You should arrive at least 30 minutes before the examination starts. This will allow time for finding a convenient parking place, bringing your materials to the examination room, making room and seating changes, and calming down. Be prepared, though, to find that the examination room is not open or ready at the designated time.

Once you have arranged the materials around you on your table, take out your morning newspaper and look cool. (Only nervous people work crossword puzzles.)

WHAT TO DO DURING THE EXAM

All of the procedures typically associated with timed, proctored, computer-graded assessment tests will be in effect when you take the PE examination.

The proctors will distribute the examination booklets and answer sheets if they are not already on your tables. However, you should not open the booklets until instructed to do so. You may read the information on the front and back covers, and you should write your name in the appropriate blank spaces.

Listen carefully to everything the proctors say. Do not ask your proctors any engineering questions. Even if they are knowledgeable in engineering, they will not be permitted to answer your questions.

Answers to questions are recorded on an answer sheet contained in the test booklet. The proctors will guide you through the process of putting your name and other biographical information on this sheet when the time comes, which will take approximately 15 minutes. You will be given the full four hours to answer questions. Time to initialize the answer sheet is not part of your four hours.

The common suggestions to "completely fill the bubbles and erase completely" apply here. NCEES provides each examinee with a mechanical pencil with HB lead. Use of ballpoint pens and felt-tip markers is prohibited for several reasons.

If you finish the exam early and there are still more than 30 minutes remaining, you will be permitted to leave the room. If you finish less than 30 minutes before the end of the exam, you may be required to remain until the end. This is done to be considerate of the people who are still working.

When you leave, you must return your exam booklet. You may not keep the exam booklet for later review.

If there are any questions that you think were flawed, in error, or unsolvable, ask a proctor for a "reporting form" on which you can submit your comments. Follow your proctor's advice in preparing this document.

WHAT ABOUT EATING AND DRINKING IN THE EXAM ROOM?

The official rule is probably the same in every state: no eating or drinking in the exam. That makes sense, for a number of reasons. Some exam sites don't want

(or don't permit) stains and messes. Others don't want crumbs to attract ants and rodents. Your table partners don't want spills or smells. Nobody wants the distractions. Your proctors can't give you a new exam booklet when the first one is ruined with coffee.

How this rule is administered varies from site to site and from proctor to proctor. Some proctors enforce the letter of law, threatening to evict you from the exam room when they see you chewing gum. Others may permit you to have bottled water, as long as you store the bottles on the floor where any spills will not harm what's on the table. No one is going to let you crack peanuts while you work on the exam, but I can't see anyone complaining about a hard candy melting away in your mouth. You'll just have to find out when you get there.

HOW TO SOLVE MULTIPLE-CHOICE QUESTIONS

When you begin each session of the exam, observe the following suggestions:

- Use only the pencil provided.

- Do not spend an inordinate amount of time on any single question. If you have not answered a question in a reasonable amount of time, make a note of it and move on.

- Set your wristwatch alarm for five minutes before the end of each four-hour session, and use that remaining time to guess at all of the remaining questions. Odds are that you will be successful with about 25% of your guesses, and these points will more than make up for the few points that you might earn by working during the last five minutes.

- Make mental notes about any questions for which you cannot find a correct response, that appears to have two correct responses, or that you believe have some technical flaw. Errors in the exam are rare, but they do occur. Such errors are usually discovered during the scoring process and discounted from the examination, so it is not necessary to tell your proctor, but be sure to mark the one best answer before moving on.

- Make sure all of your responses on the answer sheet are dark and completely fill the bubbles.

SOLVE QUESTIONS CAREFULLY

Many points are lost to carelessness. Keep the following items in mind when you are solving the end-of-chapter questions. Hopefully, these suggestions will be automatic in the exam.

[ ] Did you recheck your mathematical equations?

[ ] Do the units cancel out in your calculations?

[ ] Did you convert between radius and diameter?

[ ] Did you convert between feet and inches?

[ ] Did you convert from gage to absolute pressures?

[ ] Did you convert between kPa and Pa?

[ ] Did you recheck all data obtained from other sources, tables, and figures? (In finding the friction factor, did you enter the Moody diagram at the correct Reynolds number?)

SHOULD YOU TALK TO OTHER EXAMINEES AFTER THE EXAM?

The jury is out on this question. People react quite differently to the examination experience. Some people are energized. Most are exhausted. Some people need to unwind by talking with other examinees, describing every detail of their experience, and dissecting every examination question. Other people need lots of quiet space, and prefer to just get into a hot tub to soak and sulk. Most engineers, apparently, are in this latter category.

Since everyone who took the exam has seen it, you will not be violating your "oath of silence" if you talk about the details with other examinees. It's difficult not to ask how someone else approached a question that had you completely stumped. However, keep in mind that it is very disquieting to think you answered a question correctly, only to have someone tell you where you went wrong.

AFTER THE EXAM

Yes, there is something to do after the exam. Most people come home, throw their exam "kits" into the corner, and collapse. A week later, when they can bear to think about the experience again, they start integrating their exam kits back into their normal lives. The calculators go back into the desk, the books go back on the shelves, the $3.00 in change goes back into the piggy bank, and all of the miscellaneous stuff you brought with you to the exam is put back wherever it came from.

Here's what I suggest you do as soon as you get home, before you collapse.

[ ] Thank your spouse and children for helping you during your preparation.

[ ] Take any paperwork you received on exam day out of your pocket, purse, or wallet. Put this inside your *Civil Engineering Reference Manual*.

[ ] Reflect on any statements regarding exam secrecy to which you signed your agreement in the exam.

[ ] Visit the PPI website and complete the after-exam survey to help PPI improve the quality of its service and products.

[ ] If you participated in a PPI Passing Zone, log on one last time to thank the instructors. (Passing Zones remain posted for a week after the exam.)

[ ] Call your employer and tell him/her that you need to take a mental health day off on Monday.

A few days later, when you can face the world again, do the following.

[ ] Make notes about anything you would do differently if you had to take the exam over again.

[ ] Consolidate all of your application paperwork, correspondence to/from your state, and any paperwork that you received on exam day.

[ ] If you took a live review course, call the instructor (or write a note) to say "Thanks."

[ ] Visit the Engineering Exam Forum part of PPI's website and see what other people are saying about the exam you took.

[ ] Return any books you borrowed.

[ ] Write thank-you notes to all of the people who wrote letters of recommendation or reference for you.

[ ] Find and read the chapter in this book that covers ethics. There were no ethics questions on your PE exam, but it doesn't make any difference. Ethical behavior is expected of a PE in any case. Spend a few minutes reflecting on how your performance (obligations, attitude, presentation, behavior, appearance, etc.) might be about to change once you are licensed. Consider how you are going to be a role model for others around you.

[ ] Put all of your review books, binders, and notes someplace where they will be out of sight.

FINALLY

By the time you've "undone" all of your preparations, you might have thought of a few things that could help future examinees. If you have any sage comments about how to prepare, any suggestions about what to do in or bring to the exam, any comments on how to improve this book, or any funny anecdotes about your experience, I hope you will share these with me. By this time, you'll be the "expert," and I'll be your biggest fan.

AND THEN THERE'S THE WAIT ...

Waiting for the exam results is its own form of mental torture.

Yes, I know the exam is 100% multiple-choice, and grading should be almost instantaneous. But, you are going to wait, nevertheless. There are many reasons for the delay.

Although the actual machine grading "only takes seconds," consider the following facts: (a) NCEES prepares multiple exams for each administration, in case one becomes unusable (i.e., is inappropriately released) before the exam date. (b) Since the actual version of the exam used is not known until after it is finally given, the cut-score determination occurs after the exam date.

I wouldn't be surprised to hear that NCEES receives dozens, if not hundreds, of claims from well-meaning examinees who were 100% certain that the exams they took were fatally flawed to some degree—that there wasn't a correct answer for such-and-such question—that there were two answers for such-and-such question—or even, perhaps, that such-and-such question was missing from their exam booklet altogether. Each of these claims must be considered as a potential adjustment to the cut-score.

Then, the exams must actually be graded. Since grading nearly 100,000 exams (counting all the FE and PE exams) requires specialized equipment, software, and training not normally possessed by the average employee, as well as time to do the work (also not normally possessed by the average employee), grading is invariably outsourced.

Outsourced grading cannot begin until all of the states have returned their score sheets to NCEES and NCEES has sorted, separated, organized, and consolidated the score sheets into whatever "secret sauce sequence" is best.

During grading, some of the score sheets "pop out" with any number of abnormalities that demand manual scoring.

After the individual exams are scored, the results are analyzed in a variety of ways. Some of the analysis looks at passing rates by such delineators as degree, major, university, site, and state. Part of the analysis looks for similarities between physically adjacent examinees (to look for cheating). Part of the analysis looks for exam sites that have statistically abnormal group performance. And, some of the analysis looks for exam questions that have a disproportionate fraction of successful or unsuccessful examinees. Anyway, you get the idea: It's not merely putting your exam sheet in an electronic reader. All of these steps have to be completed for 100% of the examinees before any results can go out.

Once NCEES has graded your test and notified your state, when you hear about it depends on when the work is done by your state. Some states have to approve the results at a board meeting; others prepare the certificates before sending out notifications. Some states are more computerized than others. Some states have 50 examinees, while others have 10,000. Some states are shut down by blizzards and hurricanes; others are administratively challenged—understaffed, inadequately trained, or over budget.

There is no pattern to the public release of results. None. The exam results are not released to all states simultaneously. (The states with the fewest examinees often receive their results soonest.) They are not released by discipline. They are not released alphabetically by state or examinee name. The people who failed are not notified first (or last). Your coworker might receive his or her notification today, and you might be waiting another three weeks for yours.

Some states post the names of the successful examinees on their official state websites before the results go out. Others update their websites after the results go out. Some states don't list much of anything on their websites.

Remember, too, that the size or thickness of the envelope you receive from your state does not mean anything. Some states send a big congratulations package and certificate. Others send a big package with a new application to repeat the exam. Some states send a postcard. Some send a one-page letter. Some states simply send you an invoice for your license fees. (Ahh, what a welcome bill!) You just have to open it to find out.

Check the Engineering Exam Forum on the PPI website regularly to find out which states have released their results. You will find many other anxious examinees there. And any number of humorous conspiracy theories and rumors.

While you are waiting, I hope you will become a "Forum" regular. Log on often and help other examinees by sharing your knowledge, experiences, and wisdom. And, if you hear any good jokes at work, I hope you will share them as well.

AND WHEN YOU PASS ...

[ ] Celebrate.

[ ] Notify the people who wrote letters of recommendation or reference for you.

[ ] Read "FAQs about What Happens After You Pass the Exam" on PPI's website.

[ ] Ask your employer for a raise.

[ ] Tell the folks at PPI (who have been rootin' for you all along) the good news.

Table 4 *Schedule for Self-Study*

chapter number	subject	date to start	date to finish
1	Systems of Units		
2	Drawing		
3	Algebra		
4	Linear Algebra		
5	Vectors		
6	Trigonometry		
7	Analytic Geometry		
8	Differential Calculus		
9	Integral Calculus		
10	Differential Equations		
11	Probability and Statistics		
12	Numerical Analysis		
13	Energy, Work, and Power		
14	Fluid Properties		
15	Fluid Statics		
16	Fluid Flow Parameters		
17	Fluid Dynamics		
18	Hydraulic Machines		
19	Open Channel Flow		
20	Meterology, Climatology, and Hydrology		
21	Groundwater		
22	Inorganic Chemistry		
23	Organic Chemistry		
24	Combustion and Incineration		
25	Water Quality Supply and Testing		
26	Water Supply Treatment and Distribution		
27	Biochemistry, Biology, and Bacteriology		
28	Wastewater Quantity and Quality		
29	Wastewater Treatment		
30	Activated Sludge and Sludge Processing		
31	Municipal Solid Waste		
32	Pollutants in the Environment		
33	Disposition of Hazardous Materials		
34	Environmental Remediation		
35	Soil Properties and Testing		
36	Shallow Foundations		
37	Rigid Retaining Walls		
38	Piles and Deep Foundations		
39	Excavations		
40	Special Soil Topics		
41	Determinate Statics		
42	Properties of Areas		
43	Material Properties and Testing		
44	Strength of Materials		

(continued)

Table 4 Schedule for Self-Study (continued)

chapter number	subject	date to start	date to finish
45	Basic Elements of Design		
46	Structural Analysis I		
47	Structural Analysis II		
48	Properties of Concrete and Steel		
49	Concrete Proportioning and Mixing		
50	Reinforced Concrete: Beams		
51	Reinforced Concrete: Slabs		
52	Reinforced Concrete: Short Columns		
53	Reinforced Concrete: Long Columns		
54	Reinforced Concrete: Walls		
55	Reinforced Concrete: Footings		
56	Pretensioned Concrete		
57	Composite Concrete and Steel Bridge Girders		
58	Structural Steel: Introduction		
59	Structural Steel: Beams		
60	Structural Steel: Tension Members		
61	Structural Steel: Compression Members		
62	Structural Steel: Beam-Columns		
63	Structural Steel: Built-Up Sections		
64	Structural Steel: Composite Members for Buildings		
65	Structural Steel: Connectors		
66	Structural Steel: Welding		
67	Properties of Masonry		
68	Masonry Walls		
69	Masonry Columns		
70	Properties of Solid Bodies		
71	Kinematics		
72	Kinetics		
73	Roads and Highways: Capacity Analysis		
74	Vehicle Dynamics and Accident Analysis		
75	Flexible Pavement Design		
76	Rigid Pavement Design		
77	Plane Surveying		
78	Horizontal, Vertical, Spiral, and Compound Curves		
79	Construction Earthwork		
80	Construction Staking and Layout		
81	Building Codes and Materials Testing		
82	Construction and Jobsite Safety		
83	Electrical Systems and Equipment		
84	Instrumentation and Measurement		
85	Project Management, Budgeting, and Scheduling		
86	Engineering Economic Analysis		
87	Engineering Law		
88	Engineering Ethics		
89	Engineering Licensing in the United States		

Topic I: Background and Support

Chapter

1 Systems of Units

1. INTRODUCTION

The purpose of this chapter is to eliminate some of the confusion regarding the many units available for each engineering variable. In particular, an effort has been made to clarify the use of the so-called English systems, which for years have used the *pound* unit both for force and mass—a practice that has resulted in confusion even for those familiar with it.

2. COMMON UNITS OF MASS

The choice of a mass unit is the major factor in determining which system of units will be used in solving a problem. It is obvious that one will not easily end up with a force in pounds if the rest of the problem is stated in meters and kilograms. Actually, the choice of a mass unit determines more than whether a conversion factor will be necessary to convert from one system to another (e.g., between SI and English units). An inappropriate choice of a mass unit may actually require a conversion factor *within* the system of units.

The common units of mass are the gram, pound, kilogram, and slug.[1] There is nothing mysterious about these units. All represent different quantities of matter,

---

[1] Normally, one does not distinguish between a unit and a multiple of that unit, as is done here with the gram and the kilogram. However, these two units actually are bases for different consistent systems.

as Fig. 1.1 illustrates. In particular, note that the pound and slug do not represent the same quantity of matter.[2]

Figure 1.1 *Common Units of Mass*

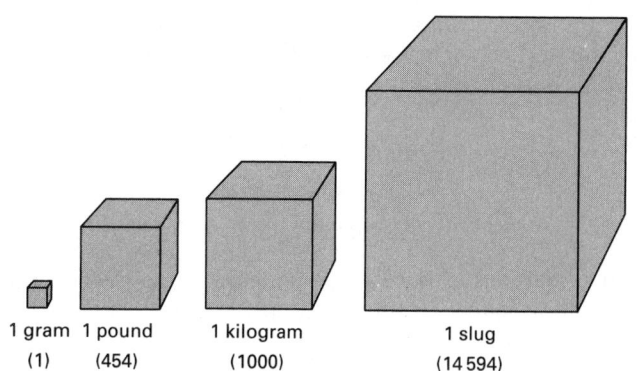

1 gram   1 pound   1 kilogram   1 slug
 (1)   (454)   (1000)   (14 594)

3. MASS AND WEIGHT

In SI, *kilograms* are used for mass and *newtons* for weight (force). The units are different, and there is no confusion between the variables. However, for years the term *pound* has been used for both mass and weight. This usage has obscured the distinction between the two: mass is a constant property of an object; weight varies with the gravitational field. Even the conventional use of the abbreviations *lbm* and *lbf* (to distinguish between pounds-mass and pounds-force) has not helped eliminate the confusion.

It is true that an object with a mass of one pound will have an earthly weight of one pound, but this is true only on the earth. The weight of the same object will be much less on the moon. Therefore, care must be taken when working with mass and force in the same problem.

The relationship that converts mass to weight is familiar to every engineering student.

$$W = mg \qquad \text{1.1}$$

Equation 1.1 illustrates that an object's weight will depend on the local acceleration of gravity as well as the object's mass. The mass will be constant, but gravity will depend on location. Mass and weight are not the same.

---

[2] A slug is equal to 32.1740 pounds-mass.

4. ACCELERATION OF GRAVITY

Gravitational acceleration on the earth's surface is usually taken as 32.2 ft/sec² or 9.81 m/s². These values are rounded from the more exact standard values of 32.1740 ft/sec² and 9.8066 m/s². However, the need for greater accuracy must be evaluated on a problem-by-problem basis. Usually, three significant digits are adequate, since gravitational acceleration is not constant anyway but is affected by location (primarily latitude and altitude) and major geographical features.

The term *standard gravity*, g_0, is derived from the acceleration at essentially any point at sea level and approximately 45° N latitude. If additional accuracy is needed, the gravitational acceleration can be calculated from Eq. 1.2. This equation neglects the effects of large land and water masses. ϕ is the latitude in degrees.

$$g_{\text{surface}} = g'\left(1 + (5.305 \times 10^{-3})\sin^2\phi\right.$$
$$\left. - (5.9 \times 10^{-6})\sin^2 2\phi\right) \qquad 1.2$$
$$g' = 32.0881 \text{ ft/sec}^2$$
$$= 9.78045 \text{ m/s}^2$$

If the effects of the earth's rotation are neglected, the gravitational acceleration at an altitude h above the earth's surface is given by Eq. 1.3. R_e is the earth's radius.

$$g_h = g_{\text{surface}}\left(\frac{R_e}{R_e + h}\right)^2 \qquad 1.3$$
$$R_e = 3960 \text{ mi}$$
$$= 6.37 \times 10^6 \text{ m}$$

5. CONSISTENT SYSTEMS OF UNITS

A set of units used in a calculation is said to be *consistent* if no conversion factors are needed.[3] For example, a moment is calculated as the product of a force and a lever arm length.

$$M = Fd \qquad 1.4$$

A calculation using Eq. 1.4 would be consistent if M was in newton-meters, F was in newtons, and d was in meters. The calculation would be inconsistent if M was in ft-kips, F was in kips, and d was in inches (because a conversion factor of 1/12 would be required).

The concept of a consistent calculation can be extended to a system of units. A *consistent system of units* is one in which no conversion factors are needed for any calculation. For example, Newton's second law of motion can be written without conversion factors. Newton's second law simply states that the force required to accelerate an object is proportional to the acceleration of the object. The constant of proportionality is the object's mass.

$$F = ma \qquad 1.5$$

[3]The terms *homogeneous* and *coherent* are also used to describe a consistent set of units.

Notice that Eq. 1.5 is $F = ma$, not $F = Wa/g$ or $F = ma/g_c$. Equation 1.5 is consistent: it requires no conversion factors. This means that in a consistent system where conversion factors are not used, once the units of m and a have been selected, the units of F are fixed. This has the effect of establishing units of work and energy, power, fluid properties, and so on.

It should be mentioned that the decision to work with a consistent set of units is desirable but unnecessary, depending on tradition and environment. Problems in fluid flow and thermodynamics are routinely solved in the United States with inconsistent units. This causes no more of a problem than working with inches and feet when calculating a moment. It is necessary only to use the proper conversion factors.

6. THE ENGLISH ENGINEERING SYSTEM

Through common and widespread use, pounds-mass (lbm) and pounds-force (lbf) have become the standard units for mass and force in the *English Engineering System*. (The English Engineering System is used in this book.)

There are subjects in the United States where the practice of using pounds for mass is firmly entrenched. For example, most thermodynamics, fluid flow, and heat transfer problems have traditionally been solved using the units of lbm/ft³ for density, Btu/lbm for enthalpy, and Btu/lbm-°F for specific heat. Unfortunately, some equations contain both lbm-related and lbf-related variables, as does the steady flow conservation of energy equation, which combines enthalpy in Btu/lbm with pressure in lbf/ft².

The units of pounds-mass and pounds-force are as different as the units of gallons and feet, and they cannot be canceled. A mass conversion factor, g_c, is needed to make the equations containing lbf and lbm dimensionally consistent. This factor is known as the *gravitational constant* and has a value of 32.1740 lbm-ft/lbf-sec². The numerical value is the same as the standard acceleration of gravity, but g_c is not the local gravitational acceleration, g.[4] g_c is a conversion constant, just as 12.0 is the conversion factor between feet and inches.

The English Engineering System is an inconsistent system as defined according to Newton's second law. $F = ma$ cannot be written if lbf, lbm, and ft/sec² are the units used. The g_c term must be included.

$$F \text{ in lbf} = \frac{(m \text{ in lbm})\left(a \text{ in } \dfrac{\text{ft}}{\text{sec}^2}\right)}{g_c \text{ in } \dfrac{\text{lbm-ft}}{\text{lbf-sec}^2}} \qquad 1.6$$

It is important to note in Eq. 1.6 that g_c does more than "fix the units." Since g_c has a numerical value of

[4]It is acceptable (and recommended) that g_c be rounded to the same number of significant digits as g. Therefore, a value of 32.2 for g_c would typically be used.

32.174, it actually changes the calculation numerically. A force of 1.0 pound will not accelerate a 1.0-pound mass at the rate of 1.0 ft/sec^2.

In the English Engineering System, work and energy are typically measured in ft-lbf (mechanical systems) or in British thermal units, Btu (thermal and fluid systems). One Btu is equal to 778.17 ft-lbf.

Example 1.1

Calculate the weight in lbf of a 1.00 lbm object in a gravitational field of 27.5 ft/sec^2.

Solution

From Eq. 1.6,

$$F = \frac{ma}{g_c} = \frac{(1.00 \text{ lbm}) \left(27.5 \dfrac{\text{ft}}{\text{sec}^2}\right)}{32.2 \dfrac{\text{lbm-ft}}{\text{lbf-sec}^2}}$$

$$= 0.854 \text{ lbf}$$

7. OTHER FORMULAS AFFECTED BY INCONSISTENCY

It is not a significant burden to include g_c in a calculation, but it may be difficult to remember when g_c should be used. Knowing when to include the gravitational constant can be learned through repeated exposure to the formulas in which it is needed, but it is safer to carry the units along in every calculation.

The following is a representative (but not exhaustive) listing of formulas that require the g_c term. In all cases, it is assumed that the standard English Engineering System units will be used.

- kinetic energy

$$E = \frac{m\text{v}^2}{2g_c} \quad \text{(in ft-lbf)} \qquad 1.7$$

- potential energy

$$E = \frac{mgz}{g_c} \quad \text{(in ft-lbf)} \qquad 1.8$$

- pressure at a depth

$$p = \frac{\rho g h}{g_c} \quad \text{(in lbf/ft}^2) \qquad 1.9$$

Example 1.2

A rocket with a mass of 4000 lbm travels at 27,000 ft/sec. What is its kinetic energy in ft-lbf?

Solution

From Eq. 1.7,

$$E_k = \frac{m\text{v}^2}{2g_c} = \frac{(4000 \text{ lbm}) \left(27,000 \dfrac{\text{ft}}{\text{sec}}\right)^2}{(2)\left(32.2 \dfrac{\text{lbm-ft}}{\text{lbf-sec}^2}\right)}$$

$$= 4.53 \times 10^{10} \text{ ft-lbf}$$

8. WEIGHT AND WEIGHT DENSITY

Weight is a force exerted on an object due to its placement in a gravitational field. If a consistent set of units is used, Eq. 1.1 can be used to calculate the weight of a mass. In the English Engineering System, however, Eq. 1.10 must be used.

$$W = \frac{mg}{g_c} \qquad 1.10$$

Both sides of Eq. 1.10 can be divided by the volume of an object to derive the *weight density*, γ, of the object. Equation 1.11 illustrates that the weight density (in lbf/ft^3) can also be calculated by multiplying the mass density (in lbm/ft^3) by g/g_c. Since g and g_c usually have the same numerical values, the only effect of Eq. 1.12 is to change the units of density.

$$\frac{W}{V} = \left(\frac{m}{V}\right)\left(\frac{g}{g_c}\right) \qquad 1.11$$

$$\gamma = \frac{W}{V} = \left(\frac{m}{V}\right)\left(\frac{g}{g_c}\right) = \frac{\rho g}{g_c} \qquad 1.12$$

Weight does not occupy volume. Only mass has volume. The concept of weight density has evolved to simplify certain calculations, particularly fluid calculations. For example, pressure at a depth is calculated from Eq. 1.13. (Compare this with Eq. 1.9.)

$$p = \gamma h \qquad 1.13$$

9. THE ENGLISH GRAVITATIONAL SYSTEM

Not all English systems are inconsistent. Pounds can still be used as the unit of force as long as pounds are not used as the unit of mass. Such is the case with the consistent *English Gravitational System*.

If acceleration is given in ft/sec^2, the units of mass for a consistent system of units can be determined from Newton's second law. The combination of units in Eq. 1.14 is known as a *slug*. g_c is not needed at all since this system is consistent. It would be needed only to convert slugs to another mass unit.

$$\text{units of } m = \frac{\text{units of } F}{\text{units of } a}$$

$$= \frac{\text{lbf}}{\dfrac{\text{ft}}{\text{sec}^2}} = \frac{\text{lbf-sec}^2}{\text{ft}} \qquad 1.14$$

Slugs and pounds-mass are not the same, as Fig. 1.1 illustrates. However, both are units for the same quantity: mass. Equation 1.15 will convert between slugs and pounds-mass.

$$\text{no. of slugs} = \frac{\text{no. of lbm}}{g_c} \qquad 1.15$$

It is important to recognize that the number of slugs is not derived by dividing the number of pounds-mass by the local gravity. g_c is used regardless of the local gravity. The conversion between feet and inches is not dependent on local gravity; neither is the conversion between slugs and pounds-mass.

Since the English Gravitational System is consistent, Eq. 1.16 can be used to calculate weight. Notice that the local gravitational acceleration is used.

$$W \text{ in lbf} = (m \text{ in slugs}) \left(g \text{ in } \frac{\text{ft}}{\text{sec}^2} \right) \qquad 1.16$$

10. THE ABSOLUTE ENGLISH SYSTEM

The obscure *Absolute English System* takes the approach that mass must have units of pounds-mass (lbm) and the units of force can be derived from Newton's second law. The units for F cannot be simplified any more than they are in Eq. 1.17. This particular combination of units is known as a *poundal*.[5] A poundal is not the same as a pound.

$$\begin{aligned} \text{units of } F &= (\text{units of } m)(\text{units of } a) \\ &= (\text{lbm}) \left(\frac{\text{ft}}{\text{sec}^2} \right) \\ &= \frac{\text{lbm-ft}}{\text{sec}^2} \qquad 1.17 \end{aligned}$$

Poundals have not seen widespread use in the United States. The English Gravitational System (using slugs for mass) has greatly eclipsed the Absolute English System in popularity. Both are consistent systems, but there seems to be little need for poundals in modern engineering.

Figure 1.2 *Common Force Units*

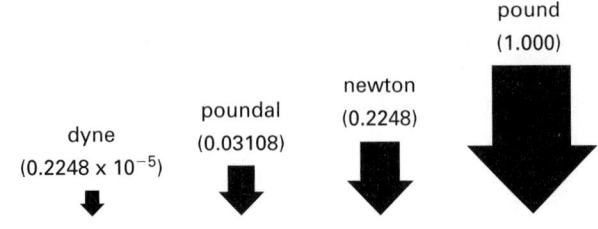

dyne
(0.2248 x 10⁻⁵)

poundal
(0.03108)

newton
(0.2248)

pound
(1.000)

---

[5]A poundal is equal to 0.03108 pounds-force.

11. METRIC SYSTEMS OF UNITS

Strictly speaking, a *metric system* is any system of units that is based on meters or parts of meters. This broad definition includes *mks systems* (based on meters, kilograms, and seconds) as well as *cgs systems* (based on centimeters, grams, and seconds).

Metric systems avoid the pounds-mass versus pounds-force ambiguity in two ways. First, a unit of weight is not established at all. All quantities of matter are specified as mass. Second, force and mass units do not share a common name.

The term *metric system* is not explicit enough to define which units are to be used for any given variable. For example, within the cgs system there is variation in how certain electrical and magnetic quantities are represented (resulting in the ESU and EMU systems). Also, within the mks system, it is common practice in some industries to use kilocalories as the unit of thermal energy, while the SI unit for thermal energy is the joule. Thus, there is a lack of uniformity even within the metricated engineering community.[6]

The "metric" parts of this book use SI, which is the most developed and codified of the so-called metric systems.[7] There will be occasional variances with local engineering custom, but it is difficult to anticipate such variances within a book that must itself be consistent.[8]

12. THE cgs SYSTEM

The *cgs system* is used widely by chemists and physicists. It is named for the three primary units used to construct its derived variables: the centimeter, the gram, and the second.

When Newton's second law is written in the cgs system, the following combination of units results.

$$\begin{aligned} \text{units of force} &= (m \text{ in g}) \left(a \text{ in } \frac{\text{cm}}{\text{s}^2} \right) \\ &= \text{g·cm/s}^2 \qquad 1.18 \end{aligned}$$

This combination of units for force is known as a *dyne*. Energy variables in the cgs system have units of dyne·cm or, equivalently, g·cm²/s². This combination is known as an *erg*. There is no uniformly accepted unit of power in the cgs system, although calories per second is frequently used.

---

[6]In the "field test" of the metric system conducted over the past 200 years, other conventions are to use kilograms-force (kgf) instead of newtons and kgf/cm² for pressure (instead of pascals).
[7]SI units are an outgrowth of the *General Conference of Weights and Measures*, an international treaty organization that established the *Système International d'Unités (International System of Units)* in 1960. The United States subscribed to this treaty in 1975.
[8]Conversion to pure SI units is essentially complete in Australia, Canada, New Zealand, and South Africa. The use of nonstandard metric units is more common among European civil engineers, who have had little need to deal with the inertial properties of mass. However, even the American Society of Civil Engineers declared its support for SI units in 1985.

The fundamental volume unit in the cgs system is the cubic centimeter (cc). Since this is the same volume as one thousandth of a liter, units of milliliters (mL) are also used.

13. SI UNITS (THE mks SYSTEM)

SI units comprise an *mks system* (so named because it uses the meter, kilogram, and second as base units). All other units are derived from the base units, which are completely listed in Table 1.1. This system is fully consistent, and there is only one recognized unit for each physical quantity (variable).

Table 1.1 *SI Base Units*

quantity	name	symbol
length	meter	m
mass	kilogram	kg
time	second	s
electric current	ampere	A
temperature	kelvin	K
amount of substance	mole	mol
luminous intensity	candela	cd

Two types of units are used: base units and derived units. The *base units* are dependent only on accepted standards or reproducible phenomena. The *derived units* (Tables 1.2 and 1.3) are made up of combinations of base units. The old *supplementary units* (Table 1.4) were classified as derived units in 1995.

Table 1.2 *Some SI Derived Units with Special Names*

quantity	name	symbol	expressed in terms of other units
frequency	hertz	Hz	$1/s$
force	newton	N	$kg \cdot m/s^2$
pressure, stress	pascal	Pa	N/m^2
energy, work, quantity of heat	joule	J	$N \cdot m$
power, radiant flux	watt	W	J/s
quantity of electricity, electric charge	coulomb	C	
electric potential, potential difference, electromotive force	volt	V	W/A
electric capacitance	farad	F	C/V
electric resistance	ohm	Ω	V/A
electric conductance	siemens	S	A/V
magnetic flux	weber	Wb	$V \cdot s$
magnetic flux density	tesla	T	Wb/m^2
inductance	henry	H	Wb/A
luminous flux	lumen	lm	
illuminance	lux	lx	lm/m^2

In addition, there is a set of non-SI units that may be used. This concession is primarily due to the significance and widespread acceptance of these units. Use of the non-SI units listed in Table 1.5 will usually create an inconsistent expression requiring conversion factors.

Table 1.3 *Some SI Derived Units*

quantity	description	symbol
area	square meter	m^2
volume	cubic meter	m^3
speed—linear	meter per second	m/s
—angular	radian per second	rad/s
acceleration—linear	meter per second squared	m/s^2
—angular	radian per second squared	rad/s^2
density, mass density	kilogram per cubic meter	kg/m^3
concentration (of amount of substance)	mole per cubic meter	mol/m^3
specific volume	cubic meter per kilogram	m^3/kg
luminance	candela per square meter	cd/m^2
absolute viscosity	pascal second	$Pa \cdot s$
kinematic viscosity	square meters per second	m^2/s
moment of force	newton meter	$N \cdot m$
surface tension	newton per meter	N/m
heat flux density, irradiance	watt per square meter	W/m^2
heat capacity, entropy	joule per kelvin	J/K
specific heat capacity, specific entropy	joule per kilogram kelvin	$J/kg \cdot K$
specific energy	joule per kilogram	J/kg
thermal conductivity	watt per meter kelvin	$W/m \cdot K$
energy density	joule per cubic meter	J/m^3
electric field strength	volt per meter	V/m
electric charge density	coulomb per cubic meter	C/m^3
surface density of charge, flux density	coulomb per square meter	C/m^2
permittivity	farad per meter	F/m
current density	ampere per square meter	A/m^2
magnetic field strength	ampere per meter	A/m
permeability	henry per meter	H/m
molar energy	joule per mole	J/mol
molar entropy, molar heat capacity	joule per mole kelvin	$J/mol \cdot K$
radiant intensity	watt per steradian	W/sr

Table 1.4 *SI Supplementary Units*[a]

quantity	name	symbol
plane angle	radian	rad
solid angle	steradian	sr

[a]classified as derived units in 1995.

The SI unit of force can be derived from Newton's second law. This combination of units for force is known as a *newton*.

$$\text{units of force} = (m \text{ in kg}) \left(a \text{ in } \frac{m}{s^2} \right)$$
$$= kg \cdot m/s^2 \qquad \textit{1.19}$$

Energy variables in SI units have units of N·m or, equivalently, kg·m^2/s^2. Both of these combinations are known as a *joule*. The units of power are joules per second, equivalent to a *watt*.

Example 1.3

A 10 kg block hangs from a cable. What is the tension in the cable? (Standard gravity equals 9.81 m/s^2.)

Solution

$$F = mg$$
$$= (10 \text{ kg}) \left(9.81 \text{ } \frac{\text{m}}{\text{s}^2}\right)$$
$$= 98.1 \text{ kg·m/s}^2 \quad (98.1 \text{ N})$$

Example 1.4

A 10 kg block is raised vertically 3 m. What is the change in potential energy?

Solution

$$\Delta E_p = mg\Delta h$$
$$= (10 \text{ kg}) \left(9.81 \text{ } \frac{\text{m}}{\text{s}^2}\right)(3 \text{ m})$$
$$= 294 \text{ kg·m}^2/\text{s}^2 \quad (294 \text{ J})$$

14. RULES FOR USING SI UNITS

In addition to having standardized units, the set of SI units also has rigid syntax rules for writing the units and combinations of units. Each unit is abbreviated with a specific symbol. The following rules for writing and combining these symbols should be adhered to.

- The expressions for derived units in symbolic form are obtained by using the mathematical signs of multiplication and division. For example, units of velocity are m/s. Units of torque are N·m (not N-m or Nm).

- Scaling of most units is done in multiples of 1000.

- The symbols are always printed in roman type, regardless of the type used in the rest of the text. The only exception to this is in the use of the symbol for liter, where the use of the lower case "el" (1) may be confused with the numeral one (1). In this case, "liter" should be written out in full, or the script ℓ or L used. (L is used in this book.)

Table 1.5 *Acceptable Non-SI Units*

quantity	unit name	symbol or abbreviation	relationship to SI unit
area	hectare	ha	1 ha = 10 000 m^2
energy	kilowatt-hour	kW·h	1 kW·h = 3.6 MJ
mass	metric ton[a]	t	1 t = 1000 kg
plane angle	degree (of arc)	°	1° = 0.017 453 rad
speed of rotation	revolution per minute	r/min	1 r/min = 2π/60 rad/s
temperature interval	degree Celsius	°C	1°C = 1K
time	minute	min	1 min = 60 s
	hour	h	1 h = 3600 s
	day (mean solar)	d	1 d = 86 400 s
	year (calendar)	a	1 a = 31 536 000 s
velocity	kilometer per hour	km/h	1 km/h = 0.278 m/s
volume	liter[b]	L	1 L = 0.001 m^3

[a]The international name for metric ton is *tonne*. The metric ton is equal to the *megagram* (Mg).
[b]The international symbol for liter is the lowercase l, which can be easily confused with the numeral 1. Several English-speaking countries have adopted the script ℓ or uppercase L (as does this book) as a symbol for liter in order to avoid any misinterpretation.

- Symbols are not pluralized: 1 kg, 45 kg (not 45 kgs).

- A period after a symbol is not used, except when the symbol occurs at the end of a sentence.

- When symbols consist of letters, there is always a full space between the quantity and the symbols: 45 kg (not 45kg). However, for planar angle designations, no space is left: 32°C (not 32° C or 32 °C); or 42° 12′ 45″ (not 42 ° 12 ′ 45 ″).

- All symbols are written in lowercase, except when the unit is derived from a proper name: m for meter; s for second; A for ampere, Wb for weber, N for newton, W for watt.

- Prefixes are printed without spacing between the prefix and the unit symbol (e.g., km is the symbol for kilometer).

- In text, when no number is involved, the unit should be spelled out. Example: Carpet is sold by the square meter, not by the m^2.

- Where a decimal fraction of a unit is used, a zero should always be placed before the decimal marker: 0.45 kg (not .45 kg). This practice draws attention to the decimal marker and helps avoid errors of scale.

Table 1.6 SI Prefixes[a]

prefix	symbol	value
exa	E	10^{18}
peta	P	10^{15}
tera	T	10^{12}
giga	G	10^{9}
mega	M	10^{6}
kilo	k	10^{3}
hecto	h	10^{2}
deka (or "deca")	da	10^{1}
deci	d	10^{-1}
centi	c	10^{-2}
milli	m	10^{-3}
micro	μ	10^{-6}
nano	n	10^{-9}
pico	p	10^{-12}
femto	f	10^{-15}
atto	a	10^{-18}

[a]There is no "B" (billion) prefix. In fact, the word "billion" means 10^{9} in the United States but 10^{12} in most other countries. This unfortunate ambiguity is handled by avoiding the use of the term billion.

- A practice in some countries is to use a comma as a decimal marker, while the practice in North America, the United Kingdom, and some other countries is to use a period (or dot) as the decimal marker. Furthermore, in some countries that use the decimal comma, a dot is frequently used to divide long numbers into groups of three. Because of these differing practices, spaces must be used instead of commas to separate long lines of digits into easily readable blocks of three digits with respect to the decimal marker: 32 453.246 072 5. A space (half-space preferred) is optional with a four-digit number: 1 234 or 1234.

- Some confusion may arise with the word "tonne" (1000 kg). When this word occurs in French text of Canadian origin, the meaning may be a ton of 2000 pounds.

15. PRIMARY DIMENSIONS

Regardless of the system of units chosen, each variable representing a physical quantity will have the same *primary dimensions*. For example, velocity may be expressed in miles per hour (mph) or meters per second (m/s), but both units have dimensions of length per unit time. Length and time are two of the primary dimensions, as neither can be broken down into more basic dimensions. The concept of primary dimensions is useful when converting little-used variables between different systems of units, as well as in correlating experimental results (i.e., dimensional analysis).

There are three different sets of primary dimensions in use.[9] In the $ML\theta T$ system, the primary dimensions are mass (M), length (L), time (θ), and temperature (T). Notice that all symbols are uppercase. In order to avoid confusion between time and temperature, the Greek letter theta is used for time.[10]

All other physical quantities can be derived from these primary dimensions.[11] For example, work in SI units has units of N·m. Since a newton is a kg·m/s², the primary dimensions of work are ML^2/θ^2. The primary dimensions for many important engineering variables are shown in Table 1.7. If it is more convenient to stay with traditional English units, it may be more desirable to work in the $FML\theta TQ$ system (sometimes called the *engineering dimensional system*). This system adds the primary dimensions of force (F) and heat (Q). Thus, work (ft-lbf in the English system) has the primary dimensions of FL. (Compare this with the primary dimensions for work in the $ML\theta T$ system.) Thermodynamic variables are similarly simplified.

Dimensional analysis will be more conveniently carried out when one of the four-dimension systems ($ML\theta T$ or $FL\theta T$) is used. Whether the $ML\theta T$, $FL\theta T$, or $FML\theta TQ$ system is used depends on what is being derived and who will be using it, and whether or not a consistent set of variables is desired. Conversion constants such as g_c and J will almost certainly be required if the $ML\theta T$ system is used to generate variables for use in the English systems. It is also much more convenient to use the $FML\theta TQ$ system when working in the fields of thermodynamics, fluid flow, heat transfer, and so on.

16. DIMENSIONLESS GROUPS

A *dimensionless group* is derived as a ratio of two forces or other quantities. Considerable use of dimensionless groups is made in certain subjects, notably fluid mechanics and heat transfer. For example, the Reynolds number, Mach number, and Froude number are used to distinguish between distinctly different flow regimes in pipe flow, compressible flow, and open channel flow, respectively.

Table 1.8 contains information about the most common dimensionless groups used in fluid mechanics and heat transfer.

---

[9]One of these, the $FL\theta T$ system, is not discussed here but appears in Table 1.7.

[10]This is the most common usage. There is a lack of consistency in the engineering world about the symbols for the primary dimensions in dimensional analysis. Some writers use t for time instead of θ. Some use H for heat instead of Q. And, in the worst mix-up of all, some have reversed the use of T and θ.

[11]A *primary dimension* is the same as a *base unit* in the SI set of units. The SI units add several other base units, as shown in Table 1.1, to deal with variables that are difficult to derive in terms of the four primary base units.

Table 1.7 *Dimensions of Common Variables*

variable	dimensional system		
	$ML\theta T$	$FL\theta T$	$FML\theta TQ$
mass (m)	M	$F\theta^2/L$	M
force (F)	ML/θ^2	F	F
length (L)	L	L	L
time (θ)	θ	θ	θ
temperature (T)	T	T	T
work (W)	ML^2/θ^2	FL	FL
heat (Q)	ML^2/θ^2	FL	Q
acceleration (a)	L/θ^2	L/θ^2	L/θ^2
frequency (N)	$1/\theta$	$1/\theta$	$1/\theta$
area (A)	L^2	L^2	L^2
coefficient of			
thermal expansion (β)	$1/T$	$1/T$	$1/T$
density (ρ)	M/L^3	$F\theta^2/L^4$	M/L^3
dimensional constant (g_c)	1.0	1.0	$ML/\theta^2 F$
specific heat at			
constant pressure (c_p);			
at constant volume (c_v)	$L^2/\theta^2 T$	$L^2/\theta^2 T$	Q/MT
heat transfer			
coefficient (h);			
overall (U)	$M/\theta^3 T$	$F/\theta LT$	$Q/\theta L^2 T$
power (P)	ML^2/θ^3	FL/θ	FL/θ
heat flow rate $(\dot{Q})$	ML^2/θ^3	FL/θ	Q/θ
kinematic viscosity (ν)	L^2/θ	L^2/θ	L^2/θ
mass flow rate $(\dot{m})$	M/θ	$F\theta/L$	M/θ
mechanical equivalent			
of heat (J)	$-$	$-$	FL/Q
pressure (p)	$M/L\theta^2$	F/L^2	F/L^2
surface tension (σ)	M/θ^2	F/L	F/L
angular velocity (ω)	$1/\theta$	$1/\theta$	$1/\theta$
volumetric flow rate			
$(\dot{m}/\rho = \dot{V})$	L^3/θ	L^3/θ	L^3/θ
conductivity (k)	$ML/\theta^3 T$	$F/\theta T$	$Q/L\theta T$
thermal diffusivity (α)	L^2/θ	L^2/θ	L^2/θ
velocity (v)	L/θ	L/θ	L/θ
viscosity, absolute (μ)	$M/L\theta$	$F\theta/L^2$	$F\theta/L^2$
volume (V)	L^3	L^3	L^3

17. LINEAL AND BOARD FOOT MEASUREMENTS

The term *lineal* is often mistaken as a typographical error for *linear*. Although "lineal" has its own specific meaning slightly different from "linear," the two are often used interchangeably by engineers.[12] The adjective *lineal* is often encountered in the building trade (e.g., 12 lineal feet of lumber), where the term is used to distinguish it from board feet measurement.

A *board foot* (abbreviated bd-ft) is not a measure of length. Rather, it is a measure of volume used with lumber. Specifically, a board foot is equal to 144 in³ (2.36×10^{-3} m³). The name is derived from the volume of a board 1 foot square and 1 inch thick. In that sense,

---

[12] *Lineal* is best used when discussing a line of succession (e.g., a lineal descendant of a particular person). *Linear* is best used when discussing length (e.g., a linear dimension of a room).

it is parallel in concept to the acre-foot. Since lumber cost is directly related to lumber weight and volume, the board foot unit is used in determining the overall lumber cost.

18. DIMENSIONAL ANALYSIS

Dimensional analysis is a means of obtaining an equation that describes some phenomenon without understanding the mechanism of the phenomenon. The most serious limitation is the need to know beforehand which variables influence the phenomenon. Once these are known or assumed, dimensional analysis can be applied by a routine procedure.

The first step is to select a system of primary dimensions. (See Sec. 15.) Usually the $ML\theta T$ system is used, although this choice may require the use of g_c and J in the final results. The dimensional formulas and symbols for variables most frequently encountered are given in Table 1.7.

The second step is to write a functional relationship between the dependent variable and the independent variable, x_i.

$$y = f(x_1, x_2, \ldots, x_m) \qquad 1.20$$

This function can be expressed as an exponentiated series. The $C_1, a_i, b_i \ldots, z_i$ in Eq. 1.21 are unknown constants.

$$y = C_1 x_1^{a_1} x_2^{b_1} x_3^{c_1} \ldots x_m^{z_1} + C_2 x_1^{a_2} x_2^{b_2} x_3^{c_2} \ldots x_m^{z_2} + \cdots \quad 1.21$$

The key to solving Eq. 1.21 is that each term on the right-hand side must have the same dimensions as y. Simultaneous equations are used to determine some of the a_i, b_i, c_i, and z_i. Experimental data are required to determine the C_i and remaining exponents. In most analyses, it is assumed that the $C_i = 0$ for $i \geq 2$.

Since this method requires working with m different variables and n different independent dimensional quantities (such as M, L, θ, and T), an easier method is desirable. One simplification is to combine the m variables into dimensionless groups called *pi-groups*. (See Table 1.8.)

If these dimensionless groups are represented by $\pi_1, \pi_2, \pi_3, \ldots, \pi_k$, the equation expressing the relationship between the variables is given by the *Buckingham π-theorem*.

$$f(\pi_1, \pi_2, \pi_3 \ldots, \pi_k) = 0 \qquad 1.22$$

$$k = m - n \qquad 1.23$$

The dimensionless pi-groups are usually found from the m variables according to an intuitive process.

Table 1.8 *Common Dimensionless Groups*

name	symbol	formula	interpretation
Biot number	Bi	$\dfrac{hL}{k_s}$	$\dfrac{\text{surface conductance}}{\text{internal conduction of solid}}$
Cauchy number	Ca	$\dfrac{\text{v}^2}{\dfrac{B_s}{\rho}} = \dfrac{\text{v}^2}{a^2}$	$\dfrac{\text{inertia force}}{\text{compressive force}} = \text{Mach number}^2$
Eckert number	Ec	$\dfrac{\text{v}^2}{2c_p\Delta T}$	$\dfrac{\text{temperature rise due to energy conversion}}{\text{temperature difference}}$
Eötvös number	Eo	$\dfrac{\rho g L^2}{\sigma}$	$\dfrac{\text{buoyancy}}{\text{surface tension}}$
Euler number	Eu	$\dfrac{\Delta p}{\rho \text{v}^2}$	$\dfrac{\text{pressure force}}{\text{inertia force}}$
Fourier number	Fo	$\dfrac{kt}{\rho c_p L^2} = \dfrac{\alpha t}{L^2}$	$\dfrac{\text{rate of conduction of heat}}{\text{rate of storage of energy}}$
Froude number[a]	Fr	$\dfrac{\text{v}^2}{gL}$	$\dfrac{\text{inertia force}}{\text{gravity force}}$
Graetz number[a]	Gz	$\left(\dfrac{D}{L}\right)\left(\dfrac{\text{v}\rho c_p D}{k}\right)$	$\dfrac{(\text{Re})(\text{Pr})}{L/D}$ $\dfrac{\text{heat transfer by convection in entrance region}}{\text{heat transfer by conduction}}$
Grashof number[a]	Gr	$\dfrac{g\beta\Delta T L^3}{\nu^2}$	$\dfrac{\text{buoyancy force}}{\text{viscous force}}$
Knudsen number	Kn	$\dfrac{\lambda}{L}$	$\dfrac{\text{mean free path of molecules}}{\text{characteristic length of object}}$
Lewis number[a]	Le	$\dfrac{\alpha}{D_c}$	$\dfrac{\text{thermal diffusivity}}{\text{molecular diffusivity}}$
Mach number	M	$\dfrac{\text{v}}{a}$	$\dfrac{\text{macroscopic velocity}}{\text{speed of sound}}$
Nusselt number	Nu	$\dfrac{hL}{k}$	$\dfrac{\text{temperature gradient at wall}}{\text{overall temperature difference}}$
Péclet number	Pé	$\dfrac{\text{v}\rho c_p D}{k}$	$(\text{Re})(\text{Pr})$ $\dfrac{\text{heat transfer by convection}}{\text{heat transfer by conduction}}$
Prandtl number	Pr	$\dfrac{\mu c_p}{k} = \dfrac{\nu}{\alpha}$	$\dfrac{\text{diffusion of momentum}}{\text{diffusion of heat}}$
Reynolds number	Re	$\dfrac{\rho \text{v} L}{\mu} = \dfrac{\text{v}L}{\nu}$	$\dfrac{\text{inertia force}}{\text{viscous force}}$
Schmidt number	Sc	$\dfrac{\mu}{\rho D_c} = \dfrac{\nu}{D_c}$	$\dfrac{\text{diffusion of momentum}}{\text{diffusion of mass}}$
Sherwood number[a]	Sh	$\dfrac{k_D L}{D_c}$	$\dfrac{\text{mass diffusivity}}{\text{molecular diffusivity}}$
Stanton number	St	$\dfrac{h}{\text{v}\rho c_p} = \dfrac{h}{c_p G}$	$\dfrac{\text{heat transfer at wall}}{\text{energy transported by stream}}$
Stokes number	Sk	$\dfrac{\Delta p L}{\mu \text{v}}$	$\dfrac{\text{pressure force}}{\text{viscous force}}$
Strouhal number[a]	Sl	$\dfrac{L}{t\text{v}} = \dfrac{L\omega}{\text{v}}$	$\dfrac{\text{frequency of vibration}}{\text{characteristic frequency}}$
Weber number	We	$\dfrac{\rho \text{v}^2 L}{\sigma}$	$\dfrac{\text{inertia force}}{\text{surface tension force}}$

[a]Multiple definitions exist.

Background and Support

Example 1.5

A solid sphere rolls down a submerged incline. Find an equation for the velocity, v.

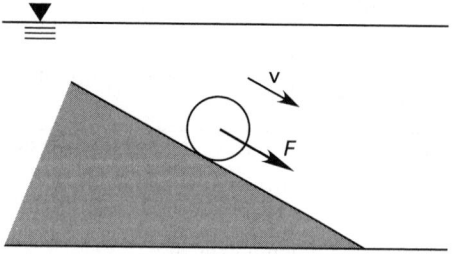

Solution

Assume that the velocity depends on the force, F, due to gravity, the diameter of the sphere, D, the density of the fluid, ρ, and the viscosity of the fluid, μ.

$$v = f(F, D, \rho, \mu) = C F^a D^b \rho^c \mu^d$$

This equation can be written in terms of the primary dimensions of the variables.

$$\frac{L}{\theta} = C \left(\frac{ML}{\theta^2}\right)^a L^b \left(\frac{M}{L^3}\right)^c \left(\frac{M}{L\theta}\right)^d$$

Since L on the left-hand side has an implied exponent of one, a necessary condition is

$$1 = a + b - 3c - d \quad (L)$$

Similarly, the other necessary conditions are

$$-1 = -2a - d \quad (\theta)$$
$$0 = a + c + d \quad (M)$$

Solving simultaneously yields

$$b = -1$$
$$c = a - 1$$
$$d = 1 - 2a$$
$$v = C F^a D^{-1} \rho^{a-1} \mu^{1-2a}$$
$$= C \left(\frac{\mu}{D\rho}\right) \left(\frac{F\rho}{\mu^2}\right)^a$$

C and a must be determined experimentally.

2 Engineering Drawing Practice[1]

Figure 2.1 *Intersecting and Non-Intersecting Lines*

(a) intersecting lines   (b) non-intersecting lines

1. NORMAL VIEWS OF LINES AND PLANES

A *normal view* of a line is a perpendicular projection of the line onto a viewing plane parallel to the line. In the normal view, all points of the line are equidistant from the observer. Therefore, the true length of a line is viewed and can be measured.

Generally, however, a line will be viewed from an oblique position and will appear shorter than it actually is. The normal view can be constructed by drawing an auxiliary view (see Sec. 5) from the orthographic view.[2]

Similarly, a normal view of a plane figure is a perpendicular projection of the figure onto a viewing plane parallel to the plane of the figure. All points of the plane are equidistant from the observer. Therefore, the true size and shape of any figure in the plane can be determined.

2. INTERSECTING AND PERPENDICULAR LINES

A single orthographic view is not sufficient to determine whether two lines intersect. However, if two or more views show the lines as having the same common point (i.e., crossing at the same position in space), then the lines intersect. In Fig. 2.1, the subscripts F and T refer to front and top views, respectively.

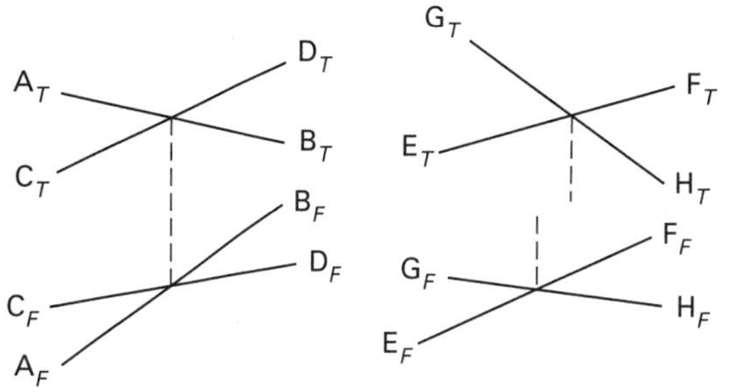

According to the *perpendicular line principle*, two perpendicular lines appear perpendicular only in a normal view of either one or both of the lines. Conversely, if two lines appear perpendicular in any view, the lines are perpendicular only if the view is a normal view of one or both of the lines.

3. TYPES OF VIEWS

Objects can be illustrated in several different ways depending on the number of views, the angle of observation, and the degree of artistic latitude taken for the purpose of simplifying the drawing process.[3] Table 2.1 categorizes the types of views.

Table 2.1 *Types of Views of Objects*

orthographic views
  principal views
  auxiliary views
  oblique views
    cavalier projection
    cabinet projection
    clinographic projection
    axonometric views
      isometric
      dimetric
      trimetric
perspective views
  parallel perspective
  angular perspective

---

[1]This chapter is not meant to show "how to do it" as much as it is to present the conventions and symbols of engineering drawing.
[2]The technique for constructing a normal view is covered in engineering drafting texts.

[3]The omission of perspective from a drawing is an example of a step taken to simplify the drawing process.

The different types of views are easily distinguished by their *projectors* (i.e., projections of parallel lines on the object). For a cube, there are three sets of projectors corresponding to the three perpendicular axes. In an *orthographic (orthogonal) view*, the projectors are parallel. In a *perspective (central) view*, some or all of the projectors converge to a point.

Figure 2.2 *Orthographic and Perspective Views of a Block*

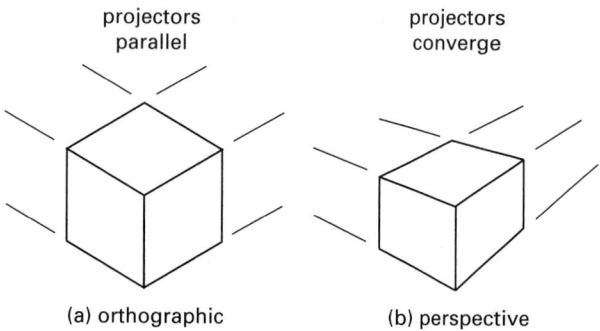

projectors parallel

projectors converge

(a) orthographic (b) perspective

4. PRINCIPAL (ORTHOGRAPHIC) VIEWS

In a *principal view* (also known as a *planar view*), one of the sets of projectors is normal to the view. That is, one of the planes of the object is seen in a normal view. The other two sets of projectors are orthogonal and are usually oriented horizontally and vertically on the paper. Because background details of an object may not be visible in a principal view, it is necessary to have at least three principal views to completely illustrate a symmetrical object. At most, six principal views will be needed to illustrate complex objects.

The relative positions of the six views have been standardized and are shown in Fig. 2.3, which also defines the *width* (also known as *depth*), *height*, and *length* of the object. The views that are not needed to illustrate features or provide dimensions (i.e., *redundant views*) can be omitted. The usual combination selected consists of the top, front, and right side views.

Figure 2.3 *Positions of Standard Orthographic Views*

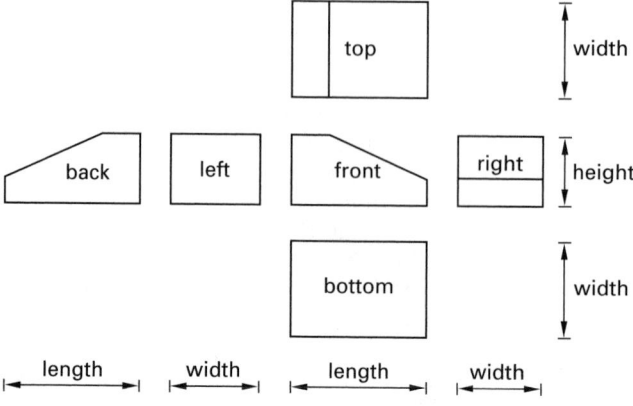

top

back left front right

bottom

width

height

width

length width length width

It is common to refer to the front, side, and back views as *elevations* and to the top and bottom views as *plan views*. These terms are not absolute since any plane can be selected as the front.

5. AUXILIARY (ORTHOGRAPHIC) VIEWS

An *auxiliary view* is needed when an object has an inclined plane or curved feature or when there are more details than can be shown in the six principal views. As with the other orthographic views, the auxiliary view is a normal (face-on) view of the inclined plane.

Figure 2.4 *Auxiliary View*

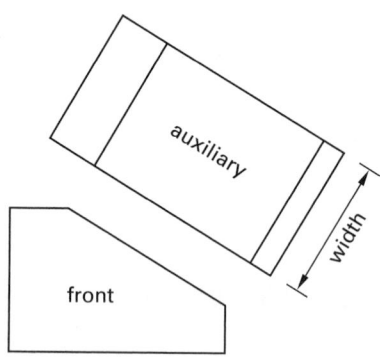

auxiliary

width

front

The projectors in an auxiliary view are perpendicular to only one of the directions in which a principal view is observed. Accordingly, only one of the three dimensions of width, height, and depth can be measured (scaled). In a *profile auxiliary view*, the object's width can be measured. In a *horizontal auxiliary view (auxiliary elevation)*, the object's height can be measured. In a *frontal auxiliary view*, the depth of the object can be measured.

6. OBLIQUE (ORTHOGRAPHIC) VIEWS

If the object is turned so that three principal planes are visible, it can be completely illustrated by a single *oblique view*.[4] In an oblique view, the direction from which the object is observed is not (necessarily) parallel to any of the directions from which principal and auxiliary views are observed.

In two common methods of oblique illustration, one of the view planes coincides with an orthographic view plane. Two of the drawing axes are at right angles to each other; one of these is vertical, and the other (the *oblique axis*) is oriented at 30° or 45° (originally chosen to coincide with standard drawing triangles). The ratio of scales used for the horizontal, vertical, and oblique axes can be 1:1:1 or 1:1:$\frac{1}{2}$. The latter ratio helps to overcome the visual distortion due to the absence of perspective in the oblique direction.

---

[4]Oblique views are not unique in this capability—perspective drawings share it. Oblique and perspective drawings are known as *pictorial drawings* because they give depth to the object by illustrating it in three dimensions.

Cavalier (45° oblique axis and 1:1:1 scale ratio) and *cabinet* (45° oblique axis and 1:1:$\frac{1}{2}$ scale ratio) *projections* are the two common types of oblique views that incorporate one of the orthographic views. If an angle of 9.5° is used (as in illustrating crystalline lattice structures), the technique is known as *clinographic projection*.

Figure 2.5 *Cavalier and Cabinet Oblique Drawings*

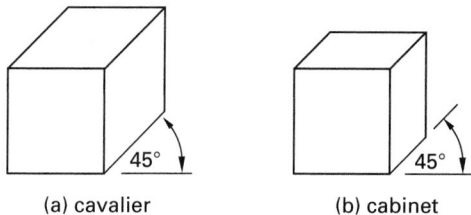

(a) cavalier   (b) cabinet

7. AXONOMETRIC (ORTHOGRAPHIC OBLIQUE) VIEWS

In axonometric views, the view plane is not parallel to any of the principal orthographic planes. Axonometric views and axonometric drawings are not the same. In a *view (projection)*, one or more of the face lengths is foreshortened. In a *drawing*, the lengths are drawn full length, resulting in a distorted illustration. Table 2.2 lists the proper ratios that should be observed.

Table 2.2 *Axonometric Foreshortening*

view	projector intersection angles	proper ratio of sides
isometric	120°, 120°, 120°	0.82:0.82:0.82
dimetric	131°25′, 131°25′, 97°10′	1:1:$\frac{1}{2}$
	103°38′, 103°38′, 152°44′	$\frac{3}{4}$:$\frac{3}{4}$:1
trimetric	102°28′, 144°16′, 113°16′	1:$\frac{2}{3}$:$\frac{7}{8}$
	138°14′, 114°46′, 107°	1:$\frac{3}{4}$:$\frac{7}{8}$

In an *isometric view*, the three projectors intersect at equal angles (120°) with the plane. This simplifies construction with standard 30° drawing triangles. All of the faces are foreshortened an equal amount, to $\sqrt{2/3}$, or approximately 81.6% of the true length. In a *dimetric view*, two of the projectors intersect at equal angles, and only two of the faces are equally reduced in length. In a *trimetric view*, all three intersection angles are different, and all three faces are reduced different amounts.

8. PERSPECTIVE VIEWS

In a *perspective view*, one or more sets of projectors converge to a fixed point known as the *center of vision*. In the *parallel perspective*, all vertical lines remain vertical in the picture; all horizontal frontal lines remain horizontal. Therefore, one face is parallel to the observer and only one set of projectors converges. In the *angular perspective*, two sets of projectors converge. In the little-used *oblique perspective*, all three sets of projectors converge.

Figure 2.6 *Types of Axonometric Views*

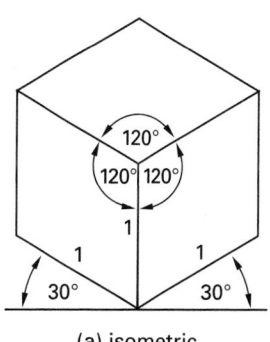

(a) isometric

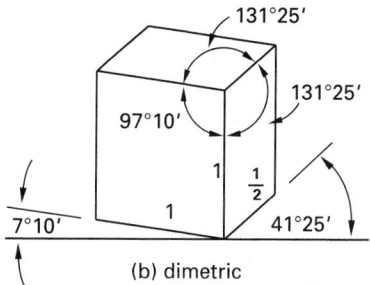

(b) dimetric

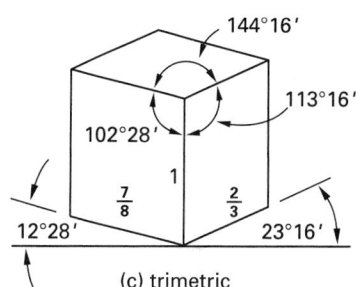

(c) trimetric

Figure 2.7 *Types of Perspective Views*

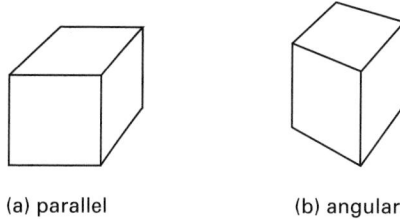

(a) parallel   (b) angular

9. SECTIONS

The term *section* is an imaginary cut taken through an object to reveal the shape or interior construction.[5] Figure 2.8 illustrates the standard symbol for a *sectioning cut* and the resulting sectional view. Section arrows are perpendicular to the cutting plane and indicate the viewing direction.

---

[5]The term *section* is also used to mean a *cross section*—a slice of finite but negligible thickness that is taken from an object to show the cross section or interior construction at the plane of the slice.

Figure 2.8 *Sectioning Cut Symbol and Sectional View*

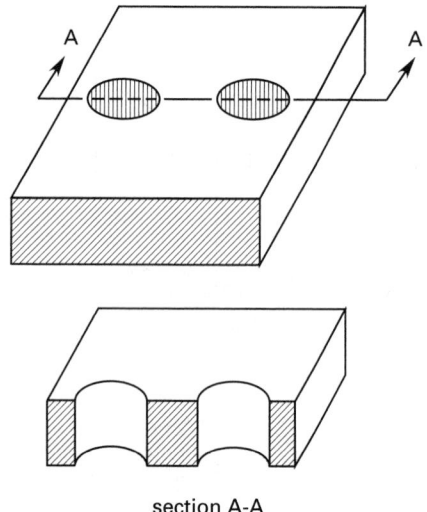

section A-A

10. TOLERANCES

The *tolerance* for a dimension is the total permissible variation or difference between the acceptable limits. The tolerance for a dimension can be specified in two ways: either as a general rule in the title block (e.g., ± 0.001 in unless otherwise specified) or as specific limits given with each dimension (e.g., 2.575 in ± 0.005 in).

11. SURFACE FINISH

ANSI B46.1 specifies surface finish by a combination of parameters.[6] The basic symbol for designating these factors is shown in Fig. 2.9. In the symbol, A is the maximum *roughness height index*, B is the optional minimum *roughness height*, C is the peak-to-valley *waviness height*, D is the optional peak-to-valley *waviness spacing (width)* rating, E is the optional *roughness width cutoff (roughness sampling length)*, F is the *lay*, and G is the *roughness width*. Unless minimums are specified, all parameters are maximum allowable values, and all lesser values are permitted.

Since the roughness varies, the waviness height is an arithmetic average within a sampled square, and the designation A is known as the *roughness weight*, R_a.[7] Values are normally given in microns (μm) or microinches (μin) in SI or customary U.S. units, respectively. A value for the roughness width cutoff of 0.80 mm (0.03 in) is assumed when E is not specified. Other standard values in common use are 0.25 mm (0.010 in) and 0.08 mm (0.003 in). The lay symbol, F, can be = (parallel to indicated surface), $\perp$ (perpendicular), C (circular), M (multidirectional), P (pitted), R (radial), or X (crosshatch).

If a small circle is placed at the A position, no machining is allowed and only cast, forged, die-cast, injection-molded, and other unfinished surfaces are acceptable.

[6]Specification does not indicate appearance (i.e., color, luster) or performance (i.e., hardness, corrosion resistance, microstructure).
[7]The symbol R_a is the same as the AA (arithmetic average) and CLA (centerline average) terms used in other (and earlier) standards.

Figure 2.9 *Surface Finish Designations*

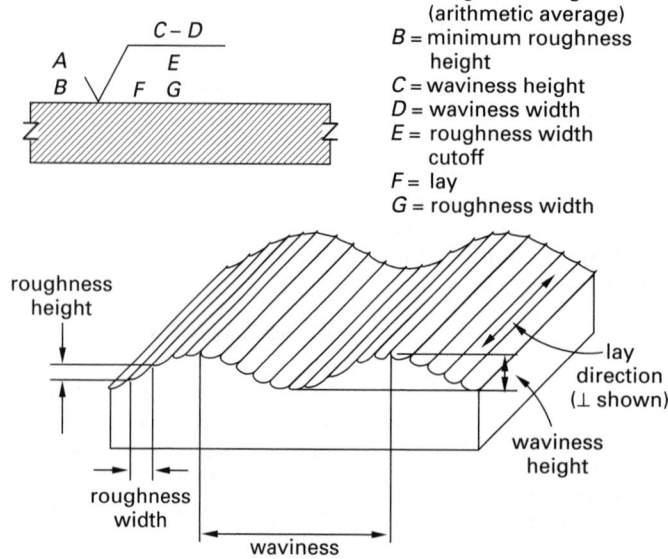

A = roughness height (arithmetic average)
B = minimum roughness height
C = waviness height
D = waviness width
E = roughness width cutoff
F = lay
G = roughness width

3 Algebra

1. INTRODUCTION

Engineers working in design and analysis encounter mathematical problems on a daily basis. Although algebra and simple trigonometry are often sufficient for routine calculations, there are many instances when certain advanced subjects are needed. This chapter and the following, in addition to supporting the calculations used in other chapters, consolidate the mathematical concepts most often needed by engineers.

2. SYMBOLS USED IN THIS BOOK

Many symbols, letters, and Greek characters are used to represent variables in the formulas used throughout this book. These symbols and characters are defined in the nomenclature section of each chapter. However, some of the other symbols in this book are listed in Table 3.2.

3. GREEK ALPHABET

Table 3.1 The Greek Alphabet

A	α	alpha	N	ν	nu
B	β	beta	Ξ	ξ	xi
Γ	γ	gamma	O	o	omicron
Δ	δ	delta	Π	π	pi
E	ϵ	epsilon	P	ρ	rho
Z	ζ	zeta	Σ	σ	sigma
H	η	eta	T	τ	tau
Θ	θ	theta	Υ	υ	upsilon
I	ι	iota	Φ	ϕ	phi
K	κ	kappa	X	χ	chi
Λ	λ	lambda	Ψ	ψ	psi
M	μ	mu	Ω	ω	omega

4. TYPES OF NUMBERS

The *numbering system* consists of three types of numbers: real, imaginary, and complex. *Real numbers*, in turn, consist of rational numbers and irrational numbers. *Rational real numbers* are numbers that can be written as the ratio of two integers (e.g., 4, $^2/_5$, and $^1/_3$).[1] *Irrational real numbers* are nonterminating, nonrepeating numbers that cannot be expressed as the ratio of two integers (e.g., π and $\sqrt{2}$). Real numbers can be positive or negative.

Imaginary numbers are square roots of negative numbers. The symbols i and j are both used to represent the square root of -1.[2] For example, $\sqrt{-5} = \sqrt{5}\sqrt{-1} = \sqrt{5}i$. *Complex numbers* consist of combinations of real and imaginary numbers (e.g., $3 - 7i$).

5. SIGNIFICANT DIGITS

The significant digits in a number include the leftmost, nonzero digits to the rightmost digit written. Final answers from computations should be rounded off to the number of decimal places justified by the data. The answer can be no more accurate than the least accurate number in the data. Of course, rounding should be done on final calculation results only. It should not be done on interim results.

---

[1] Notice that 0.3333333 is a nonterminating number, but as it can be expressed as a ratio of two integers (i.e., $^1/_3$), it is a rational number.

[2] The symbol j is used to represent the square root of -1 in electrical calculations to avoid confusion with the current variable, i.

Table 3.2 *Symbols Used in This Book*

symbol	name	use	example
$\sum$	sigma	series summation	$\sum_{i=1}^{3} x_i = x_1 + x_2 + x_3$
π	pi	$3.1415927\ldots$	$p = \pi D$
e	base of natural logs	$2.71828\ldots$	
$\prod$	pi	series multiplication	$\prod_{i=1}^{3} x_i = x_1 x_2 x_3$
Δ	delta	change in quantity	$\Delta h = h_2 - h_1$
$-$	over bar	average value	$\overline{x}$
$\cdot$	over dot	per unit time	$\dot{m} = $ mass flowing per second
$!$	factorial[a]		$x! = x(x-1)(x-2)\cdots(2)(1)$
$\vert\ \vert$	absolute value[b]		$\vert -3\vert = +3$
$\approx$	approximately equal to		$x \approx 1.5$
$\equiv$	equivalent to		$a + bi \equiv re^{i\theta}$
$\propto$	proportional to		$x \propto y$
∞	infinity		$x \to \infty$
log	base 10 logarithm		$\log(5.74)$
ln	natural logarithm		$\ln(5.74)$
exp	exponential power		$\exp(x) = e^x$
rms	root-mean-square	$\sqrt{\dfrac{1}{n}\sum_{i=1}^{n} x_i^2}$	V_{rms}
$\angle$	phasor or angle		$\angle 53°$

[a] *Zero factorial* ($0!$) is frequently encountered in the form of $(n-n)!$ when calculating permutations and combinations. Zero factorial is defined as 1.
[b] The notation $\text{abs}(x)$ is also used to indicate the absolute value.

Table 3.3 *Examples of Significant Digits*

number as written	number of significant digits	implied range
341	3	340.5 to 341.5
34.1	3	34.05 to 34.15
0.00341	3	0.003405 to 0.003415
341×10^7	3	340.5×10^7 to 341.5×10^7
3.41×10^{-2}	3	3.405×10^{-2} to 3.415×10^{-2}
3410	3	3405 to 3415
3410[a]	4	3409.5 to 3410.5
341.0	4	340.95 to 341.05

[a] It is permitted to write "3410." to distinguish the number from its 3-significant digit form, although this is rarely done.

There are two ways that significant digits can affect calculations. For the operations of multiplication and division, the final answer is rounded to the number of significant digits in the least significant multiplicand, divisor, or dividend. So, $2.0 \times 13.2 = 26$ since the first multiplicand (2.0) has two significant digits only.

For the operations of addition and subtraction, the final answer is rounded to the position of the least significant digit in the addenda, minuend, or subtrahend. So, $2.0 + 13.2 = 15.2$ because both addenda are significant to the tenth's position; but $2 + 13.4 = 15$ since the 2 is significant only in the ones' position.

The multiplication rule should not be used for addition or subtraction, as this can result in strange answers. For example, it would be incorrect to round $1700 + 0.1$ to 2000 simply because 0.1 has only one significant digit.

6. EQUATIONS

An *equation* is a mathematical statement of equality, such as $5 = 3 + 2$. *Algebraic equations* are written in terms of *variables*. In the equation $y = x^2 + 3$, the value of variable y depends on the value of variable x. Therefore, y is the *dependent variable* and x is the *independent variable*. The dependency of y on x is clearer when the equation is written in *functional form*: $y = f(x)$.

A *parametric equation* uses one or more independent variables (*parameters*) to describe a function.[3] For

[3] As used in this section, there is no difference between a parameter and an independent variable. However, the term *parameter* is also used as a descriptive measurement that determines or characterizes the form, size, or content of a function. For example, the radius is a parameter of a circle, and mean and variance are parameters of a probability distribution. Once these parameters are specified, the function is completely defined.

example, the parameter θ can be used to write the parametric equations of a unit circle.

$$x = \cos\theta \qquad \textit{3.1}$$
$$y = \sin\theta \qquad \textit{3.2}$$

A unit circle can also be described by a *nonparametric equation*.[4]

$$x^2 + y^2 = 1 \qquad \textit{3.3}$$

7. FUNDAMENTAL ALGEBRAIC LAWS

Algebra provides the rules that allow complex mathematical relationships to be expanded or condensed. Algebraic laws may be applied to complex numbers, variables, and real numbers. The general rules for changing the form of a mathematical relationship are given as follows.

- commutative law for addition:
$$A + B = B + A \qquad \textit{3.4}$$
- commutative law for multiplication:
$$AB = BA \qquad \textit{3.5}$$
- associative law for addition:
$$A + (B + C) = (A + B) + C \qquad \textit{3.6}$$
- associative law for multiplication:
$$A(BC) = (AB)C \qquad \textit{3.7}$$
- distributive law:
$$A(B + C) = AB + AC \qquad \textit{3.8}$$

8. POLYNOMIALS

A *polynomial* is a rational expression—usually the sum of several variable terms known as *monomials*—that does not involve division. The *degree of the polynomial* is the highest power to which a variable in the expression is raised. The following *standard polynomial forms* are useful when trying to find the roots of an equation.

$$(a + b)(a - b) = a^2 - b^2 \qquad \textit{3.9}$$
$$(a \pm b)^2 = a^2 \pm 2ab + b^2 \qquad \textit{3.10}$$
$$(a \pm b)^3 = a^3 \pm 3a^2b + 3ab^2 \pm b^3 \qquad \textit{3.11}$$
$$(a^3 \pm b^3) = (a \pm b)(a^2 \mp ab + b^2) \qquad \textit{3.12}$$
$$(a^n - b^n) = (a - b)(a^{n-1} + a^{n-2}b + a^{n-3}b^2 + \cdots$$
$$+ b^{n-1}) \quad [n \text{ is any positive integer}] \quad \textit{3.13}$$
$$(a^n + b^n) = (a + b)(a^{n-1} - a^{n-2}b + a^{n-3}b^2 - \cdots$$
$$+ b^{n-1}) \quad [n \text{ is any positive odd integer}]$$
$$\textit{3.14}$$

The *binomial theorem* defines a polynomial of the form $(a + b)^n$.

$$(a + b)^n = \underset{[i=0]}{a^n} + \underset{[i=1]}{na^{n-1}b} + \underset{[i=2]}{C_2a^{n-2}b^2} + \cdots$$
$$+ C_ia^{n-i}b^i + \cdots + nab^{n-1} + b^n \qquad \textit{3.15}$$
$$C_i = \frac{n!}{i!(n-i)!} \qquad [i = 0,1,2,\ldots,n] \qquad \textit{3.16}$$

----

[4]Since only the coordinate variables are used, this equation is also said to be in *Cartesian equation form*.

The coefficients of the expansion can be determined quickly from *Pascal's triangle*, Fig. 3.1. Notice that each entry is the sum of the two entries directly above it.

Figure 3.1 *Pascal's Triangle*

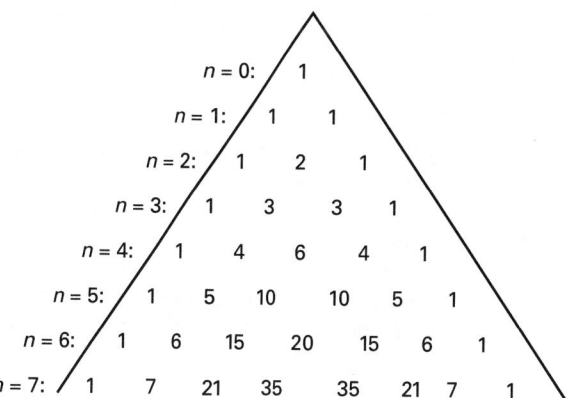

The values $r_1, r_2, \ldots, r_n$ of the independent variable x that satisfy a polynomial equation $f(x) = 0$ are known as *roots* or *zeros* of the polynomial. A polynomial of degree n with real coefficients will have at most n real roots, although they need not all be distinctly different.

9. ROOTS OF QUADRATIC EQUATIONS

A *quadratic equation* is an equation of the general form $ax^2 + bx + c = 0$ $[a \neq 0]$. The *roots*, x_1 and x_2, of the equation are the two values of x that satisfy it.

$$x_1, x_2 = \frac{-b \pm \sqrt{b^2 - 4ac}}{2a} \qquad \textit{3.17}$$
$$x_1 + x_2 = -\frac{b}{a} \qquad \textit{3.18}$$
$$x_1x_2 = \frac{c}{a} \qquad \textit{3.19}$$

The types of roots of the equation can be determined from the *discriminant* (i.e., the quantity under the radical in Eq. 3.17).

- If $(b^2 - 4ac) > 0$, the roots are real and unequal.
- If $(b^2 - 4ac) = 0$, the roots are real and equal. This is known as a *double root*.
- If $(b^2 - 4ac) < 0$, the roots are complex and unequal.

10. ROOTS OF GENERAL POLYNOMIALS

It is more difficult to find roots of cubic and higher-degree polynomials because few general techniques exist.

- *inspection*: Finding roots by inspection is equivalent to making reasonable guesses about the roots and substituting into the polynomial.

- *graphing*: If the value of a polynomial $f(x)$ is calculated and plotted for different values of x, an approximate value of a root can be determined as the value of x at which the plot crosses the x-axis.

- *numerical methods*: If an approximate value of a root is known, numerical methods (bisection method, Newton's method, etc.) can be used to refine the value. The more efficient techniques are too complex to be performed by hand.

- *factoring*: If at least one root (say, $x = r$) of a polynomial $f(x)$ is known, the quantity $(x - r)$ can be factored out of $f(x)$ by long division. The resulting quotient will be lower by one degree, and the remaining roots may be easier to determine. This method is particularly applicable if the polynomial is in one of the standard forms presented in Sec. 8.

- *special cases*: Certain polynomial forms can be simplified by substitution or solved by standard formulas if they are recognized as being special cases. (The standard solution to the quadratic equation is such a special case.) For example, $ax^4 + bx^2 + c = 0$ can be reduced to a polynomial of degree 2 if the substitution $u = x^2$ is made.

11. EXTRANEOUS ROOTS

With simple equalities, it may appear possible to derive roots by basic algebraic manipulations.[5] However, multiplying each side of an equality by a power of a variable may introduce *extraneous roots*. Such roots do not satisfy the original equation even though they are derived according to the rules of algebra. Checking a calculated root is always a good idea, but is particularly necessary if the equation has been multiplied by one of its own variables.

Example 3.1

Use algebraic operations to determine a value that satisfies the following equation. Determine if the value is a valid or extraneous root.

$$\sqrt{x - 2} = \sqrt{x} + 2$$

Solution

Square both sides.

$$x - 2 = x + 4\sqrt{x} + 4$$

Subtract x from each side, and combine the constants.

$$4\sqrt{x} = -6$$

---

[5]In this sentence, *equality* means a combination of two expressions containing an equal sign. Any two expressions can be linked in this manner, even those that are not actually equal. For example, the expressions for two non-intersecting ellipses can be equated even though there is no intersection point. Finding extraneous roots is more likely when the underlying equality is false to begin with.

Solve for x.

$$x = \left(\tfrac{-6}{4}\right)^2 = \tfrac{9}{4}$$

Substitute $x = 9/4$ into the original equation.

$$\sqrt{\tfrac{9}{4} - 2} = \sqrt{\tfrac{9}{4}} + 2$$
$$\tfrac{1}{2} = \tfrac{7}{2}$$

Since the equality is not established, $x = {}^9\!/_4$ is an extraneous root.

12. DESCARTES' RULE OF SIGNS

Descartes' rule of signs determines the maximum number of positive (and negative) real roots that a polynomial will have by counting the number of sign reversals (i.e., changes in sign from one term to the next) in the polynomial. The polynomial $f(x)$ must have real coefficients and must be arranged in terms of descending powers of x.

- The number of positive roots of the polynomial equation $f(x) = 0$ will not exceed the number of sign reversals.

- The difference between the number of sign reversals and the number of positive roots is an even number.

- The number of negative roots of the polynomial equation $f(x) = 0$ will not exceed the number of sign reversals in the polynomial $f(-x)$.

- The difference between the number of sign reversals in $f(-x)$ and the number of negative roots is an even number.

Example 3.2

Determine the possible numbers of positive and negative roots that satisfy the following polynomial equation.

$$4x^5 - 5x^4 + 3x^3 - 8x^2 - 2x + 3 = 0$$

Solution

There are four sign reversals, so up to four positive roots exist. To keep the difference between the number of positive roots and the number of sign reversals an even number, the number of positive real roots is limited to zero, two, and four.

Substituting $-x$ for x in the polynomial results in

$$-4x^5 - 5x^4 - 3x^3 - 8x^2 + 2x + 3 = 0$$

There is only one sign reversal, so the number of negative roots cannot exceed one. There must be exactly one negative real root in order to keep the difference to an even number (zero in this case).

13. RULES FOR EXPONENTS AND RADICALS

In the expression $b^n = a$, b is known as the *base* and n is the *exponent* or *power*. In Eqs. 3.20 through 3.33, a, b, m, and n are any real numbers with limitations listed.

$$b^0 = 1 \qquad [b \neq 0] \qquad 3.20$$

$$b^1 = b \qquad 3.21$$

$$b^{-n} = \frac{1}{b^n} = \left(\frac{1}{b}\right)^n \qquad [b \neq 0] \qquad 3.22$$

$$\left(\frac{a}{b}\right)^n = \frac{a^n}{b^n} \qquad [b \neq 0] \qquad 3.23$$

$$(ab)^n = a^n b^n \qquad 3.24$$

$$b^{m/n} = \sqrt[n]{b^m} = \left(\sqrt[n]{b}\right)^m \qquad 3.25$$

$$(b^n)^m = b^{nm} \qquad 3.26$$

$$b^m b^n = b^{m+n} \qquad 3.27$$

$$\frac{b^m}{b^n} = b^{m-n} \qquad [b \neq 0] \qquad 3.28$$

$$\sqrt[n]{b} = b^{1/n} \qquad 3.29$$

$$\left(\sqrt[n]{b}\right)^n = \left(b^{1/n}\right)^n = b \qquad 3.30$$

$$\sqrt[n]{ab} = \sqrt[n]{a}\,\sqrt[n]{b} = a^{1/n} b^{1/n} = (ab)^{1/n} \qquad 3.31$$

$$\sqrt[n]{\frac{a}{b}} = \frac{\sqrt[n]{a}}{\sqrt[n]{b}} = \left(\frac{a}{b}\right)^{1/n} \qquad [b \neq 0] \qquad 3.32$$

$$\sqrt[m]{\sqrt[n]{b}} = \sqrt[mn]{b} = b^{1/mn} \qquad 3.33$$

14. LOGARITHMS

Logarithms can be considered to be exponents. For example, the exponent n in the expression $b^n = a$ is the logarithm of a to the base b. Therefore, the two expressions $\log_b a = n$ and $b^n = a$ are equivalent.

The base for *common logs* is 10. Usually, "log" will be written when common logs are desired, although "$\log_{10}$" appears occasionally. The base for *natural (Napierian) logs* is $2.71828\ldots$, a number which is given the symbol e. When natural logs are desired, usually "ln" will be written, although "$\log_e$" is also used.

Most logarithms will contain an integer part (the *characteristic*) and a decimal part (the *mantissa*). The common and natural logarithms of any number less than one are negative. If the number is greater than one, its common and natural logarithms are positive. Although the logarithm may be negative, the mantissa is always positive. For negative logarithms, the characteristic is found by expressing the logarithm as the sum of a negative characteristic and a positive mantissa.

For common logarithms of numbers greater than one, the characteristics will be positive and equal to one less than the number of digits in front of the decimal. If the number is less than one, the characteristic will be negative and equal to one more than the number of zeros immediately following the decimal point.

If a negative logarithm is to be used in a calculation, it must first be converted to *operational form* by adding the characteristic and mantissa. The operational form should be used in all calculations and is the form displayed by scientific calculators.

The logarithm of a negative number is a complex number.

Example 3.3

Use logarithm tables to determine the operational form of $\log_{10}(0.05)$.

Solution

Since the number is less than one and there is one leading zero, the characteristic is found by observation to be -2. From a book of logarithm tables, the mantissa of 5.0 is 0.699. Two ways of combining the mantissa and characteristic are possible.

method 1: $\overline{2}.699$

method 2: $8.699 - 10$

The operational form of this logarithm is $-2 + 0.699 = -1.301$.

15. LOGARITHM IDENTITIES

Prior to the widespread availability of calculating devices, logarithm identities were used to solve complex calculations by reducing the solution method to table look-up, addition, and subtraction. Logarithm identities are still useful in simplifying expressions containing exponentials and other logarithms. In Eqs. 3.34 through 3.45, $a \neq 1$, $b \neq 1$, $x > 0$, and $y > 0$.

$$\log_b(b) = 1 \qquad 3.34$$

$$\log_b(1) = 0 \qquad 3.35$$

$$\log_b(b^n) = n \qquad 3.36$$

$$\log(x^a) = a \log(x) \qquad 3.37$$

$$\log\left(\sqrt[n]{x}\right) = \log\left(x^{1/n}\right) = \frac{\log(x)}{n} \qquad 3.38$$

$$b^{n \log_b(x)} = x^n = \text{antilog}[n \log_b(x)] \qquad 3.39$$

$$b^{\log_b(x)/n} = x^{1/n} \qquad 3.40$$

$$\log(xy) = \log(x) + \log(y) \qquad 3.41$$

$$\log\left(\frac{x}{y}\right) = \log(x) - \log(y) \qquad 3.42$$

$$\log_a(x) = \log_b(x) \log_a(b) \qquad 3.43$$

$$\ln(x) = \log_{10}(x) \ln(10) \approx 2.3026 \log_{10}(x) \qquad 3.44$$

$$\log_{10}(x) = \ln(x) \log_{10}(e) \approx 0.4343 \ln(x) \qquad 3.45$$

Example 3.4

The surviving fraction, x, of a radioactive isotope is given by $x = e^{-0.005t}$. For what value of t will the surviving percentage be 7%?

Solution

$$x = 0.07 = e^{-0.005t}$$

Take the natural log of both sides.

$$\ln(0.07) = \ln\left(e^{-0.005t}\right)$$

From 3.36, $\ln e^x = x$. Therefore,

$$-2.66 = -0.005t$$

$$t = 532$$

16. PARTIAL FRACTIONS

The method of *partial fractions* is used to transform a proper polynomial fraction of two polynomials into a sum of simpler expressions, a procedure known as *resolution*.[6,7] The technique can be considered to be the act of "unadding" a sum to obtain all of the addends.

Suppose $H(x)$ is a proper polynomial fraction of the form $P(x)/Q(x)$. The object of the resolution is to determine the partial fractions u_1/v_1, u_2/v_2, etc., such that

$$H(x) = \frac{P(x)}{Q(x)} = \frac{u_1}{v_1} + \frac{u_2}{v_2} + \frac{u_3}{v_3} + \cdots \qquad 3.46$$

The form of the denominator polynomial $Q(x)$ will be the main factor in determining the form of the partial fractions. The task of finding the u_i and v_i is simplified by categorizing the possible forms of $Q(x)$.

case 1: $Q(x)$ factors into n different linear terms.

$$Q(x) = (x - a_1)(x - a_2)\cdots(x - a_n) \qquad 3.47$$

Then,

$$H(x) = \sum_{i=1}^{n} \frac{A_i}{x - a_i} \qquad 3.48$$

case 2: $Q(x)$ factors into n identical linear terms.

$$Q(x) = (x - a)(x - a)\cdots(x - a) \qquad 3.49$$

Then,

$$H(x) = \sum_{i=1}^{n} \frac{A_i}{(x - a)^i} \qquad 3.50$$

---

[6]To be a *proper polynomial fraction*, the degree of the numerator must be less than the degree of the denominator. If the polynomial fraction is improper, the denominator can be divided into the numerator to obtain whole and fractional polynomials. The method of partial fractions can then be used to reduce the fractional polynomial.

[7]This technique is particularly useful for calculating integrals and inverse Laplace transforms in subsequent chapters.

case 3: $Q(x)$ factors into n different quadratic terms, $x^2 + p_i x + q_i$. Then,

$$H(x) = \sum_{i=1}^{n} \frac{A_i x + B_i}{x^2 + p_i x + q_i} \qquad 3.51$$

case 4: $Q(x)$ factors into n identical quadratic terms, $x^2 + px + q$. Then,

$$H(x) = \sum_{i=1}^{n} \frac{A_i x + B_i}{(x^2 + px + q)^i} \qquad 3.52$$

Once the general forms of the partial fractions have been determined from inspection, the *method of undetermined coefficients* is used. The partial fractions are all cross-multiplied to obtain $Q(x)$ as the denominator, and the coefficients are found by equating $P(x)$ and the cross-multiplied numerator.

Example 3.5

Resolve $H(x)$ into partial fractions.

$$H(x) = \frac{x^2 + 2x + 3}{x^4 + x^3 + 2x^2}$$

Solution

Here, $Q(x) = x^4 + x^3 + 2x^2$, which factors into $x^2(x^2 + x + 2)$. This is a combination of cases 2 and 3.

$$H(x) = \frac{A_1}{x} + \frac{A_2}{x^2} + \frac{A_3 + A_4 x}{x^2 + x + 2}$$

Cross-multiplying to obtain a common denominator yields

$$\frac{(A_1 + A_4)x^3 + (A_1 + A_2 + A_3)x^2 + (2A_1 + A_2)x + 2A_2}{x^4 + x^3 + 2x^2}$$

Since the original numerator is known, the following simultaneous equations result.

$$A_1 + A_4 = 0$$

$$A_1 + A_2 + A_3 = 1$$

$$2A_1 + A_2 = 2$$

$$2A_2 = 3$$

The solutions are $A_1 = 0.25$, $A_2 = 1.5$, $A_3 = -0.75$, and $A_4 = -0.25$.

$$H(x) = \frac{1}{4x} + \frac{3}{2x^2} - \frac{x + 3}{(4)(x^2 + x + 2)}$$

17. SIMULTANEOUS LINEAR EQUATIONS

A *linear equation* with n variables is a polynomial of degree 1 describing a geometric shape in n-space. A *homogeneous linear equation* is one that has no constant term, and a *nonhomogeneous linear equation* has a constant term.

A solution to a set of simultaneous linear equations represents the intersection point of the geometric shapes in n-space. For example, if the equations are limited to two variables (e.g., $y = 4x - 5$), they describe straight lines. The solution to two simultaneous linear equations in 2-space is the point where the two lines intersect. The set of the two equations is said to be a *consistent system* when there is such an intersection.[8]

Simultaneous equations do not always have unique solutions, and some have none at all. In addition to crossing in 2-space, lines can be parallel or they can be the same line expressed in a different equation format (i.e., dependent equations). In some cases, parallelism and dependency can be determined by inspection. In most cases, however, matrix and other advanced methods must be used to determine whether a solution exists. A set of linear equations with no simultaneous solution is known as an *inconsistent system*.

Several methods exist for solving linear equations simultaneously by hand.[9]

- *graphing*: The equations are plotted and the intersection point is read from the graph. This method is possible only with two-dimensional problems.

- *substitution*: An equation is rearranged so that one variable is expressed as a combination of the other variables. The expression is then substituted into the remaining equations wherever the selected variable appears.

- *reduction*: All terms in the equations are multiplied by constants chosen to eliminate one or more variables when the equations are added or subtracted. The remaining sum can then be solved for the other variables. This method is also known as *eliminating the unknowns*.

- *Cramer's rule*: This method is covered in Ch. 4.

Example 3.6

Solve the following set of linear equations by (a) substitution and (b) reduction.

$$2x + 3y = 12 \quad \text{[Eq. 1]}$$
$$3x + 4y = 8 \quad \text{[Eq. 2]}$$

[8]A homogeneous system always has at least one solution: the *trivial solution*, in which all variables have a value of zero.
[9]Other methods exist, but they require a computer.

Solution

(a) From Eq. 1, solve for variable x.

$$x = 6 - 1.5y \quad \text{[Eq. 3]}$$

Substitute $6 - 1.5y$ into Eq. 2 wherever x appears.

$$(3)(6 - 1.5y) + 4y = 8$$
$$18 - 4.5y + 4y = 8$$
$$y = 20$$

Substitute 20 for y in Eq. 3.

$$x = 6 - (1.5)(20) = -24$$

The solution $(-24, 20)$ should be checked to verify that it satisfies both original equations.

(b) Eliminate variable x by multiplying Eq. 1 by 3 and Eq. 2 by 2.

$$3 \times \text{Eq. 1: } 6x + 9y = 36 \quad \text{[Eq. 1']}$$
$$2 \times \text{Eq. 2: } 6x + 8y = 16 \quad \text{[Eq. 2']}$$

Subtract Eq. 2' from Eq. 1'.

$$y = 20 \quad \text{[Eq. 1' - Eq. 2']}$$

Substitute $y = 20$ into Eq. 1'.

$$6x + (9)(20) = 36$$
$$x = -24$$

The solution $(-24, 20)$ should be checked to verify that it satisfies both original equations.

18. COMPLEX NUMBERS

A *complex number*, $\mathbf{Z}$, is a combination of real and imaginary numbers. When expressed as a sum (e.g., $a + bi$), the complex number is said to be in *rectangular* or *trigonometric form*. The complex number can be plotted on the real-imaginary coordinate system known as the *complex plane*, as illustrated in Fig. 3.2.

Figure 3.2 A Complex Number in the Complex Plane

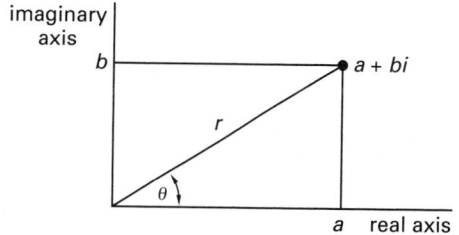

The complex number $\mathbf{Z} = a + bi$ can also be expressed in *exponential form*.[10] The quantity r is known as the *modulus* of $\mathbf{Z}$; θ is the *argument*.

$$a + bi \equiv re^{i\theta} \qquad 3.53$$

$$r = \text{mod}(\mathbf{Z}) = \sqrt{a^2 + b^2} \qquad 3.54$$

$$\theta = \arg(\mathbf{Z}) = \arctan\left(\frac{b}{a}\right) \qquad 3.55$$

Similarly, the *phasor form* (also known as the *polar form*) is

$$\mathbf{Z} = r\angle\theta \qquad 3.56$$

The *rectangular form* can be determined from r and θ.

$$a = r\cos\theta \qquad 3.57$$

$$b = r\sin\theta \qquad 3.58$$

$$\mathbf{Z} = a + bi = r\cos\theta + ir\sin\theta$$
$$= r(\cos\theta + i\sin\theta) \qquad 3.59$$

The *cis form* is a shorthand method of writing a complex number in rectangular (trigonometric) form.

$$a + bi = r(\cos\theta + i\sin\theta) = r\,\text{cis}\,\theta \qquad 3.60$$

Euler's equation (Eq. 3.61) expresses the equality of complex numbers in exponential and trigonometric form.

$$e^{i\theta} = \cos\theta + i\sin\theta \qquad 3.61$$

Related expressions are

$$e^{-i\theta} = \cos\theta - i\sin\theta \qquad 3.62$$

$$\cos\theta = \frac{e^{i\theta} + e^{-i\theta}}{2} \qquad 3.63$$

$$\sin\theta = \frac{e^{i\theta} - e^{-i\theta}}{2i} \qquad 3.64$$

Example 3.7

What is the exponential form of the complex number $\mathbf{Z} = 3 + 4i$?

Solution

$$r = \sqrt{a^2 + b^2} = \sqrt{3^2 + 4^2} = \sqrt{25} = 5$$

$$\theta = \arctan\left(\frac{b}{a}\right) = \arctan\left(\frac{4}{3}\right) = 0.927 \text{ rad}$$

$$\mathbf{Z} = re^{i\theta} = 5e^{i(0.927)}$$

19. OPERATIONS ON COMPLEX NUMBERS

Most algebraic operations (addition, multiplication, exponentiation, etc.) work with complex numbers, but notable exceptions are the inequality operators. The concept of one complex number being less than or greater than another complex number is meaningless.

---

[10]The terms *polar form*, *phasor form*, and *exponential form* are all used somewhat interchangeably.

When adding two complex numbers, real parts are added to real parts, and imaginary parts are added to imaginary parts.

$$(a_1 + ib_1) + (a_2 + ib_2) = (a_1 + a_2) + i(b_1 + b_2) \qquad 3.65$$

$$(a_1 + ib_1) - (a_2 + ib_2) = (a_1 - a_2) + i(b_1 - b_2) \qquad 3.66$$

Multiplication of two complex numbers in rectangular form is accomplished by the use of the algebraic distributive law, remembering that $i^2 = -1$.

Division of complex numbers in rectangular form requires use of the *complex conjugate*. The complex conjugate of the complex number $(a + bi)$ is $(a - bi)$. By multiplying the numerator and the denominator by the complex conjugate, the denominator will be converted to the real number $a^2 + b^2$. This technique is known as *rationalizing* the denominator and is illustrated in Ex. 3.8(c).

Multiplication and division are often more convenient when the complex numbers are in exponential or phasor forms, as Eqs. 3.67 and 3.68 show.

$$(r_1 e^{i\theta_1})(r_2 e^{i\theta_2}) = r_1 r_2 e^{i(\theta_1 + \theta_2)} \qquad 3.67$$

$$\frac{r_1 e^{i\theta_1}}{r_2 e^{i\theta_2}} = \left(\frac{r_1}{r_2}\right) e^{i(\theta_1 - \theta_2)} \qquad 3.68$$

Taking powers and roots of complex numbers requires *de Moivre's theorem*, Eqs. 3.69 and 3.70.

$$\mathbf{Z}^n = (re^{i\theta})^n = r^n e^{in\theta} \qquad 3.69$$

$$\sqrt[n]{\mathbf{Z}} = (re^{i\theta})^{1/n} = \sqrt[n]{r}\,e^{i(\theta + k(360°)/n)}$$
$$[k = 0, 1, 2, \ldots n - 1] \qquad 3.70$$

Example 3.8

Perform the following complex arithmetic.

(a) $(3 + 4i) + (2 + i)$

(b) $(7 + 2i)(5 - 3i)$

(c) $\dfrac{2 + 3i}{4 - 5i}$

Solution

(a)
$$(3 + 4i) + (2 + i) = (3 + 2) + (4 + 1)i$$
$$= 5 + 5i$$

(b)
$$(7 + 2i)(5 - 3i) = (7)(5) - (7)(3i) + (2i)(5)$$
$$- (2i)(3i)$$
$$= 35 - 21i + 10i - 6i^2$$
$$= 35 - 21i + 10i - (6)(-1)$$
$$= 41 - 11i$$

(c) Multiply the numerator and denominator by the complex conjugate of the denominator.

$$\frac{2+3i}{4-5i} = \frac{(2+3i)(4+5i)}{(4-5i)(4+5i)}$$

$$= \frac{-7+22i}{(4)^2+(5)^2}$$

$$= \left(\frac{-7}{41}\right) + i\left(\frac{22}{41}\right)$$

20. LIMITS

A *limit* (*limiting value*) is the value a function approaches when an independent variable approaches a target value. For example, suppose the value of $y = x^2$ is desired as x approaches 5. This could be written as

$$\lim_{x \to 5} x^2 \qquad\qquad 3.71$$

The power of limit theory is wasted on simple calculations such as this but is appreciated when the function is undefined at the target value. The object of limit theory is to determine the limit without having to evaluate the function at the target. The general case of a limit evaluated as x approaches the target value a is written as

$$\lim_{x \to a} f(x) \qquad\qquad 3.72$$

It is not necessary for the actual value $f(a)$ to exist for the limit to be calculated. The function $f(x)$ may be undefined at point a. However, it is necessary that $f(x)$ be defined on both sides of point a for the limit to exist. If $f(x)$ is undefined on one side, or if $f(x)$ is discontinuous at $x = a$ (as in Fig. 3.3(c) and 3.3(d)), the limit does not exist.

Figure 3.3 *Existence of Limits*

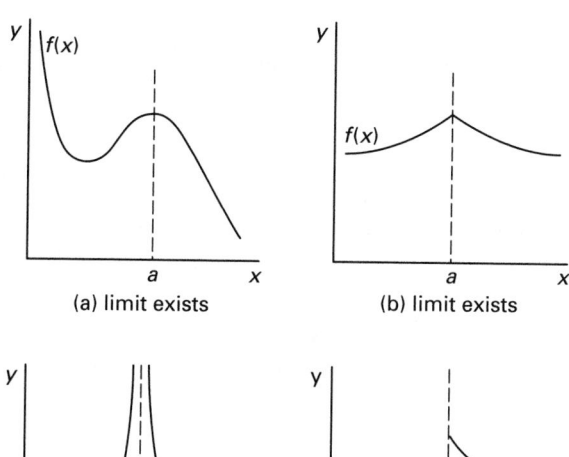

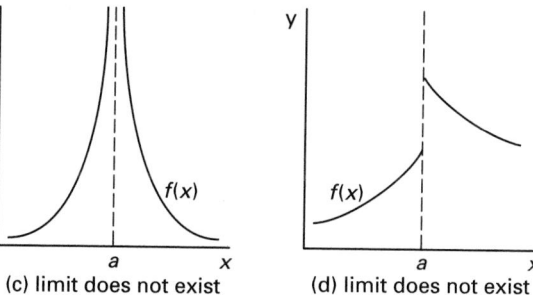

(a) limit exists (b) limit exists

(c) limit does not exist (d) limit does not exist

The following theorems can be used to simplify expressions when calculating limits.

$$\lim_{x \to a} x = a \qquad\qquad 3.73$$

$$\lim_{x \to a} (mx + b) = ma + b \qquad\qquad 3.74$$

$$\lim_{x \to a} b = b \qquad\qquad 3.75$$

$$\lim_{x \to a} (kF(x)) = k \lim_{x \to a} F(x) \qquad\qquad 3.76$$

$$\lim_{x \to a} \left(F_1(x) \begin{Bmatrix} + \\ - \\ \times \\ \div \end{Bmatrix} F_2(x) \right)$$

$$= \lim_{x \to a} (F_1(x)) \begin{Bmatrix} + \\ - \\ \times \\ \div \end{Bmatrix} \lim_{x \to a} (F_2(x)) \qquad 3.77$$

The following identities can be used to simplify limits of trigonometric expressions.

$$\lim_{x \to 0} \sin x = 0 \qquad\qquad 3.78$$

$$\lim_{x \to 0} \left(\frac{\sin x}{x} \right) = 1 \qquad\qquad 3.79$$

$$\lim_{x \to 0} \cos x = 1 \qquad\qquad 3.80$$

The following standard methods (tricks) can be used to determine limits.

- If the limit is taken to infinity, all terms can be divided by the largest power of x in the expression. This will leave at least one constant. Any quantity divided by a power of x vanishes as x approaches infinity.

- If the expression is a quotient of two expressions, any common factors should be eliminated from the numerator and denominator.

- *L'Hôpital's rule*, Eq. 3.81, should be used when the numerator and denominator of the expression both approach zero or both approach infinity.[11] $P^k(x)$ and $Q^k(x)$ are the kth derivatives of the functions $P(x)$ and $Q(x)$, respectively.[12] (L'Hôpital's rule can be applied repeatedly as required.)

$$\lim_{x \to a} \left(\frac{P(x)}{Q(x)} \right) = \lim_{x \to a} \left(\frac{P^k(x)}{Q^k(x)} \right) \qquad 3.81$$

---

[11]L'Hôpital's rule should not be used when only the denominator approaches zero. In that case, the limit approaches infinity regardless of the numerator.
[12]The calculation of derivatives is covered in Ch. 8.

Example 3.9

Evaluate the following limits.

(a) $\lim\limits_{x\to 3}\left(\dfrac{x^3-27}{x^2-9}\right)$

(b) $\lim\limits_{x\to\infty}\left(\dfrac{3x-2}{4x+3}\right)$

(c) $\lim\limits_{x\to 2}\left(\dfrac{x^2+x-6}{x^2-3x+2}\right)$

Solution

(a) Factor the numerator and denominator. (L'Hôpital's rule can also be used.)

$$\lim_{x\to 3}\left(\frac{x^3-27}{x^2-9}\right)=\lim_{x\to 3}\left(\frac{(x-3)(x^2+3x+9)}{(x-3)(x+3)}\right)$$

$$=\lim_{x\to 3}\left(\frac{x^2+3x+9}{x+3}\right)=\frac{(3)^2+(3)(3)+9}{3+3}$$

$$=\frac{9}{2}$$

(b) Divide through by the largest power of x. (L'Hôpital's rule can also be used.)

$$\lim_{x\to\infty}\left(\frac{3x-2}{4x+3}\right)=\lim_{x\to\infty}\left(\frac{3-\dfrac{2}{x}}{4+\dfrac{3}{x}}\right)$$

$$=\frac{3-\dfrac{2}{\infty}}{4+\dfrac{3}{\infty}}$$

$$=\frac{3-0}{4+0}=\frac{3}{4}$$

(c) Use L'Hôpital's rule. (Factoring can also be used.) Take the first derivative of the numerator and denominator.

$$\lim_{x\to 2}\left(\frac{x^2+x-6}{x^2-3x+2}\right)=\lim_{x\to 2}\left(\frac{2x+1}{2x-3}\right)$$

$$=\frac{(2)(2)+1}{(2)(2)-3}$$

$$=\frac{5}{1}=5$$

21. SEQUENCES AND PROGRESSIONS

A *sequence* is an ordered *progression* of numbers, a_i, such as $1,4,9,16,25,\ldots$ The *terms* in a sequence can be all positive, all negative, or of alternating signs. a_n is known as the *general term* of the sequence.

$$\{A\}=\{a_1,a_2,a_3,\ldots,a_n\} \qquad 3.82$$

A sequence is said to *diverge* (i.e., be *divergent*) if the terms approach infinity or if the terms fail to approach any finite value, and is said to *converge* (i.e.,

be *convergent*) if the terms approach any finite value (including zero). That is, the sequence converges if the limit defined by Eq. 3.83 exists.

$$\lim_{n\to\infty}a_n \begin{cases} \text{converges if } L \text{ is finite}\\ \text{diverges if } L \text{ is infinite}\\ \text{or does not exist} \end{cases} \qquad 3.83$$

The main task associated with a sequence is determining the next (or the general) term. If several terms of a sequence are known, the next (unknown) term must usually be found by intuitively determining the pattern of the sequence. In some cases, though, the method of *Rth-order differences* can be used to determine the next term. This method consists of subtracting each term from the following term to obtain a set of differences. If the differences are not all equal, the next order of differences can be calculated.

Example 3.10

What is the general term of the sequence A?

$$\{A\}=\left\{3,\frac{9}{2},\frac{27}{6},\frac{81}{24},\ldots\right\}$$

Solution

The solution is purely intuitive. The numerator is recognized as a power series based on the number 3. The denominator is recognized as the factorial sequence. The general term is

$$a_n=\frac{3^n}{n!}$$

Example 3.11

Find the sixth term in the sequence $\{7,16,29,46,67,a_6\}$.

Solution

The sixth term is not intuitively obvious, so the method of Rth-order differences is tried. The pattern is not obvious from the first order differences, but the second order differences are all 4.

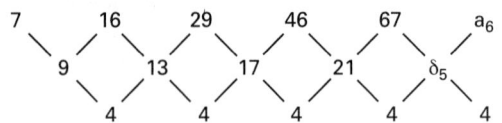

$$\delta_5-21=4$$

$$\delta_5=25$$

$$a_6-67=\delta_5=25$$

$$a_6=92$$

Example 3.12

Does the sequence with general term e^n/n converge or diverge?

Solution

See if the limit exists.

$$\lim_{n \to \infty} \left(\frac{e^n}{n} \right) = \frac{\infty}{\infty}$$

Since ∞/∞ is inconclusive, apply L'Hôpital's rule. Take the derivative of both the numerator and the denominator with respect to n.

$$\lim_{n \to \infty} \left(\frac{e^n}{1} \right) = \frac{\infty}{1}$$
$$= \infty$$

The sequence diverges.

22. STANDARD SEQUENCES

There are four standard sequences: the geometric, arithmetic, harmonic, and p-sequences.

- *geometric sequence*: The geometric sequence converges for $-1 < r \le 1$ and diverges otherwise. a is known as the *first term*; r is known as the *common ratio*.

$$a_n = ar^{n-1} \qquad \left[\begin{array}{l} a \text{ is a constant} \\ n = 1, 2, 3, \ldots, \infty \end{array} \right] \qquad 3.84$$

   example: $\{1, 2, 4, 8, 16, 32\}$ $(a = 1, r = 2)$

- *arithmetic sequence*: The arithmetic sequence always diverges.

$$a_n = a + (n-1)d \qquad \left[\begin{array}{l} a \text{ and } d \text{ are constants} \\ n = 1, 2, 3, \ldots, \infty \end{array} \right] \quad 3.85$$

   example: $\{2, 7, 12, 17, 22, 27\}$ $(a = 2, d = 5)$

- *harmonic sequence*: The harmonic sequence always converges.

$$a_n = \frac{1}{n} \qquad [n = 1, 2, 3, \ldots, \infty] \qquad 3.86$$

   example: $\{1, {}^1\!/_2, {}^1\!/_3, {}^1\!/_4, {}^1\!/_5, {}^1\!/_6\}$

- *p-sequence*: The p-sequence converges if $p \ge 0$ and diverges if $p < 0$. (Notice that this is different than the p-series.)

$$a_n = \frac{1}{n^p} \qquad [n = 1, 2, 3, \ldots, \infty] \qquad 3.87$$

   example: $\{1, {}^1\!/_4, {}^1\!/_9, {}^1\!/_{16}, {}^1\!/_{25}, {}^1\!/_{36}\}$ $(p = 2)$

23. SERIES

A *series* is the sum of terms in a sequence. There are two types of series. A *finite series* has a finite number of terms, and an *infinite series* has an infinite number of terms.[13] The main tasks associated with series are determining the sum of the terms and whether the series converges. A series is said to *converge* (be *convergent*) if the sum, S_n, of its term exists.[14] A finite series is always convergent.

The performance of a series based on standard sequences (defined in Sec. 22) is well known.

- *geometric series*:

$$S_n = \sum_{i=1}^{n} ar^{i-1} = \frac{a(1 - r^n)}{1 - r} \qquad \text{[finite]} \qquad 3.88$$

$$S_n = \sum_{i=1}^{\infty} ar^{i-1} = \frac{a}{1 - r} \qquad \left[\begin{array}{c} \text{infinite} \\ -1 < r < 1 \end{array} \right] \qquad 3.89$$

- *arithmetic series*: The infinite series diverges unless $a = d = 0$.

$$S_n = \sum_{i=1}^{n} (a + (i-1)d) = \frac{n(2a + (n-1)d)}{2}$$
$$\text{[finite]} \qquad 3.90$$

- *harmonic series*: The infinite series diverges.

- *p-series*: The infinite series diverges if $p \le 1$. The infinite series converges if $p > 1$. (Notice that this is different than the p-sequence.)

24. TESTS FOR SERIES CONVERGENCE

It is obvious that all *finite series* (i.e., series having a finite number of terms) converge. That is, the sum, S_n, defined by Eq. 3.91 exists.

$$S_n = \sum_{i=1}^{n} a_i \qquad 3.91$$

Convergence of an infinite series can be determined by taking the limit of the sum. If the limit exists, the series converges; otherwise, it diverges.

$$\lim_{n \to \infty} S_n = \lim_{n \to \infty} \sum_{i=1}^{n} a_i \qquad 3.92$$

In most cases, the expression for the general term a_n will be known, but there will be no simple expression for the sum S_n. Therefore, Eq. 3.92 cannot be used to determine convergence. It is helpful, but not conclusive, to look at the limit of the general term. If L, as defined in Eq. 3.93, is nonzero, the series diverges. If L

---

[13]The term *infinite series* does not imply the sum is infinite.
[14]This is different from the definition of convergence for a sequence where only the last term was evaluated.

equals zero, the series may either converge or diverge. Additional testing is needed in that case.

$$\lim_{n \to \infty} a_n \begin{cases} = 0 & \text{inconclusive} \\ \neq 0 & \text{diverges} \end{cases} \qquad 3.93$$

Two tests can be used independently or after Eq. 3.93 has proven inconclusive: the ratio and comparison tests. The *ratio test* calculates the limit of the ratio of two consecutive terms.

$$\lim_{n \to \infty} \frac{a_{n+1}}{a_n} \begin{cases} < 1 & \text{converges} \\ = 1 & \text{inconclusive} \\ > 1 & \text{diverges} \end{cases} \qquad 3.94$$

The *comparison test* is an indirect method of determining convergence of an unknown series. It compares a standard series (geometric and *p*-series are commonly used) against the unknown series. If all terms in a positive standard series are smaller than the terms in the unknown series and the standard series diverges, the unknown series must also diverge. Similarly, if all terms in the standard series are larger than the terms in the unknown series and the standard series converges, then the unknown series also converges.

In mathematical terms, if A and B are both series of positive terms such that $a_n < b_n$ for all values of n, then (a) B diverges if A diverges, and (b) A converges if B converges.

Example 3.13

Does the infinite series A converge or diverge?

$$A = 3 + \frac{9}{2} + \frac{27}{6} + \frac{81}{24} + \cdots$$

Solution

The general term was found in Ex. 3.10 to be

$$a_n = \frac{3^n}{n!}$$

Since limits of factorials are not easily determined, use the ratio test.

$$\lim_{n \to \infty} \left(\frac{a_{n+1}}{a_n} \right) = \lim_{n \to \infty} \left(\frac{\frac{3^{n+1}}{(n+1)!}}{\frac{3^n}{n!}} \right) = \lim_{n \to \infty} \left(\frac{3}{n+1} \right)$$

$$= \frac{3}{\infty} = 0$$

Since the limit is less than 1, the infinite series converges.

Example 3.14

Does the infinite series A converge or diverge?

$$A = 2 + \frac{3}{4} + \frac{4}{9} + \frac{5}{16} + \cdots$$

Solution

By observation, the general term is

$$a_n = \frac{1+n}{n^2}$$

The general term can be expanded by partial fractions to

$$a_n = \frac{1}{n} + \frac{1}{n^2}$$

However, $1/n$ is the harmonic series. Since the harmonic series is divergent and this series is larger than the harmonic series (by the term $1/n^2$), this series also diverges.

25. SERIES OF ALTERNATING SIGNS[15]

Some series contain both positive and negative terms. The ratio and comparison tests can both be used to determine if a series with alternating signs converges. If a series containing all positive terms converges, then the same series with some negative terms also converges. Therefore, the all-positive series should be tested for convergence. If the all-positive series converges, the original series is said to be *absolutely convergent*. (If the all-positive series diverges and the original series converges, the original series is said to be *conditionally convergent*.)

Alternatively, the ratio test can be used with the absolute value of the ratio. The same criteria apply.

$$\lim_{n \to \infty} \left| \frac{a_{n+1}}{a_n} \right| \begin{cases} < 1 & \text{converges} \\ = 1 & \text{inconclusive} \\ > 1 & \text{diverges} \end{cases} \qquad 3.95$$

---

[15]This terminology is commonly used even though it is not necessary that the signs strictly alternate.

4 Linear Algebra

1. MATRICES

A *matrix* is an ordered set of *entries* (*elements*) arranged rectangularly and set off by brackets.[1] The entries can be variables or numbers. A matrix by itself has no particular value—it is merely a convenient method of representing a set of numbers.

The size of a matrix is given by the number of rows and columns, and the nomenclature $m \times n$ is used for a matrix with m rows and n columns. For a *square matrix*, the number of rows and columns will be the same, a quantity known as the *order* of the matrix.

Bold uppercase letters are used to represent matrices, while lowercase letters represent the entries. For example, a_{23} would be the entry in the second row and third column of matrix $\mathbf{A}$.

$$\mathbf{A} = \begin{bmatrix} a_{11} & a_{12} & a_{13} \\ a_{21} & a_{22} & a_{23} \\ a_{31} & a_{32} & a_{33} \end{bmatrix}$$

A *submatrix* is the matrix that remains when selected rows or columns are removed from the original matrix.[2] For example, for matrix $\mathbf{A}$, the submatrix remaining after the second row and second column have been removed is

$$\begin{bmatrix} a_{11} & a_{13} \\ a_{31} & a_{33} \end{bmatrix}$$

[1] The term *array* is synonymous with *matrix*, although the former is more likely to be used in computer applications.
[2] By definition, a matrix is a submatrix of itself.

An *augmented matrix* results when the original matrix is extended by repeating one or more of its rows or columns or by adding rows and columns from another matrix. For example, for the matrix $\mathbf{A}$, the augmented matrix created by repeating the first and second columns is

$$\left[\begin{array}{ccc|cc} a_{11} & a_{12} & a_{13} & a_{11} & a_{12} \\ a_{21} & a_{22} & a_{23} & a_{21} & a_{22} \\ a_{31} & a_{32} & a_{33} & a_{31} & a_{32} \end{array}\right]$$

2. SPECIAL TYPES OF MATRICES

Certain types of matrices are given special designations.

- *cofactor matrix*: the matrix that is formed when every entry is replaced by the cofactor (see Sec. 4) of that entry

- *column matrix*: a matrix with only one column

- *complex matrix*: a matrix with complex number entries

- *diagonal matrix*: a square matrix with all zero entries except for the a_{ij} for which $i = j$

- *echelon matrix*: a matrix in which the number of zeros preceding the first nonzero entry of a row increases row by row until only zero rows remain. A *row-reduced echelon matrix* is an echelon matrix in which the first nonzero entry in each row is a 1 and all other entries in the columns are zero.

- *identity matrix*: a diagonal (square) matrix with all nonzero entries equal to 1, usually designated as $\mathbf{I}$, having the property that $\mathbf{AI} = \mathbf{IA} = \mathbf{A}$

- *null matrix*: the same as a zero matrix

- *row matrix*: a matrix with only one row

- *scalar matrix*:[3] a diagonal (square) matrix with all diagonal entries equal to some scalar k

- *singular matrix*: a matrix whose determinant is zero (see Sec. 10)

[3] Although the term *complex matrix* means a matrix with complex entries, the term *scalar matrix* means more than a matrix with scalar entries.

- *skew symmetric matrix*: a square matrix whose transpose (see Sec. 9) is equal to the negative of itself (i.e., $\mathbf{A} = -\mathbf{A}^t$)

- *square matrix*: a matrix with the same number of rows and columns (i.e., $m = n$)

- *symmetric(al) matrix*: a square matrix whose transpose is equal to itself (i.e., $\mathbf{A}^t = \mathbf{A}$), which occurs only when $a_{ij} = a_{ji}$

- *triangular matrix*: a square matrix with zeros in all positions above or below the diagonal

- *unit matrix*: the same as the identity matrix

- *zero matrix*: a matrix with all zero entries

$$\begin{bmatrix} 9 & 0 & 0 & 0 \\ 0 & -6 & 0 & 0 \\ 0 & 0 & 1 & 0 \\ 0 & 0 & 0 & 5 \end{bmatrix} \quad \begin{bmatrix} 2 & 18 & 2 & 18 \\ 0 & 0 & 1 & 9 \\ 0 & 0 & 0 & 9 \\ 0 & 0 & 0 & 0 \end{bmatrix} \quad \begin{bmatrix} 1 & 9 & 0 & 0 \\ 0 & 0 & 1 & 0 \\ 0 & 0 & 0 & 1 \\ 0 & 0 & 0 & 0 \end{bmatrix}$$

(a) diagonal   (b) echelon   (c) row-reduced echelon

$$\begin{bmatrix} 1 & 0 & 0 & 0 \\ 0 & 1 & 0 & 0 \\ 0 & 0 & 1 & 0 \\ 0 & 0 & 0 & 1 \end{bmatrix} \quad \begin{bmatrix} 2 & 0 & 0 & 0 \\ 7 & 6 & 0 & 0 \\ 9 & 1 & 1 & 0 \\ 8 & 0 & 4 & 5 \end{bmatrix} \quad \begin{bmatrix} 3 & 0 & 0 & 0 \\ 0 & 3 & 0 & 0 \\ 0 & 0 & 3 & 0 \\ 0 & 0 & 0 & 3 \end{bmatrix}$$

(d) identity   (e) triangular   (f) scalar

Figure 4.1 *Examples of Special Matrices*

3. ROW EQUIVALENT MATRICES

A matrix $\mathbf{B}$ is said to be *row equivalent* to a matrix $\mathbf{A}$ if it is obtained by a finite sequence of *elementary row operations* on $\mathbf{A}$:

- interchanging the ith and jth rows

- multiplying the ith row by a nonzero scalar

- replacing the ith row by the sum of the original ith row and k times the jth row

However, two matrices that are row equivalent as defined do not necessarily have the same determinants. (See Sec. 5.)

Gauss-Jordan elimination is the process of using these elementary row operations to row-reduce a matrix to echelon or row-reduced echelon forms, as illustrated in Ex. 4.8. When a matrix has been converted to a row-reduced echelon matrix, it is said to be in *row canonical form*. Thus, the terms *row-reduced echelon form* and *row canonical form* are synonymous.

4. MINORS AND COFACTORS

Minors and cofactors are determinants of submatrices associated with particular entries in the original square matrix. The *minor* of entry a_{ij} is the determinant of a submatrix resulting from the elimination of the single row i and the single column j. For example, the minor corresponding to entry a_{12} in a 3×3 matrix $\mathbf{A}$ is the determinant of the matrix created by eliminating row 1 and column 2.

$$\text{minor of } a_{12} = \begin{vmatrix} a_{21} & a_{23} \\ a_{31} & a_{33} \end{vmatrix} \qquad 4.1$$

The *cofactor* of entry a_{ij} is the minor of a_{ij} multiplied by either $+1$ or -1, depending on the position of the entry. (That is, the cofactor either exactly equals the minor or it differs only in sign.) The sign is determined according to the following positional matrix.[4]

$$\begin{bmatrix} +1 & -1 & +1 & \cdots \\ -1 & +1 & -1 & \cdots \\ +1 & -1 & +1 & \cdots \\ \vdots & \vdots & \vdots & \end{bmatrix}$$

For example, the cofactor of entry a_{12} in matrix $\mathbf{A}$ (described in Sec. 4) is

$$\text{cofactor of } a_{12} = -\begin{vmatrix} a_{21} & a_{23} \\ a_{31} & a_{33} \end{vmatrix} \qquad 4.2$$

Example 4.1

What is the cofactor corresponding to the -3 entry in the following matrix?

$$\mathbf{A} = \begin{bmatrix} 2 & 9 & 1 \\ -3 & 4 & 0 \\ 7 & 5 & 9 \end{bmatrix}$$

Solution

The minor's submatrix is created by eliminating the row and column of the -3 entry.

$$\mathbf{M} = \begin{bmatrix} 9 & 1 \\ 5 & 9 \end{bmatrix}$$

The minor is the determinant of $\mathbf{M}$.

$$|\mathbf{M}| = (9)(9) - (5)(1) = 76$$

The sign corresponding to the -3 position is negative. Therefore, the cofactor is -76.

---

[4]The sign of the cofactor a_{ij} is positive if $(i + j)$ is even and is negative if $(i + j)$ is odd.

5. DETERMINANTS

A *determinant* is a scalar calculated from a square matrix. The determinant of matrix $\mathbf{A}$ can be represented as $D\{\mathbf{A}\}$, Det $(\mathbf{A})$, $\Delta \mathbf{A}$, or $|\mathbf{A}|$.[5] The following rules can be used to simplify the calculation of determinants.

- If $\mathbf{A}$ has a row or column of zeros, the determinant is zero.

- If $\mathbf{A}$ has two identical rows or columns, the determinant is zero.

- If $\mathbf{B}$ is obtained from $\mathbf{A}$ by adding a multiple of a row (column) to another row (column) in $\mathbf{A}$, then $|\mathbf{B}| = |\mathbf{A}|$.

- If $\mathbf{A}$ is triangular, the determinant is equal to the product of the diagonal entries.

- If $\mathbf{B}$ is obtained from $\mathbf{A}$ by multiplying one row or column in $\mathbf{A}$ by a scalar k, then $|\mathbf{B}| = k|\mathbf{A}|$.

- If $\mathbf{B}$ is obtained from the $n \times n$ matrix $\mathbf{A}$ by multiplying by the scalar matrix k, then $|k\mathbf{A}| = k^n|\mathbf{A}|$.

- If $\mathbf{B}$ is obtained from $\mathbf{A}$ by switching two rows or columns in $\mathbf{A}$, then $|\mathbf{B}| = -|\mathbf{A}|$.

Calculation of determinants is laborious for all but the smallest or simplest of matrices. For a 2×2 matrix, the formula used to calculate the determinant is easy to remember.

$$\mathbf{A} = \begin{bmatrix} a & b \\ c & d \end{bmatrix}$$

$$|\mathbf{A}| = \begin{vmatrix} a & b \\ c & d \end{vmatrix} = ad - bc \qquad 4.3$$

Two methods are commonly used for calculating the determinant of 3×3 matrices by hand. The first uses an augmented matrix constructed from the original matrix and the first two columns (as presented in Sec. 1).[6] The determinant is calculated as the sum of the products in the left-to-right downward diagonals less the sum of the products in the left-to-right upward diagonals.

$$\mathbf{A} = \begin{bmatrix} a & b & c \\ d & e & f \\ g & h & i \end{bmatrix}$$

$$\text{augmented } \mathbf{A} = \begin{bmatrix} a & b & c & a & b \\ d & e & f & d & e \\ g & h & i & g & h \end{bmatrix} \qquad 4.4$$

$$|\mathbf{A}| = aei + bfg + cdh - gec - hfa - idb \qquad 4.5$$

[5] The vertical bars should not be confused with the square brackets used to set off a matrix, nor with absolute value.
[6] It is not actually necessary to construct the augmented matrix, but doing so helps avoid errors.

The second method of calculating the determinant is somewhat slower than the first for a 3×3 matrix but illustrates the method that must be used to calculate determinants of 4×4 and larger matrices. This method is known as *expansion by cofactors*. One row (column) is selected as the base row (column). The selection is arbitrary, but the number of calculations required to obtain the determinant can be minimized by choosing the row (column) with the most zeros. The determinant is equal to the sum of the products of the entries in the base row (column) and their corresponding cofactors.

$$\mathbf{A} = \begin{bmatrix} a & b & c \\ d & e & f \\ g & h & i \end{bmatrix}$$

$$|\mathbf{A}| = a\begin{vmatrix} e & f \\ h & i \end{vmatrix} - d\begin{vmatrix} b & c \\ h & i \end{vmatrix} + g\begin{vmatrix} b & c \\ e & f \end{vmatrix} \qquad 4.6$$

Example 4.2

Calculate the determinant of matrix $\mathbf{A}$ (a) by cofactor expansion, and (b) by the augmented matrix method.

$$\mathbf{A} = \begin{bmatrix} 2 & 3 & -4 \\ 3 & -1 & -2 \\ 4 & -7 & -6 \end{bmatrix}$$

Solution

(a) Since there are no zero entries, it does not matter which row or column is chosen as the base. Choose the first column as the base.

$$|\mathbf{A}| = 2\begin{vmatrix} -1 & -2 \\ -7 & -6 \end{vmatrix} - 3\begin{vmatrix} 3 & -4 \\ -7 & -6 \end{vmatrix} + 4\begin{vmatrix} 3 & -4 \\ -1 & -2 \end{vmatrix}$$

$$= (2)(6-14) - (3)(-18-28) + (4)(-6-4)$$

$$= 82$$

(b)

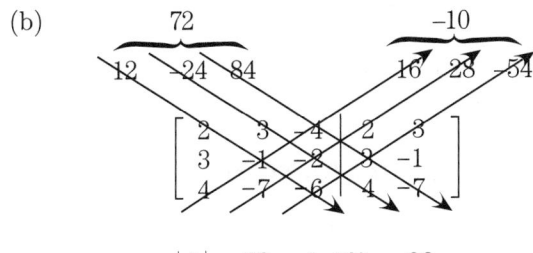

$$|\mathbf{A}| = 72 - (-10) = 82$$

6. MATRIX ALGEBRA[7]

Matrix algebra differs somewhat from standard algebra.

- *equality*: Two matrices, $\mathbf{A}$ and $\mathbf{B}$, are equal only if they have the same numbers of rows and columns *and* if all corresponding entries are equal.

- *inequality*: The $>$ and $<$ operators are not used in matrix algebra.

[7] Since matrices are used to simplify the presentation and solution of sets of linear equations, matrix algebra is also known as *linear algebra*.

- *commutative law of addition*:

$$\mathbf{A} + \mathbf{B} = \mathbf{B} + \mathbf{A} \qquad 4.7$$

- *associative law of addition*:

$$\mathbf{A} + (\mathbf{B} + \mathbf{C}) = (\mathbf{A} + \mathbf{B}) + \mathbf{C} \qquad 4.8$$

- *associative law of multiplication*:

$$(\mathbf{AB})\mathbf{C} = \mathbf{A}(\mathbf{BC}) \qquad 4.9$$

- *left distributive law*:

$$\mathbf{A}(\mathbf{B} + \mathbf{C}) = \mathbf{AB} + \mathbf{AC} \qquad 4.10$$

- *right distributive law*:

$$(\mathbf{B} + \mathbf{C})\mathbf{A} = \mathbf{BA} + \mathbf{CA} \qquad 4.11$$

- *scalar multiplication*:

$$k(\mathbf{AB}) = (k\mathbf{A})\mathbf{B} = \mathbf{A}(k\mathbf{B}) \qquad 4.12$$

It is important to recognize that, except for trivial and special cases, matrix multiplication is not commutative. That is,

$$\mathbf{AB} \neq \mathbf{BA}$$

7. MATRIX ADDITION AND SUBTRACTION

Addition and subtraction of two matrices is possible only if both matrices have the same numbers of rows and columns (i.e., order). They are accomplished by adding or subtracting the corresponding entries of the two matrices.

8. MATRIX MULTIPLICATION

A matrix can be multiplied by a scalar, an operation known as *scalar multiplication*, in which case all entries of the matrix are multiplied by that scalar. For example, for the 2×2 matrix $\mathbf{A}$,

$$k\mathbf{A} = \begin{bmatrix} ka_{11} & ka_{12} \\ ka_{21} & ka_{22} \end{bmatrix}$$

A matrix can be multiplied by another matrix, but only if the left-hand matrix has the same number of columns as the right-hand matrix has rows. *Matrix multiplication* occurs by multiplying the elements in each left-hand matrix row by the entries in each right-hand matrix column, adding the products, and placing the sum at the intersection point of the participating row and column.

Matrix division can only be accomplished by multiplying by the inverse of the denominator matrix. There is no specific division operation in matrix algebra.

Example 4.3

Determine the product matrix $\mathbf{C}$.

$$\mathbf{C} = \begin{bmatrix} 1 & 4 & 3 \\ 5 & 2 & 6 \end{bmatrix} \begin{bmatrix} 7 & 12 \\ 11 & 8 \\ 9 & 10 \end{bmatrix}$$

Solution

The left-hand matrix has three columns, and the right-hand matrix has three rows. Therefore, the two matrices can be multiplied.

The first row of the left-hand matrix and the first column of the right-hand matrix are worked with first. The corresponding entries are multiplied and the products are summed.

$$c_{11} = (1)(7) + (4)(11) + (3)(9) = 78$$

The intersection of the top row and left column is the entry in the upper left-hand corner of the matrix $\mathbf{C}$.

The remaining entries are calculated similarly.

$$c_{12} = (1)(12) + (4)(8) + (3)(10) = 74$$
$$c_{21} = (5)(7) + (2)(11) + (6)(9) = 111$$
$$c_{22} = (5)(12) + (2)(8) + (6)(10) = 136$$

The product matrix is

$$\mathbf{C} = \begin{bmatrix} 78 & 74 \\ 111 & 136 \end{bmatrix}$$

9. TRANSPOSE

The *transpose*, $\mathbf{A}^t$, of an $m \times n$ matrix $\mathbf{A}$ is an $n \times m$ matrix constructed by taking the ith row and making it the ith column. The diagonal is unchanged. For example,

$$\mathbf{A} = \begin{bmatrix} 1 & 6 & 9 \\ 2 & 3 & 4 \\ 7 & 1 & 5 \end{bmatrix}$$

$$\mathbf{A}^t = \begin{bmatrix} 1 & 2 & 7 \\ 6 & 3 & 1 \\ 9 & 4 & 5 \end{bmatrix}$$

Transpose operations have the following characteristics.

$$(\mathbf{A}^t)^t = \mathbf{A} \qquad 4.13$$
$$(k\mathbf{A})^t = k(\mathbf{A}^t) \qquad 4.14$$
$$\mathbf{I}^t = \mathbf{I} \qquad 4.15$$
$$(\mathbf{AB})^t = \mathbf{B}^t \mathbf{A}^t \qquad 4.16$$
$$(\mathbf{A} + \mathbf{B})^t = \mathbf{A}^t + \mathbf{B}^t \qquad 4.17$$
$$|\mathbf{A}^t| = |\mathbf{A}| \qquad 4.18$$

10. SINGULARITY AND RANK

A *singular matrix* is one whose determinant is zero. Similarly, a *nonsingular* matrix is one whose determinant is nonzero.

The *rank* of a matrix is the maximum number of linearly independent row or column vectors.[8] A matrix has rank r if it has at least one nonsingular square submatrix of order r but has no nonsingular square submatrix of order more than r. While the submatrix must be square (in order to calculate the determinant), the original matrix need not be.

The rank of an $m \times n$ matrix will be, at most, the smaller of m and n. The rank of a null matrix is zero. The ranks of a matrix and its transpose are the same. If a matrix is in echelon form, the rank will be equal to the number of rows containing at least one nonzero entry. For a 3×3 matrix, the rank can either be 3 (if it is nonsingular), 2 (if any one of its 2×2 submatrices is nonsingular), 1 (if it and all 2×2 submatrices are singular), or 0 (if it is null).

The determination of rank is laborious if done by hand. Either the matrix is reduced to echelon form by using elementary row operations, or exhaustive enumeration is used to create the submatrices and many determinants are calculated. If a matrix has more rows than columns and row-reduction is used, the work required to put the matrix in echelon form can be reduced by working with the transpose of the original matrix.

Example 4.4

What is the rank of matrix $\mathbf{A}$?

$$\mathbf{A} = \begin{bmatrix} 1 & -2 & -1 \\ -3 & 3 & 0 \\ 2 & 2 & 4 \end{bmatrix}$$

Solution

Matrix $\mathbf{A}$ is singular because $|\mathbf{A}| = 0$. However, there is at least one 2×2 nonsingular submatrix:

$$\begin{vmatrix} 1 & -2 \\ -3 & 3 \end{vmatrix} = (1)(3) - (-3)(-2) = -3$$

Therefore, the rank is 2.

Example 4.5

Determine the rank of matrix $\mathbf{A}$ by reducing it to echelon form.

$$\mathbf{A} = \begin{bmatrix} 7 & 4 & 9 & 1 \\ 0 & 2 & -5 & 3 \\ 0 & 4 & -10 & 6 \end{bmatrix}$$

Solution

By inspection, the matrix can be row-reduced by subtracting two times the second row from the third row. The matrix cannot be further reduced. Since there are two nonzero rows, the rank is 2.

$$\begin{bmatrix} 7 & 4 & 9 & 1 \\ 0 & 2 & -5 & 3 \\ 0 & 0 & 0 & 0 \end{bmatrix}$$

11. CLASSICAL ADJOINT

The *classical adjoint* is the transpose of the cofactor matrix. The resulting matrix can be designated as $\mathbf{A}_{\text{adj}}$, adj$\{\mathbf{A}\}$ or $\mathbf{A}^{\text{adj}}$.

Example 4.6

What is the classical adjoint of matrix $\mathbf{A}$?

$$\mathbf{A} = \begin{bmatrix} 2 & 3 & -4 \\ 0 & -4 & 2 \\ 1 & -1 & 5 \end{bmatrix}$$

Solution

The matrix of cofactors is

$$\begin{bmatrix} -18 & 2 & 4 \\ -11 & 14 & 5 \\ -10 & -4 & -8 \end{bmatrix}$$

The transpose of the matrix of cofactors is

$$\mathbf{A}_{\text{adj}} = \begin{bmatrix} -18 & -11 & -10 \\ 2 & 14 & -4 \\ 4 & 5 & -8 \end{bmatrix}$$

12. INVERSE

The product of a matrix $\mathbf{A}$ and its inverse, $\mathbf{A}^{-1}$, is the identity matrix, $\mathbf{I}$. Only square matrices have inverses, but not all square matrices are invertible. A matrix has an inverse if and only if it is nonsingular (i.e., its determinant is nonzero).

$$\mathbf{A}\mathbf{A}^{-1} = \mathbf{A}^{-1}\mathbf{A} = \mathbf{I} \qquad 4.19$$
$$(\mathbf{A}\mathbf{B})^{-1} = \mathbf{B}^{-1}\mathbf{A}^{-1} \qquad 4.20$$

The inverse of a 2×2 matrix is easily determined by formula.

$$\mathbf{A} = \begin{bmatrix} a & b \\ c & d \end{bmatrix}$$

$$\mathbf{A}^{-1} = \frac{\begin{bmatrix} d & -b \\ -c & a \end{bmatrix}}{|\mathbf{A}|} \qquad 4.21$$

---

[8]The *row rank* and *column rank* are the same.

For a 3×3 or larger matrix, the inverse is determined by dividing every entry in the classical adjoint by the determinant of the original matrix.

$$\mathbf{A}^{-1} = \frac{\mathbf{A}_{adj}}{|\mathbf{A}|} \qquad 4.22$$

Example 4.7

What is the inverse of matrix $\mathbf{A}$?

$$\mathbf{A} = \begin{bmatrix} 4 & 5 \\ 2 & 3 \end{bmatrix}$$

Solution

The determinant is calculated as

$$|\mathbf{A}| = (4)(3) - (2)(5) = 2$$

Using Eq. 4.22, the inverse is

$$\mathbf{A}^{-1} = \frac{\begin{bmatrix} 3 & -5 \\ -2 & 4 \end{bmatrix}}{2} = \begin{bmatrix} \frac{3}{2} & -\frac{5}{2} \\ -1 & 2 \end{bmatrix}$$

Check.

$$\mathbf{A}\mathbf{A}^{-1} = \begin{bmatrix} 4 & 5 \\ 2 & 3 \end{bmatrix} \begin{bmatrix} \frac{3}{2} & -\frac{5}{2} \\ -1 & 2 \end{bmatrix} = \begin{bmatrix} 6-5 & -10+10 \\ 3-3 & -5+6 \end{bmatrix}$$

$$= \begin{bmatrix} 1 & 0 \\ 0 & 1 \end{bmatrix} = \mathbf{I} \qquad [\text{OK}]$$

13. WRITING SIMULTANEOUS LINEAR EQUATIONS IN MATRIX FORM

Matrices are used to simplify the presentation and solution of sets of simultaneous linear equations. For example, the following three methods of presenting simultaneous linear equations are equivalent:

$$a_{11}x_1 + a_{12}x_2 = b_1$$
$$a_{21}x_1 + a_{22}x_2 = b_2$$

$$\begin{bmatrix} a_{11} & a_{12} \\ a_{21} & a_{22} \end{bmatrix} \begin{bmatrix} x_1 \\ x_2 \end{bmatrix} = \begin{bmatrix} b_1 \\ b_2 \end{bmatrix}$$

$$\mathbf{AX} = \mathbf{B}$$

In the second and third representations, $\mathbf{A}$ is known as the *coefficient matrix*, $\mathbf{X}$ as the *variable matrix*, and $\mathbf{B}$ as the *constant matrix*.

Not all systems of simultaneous equations have solutions, and those that do may not have unique solutions.

The existence of a solution can be determined by calculating the determinant of the coefficient matrix. These rules are summarized in Table 4.1.

- If the system of linear equations is homogeneous (i.e., $\mathbf{B}$ is a zero matrix) and $|\mathbf{A}|$ is zero, there are an infinite number of solutions.

- If the system is homogeneous and $|\mathbf{A}|$ is nonzero, only the trivial solution exists.

- If the system of linear equations is nonhomogeneous (i.e., $\mathbf{B}$ is not a zero matrix) and $|\mathbf{A}|$ is nonzero, there is a unique solution to the set of simultaneous equations.

- If $|\mathbf{A}|$ is zero, a nonhomogeneous system of simultaneous equations may still have a solution. The requirement is that the determinants of all substitutional matrices (see Sec. 14) are zero, in which case there will be an infinite number of solutions. Otherwise, no solution exists.

Table 4.1 *Solution Existence Rules for Simultaneous Equations*

	$\mathbf{B} = 0$	$\mathbf{B} \neq 0$
$|\mathbf{A}| = 0$	infinite number of solutions (linearly dependent equations)	either an infinite number of solutions or no solution at all
$|\mathbf{A}| \neq 0$	trivial solution only $(x_i = 0)$	unique nonzero solution

14. SOLVING SIMULTANEOUS LINEAR EQUATIONS

Gauss-Jordan elimination can be used to obtain the solution to a set of simultaneous linear equations. The coefficient matrix is augmented by the constant matrix. Then, elementary row operations are used to reduce the coefficient matrix to canonical form. All of the operations performed on the coefficient matrix are performed on the constant matrix. The variable values that satisfy the simultaneous equations will be the entries in the constant matrix when the coefficient matrix is in canonical form.

Determinants are used to calculate the solution to linear simultaneous equations through a procedure known as *Cramer's rule*.

The procedure is to calculate determinants of the original coefficient matrix $\mathbf{A}$ and of the n matrices resulting from the systematic replacement of a column in $\mathbf{A}$ by the constant matrix $\mathbf{B}$. For a system of three equations in three unknowns, there are three substitutional matrices, $\mathbf{A}_1$, $\mathbf{A}_2$, and $\mathbf{A}_3$, as well as the original coefficient matrix, for a total of four matrices whose determinants must be calculated.

The values of the unknowns that simultaneously satisfy all of the linear equations are

$$x_1 = \frac{|\mathbf{A}_1|}{|\mathbf{A}|} \qquad 4.23$$

$$x_2 = \frac{|\mathbf{A}_2|}{|\mathbf{A}|} \qquad 4.24$$

$$x_3 = \frac{|\mathbf{A}_3|}{|\mathbf{A}|} \qquad 4.25$$

Example 4.8

Use Gauss-Jordan elimination to solve the following system of simultaneous equations.

$$2x + 3y - 4z = 1$$
$$3x - y - 2z = 4$$
$$4x - 7y - 6z = -7$$

Solution

The augmented matrix is created by appending the constant matrix to the coefficient matrix.

$$\begin{bmatrix} 2 & 3 & -4 & | & 1 \\ 3 & -1 & -2 & | & 4 \\ 4 & -7 & -6 & | & -7 \end{bmatrix}$$

Elementary row operations are used to reduce the coefficient matrix to canonical form. For example, two times the first row is subtracted from the third row. This step obtains the 0 needed in the a_{31} position.

$$\begin{bmatrix} 2 & 3 & -4 & | & 1 \\ 3 & -1 & -2 & | & 4 \\ 0 & -13 & 2 & | & -9 \end{bmatrix}$$

This process continues until the following form is obtained.

$$\begin{bmatrix} 1 & 0 & 0 & | & 3 \\ 0 & 1 & 0 & | & 1 \\ 0 & 0 & 1 & | & 2 \end{bmatrix}$$

$x = 3$, $y = 1$, and $z = 2$ satisfy this system of equations.

Example 4.9

Use Cramer's rule to solve the following system of simultaneous equations.

$$2x + 3y - 4z = 1$$
$$3x - y - 2z = 4$$
$$4x - 7y - 6z = -7$$

Solution

The determinant of the coefficient matrix is

$$|\mathbf{A}| = \begin{vmatrix} 2 & 3 & -4 \\ 3 & -1 & -2 \\ 4 & -7 & -6 \end{vmatrix} = 82$$

The determinants of the substitutional matrices are

$$|\mathbf{A}_1| = \begin{vmatrix} 1 & 3 & -4 \\ 4 & -1 & -2 \\ -7 & -7 & -6 \end{vmatrix} = 246$$

$$|\mathbf{A}_2| = \begin{vmatrix} 2 & 1 & -4 \\ 3 & 4 & -2 \\ 4 & -7 & -6 \end{vmatrix} = 82$$

$$|\mathbf{A}_3| = \begin{vmatrix} 2 & 3 & 1 \\ 3 & -1 & 4 \\ 4 & -7 & -7 \end{vmatrix} = 164$$

The values of x, y, and z that satisfy the linear equations are

$$x = \frac{246}{82} = 3$$
$$y = \frac{82}{82} = 1$$
$$z = \frac{164}{82} = 2$$

15. EIGENVALUES AND EIGENVECTORS

Eigenvalues and eigenvectors (also known as *characteristic values* and *characteristic vectors*) of a square matrix $\mathbf{A}$ are the scalars k and matrices $\mathbf{X}$ such that

$$\mathbf{AX} = k\mathbf{X} \qquad 4.26$$

The scalar k is an eigenvalue of $\mathbf{A}$ if and only if the matrix $(k\mathbf{I} - \mathbf{A})$ is singular; that is, if $|k\mathbf{I} - \mathbf{A}| = 0$. This equation is called the *characteristic equation* of the matrix $\mathbf{A}$. When expanded, the determinant is called the *characteristic polynomial*. The method of using the characteristic polynomial to find eigenvalues and eigenvectors is illustrated in Ex. 4.10.

If all of the eigenvalues are unique (i.e., nonrepeating), then Eq. 4.27 is valid.

$$[k\mathbf{I} - \mathbf{A}]\mathbf{X} = 0 \qquad 4.27$$

Example 4.10

Find the eigenvalues and nonzero eigenvectors of the matrix $\mathbf{A}$.

$$\mathbf{A} = \begin{bmatrix} 2 & 4 \\ 6 & 4 \end{bmatrix}$$

Solution

$$k\mathbf{I} - A = \begin{bmatrix} k & 0 \\ 0 & k \end{bmatrix} - \begin{bmatrix} 2 & 4 \\ 6 & 4 \end{bmatrix} = \begin{bmatrix} k-2 & -4 \\ -6 & k-4 \end{bmatrix}$$

The characteristic polynomial is found by setting the determinant $|k\mathbf{I} - \mathbf{A}|$ equal to zero.

$$(k-2)(k-4) - (-6)(-4) = 0$$
$$k^2 - 6k - 16 = (k-8)(k+2) = 0$$

The roots of the characteristic polynomial are $k = +8$ and $k = -2$. These are the eigenvalues of $\mathbf{A}$.

Substituting $k = 8$,

$$k\mathbf{I} - \mathrm{A} = \begin{bmatrix} 8-2 & -4 \\ -6 & 8-4 \end{bmatrix} = \begin{bmatrix} 6 & -4 \\ -6 & 4 \end{bmatrix}$$

This can be interpreted as the linear equation $6x_1 - 4x_2 = 0$. The values of x that satisfy this equation define the eigenvector. An eigenvector $\mathbf{X}$ associated with the eigenvalue $+8$ is

$$\mathbf{X} = \begin{bmatrix} x_1 \\ x_2 \end{bmatrix} = \begin{bmatrix} 4 \\ 6 \end{bmatrix}$$

All other eigenvectors for this eigenvalue are multiples of $\mathbf{X}$. Normally $\mathbf{X}$ is reduced to smallest integers.

$$\mathbf{X} = \begin{bmatrix} 2 \\ 3 \end{bmatrix}$$

Similarly, the eigenvector associated with the eigenvalue -2 is

$$\mathbf{X} = \begin{bmatrix} x_1 \\ x_2 \end{bmatrix} = \begin{bmatrix} +4 \\ -4 \end{bmatrix}$$

Reducing this to smallest integers gives

$$\mathbf{X} = \begin{bmatrix} +1 \\ -1 \end{bmatrix}$$

5 Vectors

1. INTRODUCTION

A physical property or quantity can be a scalar, vector, or tensor. A *scalar* has only magnitude. Knowing its value is sufficient to define a scalar. Mass, enthalpy, density, and speed are examples of scalars.

Force, momentum, displacement, and velocity are examples of vectors. A *vector* is a directed straight line with a specific magnitude. Thus, a vector is specified completely by its direction (consisting of the vector's *angular orientation* and its *sense*) and magnitude. A vector's *point of application* (*terminal point*) is not needed to define the vector.[1] Two vectors with the same direction and magnitude are said to be *equal vectors* even though their *lines of action* may be different.[2]

A vector can be designated by a boldface variable (as in this book) or as a combination of the variable and some other symbol. For example, the notations $\mathbf{V}$, $\overline{V}$, $\hat{V}$, $\vec{V}$, and $\underline{V}$ are used by different authorities to represent vectors. The magnitude of a vector can be designated by either $|\mathbf{V}|$ or V (not bold).

Stress, dielectric constant, and magnetic susceptibility are examples of tensors. A *tensor* has magnitude in a specific direction but the direction is not unique. Tensors are frequently associated with *anisotropic materials* that have different properties in different directions. A tensor in three-dimensional space is defined by nine components, compared with the three that are required

to define vectors. These components are written in matrix form. Stress, σ, at a point, for example, would be defined by the following tensor matrix.

$$\sigma = \begin{pmatrix} \sigma_{xx} & \sigma_{xy} & \sigma_{xz} \\ \sigma_{yx} & \sigma_{yy} & \sigma_{yz} \\ \sigma_{zx} & \sigma_{zy} & \sigma_{zz} \end{pmatrix}$$

2. VECTORS IN n-SPACE

In some cases, a vector, $\mathbf{V}$, will be designated by its two endpoints in n-dimensional vector space. A common example is three-dimensional force-space. Usually, one of the points will be the origin, in which case the vector is said to be "based at the origin," "origin-based," or "zero-based."[3] If one of the endpoints is the origin, specifying a terminal point P would represent a force directed from the origin to point P.

If a coordinate system is superimposed on the vector space, a vector can be specified in terms of the n coordinates of its two endpoints. The magnitude of the vector $\mathbf{V}$ is the distance in vector space between the two points, as given by Eq. 5.1. Similarly, the direction is defined by the angle the vector makes with one of the axes.

$$|\mathbf{V}| = \sqrt{(x_2 - x_1)^2 + (y_2 - y_1)^2} \qquad 5.1$$

$$\phi = \arctan\left(\frac{y_2 - y_1}{x_2 - x_1}\right) \qquad 5.2$$

Figure 5.1 *Vector in Two-Dimensional Space*

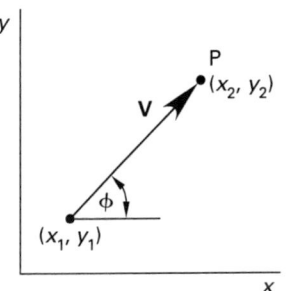

---

[1] A vector that is constrained to act at or through a certain point is a *bound vector (fixed vector)*. A *sliding vector (transmissible vector)* can be applied anywhere along its line of action. A *free vector* is not constrained and can be applied at any point in space.
[2] A distinction is sometimes made between equal vectors and equivalent vectors. *Equivalent vectors* produce the same effect but are not necessarily equal.

[3] Any vector directed from P_1 to P_2 can be transformed into a zero-based vector by subtracting the coordinates of point P_1 from the coordinates of terminal point P_2. The transformed vector will be equivalent to the original vector.

The *components* of a vector are the projections of the vector on the coordinate axes. (For a zero-based vector, the components and the coordinates of the endpoint are the same.) Simple trigonometric principles are used to resolve a vector into its components. A vector constructed from its components is known as a *resultant vector*.

$$V_x = |\mathbf{V}| \cos \phi_x \qquad 5.3$$
$$V_y = |\mathbf{V}| \cos \phi_y \qquad 5.4$$
$$V_z = |\mathbf{V}| \cos \phi_z \qquad 5.5$$
$$|\mathbf{V}| = \sqrt{V_x^2 + V_y^2 + V_z^2} \qquad 5.6$$

Figure 5.2 *Direction Angles of a Vector*

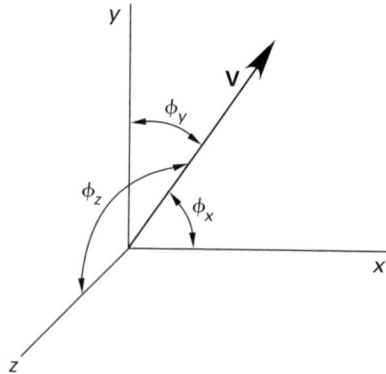

In Eqs. 5.3 through 5.5, ϕ_x, ϕ_y, and ϕ_z are the *direction angles*—the angles between the vector and the x-, y-, and z-axis, respectively. The cosines of these angles are known as *direction cosines*. The sum of the squares of the direction cosines is equal to 1.

$$\cos^2 \phi_x + \cos^2 \phi_y + \cos^2 \phi_z = 1 \qquad 5.7$$

3. UNIT VECTORS

Unit vectors are vectors with unit magnitudes (i.e., magnitudes of 1). They are represented in the same notation as other vectors. (Unit vectors in this book are written in boldface type.) Although they can have any direction, the standard unit vectors (the *Cartesian unit vectors* $\mathbf{i}$, $\mathbf{j}$, and $\mathbf{k}$) have the directions of the x-, y-, and z-coordinate axes and constitute the *Cartesian triad*.

A vector $\mathbf{V}$ can be written in terms of unit vectors and its components.

$$\mathbf{V} = |\mathbf{V}|\mathbf{a} = V_x\mathbf{i} + V_y\mathbf{j} + V_z\mathbf{k} \qquad 5.8$$

Figure 5.3 *Cartesian Unit Vectors*

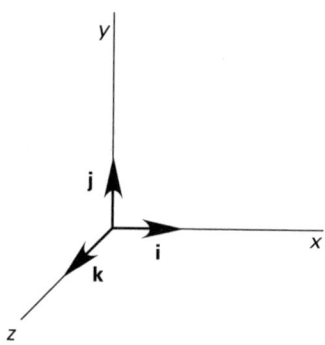

The unit vector, $\mathbf{a}$, has the same direction as the vector $\mathbf{V}$ but has a length of 1. This unit vector is calculated by dividing the original vector, $\mathbf{V}$, by its magnitude, $|\mathbf{V}|$.

$$\mathbf{a} = \frac{\mathbf{V}}{|\mathbf{V}|} = \frac{V_x\mathbf{i} + V_y\mathbf{j} + V_z\mathbf{k}}{\sqrt{V_x^2 + V_y^2 + V_z^2}} \qquad 5.9$$

4. VECTOR REPRESENTATION

The most common method of representing a vector is by writing it in *rectangular form*—a vector sum of its orthogonal components. In rectangular form, each of the orthogonal components has the same units as the resultant vector.

$$\mathbf{A} \equiv A_x\mathbf{i} + A_y\mathbf{j} + A_z\mathbf{k} \quad \text{[three dimensions]}$$

However, the vector is also completely defined by its magnitude and associated angle. These two quantities can be written together in *phasor form*, sometimes referred to as *polar form*.

$$\mathbf{A} \equiv |\mathbf{A}| \angle \phi = A \angle \phi$$

5. CONVERSION BETWEEN SYSTEMS

The choice of the $\mathbf{ijk}$ triad may be convenient but is arbitrary. A vector can be expressed in terms of another set of unit vectors, $\mathbf{uvw}$.

$$\mathbf{V} = V_x\mathbf{i} + V_y\mathbf{j} + V_z\mathbf{k} = V_x'\mathbf{u} + V_y'\mathbf{v} + V_z'\mathbf{w} \qquad 5.10$$

The two representations are related.

$$V_x' = \mathbf{V} \cdot \mathbf{u} = (\mathbf{i} \cdot \mathbf{u})V_x + (\mathbf{j} \cdot \mathbf{u})V_y + (\mathbf{k} \cdot \mathbf{u})V_z \qquad 5.11$$
$$V_y' = \mathbf{V} \cdot \mathbf{v} = (\mathbf{i} \cdot \mathbf{v})V_x + (\mathbf{j} \cdot \mathbf{v})V_y + (\mathbf{k} \cdot \mathbf{v})V_z \qquad 5.12$$
$$V_z' = \mathbf{V} \cdot \mathbf{w} = (\mathbf{i} \cdot \mathbf{w})V_x + (\mathbf{j} \cdot \mathbf{w})V_y + (\mathbf{k} \cdot \mathbf{w})V_z \qquad 5.13$$

Equations 5.11 through 5.13 can be expressed in matrix form. The dot products are known as the *coefficients of transformation*, and the matrix containing them is the *transformation matrix*.

$$\begin{pmatrix} V'_x \\ V'_y \\ V'_z \end{pmatrix} = \begin{pmatrix} \mathbf{i} \cdot \mathbf{u} & \mathbf{j} \cdot \mathbf{u} & \mathbf{k} \cdot \mathbf{u} \\ \mathbf{i} \cdot \mathbf{v} & \mathbf{j} \cdot \mathbf{v} & \mathbf{k} \cdot \mathbf{v} \\ \mathbf{i} \cdot \mathbf{w} & \mathbf{j} \cdot \mathbf{w} & \mathbf{k} \cdot \mathbf{w} \end{pmatrix} \begin{pmatrix} V_x \\ V_y \\ V_z \end{pmatrix} \quad 5.14$$

6. VECTOR ADDITION

Addition of two vectors by the *polygon method* is accomplished by placing the tail of the second vector at the head (tip) of the first. The sum (i.e., the *resultant vector*) is a vector extending from the tail of the first vector to the head of the second. Alternatively, the two vectors can be considered as the two sides of a parallelogram, while the sum represents the diagonal. This is known as addition by the *parallelogram method*.

Figure 5.4 *Addition of Two Vectors*

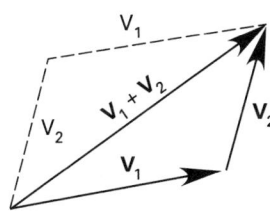

The components of the resultant vector are the sums of the components of the added vectors ($V_{1x} + V_{2x}$, $V_{1y} + V_{2y}$, $V_{1z} + V_{2z}$).

Vector addition is both commutative and associative.

$$\mathbf{V}_1 + \mathbf{V}_2 = \mathbf{V}_2 + \mathbf{V}_1 \quad 5.15$$
$$\mathbf{V}_1 + (\mathbf{V}_2 + \mathbf{V}_3) = (\mathbf{V}_1 + \mathbf{V}_2) + \mathbf{V}_3 \quad 5.16$$

7. MULTIPLICATION BY A SCALAR

A vector, $\mathbf{V}$, can be multiplied by a scalar, c. If the original vector is represented by its components, each of the components is multiplied by c.

$$c\mathbf{V} = c|\mathbf{V}|\mathbf{a} = cV_x\mathbf{i} + cV_y\mathbf{j} + cV_z\mathbf{k} \quad 5.17$$

Scalar multiplication is distributive.

$$c(\mathbf{V}_1 + \mathbf{V}_2) = c\mathbf{V}_1 + c\mathbf{V}_2 \quad 5.18$$

8. VECTOR DOT PRODUCT

The *dot product (scalar product)*, $\mathbf{V}_1 \cdot \mathbf{V}_2$, of two vectors is a scalar that is proportional to the length of the projection of the first vector onto the second vector.[4]

---

[4]The dot product is also written in parentheses without a dot, that is, $(\mathbf{V}_1\mathbf{V}_2)$.

Figure 5.5 *Vector Dot Product*

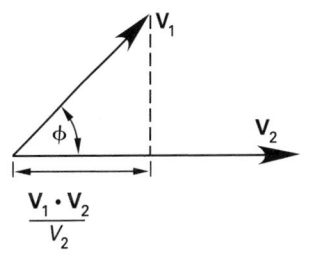

The dot product is commutative and distributive.

$$\mathbf{V}_1 \cdot \mathbf{V}_2 = \mathbf{V}_2 \cdot \mathbf{V}_1 \quad 5.19$$
$$\mathbf{V}_1 \cdot (\mathbf{V}_2 + \mathbf{V}_3) = \mathbf{V}_1 \cdot \mathbf{V}_2 + \mathbf{V}_1 \cdot \mathbf{V}_3 \quad 5.20$$

The dot product can be calculated in two ways, as Eq. 5.21 indicates. ϕ is limited to 180° and is the angle between the two vectors.

$$\mathbf{V}_1 \cdot \mathbf{V}_2 = |\mathbf{V}_1| |\mathbf{V}_2| \cos \phi$$
$$= V_{1x}V_{2x} + V_{1y}V_{2y} + V_{1z}V_{2z} \quad 5.21$$

When Eq. 5.21 is solved for the angle between the two vectors, ϕ, it is known as the *Cauchy-Schwartz theorem*.

$$\cos \phi = \frac{V_{1x}V_{2x} + V_{1y}V_{2y} + V_{1z}V_{2z}}{|\mathbf{V}_1| |\mathbf{V}_2|} \quad 5.22$$

The dot product can be used to determine whether a vector is a unit vector and to show that two vectors are orthogonal (perpendicular). For two non-null orthogonal vectors,

$$\mathbf{V}_1 \cdot \mathbf{V}_2 = 0 \quad 5.23$$

For any unit vector, $\mathbf{u}$,

$$\mathbf{u} \cdot \mathbf{u} = 1 \quad 5.24$$

Equations 5.23 and 5.24 can be extended to the Cartesian unit vectors.

$$\mathbf{i} \cdot \mathbf{i} = 1 \quad 5.25$$
$$\mathbf{j} \cdot \mathbf{j} = 1 \quad 5.26$$
$$\mathbf{k} \cdot \mathbf{k} = 1 \quad 5.27$$
$$\mathbf{i} \cdot \mathbf{j} = 0 \quad 5.28$$
$$\mathbf{i} \cdot \mathbf{k} = 0 \quad 5.29$$
$$\mathbf{j} \cdot \mathbf{k} = 0 \quad 5.30$$

Example 5.1

What is the angle between the zero-based vectors $\mathbf{V}_1 = (-\sqrt{3}, 1)$ and $\mathbf{V}_2 = (2\sqrt{3}, 2)$?

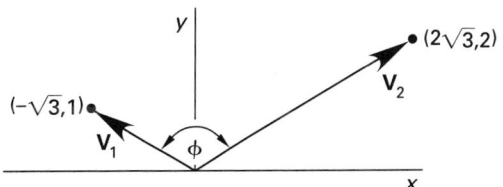

Solution

From Eq. 5.22,

$$\cos\phi = \frac{V_{1x}V_{2x} + V_{1y}V_{2y}}{|\mathbf{V}_1||\mathbf{V}_2|} = \frac{V_{1x}V_{2x} + V_{1y}V_{2y}}{\sqrt{V_{1x}^2 + V_{1y}^2}\sqrt{V_{2x}^2 + V_{2y}^2}}$$

$$= \frac{(-\sqrt{3})(2\sqrt{3}) + (1)(2)}{\sqrt{(-\sqrt{3})^2 + (1)^2}\sqrt{(2\sqrt{3})^2 + (2)^2}}$$

$$= \frac{-4}{8} = -\frac{1}{2}$$

$$\phi = \arccos\left(-\frac{1}{2}\right) = 120°$$

9. VECTOR CROSS PRODUCT

The *cross product (vector product)*, $\mathbf{V}_1 \times \mathbf{V}_2$, of two vectors is a vector that is orthogonal (perpendicular) to the plane of the two vectors.[5] The unit vector representation of the cross product can be calculated as a third-order determinant.

$$\mathbf{V}_1 \times \mathbf{V}_2 = \begin{vmatrix} \mathbf{i} & V_{1x} & V_{2x} \\ \mathbf{j} & V_{1y} & V_{2y} \\ \mathbf{k} & V_{1z} & V_{2z} \end{vmatrix} \qquad 5.31$$

The direction of the cross-product vector corresponds to the direction a right-hand screw would progress if vectors $\mathbf{V}_1$ and $\mathbf{V}_2$ are placed tail-to-tail in the plane they define and $\mathbf{V}_1$ is rotated into $\mathbf{V}_2$. The direction can also be found from the *right-hand rule*.

The magnitude of the cross product can be determined from Eq. 5.32, in which ϕ is the angle between the two vectors and is limited to 180°. The magnitude corresponds to the area of a parallelogram that has $\mathbf{V}_1$ and $\mathbf{V}_2$ as two of its sides.

$$|\mathbf{V}_1 \times \mathbf{V}_2| = |\mathbf{V}_1||\mathbf{V}_2|\sin\phi \qquad 5.32$$

Figure 5.6 *Vector Cross Product*

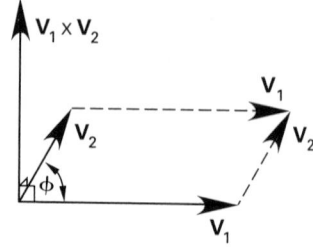

Vector cross multiplication is distributive but not commutative.

$$\mathbf{V}_1 \times \mathbf{V}_2 = -(\mathbf{V}_2 \times \mathbf{V}_1) \qquad 5.33$$
$$c(\mathbf{V}_1 \times \mathbf{V}_2) = (c\mathbf{V}_1) \times \mathbf{V}_2 = \mathbf{V}_1 \times (c\mathbf{V}_2) \qquad 5.34$$
$$\mathbf{V}_1 \times (\mathbf{V}_2 + \mathbf{V}_3) = \mathbf{V}_1 \times \mathbf{V}_2 + \mathbf{V}_1 \times \mathbf{V}_3 \qquad 5.35$$

---

[5]The cross product is also written in square brackets without a cross, that is, $[\mathbf{V}_1\mathbf{V}_2]$.

If the two vectors are parallel, their cross product will be zero.

$$\mathbf{i} \times \mathbf{i} = \mathbf{j} \times \mathbf{j} = \mathbf{k} \times \mathbf{k} = 0 \qquad 5.36$$

Equation 5.36 can be extended to the unit vectors.

$$\mathbf{i} \times \mathbf{j} = -\mathbf{j} \times \mathbf{i} = \mathbf{k} \qquad 5.37$$
$$\mathbf{j} \times \mathbf{k} = -\mathbf{k} \times \mathbf{j} = \mathbf{i} \qquad 5.38$$
$$\mathbf{k} \times \mathbf{i} = -\mathbf{i} \times \mathbf{k} = \mathbf{j} \qquad 5.39$$

Example 5.2

Find a unit vector orthogonal to $\mathbf{V}_1 = \mathbf{i} - \mathbf{j} + 2\mathbf{k}$ and $\mathbf{V}_2 = 3\mathbf{j} - \mathbf{k}$.

Solution

The cross product is a vector orthogonal to $\mathbf{V}_1$ and $\mathbf{V}_2$.

$$\mathbf{V}_1 \times \mathbf{V}_2 = \begin{vmatrix} \mathbf{i} & 1 & 0 \\ \mathbf{j} & -1 & 3 \\ \mathbf{k} & 2 & -1 \end{vmatrix}$$
$$= -5\mathbf{i} + \mathbf{j} + 3\mathbf{k}$$

Check to see whether this is a unit vector.

$$|\mathbf{V}_1 \times \mathbf{V}_2| = \sqrt{(-5)^2 + (1)^2 + (3)^2} = \sqrt{35}$$

Since its length is $\sqrt{35}$, the vector must be divided by $\sqrt{35}$ to obtain a unit vector.

$$\mathbf{a} = \frac{-5\mathbf{i} + \mathbf{j} + 3\mathbf{k}}{\sqrt{35}}$$

10. MIXED TRIPLE PRODUCT

The *mixed triple product (triple scalar product* or just *triple product)* of three vectors is a scalar quantity representing the volume of a parallelepiped with the three vectors making up the sides. It is calculated as a determinant. Since Eq. 5.40 can be negative, the absolute value must be used to obtain the volume in that case.

$$\mathbf{V}_1 \cdot (\mathbf{V}_2 \times \mathbf{V}_3) = \begin{vmatrix} V_{1x} & V_{1y} & V_{1z} \\ V_{2x} & V_{2y} & V_{2z} \\ V_{3x} & V_{3y} & V_{3z} \end{vmatrix} \qquad 5.40$$

Figure 5.7 *Vector Mixed Triple Product*

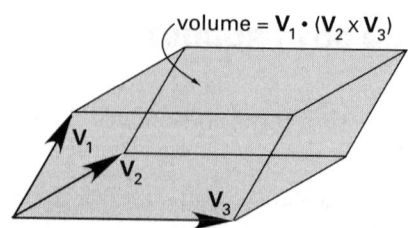

$$\text{volume} = \mathbf{V}_1 \cdot (\mathbf{V}_2 \times \mathbf{V}_3)$$

The mixed triple product has the property of *circular permutation*, as defined by Eq. 5.41.

$$\mathbf{V}_1 \cdot (\mathbf{V}_2 \times \mathbf{V}_3) = (\mathbf{V}_1 \times \mathbf{V}_2) \cdot \mathbf{V}_3 \qquad 5.41$$

11. VECTOR TRIPLE PRODUCT

The *vector triple product* is a vector defined by Eq. 5.42. The quantities in parentheses on the right-hand side are scalars.

$$\mathbf{V}_1 \times (\mathbf{V}_2 \times \mathbf{V}_3) = (\mathbf{V}_1 \cdot \mathbf{V}_3)\mathbf{V}_2 - (\mathbf{V}_1 \cdot \mathbf{V}_2)\mathbf{V}_3 \quad 5.42$$

12. VECTOR FUNCTIONS

A vector can be a function of another parameter. For example, a vector $\mathbf{V}$ is a function of variable t when its $V_x, V_y,$ and V_z are functions of t.

$$\mathbf{V}(t) = (2t - 3)\mathbf{i} + (t^2 + 1)\mathbf{j} + (-7t + 5)\mathbf{k} \qquad 5.43$$

When the functions of t are differentiated (or integrated) with respect to t, the vector itself is differentiated (integrated).[6] (Chapters 8 and 9 cover differentiation and integration.)

$$\frac{d\mathbf{V}(t)}{dt} = \left(\frac{dV_x}{dt}\right)\mathbf{i} + \left(\frac{dV_y}{dt}\right)\mathbf{j} + \left(\frac{dV_z}{dt}\right)\mathbf{k} \qquad 5.44$$

Similarly, the integral of the vector is

$$\int \mathbf{V}(t)\, dt = \mathbf{i}\int V_x\, dt + \mathbf{j}\int V_y\, dt + \mathbf{k}\int V_z\, dt \qquad 5.45$$

---

[6]This is particularly valuable when converting among position, velocity, and acceleration vectors.

6 Trigonometry

1. DEGREES AND RADIANS

Degrees and *radians* are two units for measuring angles. One complete circle is divided into 360 degrees (written 360°) or 2π radians (abbreviated *rad*).[1] The conversions between degrees and radians are

multiply	by	to obtain
radians	$\dfrac{180}{\pi}$	degrees
degrees	$\dfrac{\pi}{180}$	radians

The number of radians in an angle θ corresponds to two times the area within a circular sector with arc length θ and a radius of one. Alternatively, the area of a sector with central angle θ radians is $\theta/2$ for a *unit circle* (i.e., a circle with a radius of one unit).

Figure 6.1 *Radians and Area of Unit Circle*

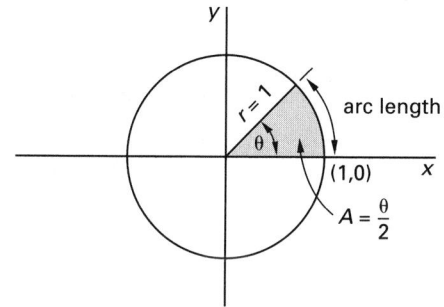

[1]The abbreviation *rad* is also used to represent *radiation absorbed dose*, a measure of radiation exposure.

2. PLANE ANGLES

A *plane angle* (usually referred to as just an *angle*) consists of two intersecting lines and an intersection point known as the *vertex*. The angle can be referred to by a capital letter representing the vertex (e.g., B in Fig. 6.2), a Greek letter representing the angular measure (e.g., β), or by three capital letters, where the middle letter is the vertex and the other two letters are two points on different lines, and either the symbol $\angle$ or $\measuredangle$ (e.g., $\measuredangle$ABC).

Figure 6.2 *Angle*

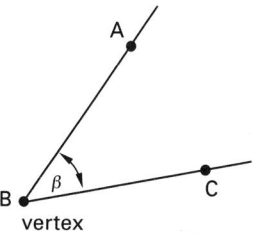

The angle between two intersecting lines generally is understood to be the smaller angle created.[2] Angles have been classified as follows.

- *acute angle*: an angle less than 90° ($\pi/2$ rad)

- *obtuse angle*: an angle more than 90° ($\pi/2$ rad) but less than 180° (π rad)

- *reflex angle*: an angle more than 180° (π rad) but less than 360° (2π rad)

- *related angle*: an angle that differs from another by some multiple of 90° ($\pi/2$ rad)

- *right angle*: an angle equal to 90° ($\pi/2$ rad)

- *straight angle*: an angle equal to 180° (π rad), that is, a straight line

Complementary angles are two angles whose sum is 90° ($\pi/2$ rad). *Supplementary angles* are two angles whose sum is 180° (π rad). *Adjacent angles* share a common vertex and one (the interior) side. Adjacent angles are supplementary only if their exterior sides form a straight line.

[2]In books on geometry, the term *ray* is used instead of *line*.

Vertical angles are the two angles with a common vertex and with sides made up by two intersecting straight lines. Vertical angles are equal.

Figure 6.3 *Vertical Angles*

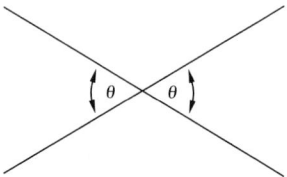

Angle of elevation and *angle of depression* are surveying terms referring to the angle above and below the horizontal plane of the observer, respectively.

3. TRIANGLES

A *triangle* is a three-sided closed polygon with three angles whose sum is 180° (π rad). Triangles are identified by their vertices and the symbol Δ (e.g., ΔABC in Fig. 6.4). A side is designated by its two endpoints (e.g., AB in Fig. 6.4) or by a lowercase letter corresponding to the capital letter of the opposite vertex (e.g., c).

In *similar triangles*, the corresponding angles are equal and the corresponding sides are in proportion. (Since there are only two independent angles in a triangle, showing that two angles of one triangle are equal to two angles of the other triangle is sufficient to show similarity.) The symbol for similarity is $\sim$. In Fig. 6.4, $\Delta ABC \sim \Delta DEF$ (i.e., ΔABC is similar to ΔDEF).

Figure 6.4 *Similar Triangles*

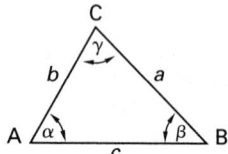 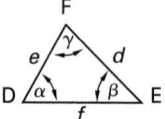

4. RIGHT TRIANGLES

A *right triangle* is a triangle in which one of the angles is 90° ($\pi/2$ rad). The remaining two angles are complementary. If one of the acute angles is chosen as the reference, the sides forming the right angle are known as the *adjacent side*, x, and the *opposite side*, y. The longest side is known as the *hypotenuse*, r. The *Pythagorean theorem* relates the lengths of these sides.

$$x^2 + y^2 = r^2 \hspace{2cm} 6.1$$

In certain cases, the lengths of unknown sides of right triangles can be determined by inspection.[3] This occurs when the lengths of the sides are in the ratios of 3:4:5, 1:1:$\sqrt{2}$, 1:$\sqrt{3}$:2, and 5:12:13.

Figure 6.5 *3:4:5 Right Triangle*

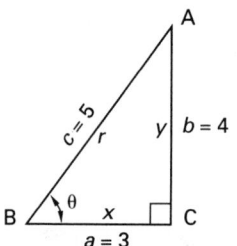

5. CIRCULAR TRANSCENDENTAL FUNCTIONS

The *circular transcendental functions* (usually referred to as the *transcendental functions*, *trigonometric functions*, or *functions of an angle*) are calculated from the sides of a right triangle. Equations 6.2 through 6.7 refer to Fig. 6.5.

$$\text{sine}: \sin\theta = \frac{y}{r} = \frac{\text{opposite}}{\text{hypotenuse}} \hspace{1cm} 6.2$$

$$\text{cosine}: \cos\theta = \frac{x}{r} = \frac{\text{adjacent}}{\text{hypotenuse}} \hspace{1cm} 6.3$$

$$\text{tangent}: \tan\theta = \frac{y}{x} = \frac{\text{opposite}}{\text{adjacent}} \hspace{1cm} 6.4$$

$$\text{cotangent}: \cot\theta = \frac{x}{y} = \frac{\text{adjacent}}{\text{opposite}} \hspace{1cm} 6.5$$

$$\text{secant}: \sec\theta = \frac{r}{x} = \frac{\text{hypotenuse}}{\text{adjacent}} \hspace{1cm} 6.6$$

$$\text{cosecant}: \csc\theta = \frac{r}{y} = \frac{\text{hypotenuse}}{\text{opposite}} \hspace{1cm} 6.7$$

---

[3]These cases are almost always contrived examples. There is nothing intrinsic in nature to cause the formation of triangles with these proportions.

Three of the transcendental functions are reciprocals of the others. Notice that while the tangent and cotangent functions are reciprocals of each other, the sine and cosine functions are not.

$$\cot \theta = \frac{1}{\tan \theta} \qquad 6.8$$

$$\sec \theta = \frac{1}{\cos \theta} \qquad 6.9$$

$$\csc \theta = \frac{1}{\sin \theta} \qquad 6.10$$

The trigonometric functions correspond to the lengths of various line segments in a right triangle with a unit hypotenuse. Figure 6.6 shows such a triangle inscribed in a unit circle.

Figure 6.6 *Trigonometric Functions in a Unit Circle*

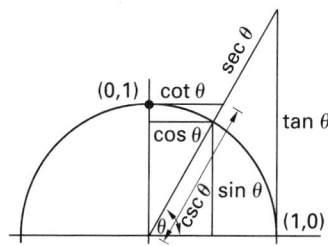

6. SMALL ANGLE APPROXIMATIONS

When an angle is very small, the hypotenuse and adjacent sides are essentially equal in length and certain approximations can be made. (The angle θ must be expressed in radians in Eqs. 6.11 and 6.12.)

$$\sin \theta \approx \tan \theta \approx \theta \Big|_{\theta < 10° \ (0.175 \ \text{rad})} \qquad 11$$

$$\cos \theta \approx 1 \Big|_{\theta < 5° \ (0.0873 \ \text{rad})} \qquad 12$$

7. GRAPHS OF THE FUNCTIONS

Figure 6.7 illustrates the periodicity of the sine, cosine, and tangent functions.[4]

Figure 6.7 *Graphs of Sine, Cosine, and Tangent Functions*

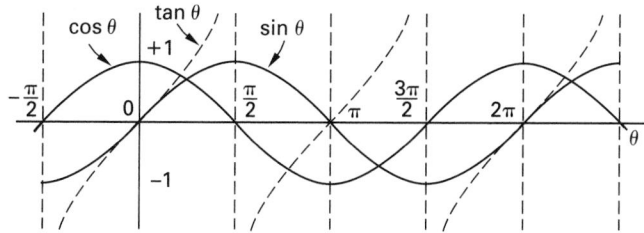

[4]The remaining functions, being reciprocals of these three functions, are also periodic.

8. SIGNS OF THE FUNCTIONS

Table 6.1 shows how the sine, cosine, and tangent functions vary in sign with different values of θ. All three functions are positive for angles $0° \leq \theta \leq 90°$ ($0 \leq \theta \leq \pi/2$ rad), but only the sine is positive for angles $90° < \theta \leq 180°$ ($\pi/2$ rad $< \theta \leq \pi$ rad). The concept of quadrants is used to summarize the signs of the functions: angles up to $90°$ ($\pi/2$ rad) are in quadrant I, between $90°$ and $180°$ ($\pi/2$ and π rad) are in quadrant II, and so on.

Table 6.1 *Signs of the Functions by Quadrant*

function	quadrant			
	I	II	III	IV
sine	+	+	−	−
cosine	+	−	−	+
tangent	+	−	+	−

9. FUNCTIONS OF RELATED ANGLES

Figure 6.7 shows that the sine, cosine, and tangent curves are symmetrical with respect to the horizontal axis. Furthermore, portions of the curves are symmetrical with respect to a vertical axis. The values of the sine and cosine functions repeat every $360°$ (2π rad), and the absolute values repeat every $180°$ (π rad). This can be written as

$$\sin(\theta + 180°) = -\sin \theta \qquad 6.13$$

Similarly, the tangent function repeats every $180°$ (π rad), and its absolute value repeats every $90°$ ($\pi/2$ rad).

Table 6.2 summarizes the functions of the related angles.

Table 6.2 *Functions of Related Angles*

$f(\theta)$	$-\theta$	$90° - \theta$	$90° + \theta$	$180° - \theta$	$180° + \theta$
sin	$-\sin\theta$	$\cos\theta$	$\cos\theta$	$\sin\theta$	$-\sin\theta$
cos	$\cos\theta$	$\sin\theta$	$-\sin\theta$	$-\cos\theta$	$-\cos\theta$
tan	$-\tan\theta$	$\cot\theta$	$-\cot\theta$	$-\tan\theta$	$\tan\theta$

10. TRIGONOMETRIC IDENTITIES

There are many relationships between trigonometric functions. For example, Eqs. 6.14 through 6.16 are well known.

$$\sin^2 \theta + \cos^2 \theta = 1 \qquad 6.14$$

$$1 + \tan^2 \theta = \sec^2 \theta \qquad 6.15$$

$$1 + \cot^2 \theta = \csc^2 \theta \qquad 6.16$$

Other relatively common identities are listed as follows.[5]

- *double-angle formulas:*

$$\sin 2\theta = 2\sin\theta\cos\theta = \frac{2\tan\theta}{1+\tan^2\theta} \qquad 6.17$$

$$\cos 2\theta = \cos^2\theta - \sin^2\theta = 1 - 2\sin^2\theta$$
$$= 2\cos^2\theta - 1 = \frac{1-\tan^2\theta}{1+\tan^2\theta} \qquad 6.18$$

$$\tan 2\theta = \frac{2\tan\theta}{1-\tan^2\theta} \qquad 6.19$$

$$\cot 2\theta = \frac{\cot^2\theta - 1}{2\cot\theta} \qquad 6.20$$

- *two-angle formulas:*

$$\sin(\theta \pm \phi) = \sin\theta\cos\phi \pm \cos\theta\sin\phi \qquad 6.21$$

$$\cos(\theta \pm \phi) = \cos\theta\cos\phi \mp \sin\theta\sin\phi \qquad 6.22$$

$$\tan(\theta \pm \phi) = \frac{\tan\theta \pm \tan\phi}{1 \mp \tan\theta\tan\phi} \qquad 6.23$$

$$\cot(\theta \pm \phi) = \frac{\cot\phi\cot\theta \mp 1}{\cot\phi \pm \cot\theta} \qquad 6.24$$

- *half-angle formulas* ($\theta < 180°$):

$$\sin\frac{\theta}{2} = \sqrt{\frac{1-\cos\theta}{2}} \qquad 6.25$$

$$\cos\frac{\theta}{2} = \sqrt{\frac{1+\cos\theta}{2}} \qquad 6.26$$

$$\tan\frac{\theta}{2} = \sqrt{\frac{1-\cos\theta}{1+\cos\theta}} = \frac{\sin\theta}{1+\cos\theta} = \frac{1-\cos\theta}{\sin\theta} \qquad 6.27$$

- *miscellaneous formulas* ($\theta < 90°$):

$$\sin\theta = 2\sin\left(\frac{\theta}{2}\right)\cos\left(\frac{\theta}{2}\right) \qquad 6.28$$

$$\sin\theta = \sqrt{\frac{1-\cos 2\theta}{2}} \qquad 6.29$$

$$\cos\theta = \cos^2\left(\frac{\theta}{2}\right) - \sin^2\left(\frac{\theta}{2}\right) \qquad 6.30$$

$$\cos\theta = \sqrt{\frac{1+\cos 2\theta}{2}} \qquad 6.31$$

$$\tan\theta = \frac{2\tan\left(\frac{\theta}{2}\right)}{1-\tan^2\left(\frac{\theta}{2}\right)}$$
$$= \frac{2\sin\left(\frac{\theta}{2}\right)\cos\left(\frac{\theta}{2}\right)}{\cos^2\left(\frac{\theta}{2}\right) - \sin^2\left(\frac{\theta}{2}\right)} \qquad 6.32$$

$$\tan\theta = \sqrt{\frac{1-\cos 2\theta}{1+\cos 2\theta}}$$
$$= \frac{\sin 2\theta}{1+\cos 2\theta} = \frac{1-\cos 2\theta}{\sin 2\theta} \qquad 6.33$$

$$\cot\theta = \frac{\cot^2\left(\frac{\theta}{2}\right) - 1}{2\cot\left(\frac{\theta}{2}\right)}$$
$$= \frac{\cos^2\left(\frac{\theta}{2}\right) - \sin^2\left(\frac{\theta}{2}\right)}{2\sin\left(\frac{\theta}{2}\right)\cos\left(\frac{\theta}{2}\right)} \qquad 6.34$$

$$\cot\theta = \sqrt{\frac{1+\cos 2\theta}{1-\cos 2\theta}}$$
$$= \frac{1+\cos 2\theta}{\sin 2\theta} = \frac{\sin 2\theta}{1-\cos 2\theta} \qquad 6.35$$

11. INVERSE TRIGONOMETRIC FUNCTIONS

Finding an angle from a known trigonometric function is a common operation known as an *inverse trigonometric operation*. The inverse function can be designated by adding "inverse," "arc-," or the superscript -1 to the name of the function. For example,

$$\text{inverse } \sin(0.5) = \arcsin(0.5) = \sin^{-1}(0.5) = 30°$$

12. HYPERBOLIC TRANSCENDENTAL FUNCTIONS

Hyperbolic transcendental functions (normally referred to as *hyperbolic functions*) are specific equations containing combinations of the terms e^θ and $e^{-\theta}$. These combinations appear regularly in certain types of problems (e.g., analysis of cables and heat transfer from fins) and are given specific names and symbols to simplify presentation.[6]

$$\text{hyperbolic sine}: \sinh\theta = \frac{e^\theta - e^{-\theta}}{2} \qquad 6.36$$

$$\text{hyperbolic cosine}: \cosh\theta = \frac{e^\theta + e^{-\theta}}{2} \qquad 6.37$$

---

[5]It is an idiosyncrasy that these formulas are conventionally referred to as *formulas*, not *identities*.

[6]The hyperbolic sine and cosine functions are pronounced (by some) as "sinch" and "cosh," respectively.

hyperbolic tangent : $\tanh\theta = \dfrac{e^\theta - e^{-\theta}}{e^\theta + e^{-\theta}} = \dfrac{\sinh\theta}{\cosh\theta}$ **6.38**

hyperbolic cotangent : $\coth\theta = \dfrac{e^\theta + e^{-\theta}}{e^\theta - e^{-\theta}} = \dfrac{\cosh\theta}{\sinh\theta}$ **6.39**

hyperbolic secant : $\operatorname{sech}\theta = \dfrac{2}{e^\theta + e^{-\theta}} = \dfrac{1}{\cosh\theta}$ **6.40**

hyperbolic cosecant : $\operatorname{csch}\theta = \dfrac{2}{e^\theta - e^{-\theta}} = \dfrac{1}{\sinh\theta}$ **6.41**

Hyperbolic functions cannot be related to a right triangle, but they are related to a rectangular (equilateral) hyperbola, as shown in Fig. 6.8. The shaded area has a value of $\theta/2$ and is sometimes given the units of *hyperbolic radians*.

$$\sinh\theta = \frac{y}{a} \qquad \textbf{6.42}$$

$$\cosh\theta = \frac{x}{a} \qquad \textbf{6.43}$$

$$\tanh\theta = \frac{y}{x} \qquad \textbf{6.44}$$

$$\coth\theta = \frac{x}{y} \qquad \textbf{6.45}$$

$$\operatorname{sech}\theta = \frac{a}{x} \qquad \textbf{6.46}$$

$$\operatorname{csch}\theta = \frac{a}{y} \qquad \textbf{6.47}$$

Figure 6.8 *Equilateral Hyperbola and Hyperbolic Functions*

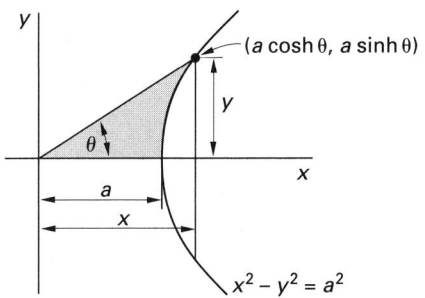

13. HYPERBOLIC IDENTITIES

The hyperbolic identities are different from the standard trigonometric identities. Some of the most important identities are presented as follows.

$$\cosh^2\theta - \sinh^2\theta = 1 \qquad \textbf{6.48}$$

$$1 - \tanh^2\theta = \operatorname{sech}^2\theta \qquad \textbf{6.49}$$

$$1 - \coth^2\theta = -\operatorname{csch}^2\theta \qquad \textbf{6.50}$$

$$\cosh\theta + \sinh\theta = e^\theta \qquad \textbf{6.51}$$

$$\cosh\theta - \sinh\theta = e^{-\theta} \qquad \textbf{6.52}$$

$$\sinh(\theta \pm \phi) = \sinh\theta\cosh\phi \pm \cosh\theta\sinh\phi \qquad \textbf{6.53}$$

$$\cosh(\theta \pm \phi) = \cosh\theta\cosh\phi \pm \sinh\theta\sinh\phi \qquad \textbf{6.54}$$

$$\tanh(\theta \pm \phi) = \frac{\tanh\theta \pm \tanh\phi}{1 \pm \tanh\theta\tanh\phi} \qquad \textbf{6.55}$$

14. GENERAL TRIANGLES

A *general triangle* (also known as an *oblique triangle*) is one that is not specifically a right triangle, as shown in Fig. 6.9. Equation 6.56 calculates the area of a general triangle.[7]

$$\text{area} = \tfrac{1}{2}ab\sin C = \tfrac{1}{2}bc\sin A = \tfrac{1}{2}ca\sin B \qquad \textbf{6.56}$$

The *law of sines* (Eq. 6.57) relates the sides and the sines of the angles.

$$\frac{\sin A}{a} = \frac{\sin B}{b} = \frac{\sin C}{c} \qquad \textbf{6.57}$$

The *law of cosines* relates the cosine of an angle to an opposite side. (Equation 6.58 can be extended to the two remaining sides.)

$$a^2 = b^2 + c^2 - 2bc\cos A \qquad \textbf{6.58}$$

Figure 6.9 *General Triangle*

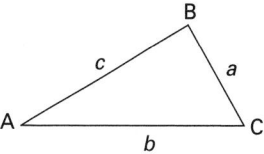

The *law of tangents* relates the sum and difference of two sides. (Equation 6.59 can be extended to the two remaining sides.)

$$\frac{a-b}{a+b} = \frac{\tan\left(\dfrac{A-B}{2}\right)}{\tan\left(\dfrac{A+B}{2}\right)} \qquad \textbf{6.59}$$

15. SPHERICAL TRIGONOMETRY

A *spherical triangle* is a triangle that has been drawn on the surface of a sphere, as shown in Fig. 6.10. The *trihedral angle* O–ABC is formed when the vertices A, B, and C are joined to the center of the sphere. The *face angles* (BOC, COA, and AOB in Fig. 6.10) are used to measure the sides (a, b, and c in Fig. 6.10). The *vertex angles* are A, B, and C. Thus, angles are used to measure both vertex angles and sides.

---

[7]Other methods of calculating the area of a general triangle are given in Chap. 7.

Figure 6.10 *Spherical Triangle ABC*

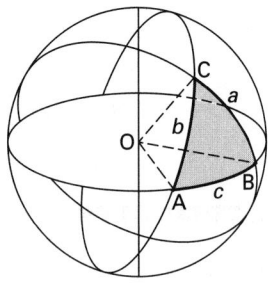

The following rules are valid for spherical triangles for which each side and angle is less than 180°.

- The sum of the three vertex angles is greater than 180° and less than 540°.

$$180° < A + B + C < 540° \qquad 6.60$$

- The sum of any two sides is greater than the third side.

- The sum of the three sides is less than 360°.

$$0° < a + b + c < 360° \qquad 6.61$$

- If the two sides are equal, the corresponding angles opposite are equal, and the converse is also true.

- If two sides are unequal, the corresponding angles opposite are unequal. The greater angle is opposite the greater side.

The *spherical excess*, ϵ, is the amount by which the sum of the vertex angles exceeds 180°. The *spherical defect*, d, is the amount by which the sum of the sides differs from 360°.

$$\epsilon = A + B + C - 180° \qquad 6.62$$
$$d = 360° - (a + b + c) \qquad 6.63$$

There are many trigonometric identities that define the relationships between angles in a spherical triangle. Some of the more common identities are presented as follows.

- *law of sines*:

$$\frac{\sin A}{\sin a} = \frac{\sin B}{\sin b} = \frac{\sin C}{\sin c} \qquad 6.64$$

- *first law of cosines*:

$$\cos a = \cos b \cos c + \sin b \sin c \cos A \qquad 6.65$$

- *second law of cosines*:

$$\cos A = -\cos B \cos C + \sin B \sin C \cos a \qquad 6.66$$

16. SOLID ANGLES

A *solid angle*, ω, is a measure of the angle subtended at the vertex of a cone. The solid angle has units of *steradians* (abbreviated *sr*). A steradian is the solid angle subtended at the center of a unit sphere (i.e., a sphere with a radius of one) by a unit area on its surface. Since the surface area of a sphere of radius r is r^2 times the surface area of a unit sphere, the solid angle is equal to the area cut out by the cone divided by r^2. (The surface area of a spherical segment is given in App. 7.B.)

$$\omega = \frac{\text{surface area}}{r^2} \qquad 6.67$$

Figure 6.11 *Solid Angle*

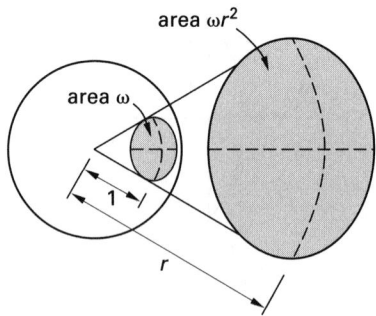

7 Analytic Geometry

1. MENSURATION OF REGULAR SHAPES

The dimensions, perimeter, area, and other geometric properties constitute the *mensuration* (i.e., the measurements) of a geometric shape. Appendices 7.A and 7.B contain formulas and tables used to calculate these properties.

Example 7.1

In the study of open channel fluid flow, the hydraulic radius is defined as the ratio of flow area to wetted perimeter. What is the hydraulic radius of a 6 in diameter pipe filled to a depth of 2 in?

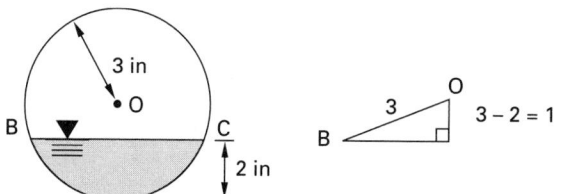

Solution

Points O, B, and C constitute a circular segment and are used to find the central angle of the circular segment.

$$\left(\tfrac{1}{2}\right)(\angle BOC) = \arccos\left(\tfrac{1}{3}\right) = 70.53°$$

$$\phi = \angle BOC = (2)(70.53°) = 141.06° = 2.462 \text{ rad}$$

From App. 7.A, the area in flow and arc length are

$$A = \tfrac{1}{2}r^2(\phi - \sin\phi)$$
$$= \left(\tfrac{1}{2}\right)(3 \text{ in})^2\left(2.462 \text{ rad} - \sin(2.462 \text{ rad})\right)$$
$$= 8.251 \text{ in}^2$$

$$s = r\phi$$
$$= (3 \text{ in})(2.462 \text{ rad}) = 7.386 \text{ in}$$

The hydraulic radius is

$$r_h = \frac{A}{s}$$
$$= \frac{8.251 \text{ in}^2}{7.386 \text{ in}} = 1.12 \text{ in}$$

2. AREAS WITH IRREGULAR BOUNDARIES

Areas of sections with irregular boundaries (such as creek banks) cannot be determined precisely, and approximation methods must be used. If the irregular side can be divided into a series of cells of equal width, either the trapezoidal rule or Simpson's rule can be used.

Figure 7.1 *Irregular Areas*

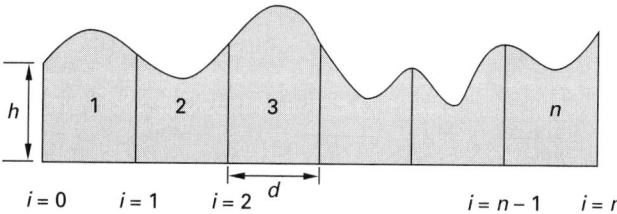

If the irregular side of each cell is fairly straight, the *trapezoidal rule* is appropriate.

$$A = \left(\frac{d}{2}\right)\left(h_0 + h_n + 2\sum_{i=1}^{n-1} h_i\right) \qquad 7.1$$

If the irregular side of each cell is curved (parabolic), *Simpson's rule* should be used. (*n* must be even to use Simpson's rule.)

$$A = \left(\frac{d}{3}\right)\left(h_0 + h_n + 4\sum_{\substack{i \text{ odd} \\ i=1}}^{n-1} h_i + 2\sum_{\substack{i \text{ even} \\ i=2}}^{n-2} h_i\right) \quad \textit{7.2}$$

3. GEOMETRIC DEFINITIONS

The following terms are used in this book to describe the relationship or orientation of one geometric figure to another.

- *abscissa*: the horizontal coordinate, typically designated as x in a rectangular coordinate system

- *asymptote*: a straight line that is approached but not intersected by a curved line

- *asymptotic*: approaching the slope of another line; attaining the slope of another line in the limit

- *center*: a point equidistant from all other points

- *collinear*: falling on the same line

- *concave*: curved inward (in the direction indicated)[1]

- *convex*: curved outward (in the direction indicated)

- *convex hull*: a closed figure whose surface is convex everywhere

- *coplanar*: falling on the same plane

- *inflection point*: a point where the second derivative changes sign or the curve changes from concave to convex. (Also known as a *point of contraflexure*.)

- *locus of points*: a set or collection of points having some common property and being so infinitely close together as to be indistinguishable from a line

- *node*: a point on a line from which other lines enter or leave

- *normal*: rotated 90°; being at right angles

- *ordinate*: the vertical coordinate, typically designated as y in a rectangular coordinate system

- *orthogonal*: rotated 90°; being at right angles

---

[1]This is easily remembered since one must go inside to explore a cave.

- *saddle point*: a point in three-dimensional space where all adjacent points are higher in one direction (the direction of the saddle) and lower in an orthogonal direction (the direction of the sides)

- *tangent*: having equal slopes at a common point

Figure 7.2 *Geometric Definitions*

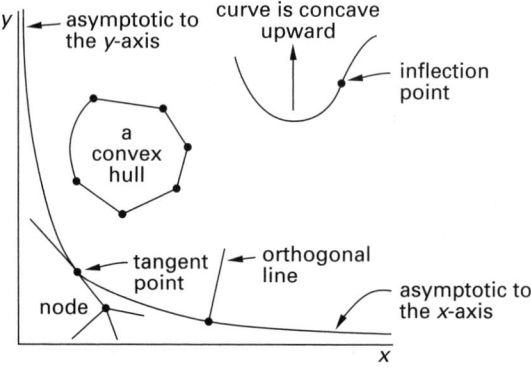

4. CONCAVE CURVES

Concavity is a term that is applied to curved lines. A *concave up curve* is one whose function's first derivative increases continuously from negative to positive values. Straight lines drawn tangent to concave up curves are all below the curve. The graph of such a function may be thought of as being able to "hold water."

The first derivative of a *concave down curve* decreases continuously from positive to negative. A graph of a concave down function may be thought of as "spilling water." See Fig. 7.3.

Figure 7.3 *Concave Curves*

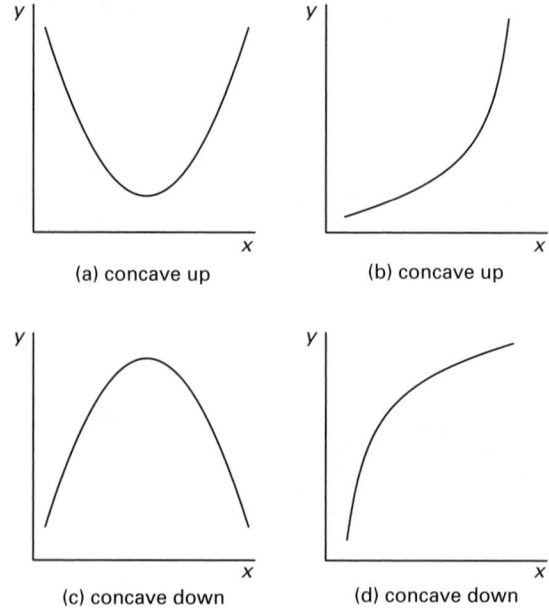

5. CONVEX REGIONS

Convexity is a term that is applied to sets and regions.[2] It plays an important role in many mathematics subjects. A set or multidimensional region is *convex* if it contains the line segment joining any two of its points; that is, if a straight line is drawn connecting any two points in a convex region, that line will lie entirely within the region. For example, the interior of a parabola is a convex region, as is a solid sphere. The *void* or *null region* (i.e., an empty set of points), single points, and straight lines are convex sets. A convex region bounded by separate, connected line segments is known as a *convex hull*. (See Fig. 7.4.)

Within a convex region, a local maximum is also the global maximum. Similarly, a local minimum is also the global minimum. (See Sec. 8.3.) The intersection of two convex regions is also convex.

Figure 7.4 *Convexity*

 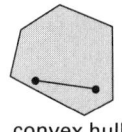 

convex region convex hull nonconvex region

6. CONGRUENCY

Congruence in geometric figures is analogous to *equality* in algebraic expressions. Congruent line segments are segments that have the same length. Congruent angles have the same angular measure. Congruent triangles have the same vertex angles and side lengths.

In general, *congruency*, indicated by the symbol $\cong$, means that there is one-to-one correspondence between all points on two objects. This correspondence is defined by the *mapping function* or *isometry*, which can be a translation, rotation, or reflection. Since the identity function is a valid mapping function, every geometric shape is congruent to itself.

Two congruent objects can be in different spaces. For example, a triangular area in three-dimensional space can be mapped into a triangle in two-dimensional space.

---

[2]It is tempting to define regions that fail the convexity test as being "concave." However, it is more proper to define such regions as "nonconvex." In any case, it is important to recognize that convexity depends on the reference point: An observer within a sphere will see the spherical boundary as convex; an observer outside the sphere may see the boundary as nonconvex.

7. COORDINATE SYSTEMS

The manner in which a geometric figure is described depends on the coordinate system that is used. The three-dimensional *rectangular coordinate system* (also known as the *Cartesian coordinate system*) with its x-, y-, and z-coordinates is the most commonly used in engineering. Table 7.1 summarizes the components needed to specify a point in the various coordinate systems. Figure 7.5 illustrates the use of and conversion between the coordinate systems.

Table 7.1 *Components of Coordinate Systems*

name	dimensions	components
rectangular	2	x, y
rectangular	3	x, y, z
polar	2	r, θ
cylindrical	3	r, θ, z
spherical	3	r, θ, ϕ

8. CURVES

A *curve* (commonly called a *line*) is a function over a finite or infinite range of the independent variable. When a curve is drawn in two- or three-dimensional space, it is known as a *graph of the curve*. It may or may not be possible to describe the curve mathematically. The *degree of a curve* is the highest exponent in the function. For example, Eq. 7.3 is a fourth-degree curve.

$$f(x) = 2x^4 + 7x^3 + 6x^2 + 3x + 9 = 0 \qquad 7.3$$

An *ordinary cycloid* (Fig. 7.6) is a curve traced out by a point on the rim of a wheel that rolls without slipping. It is described in parametric form by Eqs. 7.4 and 7.5 and in rectangular form by Eq. 7.6. In Eqs. 7.4 through 7.6, using the minus sign results in a cusp at the origin; using the plus sign results in a vertex (trough) at the origin.

$$x = r(\theta \pm \sin\theta) \qquad 7.4$$
$$y = r(1 \pm \cos\theta) \qquad 7.5$$
$$x = r\arccos\left(\frac{r-y}{r}\right) \pm \sqrt{2ry - y^2} \qquad 7.6$$

An *epicycloid* is a curve generated by a point on the rim of a wheel that rolls on the outside of a circle. A *hypocycloid* is a curve generated by a point on the rim of a wheel that rolls on the inside of a circle. The equation of a hypocycloid of four cusps is

$$x^{2/3} + y^{2/3} = r^{2/3} \quad \text{[4 cusps]} \qquad 7.7$$

Figure 7.5 *Different Coordinate Systems*

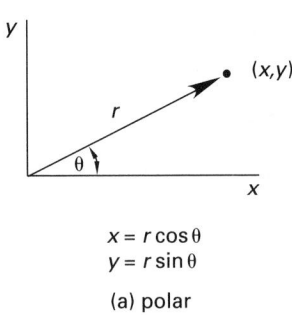

$$x = r \cos \theta$$
$$y = r \sin \theta$$

(a) polar

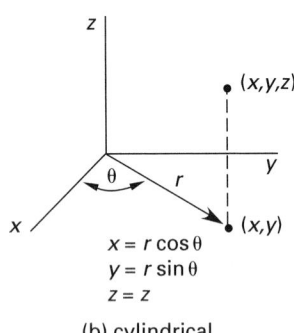

$$x = r \cos \theta$$
$$y = r \sin \theta$$
$$z = z$$

(b) cylindrical

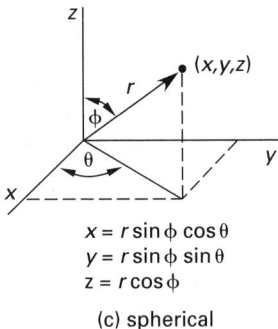

$$x = r \sin \phi \cos \theta$$
$$y = r \sin \phi \sin \theta$$
$$z = r \cos \phi$$

(c) spherical

Figure 7.6 *Cycloid (cusp at origin shown)*

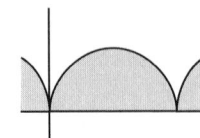

9. SYMMETRY OF CURVES

Two points, P and Q, are symmetrical with respect to a line if the line is a perpendicular bisector of the line segment PQ. If the graph of a curve is unchanged when y is replaced with $-y$, the curve is symmetrical with respect to the x-axis. If the curve is unchanged when x is replaced with $-x$, the curve is symmetrical with respect to the y-axis.

Figure 7.7 *Waveform Symmetry*

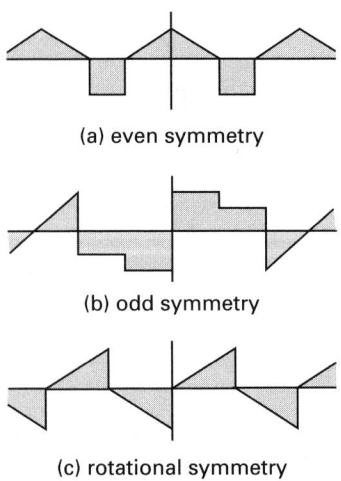

(a) even symmetry

(b) odd symmetry

(c) rotational symmetry

Repeating waveforms can be symmetrical with respect to the y-axis. A curve $f(x)$ is said to have *even symmetry* if $f(x) = f(-x)$. (Alternatively, $f(x)$ is said to be a *symmetrical function*.) With even symmetry, the function to the left of $x = 0$ is a reflection of the function to the right of $x = 0$. (In effect, the y-axis is a mirror.) The cosine curve is an example of a curve with even symmetry.

A curve is said to have *odd symmetry* if $f(x) = -f(-x)$. (Alternatively, $f(x)$ is said to be an *asymmetrical function*.)[3] The sine curve is an example of a curve with odd symmetry.

A curve is said to have *rotational symmetry* (*half-wave symmetry*) if $f(x) = -f(x + \pi)$.[4] Curves of this type are identical except for a sign reversal on alternate half-cycles.

Table 7.2 describes the type of function resulting from the combination of two functions.

Table 7.2 *Combinations of Functions*

	operation			
	$+$	$-$	$\times$	$\div$
$f_1(x)$ even, $f_2(x)$ even	even	even	even	even
$f_1(x)$ odd, $f_2(x)$ odd	odd	odd	even	even
$f_1(x)$ even, $f_2(x)$ odd	neither	neither	odd	odd

10. STRAIGHT LINES

Figure 7.8 illustrates a straight line in two-dimensional space. The *slope* of the line is m, the *y-intercept* is b, and the *x-intercept* is a. The equation of the line

---

[3] The semantics of "even symmetry" and "asymmetrical function" are contradictory.
[4] The symbol π represents half of a full cycle of the waveform, not the value 3.141 . . . .

can be represented in several forms, and the procedure for finding the equation depends on the form chosen to represent the line. In general, the procedure involves substituting one or more known points on the line into the equation in order to determine the coefficients.

- *general form:*

$$Ax + By + C = 0 \qquad 7.8$$

$$A = -mB \qquad 7.9$$

$$B = \frac{-C}{b} \qquad 7.10$$

$$C = -aA = -bB \qquad 7.11$$

- *slope-intercept form:*

$$y = mx + b \qquad 7.12$$

$$m = \frac{-A}{B} = \tan\theta = \frac{y_2 - y_1}{x_2 - x_1} \qquad 7.13$$

$$b = \frac{-C}{B} \qquad 7.14$$

$$a = \frac{-C}{A} \qquad 7.15$$

- *point-slope form:*

$$y - y_1 = m(x - x_1) \qquad 7.16$$

- *intercept form:*

$$\frac{x}{a} + \frac{y}{b} = 1 \qquad 7.17$$

- *two-point form:*

$$\frac{y - y_1}{x - x_1} = \frac{y_2 - y_1}{x_2 - x_1} \qquad 7.18$$

- *normal form:*

$$x\cos\beta + y\sin\beta - d = 0 \qquad 7.19$$

(d and β are constants; x and y are variables.)

- *polar form:*

$$r = \frac{d}{\cos(\beta - \alpha)} \qquad 7.20$$

(d and β are constants; r and α are variables.)

Figure 7.8 *Straight Line*

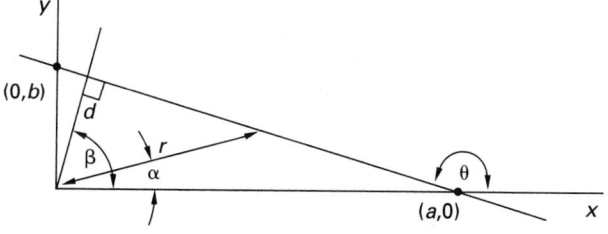

11. DIRECTION NUMBERS, ANGLES, AND COSINES

Given a directed line from (x_1, y_1, z_1) to (x_2, y_2, z_2), the *direction numbers* are

$$L = x_2 - x_1 \qquad 7.21$$

$$M = y_2 - y_1 \qquad 7.22$$

$$N = z_2 - z_1 \qquad 7.23$$

The distance between two points is

$$d = \sqrt{L^2 + M^2 + N^2} \qquad 7.24$$

The *direction cosines* are

$$\cos\alpha = \frac{L}{d} \qquad 7.25$$

$$\cos\beta = \frac{M}{d} \qquad 7.26$$

$$\cos\gamma = \frac{N}{d} \qquad 7.27$$

Note that

$$\cos^2\alpha + \cos^2\beta + \cos^2\gamma = 1 \qquad 7.28$$

The *direction angles* are the angles between the axes and the lines. They are found from the inverse functions of the direction cosines.

$$\alpha = \arccos\left(\frac{L}{d}\right) \qquad 7.29$$

$$\beta = \arccos\left(\frac{M}{d}\right) \qquad 7.30$$

$$\gamma = \arccos\left(\frac{N}{d}\right) \qquad 7.31$$

The direction cosines can be used to write the equation of the straight line in terms of the unit vectors. The line **R** would be defined as

$$\mathbf{R} \equiv d(\mathbf{i}\cos\alpha + \mathbf{j}\cos\beta + \mathbf{k}\cos\gamma) \qquad 7.32$$

Similarly, the line may be written in terms of its direction numbers.

$$\mathbf{R} = L\mathbf{i} + M\mathbf{j} + N\mathbf{k} \qquad 7.33$$

Example 7.2

A line passes through the points (4,7,9) and (0,1,6). Write the equation of the line in terms of its (a) direction numbers and (b) direction cosines.

Solution

(a) The direction numbers are

$$L = 4 - 0 = 4$$
$$M = 7 - 1 = 6$$
$$N = 9 - 6 = 3$$

Using Eq. 7.33,

$$\mathbf{R} = 4\mathbf{i} + 6\mathbf{j} + 3\mathbf{k}$$

(b) The distance between the two points is

$$d = \sqrt{(4)^2 + (6)^2 + (3)^2} = 7.81$$

The line in terms of its direction cosines is

$$\mathbf{R} = \frac{4\mathbf{i} + 6\mathbf{j} + 3\mathbf{k}}{7.81}$$
$$= 0.512\mathbf{i} + 0.768\mathbf{j} + 0.384\mathbf{k}$$

12. INTERSECTION OF TWO LINES

The intersection of two lines is a point. The location of the intersection point can be determined by setting the two equations equal and solving them in terms of a common variable. Alternatively, Eqs. 7.34 and 7.35 can be used to calculate the coordinates of the intersection point.

$$x = \frac{B_2 C_1 - B_1 C_2}{A_2 B_1 - A_1 B_2} \qquad 7.34$$

$$y = \frac{A_1 C_2 - A_2 C_1}{A_2 B_1 - A_1 B_2} \qquad 7.35$$

13. PLANES

A *plane* in three-dimensional space is completely determined by one of the following:

- three noncollinear points
- two nonparallel vectors $\mathbf{V}_1$ and $\mathbf{V}_2$ and their intersection point P_0
- a point P_0 and a vector, $\mathbf{N}$, normal to the plane (i.e., the *normal vector*)

Figure 7.9 *Plane in Three-Dimensional Space*

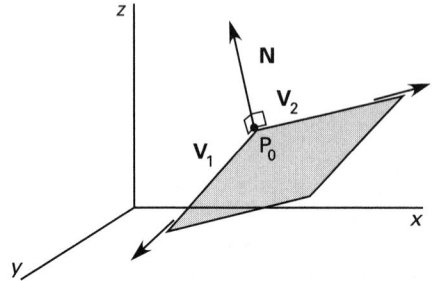

The plane can be specified mathematically in one of two ways: in rectangular form or as a parametric equation. The general form is

$$A(x - x_0) + B(y - y_0) + C(z - z_0) = 0 \qquad 7.36$$

x_0, y_0, and z_0 are the coordinates of the intersection point of any two vectors in the plane. The coefficients A, B, and C are the same as the coefficients of the normal vector, $\mathbf{N}$.

$$\mathbf{N} = \mathbf{V}_1 \times \mathbf{V}_2 = A\mathbf{i} + B\mathbf{j} + C\mathbf{k} \qquad 7.37$$

Equation 7.36 can be simplified as follows.

$$Ax + By + Cz + D = 0 \qquad 7.38$$
$$D = -(Ax_0 + By_0 + Cz_0) \qquad 7.39$$

The following procedure can be used to determine the equation of a plane from three noncollinear points, P_1, P_2, and P_3, or from a normal vector and a single point.

step 1: (If the normal vector is known, go to step 3.) Determine the equations of the vectors $\mathbf{V}_1$ and $\mathbf{V}_2$ from two pairs of the points. For example, determine $\mathbf{V}_1$ from points P_1 and P_2, and determine $\mathbf{V}_2$ from P_1 and P_3. Express the vectors in the form $A\mathbf{i} + B\mathbf{j} + C\mathbf{k}$.

$$\mathbf{V}_1 = (x_2 - x_1)\mathbf{i} + (y_2 - y_1)\mathbf{j}$$
$$+ (z_2 - z_1)\mathbf{k} \qquad 7.40$$
$$\mathbf{V}_2 = (x_3 - x_1)\mathbf{i} + (y_3 - y_1)\mathbf{j}$$
$$+ (z_3 - z_1)\mathbf{k} \qquad 7.41$$

step 2: Find the normal vector, $\mathbf{N}$, as the cross product of the two vectors.

$$\mathbf{N} = \mathbf{V}_1 \times \mathbf{V}_2$$
$$= \begin{vmatrix} \mathbf{i} & (x_2 - x_1) & (x_3 - x_1) \\ \mathbf{j} & (y_2 - y_1) & (y_3 - y_1) \\ \mathbf{k} & (z_2 - z_1) & (z_3 - z_1) \end{vmatrix} \qquad 7.42$$

step 3: Write the general equation of the plane in rectangular form (Eq. 7.36) using the coefficients A, B, and C from the normal vector and any one of the three points as P_0.

The parametric equations of a plane also can be written as a linear combination of the components of two vectors in the plane. Referring to Fig. 7.9, the two known vectors are

$$\mathbf{V}_1 = V_{1x}\mathbf{i} + V_{1y}\mathbf{j} + V_{1z}\mathbf{k} \qquad 7.43$$
$$\mathbf{V}_2 = V_{2x}\mathbf{i} + V_{2y}\mathbf{j} + V_{2z}\mathbf{k} \qquad 7.44$$

If s and t are scalars, the coordinates of each point in the plane can be written as Eqs. 7.45 through 7.47. These are the parametric equations of the plane.

$$x = x_0 + sV_{1x} + tV_{2x} \qquad 7.45$$
$$y = y_0 + sV_{1y} + tV_{2y} \qquad 7.46$$
$$z = z_0 + sV_{1z} + tV_{2z} \qquad 7.47$$

Example 7.3

The following points are coplanar.

$$P_1 = (2, 1, -4)$$
$$P_2 = (4, -2, -3)$$
$$P_3 = (2, 3, -8)$$

Determine the equation of the plane in (a) general form and (b) parametric form.

Solution

(a) Use the first two points to find a vector, $\mathbf{V}_1$.

$$\mathbf{V}_1 = (x_2 - x_1)\mathbf{i} + (y_2 - y_1)\mathbf{j} + (z_2 - z_1)\mathbf{k}$$
$$= (4 - 2)\mathbf{i} + (-2 - 1)\mathbf{j} + (-3 - (-4))\mathbf{k}$$
$$= 2\mathbf{i} - 3\mathbf{j} + 1\mathbf{k}$$

Similarly, use the first and third points to find $\mathbf{V}_2$.

$$\mathbf{V}_2 = (x_3 - x_1)\mathbf{i} + (y_3 - y_1)\mathbf{j} + (z_3 - z_1)\mathbf{k}$$
$$= (2 - 2)\mathbf{i} + (3 - 1)\mathbf{j} + (-8 - (-4))\mathbf{k}$$
$$= 0\mathbf{i} + 2\mathbf{j} - 4\mathbf{k}$$

From Eq. 7.42, determine the normal vector as a determinant.

$$\mathbf{N} = \begin{vmatrix} \mathbf{i} & 2 & 0 \\ \mathbf{j} & -3 & 2 \\ \mathbf{k} & 1 & -4 \end{vmatrix}$$

Expand the determinant across the top row.

$$\mathbf{N} = \mathbf{i}(12 - 2) - 2(-4\mathbf{j} - 2\mathbf{k})$$
$$= 10\mathbf{i} + 8\mathbf{j} + 4\mathbf{k}$$

The rectangular form of the equation of the plane uses the same constants as in the normal vector. Use the first point and write the equation of the plane in the form of Eq. 7.36.

$$(10)(x - 2) + (8)(y - 1) + (4)(z + 4) = 0$$

The three constant terms can be combined by using Eq. 7.39.

$$D = -\big((10)(2) + (8)(1) + (4)(-4)\big) = -12$$

The equation of the plane is

$$10x + 8y + 4z - 12 = 0$$

(b) The parametric equations based on the first point and for any values of s and t are

$$x = 2 + 2s + 0t$$
$$y = 1 - 3s + 2t$$
$$z = -4 + 1s - 4t$$

The scalars s and t are not unique. Two of the three coordinates can also be chosen as the parameters. Dividing the rectangular form of the plane's equation by 4 to isolate z results in an alternate set of parametric equations.

$$x = x$$
$$y = y$$
$$z = 3 - 2.5x - 2y$$

14. DISTANCES BETWEEN GEOMETRIC FIGURES

The smallest distance, d, between various geometric figures is given by the following equations.

- between two points in (x, y, z) format:

$$d = \sqrt{(x_2 - x_1)^2 + (y_2 - y_1)^2 + (z_2 - z_1)^2} \qquad 7.48$$

- between a point (x_0, y_0) and a line $Ax + By + C = 0$:

$$d = \frac{|Ax_0 + By_0 + C|}{\sqrt{A^2 + B^2}} \qquad 7.49$$

- between a point (x_0, y_0, z_0) and a plane $Ax + By + Cz + D = 0$:

$$d = \frac{|Ax_0 + By_0 + Cz_0 + D|}{\sqrt{A^2 + B^2 + C^2}} \qquad 7.50$$

- between two parallel lines $Ax + By + C = 0$:

$$d = \left| \frac{|C_2|}{\sqrt{A_2^2 + B_2^2}} - \frac{|C_1|}{\sqrt{A_1^2 + B_1^2}} \right| \qquad 7.51$$

Example 7.4

What is the minimum distance between the line $y = 2x + 3$ and the origin $(0, 0)$?

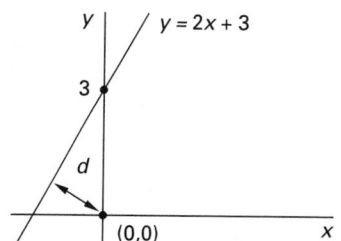

Solution

Put the equation in general form.

$$Ax + By + C = 2x - y + 3 = 0$$

Use Eq. 7.49 with $(x, y) = (0, 0)$.

$$d = \frac{|Ax + By + C|}{\sqrt{A^2 + B^2}} = \frac{(2)(0) + (-1)(0) + 3}{\sqrt{(2)^2 + (-1)^2}}$$

$$= \frac{3}{\sqrt{5}}$$

15. ANGLES BETWEEN GEOMETRIC FIGURES

The angle, ϕ, between various geometric figures is given by the following equations.

- between two lines in $Ax + By + C = 0$, $y = mx + b$, or direction angle formats:

$$\phi = \arctan\left(\frac{A_1 B_2 - A_2 B_1}{A_1 A_2 + B_1 B_2}\right) \qquad 7.52$$

$$= \arctan\left(\frac{m_2 - m_1}{1 + m_1 m_2}\right) \qquad 7.53$$

$$= |\arctan(m_1) - \arctan(m_2)| \qquad 7.54$$

$$= \arccos\left(\frac{L_1 L_2 + M_1 M_2 + N_1 N_2}{d_1 d_2}\right) \qquad 7.55$$

$$= \arccos(\cos\alpha_1 \cos\alpha_2 + \cos\beta_1 \cos\beta_2$$
$$+ \cos\gamma_1 \cos\gamma_2) \qquad 7.56$$

If the lines are parallel, then $\phi = 0$.

$$\frac{A_1}{A_2} = \frac{B_1}{B_2} \qquad 7.57$$

$$m_1 = m_2 \qquad 7.58$$

$$\alpha_1 = \alpha_2; \ \beta_1 = \beta_2; \ \gamma_1 = \gamma_2 \qquad 7.59$$

If the lines are perpendicular, then $\phi = 90°$.

$$A_1 A_2 = -B_1 B_2 \qquad 7.60$$

$$m_1 = \frac{-1}{m_2} \qquad 7.61$$

$$\alpha_1 + \alpha_2 = \beta_1 + \beta_2 = \gamma_1 + \gamma_2 = 90° \qquad 7.62$$

- between two planes in $A\mathbf{i} + B\mathbf{j} + C\mathbf{k} = 0$ format, the coefficients A, B, and C are the same as the coefficients for the normal vector. (See Eq. 7.37.) ϕ is equal to the angle between the two normal vectors.

$$\cos\phi = \frac{|A_1 A_2 + B_1 B_2 + C_1 C_2|}{\sqrt{A_1^2 + B_1^2 + C_1^2}\sqrt{A_2^2 + B_2^2 + C_2^2}} \qquad 7.63$$

Example 7.5

Use Eqs. 7.52, 7.53, and 7.54 to find the angle between the lines.

$$y = -0.577x + 2$$
$$y = +0.577x - 5$$

Solution

Write both equations in general form.

$$-0.577x - y + 2 = 0$$
$$0.577x - y - 5 = 0$$

(a) From Eq. 7.52,

$$\phi = \arctan\left(\frac{A_1 B_2 - A_2 B_1}{A_1 A_2 + B_1 B_2}\right)$$

$$= \arctan\left(\frac{(-0.577)(-1) - (0.577)(-1)}{(-0.577)(0.577) + (-1)(-1)}\right) = 60°$$

(b) Use Eq. 7.53.

$$\phi = \arctan\left(\frac{m_2 - m_1}{1 + m_1 m_2}\right)$$

$$= \arctan\left(\frac{0.577 - (-0.577)}{1 + (0.577)(-0.577)}\right) = 60°$$

(c) Use Eq. 7.54.

$$\phi = |\arctan(m_1) - \arctan(m_2)|$$

$$= |\arctan(-0.577) - \arctan(0.577)|$$

$$= |-30° - 30°| = 60°$$

16. CONIC SECTIONS

A *conic section* is any one of several curves produced by passing a plane through a cone as shown in Fig. 7.10. If α is the angle between the vertical axis and the cutting plane and β is the cone generating angle, Eq. 7.64 gives the *eccentricity*, ϵ, of the conic section. Values of the eccentricity are given in Fig. 7.10.

$$\epsilon = \frac{\cos\alpha}{\cos\beta} \qquad 7.64$$

All conic sections are described by second-degree polynomials (i.e., are *quadratic equations*) of the following form.[5]

$$Ax^2 + Bxy + Cy^2 + Dx + Ey + F = 0 \qquad 7.65$$

[5]One or more straight lines are produced when the cutting plane passes through the cone's vertex. Straight lines can be considered to be quadratic functions without second-degree terms.

Figure 7.10 *Conic Sections Produced by Cutting Planes*

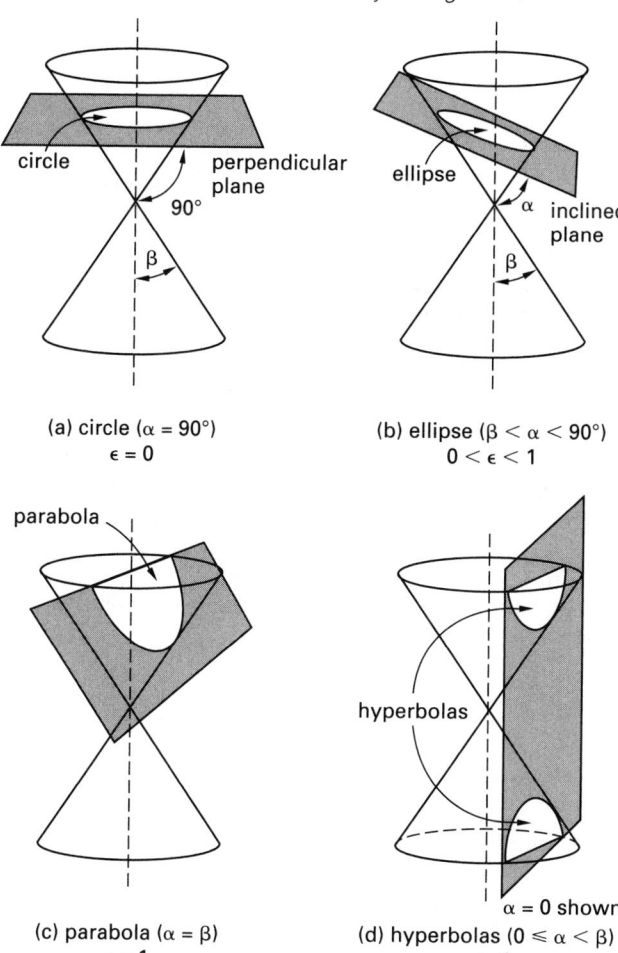

(a) circle ($\alpha = 90°$)
$\epsilon = 0$

(b) ellipse ($\beta < \alpha < 90°$)
$0 < \epsilon < 1$

(c) parabola ($\alpha = \beta$)
$\epsilon = 1$

(d) hyperbolas ($0 \leq \alpha < \beta$)
$\alpha = 0$ shown
$\epsilon > 1$

This is the *general form*, which allows the figure axes to be at any angle relative to the coordinate axes. The *standard forms* presented in the following sections pertain to figures whose axes coincide with the coordinate axes, thereby eliminating certain terms of the general equation.

Figure 7.11 can be used to determine which conic section is described by the quadratic function. The quantity $B^2 - 4AC$ is known as the *discriminant*. Figure 7.11 determines only the type of conic section; it does not determine whether the conic section is degenerate (e.g., a circle with a negative radius).

Example 7.6

What geometric figures are described by the following equations?

(a) $4y^2 - 12y + 16x + 41 = 0$

(b) $x^2 - 10xy + y^2 + x + y + 1 = 0$

(c) $x^2 + 4y^2 + 2x - 8y + 1 = 0$

(d) $x^2 + y^2 - 6x + 8y + 20 = 0$

Figure 7.11 *Determining Conic Sections from Quadratic Equations*

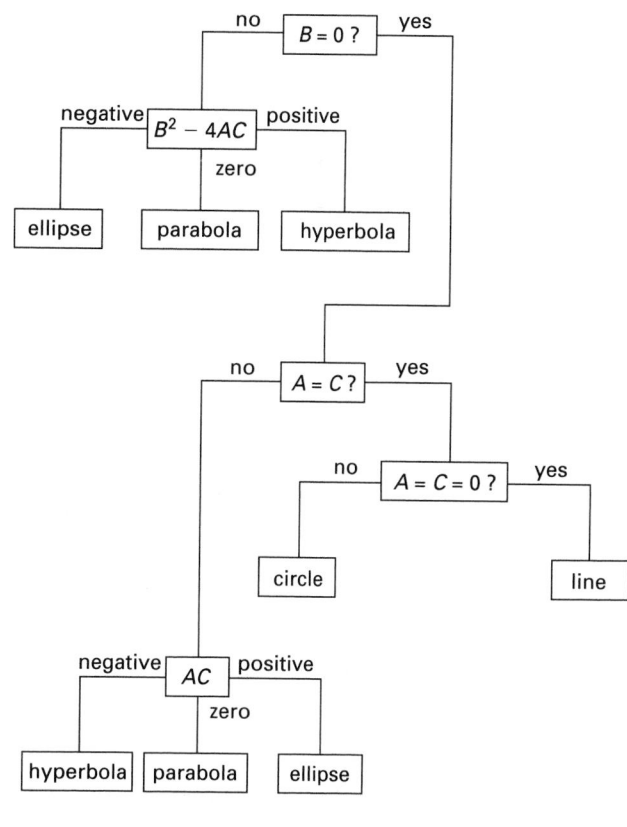

Solution

(a) Referring to Fig. 7.11, $B = 0$ since there is no xy term, $A = 0$ since there is no x^2 term, and $AC = (0)(4) = 0$. This is a parabola.

(b) $B \neq 0$; $B^2 - 4AC = (-10)^2 - (4)(1)(1) = +96$. This is a hyperbola.

(c) $B = 0$; $A \neq C$; $AC = (1)(4) = +4$. This is an ellipse.

(d) $B = 0$; $A = C$; $A = C = 1 \ (\neq 0)$. This is a circle.

17. CIRCLE

The general form of the equation of a circle is

$$Ax^2 + Ay^2 + Dx + Ey + F = 0 \qquad 7.66$$

The *center-radius form* of the equation of a circle with radius r and center at (h, k) is

$$(x - h)^2 + (y - k)^2 = r^2 \qquad 7.67$$

Figure 7.12 *Circle*

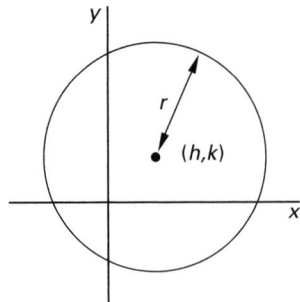

Figure 7.13 *Parabola*

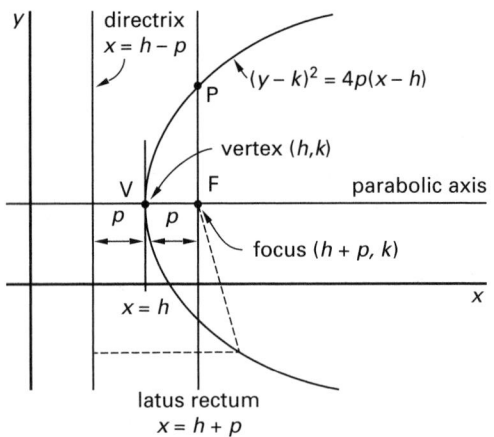

The two forms can be converted by use of Eqs. 7.68 through 7.70.

$$h = \frac{-D}{2A} \qquad 7.68$$

$$k = \frac{-E}{2A} \qquad 7.69$$

$$r^2 = \frac{D^2 + E^2 - 4AF}{4A^2} \qquad 7.70$$

If the right-hand side of Eq. 7.70 is positive, the figure is a circle. If it is zero, the circle shrinks to a point. If the right-hand side is negative, the figure is imaginary. A *degenerate circle* is one in which the right-hand side is less than or equal to zero.

18. PARABOLA

A *parabola* is the locus of points equidistant from the *focus* (point F in Fig. 7.13) and a line called the *directrix*. A parabola is symmetric with respect to its *parabolic axis*. The line normal to the parabolic axis and passing through the focus is known as the *latus rectum*. The eccentricity of a parabola is 1.

There are two common types of parabolas in the Cartesian plane—those that open right and left, and those that open up and down. Equation 7.65 is the general form of the equation of a parabola. With Eq. 7.71, the parabola points to the right if $CD > 0$ and to the left if $CD < 0$. With Eq. 7.72, the parabola points up if $AE > 0$ and down if $AE < 0$.

$$Cy^2 + Dx + Ey + F = 0 \Big|_{\substack{C,D \neq 0 \\ \text{opens horizontally}}} \qquad 7.71$$

$$Ax^2 + Dx + Ey + F = 0 \Big|_{\substack{A,E \neq 0 \\ \text{opens vertically}}} \qquad 7.72$$

The *standard form* of the equation of a parabola with vertex at (h, k), focus at $(h + p, k)$, and directrix at $x = h - p$ and that opens to the right or left is given by Eq. 7.73. The parabola opens to the right (points to the left) if $p > 0$ and opens to the left (points to the right) if $p < 0$.

$$(y - k)^2 = 4p(x - h) \Big|_{\text{opens horizontally}} \qquad 7.73$$

$$y^2 = 4px \Big|_{\substack{\text{vertex at origin} \\ h=k=0}} \qquad 7.74$$

The *standard form* of the equation of a parabola with vertex at (h, k), focus at $(h, k + p)$, and directrix at $y = k - p$ and that opens up or down is given by Eq. 7.75. The parabola opens up (points down) if $p > 0$ and opens down (points up) if $p < 0$.

$$(x - h)^2 = 4p(y - k) \Big|_{\text{opens vertically}} \qquad 7.75$$

$$x^2 = 4py \Big|_{\text{vertex at origin}} \qquad 7.76$$

The general and vertex forms of the equations can be reconciled with Eqs. 7.77 through 7.79. Whether the first or second forms of these equations are used depends on whether the parabola opens horizontally or vertically (i.e., whether $A = 0$ or $C = 0$), respectively.

$$h = \begin{cases} \dfrac{E^2 - 4CF}{4CD} & \text{[opens horizontally]} \\[2ex] \dfrac{-D}{2A} & \text{[opens vertically]} \end{cases} \qquad 7.77$$

$$k = \begin{cases} \dfrac{-E}{2C} & \text{[opens horizontally]} \\[2ex] \dfrac{D^2 - 4AF}{4AE} & \text{[opens vertically]} \end{cases} \qquad 7.78$$

$$p = \begin{cases} \dfrac{-D}{4C} & \text{[opens horizontally]} \\[2ex] \dfrac{-E}{4A} & \text{[opens vertically]} \end{cases} \qquad 7.79$$

19. ELLIPSE

An *ellipse* has two foci separated along the *major axis* by a distance $2c$. The line perpendicular to the major axis passing through the center of the ellipse is the *minor axis*. The lines perpendicular to the major axis passing through the foci are the *latus recta*. The distance between the two vertices is $2a$. The ellipse is the locus of points such that the sum of the distances from the two foci is $2a$. Referring to Fig. 7.14,

$$F_1 P + P F_2 = 2a \qquad 7.80$$

Figure 7.14 *Ellipse*

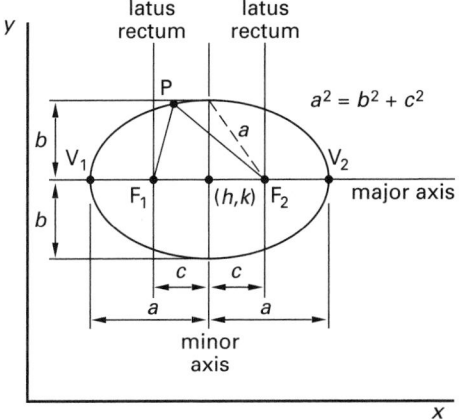

Equation 7.81 is the standard equation for an ellipse with axes parallel to the coordinate axes. (Equation 7.65 is the general form.) F is not independent of A, C, D, and E for the ellipse.

$$Ax^2 + Cy^2 + Dx + Ey + F = 0 \Big|_{\substack{AC>0 \\ A \neq C}} \qquad 7.81$$

Equation 7.82 gives the standard form of the equation of an ellipse centered at (h, k). Distances a and b are known as the *semimajor distance* and *semiminor distance*, respectively.

$$\frac{(x-h)^2}{a^2} + \frac{(y-k)^2}{b^2} = 1 \qquad 7.82$$

The distance between the two foci is $2c$.

$$2c = 2\sqrt{a^2 - b^2} \qquad 7.83$$

The *aspect ratio* of the ellipse is

$$\text{aspect ratio} = \frac{a}{b} \qquad 7.84$$

The *eccentricity*, ϵ, of the ellipse is always less than 1. If the eccentricity is zero, the figure is a circle (another form of a *degenerative ellipse*).

$$\epsilon = \frac{\sqrt{a^2 - b^2}}{a} < 1 \qquad 7.85$$

The standard and center forms of the equations of an ellipse can be reconciled by using Eqs. 7.86 through 7.89.

$$h = \frac{-D}{2A} \qquad 7.86$$

$$k = \frac{-E}{2C} \qquad 7.87$$

$$a = \sqrt{C} \qquad 7.88$$

$$b = \sqrt{A} \qquad 7.89$$

20. HYPERBOLA

A *hyperbola* has two foci separated along the *transverse axis* by a distance $2c$. Lines perpendicular to the transverse axis passing through the foci are the *conjugate axes*. The distance between the two vertices is $2a$, and the distance along a conjugate axis passing through each vertex between two points on the asymptotes is $2b$. The hyperbola is the locus of points such that the difference in distances from the two foci is $2a$. Referring to Fig. 7.15,

$$F_2 P - P F_1 = 2a \qquad 7.90$$

Figure 7.15 *Hyperbola*

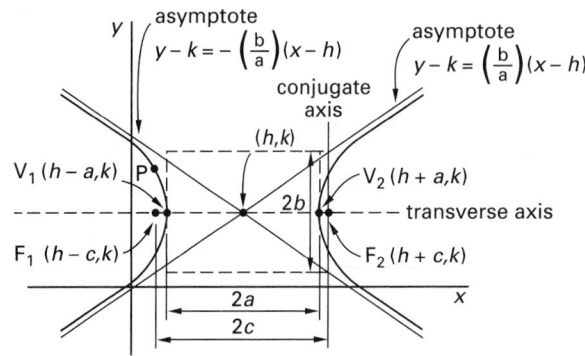

Equation 7.91 is the standard equation of a hyperbola. Coefficients A and C have opposite signs.

$$Ax^2 + Cy^2 + Dx + Ey + F = 0 \big|_{AC < 0} \qquad 7.91$$

Equation 7.92 gives the standard form of the equation of a hyperbola centered at (h, k) and opening to the left and right.

$$\frac{(x-h)^2}{a^2} - \frac{(y-k)^2}{b^2} = 1 \bigg|_{\text{opens horizontally}} \qquad 7.92$$

Equation 7.93 gives the standard form of the equation of a hyperbola centered at (h, k) and opening up and down.

$$\frac{(y-k)^2}{a^2} - \frac{(x-h)^2}{b^2} = 1 \bigg|_{\text{opens vertically}} \qquad 7.93$$

The distance between the two foci is $2c$.

$$2c = 2\sqrt{a^2 + b^2} \qquad 7.94$$

The *eccentricity*, ϵ, of the hyperbola is calculated from Eq. 7.95 and is always greater than 1.

$$\epsilon = \frac{c}{a} = \frac{\sqrt{a^2 + b^2}}{a} > 1 \qquad 7.95$$

The hyperbola is asymptotic to the lines given by Eqs. 7.96 and 7.97.

$$y = \pm\frac{b}{a}(x - h) + k \Big|_{\text{opens horizontally}} \qquad 7.96$$

$$y = \pm\frac{a}{b}(x - h) + k \Big|_{\text{opens vertically}} \qquad 7.97$$

For a *rectangular (equilateral) hyperbola*, the asymptotes are perpendicular, $a = b$, $c = \sqrt{2}a$, and the eccentricity is $\epsilon = \sqrt{2}$. If the hyperbola is centered at the origin (i.e., $h = k = 0$), then the equations are $x^2 - y^2 = a^2$ (opens horizontally) and $y^2 - x^2 = a^2$ (opens vertically).

If the asymptotes are the x- and y-axes, the equation of the hyperbola is simply

$$xy = \pm\frac{a^2}{2} \qquad 7.98$$

The general and center forms of the equations of a hyperbola can be reconciled by using Eqs. 7.99 through 7.103. Whether the hyperbola opens left and right or up and down depends on whether M/A or M/C is positive, respectively, where M is defined by Eq. 7.99.

$$M = \frac{D^2}{4A} + \frac{E^2}{4C} - F \qquad 7.99$$

$$h = \frac{-D}{2A} \qquad 7.100$$

$$k = \frac{-E}{2C} \qquad 7.101$$

$$a = \begin{cases} \sqrt{-C} & \text{[opens horizontally]} \\ \sqrt{-A} & \text{[opens vertically]} \end{cases} \qquad 7.102$$

$$b = \begin{cases} \sqrt{A} & \text{[opens horizontally]} \\ \sqrt{C} & \text{[opens vertically]} \end{cases} \qquad 7.103$$

21. SPHERE

Equation 7.104 is the general equation of a sphere. The coefficient A cannot be zero.

$$Ax^2 + Ay^2 + Az^2 + Bx + Cy + Dz + E = 0 \qquad 7.104$$

Equation 7.105 gives the standard form of the equation of a sphere centered at (h, k, l) with radius r.

$$(x - h)^2 + (y - k)^2 + (z - l)^2 = r^2 \qquad 7.105$$

The general and center forms of the equations of a sphere can be reconciled by using Eqs. 7.106 through 7.109.

$$h = \frac{-B}{2A} \qquad 7.106$$

$$k = \frac{-C}{2A} \qquad 7.107$$

$$l = \frac{-D}{2A} \qquad 7.108$$

$$r = \sqrt{\frac{B^2 + C^2 + D^2}{4A^2} - \frac{E}{A}} \qquad 7.109$$

Planes tangent to spheres and other solids are covered in Chap. 8, Sec. 7.

22. HELIX

A *helix* is a curve generated by a point moving on, around, and along a cylinder such that the distance the point moves parallel to the cylindrical axis is proportional to the angle of rotation about that axis. For a cylinder of radius r, Eqs. 7.110 through 7.112 define the three-dimensional positions of points along the helix. The quantity $2\pi k$ is the *pitch* of the helix.

$$x = r\cos\theta \qquad 7.110$$
$$y = r\sin\theta \qquad 7.111$$
$$z = k\theta \qquad 7.112$$

Figure 7.16 *Helix*

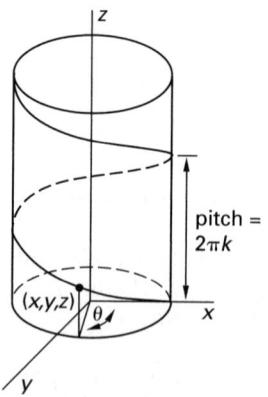

8 Differential Calculus

1. DERIVATIVE OF A FUNCTION

In most cases, it is possible to transform a continuous function, $f(x_1, x_2, x_3, \ldots)$, of one or more independent variables into a derivative function.[1] In simple cases, the *derivative* can be interpreted as the slope (tangent or rate of change) of the curve described by the original function. Since the slope of the curve depends on x, the derivative function will also depend on x. The derivative, $f'(x)$, of a function $f(x)$ is defined mathematically by Eq. 8.1. However, limit theory is seldom needed to actually calculate derivatives.

$$f'(x) = \lim_{\triangle x \to 0} \frac{\triangle f(x)}{\triangle x} \qquad 8.1$$

The derivative of a function $f(x)$, also known as the *first derivative*, is written in various ways, including

$$f'(x), \frac{df(x)}{dx}, \frac{df}{dx}, \mathbf{D}f(x), \mathbf{D}_x f(x), \dot{f}(x), sf(s)$$

A *second derivative* may exist if the derivative operation is performed on the first derivative—that is, a derivative is taken of a derivative function. This is written as

$$f''(x), \frac{d^2 f(x)}{dx^2}, \frac{d^2 f}{dx^2}, \mathbf{D}^2 f(x), \mathbf{D}_x^2 f(x), \ddot{f}(x), s^2 f(s)$$

---

[1]A function, $f(x)$, of one independent variable, x, is used in this section to simplify the discussion. Although the derivative is taken with respect to x, the independent variable can be anything.

A *regular* (*analytic* or *holomorphic*) *function* possesses a derivative. A point at which a function's derivative is undefined is called a *singular point*, as Fig. 8.1 illustrates.

Figure 8.1 *Derivatives and Singular Points*

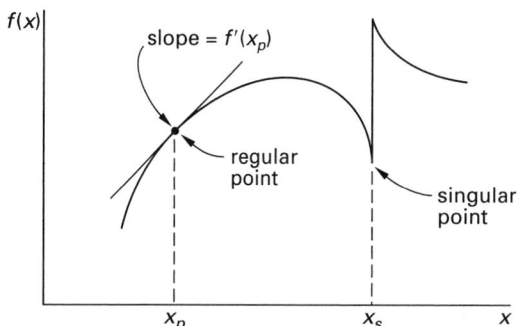

2. ELEMENTARY DERIVATIVE OPERATIONS

Equations 8.2 through 8.5 summarize the elementary derivative operations on polynomials and exponentials. Equations 8.2 and 8.3 are particularly useful. (a, n, and k represent constants. $f(x)$ and $g(x)$ are functions of x.)

$$\mathbf{D}k = 0 \qquad 8.2$$

$$\mathbf{D}x^n = nx^{n-1} \qquad 8.3$$

$$\mathbf{D}\ln x = \frac{1}{x} \qquad 8.4$$

$$\mathbf{D}e^{ax} = ae^{ax} \qquad 8.5$$

Equations 8.6 through 8.17 summarize the elementary derivative operations on transcendental (trigonometric) functions.

$$\mathbf{D}\sin x = \cos x \qquad 8.6$$

$$\mathbf{D}\cos x = -\sin x \qquad 8.7$$

$$\mathbf{D}\tan x = \sec^2 x \qquad 8.8$$

$$\mathbf{D}\cot x = -\csc^2 x \qquad 8.9$$

$$\mathbf{D}\sec x = \sec x \tan x \qquad 8.10$$

$$\mathbf{D}\csc x = -\csc x \cot x \qquad 8.11$$

$$\mathbf{D}\arcsin x = \frac{1}{\sqrt{1 - x^2}} \qquad 8.12$$

$$\mathbf{D}\arccos x = -\mathbf{D}\arcsin x \qquad 8.13$$

$$\mathbf{D}\arctan x = \frac{1}{1+x^2} \qquad 8.14$$

$$\mathbf{D}\operatorname{arccot} x = -\mathbf{D}\arctan x \qquad 8.15$$

$$\mathbf{D}\operatorname{arcsec} x = \frac{1}{x\sqrt{x^2-1}} \qquad 8.16$$

$$\mathbf{D}\operatorname{arccsc} x = -\mathbf{D}\operatorname{arcsec} x \qquad 8.17$$

Equations 8.18 through 8.23 summarize the elementary derivative operations on hyperbolic transcendental functions. Derivatives of hyperbolic functions are not completely analogous to those of the regular transcendental functions.

$$\mathbf{D}\sinh x = \cosh x \qquad 8.18$$

$$\mathbf{D}\cosh x = \sinh x \qquad 8.19$$

$$\mathbf{D}\tanh x = \operatorname{sech}^2 x \qquad 8.20$$

$$\mathbf{D}\coth x = -\operatorname{csch}^2 x \qquad 8.21$$

$$\mathbf{D}\operatorname{sech} x = -\operatorname{sech} x \tanh x \qquad 8.22$$

$$\mathbf{D}\operatorname{csch} x = -\operatorname{csch} x \coth x \qquad 8.23$$

Equations 8.24 through 8.29 summarize the elementary derivative operations on functions and combinations of functions.

$$\mathbf{D}kf(x) = k\mathbf{D}f(x) \qquad 8.24$$

$$\mathbf{D}(f(x) \pm g(x)) = \mathbf{D}f(x) \pm \mathbf{D}g(x) \qquad 8.25$$

$$\mathbf{D}(f(x)\cdot g(x)) = f(x)\mathbf{D}g(x) + g(x)\mathbf{D}f(x) \qquad 8.26$$

$$\mathbf{D}\left(\frac{f(x)}{g(x)}\right) = \frac{g(x)\mathbf{D}f(x) - f(x)\mathbf{D}g(x)}{(g(x))^2} \qquad 8.27$$

$$\mathbf{D}\left(f(x)\right)^n = n\left(f(x)\right)^{n-1}\mathbf{D}f(x) \qquad 8.28$$

$$\mathbf{D}f(g(x)) = \mathbf{D}_g f(g)\mathbf{D}_x g(x) \qquad 8.29$$

Example 8.1

What is the slope at $x = 3$ of the curve $f(x) = x^3 - 2x$?

Solution

The derivative function found from Eq. 8.3 determines the slope.

$$f'(x) = 3x^2 - 2$$

The slope at $x = 3$ is

$$f'(3) = (3)(3)^2 - 2 = 27 - 2 = 25$$

Example 8.2

What are the derivatives of the following functions?

(a) $f(x) = 5\sqrt[3]{x^5}$

(b) $f(x) = \sin x \cos^2 x$

(c) $f(x) = \ln(\cos e^x)$

Solution

(a) Using Eqs. 8.3 and 8.24,

$$f'(x) = 5\mathbf{D}\sqrt[3]{x^5} = 5\mathbf{D}\left((x^5)^{\frac{1}{3}}\right)$$

$$= (5)\left(\tfrac{1}{3}\right)(x^5)^{-\frac{2}{3}}\mathbf{D}x^5$$

$$= (5)\left(\tfrac{1}{3}\right)(x^5)^{-\frac{2}{3}}(5)(x^4)$$

$$= \frac{25x^{\frac{2}{3}}}{3}$$

(b) Using Eq. 8.26,

$$f'(x) = \sin x\,\mathbf{D}\cos^2 x + \cos^2 x\,\mathbf{D}\sin x$$

$$= (\sin x)(2\cos x)(\mathbf{D}\cos x) + \cos^2 x \cos x$$

$$= (\sin x)(2\cos x)(-\sin x) + \cos^2 x \cos x$$

$$= -2\sin^2 x \cos x + \cos^3 x$$

(c) Using Eq. 8.29,

$$f'(x) = \left(\frac{1}{\cos e^x}\right)\mathbf{D}\cos e^x$$

$$= \left(\frac{1}{\cos e^x}\right)(-\sin e^x)\mathbf{D}e^x$$

$$= \left(\frac{-\sin e^x}{\cos e^x}\right)e^x$$

$$= -e^x \tan e^x$$

3. CRITICAL POINTS

Derivatives are used to locate the local *critical points* of functions of one variable—that is, *extreme points* (also known as *maximum* and *minimum* points) as well as the *inflection points* (*points of contraflexure*). The plurals *extrema*, *maxima*, and *minima* are used without the word "points." These points are illustrated in Fig. 8.2. There is usually an inflection point between two adjacent local extrema.

Figure 8.2 *Extreme and Inflection Points*

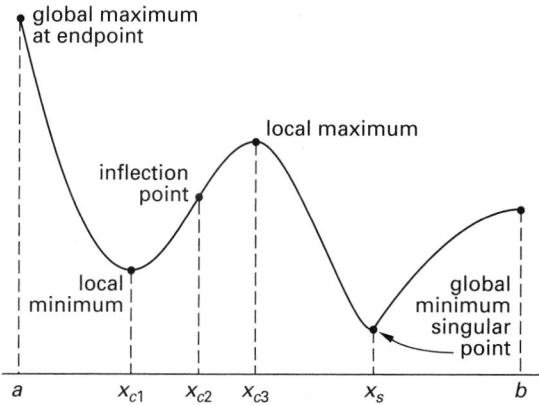

The first derivative is calculated to determine the locations of the critical points. The second derivative is calculated to determine whether a critical point is a local maximum, minimum, or inflection point, according to the following conditions. With this method, no distinction is made between local and global extrema. Therefore, the extrema should be compared with the function values at the endpoints of the interval, as illustrated in Ex. 8.3.[2] Note that $f'(x) \neq 0$ at an inflection point.

$$f'(x_c) = 0 \text{ at any extreme point, } x_c \qquad \textit{8.30}$$

$$f''(x_c) < 0 \text{ at a maximum point} \qquad \textit{8.31}$$

$$f''(x_c) > 0 \text{ at a minimum point} \qquad \textit{8.32}$$

$$f''(x_c) = 0 \text{ at an inflection point} \qquad \textit{8.33}$$

Example 8.3

Find the global extrema of the function $f(x)$ on the interval $[-2, +2]$.

$$f(x) = x^3 + x^2 - x + 1$$

Solution

The first derivative is

$$f'(x) = 3x^2 + 2x - 1$$

Since the first derivative is zero at extreme points, set $f'(x)$ equal to zero and solve for the roots of the quadratic equation.

$$3x^2 + 2x - 1 = (3x - 1)(x + 1) = 0$$

The roots are $x_1 = {}^1\!/_3$, $x_2 = -1$. These are the locations of the two extrema.

---

[2]It is also necessary to check the values of the function at singular points (i.e., points where the derivative does not exist).

The second derivative is

$$f''(x) = 6x + 2$$

Substituting x_1 and x_2 into $f''(x)$,

$$f''(x_1) = (6)\left(\tfrac{1}{3}\right) + 2 = 4$$

$$f''(x_2) = (6)(-1) + 2 = -4$$

Therefore, x_1 is a local minimum point (because $f''(x)$ is positive), and x_2 is a local maximum point (because $f''(x)$ is negative). The inflection point between these two extrema is found by setting $f''(x)$ equal to zero.

$$f''(x) = 6x + 2 = 0 \text{ or } x = -\tfrac{1}{3}$$

Since the question asked for the global extreme points, it is necessary to compare the values of $f(x)$ at the local extrema with the values at the endpoints.

$$f(-2) = -1$$

$$f(-1) = 2$$

$$f\left(\tfrac{1}{3}\right) = 22/27$$

$$f(2) = +11$$

Therefore, the actual global extrema are the endpoints.

4. DERIVATIVES OF PARAMETRIC EQUATIONS

The derivative of a function $f(x_1, x_2, \ldots x_n)$ can be calculated from the derivatives of the parametric equations $f_1(s), f_2(s), \ldots, f_n(s)$. The derivative will be expressed in terms of the parameter, s, unless the derivatives of the parametric equations can be expressed explicitly in terms of the independent variables.

Example 8.4

A circle is expressed parametrically by the equations

$$x = 5\cos\theta$$

$$y = 5\sin\theta$$

Express the derivative dy/dx (a) as a function of the parameter θ and (b) as a function of x and y.

Solution

(a) Taking the derivative of each parametric equation with respect to θ,

$$\frac{dx}{d\theta} = -5\sin\theta$$

$$\frac{dy}{d\theta} = 5\cos\theta$$

Then,

$$\frac{dy}{dx} = \frac{\dfrac{dy}{d\theta}}{\dfrac{dx}{d\theta}} = \frac{5\cos\theta}{-5\sin\theta} = -\cot\theta$$

(b) The derivatives of the parametric equations are closely related to the original parametric equations.

$$\frac{dx}{d\theta} = -5\sin\theta = -y$$

$$\frac{dy}{d\theta} = 5\cos\theta = x$$

$$\frac{dy}{dx} = \frac{\dfrac{dy}{d\theta}}{\dfrac{dx}{d\theta}} = \frac{-x}{y}$$

5. PARTIAL DIFFERENTIATION

Derivatives can be taken with respect to only one independent variable at a time. For example, $f'(x)$ is the derivative of $f(x)$ and is taken with respect to the independent variable x. If a function, $f(x_1, x_2, x_3, \ldots)$, has more than one independent variable, a *partial derivative* can be found, but only with respect to one of the independent variables. All other variables are treated as constants. Symbols for a partial derivative of f taken with respect to variable x are $\partial f/\partial x$ and $f_x(x,y)$.

The geometric interpretation of a partial derivative $\partial f/\partial x$ is the slope of a line tangent to the surface (a sphere, ellipsoid, etc.) described by the function when all variables except x are held constant. In three-dimensional space with a function described by $z = f(x,y)$, the partial derivative $\partial f/\partial x$ (equivalent to $\partial z/\partial x$) is the slope of the line tangent to the surface in a plane of constant y. Similarly, the partial derivative $\partial f/\partial y$ (equivalent to $\partial z/\partial y$) is the slope of the line tangent to the surface in a plane of constant x.

Example 8.5

What is the partial derivative $\partial z/\partial x$ of the following function?

$$z = 3x^2 - 6y^2 + xy + 5y - 9$$

Solution

The partial derivative with respect to x is found by considering all variables other than x to be constants.

$$\frac{\partial z}{\partial x} = 6x - 0 + y + 0 - 0 = 6x + y$$

Example 8.6

A surface has the equation $x^2 + y^2 + z^2 - 9 = 0$. What is the slope of a line that lies in a plane of constant y and is tangent to the surface at $(x, y, z) = (1, 2, 2)$?[3]

---

[3]Although only implied, it is required that the point actually be on the surface (i.e., it must satisfy the equation $f(x, y, z) = 0$).

Solution

Solve for the dependent variable. Then, consider variable y to be a constant.

$$z = \sqrt{9 - x^2 - y^2}$$

$$\frac{\partial z}{\partial x} = \frac{\partial(9 - x^2 - y^2)^{\frac{1}{2}}}{\partial x}$$

$$= \left(\tfrac{1}{2}\right)(9 - x^2 - y^2)^{-\frac{1}{2}}\left(\frac{\partial(9 - x^2 - y^2)}{\partial x}\right)$$

$$= \left(\tfrac{1}{2}\right)(9 - x^2 - y^2)^{-\frac{1}{2}}(-2x)$$

$$= \frac{-x}{\sqrt{9 - x^2 - y^2}}$$

At the point $(1, 2, 2)$, $x = 1$ and $y = 2$.

$$\left.\frac{\partial z}{\partial x}\right|_{(1,2,2)} = \frac{-1}{\sqrt{9 - (1)^2 - (2)^2}} = -\tfrac{1}{2}$$

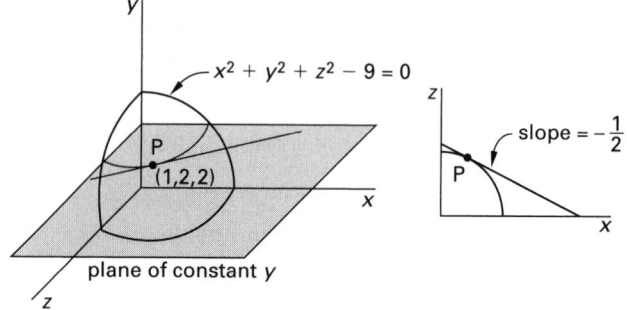

6. IMPLICIT DIFFERENTIATION

When a relationship between n variables cannot be manipulated to yield an explicit function of $n - 1$ independent variables, that relationship implicitly defines the nth variable. Finding the derivative of the implicit variable with respect to any other independent variable is known as *implicit differentiation*.

An implicit derivative is the quotient of two partial derivatives. The two partial derivatives are chosen so that dividing one by the other eliminates a common differential. For example, if z cannot be explicitly extracted from $f(x, y, z) = 0$, the partial derivatives $\partial z/\partial x$ and $\partial z/\partial y$ can still be found as follows.

$$\frac{\partial z}{\partial x} = \frac{-\dfrac{\partial f}{\partial x}}{\dfrac{\partial f}{\partial z}} \qquad 8.34$$

$$\frac{\partial z}{\partial y} = \frac{-\dfrac{\partial f}{\partial y}}{\dfrac{\partial f}{\partial z}} \qquad 8.35$$

Example 8.7

Find the derivative dy/dx of

$$f(x, y) = x^2 + xy + y^3$$

Solution

Implicit differentiation is required because x cannot be extracted from $f(x, y)$.

$$\frac{\partial f}{\partial x} = 2x + y$$

$$\frac{\partial f}{\partial y} = x + 3y^2$$

$$\frac{dy}{dx} = \frac{-\dfrac{\partial f}{\partial x}}{\dfrac{\partial f}{\partial y}} = \frac{-(2x + y)}{x + 3y^2}$$

Example 8.8

Solve Ex. 8.6 using implicit differentiation.

Solution

$$f(x, y, z) = x^2 + y^2 + z^2 - 9 = 0$$

$$\frac{\partial f}{\partial x} = 2x$$

$$\frac{\partial f}{\partial z} = 2z$$

$$\frac{\partial z}{\partial x} = \frac{-\dfrac{\partial f}{\partial x}}{\dfrac{\partial f}{\partial z}} = \frac{-2x}{2z} = -\frac{x}{z}$$

At the point $(1, 2, 2)$,

$$\frac{\partial z}{\partial x} = -\frac{1}{2}$$

7. TANGENT PLANE FUNCTION

Partial derivatives can be used to find the equation of a plane tangent to a three-dimensional surface defined by $f(x, y, z) = 0$ at some point, P_0.

$$T(x_0, y_0, z_0) = (x - x_0)\frac{\partial f(x, y, z)}{\partial x}\Big|_{P_0}$$

$$+ (y - y_0)\frac{\partial f(x, y, z)}{\partial y}\Big|_{P_0}$$

$$+ (z - z_0)\frac{\partial f(x, y, z)}{\partial z}\Big|_{P_0}$$

$$= 0 \qquad\qquad 8.36$$

The coefficients of x, y, and z are the same as the coefficients of $\mathbf{i}$, $\mathbf{j}$, and $\mathbf{k}$ of the normal vector at point P_0. (See Secs. 7.13 and 8.10.)

Example 8.9

What is the equation of the plane that is tangent to the surface defined by $f(x, y, z) = 4x^2 + y^2 - 16z = 0$ at the point $(2, 4, 2)$?

Solution

First, calculate the partial derivatives and substitute the coordinates of the point.

$$\frac{\partial f(x, y, z)}{\partial x}\Big|_{P_0} = 8x|_{(2,4,2)} = (8)(2) = 16$$

$$\frac{\partial f(x, y, z)}{\partial y}\Big|_{P_0} = 2y|_{(2,4,2)} = (2)(4) = 8$$

$$\frac{\partial f(x, y, z)}{\partial z}\Big|_{P_0} = -16|_{(2,4,2)} = -16$$

$$T(2, 4, 2) = (16)(x - 2) + (8)(y - 4) - (16)(z - 2)$$

$$= 2x + y - 2z - 4$$

Next, substitute into Eq. 8.36.

$$2x + y - 2z - 4 = 0$$

8. GRADIENT VECTOR

The slope of a function is the change in one variable with respect to a distance in a chosen direction. Usually, the direction is parallel to a coordinate axis. However, the maximum slope at a point on a surface may not be in a direction parallel to one of the coordinate axes.

The *gradient vector function* $\nabla f(x, y, z)$ (pronounced "del f") gives the maximum rate of change of the function $f(x, y, z)$.

$$\nabla f(x, y, z) = \left(\frac{\partial f(x, y, z)}{\partial x}\right)\mathbf{i} + \left(\frac{\partial f(x, y, z)}{\partial y}\right)\mathbf{j}$$

$$+ \left(\frac{\partial f(x, y, z)}{\partial z}\right)\mathbf{k} \qquad 8.37$$

Example 8.10

A two-dimensional function is defined as

$$f(x, y) = 2x^2 - y^2 + 3x - y$$

(a) What is the gradient vector for this function? (b) What is the direction of the line passing through the point $(1, -2)$ that has a maximum slope? (c) What is the maximum slope at the point $(1, -2)$?

Solution

(a) It is necessary to calculate two partial derivatives in order to use Eq. 8.37.

$$\frac{\partial f(x,y)}{\partial x} = 4x + 3$$

$$\frac{\partial f(x,y)}{\partial y} = -2y - 1$$

$$\nabla f(x,y) = (4x+3)\mathbf{i} + (-2y-1)\mathbf{j}$$

(b) The direction of the line passing through $(1,-2)$ with maximum slope is found by inserting $x = 1$ and $y = -2$ into the gradient vector function.

$$\mathbf{V} = \big((4)(1)+3\big)\mathbf{i} + \big((-2)(-2)-1\big)\mathbf{j}$$
$$= 7\mathbf{i} + 3\mathbf{j}$$

(c) The magnitude of the slope is

$$|\mathbf{V}| = \sqrt{(7)^2 + (3)^2} = \sqrt{58} = 7.62$$

9. DIRECTIONAL DERIVATIVE

Unlike the gradient vector (covered in Sec. 8), which calculates the maximum rate of change of a function, the *directional derivative*, indicated by $\nabla_u f(x,y,z)$, $D_u f(x,y,z)$, or $f'_u(x,y,z)$, gives the rate of change in the direction of a given vector, $\mathbf{u}$ or $\mathbf{U}$. The subscript u implies that the direction vector is a unit vector, but it does not need to be, as only the direction cosines are calculated from it.

$$\nabla_u f(x,y,z) = \left(\frac{\partial f(x,y,z)}{\partial x}\right)\cos\alpha + \left(\frac{\partial f(x,y,z)}{\partial y}\right)\cos\beta$$
$$+ \left(\frac{\partial f(x,y,z)}{\partial z}\right)\cos\gamma \qquad \textit{8.38}$$

$$\mathbf{U} = U_x\mathbf{i} + U_y\mathbf{j} + U_z\mathbf{k} \qquad \textit{8.39}$$

$$\cos\alpha = \frac{U_x}{|\mathbf{U}|} = \frac{U_x}{\sqrt{U_x^2 + U_y^2 + U_z^2}} \qquad \textit{8.40}$$

$$\cos\beta = \frac{U_y}{|\mathbf{U}|} \qquad \textit{8.41}$$

$$\cos\gamma = \frac{U_z}{|\mathbf{U}|} \qquad \textit{8.42}$$

Example 8.11

What is the rate of change of $f(x,y) = 3x^2 + xy - 2y^2$ at the point $(1,-2)$ in the direction $4\mathbf{i} + 3\mathbf{j}$?

Solution

The direction cosines are given by Eqs. 8.40 and 8.41.

$$\cos\alpha = \frac{U_x}{|\mathbf{U}|} = \frac{4}{\sqrt{(4)^2 + (3)^2}} = \tfrac{4}{5}$$

$$\cos\beta = \frac{U_y}{|\mathbf{U}|} = \tfrac{3}{5}$$

The partial derivatives are

$$\frac{\partial f(x,y)}{\partial x} = 6x + y$$

$$\frac{\partial f(x,y)}{\partial y} = x - 4y$$

The directional derivative is given by Eq. 8.38.

$$\nabla_u f(x,y) = \left(\tfrac{4}{5}\right)(6x+y) + \left(\tfrac{3}{5}\right)(x-4y)$$

Substituting the given values of $x = 1$ and $y = -2$,

$$\nabla_u f(1,-2) = \left(\tfrac{4}{5}\right)\big((6)(1)-2\big) + \left(\tfrac{3}{5}\right)\big(1-(4)(-2)\big)$$
$$= \tfrac{43}{5} = 8.6$$

10. NORMAL LINE VECTOR

Partial derivatives can be used to find the vector normal to a three-dimensional surface defined by $f(x,y,z) = 0$ at some point P_0. Notice that the coefficients of $\mathbf{i}$, $\mathbf{j}$, and $\mathbf{k}$ are the same as the coefficients of x, y, and z calculated for the equation of the tangent plane at point P_0. (See Secs. 7.13 and 8.7.)

$$\mathbf{N} = \left.\frac{\partial f(x,y,z)}{\partial x}\right|_{P_0}\mathbf{i} + \left.\frac{\partial f(x,y,z)}{\partial y}\right|_{P_0}\mathbf{j}$$
$$+ \left.\frac{\partial f(x,y,z)}{\partial z}\right|_{P_0}\mathbf{k} \qquad \textit{8.43}$$

Example 8.12

What is the vector normal to the surface of $f(x,y,z) = 4x^2 + y^2 - 16z = 0$ at the point $(2,4,2)$?

Solution

The equation of the tangent plane at this point was calculated in Ex. 8.9 to be

$$T(2,4,2) = 2x + y - 2z - 4 = 0$$

A vector normal to the tangent plane through this point is

$$\mathbf{N} = 2\mathbf{i} + \mathbf{j} - 2\mathbf{k}$$

11. DIVERGENCE OF A VECTOR FIELD

The *divergence*, div $\mathbf{F}$, of a vector field $\mathbf{F}(x, y, z)$ is a scalar function defined by Eqs. 8.44 through 8.46.[4] The divergence of $\mathbf{F}$ can be interpreted as the *accumulation* of flux (i.e., a flowing substance) in a small region (i.e., at a point). One of the uses of the divergence is to determine whether flow (represented in direction and magnitude by $\mathbf{F}$) is compressible. Flow is incompressible if $div\mathbf{F} = 0$, since the substance is not accumulating.

$$\mathbf{F} = P(x, y, z)\mathbf{i} + Q(x, y, z)\mathbf{j} + R(x, y, z)\mathbf{k} \quad 8.44$$

$$div\mathbf{F} = \frac{\partial P}{\partial x} + \frac{\partial Q}{\partial y} + \frac{\partial R}{\partial z} \quad 8.45$$

It may be easier to calculate the divergence from Eq. 8.46.

$$div\mathbf{F} = \nabla \cdot \mathbf{F} \quad 8.46$$

The vector del operator, ∇, is defined as

$$\nabla = \frac{\partial}{\partial x}\mathbf{i} + \frac{\partial}{\partial y}\mathbf{j} + \frac{\partial}{\partial z}\mathbf{k} \quad 8.47$$

If there is no divergence, then the dot product calculated in Eq. 8.46 is zero.

Example 8.13

Calculate the divergence of the following vector function.

$$\mathbf{F}(x, y, z) = xz\mathbf{i} + e^x y\mathbf{j} + 7x^3 y\mathbf{k}$$

Solution

From Eq. 8.45,

$$div\ \mathbf{F} = \frac{\partial}{\partial x}(xz) + \frac{\partial}{\partial y}(e^x y) + \frac{\partial}{\partial z}(7x^3 y)$$

$$= z + e^x + 0 = z + e^x$$

12. CURL OF A VECTOR FIELD

The *curl*, curl $\mathbf{F}$, of a vector field $\mathbf{F}(x, y, z)$ is a vector field defined by Eqs. 8.51 and 8.52. The curl $\mathbf{F}$ can be interpreted as the *vorticity* per unit area of flux (i.e., a flowing substance) in a small region (i.e., at a point). One of the uses of the curl is to determine whether flow (represented in direction and magnitude by $\mathbf{F}$) is rotational. Flow is irrotational if curl $\mathbf{F} = 0$. [5]

$$\mathbf{F} = P(x, y, z)\mathbf{i} + Q(x, y, z)\mathbf{j} + R(x, y, z)\mathbf{k} \quad 8.48$$

---

[4] Notice that a bold letter, $\mathbf{F}$, is used to indicate that the vector is a function of x, y, and z.

[5] If the velocity vector is $\mathbf{V}$, then the *vorticity* is

$$\boldsymbol{\omega} = \nabla \times V = \omega_x i + \omega_y j + \omega_z k \quad 8.49$$

The *circulation* (defined in Chap. 17) is the line integral of the velocity $\mathbf{V}$ along a closed curve.

$$\Gamma = \oint V \cdot ds = \oint \boldsymbol{\omega} \cdot dA \quad 8.50$$

$$\text{curl } \mathbf{F} = \left(\frac{\partial R}{\partial y} - \frac{\partial Q}{\partial z}\right)\mathbf{i} + \left(\frac{\partial P}{\partial z} - \frac{\partial R}{\partial x}\right)\mathbf{j}$$

$$+ \left(\frac{\partial Q}{\partial x} - \frac{\partial P}{\partial y}\right)\mathbf{k} \quad 8.51$$

It may be easier to calculate the curl from Eq. 8.52. (The vector del operator, ∇, was defined in Eq. 8.47.)

$$\text{curl } \mathbf{F} = \nabla \times \mathbf{F}$$

$$= \begin{vmatrix} \mathbf{i} & \mathbf{j} & \mathbf{k} \\ \dfrac{\partial}{\partial x} & \dfrac{\partial}{\partial y} & \dfrac{\partial}{\partial z} \\ P(x, y, z) & Q(x, y, z) & R(x, y, z) \end{vmatrix} \quad 8.52$$

Example 8.14

Calculate the curl of the following vector function.

$$\mathbf{F}(x, y, z) = 3x^2\mathbf{i} + 7e^x y\mathbf{j}$$

Solution

Using Eq. 8.52,

$$\text{curl } \mathbf{F} = \begin{vmatrix} \mathbf{i} & \mathbf{j} & \mathbf{k} \\ \dfrac{\partial}{\partial x} & \dfrac{\partial}{\partial y} & \dfrac{\partial}{\partial z} \\ 3x^2 & 7e^x y & 0 \end{vmatrix}$$

Expand the determinant across the top row.

$$\mathbf{i}\left(\frac{\partial}{\partial y}(0) - \frac{\partial}{\partial z}(7e^x y)\right) - \mathbf{j}\left(\frac{\partial}{\partial x}(0) - \frac{\partial}{\partial z}(3x^2)\right)$$

$$+ \mathbf{k}\left(\frac{\partial}{\partial x}(7e^x y) - \frac{\partial}{\partial y}(3x^2)\right)$$

$$= \mathbf{i}(0 - 0) - \mathbf{j}(0 - 0) + \mathbf{k}(7e^x y - 0) = 7e^x y\mathbf{k}$$

13. TAYLOR'S FORMULA

Taylor's formula (series) can be used to expand a function around a point (i.e., approximate the function at one point based on the function's value at another point). The approximation consists of a series, each term composed of a derivative of the original function and a polynomial. Using Taylor's formula requires that the original function be continuous in the interval $[a, b]$ and have the required number of derivatives. To expand a function, $f(x)$, around a point, a, in order to obtain $f(b)$, Taylor's formula is[6]

$$f(b) = f(a) + \frac{f'(a)}{1!}(b - a) + \frac{f''(a)}{2!}(b - a)^2 + \cdots$$

$$+ \frac{f^n(a)}{n!}(b - a)^n + R_n(b) \quad 8.53$$

---

[6] If $a = 0$, Eq. 8.53 is known as the *Maclaurin series*.

In Eq. 8.53, the expression f^n designates the nth derivative of the function $f(x)$. To be a useful approximation, point a must satisfy two requirements: It must be relatively close to point b, and the function and its derivatives must be known or easy to calculate. The last term, $R_n(b)$, is the uncalculated remainder after n derivatives. It is the difference between the exact and approximate values. By using enough terms, the remainder can be made arbitrarily small. That is, $R_n(b)$ approaches zero as n approaches infinity.

It can be shown that the remainder term can be calculated from Eq. 8.54, where c is some number in the interval $[a, b]$. With certain functions, the constant c can be completely determined. In most cases, however, it is possible only to calculate an upper bound on the remainder from Eq. 8.55. M_n is the maximum (positive) value of $f^{(n+1)}(x)$ on the interval $[a, b]$.

$$R_n(b) = \frac{f^{n+1}(c)}{(n+1)!}(b-a)^{n+1} \qquad \text{8.54}$$

$$|R_n(b)| \leq M_n \frac{\left|(b-a)^{n+1}\right|}{(n+1)!} \qquad \text{8.55}$$

14. COMMON SERIES APPROXIMATIONS

Taylor's formulas can be used (by expanding about $a = 0$) to derive the following series approximations.

$$\sin x \approx x - \frac{x^3}{3!} + \frac{x^5}{5!} - \frac{x^7}{7!} + \cdots$$
$$+ (-1)^n \frac{x^{2n+1}}{(2n+1)!} \qquad \text{8.56}$$

$$\cos x \approx 1 - \frac{x^2}{2!} + \frac{x^4}{4!} - \frac{x^6}{6!} + \cdots + (-1)^n \frac{x^{2n}}{(2n)!} \qquad \text{8.57}$$

$$\sinh x \approx x + \frac{x^3}{3!} + \frac{x^5}{5!} + \frac{x^7}{7!} + \cdots + \frac{x^{2n+1}}{(2n+1)!} \qquad \text{8.58}$$

$$\cosh x \approx 1 + \frac{x^2}{2!} + \frac{x^4}{4!} + \frac{x^6}{6!} + \cdots + \frac{x^{2n}}{(2n)!} \qquad \text{8.59}$$

$$e^x \approx 1 + x + \frac{x^2}{2!} + \frac{x^3}{3!} + \cdots + \frac{x^n}{n!} \qquad \text{8.60}$$

$$\ln(1+x) \approx x - \frac{x^2}{2} + \frac{x^3}{3} - \frac{x^4}{4} + \cdots + (-1)^{n+1}\frac{x^n}{n} \qquad \text{8.61}$$

$$\frac{1}{1-x} \approx 1 + x + x^2 + x^3 + \cdots + x^n \qquad \text{8.62}$$

9

Integral Calculus

1. INTEGRATION

Integration is the inverse operation of differentiation. For that reason, *indefinite integrals* are sometimes referred to as *antiderivatives*.[1] Although expressions can be functions of several variables, integrals can only be taken with respect to one variable at a time. The *differential term* (dx in Eq. 9.1) indicates that variable. In Eq. 9.1, the function $f'(x)$ is the *integrand*, and x is the variable of integration.

$$\int f'(x)dx = f(x) + C \qquad 9.1$$

While most of a function, $f(x)$, can be "recovered" through integration of its derivative, $f'(x)$, a constant term will be lost. This is because the derivative of a constant term vanishes (i.e., is zero), leaving nothing to recover from. A *constant of integration*, C, is added to the integral to recognize the possibility of such a term.

2. ELEMENTARY OPERATIONS

Equations 9.2 through 9.8 summarize the elementary integration operations on polynomials and exponentials.[2]

[1] The difference between an indefinite and definite integral (covered in Sec. 7) is simple: An *indefinite integral* is a function, while a *definite integral* is a number.
[2] More extensive listings, known as *tables of integrals*, are widely available.

Equations 9.2 and 9.3 are particularly useful. (C and k represent constants. $f(x)$ and $g(x)$ are functions of x.)

$$\int k\,dx = kx + C \qquad 9.2$$

$$\int x^m\,dx = \frac{x^{m+1}}{m+1} + C \qquad [m \neq -1] \qquad 9.3$$

$$\int \frac{1}{x}dx = \ln|x| + C \qquad 9.4$$

$$\int e^{kx}dx = \frac{e^{kx}}{k} + C \qquad 9.5$$

$$\int xe^{kx}dx = \frac{e^{kx}(kx-1)}{k^2} + C \qquad 9.6$$

$$\int k^{ax}dx = \frac{k^{ax}}{a\ln k} + C \qquad 9.7$$

$$\int \ln x\,dx = x\ln x - x + C \qquad 9.8$$

Equations 9.9 through 9.20 summarize the elementary integration operations on transcendental functions.

$$\int \sin x\,dx = -\cos x + C \qquad 9.9$$

$$\int \cos x\,dx = \sin x + C \qquad 9.10$$

$$\int \tan x\,dx = \ln|\sec x| + C$$
$$= -\ln|\cos x| + C \qquad 9.11$$

$$\int \cot x\,dx = \ln|\sin x| + C \qquad 9.12$$

$$\int \sec x\,dx = \ln|(\sec x + \tan x)| + C$$
$$= \ln\left|\tan\left(\frac{x}{2}+\frac{\pi}{4}\right)\right| + C \qquad 9.13$$

$$\int \csc x\,dx = \ln|(\csc x - \cot x)| + C$$
$$= \ln\left|\tan\frac{x}{2}\right| + C \qquad 9.14$$

$$\int \frac{dx}{k^2+x^2} = \frac{1}{k}\arctan\frac{x}{k} + C \qquad 9.15$$

$$\int \frac{dx}{\sqrt{k^2-x^2}} = \arcsin\frac{x}{k} + C \quad [k^2 > x^2] \qquad 9.16$$

$$\int \frac{dx}{x\sqrt{x^2-k^2}} = \frac{1}{k}\text{arcsec}\frac{x}{k} + C \quad [x^2 > k^2] \qquad 9.17$$

$$\int \sin^2 x\,dx = \tfrac{1}{2}x - \tfrac{1}{4}\sin 2x + C \qquad 9.18$$

$$\int \cos^2 x \, dx = \tfrac{1}{2}x + \tfrac{1}{4}\sin 2x + C \qquad \textbf{9.19}$$

$$\int \tan^2 x \, dx = \tan x - x + C \qquad \textbf{9.20}$$

Equations 9.21 through 9.26 summarize the elementary integration operations on hyperbolic transcendental functions. Integrals of hyperbolic functions are not completely analogous to those of the regular transcendental functions.

$$\int \sinh x \, dx = \cosh x + C \qquad \textbf{9.21}$$

$$\int \cosh x \, dx = \sinh x + C \qquad \textbf{9.22}$$

$$\int \tanh x \, dx = \ln|\cosh x| + C \qquad \textbf{9.23}$$

$$\int \coth x \, dx = \ln|\sinh x| + C \qquad \textbf{9.24}$$

$$\int \operatorname{sech} x \, dx = \arctan(\sinh x) + C \qquad \textbf{9.25}$$

$$\int \operatorname{csch} x \, dx = \ln\left|\tanh\left(\frac{x}{2}\right)\right| + C \qquad \textbf{9.26}$$

Equations 9.27 through 9.31 summarize the elementary integration operations on functions and combinations of functions.

$$\int k f(x) \, dx = k \int f(x) \, dx \qquad \textbf{9.27}$$

$$\int (f(x) + g(x)) \, dx = \int f(x) \, dx + \int g(x) \, dx \qquad \textbf{9.28}$$

$$\int \frac{f'(x)}{f(x)} \, dx = \ln|f(x)| + C \qquad \textbf{9.29}$$

$$\int f(x) \, dg(x) = f(x) \int dg(x) - \int g(x) \, df(x) + C$$

$$= f(x)g(x) - \int g(x) \, df(x) + C \qquad \textbf{9.30}$$

Example 9.1

Find the integral with respect to x of

$$3x^2 + \tfrac{1}{3}x - 7 = 0$$

Solution

This is a polynomial function, and Eq. 9.3 can be applied to each of the three terms.

$$\int \left(3x^2 + \tfrac{1}{3}x - 7\right) dx = x^3 + \tfrac{1}{6}x^2 - 7x + C$$

3. INTEGRATION BY PARTS

Equation 9.30, repeated here, is known as *integration by parts*. $f(x)$ and $g(x)$ are functions. The use of this method is illustrated by Ex. 9.2.

$$\int f(x) \, dg(x) = f(x)g(x) - \int g(x) \, df(x) + C \qquad \textbf{9.31}$$

Example 9.2

Find the following integral.

$$\int x^2 e^x \, dx$$

Solution

$x^2 e^x$ is factored into two parts so that integration by parts can be used.

$$f(x) = x^2$$
$$dg(x) = e^x \, dx$$
$$df(x) = 2x \, dx$$
$$g(x) = \int dg(x) = \int e^x \, dx = e^x$$

From Eq. 9.31, disregarding the constant of integration (which cannot be evaluated),

$$\int f(x) \, dg(x) = f(x)g(x) - \int g(x) \, df(x)$$

$$\int x^2 e^x \, dx = x^2 e^x - \int e^x (2x) \, dx$$

The second term is also factored into two parts, and integration by parts is used again. This time,

$$f(x) = x$$
$$dg(x) = e^x \, dx$$
$$df(x) = dx$$
$$g(x) = \int dg(x) = \int e^x \, dx = e^x$$

From Eq. 9.31,

$$\int 2x e^x \, dx = 2 \int x e^x \, dx$$

$$= (2)\left(x e^x - \int e^x \, dx\right)$$

$$= (2)(x e^x - e^x)$$

Then, the complete integral is

$$\int x^2 e^x \, dx = x^2 e^x - (2)(x e^x - e^x) + C$$

$$= e^x (x^2 - 2x + 2) + C$$

4. SEPARATION OF TERMS

Equation 9.28 shows that the integral of a sum of terms is equal to a sum of integrals. This technique is known

Background and Support

as *separation of terms*. In many cases, terms are easily separated. In other cases, the technique of *partial fractions* (see Sec. 3.16) can be used to obtain individual terms. These techniques are illustrated by Exs. 9.3 and 9.4.

Example 9.3

Find the following integral.

$$\int \frac{(2x^2 + 3)^2}{x} dx$$

Solution

$$\int \frac{(2x^2 + 3)^2}{x} dx = \int \frac{4x^4 + 12x^2 + 9}{x} dx$$

$$= \int \left(4x^3 + 12x + \frac{9}{x} \right) dx$$

$$= x^4 + 6x^2 + 9 \ln|x| + C$$

Example 9.4

Find the following integral.

$$\int \frac{3x + 2}{3x - 2} dx$$

Solution

The integrand is larger than 1, so use long division to simplify it.

$$\begin{array}{r} 1 \text{ rem } 4 \qquad \left(1 + \dfrac{4}{3x-2} \right) \\ 3x - 2 \overline{\smash{\big)}\ 3x+2} \\ \underline{3x-2} \\ 4 \text{ remainder} \end{array}$$

$$\int \frac{3x + 2}{3x - 2} dx = \int \left(1 + \frac{4}{3x - 2} \right) dx$$

$$= \int dx + \int \frac{4}{3x - 2} dx$$

$$= x + \tfrac{4}{3} \ln|(3x - 2)| + C$$

5. DOUBLE AND HIGHER-ORDER INTEGRALS

A function can be successively integrated. (This is analogous to successive differentiation.) A function that is integrated twice is known as a *double integral*; if integrated three times, it is a *triple integral*; and so on. Double and triple integrals are used to calculate areas and volumes, respectively.

The successive integrations do not need to be with respect to the same variable. Variables not included in the integration are treated as constants.

There are several notations used for a multiple integral, particularly when the product of length differentials represents a differential area or volume. A double integral (i.e., two successive integrations) can be represented by one of the following notations.

$$\iint f(x, y) dx\, dy, \quad \int_{R^2} f(x, y) dx\, dy,$$

$$\text{or} \int \int_{R^2} f(x, y) dA$$

A triple integral can be represented by one of the following notations.

$$\iiint f(x, y, z) dx\, dy\, dz,$$

$$\int_{R^3} f(x, y, z) dx\, dy\, dz,$$

$$\text{or} \quad \iiint_{R^3} f(x, y, z) dV$$

Example 9.5

Find the following double integral.

$$\iint (x^2 + y^3 x) dx\, dy$$

Solution

$$\int (x^2 + y^3 x) dx = \tfrac{1}{3} x^3 + \tfrac{1}{2} y^3 x^2 + C_1$$

$$\int \left(\tfrac{1}{3} x^3 + \tfrac{1}{2} y^3 x^2 + C_1 \right) dy = \tfrac{1}{3} y x^3 + \tfrac{1}{8} y^4 x^2 + C_1 y + C_2$$

So,

$$\iint (x^2 + y^3 x) dx\, dy = \tfrac{1}{3} y x^3 + \tfrac{1}{8} y^4 x^2 + C_1 y + C_2$$

6. INITIAL VALUES

The constant of integration, C, can be found only if the value of the function $f(x)$ is known for some value of x_0. The value $f(x_0)$ is known as an *initial value*. To completely define a function, as many initial values, $f(x_0), f'(x_0), f''(x_0)$, and so on, as there are integrations are needed.

Example 9.6

It is known that $f(x) = 4$ when $x = 2$ (i.e., the initial condition is $f(2) = 4$). Find the original function.

$$\int (3x^3 - 7x) dx$$

Solution

The function is

$$f(x) = \int (3x^3 - 7x)dx = \tfrac{3}{4}x^4 - \tfrac{7}{2}x^2 + C$$

Substituting the initial value determines that C equals 6.

$$4 = \left(\tfrac{3}{4}\right)(2)^4 - \left(\tfrac{7}{2}\right)(2)^2 + C$$
$$4 = 12 - 14 + C$$
$$C = 6$$

The function is

$$f(x) = \tfrac{3}{4}x^4 - \tfrac{7}{2}x^2 + 6$$

7. DEFINITE INTEGRALS

A *definite integral* is restricted to a specific range of the independent variable. (Unrestricted integrals of the types shown in all preceding examples are known as *indefinite integrals*.) A definite integral restricted to the region bounded by *lower* and *upper limits* (also known as *bounds*), x_1 and x_2, is written as

$$\int_{x_1}^{x_2} f(x)dx$$

Equation 9.32 indicates how definite integrals are evaluated. It is known as the *fundamental theorem of calculus*.

$$\int_{x_1}^{x_2} f'(x)dx = f(x)\Big|_{x_1}^{x_2} = f(x_2) - f(x_1) \qquad 9.32$$

A common use of a definite integral is the calculation of work performed by a force, F, that moves from position x_1 to x_2.

$$W = \int_{x_1}^{x_2} F\,dx \qquad 9.33$$

Example 9.7

Evaluate the following definite integral.

$$\int_{\pi/4}^{\pi/3} \sin x\,dx$$

Solution

From Eq. 9.32,

$$\int_{\pi/4}^{\pi/3} \sin x\,dx = \Big[-\cos x\Big]_{\pi/4}^{\pi/3}$$
$$= -\cos\tfrac{\pi}{3} - \left(-\cos\tfrac{\pi}{4}\right)$$
$$= -0.5 - (-0.707) = 0.207$$

8. AVERAGE VALUE

The average value of a function $f(x)$ that is integrable over the interval $[a, b]$ is

$$\text{average value} = \frac{1}{b-a}\int_a^b f(x)dx \qquad 9.34$$

9. AREA

Equation 9.35 calculates the area, A, bounded by $x = a$, $x = b$, $f_1(x)$ above and $f_2(x)$ below. ($f_2(x) = 0$ if the area is bounded by the x-axis.) This is illustrated in Fig. 9.1.

$$A = \int_a^b (f_1(x) - f_2(x))dx \qquad 9.35$$

Figure 9.1 *Area Between Two Curves*

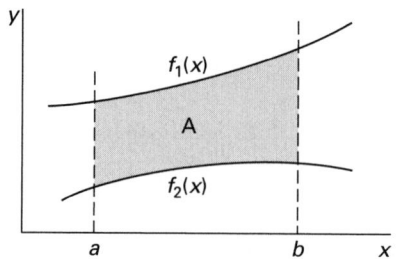

Example 9.8

Find the area between the x-axis and the parabola $y = x^2$ in the interval $[0,4]$.

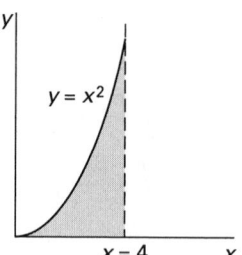

Solution

Referring to Eq. 9.35,

$$f_1(x) = x^2$$
$$f_2(x) = 0$$
$$A = \int_a^b (f_1(x) - f_2(x))dx = \int_0^4 x^2 dx$$
$$= \left[\frac{x^3}{3}\right]_0^4 = 64/3$$

10. ARC LENGTH

Equation 9.36 gives the length of a curve defined by $f(x)$ whose derivative exists in the interval $[a, b]$.

$$\text{length} = \int_a^b \sqrt{1 + (f'(x))^2}\, dx \qquad 9.36$$

11. PAPPUS' THEOREMS[3]

The first and second theorems of Pappus are:[4]

- *First Theorem*: Given a curve, C, that does not intersect the y-axis, the area of the *surface of revolution* generated by revolving C around the y-axis is equal to the product of the length of the curve and the circumference of the circle traced by the centroid of curve C.

$$A = \text{length} \times \text{circumference}$$
$$= \text{length} \times 2\pi \times \text{radius} \qquad 9.37$$

- *Second Theorem*: Given a plane region, R, that does not intersect the y-axis, the *volume of revolution* generated by revolving R around the y-axis is equal to the product of the area and the circumference of the circle traced by the centroid of area R.

$$V = \text{area} \times \text{circumference}$$
$$= \text{area} \times 2\pi \times \text{radius} \qquad 9.38$$

12. SURFACE OF REVOLUTION

The surface area obtained by rotating $f(x)$ about the x-axis is

$$A = 2\pi \int_{x=a}^{x=b} f(x)\sqrt{1 + (f'(x))^2}\, dx \qquad 9.39$$

The surface area obtained by rotating $f(y)$ about the y-axis is

$$A = 2\pi \int_{y=c}^{y=d} f(y)\sqrt{1 + (f'(y))^2}\, dy \qquad 9.40$$

Example 9.9

The curve $f(x) = \frac{1}{2}x$ over the region $x = [0, 4]$ is rotated about the x-axis. What is the surface of revolution?

---

[3]This section is an introduction to surfaces and volumes of revolution but does not involve integration.
[4]Some authorities call the first theorem the second and vice versa.

Solution

The surface of revolution is

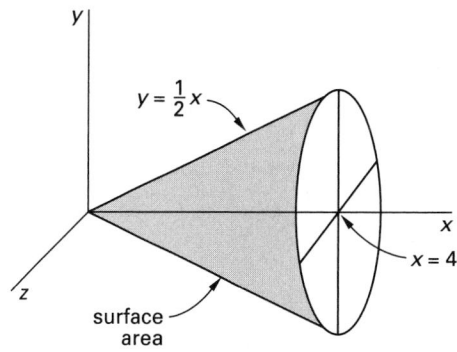

Since $f(x) = \frac{1}{2}x$, $f'(x) = \frac{1}{2}$. From Eq. 9.39, the area is

$$A = 2\pi \int_{x=a}^{x=b} f(x)\sqrt{1 + (f'(x))^2}\, dx$$

$$= 2\pi \int_0^4 \frac{1}{2}x\sqrt{1 + \left(\frac{1}{2}\right)^2}\, dx$$

$$= \frac{\sqrt{5}}{2}\pi \int_0^4 x\, dx$$

$$= \frac{\sqrt{5}}{2}\pi \left[\frac{x^2}{2}\right]_0^4$$

$$= \frac{\sqrt{5}}{2}\pi \left(\frac{(4)^2 - (0)^2}{2}\right)$$

$$= 4\sqrt{5}\pi$$

13. VOLUME OF REVOLUTION

The volume obtained by rotating $f(x)$ about the x-axis is given by Eq. 9.41. $f^2(x)$ is the square of the function, not the second derivative. Equation 9.41 is known as the *method of discs*.

$$V = \pi \int_{x=a}^{x=b} f^2(x)\, dx \qquad 9.41$$

The volume obtained by rotating $f(x)$ about the y-axis can be found from Eq. 9.41 (i.e., using the method of discs) by rewriting the limits and equation in terms of y, or alternatively, the *method of shells* can be used, resulting in the second form of Eq. 9.42.

$$V = 2\pi \int_{y=c}^{y=d} f^2(y)\, dy = \pi \int_{x=a}^{x=b} xf(x)\, dx \qquad 9.42$$

Example 9.10

The curve $f(x) = x^2$ over the region $x = [0, 4]$ is rotated about the x-axis. What is the volume of revolution?

Solution

The volume of revolution is

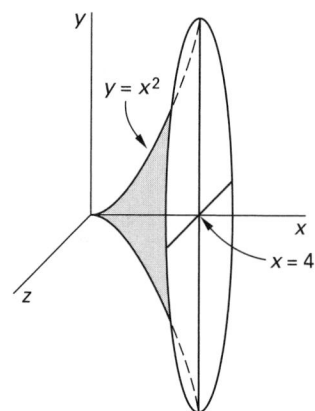

$$V = \pi \int_a^b f^2(x)dx = \pi \int_0^4 (x^2)^2 dx$$

$$= \pi \left[\frac{x^5}{5} \right]_0^4 = \pi \left(\frac{1024}{5} - 0 \right) = 204.8\pi$$

14. MOMENTS OF A FUNCTION

The *first moment of a function* is a concept used in finding centroids and centers of gravity. Equations 9.43 and 9.44 are for one- and two-dimensional problems, respectively. It is the exponent of x (1 in this case) that gives the moment its name.

$$\text{first moment} = \int x f(x)dx \qquad 9.43$$

$$\text{first moment} = \iint x f(x,y)dx\,dy \qquad 9.44$$

The *second moment of a function* is a concept used in finding moments of inertia with respect to an axis. Equations 9.45 and 9.46 are for two- and three-dimensional problems, respectively. Second moments with respect to other axes are analogous.

$$(\text{second moment})_x = \iint y^2 f(x,y)dy\,dx \qquad 9.45$$

$$(\text{second moment})_x = \iiint (y^2 + z^2)f(x,y,z)dy\,dz\,dx$$
$$9.46$$

15. FOURIER SERIES

Any periodic waveform can be written as the sum of an infinite number of sinusoidal terms, known as *harmonic terms* (i.e., an infinite series). Such a sum of terms is known as a *Fourier series*, and the process of finding the terms is *Fourier analysis*. (Extracting the

original waveform from the series is known as *Fourier inversion*.) Since most series converge rapidly, it is possible to obtain a good approximation to the original waveform with a limited number of sinusoidal terms.

Fourier's theorem is Eq. 9.47.[5] The object of a Fourier analysis is to determine the coefficients a_n and b_n. The constant a_0 can often be determined by inspection since it is the average value of the waveform.

$$f(t) = a_0 + a_1 \cos \omega t + a_2 \cos 2\omega t + \cdots$$
$$+ b_1 \sin \omega t + b_2 \sin 2\omega t + \cdots \qquad 9.47$$

ω is the *natural (fundamental) frequency* of the waveform. It depends on the actual waveform period, T.

$$\omega = \frac{2\pi}{T} \qquad 9.48$$

To simplify the analysis, the time domain can be normalized to the radian scale. The normalized scale is obtained by dividing all frequencies by ω. Then the Fourier series becomes

$$f(t) = a_0 + a_1 \cos t + a_2 \cos 2t + \cdots$$
$$+ b_1 \sin t + b_2 \sin 2t + \cdots \qquad 9.49$$

The coefficients a_n and b_n are found from the following relationships.

$$a_0 = \frac{1}{2\pi} \int_0^{2\pi} f(t)dt$$

$$= \frac{1}{T} \int_0^T f(t)dt \qquad 9.50$$

$$a_n = \frac{1}{\pi} \int_0^{2\pi} f(t) \cos nt\, dt$$

$$= \frac{2}{T} \int_0^T f(t) \cos nt\, dt \qquad [n \geq 1] \qquad 9.51$$

$$b_n = \frac{1}{\pi} \int_0^{2\pi} f(t) \sin nt\, dt$$

$$= \frac{2}{T} \int_0^T f(t) \sin nt\, dt \qquad [n \geq 1] \qquad 9.52$$

While Eqs. 9.51 and 9.52 are always valid, the work of integrating and finding a_n and b_n can be greatly simplified if the waveform is recognized as being symmetrical. Table 9.1 summarizes the simplifications.

---

[5]The independent variable used in this section is t, since Fourier analysis is most frequently used in the time domain.

Table 9.1 Fourier Analysis Simplifications for Symmetrical Waveforms

	even symmetry $f(-t) = f(t)$	odd symmetry $f(-t) = -f(t)$
full-wave symmetry* $f(t + 2\pi) = f(t)$ $|A_2| = |A_1|$ $|A_{\text{total}}| = |A_1|$ *any repeating wave form	$b_n = 0$ [all n] $a_n = \frac{1}{\pi}\int_0^{2\pi} f(t)\cos nt\, dt$ [all n]	$a_0 = 0$ $a_n = 0$ [all n] $b_n = \frac{1}{\pi}\int_0^{2\pi} f(t)\sin nt\, dt$ [all n]
half-wave symmetry* $f(t + \pi) = -f(t)$ $|A_2| = |A_1|$ $|A_{\text{total}}| = 2|A_1|$ *same as rotational symmetry	$a_n = 0$ [even n] $b_n = 0$ [all n] $a_n = \frac{2}{\pi}\int_0^{\pi} f(t)\cos nt\, dt$ [odd n]	$a_0 = 0$ $a_n = 0$ [all n] $b_n = 0$ [even n] $b_n = \frac{2}{\pi}\int_0^{\pi} f(t)\sin nt\, dt$ [odd n]
quarter-wave symmetry $f(t + \pi) = -f(t)$ $|A_2| = |A_1|$ $|A_{\text{total}}| = 4|A_1|$	$a_0 = 0$ $a_n = 0$ [even n] $b_n = 0$ [all n] $a_n = \frac{4}{\pi}\int_0^{\frac{\pi}{2}} f(t)\cos nt\, dt$ [odd n]	$a_0 = 0$ $a_n = 0$ [all n] $b_n = 0$ [even n] $b_n = \frac{4}{\pi}\int_0^{\frac{\pi}{2}} f(t)\sin nt\, dt$ [odd n]

Solution

From Eq. 9.50,

$$a_0 = \frac{1}{2\pi}\int_0^\pi (1)\, dt + \frac{1}{2\pi}\int_\pi^{2\pi} (0)\, dt = \tfrac{1}{2}$$

This value of $1/2$ corresponds to the average value of $f(t)$. It could have been found by observation.

$$a_1 = \frac{1}{\pi}\int_0^\pi (1)\cos t\, dt + \frac{1}{\pi}\int_\pi^{2\pi}(0)\cos t\, dt$$
$$= \frac{1}{\pi}\Big[\sin t\Big]_0^\pi + 0 = 0$$

In general,

$$a_n = \frac{1}{\pi}\left[\frac{\sin nt}{n}\right]_0^\pi = 0$$
$$b_1 = \frac{1}{\pi}\int_0^\pi (1)\sin t\, dt + \frac{1}{\pi}\int_\pi^{2\pi}(0)\sin t\, dt$$
$$= \frac{1}{\pi}\Big[-\cos t\Big]_0^\pi = \frac{2}{\pi}$$

In general,

$$b_n = \frac{1}{\pi}\left[\frac{-\cos nt}{n}\right]_0^\pi = \begin{cases} 0 & \text{for } n \text{ even} \\ \dfrac{2}{\pi n} & \text{for } n \text{ odd} \end{cases}$$

The series is

$$f(t) = \tfrac{1}{2} + \frac{2}{\pi}\left[\sin t + \tfrac{1}{3}\sin 3t + \tfrac{1}{5}\sin 5t + \cdots\right]$$

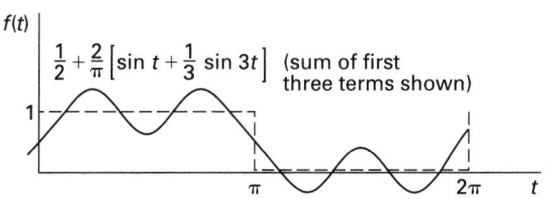

16. FAST FOURIER TRANSFORMS

Many mathematical operations are needed to implement a true Fourier transform. While the terms of a Fourier series might be slowly derived by integration, a faster method is needed to analyze real-time data. The *fast Fourier transform* (FFT) is a computer algorithm implemented in *spectrum analyzers (signal analyzers or FFT analyzers)* and replaces integration and multiplication operations with table look-ups and additions.[6]

---

[6] *Spectrum analysis*, also known as *frequency analysis, signature analysis,* and *time-series analysis,* develops a relationship (usually graphical) between some property (e.g., amplitude or phase shift) versus frequency.

Example 9.11

Find the first four terms of a Fourier series that approximates the repetitive step function illustrated.

$$f(t) = \begin{cases} 1 & 0 < t < \pi \\ 0 & \pi < t < 2\pi \end{cases}$$

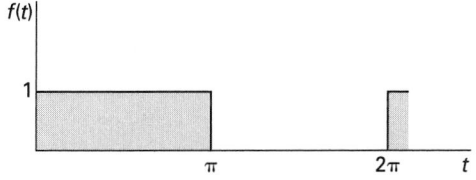

Since the complexity of the transform is reduced, the transformation occurs more quickly, enabling efficient analysis of waveforms with little or no periodicity.[7]

Using a spectrum analyzer requires choosing the frequency band (e.g., 0 to 20 kHz) to be monitored. (This step automatically selects the sampling period. The lower the frequencies sampled, the longer the sampling period.) If they are not fixed by the analyzer, the numbers of time-dependent input variable samples (e.g., 1024) and frequency-dependent output variable values (e.g., 400) are chosen.[8] There are half as many frequency lines as data points because each line contains two pieces of information—real (amplitude) and imaginary (phase). The *resolution* of the resulting frequency analysis is

$$\text{resolution} = \frac{\text{frequency bandwidth}}{\text{no. of output variable values}} \qquad 9.53$$

17. INTEGRAL FUNCTIONS

Integrals that cannot be evaluated as finite combinations of elementary functions are called *integral functions*. These functions are evaluated by series expansion. Some of the more common functions are listed as follows.[9,10]

- *integral sine function:*

$$\text{Si}(x) = \int_0^x \frac{\sin x}{x} dx$$
$$= x - \frac{x^3}{3 \cdot 3!} + \frac{x^5}{5 \cdot 5!} - \frac{x^7}{7 \cdot 7!} + \cdots \qquad 9.54$$

- *integral cosine function:*

$$\text{Ci}(x) = \int_{-\infty}^x \frac{\cos x}{x} dx = -\int_x^\infty \frac{\cos x}{x} dx$$
$$= C_E + \ln\ x - \frac{x^2}{2 \cdot 2!} + \frac{x^4}{4 \cdot 4!} - \cdots \qquad 9.55$$

- *integral exponential function:*

$$\text{Ei}(x) = \int_{-\infty}^x \frac{e^x}{x} dx = -\int_{-x}^\infty \frac{e^{-x}}{x} dx$$
$$= C_E + \ln\ x + x + \frac{x^2}{2 \cdot 2!} + \frac{x^3}{3 \cdot 3!} + \cdots \qquad 9.56$$

- *error function:*

$$\text{erf}(x) = \frac{2}{\sqrt{\pi}} \int_0^x e^{-x^2} dx$$
$$= \left(\frac{2}{\sqrt{\pi}} \right) \left(\frac{x}{1 \cdot 0!} - \frac{x^3}{3 \cdot 1!} + \frac{x^5}{5 \cdot 2!} - \frac{x^7}{7 \cdot 3!} + \cdots \right) \qquad 9.57$$

---

[7]Hours and days of manual computations are compressed into milliseconds.

[8]Two samples per time-dependent cycle (at the maximum frequency) is the lower theoretical limit for sampling, but the practical minimum rate is approximately 2.5 samples per cycle. This will ensure that *alias components* (i.e., low-level frequency signals) do not show up in the frequency band of interest.

[9]Other integral functions include the Fresnel integral, gamma function, and elliptic integral.

[10]C_E in Eqs. 9.55 and 9.56 is *Euler's constant.*

$$C_E = \int_{+\infty}^0 e^{-x} \ln x\, dx$$
$$= \lim_{m \to \infty} \left(1 + \tfrac{1}{2} + \tfrac{1}{3} + \cdots + \frac{1}{m} - \ln m \right)$$
$$= 0.577215665$$

10 Differential Equations

than one) or is embedded in another function (e.g., y embedded in $\sin y$ or e^y), the equation is said to be *nonlinear*.

Each term of a *homogeneous differential equation* contains either the function (y) or one of its derivatives—that is, the sum of derivative terms is equal to zero. In a *nonhomogeneous differential equation*, the sum of derivative terms is equal to a nonzero *forcing function* of the independent variable (e.g., $g(x)$). In order to solve a nonhomogeneous equation, it is often necessary to solve the homogeneous equation first. The homogeneous equation corresponding to a nonhomogeneous equation is known as a *reduced equation* or *complementary equation*.

The following examples illustrate the types of differential equations.

- $y' - 7y = 0$ homogeneous, first-order linear, with constant coefficients

- $y'' - 2y' + 8y = \sin 2x$ nonhomogeneous, second-order linear, with constant coefficients

- $y'' - (x^2 - 1)y^2 = \sin 4x$ nonhomogeneous, second-order, nonlinear

An *auxiliary equation* (also called the *characteristic equation*) can be written for a homogeneous linear differential equation with constant coefficients, regardless of order. This auxiliary equation is simply the polynomial formed by replacing all derivatives with variables raised to the power of their respective derivatives.

The purpose of solving a differential equation is to derive an expression for the function in terms of the independent variable. The expression does not need to be explicit in the function, but there can be no derivatives in the expression. Since, in the simplest cases, solving a differential equation is equivalent to finding an indefinite integral, it is not surprising that *constants of integration* must be evaluated from knowledge of how the system behaves. Additional data are known as *initial*

1. TYPES OF DIFFERENTIAL EQUATIONS

A *differential equation* is a mathematical expression combining a function (e.g., $y = f(x)$) and one or more of its derivatives. The *order* of a differential equation is the highest derivative in it. *First-order differential equations* contain only first derivatives of the function, *second-order differential equations* contain second derivatives (and may contain first derivatives as well), and so on.

A *linear differential equation* can be written as a sum of products of multipliers of the function and its derivatives. If the multipliers are scalars, the differential equation is said to have *constant coefficients*. If the function or one of its derivatives is raised to some power (other *values,* and any problem that includes them is known as an *initial value problem.*[1]

Most differential equations require lengthy solutions and are not efficiently solved by hand. However, several types are fairly simple and are presented in this chapter.

Example 10.1

Write the complementary differential equation for the following nonhomogeneous differential equation.

$$y'' + 6y' + 9y = e^{-14x} \sin 5x$$

Solution

The complementary equation is found by eliminating the forcing function, $e^{-14x} \sin 5x$.

$$y'' + 6y' + 9y = 0$$

Example 10.2

Write the auxiliary equation to the following differential equation.

$$y'' + 4y' + y = 0$$

Solution

Replacing each derivative with a polynomial term whose degree equals the original order, the auxiliary equation is

$$r^2 + 4r + 1 = 0$$

2. HOMOGENEOUS, FIRST-ORDER LINEAR DIFFERENTIAL EQUATIONS WITH CONSTANT COEFFICIENTS

A homogeneous, first-order linear differential equation with constant coefficients has the general form of Eq. 10.1.

$$y' + ky = 0 \qquad 10.1$$

The auxiliary equation is $r + k = 0$ and has a root of $r = -k$. Equation 10.2 is the solution.

$$y = Ae^{rx} = Ae^{-kx} \qquad 10.2$$

---

[1] The term *initial* implies that time is the independent variable. While this may explain the origin of the term, initial value problems are not limited to the time domain. A *boundary value problem* is similar, except that the data come from different points. For example, additional data in the form $y(x_0)$ and $y'(x_0)$ or $y(x_0)$ and $y'(x_1)$ that need to be simultaneously satisfied constitute an initial value problem. Data of the form $y(x_0)$ and $y(x_1)$ constitute a boundary value problem. Until solved, it is difficult to know whether a boundary value problem has no, one, or more than one solution.

If the initial condition is known to be $y(0) = y_0$, the solution is

$$y = y_0 e^{-kx} \qquad 10.3$$

3. FIRST-ORDER LINEAR DIFFERENTIAL EQUATIONS

A first-order linear differential equation has the general form of Eq. 10.4. $p(x)$ and $g(x)$ can be constants or any function of x (but not of y). However, if $p(x)$ is a constant and $g(x)$ is zero, it is easier to solve the equation as shown in Sec. 2.

$$y' + p(x)y = g(x) \qquad 10.4$$

The *integrating factor* (which is actually a function) to this differential equation is

$$u(x) = \exp\left[\int p(x)dx\right] \qquad 10.5$$

The closed-form solution to Eq. 10.4 is

$$y = \frac{1}{u(x)}\left[\int u(x)g(x)dx + C\right] \qquad 10.6$$

For the special case where $p(x)$ and $g(x)$ are both constants, Eq. 10.4 becomes

$$y' + ay = b \qquad 10.7$$

If the initial condition is $y(0) = y_0$, the solution to Eq. 10.7 is

$$y = \left(\frac{b}{a}\right)(1 - e^{-ax}) + y_0 e^{-ax} \qquad 10.8$$

Example 10.3

Find a solution to the following differential equation.

$$y' - y = 2xe^{2x} \qquad y(0) = 1$$

Solution

This is a first-order linear equation with $p(x) = -1$ and $g(x) = 2xe^{2x}$. The integrating factor is

$$u(x) = \exp\left[\int p(x)dx\right] = \exp\left[\int -1dx\right] = e^{-x}$$

The solution is given by Eq. 10.6.

$$y = \frac{1}{u(x)}\left[\int u(x)g(x)dx + C\right]$$
$$= \frac{1}{e^{-x}}\left[\int e^{-x}2xe^{2x}dx + C\right]$$
$$= e^x\left[2\int xe^x dx + C\right]$$
$$= e^x[2xe^x - 2e^x + C]$$
$$= e^x[2e^x(x-1) + C]$$

From the initial condition,

$$y(0) = 1$$
$$e^0\big((2)(e^0)(0-1) + C\big) = 1$$
$$1\big((2)(1)(-1) + C\big) = 1$$

Therefore, $C = 3$. The complete solution is

$$y = e^x\big(2e^x(x-1) + 3\big)$$

4. FIRST-ORDER SEPARABLE DIFFERENTIAL EQUATIONS

First-order separable differential equations can be placed in the form of Eq. 10.9. For clarity, y' is written as dy/dx.

$$m(x) + n(y)\frac{dy}{dx} = 0 \qquad 10.9$$

Equation 10.9 can be placed in the form of Eq. 10.10, both sides of which are easily integrated. An initial value will establish the constant of integration.

$$m(x)dx = -n(y)dy \qquad 10.10$$

5. FIRST-ORDER EXACT DIFFERENTIAL EQUATIONS

A *first-order exact differential equation* has the form

$$f_x(x,y) + f_y(x,y)y' = 0 \qquad 10.11$$

Notice that $f_x(x,y)$ is the exact derivative of $f(x,y)$ with respect to x, and $f_y(x,y)$ is the exact derivative of $f(x,y)$ with respect to y. The solution is

$$f(x,y) - C = 0 \qquad 10.12$$

6. HOMOGENEOUS, SECOND-ORDER LINEAR DIFFERENTIAL EQUATIONS WITH CONSTANT COEFFICIENTS

Homogeneous second-order linear differential equations with constant coefficients have the form of Eq. 10.13. They are most easily solved by finding the two roots of the auxiliary equation (Eq. 10.14).

$$y'' + k_1 y' + k_2 y = 0 \qquad 10.13$$

$$r^2 + k_1 r + k_2 = 0 \qquad 10.14$$

There are three cases. If the two roots of Eq. 10.14 are real and different, the solution is

$$y = A_1 e^{r_1 x} + A_2 e^{r_2 x} \qquad 10.15$$

If the two roots are real and the same, the solution is

$$y = A_1 e^{rx} + A_2 x e^{rx} \qquad 10.16$$

$$r = \frac{-k_1}{2} \qquad 10.17$$

If the two roots are imaginary, they will be of the form $(\alpha + i\omega)$ and $(\alpha - i\omega)$, and the solution is

$$y = A_1 e^{\alpha x}\cos\omega x + A_2 e^{\alpha x}\sin\omega x \qquad 10.18$$

In all three cases, A_1 and A_2 must be found from the two initial conditions.

Example 10.4

Solve the following differential equation.

$$y'' + 6y' + 9y = 0$$

$$y(0) = 0 \qquad y'(0) = 1$$

Solution

The auxiliary equation is

$$r^2 + 6r + 9 = 0$$
$$(r+3)(r+3) = 0$$

The roots to the auxiliary equation are $r_1 = r_2 = -3$. Therefore, the solution has the form of Eq. 10.16.

$$y = A_1 e^{-3x} + A_2 x e^{-3x}$$

The first initial condition is

$$y(0) = 0$$
$$A_1 e^0 + A_2(0)e^0 = 0$$
$$A_1 + 0 = 0$$
$$A_1 = 0$$

To use the second initial condition, the derivative of the equation is needed. Making use of the known fact that $A_1 = 0$,

$$y' = \frac{d}{dx}\left(A_2 x e^{-3x}\right) = -3A_2 x e^{-3x} + A_2 e^{-3x}$$

Using the second initial condition,

$$y'(0) = 1$$

$$-3A_2(0)e^0 + A_2 e^0 = 1$$

$$0 + A_2 = 1$$

$$A_2 = 1$$

The solution is

$$y = xe^{-3x}$$

7. NONHOMOGENEOUS DIFFERENTIAL EQUATIONS

A nonhomogeneous equation has the form of Eq. 10.19. $f(x)$ is known as the *forcing function*.

$$y'' + p(x)y' + q(x)y = f(x) \qquad \textbf{10.19}$$

The solution to Eq. 10.19 is the sum of two equations. The *complementary solution*, y_c, solves the complementary (i.e., homogeneous) problem. The *particular solution*, y_p, is any specific solution to the nonhomogeneous Eq. 10.19 that is known or can be found. Initial values are used to evaluate any unknown coefficients in the complementary solution *after* y_c and y_p have been combined. (The particular solution will not have any unknown coefficients.)

$$y = y_c + y_p \qquad \textbf{10.20}$$

Two methods are available for finding a particular solution. The *method of undetermined coefficients*, as presented here, can be used only when $p(x)$ and $q(x)$ are constant coefficients and $f(x)$ takes on one of the forms in Table 10.1.

The particular solution can be read from Table 10.1 if the forcing function is of one of the forms given. Of course, the coefficients A_i and B_i are not known—these are the *undetermined coefficients*. The exponent s is the smallest nonnegative number (and will be 0, 1, or 2), which ensures that no term in the particular solution, y_p, is also a solution to the complementary equation, y_c. s must be determined prior to proceeding with the solution procedure.

Table 10.1 *Particular Solutions**

form of $f(x)$	form of y_p
$P_n(x) = a_0 x^n + a_1 x^{n-1}$ $+ \cdots + a_n$	$x^s(A_0 x^n + A_1 x^{n-1} + \cdots + A_n)$
$P_n(x)e^{\alpha x}$	$x^s(A_0 x^n + A_1 x^{n-1} + \cdots + A_n)e^{\alpha x}$
$P_n(x)e^{\alpha x} \begin{Bmatrix} \sin \omega x \\ \cos \omega x \end{Bmatrix}$	$x^s[(A_0 x^n + A_1 x^{n-1} + \cdots + A_n)e^{\alpha x}\cos \omega x$ $+(B_0 x^n + B_1 x^{n-1} + \cdots + B_n)e^{\alpha x}\sin \omega x]$

*$P_n(x)$ is a polynomial of degree n.

Once y_p (including s) is known, it is differentiated to obtain y_p' and y_p'', and all three functions are substituted into the original nonhomogeneous equation. The resulting equation is rearranged to match the forcing function, $f(x)$, and the unknown coefficients are determined, usually by solving simultaneous equations.

If the forcing function, $f(x)$, is more complex than the forms shown in Table 10.1, or if either $p(x)$ or $q(x)$ is a function of x, the method of *variation of parameters* should be used. This complex and time-consuming method is not covered in this book.

Example 10.5

Solve the following nonhomogeneous differential equation.

$$y'' + 2y' + y = e^x \cos x$$

Solution

step 1: Find the solution to the complementary (homogeneous) differential equation.

$$y'' + 2y' + y = 0$$

Since this is a differential equation with constant coefficients, write the auxiliary equation.

$$r^2 + 2r + 1 = 0$$

The auxiliary equation factors in $(r+1)^2 = 0$ with two identical roots at $r = -1$. Therefore, the solution to the homogeneous differential equation is

$$y_c(x) = C_1 e^{-x} + C_2 x e^{-x}$$

step 2: Use Table 10.1 to determine the form of a particular solution. Since the forcing function has the form $P_n(x)e^{\alpha x} \cos \omega x$ with $P_n(x) = 1$ (equivalent to $n = 0$), $\alpha = 1$, and $\omega = 1$, the particular solution has the form

$$y_p(x) = x^s(Ae^x \cos x + Be^x \sin x)$$

step 3: Determine the value of s. Check to see if any of the terms in $y_p(x)$ will themselves solve the homogeneous equation. Try $Ae^x \cos x$ first.

$$\frac{d}{dx}(Ae^x \cos x) = Ae^x \cos x - Ae^x \sin x$$

$$\frac{d^2}{dx^2}(Ae^x \cos x) = -2Ae^x \sin x$$

Substitute these quantities into the homogeneous equation.

$$y'' + 2y' + y = 0$$

$$-2Ae^x \sin x + 2Ae^x \cos x$$

$$-2Ae^x \sin x + Ae^x \cos x = 0$$

$$3Ae^x \cos x - 4Ae^x \sin x = 0$$

Disregarding the trivial ($A = 0$) solution, $Ae^x \cos x$ does not solve the homogeneous equation.

Next, try $Be^x \sin x$.

$$\frac{d}{dx}\left(Be^x \sin x\right) = Be^x \cos x + Be^x \sin x$$

$$\frac{d^2}{dx^2}\left(Be^x \sin x\right) = 2Be^x \cos x$$

Substitute these quantities into the homogeneous equation.

$$y'' + 2y' + y = 0$$
$$2Be^x \cos x + 2Be^x \cos x$$
$$+2Be^x \sin x + Be^x \sin x = 0$$
$$3Be^x \sin x + 4Be^x \cos x = 0$$

Disregarding the trivial ($B = 0$) case, $Be^x \sin x$ does not solve the homogeneous equation.

Since none of the terms in $y_p(x)$ solve the homogeneous equation, $s = 0$, and a particular solution has the form

$$y_p(x) = Ae^x \cos x + Be^x \sin x$$

step 4: Use the method of unknown coefficients to determine A and B in the particular solution. Drawing on the previous steps, substitute the quantities derived from the particular solution into the nonhomogeneous equation.

$$y'' + 2y' + y = e^x \cos x$$
$$-2Ae^x \sin x + 2Be^x \cos x$$
$$+2Ae^x \cos x - 2Ae^x \sin x$$
$$+2Be^x \cos x + 2Be^x \sin x$$
$$+Ae^x \cos x + Be^x \sin x = e^x \cos x$$

Combining terms,

$$(-4A + 3B)e^x \sin x + (3A + 4B)e^x \cos x$$
$$= e^x \cos x$$

Equating the coefficients of like terms on either side of the equal sign results in the following simultaneous equations.

$$-4A + 3B = 0$$
$$3A + 4B = 1$$

The solution to these equations is

$$A = \tfrac{3}{25}$$
$$B = \tfrac{4}{25}$$

A particular solution is

$$y_p(x) = \left(\tfrac{3}{25}\right)(e^x \cos x) + \left(\tfrac{4}{25}\right)(e^x \sin x)$$

step 5: Write the general solution.

$$y(x) = y_c(x) + y_p(x)$$
$$= C_1 e^{-x} + C_2 x e^{-x} + \left(\tfrac{3}{25}\right)(e^x \cos x)$$
$$+ \left(\tfrac{4}{25}\right)(e^x \sin x)$$

The values of C_1 and C_2 would be determined at this time if initial conditions were known.

8. NAMED DIFFERENTIAL EQUATIONS

Some differential equations with specific forms are named after the individuals who developed solution techniques for them.

- *Bessel equation of order ν:*

$$x^2 y'' + xy' + (x^2 - \nu^2)y = 0 \qquad 10.21$$

- *Cauchy equation:*

$$a_0 x^n \frac{d^n y}{dx^n} + a_1 x^{n-1} \frac{d^{n-1}y}{dx^{n-1}} + \cdots$$
$$+ a_{n-1} x \frac{dy}{dx} + a_n y = f(x) \qquad 10.22$$

- *Euler equation:*

$$x^2 y'' + \alpha x y' + \beta y = 0 \qquad 10.23$$

- *Gauss' hypergeometric equation:*

$$x(1-x)y'' + \big(c - (a+b+1)x\big)y' - aby = 0 \quad 10.24$$

- *Legendre equation of order λ:*

$$(1 - x^2)y'' - 2xy' + \lambda(\lambda+1)y = 0 \quad [-1 < x < 1]$$
$$10.25$$

9. LAPLACE TRANSFORMS

Traditional methods of solving nonhomogeneous differential equations by hand are usually difficult and/or time consuming. *Laplace transforms* can be used to reduce many solution procedures to simple algebra.

Every mathematical function, $f(t)$, for which Eq. 10.26 exists has a Laplace transform, written as $\mathcal{L}(f)$ or $F(s)$. The transform is written in the s-domain, regardless of the independent variable in the original function.[2] (The variable s is equivalent to a derivative operator, although it may be handled in the equations as a simple variable.) Equation 10.26 converts a function into a Laplace transform.

$$\mathcal{L}(f(t)) = F(s) = \int_0^\infty e^{-st} f(t) dt \qquad 10.26$$

Equation 10.26 is not often needed because tables of transforms are readily available. (Appendix 10.A contains some of the most common transforms.)

Extracting a function from its transform is the *inverse Laplace transform* operation. Although other methods exist, this operation is almost always done by finding the transform in a set of tables.[3]

$$f(t) = \mathcal{L}^{-1}(F(s)) \qquad 10.27$$

Example 10.6

Find the Laplace transform of the following function.

$$f(t) = e^{at} \qquad [s > a]$$

Solution

Applying Eq. 10.26,

$$\mathcal{L}(e^{at}) = \int_0^\infty e^{-st} e^{at} dt = \int_0^\infty e^{-(s-a)t} dt$$

$$= -\left[\frac{e^{-(s-a)t}}{s-a}\right]_0^\infty = \frac{1}{s-a} \qquad [s > a]$$

10. STEP AND IMPULSE FUNCTIONS

Many forcing functions are sinusoidal or exponential in nature; others, however, can only be represented by a step or impulse function. A *unit step function*, u_t, is a function describing the disturbance of magnitude 1 that is not present before time t but is suddenly there after time t. A step of magnitude 5 at time $t = 3$ would be represented as $5u_3$. (The notation $5u(t-3)$ is used in some books.)

The *unit impulse function*, δ_t, is a function describing a disturbance of magnitude 1 that is applied and removed so quickly as to be instantaneous. An impulse of magnitude 5 at time 3 would be represented by $5\delta_3$. (The notation $5\delta(t-3)$ is used in some books.)

---

[2]It is traditional to write the original function as a function of the independent variable t rather than x. However, Laplace transforms are not limited to functions of time.
[3]Other methods include integration in the complex plane, convolution, and simplification by partial fractions.

Example 10.7

What is the notation for a forcing function of magnitude 6 that is applied at $t = 2$ and completely removed at $t = 7$?

Solution

The notation is $f(t) = 6(u_2 - u_7)$.

Example 10.8

Find the Laplace transform of u_0, a unit step at $t = 0$.

$$f(t) = 0 \text{ for } t < 0$$
$$f(t) = 1 \text{ for } t \geq 0$$

Solution

Since the Laplace transform is an integral that starts at $t = 0$, the value of $f(t)$ prior to $t = 0$ is irrelevant.

$$\mathcal{L}(u_0) = \int_0^\infty e^{-st}(1) dt = -\left[\frac{e^{-st}}{s}\right]_0^\infty$$

$$= 0 - \left(\frac{-1}{s}\right) = \frac{1}{s}$$

11. ALGEBRA OF LAPLACE TRANSFORMS

Equations containing Laplace transforms can be simplified by applying the following principles.

- *linearity theorem*: (c is a constant.)
$$\mathcal{L}(cf(t)) = c\mathcal{L}(f(t)) = cF(s) \qquad 10.28$$

- *superposition theorem*: ($f(t)$ and $g(t)$ are different functions.)
$$\mathcal{L}(f(t)) \pm g(t)) = \mathcal{L}(f(t)) \pm \mathcal{L}(g(t))$$
$$= F(s) \pm G(s) \qquad 10.29$$

- *time-shifting theorem (delay theorem)*:
$$\mathcal{L}(f(t-b)u_b) = e^{-bs} F(s) \qquad 10.30$$

- *Laplace transform of a derivative*:
$$\mathcal{L}(f^n(t)) = -f^{n-1}(0) - sf^{n-2}(0) - \cdots$$
$$- s^{n-1} f(0) + s^n F(s) \qquad 10.31$$

- *other properties*:
$$\mathcal{L}\left(\int_0^t f(u) du\right) = \left(\frac{1}{s}\right) F(s) \qquad 10.32$$

$$\mathcal{L}(tf(t)) = -\frac{dF}{ds} \qquad 10.33$$

$$\mathcal{L}\left(\frac{1}{t}f(t)\right) = \int_s^\infty F(u) du \qquad 10.34$$

12. CONVOLUTION INTEGRAL

A complex Laplace transform, $F(s)$, will often be recognized as the product of two other transforms, $F_1(s)$ and $F_2(s)$, whose corresponding functions $f_1(t)$ and $f_2(t)$ are known. Unfortunately, Laplace transforms cannot be computed with ordinary multiplication. That is, $f(t) \neq f_1(t)f_2(t)$ even though $F(s) = F_1(s)F_2(s)$.

However, it is possible to extract $f(t)$ from its *convolution*, $h(t)$, as calculated from either of the *convolution integrals* in Eq. 10.35. This process is demonstrated in Ex. 10.9. χ is a dummy variable.

$$
\begin{aligned}
f(t) &= \mathcal{L}^{-1}(F_1(s)F_2(s)) \\
&= \int_0^t f_1(t-\chi)f_2(\chi)d\chi \\
&= \int_0^t f_1(\chi)f_2(t-\chi)d\chi \quad\quad \textit{10.35}
\end{aligned}
$$

Example 10.9

Use the convolution integral to find the inverse transform of

$$ F(s) = \frac{3}{s^2(s^2+9)} $$

Solution

$F(s)$ can be factored as

$$ F_1(s)F_2(s) = \left(\frac{1}{s^2}\right)\left(\frac{3}{s^2+9}\right) $$

Since the inverse transforms of $F_1(s)$ and $F_2(s)$ are $f_1(t) = t$ and $f_2(t) = \sin 3t$, respectively, the convolution integral from Eq. 10.35 is

$$
\begin{aligned}
f(t) &= \int_0^t (t-\chi)\sin 3\chi \ d\chi \\
&= \int_0^t (t\,\sin 3\chi - \chi\,\sin 3\chi)d\chi \\
&= t\int_0^t \sin 3\chi \ d\chi - \int_0^t \chi\,\sin 3\chi \ d\chi
\end{aligned}
$$

Expand using integration by parts.

$$
\begin{aligned}
f(t) &= -\tfrac{1}{3}t\cos 3\chi + \tfrac{1}{3}\chi\cos 3\chi - \tfrac{1}{9}\sin 3\chi \Big|_0^t \\
&= \frac{3t - \sin 3t}{9}
\end{aligned}
$$

13. USING LAPLACE TRANSFORMS

Any nonhomogeneous linear differential equation with constant coefficients can be solved with the following procedure, which reduces the solution to simple algebra.

A complete table of transforms simplifies or eliminates step 5.

step 1: Put the differential equation in standard form (i.e., isolate the y'' term).

$$ y'' + k_1 y' + k_2 y = f(t) \quad\quad \textit{10.36} $$

step 2: Take the Laplace transform of both sides. Use the linearity and superposition theorems, Eqs. 10.28 and 10.29.

$$ \mathcal{L}(y'') + k_1\mathcal{L}(y') + k_2\mathcal{L}(y) = \mathcal{L}(f(t)) \quad\quad \textit{10.37} $$

step 3: Use Eqs. 10.38 and 10.39 to expand the equation. (These are specific forms of Eq. 10.31.) Use a table to evaluate the transform of the forcing function.

$$ \mathcal{L}(y'') = s^2\mathcal{L}(y) - sy(0) - y'(0) \quad\quad \textit{10.38} $$
$$ \mathcal{L}(y') = s\mathcal{L}(y) - y(0) \quad\quad \textit{10.39} $$

step 4: Use algebra to solve for $\mathcal{L}(y)$.

step 5: If needed, use partial fractions (see Ch. 3) to simplify the expression for $\mathcal{L}(y)$.

step 6: Take the inverse transform to find $y(t)$.

$$ y(t) = \mathcal{L}^{-1}(\mathcal{L}(y)) \quad\quad \textit{10.40} $$

Example 10.10

Find $y(t)$ for the following differential equation.

$$
\begin{aligned}
y'' + 2y' + 2y &= \cos t \\
y(0) = 1 \quad\quad y'(0) &= 0
\end{aligned}
$$

Solution

step 1: The equation is already in standard form.

step 2: $\mathcal{L}(y'') + 2\mathcal{L}(y') + 2\mathcal{L}(y) = \mathcal{L}(\cos t)$

step 3: Use Eqs. 10.38 and 10.39. Use App. 10.A to find the transform of $\cos t$.

$$
\begin{aligned}
&s^2\mathcal{L}(y) - sy(0) - y'(0) + 2s\mathcal{L}(y) - 2y(0) + 2\mathcal{L}(y) \\
&= \frac{s}{s^2+1}
\end{aligned}
$$

But, $y(0) = 1$ and $y'(0) = 0$.

$$ s^2\mathcal{L}(y) - s + 2s\mathcal{L}(y) - 2 + 2\mathcal{L}(y) = \frac{s}{s^2+1} $$

step 4: Combine terms and solve for $\mathcal{L}(y)$.

$$ \mathcal{L}(y)(s^2+2s+2) - s - 2 = \frac{s}{s^2+1} $$

$$
\begin{aligned}
\mathcal{L}(y) &= \frac{\dfrac{s}{s^2+1} + s + 2}{s^2+2s+2} \\
&= \frac{s^3 + 2s^2 + 2s + 2}{(s^2+1)(s^2+2s+2)}
\end{aligned}
$$

step 5: Expand the expression for $\mathcal{L}(y)$ by partial fractions. (Refer to Ch. 3.)

$$\mathcal{L}(y) = \frac{(s^3 + 2s^2 + 2s + 2)}{(s^2 + 1)(s^2 + 2s + 2)}$$

$$= \frac{A_1 s + B_1}{s^2 + 1} + \frac{A_2 s + B_2}{s^2 + 2s + 2}$$

$$= \frac{\begin{aligned}s^3(A_1 + A_2) + s^2(2A_1 + B_1 + B_2) \\ + s(2A_1 + 2B_1 + A_2) + (2B_1 + B_2))\end{aligned}}{(s^2 + 1)(s^2 + 2s + 2)}$$

The following simultaneous equations result.

$$\begin{array}{ccccccc} A_1 & + & A_2 & & & & = 1 \\ 2A_1 & & & + & B_1 & + & B_2 = 2 \\ 2A_1 & + & A_2 & + & 2B_1 & & = 2 \\ & & & & 2B_1 & + & B_2 = 2 \end{array}$$

These equations have the solutions $A_1 = \frac{1}{5}$, $A_2 = \frac{4}{5}$, $B_1 = \frac{2}{5}$, and $B_2 = \frac{6}{5}$.

step 6: Refer to App. 10.A and take the inverse transforms. (The numerator of the second term is rewritten from $(4s + 6)$ to $((4s + 4) + 2)$.

$$y = \mathcal{L}^{-1}(\mathcal{L}(y))$$

$$= \mathcal{L}^{-1}\left(\frac{\left(\frac{1}{5}\right)(s+2)}{s^2 + 1} + \frac{\left(\frac{1}{5}\right)(4s+6)}{s^2 + 2s + 2} \right)$$

$$= \left(\tfrac{1}{5}\right)\left(\mathcal{L}^{-1}\left(\frac{s}{s^2+1}\right) + 2\mathcal{L}^{-1}\left(\frac{1}{s^2+1}\right) \right.$$

$$+ 4\mathcal{L}^{-1}\left(\frac{s - (-1)}{\left(s - (-1)\right)^2 + 1} \right)$$

$$\left. + 2\mathcal{L}^{-1}\left(\frac{1}{\left(s - (-1)\right)^2 + 1} \right) \right)$$

$$= \tfrac{1}{5}(\cos t + 2\,\sin t + 4e^{-t}\cos t + 2e^{-t}\sin t)$$

14. THIRD- AND HIGHER-ORDER LINEAR DIFFERENTIAL EQUATIONS WITH CONSTANT COEFFICIENTS

The solutions of third- and higher-order linear differential equations with constant coefficients are extensions of the solutions for second-order equations of this type. Specifically, if an equation is homogeneous, the auxiliary equation is written and its roots are found. If the equation is nonhomogeneous, Laplace transforms can be used to simplify the solution.

Consider the following homogeneous differential equation with constant coefficients.

$$y^n + k_1 y^{n-1} + \cdots + k_{n-1}y' + k_n y = 0 \qquad 10.41$$

The auxiliary equation to Eq. 10.41 is

$$r^n + k_1 r^{n-1} + \cdots + k_{n-1}r + k_n = 0 \qquad 10.42$$

For each real and distinct root r, the solution contains the term

$$y = Ae^{rx} \qquad 10.43$$

For each real root r that repeats m times, the solution contains the term

$$y = \left(A_1 + A_2 x + A_3 x^2 + \cdots + A_m x^{m-1} \right) e^{rx} \qquad 10.44$$

For each pair of complex roots of the form $r = \alpha \pm i\omega$, the solution contains the terms

$$y = e^{\alpha x}(A_1 \sin \omega x + A_2 \cos \omega x) \qquad 10.45$$

15. APPLICATION: ENGINEERING SYSTEMS

There is a wide variety of engineering systems (mechanical, electrical, fluid flow, heat transfer, and so on) whose behavior is described by linear differential equations with constant coefficients.

16. APPLICATION: MIXING

A typical mixing problem involves a liquid-filled tank. The liquid may initially be pure or contain some solute. Liquid (either pure or as a solution) enters the tank at a known rate. A drain may be present to remove thoroughly mixed liquid. The concentration of the solution (or, equivalently, the amount of solute in the tank) at some given time is generally unknown.

Figure 10.1 *Fluid Mixture Problem*

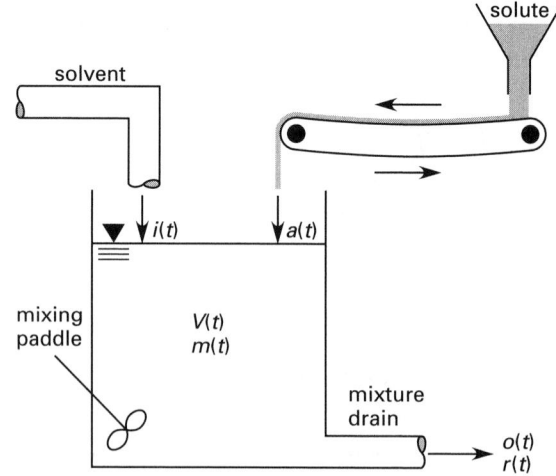

If $m(t)$ is the mass of solute in the tank at time t, the rate of solute change will be $m'(t)$. If the solute is being added at the rate of $a(t)$ and being removed at the rate of $r(t)$, the rate of change is

$$m'(t) = \text{rate of addition} - \text{rate of removal}$$

$$= a(t) - r(t) \qquad 10.46$$

The rate of solute addition $a(t)$ must be known and, in fact, may be constant. However, $r(t)$ depends on the concentration, $c(t)$, of the mixture and volumetric flow rates at time t. If $o(t)$ is the volumetric flow rate out of the tank, then

$$r(t) = c(t)o(t) \qquad 10.47$$

However, the concentration depends on the mass of solute in the tank at time t. Recognizing that the volume, $V(t)$, of the liquid in the tank may be changing with time,

$$c(t) = \frac{m(t)}{V(t)} \qquad 10.48$$

The differential equation describing this problem is

$$m'(t) = a(t) - \frac{m(t)o(t)}{V(t)} \qquad 10.49$$

Example 10.11

A tank contains 100 gal of pure water at the beginning of an experiment. Pure water flows into the tank at a rate of 1 gal/min. Brine containing $1/4$ lbm of salt per gallon enters the tank from a second source at a rate of 1 gal/min. A perfectly mixed solution drains from the tank at a rate of 2 gal/min. How much salt is in the tank 8 min after the experiment begins?

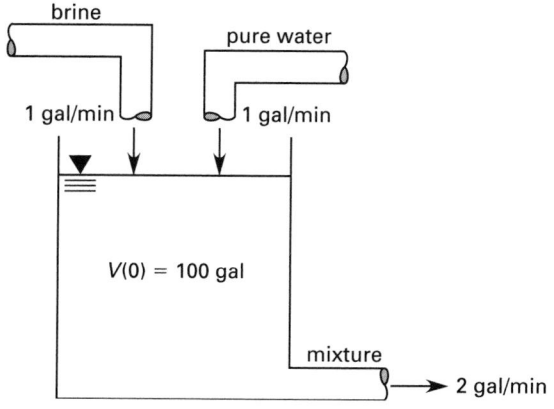

Solution

Let $m(t)$ be the mass of salt in the tank at time t. 0.25 lbm of salt enters the tank per minute (i.e., $a(t) = 0.25$ lbm/min). The salt removal rate depends on the concentration in the tank. That is,

$$r(t) = o(t)c(t)$$

$$= \left(2 \; \frac{\text{gal}}{\text{min}}\right)\left(\frac{m(t)}{100 \; \text{gal}}\right) = \left(0.02 \; \frac{1}{\text{min}}\right)m(t)$$

From Eq. 10.46, the rate of change of salt in the tank is

$$m'(t) = a(t) - r(t)$$

$$= 0.25 \; \frac{\text{lbm}}{\text{min}} - \left(0.02 \; \frac{1}{\text{min}}\right)m(t)$$

$$m'(t) + \left(0.02 \; \frac{1}{\text{min}}\right)m(t) = 0.25 \; \text{lbm/min}$$

This is a first-order linear differential equation of the form of Eq. 10.7. Since the initial condition is $m(0) = 0$, the solution is

$$m(t) = \left(\frac{0.25 \; \dfrac{\text{lbm}}{\text{min}}}{0.02 \; \dfrac{1}{\text{min}}}\right)\left(1 - e^{-\left(0.02 \; \frac{1}{\text{min}}\right)t}\right)$$

$$= (12.5 \; \text{lbm})\left(1 - e^{-\left(0.01 \; \frac{1}{\text{min}}\right)t}\right)$$

At $t = 8$,

$$m(t) = (12.5 \; \text{lbm})\left(1 - e^{-\left(0.02 \; \frac{1}{\text{min}}\right)(8 \; \text{min})}\right)$$

$$= (12.5 \; \text{lbm})(1 - 0.852)$$

$$= 1.85 \; \text{lbm}$$

17. APPLICATION: EXPONENTIAL GROWTH AND DECAY

Equation 10.50 describes the behavior of a substance whose quantity, $m(t)$, changes at a rate proportional to the quantity present. The constant of proportionality, k, will be negative for decay (e.g., radioactive decay) and positive for growth (e.g., compound interest).

$$m'(t) = km(t) \qquad 10.50$$
$$m'(t) - km(t) = 0 \qquad 10.51$$

If the initial quantity of substance is $m(0) = m_0$, Eq. 10.51 has the solution

$$m(t) = m_0 e^{kt} \qquad 10.52$$

If $m(t)$ is known for some time t, the constant of proportionality is

$$k = \left(\frac{1}{t}\right)\ln\left(\frac{m(t)}{m_0}\right) \qquad 10.53$$

For the case of a decay, the *half-life*, $t_{1/2}$, is the time at which only half of the substance remains. The relationship between k and $t_{1/2}$ is

$$kt_{1/2} = \ln\left(\tfrac{1}{2}\right) = -0.693 \qquad 10.54$$

18. APPLICATION: EPIDEMICS

During an epidemic in a population of n people, the density of sick (contaminated, contagious, affected, etc.) individuals is $\rho_s(t) = s(t)/n$, where $s(t)$ is the number of sick individuals at time t. Similarly, the density of well (uncontaminated, unaffected, susceptible, etc.) individuals is $\rho_w(t) = w(t)/n$, where $w(t)$ is the number of well individuals. Assuming there is no quarantine, the population size is constant, individuals move about freely, and sickness does not limit the activities of individuals, the rate of contagion, $\rho_s'(t)$, will be $k\rho_s(t)\rho_w(t)$, where k is a proportionality constant.

$$\rho_s'(t) = k\rho_s(t)\rho_w(t) = k\rho_s(t)(1 - \rho_s(t)) \qquad 10.55$$

This is a separable differential equation with the solution

$$\rho_s(t) = \frac{\rho_s(0)}{\rho_s(0) + (1 - \rho_s(0))e^{-kt}} \qquad 10.56$$

19. APPLICATION: SURFACE TEMPERATURE

Newton's law of cooling states that the surface temperature, T, of a cooling object changes at a rate proportional to the difference between the surface and ambient temperatures. The constant k is a positive number.

$$T'(t) = -k(T(t) - T_{\text{ambient}}) \quad [k > 0] \qquad 10.57$$

$$T'(t) + kT(t) - kT_{\text{ambient}} = 0 \quad [k > 0] \qquad 10.58$$

This first-order linear differential equation with constant coefficients has the following solution (from Eq. 10.8).

$$T(t) = T_{\text{ambient}} + (T(0) - T_{\text{ambient}})e^{-kt} \qquad 10.59$$

If the temperature is known at some time t, the constant k can be found from Eq. 10.60.

$$k = \left(\frac{-1}{t}\right) \ln\left(\frac{T(t) - T_{\text{ambient}}}{T(0) - T_{\text{ambient}}}\right) \qquad 10.60$$

20. APPLICATION: EVAPORATION

The mass of liquid evaporated from a liquid surface is proportional to the exposed surface area. Since quantity, mass, and remaining volume are all proportional, the differential equation is

$$\frac{dV}{dt} = -kA \qquad 10.61$$

For a spherical drop of radius r, Eq. 10.61 reduces to

$$\frac{dr}{dt} = -k \qquad 10.62$$

$$r(t) = r(0) - kt \qquad 10.63$$

For a cube with sides of length s, Eq. 10.61 reduces to

$$\frac{ds}{dt} = -2k \qquad 10.64$$

$$s(t) = s(0) - 2kt \qquad 10.65$$

11 Probability and Statistical Analysis of Data

1. SET THEORY

A *set* (usually designated by a capital letter) is a population or collection of individual items known as *elements* or *members*. The *null set*, $\emptyset$, is empty (i.e., contains no members). If A and B are two sets, A is a *subset* of B if every member in A is also in B. A is a *proper subset* of B if B consists of more than the elements in A. These relationships are denoted

$$A \subseteq B \qquad \text{[subset]}$$
$$A \subset B \qquad \text{[proper subset]}$$

The *universal set, U,* is one from which other sets draw their members. If A is a subset of U, then A' (also designated as A^{-1}, $\tilde{A}$, $-A$, and $\overline{A}$) is the *complement* of A and consists of all elements in U that are not in A. This is illustrated in a *Venn diagram* in Fig. 11.1(a).

The *union of two sets*, denoted by $A \cup B$ and shown in Fig. 11.1(b), is the set of all elements that are either in A or B or both. The *intersection of two sets*, denoted by $A \cap B$ and shown in Fig. 11.1(c), is the set of all elements that belong to both A and B. If $A \cap B = \emptyset$, A and B are said to be *disjoint sets*.

Figure 11.1 *Venn Diagrams*

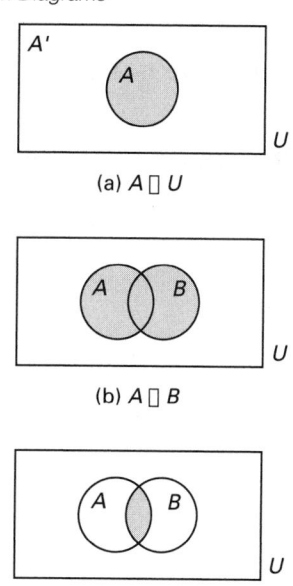

(a) $A \subseteq U$

(b) $A \cup B$

(c) $A \cap B$

If A, B, and C are subsets of the universal set, the following laws apply.

- *identity laws*:

$$A \cup \emptyset = A \qquad \qquad 11.1$$
$$A \cup U = U \qquad \qquad 11.2$$
$$A \cap \emptyset = \emptyset \qquad \qquad 11.3$$
$$A \cap U = A \qquad \qquad 11.4$$

- *idempotent laws*:

$$A \cup A = A \qquad \qquad 11.5$$
$$A \cap A = A \qquad \qquad 11.6$$

- *complement laws*:

$$A \cup A' = U \qquad \qquad 11.7$$
$$(A')' = A \qquad \qquad 11.8$$
$$A \cap A' = \emptyset \qquad \qquad 11.9$$
$$U' = \emptyset \qquad \qquad 11.10$$

- *commutative laws*:
$$A \cup B = B \cup A \qquad \textit{11.11}$$
$$A \cap B = B \cap A \qquad \textit{11.12}$$

- *associative laws*:
$$(A \cup B) \cup C = A \cup (B \cup C) \qquad \textit{11.13}$$
$$(A \cap B) \cap C = A \cap (B \cap C) \qquad \textit{11.14}$$

- *distributive laws*:
$$A \cup (B \cap C) = (A \cup B) \cap (A \cup C) \qquad \textit{11.15}$$
$$A \cap (B \cup C) = (A \cap B) \cup (A \cap C) \qquad \textit{11.16}$$

- *de Morgan's laws*:
$$(A \cup B)' = A' \cap B' \qquad \textit{11.17}$$
$$(A \cap B)' = A' \cup B' \qquad \textit{11.18}$$

2. COMBINATIONS OF ELEMENTS

There are a finite number of ways in which n elements can be combined into distinctly different groups of r items. For example, suppose a farmer has a hen, a rooster, a duck, and a cage that holds only two birds. The possible *combinations* of three birds taken two at a time are (hen, rooster), (hen, duck), and (rooster, duck). The birds in the cage will not remain stationary, and the combination (rooster, hen) is not distinctly different from (hen, rooster). That is, the groups are not *order conscious*.

The number of combinations of n items taken r at a time is written $C(n,r)$, C_r^n, $_nC_r$, or $\binom{n}{r}$ (pronounced "n choose r") and given by Eq. 11.19. It is sometimes referred to as the *binomial coefficient*.

$$\binom{n}{r} = C(n,r) = \frac{n!}{(n-r)!\,r!} \qquad \text{[for } r \le n] \quad \textit{11.19}$$

Example 11.1

Six people are on a sinking yacht. There are four life jackets. How many combinations of survivors are there?

Solution

The groups are not order-conscious. From Eq. 11.19,

$$
\begin{aligned}
C(6,4) &= \frac{6!}{(6-4)!\,4!} = \frac{6 \cdot 5 \cdot 4 \cdot 3 \cdot 2 \cdot 1}{(2 \cdot 1)(4 \cdot 3 \cdot 2 \cdot 1)} \\
&= 15
\end{aligned}
$$

3. PERMUTATIONS

An order-conscious subset of r items taken from a set of n items is the *permutation* $P(n,r)$, also written P_r^n and $_nP_r$. The permutation is order conscious because the arrangement of two items (say a_i and b_i) as $a_i b_i$

is different from the arrangement $b_i a_i$. The number of permutations is

$$P(n,r) = \frac{n!}{(n-r)!} \qquad \text{[for } r \le n] \quad \textit{11.20}$$

If groups of the entire set of n items are being enumerated, the number of permutations of n items taken n at a time is

$$P(n,n) = \frac{n!}{(n-n)!} = \frac{n!}{0!} = n! \qquad \textit{11.21}$$

A *ring permutation* is a special case of n items taken n at a time. There is no identifiable beginning or end, and the number of permutations is divided by n.

$$P_{\text{ring}}(n,n) = \frac{P(n,n)}{n} = (n-1)! \qquad \textit{11.22}$$

Example 11.2

A pianist knows four pieces but will have enough stage time to play only three of them. Pieces played in a different order constitute a different program. How many different programs can be arranged?

Solution

The groups are order conscious. From Eq. 11.20,

$$P(4,3) = \frac{4!}{(4-3)!} = \frac{4 \cdot 3 \cdot 2 \cdot 1}{1} = 24$$

Example 11.3

Seven diplomats from different countries enter a circular room. The only furnishings are seven chairs arranged around a circular table. How many ways are there of arranging the diplomats?

Solution

All seven diplomats must be seated, so the groups are permutations of seven objects taken seven at a time. Since there is no head chair, the groups are ring permutations. From Eq. 11.22,

$$P_{\text{ring}}(7,7) = (7-1)! = 6 \cdot 5 \cdot 4 \cdot 3 \cdot 2 \cdot 1 = 720$$

4. PROBABILITY THEORY

The act of conducting an experiment (trial) or taking a measurement is known as *sampling*. *Probability theory* determines the relative likelihood that a particular event will occur. An *event, e*, is one of the possible outcomes of the *trial*. Taken together, all of the possible events constitute a finite *sample space*, $E = [e_1, e_2, \ldots, e_n]$. The trial is drawn from the *population* or *universe*. Populations can be finite or infinite in size.

Events can be numerical or nonnumerical, discrete or continuous, and dependent or independent. An example of a nonnumerical event is getting tails on a coin toss. The roll of a die is a discrete numerical event. The measured diameter of a bolt produced from an automatic screw machine is a numerical event. Since the diameter can (within reasonable limits) take on any value, its measured value is a continuous numerical event.

An event is *independent* if its outcome is unaffected by previous outcomes (i.e., previous runs of the experiment) and *dependent* otherwise. Whether or not an event is independent depends on the population size and how the sampling is conducted. Sampling (a trial) from an infinite population is implicitly independent. When the population is finite, *sampling with replacement* produces independent events, while *sampling without replacement* changes the population and produces dependent events.

The terms *success* and *failure* are loosely used in probability theory to designate obtaining and not obtaining, respectively, the tested-for condition. "Failure" is not the same as a *null event* (i.e., one that has a zero probability of occurrence).

The *probability* of event e_1 occurring is designated as $p\{e_1\}$ and is calculated as the ratio of the total number of ways the event can occur to the total number of outcomes in the sample space.

Example 11.4

There are 380 students in a rural school—200 girls and 180 boys. One student is chosen at random and is checked for gender and height. (a) Define and categorize the population. (b) Define and categorize the sample space. (c) Define the trials. (d) Define and categorize the events. (e) In determining the probability that the student chosen is a boy, define success and failure. (f) What is the probability that the student is a boy?

Solution

(a) The population consists of 380 students and is finite.

(b) In determining the gender of the student, the sample space consists of the two outcomes $E = [\text{girl, boy}]$. This sample space is nonnumerical and discrete. In determining the height, the sample space consists of a range of values and is numerical and continuous.

(c) The trial is the actual sampling (i.e., the determination of gender and height).

(d) The events are the outcomes of the trials (i.e., the gender and height of the student). These events are independent if each student returns to the population prior to the random selection of the next student; otherwise, the events are dependent.

(e) The event is a success if the student is a boy and is a failure otherwise.

(f) From the definition of probability,

$$p\{\text{boy}\} = \frac{\text{no. of boys}}{\text{no. of students}} = \frac{180}{380} = \frac{9}{19} = 0.47$$

5. JOINT PROBABILITY

Joint probability rules specify the probability of a combination of events. If n mutually exclusive events from the set E have probabilities $p\{e_i\}$, the probability of any one of these events occurring in a given trial is the sum of the individual probabilities. Notice that the events in Eq. 11.23 come from a single sample space and are linked by the word *or*.

$$p\{e_1 \text{ or } e_2 \text{ or } \ldots \text{ or } e_k\} = p\{e_1\} + p\{e_2\} + \ldots + p\{e_k\}$$
$$11.23$$

Given two independent sets of events, E and G, Eq. 11.24 gives the probability that either event e_i or g_i, or both, will occur. Notice that the events in Eq. 11.24 come from two different sample spaces and are linked by the word *or*.

$$p\{e_i \text{ or } g_i\} = p\{e_i\} + p\{g_i\} - p\{e_i\}p\{g_i\} \qquad 11.24$$

Given two independent sets of events, E and G, Eq. 11.25 gives the probability that events e_i and g_i will both occur. Notice that the events in Eq. 11.25 come from two different sample spaces and are linked by the word *and*.

$$p\{e_i \text{ and } g_i\} = p\{e_i\}p\{g_i\} \qquad 11.25$$

Example 11.5

A bowl contains five white balls, two red balls, and three green balls. What is the probability of getting either a white ball or a red ball in one draw from the bowl?

Solution

The two possible events are mutually exclusive and come from the same sample space, so Eq. 11.23 can be used.

$$p\{\text{white or red}\} = p\{\text{white}\} + p\{\text{red}\} = \frac{5}{10} + \frac{2}{10} = \frac{7}{10}$$

Example 11.6

One bowl contains five white balls, two red balls, and three green balls. Another bowl contains three yellow balls and seven black balls. What is the probability of getting a red ball from the first bowl and a yellow ball from the second bowl in one draw from each bowl?

Solution

The two trials are independent, so Eq. 11.25 can be used.

$$p\{\text{red and yellow}\} = p\{\text{red}\}p\{\text{yellow}\}$$
$$= \left(\frac{2}{10}\right)\left(\frac{3}{10}\right) = \frac{6}{100}$$

6. COMPLEMENTARY PROBABILITIES

The probability of an event occurring is equal to one minus the probability of the event not occurring. This is known as *complementary probability*.

$$p\{e_i\} = 1 - p\{\text{not } e_i\} \qquad 11.26$$

Equation 11.26 can be used to simplify some probability calculations. Specifically, calculation of the probability of numerical events being "greater than" or "less than" or quantities being "at least" a certain number can often be simplified by calculating the probability of complementary event.

Example 11.7

A fair coin is tossed five times.[1] What is the probability of getting at least one tail?

Solution

The probability of getting at least one tail in five tosses could be calculated as

$$p\{\text{at least 1 tail}\} = p\{1 \text{ tail}\} + p\{2 \text{ tails}\}$$
$$+ p\{3 \text{ tails}\} + p\{4 \text{ tails}\}$$
$$+ p\{5 \text{ tails}\}$$

However, it is easier to calculate the complementary probability of getting no tails (i.e., getting all heads).

$$p\{\text{at least 1 tail}\} = 1 - p\{0 \text{ tails}\}$$
$$= 1 - (0.5)^5 = 0.96875$$

7. CONDITIONAL PROBABILITY

Given two dependent sets of events, E and G, the probability that event e_k will occur given the fact that the dependent event g has already occurred is written as $p\{e_k|g\}$ and given by *Bayes' theorem*, Eq. 11.27.

$$p\{e_k|g\} = \frac{p\{e_k \text{ and } g\}}{p\{g\}} = \frac{p\{g|e_k\}p\{e_k\}}{\sum_{i=1}^{n} p\{g|e_i\}p\{e_i\}} \qquad 11.27$$

---

[1]It makes no difference whether one coin is tossed five times or five coins are each tossed once.

8. PROBABILITY DENSITY FUNCTIONS

A *density function* is a nonnegative function whose integral taken over the entire range of the independent variable is unity. A *probability density function* (PDF) is a mathematical formula that gives the probability of a discrete numerical event. A *numerical event* is an occurrence that can be described (usually) by an integer. For example, 27 cars passing through a bridge toll booth in an hour is a discrete numerical event.

A probability density function, $f(x)$, gives the probability that discrete event x will occur. That is, $p\{x\} = f(x)$. Important discrete probability density functions are the binomial, hypergeometric, and Poisson distributions.

Figure 11.2 *Probability Density Function*

9. BINOMIAL DISTRIBUTION

The *binomial probability density function* is used when all outcomes can be categorized as either successes or failures. The probability of success in a single trial is $\hat{p}$, and the probability of failure is the complement, $\hat{q} = 1 - \hat{p}$. The population is assumed to be infinite in size so that sampling does not change the values of $\hat{p}$ and $\hat{q}$. (The binomial distribution can also be used with finite populations when sampling with replacement.)

Equation 11.28 gives the probability of x successes in n independent *successive trials*. The quantity $\binom{n}{x}$ is the binomial coefficient, identical to the number of combinations of n items taken x at a time.

$$p\{x\} = f(x) = \binom{n}{x} \hat{p}^x \hat{q}^{n-x} \qquad 11.28$$
$$\binom{n}{x} = \frac{n!}{(n-x)!x!} \qquad 11.29$$

Equation 11.28 is a true (discrete) distribution, taking on values for each integer value up to n. The mean, μ, and variance, σ^2 (see Secs. 18 and 19), of this distribution are

$$\mu = n\hat{p} \qquad 11.30$$
$$\sigma^2 = n\hat{p}\hat{q} \qquad 11.31$$

Example 11.8

Five percent of a large batch of high-strength steel bolts purchased for bridge construction are defective. (a) If seven bolts are randomly sampled, what is the probability that exactly three will be defective? (b) What is the probability that two or more bolts will be defective?

Solution

(a) The bolts are either defective or not, so the binomial distribution can be applied.

$$\hat{p} = 0.05 \quad [\text{success} = \text{defective}]$$

$$\hat{q} = 1 - 0.05 = 0.95 \quad [\text{failure} = \text{not defective}]$$

From Eq. 11.28,

$$p\{3\} = f(3) = \binom{7}{3} \hat{p}^3 \hat{q}^{7-3}$$

$$= \left(\frac{7 \cdot 6 \cdot 5 \cdot 4 \cdot 3 \cdot 2 \cdot 1}{4 \cdot 3 \cdot 2 \cdot 1 \cdot 3 \cdot 2 \cdot 1} \right) (0.05)^3 (0.95)^4$$

$$= 0.00356$$

(b) The probability that two or more bolts will be defective could be calculated as

$$p\{x \geq 2\} = p\{2\} + p\{3\} + p\{4\} + p\{5\} + p\{6\} + p\{7\}$$

This method would require six probability calculations. It is easier to use the complement of the desired probability.

$$p\{x \geq 2\} = 1 - p\{x \leq 1\} = 1 - (p\{0\} + p\{1\})$$

$$p\{0\} = \binom{7}{0} (0.05)^0 (0.95)^7 = (0.95)^7$$

$$p\{1\} = \binom{7}{1} (0.05)^1 (0.95)^6 = (7)(0.05)(0.95)^6$$

$$p\{x \geq 2\} = 1 - ((0.95)^7 + (7)(0.05)(0.95)^6)$$

$$= 1 - (0.6983 + 0.2573)$$

$$= 0.0444$$

10. HYPERGEOMETRIC DISTRIBUTION

Probabilities associated with sampling from a finite population without replacement are calculated from the *hypergeometric distribution*. If a population of finite size M contains K items with a given characteristic (e.g., red color, defective construction), then the probability of finding x items with that characteristic in a sample of n items is

$$p\{x\} = f(x) = \frac{\binom{K}{x} \binom{M-K}{n-x}}{\binom{M}{n}} \quad [\text{for } x \leq n] \quad \textbf{11.32}$$

11. MULTIPLE HYPERGEOMETRIC DISTRIBUTION

Sampling without replacement from finite populations containing several different types of items is handled by the *multiple hypergeometric distribution*. If a population of finite size M contains K_i items of type i (such that $\sum K_i = M$), the probability of finding x_1 items of type 1, x_2 items of type 2, and so on, in a sample size n (such that $\sum x_i = n$) is

$$p\{x_1, x_2, x_3, \ldots\} = \frac{\binom{K_1}{x_1} \binom{K_2}{x_2} \binom{K_3}{x_3} \cdots}{\binom{M}{n}} \quad \textbf{11.33}$$

12. POISSON DISTRIBUTION

Certain events occur relatively infrequently but at a relatively regular rate. The probability of such an event occurring is given by the *Poisson distribution*. Suppose an event occurs, on the average, λ times per period. The probability that the event will occur x times per period is

$$p\{x\} = f(x) = \frac{e^{-\lambda} \lambda^x}{x!} \quad [\lambda > 0] \quad \textbf{11.34}$$

λ is both the mean and the variance of the Poisson distribution.

Example 11.9

The number of customers arriving at a hamburger stand in the next period is a Poisson distribution having a mean of eight. What is the probability that exactly six customers will arrive in the next period?

Solution

$\lambda = 8$ and $x = 6$. From Eq. 11.34,

$$p\{6\} = \frac{e^{-\lambda} \lambda^x}{x!} = \frac{e^{-8} (8)^6}{6!} = 0.122$$

13. CONTINUOUS DISTRIBUTION FUNCTIONS

Most numerical events are *continuously distributed* and are not constrained to discrete or integer values. For example, the resistance of a 10% 1 Ω resistor may be any value between 0.9 and 1.1 Ω. The probability of an exact numerical event is zero for continuously distributed variables. That is, there is no chance that a numerical event will be *exactly* x.[2] It is possible to determine only the probability that a numerical event will be less than x, greater than x, or between the values of x_1 and x_2, but not exactly equal to x.

---

[2]It is important to understand the rationale behind this statement. Since the variable can take on any value and has an infinite number of significant digits, we can infinitely continue to increase the precision of the value. For example, the probability is zero that a resistance will be exactly 1 Ω because the resistance is really 1.03 or 1.0260008 or 1.02600080005, and so on.

While an expression, $f(x)$, for a probability density function can be written, it is used to derive the *continuous distribution function* (CDF), $F(x_0)$, which gives the probability of numerical event x_0 or less occurring.

$$p\{X < x_0\} = F(x_0) = \int_0^{x_0} f(x)dx \qquad 11.35$$

$$f(x) = \frac{dF(x)}{dx} \qquad 11.36$$

Figure 11.3 *Continuous Distribution Function*

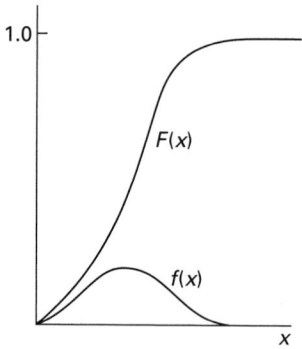

14. EXPONENTIAL DISTRIBUTION

The continuous *exponential distribution* is given by its probability density and continuous distribution functions.

$$f(x) = \lambda e^{-\lambda x} \qquad 11.37$$

$$p\{X < x\} = F(x) = 1 - e^{-\lambda x} \qquad 11.38$$

The mean and variance of the exponential distribution are

$$\mu = \frac{1}{\lambda} \qquad 11.39$$

$$\sigma^2 = \frac{1}{\lambda^2} \qquad 11.40$$

15. NORMAL DISTRIBUTION

The *normal distribution (Gaussian distribution)* is a symmetrical distribution commonly referred to as the *bell-shaped curve*, which represents the distribution of outcomes of many experiments, processes, and phenomena. The probability density and continuous distribution functions for the normal distribution with mean μ and variance σ^2 are

$$f(x) = \frac{e^{-\frac{1}{2}\left(\frac{x-\mu}{\sigma}\right)^2}}{\sigma\sqrt{2\pi}} \qquad [-\infty < x < +\infty] \qquad 11.41$$

$$p\{\mu < X < x_0\} = F(x_0)$$
$$= \frac{1}{\sigma\sqrt{2\pi}} \int_0^{x_0} e^{-\frac{1}{2}\left(\frac{x-\mu}{\sigma}\right)^2} dx \qquad 11.42$$

Figure 11.4 *Normal Distribution*

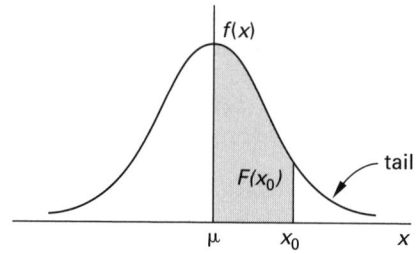

Since $f(x)$ is difficult to integrate, Eq. 11.41 is seldom used directly, and a *standard normal table* (see App. 11.A) is used instead. The standard normal table is based on a normal distribution with a mean of zero and a standard deviation of 1. Since the range of values from an experiment or phenomenon will not generally correspond to the standard normal table, a value, x_0, must be converted to a *standard normal value*, z. In Eq. 11.43, μ and σ are the mean and standard deviation, respectively, of the distribution from which x_0 comes. For all practical purposes, all normal distributions are completely bounded by $\mu \pm 3\sigma$.

$$z = \frac{x_0 - \mu}{\sigma} \qquad 11.43$$

Numbers in the standard normal table given by App. 11.A are the probabilities of the normalized x being between zero and z and represent the areas under the curve up to point z. When x is less than μ, z will be negative. However, the curve is symmetrical, so the table value corresponding to positive z can be used. The probability of x being greater than z is the complement of the table value. The curve area past point z is known as the *tail of the curve*.

The *error function*, $\mathrm{erf}(x)$, and its complement, $\mathrm{erfc}(x)$, are defined by Eqs. 11.44 and 11.45. The error function is used to determine the probable error of a measurement.

$$\mathrm{erf}(x_0) = \frac{2}{\sqrt{\pi}} \int_0^{x_0} e^{-x^2} dx \qquad 11.44$$

$$\mathrm{erfc}(x_0) = 1 - \mathrm{erf}(x_0) \qquad 11.45$$

Example 11.10

The mass, m, of a particular hand-laid fiberglass (Fibreglas^{TM}) part is normally distributed with a mean of 66 kg and a standard deviation of 5 kg. (a) What percent of the parts will have a mass less than 72 kg? (b) What percent of the parts will have a mass in excess of 72 kg? (c) What percent of the parts will have a mass between 61 and 72 kg?

Solution

(a) The value of 72 kg must be normalized by using Eq. 11.43. The standard normal variable is

$$z = \frac{x - \mu}{\sigma} = \frac{72 \text{ kg} - 66 \text{ kg}}{5 \text{ kg}} = 1.2$$

Reading from App. 11.A, the area under the normal curve is 0.3849. This represents the probability of the mass, m, being between 66 kg and 72 kg (i.e., z being between 0 and 1.2). However, the probability of the mass being less than 66 kg is also needed. Since the curve is symmetrical, this probability is 0.5. Therefore,

$$p\{m < 72 \text{ kg}\} = p\{z < 1.2\} = 0.5 + 0.3849 = 0.8849$$

(b) The probability of the mass exceeding 72 kg is the area under the tail past point z.

$$p\{m > 72 \text{ kg}\} = p\{z > 1.2\} = 0.5 - 0.3849 = 0.1151$$

(c) The standard normal variable corresponding to $m = 61$ kg is

$$z = \frac{x - \mu}{\sigma} = \frac{61 \text{ kg} - 66 \text{ kg}}{5 \text{ kg}} = -1$$

Since the two masses are on opposite sides of the mean, the probability will have to be determined in two parts.

$$p\{61 < m < 72\} = p\{61 < m < 66\} + p\{66 < m < 72\}$$
$$= p\{-1 < z < 0\} + p\{0 < z < 1.2\}$$
$$= 0.3413 + 0.3849 = 0.7262$$

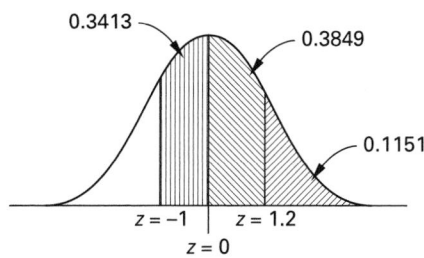

16. APPLICATION: RELIABILITY

Introduction

Reliability, $R\{t\}$, is the probability that an item will continue to operate satisfactorily up to time t. The *bathtub distribution*, Fig. 11.5, is often used to model the probability of failure of an item (or, the number of failures from a large population of items) as a function of time. Items initially fail at a high rate, a phenomenon known as *infant mortality*. For the majority of the operating time, known as the *steady-state operation*, the failure rate is constant (i.e., is due to random causes).

After a long period of time, the items begin to deteriorate and the failure rate increases. (No mathematical distribution describes all three of these phases simultaneously.)

Figure 11.5 *Bathtub Reliability Curve*

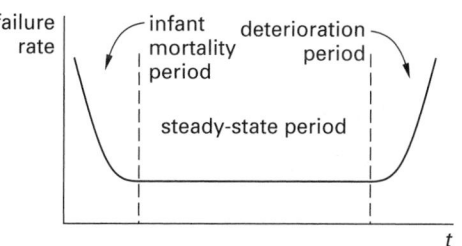

The *hazard function*, $z\{t\}$, represents the *conditional probability of failure*—the probability of failure in the next time interval, given that no failure has occurred thus far.[3]

$$z\{t\} = \frac{f(t)}{R(t)} = \frac{\dfrac{dF(t)}{dt}}{1 - F(t)} \qquad 11.46$$

Exponential Reliability

Steady-state reliability is often described by the *negative exponential distribution*. This assumption is appropriate whenever an item fails only by random causes and does not experience deterioration during its life. The parameter λ is related to the *mean time to failure* (MTTF) of the item.[4]

$$R\{t\} = e^{-\lambda t} = e^{-t/\text{MTTF}} \qquad 11.47$$

$$\lambda = \frac{1}{\text{MTTF}} \qquad 11.48$$

Equation 11.47 and the exponential continuous distribution function, Eq. 11.38, are complementary.

$$R\{t\} = 1 - F(t) = 1 - (1 - e^{-\lambda t}) = e^{-\lambda t} \qquad 11.49$$

The hazard function for the negative exponential distribution is

$$z\{t\} = \lambda \qquad 11.50$$

Thus, the hazard function for exponential reliability is constant and does not depend on t (i.e., on the age of the item). In other words, the expected future life of an item is independent of the previous history (length of operation). This lack of memory is consistent with the assumption that only random causes contribute to failure during steady-state operations. And since random causes are unlikely discrete events, their probability of

---

[3]The symbol $z\{t\}$ is traditionally used for the hazard function and is not related to the standard normal variable.

[4]The term "mean time *between* failures" is improper. However, the term *mean time before failure* (MTBF) is acceptable.

occurrence can be represented by a Poisson distribution with mean λ. That is, the probability of having x failures in any given period is

$$p\{x\} = \frac{e^{-\lambda}\lambda^x}{x!} \qquad 11.51$$

Serial System Reliability

In the analysis of system reliability, the binary variable X_i is defined as one if item i operates satisfactorily and zero otherwise. Similarly, the binary variable Φ is one only if the entire system operates satisfactorily. Thus, Φ will depend on a *performance function* containing the X_i.

A *serial system* is one for which all items must operate correctly for the system to operate. Each item has its own reliability, R_i. For a serial system of n items, the performance function is

$$\Phi = X_1 X_2 X_3 \cdots X_n = \min(X_i) \qquad 11.52$$

The probability of a serial system operating correctly is

$$p\{\Phi = 1\} = R_{\text{serial system}} = R_1 R_2 R_3 \cdots R_n \qquad 11.53$$

Parallel System Reliability

A *parallel system* with n items will fail only if all n items fail. Such a system is said to be *redundant* to the nth degree. Using redundancy, a highly reliable system can be produced from components with relatively low individual reliabilities.

The performance function of a redundant system is

$$\Phi = 1 - (1 - X_1)(1 - X_2)(1 - X_3) \cdots (1 - X_n)$$
$$= \max(X_i) \qquad 11.54$$

The reliability of the parallel system is

$$R = p\{\Phi = 1\} = 1 - (1 - R_1)(1 - R_2)(1 - R_3) \cdots (1 - R_n) \qquad 11.55$$

Example 11.11

The reliability of an item is exponentially distributed with mean time to failure (MTTF) of 1000 hr. What is the probability that the item will not have failed after 1200 hr of operation?

Solution

The probability of not having failed before time t is the reliability. From Eqs. 11.48 and 11.49,

$$\lambda = \frac{1}{\text{MTTF}} = \frac{1}{1000 \text{ hr}} = 0.001 \text{ hr}^{-1}$$

$$R\{1200\} = e^{-\lambda t} = e^{(-0.001 \text{ hr}^{-1})(1200 \text{ hr})} = 0.3$$

Example 11.12

What are the reliabilities of the following systems?

(a)

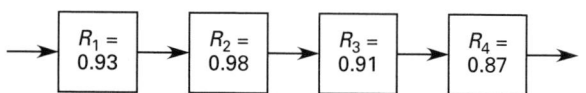

(b)

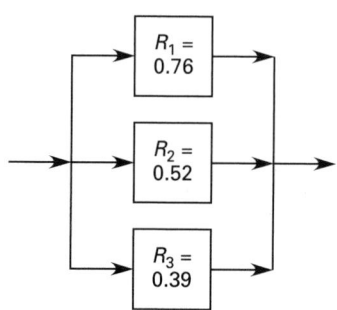

Solution

(a) This is a serial system. From Eq. 11.53,

$$R = R_1 R_2 R_3 R_4 = (0.93)(0.98)(0.91)(0.87)$$
$$= 0.72$$

(b) This is a parallel system. From Eq. 11.55,

$$R = 1 - (1 - R_1)(1 - R_2)(1 - R_3)$$
$$= 1 - (1 - 0.76)(1 - 0.52)(1 - 0.39) = 0.93$$

17. ANALYSIS OF EXPERIMENTAL DATA

Experiments can take on many forms. An experiment might consist of measuring the mass of one cubic foot of concrete, or measuring the speed of a car on a roadway. Generally, such experiments are performed more than once to increase the precision and accuracy of the results.

Both systematic and random variations in the process being measured will cause the observations to vary, and the experiment would not be expected to yield the same result each time it was performed. Eventually, a collection of experimental outcomes (observations) will be available for analysis.

The *frequency distribution* is a systematic method for ordering the observations from small to large, according to some convenient numerical characteristic. The *step interval* should be chosen so that the data are presented in a meaningful manner. If there are too many intervals, many of them will have zero frequencies; if there are too few intervals, the frequency distribution will have little value. Generally, 10 to 15 intervals are used.

Once the frequency distribution is complete, it can be represented graphically as a *histogram*. The procedure in drawing a histogram is to mark off the interval limits (also known as *class limits*) on a number line and then draw contiguous bars with lengths that are proportional to the frequencies in the intervals and that are centered on the midpoints of their respective intervals. The continuous nature of the data can be depicted by a *frequency polygon*. The number or percentage of observations that occur up to and including some value can be shown in a *cumulative frequency table*.

Example 11.13

The number of cars that travel through an intersection between 12 noon and 1 p.m. is measured for 30 consecutive working days. The results of the 30 observations are

79, 66, 72, 70, 68, 66, 68, 76, 73, 71, 74, 70, 71, 69, 67, 74, 70, 68, 69, 64, 75, 70, 68, 69, 64, 69, 62, 63, 63, 61

(a) What are the frequency and cumulative distributions? (Use a distribution interval of two cars per hour.) (b) Draw the histogram. (Use a cell size of two cars per hour.) (c) Draw the frequency polygon. (d) Graph the cumulative frequency distribution.

Solution

(a)

cars per hour	frequency	cumulative frequency	cumulative percent
60–61	1	1	3
62–63	3	4	13
64–65	2	6	20
66–67	3	9	30
68–69	8	17	57
70–71	6	23	77
72–73	2	25	83
74–75	3	28	93
76–77	1	29	97
78–79	1	30	100

(b)

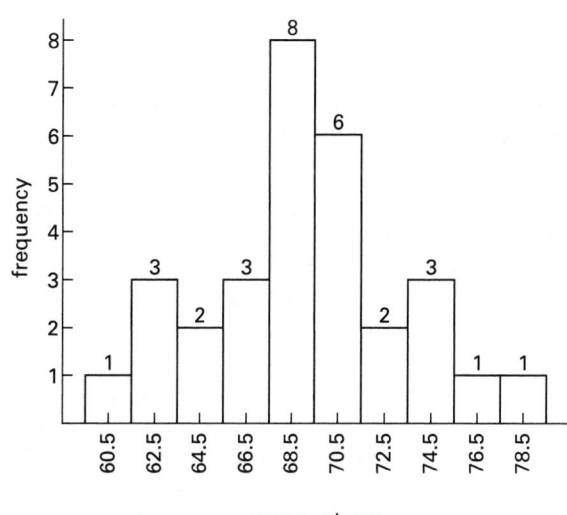

(c)

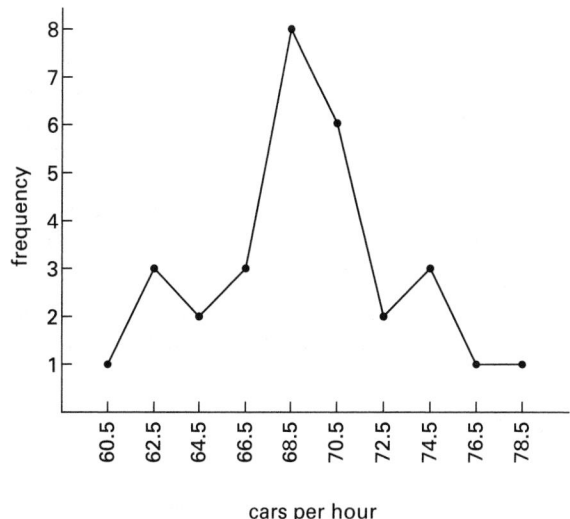

(d)

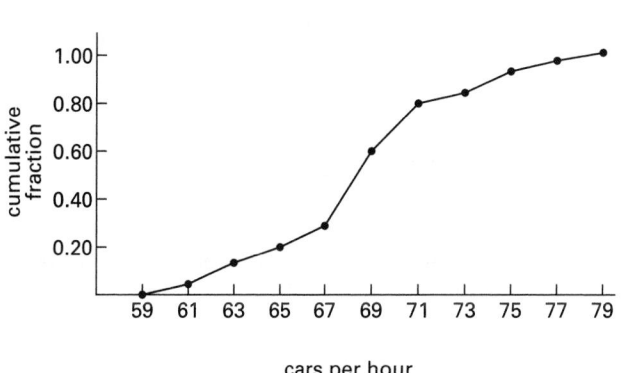

18. MEASURES OF CENTRAL TENDENCY

It is often unnecessary to present the experimental data in their entirety, either in tabular or graphical form. In such cases, the data and distribution can be represented by various parameters. One type of parameter is a measure of *central tendency*. Mode, median, and mean are measures of central tendency.

The *mode* is the observed value that occurs most frequently. The mode may vary greatly between series of observations. Therefore, its main use is as a quick measure of the central value since little or no computation is required to find it. Beyond this, the usefulness of the mode is limited.

The *median* is the point in the distribution that partitions the total set of observations into two parts containing equal numbers of observations. It is not influenced by the extremity of scores on either side of the distribution. The median is found by counting up (from either end of the frequency distribution) until half of the observations have been accounted for.

For even numbers of observations, the median is estimated as some value (i.e., the average) between the two center observations.

Similar in concept to the median are *percentiles (percentile ranks), quartiles,* and *deciles.* The median could also have been called the *50th percentile* observation. Similarly, the 80th percentile would be the observed value (e.g., the number of cars per hour) for which the cumulative frequency was 80%. The quartile and decile points on the distribution divide the observations or distribution into segments of 25% and 10%, respectively.

The *arithmetic mean* is the arithmetic average of the observations. The sample mean, $\overline{x}$, can be used as an unbiased estimator of the population mean, μ. The *mean* may be found without ordering the data (as was necessary to find the mode and median). The mean can be found from the following formula.

$$\overline{x} = \left(\frac{1}{n}\right)(x_1 + x_2 + \cdots + x_n) = \frac{\sum x_i}{n} \qquad 11.56$$

The *geometric mean* is used occasionally when it is necessary to average ratios. The geometric mean is calculated as

$$\text{geometric mean} = \sqrt[n]{x_1 x_2 x_3 \cdots x_n} \quad [x_i > 0] \qquad 11.57$$

The *harmonic mean* is defined as

$$\text{harmonic mean} = \frac{n}{\dfrac{1}{x_1} + \dfrac{1}{x_2} + \cdots + \dfrac{1}{x_n}} \qquad 11.58$$

The *root-mean-squared* (rms) *value* of a series of observations is defined as

$$x_{\text{rms}} = \sqrt{\frac{\sum x_i^2}{n}} \qquad 11.59$$

Example 11.14

Find the mode, median, and arithmetic mean of the distribution represented by the data given in Ex. 11.13.

Solution

First, resequence the observations in increasing order.

61, 62, 63, 63, 64, 64, 66, 66, 67, 68, 68, 68, 68, 69, 69, 69, 69, 70, 70, 70, 70, 71, 71, 72, 73, 74, 74, 75, 76, 79

The mode is the interval 68–69, since this interval has the highest frequency. If 68.5 is taken as the interval center, then 68.5 would be the mode.

The 15th and 16th observations are both 69, so the median is

$$\frac{69 + 69}{2} = 69$$

The mean can be found from the raw data or from the grouped data using the interval center as the assumed observation value. Using the raw data,

$$\overline{x} = \frac{\sum x}{n} = \frac{2069}{30} = 68.97$$

19. MEASURES OF DISPERSION

The simplest statistical parameter that describes the variability in observed data is the *range.* The range is found by subtracting the smallest value from the largest. Since the range is influenced by extreme (low probability) observations, its use as a measure of variability is limited.

The *standard deviation* is a better estimate of variability because it considers every observation. That is, N in Eq. 11.60 is the total population size, not the sample size, n.

$$\sigma = \sqrt{\frac{\sum (x_i - \mu)^2}{N}} = \sqrt{\frac{\sum x_i^2}{N} - \mu^2} \qquad 11.60$$

The standard deviation of a sample (particularly a small sample) is a biased (i.e., is not a good) estimator of the population standard deviation. An *unbiased estimator* of the population standard deviation is the *sample standard deviation, s.*[5]

$$s = \sqrt{\frac{\sum (x_i - \overline{x})^2}{n - 1}} = \sqrt{\frac{\sum x_i^2 - \dfrac{(\sum x_i)^2}{n}}{n - 1}} \qquad 11.61$$

---

[5]There is a subtle yet significant difference between *standard deviation of the sample,* σ (obtained from Eq. 11.60 for a finite sample drawn from a larger population) and the *sample standard deviation, s* (obtained from Eq. 11.61). While σ can be calculated, it has no significance or use as an estimator. It is true that the difference between σ and s approaches zero when the sample size, n, is large, but this convergence does nothing to legitimize the use of σ as an estimator of the true standard deviation. (Some people say "large" is 30, others say 50 or 100.)

If the sample standard deviation, s, is known, the standard deviation of the sample, σ_{sample}, can be calculated.

$$\sigma_{\text{sample}} = s\sqrt{\frac{n-1}{n}} \qquad 11.62$$

The *variance* is the square of the standard deviation. Since there are two standard deviations, there are two variances. The *variance of the sample* is σ^2, and the *sample variance* is s^2.

The *relative dispersion* is defined as a measure of dispersion divided by a measure of central tendency. The *coefficient of variation* is a relative dispersion calculated from the sample standard deviation and the mean.

$$\text{coefficient of variation} = \frac{s}{\overline{x}} \qquad 11.63$$

Skewness is a measure of a frequency distribution's lack of symmetry.

$$\text{skewness} = \frac{\overline{x} - \text{mode}}{s}$$

$$\approx \frac{3(\overline{x} - \text{median})}{s} \qquad 11.64$$

Example 11.15

For the data given in Ex. 11.13, calculate (a) the sample range, (b) the standard deviation of the sample, (c) an unbiased estimator of the population standard deviation, (d) the variance of the sample, and (e) the sample variance.

Solution

$$\sum x_i = 2069$$

$$\left(\sum x_i\right)^2 = (2069)^2 = 4{,}280{,}761$$

$$\sum x_i^2 = 143{,}225$$

$$n = 30$$

$$\overline{x} = \frac{2069}{30} = 68.967$$

(a) $\quad R = x_{\max} - x_{\min} = 79 - 61 = 18$

(b) $\quad \sigma = \sqrt{\dfrac{\sum x_i^2}{n} - (\overline{x})^2} = \sqrt{\dfrac{143{,}225}{30} - \left(\dfrac{2069}{30}\right)^2}$

$\qquad = \sqrt{17.766} = 4.215$

(c) $\quad s = \sqrt{\dfrac{\sum x_i^2 - \dfrac{\left(\sum x_i\right)^2}{n}}{n-1}}$

$\qquad = \sqrt{\dfrac{143{,}225 - \dfrac{4{,}280{,}761}{30}}{29}}$

$\qquad = \sqrt{18.378} = 4.287$

(d) $\quad \sigma^2 = 17.77$

(e) $\quad s^2 = 18.38$

20. CENTRAL LIMIT THEOREM

Measuring a sample of n items from a population with mean μ and standard deviation σ is the general concept of an experiment. The sample mean, $\overline{x}$, is one of the parameters that can be derived from the experiment. This experiment can be repeated k times, yielding a set of averages $(\overline{x}_1, \overline{x}_2, \ldots, \overline{x}_k)$. The k numbers in the set themselves represent samples from distributions of averages. The average of averages, $\overline{\overline{x}}$, and sample standard deviation of averages, $s_{\overline{x}}$ (known as the *standard error of the mean*), can be calculated.

The *central limit theorem* defines the mean of the sample averages. The theorem can be stated in several ways, but the essential elements are the following points.

1. The averages, $\overline{x}_i$, are normally distributed variables, even if the original data from which they are calculated are not normally distributed.

2. The grand average, $\overline{\overline{x}}$ (i.e., the average of the averages), approaches and is an unbiased estimator of μ.

$$\mu \approx \overline{\overline{x}} \qquad 11.65$$

The standard deviation of the original distribution, σ, is much larger than the standard error of the mean.

$$\sigma \approx \sqrt{n}\, s_{\overline{x}} \qquad 11.66$$

21. CONFIDENCE LEVEL

The results of experiments are seldom correct 100% of the time. Recognizing this, researchers accept a certain probability of being wrong. In order to minimize this probability, the experiment is repeated several times. The number of repetitions depends on the desired level of confidence in the results.

If the results have a 5% probability of being wrong, the *confidence level*, C, is 95% that the results are correct, in which case the results are said to be *significant*. If the results have only a 1% probability of being wrong, the confidence level is 99%, and the results are said to be *highly significant*. Other confidence levels (90%, 99.5%, etc.) are used as appropriate.

22. APPLICATION: CONFIDENCE LIMITS

As a consequence of the central limit theorem, sample means of n items taken from a normal distribution with mean μ and standard deviation σ will be normally distributed with mean μ and variance σ^2/n. Thus, the probability that any given average, $\overline{x}$, exceeds some value, L, is

$$p\{\overline{x} > L\} = p\left\{ z > \left| \frac{L - \mu}{\frac{\sigma}{\sqrt{n}}} \right| \right\} \qquad 11.67$$

L is the *confidence limit* for the confidence level $1 - p\{\overline{x} > L\}$ (expressed as a percent). Values of z are read directly from the standard normal table. As an example, $z = 1.645$ for a 95% confidence level since only 5% of the curve is above that z in the upper tail. Similar values are given in Table 11.1. This is known as a *one-tail confidence limit* because all of the probability is given to one side of the variation.

Table 11.1 *Values of z for Various Confidence Levels*

confidence level, C	one-tail limit z	two-tail limit z
90%	1.28	1.645
95%	1.645	1.96
97.5%	1.96	2.17
99%	2.33	2.575
99.5%	2.575	2.81
99.75%	2.81	3.00

With *two-tail confidence limits*, the probability is split between the two sides of variation. There will be upper and lower confidence limits, UCL and LCL, respectively.

$$p\{\text{LCL} < \overline{x} < \text{UCL}\} = p\left\{ \frac{\text{LCL} - \mu}{\frac{\sigma}{\sqrt{n}}} < z < \frac{\text{UCL} - \mu}{\frac{\sigma}{\sqrt{n}}} \right\} \qquad 11.68$$

23. APPLICATION: BASIC HYPOTHESIS TESTING

A *hypothesis test* is a procedure that answers the question, "Did these data come from [a particular type of] distribution?" There are many types of tests, depending on the distribution and parameter being evaluated. The simplest hypothesis test determines whether an average value obtained from n repetitions of an experiment could have come from a population with known mean, μ, and standard deviation, σ. A practical application of this question is whether a manufacturing process has changed from what it used to be or should be. Of course, the answer (i.e., "yes" or "no") cannot be given with absolute certainty—there will be a confidence level associated with the answer.

The following procedure is used to determine whether the average of n measurements can be assumed (with a given confidence level) to have come from a known population.

step 1: Assume random sampling from a normal population.

step 2: Choose the desired confidence level, C.

step 3: Decide on a one-tail or two-tail test. If the hypothesis being tested is that the average has or has not *increased* or *decreased*, choose a one-tail test. If the hypothesis being tested is that the average has or has not *changed*, choose a two-tail test.

step 4: Use Table 11.1 or the standard normal table to determine the z-value corresponding to the confidence level and number of tails.

step 5: Calculate the actual standard normal variable, z'.

$$z' = \frac{\overline{x} - \mu}{\frac{\sigma}{\sqrt{n}}} \qquad 11.69$$

step 6: If $z' \geq z$, the average can be assumed (with confidence level C) to have come from a different distribution.

Example 11.16

When it is operating properly, a cement plant has a daily production rate that is normally distributed with a mean of 880 tons/day and a standard deviation of 21 tons/day. During an analysis period, the output is measured on 50 consecutive days, and the mean output is found to be 871 tons/day. With a 95% confidence level, determine whether the plant is operating properly.

Solution

step 1: Given.

step 2: $C = 0.95$ is given.

step 3: Since a specific direction in the variation is not given (i.e., the example does not ask whether the average has decreased), use a two-tail hypothesis test.

step 4: From Table 11.1, $z = 1.96$.

step 5: From Eq. 11.69,

$$z' = \left| \frac{\overline{x} - \mu}{\frac{\sigma}{\sqrt{n}}} \right| = \left| \frac{871 - 880}{\frac{21}{\sqrt{50}}} \right| = 3.03$$

Since $3.03 > 1.96$, the distributions are not the same. There is at least a 95% probability that the plant is not operating correctly.

24. APPLICATION: STATISTICAL PROCESS CONTROL

All manufacturing processes contain variation due to random and nonrandom causes. Random variation cannot be eliminated. *Statistical process control* (SPC) is the act of monitoring and adjusting the performance of a process to detect and eliminate nonrandom variation.

Statistical process control is based on taking regular (hourly, daily, etc.) samples of n items and calculating the mean, $\overline{x}$, and range, R, of the sample. To simplify the calculations, the range is used as a measure of the dispersion. These two parameters are graphed on their respective x-bar and R-control charts.[6] Confidence limits are drawn at $\pm 3\sigma/\sqrt{n}$. From a statistical standpoint, the control chart tests a hypothesis each time a point is plotted. When a point falls outside these limits, there is a 99.75% probability that the process is out of control. Until a point exceeds the control limits, no action is taken.[7]

Figure 11.6 *Typical Statistical Process Control Charts*

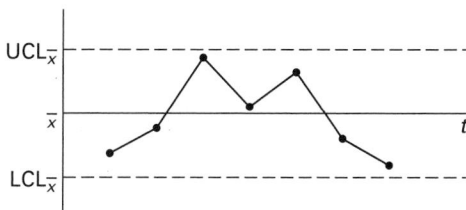

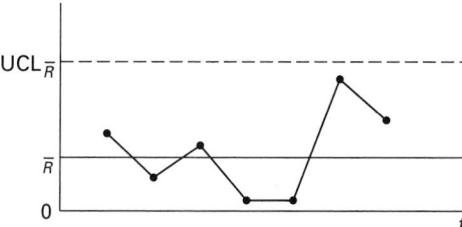

25. MEASURES OF EXPERIMENTAL ADEQUACY

An experiment is said to be *accurate* if it is unaffected by experimental error. In this case, *error* is not synonymous with *mistake*, but rather includes all variations not within the experimenter's control.

For example, suppose a gun is aimed at a point on a target and five shots are fired. The mean distance from the point of impact to the sight in point is a measure of the alignment accuracy between the barrel and sights. The difference between the actual value and the experimental value is known as *bias*.

Precision is not synonymous with accuracy. Precision is concerned with the repeatability of the experimental results. If an experiment is repeated with identical results, the experiment is said to be precise.

The average distance of each impact from the centroid of the impact group is a measure of the precision of the experiment. Thus, it is possible to have a highly precise experiment with a large bias.

Most of the techniques applied to experiments in order to improve the accuracy of the experimental results (e.g., repeating the experiment, refining the experimental methods, or reducing variability) actually increase the precision.

Sometimes the word *reliability* is used with regard to the precision of an experiment. Thus, a "reliable estimate" is used in the same sense as a "precise estimate."

Stability and *insensitivity* are synonymous terms. A stable experiment will be insensitive to minor changes in the experimental parameters. For example, suppose the centroid of a bullet group is 2.1 in from the target point at 65°F and 2.3 in away at 80°F. The sensitivity of the experiment to temperature change would be

$$\text{sensitivity} = \frac{\Delta x}{\Delta T} = \frac{2.3 \text{ in} - 2.1 \text{ in}}{80°F - 65°F} = 0.0133 \text{ in/°F}$$

26. LINEAR REGRESSION

If it is necessary to draw a straight line ($y = mx + b$) through n data points $(x_1, y_1), (x_2, y_2), \ldots, (x_n, y_n)$, the following method based on the *method of least squares* can be used.

step 1: Calculate the following nine quantities.

$$\sum x_i \quad \sum x_i^2 \quad \left(\sum x_i\right)^2 \quad \overline{x} = \frac{\sum x_i}{n} \quad \sum x_i y_i$$

$$\sum y_i \quad \sum y_i^2 \quad \left(\sum y_i\right)^2 \quad \overline{y} = \frac{\sum y_i}{n}$$

step 2: Calculate the slope, m, of the line.

$$m = \frac{n\sum(x_i y_i) - (\sum x_i)(\sum y_i)}{n\sum x_i^2 - (\sum x_i)^2} \qquad 11.70$$

step 3: Calculate the y-intercept, b.

$$b = \overline{y} - m\overline{x} \qquad 11.71$$

step 4: To determine the goodness of fit, calculate the correlation coefficient, r.

$$r = \frac{n\sum(x_i y_i) - (\sum x_i)(\sum y_i)}{\sqrt{(n\sum x_i^2 - (\sum x_i)^2)(n\sum y_i^2 - (\sum y_i)^2)}} \qquad 11.72$$

If m is positive, r will be positive; if m is negative, r will be negative. As a general rule, if the absolute value of r exceeds 0.85, the fit is good; otherwise, the fit is poor. r equals 1.0 if the fit is a perfect straight line.

---

[6]Other charts (e.g., the *sigma chart*, *p-chart*, and *c-chart*) are less common but are used as required.

[7]Other indications that a correction may be required are seven measurements on one side of the average and seven consecutively increasing measurements. Rules such as these detect shifts and trends.

A low value of r does not eliminate the possibility of a nonlinear relationship existing between x and y. It is possible that the data describe a parabolic, logarithmic, or other nonlinear relationship. (Usually this will be apparent if the data are graphed.) It may be necessary to convert one or both variables to new variables by taking squares, square roots, cubes, or logarithms, to name a few of the possibilities, in order to obtain a linear relationship. The apparent shape of the line through the data will give a clue to the type of variable transformation that is required. The curves in Fig. 11.7 may be used as guides to some of the simpler variable transformations.

Figure 11.7 *Nonlinear Data Curves*

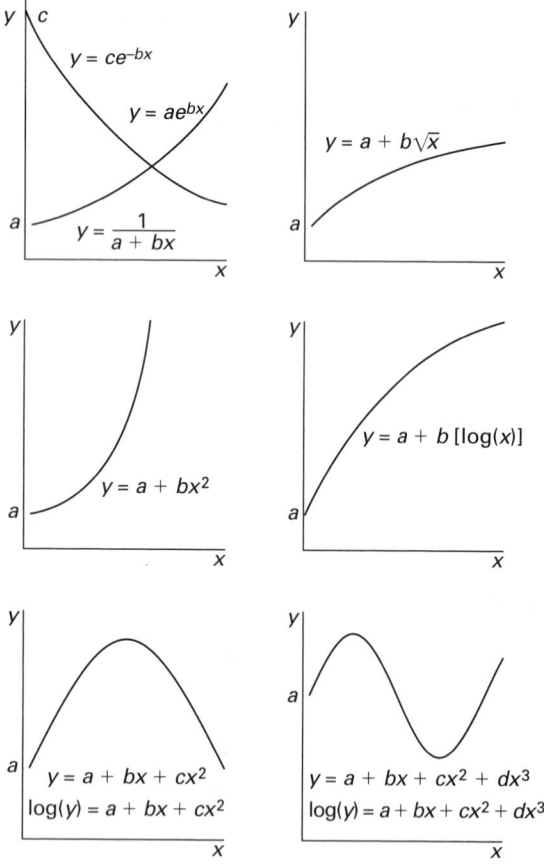

Figure 11.8 illustrates several common problems encountered in trying to fit and evaluate curves from experimental data. Figure 11.8(a) shows a graph of clustered data with several extreme points. There will be moderate correlation due to the weighting of the extreme points, although there is little actual correlation at low values of the variables. The extreme data should be excluded, or the range should be extended by obtaining more data.

Figure 11.8(b) shows that good correlation exists in general, but extreme points are missed and the overall correlation is moderate. If the results within the small linear range can be used, the extreme points should be

excluded. Otherwise, additional data points are needed, and curvilinear relationships should be investigated.

Figure 11.8 *Common Regression Difficulties*

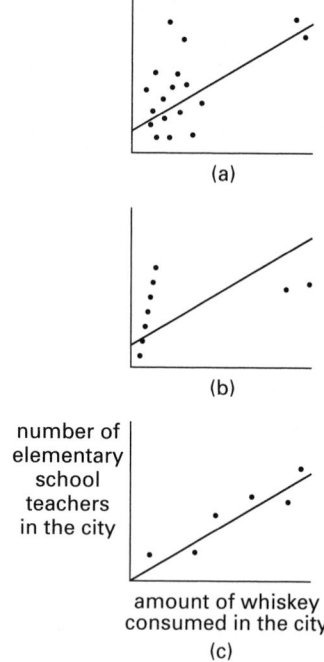

Figure 11.8(c) illustrates the problem of drawing conclusions of cause and effect. There may be a predictable relationship between variables, but that does not imply a cause and effect relationship. In the case shown, both variables are functions of a third variable, the city population. But there is no direct relationship between the plotted variables.

Example 11.17

An experiment is performed in which the dependent variable y is measured against the independent variable x. The results are as follows.

x	y
1.2	0.602
4.7	5.107
8.3	6.984
20.9	10.031

(a) What is the least squares straight line equation that represents this data? (b) What is the correlation coefficient?

Solution

(a)
$$\sum x_i = 35.1$$
$$\sum y_i = 22.72$$

$$\sum x_i^2 = 529.23$$

$$\sum y_i^2 = 175.84$$

$$\left(\sum x_i\right)^2 = 1232.01$$

$$\left(\sum y_i\right)^2 = 516.38$$

$$\overline{x} = 8.775$$

$$\overline{y} = 5.681$$

$$\sum x_i y_i = 292.34$$

$$n = 4$$

From Eq. 11.70, the slope is

$$m = \frac{(4)(292.34) - (35.1)(22.72)}{(4)(529.23) - (35.1)^2} = 0.42$$

From Eq. 11.71, the y-intercept is

$$b = 5.681 - (0.42)(8.775) = 2.0$$

The equation of the line is

$$y = 0.42x + 2.0$$

(b) From Eq. 11.72, the correlation coefficient is

$$r = \frac{(4)(292.34) - (35.1)(22.72)}{\sqrt{\begin{array}{c}((4)(529.23) - 1232.01)\\ \times \left((4)(175.84) - 516.38\right)\end{array}}} = 0.914$$

Example 11.18

Repeat Ex. 11.17 assuming the relationship between the variables is nonlinear.

Solution

The first step is to graph the data. Since the graph has the appearance of the fourth case, it can be assumed that the relationship between the variables has the form of $y = a + b[\log(x)]$. Therefore, the variable change $z = \log(x)$ is made, resulting in the following set of data.

z	y
0.0792	0.602
0.672	5.107
0.919	6.984
1.32	10.031

If the regression analysis is performed on this set of data, the resulting equation and correlation coefficient are

$$y = 7.599z + 0.000247$$

$$r = 0.999$$

This is a very good fit. The relationship between the variable x and y is approximately

$$y = 7.599\log(x) + 0.000247$$

12 Numerical Analysis

1. NUMERICAL METHODS

Although the roots of second-degree polynomials are easily found by a variety of methods (by factoring, completing the square, or using the quadratic equation), easy methods of solving cubic and higher-order equations exist only for specialized cases. However, cubic and higher-order equations occur frequently in engineering, and they are difficult to factor. Trial and error solutions, including graphing, are usually satisfactory for finding only the general region in which the root occurs.

Numerical analysis is a general subject that covers, among other things, iterative methods for evaluating roots to equations. The most efficient numerical methods are too complex to present and, in any case, work by hand. However, some of the simpler methods are presented here. Except in critical problems that must be solved in real time, a few extra calculator or computer iterations will make no difference.[1]

2. FINDING ROOTS: BISECTION METHOD

The *bisection method* is an iterative method that "brackets" (also known as "straddles") an interval containing the *root* or *zero* of a particular equation.[2] The size of the interval is halved after each iteration. As the method's name suggests, the best estimate of the root after any iteration is the midpoint of the interval. The maximum error is half the interval length. The procedure continues until the size of the maximum error is "acceptable."[3]

---

[1]Most advanced hand-held calculators now have "root finder" functions that use numerical methods to iteratively solve equations.

[2]The equation does not have to be a pure polynomial. The bisection method requires only that the equation be defined and determinable at all points in the interval.

[3]The bisection method is not a closed method. Unless the root actually falls on the midpoint of one iteration's interval, the method continues indefinitely. Eventually, the magnitude of the maximum error is small enough not to matter.

The disadvantages of the bisection method are (a) the slowness in converging to the root, (b) the need to know the interval containing the root before starting, and (c) the inability to determine the existence of or find other real roots in the starting interval.

The bisection method starts with two values of the independent variable, $x = L_0$ and $x = R_0$, which straddle a root. Since the function passes through zero at a root, $f(L_0)$ and $f(R_0)$ will have opposite signs. The following algorithm describes the remainder of the bisection method.

Let n be the iteration number. Then, for $n = 0, 1, 2, \ldots$, perform the following steps until sufficient accuracy is attained.

step 1: Set $m = \frac{1}{2}(L_n + R_n)$.

step 2: Calculate $f(m)$.

step 3: If $f(L_n)f(m) \leq 0$, set $L_{n+1} = L_n$ and $R_{n+1} = m$. Otherwise, set $L_{n+1} = m$ and $R_{n+1} = R_n$.

step 4: $f(x)$ has at least one root in the interval (L_{n+1}, R_{n+1}). The estimated value of that root, x^*, is

$$x^* \approx \frac{1}{2}(L_{n+1} + R_{n+1})$$

The maximum error is $\frac{1}{2}(R_{n+1} - L_{n+1})$.

Example 12.1

Use two iterations of the bisection method to find a root of

$$f(x) = x^3 - 2x - 7$$

Solution

The first step is to find L_0 and R_0, which are the values of x that straddle a root and have opposite signs. A table can be made and values of $f(x)$ calculated for random values of x.

x	-2	-1	0	$+1$	$+2$	$+3$
$f(x)$	-11	-6	-7	-8	-3	$+14$

Since $f(x)$ changes sign between $x = 2$ and $x = 3$, $L_0 = 2$ and $R_0 = 3$. First iteration, $n = 0$:

$$m = \left(\tfrac{1}{2}\right)(2 + 3) = 2.5$$
$$f(2.5) = (2.5)^3 - (2)(2.5) - 7 = 3.625$$

Since $f(2.5)$ is positive, a root must exist in the interval $(2, 2.5)$. Therefore, $L_1 = 2$ and $R_1 = 2.5$. At this point, the best estimate of the root is

$$x^* \approx \left(\tfrac{1}{2}\right)(2 + 2.5) = 2.25$$

The maximum error is $\left(\tfrac{1}{2}\right)(2.5 - 2) = 0.25$.

Second iteration, $n = 1$:

$$m = \left(\tfrac{1}{2}\right)(2 + 2.5) = 2.25$$
$$f(2.25) = (2.25)^3 - (2)(2.25) - 7 = -0.1094$$

Since $f(2.25)$ is negative, a root must exist in the interval $(2.25, 2.5)$. Therefore, $L_2 = 2.25$ and $R_2 = 2.5$. The best estimate of the root is

$$x^* \approx \left(\tfrac{1}{2}\right)(2.25 + 2.5) = 2.375$$

The maximum error is $\left(\tfrac{1}{2}\right)(2.5 - 2.25) = 0.125$.

3. FINDING ROOTS: NEWTON'S METHOD

Many other methods have been developed to overcome one or more of the disadvantages of the bisection method. These methods have their own disadvantages.[4]

Newton's method is a particular form of *fixed-point iteration*. In this sense, "fixed point" is often used as a synonym for "root" or "zero." However, fixed-functions with the characteristic property $x = g(x)$ such that the limit of $g(x)$ is the fixed point (i.e., is the root).

All fixed-point techniques require a starting point. Preferably, the starting point will be close to the actual root.[5] And, while Newton's method converges quickly, it requires the function to be continuously differentiable.

Newton's method algorithm is simple. At each iteration ($n = 0$, 1, 2, etc.), Eq. 12.1 estimates the root. The maximum error is determined by looking at how much the estimate changes after each iteration. If the change between the previous and current estimates (representing the magnitude of error in the estimate) is too large, the current estimate is used as the independent variable for the subsequent iteration.[6]

$$x_{n+1} = g(x_n) = x_n - \frac{f(x_n)}{f'(x_n)} \qquad 12.1$$

---

[4]The *regula falsi (false position) method* converges faster than the bisection method but is unable to specify a small interval containing the root. The *secant method* is prone to round-off errors and gives no indication of the remaining distance to the root.

[5]Theoretically, the only penalty for choosing a starting point too far away from the root will be a slower convergence to the root.

[6]Actually, the theory defining the maximum error is more definite than this. For example, for a large enough value of n, the error decreases approximately linearly. Therefore, the consecutive values of x_n converge linearly to the root as well.

Example 12.2

Solve Ex. 12.1 using two iterations of Newton's method. Use $x_0 = 2$.

Solution

The function and its first derivative are

$$f(x) = x^3 - 2x - 7$$
$$f'(x) = 3x^2 - 2$$

First iteration, $n = 0$:

$$x_0 = 2$$
$$f(x_0) = f(2) = (2)^3 - (2)(2) - 7 = -3$$
$$f'(x_0) = f'(2) = (3)(2)^2 - 2 = 10$$
$$x_1 = x_0 - \frac{f(x_0)}{f'(x_0)} = 2 - \frac{-3}{10} = 2.3$$

Second iteration, $n = 1$:

$$x_1 = 2.3$$
$$f(x_1) = (2.3)^3 - (2)(2.3) - 7 = 0.567$$
$$f'(x_1) = (3)(2.3)^2 - 2 = 13.87$$
$$x_2 = x_1 - \frac{f(x_1)}{f'(x_1)} = 2.3 - \frac{0.567}{13.87} = 2.259$$

4. NONLINEAR INTERPOLATION: LAGRANGIAN INTERPOLATING POLYNOMIAL

Interpolating between two points of known data is common in engineering. Primarily due to its simplicity and speed, straight-line interpolation is used most often. Even if more than two points on the curve are explicitly known, they are not used. Since straight-line interpolation ignores all but two of the points on the curve, it ignores the effects of curvature.

A more powerful technique that accounts for the curvature is the *Lagrangian interpolating polynomial*. This method uses an nth degree parabola (polynomial) as the interpolating curve.[7] This method requires that $f(x)$ be continuous and real-valued on the interval $[x_0, x_n]$ and that $n + 1$ values of $f(x)$ are known corresponding to $x_0, x_1, x_2, \ldots, x_n$.

The procedure for calculating $f(x)$ at some intermediate point x^* starts by calculating the Lagrangian interpolating polynomial for each known point.

$$L_k(x^*) = \prod_{\substack{i=0 \\ i \neq k}}^{n} \frac{x^* - x_i}{x_k - x_i} \qquad 12.2$$

---

[7]The Lagrangian interpolating polynomial reduces to straight-line interpolation if only two points are used.

The value of $f(x)$ at x^* is calculated from Eq. 12.3.

$$f(x^*) = \sum_{k=0}^{n} f(x_k)L_k(x^*) \qquad 12.3$$

The Lagrangian interpolating polynomial has two primary disadvantages. The first is that a large number of additions and multiplications are needed.[8] The second is that the method does not indicate how many interpolating points should be (or should have been) used. Other interpolating methods have been developed that overcome these disadvantages.[9]

Example 12.3

A real-valued function has the following values.

$$f(1) = 3.5709$$
$$f(4) = 3.5727$$
$$f(6) = 3.5751$$

Use the Lagrangian interpolating polynomial to determine the value of the function at 3.5.

Solution

The procedure for applying Eq. 12.2, the Lagrangian interpolating polynomial, is illustrated in tabular form. Notice that the term corresponding to $i = k$ is omitted from the product.

$$
\begin{array}{ccc}
& i=0 \quad & i=1 \quad & i=2 \\
k=0: L_0(3.5) = \left(\dfrac{3.5-1}{1-1}\right)\left(\dfrac{3.5-4}{1-4}\right)\left(\dfrac{3.5-6}{1-6}\right)
\end{array}
$$
$$= 0.08333$$

$$
k=1: L_1(3.5) = \left(\dfrac{3.5-1}{4-1}\right)\left(\dfrac{3.5-4}{4-4}\right)\left(\dfrac{3.5-6}{4-6}\right)
$$
$$= 1.04167$$

$$
k=2: L_2(3.5) = \left(\dfrac{3.5-1}{6-1}\right)\left(\dfrac{3.5-4}{6-4}\right)\left(\dfrac{3.5-6}{6-6}\right)
$$
$$= -0.12500$$

Equation 12.3 is used to calculate the estimate.

$$f(3.5) = (3.5709)(0.08333) + (3.5727)(1.04167)$$
$$+ (3.5751)(-0.12500)$$
$$= 3.57225$$

---

[8]As with the numerical methods for finding roots previously discussed, the number of calculations probably will not be an issue if the work is performed by a calculator or computer.

[9]Other common methods for performing interpolation include the *Newton form* and *divided difference table*.

5. NONLINEAR INTERPOLATION: NEWTON'S INTERPOLATING POLYNOMIAL

Newton's form of the interpolating polynomial is more efficient than the Lagrangian method of interpolating between known points.[10] Given $n + 1$ known points for $f(x)$, the *Newton form of the interpolating polynomial* is

$$f(x^*) = \sum_{i=0}^{n} \left(f[x_0, x_1, \ldots, x_i] \prod_{j=0}^{i-1} (x^* - x_j) \right) \qquad 12.4$$

$f[x_0, x_1, \ldots, x_i]$ is known as the ith *divided difference*.

$$f[x_0, x_1, \ldots, x_i] = \sum_{k=0}^{i} \left(\frac{f(x_k)}{(x_k - x_0) \cdots (x_k - x_{k-1})} \right.$$
$$\left. \times (x_k - x_{k+1}) \cdots (x_k - x_i) \right)$$
$$12.5$$

It is necessary to define the following two terms.

$$f[x_0] = f(x_0) \qquad 12.6$$
$$\prod (x^* - x_j) = 1 \qquad [i = 0] \qquad 12.7$$

Example 12.4

Repeat Ex. 12.3 using Newton's form of the interpolating polynomial.

Solution

Since there are $n + 1 = 3$ data points, $n = 2$. Evaluate the terms for $i = 0$ to 2.

$i = 0$:

$$f[x_0] \prod_{j=0}^{-1} (x^* - x_j) = f(x_0)(1) = f(x_0)$$

$i = 1$:

$$f[x_0, x_1] \prod_{j=0}^{0} (x^* - x_j) = f[x_0, x_1](x^* - x_0)$$
$$f[x_0, x_1] = \frac{f(x_0)}{x_0 - x_1} + \frac{f(x_1)}{x_1 - x_0}$$

$i = 2$:

$$f[x_0, x_1, x_2] \prod_{j=0}^{1} (x^* - x_j) = f[x_0, x_1, x_2](x^* - x_0)$$
$$\times (x^* - x_1)$$
$$f[x_0, x_1, x_2] = \frac{f(x_0)}{(x_0 - x_1)(x_0 - x_2)}$$
$$+ \frac{f(x_1)}{(x_1 - x_0)(x_1 - x_2)}$$
$$+ \frac{f(x_2)}{(x_2 - x_0)(x_2 - x_1)}$$

---

[10]In this case, "efficiency" relates to the ease in adding new known points without having to repeat all previous calculations.

Substitute known values.

$$f(3.5) = 3.5709 + \left(\frac{3.5709}{1-4} + \frac{3.5727}{4-1} \right)(3.5-1)$$
$$+ \left(\frac{3.5709}{(1-4)(1-6)} + \frac{3.5727}{(4-1)(4-6)} \right.$$
$$+ \left. \frac{3.5751}{(6-1)(6-4)} \right)$$
$$\times (3.5-1)(3.5-4)$$
$$= 3.57225$$

This answer is the same as that determined in Ex. 12.3.

13

Energy, Work, and Power

Symbols

δ	displacement	ft	m
η	efficiency	–	–
θ	angular position	rad	rad
ρ	mass density	lbm/ft^3	kg/m^3
v	specific volume	ft^3/lbm	m^3/kg
ϕ	angle	deg	deg
ω	angular velocity	rad/sec	rad/s

Subscripts

f	frictional
p	constant pressure
v	constant volume

Nomenclature

c	specific heat	Btu/lbm-°F	J/kg·°C
C	molar specific heat	Btu/lbmol-°F	J/kmol·°C
E	energy	ft-lbf	J
F	force	lbf	N
g	gravitational acceleration	ft/sec^2	m/s^2
g_c	gravitational constant	lbm-ft/lbf-sec²	n.a.
h	height	ft	m
I	mass moment of inertia	$lbm\text{-}ft^2$	$kg\cdot m^2$
J	Joule's constant	ft-lbf/Btu	n.a.
k	spring constant	lbf/ft	N/m
m	mass	lbm	kg
MW	molecular weight	lbm/lbmol	kg/kmol
p	pressure	lbf/ft^2	Pa
P	power	ft-lbf/sec	W
q	heat	Btu/lbm	J/kg
Q	heat	Btu	J
r	radius	ft	m
s	distance	ft	m
t	time	sec	s
T	temperature	°F	°C
T	torque	ft-lbf	N·m
u	specific energy	ft-lbf/lbm	J/kg
U	internal energy	Btu	J
v	velocity	ft/sec	m/s
W	work	ft-lbf	J
x	displacement	ft	m

1. ENERGY OF A MASS

The *energy* of a mass represents the capacity of the mass to do work. Such energy can be stored and released. There are many forms that it can take, including mechanical, thermal, electrical, and magnetic energies. Energy is a positive, scalar quantity (although the change in energy can be either positive or negative).

The total energy of a body can be calculated from its mass, m, and its *specific energy*, u (i.e., the energy per unit mass).[1]

$$E = mu \qquad 13.1$$

Typical units of mechanical energy are foot-pounds and joules. (A joule is equivalent to the units of N·m and kg·m²/s².) In traditional English-unit countries, the *British thermal unit* (Btu) is used for thermal energy, whereas the kilocalorie (kcal) is still used in some applications in SI countries. *Joule's constant* or the *Joule equivalent* (778.26 ft-lbf/Btu, usually shortened to 778, three significant digits) is used to convert between English mechanical and thermal energy units.

$$\text{energy in Btu} = \frac{\text{energy in ft-lbf}}{J} \qquad 13.2$$

Two other units of large amounts of energy are the therm and the quad. A *therm* is 10^5 Btu (1.055×10^8 J). A *quad* is equal to a quadrillion (10^{15}) Btu. This is 1.055×10^{18} J, roughly the energy contained in 200 million barrels of oil.

---

[1]The use of symbols E and U for energy is not consistent in the engineering field.

2. LAW OF CONSERVATION OF ENERGY

The *law of conservation of energy* says that energy cannot be created or destroyed. However, energy can be converted into different forms. Therefore, the sum of all energy forms is constant.

$$\sum E = \text{constant} \qquad 13.3$$

3. WORK

Work, W, is the act of changing the energy of a particle, body, or system. For a mechanical system, *external work* is work done by an external force, whereas *internal work* is done by an internal force. Work is a signed, scalar quantity. Typical units are inch-pounds, foot-pounds, and joules. Mechanical work is seldom expressed in British thermal units or kilocalories.

For a mechanical system, work is positive when a force acts in the direction of motion and helps a body move from one location to another. Work is negative when a force acts to oppose motion. (Friction, for example, always opposes the direction of motion and can do only negative work.) The work done on a body by more than one force can be found by superposition.

From a thermodynamic standpoint, work is positive if a particle or body does work on its surroundings. Work is negative if the surroundings do work on the object. (Thus, blowing up a balloon represents negative work to the balloon.) Although this may be a difficult concept, it is consistent with the conservation of energy, since the sum of negative work and the positive energy increase is zero (i.e., no net energy change in the system).[2]

The work performed by a variable force or torque is calculated from the dot products of Eqs. 13.4 and 13.5,

$$W_{\text{variable force}} = \int \mathbf{F} \cdot d\mathbf{s} \quad \text{[linear systems]} \qquad 13.4$$

$$W_{\text{variable torque}} = \int \mathbf{T} \cdot d\boldsymbol{\theta} \quad \text{[rotational systems]} \quad 13.5$$

The work done by a force or torque of constant magnitude is

$$W_{\text{constant force}} = \mathbf{F} \cdot \mathbf{s} = Fs\cos\phi \quad \text{[linear systems]} \quad 13.6$$

$$W_{\text{constant torque}} = \mathbf{T} \cdot \boldsymbol{\theta} = Fr\theta\cos\phi$$
$$\text{[rotational systems]} \quad 13.7$$

The nonvector forms, Eqs. 13.6 and 13.7, illustrate that only the component of force or torque in the direction of motion contributes to work.

Figure 13.1 *Work of a Constant Force*

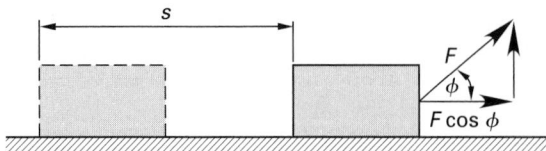

---

[2]This is just a partial statement of the *first law of thermodynamics*.

Common applications of the work done by a constant force are frictional work and gravitational work. The work to move an object a distance s against a frictional force of F_f is

$$W_{\text{friction}} = F_f s \qquad 13.8$$

The work done against gravity when a mass m changes in elevation from h_1 to h_2 is

$$W_{\text{gravity}} = mg(h_2 - h_1) \qquad \text{[SI]} \qquad 13.9(a)$$

$$W_{\text{gravity}} = \left(\frac{mg}{g_c}\right)(h_2 - h_1) \quad \text{[U.S.]} \quad 13.9(b)$$

The work done by or on a *linear spring* whose length or deflection changes from δ_1 to δ_2 is given by Eq. 13.10.[3] It does not make any difference whether the spring is a compression spring or an extension spring.

$$W_{\text{spring}} = \tfrac{1}{2}k\left(\delta_2^2 - \delta_1^2\right) \qquad 13.10$$

Example 13.1

A lawn mower engine is started by pulling a cord wrapped around a sheave. The sheave radius is 8.0 cm. The cord is wrapped around the sheave two times. If a constant tension of 90 N is maintained in the cord during starting, what work is done?

Solution

The starting torque on the engine is

$$T = Fr = (90 \text{ N})\left(\frac{8 \text{ cm}}{100 \; \frac{\text{cm}}{\text{m}}}\right)$$
$$= 7.2 \text{ N·m}$$

The cord wraps around the sheave $(2)(2\pi) = 12.6$ rad. From Eq. 13.7, the work done by a constant torque is

$$W = T\theta = (7.2 \text{ N·m})(12.6 \text{ rad})$$
$$= 90.7 \text{ J}$$

Example 13.2

A 200 lbm crate is pushed 25 ft at constant velocity across a warehouse floor. There is a frictional force of 60 lbf between the crate and floor. What work is done by the frictional force on the crate?

Solution

From Eq. 13.8,

$$W_{\text{friction}} = F_f s = (60 \text{ lbf})(25 \text{ ft})$$
$$= 1500 \text{ ft-lbf}$$

---

[3]A *linear spring* is one for which the linear relationship $F = kx$ is valid.

4. POTENTIAL ENERGY OF A MASS

Potential energy (gravitational energy) is a form of mechanical energy possessed by a body due to its relative position in a gravitational field. Potential energy is lost when the elevation of a body decreases. The lost potential energy usually is converted to kinetic energy or heat.

$$E_{\text{potential}} = mgh \quad \text{[SI]} \qquad \textbf{13.11(a)}$$

$$E_{\text{potential}} = \frac{mgh}{g_c} \quad \text{[U.S.]} \qquad \textbf{13.11(b)}$$

In the absence of friction and other nonconservative forces, the change in potential energy of a body is equal to the work required to change the elevation of the body.

$$W = \Delta E_{\text{potential}} \qquad \textbf{13.12}$$

5. KINETIC ENERGY OF A MASS

Kinetic energy is a form of mechanical energy associated with a moving or rotating body. The kinetic energy of a body moving with instantaneous linear velocity v is

$$E_{\text{kinetic}} = \frac{1}{2}m\text{v}^2 \quad \text{[SI]} \qquad \textbf{13.13(a)}$$

$$E_{\text{kinetic}} = \frac{m\text{v}^2}{2g_c} \quad \text{[U.S.]} \qquad \textbf{13.13(b)}$$

According to the *work-energy principle* (see Sec. 13.9), the kinetic energy is equal to the work necessary to initially accelerate a stationary body, or to bring a moving body to rest.

$$W = \Delta E_{\text{kinetic}} \qquad \textbf{13.14}$$

A body can also have rotational kinetic energy.

$$E_{\text{rotational}} = \frac{1}{2}I\omega^2 \quad \text{[SI]} \qquad \textbf{13.15(a)}$$

$$E_{\text{rotational}} = \frac{I\omega^2}{2g_c} \quad \text{[U.S.]} \qquad \textbf{13.15(b)}$$

Example 13.3

A solid disk flywheel ($I = 200$ kg·m²) is rotating with a speed of 900 rpm. What is its rotational kinetic energy?

Solution

The angular rotational velocity is

$$\omega = \frac{(900 \text{ rpm})\left(2\pi \ \dfrac{\text{rad}}{\text{rev}}\right)}{60 \ \dfrac{\text{s}}{\text{min}}} = 94.25 \text{ rad/s}$$

From Eq. 13.15, the rotational kinetic energy is

$$E = \frac{1}{2}I\omega^2 = \left(\frac{1}{2}\right)(200 \text{ kg·m}^2)\left(94.25 \ \frac{\text{rad}}{\text{s}}\right)^2$$

$$= 888 \times 10^3 \text{ J} \quad (888 \text{ kJ})$$

6. SPRING ENERGY

A spring is an energy storage device, since the spring has the ability to perform work. In a perfect spring, the amount of energy stored is equal to the work required to compress the spring initially. The stored spring energy does not depend on the mass of the spring. Given a spring with spring constant (stiffness) k, the *spring energy* is

$$E_{\text{spring}} = \frac{1}{2}k\delta^2 \qquad \textbf{13.16}$$

Example 13.4

A body of mass m falls from height h onto a massless, simply supported beam. The mass adheres to the beam. If the beam has a lateral stiffness k, what will be the deflection, δ, of the beam?

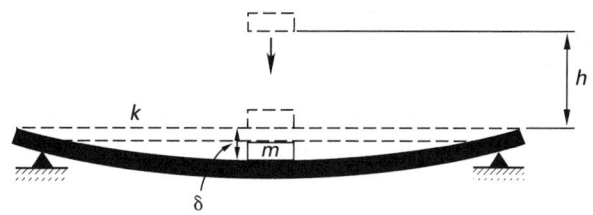

Solution

The initial energy of the system consists of only the potential energy of the body. Using consistent units, the change in potential energy is

$$E = mg(h + \delta) \quad \text{[consistent units]}$$

All of this energy is stored in the spring. Therefore,

$$\frac{1}{2}k\delta^2 = mg(h + \delta)$$

Solving for the deflection,

$$\delta = \frac{mg \pm \sqrt{mg(2hk + mg)}}{k}$$

7. PRESSURE ENERGY OF A MASS

Since work is done in increasing the pressure of a system (i.e., it takes work to blow up a balloon), mechanical energy can be stored in pressure form. This is known as *pressure energy, static energy, flow energy, flow work,* and *p-V work (energy).* For a system of pressurized mass m, the pressure energy is

$$E_{\text{flow}} = \frac{mp}{\rho} = mpv \quad [v = \text{specific volume}] \qquad \textbf{13.17}$$

8. INTERNAL ENERGY OF A MASS

The total internal energy, usually given the symbol U, of a body increases when the body's temperature increases.[4] In the absence of any work done on or by the body, the change in internal energy is equal to the heat flow, Q, into the body. Q is positive if the heat flow is into the body and negative otherwise.

$$U_2 - U_1 = Q \qquad \text{13.18}$$

The property of internal energy is encountered primarily in thermodynamics problems. Typical units are British thermal units, joules, and kilocalories.

An increase in internal energy is needed to cause a rise in temperature. Different substances differ in the quantity of heat needed to produce a given temperature increase. The ratio of heat, Q, required to change the temperature of a mass, m, by an amount ΔT is called the *specific heat* (*heat capacity*) of the substance, c.

Because specific heats of solids and liquids are slightly temperature dependent, the mean specific heats are used when evaluating processes covering a large temperature range.

$$Q = mc\,\Delta T \qquad \text{13.19}$$
$$c = \frac{Q}{m\Delta T} \qquad \text{13.20}$$

The lowercase c implies that the units are Btu/lbm-°F or J/kg·°C. Typical values of specific heat are given in Table 13.1. The *molar specific heat*, designated by the symbol C, has units of Btu/lbmol-°F or J/kmol·°C.

$$C = (\text{MW}) \times \text{c} \qquad \text{13.21}$$

For gases, the specific heat depends on the type of process during which the heat exchange occurs. Specific heats for constant-volume and constant-pressure processes are designated by c_v and c_p, respectively.

$$Q = mc_{\text{v}}\Delta T \qquad \text{[constant-volume process]} \qquad \text{13.22}$$
$$Q = mc_p\Delta T \qquad \text{[constant-pressure process]} \qquad \text{13.23}$$

Approximate values of c_p and c_v for solids and liquids are essentially the same. However, the designation c_p is often encountered for solids and liquids.

9. WORK-ENERGY PRINCIPLE

Since energy can neither be created nor destroyed, external work performed on a conservative system goes into changing the system's total energy. This is known as the *work-energy principle* (or *principle of work and energy*).

$$W = \Delta E = E_2 - E_1 \qquad \text{13.24}$$

---

[4]The *thermal energy*, represented by the body's enthalpy, is the sum of internal and pressure energies.

Table 13.1 *Approximate Specific Heats of Selected Liquids and Solids*[a]

substance	c_p Btu/lbm-°F	c_p kJ/kg·°C
aluminum, pure	0.23	0.96
aluminum, 2024-T4	0.2	0.84
ammonia	1.16	4.86
asbestos	0.20	0.84
benzene	0.41	1.72
brass, red	0.093	0.39
bronze	0.082	0.34
concrete	0.21	0.88
copper, pure	0.094	0.39
Freon-12	0.24	1.00
gasoline	0.53	2.20
glass	0.18	0.75
gold, pure	0.031	0.13
ice	0.49	2.05
iron, pure	0.11	0.46
iron, cast (4% C)	0.10	0.42
lead, pure	0.031	0.13
magnesium, pure	0.24	1.00
mercury	0.033	0.14
oil, light hydrocarbon	0.5	2.09
silver, pure	0.06	0.25
steel, 1010	0.10	0.42
steel, stainless 301	0.11	0.46
tin, pure	0.055	0.23
titanium, pure	0.13	0.54
tungsten, pure	0.032	0.13
water	1.0	4.19
wood (typical)	0.6	2.50
zinc, pure	0.088	0.37

(Multiply Btu/lbm-°F by 4.1868 to obtain kJ/kg·°C.)
[a]Values in cal/g·°C are the same as Btu/lbm-°F. Values in kJ/kg·°C are the same as kJ/kg·K.

Generally, the term *work-energy principle* is limited to use with mechanical energy problems (i.e., conversion of work into kinetic or potential energies). When energy is limited to kinetic energy, the work-energy principle is a direct consequence of Newton's second law but is valid for only inertial reference systems.

By directly relating forces, displacements, and velocities, the work-energy principle introduces some simplifications into many mechanical problems.

- It is not necessary to calculate or know the acceleration of a body to calculate the work performed on it.

- Forces that do not contribute to work (e.g., are normal to the direction of motion) are irrelevant.

- Only scalar quantities are involved.

- It is not necessary to individually analyze the particles or component parts in a complex system.

Example 13.5

A 4000 kg elevator starts from rest, accelerates uniformly to a constant speed of 2.0 m/s, and then decelerates uniformly to a stop 20 m above its initial position. Neglecting friction and other losses, what work was done on the elevator?

Solution

By the work-energy principle, the work done on the elevator is equal to the change in the elevator's energy. Since the initial and final kinetic energies are zero, the only mechanical energy change is the potential energy change.

Taking the initial elevation of the elevator as the reference (i.e., $h_1 = 0$),

$$W = E_{2,\text{potential}} - E_{1,\text{potential}} = mg(h_2 - h_1)$$

$$= (4000 \text{ kg})\left(9.81 \frac{\text{m}}{\text{s}^2}\right)(20 \text{ m}) = 785 \times 10^3 \text{ J} \quad (785 \text{ kJ})$$

10. CONVERSION BETWEEN ENERGY FORMS

Conversion of one form of energy into another does not violate the conservation of energy law. However, most problems involving conversion of energy are really just special cases of the work-energy principle. For example, consider a falling body that is acted upon by a gravitational force. The conversion of potential energy into kinetic energy can be interpreted as equating the work done by the constant gravitational force to the change in kinetic energy.

In general terms, *Joule's law* states that one energy form can be converted without loss into another. There are two specific formulations of Joule's law. As related to electricity, $P = I^2R = V^2/R$ is the common formulation of Joule's law. As related to thermodynamics and ideal gases, Joule's law states that "the change in internal energy of an ideal gas is a function of the temperature change, not of the volume." This latter form can also be stated more formally as "at constant temperature, the internal energy of a gas approaches a finite value that is independent of the volume as the pressure goes to zero."

Example 13.6

A 2.0 lbm projectile is launched straight up with an initial velocity of 700 ft/sec. Neglecting air friction, calculate the (a) kinetic energy immediately after launch, (b) kinetic energy at maximum height, (c) potential energy at maximum height, (d) total energy at an elevation where the velocity has dropped to 300 ft/sec, and (e) maximum height attained.

Solution

(a) From Eq. 13.13, the kinetic energy is

$$E_{\text{kinetic}} = \frac{m\text{v}^2}{2g_c} = \frac{(2 \text{ lbm})\left(700 \frac{\text{ft}}{\text{sec}}\right)^2}{(2)\left(32.2 \frac{\text{lbm-ft}}{\text{sec}^2\text{-lbf}}\right)}$$

$$= 15{,}217 \text{ ft-lbf}$$

(b) The velocity is zero at the maximum height. Therefore, the kinetic energy is zero.

(c) At the maximum height, all of the kinetic energy has been converted into potential energy. Therefore, the potential energy is 15,217 ft-lbf.

(d) Although some of the kinetic energy has been transformed into potential energy, the total energy is still 15,217 ft-lbf.

(e) Since all of the kinetic energy has been converted into potential energy, the maximum height can be found from Eq. 13.11.

$$E_{\text{potential}} = \frac{mgh}{g_c}$$

$$15{,}217 \text{ ft-lbf} = \frac{(2 \text{ lbm})\left(32.2 \frac{\text{ft}}{\text{sec}^2}\right)h}{32.2 \frac{\text{lbm-ft}}{\text{sec}^2\text{-lbf}}}$$

$$h = 7609 \text{ ft}$$

Example 13.7

A 4500 kg ore car rolls down an incline and passes point A traveling at 1.2 m/s. The ore car is stopped by a spring bumper that compresses 0.6 m. A constant friction force of 220 N acts on the ore car at all times. What spring constant is required?

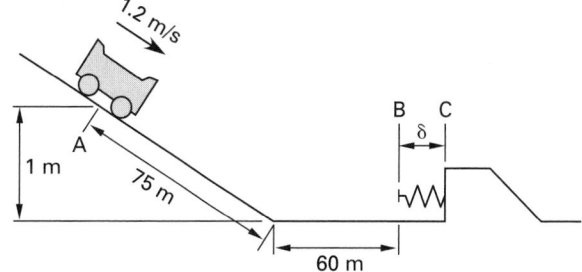

Solution

The car's total energy at point A is the sum of the kinetic and potential energies.

$$E_{\text{total,A}} = E_{\text{kinetic}} + E_{\text{potential}}$$

$$= \tfrac{1}{2}m\text{v}^2 + mgh$$

$$= \left(\tfrac{1}{2}\right)(4500 \text{ kg})\left(1.2 \frac{\text{m}}{\text{s}}\right)^2$$

$$+ (4500 \text{ kg})\left(9.81 \frac{\text{m}}{\text{s}^2}\right)(1 \text{ m})$$

$$= 47\,385 \text{ J}$$

At point B, the potential energy has been converted into additional kinetic energy. However, except for friction, the total energy is the same as at point A. Since the frictional force does negative work, the total energy remaining at point B is

$$E_{\text{total,B}} = E_{\text{total,A}} - W_{\text{friction}}$$
$$= 47\,385 \text{ J} - (220 \text{ N})(75 \text{ m} + 60 \text{ m})$$
$$= 17\,685 \text{ J}$$

At point C, the maximum compression point, the remaining energy has gone into compressing the spring a distance $\delta = 0.6$ m and performing a small amount of frictional work.

$$E_{\text{total,B}} = E_{\text{total,C}} = W_{\text{spring}} + W_{\text{friction}}$$
$$= \tfrac{1}{2}k\delta^2 + F_f\delta$$
$$17\,685 \text{ J} = \left(\tfrac{1}{2}\right)(k)(0.6 \text{ m})^2 + (220 \text{ N})(0.6 \text{ m})$$

The spring constant can be determined directly.

$$k = 97\,520 \text{ N/m} \quad (97.5 \text{ kN/m})$$

11. POWER

Power is the amount of work done per unit time. It is a scalar quantity. (Although power is calculated from two vectors, the vector dot-product operation is seldom needed.)

$$P = \frac{W}{\Delta t} \qquad \text{13.25}$$

For a body acted upon by a force or torque, the instantaneous power can be calculated from the velocity.

$$P = F\text{v} \quad \text{[linear systems]} \qquad \text{13.26}$$
$$P = T\omega \quad \text{[rotational systems]} \qquad \text{13.27}$$

For a fluid flowing at a rate of $\dot{m}$, the unit of time is already incorporated into the flow rate (e.g., lbm/sec). If the fluid experiences a specific energy change of Δu, the power generated or dissipated will be

$$P = \dot{m}\Delta u \qquad \text{13.28}$$

Typical basic units of power are ft-lbf/sec and watts (J/s), although *horsepower* is widely used. Table 13.2 can be used to convert units of power.

Table 13.2 *Useful Power Conversion Formulas*

$$1 \text{ hp} = 550 \text{ ft-lbf/sec}$$
$$= 33{,}000 \text{ ft-lbf/min}$$
$$= 0.7457 \text{ kW}$$
$$= 0.7068 \text{ Btu/sec}$$

$$1 \text{ kW} = 737.6 \text{ ft-lbf/sec}$$
$$= 44{,}250 \text{ ft-lbf/min}$$
$$= 1.341 \text{ hp}$$
$$= 0.9483 \text{ Btu/sec}$$

$$1 \text{ Btu/sec} = 778.26 \text{ ft-lbf/sec}$$
$$= 46{,}680 \text{ ft-lbf/min}$$
$$= 1.415 \text{ hp}$$

Example 13.8

When traveling at 100 km/h, a car supplies a constant horizontal force of 50 N to the hitch of a trailer. What tractive power (in horsepower) is required for the trailer alone?

Solution

From Eq. 13.26, the power being generated is

$$P = F\text{v} = \frac{(50 \text{ N})\left(100 \text{ } \frac{\text{km}}{\text{h}}\right)\left(1000 \text{ } \frac{\text{m}}{\text{km}}\right)}{\left(60 \text{ } \frac{\text{s}}{\text{min}}\right)\left(60 \text{ } \frac{\text{min}}{\text{h}}\right)\left(1000 \text{ } \frac{\text{W}}{\text{kW}}\right)}$$
$$= 1.389 \text{ kW}$$

Using a conversion from Table 13.2, the horsepower is

$$P = \left(1.341 \text{ } \frac{\text{hp}}{\text{kW}}\right)(1.389 \text{ kW})$$
$$= 1.86 \text{ hp}$$

12. EFFICIENCY

For energy-using systems (such as cars, electrical motors, elevators, etc.), the *energy-use efficiency*, η, of a system is the ratio of an ideal property to an actual property. The property used is commonly work, power, or, for thermodynamic problems, heat. When the rate of work is constant, either work or power can be used to calculate the efficiency. Otherwise, power should be used. Except in rare instances, the numerator and denominator of the ratio must have the same units.[5]

$$\eta = \frac{P_{\text{ideal}}}{P_{\text{actual}}} \qquad [P_{\text{actual}} \geq P_{\text{ideal}}] \qquad \textbf{13.29}$$

For energy-producing systems (such as electrical generators, prime movers, and hydroelectric plants), the *energy-production efficiency* is

$$\eta = \frac{P_{\text{actual}}}{P_{\text{ideal}}} \qquad [P_{\text{ideal}} \geq P_{\text{actual}}] \qquad \textbf{13.30}$$

The efficiency of an *ideal machine* is 1.0 (100%). However, all *real machines* have efficiencies of less than 1.0.

---

[5]The *energy-efficiency ratio* used to evaluate refrigerators, air conditioners, and heat pumps, for example, has units of Btu per watt-hour (Btu/W-hr).

Topic II: Water Resources

14 Fluid Properties

Nomenclature

a	speed of sound	ft/sec	m/s
A	area	ft^2	m^2
d	diameter	ft	m
E	bulk modulus	lbf/ft^2	Pa
F	force	lbf	N
g	gravitational acceleration	ft/sec^2	m/s^2
g_c	gravitational conversion constant	lbm-ft/lbf-sec^2	n.a.
h	height	ft	m
k	ratio of specific heats	–	–
L	length	ft	m
M	Mach number	–	–
M	molar concentration	lbmol/ft^3	kmol/m^3
MW	molecular weight	lbm/lbmol	kg/kmol
n	number of moles	–	–
p	pressure	lbf/ft^2	Pa
r	radius	ft	m
R	specific gas constant	ft-lbf/lbm-°R	J/kg·K
R^*	universal gas constant	ft-lbf/lbm-°R	J/kmol·K
SG	specific gravity	–	–
T	absolute temperature	°R	K
v	velocity	ft/sec	m/s
V	volume	ft^3	m^3
VI	viscosity index	various	various
x	mole fraction	–	–
y	distance	ft	m
Z	compressibility factor	–	–

Symbols

β	compressibility	ft^2/lbf	Pa^{-1}
β	contact angle	deg	deg
γ	specific weight	lbf/ft^3	n.a.
μ	absolute viscosity	lbf-sec/ft^2	Pa·s
ν	kinematic viscosity	ft^2/sec	m^2/s
π	osmotic pressure	lbf/ft^2	Pa
ρ	density	lbm/ft^3	kg/m^3
σ	surface tension	lbf/ft	N/m
τ	shear stress	lbf/ft^2	Pa
υ	specific volume	ft^3/lbm	m^3/kg

Subscripts

c	critical
o	original
p	constant pressure
T	constant temperature

1. CHARACTERISTICS OF A FLUID

Liquids and gases can both be categorized as fluids, although this chapter is primarily concerned with incompressible liquids. There are certain characteristics shared by all fluids, and these characteristics can be used, if necessary, to distinguish between liquids and gases.[1]

- *Compressibility*: Liquids are only slightly compressible and are assumed to be incompressible for most purposes. Gases are highly compressible.

- *Shear resistance*: Liquids and gases cannot support shear, and they deform continuously to minimize applied shear forces.

- *Shape and volume*: As a consequence of their inability to support shear forces, liquids and gases take on the shapes of their containers. Only liquids have free surfaces. Liquids have fixed volumes, regardless of their container volumes, and these volumes are not significantly affected by temperature and pressure. Unlike liquids, gases take

---

[1]The differences between liquids and gases become smaller as temperature and pressure are increased. Gas and liquid properties become the same at the critical temperature and pressure.

on the volumes of their containers. If allowed to do so, gas densities will change as temperature and pressure are varied.

- *Resistance to motion*: Due to viscosity, liquids resist instantaneous changes in velocity, but the resistance stops when liquid motion stops. Gases have very low viscosities.

- *Molecular spacing*: Molecules in liquids are relatively close together and are held together with strong forces of attraction. Liquid molecules have low kinetic energy. The distance each liquid molecule travels between collisions is small. In gases, the molecules are relatively far apart and the attractive forces are weak. Kinetic energy of the molecules is high. Gas molecules travel larger distances between collisions.

- *Pressure*: The pressure at a point in a fluid is the same in all directions. Pressure exerted by a fluid on a solid surface (e.g., container wall) is always normal to that surface.

2. TYPES OF FLUIDS

For computational convenience, fluids are generally divided into two categories: ideal fluids and real fluids. *Ideal fluids* are assumed to have no viscosity (and hence, no resistance to shear), be incompressible, and have uniform velocity distributions when flowing. In an ideal fluid, there is no friction between moving layers of fluid, and there are no eddy currents or turbulence.

Real fluids exhibit finite viscosities and non-uniform velocity distributions, are compressible, and experience friction and turbulence in flow. Real fluids are further divided into *Newtonian fluids* and *non-Newtonian fluids*, depending on their viscous behavior. The differences between Newtonian and the various types of non-Newtonian fluids are described in Sec. 9.

For convenience, most fluid problems assume real fluids with Newtonian characteristics. This is an appropriate assumption for water, air, gases, steam, and other simple fluids (alcohol, gasoline, acid solutions, etc.). However, slurries, pastes, gels, suspensions, and polymer/electrolyte solutions may not behave according to simple fluid relationships.

Figure 14.1 *Types of Fluids*

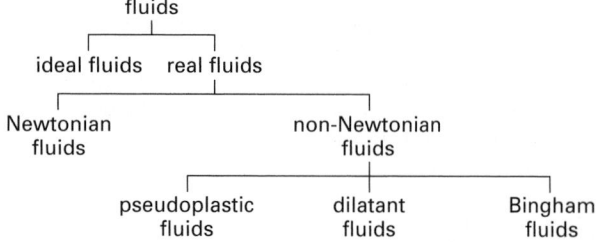

3. FLUID PRESSURE AND VACUUM

In the English system, fluid pressure is measured in pounds per square inch (lbf/in^2 or psi) and pounds per square foot (lbf/ft^2 or psf), although tons (2000 pounds) per square foot (tsf) are occasionally used. In SI units, pressure is measured in pascals (Pa). Because a pascal is very small, kilopascals (kPa) are usually used. Other units of pressure are bars; millibars; atmospheres; inches and feet of water; millimeters, centimeters, and inches of mercury; and torrs.

Figure 14.2 *Relative Sizes of Pressure Units*

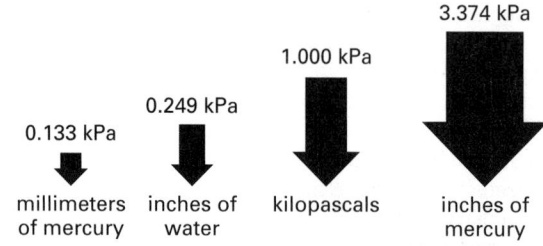

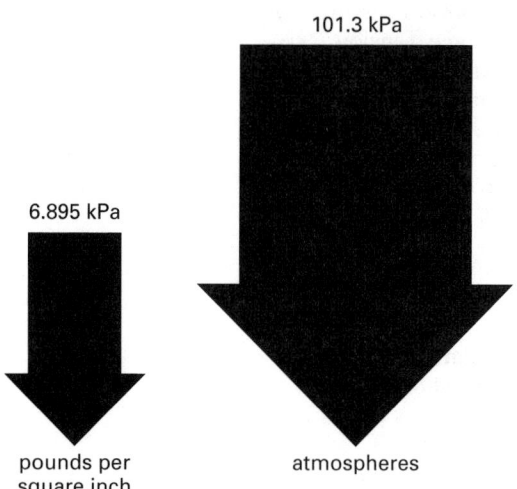

Fluid pressures are measured with respect to two pressure references: zero pressure and atmospheric pressure. Pressures measured with respect to a true zero pressure reference are known as *absolute pressures*. Pressures measured with respect to atmospheric pressure are known as *gage pressures*.[2] Most pressure gauges read the excess of the test pressure over atmospheric pressure (i.e., the gage pressure). To distinguish between these two pressure measurements, the letters "a" and "g" are traditionally added to the the unit symbols in the English unit system (e.g., 14.7 psia and 4015 psfg). For SI units, the actual words "gauge" and "absolute" can be added to the measurement (e.g., 25.1 kPa absolute). Alternatively, the pressure is assumed to be absolute unless the "g" is used (e.g., 15 Pag).

---

[2]The spelling *gage* persists even though pressures are measured with *gauges*. In some countries, the term *meter pressure* is used instead of gage pressure.

Absolute and gage pressures are related by Eq. 14.1. It should be mentioned that $p_{atmospheric}$ in Eq. 14.1 is the actual atmospheric pressure existing when the gage measurement is taken. It is not standard atmospheric pressure, unless that pressure is implicitly or explicitly applicable. Also, since a barometer measures atmospheric pressure, *barometric pressure* is synonymous with atmospheric pressure. Table 14.1 lists standard atmospheric pressure in various units.

$$p_{absolute} = p_{gage} + p_{atmospheric} \qquad 14.1$$

Table 14.1 Standard Atmospheric Pressure

1.000 atm	(atmosphere)
14.696 psia	(pounds per square inch absolute)
2116.2 psfa	(pounds per square foot absolute)
407.1 in wg	(inches of water, inches water gage)
33.93 ft wg	(feet of water, feet water gage)
29.921 in Hg	(inches of mercury)
760.0 mm Hg	(millimeters of mercury)
760.0 torr	
1.013 bars	
1013 millibars	
1.013×10^5 Pa	(pascals)
101.3 kPa	(kilopascals)

A *vacuum* measurement is implicitly a pressure below atmospheric (i.e., a negative gage pressure). It must be assumed that any measured quantity given as a vacuum is a quantity to be subtracted from the atmospheric pressure. Thus, when a condenser is operating with a vacuum of 4.0 in Hg (4 in of mercury), the absolute pressure is approximately 29.92 in Hg − 4.0 in Hg = 25.92 in Hg. Vacuums are generally stated as positive numbers.

$$p_{absolute} = p_{atmospheric} - p_{vacuum} \qquad 14.2$$

A difference in two pressures may be reported with units of psid (i.e., a differential in psi).

4. DENSITY

The *density*, ρ, of a fluid is its mass per unit volume.[3] In SI units, density is measured in kg/m^3. In a consistent English system, density is measured in $slugs/ft^3$, even though fluid density is traditionally reported in lbm/ft^3.

$$\rho = \text{fluid density} \qquad 14.3$$

The density of a fluid in a liquid form is usually given, known in advance, or easily obtained from tables in any one of a number of sources. Most English fluid data are reported on a per pound basis, and the data included in

---

[3]Mass is an absolute property of a substance. Weight is not absolute, since it depends on the local gravity. Some fluids books continue to use γ as the symbol for weight density. The equations using γ that result (such as Bernoulli's equation) cannot be used with SI data, since the equations are not consistent. Thus, engineers end up with two different equations for the same thing.

this book follow that tradition. To make the conversion from pounds to slugs, divide by g_c as an implied step whenever using pound-basis fluid data.

$$\rho_{slugs} = \frac{\rho_{lbm}}{g_c} \qquad 14.4$$

The density of an ideal gas can be found from the specific gas constant and the ideal gas law.

$$\rho = \frac{p}{RT} \qquad 14.5$$

Table 14.2 Approximate Room-Temperature Densities of Common Fluids

fluid	lbm/ft^3	kg/m^3
air (STP)	0.0807	1.29
air (70°F, 1 atm)	0.075	1.20
alcohol	49.3	790
ammonia	38	602
gasoline	44.9	720
glycerin	78.8	1260
mercury	848	13 600
water	62.4	1000

(Multiply lbm/ft^3 by 16.01 to obtain kg/m^3.)

Example 14.1

The density of water is typically taken as 62.4 lbm/ft^3 for engineering problems where greater accuracy is not required. What is the value in (a) $slugs/ft^3$ and (b) kg/m^3?

Solution

(a) Equation 14.4 can be used to calculate the slug-density of water.

$$\rho = \frac{\rho_{lbm}}{g_c} = \frac{62.4 \, \frac{lbm}{ft^3}}{32.2 \, \frac{lbm\text{-}ft}{lbf\text{-}sec^2}} = 1.94 \, lbf\text{-}sec^2/ft\text{-}ft^3$$

$$= 1.94 \, slugs/ft^3$$

(b) The conversion between lbm/ft^3 and kg/m^3 is approximately 16.0, derived as follows.

$$\rho = \left(62.4 \, \frac{lbm}{ft^3}\right) \left(\frac{35.31 \, \frac{ft^3}{m^3}}{2.205 \, \frac{lbm}{kg}} \right)$$

$$= \left(62.4 \, \frac{lbm}{ft^3}\right) \left(16.01 \, \frac{kg\text{-}ft^3}{m^3\text{-}lbm}\right) = 999 \, kg/m^3$$

In SI problems, it is common to take the density of water as 1000 kg/m^3.

5. SPECIFIC VOLUME

Specific volume, v, is the volume occupied by a unit mass of fluid.[4] Since specific volume is the reciprocal of density, typical units will be ft^3/lbm, $ft^3/lbmole$, or m^3/kg.[5]

$$v = \frac{1}{\rho} \qquad 14.6$$

6. SPECIFIC GRAVITY

Specific gravity (SG) is a dimensionless ratio of a fluid's density to some standard reference density.[6] For liquids and solids, the reference is the density of pure water. There is some variation in this reference density, however, since the temperature at which the water density is evaluated is not standardized. Temperatures of $39.2°F$ ($4°C$), $60°F$, and $70°F$ have been reported.[7]

Fortunately, the density of water is the same to three significant digits over the normal ambient temperature range: 62.4 lbm/ft^3 or 1000 kg/m^3. However, to be precise, the temperature of both the fluid and water should be specified (e.g., "... the specific gravity of the $20°C$ fluid is 1.05 referred to $4°C$ water ...").

$$SG_{liquid} = \frac{\rho_{liquid}}{\rho_{water}} \qquad 14.7$$

Since the SI density of water is very nearly 1.000 g/cm^3 (1000 kg/m^3), the numerical values of density in g/cm^3 and specific gravity are the same. Such is not the case with English units.

The standard reference used to calculate the specific gravity of gases is the density of air. Since the density of a gas depends on temperature and pressure, both must be specified for the gas and air (i.e., two temperatures and two pressures must be specified). While STP (standard temperature and pressure) conditions are commonly specified, they are not universal.[8] Table 14.3 lists several common sets of standard conditions.

$$SG_{gas} = \frac{\rho_{gas}}{\rho_{air}} \qquad 14.8$$

Table 14.3 *Commonly Quoted Values of Standard Temperature and Pressure*

system	temperature	pressure
SI	273.15K	101.325 kPa
scientific	0.0°C	760 mm Hg
U.S. engineering	32°F	14.696 psia
natural gas industry (U.S.)	60°F	14.65, 14.73, or 15.025 psia
natural gas industry (Canada)	60°F	14.696 psia

If it is known or implied that the temperature and pressure of the air and gas are the same, the specific gravity of the gas will be equal to the ratio of molecular weights and the inverse ratio of specific gas constants. The density of air evaluated at STP is listed in Table 14.2. At $70°F$ ($21.1°C$) and 1.0 atm, the density is approximately 0.075 lbm/ft^3 (1.20 kg/m^3).

$$SG_{gas} = \frac{MW_{gas}}{MW_{air}} = \frac{MW_{gas}}{29.0}$$

$$= \frac{R_{air}}{R_{gas}} = \frac{53.3 \ \frac{ft\text{-}lbf}{lbm\text{-}°R}}{R_{gas}} \qquad 14.9$$

Specific gravities of petroleum liquids and aqueous solutions (of acid, antifreeze, salts, etc.) can be determined by use of a *hydrometer*. In its simplest form, a hydrometer is constructed as a graduated scale weighted at one end so it will float vertically. The height at which the hydrometer floats depends on the density of the fluid, and the graduated scale can be calibrated directly in specific gravity.[9]

Figure 14.3 *Hydrometer*

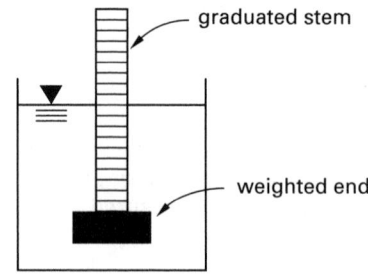

There are two standardized hydrometer scales (i.e., methods for calibrating the hydrometer stem).[10] Both state specific gravity in degrees, although temperature is not being measured. The *American Petroleum Institute* (API) scale (°API) may be used with all liquids,

---

[4]Care must be taken to distinguish between the symbol upsilon, v, used for specific volume, and italic roman "vee," v, used for velocity in many engineering textbooks.

[5]Units of $ft^3/slug$ are also possible, but this combination of units is almost never encountered.

[6]The symbols S.G., sp.gr., S, and G are also used. In fact, petroleum engineers in the United States use γ, a symbol that civil engineers use for specific weight. There is no standard engineering symbol for specific gravity.

[7]Density of liquids is sufficiently independent of pressure to make consideration of pressure in specific gravity calculations unnecessary.

[8]The abbreviation "SC" (standard conditions) is interchangeable with "STP."

[9]This is a direct result of the buoyancy principle of Archimedes.

[10]In addition to °Be and °API mentioned in this chapter, the *Twaddell scale* (°Tw) is used in chemical processing, the *Brix* and *Balling's scales* are used in the sugar industry, and the *Salometer scale* is used to measure salt (NaCl and $CaCl_2$) solutions.

not only with oils or other hydrocarbons. For the specific gravity value, a standard reference temperature of 60°F (15.6°C) is implied for both the liquid and the water.

$$°API = \frac{141.5}{SG} - 131.5 \qquad \textit{14.10}$$

$$SG = \frac{141.5}{°API + 131.5} \qquad \textit{14.11}$$

The *Baumé scale* (°Be) has been used in the past. It is somewhat confusing because there are actually two Baumé scales—one for liquids heavier than water and another for liquids lighter than water. (There is also a discontinuity in the scales at SG = 1.00.) As with the API scale, the specific gravity value assumes 60°F is the standard temperature for both scales.

$$SG = \frac{140.0}{130.0 + °Be} \qquad [SG \leq 1.00] \qquad \textit{14.12}$$

$$SG = \frac{145.0}{145.0 - °Be} \qquad [SG \geq 1.00] \qquad \textit{14.13}$$

Example 14.2

Determine the specific gravity of carbon dioxide gas (molecular weight = 44) at 66°C (150°F) and 138 kPa (20 psia) using STP air as a reference.

SI Solution

$$R^* = 8314 \text{ J/kmol·K}$$

$$R = \frac{R^*}{MW} = \frac{8314 \dfrac{\text{J}}{\text{kmol·K}}}{44 \dfrac{\text{kg}}{\text{kmol}}} = 189.0 \text{ J/kg·K}$$

$$\rho = \frac{p}{RT} = \frac{1.38 \times 10^5 \text{ Pa}}{\left(189.0 \dfrac{\text{J}}{\text{kg·K}}\right)(66°C + 273)}$$

$$= 2.15 \text{ kg/m}^3$$

$$SG = \frac{2.15 \dfrac{\text{kg}}{\text{m}^3}}{1.29 \dfrac{\text{kg}}{\text{m}^3}} = 1.67$$

Customary U.S. Solution

Since the conditions of the carbon dioxide and air are different, Eq. 14.9 cannot be used. It is necessary to calculate the density of the carbon dioxide from Eq. 14.5. The specific gas constant of carbon dioxide is 35.1 ft-lbf/lbm-°R (App. 24.B). The density is

$$\rho = \frac{p}{RT} = \frac{\left(20 \dfrac{\text{lbf}}{\text{in}^2}\right)\left(144 \dfrac{\text{in}^2}{\text{ft}^2}\right)}{\left(35.1 \dfrac{\text{ft-lbf}}{\text{lbm-°R}}\right)(150°F + 460)}$$

$$= 0.135 \text{ lbm/ft}^3$$

From Table 14.2, the density of STP air is 0.0807 lbm/ft^3. From Eq. 14.8, the specific gravity of carbon dioxide at the conditions given is

$$SG = \frac{\rho_{gas}}{\rho_{air}} = \frac{0.135 \dfrac{\text{lbm}}{\text{ft}^3}}{0.0807 \dfrac{\text{lbm}}{\text{ft}^3}} = 1.67$$

7. SPECIFIC WEIGHT

Specific weight, γ, is the weight of fluid per unit volume. The use of specific weight is most often encountered in civil engineering works from the United States, where it is commonly called "density." Mechanical and chemical engineers seldom encounter the term. The usual units of specific weight are lbf/ft^3.[11] Specific weight is not an absolute property of a fluid, since it depends not only on the fluid but on the local gravitational field as well.

$$\gamma = g\rho \qquad [SI] \qquad \textit{14.14(a)}$$

$$\gamma = \rho \times \frac{g}{g_c} \qquad [U.S.] \qquad \textit{14.14(b)}$$

If the gravitational acceleration is 32.2 ft/sec^2, as it is almost everywhere on the earth, the specific weight in lbf/ft^3 will be numerically equal to the density in lbm/ft^3. This is illustrated in Ex. 14.3.

Example 14.3

What is the sea level ($g = 32.2$ ft/sec^2) specific weight (in lbf/ft^3) of liquids with densities of (a) 1.95 slug/ft^3, and (b) 58.3 lbm/ft^3?

Solution

(a) Equation 14.14(a) can be used with any consistent set of units, including densities involving slugs.

$$\gamma = g\rho = \left(32.2 \dfrac{\text{ft}}{\text{sec}^2}\right)\left(1.95 \dfrac{\text{slug}}{\text{ft}^3}\right)$$

$$= \left(32.2 \dfrac{\text{ft}}{\text{sec}^2}\right)\left(1.95 \dfrac{\text{lbf-sec}^2}{\text{ft-ft}^3}\right) = 62.8 \text{ lbf/ft}^3$$

(b) From Eq. 14.14(b),

$$\gamma = \rho \times \frac{g}{g_c}$$

$$= \left(58.3 \dfrac{\text{lbm}}{\text{ft}^3}\right)\left(\dfrac{32.2 \dfrac{\text{ft}}{\text{sec}^2}}{32.2 \dfrac{\text{lbm-ft}}{\text{lbf-sec}^2}}\right) = 58.3 \text{ lbf/ft}^3$$

8. MOLE FRACTION

Mole fraction is an important parameter in many practical engineering problems, particularly in chemistry

---

[11]Notice that the units are lbf/ft^3, not lbm/ft^3. Pound-mass (lbm) is a mass unit, not a weight (force) unit.

and chemical engineering. The composition of a fluid consisting of two or more distinctly different substances, A, B, C, and so on, can be described by the mole fractions, x_A, x_B, x_C, and so on, of each substance.[12] The mole fraction of component A is the number of moles of that component, n_A, divided by the total number of moles in the combined fluid.

$$x_A = \frac{n_A}{n_A + n_B + n_C + \cdots} \qquad 14.15$$

Mole fraction is a number between 0 and 1.000. *Mole percent* is the mole fraction multiplied by 100, expressed in percent.

9. VISCOSITY

The *viscosity* of a fluid is a measure of that fluid's resistance to flow when acted upon by an external force such as a pressure differential or gravity. Some fluids, such as heavy oils, jellies, and syrups, are very viscous. Other fluids, such as water, lighter hydrocarbons, and gases, are not as viscous.

Most viscous liquids will flow more easily when their temperatures are raised. However, the behavior of a fluid when temperature, pressure, or stress is varied will depend on the type of fluid. The different types of fluids can be determined with a *sliding plate viscometer* test.[13]

---

[12]There are other methods of specifying the composition. The subjects of moles, mole fraction, gravimetric fraction, and volumetric fraction are covered in Ch. 22.

[13]This test is conceptually simple but is not always practical, since the liquid leaks out between the plates. In research work with liquids, it is common to determine viscosity with a *concentric cylinder viscometer*, also known as a *cup-and-bob viscometer*. Viscosities of perfect gases can be predicted by the kinetic theory of gases.

Viscosity can also be measured by a *Saybolt viscometer*, which is essentially a container that allows a given quantity of fluid to leak out through one of two different-sized orifices. The more viscous the fluid, the more time will be required for the fluid to leak out. *Saybolt Seconds Universal* (SSU) and *Saybolt Seconds Furol* (SSF) are scales of such viscosity measurement based on the smaller and larger orifices, respectively. Seconds can be converted (empirically) to viscosity in other units.

The following relations are approximate conversions between SSU, stokes, and poise.

SSU < 100 sec:
$$\nu_{\text{stokes}} = (0.00226)(\text{SSU}) - \frac{1.95}{\text{SSU}}$$

$$\mu_{\text{poise}} = (\text{SG})(\nu_{\text{stokes}})$$

SSU > 100 sec:
$$\nu_{\text{stokes}} = (0.00220)(\text{SSU}) - \frac{1.35}{\text{SSU}}$$

$$\mu_{\text{poise}} = (\text{SG})(\nu_{\text{stokes}})$$

Consider two plates of area A separated by a fluid with thickness y_0, as shown in Fig. 14.4. The bottom plate is fixed, and the top plate is kept in motion at a constant velocity, v_0, by a force, F.

Experiments with Newtonian fluids have shown that the force, F, required to maintain the velocity, v_0, is proportional to the velocity and the area and is inversely proportional to the separation of the plates. That is,

$$\frac{F}{A} \propto \frac{dv}{dy} \qquad 14.16$$

Figure 14.4 *Sliding Plate Viscometer*

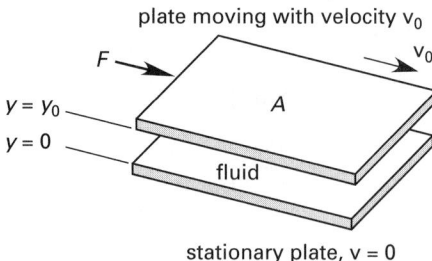

The constant of proportionality needed to make Eq. 14.16 an equality is the *absolute viscosity*, μ, also known as the *coefficient of viscosity*.[14] The reciprocal of absolute viscosity, $1/\mu$, is known as the *fluidity*.

$$\frac{F}{A} = \mu \frac{dv}{dy} \qquad 14.17$$

F/A is the *fluid shear stress*, τ. The quantity dv/dy (v_0/y_0) is known by various names, including *rate of strain, shear rate, velocity gradient*, and *rate of shear formation*. Equation 14.17 is known as *Newton's law of viscosity*, from which Newtonian fluids get their name.[15] Equation 14.18 is simply the equation of a straight line.

$$\tau = \mu \frac{dv}{dy} \qquad 14.18$$

Not all fluids are Newtonian (although most common fluids are), and Eq. 14.17 is not universally applicable. Figure 14.5 illustrates how differences in fluid shear stress behavior (at constant temperature and pressure) can be used to define Bingham, pseudoplastic, and dilatant fluids, as well as Newtonian fluids.

Gases, water, alcohol, and benzene are examples of *Newtonian fluids*. In fact, all liquids with a simple chemical formula are Newtonian. Also, most solutions of simple compounds, such as sugar and salt, are Newtonian. For

---

[14]Another name for absolute viscosity is *dynamic viscosity*. The name *absolute viscosity* is preferred, if for no other reason than to avoid confusion with *kinematic viscosity*.

[15]Sometimes Eq. 14.18 is written with a minus sign to compare viscous behavior with other behavior. However, the direction of positive shear stress is arbitrary.

a highly viscous fluid, the straight line (see Fig. 14.5) will be closer to the τ axis (i.e., the slope will be higher). For low-viscosity fluids, the straight line will be closer to the dv/dy axis (i.e., the slope will be lower).

Pseudoplastic fluids (muds, motor oils, polymer solutions, natural gums, and most slurries) exhibit viscosities that decrease with an increasing velocity gradient. Such fluids present no serious pumping problems.

Bingham fluids (Bingham plastics), typified by toothpaste, jellies, bread dough, and some slurries, are capable of indefinitely resisting a small shear stress but move easily when the stress becomes large—that is, Bingham fluids become pseudoplastic when the stress increases.

Dilatant fluids are rare but include clay slurries, various starches, some paints, milk chocolate with nuts, and other candy compounds. They exhibit viscosities that increase with increasing agitation (i.e., with increasing velocity gradients), but they return rapidly to their normal viscosity after the agitation ceases. Pump selection is critical for dilatant fluids because these fluids can become almost solid if the shear rate is high enough.

Plastic materials, such as tomato catsup, behave similarly to pseudoplastic fluids once movement begins—their viscosities decrease with agitation. However, a finite force must be applied before any fluid movement occurs.

Figure 14.5 *Shear Stress Behavior for Different Types of Fluids*

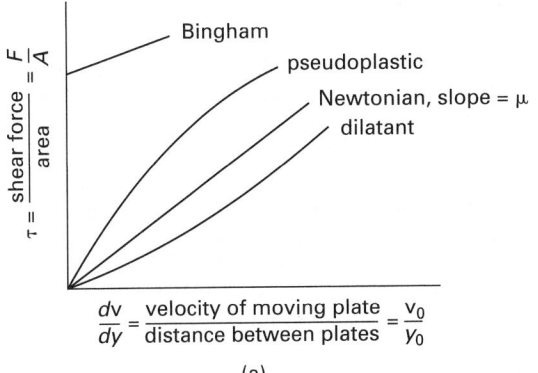

(a)

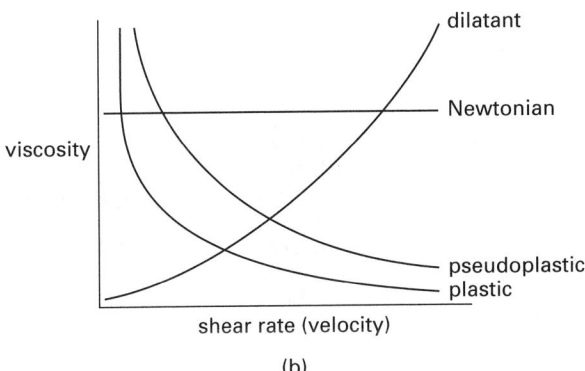

(b)

Viscosity can also change with time (all other conditions being constant). If viscosity decreases with time, the fluid is said to be a *thixotropic fluid*. If viscosity increases (usually up to a finite value) with time, the fluid is a *rheopectic fluid*. Viscosity does not change in time-independent fluids. *Colloidal materials*, such as gelatinous compounds, lotions, shampoos, and low-temperature solutions of soaps in water and oil, behave like *thixotropic liquids*—their viscosities decrease as the rate of shear is increased. However, viscosity does not return to its original state after the agitation ceases.

Molecular cohesion is the dominating cause of viscosity in liquids. As the temperature of a liquid increases, these cohesive forces decrease, resulting in a decrease in viscosity.

In gases, the dominant cause of viscosity is random collisions between gas molecules. This molecular agitation increases with increases in temperature. Therefore, viscosity in gases increases with temperature.

Although viscosity of liquids increases slightly with pressure, the increase is insignificant over moderate pressure ranges. Therefore, the absolute viscosity of both gases and liquids is usually considered to be essentially independent of pressure.[16]

The units of absolute viscosity, as derived from Eq. 14.18, are lbf-sec/ft². Such units are actually used in the English engineering system.[17] Another common unit used throughout the world is the *poise* (abbreviated P), equal to a dyne·s/cm². These dimensions are the same primary dimensions (see Ch. 1) as in the English system, $F\theta/L^2$ or $M/L\theta$, and are functionally the same as a g/cm·s. Since the poise is a large unit, the *centipoise* (abbreviated cP) scale is generally used. By coincidence, the viscosity of pure water at room temperature is approximately 1 cP.

Absolute viscosity is measured in pascal-seconds (Pa·s) in SI units.

Example 14.4

A liquid ($\mu = 5.2 \times 10^{-5}$ lbf-sec/ft²) is flowing in a rectangular duct. The equation of the symmetrical velocity (in ft/sec) is approximately $v = 3y^{0.7}$ ft/sec, where y is in inches. (a) What is the velocity gradient at $y = 3.0$ in from the duct wall? (b) What is the shear stress in the fluid at that point?

Solution

(a) The velocity is not a linear function of y, so dv/dy must be calculated as a derivative.

$$\frac{dv}{dy} = \frac{d}{dy}\left(3y^{0.7}\right)$$

$$= (3)\left(0.7y^{-0.3}\right) = 2.1y^{-0.3}$$

---

[16]This is not true for kinematic viscosity, however.

[17]Units of lbm/ft-sec are also used for absolute viscosity in the English system, although it is difficult to see how such units are derived from the sliding-plate test. These units are obtained by dividing lbf-sec/ft² units by g_c.

At $y = 3$ in,

$$\frac{dv}{dy} = (2.1)(3)^{-0.3} = 1.51 \text{ ft/sec-in}$$

(b) From Eq. 14.18, the shear stress is

$$\tau = \mu \frac{dv}{dy}$$

$$= \left(5.2 \times 10^{-5} \, \frac{\text{lbf-sec}}{\text{ft}^2}\right)\left(1.51 \, \frac{\text{ft}}{\text{sec-in}}\right)\left(12 \, \frac{\text{in}}{\text{ft}}\right)$$

$$= 9.42 \times 10^{-4} \text{ lbf/ft}^2$$

10. KINEMATIC VISCOSITY

Another quantity with the name *viscosity* is the ratio of absolute viscosity to mass density. This combination of variables, known as *kinematic viscosity, ν*, appears sufficiently often in fluids and other problems as to warrant its own symbol and name. Thus, kinematic viscosity is the name given to a frequently occurring combination of variables.

$$\nu = \frac{\mu}{\rho} \qquad \text{[SI]} \qquad \textit{14.19(a)}$$

$$\nu = \frac{\mu g_c}{\rho} \qquad \text{[U.S.]} \qquad \textit{14.19(b)}$$

The primary dimensions (see Ch. 1) of kinematic viscosity are L^2/θ. Typical units are ft^2/sec and cm^2/s (the *stoke*, St). It is also common to give kinematic viscosity in *centistokes*, cSt. The SI units of kinematic viscosity are m^2/s.

It is essential that consistent units be used with Eq. 14.19(a). The following sets of units are consistent:

$$\text{ft}^2/\text{sec} = \frac{\text{lbf-sec/ft}^2}{\text{slugs/ft}^3}$$

$$\text{m}^2/\text{s} = \frac{\text{Pa·s}}{\text{kg/m}^3}$$

$$\text{St (stoke)} = \frac{\text{P (poise)}}{\text{g/cm}^3}$$

$$\text{cSt (centistokes)} = \frac{\text{cP (centipoise)}}{\text{g/cm}^3}$$

Unlike absolute viscosity, kinematic viscosity is greatly dependent on both temperature and pressure, since these variables affect the density of the fluid. Referring to Eq. 14.19, even if absolute viscosity is independent of temperature or pressure, the change in density will change the kinematic viscosity.

11. VISCOSITY CONVERSIONS

The most common units of absolute and kinematic viscosity are listed in Table 14.4.

Table 14.4 *Common Viscosity Units*

	absolute (μ)	kinematic (ν)
English	lbf-sec/ft^2 (slug/ft-sec)	ft^2/sec
conventional metric	dyne·s/cm^2 (poise)	cm^2/s (stoke)
SI	Pa·s (N·s/m^2)	m^2/s

Table 14.5 contains conversions between the various viscosity units.

Table 14.5 *Viscosity Conversions*[a]

multiply	by	to obtain
absolute viscosity, μ		
dyne·s/cm^2	0.10	Pa·s
lbf-sec/ft^2	478.8	P
lbf-sec/ft^2	47,880	cP
lbf-sec/ft^2	47.88	Pa·s
slug/ft-sec	47.88	Pa·s
lbm/ft-sec	1.488	Pa·s
cP	1.0197×10^{-4}	kgf·s/m^2
cP	2.0885×10^{-5}	lbf-sec/ft^2
cP	0.001	Pa·s
Pa·s	0.020885	lbf-sec/ft^2
Pa·s	1000	cP
kinematic viscosity, ν		
ft^2/sec	92,903	cSt
ft^2/sec	0.092903	m^2/s
m^2/s	10.7639	ft^2/sec
m^2/s	1×10^6	cSt
cSt	1×10^{-6}	m^2/sec
cSt	1.0764×10^{-5}	ft^2/sec
absolute viscosity to kinematic viscosity		
cP	$1/\rho$ in g/cm^3	cSt
cP	$6.7195 \times 10^{-4}/\rho$ in lbm/ft^3	ft^2/sec
lbf-sec/ft^2	$32.174/\rho$ in lbm/ft^3	ft^2/sec
kgf·s/m^2	$9.807/\rho$ in kg/m^3	m^2/s
Pa·s	$1000/\rho$ in g/cm^3	cSt
kinematic viscosity to absolute viscosity		
cSt	ρ in g/cm^3	cP
cSt	$1.6 \times 10^{-5} \times \rho$ in g/cm^3	Pa·s
m^2/s	$0.10197 \times \rho$ in kg/m^3	kgf·s/m^2
m^2/s	$1000 \times \rho$ in g/cm^3	Pa·s
ft^2/sec	$0.031081 \times \rho$ in lbm/ft^3	lbf-sec/ft^2
ft^2/sec	$1488.2 \times \rho$ in lbm/ft^3	cP

[a]cP: centipoise; cSt: centistoke; P: poise

Example 14.5

Water at 60°F has a specific gravity of 0.999 and a kinematic viscosity of 1.12 cSt. What is the absolute viscosity in lbf-sec/ft²?

Solution

The density of a liquid expressed in g/cm³ is numerically equal to its specific gravity.

$$\rho = 0.999 \text{ g/cm}^3$$

The centistoke (cSt) is a measure of kinematic viscosity. Kinematic viscosity is converted first to the absolute viscosity units of centipoise. From Table 14.5,

$$\mu_{cP} = \nu_{cSt}\rho_{g/cm^3}$$
$$= (1.12)(0.999) = 1.119 \text{ cP}$$

Next, centipoises are converted to lbf-sec/ft².

$$\mu_{lbf\text{-}sec/ft^2} = \mu_{cP}\left(2.0885 \times 10^{-5}\right)$$
$$= (1.119)\left(2.0885 \times 10^{-5}\right)$$
$$= 2.34 \times 10^{-5} \text{ lbf-sec/ft}^2$$

12. VISCOSITY INDEX

Viscosity index (VI) is a measure of a fluid's sensitivity to changes in viscosity with changes in temperature. It has traditionally been applied to crude and refined oils through use of a 100-point scale.[18] The viscosity is measured at two temperatures: 100°F and 210°F (38°C and 99°C). These viscosities are converted into a viscosity index in accordance with standard ASTM D2270.

13. VAPOR PRESSURE

Molecular activity in a liquid will allow some of the molecules to escape the liquid surface. Strictly speaking, a small portion of the liquid vaporizes. Molecules of the vapor also condense back into the liquid. The vaporization and condensation at constant temperature are equilibrium processes. The equilibrium pressure exerted by these free molecules is known as the *vapor pressure* or *saturation pressure*. (Vapor pressure does not include the pressure of other substances in the mixture.)

Some liquids, such as propane, butane, ammonia, and Freon, have significant vapor pressures at normal temperatures. Liquids near their boiling points or that

vaporize easily are said to be *volatile liquids*.[19] Other liquids, such as mercury, have insignificant vapor pressures at the same temperature. Liquids with low vapor pressures are used in accurate barometers.

The tendency toward vaporization is dependent on the temperature of the liquid. *Boiling* occurs when the liquid temperature is increased to the point that the vapor pressure is equal to the local ambient pressure. Thus, a liquid's boiling temperature depends on the local ambient pressure as well as on the liquid's tendency to vaporize.

Vapor pressure is usually considered to be a nonlinear function of temperature only. It is possible to derive correlations between vapor pressure and temperature, and such correlations usually involve a logarithmic transformation of vapor pressure.[20] Vapor pressure can also be graphed against temperature in a (logarithmic) *Cox chart* when values are needed over larger temperature extremes. Although there is also some variation with external pressure, the external pressure effect is negligible under normal conditions.

Typical values of vapor pressure are given in Table 14.6.

Table 14.6 *Typical Vapor Pressures*

fluid	lbf/ft², 68°F	kPa, 20°C
mercury	0.00362	0.000173
turpentine	1.115	0.0534
water	48.9	2.34
ethyl alcohol	122.4	5.86
ether	1231	58.9
butane	4550	218
Freon-12	12,200	584
propane	17,900	855
ammonia	18,550	888

14. OSMOTIC PRESSURE

Osmosis is a special case of diffusion in which molecules of the *solvent* move under pressure from one fluid to another (i.e., from the *solvent* to the *solution*) in one direction only, usually through a *semipermeable membrane*.[21] Osmosis continues until sufficient solvent has passed through the membrane to make the activity (or solvent pressure) of the solution equal to that of the solvent.[22] The pressure at equilibrium is known as the *osmotic pressure*, π.

---

[18]Use of the *viscosity index* has been adopted by other parts of the chemical process industry (CPI), including in the manufacture of solvents, polymers, and other synthetics. The 100-point scale may be exceeded (on both ends) for these uses. Refer to standard ASTM D2270 for calculating extreme values of the viscosity index.

[19]Because a liquid that vaporizes easily has an aroma, the term *aromatic liquid* is also occasionally used.
[20]The *Clausius-Clapeyron equation* and *Antoine equation* are two such logarithmic correlations of vapor pressure with temperature.
[21]A semipermeable membrane will be impermeable to the solute but permeable for the solvent.
[22]Two solutions in equilibrium (i.e., whose activities are equal) are said to be in *isopiestic equilibrium*.

Figure 14.6 illustrates an *osmotic pressure apparatus*. The fluid column can be interpreted as the result of an osmotic pressure that has developed through diffusion into the solution. The fluid column will continue to increase in height until equilibrium is reached. Alternatively, the fluid column can be adjusted so that the solution pressure just equals the osmotic pressure that would develop otherwise, in order to prevent the flow of solvent. For the arrangement in Fig. 14.6, the osmotic pressure can be calculated from the difference in fluid level heights.

$$\pi = \rho g h \qquad \text{[SI]} \qquad 14.20(a)$$

$$\pi = \frac{\rho g h}{g_c} \qquad \text{[U.S.]} \qquad 14.20(b)$$

Figure 14.6 *Osmotic Pressure Apparatus*

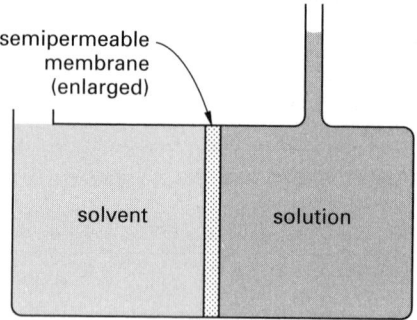

In dilute solutions, osmotic pressure obeys the ideal gas law. The solute acts like a gas in exerting pressure against the membrane. The solvent exerts no pressure since it can pass through. In Eq. 14.21, M is the molarity (concentration). Consistent units must be used.

$$\pi = MR^*T \qquad 14.21$$

Example 14.6

An aqueous solution is in isopiestic equilibrium with a 0.1 molarity sucrose solution at 22°C. What is the osmotic pressure?

Solution

Referring to Eq. 14.21,

$$M = 0.1 \text{mol/L of solution}$$

$$R^* = 0.0821 \text{atm·L/mol·K}$$

$$T = 22°C + 273 = 295K$$

$$\pi = MR^*T = \left(0.1\ \frac{\text{mol}}{\text{L}}\right)\left(0.0821\ \frac{\text{atm·L}}{\text{mol·K}}\right)(295K)$$

$$= 2.42 \text{ atm}$$

15. SURFACE TENSION

The membrane or "skin" that seems to form on the free surface of a fluid is due to the intermolecular cohesive forces and is known as *surface tension*, σ. Surface tension is the reason that insects are able to sit on water and a needle is able to float on it. Surface tension also causes bubbles and droplets to take on a spherical shape, since any other shape would have more surface area per unit volume.

Data on the surface tension of liquids is important in determining the performance of heat-, mass-, and momentum-transfer equipment, including heat transfer devices.[23] Surface tension data is needed to calculate the nucleate boiling point (i.e., the initiation of boiling) of liquids in a pool (using the *Rohsenow equation*) and the maximum heat flux of boiling liquids in a pool (using the *Zuber equation*).

Surface tension can be interpreted as the tension between two points a unit distance apart on the surface or as the amount of work required to form a new unit of surface area in an apparatus similar to that shown in Fig. 14.7. Typical units of surface tension are lbf/ft (ft-lbf/ft^2), dyne/cm, and N/m.

Figure 14.7 *Wire Frame for Stretching a Film*

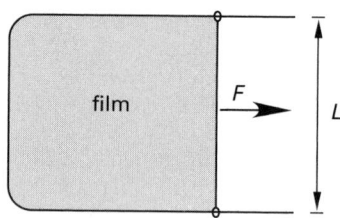

The apparatus shown in Fig. 14.7 consists of a wire frame with a sliding side that has been dipped in a liquid to form a film. Surface tension is determined by measuring the force necessary to keep the sliding side stationary, against the surface tension pull of the film.[24] (The film does not act like a spring, since the force, F, does not increase as the film is stretched.) Since the film has two surfaces (i.e., two surface tensions), the surface tension is

$$\sigma = \frac{F}{2L} \qquad 14.22$$

Alternatively, surface tension can also be determined by measuring the force required to pull a wire ring out of the liquid, as shown in Fig. 14.8.[25] Since the ring's inner and outer sides are in contact with the liquid,

---

[23]Surface tension plays a role in processes involving dispersion, emulsion, flocculation, foaming, and solubilization. It is not surprising that surface tension data are particularly important in determining the performance of equipment in the chemical process industry (CPI), such as distillation columns, packed towers, wetted-wall columns, strippers, and phase-separation equipment.

[24]The force includes the weight of the sliding side wire if the frame is oriented vertically, with gravity acting on the sliding side wire to stretch the film.

[25]This apparatus is known as a *Du Nouy torsion balance*. The ring is made of platinum with a diameter of 4.00 cm.

the wetted perimeter is twice the circumference. The surface tension is

$$\sigma = \frac{F}{4\pi r} \qquad 14.23$$

Figure 14.8 Du Nouy Ring Surface Tension Apparatus

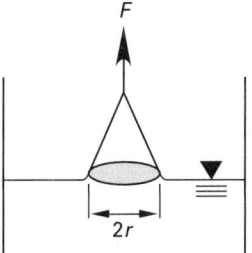

Surface tension depends slightly on the gas in contact with the free surface. Surface tension values are usually quoted for air contact. Typical values of surface tension are listed in Table 14.7.

Table 14.7 Approximate Values of Surface Tension (air contact)

fluid	lbf/ft, 68°F	N/m, 20°C
n-octane	0.00149	0.0217
ethyl alcohol	0.00156	0.0227
acetone	0.00162	0.0236
kerosene	0.00178	0.0260
carbon tetrachloride	0.00185	0.0270
turpentine	0.00186	0.0271
toluene	0.00195	0.0285
benzene	0.00198	0.0289
olive oil	0.0023	0.034
glycerin	0.00432	0.0631
water	0.00499	0.0728
mercury	0.0356	0.519

(Multiply lbf/ft by 14.59 to obtain N/m.)
(Multiply dyne/cm by 0.001 to obtain N/m.)

At temperatures below freezing, the substance will be a solid, so surface tension is a moot point. As the temperature of a liquid is raised, the surface tension decreases because the cohesive forces decrease. Surface tension is zero at a substance's critical temperature. If a substance's critical temperature is known, the *Othmer correlation*, Eq. 14.24, can be used to determine the surface tension at one temperature from the surface tension at another temperature.[26]

$$\sigma_2 = \sigma_1 \left(\frac{T_c - T_2}{T_c - T_1} \right)^{11/9} \qquad 14.24$$

---

[26]An extensive listing of critical temperatures and surface tension data was presented in "633 Organic Chemicals: Surface Tension Data," *Chemical Engineering Magazine*, March 1991, pp. 140–150.

Surface tension is the reason that the pressure on the inside of bubbles and droplets is greater than on the outside. Equation 14.25 gives the relationship between the surface tension in a hollow bubble surrounded by a gas and the difference between the inside and outside pressures. For a spherical droplet or a bubble in a liquid, where in both cases there is only one surface in tension, the surface tension is twice as large. (r is the radius of the bubble or droplet.)

$$\sigma_{\text{bubble}} = \frac{r(p_{\text{inside}} - p_{\text{outside}})}{4} \qquad 14.25$$

$$\sigma_{\text{droplet}} = \frac{r(p_{\text{inside}} - p_{\text{outside}})}{2} \qquad 14.26$$

16. CAPILLARY ACTION

Capillary action (capillarity) is the name given to the behavior of a liquid in a thin-bore tube. Capillary action is caused by surface tension between the liquid and a vertical solid surface.[27] In the case of liquid water in a glass tube, the adhesive forces between the liquid molecules and the surface are greater than (i.e., dominate) the cohesive forces between the water molecules themselves.[28] The adhesive forces cause the water to attach itself to and climb a solid vertical surface. It can be said that the water "reaches up and tries to wet as much of the interior surface as it can." In so doing, the water rises above the general water surface level. This is illustrated in Fig. 14.9.

Figure 14.9 Capillarity of Liquids

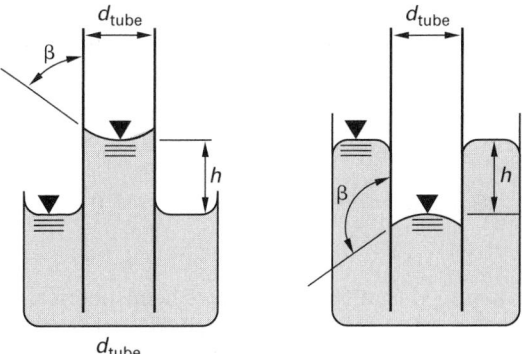

(a) adhesive force dominates   (b) cohesive force dominates

Figure 14.9 also illustrates that the same surface tension forces that keep a droplet spherical are at work on the surface of the liquid in the tube. The curved liquid surface, known as the *meniscus*, can be considered to be an incomplete droplet. If the inside diameter of the tube is less than approximately 0.1 in (2.5 mm), the meniscus is essentially hemispherical, and $r_{\text{meniscus}} = r_{\text{tube}}$.

For a few other liquids, such as mercury, the molecules have a strong affinity for each other (i.e., the cohesive

---

[27]In fact, observing the rise of liquid in a capillary tube is another method of determining the surface tension of a liquid.
[28]*Adhesion* is the attractive force between molecules of different substances. *Cohesion* is the attractive force between molecules of the same substance.

forces dominate). The liquid avoids contact with the tube surface. In such liquids, the meniscus in the tube will be below the general surface level.

Table 14.8 Contact Angles, β

materials	angle
mercury–glass	140°
water–paraffin	107°
water–silver	90°
kerosene–glass	26°
glycerin–glass	19°
water–glass	0°
ethyl alcohol–glass	0°

The *angle of contact*, β, is an indication of whether adhesive or cohesive forces dominate. For contact angles less than 90°, adhesive forces dominate. For contact angles greater than 90°, cohesive forces dominate.

Equation 14.27 can be used to predict the capillary rise in a small-bore tube. Surface tension and contact angles can be obtained from Tables 14.7 and 14.8, respectively.

$$h = \frac{4\sigma \cos \beta}{\rho d_{\text{tube}}\, g} \qquad \text{[SI]} \quad 14.27(a)$$

$$h = \left(\frac{4\sigma \cos \beta}{\rho d_{\text{tube}}} \right) \times \left(\frac{g_c}{g} \right) \qquad \text{[U.S.]} \quad 14.27(b)$$

$$\sigma = \frac{h\rho d_{\text{tube}}\, g}{4 \cos \beta} \qquad \text{[SI]} \quad 14.28(a)$$

$$\sigma = \left(\frac{h\rho d_{\text{tube}}}{4 \cos \beta} \right) \times \left(\frac{g}{g_c} \right) \qquad \text{[U.S.]} \quad 14.28(b)$$

$$r_{\text{meniscus}} = \frac{d_{\text{tube}}}{2 \cos \beta} \qquad 14.29$$

If it is assumed that the meniscus is hemispherical, then $r_{\text{meniscus}} = r_{\text{tube}}$, $\beta = 0°$, and $\cos \beta = 1.0$, and the above equations can be simplified. (Such an assumption can only be made when the diameter of the capillary tube is less than 0.1 in.)

Example 14.7

To what height will 68°F (20°C) ethyl alcohol rise in a 0.005 in (0.127 mm) internal diameter glass capillary tube? The density of the alcohol is 49 lbm/ft^3 (790 kg/m^3).

SI Solution

$$\sigma = 0.0227 \text{ N/m}$$

$$\beta = 0°$$

The acceleration due to gravity is 9.81 m/s^2. From Eq. 14.27, the height is

$$h = \frac{4\sigma \cos \beta}{\rho d_{\text{tube}}\, g}$$

$$= \frac{(4)\left(0.0227\, \dfrac{\text{N}}{\text{m}} \right)(1.0)\left(1000\, \dfrac{\text{mm}}{\text{m}} \right)}{\left(790\, \dfrac{\text{kg}}{\text{m}^3} \right)(0.127 \text{ mm})\left(9.81\, \dfrac{\text{m}}{\text{sec}^2} \right)}$$

$$= 0.0923 \text{ m}$$

Customary U.S. Solution

From Tables 14.7 and 14.8, respectively, the surface tension and contact angle are

$$\sigma = 0.00156 \text{ lbf/ft}$$

$$\beta = 0°$$

The acceleration due to gravity is 32.2 ft/sec^2. From Eq. 14.27, the height is

$$h = \frac{4\sigma \cos \beta g_c}{\rho d_{\text{tube}}\, g}$$

$$= \frac{(4)\left(0.00156\, \dfrac{\text{lbf}}{\text{ft}} \right)(1.0)\left(32.2\, \dfrac{\text{lbm-ft}}{\text{lbf-sec}^2} \right)\left(12\, \dfrac{\text{in}}{\text{ft}} \right)}{\left(49\, \dfrac{\text{lbm}}{\text{ft}^3} \right)(0.005 \text{ in})\left(32.2\, \dfrac{\text{ft}}{\text{sec}^2} \right)}$$

$$= 0.306 \text{ ft}$$

17. COMPRESSIBILITY[29]

Compressibility (also known as the *coefficient of compressibility*), β, is the fractional change in the volume of a fluid per unit change in pressure in a constant-temperature process.[30] Typical units are 1/psi, 1/psf, 1/atm, and 1/kPa. It is the reciprocal of the bulk modulus, a quantity that is more commonly tabulated than compressibility.

$$\beta = \frac{-\dfrac{\Delta V}{V_0}}{\Delta p} = \frac{1}{E} \qquad 14.30$$

Compressibility can also be written in terms of partial derivatives.

$$\beta = \left(\frac{-1}{V_0} \right)\left(\frac{\partial V}{\partial p} \right)_T = \left(\frac{1}{\rho_0} \right)\left(\frac{\partial \rho}{\partial p} \right)_T \qquad 14.31$$

---

[29]Compressibility should not be confused with the *thermal coefficient of expansion*, $(1/V_0)\,(\partial V/\partial T)_p$, which is the fractional change in volume per unit temperature change in a constant-pressure process (with units of 1°F or 1°C), or the dimensionless *compressibility factor*, Z, used with the ideal gas law.

[30]Other symbols used for compressibility are c, C, and K. Equation 14.30 is written with a negative sign to show that volume decreases as pressure increases.

Compressibility changes only slightly with temperature. The small compressibility of liquids is typically considered to be insignificant, giving rise to the common understanding that liquids are incompressible.

The density of a compressible fluid depends on the fluid's pressure. For small changes in pressure, the density at one pressure can be calculated from the density at another pressure from Eq. 14.32.

$$\rho_2 \approx \rho_1 \left(1 + \beta(p_2 - p_1)\right) \qquad \textit{14.32}$$

Gases, of course, are easily compressed. The compressibility of an ideal gas depends on its pressure, p, its ratio of specific heats, k, and the nature of the process.[31] Depending on the process, the compressibility may be known as *isothermal compressibility* or *(adiabatic) isentropic compressibility*. Of course, compressibility is zero for constant-volume processes and is infinite (or undefined) for constant-pressure processes.

$$\beta_T = \frac{1}{p} \qquad \text{[isothermal ideal gas processes]} \qquad \textit{14.33}$$

$$\beta_s = \frac{1}{kp} \qquad \text{[adiabatic ideal gas processes]} \qquad \textit{14.34}$$

Table 14.9 Approximate Compressibility of Common Liquids at 1 atm

liquid	temperature	β, 1/psi	β, 1/atm
mercury	32°F	0.027×10^{-5}	0.39×10^{-5}
glycerin	60°F	0.16×10^{-5}	2.4×10^{-5}
water	60°F	0.33×10^{-5}	4.9×10^{-5}
ethyl alcohol	32°F	0.68×10^{-5}	10×10^{-5}
chloroform	32°F	0.68×10^{-5}	10×10^{-5}
gasoline	60°F	1.0×10^{-5}	15×10^{-5}
hydrogen	20K	11×10^{-5}	160×10^{-5}
helium	2.1K	48×10^{-5}	700×10^{-5}

(Multiply 1/psi by 0.14504 to obtain 1/kPa.)
(Multiply 1/psi by 14.696 to obtain 1/atm.)

Example 14.8

Water at 68°F (20°C) and 1 atm has a density of 62.3 lbm/ft³ (997 kg/m³). What is the new density if the pressure is isothermally increased from 14.7 psi (100 kPa) to 400 psi (2760 kPa)? Assume that the bulk modulus has a constant value of 320,000 psi (2.2×10^6 kPa).

SI Solution

Compressibility is the reciprocal of the bulk modulus.

$$\beta = \frac{1}{E} = \frac{1}{2.2 \times 10^6 \text{ kPa}} = 4.55 \times 10^{-7} \text{ 1/kPa}$$

All other information needed to use Eq. 14.32 is provided.

$$
\begin{aligned}
\rho_2 &= \rho_1 \left(1 + \beta(p_2 - p_1)\right) \\
&= \left(997 \ \frac{\text{kg}}{\text{m}^3}\right) \left(1 + \left(4.55 \times 10^{-7} \frac{1}{\text{kPa}}\right) \right. \\
&\quad \left. \times (2760 \text{ kPa} - 100 \text{ kPa})\right) \\
&= 998.2 \text{ kg/m}^3
\end{aligned}
$$

Customary U.S. Solution

Compressibility is the reciprocal of the bulk modulus.

$$\beta = \frac{1}{E} = \frac{1}{320,000 \ \frac{\text{lbf}}{\text{in}^2}} = 0.3125 \times 10^{-5} \text{ in}^2/\text{lbf}$$

All other information needed to use Eq. 14.32 is provided.

$$
\begin{aligned}
\rho_2 &= \rho_1 \left(1 + \beta(p_2 - p_1)\right) \\
&= \left(62.3 \ \frac{\text{lbm}}{\text{ft}^3}\right) \left(1 + \left(0.3125 \times 10^{-5} \ \frac{\text{in}^2}{\text{lbf}}\right) \right. \\
&\quad \left. \times \left(400 \ \frac{\text{lbf}}{\text{in}^2} - 14.7 \ \frac{\text{lbf}}{\text{in}^2}\right)\right) \\
&= 62.38 \text{ lbm/ft}^3
\end{aligned}
$$

18. BULK MODULUS

The *bulk modulus*, E, of a fluid is analogous to the modulus of elasticity of a solid.[32] Typical units are psi, atm, and kPa. The term Δp in Eq. 14.35 represents an increase in stress. The term $\Delta V/V_0$ is a *volumetric strain*. Analogous to Hooke's law describing elastic formation, the *bulk modulus* of a fluid (liquid or gas) is given by Eq. 14.35.

$$E = \frac{\text{stress}}{\text{strain}} = \frac{-\Delta p}{\dfrac{\Delta V}{V_0}} \qquad \textit{14.35(a)}$$

$$E = -V_0 \left(\frac{\partial p}{\partial V}\right)_T \qquad \textit{14.35(b)}$$

The term *secant bulk modulus* is associated with Eq. 14.35(a) (using finite differences), while the terms *tangent bulk modulus* and *point bulk modulus* are associated with Eq. 14.35(b) (using partial derivatives).

The bulk modulus is the reciprocal of compressibility.

$$E = \frac{1}{\beta} \qquad \textit{14.36}$$

---

[31]For air, $k = 1.4$.

[32]To distinguish it from the modulus of elasticity, the bulk modulus is represented by the symbol B when dealing with solids.

Water Resources

The bulk modulus changes only slightly with temperature. Water's bulk modulus is usually taken as 300,000 psi (2.1×10^6 kPa) unless greater accuracy is required, in which case Table 14.10 or App. 14.A can be used.

Table 14.10 Approximate Bulk Modulus of Water

pressure (psi)	32°F	68°F	120°F	200°F	300°F
	(thousands of psi)				
15	292	320	332	308	–
1500	300	330	340	319	218
4500	317	348	362	338	271
15,000	380	410	420	405	350

(Multiply psi by 6.8948 to obtain kPa.)

Reprinted with permission from Victor L. Streeter, *Handbook of Fluid Dynamics,* © 1961, by McGraw-Hill Book Company.

19. SPEED OF SOUND

The *speed of sound* (*acoustical velocity* or *sonic velocity*), a, in a fluid is a function of its bulk modulus (or, equivalently, of its compressibility).[33] Equation 14.37 gives the speed of sound through a liquid.

$$a = \sqrt{\frac{E}{\rho}} = \sqrt{\frac{1}{\beta\rho}} \quad \text{[SI]} \quad 14.37(a)$$

$$a = \sqrt{\frac{Eg_c}{\rho}} = \sqrt{\frac{g_c}{\beta\rho}} \quad \text{[U.S.]} \quad 14.37(b)$$

Equation 14.38 gives the speed of sound in an ideal gas. The temperature, T, must be in degrees absolute (i.e., °R or K). For air, the ratio of specific heats is $k = 1.40$, the molecular weight is 28.967, and the universal gas constant is $R^* = 1545.4$ ft-lbf/lbmol-°R (8314.57 J/kmol·K).

$$a = \sqrt{\frac{E}{\rho}} = \sqrt{\frac{kp}{\rho}}$$

$$= \sqrt{kRT} = \sqrt{\frac{kR^*T}{MW}} \quad \text{[SI]} \quad 14.38(a)$$

$$a = \sqrt{\frac{Eg_c}{\rho}} = \sqrt{\frac{kg_cp}{\rho}}$$

$$= \sqrt{kg_cRT} = \sqrt{\frac{kg_cR^*T}{MW}} \quad \text{[U.S.]} \quad 14.38(b)$$

Since k and R are constant for an ideal gas, the speed of sound is a function of temperature only. Equation 14.39 can be used to calculate the new speed of sound when temperature is varied.

$$\frac{a_1}{a_2} = \sqrt{\frac{T_1}{T_2}} \quad 14.39$$

The *Mach number* of an object is the ratio of the object's speed to the speed of sound in the medium through which it is traveling.

$$M = \frac{v}{a} \quad 14.40$$

The term *subsonic travel* implies $M < 1$.[34] Similarly, *supersonic travel* implies $M > 1$, but usually $M < 5$. Travel above $M = 5$ is known as *hypersonic travel.* Travel in the transition region between subsonic and supersonic (i.e., $0.8 < M < 1.2$) is known as *transonic travel.* A *sonic boom* (a shock-wave phenomenon) occurs when an object travels at supersonic speed.

Example 14.9

What is the speed of sound in 150°F (66°C) water? The density is 61.2 lbm/ft^3 (980 kg/m^3), and the bulk modulus is 328,000 psi (2.26×10^6 kPa).

SI Solution

$$a = \sqrt{\frac{(2.26 \times 10^6 \text{ kPa})\left(1000 \dfrac{\text{Pa}}{\text{kPa}}\right)}{980 \dfrac{\text{kg}}{\text{m}^3}}} = 1519 \text{ m/s}$$

Customary U.S. Solution

From Eq. 14.37,

$$a = \sqrt{\frac{Eg_c}{\rho}}$$

$$= \sqrt{\frac{\left(328,000 \dfrac{\text{lbf}}{\text{in}^2}\right)\left(144 \dfrac{\frac{\text{lbf}}{\text{ft}^2}}{\frac{\text{lbf}}{\text{in}^2}}\right)\left(32.2 \dfrac{\text{lbm-ft}}{\text{lbf-sec}^2}\right)}{61.2 \dfrac{\text{lbm}}{\text{ft}^3}}}$$

$$= 4985 \text{ ft/sec}$$

Example 14.10

What is the speed of sound in 150°F (66°C) air at standard atmospheric pressure?

SI Solution

$$R = \frac{8314.57 \dfrac{\text{J}}{\text{kmol·K}}}{28.967 \dfrac{\text{kg}}{\text{kmol}}} = 287.03 \text{ J/kg·K}$$

$$T = 66°C + 273 = 339\text{K}$$

$$a = \sqrt{kRT} = \sqrt{(1.4)\left(287.03 \dfrac{\text{J}}{\text{kg·K}}\right)(339\text{K})}$$

$$= 369 \text{ m/s}$$

---

[33]The symbol c is also used for the sonic velocity.

[34]In the language of compressible fluid flow, this is known as the *subsonic flow regime.*

Customary U.S. Solution

The specific gas constant, R, for air is

$$R = \frac{R^*}{MW} = \frac{1545.4 \frac{\text{ft-lbf}}{\text{lbmol-}^\circ\text{R}}}{28.967 \frac{\text{lbm}}{\text{lbmol}}}$$

$$= 53.35 \text{ ft-lbf/lbm-}^\circ\text{R}$$

The absolute temperature is

$$T = 150^\circ\text{F} + 460 = 610^\circ\text{R}$$

From Eq. 14.38(b),

$$a = \sqrt{kg_cRT}$$

$$= \sqrt{(1.4)\left(32.2 \frac{\text{ft-lbm}}{\text{lbf-sec}^2}\right)\left(53.35 \frac{\text{ft-lbf}}{\text{lbm-}^\circ\text{R}}\right)(610^\circ\text{R})}$$

$$= 1211 \text{ ft/sec}$$

20. PROPERTIES OF SOLUTIONS

There are few convenient ways of predicting the properties of nonreacting, nonvolatile organic and aqueous solutions (acids, brines, alcohol mixtures, coolants, etc.) and mixtures from the individual properties of the components.

Volumes of two combining organic liquids (e.g., acetone and chloroform) are essentially additive. The volume change upon mixing will seldom be more than a few tenths of a percent. The volume change in aqueous solutions is often slightly greater, but is still limited to a few percent (e.g., 3% for some solutions of methanol and water).

Thus, the specific gravity (density, specific weight, etc.) can be considered to be a volumetric weighting of the individual specific gravities. Most times, however, the specific gravity of a known solution must be calculated from known data regarding one of the various density scales (see Sec. 6) or determined through research.

Most other important fluid properties of aqueous solutions, such as viscosity, compressibility, surface tension, and vapor pressure, have been measured and are usually determined through research.[35] It is important to be aware of the operating conditions of the solution. Data for one concentration or condition should not be used for another concentration or condition.

---

[35] There is no substitute for a complete fluid properties data book.

15 Fluid Statics

Nomenclature

a	acceleration	ft/sec^2	m/s^2
A	area	ft^2	m^2
b	base length	ft	m
d	diameter	ft	m
e	eccentricity	ft	m
F	force	lbf	N
FS	factor of safety	–	–
g	gravitational acceleration	ft/sec^2	m/s^2
g_c	gravitational conversion constant (32.2)	lbm-ft/lbf-sec^2	n.a.
h	height	ft	m
I	moment of inertia	ft^4	m^4
J	polar moment of inertia	ft^4	m^4
k	radius of gyration	ft	m
k	ratio of specific heats	–	–
L	length	ft	m
m	mass	lbm	kg
M	mechanical advantage	–	–
M	moment	ft-lbf	N·m
n	polytropic exponent	–	–
N	normal force	lbf	N
p	pressure	lbf/ft^2	Pa
r	radius	ft	m
R	resultant force	lbf	N
R	specific gas constant	ft-lbf/lbm-°R	J/kg·K
SG	specific gravity	–	–
t	wall thickness	ft	m
T	temperature	°R	K
v	velocity	ft/sec	m/s
V	volume	ft^3	m^3
W	weight	lbf	N
x	distance	ft	m
x	fraction	–	–
y	distance	ft	m

Symbols

γ	specific weight	lbf/ft^3	n.a.
η	efficiency	–	–
θ	angle	deg	deg
μ	coefficient of friction	–	–
ρ	density	lbm/ft^3	kg/m^3
σ	stress	lbf/ft^2	Pa
υ	specific volume	$\text{ft}^3\text{/lbm}$	$\text{m}^3\text{/kg}$
ω	angular velocity	rad/sec	rad/s

Subscripts

a	atmospheric
b	buoyant
bg	between CB and CG
c	centroidal
f	frictional
F	force
h	hoop
l	lever or longitudinal
m	manometer fluid, mercury, or metacentric
p	plunger
r	ram
R	resultant
t	tank
v	vapor or vertical
w	water

1. PRESSURE-MEASURING DEVICES

There are many devices for measuring and indicating fluid pressure. Some devices measure gage pressure; others measure absolute pressure. The effects of nonstandard atmospheric pressure and nonstandard gravitational acceleration must be determined, particularly for devices relying on columns of liquid to indicate pressure. Table 15.1 lists the common types of devices and the ranges of pressure appropriate for each.

Table 15.1 *Common Pressure-Measuring Devices*

device	approximate range (in atm)
water manometer	0–0.1
mercury barometer	0–1
mercury manometer	0.001–1
metallic diaphragm	0.01–200
transducer	0.001–700
Bourdon pressure gauge	1–3000
Bourdon vacuum gauge	0.1–1

The *Bourdon pressure gauge* is the most common pressure-indicating device. This mechanical device consists of a coiled hollow tube that tends to straighten out (i.e., unwind) when the tube is subjected to an internal pressure. The degree to which the coiled tube unwinds depends on the difference between the internal and external pressures. A Bourdon gauge directly indicates *gage pressure*. Extreme accuracy is generally not a characteristic of Bourdon gauges.

Figure 15.1 *C-Bourdon Pressure Gauge*

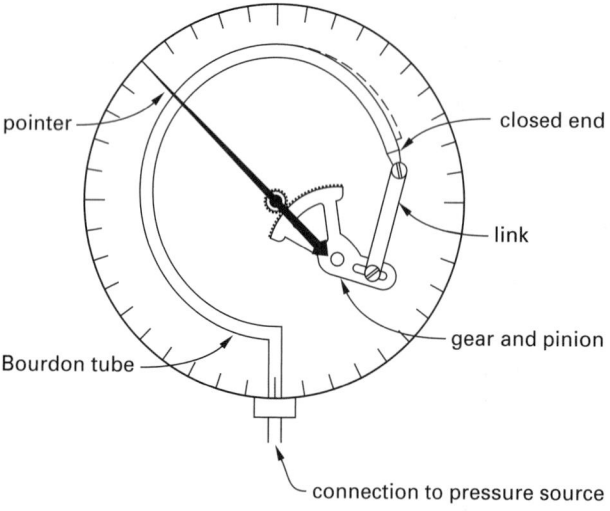

The *barometer* is a common device for measuring the absolute pressure of the atmosphere.[1] It is constructed by filling a long tube open at one end with mercury (or alcohol, or some other liquid) and inverting the tube so that the open end is below the level of a mercury-filled container. If the vapor pressure of the mercury in the tube is neglected, the fluid column will be supported only by the atmospheric pressure transmitted through the container fluid at the lower, open end.

In non-SI installations, gauges are always calibrated in psi, unless the dial is marked "altitude" (measuring in feet of water) or "vacuum" (measuring in inches of mercury) on its face. To avoid confusion, the gauge dial will be clearly marked if other units are indicated.

---

[1]A barometer can be used to measure the pressure inside any vessel. However, the barometer must be completely enclosed in the vessel, which may not be possible. Also, it is difficult to read a barometer enclosed within a tank.

Strain gages, diaphragm gauges, quartz-crystal transducers and other devices using the *piezoelectric effect* are also used to measure stress and pressure, particularly when pressure fluctuates quickly (e.g., as in a rocket combustion chamber). With these devices, calibration is required to interpret pressure from voltage generation or changes in resistance, capacitance, or inductance. These devices are generally unaffected by atmospheric pressure or gravitational acceleration.

Manometers (U-tube manometers) can also be used to indicate small pressure differences, and for this purpose they provide great accuracy. (Manometers are not suitable for measuring pressures much larger than 10 psi (70 kPa), however.) A difference in manometer fluid surface heights is converted into a pressure difference. If one end of a manometer is open to the atmosphere, the manometer indicates gage pressure. It is theoretically possible, but impractical, to have a manometer indicate absolute pressure, since one end of the manometer would have to be exposed to a perfect vacuum.

A *static pressure tube (piezometer tube)* is a variation of the manometer. It is a simple method of determining the static pressure in a pipe or other vessel, regardless of fluid motion in the pipe. A vertical transparent tube is connected to a hole in the pipe wall.[2] (None of the tube projects into the pipe.) The static pressure will force the contents of the pipe up into the tube. The height of the contents will be an indication of gage pressure in the pipe.

Figure 15.2 *Static Pressure Tube*

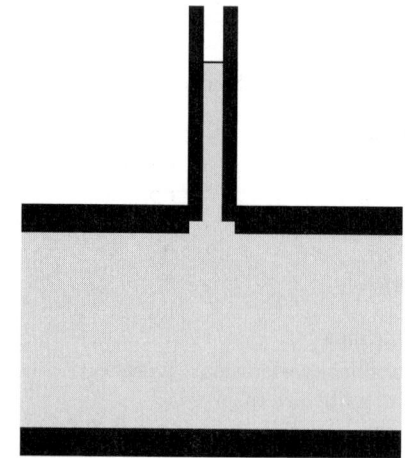

The device used to measure the pressure should not be confused with the method used to obtain exposure to the pressure. For example, a static pressure *tap* in a pipe is merely a hole in the pipe wall. A Bourdon gauge, manometer, or transducer can then be used with the tap to indicate pressure.

---

[2]Where greater accuracy is required, multiple holes may be drilled around the circumference of the pipe and connected through a manifold (*piezometer ring*) to the pressure-measuring device.

Tap holes are generally $1/8$ to $1/4$ in in diameter, drilled at right angles to the wall, and smooth and flush with the pipe wall. No part of the gauge or connection projects into the pipe. The tap holes should be at least 5 to 10 pipe diameters downstream from any source of turbulence (e.g., a bend, fitting, or valve).

2. MANOMETERS

Figure 15.3 illustrates a simple U-tube manometer used to measure the difference in pressure between two vessels. When both ends of the manometer are connected to pressure sources, the name *differential manometer* is used. If one end of the manometer is open to the atmosphere, the name *open manometer* is used.[3] The open manometer implicitly measures gage pressures.

Figure 15.3 Simple U-Tube Manometer

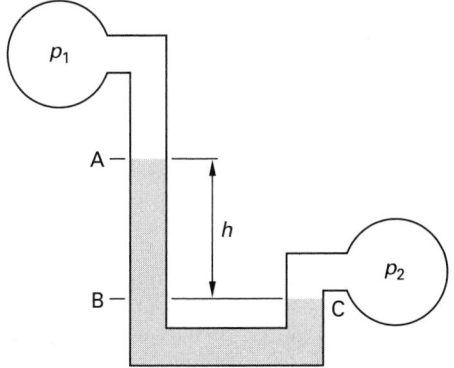

Since the pressure at point B in Fig. 15.3 is the same as at point C, the pressure differential produces the vertical fluid column of height h. In Eq. 15.2, A is the area of the tube. Equations 15.2 and 15.3 assume consistent density units. In the absence of any capillary action, the inside diameters of the manometer tubes are irrelevant.

$$F_{\text{net}} = \text{weight of fluid column} \qquad 15.1$$

$$(p_2 - p_1) \times A = \rho_m g h \times A \qquad 15.2$$

$$p_2 - p_1 = \rho_m g h \qquad [\text{SI}] \qquad 15.3$$

In countries that do not use SI units, densities are commonly quoted in pounds per cubic foot. In that case, Eq. 15.3 can be written as

$$p_2 - p_1 = \rho_m \times \frac{g}{g_c} \times h$$
$$= \gamma_m h \qquad [\text{U.S.}] \qquad 15.4$$

The quantity g/g_c has a value of 1.0 lbf/lbm in almost all cases, and thus γ_m is numerically equal to ρ_m with units of lbf/ft^3.

---

[3]If one of the manometer legs is inclined, the term *inclined manometer* or *draft gauge* is used. Although only the vertical distance between the manometer fluid surfaces should be used to calculate the pressure difference, with small pressure differences it may be more accurate to read the inclined distance (which is larger than the vertical distance) and compute the vertical distance from the angle of inclination.

Equations 15.3 and 15.4 assume that the manometer fluid height is small, or that only low-density gases fill the tubes above the manometer fluid. If a high-density fluid (such as water) is present above the measuring fluid, or if the columns h_1 or h_2 are very long, corrections will be necessary.

Figure 15.4 Manometer Requiring Corrections

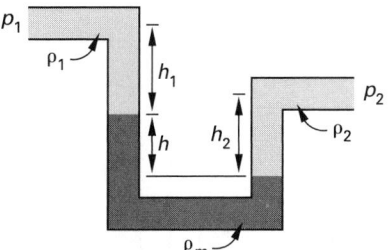

Fluid column h_2 "sits on top" of the manometer fluid, forcing the manometer fluid to the left. This increase must be subtracted out. Similarly, the column h_1 restricts the movement of the manometer fluid. The observed measurement must be increased to correct for this restriction.

$$p_2 - p_1 = g(\rho_m h + \rho_1 h_1 - \rho_2 h_2) \qquad [\text{SI}] \qquad 15.5(a)$$

$$p_2 - p_1 = \frac{g}{g_c} \times (\rho_m h + \rho_1 h_1 - \rho_2 h_2)$$
$$= \gamma_m h + \gamma_1 h_1 - \gamma_2 h_2 \qquad [\text{U.S.}] \qquad 15.5(b)$$

When a manometer is used to measure the pressure difference across an orifice or other fitting where the same liquid exists in both manometer sides (as in Fig. 15.5), it is not necessary to correct the manometer reading for all of the liquid present above the manometer fluid. This is because parts of the correction for both sides of the manometer are the same. Thus, the distance y in Fig. 15.5 is an irrelevant distance.

Figure 15.5 Irrelevant Distance

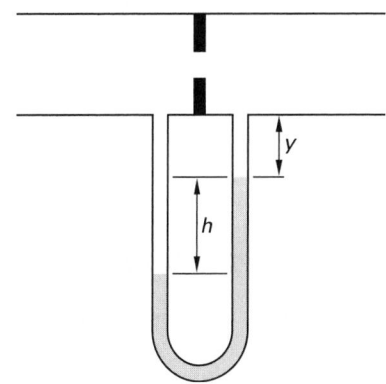

Manometer tubes are generally large enough in diameter to avoid significant capillary effects. Corrections for capillarity are seldom necessary.

Example 15.1

The pressure at the bottom of a tank of water (62.4 lbm/ft³; $\rho = 998$ kg/m³) is measured with a mercury manometer. (The density of mercury is 848 lbm/ft³; 13 575 kg/m³.) What is the gage pressure at the bottom of the water tank?

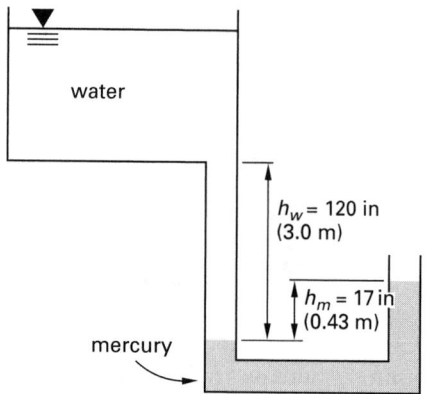

SI Solution

From Eq. 15.5(a),

$$\Delta p = g(\rho_m h_m - \rho_w h_w)$$

$$= \left(9.81 \ \frac{\text{m}}{\text{s}^2}\right)\left(\left(13\,575 \ \frac{\text{kg}}{\text{m}^3}\right)(0.43 \ \text{m})\right.$$

$$\left. - \left(998 \ \frac{\text{kg}}{\text{m}^3}\right)(3.0 \ \text{m})\right)$$

$$= 27\,892 \ \text{Pa} \quad (27.9 \ \text{kPa gage})$$

Customary U.S. Solution

From Eq. 15.5(b),

$$\Delta p = \gamma_m h_m - \gamma_w h_w$$

$$= \frac{\left(848 \ \frac{\text{lbf}}{\text{ft}^3}\right)(17 \ \text{in}) - \left(62.4 \ \frac{\text{lbf}}{\text{ft}^3}\right)(120 \ \text{in})}{\left(12 \ \frac{\text{in}}{\text{ft}}\right)^3}$$

$$= 4.01 \ \text{lbf/in}^2 \quad (\text{psig})$$

3. HYDROSTATIC PRESSURE

Hydrostatic pressure is the pressure a fluid exerts on an immersed object or container walls.[4] Pressure is equal to the force per unit area of surface.

$$p = \frac{F}{A} \qquad \textbf{\textit{15.6}}$$

---

[4]The term *hydrostatic* is used with all fluids, not only with water.

Hydrostatic pressure in a stationary, incompressible fluid behaves according to the following characteristics.

- Pressure is a function of vertical depth (and density) only. The pressure will be the same at two points with identical depths.

- Pressure varies linearly with (vertical) depth.

- Pressure is independent of an object's area and size and the weight (mass) of water above the object. Figure 15.6 illustrates the *hydrostatic paradox*. The pressures at depth h are the same in all four columns because pressure depends on depth, not volume.

Figure 15.6 *Hydrostatic Paradox*

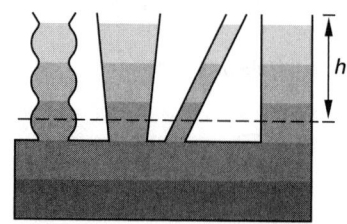

- Pressure at a point has the same magnitude in all directions (*Pascal's law*). Thus, pressure is a scalar quantity.

- Pressure is always normal to a surface, regardless of the surface's shape or orientation. (This is a result of the fluid's inability to support shear stress.)

- The resultant of the pressure distribution acts through the *center of pressure*.

4. FLUID HEIGHT EQUIVALENT TO PRESSURE

Pressure varies linearly with depth. The relationship between pressure and depth (i.e., the *hydrostatic head*) for an incompressible fluid is given by Eq. 15.7.

$$p = \rho g h \qquad \text{[SI]} \qquad \textbf{\textit{15.7(a)}}$$

$$p = \frac{\rho g h}{g_c} = \gamma h \qquad \text{[U.S.]} \qquad \textbf{\textit{15.7(b)}}$$

Since ρ and g are constants, Eq. 15.7 shows that p and h are linearly related. Knowing one determines the other.[5] For example, the height of a fluid column needed to produce a pressure is

$$h = \frac{p}{\rho g} \qquad \text{[SI]} \qquad \textbf{\textit{15.8(a)}}$$

$$h = \frac{p g_c}{\rho g} = \frac{p}{\gamma} \qquad \text{[U.S.]} \qquad \textbf{\textit{15.8(b)}}$$

---

[5]In fact, pressure and height of a fluid column can be used interchangeably. The height of a fluid column is known as *head*. For example: "The fan developed a static head of 3 inches of water," or "The pressure head at the base of the water tank was 8 meters." When the term "head" is used, it is essential to specify the fluid.

Table 15.2 lists six important fluid height equivalents that many engineers commit to memory.[6]

Table 15.2 Approximate Fluid Height Equivalents at 68° F (20° C)

liquid	height equivalents	
water	0.0361 psi/in	27.70 in/psi
water	62.4 psf/ft	0.01603 ft/psf
water	9.81 kPa/m	0.1019 m/kPa
water	0.4329 psi/ft	2.31 ft/psi
mercury	0.491 psi/in	2.036 in/psi
mercury	133.3 kPa/m	0.00750 m/kPa

A barometer is an example of the measurement of pressure by the height of a fluid column. If the vapor pressure of the barometer liquid is neglected, the atmospheric pressure will be given by Eq. 15.9.

$$p_a = \rho g h \qquad \text{[SI]} \qquad \textbf{15.9(a)}$$

$$p_a = \frac{\rho g h}{g_c} = \gamma h \qquad \text{[U.S.]} \qquad \textbf{15.9(b)}$$

If the vapor pressure of the barometer liquid is significant (as it would be with alcohol or water), the vapor pressure effectively reduces the height of the fluid column, as Eq. 15.10 illustrates.

$$p_a - p_v = \rho g h \qquad \text{[SI]} \qquad \textbf{15.10(a)}$$

$$p_a - p_v = \frac{\rho g h}{g_c} = \gamma h \qquad \text{[U.S.]} \qquad \textbf{15.10(b)}$$

Example 15.2

A vacuum pump is used to drain a flooded mine shaft of 68°F (20°C) water.[7] The vapor pressure of water at this temperature is 0.34 psi (2.34 kPa). The pump is incapable of lifting the water higher than 400 in (10.16 m). What is the atmospheric pressure?

SI Solution

From Eq. 15.10,

$$p_a = p_v + \rho g h$$

$$= 2.34 \text{ kPa} + \frac{\left(998 \ \frac{\text{kg}}{\text{m}^3}\right)\left(9.81 \ \frac{\text{m}}{\text{s}^2}\right)(10.16 \text{ m})}{1000 \ \frac{\text{Pa}}{\text{kPa}}}$$

$$= 101.8 \text{ kPa}$$

(alternate SI solution, using Table 15.2)

$$p_a = p_v + \rho g h$$

$$= 2.34 \text{ kPa} + \left(9.8 \ \frac{\text{kPa}}{\text{m}}\right)(10.16 \text{ m})$$

$$= 101.9 \text{ kPa}$$

Customary U.S. Solution

From Table 15.2, the height equivalent of water is approximately 0.0361 psi/in. Notice that psi/in is the same as lbf/in^3, the units of γ. From Eq. 15.10, the atmospheric pressure is

$$p_a = p_v + \rho g h = p_v + \gamma h$$

$$= 0.34 \ \frac{\text{lbf}}{\text{in}^2} + \left(0.0361 \ \frac{\text{lbf}}{\text{in}^3}\right)(400 \text{ in})$$

$$= 14.78 \text{ lbf/in}^2 \text{ (psia)}$$

5. MULTIFLUID BAROMETERS

It is theoretically possible to fill a barometer tube with several different immiscible fluids.[8] Upon inversion, the fluids will separate, leaving the most dense fluid at the bottom and the least dense fluid at the top. All of the fluids will contribute, by superposition, to the balance between the external atmospheric pressure and the weight of the fluid column.

$$p_a - p_v = g \sum \rho_i h_i \qquad \text{[SI]} \qquad \textbf{15.11(a)}$$

$$p_a - p_v = \frac{g}{g_c} \sum \rho_i h_i = \sum \gamma_i h_i \qquad \text{[U.S.]} \qquad \textbf{15.11(b)}$$

The pressure at any intermediate point within the fluid column is found by starting at a location where the pressure is known, and then adding or subtracting $\rho g h$ to get to the point where the pressure is needed. Usually, the known pressure will be the atmospheric pressure located in the barometer barrel at the level (elevation) of the fluid outside of the barometer.

Example 15.3

Neglecting vapor pressure, what is the pressure of the air (at point E) in the container shown? The external pressure is 1.0 atm.

---

[6]Of course, these values are recognized to be the approximate specific weights of the liquids.
[7]A reciprocating or other direct-displacement pump would be a better choice to drain a mine.

[8]In practice, barometers are never constructed this way. This theory is more applicable to a category of problems dealing with up-ended containers, as illustrated in Ex. 15.3.

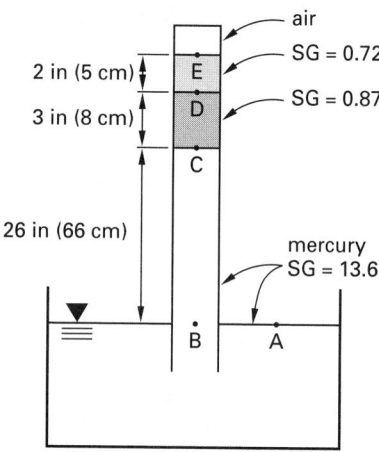

SI Solution

$$p_E = p_{\text{atm}} - g\rho_{\text{water}} \sum (\text{SG}_i) h_i$$

$$= 101\,300 \text{ Pa} - \left(9.81 \frac{\text{m}}{\text{s}^2}\right) \left(1000 \frac{\text{kg}}{\text{m}^3}\right)$$

$$\times \left((0.66 \text{ m})(13.6) + (0.08 \text{ m})(0.87)\right.$$

$$\left. + (0.05 \text{ m})(0.72)\right)$$

$$= 12\,210 \text{ Pa} \quad (12.2 \text{ kPa})$$

Customary U.S. Solution

The pressure at point B is the same as the pressure at point A—1.0 atm. The density of mercury is $13.6 \times 0.0361 = 0.491$ lbm/in^3. The pressure at point C is

$$p_C = 14.7 \text{ psia} - (26 \text{ in}) \left(0.491 \frac{\frac{\text{lbf}}{\text{in}^2}}{\text{in}}\right)$$

$$= 1.93 \text{ lbf/in}^2 \text{ (psia)}$$

Similarly, the pressure at point E (and anywhere within the captive air) is

$$p_E = 14.7 \frac{\text{lbf}}{\text{in}^2} - (26 \text{ in}) \left(0.491 \frac{\text{lbf}}{\text{in}^3}\right)$$

$$- (3 \text{ in})(0.87) \left(0.0361 \frac{\text{lbf}}{\text{in}^3}\right)$$

$$- (2 \text{ in})(0.72) \left(0.0361 \frac{\text{lbf}}{\text{in}^3}\right)$$

$$= 1.79 \text{ lbf/in}^2 \text{ (psia)}$$

6. PRESSURE ON A HORIZONTAL PLANE SURFACE

The pressure on a horizontal plane surface is uniform over the surface because the depth of the fluid is uniform. The resultant of the pressure distribution acts

through the center of pressure of the surface, which corresponds to the centroid of the surface.

Figure 15.7 *Hydrostatic Pressure on a Horizontal Plane Surface*

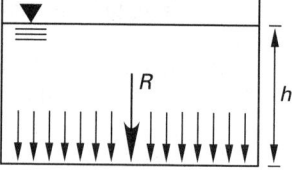

The uniform pressure at depth h is given by Eq. 15.12.[9]

$$p = \rho g h \qquad \text{[SI]} \qquad 15.12(a)$$

$$p = \frac{\rho g h}{g_c} = \gamma h \qquad \text{[U.S.]} \qquad 15.12(b)$$

The total vertical force on the horizontal plane of area A is given by Eq. 15.13.

$$R = pA \qquad 15.13$$

It is tempting, but not always correct, to calculate the vertical force on a submerged surface as the weight of the fluid above it. Such an approach works only when there is no change in the cross-sectional area of the fluid above the surface. This is a direct result of the *hydrostatic paradox*. Figure 15.8 illustrates two containers with the same pressure distribution (force) on their bottom surfaces.

Figure 15.8 *Two Containers with the Same Pressure Distribution*

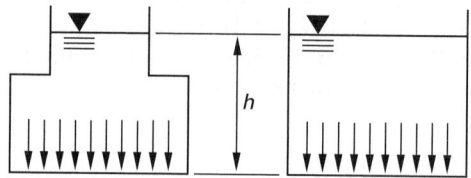

7. PRESSURE ON A RECTANGULAR VERTICAL PLANE SURFACE

The pressure on a vertical rectangular plane surface increases linearly with depth. The pressure distribution will be triangular, as in Fig. 15.9(a), if the plane surface extends to the surface; otherwise, the distribution will be trapezoidal, as in Fig. 15.9(b).

---

[9]The phrase *pressure at a depth* is universally understood to mean the *gage pressure*, as given by Eq. 15.12.

Figure 15.9 *Hydrostatic Pressure on a Vertical Plane Surface*

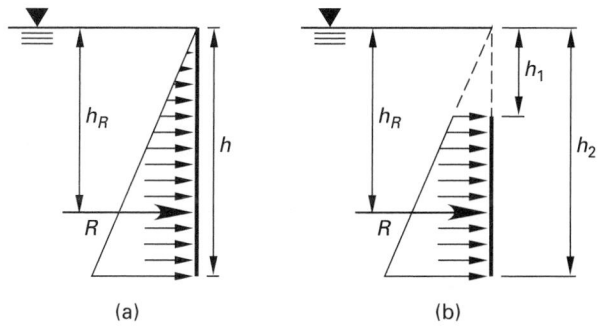

(a)   (b)

The resultant force is calculated from the *average pressure*.

$$\overline{p} = \tfrac{1}{2}(p_1 + p_2) \qquad\qquad 15.14$$

$$= \tfrac{1}{2}\rho g(h_1 + h_2) \qquad \text{[SI]} \qquad 15.15(a)$$

$$\overline{p} = \frac{\tfrac{1}{2}\rho g(h_1 + h_2)}{g_c} = \tfrac{1}{2}\gamma(h_1 + h_2) \quad \text{[U.S.]} \quad 15.15(b)$$

$$R = \overline{p}A \qquad\qquad 15.16$$

Although the resultant is calculated from the average depth, it does not act at the average depth. The resultant of the pressure distribution passes through the centroid of the pressure distribution. For the triangular distribution of Fig. 15.9(a), the resultant is located at a depth of $h_R = \,^2\!/_3 h$. For the more general case of Fig. 15.9(b), the resultant is located from Eq. 15.17.

$$h_R = \tfrac{2}{3}\left(h_1 + h_2 - \frac{h_1 h_2}{h_1 + h_2}\right) \qquad 15.17$$

8. PRESSURE ON A RECTANGULAR INCLINED PLANE SURFACE

The average pressure and resultant force on an inclined rectangular plane surface are calculated in much the same fashion as for the vertical plane surface. The pressure varies linearly with depth. The resultant is calculated from the average pressure, which, in turn, depends on the average depth.

Figure 15.10 *Hydrostatic Pressure on an Inclined Rectangular Plane Surface*

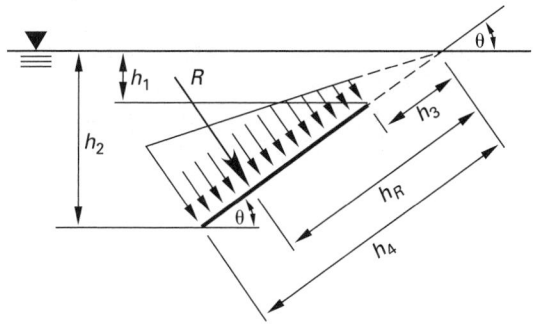

The average pressure and resultant are given by Eqs. 15.18 through 15.21.

$$\overline{p} = \tfrac{1}{2}(p_1 + p_2) \qquad\qquad 15.18$$

$$\overline{p} = \tfrac{1}{2}\rho g(h_1 + h_2) \qquad \text{[SI]} \qquad 15.19$$

$$\overline{p} = \tfrac{1}{2}\rho g(h_3 + h_4)\,\sin\theta \qquad \text{[SI]} \quad 15.20(a)$$

$$\overline{p} = \frac{\tfrac{1}{2}\rho g(h_1 + h_2)}{g_c} = \frac{\tfrac{1}{2}\rho g(h_3 + h_4)\,\sin\theta}{g_c}$$

$$= \tfrac{1}{2}\gamma(h_1 + h_2) = \tfrac{1}{2}\gamma(h_3 + h_4)\,\sin\theta \quad \text{[U.S.]} \; 15.20(b)$$

$$R = \overline{p}A \qquad\qquad 15.21$$

As with the vertical plane surface, the resultant acts at the centroid of the pressure distribution, not at the average depth. Equation 15.17 is rewritten in terms of inclined depths.[10]

$$h_R = \left(\frac{\tfrac{2}{3}}{\sin\theta}\right)\left(h_1 + h_2 - \frac{h_1 h_2}{h_1 + h_2}\right) \qquad 15.22$$

$$h_R = \left(\tfrac{2}{3}\right)\left(h_3 + h_4 - \frac{h_3 h_4}{h_3 + h_4}\right) \qquad 15.23$$

Example 15.4

The tank shown is filled with water ($\rho = 62.4$ lbm/ft^3; $\rho = 1000$ kg/m^3). (a) What is the total force on a 1 ft (1 m) width of the inclined portion of the wall?[11] (b) At what depth (vertical distance) is the resultant force located?

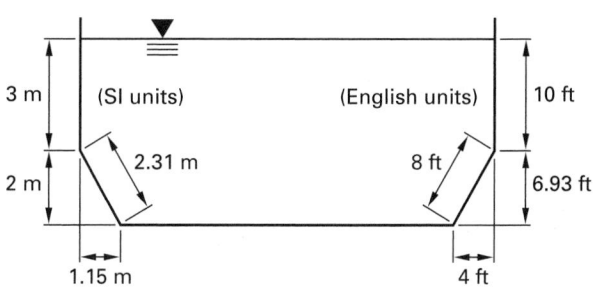

---

[10]Notice that h_R is an inclined distance. If a vertical distance is wanted, it must usually be calculated from h_R and $\sin\theta$. Equation 15.22 can be derived simply by dividing Eq. 15.17 by $\sin\theta$.
[11]Since the width of the tank (the distance into and out of the illustration) is unknown, it is common to calculate the pressure or force per unit width of tank wall. This is the same as calculating the pressure on a 1 ft (1 m) wide section of wall.

SI Solution

(a) The depth of the tank bottom is

$$h_2 = 3 \text{ m} + 2 \text{ m} = 5 \text{ m}$$

From Eq. 15.19, the average gage pressure on the inclined section is

$$\overline{p} = \left(\tfrac{1}{2}\right)\left(1000 \; \frac{\text{kg}}{\text{m}^3}\right)\left(9.81 \; \frac{\text{m}}{\text{s}^2}\right)(3 \text{ m} + 5 \text{ m})$$

$$= 39\,240 \text{ Pa (gage)}$$

The total force on a 1 m section of wall is

$$R = \overline{p}A = (39\,240 \text{ Pa})(2.31 \text{ m})(1 \text{ m})$$

$$= 90\,644 \text{ N} \quad (90.6 \text{ kN})$$

(b) θ must be known to determine h_R.

$$\theta = \arctan\left(\frac{2 \text{ m}}{1.15 \text{ m}}\right) = 60°$$

The location of the resultant can be calculated from Eq. 15.23 once h_3 and h_4 are known.

$$h_3 = \frac{3 \text{ m}}{\sin 60°} = 3.464 \text{ m}$$

$$h_4 = \frac{5 \text{ m}}{\sin 60°} = 5.774 \text{ m}$$

$$h_R = \left(\tfrac{2}{3}\right)\left(3.464 + 5.774 - \frac{(3.464)(5.774)}{3.464 + 5.774}\right)$$

$$= 4.715 \text{ m} \quad \text{[inclined]}$$

The vertical depth at which the resultant acts is

$$h = h_R \sin\theta = (4.715 \text{ m})(\sin 60°)$$

$$= 4.08 \text{ m} \quad \text{[vertical]}$$

Customary U.S. Solution

(a) The water density is given in traditional U.S. mass units. The specific weight, γ, is

$$\gamma = \frac{\rho g}{g_c} = \frac{\left(62.4 \; \dfrac{\text{lbm}}{\text{ft}^3}\right)\left(32.2 \; \dfrac{\text{ft}}{\text{sec}^2}\right)}{32.2 \; \dfrac{\text{lbm-ft}}{\text{lbf-sec}^2}}$$

$$= 62.4 \text{ lbf/ft}^3$$

The depth of the tank bottom is

$$h_2 = 10 \text{ ft} + 6.93 \text{ ft} = 16.93 \text{ ft}$$

From Eq. 15.19, the average pressure on the inclined section is

$$\overline{p} = \left(\tfrac{1}{2}\right)\left(62.4 \; \frac{\text{lbf}}{\text{ft}^3}\right)(10 \text{ ft} + 16.93 \text{ ft})$$

$$= 840.2 \text{ lbf/ft}^2 \text{ (gage)}$$

The total force on a 1 ft section of wall is

$$R = \overline{p}A = \left(840.2 \; \frac{\text{lbf}}{\text{ft}^2}\right)(8 \text{ ft})(1 \text{ ft})$$

$$= 6722 \text{ lbf}$$

(b) θ must be known to determine h_R.

$$\theta = \arctan\left(\frac{6.93 \text{ ft}}{4 \text{ ft}}\right) = 60°$$

The location of the resultant can be calculated from Eq. 15.23 once h_3 and h_4 are known.

$$h_3 = \frac{10 \text{ ft}}{\sin 60°} = 11.55 \text{ ft}$$

$$h_4 = \frac{16.93 \text{ ft}}{\sin 60°} = 19.55 \text{ ft}$$

From Eq. 15.23,

$$h_R = \left(\tfrac{2}{3}\right)\left(11.55 + 19.55 - \frac{(11.55)(19.55)}{11.55 + 19.55}\right)$$

$$= 15.89 \text{ ft} \quad \text{[inclined]}$$

The vertical depth at which the resultant acts is

$$h = h_R \sin\theta = (15.89 \text{ ft})(\sin 60°)$$

$$= 13.76 \text{ ft} \quad \text{[vertical]}$$

9. PRESSURE ON A GENERAL PLANE SURFACE

Figure 15.11 illustrates a nonrectangular plane surface that may or may not extend to the liquid surface and that may or may not be inclined, as shown in Fig. 15.10. As with other regular surfaces, the resultant force depends on the average pressure and acts through the center of pressure (CP). The average pressure is calculated from the depth of the surface's centroid (CG).

$$\overline{p} = \rho g h_c \sin\theta \quad \text{[SI]} \quad \textit{15.24(a)}$$

$$\overline{p} = \frac{\rho g h_c \sin\theta}{g_c} = \gamma h_c \sin\theta \quad \text{[U.S.]} \quad \textit{15.24(b)}$$

$$R = \overline{p}A \quad \textit{15.25}$$

Figure 15.11 General Plane Surface

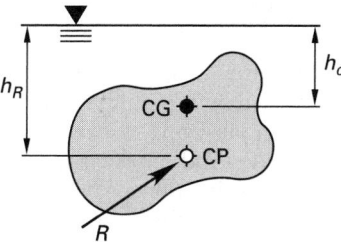

The resultant force acts at depth h_R normal to the plane surface. I_c in Eq. 15.26 is the centroidal area moment of inertia, with dimensions of L^4 (length^4) about an axis parallel to the surface. (Appendix 42.A contains equations for the moment of inertia of common shapes.) Both h_c and h_R are measured parallel to the plane surface. That is, if the plane surface is inclined, h_c and h_R are inclined distances.

$$h_R = h_c + \frac{I_c}{A\, h_c} \qquad \textit{15.26}$$

Example 15.5

The top edge of a vertical circular observation window in a submarine is located 4.0 ft (1.25 m) below the surface of the water. The window is 1.0 ft (0.3 m) in diameter. The density of the water is 62.4 lbm/ft^3 (1000 kg/m^3). Neglect the salinity of the water. (a) What is the resultant force on the window? (b) At what depth does the resultant force act?

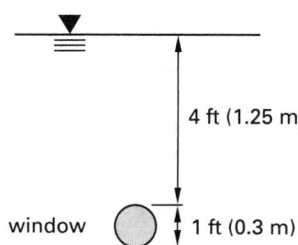

window   1 ft (0.3 m)

4 ft (1.25 m)

SI Solution

(a) The radius of the window is

$$r = \frac{0.3\text{ m}}{2} = 0.15\text{ m}$$

$$h_c = 1.25\text{ m} + 0.15\text{ m} = 1.4\text{ m}$$

$$A = \pi(0.15\text{ m})^2 = 0.0707\text{ m}^2$$

$$\bar{p} = \left(1000\ \frac{\text{kg}}{\text{m}^3}\right)\left(9.81\ \frac{\text{m}}{\text{s}^2}\right)(1.4\text{ m})$$

$$= 13\,734\text{ Pa (gage)}$$

$$R = (13\,734\text{ Pa})(0.0707\text{ m}^2) = 971\text{ N}$$

(b)
$$I_c = \left(\frac{\pi}{4}\right)(0.15\text{ m})^4 = 3.976 \times 10^{-4}\text{ m}^4$$

$$h_R = 1.4\text{ m} + \frac{3.976 \times 10^{-4}\text{ m}^4}{(0.0707\text{ m}^2)(1.4\text{ m})}$$

$$= 1.404\text{ m}$$

Customary U.S. Solution

(a) The radius of the window is

$$r = \frac{1\text{ ft}}{2} = 0.5\text{ ft}$$

The depth at which the centroid of the circular window is located is

$$h_c = 4.0\text{ ft} + 0.5\text{ ft} = 4.5\text{ ft}$$

The area of the circular window is

$$A = \pi r^2 = \pi(0.5\text{ ft})^2$$

$$= 0.7854\text{ ft}^2$$

As in Ex. 15.4, the resultant force will be calculated using γ as the weight density.

The average pressure is

$$\bar{p} = \gamma h_c = \left(62.4\ \frac{\text{lbf}}{\text{ft}^3}\right)(4.5\text{ ft})$$

$$= 280.8\text{ lbf/ft}^2\text{ (psfg)}$$

The resultant is calculated from Eq. 15.21.

$$R = \bar{p}A = \left(280.8\ \frac{\text{lbf}}{\text{ft}^2}\right)(0.7854\text{ ft}^2)$$

$$= 220.5\text{ lbf}$$

(b) The centroidal area moment of inertia of a circle is

$$I_c = \frac{\pi}{4}r^4 = \left(\frac{\pi}{4}\right)(0.5\text{ ft})^4$$

$$= 0.049\text{ ft}^4$$

From Eq. 15.26, the depth at which the resultant force acts is

$$h_R = 4.5\text{ ft} + \frac{0.049\text{ ft}^4}{(0.7854\text{ ft}^2)(4.5\text{ ft})}$$

$$= 4.514\text{ ft}$$

10. SPECIAL CASES: VERTICAL SURFACES

Several simple wall shapes and configurations recur frequently. Figure 15.12 indicates the location of their hydrostatic pressure resultants (*centers of pressure*). In all cases, the surfaces are vertical and extend to the liquid's surface.

Figure 15.12 *Centers of Pressure for Common Configurations*

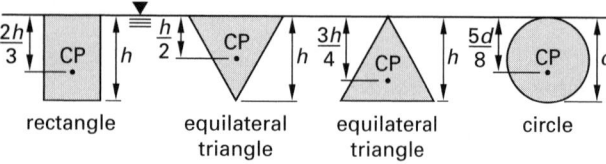

11. FORCES ON CURVED AND COMPOUND SURFACES

Figure 15.13 illustrates a curved surface. The resultant force acting on such a curved surface is not difficult to determine, although the x- and y-components of the resultant usually must be calculated first. The magnitude and direction of the resultant are found by conventional methods.

$$R = \sqrt{R_x^2 + R_y^2} \qquad 15.27$$

$$\theta = \arctan\left(\frac{R_y}{R_x}\right) \qquad 15.28$$

The horizontal component of the resultant hydrostatic force is found in the same manner as for a vertical plane surface.

Figure 15.13 *Pressure Distributions on a Curved Surface*

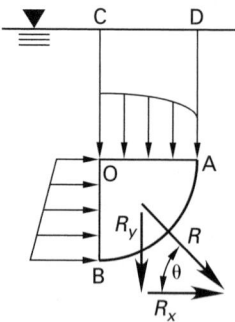

The fact that the surface is curved does not affect the calculation of the horizontal force. In Fig. 15.13, the horizontal pressure distribution on curved surface BA is the same as the horizontal pressure distribution on imaginary projected surface BO.

The vertical component of force on the curved surface is most easily calculated as the weight of the liquid above it.[12] In Fig. 15.13, the vertical component of force on

---

[12] Calculating the vertical force component is not in conflict with the hydrostatic paradox as long as the cross-sectional area of liquid above the curved surface does not decrease between the curved surface and the liquid's free surface. If there is a change in the cross-sectional area, the vertical component of force is equal to the weight of fluid in an unchanged cross-sectional area (i.e., the equivalent area).

the curved surface BA is the weight of liquid within the area ABCD, with a vertical line of action passing through the centroid of the area ABCD.

Figure 15.14 illustrates a curved surface with no liquid above it. However, it is not difficult to show that the resultant force acting upward on the curved surface HG is equal in magnitude (and opposite in direction) to the force that would be acting downward due to the missing area EFGH. Such an imaginary area used to calculate hydrostatic pressure is known as an *equivalent area*.

Figure 15.14 *Equivalent Area*

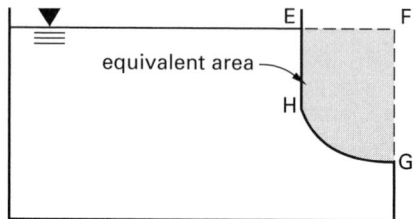

Example 15.6

What is the total force on a 1 ft section of the wall in Ex. 15.4?

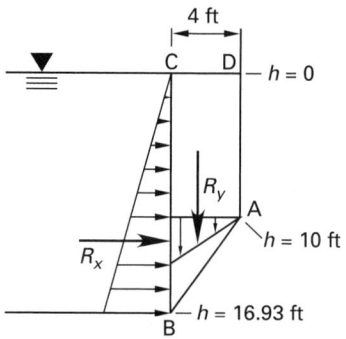

Solution

The average depth is

$$\overline{h} = \left(\tfrac{1}{2}\right)(0 + 16.93 \text{ ft}) = 8.465 \text{ ft}$$

The average pressure and horizontal component of the resultant on a 1 ft section of wall are

$$\overline{p} = \gamma\overline{h} = \left(62.4 \ \frac{\text{lbf}}{\text{ft}^3}\right)(8.465 \text{ ft})$$

$$= 528.2 \text{ lbf/ft}^2 \text{ (psfg)}$$

$$R_x = \overline{p}A = \left(528.2 \ \frac{\text{lbf}}{\text{ft}^2}\right)(16.93 \text{ ft})(1 \text{ ft})$$

$$= 8942 \text{ lbf}$$

The volume of a 1 ft section of area ABCD is

$$V_{\text{ABCD}} = (1 \text{ ft})\big((4 \text{ ft})(10 \text{ ft}) + \left(\tfrac{1}{2}\right)(4 \text{ ft})(6.93 \text{ ft})\big)$$

$$= 53.86 \text{ ft}^3$$

The vertical component is

$$R_y = \gamma V = \left(62.4 \; \frac{\text{lbf}}{\text{ft}^3}\right)(53.86 \; \text{ft}^3)$$
$$= 3361 \; \text{lbf}$$

The total resultant force is

$$R = \sqrt{(8942 \; \text{lbf})^2 + (3361 \; \text{lbf})^2}$$
$$= 9553 \; \text{lbf}$$

12. TORQUE ON A GATE

When an openable gate or door is submerged in such a manner as to have unequal depths of liquid on either of its sides, or when there is no liquid present on one side of a gate or door, the hydrostatic pressure will act to either open or close the door. If the gate does not open, this pressure is resisted, usually by a latching mechanism on the gate itself.[13] The magnitude of the resisting latch force can be determined from the *hydrostatic torque (hydrostatic moment)* acting on the gate. The moment is almost always taken with respect to the gate hinges.

Figure 15.15 *Torque on a Hinge*

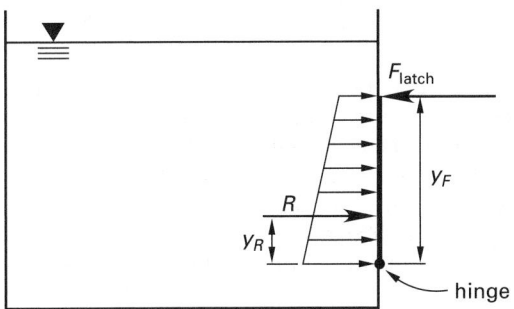

The applied moment is calculated as the product of the resultant force on the gate and the distance from the hinge to the resultant on the gate. This applied moment is balanced by the resisting moment, calculated as the latch force times the separation of the latch and hinge.

$$M_{\text{applied}} = M_{\text{resisting}} \qquad 15.29$$
$$R \times y_R = F_{\text{latch}} \times y_F \qquad 15.30$$

13. HYDROSTATIC FORCES ON A DAM

The techniques presented in the preceding sections are applicable to dams. That is, the horizontal force on the dam face can be found as in Ex. 15.6, regardless of inclination or curvature of the dam face. The vertical

[13]Any contribution to resisting force from stiff hinges or other sources of friction is typically neglected.

force on the dam face is calculated as the weight of the water above the dam face. Of course, the vertical force is zero if the dam face is vertical.

Figure 15.16 illustrates a typical dam, defining its *heel, toe,* and *crest.* x_{CG}, the distance to the dam's center of gravity, is not shown.

Figure 15.16 *Dam*

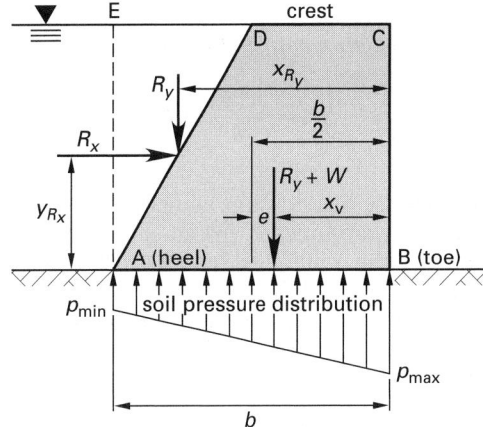

However, there are other considerations for gravity dams.[14] Most notably, the dam must not tip over or slide away due to the hydrostatic pressure. Furthermore, the pressure distribution within the soil under the dam is not uniform, and soil loading must not be excessive.

The *overturning moment* is a measure of the horizontal pressure's tendency to tip the dam over, pivoting it about the toe of the dam (point B in Fig. 15.16). (Usually, moments are calculated with respect to the pivot point.) The overturning moment is calculated as the product of the horizontal component of hydrostatic pressure (i.e., the x-component of the resultant) and the vertical distance between the toe and the line of action of the force.

$$M_{\text{overturning}} = R_x \times y_{R_x} \qquad 15.31$$

In most configurations, the overturning is resisted jointly by moments from the dam's own weight, W, and the vertical component of the resultant, R_y (i.e., the weight of area EAD in Fig. 15.16).[15]

$$M_{\text{resisting}} = (R_y \times x_{R_y}) + (W \times x_{\text{CG}}) \qquad 15.32$$

The *factor of safety against overturning* is

$$(\text{FS})_{\text{overturning}} = \frac{M_{\text{resisting}}}{M_{\text{overturning}}} \qquad 15.33$$

[14]A *gravity dam* is one that is held in place and orientation by its own mass (weight) and the friction between its base and the ground.
[15]The density of concrete or masonry with steel reinforcing is usually taken to be approximately 150 lbm/ft³ (2400 kg/m³).

In addition to causing the dam to tip over, the horizontal component of hydrostatic force will also cause the dam to tend to slide along the ground. This tendency is resisted by the frictional force between the dam bottom and soil. The frictional force, F_f, is calculated as the product of the *normal force*, N, and the *coefficient of static friction*, μ.

$$F_f = \mu_{\text{static}} \, N = \mu_{\text{static}}(W + R_y) \qquad 15.34$$

The *factor of safety against sliding* is

$$(\text{FS})_{\text{sliding}} = \frac{F_f}{R_x} \qquad 15.35$$

The soil pressure distribution beneath the dam is usually assumed to vary linearly from a minimum to a maximum value. (The minimum value must be greater than zero, since soil cannot be in a state of tension. The maximum pressure should not exceed the allowable soil pressure.) Equation 15.36 predicts the minimum and maximum soil pressures.

$$p_{\max}, p_{\min} = \left(\frac{R_y + W}{b}\right)\left(1 \pm \frac{6e}{b}\right) \quad \begin{bmatrix}\text{per unit} \\ \text{width}\end{bmatrix} \quad 15.36$$

The *eccentricity*, e, in Eq. 15.36 is the distance between the mid-length of the dam and the line of action of the total vertical force, $W + R_y$. (The eccentricity must be less than $b/6$ for the entire base to be in compression.) Notice that distances x_{v} and x_{CG} are different.

$$e = \frac{b}{2} - x_{\text{v}} \qquad 15.37$$

$$x_{\text{v}} = \frac{M_{\text{resisting}}}{R_y + W} \qquad 15.38$$

14. PRESSURE DUE TO SEVERAL IMMISCIBLE LIQUIDS

Figure 15.17 illustrates the non-uniform pressure distribution due to two immiscible liquids (e.g., oil on top and water below).

Figure 15.17 *Pressure Distribution from Two Immiscible Liquids*

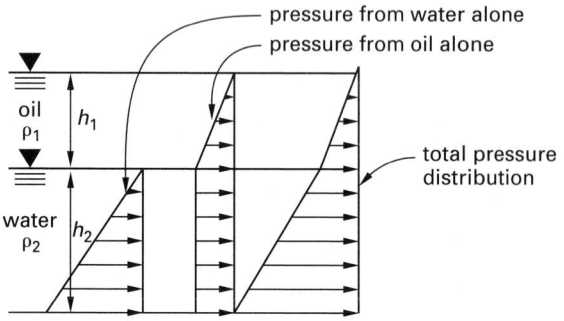

The pressure due to the upper liquid (oil), once calculated, serves as a *surcharge* to the liquid below (water). The pressure at the tank bottom is given by Eq. 15.39. (The principle can be extended to three or more immiscible liquids as well.)

$$p_{\text{bottom}} = \rho_1 g h_1 + \rho_2 g h_2 \qquad [\text{SI}] \qquad 15.39(a)$$

$$p_{\text{bottom}} = \frac{\rho_1 g h_1}{g_c} + \frac{\rho_2 g h_2}{g_c}$$
$$= \gamma_1 h_1 + \gamma_2 h_2 \qquad [\text{U.S.}] \qquad 15.39(b)$$

15. PRESSURE FROM COMPRESSIBLE FLUIDS

Fluid density, thus far, has been assumed to be independent of pressure. In reality, even "incompressible" liquids are slightly compressible. Sometimes, the effect of this compressibility cannot be neglected.

The familiar $p = \rho g h$ equation is a special case of Eq. 15.40. (It is assumed that $h_2 > h_1$. The minus sign in Eq. 15.40 indicates that pressure decreases as elevation (height) increases.)

$$\int_{p_1}^{p_2} \frac{dp}{\rho g} = -(h_2 - h_1) \qquad [\text{SI}] \qquad 15.40(a)$$

$$\int_{p_1}^{p_2} \frac{g_c \, dp}{\rho g} = -(h_2 - h_1) \qquad [\text{U.S.}] \qquad 15.40(b)$$

If the fluid is a perfect gas, and if compression is an isothermal (i.e., constant temperature) process, the relationship between pressure and density is given by Eq. 15.41. The isothermal assumption is appropriate, for example, for the earth's *stratosphere* (i.e., above 35,000 ft or 11 000 m), where the temperature is assumed to be constant at approximately $-67°F$ ($-55°C$).

$$pv = \frac{p}{\rho} = RT = \text{constant} \qquad 15.41$$

In the isothermal case, Eq. 15.41 can be rewritten as Eq. 15.42. (For air, $R = 53.35$ ft-lbf/lbm-°R; $R = 287.03$ J/kg·K.) Of course, the temperature, T, must be in degrees absolute (i.e., in °R or K). Equation 15.42 is known as the *barometric height relationship*[16] because knowledge of atmospheric temperature and the pressures at two points is sufficient to determine the height difference between the two points.

$$h_2 - h_1 = \left(\frac{RT}{g}\right) \ln\left(\frac{p_1}{p_2}\right) \qquad [\text{SI}] \qquad 15.42(a)$$

$$h_2 - h_1 = \left(\frac{g_c RT}{g}\right) \ln\left(\frac{p_1}{p_2}\right) \qquad [\text{U.S.}] \qquad 15.42(b)$$

---

[16]You may recognize this as being equivalent to the work done in an isothermal compression process. The elevation (height) difference $h_2 - h_1$ (with units of feet) can be interpreted as the work done per unit mass during compression (with units of ft-lbf/lbm).

The pressure at an elevation (height) h_2 in a layer of perfect gas that has been isothermally compressed is given by Eq. 15.43.

$$p_2 = p_1 e^{g(h_1 - h_2)/RT} \qquad \text{[SI]} \qquad 15.43(a)$$

$$p_2 = p_1 e^{g(h_1 - h_2)/g_c RT} \qquad \text{[U.S.]} \qquad 15.43(b)$$

If the fluid is a perfect gas, and if compression is an *adiabatic process*, the relationship between pressure and density is given by Eq. 15.44[17] where k is the *ratio of specific heats*, a property of the gas. ($k = 1.4$ for air, hydrogen, oxygen, and carbon monoxide, among others.)

$$pv^k = p \left(\frac{1}{\rho} \right)^k = \text{constant} \qquad 15.44$$

The following three equations apply to adiabatic compression of an ideal gas.

$$h_2 - h_1 = \left(\frac{k}{k-1} \right) \left(\frac{RT_1}{g} \right) \left(1 - \left(\frac{p_2}{p_1} \right)^{k-1/k} \right)$$
$$\text{[SI]} \qquad 15.45(a)$$

$$h_2 - h_1 = \left(\frac{k}{k-1} \right) \left(\frac{g_c}{g} \right) RT_1 \left(1 - \left(\frac{p_2}{p_1} \right)^{k-1/k} \right)$$
$$\text{[U.S.]} \qquad 15.45(b)$$

$$p_2 = p_1 \left(1 - \left(\frac{k-1}{k} \right) \left(\frac{g}{RT_1} \right) (h_2 - h_1) \right)^{k/k-1}$$
$$\text{[SI]} \qquad 15.46(a)$$

$$p_2 = p_1 \left(1 - \left(\frac{k-1}{k} \right) \left(\frac{g}{g_c} \right) \left(\frac{h_2 - h_1}{RT_1} \right) \right)^{k/k-1}$$
$$\text{[U.S.]} \qquad 15.46(b)$$

$$T_2 = T_1 \left(1 - \left(\frac{k-1}{k} \right) \left(\frac{g}{RT_1} \right) (h_2 - h_1) \right)$$
$$\text{[SI]} \qquad 15.47(a)$$

$$T_2 = T_1 \left(1 - \left(\frac{k-1}{k} \right) \left(\frac{g}{g_c} \right) \left(\frac{h_2 - h_1}{RT_1} \right) \right)$$
$$\text{[U.S.]} \qquad 15.47(b)$$

The three adiabatic compression equations can be used for the more general *polytropic compression* case simply by substituting the *polytropic exponent*, n, for

k.[18] Unlike the ratio of specific heats, the polytropic exponent is a function of the process, not the gas. The polytropic compression assumption is appropriate for the earth's *troposphere*.[19] Assuming a linear decrease in temperature with altitude of $-0.00356°F/ft$ ($-0.00649°$ C/m), a polytropic exponent of $n = 1.235$ can be derived.

Example 15.7

The air pressure and temperature at sea level are 1.0 standard atmosphere and $68°F$ ($20°C$), respectively. Assume polytropic compression with $n = 1.235$. What is the pressure at an altitude of 5000 ft (1525 m)?

SI Solution

The absolute temperature of the air is $20°C + 273 = 293K$. From Eq. 15.46, the pressure at 1525 m altitude is

$$p_2 = (1.0 \text{ atm}) \left(1 - \left(\frac{1.235 - 1}{1.235} \right) \left(9.81 \, \frac{\text{m}}{\text{s}^2} \right) \right.$$
$$\left. \times \left(\frac{1525 \text{ m}}{\left(287.03 \, \frac{\text{J}}{\text{kg·K}} \right) (293\text{K})} \right) \right)^{1.235/1.235 - 1}$$
$$= 0.834 \text{ atm}$$

Customary U.S. Solution

The absolute temperature of the air is $68°F + 460 = 528°R$. From Eq. 15.46 (substituting $k = n = 1.235$ for polytropic compression), the pressure at 5000 ft altitude is

$$p_2 = (1.0 \text{ atm}) \left(1 - \left(\frac{1.235 - 1}{1.235} \right) \left(\frac{32.2 \, \frac{\text{ft}}{\text{sec}^2}}{32.2 \, \frac{\text{lbm-ft}}{\text{lbf-sec}^2}} \right) \right.$$
$$\left. \times \left(\frac{5000 \text{ ft}}{\left(53.35 \, \frac{\text{ft-lbf}}{\text{lbm-°R}} \right) (528°\text{R})} \right) \right)^{1.235/(1.235 - 1)}$$
$$= 0.835 \text{ atm}$$

---

[17]There is no heat or energy transfer to or from the ideal gas in an adiabatic process. However, this is not the same as an isothermal process.

[18]Actually, polytropic compression is the general process. Isothermal compression is a special case ($n = 1$) of the polytropic process, as is adiabatic compression ($n = k$).

[19]The *troposphere* is the part of the earth's atmosphere we live in and where most atmospheric disturbances occur. The *stratosphere*, starting at approximately 35,000 ft (11 000 m), is cold, clear, dry, and still. Between the troposphere and the stratosphere is the *tropopause*, a transition layer that contains most of the atmosphere's dust and moisture. Temperature actually increases with altitude in the stratosphere and decreases with altitude in the troposphere, but is constant in the tropopause.

16. EXTERNALLY PRESSURIZED LIQUIDS

If the gas above a liquid in a closed tank is pressurized to a gage pressure of p_t, this pressure will add to the hydrostatic pressure anywhere in the fluid. The pressure at the tank bottom illustrated in Fig. 15.18 is given by Eq. 15.48.

$$p_{\text{bottom}} = p_t + \rho g h \qquad \text{[SI]} \quad 15.48(a)$$

$$p_{\text{bottom}} = p_t + \frac{\rho g h}{g_c} = p_t + \gamma h \qquad \text{[U.S.]} \quad 15.48(b)$$

Figure 15.18 Externally Pressurized Liquid

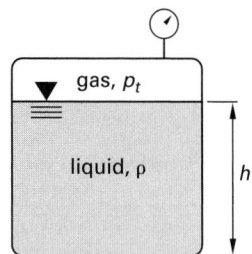

17. HYDRAULIC RAM

A *hydraulic ram (hydraulic jack, hydraulic press, fluid press,* etc.) is illustrated in Fig. 15.19. This is a force-multiplying device. A force, F_p, is applied to the *plunger*, and a useful force, F_r, appears at the *ram*. Even though the pressure in the hydraulic fluid is the same on the ram and plunger, the forces at the two cylinders will be proportional to their respective cross-sectional areas.[20]

Figure 15.19 Hydraulic Ram

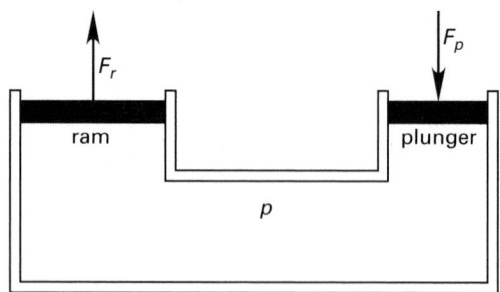

Since the pressure is the same everywhere, Eq. 15.49 can be solved for p for the ram and plunger.

$$F = pA = p\pi r^2 \qquad 15.49$$

$$p_p = p_r \qquad 15.50$$

$$\frac{F_p}{A_p} = \frac{F_r}{A_r} \qquad 15.51$$

$$\frac{F_p}{d_p^2} = \frac{F_r}{d_r^2} \qquad 15.52$$

---

[20] *Pascal's law*: Pressure in a liquid is the same in all directions.

Small, manually actuated hydraulic rams usually have a lever handle to increase the mechanical advantage of the ram from 1.0 to M, as illustrated in Fig. 15.20. In most cases, the pivot and connection mechanism will not be frictionless, and some of the applied force will be used to overcome the friction. This friction loss is accounted for by a *lever efficiency* or *lever effectiveness*, η.

$$M = \frac{L_1}{L_2} \qquad 15.53$$

$$F_p = \eta M F_l \qquad 15.54$$

$$\frac{\eta M F_l}{A_p} = \frac{F_r}{A_r} \qquad 15.55$$

Figure 15.20 Hydraulic Ram with Mechanical Advantage

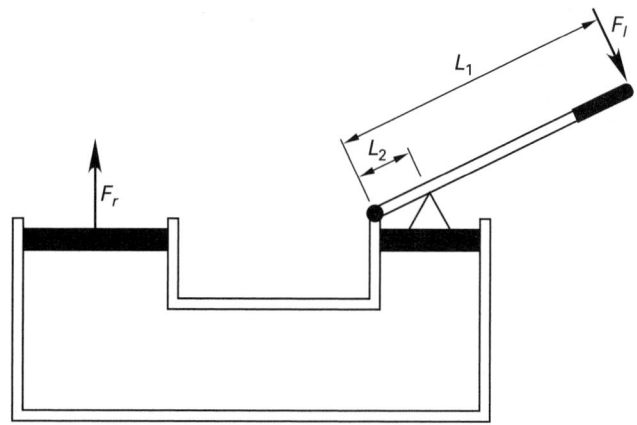

18. BUOYANCY

Buoyant force is an upward force that acts on all objects that are partially or completely submerged in a fluid. The fluid can be a liquid, as in the case of a ship floating at sea, or the fluid can be a gas, as in a balloon floating in the atmosphere.

There is a buoyant force on all submerged objects, not just those that are stationary or ascending. There will be, for example, a buoyant force on a rock sitting at the bottom of a pond. There will also be a buoyant force on a rock sitting exposed on the ground, since the rock is "submerged" in air. A buoyant force due to displaced air also exists, although it may be insignificant, in the case of partially exposed floating objects such as icebergs.

Buoyant force always acts to counteract an object's weight (i.e., buoyancy acts against gravity). The magnitude of the buoyant force is predicted from *Archimedes' principle (the buoyancy theorem)*: The buoyant force on a submerged object is equal to the weight of the displaced fluid.[21] An equivalent statement of Archimedes' principle is: A floating object displaces liquid equal in weight to its own weight.

---

[21] The volume term in Eq. 15.56 is the total volume of the object only in the case of complete submergence.

$$F_{\text{buoyant}} = \rho g V_{\text{displaced}} \qquad \text{[SI]} \quad \textbf{15.56(a)}$$

$$F_{\text{buoyant}} = \frac{\rho g V_{\text{displaced}}}{g_c} = \gamma V_{\text{displaced}} \quad \text{[U.S.]} \ \textbf{15.56(b)}$$

In the case of stationary (i.e., not moving vertically) floating or submerged objects, the buoyant force and object weight are in equilibrium. If the forces are not in equilibrium, the object will rise or fall until equilibrium is reached. That is, the object will sink until its remaining weight is supported by the bottom, or it will rise until the weight of displaced liquid is reduced by breaking the surface.[22]

The specific gravity (SG) of an object submerged in water can be determined from its dry and submerged weights. Neglecting the buoyancy of any surrounding gases,

$$\text{SG} = \frac{W_{\text{dry}}}{W_{\text{dry}} - W_{\text{submerged}}} \qquad \textbf{15.57}$$

Figure 15.21 illustrates an object floating partially exposed in a liquid. Neglecting the insignificant buoyant force from the displaced air (or other gas), the fractions of volume exposed and submerged are easily determined.

Figure 15.21 *Partially Submerged Object*

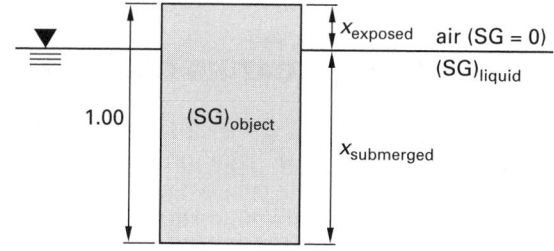

$$x_{\text{submerged}} = \frac{\rho_{\text{object}}}{\rho_{\text{liquid}}} = \frac{(\text{SG})_{\text{object}}}{(\text{SG})_{\text{liquid}}} \qquad \textbf{15.58}$$

$$x_{\text{exposed}} = 1 - x_{\text{submerged}} \qquad \textbf{15.59}$$

Figure 15.22 illustrates a somewhat more complicated situation—that of an object floating at the interface between two liquids of different densities. The fractions of immersion in each liquid are given by the following equations.

$$x_1 = \frac{(\text{SG})_2 - (\text{SG})_{\text{object}}}{(\text{SG})_2 - (\text{SG})_1} \qquad \textbf{15.60}$$

$$x_2 = 1 - x_1 = \frac{(\text{SG})_{\text{object}} - (\text{SG})_1}{(\text{SG})_2 - (\text{SG})_1} \qquad \textbf{15.61}$$

---

[22]An object can also stop rising or falling due to a change in the fluid's density. The buoyant force will increase with increasing depth in the ocean due to an increase in density at great depths. The buoyant force will decrease with increasing altitude in the atmosphere due to a decrease in density at great heights.

Figure 15.22 *Object Floating in Two Liquids*

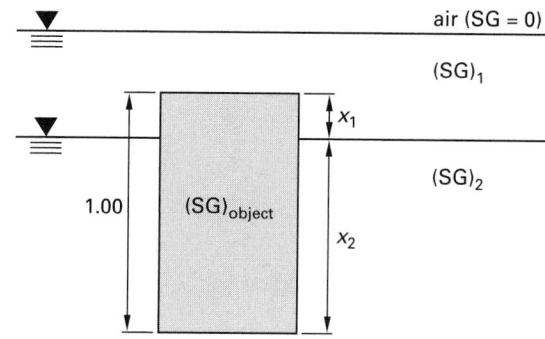

A more general case of a floating object is shown in Fig. 15.23. Situations of this type are easily evaluated by equating the object's weight with the sum of the buoyant forces.

In the case of Fig. 15.23 (with two liquids), the following relationships apply.

$$(\text{SG})_{\text{object}} = x_1(\text{SG})_1 + x_2(\text{SG})_2 \qquad \textbf{15.62}$$

$$(\text{SG})_{\text{object}} = (1 - x_0 - x_2)(\text{SG})_1 + x_2(\text{SG})_2 \quad \textbf{15.63}$$

$$(\text{SG})_{\text{object}} = x_1(\text{SG})_1 + (1 - x_0 - x_1)(\text{SG})_2 \quad \textbf{15.64}$$

Figure 15.23 *General Two-Liquid Buoyancy Problem*

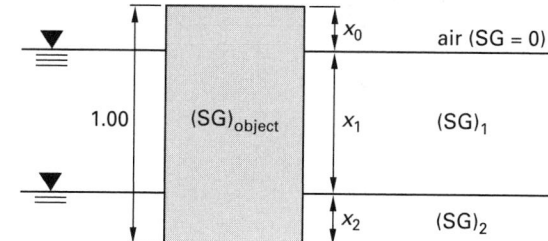

Example 15.8

An empty polyethylene telemetry balloon and payload have a mass of 500 lbm (225 kg). The balloon is filled with helium when the atmospheric conditions are 60°F (15.6°C) and 14.8 psia (102 kPa). The specific gas constant of helium is 2079 J/kg·K (386.3 ft-lbf/lbm-°R). What volume of helium is required for lift-off from a sea-level platform?

SI Solution

$$\rho_{\text{air}} = \frac{p}{RT} = \frac{1.02 \times 10^5 \ \text{Pa}}{\left(287.03 \ \dfrac{\text{J}}{\text{kg·K}}\right)(15.6°\text{C} + 273)}$$

$$= 1.231 \ \text{kg/m}^3$$

$$\rho_{\text{helium}} = \frac{1.02 \times 10^5 \text{ Pa}}{\left(2079 \ \dfrac{\text{J}}{\text{kg·K}}\right)(288.6\text{K})} = 0.17 \text{ kg/m}^3$$

$$m = 225 \text{ kg} + \left(0.17 \ \frac{\text{kg}}{\text{m}^3}\right) V_{\text{He}}$$

$$m_b = \left(1.231 \ \frac{\text{kg}}{\text{m}^3}\right) V_{\text{He}}$$

$$225 \text{ kg} + \left(0.17 \ \frac{\text{kg}}{\text{m}^3}\right) V_{\text{He}} = \left(1.231 \ \frac{\text{kg}}{\text{m}^3}\right) V_{\text{He}}$$

$$V_{\text{He}} = 212.1 \text{ m}^3$$

Customary U.S. Solution

The gas densities are

$$\rho_{\text{air}} = \frac{p}{RT} = \frac{\left(14.8 \ \dfrac{\text{lbf}}{\text{in}^2}\right)\left(144 \ \dfrac{\text{in}^2}{\text{ft}^2}\right)}{\left(53.35 \ \dfrac{\text{ft-lbf}}{\text{lbm-°R}}\right)(60°\text{F} + 460)}$$

$$= 0.07682 \text{ lbm/ft}^3$$

$$\gamma_{\text{air}} = \rho \times \frac{g}{g_c} = 0.07682 \text{ lbf/ft}^3$$

$$\rho_{\text{helium}} = \frac{\left(14.8 \ \dfrac{\text{lbf}}{\text{in}^2}\right)\left(144 \ \dfrac{\text{in}^2}{\text{ft}^2}\right)}{\left(386.3 \ \dfrac{\text{ft-lbf}}{\text{lbm-°R}}\right)(520°\text{R})}$$

$$= 0.01061 \text{ lbm/ft}^3$$

$$\gamma_{\text{helium}} = 0.01061 \text{ lbf/ft}^3$$

The total weight of the balloon, payload, and helium is

$$W = 500 \text{ lbf} + \left(0.01061 \ \frac{\text{lbf}}{\text{ft}^3}\right) V_{\text{He}}$$

The buoyant force is the weight of the displaced air. Neglecting the payload volume, the displaced air volume is the same as the helium volume.

$$F_b = \left(0.07682 \ \frac{\text{lbf}}{\text{ft}^3}\right) V_{\text{He}}$$

At lift-off, the weight of the balloon is just equal to the buoyant force.

$$W = F_b$$

$$500 \text{ lbf} + \left(0.01061 \ \frac{\text{lbf}}{\text{ft}^3}\right) V_{\text{He}} = \left(0.07682 \ \frac{\text{lbf}}{\text{ft}^3}\right) V_{\text{He}}$$

$$V_{\text{He}} = 7552 \text{ ft}^3$$

19. BUOYANCY OF SUBMERGED PIPELINES

Whenever possible, submerged pipelines for river crossings should be completely buried at a level below river scour. This will reduce or eliminate loads and movement due to flutter, scour and fill, drag, collisions, and buoyancy. Submerged pipelines should cross at right angles to the river. For maximum flexibility, ductility, and weighting, pipelines should be made of thick-walled mild steel.

Submerged pipelines should be weighted to achieve a minimum of 20% negative buoyancy (i.e., an average density of 1.2 times the environment, approximately 72 lbm/ft³ or 1200 kg/m³). Metal or concrete clamps can be used for this purpose, as well as concrete coatings. Thick steel clamps have the advantage of a smaller lateral exposed area (resulting in less drag from river flow), while brittle concrete coatings are sensitive to pipeline flutter and temperature fluctuations.

Due to the critical nature of many pipelines and the difficulty in accessing submerged pipelines for repair, it is common to provide a parallel auxiliary line. The auxiliary and main lines are provided with crossover and mainline valves, respectively, on high ground at both sides of the river to permit either or both lines to be used.

20. INTACT STABILITY: STABILITY OF FLOATING OBJECTS

A stationary object is said to be in *static equilibrium*. However, an object in static equilibrium is not necessarily stable. For example, a coin balanced on edge is in static equilibrium, but it will not return to the balanced position if it is disturbed. An object is said to be *stable* (i.e., in *stable equilibrium*) if it tends to return to the equilibrium position when slightly displaced.

Stability of floating and submerged objects is known as *intact stability*.[23] There are two forces acting on a stationary floating object: the buoyant force and the object's weight. The buoyant force acts upward through the centroid of the displaced volume (not the object's volume). This centroid is known as the *center of buoyancy*. The gravitational force on the object (i.e., the object's weight) acts downward through the entire object's center of gravity.

For a totally submerged object (as in the balloon and submarine shown in Fig. 15.24) to be stable, the center of buoyancy must be above the center of gravity. The object will be stable because a *righting moment* will be created if the object tips over, since the center of buoyancy will move outward from the center of gravity.

[23]The subject of intact stability, being a part of naval architecture curriculum, is not covered extensively in most fluids books. However, it is covered extensively in basic ship design and naval architecture books.

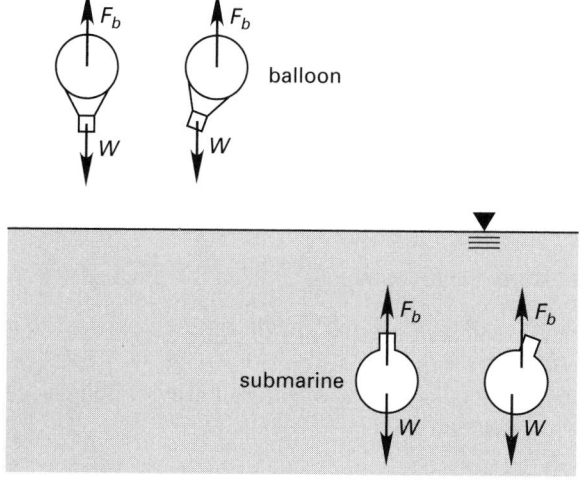

Figure 15.24 *Stability of a Submerged Object*

This righting couple exists when the extension of the buoyant force, F_b, intersects line O-O above the center of gravity at M, the *metacenter*. If M lies below the center of gravity, an overturning couple will exist. The distance between the center of gravity and the metacenter is called the *metacentric height*, and it is reasonably constant for heel angles less than 10°. Also, for angles less than 10°, the center of buoyancy follows a locus for which the metacenter is the instantaneous center.

The metacentric height is one of the most important and basic parameters in ship design. It determines the ship's ability to remain upright as well as the ship's roll and pitch characteristics.

The stability criterion is different for partially submerged objects (e.g., surface ships). For partially submerged objects to be stable, the *metacenter* must be above the center of gravity. If the vessel shown in Fig. 15.25 heels (i.e., lists or rolls), the location of the center of gravity of the object does not change.[24] However, the center of buoyancy shifts to the centroid of the new submerged section 123. The centers of buoyancy and gravity are no longer in line. The righting couple resists further overturning.

"Acceptable" minimum values of the metacentric height have been established from experience, and these depend on the ship type and class. For example, many submarines are required to have a metacentric height of 1 ft (0.3 m) when surfaced. This will increase to approximately 3.5 ft (1.2 m) for some of the largest surface ships. If an acceptable metacentric height is not achieved initially, the center of gravity must be lowered or the keel depth increased. The beam width can also be increased slightly to increase the waterplane moment of inertia.

For a surface vessel rolling through an angle less than approximately 10°, the distance between the vertical center of gravity and the metacenter can be found from Eq. 15.65. Variable I is the centroidal area moment of inertia of the original waterline (free surface) cross section about a longitudinal (fore and aft) waterline axis; V is the displaced volume.

If the distance, y_{bg}, separating the centers of buoyancy and gravity is known, Eq. 15.65 can be solved for the metacentric height. y_{bg} is positive when the center of gravity is above the center of buoyancy. This is the normal case. Otherwise, y_{bg} is negative.

$$y_{bg} + h_m = \frac{I}{V} \qquad \textit{15.65}$$

The righting moment (also known as the *restoring moment*) is the stabilizing moment exerted when the ship rolls. Values of the righting moment are typically specified with units of foot-tons (MN·m).

$$M_{\text{righting}} = h_m \gamma_w V_{\text{displaced}} \sin \theta \qquad \textit{15.66}$$

The transverse (roll) and longitudinal (pitch) *periods* also depend on the metacentric height. The roll characteristics are found from the differential equation formed by equating the righting moment to the product of the ship's transverse mass moment of inertia and the angular acceleration. Larger metacentric heights result in lower roll periods. If k is the radius of gyration about the roll axis, the roll period is

$$T_{\text{roll}} = \frac{2\pi k}{\sqrt{g h_m}} \qquad \textit{15.67}$$

Figure 15.25 *Stability of a Partially Submerged Floating Object*

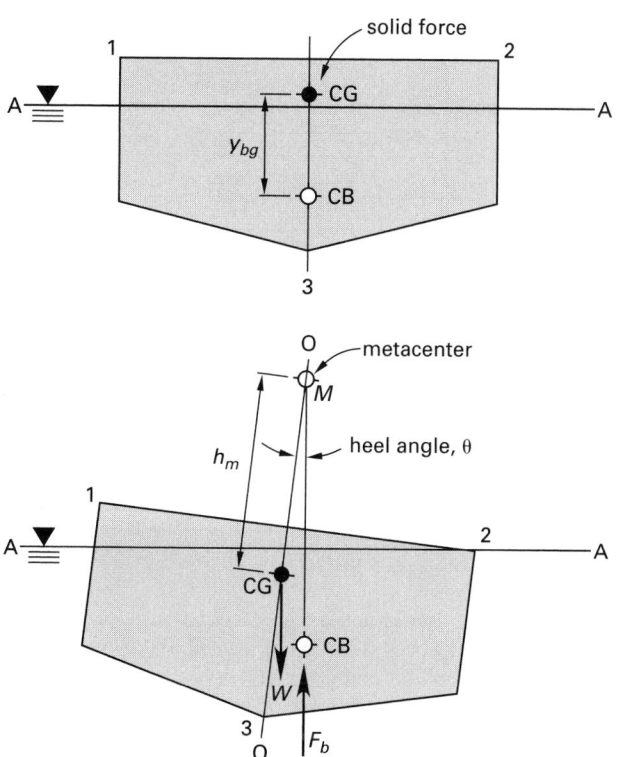

---

[24]The verbs *roll*, *list* and *heel* are synonymous.

The roll and pitch periods must be adjusted for the appropriate level of crew and passenger comfort. A "beamy" ice-breaking ship will have a metacentric height much larger than normally required for intact stability, resulting in a low, nauseating, roll period. The designer of a passenger ship, however, would have to decrease the intact stability (i.e., decrease the metacentric height) in order to achieve an acceptable ride characteristic. This requires a metacentric height that is less than approximately 6% of the beam length.

Example 15.9

A 600,000 lbm (280 000 kg) rectangular barge has external dimensions of 24 ft width, 98 ft length, and 12 ft height (7 m × 30 m × 3.6 m). It floats in seawater ($\gamma_w = 64.0$ lbf/ft^3; $\rho_w = 1024$ kg/m^3). The center of gravity is 7.8 ft (2.4 m) from the top of the barge as loaded. Find (a) the location of the center of buoyancy when the barge is floating on an even keel, and (b) the approximate location of the metacenter when the barge experiences a 5° heel.

SI Solution

(a) Refer to the following diagram. Let dimension y represent the depth of the submerged barge.

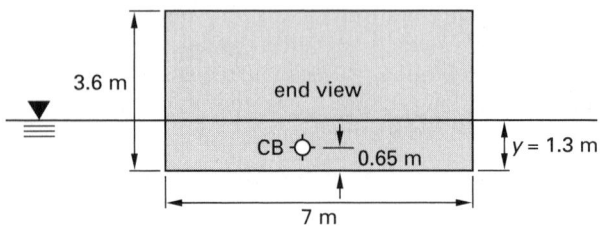

From Archimedes' principle, the buoyant force equals the weight of the barge. This, in turn, equals the weight of the displaced seawater.

$$F_b = W = V\gamma_w = V\rho_w g$$

$$(280\,000 \text{ kg})\left(9.81 \frac{\text{m}}{\text{s}^2}\right) = y(7 \text{ m})(30 \text{ m})\left(1024 \frac{\text{kg}}{\text{m}^3}\right)$$
$$\times \left(9.81 \frac{\text{m}}{\text{s}^2}\right)$$
$$y = 1.30 \text{ m}$$

The center of buoyancy is located at the centroid of the submerged cross section. When floating on an even keel, the submerged cross section is rectangular with a height of 1.30 m. The height of the center of buoyancy above the keel is

$$\frac{1.30 \text{ m}}{2} = 0.65 \text{ m}$$

(b) While the location of the new center of buoyancy can be determined, the location of the metacenter does

not change significantly for small angles of heel. Therefore, for approximate calculations, the angle of heel is not significant.

The area moment of inertia of the longitudinal waterline cross section is

$$I = \frac{Lw^3}{12} = \frac{(30 \text{ m})(7 \text{ m})^3}{12}$$
$$= 858 \text{ m}^4$$

The submerged volume is

$$V = (1.3 \text{ m})(7 \text{ m})(30 \text{ m}) = 273 \text{ m}^3$$

The distance between the center of gravity and the center of buoyancy is

$$y_{bg} = 3.6 \text{ m} - 2.4 \text{ m} - 0.65 \text{ m} = 0.55 \text{ m}$$

The metacentric height measured above the center of gravity is

$$h_m = \frac{I}{V} - y_{bg}$$
$$= \frac{858 \text{ m}^4}{273 \text{ m}^3} - 0.55 \text{ m} = 2.6 \text{ m}$$

Customary U.S. Solution

(a) Refer to the following diagram. Let dimension y represent the depth of the submerged barge.

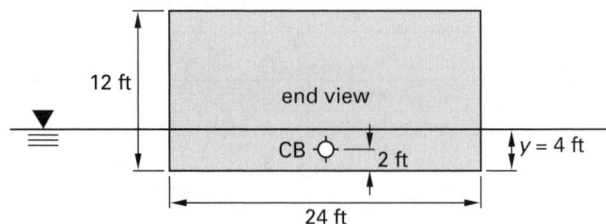

From Archimedes' principle, the buoyant force equals the weight of the barge. This, in turn, equals the weight of the displaced seawater.

$$F_b = W = V\gamma_w$$

$$600,000 \text{ lbf} = y(24 \text{ ft})(98 \text{ ft})\left(64 \frac{\text{lbf}}{\text{ft}^3}\right)$$
$$y = 4.00 \text{ ft}$$

The center of buoyancy is located at the centroid of the submerged cross section. When floating on an even keel, the submerged cross section is rectangular with a height of 4.00 ft. The height of the center of buoyancy above the keel is

$$\frac{4.00 \text{ ft}}{2} = 2.00 \text{ ft}$$

(b) While the location of the new center of buoyancy can be determined, the location of the metacenter does not change significantly for small angles of heel. Therefore, for approximate calculations, the angle of heel is not significant.

The area moment of inertia of the longitudinal waterline cross section is

$$I = \frac{Lw^3}{12} = \frac{(98 \text{ ft})(24 \text{ ft})^3}{12}$$
$$= 112{,}900 \text{ ft}^4$$

The submerged volume is

$$V = (4 \text{ ft})(24 \text{ ft})(98 \text{ ft}) = 9408 \text{ ft}^3$$

The distance between the center of gravity and the center of buoyancy is

$$y_{bg} = 12 \text{ ft} - 7.8 \text{ ft} - 2.0 \text{ ft} = 2.2 \text{ ft}$$

The metacentric height measured above the center of gravity is

$$h_m = \frac{I}{V} - y_{bg}$$
$$= \frac{112{,}900 \text{ ft}^4}{9408 \text{ ft}^3} - 2.2 \text{ ft}$$
$$= 9.8 \text{ ft}$$

21. FLUID MASSES UNDER EXTERNAL ACCELERATION

Up to this point, fluid masses have been stationary, and gravity has been the only force acting on them. If a fluid mass is subjected to an external acceleration (moved sideways, rotated, etc.), an additional force will be introduced. This force will change the equilibrium position of the fluid surface as well as the hydrostatic pressure distribution.

Figure 15.26 illustrates a liquid mass subjected to constant accelerations in the vertical and/or horizontal directions. (a_y is negative if the acceleration is downward.) The surface is inclined at the angle predicted by Eq. 15.68. The planes of equal hydrostatic pressure beneath the surface are also inclined at the same angle.[25]

$$\theta = \arctan\left(\frac{a_x}{a_y + g}\right) \qquad 15.68$$

$$p = \rho g h \left(1 + \frac{a_y}{g}\right) \qquad \text{[SI]} \qquad 15.69(a)$$

$$p = \left(\frac{\rho g h}{g_c}\right)\left(1 + \frac{a_y}{g}\right) = \gamma h \left(1 + \frac{a_y}{g}\right)$$
$$\text{[U.S.]} \qquad 15.69(b)$$

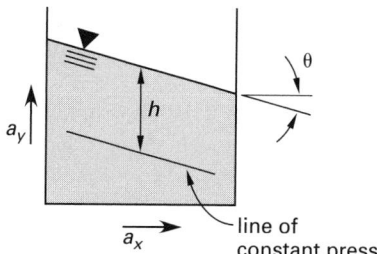

Figure 15.26 *Fluid Mass Under Constant Linear Acceleration*

Figure 15.27 illustrates a fluid mass rotating at constant angular velocity, ω in rad/sec.[26] The resulting surface is parabolic in shape. The elevation of the fluid surface at point A at distance r from the axis of rotation is given by Eq. 15.71. The distance h in Fig. 15.27 is measured from the lowest fluid elevation during rotation. h is not measured from the original elevation of the stationary fluid.

$$\theta = \arctan\left(\frac{\omega^2 r}{g}\right) \qquad 15.70$$

$$h = \frac{(\omega r)^2}{2g} = \frac{\text{v}^2}{2g} \qquad 15.71$$

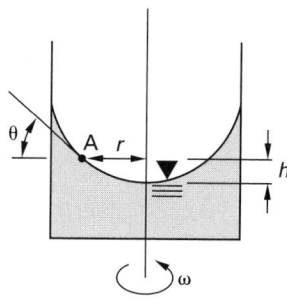

Figure 15.27 *Rotating Fluid Mass*

---

[25]Once the orientation of the surface is known, the pressure distribution can be determined without considering the acceleration. The hydrostatic pressure at a point depends only on the height of the liquid above that point. The acceleration affects that height but does not change the $p = \rho g h$ relationship.

[26]Even though the rotational speed is not increasing, the fluid mass experiences a constant *centripetal acceleration* radially outward from the axis of rotation.

16 Fluid Flow Parameters

Nomenclature

A	area	ft^2	m^2
C	correction factor	–	–
d	depth	ft	m
D	diameter	ft	m
E	specific energy	ft-lbf/lbm	J/kg
g	gravitational acceleration	ft/sec^2	m/s^2
g_c	gravitational conversion constant	lbm-ft/lbf-sec^2	n.a.
G	mass flow rate per unit area	lbm/ft^2-sec	kg/m$^2\cdot$s
h	height, or head	ft	m
L	length	ft	m
p	pressure	lbf/ft^2	Pa
r	radius	ft	m
Re	Reynolds number	–	–
s	wetted perimeter	ft	m
v	velocity	ft/sec	m/s
$\dot{V}$	volumetric flow rate	ft^3/sec	m^3/s
y	distance	ft	m
z	elevation	ft	m

Symbols

α	angle	rad	rad
θ	time	sec	s
μ	absolute viscosity	lbf-sec/ft^2	Pa·s
ν	kinematic viscosity	ft^2/sec	m^2/s
ρ	density	lbm/ft^3	kg/m^3
ϕ	angle	rad	rad

Subscripts

e	equivalent
h	hydraulic
i	impact or inner
o	outer
p	pressure
r	radius
s	static
t	total
v	velocity
z	potential

1. INTRODUCTION TO FLUID ENERGY UNITS

Several important fluids and thermodynamics equations, such as Bernoulli's equation and the steady-flow energy equation, are special applications of the *conservation of energy* concept. However, it is not always obvious how some formulations of these equations can be termed "energy." For example, elevation (z), with units of feet, is often called *gravitational energy*.[1]

Since every fluids problem is different, energy must be expressed per unit mass (i.e., *specific energy*). With SI formulations, the choice of units is unambiguous: J/kg is the only choice.

$$\frac{\text{J}}{\text{kg}} = \frac{\text{N}\cdot\text{m}}{\text{kg}} = \frac{\text{m}^2}{\text{s}^2} \qquad \textit{16.1}$$

With problems formulated in English units, specific energy units will be ft-lbf/lbm.[2] (If the consistent set of units ft-lbf/slug is chosen, the same equations can be used with SI and consistent English units, since the primary dimensions (see Ch. 1) are the same, that is, L^2/θ^2.)

$$\frac{\text{ft-lbf}}{\text{slug}} = \frac{\text{ft}^2}{\text{sec}^2} \qquad \textit{16.2}$$

The gravitational conversion constant, g_c, must be used if the units ft-lbf/lbm are used for specific fluid energy.

In many cases, the ratio g/g_c appears in equations. Since g and g_c have the same numerical value in most

---

[1]Foot and ft-lbf/lbm may be thought of as one and the same. Certainly, the set of units ft-lbf/lbm represents energy per unit mass. Unfortunately, lbf and lbm do not really cancel out to yield ft.

[2]Btu could be used for energy, as is common in thermodynamics problems, instead of ft-lbf. However, this is almost never done in fluids problems.

instances, this ratio affects the units without affecting the calculation. Because of this, some engineers omit the term g/g_c entirely. While it is easy to justify such a practice, the resulting equations are not dimensionally consistent.[3]

2. KINETIC ENERGY

Energy is required to accelerate a stationary body. Thus, a moving mass of fluid possesses more energy than an identical, stationary mass. The energy difference is the *kinetic energy* of the fluid.[4] If the kinetic energy is evaluated per unit mass, the term *specific kinetic energy* is used. Equation 16.3 gives the specific kinetic energy corresponding to a fluid flow with uniform (i.e., turbulent) velocity, v.

$$E_{\text{v}} = \frac{\text{v}^2}{2} \qquad \text{[SI]} \qquad \textit{16.3(a)}$$

$$E_{\text{v}} = \frac{\text{v}^2}{2g_c} \qquad \text{[U.S.]} \qquad \textit{16.3(b)}$$

The units of specific kinetic energy in consistent units are clearly m^2/s^2 or ft^2/sec^2, which coincide with Eqs. 16.1 and 16.2 as representing energy per consistent mass unit. The units in traditional English units are

$$\frac{\left(\dfrac{\text{ft}}{\text{sec}}\right)^2}{\dfrac{\text{lbm-ft}}{\text{lbf-sec}^2}} = \frac{\text{ft-lbf}}{\text{lbm}} \qquad \textit{16.4}$$

3. POTENTIAL ENERGY

Work is performed in elevating a body. Thus, a mass of fluid at a high elevation will have more energy than an identical mass of fluid at a lower elevation. The energy difference is the *potential energy* of the fluid.[5] Like kinetic energy, potential energy is usually expressed per unit mass. Equation 16.5 gives the potential energy of fluid at an elevation z.

$$E_z = zg \qquad \text{[SI]} \qquad \textit{16.5(a)}$$

$$E_z = \frac{zg}{g_c} \qquad \text{[U.S.]}[6] \qquad \textit{16.5(b)}$$

The units of potential energy in a consistent system are again m^2/s^2 or ft^2/sec^2. The units in a traditional English system are ft-lbf/lbm.

---

[3]More than being dimensionally inconsistent, the resulting equations will be numerically incorrect in any nonstandard (other than one gravity) gravitational field.

[4]The terms *velocity energy* and *dynamic energy* are used less often.

[5]The term *gravitational energy* is also used.

[6]Since $g = g_c$ (numerically), it is tempting to write $E_z = z$. In fact, many engineers in the United States do just that.

z is the elevation of the fluid. The reference point (i.e., zero elevation point) is entirely arbitrary and can be chosen for convenience. This is because potential energy always appears in a difference equation (i.e., ΔE_z), and the reference point cancels out.

4. PRESSURE ENERGY

Work is performed and energy is expended when a substance is compressed. Thus, a mass of fluid at a high pressure will have more energy than an identical mass of fluid at a lower pressure. The energy difference is the *pressure energy* of the fluid.[7] Pressure energy is usually found in equations along with kinetic and potential energies, and is expressed as energy per unit mass. Equation 16.6 gives the pressure energy of fluid at pressure p.

$$E_p = \frac{p}{\rho} \qquad \textit{16.6}$$

The consistent SI units of pressure energy are

$$\frac{\text{Pa·m}^3}{\text{kg}} = \frac{\text{N·m}}{\text{kg}} = \frac{\text{J}}{\text{kg}} \qquad \textit{16.7}$$

The units of pressure energy in the consistent English system are

$$\frac{\text{lbf-ft}^3}{\text{ft}^2\text{-slug}} = \frac{\text{ft-lbf}}{\text{slug}} \qquad \textit{16.8}$$

The units of pressure energy in the traditional English system are

$$\frac{\text{lbf-ft}^3}{\text{ft}^2\text{-lbm}} = \frac{\text{ft-lbf}}{\text{lbm}} \qquad \textit{16.9}$$

5. BERNOULLI EQUATION

The *Bernoulli equation* is an energy conservation equation based on several reasonable assumptions. The equation assumes the following.

- The fluid is incompressible.

- There is no fluid friction.

- Changes in thermal energy are negligible.[8]

The Bernoulli equation states that the *total energy* of a fluid flowing without friction losses in a pipe is constant.[9] The total energy possessed by the fluid is the

---

[7]The terms *static energy* and *flow energy* are also used. The name *flow energy* results from the need to push (pressurize) a fluid to get it to flow through a pipe. However, flow energy and kinetic energy are not the same.

[8]In thermodynamics, the fluid flow is said to be *adiabatic*.

[9]Strictly speaking, this is the *total specific energy*, since the energy is per unit mass. However, the word "specific," being understood, is seldom used. Of course, "the total energy of the system" means something else and requires knowing the fluid mass in the system.

sum of its pressure, kinetic, and potential energies. Drawing on Eqs. 16.3, 16.5, and 16.6, the Bernoulli equation is written as

$$E_t = E_p + E_v + E_z \qquad \textit{16.10}$$

$$E_t = \frac{p}{\rho} + \frac{v^2}{2} + zg \qquad \text{[SI]} \qquad \textit{16.11(a)}$$

$$E_t = \frac{p}{\rho} + \frac{v^2}{2g_c} + \frac{zg}{g_c} \qquad \text{[U.S.]} \qquad \textit{16.11(b)}$$

Equation 16.11 is valid for both laminar and turbulent flows. It can also be used for gases and vapors if the incompressibility assumption is valid.[10]

The quantities known as *total head*, h_t, and *total pressure*, p_t, can be calculated from total energy.

$$h_t = \frac{E_t}{g} \qquad \text{[SI]} \qquad \textit{16.12(a)}$$

$$h_t = E_t \times \frac{g_c}{g} \qquad \text{[U.S.]} \qquad \textit{16.12(b)}$$

$$p_t = \rho g h_t \qquad \text{[SI]} \qquad \textit{16.13(a)}$$

$$p_t = \rho h_t \times \frac{g}{g_c} \qquad \text{[U.S.]} \qquad \textit{16.13(b)}$$

Example 16.1

A pipe draws water from the bottom of a reservoir and discharges it freely at point C, 100 ft (30 m) below the surface. The flow is frictionless. (a) What is the total specific energy at an elevation 50 ft (15 m) below the water surface (i.e., point B)? (b) What is the velocity at point C?

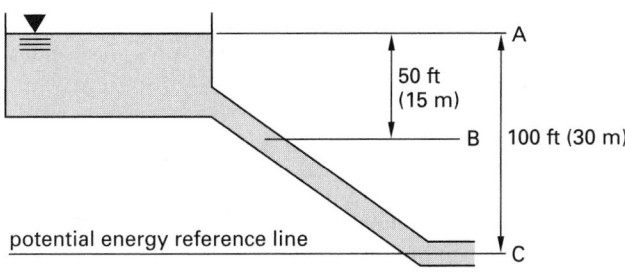

SI Solution

(a) At point A, the velocity and gage pressure are both zero. Therefore, the total energy consists only of potential energy. Point C is chosen as the reference ($z = 0$) elevation.

$$E_A = z_A g = (30 \text{ m}) \left(9.81 \, \frac{\text{m}}{\text{s}^2} \right)$$
$$= 294.3 \text{ m}^2/\text{s}^2 \text{ (J/kg)}$$

At point B, the fluid is moving and possesses kinetic energy. The fluid is also under hydrostatic pressure and possesses pressure energy. These energy forms have come at the expense of potential energy. (This is a direct result of the Bernoulli equation.) Also, the flow is frictionless. Thus, there is no net change in the total energy between points A and B.

$$E_B = E_A = 294.3 \text{ m}^2/\text{s}^2 \text{ (J/kg)}$$

(b) At point C, the gage pressure and pressure energy are again zero, since the discharge is at atmospheric pressure. The potential energy is zero, since $z = 0$. The total energy of the system has been converted to kinetic energy. From Eq. 16.11,

$$E_t = 294.3 \, \frac{\text{m}^2}{\text{s}^2} = 0 + \frac{v^2}{2} + 0$$

$$v = 24.3 \text{ m/s}$$

Customary U.S. Solution

(a) $\quad E_A = \dfrac{z_A g}{g_c} = \dfrac{(100 \text{ ft}) \left(32.2 \, \dfrac{\text{ft}}{\text{sec}^2} \right)}{32.2 \, \dfrac{\text{lbm-ft}}{\text{lbf-sec}^2}}$

$$= 100 \text{ ft-lbf/lbm}$$

$$E_B = E_A = 100 \text{ ft-lbf/lbm}$$

(b) $\quad E_t = 100 \, \dfrac{\text{ft-lbf}}{\text{lbm}} = 0 + \dfrac{v^2}{2g_c} + 0$

$$v^2 = (2) \left(32.2 \, \frac{\text{lbm-ft}}{\text{lbf-sec}^2} \right) \left(100 \, \frac{\text{ft-lbf}}{\text{lbm}} \right)$$

$$= 6440 \text{ ft}^2/\text{sec}^2$$

$$v = 80.2 \text{ ft/sec}$$

Example 16.2

Water (62.4 lbm/ft³; 1000 kg/m³) is pumped up a hillside into a reservoir. The pump discharges water at the rate of 6 ft/sec (2 m/s) and pressure of 150 psig (1000 kPa). Disregarding friction, what is the maximum elevation (above the centerline of the pump's discharge) of the reservoir's water surface?

SI Solution

At the centerline of the pump's discharge, the potential energy is zero. The pressure and velocity energies are

$$E_p = \frac{p}{\rho} = \frac{(1000 \text{ kPa}) \left(1000 \, \dfrac{\text{Pa}}{\text{kPa}} \right)}{1000 \, \dfrac{\text{kg}}{\text{m}^3}} = 1000 \text{ J/kg}$$

$$E_v = \frac{v^2}{2} = \frac{\left(2 \, \dfrac{\text{m}}{\text{s}} \right)^2}{2} = 2 \text{ J/kg}$$

---

[10]A gas or vapor can be considered to be incompressible as long as its pressure does not change by more than 10% between the entrance and exit, and its velocity is less than Mach 0.3 everywhere.

The total energy at the pump's discharge is

$$E_{t,1} = E_p + E_v$$

$$= 1000 \ \frac{J}{kg} + 2 \ \frac{J}{kg} = 1002 \ J/kg$$

Since the flow is frictionless, the same energy is possessed by the water at the reservoir's surface. Since the velocity and gage pressure at the surface are zero, all of the available energy has been converted to potential energy.

$$E_{t,2} = E_{t,1}$$

$$z_2 g = 1002 \ J/kg$$

$$z_2 = \frac{E_{t,2}}{g} = \frac{1002 \ \dfrac{J}{kg}}{9.81 \ \dfrac{m}{s^2}} = 102.1 \ m$$

Notice that the volumetric flow rate of the water is not relevant since the water velocity was known. Similarly, the pipe size is not needed.

Customary U.S. Solution

$$E_p = \frac{p}{\rho} = \frac{\left(150 \ \dfrac{lbf}{in^2}\right)\left(144 \ \dfrac{in^2}{ft^2}\right)}{62.4 \ \dfrac{lbm}{ft^3}}$$

$$= 346.15 \ ft\text{-}lbf/lbm$$

$$E_v = \frac{v^2}{2g_c} = \frac{\left(6 \ \dfrac{ft}{sec}\right)^2}{(2)\left(32.2 \ \dfrac{lbm\text{-}ft}{lbf\text{-}sec^2}\right)}$$

$$= 0.56 \ ft\text{-}lbf/lbm$$

$$E_{t,1} = E_p + E_v$$

$$= 346.15 \ \frac{ft\text{-}lbf}{lbm} + 0.56 \ \frac{ft\text{-}lbf}{lbm} = 346.71 \ ft\text{-}lbf/lbm$$

$$E_{t,2} = E_{t,1}$$

$$\frac{z_2 g}{g_c} = 346.71 \ ft\text{-}lbf/lbm$$

$$z_2 = \frac{E_{t,2} g_c}{g} = \frac{\left(346.71 \ \dfrac{ft\text{-}lbf}{lbm}\right)\left(32.2 \ \dfrac{lbm\text{-}ft}{lbf\text{-}sec^2}\right)}{32.2 \ \dfrac{ft}{sec^2}}$$

$$= 346.71 \ ft$$

6. PITOT TUBE

A *pitot tube* (also known as an *impact tube* or *stagnation tube*) is simply a hollow tube that is placed longitudinally in the direction of fluid flow, allowing the flow to enter one end at the fluid's *velocity of approach*. It is used to measure velocity of flow and finds uses in both subsonic and supersonic applications.

Figure 16.1 *Pitot Tube*

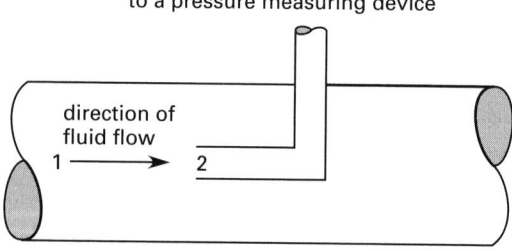

When the fluid enters the pitot tube, it is forced to come to a stop (at the *stagnation point*), and the velocity energy is transformed into pressure energy. If the fluid is a low-velocity gas, the stagnation is assumed to occur without compression heating of the gas. If there is no friction (the common assumption), the process is said to be adiabatic.

Bernoulli's equation can be used to predict the static pressure at the stagnation point. Since the velocity of the fluid within the pitot tube is zero, the upstream velocity can be calculated if the static and stagnation pressures are known.

$$\frac{p_1}{\rho} + \frac{v_1^2}{2} = \frac{p_2}{\rho} \qquad \qquad 16.14$$

$$v_1 = \sqrt{\frac{2(p_2 - p_1)}{\rho}} \qquad \text{[SI]} \quad 16.15(a)$$

$$v_1 = \sqrt{\frac{2g_c(p_2 - p_1)}{\rho}} \qquad \text{[U.S.]} \quad 16.15(b)$$

In reality, both friction and heating occur, and the fluid may be compressible. These errors are taken care of by a correction factor known as the *impact factor*, C_i, which is applied to the derived velocity. C_i is usually very close to 1.00 (e.g., 0.99 or 0.995).

$$v_{actual} = C_i \ v_{indicated}$$

Since accurate measurements of fluid velocity are dependent on one-dimensional fluid flow, it is essential that any obstructions or pipe bends be more than ten pipe diameters upstream from the pitot tube.

7. IMPACT ENERGY

Impact energy, E_i (also known as *stagnation energy* and *total energy*), is the sum of the kinetic and pressure energy terms.[11] Equation 16.17 is applicable to liquids

---

[11]It is confusing to label Eq. 16.16 *total* when the gravitational energy term has been omitted. However, the reference point for gravitational energy is arbitrary, and in this application the reference coincides with the centerline of the fluid flow. In truth, the effective pressure developed in a fluid which has been brought to rest adiabatically does not depend on the elevation or altitude of the fluid. This situation is seldom ambiguous. The application will determine which definition of total head or total energy is intended.

and gases flowing with velocities less than approximately Mach 0.3.

$$E_i = E_p + E_v \qquad 16.16$$

$$E_i = \frac{p}{\rho} + \frac{v^2}{2} \qquad \text{[SI]} \qquad 16.17(a)$$

$$E_i = \frac{p}{\rho} + \frac{v^2}{2g_c} \qquad \text{[U.S.]} \qquad 16.17(b)$$

Impact head, h_i, is calculated from the impact energy in a manner analogous to Eq. 16.12. Impact head represents the height the liquid will rise in a piezometer-pitot tube when the liquid has been brought to rest (i.e., stagnated) in an adiabatic manner. Such a case is illustrated in Fig. 16.2. If a gas or high-velocity, high-pressure liquid is flowing, it will be necessary to use a mercury manometer or pressure gauge to measure stagnation head.

Figure 16.2 *Pitot Tube-Piezometer Apparatus*

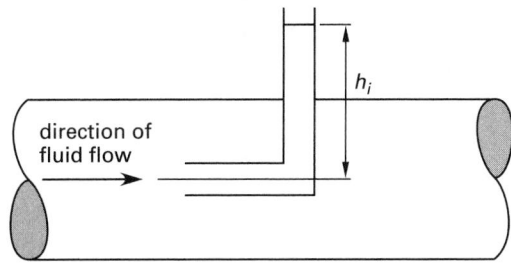

Example 16.3

The static pressure of air $(0.075 \text{ lbm/ft}^3; 1.2 \text{ kg/m}^3)$ flowing in a pipe is measured by a precision gauge to be 10.00 psig (68.95 kPa). A pitot tube-manometer indicates 20.6 in (0.52 m) of mercury. The density of mercury is 0.491 lbm/in^3 (13 600 kg/m^3). Losses are insignificant. What is the velocity of the air in the pipe?

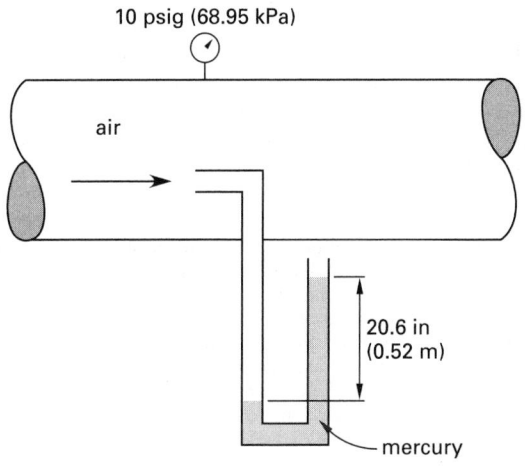

SI Solution

From Table 14.2, the density of mercury is 13 600 kg/m^3. The impact pressure is

$$p_i = \rho g h$$

$$= \left(13\,600 \; \frac{\text{kg}}{\text{m}^3}\right)\left(9.81 \; \frac{\text{m}}{\text{s}^2}\right)\left(\frac{0.52 \text{ m}}{1000 \; \frac{\text{Pa}}{\text{kPa}}}\right)$$

$$= 69.38 \text{ kPa}$$

From Eq. 16.15, the velocity is

$$v = \sqrt{\frac{2(p_i - p_s)}{\rho}}$$

$$= \sqrt{\frac{(2)(69.38 \text{ kPa} - 68.95 \text{ kPa})\left(1000 \; \frac{\text{Pa}}{\text{kPa}}\right)}{1.2 \; \frac{\text{kg}}{\text{m}^3}}}$$

$$= 26.8 \text{ m/s}$$

Customary U.S. Solution

The pitot tube measures impact (stagnation) pressure. The impact pressure could be calculated from $p = \gamma h$. Alternatively, from Table 15.2, the fluid height equivalent (specific weight) of mercury is 0.491 psi/in. Therefore, the impact pressure is

$$p_i = (20.6 \text{ in})\left(0.491 \; \frac{\text{psi}}{\text{in}}\right) = 10.11 \text{ psig}$$

Since impact pressure is the sum of the static and kinetic (velocity) pressures, the kinetic pressure is

$$p_v = p_i - p_s$$
$$= 10.11 \text{ psig} - 10.00 \text{ psig} = 0.11 \text{ psi}$$

The velocity is calculated from Eq. 16.15.

$$v = \sqrt{\frac{2g_c(p_i - p_s)}{\rho}}$$

$$= \sqrt{\frac{(2)\left(32.2 \; \frac{\text{lbm-ft}}{\text{lbf-sec}^2}\right)\left(0.11 \; \frac{\text{lbf}}{\text{in}^2}\right)\left(144 \; \frac{\text{in}^2}{\text{ft}^2}\right)}{0.075 \; \frac{\text{lbm}}{\text{ft}^3}}}$$

$$= 117 \text{ ft/sec}$$

8. HYDRAULIC RADIUS

The *hydraulic radius* is defined as the area in flow divided by the *wetted perimeter*.[12] (The hydraulic radius is not the same as the radius of a pipe.) The area in flow is the cross-sectional area of the fluid flowing. When a fluid is flowing under pressure in a pipe (i.e., *pressure flow* in a *pressure conduit*), the area in flow will be the internal area of the pipe. However, the fluid may not completely fill the pipe and may flow simply because of a sloped surface (i.e., *gravity flow* or *open channel flow*).

The wetted perimeter is the length of the line representing the interface between the fluid and the pipe or channel. It does not include the *free surface* length (i.e., the interface between fluid and atmosphere).

$$r_h = \frac{\text{area in flow}}{\text{wetted perimeter}} = \frac{A}{s} \qquad 16.18$$

Consider a circular pipe flowing completely full. The area in flow is πr^2. The wetted perimeter is the entire circumference, $2\pi r$. The hydraulic radius is

$$r_{h,\text{pipe}} = \frac{\pi r^2}{2\pi r} = \frac{r}{2} = \frac{D}{4} \qquad 16.19$$

The hydraulic radius of a pipe flowing half full is also $r/2$, since the flow area and wetted perimeter are both halved. However, it is time-consuming to calculate the hydraulic radius for pipe flow at any intermediate depth, due to the difficulty in evaluating the flow area and wetted perimeter. Appendix 16.A greatly simplifies such calculations.

Example 16.4

A pipe (internal diameter = 6 units) carries water with a depth of 2 units flowing under the influence of gravity. (a) Calculate the hydraulic radius analytically. (b) Verify the result by using App. 16.A.

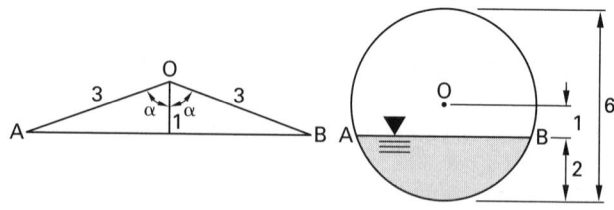

Solution

(a) The equations for a circular segment must be used. The radius is $6/2 = 3$.

---

[12]The hydraulic radius can also be calculated as one-fourth of the hydraulic diameter of the pipe or channel, as will be subsequently shown. That is, $r_h = {}^1/_4 D_h$.

Points A, O, and B are used to find the central angle of the circular segment.

$$\phi = 2\alpha = 2\left(\arccos \tfrac{1}{3}\right)$$
$$= (2)(70.53°) = 141.06°$$

ϕ must be expressed in radians.

$$\phi = 2\pi\left(\frac{141.06°}{360°}\right) = 2.46 \text{ rad}$$

The area of the circular segment (i.e., the area in flow) is

$$A = \tfrac{1}{2}r^2(\phi - \sin\phi) \quad [\phi \text{ in radians}]$$
$$= (0.5)(3)^2(2.46 \text{ rad} - \sin(2.46 \text{ rad}))$$
$$= 8.235 \text{ units}^2$$

The arc length (i.e., the wetted perimeter) is

$$s = r\phi = (3)(2.46 \text{ rad}) = 7.38 \text{ units}$$

The hydraulic radius is

$$r_h = \frac{A}{s} = \frac{8.235 \text{ units}^2}{7.38 \text{ units}} = 1.12 \text{ units}$$

(b) The ratio d/D is needed to use App. 16.A.

$$\frac{d}{D} = \frac{2 \text{ units}}{6 \text{ units}} = 0.333$$

From App. 16.A,

$$\frac{r_h}{D} \approx 0.186$$
$$r_h = (0.186)(6 \text{ units}) = 1.12 \text{ units}$$

9. HYDRAULIC DIAMETER

Many fluid, thermodynamic, and heat transfer processes are dependent on the physical length of an object. The general name for this controlling variable is *characteristic dimension*. The characteristic dimension in evaluating fluid flow is the *hydraulic diameter* (also known as the *equivalent hydraulic diameter*).[13] The hydraulic diameter for a full-flowing pipe is simply its inside diameter. The hydraulic diameters of other cross sections in flow are given in Table 16.1. If the hydraulic radius is known, it can be used to calculate the hydraulic diameter.

$$D_h = 4r_h \qquad 16.20$$

---

[13]The engineering community is very inconsistent, but the three terms—hydraulic depth, hydraulic diameter, and equivalent diameter—do not have the same meanings. Hydraulic depth (flow area divided by exposed surface width) is a characteristic length used in Froude number and other open channel flow calculations. Hydraulic diameter (four times the area in flow divided by the wetted surface) is a characteristic length used in Reynolds number and friction loss calculations. Equivalent diameter $1.3(ab)^{0.625}/(a+b)^{0.25}$ is the diameter of a round duct or pipe that will have the same friction loss per unit length as a rectangular duct. Unfortunately, these terms are often used interchangeably.

Table 16.1 *Hydraulic Diameters for Common Conduit Shapes*

conduit cross section	D_h
flowing full	
circle	D
annulus (outer diameter D_o, inner diameter D_i)	$D_o - D_i$
square (side L)	L
rectangle (sides L_1 and L_2)	$\dfrac{2L_1 L_2}{L_1 + L_2}$
flowing partially full	
half-filled circle (diameter D)	D
rectangle (h deep, L wide)	$\dfrac{4hL}{L + 2h}$
wide, shallow stream (h deep)	$4h$
triangle (h deep, L broad, s side)	$\dfrac{hL}{s}$
trapezoid (h deep, a wide at top, b wide at bottom, s side)	$\dfrac{2h(a+b)}{b + 2s}$

Example 16.5

Determine the hydraulic diameter and hydraulic radius for the open trapezoidal channel shown.

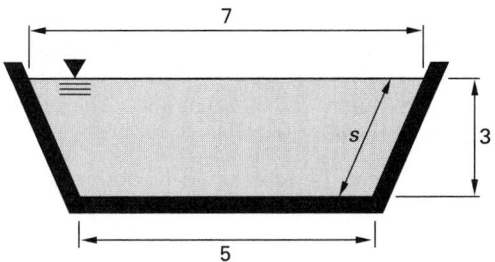

Solution

The batter of the inclined walls is $(7-5)/2$ walls $= 1$.

$$s = \sqrt{(3)^2 + (1)^2} = 3.16$$

Using Table 16.1,

$$D_h = \frac{2h(a+b)}{b+2s} = \frac{(2)(3)(7+5)}{5 + (2)(3.16)} = 6.36$$

From Eq. 16.20,

$$r_h = \frac{D_h}{4} = \frac{6.36}{4} = 1.59$$

10. REYNOLDS NUMBER

The *Reynolds number*, Re, is a dimensionless number interpreted as the ratio of inertial forces to viscous forces in the fluid.[14]

$$\text{Re} = \frac{\text{inertial forces}}{\text{viscous forces}} \qquad 16.21$$

The inertial forces are proportional to the flow diameter, velocity, and fluid density. (Increasing these variables will increase the momentum of the fluid in flow.) The viscous force is represented by the fluid's absolute viscosity, μ. Thus, the Reynolds number is calculated as

$$\text{Re} = \frac{D_h \text{v} \rho}{\mu} \qquad \text{[SI]} \qquad 16.22(a)$$

$$\text{Re} = \frac{D_h \text{v} \rho}{g_c \mu} \qquad \text{[U.S.]} \qquad 16.22(b)$$

Since μ/ρ is defined as the *kinematic viscosity*, ν, Eq. 16.22 can be simplified.[15]

$$\text{Re} = \frac{D_h \text{v}}{\nu} \qquad 16.23$$

Occasionally, the *mass flow rate per unit area*, $G = \rho \text{v}$, will be known. This variable expresses the quantity of fluid flowing in kg/m²·s or lbm/ft²-sec.

$$\text{Re} = \frac{D_h G}{\mu} \qquad \text{[SI]} \qquad 16.24(a)$$

$$\text{Re} = \frac{D_h G}{g_c \mu} \qquad \text{[U.S.]} \qquad 16.24(b)$$

11. LAMINAR FLOW

Laminar flow gets its name from the word *laminae* (layers). If all of the fluid particles move in paths parallel to the overall flow direction (i.e., in layers), the flow is said to be *laminar*. (The terms *viscous flow* and *streamline flow* are also used.) This occurs in pipeline flow when the Reynolds number is less than (approximately) 2100. Laminar flow is typical when the flow channel is small, the velocity is low, and the fluid is viscous. Viscous forces are dominant in laminar flow.

In laminar flow, a stream of dye inserted in the flow will continue from the source in a continuous, unbroken line with very little mixing of the dye and surrounding liquid. The fluid particle paths coincide with imaginary *streamlines*. (Streamlines and velocity vectors are always tangent to each other.) A "bundle" of these streamlines (i.e., a *streamtube*) constitutes a complete fluid flow.

---

[14]Engineering authors are not in agreement about the symbol for the Reynolds number. In addition to Re (used in this book), engineers commonly use **Re**, R, $\Re$, N_{Re}, and N_R.

[15]This simplification implies a caveat as well. If the viscosity is known or is given in a problem, the units must be used to determine if this viscosity is μ or ν.

Water Resources

12. TURBULENT FLOW

A fluid is said to be in *turbulent flow* if the Reynolds number is greater than (approximately) 4000. (This is the most common situation.) Turbulent flow is characterized by a three-dimensional movement of the fluid particles superimposed on the overall direction of motion. A stream of dye injected into a turbulent flow will quickly disperse and uniformly mix with the surrounding flow. Inertial forces dominate in turbulent flow. At very high Reynolds numbers, the flow is said to be *fully turbulent*.

13. CRITICAL FLOW

The flow is said to be in a *critical zone* or *transition region* when the Reynolds number is between 2100 and 4000. These numbers are known as the lower and upper *critical Reynolds numbers* for fluid flow, respectively. (Critical Reynolds numbers for other processes are different.) It is difficult to design for the transition region, since fluid behavior is not consistent and few processes operate in the critical zone. In the event a critical zone design is required, the conservative assumption of turbulent flow will result in the greatest value of friction loss.

14. FLUID VELOCITY DISTRIBUTION IN PIPES

With laminar flow, the viscous effects make some fluid particles adhere to the pipe wall. The closer a particle is to the pipe wall, the greater the tendency will be for the fluid to adhere to the pipe wall. The following statements characterize laminar flow.

- The velocity distribution is parabolic.

- The velocity is zero at the pipe wall.

- The velocity is maximum at the center and equal to twice the average velocity.

$$v_{ave} = \frac{\dot{V}}{A} = \frac{v_{max}}{2} \quad \text{[laminar]} \qquad 16.25$$

With turbulent flow, there is generally no distinction made between the velocities of particles near the pipe wall and particles at the pipe centerline.[16] All of the fluid particles are assumed to have the same velocity. This velocity is known as the *average* or *bulk velocity*. It can be calculated from the volume flowing.

$$v_{ave} = \frac{\dot{V}}{A} \quad \text{[turbulent]} \qquad 16.26$$

---

[16]This disregards the *boundary layer*, a thin layer near the pipe wall, where the velocity goes from zero to v_{ave}.

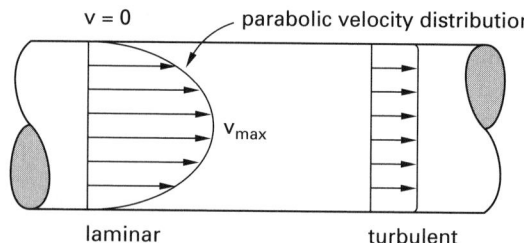

Figure 16.3 *Laminar and Turbulent Velocity Distributions*

In actuality, no flow is completely turbulent, and there is a difference between the *centerline velocity* and the average velocity. The error decreases as the Reynolds number increases. The ratio of v_{ave}/v_{max} starts at approximately 0.75 for Re = 4000 and increases to approximately 0.86 at Re = 10^6. Most problems ignore the difference between v_{ave} and v_{max}, but care should be taken when a centerline measurement (as from a pitot tube) is used to evaluate the average velocity.

For turbulent flow (Re $\approx 10^5$) in a smooth, circular pipe of radius r_o, the velocity at a radial distance r from the centerline is given by the *$1/7$-power law*:

$$v_r = v_{max} \left(\frac{r_o - r}{r_o} \right)^{1/7} \quad \text{[turbulent flow]} \qquad 16.27$$

The fluid's *velocity profile*, given by Eq. 16.27, is valid in smooth pipes up to a Reynolds number of approximately 100,000. Above that, up to a Reynolds number of approximately 400,000, an exponent of $1/8$ fits experimental data better. For rough pipes, the exponent is larger (e.g., $1/5$).

The ratio of the average velocity to maximum velocity is known as the *pipe coefficient* or *pipe factor*. Considering all the other coefficients used in pipe flow, these names are somewhat vague and ambiguous. Therefore, they are not in widespread use.

Equation 16.27 can be integrated to determine the average velocity.

$$v_{ave} = \left(\frac{49}{60} \right) v_{max} = 0.817 v_{max} \quad \text{[turbulent flow]} \qquad 16.28$$

When the flow is laminar, the velocity profile within a pipe will be parabolic and of the form of Eq. 16.29. (Equation 16.27 is for turbulent flow and does not describe a parabolic velocity profile.) The velocity at a radial distance r from the centerline in a pipe of radius r_o is

$$v_r = v_{max} \left(\frac{r_o^2 - r^2}{r_o^2} \right) \quad \text{[laminar flow]} \qquad 16.29$$

When the velocity profile is parabolic, the flow rate and pressure drop can easily be determined. The average velocity is half of the maximum velocity given in the velocity profile equation.

$$v_{ave} = \tfrac{1}{2} v_{max} \quad \text{[laminar flow]} \qquad 16.30$$

The average velocity is used to determine the flow quantity and friction loss. The friction loss is determined by traditional means.

$$\dot{V} = A\mathrm{v}_{\mathrm{ave}} \qquad 16.31$$

The kinetic energy of laminar flow can be found by integrating the velocity profile equation, resulting in Eq. 16.32.

$$E_{\mathrm{v}} = \mathrm{v}_{\mathrm{ave}}^2 \qquad \text{[laminar flow]} \qquad \text{[SI]} \qquad 16.32(a)$$

$$E_{\mathrm{v}} = \frac{\mathrm{v}_{\mathrm{ave}}^2}{g_c} \qquad \text{[laminar flow]} \qquad \text{[U.S.]} \quad 16.32(b)$$

15. ENERGY GRADE LINE

The *energy grade line* (EGL) is a graph of the total energy (total specific energy) along a length of pipe.[17] In a frictionless pipe without pumps or turbines, the total specific energy is constant, and the EGL will be horizontal. (This is a restatement of the Bernoulli equation.)

$$\text{elevation of EGL} = h_p + h_{\mathrm{v}} + h_z \qquad 16.33$$

The *hydraulic grade line* (HGL) is the graph of the sum of the pressure and gravitational heads, plotted as a position along the pipeline. Since the pressure head can increase at the expense of the velocity head, the HGL can increase in elevation if the flow area is increased.

$$\text{elevation of HGL} = h_p + h_z \qquad 16.34$$

The difference between the EGL and the HGL is the velocity head, h_{v}, of the fluid.

$$h_{\mathrm{v}} = \text{elevation of EGL} - \text{elevation of HGL} \qquad 16.35$$

The following rules apply to these grade lines in a frictionless environment, in a pipe flowing full (i.e., under pressure), without pumps or turbines.

- The EGL is always horizontal.
- The HGL is always equal to or below the EGL.
- For still (v = 0) fluid at a free surface, EGL = HGL (i.e., the EGL coincides with the fluid surface in a reservoir).
- If flow velocity is constant (i.e., flow in a constant-area pipe), the HGL will be horizontal and parallel to the EGL, regardless of pipe orientation or elevation.
- When the flow area decreases, the HGL decreases.
- When the flow area increases, the HGL increases.
- In a free jet (i.e., a stream of water from a hose), the HGL coincides with the jet elevation, following a parabolic path.

---

[17]The term *energy line* (EL) is also used.

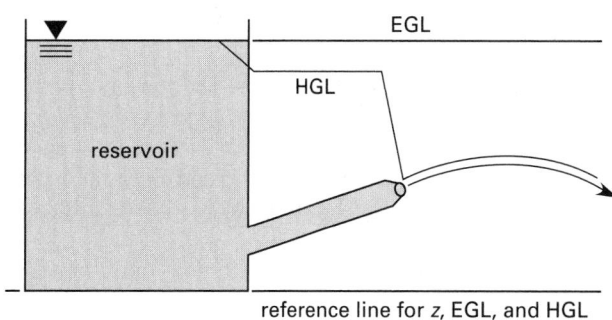

Figure 16.4 *Energy and Hydraulic Grade Lines Without Friction*

reference line for z, EGL, and HGL

16. SPECIFIC ENERGY

Specific energy is a term that is used primarily with open channel flow. It is the total energy with respect to the channel bottom, consisting of pressure and velocity energy terms only.

$$E_{\mathrm{specific}} = E_p + E_{\mathrm{v}} \qquad 16.36$$

Since the channel bottom is chosen as the reference elevation ($z = 0$) for gravitational energy, there is no contribution by gravitational energy to specific energy.

$$E_{\mathrm{specific}} = \frac{p}{\rho} + \frac{\mathrm{v}^2}{2} \qquad \text{[SI]} \qquad 16.37(a)$$

$$E_{\mathrm{specific}} = \frac{p}{\rho} + \frac{\mathrm{v}^2}{2g_c} \qquad \text{[U.S.]} \quad 16.37(b)$$

However, since p is the hydrostatic pressure at the channel bottom due to a fluid depth, d, p/ρ can be interpreted as the depth of the fluid.

$$E_{\mathrm{specific}} = d + \frac{\mathrm{v}^2}{2g} \qquad 16.38$$

Specific energy is constant when the flow depth and width are constant (i.e., *uniform flow*). A change in channel width will cause a change in flow depth, and since width is not part of the equation for specific energy, there will be a corresponding change in specific energy. There are other ways that specific energy can decrease, also.[18]

17. PIPE MATERIALS AND SIZES

Many materials are used for pipes. The material used depends on the application. Water supply distribution, wastewater collection, and air conditioning refrigerant lines all place different demands on pipe material performance. Pipe materials are chosen on the basis of strength to withstand internal pressures, strength

---

[18]Specific energy changes dramatically in a *hydraulic jump* or *hydraulic drop*.

to withstand external loads from backfill and traffic, smoothness, corrosion resistance, chemical inertness, cost, and other factors.

The following are characteristics of the major types of commercial pipe materials that are in use.

- *asbestos cement*: immune to electrolysis and corrosion, light in weight but weak structurally; environmentally limited
- *concrete*: durable, water-tight, low maintenance, smooth interior
- *copper and brass*: used primarily for water, condensate, and refrigerant lines; in some cases, easily bent by hand, good thermal conductivity
- *ductile cast iron*: long-lived, strong, impervious, heavy, scour-resistant, but costly
- *plastic* (PVC and ABS):[19] chemically inert, resistant to corrosion, very smooth, lightweight, low cost
- *steel*: high strength, ductile, resistant to shock, very smooth interior, but susceptible to corrosion
- *vitrified clay*: resistant to corrosion, acids (e.g., hydrogen sulfide from septic sewage), scour, and erosion

Table 16.2 lists recommendations for pipe materials in several common applications.

Table 16.2 Recommended Pipe Materials by Application

service	pipe material
most refrigerants (suction, liquid, and hot gas lines)	hard copper tubing (type L);[a] standard wall steel pipe, lap-welded or seamless
chilled water	hard copper tubing; plain (black) or galvanized steel pipe[b]
condenser or quad make-up water	hard copper tubing; plain or galvanized steel pipe[b]
steam or condensate	hard copper tubing; steel pipe[b]
hot water	hard copper tubing; steel pipe

[a]Soft copper may be used for $1/4$ in (6.3 mm) and $3/8$ in (9.5 mm) (outside diameter) with wall thicknesses of 0.30 in (7.6 mm) and 0.32 in (8.1 mm), respectively. Soft copper refrigeration lines are commonly used up to $1^3/8$ in (35 mm) (outside diameter). Mechanical joints should not be used with soft copper tubing larger than $7/8$ in (22 mm).

[b]Standard wall steel pipe or type-M hard copper tubing are usually satisfactory for air conditioning applications. However, the pressure rating of the pipe material should be checked at the design temperature.

The required wall thickness of a pipe is proportional to the pressure the pipe must carry. However, not all pipes operate at high pressures. Therefore, pipes and tubing may be available in different wall thicknesses (*schedules,*

[19]PVC: polyvinyl chloride; ABS: acrylonitrile-butadiene-styrene.

series, or *types*). Steel pipe, for example, is available in schedules 40, 80, and others.[20]

For initial estimates, the approximate schedule of steel pipe can be calculated from Eq. 16.39. p is the operating pressure in psig; S is the allowable stress in the pipe material; and E is the *joint efficiency*, also known as the *joint quality factor* (typically 1.00 for seamless pipe, 0.85 for electric resistance-welded pipe, 0.80 for electric fusion-welded pipe, and 0.60 for furnace butt-welded pipe). For seamless carbon steel (A53) pipe used below 650°F (340°C), the allowable stress is approximately 12,000 to 15,000 psi. So, with butt-welded joints, a value of 6500 psi is often used for the product SE.

$$\text{schedule} \approx \frac{1000p}{SE} \qquad 16.39$$

Steel pipe is available in black (i.e., plain) and galvanized (inside, outside, or both) varieties. Steel pipe is manufactured in plain-carbon and stainless varieties. AISI 316 stainless is particularly corrosion resistant.

The actual dimensions of some pipes (concrete, clay, some cast iron, etc.) coincide with their *nominal dimensions*. For example, a 12 in concrete pipe has an inside diameter of 12 in, and no further refinement is needed. However, some pipes and tubing (e.g., steel pipe, copper and brass tubing, and some cast iron) are called out by a nominal diameter that has nothing to do with the internal diameter of the pipe. For example, a 16 in schedule-40 steel pipe has an actual inside diameter of 15 in. In some cases, the nominal size does not coincide with the external diameter, either.

PVC (polyvinyl chloride) pipe is used extensively as water and sewer pipe due to its combination of strength, ductility, and corrosion resistance. Manufactured lengths are approximately 10 to 13 ft (3 to 3.9 m) for sewer pipe and 20 ft (6 m) for water pipe, with integral gasketed joints or solvent-weld bells. Infiltration is very low (less than 50 gal/in-mile-day), even in the wettest environments. The low Manning's roughness constant (0.009 typical) allows PVC sewer pipe to be used with flatter grades or smaller diameters. PVC pipe is resistant to corrosive soils and sewerage gases and is generally resistant to abrasion from pipe-cleaning tools.

It is essential that tables of pipe sizes, such as App. 16.B, be used when working problems involving steel and copper pipes since there is no other way to obtain the inside diameters of such pipes.[21]

[20]Other schedules of steel pipe, such as 30, 60, 120, etc., also exist, but in limited sizes, as Table 16.3 indicates. Schedule-40 pipe roughly corresponds to the standard weight (S) designation used in the past. Schedule-80 roughly corresponds to the extra-strong (X) designation. There is no uniform replacement for double-extra-strong (XX) pipe.

[21]It is a characteristic of standard steel pipes that the schedule number does not affect the outside diameter of the pipe. An 8 in schedule-40 pipe has the same exterior dimensions as an 8 in schedule-80 pipe. However, the interior flow area will be less for the schedule-80 pipe.

Table 16.3 *Dimensions of Commercial Steel Pipe[a] (English Units)*

nominal diameter	outside diameter	schedule									
		10	20	30	40	60	80	100	120	140	160
		wall thickness (in)									
$\frac{1}{2}$	0.840	....	....	....	0.109	....	0.147	....	....	....	0.187
$\frac{3}{4}$	1.05	....	....	....	0.113	....	0.154	....	....	....	0.218
1	1.315	....	....	....	0.133	....	0.179	....	....	....	0.250
$1\frac{1}{4}$	1.660	....	....	....	0.140	....	0.191	....	....	....	0.250
$1\frac{1}{2}$	1.900	....	....	....	0.145	....	0.200	....	....	....	0.281
2	2.375	....	....	....	0.154	....	0.218	....	....	....	0.343
$2\frac{1}{2}$	2.875	....	....	....	0.203	....	0.276	....	....	....	0.375
3	3.500	....	....	....	0.216	....	0.300	....	....	....	0.437
$3\frac{1}{2}$	4.000	....	....	....	0.226	....	0.318	....	....	....	....
4	4.500	....	....	....	0.237	....	0.337	....	0.437	....	0.531
5	5.563	....	....	....	0.258	....	0.375	....	0.500	....	0.625
6	6.625	....	....	....	0.280	....	0.432	....	0.562	....	0.718
8	8.625	....	0.250	0.277	0.322	0.406	0.500	0.593	0.718	0.812	0.906
10	10.75	....	0.250	0.307	0.365	0.500	0.593	0.718	0.843	1.000	1.125
12	12.75	....	0.250	0.330	0.406	0.562	0.687	0.843	1.000	1.125	1.312
14	14.00	0.250	0.312	0.375	0.437	0.593	0.750	0.937	1.062	1.250	1.406
16	16.00	0.250	0.312	0.375	0.500	0.656	0.843	1.031	1.218	1.437	1.562
18	18.00	0.250	0.312	0.437	0.562	0.718	0.937	1.156	1.343	1.562	1.750
20	20.00	0.250	0.375	0.500	0.593	0.812	1.031	1.250	1.500	1.750	1.937
24	24.00	0.250	0.375	0.562	0.687	0.937	1.218	1.500	1.750	2.062	2.312

[a] Also, see Apps. 16.B and 16.C.

18. MANUFACTURED PIPE STANDARDS

There are many different standards governing pipe diameters and wall thicknesses. A pipe's nominal outside diameter is rarely sufficient to determine the internal dimensions of the pipe. A manufacturing specification and class or category are usually needed to completely specify pipe dimensions.

Cast iron and ductile iron (CI/DI) pipes are produced to AWWA C151 standards. Gasketed PVC sewer pipe up to 15 in inside diameter are produced to ASTM D3034 standards. Gasketed sewer PVC pipe from 18 in to 28 in are produced to ASTM F679 standards. PVC pressure pipe for water distribution is manufactured to AWWA C900 standards. Truss pipe is manufactured to ASTM D2680 standards. Reinforced concrete pipe (RCP) is manufactured to ASTM C76 standards.

19. PAINTS, COATINGS, AND LININGS

Various materials are used to protect steel and ductile iron pipes against rust and other forms of corrosion. *Red primer* is a shop-applied rust inhibiting primer applied to prevent short-term rust prior to shipment and the application of subsequent coatings. *Asphaltic coating* ("tar" coating) is applied to the exterior of underground pipes. *Bituminous coating* refers to a similar coating made from tar pitch. Both asphaltic and bituminous coatings should be completely removed or sealed with a synthetic resin prior to the pipe being finish-coated, since their oils may bleed through otherwise.

Though bituminous materials (i.e., asphaltic materials) continue to be cost effective, epoxy-based products are now extensively used. Epoxy products are delivered as a two-part formulation (a polyamide resin and liquid chemical hardener) that is mixed together prior to application. *Coal tar epoxy*, also referred to as *epoxy coal tar*, a generic name, sees frequent use in pipes exposed to high humidity, seawater, other salt solutions, and crude oil. Though suitable for coating steel penstocks of hydroelectric installations, coal tar epoxy is generally not suitable for potable water delivery systems. Though it is self-priming, appropriate surface preparation is required for adequate adhesion. It has a density of 1.9 to 2.3 lbm/gal (230 to 280 g/L).

Figure 16.5 *Types of Valves*

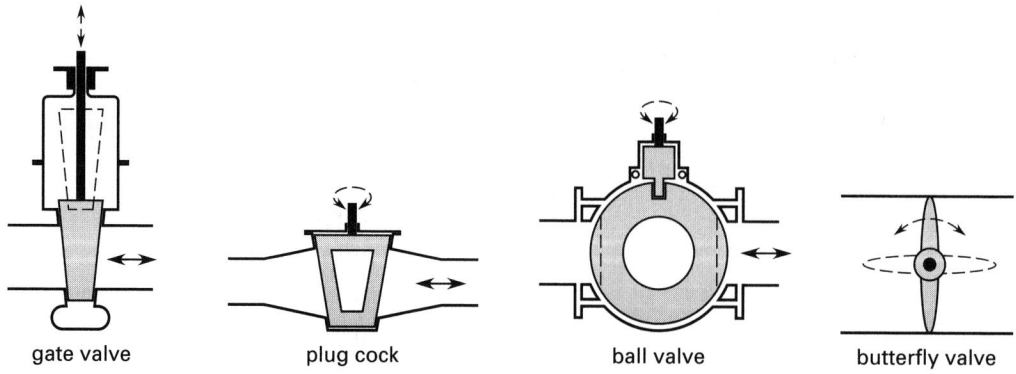

gate valve   plug cock   ball valve   butterfly valve

(a) valves for shut-off service

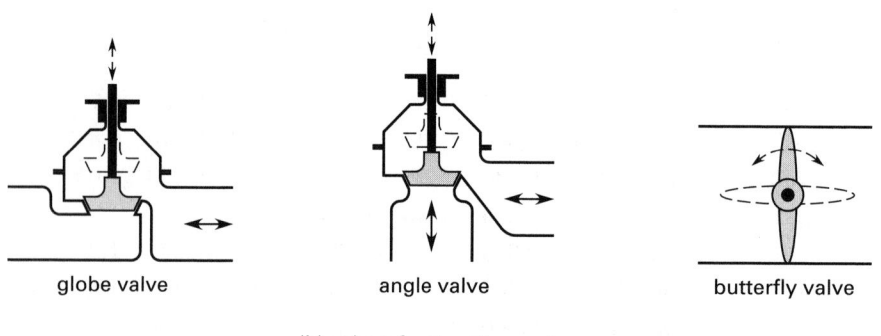

globe valve   angle valve   butterfly valve

(b) valves for throttle service

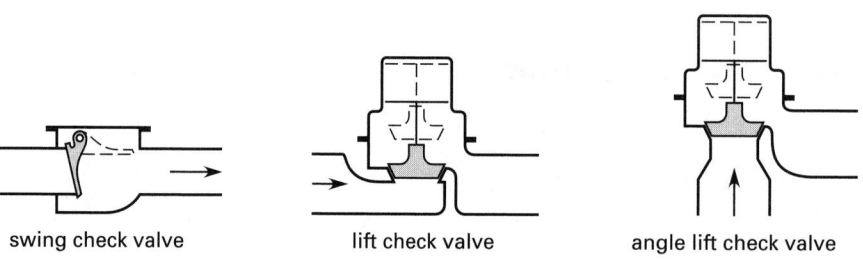

swing check valve   lift check valve   angle lift check valve

(c) valves for antireversal service

Table 16.4 *Characteristics of Common Valve Types*

valve type	fluid condition	switching frequency	pressure drop (fully open)	control response	maximum pressure, atm	maximum temperature, °C
ball	clean	low	low	very poor	160	300
butterfly	clean	low	low	poor	50	400
diaphragm	clean to slurried	very high	low to medium	very good	16	150
gate	clean	low	low	very poor	50	400
globe	clean	high	medium to high	very good	80	300
plug	clean	low	low	very poor	160	300

Water Resources

20. TYPES OF VALVES

Valves used for *shutoff service* (e.g., gate, plug, ball, and butterfly valves) are used fully open or fully closed. *Gate valves* offer minimum resistance to flow. They are used in clean fluid and slurry services when valve operation is infrequent. Many turns of the handwheels are required to raise or lower their gates. *Plug valves* provide for tight shutoff. A 90° turn of their handles is sufficient to rotate the plugs fully open or closed. *Eccentric plug valves*, in which the plug rotates out of the fluid path when open, are among the most common wastewater valves. *Plug cock valves* have a hollow passageway in their plugs through which fluid can flow. Both eccentric plug valves and plug cock valves are referred to as "plug valves." *Ball valves* offer an unobstructed flow path and tight shutoff. They are often used with slurries and viscous fluids, as well as with cryogenic fluids. A 90° turn of their handles rotates the balls fully open or closed. *Butterfly valves* (when specially designed with appropriate seats) can be used for shutoff operation. They are particularly applicable to large flows of low-pressure (vacuum up to 150 psig (1 MPa)) gases or liquids. Their straight-through, open-disk design results in minimal solids build-up and low pressure drops.

Other valve types (e.g., globe, needle, Y-, angle, and butterfly valves) are more suitable for *throttling service*. *Globe valves* provide positive shutoff and precise metering on clean fluids. However, since the seat is parallel to the direction of flow and the fluid makes two right-angle turns, there is substantial resistance and pressure drop through them, as well as relatively fast erosion of the seat. Globe valves are intended for frequent operation. *Needle valves* are similar to globe valves, except that the plug is a tapered, needle-like cone. Needle valves provide accurate metering of small flows of clean fluids. Needle valves are applicable to cryogenic fluids. *Y-valves* are similar to globe valves in operation, but their seats are inclined to the direction of flow, offering more of a straight-through passage and unobstructed flow than the globe valve. *Angle valves* are essentially globe valves where the fluid makes a 90° turn. They can be used for throttling and shut-off of clean or viscous fluids and slurries. *Butterfly valves* are often used for throttling services with the same limitations and benefits as those listed for shutoff use.

Other valves are of the *check (nonreverse-flow)* variety. These react automatically to changes in pressure to prevent reversals of flow. Special check valves can also prevent excess flow. Figure 16.5 illustrates *swing*, *lift*, and *angle lift check valves*.

17 Fluid Dynamics

Nomenclature

a	length	ft	m
a	speed of sound	ft/sec	m/s
A	area	ft^2	m^2
C	coefficient	–	–
C	Hazen-Williams coefficient	–	–
d	diameter	in	cm
D	diameter	ft	m
E	bulk modulus	lbf/ft^2	Pa
E	specific energy	ft-lbf/lbm	J/kg
f	Darcy friction factor	–	–
f	fraction split	–	–
F	force	lbf	N
Fr	Froude number	–	–
g	gravitational acceleration	ft/sec^2	m/s^2
g_c	gravitational conversion constant	lbm-ft/lbf-sec^2	n.a.
G	mass flow rate per unit area	lbm/ft^2-sec	kg/m^2·s
h	height or head	ft	m
I	impulse	lbf-sec	N·s
k	ratio of specific heats	–	–
K	minor loss coefficient	–	–
l	length	ft	m
L	length	ft	m
m	mass	lbm	kg
$\dot{m}$	mass flow rate	lbm/sec	kg/s
MW	molecular weight	lbm/lbmol	kg/kmol
n	Manning roughness constant	–	–
n	flow rate exponent	–	–
p	pressure	lbf/ft^2	Pa
P	momentum	lbm-ft/sec	kg·m/s

P	power	ft-lbf/sec	W
Q	flow rate	gal/min	n.a.
r	radius	ft	m
rpm	rotational speed	rev/min	rev/min
R	resultant force	lbf	N
R^*	universal gas constant	ft-lbf/ lbmol-°R	J/ kmol·K
Re	Reynolds number	–	–
SG	specific gravity	–	–
t	thickness	ft	m
t	time	sec	s
T	absolute temperature	°R	K
u	x-component of velocity	ft/sec	m/s
v	y-component of velocity	ft/sec	m/s
v	velocity	ft/sec	m/s
V	volume	ft^3	m^3
$\dot{V}$	volumetric flow rate	ft^3/sec	m^3/s
W	work	ft-lbf	J
We	Weber number	–	–
WHP	water horsepower	hp	n.a.
x	x-coordinate of position	ft	m
y	y-coordinate of position	ft	m
Y	expansion factor	–	–
z	elevation	ft	m

Symbols

β	diameter ratio	–	–
γ	specific weight	lbf/ft^3	N/m^3
Γ	circulation	ft^2/sec	m^2/s
ϵ	specific roughness	ft	m
η	efficiency	–	–
η	non-Newtonian viscosity	lbf-sec/ft^2	Pa·s
θ	angle	deg	deg
μ	absolute viscosity	lbf-sec/ft^2	Pa·s
ν	kinematic viscosity	ft^2/sec	m^2/s
ρ	density	lbm/ft^3	kg/m^3
σ	surface tension	lbf/ft	N/m
τ	shear stress	lbf/ft^2	Pa
υ	specific volume	ft^3/lbm	m^3/kg
ϕ	angle	deg	deg
Φ	stream potential	–	–
ψ	sphericity	–	–
Ψ	stream function	–	–
ω	angular velocity	rad/sec	rad/s

Subscripts

A	added (by pump)
b	blade or buoyant
c	contraction
d	discharge
D	drag
e	equivalent
E	extracted (by turbine)
f	friction or flow
i	inside
I	instrument
L	lift
m	minor, model, or manometer fluid
o	orifice or outside
p	pressure or prototype
r	ratio
s	static
t	total, tank, or theoretical
v	velocity
va	velocity of approach
z	potential

1. HYDRAULICS AND HYDRODYNAMICS

This chapter investigates fluid moving through pipes, measurements with venturis and orifices, and other motion-related topics such as model theory, lift and drag, and pumps. In a strict interpretation, any fluid-related phenomenon that is not hydro*statics* should be hydro*dynamics*. However, tradition has separated the study of moving fluids into the fields of hydraulics and hydrodynamics.

In a general sense, *hydraulics* is the study of the practical laws of fluid flow and resistance in pipes and open channels. Hydraulic formulas are often developed from experimentation, empirical factors, and curve fitting, without an attempt to justify why the fluid behaves the way it does.

On the other hand, *hydrodynamics* is the study of fluid behavior based on theoretical considerations. Hydrodynamicists start with Newton's laws of motion and try to develop models of fluid behavior. Models developed in this manner are complicated greatly by the inclusion of viscous friction and compressibility. Therefore, hydrodynamic models assume a perfect fluid with constant density and zero viscosity. The conclusions reached by hydrodynamicists can differ greatly from those reached by hydraulicians.[1]

2. CONSERVATION OF MASS

Fluid mass is always conserved in fluid systems, regardless of the pipeline complexity, orientation of the flow, or which fluid is flowing. This single concept is often sufficient to solve simple fluid problems.

$$\dot{m}_1 = \dot{m}_2 \qquad 17.1$$

When applied to fluid flow, the conservation of mass law is known as the *continuity equation*.

$$\rho_1 A_1 v_1 = \rho_2 A_2 v_2 \qquad 17.2$$

If the fluid is incompressible, then $\rho_1 = \rho_2$.

$$A_1 v_1 = A_2 v_2 \qquad 17.3$$

$$\dot{V}_1 = \dot{V}_2 \qquad 17.4$$

Various units and symbols are used for *volumetric flow rate*. (Though this book uses $\dot{V}$, the symbol Q is often used when the flow rate is expressed in gallons.)

---

[1]Perhaps the most disparate conclusion is *D'Alembert's paradox*. In 1744, D'Alembert derived theoretical results "proving" that there is no resistance to bodies moving through an ideal (non-viscous) fluid.

MGD (millions of gallons per day) and mgpcd (millions of gallons per capita day) are units commonly used in municipal water works problems. MMSCFD (millions of standard cubic feet per day) may be used to express gas flows.

Calculation of flow rates is often complicated by the interdependence between flow rate and friction loss. Each affects the other. Hence, many pipe flow problems must be solved iteratively. Usually, a reasonable friction factor is assumed and is used to calculate an initial flow rate. The flow rate establishes the flow velocity, from which a revised friction factor can be determined.

3. TYPICAL VELOCITIES IN PIPES

Fluid friction in pipes is kept at acceptable levels by maintaining reasonable fluid velocities. Table 17.1 lists typical maximum fluid velocities. Higher velocities may be observed in practice, but only with a corresponding excessive increase in friction and pumping power.

Table 17.1 *Typical Fluid Velocities*

fluid and application	ft/sec	m/s
water: city service	7 (2–5 typ)	2.1 (0.6–1.5 typ)
water: boiler feed	15 (8–10 typ)	4.5 (2.4–3.0 typ)
air: compressor suction	75–200	23–60
air: compressor discharge	100–250	30–75
refrigerant: suction	15–35	4.5–11
refrigerant: discharge	35–60	11–18
steam, saturated: heating	65–100	20–30
steam, saturated: miscellaneous	100–200	30–60
steam, superheated: turbine feed	160–250	50–75

4. STREAM POTENTIAL AND STREAM FUNCTION

An application of hydrodynamic theory is the derivation of the stream function from stream potential. The *stream potential function (velocity potential function)*, Φ, is the algebraic sum of the component velocity potential functions.[2]

$$\Phi = \Phi_x(x, y) + \Phi_y(x, y) \qquad 17.5$$

The velocity component of the resultant in the x-direction is

$$u = \frac{\partial \Phi}{\partial x} \qquad 17.6$$

---

[2]The two-dimensional derivation of the stream function can be extended to three dimensions, if necessary. The stream function can also be expressed in the cylindrical coordinate system.

The velocity component of the resultant in the y-direction is

$$v = \frac{\partial \Phi}{\partial y} \qquad 17.7$$

The total derivative of the stream potential function is

$$d\Phi = \frac{\partial \Phi}{\partial x}dx + \frac{\partial \Phi}{\partial y}dy$$
$$= u\, dx + v\, dy \qquad 17.8$$

An *equipotential line* is a line along which the function Φ is constant (i.e., $d\Phi = 0$). The slope of the equipotential line is derived from Eq. 17.8.

$$\left.\frac{dy}{dx}\right|_{\text{equipotential}} = -\frac{u}{v} \qquad 17.9$$

For flow through a porous, permeable medium, pressure will be constant along equipotential lines (i.e., along lines of constant Φ). However, for an ideal, non-viscous fluid flowing in a frictionless environment, Φ has no physical significance. Even though Φ has a theoretical basis, it does not coincide with any measurable physical quantity.

Figure 17.1 *Equipotential Lines and Streamlines*

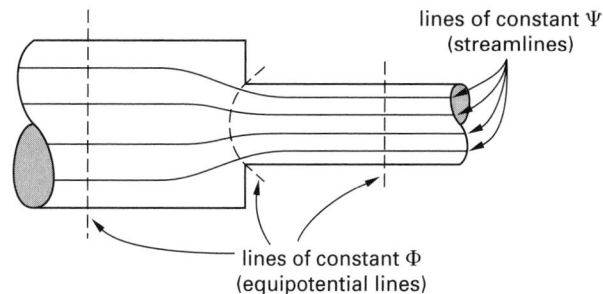

lines of constant Ψ (streamlines)

lines of constant Φ (equipotential lines)

The *stream function (Lagrange stream function)*, $\Psi(x, y)$, defines the direction of flow at a point.

$$u = \frac{\partial \Psi}{\partial y} \qquad 17.10$$

$$v = -\frac{\partial \Psi}{\partial x} \qquad 17.11$$

The stream function can also be written in total derivative form.

$$d\Psi = \frac{\partial \Psi}{\partial x}dx + \frac{\partial \Psi}{\partial y}dy$$
$$= -v\, dx + u\, dy \qquad 17.12$$

The stream function, $\Psi(x, y)$, satisfies Eq. 17.12. For a given streamline, $d\Psi = 0$, and each streamline is a line representing a constant value of Ψ. A streamline is perpendicular to an equipotential line.

$$\left.\frac{dy}{dx}\right|_{\text{streamline}} = \frac{v}{u} \qquad 17.13$$

Example 17.1

The stream potential function for water flowing through a particular valve is

$$\Phi = 3xy - 2y$$

What is the stream function, Ψ?

Solution

First, work with Φ to obtain u and v.

$$u = \frac{\partial \Phi}{\partial x} = \frac{\partial(3xy - 2y)}{\partial x} = 3y$$

$$v = \frac{\partial \Phi}{\partial y} = 3x - 2$$

u and v are also related to the stream function, Ψ. From Eq. 17.10,

$$u = \frac{\partial \Psi}{\partial y}$$

$$\partial \Psi = u \, \partial y$$

$$\Psi = \int 3y \, dy = \tfrac{3}{2}y^2 + \text{some function of } x + C_1$$

Similarly, from Eq. 17.11,

$$v = -\frac{\partial \Psi}{\partial x}$$

$$\partial \Psi = -v \, \partial x$$

$$\Psi = -\int (3x - 2) \, dx$$

$$= 2x - \tfrac{3}{2}x^2 + \text{some function of } y + C_2$$

Ψ is found by superposition of these two results.

$$\Psi = \tfrac{3}{2}y^2 + 2x - \tfrac{3}{2}x^2 + C$$

5. HEAD LOSS DUE TO FRICTION

The original Bernoulli equation was based on an assumption of frictionless flow. In actual practice, friction occurs during fluid flow. This friction "robs" the fluid of energy, so that the fluid at the end of a pipe section has less energy than it does at the beginning.[3]

$$E_1 > E_2 \qquad\qquad 17.14$$

---

[3]The friction generates minute amounts of heat. The heat is lost to the surroundings.

Most formulas for calculating friction loss use the symbol h_f to represent the *head loss due to friction*.[4] This loss is added into the original Bernoulli equation to restore the equality. Of course, the units of h_f must be the same as the units for the other terms in the Bernoulli equation. (See Eq. 17.23.) If the Bernoulli equation is written in terms of energy, the units will be ft-lbf/lbm or J/kg.

$$E_1 = E_2 + E_f \qquad\qquad 17.15$$

Consider the constant-diameter, horizontal pipe in Fig. 17.2. An incompressible fluid is flowing at a steady rate. Since the elevation of the pipe does not change, the potential energy is constant. Since the pipe has a constant area, the kinetic energy (velocity) is constant. Therefore, the friction energy loss must show up as a decrease in pressure energy. Since the fluid is incompressible, this can only occur if the pressure decreases in the direction of flow.

Figure 17.2 *Pressure Drop in a Pipe*

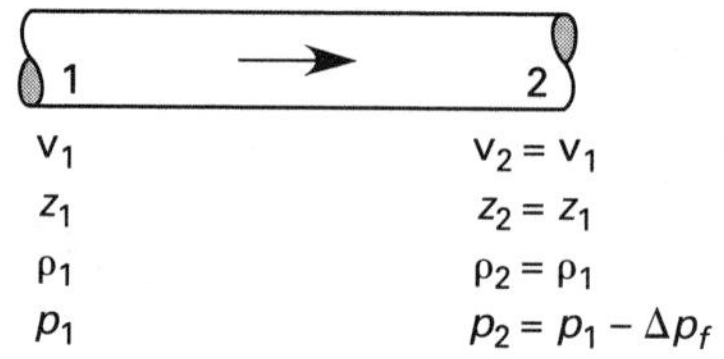

v_1 $v_2 = v_1$
z_1 $z_2 = z_1$
ρ_1 $\rho_2 = \rho_1$
p_1 $p_2 = p_1 - \Delta p_f$

6. RELATIVE ROUGHNESS

It is intuitive that pipes with rough inside surfaces will experience greater friction losses than smooth pipes.[5] *Specific roughness*, ϵ, is a parameter that measures the average size of imperfections inside the pipe. Table 17.2 lists values of ϵ for common pipe materials. (Also, see App. 17.A.)

Table 17.2 *Values of Specific Roughness for Common Pipe Materials*

material	ϵ ft	ϵ m
plastic (PVC, ABS)	0.000005	1.5×10^{-6}
copper and brass	0.000005	1.5×10^{-6}
steel	0.0002	6.0×10^{-5}
plain cast iron	0.0008	2.4×10^{-4}
concrete	0.004	1.2×10^{-3}

---

[4]Other names and symbols for this friction loss are *friction head loss* (h_L), *lost work* (LW), *friction heating* ($\mathcal{F}$), *skin friction loss* (F_f), and *pressure drop due to friction* (Δp_f). All terms and symbols mean basically the same thing, although the units may need to be converted.

[5]Surprisingly, this intuitive statement is valid only for turbulent flow. The roughness does not (ideally) affect the friction loss for laminar flow.

However, an imperfection the size of a sand grain will have much more effect in a small-diameter hydraulic line than in a large-diameter sewer. Therefore, the *relative roughness*, ϵ/D, is a better indicator of pipe roughness. Both ϵ and D have units of length (e.g., feet or meters), and the relative roughness is dimensionless.

7. FRICTION FACTOR

The *Darcy friction factor*, f, is one of the parameters used to calculate friction loss.[6] The friction factor is not constant, but decreases as the Reynolds number (fluid velocity) increases, up to a certain point known as *fully turbulent flow* (or *rough-pipe flow*). Once the flow is fully turbulent, the friction factor remains constant and depends only on the relative roughness and not on the Reynolds number. For very smooth pipes, fully turbulent flow is achieved only at very high Reynolds numbers.

Figure 17.3 *Friction Factor as a Function of Reynolds Number*

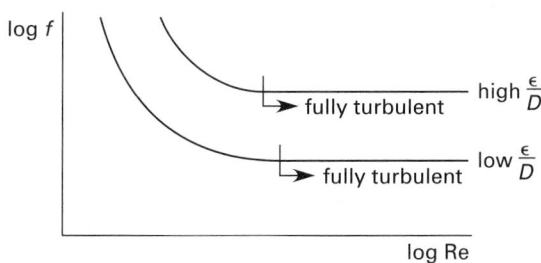

The friction factor is not dependent on the material of the pipe but is affected by the roughness. For example, for a given Reynolds number, the friction factor will be the same for any smooth pipe material (glass, plastic, smooth brass and copper, etc.).

The friction factor is determined from the relative roughness, ϵ/D, and the Reynolds number, Re, by various methods. These methods include explicit and implicit equations, the Moody diagram, and tables. The values obtained are based on experimentation, primarily the work of J. Nikuradse in the early 1930s.

When a moving fluid initially encounters a parallel surface (as when a moving gas encounters a flat plate or when a fluid first enters the mouth of a pipe), the flow will generally not be turbulent, even for very rough surfaces. The flow will be laminar for a certain *critical distance* before becoming turbulent.

---

[6]There are actually two friction factors: the Darcy friction factor and the *Fanning friction factor*, f_{Fanning}, also known as the *skin friction coefficient* and *wall shear stress factor*. Both factors are in widespread use, sharing the same symbol, f. Civil and (most) mechanical engineers use the Darcy friction factor. The Fanning friction factor is encountered more often in the chemical industry. One can be derived from the other: $f_{\text{Darcy}} = 4f_{\text{Fanning}}$.

Friction Factors for Laminar Flow

The easiest method of obtaining the friction factor for laminar flow (Re < 2100) is to calculate it. Equation 17.16 illustrates that roughness is not a factor in determining the frictional loss in ideal laminar flow.

$$f = \frac{64}{\text{Re}} \qquad \textit{17.16}$$

Friction Factors for Turbulent Flow: by Formula

One of the earliest attempts to predict the friction factor for turbulent flow in smooth pipes resulted in the *Blasius equation* (claimed "valid" for 3000 < Re < 100,000).

$$f = \frac{0.316}{\text{Re}^{0.25}} \qquad \textit{17.17}$$

The *Nikuradse equation* can also be used to determine the friction factor for smooth pipes (i.e., when $\epsilon/D = 0$). Unfortunately, this equation is implicit in f and must be solved iteratively.

$$\frac{1}{\sqrt{f}} = 2.0 \ \log_{10}\left(\text{Re}\sqrt{f}\right) - 0.80 \qquad \textit{17.18}$$

The *Karman-Nikuradse equation* predicts the fully turbulent friction factor (i.e., when Re is very large).

$$\frac{1}{\sqrt{f}} = 1.74 - 2 \ \log_{10}\left(\frac{2\epsilon}{D}\right) \qquad \textit{17.19}$$

The most widely known method of calculating the friction factor for any pipe roughness and Reynolds number is another implicit formula, the *Colebrook equation*. Most other equations are variations of this equation. (Notice that the relative roughness, ϵ/D, is used to calculate f.)

$$\frac{1}{\sqrt{f}} = -2 \ \log_{10}\left(\frac{\frac{\epsilon}{D}}{3.7} + \frac{2.51}{\text{Re}\sqrt{f}}\right) \qquad \textit{17.20}$$

A suitable approximation would appear to be the Swamee-Jain equation, which claims to have less than 1% error (as measured against the Colebrook equation) for relative roughnesses between 0.000001 and 0.01, and for Reynolds numbers between 5000 and 100,000,000.[7] Even with a 1% error, this equation produces more accurate results than can be read from the Moody diagram.

$$f = \frac{0.25}{\left(\log_{10}\left(\frac{\frac{\epsilon}{D}}{3.7} + \frac{5.74}{\text{Re}^{0.9}}\right)\right)^2} \qquad \textit{17.21}$$

---

[7]*ASCE Hydraulic Division Journal*, Vol. 102, May 1976, p. 657. This is not the only explicit approximation to the Colebrook equation in existence.

Friction Factors for Turbulent Flow: by Moody Chart

The *Moody friction factor chart*, Fig. 17.4, presents the friction factor graphically as a function of Reynolds number and relative roughness. There are different lines for selected discrete values of relative roughness. Due to the complexity of this graph, it is easy to mislocate the Reynolds number or use the wrong curve. Nevertheless, the Moody chart remains the most common method of obtaining the friction factor.

Friction Factors for Turbulent Flow: by Table

Appendix 17.B (based on the Colebrook equation), or a similar table, will usually be the most convenient method of obtaining friction factors for turbulent flow.

Example 17.2

Determine the friction factor for a Reynolds number of Re = 400,000 and a relative roughness of $\epsilon/D = 0.004$ using (a) the Moody diagram, (b) App. 17.B, and (c) the Swamee-Jain approximation. (d) Check the table value of f with the Colebrook equation.

Solution

(a) From Fig. 17.4, the friction factor is approximately 0.028.

(b) Appendix 17.B lists the friction factor as 0.0287.

(c) From Eq. 17.21,

$$f = \frac{0.25}{\left(\log_{10}\left(\dfrac{0.004}{3.7} + \dfrac{5.74}{(400{,}000)^{0.9}}\right)\right)^2}$$

$$= 0.0288$$

(d) From Eq. 17.20,

$$\frac{1}{\sqrt{0.0287}} = -2 \log_{10}\left(\frac{0.004}{3.7} + \frac{2.51}{(400{,}000)\sqrt{0.0287}}\right)$$

$$5.903 = 5.903$$

Figure 17.4 *Moody Friction Factor Chart*

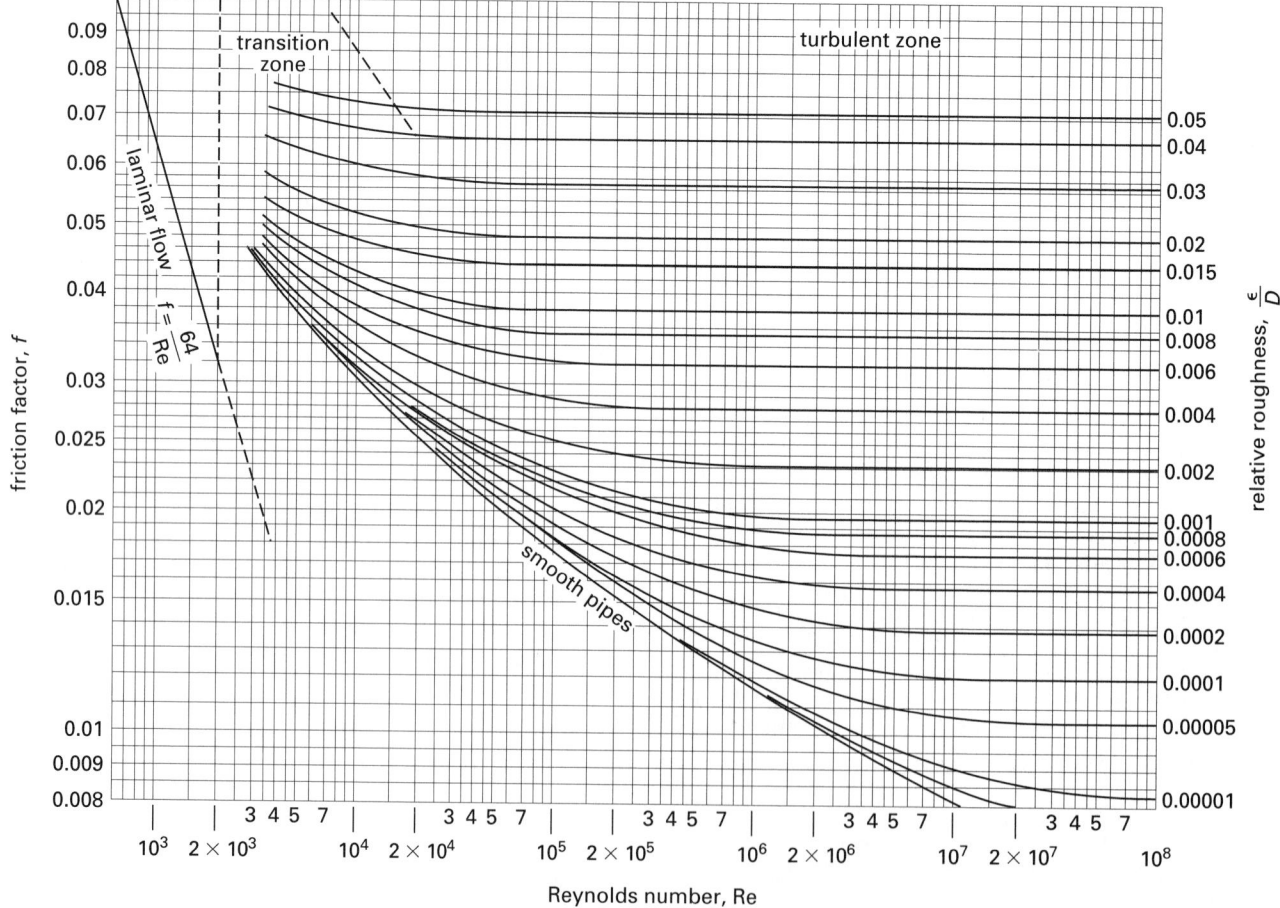

8. ENERGY LOSS DUE TO FRICTION: LAMINAR FLOW

Two methods are available for calculating the frictional energy loss for fluids experiencing laminar flow. The most common is the *Darcy equation* (also known as the *Weisbach equation* or the *Darcy-Weisbach equation*), which can be used for both laminar and turbulent flow.[8] One of the advantages to using the Darcy equation is that the assumption of laminar flow does not need to be confirmed if f is known.

$$h_f = \frac{fLv^2}{2Dg} \qquad 17.22$$

$$E_f = h_f g = \frac{fLv^2}{2D} \qquad \text{[SI]} \quad 17.23(a)$$

$$E_f = h_f \times \left(\frac{g}{g_c}\right) = \frac{fLv^2}{2Dg_c} \qquad \text{[U.S.]} \quad 17.23(b)$$

If the flow is truly laminar and the fluid is flowing in a circular pipe, then the *Hagen-Poiseuille equation* can be used.

$$E_f = \frac{32\mu vL}{D^2\rho} \qquad \text{[SI]} \quad 17.24(a)$$

$$E_f = \frac{32\mu vLg_c}{D^2\rho} \qquad \text{[U.S.]} \quad 17.24(b)$$

An alternate form of the Hagen-Poiseuille equation substitutes $\dot{V}/A$ for v.

$$E_f = \frac{128\mu\dot{V}L}{\pi D^4\rho} \qquad \text{[SI]} \quad 17.25(a)$$

$$E_f = \frac{128\mu\dot{V}Lg_c}{\pi D^4\rho} \qquad \text{[U.S.]} \quad 17.25(b)$$

If necessary, h_f can be converted to an actual pressure drop in psi or Pa by multiplying by the fluid density.

$$\Delta p = h_f \times \rho g \qquad \text{[SI]} \quad 17.26(a)$$

$$\Delta p = h_f \times \rho\left(\frac{g}{g_c}\right) \qquad \text{[U.S.]} \quad 17.26(b)$$

Values of the Darcy friction factor, f, are often quoted for new, clean pipe. The friction head losses and pumping power requirements calculated from these values are minimal values. Depending on the nature of the service, scale and impurity buildup within pipes may decrease the pipe diameters over time. Since the frictional loss is proportional to the fifth power of the diameter, such diameter decreases can produce dramatic increases in the friction loss.

$$\frac{h_{f,scaled}}{h_{f,new}} = \left(\frac{D_{\text{new}}}{D_{\text{scaled}}}\right)^5 \qquad 17.27$$

[8]The difference is that the friction factor can be derived by hydrodynamics: $f = 64/\text{Re}$. For turbulent flow, f is empirical.

Equation 17.27 accounts only for the decrease in diameter. Any increase in roughness (i.e., friction factor) will produce a proportional increase in friction loss.

Because the "new, clean" condition is transitory in most applications, an uprating factor of 10 to 30% is often applied to either the friction factor, f, or the head loss, h_f. Of course, even larger increases should be considered when extreme fouling is expected.

Another approach eliminates the need to estimate the scaled pipe diameter. This simplistic approach multiplies the initial friction loss by a factor based on the age of the pipe. For example, for schedule-40 pipe 4 to 10 in (10 to 25 cm) in diameter, the multipliers of 1.4, 2.2, and 5.0 have been proposed for pipe ages of 5, 10, and 20 years, respectively. For larger pipes, the corresponding multipliers are 1.3, 1.6, and 2.0. Obviously, use of these values should be based on a clear understanding of the method's limitations.

9. ENERGY LOSS DUE TO FRICTION: TURBULENT FLOW

The *Darcy equation* is used almost exclusively to calculate the head loss due to friction for turbulent flow.

$$h_f = \frac{fLv^2}{2Dg} \qquad 17.28$$

The head loss can be converted to pressure drop.

$$\Delta p = h_f \times \rho g \qquad \text{[SI]} \quad 17.29(a)$$

$$\Delta p = h_f \times \rho\left(\frac{g}{g_c}\right) \qquad \text{[U.S.]} \quad 17.29(b)$$

In problems where the pipe size is unknown, it will be impossible to obtain an accurate initial value of the friction factor, f (since f depends on velocity). In such problems, an iterative solution will be necessary.

It is not uncommon for civil engineers to use the *Hazen-Williams equation* to calculate head loss. This method requires knowledge of the Hazen-Williams *roughness coefficient*, C, values of which are widely tabulated.[9] (See App. 17.A.) The advantage of using this equation is that C does not depend on the Reynolds number.

$$h_{f,\text{feet}} = \frac{3.022 v_{\text{ft/sec}}^{1.85} L_{\text{ft}}}{C^{1.85} D^{1.17}} \qquad \text{[U.S.]} \quad 17.30$$

Or, in terms of other units,

$$h_{f,\text{feet}} = \frac{10.44 L_{\text{ft}} \dot{V}_{\text{gpm}}^{1.85}}{C^{1.85} d_{\text{inches}}^{4.87}} \qquad \text{[U.S.]} \quad 17.31$$

[9]An approximate value of $C = 140$ is often chosen for initial calculations for new water pipe. $C = 100$ is more appropriate for water pipe that has been in service for some time. For sludge, C values are 20 to 40% lower than the equivalent water pipe values.

The Hazen-Williams equation is empirical and is not dimensionally homogeneous. It is taken as a matter of faith that the units of h_f are feet.

The Hazen-Williams equation should be used only for turbulent flow. It gives good results for liquids that have kinematic viscosities around 1.2×10^{-5} ft^2/sec (1.1×10^{-6} m^2/s), which corresponds to the viscosity of 60°F (16°C) water. At extremely high and low temperatures, the Hazen-Williams equation can be 20% or more in error for water.

Example 17.3

50°F water is pumped through 1000 ft of 4 in, schedule-40 welded steel pipe at the rate of 300 gpm. What friction loss (in ft-lbf/lbm) is predicted by the Darcy equation?

Solution

First, it is necessary to collect data on the pipe and water. The fluid viscosity, pipe dimensions, and other parameters can be found from the appendices.

$$\nu = 1.41 \times 10^{-5} \text{ ft}^2/\text{sec} \quad [\text{App. 14.A}]$$

$$\epsilon = 0.0002 \text{ ft} \quad [\text{App. 17.A}]$$

$$D = 0.3355 \text{ ft} \quad [\text{App. 16.B}]$$

$$A = 0.0884 \text{ ft}^2 \quad [\text{App. 16.B}]$$

The flow quantity is converted from gallons per minute to cubic feet per second.

$$\dot{V} = (300 \text{ gpm}) \left(0.002228 \frac{\frac{\text{ft}^3}{\text{sec}}}{\text{gpm}} \right) = 0.6684 \text{ ft}^3/\text{sec}$$

The velocity is

$$\text{v} = \frac{\dot{V}}{A} = \frac{0.6684 \frac{\text{ft}^3}{\text{sec}}}{0.0884 \text{ ft}^2} = 7.56 \text{ ft/sec}$$

The Reynolds number is

$$\text{Re} = \frac{D\text{v}}{\nu} = \frac{(0.3355 \text{ ft}) \left(7.56 \frac{\text{ft}}{\text{sec}} \right)}{1.41 \times 10^{-5} \frac{\text{ft}^2}{\text{sec}}}$$

$$= 1.8 \times 10^5$$

The relative roughness is

$$\frac{\epsilon}{D} = \frac{0.0002}{0.3355} = 0.0006$$

From the friction factor table (or the Moody friction factor chart), $f = 0.0195$.

Equation 17.23(b) is used to calculate the friction loss.

$$E_f = h_f \times \left(\frac{g}{g_c} \right) = \frac{fL\text{v}^2}{2Dg_c}$$

$$= \frac{(0.0195)(1000 \text{ ft}) \left(7.56 \frac{\text{ft}}{\text{sec}} \right)^2}{(2)(0.3355 \text{ ft}) \left(32.2 \frac{\text{lbm-ft}}{\text{lbf-sec}^2} \right)}$$

$$= 51.6 \text{ ft-lbf/lbm}$$

Example 17.4

Calculate the head loss due to friction for the pipe in Ex. 17.3 using the Hazen-Williams formula. Assume $C = 100$.

Solution

Substituting the parameters derived in Ex. 17.3 into Eq. 17.30,

$$h_f = \frac{(3.022) \left(7.56 \frac{\text{ft}}{\text{sec}} \right)^{1.85} (1000 \text{ ft})}{(100)^{1.85} (0.3355 \text{ ft})^{1.17}} = 91.3 \text{ ft}$$

Alternatively, the given data can be substituted directly into Eq. 17.31.

$$h_f = \frac{(10.44)(1000 \text{ ft})(300 \text{ gpm})^{1.85}}{(100)^{1.85}(4.026 \text{ in})^{4.87}} = 90.3 \text{ ft}$$

10. FRICTION LOSS FOR WATER FLOW IN STEEL PIPES

Friction loss and velocity for water flowing through steel pipe (as well as for other liquids and other pipe materials) in table and chart form are widely available. (Appendix 17.C is an example of such a table.) These tables and charts are unable to compensate for the effects of fluid temperature and different pipe roughness. Unfortunately, the assumptions made in developing the tables and charts are seldom listed. Another disadvantage is that the values can be read to only a few significant figures.

Since water's specific volume is essentially constant within the normal temperature range, tables and charts can be used to determine water velocity. Friction loss data, however, should be considered accurate to only ±20%. Alternatively, a 20% safety margin should be established in choosing pumps and motors.

Tables and charts almost always give the friction loss per 100 ft or 10 m of pipe. The pressure drop is proportional to the length, so the value read can be scaled for other pipe lengths. Flow velocity is independent of pipe length.

11. FRICTION LOSS IN NONCIRCULAR DUCTS

The frictional energy loss by a fluid flowing in a rectangular, annular, or other noncircular duct can be calculated from the Darcy equation by using the *hydraulic diameter*, D_h, in place of the diameter variable, D.[10] The friction factor, f, is determined in any of the conventional manners.

12. FRICTION LOSS FOR STEAM AND GASES

The Darcy equation can be used to calculate the frictional energy loss for all incompressible liquids, not just for water. Alcohol, gasoline, fuel oil, and refrigerants, for example, are all handled well, since the effect of viscosity is considered in determining the friction factor, f.[11]

In fact, the Darcy equation is commonly used with noncondensing vapors and compressed gases, such as air, nitrogen, and steam.[12] In such cases, reasonable accuracy will be achieved as long as the fluid is not moving too fast (i.e., less than Mach 0.3) and is incompressible. The fluid is assumed to be incompressible if the pressure (or density) change along the section of interest is less than 10% of the starting pressure.

If possible, it is preferred to base all calculations on the average properties of the fluid.[13] Specifically, the fluid velocity would normally be calculated as

$$\text{v} = \frac{\dot{m}}{\rho_{\text{ave}} A} \qquad 17.32$$

However, the average density of a gas depends on the average pressure, which is unknown at the start of a problem. The solution is to write the Reynolds number and Darcy equation in terms of the constant mass flow rate per unit area, G, instead of velocity, v, which varies.

$$G = \text{v}_{\text{ave}} \rho_{\text{ave}} \qquad 17.33$$

$$\text{Re} = \frac{DG}{\mu} \qquad \text{[SI]} \qquad 17.34(a)$$

$$\text{Re} = \frac{DG}{g_c \mu} \qquad \text{[U.S.]} \qquad 17.34(b)$$

---

[10]Although it is used for both, this approach is better suited for turbulent flow than for laminar flow. Also, the accuracy of this method decreases as the flow area becomes more noncircular. The friction drop in long, narrow slits is poorly predicted, for example. However, there is no other convenient method of predicting friction drop. Experimentation should be used with a particular flow geometry if extreme accuracy is required.

[11]Since viscosity is not an explicit factor in the formula, it should be obvious that the Hazen-Williams equation is primarily used for water.

[12]Use of the Darcy equation is limited only by the availability of the viscosity data needed to calculate the Reynolds number.

[13]Of course, the entrance (or exit) conditions can be used if great accuracy is not needed.

$$\Delta p_f = p_1 - p_2 = \rho_{\text{ave}} h_f g = \frac{fLG^2}{2D\rho_{\text{ave}}} \qquad \text{[SI]} \qquad 17.35(a)$$

$$\Delta p_f = p_1 - p_2 = \gamma_{\text{ave}} h_f = \rho_{\text{ave}} h_f \frac{g}{g_c}$$

$$= \frac{fLG^2}{2D\rho_{\text{ave}} g_c} \qquad \text{[U.S.]} \qquad 17.35(b)$$

Assuming a perfect gas with a molecular weight of MW, the ideal gas law can be used to calculate ρ_{ave} from the absolute temperature, T, and $p_{\text{ave}} = (p_1 + p_2)/2$.

$$p_1^2 - p_2^2 = \frac{fLG^2 R^* T}{D(\text{MW})} \qquad \text{[SI]} \qquad 17.36(a)$$

$$p_1^2 - p_2^2 = \frac{fLG^2 R^* T}{D g_c (\text{MW})} \qquad \text{[U.S.]} \qquad 17.36(b)$$

To summarize, use the following guidelines when working with compressible gases or vapors flowing in a pipe or duct. (a) If the pressure drop, based on the entrance pressure, is less than 10%, the fluid can be assumed to be incompressible, and the gas properties can be evaluated at any point known along the pipe. (b) If the pressure drop is between 10% and 40%, use of the midpoint properties will yield reasonably accurate friction losses. (c) If the pressure drop is greater than 40%, the pipe can be divided into shorter sections and the losses calculated for each section, or exact calculations based on compressible flow theory must be made.

Calculating a friction loss for steam flow can be frustrating if steam viscosity data are unavailable. Generally, the steam viscosities listed in compilations of heat transfer data are sufficiently accurate. Various empirical methods are also in use. For example, the *Babcock formula* (Eq. 17.37) for pressure drop when steam with a specific volume of v flows in a pipe of diameter d is

$$\Delta p_{\text{psi}} = (0.470) \left(\frac{d_{\text{in}} + 3.6}{d_{\text{in}}^6} \right) (\dot{m}_{\text{lbm/sec}})^2 L_{\text{ft}} v_{\text{ft}^3/\text{lbm}} \qquad 17.37$$

Use of empirical formulas is not limited to steam. Theoretical formulas (e.g., the *complete isothermal flow equation*) and specialized empirical formulas (e.g., the *Weymouth*, *Panhandle*, and *Spitzglass formulas*) have been developed, particularly by the gas pipeline industry. Each of these provides reasonable accuracy within their operating limits. However, none should be used without knowing the assumptions and operational limitations that were used in their derivations.

Example 17.5

0.0011 kg/s of 25°C nitrogen gas flows isothermally through a 175 m section of smooth tubing (inside diameter = 0.012 m). The viscosity of the nitrogen is 1.8×10^{-5} Pa·s. The pressure of the nitrogen is 200 kPa originally. At what pressure is the nitrogen delivered?

SI Solution

The flow area of the pipe is

$$A = \frac{\pi}{4}D^2 = \frac{\pi(0.012 \text{ m})^2}{4} = 1.131 \times 10^{-4} \text{ m}^2$$

The mass flow rate per unit area is

$$G = \frac{\dot{m}}{A} = \frac{0.0011 \frac{\text{kg}}{\text{s}}}{1.131 \times 10^{-4} \text{ m}^2} = 9.73 \text{ kg/m}^2\text{·s}$$

The Reynolds number is

$$\text{Re} = \frac{DG}{\mu} = \frac{(0.012 \text{ m})\left(9.73 \frac{\text{kg}}{\text{m}^2\text{·s}}\right)}{1.8 \times 10^{-5} \text{ Pa·s}} = 6487$$

The flow is turbulent, and the pipe is said to be smooth. Therefore, the friction factor is interpolated (from App. 17.B) as 0.0347.

Since two atoms of nitrogen form a molecule of nitrogen gas, the molecular weight of nitrogen is twice the atomic weight, or 28.0 kg/kmol. The temperature must be in degrees absolute: $T = 25°C + 273 = 298K$. The universal gas constant is 8314.3 J/kmol·K.

From Eq. 17.36, the final pressure is

$$p_2^2 = p_1^2 - \frac{fLG^2R^*T}{D(MW)}$$

$$= (200{,}000 \text{ Pa})^2$$

$$- \frac{(0.0347)(175 \text{ m})\left(9.73 \frac{\text{kg}}{\text{m}^2\text{·s}}\right)^2}{\times \left(8314.3 \frac{\text{J}}{\text{kmol·K}}\right)(298K)}{(0.012 \text{ m})\left(28 \frac{\text{kg}}{\text{kmol}}\right)}$$

$$= 4 \times 10^{10} \text{ Pa}^2 - 4.24 \times 10^9 \text{ Pa}^2 = 3.576 \times 10^{10} \text{ Pa}^2$$

$$p_2 = \sqrt{3.576 \times 10^{10} \text{ Pa}^2} = 1.89 \times 10^5 \text{ Pa} \quad (189 \text{ kPa})$$

The percentage drop in pressure should not be more than 10%.

$$\frac{200 \text{ kPa} - 189 \text{ kPa}}{200 \text{ kPa}} = 0.055 \ (5.5\%) \quad [\text{OK}]$$

Example 17.6

Superheated steam at 140 psi and 500°F enters a 200 ft long steel pipe with an internal diameter of 3.826 in. The pipe is insulated so that there is no heat loss. (a) Use the Babcock formula to determine the maximum velocity and mass flow rate such that the steam does not experience more than a 10% drop in pressure. (b) Verify the velocity by calculating the pressure drop with the Darcy equation.

Solution

(a) From superheated steam tables, the specific volume of the steam is 3.954 ft³/lbm. The maximum pressure drop is 10% of 140 psi or 14 psi.

From Eq. 17.37,

$$\Delta p_{\text{psi}} = (0.470)\left(\frac{d_{\text{in}} + 3.6}{d_{\text{in}}^6}\right)(\dot{m}_{\text{lbm/sec}})^2 L_{\text{ft}} v$$

$$14 \text{ psi} = (0.470)\left(\frac{3.826 \text{ in} + 3.6}{(3.826 \text{ in})^6}\right)\dot{m}^2$$

$$\times (200 \text{ ft})\left(3.954 \frac{\text{ft}^3}{\text{lbm}}\right)$$

$$\dot{m} = 3.99 \text{ lbm/sec} \ (4 \text{ lbm/sec})$$

$$v = \frac{Q}{A} = \frac{\dot{m}}{\rho A} = \frac{\dot{m}v}{A}$$

$$= \frac{\left(4 \frac{\text{lbm}}{\text{sec}}\right)\left(3.954 \frac{\text{ft}^3}{\text{lbm}}\right)}{\left(\frac{\pi}{4}\right)\left(\frac{3.826 \text{ in}}{12 \frac{\text{in}}{\text{ft}}}\right)^2}$$

$$= 198 \text{ ft/sec}$$

(b) Assume a Darcy friction factor of 0.02 (typical for turbulent flow in steel pipe). The steam flow velocity is

$$h_f = \frac{fLv^2}{2Dg}$$

$$= \frac{(0.02)(200 \text{ ft})\left(198 \frac{\text{ft}}{\text{sec}}\right)^2}{(2)\left(\frac{3.826 \text{ in}}{12 \frac{\text{in}}{\text{ft}}}\right)\left(32.2 \frac{\text{ft}}{\text{sec}^2}\right)}$$

$$= 7637 \text{ ft of steam}$$

$$\Delta p = \rho h_f \times \left(\frac{g}{g_c}\right) = \left(\frac{h_f}{v}\right) \times \left(\frac{g}{g_c}\right)$$

$$= \left(\frac{7637 \text{ ft}}{\left(3.954 \frac{\text{ft}^3}{\text{lbm}}\right)\left(144 \frac{\text{in}^2}{\text{ft}^2}\right)}\right)$$

$$\times \left(\frac{32.2 \frac{\text{ft}}{\text{sec}^2}}{32.2 \frac{\text{ft-lbm}}{\text{lbf-sec}^2}}\right)$$

$$= 13.4 \text{ lbf/in}^2 \ (\text{psi})$$

13. EFFECT OF VISCOSITY ON HEAD LOSS

Friction loss in a pipe is affected by the fluid viscosity. For both laminar and turbulent flow, viscosity is considered when the Reynolds number is calculated. When viscosities substantially increase without a corresponding decrease in flow rate, two things usually happen: (a) the friction loss greatly increases, and (b) the flow becomes laminar.

It is sometimes necessary to estimate head loss for a new fluid viscosity based on head loss at an old fluid viscosity. The estimation procedure used depends on the flow regimes for the new and old fluids.

For laminar flow, the friction factor is directly proportional to the viscosity. If the flow is laminar for both fluids, the ratio of new-to-old head losses will be equal to the ratio of new-to-old viscosities. Thus, if a flow is already known to be laminar at one viscosity and the fluid viscosity increases, a simple ratio will define the new friction loss.

If both flows are fully turbulent, the friction factor will not change. If flow is fully turbulent and the viscosity decreases, the Reynolds number will increase. Theoretically, this will have no effect on the friction loss.

There are no analytical ways of estimating the change in friction loss when the flow regime changes between laminar and turbulent or between semiturbulent and fully turbulent. Various graphical methods are used, particularly by the pump industry, for calculating power requirements.

14. FRICTION LOSS WITH SLURRIES AND NON-NEWTONIAN FLUIDS

A *slurry* is a mixture of a liquid (usually water) and a solid (e.g., coal, paper pulp, foodstuffs). The liquid is generally used as the transport mechanism (i.e., the *carrier*) for the solid.

Friction loss calculations for slurries vary in sophistication depending on what information is available. In many cases, only the slurry's specific gravity is known. In that case, use is made of the fact that friction loss can be reasonably predicted by multiplying the friction loss based on the pure carrier (e.g., water) by the specific gravity of the slurry.

Another approach is possible if the density and viscosity in the operating range are known. The traditional Darcy equation (Eq. 17.28) and Reynolds number can be used for thin slurries as long as the flow velocity is high enough to keep solids from settling. (Settling is more of a concern for laminar flow. With turbulent flow, the direction of velocity components fluctuates, assisting the solids to remain in suspension.)

The most analytical approach to slurries or other non-Newtonian fluids requires laboratory-derived rheological data. Non-Newtonian viscosity (η, in Pa·s) is fitted to data of the shear rate (dv/dy, in s^{-1}) according to two common models: the power-law model and the Bingham-plastic model. (These two models are applicable to both laminar and turbulent flow, although each has its advantages and disadvantages.)

The *power-law model* has two empirical constants, m and n, that must be determined.

$$\eta = m \left(\frac{dv}{dy} \right)^{n-1} \qquad \text{17.38}$$

The *Bingham-plastic model* also requires finding two empirical constants: the yield stress τ_0 (in Pa) and the Bingham-plastic limiting viscosity μ_∞ (in Pa·s).

$$\eta = \frac{\tau_0}{\dfrac{dv}{dy}} + \mu_\infty \qquad \text{17.39}$$

Once m and n (or τ_0 and μ_∞) have been determined, the friction factor is determined from one of various models (e.g., Buckingham-Reiner, Dodge-Metzner, Metzner-Reed, Hanks-Ricks, Darby, or Hanks-Dadia). Specialized texts and articles cover these models in greater detail. The friction loss is calculated from the traditional Moody equation.

15. MINOR LOSSES

In addition to the frictional energy lost due to viscous effects, friction losses also result from fittings in the line, changes in direction, and changes in flow area. These losses are known as *minor losses* or *local losses*, since they are usually much smaller in magnitude than the pipe wall frictional loss.[14] Two methods are used to calculate minor losses: equivalent lengths and loss coefficients.

With the *method of equivalent lengths*, each fitting or other flow variation is assumed to produce friction equal to the pipe wall friction from an *equivalent length* of pipe. For example, a 2 in globe valve may produce the same amount of friction as 54 ft (its equivalent length) of 2 in pipe. The equivalent lengths for all minor losses are added to the pipe length term, L, in the Darcy equation. This method can be used with all liquids, but it is generally limited to turbulent flow.

$$L_t = L + \sum L_e \qquad \text{17.40}$$

Equivalent lengths are simple to use, but the method depends on having a table of equivalent length values. The actual value for a fitting will depend on the fitting manufacturer, as well as the fitting material (e.g., brass, cast iron, or steel) and the method of attachment (e.g.,

---

[14]Example and practice problems often include the instruction to "Ignore minor losses." In some industries, valves are considered to be "components," not fittings. In such cases, instructions to "Ignore minor losses in fittings" would be ambiguous, since minor losses in valves would be included in the calculations. However, this interpretation is rare in examples and practice problems.

weld, thread, or flange).[15] Because of these many variations, it may be necessary to use a "generic table" of equivalent lengths during the initial design stages. (See Table 17.3 and App. 17.D.)

Table 17.3 *Typical Equivalent Lengths*
(schedule-40, screwed steel fittings)

	pipe size		
	1 in	2 in	4 in
fitting type	equivalent length, ft		
angle valve	17.0	18.0	18.0
coupling or union	0.29	0.45	0.65
gate valve	0.84	1.5	2.5
globe valve	29.0	54.0	110.0
long radius 90° elbow	2.7	3.6	4.6
regular 45° elbow	1.3	2.7	5.5
regular 90° elbow	5.2	8.5	13.0
swing check valve	11.0	19.0	38.0
tee, flow through line (run)	3.2	7.7	17.0
tee, flow through stem	6.6	12.0	21.0
180° return bend	5.2	8.5	13.0

An alternative method of calculating the minor loss for a fitting is to use the *method of loss coefficients*. Each fitting has a *loss coefficient*, K, associated with it, which, when multiplied by the kinetic energy, gives the loss. Thus, a loss coefficient is the minor loss expressed in fractions (or multiples) of the velocity head.

$$ h_m = K h_v \qquad 17.41 $$

The loss coefficient for any minor loss can be calculated if the equivalent length is known. However, there is no advantage to using one method over the other, except for consistency in calculations.

$$ K = \frac{f L_e}{D} \qquad 17.42 $$

Exact friction loss coefficients for bends, fittings, and valves are unique to each manufacturer. Furthermore, except for contractions, enlargements, exits, and entrances, the coefficients decrease fairly significantly (according to the fourth power of the diameter ratio) with increases in valve size. Therefore, a single K value is seldom applicable to an entire family of valves. Nevertheless, generic tables and charts have been developed. These compilations can be used for initial estimates as long as the general nature of the data is recognized.

[15]In the language of pipe fittings, a *threaded fitting* is known as a *screwed fitting*, even though no screws are used.

Table 17.4 *Typical Loss Coefficients*[a]

device	K
angle valve	5
bend, close return	2.2
butterfly valve[b]	
2 to 8 in	$45 f_t$
10 to 14 in	$35 f_t$
16 to 24 in	$25 f_t$
check valve, swing, fully open	2.3
corrugated bends	1.3 to 1.6 times value for smooth bend
standard 90° elbow	0.9
long radius 90° elbow	0.6
45° elbow	0.42
gate valve	
fully open	0.19
$1/4$ closed	1.15
$1/2$ closed	5.6
$3/4$ closed	24
globe valve	10
meter	
disk or wobble	3.4 to 10
rotary (star or cog-wheel piston)	10
reciprocating piston	15
turbine wheel (double flow)	5 to 7.5
tee, standard	1.8

[a]The actual loss coefficient will usually depend on the size of the valve. Average values are given.
[b]Loss coefficients for butterfly valves are calculated from the friction factors for the pipes with complete turbulent flow.

Loss coefficients for specific fittings and valves must be known in order to be used. They cannot be derived theoretically. However, the loss coefficients for certain changes in flow area can be calculated from the following equations.[16]

- *sudden enlargements:* (D_1 is the smaller of the two diameters)

$$ K = \left(1 - \left(\frac{D_1}{D_2} \right)^2 \right)^2 \qquad 17.43 $$

- *sudden contractions:* (D_1 is the smaller of the two diameters)

$$ K = \frac{1}{2} \left(1 - \left(\frac{D_1}{D_2} \right)^2 \right) \qquad 17.44 $$

- *pipe exit:* (projecting exit, sharp-edged or rounded)

$$ K = 1.0 \qquad 17.45 $$

[16]No attempt is made to imply great accuracy with these equations. Correlation between actual and theoretical losses is fair.

- *pipe entrance:*

  reentrant: $K = 0.78$

  sharp-edged: $K = 0.50$

  rounded:

$\dfrac{\text{bend radius}}{D}$	K
0.02	0.28
0.04	0.24
0.06	0.15
0.10	0.09
0.15	0.04

- *tapered diameter changes:*

$$\beta = \frac{\text{small diameter}}{\text{large diameter}} = \frac{D_1}{D_2}$$

$$\phi = \text{wall-to-horizontal angle}$$

enlargement, $\phi \le 22°$:

$$K = 2.6 \, \sin\phi \, (1 - \beta^2)^2 \qquad 17.46$$

enlargement, $\phi > 22°$:

$$K = (1 - \beta^2)^2 \qquad 17.47$$

contraction, $\phi \le 22°$:

$$K = 0.8 \, \sin\phi \, (1 - \beta^2) \qquad 17.48$$

contraction, $\phi > 22°$:

$$K = 0.5 \, \sqrt{\sin\phi} \, (1 - \beta^2) \qquad 17.49$$

Example 17.7

Determine the total equivalent length of the piping system shown. The pipeline contains one gate valve, five regular 90° elbows, one tee (flow through the run), and 228 ft of straight pipe. All fittings are 1 in screwed steel pipe. Disregard entrance and exit losses.

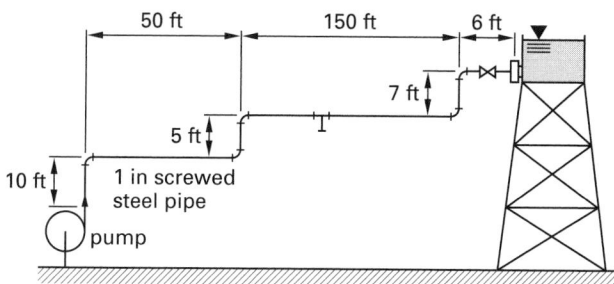

Solution

From Table 17.3, the individual and total equivalent lengths are

1	gate valve	1×0.84	=	0.84
5	regular elbows	5×5.2	=	26.0
1	tee run	1×3.2	=	3.2
	straight pipe			228.0
		total L_t	=	258.0 ft

16. VALVE FLOW COEFFICIENTS

Valve flow capacities depend on the geometry of the inside of the valve. The *flow coefficient*, C_v, for a valve (particularly a control valve) relates the flow quantity (in gallons per minute) of a fluid with specific gravity to the pressure drop (in pounds per square inch). (The flow coefficient for a valve is not the same as the coefficient of flow for an orifice or venturi meter.) As Eq. 17.50 shows, the flow coefficient is not dimensionally homogeneous and is specifically limited to English units.

$$Q_{\text{gpm}} = C_v \sqrt{\frac{\Delta p_{\text{psi}}}{\text{SG}}} \qquad 17.50$$

When selecting a control valve for a particular application, the value of C_v is first calculated. Depending on the application and installation, C_v may be further modified by dividing by *piping geometry* and *Reynolds number factors*. (These additional procedures are often specified by the valve manufacturer.) Then, a valve with the required value of C_v is selected.

Although the flow coefficient concept is generally limited to control valves, its use can be extended to all fittings and valves. The relationship between C_v and the loss coefficient, K, is

$$C_v = \frac{29.9 d_{\text{in}}^2}{\sqrt{K}} \qquad 17.51$$

17. SHEAR STRESS IN CIRCULAR PIPES

Shear stress in fluid always acts to oppose the motion of the fluid. (That is the reason the term *frictional force* is occasionally used.) Shear stress for a fluid in laminar flow can always be calculated from the basic definition of absolute viscosity.

$$\tau = \mu \, \frac{dv}{dy} \qquad 17.52$$

In the case of the flow in a circular pipe, dr can be substituted for dy in the expression for *shear rate (velocity gradient)*, dv/dy.

$$\tau = \mu \, \frac{dv}{dr} \qquad 17.53$$

Equation 17.54 calculates the shear stress between fluid layers a distance r from the pipe centerline from the pressure drop across a length L of the pipe.[17] Equation 17.54 is valid for both laminar and turbulent flows.

$$\tau = \frac{(p_1 - p_2)r}{2L} \quad [r \le \tfrac{D}{2}] \qquad 17.54$$

---

[17]In highly turbulent flow, shear stress is not caused by viscous effects but rather by momentum effects. Equation 17.54 is derived from a shell momentum balance. Such an analysis requires the concept of *momentum flux*. In a circular pipe with laminar flow, momentum flux is maximum at the pipe wall, zero at the flow centerline, and varies linearly in between.

The quantity $(p_1 - p_2)$ can be calculated from the Darcy equation (Eq. 17.28). If v is the average flow velocity, the shear stress at the wall (where $r = D/2$) is

$$\tau_{\text{wall}} = \frac{f\rho v^2}{8} \qquad \text{[SI]} \qquad 17.55(a)$$

$$\tau_{\text{wall}} = \frac{f\rho v^2}{8g_c} \qquad \text{[U.S.]} \qquad 17.55(b)$$

Equation 17.54 can be rearranged somewhat to give the relationship between the pressure gradient and the shear stress at the wall.

$$\frac{dp}{dL} = \frac{4\tau_{\text{wall}}}{D} \qquad 17.56$$

Equation 17.55 can be combined with the Hagen-Poiseuille equation (Eq. 17.24) if the flow is laminar. (v in Eq. 17.57 is the average velocity of fluid flow.)

$$\tau = \frac{16\mu v r}{D^2} \qquad [r \le \tfrac{D}{2}] \qquad 17.57$$

At the pipe wall, $r = D/2$ and the shear stress is maximum. Therefore,

$$\tau_{\text{wall}} = \frac{8\mu v}{D} \qquad 17.58$$

Figure 17.5 *Shear Stress Distribution in a Circular Pipe*

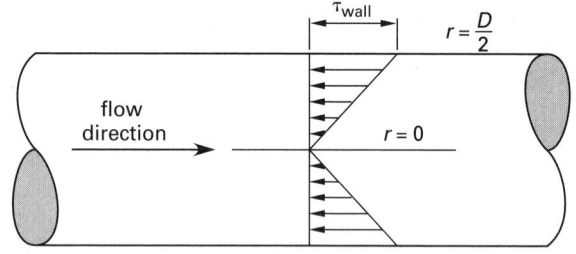

18. INTRODUCTION TO PUMPS AND TURBINES[18]

A *pump* adds energy to the fluid flowing through it. The amount of energy that a pump puts into the fluid stream can be determined by the difference between the total energy on either side of the pump. In most situations, a pump will add primarily pressure energy. The specific energy added (a positive number) on a per-unit mass basis (i.e., ft-lbf/lbm or J/kg) is given by Eq. 17.59.

$$E_A = E_{t,2} - E_{t,1} \qquad 17.59$$

---

[18]Pumps and turbines are covered in greater detail in Chap. 18.

Figure 17.6 *Pump and Turbine Representation*

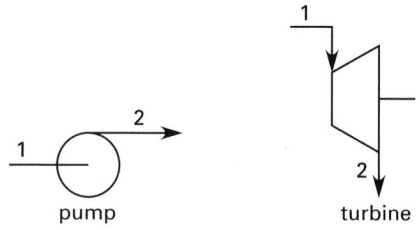

The *head added* by a pump is

$$h_A = \frac{E_A}{g} \qquad \text{[SI]} \qquad 17.60(a)$$

$$h_A = \frac{E_A g_c}{g} \qquad \text{[U.S.]} \qquad 17.60(b)$$

The specific energy added by a pump can also be calculated from the input power if the mass flow rate is known. The input power to the pump will be the output power of the electric motor or engine driving the pump.

$$E_A = \frac{\left(1000 \, \frac{\text{W}}{\text{kW}}\right)(P_{\text{kW,input}})(\eta_{\text{pump}})}{\dot{m}} \qquad \text{[SI]} \qquad 17.61(a)$$

$$E_A = \frac{\left(550 \, \frac{\text{ft-lbf}}{\text{sec-hp}}\right)(P_{\text{hp,input}})(\eta_{\text{pump}})}{\dot{m}} \qquad \text{[U.S.]} \qquad 17.61(b)$$

The *water horsepower* (WHP, also known as the *hydraulic horsepower* and *theoretical horsepower*) is the amount of power actually entering the fluid.

$$\text{WHP} = (P_{\text{hp,input}})\eta_{\text{pump}} \qquad 17.62$$

A *turbine* extracts energy from the fluid flowing through it. As with a pump, the energy extraction can be obtained by evaluating the Bernoulli equation on both sides of the turbine and taking the difference. The energy extracted (a positive number) on a per-unit mass basis is given by Eq. 17.63.

$$E_E = E_{t,1} - E_{t,2} \qquad 17.63$$

19. EXTENDED BERNOULLI EQUATION

The original Bernoulli equation assumes frictionless flow and does not consider the effects of pumps and turbines. When friction is present and when there are minor losses such as fittings and other energy-related devices in a pipeline, the energy balance is affected. The *extended Bernoulli equation* takes these additional factors into account.

$$(E_p + E_v + E_z)_1 + E_A$$
$$= (E_p + E_v + E_z)_2 + E_E + E_f + E_m \qquad 17.64$$

$$\frac{p_1}{\rho} + \frac{v_1^2}{2} + z_1 g + E_A$$

$$= \frac{p_2}{\rho} + \frac{v_2^2}{2} + z_2 g + E_E + E_f + E_m \quad \text{[SI]} \quad 17.65(a)$$

$$\frac{p_1}{\rho} + \frac{v_1^2}{2g_c} + \frac{z_1 g}{g_c} + E_A$$

$$= \frac{p_2}{\rho} + \frac{v_2^2}{2g_c} + \frac{z_2 g}{g_c} + E_E + E_f + E_m \quad \text{[U.S.]} \quad 17.65(b)$$

As defined, E_A, E_E, and E_f are all positive terms. None of the terms in Eq. 17.64 is negative.

The concepts of sources and sinks can be used to decide whether the friction, pump, and turbine terms appear on the left or right side of the Bernoulli equation. An *energy source* puts energy into the system. The incoming fluid and a pump contribute energy to the system. An *energy sink* removes energy from the system. The leaving fluid, friction, and a turbine remove energy from the system. In an energy balance, all energy must be accounted for, and the energy sources just equal the energy sinks.

$$\sum E_{\text{sources}} = \sum E_{\text{sinks}} \qquad 17.66$$

Therefore, the energy added by a pump always appears on the entrance side of the Bernoulli equation. Similarly, the frictional energy loss always appears on the discharge side.

20. ENERGY AND HYDRAULIC GRADE LINES WITH FRICTION[19]

The *energy grade line* (EGL, also known as *total energy line*) is a graph of the total energy versus position in a pipeline. Since a pitot tube measures total (stagnation) energy, EGL will always coincide with the elevation of a pitot-piezometer fluid column. When friction is present, the EGL will always slope down, in the direction of flow. Figure 17.7 illustrates the EGL for a complex pipe network. The difference between EGL$_{\text{frictionless}}$ and EGL$_{\text{with friction}}$ is the energy loss due to friction.

Notice that the EGL line in Figure 17.7 is discontinuous at point 2, since the friction in pipe section B-C cannot be portrayed without disturbing the spatial correlation of points in the figure. Since the friction loss is proportional to v^2, the slope is steeper when the fluid velocity increases (i.e., when the pipe decreases in flow area), as it does in section D-E. Disregarding air friction, the EGL becomes horizontal at point 6 when the fluid becomes a free jet.

The *hydraulic grade line* (HGL) is a graph of the sum of pressure and potential energies versus position in the pipeline. (That is, the EGL and HGL differ by the kinetic energy.) The HGL will always coincide with

[19]The energy grade line and hydraulic grade line were covered in Ch. 16 for the frictionless case.

Figure 17.7 Energy and Hydraulic Grade Lines

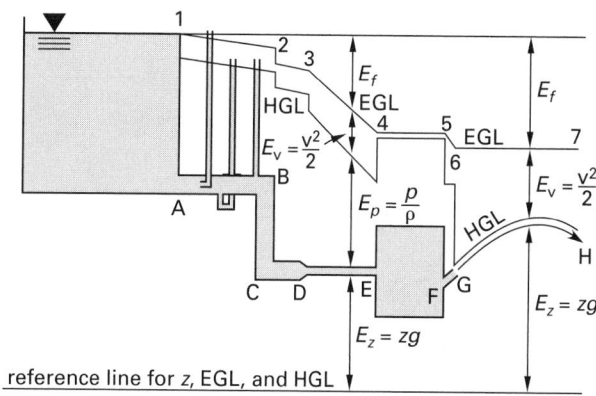

reference line for z, EGL, and HGL

the height of the fluid column in a static piezometer tube. The reference point for elevation is arbitrary, and the pressure energy is usually referenced to atmospheric pressure. Thus, the pressure energy, E_p, for a free jet will be zero, and the HGL will consist only of the potential energy, as shown in section G-H.

The easiest way to draw the energy and hydraulic grade lines is to start with the EGL. The EGL can be drawn simply by recognizing that the rate of divergence from the horizontal EGL$_{\text{frictionless}}$ line is proportional to v^2. Then, since EGL and HGL differ by the velocity head, the HGL can be drawn parallel to the EGL when the pipe diameter is constant. The larger the pipe diameter, the closer the two lines will be.

The EGL for a pump will increase in elevation by E_A across the pump. (The actual energy "path" taken by the fluid is unknown, and a dotted line is used to indicate a lack of knowledge about what really happens in the pump.) The placement of the HGL for a pump will depend on whether the pump increases the fluid velocity and elevation as well as the fluid pressure. In most cases, only the pressure will be increased. Figure 17.8 illustrates the HGL for the case of a pressure increase only.

Figure 17.8 EGL and HGL for a Pump

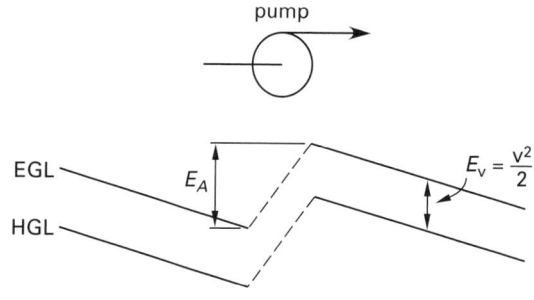

The EGL and HGL for minor losses (fittings, contractions, expansions, etc.) are shown in Fig. 17.9.

Figure 17.9 EGL and HGL for Minor Losses

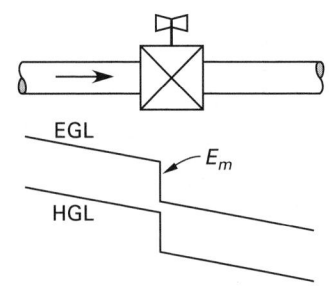

(a) valve, fitting, or obstruction

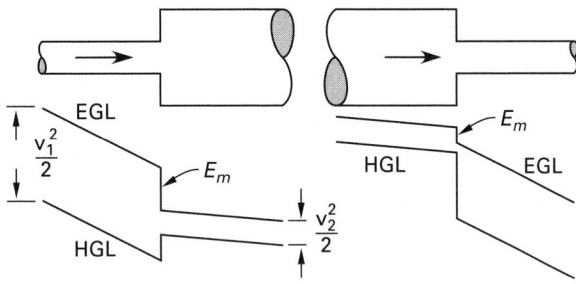

(b) sudden enlargement   (c) sudden contraction

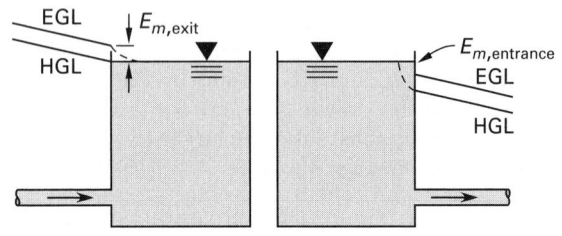

(d) transition to reservoir   (e) transition to pipeline

21. DISCHARGE FROM TANKS

The velocity of a jet issuing from an orifice in a tank can be determined by comparing the total energies at the free fluid surface and the jet itself. At the fluid surface, $p_1 = 0$ (atmospheric) and $v_1 = 0$. (v_1 is known as the *velocity of approach*.) The only energy the fluid has is potential energy. At the jet, $p_2 = 0$. All of the potential energy difference ($z_1 - z_2$) has been converted to kinetic energy. The theoretical velocity of the jet can be derived from the Bernoulli equation. Equation 17.67 is known as the equation for *Torricelli's speed of efflux*.

$$v_t = \sqrt{2gh} \qquad 17.67$$
$$h = z_1 - z_2 \qquad 17.68$$

The actual jet velocity is affected by the orifice geometry. The *coefficient of velocity*, C_v, is an empirical factor that accounts for the friction and turbulence at the orifice. Typical values of C_v are given in Table 17.5.

$$v_o = C_v \sqrt{2gh} \qquad 17.69$$

$$C_v = \frac{\text{actual velocity}}{\text{theoretical velocity}} = \frac{v_o}{v_t} \qquad 17.70$$

Figure 17.10 Discharge from a Tank

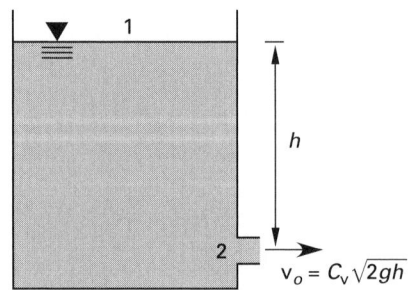

The specific energy loss due to turbulence and friction at the orifice is calculated as a multiple of the jet's kinetic energy.

$$E_f = \left(\frac{1}{C_v^2} - 1\right)\left(\frac{v_o^2}{2}\right) = \left(1 - C_v^2\right)gh \quad \text{[SI]} \quad 17.71(a)$$

$$E_f = \left(\frac{1}{C_v^2} - 1\right)\left(\frac{v_o^2}{2g_c}\right) = \left(1 - C_v^2\right)h \times \left(\frac{g}{g_c}\right)$$
$$\text{[U.S.]} \quad 17.71(b)$$

The total head producing discharge (*effective head*) is the difference in elevations that would produce the same velocity from a frictionless orifice.

$$h_{\text{effective}} = C_v^2 h \qquad 17.72$$

The orifice guides quiescent water from the tank into the jet geometry. Unless the orifice is very smooth and the transition is gradual, momentum effects will continue to cause the jet to contract after it has passed through. The velocity calculated from Eq. 17.69 is usually assumed to be the velocity at the *vena contracta*, the section of smallest cross-sectional area.

Figure 17.11 Vena Contracta of a Fluid Jet

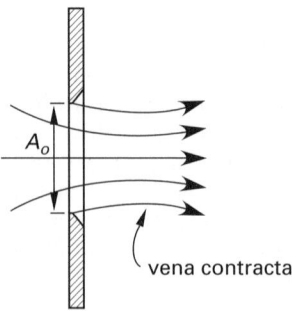

For a thin plate or sharp-edged orifice, the vena contracta is often assumed to be located approximately one half an orifice diameter past the orifice, although the distance can vary from $0.3D_o$ to $0.8D_o$. The area of the vena contracta can be calculated from the orifice area

Table 17.5 Approximate Orifice Coefficients for Turbulent Water

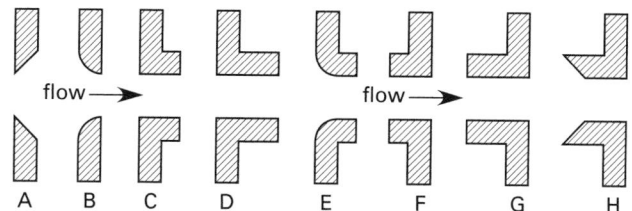

illustration	description	C_d	C_c	C_v
A	sharp-edged	0.62	0.63	0.98
B	round-edged	0.98	1.00	0.98
C	short tube[a] (fluid separates from walls)	0.61	1.00	0.61
D	sharp tube (no separation)	0.82	1.00	0.82
E	short tube with rounded entrance	0.97	0.99	0.98
F	reentrant tube, length less than one-half of pipe diameter	0.54	0.55	0.99
G	reentrant tube, length 2 to 3 pipe diameters	0.72	1.00	0.72
H	Borda	0.51	0.52	0.98
(none)	smooth, well-tapered nozzle	0.98	0.99	0.99

[a]A short tube has a length less than 2 to 3 diameters.

and the *coefficient of contraction*, C_c. For water flowing with a high Reynolds number through a small sharp-edged orifice, the contracted area is approximately 61% to 63% of the orifice area.

$$A_{\text{vena contracta}} = C_c A_o \qquad 17.73$$

$$C_c = \frac{\text{area of vena contracta}}{\text{orifice area}} \qquad 17.74$$

The theoretical discharge rate from a tank is $\dot{V} = A_o\sqrt{2gh}$. However, this relationship needs to be corrected for friction and contraction by multiplying by C_v and C_c. The *coefficient of discharge*, C_d, is the product of the coefficients of velocity and contraction.

$$\dot{V} = C_c v_o A_o = C_d v_t A_o = C_d A_o \sqrt{2gh} \qquad 17.75$$

$$C_d = C_v C_c$$

$$= \frac{\text{actual discharge}}{\text{theoretical discharge}} \qquad 17.76$$

22. DISCHARGE FROM PRESSURIZED TANKS

If the gas or vapor above the liquid in a tank is at gage pressure p, and the discharge is to atmospheric pressure, the head causing discharge will be

$$h = z_1 - z_2 + \frac{p}{\rho g} \qquad \text{[SI]} \qquad 17.77(a)$$

$$h = z_1 - z_2 + \left(\frac{p}{\rho}\right) \times \left(\frac{g_c}{g}\right) \qquad \text{[U.S.]} \qquad 17.77(b)$$

The discharge velocity can be calculated from Eq. 17.69 using the increased discharge head.

Figure 17.12 Discharge from a Pressurized Tank

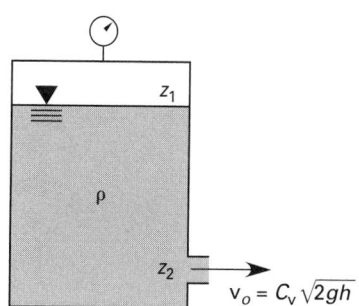

23. COORDINATES OF A FLUID STREAM

Fluid discharged from an orifice in a tank gets its initial velocity from the conversion of potential energy. After discharge, no additional energy conversion occurs, and all subsequent velocity changes are due to external forces.

Figure 17.13 Coordinates of a Fluid Stream

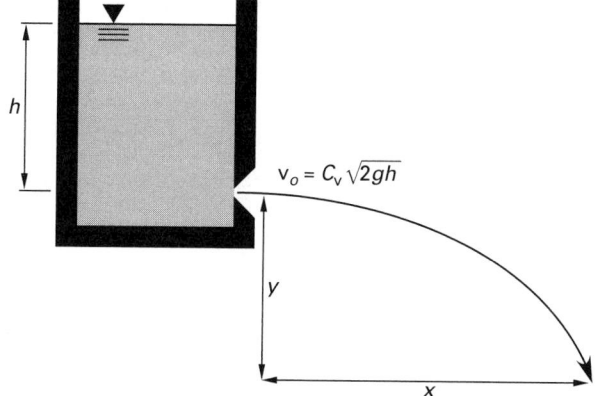

In the absence of air friction (drag), there are no retarding or accelerating forces in the x-direction on the fluid stream. The x-component of velocity is constant. Projectile motion equations (Ch. 72) can be used to predict the path of the fluid stream.

$$v_x = v_o \qquad 17.78$$

$$x = v_o t = v_o \sqrt{\frac{2y}{g}} = 2C_v \sqrt{hy} \qquad 17.79$$

After discharge, the fluid stream is acted upon by a constant gravitational acceleration. The y-component of velocity is zero at discharge but increases linearly with time.

$$v_y = gt \qquad 17.80$$

$$y = \frac{gt^2}{2} = \frac{gx^2}{2v_o^2} = \frac{x^2}{4hC_v^2} \qquad 17.81$$

24. TIME TO EMPTY A TANK

If the fluid in an open or vented tank is not replenished at the rate of discharge, the static head forcing discharge through the orifice will decrease with time. If the tank has a varying cross section, A_t, Eq. 17.82 specifies the basic relationship between the change in elevation and elapsed time. (The negative sign indicates that z decreases as t increases.)

$$\dot{V}dt = -A_t dz \qquad 17.82$$

If A_t can be expressed as a function of h, Eq. 17.83 can be used to determine the time to lower the fluid elevation from z_1 to z_2.

$$t = \int_{z_1}^{z_2} \frac{-A_t dz}{C_d A_o \sqrt{2gz}} \qquad 17.83$$

For a tank with a constant cross-sectional area, A_t, the time required to lower the fluid elevation is

$$t = \frac{2A_t(\sqrt{z_1} - \sqrt{z_2})}{C_d A_o \sqrt{2g}} \qquad 17.84$$

If a tank is replenished at a rate of $\dot{V}_{in}$, Eq. 17.85 can be used to calculate the discharge time. If the tank is replenished at a rate greater than the discharge rate, t in Eq. 17.85 will represent the time to raise the fluid level from z_1 to z_2.

$$t = \int_{z_1}^{z_2} \frac{A_t dz}{(C_d A_o \sqrt{2gz}) - \dot{V}_{in}} \qquad 17.85$$

Example 17.8

A tank 15 ft in diameter discharges $150°$F water ($\rho = 61.20$ lbm/ft^3) through a sharp-edged 1.0 in diameter

orifice ($C_d = 0.62$) in the bottom. The original water depth is 12 ft. The tank is continually pressurized to 50 psig. What is the time to empty the tank?

Solution

The area of the orifice is

$$A_o = \frac{\pi D^2}{4} = \frac{\pi (1 \text{ in})^2}{(4)\left(12 \frac{\text{in}}{\text{ft}}\right)^2} = 0.00545 \text{ ft}^2$$

The tank area constant with respect to z is

$$A_t = \frac{\pi D^2}{4} = \frac{\pi (15 \text{ ft})^2}{4} = 176.7 \text{ ft}^2$$

The total initial head includes the effect of the pressurization. Use Eq. 17.77.

$$h_1 = 12 \text{ ft} + \left(\frac{\left(50 \frac{\text{lbf}}{\text{in}^2}\right)\left(144 \frac{\text{in}^2}{\text{ft}^2}\right)}{61.2 \frac{\text{lbm}}{\text{ft}^3}}\right)\left(\frac{32.2 \frac{\text{lbm-ft}}{\text{lbf-sec}^2}}{32.2 \frac{\text{ft}}{\text{sec}^2}}\right)$$

$$= 12 \text{ ft} + 117.6 \text{ ft} = 129.6 \text{ ft}$$

When the fluid has reached the level of the orifice, the fluid potential head will be zero, but the pressurization will remain.

$$h_2 = 117.6 \text{ ft}$$

The time to empty the tank is given by Eq. 17.84.

$$t = \frac{(2)(176.7 \text{ ft}^2)\left(\sqrt{129.6 \text{ ft}} - \sqrt{117.6 \text{ ft}}\right)}{(0.62)(0.00545 \text{ ft}^2)\sqrt{(2)\left(32.2 \frac{\text{ft}}{\text{sec}^2}\right)}}$$

$$= 7036 \text{ sec}$$

25. DISCHARGE FROM LARGE ORIFICES

When an orifice diameter is large compared with the discharge head, the jet velocity at the top edge of the orifice will be less than the velocity at the bottom edge. Since the velocity is related to the square root of the head, the distance used to calculate the effective jet velocity should be measured from the fluid surface to a point above the centerline of the orifice.

This correction is generally neglected, however, since it is small for heads of more than twice the orifice diameter. Furthermore, if an orifice is intended to work regularly with small heads, the orifice should be calibrated in place. The discrepancy can then be absorbed into the discharge coefficient, C_d.

26. CULVERTS

A *culvert* is a water path (usually a large diameter pipe) used to channel water around or through an obstructing feature. In most instances, a culvert is used to restore a natural water path obstructed by a manufactured feature. For example, when a road is built across (perpendicular to) a natural ravine or arroyo, a culvert can be used to channel water under the road.

Figure 17.14 Simple Pipe Culvert

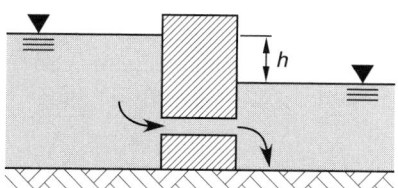

Because culverts usually operate only partially full and with low heads, Torricelli's equation does not apply. Therefore, most culvert designs are empirical. However, if the entrance and exit of a culvert are both submerged, the culvert will flow full, and the discharge will be independent of the barrel slope. Equation 17.86 can be used to calculate the discharge.

$$\dot{V} = C_d A \sqrt{2gh} \qquad 17.86$$

If the culvert is long (more than 60 ft or 20 m), or if the entrance is not gradual, the available energy will be divided between friction and velocity heads. The effective head used in Eq. 17.86 should be

$$h_{\text{effective}} = h - h_{f,\text{barrel}} - h_{m,\text{entrance}} \qquad 17.87$$

The friction loss in the barrel can be found in the usual manner, from either the Darcy equation or the Hazen-Williams equation. The entrance loss is calculated using the standard method of loss coefficients. Representative values of the loss coefficient, K, are given in Table 17.6. Since the fluid velocity is not initially known but is needed to find the friction factor, a trial-and-error solution will be necessary.

Table 17.6 Representative Loss Coefficients for Culvert Entrances

entrance	K
smooth and gradual transition	0.08
flush vee or bell shape	0.10
projecting vee or bell shape	0.15
flush, square-edged	0.50
projecting, square-edged	0.90

27. SIPHONS

A *siphon* is a bent or curved tube that carries fluid from a container at a high elevation to another container at a lower elevation. Normally, it would not seem difficult

to have a fluid flow to a lower elevation. However, the fluid seems to flow "uphill" in a siphon. Figure 17.15 illustrates a siphon.

Figure 17.15 Siphon

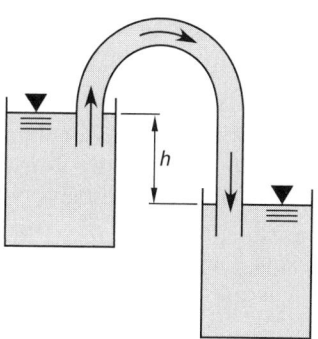

Starting a siphon requires the tube to be completely filled with liquid. Then, since the fluid weight is greater in the longer arm than in the shorter arm, the fluid in the longer arm "falls" out of the siphon, "pulling" more liquid into the shorter arm and over the bend.

Operation of a siphon is essentially independent of atmospheric pressure. The theoretical discharge is the same as predicted by the Torricelli equation. A correction for discharge is necessary, but little data is available on typical values of C_d. Therefore, siphons should be tested and calibrated in place.

$$\dot{V} = C_d A \mathrm{v} = C_d A \sqrt{2gh} \qquad 17.88$$

28. SERIES PIPE SYSTEMS

A system of pipes in series consists of two or more lengths of different-diameter pipes connected together. In the case of the series pipe from a reservoir discharging to the atmosphere shown in Fig. 17.16, the available head will be split between the velocity head and the friction loss.

$$h = h_{\mathrm{v}} + h_f \qquad 17.89$$

Figure 17.16 Series Pipe System

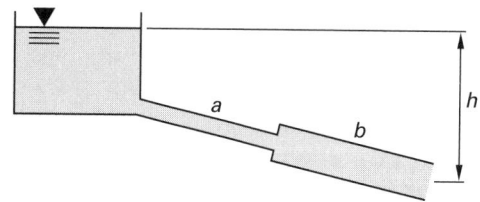

If the flow rate or velocity in any part of the system is known, the friction loss can easily be found as the sum of the friction losses in the individual sections. The solution is somewhat more simple than first appears, since the velocity of all sections can be written in terms of only one velocity.

$$h_{f,t} = h_{f,a} + h_{f,b} \qquad 17.90$$
$$A_a \mathrm{v}_a = A_b \mathrm{v}_b \qquad 17.91$$

If neither the velocity nor the flow quantity is known, a trial-and-error solution will be required, since a friction factor must be known to calculate h_f. A good starting point is to assume fully turbulent flow.

When velocity and flow rate are both unknown, the following procedure using the Darcy friction factor can be used.[20]

step 1: Calculate the relative roughness, ϵ/D, for each section. Use the Moody diagram to determine f_a and f_b for fully turbulent flow (i.e., the horizontal portion of the curve).

step 2: Write all of the velocities in terms of one unknown velocity.

$$\dot{V}_a = \dot{V}_b \qquad 17.92$$

$$v_b = \left(\frac{A_a}{A_b}\right)v_a \qquad 17.93$$

step 3: Write the total friction loss in terms of the unknown velocity.

$$
\begin{aligned}
h_{f,t} &= \frac{f_a L_a v_a^2}{2D_a g} + \left(\frac{f_b L_b}{2D_b g}\right)\left(\frac{A_a}{A_b}\right)^2 v_a^2 \\
&= \left(\frac{v_a^2}{2g}\right)\left(\left(\frac{f_a L_a}{D_a}\right) + \left(\frac{f_b L_b}{D_b}\right)\left(\frac{A_a}{A_b}\right)^2\right)
\end{aligned}
$$

$$17.94$$

step 4: Solve for the unknown velocity using the Bernoulli equation between the free reservoir surface ($p = 0$, $v = 0$, $z = h$) and the discharge point ($p = 0$, if free discharge; $z = 0$). Include pipe friction, but disregard minor losses for convenience.

$$
\begin{aligned}
h &= \frac{v_b^2}{2g} + h_{f,t} \\
&= \left(\frac{v_a^2}{2g}\right)\left(\left(\frac{A_a}{A_b}\right)^2\left(1 + \frac{f_b L_b}{D_b}\right) + \frac{f_a L_a}{D_a}\right)
\end{aligned}
$$

$$17.95$$

step 5: Using the value of v_a, calculate v_b. Calculate the Reynolds number and check the values of f_a and f_b from step 4. Repeat steps 3 and 4 if necessary.

29. PARALLEL PIPE SYSTEMS

A *pipe loop* is a set of two pipes placed in parallel, both originating and terminating at the same junction. Adding a second pipe in parallel with a first is a standard method of increasing the capacity of a line.

---

[20]If Hazen-Williams constants are given for the pipe sections, the procedure for finding the unknown velocities is similar, although considerably more difficult since v^2 and $v^{1.85}$ cannot be combined. A first approximation, however, can be obtained by replacing $v^{1.85}$ in the Hazen-Williams equation for friction loss. A trial and error method can then be used to find velocity.

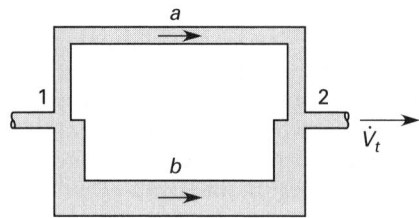

Figure 17.17 *Parallel Pipe System*

There are three principles that govern the distribution of flow between the two branches.

- The flow divides in such a manner as to make the head loss in each branch the same.

$$h_{f,a} = h_{f,b} \qquad 17.96$$

- The head loss between the junctions 1 and 2 is the same as the head loss in branches a and b.

$$h_{f,1-2} = h_{f,a} = h_{f,b} \qquad 17.97$$

- The total flow rate is the sum of the flow rates in the two branches.

$$\dot{V}_t = \dot{V}_a + \dot{V}_b \qquad 17.98$$

If the pipe diameters are known, Eq. 17.96 and Eq. 17.98 can be solved simultaneously for the branch velocities. In such problems, it is common to neglect minor losses, the velocity head, and the variation in the friction factor, f, with velocity.

If the parallel system has only two branches, the unknown branch flows can be determined by solving Eqs. 17.99 and 17.101 simultaneously.

$$\frac{f_a L_a v_a^2}{2D_a g} = \frac{f_b L_b v_b^2}{2D_b g} \qquad 17.99$$

$$\dot{V}_a + \dot{V}_b = \dot{V}_t \qquad 17.100$$

$$\left(\frac{\pi}{4}\right)(D_a^2 v_a + D_b^2 v_b) = \dot{V}_t \qquad 17.101$$

However, if the parallel system has three or more branches, it is easier to use the following iterative procedure. This procedure can be used for problems (a) where the flow rate is unknown but the pressure drop between the two junctions is known, or (b) where the total flow rate is known but the pressure drop and velocity are both unknown. In both cases, the solution iteratively determines the friction coefficients (f).

step 1: Solve the friction head loss (h_f) expression (either Darcy or Hazen-Williams) for velocity in each branch. If the pressure drop is known, first convert it to friction head loss.

$$v = \sqrt{\frac{2Dgh_f}{fL}} \quad \text{[Darcy]} \qquad 17.102$$

$$v = \frac{0.355 C D^{0.63} h_f^{0.54}}{L^{0.54}} \quad \text{[Hazen-Williams; SI]} \quad \textit{17.103(a)}$$

$$v = \frac{0.550 C D^{0.63} h_f^{0.54}}{L^{0.54}} \quad \text{[Hazen-Williams; U.S.]} \quad \textit{17.103(b)}$$

step 2: Solve for the flow rate in each branch. If they are unknown, friction factors, f, must be assumed for each branch. The fully turbulent assumption provides a good initial estimate. (The value of k' will be different for each branch.)

$$\dot{V} = Av = A\sqrt{\frac{2Dgh_f}{fL}}$$
$$= k'\sqrt{h_f} \quad \text{[Darcy]} \quad \textit{17.104}$$

step 3: Write the expression for the conservation of flow. Calculate the friction head loss from the total flow rate. For example, for a three-branch system,

$$\dot{V}_t = \dot{V}_1 + \dot{V}_2 + \dot{V}_3$$
$$= (k'_1 + k'_2 + k'_3)\sqrt{h_f} \quad \textit{17.105}$$

step 4: Check the assumed values of the friction factor. Repeat as necessary.

Example 17.9

3.0 ft^3/sec of water enter the parallel pipe network shown. All pipes are schedule-40 steel with the nominal sizes shown. Minor losses are insignificant. What is the total friction head loss between junctions A and B?

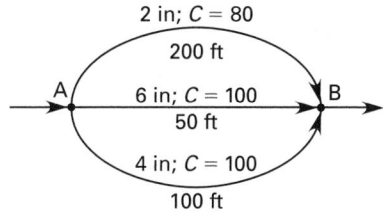

Solution

step 1: Collect the pipe dimensions.

	2 in	4 in	6 in
flow area	0.0233 ft^2	0.0884 ft^2	0.2006 ft^2
diameter	0.1723 ft	0.3355 ft	0.5054 ft

Follow the procedure given in Sec. 29. Since the Hazen-Williams loss coefficients are given for each branch, the Hazen-Williams friction loss equation must be used.

$$h_f = \frac{3.022 v^{1.85} L}{C^{1.85} D^{1.165}}$$

$$v = \frac{0.550 C D^{0.63} h_f^{0.54}}{L^{0.54}}$$

The velocity (expressed in ft/sec) in the 2 in pipe branch is

$$v_{2\ in} = \frac{(0.550)(80)(0.1723\ \text{ft})^{0.63} h_f^{0.54}}{(200\ \text{ft})^{0.54}}$$
$$= 0.831 h_f^{0.54}$$

The velocities in the other two branches are

$$v_{6\ in} = 4.327 h_f^{0.54}$$
$$v_{4\ in} = 2.299 h_f^{0.54}$$

step 2: The flow rates are

$$\dot{V} = Av$$
$$\dot{V}_{2\ in} = (0.0233\ \text{ft}^2)(0.831) h_f^{0.54}$$
$$= 0.0194 h_f^{0.54}$$
$$\dot{V}_{6\ in} = (0.2006\ \text{ft}^2)(4.327) h_f^{0.54}$$
$$= 0.8680 h_f^{0.54}$$
$$\dot{V}_{4\ in} = (0.0884\ \text{ft}^2)(2.299) h_f^{0.54}$$
$$= 0.2032 h_f^{0.54}$$

step 3: $\dot{V}_t = \dot{V}_{2\ in} + \dot{V}_{6\ in} + \dot{V}_{4\ in}$

$$3\ \text{ft}^3/\text{sec} = 0.0194 h_f^{0.54} + 0.8680 h_f^{0.54} + 0.2032 h_f^{0.54}$$

The friction head loss is the same in all parallel branches.

$$3\ \text{ft}^3/\text{sec} = (0.0194 + 0.8680 + 0.2032) h_f^{0.54}$$
$$h_f = 6.5\ \text{ft}$$

30. MULTIPLE RESERVOIR SYSTEMS

In the *three-reservoir problem*, there are many possible choices for the unknown quantity (pipe length, diameter, head, flow rate, etc.). In all but the simplest cases, the solution technique is by trial and error based on conservation of mass and energy.

Figure 17.18 Three-Reservoir System

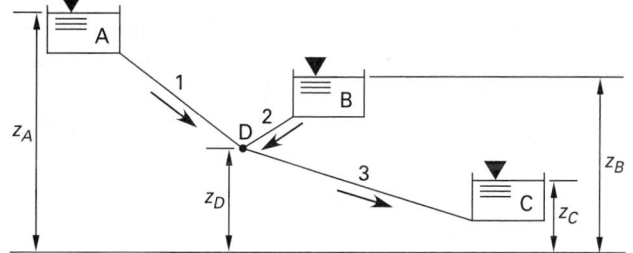

For simplification, velocity heads and minor losses are usually insignificant and can be neglected. However, the presence of a pump in any of the lines must be included in the solution procedure. This is most easily done by adding the pump head to the elevation of the reservoir feeding the pump. If the pump head is not known or depends on the flow rate, it must be determined iteratively as part of the solution procedure.

Case 1: Given all lengths, diameters, and elevations, find all flow rates.

Although an analytical solution method is possible, this type of problem is easily solved iteratively. The following procedure makes an initial estimate of a flow rate and uses it to calculate p_D. Since this method may not converge if the initial estimate of $\dot{V}_1$ is significantly in error, it is helpful to use other information (e.g., normal pipe velocities) to obtain the initial estimate. An alternative procedure is simply to make several estimates of p_D and calculate the corresponding values of flow rate.

step 1: Assume a reasonable value for $\dot{V}_1$. Calculate the corresponding friction loss, $h_{f,1}$. Use the Bernoulli equation to find the corresponding value of p_D. Disregard minor losses and velocity head.

$$v_1 = \frac{\dot{V}_1}{A_1} \qquad \qquad 17.106$$

$$z_A = z_D + \frac{p_D}{\gamma} + h_{f,1} \qquad \qquad 17.107$$

step 2: Use the value of p_D to calculate $h_{f,2}$. Use the friction loss to determine v_2. Use v_2 to determine $\dot{V}_2$.

$$z_B = z_D + \frac{p_D}{\gamma} \pm h_{f,2} \qquad \qquad 17.108$$

$$\dot{V}_2 = v_2 A_2 \qquad \qquad 17.109$$

If flow is out of reservoir B, $h_{f,2}$ should be added. If $z_D + (p_D/\gamma) > z_B$, flow will be into reservoir B. In this case, $h_{f,2}$ should be subtracted.

step 3: Similarly, use the value of p_D to calculate $h_{f,3}$. Use the friction loss to determine v_3. Use v_3 to determine $\dot{V}_3$.

$$z_C = z_D + \frac{p_D}{\gamma} - h_{f,3} \qquad \qquad 17.110$$

$$\dot{V}_3 = v_3 A_3 \qquad \qquad 17.111$$

step 4: Check that $\dot{V}_1 \pm \dot{V}_2 = \dot{V}_3$. If it does not, repeat steps 1 through 4. After the second iteration, plot $\dot{V}_1 \pm \dot{V}_2 - \dot{V}_3$ versus $\dot{V}_1$. Interpolate or extrapolate the value of $\dot{V}_1$ that makes the difference zero.

Case 2: Given $\dot{V}_1$ and all lengths, diameters, and elevations except z_C, find z_C.

step 1: Calculate v_1.

$$v_1 = \frac{\dot{V}_1}{A_1} \qquad \qquad 17.112$$

step 2: Calculate the corresponding friction loss, $h_{f,1}$. Use the Bernoulli equation to find the corresponding value of p_D. Disregard minor losses and velocity head.

$$z_A = z_D + \frac{p_D}{\gamma} + h_{f,1} \qquad \qquad 17.113$$

step 3: Use the value of p_D to calculate $h_{f,2}$. Use the friction loss to determine v_2. Use v_2 to determine $\dot{V}_2 A_2$.

$$z_B = z_D + \frac{p_D}{\gamma} \pm h_{f,2} \qquad \qquad 17.114$$

$$\dot{V}_2 = v_2 A_2 \qquad \qquad 17.115$$

If flow is out of reservoir B, $h_{f,2}$ should be added. If $z_D + (p_D/\gamma) > z_B$, flow will be into reservoir B. In this case, $h_{f,2}$ should be subtracted.

step 4: $\dot{V}_3 = \dot{V}_1 \pm \dot{V}_2 \qquad \qquad 17.116$

step 5: $v_3 = \dfrac{\dot{V}_3}{A_3} \qquad \qquad 17.117$

step 6: Calculate $h_{f,3}$.

step 7: $z_C = z_D + \dfrac{p_D}{\gamma} - h_{f,3} \qquad \qquad 17.118$

Case 3: Given $\dot{V}_1$, all lengths, all elevations, and all diameters except D_3, find D_3.

steps 1 through 4: Repeat steps 1 through 4 from case 2.

step 5: Calculate $h_{f,3}$ from

$$z_C = z_D + \frac{p_D}{\gamma} - h_{f,3} \qquad \qquad 17.119$$

step 6: Calculate D_3 from $h_{f,3}$.

Case 4: Given all lengths, diameters, and elevations except z_D, find all flow rates.

step 1: Calculate the head loss between each reservoir and junction D. Combine as many terms as possible into constant k'.

$$\dot{V} = Av = A\sqrt{\frac{2Dgh_f}{f_L}}$$

$$= k'\sqrt{h_f} \quad \text{[Darcy]} \qquad \qquad 17.120$$

$$\dot{V} = Av = \frac{A(0.550)CD^{0.63}h_f^{0.54}}{L^{0.54}}$$

$$= k'(h_f)^{0.54} \quad \text{[Hazen-Williams, U.S.]} \qquad \qquad 17.121$$

step 2: Assume that the flow direction in all three pipes is toward junction D. Write the conservation equation for junction D.

$$\dot{V}_{D,t} = \dot{V}_1 + \dot{V}_2 + \dot{V}_3 = 0 \qquad 17.122$$

$$k_1'\sqrt{h_{f,1}} + k_2'\sqrt{h_{f,2}} + k_3'\sqrt{h_{f,3}} = 0 \quad [\text{Darcy}] \qquad 17.123$$

$$k_1'(h_{f,1})^{0.54} + k_2'(h_{f,2})^{0.54} + k_3'(h_{f,3})^{0.54} = 0 \quad [\text{Hazen-Williams}] \qquad 17.124$$

step 3: Write the Bernoulli equation between each reservoir and junction D. Since $p_A = p_B = p_C = 0$, and $v_A = v_B = v_C = 0$, the friction loss in branch 1 is

$$h_{f,1} = z_A - z_D - \frac{p_D}{\gamma} \qquad 17.125$$

However, z_D and p_D can be combined since they are related constants in any particular situation. Define R_D as

$$R_D = z_D + \frac{p_D}{\gamma} \qquad 17.126$$

Then, the friction head losses in the branches are

$$h_{f,1} = z_A - R_D \qquad 17.127$$
$$h_{f,2} = z_B - R_D \qquad 17.128$$
$$h_{f,3} = z_C - R_D \qquad 17.129$$

step 4: Assume a value for R_D. Calculate the corresponding h_f values. Use Eq. 17.120 to find $\dot{V}_1$, $\dot{V}_2$, and $\dot{V}_3$. Calculate the corresponding $\dot{V}_t$ value. Repeat until $\dot{V}_t$ converges to zero. It is not necessary to calculate p_D or z_D once all of the flow rates are known.

31. PIPE NETWORKS

Network flows in a *multiloop system* cannot be determined by any closed-form equation. Most real-world problems involving multiloop systems are analyzed on a computer. Computer programs are based on the *Hardy Cross method*, which can also be performed manually when there are only a few loops. In this method, flows in all of the branches are first assumed, and adjustments are made in consecutive iterations to the assumed flow.

Figure 17.19 *Multiloop System*

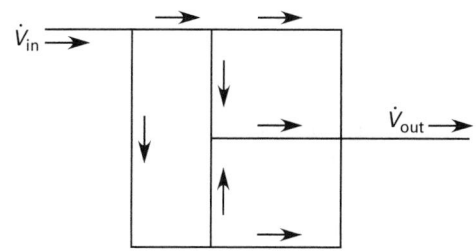

The Hardy Cross method is based on the following principles.

Principle 17.1: Conservation: The flows entering a junction equal the flows leaving the junction.

Principle 17.2: The algebraic sum of head losses around any closed loop is zero.

The friction head loss has the form $h_f = K'\dot{V}^n$, with h_f having units of feet. (Note that k' used in Eq. 17.104 is equal to $\sqrt{1/K'}$.) The Darcy friction factor, f, is usually assumed to be the same in all parts of the network. For a Darcy head loss, the exponent is $n = 2$. For a Hazen-Williams loss, $n = 1.85$.

- For $\dot{V}$ in ft^3/sec, L in feet, and D in feet, the friction coefficient is

$$K' = \frac{0.02517fL}{D^5} \quad [\text{Darcy}] \qquad 17.130$$

$$K' = \frac{4.727L}{D^{4.8655}C^{1.85}} \quad [\text{Hazen-Williams}] \qquad 17.131$$

- For $\dot{V}$ in gal/min, L in feet, and d in inches, the friction coefficient is

$$K' = \frac{0.03109fL}{d^5} \quad [\text{Darcy}] \qquad 17.132$$

$$K' = \frac{10.44L}{d^{4.8655}C^{1.85}} \quad [\text{Hazen-Williams}] \qquad 17.133$$

- For $\dot{V}$ in gal/min, L in feet, and D in feet, the friction coefficient is

$$K' = \frac{1.251 \times 10^{-7}fL}{D^5} \quad [\text{Darcy}] \qquad 17.134$$

$$K' = \frac{5.862 \times 10^{-5}L}{D^{4.8655}C^{1.85}} \quad [\text{Hazen-Williams}] \qquad 17.135$$

- For $\dot{V}$ in MGD (millions of gallons per day), L in feet, and D in feet, the friction coefficient is

$$K' = \frac{0.06026fL}{D^5} \quad [\text{Darcy}] \qquad 17.136$$

$$K' = \frac{10.59L}{D^{4.8655}C^{1.85}} \quad [\text{Hazen-Williams}] \qquad 17.137$$

If $\dot{V}_a$ is the assumed flow in a pipe, the true value $(\dot{V})$ can be calculated from the difference (correction), δ.

$$\dot{V} = \dot{V}_a + \delta \qquad \qquad 17.138$$

The friction loss term for the assumed value and its correction can be expanded as a series. Since the correction is small, higher order terms can be omitted.

$$h_f = K'(\dot{V}_a + \delta)^n$$
$$\approx K'\dot{V}_a^n + nK'\delta\dot{V}_a^{n-1} \qquad 17.139$$

From principle 17.2, the sum of the friction drops is zero around a loop. The correction, δ, is the same for all pipes in the loop and can be taken out of the summation. Since the loop closes on itself, all elevations can be omitted.

$$\sum h_f = \sum K'\dot{V}_a^n + n\delta \sum K'\dot{V}_a^{n-1} = 0 \qquad 17.140$$

This equation can be solved for δ.

$$\delta = \frac{-\sum K'\dot{V}_a^n}{n\sum |K'\dot{V}_a^{n-1}|}$$
$$= -\frac{\sum h_f}{n\sum \left|\dfrac{h_f}{\dot{V}_a}\right|} \qquad 17.141$$

The Hardy Cross procedure is as follows.

step 1: Determine the value of n. For a Darcy head loss, the exponent is $n = 2$. For a Hazen-Williams loss, $n = 1.85$.

step 2: Arbitrarily select a positive direction (e.g., clockwise).

step 3: Label all branches and junctions in the network.

step 4: Separate the network into independent loops such that each branch is included in at least one loop.

step 5: Calculate K' for each branch in the network.

step 6: Assume consistent and reasonable flow rates and directions for each branch in the network.

step 7: Calculate the correction, δ, for each independent loop. (The numerator is the sum of head losses around the loop, taking signs into consideration.) It is not necessary for the loop to be at the same elevation everywhere. Disregard elevations. Since the loop closes on itself, all elevations can be omitted.

step 8: Apply the correction, δ, to each branch in the loop. The correction must be applied in the same sense to each branch in the loop. If clockwise has been taken as the positive

direction, then δ is added to clockwise flows and subtracted from counterclockwise flows.

step 9: Repeat steps 7 and 8 until the correction is sufficiently small.

Example 17.10

A two-loop pipe network is shown. All junctions are at the same elevation. The Darcy friction factor is 0.02 for all pipes in the network. For convenience, use the nominal pipe sizes shown in the figure. Determine the flows in all branches.

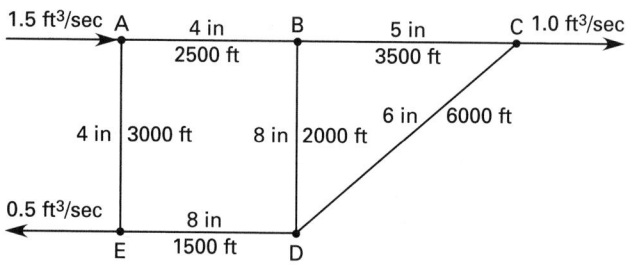

Solution

step 1: The Darcy friction factor is given, so $n = 2$.

step 2: Select clockwise as the positive direction.

step 3: Use the junction letters in the illustration.

step 4: Two independent loops are needed. Work with loops ABDE and BCD. (Loop ABCDE could also be used but would be more complex than loop BCD.)

step 5: Work with branch AB.

$$D = \frac{4 \text{ in}}{12 \dfrac{\text{in}}{\text{ft}}} = 0.3333 \text{ ft}$$

Use Eq. 17.130.

$$K'_{AB} = \frac{0.0252fL}{D^5}$$
$$= \frac{(0.0252)(0.02)(2500 \text{ ft})}{(0.3333 \text{ ft})^5}$$
$$= 306.2$$

Similarly,

$$K'_{BC} = 140.5$$
$$K'_{DC} = 96.8$$
$$K'_{BD} = 7.7$$
$$K'_{ED} = 5.7$$
$$K'_{AE} = 367.4$$

step 6: Assume the direction and flow rates shown.

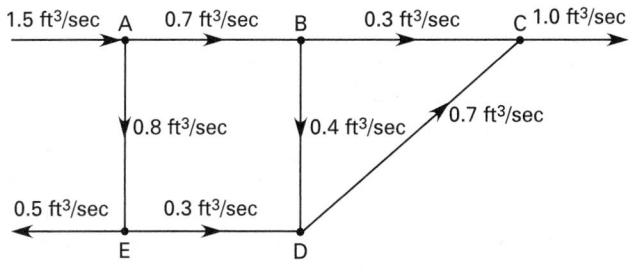

step 7: Use Eq. 17.141.

$$\delta = \frac{-\sum K'\dot{V}_a^n}{n\sum |K'\dot{V}_a^{n-1}|}$$

$$\delta_{\text{ABDE}} = \frac{\begin{array}{c} -\left((306.2)(0.7)^2 + (7.7)(0.4)^2 \right. \\ \left. -(5.7)(0.3)^2 - (367.4)(0.8)^2\right) \end{array}}{\begin{array}{c} (2)\big((306.2)(0.7) + (7.7)(0.4) \\ +(5.7)(0.3) + (367.4)(0.8)\big) \end{array}}$$

$$= 0.08 \text{ ft}^3/\text{sec}$$

$$\delta_{\text{BCD}} = \frac{-\left((140.5)(0.3)^2 - (96.8)(0.7)^2 - (7.7)(0.4)^2\right)}{(2)\big((140.5)(0.3) + (96.8)(0.7) + (7.7)(0.4)\big)}$$

$$= 0.16 \text{ ft}^3/\text{sec}$$

step 8: The corrected flows are

$$\dot{V}_{\text{AB}} = 0.7\,\frac{\text{ft}^3}{\text{sec}} + 0.08\,\frac{\text{ft}^3}{\text{sec}} = 0.78 \text{ ft}^3/\text{sec}$$

$$\dot{V}_{\text{BC}} = 0.3\,\frac{\text{ft}^3}{\text{sec}} + 0.16\,\frac{\text{ft}^3}{\text{sec}} = 0.46 \text{ ft}^3/\text{sec}$$

$$\dot{V}_{\text{DC}} = 0.7\,\frac{\text{ft}^3}{\text{sec}} - 0.16\,\frac{\text{ft}^3}{\text{sec}} = 0.54 \text{ ft}^3/\text{sec}$$

$$\dot{V}_{\text{BD}} = 0.4\,\frac{\text{ft}^3}{\text{sec}} + 0.08\,\frac{\text{ft}^3}{\text{sec}} - 0.16\,\frac{\text{ft}^3}{\text{sec}}$$

$$= 0.32 \text{ ft}^3/\text{sec}$$

$$\dot{V}_{\text{ED}} = 0.3\,\frac{\text{ft}^3}{\text{sec}} - 0.08\,\frac{\text{ft}^3}{\text{sec}} = 0.22 \text{ ft}^3/\text{sec}$$

$$\dot{V}_{\text{AE}} = 0.8\,\frac{\text{ft}^3}{\text{sec}} - 0.08\,\frac{\text{ft}^3}{\text{sec}} = 0.72 \text{ ft}^3/\text{sec}$$

32. FLOW MEASURING DEVICES

A device that measures flow can be calibrated to indicate either velocity or volumetric flow rate. There are many methods available to obtain the flow rate. Some are indirect, requiring the use of transducers and solid-state electronics, and others can be evaluated using the Bernoulli equation. Some are more appropriate for one variety of fluid than others, and some are limited to specific ranges of temperature and pressure.

Table 17.7 categorizes a few common flow measurement methods. Many other methods and variations thereof exist, particularly for specialized industries. Some of the methods listed are so basic that only a passing mention will be made of them. Others, particularly those that can be analyzed with energy and mass conservation laws, will be covered in greater detail in subsequent sections.

Table 17.7 Flow Measuring Devices

I. direct (primary) measurements
    positive-displacement meters
    volume tanks
    weight and mass scales

II. indirect (secondary) measurements
    obstruction meters
      – flow nozzles
      – orifice plate meters
      – variable-area meters
      – venturi meters
    velocity probes
      – direction sensing probes
      – pitot-static meters
      – pitot tubes
      – static pressure probes
    miscellaneous methods
      – hot-wire meters
      – magnetic flow meters
      – mass flow meters
      – sonic flow meters
      – turbine and propeller meters

The utility meters used to measure gas and water usage are examples of *displacement meters*. Such devices are cyclical, fixed-volume devices with counters to record the numbers of cycles. Displacement devices are generally unpowered, drawing on only the pressure energy to overcome mechanical friction. Most configurations for positive-displacement pumps (e.g., reciprocating piston, helical screw, and nutating disk) have also been converted to measurement devices.

The venturi nozzle, orifice plate, and flow nozzle are examples of *obstruction meters*. These devices rely on a decrease in static pressure to measure the flow velocity. One disadvantage of these devices is that the pressure drop is proportional to the square of the velocity, limiting the range over which any particular device can be used.

An obstruction meter that somewhat overcomes the velocity range limitation is the *variable-area meter*, also known as a *rotameter*, illustrated in Fig. 17.20.[21] This

[21] The rotameter has its own disadvantages, however. It must be installed vertically; the fluid cannot be opaque; and it is more difficult to manufacture for use with high-temperature, high-pressure fluids.

device consists of a float (which is actually more dense than the fluid) and a transparent sight tube. With proper design, the effects of fluid density and viscosity can be minimized. The sight glass can be directly calibrated in volumetric flow rate, or the height of the float above the zero position can be used in a volumetric calculation.

Figure 17.20 Variable-Area Rotameter

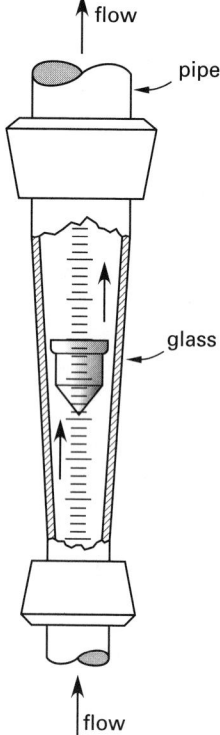

It is necessary to be able to measure static pressures in order to use obstruction meters and pitot-static tubes. In some cases, a *static pressure probe* is used. Figure 17.21 illustrates a simplified static pressure probe. In practice, such probes are sensitive to burrs and irregularities in the tap openings, orientation to the flow (i.e., *yaw*), and interaction with the pipe walls and other probes. A *direction-sensing probe* overcomes some of these problems.

Figure 17.21 Simple Static Pressure Probe

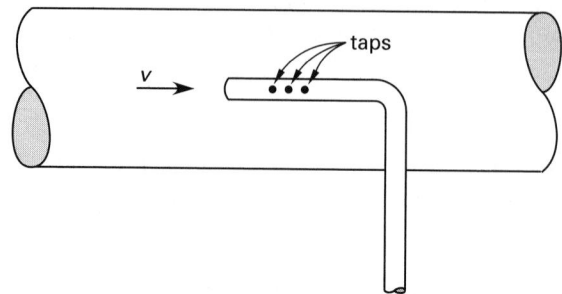

A weather station *anemometer* used to measure wind velocity is an example of a simple *turbine meter*. Similar devices are used to measure the speed of a stream or river, in which case the name *current meter* may be used. Turbine meters are further divided into cup-type meters and propeller-type meters, depending on the orientation of the turbine axis relative to the flow direction. (The turbine axis and flow direction are parallel for propeller-type meters; they are perpendicular for cup-type meters.) Since the wheel motion is proportional to the flow velocity, the velocity is determined by counting the number of revolutions made by the wheel per unit time.

A more sophisticated turbine flowmeter uses a reluctance-type pickup coil to detect wheel motion. The permeability of a magnetic circuit changes each time a wheel blade passes the pole of a permanent magnet in the meter body. This change is detected to indicate velocity or flow rate.

A *hot-wire anemometer* measures velocity by determining the cooling effect of fluid (usually a gas) flowing over an electrically heated tungsten, platinum, or nickel wire. Cooling is primarily by convection; radiation and conduction are neglected. Circuitry can be used either to keep the current constant (in which case, the changing resistance is measured) or to keep the temperature constant (in which case, the changing current is measured). Additional circuitry can be used to compensate for thermal lag if the velocity changes rapidly.

A voltage proportional to the velocity will be generated when a conductor passes through a magnetic field.[22] This characteristic can be used to measure flow velocity if the fluid is electrically conductive. *Magnetic flowmeters* are ideal for measuring the flow of liquid metals, but variations of the device can also be used when the fluid is only slightly conductive. In some cases, precise quantities of conductive ions can be added to the fluid to permit measurement by this method.

In an *ultrasonic flowmeter*, two electric or magnetic transducers are placed a short distance apart on the outside of the pipe. One transducer serves as a transmitter of ultrasonic waves; the other transducer is a receiver. As an ultrasonic wave travels from the transmitter to the receiver, its velocity will be increased (or decreased) by the relative motion of the fluid. The phase shift between the fluid-carried waves and the waves passing through a stationary medium can be measured and converted to fluid velocity.

33. PITOT-STATIC GAUGE

Measurements from pitot tubes are used to determine total (stagnation) energy. Piezometer tubes and wall taps are used to measure static pressure energy. The difference between the total and static energies is the kinetic energy of the flow. Figure 17.22 illustrates a

---

[22] The magnitude of this induced voltage is predicted by *Faraday's law.*

comparative method of directly measuring the velocity head for an incompressible fluid.

$$\frac{v^2}{2} = \frac{p_t - p_s}{\rho} = hg \qquad \text{[SI]} \quad \textit{17.142(a)}$$

$$\frac{v^2}{2g_c} = \frac{p_t - p_s}{\rho} = h \times \left(\frac{g}{g_c}\right) \quad \text{[U.S.]} \quad \textit{17.142(b)}$$

$$v = \sqrt{2gh} \qquad \textit{17.143}$$

Figure 17.22 *Comparative Velocity Head Measurement*

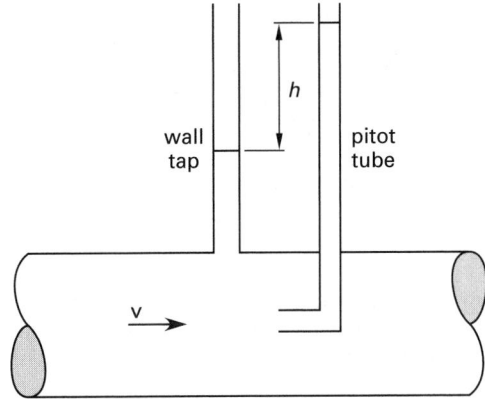

The pitot tube and static pressure tap shown in Fig. 17.22 can be combined into a *pitot-static gauge*. In a pitot-static gauge, one end of the manometer is acted upon by the static pressure (also referred to as the *transverse pressure*). The other end of the manometer experiences the total pressure. The difference in elevations of the manometer fluid columns is the velocity head. This distance must be corrected if the density of the flowing fluid is significant.

$$\frac{v^2}{2} = \frac{p_t - p_s}{\rho} = \frac{h(\rho_m - \rho)g}{\rho} \qquad \text{[SI]} \quad \textit{17.144(a)}$$

$$\frac{v^2}{2g_c} = \frac{p_t - p_s}{\rho} = \frac{h(\rho_m - \rho)}{\rho} \times \left(\frac{g}{g_c}\right) \quad \text{[U.S.]} \quad \textit{17.144(b)}$$

$$v = \sqrt{\frac{2gh(\rho_m - \rho)}{\rho}} \qquad \textit{17.145}$$

Another correction, which is seldom made, is to multiply the velocity calculated from Eq. 17.145 by C_I, the *coefficient of the instrument*. Since the flow past the pitot-static tube is slightly faster than the free-fluid velocity, the static pressure measured will be slightly lower than the true value. This makes the indicated velocity slightly higher than the true value. C_I, a number close to but less than 1.0, corrects for this.

Figure 17.23 *Pitot-Static Gauge*

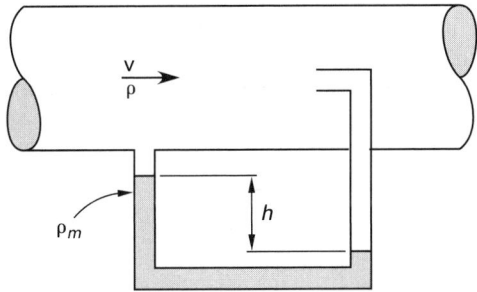

A pitot-static tube indicates the velocity at only one point in a pipe. If the flow is laminar, and if the pitot-static tube is in the center of the pipe, v_{max} will be determined. The average velocity, however, will be only half the maximum value.

Pitot tube measurements are sensitive to the condition of the opening and errors in installation alignment. The *yaw angle* (i.e., the acute angle between the pitot tube axis and the flow streamline) should be zero.

Example 17.11

Water ($62.4\ \text{lbm/ft}^3$; $\rho = 1000\ \text{kg/m}^3$) is flowing through a pipe. A pitot-static gauge registers 3.0 in (0.076 m) of mercury. What is the velocity of the water in the pipe?

SI Solution

The density of mercury is $\rho = 13\,580\ \text{kg/m}^3$. The velocity can be calculated directly from Eq. 17.145.

$$v = \sqrt{\frac{(2)\left(9.81\ \dfrac{m}{s^2}\right)(0.076\ m) \times \left(13\,580\ \dfrac{kg}{m^3} - 1000\ \dfrac{kg}{m^3}\right)}{1000\ \dfrac{kg}{m^3}}}$$

$$= 4.33\ \text{m/s}$$

Customary U.S. Solution

The density of mercury is $\rho = 848.6\ \text{lbm/ft}^3$.

From Eq. 17.145,

$$v = \sqrt{\frac{(2)\left(32.2\ \dfrac{ft}{sec^2}\right)\left((3\ in)\left(\dfrac{1\ ft}{12\ in}\right)\right) \times \left(848.6\ \dfrac{lbm}{ft^3} - 62.4\ \dfrac{lbm}{ft^3}\right)}{62.4\ \dfrac{lbm}{ft^3}}}$$

$$= 14.24\ \text{ft/sec}$$

34. VENTURI METER

Figure 17.24 illustrates a simple *venturi*. (Sometimes the venturi is called a *converging-diverging nozzle.*) This flow measuring device can be inserted directly into a pipeline. Since the diameter changes are gradual, there is very little friction loss.[23] Static pressure measurements are taken at the throat and upstream of the diameter change. These measurements are traditionally made by manometer.

Figure 17.24 *Venturi Meter*

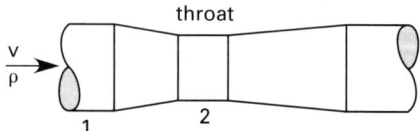

The analysis of *venturi meter* performance is relatively simple. The traditional derivation of upstream velocity starts by assuming a horizontal orientation and frictionless, incompressible, and turbulent flow. Then the Bernoulli equation is written for points 1 and 2. Equation 17.146 shows that the static pressure decreases as the velocity increases. This is known as the *venturi effect*.

$$\frac{v_1^2}{2} + \frac{p_1}{\rho} = \frac{v_2^2}{2} + \frac{p_2}{\rho} \qquad \text{[SI]} \qquad 17.146(a)$$

$$\frac{v_1^2}{2g_c} + \frac{p_1}{\rho} = \frac{v_2^2}{2g_c} + \frac{p_2}{\rho} \qquad \text{[U.S.]} \qquad 17.146(b)$$

The two velocities are related by the continuity equation.

$$A_1 v_1 = A_2 v_2 \qquad 17.147$$

Combining Eqs. 17.146 and 17.147 and eliminating the unknown v_1 produces an expression for the throat velocity. Also, a *coefficient of velocity* is used to account for the small effect of friction. (C_v is very close to 1.0, usually 0.98 or 0.99.)

$$v_2 = C_v v_{2,\text{ideal}}$$

$$= \left(\frac{C_v}{\sqrt{1 - \left(\frac{A_2}{A_1}\right)^2}} \right) \sqrt{\frac{2(p_1 - p_2)}{\rho}} \qquad \text{[SI]} \qquad 17.148(a)$$

$$v_2 = \left(\frac{C_v}{\sqrt{1 - \left(\frac{A_2}{A_1}\right)^2}} \right) \sqrt{\frac{2g_c(p_1 - p_2)}{\rho}} \qquad \text{[U.S.]} \qquad 17.148(b)$$

The *velocity of approach factor*, F_{va}, also known as the *meter constant*, is the reciprocal of the denominator of the first term of Eq. 17.148. The *beta ratio* can be incorporated into the formula for F_{va}.

$$\beta = \frac{D_2}{D_1} \qquad 17.149$$

---

[23]The actual friction loss is approximately 10% of the pressure difference $p_1 - p_2$.

$$F_{\text{va}} = \frac{1}{\sqrt{1 - \left(\frac{A_2}{A_1}\right)^2}} = \frac{1}{\sqrt{1 - \beta^4}} \qquad 17.150$$

If a manometer is used to measure the pressure difference directly, Eq. 17.148 can be rewritten in terms of the manometer fluid reading.

$$v_2 = \left(\frac{C_v}{\sqrt{1 - \beta^4}} \right) \sqrt{\frac{2g(\rho_m - \rho)h}{\rho}}$$

$$= C_v F_{\text{va}} \sqrt{\frac{2g(\rho_m - \rho)h}{\rho}} \qquad 17.151$$

Figure 17.25 *Venturi Meter with Manometer*

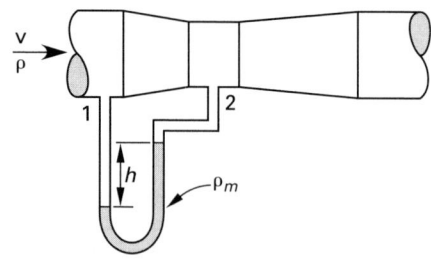

The flow rate through a venturi meter can be calculated from the throat area. There is an insignificant amount of contraction of the flow as it passes through the throat, and the *coefficient of contraction* is seldom encountered in venturi meter work. The *coefficient of discharge* ($C_d = C_c C_v$) is quoted, nevertheless. Values of C_d range from slightly less than 0.90 to over 0.99, depending on the Reynolds number. C_d is seldom less than 0.95 for turbulent flow.

$$\dot{V} = C_d A_2 v_{2,\text{ideal}} \qquad 17.152$$

The product $C_d F_{\text{va}}$ is known as the *coefficient of flow* or *flow coefficient*, not to be confused with the coefficient of discharge.[24] This factor is used for convenience, since it combines the losses with the meter constant.

$$C_f = C_d F_{\text{va}} = \frac{C_d}{\sqrt{1 - \beta^4}} \qquad 17.153$$

$$\dot{V} = C_f A_2 \sqrt{\frac{2g(\rho_m - \rho)h}{\rho}} \qquad 17.154$$

Figure 17.26 *Typical Venturi Meter Discharge Coefficients (Long Radius Venturi Meter)*

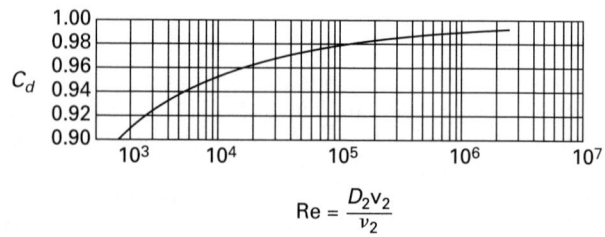

---

[24]Some writers use the symbol K for the flow coefficient.

35. ORIFICE METER

The *orifice meter* (or *orifice plate*) is used more frequently than the venturi meter to measure flow rates in small pipes. It consists of a thin or sharp-edged plate with a central, round hole through which the fluid flows. Such a plate is easily clamped between two flanges in an existing pipeline.

While (for small pipes) the orifice meter may consist of a thin plate without significant thickness, various types of bevels and rounded edges are also used with thicker plates. There is no significant difference in the analysis procedure between "flat plate," "sharp-edged," or "square-edged" orifice meters. Any effect that the orifice edges have is accounted for in the discharge and flow coefficient correlations. Similarly, the direction of the bevel will affect the coefficients, but not the analysis method.

As with the venturi meter, pressure taps are used to obtain the static pressure upstream of the orifice plate and at the *vena contracta* (i.e., at the point of minimum pressure).[25] A differential manometer connected to the two taps conveniently indicates the difference in static pressures.

Figure 17.27 *Orifice Meter with Differential Manometer*

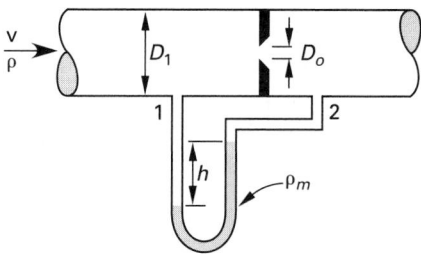

The derivation of the governing equations for an orifice meter is similar to that of the venturi meter. (The obvious falsity of assuming frictionless flow through the orifice is corrected by the coefficient of discharge.) The major difference is that the coefficient of contraction is taken into consideration in writing the mass continuity equation, since the pressure is measured at the vena contracta, not at the orifice.

$$A_2 = C_c A_o \qquad 17.155$$

[25]Calibration of the orifice meter is sensitive to tap placement. Upstream taps are placed between one-half and two pipe diameters upstream from the orifice. (An upstream distance of one pipe diameter is often quoted and used.) There are three tap-placement options: flange, vena contracta, and standardized. Flange taps are used with prefabricated orifice meters that are inserted (by flange bolting) in pipes. If the location of the vena contracta is known, a tap can be placed there. However, the location of the vena contracta depends on the diameter ratio $\beta = D_o/D$ and varies from approximately 0.4 to 0.7 pipe diameters downstream. Due to the difficulty of locating the vena contracta, the standardized $1D$-$\frac{1}{2}D$ configuration is often used. The upstream tap is one diameter before the orifice; the downstream tap is one-half diameter after the orifice. Since approaching flow should be stable and uniform, care must be taken not to install the orifice meter less than approximately five diameters after a bend or elbow.

$$v_o = \left(\frac{C_v}{\sqrt{1 - \left(\frac{C_c A_o}{A_1} \right)^2}} \right) \sqrt{\frac{2(p_1 - p_2)}{\rho}} \quad \text{[SI]} \quad 17.156(a)$$

$$v_o = \left(\frac{C_v}{\sqrt{1 - \left(\frac{C_c A_o}{A_1} \right)^2}} \right) \sqrt{\frac{2g_c(p_1 - p_2)}{\rho}} \quad \text{[U.S.]} \quad 17.156(b)$$

If a manometer is used to indicate the differential pressure $p_1 - p_2$, the velocity at the vena contracta can be calculated from Eq. 17.157.

$$v_o = \left(\frac{C_v}{\sqrt{1 - \left(\frac{C_c A_o}{A_1} \right)^2}} \right) \sqrt{\frac{2g(\rho_m - \rho)h}{\rho}} \qquad 17.157$$

Although the orifice meter is simpler and less expensive than a venturi meter, its discharge coefficient is much less than that of a venturi meter. C_d usually ranges from 0.55 to 0.75, with values of 0.60 and 0.61 often being quoted. (The coefficient of contraction has a large effect, since $C_d = C_v C_c$.) Also, its pressure recovery is poor (i.e., there is a permanent pressure reduction), and it is susceptible to inaccuracies from wear and abrasion.[26]

The velocity of approach factor, F_{va}, for an orifice meter is defined differently than for a venturi meter, since it takes into consideration the contraction of the flow. However, the velocity of approach factor is still combined with the coefficient of discharge into the flow coefficient, C_f. Figure 17.28 illustrates how the flow coefficient varies with the area ratio and the Reynolds number.

$$F_{va} = \frac{1}{\sqrt{1 - \left(\frac{C_c A_o}{A_1} \right)^2}} \qquad 17.158$$

$$C_f = C_d F_{va} \qquad 17.159$$

[26]The actual permanent pressure loss varies from 40 to 90% of the differential pressure. The loss depends on the diameter ratio $\beta = D_o/D_1$, and is not particularly sensitive to the Reynolds number for turbulent flow. For $\beta = 0.5$, the loss is 73% of the measured pressure difference, $p_1 - p_2$. This decreases to approximately 56% of the pressure difference when $\beta = 0.65$ and to 38% when $\beta = 0.8$. For any diameter ratio, the pressure drop coefficient, K, in multiples of the orifice velocity head is

$$K = \frac{1 - \beta^2}{C_f^2}$$

Figure 17.28 *Typical Flow Coefficients for Orifice Plates*

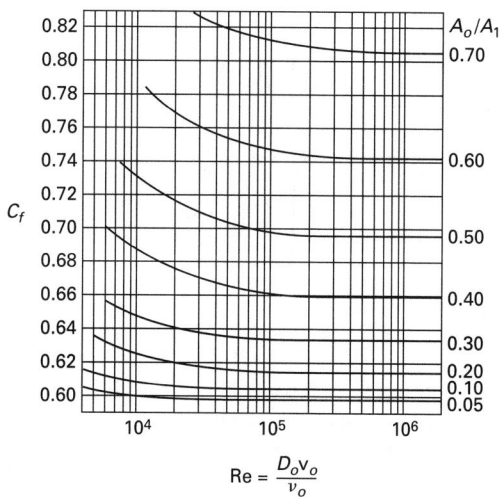

$$\text{Re} = \frac{D_o v_o}{\nu_o}$$

The flow rate through an orifice meter is given by Eq. 17.160.

$$\dot{V} = C_f A_o \sqrt{\frac{2g(\rho_m - \rho)h}{\rho}} = C_f A_o \sqrt{\frac{2(p_1 - p_2)}{\rho}}$$

$$[\text{SI}] \quad \textbf{17.160(a)}$$

$$\dot{V} = C_f A_o \sqrt{\frac{2g(\rho_m - \rho)h}{\rho}} = C_f A_o \sqrt{\frac{2g_c(p_1 - p_2)}{\rho}}$$

$$[\text{U.S.}] \quad \textbf{17.160(b)}$$

Example 17.12

150°F water ($\rho = 61.2$ lbm/ft^3) flows in an 8 in schedule-40 steel pipe at the rate of 2.23 ft^3/sec. A sharp-edged orifice with a 7 in diameter hole is placed in the line. A mercury differential manometer is used to record the pressure difference. If the orifice has a flow coefficient, C_f, of 0.62, what deflection in inches of mercury is observed? (Mercury has a density of 848.6 lbm/ft^3.)

Solution

The orifice area is

$$A_o = \frac{\pi}{4} D_o^2 = \left(\frac{\pi}{4}\right)\left(\frac{7 \text{ in}}{12 \frac{\text{in}}{\text{ft}}}\right)^2$$

$$= 0.2673 \text{ ft}^2$$

Equation 17.160 is solved for h.

$$h = \frac{\dot{V}^2 \rho}{2g C_f^2 A_o^2 (\rho_m - \rho)}$$

$$= \frac{\left(2.23 \frac{\text{ft}^3}{\text{sec}}\right)^2 \left(61.2 \frac{\text{lbm}}{\text{ft}^3}\right)\left(12 \frac{\text{in}}{\text{ft}}\right)}{(2)\left(32.2 \frac{\text{ft}}{\text{sec}^2}\right)(0.62)^2}$$

$$\times (0.2673 \text{ ft}^2)^2 \left(848.6 \frac{\text{lbm}}{\text{ft}^3} - 61.2 \frac{\text{lbm}}{\text{ft}^3}\right)$$

$$= 2.62 \text{ in}$$

36. FLOW NOZZLE

A typical flow nozzle is illustrated in Fig. 17.29. This device consists only of a converging section. It is somewhat between an orifice plate and a venturi meter in performance, possessing some of the advantages and disadvantages of each. The venturi performance equations can be used for the flow nozzle.

Figure 17.29 *Flow Nozzle*

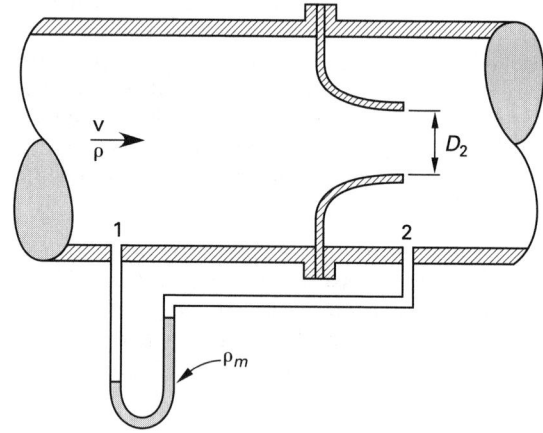

The geometry of the nozzle entrance is chosen to prevent separation of the fluid from the wall. The converging portion and the subsequent parallel section keep the coefficients of velocity and contraction close to 1.0. However, the absence of a diffuser section disrupts the orderly return of fluid to its original condition. The permanent pressure drop is more similar to that of the orifice meter than the venturi meter.

Since the nozzle geometry greatly affects the performance, values of C_d and C_f have been established for only a limited number of specific proprietary nozzles.[27]

37. FLOW MEASUREMENTS OF COMPRESSIBLE FLUIDS

Volume measurements of compressible fluids (i.e., gases) are not very meaningful. The volume of a gas will depend on its temperature and pressure. For that reason, flow quantities of gases discharged should be stated as mass flow rates.

$$\dot{m} = \rho_2 A_2 v_2 \qquad \textbf{17.161}$$

Equation 17.161 requires that the velocity and area be measured at the same point. More important, the density must be measured at that point, as well. However, it is common practice in flow measurement work to use the density of the upstream fluid at position 1. (Note that this is not the stagnation density.)

---

[27]Some of the proprietary nozzles for which detailed performance data exist are the ASME long-radius nozzle (low-β and high-β series) and the International Standards Association (ISA) nozzle (German standard nozzle).

The significant error introduced by this simplification is corrected by the use of an *expansion factor*, Y. For venturi meters and flow nozzles, values of the expansion factor are generally calculated theoretical values. Values of Y are determined experimentally for orifice plates.

$$\dot{m} = Y \rho_1 A_2 \mathrm{v}_2 \qquad \textit{17.162}$$

Derivation of the theoretical formula for the expansion factor for venturi meters and flow nozzles is based on thermodynamic principles and an assumption of adiabatic flow.

$$Y = \sqrt{\frac{(1-\beta^4)\left(\left(\dfrac{p_2}{p_1}\right)^{2/k} - \left(\dfrac{p_2}{p_1}\right)^{(k+1)/k}\right)}{\left(\dfrac{k-1}{k}\right)\left(1-\dfrac{p_2}{p_1}\right)\left(1-\beta^4\left(\dfrac{p_2}{p_1}\right)^{2/k}\right)}}$$
$$\textit{17.163}$$

Figure 17.30 *Approximate Expansion Factors*

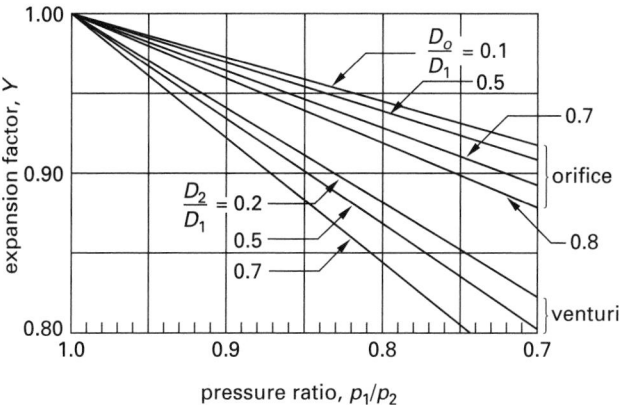

Transport Processes, 3rd ed., by Geankoplis, C. J., © 1993. Reprinted by permission of Prentice-Hall, Inc., Upper Saddle River, NJ.

Once the expansion factor is known, it can be used with the standard flow rate ($\dot{V}$) equations for venturi meters, flow nozzles, and orifice meters. For example, for a venturi meter, the mass flow rate would be calculated from Eq. 17.164.

$$\dot{m} = Y\dot{m}_{\text{ideal}} = \left(\frac{YC_dA_2}{\sqrt{1-\beta^4}}\right)\sqrt{2\rho_1(p_1-p_2)}$$
$$\text{[SI]} \qquad \textit{17.164(a)}$$

$$\dot{m} = Y\dot{m}_{\text{ideal}} = \left(\frac{YC_dA_2}{\sqrt{1-\beta^4}}\right)\sqrt{2g_c\rho_1(p_1-p_2)}$$
$$\text{[U.S.]} \qquad \textit{17.164(b)}$$

38. IMPULSE-MOMENTUM PRINCIPLE

(The convention of this section is to make F and x positive when they are directed toward the right. F and y are positive when directed upward. Also, the fluid is assumed to flow horizontally from left to right, and it has no initial y-component of velocity.)

The *momentum*, $\mathbf{P}$ (also known as *linear momentum* to distinguish it from *angular momentum*, which is not considered here), of a moving object is a vector quantity defined as the product of the object's mass and velocity.[28]

$$\mathbf{P} = m\mathbf{v} \qquad \text{[SI]} \qquad \textit{17.165(a)}$$

$$\mathbf{P} = \frac{m\mathbf{v}}{g_c} \qquad \text{[U.S.]} \qquad \textit{17.165(b)}$$

The *impulse*, I, of a constant force is calculated as the product of the force's magnitude and the length of time the force is applied.

$$\mathbf{I} = \mathbf{F}\Delta t \qquad \textit{17.166}$$

The *impulse-momentum principle* states that the impulse applied to a body is equal to the change in momentum. (This is also known as the *law of conservation of momentum*, even though fluid momentum is not always conserved.) Equation 17.167 is one way of stating Newton's second law.

$$\mathbf{I} = \Delta\mathbf{P} \qquad \textit{17.167}$$

$$F\Delta t = m\Delta\mathrm{v} = m(\mathrm{v}_2-\mathrm{v}_1) \qquad \text{[SI]} \qquad \textit{17.168(a)}$$

$$F\Delta t = \frac{m\Delta\mathrm{v}}{g_c} = \frac{m(\mathrm{v}_2-\mathrm{v}_1)}{g_c} \qquad \text{[U.S.]} \qquad \textit{17.168(b)}$$

For fluid flow, there is a mass flow rate, $\dot{m}$, but no mass per se. Since $\dot{m} = m/\Delta t$, the impulse-momentum equation can be rewritten as follows.

$$F = \dot{m}\Delta\mathrm{v} \qquad \text{[SI]} \qquad \textit{17.169(a)}$$

$$F = \frac{\dot{m}\Delta\mathrm{v}}{g_c} \qquad \text{[U.S.]} \qquad \textit{17.169(b)}$$

Equation 17.169 calculates the constant force required to accelerate or retard a fluid stream. This would occur when fluid enters a reduced or enlarged flow area. If the flow area decreases, for example, the fluid will be accelerated by a wall force up to the new velocity. Ultimately, this force must be resisted by the pipe supports.

As Eq. 17.169 illustrates, fluid momentum is not always conserved, since it is generated by the external force, F. Examples of external forces are gravity (considered zero for horizontal pipes), gage pressure, friction, and turning forces from walls and vanes. Only if these external forces are absent is fluid momentum conserved.

Since force is a vector, it can be resolved into its x- and y-components of force.

$$F_x = \dot{m}\Delta\mathrm{v}_x \qquad \text{[SI]} \qquad \textit{17.170(a)}$$

$$F_x = \frac{\dot{m}\Delta\mathrm{v}_x}{g_c} \qquad \text{[U.S.]} \qquad \textit{17.170(b)}$$

$$F_y = \dot{m}\Delta\mathrm{v}_y \qquad \text{[SI]} \qquad \textit{17.171(a)}$$

$$F_y = \frac{\dot{m}\Delta\mathrm{v}_y}{g_c} \qquad \text{[U.S.]} \qquad \textit{17.171(b)}$$

---

[28]The symbol B is also used for momentum. In many texts, however, momentum is given no symbol at all.

If the flow is initially at velocity v but is directed through an angle θ with respect to the original direction, the x- and y-components of velocity can be calculated from Eqs. 17.172 and 17.173.

$$\Delta v_x = v(\cos\theta - 1) \qquad 17.172$$

$$\Delta v_y = v\sin\theta \qquad 17.173$$

Since F and v are vector quantities and Δt and m are scalars, F must have the same direction as $v_2 - v_1$. This provides an intuitive method of determining the direction in which the force acts. Essentially, one needs to ask, "In which direction must the force act in order to push the fluid stream into its new direction?" (The force, F, is the force on the fluid. The force on the pipe walls or pipe supports has the same magnitude but is opposite in direction.)

Figure 17.31 *Force on a Confined Fluid Stream*

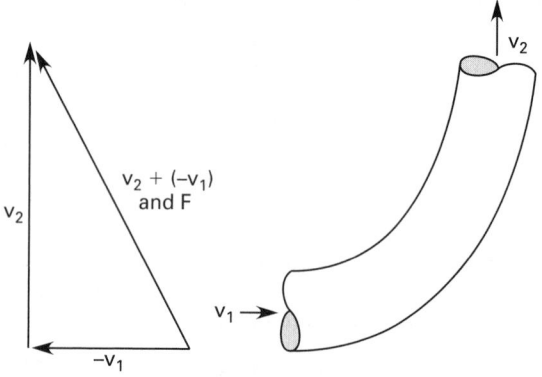

If a jet is completely stopped by a flat plate placed perpendicular to its flow, then $\theta = 90°$ and $\Delta v_x = -v$. If a jet is turned around so that it ends up returning to where it originated, then $\theta = 180°$ and $\Delta v_x = -2v$. Notice that a positive Δv indicates an increase in velocity. A negative Δv indicates a decrease in velocity.

39. JET PROPULSION

A basic application of the impulse-momentum principle is the analysis of jet propulsion. Air enters a jet engine and is mixed with a small amount of jet fuel. The air and fuel mixture is compressed and ignited, and the exhaust products leave the engine at a greater velocity than was possessed by the original air. The change in momentum of the air produces a force on the engine.

The governing equation for a jet engine is Eq. 17.174. In the special case of VTOL (vertical takeoff and landing) aircraft, there will also be a y-component of force. The mass of the jet fuel is small compared with the air mass, and the fuel mass is commonly disregarded.

$$F_x = \dot{m}(v_2 - v_1) \qquad 17.174$$

$$F_x = \dot{V}_2\rho_2 v_2 - \dot{V}_1\rho_1 v_1 \qquad [\text{SI}] \qquad 17.175(a)$$

$$F_x = \frac{\dot{V}_2\rho_2 v_2 - \dot{V}_1\rho_1 v_1}{g_c} \qquad [\text{U.S.}] \qquad 17.175(b)$$

Figure 17.32 *Jet Engine*

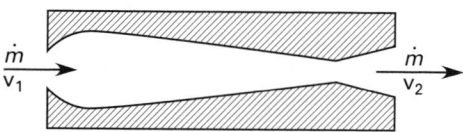

40. OPEN JET ON A VERTICAL FLAT PLATE

Figure 17.33 illustrates an open jet on a vertical flat plate. The fluid approaches the plate with no vertical component of velocity; it leaves the plate with no horizontal component of velocity. (This is another way of saying there is no splash-back.) Thus, all of the velocity in the x-direction is canceled. (The minus sign in Eq. 17.176 indicates that the force is opposite the initial velocity direction.)

$$\Delta v = -v \qquad 17.176$$

$$F_x = -\dot{m}v \qquad [\text{SI}] \qquad 17.177(a)$$

$$F_x = \frac{-\dot{m}v}{g_c} \qquad [\text{U.S.}] \qquad 17.177(b)$$

Figure 17.33 *Jet on a Vertical Plate*

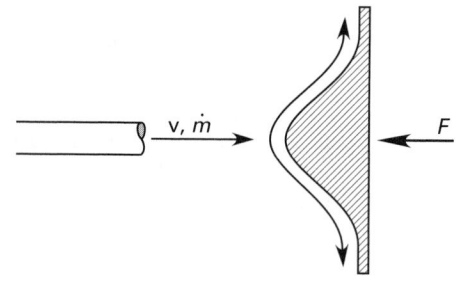

Since the flow is divided, half going up and half going down, the net velocity change in the y-direction is zero. There is no force in the y-direction on the fluid.

41. OPEN JET ON A HORIZONTAL FLAT PLATE

If a jet of fluid is directed upward, its velocity will decrease due to the effect of gravity. The force exerted on the fluid by the plate will depend on the fluid velocity at the plate surface, v_y, not the original jet velocity, v_o. All of this velocity is canceled. Since the flow divides evenly in both horizontal directions ($\Delta v_x = 0$), there is no force component in the x-direction.

$$v_y = \sqrt{v_o^2 - 2gh} \qquad 17.178$$

$$\Delta v_y = -\sqrt{v_o^2 - 2gh} \qquad 17.179$$

$$F_y = -\dot{m}\sqrt{v_o^2 - 2gh} \qquad [\text{SI}] \qquad 17.180(a)$$

$$F_y = \frac{-\dot{m}\sqrt{v_o^2 - 2gh}}{g_c} \qquad [\text{U.S.}] \qquad 17.180(b)$$

Figure 17.34 *Open Jet on a Horizontal Plate*

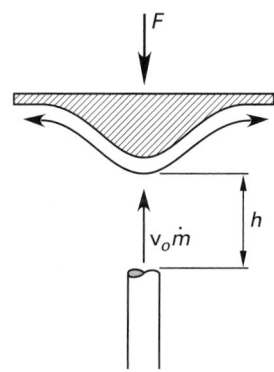

42. OPEN JET ON AN INCLINED PLATE

An open jet will be diverted both up and down (but not laterally) a stationary, inclined plate, as shown in Fig. 17.35. In the absence of friction, the velocity in each diverted flow will be v, the same as in the approaching jet. The fractions f_1 and f_2 of the jet that are diverted up and down can be found from Eq. 17.181 through Eq. 17.184.

$$f_1 = \frac{1 + \cos\theta}{2} \qquad 17.181$$

$$f_2 = \frac{1 - \cos\theta}{2} \qquad 17.182$$

$$f_1 - f_2 = \cos\theta \qquad 17.183$$

$$f_1 + f_2 = 1.0 \qquad 17.184$$

Figure 17.35 *Open Jet on an Inclined Plate*

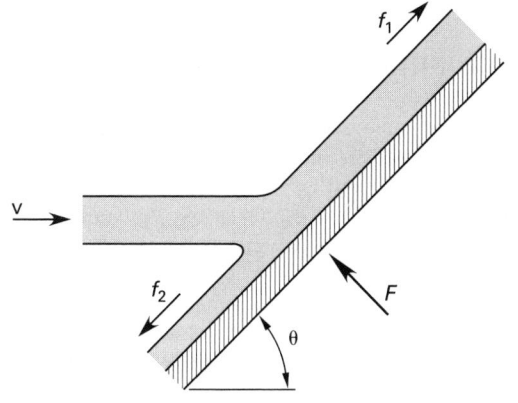

If the flow along the plate is frictionless, there will be no force component parallel to the plate. The force perpendicular to the plate is given by Eq. 17.185.

$$F = \dot{m}\mathrm{v}\sin\theta \qquad [\text{SI}] \qquad 17.185(a)$$

$$F = \frac{\dot{m}\mathrm{v}\sin\theta}{g_c} \qquad [\text{U.S.}] \qquad 17.185(b)$$

43. OPEN JET ON A SINGLE STATIONARY BLADE

Figure 17.36 illustrates a fluid jet being turned through an angle θ by a stationary blade (also called a *vane*). It is common to assume that $|\mathrm{v}_2| = |\mathrm{v}_1|$, although this will not be strictly true if friction between the blade and fluid is considered. Since the fluid is both retarded (in the x-direction) and accelerated (in the y-direction), there will be two components of force on the fluid.

$$\Delta\mathrm{v}_x = \mathrm{v}_2\cos\theta - \mathrm{v}_1 \qquad 17.186$$

$$\Delta\mathrm{v}_y = \mathrm{v}_2\sin\theta \qquad 17.187$$

$$F_x = \dot{m}\,(\mathrm{v}_2\cos\theta - \mathrm{v}_1) \qquad [\text{SI}] \qquad 17.188(a)$$

$$F_x = \frac{\dot{m}\,(\mathrm{v}_2\cos\theta - \mathrm{v}_1)}{g_c} \qquad [\text{U.S.}] \qquad 17.188(b)$$

$$F_y = \dot{m}\mathrm{v}_2\sin\theta \qquad [\text{SI}] \qquad 17.189(a)$$

$$F_y = \frac{\dot{m}\mathrm{v}_2\sin\theta}{g_c} \qquad [\text{U.S.}] \qquad 17.189(b)$$

$$F = \sqrt{F_x^2 + F_y^2} \qquad 17.190$$

Figure 17.36 *Open Jet on a Stationary Blade*

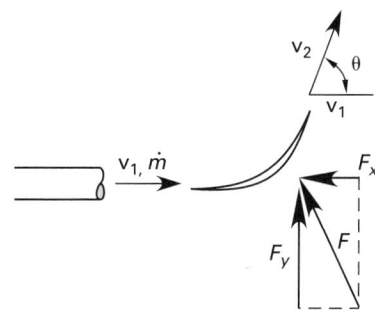

44. OPEN JET ON A SINGLE MOVING BLADE

If a blade is moving away at velocity v_b from the source of the fluid jet, only the *relative velocity difference* between the jet and blade produces a momentum change. Furthermore, not all of the fluid jet overtakes the moving blade. The equations used for the single stationary blade can be used by substituting $(\mathrm{v} - \mathrm{v}_b)$ for v and by using the effective mass flow rate, $\dot{m}_{\text{eff}}$.

$$\Delta\mathrm{v}_x = (\mathrm{v} - \mathrm{v}_b)(\cos\theta - 1) \qquad 17.191$$

$$\Delta\mathrm{v}_y = (\mathrm{v} - \mathrm{v}_b)(\sin\theta) \qquad 17.192$$

$$\dot{m}_{\text{eff}} = \left(\frac{\mathrm{v} - \mathrm{v}_b}{\mathrm{v}}\right)\dot{m} \qquad 17.193$$

$$F_x = \dot{m}_{\text{eff}}(\mathrm{v} - \mathrm{v}_b)(\cos\theta - 1) \qquad [\text{SI}] \qquad 17.194(a)$$

$$F_x = \frac{\dot{m}_{\text{eff}}(\mathrm{v} - \mathrm{v}_b)(\cos\theta - 1)}{g_c} \qquad [\text{U.S.}] \qquad 17.194(b)$$

$$F_y = \dot{m}_{\text{eff}}(\mathrm{v} - \mathrm{v}_b)(\sin\theta) \qquad [\text{SI}] \qquad 17.195(a)$$

$$F_y = \frac{\dot{m}_{\text{eff}}(\mathrm{v} - \mathrm{v}_b)(\sin\theta)}{g_c} \qquad [\text{U.S.}] \qquad 17.195(b)$$

Figure 17.37 *Open Jet on a Moving Blade*

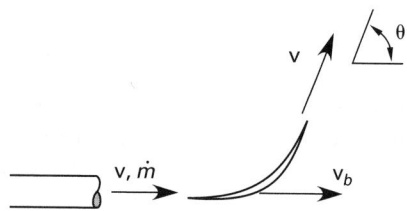

45. OPEN JET ON A MULTIPLE-BLADED WHEEL

An *impulse turbine* consists of a series of blades (buckets or vanes) mounted around a wheel. The tangential velocity of the blades is approximately parallel to the jet. The effective mass flow rate, $\dot{m}_{\text{eff}}$, used in calculating the reaction force is the full discharge rate, since when one blade moves away from the jet, other blades will have moved into position. Thus, all of the fluid discharged is captured by the blades. Equations 17.194 and 17.195 are applicable if the total flow rate is used.

$$v_b = \frac{\text{rpm} \times 2\pi r}{60} = \omega r \qquad 17.196$$

Figure 17.38 *Impulse Turbine*

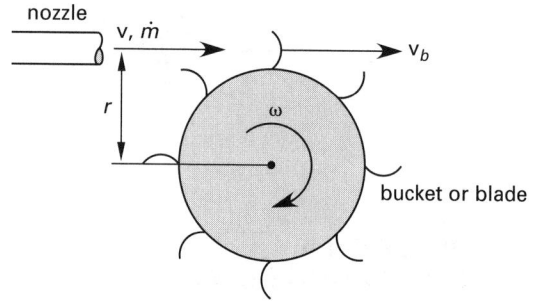

46. IMPULSE TURBINE POWER

The total power potential of a fluid jet can be calculated from the kinetic energy of the jet and the mass flow rate.[29] (This neglects the pressure energy, which is small by comparison.)

$$P_{\text{jet}} = \frac{\dot{m}v^2}{2} \qquad \text{[SI]} \qquad 17.197(a)$$

$$P_{\text{jet}} = \frac{\dot{m}v^2}{2g_c} \qquad \text{[U.S.]} \qquad 17.197(b)$$

The power transferred from a fluid jet to the blades of a turbine is calculated from the x-component of force on the blades. The y-of force does no work.

---

[29]The full jet discharge is used in this section. If only a single blade is involved, the effective mass flow rate, $\dot{m}_{\text{eff}}$, must be used.

$$P = F_x v_b \qquad 17.198$$

$$P = \dot{m}v_b\,(v - v_b)(1 - \cos\theta) \qquad \text{[SI]} \qquad 17.199(a)$$

$$P = \frac{\dot{m}v_b\,(v - v_b)(1 - \cos\theta)}{g_c} \qquad \text{[U.S.]} \; 17.199(b)$$

The maximum theoretical blade velocity is the velocity of the jet: $v_b = v$. This is known as the *runaway speed* and can only occur when the turbine is unloaded. If Eq. 17.199 is maximized with respect to v_b, however, the maximum power will be found to occur when the blade is traveling at half of the jet velocity: $v_b = v/2$. The power (force) is also affected by the deflection angle of the blade. Power is maximized when $\theta = 180°$. Figure 17.39 illustrates the relationship between power and the variables θ and v_b.

Figure 17.39 *Turbine Power*

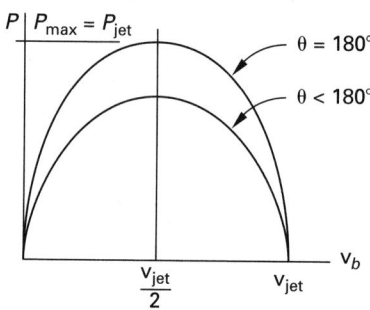

Putting $\theta = 180°$ and $v_b = v/2$ into Eq. 17.199 results in $P_{\text{max}} = \dot{m}v^2/2$, which is the same as P_{jet} in Eq. 17.197. If the machine is 100% efficient, 100% of the jet power can be transferred to the machine.

47. CONFINED STREAMS IN PIPE BENDS

As presented in Sec. 38, momentum can also be changed by pressure forces. Such is the case when fluid enters a pipe fitting or bend. Since the fluid is confined, the forces due to static pressure must be included in the analysis. (The effects of gravity and friction are neglected.)

$$F_x = p_2 A_2 \cos\theta - p_1 A_1 + \dot{m}(v_2 \cos\theta - v_1) \qquad \text{[SI]} \qquad 17.200(a)$$

$$F_x = p_2 A_2 \cos\theta - p_1 A_1 + \frac{\dot{m}(v_2 \cos\theta - v_1)}{g_c} \qquad \text{[U.S.]} \; 17.200(b)$$

$$F_y = (p_2 A_2 + \dot{m}v_2)\sin\theta \qquad \text{[SI]} \qquad 17.201(a)$$

$$F_y = \left(p_2 A_2 + \frac{\dot{m}v_2}{g_c}\right)\sin\theta \qquad \text{[U.S.]} \qquad 17.201(b)$$

Figure 17.40 *Pipe Bend*

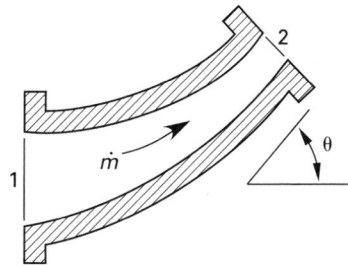

Example 17.13

60°F water ($\rho = 62.4$ lbm/ft^3) at 40 psig enters a 12 in ×
8 in reducing elbow at 8 ft/sec and is turned through
an angle of 30°. Water leaves 26 in higher in elevation.
(a) What is the resultant force exerted on the water by
the elbow? (b) What other forces should be considered
in the design of supports for the fitting?

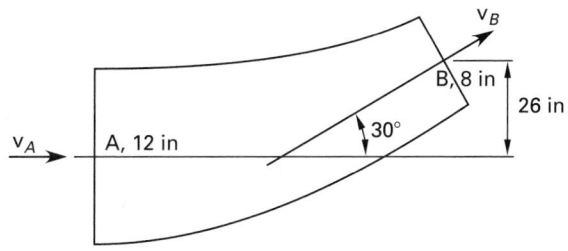

Solution

(a) The velocity and pressure at point B are both needed.
The velocity is easily calculated from the continuity
equation.

$$A_A = \frac{\pi}{4}D_A^2 = \left(\frac{\pi}{4}\right)\left(\frac{12 \text{ in}}{12 \frac{\text{in}}{\text{ft}}}\right)^2 = 0.7854 \text{ ft}^2$$

$$A_B = \left(\frac{\pi}{4}\right)\left(\frac{8}{12}\right)^2 = 0.3491 \text{ ft}^2$$

$$v_B = \frac{v_A A_A}{A_B}$$

$$= \left(8 \frac{\text{ft}}{\text{sec}}\right)\left(\frac{0.7854 \text{ ft}^2}{0.3491 \text{ ft}^2}\right) = 18 \text{ ft/sec}$$

$$p_A = \left(40 \frac{\text{lbf}}{\text{in}^2}\right)\left(144 \frac{\text{in}^2}{\text{ft}^2}\right) = 5760 \text{ lbf/ft}^2$$

The Bernoulli equation is used to calculate p_B. (Notice
that gage pressures are used. Absolute pressures could
also be used, but the addition of p_{atm}/ρ to both sides
of the Bernoulli equation would not affect p_B.)

$$\frac{5760 \frac{\text{lbf}}{\text{ft}^2}}{62.4 \frac{\text{lbm}}{\text{ft}^3}} + \frac{\left(8 \frac{\text{ft}}{\text{sec}}\right)^2}{(2)\left(32.2 \frac{\text{lbm-ft}}{\text{lbf-sec}^2}\right)}$$

$$= \frac{p_B}{62.4} + \frac{(18)^2}{(2)(32.2)} + \left(\frac{26 \text{ in}}{12 \frac{\text{in}}{\text{ft}}}\right) \times \left(\frac{g}{g_c}\right)$$

$$p_B = 5373 \text{ lbf/ft}^2$$

The mass flow rate is

$$\dot{m} = \dot{V}\rho = vA\rho$$

$$= \left(8 \frac{\text{ft}}{\text{sec}}\right)(0.7854 \text{ ft}^2)\left(62.4 \frac{\text{lbm}}{\text{ft}^3}\right)$$

$$= 392.1 \text{ lbm/sec}$$

From Eq. 17.200,

$$F_x = \left(5373 \frac{\text{lbf}}{\text{ft}^2}\right)(0.3491 \text{ ft}^2)(\cos 30°)$$

$$- \left(5760 \frac{\text{lbf}}{\text{ft}^2}\right)(0.7854 \text{ ft}^2)$$

$$+ \frac{\left(392.1 \frac{\text{lbm}}{\text{sec}}\right)\left(\left(18 \frac{\text{ft}}{\text{sec}}\right)(\cos 30°) - 8 \frac{\text{ft}}{\text{sec}}\right)}{32.2 \frac{\text{lbm-ft}}{\text{lbf-sec}^2}}$$

$$= -2807 \text{ lbf}$$

From Eq. 17.201,

$$F_y = \left((5373)(0.3491) + \frac{(392.1)(18)}{32.2}\right)(\sin 30°)$$

$$= 1047 \text{ lbf}$$

The resultant force on the water is

$$R = \sqrt{F_x^2 + F_y^2} = \sqrt{(-2807 \text{ lbf})^2 + (1047 \text{ lbf})^2}$$

$$= 2996 \text{ lbf}$$

(b) In addition to counteracting the resultant force, R,
the support should be designed to carry the weight of
the elbow and the water in it. Also, the support must
carry a part of the pipe and water weight tributary to
the elbow.

48. WATER HAMMER

Water hammer in a long pipe is an increase in fluid pressure caused by a sudden velocity decrease. The sudden
velocity decrease will usually be caused by a valve closing. Analysis of the water hammer phenomenon can

take two approaches, depending on whether or not the pipe material is assumed to be elastic.

If the pipe material is assumed to be inelastic (i.e., rigid pipe), the time required for the water hammer shock wave to travel from the suddenly closed valve to a point of interest depends only on the velocity of sound in the fluid (a) and the distance (L) between the two points. This is also the time required to bring all of the fluid in the pipe to rest.

$$t = \frac{L}{a} \qquad 17.202$$

When the water hammer shock wave reaches the original source of water, the pressure wave will dissipate. A rarefaction wave (at the pressure of the water source) will return at velocity a to the valve. The time for the compression shock wave to travel to the source and the rarefaction wave to return to the valve is given by Eq. 17.203. This is also the length of time that the pressure is constant at the valve.

$$t = \frac{2L}{a} \qquad 17.203$$

The fluid pressure increase resulting from the shock wave is calculated by equating the kinetic energy change of the fluid with the average pressure during the compression process. The pressure increase is independent of the length of pipe. If the velocity is decreased by an amount Δv instantaneously, the increase in pressure will be

$$\Delta p = \rho a \Delta v \qquad \text{[SI]} \qquad 17.204(a)$$

$$\Delta p = \frac{\rho a \Delta v}{g_c} \qquad \text{[U.S.]} \qquad 17.204(b)$$

It is interesting that the pressure increase at the valve depends on Δv but not on the actual length of time it takes to close the valve, as long as the valve is closed when the wave returns to it. Therefore, there is no difference in pressure buildups at the valve for an "instantaneous closure," "rapid closure," or "sudden closure."[30] It is only necessary for the closure to occur rapidly. Because of this, some authorities recommend that the valve closure time be at least ten times the wave return duration calculated in Eq. 17.203.

Having a very long pipe is equivalent to assuming an instantaneous closure. When the pipe is long, the time for the shock wave to travel round-trip is much longer than the time to close the valve. Thus, the valve will be closed when the rarefaction wave returns to the valve.

If the pipe is short, it will be difficult to close the valve before the rarefaction wave returns to the valve. With a short pipe, the pressure buildup will be less than is predicted by Eq. 17.204. (Having a short pipe is equivalent to the case of "slow closure.") The actual pressure history is complex, and no simple method exists for calculating the pressure buildup in short pipes.

---

[30]The pressure elsewhere along the pipe, however, will be lower for slow closures than for instantaneous closures.

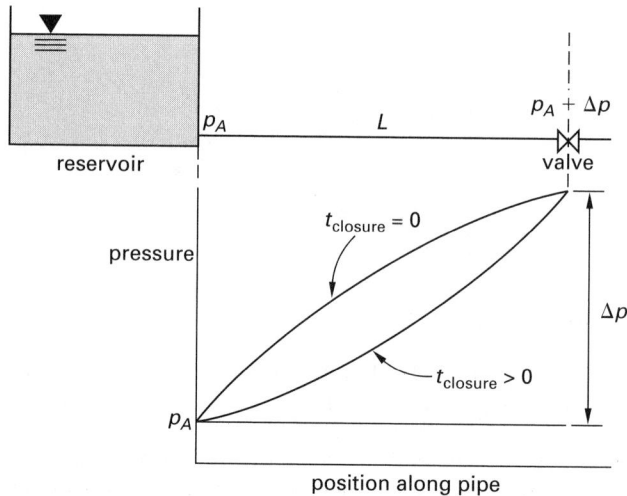

Figure 17.41 Water Hammer

Installing a *surge tank, accumulator, slow-closing valve* (e.g., a gate valve), or *pressure-relief valve* in the line will protect against water hammer damage. The surge tank (or *surge chamber*) is an open tank or reservoir. Since the water is unconfined, large pressure buildups do not occur. An accumulator is a closed tank that is partially filled with air. Since the air is much more compressible than the water, it will be compressed by the water hammer shock wave. The energy of the shock wave is dissipated when the air is compressed.

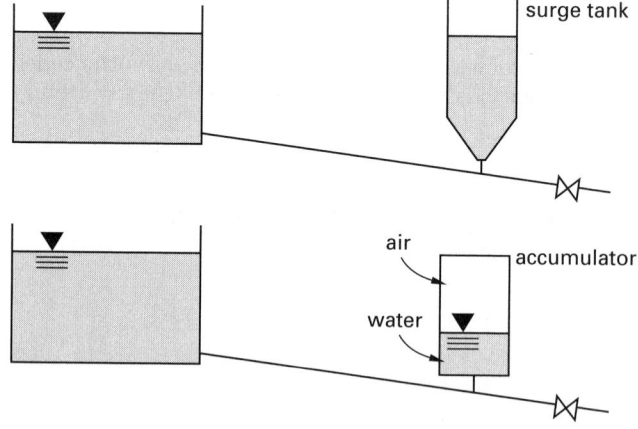

Figure 17.42 Water Hammer Protective Devices

If the pipe material is elastic (the typical assumption for steel and plastic), the previous analysis of water hammer effects (t and Δp) is still valid. However, the calculation of the speed of sound in water must account for the elasticity of the pipe material. This is accomplished by using Eq. 17.205 for the modulus of elasticity (bulk modulus) when calculating the speed of sound, a. (In Eq. 17.205, t is the pipe wall thickness and D is the pipe inside diameter.) At room temperature, the modulus of elasticity of ductile steel is approximately

2.9×10^7 lbf/in^2 (200 GPa); for ductile cast iron, it is $2.2\text{-}2.5 \times 10^7$ lbf/in^2 (150-170 GPa).

$$E = \frac{E_{\text{water}} \, t_{\text{pipe}} \, E_{\text{pipe}}}{t_{\text{pipe}} \, E_{\text{pipe}} + D_{\text{pipe}} E_{\text{water}}} \qquad \textit{17.205}$$

Equation 17.205 indicates that E (and, hence, the effect of water hammer) can be reduced by using a larger diameter pipe. The size of the valve does not affect the wave velocity.

Example 17.14

Water ($\rho = 1000$ kg/m^3, $E = 2 \times 10^9$ Pa) is flowing at 4 m/s through a long length of 4 in schedule-40 steel pipe ($D_i = 0.102$ m, $t = 0.00602$ m, $E = 2 \times 10^{11}$ Pa) when a valve suddenly closes completely. What is the theoretical increase in pressure?

Solution

From Eq. 17.205, the modulus of elasticity to be used in calculating the speed of sound is

$$E = \frac{(2 \times 10^9 \text{ Pa})(0.00602 \text{ m})(2 \times 10^{11} \text{ Pa})}{(0.00602 \text{ m})(2 \times 10^{11} \text{ Pa})}$$
$$+ (0.102 \text{ m})(2 \times 10^9 \text{ Pa})$$

$$= 1.71 \times 10^9 \text{ Pa}$$

The speed of sound in the pipe is

$$a = \sqrt{\frac{E}{\rho}} = \sqrt{\frac{1.71 \times 10^9 \text{ Pa}}{1000 \, \dfrac{\text{kg}}{\text{m}^3}}}$$

$$= 1308 \text{ m/s}$$

From Eq. 17.204, the pressure increase is

$$\Delta p = \rho a \Delta v = \left(1000 \, \frac{\text{kg}}{\text{m}^3}\right)\left(1308 \, \frac{\text{m}}{\text{s}}\right)\left(4 \, \frac{\text{m}}{\text{s}}\right)$$

$$= 5.23 \times 10^6 \text{ Pa}$$

49. LIFT

Lift is an upward force that is exerted on an object (flat plate, airfoil, rotating cylinder, etc.) as the object passes through a fluid. Lift combines with drag to form the resultant force on the object, as shown in Fig. 17.43.

Figure 17.43 *Lift and Drag on an Airfoil*

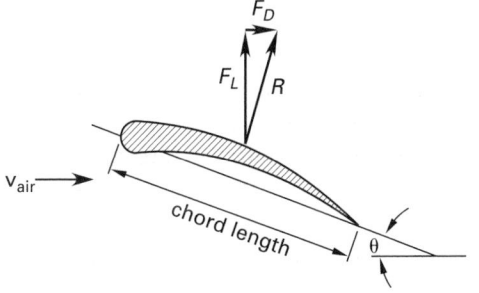

The generation of lift from air flowing over an airfoil is predicted by Bernoulli's equation. Air molecules must travel a longer distance over the top surface of the airfoil than over the lower surface, and, therefore, they travel faster over the top surface. Since the total energy of the air is constant, the increase in kinetic energy comes at the expense of pressure energy. The static pressure on the top of the airfoil is reduced, and a net upward force is produced.

Within practical limits, the lift produced can be increased at lower speeds by increasing the curvature of the wing. This increased curvature is achieved by the use of *flaps*. When a plane is traveling slowly (e.g., during take-off or landing), its flaps are extended to create the lift needed.

Figure 17.44 *Use of Flaps in an Airfoil*

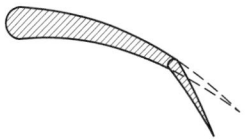

The lift produced can be calculated from Eq. 17.206, whose use is not limited to airfoils.

$$F_L = \frac{C_L A \rho v^2}{2} \qquad \text{[SI]} \qquad \textit{17.206(a)}$$

$$F_L = \frac{C_L A \rho v^2}{2 g_c} \qquad \text{[U.S.]} \qquad \textit{17.206(b)}$$

The dimensions of an airfoil or wing are frequently given in terms of chord length and aspect ratio. The *chord length* is the front-to-back dimension of the airfoil. The *aspect ratio* is the ratio of the *span* (wing length) to chord length. The area, A, in Eq. 17.206 is the airfoil's area projected onto the plane of the chord. Thus, for a rectangular airfoil, $A = $ chord $\times$ span.

The dimensionless *coefficient of lift*, C_L, is used to measure the effectiveness of the airfoil. The coefficient of lift depends on the shape of the airfoil and the Reynolds number. No simple relationship can be given for calculating the coefficient of lift for airfoils, but the theoretical coefficient of lift for a thin plate in two-dimensional flow at a low angle of attack, θ, is given by Eq. 17.207. Actual airfoils are able to achieve only 80 to 90% of this theoretical value.

$$C_L = 2\pi \sin \theta \qquad \textit{17.207}$$

The coefficient of lift for an airfoil cannot be increased without limit merely by increasing θ. Eventually, the *stall angle* is reached, at which point the coefficient of lift decreases dramatically.

Figure 17.45 *Typical Plot of Lift Coefficient*

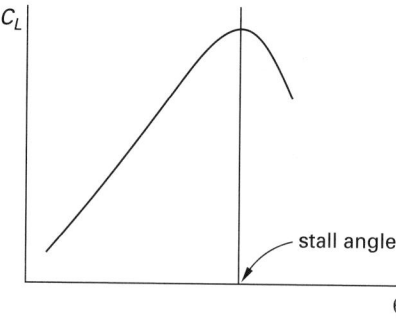

50. CIRCULATION

A theoretical concept for calculating the lift generated by an object (an airfoil, propeller, turbine blade, etc.) is *circulation*. Circulation, Γ, is defined by Eq. 17.208.[31] Its units are length²/time.

$$\Gamma = \oint \text{v} \cos\theta \; dl \qquad \qquad 17.208$$

Figure 17.46 illustrates an arbitrary closed curve drawn around a point (or body) in steady flow. The tangential components of velocity, v, at all points around the curve are $\text{v}\cos\theta$. It is a fundamental theorem that circulation has the same value for every closed curve that can be drawn around a body.

Figure 17.46 *Circulation Around a Point*

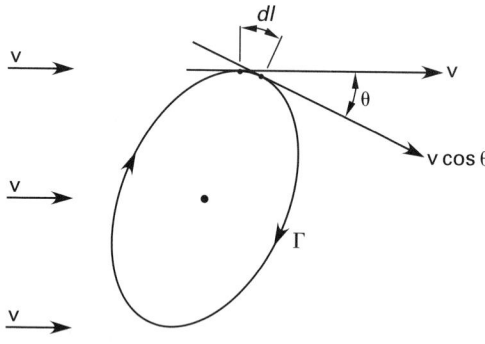

Lift on a body traveling with relative velocity v through a fluid of density ρ can be calculated from the circulation by using Eq. 17.209.[32]

$$F_L = \rho \text{v} \Gamma \times \text{chord length} \qquad 17.209$$

There is no actual circulation of air "around" an airfoil, but this mathematical concept can be used, nevertheless. However, since the flow of air "around" an airfoil is

---

[31] Equation 17.208 is analogous to the calculation of work being done by a constant force moving around a curve. If the force makes an angle of θ with the direction of motion, the work done as the force moves a distance dl around a curve is $W = \oint F \cos\theta dl$. In calculating circulation, velocity takes the place of force.
[32] U is the traditional symbol of velocity in circulation studies.

not symmetrical in path or velocity, experimental determination of C_L is favored over theoretical calculations of circulation.

51. LIFT FROM ROTATING CYLINDERS

When a cylinder is placed transversely to an airflow traveling at velocity v_∞, the velocity at a point on the surface of the cylinder is $2\text{v}_\infty \sin\theta$. Since the flow is symmetrical, however, no lift is produced.

Figure 17.47 *Flow Over a Cylinder*

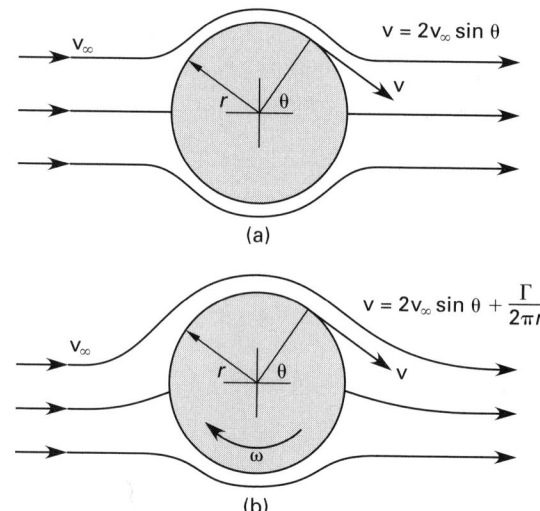

If the cylinder with radius r is rotating at ω rad/sec while it moves with a relative velocity v_∞ through the air, the *Kutta-Joukowsky result* (theorem) can be used to calculate the lift per unit length of cylinder.[33] This is known as the *Magnus effect*.

$$F_L(\text{per unit length}) = \rho \text{v}_\infty \Gamma \qquad \text{[SI]} \qquad 17.210(a)$$

$$F_L = \frac{\rho \text{v}_\infty \Gamma}{g_c} \qquad \text{[U.S.]} \qquad 17.210(b)$$

$$\Gamma = 2\pi r^2 \omega \qquad \qquad 17.211$$

Equation 17.210 assumes that there is no slip (i.e., that the air drawn around the cylinder by rotation moves at ω), and in that ideal case, the maximum coefficient of lift is 4π. Practical rotating devices, however, seldom achieve a coefficient of lift in excess of 9 or 10, and even then, the power expenditure is excessive.

52. DRAG

Drag is a frictional force that acts parallel but opposite to the direction of motion. It combines with the lift (acting perpendicular to the direction of motion) to produce a resultant force on the object. The total drag

---

[33] A similar analysis can be used to explain why a pitched baseball curves. The rotation of the ball produces a force that changes the path of the ball as it travels.

force is made up of *skin friction* and *pressure drag* (also known as *form drag*). These components, in turn, can be subdivided and categorized into *wake drag, induced drag*, and *profile drag*.

Figure 17.48 *Components of Total Drag*

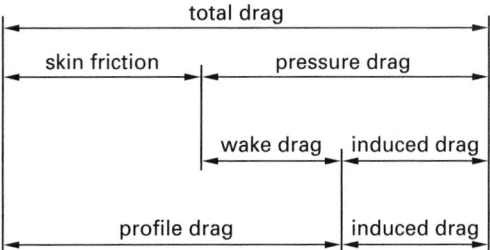

Most aeronautical engineering books contain descriptions of these drag terms. However, the difference between the situations where either skin friction drag or pressure drag predominates is illustrated in Fig. 17.49.

Figure 17.49 *Extreme Cases of Pressure Drag and Skin Friction*

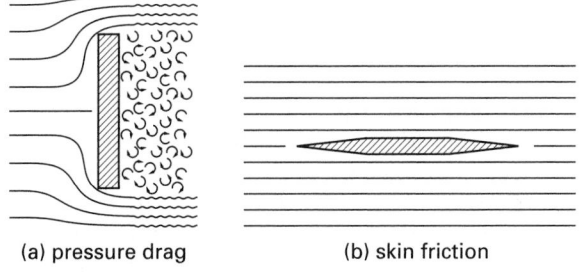

(a) pressure drag   (b) skin friction

Total drag is most easily calculated from the dimensionless *drag coefficient*, C_D. It can be shown by dimensional analysis that the drag coefficient depends only on the Reynolds number.

$$F_D = \frac{C_D A \rho \mathrm{v}^2}{2} \qquad \text{[SI]} \qquad \textit{17.212(a)}$$

$$F_D = \frac{C_D A \rho \mathrm{v}^2}{2 g_c} \qquad \text{[U.S.]} \qquad \textit{17.212(b)}$$

In most cases, the area, A, in Eq. 17.212 is the projected area (i.e., the *frontal area*) normal to the stream. This is appropriate for spheres, cylinders, and automobiles. In a few cases (e.g., for airfoils and flat plates), the area is a projection of the object onto a plane parallel to the stream.

Typical drag coefficients for production cars vary from approximately 0.35 to approximately 0.55, with most modern cars being nearer the lower end. By comparison, other low-speed drag coefficients are approximately 0.05 (aircraft wing), 0.10 (sphere in turbulent flow), and 1.2 (flat plate).

Aero horsepower is a term used by automobile manufacturers to designate the power required to move a car horizontally at 50 mi/hr (80.5 km/h) against the drag force. Aero horsepower varies from approximately 7 hp (5.2 kW) for a streamlined subcompact car to approximately 100 hp (75 kW) for a box-shaped truck.

53. DRAG ON SPHERES AND DISKS

The drag coefficient varies linearly with the Reynolds number for laminar flow around a sphere or disk. In this region, the drag is almost entirely due to skin friction. For Reynolds numbers below approximately 0.4, experiments have shown that the drag coefficient can be calculated from Eq. 17.213.[34] In calculating the Reynolds number, the sphere or disk diameter should be used as the characteristic dimension.

$$C_D = \frac{24}{\mathrm{Re}} \qquad \textit{17.213}$$

Substituting this value of C_D into Eq. 17.212 results in *Stokes' law*, which is applicable to slow motion (ascent or descent) of spherical particles and bubbles traveling at velocity v through a fluid. Stokes' law is based on the assumptions that (a) flow is laminar, (b) Newton's law of viscosity is valid, and (c) all higher-order velocity terms (v^2, etc.) are negligible.

$$F_D = 3\pi\mu\mathrm{v}D \qquad \textit{17.214}$$

The drag coefficients for disks and spheres operating outside the region covered by Stokes' law have been determined experimentally. In the turbulent region, pressure drag is predominant. Figure 17.50 can be used to obtain approximate values for C_D.

Figure 17.50 *Drag Coefficients for Spheres and Circular Flat Disks*

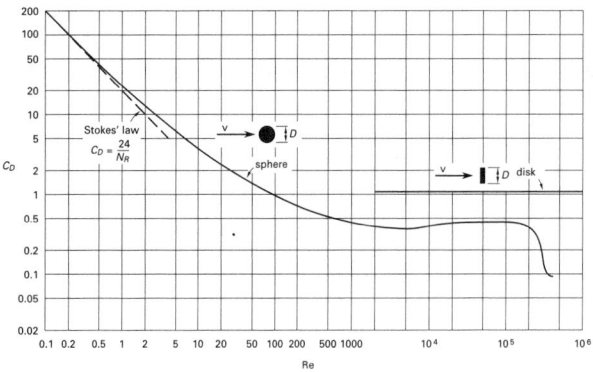

Figure 17.50 shows that there is a dramatic drop in the drag coefficient around Re $= 10^5$. The explanation

---

[34]Some sources report that the region in which Stokes' law applies extends to Re = 1.0.

for this is that the point of separation of the boundary layer shifts, decreasing the width of the wake. Since the drag force is primarily pressure drag at higher Reynolds numbers, a reduction in the wake reduces the pressure drag. Thus, anything that can be done to a sphere (scuffing or wetting a baseball, dimpling a golf ball, etc.) to induce a smaller wake will reduce the drag. There can be no shift in the boundary layer separation point for a thin disk, since the disk has no depth in the direction of flow. Therefore, the drag coefficient remains the same at all turbulent Reynolds numbers.

Figure 17.51 *Turbulent Flow Around a Sphere at Various Reynolds Numbers*

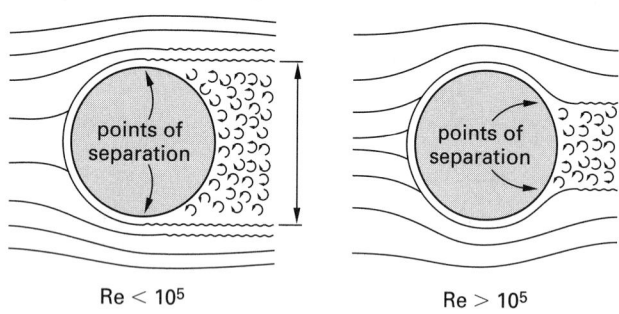

54. TERMINAL VELOCITY

The velocity of an object falling through a fluid will continue to increase until the drag force equals the net downward force (i.e., the weight less the buoyant force). The maximum velocity attained is known as the *terminal velocity*. At terminal velocity,

$$F_D = mg - F_b \qquad \text{[SI]} \qquad 17.215(a)$$

$$F_D = m \times \left(\frac{g}{g_c}\right) - F_b \qquad \text{[U.S.]} \qquad 17.215(b)$$

If the drag coefficient is known, the terminal velocity can be calculated from Eq. 17.216. For small, heavy objects falling in air, the buoyant force can be neglected.

$$\mathrm{v} = \sqrt{\frac{2(mg - F_b)}{C_D A \rho_{\text{fluid}}}} = \sqrt{\frac{2Vg(\rho_{\text{object}} - \rho_{\text{fluid}})}{C_D A \, \rho_{\text{fluid}}}} \qquad 17.216$$

For a sphere of diameter D, the terminal velocity is

$$\mathrm{v} = \sqrt{\frac{4Dg(\rho_{\text{sphere}} - \rho_{\text{fluid}})}{3C_D \rho_{\text{fluid}}}} \qquad 17.217$$

If the spherical particle is very small, Stokes' law may apply. In that case, the terminal velocity can be calculated from Eq. 17.218.

$$\mathrm{v} = \frac{D^2 g(\rho_{\text{sphere}} - \rho_{\text{fluid}})}{18\,\mu} \qquad \text{[SI]} \qquad 17.218(a)$$

$$\mathrm{v} = \frac{D^2 (\rho_{\text{sphere}} - \rho_{\text{fluid}})}{18\,\mu} \times \left(\frac{g}{g_c}\right) \qquad \text{[U.S.]} \qquad 17.218(b)$$

55. NONSPHERICAL PARTICLES

Only the most simple bodies can be modeled as spheres. One method of overcoming the complexity of dealing with the flow of real particles is to correlate performance with *sphericity*. For a particle and a sphere with the same volume, the sphericity is defined by Eq. 17.219. Sphericity will always be less than or equal to 1.0.

$$\psi = \text{sphericity} = \frac{A_{\text{sphere}}}{A_{\text{particle}}} \qquad 17.219$$

Figure 17.52 *Drag Coefficients for Nonspherical Particles*

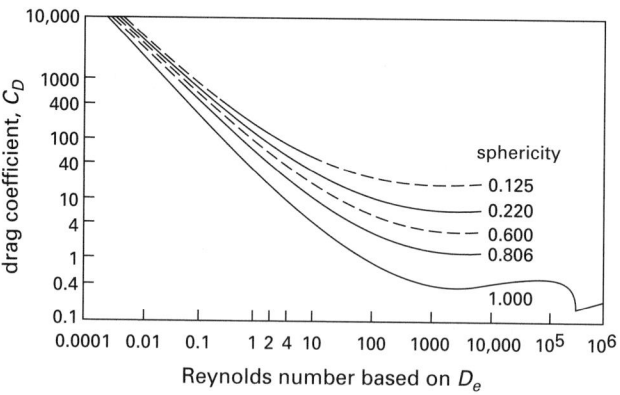

Reprinted with permission from George Granger Brown, et al., *Unit Operations*, © 1950, by John Wiley & Sons, Inc.

Another parameter that can be used to describe the deviation from ideal spherical behavior is the ratio of equivalent to average diameters, D_e/D_{ave}. The *average diameter* can be determined by screening a sample of the particles and evaluating the size distribution. The *equivalent diameter* is the diameter of a sphere having the same volume as the particle.

56. FLOW AROUND A CYLINDER

The characteristic drag coefficient plot for cylinders placed normal to the fluid flow is similar to the plot for spheres. The plot shown in Fig. 17.53 is for infinitely long cylinders, since there is additional wake drag at the cylinder ends. In calculating the Reynolds number, the cylinder diameter should be used as the characteristic dimension.

Figure 17.53 *Drag Coefficient for a Cylinder*

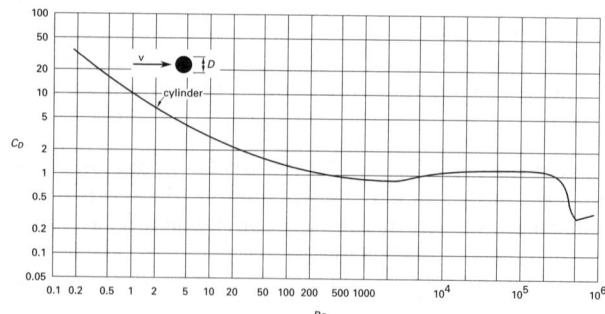

Table 17.8 *Sphericity and D_e/D_{ave} Ratios for Nonspherical Particles*

shape	sphericity	D_e/D_{ave}
sphere	1.00	1.00
octahedron	0.817	0.965
cube	0.806	1.24
prisms		
$a \times a \times 2a$	0.767	1.564
$a \times 2a \times 2a$	0.761	0.985
$a \times 2a \times 3a$	0.725	1.127
cylinders		
$h = 2r$	0.874	1.135
$h = 3r$	0.860	1.31
$h = 10r$	0.691	1.96
$h = 20r$	0.580	2.592
disks		
$h = 1.33r$	0.858	1.00
$h = r$	0.827	0.909
$h = r/3$	0.594	0.630
$h = r/10$	0.323	0.422
$h = r/15$	0.254	0.368

Reprinted with permission from George Granger Brown, et al., *Unit Operations*, © 1950, by John Wiley & Sons, Inc.

Example 17.15

A 50 ft (15 m) high flagpole is constructed of a uniformly smooth cylinder 10 in (25 cm) in diameter. The surrounding air is at 40°F (0°C) and 14.6 psia (100 kPa). What is the total drag force on the flagpole in a 30 mi/hr (50 km/h) gust? (Neglect variations in the wind speed with height above the ground.)

SI Solution

At 0°C, the absolute viscosity of air is

$$\mu_{p,0°C} = 1.709 \times 10^{-5} \text{ Pa·s} \quad [\text{App. 14.E}]$$

The density of the air is

$$\rho_p = \frac{p}{RT} = \frac{(100 \text{ kPa}) \left(1000 \frac{\text{Pa}}{\text{kPa}}\right)}{\left(287 \frac{\text{J}}{\text{kg·K}}\right)(0°\text{C} + 273)}$$

$$= 1.276 \text{ kg/m}^3$$

The kinematic viscosity is

$$\nu_p = \frac{\mu}{\rho} = \frac{1.709 \times 10^{-5} \text{ Pa·s}}{1.276 \frac{\text{kg}}{\text{m}^3}}$$

$$= 1.339 \times 10^{-5} \text{ m}^2/\text{s}$$

The wind speed is

$$v = \frac{\left(50 \frac{\text{km}}{\text{h}}\right)\left(1000 \frac{\text{m}}{\text{km}}\right)}{3600 \frac{\text{s}}{\text{h}}} = 13.89 \text{ m/s}$$

The characteristic dimension of the flagpole is its diameter. The Reynolds number is

$$\text{Re} = \frac{Lv}{\nu} = \frac{Dv}{\nu} = \frac{\left(\frac{25 \text{ cm}}{100 \frac{\text{cm}}{\text{m}}}\right)\left(13.89 \frac{\text{m}}{\text{s}}\right)}{1.339 \times 10^{-5} \frac{\text{m}^2}{\text{s}}}$$

$$= 2.59 \times 10^5$$

From Fig. 17.53 for this Reynolds number, the drag coefficient is approximately 1.2.

The frontal area of the flagpole is

$$A = DL = \frac{(25 \text{ cm})(15 \text{ m})}{100 \frac{\text{cm}}{\text{m}}}$$

$$= 3.75 \text{ m}^2$$

From Eq. 17.212(a), the drag on the flagpole is

$$F_D = \frac{C_D A \rho v^2}{2}$$

$$= \frac{(1.2)(3.75 \text{ m}^2)\left(1.276 \frac{\text{kg}}{\text{m}^3}\right)\left(13.89 \frac{\text{m}}{\text{s}}\right)^2}{2}$$

$$= 554 \text{ N}$$

Customary U.S. Solution

At 40°F, the absolute viscosity of air is

$$\mu_{p,40°F} = 3.62 \times 10^{-7} \text{ lbf-sec/ft}^2 \quad [\text{App. 14.D}]$$

The density of the air is

$$\rho_p = \frac{p}{RT} = \frac{\left(14.6 \frac{\text{lbf}}{\text{in}^2}\right)\left(144 \frac{\text{in}^2}{\text{ft}^2}\right)}{\left(53.3 \frac{\text{ft-lbf}}{\text{lbm-°R}}\right)(40°\text{F} + 460)}$$

$$= 0.07889 \text{ lbm/ft}^3$$

The kinematic viscosity is

$$\nu_p = \frac{\mu g_c}{\rho} = \frac{\left(3.62 \times 10^{-7} \frac{\text{lbf-sec}}{\text{ft}^2}\right)\left(32.2 \frac{\text{ft-lbm}}{\text{lbf-sec}^2}\right)}{0.07889 \frac{\text{lbm}}{\text{ft}^3}}$$

$$= 1.478 \times 10^{-4} \text{ ft}^2/\text{sec}$$

The wind speed is

$$v = \frac{\left(30 \frac{mi}{hr}\right)\left(5280 \frac{ft}{mi}\right)}{3600 \frac{sec}{hr}} = 44 \text{ ft/sec}$$

The characteristic dimension of the flagpole is its diameter. The Reynolds number is

$$Re = \frac{Lv}{\nu} = \frac{Dv}{\nu} = \frac{\left(\frac{10 \text{ in}}{12 \frac{in}{ft}}\right)\left(44 \frac{ft}{sec}\right)}{1.478 \times 10^{-4} \frac{ft^2}{sec}}$$

$$= 2.48 \times 10^5$$

From Fig. 17.53 for this Reynolds number, the drag coefficient is approximately 1.2.

The frontal area of the flagpole is

$$A = DL = \frac{(10 \text{ in})(50 \text{ ft})}{12 \frac{in}{ft}} = 41.67 \text{ ft}^2$$

From Eq. 17.212(b), the drag on the flagpole is

$$F_D = \frac{C_D A \rho v^2}{2g_c}$$

$$= \frac{(1.2)(41.67 \text{ ft}^2)\left(0.07889 \frac{lbm}{ft^3}\right)\left(44 \frac{ft}{sec}\right)^2}{(2)\left(32.2 \frac{ft\text{-}lbm}{lbf\text{-}sec^2}\right)}$$

$$= 118.6 \text{ lbf}$$

57. FLOW OVER A PARALLEL FLAT PLATE

The drag experienced by a flat plate oriented parallel to the direction of flow is almost totally skin friction drag. Prandtl's *boundary layer theory* can be used to evaluate the frictional effects. Such an analysis shows that the shape of the boundary layer is a function of the Reynolds number. The characteristic dimension used in calculating the Reynolds number is the chord length, L (i.e., the dimension of the plate parallel to the flow).

Figure 17.54 *Flow Over a Parallel Flat Plate*
(one side)

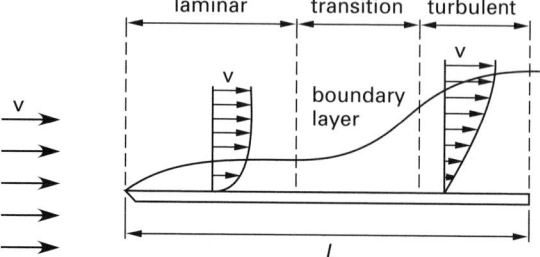

When skin friction predominates, it is common to use the symbol C_f (i.e., the *skin friction coefficient*) for the drag coefficient. For laminar flow over a smooth, flat plate, the drag coefficient based on the boundary layer theory is given by Eq. 17.220, which is known as the *Blasius solution*.[35] The critical Reynolds number for laminar flow is often reported to be 530,000. However, the transition region between laminar flow and turbulent flow actually occupies the Reynolds number range of 100,000 to 1,000,000.

$$C_f = \frac{1.328}{\sqrt{Re}} \qquad \textit{17.220}$$

Prandtl reported that the skin friction coefficient for turbulent flow is

$$C_f = \frac{0.455}{(\log Re)^{2.58}} \qquad \textit{17.221}$$

The drag force is calculated from Eq. 17.222. The factor 2 is used because there is friction on two sides of the flat plate.

$$F_D = 2\left(\frac{C_f A \rho v^2}{2}\right) = C_f A \rho v^2 \qquad \text{[SI]} \quad \textit{17.222(a)}$$

$$F_D = 2\left(\frac{C_f A \rho v^2}{2g_c}\right) = \frac{C_f A \rho v^2}{g_c} \qquad \text{[U.S.]} \quad \textit{17.222(b)}$$

58. SIMILARITY

Similarity considerations between a *model* (subscript m) and a full-sized object (subscript p, for *prototype*) imply that the model can be used to predict the performance of the prototype. Such a model is said to be *mechanically similar* to the prototype.

Complete *mechanical similarity* requires both geometric and dynamic similarity.[36] *Geometric similarity* means that the model is true to scale in length, area, and volume. The *model scale (length ratio)* is defined as

$$L_r = \frac{\text{size of model}}{\text{size of prototype}} \qquad \textit{17.223}$$

The area and volume ratios are based on the model scale.

$$\frac{A_m}{A_p} = (L_r)^2 \qquad \textit{17.224}$$

$$\frac{V_m}{V_p} = (L_r)^3 \qquad \textit{17.225}$$

---

[35] Other correlations substitute the coefficient 1.44, the original value calculated by Prandtl, for 1.328 in Eq. 17.220. The Blasius solution is considered to be more accurate.
[36] Complete mechanical similarity also requires kinematic and thermal similarity, which are not discussed in this book.

Dynamic similarity means that the ratios of all types of forces are equal for the model and the prototype. These forces result from inertia, gravity, viscosity, elasticity (i.e., fluid compressibility), surface tension, and pressure.

The number of possible ratios of forces is large. For example, the ratios of viscosity/inertia, inertia/gravity, and inertia/surface tension are only three of the ratios of forces that must match for every corresponding point on the model and prototype. Fortunately, some force ratios can be neglected because the forces are negligible or are self-canceling.

In some cases, the geometric scale may be deliberately distorted. For example, with scale models of harbors and rivers, the water depth might be only a fraction of an inch if the scale is followed loyally. Not only will the surface tension be excessive, but the shifting of the harbor or river bed may not be properly observed. Therefore, the vertical scale is chosen to be different from the horizontal scale. Such models are known as *distorted models*. Experience is needed to interpret observations made of distorted models.

The following sections deal with the most common similarity problems, but not all. For example, similarity of steady laminar flow in a horizontal pipe requires the Stokes number, similarity of high-speed (near-sonic) aircraft requires the Mach number, and similarity of capillary rise in a tube requires the *Eötvös number* $(g\rho L^2/\sigma)$.

59. VISCOUS AND INERTIAL FORCES DOMINATE

Consider the testing of a completely submerged object, such as the items listed in Table 17.9. Surface tension will be negligible. The fluid can be assumed to be incompressible for low velocity. Gravity does not change the path of the fluid particles significantly during the passage of the object.

Table 17.9 *Cases with Reynolds Number Similarity*

aircraft (subsonic)
airfoils (subsonic)
closed-pipe flow (turbulent)
drainage through tank orifices
fans
flow meters
open channel flow (without wave action)
pumps
submarines
torpedoes
turbines

Only inertial, viscous, and pressure forces are significant. Because they are the only forces that are acting, these three forces are in equilibrium. Since they are in equilibrium, knowing any two forces will define the third force. This third force is dependent and can be

omitted from the similarity analysis. For submerged objects, pressure is traditionally chosen as the dependent force.

The dimensionless ratio of the inertial forces to the viscous forces is the Reynolds number. Equating the model's and prototype's Reynolds numbers will ensure similarity.[37]

$$\text{Re}_m = \text{Re}_p \qquad 17.226$$

$$\frac{L_m \text{v}_m}{\nu_m} = \frac{L_p \text{v}_p}{\nu_p} \qquad 17.227$$

If the model is tested in the same fluid and at the same temperature in which the prototype is expected to operate, setting the Reynolds numbers equal is equivalent to setting $L_m \text{v}_m = L_p \text{v}_p$.

Example 17.16

A $^1/_{30}$ size scale model of a helicopter fuselage is tested in a wind tunnel at 120 mi/hr (190 km/h). The conditions in the wind tunnel are 50 psia and 100°F (350 kPa and 50°C). What is the corresponding speed of a prototype traveling in 14.0 psia and 40°F still air (100 kPa, 0°C)?

SI Solution

From App. 14.E,

$$\mu_{\text{p},0°\text{C}} = 1.709 \times 10^{-5} \text{ Pa·s}$$
$$\mu_{\text{m},50°\text{C}} = 1.951 \times 10^{-5} \text{ Pa·s}$$

The densities of air at the two conditions are

$$\rho_p = \frac{p}{RT} = \frac{(100 \text{ kPa})\left(1000 \frac{\text{Pa}}{\text{kPa}}\right)}{\left(287 \frac{\text{J}}{\text{kg·K}}\right)(0°\text{C} + 273)}$$
$$= 1.276 \text{ kg/m}^3$$

$$\rho_m = \frac{p}{RT} = \frac{(350 \text{ kPa})\left(1000 \frac{\text{Pa}}{\text{kPa}}\right)}{\left(287 \frac{\text{J}}{\text{kg·K}}\right)(50°\text{C} + 273)}$$
$$= 3.776 \text{ kg/m}^3$$

---

[37]An implied assumption is that the drag coefficients are the same for the model and the prototype. In the case of pipe flow, it is assumed that flow will be in the turbulent region with the same relative roughness.

The kinematic viscosities are

$$\nu_p = \frac{\mu}{\rho} = \frac{1.709 \times 10^{-5} \text{ Pa·s}}{1.276 \ \frac{\text{kg}}{\text{m}^3}}$$

$$= 1.339 \times 10^{-5} \text{ m}^2/\text{s}$$

$$\nu_m = \frac{\mu}{\rho} = \frac{1.951 \times 10^{-5} \text{ Pa·s}}{3.776 \ \frac{\text{kg}}{\text{m}^3}}$$

$$= 5.167 \times 10^{-6} \text{ m}^2/\text{s}$$

From Eq. 17.227,

$$v_p = v_m \left(\frac{L_m}{L_p}\right)\left(\frac{\nu_p}{\nu_m}\right)$$

$$= \frac{\left(190 \ \frac{\text{km}}{\text{h}}\right)\left(\frac{1}{30}\right)\left(1.339 \times 10^{-5} \ \frac{\text{m}^2}{\text{s}}\right)}{5.167 \times 10^{-6} \ \frac{\text{m}^2}{\text{s}}}$$

$$= 16.4 \text{ km/h}$$

Customary U.S. Solution

Since surface tension and gravitational forces on the air particles are insignificant and since the flow velocities are low, viscous and inertial forces dominate. The Reynolds numbers of the model and prototype are equated.

The kinematic viscosity of air must be evaluated at the respective temperatures and pressures. As kinematic viscosity tables for air are not readily available, the viscosity must be calculated. Although kinematic viscosity depends on the temperature and pressure, absolute viscosity is essentially independent of pressure. From App. 14.D,

$$\mu_{p,40°F} = 3.62 \times 10^{-7} \text{ lbf-sec/ft}^2$$

$$\mu_{m,100°F} = 3.96 \times 10^{-7} \text{ lbf-sec/ft}^2$$

The densities of air at the two conditions are

$$\rho_p = \frac{p}{RT} = \frac{\left(14.0 \ \frac{\text{lbf}}{\text{in}^2}\right)\left(144 \ \frac{\text{in}^2}{\text{ft}^2}\right)}{\left(53.3 \ \frac{\text{ft-lbf}}{\text{lbm-°R}}\right)(40°F + 460)}$$

$$= 0.0756 \text{ lbm/ft}^3$$

$$\rho_m = \frac{p}{RT} = \frac{\left(50.0 \ \frac{\text{lbf}}{\text{in}^2}\right)\left(144 \ \frac{\text{in}^2}{\text{ft}^2}\right)}{\left(53.3 \ \frac{\text{ft-lbf}}{\text{lbm-°R}}\right)(100°F + 460)}$$

$$= 0.2412 \text{ lbm/ft}^3$$

The kinematic viscosities are

$$\nu_p = \frac{\mu g_c}{\rho} = \frac{\left(3.62 \times 10^{-7} \ \frac{\text{lbf-sec}}{\text{ft}^2}\right)g_c}{0.0756 \ \frac{\text{lbm}}{\text{ft}^3}}$$

$$= 4.79 \times 10^{-6} \times g_c \text{ ft}^2/\text{sec}$$

$$\nu_m = \frac{\mu g_c}{\rho} = \frac{\left(3.96 \times 10^{-7} \ \frac{\text{lbf-sec}}{\text{ft}^2}\right)g_c}{0.2412 \ \frac{\text{lbm}}{\text{ft}^3}}$$

$$= 1.64 \times 10^{-6} \times g_c \text{ ft}^2/\text{sec}$$

From Eq. 17.227,

$$v_p = v_m \left(\frac{L_m}{L_p}\right)\left(\frac{\nu_p}{\nu_m}\right)$$

$$= \frac{\left(120 \ \frac{\text{mi}}{\text{hr}}\right)\left(\frac{1}{30}\right)\left(4.79 \times 10^{-6} g_c \ \frac{\text{ft}^2}{\text{sec}}\right)}{\left(1.64 \times 10^{-6} g_c \ \frac{\text{ft}^2}{\text{sec}}\right)}$$

$$= 11.7 \text{ mi/hr}$$

60. INERTIAL AND GRAVITATIONAL FORCES DOMINATE

Table 17.10 lists the cases when elasticity and surface tension forces can be neglected but gravitational forces cannot. Omitting these two forces from the similarity calculations leaves pressure, inertia, viscosity, and gravity forces, which are in equilibrium. Pressure is chosen as the dependent force and is omitted from the analysis.

Table 17.10 Cases with Froude Number Similarity

> bow waves from ships
> flow over spillways
> flow over weirs
> motion of a fluid jet
> open channel flow with varying surface levels
> oscillatory wave action
> seaplane hulls
> surface ships
> surface wave action
> surge and flood waves

There are only two possible combinations of the remaining three forces. The ratio of inertial to viscous forces is the Reynolds number. The ratio of the inertial forces to the gravitational forces is the *Froude number*, Fr.[38] The Froude number is used when gravitational forces are significant, such as in wave motion produced by a ship or seaplane hull.

$$\text{Fr} = \frac{\text{v}^2}{Lg} \qquad 17.228$$

Thus, similarity is ensured when Eqs. 17.229 and 17.230 are satisfied.

$$\text{Re}_m = \text{Re}_p \qquad 17.229$$

$$\text{Fr}_m = \text{Fr}_p \qquad 17.230$$

As an alternative, Eqs. 17.229 and 17.230 can be solved simultaneously. This results in the following requirement for similarity, which indicates that it is necessary to test the model in a manufactured liquid with a specific viscosity.

$$\frac{\nu_m}{\nu_p} = \left(\frac{L_m}{L_p}\right)^{3/2} = (L_r)^{3/2} \qquad 17.231$$

Sometimes it is not possible to satisfy Eqs. 17.229 and 17.230. This occurs when a model fluid viscosity is called for that is not available. If only one of the equations is satisfied, the model is said to be *partially similar*. In such a case, corrections based on other factors are used.

Another problem with trying to achieve similarity in open channel flow problems is the need to scale surface drag. It can be shown that the ratio of Manning's roughness constants is given by Eq. 17.232. In some cases, it may not be possible to create a surface smooth enough to satisfy this requirement.

$$n_r = (L_r)^{1/6} \qquad 17.232$$

61. SURFACE TENSION FORCE DOMINATES

Table 17.11 lists some of the cases where surface tension is the predominant force. Such cases can be handled by equating the Weber numbers, We, of the model and prototype.[39] (The *Weber number* is the ratio of inertial force to surface tension.)

$$\text{We} = \frac{\text{v}^2 L \rho}{\sigma} \qquad 17.233$$

$$\text{We}_m = \text{We}_p \qquad 17.234$$

Table 17.11 Cases with Weber Number Similarity

air entrainment
bubbles
droplets
waves

---

[38]There are two definitions of the Froude number. Dimensional analysis determines the Froude number as Eq. 17.228 (v^2/Lg), a form that is used in model similitude. However, in open channel flow studies performed by civil engineers, the Froude number is taken as the square root of Eq. 17.228. Whether the derived form or its square root is used can sometimes be determined from the application. If the Froude number is squared (e.g., as in $dE/dx = 1 - \text{Fr}^2$), the square root form is probably needed. In similarity problems, it doesn't make any difference which definition is used.

[39]There are two definitions of the Weber number. The alternate definition is the square root of Eq. 17.233. In similarity problems, it does not make any difference which definition of the Weber number is used.

18 Hydraulic Machines

Nomenclature

C	coefficient	–	–
D	diameter	ft	m
E	specific energy	ft-lbf/lbm	J/kg
f	Darcy friction factor	–	–
f	frequency	Hz	Hz
g	gravitational acceleration	ft/sec^2	m/s^2
g_c	gravitational constant	lbm-ft/ lbf-sec^2	n.a.
h	height or head	ft	m
h_{ac}	acceleration head	ft	m
K	dimensionless factor	–	–
L	length	ft	m
m	mass	lbm	kg
$\dot{m}$	mass flow rate	lbm/sec	kg/s
n	dimensionless exponent	–	–
n	rotational speed	rpm	rpm
NPSHA	net positive suction head available	ft	m
NPSHR	net positive suction head required	ft	m
p	pressure	lbf/ft^2	Pa
P	power	ft-lbf/sec	W
Q	volumetric flow rate	gal/min	L/s
r	radius	ft	m
SA	suction specific speed available	rpm	rpm
SG	specific gravity	–	–
t	time	sec	s
T	transmitted torque	ft-lbf	N·m
v	velocity	ft/sec	m/s
$\dot{V}$	volumetric flow rate	ft^3/sec	m^3/s
W	work	ft-lbf	kW·h
WHP	water horsepower	hp	n.a.
WkW	water kilowatts	n.a.	kW
z	elevation	ft	m

Symbols

γ	specific weight	lbf/ft^3	n.a.
η	efficiency	–	–
θ	angle	deg	deg
ν	kinematic viscosity	ft^2/sec	m^2/s
ρ	density	lbm/ft^3	kg/m^3
σ	cavitation number	–	–
ω	angular velocity	rad/sec	rad/s

Subscripts

atm	atmospheric
A	added (by pump)
b	blade
cr	critical
d	discharge
f	friction
i	inlet
j	jet
m	motor
n	nozzle
o	outlet
p	pressure or pump
s	suction or specific
ss	suction specific
t	total or tangential
th	theoretical
v	velocity or volumetric
vp	vapor pressure
z	potential

1. HYDRAULIC MACHINES

Pumps and turbines are the two basic types of hydraulic machines discussed in this chapter. Pumps convert mechanical energy into fluid energy, increasing the energy possessed by the fluid. Turbines convert fluid energy into mechanical energy, extracting energy from the fluid.

2. TYPES OF PUMPS

Pumps can be classified according to the method by which pumping energy is transferred to the fluid. This classification separates pumps into positive displacement pumps and kinetic pumps.

The most common types of *positive displacement pumps* are *reciprocating action pumps* (which use pistons, plungers, diaphragms, or bellows) and *rotary action pumps* (using vanes, screws, lobes, or progressing cavities). Such pumps discharge a fixed volume for each stroke or revolution. Energy is added intermittently to the fluid.

Kinetic pumps transform fluid kinetic energy into fluid static pressure energy. The pump imparts the kinetic energy; the pump mechanism or housing is constructed in a manner that causes the transformation. *Jet pumps* and *ejector pumps* fall into the kinetic pump category, but centrifugal pumps are the primary examples.

In the operation of a *centrifugal pump*, liquid flowing into the *suction side* (the *inlet*) is captured by the *impeller* and thrown to the outside of the pump casing. Within the casing, the velocity imparted to the fluid by the impeller is converted into pressure energy. The fluid leaves the pump through the *discharge line* (the *exit*). It is a characteristic of most centrifugal pumps that the fluid is turned approximately 90° from the original flow direction.

Table 18.1 *Generalized Characteristics of Positive Displacement and Kinetic Pumps*

characteristic	positive displacement pumps	kinetic pumps
flow rate	low	high
pressure rise per stage	high	low
constant quantity over operating range	flow rate	pressure rise
self-priming	yes	no
discharge stream	pulsing	steady
works with high viscosity fluids	yes	no

3. POSITIVE DISPLACEMENT RECIPROCATING PUMPS

Reciprocating positive displacement (PD) pumps are used with viscous fluids and slurries (up to about 8000 SSU), when the fluid is sensitive to shear, and when a high discharge pressure is required.[1] By entrapping a volume of liquid in the cylinder, reciprocating pumps provide a fixed-displacement volume per cycle. They are self-priming and inherently leak-free. Within the pressure limits of the line and pressure relief valve and the current capacity of the motor circuit, reciprocating pumps can provide an infinite discharge pressure.[2]

There are three main types of reciprocating pumps: power, direct-acting, and diaphragm. A *power pump* is a *cylinder-operated pump*. It can be single-acting or double-acting. A *single-acting pump* discharges liquid (or takes suction) only on one side of the piston, and there is only one transfer operation per crankshaft revolution. A *double-acting pump* discharges from both sides, and there are two transfers per revolution of the crank.

Traditional reciprocating pumps with pistons and rods can be either single-acting or double-acting and are suitable up to approximately 2000 psi (14 MPa). *Plunger pumps* are only single-acting and are suitable up to approximately 10,000 psi (70 MPa).

Simplex pumps have one cylinder, *duplex pumps* have two cylinders, *triplex pumps* have three cylinders, and so forth. *Direct-acting pumps* (sometimes referred to as *steam pumps*) are always double-acting. They use steam or unburned fuel gas as a motive fluid.

PD pumps are limited by both their NPSHR characteristics, acceleration head, and (for rotary pumps) slip.[3] Because the flow is unsteady, a certain amount of energy, the *acceleration head* (h_{ac}), is required to accelerate the fluid flow each stroke or cycle. If the acceleration head is too large, the NPSHR requirements may not be attainable. Acceleration head can be reduced by increasing the pipe diameter, shortening the suction piping, decreasing the pump speed, or placing a *pulsation damper* (*stabilizer*) in the suction line.[4]

Generally, friction losses with pulsating flow are calculated based on the maximum velocity attained by the fluid. Since this is difficult to determine, the maximum velocity can be approximated by multiplying the average velocity (calculated from the rated capacity) by the factors in Table 18.2.

When the suction line is "short," the acceleration head can be calculated from the length of the suction line, the average velocity in the line, and the rotational speed.[5]

---

[1] For viscosities of Saybolt seconds universal (SSU) > 240, multiply SSU viscosity by 0.216 to get viscosity in centistokes.

[2] For this reason, a relief valve should be included in every installation of positive displacement pumps. Rotary pumps typically have integral relief valves, but external relief valves are often installed to provide easier adjusting, cleaning, and checking.

[3] Manufacturers of PD pumps prefer the term *net positive inlet pressure* (NPIP) to NPSH. NPIPA corresponds to NPSHA; NPIPR corresponds to NPSHR. Pressure and head are related by $p = \gamma h$.

[4] Pulsation dampers are not needed with rotary-action PD pumps, as the discharge is essentially constant.

[5] With a properly designed pulsation damper, the effective length of the suction line is reduced to approximately 10 pipe diameters.

In Eq. 18.1, C and K are dimensionless factors. K represents the relative compressibility of the liquid. (Typical values are 1.4 for hot water; 1.5 for amine, glycol, and cold water; and 2.5 for hot oil.) Values of C are given in Table 18.3.

$$h_{ac} = \left(\frac{C}{K}\right)\left(\frac{L_{suction}v_{ave}n_{rpm}}{g}\right) \qquad \textbf{18.1}$$

Table 18.2 *Typical* v_{max}/v_{ave} *Velocity Ratios*[a,b]

pump type	single-acting	double-acting
simplex	3.2	2.0
duplex	1.6	1.3
triplex	1.1	1.1
quadriplex	1.1	1.1
quintuplex and up	1.05	1.05

[a]Without stabilization. With properly sized stabilizers, use 1.05 to 1.1 for all cases.
[b]Multiply the values by 1.3 for metering pumps where lost fluid motion is relied on for capacity control.

Table 18.3 *Typical Acceleration Head C-Values*[a]

pump type	single-acting	double-acting
simplex	0.4	0.2
duplex	0.2	0.115
triplex	0.066	0.066
quadriplex	0.040	0.040
quintuplex and up	0.028	0.028

[a]Typical values for common connecting rod lengths and crank radii.

4. ROTARY PUMPS

Rotary pumps are *positive displacement* (PD) pumps that move fluid by means of screws, progressing cavities, gears, lobes, or vanes turning within a fixed casing (the *stator*). Rotary pumps are useful for high viscosities (up to 4×10^6 SSU for screw pumps). The rotation creates a cavity of fixed volume near the pump input; atmospheric or external pressure forces the liquid into that cavity. Near the outlet, the cavity is collapsed, forcing the liquid out. Figure 18.1 illustrates the external circumferential piston rotary pump.

Discharge from rotary pumps is relatively smooth. Acceleration head is negligible. Pulsation dampers and suction stabilizers are not required.

Slip in rotary pumps is the amount (sometimes expressed as a percentage) of each rotational fluid volume that "leaks" back to the suction line on each revolution. Slip reduces pump capacity. It is a function of clearance, differential pressure, and viscosity. Slip is proportional

to the third power of the clearance between the rotating element and the casing. Slip decreases with increases in viscosity; it increases linearly with increases in differential pressure. Slip is not affected by rotational speed. The *volumetric efficiency* is defined by Eq. 18.2. Figure 18.2 illustrates the relationship between flow rate, speed, slip, and differential pressure.

$$\eta_v = \frac{Q_{actual}}{Q_{ideal}} = \frac{Q_{ideal} - Q_{slip}}{Q_{ideal}} \qquad \textbf{18.2}$$

Figure 18.1 *External Circumferential Piston Rotary Pump*

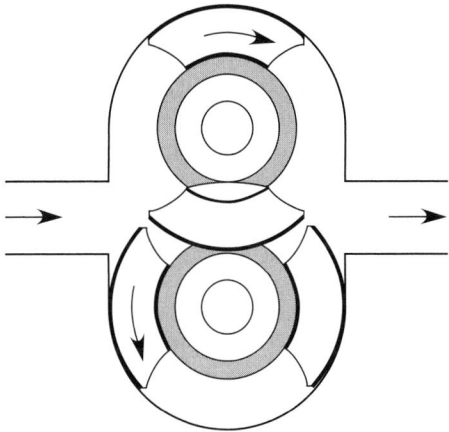

Figure 18.2 *Slip in Rotary Pumps*

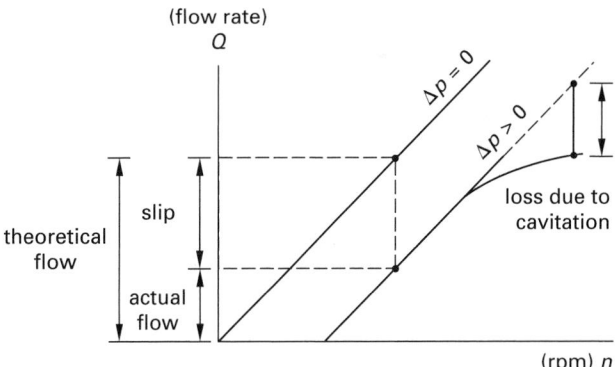

Except for screw pumps, rotary pumps are generally not used for handling abrasive fluids or materials with suspended solids. Few rotary pumps are suitable when variable flow is required.

5. DIAPHRAGM PUMPS

Hydraulically operated *diaphragm pumps* have a diaphragm that completely separates the pumped fluid from the rest of the pump. A reciprocating plunger pressurizes and moves a hydraulic fluid that, in turn, flexes the diaphragm. Single-ball check valves in the suction and discharge lines determine the direction of flow during both phases of the diaphragm action.

Metering is a common application of diaphragm pumps. They have no packing and are essentially leak-proof. This makes them ideal when fugitive emissions are undesirable. Diaphragm pumps are suitable for pumping a wide range of materials, from liquefied gases to coal slurries, though the upper viscosity limit is approximately 3500 SSU. Within the limits of their reactivities, hazardous and reactive materials can also be handled.

Diaphragm pumps are limited by capacity, suction pressure, and discharge pressure and temperature. Because of their construction and size, most diaphragm pumps are limited to discharge pressures of 5000 psi (35 MPa) or less, and most high-capacity pumps are limited to 2000 psi (14 MPa). Suction pressures are similarly limited to 5000 psi (35 MPa). A minimum pressure of 3 to 9 psi (20 to 60 kPa) is often quoted as the minimum liquid-side pressure for metering applications.

The discharge is inherently pulsating, and the dampers or stabilizers are often used. (The acceleration head term is required when calculating NPSHR.) The discharge can be smoothed out somewhat by using two or three (i.e., duplex or triplex) plungers.

Diaphragms are commonly manufactured from stainless steel (type 316) and polytetrafluorethylene (PTFE) or other elastomers. PTFE diaphragms are suitable in the range of $-50°F$ to $300°F$ ($-45°C$ to $150°C$), while metal diaphragms (and some ketone resin diaphragms) are used up to approximately $400°F$ ($200°C$), with life expectancy being reduced at higher temperatures. Although most diaphragm pumps usually operate below 200 spm (strokes per minute), diaphragm life will be improved by limiting the maximum speed to 100 spm.

6. TYPES OF CENTRIFUGAL PUMPS

Centrifugal pumps can be classified according to the way their impellers impart energy to the fluid. Each category of pump is suitable for a different application and (specific) speed range. Figure 18.3 illustrates a typical centrifugal pump and its schematic symbol.

Figure 18.3 *Centrifugal Pump and Symbol*

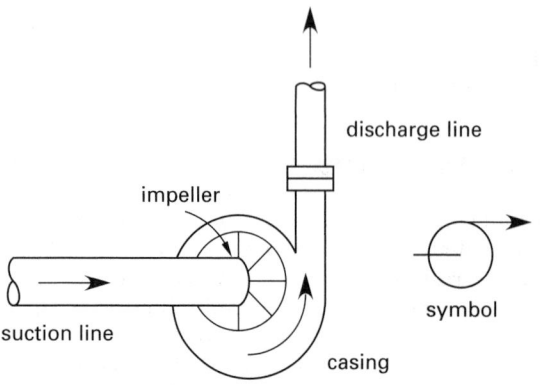

Radial-flow impellers impart energy primarily by centrifugal force. Liquid enters the impeller at the hub and flows radially to the outside of the casing. Radial-flow pumps are suitable for adding high pressure at low fluid flow rates. *Axial-flow impellers* impart energy to the fluid by acting as compressors. Fluid enters and exits along the axis of rotation. Axial-flow pumps are suitable for adding low pressures at high fluid flow rates.[6]

Figure 18.4 *Radial and Axial Flow Impellers*

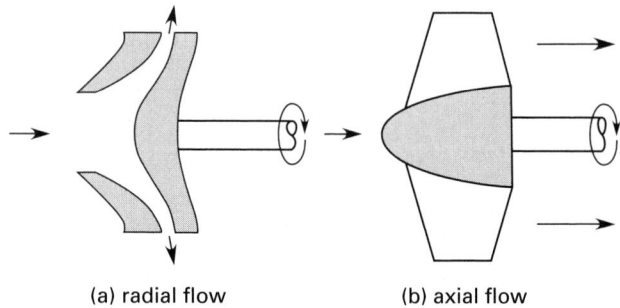

(a) radial flow   (b) axial flow

Radial-flow pumps can be designed for either single- or double-suction operation. In a *single-suction pump*, fluid enters from only one side of the impeller. In a *double-suction pump*, fluid enters from both sides of the impeller.[7] (That is, the impeller is two-sided.) Operation is similar to having two single-suction pumps in parallel.

A *multiple-stage pump* consists of two or more impellers within a single casing. The discharge of one stage feeds the input of the next stage, and operation is similar to having several pumps in series. In this manner, higher heads are achieved than would be possible with a single impeller.

The *circular blade pitch* is the impeller's circumference divided by the number of impeller vanes. The impeller *tip speed* (not to be confused with the specific and suction specific speeds) is easily calculated from the impeller diameter and rotational speed. The impeller "tip speed" is actually the tangential velocity at the periphery. Tip speed is typically somewhat less than 1000 ft/sec (300 m/s).

$$v_{\text{tip}} = \frac{\pi D n}{60 \frac{\text{sec}}{\text{min}}} = \frac{D\omega}{2} \qquad 18.3$$

7. SEWAGE PUMPS

The primary consideration in choosing a pump for sewage and large solids is resistance to clogging. Centrifugal pumps should always be the single-suction type with

---

[6]There is a third category of centrifugal pumps known as *mixed flow pumps*. Mixed flow pumps have operational characteristics between those of radial flow and axial flow pumps.

[7]The double-suction pump can handle a greater fluid flow rate than a single-suction pump with the same specific speed. Also, the double-suction pump will have a lower NPSHR.

nonclog, open impellers. (Double-suction pumps are prone to clogging because rags catch and wrap around the shaft extending through the impeller eye.) Clogging can be further minimized by limiting the number of impeller blades to two or three, providing for large passageways, and using a bar screen ahead of the pump.

Though made of heavy construction, nonclog pumps are constructed for ease of cleaning and repair. Horizontal pumps usually have a split casing, half of which can be removed for maintenance. A hand-sized cleanout opening may also be built into the casing. Although designed for long life, a sewage pump should normally be used with a grit chamber for prolonged bearing life.

The solids-handling capacity of a pump may be specified in terms of the largest sphere that can pass through it without clogging, usually about 80% of the inlet diameter. For example, a wastewater pump with a 6 in (150 mm) inlet should be able to pass a 4 in (100 mm) sphere. The pump must be capable of handling spheres with diameters slightly larger than the bar screen spacing.

Figure 18.5 shows a simplified wastewater pump installation. Not shown are instrumentation and water level measurement devices, baffles, lighting, drains for the dry well, electrical power, pump lubrication equipment, and access ports. (Totally submerged pumps do not require dry wells. However, such pumps without dry wells are more difficult to access, service, and repair.)

Figure 18.5 *Typical Wastewater Pump Installation (greatly simplified)*

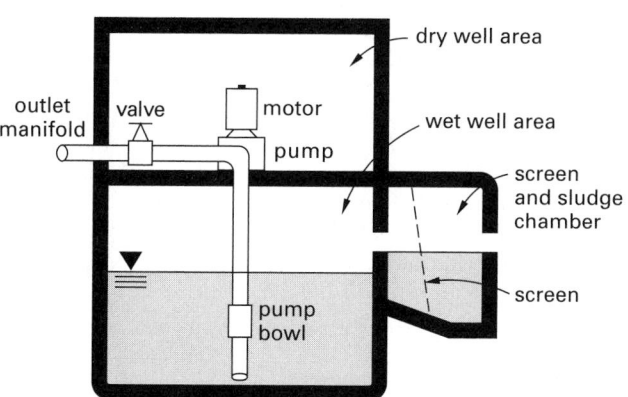

The multiplicity and redundancy of pumping equipment is not apparent from Fig. 18.5. The number of pumps used in a wastewater installation largely depends on the expected demand, pump capacity, and design criteria for backup operation. Although there may be state and federal regulations affecting the design, it is considered good practice to install pumps in sets of two, with a third backup pump being available for each set of pumps that performs the same function. The number of pumps and their capacities should be able to handle the peak flow when one pump in the set is out of service.

8. SLUDGE PUMPS AND GRAVITY FLOW

Centrifugal and reciprocating pumps are extensively used for pumping sludge. Progressive cavity screw impeller pumps are also used.

As described in Sec. 28, the pumping power is proportional to the specific gravity. Accordingly, pumping power for dilute and well-digested sludges is typically only 10 to 25% higher than for water. However, most sludges are non-Newtonian fluids, often flow in a laminar mode, and have characteristics that may change with the season. Also, sludge characteristics change greatly during the pumping cycle. Therefore, engineering judgment and rules of thumb are often important in choosing sludge pumps. For example, a general rule is to choose sludge pumps capable of developing at least 50 to 100% excess head.

One method of determining the required pumping power is to multiply the power required for pumping pure water by a numerical factor. Empirical data is the best method of selecting this factor. Choice of initial values is a matter of judgment. Guidelines are listed in Table 18.4.

Table 18.4 *Pumping Power Multiplicative Factors*

solids concentration	digested sludge	untreated, primary, and concentrated sludge
0%	1.0	1.0
2%	1.2	1.4
4%	1.3	2.5
6%	1.7	4.1
8%	2.2	7.0
10%	3.0	10.0

Derived from *Wastewater Engineering: Treatment, Disposal, Reuse*, 3rd ed., by Metcalf & Eddy, et al., © 1991, with permission from The McGraw-Hill Companies.

Generally, sludge will thin out during a pumping cycle. The most dense sludge components will be pumped first, with more watery sludge appearing at the end of the pumping cycle. With a constant power input, the reduction in load at the end of pumping cycles may cause centrifugal pumps to operate far from the desired operating point and experience overload failures. The operating point should be evaluated with high-, medium-, and low-density sludges.

To avoid cavitation, sludge pumps should always be under a positive suction head of at least 4 ft (1.2 m), and suction lifts should be avoided. The minimum diameters of suction and discharge lines for pumped sludge are typically 6 in (150 mm) and 4 in (100 mm), respectively.

Not all sludge is moved by pump action. Some installations rely on gravity flow to move sludge. The minimum diameter of sludge gravity transfer lines is typically 8 in (200 mm), and the recommended minimum slope is 3%.

To avoid clogging due to settling, flow should be above the transition from laminar to turbulent flow, known as the *critical velocity*. The critical velocity for most sludges is approximately 3.5 ft/sec (1.1 m/s). Velocities of 5 to 8 ft/sec (1.5 to 2.4 m/s) are typical and adequate.

9. TERMINOLOGY OF HYDRAULIC MACHINES

A pump will always have an inlet (designated the *suction*) and an outlet (designated the *discharge*). The subscripts s and d refer to the inlet and outlet of the pump, not of the pipeline.

All of the terms that are discussed in this section are *head* terms and, as such, have units of length. When working with hydraulic machines, it is common to hear such phrases as "a pressure head of 50 feet" and "a static discharge head of 15 meters" The term *head* is often substituted for pressure or pressure drop. Any head term (*pressure head, atmospheric head, vapor pressure head*, etc.) can be calculated from pressure by using Eq. 18.4.[8]

$$h = \frac{p}{\gamma} \qquad\qquad 18.4$$

$$h = \frac{p}{g\rho} \qquad \text{[SI]} \qquad 18.5(a)$$

$$h = \left(\frac{p}{\rho}\right) \times \left(\frac{g_c}{g}\right) \qquad \text{[U.S.]} \qquad 18.5(b)$$

Some of the terms used in the description of pipelines appear to be similar and may be initially confusing (e.g., suction head and total suction head). The following general rules will help to clarify the meanings.[9]

Rule 18.1: The word *suction* or *discharge* limits the quantity to the suction line or discharge line, respectively. The absence of either word implies that both the suction and discharge lines are included. Example: discharge head.

Rule 18.2: The word *static* means that static head only is included (not velocity head, friction head, etc.). Example: static suction head.

Rule 18.3: The word *total* means that static head, velocity head, and friction head are all included. (Note that total does not mean the combination of suction and discharge.) Example: total suction head.

The following terms are commonly encountered.

---

[8]Equation 18.5 can be used to define *pressure head, atmospheric head*, and *vapor pressure head*, whose meanings and derivations should be obvious.
[9]The term *dynamic* is not as consistently applied as are the terms described in the rules. In particular, it is not clear whether friction head is to be included with the velocity head.

- *friction head* (h_f): The head required to overcome resistance to flow in the pipes, fittings, valves, entrances, and exits.

$$h_f = \frac{fLv^2}{2Dg} \qquad\qquad 18.6$$

- *velocity head* (h_v): The specific kinetic energy of the fluid. Also known as *dynamic head*.

$$h_v = \frac{v^2}{2g} \qquad\qquad 18.7$$

- *static suction head* ($h_{z(s)}$): The vertical distance above the centerline of the pump inlet to the free level of the fluid source. If the free level of the fluid is below the pump inlet, h_z will be negative and is known as *static suction lift*.

Figure 18.6 Static Suction Lift

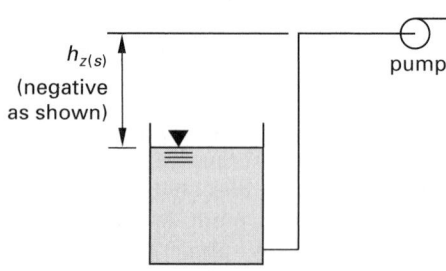

- *static discharge head* ($h_{z(d)}$): The vertical distance above the centerline of the pump inlet to the point of free discharge or surface level of the discharge tank.

The ambiguous term *effective head* is not commonly used when discussing hydraulic machines, but when used, the term most closely means *net head* (i.e., starting head less losses). Consider a hydroelectric turbine that is fed by water with a static head of H. After frictional and other losses, the net head acting on the turbine will be less than H. The turbine output will coincide with an ideal turbine being acted upon by the net or effective head. Similarly, the actual increase in pressure across a pump will be the effective head (i.e., the head net of internal losses and geometric effects).

Figure 18.7 Static Discharge Head

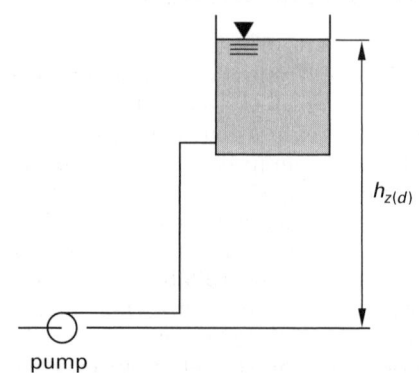

Example 18.1

Write the symbolic equations for the following terms: (a) total suction head, (b) total discharge head, and (c) total head.

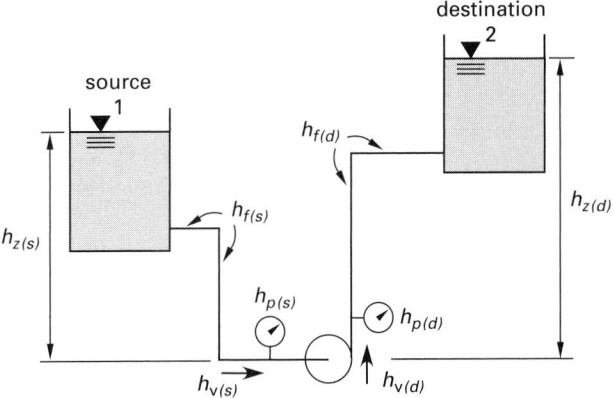

Solution

(a) The total suction head at the pump inlet is the sum of static (pressure) head and velocity head at the pump suction.

$$h_{t(s)} = h_{p(s)} + h_{v(s)}$$

Total suction head can also be calculated from the conditions existing at the source (1), in which case suction line friction would also be considered. (With an open reservoir, $h_{p(1)}$ will be zero if gage pressures are used, and $h_{v(1)}$ will be zero if the source is large.)

$$h_{t(s)} = h_{p(1)} + h_{z(s)} + h_{v(1)} - h_{f(s)}$$

(b) The total discharge head at the pump outlet is the sum of the static (pressure) and velocity heads at the pump outlet. Friction head is zero since the fluid has not yet traveled through any length of pipe when it is discharged.

$$h_{t(d)} = h_{p(d)} + h_{v(d)}$$

The total discharge head can also be evaluated at the destination (2) if the friction head, $h_{f(d)}$, between the discharge and the destination is known. (With an open reservoir, $h_{p(2)}$ will be zero if gage pressures are used, and $h_{v(2)}$ will be zero if the destination is large.)

$$h_{t(d)} = h_{p(2)} + h_{z(d)} + h_{v(2)} + h_{f(d)}$$

(c) The total head added by the pump is the total discharge head less the total suction head. Assuming suction from and discharge to reservoirs exposed to the atmosphere and assuming negligible reservoir velocities,

$$h_t = h_A = h_{t(d)} - h_{t(s)} \approx h_{z(d)} - h_{z(s)} + h_{f(d)} + h_{f(s)}$$

10. PUMPING POWER

The energy (head) added by a pump can be determined from the difference in total energy on either side of the pump. Writing the Bernoulli equation for the discharge and suction conditions produces Eq. 18.8, an equation for the *total dynamic head*, often abbreviated TDH.

$$h_A = h_t = h_{t(d)} - h_{t(s)} \qquad \text{18.8}$$

$$h_A = \frac{p_d - p_s}{\rho g} + \frac{v_d^2 - v_s^2}{2g} + z_d - z_s \qquad \text{[SI]} \quad \text{18.9(a)}$$

$$h_A = \frac{(p_d - p_s)g_c}{\rho g} + \frac{v_d^2 - v_s^2}{2g} + z_d - z_s \qquad \text{[U.S.]} \quad \text{18.9(b)}$$

In most applications, the change in velocity and potential heads is either zero or small in comparison to the increase in pressure head. Equation 18.9 then reduces to Eq. 18.10.

$$h_A = \frac{p_d - p_s}{\rho g} \qquad \text{[SI]} \quad \text{18.10(a)}$$

$$h_A = \left(\frac{p_d - p_s}{\rho}\right) \times \left(\frac{g_c}{g}\right) \qquad \text{[U.S.]} \quad \text{18.10(b)}$$

It is important to recognize that the variables in Eqs. 18.9 and 18.10 refer to the conditions at the pump's immediate inlet and discharge, not to the distant ends of the suction and discharge lines. However, the total dynamic head added by a pump can be calculated in another way. For example, for a pump raising water from one open reservoir to another, the total dynamic head would consist of the total elevation rise, the velocity head (often negligible), and the friction losses in the suction and discharge lines.

The head added by the pump can also be calculated from the impeller and fluid speeds. Equation 18.11 is useful for radial- and mixed-flow pumps for which the incoming fluid has little or no rotational velocity component (i.e., up to a specific speed of approximately 2000 U.S. or 40 SI). In Eq. 18.11, $v_{impeller}$ is the tangential impeller velocity at the radius being considered, and v_{fluid} is the average tangential velocity imparted to the fluid by the impeller. The impeller efficiency, $\eta_{impeller}$, is typically 0.85 to 0.95. This is much higher than the total pump efficiency (Sec. 11) because it does not include mechanical and fluid friction losses.

$$h_A = \frac{\eta_{impeller} v_{impeller} v_{fluid}}{g} \qquad \text{18.11}$$

The pumping power depends on the head added (h_A) and the mass flow rate. For example, the product $\dot{m} h_A$ has the units of foot-pounds per second (in customary U.S. units), which can be easily converted to horsepower. Pump output power is known as *hydraulic power* or *water power*. Hydraulic power is the net power actually transferred to the fluid.

Horsepower is the unit of power in the United States and other non-SI countries, which gives rise to the terms *hydraulic horsepower* and *water horsepower*, WHP. Various relationships for finding the hydraulic horsepower are given in Table 18.5.

The unit of power in SI units is the watt (kilowatt). Table 18.6 can be used to determine *hydraulic kilowatts*, WkW.

Table 18.5 *Hydraulic Horsepower Equations*[a]

	Q (gal/min)	$\dot{m}$ (lbm/sec)	$\dot{V}$ (ft^3/sec)
h_A in feet	$\dfrac{h_A Q(SG)}{3956}$	$\left(\dfrac{h_A \dot{m}}{550}\right)\times\left(\dfrac{g}{g_c}\right)$	$\dfrac{h_A \dot{V}(SG)}{8.814}$
Δp in psi[b]	$\dfrac{\Delta p Q}{1714}$	$\left(\dfrac{\Delta p \dot{m}}{(238.3)(SG)}\right)\times\left(\dfrac{g}{g_c}\right)$	$\dfrac{\Delta p \dot{V}}{3.819}$
Δp in psf[b]	$\dfrac{\Delta p Q}{2.468\times 10^5}$	$\left(\dfrac{\Delta p \dot{m}}{(34,320)(SG)}\right)\times\left(\dfrac{g}{g_c}\right)$	$\dfrac{\Delta p \dot{V}}{550}$
W in $\dfrac{\text{ft-lbf}}{\text{lbm}}$	$\dfrac{W Q(SG)}{3956}$	$\dfrac{W \dot{m}}{550}$	$\dfrac{W \dot{V}(SG)}{8.814}$

(Multiply horsepower by 0.7457 to obtain kilowatts.)
[a]Table 18.5 is based on $\rho_{\text{water}} = 62.4\,\text{lbm/ft}^3$ and $g = 32.2\,\text{ft/sec}^2$.
[b]Velocity head changes must be included in Δp.

Table 18.6 *Hydraulic Kilowatt Equations*[a]

	Q (L/s)	$\dot{m}$ (kg/s)	$\dot{V}$ (m^3/s)
h_A in meters	$\dfrac{(9.81)h_A Q(SG)}{1000}$	$\dfrac{(9.81)h_A \dot{m}}{1000}$	$(9.81)h_A \dot{V}(SG)$
Δp in kPa[b]	$\dfrac{\Delta p Q}{1000}$	$\dfrac{\Delta p \dot{m}}{1000(SG)}$	$\Delta p \dot{V}$
W in $\dfrac{\text{J}}{\text{kg}}$[b]	$\dfrac{W Q(SG)}{1000}$	$\dfrac{W \dot{m}}{1000}$	$W \dot{V}(SG)$

(Multiply kilowatts by 1.341 to obtain horsepower.)
[a]Table 18.6 is based on $\rho_{\text{water}} = 1000\,\text{kg/m}^3$ and $g = 9.81\,\text{m/s}^2$.
[b]Velocity head changes must be included in Δp.

Example 18.2

A pump adds 550 ft of pressure head to 100 lbm/sec of water. (a) Complete the following table of performance data. (b) What is the hydraulic power in horsepower and kilowatts? (Assume $\rho = 62.4\,\text{lbm/ft}^3$ or 1000 kg/m^3, and $g = 9.81\,\text{m/s}^2$.)

item	customary U.S.	SI
$\dot{m}$	100 lbm/sec	___ kg/s
h	550 ft	___ m
Δp	___ lbf/ft^2	___ kPa
$\dot{V}$	___ ft^3/sec	___ m^3/s
W	___ ft-lbf/lbm	___ J/kg
P	___ hp	___ kW

Solution

(a) Work initially with the customary U.S. data.

$$\Delta p = \rho h \times \left(\frac{g}{g_c}\right) = \left(62.4\,\frac{\text{lbm}}{\text{ft}^3}\right)(550\,\text{ft})\times\left(\frac{g}{g_c}\right)$$
$$= 34{,}320\,\text{lbf/ft}^2$$

$$\dot{V} = \frac{\dot{m}}{\rho} = \frac{100\,\dfrac{\text{lbm}}{\text{sec}}}{62.4\,\dfrac{\text{lbm}}{\text{ft}^3}} = 1.603\,\text{ft}^3/\text{sec}$$

$$W = h \times \left(\frac{g}{g_c}\right) = 550\,\text{ft-lbf/lbm}$$

Now, convert to SI units.

$$\dot{m} = \frac{100\,\dfrac{\text{lbm}}{\text{sec}}}{2.201\,\dfrac{\text{lbm}}{\text{kg}}} = 45.43\,\text{kg/s}$$

$$h = \frac{550\,\text{ft}}{3.281\,\dfrac{\text{ft}}{\text{m}}} = 167.6\,\text{m}$$

$$\Delta p = \left(34{,}320\,\frac{\text{lbf}}{\text{ft}^2}\right)\left(\frac{1}{144\,\dfrac{\text{in}^2}{\text{ft}^2}}\right)\left(6.895\,\frac{\text{kPa}}{\dfrac{\text{lbf}}{\text{in}^2}}\right)$$
$$= 1643\,\text{kPa}$$

$$\dot{V} = \left(1.603\,\frac{\text{ft}^3}{\text{sec}}\right)\left(0.0283\,\frac{\text{m}^3}{\text{ft}^3}\right) = 0.0454\,\text{m}^3/\text{s}$$

$$W = \left(550\,\frac{\text{ft-lbf}}{\text{lbm}}\right)\left(1.356\,\frac{\text{J}}{\text{ft-lbf}}\right)\left(2.201\,\frac{\text{lbm}}{\text{kg}}\right)$$
$$= 1642\,\text{J/kg}$$

(b) From Table 18.5, the hydraulic horsepower is

$$\text{WHP} = \frac{h_A \dot{m}}{550}\times\left(\frac{g}{g_c}\right)$$
$$= \left(\frac{(550\,\text{ft})\left(100\,\dfrac{\text{lbm}}{\text{sec}}\right)}{550\,\dfrac{\text{ft-lbf}}{\text{hp-sec}}}\right)\times\left(\frac{g}{g_c}\right)$$
$$= 100\,\text{hp}$$

From Table 18.6, the power is

$$\text{WkW} = \frac{\Delta p \dot{m}}{(1000)(SG)} = \frac{(1643\,\text{kPa})\left(45.43\,\dfrac{\text{kg}}{\text{s}}\right)}{\left(1000\,\dfrac{\text{W}}{\text{kW}}\right)(1.0)}$$
$$= 74.6\,\text{kW}$$

11. PUMPING EFFICIENCY

Hydraulic power is the net energy actually transferred to the fluid per unit time. The input power delivered by the motor to the pump is known as the *brake pump power*. (For example, the term *brake horsepower*, bhp, is commonly used.) Due to frictional losses between the fluid and the pump and mechanical losses in the pump itself, the brake pump power will be greater than the hydraulic power.

The ratio of hydraulic power to brake pump power is the pump efficiency, η_p. (Figure 18.8 gives typical pump efficiencies as a function of the pump's specific speed. The difference between the brake and hydraulic powers is known as the *friction power* (or *friction horsepower*).

Figure 18.8 *Average Pump Efficiency Versus Specific Speed*

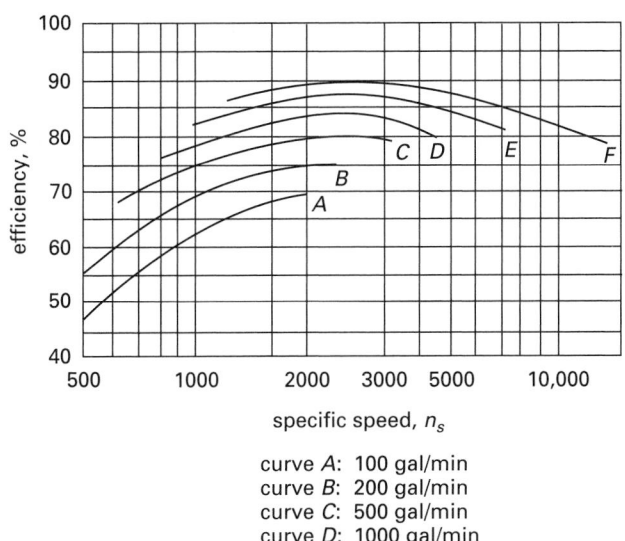

curve *A*:   100 gal/min
curve *B*:   200 gal/min
curve *C*:   500 gal/min
curve *D*: 1000 gal/min
curve *E*: 3000 gal/min
curve *F*: 10,000 gal/min

$$\text{brake pump power} = \frac{\text{hydraulic power}}{\eta_p} \qquad 18.12$$

$$\begin{matrix}\text{friction}\\\text{power}\end{matrix} = \begin{matrix}\text{brake}\\\text{pump}\\\text{power}\end{matrix} - \begin{matrix}\text{hydraulic}\\\text{power}\end{matrix} \qquad 18.13$$

Pumping efficiency is not constant for any specific pump; rather, it depends on the operating point (Sec. 23) and the speed of the pump. A specific pump's characteristic efficiency curves will be published by its manufacturer.

With pump characteristic curves given by the manufacturer, the efficiency is not determined from the intersection of the system curve (Sec. 22) and the efficiency curve. Rather, the efficiency is a function of only the flow rate. Therefore, the operating efficiency is read from the efficiency curve directly above or below the operating point (Sec. 23).

Figure 18.9 *Typical Centrifugal Pump Efficiency Curves*

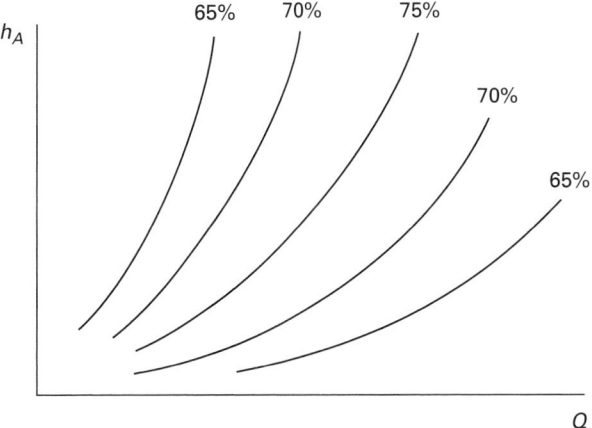

Unfortunately, efficiency curves published by a manufacturer may not be representative of the actual installed efficiency. The manufacturer's efficiency may not include such losses in the suction elbow, discharge diffuser, couplings, bearing frame, seals, or pillow blocks. Up to 15% of the motor horsepower may be lost to these factors. Therefore, the manufacturer should be requested to provide the pump's installed *wire-to-water efficiency* (i.e., the fraction of the electrical power drawn that is converted to hydraulic power).

The pump must be driven by an engine or motor.[10] The power delivered to the motor is greater than the power delivered to the pump, as accounted for by the motor efficiency, η_m. If the pump motor is electrical, its input power requirements will be stated in kilowatts.[11]

$$\begin{matrix}\text{motor}\\\text{input}\\\text{power}\end{matrix} = \frac{\text{brake pump power}}{\eta_m} \qquad 18.14$$

$$\begin{pmatrix}\text{motor}\\\text{input}\\\text{power}\end{pmatrix}_{kW} = \frac{0.7457 \times (\text{brake pump power})_{hp}}{\eta_m} \qquad 18.15$$

The *overall efficiency* of the pump installation is the product of the pump and motor efficiencies.

$$\eta = \eta_p \eta_m = \frac{\text{hydraulic horsepower}}{\text{motor horsepower}} \qquad 18.16$$

12. COST OF ELECTRICITY

The power utilization of pump motors is usually measured in kilowatts. The kilowatt usage represents the rate that energy is transferred by the pump motor. The total amount of work, W, done by the pump motor is found by multiplying the rate of energy usage by the length of time the pump is in operation.

$$W = Pt \qquad 18.17$$

---

[10]The source of power is sometimes called the *prime mover*.
[11]A *watt* is a joule per second.

Although the units horsepower-hours are occasionally encountered, it is more common to measure electrical work in *kilowatt-hours* (kW-hr). Accordingly, the cost of electrical energy is stated per kW-hr (e.g., $0.10 per kW-hr).

$$\text{cost} = \frac{W_{\text{kW-hr}} \times \text{cost per kW-hr}}{\eta_m} \qquad \textit{18.18}$$

13. STANDARD MOTOR SIZES AND SPEEDS

An effort should be made to specify standard motor sizes when selecting the source of pumping power. Table 18.7 lists NEMA (National Electrical Manufacturers Association) standard motor sizes by horsepower *nameplate rating*.[12] The rated horsepower is the maximum power the motor can provide without incurring damage. Other motor sizes may also be available by special order.

Table 18.7 *NEMA Standard Motor Sizes (brake horsepower)*

$\frac{1}{8}$,	$\frac{1}{6}$,	$\frac{1}{4}$,	$\frac{1}{3}$				
0.5,	0.75,	1,	1.5,	2,	3,	5,	7.5
10,	15,	20,	25,	30,	40,	50,	60
75,	100,	125,	150,	200,	250,	300,	350
400,	450,	500,	600,	700,	800,	900,	1000
1250,	1500,	1750,	2000,	2250,	2500,	2750,	3000
3500,	4000,	4500,	5000				

Larger horsepower motors are usually three-phase induction motors. The *synchronous speed, n* in rpm, of such motors is the speed of the rotating field, which depends on the number of poles per stator phase and the frequency, f. The number of poles must be an even number. The frequency is typically 60 Hz (hertz, previously cycles per second), as in the United States, or 50 Hz, as in European countries. Table 18.8 lists common synchronous speeds.

$$n = \frac{120 \times f}{\text{no. of poles}} \qquad \textit{18.19}$$

Table 18.8 *Common Synchronous Speeds*

number of poles	n (rpm) 60 Hz	50 Hz
2	3600	3000
4	1800	1500
6	1200	1000
8	900	750
10	720	600
12	600	500
14	514	428
18	400	333
24	300	250
48	150	125

[12]The nameplate rating gets its name from the information stamped on the motor's identification plate. Besides the horsepower rating, other nameplate data used to classify the motor are the service class, voltage, full-load current, speed, number of phases, frequency of the current, and maximum ambient temperature (or, for older motors, the motor temperature rise).

Induction motors do not run at their synchronous speeds when loaded. Rather, they run at slightly less than synchronous speed. The deviation is known as the *slip*. Slip is typically around 4% and is seldom greater than 10% at full load for motors in the 1 to 75 hp range.

$$\begin{array}{l}\text{slip} \\ \text{(in rpm)}\end{array} = \text{synchronous speed} - \text{actual speed} \qquad \textit{18.20}$$

$$\begin{array}{l}\text{percent} \\ \text{slip}\end{array} = 100\% \times \frac{\begin{array}{c}\text{synchronous speed} \\ - \text{ actual speed}\end{array}}{\text{synchronous speed}} \qquad \textit{18.21}$$

Induction motors may also be specified in terms of their kVA (kilovolt-amp) ratings. The kVA rating is not the same as the power in kilowatts, although one can be derived from the other if the motor's power factor is known. Such power factors typically range from 0.8 to 0.9, depending on the installation and motor size.

$$\text{kVA rating} = \frac{\text{motor power in kW}}{\text{power factor}} \qquad \textit{18.22}$$

Example 18.3

A pump driven by an electrical motor moves 25 gal/min of water from reservoir A to reservoir B, lifting the water a total of 245 ft. The efficiencies of the pump and motor are 64% and 84%, respectively. Electricity costs $0.08/kW-hr. Neglect velocity head, friction, and minor losses. (a) What size motor is required? (b) How much does it cost to operate the pump for 6 hr?

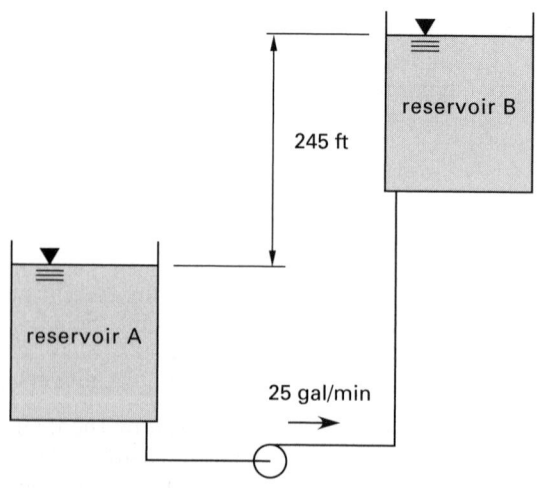

Solution

(a) The head added is 245 ft.

From Table 18.5, the motor power required is

$$P = \frac{h_A \, Q(\text{SG})}{3956\eta_p} = \frac{(245 \text{ ft}) \left(25 \, \frac{\text{gal}}{\text{min}} \right) (1.0)}{(3956)(0.64)}$$

$$= 2.42 \text{ hp}$$

From Table 18.7, select a 3 hp motor.

(b) From Eqs. 18.17 and 18.18,

$$\text{cost} = \text{cost per kW-hr} \times \left(\frac{Pt}{\eta_m} \right)$$

$$= \frac{\left(0.08 \, \frac{\$}{\text{kW-hr}} \right) \left(0.7457 \, \frac{\text{kW}}{\text{hp}} \right) (2.42 \text{ hp}) \, (6 \text{ hr})}{0.84}$$

$$= \$1.03$$

Notice that the developed power, not the motor's rated power, is used.

14. PUMP SHAFT LOADING

The torque on a pump or motor shaft, brake power, and speed are all related. The power used in Eqs. 18.23 through 18.25 can be either the brake (shaft) power or the hydraulic power developed.

$$T_{\text{in-lbf}} = \frac{63{,}025 \times P_{\text{hp}}}{n} \qquad \textit{18.23}$$

$$T_{\text{ft-lbf}} = \frac{5252 \times P_{\text{hp}}}{n} \qquad \textit{18.24}$$

$$T_{\text{N·m}} = \frac{9549 \times P_{\text{kW}}}{n} \qquad \textit{18.25}$$

The actual (developed) torque can be calculated from the change in momentum of the fluid flow. For radial impellers, the fluid enters through the eye and is turned $90°$. The direction change is related to the increase in momentum and the shaft torque. When fluid is introduced axially through the eye of the impeller, the tangential velocity at the inlet (eye), $v_{t,i}$, is zero.[13]

$$T_{\text{actual}} = \left(\frac{\dot{m}}{g_c} \right) \left(v_{t(d)} r_{\text{impeller}} - v_{t(i)} r_{\text{eye}} \right) \qquad \textit{18.26}$$

Centrifugal pumps may be driven directly from a motor, or a speed changer may be used. Rotary pumps generally require a speed reduction. *Gear motors* have integral speed reducers. V-belt drives are widely used because of their initial low cost, although timing belts and chains can be used in some applications.

When a belt or chain is used, the pump's and motor's maximum overhung loads must be checked. This is particularly important for high-power, low-speed applications (such as rotary pumps). *Overhung load* is the side

---

[13]The tangential component of fluid velocity is sometimes referred to as the *velocity of whirl*.

load (force) put on shafts and bearings. The overhung load is calculated from Eq. 18.27. The empirical factor K is 1.0 for chain drives, 1.25 for timing belts, and 1.5 for V-belts.

$$\text{overhung load} = \frac{2KT}{D_{\text{sheave}}} \qquad \textit{18.27}$$

If a direct drive cannot be used and the overhung load is excessive, the installation can incorporate a jack shaft or outboard bearing.

Example 18.4

A centrifugal pump delivers 275 lbm/sec (125 kg/s) of water while turning at 850 rpm. The impeller has straight radial vanes and an outside diameter of 10 in (25.4 cm). Water enters the impeller through the eye. The driving motor delivers 30 hp (22 kW). What are the (a) theoretical torque, (b) pump efficiency, and (c) total developed dynamic head?

SI Solution

(a) From Eq. 18.3, the impeller's tangential velocity is

$$v_t = \frac{\pi D n}{60 \, \frac{\text{s}}{\text{min}}} = \frac{\pi (25.4 \text{ cm})(850 \text{ rpm})}{\left(60 \, \frac{\text{s}}{\text{min}} \right) \left(100 \, \frac{\text{cm}}{\text{m}} \right)}$$

$$= 11.3 \text{ m/s}$$

Since water enters axially, the incoming water has no tangential component. From Eq. 18.26, the developed torque is

$$T = \dot{m} v_{t(d)} r_{\text{impeller}}$$

$$= \frac{\left(125 \, \frac{\text{kg}}{\text{s}} \right) \left(11.3 \, \frac{\text{m}}{\text{s}} \right) \left(\frac{25.4 \text{ cm}}{2} \right)}{100 \, \frac{\text{cm}}{\text{m}}}$$

$$= 179.4 \text{ N·m}$$

(b) From Eq. 18.25, the developed power is

$$P_{\text{kW}} = \frac{n T_{\text{N·m}}}{9549} = \frac{(850 \text{ rpm})(179.4 \text{ N·m})}{9549}$$

$$= 15.97 \text{ kW}$$

The pump efficiency is

$$\eta_p = \frac{P_{\text{developed}}}{P_{\text{input}}}$$

$$= \frac{15.97 \text{ kW}}{22 \text{ kW}} = 0.726 \quad (72.6\%)$$

(c) From Table 18.6, the total dynamic head is

$$h_A = \frac{P_{\text{kW}}\left(1000\ \dfrac{\text{W}}{\text{kW}}\right)}{\dot{m}\left(9.81\ \dfrac{\text{m}}{\text{s}^2}\right)}$$

$$= \frac{(15.97\ \text{kW})\left(1000\ \dfrac{\text{W}}{\text{kW}}\right)}{\left(125\ \dfrac{\text{kg}}{\text{s}}\right)\left(9.81\ \dfrac{\text{m}}{\text{s}^2}\right)}$$

$$= 13.0\ \text{m}$$

Customary U.S. Solution

(a) From Eq. 18.3, the impeller's tangential velocity is

$$v_t = \frac{\pi D n}{60\ \dfrac{\text{sec}}{\text{min}}} = \frac{\pi(10\ \text{in})(850\ \text{rpm})}{\left(60\ \dfrac{\text{sec}}{\text{min}}\right)\left(12\ \dfrac{\text{in}}{\text{ft}}\right)}$$

$$= 37.08\ \text{ft/sec}$$

Since water enters axially, the incoming water has no tangential component. From Eq. 18.26, the developed torque is

$$T = \frac{\dot{m}\, v_{t(d)}\, r_{\text{impeller}}}{g_c}$$

$$= \frac{\left(275\ \dfrac{\text{lbm}}{\text{sec}}\right)\left(37.08\ \dfrac{\text{ft}}{\text{sec}}\right)\left(\dfrac{10\ \text{in}}{2}\right)}{\left(32.2\ \dfrac{\text{lbm-ft}}{\text{lbf-sec}^2}\right)\left(12\ \dfrac{\text{in}}{\text{ft}}\right)}$$

$$= 131.9\ \text{ft-lbf}$$

(b) From Eq. 18.24, the developed power is

$$P_{\text{hp}} = \frac{T_{\text{ft-lbf}}\, n}{5252}$$

$$= \frac{(131.9\ \text{ft-lbf})(850\ \text{rpm})}{5252} = 21.35\ \text{hp}$$

The pump efficiency is

$$\eta_p = \frac{P_{\text{developed}}}{P_{\text{input}}}$$

$$= \frac{21.35\ \text{hp}}{30\ \text{hp}} = 0.712 \quad (71.2\%)$$

(c) From Table 18.5, the total dynamic head is

$$h_A = \left(\frac{\left(550\ \dfrac{\text{ft-lbf}}{\text{hp-sec}}\right)(P_{\text{hp}})}{\dot{m}}\right) \times \left(\frac{g_c}{g}\right)$$

$$= \left(\frac{\left(550\ \dfrac{\text{ft-lbf}}{\text{hp-sec}}\right)(21.35\ \text{hp})}{275\ \dfrac{\text{lbm}}{\text{sec}}}\right)$$

$$\times \left(\frac{32.2\ \dfrac{\text{lbm-ft}}{\text{lbf-sec}^2}}{32.2\ \dfrac{\text{ft}}{\text{sec}^2}}\right)$$

$$= 42.7\ \text{ft}$$

15. SPECIFIC SPEED

The capacity and efficiency of a centrifugal pump are partially governed by the impeller design. For a desired flow rate and added head, there will be one optimum impeller design. The quantitative index used to optimize the impeller design is known as *specific speed*, n_s, also known as *impeller specific speed*. Table 18.9 lists the impeller designs that are appropriate for different specific speeds.[14]

Table 18.9 Specific Speed versus Impeller Design

approximate range of specific speed (rpm)		
customary U.S. units	SI units	impeller type
500 to 1000	10 to 20	radial vane
2000 to 3000	40 to 60	Francis (mixed) vane
4000 to 7000	80 to 140	mixed flow
9000 and above	180 and above	axial flow

Highest heads per stage are developed at low specific speeds. However, for best efficiency, specific speed should be greater than 650 (13 in SI units). If the specific speed for a given set of conditions drops below 650 (13), a multiple-stage pump should be selected.[15]

---

[14]Specific speed is useful for more than just selecting an impeller type. Maximum suction lift, pump efficiency, and net positive suction head required (NPSHR) can be correlated with specific speed.

[15]Advances in *partial emission, forced vortex centrifugal pumps* allow operation down to specific speeds of 150 (3 in SI). Such pumps have been used for low-flow, high-head applications, such as high-pressure petrochemical cracking processes.

Specific speed is a function of a pump's capacity, head, and rotational speed at peak efficiency, as shown in Eq. 18.28. For a given pump and impeller configuration, the specific speed remains essentially constant over a range of flow rates and heads.

While specific speed is not dimensionless, the units are meaningless. Specific speed may be assigned units of rpm, but most often it is expressed simply as a pure number. (Q or $\dot{V}$ in Eq. 18.28 is half of the full flow rate for double-suction pumps.)

$$n_s = \frac{n\sqrt{\dot{V}}}{(h_A)^{0.75}} \qquad \text{[SI]} \qquad \textit{18.28(a)}$$

$$n_s = \frac{n\sqrt{Q}}{(h_A)^{0.75}} \qquad \text{[U.S.]} \qquad \textit{18.28(b)}$$

The numerical range of acceptable performance for each impeller type is redefined when SI units are used. The SI specific speed is obtained by dividing the customary U.S. specific speed by 51.66.

A common definition of specific speed is the speed (in rpm) at which a *homologous pump* would have to turn in order to deliver one gallon per minute at one foot total added head.[16] This definition is implicit to Eq. 18.28 but is not very useful otherwise.

Specific speed can be used to determine the type of impeller needed. Once a pump is selected, its specific speed and Eq. 18.28 can be used to determine other operational parameters (e.g., maximum rotational speed). Specific speed can be used with Fig. 18.8 to obtain an approximate pump efficiency.

Example 18.5

A centrifugal pump powered by a direct-drive induction motor is needed to discharge 150 gal/min against a 300 ft total head when turning at the fully loaded speed of 3500 rpm. What type of pump should be selected?

Solution

From Eq. 18.28, the specific speed is

$$n_s = \frac{n\sqrt{Q}}{(h_A)^{0.75}} = \frac{3500\sqrt{150}}{(300)^{0.75}} = 595$$

---

[16]*Homologous pumps* are geometrically similar. This means that each pump is a scaled up or down version of the others. Such pumps are said to belong to a *homologous family*.

From Table 18.9, the pump should be a radial vane type. However, pumps achieve their highest efficiencies when specific speed exceeds 650. (See Fig. 18.8.) To increase the specific speed, the rotational speed can be increased, or the total added head can be decreased. Since the pump is direct-driven and 3600 rpm is the maximum speed for induction motors (see Table 18.8), the total added head should be divided evenly between two stages, or two pumps should be used in series.

In a two-stage system, the specific speed would be

$$n_s = \frac{3500\sqrt{150}}{(150)^{0.75}} = 1000$$

This is satisfactory for a radial vane pump.

Example 18.6

An induction motor turning at 1200 rpm is to be selected to drive a single-stage, single-suction centrifugal water pump through a direct drive. The total dynamic head added by the pump is 26 ft. The flow rate is 900 gal/min. What size motor should be selected?

Solution

The specific speed is

$$n_s = \frac{n\sqrt{Q}}{(h_A)^{0.75}} = \frac{1200\sqrt{900}}{(26)^{0.75}}$$
$$= 3127$$

From Fig. 18.8, the pump efficiency will be approximately 82%.

From Table 18.5, the minimum motor horsepower is

$$P_{hp} = \frac{h_A Q(\text{SG})}{\left(3956 \dfrac{\text{ft-gal}}{\text{hp-min}}\right)\eta_p}$$

$$= \frac{(26 \text{ ft})\left(900 \dfrac{\text{gal}}{\text{min}}\right)(1.0)}{\left(3956 \dfrac{\text{ft-gal}}{\text{hp-min}}\right)(0.82)}$$

$$= 7.2 \text{ hp}$$

From Table 18.6, select a 7.5 hp motor.

Example 18.7

A single-stage pump driven by a 3600 rpm motor is currently delivering 150 gal/min. The total dynamic head is 430 ft. What would be the approximate increase in efficiency per stage if the single-stage pump is replaced by a double-stage pump?

Solution

The specific speed is

$$n_s = \frac{n\sqrt{Q}}{(h_A)^{0.75}} = \frac{3600\sqrt{150}}{(430)^{0.75}}$$
$$= 467$$

From Fig. 18.8, the approximate efficiency is 45%.

In a two-stage pump, each stage adds half of the head. The specific speed per stage would be

$$n_s = \frac{n\sqrt{Q}}{(h_A)^{0.75}} = \frac{3600\sqrt{150}}{\left(\dfrac{430}{2}\right)^{0.75}}$$

$$= 785$$

From Fig. 18.8, the efficiency for this configuration is approximately 60%.

The increase in stage efficiency is $60\% - 45\% = 15\%$. Whether or not the cost of multistaging is worthwhile in this low-volume application would have to be determined. The overall efficiency of the two-stage pump may actually be lower than the efficiency of the one-stage pump.

16. CAVITATION

Cavitation is the spontaneous vaporization of the fluid, resulting in a degradation of pump performance. If the fluid pressure is less than the vapor pressure, small pockets of vapor will form. These pockets usually form only within the pump itself, although cavitation slightly upstream within the suction line is also possible. As the vapor pockets reach the surface of the impeller, the local high fluid pressure collapses them. Noise, vibration, impeller pitting, and structural damage to the pump casing are manifestations of cavitation.

Cavitation can be caused by any of the following conditions.

- discharge head far below the pump head at peak efficiency
- high suction lift or low suction head
- excessive pump speed
- high liquid temperature (i.e., high vapor pressure)

17. NET POSITIVE SUCTION HEAD

The occurrence of cavitation is predictable. Cavitation will occur when the net pressure in the fluid drops below the vapor pressure. This criterion is commonly stated in terms of head: Cavitation occurs when the available head is less than the required head for satisfactory operation.

$$\text{available head} < \text{required head} \quad \begin{bmatrix} \text{criterion for} \\ \text{cavitation} \end{bmatrix}$$

The minimum fluid energy required at the pump inlet for satisfactory operation (i.e., the required head) is known as the *net positive suction head required*, NPSHR.[17] NPSHR is a function of the pump and will

---

[17]If NPSHR (a head term) is multiplied by the fluid specific weight, it is known as the *net inlet pressure required*, NIPR. Similarly, NPSHA can be converted to NIPA.

be given by the pump manufacturer as part of the pump performance data.[18] NPSHR is dependent on the flow rate. However, if NPSHR is known for one flow rate, it can be determined for another flow rate from Eq. 18.29.

$$\frac{\text{NPSHR}_2}{\text{NPSHR}_1} = \left(\frac{Q_2}{Q_1}\right)^2 \qquad \textit{18.29}$$

Net positive suction head available, NPSHA, is the actual fluid energy at the inlet. There are two different methods for calculating NPSHA, both of which are correct and yield identical answers. Equation 18.30(a) is based on the conditions at the fluid surface at the top of an open fluid source (e.g., tank or reservoir). There is a potential energy term but no kinetic energy term. Equation 18.30(b) is based on the conditions at the immediate entrance (suction, subscript s) to the pump. At that point, some of the potential head has been converted to velocity head. Frictional losses are implicitly part of the reduced pressure head, as is the atmospheric pressure head. Since pressure head, $h_{p(s)}$, is absolute, it includes the atmospheric pressure head, and the effect of higher altitudes is explicit in Eq. 18.30(a) and implicit in Eq. 18.30(b). If the source was pressurized instead of being open to the atmosphere, the pressure head would replace h_{atm} in Eq. 18.30(a) but would be implicit in $h_{p(s)}$ in Eq. 18.30(b).

$$\text{NPSHA} = h_{atm} + h_{z(s)} - h_{f(s)} - h_{vp} \qquad \textit{18.30(a)}$$

$$\text{NPSHA} = h_{p(s)} + h_{v(s)} - h_{vp} \qquad \textit{18.30(b)}$$

The net positive suction head available (NPSHA) for most positive displacement pumps includes a term for acceleration head.[19]

$$\text{NPSHA} = h_{atm} + h_{z(s)} - h_{f(s)} - h_{vp} - h_{ac} \qquad \textit{18.31}$$

$$\text{NPSHA} = h_{p(s)} + h_{v(s)} - h_{vp} - h_{ac} \qquad \textit{18.32}$$

If NPSHA is less than NPSHR, the fluid will cavitate. The criterion for cavitation is given by Eq. 18.33. (In practice, it is desirable to have a safety margin.)

$$\text{NPSHA} < \text{NPSHR} \qquad [\text{criterion for cavitation}] \qquad \textit{18.33}$$

Example 18.8

2.0 ft^3/sec (56 L/s) of 60°F (16°C) water are pumped from an elevated feed tank to an open reservoir through 6 in (15.2 cm), schedule-40 steel pipe, as shown. The friction loss for the piping and fittings in the suction line is 2.6 ft (0.9 m). The friction loss for the piping and fittings in the discharge line is 13 ft (4.3 m). The

---

[18]It is also possible to calculate NPSHR from other information, such as suction specific speed. However, this still depends on information provided by the manufacturer.

[19]The friction loss and the acceleration are both maximum values, but they do not occur in phase. Combining them is conservative.

atmospheric pressure is 14.7 psia (101 kPa). What is the NPSHA?

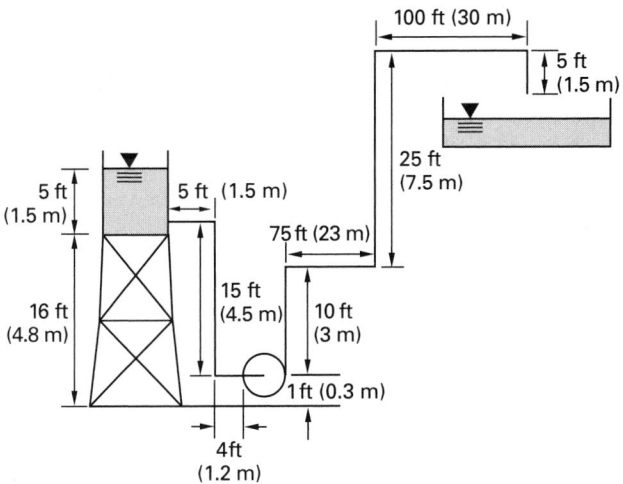

SI Solution

The density of water is approximately 1000 kg/m³. The atmospheric head is

$$h_{\text{atm}} = \frac{p}{\rho g} = \frac{(101 \text{ kPa})\left(1000 \dfrac{\text{Pa}}{\text{kPa}}\right)}{\left(1000 \dfrac{\text{kg}}{\text{m}^3}\right)\left(9.81 \dfrac{\text{m}}{\text{s}^2}\right)}$$

$$= 10.3 \text{ m}$$

For 16°C water, the vapor pressure is approximately 0.01818 bars. The vapor pressure head is

$$h_{\text{vp}} = \frac{p}{\rho g} = \frac{(0.01818 \text{ bar})\left(1 \times 10^5 \dfrac{\text{Pa}}{\text{bar}}\right)}{\left(1000 \dfrac{\text{kg}}{\text{m}^3}\right)\left(9.81 \dfrac{\text{m}}{\text{s}^2}\right)}$$

$$= 0.2 \text{ m}$$

From Eq. 18.30(a), the NPSHA is

$$\text{NPSHA} = h_{\text{atm}} + h_{z(s)} - h_{f(s)} - h_{\text{vp}}$$
$$= 10.3 \text{ m} + 1.5 \text{ m} + 4.8 \text{ m} - 0.3 \text{ m}$$
$$\quad - 0.9 \text{ m} - 0.2 \text{ m}$$
$$= 15.2 \text{ m}$$

Customary U.S. Solution

The specific weight of water is approximately 62.4 lbf/ft³. The atmospheric head is

$$h_{\text{atm}} = \frac{p}{\gamma} = \frac{\left(14.7 \dfrac{\text{lbf}}{\text{in}^2}\right)\left(144 \dfrac{\text{in}^2}{\text{ft}^2}\right)}{62.4 \dfrac{\text{lbf}}{\text{ft}^3}}$$

$$= 33.9 \text{ ft}$$

From App. 14.A, for 60°F water, the vapor pressure head is 0.59 ft. Use 0.6 ft.

From Eq. 18.30(a), the NPSHA is

$$\text{NPSHA} = h_{\text{atm}} + h_{z(s)} - h_{f(s)} - h_{\text{vp}}$$
$$= 33.9 \text{ ft} + 5 \text{ ft} + 16 \text{ ft} - 1 \text{ ft} - 2.6 \text{ ft} - 0.6 \text{ ft}$$
$$= 50.7 \text{ ft}$$

18. PREVENTING CAVITATION

Cavitation is eliminated by increasing NPSHA or decreasing NPSHR. NPSHA can be increased by:

- increasing the height of the fluid source
- lowering the pump
- reducing friction and minor losses by shortening the suction line or using a larger pipe size
- reducing the temperature of the fluid at the pump entrance
- pressurizing the fluid supply tank
- reducing the flow rate or velocity (i.e., reducing the pump speed)

NPSHR can be reduced by:

- placing a throttling valve or restriction in the discharge line[20]
- using an oversized pump
- using a double-suction pump
- using an impeller with a larger eye
- using an inducer

High NPSHR applications, such as boiler feed pumps needing 150 to 250 ft (50 to 80 m), should use one or more booster pumps in front of each high-NPSHR pump. Such booster pumps are typically single-stage, double-suction pumps running at low speed. Their NPSHR can be 25 ft (8 m) or less.

Note that throttling the input line to a pump and venting or evacuating the receiving tank both increase cavitation. Throttling the input line increases the friction head and decreases NPSHA. Evacuating the receiving tank increases the flow rate, increasing NPSHR while simultaneously increasing the friction head and reducing NPSHA.

19. CAVITATION COEFFICIENT

The *cavitation coefficient* (or *cavitation number*), σ, is a dimensionless number that can be used in modeling and extrapolating experimental results. The actual cavitation coefficient is compared with the critical cavitation number obtained experimentally. If the actual cavitation number is less than the critical cavitation number,

---

[20]This will increase the total head (h_A) added by the pump, thereby reducing the pump's output and driving the pump's operating point into a region of lower NPSHR.

cavitation will occur. Absolute pressure must be used in calculating σ.

$$\sigma = \frac{2(p - p_{vp})}{\rho v^2} = \frac{\text{NPSHA}}{h_A} \quad \text{[SI]} \qquad 18.34(a)$$

$$\sigma = \frac{2g_c(p - p_{vp})}{\rho v^2} = \frac{\text{NPSHA}}{h_A} \quad \text{[U.S.]} \qquad 18.34(b)$$

$$\sigma < \sigma_{cr} \quad \text{[criterion for cavitation]} \qquad 18.35$$

The two forms of Eq. 18.34 yield slightly different results. The first form is essentially the ratio of the net pressure available for collapsing a vapor bubble to the velocity pressure creating the vapor. It is useful in model experiments. The second form is applicable to tests of production model pumps.

20. SUCTION SPECIFIC SPEED

The formula for *suction specific speed*, n_{ss}, can be derived by substituting NPSHR for total head in the expression for specific speed. Q is halved for double-suction pumps.

$$n_{ss} = \frac{n\sqrt{\dot{V}}}{(\text{NPSHR in m})^{0.75}} \quad \text{[SI]} \qquad 18.36(a)$$

$$n_{ss} = \frac{n\sqrt{Q}}{(\text{NPSHR in ft})^{0.75}} \quad \text{[U.S.]} \qquad 18.36(b)$$

Suction specific speed is an index of the suction characteristics of the impeller. Ideally, it should be approximately 8500 (165 in SI) for both single- and double-suction pumps. This assumes the pump is operating at or near its point of optimum efficiency.

Suction specific speed can be used to determine the maximum recommended operating speed by substituting 8500 (165 in SI) for n_{ss} in Eq. 18.36 and solving for n.

If the suction specific speed is known, it can be used to determine the NPSHR. If the pump is known to be operating at or near its optimum efficiency, an approximate NPSHR value can be found by substituting 8500 (165 in SI) for n_{ss} in Eq. 18.36 and solving for NPSHR.

Suction specific speed available, SA, is obtained when NPSHA is substituted for total head in the expression for specific speed. The suction specific speed available must be less than the suction specific speed required to prevent cavitation.[21]

21. PUMP PERFORMANCE CURVES

For a given impeller diameter and constant speed, the head added will decrease as the flow rate increases. This

can be shown graphically on the *pump performance curve (pump curve)* supplied by the pump manufacturer. Other operating characteristics (e.g., power requirement, NPSHR, and efficiency) also vary with flow rate, and these are usually plotted on a common graph, as shown in Fig. 18.10.[22] Manufacturers' pump curves show performance over a limited number of calibration speeds. If an operating point is outside the range of published curves, the affinity laws can be used to estimate the speed at which the pump gives the required performance.

Figure 18.10 *Pump Performance Curves*

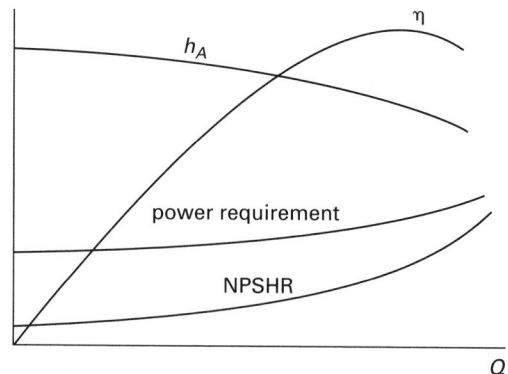

On the pump curve, the *shutoff point* (also known as *churn*) corresponds to a closed discharge valve (i.e., zero flow); the *rated point* is where the pump operates with rated 100% of capacity and head; the *overload point* corresponds to 65% of the rated head.

Figure 18.10 is for a pump with a fixed impeller diameter and rotational speed. The characteristics of a pump operated over a range of speed or for different impeller diameters are illustrated in Fig. 18.11.

22. SYSTEM CURVES

A *system curve* (or *system performance curve*) is a plot of the static and friction energy losses experienced by the fluid for different flow rates. Unlike the pump curve, which depends only on the pump, the system curve depends only on the configuration of the suction and discharge lines. (The following equations assume equal pressures at the fluid source and destination surfaces, which is the case for pumping from one atmospheric reservoir to another. The velocity head is insignificant and is disregarded.)

$$h_A = h_z + h_f \qquad 18.37$$

$$h_z = h_{z(d)} - h_{z(s)} \qquad 18.38$$

$$h_f = h_{f(s)} + h_{f(d)} \qquad 18.39$$

---

[21]Since speed and flow rate are constants, this is another way of saying NPSHA must equal or exceed NPSHR.

[22]The term *pump curve* is commonly used to designate the h_A versus Q characteristics, whereas *pump characteristics curve* implies all of the pump data.

Figure 18.11 *Centrifugal Pump Characteristics Curves*

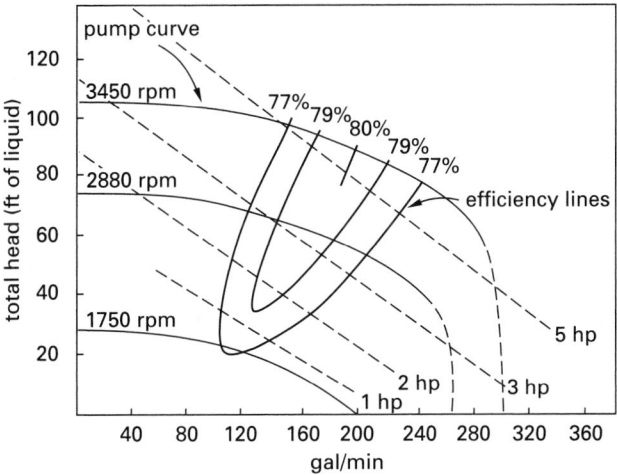

(a) variable speed

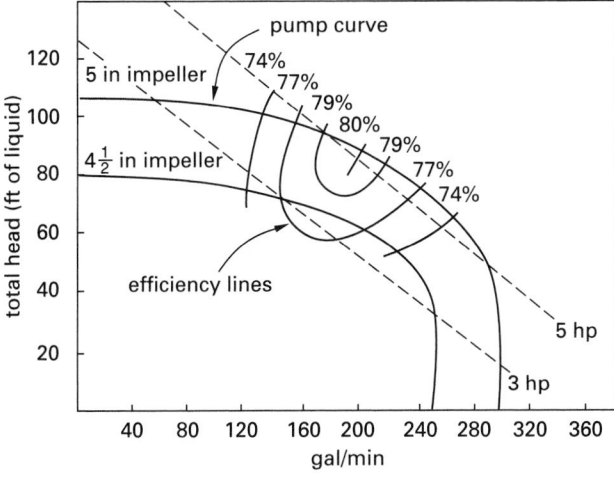

(b) variable impeller diameter

Figure 18.12 *System Curve*

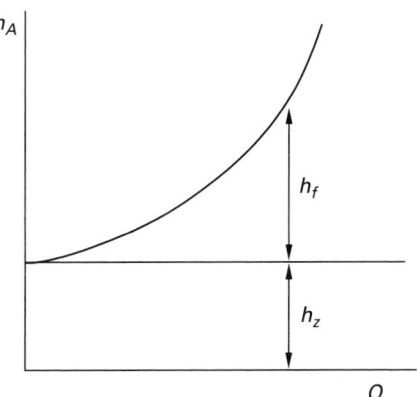

23. OPERATING POINT

The intersection of the pump curve and the system curve determines the *operating point*, as shown in Fig. 18.13. The operating point defines the system head and system flow rate.

Figure 18.13 *Extreme Operating Points*

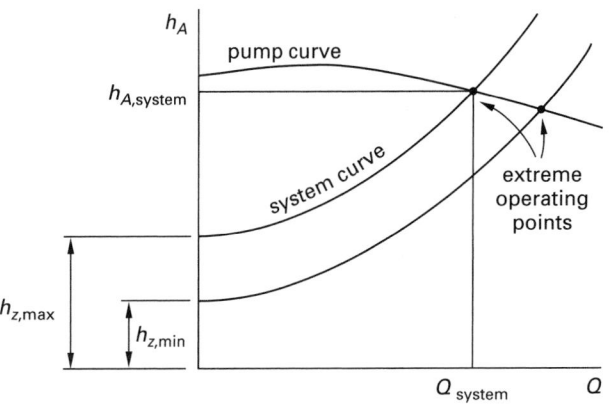

When selecting a pump, the system curve is plotted on manufacturers' pump curves for different speeds and/or impeller diameters (i.e., Fig. 18.11). There will be several possible operating points corresponding to the various pump curves shown. Generally, the design operating point should be close to the highest pump efficiency. This, in turn, will determine speed and impeller diameter.

In many systems, the static head will vary as the source reservoir is drained or as the destination reservoir fills. The system head is then defined by a pair of matching system friction curves intersecting the pump curve. The two intersection points are the *extreme operating points*—the maximum and minimum capacity requirements.

After a pump is installed, it may be desired to change the operating point. This can be done without replacing the pump by placing a throttling valve in the discharge line. The operating point can then be moved along the

---

If the fluid reservoirs are large, or if the fluid reservoir levels are continually replenished, the net static suction head $(h_{z(1)} - h_{z(2)})$ will be constant for all flow rates. The friction loss, h_f, varies with v^2 (and hence with Q^2) in the Darcy friction formula. This makes it easy to find friction losses for other flow rates (subscript 2) once one friction loss (subscript 1) is known.[23]

$$\frac{h_{f,1}}{h_{f,2}} = \left(\frac{Q_1}{Q_2}\right)^2 \qquad \textit{18.40}$$

Figure 18.12 illustrates a system curve with a negative suction head (i.e., a fluid source below the fluid destination). The system curve is shifted upward, intercepting the vertical axis at some positive value of h_A.

---

[23]Equation 18.40 implicitly assumes that the friction factor, f, is constant. This may be true over a limited range of flow rates, but it is not true over large ranges unless the Hazen-Williams friction loss equation is being used. Nevertheless, Eq. 18.40 is often used to quickly construct preliminary versions of the system curve.

Water Resources

pump curve by opening or closing the valve, as illustrated in Fig. 18.14. (A throttling valve should never be placed in the suction line since that would reduce NPSHA.)

Figure 18.14 *Throttling the Discharge*

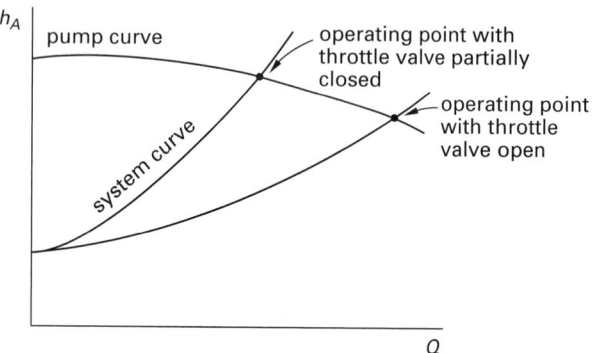

24. PUMPS IN PARALLEL

Parallel operation is obtained by having two pumps discharging into a common header. This type of connection is advantageous when the system demand varies greatly and when high reliability is required. A single pump providing total flow would have to operate far from its optimum efficiency at one point or another. With two pumps in parallel, one can be shut down during low demand. This allows the remaining pump to operate close to its optimum efficiency point.

Figure 18.15 illustrates that parallel operation increases the capacity of the system while maintaining the same total head.

Figure 18.15 *Pumps Operating in Parallel*

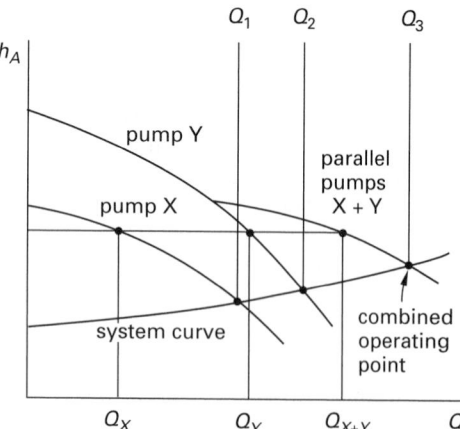

The performance curve for a set of pumps in parallel can be plotted by adding the capacities of the two pumps at various heads. Capacity does not increase at heads above the maximum head of the smaller pump.

Furthermore, a second pump will operate only when its discharge head is greater than the discharge head of the pump already running.

When the parallel performance curve is plotted with the system head curve, the operating point is the intersection of the system curve with the X + Y curve. With pump X operating alone, the capacity is given by Q_1. When pump Y is added, the capacity increases to Q_3 with a slight increase in total head.

25. PUMPS IN SERIES

Series operation is achieved by having one pump discharge into the suction of the next. This arrangement is used primarily to increase the discharge head, although a small increase in capacity also results.

The performance curve for a set of pumps in series can be plotted by adding the heads of the two pumps at various capacities.

Figure 18.16 *Pumps Operating in Series*

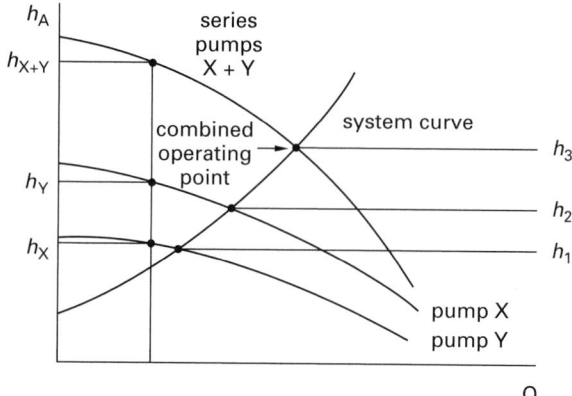

26. AFFINITY LAWS

Most parameters (impeller diameter, speed, and flow rate) determining a specific pump's performance can vary. If the impeller diameter is held constant and the speed is varied, the following ratios are maintained with no change in efficiency.

$$\frac{Q_2}{Q_1} = \frac{n_2}{n_1} \qquad 18.41$$

$$\frac{h_2}{h_1} = \left(\frac{n_2}{n_1}\right)^2 = \left(\frac{Q_2}{Q_1}\right)^2 \qquad 18.42$$

$$\frac{P_2}{P_1} = \left(\frac{n_2}{n_1}\right)^3 = \left(\frac{Q_2}{Q_1}\right)^3 \qquad 18.43$$

If the speed is held constant and the impeller size is varied,

$$\frac{Q_2}{Q_1} = \frac{D_2}{D_1} \qquad 18.44$$

$$\frac{h_2}{h_1} = \left(\frac{D_2}{D_1}\right)^2 \qquad 18.45$$

$$\frac{P_2}{P_1} = \left(\frac{D_2}{D_1}\right)^3 \qquad 18.46$$

The affinity laws are based on the assumption that the efficiencies of the two pumps are the same. In reality, larger pumps are somewhat more efficient than smaller pumps. Therefore, extrapolations to greatly different sized pumps should be avoided. Equation 18.47 can be used to estimate the efficiency of a different sized pump. The dimensionless exponent varies from 0 to approximately 0.26, with 0.2 being a typical value.

$$\frac{1 - \eta_{\text{smaller}}}{1 - \eta_{\text{larger}}} = \left(\frac{D_{\text{larger}}}{D_{\text{smaller}}} \right)^n \qquad 18.47$$

Example 18.9

A pump operating at 1770 rpm delivers 500 gal/min against a total head of 200 ft. Changes in the piping system have increased the total head to 375 ft. At what speed should this pump be operated to achieve this new head at the same efficiency?

Solution

From Eq. 18.42,

$$n_2 = n_1 \sqrt{\frac{h_2}{h_1}} = 1770 \text{ rpm} \sqrt{\frac{375 \text{ ft}}{200 \text{ ft}}}$$
$$= 2424 \text{ rpm}$$

Example 18.10

A pump is required to pump 500 gal/min while providing a total dynamic head of 425 ft. The hydraulic system has no static head change. Only the 1750 rpm performance curve is known for the pump. At what speed must the pump be turned to achieve the desired performance with no change in efficiency or impeller size?

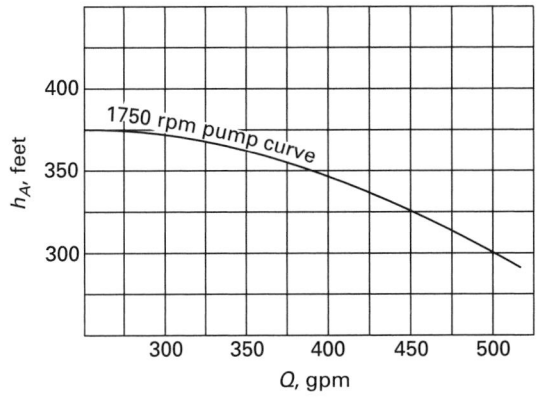

Solution

A flow of 500 gal/min with a head of 425 ft does not correspond to any point on the 1750 rpm curve.

From Eq. 18.42, the quantity h/Q^2 is constant.

$$\frac{h}{Q^2} = \frac{425 \text{ ft}}{\left(500 \dfrac{\text{gal}}{\text{min}} \right)^2} = 1.7 \times 10^{-3} \quad \text{[mixed units]}$$

In order to use the affinity laws, the operating point on the 1750 rpm curve must be determined. Random values of Q are chosen and the corresponding values of h are determined such that the ratio h/Q^2 is unchanged.

Q	h
475	383
450	344
425	307
400	272

These points are plotted and connected to draw the system curve. The intersection of the system and 1750 rpm pump curve at 440 gal/min defines the operating point at that speed. From Eq. 18.41,

$$n_2 = \frac{n_1 Q_2}{Q_1} = \frac{(1750 \text{ rpm}) \left(500 \dfrac{\text{gal}}{\text{min}} \right)}{440 \dfrac{\text{gal}}{\text{min}}}$$

$$= 1989 \text{ rpm}$$

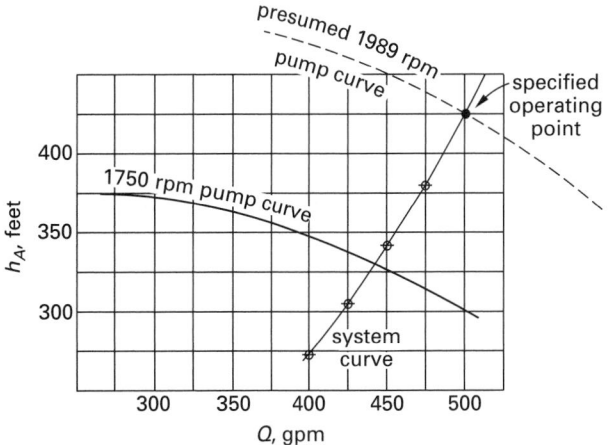

27. PUMP SIMILARITY

The performance of one pump can be used to predict the performance of a *dynamically similar (homologous) pump*. This can be done by using Eqs. 18.48 through 18.53.

$$\frac{n_1 D_1}{\sqrt{h_1}} = \frac{n_2 D_2}{\sqrt{h_2}} \qquad 18.48$$

$$\frac{Q_1}{D_1^2 \sqrt{h_1}} = \frac{Q_2}{D_2^2 \sqrt{h_2}} \qquad 18.49$$

$$\frac{P_1}{\rho_1 D_1^2 h_1^{1.5}} = \frac{P_2}{\rho_2 D_2^2 h_2^{1.5}} \qquad 18.50$$

$$\frac{Q_1}{n_1 D_1^3} = \frac{Q_2}{n_2 D_2^3} \qquad 18.51$$

$$\frac{P_1}{\rho_1 n_1^3 D_1^5} = \frac{P_2}{\rho_2 n_2^3 D_2^5} \qquad 18.52$$

$$\frac{n_1 \sqrt{Q_1}}{h_1^{0.75}} = \frac{n_2 \sqrt{Q_2}}{h_2^{0.75}} \qquad 18.53$$

These *similarity laws* assume that both pumps:

- operate in the turbulent region
- have the same pump efficiency
- operate at the same percentage of wide-open flow

Similar pumps also will have the same specific speed and cavitation number.

As with the affinity laws, these relationships assume that the efficiencies of the larger and smaller pumps are the same. In reality, larger pumps will be more efficient than smaller pumps. Therefore, extrapolations to much larger or much smaller sizes should be avoided.

Example 18.11

A 6 in pump operating at 1770 rpm discharges 1500 gal/min of cold water (SG = 1.0) against an 80 ft head at 85% efficiency. A homologous 8 in pump operating at 1170 rpm is being considered as a replacement. (a) What total head and capacity can be expected from the new pump? (b) What would be the new horsepower requirement?

Solution

(a) From Eq. 18.48,

$$h_2 = \left(\frac{D_2 n_2}{D_1 n_1}\right)^2 h_1 = \left(\frac{(8 \text{ in})(1170 \text{ rpm})}{(6 \text{ in})(1770 \text{ rpm})}\right)^2 (80 \text{ ft})$$

$$= 62.14 \text{ ft}$$

From Eq. 18.51,

$$Q_2 = \left(\frac{n_2 D_2^3}{n_1 D_1^3}\right) Q_1$$

$$= \left(\frac{(1170 \text{ rpm})(8 \text{ in})^3}{(1770 \text{ rpm})(6 \text{ in})^3}\right)\left(1500 \frac{\text{gal}}{\text{min}}\right)$$

$$= 2350.3 \text{ gal/min}$$

(b) From Table 18.5, the hydraulic horsepower is

$$\text{WHP}_2 = \frac{h_2 Q_2 (\text{SG})}{3956} = \frac{(62.14 \text{ ft})\left(2350.3 \frac{\text{gal}}{\text{min}}\right)(1.0)}{3956 \frac{\text{ft-gal}}{\text{hp-min}}}$$

$$= 36.92 \text{ hp}$$

From Eq. 18.47,

$$\eta_{\text{larger}} = 1 - \frac{1 - \eta_{\text{smaller}}}{\left(\dfrac{D_{\text{larger}}}{D_{\text{smaller}}}\right)^{0.2}}$$

$$= 1 - \frac{1 - 0.85}{\left(\dfrac{8 \text{ in}}{6 \text{ in}}\right)^{0.2}} = 0.858$$

$$\text{BHP}_2 = \frac{\text{WHP}_2}{\eta_p} = \frac{36.92 hp}{0.858} = 43.0 \text{ hp}$$

28. PUMPING LIQUIDS OTHER THAN COLD WATER

Many pump parameters are determined from tests with cold, clear water at 85°F (29°C). The following guidelines can be used when pumping water at other temperatures or when pumping other fluids.

- Head developed is independent of the liquid's specific gravity. Pump performance curves from tests with water can be used with other Newtonian fluids (e.g., gasoline, alcohol, and aqueous solutions) having similar viscosities.

- Head, flow rate, and efficiency are all reduced when pumping highly viscous non-Newtonian fluids. No exact method exists for determining the reduction factors, other than actual tests of an installation using both fluids. Some sources have published charts of correction factors based on tests over limited viscosity and size ranges.[24]

- The hydraulic horsepower depends on the specific gravity of the fluid. If the pump characteristic curve is used to find the operating point, multiply the horsepower reading by the specific gravity. Tables 18.5 and 18.6 incorporate the specific gravity term in the calculation of hydraulic power where required.

- Efficiency is not affected by changes in temperature that cause only the specific gravity to change.

- Efficiency is nominally affected by changes in temperature that cause the viscosity to change. Equation 18.54 is an approximate relationship suggested by the Hydraulics Institute when extrapolating the efficiency (in decimal form) from cold water to hot water. n is an experimental exponent established by the pump manufacturer, generally in the range of 0.05 to 0.1.

$$\eta_{\text{hot}} = 1 - (1 - \eta_{\text{cold}})\left(\frac{\nu_{\text{hot}}}{\nu_{\text{cold}}}\right)^n \qquad \textit{18.54}$$

- NPSHR is not significantly affected by minor variations in the water temperature.

- When hydrocarbons are pumped, the NPSHR determined from cold water can usually be reduced. This reduction is apparently due to the slow vapor release of complex organic liquids. If the hydrocarbon's vapor pressure at the pumping temperature is known, Fig. 18.17 will give the percentage of the cold-water NPSHR.

---

[24] A chart published by the Hydraulics Institute is widely distributed.

Figure 18.17 *Hydrocarbon NPSHR Correction Factor*

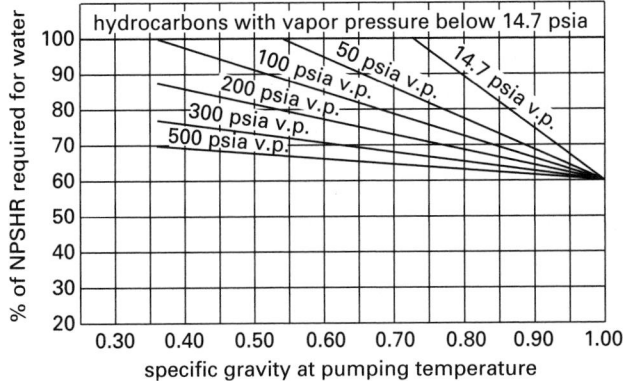

Figure 18.18 *Typical Hydroelectric Plant*

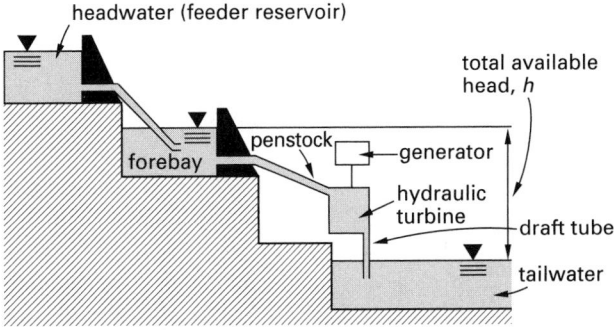

• Pumping many fluids requires expertise that goes far beyond simply extrapolating parameters in proportion to the fluid's specific gravity. Such special cases include pumping liquids containing abrasives, liquids that solidify, highly corrosive liquids, liquids with vapor or gas, highly viscous fluids, paper stock, and hazardous fluids.

Example 18.12

A centrifugal pump has an NIPR (NPSHR) of 12 psi based on cold water. $10°F$ isobutane has a specific gravity of 0.60 and a vapor pressure of 15 psia. What NPSHR should be used with $10°F$ liquid isobutane?

Solution

From Fig. 18.17, the intersection of a specific gravity of 0.60 and 15 psia is above the horizontal 100% line. The full NPSHR of 12 psi should be used.

29. TURBINE SPECIFIC SPEED

Like centrifugal pumps, turbines are classified according to the manner in which the impeller extracts energy from the fluid flow. This is measured by the turbine-specific speed equation, which is different from the equation used to calculate specific speed for pumps.

$$n_s = \frac{n\sqrt{P \text{ in kW}}}{(h_t)^{1.25}} \qquad \text{[SI]} \qquad 18.55(a)$$

$$n_s = \frac{n\sqrt{P \text{ in hp}}}{(h_t)^{1.25}} \qquad \text{[U.S.]} \qquad 18.55(b)$$

30. HYDROELECTRIC GENERATING PLANTS

In a typical hydroelectric generating plant using reaction turbines, the turbine is generally housed in a *powerhouse*, with water conducted to the turbine through the *penstock* piping. Water originates in a reservoir, dam, or *forebay* (in the instance where the reservoir is a long distance from the turbine).

After the water passes through the turbine, it is discharged through the draft tube to the receiving reservoir, known as the *tailwater*. The *draft tube* is used to keep the turbine up to 15 ft (5 m) above the tailwater surface, while still being able to extract the total available head. If a draft tube is not employed, water may be returned to the tailwater by way of a channel known as the *tail race*. The turbine, draft tube, and all related parts comprise what is known as the *setting*.

When a forebay is not part of the generating plant's design, it will be desirable to provide a *surge chamber* in order to relieve the effects of rapid changes in flow rate. In the case of a sudden power demand, the surge chamber would provide an immediate source of water, without waiting for a contribution from the feeder reservoir.

Similarly, in the case of a sudden decrease in discharge through the turbine, the excess water would surge back into the surge chamber.

31. IMPULSE TURBINES

An *impulse turbine* consists of a rotating shaft (called a *turbine runner*) on which buckets or blades are mounted. (This is commonly called a *Pelton wheel.*)[25] A jet of water (or other fluid) hits the buckets and causes the turbine to rotate. The kinetic energy of the jet is converted into rotational kinetic energy. The jet is essentially at atmospheric pressure.

Impulse turbines are generally employed where the available head is very high, above 800 to 1600 ft (250 to 500 m). (There is no exact value for the critical head, hence the range. What is important is that impulse turbines are *high-head turbines*.) Efficiencies are in the range of 80 to 90%, with the higher efficiencies being associated with turbines having two or more jets per runner.

---

[25]In a Pelton wheel turbine, the spoon-shaped buckets are divided into two halves, with a ridge between the halves. Half of the water is thrown to each side of the bucket. A Pelton wheel is known as a *tangential turbine (tangential wheel)* because the centerline of the jet is directed at the centers of the buckets.

Figure 18.19 *Impulse Turbine Installation*

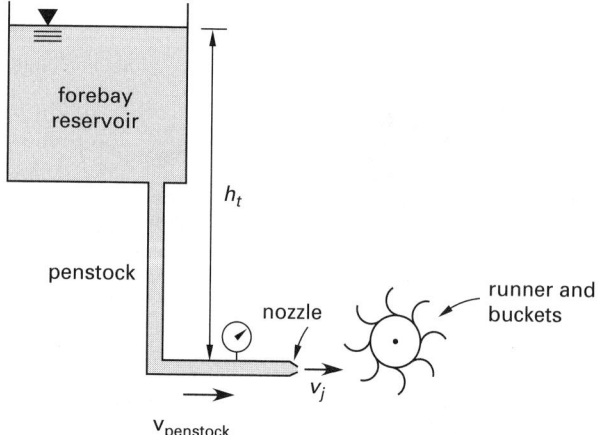

The total available head in this installation is h_t, but not all of this energy can be extracted. Some of the energy is lost to friction in the penstock. Minor losses also occur, but these small losses are usually disregarded. In the penstock, immediately before entering the nozzle, the remaining head is divided between the pressure head and the velocity head.[26]

$$h' = h_t - h_f = (h_p + h_v)_{\text{penstock}} \qquad \textbf{18.56}$$

Another loss, h_n, occurs in the nozzle itself. The head remaining to turn the turbine is

$$h'' = h' - h_n$$
$$= h_t - h_f - h_n \qquad \textbf{18.57}$$

h_f is calculated from either the Darcy or the Hazen-Williams equation. The nozzle loss is calculated from the *nozzle coefficient*, C_v.

$$h_n = h'(1 - C_v^2) \qquad \textbf{18.58}$$

The total head at the nozzle exit is converted into velocity head according to the Torricelli equation.

$$v_j = \sqrt{2gh''} = C_v\sqrt{2gh'} \qquad \textbf{18.59}$$

In order to maximize power output, buckets are usually designed to partially reverse the direction of the water jet flow. The forces on the turbine buckets can be found from the impulse-momentum equations. If the water is turned through an angle θ and the wheel's *tangential velocity* is v_b, the energy transmitted by each unit mass of water to the turbine runner is[27]

$$E = v_b(v_j - v_b)(1 - \cos\theta) \qquad \text{[SI]} \qquad \textbf{18.60(a)}$$

$$E = \left(\frac{v_b(v_j - v_b)}{g_c}\right)(1 - \cos\theta) \quad \text{[U.S.]} \qquad \textbf{18.60(b)}$$

$$v_b = \frac{2\pi n r}{60} = \omega r \qquad \textbf{18.61}$$

[26]Care must be taken to distinguish between the conditions existing in the penstock, the nozzle throat, and the jet itself. The velocity in the nozzle throat and jet will be the same, but this is different from the penstock velocity. Similarly, the pressure in the jet is zero, although it is nonzero in the penstock.
[27]$\theta = 180°$ would be ideal. However, the actual angle is limited to approximately 165° to keep the deflected jet out of the way of the incoming jet.

The theoretical *turbine power* is found by multiplying Eq. 18.60 by the mass flow rate. The actual power will be less than the theoretical output. Typical turbine efficiencies range between 80 and 90%. For a mass flow rate in kg/s, the theoretical power (in kilowatts) will be

$$P_{\text{th}} = \frac{\dot{m}E}{1000} \qquad \text{[SI]} \qquad \textbf{18.62(a)}$$

For a mass flow rate in lbm/sec, the theoretical horsepower will be

$$P_{\text{th}} = \frac{\dot{m}E}{550} \qquad \text{[U.S.]} \qquad \textbf{18.62(b)}$$

Example 18.13

A Pelton wheel impulse turbine develops 100 hp (brake) while turning at 500 rpm. The water is supplied from a penstock with an internal area of 0.3474 ft². The water subsequently enters a nozzle with a reduced flow area. The total head is 200 ft before nozzle loss. The turbine efficiency is 80%, and the nozzle coefficient, C_v, is 0.95. Disregard penstock friction losses. What are the (a) flow rate (in ft³/sec), (b) area of the jet, and (c) pressure head in the penstock just before the nozzle?

Solution

(a) From Eq. 18.59 with $h' = h_t = 200$ ft, the jet velocity is

$$v_j = C_v\sqrt{2gh'} = 0.95\sqrt{(2)\left(32.2\ \frac{\text{ft}}{\text{sec}^2}\right)(200\ \text{ft})}$$
$$= 107.8\ \text{ft/sec}$$

From Eq. 18.58, the nozzle loss is

$$h_n = h'(1 - C_v^2) = (200\ \text{ft})\left(1 - (0.95)^2\right)$$
$$= 19.5\ \text{ft}$$

From Table 18.5, the flow rate is

$$\dot{V} = \frac{(8.814)(P)}{h_A(\text{SG})\eta} = \frac{(8.814)(100\ \text{hp})}{(200\ \text{ft - }19.5\ \text{ft})(1)(0.8)}$$
$$= 6.104\ \text{ft}^3/\text{sec}$$

(b) The jet area is

$$A_j = \frac{\dot{V}}{v_j} = \frac{6.104\ \dfrac{\text{ft}^3}{\text{sec}}}{107.8\ \dfrac{\text{ft}}{\text{sec}}} = 0.0566\ \text{ft}^2$$

(c) The velocity in the penstock is

$$v_{\text{penstock}} = \frac{\dot{V}}{A} = \frac{6.104\ \dfrac{\text{ft}^3}{\text{sec}}}{0.3474\ \text{ft}^2}$$
$$= 17.57\ \text{ft/sec}$$

The pressure head in the penstock is

$$h_p = h' - h_v = h' - \frac{v^2}{2g}$$

$$= 200 \text{ ft} - \frac{\left(17.57 \frac{\text{ft}}{\text{sec}}\right)^2}{(2)\left(32.2 \frac{\text{ft}}{\text{sec}^2}\right)}$$

$$= 195.2 \text{ ft}$$

32. REACTION TURBINES

Reaction turbines (also known as *Francis turbines* or *radial-flow turbines*) are essentially centrifugal pumps operating in reverse. They are used when the total available head is small, typically below 600 to 800 ft. However, their energy conversion efficiency is higher than that of impulse turbines, typically in the 85 to 95% range.

In a reaction turbine, water enters the turbine housing with a pressure greater than atmospheric pressure. The water completely surrounds the turbine runner (impeller) and continues through the draft tube. There is no vacuum or air pocket between the turbine and the tailwater.

All of the power, affinity, and similarity relationships used with centrifugal pumps can be used with reaction turbines.

Figure 18.20 *Reaction Turbine*

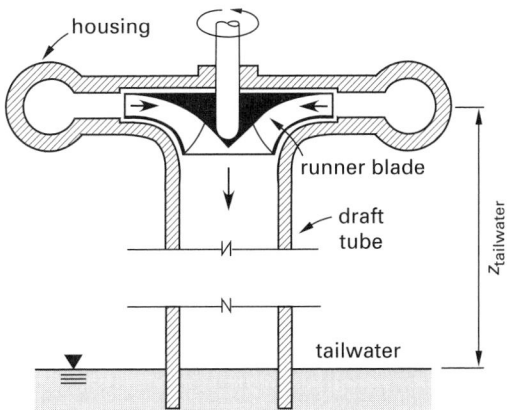

Example 18.14

A reaction turbine with a draft tube develops 500 hp (brake) when 50 ft³/sec of water flow through it. Water enters the turbine at 20 ft/sec with a 100 ft pressure head. The elevation of the turbine above the tailwater level is 10 ft. Disregarding friction, what are the (a) total available head and (b) turbine efficiency?

Solution

(a) The available head is the difference between the forebay and tailwater elevations. The tailwater depression is known, but the height of the forebay above the turbine is not known. At the turbine entrance, this unknown potential energy has been converted to pressure and velocity head. Therefore, the total available head (exclusive of friction) is

$$h_t = z_{\text{forebay}} - z_{\text{tailwater}}$$

$$= h_p + h_v - z_{\text{tailwater}}$$

$$= 100 \text{ ft} + \frac{\left(20 \frac{\text{ft}}{\text{sec}}\right)^2}{(2)\left(32.2 \frac{\text{ft}}{\text{sec}^2}\right)} - (-10 \text{ ft})$$

$$= 116.2 \text{ ft}$$

(b) From Table 18.5, the theoretical hydraulic horsepower is

$$P_{\text{th}} = \frac{h_A \dot{V}(\text{SG})}{8.814} = \frac{(116.2 \text{ ft})\left(50 \frac{\text{ft}^3}{\text{sec}}\right)(1.0)}{8.814}$$

$$= 659.2 \text{ hp}$$

The efficiency of the turbine is

$$\eta = \frac{P_{\text{brake}}}{P_{\text{th}}} = \frac{500 \text{ hp}}{659.2 \text{ hp}} = 0.758 \quad (75.8\%)$$

33. TYPES OF REACTION TURBINES

Each of the three types of turbines is associated with a range of specific speeds.

- Axial-flow reaction turbines are used for low heads, high rotational speeds, and large flow rates. These propeller turbines operate with specific speeds in the 80 to 200 range (300 to 760 in SI). Their best efficiencies, however, are produced with specific speeds between 120 and 160 (460 and 610 in SI).

- For mixed-flow reaction turbines, the specific speed varies from 10 to 100 (38 and 380 in SI). Best efficiencies are found in the 40 to 60 (150 and 230 in SI) range with heads below 600 to 800 ft (180 to 240 m).

- Radial-flow reaction turbines have the lowest flow rates and specific speeds but are used when heads are high. These turbines have specific speeds below 5 (19 in SI).

34. AXIAL-FLOW TURBINES

Figure 18.21 illustrates an *axial-flow turbine*, also known as a *propeller turbine*. Analysis of performance is similar to that of reaction and impulse turbines.

Figure 18.21 *Axial-Flow Turbine*

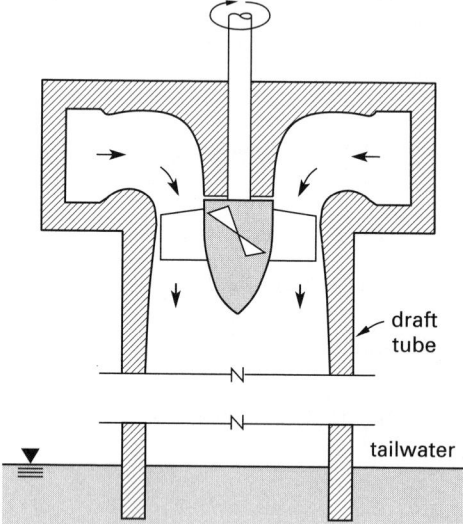

19 Open Channel Flow

Water Resources

Nomenclature

A	area	ft^2	m^2
b	weir or channel width	ft	m
C	coefficient	ft$^{1/2}$/sec	m$^{1/2}$/s
d	depth of flow	ft	m
d	diameter	in	cm
D	diameter	ft	m
E	specific energy	ft	m
Fr	Froude number	–	–
g	acceleration due to gravity	ft/sec^2	m/s^2
g_c	gravitational constant	ft-lbm/lbf-sec^2	–
h	head	ft	m
H	total hydraulic head	ft	m
k	minor loss coefficient	–	–
K	conveyance	ft^3/sec	m^3/s
K'	modified conveyance	–	–
L	channel length	ft	m
m	cotangent of side slope angle	–	–
n	Manning roughness coefficient	–	–
N	number of end contractions	–	–
p	pressure	lbf/ft^2	Pa
P	wetted perimeter	ft	m
Q	flow quantity	ft^3/sec	m^3/s
R	hydraulic radius	ft	m
S	slope of energy line (energy gradient)	–	–
S_0	channel slope	–	–
T	width of surface	ft	m
v	velocity	ft/sec	m/s
w	channel width	ft	m
x	distance	ft	m
Y	weir height	ft	m
z	height above datum	ft	m

Symbols

α	velocity-head coefficient	–	–
γ	specific weight	lbf/ft^3	–
ρ	density	lbm/ft^3	kg/m^3
θ	angle	deg	deg

Subscripts

b	brink
c	critical or composite
d	discharge
e	entrance
f	friction
h	hydraulic
n	normal
o	channel or culvert barrel
s	spillway
t	total
w	weir

1. INTRODUCTION

An *open channel* is a fluid passageway that allows part of the fluid to be exposed to the atmosphere. This type of channel includes natural waterways, canals, culverts, flumes, and pipes flowing under the influence of gravity (as opposed to pressure conduits, which always flow full). A *reach* is a straight section of open channel with uniform shape, depth, slope, and flow quantity.

There are difficulties in evaluating open channel flow. The unlimited geometric cross sections and variations in roughness have contributed to a relatively small number of scientific observations upon which to estimate the required coefficients and exponents. Therefore, the analysis of open channel flow is more empirical and less exact than that of pressure conduit flow. This lack of precision, however, is more than offset by the percentage error in runoff calculations that generally precede the channel calculations.

Flow can be categorized on the basis of the channel material, for example, concrete or metal pipe or earth material. Except for a short discussion of erodible canals in Sec. 36, this chapter assumes the channel is nonerodible.

2. TYPES OF FLOW

Flow in open channels is almost always turbulent; laminar flow will occur only in very shallow channels or at very low fluid velocities. However, within the turbulent category are many somewhat confusing categories of flow. Flow can be a function of time and location. If the flow quantity (volume per unit of time across an area in flow) is invariant, it is said to be *steady flow*. (Flow that varies with time, such as stream flow during a storm, known as *varied flow*, is not covered in this chapter.) If the flow cross section does not depend on the location along the channel, it is said to be *uniform flow*. Steady flow can also be *nonuniform flow*, as in the case of a river with a varying cross section or on a steep slope. Furthermore, uniform channel construction does not ensure uniform flow, as will be seen in the case of hydraulic jumps.

Table 19.1 summarizes some of the more common categories and names of steady open channel flow. All of the subcategories are based on variations in depth and flow area with respect to location along the channel.

Table 19.1 *Categories of Steady Open Channel Flow*

subcritical flow (tranquil flow)
    uniform flow
        normal flow
    nonuniform flow
        accelerating flow
        decelerating flow (retarded flow)
critical flow
supercritical flow (rapid flow, shooting flow)
    uniform flow
        normal flow
    nonuniform flow
        accelerating flow
        decelerating flow

3. MINIMUM VELOCITIES

The minimum permissible velocity in a sewer or other nonerodible channel is the lowest that prevents sedimentation and plant growth. Velocities of 2 to 3 ft/sec (0.6 to 0.9 m/s) keep all but the heaviest silts in suspension. 2.5 ft/sec (0.75 m/s) is considered the minimum to prevent plant growth.

4. VELOCITY DISTRIBUTION

Due to the adhesion between the wetted surface of the channel and the water, the velocity will not be uniform across the area in flow. The velocity term used in this chapter is the *mean velocity*. The mean velocity, when multiplied by the flow area, gives the flow quantity.

$$Q = A\text{v} \qquad\qquad 19.1$$

The location of the mean velocity depends on the distribution of velocities in the waterway, which is generally quite complex. The procedure for measuring the velocity of a channel (called *stream gauging*) involves measuring the average channel velocity at multiple locations across the channel width. These subaverage velocities are then themselves averaged to give a grand average (mean) flow velocity.

Figure 19.1 *Velocity Distribution in an Open Channel*

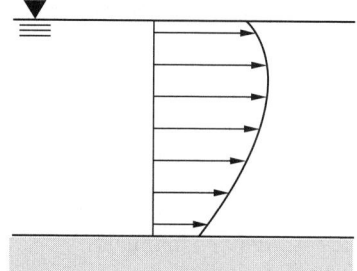

5. PARAMETERS USED IN OPEN CHANNEL FLOW

The *hydraulic radius* is the ratio of the area in flow to the wetted perimeter.[1]

$$R = \frac{A}{P} \qquad\qquad 19.2$$

For a circular channel flowing either full or half-full, the hydraulic radius is one-fourth of the equivalent diameter, $D_e/4$. The hydraulic radii of other channel shapes are easily calculated from the basic definition. Table 19.2 summarizes parameters for the basic shapes. For very wide channels such as rivers, the hydraulic radius is approximately equal to the depth.

---

[1]The hydraulic radius is also referred to as the *hydraulic mean depth*. However, this name is easily confused with "mean depth" and "hydraulic depth," both of which have different meanings. Therefore, the term "hydraulic mean depth" is not used in this chapter.

Table 19.2 *Hydraulic Parameters of Basic Channel Sections*

section	area, A	wetted perimeter, P	hydraulic radius, R
rectangle	dw	$2d + w$	$\dfrac{dw}{w + 2d}$
trapezoid	$\left(b + \dfrac{d}{\tan\theta}\right)d$	$b + 2\left(\dfrac{d}{\sin\theta}\right)$	$\dfrac{bd\sin\theta + d^2\cos\theta}{b\sin\theta + 2d}$
triangle	$\dfrac{d^2}{\tan\theta}$	$\dfrac{2d}{\sin\theta}$	$\dfrac{d\cos\theta}{2}$
circle	$\frac{1}{8}(\theta - \sin\theta)D^2$ [θ in radians]	$\frac{1}{2}\theta D$ [θ in radians]	$\frac{1}{4}\left(1 - \dfrac{\sin\theta}{\theta}\right)D$ [θ in radians]

The *hydraulic depth* is the ratio of the area in flow to the width of the channel at the fluid surface.[2]

$$D_h = \frac{A}{w} \qquad \text{19.3}$$

The uniform flow *section factor* represents a frequently occurring variable group. The section factor is often evaluated against depth of flow when working with discharge from irregular cross sections.

$$\text{section factor} = AR^{2/3} \quad \text{[general uniform flow]} \quad \text{19.4}$$

$$\text{section factor} = A\sqrt{D_h} \quad \text{[critical flow only]} \quad \text{19.5}$$

The *slope*, S, is the slope of the energy line. In general, the slope can be calculated from the Bernoulli equation as the energy loss per unit length of channel. For small slopes typical of almost all natural waterways, the channel length and horizontal run are essentially identical.

---

[2]For a rectangular channel, $D_h = d$.

$$S = \frac{dE}{dL} \qquad \text{19.6}$$

If the flow is uniform, the slope of the energy line will parallel the water surface and channel bottom, and the *energy gradient* will equal the *geometric slope*, S_0.

$$S_0 = \frac{\Delta z}{L} \quad \text{[uniform flow]} \quad \text{19.7}$$

Any open channel performance equation can be written using the geometric slope, S_0, instead of the hydraulic slope, S, but only under the condition of uniform flow.

In most problems, the slope is a function of the terrain and is known. However, it may be necessary to calculate the slope that results in some other specific parameter. The slope that produces flow at some normal depth, d, is called the *normal slope*. The slope that produces flow at some critical depth, d_c, is called the *critical slope*. Both are determined by solving the Manning equation for slope.

6. GOVERNING EQUATIONS FOR UNIFORM FLOW

Since water is incompressible, the continuity equation is

$$A_1 \text{v}_1 = A_2 \text{v}_2 \qquad \textit{19.8}$$

The most common equation used to calculate the flow velocity in open channels is the 1768 *Chezy equation*.[3]

$$\text{v} = C\sqrt{RS} \qquad \textit{19.9}$$

Various methods for evaluating the *Chezy coefficient*, C, or "Chezy's C," have been proposed.[4] If the channel is small and very smooth, Chezy's own formula can be used. The friction factor, f, is dependent on the Reynolds number and can be found in the usual manner from the Moody diagram.

$$C = \sqrt{\frac{8g}{f}} \qquad \textit{19.10}$$

If the channel is large and the flow is fully turbulent, then the friction loss will not depend so much on the Reynolds number as on the channel roughness. The 1888 Manning formula is frequently used to evaluate the constant C.[5] Notice that the value of C depends only on the channel roughness and geometry. (The conversion constant 1.49 in Eq. 19.11(b) is reported as 1.486 by some authorities. 1.486 is the correct SI-to-English conversion, but it is doubtful whether this equation warrants four significant digits.)

$$C = \left(\frac{1.00}{n}\right) R^{1/6} \qquad \text{[SI]} \qquad \textit{19.11(a)}$$

$$C = \left(\frac{1.49}{n}\right) R^{1/6} \qquad \text{[U.S.]} \qquad \textit{19.11(b)}$$

n is the *Manning roughness coefficient* (*Manning constant*). Typical values of Manning's n are given in App. 19.A. Judgment is needed in selecting values since tabulated values often differ by as much as 30%. More important to recognize for sewer work is the layer of slime that often coats the sewer walls. Since the slime characteristics can change with location in the sewer, there can be variations in Manning's roughness coefficient along the sewer length.

Independent of these factors, the value of n also depends on the depth of flow, leading to a value (n_{full}) specifically intended for use with full flow. (It is seldom clear from tabulations such as App. 19.A whether

[3]Pronounced "Shay'-zee." This equation does not appear to be dimensionally consistent. However, the coefficient C is not a pure number. Rather, it has units of (length)$^{1/2}$/time (i.e., (acceleration)$^{1/2}$).
[4]Other methods of evaluating C include the *Kutter equation* (also known as the *G.K. formula*) and the *Bazin formula*. These methods are interesting from a historical viewpoint, but both have been replaced by the Manning equation.
[5]This equation was originally proposed in 1868 by Gaukler and again in 1881 by Hagen, both working independently. For some reason, the Frenchman Flamant attributed the equation to an Irishman, R. Manning. In Europe and many other places, the Manning equation may be known as the *Strickler equation*.

the values are for full flow or general use.) The variation in n can be taken into consideration using *Camp's correction*, shown in App. 19.C. However, this degree of sophistication cannot be incorporated into an analysis problem unless a specific value of n is known for a specific depth of flow.

Combining Eqs. 19.9 and 19.11 produces the *Manning equation*, also known as the *Chezy-Manning equation*.

$$\text{v} = \left(\frac{1.00}{n}\right) R^{2/3}\sqrt{S} \qquad \text{[SI]} \qquad \textit{19.12(a)}$$

$$\text{v} = \left(\frac{1.49}{n}\right) R^{2/3}\sqrt{S} \qquad \text{[U.S.]} \qquad \textit{19.12(b)}$$

All of the coefficients and constants in the Manning equation may be combined into the *conveyance*, K.

$$Q = \text{v}A = \left(\frac{1.00}{n}\right) A R^{2/3}\sqrt{S}$$
$$= K\sqrt{S} \qquad \text{[SI]} \qquad \textit{19.13(a)}$$

$$Q = \text{v}A = \left(\frac{1.49}{n}\right) A R^{2/3}\sqrt{S}$$
$$= K\sqrt{S} \qquad \text{[U.S.]} \qquad \textit{19.13(b)}$$

Example 19.1

A rectangular channel on a 0.002 slope is constructed of finished concrete. The channel is 8 ft (2.4 m) wide. Water flows at a depth of 5 ft (1.5 m). What is the flow rate?

SI Solution

The hydraulic radius is

$$R = \frac{A}{P} = \frac{(2.4 \text{ m})(1.5 \text{ m})}{1.5 \text{ m} + 2.4 \text{ m} + 1.5 \text{ m}}$$
$$= 0.67 \text{ m}$$

From App. 19.A, the roughness coefficient for finished concrete is 0.012. The Manning coefficient is determined by Eq. 19.11(a).

$$C = \left(\frac{1.00}{n}\right) R^{1/6} = \left(\frac{1.00}{0.012}\right)(0.67 \text{ m})^{1/6}$$
$$= 77.9$$

The discharge is

$$Q = \text{v}A = C\sqrt{RS}A$$
$$= \left(77.9 \,\frac{\sqrt{\text{m}}}{\text{s}}\right)\left(\sqrt{(0.67 \text{ m})(0.002)}\right)(1.5 \text{ m})(2.4 \text{ m})$$
$$= 10.3 \text{ m}^3/\text{s}$$

Customary U.S. Solution

The hydraulic radius is

$$R = \frac{A}{P} = \frac{(8 \text{ ft})(5 \text{ ft})}{5 \text{ ft} + 8 \text{ ft} + 5 \text{ ft}}$$
$$= 2.22 \text{ ft}$$

From App. 19.A, the roughness coefficient for finished concrete is 0.012. The Manning coefficient is determined by Eq. 19.11(b).

$$C = \left(\frac{1.49}{n}\right) R^{1/6}$$
$$= \left(\frac{1.49}{0.012}\right)(2.22 \text{ ft})^{1/6}$$
$$= 141.8$$

The discharge is

$$Q = vA = C\sqrt{RS}A$$
$$= \left(141.8 \frac{\sqrt{\text{ft}}}{\text{sec}}\right)\left(\sqrt{(2.22 \text{ ft})(0.002)}\right)(8 \text{ ft})(5 \text{ ft})$$
$$= 377.9 \text{ ft}^3/\text{sec}$$

7. VARIATIONS IN THE MANNING CONSTANT

The Manning coefficient, n, actually varies with depth. For most calculations, however, n is assumed to be constant. The accuracy of other parameters used in open-flow calculations often does not warrant considering the variation of n with depth, and the choice to use a constant or varying n value is left to the individual designer.

If it is desired to acknowledge variations in n with respect to depth, it is expedient to use tables or graphs of hydraulic elements prepared for that purpose. Table 19.3 lists such hydraulic elements under the assumption that n varies. (Appendix 19.C can be used for both varying and constant n.)

Table 19.3 Circular Channel Ratios (varying n)

$\frac{d}{D}$	$\frac{Q}{Q_{\text{full}}}$	$\frac{v}{v_{\text{full}}}$
0.1	0.02	0.31
0.2	0.07	0.48
0.3	0.14	0.61
0.4	0.26	0.71
0.5	0.41	0.80
0.6	0.56	0.88
0.7	0.72	0.95
0.8	0.87	1.01
0.9	0.99	1.04
0.95	1.02	1.03
1.00	1.00	1.00

Example 19.2

2.5 ft³/sec (0.07 m³/s) of water flow in a 20 in (0.5 m) sewer line ($n = 0.015$, $S = 0.001$). The Manning coefficient, n, varies with depth. Flow is uniform and steady. What are the velocity and depth?

SI Solution

The hydraulic radius is

$$R = \frac{D}{4} = \frac{0.5 \text{ m}}{4} = 0.125 \text{ m}$$

From Eq. 19.12(a),

$$v_{\text{full}} = \left(\frac{1.00}{n}\right) R^{2/3}\sqrt{S}$$
$$= \left(\frac{1.00}{0.015}\right)(0.125 \text{ m})^{2/3}\sqrt{0.001}$$
$$= 0.53 \text{ m/s}$$

If the pipe were flowing full, it would carry Q_{full}.

$$Q_{\text{full}} = v_{\text{full}}A$$
$$= \left(0.53 \frac{\text{m}}{\text{s}}\right)\left(\frac{\pi}{4}\right)(0.5 \text{ m})^2$$
$$= 0.10 \text{ m}^3/\text{s}$$

$$\frac{Q}{Q_{\text{full}}} = \frac{0.07 \frac{\text{m}^3}{\text{s}}}{0.10 \frac{\text{m}^3}{\text{s}}} = 0.7$$

From App. 19.C, $d/D = 0.68$ and $v/v_{\text{full}} = 0.94$.

$$v = (0.94)\left(0.53 \frac{\text{m}}{\text{s}}\right) = 0.50 \text{ m/s}$$
$$d = (0.68)(0.5 \text{ m}) = 0.34 \text{ m}$$

Customary U.S. Solution

The hydraulic radius is

$$R = \frac{D}{4} = \frac{\frac{20 \text{ in}}{12 \frac{\text{in}}{\text{ft}}}}{4} = 0.417 \text{ ft}$$

From Eq. 19.12(b),

$$v_{\text{full}} = \left(\frac{1.49}{n}\right) R^{2/3}\sqrt{S}$$
$$= \left(\frac{1.49}{0.015}\right)(0.417 \text{ ft})^{2/3}\sqrt{0.001}$$
$$= 1.75 \text{ ft/sec}$$

If the pipe were flowing full, it would carry Q_{full}.

$$Q_{\text{full}} = v_{\text{full}}A$$
$$= \left(1.75 \frac{\text{ft}}{\text{sec}}\right)\left(\frac{\pi}{4}\right)\left(\frac{20 \text{ in}}{12 \frac{\text{in}}{\text{ft}}}\right)^2$$
$$= 3.83 \text{ ft}^3/\text{sec}$$

$$\frac{Q}{Q_{\text{full}}} = \frac{2.5 \frac{\text{ft}^3}{\text{sec}}}{3.83 \frac{\text{ft}^3}{\text{sec}}} = 0.65$$

From App. 19.C, $d/D = 0.66$ and $v/v_{full} = 0.92$.

$$v = (0.92)\left(1.75 \frac{\text{ft}}{\text{sec}}\right) = 1.61 \text{ ft/sec}$$

$$d = (0.66)(20 \text{ in}) = 13.2 \text{ in}$$

8. HAZEN-WILLIAMS VELOCITY

The empirical Hazen-Williams open channel velocity equation was developed in the early 1920s. It is still occasionally used in the United States for sizing gravity sewers. It is applicable to water flows at reasonably high Reynolds numbers and is based on sound dimensional analysis. However, the constants and exponents were developed experimentally.

The equation uses the Hazen-Williams constant, C, to characterize the roughness of the channel. Since the equation is used only for water within "normal" ambient conditions, the effects of temperature, pressure, and viscosity are disregarded. The primary advantage of this approach is that the constant, C, depends only on the roughness, not on the fluid characteristics. This is also the method's main disadvantage, since professional judgment is required in choosing the value of C.

$$v = 0.85 C R^{0.63} S_0^{0.54} \quad \text{[SI]} \quad \textbf{19.14(a)}$$

$$v = 1.318 C R^{0.63} S_0^{0.54} \quad \text{[U.S.]} \quad \textbf{19.14(b)}$$

9. NORMAL DEPTH

When the depth of flow is constant along the length of the channel (i.e., the depth is neither increasing nor decreasing), the flow is said to be *uniform*. The depth of flow in that case is known as the *normal depth*, d_n. If the normal depth is known, it can be compared with the actual depth of flow to determine if the flow is uniform.[6]

The difficulty with which the normal depth is calculated depends on the cross section of the channel. If the width is very large compared to the depth, the flow cross section will essentially be rectangular and the Manning equation can be used. (Equation 19.15 assumes that the hydraulic radius equals the normal depth.)

$$d_n = \left(\frac{nQ}{w\sqrt{S}}\right)^{3/5} \quad [w \gg d_n] \quad \text{[SI]} \quad \textbf{19.15(a)}$$

$$d_n = 0.788\left(\frac{nQ}{w\sqrt{S}}\right)^{3/5} \quad [w \gg d_n] \quad \text{[U.S.]} \quad \textbf{19.15(b)}$$

Normal depth in circular channels can be calculated directly only under limited conditions. If the circular channel is flowing full, the normal depth is the inside pipe diameter.

---

[6]Normal depth is a term that applies only to uniform flow. The two alternate depths that can occur in nonuniform flow are not normal depths.

$$D = d_n = 1.548\left(\frac{nQ}{\sqrt{S}}\right)^{3/8} \quad \text{[full]} \quad \text{[SI]} \quad \textbf{19.16(a)}$$

$$D = d_n = 1.335\left(\frac{nQ}{\sqrt{S}}\right)^{3/8} \quad \text{[full]} \quad \text{[U.S.]} \quad \textbf{19.16(b)}$$

If a circular channel is flowing half full, the normal depth is half of the inside pipe diameter.

$$D = 2d_n = 2.008\left(\frac{nQ}{\sqrt{S}}\right)^{3/8} \quad \text{[half full]}$$
$$\text{[SI]} \quad \textbf{19.17(a)}$$

$$D = 2d_n = 1.731\left(\frac{nQ}{\sqrt{S}}\right)^{3/8} \quad \text{[half full]}$$
$$\text{[U.S.]} \quad \textbf{19.17(b)}$$

For other cases of uniform flow (trapezoidal, triangular, etc.), it is more difficult to determine normal depth. Various researchers have prepared tables and figures to assist in the calculations. For example, Table 19.3 is derived from App. 19.C and can be used for circular channels flowing other than full or half full.

In the absence of tables or figures, trial-and-error solutions are required. The appropriate expressions for the flow area and hydraulic radius are used in the Manning equation. Trial values are used in conjunction with graphical techniques, linear interpolation, or extrapolation to determine the normal depth. The Manning equation is solved for flow rate with various assumed values of d_n. The calculated value is compared to the actual known flow quantity, and the normal depth is approached iteratively.

For a rectangular channel whose width is small compared to the depth, the hydraulic radius and area in flow are

$$R = \frac{wd_n}{w + 2d_n} \quad\quad \textbf{19.18}$$

$$A = wd_n \quad\quad \textbf{19.19}$$

$$Q = \left(\frac{1.00}{n}\right)(wd_n)\left(\frac{wd_n}{w + 2d_n}\right)^{2/3}\sqrt{S} \quad \text{[rectangular]}$$
$$\text{[SI]} \quad \textbf{19.20(a)}$$

$$Q = \left(\frac{1.49}{n}\right)(wd_n)\left(\frac{wd_n}{w + 2d_n}\right)^{2/3}\sqrt{S} \quad \text{[rectangular]}$$
$$\text{[U.S.]} \quad \textbf{19.20(b)}$$

For a trapezoidal channel with exposed surface width w, base width b, side length s, and normal depth of flow d_n, the hydraulic radius and area in flow are

$$R = \frac{d_n(b + w)}{2(b + 2s)} \quad \text{[trapezoidal]} \quad\quad \textbf{19.21}$$

$$A = \frac{d_n(w + b)}{2} \quad \text{[trapezoidal]} \quad\quad \textbf{19.22}$$

For a symmetrical triangular channel with exposed surface width w, side slope 1:z (vertical:horizontal), and normal depth of flow d_n, the hydraulic radius and area in flow are

$$R = \frac{zd_n}{2\sqrt{1+z^2}} \qquad \textit{19.23}$$

$$A = zd_n^2 \qquad \textit{19.24}$$

10. ENERGY AND FRICTION RELATIONSHIPS

Bernoulli's equation is an expression for the conservation of energy along a fluid streamline. The Bernoulli equation can also be written for two points along the bottom of an open channel.

$$\frac{p_1}{\rho g} + \frac{v_1^2}{2g} + z_1 = \frac{p_2}{\rho g} + \frac{v_2^2}{2g} + z_2 + h_f \quad \text{[SI]} \qquad \textit{19.25(a)}$$

$$\frac{p_1}{\gamma} + \frac{v_1^2}{2g} + z_1 = \frac{p_2}{\gamma} + \frac{v_2^2}{2g} + z_2 + h_f \quad \text{[U.S.]} \qquad \textit{19.25(b)}$$

However, $p/\rho g = d$.

$$d_1 + \frac{v_1^2}{2g} + z_1 = d_2 + \frac{v_2^2}{2g} + z_2 + h_f \qquad \textit{19.26}$$

And since $d_1 = d_2$ and $v_1 = v_2$ for uniform flow at the bottom of a channel,

$$h_f = z_1 - z_2 \qquad \textit{19.27}$$

$$S_0 = \frac{z_1 - z_2}{L} \qquad \textit{19.28}$$

The channel slope, S_0, and the hydraulic energy gradient, S, are numerically the same for uniform flow. Therefore, the total friction loss along a channel is

$$h_f = LS \qquad \textit{19.29}$$

Combining Eq. 19.29 with the Manning equation results in a method for calculating friction loss.

$$h_f = \frac{Ln^2v^2}{R^{4/3}} \qquad \text{[SI]} \qquad \textit{19.30(a)}$$

$$h_f = \frac{Ln^2v^2}{2.208R^{4/3}} \qquad \text{[U.S.]} \qquad \textit{19.30(b)}$$

Example 19.3

The velocities upstream and downstream, v_1 and v_2, of a 12 ft (4.0 m) wide sluice gate are both unknown. The upstream and downstream depths are 6 ft (2.0 m) and 2 ft (0.6 m), respectively. Flow is uniform and steady. What is the downstream velocity, v_2?

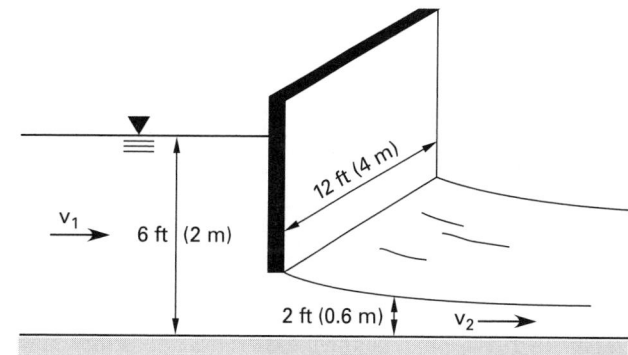

SI Solution

Since the channel bottom is essentially level on either side of the gate, $z_1 = z_2$. Bernoulli's equation reduces to

$$d_1 + \frac{v_1^2}{2g} = d_2 + \frac{v_2^2}{2g}$$

$$2 \text{ m} + \frac{v_1^2}{2g} = 0.6 \text{ m} + \frac{v_2^2}{2g}$$

v_1 and v_2 are related by continuity.

$$Q_1 = Q_2$$
$$A_1v_1 = A_2v_2$$
$$(2 \text{ m})(4 \text{ m})v_1 = (0.6 \text{ m})(4 \text{ m})v_2$$
$$v_1 = 0.3v_2$$

Substituting the expression for v_1 into the Bernoulli equation gives

$$2 \text{ m} + \frac{(0.3)^2v_2^2}{(2)\left(9.81 \dfrac{\text{m}}{\text{s}^2}\right)} = 0.6 \text{ m} + \frac{v_2^2}{(2)\left(9.81 \dfrac{\text{m}}{\text{s}^2}\right)}$$

$$2 \text{ m} + 0.004587v_2^2 = 0.6 \text{ m} + 0.050968v_2^2$$

$$v_2 = 5.5 \text{ m/s}$$

Customary U.S. Solution

Since the channel bottom is essentially level on either side of the gate, $z_1 = z_2$. Bernoulli's equation reduces to

$$d_1 + \frac{v_1^2}{2g} = d_2 + \frac{v_2^2}{2g}$$

$$6 \text{ ft} + \frac{v_1^2}{2g} = 2 \text{ ft} + \frac{v_2^2}{2g}$$

v_1 and v_2 are related by continuity.

$$Q_1 = Q_2$$
$$A_1v_1 = A_2v_2$$
$$(6 \text{ ft})(12 \text{ ft})v_1 = (2 \text{ ft})(12 \text{ ft})v_2$$
$$v_1 = \frac{v_2}{3}$$

Substituting the expression for v_1 into the Bernoulli equation gives

$$6 \text{ ft} + \frac{v_2^2}{(3)^2(2)\left(32.2 \frac{\text{ft}}{\text{sec}^2}\right)} = 2 \text{ ft} + \frac{v_2^2}{(2)\left(32.2 \frac{\text{ft}}{\text{sec}^2}\right)}$$

$$6 \text{ ft} + 0.00173v_2^2 = 2 \text{ ft} + 0.0155v_2^2$$

$$v_2 = 17.0 \text{ ft/sec}$$

Example 19.4

In Ex. 19.1, the open channel experiencing normal flow had the following characteristics: $S = 0.002$, $n = 0.012$, $v = 9.447$ ft/sec (2.9 m/s), and $R = 2.22$ ft (0.68 m). What is the energy loss per 1000 ft (100 m)?

SI Solution

There are two methods for finding the energy loss. From Eq. 19.29,

$$h_f = LS = (100 \text{ m})(0.002)$$
$$= 0.2 \text{ m}$$

From Eq. 19.30(a),

$$h_f = \frac{Ln^2v^2}{R^{4/3}}$$

$$= \frac{(100 \text{ m})(0.012)^2\left(2.9 \frac{\text{m}}{\text{s}}\right)^2}{(0.68 \text{ m})^{4/3}}$$

$$= 0.2 \text{ m}$$

Customary U.S. Solution

There are two methods for finding the energy loss. From Eq. 19.29,

$$h_f = LS = (1000 \text{ ft})(0.002)$$
$$= 2 \text{ ft}$$

From Eq. 19.30(b),

$$h_f = \frac{Ln^2v^2}{2.208R^{4/3}}$$

$$= \frac{(1000 \text{ ft})(0.012)^2\left(9.447 \frac{\text{ft}}{\text{sec}}\right)^2}{(2.208)(2.22 \text{ ft})^{4/3}}$$

$$= 2 \text{ ft}$$

11. SIZING TRAPEZOIDAL AND RECTANGULAR CHANNELS

Trapezoidal and rectangular cross sections are commonly used for artificial surface channels. The flow through a trapezoidal channel is easily determined from the Manning equation when the cross section is known. However, when the cross section or uniform depth is unknown, a trial-and-error solution is required.

Figure 19.2 *Trapezoidal Cross Section*

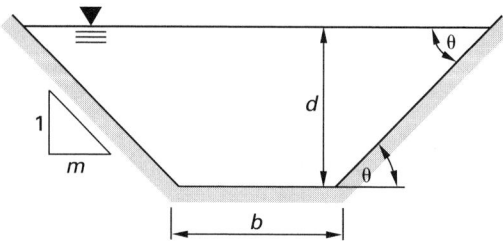

For such problems involving rectangular and trapezoidal channels, it is common to calculate and plot the *conveyance*, K (or alternatively, the product Kn), against depth. For trapezoidal sections, it is particularly convenient to write the uniform flow, Q, in terms of a modified conveyance, K'. b is the base width of the channel, d is the depth of flow, and m is the cotangent of the side slope angle. m and the ratio d/b are treated as independent variables. Values of K' are tabulated in App. 19.F.

$$Q = \frac{K'b^{8/3}\sqrt{S_0}}{n} \qquad 19.31$$

$$K' = \left(\frac{\left(1 + m\left(\frac{d}{b}\right)\right)^{5/3}}{\left(1 + 2\left(\frac{d}{b}\right)\sqrt{1+m^2}\right)^{2/3}}\right)\left(\frac{d}{b}\right)^{5/3}$$

$$\text{[SI]} \quad 19.32(a)$$

$$K' = \left(\frac{1.49\left(1 + m\left(\frac{d}{b}\right)\right)^{5/3}}{\left(1 + 2\left(\frac{d}{b}\right)\sqrt{1+m^2}\right)^{2/3}}\right)\left(\frac{d}{b}\right)^{5/3}$$

$$\text{[U.S.]} \quad 19.32(b)$$

$$m = \cot\theta \qquad 19.33$$

For any fixed value of m, enough values of K' are calculated over a reasonable range of the d/b ratio ($0.05 < d/b < 0.5$) to define a curve. Then, given specific values of Q, n, S_0, and b, the value of K' can be calculated from the expression for Q. The graph is used to determine the ratio d/b, giving the depth of uniform flow, d, since b is known.

When the ratio of d/b is very small (less than 0.02), it is satisfactory to consider the trapezoidal channel as a wide rectangular channel with area $A = bd$.

12. MOST EFFICIENT CROSS SECTION

The most efficient open channel cross section will maximize the flow for a given Manning coefficient, slope, and flow area. Accordingly, the Manning equation requires that the hydraulic radius be maximum. For a given flow area, the wetted perimeter will be minimum.

Semicircular cross sections have the smallest wetted perimeter; therefore, the cross section with the highest efficiency is the semicircle. Although such a shape can be constructed with concrete, it cannot be used with earth channels.

The most efficient cross section is also generally assumed to minimize construction cost. This is true only in the most simplified cases, however, since the labor and material costs of excavation and formwork must be considered. Rectangular and trapezoidal channels are much easier to form than semicircular channels. So in this sense the "least efficient" (i.e, most expensive) cross section (i.e., semicircular) is also the "most efficient."

The most efficient rectangle is one having depth equal to one-half of the width (i.e., is one-half of a square).

$$d = \frac{w}{2} \quad \text{[most efficient rectangle]} \qquad 19.34$$

$$A = dw = \frac{w^2}{2} = 2d^2 \qquad 19.35$$

$$P = d + w + d = 2w = 4d \qquad 19.36$$

$$R = \frac{w}{4} = \frac{d}{2} \qquad 19.37$$

The most efficient trapezoidal channel is always one in which the flow depth is twice the hydraulic radius. If the side slope is adjustable, the sides of the most efficient trapezoid should be inclined at 60° from the horizontal. Since the surface width will be equal to twice the sloping side length, the most efficient trapezoidal channel will be half of a regular hexagon (i.e., three adjacent equilateral triangles of side length $2d/\sqrt{3}$). If the side slope is any other angle, only the $d = 2R$ criterion is applicable.

$$d = 2R \quad \text{[most efficient trapezoid]} \qquad 19.38$$

$$b = \frac{2d}{\sqrt{3}} \qquad 19.39$$

$$A = \sqrt{3}d^2 \qquad 19.40$$

$$P = 3b = 2\sqrt{3}d \quad \text{[most efficient]} \qquad 19.41$$

$$R = \frac{d}{2} \qquad 19.42$$

A semicircle with its center at the middle of the water surface can always be inscribed in a cross section with maximum efficiency.

Figure 19.3 Circles Inscribed in Efficient Channels

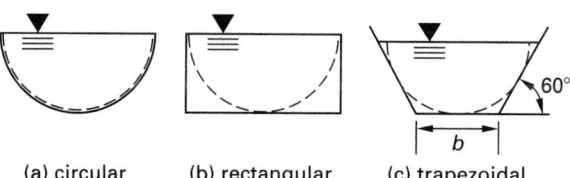

    (a) circular   (b) rectangular   (c) trapezoidal

Example 19.5

A rubble masonry open channel is being designed to carry 500 ft³/sec (14 m³/s) of water on a 0.0001 slope. Using $n = 0.017$, find the most efficient dimensions for a rectangular channel.

SI Solution

Let the depth and width be d and w, respectively. For an efficient rectangle, $d = w/2$.

$$A = dw = \left(\frac{w}{2}\right)w = \frac{w^2}{2}$$

$$P = d + w + d = \frac{w}{2} + w + \frac{w}{2} = 2w$$

$$R = \frac{A}{P} = \frac{\frac{w^2}{2}}{2w} = \frac{w}{4}$$

Using Eq. 19.13(a),

$$Q = \left(\frac{1.00}{n}\right)AR^{2/3}\sqrt{S}$$

$$14 \text{ m}^3/\text{s} = \left(\frac{1.00}{0.017}\right)\left(\frac{w^2}{2}\right)\left(\frac{w}{4}\right)^{2/3}\sqrt{0.0001}$$

$$14 \text{ m}^3/\text{s} = 0.1167w^{8/3}$$

$$w = 6.02 \text{ m}$$

$$d = \frac{w}{2} = \frac{6.02 \text{ m}}{2} = 3.01 \text{ m}$$

Customary U.S. Solution

Let the depth and width be d and w, respectively. For an efficient rectangle, $d = w/2$.

$$A = dw = \left(\frac{w}{2}\right)w = \frac{w^2}{2}$$

$$P = d + w + d = \frac{w}{2} + w + \frac{w}{2} = 2w$$

$$R = \frac{A}{P} = \frac{\frac{w^2}{2}}{2w} = \frac{w}{4}$$

Using Eq. 19.13(b),

$$Q = \left(\frac{1.49}{n}\right)AR^{2/3}\sqrt{S}$$

$$500 \text{ ft}^3/\text{sec} = \left(\frac{1.49}{0.017}\right)\left(\frac{w^2}{2}\right)\left(\frac{w}{4}\right)^{2/3}\sqrt{0.0001}$$

$$500 \text{ ft}^3/\text{sec} = 0.1739w^{8/3}$$

$$w = 19.82 \text{ ft}$$

$$d = \frac{w}{2} = \frac{19.82 \text{ ft}}{2} = 9.91 \text{ ft}$$

13. ANALYSIS OF NATURAL WATERCOURSES

Natural watercourses do not have uniform paths or cross sections. This complicates their analysis considerably. Frequently, analyzing the flow from a river is a matter of making the most logical assumptions. Many evaluations can be solved with a reasonable amount of error.

As was seen in Eq. 19.30, the friction loss (and hence the hydraulic gradient) depends on the square of the roughness coefficient. Therefore, an attempt must be made to evaluate the roughness constant as accurately as possible. If the channel consists of a river with flood plains, it should be treated as parallel channels. The flow from each subdivision can be calculated independently and the separate values added to obtain the total flow. (The common interface between adjacent subdivisions is not included in the wetted perimeter.) Alternatively, a composite value of the roughness coefficient, n_c, can be approximated from the individual values of n and the corresponding wetted perimeters.

$$n_c = \left(\frac{\sum P_i (n_i)^{3/2}}{\sum P_i} \right)^{2/3} \qquad \textbf{19.43}$$

Figure 19.4 *River with Flood Plain*

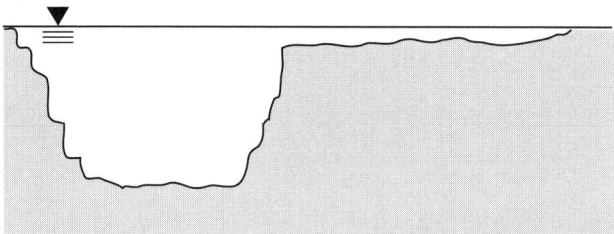

If the channel is divided by an island into two channels, some combination of flows will usually be known. For example, if the total flow, Q, is known, Q_1 and Q_2 may be unknown. If the slope is known, Q_1 and Q_2 may be known. Iterative trial-and-error solutions are often required.

Figure 19.5 *Divided Channel*

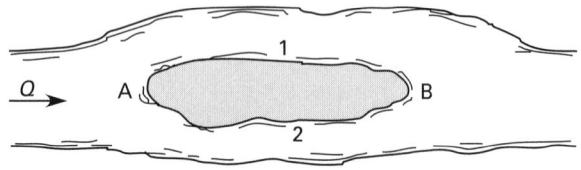

Since the elevation drop $(z_A - z_B)$ between points A and B is the same regardless of flow path,

$$S_1 = \frac{z_A - z_B}{L_1} \qquad \textbf{19.44}$$

$$S_2 = \frac{z_A - z_B}{L_2} \qquad \textbf{19.45}$$

Once the slopes are known, initial estimates Q_1 and Q_2 can be calculated from Eq. 19.13. The sum of Q_1 and Q_2 will probably not be the same as the given flow quantity, Q. In that case, Q should be prorated according to the ratios of Q_1 and Q_2 to $Q_1 + Q_2$.

If the lengths L_1 and L_2 are the same or almost so, the Manning equation may be solved for the slope by writing Eq. 19.46.

$$Q = Q_1 + Q_2$$
$$= \left(\left(\frac{A_1}{n_1} \right) (R_1)^{2/3} + \left(\frac{A_2}{n_2} \right) (R_2)^{2/3} \right) \sqrt{S}$$
$$\text{[SI]} \qquad \textbf{19.46(a)}$$

$$Q = Q_1 + Q_2$$
$$= 1.49 \left(\left(\frac{A_1}{n_1} \right) (R_1)^{2/3} + \left(\frac{A_2}{n_2} \right) (R_2)^{2/3} \right) \sqrt{S}$$
$$\text{[U.S.]} \qquad \textbf{19.46(b)}$$

Equation 19.46 yields only a rough estimate of the flow quantity, as the geometry and roughness of a natural channel changes considerably along its course.

14. FLOW MEASUREMENT WITH WEIRS

A *weir* is an obstruction in an open channel over which flow occurs. Although a dam spillway is a specific type of weir, most weirs are intended specifically for flow measurement.

Measurement weirs consist of a vertical flat plate with sharp edges. Because of their construction, they are called *sharp-crested weirs*. Sharp-crested weirs are most frequently rectangular, consisting of a straight, horizontal crest. However, weirs may also have trapezoidal and triangular openings.

For any given width of weir opening (referred to as the *weir length*), the discharge will be a function of the head over the weir. The head (or sometimes surface elevation) can be determined by a standard *staff gauge* mounted adjacent to the weir.

The full channel flow usually goes over the weir. However, it is also possible to divert a small portion of the total flow through a measurement channel. The full channel flow rate can be extrapolated from a knowledge of the split fractions.

If a rectangular weir is constructed with an opening width less than the channel width, the falling liquid sheet (called the *nappe*) decreases in width as it falls. Because of this *contraction* of the nappe, these weirs are known as *contracted weirs*, although it is the nappe that is actually contracted, not the weir. If the opening of the weir extends the full channel width, the weir is known as a *suppressed weir*, since the contractions are suppressed.

Water Resources

Figure 19.6 *Contracted and Suppressed Weirs*

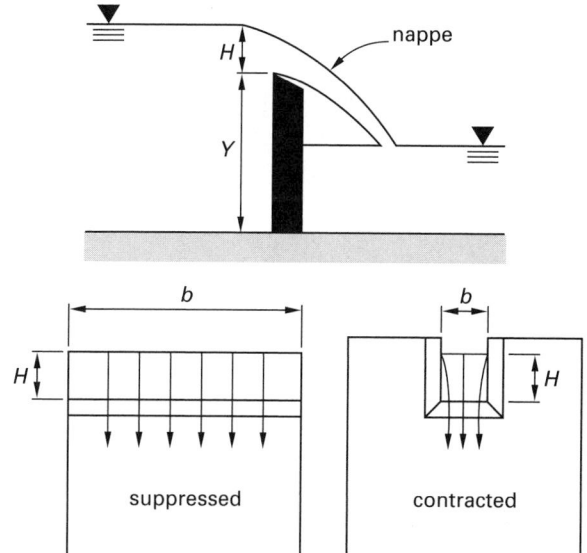

The derivation of an expression for the quantity flowing over a weir is dependent on many simplifying assumptions. The basic weir equation (Eq. 19.47 or 19.48) is, therefore, an approximate result requiring correction by experimental coefficients.

If it is assumed that the contractions are suppressed, upstream velocity is uniform, flow is laminar over the crest, nappe pressure is zero, the nappe is fully ventilated, and viscosity, turbulence, and surface tension effects are negligible, then the following equation may be derived from the Bernoulli equation.

$$Q = \tfrac{2}{3}b\sqrt{2g}\left(\left(H + \frac{\text{v}_1^2}{2g}\right)^{3/2} - \left(\frac{\text{v}_1^2}{2g}\right)^{3/2}\right) \qquad \textbf{19.47}$$

If the velocity of approach, v_1, is negligible, then

$$Q = \tfrac{2}{3}b\sqrt{2g}H^{3/2} \qquad \textbf{19.48}$$

Equation 19.48 must be corrected for all of the assumptions made, primarily for a nonuniform velocity distribution. This is done by introducing an empirical discharge coefficient, C_1. Equation 19.49 is known as the *Francis weir equation.*

$$Q = \tfrac{2}{3}C_1 b\sqrt{2g}H^{3/2} \qquad \textbf{19.49}$$

Many investigations have been done to evaluate C_1 analytically. Perhaps the most widely known is the coefficient formula developed by *Rehbock.*[7]

---

[7]There is much variation in how different investigators calculate the discharge coefficient, C_1. For ratios of H/b less than 5, $C_1 = 0.622$ gives a reasonable value. With the questionable accuracy of some of the other variables used in open channel flow problems, the pursuit of greater accuracy is of dubious value.

$$C_1 = \left(0.6035 + 0.0813\left(\frac{H}{Y}\right) + \frac{0.000295}{Y}\right)$$
$$\times \left(1 + \frac{0.00361}{H}\right)^{3/2} \qquad \text{[U.S. only]}$$
$$\approx 0.602 + 0.083\left(\frac{H}{Y}\right) \qquad \text{[U.S. and SI]} \qquad \textbf{19.50}$$

When $H/Y < 0.2$, C_1 approaches 0.61 to 0.62. In most cases, a value in this range is adequate. Other constants (i.e., $^2/_3$ and $\sqrt{2g}$, can be taken out of Eq. 19.49. In that case,

$$Q \approx 1.84bh^{3/2} \qquad \text{[SI]} \qquad \textbf{19.51(a)}$$
$$Q \approx 3.33bh^{3/2} \qquad \text{[U.S.]} \qquad \textbf{19.51(b)}$$

If the contractions are not suppressed (i.e., one or both sides do not extend to the channel sides) then the actual width, b, should be replaced with the *effective width*. In Eq. 19.52, N is 1 if one side is contracted and N is 2 if there are two end contractions.

$$b_{\text{effective}} = b_{\text{actual}} - 0.1NH \qquad \textbf{19.52}$$

A *submerged rectangular weir* requires a more complex analysis because of the difficulty in measuring H and because the discharge depends on both the upstream and downstream depths. The following equation, however, may be used with little difficulty.

$$Q_{\text{submerged}} = Q_{\text{free flow}}\left(1 - \left(\frac{H_{\text{downstream}}}{H_{\text{upstream}}}\right)^{3/2}\right)^{0.385} \qquad \textbf{19.53}$$

Equation 19.53 is used by first finding the flow rate, Q, from Eq. 19.49 and then correcting it with the bracketed quantity.

Figure 19.7 *Submerged Weir*

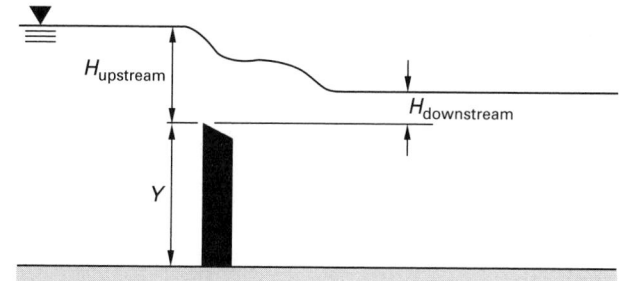

Example 19.6

The crest of a sharp-crested, rectangular weir with two contractions is 2.5 ft (1.0 m) high above the channel bottom. The crest is 4 ft (1.6 m) long. A 4 in (100 mm) head exists over the weir. What is the velocity of approach?

SI Solution

The number of contractions, N, is 2. From Eq. 19.52, the effective width is

$$H = \frac{100 \text{ mm}}{1000 \frac{\text{mm}}{\text{m}}} = 0.1 \text{ m}$$

$$b_{\text{effective}} = b_{\text{actual}} - 0.1NH = 1.6 \text{ m} - (0.1)(2)(0.1 \text{ m})$$
$$= 1.58 \text{ m}$$

$$C_1 \approx 0.602 + 0.083 \left(\frac{H}{Y}\right)$$
$$= 0.602 + (0.083)\left(\frac{0.1}{1}\right) = 0.61$$

From Eq. 19.49, the flow is

$$Q = \tfrac{2}{3} C_1 b \sqrt{2g} H^{3/2}$$
$$= \left(\tfrac{2}{3}\right)(0.61)(1.58 \text{ m})\sqrt{(2)\left(9.81 \frac{\text{m}}{\text{s}^2}\right)}(0.10 \text{ m})^{3/2}$$
$$= 0.090 \text{ m}^3/\text{s}$$

$$\text{v} = \frac{Q}{A}$$
$$= \frac{0.090 \frac{\text{m}^3}{\text{s}}}{(1.6 \text{ m})(1.0 \text{ m} + 0.1 \text{ m})}$$
$$= 0.05 \text{ m/s}$$

Customary U.S. Solution

The number of contractions, N, is 2. From Eq. 19.52, the effective width is

$$H = \frac{4 \text{ in}}{12 \frac{\text{in}}{\text{ft}}} = 0.333 \text{ ft}$$

$$b_{\text{effective}} = b_{\text{actual}} - 0.1NH = 4 \text{ ft} - (0.1)(2)(0.333 \text{ ft})$$
$$= 3.93 \text{ ft}$$

The Rehbock coefficient (Eq. 19.50) is

$$C_1 = \left(0.6035 + 0.0813 \left(\frac{H}{Y}\right)\right.$$
$$\left. + \left(\frac{0.000295}{Y}\right)\right)\left(1 + \frac{0.00361}{H}\right)^{3/2}$$
$$= \left(0.6035 + (0.0813)\left(\frac{0.333 \text{ ft}}{2.5 \text{ ft}}\right) + \frac{0.000295}{2.5 \text{ ft}}\right)$$
$$\times \left(1 + \frac{0.00361}{0.333 \text{ ft}}\right)^{3/2}$$
$$= 0.624$$

From Eq. 19.49, the flow is

$$Q = \tfrac{2}{3} C_1 b \sqrt{2g} H^{3/2}$$
$$= \left(\tfrac{2}{3}\right)(0.624)(3.93 \text{ ft})\sqrt{(2)\left(32.2 \frac{\text{ft}}{\text{sec}^2}\right)}(0.333 \text{ ft})^{3/2}$$
$$= 2.52 \frac{\text{ft}^3}{\text{sec}}$$

$$\text{v} = \frac{Q}{A}$$
$$= \frac{2.52 \frac{\text{ft}^3}{\text{sec}}}{(4 \text{ ft})(2.5 \text{ ft} + 0.333 \text{ ft})}$$
$$= 0.222 \text{ ft/sec}$$

15. TRIANGULAR WEIRS

Triangular weirs (*V-notch weirs*) should be used when small flow rates are to be measured. The flow coefficient over a triangular weir depends on the notch angle, θ, but generally varies from 0.58 to 0.61. For a 90° weir, $C_2 \approx 0.593$.

$$Q = C_2 \left(\frac{8}{15}\right)\tan\left(\frac{\theta}{2}\right)\sqrt{2g} H^{5/2} \qquad \textit{19.54}$$

$$Q \approx 1.4 H^{2.5} \quad [\text{90° weir}] \qquad [\text{SI}] \quad \textit{19.55(a)}$$

$$Q \approx 2.5 H^{2.5} \quad [\text{90° weir}] \qquad [\text{U.S.}] \quad \textit{19.55(b)}$$

Figure 19.8 *Triangular Weir*

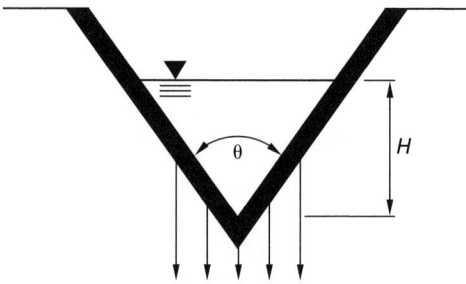

16. TRAPEZOIDAL WEIRS

A *trapezoidal weir* is essentially a rectangular weir with a triangular weir on either side. If the angle of the sides from the vertical is approximately 14° (i.e., 4 vertical and 1 horizontal), the weir is known as a *Cipoletti weir*. The discharge from the triangular ends of a Cipoletti weir approximately make up for the contractions that would reduce the flow over a rectangular weir. Therefore, no correction is theoretically necessary. This is not completely accurate, and for this reason, Cipoletti weirs are not used where great accuracy is required. The discharge is

$$Q = \tfrac{2}{3} C_d b \sqrt{2g} H^{3/2} \qquad \textit{19.56}$$

The average value of the discharge coefficient is 0.63. The discharge from a Cipoletti weir is given by Eq. 19.57.

$$Q = 1.86bH^{3/2} \qquad \text{[SI]} \qquad 19.57(a)$$

$$Q = 3.367bH^{3/2} \qquad \text{[U.S.]} \qquad 19.57(b)$$

Figure 19.9 *Trapezoidal Weir*

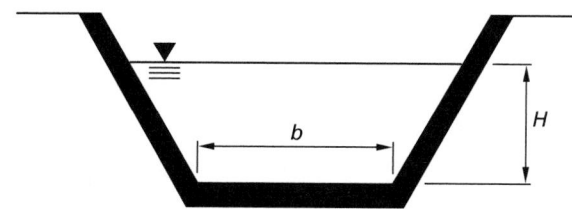

17. BROAD-CRESTED WEIRS AND SPILLWAYS

Most weirs used for flow measurement are sharp-crested. However, the flow over spillways, broad-crested weirs, and similar features can be calculated from Eq. 19.49 even though flow measurement is not the primary function of the feature. (A weir is broad-crested if the weir thickness is greater than half of the head, H.)

A *spillway* (*overflow spillway*) is designed for a capacity based on the dam's inflow hydrograph, turbine capacity, and storage capacity. Spillways frequently have a cross section known as an *ogee*, which closely approximates the underside of a nappe from a sharp-crested weir. This cross section minimizes the cavitation that is likely to occur if the water surface breaks contact with the spillway due to upstream heads that are higher than designed for.[8]

Discharge from an overflow spillway is derived in the same manner as for a weir. Equation 19.58 can be used for broad-crested weirs ($C_1 = 0.5$ to 0.57) and ogee spillways ($C_1 = 0.60$ to 0.75).

$$Q = \tfrac{2}{3}C_1 b\sqrt{2g}H^{3/2} \qquad 19.58$$

The *Horton equation* (Eq. 19.59) for broad-crested weirs combines all of the coefficients into a spillway (weir) coefficient and adds the velocity of approach to the upstream head. The Horton coefficient, C_{Horton}, is specific to the Horton equation.

$$Q = C_{\text{Horton}}b\left(H + \frac{\mathrm{v}^2}{2g}\right)^{3/2} \qquad 19.59$$

If the velocity of approach is insignificant, the discharge is

$$Q = C_s bH^{3/2} \qquad 19.60$$

---

[8]Cavitation and separation will normally not occur as long as the actual head, H, is less than twice the design value. The shape of the ogee spillway will be a function of the design head.

C_s is a *spillway coefficient*, which varies from about 3.3 to 3.98 $\text{ft}^{0.5}/\text{sec}$ (1.8 to 2.2 $\text{m}^{0.5}/\text{s}$) for ogee spillways. 3.97 $\text{ft}^{0.5}/\text{sec}$ (2.2 $\text{m}^{0.5}/\text{s}$) is frequently used for first approximations. (Notice that C_s and C_1 differ by a factor of about 5 and cannot easily be mistaken for each other.) For broad-crested weirs, C_s varies between 2.63 and 3.33 $\text{ft}^{0.5}/\text{sec}$ (1.45 and 1.84 $\text{m}^{0.5}/\text{s}$). (Use 3.33 $\text{ft}^{0.5}/\text{sec}$ (1.84 $\text{m}^{0.5}/\text{s}$) for initial estimates.) C_s increases as the upstream design head above the spillway top, H, increases, and the larger values apply to the higher heads.

Broad-crested weirs and spillways should be calibrated to obtain greater accuracy in predicting flow rates.

Scour protection is usually needed at the toe of a spillway to protect the area exposed to a hydraulic jump. This protection usually takes the form of an extended horizontal or sloping apron. Other measures, however, are needed if the tailwater exhibits large variations in depth.

18. PROPORTIONAL WEIRS

The *proportional weir* (*Sutro weir*) is used in water level control because it demonstrates a linear relationship between Q and H. Figure 19.10 illustrates a proportional weir whose sides are hyperbolic in shape.

$$Q = C_d K\left(\frac{\pi}{2}\right)\sqrt{2g}H \qquad 19.61$$

$$K = 2x\sqrt{y} \qquad 19.62$$

Figure 19.10 *Proportional Weir*

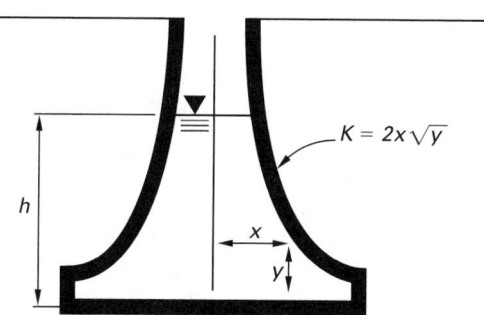

19. FLOW MEASUREMENT WITH PARSHALL FLUMES

The Parshall flume is widely used for measuring open channel wastewater flows. It performs well when head losses must be kept to a minimum and when there are high amounts of suspended solids.

The Parshall flume is constructed with a converging upstream section, a throat, and a diverging downstream section. The walls of the flume are vertical, but the floor of the throat section drops. The length, width, and height of the flume are essentially predefined by the anticipated flow rate.[9]

---

[9]This chapter does not attempt to design the Parshall flume, only to predict flow rates through its use.

Figure 19.11 Parshall Flume

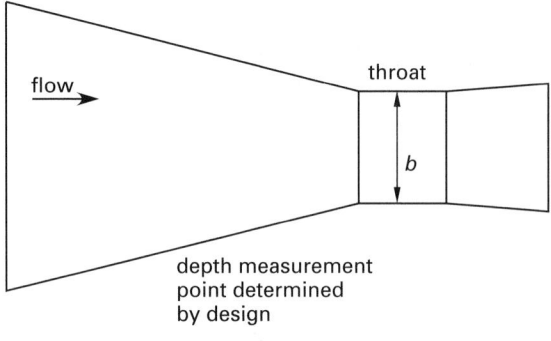

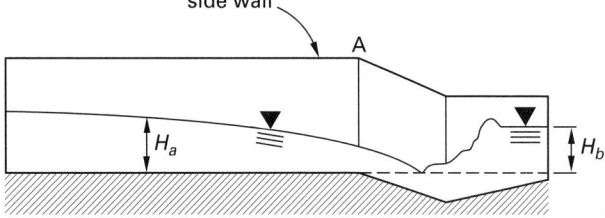

The throat geometry in a Parshall flume has been designed to force the occurrence of critical flow (Sec. 25) at that point. Following the critical section is a short length of supercritical flow followed by a hydraulic jump (Sec. 33). This design eliminates any dead water region where debris and silt can accumulate (as are common with flat-topped weirs).

The discharge relationship for a Parshall flume is given by Eq. 19.63 for submergence ratios of H_b/H_a up to 0.7. Above 0.7, the true discharge is less than predicted by Eq. 19.63. Values of K are given in Table 19.4, although using a value of 4.0 is accurate for most purposes.

$$Q = KbH_a^n \qquad \textbf{19.63}$$

$$n = 1.522b^{0.026} \qquad \textbf{19.64}$$

Above a certain tailwater height, the Parshall flume no longer operates in the *free-flow mode*. Rather, it operates in a *submerged mode*. A very high tailwater reduces the flow rate through the flume. Equation 19.63 predicts the flow rate with reasonable accuracy, however, even for 50 to 80% submergence (calculated as H_b/H_a). For large submergence, the tailwater height must be known and a different analysis method must be used.

Table 19.4 *Parshall Flume K-Values*

b, ft (m)	K
0.25 (0.075)	3.97
0.50 (0.15)	4.12
0.75 (0.225)	4.09
1.0 (0.3)	4.00
1.5 (0.45)	4.00
2.0 (0.6)	4.00
3.0 (0.9)	4.00
4.0 (1.2)	4.00

(Multiply ft by 0.3 to obtain *m*.)

20. UNIFORM AND NONUNIFORM STEADY FLOW

Steady flow is constant-volume flow. However, the flow may be uniform or nonuniform (varied) in depth. There may be significant variations over long and short distances without any change in the flow rate.

Figure 19.12 illustrates the three definitions of "slope" existing for open channel flow. These three slopes are the slope of the channel bottom, the slope of the water surface, and the slope of the energy gradient line.

Figure 19.12 Slopes Used in Open Channel Flow

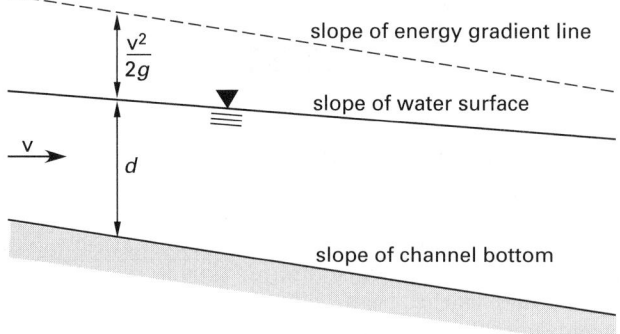

Under conditions of uniform flow, all of these three slopes are equal since the flow quantity and flow depth are constant along the length of flow.[10] With nonuniform flow, however, the flow velocity and depth vary along the length of channel and the three slopes are not necessarily equal.

If water is introduced down a path with a steep slope (as after flowing over a spillway), the effect of gravity will cause the velocity to increase. As the velocity increases, the depth decreases in accordance with the continuity of flow equation. The downward velocity is opposed by friction. Because the gravitational force is constant but friction varies with the square of velocity, these two forces eventually become equal. When equal, the velocity stops increasing, the depth stops decreasing, and the flow becomes uniform. Until they become equal, however, the flow is nonuniform (varied).

21. SPECIFIC ENERGY

The total head possessed by a fluid is given by the Bernoulli equation.

$$\frac{p}{\rho g} + \frac{v^2}{2g} + z \qquad \text{[SI]} \qquad \textbf{19.65(a)}$$

$$\frac{p}{\gamma} + \frac{v^2}{2g} + z \qquad \text{[U.S.]} \qquad \textbf{19.65(b)}$$

---

[10]As a simplification, this chapter deals only with channels of constant width. If the width is varied, changes in flow depth may not coincide with changes in flow quantity.

Specific energy, E, is defined as the total head with respect to the channel bottom. In this case, $z = 0$ and $p/\gamma = d$.

$$E = d + \frac{v^2}{2g} \qquad \textit{19.66}$$

Equation 19.66 is not meant to imply that the potential energy is an unimportant factor in open channel flow problems. The concept of specific energy is used for convenience only, and it should be clear that the Bernoulli equation is still the valid energy conservation equation.

In uniform flow, total head decreases due to the frictional effects, but specific energy is constant. In nonuniform flow, total head decreases, but specific energy may increase or decrease.

Since $v = Q/A$, Eq. 19.66 can be written as

$$E = d + \frac{Q^2}{2gA^2} \quad \text{[general]} \qquad \textit{19.67}$$

For a rectangular channel, the velocity can be written in terms of the width and flow depth.

$$v = \frac{Q}{A} = \frac{Q}{wd} \qquad \textit{19.68}$$

The specific energy equation for a rectangular channel is given by Eq. 19.69 and shown in Fig. 19.13.

$$E = d + \frac{Q^2}{2g(wd)^2} \quad \text{[rectangular]} \qquad \textit{19.69}$$

Figure 19.13 *Specific Energy Diagram*

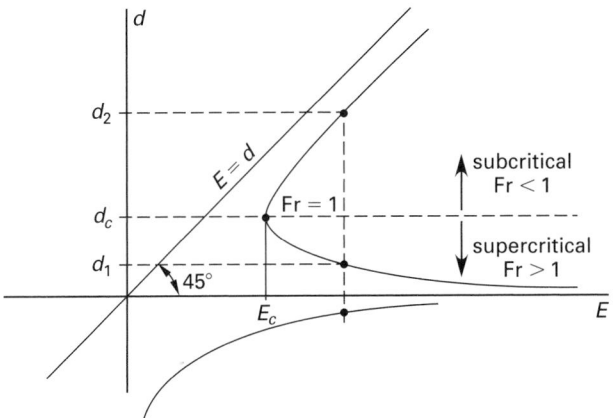

22. SPECIFIC FORCE

The *specific force* of a general channel section is the total force per unit weight acting on the water. Equivalently, specific force is the total force that a submerged object would experience. In Eq. 19.70, $\overline{d}$ is the distance from the free surface to the centroid of the flowing area cross section, A.

$$\frac{F}{g\rho} = \frac{Q^2}{gA} + \overline{d}A \qquad \text{[SI]} \qquad \textit{19.70(a)}$$

$$\frac{F}{\gamma} = \frac{Q^2}{gA} + \overline{d}A \qquad \text{[U.S.]} \qquad \textit{19.70(b)}$$

The first term in Eq. 19.70 represents the momentum flow through the channel per unit time and per unit mass of water. The second term is the pressure force per unit mass of water. Graphs of specific force and specific energy are similar in appearance and predict equivalent results for the critical and alternate depths.

23. CHANNEL TRANSITIONS

Sudden changes in channel width or bottom elevation are known as *channel transitions*. (Contractions in width are not covered in this chapter.) For sudden vertical steps in channel bottom, the Bernoulli equation, written in terms of the specific energy, is used to predict the flow behavior (i.e., the depth).

$$E_1 + z_1 = E_2 + z_2 \qquad \textit{19.71}$$
$$E_1 - E_2 = z_2 - z_1 \qquad \textit{19.72}$$

The maximum possible change in bottom elevation without affecting the energy equality occurs when the depth of flow over the step is equal to the critical depth (d_2 equals d_c). (See Sec. 25.)

24. ALTERNATE DEPTHS

Since the area depends on the depth, fixing the channel shape and slope and assuming a depth will determine the flow rate, Q, as well as the specific energy. Since Eq. 19.69 is a cubic equation, there are three values of depth of flow, d, that will satisfy it. One of them is negative, as Fig. 19.13 shows. Since depth cannot be negative, that value can be discarded. The two remaining values are known as *alternate depths*.

For a given flow rate, the two alternate depths have the same energy. One represents a high velocity with low depth; the other represents a low velocity with high depth. The former is called *supercritical (rapid) flow*; the latter is called *subcritical (tranquil) flow*.

The Bernoulli equation cannot predict which of the two alternate depths will occur for any given flow quantity. The concept of *accessibility* is required to evaluate the two depths. Specifically, the upper and lower limbs of the energy curve are not accessible from each other unless there is a local restriction in the flow.

Energy curves can be drawn for different flow quantities, as shown in Fig. 19.14 for flow quantities Q_A and Q_B. Suppose that flow is initially at point 1. (Since the flow is on the upper limb, the flow is initially subcritical.) If there is a step up in the channel bottom, Eq. 19.72 predicts that the specific energy will decrease.

Figure 19.14 *Specific Energy Curve Families*

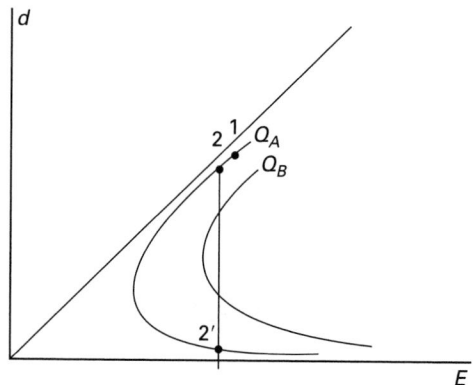

However, the flow cannot arrive at point 2′ without the flow quantity changing (i.e., going through a specific energy curve for a different flow quantity). Therefore, point 2′ is not accessible from point 1 without going through point 2 first.[11]

If the flow is well up on the top limb of the specific energy curve (as it is in Ex. 19.7), the water level will drop only slightly. Since the upper limb is asymptotic to a 45° diagonal line, any change in specific energy will result in almost the same change in depth.[12] Therefore, the surface level will remain almost the same.

$$\Delta d \approx \Delta E \quad \text{[fully subcritical]} \qquad 19.73$$

However, if the initial point on the limb is close to the critical point (i.e., the nose of the curve), then a small change in the specific energy (such as might be caused by a small variation in the channel floor) will cause a large change in depth. That is why severe turbulence commonly occurs near points of critical flow.

Example 19.7

4 ft/sec (1.2 m/s) of water flows in a 7 ft (2.1 m) wide, 6 ft (1.8 m) deep open channel. The flow encounters a 1.0 ft (0.3 m) step in the channel bottom. What is the depth of flow above the step?

---

[11]Actually, specific energy curves are typically plotted for flow per unit width, $q = Q/w$. If that is the case, a jump from one limb to the other could take place if the width were allowed to change as well as depth.
[12]A rise in the channel bottom does not always produce a drop in the water surface. Only if the flow is initially subcritical will the water surface drop upon encountering a step. The water surface will rise if the flow is initially supercritical.

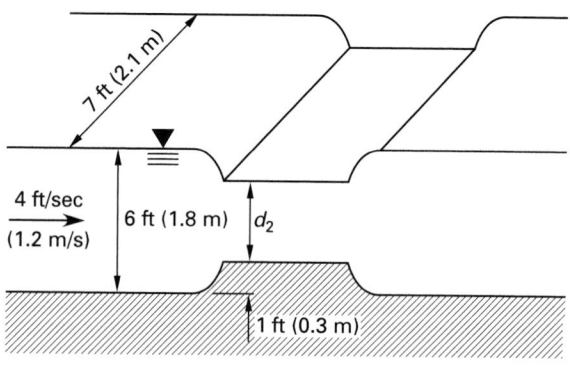

SI Solution

The initial specific energy is found from Eq. 19.66.

$$E_1 = d + \frac{v^2}{2g}$$

$$= 1.8 \text{ m} + \frac{\left(1.2 \; \frac{m}{s}\right)^2}{(2)\left(9.81 \; \frac{m}{s^2}\right)} = 1.87 \text{ m}$$

From Eq. 19.72, the specific energy above the step is

$$E_2 = E_1 + z_1 - z_2 = 1.87 \text{ m} + 0 - 0.3 \text{ m}$$
$$= 1.57 \text{ m}$$

The quantity flowing is

$$Q = Av = (2.1 \text{ m})(1.8 \text{ m})\left(1.2 \; \frac{m}{s}\right)$$
$$= 4.54 \text{ m}^3/\text{s}$$

Substituting Q into Eq. 19.69 gives

$$E = d + \frac{Q^2}{2g(wd)^2}$$

$$1.57 \text{ m} = d_2 + \frac{\left(4.54 \; \frac{m^3}{s}\right)^2}{(2)\left(9.81 \; \frac{m}{s^2}\right)(2.1 \text{ m})^2(d_2)^2}$$

By trial and error or a calculator's equation solver, the alternate depths are $d_2 = 0.46$ m, 1.46 m.

Since the 0.46 m depth is not accessible from the initial depth of 1.8 m, the depth over the step is 1.5 m. The drop in the water level is

$$1.8 \text{ m} - (1.46 \text{ m} + 0.3 \text{ m}) = 0.04 \text{ m}$$

Customary U.S. Solution

The initial specific energy is found from Eq. 19.66.

$$E_1 = d + \frac{v^2}{2g}$$

$$= 6 \text{ ft} + \frac{\left(4 \frac{\text{ft}}{\text{sec}}\right)^2}{(2)\left(32.2 \frac{\text{ft}}{\text{sec}^2}\right)} = 6.25 \text{ ft}$$

From Eq. 19.72, the specific energy over the step is

$$E_2 = E_1 + z_1 - z_2 = 6.25 \text{ ft} + 0 - 1 \text{ ft}$$
$$= 5.25 \text{ ft}$$

The quantity flowing is

$$Q = Av = (7 \text{ ft})(6 \text{ ft})\left(4 \frac{\text{ft}}{\text{sec}}\right)$$
$$= 168 \text{ ft}^3/\text{sec}$$

Substituting Q into Eq. 19.69 gives

$$E = d + \frac{Q^2}{2g(wd)^2}$$

$$5.25 \text{ ft} = d_2 + \frac{\left(168 \frac{\text{ft}^3}{\text{sec}}\right)^2}{(2)\left(32.2 \frac{\text{ft}}{\text{sec}^2}\right)(7 \text{ ft})^2(d_2)^2}$$

By trial and error or a calculator's equation solver, the alternate depths are $d_2 = 1.6$ ft, 4.9 ft.

Since the 1.6 ft depth is not accessible from the initial depth of 6 ft, the depth over the step is 4.9 ft. The drop in the water level is

$$6 \text{ ft} - (4.9 \text{ ft} + 1 \text{ ft}) = 0.1 \text{ ft}$$

25. CRITICAL FLOW AND CRITICAL DEPTH IN RECTANGULAR CHANNELS

There is one depth, known as the *critical depth*, that minimizes the energy of flow. (The depth is not minimized, however.) The critical depth for a given flow depends on the shape of the channel.

For a rectangular channel, if Eq. 19.69 is differentiated with respect to depth in order to minimize the specific energy, Eq. 19.74 results.

$$d_c^3 = \frac{Q^2}{gw^2} \quad \text{[rectangular]} \qquad \textit{19.74}$$

Geometrical and analytical methods can be used to correlate the critical depth and the minimum specific energy.

$$d_c = \tfrac{2}{3} E_c \qquad \textit{19.75}$$

For a rectangular channel, $Q = d_c w v_c$. Substituting this into Eq. 19.74 produces an equation for the *critical velocity*.

$$v_c = \sqrt{gd_c} \qquad \textit{19.76}$$

The expression for critical velocity also coincides with the expression for the velocity of a low-amplitude *surface wave* (*surge wave*) moving in a liquid of depth d_c. Since surface disturbances are transmitted as ripples upstream (and downstream) at velocity v_c, it is apparent that a surge wave will be stationary in a channel moving at the critical velocity. Such motionless waves are known as *standing waves*.

If the flow velocity is less than the surge wave velocity (for the actual depth), then a ripple can make its way upstream. If the flow velocity exceeds the surge wave velocity, the ripple will be swept downstream.

Example 19.8

500 ft³/sec (14 m³/s) of water flow in a 20 ft (6 m) wide rectangular channel. What are the (a) critical depth and (b) critical velocity?

SI Solution

(a) From Eq. 19.74, the critical depth is

$$d_c^3 = \frac{Q^2}{gw^2}$$

$$= \frac{\left(14 \frac{\text{m}^3}{\text{s}}\right)^2}{\left(9.81 \frac{\text{m}}{\text{s}^2}\right)(6 \text{ m})^2}$$

$$d_c = 0.822 \text{ m}$$

(b) From Eq. 19.76, the critical velocity is

$$v_c = \sqrt{gd_c} = \sqrt{\left(9.81 \frac{\text{m}}{\text{s}^2}\right)(0.822 \text{ m})}$$

$$= 2.84 \text{ m/s}$$

Customary U.S. Solution

(a) From Eq. 19.74, the critical depth is

$$d_c^3 = \frac{Q^2}{gw^2}$$

$$= \frac{\left(500 \frac{\text{ft}^3}{\text{sec}}\right)^2}{\left(32.2 \frac{\text{ft}}{\text{sec}^2}\right)(20 \text{ ft})^2}$$

$$d_c = 2.687 \text{ ft}$$

(b) From Eq. 19.76, the critical velocity is

$$v_c = \sqrt{gd_c} = \sqrt{\left(32.2 \frac{\text{ft}}{\text{sec}^2}\right)(2.687 \text{ ft})}$$

$$= 9.30 \text{ ft/sec}$$

Water Resources

26. CRITICAL FLOW AND CRITICAL DEPTH IN NONRECTANGULAR CHANNELS

For nonrectangular shapes (including trapezoidal channels), the critical depth can be found by trial and error from the following equation in which T is the surface width. To use Eq. 19.77, assume trial values of the critical depth, use them to calculate dependent quantities in the equation, and then verify the equality.

$$\frac{Q^2}{g} = \frac{A^3}{T} \quad \text{[nonrectangular]} \qquad 19.77$$

Equation 19.77 is particularly difficult to use with circular channels. Appendix 19.D is a convenient method of determining critical depth in circular channels.

27. FROUDE NUMBER

The dimensionless Froude number, Fr, is a convenient index of the flow regime. It can be used to determine whether the flow is subcritical or supercritical. L is the *characteristic length*, also referred to as the *characteristic (length) scale*, hydraulic depth, mean hydraulic depth, and others, depending on the channel configuration. d is the depth corresponding to velocity v. For circular channels flowing half full, $L = \pi D/8$. For a rectangular channel, $L = d$. For trapezoidal and semi-circular channels and in general, L is the area in flow divided by the top width, T.

$$\text{Fr} = \frac{\text{v}}{\sqrt{gL}} \qquad 19.78$$

When the Froude number is less than one, the flow is subcritical; that is, the depth of flow is greater than the critical depth, and the velocity is less than the critical velocity.

For convenience, the Froude number can be written in terms of the flow rate per average unit width.

$$\text{Fr} = \frac{\dfrac{Q}{b}}{\sqrt{gd^3}} \quad \text{[rectangular]} \qquad 19.79$$

$$\text{Fr} = \frac{\dfrac{Q}{b_{\text{ave}}}}{\sqrt{g\left(\dfrac{A}{b_{\text{ave}}}\right)^3}} \quad \text{[nonrectangular]} \qquad 19.80$$

When the Froude number is greater than one, the flow is supercritical. The depth is less than critical depth, and the flow velocity is greater than the critical velocity.

When the Froude number is equal to one, the flow is critical.[13]

The Froude number has another form. Dimensional analysis determines it to be v^2/gL, a form that is also

---

[13]The similarity of the Froude number to the Mach number used to classify gas flows is more than coincidental. Both bodies of knowledge employ similar concepts.

used in analyzing similarity of models. Whether the derived form or the square root form is used can sometimes be determined by observing the form of the intended application. If the Froude number is squared (as it is in Eq. 19.81), then the square root form is probably intended. For open channel flow, the Froude number is always the square root of the derived form.

28. PREDICTING OPEN CHANNEL FLOW BEHAVIOR

Upon encountering a variation in the channel bottom, the behavior of an open channel flow is dependent on whether the flow is initially subcritical or supercritical. Open channel flow is governed by Eq. 19.81 in which the Froude number is the primary independent variable.

$$\frac{dd}{dx}(1 - \text{Fr}^2) + \frac{dz}{dx} = 0 \qquad 19.81$$

The quantity dd/dx is the slope of the surface (i.e., it is the derivative of the depth with respect to the channel length). The quantity dz/dx is the slope of the channel bottom.

For an upward step, $dz/dx > 0$. If the flow is initially subcritical (i.e., Fr < 1), then Eq. 19.81 requires that $dd/dx < 0$, a drop in depth.

This logic can be repeated for other combinations of the terms. Table 19.5 lists the various behaviors of open channel flow surface levels based on Eq. 19.81.

Table 19.5 *Surface Level Change Behavior*

initial flow	step up	step down
subcritical	surface drops	surface rises
supercritical	surface rises	surface drops

If $dz/dx = 0$ (i.e., a horizontal slope), then either the depth must be constant or the Froude number must be unity. The former case is obvious. The latter case predicts critical flow. Such critical flow actually occurs where the slope is horizontal over broad-crested weirs (see Fig. 19.17) and at the top of a rounded spillway. Since broad-crested weirs and spillways produce critical flow, they represent a class of controls on flow.

29. OCCURRENCES OF CRITICAL FLOW

The critical depth not only minimizes the energy of flow, but also maximizes the quantity flowing for a given cross section and slope. Critical flow is generally quite turbulent because of the large changes in energy that occur with small changes in elevation and depth. Critical depth flow is often characterized by successive water surface undulations over a very short stretch of channel.

For any given discharge and cross section, there is a unique slope that will produce and maintain flow at critical depth. Once d_c is known, this critical slope can be found from the Manning equation. In all of the instances of critical depth, Eq. 19.76 can be used to calculate the actual velocity.

Figure 19.15 *Occurrence of Critical Depth*

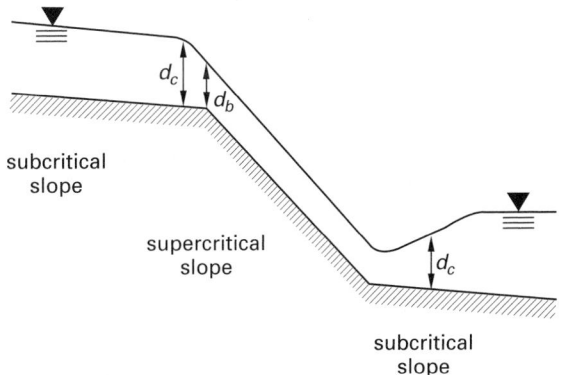

Critical depth occurs at free outfall from a channel of mild slope. The occurrence is at the point of curvature inversion, just upstream from the brink. For mild slopes, the *brink depth* is approximately

$$d_b = 0.715d_c \qquad\qquad \textbf{19.82}$$

Figure 19.16 *Free Outfall*

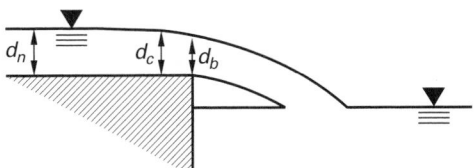

Critical flow can occur across a broad-crested weir, as shown in Fig. 19.17.[14] With no obstruction to hold the water, it falls from the normal depth to the critical depth, but it can fall no more than that because there is no source to increase the specific energy (to increase the velocity). This is not a contradiction of the previous free outfall case where the brink depth is less than the critical depth. The flow curvatures in free outfall are a result of the constant gravitational acceleration.

Figure 19.17 *Broad-Crested Weir*

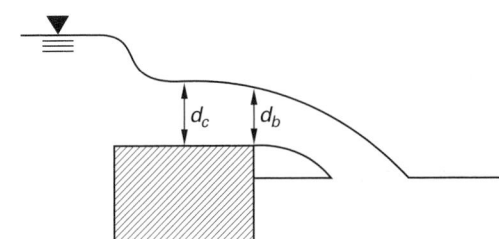

---

[14]Figure 19.17 is an example of a *hydraulic drop*, the opposite of a hydraulic jump. A hydraulic drop can be recognized by the sudden decrease in depth over a short length of channel.

Critical depth can also occur when a channel bottom has been raised sufficiently to choke the flow. A raised channel bottom is essentially a broad-crested weir.

Figure 19.18 *Raised Channel Bottom with Choked Flow*

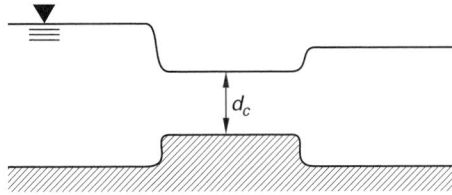

Example 19.9

At a particular point in an open rectangular channel ($n = 0.013$, $S = 0.002$, and $w = 10$ ft (3 m)), the flow is 250 ft^3/sec (7 m^3/s) and the depth is 4.2 ft (1.3 m).

(a) Is the flow tranquil, critical, or rapid?

(b) What is the normal depth?

(c) If the channel ends in a free outfall, what is the brink depth?

SI Solution

(a) From Eq. 19.74, the critical depth is

$$d_c = \left(\frac{Q^2}{gw^2}\right)^{1/3}$$

$$= \left(\frac{\left(7\ \frac{m^3}{s}\right)^2}{(3\ m)^2\left(9.81\ \frac{m}{s^2}\right)}\right)^{1/3}$$

$$= 0.82\ m$$

Since the actual depth exceeds the critical depth, the flow is tranquil.

(b) From Eq. 19.13,

$$Q = \left(\frac{1.00}{n}\right)AR^{2/3}\sqrt{S}$$

$$R = \frac{A}{P}$$

$$= \frac{d_n(3\ m)}{2d_n + 3\ m}$$

Substitute the expression for R into Eq. 19.13 and solve for d_n.

$$7\ \frac{m^3}{s} = \left(\frac{1.00}{0.013}\right)d_n(3\ m)\left(\frac{d_n(3\ m)}{2d_n + 3\ m}\right)^{2/3}\sqrt{0.002}$$

By trial and error or a calculator's equation solver, $d_n = 0.97$ m. Since the actual and normal depths are different, the flow is nonuniform.

(c) From Eq. 19.82, the brink depth is

$$d_b = 0.715 d_c = (0.715)(0.82 \text{ m})$$
$$= 0.59 \text{ m}$$

Customary U.S. Solution

(a) From Eq. 19.74, the critical depth is

$$d_c = \left(\frac{Q^2}{gw^2}\right)^{1/3}$$
$$= \left(\frac{\left(250 \ \frac{\text{ft}^3}{\text{sec}}\right)^2}{\left(32.2 \ \frac{\text{ft}}{\text{sec}^2}\right)(10 \text{ ft})^2}\right)^{1/3}$$
$$= 2.69 \text{ ft}$$

Since the actual depth exceeds the critical depth, the flow is tranquil.

(b) From Eq. 19.13,

$$Q = \left(\frac{1.49}{n}\right) A R^{2/3} \sqrt{S}$$
$$R = \frac{A}{P}$$
$$= \frac{d_n(10 \text{ ft})}{2d_n + 10 \text{ ft}}$$

Substitute the expression for R into Eq. 19.13 and solve for d_n.

$$250 \ \frac{\text{ft}^3}{\text{sec}} = \left(\frac{1.49}{0.013}\right) d_n(10 \text{ ft}) \left(\frac{d_n(10 \text{ ft})}{2d_n + 10 \text{ ft}}\right)^{2/3} \sqrt{0.002}$$

By trial and error or a calculator's equation solver, $d_n = 3.1$ ft. Since the actual and normal depths are different, the flow is nonuniform.

(c) From Eq. 19.82, the brink depth is

$$d_b = 0.715 d_c$$
$$= (0.715)(2.69 \text{ ft})$$
$$= 1.92 \text{ ft}$$

30. CONTROLS ON FLOW

If flow is subcritical, then a disturbance downstream will be able to affect the upstream conditions. Since the flow velocity is less than the critical velocity, a ripple will be able to propagate upstream to signal a change in the downstream conditions. Any object downstream that affects the flow rate, velocity, or depth upstream is known as a *downstream control*.

If a flow is supercritical, then a downstream obstruction will have no effect upstream, since disturbances cannot propagate upstream faster than the flow velocity. The only effect on supercritical flow is from an upstream obstruction. Such an obstruction is said to be an *upstream control*.

Figure 19.19 *Control on Flow*

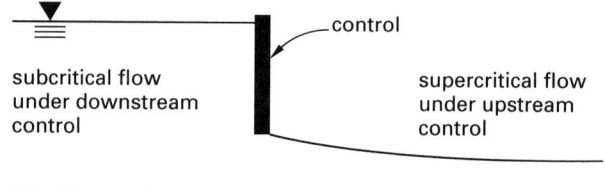

In general, any feature that affects depth and discharge rates is known as a *control on flow*. Controls may consist of constructed control structures (weirs, gates, sluices, etc.), forced flow through critical depth (as in a free outfall), sudden changes of slope (which forces a hydraulic jump or hydraulic drop to the new normal depth), or free flow between reservoirs of different surface elevations. A downstream control may also be an upstream control, as Fig. 19.19 shows.

31. FLOW CHOKING

A channel feature that causes critical flow to occur is known as a *choke*, and the corresponding flow past the feature and downstream is known as *choked flow*.

In the case of vertical transitions (i.e., upward or downward steps in the channel bottom), choked flow will occur when the step size is equal to the difference between the upstream specific energy and the critical flow energy.

$$\Delta z = E_1 - E_c \quad \text{[choked flow]} \qquad \textbf{19.83}$$

In the case of a rectangular channel, the maximum variation in channel bottom will be

$$\Delta z = E_1 - \left(d_c + \frac{\text{v}_c^2}{2g}\right)$$
$$= E_1 - \tfrac{3}{2} d_c \qquad \textbf{19.84}$$

The flow downstream from a choke point can be subcritical or supercritical, depending on the downstream conditions. If there is a downstream control, such as a sluice gate, the flow downstream will be subcritical. If there is additional gravitational acceleration (as with flow down the side of a dam spillway), then the flow will be supercritical.

32. VARIED FLOW

Accelerated flow occurs in any channel where the actual slope exceeds the friction loss per foot.

$$S_0 > \frac{h_f}{L} \qquad \textbf{19.85}$$

Retarded flow occurs when the actual slope is less than the unit friction loss.

$$S_0 < \frac{h_f}{L} \qquad \textit{19.86}$$

Figure 19.20 *Varied Flow*

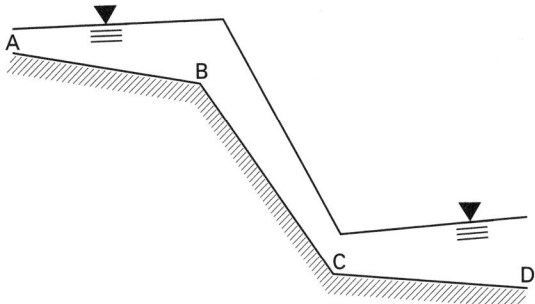

In sections AB and CD of Fig. 19.20, the slopes are less than the energy gradient, so the flows are retarded. In section BC, the slope is greater than the energy gradient, so the velocity increases (i.e., the flow is accelerated). If section BC were long enough, the friction loss would eventually become equal to the accelerating energy and the flow would become uniform.

The distance between points 1 and 2 with two known depths in accelerated or retarded flow can be determined from the average velocity. Equations 19.87 and 19.88 assume that the friction losses are the same for varied flow as for uniform flow.

$$S_{\text{ave}} = \left(\frac{n v_{\text{ave}}}{(R_{\text{ave}})^{2/3}} \right)^2 \quad \text{[SI]} \qquad \textit{19.87(a)}$$

$$S_{\text{ave}} = \left(\frac{n v_{\text{ave}}}{1.49(R_{\text{ave}})^{2/3}} \right)^2 \quad \text{[U.S.]} \qquad \textit{19.87(b)}$$

$$v_{\text{ave}} = \tfrac{1}{2}(v_1 + v_2) \qquad \textit{19.88}$$

S is the slope of the energy gradient from Eq. 19.87, not the channel slope S_0. The usual method of finding the *depth profile* is to start at a point in the channel where d_2 and v_2 are known. Then, assume a depth d_1, find v_1 and S, and solve Eq. 19.89 for L. Repeat as needed.

$$L = \frac{\left(d_1 + \frac{v_1^2}{2g} \right) - \left(d_2 + \frac{v_2^2}{2g} \right)}{S - S_0}$$
$$= \frac{E_1 - E_2}{S - S_0} \qquad \textit{19.89}$$

In Eqs. 19.88 and 19.89, d_1 is always the smaller of the two depths.

Example 19.10

How far from the point described in Ex. 19.9 will the depth be 4 ft (1.2 m)?

SI Solution

The difference between 1.3 m and 1.2 m is small, so a one-step calculation will probably be sufficient.

$$d_1 = 1.2 \text{ m}$$

$$v_1 = \frac{Q}{A} = \frac{7 \, \frac{\text{m}^3}{\text{s}}}{(1.2 \text{ m})(3 \text{ m})} = 1.94 \text{ m/s}$$

$$E_1 = d_1 + \frac{v_1^2}{2g} = 1.2 \text{ m} + \frac{\left(1.94 \, \frac{\text{m}}{\text{s}} \right)^2}{(2) \left(9.81 \, \frac{\text{m}}{\text{s}^2} \right)}$$
$$= 1.39 \text{ m}$$

$$R_1 = \frac{A_1}{P_1} = \frac{(1.2 \text{ m})(3 \text{ m})}{1.2 \text{ m} + 3 \text{ m} + 1.2 \text{ m}} = 0.67 \text{ m}$$

$$d_2 = 1.3 \text{ m}$$

$$v_2 = \frac{7 \, \frac{\text{m}^3}{\text{s}}}{(1.3 \text{ m})(3 \text{ m})} = 1.79 \text{ m/s}$$

$$E_2 = d_2 + \frac{v_2^2}{2g} = 1.3 \text{ m} + \frac{\left(1.79 \, \frac{\text{m}}{\text{s}} \right)^2}{(2) \left(9.81 \, \frac{\text{m}}{\text{s}^2} \right)} = 1.46 \text{ m}$$

$$R_2 = \frac{A_2}{P_2} = \frac{(1.3 \text{ m})(3 \text{ m})}{1.3 \text{ m} + 3 \text{ m} + 1.3 \text{ m}} = 0.70 \text{ m}$$

$$v_{\text{ave}} = \tfrac{1}{2}(v_1 + v_2) = \left(\tfrac{1}{2} \right) \left(1.94 \, \frac{\text{m}}{\text{s}} + 1.79 \, \frac{\text{m}}{\text{s}} \right)$$
$$= 1.865 \text{ m/s}$$

$$R_{\text{ave}} = \tfrac{1}{2}(R_1 + R_2) = \left(\tfrac{1}{2} \right) (0.70 \text{ m} + 0.67 \text{ m})$$
$$= 0.685 \text{ m}$$

From Eq. 19.87,

$$S = \left(\frac{n v_{\text{ave}}}{(R_{\text{ave}})^{2/3}} \right)^2$$
$$= \left(\frac{(0.013) \left(1.865 \, \frac{\text{m}}{\text{s}} \right)}{(1.00)(0.685 \text{ m})^{2/3}} \right)^2$$
$$= 0.000973$$

From Eq. 19.89,

$$L = \frac{E_1 - E_2}{S - S_0} = \frac{1.39 \text{ m} - 1.46 \text{ m}}{0.000973 - 0.002}$$
$$= 68.2 \text{ m}$$

Customary U.S. Solution

The difference between 4 ft and 4.2 ft is small, so a one-step calculation will probably be sufficient.

$$d_1 = 4 \text{ ft}$$

$$v_1 = \frac{Q}{A} = \frac{250 \frac{\text{ft}^3}{\text{sec}}}{(4 \text{ ft})(10 \text{ ft})} = 6.25 \text{ ft/sec}$$

$$E_1 = d_1 + \frac{v_1^2}{2g}$$

$$= 4 \text{ ft} + \frac{\left(6.25 \frac{\text{ft}}{\text{sec}}\right)^2}{(2)\left(32.2 \frac{\text{ft}}{\text{sec}^2}\right)}$$

$$= 4.607 \text{ ft}$$

$$R_1 = \frac{A_1}{P_1} = \frac{(4 \text{ ft})(10 \text{ ft})}{4 \text{ ft} + 10 \text{ ft} + 4 \text{ ft}}$$

$$= 2.22 \text{ ft}$$

$$d_2 = 4.2 \text{ ft}$$

$$v_2 = \frac{250 \frac{\text{ft}^3}{\text{sec}}}{(4.2 \text{ ft})(10 \text{ ft})} = 5.95 \text{ ft/sec}$$

$$E_2 = d_2 + \frac{v_2^2}{2g} = 4.2 \text{ ft} + \frac{\left(5.95 \frac{\text{ft}}{\text{sec}}\right)^2}{(2)\left(32.2 \frac{\text{ft}}{\text{sec}^2}\right)}$$

$$= 4.75 \text{ ft}$$

$$R_2 = \frac{A_2}{P_2} = \frac{(4.2 \text{ ft})(10 \text{ ft})}{4.2 \text{ ft} + 10 \text{ ft} + 4.2 \text{ ft}} = 2.28 \text{ ft}$$

$$v_{\text{ave}} = \tfrac{1}{2}(v_1 + v_2)$$

$$= \left(\tfrac{1}{2}\right)\left(6.25 \frac{\text{ft}}{\text{sec}} + 5.95 \frac{\text{ft}}{\text{sec}}\right) = 6.1 \text{ ft/sec}$$

$$R_{\text{ave}} = \tfrac{1}{2}(R_1 + R_2)$$

$$= \left(\tfrac{1}{2}\right)(2.22 \text{ ft} + 2.28 \text{ ft}) = 2.25 \text{ ft}$$

From Eq. 19.87,

$$S = \left(\frac{n v_{\text{ave}}}{1.49(R_{\text{ave}})^{2/3}}\right)^2$$

$$= \left(\frac{(0.013)\left(6.1 \frac{\text{ft}}{\text{sec}}\right)}{(1.49)(2.25 \text{ ft})^{2/3}}\right)^2$$

$$= 0.000965$$

From Eq. 19.89,

$$L = \frac{E_1 - E_2}{S - S_0}$$

$$= \frac{4.607 \text{ ft} - 4.75 \text{ ft}}{0.000965 - 0.002} = 138 \text{ ft}$$

33. HYDRAULIC JUMP

If water is introduced at high (supercritical) velocity to a section of slow-moving (subcritical) flow (as in Fig. 19.21), the velocity will be reduced rapidly over a short length of channel. The abrupt rise in the water surface is known as a *hydraulic jump*. The increase in depth is always from below the critical depth to above the critical depth.[15] The depths on either side of the hydraulic jump are known as *conjugate depths*. The conjugate depths and the relationship between them are as follows.

$$d_1 = -\tfrac{1}{2}d_2 + \sqrt{\frac{2v_2^2 d_2}{g} + \frac{d_2^2}{4}} \quad \left[\begin{array}{c}\text{rectangular}\\\text{channels}\end{array}\right] \quad \textit{19.90}$$

$$d_2 = -\tfrac{1}{2}d_1 + \sqrt{\frac{2v_1^2 d_1}{g} + \frac{d_1^2}{4}} \quad \left[\begin{array}{c}\text{rectangular}\\\text{channels}\end{array}\right] \quad \textit{19.91}$$

$$\frac{d_2}{d_1} = \tfrac{1}{2}\left(\sqrt{1 + 8(\text{Fr}_1)^2} - 1\right) \quad \left[\begin{array}{c}\text{rectangular}\\\text{channels}\end{array}\right] \quad \textit{19.92(a)}$$

$$\frac{d_1}{d_2} = \tfrac{1}{2}\left(\sqrt{1 + 8(\text{Fr}_2)^2} - 1\right) \quad \left[\begin{array}{c}\text{rectangular}\\\text{channels}\end{array}\right] \quad \textit{19.92(b)}$$

Figure 19.21 *Conjugate Depths*

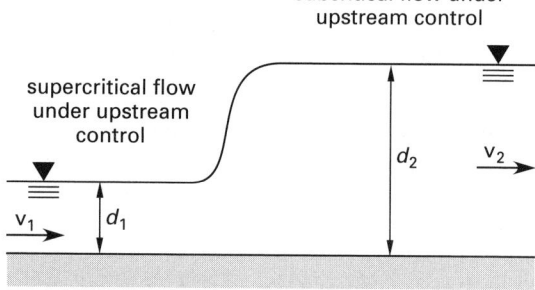

If the depths d_1 and d_2 are known, then the upstream velocity can be found from Eq. 19.93.

$$v_1^2 = \left(\frac{g d_2}{2 d_1}\right)(d_1 + d_2) \quad \left[\begin{array}{c}\text{rectangular}\\\text{channels}\end{array}\right] \quad \textit{19.93}$$

Conjugate depths are not the same as alternate depths. Alternate depths are derived from the conservation of energy equation (i.e., a variation of the Bernoulli equation). Conjugate depths (calculated in Eqs. 19.90 through 19.92) are derived from a conservation of momentum equation. Conjugate depths are calculated only when there has been an abrupt energy loss such as occurs in a hydraulic jump or drop.

Hydraulic jumps have practical applications in the design of stilling basins. Stilling basins are designed to intentionally reduce energy of flow through hydraulic jumps. In the case of a concrete apron at the bottom of

---

[15]This provides a way of determining if a hydraulic jump can occur in a channel. If the original depth is above the critical depth, the flow is already subcritical. Therefore, a hydraulic jump cannot form. Only a hydraulic drop could occur.

a dam spillway, the apron friction is usually low, and the water velocity will decrease only gradually. However, supercritical velocities can be reduced to much slower velocities by having the flow cross a series of baffles on the channel bottom.

The specific energy lost in the jump is the energy lost per pound of water flowing.

$$\Delta E = \left(d_1 + \frac{v_1^2}{2g}\right) - \left(d_2 + \frac{v_2^2}{2g}\right) \approx \frac{(d_2 - d_1)^3}{4d_1 d_2} \quad \textit{19.94}$$

Evaluation of hydraulic jumps in stilling basins starts by determining the depth at the toe. The depth of flow at the toe of a spillway is found from an energy balance. Neglecting friction, the total energy at the toe equals the total upstream energy before the spillway. The total upstream energy before the spillway is

$$E_{\text{upstream}} = E_{\text{toe}} \quad \textit{19.95}$$

$$y_{\text{crest}} + H + \frac{v^2}{2g} = d_{\text{toe}} + \frac{v_{\text{toe}}^2}{2g} \quad \textit{19.96}$$

The upstream velocity, v, is the velocity before the spillway (which is essentially zero), not the velocity over the brink. If the brink depth is known, the velocity over the brink can be used with the continuity equation to calculate the upstream velocity, but the velocity over the brink should not be used with H to determine total energy since $d_{\text{brink}} \neq H$.

Figure 19.22 *Total Energy Upstream of a Spillway*

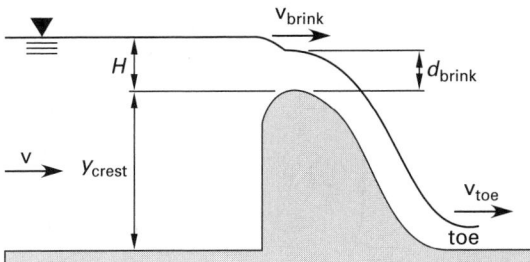

If water is drained quickly from the apron so that the tailwater depth is small or zero, no hydraulic jump will form. This is because the tailwater depth is already less than the critical depth.

A hydraulic jump will form along the apron at the bottom of the spillway when the actual tailwater depth equals the conjugate depth d_2 corresponding to the depth at the toe. That is, the jump is located at the toe when $d_2 = d_{\text{tailwater}}$, where d_2 and $d_1 = d_{\text{toe}}$ are conjugate depths. The tailwater and toe depths are implicitly the conjugate depths. This is shown in Fig. 19.23(a) and is the proper condition for energy dissipation in a stilling basin.

When the actual tailwater depth is less than the conjugate depth d_2 corresponding to d_{toe}, but still greater than the critical depth, flow will continue along the

apron until the depth increases to conjugate depth d_1 corresponding to the actual tailwater depth. This is shown in Fig. 19.23(b). A hydraulic jump will form at that point to increase the depth to the tailwater depth. (Another way of saying this is that the hydraulic jump moves downstream from the toe.) This is an undesirable condition, since the location of the jump is often inadequately protected from scour.

If the tailwater depth is greater than the conjugate depth corresponding to the depth at the toe, as in Fig. 19.23(c), the hydraulic jump may occur up on the spillway, or it may be completely submerged (i.e., it will not occur at all).

Figure 19.23 *Hydraulic Jump to Reach Tailwater Level*

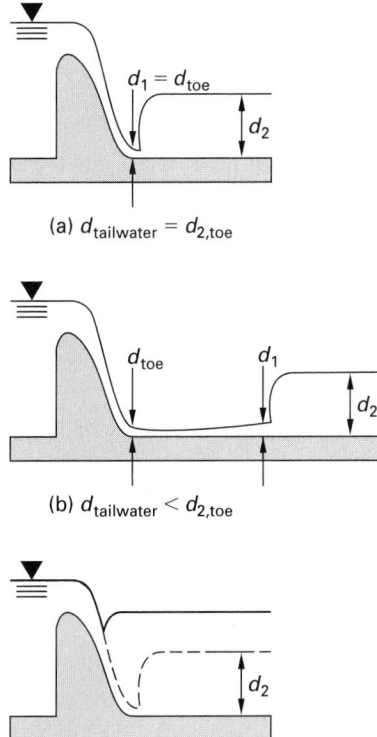

Example 19.11

A hydraulic jump is produced at a point in a 10 ft (3 m) wide channel where the depth is 1 ft (0.3 m). The flow rate is 200 ft³/sec (5.7 m³/s). (a) What is the depth after the jump? (b) What is the total power dissipated?

SI Solution

(a)
$$v_1 = \frac{Q}{A} = \frac{5.7 \, \frac{\text{m}^3}{\text{s}}}{(3 \text{ m})(0.3 \text{ m})}$$

$$= 6.33 \text{ m/s}$$

From Eq. 19.91,

$$d_2 = -\tfrac{1}{2}d_1 + \sqrt{\frac{2v_1^2 d_1}{g} + \frac{d_1^2}{4}}$$

$$= -\left(\tfrac{1}{2}\right)(0.3 \text{ m})$$

$$+ \sqrt{\frac{(2)\left(6.33 \frac{\text{m}}{\text{s}}\right)^2 (0.3 \text{ m})}{9.81 \frac{\text{m}}{\text{s}^2}} + \frac{(0.3 \text{ m})^2}{4}}$$

$$= 1.42 \text{ m}$$

(b) The mass flow rate is

$$\dot{m} = \left(5.7 \frac{\text{m}^3}{\text{s}}\right)\left(1000 \frac{\text{kg}}{\text{m}^3}\right)$$

$$= 5700 \text{ kg/s}$$

The velocity after the jump is

$$v_2 = \frac{Q}{A_2} = \frac{5.7 \frac{\text{m}^3}{\text{s}}}{(3 \text{ m})(1.42 \text{ m})}$$

$$= 1.33 \text{ m/s}$$

From Eq. 19.94, the change in specific energy is

$$\Delta E = \left(d_1 + \frac{v_1^2}{2g}\right) - \left(d_2 + \frac{v_2^2}{2g}\right)$$

$$= \left(0.3 \text{ m} + \frac{\left(6.33 \frac{\text{m}}{\text{s}}\right)^2}{(2)\left(9.81 \frac{\text{m}}{\text{s}^2}\right)}\right)$$

$$- \left(1.423 \text{ m} + \frac{\left(1.33 \frac{\text{m}}{\text{s}}\right)^2}{(2)\left(9.81 \frac{\text{m}}{\text{s}^2}\right)}\right)$$

$$= 0.83 \text{ m}$$

The total power dissipated is

$$P = \dot{m}g\Delta E$$

$$= \left(5700 \frac{\text{kg}}{\text{s}}\right)\left(9.81 \frac{\text{m}}{\text{s}^2}\right)\left(\frac{0.83 \text{ m}}{1000 \frac{\text{W}}{\text{kW}}}\right)$$

$$= 46.4 \text{ kW}$$

Customary U.S. Solution

(a) $$v_1 = \frac{Q}{A} = \frac{200 \frac{\text{ft}^3}{\text{sec}}}{(10 \text{ ft})(1 \text{ ft})}$$

$$= 20 \text{ ft/sec}$$

From Eq. 19.91,

$$d_2 = -\tfrac{1}{2}d_1 + \sqrt{\frac{2v_1^2 d_1}{g} + \frac{d_1^2}{4}}$$

$$= -\left(\tfrac{1}{2}\right)(1 \text{ ft}) + \sqrt{\frac{(2)\left(20 \frac{\text{ft}}{\text{sec}}\right)^2 (1 \text{ ft})}{32.2 \frac{\text{ft}}{\text{sec}^2}} + \frac{(1 \text{ ft})^2}{4}}$$

$$= 4.51 \text{ ft}$$

(b) The mass flow rate is

$$\dot{m} = \left(200 \frac{\text{ft}^3}{\text{sec}}\right)\left(62.4 \frac{\text{lbm}}{\text{ft}^3}\right)$$

$$= 12{,}480 \text{ lbm/sec}$$

The velocity after the jump is

$$v_2 = \frac{Q}{A_2} = \frac{200 \frac{\text{ft}^3}{\text{sec}}}{(10 \text{ ft})(4.51 \text{ ft})}$$

$$= 4.43 \text{ ft/sec}$$

From Eq. 19.94, the change in specific energy is

$$\Delta E = \left(d_1 + \frac{v_1^2}{2g}\right) - \left(d_2 + \frac{v_2^2}{2g}\right)$$

$$= \left(1 \text{ ft} + \frac{\left(20 \frac{\text{ft}}{\text{sec}}\right)^2}{(2)\left(32.2 \frac{\text{ft}}{\text{sec}^2}\right)}\right)$$

$$- \left(4.51 \text{ ft} + \frac{\left(4.43 \frac{\text{ft}}{\text{sec}}\right)^2}{(2)\left(32.2 \frac{\text{ft}}{\text{sec}^2}\right)}\right)$$

$$= 2.4 \text{ ft}$$

The total power dissipated is given by Eq. 13.28.

$$P = \frac{\dot{m}g\Delta E}{g_c}$$

$$= \frac{\left(12{,}480 \frac{\text{lbm}}{\text{sec}}\right)\left(32.2 \frac{\text{ft}}{\text{sec}^2}\right)(2.4 \text{ ft})}{\left(32.2 \frac{\text{ft-lbm}}{\text{lbf-sec}^2}\right)\left(550 \frac{\text{ft-lbf}}{\text{hp-sec}}\right)}$$

$$= 54.5 \text{ hp}$$

34. LENGTH OF HYDRAULIC JUMP

For practical stilling basin design, it is helpful to have an estimate of the length of the hydraulic jump. Lengths of hydraulic jumps are difficult to measure because of the difficulty in defining the endpoints of the jumps. However, the length of the jump, L, varies within the limits of $5 < L/d_2 < 6.5$, in which d_2 is the conjugate depth after the jump. Where greater accuracy is warranted, Table 19.6 can be used. This table correlates the length of the jump to the upstream Froude number.

Table 19.6 *Approximate Lengths of Hydraulic Jumps*

$N_{\text{Fr},1}$	$\dfrac{L}{d_2}$
3	5.25
4	5.8
5	6.0
6	6.1
7	6.15
8	6.15

35. HYDRAULIC DROP

A *hydraulic drop* is the reverse of a hydraulic jump. If water is introduced at low (subcritical) velocity to a section of fast-moving (supercritical) flow, the velocity will be increased rapidly over a short length of channel. The abrupt drop in the water surface is known as a hydraulic drop. The decrease in depth is always from above the critical depth to below the critical depth.

Water flowing over a spillway and down a long, steep chute typically experiences a hydraulic drop, with critical depth occurring just before the brink. This is illustrated in Figs. 19.17 and 19.22.

The depths on either side of the hydraulic drop are the *conjugate depths*, which are determined from Eqs. 19.90 and 19.91. The equations for calculating specific energy and power changes are the same for hydraulic jumps and drops.

36. ERODIBLE CHANNELS

Given an appropriate value of the Manning coefficient, the analysis of channels constructed of erodible materials is similar to that for concrete or pipe channels.

However, for design problems, maximum velocities and permissible side slopes must also be considered. The present state of knowledge is not sufficiently sophisticated to allow for precise designs. The usual uniform flow equations are insufficient because the stability of erodible channels is dependent on the properties of the channel material rather than on the hydraulics of flow. Two methods of design exist: (a) the tractive force method and (b) the simpler maximum permissible velocity method. Maximum velocities that should be used with erodible channels are given in Table 19.7.

Table 19.7 *Suggested Maximum Velocities*

soil type or lining (earth; no vegetation)	maximum permissible velocities (ft/sec)		
	clear water	water carrying fine silts	water carrying sand and gravel
fine sand (noncolloidal)	1.5	2.5	1.5
sandy loam (noncolloidal)	1.7	2.5	2.0
silt loam (noncolloidal)	2.0	3.0	2.0
ordinary firm loam	2.5	3.5	2.2
volcanic ash	2.5	3.5	2.0
fine gravel	2.5	5.0	3.7
stiff clay (very colloidal)	3.7	5.0	3.0
graded, loam to cobbles (noncolloidal)	3.7	5.0	5.0
graded, silt to cobbles (colloidal)	4.0	5.5	5.0
alluvial silts (noncolloidal)	2.0	3.5	2.0
alluvial silts (colloidal)	3.7	5.0	3.0
coarse gravel (noncolloidal)	4.0	6.0	6.5
cobbles and shingles	5.0	5.5	6.5
shales and hard pans	6.0	6.0	5.0

(Multiply ft/sec by 0.3 to obtain m/s.)

Source: Special Committee on Irrigation Research, ASCE, 1926.

The sides of the channel should not have a slope exceeding the natural angle of repose for the material used. Although there are other factors that determine the maximum permissible side slope, Table 19.8 lists some guidelines.

Table 19.8 *Recommended Side Slopes*

type of channel	side slope (horizontal:vertical)
firm rock	vertical to $\frac{1}{4}$:1
concrete-lined stiff clay	$\frac{1}{2}$:1
fissured rock	$\frac{1}{2}$:1
firm earth with stone lining	1:1
firm earth, large channels	1:1
firm earth, small channels	$1\frac{1}{2}$:1
loose, sandy earth	2:1
sandy, porous loam	3:1

37. CULVERTS[16]

A *culvert* is a pipe that carries water under or through some feature (usually a road or highway) that would otherwise block the flow of water. For example, highways are often built at right angles to ravines draining hillsides and other watersheds. Culverts under the highway keep the construction fill from blocking the natural runoff.

Culverts are classified according to which of their ends controls the discharge capacity: inlet control or outlet control. If water can flow through and out of the culvert faster than it can enter, the culvert is under *inlet control*. If water can flow into the culvert faster than it can flow through and out, the culvert is under *outlet control*. Culverts under inlet control will always flow partially full. Culverts under outlet control can flow either partially full or full.

The culvert length is one of the most important factors in determining whether the culvert flows full. A culvert may be known as "hydraulically long" if it runs full and "hydraulically short" if it does not.[17]

All culvert design theory is closely dependent on energy conservation. However, due to the numerous variables involved, no single formula or procedure can be used to design a culvert. Culvert design is often an empirical, trial-and-error process. Figure 19.24 illustrates some of the important variables that affect culvert performance.

Figure 19.24 Flow Profiles in Culvert Design

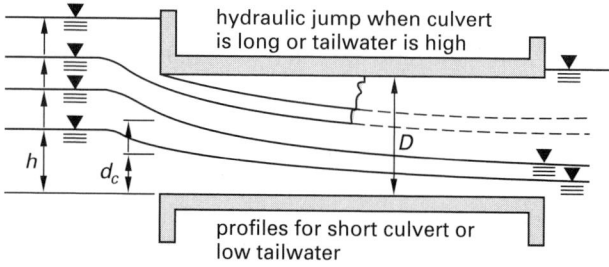

A culvert can operate with its entrance partially or totally submerged. Similarly, the exit can be partially or totally submerged, or it can have free outfall. The upstream head, h, is the water surface level above the lowest part of the culvert barrel, known as the *invert*.[18]

In Fig. 19.24, the three lowermost surface level profiles are of the type that would be produced with inlet control. Such a situation can occur if the culvert is short and the slope is steep. Flow at the entrance is critical as the water falls over the brink. Since critical flow occurs, the flow is choked and the inlet controls the flow rate.

Downstream variations cannot be transmitted past the critical section.

If the tailwater covers the culvert exit completely (i.e., a submerged exit), the culvert will be full at that point, even though the inlet control forces the culvert to be only partially full at the inlet. The transition from partially full to totally full occurs in a hydraulic jump, the location of which depends on the flow resistance and water levels. If the flow resistance is very high, or if the headwater and tailwater levels are high enough, the jump will occur close to or at the entrance.

If the flow in a culvert is full for its entire length, then the flow is under outlet control. The discharge will be a function of the differences in tailwater and headwater levels, as well as the flow resistance along the barrel length.

38. DETERMINING TYPE OF CULVERT FLOW

For convenience, culvert flow is classified into six different types on the basis of the type of control, the steepness of the barrel, the relative tailwater and headwater heights, and in some cases, the relationship between critical depth and culvert size. These parameters are quantified through the use of the ratios in Table 19.9.[19] The six types are illustrated in Fig. 19.25. Identification of the type of flow beyond the guidelines in Table 19.9 requires a trial-and-error procedure.

In the following cases, several variables appear repeatedly. C_d is the discharge coefficient, a function of the barrel inlet geometry. Orifice data can be used to approximate the discharge coefficient when specific information is unavailable. v_1 is the average velocity of the water approaching the culvert entrance and is often insignificant. The velocity-head coefficient, α, also called the *Coriolis coefficient*, accounts for a nonuniform distribution of velocities over the channel section. However, it represents only a second-order correction and is normally neglected (i.e., assumed equal to 1.0). d_c is the critical depth, which may not correspond to the actual depth of flow. (It must be calculated from the flow conditions.) h_f is the friction loss in the identified section. For culverts flowing full, the friction loss can be found in the usual manner developed for pipe flow: from the Darcy formula and the Moody friction factor chart. For partial flow, the Manning equation and its variations (e.g., Eq. 19.30) can also be used. The Manning equation is particularly useful since it eliminates the need for trial-and-error solutions. The friction head loss between sections 1 and 2, for example, can be calculated from Eq. 19.97.

$$h_{f,1-2} = \frac{LQ^2}{K_1 K_2} \qquad 19.97$$

---

[16]The methods of culvert flow analysis in this chapter are based on *Measurement of Peak Discharge at Culverts by Indirect Methods*, U.S. Department of the Interior (1968).
[17]Proper design of culvert entrances can reduce the importance of length on culvert filling.
[18]The highest part of the culvert barrel is known as the *soffit* or *crown*.

[19]The six cases presented here do not exhaust the various possibilities for entrance and exit control. Culvert design is complicated by this multiplicity of possible flows. Since only the easiest problems can be immediately categorized as one of the six cases, each situation needs to be carefully evaluated.

$$K = \left(\frac{1.00}{n}\right) R^{2/3} A \quad \text{[SI]} \quad \textit{19.98(a)}$$

$$K = \left(\frac{1.49}{n}\right) R^{2/3} A \quad \text{[U.S.]} \quad \textit{19.98(b)}$$

The total hydraulic head available, H, is divided between the velocity head in the culvert, the entrance loss (if considered), and the friction.

$$H = \frac{v^2}{2g} + k_e\left(\frac{v^2}{2g}\right) + \frac{v^2 n^2 L}{R^{4/3}} \quad \text{[SI]} \quad \textit{19.99(a)}$$

$$H = \frac{v^2}{2g} + k_e\left(\frac{v^2}{2g}\right) + \frac{v^2 n^2 L}{2.21 R^{4/3}} \quad \text{[U.S.]} \quad \textit{19.99(b)}$$

Equation 19.99 can be solved directly for the velocity. Equation 19.100 is valid for culverts of any shape.

$$v = \sqrt{\frac{H}{\dfrac{1 + k_e}{2g} + \dfrac{n^2 L}{R^{4/3}}}} \quad \text{[SI]} \quad \textit{19.100(a)}$$

$$v = \sqrt{\frac{H}{\dfrac{1 + k_e}{2g} + \dfrac{n^2 L}{2.21 R^{4/3}}}} \quad \text{[U.S.]} \quad \textit{19.100(b)}$$

Table 19.9 *Culvert Flow Classification Parameters*

flow type	$\dfrac{h_1 - z}{D}$	$\dfrac{h_4}{h_c}$	$\dfrac{h_4}{D}$	culvert slope	barrel flow	location of control	kind of control
1	< 1.5	< 1.0	≤ 1.0	steep	partial	inlet	critical depth
2	< 1.5	< 1.0	≤ 1.0	mild	partial	outlet	critical depth
3	< 1.5	> 1.0	≤ 1.0	mild	partial	outlet	backwater
4	> 1.0		≥ 1.0	any	full	outlet	backwater
5	≥ 1.5		≤ 1.0	any	partial	inlet	entrance geometry
6	≥ 1.5		≤ 1.0	any	full	outlet	entrance and barrel geometry

Figure 19.25 *Culvert Flow Classifications*

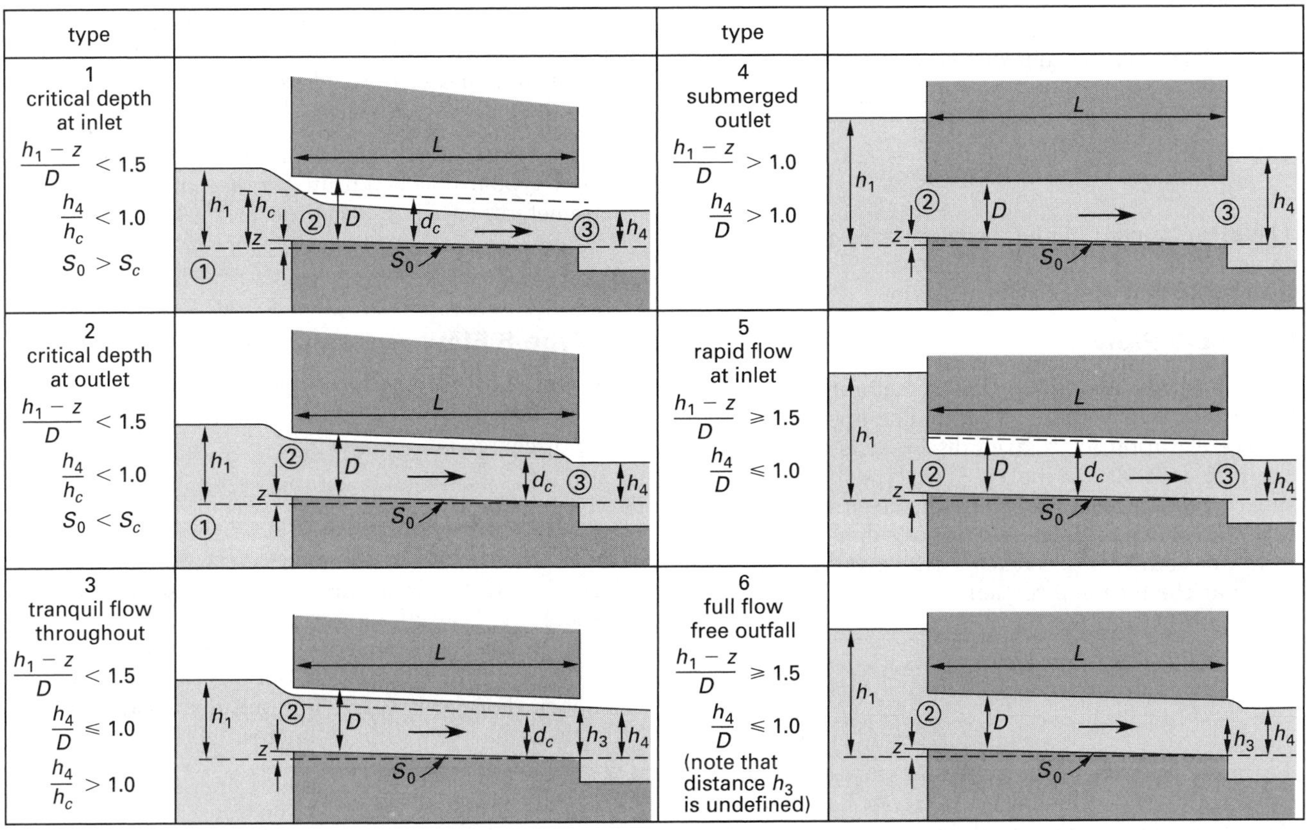

Table 19.10 *Minor Entrance Loss Coefficients*

k_e	condition of entrance
0.08	smooth, tapered
0.10	flush concrete groove
0.10	flush concrete bell
0.15	projecting concrete groove
0.15	projecting concrete bell
0.50	flush, square-edged
0.90	projecting, square-edged

A. Type-1 Flow

Water passes through the critical depth near the culvert entrance, and the culvert flows partially full. The slope of the culvert barrel is greater than the critical slope, and the tailwater elevation is less than the elevation of the water surface at the control section.

The discharge is

$$Q = C_d A_c \sqrt{2g\left(h_1 - z + \frac{\alpha v_1^2}{2g} - d_c - h_{f,1\text{-}2}\right)} \qquad 19.101$$

The area, A, used in the discharge equation is not the culvert area since the culvert does not flow full. A_c is the area in flow at the critical section.

B. Type-2 Flow

As in type-1 flow, flow passes through the critical depth at the culvert outlet, and the barrel flows partially full. The slope of the culvert is less than critical, and the tailwater elevation does not exceed the elevation of the water surface at the control section.

$$Q = C_d A_c \sqrt{2g\left(h_1 + \frac{\alpha v_1^2}{2g} - d_c - h_{f,1\text{-}2} - h_{f,2\text{-}3}\right)}$$
$$19.102$$

The area, A, used in the discharge equation is not the culvert area since the culvert does not flow full. A_c is the area in flow at the critical section.

C. Type-3 Flow

When backwater is the controlling factor in culvert flow, the critical depth cannot occur. The upstream water-surface elevation for a given discharge is a function of the height of the tailwater. For type-3 flow, flow is subcritical for the entire length of the culvert, with the flow being partial. The outlet is not submerged, but the tailwater elevation does exceed the elevation of critical depth at the terminal section.

$$Q = C_d A_3 \sqrt{2g\left(h_1 + \frac{\alpha v_1^2}{2g} - h_3 - h_{f,1\text{-}2} - h_{f,2\text{-}3}\right)}$$
$$19.103$$

The area, A, used in the discharge equation is not the culvert area since the culvert does not flow full. A_3 is the area in flow at numbered section 3 (i.e., the exit).

D. Type-4 Flow

As in type-3 flow, the backwater elevation is the controlling factor in this case. Critical depth cannot occur, and the upstream water surface elevation for a given discharge is a function of the tailwater elevation. Discharge is independent of barrel slope. The culvert is submerged at both the headwater and the tailwater. No differentiation between low head and high head is made for this case. If the velocity head at section 1 (the entrance), the entrance friction loss, and the exit friction loss are neglected, the discharge can be calculated. A_o is the culvert area.

$$Q = C_d A_o \sqrt{2g\left(\frac{h_1 - h_4}{1 + \dfrac{29 C_d^2 n^2 L}{R^{4/3}}}\right)} \qquad 19.104$$

The complicated term in the denominator corrects for friction. For rough estimates and for culverts less than 50 ft long, the friction loss can be ignored.

$$Q = C_d A_o \sqrt{2g(h_1 - h_4)} \qquad 19.105$$

E. Type-5 Flow

Partially full flow under a high head is classified as type-5 flow. The flow pattern is similar to the flow downstream from a sluice gate, with rapid flow near the entrance. Usually, type-5 flow requires a relatively square entrance that causes contraction of the flow area to less than the culvert area. In addition, the barrel length, roughness, and bed slope must be sufficient to keep the velocity high throughout the culvert.

It is difficult to distinguish in advance between type-5 and type-6 flow. Within a range of the important parameters, either flow can occur.[20] A_o is the culvert area.

$$Q = C_d A_o \sqrt{2g(h_1 - z)} \qquad 19.106$$

F. Type-6 Flow

Type-6 flow, like type-5 flow, is considered a high-head flow. The culvert is full under pressure with free outfall. The discharge is

$$Q = C_d A_o \sqrt{2g(h_1 - h_3 - h_{f,2\text{-}3})} \qquad 19.107$$

Equation 19.107 is inconvenient because h_3 (the true piezometric head at the outfall) is difficult to evaluate without special graphical aids. The actual hydraulic head driving the culvert flow is a function of the Froude number. For conservative first approximations, h_3 can be taken as the barrel diameter. This will give the

---

[20]If the water surface ever touches the top of the culvert, the passage of air in the culvert will be prevented and the culvert will flow full everywhere. This is type-6 flow.

minimum hydraulic head. In reality, h_3 varies from somewhat less than half the barrel diameter to the full diameter.

If h_3 is taken as the barrel diameter, the total hydraulic head ($H = h_1 - h_3$) will be split between the velocity head and friction. In that case, Eq. 19.100 can be used to calculate the velocity. The discharge is easily calculated from Eq. 19.108.[21] A_o is the culvert area.

$$Q = A_o \mathrm{v} \qquad\qquad 19.108$$

Example 19.12

Size a square culvert with an entrance fluid level 5 ft above the barrel top and a free exit to operate with the following characteristics.

$$\text{slope} = 0.01$$
$$\text{length} = 250 \text{ ft}$$
$$\text{capacity} = 45 \text{ ft}^3/\text{sec}$$
$$n = 0.013$$

Solution

Since the h_1 dimension is measured from the culvert invert, it is difficult to classify the type of flow at this point. However, either type 5 or type 6 is likely since the head is high.

step 1: Assume a trial culvert size. Select a square opening with 1.0 ft sides.

step 2: Calculate the flow assuming case 5 (entrance control). The entrance will act like an orifice.

$$A_o = (1 \text{ ft})(1 \text{ ft}) = 1 \text{ ft}^2$$

$$\begin{aligned} H &= h_1 - z \\ &= \big(5 \text{ ft} + 1 \text{ ft} + (0.01)(250 \text{ ft})\big) - (0.01)(250 \text{ ft}) \\ &= 6 \text{ ft} \end{aligned}$$

C_d is approximately 0.62 for square-edged openings with separation from the wall (see Table 17.5). From Eq. 19.106,

$$Q = (0.62)(1 \text{ ft}^2)\sqrt{(2)\left(32.2\,\frac{\text{ft}}{\text{sec}^2}\right)(6 \text{ ft})}$$

$$= 12.2 \text{ ft}^3/\text{sec}$$

Since this size has insufficient capacity, try a larger culvert. Choose a square opening with 2.0 ft sides.

$$A_o = (2 \text{ ft})(2 \text{ ft}) = 4 \text{ ft}^2$$

$$H = h_1 - z = 5 \text{ ft} + 2 \text{ ft} = 7 \text{ ft}$$

$$Q = C_d A_o \sqrt{2g(h_1 - z)}$$

$$= (0.62)(4 \text{ ft}^2)\sqrt{(2)\left(32.2\,\frac{\text{ft}}{\text{sec}^2}\right)(7 \text{ ft})}$$

$$= 52.7 \text{ ft}^3/\text{sec}$$

step 3: Begin checking the entrance control assumption by calculating the maximum hydraulic radius. The upper surface of the culvert is not wetted because the flow is entrance controlled. The hydraulic radius is maximum at the entrance.

$$R = \frac{A_o}{P}$$

$$= \frac{4 \text{ ft}^2}{2 \text{ ft} + 2 \text{ ft} + 2 \text{ ft}}$$

$$= 0.667 \text{ ft}$$

step 4: Calculate the velocity using the Manning equation for open channel flow. Since the hydraulic radius is maximum, the velocity will also be maximum.

$$\mathrm{v} = \frac{1.49}{n} R^{2/3}\sqrt{S}$$

$$= \left(\frac{1.49}{0.013}\right)(0.667 \text{ ft})^{2/3}\sqrt{0.01}$$

$$= 8.75 \text{ ft/sec}$$

step 5: Calculate the normal depth, d_n.

$$d_n = \frac{Q}{\mathrm{v}w} = \frac{45\,\dfrac{\text{ft}^3}{\text{sec}}}{\left(8.75\,\dfrac{\text{ft}}{\text{sec}}\right)(2 \text{ ft})}$$

$$= 2.57 \text{ ft}$$

Since the normal depth is greater than the culvert size, the culvert will flow full under pressure. (It was not necessary to calculate the critical depth since the flow is implicitly subcritical.) The entrance control assumption was, therefore, not valid for this size culvert.[22] At this point, two things can be done: A larger culvert can be chosen if entrance control is desired, or the solution can continue by checking to see if the culvert has the required capacity as a pressure conduit.

step 6: Check the capacity as a pressure conduit. H is the total available head.

$$\begin{aligned} H &= h_1 - h_3 \\ &= \big(5 \text{ ft} + 2 \text{ ft} + (0.01)(250 \text{ ft})\big) - 2 \text{ ft} \\ &= 7.5 \text{ ft} \end{aligned}$$

---

[21]Equation 19.108 does not include the discharge coefficient. Velocity, v, when calculated from Eq. 19.100, is implicitly the velocity in the barrel.

[22]If the normal depth had been less than the barrel diameter, it would still be necessary to determine the critical depth of flow. If the normal depth was less than the critical depth, the entrance control assumption would have been valid.

step 7: Since the pipe is flowing full, the hydraulic radius is

$$R = \frac{A}{P} = \frac{4 \text{ ft}^2}{8 \text{ ft}} = 0.5 \text{ ft}$$

step 8: Equation 19.100 can be used to calculate the flow velocity. Since the culvert has a square-edged entrance, a loss coefficient of $k_e = 0.5$ is used. However, this does not greatly affect the velocity.

$$v = \sqrt{\frac{H}{\dfrac{1 + k_e}{2g} + \dfrac{n^2 L}{2.21 R^{4/3}}}}$$

$$= \sqrt{\frac{7.5 \text{ ft}}{\dfrac{1 + 0.5}{(2)\left(32.2 \dfrac{\text{ft}}{\text{sec}}\right)} + \dfrac{(0.013)^2 (250 \text{ ft})}{(2.21)(0.5 \text{ ft})^{4/3}}}}$$

$$= 10.24 \text{ ft/sec}$$

step 9: Check the capacity.

$$Q = v A_o$$
$$= \left(10.24 \frac{\text{ft}}{\text{sec}}\right)(4 \text{ ft}^2)$$
$$= 40.96 \text{ ft}^3/\text{sec}$$

The culvert size is not acceptable since its discharge under the maximum head does not have a capacity of 45 ft^3/sec.

step 10: Repeat from step 2, trying a larger-size culvert. With a 2.5 ft side, the following values are obtained.

$$A_o = (2.5 \text{ ft})(2.5 \text{ ft}) = 6.25 \text{ ft}^2$$
$$H = 5 \text{ ft} + 2.5 \text{ ft} = 7.5 \text{ ft}$$
$$Q = (0.62)(6.25 \text{ ft}^2)$$
$$\times \sqrt{(2)\left(32.2 \frac{\text{ft}}{\text{sec}^2}\right)(7.5 \text{ ft})}$$
$$= 85.2 \text{ ft}^3/\text{sec}$$

$$R = \frac{6.25 \text{ ft}^2}{7.5 \text{ ft}} = 0.833 \text{ ft}$$

$$\frac{v}{-} = \left(\frac{1.49}{0.013}\right)(0.833 \text{ ft})^{2/3}\sqrt{0.01}$$
$$= 10.12 \text{ ft/sec}$$

$$d_n = \frac{\left(45 \dfrac{\text{ft}^3}{\text{sec}}\right)}{\left(10.12 \dfrac{\text{ft}}{\text{sec}}\right)(2.5 \text{ ft})}$$
$$= 1.78 \text{ ft}$$

step 11: Calculate the critical depth. For rectangular channels, Eq. 19.74 can be used.

$$d_c = \left(\frac{Q^2}{gw^2}\right)^{1/3}$$

$$= \left(\frac{\left(45 \dfrac{\text{ft}^3}{\text{sec}}\right)^2}{\left(32.2 \dfrac{\text{ft}}{\text{sec}^2}\right)(2.5 \text{ ft})^2}\right)^{1/3}$$

$$= 2.16 \text{ ft}$$

Since the normal depth is less than the critical depth, the flow is supercritical. The entrance control assumption was correct for the culvert. The culvert has sufficient capacity to carry 45 ft^3/sec.

39. CULVERT DESIGN

Designing a culvert is somewhat easier than culvert analysis because of common restrictions placed on designers and the flexibility to change almost everything else. For example, culverts may be required to (a) never be more than 50% full (deep), (b) always be under inlet control, or (c) always operate with some minimum head (above the centerline or crown). In the absence of any specific guidelines, a culvert may be designed using the following procedure.

step 1: Determine the required flow rate.

step 2: Determine all water surface elevations, lengths, and other geometric characteristics.

step 3: Determine the material to be used for the culvert and its roughness.

step 4: Assume type 1 flow (inlet control).

step 5: Select a trial diameter.

step 6: Assume a reasonable slope.

step 7: Position the culvert entrance such that the ratio of headwater depth (inlet to water surface) to culvert diameter is 1:2 to 1:2.5.

step 8: Calculate the flow. Repeat steps 5 through 7 until the capacity is adequate.

step 9: Determine the location of the outlet. Check for outlet control. If the culvert is outlet controlled, repeat steps 5 through 7 using a different flow model.

step 10: Calculate the discharge velocity. Specify riprap, concrete, or other protection to prevent erosion at the outlet.

40. CORRUGATED METAL PIPE

Corrugated metal pipe (CMP, also known as *corrugated steel pipe*) is frequently used for culverts. Pipe is made from corrugated sheets of galvanized steel that are rolled and riveted together along a longitudinal seam. Aluminized steel may also be used in certain ranges of soil pH. Standard round pipe diameters range from 8 to 96 in (200 to 2450 mm). Metric dimensions of standard diameters are usually rounded to the nearest 25 or 50 mm (e.g., a 42 in culvert would be specified as a 1050 mm culvert, not 1067 mm).

Larger and noncircular culverts can be created out of curved steel plate. Standard section lengths are 10 to 20 ft (3 to 6 m). Though most corrugations are transverse (i.e., annular), helical corrugations are also used. Metal gages of 8, 10, 12, 14, and 16 are commonly used, depending on the depth of burial.

The most common corrugated steel pipe has transverse corrugations that are $1/2$ in (13 mm) deep and $2^2/3$ in (68 mm) from crest to crest. These are referred to as "$2^1/2$ inch" or "68×13" corrugations. For larger culverts, corrugations with a 2 in (25 mm) depth and 3, 5, and 6 in (76, 125, or 152 mm) pitches are used. Plate-based products using 6 in by 2 in (152 mm by 51 mm) corrugations are known as *structural plate corrugated steel pipe* (SPCSP) and *multiplate* after the trade-named product "Multi-Plate^{TM}."

The flow area for circular culverts is based on the nominal culvert diameter, regardless of the gage of the plate metal used to construct the pipe. Flow area is calculated to (at most) three significant digits.

A Hazen-Williams coefficient, C, of 60 is typically used with all sizes of corrugated pipe. Values of C and Manning's constant, n, for corrugated pipe are generally not affected by age. *Design Charts for Open Channel Flow* (U.S. Department of Transportation, 1979) recommends a Manning constant of $n = 0.024$ for all cases. The U.S. Department of the Interior recommends the following values. For standard ($2^2/3$ in by $1/2$ in or 68 mm by 13 mm) corrugated pipe with the diameters given: 12 in (457 mm), 0.027; 24 in (610 mm), 0.025; 36 to 48 in (914 to 1219 mm), 0.024; 60 to 84 in (1524 to 2134 mm), 0.023; 96 in (2438 mm), 0.022. For (6 in by 2 in or 152 mm by 51 mm) multiplate construction with the diameters given: 5 to 6 ft (1.5 to 1.8 m), 0.034; 7 to 8 ft (2.1 to 2.4 m), 0.033; 9 to 11 ft (2.7 to 3.3 m), 0.032; 12 to 13 ft (3.6 to 3.9 m), 0.031; 14 to 15 ft (4.2 to 4.5 m), 0.030; 16 to 18 ft (4.8 to 5.4 m), 0.029; 19 to 20 ft (5.8 to 6.0 m), 0.028; 21 to 22 ft (6.3 to 6.6 m), 0.027.

If the inside of the corrugated pipe has been asphalted completely smooth $360°$ circumferentially, Manning's n ranges from 0.009 to 0.011. For culverts with 40% asphalted inverts, $n = 0.019$. For other percentages of paved invert, the resulting value is proportional to the percentage and the values normally corresponding to that diameter pipe. For field-bolted corrugated metal pipe arches, $n = 0.025$.

It is also possible to calculate the Darcy friction loss if the corrugation depth, 0.5 in (13 mm) for standard corrugations and 2.0 in (51 mm) for multiplate, is taken as the specific roughness.

Water Resources

20 Meteorology, Climatology, and Hydrology

Nomenclature

A	area	ft^2	m^2
A_d	drainage area	ac	km^2
C	coefficient	–	–
C	rational runoff coefficient	–	–
CN	curve number	–	–
d	distance between stations	mi	km
E	evaporation	in/day	cm/day
F	factor	–	–
F	frequency of occurrence	1/yr	1/yr
F	infiltration	in	cm
H	elevation difference	ft	m
I	rainfall intensity	in/hr	cm/h
I_a	initial abstraction	in	cm
Imp	imperviousness	%	%
K	coefficient	–	–
L	length	ft	m
M	order number	–	–
n	Manning roughness coefficient	–	–
n	number of years	yr	yr
n_y	number of years of streamflow data	yr	yr
N	normal annual precipitation	in	cm
N	number	–	–
N	time from peak to end of runoff	hr	h
P	precipitation	in	cm
q	runoff	ft^3/mi^2-in	m^3/km^2·cm
Q	flow quantity	ft^3/sec	m^3/s
S_{decimal}	slope	ft/ft	m/m
S_{percent}	slope	%	%
S	storage capacity	in	cm
t	time	min	min
t_c	time to concentration	min	min
t_p	time from start of storm to peak runoff	hr	h
t_R	rainstorm duration	hr	h
v	flow velocity	ft/sec	m/s
V	volume	ft^3	m^3
W	width of unit hydrograph	min	min

Subscripts

c	concentration
d	drainage
n	period n
o	overland
p	peak, pond, or pan
R	rain (storm) or reservoir
t	time
u	unit

1. HYDROLOGIC CYCLE

The *hydrologic cycle* is the full "life cycle" of water. The cycle begins with *precipitation*, which encompasses all of the hydrometeoric forms, including rain, snow, sleet, and hail from a storm. Precipitation can (a) fall on vegetation and structures and evaporate back into the atmosphere, (b) be absorbed into the ground and either make its way to the water table or be absorbed by plants after which it evapotranspires back into the atmosphere, or (c) travel as surface water to a depression, watershed, or creek from which it either evaporates back into the atmosphere, infiltrates into the ground water system, or flows off in streams and rivers to an ocean or lakes. The cycle is completed when lake and ocean water evaporates into the atmosphere.

The *water balance equation* is the application of conservation to the hydrologic cycle.

$$\begin{aligned} \text{total precipitation} = {}& \text{net change in surface water removed} \\ &+ \text{net change in ground water removed} \\ &+ \text{evapotranspiration} \\ &+ \text{interception evaporization} \\ &+ \text{net increase in surface water storage} \\ &+ \text{net increase in ground water storage} \end{aligned}$$

20.1

The total amount of water that is intercepted (and which subsequently evaporates) and absorbed into ground water before runoff begins is known as the *initial abstraction*. Even after runoff begins, the soil continues to absorb some infiltrated water. Initial abstraction and infiltration do not contribute to surface runoff. Equation 20.1 can be restated as Eq. 20.2.

$$\begin{aligned} \text{total precipitation} = {}& \text{initial abstraction} \\ &+ \text{infiltration} \\ &+ \text{surface runoff} \end{aligned}$$

20.2

2. STORM CHARACTERISTICS

Storm rainfall characteristics include the duration, total volume, intensity, and areal distribution of a storm. Storms are also characterized by their recurrence intervals. (See Sec. 6.)

The duration of storms is measured in hours and days. The volume of rainfall is simply the total quantity of precipitation dropping on the watershed. Average rainfall intensity is the volume divided by the duration of the storm. Average rainfall can be considered to be generated by an equivalent theoretical storm that drops the same volume of water uniformly and constantly over the entire watershed area.

A storm *hyetograph* is the instantaneous rainfall intensity measured as a function of time. Hyetographs are usually bar graphs showing constant rainfall intensities over short periods of time. Hyetograph data can be reformulated as a *cumulative rainfall curve*, also known as a rainfall *mass curve*.

3. PRECIPITATION DATA

Precipitation data on rainfall can be collected in a number of ways, but use of an open precipitation rain gauge is quite common. This type of gauge measures only the volume of rain collected between readings, usually 24 hr.

The *average precipitation* over a specific area can be found from station data in several ways.

Method 1: If the stations are uniformly distributed over a flat site, their precipitations can be averaged. This also requires that the individual precipitation records not vary too much from the mean.

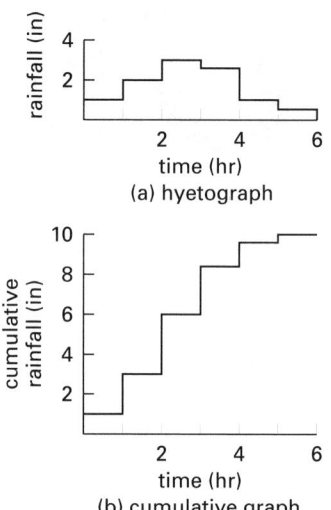

Figure 20.1 *Storm Hyetograph and Cumulative Rainfall Curves*

Method 2: The *Thiessen method* calculates the average by weighting station measurements by the area of the assumed watershed for each station. These assumed watershed areas are found by drawing dotted lines between all stations and bisecting these dotted lines with solid lines (which are extended outward until they connect with other solid lines). The solid lines will form a polygon whose area is the assumed watershed area.

Method 3: The *isohyetal method* requires plotting lines of constant precipitation (*isohyets*) and weighting the isohyet values by the areas enclosed by the lines. This method is the most accurate of the three methods, as long as there are enough observations to permit drawing of isohyets. Station data are used to draw isohyets, but they are not used in the calculation of average rainfall.

4. ESTIMATING UNKNOWN PRECIPITATION

If a precipitation measurement at a location is unknown, it may still be possible to estimate the value by one of the following procedures.

Method 1: Choose three stations close to and evenly spaced around the location with missing data. If the normal annual precipitations at the three sites do not vary more than 10% from the missing station's normal annual precipitation, the rainfall can be estimated as the arithmetic mean of the three neighboring stations' precipitations for the period in question.

Method 2: If the precipitation difference between locations is more than 10%, the *normal-ratio method* can be used. In Eq. 20.3, P_x is the precipitation at the missing station; N_x is the long-term normal precipitation at the missing station; P_A, P_B, and P_C are the precipitations at known stations; and N_A, N_B, and N_C are the long-term normal precipitations at the known stations.

$$P_x = \tfrac{1}{3}\left(\left(\frac{N_x}{N_A}\right)P_A + \left(\frac{N_x}{N_B}\right)P_B + \left(\frac{N_x}{N_C}\right)P_C\right) \quad 20.3$$

Method 3: Use data from stations in the four nearest quadrants (north, south, east, and west of the unknown station) and weight the data with the inverse squares of the distance between the stations. In Eq. 20.4, P_x, P_A, P_B, P_C, and P_D are defined as in Method 2, and d_{A-x}, d_{B-x}, d_{C-x}, and d_{D-x} are the distances between stations A and x, B and x, and so on, respectively.

$$P_x = \frac{\dfrac{P_A}{d_{A-x}^2} + \dfrac{P_B}{d_{B-x}^2} + \dfrac{P_C}{d_{C-x}^2} + \dfrac{P_D}{d_{D-x}^2}}{\dfrac{1}{d_{A-x}^2} + \dfrac{1}{d_{B-x}^2} + \dfrac{1}{d_{C-x}^2} + \dfrac{1}{d_{D-x}^2}} \qquad \textit{20.4}$$

5. TIME OF CONCENTRATION

Time of concentration, t_c, is defined as the time of travel from the hydraulically most remote (timewise) point in the watershed to the watershed outlet or other design point. For points (e.g., manholes) along storm drains being fed from a watershed, time of concentration is taken as the largest combination of overland flow time (sheet flow), swale or ditch flow (shallow concentrated flow), and storm drain, culvert, or channel time. It is unusual for time of concentration to be less than 0.1 hr (6 min) when using the NRCS method or less than 10 min when using the rational method.

$$t_c = t_{\text{sheet}} + t_{\text{shallow}} + t_{\text{channel}} \qquad \textit{20.5}$$

The Natural Resources Conservation Service (NRCS) specifies using the *Manning kinematic equation* (Overton and Meadows 1976 formulation) for calculating *sheet flow* travel time over distances less than 300 ft (100 m). In Eq. 20.6, n is the Manning roughness coefficient for sheet flow (Table 20.1), P_2 is the 2 yr, 24 hr rainfall in inches, and S is the slope of the hydraulic grade line in ft/ft.

$$t_{\text{sheet flow}} = \frac{0.007(nL_o)^{0.8}}{\sqrt{P_2}(S_{\text{decimal}})^{0.4}} \qquad \textit{20.6}$$

After about 300 ft (100 m), the flow usually becomes shallow concentrated flow (swale, ditch flow). Travel time is calculated as L/v. Velocity can be found from the Manning equation if the flow geometry is well defined, but must be determined from other correlations otherwise, such as those specified by the NRCS in Eqs. 20.7 and 20.8.

$$v_{\text{shallow,ft/sec}} = 16.1345\sqrt{S_{\text{decimal}}} \quad \text{[unpaved]} \quad \textit{20.7}$$

$$v_{\text{shallow,ft/sec}} = 20.3282\sqrt{S_{\text{decimal}}} \quad \text{[paved]} \quad \textit{20.8}$$

Storm drain (channel) time is found by dividing the storm drain length by the actual or an assumed channel velocity. Storm drain velocity is found from either the Manning or the Hazen-Williams equation. Since size and velocity are related, an iterative trial-and-error solution is generally required.[1]

---

[1]If the pipe or channel size is known, the velocity can be found from $Q = Av$. If the pipe size is not known, the area will have to be estimated. In that case, one might as well estimate velocity instead. 5 ft/sec is a reasonable flow velocity for open channel flow. The minimum velocity for a *self-cleansing pipe* is 2 ft/sec.

Table 20.1 *Manning Roughness Coefficient for Sheet Flow*

surface	n
smooth surfaces (concrete, asphalt, gravel, or bare soil)	0.011
fallow (no residue cover)	0.05
cultivated soils	
residue cover $\leq$ 20%	0.06
residue cover > 20%	0.17
grasses	
short prairie grass	0.15
dense grass[a]	0.24
Bermuda grass	0.41
range, natural	0.13
woods[b]	
light underbrush	0.40
dense underbrush	0.80

[a]Includes species such as weeping lovegrass, bluegrass, buffalo grass, blue gramma grass, and native grass mixtures.
[b]When selecting a value of n, consider the cover to a height of about 0.1 ft (3 cm). This is the only part of the plant that will obstruct sheet flow.

Reprinted from *Urban Hydrology for Small Watersheds*, Technical Release TR 55, United States Department of Agriculture, Natural Resources Conservation Service, Table 3-1, after Engman (1986).

There are a variety of methods available for estimating time of concentration. Early methods include Kirpich (1940), California Culverts Practice (1942), Hathaway (1945), and Izzard (1946). More recent methods include those from the Federal Aviation Administration, or FAA (1970), the kinematic wave formulas of Morgali (1965) and Aron (1973), the NRCS lag equation (1975), and NRCS average velocity charts (1975). Estimates of time of concentration from these methods can vary by as much as 100%. These differences carry over into estimates of peak flow, hence the need to carefully determine the validity of any method used.

The distance L_o in the various equations that follow is the longest distance to the collection point, as shown in Fig. 20.2.

Figure 20.2 *Overland Flow Distances*

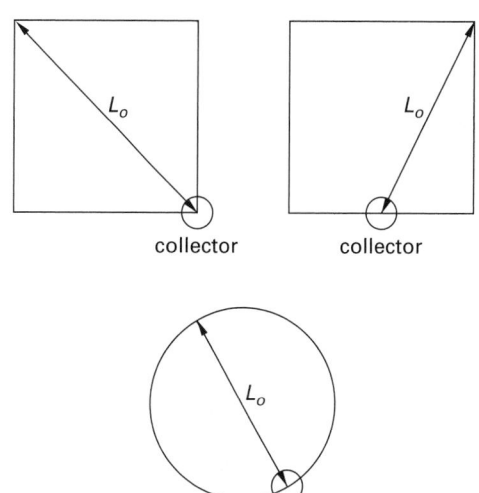

For irregularly shaped drainage areas, it may be necessary to evaluate several alternative overland flow distances. For example, Fig. 20.3 shows a drainage area with a long tongue. Although the tongue area contributes little to the drainage area, it does lengthen the overland flow time. Depending on the intensity-duration-frequency curve, the longer overland flow time (resulting in a lower rainfall intensity) may offset the increase in area due to the tongue. Therefore, two runoffs need to be compared, one ignoring and the other including the tongue.

Figure 20.3 *Irregular Drainage Area*

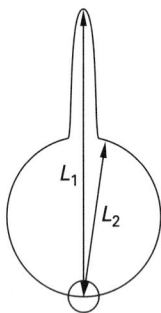

The *FAA formula*, Eq. 20.9, was developed from airfield drainage data collected by the Army Corp of Engineers. However, it has been widely used for urbanized areas. C is the rational method runoff coefficient. The slope in Eq. 20.9 is in percent. L_o is in ft.

$$t_c = \frac{(1.8)(1.1 - C)\sqrt{L_{o,\text{ft}}}}{S_{\text{percent}}^{1/3}} \qquad 20.9$$

The *kinematic wave formula* is particularly accurate for uniform planar homogenous areas (e.g., paved areas such as parking lots and streets). It requires iteration, since both intensity and time of concentration are unknown. In Eq. 20.10, L_o is in ft, n is the Manning roughness coefficient, I is the intensity in in/hr, and S_{decimal} is the slope in ft/ft. Recommended values of n for this application are somewhat different than for open channel flow and are: smooth impervious surfaces, 0.035; smooth bare-packed soil, free of stones, 0.05; poor grass, moderately bare surface, 0.10; pasture or average grass cover, 0.20; dense grass or forest, 0.40. Equation 20.10 is solved iteratively since the intensity depends on the time to concentration.

$$t_{c,\text{min}} = \frac{0.94 L_{o,\text{ft}}^{0.6} n^{0.6}}{I^{0.4} S_{\text{decimal}}^{0.3}} \qquad 20.10$$

The NRCS *lag equation* was developed from observations of agricultural watersheds where overland flow paths are poorly defined and channel flow is absent. However, it has been adapted to small urban watersheds under 2000 ac. The equation performs reasonably well for areas that are completely paved, as well. Correction factors are used to account for channel improvement and impervious areas. L_o is in ft, CN is the NRCS runoff curve number, S_{in} is the potential maximum retention in the watershed after runoff begins in inches,

and S_{percent} is the average slope in percent. The factor 1.67 converts the watershed lag time to the time of concentration. Since the formula overestimates time for mixed areas, different adjustment factors have been proposed. The NRCS lag equation performs poorly when channel flow is a significant part of the time of concentration.

$$t_{c,\text{min}} = 1.67 t_{\text{watershed lag time}}$$

$$= \frac{100 L_{o,\text{ft}}^{0.8}(S_{\text{in}} + 1)^{0.7}}{1900\sqrt{S_{\text{percent}}}}$$

$$= \frac{100 L_{o,\text{ft}}^{0.8}\left(\dfrac{1000}{\text{CN}} - 9\right)^{0.7}}{1900\sqrt{S_{\text{percent}}}} \qquad 20.11$$

If the velocity of runoff water is known, the time of concentration can be easily determined. Charts of average velocity as functions of watercourse slope and surface cover have been published by the NRCS. The charts are best suited for flow paths of at least several hundred feet (at least 70 m). The time of concentration is easily determined from these charts as

$$t_c = \frac{\sum L_{o,\text{ft}}}{\text{v}\left(60\,\dfrac{\text{sec}}{\text{min}}\right)} \qquad 20.12$$

Figure 20.4 *NRCS Average Velocity Chart for Overland Flow Travel Time*

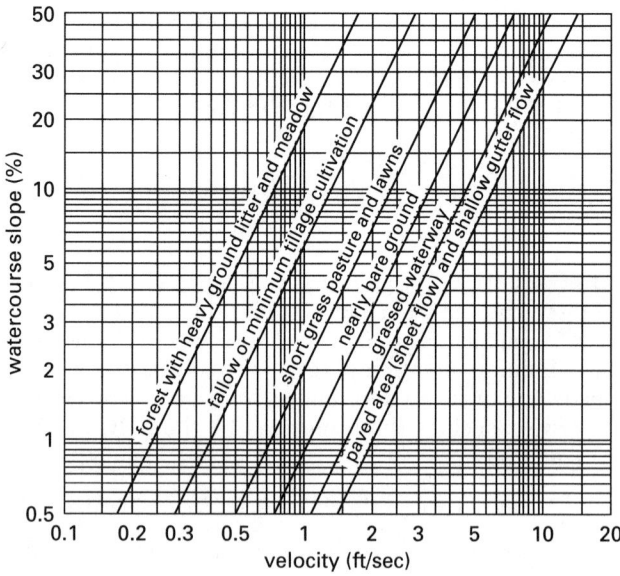

Reprinted from NRCS TR-55-1975. TR-55-1986 contains a similar graph for shallow-concentrated flow over paved and unpaved surfaces, but it does not contain this particular graph.

6. RAINFALL INTENSITY

Effective design of a surface feature depends on its geographical location and required degree of protection. Once the location of the feature is known, the *design storm* (or *design flood*) must be determined based on some probability of recurrence.

Rainfall intensity is the amount of precipitation per hour. The instantaneous intensity changes throughout the storm. However, it may be averaged over short time intervals or over the entire storm duration. Average intensity will be low for most storms, but it can be high for some. These high-intensity storms can be expected infrequently, say, every 20, 50, or 100 years. The average number of years between storms of a given intensity is known as the *frequency of occurrence* or *recurrence interval*.

In general, the design storm may be specified by its recurrence interval (e.g., "100-year storm"), its annual probability of occurrence (e.g., "1% storm"), or a nickname (e.g., "century storm"). A 1% storm is a storm that would be exceeded in severity only once every hundred years on the average.

The average intensity of a storm over a time period t can be calculated from Eq. 20.13 (and similar correlations). (When using the rational method described in Sec. 15, t is the time of concentration.) In the United States, it is understood that the units of intensity calculated using Eq. 20.13 will be in in/hr.

$$I = \frac{K'F^a}{(t+b)^c} \qquad 20.13$$

K', F, a, b, and c are constants that depend on the conditions, recurrence interval, and location of a storm. For many reasons, these constants may be unavailable. The *Steel formula* is a simplification of Eq. 20.13.

$$I = \frac{K}{t_c + b} \qquad 20.14$$

Values of the constants K and b in Eq. 20.14 are not difficult to obtain once the intensity-duration-frequency curve is established. Although a logarithmic transformation could be used to convert the data to straight-line form, an easier method exists. This method starts by taking the reciprocal of Eq. 20.14 and converting the equation to a straight line.

$$\frac{1}{I} = \frac{t_c + b}{K}$$

$$= \frac{t_c}{K} + \frac{b}{K}$$

$$= C_1 t_c + C_2 \qquad 20.15$$

Once C_1 and C_2 have been found, K and b can be calculated.

$$K = \frac{1}{C_1} \qquad 20.16$$

$$b = \frac{C_2}{C_1} \qquad 20.17$$

Published values of K and b can be obtained from compilations, but these values are suitable only for very rough estimates. Table 20.2 is typical of some of this general data.

The total rainfall can be calculated from the duration.

$$P = It \qquad 20.18$$

Figure 20.5 *Steel Formula Rainfall Regions*

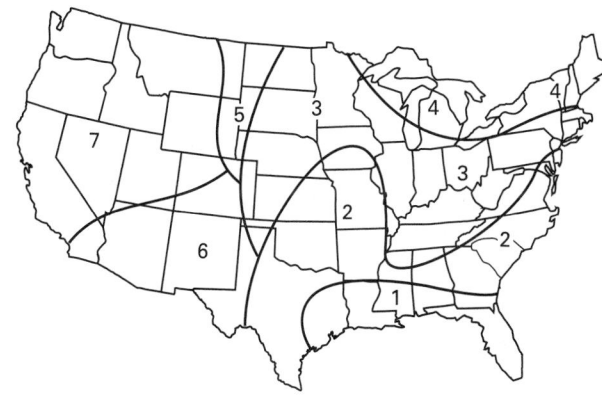

Table 20.2 *Steel Formula Coefficients (for intensities of in/hr)*

frequency in years	coefficients	region 1	2	3	4	5	6	7
2	K	206	140	106	70	70	68	32
	b	30	21	17	13	16	14	11
4	K	247	190	131	97	81	75	48
	b	29	25	19	16	13	12	12
10	K	300	230	170	111	111	122	60
	b	36	29	23	16	17	23	13
25	K	327	260	230	170	130	155	67
	b	33	32	30	27	17	26	10
50	K	315	350	250	187	187	160	65
	b	28	38	27	24	25	21	8
100	K	367	375	290	220	240	210	77
	b	33	36	31	28	29	26	10

(Multiply in/hr by 2.54 to obtain cm/h.)

Rainfall data can be compiled into *intensity-duration-frequency curves (IDF curves)* similar to those in Fig. 20.6.

Figure 20.6 *Typical Intensity-Duration-Frequency Curves*

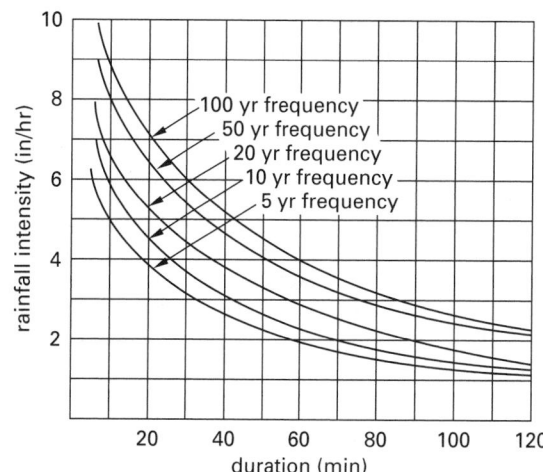

Example 20.1

A storm has an intensity given by

$$I = \frac{100}{t_c + 10}$$

15 min are required for runoff from the farthest corner of a 5 ac watershed to reach a discharge culvert. What is the design intensity?

Solution

The intensity is

$$I = \frac{100}{t_c + 10} = \frac{100 \; \frac{\text{in-min}}{\text{hr}}}{15 \; \text{min} + 10 \; \text{min}}$$

$$= 4 \; \text{in/hr}$$

7. FLOODS

A *flood* occurs when more water arrives than can be drained away. When a watercourse (i.e., a creek or river) is too small to contain the flow, the water overflows the banks.

The flooding may be categorized as nuisance, damaging, or devastating. *Nuisance floods* result in inconveniences such as wet feet, tire spray, and soggy lawns. *Damaging floods* soak flooring, carpeting, and first-floor furniture. *Devastating floods* wash buildings, vehicles, and livestock downstream.

Although rain causes flooding, large storms do not always cause floods. The size of a flood depends not only on the amount of rainfall, but also on the conditions within the watershed before and during the storm. Runoff will occur only when the rain falls on a very wet watershed that is unable to absorb additional water, or when a very large amount of rain falls on a dry watershed faster than it can be absorbed.

Specific terms are sometimes used to designate the degree of protection required. For example, the *probable maximum flood* (PMF) is a hypothetical flood that can be expected to occur as a result of the most severe combination of critical meteorologic and hydrologic conditions possible within a region.

Designing for the *probable maximum precipitation* (PMP) or *probable maximum flood* (PMF) is very conservative and usually uneconomical since the recurrence interval for these events exceeds 100 yr and may even approach 1000 yr. Designing for 100 yr floods and floods with even lower recurrence intervals is more common. (100 yr floods are not necessarily caused by 100 yr storms.)

The *design flood* or *design basis flood* (DBF) depends on the site. It is the flood that is adopted as the basis for design of a particular project. The DBF is usually determined from economic considerations, or it is specified as part of the contract document.

The *standard flood* or *standard project flood* (SPF) is a flood that can be selected from the most severe combinations of meteorological and hydrological conditions reasonably characteristic of the region, excluding extremely rare combinations of events. SPF volumes are commonly 40 to 60% of the PMF volumes.

The probability that a flooding event in any year will equal a design basis flood with a recurrence interval of F is

$$p\{F\text{event in one year}\} = \frac{1}{F} \qquad \text{20.19}$$

The probability of an F event occurring in n years is

$$p\{F\text{event in }n\text{ years}\} = 1 - \left(1 - \frac{1}{F}\right)^n \qquad \text{20.20}$$

Planning for a 1% flood has proven to be a good compromise between not doing enough and spending too much. Although the 1% flood is a common choice for the design basis flood, shorter recurrence intervals are often used, particularly in low-value areas such as cropland. For example, a 5 yr value can be used in residential areas, a 10 yr value in business sections, and a 15 yr value for high-value districts where flooding will result in more extensive damage. The ultimate choice of recurrence interval, however, must be made on the basis of economic considerations and trade-offs.

Example 20.2

A wastewater treatment plant has been designed to be in use for 40 yr. What is the probability that a 1% flood will occur within the useful lifetime of the plant?

Solution

Use Eq. 20.20.

$$p\{F\text{event in }n\text{ years}\} = 1 - \left(1 - \frac{1}{F}\right)^n$$

$$p\{100 \text{ yr flood in 40 years}\} = 1 - \left(1 - \frac{1}{100}\right)^{40}$$

$$= 0.33 \; (33\%)$$

8. TOTAL SURFACE RUNOFF FROM STREAM HYDROGRAPH

After a rain, runoff and groundwater increases stream flow. A plot of the stream discharge versus time is known as a *hydrograph*. Hydrograph periods may be very short (e.g., hours) or very long (e.g., days, weeks, or months). A typical hydrograph is shown in Fig. 20.7. The *time base* is the length of time that the stream flow exceeds the original *base flow*. The flow rate increases on the *rising limb* (*concentration curve*) and decreases on the *falling limb* (*recession curve*).

Figure 20.7 Stream Hydrograph

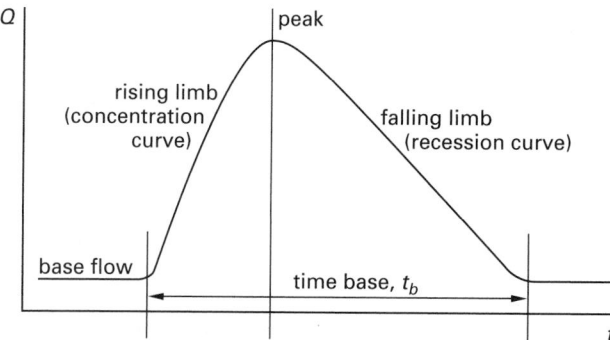

9. HYDROGRAPH SEPARATION

The stream discharge consists of both runoff and groundwater (subsurface) flows. Since culverts do not have to be designed to carry groundwater flow, a procedure known as *hydrograph separation* or *hydrograph analysis* is necessary to separate runoff (*surface flow, net flow,* or *overland flow*) and groundwater (*base flow*).[2]

There are many methods of separating base flow from overland flow. Most of the methods are somewhat arbitrary. Three methods that are easily carried out manually are presented here.

Method 1: In the *straight-line method*, a horizontal line is drawn from the start of the rising limb to the falling limb. All of the flow under the horizontal line is considered base flow. This assumption is not theoretically accurate, but the error can be small. This method is illustrated in Fig. 20.8.

Figure 20.8 Straight-Line Method

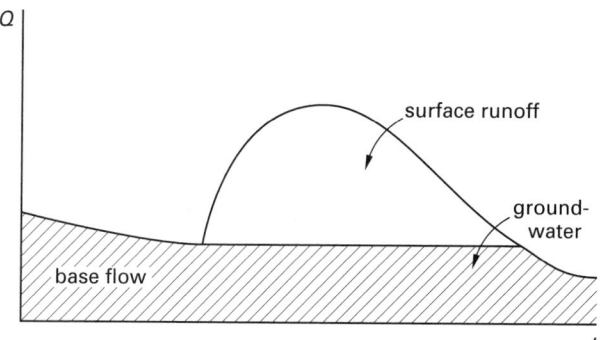

Method 2: In the *fixed-base method*, the base flow existing before the storm is projected graphically down to a point directly under the peak of the hydrograph. Then, a straight line is used to connect the projection

---

[2]The total rain dropped by a storm is the *gross rain*. The rain that actually appears as immediate runoff can be called *surface runoff, overland flow, surface flow,* and *net rain*. The water that is absorbed by the soil and that does not contribute to the surface runoff can be called *base flow, groundwater, infiltration,* and *dry weather flow*.

---

to the falling limb. The duration of the recession limb is determined by inspection, or it can be calculated from correlations with the drainage area.

Figure 20.9 Fixed-Base Method

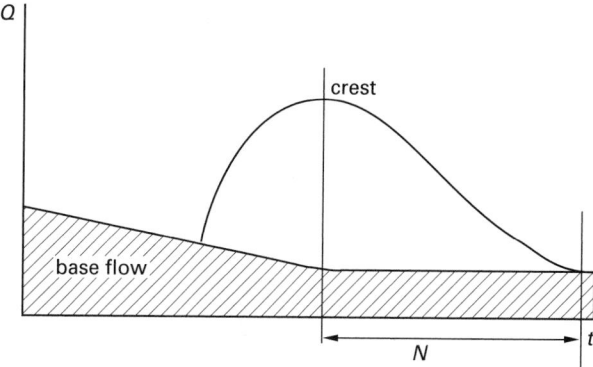

Method 3: The *variable-slope method* recognizes that the shape of the base flow curve before the storm will probably match the shape of the base flow curve after the storm. The groundwater curve after the storm is projected back under the hydrograph to a point under the inflection point of the falling limb. The separation line under the rising limb is drawn arbitrarily.

Figure 20.10 Variable-Slope Method

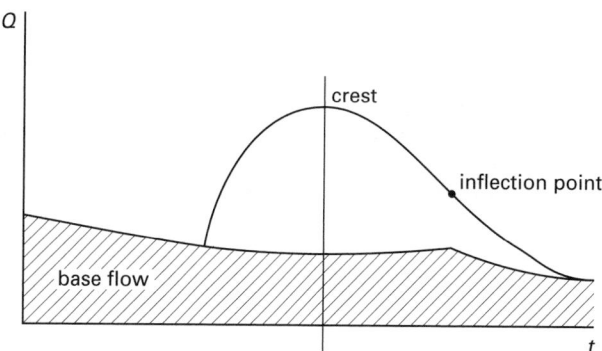

Once the base flow is separated out, the hydrograph of surface runoff will have the approximate appearance of Fig. 20.11.

Figure 20.11 Overland Flow Hydrograph

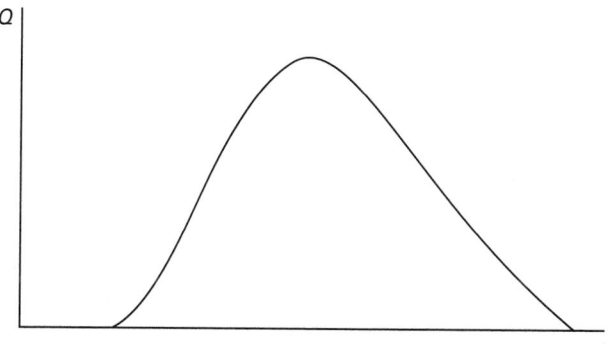

10. UNIT HYDROGRAPH

Once the overland flow hydrograph for a watershed has been developed, the total runoff (i.e., "excess rainfall," "direct runoff") volume, V, from the storm can be found as the area under the curve. Although this can be found by integration, planimetry, or computer methods, it is often sufficiently accurate to approximate the hydrograph with a histogram and to sum the areas of the rectangles.

Figure 20.12 *Hydrograph Histogram*

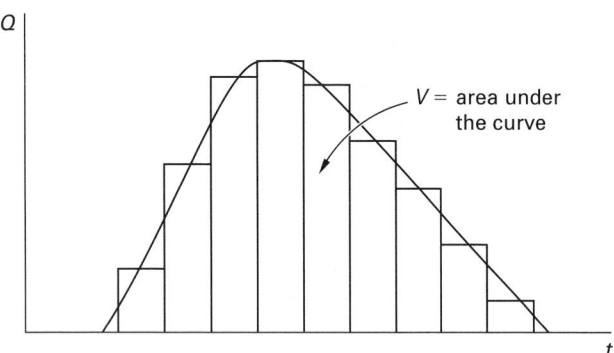

Since the area of the watershed is known, the average depth, P_{ave}, of the excess precipitation can be calculated. (Consistent units must be used in Eq. 20.21.)

$$V = A_d P_{ave,excess} \qquad \text{20.21}$$

A *unit hydrograph* (UH) is developed by dividing every point on the overland flow hydrograph by the average excess precipitation, $P_{ave,excess}$. This is a hydrograph of a storm dropping 1 in (1 cm) of excess precipitation (runoff) evenly on the entire watershed. Units of the unit hydrograph are in/in (cm/cm). Figure 20.13 shows how a unit hydrograph compares to its surface runoff hydrograph.

Figure 20.13 *Unit Hydrograph*

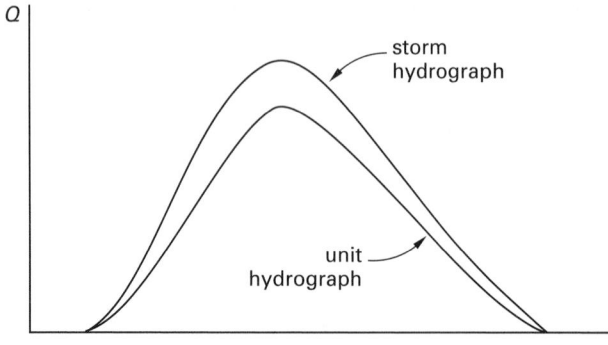

Once a unit hydrograph has been developed from historical data of a particular storm volume, it can be used for other storm volumes. Such application is based on several assumptions: (a) All storms in the watershed

have the same duration, (b) the time base is constant for all storms, (c) the shape of the rainfall curve is the same for all storms, and (d) only the total amount of rainfall varies from storm to storm.

The hydrograph of a storm producing more or less than 1 in (1 cm) of rain is found by multiplying all ordinates on the unit hydrograph by the total precipitation of the storm.

A unit hydrograph can be used to predict the runoff for storms that have durations somewhat different than the storms used to develop the unit hydrograph. Generally, storm durations differing by up to ±25% are considered to be equivalent.

Example 20.3

After a 2 hr storm, a station downstream from a 45 mi^2 (115 km^2) drainage watershed records a peak discharge of 9400 ft^3/sec (250 m^3/s) and a total runoff of 3300 ac-ft (4×10^6 m^3). (a) What is the unit hydrograph peak discharge? (b) What would be the peak runoff and design flood volume if a 2 hr storm dropped 2.5 in (6 cm) net precipitation?

SI Solution

(a) Use Eq. 20.21 to find the average precipitation for the drainage watershed.

$$P_{ave} = \frac{V}{A_d} = \frac{4 \times 10^6 \text{ m}^3}{(115 \text{ km}^2)\left(1000 \dfrac{\text{m}}{\text{km}}\right)^2}$$
$$= 0.0348 \text{ m} \quad (3.5 \text{ cm})$$

The ordinates for the unit hydrograph are found by dividing every point on the 3.5 cm hydrograph by 3.5 cm. Therefore the peak discharge for the unit hydrograph is

$$Q_{p,unit} = \frac{250 \dfrac{\text{m}^3}{\text{s}}}{3.5 \text{ cm}}$$
$$= 71.4 \text{ m}^3/\text{s·cm}$$

(b) The hydrograph for a storm producing more than 1 cm of rain is found by multiplying the ordinates of the unit hydrograph by the total precipitation. For a 6 cm storm, the peak discharge is

$$Q_p = \left(71.4 \dfrac{\text{m}^3}{\text{s·cm}}\right)(6 \text{ cm})$$
$$= 429 \text{ m}^3/\text{s}$$

To find the design flood volume, first use Eq. 20.21 to find the unit hydrograph total volume.

$$V = P A_d = \left(1 \dfrac{\text{cm}}{\text{cm}}\right)\left(\dfrac{1 \text{ m}}{100 \text{ cm}}\right)(115 \text{ km}^2)\left(1000 \dfrac{\text{m}}{\text{km}}\right)^2$$
$$= 1.15 \times 10^6 \text{ m}^3/\text{cm}$$

The design flood volume for a 6 cm storm is

$$V = (6 \text{ cm}) \left(1.15 \times 10^6 \frac{\text{m}^3}{\text{cm}} \right)$$

$$= 6.9 \times 10^6 \text{ m}^3$$

Customary U.S. Solution

(a) Use Eq. 20.21 to find the average precipitation for the drainage watershed.

$$P_{\text{ave}} = \frac{V}{A_d}$$

$$= \frac{(3300 \text{ ac-ft}) \left(43{,}560 \frac{\text{ft}^2}{\text{ac}} \right) \left(12 \frac{\text{in}}{\text{ft}} \right)}{(45 \text{ mi}^2) \left(5280 \frac{\text{ft}}{\text{mi}} \right)^2}$$

$$= 1.375 \text{ in}$$

The ordinates for the unit hydrograph are found by dividing every point on the 1.375 in hydrograph by 1.375 in. Therefore, the peak discharge for the unit hydrograph is

$$Q_{p,\text{unit}} = \frac{9400 \frac{\text{ft}^3}{\text{sec}}}{1.375 \text{ in}}$$

$$= 6836 \text{ ft}^3/\text{sec-in}$$

(b) The hydrograph for a storm producing more than 1 in of rain is found by multiplying the ordinates of the unit hydrograph by the total precipitation. For a 2.5 in storm, the peak discharge is

$$Q_p = \left(6836 \frac{\text{ft}^3}{\text{sec-in}} \right) (2.5 \text{ in})$$

$$= 17{,}091 \text{ ft}^3/\text{sec}$$

To find the design flood volume, first use Eq. 20.21 to find the unit hydrograph total volume.

$$V = PA_d = \left(1 \frac{\text{in}}{\text{in}} \right) (45 \text{ mi}^2) \left(\frac{640 \frac{\text{ac}}{\text{mi}^2}}{12 \frac{\text{in}}{\text{ft}}} \right)$$

$$= 2399 \text{ ac-ft/in}$$

The design flood volume for the 2.5 in storm is

$$V = (2.5 \text{ in}) \left(2399 \frac{\text{ac-ft}}{\text{in}} \right) = 5996 \text{ ac-ft}$$

Example 20.4

A 6 hr storm rains on a 25 mi² (65 km²) drainage watershed. Records from a stream gauging station draining the watershed are shown. (a) Construct the unit hydrograph for the 6 hr storm. (b) Find the runoff rate at $t = 15$ hr from a two-storm system if the first storm drops 2 in (5 cm) starting at $t = 0$ and the second storm drops 5 in (12 cm) starting at $t = 12$ hr.

t (hr)	Q (ft³/sec)	Q (m³/s)
0	0	0
3	400	10
6	1300	35
9	2500	70
12	1700	50
15	1200	35
18	800	20
21	600	15
24	400	10
27	300	10
30	200	5
33	100	3
36	0	0
Totals	9500	263

SI Solution

(a) Plot the stream gauging data.

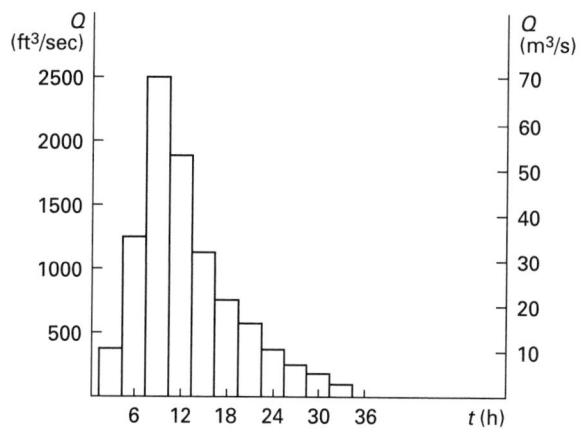

The total runoff is the total area under the curve, found by multiplying the total runoff by the rainfall data interval. This is equivalent to summing the areas of each rectangle of the histogram. Each of the histogram bars is 3 h "wide."

$$V = \left(263 \frac{\text{m}^3}{\text{s}} \right) (3 \text{ h}) \left(3600 \frac{\text{s}}{\text{h}} \right) = 2.8404 \times 10^6 \text{ m}^3$$

The watershed drainage area is

$$A_d = (65 \text{ km}^2) \left(1000 \frac{\text{m}}{\text{km}} \right)^2$$

$$= 65 \times 10^6 \text{ m}^2$$

The average precipitation is calculated from Eq. 20.21.

$$P = \frac{V}{A_d}$$
$$= \frac{(2.8404 \times 10^6 \text{ m}^3) \left(100 \frac{\text{cm}}{\text{m}}\right)}{65 \times 10^6 \text{ m}^2}$$
$$= 4.37 \text{ cm}$$

The unit hydrograph has the same shape as the actual hydrograph with all ordinates reduced by a factor of 4.37.

(b) To find the flow at 15 hr, add the contributions from each storm. For the 5 cm storm, the contribution is the 15 hr runoff multiplied by its scaling factors; for the 12 cm storm, the contribution is the 15 hr − 12 hr = 3 hr runoff multiplied by its scaling factors.

$$Q = \frac{(5 \text{ cm}) \left(35 \frac{\text{m}^3}{\text{s}}\right)}{4.37 \text{ cm}} + \frac{(12 \text{ cm}) \left(10 \frac{\text{m}^3}{\text{s}}\right)}{4.37 \text{ cm}}$$
$$= 67.5 \text{ m}^3/\text{s}$$

Customary U.S. Solution

(a) Plot the stream gauging data.

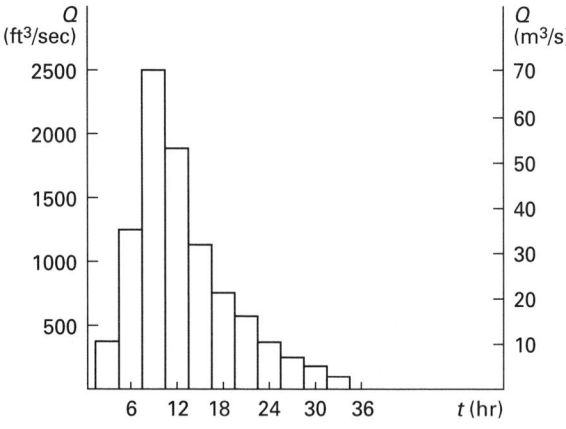

The total runoff is the area under the curve, found by multiplying the total runoff by the rainfall data interval. This is equivalent to summing the areas of each rectangle of the histogram. Each of the histogram bars is 3 hr "wide."

$$V = \left(9500 \frac{\text{ft}^3}{\text{sec}}\right) (3 \text{ hr}) \left(3600 \frac{\text{sec}}{\text{hr}}\right)$$
$$= 1.026 \times 10^8 \text{ ft}^3$$

The watershed drainage area is

$$A_d = (25 \text{ mi}^2) \left(5280 \frac{\text{ft}}{\text{mi}}\right)^2$$
$$= 6.97 \times 10^8 \text{ ft}^2$$

The average precipitation is calculated from Eq. 20.21.

$$P = \frac{V}{A_d}$$
$$= \frac{(1.026 \times 10^8 \text{ ft}^3) \left(12 \frac{\text{in}}{\text{ft}}\right)}{6.97 \times 10^8 \text{ ft}^2}$$
$$= 1.766 \text{ in}$$

The unit hydrograph has the same shape as the actual hydrograph with all ordinates reduced by a factor of 1.766.

(b) To find the flow at $t = 15$ hr, add the contributions from each storm. For the 2 in storm, the contribution is the 15 hr runoff multiplied by its scaling factors; for the 5 in storm, the contribution is the 15 hr − 12 hr = 3 hr runoff multiplied by its scaling factors.

$$Q = \frac{(2 \text{ in}) \left(1200 \frac{\text{ft}^3}{\text{sec}}\right)}{1.766 \text{ in}} + \frac{(5 \text{ in}) \left(400 \frac{\text{ft}^3}{\text{sec}}\right)}{1.766 \text{ in}}$$
$$= 2492 \text{ ft}^3/\text{sec}$$

11. NRCS SYNTHETIC UNIT HYDROGRAPH

If a watershed is ungauged such that no historical records are available to produce a unit hydrograph, the *synthetic hydrograph* can still be reasonably approximated. The process of developing a synthetic hydrograph is known as *hydrograph synthesis*.

Pioneering work was done in 1938 by Snyder, who based his analysis on Appalachian highland watersheds with areas of 10 to 10,000 mi^2. Snyder's work has been largely replaced by more sophisticated analyses, including the NRCS methods.

The U.S. Natural Resources Conservation Service (NRCS), previously known as the Soil Conservation Service (SCS), developed a synthetic unit hydrograph based on the *curve number*, CN. The method was originally intended for use with rural watersheds up to 2000 ac, but it appears to be applicable for urban conditions up to 4000 to 5000 ac.

In order to draw the NRCS synthetic unit hydrograph, it is necessary to calculate the time to peak flow (t_p) and the peak discharge (Q_p). Provisions for calculating both of these parameters are included in the method.

$$t_p = 0.5t_R + t_1 \qquad \qquad 20.22$$

t_R in Eq. 20.22 is the storm duration (i.e., of the rainfall). t_1 in Eq. 20.22 is the *lag time* (i.e., the time from the centroid of the rainfall distribution to the peak discharge). Lag time can be determined from correlations with geographical region and drainage area or calculated from Eq. 20.23. Although Eq. 20.23 was developed for natural watersheds, limited studies of urban watersheds indicate that it does not change significantly for urbanized watersheds. S in Eq. 20.23 is the soil water storage capacity in inches, computed as a function of the curve number. (See Eq. 20.43.)

$$t_{1,\text{hr}} = \frac{L_{o,\text{ft}}^{0.8}(S+1)^{0.7}}{1900\sqrt{S_{\text{percent}}}} \qquad 20.23$$

The peak runoff is calculated as

$$Q_p = \frac{0.756A_{d,\text{ac}}}{t_p} \qquad 20.24$$

$$Q_p = \frac{484A_{d,\text{mi}^2}}{t_p} \qquad 20.25$$

Q_p and t_p only contribute one point to the construction of the unit hydrograph. To construct the remainder, Table 20.3 must be used. Using time as the independent variable, selections of time (different from t_p) are arbitrarily made, and the ratio t/t_p is calculated. The curve is then used to obtain the ratio of Q_t/Q_p.

12. NRCS SYNTHETIC UNIT TRIANGULAR HYDROGRAPH

The NRCS unit triangular hydrograph is shown in Fig. 20.14. It is found from the peak runoff, the time to peak, and the duration of runoff. Peak runoff, Q_p, is found from Eq. 20.24 or 20.25. Time to peak and duration are correlated with the time of concentration. These are generalizations that apply to specific storm and watershed types.

Equation 20.26 can be used to estimate the time to peak.

$$t_p = 0.5t_R + t_1 \qquad 20.26$$
$$t_1 \approx 0.6t_c \qquad 20.27$$

Alternatively, the time to peak has been roughly correlated to the time of concentration.

$$t_p = 0.67t_c \qquad 20.28$$

The total duration of the unit hydrograph is the sum of time to peak and length of recession limb, assumed to be $1.67t_p$.

$$t_b = t_p + t_{\text{recession}} = t_p + 1.67t_p = 2.67t_p \qquad 20.29$$

Table 20.3 NRCS Dimensionless Unit Hydrograph and Mass Curve Ratios[a]

time ratios (t/t_p)	discharge ratios (Q/Q_p)	cumulative mass curve fraction
0.0	0.000	0.000
0.1	0.030	0.001
0.2	0.100	0.006
0.3	0.190	0.012
0.4	0.310	0.035
0.5	0.470	0.065
0.6	0.660	0.107
0.7	0.820	0.163
0.8	0.930	0.228
0.9	0.990	0.300
1.0	1.000	0.375
1.1	0.990	0.450
1.2	0.930	0.522
1.3	0.860	0.589
1.4	0.780	0.650
1.5	0.680	0.700
1.6	0.560	0.751
1.7	0.460	0.790
1.8	0.390	0.822
1.9	0.330	0.849
2.0	0.280	0.871
2.2	0.207	0.908
2.4	0.147	0.934
2.6	0.107	0.953
2.8	0.077	0.967
3.0	0.055	0.977
3.2	0.040	0.984
3.4	0.029	0.989
3.6	0.021	0.993
3.8	0.015	0.995
4.0	0.011	0.997
4.5	0.005	0.999
5.0	0.000	1.000

[a]Reprinted from NRCS, 1969.

Figure 20.14 NRCS Synthetic Unit Triangular Hydrograph

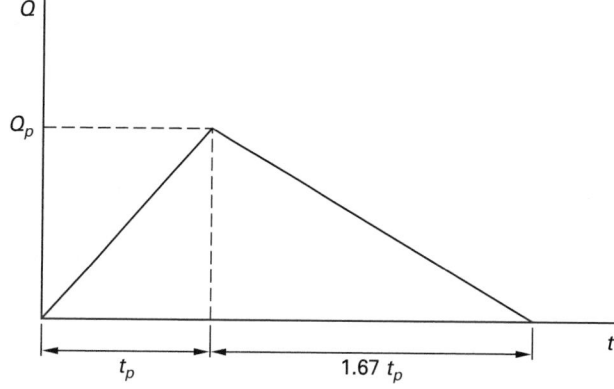

Figure 20.15 *Espey Watershed Conveyance Factor*

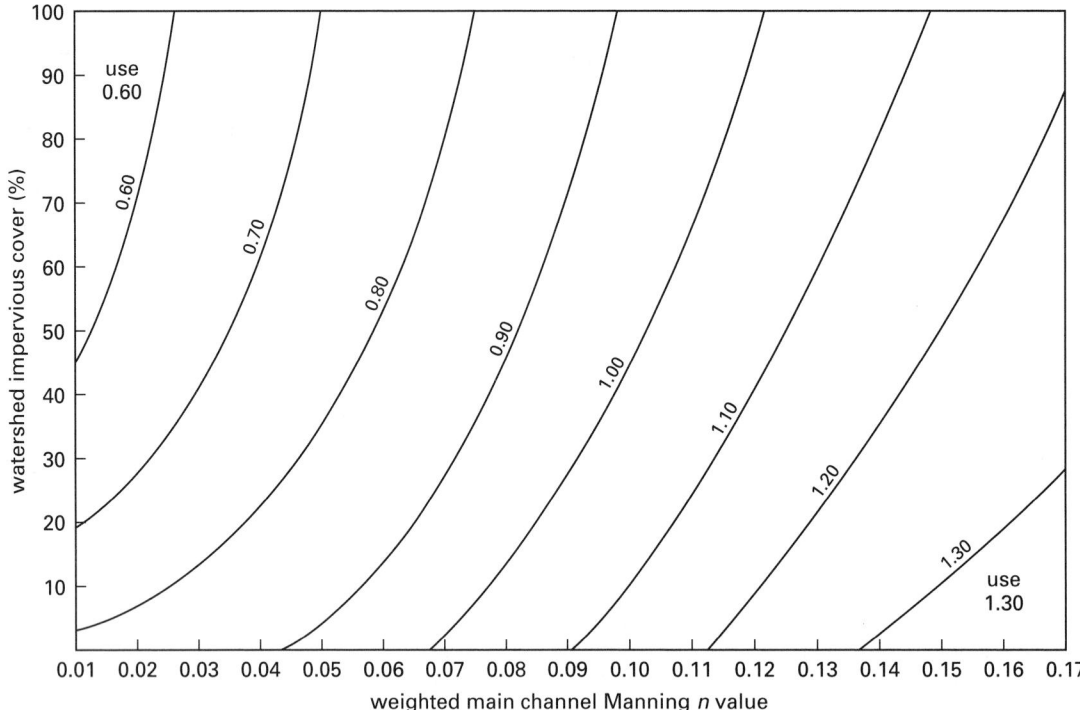

Reprinted from *Recommended Hydrologic Procedures for Computing Urban Runoff from Small Watersheds in Pennsylvania*, Commonwealth of Pennsylvania, Fig. 6-3, 1982, Department of Environmental Resources.

13. ESPEY SYNTHETIC UNIT HYDROGRAPH

The *Espey method* (1978) calculates the time to peak (t_p in min), peak discharge (Q_p in ft³/sec), total hydrograph base (t_b in min), and the hydrograph widths at 50% and 75% of the peak discharge rates (W_{50} and W_{75} in min). These values depend on the watershed area (A, in mi²), main channel flow path length (L in ft), slope (S in ft/ft), roughness, and percent imperviousness (Imp). ϕ is a dimensionless watershed conveyance factor ($0.6 < \phi < 1.3$) that depends on the percent imperviousness and weighted main channel Manning roughness coefficient, n. ϕ is found graphically from Fig. 20.15.

$$t_p = 3.1 L^{0.23} S_{\text{decimal}}{}^{-0.25} (\text{Imp}_{\text{percent}})^{-0.18} \phi^{1.57} \quad 20.30$$

$$Q_p = (31.62 \times 10^3) A^{0.96} t_p^{-1.07} \quad 20.31$$

$$t_b = (125.89 \times 10^3) A Q_p^{-0.95} \quad 20.32$$

$$W_{50} = (16.22 \times 10^3) A^{0.93} Q_p^{-0.92} \quad 20.33$$

$$W_{75} = (3.24 \times 10^3) A^{0.79} Q_p^{-0.78} \quad 20.34$$

To use this method, the geometric slope, S, used in Eq. 20.30 is specifically calculated from Eq. 20.35. H is the difference in elevation of points A and B. Point A is the channel bottom a distance $0.2L$ downstream from the upstream watershed boundary. Point B is the channel bottom at the downstream watershed boundary.

$$S_{\text{decimal}} = \frac{H}{0.8L} \quad 20.35$$

The unit hydrograph is drawn by manually "fitting" a smooth curve over the seven computed points. The widths of W_{50} and W_{75} are allocated in a 1:2 ratio to the rising and falling hydrograph limbs, respectively. After the curve is drawn, it is adjusted to be a unit hydrograph. The resulting curve is sometimes referred to as an *Espey 10-minute unit hydrograph*.

Figure 20.16 *Espey Synthetic Hydrograph*

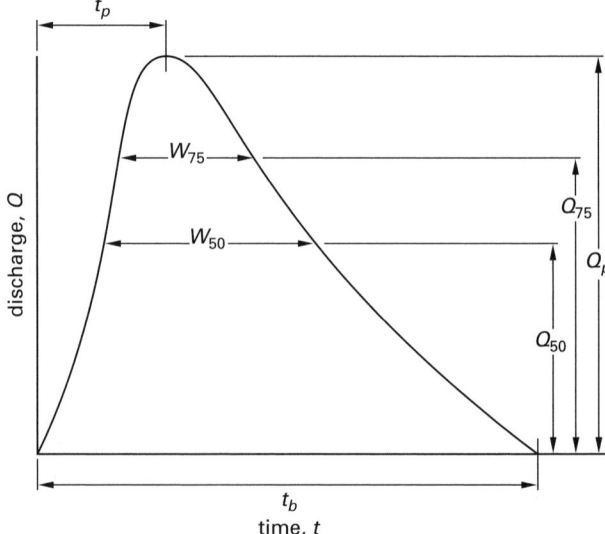

14. HYDROGRAPH SYNTHESIS

If a storm's duration is not the same or close to the hydrograph base length, the unit hydrograph cannot be used to predict runoff. For example, the runoff from a 6 hr storm cannot be predicted from a unit hydrograph derived from a 2 hr storm. However, the technique of hydrograph synthesis can be used to construct the hydrograph of the longer storm from the unit hydrograph of a shorter storm.

A. Lagging Storm Method

If a unit hydrograph for a storm of duration t_R is available, the *lagging storm method* can be used to construct the hydrograph of a storm whose duration is a whole multiple of t_R. For example, a 6 hr storm hydrograph can be constructed from a 2 hr unit hydrograph.

Let the whole multiple number be n. To construct the longer hydrograph, draw n unit hydrographs, each separated by time t_R. Then add the ordinates to obtain a hydrograph for an $n t_R$ duration storm. Since the total rainfall from this new hydrograph is n inches (having been constructed from n unit hydrographs), the curve will have to be reduced (i.e., divided) by n everywhere to produce a unit hydrograph.

Figure 20.17 Lagging Storm Method

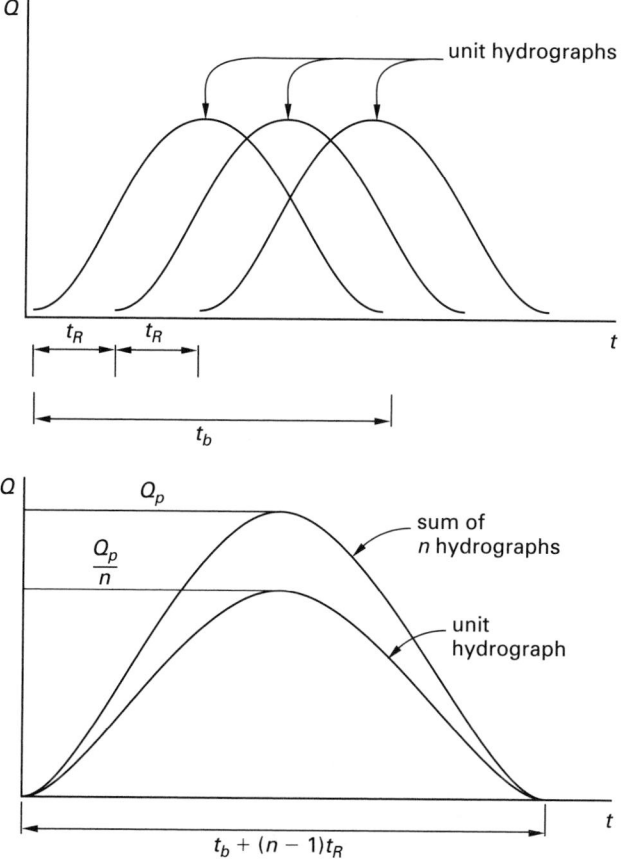

B. S-Curve Method

The *S-curve method* can be used to construct hydrographs from unit hydrographs with longer or shorter durations, even when the storm durations are not multiples. This method begins by adding the ordinates of many unit hydrographs, each lagging the other by time t_R, the duration of the storm that produced the unit hydrograph. After a sufficient number of lagging unit hydrographs have been added together, the accumulation will level off and remain constant. At that point, the lagging can be stopped. The resulting accumulation is known as an S-curve.

Figure 20.18 Constructing the S-Curve

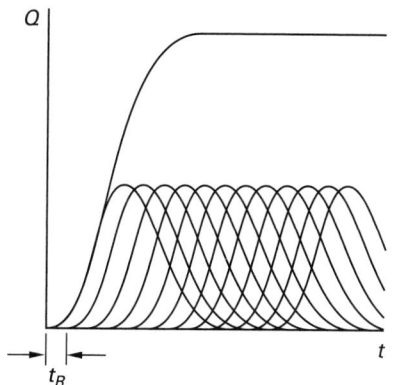

If two S-curves are drawn, one lagging the other by time $t_{R'}$, the area between the two curves represents a hydrograph area for a storm of duration t'_R. The differences between the two curves can be plotted and scaled to a unit hydrograph by multiplying by the ratio of (t_R/t'_R).

15. PEAK RUNOFF FROM THE RATIONAL METHOD

Although total runoff volume is required for reservoir and dam design, the instantaneous peak runoff is needed to size culverts and storm drains.

The *rational formula* ("method," "equation," etc.), Eq. 20.36, for peak discharge has been in widespread use in the United States since the early 1900s.[3] It is applicable to small areas (i.e., less than several hundred acres or so), but is seldom used for areas greater than 1 to 2 mi^2. The intensity used in Eq. 20.36 depends on the time of concentration and the degree of protection desired (i.e., the recurrence interval).[4]

$$Q_p = CIA_d \qquad \textit{20.36}$$

---

[3]In Great Britain, the rational equation is known as the *Lloyd-Davies equation*.

[4]When using intensity-duration-frequency curves to size storm sewers, culverts, and other channels, it is assumed that the frequencies and probabilities of flood damage and storms are identical. This is not generally true, but the assumption is usually made anyway.

Figure 20.19 *Using S-Curves to Construct a t' Hydrograph*

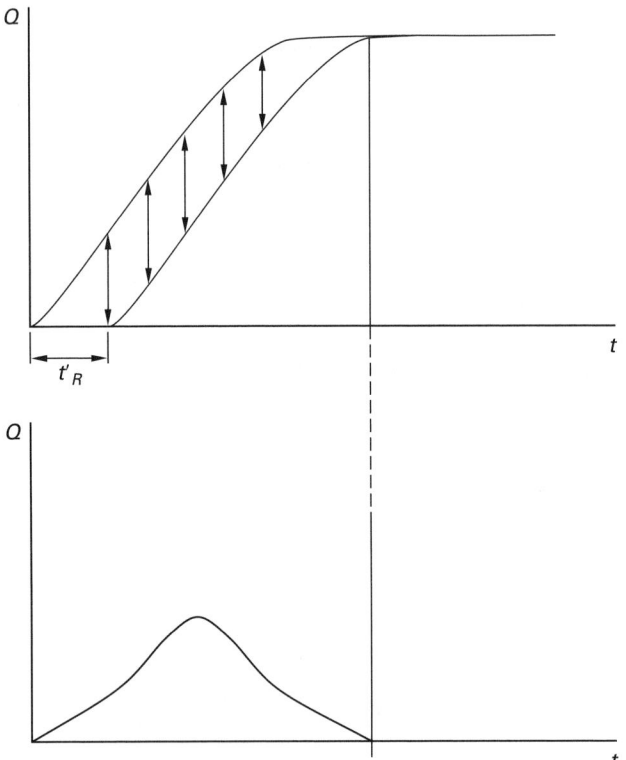

intensity after 25 min is 3.9 in/hr. (a) Calculate the time to concentration. (b) Use the rational method to calculate the peak flow.

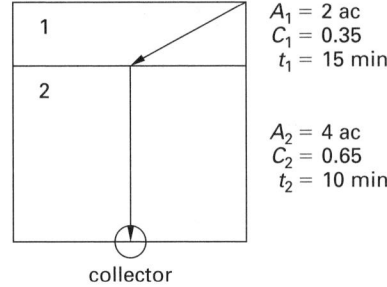

$A_1 = 2$ ac
$C_1 = 0.35$
$t_1 = 15$ min

$A_2 = 4$ ac
$C_2 = 0.65$
$t_2 = 10$ min

collector

Solution

(a) The overland flow time is given for both areas. The time for water from the farthest corner to reach the collector is

$$t_c = 15 \text{ min} + 10 \text{ min} = 25 \text{ min}$$

(b) The runoff coefficients are given for each area. Since we want to size the pipe carrying the total runoff, the coefficients are weighted by their respective contributing areas.

$$C = \frac{(2 \text{ ac})(0.35) + (4 \text{ ac})(0.65)}{2 \text{ ac} + 4 \text{ ac}}$$
$$= 0.55$$

The intensity after 25 min was given as 3.9 in/hr.

The total area is 4 ac + 2 ac = 6 ac.

The peak flow is found from Eq. 20.36.

$$Q_p = CIA_d = (0.55)\left(3.9 \ \frac{\text{in}}{\text{hr}}\right)(6 \text{ ac})$$
$$= 12.9 \text{ ac-in/hr} \ (\text{ft}^3/\text{sec})$$

Since A_d is in acres, Q_p is in ac-in/hr. However, Q_p is taken as ft^3/sec since the conversion factor between these two units is 1.008.

Typical values of C coefficients are found in App. 20.A. If more than one area contributes to the runoff, the coefficient is weighted by the areas.

Accurate values of the C coefficient depend not only on the surface cover and soil type, but also on the recurrence interval, antecedent moisture content, rainfall intensity, drainage area, slope, and fraction of imperviousness. These factors have been investigated and quantified by Rossmiller (1981), who correlated these effects with the NRCS curve number. The Schaake, et. al. (1967), equation developed at Johns Hopkins University was intended for use in urban areas, correlating the impervious fraction and slope.

$$C = 0.14 + 0.65(\text{Imp}_{\text{decimal}}) + 0.05S_{\text{percent}} \qquad 20.37$$

The rational method assumes that rainfall occurs at a constant rate. If this is true, then the peak runoff will occur when the entire drainage area is contributing to surface runoff, which will occur at t_c. Other assumptions include (a) the recurrence interval of the peak flow is the same as for the design storm, (b) the runoff coefficient is constant, and (c) the rainfall is spatially uniform over the drainage area.

Example 20.5

Two adjacent fields, as shown, contribute runoff to a collector whose capacity is to be determined. The storm

Example 20.6

Three watersheds contribute runoff to a storm drain. The watersheds have the following characteristics.

watershed	area (ac)	overland flow time (min)	C
A	10	20	0.3
B	2	5	0.7
C	15	25	0.4

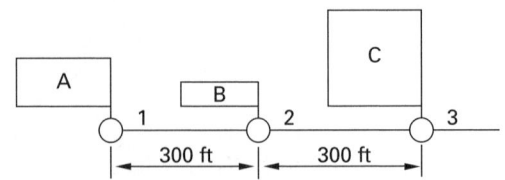

The instantaneous rainfall intensity (in/hr) for the area and specified storm frequency is

$$I = \frac{115}{t_c + 15}$$

The collector inlets are 300 ft apart. The pipe slope is 0.009. The Manning roughness coefficient is 0.015. Assume the flow velocity is 5 ft/sec in all sections. The storm duration is long enough to permit all three areas to contribute to the combined flow through section 3. (a) What should be the pipe size in section 2? (b) What is the maximum flow through section 3?

Solution

The maximum flow occurs when the overland flow from both areas A and B have reached the second manhole. The time of concentration for area A is the time of concentration to the first manhole plus the travel time in pipe section 1.

$$t_c = 20 \text{ min} + \frac{\dfrac{300 \text{ ft}}{5 \dfrac{\text{ft}}{\text{sec}}}}{60 \dfrac{\text{sec}}{\text{min}}}$$

$$= 21 \text{ min}$$

The time of concentration for area B is 5 min. The pipe in section 2 should be sized for flow at 21 min.

$$I = \frac{115}{t_c + 15} = \frac{115}{21 + 15}$$

$$= 3.19 \text{ in/hr}$$

Use the sum of the CA values in Eq. 20.36.

$$\sum CA = (0.3)(10 \text{ ac}) + (0.7)(2 \text{ ac})$$

$$= 4.4 \text{ ac}$$

Use the rational formula, Eq. 20.36.

$$Q = \sum CA_d I = (4.4 \text{ ac}) \left(3.19 \frac{\text{in}}{\text{hr}} \right)$$

$$= 14.0 \text{ ac-in/hr } (\text{ft}^3/\text{sec})$$

This value of Q should be used to design section 2 of the pipe. If the pipe is assumed to flow full, from Eq. 19.16,

$$D = 1.33 \left(\frac{nQ}{\sqrt{S_{\text{decimal}}}} \right)^{3/8}$$

$$= (1.33) \left(\frac{(0.015) \left(14.0 \dfrac{\text{ft}^3}{\text{sec}} \right)}{\sqrt{0.009}} \right)^{3/8}$$

$$= 1.79 \text{ ft} \quad [\text{round to 2.0 ft}]$$

The maximum flow through section 3 occurs when the overland flow from plots A, B, and C reach the third manhole.

For plot A,

$$t_c = 20 \text{ min} + 1 \text{ min} + 1 \text{ min} = 22 \text{ min}$$

For plot B,

$$t_c = 5 \text{ min} + 1 \text{ min} = 6 \text{ min}$$

For plot C,

$$t_c = 25 \text{ min}$$

The maximum runoff will occur 25 min after the start of the storm.

$$I = \frac{115}{25 + 15} = 2.875 \text{ in/hr}$$

The sum of the CA values is

$$\sum CA = (0.3)(10 \text{ ac}) + (0.7)(2 \text{ ac}) + (0.4)(15 \text{ ac})$$

$$= 10.4 \text{ ac}$$

$$Q = \sum CA_d I = (10.4 \text{ ac}) \left(2.875 \frac{\text{in}}{\text{hr}} \right)$$

$$= 29.9 \text{ ac-in/hr } (\text{ft}^3/\text{sec})$$

16. NRCS CURVE NUMBER

Several methods of calculating total and peak runoff have been developed over the years by the U.S. Natural Resources Conservation Service. These methods have generally been well correlated with actual experience, and the NRCS methods have become dominant in the United States.

The NRCS methods classify the land use and soil type by a single parameter called the *curve number*, CN. This method can be used for any size homogeneous watershed with a known percentage of imperviousness. If the watershed varies in soil type or in cover, it generally should be divided into regions to be analyzed separately. A composite curve number can be calculated by weighting the curve number for each region by its area. Alternatively, the runoffs from each region can be calculated separately and added.

The NRCS method of using precipitation records and an assumed distribution of rainfall to construct a synthetic storm is based on several assumptions. First, a type II storm is assumed. Type I storms, which drop most of their precipitation early, are applicable to Hawaii, Alaska, and the coastal side of the Sierra Nevada and Cascade mountains in California, Oregon, and Washington. Type II distributions are typical of the rest of the United States, Puerto Rico, and the Virgin Islands.

This method assumes that initial abstraction (depression storage, evaporation, and interception losses) is equal to 20% of the storage capacity.

$$I_a = 0.2S \qquad \qquad 20.38$$

For there to be any runoff at all, the gross rain must equal or exceed the initial abstraction.

$$P \geq I_a \qquad 20.39$$

The storage capacity must be great enough to absorb the initial abstraction plus the infiltration.

$$S \geq I_a + F \qquad 20.40$$

The following steps constitute the NRCS method.

step 1: Classify the soil into a *hydrologic soil group* (HSG) according to its infiltration rate. Soil is classified into HSG A (low runoff potential) through D (high runoff potential).

Group A: High infiltration rates (> 0.30 in/hr) even if thoroughly saturated; chiefly deep sands and gravels with good drainage and high moisture transmission. In urbanized areas, this category includes sand, loamy sand, and sandy loam.

Group B: Moderate infiltration rates if thoroughly wetted (0.15 to 0.30 in/hr), moderate rates of moisture transmission, and consisting chiefly of coarse to moderately fine textures. In urbanized areas, this category includes silty loam and loam.

Group C: Slow infiltration rates (0.05 to 0.15 in/hr) if thoroughly wetted, and slow moisture transmission; soils having moderately fine to fine textures or that impede the downward movement of water. In urbanized areas, this category includes sandy clay loam.

Group D: Very slow infiltration rates (less than 0.05 in/hr) if thoroughly wetted, very slow water transmission, and consisting primarily of clay soils with high potential for swelling; soils with permanent high water tables; or soils with an impervious layer near the surface. In urbanized areas, this category includes clay loam, silty clay loam, sandy clay, silty clay, and clay.

(Note that as a result of urbanization, the underlying soil may be disturbed or covered by a new layer. The original classification will no longer be applicable, and the "urbanized" soil HSGs are applicable.)

step 2: Determine the preexisting soil conditions. The soil condition is classified into *antecedent runoff conditions* (ARC) I through III.[5] Generally, "average" conditions (ARC II) are assumed.

ARC I: Dry soils, prior to or after plowing or cultivation, or after periods without rain.

ARC II: Typical conditions existing before maximum annual flood.

ARC III: Saturated soil due to heavy rainfall (or light rainfall with freezing temperatures) during 5 days prior to storm.

step 3: Classify *cover type* and hydrologic condition of the soil-cover complex. For pasture, range, row crops, arid, and semiarid lands, the NRCS method includes additional tables to classify the cover and hydrologic conditions. In order to use these tables, it is necessary to characterize the surface coverage. The condition is "good" if it is lightly grazed or has plant cover over 75% or more of its area. The condition is "fair" if plant coverage is 50 to 75% or not heavily grazed. The condition is "poor" if the area is heavily grazed, has no mulch, or has plant cover over less than 50% of the area.

step 4: Use Table 20.4 or Table 20.5 to determine the *curve number*, CN, corresponding to the soil classification for ARC II.

step 5: If the soil is ARC I or ARC III, convert the curve number from step 4 using Eq. 20.41 or 20.42 and rounding up.

$$\mathrm{CN_I} = \frac{4.2\,\mathrm{CN_{II}}}{10 - 0.058\,\mathrm{CN_{II}}} \qquad 20.41$$

$$\mathrm{CN_{III}} = \frac{23\,\mathrm{CN_{II}}}{10 + 0.13\,\mathrm{CN_{II}}} \qquad 20.42$$

step 6: If any significant fraction of the watershed is impervious (i.e., CN = 98), or if the watershed consists of areas with different curve numbers, calculate the composite curve number by weighting by the runoff areas (same as weighting by the impervious and pervious fractions). If the watershed's impervious fraction is different from the value implicit in step 3's classification, or if the impervious area is not connected directly to a storm drainage system, the NRCS method includes direct and graphical adjustments.

step 7: Estimate the time of concentration of the watershed.[6] (See Sec. 5.)

step 8: Determine the *gross (total) rainfall*, P, from the storm. (See Eq. 20.18.) To do this, it is necessary to assume the storm length and recurrence interval. It is a characteristic of

---

[5]The *antecedent runoff condition* (ARC) may also be referred to as the *antecedent moisture condition* (AMC).

[6]The NRCS method uses T_c, not t_c, as the symbol for time of concentration.

Table 20.4 *Runoff Curve Numbers of Urban Areas (ARC II)*

cover description		curve numbers for hydrologic soil			
cover type and hydrologic condition	average percent impervious area	group A	group B	group C	group D
fully developed urban areas (vegetation established)					
open space (lawns, parks, golf courses, cemeteries, etc.)					
poor condition (grass cover < 50%)		68	79	86	89
fair condition (grass cover 50 to 75%)		49	69	79	84
good condition (grass cover > 75%)		39	61	74	80
impervious areas					
paved parking lots, roofs, driveways, etc. (excluding right-of-way)		98	98	98	98
streets and roads					
paved; curbs and storm sewers (excluding right-of-way)		98	98	98	98
paved; open ditches (including right-of-way)		83	89	92	93
gravel (including right-of-way)		76	85	89	91
dirt (including right-of-way)		72	82	87	89
western desert urban areas					
natural desert landscaping (pervious areas only)		63	77	85	88
artificial desert landscaping (impervious weed barrier, desert shrub with 1 to 2 in sand or gravel mulch and basin borders)		96	96	96	96
urban districts					
commercial and business	85	89	92	94	95
industrial	72	81	88	91	93
residential districts by average lot size					
$\frac{1}{8}$ acre or less (townhouses)	65	77	85	90	92
$\frac{1}{4}$ acre	38	61	75	83	87
$\frac{1}{3}$ acre	30	57	72	81	86
$\frac{1}{2}$ acre	25	54	70	80	85
1 acre..	20	51	68	79	84
2 acres...	12	46	65	77	82
developing urban areas					
newly graded areas (pervious areas only, no vegetation)		77	86	91	94

Reprinted from *Urban Hydrology for Small Watersheds*, Technical Release TR 55, United States Department of Agriculture, Natural Resources Conservation Service, Table 2-2a, 1986.

Table 20.5 *Runoff Curve Numbers for Cultivated Agricultural Lands (ARC II)*

| | cover description | | curve numbers for hydrologic soil | | | |
cover type	treatment[a]	hydrologic condition[b]	group A	group B	group C	group D
fallow	bare soil	–	77	86	91	94
	crop residue cover (CR)	poor	76	85	90	93
		good	74	83	88	90
row crops	straight row (SR)	poor	72	81	88	91
		good	67	78	85	89
	SR + CR	poor	71	80	87	90
		good	64	75	82	85
	contoured (C)	poor	70	79	84	88
		good	65	75	82	86
	C + CR	poor	69	78	83	87
		good	64	74	81	85
	contoured and terraced (C&T)	poor	66	74	80	82
		good	62	71	78	81
	C&T + CR	poor	65	73	79	81
		good	61	70	77	80
small grain	SR	poor	65	76	84	88
		good	63	75	83	87
	SR + CR	poor	64	75	83	86
		good	60	72	80	84
	C	poor	63	74	82	85
		good	61	73	81	84
	C + CR	poor	62	73	81	84
		good	60	72	80	83
	C&T	poor	61	72	79	82
		good	59	70	78	81
	C&T + CR	poor	60	71	78	81
		good	58	69	77	80
close-seeded or broadcast legumes or rotation meadow	SR	poor	66	77	85	89
		good	58	72	81	85
	C	poor	64	75	83	85
		good	55	69	78	83
	C&T	poor	63	73	80	83
		good	51	67	76	80

[a] *Crop residue cover* applies only if residue is on at least 5% of the surface throughout the year.
[b]Hydrologic condition is based on a combination of factors that affect infiltration and runoff, including (a) density and canopy of vegetative areas, (b) amount of year-round cover, (c) amount of grass or close-seeded legumes in rotations, (d) percent of residue cover on the land surface (good ≥ 20%), and (e) degree of surface roughness.
   Poor: Factors impair infiltration and tend to increase runoff.
   Good: Factors encourage average and better-than-average infiltration and tend to decrease runoff.
Reprinted from *Urban Hydrology for Small Watersheds*, Technical Release TR 55, United States Department of Agriculture, Natural Resources Conservation Service, Table 2-b, 1986.

the NRCS methods to use a 24 hr storm. Maps from the U.S. Weather Bureau can be used to read gross point rainfalls for storms with frequencies from 1 to 100 years.[7]

step 9: Multiply the gross rain point value from step 8 by a factor from Fig. 20.20 to make the gross rain representative of larger areas. This is the *areal rain*.

Figure 20.20 *Point to Areal Rain Conversion Factors*

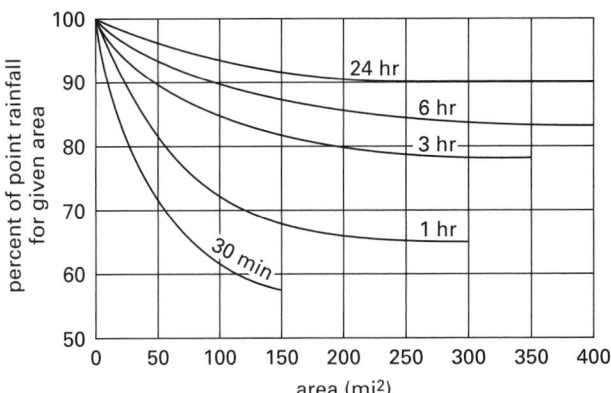

step 10: The NRCS method assumes that infiltration follows an exponential decay curve with time. Storage capacity of the soil (i.e., the potential maximum retention after runoff begins), S, is calculated from the curve number by using Eq. 20.43.

$$S = \frac{1000}{CN} - 10 \qquad 20.43$$

step 11: Calculate the total runoff (net rain, precipitation excess, etc.), Q, in inches from the areal rain. Equation 20.44 subtracts losses from interception, storm period evaporation, depression storage, and infiltration from the *gross rain* to obtain the *net rain*. (Equation 20.44 can be derived from the water balance equation, Eq. 20.2.)

$$Q_{\text{inches}} = \frac{(P_g - I_a)^2}{P_g - I_a + S} = \frac{(P_g - 0.2S)^2}{P_g + 0.8S} \qquad 20.44$$

17. NRCS GRAPHICAL PEAK DISCHARGE METHOD

Two NRCS methods are available for calculating the peak discharge. When a full hydrograph is not needed, the so-called graphical method can be used. If a hydrograph is needed, the tabular method can be used.[8]

---

[7] *Rainfall Frequency Atlas of the United States*, U.S. Weather Bureau, Technical Paper 40.

[8] The graphical and tabular methods both rely on graphs and tables that are contained in TR-55. These graphs and tables are not included in this chapter. The NRCS tabular method is also not described in this book.

The graphical method is applicable when (a) CN > 40, (b) 0.1 hr < t_c < 10 hr, (c) the watershed is relatively homogeneous or uniformly mixed, (d) all streams have the same time of concentration, and (e) there is no interim storage along the stream path.

Equation 20.45 is the NRCS peak discharge equation. q_p is the peak discharge (ft^3/sec), q_u is the unit peak discharge (cubic feet per square mile per inch of runoff, csm/in), A_{mi^2} is the drainage area in mi^2, Q_{inches} is the runoff (in), and F_p is a pond and swamp adjustment factor.

$$q_p = q_u A_{\text{mi}^2} Q_{\text{inches}} F_p \qquad 20.45$$

The graphical method begins by obtaining the design 24 hr rainfall, P, for the area. Total runoff, Q, and the curve number, CN, are determined as in Sec. 16. Then, the initial abstraction, I_a, is obtained from Table 20.6, and the ratio I_a/P is calculated. Next, NRCS curves are entered with t_c and the I_a/P ratio to determine the runoff in cubic feet per square mile per inch of runoff (csm/in). The appropriate pond and swamp adjustment factor is selected from the NRCS literature. ($F_p = 1$ if there are no ponds or swamps.)

Table 20.6 *Initial Abstraction versus Curve Number*

curve number	I_a (in)	curve number	I_a (in)
40	3.000	70	0.857
41	2.878	71	0.817
42	2.762	72	0.778
43	2.651	73	0.740
44	2.545	74	0.703
45	2.444	75	0.667
46	2.348	76	0.632
47	2.255	77	0.597
48	2.167	78	0.564
49	2.082	79	0.532
50	2.000	80	0.500
51	1.922	81	0.469
52	1.846	82	0.439
53	1.774	83	0.410
54	1.704	84	0.381
55	1.636	85	0.353
56	1.571	86	0.326
57	1.509	87	0.299
58	1.448	88	0.273
59	1.390	89	0.247
60	1.333	90	0.222
61	1.279	91	0.198
62	1.226	92	0.174
63	1.175	93	0.151
64	1.125	94	0.128
65	1.077	95	0.105
66	1.030	96	0.083
67	0.985	97	0.062
68	0.941	98	0.041
69	0.899		

(Multiply in by 2.54 to obtain cm.)

Reprinted from *Urban Hydrology for Small Watersheds*, Technical Release TR 55, United Stated Department of Agriculture, Natural Resources Conservation Service, Table 4-1, 1986.

18. RESERVOIR SIZING: MODIFIED RATIONAL METHOD

An effective method of preventing flooding is to store surface runoff temporarily. After the storm is over, the stored water can be gradually released. An *impounding reservoir* (*retention watershed* or *detention watershed*) is a watershed used to store excess flow from a stream or river. The stored water is released when the stream flow drops below the minimum level that is needed to meet water demand. The *impoundment depth* is the design depth. Finding the impoundment depth is equivalent to finding the design storage capacity of the reservoir.

The purpose of a *reservoir sizing* (*reservoir yield*) analysis is to determine the proper size of a reservoir or dam, or to evaluate the ability of an existing reservoir to meet water demands.

The volume of a reservoir needed to hold streamflow from a storm is simply the total area of the hydrograph. Similarly, when comparing two storms, the incremental volume needed is simply the difference in the areas of their two hydrographs.

Poertner's (1974) *modified rational method* can be used to design detention storage facilities for small areas (up to 20 ac). A trapezoidal hydrograph is constructed with the peak flow calculated from the Rational equation. The total hydrograph base (i.e., the total duration of surface runoff) is the storm duration plus the time to concentration. The durations of the rising and falling limbs are both taken as t_c.

Figure 20.21 *Modified Rational Method Hydrograph*

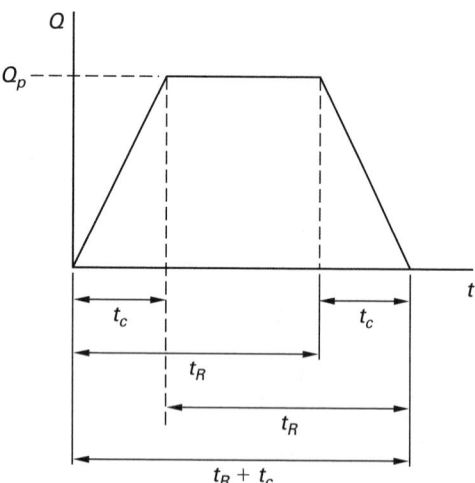

To size a detention watershed using this method, a trapezoidal hydrograph is drawn for several storm durations greater than $2t_c$ (e.g., 10, 15, 20, and 30 min). For each of the storm durations, the total rainfall is calculated as the hydrograph area. The difference between the total rainfall and the total released water (calculated as the product of the allowable release rate and the storm duration) is the required storage capacity.

Detention watershed sizing for areas larger than 20 ac should be accomplished using full hydrograph methods in combination with reservoir routing.

19. RESERVOIR SIZING: NONSEQUENTIAL DROUGHT METHOD

The *nonsequential drought method* is somewhat complex, but it has the advantage of giving an estimate of the required reservoir size, rather than merely evaluating a trial size. In the absence of synthetic drought information, it is first necessary to develop intensity-duration-frequency curves from stream flow records.

step 1: Choose a duration. Usually, the first duration used will be 7 days, although choosing 15 days will not introduce too much error.

step 2: Search the streamflow records to find the smallest flow during the duration chosen. (The first time through, for example, find the smallest discharge totaled over any 7 days.) The days do not have to be sequential.

step 3: Continue searching the discharge records to find the next smallest discharge over the number of days in the period. Continue searching and finding the smallest discharges (which gradually increase) until all of the days in the record have been used up. Do not use the same day more than once.

step 4: Give the values of smallest discharge order numbers; that is, give $M = 1$ to the smallest discharge, $M = 2$ to the next smallest, etc.

step 5: For each observation, calculate the recurrence interval as

$$F = \frac{n_y}{M} \qquad 20.46$$

n_y is the number of years of streamflow data that was searched to find the smallest discharges.

step 6: Plot the points as discharge on the y-axis versus F in years on the x-axis. Draw a reasonably continuous curve through the points.

step 7: Return to step 1 for the next duration. Repeat for all of the following durations: 7, 15, 30, 60, 120, 183, and 365 days.

Figure 20.22 *Sample Family of Synthetic Inflow Curves*

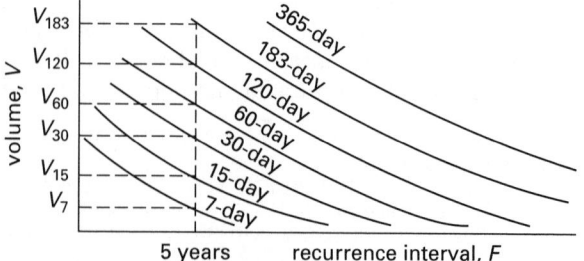

A synthetic drought can be constructed for any recurrence interval. For example, if a 5 year drought is to be planned for, the discharges V_7, V_{15}, V_{30}, ..., V_{365} are read from the appropriate curves for $F = 5$ yr.

The next step is to plot the reservoir *mass diagram* (also known as a *Rippl diagram*). This is a plot of the cumulative net volume—a simultaneous plot of the cumulative demand (known as *draft*) and cumulative inflow. The mass diagram is used to graphically determine the reservoir storage requirements (i.e., size).

As long as the slopes of the cumulative demand and inflow lines are equal, the water reserve in the reservoir will not change. When the slope of the inflow is less than the slope of the demand, the inflow cannot by itself satisfy the community's water needs, and the reservoir is drawn down to make up the difference. A peak followed by a trough is, therefore, a drought condition.

If the reservoir is to be sized so that the community will not run dry during a drought, the required capacity is the maximum separation between two parallel lines (pseudo-demand lines with slopes equal to the demand rate) drawn tangent to a peak and a subsequent trough. If the mass diagram covers enough time so that multiple droughts are present, the largest separation between peaks and subsequent troughs represents the capacity.

In order for the reservoir to supply enough water during a drought condition, the reservoir must be full prior to the start of the drought. This fact is not represented when the mass diagram is drawn, hence the need to draw a pseudo-demand line parallel to the peak.

After a drought equal to the capacity of the reservoir, the reservoir will again be empty. At the trough, however, the reservoir begins to fill up again. When the cumulative excess exceeds the reservoir capacity, the reservoir will have to "spill" (i.e., release) water. This occurs when the cumulative inflow line crosses the prior peak's pseudo-demand line as shown in Fig. 20.23.

A *flood-control dam* is built to keep water in and must be sized so that water is not spilled. The mass diagram can still be used, but the maximum separation between troughs and subsequent peaks (not peaks followed by troughs) is the required capacity.

Figure 20.23 *Reservoir Mass Diagram (Rippl Diagram)*

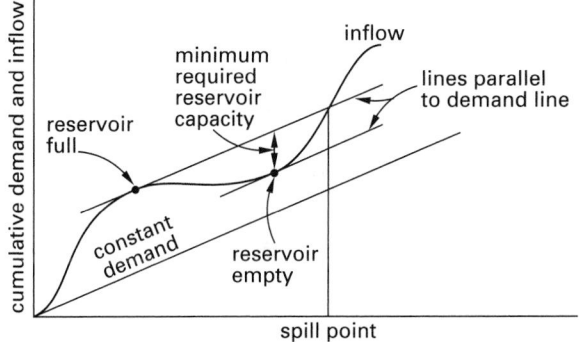

20. RESERVOIR SIZING: RESERVOIR ROUTING

Reservoir routing is the process by which the outflow hydrograph (i.e., the outflow over time) of a reservoir is determined from the inflow hydrograph (i.e., the inflow over time), the initial storage, and other characteristics of the reservoir. The simplest method is to keep track of increments in inflow, storage, and outflow period by period in a tabular simulation. This is the basis of the *storage indication method*, which is basically a bookkeeping process. The validity of this method is dependent on choosing time increments that are as small as possible.

step 1: Determine the starting storage volume, V_n. If the starting volume is zero or considerably different from the average steady-state storage, a large number of iterations will be required before the simulation reaches its steady-state results. A convergence criterion should be determined before the simulation begins.

step 2: For the next iteration, determine the inflow, discharge, evaporation, and seepage. The starting storage volume for the next iteration is found by solving Eq. 20.47.

$$V_{n+1} = V_n + (\text{inflow})_n - (\text{discharge})_n - (\text{seepage})_n - (\text{evaporation})_n \quad \textit{20.47}$$

Repeat step 2 as many times as necessary.

Loss due to seepage is generally very small compared to inflow and discharge, and it is often neglected. Reservoir *evaporation* can be estimated from analytical relationships or by evaluating data from *evaporation pans*. Pan data is extended to reservoir evaporation by the pan coefficient formula. In Eq. 20.48, the summation is taken over the number of days in the simulation period. Units of evaporation are typically in/day. A typical value of C_p is 0.7.

$$E_R = \sum C_p E_p \quad \textit{20.48}$$

Inflow can be taken from actual past history. However, since it is unlikely that history will repeat exactly, the next method is preferred.

21. RESERVOIR ROUTING: STOCHASTIC SIMULATION

The *stochastic simulation* method of reservoir routing is the same as tabular simulation except for the method of determining the inflow. This description uses a *Monte Carlo simulation* technique that is dependent on enough historical data to establish a cumulative inflow distribution. A Monte Carlo simulation is suitable if long periods are to be simulated. If short periods are to be simulated, the simulation should be performed several times and the results averaged.

step 1: Tabulate or otherwise determine a frequency distribution of inflow quantities.

step 2: Form a cumulative distribution of inflow quantities.

step 3: Multiply the cumulative x-axis (which runs from 0 to 1) by 100 or 1000, depending on the accuracy needed.

step 4: Generate random numbers between 0 and 100 (or 0 and 1000). Use of a random number table (App. 20.B) is adequate for hand simulation.

step 5: Locate the inflow quantity corresponding to the random number from the cumulative distribution x-axis.

Example 20.7

A well-monitored stream has been observed for 50 yr and has the following frequency distribution of total annual discharges.

discharge (units)	frequency (yr)	% of time
0 to 0.5	5	0.10
0.5 to 1.0	21	0.42
1.0 to 1.5	17	0.34
1.5 to 2.0	7	0.14

It is proposed that a dam be placed across the stream to create a reservoir with a capacity of 1.8 units. The reservoir is to support a town that will draw 1.2 units per year. Use stochastic simulation to simulate 10 yr of reservoir operation, assuming it starts filled with 1.5 units.

Solution

step 1: The frequency distribution is given.

steps 2 and 3: Make a table to form the cumulative distribution with multiplied frequencies.

discharge	cumulative frequency	cumulative frequency × 100
0 to 0.5	0.10	10
0.5 to 1.0	0.52	52
1.0 to 1.5	0.86	86
1.5 to 2.0	1.00	100

step 4: From App. 20.B, choose ten 2-digit numbers. The starting point within the table is arbitrary, but the numbers must come sequentially from a row or column. Use the first row for this simulation.

78, 46, 68, 33, 26, 96, 58, 98, 87, 27

step 5: For the first year, the random number is 78. Since 78 is greater than 52 but less than 86, the inflow is in the 1.0 to 1.5 unit range. The midpoint of this range is taken as the inflow, which would be 1.25. The reservoir volume after the first year would be $1.5 + 1.25 - 1.20 = 1.55$. The remaining years can be similarly simulated.

year	starting volume	+ inflow	− usage	= ending volume	+ spill
1	1.5	1.25	1.2	1.55	
2	1.55	0.75	1.2	1.1	
3	1.1	1.25	1.2	1.15	
4	1.15	0.75	1.2	0.7	
5	0.7	0.75	1.2	0.25	
6	0.25	1.75	1.2	0.8	
7	0.8	1.25	1.2	0.85	
8	0.85	1.75	1.2	1.4	
9	1.4	1.75	1.2	1.95	0.15
10	1.8	0.75	1.2	1.35	

No shortages are experienced; one spill is required.

22. STORMWATER/WATERSHED MANAGEMENT MODELING

Flood routing and *channel routing* are terms used to describe the passage of a flood or runoff wave through a complex system. Due to flow times, detention, and processing, the wave front appears at different points in a system at different times. The ability to simulate flood, channel, and reservoir routing is particularly useful when evaluating competing features (e.g., treatment, routing, storage options).

The interaction of stormwater features (e.g., watersheds, sewers and storm drains, detention watersheds, treatment facilities) during and after a storm is complex. In addition to continuity considerations, the interaction is affected by hydraulic considerations (e.g., the characteristics of the flow paths) and topography (e.g., the elevation changes from point to point).

Many computer programs have been developed to predict the performance of such complex systems. These simulation programs vary considerably in complexity, degree of hydraulic detail, length of simulation interval, and duration of simulation study. The Environmental Protection Agency (EPA) Stormwater Management Model is a well-known micro-scale model, particularly well-suited to areas with a high impervious fraction. The Army Corp of Engineers' STORM model is a macro-scale model. The NRCS TR-20 program is consistent with other NRCS methodology.

23. FLOOD CONTROL CHANNEL DESIGN

For many years, the traditional method of flood control was *channelization*, converting a natural stream to a uniform channel cross section. However, this method does not always work as intended and may fail at critical moments. Some of the reasons for reduced capacities are sedimentation, increased flow resistance, and inadequate maintenance.

The design of artificial channels is often based on clear-water hydraulics without sediment. However, the capacity of such channels is greatly affected by *sedimentation*. Silting and sedimentation can double or triple the Manning roughness coefficient. Every large flood carries appreciable amounts of bed load, significantly increasing the composite channel roughness. Also, when the channel is unlined, scour and erosion can significantly change the channel cross-sectional area.

To minimize cost and right-of-way requirements, shallow supercritical flow is often intended when the channel is designed. However, supercritical flow can occur only with low bed and side roughness. When the roughness increases during a flood, the flow shifts back to deeper, slower-moving subcritical flows.

Debris carried downstream by floodwaters can catch on bridge pilings and culverts, obstructing the flow. Actual flood profiles can resemble a staircase consisting of a series of backwater pools behind obstructed bridges, rather than a uniformly sloping surface. In such situations, the most effective flood control method may be replacement of bridges or improved emergency maintenance procedures to remove debris.

Vegetation and other debris that collects during dry periods in the flow channel reduces the flow capacity. Maintenance of channels to eliminate the vegetation is often haphazard.

Lately, increased emphasis in flood control channel design programs has been on "creek protection" and the long-term management of natural channels. Elements of such programs include (a) excavation of a low-flow channel within the natural channel, (b) periodic intervention to keep the channel clear, (c) establishment of a wide floodplain terrace along the channel banks, incorporating wetlands, vegetation, and public-access paths, and (d) planting riparian vegetation and trees along the terrace to slow bank erosion and provide a continuous corridor for wildlife.

Water Resources

21 Groundwater

Nomenclature

A	area	ft^2	m^2
A	availability	–	–
b	aquifer width	ft	m
C	concentration	ppm	ppm
C	constant	various	various
D	diameter	ft	m
e	void ratio	–	–
f	infiltration	in/hr	cm/h
F	cumulative infiltration	in	cm
FS	factor of safety	–	–
g	acceleration of gravity	ft/sec^2	m/s^2
g_c	gravitational constant	ft-lbm/ lbf-sec^2	–
H	total hydraulic head	ft	m
i	hydraulic gradient	ft/ft	m/m
k	intrinsic permeability	ft^2	m^2
K	hydraulic conductivity	$\text{ft}^3/$ day-ft^2	$\text{m}^3/$ day·m^2
L	length	ft	m
m	mass	lbm	kg
n	porosity	–	–
N	neutral stress coefficient	–	–
N	quantity (number of)	–	–
p	pressure	lbf/ft^2	Pa
q	specific discharge	ft/sec	m/s
Q	flow quantity	ft^3/sec	m^3/s
r	radial distance from well	ft	m
Re	Reynolds number	–	–
s	drawdown	ft	m
S	storage constant	–	–

S_r	specific retention	–	–
S_y	specific yield	–	–
T	transmissivity	$\text{ft}^3/$ day-ft	$\text{m}^3/$ day·m
u	well function argument	–	–
U	uplift force	lbf	N
v	flow velocity	ft/sec	m/s
V	volume	ft^3	m^3
w	moisture content	–	–
$W(u)$	well function	–	–
y	aquifer thickness after drawdown	ft	m
Y	original aquifer phreatic zone thickness	ft	m
z	distance below datum	–	–

Symbols

γ	specific weight	lbf/ft^3	–
μ	absolute viscosity	lbf-sec/ft^2	Pa·s
ν	kinematic viscosity	ft^2/sec	m^2/s
ρ	density	lbm/ft^3	kg/m^3

Subscripts

c	equilibrium
e	effective
f	flow
o	at well
p	equipotential
r	radius
s	solid
t	total
u	uniformity or uplift
v	void
w	water

1. AQUIFERS

Underground water, also known as *subsurface water*, is contained in saturated geological formations known as *aquifers*. Aquifers are divided into two zones by the water table surface. The *vadose zone* is above the elevation of the water table. Pores in the vadose zone may be either saturated, partially saturated, or empty. The *phreatic zone* is below the elevation of the water table. Pores are always saturated in the phreatic zone.

An aquifer whose water surface is at atmospheric pressure and that can rise or fall with changes in volume is a *free aquifer*, also known as an *unconfined aquifer*. If a well is drilled into an unconfined aquifer, the water level in the well will correspond to the water table. Such a well is known as a *gravity well*.

An aquifer that is bounded on all extents is known as a *confined aquifer*. The water in confined aquifers may be under pressure. If a well is drilled into such an aquifer, the water in the well will rise to a height corresponding to the hydrostatic pressure. The *piezometric height* of the rise is

$$H = \frac{p}{\rho g} \qquad \text{[SI]} \qquad 21.1(a)$$

$$H = \frac{p}{\gamma} = \frac{p}{\rho} \times \frac{g_c}{g} \qquad \text{[U.S.]} \qquad 21.1(b)$$

If the confining pressure is high enough, the water will be expelled from the surface, and the source is known as an *artesian well*.

2. AQUIFER CHARACTERISTICS

Soil moisture content (water content), w, can be determined by oven drying a sample of soil and measuring the change in mass.[1] The water content is the ratio of the mass of water to the mass of solids, expressed as a percentage. The water content can also be determined with a *tensiometer*, which measures the vapor pressure of the moisture in the soil.

$$w = \frac{m_w}{m_s} = \frac{m_t - m_s}{m_s} \qquad 21.2$$

The *porosity*, n, of the aquifer is the percentage of void volume to total volume.[2]

$$n = \frac{V_v}{V_t} = \frac{V_t - V_s}{V_t} \qquad 21.3$$

The *void ratio*, e, is

$$e = \frac{V_v}{V_s} = \frac{V_t - V_s}{V_s} \qquad 21.4$$

Void ratio and porosity are related.

$$e = \frac{n}{1 - n} \qquad 21.5$$

Some pores and voids are dead ends or are too small to contribute to seepage. Only the *effective porosity*, n_e, 95 to 98% of the total porosity, contributes to groundwater flow.

The *hydraulic gradient*, i, is the change in hydraulic head over a particular distance. The hydraulic head at a point is determined as the piezometric head at observation wells.

$$i = \frac{\Delta H}{L} \qquad 21.6$$

3. PERMEABILITY

The flow of a liquid through a permeable medium is affected by both the fluid and the medium. The effects of the medium (independent of the fluid properties) are characterized by the *intrinsic permeability* (*specific permeability*), k. Intrinsic permeability has dimensions of length squared. The *darcy* has been widely accepted as the unit of intrinsic permeability. One darcy is 0.987×10^{-8} cm^2.

For studies involving the flow of water through an aquifer, effects of intrinsic permeability and the water are combined into the *hydraulic conductivity*, also known as the *coefficient of permeability* or simply the *permeability*, K. Hydraulic conductivity can be determined from a number of water-related tests.[3] It has units of volume per unit area per unit time, which is equivalent to length divided by time (i.e., units of velocity). Volume may be expressed as cubic feet and cubic meters, or gallons and liters.

$$K = \frac{k\gamma}{\mu} \qquad 21.7$$

For many years in the United States, hydraulic conductivity was specified in *Meinzer units* (gallons per day per square foot). To avoid confusion related to multiple definitions and ambiguities in these definitions, hydraulic conductivity is now often specified in units of ft/day (m/day).

The coefficient of permeability is proportional to the square of the mean particle diameter.

$$K = C D_{\text{mean}}^2 \qquad 21.8$$

Hazen's empirical formula can be used to calculate an approximate coefficient of permeability for clean, uniform sands. D_{10} is the *effective* size in mm (i.e., the size for which 10% of the distribution is finer).

$$K_{\text{cm/s}} \approx C(D_{10,\text{mm}})^2 \quad [0.1 \text{ mm} \leq D_{10,\text{mm}} \leq 3.0 \text{ mm}]$$
$$21.9$$

The coefficient C is 40 to 80 for very fine sand (poorly sorted) or fine sand with appreciable fines; 80 to 120 for medium sand (well sorted) or coarse sand (poorly sorted); and 120 to 150 for coarse sand (well sorted and clean).

4. DARCY'S LAW

Movement of groundwater through an aquifer is given by *Darcy's law*, Eq. 21.10.[4] The hydraulic gradient may be specified in either ft/ft (m/m) or ft/mi (m/km), depending on the units of area used.

$$Q = -K i A_{\text{gross}} = -v_e A_{\text{gross}} \qquad 21.10$$

---

[1]It is common in civil engineering to use the term "weight" in place of mass. For example, the *water content* would be defined as the ratio of the weight of water to the weight of solids, expressed as a percentage.

[2]The symbol θ is sometimes used for porosity.

[3]Permeability can be determined from constant-head permeability tests (sands), falling-head permeability tests (fine sands and silts), consolidation tests (clays), and field tests of wells (in situ gravels and sands).

[4]The negative sign in Darcy's law accounts for the fact that flow is in the direction of decreasing head. That is, the hydraulic gradient is negative in the direction of flow. The negative sign is omitted in the remainder of this chapter, or appropriate equations are rearranged to be positive.

Table 21.1 *Typical Permeabilities*

	k (cm^2)	k (m^2)	k (darcys)	K (cm/sec)	K (gal/day-ft^2)
gravel	10^{-5}–10^{-3}	10^{-1}–10	10^3–10^5	0.5–50	10^4–10^6
gravelly sand	10^{-5}	10^{-1}	10^3	0.5	10^4
clean sand	10^{-6}	10^{-2}	10^2	0.05	10^3
sandstone	10^{-8}	10^{-3}	10	0.005	10^2
dense shale or limestone	10^{-9}	10^{-5}	10^{-1}	0.000 05	1
granite or quartzite	10^{-11}	10^{-7}	10^{-3}	0.000 000 5	10^{-2}
clay	10^{-11}	10^{-7}	10^{-3}	0.000 000 5	10^{-2}

Multiply gal/day-ft^2 by 0.1337 to obtain ft^3/day-ft^2.
Multiply darcys by 0.987×10^{-8} to obtain cm^2.
Multiply darcys by 0.987×10^{-12} to obtain m^2.
Multiply cm^2 by 10^{-4} to obtain m^2.
Multiply ft^2 by 9.4135×10^{10} to obtain darcys.
Multiply m^2 by 10^4 to obtaim cm^2.
Multiply gal/day-ft^2 by 4.716×10^{-5} to obtain cm/sec.

The *specific discharge* is the same as the *effective velocity*, v_e.

$$q = \frac{Q}{A_{\text{gross}}} = Ki = v_e \qquad 21.11$$

Darcy's law is applicable only when the Reynolds number is less than 1. Significant deviations have been noted when the Reynolds number is even as high as 2. In Eq. 21.12, D_{mean} is the mean grain diameter.

$$\text{Re} = \frac{\rho q D_{\text{mean}}}{\mu} = \frac{q D_{\text{mean}}}{\nu} \qquad 21.12$$

5. TRANSMISSIVITY

Transmissivity (also known as the *coefficient of transmissivity*) is an index for the rate of groundwater movement. The transmissivity of flow from a saturated aquifer of thickness Y and width b is given by Eq. 21.13. The thickness, Y, of a confined aquifer is the difference in elevations of the bottom and top of the saturated formation. For permeable soil, the thickness, Y, is the difference in elevations of the impermeable bottom and the water table.

$$T = KY \qquad 21.13$$

Combining Eqs. 21.10 and 21.13 gives

$$Q = bTi \qquad 21.14$$

6. SPECIFIC YIELD, RETENTION, AND CAPACITY

The dimensionless *storage constant (storage coefficient)*, S, of a confined aquifer is the change in aquifer water volume per unit surface area of the aquifer per unit change in head. That is, the storage constant is the amount of water that is removed from a column of the aquifer 1 ft^2 (1 m^2) in plan area when the water table drops 1 ft (1 m). For unconfined aquifers, the storage coefficient is virtually the same as the specific yield. Various methods have been proposed for calculating the storage coefficient directly from properties of the rock and water. It can also be determined from unsteady flow analysis of wells. (See Sec. 12.)

The *specific yield*, S_y, is the water yielded when water-bearing material drains by gravity. It is the volume of water removed per unit area when a drawdown of one length unit is experienced. A time period may be given for different values of specific yield.

$$S_y = \frac{V_{\text{yielded}}}{V_{\text{total}}} \qquad 21.15$$

The *specific retention* is the volume of water that, after being saturated, will remain in the aquifer against the pull of gravity.

$$S_r = \frac{V_{\text{retained}}}{V_{\text{total}}} = n - S_y \qquad 21.16$$

The *specific capacity* of an aquifer is the discharge rate divided by the drawdown.

$$\text{specific capacity} = \frac{Q}{s} \qquad 21.17$$

7. DISCHARGE VELOCITY AND SEEPAGE VELOCITY

The *pore velocity* (*linear velocity, flow front velocity,* and *seepage velocity*) is given by Eq. 21.18.[5]

$$v_{pore} = \frac{Q}{A_{net}} = \frac{Q}{nA_{gross}} = \frac{Q}{nbY} = \frac{Ki}{n} \qquad 21.18$$

The gross cross-sectional area in flow depends on the aquifer dimensions. However, water can only flow through voids (pores) in the aquifer. If the gross cross-sectional area is known, it will be necessary to multiply by the porosity to reduce the area in flow. (The hydraulic conductivity is not affected.)

The *effective velocity* (also known as the *apparent velocity, Darcy velocity, Darcian velocity, Darcy flux, discharge velocity, specific discharge, superficial velocity, face velocity,* and *approach velocity*) through a porous medium is the velocity of flow averaged over the gross aquifer cross-sectional area.

$$v_e = nv_{pore} = \frac{Q}{A_{gross}} = \frac{Q}{bY} = Ki \qquad 21.19$$

Contaminants introduced into an acquifer will migrate from place to place relative to the surface at the effective velocity given by Eq. 21.19. The pore velocity (Eq. 21.18) should not be used to determine the overland time taken and distance moved by a contaminant.

8. FLOW DIRECTION

Flow direction will be from an area with a high piezometric head (as determined from observation wells) to an area of low piezometric head. Piezometric head is assumed to vary linearly between points of known head. Similarly, all points along a line joining two points with the same piezometric head can be considered to have the same head.

9. WELLS

Water in aquifers can be extracted from *gravity wells*. However, *monitor wells* may also be used to monitor the quality and quantity of water in an aquifer. *Relief wells* are used to dewater soil.

Wells may be dug, bored, driven, jetted, or drilled in a number of ways, depending on the aquifer material and the depth of the well. Wells deeper than 100 ft (30 m) are usually drilled. After construction, the well is *developed*, which includes the operations of removing any fine sand and mud. Production is *stimulated* by increasing the production rate. The fractures in the rock surrounding the well are increased in size by injecting high-pressure water or using similar operations.

---

[5]Terms used to describe pore and effective velocities are not used consistently.

Well equipment is *sterilized* by use of chlorine or other disinfectants. The strength of any chlorine solution used to disinfect well equipment should not be less than 100 ppm by weight (i.e., 100 kg of chlorine per 10^6 kg of water). Calcium hypochlorite (which contains 65% available chlorine) and sodium hypochlorite (which contains 12.5% available chlorine) are commonly used for this purpose. The mass of any chlorine-supplying compound with fractional availability A required to produce V gallons of disinfectant with concentration C is

$$m_{lbm} = (8.33 \times 10^{-6})V_{gal} \left(\frac{C_{ppm}}{A_{decimal}} \right) \qquad 21.20$$

Figure 21.1 illustrates a typical water-supply well. Water is removed from the well through the *riser pipe* (*eductor pipe*). Water enters the well through a perforated or slotted casing known as the *screen*. The required *open area* depends on the flow rate and is limited by the maximum permissible entrance velocity that will not lift grains larger than a certain size. Table 21.2 recommends maximum flow velocities as functions of the grain diameter. A safety factor of 1.5 to 2.0 is also used to account for the fact that parts of the screen may become blocked. As a general rule, the openings should also be smaller than D_{50} (i.e., smaller than 50% of the screen material particles).

Table 21.2 *Lifting Velocity of Water*[a]

grain diameter (mm)	maximum water velocity	
	(ft/sec)	(m/s)
0 to 0.25	0.10	0.030
0.25 to 0.50	0.22	0.066
0.50 to 1.00	0.33	0.10
1.00 to 2.00	0.56	0.17
2.00 to 4.00	2.60	0.78

(Multiply ft/sec by 0.3 to obtain m/s.)
[a]Spherical particles with specific gravity of 2.6.

Screens do not necessarily extend the entire length of the well. The required screen length can be determined from the total amount of open area required and the open area per unit length of casing. For confined aquifers, screens are usually installed in the middle 70 to 80% of the well. For unconfined aquifers, screens are usually installed in the lower 30 to 40% of the well.

It is desirable to use screen openings as large as possible to reduce entrance friction losses. Larger openings can be tolerated if the well is surrounded by a *gravel pack* to prevent fine material from entering the well. Gravel packs are generally required in soils where the D_{90} size (i.e., the sieve size retaining 90% of the soil) is less than 0.01 in (0.25 mm) and when the well goes through layers of sand and clay.

Figure 21.1 *Typical Gravity Well*

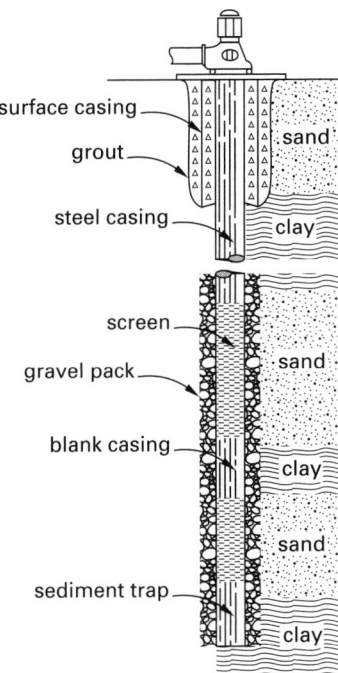

10. DESIGN OF GRAVEL SCREENS AND POROUS FILTERS

Gravel and other porous materials may be used as filters as long as their voids are smaller than the particles to be excluded. Actually, only the largest 15% of the particles need to be filtered out, since the agglomeration of these particles will themselves create even smaller openings, and so on.

In specifying the opening sizes for screens, perforated pipe, and fabric filters, the following criteria are in widespread use as the basis for filter design.

$$D_{\text{opening,filter}} \leq D_{85,\text{soil}} \quad \text{[screen filters]}$$
21.21

$$[\text{filtering criterion}]^6 D_{15,\text{filter}} \leq 5D_{85,\text{soil}} \quad \text{[filter beds]}$$
21.22

$$[\text{permeability criterion}]^7 D_{15,\text{filter}} \leq 5D_{15,\text{soil}} \quad \text{[filter beds]}$$
21.23

As stated in Sec. 9, the screen openings should also be smaller than D_{50} (i.e., smaller than 50% of the screen material particles).

The *coefficient of uniformity*, C_u, is used to determine whether the filter particles are properly graded. Only uniform materials ($C_u < 2.5$) and well-graded materials ($2.5 < C < 6$) are suitable for use as filters.

$$C_u = \frac{D_{60}}{D_{10}}$$
21.24

---

[6]Some authorities replace "5" with "9."
[7]Some authorities replace "5" with "4."

11. WELL DRAWDOWN IN AQUIFERS

An aquifer with a well is shown in Fig. 21.2. Once pumping begins, the water table will be lowered in the vicinity of the well. The resulting water table surface is referred to as a *cone of depression*. The decrease in water level at some distance r from the well is known as the *drawdown*, s_r. The drawdown at the well is denoted as s_o.

Figure 21.2 *Well Drawdown in an Unconfined Aquifer*

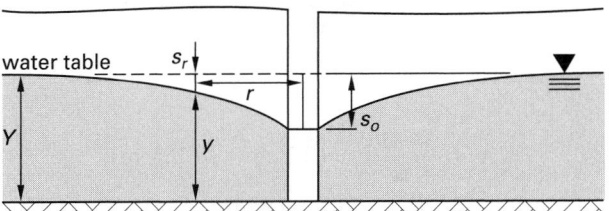

If the drawdown is small with respect to the aquifer phreatic zone thickness, Y, and the well completely penetrates the aquifer, the equilibrium (steady-state) well discharge is given by the *Dupuit equation*, Eq. 21.25. Equation 21.25 can only be used to determine the equilibrium flow rate a "long time" after pumping has begun.

$$Q = \frac{\pi K(y_1^2 - y_2^2)}{\ln\left(\frac{r_1}{r_2}\right)}$$
21.25

In Eq. 21.25, y_1 and y_2 are the aquifer depths at radial distances r_1 and r_2, respectively, from the well. y_1 can also be taken as the original aquifer depth, Y, if r_1 is the well's *radius of influence*, the distance at which the well has no effect on the water table level.

In very thick unconfined aquifers where the drawdown is negligible compared to the aquifer thickness, or in confined aquifers where there is no cone of depression at all, $y_1 + y_2$ is essentially equal to $2Y$. Then, since $y_2 - y_1 = s_1 - s_2$, $y_2^2 - y_1^2 = (y_2 - y_1)(y_2 + y_1) = 2Y(s_1 - s_2)$. Using this equality and Eq. 21.13, Eq. 21.25 can be written as

$$Q = \frac{2\pi T(s_2 - s_1)}{\ln\left(\frac{r_1}{r_2}\right)}$$
21.26

For an artesian well fed by a confined aquifer of thickness Y, the discharge is given by the *Thiem equation*.

$$Q = \frac{2\pi K Y(y_1 - y_2)}{\ln\left(\frac{r_1}{r_2}\right)}$$
21.27

The rate (or amount in some period) of water that can be extracted without experiencing some undesirable result is known as the *safe yield*. Undesirable results include deteriorations in the water quality, large pump lifts, and infringement on the water rights of others. Extractions in excess of the safe yield are known as *overdrafts*.

Example 21.1

A 9 in (25 cm) diameter well is pumped at the rate of 50 gal/min (0.2 m³/min). The aquifer is 100 ft (30 m) thick. After some time, the well sides cave in and are replaced with an 8 in (20 cm) diameter tube. The drawdown is 6 ft (2 m). The water table recovers its original thickness 2500 ft (750 m) from the well. What will be the steady flow from the new well?

SI Solution

$$r_2 = \frac{25 \text{ cm}}{2} = 12.5 \text{ cm} \quad (0.125 \text{ m})$$

$$y_2 = 30 \text{ m} - 2 \text{ m} = 28 \text{ m}$$

Rearrange Eq. 21.25 to determine hydraulic conductivity from the given information.

$$K = \frac{Q \ln\left(\dfrac{r_1}{r_2}\right)}{\pi(y_1^2 - y_2^2)}$$

$$= \frac{\left(0.2 \dfrac{\text{m}^3}{\text{min}}\right) \ln\left(\dfrac{750 \text{ m}}{0.125 \text{ m}}\right)}{\pi\left((30 \text{ m})^2 - (28 \text{ m})^2\right)}$$

$$= 0.004774 \text{ m/min} \ (0.007957 \text{ cm/s})$$

For the relined well,

$$r_2 = \frac{20 \text{ cm}}{2} = 10 \text{cm} \quad (0.10 \text{ m})$$

Use Eq. 21.25 to solve for the new flow rate.

$$Q = \frac{\pi K(y_1^2 - y_2^2)}{\ln\left(\dfrac{r_1}{r_2}\right)}$$

$$= \pi\left(0.004774 \ \frac{\text{m}}{\text{min}}\right)\left(\frac{(30 \text{ m})^2 - (28 \text{ m})^2}{\ln\left(\dfrac{750 \text{ m}}{0.10 \text{ m}}\right)}\right)$$

$$= 0.195 \text{ m}^3/\text{min}$$

Customary U.S. Solution

$$r_2 = \frac{9 \text{ in}}{(2)\left(12 \dfrac{\text{in}}{\text{ft}}\right)} = 0.375 \text{ ft}$$

$$y_2 = 100 \text{ ft} - 6 \text{ ft} = 94 \text{ ft}$$

$$Q = \left(50 \ \frac{\text{gal}}{\text{min}}\right)\left(0.002228 \ \frac{\text{ft}^3\text{-min}}{\text{sec-gal}}\right)$$

$$= 0.1114 \text{ ft}^3/\text{sec}$$

Rearrange Eq. 21.25 to find the hydraulic conductivity.

$$K = \frac{Q \ln\left(\dfrac{r_1}{r_2}\right)}{\pi(y_1^2 - y_2^2)}$$

$$= \frac{\left(50 \ \dfrac{\text{gal}}{\text{min}}\right)\left(1440 \ \dfrac{\text{min}}{\text{day}}\right) \ln\left(\dfrac{2500 \text{ ft}}{0.375 \text{ ft}}\right)}{\pi\left((100 \text{ ft})^2 - (94 \text{ ft})^2\right)}$$

$$= 173.4 \text{ gal/day-ft}^2$$

For the relined well,

$$r_2 = \frac{8 \text{ in}}{(2)\left(12 \dfrac{\text{in}}{\text{ft}}\right)} = 0.333 \text{ ft}$$

Use Eq. 21.25 to solve for the new flow rate.

$$Q = \frac{\pi K(y_1^2 - y_2^2)}{\ln\left(\dfrac{r_1}{r_2}\right)}$$

$$= \frac{\pi\left(173.4 \ \dfrac{\text{gal}}{\text{day-ft}^2}\right)\left((100 \text{ ft})^2 - (94 \text{ ft})^2\right)}{\ln\left(\dfrac{2500 \text{ ft}}{0.333 \text{ ft}}\right)}$$

$$= 71{,}042 \text{ gal/day}$$

12. UNSTEADY FLOW

When pumping first begins, the removed water also comes from the aquifer above the equilibrium cone of depression. Therefore, Eq. 21.26 cannot be used, and a nonequilibrium analysis is required.

Nonequilibrium solutions to well problems have been formulated in terms of dimensionless numbers. For small drawdowns compared with the initial thickness of the aquifer, the *Theis equation* for the drawdown at a distance r from the well and after pumping for time t is

$$s_{r,t} = \left(\frac{Q}{4\pi KY}\right) W(u)$$

$$= \left(\frac{Q}{4\pi T}\right) W(u) \qquad 21.28$$

$W(u)$ is a dimensionless *well function*. Though it is possible to obtain $W(u)$ from u, extracting u from $W(u)$ is more difficult. The relationship between u and $W(u)$ is often given in tabular or graphical forms. In Eq. 21.30, S is the aquifer *storage constant*.

$$W(u) = -0.577216 - \ln u + u - \frac{u^2}{(2)(2!)}$$

$$+ \frac{u^3}{(3)(3!)} - \frac{u^4}{(4)(4!)} + \cdots \qquad 21.29$$

$$u = \frac{r^2 S}{4KYt} = \frac{r^2 S}{4Tt} \qquad 21.30$$

Accordingly, for any two different times in the pumping cycle,

$$s_1 - s_2 = y_2 - y_1 = \left(\frac{Q}{4\pi KY}\right)\left(W(u_1) - W(u_2)\right) \quad \textbf{\textit{21.31}}$$

If $u < 0.01$, then *Jacob's equation* can be used.

$$s_{r,t} = \left(\frac{Q}{4\pi T}\right)\ln\left(\frac{2.25Tt}{r^2 S}\right) \qquad \textbf{\textit{21.32}}$$

13. PUMPING POWER

Various types of pumps are used in wells. Problems with excessive suction lift are avoided by the use of submersible pumps.

Pumping power can be determined from the hydraulic (water) power equations presented in Ch. 18. The total head is the sum of static lift, velocity head, drawdown, pipe friction, and minor entrance losses from the casing, strainer, and screen. The Hazen-Williams equation is commonly used with a coefficient of $C = 100$ to determine the pipe friction.

14. FLOW NETS

Groundwater seepage is from locations of high hydraulic head to locations of lower hydraulic head. Relatively complex two-dimensional problems may be evaluated using a graphical technique that shows the decrease in hydraulic head along the flow path. The resulting graphic representation of pressure and flow path is called a *flow net*.

The flow net concept as discussed here is limited to cases where the flow is steady, two-dimensional, incompressible, and through a homogeneous medium, and where the liquid has a constant viscosity. This is the ideal case of groundwater seepage.

Flow nets are constructed from streamlines and equipotential lines. *Streamlines (flow lines)* show the path taken by the seepage. *Equipotential lines* are contour lines of constant driving (differential) hydraulic head. (This head does not include static head, which varies with depth.)

The object of a graphical flow net solution is to construct a network of flow paths (outlined by the streamlines) and equal pressure drops (bordered by equipotential lines). No fluid flows across streamlines, and a constant amount of fluid flows between any two streamlines.

Flow nets are constructed according to the following rules.

Rule 21.1: Streamlines enter and leave pervious surfaces perpendicular to those surfaces.

Rule 21.2: Streamlines approach the line of seepage (above which there is no hydrostatic pressure) asymptotically to (i.e., parallel but gradually approaching) that surface.

Rule 21.3: Streamlines are parallel to but cannot touch impervious surfaces that are streamlines.

Rule 21.4: Streamlines are parallel to the flow direction.

Rule 21.5: Equipotential lines are drawn perpendicular to streamlines such that the resulting cells are approximately square and the intersections are 90° angles. Theoretically, it should be possible to draw a perfect circle within each cell that touches all four boundaries, even though the cell is not actually square.

Table 21.3 *Well Function W(u) for Various Values of u*

u	1.0	2.0	3.0	4.0	5.0	6.0	7.0	8.0	9.0
$\times 1$	0.219	0.049	0.013	0.0038	0.0011	0.00036	0.00012	0.000038	0.000012
$\times 10^{-1}$	1.82	1.22	0.91	0.70	0.56	0.45	0.37	0.31	0.26
$\times 10^{-2}$	4.04	3.35	2.96	2.68	2.47	2.30	2.15	2.03	1.92
$\times 10^{-3}$	6.33	5.64	5.23	4.95	4.73	4.54	4.39	4.26	4.14
$\times 10^{-4}$	8.63	7.94	7.53	7.25	7.02	6.84	6.69	6.55	6.44
$\times 10^{-5}$	10.94	10.24	9.84	9.55	9.33	9.14	8.99	8.86	8.74
$\times 10^{-6}$	13.24	12.55	12.14	11.85	11.63	11.45	11.29	11.16	11.04
$\times 10^{-7}$	15.54	14.85	14.44	14.15	13.93	13.75	13.60	13.46	13.34
$\times 10^{-8}$	17.84	17.15	16.74	16.46	16.23	16.05	15.90	15.76	15.65
$\times 10^{-9}$	20.15	19.45	19.05	18.76	18.54	18.35	18.20	18.07	17.95
$\times 10^{-10}$	22.45	21.76	21.35	21.06	20.84	20.66	20.50	20.37	20.25
$\times 10^{-11}$	24.75	24.06	23.65	23.36	23.14	22.96	22.81	22.67	22.55
$\times 10^{-12}$	27.05	26.36	25.96	25.67	25.44	25.26	25.11	24.97	24.86
$\times 10^{-13}$	29.36	28.66	28.26	27.97	27.75	27.56	27.41	27.28	27.16
$\times 10^{-14}$	31.66	30.97	30.56	30.27	30.05	29.87	29.71	29.58	29.46
$\times 10^{-15}$	33.96	33.27	32.86	32.58	32.35	32.17	32.02	31.88	31.76

Reprinted from L. K. Wenzel, "Methods for Determining Permeability of Water Bearing Materials with Special Reference to Discharging Well Methods," U.S. Geological Survey, 1942, Water-Supply Paper 887.

Rule 21.6: Equipotential lines enter and leave impervious surfaces perpendicular to those surfaces.

Many flow nets with differing degrees of detail can be drawn, and all will be more or less correct. Generally, three to five streamlines are sufficient for initial graphical evaluations. The size of the cells is determined by the number of intersecting streamlines and equipotential lines. As long as the rules are followed, the ratio of stream flow channels to equipotential drops will be approximately constant regardless of whether the grid is coarse or fine.

Figure 21.3 shows flow nets for several common cases. A careful study of the flow nets will help to clarify the rules and conventions previously listed.

15. SEEPAGE FROM FLOW NETS

Once a flow net is drawn, it can be used to calculate the seepage. First, the number of flow channels, N_f, between the streamlines is counted. Then, the number of equipotential drops, N_p, between equipotential lines is counted. The total hydraulic head, H, is determined as a function of the water surface levels.

$$Q = KH \left(\frac{N_f}{N_p} \right) \qquad \begin{bmatrix} \text{per unit} \\ \text{width} \end{bmatrix} \qquad 21.33$$

$$H = H_1 - H_2 \qquad 21.34$$

16. HYDROSTATIC PRESSURE ALONG FLOW PATH

The hydrostatic pressure at equipotential drop j (counting the last line on the downstream side as zero) in a flow net with a total of N_p equipotential drops is given by Eq. 21.35. If the point being investigated is along the bottom of a dam or other structure, the hydrostatic pressure is referred to as *uplift pressure*. Since uplift due to velocity is negligible, all uplift is due to the neutral pore pressure. Therefore, *neutral pressure* is synonymous with uplift pressure.

$$p_u = \left(\frac{j}{N_p} \right) H \gamma_w \qquad 21.35$$

Figure 21.3 *Typical Flow Nets*

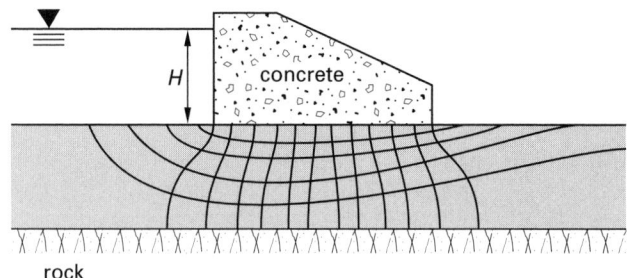

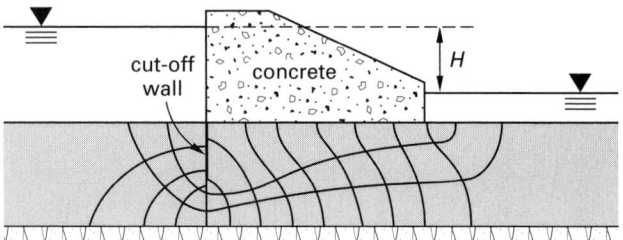

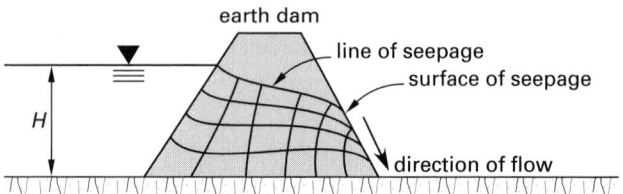

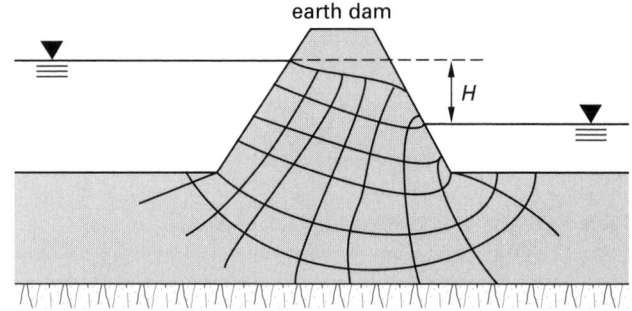

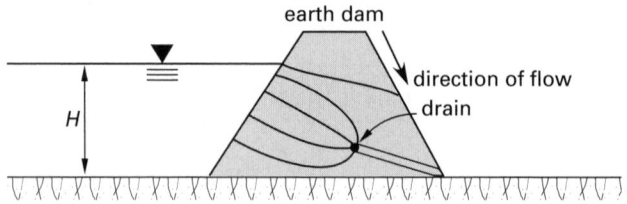

If the point being investigated is located a distance z below the datum, then the pressure at the point is given by Eq. 21.36. If the point being investigated is above the datum, z is negative.

$$p_u = \left(\left(\frac{j}{N_p} \right) H + z \right) \gamma_w \qquad 21.36$$

The actual *uplift force* on a surface of area A is given by Eq. 21.37. N is the *neutral stress coefficient*, defined as the fraction of the surface that is exposed to the hydrostatic pressure. For soil, $N \approx 1$. However, for fractured rock, concrete, porous limestone, or sandstone, it varies from 0.5 to 1.0. For impervious materials such as marble or granite, it can be as low as 0.1.

$$U = N p_u A \qquad 21.37$$

As Table 21.1 indicates, some clay layers are impervious enough to prohibit the passing of water. However, the hydrostatic uplift pressure below a clay layer can still be transmitted through the clay layer. The transfer geometry for thick layers is complex, making estimates of the amount of uplift force difficult. Upward forces on an impervious layer due to hydrostatic pressure below is known as *heave*. A factor of safety against heave of at least 1.5 is desirable.

$$(\text{FS})_\text{heave} = \frac{\text{downward pressure}}{\text{uplift pressure}} \qquad 21.38$$

17. INFILTRATION

Aquifers can be recharged (refilled) in a number of natural and artificial ways. The *Horton equation* gives a lower bound on the *infiltration capacity* (in inches of rainfall per unit time), which is the capacity of an aquifer to absorb water from a saturated source above as a function of time. When the rainfall supply exceeds the infiltration capacity, infiltration decreases exponentially over time. In Eq. 21.39, f_c is the final or equilibrium capacity, f_0 is the initial infiltration capacity, and f_t is the infiltration capacity at time t. k is an infiltration decay constant with dimensions of time^{-1} determined from a *double-ring infiltration test*.

$$f_t = f_c + (f_0 - f_c)e^{-kt} \qquad 21.39$$

The cumulative infiltration will not correspond to the increase in water table elevation due to the effects of porosity. However, the increase in water table elevation can be determined from a knowledge of the aquifer properties.

Infiltration rates for intermediate, silty soils are approximately 50% of those for sand. Clay infiltration rates are approximately 10% of those for sand. The actual infiltration rate at any time is equal (in the Horton model) to the smaller of the rainfall intensity, $i(t)$, and the instantaneous capacity.

$$f = i \quad [i \le f_t] \qquad 21.40$$
$$f = f_t \quad [i > f_t] \qquad 21.41$$

The cumulative infiltration over time is

$$F_t = f_c t + \left(\frac{f_0 - f_c}{k} \right)(1 - e^{-kt}) \qquad 21.42$$

The average infiltration rate is found by dividing Eq. 21.42 by the duration of infiltration.

$$\overline{f} = \frac{F_t}{t_\text{infiltration}} \qquad 21.43$$

22 Inorganic Chemistry

Nomenclature

A	atomic weight	lbm/lbmol	kg/kmol
EW	equivalent weight	lbm/lbmol	kg/kmol
F	formality	n.a.	FW/L
FW	formula weight	lbm/lbmol	kg/kmol
H	enthalpy	Btu/lbmol	kcal/mol
H	Henry's law constant	1/atm	1/atm
k	reaction rate constant	1/min	1/min
K	equilibrium constant	–	–
m	mass	lbm	kg
m	molality	n.a.	mol/1000 g
M	molarity	n.a.	mol/L
MW	molecular weight	lbm/lbmol	kg/kmol
n	number of moles	–	–
N	normality	n.a.	GEW/L
N_A	Avogadro's number	–	1/mol
p	pressure	lbf/ft^2	Pa
R^*	universal gas constant	ft-lbf/ lbmol-°R	kJ/kmol·K
T	temperature	°R	K
v	rate of reaction	–	mol/L·s
V	volume	ft^3	m^3
x	gravimetric fraction	–	–
x	mole fraction	–	–
x	relative abundance	–	–
X	fraction ionized	–	–
Z	atomic number	–	–

Subscripts

a	acid
A	Avogadro
b	base
eq	equilibrium
f	formation
r	reaction
sp	solubility product
t	total

1. ATOMIC STRUCTURE

An *element* is a substance that cannot be decomposed into simpler substances during ordinary chemical reactions.[1] An *atom* is the smallest subdivision of an element that can take part in a chemical reaction. A *molecule* is the smallest subdivision of an element or compound that can exist in a natural state.

The atomic nucleus consists of neutrons and protons, known as *nucleons.* The masses of neutrons and protons are essentially the same—one *atomic mass unit*, amu. (One amu is exactly $1/12$ of the mass of an atom of carbon-12, approximately equal to 1.66×10^{-27} kg.[2]) The *relative atomic weight* or *atomic weight*, A, of an atom is approximately equal to the number of protons

---

[1] Atoms of an element can be decomposed into subatomic particles in nuclear reactions.
[2] Until 1961, the atomic mass unit was defined as $1/16$ of the mass of one atom of oxygen-16. The carbon-12 reference has now been universally adopted.

and neutrons in the nucleus.[3] The *atomic number*, Z, of an atom is equal to the number of protons in the nucleus.

The atomic number and atomic weight of an element E are written in symbolic form as $_Z E^A$, E_Z^A, or $_Z^A E$. For example, carbon is the sixth element; radioactive carbon has an atomic mass of 14. Therefore, the symbol for carbon-14 is C_6^{14}. Since the atomic number is superfluous if the chemical symbol is given, the atomic number can be omitted (e.g., C^{14}).

2. ISOTOPES

Although an element can have only a single atomic number, atoms of that element can have different atomic weights. Many elements possess *isotopes*. The nuclei of isotopes differ from one another only in the number of neutrons. Isotopes behave the same way chemically.[4] Therefore, isotope separation must be done physically (e.g., by centrifugation or gaseous diffusion) rather than chemically.

Hydrogen has three isotopes. H_1^1 is *normal hydrogen* with a single proton nucleus. H_1^2 is known as *deuterium* (*heavy hydrogen*), with a nucleus of a proton and neutron. (This nucleus is known as a *deuteron*.) Finally, H_1^3 (*tritium*) has two neutrons in the nucleus. While normal hydrogen and deuterium are stable, tritium is radioactive. Many elements have more than one stable isotope. Tin, for example, has ten.

The *relative abundance*, x_i, of an isotope i is equal to the fraction of that isotope in a naturally occurring sample of the element. The *chemical atomic weight* is the weighted average of the isotope weights.

$$A_{\text{average}} = x_1 A_1 + x_2 A_2 + \cdots \qquad 22.1$$

3. PERIODIC TABLE

The *periodic table* (App. 22.B) is organized around the *periodic law:* The properties of the elements depend on the atomic structure and vary with the atomic number in a systematic way. Elements are arranged in order of increasing atomic numbers from left to right. Adjacent elements in horizontal rows differ decidedly in both physical and chemical properties. However, elements in the same column have similar properties. Graduations in properties, both physical and chemical, are most pronounced in the *periods* (i.e., the horizontal rows).

The vertical columns are known as *groups*, numbered in Roman numerals. Elements in a group are called *cogeners*. Each vertical group except 0 and VIII has A

and B subgroups (*families*). The elements of a family resemble each other more than they resemble elements in the other family of the same group. Graduations in properties are definite but less pronounced in vertical families. The trend in any family is toward more *metallic properties* as the atomic weight increases.

Metals (elements at the left end of the periodic chart) have low electron affinities, are reducing agents, form positive ions, and have positive oxidation numbers. They have high electrical conductivities, luster, generally high melting points, ductility, and malleability.

Nonmetals (elements at the right end of the periodic chart) have high electron affinities, are oxidizing agents, form negative ions, and have negative oxidation numbers. *Nonmetals* are poor electrical conductors, have little or no luster, and form brittle solids.

The *metalloids* (e.g., boron, silicon, germanium, arsenic, antimony, tellurium, and polonium) have characteristics of both metals and nonmetals. Electrically, they are semiconductors.

Elements in the periodic table are often categorized into the following groups.

- *actinides:* same as actinons
- *actinons:* elements 90 to 103[5]
- *alkali metals:* group IA
- *alkaline earth metals:* group IIA
- *halogens:* group VIIA
- *heavy metals:* metals near the center of the chart
- *inner transition elements:* same as transition metals
- *lanthanides:* same as lanthanons
- *lanthanons:* elements 58 to 71[6]
- *light metals:* elements in the first two groups
- *metals:* everything except the nonmetals
- *metalloids:* elements along the dark line in the chart separating metals and nonmetals
- *noble gases:* group 0
- *nonmetals:* elements 2, 5 to 10, 14 to 18, 33 to 36, 52 to 54, 85, and 86
- *rare earths:* same as lanthanons
- *transition elements:* same as transition metals
- *transition metals:* all B families and group VIII[7]

---

[3]The term *weight* is used even though all chemical calculations involve mass. The atomic weight of an atom includes the mass of the electrons. Published *chemical atomic weights* of elements are averages of all the atomic weights of stable isotopes, taking into consideration the relative abundances of the isotopes.

[4]There are slight differences, known as *isotope effects*, in the chemical behavior of isotopes. These effects usually influence only the rate of reaction, not the kind of reaction.

[5]The *actinons* resemble element 89, *actinium*. Therefore, element 89 is sometimes included as an actinon.

[6]The *lanthanons* resemble element 57, *lanthanum*. Therefore element 57 is sometimes included as a lanthanon.

[7]The *transition metals* are elements whose electrons occupy the d sublevel. They can have various oxidation numbers, including +2, +3, +4, +6, and +7.

4. OXIDATION NUMBER

The *oxidation number* (*oxidation state*) is an electrical charge assigned by a set of prescribed rules. It is actually the charge assuming all bonding is ionic. The sum of the oxidation numbers equals the net charge. For monoatomic ions, the oxidation number is equal to the charge. The oxidation numbers of some common ions and radicals are given in Table 22.1.

For atoms in a free-state molecule, the oxidation number is zero. Hydrogen gas is a diatomic molecule, H_2. Thus, the oxidation number of the hydrogen molecule, H_2, is zero. The same is true for the atoms in O_2, N_2, Cl_2, and so on. Also, the sum of all the oxidation numbers of atoms in a neutral molecule is zero.

For a charged *radical* (a group of atoms that combine as a single unit), the net oxidation number is equal to the charge on the radical.

5. COMPOUNDS

Combinations of elements are known as *compounds*. *Binary compounds* contain two elements; *ternary (tertiary) compounds* contain three elements. A *chemical formula* is a representation of the relative numbers of each element in the compound. For example, the formula $CaCl_2$ shows that there are one calcium atom and two chlorine atoms in one molecule of calcium chloride.

Generally, the numbers of atoms are reduced to their lowest terms. However, there are exceptions. For example, acetylene is C_2H_2, and hydrogen peroxide is H_2O_2.

6. FORMATION OF COMPOUNDS

Compounds form according to the *law of definite (constant) proportions*: A pure compound is always composed of the same elements combined in a definite proportion by mass. For example, common table salt is always NaCl. It is not sometimes NaCl and other times Na_2Cl or $NaCl_3$ (which do not exist, in any case).

Furthermore, compounds form according to the *law of (simple) multiple proportions*: When two elements combine to form more than one compound, the masses of one element that combine with the same mass of the other are in the ratios of small integers.

In order to evaluate whether a compound formula is valid, it is necessary to know the *oxidation numbers* of the interacting atoms. Although some atoms have more than one possible oxidation number, most do not.

Table 22.1 *Oxidation Numbers of Selected Atoms and Charge Numbers of Radicals*

name	symbol	oxidation or charge number
acetate	$C_2H_3O_2$	-1
aluminum	Al	$+3$
ammonium	NH_4	$+1$
barium	Ba	$+2$
borate	BO_3	-3
boron	B	$+3$
bromine	Br	-1
calcium	Ca	$+2$
carbon	C	$+4, -4$
carbonate	CO_3	-2
chlorate	ClO_3	-1
chlorine	Cl	-1
chlorite	ClO_2	-1
chromate	CrO_4	-2
chromium	Cr	$+2, +3, +6$
copper	Cu	$+1, +2$
cyanide	CN	-1
dichromate	Cr_2O_7	-2
fluorine	F	-1
gold	Au	$+1, +3$
hydrogen	H	$+1$
hydroxide	OH	-1
hypochlorite	ClO	-1
iron	Fe	$+2, +3$
lead	Pb	$+2, +4$
lithium	Li	$+1$
magnesium	Mg	$+2$
mercury	Hg	$+1, +2$
nickel	Ni	$+2, +3$
nitrate	NO_3	-1
nitrite	NO_2	-1
nitrogen	N	$-3, +1, +2, +3, +4, +5$
oxygen	O	-2 (-1 in peroxides)
perchlorate	ClO_4	-1
permanganate	MnO_4	-1
phosphate	PO_4	-3
phosphorus	P	$-3, +3, +5$
potassium	K	$+1$
silicon	Si	$+4, -4$
silver	Ag	$+1$
sodium	Na	$+1$
sulfate	SO_4	-2
sulfite	SO_3	-2
sulfur	S	$-2, +4, +6$
tin	Sn	$+2, +4$
zinc	Zn	$+2$

The sum of the oxidation numbers must be zero if a neutral compound is to form. For example, H_2O is a valid compound since the two hydrogen atoms have a total positive oxidation number of $2 \times 1 = +2$. The oxygen ion has an oxidation number of -2. These oxidation numbers sum to zero.

On the other hand, $NaCO_3$ is not a valid compound formula. The sodium (Na) ion has an oxidation number of $+1$. However, the carbonate radical has a *charge number* of -2. The correct sodium carbonate molecule is Na_2CO_3.

7. MOLES

The *mole* is a measure of the quantity of an element or compound. Specifically, a mole of an element will have a mass equal to the element's atomic (or molecular) weight. The three main types of moles are based on mass being measured in grams, kilograms, and pounds.[8] Obviously, a gram-based mole of carbon (12.0 grams) is not the same quantity as a pound-based mole of carbon (12.0 pounds). Although "mol" is understood in SI countries to mean a gram-mole, the term *mole* is ambiguous, and the units mol (gmol), kmol (kgmol), or lbmol must be specified, or the type of mole must be spelled out.[9]

One gram-mole of any substance has a number of particles (atoms, molecules, ions, electrons, etc.) equal to 6.022×10^{23}, *Avogadro's number*, N_A. A pound-mole contains approximately 454 times the number of particles in a gram-mole.

Avogadro's law (hypothesis) holds that equal volumes of all gases at the same temperature and pressure contain equal numbers of gas molecules. Specifically, at standard scientific conditions (1.0 atm and 0°C), one gram-mole of any gas contains 6.022×10^{23} molecules and occupies 22.4 L. Of course, a pound-mole occupies 454 times that volume (359 ft^3).

"Molar" is used as an adjective when describing properties of a mole. For example, a *molar volume* is the volume of a mole.

Example 22.1

How many moles are in 0.01 g of gold ($A = 196.97$; $Z = 79$)?

Solution

The number of gram-moles of gold present is

$$n = \frac{m}{A} = \frac{0.01 \text{ g}}{196.97 \frac{\text{g}}{\text{mol}}} = 5.077 \times 10^{-5} \text{ mol}$$

---

[8]Theoretically, a slug-mole could be defined, but it is not used.
[9]There are also variations on the presentation of these units, such as g mol, gmole, g-mole, kmole, kg-mol, lb-mole, pound-mole, and p-mole. In most cases, the intent is clear.

8. FORMULA AND MOLECULAR WEIGHTS

The *formula weight*, FW, of a molecule (compound) is the sum of the atomic weights of all elements in the molecule. The *molecular weight*, MW, is generally the same as the formula weight. The units of molecular weight are actually g/mol, kg/kmol, or lb/lbmol. However, units are sometimes omitted because weights are relative. For example,

$$CaCO_3: FW = MW = 40.1 + 12 + 3 \times 16 = 100.1$$

An *ultimate analysis* (which determines how much of each element is present in a compound) will not determine the molecular formula. It will determine only the relative proportions of each element. Therefore, except for hydrated molecules and other linked structures, the molecular weight will be an integer multiple of the formula weight.

For example, an ultimate analysis of hydrogen peroxide (H_2O_2) will show that the compound has one oxygen atom for each hydrogen atom. In this case, the formula would be assumed to be HO and the formula weight would be approximately 17, although the actual molecular weight is 34.

For *hydrated molecules* (e.g., $FeSO_4 \cdot 7H_2O$), the mass of the *water of hydration* (also known as the *water of crystallization*) is included in the formula and in the molecular weight.

9. EQUIVALENT WEIGHT

The *equivalent weight* (i.e., an *equivalent*) is the amount of substance (in grams) that supplies one gram-mole (i.e., 6.022×10^{23}) of reacting units. For acid-base reactions an acid equivalent supplies one gram-mole of H^+ ions. A base equivalent supplies one gram-mole of OH^- ions. In oxidation-reduction reactions, an equivalent of a substance gains or loses a gram-mole of electrons. Similarly, in electrolysis reactions an equivalent weight is the weight of substance that either receives or donates one gram-mole of electrons at an electrode.

The equivalent weight can be calculated as the molecular weight divided by the change in oxidation number experienced in a chemical reaction. A substance can have several equivalent weights.

$$EW = \frac{MW}{\Delta \text{ oxidation number}} \qquad 22.2$$

Example 22.2

What are the equivalent weights of the following compounds?

(a) Al in the reaction

$$Al^{+++} + 3e^- \longrightarrow Al$$

(b) H_2SO_4 in the reaction

$$H_2SO_4 + H_2O \longrightarrow 2H^+ + SO_4^{-2} + H_2O$$

(c) NaOH in the reaction

$$NaOH + H_2O \longrightarrow Na^+ + OH^- + H_2O$$

Solution

(a) The atomic weight of aluminum is approximately 27. Since the change in the oxidation number is 3, the equivalent weight is $27/3 = 9$.

(b) The molecular weight of sulfuric acid is approximately 98. Since the acid changes from a neutral molecule to ions with two charges each, the equivalent weight is $98/2 = 49$.

(c) Sodium hydroxide has a molecular weight of approximately 40. The originally neutral molecule goes to a singly charged state. Therefore, the equivalent weight is $40/1 = 40$.

10. GRAVIMETRIC FRACTION

The *gravimetric fraction*, x_i, of an element i in a compound is the fraction by weight of that element in the compound. The gravimetric fraction is found from an *ultimate analysis* (also known as a *gravimetric analysis*) of the compound.

$$x_i = \frac{m_i}{m_1 + m_2 + \cdots + m_i + \cdots + m_n} = \frac{m_i}{m_t} \quad 22.3$$

The *percentage composition* is the gravimetric fraction converted to percentage.

$$\%\text{composition} = x_i \times 100\% \quad 22.4$$

If the gravimetric fractions are known for all elements in a compound, the *combining weights* of each element can be calculated. (The term *weight* is used even though mass is the traditional unit of measurement.)

$$m_i = x_i m_t \quad 22.5$$

11. EMPIRICAL FORMULA DEVELOPMENT

It is relatively simple to determine the *empirical formula* of a compound from the atomic and combining weights of elements in the compound. The empirical formula gives the relative number of atoms (i.e., the formula weight is calculated from the empirical formula).

step 1: Divide the gravimetric fractions (or percentage compositions) by the atomic weight of each respective element.

step 2: Determine the smallest ratio from step 1.

step 3: Divide all of the ratios from step 1 by the smallest ratio.

step 4: Write the chemical formula using the results from step 3 as the numbers of atoms. Multiply through as required to obtain all integer numbers of atoms.

Example 22.3

A clear liquid is analyzed, and the following percentage compositions are recorded: carbon, 37.5%; hydrogen, 12.5%; oxygen, 50%. What is the chemical formula for the liquid?

Solution

step 1: Divide the percentage compositions by the atomic weight.

$$C: \quad \frac{37.5}{12} = 3.125$$

$$H: \quad \frac{12.5}{1} = 12.5$$

$$O: \quad \frac{50}{16} = 3.125$$

steps 2 and 3: The smallest ratio is 3.125. Divide all ratios by 3.125.

$$C: \quad \frac{3.125}{3.125} = 1$$

$$H: \quad \frac{12.5}{3.125} = 4$$

$$O: \quad \frac{3.125}{3.125} = 1$$

step 4: The empirical formula is CH_4O.

If it had been known that the liquid behaved chemically as though it had a hydroxyl (OH) radical, the formula would have been written as CH_3OH. This is recognized as methyl alcohol.

12. CHEMICAL REACTIONS

During chemical reactions, bonds between atoms are broken and new bonds are usually formed. The starting substances are known as *reactants*; the ending substances are known as *products*. In a chemical reaction, reactants are either converted to simpler products or synthesized into more complex compounds. There are four common types of reactions.

- *direct combination* (or *synthesis*): This is the simplest type of reaction where two elements or compounds combine directly to form a compound.

$$2H_2 + O_2 \longrightarrow 2H_2O$$
$$SO_2 + H_2O \longrightarrow H_2SO_3$$

- *decomposition* (or *analysis):* Bonds within a compound are disrupted by heat or other energy to produce simpler compounds or elements.

$$2HgO \longrightarrow 2Hg + O_2$$
$$H_2CO_3 \longrightarrow H_2O + CO_2$$

- *single displacement* (or *replacement):* This type of reaction has one element and one compound as reactants.

$$2Na + 2H_2O \longrightarrow 2NaOH + H_2$$
$$2KI + Cl_2 \longrightarrow 2KCl + I_2$$

- *double displacement* (or *replacement*[10]*):* These are reactions with two compounds as reactants and two compounds as products.

$$AgNO_3 + NaCl \longrightarrow AgCl + NaNO_3$$
$$H_2SO_4 + ZnS \longrightarrow H_2S + ZnSO_4$$

13. BALANCING CHEMICAL EQUATIONS

The coefficients in front of element and compound symbols in chemical reaction equations are the numbers of molecules or moles taking part in the reaction. (For gaseous reactants and products, the coefficients also represent the numbers of volumes. This is a direct result of Avogadro's hypothesis that equal numbers of molecules in the gas phase occupy equal volumes under the same conditions.)[11]

Since atoms cannot be changed in a normal chemical reaction (i.e., mass is conserved), the numbers of each element must match on both sides of the equation. When the numbers of each element match, the equation is said to be "balanced." The total atomic weights on both sides of the equation will be equal when the equation is balanced.

Balancing simple chemical equations is largely a matter of deductive trial and error. More complex reactions require use of oxidation numbers.

Example 22.4

Balance the following reaction equation.

$$Al + H_2SO_4 \longrightarrow Al_2(SO_4)_3 + H_2$$

---

[10]Another name for replacement is *metathesis*.
[11]When water is part of the reaction, the interpretation that the coefficients are volumes is valid only if the reaction takes place at a high enough temperature to vaporize the water.

Solution

As written, the reaction is not balanced. For example, there is one aluminum on the left, but there are two on the right. The starting element in the balancing procedure is chosen somewhat arbitrarily.

step 1: Since there are two aluminums on the right, multiply Al by 2.

$$2Al + H_2SO_4 \longrightarrow Al_2(SO_4)_3 + H_2$$

step 2: Since there are three sulfate radicals (SO_4) on the right, multiply H_2SO_4 by 3.

$$2Al + 3H_2SO_4 \longrightarrow Al_2(SO_4)_3 + H_2$$

step 3: Now there are six hydrogens on the left, so multiply H_2 by 3 to balance the equation.

$$2Al + 3H_2SO_4 \longrightarrow Al_2(SO_4)_3 + 3H_2$$

14. STOICHIOMETRIC REACTIONS

Stoichiometry is the study of the proportions in which elements and compounds react and are formed. A *stoichiometric reaction* (also known as a *perfect reaction* or an *ideal reaction*) is one in which just the right amounts of reactants are present. After the reaction stops, there are no unused reactants.

Stoichiometric problems are known as *weight and proportion problems* because their solutions use simple ratios to determine the masses of reactants required to produce given masses of products, or vice versa. The procedure for solving these problems is essentially the same regardless of the reaction.

step 1: Write and balance the chemical equation.
step 2: Determine the atomic (molecular) weight of each element (compound) in the equation.
step 3: Multiply the atomic (molecular) weights by their respective coefficients and write the products under the formulas.
step 4: Write the given mass data under the weights determined in step 3.
step 5: Fill in the missing information by calculating simple ratios.

Example 22.5

Caustic soda (NaOH) is made from sodium carbonate (Na_2CO_3) and slaked lime ($Ca(OH)_2$) according to the given reaction. How many kilograms of caustic soda can be made from 2000 kg of sodium carbonate?

Solution

	Na_2CO_3	$+ Ca(OH)_2$	$\longrightarrow 2NaOH$	$+ CaCO_3$
molecular weights	106	74	2×40	100
given data	2000 kg		X kg	

The simple ratio used is

$$\frac{NaOH}{Na_2CO_3} = \frac{80}{106} = \frac{X}{2000}$$

Solving for the unknown mass, $X = 1509$ kg.

15. NONSTOICHIOMETRIC REACTIONS

In many cases, it is not realistic to assume a stoichiometric reaction because an excess of one or more reactants is necessary to assure that all of the remaining reactants take part in the reaction. Combustion is an example where the stoichiometric assumption is, more often than not, invalid. Excess air is generally needed to ensure that all of the fuel is burned.

With nonstoichiometric reactions, the reactant that is used up first is called the *limiting reactant*. The amount of product will be dependent on (limited by) the limiting reactant.

The *theoretical yield* or *ideal yield* of a product is the maximum amount of product per unit amount of limiting reactant that can be obtained from a given reaction if the reaction goes to completion. The *percentage yield* is a measure of the efficiency of the actual reaction.

$$\text{percentage yield} = \frac{\text{actual yield} \times 100\%}{\text{theoretical yield}} \qquad 22.6$$

16. SOLUTIONS OF GASES IN LIQUIDS

Henry's law states that the amount (i.e., mole fraction) of a slightly soluble gas dissolved in a liquid is proportional to the partial pressure of the gas. This law applies separately to each gas to which the liquid is exposed, as if each gas were present alone. The algebraic form of Henry's law is given by Eq. 22.7, in which H is the *Henry's law constant* in mole fractions/atmosphere.

$$p_i = Hx_i \qquad 22.7$$

Generally, the solubility of gases in liquids decreases with increasing temperature.

The volume of gas absorbed at a partial pressure of 1 atm and 0°C is known as the *absorption coefficient*. Typical absorption coefficients for solutions in water are: H_2, 0.017 1/L; He, 0.009 1/L; N_2, 0.015 1/L; O_2, 0.028 1/L; CO, 0.025 1/L; and CO_2, 0.88 1/L.

The amount of gas dissolved in a liquid varies with the temperature of the liquid and the concentration of dissolved salts in the liquid. Appendix 22.D lists the saturation values of dissolved oxygen in water at various temperatures and for various amounts of chloride ion (also referred to as *salinity*).

Example 22.6

At 20°C and 1 atm, 1 L of water will absorb 0.043 g of oxygen or 0.019 g of nitrogen. Atmospheric air is 20.9%

oxygen by volume, and the remainder is assumed to be nitrogen. What masses of oxygen and nitrogen will be absorbed by 1 L of water exposed to 20°C air at 1 atm?

Solution

Since partial pressure is volumetrically weighted,

$$m_{oxygen} = (0.209)\left(0.043 \; \frac{g}{L}\right) = 0.009 \; g/L$$

$$m_{nitrogen} = (1.000 - 0.209)\left(0.019 \; \frac{g}{L}\right) = 0.015 \; g/L$$

Example 22.7

At an elevation of 4000 ft, the barometric pressure is 660 mm Hg. What is the dissolved oxygen concentration of 18°C water with a 800 mg/L chloride concentration at that elevation?

Solution

From App. 22.D, oxygen's saturation concentration for 18°C water corrected for a 800 mg/L chloride concentration is

$$C_s = 9.5 \; \frac{mg}{L} - (8)\left(0.009 \; \frac{mg}{L}\right) = 9.4 \; mg/L$$

Use the appendix footnote to correct for the barometric pressure.

$$C'_s = \left(9.4 \; \frac{mg}{L}\right)\left(\frac{660 \; mm - 16 \; mm}{760 \; mm - 16 \; mm}\right)$$

$$= 8.1 \; mg/L$$

17. SOLUTIONS OF SOLIDS IN LIQUIDS

When a solid is added to a liquid, the solid is known as the *solute* and the liquid is known as the *solvent*.[12] If the dispersion of the solute throughout the solvent is at the molecular level, the mixture is known as a *solution*. If the solute particles are larger than molecules, the mixture is known as a *suspension*.[13]

In some solutions, the solvent and solute molecules bond loosely together. This loose bonding is known as *solvation*. If water is the solvent, the bonding process is also known as *aquation* or *hydration*.

The solubility of most solids in liquid solvents usually increases with increasing temperature. Pressure has very little effect on the solubility of solids in liquids.

---

[12]The term *solvent* is often associated with volatile liquids, but the term is more general than that. (A *volatile liquid* evaporates rapidly and readily at normal temperatures.) Water is the solvent in aqueous solutions.
[13]An *emulsion* is not a mixture of a solid in a liquid. It is a mixture of two immiscible liquids.

When the solvent has dissolved as much solute as it can, it is a *saturated solution*.[14] Adding more solute to an already saturated solution will cause the excess solute to settle to the bottom of the container, a process known as *precipitation*. Other changes (in temperature, concentration, etc.) can be made to cause precipitation from saturated and unsaturated solutions.

18. UNITS OF CONCENTRATION

There are many units of concentration to express solution strengths.

F— formality: The number of gram formula weights (i.e., molecular weights in grams) per liter of solution.

m— molality: The number of gram-moles of solute per 1000 grams of solvent. A "molal" solution contains 1 gram-mole per 1000 grams of solvent.

M— molarity: The number of gram-moles of solute per liter of solution. A "molar" (i.e., 1 M) solution contains 1 gram-mole per liter of solution. Molarity is related to normality: $N = M \times \Delta$ oxidation number.

N— normality: The number of gram equivalent weights of solute per liter of solution. A solution is "normal" (i.e., 1 N) if there is exactly one gram equivalent weight per liter of solution.

x— mole fraction: The number of moles of solute divided by the number of moles of solvent and all solutes.

meq/L— milligram equivalent weights of solute *per liter* of solution: calculated by multiplying normality by 1000 or dividing concentration in mg/L by equivalent weight.

mg/L— milligrams per liter: The number of milligrams of solute per liter of solution. Same as ppm for solutions of water.

ppm— parts per million: The number of pounds (or grams) of solute per million pounds (or grams) of solution. Same as mg/L for solutions of water.

For compounds whose molecules do not dissociate in solution (e.g., table sugar), there is no difference between molarity and formality. There is a difference, however, for compounds that dissociate into ions (e.g., table salt). Consider a solution derived from 1 gmol of magnesium nitrate $Mg(NO_3)_2$ in enough water to bring the volume to 1 L. The formality is 1.0 (i.e., the solution is 1.0 formal). However, 3 moles of ions will be produced—1 mole of Mg^{++} ions and 2 moles of NO_3^-

ions. Therefore, molarity is 1.0 for the magnesium ion and 2.0 for the nitrate ion.

The use of formality avoids the ambiguity in specifying concentrations for ionic solutions. Also, the use of formality avoids the problem of determining a molecular weight when there are no discernible molecules (e.g., as in a crystalline solid such as NaCl). However, in their quest for uniformity in nomenclature, most modern chemists do not make the distinction between molarity and formality, and molarity is used as if it were formality.

Example 22.8

A solution is made by dissolving 0.353 g of $Al_2(SO_4)_3$ in 730 g of water. Assuming 100% ionization, what is the concentration expressed as normality, molarity, and mg/L?

Solution

The molecular weight of $Al_2(SO_4)_3$ is

$$MW = (2)(26.98) + (3)(32.06 + (4)(16)) = 342.14$$

The equivalent weight is

$$EW = \frac{342.14}{6} = 57.02$$

The number of gram equivalent weights used is

$$\frac{0.353}{57.02} = 6.19 \times 10^{-3} \text{ GEW}$$

The number of liters of solution (same as the solvent volume if the small amount of solute is neglected) is 0.73.

The normality is

$$N = \frac{6.19 \times 10^{-3} \text{ GEW}}{0.73 \text{ L}} = 8.48 \times 10^{-3}$$

The number of moles of solute used is 0.353 g/342.14 g/mol = 1.03 $\times 10^{-3}$ mol.

The molarity is

$$M = \frac{1.03 \times 10^{-3} \text{ mol}}{0.73 \text{ L}} = 1.41 \times 10^{-3}$$

The number of milligrams is

$$\frac{0.353 \text{ g}}{0.001 \frac{g}{mg}} = 353 \text{ mg}$$

$$mg/L = \frac{353 \text{ mg}}{0.73 \text{ L}} = 483.6$$

19. pH AND pOH

A measure of the strength of an acid or base is the number of hydrogen or hydroxide ions in a liter of solution.

---

[14]Under certain circumstances, a *supersaturated solution* can exist for a limited amount of time.

Since these are very small numbers, a logarithmic scale is used.

$$pH = -\log_{10}[H^+] = \log_{10}\left(\frac{1}{[H^+]}\right) \qquad 22.8$$

$$pOH = -\log_{10}[OH^-] = \log_{10}\left(\frac{1}{[OH^-]}\right) \qquad 22.9$$

The quantities $[H^+]$ and $[OH^-]$ in square brackets are the *ionic concentrations* in moles of ions per liter. The number of moles can be calculated from Avogadro's law by dividing the actual number of ions per liter by 6.022×10^{23}. Alternatively, for a partially ionized compound, X, in a solution of known molarity, M, the ionic concentration is

$$[X] = (\text{fraction ionized}) \times M \qquad 22.10$$

A *neutral solution* has a pH of 7.[15] Solutions with a pH below 7 are acidic; the smaller the pH, the more acidic the solution. Solutions with a pH above 7 are basic.

The relationship between pH and pOH is

$$pH + pOH = 14 \qquad 22.11$$

Example 22.9

A 4.2% ionized 0.01M ammonia solution is prepared from ammonium hydroxide (NH_4OH). Calculate the pH, pOH, and concentrations of H^+ and OH^-.

Solution

From Eq. 22.10,

$$[OH^-] = (\text{fraction ionized}) \times M = (0.042)(0.01)$$
$$= 4.2 \times 10^{-4} \text{ mol/L}$$

From Eq. 22.9,

$$pOH = -\log[OH^-] = -\log(4.2 \times 10^{-4})$$
$$= 3.38$$

From Eq. 22.11,

$$pH = 14 - pOH = 14 - 3.38$$
$$= 10.62$$

The $[H^+]$ ionic concentration can be extracted from the definition of pH.

$$[H^+] = 10^{-pH} = 10^{-10.62}$$
$$= 2.4 \times 10^{-11} \text{ mol/L}$$

---

[15]The pH of a neutral solution depends on the temperature. At 25°C, the pH is 7. When the temperature is higher (lower) than 25°C, the pH will be less (greater) than 7.

20. BUFFERS

A *buffer solution* resists changes in acidity and maintains a relatively constant pH when a small amount of an acid or base is added to it. Buffers are usually combinations of weak acids and their salts. A buffer is most effective when the acid and salt concentrations are equal.

21. NEUTRALIZATION

Acids and bases neutralize each other to form water.

$$H^+ + OH^- \longrightarrow H_2O$$

Assuming 100% ionization of the solute, the volumes, V, required for complete neutralization can be calculated from the normalities, N, or the molarities, M.

$$V_{\text{base}} N_{\text{base}} = V_{\text{acid}} N_{\text{acid}} \qquad 22.12$$

$$V_{\text{base}} M_{\text{base}} \Delta_{\text{base charge}} = V_{\text{acid}} M_{\text{acid}} \Delta_{\text{acid charge}} \qquad 22.13$$

22. REVERSIBLE REACTIONS

Reversible reactions are capable of going in either direction and do so to varying degrees (depending on the concentrations and temperature) simultaneously. These reactions are characterized by the simultaneous presence of all reactants and all products. For example, the chemical equation for the exothermic formation of ammonia from nitrogen and hydrogen is

$$N_2 + 3H_2 \longleftrightarrow 2NH_3 + \Delta H = -24.5 \text{ kcal}$$

At *chemical equilibrium*, reactants and products are both present. However, the concentrations of the reactants and products do not change after equilibrium is reached.

23. LE CHÂTELIER'S PRINCIPLE

Le Châtelier's principle predicts the direction in which a reversible reaction at equilibrium will go when some condition (e.g., temperature, pressure, concentration) is "stressed" (i.e., changed). The principle says that when an equilibrium state is stressed by a change, a new equilibrium is formed that reduces that stress.

Consider the formation of ammonia from nitrogen and hydrogen. When the reaction proceeds in the forward direction, energy in the form of heat is released and the temperature increases. If the reaction proceeds in the reverse direction, heat is absorbed and the temperature decreases. If the system is stressed by increasing the temperature, the reaction will proceed in the reverse direction because that direction absorbs heat and reduces the temperature.

For reactions that involve gases, the reaction equation coefficients can be interpreted as volumes. In the nitrogen-hydrogen reaction, four volumes combine to form two volumes. If the equilibrium system is stressed by increasing the pressure, then the forward reaction will occur because this direction reduces the volume and pressure.[16]

If the concentration of any substance is increased, the reaction proceeds in a direction away from the substance with the increase in concentration. (For example, an increase in the concentration of the reactants shifts the equilibrium to the right, increasing the amount of products formed.)

The *common ion effect* is a special case of Le Châtelier's principle. If a salt containing a common ion is added to a solution of a weak acid, almost all of the salt will dissociate, adding large quantities of the common ion to the solution. Ionization of the acid will be greatly suppressed, a consequence of the need to have an unchanged equilibrium constant.

24. IRREVERSIBLE REACTION KINETICS

The rate at which a compound is formed or used up in an irreversible (one-way) reaction is known as the *rate of reaction*, also known as the *speed of reaction, reaction velocity*, and so on. The rate, v, is the change in concentration per unit time, usually measured in mol/L·s.

$$v = \text{change in concentration/time} \qquad 22.14$$

According to the *law of mass action*, the rate of reaction varies with the concentrations of the reactants and products. Specifically, the rate is proportional to the molar concentrations (i.e., the molarities). The rate of the formation or a conversion of substance A is represented in various forms, such as r_A, dA/dt, and $d[A]/dt$, where the variable A or [A] can represent either the mass or the concentration of substance A. Substance A can be either a pure element or a compound.

The rate of reaction is generally not affected by pressure, but does depend on five other factors.

- *type of substances in the reaction:* Some substances are more reactive than others.
- *exposed surface area:* The rate of reaction is proportional to the amount of contact between the reactants.
- *concentrations:* The rate of reaction increases with increases in concentration.
- *temperature:* The rate of reaction approximately doubles with every 10°C increase in temperature.
- *catalysts:* If a catalyst is present, the rate of reaction increases. However, the equilibrium point is

not changed. (A catalyst is a substance that increases the reaction rate without being consumed in the reaction.)

25. ORDER OF THE REACTION

The *order of the reaction* is the total number of reacting molecules in or before the slowest step in the mechanism.[17] The order must be determined experimentally. However, for an irreversible elementary reaction, the order is usually assumed from the stoichiometric reaction equation as the sum of the combining coefficients for the reactants.[18,19] For example, for the reaction $mA + nB \rightarrow pC$, the overall order of the forward reaction is assumed to be $m + n$.

Many reactions (e.g., dissolving metals in acid or the evaporation of condensed materials) have *zero-order reaction rates*. These reactions do not depend on the concentrations or temperature at all, but rather, are affected by other factors such as the availability of reactive surfaces or the absorption of radiation. The formation (conversion) rate of a compound in a zero-order reaction is constant. That is, $dA/dt = -k_0$. k_0 is known as the *reaction rate constant*. (The subscript "0" refers to the zero-order.) Since the concentration (amount) of the substance decreases with time, dA/dt is negative. Since the negative sign is explicit in rate equations, the reaction rate constant is always considered to be a positive number.

Table 22.2 contains reaction rate and half-life equations for various types of low-order reactions. Once a reaction rate equation is known, it can be integrated to obtain an expression for the concentration (mass) of the substance at various times. The time for half of the substance to be formed (or converted) is the *half-life, $t_{1/2}$*.

Example 22.10

Nitrogen pentoxide decomposes according to the following first-order reaction.

$$N_2O_5 \rightarrow 2NO_2 + \tfrac{1}{2}O_2$$

At a particular temperature, the decomposition of nitrogen pentoxide is 85% complete at the end of 11 min. The reaction rate constant is to be determined.

---

[17]This definition is valid for elementary reactions. For complex reactions, the order is an empirical number that need not be an integer.
[18]The overall order of the reaction is the sum of the orders with respect to the individual reactants. For example, in the reaction $2NO + O_2 \rightarrow 2NO_2$, the reaction is second order with respect to NO, first order with respect to O_2, and third order overall.
[19]In practice, the order of the reaction must be known, given, or determined experimentally. It is not always equal to the sum of the combining coefficients for the reactants. For example, in the reaction $H_2 + I_2 \rightarrow 2HI$, the overall order of the reaction is indeed 2, as expected. However, in the reaction $H_2 + Br_2 \rightarrow 2HBr$, the overall order is found experimentally to be 3/2, even though the two reactions have the same stoichiometry, and despite the similarities of iodine and bromine.

---

[16]The exception to this rule is the addition of an inert gas to a gaseous equilibrium system. Although there is an increase in total pressure, the position of the equilibrium is not affected.

Table 22.2 *Reaction Rates and Half-Life Equations*

reaction	order	rate equation	integrated forms
$A \rightarrow B$	zero	$\dfrac{d[A]}{dt} = -k_0$	$[A] = [A]_0 - k_0 t$ $\qquad t_{1/2} = \dfrac{[A]_0}{2k_0}$
$A \rightarrow B$	first	$\dfrac{d[A]}{dt} = -k_1[A]$	$\ln \dfrac{[A]}{[A]_0} = k_1 t$ $\qquad t_{1/2} = \dfrac{1}{k_1}\ln 2$
$A + A \rightarrow P$	second, type I	$\dfrac{d[A]}{dt} = -k_2[A]^2$	$\dfrac{1}{[A]} - \dfrac{1}{[A]_0} = k_2 t$ $\qquad t_{1/2} = \dfrac{1}{k_2[A]_0}$
$aA + bB \rightarrow P$	second, type II	$\dfrac{d[A]}{dt} = -k_2[A][B]$	$\ln \dfrac{[A]_0 - [B]}{[B]_0 - \left(\dfrac{b}{a}\right)[X]} = \ln \dfrac{[A]}{[B]}$ $= \left(\dfrac{b[A]_0 - a[B]_0}{a}\right)k_2 t + \ln \dfrac{[A]_0}{[B]_0}$ $t_{1/2} = \left[\dfrac{a}{k_2(b[A]_0 - a[B]_0)}\right]$ $\times \ln\left[\dfrac{a[B]_0}{2a[B]_0 - b[A]_0}\right]$

Solution

The reaction is given as first order. Use the integrated reaction rate equation from Table 22.2. Since the decomposition reaction is 85% complete, the surviving fraction is 15% (0.15).

$$\ln\left(\frac{[A]}{[A]_0}\right) = k_1 t$$
$$\ln(0.15) = k(11 \text{ min})$$
$$k = -0.172 \text{ } 1/\text{min} \quad (0.172 \text{ } 1/\text{min})$$

(The rate constant is considered to be a positive number.)

26. REVERSIBLE REACTION KINETICS

Consider the following reversible reaction.

$$aA + bB \longleftrightarrow cC + dD \qquad 22.15$$

In Eqs. 22.16 and 22.17, the *reaction rate constants* are k_{forward} and k_{reverse}. The order of the forward reaction is $a + b$; the order of the reverse reaction is $c + d$.

$$v_{\text{forward}} = k_{\text{forward}}[A]^a[B]^b \qquad 22.16$$
$$v_{\text{reverse}} = k_{\text{reverse}}[C]^c[D]^d \qquad 22.17$$

At equilibrium, the forward and reverse speeds of reaction are equal.

$$v_{\text{forward}} = v_{\text{reverse}}\big|_{\text{equilibrium}} \qquad 22.18$$

27. EQUILIBRIUM CONSTANT

For reversible reactions, the *equilibrium constant*, K, is proportional to the ratio of the reverse rate of reaction to the forward rate of reaction.[20] Except for catalysis, the equilibrium constant depends on the same factors affecting the reaction rate. For the complex reversible reaction given by Eq. 22.15, the equilibrium constant is given by the *law of mass action*.

$$K = \frac{[C]^c[D]^d}{[A]^a[B]^b} = \frac{k_{\text{forward}}}{k_{\text{reverse}}} \qquad 22.19$$

If any of the reactants or products are in pure solid or pure liquid phases, their concentrations are omitted from the calculation of the equilibrium constant. For example, in weak aqueous solutions, the concentration of water, H_2O, is very large and essentially constant; therefore, that concentration is omitted.

[20]The symbols K_c (in molarity units) and K_{eq} are occasionally used for the equilibrium constant.

For gaseous reactants and products, the concentrations (i.e., the numbers of atoms) will be proportional to the partial pressures. Therefore, an equilibrium constant can be calculated directly from the partial pressures and is given the symbol K_p. For example, for the formation of ammonia gas from nitrogen and hydrogen, the equilibrium constant is

$$K_p = \frac{[p_{NH_3}]^2}{[p_{N_2}][p_{H_2}]^3} \qquad \textit{22.20}$$

K and K_p are not numerically the same, but they are related by Eq. 22.21. Δn is the number of moles of products minus the number of moles of reactants.

$$K_p = K(R^*T)^{\Delta n} \qquad \textit{22.21}$$

Example 22.11

A particularly weak solution of acetic acid $(HC_2H_3O_2)$ in water has the ionic concentrations (in mol/L) given. What is the equilibrium constant?

$$HC_2H_3O_2 + H_2O \longleftrightarrow H_3O^+ + C_2H_3O_2^-$$
$$[HC_2H_3O_2] = 0.09866$$
$$[H_2O] = 55.5555$$
$$[H_3O^+] = 0.00134$$
$$[C_2H_3O_2^-] = 0.00134$$

Solution

The concentration of the water molecules is not included in the calculation of the equilibrium or ionization constant. Therefore, the equilibrium constant is

$$\begin{aligned} K = K_a &= \frac{[H_3O^+][C_2H_3O_2^-]}{[HC_2H_3O_2]} \\ &= \frac{(0.00134)(0.00134)}{0.09866} = 1.82 \times 10^{-5} \end{aligned}$$

28. IONIZATION CONSTANT

The equilibrium constant for a weak solution is essentially constant and is known as the *ionization constant* (also known as a *dissociation constant*). For weak acids, the symbol K_a and name *acid constant* are used. For weak bases, the symbol K_b and the name *base constant* are used. For example, for the ionization of hydrocyanic acid,

$$HCN \longleftrightarrow H^+ + CN^-$$
$$K_a = \frac{[H^+][CN^-]}{[HCN]}$$

Pure water is itself a very weak electrolyte and ionizes only slightly.

$$2H_2O \longleftrightarrow H_3O^+ + OH^- \qquad \textit{22.22}$$

At equilibrium, the ionic concentrations are equal.

$$[H_3O^+] = 10^{-7}$$
$$[OH^-] = 10^{-7}$$

From Eq. 22.19, the ionization constant (*ion product*) for pure water is

$$\begin{aligned} K_w = K_{a,\text{water}} &= [H_3O^+][OH^-] = (10^{-7})(10^{-7}) \\ &= 10^{-14} \qquad \textit{22.23} \end{aligned}$$

If the molarity, M, and fraction ionization, X, are known, the ionization constant can be calculated from Eq. 22.24.

$$K_{\text{ionization}} = \frac{MX^2}{1-X} \qquad \textit{22.24}$$

The reciprocal of the ionization constant is the *stability constant* (*overall stability constant*), also known as the *formation constant*. Stability constants are used to describe complex ions that dissociate readily.

Table 22.3 *Approximate Ionization Constants*

substance	0°C	5°C	10°C	15°C	20°C	25°C
Ca(OH)$_2$						3.74×10^{-3}
HClO	2.0×10^{-8}	2.3×10^{-8}	2.6×10^{-8}	3.0×10^{-8}	3.3×10^{-8}	3.7×10^{-8}
HC$_2$H$_3$O$_2$	1.67×10^{-5}	1.70×10^{-5}	1.73×10^{-5}	1.75×10^{-5}	1.75×10^{-5}	1.75×10^{-5}
HBrO					$\approx 2 \times 10^{-9}$	
H$_2$CO$_3$ (K_1)	2.6×10^{-7}	3.04×10^{-7}	3.44×10^{-7}	3.81×10^{-7}	4.16×10^{-7}	4.45×10^{-7}
HClO$_2$					$\approx 1.1 \times 10^{-2}$	
NH$_3$	1.37×10^{-5}	1.48×10^{-5}	1.57×10^{-5}	1.65×10^{-5}	1.71×10^{-5}	1.77×10^{-5}
NH$_4$OH						1.79×10^{-5}
water^a	14.9435	14.7338	14.5346	14.3463	14.1669	13.9965

$^a - \log_{10}(K)$ given

Example 22.12

A 0.1 molar (0.1M) acetic acid solution is 1.34% ionized. Find the (a) hydrogen ion concentration, (b) acetate ion concentration, (c) un-ionized acid concentration, and (d) ionization constant.

Solution

(a) From Eq. 22.10, the hydrogen ion concentration is

$$[H_3O^+] = \text{(fraction ionized)}\,\text{(molarity)}$$
$$= (0.0134)(0.1) = 0.00134 \text{ mol/L}$$

(b) Since every hydronium ion has a corresponding acetate ion, the acetate and hydronium ion concentrations are the same.

$$[C_2H_3O_2^-] = [H_3O^+] = 0.00134 \text{ mol/L}$$

(c) The concentration of un-ionized acid can be derived from Eq. 22.10.

$$[HC_2H_3O_2] = \text{(fraction not ionized)}\,\text{(molarity)}$$
$$= (1 - 0.0134)(0.1) = 0.09866 \text{ mol/L}$$

(d) The ionization constant is calculated from Eq. 22.24.

$$K_a = \frac{\text{(molarity)}\,\text{(fraction ionized)}^2}{1 - \text{fraction ionized}}$$
$$= \frac{(0.1)(0.0134)^2}{1 - 0.0134} = 1.82 \times 10^{-5}$$

Example 22.13

The ionization constant for acetic acid is 1.82×10^{-5}. What is the hydrogen ion concentration for a 0.2M solution?

Solution

From Eq. 22.24,

$$K_a = \frac{MX^2}{1 - X}$$
$$1.82 \times 10^{-5} = \frac{0.2X^2}{1 - X}$$

Since acetic acid is a weak acid, X is known to be small. Therefore, the computational effort can be reduced by assuming that $1 - X \approx 1$.

$$1.82 \times 10^{-5} = 0.2X^2$$
$$X = 9.49 \times 10^{-3}$$

From Eq. 22.10, the concentration of the hydrogen ion is

$$[H_3O^+] = XM = (9.49 \times 10^{-3})(0.2)$$
$$= 1.9 \times 10^{-3} \text{ mol/L}$$

Example 22.14

The ionization constant for acetic acid $(HC_2H_3O_2)$ is 1.82×10^{-5}. What is the hydrogen ion concentration of a solution with 0.1 mole of 80% ionized ammonium acetate $(NH_4C_2H_3O_2)$ in 1 L of 0.1M acetic acid?

Solution

The acetate ion $(C_2H_3O_2^-)$ is a *common ion*, since it is supplied by both the acetic acid and the ammonium acetate. Both sources contribute to the ionic concentration. However, the ammonium acetate's contribution dominates. Since the acid dissociates into an equal number of hydrogen and acetate ions,

$$[C_2H_3O_2^-]_{\text{total}} = [C_2H_3O_2^-]_{\text{acid}}$$
$$+ [C_2H_3O_2^-]_{\text{ammonium acetate}}$$
$$= [H_3O^+] + (0.8)(0.1)$$
$$\approx (0.8)(0.1) = 0.08$$

As a result of the common ion effect and Le Châtelier's law, the acid's dissociation is essentially suppressed by the addition of the ammonium acetate. The concentration of un-ionized acid is

$$[HC_2H_3O_2] = 0.1 - [H_3O^+]$$
$$\approx 0.1$$

The ionization constant is unaffected by the number of sources of the acetate ion.

$$K_a = \frac{[H_3O^+][C_2H_3O_2^-]}{[HC_2H_3O_2]}$$
$$1.82 \times 10^{-5} = \frac{[H_3O^+] \times 0.08}{0.1}$$
$$[H_3O^+] = 2.3 \times 10^{-5} \text{ mol/L}$$

29. IONIZATION CONSTANTS FOR POLYPROTIC ACIDS

A polyprotic acid has as many ionization constants as it has acidic hydrogen atoms. For oxyacids, each successive ionization constant is approximately 10^5 times smaller than the preceding one. For example, phosphoric acid (H_3PO_4) has three ionization constants:

$$K_1 = 7.1 \times 10^{-3} \quad (H_3PO_4)$$
$$K_2 = 6.3 \times 10^{-8} \quad (H_2PO_4^-)$$
$$K_3 = 4.4 \times 10^{-13} \quad (HPO_4^{-2})$$

30. SOLUBILITY PRODUCT

When an ionic solid is dissolved in a solvent, it dissociates. For example, consider the ionization of silver chloride in water.

$$AgCl(s) \longleftrightarrow Ag^+(aq) + Cl^-(aq)$$

If the equilibrium constant is calculated, the terms for pure solids and liquids (in this case, [AgCl] and [H_2O]) are omitted. Thus, the *solubility product*, K_{sp}, consists only of the ionic concentrations. As with the general case of ionization constants, the solubility product for slightly soluble solutes is essentially constant at a standard value.

$$K_{sp} = [\text{Ag}^+][\text{Cl}^-] \qquad \textbf{22.25}$$

When the product of terms exceeds the standard value of the solubility product, solute will precipitate out until the product of the remaining ion concentrations attains the standard value. If the product is less than the standard value, the solution is not saturated.

The solubility products of nonhydrolyzing compounds are relatively easy to calculate. (Example 22.14 illustrates a method.) Such is the case for chromates (CrO_4^{-2}), halides (F^-, Cl^-, Br^-, I^-), sulfates (SO_4^{-2}), and iodates (IO_3^-). However, compounds that hydrolyze must be treated differently. The method used in Ex. 22.14 cannot be used for hydrolyzing compounds.

Example 22.15

At a particular temperature, it takes 0.038 grams of lead sulfate ($PbSO_4$, molecular weight = 303.25) per liter of water to prepare a saturated solution. What is the solubility product of lead sulfate if all of the lead sulfate ionizes?

Solution

Sulfates are not one of the hydrolyzing ions. Therefore, the solubility product can be calculated from the concentrations.

Since 1 L of water has a mass of 1 kg, the number of moles of lead sulfate dissolved per saturated liter of solution is

$$n = \frac{\text{m}}{\text{MW}} = \frac{0.038 \text{ g}}{303.25 \, \frac{\text{g}}{\text{mol}}}$$

$$= 1.25 \times 10^{-4} \text{ mol}$$

Lead sulfate ionizes according to the following reaction.

$$\text{PbSO}_4(s) \longleftrightarrow \text{Pb}^{+2}(aq) + \text{SO}_4^-(aq) \quad \text{[in water]}$$

Since all of the lead sulfate ionizes, the number of moles of each ion is the same as the number of moles of lead sulfate. Therefore,

$$K_{sp} = [\text{Pb}^{+2}][\text{SO}_4^{-2}] = (1.25 \times 10^{-4})(1.25 \times 10^{-4})$$

$$= 1.56 \times 10^{-8}$$

31. ENTHALPY OF FORMATION

Enthalpy, H, is the potential energy that a substance possesses by virtue of its temperature, pressure, and phase.[21] The *enthalpy of formation* (*heat of formation*), ΔH_f, of a compound is the energy absorbed during the formation of 1 gmol of the compound from the elements.[22] The enthalpy of formation is assigned a value of zero for elements in their free states at 25°C and 1 atm. This is the so-called *standard state* for enthalpies of formation.

Table 22.4 contains enthalpies of formation for some common elements and compounds. The enthalpy of formation depends on the temperature and phase of the compound. A standard temperature of 25°C is used in most tables of enthalpies of formation.[23] Compounds are solid (*s*) unless indicated to be gaseous (*g*) or liquid (*l*). Some aqueous (*aq*) values are also encountered.

32. ENTHALPY OF REACTION

The *enthalpy of reaction* (*heat of reaction*), ΔH_r, is the energy absorbed during a chemical reaction under constant volume conditions. It is found by summing the enthalpies of formation of all products and subtracting the sum of enthalpies of formation of all reactants. This is essentially a restatement of the energy conservation principle and is known as *Hess' law of energy summation*.

$$\Delta H_r = \sum \Delta H_{f,\text{products}} - \sum \Delta H_{f,\text{reactants}} \qquad \textbf{22.26}$$

Reactions that give off energy (i.e., have negative enthalpies of reaction) are known as *exothermic reactions*. Many (but not all) exothermic reactions begin spontaneously. On the other hand, *endothermic reactions* absorb energy and require heat or electrical energy to begin.

Example 22.16

Using enthalpies of formation, calculate the heat of stoichiometric combustion (standardized to 25°C) of gaseous methane (CH_4) and oxygen.

Solution

The balanced chemical equation for the stoichiometric combustion of methane is

$$\text{CH}_4 + 2\text{O}_2 \longrightarrow 2\text{H}_2\text{O} + \text{CO}_2$$

---

[21] The older term *heat* is rarely encountered today.
[22] The symbol H is used to denote molar enthalpies. The symbol h is used for specific enthalpies (i.e., energy per kilogram or per pound).
[23] It is possible to correct the enthalpies of formation to account for other reaction temperatures.

Table 22.4 *Standard Enthalpies of Formation*
(kcal/mol at 25° C)

element/compound	ΔH_f
Al (s)	0.00
Al_2O_3	−399.09
C (graphite)	0.00
C (diamond)	0.45
C (g)	171.70
CO (g)	−26.42
CO_2 (g)	−94.05
CH_4 (g)	−17.90
C_2H_2 (g)	54.19
C_2H_4 (g)	12.50
C_2H_6 (g)	−20.24
CCl_4 (g)	−25.5
$CHCl_4$ (g)	−24
CH_2Cl_2 (g)	−21
CH_3Cl (g)	−19.6
CS_2 (g)	27.55
COS (g)	−32.80
$(CH_3)_2S$ (g)	−8.98
CH_3OH (g)	−48.08
C_2H_5OH (g)	−56.63
$(CH_3)_2O$ (g)	−44.3
C_3H_6 (g)	9.0
C_6H_{12} (g)	−29.98
C_6C_{10} (g)	−1.39
C_6H_6 (g)	19.82
Fe (s)	0.00
Fe (g)	99.5
Fe_2O_3 (s)	−196.8
Fe_3O_4 (s)	−267.8
H_2 (g)	0.00
H_2O (g)	−57.80
H_2O (l)	−68.32
H_2O_2 (g)	−31.83
H_2S (g)	−4.82
N_2 (g)	0.00
NO (g)	21.60
NO_2 (g)	8.09
NO_3 (g)	13
NH_3 (g)	−11.04
O_2 (g)	0.00
O_3 (g)	34.0
S (s)	0.00
SO_2 (g)	−70.96
SO_3 (g)	−94.45

(Multiply kcal/mol by 1800/MW to obtain Btu/lbm.)

Using enthalpies of formation from Table 22.4 in Eq. 22.13, the enthalpy of reaction per mole of methane is

$$\Delta H_r = 2\Delta H_{f,H_2O} + \Delta H_{f,CO_2} - \Delta H_{f,CH_4}$$
$$- 2\Delta H_{f,O_2}$$
$$= (2)\left(-57.80\ \frac{kcal}{mol}\right) + \left(-94.05\ \frac{kcal}{mol}\right)$$
$$- \left(-17.90\ \frac{kcal}{mol}\right) - (2)(0)$$
$$= -191.75\ \text{kcal/mol } CH_4 \quad \text{[exothermic]}$$

Notice that the enthalpy of formation of oxygen gas (its free-state configuration) is zero.

Using the footnote to Table 22.4, this value can be converted to Btu/lbm. The molecular weight of methane is

$$MW_{CH_4} = 12 + (4)(1) = 16$$

$$\text{higher heating value} = \frac{\left(191.75\ \frac{kcal}{mol}\right)(1800)}{16}$$
$$= 21{,}570\ \text{Btu/lbm}$$

33. CORROSION

Corrosion is an undesirable degradation of a material resulting from a chemical or physical reaction with the environment. Conditions within the crystalline structure can accentuate or retard corrosion. The main types of corrosion are listed in subsequent sections. Corrosion rates are reported in units of mils per year (mpy) and micrometers per year (μm/y).

34. UNIFORM ATTACK CORROSION

Uniform rusting of steel and oxidation of aluminum over entire exposed surfaces are examples of *uniform attack corrosion*. Uniform attack is usually prevented by the use of paint, plating, and other protective coatings.

35. INTERGRANULAR CORROSION

Some metals are particularly sensitive to *intergranular corrosion*, IGC—selective or localized attack at metal-grain boundaries. For example, the Cr_2O_3 oxide film on stainless steel contains numerous imperfections at grain boundaries, and these boundaries can be attacked and enlarged by chlorides.

Intergranular corrosion may occur after a metal has been heated, in which case it may be known as *weld decay*. In the case of type 304 austenitic stainless steels, heating to 930 to 1300°F (500 to 700°C) in a welding process causes chromium carbides to precipitate out, reducing the corrosion resistance.[24] Reheating to 1830 to 2010°F (1000 to 1100°C) followed by rapid cooling will redissolve the chromium carbides and restore corrosion resistance.

36. PITTING

Pitting is a localized perforation on the surface. It can occur even where there is little or no other visible damage. Chlorides and other halogens (e.g., HF and HCl) in the presence of water foster pitting in passive alloys, especially in stainless steels and aluminum alloys.

37. CONCENTRATION-CELL CORROSION

Concentration-cell corrosion (also known as *crevice corrosion* and *intergranular attack*, IGA) occurs when a metal is in contact with different electrolyte concentrations. It usually occurs in crevices, between two assembled parts, under riveted joints, or where there are scale and surface deposits that create stagnant areas in a corrosive medium.

38. EROSION CORROSION

Erosion corrosion is the deterioration of metals buffeted by the entrained solids in a corrosive medium.

39. SELECTIVE LEACHING

Selective leaching is the dealloying process in which one of the alloy ingredients is removed from the solid solution. This occurs because the lost ingredient has a lower corrosion resistance than the remaining ingredient. *Dezincification* is the classic case where zinc is selectively destroyed in brass. Other examples are the dealloying of nickel from copper-nickel alloys, iron from steel, and aluminum from copper-aluminum alloys.

40. HYDROGEN EMBRITTLEMENT

Hydrogen damage occurs when hydrogen gas diffuses through and decarburizes steel (i.e., reacts with carbon to form methane). *Hydrogen embrittlement* (also known as *caustic embrittlement*) is hydrogen damage from hydrogen produced by caustic corrosion.

41. GALVANIC ACTION

Galvanic action (*galvanic corrosion* or *two-metal corrosion*) results from the difference in oxidation potentials

---

[24] *Austenitic stainless steels* are the 300 series. They consist of chromium nickel alloys with up to 8% nickel. They are not hardenable by heat treatment, are nonmagnetic, and offer the greatest resistance to corrosion. *Martensitic stainless steels* are hardenable and magnetic. *Ferritic stainless steels* are magnetic and not hardenable.

of metallic ions. The greater the difference in oxidation potentials, the greater will be the galvanic corrosion. If two metals with different oxidation potentials are placed in an electrolytic medium (e.g., seawater), a galvanic cell will be created. The metal with the higher potential (i.e., the more "active" metal) will act as an anode and will corrode. The metal with the lower potential (the more "noble" metal), being the cathode, will be unchanged. In one extreme type of intergranular corrosion known as *exfoliation*, open endgrains separate into layers.

Metals are often classified according to their positions in the *galvanic series* listed in Table 22.5. As would be expected, the metals in this series are in approximately the same order as their half-cell potentials. However, alloys and proprietary metals are also included in the series.

Table 22.5 Galvanic Series in Seawater (top to bottom anodic (sacrificial, active) to cathodic (noble, passive))

magnesium
zinc
Alclad 3S
cadmium
2024 aluminum alloy
low-carbon steel
cast iron
stainless steels (active)
no. 410
no. 430
no. 404
no. 316
Hastelloy A
lead
lead-tin alloys
tin
nickel
brass (copper-zinc)
copper
bronze (copper-tin)
90/10 copper-nickel
70/30 copper-nickel
Inconel
silver solder
silver
stainless steels (passive)
Monel metal
Hastelloy C
titanium
graphite
gold

Precautionary measures can be taken to inhibit or eliminate galvanic action when use of dissimilar metals is unavoidable.

- Use dissimilar metals that are close neighbors in the galvanic series.

- Use *sacrificial anodes*. In marine saltwater applications, sacrificial zinc plates can be used.

- Use protective coatings, oxides, platings, or inert spacers to reduce or eliminate the access of corrosive environments to the metals.[25]

42. STRESS CORROSION

When subjected to sustained surface tensile stresses (including low residual stresses from manufacturing) in corrosive environments, certain metals exhibit catastrophic *stress corrosion* cracking, SCC. When the stresses are cyclic, this type of corrosion is called *corrosion fatigue*, which can lead to fatigue failures well below normal yield stresses.

Stress corrosion occurs because the more highly stressed grains (at the crack tip) are slightly more anodic than neighboring grains with lower stresses. Although intergranular cracking (at grain boundaries) is more common, corrosion cracking may be *intergranular* (between the grains), *transgranular* (through the grains), or a combination of the two, depending on the alloy. Cracks propagate, often with extensive branching, until failure occurs.

The precautionary measures that can be taken to inhibit or eliminate stress corrosion are as follows.

- Avoid using metals that are susceptible to stress corrosion. These include austenitic stainless steels without heat treatment in seawater; certain tempers of the aluminum alloys 2124, 2219, 7049, and 7075 in seawater; and copper alloys exposed to ammonia.

- Protect open-grain surfaces from the environment. For example, press-fitted parts in drilled holes can be assembled with wet zinc chromate paste. Also, weldable aluminum can be "buttered" with pure aluminum rod.

- Stress-relieve by annealing heat treatment after welding or cold working.

43. FRETTING CORROSION

Fretting corrosion occurs when two highly loaded members have a common surface at which rubbing and sliding take place. The phenomenon is a combination of wear and chemical corrosion. Metals that depend on a film of surface oxide for protection, such as aluminum and stainless steel, are especially susceptible.

Fretting corrosion can be reduced by the following methods.

- Lubricate the rubbing surfaces.

- Seal the surfaces.

- Reduce vibration and movement.

44. CAVITATION CORROSION

Cavitation is the formation and collapse of minute bubbles of vapor in liquids. It is caused by a combination of reduced pressure and increased velocity in the fluid. In effect, very small amounts of the fluid vaporize (i.e., boil) and almost immediately condense. The repeated collapse of the bubbles hammers and work-hardens the surface.

When the surface work-hardens, it becomes brittle. Small amounts of the surface flake away, and the surface becomes pitted. This is known as *cavitation corrosion*. Eventually, the entire piece may work-harden and become brittle, leading to structural failure.

---

[25]While cadmium, nickel, chromium, and zinc are often used as protective deposits on steel, porosities in the surfaces can act as small galvanic cells, resulting in invisible subsurface corrosion.

Water Resources

23 Organic Chemistry

1. INTRODUCTION TO ORGANIC CHEMISTRY

Organic chemistry deals with the formation and reaction of compounds of carbon, many of which are produced by living organisms. Organic compounds typically have one or more of the following characteristics.

- They are insoluble in water.[1]

- They are soluble in concentrated acids.

- They are relatively nonionizing.

- They are unstable at high temperatures.

The method of naming organic compounds was standardized in 1930 at the International Union Chemistry meeting in Belgium. Names conforming to the established guidelines are known as *IUC names* or *IUPAC names*.[2]

2. FUNCTIONAL GROUPS

Certain combinations of atoms occur repeatedly in organic compounds and remain intact during reactions. Such combinations are called *functional groups*. For example, the radical OH^- is known as a *hydroxyl group*. Table 23.1 contains the most important functional groups.

3. FAMILIES OF ORGANIC COMPOUNDS

For convenience, organic compounds are categorized into families. Compounds within each family have similar structures, being based on similar combinations of groups. For example, all alcohols have the structure [R]-OH, where [R] is any alkyl group and -OH is the hydroxyl group.

Table 23.1 Functional Groups

name	standard symbol	formula	number of single bonding sites
aldehyde		CHO	1
alkyl	[R]	C_nH_{2n+1}	1
alkoxy	[RO]	$C_nH_{2n+1}O$	1
amine (amino, $n = 2$)		NH_n	$3 - n\ [n = 0, 1, 2]$
aryl (benzene ring)	[Ar]	C_6H_5	1
carbinol		COH	3
carbonyl (keto)	[CO]	CO	2
carboxyl		COOH	1
ester		COO	1
ether		O	2
halogen (halide)	[X]	Cl, Br, I, or F	1
hydroxyl		OH	1
nitrile		CN	1
nitro		NO_2	1

Families of compounds can be further subdivided into subfamilies. For example, the hydrocarbons are classified into alkanes (single carbon-carbon bond), alkenes (double carbon-carbon bond), and alkynes (triple carbon-carbon bond).[3]

Table 23.2 contains the most common organic families.

4. SYMBOLIC REPRESENTATION

The nature and structure of organic groups and families cannot be explained fully without showing the types of bonds between the elements. Figure 23.1 illustrates the symbolic representation of some of the functional groups and families.

5. FORMATION OF ORGANIC COMPOUNDS

There are usually many ways of producing an organic compound. The types of reactions contained in this section deal only with the interactions between the organic families. The following processes are referred to.

---

[1]This is especially true for hydrocarbons. However, many organic compounds containing oxygen are water soluble. The sugar family is an example of water-soluble compounds.

[2]IUPAC stands for *International Union of Pure and Applied Chemistry*.

[3]Hydrocarbons with two double carbon-carbon bonds are known as *dienes*.

Table 23.2 *Families of Organic Compounds*

family	structure[a]	example
acids		
carboxylic acids	[R]-COOH	acetic acid ($(CH_3)COOH$)
fatty acids	[Ar]-COOH	benzoic acid (C_6H_5COOH)
alcohols		
aliphatic	[R]-OH	methanol (CH_3OH)
aromatic	[Ar]-[R]-OH	benzyl alcohol ($C_6H_5CH_2OH$)
aldehydes	[R]-CHO	formaldehyde (HCHO)
alkyl halides		
(haloalkane)	[R]-[X]	chloromethane (CH_3Cl)
amides	[R]-CO-NH$_n$	β-methylbutyramide ($C_4H_9CONH_2$)
amines	[R]$_{3-n}$-NH$_n$	methylamine (CH_3NH_2)
	[Ar]$_{3-n}$-NH$_n$	aniline ($C_6H_5NH_2$)
primary amines	$n = 2$	
secondary amines	$n = 1$	
tertiary amines	$n = 0$	
amino acids	CH-[R]-(NH$_2$)COOH	glycine ($CH_2(NH_2)COOH$)
anhydrides	[R]-CO-O-CO-[R']	acetic anhydride ($(CH_3CO)_2O$)
arene (aromatics)	ArH $= C_nH_{2n-6}$	benzene (C_6H_6)
aryl halides	[Ar]-[X]	fluorobenzene (C_6H_5F)
carbohydrates	$C_x(H_2O)_y$	dextrose ($C_6H_{12}O_6$)
sugars		
polysaccharides		
esters	[R]-COO-[R']	methyl acetate (CH_3COOCH_3)
ethers	[R]-O-[R]	diethyl ether ($C_2H_5OC_2H_5$)
	[Ar]-O-[R]	methyl phenyl ether ($CH_3OC_6H_5$)
	[Ar]-O-[Ar]	diphenyl ether ($C_6H_5OC_6H_5$)
glycols	$C_nH_{2n}(OH)_2$	ethylene glycol ($C_2H_4(OH)_2$)
hydrocarbons		
alkanes[b]	RH $= C_nH_{2n+2}$	octane (C_8H_{18})
saturated hydrocarbons		
cycloalkanes (cycloparaffins)		
	C_nH_{2n}	cyclohexane (C_6H_{12})
alkenes[c]	C_nH_{2n}	ethylene (C_2H_4)
unsaturated hydrocarbons		
cycloalkenes	C_nH_{2n-2}	cyclohexene (C_6H_{10})
alkynes	C_nH_{2n-2}	acetylene (C_2H_2)
unsaturated hydrocarbons		
ketones	[R]-[CO]-[R]	acetone ($(CH_3)_2CO$)
nitriles	[R]-CN	acetonitrile (CH_3CN)
phenols	[Ar]-OH	phenol (C_6H_5OH)

[a]See Table 23.1 for definitions of [R], [Ar], [X], and [CO].
[b]Alkanes are also known as the *paraffin series* and *methane series*.
[c]Alkenes are also known as the *olefin series*.

- *oxidation:* replacement of a hydrogen atom with a hydroxyl group

- *reduction:* replacement of a hydroxyl group with a hydrogen atom

- *hydrolysis:* addition of one or more water molecules

- *dehydration:* removal of one or more water molecules

Figure 23.1 *Representation of Functional Groups and Families*

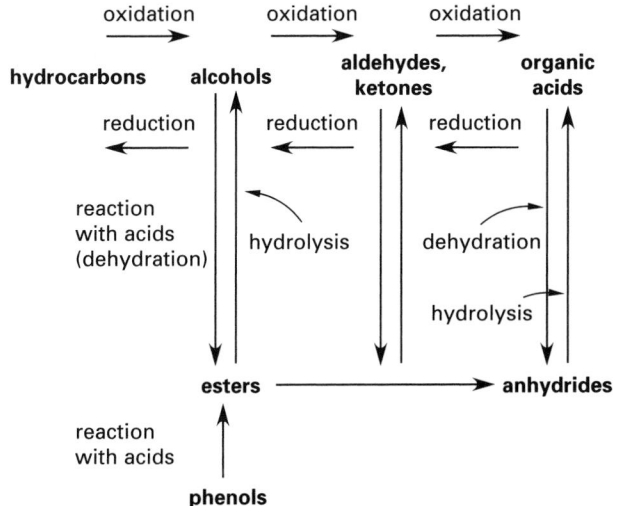

— CH $\parallel$ O or — C $\overset{H}{\underset{O}{<}}$	aldehyde group	R — C $\overset{H}{\underset{O}{<}}$ aldehyde
— NH$_2$	amino group	R — NH$_2$ amine
(benzene ring)	aryl group (benzene ring)	⬡—X aryl halide
— C — $\parallel$ O or — C $\overset{R}{\underset{O}{<}}$	carbonyl (keto) group	R — C $\overset{R}{\underset{O}{<}}$ ketone
— C — OH $\parallel$ O or — C $\overset{O-H}{\underset{O}{<}}$	carboxyl group	R — C $\overset{O-H}{\underset{O}{<}}$ organic acid
— C — OR $\parallel$ O or — C $\overset{O-R}{\underset{O}{<}}$	ester group	R — C $\overset{O-R}{\underset{O}{<}}$ ester
— OH	hydroxyl group	R — OH alcohol

Table 23.3 *Synthesis Routes for Various Classes of Organic Compounds*

acids
  oxidation of primary alcohols
  oxidation of ketones
  oxidation of aldehydes
  hydrolysis of esters
alcohols
  oxidation of hydrocarbons
  reduction of aldehydes
  reduction of organic acids
  hydrolysis of esters
  hydrolysis of alkyl halides
  hydrolysis of alkenes (aromatic hydrocarbons)
aldehydes
  oxidation of primary and tertiary alcohols
  oxidation of esters
  reduction of organic acids
amides
  replacement of hydroxyl group in an acid
    with an amino group
anhydrides
  dehydration of organic acids (withdrawal of
    one water molecule from two acid molecules)
carbohydrates
  oxidation of alcohols
esters
  reaction of acids with alcohols (*ester alcohols*)[a]
  reaction of acids with phenols (*ester phenols*)
  dehydration of alcohols
  dehydration of organic acids
ethers
  dehydration of alcohol
hydrocarbons
  alkanes: reduction of alcohols and organic acids
       hydrogenation of alkenes
  alkenes: dehydration of alcohols
       dehydrogenation of alkanes
ketones
  oxidation of secondary and tertiary alcohols
  reduction of organic acids
phenols
  hydrolysis of aryl halides

[a]The reaction of an organic acid with an alcohol is called *esterification*.

Figure 23.2 *Reactions Between Organic Compounds*

hydrocarbons —oxidation→ alcohols —oxidation→ aldehydes, ketones —oxidation→ organic acids

←reduction ←reduction ←reduction

reaction with acids (dehydration) | hydrolysis | dehydration | hydrolysis

esters ——→ anhydrides

reaction with acids

phenols

Combustion and Incineration

Nomenclature

A	area	ft^2	m^2
B	volumetric fraction	–	–
c	specific heat	Btu/lbm-°F	kJ/kg·°C
d	diameter	in	cm
D	diameter	ft	m
D	draft	in wg	kPa
g	acceleration of gravity	ft/sec^2	m/s^2
g_c	gravitational constant	$ft\text{-}lbm/lbf\text{-}sec^2$	–
G	gravimetric fraction	–	–
h	enthalpy	Btu/lbm	kJ/kg
h	head	ft	m
H	height	ft	m
HHV	higher heating value	Btu/lbm	kJ/kg
HV	heating value	Btu/lbm	kJ/kg
J	gravimetric air-fuel ratio	–	–
K	volumetric air-fuel ratio	–	–
L	length	ft	m
LHV	lower heating value	Btu/lbm	kJ/kg
m	mass	lbm	kg
M	fraction moisture	–	–
ON	octane number	–	–
p	pressure	lbf/ft^2	Pa
P	power	Btu/sec	kW
PN	performance number	–	–
q	heat loss	Btu/lbm	kJ/kg
Q	flow rate	ft^3/sec	m^3/s
R	ratio	lbm/lbm	kg/kg
R	specific gas constant	ft-lbf/lbm-°R	kJ/kg·K
T	temperature	°F	°C
v	velocity	ft/sec	m/s

Symbols

γ	specific weight	lbf/ft^3	–
η	efficiency	–	–
ω	humidity ratio	–	–

Subscripts

a/f	air/fuel
fg	vaporization
g	gas
i	initial
p	constant pressure
SE	stack effect

1. HYDROCARBONS

With the exception of sulfur and related compounds, most fuels are hydrocarbons. Hydrocarbons are further categorized into subfamilies such as *alkynes* (C_nH_{2n-2}, such as acetylene C_2H_2), *alkenes* (C_nH_{2n}, such as ethylene C_2H_4), and *alkanes* (C_nH_{2n+2}, such as octane C_8H_{18}). The alkynes and alkenes are referred to as *unsaturated hydrocarbons*, while the alkanes are referred to as *saturated hydrocarbons*. The alkanes are also known

Table 24.1 *Approximate Specific Heats (at constant pressure) of Gases*
(c_p *in Btu/lbm-°R; at 1 atm*)

gas	temperature (°R)							
	500	1000	1500	2000	2500	3000	4000	5000
air	0.240	0.249	0.264	0.277	0.286	0.294	0.302	–
carbon dioxide	0.196	0.251	0.282	0.302	0.314	0.322	0.332	0.339
carbon monoxide	0.248	0.257	0.274	0.288	0.298	0.304	0.312	0.316
hydrogen	3.39	3.47	3.52	3.63	3.77	3.91	4.14	4.30
nitrogen	0.248	0.255	0.270	0.284	0.294	0.301	0.310	0.315
oxygen	0.218	0.236	0.253	0.264	0.271	0.276	0.286	0.294
sulfur dioxide	0.15	0.16	0.18	0.19	0.20	0.21	0.23	–
water vapor	0.444	0.475	0.519	0.566	0.609	0.645	0.696	0.729

(Multiply Btu/lbm-°R by 4.187 to obtain kJ/kg·K.)

as the *paraffin series* and *methane series*. The alkenes are subdivided into the chain-structured *olefin series* and the ring-structured *naphthalene series*. *Aromatic hydrocarbons* (C_nH_{2n-6}, such as benzene C_6H_6) constitute another subfamily. Names for common hydrocarbon compounds are listed in App. 24.A.

2. CRACKING OF HYDROCARBONS

Cracking is the process of splitting hydrocarbon molecules into smaller molecules. For example, alkane molecules crack into a smaller member of the alkane subfamily and a member of the alkene subfamily. Cracking is used to obtain lighter hydrocarbons (such as those used in gasoline) from heavy hydrocarbons (e.g., crude oil).

Cracking can proceed under the influence of high temperatures (*thermal cracking*) or catalysts (*catalytic cracking* or "cat cracking"). Since (from Le Châtelier's principle) cracking at high pressure favors recombination, catalytic cracking is performed at pressures near atmospheric. Catalytic cracking also produces gasolines with better antiknock properties than does thermal cracking.

3. FUEL ANALYSIS

Fuels analyses are reported as either percentages by weight (for liquid and solid fuels) or percentages by volume (for gaseous fuels). Percentages by weight are known as *gravimetric analyses*, while percentages by volume are known as *volumetric analyses*. An *ultimate analysis* is a special type of gravimetric analysis in which the constituents are reported by atomic species rather than by compound. In an ultimate analysis, combined hydrogen from moisture in the fuel is added to hydrogen from the combustive compounds. (See Sec. 6.)

A *proximate analysis* (not "approximate") gives the gravimetric fraction of moisture, volatile matter, fixed carbon, and ash. Sulfur may be combined with the ash or may be specified separately.

Gas analyses are typically specified as volumetric fractions. For a gas in a mixture, its *volumetric fraction* (i.e., its volumetric percentage) is the same as its *mole fraction* and *partial pressure fraction*.

A volumetric fraction can be converted to a gravimetric fraction by multiplying by the molecular weight and then dividing by the sum of the products of all the volumetric fractions and molecular weights.

4. WEIGHTING OF THERMODYNAMIC PROPERTIES

Many gaseous fuels (and all gaseous combustion products) are mixtures of different compounds. Some thermodynamic properties of mixtures are gravimetrically weighted, while others are volumetrically weighted. Specific heat, specific gas constant, enthalpy, internal energy, and entropy are gravimetrically weighted. For gases, molecular weight, density, and all molar properties are volumetrically weighted.[1]

When a compound experiences a large temperature change, the thermodynamic properties should be evaluated at the average temperature. Table 24.1 can be used to find the specific heat of gases at various temperatures.

5. STANDARD CONDITIONS

Though "standard conditions" usually means 70°F (21°C) and 1 atm pressure, *standard temperature and pressure*, STP, for manufactured fuel gases is 60°F (16°C) and 1 atm pressure. Since this convention is not

---

[1] For gases, molar properties include molar specific heats, enthalpy per mole, and internal energy per mole.

well standardized, the actual temperature and pressure should be stated.[2]

Some combustion equipment (e.g., particulate collectors) operates within a narrow range of temperatures and pressures. These conditions are referred to as *normal temperature and pressure*, NTP.

6. MOISTURE

If an ultimate analysis of a solid or liquid fuel is given, all of the oxygen is assumed to be in the form of free water.[3] The amount of hydrogen combined as free water is assumed to be one-eighth of the oxygen weight.[4] All remaining hydrogen, known as the *available hydrogen*, is assumed to be combustible.

$$G_{H,combined} = \frac{G_O}{8} \qquad 24.1$$

$$G_{H,available} = G_{H,total} - \frac{G_O}{8} \qquad 24.2$$

For coal, the *"bed" moisture level* refers to the moisture level when the coal is mined. The terms *dry* and *as fired* are often used in commercial coal specifications. The "as fired" condition corresponds to a specific moisture content when placed in the furnace. The "as fired" heating value should be used, since the moisture actually decreases the combustion heat. The relationship between the two heating values is given by Eq. 24.3, where M is the moisture content from a proximate analysis.

$$HV_{as\ fired} = HV_{dry}(1 - M) \qquad 24.3$$

Moisture in fuel is undesirable because it increases fuel weight (transportation costs) and decreases available combustion heat.[5]

7. ASH AND MINERAL MATTER

Mineral matter is the noncombustible material in a fuel. *Ash* is the residue remaining after combustion. Ash may contain some combustible carbon. The two terms ("mineral matter" and "ash") are often used interchangeably when reporting fuel analyses.

Ash may also be categorized according to where it is recovered. Dry and wet *bottom ashes* are recovered from *ash pits*. (As little as 10% of the total ash content may be recovered in the ash pit.) *Flyash* is carried out of the boiler by the flue gas. Flyash can be deposited on walls and heat transfer surfaces. It will be discharged from the stack if not captured. *Economizer ash* and *air heater ash* are recovered from the devices the ash is named after.

The finely powdered ash that covers combustion grates protects them from high temperatures.[6] If the ash has a low (i.e., below 2200°F; 1200°C) fusion temperature (melting point), it may form *clinkers* in the furnace and/or *slag* in other high-temperature areas. In extreme cases, it can adhere to the surfaces. Ashes with high melting (fusion) temperatures (i.e., above 2600°F; 1430°C) are known as *refractory ashes*. The T_{250} *temperature* is used as an index of slagging tendencies of an ash. This is the temperature at which the slag becomes molten with a viscosity of 250 poise. Slagging will be experienced when the T_{250} temperature is exceeded.

The actual melting point depends on the ash composition. Ash is primarily a mixture of silica (SiO_2), alumina (Al_2O_3), and ferric oxide (Fe_2O_3).[7] The relative proportions of each will determine the melting point, with lower melting points resulting from high amounts of ferric oxide and calcium oxide. The melting points of pure alumina and pure silica are in the 2700 to 2800°F (1480 to 1540°C) range.

Coal ash is either of a bituminous type or lignite type. Bituminous-type ash (from midwestern and eastern coals) contains more ferric oxide than lime and magnesia. Lignite-type ash (from western coals) contains more lime and magnesia than ferric oxide.

8. SULFUR

Several forms of sulfur are present in coal and fuel oils. *Pyritic sulfur* (FeS_2) is the primary form. *Organic sulfur* is combined with hydrogen and carbon in other compounds. *Sulfate sulfur* is iron sulfate and gypsum ($CaSO_4 \cdot 2H_2O$). Sulfur in elemental, organic, and pyritic forms oxidizes to sulfur dioxide. *Sulfur trioxide* can be formed under certain conditions. Sulfur trioxide combines with water to form sulfuric acid and is a major source of boiler/stack corrosion and pollution.

$$SO_3 + H_2O \longrightarrow H_2SO_4$$

9. WOOD

Wood is not an industrial fuel, though it may be used in small quantities in developing countries. Most woods have heating values around 8300 Btu/lbm (19 MJ/kg), with specific values depending on the species and moisture content. Variations in wood properties are so great that generalized properties are meaningless.

---

[2]Both of these are different from the standard temperature and pressure used for scientific work. The standard temperature in that case is 32°F (0°C).

[3]This assumes that none of the oxygen is in the form of carbonates.

[4]The value of $1/8$ follows directly from the combustion reaction of hydrogen and oxygen.

[5]A moisture content up to 5% is reported to be beneficial in some mechanically fired boilers.

[6]Some boiler manufacturers rely on the thermal protection the ash provides. For example, coal burned in cyclone boilers should have a minimum ash content of 7% to cover and protect the cyclone barrel tubes. Boiler wear and ash carryover will increase with lower ash contents.

[7]Calcium oxide (CaO), magnesium oxide ("magnesia," MgO), titanium oxide ("titania," TiO_2), ferrous oxide (FeO), and alkalies (Na_2O and K_2O) may be present in smaller amounts.

10. WASTE FUELS

Waste fuels are increasingly being burned or incinerated in industrial boilers and furnaces. Such fuels include digester and landfill gases, waste process gases, flammable waste liquids, and volatile organic compounds (VOCs) such as benzene, toluene, xylene, ethanol, and methane. Other waste fuels include oil shale, tar sands, green wood, seed and rice hulls, biomass refuse, peat, tire shreddings, and shingle/roofing waste.

The term *refuse-derived fuels*, RDF, is used to describe fuel produced from municipal waste. After separation (removal of glass, plastics, metals, corrugated cardboard, etc.), the waste is merely pushed into the combustion chamber. If the waste is to be used elsewhere, it is compressed and baled.

The heating value of RDF depends on the moisture content and fraction of combustible material. For RDFs derived from typical municipal wastes, the heating value will range from 3000 to 6000 Btu/lbm (7 to 14 MJ/kg), though higher ranges (7500 to 8500 Btu/lbm (17.5 to 19.8 MJ/kg)) can be obtained by careful selection of ingredients. Pelletized RDF (containing some coal and a limestone binder) with heating values around 8000 Btu/lbm (18.6 MJ/kg) can be used as a supplemental fuel in coal-fired units.

Scrap tires are an attractive fuel source because of their high heating values—12,000 to 16,000 Btu/lbm (28 to 37 MJ/kg). To be compatible with existing coal-loading equipment, tires are chipped or shredded to 1-inch size (25 mm). Tires in this form are known as *tire-derived fuel*, TDF. Metal (from tire reinforcement) may or may not be present.

TDF has been shown capable of supplying up to 90% of a steam-generating plant's total Btu input without any deterioration in particulate emissions, pollutants, and stack opacity. In fact, compared with some low-quality coals (e.g., lignite), TDF is far superior: about 2.5 times the heating value and about 2.5 times less sulfur per Btu.

11. INCINERATION

Many toxic wastes are incinerated rather than "burned." Incineration and combustion are not the same. *Incineration* is the term used to describe a disposal process that uses combustion to render wastes ineffective (nonharmful, nontoxic, etc.). Wastes and combustible fuel are combined in a furnace, and the heat of combustion destroys the waste.[8] Wastes may themselves be combustible, though they may not be self-sustaining if the moisture content is too high.

Incinerated wastes are categorized into seven types. Type 0 is *trash* (highly combustible paper and wood, with 10% or less moisture); type 1 is *rubbish* (combustible waste with up to 25% moisture); type 2 is *refuse*

(a mixture of rubbish and garbage, with up to 50% moisture); type 3 is *garbage* (residential waste with up to 70% moisture); type 4 is animal solids and pathological wastes (85% moisture); type 5 is industrial process wastes in gaseous, liquid, and semiliquid form; and type 6 is industrial process wastes in solid and semisolid form requiring incineration in hearth, retort, or grate burning equipment.

12. COAL

Coal consists of volatile matter, fixed carbon, moisture, noncombustible mineral matter ("ash"), and sulfur. *Volatile matter* is driven off as a vapor when the coal is heated, and it is directly related to flame size. *Fixed carbon* is the combustible portion of the solid remaining after the volatile matter is driven off. Moisture is present in the coal as free water and (for some mineral compounds) as water of hydration. Sulfur, an undesirable component, contributes to heat content.

Coals are categorized into anthracitic, bituminous, and lignitic types. *Anthracite coal* is clean, dense, and hard. It is comparatively difficult to ignite but burns uniformly and smokelessly with a short flame. *Bituminous coal* varies in composition, but generally has a higher volatile content than anthracite, starts easily, and burns freely with a long flame. Smoke and soot are possible if bituminous coal is improperly fired. *Lignite coal* is a coal of woody structure, very high in moisture and with a low heating value. It normally ignites slowly due to its moisture, breaks apart when burning, and burns with little smoke or soot.

Coal is burned efficiently in a particular furnace only if it is uniform in size. Screen sizes are used to grade coal, but descriptive terms can also be used.[9] *Run-of-mine coal*, ROM, is coal as mined. *Lump coal* is in the 1 to 6 in (25 to 150 mm) range. *Nut coal* is smaller, followed by even smaller *pea coal screenings*, and *fines* (dust).

13. LOW-SULFUR COAL

Switching to low-sulfur coal is one way of meeting strict sulfur emission standards. Western and eastern low-sulfur coals have different properties.[10,11] Eastern low-sulfur coals are generally low-impact coals (that is, few changes need to be made to the power plant when switching to them). Western coals are generally high-impact coals. Properties of typical high- and low-sulfur fuels are shown in Table 24.2.

---

[8]Rotary kilns can accept waste in many forms. They are "workhorse" incinerators.

[9]The problem with descriptive terms is that one company's "pea coal" may be as small as $1/4$ in (6 mm), while another's may start at $1/2$ in (13 mm).

[10]In the United States, low-sulfur coals predominately come from the western United States ("western subbituminous"), although some come from the east ("eastern bituminous").

[11]Some parameters affected by coal type are coal preparation, firing rate, ash volume and handling, slagging, corrosion rates, dust collection and suppression, and fire and explosion prevention.

The lower sulfur content results in less boiler corrosion. However, of all the coal variables, the different ash characteristics are the most significant with regard to the steam generator components. The slagging and fouling tendencies are prime concerns.

Table 24.2 Typical Properties of High- and Low-Sulfur Coals[a]

property	high-sulfur	low-sulfur eastern	low-sulfur western
higher heating value,			
Btu/lbm	10,500	13,400	8000
(MJ/kg)	(24.4)	(31.2)	(18.6)
moisture content, %	11.7	6.9	30.4
ash content, %	11.8	4.5	6.4
sulfur content, %	3.2	0.7	0.5
slag melting			
temperature, °F	2400	2900	2900
(°C)	(1320)	(1590)	(1590)

(Multiply Btu/lbm by 2.326 to obtain kJ/kg.)
[a]All properties are "as received."

14. CLEAN COAL TECHNOLOGIES

A lot of effort has been put into developing technologies that will reduce acid rain, pollution and air toxics (NO_x and SO_2), and global warming. These technologies are loosely labeled as *clean coal technologies*, CCTs. Whether or not these technologies can be retrofitted into an existing plant or designed into a new plant depends on the economics of the process.

With *coal cleaning*, coal is ground to ultrafine sizes to remove sulfur and ash-bearing minerals.[12] However, finely ground coal creates problems in handling, storage, and dust production. The risk of fire and explosion increases. Different approaches to reducing the problems associated with transporting and storing finely ground coal include the use of dust suppression chemicals, pelletizing, transportation of coal in liquid slurry form, and pelletizing followed by reslurrying. Some of these technologies may not be suitable for retrofit into existing installations.

With *coal upgrading*, moisture is thermally removed from low-rank coal (e.g., lignite or subbituminous coal). With some technologies, sulfur and ash are also removed when the coal is upgraded.

Reduction in sulfur dioxide emissions is the goal of *SO₂ control* technologies. These technologies include conventional use of lime and limestone in *flue gas desulfurization* (FGD) systems, *furnace sorbent-injection* (FSI) and *duct sorbent-injection*. *Advanced scrubbing* is included in FGD technologies.

Redesigned burners and injectors and adjustment of the flame zone are typical types of *NOₓ control*. Use of secondary air, injection of ammonia or urea, and selective

catalytic reduction (SCR) are also effective in NO_x reduction.

Fluidized-bed combustion (FBC) reduces NO_x emissions by reducing combustion temperatures to around 1500°F (815°C). FBC is also effective in removing up to 90% of the SO_2. *Atmospheric FBC* operates at atmospheric pressure, but higher thermal efficiencies are achieved in *pressurized FBC* units operating at pressures up to 10 atm.

Integrated gasification/combined cycle (IGCC) processes are able to remove 99% of all sulfur while reducing NO_x to well below current emission standards. *Synthetic gas* (syngas) is derived from coal. Syngas has a lower heating value than natural gas, but it can be used to drive gas turbines in combined cycles or as a reactant in the production of other liquid fuels.

15. COKE

Coke, typically used in blast furnaces, is produced by heating coal in the absence of oxygen. The heavy hydrocarbons crack (i.e., the hydrogen is driven off), leaving only a carbonaceous residue containing ash and sulfur. Coke burns smokelessly. *Breeze* is coke smaller than $5/8$ in (16 mm). It is not suitable for use in blast furnaces, but steam boilers can be adapted to use it. *Char* is produced from coal in a 900°F (500°C) carbonization process. The volatile matter is removed, but there is little cracking. The process is used to solidify tars, bitumens, and some gases.

16. LIQUID FUELS

Liquid fuels are lighter hydrocarbon products refined from crude petroleum oil. They include liquefied petroleum gas (LPG), gasoline, kerosene, jet fuel, diesel fuels, and heating oils. Important characteristics of a liquid fuel are its composition, ignition temperature, flash point,[13] viscosity, and heating value.

17. FUEL OILS

In the United States, fuel oils are categorized into grades 1 through 6 according to their viscosities.[14] Viscosity is the major factor in determining firing rate and the need for preheating for pumping or atomizing prior to burning. Grades 1 and 2 can be easily pumped at ambient temperatures. In the United States, the heaviest fuel oil used is grade 6, also known as *Bunker C oil*.[15,16]

---

[12]80% or more of *micronized coal* is 44 microns or less in size.

[13]This is different from the *flash point* that is the temperature at which fuel oils generate enough vapor to sustain ignition in the presence of spark or flame.
[14]Grade 3 became obsolete in 1948. Grade 5 is also subdivided into light and heavy categories.
[15]120°F (48°C) is the optimum temperature for pumping no. 6 fuel oil. At that temperature, no. 6 oil has a viscosity of approximately 3000 SSU. Further heating is necessary to lower the viscosity to 150 to 350 SSU for atomizing.
[16]To avoid *coking* of oil, heating coils in contact with oil should not be hotter than 240°F (116°C).

Table 24.3 Typical Properties of Common Commercial Fuels

	butane	no. 1 diesel	no. 2 diesel	ethanol	gasoline	JP-4	methanol	propane
chemical formula	C_4H_{10}	–	–	C_2H_5OH	–	–	CH_3OH	C_3H_8
molecular weight	58.12	≈ 170	≈ 184	46.07	≈ 126			44.09
heating value								
higher Btu/lbm	21,240	19,240	19,110	12,800	20,260			21,646
lower Btu/lbm	19,620	18,250	18,000	11,500	18,900	18,400	9078	19,916
lower Btu/gal	102,400	133,332	138,110	76,152	116,485	123,400	60,050	81,855
latent heat of								
vaporization Btu/lbm		115	105	361	142			147
specific gravity[a]	2.01	0.876	0.920	0.794	0.68–0.74	0.8017	0.793	1.55

(Multiply Btu/lbm by 2.326 to obtain kJ/kg.)
(Multiply Btu/gal by 0.2786 to obtain MJ/m^3.)
[a] Specific gravities of propane and butane are with respect to air.

Fuel oils are also classified according to their viscosities as *distillate oils* (lighter) and *residual fuel oils* (heavier).

Like coal, fuel oils contain sulfur and ash that may cause pollution, slagging on the hot end of the boiler, and corrosion in the cold end. Table 24.4 lists typical properties of fuel oils.

Table 24.4 Typical Properties of Fuel Oils[a]

grade	specific gravity	heating value MBtu/gal[b]	heating value GJ/m^3
1	0.805	134	37.3
2	0.850	139	38.6
4	0.903	145	40.4
5	0.933	148	41.2
6	0.965	151	41.9

(Multiply MBtu/gal by 0.2786 to obtain GJ/m^3.)
[a] Actual values will vary depending on composition.
[b] One MBtu equals one thousand Btus.

18. GASOLINE

Gasoline is not a pure compound. It is a mixture of various hydrocarbons blended to give a desired flammability, volatility, heating value, and octane rating. There are numerous blends that can be used to produce gasoline.

Gasoline's heating value depends only slightly on composition. Within a variation of $1\frac{1}{2}\%$, the heating value can be taken as 20,200 Btu/lbm (47.0 MJ/kg) for regular gasoline and as 20,300 Btu/lbm (47.2 MJ/kg) for high-octane aviation fuel.

Since gasoline is a mixture of hydrocarbons, different fractions will evaporate at different temperatures. The *volatility* is the percentage of the fuel that has evaporated by a given temperature. Typical volatility specifications call for 10% at 167°F (75°C), 50% at 221°F (105°C), and 90% at 275°F (135°C). Low volatility causes difficulty in starting and poor engine performance at low temperatures.

The *octane number*, ON, is a measure of knock resistance. It is based on comparison, performed in a standardized one-cylinder engine, with the burning of isooctane and *n*-heptane. *n*-heptane, C_7H_{16}, is rated zero and produces violent knocking. Isooctane, C_8H_{18}, is rated 100 and produces relatively knock-free operation. The percentage blend by volume of these fuels that matches the performance of the gasoline is the octane rating. The *research octane number* (RON) is a measure of the fuel's antiknock characteristics while idling; the *motor octane number* (MON) applies to high-speed, high-acceleration operations. The octane rating reported for commercial gasoline is an average of the two.

The *performance number* (PN) of gasoline containing antiknock compounds (e.g., tetraethyl lead, TEL) is related to the octane number.

$$ON = 100 + \frac{PN - 100}{3} \qquad 24.4$$

19. OXYGENATED GASOLINE

In parts of the United States, gasoline is "oxygenated" during the cold winter months. The addition of *oxygenates* raises the combustion temperature, reducing carbon monoxide and unburned hydrocarbons.[17] Common oxygenates used in *reformulated gasoline* (RFG) include methyl tertiary-butyl ether (MTBE) and ethanol. Methanol, ethyl tertiary-butyl ether (ETBE) tertiary-amyl methyl ether (TAME) and tertiary-amyl ethyl ether (TAEE) may also be used. Oxygenates are added to bring the minimum oxygen level to 2 to 3% by weight.[18]

---

[17] Oxygenation may not be successful in reducing carbon dioxide. Since the heating value of the oxygenates is lower, fuel consumption of oxygenated fuels is higher. On a per-gallon (per-liter) basis, oxygenation reduces carbon dioxide. On a per-mile (per-kilometer) basis, however, oxygenation appears to increase carbon dioxide.
[18] Other specifications on gasoline during the winter months intended to reduce pollution may include maximum percentages of benzene and total aromatics, and limits on Reid vapor pressure.

Table 24.5 *Typical Properties of Common Oxygenates*

	MTBE[c]	TAME	ETBE	TAEE
specific gravity	0.744	0.740	0.770	0.791
octane	110	112	105	100
heating value (MBtu/gal)[a]	93.6			
Reid vapor pressure (psig)[b]	8	4	3	2
percent oxygen by weight	18.2	15.7	15.7	13.8
volumetric percent needed to achieve gasoline				
2.7% oxygen by weight	15.1	17.2	17.2	19.4
2.0% oxygen by weight	11.0	12.4	12.7	13.0

(Multiply MBtu/gal by 0.2786 to obtain MJ/m^3.)
[a]One MBtu equals one thousand Btus.
[b]The Reid vapor pressure is the vapor pressure when heated to $100°F$ ($38°C$).
[c]MTBE is water soluble and does not degrade. As a suspected carcinogen, it is proving to be a serious threat to water supplies.

20. DIESEL FUEL

Properties and specifications for various grades of diesel fuel oil are similar to specifications for fuel oils. Grade 1-D ("D" for diesel) is a light distillate oil for high-speed engines in service requiring frequent speed and load changes. Grade 2-D is a distillate of lower volatility for engines in industrial and heavy mobile service. Grade 4-D is for use in medium speed engines under sustained loads.

Diesel oils are specified by a *cetane number*, which is a measure of the ignition quality (ignition delay) of a fuel. Like the octane number for gasoline, the cetane number is determined by comparison with standard fuels. Cetane, $C_{16}H_{34}$, has a cetane number of 100. *n*-methylnaphthalene, $C_{11}H_{10}$, has a cetane number of zero. A cetane number of approximately 30 is required for satisfactory operation of low-speed diesel engines. High-speed engines, such as those used in cars, require a cetane number of 45 or more. The cetane number can be increased by use of such additives as amyl nitrate, ethyl nitrate, and ether.

A diesel fuel's *pour point* number refers to its viscosity. A fuel with a pour point of $10°F$ ($-12°C$) will flow freely above that temperature. A fuel with a high pour point will thicken in cold temperatures.

The *cloud point* refers to the temperature at which wax crystals cloud the fuel at lower temperatures. The cloud point should be $20°F$ ($-7°C$) or higher. Below that temperature, the engine will not run well.

21. ALCOHOL

Both methanol and ethanol can be used in internal combustion engines. *Methanol* (*methyl alcohol*) is produced from natural gas and coal, although it can also be produced from wood and organic debris. *Ethanol* (*ethyl alcohol, grain alcohol*) is distilled from grain, sugar cane, potatoes, and other agricultural products containing various amounts of sugars, starches, and cellulose.

Although methanol generally works as well as ethanol, only ethanol can be produced in large quantities from inexpensive agricultural products and by-products.

Alcohol is water soluble. The concentration of alcohol is measured by its *proof*, where 200 proof is pure alcohol. (180 proof is 90% alcohol and 10% water.)

Gasohol is a mixture of approximately 90% gasoline and 10% alcohol (generally ethanol).[19] Alcohol's heating value is less than gasoline's, so fuel consumption (per distance traveled) is higher with gasohol. Also, since alcohol absorbs moisture more readily than gasoline, corrosion of fuel tanks becomes problematic. In some engines, significantly higher percentages of alcohol may require such modifications as including larger carburetor jets, timing advances, heaters for preheating fuel in cold weather, tank lining to prevent rusting, and alcohol-resistant gaskets.

Mixtures of gasoline and alcohol can be designated by the first letter and the fraction of the alcohol. E10 is a mixture of 10% ethanol in gasoline. M85 is a blend of 85% methanol and 15% gasoline.

Alcohol is a poor substitute for diesel fuel because alcohol's cetane number is low—from -20 to $+8$. Straight injection of alcohol results in poor performance and heavy knocking.

22. GASEOUS FUELS

Various gaseous fuels are used as energy sources, but most applications are limited to natural gas and *liquefied petroleum gases*, LPGs (i.e., propane, butane, and mixtures of the two).[20,21] Natural gas is a mixture of methane (55 to 95%), higher hydrocarbons

---

[19]In fact, oxygenated gasoline may use more than 10% alcohol.
[20]A number of *manufactured gases* are of practical (and historical) interest in specific industries, including *coke-oven gas, blast-furnace gas, water gas, producer gas*, and *town gas*. However, these gases are not now in widespread use.
[21]At atmospheric pressure, propane boils at $-44°F$ ($-42°C$), while butane boils at $31°F$ ($-0.5°C$).

Water Resources

(primarily ethane), and other noncombustible gases. Typical heating values for natural gas range from 950 to 1100 Btu/ft^3 (35 to 41 MJ/m^3).

The production of *synthetic gas (syngas)* through coal gasification may be applicable to large power generating plants. The cost of gasification, though justifiable to reduce sulfur and other pollutants, is too high for syngas to become a widespread substitute for natural gas.

23. IGNITION TEMPERATURE

The *ignition temperature (autoignition temperature)* is the minimum temperature at which combustion can be sustained. It is the temperature at which more heat is generated by the combustion reaction than is lost to the surroundings, after which combustion becomes self-sustaining. For coal, the minimum ignition temperature varies from around 800°F (425°C) for bituminous varieties to 900 to 1100°F (480 to 590°C) for anthracite. For sulfur and charcoal, the ignition temperatures are approximately 470°F (240°C) and 650°F (340°C), respectively.

For gaseous fuels, the ignition temperature depends on the air/fuel ratio, temperature, pressure, and length of time the source of heat is applied. Ignition can be instantaneous or with a lag, depending on the temperature. Generalizations can be made for any gas, but the generalized temperatures will be meaningless without specifying all of these factors.

24. ATMOSPHERIC AIR

It is important to make a distinction between "air" and "oxygen." Atmospheric air is a mixture of oxygen, nitrogen, and small amounts of carbon dioxide, water vapor, argon, and other inert ("rare") gases. For the purpose of combustion calculations, all constituents except oxygen are grouped with nitrogen. It is necessary to supply 4.32 (i.e., 1/0.2315) masses of air to obtain one mass of oxygen. Similarly, it is necessary to supply 4.773 volumes of air to obtain one volume of oxygen. The average molecular weight of air is 28.97. The specific gas constant is 53.35 ft-lbf/lbm-°R (287.03 J/kg·K).

Table 24.6 *Composition of Dry Air* [a]

component	percent by weight	percent by volume
oxygen	23.15	20.95
nitrogen/inerts	76.85	79.05
ratio of nitrogen to oxygen	3.320	3.773[b]
ratio of air to oxygen	4.320	4.773

[a] Inert gases and CO$_2$ included as N$_2$.
[b] The value is also reported by various sources as 3.76, 3.78, and 3.784.

25. COMBUSTION REACTIONS

A limited number of elements appear in combustion reactions. Carbon, hydrogen, sulfur, hydrocarbons, and oxygen are the reactants. Carbon dioxide and water vapor are the main products, with carbon monoxide, sulfur dioxide, and sulfur trioxide occurring in lesser amounts. Nitrogen and excess oxygen emerge essentially unchanged from the stack.

Combustion reactions occur according to the normal chemical reaction principles. Balancing combustion reactions is usually easiest if carbon is balanced first, followed by hydrogen and then oxygen. When a gaseous fuel has several combustible gases, the volumetric fuel composition can be used as coefficients in the chemical equation.

Table 24.7 lists ideal combustion reactions. These reactions do not include any nitrogen or water vapor that are present in the combustion air.

Example 24.1

A gaseous fuel is 20% hydrogen and 80% methane by volume. What volume of oxygen is required to burn 120 volumes of fuel at the same conditions?

Solution

Write the unbalanced combustion reaction.

$$H_2 + CH_4 + O_2 \longrightarrow CO_2 + H_2O$$

Use the volumetric analysis as coefficients of the fuel.

$$0.2H_2 + 0.8CH_4 + O_2 \longrightarrow CO_2 + H_2O$$

Balance the carbons.

$$0.2H_2 + 0.8CH_4 + O_2 \longrightarrow 0.8CO_2 + H_2O$$

Balance the hydrogens.

$$0.2H_2 + 0.8CH_4 + O_2 \longrightarrow 0.8CO_2 + 1.8H_2O$$

Balance the oxygens.

$$0.2H_2 + 0.8CH_4 + 1.7O_2 \longrightarrow 0.8CO_2 + 1.8H_2O$$

For gaseous components, the coefficients correspond to the volumes. Since one (0.2 + 0.8) volume of fuel requires 1.7 volumes of oxygen, the required oxygen is

$$(1.7)(120 \text{ volumes of fuel}) = 204 \text{ volumes of oxygen}$$

Table 24.7 *Ideal Combustion Reactions*

fuel	formula	reaction equation (excluding nitrogen)
carbon (to CO)	C	$2C + O_2 \longrightarrow 2CO$
carbon (to CO_2)	C	$C + O_2 \longrightarrow CO_2$
sulfur (to SO_2)	S	$S + O_2 \longrightarrow SO_2$
sulfur (to SO_3)	S	$2S + 3O_2 \longrightarrow 2SO_3$
carbon monoxide	CO	$2CO + O_2 \longrightarrow 2CO_2$
methane	CH_4	$CH_4 + 2O_2$ $\longrightarrow CO_2 + 2H_2O$
acetylene	C_2H_2	$2C_2H_2 + 5O_2$ $\longrightarrow 4CO_2 + 2H_2O$
ethylene	C_2H_4	$C_2H_4 + 3O_2$ $\longrightarrow 2CO_2 + 2H_2O$
ethane	C_2H_6	$2C_2H_6 + 7O_2$ $\longrightarrow 4CO_2 + 6H_2O$
hydrogen	H_2	$2H_2 + O_2 \longrightarrow 2H_2O$
hydrogen sulfide	H_2S	$2H_2S + 3O_2$ $\longrightarrow 2H_2O + 2SO_2$
propane	C_3H_8	$C_3H_8 + 5O_2$ $\longrightarrow 3CO_2 + 4H_2O$
n-butane	C_4H_{10}	$2C_4H_{10} + 13O_2$ $\longrightarrow 8CO_2 + 10H_2O$
octane	C_8H_{18}	$2C_8H_{18} + 25O_2$ $\longrightarrow 16CO_2 + 18H_2O$
olefin series	C_nH_{2n}	$2C_nH_{2n} + 3nO_2$ $\longrightarrow 2nCO_2 + 2nH_2O$
paraffin series	C_nH_{2n+2}	$2C_nH_{2n+2} + (3n+1)O_2$ $\longrightarrow 2nCO_2$ $+(2n+2)H_2O$

(Multiply oxygen volumes by 3.773 to get nitrogen volumes.)

26. STOICHIOMETRIC REACTIONS

Stoichiometric quantities (*ideal quantities*) are the exact quantities of reactants that are needed to complete a combustion reaction with no reactants left over. Table 24.7 contains some of the more common chemical reactions. Stoichiometric volumes and masses can always be determined from the balanced chemical reaction equation. Table 24.8 can be used to quickly determine stoichiometric amounts for some fuels.

27. STOICHIOMETRIC AIR

Stoichiometric air (*ideal air*) is the air necessary to provide the exact amount of oxygen for complete combustion of a fuel. Stoichiometric air includes atmospheric nitrogen. For each volume of oxygen, 3.773 volumes of nitrogen pass unchanged through the reaction.[22]

---

[22]The only major change in the nitrogen gas is its increase in temperature. Dissociation of nitrogen and formation of nitrogen compounds can occur but are essentially insignificant.

Stoichiometric air can be stated in units of mass (pounds or kilograms of air) for solid and liquid fuels, and in units of volume (cubic feet or cubic meters of air) for gaseous fuels. When stated in terms of mass, the stoichiometric ratio of air to fuel masses is known as the ideal *air/fuel ratio*, $R_{a/f}$.

$$R_{a/f,\text{ideal}} = \frac{m_{\text{air,ideal}}}{m_{\text{fuel}}} \qquad 24.5$$

The ideal air/fuel ratio can always be determined from the combustion reaction equation. It can also be determined by adding the oxygen and nitrogen amounts from Table 24.8.

For fuels whose ultimate analysis is known, the approximate stoichiometric air (oxygen and nitrogen) requirement in pounds of air per pound of fuel (kilograms of air per kilogram of fuel) can be quickly calculated from Eq. 24.6.[23] All oxygen in the fuel is assumed to be free moisture. All of the reported oxygen is assumed to be locked up in the form of water. Any free oxygen (i.e., oxygen dissolved in liquid fuels) is subtracted from the oxygen requirements.

$$R_{a/f,\text{ideal}} = (34.5)\left(\frac{G_C}{3} + G_H - \frac{G_O}{8} + \frac{G_S}{8}\right)$$
$$\text{[solid fuels]} \qquad 24.6$$

For fuels consisting of a mixture of gases, Eq. 24.7 and the constants J_i from Table 24.9 can be used to quickly determine the stoichiometric air requirements.

$$R_{a/f,\text{ideal}} = \sum J_i G_i \quad \text{[gaseous fuels]} \qquad 24.7$$

For fuels consisting of a mixture of gases, the air/fuel ratio can also be expressed in volumes of air per volume of fuel.

$$\text{volumetric air/fuel ratio} = \sum K_i B_i \quad \text{[gaseous fuels]}$$
$$24.8$$

---

[23]This is a "compromise" equation. Variations in the atomic weights will affect the coefficients slightly. The coefficient 34.5 is reported as 34.43 in some older books. 34.5 is the exact value needed for carbon and hydrogen, which constitute the bulk of the fuel. 34.43 is the correct value for sulfur, but the error is small and is disregarded in this equation.

Table 24.8 Consolidated Combustion Data^{a,b,c,d}

fuel	for 1 mole of fuel — air O$_2$	N$_2$	other products CO$_2$	H$_2$O	SO$_2$	for 1 ft^3 of fuel^e — air O$_2$	N$_2$	other products CO$_2$	H$_2$O	for 1 lbm of fuel — air O$_2$	N$_2$	other products CO$_2$	H$_2$O	SO$_2$	units of fuel
C carbon	1.0	3.773	1.0							0.0833	0.3143	0.0833			moles
	379.5	1432	379.5							31.63	119.3	31.63			ft^3
	32.0	106	44.0							2.667	8.883	3.667			lbm
H$_2$ hydrogen	0.5	1.887		1.0		0.001317	0.004969		0.002635	0.248	0.9357		0.496		moles
	189.8	716.1		379.5		0.5	1.887		1.0	94.12	355.1		188.25		ft^3
	16.0	53.0		18.0		0.04216	0.1397		0.04747	7.936	26.29		8.936		lbm
S sulfur	1.0	3.773			1.0					0.03119	0.1177			0.03119	moles
	379.5	1432			379.5					11.84	44.67			11.84	ft^3
	32.0	106.0			64.06					0.998	3.306			1.998	lbm
CO carbon monoxide	0.5	1.887	1.0			0.001317	0.004969	0.002635		0.01785	0.06735	0.03570			moles
	189.8	716.1	379.5			0.5	1.887	1.0		6.774	25.56	13.55			ft^3
	16.0	53.0	44.01			0.04216	0.1397	0.1160		0.5712	1.892	1.572			lbm
CH$_4$ methane	2.0	7.546	1.0	2.0		0.00527	0.01988	0.002635	0.00527	0.1247	0.4705	0.06233	0.1247		moles
	759	2864	379.5	758		2.0	7.546	1.0	2.0	47.31	178.5	23.66	47.31		ft^3
	64.0	212.0	44.01	36.03		0.1686	0.5586	0.1160	0.0949	3.989	13.21	2.743	2.246		lbm
C$_2$H$_2$ acetylene	2.5	9.433	2.0	1.0		0.006588	0.02486	0.00527	0.002635	0.09601	0.3622	0.07681	0.03841		moles
	948.8	3580	758	379.5		2.5	9.443	2.0	1.0	36.44	137.5	29.15	14.57		ft^3
	80.0	265.0	88.02	18.02		0.2108	0.6983	0.2319	0.04747	3.072	10.18	3.380	0.6919		lbm
C$_2$H$_4$ ethylene	3.0	11.32	2.0	2.0		0.007905	0.02983	0.00527	0.00527	0.1069	0.4033	0.07129	0.07129		moles
	1139	4297	758	758		3.0	11.32	2.0	2.0	40.58	153.1	27.05	27.05		ft^3
	96.0	318.0	88.02	36.03		0.2530	0.8380	0.2319	0.0949	3.422	11.34	3.137	1.284		lbm
C$_2$H$_6$ ethane	3.5	13.21	2.0	3.0		0.009223	0.03480	0.00527	0.007905	0.1164	0.4392	0.06651	0.09977		moles
	1328	5010	758	1139		3.5	13.21	2.0	3.0	44.17	166.7	25.24	37.86		ft^3
	112.0	371.0	88.02	54.05		0.2951	0.9776	0.2319	0.1424	3.724	12.34	2.927	1.797		lbm

(Multiply lbm/ft^3 by 0.06243 to obtain kg/m^3.)

^{a} Rounding of molecular weights and air composition may introduce slight inconsistencies in the table values. This table is based on atomic weights with at least four significant digits, a ratio of 3.773 volumes of nitrogen per volume of oxygen, and 379.5 ft^3 per mole at 1 atm and 60°F.

^{b} Volumes per unit mass are at 1 atm and 60°F (16°C). To obtain volumes at other temperatures, multiply by (T°F + 460)/520 or (T°C + 273)/289.

^{c} The volume of water applies only when the combustion products are at such high temperatures that all of the water is in vapor form.

^{d} This table can be used to directly determine some SI ratios. For kg/kg ratios, the values are the same as lbm/lbm. For mol/mol, use mole/mole. For mixed units (e.g., ft^3/lbm), conversions are required.

^{e} Sulfur is not used in gaseous form.

Table 24.9 *Approximate Air/Fuel Ratio Coefficients for Components of Natural Gas*[a]

fuel component	J (gravimetric)	K (volumetric)
acetylene, C_2H_2	13.25	11.945
butane, C_4H_{10}	15.43	31.06
carbon monoxide, CO	2.463	2.389
ethane, C_2H_6	16.06	16.723
ethylene, C_2H_4	14.76	14.33
hydrogen, H_2	34.23	2.389
hydrogen sulfide, H_2S	6.074	7.167
methane, CH_4	17.20	9.556
oxygen, O_2	−4.320	−4.773
propane, C_3H_8	15.65	23.89

[a]Rounding of molecular weights and air composition may introduce slight inconsistencies in the table values. This table is based on atomic weights with at least four significant digits and a ratio of 3.773 volumes of nitrogen per volume of oxygen.

Example 24.2

Use Table 24.8 to determine the theoretical volume of 90°F (32°C) air required to burn 1 volume of 60°F (16°C) carbon monoxide to carbon dioxide.

Solution

From Table 24.8, 0.5 volumes of oxygen are required to burn 1 volume of carbon monoxide to carbon dioxide. 1.887 volumes of nitrogen accompany the oxygen. The total amount of air at the temperature of the fuel is 0.5 + 1.887 = 2.387 volumes.

This volume will expand at the higher temperature. The volume at the higher temperature is

$$V_2 = \frac{T_2 V_1}{T_1}$$

$$= \frac{(90°F + 460)(2.387 \text{ volumes})}{60°F + 460} = 2.53 \text{ volumes}$$

Example 24.3

How much air is required for the ideal combustion of (a) coal with an ultimate analysis of 93.5% carbon, 2.6% hydrogen, 2.3% oxygen, 0.9% nitrogen, and 0.7% sulfur, (b) fuel oil with a gravimetric analysis of 84% carbon, 15.3% hydrogen, 0.4% nitrogen, and 0.3% sulfur, and (c) natural gas with a volumetric analysis of 86.92% methane, 7.95% ethane, 2.81% nitrogen, 2.16% propane, and 0.16% butane?

Solution

(a) Use Eq. 24.6.

$$R_{a/f,\text{ideal}} = (34.5)\left(\frac{G_C}{3} + G_H - \frac{G_O}{8} + \frac{G_S}{8}\right)$$

$$= (34.5)\left(\frac{0.935}{3} + 0.026 - \frac{0.023}{8} + \frac{0.007}{8}\right)$$

$$= 11.58 \text{ lbm/lbm (kg/kg)}$$

(b) Use Eq. 24.6.

$$R_{a/f,\text{ideal}} = (34.5)\left(\frac{G_C}{3} + G_H + \frac{G_S}{8}\right)$$

$$= (34.5)\left(\frac{0.84}{3} + 0.153 + \frac{0.003}{8}\right)$$

$$= 14.95 \text{ lbm/lbm (kg/kg)}$$

(c) Use Eq. 24.8 and the coefficients from Table 24.9.

$$\frac{\text{volumetric}}{\text{air/fuel ratio}} = \sum K_i B_i$$

$$= (0.8692)(9.556) + (0.0795)(16.723)$$

$$+ (0.0216)(23.89) + (0.0016)(31.06)$$

$$= 10.20 \text{ ft}^3/\text{ft}^3 \quad (\text{m}^3/\text{m}^3)$$

28. INCOMPLETE COMBUSTION

Incomplete combustion occurs when there is insufficient oxygen to burn all of the hydrogen, carbon, and sulfur in the fuel. Without enough available oxygen, carbon burns to carbon monoxide.[24] Carbon monoxide in the flue gas indicates incomplete and inefficient combustion. Incomplete combustion is caused by cold furnaces, low combustion temperatures, poor air supply, smothering from improperly vented stacks, and insufficient mixing of air and fuel.

29. FLUE GAS ANALYSIS

Combustion products that pass through a furnace's exhaust system are known as *flue gases* (*stack gases*). Flue gases are almost all nitrogen.[25] (Nitrogen oxides are not present in large enough amounts to be included separately in combustion reactions.)

The actual composition of flue gases can be obtained in a number of ways, including by electronic detectors, less expensive "length-of-stain" detectors, and direct sampling with an Orsat apparatus.

The *Orsat apparatus* determines the volumetric percentages of CO_2, CO, O_2, and N_2 in a flue gas. The sampled flue gas passes through a series of chemical compounds. The first compound absorbs only CO_2, the next only O_2, and the third only CO. The unabsorbed gas is assumed to be N_2 and is found by subtracting the volumetric percentages of all other components from 100%. An Orsat analysis is a dry analysis; the percentage of water vapor is not usually determined. A wet volumetric analysis (needed to compute the dew-point temperature) can be derived if the volume of water vapor is added to the Orsat volumes.

---

[24]Toxic alcohols, ketones, and aldehydes may also be formed during incomplete combustion.
[25]This assumption is helpful in making quick determinations of the thermodynamic properties of flue gases.

Modern electronic detectors can determine free oxygen (and other gases) independently of the other gases. Because the relationship between oxygen and excess air is relatively insensitive to fuel composition, oxygen measurements are replacing standard carbon dioxide measurements in determining combustion efficiency.

30. ACTUAL AND EXCESS AIR

Complete combustion occurs when all of the fuel is burned. Usually, *excess air* is required to achieve complete combustion. Excess air is expressed as a percentage of the theoretical air requirements. Different fuel types burn more efficiently with different amounts of excess air. Coal-fired boilers need approximately 30 to 35% excess air, oil-based units need about 15%, and natural gas burners need about 10%.

The actual air/fuel ratio for dry, solid fuels with no unburned carbon can be estimated from the flue gas analysis and the fraction of carbon in the fuel.

$$R_{a/f,\text{actual}} = \frac{m_{\text{air,actual}}}{m_{\text{fuel}}} = \frac{3.04B_{N_2}\left(G_C + \frac{G_S}{1.833}\right)}{B_{CO_2} + B_{CO}} \quad 24.9$$

Too much free oxygen or too little carbon dioxide in the flue gas is indicative of excess air. The relationship between excess air and oxygen in the flue gases is not highly affected by fuel composition.[26] The relationship between excess air and the volumetric fraction of oxygen in the flue gas is given in Table 24.10.

Reducing the air/fuel ratio will have several results. (a) The furnace temperature will increase due to a reduction in cooling air. (b) The flue gas will decrease in quantity. (c) The heat loss will decrease. (d) The furnace efficiency will increase. (e) Pollutants will (usually) decrease.

With a properly adjusted furnace and good mixing, the flue gas will contain no carbon monoxide, and the amount of carbon dioxide will be maximized. The stoichiometric amount of carbon dioxide in the flue gas is known as the *ultimate* CO_2. The air/fuel mixture should be adjusted until the maximum level of carbon dioxide is attained.

Table 24.10 *Approximate Volumetric Percentage of Oxygen in Stack Gas*

	excess air							
fuel[a]	0%	1%	5%	10%	20%	50%	100%	200%
fuel oils, no. 2–6	0	0.22	1.06	2.02	3.69	7.29	10.8	14.2
natural gas	0	0.25	1.18	2.23	4.04	7.83	11.4	14.7
propane	0	0.23	1.08	2.06	3.75	7.38	10.9	14.3

[a]Values for coal are only marginally lower than the values for fuel oils.

[26]The relationship between excess air and CO_2 is much more dependent on fuel type and composition.

Example 24.4

Propane (C_3H_8) is burned completely with 20% excess air. What is the volumetric fraction of carbon dioxide in the flue gas?

Solution

The balanced chemical reaction equation is

$$C_3H_8 + 5O_2 \longrightarrow 3CO_2 + 4H_2O$$

With 20% excess air, the oxygen volume is $(1.2)(5) = 6$.

$$C_3H_8 + 6O_2 \longrightarrow 3CO_2 + 4H_2O + O_2$$

From Table 24.6, there are 3.773 volumes of nitrogen for every volume of oxygen.

$$(6)(3.773) = 22.6$$

$$C_3H_8 + 6O_2 + 22.6N_2 \longrightarrow 3CO_2 + 4H_2O + O_2 + 22.6N_2$$

For gases, the coefficients can be interpreted as volumes. The volumetric fraction of carbon dioxide is

$$B_{CO_2} = \frac{3}{3+4+1+22.6} = 0.0980 \ (9.8\%)$$

31. CALCULATIONS BASED ON FLUE GAS ANALYSIS

Equation 24.10 gives the approximate percentage (by volume) of actual excess air.

$$\text{actual excess air \% by volume} = \frac{(100\%)(B_{O_2} - 0.5B_{CO})}{0.264B_{N_2} - B_{O_2} + 0.5B_{CO}} \quad 24.10$$

The ultimate CO_2 (i.e., the maximum theoretical carbon dioxide) can be determined from Eq. 24.11.

$$\text{ultimate CO}_2\text{, \% by volume} = \frac{(100\%)B_{CO_2,\text{actual}}}{1 - 4.773B_{O_2,\text{actual}}} \quad 24.11$$

The mass ratio of dry flue gases to solid fuel is given by Eq. 24.12.

$$\frac{\text{mass of flue gas}}{\text{mass of solid fuel}} = \frac{(11B_{CO_2} + 8B_{O_2} + 7(B_{CO} + B_{N_2})) \times \left(G_C + \frac{G_S}{1.833}\right)}{3(B_{CO_2} + B_{CO})} \quad 24.12$$

Example 24.5

A sulfur-free coal has a proximate analysis of 75% carbon. The volumetric analysis of the flue gas is 80.2% nitrogen, 12.6% carbon dioxide, 6.2% oxygen, and 1.0% carbon monoxide. Find the (a) actual air/fuel ratio, (b) percentage excess air, (c) ultimate carbon dioxide, and (d) mass of flue gas per mass of fuel.

Solution

(a) Use Eq. 24.9.

$$R_{a/f,\text{actual}} = \frac{3.04 B_{N_2} G_C}{B_{CO_2} + B_{CO}}$$

$$= \frac{(3.04)(0.802)(0.75)}{0.126 + 0.01}$$

$$= 13.4 \text{ lbm air/lbm fuel (kg air/kg fuel)}$$

(b) Use Eq. 24.10.

$$\frac{\text{actual}}{\text{excess air}} = \frac{(100\%)(B_{O_2} - 0.5 B_{CO})}{0.264 B_{N_2} - B_{O_2} + 0.5 B_{CO}}$$

$$= \frac{(100\%)\big(0.062 - (0.5)(0.01)\big)}{(0.264)(0.802) - 0.062 + (0.5)(0.01)}$$

$$= 36.8\% \text{ by volume}$$

(c) Use Eq. 24.11.

$$\text{ultimate } CO_2 = \frac{(100\%) B_{CO_2,\text{actual}}}{1 - 4.773 B_{O_2,\text{actual}}}$$

$$= \frac{(100\%)(0.126)}{1 - (4.773)(0.062)}$$

$$= 17.9\% \text{ by volume}$$

(d) Use Eq. 24.12.

$$\frac{\text{mass of flue gas}}{\text{mass of solid fuel}} = \frac{\begin{pmatrix} 11 B_{CO_2} + 8 B_{O_2} \\ + 7(B_{CO} + B_{N_2}) \end{pmatrix} \times \left(G_C + \dfrac{G_S}{1.833} \right)}{3(B_{CO_2} + B_{CO})}$$

$$= \frac{\begin{pmatrix} (11)(0.126) + (8)(0.062) \\ + (7)(0.01 + 0.802) \end{pmatrix}(0.75)}{(3)(0.126 + 0.01)}$$

$$= \begin{array}{l} 13.9 \text{ lbm flue gas/lbm fuel} \\ \text{(kg flue gas/kg fuel)} \end{array}$$

32. TEMPERATURE OF FLUE GAS

The temperature of the gas at the furnace outlet—before the gas reaches any other equipment—should be approximately 550°F (300°C). Overly low temperatures mean there is too much excess air. Overly high temperatures—above 750°F (400°C)—mean that heat is being wasted to the atmosphere and indicate other problems (ineffective heat transfer surfaces, overfiring, defective combustion chamber, etc.).

The *net stack temperature* is the difference between the stack and local environment temperatures. The net stack temperature should be as low as possible without causing corrosion of the low end.

33. SMOKE

The amount of smoke can be used as an indicator of combustion cleanliness. Smoky combustion may indicate improper air/fuel ratio, insufficient draft, leaks, insufficient preheat, or misadjustment of the fuel system.

Smoke measurements are made in a variety of ways, with the standards depending on the equipment used. Photoelectric sensors in the stack may be used to continuously monitor smoke. The *smoke spot number* (SSN) and ASTM smoke scale are used with continuous stack monitors. For coal-fired furnaces, the maximum desirable smoke number is SSN 4. For grade 2 fuel oil, the SSN should be less than 1; for grade 4, SSN 4; for grades 5L, 5H, and low-sulfur residual fuels, SSN 3; for grade 6, SSN 4.

The *Ringelman scale* is a subjective method in which the smoke density is visually compared to five standardized white-black grids. Ringelman chart no. 0 is solid white; chart no. 5 is solid black. Ringelman chart no. 1, which is 20% black, is the preferred operating point for most power plants.

34. DEW POINT OF FLUE GAS MOISTURE

The *dew point* is the temperature at which the water vapor in the flue gas begins to condense in a constant pressure process. To avoid condensation and corrosion in the stack, the temperature of the flue gases must be above the dew point.

Dalton's law predicts the dew point of moisture in the flue gas. The partial pressure of the water vapor depends on the mole fraction (i.e., the volumetric fraction) of water vapor. The higher the water vapor pressure, the higher the dew point. Air entering a furnace can also contribute to moisture in the flue gas. This moisture should be added to the water vapor from combustion when calculating the mole fraction.

$$\begin{aligned} \text{partial pressure} = &\ (\text{water vapor mole fraction}) \\ &\times (\text{flue gas pressure}) \end{aligned} \qquad \textit{24.13}$$

Once the water vapor's partial pressure is known, the dew point can be found from steam tables as the saturation temperature corresponding to the partial pressure.

When there is no sulfur in the fuel, the dew point is typically around 100°F (40°C). The presence of sulfur in virtually any quantity increases the actual dew point to approximately 300°F (150°C).[27]

---

[27]The theoretical dew point is even higher—up to 350 to 400°F (175 to 200°C). For complex reasons, the theoretical value is not attained.

35. HEAT OF COMBUSTION

The *heating value* of a fuel can be determined experimentally in a *bomb calorimeter*, or it can be estimated from the fuel's chemical analysis. The *higher heating value*, HHV (or *gross heating value*), of a fuel includes the heat of vaporization (condensation) of the water vapor formed from hydrogen during combustion. The *lower heating value*, LHV (or *net heating value*), assumes that all the products of combustion remain gaseous. The LHV is generally the value to use in calculations of thermal energy generated, since the heat of vaporization is not recovered in the stack system.

Traditionally, heating values have been reported on an HHV basis for coal-fired systems but on an LHV basis for natural gas-fired combustion turbines. There is an 11% difference between HHV and LHV thermal efficiencies for gas-fired systems and a 4% difference for coal-fired systems, approximately.

The HHV can be calculated from the LHV if the enthalpy of vaporization, h_{fg}, is known at the pressure of the water vapor.[28] In Eq. 24.14, m_{water} is the mass of water produced per unit (lbm, mole, m^3, etc.) of fuel.

$$\text{HHV} = \text{LHV} + m_{water} h_{fg} \qquad \textbf{24.14}$$

Only the hydrogen that is not locked up with oxygen in the form of water is combustible. This is known as the *available hydrogen*. The correct percentage of combustible hydrogen, $G_{H,available}$, is calculated from the hydrogen and oxygen fraction. Equation 24.15 assumes that all of the oxygen is present in the form of water.

$$G_{H,available} = G_{H,total} - \frac{G_O}{8} \qquad \textbf{24.15}$$

Dulong's formula calculates the higher heating value of coals and coke with a 2 to 3% accuracy for moisture contents below approximately 10%.[29] The gravimetric or volumetric analysis percentages for each combustible element (including sulfur) are multiplied by the heating value per unit (mass or volume) from App. 24.A and summed.

$$\text{HHV}_{MJ/kg} = 32.78 G_C + 141.8 \left(G_H - \frac{G_O}{8} \right) + 9.264 G_S$$
$$\text{[SI]} \qquad \textbf{24.16(a)}$$

$$\text{HHV}_{Btu/lbm} = 14{,}093 G_C + 60{,}958 \left(G_H - \frac{G_O}{8} \right) + 3983 G_S$$
$$\text{[U.S.]} \qquad \textbf{24.16(b)}$$

---

[28] For the purpose of initial studies, the heat of vaporization is usually assumed to be 1040 Btu/lbm (2.42 kJ/kg). This corresponds to a partial pressure of approximately 1 psia (7 kPa) and a dew point of 100°F (38°C).

[29] The coefficients in Eq. 24.16 are slightly different from the coefficients originally proposed by Dulong. Equation 24.16 reflects currently accepted heating values that were unavailable when Dulong developed his formula. Equation 24.16 makes the following assumptions: (1) None of the oxygen is in carbonate form. (2) There is no free oxygen. (3) The hydrogen and carbon are not combined as hydrocarbons. (4) Carbon is amorphous, not graphitic. (5) Sulfur is not in sulfate form. (6) Sulfur burns to sulfur dioxide.

The higher heating value of gasoline can be approximated from the Baumé specific gravity.

$$\text{HHV}_{gasoline,MJ/kg} = 42.61 + 0.093(°\text{Baumé} - 10)$$
$$\text{[SI]} \qquad \textbf{24.17(a)}$$

$$\text{HHV}_{gasoline,Btu/lbm} = 18{,}320 + 40(°\text{Baumé} - 10)$$
$$\text{[U.S.]} \qquad \textbf{24.17(b)}$$

The heating value of petroleum oils (including diesel fuel) can also be approximately determined from the oil's specific gravity. The values derived from Eq. 24.18 may not exactly agree with values for specific oils because the equation does not account for refining methods and sulfur content. Equation 24.18 was originally intended for combustion at constant volume, as in a gasoline engine. However, variations in heating values for different oils are very small, and Eq. 24.18 is widely used as an approximation for all types of combustion, including constant pressure combustion in industrial boilers.

$$\text{HHV}_{fuel\ oil,MJ/kg} = 51.92 - 8.792(\text{SG})^2$$
$$\text{[SI]} \qquad \textbf{24.18(a)}$$

$$\text{HHV}_{fuel\ oil,Btu/lbm} = 22{,}320 - 3780(\text{SG})^2$$
$$\text{[U.S.]} \qquad \textbf{24.18(b)}$$

Example 24.6

A coal has an ultimate analysis of 93.9% carbon, 2.1% hydrogen, 2.3% oxygen, 0.3% nitrogen, and 1.4% ash. What are its (a) higher and (b) lower heating values in Btu/lbm?

Solution

(a) The noncombustible ash, oxygen, and nitrogen do not contribute to heating value. Some of the hydrogen is in the form of water. From Eq. 24.2, the available hydrogen is

$$G_{H,available} = G_{H,total} - \frac{G_O}{8}$$
$$= 2.1\% - \frac{2.3\%}{8} = 1.8\%$$

From App. 24.A, the higher heating values of carbon and hydrogen are 14,093 Btu/lbm and 60,958 Btu/lbm, respectively. From Eq. 24.16, the total heating value per pound of coal is

$$\text{HHV} = (0.939) \left(14{,}093 \ \frac{\text{Btu}}{\text{lbm}} \right)$$
$$+ (0.018) \left(60{,}958 \ \frac{\text{Btu}}{\text{lbm}} \right)$$
$$= 14{,}331 \ \text{Btu/lbm}$$

(b) All of the combustible hydrogen forms water vapor. The mass of water produced is equal to the hydrogen mass plus eight times as much oxygen.

$$m_{water} = m_{H,available} + m_{oxygen}$$
$$= 0.018 + (8)(0.018)$$
$$= 0.162 \text{ lbm water/lbm coal}$$

Assume that the partial pressure of the water vapor is approximately 1 psia (7 kPa). Then, from steam tables, the heat of condensation will be 1040 Btu/lbm. From Eq. 24.14, the lower heating value is approximately

$$\text{LHV} = \text{HHV} - m_{water}h_{fg}$$
$$= 14{,}331 \frac{\text{Btu}}{\text{lbm}} - \left(0.162 \frac{\text{lbm}}{\text{lbm}}\right)\left(1040 \frac{\text{Btu}}{\text{lbm}}\right)$$
$$= 14{,}163 \text{ Btu/lbm}$$

Alternatively, the lower heating value of the coal can be calculated from Eq. 24.16 by substituting the lower (net) hydrogen heating value from App. 24.A, 51,623 Btu/lbm, for the gross heating value of 60,598 Btu/lbm. This yields 14,160 Btu/lbm.

36. MAXIMUM THEORETICAL COMBUSTION (FLAME) TEMPERATURE

It can be assumed that the maximum theoretical increase in flue gas temperature will occur if all of the combustion energy is absorbed adiabatically by the combustion products. This provides a method of estimating the *maximum theoretical combustion temperature*, also sometimes called the *maximum flame temperature* or *adiabatic flame temperature*.[30]

In Eq. 24.19, the mass of the products is the sum of the fuel, oxygen, and nitrogen masses for stoichiometric combustion. The mean specific heat is a gravimetrically weighted average of the values of c_p for all combustion gases. (Since nitrogen comprises the majority of the combustion gases, the mixture's specific heat will be approximately that of nitrogen.) The heat of combustion can be found either from the lower heating value, LHV, or from a difference in air enthalpies across the furnace.

$$T_{max} = T_i + \frac{\text{lower heat of combustion}}{m_{products}} c_{p,\text{mean}} \quad \textit{24.19}$$

Actual flame temperatures are always lower than the theoretical temperature. Most fuels produce flame temperatures in the range of 3350 to 3800°F (1850 to 2100°C).

37. COMBUSTION LOSSES

A portion of the combustion energy is lost in heating the dry flue gases, dfg.[31] This is known as *dry flue gas loss.*

[30]Flame temperature is limited by the dissociation of common reaction products (CO_2, N_2, etc.). At high enough temperatures (3400 to 3800°F; 1880 to 2090°C), the endothermic dissociation process reabsorbs combustion heat and the temperature stops increasing. The temperature at which this occurs is known as the *maximum flame temperature.* This definition of flame temperature is not a function of heating values and flow rates.
[31]The abbreviation "dfg" for dry flue gas is peculiar to the combustion industry. It may not be recognized outside of that field.

In Eq. 24.20, $m_{\text{flue gas}}$ is the mass of dry flue gas per unit mass of fuel. It can be estimated from Eq. 24.12. Although the full temperature difference is used, the specific heat should be evaluated at the average temperature of the flue gas. For quick estimates, the dry flue gas can be assumed to be pure nitrogen.

$$q_1 = m_{\text{flue gas}} c_p (T_{\text{flue gas}} - T_{\text{incoming air}}) \quad \textit{24.20}$$

Heat is lost in the vapor formed during the combustion of hydrogen. In Eq. 24.21, m_{vapor} is the mass of vapor per pound of fuel. G_H is the gravimetric fraction of hydrogen in the fuel. h_g is the enthalpy of superheated steam at the flue gas temperature and the partial pressure of the water vapor. h_f is the enthalpy of saturated liquid at the air's entrance temperature.

$$q_2 = m_{\text{vapor}}(h_g - h_f) = 8.94 G_H(h_g - h_f) \quad \textit{24.21}$$

Heat is lost when it is absorbed by moisture originally in the combustion air (and by free moisture in the fuel, if any). In Eq. 24.22, $m_{\text{combustion air}}$ is the mass of combustion air per pound of fuel. ω is the humidity ratio. h'_g is the enthalpy of superheated steam at the air's entrance temperature and partial pressure of the water vapor.

$$q_3 = m_{\text{atmospheric water vapor}}(h_g - h'_g)$$
$$= \omega m_{\text{combustion air}}(h_g - h'_g) \quad \textit{24.22}$$

When carbon monoxide appears in the flue gas, potential energy is lost in incomplete combustion. The difference in the two heating values in Eq. 24.23 is 9746 Btu/lbm (24.67 MJ/kg).[32]

$$q_4 = \frac{(\text{HHV}_C - \text{HHV}_{CO})G_C B_{CO}}{B_{CO_2} + B_{CO}} \quad \textit{24.23}$$

For solid fuels, potential energy is lost in unburned carbon in the ash. (Some carbon may be carried away in the flue gas, as well.) This is known as *combustible loss* or *unburned fuel loss.* In Eq. 24.24, m_{ash} is the mass of ash produced per pound of fuel consumed. $G_{C,\text{ash}}$ is the gravimetric fraction of carbon in the ash. The heating value of carbon is 14,093 Btu/lbm (32.8 MJ/kg).

$$q_5 = \text{HHV}_C m_{\text{ash}} G_{C,\text{ash}} \quad \textit{24.24}$$

Energy is also lost through radiation from boiler surfaces. This can be calculated if enough information is known. The *radiation loss* is fairly insensitive to different firing rates, and once calculated it can be considered constant for different conditions.

Other conditions where energy can be lost include air leaks, poor pulverizer operation, excessive blowdown, steam leaks, missing or loose insulation, and excessive soot-blower operation. Losses due to these sources must be evaluated on a case-by-case basis.

[32]Obsolete values of 10,150 Btu/lbm (23.61 MJ/kg) and 10,190 Btu/lbm (23.70 MJ/kg) are still encountered for the difference in heating values between carbon and carbon monoxide.

38. COMBUSTION EFFICIENCY

The *combustion efficiency* (also referred to as *boiler efficiency, furnace efficiency,* and *thermal efficiency*) is the overall thermal efficiency of the combustion reaction. Coal and fuel oil furnaces with air heaters and economizers have 75 to 85% efficiencies. Gas furnaces have 75 to 80% efficiencies.

In Eq. 24.25, m_{steam} is the mass of steam produced per pound of fuel burned. The useful heat may also be determined from the boiler rating. One *boiler horsepower* is equal to approximately 33,475 Btu/hr (9.808 kW).[33]

$$\eta = \frac{\text{useful heat extracted}}{\text{heating value}}$$

$$= \frac{m_{\text{steam}}(h_{\text{steam}} - h_{\text{feedwater}})}{\text{HHV}} \qquad 24.25$$

Calculating the efficiency by subtracting all known losses is known as the *loss method*. Minor sources of thermal energy such as the entering air and feedwater are essentially disregarded.

$$\eta = \frac{\text{HHV} - q_1 - q_2 - q_3 - q_4 - q_5 - \text{radiation}}{\text{HHV}}$$

$$= \frac{\text{LHV} - q_1 - q_4 - q_5 - \text{radiation}}{\text{HHV}} \qquad 24.26$$

Combustion efficiency can be improved by decreasing either the temperature or the volume of the flue gas or both. Since the latent heat of moisture is a loss, and since the amount of moisture generated corresponds to the hydrogen content of the fuel, a minimum efficiency loss due to moisture formation cannot be eliminated. This minimum loss is approximately 13% for natural gas, 8% for oil, and 6% for coal.

---

[33]Boiler horsepower is sometimes equated with a gross heating rate of 44,633 Btu/hr (13.08 kW). However, this is the total incoming heating value assuming a standardized 75% combustion efficiency.

Nomenclature

Ca	calcium	mg/L CaCO$_3$
H	total hardness	mg/L CaCO$_3$
L	free lime	mg/L CaCO$_3$
M	alkalinity	mg/L CaCO$_3$
N	normality	gEW/L
O	hydroxides	mg/L CaCO$_3$
P	phenolphthalein alkalinity	mg/L CaCO$_3$
S	sulfate	mg/L CaCO$_3$
T	total alkalinity	mg/L CaCO$_3$
V	volume	mL

1. CATIONS AND ANIONS IN NEUTRAL SOLUTIONS

Equivalency concepts provide a useful check on the accuracy of water analyses. For the water to be electrically neutral, the sum of anion equivalents must equal the sum of cation equivalents.

Concentrations of dissolved compounds in water are usually expressed in mg/L, not equivalents. (See Ch. 22, Sec. 18.) However, anionic and cationic substances can be converted to their equivalent concentrations in milliequivalents per liter (meq/L) by dividing their concentrations in mg/L by their equivalent weights. Appendix 22.C is useful for this purpose.

Example 25.1

A water analysis reveals the following ionic components in solution: Ca^{++}, 29.0 mg/L; Mg^{++}, 16.4 mg/L; Na^+, 23.0 mg/L; K^+, 17.5 mg/L; HCO_3^-, 171 mg/L; SO_4^{--}, 36.0 mg/L; Cl^-, 24.0 mg/L. Verify that the analysis is reasonably accurate.

Solution

Use the equivalent weights in App. 22.C to complete the following table.

compound	concentration (mg/L)	equivalent weight	meq/L
cations			
Ca^{++}	29.0	20.0	1.45
Mg^{++}	16.4	12.2	1.34
Na^+	23.0	23.0	1.00
K^+	17.5	39.1	0.45
		total	4.24
anions			
HCO_3^-	171	61.0	2.81
SO_4^{--}	36.0	48.0	0.75
Cl^-	24.0	35.5	0.68
		total	4.24

The sums of the cation equivalents and anion equivalents are equal. The analysis is presumed to be reasonably accurate.

2. ACIDITY

Acidity is a measure of acids in solution. Acidity in surface water is caused by formation of carbonic acid (H_2CO_3) from carbon dioxide in the air (Eq. 25.1). Carbonic acid is aggressive and must be neutralized to eliminate a cause of water pipe corrosion. If the pH of water is greater than 4.5, carbonic acid ionizes to form bicarbonate (Eq. 25.2). If the pH is greater than 8.3, carbonate ions form (Eq. 25.3).

$$CO_2 + H_2O \rightarrow H_2CO_3 \qquad \text{25.1}$$

$$H_2CO_3 + H_2O \rightarrow HCO_3^- + H_3O^+ \ (\text{pH} > 4.5) \qquad \text{25.2}$$

$$HCO_3^- + H_2O \rightarrow CO_3^{--} + H_3O^+ \ (\text{pH} > 8.3) \qquad \text{25.3}$$

Measurement of acidity is done by titration with a standard basic measuring solution. Acidity in water is typically given in terms of the $CaCO_3$ equivalent that would neutralize the acid. The constant 50,000 in Eq. 25.4 is the product of the equivalent weight of $CaCO_3$ (50 g) and 1000 mg/g.

acidity (mg/L of $CaCO_3$) =
$$\frac{(V_{\text{titrant,mL}})(N_{\text{titrant}})(50{,}000)}{V_{\text{sample,mL}}}$$
25.4

The standard titration procedure for determining acidity measures the amount of titrant needed to raise the pH to 3.7 plus the amount needed to raise the pH to 8.3. The total of these two is used in Eq. 25.4. Most water samples, unless grossly polluted with industrial wastes, will exist at a pH greater than 3.7, so the titration will be a one-step process.

3. ALKALINITY

Alkalinity is a measure of the ability of a water to neutralize acids (i.e., to absorb hydrogen ions without significant pH change). The principal alkaline ions are OH^-, CO_3^{--}, and HCO_3^-. Other radicals, such as NO_3^-, also contribute to alkalinity, but their presence is rare. The measure of alkalinity is the sum of concentrations of each of the substances measured as equivalent $CaCO_3$.

The standard titration method for determining alkalinity measures the amount of acidic titrant needed to lower the pH to 8.3 plus the amount needed to lower the pH to 4.5. Therefore, alkalinity is the sum of all titratable base concentrations down to a pH of 4.5. The constant 50,000 in Eq. 25.5 is the product of the equivalent weight of $CaCO_3$ (50 g) and 1000 mg/g.

$$M\text{(mg/L of } CaCO_3) =$$
$$\frac{(V_{\text{titrant,mL}})(N_{\text{titrant}})(50{,}000)}{V_{\text{sample,mL}}}$$
25.5

Example 25.2

Water from a city well is analyzed and found to contain 20 mg/L as substance of HCO_3^- and 40 mg/L as substance of CO_3^{--}. What is the alkalinity of the water expressed as $CaCO_3$?

Solution

From App. 22.C, the equivalent weight of HCO_3^- is 61 g/mol, the equivalent weight of CO_3^{--} is 30 g/mol, and the equivalent weight of $CaCO_3$ is 50 g/mol.

$$M\text{(mg/L of } CaCO_3) = \left(20\ \frac{\text{mg}}{\text{L}}\right)\left(\frac{50\ \frac{\text{g}}{\text{mol}}}{61\ \frac{\text{g}}{\text{mol}}}\right)$$
$$+ \left(40\ \frac{\text{mg}}{\text{L}}\right)\left(\frac{50\ \frac{\text{g}}{\text{mol}}}{30\ \frac{\text{g}}{\text{mol}}}\right)$$
$$= 83.1\ \text{mg/L as } CaCO_3$$

4. INDICATOR SOLUTIONS

End points for acidity and alkalinity titrations are determined by color changes in indicator dyes that are pH sensitive. Several commonly used indicators are listed in Table 25.1.

Table 25.1 Indicator Solutions Commonly Used in Water Chemistry

indicator	titration	end point pH	color change
bromophenol blue	acidity	3.7	yellow to blue
phenolphthalein	acidity	8.3	colorless to red-violet
phenolphthalein	alkalinity	8.3	red-violet to colorless
mixed bromocresol/ green-methyl red	alkalinity	4.5	grayish to orange-red

Depending on pH, alkaline samples can contain hydroxide alone, hydroxide and carbonate, carbonate alone, carbonate and bicarbonate, or bicarbonate alone. Samples containing hydroxide or hydroxide and carbonate have a high pH, usually greater than 10. If the titration is complete at the phenolphthalein end point (i.e., the mixed bromocresol green-methyl red indicator does not change color), the alkalinity is hydroxide alone. Samples containing carbonate and bicarbonate alkalinity have a pH greater than 8.3, and the titration to the phenolphthalein end point represents stoichiometrically one-half of the carbonate alkalinity. If the volume of phenolphthalein titrant equals the volume of titrant needed to reach the mixed bromocresol green-methyl red end point, all of the alkalinity is in the form of carbonate.

If abbreviations of P for the measured phenolphthalein alkalinity and M for the total alkalinity are used, the following relationships define the possible alkalinity states of the sample. In sequence, the relationships can be used to determine the actual state of a sample.

(state I)	hydroxide $= P = M$
(state II)	hydroxide $= 2P - M$ and carbonate $= 2(M - P)$
(state III)	carbonate $= 2P = M$
(state IV)	carbonate $= 2P$ and bicarbonate $= M - 2P$
(state V)	bicarbonate $= M$

Example 25.3

A 100 mL sample is titrated for alkalinity by using 0.02 N sulfuric acid solution. To reach the phenolphthalein end point requires 3.0 mL of the acid solution, and an additional 12.0 mL is added to reach the mixed bromocresol green-methyl red end point. Calculate the (a) phenolphthalein alkalinity and (b) total alkalinity. (c) What are the ionic forms of alkalinity present?

Solution

(a) Use Eq. 25.5.

$$P = \frac{(V_{titrant,mL})(N_{titrant})(50{,}000)}{V_{sample,mL}}$$

$$= \frac{(3.0\ mL)\left(0.02\ \dfrac{gEW}{L}\right)\left(50{,}000\ \dfrac{mg}{gEW}\right)}{100\ mL}$$

$$= 30\ mg/L\ (as\ CaCO_3)$$

(b) Use Eq. 25.5.

$$M = \frac{(V_{titrant,mL})(N_{titrant})(50{,}000)}{V_{sample,mL}}$$

$$= \frac{(3.0\ mL + 12.0\ mL)}{100\ mL}$$
$$\times \left(0.02\ \dfrac{gEW}{L}\right)\left(50{,}000\ \dfrac{mg}{gEW}\right)$$

$$= 150\ mg/L\ (as\ CaCO_3)$$

(c) Test the state relationships in sequence.

(I) hydroxide $= P = M$:
$$30\ mg/L \neq 150\ mg/L$$
$$(invalid:\ P \neq M)$$

(II) hydroxide $= 2P - M$:
$$(2)\left(30\ \frac{mg}{L}\right) - 150\ \frac{mg}{L}$$
$$= 60\ \frac{mg}{L} - 150\ \frac{mg}{L} = -90\ mg/L$$
$$(invalid:\ negative\ value)$$

(III) carbonate $= 2P = M$:
$$(2)\left(30\ \frac{mg}{L}\right) \neq 150\ mg/L$$
$$(invalid:\ 2P \neq M)$$

(IV) carbonate $= 2P$:
$$(2)\left(30\ \frac{mg}{L}\right) = 60\ mg/L$$

bicarbonate $= M - 2P$:
$$150\ \frac{mg}{L} - (2)\left(30\ \frac{mg}{L}\right) = 90\ mg/L$$

The process ends at (IV) since a valid answer is obtained. It is not necessary to check relationship (V), as (IV) gives consistent results. The alkalinity is composed of 60 mg/L as $CaCO_3$ of carbonate and 90 mg/L as $CaCO_3$ of bicarbonate.

5. HARDNESS

Hardness in natural water is caused by the presence of polyvalent (but not singly charged) metallic cations. Principal cations causing hardness in water and the major anions associated with them are presented in Table 25.2. Because the most prevalent of these species are the divalent cations of calcium and magnesium, total hardness is typically defined as the sum of the concentration of these two elements and is expressed in terms of milligrams per liter as $CaCO_3$. (Hardness is occasionally expressed in units of *grains per gallon*, where 7000 grains are equal to a pound.)

Carbonate hardness is caused by cations from the dissolution of calcium or magnesium carbonate and bicarbonate in the water. Carbonate hardness is hardness that is chemically equivalent to alkalinity, where most of the alkalinity in natural water is caused by the bicarbonate and carbonate ions.

Noncarbonate hardness is caused by cations from calcium (i.e., calcium hardness) and magnesium (i.e., magnesium hardness) compounds of sulfate, chloride, or silicate that are dissolved in the water. Noncarbonate hardness is equal to the total hardness minus the carbonate hardness.

Table 25.2 *Principal Cations and Anions Indicating Hardness*

cations	anions
Ca^{++}	HCO_3^-
Mg^{++}	SO_4^{--}
Sr^{++}	Cl^-
Fe^{++}	NO_3^-
Mn^{++}	SiO_3^{--}

Hardness can be classified as shown in Table 25.3. Although high values of hardness do not present a health risk, they have an impact on the aesthetic acceptability of water for domestic use. (Hardness reacts with soap to reduce its cleansing effectiveness and to form scum on the water surface.) Where feasible, carbonate hardness in potable water should be reduced to the 25 to 40 mg/L range and total hardness reduced to the 50 to 75 mg/L range.

Table 25.3 *Relationship of Hardness Concentration to Classification*

hardness (mg/L as $CaCO_3$)	classification
0 to 60	soft
61 to 120	moderately hard
121 to 180	hard
181 to 350	very hard
> 350	saline; brackish

Water containing bicarbonate (HCO_3^-) can be heated to precipitate carbonate (CO_3^{--}) as a scale. Water used in steam-producing equipment (e.g., boilers) must be essentially hardness-free to avoid deposit of scale.

$$Ca^{++} + 2HCO_3^- + heat \rightarrow CaCO_3 \downarrow$$
$$+ CO_2 + H_2O \qquad 25.6$$
$$Mg^{++} + 2HCO_3^- + heat \rightarrow MgCO_3 \downarrow$$
$$+ CO_2 + H_2O \qquad 25.7$$

Noncarbonate hardness, also called *permanent hardness*, cannot be removed by heating. It can be removed by precipitation softening processes (typically the lime-soda ash process) or by ion exchange processes using resins selective for ions causing hardness.

Hardness is measured in the laboratory by titrating the sample using a standardized solution of ethylene-diaminetetraacetic acid (EDTA) and an indicator dye such as Eriochrome Black T. The sample is titrated at a pH of approximately 10 until the dye color changes from red to blue. The standardized solution of EDTA is usually prepared such that 1 mL of EDTA is equivalent to 1 mg/L of hardness, but Eq. 25.8 can be used to determine hardness with any strength EDTA solution.

$$H \text{ (as mg/L } CaCO_3) =$$
$$\frac{(V_{titrant,mL})(CaCO_3 \text{ equivalent of EDTA})(1000)}{V_{sample,mL}} \qquad 25.8$$

Example 25.4

A water sample is found to contain sodium (Na^+, 15 mg/L), magnesium (Mg^{++}, 70 mg/L), and calcium (Ca^{++}, 40 mg/L). What is the hardness?

Solution

Sodium is singly charged, so it does not contribute to hardness. The necessary approximate equivalent weights are found in App. 22.C.

Mg: 12.2 g/mol

Ca: 20.0 g/mol

$CaCO_3$: 50 g/mol

The equivalent hardness is

$$H = \left(70 \ \frac{mg}{L}\right) \left(\frac{50 \ \frac{g}{mol}}{12.2 \ \frac{g}{mol}}\right) + \left(40 \ \frac{mg}{L}\right) \left(\frac{50 \ \frac{g}{mol}}{20 \ \frac{g}{mol}}\right)$$
$$= 387 \text{ mg/L as } CaCO_3$$

Example 25.5

A 75 mL water sample was titrated using 8.1 mL of an EDTA solution formulated such that 1 mL of EDTA is equivalent to 0.8 mg/L of $CaCO_3$. What is the hardness of the sample, measured in terms of mg/L of $CaCO_3$?

Solution

Use Eq. 25.8.

$$H = \frac{(V_{titrant,mL})(CaCO_3 \text{ equivalent of EDTA})(1000)}{V_{sample,mL}}$$

$$= \frac{(8.1 \text{ mL}) \left(0.8 \ \frac{gEW}{L}\right) \left(1000 \ \frac{mg}{gEW}\right)}{75 \text{ mL}}$$

$$= 86.4 \text{ mg/L as } CaCO_3$$

6. HARDNESS AND ALKALINITY

Hardness is caused by multi-positive ions. Alkalinity is caused by negative ions. Both positive and negative ions are present simultaneously. Therefore, an alkaline water can also be hard.

With some assumptions and minimal information about the water composition, it is possible to determine the ions in the water from the hardness and alkalinity. For example, Fe^{++} is an unlikely ion in most water supplies, and it is often neglected. Figure 25.1 can be used to quickly deduce the compounds in the water from hardness and alkalinity.

If hardness and alkalinity (both as $CaCO_3$) are the same and there are no mono-valent cations, then there are no SO_4^{--}, Cl^-, or NO_3^- ions present. That is, there is no noncarbonate (permanent) hardness. If hardness is greater than the alkalinity, however, then noncarbonate hardness is present, and the carbonate (temporary) hardness is equal to the alkalinity. If hardness is less than the alkalinity, then all hardness is carbonate hardness, and the extra HCO_3^- comes from other sources (such as $NaHCO_3$).

Example 25.6

An analysis of a sample of water results in the following: alkalinity, 220 mg/L; hardness, 180 mg/L; Ca^{++}, 140 mg/L; OH^-, insignificant. All concentrations are expressed as $CaCO_3$. (a) What is the noncarbonate hardness? (b) What is the Mg^{++} content in mg/L as substance?

Solution

(a) Use Fig. 25.1 with $M > H$ (alkalinity greater than hardness).

$$[NaHCO_3] = M - H = 220 \ \frac{mg}{L} - 180 \ \frac{mg}{L}$$
$$= 40 \text{ mg/L as } CaCO_3$$
$$[Mg(HCO_3)_2] = H - Ca = 180 \ \frac{mg}{L} - 140 \ \frac{mg}{L}$$
$$= 40 \text{ mg/L as } CaCO_3$$
$$Ca(HCO_3)_2 = Ca = 140 \text{ mg/L as } CaCO_3$$
$$\text{carbonate hardness} = H = 180 \text{ mg/L as } CaCO_3$$

noncarbonate hardness = 0

Figure 25.1 *Hardness and Alkalinity*[a,b]

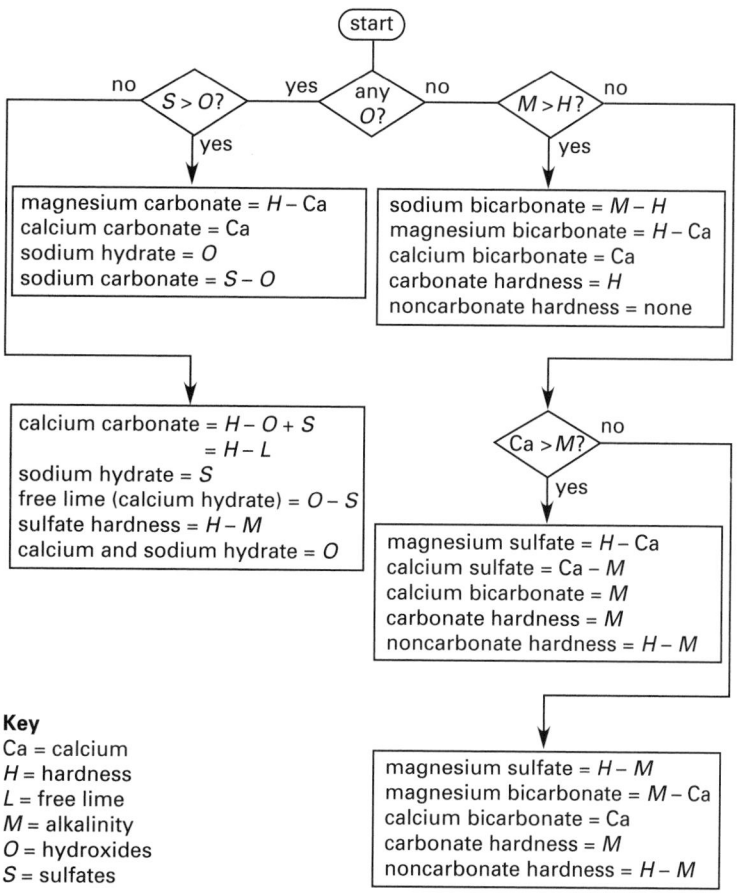

^{a} All concentrations are expressed as $CaCO_3$.
^{b} Not for use when other ionic species are present in significant quantities.

(b) The Mg^{++} ion content as $CaCO_3$ is equal to the $Mg(HCO_3)_2$ content as $CaCO_3$. Use App. 22.C to convert $CaCO_3$ equivalents to amounts as substance. The factor that converts Mg^{++} as $CaCO_3$ to Mg^{++} as substance is 4.1.

$$Mg^{++} = \frac{40 \; \frac{mg}{L}}{4.1} = 9.8 \text{ mg/L as substance}$$

7. NATIONAL PRIMARY DRINKING WATER STANDARDS

Following passage of the Safe Drinking Water Act in the United States, the Environmental Protection Agency (EPA) established minimum primary drinking water standards. These standards set limits on the amount of various substances in drinking water. Every public water supply serving at least 15 service connections or 25 or more people must ensure that its water meets these minimum standards.

Accordingly, the EPA has established the National Primary Drinking Water Standards, outlined in App. 25.A, and the National Secondary Drinking Water Standards. The primary standards establish *maximum contaminant levels* (MCL) and *maximum contaminant level goals* (MCLG) for materials that are known or suspected health hazards. The MCL is the enforceable level that the water supplier must not exceed, while the MCLG is an unenforceable health goal equal to the maximum level of a contaminant that is not expected to cause any adverse health effects over a lifetime of exposure.

8. NATIONAL SECONDARY DRINKING WATER STANDARDS

The national secondary drinking water standards, outlined in Table 25.4, are not designed to protect public health. Instead, they are intended to protect "public welfare" by providing helpful guidelines regarding the taste, odor, color, and other aesthetic aspects of drinking water.

Table 25.4 *National Secondary Drinking Water Standards (Code of Federal Regulations (CFR) Title 40, Ch. I, Part 143)*

contaminant	suggested levels	effects
aluminum	0.05 to 0.2 mg/L	discoloration of water
chloride	250 mg/L	salty taste and pipe corrosion
color	15 color units	visible tint
copper	1.0 mg/L	metallic taste and staining
corrosivity	noncorrosive	taste, staining, and corrosion
fluoride	2.0 mg/L	dental fluorosis
foaming agents	0.5 mg/L	froth, odor, and bitter taste
iron	0.3 mg/L	taste, staining, and sediment
manganese	0.05 mg/L	taste and staining
odor	3 TON[a]	"rotten egg," musty, and chemical odor
pH	6.5 to 8.5	low pH—metallic taste, and corrosion high pH—slippery feel, soda taste, and deposits
silver	0.1 mg/L	discoloration of skin and graying of eyes
sulfate	250 mg/L	salty taste and laxative effect
total dissolved solids (TDS)	500 mg/L	taste, corrosivity, and soap interference
zinc	5 mg/L	metallic taste

[a]threshold odor number

9. IRON

Even at low concentrations, iron is objectionable because it stains porcelain bathroom fixtures, causes a brown color in laundered clothing, and can be tasted. Typically, iron is a problem in groundwater pumped from anaerobic aquifers in contact with iron compounds. Soluble ferrous ions can be formed under these conditions, which, when exposed to atmospheric air at the surface or to dissolved oxygen in the water system, are oxidized to the insoluble ferric state, causing the color and staining problems mentioned. Iron determinations are made through colorimetric analysis.

10. MANGANESE

Manganese ions are similar in formation, effect, and measurement to iron ions.

11. FLUORIDE

Natural fluoride is found in groundwaters as a result of dissolution from geologic formations. Surface waters generally contain much smaller concentrations of fluoride. An absence or low concentration of ingested fluoride causes the formation of tooth enamel less resistant to decay, resulting in a high incidence of dental cavities in children's teeth. Excessive concentration of fluoride causes *fluorosis*, a brownish discoloration of dental enamel. The MCL of 4.0 mg/L established by the EPA is to prevent unsightly fluorosis.

Communities with water supplies deficient in natural fluoride chemically add fluoride during the treatment process. Since water consumption is influenced by climate, the recommended optimum concentrations listed in Table 25.5 are based on the annual average of the maximum air temperatures based on a minimum of five years of records.

Compounds commonly used as fluoride sources in water treatment are listed in Table 25.6.

Table 25.5 *Recommended Optimum Concentrations of Fluoride in Drinking Water*

average air temperature range (°F)	recommended optimum concentration (mg/L)
53.7 and below	1.2
53.8 to 58.3	1.1
58.4 to 63.8	1.0
63.9 to 70.6	0.9
70.7 to 79.2	0.8
79.3 to 90.5	0.7

Table 25.6 *Fluoridation Chemicals*

compound	formula	percentage F^- ion (%)
sodium fluoride	NaF	45
sodium silicofluoride	Na_2SiF_6	61
hydrofluosilicic acid	H_2SiF_6	79
ammonium silicofluoride[a]	$(NH_4)_2SiF_6$	64

[a]used in conjunction with chlorine disinfection where it is desired to maintain a chloramine residual in the distribution system.

Example 25.7

A liquid feeder adds a 4.0% saturated sodium fluoride solution to a water supply, increasing the fluoride concentration from the natural fluoride level of 0.4 mg/L to 1.0 mg/L. The commercial NaF powder used to prepare the NaF solution contains 45% fluoride by weight. (a) How many pounds (kilograms) of NaF are required per million gallons (liters) treated? (b) What volume of 4% NaF solution is used per million gallons (liters)?

SI Solution

(a)
$$m_{NaF} = \frac{1.0 \frac{mg}{L} - 0.4 \frac{mg}{L}}{0.45}$$
$$= 1.33 \text{ kg/ML (kg per million liters)}$$

(b) A 4% NaF solution contains $40\,000$ mg NaF per liter.

$$\left(40\,000\ \frac{\text{mg}}{\text{L}}\right)(0.45) = 18\,000\ \text{mg/L}$$

$$V = \frac{\left(1.0 \times 10^6\ \dfrac{\text{L}}{\text{ML}}\right) \times \left(1.0\ \dfrac{\text{mg}}{\text{L}} - 0.4\ \dfrac{\text{mg}}{\text{L}}\right)}{18\,000\ \dfrac{\text{mg}}{\text{L}}}$$

$$= 33.3\ \text{L/ML}$$

Customary U.S. Solution

(a) $\quad m_{\text{NaF}} = \left(\dfrac{1.0\ \dfrac{\text{mg}}{\text{L}} - 0.4\ \dfrac{\text{mg}}{\text{L}}}{0.45}\right)\left(8.345\ \dfrac{\text{lbm-L}}{\text{mg-MG}}\right)$

$$= 11.1\ \text{lbm/MG (lbm per million gallons)}$$

(b) A 4% NaF solution contains 40,000 mg NaF per liter. The concentration needs to be increased from 0.4 mg/L to 1.0 mg/L.

$$\left(40{,}000\ \frac{\text{mg}}{\text{L}}\right)(0.45) = 18{,}000\ \text{mg/L}$$

$$V = \frac{\left(1.0 \times 10^6\ \dfrac{\text{gal}}{\text{MG}}\right) \times \left(1.0\ \dfrac{\text{mg}}{\text{L}} - 0.4\ \dfrac{\text{mg}}{\text{L}}\right)}{18{,}000\ \dfrac{\text{mg}}{\text{L}}}$$

$$= 33.3\ \text{gal/MG}$$

12. PHOSPHORUS

Phosphate content is more of a concern in wastewater treatment than in supply water, although phosphorus can enter water supplies in large amounts from runoff. Excessive phosphate discharge contributes to aquatic plant (phytoplankton, algae, and macrophytes) growth and subsequent *eutrophication*. (Eutrophication is an "over-fertilization" of receiving waters.)

Phosphorus is of considerable interest in the management of lakes and reservoirs because phosphorus is a nutrient that has a major effect on aquatic plant growth. Algae normally have a phosphorus content of 0.1 to 1% of dry weight. The molar N:P ratio for ideal algae growth is 16:1.

Phosphorus exists in several forms in aquatic environments. Soluble phosphorus occurs as *orthophosphate*, as condensed *polyphosphates* (from detergents), and as various organic species. Orthophosphates ($H_2PO_4^-$, HPO_4^{--}, and PO_4^{---}) and polyphosphates (such as $Na_3(PO_3)_6$) result from the use of synthetic detergents (*syndets*). A sizable fraction of the soluble phosphorus is in the organic form, originating from the decay or excretion of nucleic acids and algal storage products. Particulate phosphorus occurs in the organic form as a part of living organisms and detritus as well as in the inorganic form of minerals such as apatite.

Phosphorus is normally measured by colorimetric or digestion methods. The results are reported in terms of mg/L of phosphorus (e.g., "mg/L of P" or "mg/L of total P"), although the tests actually measure the concentration of orthophosphate. The concentration of a particular compound is found by multiplying the concentration as P by the molecular weight of the compound and dividing by the atomic weight of phosphorus (30.97).

A substantial amount of the phosphorus that enters lakes is probably not available to aquatic plants. Bioavailable species include orthophosphates, polyphosphates, most soluble organic phosphorus, and a portion of the particulate fraction. Studies have indicated that bioavailable phosphorus generally does not exceed 60% of the total phosphorus.

In aquatic systems, phosphorus does not enter into any redox reactions, nor are any common species volatile. Therefore, lakes retain a significant portion of the entering phosphorus. The main mechanism for retaining phosphorus is simple sedimentation of particles containing the phosphorus. Particulate phosphorus can originate from the watershed (*allochthonous material*) or can be formed within the lake (*autochthonous material*).

Much of the phosphorus that enters a lake is recycled, and a large fraction of the recycling occurs at the sediment-water interface. Recycling of phosphorus is linked to iron and manganese recycling. Soluble phosphorus in the water column is removed by adsorption onto iron and manganese hydroxides, which precipitate under aerobic conditions. However, when the *hypolimnion* (i.e., the lower part of the lake that is essentially stagnant) becomes anaerobic, the iron (or manganese) is reduced, freeing up phosphorus. This is consistent with a fairly general observation that phosphorus release rates are nearly an order of magnitude higher under anaerobic conditions than under aerobic conditions.

Factors that control phosphorus recycling rates are not well understood, although it is clear that oxygen status, phosphorus speciation, temperature, and pH are important variables. Phosphorus release from sediments is usually considered to be constant (usually less than 1 mg/m^2/day under aerobic conditions).

In addition to direct regeneration from sediments, *macrophytes* (large aquatic plants) often play a significant role in phosphorus recycling. Macrophytes with highly developed root systems derive most of their phosphorus from the sediments. Regeneration to the water column can occur by excretion or through decay, effectively "pumping" phosphorus from the sediments. The internal loading generated by recycling is particularly important in shallow, eutrophic lakes. In several cases where phosphorus inputs have been reduced to control algal blooms, regeneration of phosphorus from phosphorus-rich sediments has slowed the rate of recovery.

13. NITROGEN

Compounds containing nitrogen are not abundant in virgin surface waters. However, nitrogen can reach large concentrations in ground waters that have been contaminated with barnyard runoff or that have percolated through heavily fertilized fields. Sources of surface water contamination include agricultural runoff and discharge from sewage treatment facilities.

Of greatest interest, in order of decreasing oxidation state, are nitrates (NO_3^-), nitrites (NO_2^-), ammonia (NH_3), and organic nitrogen. These three compounds are reported as *total nitrogen*, TN, with units of "mg/L of N" or "mg/L of total N." The concentration of a particular compound is found by multiplying the concentration as N by the molecular weight of the compound and dividing by the atomic weight of nitrogen (14.01).

Excessive amounts of nitrate in water can contribute to the illness in infants known as *methemoglobinemia* ("blue baby" syndrome). As with phosphorus, nitrogen stimulates aquatic plant growth.

Un-ionized ammonia is a colorless gas at standard temperature and pressure. A pungent odor is detectable at levels above 50 mg/L. Ammonia is very soluble in water at low pH.

Ammonia levels in zero-salinity surface water increase with increasing pH and temperature (see Table 25.7). At low pH and temperature, ammonia combines with water to produce ammonium ion (NH_4^+) and hydroxide (OH^-) ions. The ammonium ion is nontoxic to aquatic life and not of great concern. The un-ionized ammonia (NH_3), however, can easily cross cell membranes and have a toxic effect on a wide variety of fish. The EPA has established the following criteria for fresh- and saltwater fish.

freshwater	0.002 mg/L un-ionized ammonia
saltwater	
acute (1 hr average)	0.233 mg/L un-ionized ammonia
chronic (4 hr average)	0.035 mg/L un-ionized ammonia

Ammonia is usually measured by a distillation and titration technique or with an ammonia-selective electrode. The results are reported as ammonia nitrogen. Nitrites are measured by a colorimetric method. The results are reported as nitrite nitrogen. Nitrates are measured by ultraviolet spectrophotometry, selective electrode, or reduction methods. The results are reported as nitrate nitrogen. Organic nitrogen is determined by a digestion process that identifies organic and ammonia nitrogen combined. Organic nitrogen is found by subtracting the ammonia nitrogen value from the digestion results.

Table 25.7 *Percent of Total Ammonia Present in Toxic, Un-Ionized Form*

temp (°C)	pH								
	6.0	6.5	7.0	7.5	8.0	8.5	9.0	9.5	10.0
5	0.013	0.040	0.12	0.39	1.2	3.8	11	28	56
10	0.019	0.059	0.19	0.59	1.8	5.6	16	37	65
15	0.027	0.087	0.27	0.86	2.7	8.0	21	46	73
20	0.040	0.13	0.40	1.2	3.8	11	28	56	80
25	0.057	0.18	0.57	1.8	5.4	15	36	64	85
30	0.080	0.25	0.80	2.5	7.5	20	45	72	89

14. COLOR

Color in water is caused by substances in solution, known as *true color*, and by substances in suspension, mostly organics, known as *apparent* or *organic color*. Iron, copper, manganese, and industrial wastes all can cause color. Color is aesthetically undesirable, and it stains fabrics and porcelain bathroom fixtures.

Water color is determined by comparison with standard platinum/cobalt solutions or by spectrophotometric methods. The standard color scales range from 0 (clear) to 70. Water samples with more intense color can be evaluated using a dilution technique.

15. TURBIDITY

Turbidity is a measure of the light-transmitting properties of water and is comprised of suspended and colloidal material. Turbidity is expressed in *nephelometric turbidity units* (NTU). Viruses and bacteria become attached to these particles, where they can be protected from the bactericidal and viricidal effects of chlorine, ozone, and other disinfecting agents. The organic material included in turbidity has also been identified as a potential precursor to carcinogenic disinfection byproducts.

Turbidity in excess of 5 NTU is noticeable by visual observation. Turbidity in a typical clear lake is approximately 25 NTU, and muddy water exceeds 100 NTU. Turbidity is measured using an electronic instrument called a nephelometer, which detects light scattered by the particles when a focused light beam is shown through the sample.

16. SOLIDS

Solids present in a sample of water can be classified in several ways.

- *total solids* (TS): Total solids are the material residue left in the vessel after the evaporation of the sample at 103 to 105°C. Total solids include total suspended solids and total dissolved solids.

- *total suspended solids* (TSS): The material retained on a standard glass-fiber filter disk is defined as the suspended solids in a sample. The filter is weighed before filtration, dried at 103 to 105°C, and weighed again. The gain in weight is the amount of suspended solids. Suspended solids can also be categorized into *volatile suspended solids* (VSS) and *fixed suspended solids* (FSS).

- *total dissolved solids* (TDS): These solids are in solution and pass through the pores of the standard glass-fiber filter. Dissolved solids are determined by passing the sample through a filter, collecting the filtrate in a weighed drying dish, and evaporating the liquid at 180°C. The gain in weight represents the dissolved solids. Dissolved solids can be categorized into *volatile dissolved solids* (VDS) and *fixed dissolved solids* (FDS).

- *total volatile solids* (TVS): The residue from one of the previous determinations is ignited to constant weight in an electric muffle furnace at 550°C. The loss in weight during the ignition process represents the volatile solids.

- *total fixed solids* (TFS): The weight of solids that remain after the ignition used to determine volatile solids represents the fixed solids.

- *settleable solids*: The volume (mL/L) of settleable solids is measured by allowing a sample to settle for one hour in a graduated conical container (*Imhoff cone*).

The following relationships exist.

$$TS = TSS + TDS = TVS + TFS \qquad \text{25.9}$$
$$TSS = VSS + FSS \qquad \text{25.10}$$
$$TDS = VDS + FDS \qquad \text{25.11}$$
$$TVS = VSS + VDS \qquad \text{25.12}$$
$$TFS = FSS + FDS \qquad \text{25.13}$$

Waters with high concentrations of suspended solids are classified as *turbid waters*. Waters with high concentrations of dissolved solids often can be tasted. Therefore, a limit of 500 mg/L has been established as the suggested level for dissolved solids.

17. CHLORINE AND CHLORAMINES

Chlorine is the most common disinfectant used in water treatment. It is a strong oxidizer that deactivates microorganisms. Its oxidizing capability also makes it useful in removing soluble iron and manganese ions.

Chlorine gas in water forms *hydrochloric* and *hypochlorous acids*. At a pH greater than 9, hypochlorous acid dissociates to hydrogen and hypochlorite ions, as shown in Eq. 25.14.

$$Cl_2 + H_2O \underset{pH<4}{\overset{pH>4}{\longleftrightarrow}} HCl + HOCl \underset{pH<9}{\overset{pH>9}{\longleftrightarrow}} H^+ + OCL^- \quad \text{25.14}$$

Free chlorine, hypochlorous acid, and hypochlorite ions left in water after treatment are known as *free chlorine residuals*. Hypochlorous acid reacts with ammonia (if it is present) to form *chloramines*. Chloramines are known as *combined residuals*. Chloramines are more stable than free residuals, but their disinfecting ability is less.

$$HOCl + NH_3 \rightarrow H_2O + NH_2Cl \text{ (monochloramine)}$$
$$\text{25.15}$$

$$HOCl + NH_2Cl \rightarrow H_2O + NHCl_2 \text{ (dichloramine)}$$
$$\text{25.16}$$

$$HOCl + NHCl_2 \rightarrow H_2O + NCl_3 \text{ (trichloramine)}$$
$$\text{25.17}$$

Free and combined residual chlorine can be determined by color comparison, by titration, and with chlorine-sensitive electrodes. Color comparison is the most common field method.

18. HALOGENATED COMPOUNDS

Halogenated compounds have become a subject of concern in the treatment of water due to their potential as carcinogens. The use of chlorine as a disinfectant in water treatment generates these compounds by reacting with organic substances in the water. (For this reason there are circumstances under which chlorine may not be the most appropriate disinfectant.)

The organic substances, called *precursors*, are not in themselves harmful, but the chlorinated end products, known by the term *disinfection byproducts* (DBPs) raise serious health concerns. Typical precursors are decay byproducts such as humic and fulvic acids. Several of the DBPs contain bromine, which is found in low concentrations in most surface waters and can also occur as an impurity in commercial chlorine gas.

The DBPs can take a variety of forms depending on the precursors present, the concentration of free chlorine, the contact time, the pH, and the temperature. The most common ones are the trihalomethanes (THMs), haloacetic acids (HAAs), dihaloacetonitriles (DHANs), and various trichlorophenol isomers.

19. TRIHALOMETHANES

Only four trihalomethane (THM) compounds are normally found in chlorinated waters.

$CHCl_3$	trichloromethane (chloroform)
$CHBrCl_2$	bromodichloromethane
$CHBr_2Cl$	dibromochloromethane
$CHBr_3$	tribromomethane (bromoform)

Trihalomethanes are regulated by the EPA under the National Primary Drinking Water Standards. Because of this, there is a trend toward using treatment and disinfection processes that are designed to minimize the formation of THMs.

20. HALOACETIC ACIDS

The haloacetic acids (HAAs) exist in tri-, di-, and mono-forms, abbreviated as THAAs, DHAAs, and MHAAs. All of the haloacetic acids are toxic and are suspected or proven carcinogens. Maximum concentration limits are listed in App. 25.A for HAAs.

$C_2HCl_3O_2$	trichloroacetic acid	(TCAA)
$C_2HBrCl_2O_2$	bromodichloroacetic acid	(BDCAA)
$C_2HBr_2ClO_2$	dibromochloroacetic acid	(DBCAA)
$C_2HBr_3O_2$	tribromoacetic acid	(TBAA)
$C_2H_2Cl_2O_2$	dichloroacetic acid	(DCAA)
$C_2H_2BrClO_2$	bromochloroacetic acid	(BCAA)
$C_2H_2Br_2O_2$	dibromoacetic acid	(DBAA)
$C_2H_3ClO_2$	monochloroacetic acid	(MCAA)
$C_2H_3BrO_2$	monobromoacetic acid	(MBAA)

21. DIHALOACETONITRILES

The dihaloacetonitriles (DHANs) are formed when acetonitrile (methyl cyanide: C_2H_3N (structurally H_3CCN)) is exposed to chlorine. All are toxic and suspected carcinogens. Maximum concentration limits have not been established by the EPA for DHANs.

C_2HCl_2N	dichloroacetonitrile	(DCAN)
$C_2HBrClN$	bromochloroacetonitrile	(BCAN)
C_2HBr_2N	dibromoacetonitrile	(DBAN)

22. TRICHLOROPHENOL

Trichlorophenol can exist in six isomeric forms, with varying potential toxicities. As with all halogenated organics, they are potential carcinogens.

2,3,4-trichlorophenol	
2,3,5-trichlorophenol	
2,3,6-trichlorophenol	
2,4,5-trichlorophenol	irritant
2,4,6-trichlorophenol	fungicide, bactericide
3,4,5-trichlorophenol	

23. AVOIDANCE OF DISINFECTION BYPRODUCTS

Reduction of DBPs can best be achieved by avoiding their production in the first place. The best strategy dictates using source water with few or no precursors.

Often source water choices are limited, necessitating tailoring treatment processes to produce the desired result. This entails removing the precursors prior to the application of chlorine, applying chlorine at certain points in the treatment process that minimize production of DBPs, using disinfectants that do not produce significant DBPs, or a combination of these techniques.

Removal of precursors is achieved by preventing growth of vegetative material (algae, plankton, etc.) in the source water and by collecting source water at various depths to avoid concentrations of precursors. Oxidizers such as potassium permanganate and chlorine dioxide can often reduce the concentration of the precursors without forming the DBPs. Under some instances, application of powdered activated carbon or a pH-adjustment process can reduce the impact of chlorination.

Chlorine application should be delayed if possible until after the flocculation, coagulation, settling, and filtration processes have been completed. In this manner, turbidity and common precursors will be reduced. If it is necessary to chlorinate early to facilitate treatment processes, chlorination can be followed by dechlorination to reduce contact time. Granular activated carbon has been used to some extent to remove DBPs after they form, but the carbon needs frequent regeneration.

Alternative disinfectants include ozone, chloramines, chlorine dioxide, potassium permanganate, and ultraviolet radiation. Ozone and chloramines, singularly or together, are often used for control of THMs. However, ozone creates other DBPs, including aldehydes, hydrogen peroxide, carboxylic acids, ketones, and phenols. When ozone is used as the primary disinfectant, a secondary disinfectant such as chlorine or chloramine must be used to provide an active residual that can be measured within the distribution system.

26 Water Supply Treatment and Distribution

Nomenclature

AADF	average annual daily flow	gal/day	L/day
A	surface area	ft^2	m^2
b	width, length, distance	ft	m
C	coefficient	–	–
C	concentration	mg/L	mg/L
D	diameter	ft	m
D	dose	mg/L	mg/L
F	feed rate	lbm/day	kg/day
F	fire fighting construction factor	–	–
F	force	lbf	N
g	acceleration of gravity	ft/sec^2	m/s^2
g_c	gravitational constant	ft-lbm/ lbf-sec^2	–
G	gravimetric fraction	–	–
G	mixing velocity gradient	1/sec	1/s
h	head (depth), height	ft	m
K	mixing constant	1/sec	1/s
L	length	ft	m
m	mass	lbm	kg
M	demand multiplier	–	–
M	dose	mg/L	mg/L
n	rotational speed	rev/sec	rev/s
N	dimensionless number	–	–
P	population	thousands	thousands
P	power	hp	kW
P	purity	–	–
Q	flow rate	ft^3/sec	m^3/s
Q	flow rate	MGD	L/s
R	radius	ft	m
Re	Reynolds number	–	–
SG	specific gravity	–	–
t	time	sec	s
TSS	total suspended solids	mg/L	mg/L
v	velocity	ft/sec	m/s
v^*	overflow rate	gpd/ft^2	L/day·m^2
V	volume	ft^3	m^3

Symbols

γ	specific weight	lbf/ft^3	N/m^3
μ	absolute viscosity	lbf-sec/ft^2	N·s
ν	kinematic viscosity	ft^2/sec	m^2/s
ρ	density	lbm/ft^3	kg/m^3

Subscripts

d	detention
i	in
D	drag
o	out
P	power
Q	flow
s	settling
v	velocity

1. WATER TREATMENT PLANT LOCATION

The location chosen for a water treatment plant is influenced by many factors. The most common factors include availability of resources such as (a) local water, (b) power, and (c) sewerage services; economic factors such as (d) land cost and (e) annual taxes; and environmental factors such as (f) traffic and (g) other concerns that would be identified on an environmental impact report.

It is imperative to locate water treatment plants (a) above the flood plain and (b) where a 15 to 20 ft (4.5 to 6 m) elevation difference exists. This latter requirement will eliminate the need to pump the water between processes. Traditional treatment requires a total head of approximately 15 ft (4.5 m), whereas advanced processes such as granular activated charcoal and ozonation increase the total head required to approximately 20 ft (6 m).

The average operational life of equipment and facilities is determined by economics. However, it is unlikely that a lifetime of less than 50 yr would be economically viable. Equipment lifetimes of 25 to 30 yr are typical.

2. PROCESS INTEGRATION

The processes and sequences used in a water treatment plant depend on the characteristics of the incoming water. However, some sequences are more appropriate than others due to economic and hydraulic considerations. *Conventional filtration*, also referred to as *complete filtration*, is a term used to describe the traditional sequence of adding coagulation chemicals, flash mixing, coagulation-flocculation, sedimentation, and subsequent filtration. Coagulants, chlorine (or an alternative disinfectant), fluoride, and other chemicals are added at various points along the path, as indicated by Fig. 26.1. Conventional filtration is still the best choice when incoming water has high color, turbidity, or other impurities.

Direct filtration refers to a modern sequence of adding coagulation chemicals, flash mixing, minimal flocculation, and subsequent filtration. In direct filtration, the physical chemical reactions of flocculation occur to some extent, but special flocculation and sedimentation facilities are eliminated. This reduces the amount of sludge that has to be treated and disposed of. Direct filtration is applicable when the incoming water is of high initial quality.

In-line filtration refers to another modern sequence that starts with adding coagulation chemicals at the filter inlet pipe. Mixing occurs during the turbulent flow toward a filter, which is commonly of the pressure-filter variety. As with direct filtration, flocculation and sedimentation facilities are not used.

Table 26.1 provides guidelines based on incoming water characteristics for choosing processes required to achieve satisfactory quality.

Figure 26.1 *Chemical Application Points*

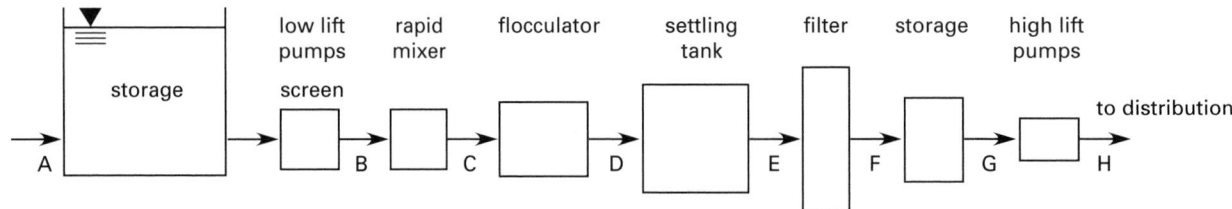

typical flow diagram of water treatment plant

category of chemicals	possible points of application							
	A	B	C	D	E	F	G	H
algicide	X				X			
disinfectant		X	X		X	X	X	X
activated carbon		X	X	X	X			
coagulants		X	X					
coagulation aids		X	X		X			
alkali								
for flocculation			X					
for corrosion control						X		
for softening			X					
acidifier			X			X		
fluoride						X		
cupric-chloramine						X		
dechlorinating agent						X		X

Note: With solids contact reactors, point C is the same as point D.

Table 26.1 Treatment Methods

incoming water quality		pretreatment				treatment					special treatments			
constituents	concentration (mg/L)	screening	prechlorination	plain settling	aeration	lime softening	coagulation and sedimentation	rapid sand filtration	slow sand filtration	postchlorination	superchlorination[a] or chlorammoniation	active carbon	special chemical treatment	salt water conversion[b]
coliform monthly average mpn[e]/100 mL	0–20									E				
	20–100			O			O	O	O	E				
	100–5000		E				E	E	O	E				
	>5000		E	O[c]			E	E		E	O			
suspended solids	0–100	O							O					
	100–200	O					E	E						
	>200	O		O[d]			E	E						
color, mg/L	20–70						O	O				O		
	>70						E	E				O		
tastes and odors	noticeable		O		O					O		E		
CaCO$_3$, mg/L	>200					E	E	E					E	
pH	<5.0–9.0<													
iron and manganese, mg/L	≤0.3		O	O										
	0.3–1.0				O		E	E	O					
	>1.0		E		E		E	E	O				O	
chloride, mg/L	0–250													
	250–500													O
	500+													E
phenolic compounds, mg/L	0–0.005						O	O			O	O		
	>0.005						E	E			O	E	O	
toxic chemicals							E	E					E	O
less critical chemicals							O	O					O	O

Note: E = essential, O = optional
[a]Superchlorination shall be followed by dechlorination.
[b]As an alternative, dilute with low chloride water.
[c]Double settling shall be provided for coliform exceeding 20,000 most probably number/100 mL[5].
[d]For extremely muddy water, presedimentation by plain settling may be provided.
[e]mpn = most probable number

3. PRETREATMENT

Preliminary treatment is a general term that usually includes all processes prior to the first flocculation operation (i.e, pretreatment, screening, presedimentation, microstraining, aeration, and chlorination). Flow measurement is usually considered to be part of the pretreatment sequence.

4. SCREENING

Screens are used to protect pumps and mixing equipment from large objects. The degree of screening required will depend on the nature of solids expected. Screens can be either manually or automatically cleaned.

5. MICROSTRAINING

Microstrainers are effective at removing 50 to 95% of the algae in incoming water. Microstrainers are constructed from woven stainless steel fabric mounted on a hollow drum that rotates at 4 to 7 rpm. Flow is usually radially outward through the drum. The accumulated filter cake is removed by backwashing.

6. ALGAE PRETREATMENT

Biological growth in water from impounding reservoirs, lakes, storage reservoirs, and settling basins can be prevented or eliminated with an *algicide* such as copper sulfate. Such growth can produce unwanted taste and odors, clog fine-mesh filters, and contribute to the buildup of slime.

Typical dosages vary from 0.54 to 5.4 lbm/ac-ft (0.2 to 2.0 mg/L), with the lower dose usually being adequate for waters that are soft or have alkalinities less than 50 mg/L as $CaCO_3$. (Units of mg/L and ppm are equivalent.) Since it is toxic, copper sulfate should not be used without considering and monitoring the effects on aquatic life (e.g., fish).

7. PRECHLORINATION

A prechlorination process was traditionally employed in most water treatment plants. Many plants have now eliminated all or part of the prechlorination due to the formation of THMs. In so doing, benefits such as algae control have been eliminated. Alternative disinfection chemicals (e.g., ozone and potassium permanganate) can be used. Otherwise, coagulant doses must be increased.

8. PRESEDIMENTATION

The purpose of presedimentation is to remove easily settled sand and grit. This can be accomplished by using pure sedimentation basins, sand and grit chambers, and various passive cyclone degritters. *Trash racks* may be integrated into sedimentation basins to remove leaves and other floating debris.

9. FLOW MEASUREMENT

Flow measurement is incorporated into the treatment process whenever the water is conditioned enough to be compatible with the measurement equipment. Flow measurement devices should not be exposed to scour from grit or highly corrosive chemicals (e.g., chlorine).

Flow measurement often takes place in a Parshall flume. Chemicals may be added in the flume to take advantage of the turbulent mixing that occurs at that point. All of the traditional fluid measurement devices are also applicable, including venturi meters, orifice plates, propeller and turbine meters, and modern passive devices such as magnetic and ultrasonic flowmeters.

10. AERATION

Aeration is used to reduce taste- and odor-causing compounds, to lower the concentration of dissolved gases (e.g., hydrogen sulfide), to increase dissolved CO_2 (i.e., recarbonation) or decrease CO_2, to reduce iron and manganese, and to increase dissolved oxygen.

Various types of aerators are used. The best transfer efficiencies are achieved when the air-water contact area is large, the air is changed rapidly, and the aeration period is long. *Force draft air injection* is common. The release depth varies from 10 to 25 ft (3 to 7.5 m). The ideal compression power required to aerate water with a simple air injector depends on the air flow rate, Q, and head (i.e., which must be greater than

the release depth), h, at the point where the air is injected. Motor-compression-distribution efficiencies are typically around 75%.

$$P_{kW} = \frac{Q_{L/s}h_m}{100} \qquad \text{[SI]} \qquad 26.1(a)$$

$$P_{hp} = \frac{Q_{cfm}h_{ft}}{528} \qquad \text{[U.S.]} \qquad 26.1(b)$$

Diffused air systems with compressed air at 5 to 10 psig (35 to 70 kPa) are the most efficient methods of aerating water. The air injection rate is 0.2 to 0.3 ft^3/gal. However, the equipment required to produce and deliver compressed air is more complex than with simple injectors. The *transfer efficiency* of a diffused air system varies with depth and bubble size. If coarse bubbles are produced, only 4 to 8% of the available oxygen will be transferred to the water. With medium-sized bubbles, the efficiency can be 6 to 15%, and it can approach 10 to 30% with fine bubble systems.

11. SEDIMENTATION PHYSICS

Water containing suspended sediment can be held in a *plain sedimentation tank* (basin) that allows the particles to settle out.[1] Settling velocity and settling time for sediment depends on the water temperature (i.e., viscosity), particle size, and particle specific gravity. (The specific gravity of sand is usually taken as 2.65.) Typical settling velocities are: gravel, 1 m/s; coarse sand, 0.1 m/s; fine sand, 0.01 m/s; and silt, 0.0001 m/s. Bacteria and colloidal particles are generally considered to be nonsettleable during the detention periods available in water treatment facilities.

Settlement time can be calculated from the settling velocity and the depth of the tank. If it is necessary to determine the settling velocity of a particle with diameter D, the following procedure can be used.[2]

step: 1 Assume a settling velocity, v_s.

step: 2 Calculate the settling Reynolds number, Re.

$$\text{Re} = \frac{v_s D}{\nu} \qquad 26.2$$

step 3(a): If Re < 1, use *Stokes' law*.

$$v_{s,m/s} = \frac{(\rho_{particle} - \rho_{water})D_m^2 g}{18\mu}$$

$$= \frac{(SG_{particle} - 1)D_m^2 g}{18\nu} \qquad \text{[SI]} \qquad 26.3(a)$$

$$v_{s,ft/sec} = \frac{(\rho_{particle} - \rho_{water})D_{ft}^2 g}{18\mu g_c}$$

$$= \frac{(SG_{particle} - 1)D_{ft}^2 g}{18\nu} \qquad \text{[U.S.]} \qquad 26.3(b)$$

---

[1]The term "plain" refers to the fact that no chemicals are used as coagulants.

[2]This calculation procedure is appropriate for the *Type I settling* that describes sand, sediment, and grit particle performance. It is not appropriate for *Type II settling*, which describes floc and other particles that grow as they settle.

step 3(b): If $1 < \text{Re} < 2000$, use Fig. 26.2, which gives theoretical settling velocities in 68°F (20°C) water for spherical particles with specific gravities of 1.05, 1.2, and 2.65. Actual settling velocities will be much less than shown because particles are not actually spherical.

Figure 26.2 *Settling Velocities, 68° F (20° C)*

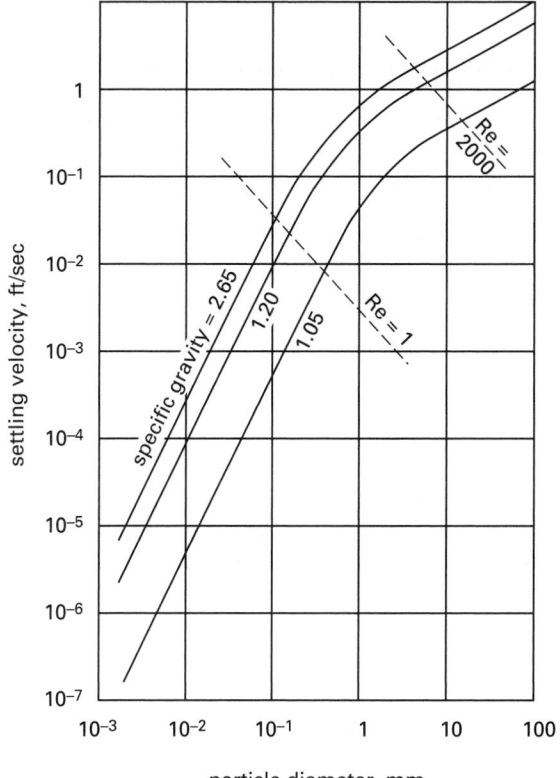

step 3(c): If $\text{Re} > 2000$, use *Newton's first law of motion* and balance the weight of the particle against the buoyant and drag forces. A spherical shape is often assumed in determining the drag coefficient, C_D. In that case, for laminar descent, use $C_D = 24/\text{Re}$.

$$v_s = \sqrt{\frac{4gD(\text{SG}_{\text{particle}} - 1)}{3C_D}} \qquad 26.4$$

12. SEDIMENTATION TANKS

Sedimentation tanks are usually concrete, rectangular, or circular in plan, and are equipped with scrapers or raking arms to periodically remove accumulated sediment. Steel should be used only for small or temporary installations. Where steel parts are unavoidable, as in the case of some rotor parts, adequate corrosion resistance is necessary.

Figure 26.3 *Sedimentation Basin*

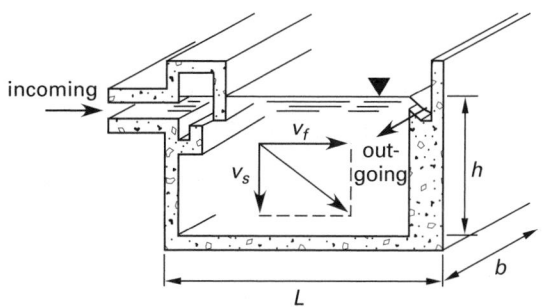

Water flows through the tank at the average *flow-through velocity*, v_f. The flow-through velocity can be found by injecting a colored dye into the tank. It should not exceed 1 ft/min (0.5 cm/sec). The time that water spends in the tank depends on the flow-through velocity and the tank length, L, typically 100 to 200 ft (30 to 60 m).

The minimum settling time depends on the tank depth, h, typically 6 to 15 ft (1.8 to 4.5 m).

$$t_{\text{settling}} = \frac{h}{v_s} \qquad 26.5$$

The time that water remains in the basin is known as the *detention time* (*retention time, detention period,* etc.). Typical detention times are 2 to 6 hr, although periods from 1 to 12 hr are used depending on the size of particles. All particles will be removed whose t_{settling} is less than t_d.

$$t_d = \frac{V_{\text{tank}}}{Q} = \frac{Ah}{Q} \qquad 26.6$$

Rectangular basins are preferred. Rectangular basins should be constructed with aspect ratios greater than 3:1, and preferably greater than 4:1. The bottom should be sloped toward the drain at no less than 1%. Multiple inlet ports along the entire inlet wall should be used. If there are fewer than four inlet ports, an inlet baffle should be provided.

Square or circular basins are appropriate only when space is limited. The slope toward the drain should be greater, typically 8%. A baffled center inlet should be provided. For radial-flow basins, the diameter is on the order of 100 ft (30 m).

Basin efficiency can approach 80% for fine sediments. Virtually all of the coarse particles are removed. Theoretically, all particles with settling velocities greater than the *overflow rate*, v^*, also known as the *surface loading* or *critical velocity*, will be removed. In Eq. 26.7, b is the tank width, typically 30 to 40 ft (9 to 12 m). At least two basins should be constructed (both normally operating) so that one can be out of service for cleaning during low-volume periods without interrupting plant operation. Therefore, Q_{filter} in Eq. 26.7 should be calculated by dividing the total plant flow by 2 or more.

The overflow rate is typically 600 to 1000 gpd/ft^2 (24 to 40 kL/day·m^2) for rectangular tanks. For square and circular basins, the range is approximately 500 to 750 gpd/ft^2 (20 to 30 kL/day·m^2).

$$v^* = \frac{Q_{\text{filter}}}{A_{\text{surface}}} = \frac{Q_{\text{filter}}}{bL} \qquad 26.7$$

In some modern designs, sedimentation has been enhanced by the installation of vertically inclined *laminar tubes* or inclined plates *(lamella plates)* at the tank bottom. These passive tubes, typically about 2 in (50 mm) in diameter, allow the particles to fall only a short distance in more turbulent water before entering the tube in which the flow is laminar. Incoming solids settle to the lower surface of the tube, slide downward, exit the tube, and settle to the sedimentation basin floor.

The *weir loading* is the daily flow rate divided by the total effluent weir length. Weir loading is commonly specified as 15,000 to 20,000 gal/day-ft (190 to 250 kL/day·m), but certainly less than 50,000 gal/day-ft (630 kL/day·m).

The accumulated sediment is referred to as *sludge*. It is removed either periodically or on a continual basis, when it has reached a concentration of 25 mg/L or is organic. Various methods of removing the sludge are used, including scrapers and pumps. The linear velocity of sludge scrapers should be 15 ft/min (7.5 cm/sec) or higher.

13. COAGULANTS

Various chemicals can be added to remove fine solids. There are two main categories of coagulating chemicals: hydrolyzing metal ions (based on either aluminum or iron) and ionic polymers. Since the chemicals work by agglomerating particles in the water to form floc, they are known as *coagulants*. *Floc* is the precipitate that forms when the coagulant allows the colloidal particles to agglomerate.

Common *hydrolyzing metal ion* coagulants are aluminum sulfate ($Al_2(SO_4)_3 \cdot (14\text{-}18)H_2O$, commonly referred to as "alum"), ferrous sulfate ($FeSO_4 \cdot 7H_2O$, sometimes referred to as "copperas"), and chlorinated copperas (a mixture of ferrous sulfate and ferric chloride).

Alum is, by far, the most common compound used in water treatment.[3] Alum provides the positive charges needed to attract and neutralize negative colloidal particles. It reacts with alkalinity in the water to form gelatinous aluminum hydroxide ($Al(OH)_3$) in the proportion of 1 mg/L alum:$^{1}\!/_{2}$ mg/L alkalinity. The aluminum hydroxide forms the nucleus for floc agglomeration.

For alum coagulation to be effective, the following requirements must be met: (a) Enough alum must be used to neutralize all of the incoming negative particles. (b) Enough alkalinity must be present to permit as complete as possible a conversion of the aluminum

sulfate to aluminum hydroxide. (c) The pH must be maintained within the effective range.

Alum reacts with natural alkalinity in the water according to the following reaction. (X is the number of waters of hydration, approximately 14.3.)

$$\begin{aligned} Al_2(SO_4)_3 \cdot XH_2O &+ 3Ca(HCO_3)_2 \\ &\rightarrow 2Al(OH)_3 \downarrow \\ &+ 3CaSO_4 + 6CO_2 + XH_2O \qquad 26.8 \end{aligned}$$

Since alum is naturally acidic, if the water is not sufficiently alkaline, lime (CaO) and soda ash ("caustic soda," Na_2CO_3) can be added for preliminary pH adjustment.[4] If there is inadequate alkalinity and lime is added, the reaction is

$$\begin{aligned} Al_2(SO_4)_3 \cdot XH_2O &+ 3Ca(OH)_2 \\ &\rightarrow 2Al(OH)_3 \downarrow \\ &+ 3CaSO_4 + XH_2O \qquad 26.9 \end{aligned}$$

Alum reacts with soda ash according to

$$\begin{aligned} Al_2(SO_4)_3 \cdot XH_2O &+ 3Na_2CO_3 + 3H_2O \\ &\rightarrow 2Al(OH)_3 \downarrow \\ &+ 3Na_2SO_4 + 3CO_2 + XH_2O \qquad 26.10 \end{aligned}$$

Alum dosage is generally 5 to 50 mg/L, depending on the turbidity. Alum floc is effective within a wide pH range of 5.5 to 8.0. However, the hydrolysis of the aluminum ion depends on pH and other factors in complex ways, and stoichiometric relationships rarely tell the entire story. Although a pH of 6–7 is a typical operating range for alum coagulation, depending on the contaminant, the pH can range as high as 9. Alum removal of chromium, for example, is effective within a pH range of 7–9.

Ferrous sulfate ($FeSO_4 \cdot 7H_2O$) reacts with slaked lime ($Ca(OH)_2$) to flocculate ferric hydroxide ($Fe(OH)_3$). This is an effective method of clarifying turbid water at higher pH.

Ferric sulfate ($Fe_2(SO_4)_3$) reacts with natural alkalinity and lime. It can also be used for color removal at low pH. At high pH, it is useful for iron and manganese removal, as well as a coagulant with precipitation softening.

If alkalinity is high, it may be necessary to use hydrochloric or sulfuric acid for pH control rather than adding lime. The iron salts are effective above a pH of 7, so they may be advantageous in alkaline waters.[5]

Polymers are chains or groups of repeating identical molecules (*mers*) with many available active adsorption sites. Their molecular weights range from several hundred to several million. Polymers can be positively

---

[3]Lime is the second most-used compound.

[4]The cost of soda ash used for pH adjustment is about three times that of lime. Also, soda ash leaves sodium ions in solution, an increasingly undesirable contaminant for people with hypertension.
[5]Optimum pH for ferric floc is approximately 7.5. However, iron salts are active over the pH range of 3.8 to 10.

charged (cationic polymers), negatively charged (anionic polymers), or neutral (nonionic polymers). The charge can vary with pH. *Organic polymers* (*polyelectrolytes*) such as starches and polysaccarides and *synthetic polymers* such as polysacylimides are used.

Polymers are useful in specialized situations, such as when particular metallic ions are to be removed. For example, conventional alum will not remove positive iron ions. However, an anionic polymer will combine with the iron. Polymers are effective in narrow ranges of turbidity and alkalinity. In some cases, it may be necessary to artificially seed the water with clay or alum (to produce floc).

14. FLOCCULATION ADDITIVES

Flocculation additives improve the coagulation efficiency by changing the floc size. Additives include *weighting agents* (e.g., bentonite clays), *adsorbents* (e.g., powdered activated carbon), and *oxidants* (chlorine). Polymers are also used in conjunction with metallic ion coagulants.

15. DOSES OF COAGULANTS AND OTHER COMPOUNDS

Feed rate, F, of a compound with purity P and fractional availability G can be calculated from the *dose equation*, Eq. 26.11. Doses in mg/L and ppm for aqueous solutions are the same. Also, each 1% by weight of concentration is equivalent to 10,000 mg/L of solution. Doses may be given in terms of grains per gallon. There are 7000 grains per pound.

$$F_{\rm kg/day} = \frac{D_{\rm mg/L} Q_{\rm ML/day}}{PG} \qquad {\rm [SI]} \qquad \textit{26.11(a)}$$

$$F_{\rm lbm/day} = \frac{D_{\rm mg/L} Q_{\rm MGD} \left(8.345 \; \frac{\rm lbm\text{-}L}{\rm mg\cdot MG} \right)}{PG}$$
$${\rm [U.S.]} \qquad \textit{26.11(b)}$$

Example 26.1

Incoming water contains 2.5 mg/L as a substance of natural alkalinity (HCO_3^-). The flow rate is 2.5 MGD (9.5 ML/day). (a) What feed rate is required if the alum dose is 7 mg/L? (b) How much lime in the form of $Ca(OH)_2$ is required to react completely with the alum?

SI Solution

(a) The dose is specifically in terms of mg/L of alum. Therefore, Eq. 26.11 can be used. Assume $P = 100\%$.

$$
\begin{aligned}
F_{\rm kg/day} &= \frac{D_{\rm mg/L} Q_{\rm ML/day}}{PG} \\
&= \frac{\left(7 \; \frac{\rm mg}{\rm L} \right) \left(9.5 \; \frac{\rm ML}{\rm day} \right)}{(1.0)(1.0)} = 66.5 \; {\rm kg/day}
\end{aligned}
$$

The amount calculated should also be adjusted for purity of the alum.

(b) From Eq. 26.8, each alum molecule reacts with six ions of alkalinity.

$$
\begin{array}{lcc}
& {\rm Al_2(SO_4)_3} + 6{\rm HCO_3^-} & \rightarrow \\
{\rm MW:} & 342 & (6)(61) \\
{\rm mg/L:} & X & 2.5
\end{array}
$$

By simple proportion, the amount of alum used to counteract the natural alkalinity is

$$X = \frac{\left(2.5 \; \frac{\rm mg}{\rm L} \right) (342)}{(6)(61)} = 2.33 \; {\rm mg/L}$$

The alum remaining is

$$7 \; \frac{\rm mg}{\rm L} - 2.33 \; \frac{\rm mg}{\rm L} = 4.7 \; {\rm mg/L}$$

From Eq. 26.9, one alum molecule reacts with three slaked lime molecules.

$$
\begin{array}{lcc}
& {\rm Al_2(SO_4)_3} + 3{\rm Ca(OH)_2} & \rightarrow \\
{\rm MW:} & 342 & (3)(74) \\
{\rm mg/L:} & 4.7 & X
\end{array}
$$

By simple proportion, the amount of lime needed is

$$X = \frac{\left(4.7 \; \frac{\rm mg}{\rm L} \right) (3)(74)}{342} = 3.05 \; {\rm mg/L}$$

Using Eq. 26.11 and assuming the lime purity is 100%,

$$
\begin{aligned}
F_{\rm kg/day} &= \frac{D_{\rm mg/L} Q_{\rm ML/day}}{PG} \\
&= \left(3.05 \; \frac{\rm mg}{\rm L} \right) \left(9.5 \; \frac{\rm ML}{\rm day} \right) \\
&= 29.0 \; {\rm kg/day}
\end{aligned}
$$

Customary U.S. Solution

(a) The dose is specifically in terms of mg/L of alum. Therefore, Eq. 26.11 can be used. Assume $P = 100\%$.

$$
\begin{aligned}
F_{\rm lbm/day} &= \frac{D_{\rm mg/L} Q_{\rm MGD} \left(8.345 \; \frac{\rm lbm\text{-}L}{\rm mg\text{-}MG} \right)}{PG} \\
&= \frac{\left(7 \; \frac{\rm mg}{\rm L} \right) (2.5 \; {\rm MGD}) \left(8.345 \; \frac{\rm lbm\text{-}L}{\rm mg\text{-}MG} \right)}{(1.0)(1.0)} \\
&= 146 \; {\rm lbm/day}
\end{aligned}
$$

The amount calculated should also be adjusted for purity of the alum.

(b) From Eq. 26.8, each alum molecule reacts with six ions of alkalinity.

$$
\begin{array}{lcc}
& {\rm Al_2(SO_4)_3} + 6{\rm HCO_3^-} & \rightarrow \\
{\rm MW:} & 342 & (6)(61) \\
{\rm mg/L:} & X & 2.5
\end{array}
$$

By simple proportion, the amount of alum used to counteract the natural alkalinity is

$$X = \frac{\left(2.5 \dfrac{\text{mg}}{\text{L}}\right)(342)}{(6)(61)} = 2.33 \text{ mg/L}$$

The alum remaining is

$$7 \frac{\text{mg}}{\text{L}} - 2.33 \frac{\text{mg}}{\text{L}} = 4.7 \text{ mg/L}$$

From Eq. 26.8, one alum molecule reacts with three lime molecules.

$$\text{Al}_2(\text{SO}_4)_3 + 3\text{Ca}(\text{OH})_2 \rightarrow$$

MW:	342	(3)(74)
mg/L:	4.7	X

By simple proportion, the amount of lime needed is

$$X = \frac{\left(4.7 \dfrac{\text{mg}}{\text{L}}\right)(3)(74)}{342} = 3.05 \text{ mg/L}$$

Using Eq. 26.11 and assuming the lime purity is 100%,

$$\begin{aligned} F_{\text{lbm/day}} &= \frac{D_{\text{mg/L}} Q_{\text{MGD}} \left(8.345 \dfrac{\text{lbm-L}}{\text{mg-MG}}\right)}{PG} \\ &= \frac{\left(3.05 \dfrac{\text{mg}}{\text{L}}\right)(2.5 \text{ MGD}) \left(8.345 \dfrac{\text{lbm-L}}{\text{mg-MG}}\right)}{(1.0)(1.0)} \\ &= 63.6 \text{ lbm/day} \end{aligned}$$

16. MIXERS AND MIXING KINETICS

Coagulants and other water treatment chemicals are added in *mixers*. If the mixer adds a coagulant for the removal of colloidal sediment, a downstream location (i.e., a tank or basin) with a reduced velocity gradient may be known as a *flocculator*.

There are two basic models: plug flow mixing and complete mixing. The *complete mixing* model is appropriate when the chemical is distributed throughout by impellers or paddles. If the basin volume is small, so that time for mixing is low, the tank is known as a *flash mixer*, *rapid mixer*, or *quick mixer*. The volume of flash mixers is seldom greater than 300 ft³ (8 m³), and flash mixer detention time is usually 30 to 60 sec. Flash mixing kinetics are described by the complete mixing model.

Flash mixers are usually concrete tanks, square in horizontal cross section, and fitted with vertical shaft impellers. The size of a mixing basin can be determined from various combinations of dimensions that satisfy the volume-flow rate relationship.

$$V = tQ \qquad\qquad 26.12$$

The detention time required for complete mixing in a tank of volume V depends on the *mixing rate constant*, K, and the incoming and outgoing concentrations.

$$t_{\text{complete}} = \frac{V}{Q} = \frac{1}{K}\left(\frac{C_i}{C_o} - 1\right) \qquad 26.13$$

The *plug flow mixing* model is appropriate when the water flows through a long narrow chamber, the chemical is added at the entrance, and there is no mechanical agitation. All of the molecules remain in the plug flow mixer for the same amount of time as they flow through. For any mixer, the maximum chemical conversion will occur with plug flow, since all of the molecules have the maximum opportunity to react. The detention time in a plug flow mixer of length L is

$$\begin{aligned} t_{\text{plug flow}} &= \frac{V}{Q} \\ &= \frac{L}{\text{v}_{\text{flow through}}} = \frac{1}{K} \ln\left(\frac{C_i}{C_o}\right) \qquad 26.14 \end{aligned}$$

17. MIXING PHYSICS

The drag force on a paddle is given by the standard fluid drag force equation. For flat plates, the coefficient of drag, C_D, is approximately 1.8.

$$F_D = \frac{C_D A \rho \text{v}_{\text{mixing}}^2}{2} \qquad\qquad \text{[SI]} \quad 26.15(a)$$

$$F_D = \frac{C_D A \rho \text{v}_{\text{mixing}}^2}{2g_c} = \frac{C_D A \gamma \text{v}_{\text{mixing}}^2}{2g}$$
$$\text{[U.S.]} \quad 26.15(b)$$

The power required is calculated from the drag force and the mixing velocity. The average *mixing velocity*, v_{mixing}, also known as the *relative paddle velocity*, is the difference in paddle and average water velocities. The mixing velocity is approximately 0.7 to 0.8 times the tip speed.

$$\text{v}_{\text{paddle,ft/sec}} = \frac{2\pi R n_{\text{rpm}}}{60 \dfrac{\text{sec}}{\text{min}}} \qquad\qquad 26.16$$

$$\text{v}_{\text{mixing}} = \text{v}_{\text{paddle}} - \text{v}_{\text{water}} \qquad\qquad 26.17$$

$$P_{\text{kW}} = \frac{F_D \text{v}_{\text{mixing}}}{1000 \dfrac{\text{W}}{\text{kW}}} \qquad\qquad 26.18$$

$$= \frac{C_D A \rho \text{v}_{\text{mixing}}^3}{2\left(1000 \dfrac{\text{W}}{\text{kW}}\right)} \qquad \text{[SI]} \quad 26.19(a)$$

$$P_{\text{hp}} = \frac{F_D \text{v}_{\text{mixing}}}{550 \dfrac{\text{ft-lbf}}{\text{hp-sec}}}$$

$$= \frac{C_D A \gamma \text{v}_{\text{mixing}}^3}{2g\left(550 \dfrac{\text{ft-lbf}}{\text{hp-sec}}\right)} \qquad \text{[U.S.]} \quad 26.19(b)$$

For slow-moving paddle mixers, the *velocity gradient*, G, varies from 20 to 75 sec^{-1} for a 15 to 30 minute mixing period. Typical units in Eq. 26.20 are ft-lbf/sec for power (multiply hp by 550 to obtain ft-lbf/sec), lbf-sec/ft² for μ, and ft³ for volume. In the SI system, power in kW is multiplied by 1000 to obtain W, viscosity is in Pa·s, and volume is in m³. (Multiply viscosity in cP by 0.001 to obtain Pa·s.)

$$G = \sqrt{\frac{P}{\mu V_{\text{tank}}}} \qquad 26.20$$

Equation 26.20 can also be used for rapid mixers, in which case the mean velocity gradient is much higher: approximately 500 to 1000 for 10 to 30 sec mixing period, or 3000 to 5000 for a 0.5 to 1.0 sec mixing period in an in-line blender configuration.

Equation 26.20 can be rearranged to calculate the power requirement. Power is typically 0.5 to 1.5 hp/MGD for rapid mixers.

$$P = \mu G^2 V_{\text{tank}} \qquad 26.21$$

The dimensionless product, Gt_d, of the velocity gradient and detention time is known as the *mixing opportunity parameter*. Typical values range from 10^4 to 10^5.

$$Gt_d = \frac{V_{\text{tank}}}{Q}\sqrt{\frac{P}{\mu V_{\text{tank}}}}$$
$$= \frac{1}{Q}\sqrt{\frac{P V_{\text{tank}}}{\mu}} \qquad 26.22$$

18. IMPELLER CHARACTERISTICS

Mixing equipment uses rotating impellers on rotating shafts. The blades of *radial-flow impellers* (paddle-type impellers, turbine impellers, etc.) are parallel to the drive shaft. *Axial-flow impellers* (propellers, pitched-blade impellers, etc.) have blades inclined with respect to the drive shaft. Axial-flow impellers are better at keeping materials (e.g., water softening chemicals) in suspension.

Figure 26.4 *Typical Axial Flow Mixing Impellers*

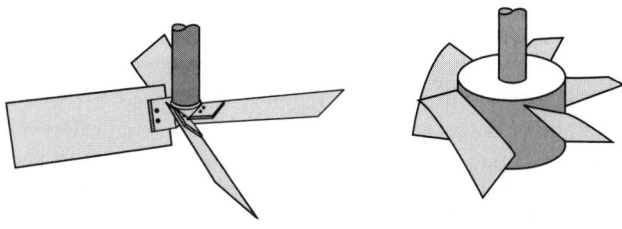

The *mixing Reynolds number* depends on the impeller diameter, D, suspension specific gravity, SG, liquid viscosity, μ, and rotational speed, n, in rps. Flow is in

transition for $10 < \text{Re} < 10{,}000$, is laminar below 10, and is turbulent above 10,000.

$$\text{Re} = \frac{D^2 n \rho}{\mu} \qquad \text{[SI]} \qquad 26.23(a)$$

$$\text{Re} = \frac{D^2 n \rho}{g_c \mu} \qquad \text{[U.S.]} \qquad 26.23(b)$$

The dimensionless *power number*, N_P, of the impeller is defined implicitly by the power required to drive the impeller. (For any given level of suspension, Eqs. 26.24 through 26.27 apply to geometrically similar impellers and turbulent flow.

$$P = N_P n^3 D^5 \rho \qquad \text{[SI]} \qquad 26.24(a)$$

$$P = \frac{N_P n^3 D^5 \rho}{g_c} \qquad \text{[U.S.]} \qquad 26.24(b)$$

$$P = \rho g Q h_{\text{v}} \qquad \text{[SI]} \qquad 26.25(a)$$

$$P = \frac{\rho g Q h_{\text{v}}}{g_c} \qquad \text{[U.S.]} \qquad 26.25(b)$$

The impeller's dimensionless *flow number*, N_Q, is defined implicitly by the *flow rate equation*.

$$Q = N_Q n D^3 \qquad 26.26$$

The velocity head can be calculated from a combination of the power and flow numbers.

$$h_{\text{v}} = \frac{N_P n^2 D^2}{N_Q g} \qquad 26.27$$

Vibration near the critical speed can be a major problem with modern, high-efficiency, high-speed impellers. Mixing speed should be well below the first critical speed of the shaft. Other important design factors include tip speed and shaft bending moment.

19. FLOCCULATION

After flash mixing, the floc is allowed to form during a 20 to 60 min period of gentle mixing. Flocculation is enhanced by the gentle agitation, but the floc disintegrates with violent agitation. During this period, the flow-through velocity should be limited to 0.5 to 1.5 ft/min (0.25 to 0.75 cm/s). The peripheral speed of mixing paddles should vary approximately from 0.5 ft/sec (0.15 m/s) for fragile, cold-water floc to 3.0 ft/sec (0.9 m/s) for warm-water floc.

Many modern designs make use of *tapered flocculation*, also known as *tapered energy*, a process in which the amount (severity) of flocculation gradually decreases as the treated water progresses through the flocculation.

Flocculation is followed by sedimentation for 2 to 8 hr (4 hr typical) in a low-velocity portion of the basin.

(The flocculation time is determined from settling column data.) A good settling process will remove 90% of the settleable solids. Poor design, usually resulting in some form of *short-circuiting* of the flow path, will reduce the effective time in which particles have to settle.

20. FLOCCULATOR-CLARIFIERS

A *flocculator-clarifier* combines mixing, flocculation, and sedimentation into a single tank. Such units are called *solid contact units* and *upflow tanks*. They are generally round in construction, with mixing and flocculation taking place near the central hub and sedimentation occurring at the periphery. Flocculator-clarifiers are most suitable when combined with softening, since the precipitated solids help seed the floc.

Typical operational characteristics of flocculator-clarifiers are as follows.

typical flocculation and mixing time	20 to 60 min
minimum detention time	1.5 to 2.0 hr
maximum weir loading	10 gpm/ft (2 L/s·m)
upflow rate	0.8 to 1.7 gpm/ft^2; 1.0 gpm/ft^2 typical (0.54 to 1.2 L/s·m^2; 0.68 L/s·m^2 typical)
maximum sludge formation rate	5% of water flow

21. SLUDGE QUANTITIES

Sludge is the watery waste that carries off the settled floc and water softening precipitates. The sludge volume produced is given by Eq. 26.28, in which G is the gravimetric fraction of solids. The gravimetric solids fraction for coagulation sludge is generally less than 0.02. For water softening sludge, it is on the order of 0.10. In Eq. 26.28, m_{sludge} can be either a specific quantity or a rate of production per unit time.

$$V_{sludge} = \frac{m_{sludge}}{\rho_{water}(SG)_{sludge}G} \qquad 26.28$$

The most accurate way to calculate the mass of sludge is to extrapolate from jar or pilot test data. There is no absolute correlation between the mass generated and other water quality measurements. However, a few generalizations are possible. (a) Each unit mass (lbm, mg/L, etc.) of alum produces 0.46 unit mass of floc.[6] (b) 100% of the reduction in suspended solids (expressed as substance) shows up as floc. (c) 100% of any supplemental flocculation aids is recovered in the sludge.

Suspended solids in water may be reported in turbidity units. There is no easy way to calculate total suspended solids (TSS) in mg/L from turbidity in NTU. The ratio

---

[6]Some researchers report other ratios, including 0.26. However, 0.456 is the stoichiometric ratio, as determined from Eq. 26.8.

---

of TSS to NTU normally varies from 1.0 to 2.0, and can be as high as 10. A value of 1 or 1.5 is generally appropriate.

The rate of dry sludge production from coagulation processes can be estimated from Eq. 26.29. M is an extra factor that accounts for the use of any miscellaneous inorganic additives such as clay.

$$m_{sludge,kg/day} =$$
$$\frac{\left(86{,}400 \dfrac{s}{day}\right)(Q_{m^3/s})\left(1000 \dfrac{L}{m^3}\right)}{\times (0.46D_{alum,mg/L} + \Delta TSS_{mg/L} + M_{mg/L})}{10^6 \dfrac{mg}{kg}}$$

$$[SI] \qquad 26.29(a)$$

$$m_{sludge,lbm/day} =$$
$$\left(8.345 \frac{lbm\text{-}L}{mg\text{-}MG}\right)$$
$$\times (Q_{MGD})(0.46D_{alum,mg/L} + \Delta TSS_{mg/L} + M_{mg/L})$$

$$[U.S.] \qquad 26.29(b)$$

After a sludge dewatering process from sludge solids constant G_1 to G_2, the resulting volume will be

$$V_{2,sludge} \approx V_{1,sludge}\left(\frac{G_1}{G_2}\right) \qquad 26.30$$

22. FILTRATION

Nonsettling floc, algae, suspended precipitates from softening, and metallic ions (iron and manganese) are removed by filtering. *Sand filters* (and in particular, rapid sand filters) are commonly used for this purpose. Sand filters are beds of gravel, sand, and other granulated materials.[7]

Although filter box heights are on the order of 10 ft (3 m) to provide for expansion and freeboard during backwashing, the almost-universal specification used before 1960 for a single-media filter was 24 to 30 in (610 to 760 mm) of sand or ground coal. Filters are usually square or nearly square in plan, and they operate with a hydraulic head of 1 to 8 ft (0.3 to 2.4 m).

Although most plants have six or more filters, there should be at least three filters so that two can be operational when one is being cleaned. Cleaning of multiple filters should be staged.

Filter operation is enhanced when the top layer of sand is slightly more coarse than the deeper sand. During backwashing, however, the finest sand rises to the top. Various *dual-layer* and *multi-layer* (also known as *dual-media* and *multi-media*) filter designs using layers of coal

---

[7]Most sand filter beds were designed when a turbidity level of 5 NTU was acceptable. With the U.S. federal MCL at 1 NTU, some states at 0.5 NTU, and planning "on the horizon" for 0.2 NTU, these early conventional filters are clearly inadequate.

("anthrafilt") and (more recently) granular activated carbon, alone or in conjunction with sand, overcome this problem. Since coal has a lower specific gravity than sand, media particle size is larger, as is pore space.

Historically, the *loading rate* (*flow rate*) was set at 2 to 3 gpm/ft^2 (1.4 to 2.0 L/s·m^2) for *rapid sand filters*. 4 to 6 gpm/ft^2 (2.7 to 4.1 L/s·m^2) is a reasonable minimum rate for dual-media filters. Multi-media filters operate at 5 to 10 gpm/ft^2 (3.4 to 6.8 L/s·m^2) and above.[8] The filter loading rate is

$$\text{loading rate} = \frac{Q}{A} \qquad 26.31$$

Filters discharge into a storage reservoir known as a *clearwell*. The *hydraulic head* (the distance between the water surfaces in the filter and clearwell) is usually 9 to 12 ft (2.7 to 3.6 m). This allows for a substantial decrease in available head prior to backwashing. Clearwell storage volume is 30 to 60% of the daily filter output, with a minimum capacity of 12 hr of the maximum daily demand so that demand can be satisfied from the clearwell while the filter is being cleaned or serviced.

23. FILTER BACKWASHING

The most common type of service needed by filters is backwashing, which is needed when the pores between the filter particles clog up. Typically, this occurs after 1 to 3 days of operation, when the head loss reaches 6 to 8 ft (1.6 to 2.4 m). There are two parameters that can trigger backwashing: head loss and turbidity. Head loss increases almost linearly with time, while turbidity remains constant for several days before suddenly increasing. The point of sudden increase is known as *breakthrough*.[9] Since head loss is more easily monitored than turbidity, it is desired to have head loss trigger the backwashing cycle.

Backwashing with filtered water pumped back through the filter from the bottom to the top expands the sand layer 30 to 50%, which dislodges trapped material. Backwashing for 3 to 5 min at 8 to 15 gpm/ft^2 (5.4 to 10 L/s·m^2) is a typical specification.[10] The head loss is reduced to approximately 1 ft (0.3 m) after washing.

Experience has shown that supplementary agitation of the filter media is necessary to prevent "caking" and "mudballs" in almost all installations. Prior to backwashing, the filter material may be expanded by an *air prewash* volume of 1 to 8 (2 to 5 typical) times the sand filter volume per minute for 2 to 10 (3 to 5 typical) min. Alternatively, turbulence in the filter material may

be encouraged during backwashing with an *air wash* or with rotating hydraulic surface jets.

During backwashing, the rate at which the water rises in the filter housing will be 1 to 3 ft/min (0.5 to 1.5 cm/s). This rise should not exceed the settling velocity of the smallest particle that is to be retained in the filter. The wash water, which is collected in troughs for disposal, constitutes approximately 1 to 5% of the total processed water. The total water used is approximately 75 to 100 gal/ft^2 (3 to 4 kL/m^2). The actual amount of backwash water is

$$V = A_{\text{filter}}(\text{rate of rise})t_{\text{backwash}} \qquad 26.32$$

The temperature of the water used in backwashing is important since the viscosity changes. 40°F (4°C) water is significantly more viscous than 70°F (21°C) water. Media particles, therefore, may be expanded to the same extent using lower upflow rates at the lower backwash temperature.

24. OTHER FILTRATION METHODS

Pressure filters operate similarly to rapid sand filters except that incoming water is pressurized up to 25 ft (0.75 m) gage. Filter rates are 2 to 4 gpm/ft^2 (1.4 to 2.7 L/s·m^2). Pressure filters are not used in large installations.

Ultrafilters are membranes that act as sieves to retain turbidity, microorganisms, and large organic molecules that are THM precursors, while allowing water, salts, and small molecules to pass through. Ultrafiltration is effective in removing particles ranging in size of 0.001 to 10 μm.[11] A pressure of 15 to 75 psig (100 to 500 kPa) is required to drive the water through the membrane.

Biofilm filtration (*biofilm process*) uses microorganisms to remove selected contaminants (e.g., aromatics and other hydrocarbons). Operation of biofilters is similar to trickling filters used in wastewater processing. Sand filter facilities are relatively easy to modify—sand is replaced with gravel in the 4 to 14 mm size range, application rates are decreased, and exposure to chlorine from incoming and backwash water is eliminated.

Slow sand filters are primarily of historical interest, though there are some similarities with modern biomethods used to remediate toxic spills. Slow sand filters operate similarly to rapid sand filters except that the exposed surface (loading) area is much larger and the flow rate is much lower (0.05 to 0.1 gpm/ft^2; 0.03 to 0.07 L/s·m^2). Slow sand filters are limited to low turbidity applications not requiring chemical treatment and where large rural areas are available to spread out the facilities. Slow sand filters must be operated as biofilm processes with a layer of biological slime that is allowed to form on the top of the filter medium.[12] Slow filter cleaning usually involves removing a few inches of sand.

---

[8]Recent testing has demonstrated that deep-bed, uniformly graded anthracite filters can operate at 10 to 15 gpm/ft^2 (7 to 10 L/s·m^2), although this requires high-efficiency preozonation and microflocculation.

[9]Technically, *breakthrough* is the point at which the turbidity rises above the MCL permitted. With low MCLs, this occurs very soon after the beginning of filter performance degradation.

[10]While the maximum may never be used, a maximum backwash rate of 20 gpm/ft^2 (14 L/s·m^2) should be provided for.

[11]A μm is the same as a micron.

[12]The biological slime is called *schmutzdecke*. The slime forms a physical barrier that traps particles.

25. ADSORPTION

Many dissolved organic molecules and inorganic ions can be removed by adsorption processes. *Adsorption* is not a straining process, but occurs when contaminants are trapped on the surface or interior of the adsorption particles.

Granular activated carbon (GAC) is considered to be the best available technology for removal of THMs and synthetic organic chemicals from water. GAC is also useful in removing compounds that contribute to taste, color, and odor. Activated carbon can be used in the form of powder (which must be subsequently removed) or granules. GAC can be integrated into the design of sand filters.[13]

Eventually, the GAC becomes saturated and must be removed and reactivated. In such dual-media filters, the reactivation interval for GAC is approximately 1 to 2 yr.

26. FLUORIDATION

Fluoridation can occur any time after filtering and can involve solid compounds or liquid solutions. Small utilities may manufacturer their own liquid solution on-site from sodium silicofluoride (Na_2SiF_6, typically 22 to 30% purity) or sodium fluoride (NaF, typically 90 to 98% purity) for use with a volumetric metering system. Larger utilities use gravimetric dry feeders with sodium silicofluoride (Na_2SiF_6, typically 98 to 99% purity) or solution feeders with fluorsilic acid (H_2SiF_6).

---

[13]For example, an 80 inch GAC layer may be placed on top of a 40 inch sand layer.

Assuming 100% ionization, the application rate, F (in pounds per day), of a compound with fluoride gravimetric fraction, G, and a fractional purity, P, needed to obtain a final concentration, C, of fluoride is

$$F_{kg/day} = \frac{C_{mg/L} Q_{L/d}}{PG} \qquad \text{[SI]} \qquad 26.33(a)$$

$$F_{lbm/day} = \frac{C_{mg/L} Q_{MGD} \left(8.345 \ \frac{lbm\text{-}L}{mg\text{-}MG} \right)}{PG} \qquad \text{[U.S.]} \qquad 26.33(b)$$

27. IRON AND MANGANESE REMOVAL

Several methods can be used to remove iron and manganese. Most involve aeration with chemical oxidation since manganese is not easily removed by aeration alone. These processes are described in Table 26.2.

28. TASTE AND ODOR CONTROL

A number of different processes affect taste and odor. Some are more effective and appropriate than others. *Microstraining*, using a 35 μm or finer metal cloth, can be used to reduce the number of algae and other organisms in the water, since these are sources of subsequent tastes and odors. Microstraining does not remove dissolved or colloidal organic material, however.

Table 26.2 Processes for Iron and Manganese Removal

processes	iron and/or manganese removed	pH required	remarks
aeration, settling, and filtration	ferrous bicarbonate	7.5	provide aeration unless incoming water contains adequate dissolved oxygen
	ferrous sulfate	8.0	
	manganous bicarbonate	10.3	
	manganous sulfate	10.0	
aeration, free residual chlorination, settling, and filtration	ferrous bicarbonate	5.0	provide aeration unless incoming water contains adequate dissolved oxygen
	manganous bicarbonate	9.0	
aeration, lime softening, settling, and filtration	ferrous bicarbonate	8.5–9.6	
	manganous bicarbonate		
aeration, coagulation, lime softening, settling, and filtration	colloidal or organic iron	8.5–9.6	require lime, and alum or iron coagulant
	colloidal or organic manganese	10.0	
ion exchange	ferrous bicarbonate	≈ 6.5	water must be devoid of oxygen
	manganous bicarbonate		iron and managanese in raw water not to exceed 2.0 mg/L
			consult manufacturers for type of ion exchange resin to be used

Activated carbon removes more tastes and odors, as well as a wide variety of chemical contaminants.

Aeration can be used when dissolved oxygen is low or when hydrogen sulfide is present. Aeration has little effect on most other tastes and odors.

Chlorination disinfects and reduces odors caused by organic matter and industrial wastes. Normally, the dosage required will be several times greater than those for ordinary disinfection, and the term *superchlorination* is used to describe applying enough chlorine to maintain an excessively large residual. Subsequent dechlorination will be required to remove the excess chlorine. Similar results will be obtained with chlorine dioxide and other disinfection products.

A quantitative odor ranking is the *threshold odor number* (TON), which is determined by adding increasing amounts of odor-free dilution water to a sample until the combined sample is virtually odor free.

$$\text{TON} = \frac{V_{\text{raw sample}} + V_{\text{dilution water}}}{V_{\text{raw sample}}}$$

A TON of 3 or less is ideal. Untreated river water usually has a TON between 6 and 24. Treated water normally has a TONs between 3 and 6. At TON of 5 and above, customers will begin to notice the taste and odor of their water.

29. PRECIPITATION SOFTENING

Precipitation softening using the *lime-soda ash process* adds lime (CaO), also known as *quicklime*, and soda ash (Na_2CO_3) to remove calcium and magnesium from hard water.[14] Granular quicklime is available with a minimum purity of 90%, and soda ash is available with a 98% purity.

Lime forms *slaked lime* (also known as *hydrated lime*), $Ca(OH)_2$, in an exothermic reaction when added to feed water. The slaked lime is delivered to the water supply as a *milk of lime* suspension.

$$CaO + H_2O \rightarrow Ca(OH)_2 + \text{heat} \qquad 26.34$$

Slaked lime reacts first with any carbon dioxide dissolved in the water, as in Eq. 26.35. No softening occurs, but the carbon dioxide demand must be satisfied before any reactions involving calcium or magnesium can occur.

carbon dioxide removal:

$$CO_2 + Ca(OH)_2 \rightarrow CaCO_3 \downarrow + H_2O \qquad 26.35$$

Lime next reacts with any carbonate hardness, precipitating calcium carbonate and magnesium hydroxide as shown in Eqs. 26.36 and 26.37. Note that removal of

carbonate hardness caused by magnesium (characterized by magnesium bicarbonate in Eq. 26.37) requires two molecules of calcium hydroxide to precipitate calcium carbonate and magnesium hydroxide.

calcium carbonate hardness removal:

$$Ca(HCO_3)_2 + Ca(OH)_2 \rightarrow 2CaCO_3 \downarrow + 2H_2O \qquad 26.36$$

magnesium carbonate hardness removal:

$$Mg(HCO_3)_2 + 2Ca(OH)_2 \rightarrow$$
$$2CaCO_3 \downarrow + Mg(OH)_2 \downarrow + 2H_2O \qquad 26.37$$

To remove noncarbonate hardness (characterized by sulfate ions in Eqs. 26.38 and 26.39), it is necessary to add soda ash and more lime. The sodium sulfate that remains in solution does not contribute to hardness, for sodium is a single-valent ion (i.e., Na^+).

magnesium noncarbonate hardness removal:

$$MgSO_4 + Ca(OH)_2 \rightarrow$$
$$Mg(OH)_2 \downarrow + CaSO_4 \qquad 26.38$$

calcium noncarbonate hardness removal:

$$CaSO_4 + Na_2CO_3 \rightarrow CaCO_3 \downarrow + Na_2SO_4 \qquad 26.39$$

The calcium ion can be effectively reduced by the lime addition shown in the previous equations, raising the pH of the water to approximately 10.3. (Thus, the lime added removes itself.) Precipitation of the magnesium ion, however, demands a higher pH and the presence of excess lime in the amount of approximately 35 mg/L of CaO or 50 mg/L of $Ca(OH)_2$ above the stoichiometric requirements. The practical limits of precipitation softening are 30 to 40 mg/L of $CaCO_3$ and 10 mg/L of $Mg(OH)_2$, both as $CaCO_3$.

After softening, the water must be recarbonated to lower its pH and to reduce its scale-forming potential. This is accomplished by bubbling carbon dioxide gas through the water.

$$Ca(OH)_2 + CO_2 \rightarrow CaCO_3 \downarrow + H_2O \qquad 26.40$$

The treatment process could be designed such that all of these reactions take place sequentially. That is, first slaked lime would be added (Eqs. 26.36 and 26.37), then the water would be recarbonated (Eq. 26.40), then excess lime would be added to raise the pH, followed by soda ash treatment and recarbonation. In fact, essentially such a sequence is used in a *double-stage process*: two chemical application points and two recarbonation points.

A *split process* can be used to reduce the amount of lime that is neutralized in recarbonation. The excess lime needed to raise the pH is added prior to the first flocculator/clarifier stage. The soda ash is added to the first stage effluent, prior to the second flocculator/clarifier

---

[14]Soda ash does not always need to be used. When lime alone is used, the process may be referred to as *lime softening*.

stage. Recarbonation, if used, is applied to the effluent of the second stage. A portion of the flow is bypassed (i.e., is not softened) and is later recombined with softened water to obtain the desired hardness.

Example 26.2

Water contains 130 mg/L of calcium bicarbonate $(Ca(HCO_3)_2)$ as $CaCO_3$. How much slaked lime $(Ca(OH)_2)$ is required to remove the hardness?

Solution

The hardness is given as a $CaCO_3$ equivalent. Therefore, the amount of slaked lime is implicitly the same: 130 mg/L as $CaCO_3$. Appendix 22.C can be used to convert the quantity to an "as substance" measurement. From App. 22.C, the conversion factor for $Ca(OH)_2$ is 1.35.

$$Ca(OH)_2: \frac{130 \frac{mg}{L}}{1.35} = 96.3 \text{ mg/L as substance}$$

Example 26.3

Water is received with the following characteristics.

total hardness	250 mg/L as $CaCO_3$
alkalinity	150 mg/L as $CaCO_3$
carbon dioxide	5 mg/L as substance

The water is to be treated with precipitation softening and recarbonation. Lime (90% pure) and soda ash (98% pure) are available.

What stoichiometric amounts (as substance) of slaked lime, soda ash, and carbon dioxide are required to reduce the hardness of the water to zero?

Solution

First, convert the carbon dioxide concentration to a $CaCO_3$ equivalent. From App. 22.C, the factor is 2.27.

$$CO_2: \left(5 \frac{mg}{L}\right)(2.27) = 11.35 \text{ mg/L as } CaCO_3$$

Since the alkalinity is less than the hardness and no hydroxides or calcium are reported, it is concluded that the carbonate hardness is equal to the alkalinity, and the noncarbonate hardness is equal to the difference in total hardness and alkalinity.

Since the alkalinity is reported as a $CaCO_3$ equivalent, the first-state treatment to remove carbon dioxide and carbonate hardness requires lime in the amount of

$$Ca(OH)_2: 11.35 \frac{mg}{L} + 150 \frac{mg}{L} =$$
$$161.35 \text{ mg/L as } CaCO_3$$

Approximately 50 mg/L are added to raise the pH so that the noncarbonate hardness can be removed.

$$Ca(OH)_2: 161.35 \frac{mg}{L} + 50 \frac{mg}{L} =$$
$$211.35 \text{ mg/L as } CaCO_3$$

Use the 90% fractional purity and App. 22.C to convert to quantity as substance.

$$Ca(OH)_2: \frac{211.35 \frac{mg}{L}}{(1.35)(0.9)} = 174.0 \text{ mg/L as substance}$$

The noncarbonate hardness is

$$250 \frac{mg}{L} - 150 \frac{mg}{L} = 100 \text{ mg/L as } CaCO_3$$

The soda ash requirement is

$$Na_2CO_3: \frac{100 \frac{mg}{L}}{(0.94)(0.98)} = 108.6 \text{ mg/L as substance}$$

The recarbonation required to lower the pH depends on the amount of excess lime added.

$$CO_2: \frac{50 \frac{mg}{L}}{2.27} = 22.03 \text{ mg/L}$$

30. ADVANTAGES AND DISADVANTAGES OF PRECIPITATION SOFTENING

Precipitation softening is relatively inexpensive for large quantities of water. Both alkalinity and total solids are reduced. The high pH and lime help disinfect the water.

However, the process produces large quantities of sludge that constitute a disposal problem. The intrinsic solubility of some of the compounds means that complete softening cannot be achieved. Flow rates and chemical feed rates must be closely monitored.

31. WATER SOFTENING BY ION EXCHANGE

In the *ion exchange process* (also known as the *zeolite process* and the *base exchange method*), water is passed through a filter bed of exchange material. Ions in the insoluble exchange material are displaced by ions in the water. The processed water leaves with zero hardness. However, since there is no need for water with zero hardness, some of the water is typically bypassed around the process.

If dissolved solids or sodium concentration of the water are issues, then ion exchange may not be suitable.

There are several types of ion exchange materials. *Green sand (glauconite)* is a natural substance that is mined and treated with manganese dioxide. Green sand is not used commercially.

Synthetic zeolites have historically been the workhorses for water softening and demineralization. Porosity through continuous-phase *gelular resins* is low, and dry contact surface areas of 500 ft^2/lbm (0.1 m^2/g) or less is common. This makes gel-based zeolites suitable only for small volumes.

Synthetic *macroporous resins* (*macroreticular resins*) are suitable for use in large-volume water processing systems, when chemical resistance is required, and when specific ions are to be removed. These are discontinuous, three-dimensional copolymer beads in a rigid-sponge type formation.[15] Each bead is made up of thousands of microspheres of the resin. Porosity is much higher than with gelular resins, and dry contact surface areas are approximately 270,000 to 320,000 ft^2/lbm (55 to 65 m^2/g).

There are four primary resin families: strong acid, strong base, weak acid, and weak base. Each family has different resistances to fouling by organic and inorganic chemicals, stabilities, and lifetimes. For example, strong acid resins can last for 20 years. Strong base resins, on the other hand, may have lifetimes of only 3 years. Each family can be used in either gelular or macroporous forms.

During operation, calcium and magnesium ions in hard water are removed according to the following reaction in which R is the zeolite anion. The resulting sodium compounds are soluble.

$$
\begin{Bmatrix} Ca \\ Mg \end{Bmatrix} \begin{Bmatrix} (HCO_3)_2 \\ SO_4 \\ Cl_2 \end{Bmatrix} + Na_2R
$$
$$
\rightarrow Na_2 \begin{Bmatrix} (HCO_3)_2 \\ SO_4 \\ Cl_2 \end{Bmatrix} + \begin{Bmatrix} Ca \\ Mg \end{Bmatrix} R \qquad \textit{26.41}
$$

Typical saturation capacities of synthetic resins are 1.0 to 1.5 meq/mL for anion exchange resins and 1.7 to 1.9 meq/mL for cation exchange resins. However, working capacities are more realistic measures than saturation capacities. The specific working capacity for zeolites is approximately 3 to 11 kilograins/ft^3 (6.9 to 25 kg/m^3) of hardness before regeneration.[16] For synthetic resins, the specific working capacity is approximately 10 to 15 kilograins/ft^3 (23 to 35 kg/m^3) of hardness.

The volume of water that can be softened per cycle (between regenerations) is

$$
V_{\text{water}} = \frac{(\text{specific working capacity})(V_{\text{exchange material}})}{\text{hardness}}
$$
$$
\textit{26.42}
$$

[15]The differences in structure between gel polymers and macroporous polymers occur during the polymerization step. Either can be obtained from the same zeolite.

[16]There are 7000 grains in a pound. 1000 grains of hardness (i.e., $1/7$ of a pound) is known as a *kilograin*. It should not be confused with a kilogram. Other conversions are 1 grain = 64.8 mg and 1 grain/gal = 17.12 mg/L.

Flow rates are typically 1 to 6 gpm/ft^3 (2 to 13 L/s·m^3) of resin volume. The flow rate across the exposed surface will depend on the geometry of the bed, but values of 3 to 15 gpm/ft^2 (2 to 10 L/s·m^2) are typical.

Example 26.4

A municipal plant processes water with a total initial hardness of 200 mg/L. The desired discharge hardness is 50 mg/L. If an ion exchange process is used, what is the bypass factor?

Solution

The water passing through the ion exchange unit is reduced to zero hardness. If x is the water fraction bypassed around the zeolite bed,

$$
(1-x)(0) + x\left(200\ \frac{\text{mg}}{\text{L}}\right) = 50\ \text{mg/L}
$$
$$
x = 0.25
$$

32. REGENERATION OF ION EXCHANGE RESINS

Ion exchange material has a finite capacity for ion removal. When the zeolite approaches saturation, it is regenerated (rejuvenated). Standard ion exchange units are regenerated when the alkalinity of their effluent increases to the *set point*. Most units that collect *crud* are operated to a pressure-drop endpoint.[17] The pressure drop through the ion exchange unit is primarily dependent on the amount of crud collected.

Regeneration of synthetic ion exchange resins is accomplished by passing a *regenerating solution* over/through the resin. Although regeneration can occur in the ion exchange unit itself, external regeneration is becoming common. This involves removing the bed contents hydraulically, backwashing to separate the components (for mixed beds), regenerating the bed components separately, washing, and then recombining and transferring the bed components back into service. For complete regeneration, a contact time of 30 to 45 minutes may be required.

Common regeneration compounds are NaCl (for water hardness removal units), H_2SO_4 (for cation exchange resins), and NaOH (for anion exchange resins). The amount of regeneration solution depends on the resin's degree of saturation. A rule of thumb is to expect to use 5 to 25 pounds of regeneration compound per cubic foot of resin (80 to 400 kilograms per cubic meter). Alternatively, dosage of the regeneration compound may be specified in terms of hardness removed (e.g., 0.4 lbm of salt per 1000 grains of hardness removed.)

The salt requirement per regeneration cycle is

$$
m_{\text{salt}} = (\text{specific working capacity})(V_{\text{exchange material}})
$$
$$
\times (\text{salt requirement}) \qquad \textit{26.43}
$$

[17]Since water has been filtered before reaching the zeolite bed, *crud* consists primarily of iron-corrosion products ranging from dissolved to particulate matter.

33. STABILIZATION

Stabilization treatment is used to eliminate or reduce the potential for scaling and corrosion in pipes after the treated water is distributed. Various factors influence the stability of water, including temperature, dissolved oxygen, dissolved solids, pH, and alkalinity. Different pipe materials are more susceptible to corrosion than others, as are combinations of pipe fitting materials.[18]

Stability is a general term used to describe the water's tendency to be corrosive. Three indices are commonly used to quantify water stability: the *Langelier index*, LI (or *Langelier saturation index*, SI), the *Ryznar index*, and the *aggressive index*. Table 26.3 describes water stability in terms of these indices.

Table 26.3 *Stability of Water*

characteristics	Langelier index	aggressive index	Ryznar index
highly aggressive	< -2	< 10	> 10
moderately aggressive	-2 to 0	10 to 12	6 to 10
nonaggressive	> 0	> 12	< 6

In Eq. 26.44, pH_{sat} is the saturation pH (i.e., the pH at $CaCO_3$ saturation), M is the alkalinity in mg/L as $CaCO_3$, and Ca is the calcium ion content in mg/L as $CaCO_3$. If the Langelier stability index is positive, the water will continue to deposit $CaCO_3$ in the water lines downstream. A corrosive water will have a negative stability index.

$$LI = pH - pH_{sat} \qquad 26.44$$

$$pH_{sat} = (pK_2 - pK_1) + pCa + pM$$
$$= -\log\left(\frac{K_2}{K_1}\right)[Ca^{++}][M] \qquad 26.45$$

K_2 is the second ionization constant for H_2CO_3, and K_1 is the solubility product constant for $CaCO_3$. The quantity $(pK_2 - pK_1)$ where $pK = -\log K$ depends in a complex manner on the total ion strength (or total dissolved solids) and the water temperature. The Ryznar index, RI, is similar.

$$RI = 2pH_{sat} - pH \qquad 26.46$$

The main stabilization processes are pH and alkalinity adjustment using lime or soda ash (if the pH is too low), and with carbon dioxide, sulfuric acid, or hydrochloric acid if the pH is too high. When lime is used for pH adjustment, care must be exercised that unreacted lime does not form *clinker* particles that can be carried out into the water distribution system.

Corrosion and scaling can also be prevented within the treatment plant by use of protective pipe linings and coatings. Various inhibitors and sequestering agents can be used.[19]

---

[18]The galvanic series is useful in predicting the likelihood of corrosion between pipe fittings of different materials.

[19]A *sequestering agent* is a compound that is useful in corrosion control. Such agents work by sequestering normally positive ions in a complex negative ion.

34. DISINFECTION

Chlorination is commonly used for disinfection. Chlorine can be added as a gas or as a liquid. (If it is added to the water as a gas, it is stored as a liquid which vaporizes around $-31°F$ ($-35°C$). Liquid chlorine is the primary form used since it is less expensive than calcium hypochlorite solid ($Ca(OCl)_2$) and sodium hypochlorite ($NaOCl$).

Chlorine is corrosive and toxic. Special safety and handling procedures must be followed with its use.

35. CHLORINATION CHEMISTRY

When chlorine gas dissolves in water, it forms hydrochloric acid (HCl) and hypochlorous acid (HOCl).

$$Cl_2 + H_2O \rightarrow HCl + HOCl \qquad 26.47$$
$$HCl \rightarrow H^+ + Cl^- \qquad 26.48$$
$$HOCl \leftrightarrow H^+ + OCl^- \qquad 26.49$$

The fraction of HOCl ionized into H^+ and OCl^- depends on the pH and can be determined from Fig. 26.5.

When calcium hypochlorite solid is added to water, the ionization reaction is

$$Ca(OCl)_2 \rightarrow Ca^{++} + 2OCl^- \qquad 26.50$$

Figure 26.5 *Ionized HOCl Fraction vs. pH*

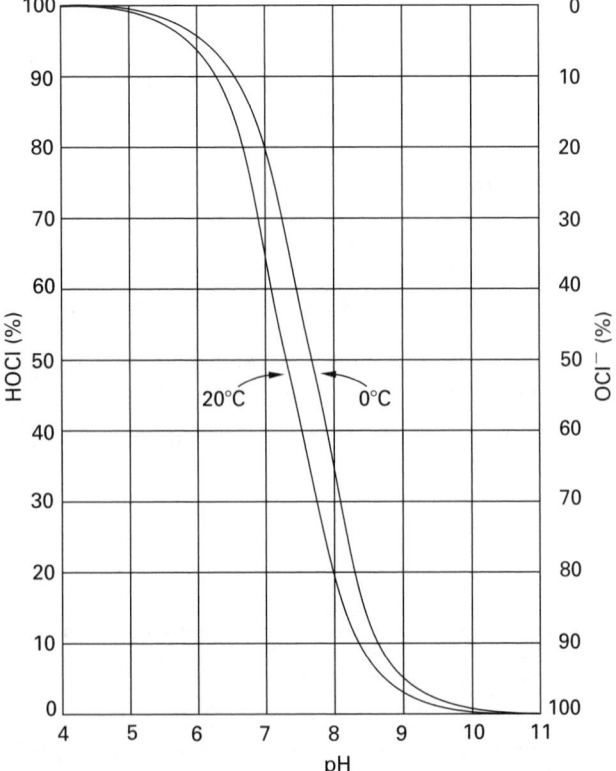

36. CHLORINE DOSE

Figure 26.6 illustrates a breakpoint chlorination curve. Basically, the concept of *breakpoint chlorination* is to continue adding chlorine until the desired quantity of free residuals appears. This cannot occur until after the demand for combined residuals has been satisfied.

Figure 26.6 Breakpoint Chlorination Curve

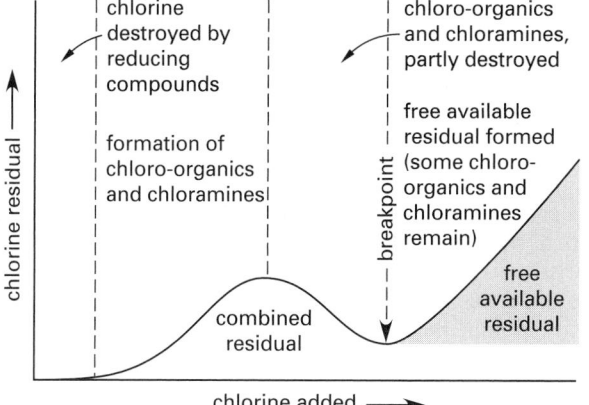

The amount of chlorine necessary for disinfection varies with the organic and inorganic material present in the water, the pH, the temperature, and the contact time. Thirty minutes of chlorine contact time are generally sufficient to deactivate giardia cysts. Satisfactory results can generally be obtained if a free chlorine residual (consisting of hypochlorous acid and hypochlorite ions) of 0.2 to 0.5 mg/L can be maintained throughout the distribution system. Combined residuals, if use is approved by health authorities, should be 1.0 to 2.0 mg/L at distant points in the distribution system.

Excess chlorine can be removed with a reducing agent, usually referred to as *dechlor*. Sodium dioxide and sodium bisulfate (metabisulfate) are used for this purpose. Aeration also reduces chlorine content, as does contact with activated charcoal.

Example 26.5

The flow rate through a treatment plant is 2 MGD (7.5 ML/day). The hypochlorite ion (OCl^-) dose is 20 mg/L as substance. The purity of calcium hypochlorite, $Ca(OCl)_2$, is 97.5% as delivered. How many pounds of calcium hypochlorite are needed to treat the water?

SI Solution

The approximate molecular weights for the components of calcium hypochlorite are

Ca:		40.1
O:	(2)(16) =	32.0
Cl:	(2)(35.5) =	70.0
	total	142.1

The fraction of available chlorine in the form of the hypochlorite ion is

$$G = \frac{32.0 + 70.0}{142.1} = 0.718$$

Use the standard dose equation with adjustments for both purity and availability.

$$F_{kg/day} = \frac{D_{mg/L}Q_{ML/day}}{PG}$$
$$= \frac{\left(20 \ \dfrac{mg}{L}\right)\left(7.5 \ \dfrac{ML}{day}\right)}{(0.975)(0.718)}$$
$$= 214 \ kg/day$$

Customary U.S. Solution

The approximate molecular weights for the components of calcium hypochlorite are

Ca:		40.1
O:	(2)(16) =	32.0
Cl:	(2)(35.5) =	70.0
	total	142.1

The fraction of available chlorine in the form of the hypochlorite ion is

$$\frac{32.0 + 70.0}{142.1} = 0.718$$

Use the standard dose equation with adjustments for both purity and availability.

$$F_{lbm/day} = \frac{D_{mg/L}Q_{MGD}\left(8.345 \ \dfrac{lbm\text{-}L}{mg\text{-}MG}\right)}{PG}$$
$$= \frac{\left(20 \ \dfrac{mg}{L}\right)(2 \ MGD)\left(8.345 \ \dfrac{lbm\text{-}L}{mg\text{-}MG}\right)}{(0.975)(0.718)}$$
$$= 477 \ lbm/day$$

37. ADVANCED OXIDATION PROCESSES

Alternatives to chlorination have become necessary since THMs were traced to the chlorination process. These alternatives are categorized as *advanced oxidation processes* (AOPs).

Both bromine and iodine have properties similar to chlorine and can be used for disinfection. They are seldom used because they are relatively costly and produce their own disinfection by-products.

Chlorine dioxide (ClO_2) is manufactured at the water treatment plant from chlorine and sodium chlorite. Its ionization by-products (chlorite and chlorate) and high cost have limited its use.

Ozone is used extensively throughout the world, though not in large quantities in the United States. Ozone is a more powerful disinfectant than chlorine. Ozone can be used alone or (in several developing technologies) in conjunction with hydrogen peroxide. Ozone is generated on-site by running high voltage electricity through dry air or pure oxygen. Gases (oxygen and unused ozone) developing during ozonation must be collected and destroyed in an ozone-destruct unit to ensure that no ozone escapes into the atmosphere.

Ultraviolet radiation is effective in disinfecting shallow (e.g., less than 10 cm) bodies of water. Its primary disadvantages are cost and the absence of any residual disinfection for downstream protection.

38. CHLORAMINATION

Before leaving the plant, treated water may be *chloraminated*. This step ensures lasting disinfection by providing a residual level of chloramines, protecting the water against bacteria regrowth as it travels through the distribution system.

39. DEMINERALIZATION AND DESALINATION

Demineralization and desalination (*salt water conversion*) are required when only brackish water supplies are available.[20] These processes are carried out in distillation, electrodialysis, ion exchange, and membrane processes.

Distillation is a process whereby the raw water is vaporized, leaving the salt and minerals behind. The water vapor is reclaimed by condensation. Distillation cannot be used to economically provide large quantities of water.

Reverse osmosis is the least costly and most attractive membrane demineralization process. A thin membrane separates two solutions of different concentrations. Pore size is smaller (0.0001 to 0.001 μm) than with ultrafilter membranes (Sec. 24), as salt ions are not permitted to pass through. Typical large-scale osmosis units operate at 150 to 500 psi (1.0 to 5.2 MPa).[21]

Nanofiltration is similar to ultrafiltration and reverse osmosis, with pore size (0.001 μm) and operating pressure (75 to 250 psig; 0.5 to 1.7 MPa) intermediate between the two. Nanofilters are commonly referred to as *softening membranes*.

In *electrodialysis*, positive and negative ions flow through selective membranes under the influence of an induced electrical current. Unlike pressure-driven filtration

processes, however, the ions (not the water molecules) pass through the membrane. The ions removed from the water form a concentrate stream that is discarded.

The *ion exchange* (Sec. 31) process is an excellent solution to demineralization and desalination.

40. WATER DEMAND

Normal water demand is specified in gallons per capita day (gpcd)—the average number of gallons used by each person each day. This is referred to as *average annual daily flow* (AADF) if the average is taken over a period of a year. Residential (i.e., domestic), commercial, industrial, and public uses all contribute to normal water demand, as do waste and unavoidable loss.[22]

As Table 26.4 illustrates, an AADF of 165 gpcd (625 Lpcd) is a typical minimum for planning purposes. If large industries are present (e.g., canning, steel making, automobile production, and electronics), then their special demand requirements must be added.

Table 26.4 *Annual Average Water Requirements*[a]

	demand	
use	gpcd	Lpcd[b]
residential	75 to 130	284 to 490
commercial and		
industrial	70 to 100	265 to 380
public	10 to 20	38 to 80
loss and waste	10 to 20	38 to 80
totals	165 to 270	625 to 1030

(Multiply gpcd by 3.79 to obtain Lpcd.)
[a]exclusive of fire fighting requirements
[b]liters per capita-day

Water demand varies with the time of day and season. Each community will have its own demand distribution curve. Table 26.5 gives typical multipliers, M, that might be used to estimate instantaneous demand from the average daily flow.

$$Q_{\text{instantaneous}} = M(\text{AADF}) \qquad 26.51$$

Table 26.5 *Typical Demand Multipliers*
(to be used with average annual daily flow)

period of usage	multiplier, M
maximum daily	1.5 to 1.8
maximum hourly	2.0 to 3.0
early morning	0.25 to 0.40
noon	1.5 to 2.0
winter	0.80
summer	1.30

[22] "Public" use includes washing streets, flushing water and sewer mains, flushing fire hydrants, filling public fountains, and fighting fires.

---

[20]In the United States, Florida and Arizona are leaders in desalinization installations. Potable water is routinely made from sea water in the Middle East.

[21]Membrane processes operate in a *fixed flux condition*. In order to keep the yield constant over time, the pressure must be constantly increased in order to compensate for the effects of fouling and compaction. Fouling by inorganic substances and biofilms is the biggest problem with membranes.

Per capita demand must be multiplied by the population to obtain the total system demand. Since a population can be expected to change in number, a supply system must be designed to handle the water demand for a reasonable amount of time into the future. Several methods can be used for estimating future demand, including mathematical, comparative, and correlative methods. Various assumptions can be made for mathematical predictions, including uniform growth rate (same as straight-line extrapolation) and constant percentage growth rate. (In the case of decreasing population, the "growth" rate would be negative.)

Economic aspects will dictate the number of years of future capacity that are installed. Excess capacities of 25% for large systems and 50% for small systems are typically built into the system initially.

41. FIRE FIGHTING DEMAND

The Insurance Services Office (ISO) has specified that the *needed fire flow* (NFF) for 1- and 2-family dwellings not exceeding 2 stories in height should be 1500 gpm (95 L/s) for house separations 10 ft (3 m) or less, 1000 gpm (63 L/s) for separations 11 to 30 ft (3.3 to 9 m), 750 gpm (47 L/s) for separations 31 to 100 ft (9 to 30 m), and 500 gpm (31 L/s) for separations over 100 ft (30 m). For any other single building, the NFF will vary between 500 gpm (31 L/s) and 12,000 gpm (760 L/s).[23] Locations with this capability are given favorable insurance ratings.

A municipality must continue to service its domestic, commercial, and industrial customers during a fire. ISO has recommended that a fire system should be able to operate with the potable water system operating at the maximum 24 hr average daily rate plus fire for a minimum of 2 hr (NFF < 3000 gpm), 3 hr (3000 gpm < NFF < 3500 gpm), and 4 hr (NFF > 4000 gpm).[24]

The needed fire flow proposed by the ISO for a particular building depends on the construction type, occupancy, height, floor area, adjacent spaces (communication), and exposure. The basic needed fire flow for a particular building is given by Eq. 26.52. The needed fire flow is obtained from Eq. 26.52 after adjusting for occupancy, exposure, and communication.

$$Q_{\text{gpm}} = 18F\sqrt{A_{\text{ft}^2}} \qquad 26.52$$

In Eq. 26.52, F is a constant that depends on construction: 1.5 for wood frame (Class 1), 1.0 for ordinary (joisted masonry) construction (Class 2), 0.8 for noncombustible (masonry) construction (Classes 3 and 4), and 0.6 for fire-resistant construction (Classes 5 and 6). The *effective area*, A, includes all stories in the building except for the basement. Special rules are used to find

the area for multistory, fire-resistant structures, buildings with various fire loadings, and buildings with sprinkler systems. Q is rounded to the nearest 250 gpm (16 L/s), but it should not be less than 500 gpm (31 L/s) or more than 6000 gpm (380 L/s) for any single-story building or any Class 3, 4, 5, or 6 building, or 8000 gpm (500 L/s) for any Class 1 or 2 building.

In estimating the water requirements for fire fighting on a population basis, the American Insurance Association has recommended Eq. 26.53. The population, P, is expressed in thousands of people.

$$Q_{\text{gpm}} = 1020\sqrt{P}\left(1 - 0.01\sqrt{P}\right) \qquad 26.53$$

Fire hydrants are generally spaced at a distance of 1000 ft (300 m) or closer. They are ordinarily located near street corners where use from four directions is possible. The actual separation of hydrants can be specified in building codes, local ordinances, and other published standards.

42. STORAGE AND DISTRIBUTION

Water is stored to provide water pressure, equalize pumping rates, equalize supply and demand over periods of high consumption, provide surge relief, and furnish water during fires and other emergencies when power is disrupted. Storage may also serve as part of the treatment process, either by providing increased detention time or by blending water supplies to obtain a desired concentration.

Several methods are used to distribute water depending on terrain, economics, and other local conditions. *Gravity distribution* is used when a lake or reservoir is located significantly higher in elevation than the population.

Distribution from *pumped storage* is the most common option when gravity distribution cannot be used. Excess water is pumped during periods of low hydraulic and electrical demands (usually at night) into elevated storage. During periods of high consumption, water is drawn from the storage. With pumped storage, pumps are able to operate at a uniform rate and near their rated capacity most of the time.

Using pumps without storage to force water directly into the mains is the least desirable option. Without storage, pumps and motors will not always be able to run in their most efficient ranges since they must operate during low, average, and peak flows. In a power outage, all water supply will be lost unless a backup power source comes online quickly or water can be obtained by gravity flow.

Water is commonly stored in surface and elevated tanks. The elevation of the water surface in the tank directly determines the distribution pressure. This elevation is controlled by an *altitude valve* that operates on the differential in pressure between the height of the water and

---

[23]A standard 1.125 in (29 mm) diameter smooth fire nozzle discharges approximately 250 gpm (16 L/s) of water.
[24]Source: *Fire Suppression Rating Schedule*, Insurance Services Office, copyright © 1980.

an adjustable spring-loaded pilot on the valve. Altitude valves are installed at ground level and, when properly adjusted, can maintain the water levels to within 4 in. Tanks must be vented to the atmosphere. Otherwise, a rapid withdrawal of water will create a vacuum that could easily cause the tank to collapse inward.

The preferred location of an elevated tank is on the opposite side of the high-consumption district from the pumping station. During periods of high water use, the district will be fed from both sides, reducing the loss of head in the mains below what would occur without elevated storage.

Equalizing the pumping rate during the day ordinarily requires storage of at least 15 to 20% of the maximum daily use. Storage for fires and emergencies is more difficult to determine. Fire storage is essentially dictated by building ordinances and insurance (i.e., economic benefits to the public). Private on-site storage volume for large industries may be dictated by local codes and ordinances.

43. WATER PIPE

Several types of pipe are used in water distribution depending on the flow rate, installed location (i.e., above or below ground), depth of installation, and surface surcharge. Pipes must have adequate strength to withstand external loads from backfill, traffic, and earth movement. They must have high burst strength to withstand internal pressure, a smooth interior surface to reduce friction, corrosion resistance, and tight joints to minimize loss. Once all other requirements have been satisfied, the choice of pipe material can be made on the basis of economics.

Asbestos-cement pipe was used extensively in the past. It has a smooth inner surface and is immune to galvanic corrosion. However, it has low flexural strength. *Concrete pipe* is durable, watertight, has a smooth interior, and requires little maintenance. It is manufactured in plain and reinforced varieties. *Cast-iron pipe* (ductile

and gray varieties) is strong, offers long life, and is impervious. However, it has a high initial cost and is heavy (i.e., difficult to transport and install). It may need to be coated on the exterior and interior to resist corrosion of various types. *Steel pipe* offers a smooth interior, high strength, and high ductility, but it is susceptible to corrosion inside and out. The exterior and interior may both need to be coated for protection. *Plastic pipe* (ABS and PVC) has a smooth interior and is chemically inert, corrosion resistant, and easily transported and installed. However, it has low strength.

44. SERVICE PRESSURE

For ordinary domestic use, the minimum water pressure at the tap should be 25 to 40 psig (170 to 280 kPa). A minimum of 60 psig (400 kPa) at the fire hydrant is usually adequate, since that allows for up to a 20 psig (140 kPa) pressure drop in fire hoses. 75 psig (500 kPa) and higher is common in commercial and industrial districts. *Pressure regulators* can be installed if delivery pressure is too high.

45. WATER MANAGEMENT

Unaccounted-for water is the potable water that is produced at the water treatment plant but is not accounted for in billing. It includes known unmetered uses (such as fire fighting and hydrant flushing) and all unknown uses (e.g., from leaks, broken meters, theft, and illegal connections).

Unaccounted-for water can be controlled with proper attention to meter selection, master metering, leak detection, quality control (installation and maintenance of meters), control of system pressures, and accurate data collection.

Since delivery is under pressure, infiltration of groundwater into pipe joints is not normally an issue with distribution.

Topic III: Environmental

Environmental

27 Biochemistry, Biology, and Bacteriology

1. MICROORGANISMS

Microorganisms occur in untreated water and represent a potential human health risk if they enter the water supply. Microorganisms include viruses, bacteria, fungi, algae, protozoa, worms, rotifers, and crustaceans.

Microorganisms are organized into three broad groups based upon their structural and functional differences. The groups are called *kingdoms*. The three kingdoms are animals (rotifers and crustaceans), plants (mosses and ferns), and *Protista* (bacteria, algae, fungi, and protozoa). Bacteria and protozoa of the kingdom *Protista* make up the major groups of microorganisms in the biological system that is used in secondary treatment of wastewater.

2. PATHOGENS

Many infectious diseases in humans or animals are caused by organisms categorized as *pathogens*. Pathogens are found in fecal wastes that are transmitted and transferred through the handling of wastewater. Pathogens will proliferate in areas where sanitary disposal of feces is not adequately practiced and where contamination of water supply from infected individuals is not properly controlled. The wastes may also be improperly discharged into surface waters, making the water *nonpotable* (unfit for drinking). Certain shellfish can become toxic when they concentrate pathogenic organisms in their tissues, increasing the toxic levels much higher than the levels in the surrounding waters.

Organisms that are considered to be pathogens include bacteria, protozoa, viruses, and helminths (worms). Table 27.1 lists potential waterborne diseases, the causative organisms, and the typical infection sources.

Not all microorganisms are considered pathogens. Some microorganisms are exploited for their usefulness in wastewater processing. Most wastewater engineering (and an increasing portion of environmental engineering) involves designing processes and operating facilities that utilize microorganisms to destroy organic and inorganic substances.

3. MICROBE CATEGORIZATION

Carbon is the basic building block for cell synthesis, and it is prevalent in large quantities in wastewater. Wastewater treatment converts carbon into microorganisms that are subsequently removed from the water by settling. Therefore, the growth of organisms that use organic material as energy is encouraged.

If a microorganism uses organic material as a carbon supply, it is *heterotrophic*. *Autotrophs* require only carbon dioxide to supply their carbon needs. Organisms that rely only on the sun for energy are called *phototrophs*. *Chemotrophs* extract energy from organic or inorganic oxidation/reduction (redox) reactions. *Organotrophs* use organic materials, while *lithotrophs* oxidize inorganic compounds.

Most microorganisms in wastewater treatment processes are bacteria. Conditions in the treatment plant are readjusted so that chemoheterotrophs predominate.

Each species of bacteria reproduces most efficiently within a limited range of temperatures. Four temperature ranges are used to classify bacteria. Those bacteria that grow best below 68°F (20°C) are called *psychrophiles* (i.e., are *psychrophilic*, also known as *cryophilic*). *Mesophiles* (i.e., *mesophilic bacteria*) grow best at temperatures in a range starting around 68°F (20°C) and ending around 113°F (45°C). Between 113°F (45°C) and 140°F (60°C), the *thermophiles* (*thermophilic bacteria*) grow best. Above 140°F (60°C), *stenothermophiles* grow best.

Table 27.1 *Potential Pathogens*

name of organism	major disease	source
Bacteria		
Salmonella typhi	typhoid fever	human feces
Salmonella paratyphi	paratyphoid fever	human feces
other *Salmonella*	salmonellosis	human/animal feces
Shigella	bacillary dysentery	human feces
Vibrio cholerae	cholera	human feces
Enteropathogenic coli	gastroenteritis	human feces
Yersinia enterocolitica	gastroenteritis	human/animal feces
Campylobacter jejuni	gastroenteritis	human/animal feces
Legionella pneumophila	acute respiratory illness	thermally enriched waters
Mycobacterium	tuberculosis	human respiratory exudates
other *Mycobacteria*	pulmonary illness	soil and water
Opportunistic bacteria	variable	natural waters
Enteric Viruses/Enteroviruses		
Polioviruses	poliomyelitis	human feces
Coxsackieviruses A	aseptic meningitis	human feces
Coxsackieviruses B	aseptic meningitis	human feces
Echoviruses	aseptic meningitis	human feces
other *Enteroviruses*	encephalitis	human feces
Reoviruses	upper respiratory and gastrointestinal illness	human/animal feces
Rotaviruses	gastroenteritis	human feces
Adenoviruses	upper respiratory and gastrointestinal illness	human feces
Hepatitis A virus	infectious hepatitis	human feces
Norwalk and related gastrointestinal viruses	gastroenteritis	human feces
Fungi		
Aspergillus	ear, sinus, lung, and skin infections	airborne spores
Candida	yeast infections	various
Protozoa		
Acanthamoeba castellani	amoebic meningoencephalitis	soil and water
Balantidium coli	balantidosis (dysentery)	human feces
Cryptosporidium[a]	cryptosporidiosis	human/animal feces
Entamoeba histolytica	amoebic dysentery	human feces
Giardia lamblia	giardiasis (gastroenteritis)	human/animal feces
Naegleria fowleri	amoebic meningoencephalitis	soil and water
Algae (blue-green)		
Anabaena flos-aquae	gastroenteritis (possible)	natural waters
Microcystis aeruginosa	gastroenteritis (possible)	natural waters
Alphanizomenon flos-aquae	gastroenteritis (possible)	natural waters
Schizothrix calciola	gastroenteritis (possible)	natural waters
Helminths (intestinal parasites/worms)		
Ascaris lumbricoides (roundworm)	digestive disturbances	ingested worm eggs
E. vericularis (pinworm)	any part of the body	ingested worm eggs
Hookworm	pneumonia, anemia	ingested worm eggs
Threadworm	abdominal pain, nausea, weight loss	ingested worm eggs
T. trichiuro (whipworm)	trichinosis	ingested worm eggs
Tapeworm	digestive disturbances	ingested worm eggs

[a]Disinfectants have little effect on *Cryptosporidia*. Most large systems now use filtration, the most effective treatment to date against *Cryptosporidia*.

Because most reactions proceed slowly at these temperatures, cells use *enzymes* to speed up the reactions and control the rate of growth. Enzymes are proteins, ranging from simple structures to complex conjugates, and are specialized for the reactions they catalyze.

The temperature ranges are qualitative and somewhat subjective. The growth range of facultative thermophiles extends from the thermophilic range into the mesophilic range. Bacteria will grow in a range of temperatures and will survive at a very large range of temperatures. *E. coli*, for example, is classified as a mesophile. It grows best at temperatures between 68°F (20°C) and 122°F (50°C) but can continue to reproduce at temperatures down to 32°F (0°C).

Nonphotosynthetic bacteria are classified into two heterotrophic and autotrophic groups depending on their sources of nutrients and energy. *Heterotrophs* use organic matter as both an energy source and a carbon source for synthesis. Heterotrophs are further subdivided into groups depending on their behavior toward free oxygen: aerobes, anaerobes, and facultative bacteria. *Obligate aerobes* require free dissolved oxygen while they decompose organic matter to gain energy for growth and reproduction. *Obligate anaerobes* oxidize organics in the complete absence of dissolved oxygen by using the oxygen bound in other compounds, such as nitrate and sulfate. *Facultative bacteria* comprise a group that uses free dissolved oxygen when available but that can also behave anaerobically in the absence of free dissolved oxygen (i.e., *anoxic conditions*). Under anoxic conditions, a group of facultative anaerobes, called *denitrifiers*, utilizes nitrites and nitrates instead of oxygen. Nitrate nitrogen is converted to nitrogen gas in the absence of oxygen. This process is called *anoxic denitrification*.

Autotrophic bacteria (*autotrophs*) oxidize inorganic compounds for energy, use free oxygen, and use carbon dioxide as a carbon source. Significant members of this group are the *Leptothrix* and *Crenothrix* families of *iron bacteria*. These have the ability to oxidize soluble ferrous iron into insoluble ferric iron. Because soluble iron is often found in well waters and iron pipe, these bacteria deserve some attention. They thrive in water pipes where dissolved iron is available as an energy source and bicarbonates are available as a carbon source. As the colonies die and decompose, they release foul tastes and odors and have the potential to cause staining of porcelain or fabrics.

4. VIRUSES

Viruses are parasitic organisms that pass through filters that retain bacteria, can only be seen with an electron microscope, and grow and reproduce only inside living cells, but they can survive outside the host. They are not cells, but particles composed of a protein sheath surrounding a nucleic-acid core. Most viruses of interest in water supply range in size from 10 to 25 nm.

The viron particles invade living cells and the viral genetic material redirects cell activities toward production of new viral particles. A large number of viruses are released to infect other cells when the infected cell dies. Viruses are host-specific, attacking only one type of organism.

There are more than 100 types of human enteric viruses. Those of interest in drinking water are *Hepatitis A*, *Norwalk*-type viruses, *Rotaviruses*, *Adenoviruses*, *Enteroviruses*, and *Reoviruses*.

5. BACTERIA

Bacteria are microscopic organisms (*microorganisms*) having round, rodlike, spiral or filamentous single-celled or noncellular bodies. Bacteria are *prokaryotes* (i.e., they lack nucleii structures). They are often aggregated into colonies. Bacteria use soluble food and reproduce through binary fission. Most bacteria are not pathogenic to humans, but they do play a significant role in the decomposition of organic material and can have an impact on the aesthetic quality of water.

6. FUNGI

Fungi are aerobic, multicellular, nonphotosynthetic, heterotrophic, eukaryotic protists. Most fungi are saprophytes that degrade dead organic matter. Fungi grow in low-moisture areas, and they are tolerant of low-pH environments. Fungi release carbon dioxide and nitrogen during the breakdown of organic material.

Fungi are obligate aerobes that reproduce by a variety of methods including fission, budding, and spore formation. They form normal cell material with one-half the nitrogen required by bacteria. In nitrogen-deficient wastewater, they may replace bacteria as the dominant species.

7. ALGAE

Algae are autotrophic, photosynthetic organisms (*photoautotrophs*) and may be either unicellular or multicellular. They take on the color of the pigment that is the catalyst for photosynthesis. In addition to chlorophyll (green), different algae have different pigments, such as carotenes (orange), phycocyanin (blue), phycoerythrin (red), fucoxanthin (brown), and xanthophylls (yellow).

Algae derive carbon from carbon dioxide and bicarbonates in water. The energy required for cell synthesis is obtained through photosynthesis. Algae utilize oxygen for respiration in the absence of light. Algae and bacteria have a symbiotic relationship in aquatic systems, with the algae producing oxygen used by the bacterial population.

In the presence of sunlight, the photosynthetic production of oxygen is greater than the amount used in respiration. At night algae use up oxygen in respiration.

Environmental

If the daylight hours exceed the night hours by a reasonable amount, there is a net production of oxygen. Excessive algal growth (*algal blooms*) can result in supersaturated oxygen conditions in the daytime and anaerobic conditions at night.

Some algae cause tastes and odors in natural water. While they are not generally considered pathogenic to humans, algae do cause turbidity, and turbidity provides a residence for microorganisms that are pathogenic.

8. PROTOZOA

Protozoa are single-celled animals that reproduce by *binary fission* (dividing in two). Most are aerobic chemoheterotrophs (i.e., *facultative heterotrophs*). Protozoa have complex digestive systems and use solid organic matter as food, including algae and bacteria. Therefore, they are desirable in wastewater effluent because they act as polishers in consuming the bacteria.

Flagellated protozoa are the smallest protozoans. The *flagella* (long hairlike strands) provide motility by a whiplike action. Amoeba move and take in food through the action of a mobile protoplasm. Free-swimming protozoa have *cilia*, small hairlike features, used for propulsion and gathering in organic matter.

9. WORMS AND ROTIFERS

Rotifers are aerobic, multicellular chemoheterotrophs. The rotifer derives its name from the apparent rotating motion of two sets of cilia on its head. The cilia provide mobility and a mechanism for catching food. Rotifers consume bacteria and small particles of organic matter.

Many *worms* are aquatic parasites. *Flatworms* of the class *Trematoda* are known as *flukes*, and the *Cestoda* are tapeworms. *Nemotodes* of public health concern are *Trichinella*, which causes trichinosis; *Necator*, which causes pneumonia; *Ascaris*, which is the common roundworm that takes up residence in the human intestine (ascariasis); and *Filaria*, which causes filariasis.

10. MOLLUSKS

Mollusks, such as mussels and clams, are characterized by a shell structure. They are aerobic chemoheterotrophs that feed on bacteria and algae. They are a source of food for fish and are not found in wastewater treatment systems to any extent, except in underloaded lagoons. Their presence is indicative of a high level of dissolved oxygen and a very low level of organic matter.

Macrofouling is a term referring to infestation of water inlets and outlets by clams and mussels. *Zebra mussels*, accidentally introduced into the United States in 1986, are particularly troublesome for several reasons. First, young mussels are microscopic, easily passing through intake screens. Second, they attach to anything, even other mussels, producing thick colonies. Third, adult mussels quickly sense some biocides, most notably those that are halogen-based (including chlorine), quickly closing and remaining closed for days or weeks.

The use of biocides in the control of zebra mussels is controversial. Chlorination is a successful treatment that is recommended with some caution since it results in increased toxicity, affecting other species and THM production. Nevertheless, an ongoing biocide program aimed at pre-adult mussels, combined with slippery polymer-based surface coatings, is probably the best approach at prevention. Once a pipe is colonized, mechanical removal by scraping or water blasting is the only practical option.

11. INDICATOR ORGANISMS

The techniques for comprehensive bacteriological examination for pathogens are complex and time-consuming. Isolating and identifying specific pathogenic microorganisms is a difficult and lengthy task. Many of these organisms require sophisticated tests that take several days to produce results. Because of these difficulties, and also because the number of pathogens relative to other microorganisms in water can be very small, *indicator organisms* are used as a measure of the quality of the water. The primary function of an indicator organism is to provide evidence of recent fecal contamination from warm-blooded animals.

Characteristics of a good indicator organism are: (a) The indicator is always present when the pathogenic organism of concern is present. It is absent in clean, uncontaminated water. (b) The indicator is present in fecal material in large numbers. (c) The indicator responds to natural environmental conditions and to treatment processes in a manner similar to the pathogens of interest. (d) The indicator is easy to isolate, identify, and enumerate. (e) The ratio of indicator to pathogen should be high. (f) The indicator and pathogen should come from the same source (e.g., gastrointestinal tract).

While there are several microorganisms that meet these criteria, *total coliform* and *fecal coliform* are the indicators generally used. *Total coliform* refers to the group of aerobic and facultatively anaerobic, gram-negative, nonspore-forming, rod-shaped bacteria that ferment lactose with gas formation within 48 hr at $662°F$ ($350°C$). This encompasses a variety of organisms, mostly of intestinal origin, including *Escherichia coli* (*E. coli*), the most numerous facultative bacterium in the feces of warm-blooded animals. Unfortunately, this group also includes *Enterobacter*, *Klebsiella*, and *Citrobacter*, which are present in wastewater but can be derived from other environmental sources such as soil and plant materials.

Fecal coliforms are a subgroup of the total coliforms that come from the intestines of warm-blooded animals. They are measured by running the standard total coliform fermentation test at an elevated temperature of 112°F (44.5°C), providing a means to distinguish false positives in the total coliform test.

Results of fermentation tests are reported as a *most probable number* (MPN) *index*. This is an index of the number of coliform bacteria that, more probably than any other number, would give the results shown by the laboratory examination; it is not an actual enumeration.

12. METABOLISM/METABOLIC PROCESSES

Metabolism is a term given to describe all chemical activities performed by a cell. The cell uses *adenosine triphosphate* (ATP) as the principle energy currency in all processes. Those processes that allow the bacterium to synthesize new cells from the energy stored within its body are said to be *anabolic*. All biochemical processes in which cells convert substrate into useful energy and waste products are said to be *catabolic*.

13. DECOMPOSITION OF WASTE

Decomposition of waste involves oxidation/reduction reactions and is classified as aerobic or anaerobic. The type of electron acceptor available for catabolism determines the type of decomposition used by a mixed culture of microorganisms. Each type of decomposition has peculiar characteristics that affect its use in waste treatment.

14. AEROBIC DECOMPOSITION

Molecular oxygen (O_2) must be present as the terminal electron acceptor in order for decomposition to proceed by aerobic oxidation. As in natural water bodies, the dissolved oxygen content is measured. When oxygen is present, it is the only terminal electron acceptor used. Hence, the chemical end products of decomposition are primarily carbon dioxide, water, and new cell material as demonstrated by Table 27.2. Odoriferous, gaseous end products are kept to a minimum. In healthy natural water systems, aerobic decomposition is the principal means of self-purification.

A wider spectrum of organic material can be oxidized by aerobic decomposition than by any other type of decomposition. Because of the large amount of energy released in aerobic oxidation, most aerobic organisms are capable of high growth rates. Consequently, there is a relatively large production of new cells in comparison with the other oxidation systems. This means that more biological sludge is generated in aerobic oxidation than in the other oxidation systems.

Aerobic decomposition is the preferred method for large quantities of dilute ($BOD_5 < 500$ mg/L) wastewater because decomposition is rapid and efficient and has a low odor potential. For high-strength wastewater ($BOD_5 > 1000$ mg/L), aerobic decomposition is not suitable because of the difficulty in supplying enough oxygen and because of the large amount of biological sludge that is produced.

15. ANOXIC DECOMPOSITION

Some microorganisms will use nitrates in the absence of oxygen needed to oxidize carbon. The end products from such *denitrification* are nitrogen gas, carbon dioxide, water, and new cell material. The amount of energy made available to the cell during denitrification is about the same as that made available during aerobic decomposition. The production of cells, although not as high as in aerobic decomposition, is relatively high.

Table 27.2 Waste Decomposition End Products

	representative end products		
substrates	aerobic decomposition	anoxic decomposition	anaerobic decomposition
proteins and other organic nitrogen compounds	amino acids ammonia → nitrites → nitrates alcohols organic acids } → $CO_2 + H_2O$	amino acids nitrates → nitrites → N_2 alcohols organic acids } → $CO_2 + H_2O$	amino acids ammonia hydrogen sulfide methane carbon dioxide alcohols organic acids
carbohydrates	alcohols fatty acids } → $CO_2 + H_2O$	alcohols fatty acids } → $CO_2 + H_2O$	carbon dioxide alcohols fatty acids
fats and related substances	fatty acids + glycerol alcohols lower fatty acids } → $CO_2 + H_2O$	fatty acids + glycerol alcohols lower fatty acids } → $CO_2 + H_2O$	fatty acids + glycerol carbon dioxide alcohols lower fatty acids

Denitrification is of importance in wastewater treatment when nitrogen must be removed. In such cases, a special treatment step is added to the conventional process for removal of carbonaceous material. One other important aspect of denitrification is in final clarification of the treated wastewater. If the final clarifier becomes anoxic, the formation of nitrogen gas will cause large masses of sludge to float to the surface and escape from the treatment plant into the receiving water. Thus, it is necessary to ensure that anoxic conditions do not develop in the final clarifier.

16. ANAEROBIC DECOMPOSITION

In order to achieve anaerobic decomposition, molecular oxygen and nitrate must not be present as terminal electron acceptors. Sulfate, carbon dioxide, and organic compounds that can be reduced serve as terminal electron acceptors. The reduction of sulfate results in the production of hydrogen sulfide (H_2S) and a group of equally odoriferous organic sulfur compounds called *mercaptans*.

The anaerobic decomposition of organic matter, also known as *fermentation*, generally is considered to be a two-step process. In the first step, complex organic compounds are fermented to low molecular weight *fatty acids* (*volatile acids*). In the second step, the organic acids are converted to methane. Carbon dioxide serves as the electron acceptor.

Anaerobic decomposition yields carbon dioxide, methane, and water as the major end products. Additional end products include ammonia, hydrogen sulfide, and mercaptans. As a consequence of these last three compounds, anaerobic decomposition is characterized by a malodorous stench.

Because only small amounts of energy are released during anaerobic oxidation, the amount of cell production is low. Thus, sludge production is correspondingly low. Wastewater treatment based on anaerobic decomposition is used to stabilize sludges produced during aerobic and anoxic decomposition.

Direct anaerobic decomposition of wastewater generally is not feasible for dilute waste. The optimum growth temperature for the anaerobic bacteria is at the upper end of the mesophilic range. Thus, to get reasonable biodegradation, the temperature of the culture must first be elevated. For dilute wastewater, this is not practical. For concentrated wastes ($BOD_5 > 1000$ mg/L), anaerobic digestion is quite appropriate.

17. FACTORS AFFECTING DISEASE TRANSMISSION

Waterborne disease transmission is influenced by the latency, persistence, and quantity (dose) of the pathogens. *Latency* is the period of time between excretion of a pathogen and its becoming infectious to a new host. *Persistence* is the length of time that a pathogen remains viable in the environment outside a human host. The *infective dose* is the number of organisms that must be ingested to result in disease.

18. AIDS

Acquired immunodeficiency syndrome (AIDS) is caused by the *human immunodeficiency virus* (HIV). HIV is present in virtually all body excretions of infected persons, and therefore, is present in wastewater. Present research, statistics, and available information indicates that the risk of contracting AIDS through contact with wastewater or working at a wastewater plant is nil. This is based on the following facts.

- HIV is relatively weak and does not remain viable for long periods of time in harsh environments (e.g., wastewater).

- HIV is quickly inactivated by alcohol, chlorine, and exposure to air. The chlorine concentration present in water used to flush the toilet would probably deactivate HIV.

- HIV that survives disinfection would be too dilute to be infectious.

- HIV replicates in white blood cells, not in the human intestinal tract, and, therefore, would not reproduce in wastewater.

- There is no evidence that HIV can be transmitted through water, air, food, or contact. HIV must enter the bloodstream directly, through a wound. It cannot enter through unbroken skin or through respiration.

- HIV is less infectious than the hepatitis virus.

- There are no reported AIDS cases linked to occupational exposure in wastewater collection and treatment.

28 Wastewater Quantity and Quality

Nomenclature

BOD	biochemical oxygen demand	mg/L	mg/L
C	concentration	mg/L	mg/L
COD	chemical oxygen demand	mg/L	mg/L
d	depth of flow	ft	m
D	oxygen deficit	mg/L	mg/L
DO	dissolved oxygen	mg/L	mg/L
DO^*	dissolved oxygen of the seed	mg/L	mg/L
K	rate constant	days^{-1}	days^{-1}
n	slope	–	–
P	population	–	–
Q	flow quantity	gal/day	L/day
r	rate of change in oxygen content	mg/L-day	mg/L·day
S	slope	ft/ft	m/m
t	time	sec	s
T	temperature	°F	°C
v	velocity	ft/sec	m/s
V	volume	mL	mL
x	distance	ft	m

Symbols

θ	temperature constant	–	–

Subscripts

0	initial
5	five-day
c	critical
d	deoxygenation
e	equivalent
f	final
i	initial
m	maximum
r	reoxygenation or river
sat	saturated
t	at time t
T	at temperature T
u	ultimate carbonaceous
w	wastewater

1. DOMESTIC WASTEWATER

Domestic wastewater refers to *sanitary wastewater* discharged from residences, commercial buildings, and institutions. Other examples of domestic wastewater sources are mobile homes, hotels, schools, offices, factories, and other commercial enterprises, excluding manufacturing facilities.

The volume of domestic wastewater varies from 50 to 250 gallons per capita day (gpcd) (190 to 950 Lpcd) depending on sewer uses. A more common range for domestic wastewater flow is 100 to 120 gpcd (380 to 450 Lpcd), which assumes that residential dwellings have major water-using appliances such as dishwashers and washing machines.

2. INDUSTRIAL WASTEWATER

Small manufacturing facilities within municipal limits ordinarily discharge *industrial wastewater* into the city sewer system only after pretreatment. In *joint processing* of wastewater, the municipality accepts responsibility for final treatment and disposal. The manufacturing plant may be required to pretreat the wastewater and equalize flow by holding it in a basin for stabilization prior to discharge to the sewer.

Uncontaminated streams (e.g., cooling water), in many cases, can be discharged into sewers directly. However, pretreatment at the industrial site is required for wastewaters having strengths and/or characteristics significantly different from sanitary wastewater.

To minimize the impact on the sewage treatment plant, consideration is given to modifications in industrial processes, segregation of wastes, flow equalization, and waste strength reduction. Modern industrial/manufacturing processes require segregation of separate waste streams for individual pretreatment, controlled mixing, and/or separate disposal. Process changes, equipment modifications, by-product recovery, and in-plant wastewater reuse can result in cost savings for both water supply and wastewater treatment.

Toxic waste streams are not generally accepted into the municipal treatment plant at all. Toxic substances require appropriate pretreatment prior to disposal by other means.

3. MUNICIPAL WASTEWATER

Municipal wastewater is the general name given to the liquid collected in sanitary sewers and routed to municipal sewage treatment plants. Many older cities have *combined sewer systems* where storm water and sanitary wastewaters are collected in the same lines. The combined flows are conveyed to the treatment plant for processing during dry weather. During wet weather, when the combined flow exceeds the plant's treatment capacity, the excess flow often bypasses the plant and is discharged directly into the watercourse.

4. WASTEWATER QUANTITY

Approximately 70 to 80% of a community's domestic and industrial water supply returns as wastewater. This water is discharged into the sewer systems, which may or may not also function as storm drains. Therefore, the nature of the return system must be known before sizing can occur.

Infiltration, due to cracks and poor joints in old or broken lines, can increase the sewer flow significantly. Infiltration per mile (kilometer) per in (mm) of pipe diameter is limited by some municipal codes to 500 gpd/mi-in (46 Lpd/km·mm). Modern piping materials and joints easily reduce the infiltration to 200 gpd/in-mi (18 Lpd/km·mm) and below. Infiltration can also be roughly estimated as 3 to 5% of the peak hourly domestic rate or as 10% of the average rate.

Inflow is another contributor to the flow in sewers. Inflow is water discharged into a sewer system from such sources as roof down spouts, yard and area drains, parking area catch basins, curb inlets, and holes in manhole covers.

Sanitary sewer sizing is commonly based on an assumed average of 100 to 125 gpcd (380 to 474 Lpcd). There will be variations in the flow over time, although the variations are not as pronounced as they are for water supply. Hourly variations are the most significant. The flow rate pattern is essentially the same from day to day. Weekend flow patterns are not significantly different from weekday flow patterns. Seasonal variation depends on the location, local industries, and infiltration.

Table 28.1 lists typical *peaking factors* (i.e., peak multipliers) for treatment plant influent volume. Due to storage in ponds, clarifiers, and sedimentation basins, these multipliers are not applicable throughout all processes in the treatment plant.

Table 28.1 Typical Variations in Wastewater Flows
(based on average annual daily flow)

flow description	typical time	peaking factor
daily average	–	1.0
daily peak	10 to 12 a.m.	2.25
daily minimum	4 to 5 a.m.	0.4
seasonal average	May, June	1.0
seasonal peak	late summer	1.25
seasonal minimum	late winter	0.9

Recommended Standards for Sewage Works ("*Ten States' Standards*," abbreviated TSS) specifies that new sanitary sewer systems should be designed for an average flow of 100 gpcd (380 Lpcd or 0.38 m³/day), which includes an allowance for normal infiltration [TSS Sec. 11.243]. However, the sewer pipe must be sized to carry the peak flow as a gravity flow. In the absence of any studies or other justifiable methods, the ratio of peak hourly flow to average flow should be calculated from the following relationship in which P is the population served in thousands of people at a particular point in the network.

$$\frac{Q_{\text{peak}}}{Q_{\text{ave}}} = \frac{18 + \sqrt{P}}{4 + \sqrt{P}} \qquad 28.1$$

Collectors (i.e., *collector sewers*, *trunks*, or *mains*) are pipes that collect wastewater from individual sources and carry it to interceptors (see Fig. 28.1). Collectors must be designed to handle the maximum hourly flow, including domestic and infiltration, as well as additional discharge from industrial plants nearby. Peak flows of 400 gpcd (1500 Lpcd) for laterals and submains flowing full and 250 gpcd (950 Lpcd) for main, trunk, and outfall sewers can be assumed for design purposes, making the peaking factors approximately 4.0 and 2.5 for submains and mains, respectively. Both of these generalizations include generous allowances for infiltration. The lower flow rates take into consideration the averaging effect of larger contributing populations and the damping (storage) effect of larger distances from the source.

Interceptors are major sewer lines receiving wastewater from collector sewers and carrying it to a treatment plant or to another interceptor.

Figure 28.1 *Classification of Sewer Lines*

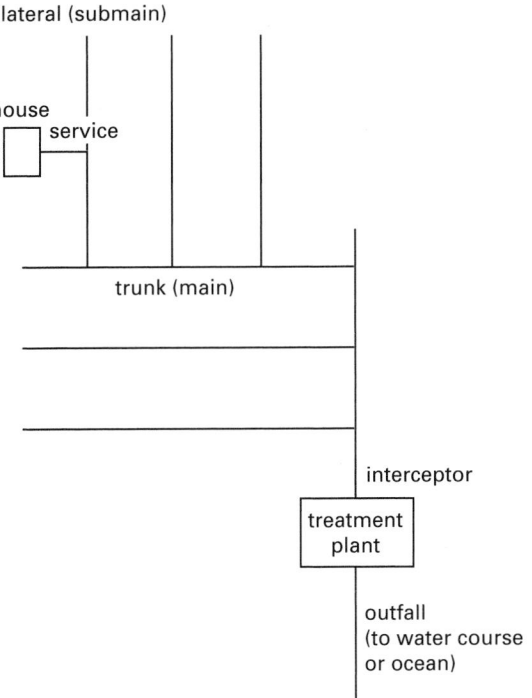

5. TREATMENT PLANT LOADING

Ideally, the quantity and organic strength of wastewater should be based on actual measurements taken throughout the year in order to account for variations that result from seasonal, climatic changes and other factors.

The *hydraulic loading* for over 75% of the sewage treatment plants in the United States is 1 MGD or less. Plants with flows less than 1 MGD are categorized as *minors*. Treatment plants handling 1 MGD or more, or serving equivalent service populations of 10,000, are categorized as *majors*.

The *organic loading* of treatment units is expressed in terms of pounds (kilograms) of biochemical oxygen demand (BOD) per day or pounds (kilograms) of solids per day. In communities where a substantial portion of the household kitchen wastes is discharged to the sewer system through garbage disposals, the average organic matter contributed by each person each day to domestic wastewater is approximately 0.24 lbm (110 g) of suspended solids and approximately 0.17 to 0.20 lbm (77 to 90 g) of BOD.

Since the average BOD of domestic waste is typically taken as 0.2 pounds per capita day (90 g per capita day), the *population equivalent* of any wastewater source, including industrial sources, can be calculated.

$$P_{\text{equivalent},1000s} = \frac{(\text{BOD}_{\text{mg/L}})(Q_{\text{ML/day}})}{90 \, \dfrac{\text{g}}{\text{person-day}}}$$

[SI] *28.2(a)*

$$P_{\text{equivalent},1000s} = \frac{(\text{BOD}_{\text{mg/L}})(Q_{\text{gal/day}}) \times \left(8.345 \, \dfrac{\text{lbm-L}}{\text{MG-mg}}\right)}{\left(10^6 \, \dfrac{\text{gal}}{\text{MG}}\right)(1000 \text{ persons})} \times (0.20 \text{ lbm/person-day})$$

[U.S.] *28.2(b)*

6. SEWER PIPE MATERIALS

In the past, sewer pipes were constructed from clay, concrete, asbestos-cement, steel, and cast iron. Due to cost, modern large-diameter sewer lines are now almost always concrete, and new small-diameter lines are generally plastic.

Concrete pipe is commonly used for gravity sewers and pressure mains. Circular pipe is used in most applications, although special shapes (e.g., arch, egg, elliptical) can be used. Concrete pipe in diameters up to 24 in is available in standard 3 and 4 ft (0.9 and 1.2 m) lengths and is usually not reinforced. *Reinforced concrete pipe* (RCP) in diameters ranging from 12 to 144 in (305 to 3660 mm) is available in lengths from 4 to 12 ft (1.2 to 3.6 m).

Concrete pipe is used for large diameter (16 in or larger) trunk and interceptor sewers. In some geographical regions, concrete pipe is used for smaller domestic sewers. However, concrete domestic lines should be selected only where stale or septic sewage is not anticipated. They should not be used with corrosive wastes or in corrosive soils.

Cast-iron pipe is particularly suited to installations where scour, high velocity waste, and high external loads are anticipated. It can be used for domestic connections, although it is more expensive than plastic. Special linings, coatings, wrappings, or encasements are required for corrosive wastes and soils.

Polyvinyl chloride (PVC) and acrylonitrile-butadiene-styrene (ABS) are two *plastic pipe* compositions that can be used for normal domestic sewage and industrial wastewater lines. They have excellent resistance to corrosive soils. However, special attention and care must be given to trench loading and pipe bedding.

ABS plastic can also be combined with concrete reinforcement for collector lines for use with corrosive domestic sewage and industrial waste. Such pipe is known as *truss pipe* due to its construction. The plastic is extruded with inner and outer web-connected pipe walls. The annular voids and the inner and outer walls are filled with lightweight concrete.

For pressure lines (i.e., *force mains*), welded steel pipe with an epoxy liner and cement-lined and coated-steel pipe are also used occasionally.

Vitrified clay pipe is resistant to acids, alkalies, hydrogen sulfide (septic sewage), erosion, and scour. Two strengths of clay pipe are available. The standard strength is suitable for pipes less than 12 in (300 mm) in diameter for any depth of cover if the "$4D/3 + 8$ in trench width" ($4D/3 + 200$ mm) rule is observed. Double-strength pipe is recommended for large pipe that is deeply trenched. Clay is seldom used for diameters greater than 36 in (910 mm).

Asbestos-cement pipe has been used in the past for both gravity and pressure sewers carrying non-septic and non-corrosive waste through non-corrosive soils. The lightweight and longer laying lengths are inherent advantages of asbestos-cement pipe. However, such pipes have fallen from favor due to the asbestos content.

7. GRAVITY AND FORCE COLLECTION SYSTEMS

Wastewater collection systems are made up of a network of discharge and flow lines, drains, inlets, valve works, and connections for transporting domestic and industrial wastewater flows to treatment facilities.

Flow through gradually sloping *gravity sewers* is the most desirable means of moving sewage since it does not require pumping energy. In some instances, though, it may be necessary to use pressurized *force mains* to carry sewage uphill or over long, flat distances.

Alternative sewers are used in some remote housing developments for domestic sewage where neither the conventional sanitary sewer nor the septic tank/leach field is acceptable. Alternative sewers use pumps to force raw or communited sewage through small-diameter plastic lines to a more distant communal treatment or collection system.

There are four categories of alternative sewer systems. In the *G-P system*, a grinder-pump pushes chopped-up but untreated wastewater through a small-diameter plastic pipe. The *STEP system* (septic-tank-effluent pumping) uses a septic tank at each house, but there is no leach field. The clear overflow goes to a pump that pushes the effluent through a small-diameter plastic sewer line operating under low pressure. The *vacuum sewer system* uses a vacuum valve at each house that periodically charges a slug of wastewater into a vacuum sewer line. The vacuum is created by a central pumping station. The *small-diameter gravity sewer* (SDG) also uses a small-diameter plastic pipe but relies on gravity instead of pumps. The pipe carries the effluent from a septic tank at each house. Costs in all four

systems can be reduced greatly by having two or more houses share septic tanks, pumps, and vacuum valves.

8. SEWER VELOCITIES

The minimum design velocity actually depends on the particulate matter size. However, 2 ft/sec (0.6 m/s) is commonly quoted as the minimum *self-cleansing velocity*, although 1.5 ft/sec (0.45 m/s) may be acceptable if the line is occasionally flushed out by peak flows [TSS Sec. 33.42].

Table 28.2 lists the approximate minimum slope needed to achieve a 2 ft/sec (0.6 m/s) flow. Slopes slightly less than those listed may be permitted (with justification) in lines where design average flow provides a depth of flow greater than 30% of the pipe diameter. Velocity greater than 10 to 15 ft/sec (3 to 4.5 m/s) requires special provisions to protect the pipe and manholes against erosion and displacement by shock hydraulic loadings.

Table 28.2 Minimum Slopes and Capacities for Sewers[a]

sewer diameter (in)	(mm)	minimum change in elevation[b] (ft/100 ft or m/100 m)	full flow discharge at minimum slope[c,d] (ft^3/sec)
8	(200)	0.40	0.771
9	(230)	0.33	0.996
10	(250)	0.28	1.17
12	(300)	0.22	1.61
14	(360)	0.17	2.23
15	(380)	0.15	2.52
16	(410)	0.14	2.90
18	(460)	0.12	3.67
21	(530)	0.10	5.05
24	(610)	0.08	6.45
27	(690)	0.067	8.08
30	(760)	0.058	9.96
36	(910)	0.046	14.4

(Multiply in by 25.4 to obtain mm.)
(Multiply ft/100 ft by 1 to obtain m/100 m.)
(Multiply ft^3/sec by 448.8 to obtain gal/min.)
(Multiply ft^3/sec by 28.32 to obtain L/s.)
(Multiply gal/min by 0.0631 to obtain L/s.)
[a]to achieve a velocity of 2 ft/sec (0.6 m/s) when flowing full
[b]as specified in Sec. 24.31 of *Recommended Standards for Sewage Works* (RSSW), also known as "*Ten States' Standards*" (TSS), published by the Health Education Service, Inc. [TSS Sec. 33.41]
[c]$n = 0.013$ assumed
[d]For any diameter in inches and $n = 0.013$, calculate the full flow as

$$Q = (0.0472)(d_{\text{in}})^{8/3}\sqrt{S}$$

9. SEWER SIZING

The Manning equation is traditionally used to size gravity sewers. Depth of flow at the design flow rate is usually less than 70 to 80% of the pipe diameter. Thus, a pipe should be sized to be able to carry the peak flow (including normal infiltration) at a depth of approximately 70% of its diameter.

Table 28.3 *Minimum Flow Velocities*

fluid	minimum velocity to keep particles in suspension ft/sec (m/s)	minimum resuspension velocity ft/sec (m/s)
raw sewage	2.5 (0.75)	3.5 (1.1)
grit tank effluent	2 (0.6)	2.5 (0.75)
primary settling tank effluent	1.5 (0.45)	2 (0.6)
mixed liquor	1.5 (0.45)	2 (0.6)
trickling filter effluent	1.5 (0.45)	2 (0.6)
secondary settling tank effluent	0.5 (0.15)	1 (0.3)

(Multiply ft/sec by 0.3 to obtain m/s.)

In general, sewers in the collection system (including laterals, interceptors, trunks, and mains) should have diameters of at least 8 in (203 mm). Building service connections can be as small as 4 in (101 mm).

10. STREET INLETS

Street inlets are required at all low points where ponding could occur, and they should be placed no more than 600 ft (180 m) apart, and a limit of 300 ft (90 m) is preferred. A common practice is to install three inlets in a sag vertical curve—one at the lowest point and one on each side with an elevation of 0.2 ft (60 mm) above the center inlet.

Depth of gutter flow is found using Manning's equation. Inlet capacities of street inlets have traditionally been calculated from semi-empirical formulas.

Grate-type street inlets (known as *gutter inlets*) less than 0.4 ft (120 mm) deep have approximate capacities given by Eq. 28.3. Equation 28.3 should also be used for combined curb-grate inlets. All dimensions are in ft. For gutter inlets, the bars should ideally be parallel to the flow and at least 1.5 ft (450 mm) long. Gutter inlets are more efficient than curb inlets, but clogging is a problem. Depressing the grating level below the street level increases the capacity.

$$Q_{\text{ft}^3/\text{sec}} = 3.0 \text{ (grate perimeter length)} \times \text{(inlet flow depth)}^{3/2} \quad 28.3$$

The capacity of a *curb inlet* (where there is an opening in the vertical plane of the gutter rather than in the horizontal plane) is given by Eq. 28.4. All dimensions are in ft. Not all curb inlets have inlet depressions. A typical curb inlet depression is 0.4 ft (130 mm).

$$Q_{\text{ft}^3/\text{sec}} = 0.7 \text{ (curb opening length)} \times \text{(inlet flow depth + curb inlet depression)}^{3/2} \quad 28.4$$

11. MANHOLES

Manholes along sewer lines should be provided at sewer line intersections and at changes in elevation, direction, size, diameter, and slope. If a sewer line is too small for a person to enter, manholes should be placed every 400 ft (120 m) to allow for cleaning. Recommended maximum spacings are 400 ft (120 m) for pipes with diameters less than 18 in (460 mm), 500 ft (150 m) for 18 to 48 in (460 to 1220 mm) pipes, and 600 to 700 ft (180 to 210 m) for larger pipes.

12. SULFIDE ATTACK

Wastewater flowing through sewers often turns septic and releases *hydrogen sulfide* gas, H_2S. The common *Thiobacillus* sulfur bacterium converts the hydrogen sulfide to sulfuric acid. The acid attacks the crown of concrete pipes that are not flowing full.

$$2H_2S + O_2 \rightarrow 2S + 2H_2O \quad 28.5$$
$$2S + 3O_2 + 2H_2O \rightarrow 2H_2SO_4 \quad 28.6$$

In warm climates, hydrogen sulfide can be generated in sanitary sewers placed along on flat grades. Hydrogen sulfide also occurs in sewers supporting food processing industries. Sulfide generation is aggravated by the use of home garbage disposals and, to a lesser extent, by recycling. Interest in nutrition and fresh foods has also added to food wastes.

Sulfide attack in partially full sewers can be prevented by maintaining the flow rate above 5 ft/sec (1.5 m/s), raising the pH above 10.4, using biocides, precipitating the sulfides, or a combination thereof. Not all methods may be possible or economical, however. In sewers, the least expensive method generally involves intermittent (e.g., biweekly during the critical months) *shock treatments* of sodium hydroxide and ferrous chloride. The sodium hydroxide is a sterilization treatment that works by raising the pH, while the ferrous chloride precipitates the sulfides.

Sulfide attack is generally not an issue in force mains because sewage is continually moving and always makes contact with all of the pipe wall.

13. WASTEWATER CHARACTERISTICS

Not all sewage flows are the same. Some sewages are stronger than others. Table 28.4 lists typical values for strong and weak domestic sewages.

14. SOLIDS

Solids in wastewater are categorized in the same manner as in water supplies. *Total solids* consist of *suspended* and *dissolved solids*. Generally, total solids constitute only a small amount of the incoming flow—less than 1/10% by mass. Therefore, wastewater fluid properties are essentially those of water. Figure 28.2 illustrates the solids categorization, along with typical percentages. Each category can be further divided into organic and inorganic groups.

Table 28.4 *Strong and Weak Domestic Sewages[a,b]*

constituent	strong	weak
solids, total	1200	350
dissolved, total	850	250
fixed	525	145
volatile	325	105
suspended, total	350	100
fixed	75	30
volatile	275	70
settleable solids (mL/L)	20	5
biochemical oxygen demand, five-day, 20°C	300	100
total organic carbon	300	100
chemical oxygen demand	1000	250
nitrogen (total as N)	85	20
organic	35	8
free ammonia	50	12
nitrites	0	0
nitrates	0	0
phosphorus (total as P)	20	6
organic	5	2
inorganic	15	4
chlorides	100	30
alkalinity (as $CaCO_3$)	200	50
grease	150	50

[a]All concentrations are in mg/L unless noted.
[b]Also see Table 29.3.

Figure 28.2 *Wastewater Solids[a]*

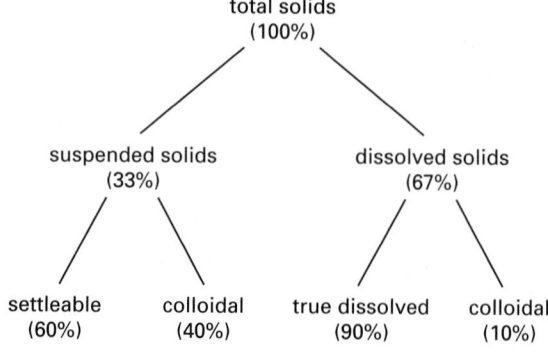

[a]typical percentages given

The amount of *volatile solids* can be used as a measure of the organic pollutants capable of affecting the oxygen content. As in water supply testing, volatile solids are measured by igniting filtered solids and measuring the decrease in mass.

Refractory solids (*refractory organics*) are solids that are difficult to remove by common wastewater treatment processes.

15. MICROBIAL GROWTH

For the large numbers and mixed cultures of microorganisms found in waste treatment systems, it is more convenient to measure biomass than numbers of organisms. This is accomplished by measuring suspended or volatile suspended solids. An expression that depicts the rate of conversion of food into biomass for wastewater treatment was developed by Monod.

Equation 28.7 calculates the rate of growth, r_g, of a bacterial culture as a function of the rate of concentration of limiting food in solution. The growth rate, X, of microorganisms, with dimensions of mass per unit volume (e.g., mg/L) is given by Eq. 28.7. μ is a specific growth rate factor with dimensions of time^{-1}.

$$r_g = \frac{dX}{dt} = \mu X \qquad 28.7$$

When one essential nutrient (referred to as a *substrate*) is present in a limited amount, the specific growth rate will increase up to a maximum value and is given by *Monod's equation*, Eq. 28.8. μ_m is the maximum specific growth rate with dimensions of time^{-1}. S is the concentration of the growth-limiting nutrient with dimensions of mass/unit volume. K_s is the *half-velocity coefficient*, the nutrient concentration at one half of the maximum growth rate, with dimensions of mass/unit volume.

$$\mu = \mu_m \left(\frac{S}{K_s + S} \right) \qquad 28.8$$

Combining Eqs. 28.7 and 28.8, the rate of growth is

$$r_g = \frac{\mu_m X S}{K_s + S} \qquad 28.9$$

The rate of substrate (nutrient) utilization is easily calculated if the *maximum yield coefficient*, Y, defined as the ratio of the mass of cells formed to the mass of nutrient consumed (with dimensions of mass/mass) is known.

$$r_{su} = \frac{dS}{dt} = \frac{-\mu_m X S}{Y(K_s + S)} \qquad 28.10$$

The maximum rate of substrate utilization per unit mass of microorganisms is

$$k = \frac{\mu_m}{Y} \qquad 28.11$$

Combining Eqs. 28.10 and 28.11,

$$\frac{dS}{dt} = \frac{-k X S}{K_s + S} \qquad 28.12$$

Figure 28.3 *Bacterial Growth Rate with Limited Nutrient*

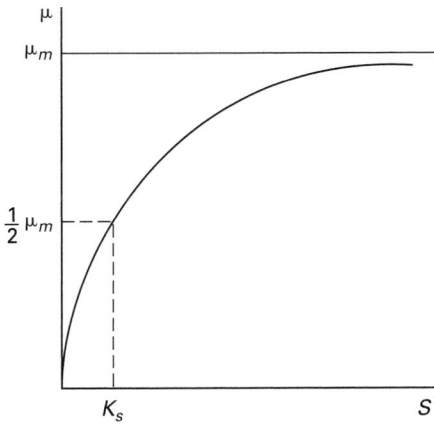

The Monod equation assumes that all microorganisms are the same age. The actual distribution of ages and other factors (death and predation of cells) decreases the rate of cell growth. The decrease is referred to as *endogenous decay* and is accounted for with an *endogenous decay coefficient*, k_d, with dimensions of time^{-1}. The rate of endogenous decay is

$$r_d = -k_d X \qquad 28.13$$

The net rate of cell growth and net specific growth rate are

$$r'_g = \frac{\mu_m X S}{K_s + S} - k_d X$$
$$= -Y r_{su} - k_d X \qquad 28.14$$
$$\mu' = \mu_m \left(\frac{S}{K_s + S} \right) - k_d \qquad 28.15$$

16. DISSOLVED OXYGEN IN WASTEWATER

A body of water exposed to the atmosphere will normally become saturated with oxygen. If the dissolved oxygen content, DO, of water is less than the saturated values (given in App. 22.D), there will be good reason to believe that the water is organically polluted.

The difference between the saturated and actual dissolved oxygen concentration is known as the *oxygen deficit*, D.

$$D = \text{DO}_{\text{sat}} - \text{DO} \qquad 28.16$$

17. REOXYGENATION

The oxygen deficit can be reduced (i.e., the dissolved oxygen concentration can be increased) only by aerating the water. This occurs naturally in free-running water (e.g., in rivers). The rate of reaeration is

$$r_r = K_r(\text{DO}_{\text{sat}} - \text{DO}) \qquad 28.17$$

K_r is the *reoxygenation (reaeration) rate constant*, which depends on the type of flow and temperature and with common units of days^{-1}. (K_r may be written as K_R and K_2 by some authorities.) Typical values of K_r (base-10) are: small stagnant ponds, 0.05 to 0.10; sluggish streams, 0.10 to 0.15; large lakes, 0.10 to 0.15; large streams, 0.15 to 0.30; swiftly flowing rivers and streams, 0.3 to 0.5; whitewater and waterfalls, 0.5 and above.

An exponential model is used to predict the oxygen deficit, D, as a function of time during a pure reoxygenation process. Reoxygenation constants are given for use with a base-10 and natural logarithmic base.

$$D_t = D_0 10^{-K_r t} \quad \text{[base-10]} \qquad 28.18$$
$$D_t = D_0 e^{-K'_r t} \quad \text{[base-}e\text{]} \qquad 28.19$$

The constant K'_r may be written as $K_{r,\text{base-}e}$ by some authorities. The constants K_r and K'_r are different but related.

$$K'_r = 2.303 K_r \qquad 28.20$$

K'_r can be approximated from the *O'Connor and Dobbins formula* for moderate to deep natural streams with low to moderate velocities. Both v and d are average values.

$$K'_{r,20°\text{C}} \approx \frac{3.93\sqrt{v_{\text{m/s}}}}{d_{\text{m}}^{1.5}} \quad \begin{bmatrix} 0.3 \text{ m} < d < 9.14 \text{ m} \\ 0.15 \text{ m/s} < v < 0.49 \text{ m/s} \end{bmatrix}$$
$$\text{[SI]} \quad 28.21(a)$$

$$K'_{r,68°\text{F}} \approx \frac{12.9\sqrt{v_{\text{ft/sec}}}}{d_{\text{ft}}^{1.5}} \quad \begin{bmatrix} 1 \text{ ft} < d < 30 \text{ ft} \\ 0.5 \text{ ft/sec} < v < 1.6 \text{ ft/sec} \end{bmatrix}$$
$$\text{[U.S.]} \quad 28.21(b)$$

An empirical formula that has been proposed for faster moving water is the *Churchill formula*.

$$K'_{r,20°\text{C}} \approx \frac{5.049(v_{\text{m/s}})^{0.969}}{d_{\text{m}}^{1.67}} \quad \begin{bmatrix} 0.61 \text{ m} < d < 3.35 \text{ m} \\ 0.55 \text{ m/s} < v < 1.52 \text{ m/s} \end{bmatrix}$$
$$\text{[SI]} \quad 28.22(a)$$

$$K'_{r,68°\text{F}} = \frac{11.61(v_{\text{ft/sec}})^{0.969}}{d_{\text{ft}}^{1.67}} \quad \begin{bmatrix} 2 \text{ ft} < d < 11 \text{ ft} \\ 1.8 \text{ ft/sec} < v < 5 \text{ ft/sec} \end{bmatrix}$$
$$\text{[U.S.]} \quad 28.22(b)$$

The variation in K'_r with temperature is given approximately by Eq. 28.23. The constant 1.024 has been reported as 1.016 by some authorities.

$$K'_{r,T} = K'_{r,20°\text{C}}(1.024)^{T-20°\text{C}} \qquad 28.23$$

Equation 28.23 is actually a special case of Eq. 28.24, which gives the relationship between values of K'_r between any two temperatures.

$$K'_{r,T_1} = K'_{r,T_2} \theta_r^{T_1 - T_2} \qquad 28.24$$

18. DEOXYGENATION

The rate of deoxygenation at time t is

$$r_{d,t} = -K_d \text{DO} \qquad 28.25$$

Environmental

The *deoxygenation rate constant*, K_d, for treatment plant effluent is approximately 0.05 to 0.10 day^{-1} and is typically taken as 0.1 day^{-1} (base-10 values). (K_d may be represented as K and K_1 by some authorities.) For highly polluted shallow streams, it can be as high as 0.25 day^{-1}. For raw sewage, it is approximately 0.15 to 0.30 day^{-1}.

K_d for other temperatures can be found from Eq. 28.26. The *temperature variation constant*, θ, is often quoted in literature as 1.047. However, this value should not be used with temperatures below 68°F (20°C). Additional research suggests that θ varies from 1.135 for temperatures between 39°F to 68°F (4°C to 20°C) up to 1.056 for temperatures between 68°F to 86°F (20°C to 30°C).

$$K'_{d,T} = K'_{d,20°C}\theta^{T-20°C} \qquad \textit{28.26}$$

Equation 28.26 is actually a special case of Eq. 28.27, which gives the relationship between values of K'_d between any two temperatures.

$$K'_{d,T_1} = K'_{d,T_2}\theta_d^{T_1-T_2} \qquad \textit{28.27}$$

19. TESTS OF WASTEWATER CHARACTERISTICS

The most common wastewater analyses used to determine characteristics of a municipal wastewater determine biochemical oxygen demand (BOD) and suspended solids (SS). BOD and flow data are basic requirements for the operation of biological treatment units. The concentration of suspended solids relative to BOD indicates the degree that organic matter is removable by primary settling. Additionally, temperature, pH, COD, alkalinity, color, grease, and the quantity of heavy metals are necessary to specify municipal wastewater characteristics.

20. BIOCHEMICAL OXYGEN DEMAND

When oxidizing organic material in water, biological organisms also remove oxygen from the water. This is typically considered to occur through oxidation of organic material to CO_2 and H_2O by microorganisms at the molecular level and is referred to as *biochemical oxygen demand* (BOD). Therefore, oxygen use is an indication of the organic waste content.

Typical values of BOD for various industrial wastewaters are given in Table 28.5. A BOD of 100 mg/L is considered to be a weak wastewater; a BOD of 200 to 250 mg/L is considered to be a medium strength wastewater; and a BOD above 300 mg/L is considered to be a strong wastewater.

Table 28.5 *Typical BOD and COD of Industrial Wastewaters*

industry/ type of waste	BOD	COD
canning		
corn	19.5 lbm/ton corn	
tomatoes	8.4 lbm/ton tomatoes	
dairy milk		
processing	1150 lbm/ton raw milk 1000 mg/L	1900 mg/L
beer brewing	1.2 lbm/barrel beer	
commercial		
laundry	1250 lbm/1000 lbm dry 700 mg/L	2400 mg/L
slaughterhouse (meat packing)	7.7 lbm/animal 1400 mg/L	2100 mg/L
papermill	121 lbm/ton pulp	
synthetic textile	1500 mg/L	3300 mg/L
chlorophenolic manufacturing	4300 mg/L	5400 mg/L
milk bottling	230 mg/L	420 mg/L
cheese production	3200 mg/L	5600 mg/L
candy production	1600 mg/L	3000 mg/L

In the past, the BOD has been determined in a lab using a traditional standardized BOD testing procedure. Since this procedure requires five days of incubating, online measurement systems have been developed to provide essentially instantaneous and continuous BOD information at wastewater plants.

The standardized BOD test consists of adding a measured amount of wastewater (which supplies the organic material) to a measured amount of dilution water (which reduces toxicity and supplies dissolved oxygen) and then incubating the mixture at a specific temperature. The standard procedure calls for a five-day incubation period at 68°F (20°C), though other temperatures are used. The measured BOD is designated as BOD$_5$ or BOD-5. The BOD of a biologically active sample at the end of the incubation period is given by Eq. 28.28.

$$\text{BOD}_5 = \frac{\text{DO}_i - \text{DO}_f}{\dfrac{V_{\text{sample}}}{V_{\text{sample}} + V_{\text{dilution}}}} \qquad \textit{28.28}$$

If more than one identical sample is prepared, the increase in BOD over time can be determined, rather than just the final BOD. Figure 28.4 illustrates a typical plot of BOD versus time. The BOD at any time t is known as the *BOD exertion*.

BOD exertion can be found from Eq. 28.29. The ultimate BOD (BOD$_u$) is the total oxygen used by *carbonaceous bacteria* if the test is run for a long period of time, usually taken as 20 days. (The symbols L, BOD$_L$, and BOD$_{\text{ult}}$ are also widely used to represent the ultimate BOD.) Use the proper form of Eq. 28.29 depending on the base of the deoxygenation rate constant.

$$\begin{aligned}\text{BOD}_t &= \text{BOD}_u(1 - 10^{-K_d t}) \\ &= \text{BOD}_u(1 - e^{-K'_d t}) \qquad \textit{28.29}\end{aligned}$$

The ultimate BOD, BOD_u, usually cannot be found from long-term studies due to the effect of nitrogen-consuming bacteria in the sample. However, if K_d is 0.1 day^{-1}, the ultimate BOD can be approximated from Eq. 28.30 derived from Eq. 28.29.

$$BOD_u \approx 1.463 BOD_5 \qquad 28.30$$

Figure 28.4 BOD Exertion

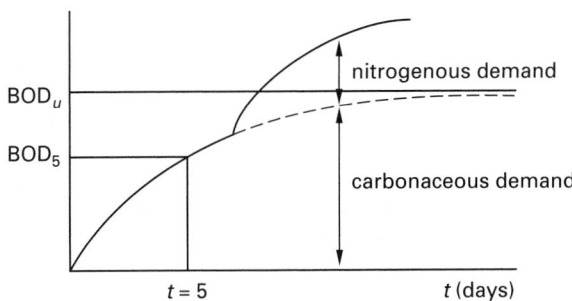

The approximate variation in first-stage BOD (either BOD_5 or BOD_u) municipal wastewater with temperature is given by Eq. 28.31.[1]

$$BOD_{T°C} = BOD_{20°C}(0.02 T_{°C} + 0.6) \qquad 28.31$$

The deviation from the expected exponential growth curve in Fig. 28.4 is due to nitrification and is considered to be *nitrogenous demand*. *Nitrification* is the use of oxygen by autotrophic bacteria. Autotrophic bacteria oxidize ammonia to nitrites and nitrates.

The number of autotrophic bacteria in a wastewater sample is initially small. Generally, 6 to 10 days are required for the autotrophic population to become sufficiently large to affect a BOD test. Therefore, the standard BOD test is terminated at five days, before the autotrophic contribution to BOD becomes significant.

Until recently, BOD has not played a significant role in the daily operations decisions of treatment plants because of the length of time it took to obtain BOD data. Other measurements, such as COD, TOC, turbidity, suspended solids, respiration rates, or flow have been used.

Example 28.1

Ten 5 mL samples of wastewater are placed in 300 mL BOD bottles and diluted to full volume. Half of the bottles are titrated immediately, and the average initial concentration of dissolved oxygen is found to be 7.9 mg/L. The remaining bottles are incubated for five days, after which the average dissolved oxygen is determined to be 4.5 mg/L. The deoxygenation rate constant is known to be 0.13 day^{-1}. What are the (a) standard BOD and (b) ultimate carbonaceous BOD?

[1] As reported in *Water-Resources Engineering*, 2nd Ed., Ray K. Linsley and Joseph B. Franzini, McGraw-Hill Book Company, 1972, New York.

Solution

(a) Use Eq. 28.28.

$$BOD_5 = \frac{DO_i - DO_f}{\dfrac{V_{sample}}{V_{sample} + V_{dilution}}}$$

$$= \frac{7.9\,\dfrac{mg}{L} - 4.5\,\dfrac{mg}{L}}{\dfrac{5\,mL}{300\,mL}}$$

$$= 204\ mg/L$$

(b) From Eq. 28.29, the ultimate BOD is

$$BOD_u = \frac{BOD_t}{1 - 10^{-K_d t}}$$

$$= \frac{204\,\dfrac{mg}{L}}{1 - 10^{(-0.13\ day^{-1})(5\ days)}}$$

$$= 263\ mg/L$$

21. SEEDED BOD

Industrial wastewater may lack sufficient microorganisms to metabolize the organic matter. In this situation, the standard BOD test will not accurately determine the amount of organic matter unless seed organisms are added.

The BOD in seeded samples is found by measuring dissolved oxygen in the seeded sample after 15 min (DO_i) and after five days (DO_f), as well as by measuring the dissolved oxygen of the seed material itself after 15 min (DO_i^*) and after five days (DO_f^*). In Eq. 28.32, x is the ratio of the volume of seed added to the sample to the volume of seed used to find DO^*.

$$BOD = \frac{DO_i - DO_f - x(DO_i^* - DO_f^*)}{\dfrac{V_{sample}}{V_{sample} + V_{dilution}}} \qquad 28.32$$

22. DILUTION PURIFICATION

Dilution purification (also known as *self-purification*) refers to the discharge of untreated or partially treated sewage into a large body of water such as a river. After the treatment plant effluent has mixed with the river water, the river experiences both deoxygenation (from the biological activity of the waste) and reaeration (from the agitation of the flow). If the body is large and is adequately oxygenated, the sewage's BOD may be satisfied without putrefaction.

When two streams merge, the characteristics of the streams are blended in the combined flow. If values of the characteristics of interest are known for the upstream flows, the blended concentration can be found as a weighted average. Equation 28.33 can be used to calculate the final temperature, dissolved oxygen, BOD, or suspended solids content immediately after two flows are mixed. (The subscript a is also used in the literature to represent the condition immediately after mixing.)

$$C_f = \frac{C_1 Q_1 + C_2 Q_2}{Q_1 + Q_2} \qquad 28.33$$

Example 28.2

6 MGD of wastewater with a dissolved oxygen concentration of 0.9 mg/L is discharged into a 50°F river flowing at 40 ft³/sec. The river is initially saturated with oxygen. What is the oxygen content of the river immediately after mixing?

Solution

Both flow rates must have the same units. Convert MGD to ft³/sec.

$$Q_w = \frac{(6 \text{ MGD}) \left(10^6 \frac{\text{gal}}{\text{MG}}\right)}{\left(7.48 \frac{\text{gal}}{\text{ft}^3}\right) \left(24 \frac{\text{hr}}{\text{day}}\right) \left(60 \frac{\text{min}}{\text{hr}}\right) \left(60 \frac{\text{sec}}{\text{min}}\right)}$$

$$= 9.28 \text{ ft}^3/\text{sec}$$

From App. 22.D, the saturated oxygen content at 50°F (10°C) is 11.3 mg/L. From Eq. 28.33,

$$C_f = \frac{C_w Q_w + C_r Q_r}{Q_w + Q_r}$$

$$= \frac{\left(0.9 \frac{\text{mg}}{\text{L}}\right) \left(9.28 \frac{\text{ft}^3}{\text{sec}}\right) + \left(11.3 \frac{\text{mg}}{\text{L}}\right) \left(40 \frac{\text{ft}^3}{\text{sec}}\right)}{9.28 \frac{\text{ft}^3}{\text{sec}} + 40 \frac{\text{ft}^3}{\text{sec}}}$$

$$= 9.34 \text{ mg/L}$$

23. RESPONSE TO DILUTION PURIFICATION

The oxygen deficit is the difference between actual and saturated oxygen concentrations, expressed in mg/L. Since reoxygenation and deoxygenation of a polluted river occur simultaneously, the oxygen deficit will increase if the reoxygenation rate is less than the deoxygenation rate. If the oxygen content goes to zero, anaerobic decomposition and putrefaction will occur.

The minimum dissolved oxygen concentration that will protect aquatic life in the river is the *dissolved oxygen standard* for the river. 4 to 6 mg/L is the generally accepted range of dissolved oxygen required to support fish populations. 5 mg/L is adequate as is evidenced in high-altitude trout lakes. However, 6 mg/L is preferable, particularly for large fish populations.

The oxygen deficit at time t is given by the *Streeter-Phelps equation*. In Eq. 28.34, BOD_u is the ultimate carbonaceous BOD of the river immediately after mixing. D_0 is the dissolved oxygen deficit immediately after mixing, which is often given the symbol D_a. (If the constants K_r and K_d are base-e constants, the base-10 exponentiation in Eq. 28.34 should be replaced with the base-e exponentiation, e^x.)

$$D_t = \left(\frac{K_d BOD_u}{K_r - K_d}\right) (10^{-K_d t} - 10^{-K_r t}) + D_0(10^{-K_r t})$$

$$28.34$$

A graph of dissolved oxygen versus time (or distance downstream) is known as a dissolved oxygen *sag curve*. The location at which the lowest dissolved oxygen concentration occurs is the *critical point*. The location of the critical point is found from the river velocity and flow time to that point.

$$x_c = v t_c \qquad 28.35$$

The time to the critical point is given by Eq. 28.36. The ratio K_r/K_d in Eq. 28.36 is known as the *self-purification constant*. (If the constants K_r and K_d are base-e constants, the base-10 logarithm in Eq. 28.36 should be replaced with the base-e logarithm, ln.)

$$t_c = \left(\frac{1}{K_r - K_d}\right)$$
$$\times \log_{10}\left(\left(\frac{K_d BOD_u - K_r D_0 + K_d D_0}{K_d BOD_u}\right)\left(\frac{K_r}{K_d}\right)\right)$$

$$28.36$$

The critical oxygen deficit can be found by substituting t_c from Eq. 28.36 into Eq. 28.34. However, it is more expedient to use Eq. 28.37.

$$D_c = \left(\frac{K_d BOD_u}{K_r}\right) 10^{-K_d t_c} \qquad 28.37$$

Figure 28.5 *Oxygen Sag Curve*

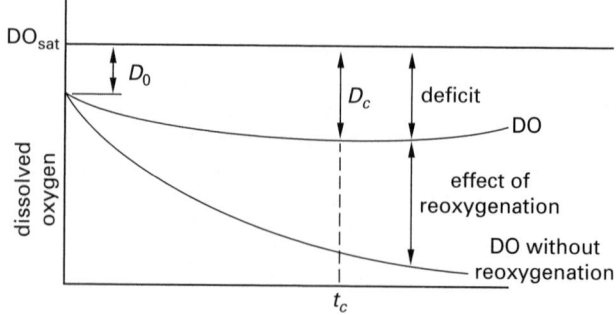

Environmental

Example 28.3

A treatment plant effluent has the following characteristics.

quantity	15 ft³/sec
$BOD_{5,20°C}$	45 mg/L BOD
DO	2.9 mg/L
temperature	24°C
$K_{d,20°C}$	0.1 day⁻¹ (base-10)

temperature variation constant for K_d, θ_d 1.047
temperature variation constant for K_r, θ_r 1.016

The outfall is located in a river carrying 120 ft³/sec of water with the following characteristics.

quantity	120 ft³/sec
velocity	0.55 ft/sec
average depth	4 ft
$BOD_{5,20°C}$	4 mg/L
DO	8.3 mg/L
temperature	16°C

When the treatment plant effluent and river are mixed, the mixture temperature persists for a long distance downstream.

(a) Determine the distance downstream where the oxygen level is at a minimum. (b) Determine if the river can support fish life at that point.

Solution

(a) Find the river conditions immediately after mixing. Use Eq. 28.33 three times.

$$C_f = \frac{C_1 Q_1 + C_2 Q_2}{Q_1 + Q_2}$$

$$BOD_{5,20°} = \frac{\left(15 \frac{ft^3}{sec}\right)\left(45 \frac{mg}{L}\right) + \left(120 \frac{ft^3}{sec}\right)\left(4 \frac{mg}{L}\right)}{15 \frac{ft^3}{sec} + 120 \frac{ft^3}{sec}}$$

$$= 8.56 \text{ mg/L}$$

$$DO = \frac{\left(15 \frac{ft^3}{sec}\right)\left(2.9 \frac{mg}{L}\right) + \left(120 \frac{ft^3}{sec}\right)\left(8.3 \frac{mg}{L}\right)}{15 \frac{ft^3}{sec} + 120 \frac{ft^3}{sec}}$$

$$= 7.7 \text{ mg/L}$$

$$T = \frac{\left(15 \frac{ft^3}{sec}\right)(24°C) + \left(120 \frac{ft^3}{sec}\right)(16°C)}{15 \frac{ft^3}{sec} + 120 \frac{ft^3}{sec}}$$

$$= 16.89°C$$

Calculate the approximate rate constants. From Eq. 28.26,

$$K_{d,T} = K_{d,20°C}(1.047)^{T-20°C}$$
$$K_{d,16.89°C} = (0.1 \text{ day}^{-1})(1.047)^{16.89°C-20°C}$$
$$= 0.0867 \text{ day}^{-1}$$

Since v = 0.55 ft/sec, use Eqs. 28.20 and 28.21 to calculate the reoxygenation constant.

$$K_{r,20°} = \frac{K'_{r,20°}}{2.303}$$
$$= \frac{12.9\sqrt{v}}{2.303 d^{1.5}}$$
$$= \frac{12.9\sqrt{0.55 \frac{ft}{sec}}}{(2.303)(4 \text{ ft})^{1.5}}$$
$$= 0.519 \text{ day}^{-1}$$

From Eq. 28.23, and using the given value of θ,

$$K_{r,T} = K_{r,20°C}(1.016)^{T-20°C}$$
$$K_{r,16.89°C} = (0.519 \text{ day}^{-1})(1.016)^{16.89°C-20°C}$$
$$= 0.494 \text{ day}^{-1}$$

Estimate BOD_u from Eq. 28.29.

$$BOD_{u,20°C} = \frac{BOD_{5,20°}}{1 - 10^{-K_d t}}$$
$$= \frac{8.56 \frac{mg}{L}}{1 - 10^{(-0.1 \text{ day}^{-1})(5 \text{ day})}}$$
$$= 12.52 \text{ mg/L}$$

From Eq. 28.31,

$$BOD_T = BOD_{20°C}(0.02T_{°C} + 0.6)$$
$$BOD_{u,16.89°C} = \left(12.52 \frac{mg}{L}\right)((0.02)(16.89°C) + 0.6)$$
$$= 11.74 \text{ mg/L}$$

Calculate the original oxygen deficit. By interpolation, from App. 22.D, the dissolved oxygen concentration at 16.89°C is approximately 9.76 mg/L.

$$D_0 = 9.76 \frac{mg}{L} - 7.7 \frac{mg}{L} = 2.06 \text{ mg/L}$$

Calculate t_c from Eq. 28.36.

$$t_c = \left(\frac{1}{K_r - K_d}\right)$$

$$\times \log_{10}\left(\left(\frac{K_d\text{BOD}_u - K_r D_0 + K_d D_0}{K_d\text{BOD}_u}\right)\left(\frac{K_r}{K_d}\right)\right)$$

$$= \left(\frac{1}{0.494 \text{ day}^{-1} - 0.0867 \text{ day}^{-1}}\right)$$

$$\times \log_{10}\left(\left(\frac{\begin{array}{c}(0.0867 \text{ day}^{-1})\left(11.74 \frac{\text{mg}}{\text{L}}\right)\\[4pt] - (0.494 \text{ day}^{-1})\left(2.06 \frac{\text{mg}}{\text{L}}\right)\\[4pt] + (0.0867 \text{ day}^{-1})\left(2.06 \frac{\text{mg}}{\text{L}}\right)\end{array}}{(0.0867 \text{ day}^{-1})\left(11.74 \frac{\text{mg}}{\text{L}}\right)}\right)\right.$$

$$\left. \times \left(\frac{0.494 \text{ day}^{-1}}{0.0867 \text{ day}^{-1}}\right)\right)$$

$$= 0.00107 \text{ days}$$

The distance downstream is

$$x_c = vt_c$$

$$= \frac{\left(0.55 \frac{\text{ft}}{\text{sec}}\right)(0.00107 \text{ days})\left(86{,}400 \frac{\text{sec}}{\text{day}}\right)}{5280 \frac{\text{ft}}{\text{mi}}}$$

$$= 0.0096 \text{ mi} \quad [\text{essentially, at the outfall}]$$

The critical oxygen deficit is found from Eq. 28.37.

$$D_c = \left(\frac{K_d\text{BOD}_u}{K_r}\right)10^{-K_d t_c}$$

$$= \left(\frac{(0.0867 \text{ day}^{-1})\left(11.74 \frac{\text{mg}}{\text{L}}\right)}{0.494 \text{ day}^{-1}}\right)$$

$$\times \left(10^{(-0.0867 \text{ day}^{-1})(0.00107 \text{ day})}\right)$$

$$= 2.06 \text{ mg/L} \quad [\text{the oxygen deficit at the outfall}]$$

(b) The saturated oxygen content at the critical point with a water temperature of $16.89°C$ is 9.76 mg/mL. The actual oxygen content is

$$\text{DO} = \text{DO}_{\text{sat}} - D_c = 9.76 \frac{\text{mg}}{\text{L}} - 2.06 \frac{\text{mg}}{\text{L}}$$

$$= 7.70 \text{ mg/L}$$

Since this is greater than 4–6 mg/L, fish life is supported.

24. CHEMICAL OXYGEN DEMAND

Chemical oxygen demand (COD) is a measure of a maximum oxidizable substance (unlike biochemical oxygen demand, which is a measure of oxygen removed only by biological organisms). Therefore, COD is a good measure of total effluent strength.

COD testing is required where there is industrial chemical pollution. In such environments, the organisms necessary to metabolize organic compounds may not exist. Furthermore, the toxicity of the wastewater may make the standard BOD test impossible to carry out. Standard COD test results are usually available in a matter of hours.

If toxicity is low, BOD and COD test results can be correlated. The BOD/COD ratio typically varies from 0.4 to 0.8. This is a wide range but, for any given treatment plant and waste type, the correlation is essentially constant. The correlation can, however, vary along the treatment path.

25. RELATIVE STABILITY

The *relative stability* test is easier to perform than the BOD test, although it is less accurate. The relative stability of an effluent is defined as the percentage of initial BOD that has been satisfied. The test consists of taking a sample of effluent and adding a small amount of methylene blue dye. The mixture is then incubated, usually at 20°C. When all oxygen has been removed from the water, anaerobic bacteria start to remove the dye. The amount of time it takes for the color to start degrading is known as the *stabilization time* or *decoloration time*. The relative stability can be found from the stabilization time by using Table 28.6.

Table 28.6 Relative Stability[a]

stabilization time (days)	relative stability[b] (%)
1/2	11
1	21
1 1/2	30
2	37
2 1/2	44
3	50
4	60
5	68
6	75
7	80
8	84
9	87
10	90
11	92
12	94
13	95
14	96
16	97
18	98
20	99

[a] incubation at 20°C
[b] calculated as $(100\%)(1 - 0.794^t)$

26. CHLORINE DEMAND AND DOSE

Chlorination destroys bacteria, hydrogen sulfide, and other compounds and substances. The *chlorine demand* (*chlorine dose*) must be determined by careful monitoring of coliform counts and free residuals since there are several ways that chlorine can be used up without producing significant disinfection. Only after uncombined (free) chlorine starts showing up is it assumed that all chemical reactions and disinfection are complete.

Chlorine demand is the amount of chlorine (or its chloramine or hypochlorite equivalent) required to leave the desired residual (usually 0.5 mg/L) 15 min after mixing. Fifteen minutes is the recommended mixing and holding time prior to discharge, since during this time nearly all pathogenic bacteria in the water will have been killed.

Typical doses for wastewater effluent depend on the application point and are widely variable, though doses rarely exceed 30 mg/L. For example, chlorine may be applied at 5 to 25 mg/L prior to primary sedimentation, 2 to 6 mg/L after sand filtration, and 3 to 15 mg/L after trickle filtration.

27. BREAKPOINT CHLORINATION

Because of their reactivities, chlorine is initially used up in the neutralization of hydrogen sulfide and the rare ferrous and manganous (Fe^{+2} and Mn^{+2}) ions. For example, hydrogen sulfide is oxidized according to Eq. 28.38. The resulting HCl, $FeCl_2$, and $MnCl_2$ ions do not contribute to disinfection. They are known as *unavailable combined residuals*.

$$H_2S + 4H_2O + 4Cl_2 \rightarrow H_2SO_4 + 8HCl \qquad 28.38$$

Ammonia nitrogen combines with chlorine to form the family of *chloramines*. Depending on the water pH, monochloramines (NH_2Cl), dichloramines ($NHCl_2$), or trichloramines (nitrogen trichloride, NCl_3) may form. Chloramines have long-term disinfection capabilities and are therefore known as *available combined residuals*. Equation 28.39 is a typical chloramine formation reaction.

$$NH_4^+ + HOCl \leftrightarrow NH_2Cl + H_2O + H^+ \qquad 28.39$$

The continued addition of chlorine after chloramines begin forming changes the pH and allows chloramine destruction to begin. Chloramines are converted to nitrogen gas (N_2) and nitrous oxide (N_2O). Equation 28.40 is a typical chloramine destruction reaction.

$$2NH_2Cl + HOCl \leftrightarrow N_2 + 3HCl + H_2O \qquad 28.40$$

The destruction of chloramines continues, with the repeated application of chlorine, until no ammonia remains in the water. The point at which all ammonia has been removed is known as the *breakpoint*.

In the *breakpoint chlorination* method, additional chlorine is added after the breakpoint in order to obtain free chlorine residuals. The free residuals have a high disinfection capacity. Typical free residuals are free chlorine (Cl_2), hypochlorous acid (HOCl), and hypochlorite ions. Equations 28.41 and 28.42 illustrate the formation of these free residuals.

$$Cl_2 + H_2O \rightarrow HCl + HOCl \qquad 28.41$$

$$HOCl \rightarrow H^+ + ClO^- \qquad 28.42$$

There are several undesirable characteristics of breakpoint chlorination. First, it may not be economical to use breakpoint chlorination unless the ammonia nitrogen has been reduced. Second, free chlorine residuals produce trihalomethanes. Third, if free residuals are prohibited to prevent trihalomethanes, the water may need to be dechlorinated using sulfur dioxide gas or sodium bisulfate. (Where small concentrations of free residuals are permitted, dechlorination may be needed only during the dry months. During winter storm months, the chlorine residuals may be adequately diluted with rain water.)

28. NITROGEN

Nitrogen occurs in water in organic, ammonia, nitrate, nitrite, and dissolved gaseous forms. Nitrogen in municipal wastewater results from human excreta, ground garbage, and industrial wastes (primarily food processing). Bacterial decomposition and the hydrolysis of urea produces ammonia, NH_3. Ammonia in water forms the ammonium ion NH_4^+, also known as *ammonia nitrogen*. Ammonia nitrogen must be removed from the waste effluent stream due to potential exertion of oxygen demand. The total of organic and ammonia nitrogen is known as *total Kjeldahl nitrogen*, or TKN. *Total nitrogen*, TN, includes TKN plus inorganic nitrates and nitrites.

Nitrification (the oxidation of ammonia nitrogen) occurs as follows.

$$NH_4^+ + 2O_2 \rightarrow NO_3^- + H_2O + 2H^+ \qquad 28.43$$

Denitrification, the reduction of nitrate nitrogen to nitrogen gas, occurs as follows.

$$2NO_3^- + \text{organic matter} \rightarrow N_2 + CO_2 + H_2O \quad 28.44$$

Ammonia nitrogen can be removed (*denitrification*) chemically from water by raising the pH level. This converts the ammonium ion back into ammonia, which can then be stripped from the water by passing large quantities of air through the water. Lime is added to provide the hydroxide for the reaction. The *ammonia stripping* reaction is

$$NH_4^+ + OH^- \rightarrow NH_3 + H_2O \qquad 28.45$$

Ammonia stripping has no effect on nitrate, which is commonly removed in an activated sludge process with a short cell detention time.

Environmental

29. ORGANIC COMPOUNDS IN WASTEWATER

Biodegradable organic matter in municipal wastewater is classified into three major categories: proteins, carbohydrates, and greases (fats).

Proteins are long strings of amino acids containing carbon, hydrogen, oxygen, nitrogen, and phosphorus.

Carbohydrates consist of sugar units containing the elements of carbon, hydrogen, and oxygen. They are identified by the presence of a saccharide ring. The rings range from simple monosaccharides to polysaccharides (long chain sugars categorized as either readily degradable starches found in potatoes, rice, corn, and other edible plants, or as cellulose found in wood and similar plant tissues).

Greases are a variety of biochemical substances that have the common property of being soluble to varying degrees in organic solvents (acetone, ether, ethanol, and hexane) while being only sparingly soluble in water. The low solubility of grease in water causes problems in pipes and tanks where it accumulates, reduces contact areas during various filtering processes, and produces a sludge that is difficult to dispose of.

Degradation of greases by microorganisms occurs at a very slow rate. A simple fat is a triglyceride composed of a glycerol unit with short-chain or long-chain fatty acids attached.

The majority of carbohydrates, fats, and proteins in wastewater are in the form of large molecules that cannot penetrate the cell membrane of microorganisms. Bacteria, in order to metabolize high molecular-weight substances, must be capable of breaking down the large molecules into diffusible fractions for assimilation into the cell. Of the organic matter in wastewater, 60 to 80% is readily available for biodegradation.

Several organic compounds such as cellulose, long-chain saturated hydrocarbons, and other complex compounds, although available as a bacterial substrate, are considered nonbiodegradable because of the time and environmental limitations of biological wastewater treatment systems.

Volatile organic chemicals (VOCs) are released in large quantities in industrial, commercial, agricultural, and household activities. The adverse health effects of VOCs include cancer and chronic effects on the liver, kidney, and nervous system. *Synthetic organic chemicals* (SOCs) and *inorganic chemicals* (IOCs) are used in agricultural and industrial processes as pesticides, insecticides, and herbicides.

Organic petroleum derivatives, detergents, pesticides, and other synthetic, organic compounds are particularly resistant to biodegradation in wastewater treatment plants, and some are toxic and inhibit the activity of microorganisms in biological treatment processes.

30. HEAVY METALS IN WASTEWATER

Metals are classified by four categories: *Dissolved metals* are those constituents of an unacidified sample that pass through a 0.45 μm membrane filter. *Suspended metals* are those constituents of an unacidified sample retained by a 0.45 μm membrane filter. *Total metals* are the concentration determined on an unfiltered sample after vigorous digestion or the sum of both dissolved and suspended fractions. *Acid-extractable metals* remain in solution after treatment of an unfiltered sample with hot dilute mineral acid.

The metal concentration in the wastewater is calculated from Eq. 28.46. The sample size is selected so that the product of the sample size (in mL) and the metal concentration (C, in mg/L) is approximately 1000.

$$C = \frac{C_{\text{digestate}} V_{\text{digested solution}}}{V_{\text{sample}}} \qquad 28.46$$

Some metals are biologically essential; others are toxic and adversely affect wastewater treatment systems and receiving waters.

31. WASTEWATER COMPOSITING

Proper sampling techniques are essential for accurate evaluation of wastewater flows. Samples should be well mixed, representative of the wastewater flow stream in composition, taken at regular time intervals, and stored properly until analysis can be performed. *Compositing* is the sampling procedure that accomplishes these goals.

Samples are collected (i.e., "grabbed," hence the name *grab sample*) at regular time intervals (e.g., every hour on the hour), stored in a refrigerator or ice chest, and then integrated to formulate the desired combination for a particular test. Flow rates are measured at each sampling to determine the wastewater flow pattern.

The total volume of the composite sample desired depends on the kinds and number of laboratory tests to be performed.

32. WASTEWATER STANDARDS

Treatment plants and industrial complexes must meet all applicable wastewater quality standards for surface waters, drinking waters, air, and effluents of various types. For example, standards for discharges from secondary treatment plants are given in terms of *maximum contaminant levels* (MCLs) for five-day BOD, suspended solids, coliform count, and pH. Some chemicals, compounds, and microbial varieties are regulated; others are unregulated but monitored. MCLs for regulated and monitored chemicals, compounds, and microbes are subject to ongoing legislation and change.

29 Wastewater Treatment: Equipment and Processes

Nomenclature

A	area	ft^2	m^2
BOD	biochemical oxygen demand	mg/L	mg/L
D	diameter	ft	m
f	Darcy friction factor	–	–
F	effective number of passes	–	–
g	acceleration of gravity	ft/sec^2	m/s^2
g_c	gravitational constant	$ft\text{-}lbm/lbf\text{-}sec^2$	–
k	Camp formula constant	–	–
k	removal rate constant	$days^{-1}$	$days^{-1}$
L	length	ft	m
L_{BOD}	BOD loading	$lbm/1000\ ft^3\text{-}day$	$kg/m^3\cdot day$
L_H	hydraulic loading	$gal/day\text{-}ft^2$	$m^3/day\cdot m^2$
m	exponent	–	–
m	mass	lbm	kg
n	exponent	–	–
P	power	hp	kW
Q	air flow quantity	ft^3/min	L/s
Q	water flow quantity	gal/day	m^3/day
R	recirculation ratio	–	–
S	BOD	mg/L	mg/L
SG	specific gravity	–	–
t	time	days	days
T	temperature	°F	°C
v	velocity	ft/sec	m/s
v^*	overflow rate	$gal/day\text{-}ft^2$	$m^3/day\cdot m^2$
V	volume	ft^3	m^3
w	empirical weighting factor	–	–
X	removable fraction of BOD	–	–
Z	depth (of filter)	ft	m

Symbols

η	BOD removal fraction	–	–
η	mechanical efficiency	–	–
θ	temperature constant	–	–
ρ	density	lbm/ft^3	kg/m^3

Subscripts

0	initial
d	detention
i	in
o	out
p	particle
ps	primary settling tank
r	recirculation
T	at temperature T
w	wastewater

1. INDUSTRIAL WASTEWATER TREATMENT

The *National Pollution Discharge Elimination System* (NPDES) places strict controls on the discharge of industrial wastewaters into municipal sewers. Any industrial wastes that would harm subsequent municipal

Environmental

treatment facilities or that would upset subsequent biological processes need to be pretreated. Manufacturing plants may also be required to equalize wastewaters by holding them in basins for stabilization prior to their discharge to the sewer. Table 29.1 lists typical limitations on industrial wastewaters.

Table 29.1 *Typical Industrial Wastewater Effluent Limitations*

characteristic	concentration[a], mg/L
COD	300–2000
BOD	100–300
oil and grease or TPH[b]	15–55
total suspended solids	15–45
pH	6.0–9.0
temperature	less than 40°C
color	2 color units
NH_3/NO_3	1.0–10
phosphates	0.2
heavy metals	0.1–5.0
surfactants	0.5–1.0 (total)
sulfides	0.01–0.1
phenol	0.1–1.0
toxic organics	1.0 total
cyanide	0.1

[a]maximum permitted at discharge
[b]total petroleum hydrocarbons

2. CESSPOOLS

A *cesspool* is a previously dug covered pit into which domestic (i.e., household) sewage is discharged. Cesspools for temporary storage can be constructed as watertight enclosures. However, most are *leaching cesspools* that allow seepage of liquid into the soil. Cesspools are rarely used today. They are acceptable for disposal for only very small volumes (e.g., from a few families).

3. SEPTIC TANKS

A *septic tank* is a simply constructed tank that holds domestic sewage while sedimentation and digestion occur. Typical detention times are 8 to 24 hr. Only 30 to 50% of the suspended solids are digested in a septic tank. The remaining solids settle, clogging the tank, and eventually must be removed. Semi-clarified effluent percolates into the surrounding soil through lateral lines placed at the *flow line* and that lead into an underground leach field. In the past, clay drainage tiles were used. The terms "tile field" and "tile bed" are still encountered even though perforated plastic pipe is now widely used.

Most septic tanks are built for use by one to three families. Larger communal tanks can be constructed for small groups (they should be designed to hold 12 to 24 hours of flow plus stored sludge). A general rule is to allow at least 30 gal (0.1 m³) of storage per person served

by the tank. Typical design parameters of domestic septic tanks are contained in Table 29.2.

Proper design of the percolation field is the key to successful operation. Soils studies are essential to ensure adequate absorption into the soil.

Table 29.2 *Typical Characteristics of Domestic Septic Tanks*

minimum capacity	
below flow line	300–500 gal (1.1–1.9 m³), plus 30 gal (0.1 m³) for each person served over 5
plan aspect ratio	1:2
minimum depth	
below flow line	3–4 ft (0.9–1.2 m)
minimum freeboard	
above flow line	1 ft (0.3 m)
tank burial depth	1–2 ft (0.3–0.6 m)
drainage field tile length	30 ft (9 m) per person
maximum drainage	
field tile length	60 ft (18 m)
minimum tile depth	1.5–2.5 ft (0.45–0.75 m)
lateral line spacing	6 ft (1.8 m)
gravel bed	0.33 ft (0.1 m) above lateral, 1–3 ft (0.3–1 m) below
soil layer below tile bed	10 ft (3 m)

Figure 29.1 *Septic Tank*

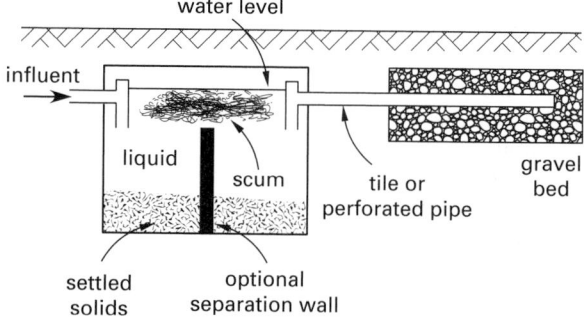

4. DISPOSAL OF SEPTAGE

Septage is the water and solid material pumped periodically from septic tanks, cesspools, or privies. In the United States, disposal of septage is controlled by federal sludge disposal regulations, and most local and state governments have their own regulations, as well. While surface and land spreading were acceptable disposal methods in the past, septage haulers are now turning to municipal treatment plants for disposal. Since septage is many times more concentrated than sewage, it can cause *shock loads* in preliminary treatment processes. Typical problems are: plugged screens and aerator inlets, reduced efficiency in grit chambers and aeration basins, and increased odors. The impact on subsequent processes is less pronounced due to the effect of dilution.

Table 29.3 Comparison of Typical Septage and Municipal Sewage[a,b]

characteristic	septage	sewage[c]
BOD	7000	220
COD	15,000	500
total solids	40,000	720
total volatile solids	25,000	365
total suspended solids	15,000	220
volatile suspended solids	10,000	165
TKN	700	40
NH_3 as N	150	25
alkalinity	1000	100
grease	8000	100
pH	6.0	–

[a] all values except pH in mg/L
[b] Also see Table 28.4.
[c] medium strength

5. WASTEWATER TREATMENT PLANTS

For traditional *wastewater treatment plants* (WWTP), *preliminary treatment* of the wastewater stream is essentially a mechanical process intended to remove large objects, rags, and wood. Heavy solids and excessive oils and grease are also eliminated. Damage to pumps and other equipment would occur without preliminary treatment.

Odor control through chlorination or ozonation, freshening of septic waste by aeration, and flow equalization in holding basins can also be loosely categorized as pretreatment processes.

After preliminary treatment, there are three "levels" of wastewater treatment: primary, secondary, and tertiary. *Primary treatment* is a mechanical (settling) process used to remove oil and most (i.e., approximately 50%) of the settleable solids. With domestic wastewater, a 25 to 35% reduction in BOD is also achieved, but BOD reduction is not the goal of primary treatment.

In the United States, secondary treatment is mandatory for all publicly owned wastewater treatment plants. *Secondary treatment* involves biological treatment in trickling filters, rotating contactors, biological beds, and activated sludge processes (covered in Chap. 30). Processing typically reduces the suspended solids and BOD content by more than 85%, volatile solids by 50%, total nitrogen by about 25%, and phosphorus by 20%.

Tertiary treatment (also known as *advanced wastewater treatment*, AWT) is targeted at specific pollutants or wastewater characteristics that have passed through previous processes in concentrations that are not allowed in the discharge. *Suspended solids* are removed by microstrainers or polishing filter beds. *Phosphorus* is removed by chemical precipitation. Aluminum and iron coagulants, as well as lime, are effective in removing phosphates. *Ammonia* can be removed by air stripping, biological denitrification, breakpoint chlorination, anion

exchange, and algae ponds. Ions from *inorganic salts* can be removed by electrodialysis and reverse osmosis. The so-called *trace organics* or *refractory substances*, *dissolved organic solids* that are resistant to biological processes, can be removed by filtering through carbon or ozonation.

6. WASTEWATER PLANT SITING CONSIDERATIONS

Wastewater plants should be located as far as possible from inhabited areas. A minimum distance of 1000 ft (300 m) for uncovered plants and lagoons is desired. Uncovered plants should be located downwind when a definite wind direction prevails. Soil conditions need to be evaluated, as does the proximity of the water table. Elevation in relationship to the need for sewage pumping (and for dikes around the site) is relevant.

The plant must be protected against flooding. One hundred-year storms are often chosen as the design flood when designing dikes and similar facilities. Distance to the outfall and possible effluent pumping need to be considered.

Table 29.4 lists the approximate acreage for sizing wastewater treatment plants. Estimates of population expansion should provide for future capacity.

Table 29.4 Treatment Plant Acreage Requirements

type of treatment	surface area required (ac/MGD)
physical-chemical plants	1.5
activated sludge plants	2
trickling filter plants	3
aerated lagoons	16
stabilization basins	20

7. PUMPS USED IN WASTEWATER PLANTS

Wastewater treatment plants should be gravity-fed wherever possible. The influent of most plants is pumped to the starting elevation, and wastewater flows through subsequent processes by gravity thereafter. However, there are still many instances when pumping is required. Table 29.5 lists pump types by application.

At least two identical pumps should be present at every location, each capable of handling the entire peak flow. Three or more pumps are suggested for flows greater than 1 MGD, and peak flow should be handled when one of the pumps is being serviced.

8. FLOW EQUALIZATION

Equalization tanks or ponds are used to smooth out variations in flow that would otherwise overload wastewater processes. Graphical or tabular techniques similar to those used in reservoir sizing can be used to size equalization ponds. In practice, up to 25% excess capacity is added as a safety factor.

Table 29.5 *Pumps Used in Wastewater Plants*

flow	flow rate gal/min (L/min)	pump type
raw sewage	<50 (190)	pneumatic ejector
	50–200 (190–760)	submersible or end-suction, nonclog centrifugal
	>200 (760)	end-suction, nonclog centrifugal
settled sewage	<500 (1900)	end-suction, nonclog centrifugal
	>500 (1900)	vertical axial or mixed-flow centrifugal
sludge, primary, thickened, or digested	–	plunger
sludge, secondary	–	end-suction, nonclog centrifugal
scum	–	plunger or recessed impeller
grit	–	recessed impeller, centrifugal, pneumatic ejector, or conveyor rake

In a pure flow equalization process, there is no settling. Mechanical aerators provide the turbulence necessary to keep the solids in suspension while providing oxygen to prevent putrefaction. For typical municipal wastewater, air is provided at the rate of 1.25 to 2.0 ft^3/min-1000 gal (0.01 to 0.015 m^3/min·m^3). Power requirements are approximately 0.02 to 0.04 hp/1000 gal (4 to 8 W/m^3).

Figure 29.2 *Equalization Volume: Mass Diagram Method*

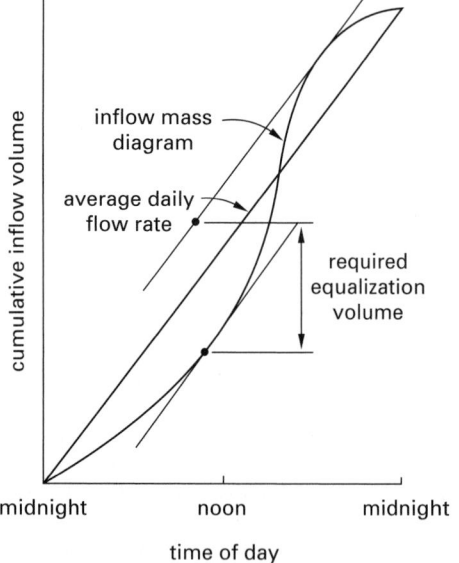

9. STABILIZATION PONDS

The term *stabilization pond* (*oxidation pond* or *stabilization lagoon*) refers to a pond used to treat organic waste by biological and physical processes. Aquatic plants, weeds, algae, and microorganisms stabilize the organic matter. The algae give off oxygen that is used by microorganisms to digest the organic matter. The microorganisms give off carbon dioxide, ammonia, and phosphates that the algae use. Even in modern times, such ponds may be necessary in remote areas (e.g., national parks and campgrounds). To keep toxic substances from leaching into the ground, ponds should not be used without strictly enforced industrial pretreatment requirements.

There are several types of stabilization ponds. *Aerobic ponds* are shallow ponds, less than 4 ft (1.2 m) in depth, where dissolved oxygen is maintained throughout the entire depth, mainly by action of photosynthesis. *Facultative ponds* have an anaerobic lower zone, a facultative middle zone, and an aerobic upper zone. The upper zone is maintained in an aerobic condition by photosynthesis and, in some cases, mechanical aeration at the surface. *Anaerobic ponds* are so deep and receive such a high organic loading that anaerobic conditions prevail throughout the entire pond depth. *Maturation ponds* (*tertiary ponds* or *polishing ponds*) are used for polishing effluent from secondary biological processes. Dissolved oxygen is furnished through photosynthesis and mechanical aeration. *Aerated lagoons* are oxygenated through the action of surface or diffused air aeration. They are often used with activated sludge processes.

10. FACULTATIVE PONDS

Facultative ponds are the most common pond type selected for small communities. Approximately 25% of the municipal wastewater treatment plants in this country use ponds, and about 90% of these are located in communities of 5000 people or fewer. Long retention times and large volumes easily handle large fluctuations in wastewater flow and strength with no significant effect on effluent quality. Also capital, operating, and maintenance costs are less than for other biological systems that provide equivalent treatment.

In a facultative pond, raw wastewater enters at the center of the pond. Suspended solids contained in the wastewater settle to the pond bottom where an anaerobic layer develops. A facultative zone develops just above the anaerobic zone. Molecular oxygen is not available in the region at all times. Generally, the zone is aerobic during the daylight hours and anaerobic during the hours of darkness.

An aerobic zone with molecular oxygen present at all times exists above the facultative zone. Some oxygen is supplied from diffusion across the pond surface, but the majority is supplied through algal photosynthesis.

Table 29.6 *Typical Characteristics of Non-Aerated Stabilization Ponds*
(See Table 29.7 for Aerated Lagoons)

characteristic	aerobic — algae-growth maximizing (high rate)	aerobic — oxygen-transfer maximizing (low rate)	facultative	facultative with surface agitation	anaerobic
size (cell)	0.5–2.5 ac (0.25–1 ha)	< 10 ac (< 4 ha)	2.5–10 ac (1–4 ha)	2.5–10 ac (1–4 ha)	0.5–2.5 ac (0.25–1 ha)
depth	1.0–1.5 ft (0.3–0.45 m)	3.0–5.0 ft (1–1.5 m)	3.0–7.0 ft (1–2 m)	3.0–8.5 ft (1–2.5 m)	8.0–15 ft (2.5–5 m)
BOD_5 loading	75–150 lbm/ac-day (80–160 kg/ha-day)	25–100 lbm/ac-day (40–120 kg/ha-day)	12–70 lbm/ac-day (15–80 kg/ha-day)	45–175 lbm/ac-day (50–200 kg/ha-day)	175–450 lbm/ac-day (200–500 kg/ha-day)
BOD_5 conversion	80–90%	80–90%	80–90%	80–90%	50–85%
detention time	4–6 days	10–40 days	7–30 days	7–20 days	20–50 days
temperature	40–85°F (5–30°C)	32–85°F (0–30°C)	32–120°F (0–50°C)	32–120°F (0–50°C)	40–120°F (5–50°C)
algal concentration	100–250 mg/L	40–100 mg/L	20–80 mg/L	5–20 mg/L	0–5 mg/L
suspended solids in effluent	150–300 mg/L	80–140 mg/L	40–100 mg/L	40–60 mg/L	80–160 mg/L
cell arrangement	series	parallel or series	parallel or series	parallel or series	series
minimum dike width			8 ft (2.5 m)		
maximum dike wall slope			1:3 (vertical:horizontal)		
minimum dike wall slope			1:4 (vertical:horizontal)		
minimum freeboard			3 ft (1 m)		

(Multiply lbm/ac-day by 1.12 to obtain kg/ha-day.)
(Multiply ft by 0.3 to obtain m.)
(Multiply ac by 0.4 to obtain ha.)
(Multiply mg/L by 1.0 to obtain g/m^3.)

General guidelines are used to design facultative ponds. Ponds may be round, square, or rectangular. Usually, there are three cells, piped to permit operation in series or in parallel. Two of the three cells should be identical, each capable of handling half of the peak design flow. The third cell should have a minimum volume of one-third of the peak design flow.

11. AERATED LAGOONS

An *aerated lagoon* is a stabilization pond that is mechanically aerated. Such lagoons are typically deeper and have shorter detention times than nonaerated ponds. In warm climates and with floating aerators, one acre can support several hundred pounds (a hundred kilograms) of BOD per day.

The basis for the design of aerated lagoons is typically the organic loading and/or detention time. The detention time will depend on the desired BOD reduction fraction, η, and the reaction constant, k. Other factors

that must be considered in the design process are solids removal requirements, oxygen requirements, temperature effects, and energy for mixing.

The BOD reduction can be calculated from the overall, first-order BOD *removal rate constant*, k_1. Typical values of $k_{1,\text{base-10}}$ (as specified by TSS) are 0.12 day^{-1} at 20°C and 0.06 day^{-1} at 1°C.

$$t_d = \frac{V}{Q} = k_{1,\text{base}-e}^{1-\eta} \qquad 29.1$$

$$k_{1,\text{base}-e} = 2.3 k_{1,\text{base}-10} \qquad 29.2$$

Aeration should maintain a minimum oxygen content of 2 mg/L at all times. Design depth is larger than for nonaerated lagoons—10 to 15 ft (3 to 4.5 m). Common design characteristics of mechanically aerated lagoons are shown in Table 29.7. Other characteristics are similar to those listed in Table 29.6.

Table 29.7 Typical Characteristics of Aerated Lagoons

aspect ratio	less than 3:1
depth	10–15 ft (3.0–4.5 m)
detention time	4–10 days
BOD loading	20–400 lbm/day-ac;
	200 lbm/day-ac typical
	(22–440 kg/day·ha;
	220 kg/day·ha typical)
operating temperature	0–38°C (21°C optimum)
typical effluent BOD	20–70 mg/L
oxygen required	0.7–1.4 times BOD removed

Table 29.8 Typical Characteristics of Grit Chambers

grit size	0.2 mm and larger
grit specific gravity	2.65
grit arrival/removal rate	0.5–5 ft³/MG
	$(4–40 \times 10^{-6} \text{ m}^3/\text{m}^3)$
depth of chamber	4–10 ft (1.2–3 m)
length	40–100 ft (12–30 m)
width	varies (not critical)
detention time	90–180 sec
horizontal velocity	0.75–1.25 ft/sec
	(0.23–0.38 m/s)

12. RACKS AND SCREENS

Trash racks or *coarse screens* with openings 2 in or larger should precede pumps to prevent clogging. *Medium screens* ($^1/_2$ in to $1^1/_2$ in openings) and *fine screens* ($^1/_{16}$ in to $^1/_8$ in) are also used to relieve the load on grit chambers and sedimentation basins. Fine screens are rare except when used with selected industrial waste processing plants. Screens in all but the smallest municipal plants are cleaned by automatic scraping arms. A minimum of two screen units is advisable.

Screen capacities and head losses are specified by the manufacturer. Although the flow velocity must be sufficient to maintain sediment in suspension, the approach velocity should be limited to 3 ft/sec (0.9 m/s) to prevent debris from being forced through the screen.

13. GRIT CHAMBERS

Abrasive *grit* can erode pumps, clog pipes, and accumulate in excessive volumes. In a *grit chamber* (also known as *grit clarifier* or *detritus tank*) the wastewater is slowed, allowing the grit to settle out but allowing the organic matter to continue through. Grit can be manually or mechanically removed with buckets or screw conveyors. A minimum of two units is needed.

Horizontal flow grit chambers are designed to keep the flow velocity as close to 1 ft/sec (0.3 m/s) as possible. If an analytical design based on settling velocity is required, the *scouring velocity* should not be exceeded. Scouring is the dislodging of particles that have already settled. Scouring velocity is not the same as settling velocity.

Scouring will be prevented if the horizontal velocity is kept below that predicted by the *Camp formula*, Eq. 29.3. SG_p is the specific gravity of the particle, typically taken as 2.65 for sand. d_p is the particle diameter. k is a dimensionless constant with typical values of 0.04 for sand and 0.06 or more for sticky, interlocking matter. The dimensionless Darcy friction factor is approximately 0.02 to 0.03. Any consistent set of units can be used with Eq. 29.3.

$$v = \sqrt{8k\left(\frac{gd_p}{f}\right)(SG_p - 1)} \qquad 29.3$$

14. AERATED GRIT CHAMBERS

An *aerated grit chamber* is a bottom-hoppered tank with a short detention time. Diffused aeration from one side of the tank rolls the water and keeps the organics in suspension while the grit drops into the hopper. The water spirals or rolls through the tank. Influent enters through the side and removes the degritted wastewater. A minimum of two units is needed.

Solids are removed by pump, screw conveyer, bucket elevator, or gravity flow. However, the grit will have a significant organic content. A grit washer or cyclone separator can be used to clean the grit.

Table 29.9 Typical Characteristics of Aerated Grit Chambers

detention time	2–5 min at peak flow; 3 min typical
air supply	
shallow tanks	1.5–5 cfm/ft length; 3 typical (0.13–0.45 m³/min·m; 0.27 typical)
deep tanks	3–8 cfm/ft length; 5 typical (0.27–0.7 m³/min·m; 0.45 typical)
grit and scum	
quantity	0.5–25 ft³/MG (2 typical) $(4–190 \times 10^{-6} \text{ m}^3/\text{m}^3)$
length:width	
ratio	2.5:1–5:1 (3:1–4:1 typical)
depth	6–15 ft (1.8–4.5 m)
length	20–60 ft (6–18 m)
width	7–20 ft (2.1–6 m)

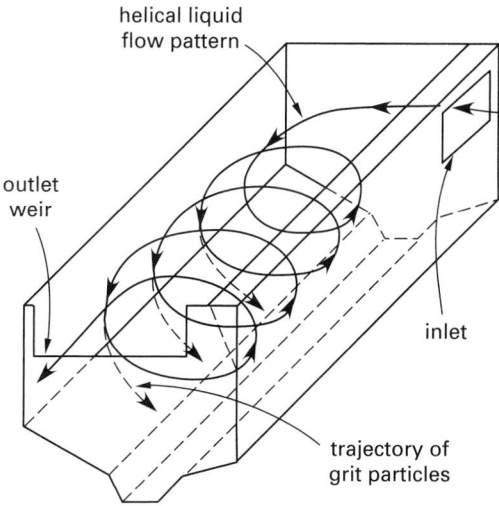

Figure 29.3 Aerated Grit Chamber

helical liquid
flow pattern

outlet
weir

inlet

trajectory of
grit particles

15. SKIMMING TANKS

If the sewage has more than 50 mg/L of floating grease or oil, a basin 8 to 10 ft (2.4 to 3 m) providing 5 to 15 min of detention time will allow the grease to rise as *floatables* (*scum*) to the surface. An aerating device will help coagulate and float grease to the surface. 40 to 80 psig (280 to 550 kPa) of air should be provided at the approximate rate of 0.01 to 0.1 ft^3/gal (0.07 to 0.7 m^3/m^3) of influent. A small fraction (e.g., 30%) of the influent may be recycled in some cases. Grease rise rates are typically 0.2 to 1 ft/min (0.06 to 0.5 m/min) depending on the degree of dispersal, amount of colloidal fines, and aeration.

Scum is mechanically removed by skimming troughs. Scum may not generally be disposed of by landfilling. It is processed with other solid wastes in anaerobic digesters. *Scum grinding* may be needed to reduce the scum to a small enough size for thorough digestion. In some cases, scum may be incinerated, although this practice may be affected by air-quality regulations.

16. SHREDDERS

Shredders (also called *comminutors*) cut waste solids to approximately $^1/_4$ in (6 mm) in size, reducing the amount of screenings that must be disposed of. Shreddings stay with the flow for later settling.

17. PLAIN SEDIMENTATION BASINS/CLARIFIERS

Plain sedimentation basins/clarifiers (i.e., basins in which no chemicals are added to encourage clarification) are similar in concept and design to those used to treat water supplies. Design characteristics for wastewater treatment sedimentation basins are listed in

Table 29.10. Since the bottom slopes slightly, the depth varies with location. Therefore, the *side water depth* is usually quoted. The *surface loading* (*surface loading rate*, *overflow rate*, or *settling rate*), v*, along with sludge storage volume, is the primary design parameter.

$$v^* = \frac{Q}{A} \qquad 29.4$$

The *detention time* (*mean residence time* or *retention period*) is

$$t_d = \frac{V}{Q} \qquad 29.5$$

The *weir loading* (*weir loading rate*) is

$$\text{weir loading} = \frac{Q}{L} \qquad 29.6$$

Table 29.10 Typical Characteristics of Clarifiers[a]

BOD reduction	20–40% (25–35% typical)
total suspended solids reduction	35–65%
bacteria reduction	50–60%
organic content of settled solids	50–75%
specific gravity of settled solids	1.2 or less
minimum settling velocity	4 ft/hr (1.2 m/hr) typical
plan shape	rectangular (or circular)
basin depth (side water depth)	6–15 ft; 10–12 ft typical (1.8–4.5 m; 3–3.6 m typical)
basin width	10–50 ft (3–15 m)
plan aspect ratio	3:1 to 5:1
basin diameter (circular only)	50–150 ft; 100 ft typical (15–45 m; 30 m typical)
minimum freeboard	1.5 ft (0.45 m)
minimum hopper wall angle	60°
detention time	1.5–2.5 hr
flow-through velocity	18 ft/hr (1.5 mm/s)
minimum flow-through time	30% of detention time
weir loading	10,000–20,000 gal/day-ft (125–250 m^3/day·m)
surface loading	400–2000 gal/day-ft^2; 800–1200 gal/day-ft^2 typical (16–80 m^3/day·m^2; 32–50 m^3/day·m^2 typical)
bottom slope to hopper	8%
inlet	baffled to prevent turbulence
scum removal	mechanical or manual

[a]See Secs. 26 and 27 for intermediate and final clarifiers, respectively.

18. CHEMICAL SEDIMENTATION BASINS/CLARIFIERS

Chemical flocculation (*clarification* or *coagulation*) operations in chemical sedimentation basins are similar to those encountered in the treatment of water supplies except that the coagulant doses are greater. Chemical precipitation may be used when plain sedimentation is insufficient, or occasionally when the stream into which the outfall discharges is running low, or when there is a large increase in sewage flow. As with water treatment, the five coagulants used most often are (a) aluminum sulfate, $Al_2(SO_4)_3$; (b) ferric chloride, $FeCl_3$; (c) ferric sulfate, $Fe_2(SO_4)_3$; (d) ferrous sulfate, $FeSO_4$; and (e) chlorinated copperas. Lime and sulfuric acid may be used to adjust the pH for proper coagulation.

Example 29.1

A primary clarifier receives 1.4 MGD of domestic waste. The clarifier has a peripheral weir, is 50 ft in diameter, and is filled to a depth of 7 ft. Determine if the clarifier has been sized properly.

Solution

The perimeter length is

$$L = \pi D = \pi(50 \text{ ft}) = 157 \text{ ft}$$

The surface area is

$$A = \frac{\pi}{4}D^2 = \left(\frac{\pi}{4}\right)(50 \text{ ft})^2 = 1963 \text{ ft}^2$$

The volume is

$$V = AZ = (1963 \text{ ft}^2)(7 \text{ ft}) = 13{,}740 \text{ ft}^3$$

The surface loading should be 400 to 2000 gal/day-ft^2.

$$v^* = \frac{Q}{A} = \frac{1.4 \times 10^6 \dfrac{\text{gal}}{\text{day}}}{1963 \text{ ft}^2}$$
$$= 713 \text{ gal/day-ft}^2 \quad [\text{OK}]$$

The detention time should be 1.5 to 2.5 hr.

$$t = \frac{V}{Q} = \frac{(13{,}740 \text{ ft}^3)\left(24 \dfrac{\text{hr}}{\text{day}}\right)}{\left(1.4 \times 10^6 \dfrac{\text{gal}}{\text{day}}\right)\left(0.1337 \dfrac{\text{ft}^3}{\text{gal}}\right)}$$
$$= 1.76 \text{ hr} \quad [\text{OK}]$$

The weir loading should be 10,000 to 20,000 gal/day-ft.

$$\frac{Q}{L} = \frac{1.4 \times 10^6 \dfrac{\text{gal}}{\text{day}}}{157 \text{ ft}}$$
$$= 8917 \text{ gal/day-ft} \quad [\text{low but probably OK}]$$

19. TRICKLING FILTERS

Trickling filters (also known as *biological beds* and *fixed media filters*) consist of beds of rounded river rocks with approximate diameters of 2 to 5 in (50 to 125 mm), wooden slats, or modern synthetic media. Wastewater from primary sedimentation processing is sprayed intermittently over the bed. The biological and microbial slime growth attached to the bed purifies the wastewater as it trickles down. The water is introduced into the filter by rotating arms that move by virtue of spray reaction (reaction-type) or motors (motor-type). The clarified water is collected by an underdrain system.

The distribution rate is sometimes given by an *SK rating*, where SK is the water depth in mm deposited per pass of the distributor. Though there is strong evidence that rotational speeds of 1 to 2 rev/hr (high SK) produces significant operational improvement, traditional distribution arms revolve at 1 to 5 rev/min (low SK).

On the average, one acre of *low-rate filter* (also referred to as *standard-rate filter*) is needed for each 20,000 people served. Trickling filters can remove 70 to 90% of the suspended solids, 65 to 85% of the BOD, and 70 to 95% of the bacteria. Although low-rate filters have rocks to a depth of 6 ft (1.8 m), most of the reduction occurs in the first few feet of bed, and organisms in the lower part of the bed may be in a near-starvation condition.

Due to the low concentration of carbonaceous material in the water near the bottom of the filter, nitrogenous bacteria produce a highly nitrified effluent from low-rate filters. With low-rate filters, the bed will periodically slough off (unload) parts of its slime coating. Therefore, sedimentation after filtering is necessary. *Filter flies* are a major problem with low-rate filters, since fly larvae are provided with an undisturbed environment in which to breed.

Since there are limits to the heights of trickling filters, longer contact times can be achieved by returning some of the collected filter water back to the filter. This is known as *recirculation* or *recycling*. Recirculation is also used to keep the filter medium from drying out and to smooth out fluctuations in the hydraulic loading.

High-rate filters are now in use in most facilities. The higher hydraulic loading flushes the bed and inhibits excess biological growth. High-rate stone filters may be only 3 to 6 ft (0.9 to 1.8 m) deep. The high rate is possible because much of the filter discharge is recirculated. With the high flow rates, fly larvae are washed out, minimizing the filter fly problem. Since the biofilm is less thick and provided with carbon-based nutrients at a high rate, the effluent is nitrified only when the filter experiences low loading.

Super high-rate filters (*oxidation towers*) using synthetic media may be up to 40 ft (12 m) tall. High-rate and super high-rate trickling filters may be used as *roughing filters*, receiving wastewater at high hydraulic or organic loading and providing intermediate treatment or the first step of a multistage biological treatment process.

A BOD balance around a filter with recirculation results in Eq. 29.7 in which S_i is the BOD applied to the filter by the diluted influent, S_{ps} is the BOD of the effluent from the primary settling tank (or the plant influent BOD if there is no primary sedimentation), and S_o is the BOD of the trickling filter effluent.

$$S_{ps} + RS_o = (1 + R)S_i \qquad 29.7$$

$$S_i = \frac{S_{ps} + RS_o}{1 + R} \qquad 29.8$$

BOD is reduced significantly in a trickling filter. Standard-rate filters produce an 80 to 85% reduction, and high-rate filters remove 65 to 80% of BOD, less because of reduced contact area and time. The *removal fraction* (*removal efficiency*) of a single-stage trickling filter is

$$\eta = \frac{S_{\text{removed}}}{S_{ps}} = \frac{S_{ps} - S_o}{S_{ps}} \qquad 29.9$$

By definition, the *recirculation ratio*, R, is zero for standard-rate filters but can be as high as 4:1 for high-rate filters.

$$R = \frac{Q_r}{Q_w} \qquad 29.10$$

Filters may be classified as high-rate based on their hydraulic loading, organic loading, or both. The *hydraulic loading* of a trickling filter is the total water flow divided by the plan area. Typical values of hydraulic loading are 25 to 100 gal/day-ft² (1 to 4 m³/day·m²) for standard filters and 250 to 1000 gal/day-ft² (10 to 40 m³/day·m²) or higher for high-rate filters.

$$L_H = \frac{Q_w + Q_r}{A} = \frac{Q_w(1 + R)}{A} \qquad 29.11$$

The *BOD loading* (*organic loading* or *surface loading*) is calculated without considering any recirculated flow. BOD loading for the filter/clarifier combination is essentially the BOD of the applied wastewater divided by the filter volume. BOD loading is typically given in pounds per 1000 ft³ per day (hence the 1000 term in Eq. 29.12.) Typical values are 5 to 25 lbm/1000 ft³-day (0.08 to 0.4 kg/m³·day) for low-rate filters and 25 to 110 lbm/1000 ft³-day (0.4 to 1.8 kg/m³·day) for high-rate filters.

$$L_{\text{BOD}} = \frac{Q_{w,\text{MGD}} S_{\text{mg/L}} \left(8.345 \, \frac{\text{lbm-L}}{\text{MG-mg}}\right)(1000)}{V_{\text{ft}^3}} \qquad 29.12$$

The *specific area* of the filter is the total surface area of the exposed filter medium divided by the total volume of the filter.

Example 29.2

A single-stage trickling filter plant processes 1.4 MGD of raw domestic waste with a BOD of 170 mg/L. The trickling filter is 90 ft in diameter and has river rock media to a depth of 7 ft. The recirculation rate is 50%, and the filter is classified as high-rate. Water passes through clarification operations both before and after the trickling filter operation. Determine if the trickling filter has been sized properly.

Solution

The filter area is

$$A = \frac{\pi}{4} D^2 = \left(\frac{\pi}{4}\right)(90 \text{ ft})^2 = 6362 \text{ ft}^2$$

The rock volume is

$$V = AZ = (6362 \text{ ft}^2)(7 \text{ ft}) = 44,534 \text{ ft}^3$$

The hydraulic load for a high-rate filter should be 250 to 1000 gal/day-ft².

$$\begin{aligned} L_H &= \frac{Q_w(1 + R)}{A} \\ &= \frac{\left(1.4 \times 10^6 \, \frac{\text{gal}}{\text{day}}\right)(1 + 0.5)}{6362 \text{ ft}^2} \\ &= 330 \text{ gal/day-ft}^2 \quad [\text{OK}] \end{aligned}$$

From Table 29.10, the primary clarification process will remove approximately 30% of the BOD. The remaining BOD is

$$\begin{aligned} \text{BOD}_i &= (1 - 0.3)\left(170 \, \frac{\text{mg}}{\text{L}}\right) \\ &= 119 \text{ mg/L} \end{aligned}$$

The BOD loading should be approximately 25 to 110 lbm/1000 ft³-day.

$$\begin{aligned} L_{\text{BOD}} &= \frac{Q_{w,\text{MGD}} \text{BOD}_{\text{mg/L}} \left(8.345 \, \frac{\text{lbm-L}}{\text{MG-mg}}\right)(1000)}{V_{\text{ft}^3}} \\ &= \frac{\begin{array}{c}\left(1.4 \, \frac{\text{MG}}{\text{day}}\right)\left(119 \, \frac{\text{mg}}{\text{L}}\right) \\ \times \left(8.345 \, \frac{\text{lbm-L}}{\text{MG-mg}}\right)(1000)\end{array}}{44,534 \text{ ft}^3} \\ &= 31.2 \text{ lbm/day-1000 ft}^3 \quad [\text{OK}] \end{aligned}$$

20. TWO-STAGE TRICKLING FILTERS

If a higher BOD and solids removal fraction is needed, two filters can be connected in series to form a *two-stage filter* system with an optional intermediate settling tank. The efficiency of the second-stage filter is considerably less than that of the first-stage filter because much of the biological food has been removed from the flow.

21. NATIONAL RESEARCH COUNCIL EQUATION

In 1946, the National Research Council (NRC) studied sewage treatment facilities at military installations. The wastewater at these facilities was stronger than typical municipal wastewater. Not surprisingly, the NRC concluded that the organic loading had a greater effect on removal efficiency than did the hydraulic loading.

If it is assumed that the biological layer and hydraulic loading are uniform, the water is at 20°C, and the filter is single-stage rock followed by a settling tank, then the *NRC equation*, Eq. 29.13, can be used to calculate the BOD removal fraction of the single-stage filter/clarifier combination. Inasmuch as installations with high BOD and low hydraulic loads were used as the basis of the studies, BOD removal efficiencies in typical municipal facilities are higher than predicted by Eq. 29.13. The value of L_{BOD} excludes recirculation returned directly from the filter outlet, which is accounted for in the value of F.

$$\eta = \frac{1}{1 + 0.0561 \sqrt{\dfrac{L_{BOD}}{F}}} \qquad \text{29.13}$$

It is important to recognize the BOD loading, L_{BOD} in Eq. 29.13, has units of lbm/day-1000 ft³. The constant 0.0085 is often encountered in place of 0.0561 in the literature. However, this value is for use with media volumes expressed in lbm/ac-ft, not in thousands of ft³.

$$\eta = \frac{1}{1 + 0.0085 \sqrt{\dfrac{L_{BOD,lbm/day}}{V_{ac\text{-}ft} F}}} \qquad \text{29.14}$$

There are a number of ways to recirculate water from the output of the trickling filters back to the filter. Water can be brought back to a wet well, to the primary settling tank, to the filter itself, or to a combination of the three. The variations in performance are not significant as long as sludge itself is not recirculated. Equation 29.13 can be used with any of the recirculation schemes.

F is the *effective number of passes* of the organic material through a filter. R is the ratio of filter discharge returned to the inlet to the raw influent. In Eq. 29.15, w is a weighting factor typically assigned a value of 0.1.

$$F = \frac{1 + R}{(1 + wR)^2} \qquad \text{29.15}$$

It is time-consuming to extract the organic loading, L_{BOD}, from Eq. 29.13 given η and R. Figure 29.4 can be used for this purpose.

Based on the NRC model, the removal fraction for a two-stage filter with an intermediate clarifier is

$$\eta_2 = \frac{1}{1 + \left(\dfrac{0.0561}{1 - \eta_1}\right) \sqrt{\dfrac{L_{BOD}}{F}}} \qquad \text{29.16}$$

In recent years, the NRC equations have fallen from favor for several reasons. These reasons include applicability only to rock media, inability to correct adequately for temperature variations, inapplicability to industrial waste, and empirical basis.

Figure 29.4 *Trickling Filter Performance*[a]

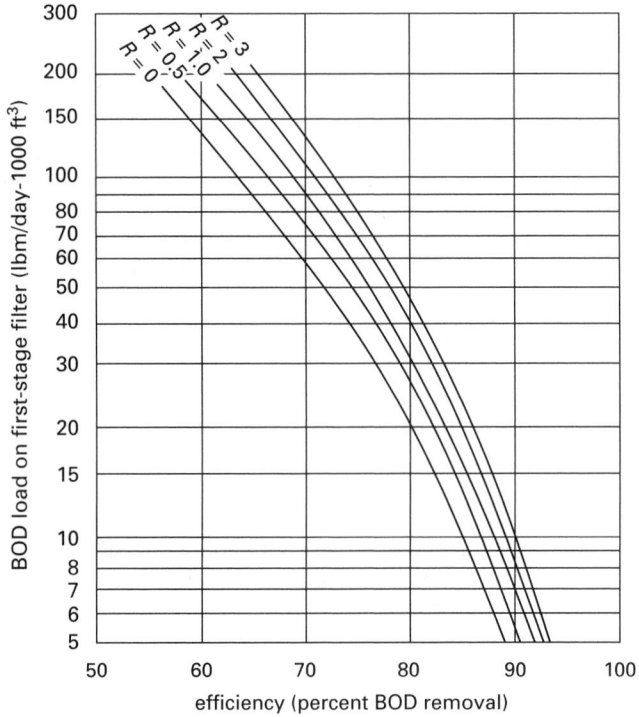

[a]NRC model, single-stage

22. VELZ EQUATION

Equation 29.17, known as the *Velz equation*, was the first semi-theoretical analysis of trickling filter performance versus depth of media. It is useful in predicting the BOD removal fraction for any generic trickling filter, including those with synthetic media. The original Velz used the term "total removable fraction of BOD," though this is understood to mean the maximum fraction of removable BOD removed, generally S_i.

Values of the *Velz decay rate*, K, an empirical rate constant, are highly dependent on the installation and operating characteristics, particularly the hydraulic loading, which is not included in the formula. Subsequent to Velz' work, the industry has developed a large database of applicable values for numerous application scenarios.

$$\frac{S_o}{S_i} = e^{-KZ} \ (\text{or } 10^{-kZ}) \qquad \text{29.17}$$

The variation in temperature was assumed to be

$$K_T = K_{20°C}(1.047)^{T-20°C} \qquad \text{29.18}$$

23. MODERN FORMULATIONS

Various researchers have built upon the *Velz equation* and developed correlations of the form of Eq. 29.19 for various situations and types of filters. (In Eq. 29.19, S_a is the specific area, not the BOD.) Each researcher has used different nomenclature and assumptions. For example, the inclusion of the effects of dilution by recirculation is far from universal. The specific form of the equation, values of constants and exponents, temperature coefficients, limitations, and assumptions are needed before such a correlation can be reliably used.

$$\frac{S_o}{S_i} = \exp\left\{ -KZS_a^m \left(\frac{A}{Q}\right)^n \right\} \qquad \textbf{29.19}$$

As an example of the difficulty in finding one model that predicts all trickling filter performance, consider the BOD removal efficiency as predicted by the *Schulze correlation* (1960) for rock media and the *Germain correlation* (1965) for synthetic media. Though both correlations have the same form (Eq. 29.20), both measured filter depths in feet, and both expressed hydraulic loadings in gal/min-ft², the Schulze correlation included recirculation while the Germain correlation did not. The treatability constant, k, was 0.51 to 0.76 day⁻¹ for the Schulze correlation and 0.088 day⁻¹ for the Germain correlation. The *media factor* exponent, n, was 0.67 for the Schulze correlation (rock media) and 0.5 for the Germain correlation (plastic media).

$$\frac{S_o}{S_i} = \exp\left\{ \frac{-kZ}{L_H^n} \right\} \qquad \textbf{29.20}$$

24. ROTATING BIOLOGICAL CONTACTORS

Rotating biological contactors, RBCs (also known as *rotating biological reactors*), consist of large-diameter plastic disks, partially immersed in wastewater, on which biofilm is allowed to grow. The disks are mounted on shafts that turn slowly. The rotation progressively wets the disks, alternately exposing the biofilm to organic material in the wastewater and to oxygen in the air. The biofilm population, since it is well oxygenated, efficiently removes organic solids from the wastewater. RBCs are primarily used for carbonaceous BOD removal, although they can also be used for nitrification or a combination of both.

The primary design criterion is hydraulic loading, not organic (BOD) loading. For a specific hydraulic loading, the BOD removal efficiency will be essentially constant, regardless of variations in BOD. Other design criteria are listed in Table 29.11.

RBC operation is more efficient when several stages are used. Recirculation is not common with RBC processes. The process can be placed in series or in parallel with existing trickling filter or activated sludge processes.

Table 29.11 Typical Characteristics of Rotating Biological Contactors

number of stages	2–4
disk diameter	10–12 ft (3–3.6 m)
immersion, percentage of area	40%
hydraulic loading secondary treatment	2–4 gal/day-ft² (0.08–0.16 m³/day·m²)
tertiary treatment (nitrification)	0.75–2 gal/day-ft² (0.03–0.08 m³/day·m²)
optimum peripheral rotational speed	1 ft/sec (0.3 m/s)
tank volume	0.12 gal/ft² (0.0049 m³/m²) of biomass area
operating temperature	13–32°C
BOD removal fraction	70–80%

Figure 29.5 Rotating Biological Contactor

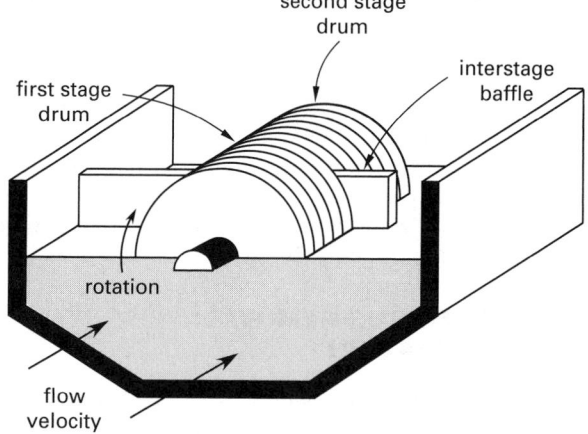

25. SAND FILTERS

For small populations, a *slow sand filter* (*intermittent sand filter*) can be used. Because of the lower flow rate, the filter area per person is higher than for a trickling filter. Roughly one acre is needed for each 1000 people. The filter is constructed as a sand bed 2 to 3 ft (0.6 to 0.9 m) deep over a 6 to 12 in (150 to 300 mm) gravel bed. The filter is alternately exposed to water from a settling tank and to air (hence the term intermittent). Straining and aerobic decomposition clean the water. Application rates are usually 2 to 2.5 gal/day-ft² (0.08 to 0.1 m³/day·m²). Up to 95% of the BOD can be satisfied in an intermittent sand filter. The filter is cleaned by removing the top layer of clogged sand.

If the water is applied continuously as a final process following secondary treatment, the filter is known as a *polishing filter* or *rapid sand filter*. The water rate of a polishing filter may be as high as 10 gal/day-ft² (0.4 m³/day·m²). Although the designs are similar to those used in water supply treatment, coarser media is used since the turbidity requirements are less stringent. Backwashing is required more frequently and is more aggressive than with water supply treatment.

26. INTERMEDIATE CLARIFIERS

Sedimentation tanks located between trickling filter stages or between a filter and subsequent aeration are known as *intermediate clarifiers* or *secondary clarifiers*. Typical characteristics of intermediate clarifiers used with trickling filter processes are: maximum overflow rate, 1500 gal/day-ft^2 (61 m^3/day·m^2); water depth, 10 to 13 ft (3 to 4 m); and maximum weir loading, 10,000 gal/day-ft (125 m^3/day·m) for plants processing 1 MGD or less and 15,000 gal/day-ft (185 m^3/day·m) for plants processing over 1 MGD.

27. FINAL CLARIFIERS

Sedimentation following secondary treatment occurs in *final clarifiers*. The purpose of final clarification is to collect sloughed-off material from trickling filter processes or to collect sludge and return it for activated sludge processes, not to reduce BOD. The depth is approximately 10 to 12 ft (3.0 to 3.7 m), the average overflow rate is 500 to 600 gal/day-ft^2 (20 to 24 m^3/day·m^2), and the maximum overflow rate is approximately 1100 gal/day-ft^2 (45 m^3/day·m^2). The maximum weir loading is the same as for intermediate clarifiers, but the lower rates are preferred. For settling following extended aeration, the overflow rate and loading should be reduced approximately 50%.

28. PHOSPHORUS REMOVAL: PRECIPITATION

Phosphorus concentrations of 5 to 15 mg/L (as P) are experienced in untreated wastewater, most of which originates from synthetic detergents and human waste. Approximately 10% of the total phosphorus is insoluble and can be removed in primary settling. The amount that is removed by absorption in conventional biological processes is small. The remaining phosphorus is soluble and must be removed by converting it into an insoluble precipitate.

Soluble phosphorus is removed by precipitation and settling. Aluminum sulfate, ferric chloride (FeCl$_3$), and lime may be used depending on the nature of the phosphorus radical. Aluminum sulfate is more desirable since lime reacts with hardness and forms large quantities of additional precipitates. (Hardness removal is not as important as in water supply treatment.) However, the process requires about 10 pounds (10 kilograms) of aluminum sulfate for each pound (kilogram) of phosphorus removed. The process also produces a chemical sludge that is difficult to dewater, handle, and dispose of.

$$Al_2(SO_4)_3 + 2PO_4 \rightleftharpoons 2AlPO_4 + 3SO_4 \qquad 29.21$$
$$FeCl_3 + PO_4 \rightleftharpoons FePO_4 + 3Cl \qquad 29.22$$

Due to the many other possible reactions the compounds can participate in, the dosage should be determined from testing. The stoichiometric chemical reactions describe how the phosphorus is removed, but they do not accurately predict the quantities of coagulants needed.

29. AMMONIA REMOVAL: AIR STRIPPING

Ammonia may be removed by either biological processing or air stripping. In the biological *nitrification and denitrification process*, ammonia is first aerobically converted to nitrite and then to nitrate (nitrification) by bacteria. Then, the nitrates are converted to nitrogen gas, which escapes (denitrification).

In the *air-stripping (ammonia-stripping)* method, lime is added to water to increase its pH to about 10. This causes the ammonium ions, NH_4^+, to change to dissolved ammonia gas, NH_3. The water is passed through a packed tower into which air is blown at high rates. The air strips the ammonia gas out of the water. Recarbonation follows to remove the excess lime.

30. CARBON ADSORPTION

Adsorption uses high surface-area activated carbon to remove organic contaminants. Adsorption can use *granular activated carbon* (GAC) in column or fluidized-bed reactors or *powdered activated carbon* (PAC) in complete-mix reactors. Activated carbon is relatively nonspecific, and it will remove a wide variety of refractory organics as well as some inorganic contaminants. It should generally be considered for organic contaminants that are nonpolar, have low solubility, or have high molecular weights.

The most common problems associated with columns are breakthrough, excessive headloss due to plugging, and premature exhaustion. *Breakthrough* occurs when the carbon becomes saturated with the target compound and can hold no more. *Plugging* occurs when biological growth blocks the spaces between carbon particles. *Premature exhaustion* occurs when large and high molecular-weight molecules block the internal pores in the carbon particles. These latter two problems can be prevented by locating the carbon columns downstream of filtration.

31. CHLORINATION

Chlorination to disinfect and deodorize is one of the final steps prior to discharge. Vacuum-type feeders are used predominantly with chlorine gas. Chlorine under vacuum is combined with wastewater to produce a chlorine solution. A *flow-pacing* chlorinator will reduce the chlorine solution feed rate when the wastewater flow decreases (e.g., at night).

The size of the *contact tank* is variable, depending on economics and other factors. An average design detention time is 30 min at average flow, some of which can occur in the plant outfall after the contact basin. Contact tanks are baffled to prevent short-circuiting that would otherwise reduce chlorination time and effectiveness.

Alternatives to disinfection by chlorine include sodium hypochlorite, ozone, ultraviolet light, bromine (as bromine chloride), chlorine dioxide, and hydrogen peroxide. All alternatives have one or more disadvantages when compared to chlorine gas.

32. DECHLORINATION

Toxicity, by-products, and strict limits on *total residual oxidants* (TROs) now make dechlorination mandatory at many installations. Sulfur dioxide (SO_2) and sodium thiosulfite ($Na_2S_2SO_3$) are the primary compounds used as dechlorinators today. Other compounds seeing limited use are sodium metabisulfate ($Na_2S_2O_5$), sodium bisulfate ($NaHSO_3$), sodium sulfite (Na_2SO_3), hydrogen peroxide (H_2O_2), and granular activated carbon.

Though reaeration was at one time thought to replace oxygen in water depleted by sulfur dioxide, this is now known to be false.

33. EFFLUENT DISPOSAL

Organic material and bacteria present in wastewater are generally not removed in their entireties, though the removal efficiency is high (e.g., better than 95% for some processes). Therefore, effluent must be discharged to large bodies of water where the remaining contaminants can be substantially diluted. Discharge to flowing surface water and oceans is the most desirable. Discharge to lakes and reservoirs should be avoided.

In some areas, *combined sewer overflow* (CSO) outfalls still channel wastewater into waterways during heavy storms when treatment plants are overworked. Such pollution can be prevented by the installation of large retention basins (*diversion chambers*), miles of tunnels, reservoirs to capture overflows for later controlled release to treatment plants, screening devices to separate solids from wastewater, swirl concentrators to capture solids for treatment while permitting the clearer portion to be chlorinated and released to waterways, and more innovative vortex solids separators.

34. WASTEWATER RECLAMATION

Treated wastewater can be used for irrigation, firefighting, road maintenance, or flushing. Public access to areas where reclaimed water is disposed of through spray-irrigation (e.g., in sod farms, fodder crops, and pasture lands) should be restricted. When the reclaimed water is to be used for irrigation of public areas, the water quality standards regarding suspended solids, fecal coliforms, and viruses should be strictly controlled.

Although it has been demonstrated that wastewater can be made potable at great expense, such practice has not gained widespread favor.

Environmental

30 Activated Sludge and Sludge Processing

Nomenclature

A	area	ft^2	m^2
BOD	biochemical oxygen demand	mg/L	mg/L
c_p	specific heat	Btu/lbm-°F	J/kg·°C
C	cost power electricity	\$/kW-hr	\$/kW·h
D	oxygen deficit	mg/L	mg/L
DO	dissolved oxygen	mg/L	mg/L
E	efficiency of waste utilization	–	–
f	BOD_5/BOD_u ratio	–	–
F	food arrival rate	mg/day	mg/day
F:M	food to microorganism ratio	day^{-1}	day^{-1}
g	acceleration of gravity	ft/sec^2	m/s^2
G	fraction of solids that are biodegradable	–	–
G	number of gravities	–	–
h	film coefficient	Btu/hr-ft^2-°F	W/m^2·K
J	Joule's constant (778)	ft-lbf/Btu	–
k	ratio of specific heats	–	–
k	thermal conductivity	Btu-ft/ hr-ft^2-°F	W/m·K
k_d	microorganism endogenous decay rate	$days^{-1}$	$days^{-1}$
K	cell yield constant (removal efficiency)	lbm/lbm	mg/mg
K_s	half-velocity coefficient	mg/L	mg/L
K_t	oxygen transfer coefficient	1/hr	1/h
L	BOD loading	lbm/ day-1000 ft^3	kg/day
L	thickness	ft	m
LHV	lower heating value	Btu/ft^3	kJ/m^3
$\dot{m}$	mass flow rate	lbm/day	kg/day
M	mass of microorganisms	lbm	mg
MLSS	total mixed liquor suspended solids	mg/L	mg/L
p	pressure	lbf/in^2	Pa
P	power	hp	kW
P_x	mass of sludge wasted	lbm/day	kg/day
q	heat transfer	Btu	kJ
Q	flow rate	ft^3/day	m^3/day
r	distance from center of rotation	ft	m
r	rate	mg/m^3-day	mg/m^3·day
R	recirculation ratio	–	–
R	specific gas constant	ft-lbf/ lbm-°R	J/kg·K
s	gravimetric fractional solids content	–	–
S	growth-limited substrate concentration	mg/L	mg/L
SG	specific gravity	–	–
SS	suspended solids	mg/L	mg/L
SVI	sludge volume index	mL/g	mL/g
t	time	days	days
T	absolute temperature	°R	K
U	overall coefficient of heat transfer	Btu/ hr-ft^2-°F	W/m^2·K
U	specific substrate utilization	$days^{-1}$	$days^{-1}$
V	volume	ft^3	m^3
$\dot{V}$	volumetric flow rate	ft^3/day	m^3/day
X	mixed liquor volatile suspended solids	mg/L	mg/L
Y	maximum yield coefficient	lbm/lbm	mg/mg

Symbols

β	oxygen saturation coefficient	–	–
η	efficiency	–	–
θ	residence/detention time	days	days
μ_m	maximum specific growth rate	$days^{-1}$	$days^{-1}$
ρ	density	lbm/ft^3	kg/m^3
ω	rotational speed	rad/sec	rad/s

Environmental

Subscripts

5	5-day
a	aeration tank
c	cell or compressor
e	effluent
i	influent diluted with recycle flow
o	influent
obs	observed
r	recirculation
s	settling tank
su	substrate utilization
t	transfer
u	ultimate
w	wasted

1. SLUDGE

Sludge is the mixture of water, organic and inorganic solids, and treatment chemicals that accumulates in settling tanks. The term is also used to refer to the dried residue (screenings, grit, filter cake, and drying bed scraping) from separation and drying processes, although the term *biosolids* is becoming more common in this regard. (The term *residuals* is also used, though this term more commonly refers to sludge from water treatment plants.)

2. ACTIVATED SLUDGE PROCESS

The *activated sludge process* is a secondary biological wastewater treatment technique in which a mixture of wastewater and sludge solids is aerated. The sludge mixture produced during this oxidation process contains an extremely high concentration of aerobic bacteria, most of which are near starvation. This condition makes the sludge an ideal medium for the destruction of any organic material in the mixture. Since the bacteria are voraciously active, the sludge is called *activated sludge*.

The well-aerated mixture of wastewater and sludge, known as *mixed liquor*, flows from the aeration tank to a secondary clarifier where the sludge solids settle out. Most of the settled sludge solids are returned to the aeration tank in order to maintain the high population of bacteria needed for rapid breakdown of the organic material. However, because more sludge is produced than is needed, some of the return sludge is diverted ("wasted") for subsequent treatment and disposal. This wasted sludge is referred to as *waste activated sludge*, WAS. The volume of sludge returned to the aeration basin is typically 20 to 30% of the wastewater flow. The liquid fraction removed from the secondary clarifier weir is chlorinated and discharged.

Though diffused aeration and mechanical aeration are the most common methods of oxygenating the mixed liquor, various methods of staging the aeration are used, each having its own characteristic ranges of operating parameters.

In a traditional activated sludge plant using conventional aeration, the wastewater is typically aerated for 6 to 8 hr in long, rectangular aeration basins. Sufficient air, about eight volumes for each volume of wastewater treated, is provided to keep the sludge in suspension. The air is injected near the bottom of the aeration tank through a system of diffusers.

Figure 30.1 *Typical Activated Sludge Plant*

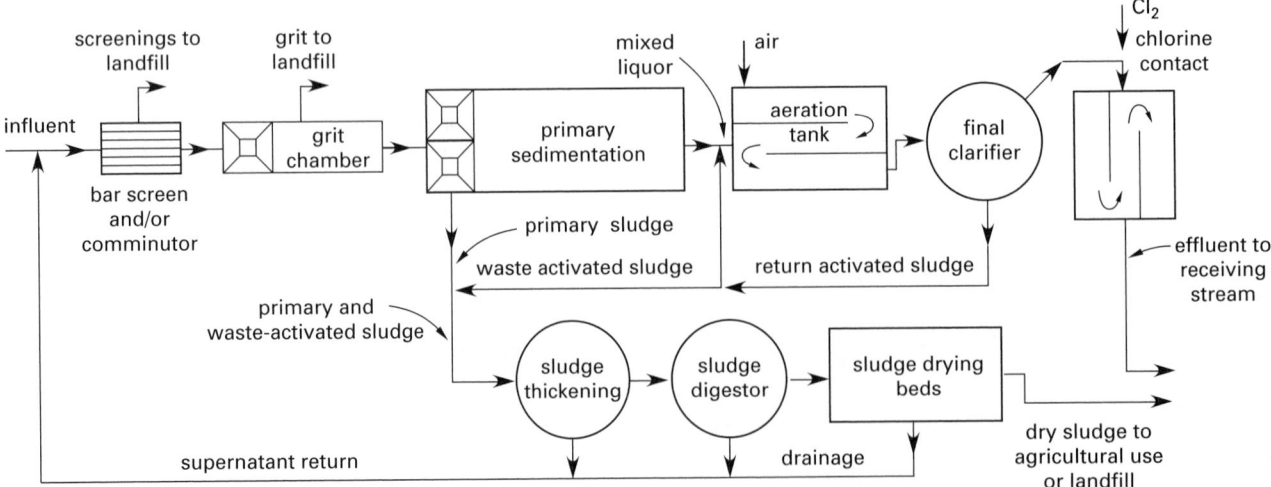

Table 30.1 *Conventional Activated Sludge Plants Typical Characteristics*[a]

BOD reduction	90–95%
effluent BOD	5–30 mg/L
effluent COD	15–90 mg/L
effluent suspended solids	5–30 mg/L
F:M	0.2–0.5
maximum aeration chamber volume	5000 ft^3 (140 m^3)
aeration chamber depth	10–15 ft (3–4.5 m)
aeration chamber width	20 ft (6 m)
length:width ratio	5:1 or greater
aeration air rate	0.5–2 ft^3 air/gal raw wastewater (3.6–14 m^3/m^3)
minimum dissolved oxygen	1 mg/L
MLSS	1000–4000 mg/L
sludge volume index	50–150
settling basin depth	15 ft (4.5 m)
settling basin detention time	4–8 hr
basin overflow rate	400–2000 gal/day-ft^2; 1000 gal/day-ft^2 typical (16–81 m^3/day·m^2; 40 m^3/day·m^2 typical)
settling basin weir loading	10,000 gal/day-ft (130 m^3/day·m)
fractional sludge recycle	0.20–0.30
frequency of sludge recycle	hourly

[a]See Table 30.2 for additional information.

3. AERATION STAGING METHODS

Small wastewater quantities can be treated with *extended aeration*. This method uses mechanical floating or fixed subsurface aerators to oxygenate the mixed liquor for 24 to 36 hr in a large lagoon. There is no primary clarification, and there is generally no sludge wasting process. Sludge is allowed to accumulate at the bottom of the lagoon for several months. Then the system is shut down and the lagoon is pumped out. Sedimentation basins are sized very small, with low overflow rates of 200 to 600 gal/day-ft^2 (8.1 to 24 m^3/day·m^2) and long retention times.

An *oxidation ditch* is a type of extended aeration system, configured in a continuous long and narrow oval. The basin contents circulate continuously to maintain mixing and aeration. Mixing occurs by brush aerators or a combination of low-head pumping and diffused aeration.

In *conventional aeration*, the influent is taken from a primary clarifier and then aerated. The amount of aeration is usually decreased as the wastewater travels along the aeration path since the BOD also decreases along the route. This is known as *tapered aeration*.

With *step-flow aeration*, aeration is constant along the length of the aeration path, but influent is introduced at various points along the path.

With *complete-mix aeration*, wastewater is added uniformly and the mixed liquor is removed uniformly over the length of the tank. The mixture is essentially uniform in composition throughout.

Units for the *contact stabilization (biosorption)* process are typically factory-built and brought to the site for installation, although permanent facilities can be built for the same process. Prebuilt units are compact but not as economical or efficient as larger plants running on the same process. The aeration tank is called a *contact tank*. The *stabilization tank* takes the sludge from the clarifier and aerates it. With this process, colloidal solids are absorbed in the activated sludge during the 30 to 90 min of aeration in the contact tank. Then, the sludge is removed by clarification and the return sludge is aerated for 3 to 6 more hours in the stabilization tank. Less time and space is required for this process because the sludge stabilization is done while the sludge is still concentrated. This method is very efficient in handling colloidal wastes.

The *high-rate aeration* method uses mechanical mixing along with aeration to decrease the aeration period and increase the BOD load per unit volume.

The *high purity oxygen aeration* method requires the use of bottled or manufactured oxygen that is introduced into closed/covered aerating tanks. Mechanical mixers are needed to take full advantage of the oxygen supply because little excess oxygen is provided. Retention times in aeration basins are longer than for other aerobic systems, producing higher concentrations of MLSS and better stabilization of sludge that is ultimately wasted. This method is applicable to high-strength sewage and industrial wastewater.

Sequencing batch reactors (SBRs) operate in a fill-and-drain sequence. The reactor is filled with influent, contents are aerated, and sludge is allowed to settle. Effluent is drawn from the basin along with some of the settled sludge, and new effluent is added to repeat the process. The "return sludge" is the sludge that remains in the basin.

Figure 30.2 *Methods of Aeration*

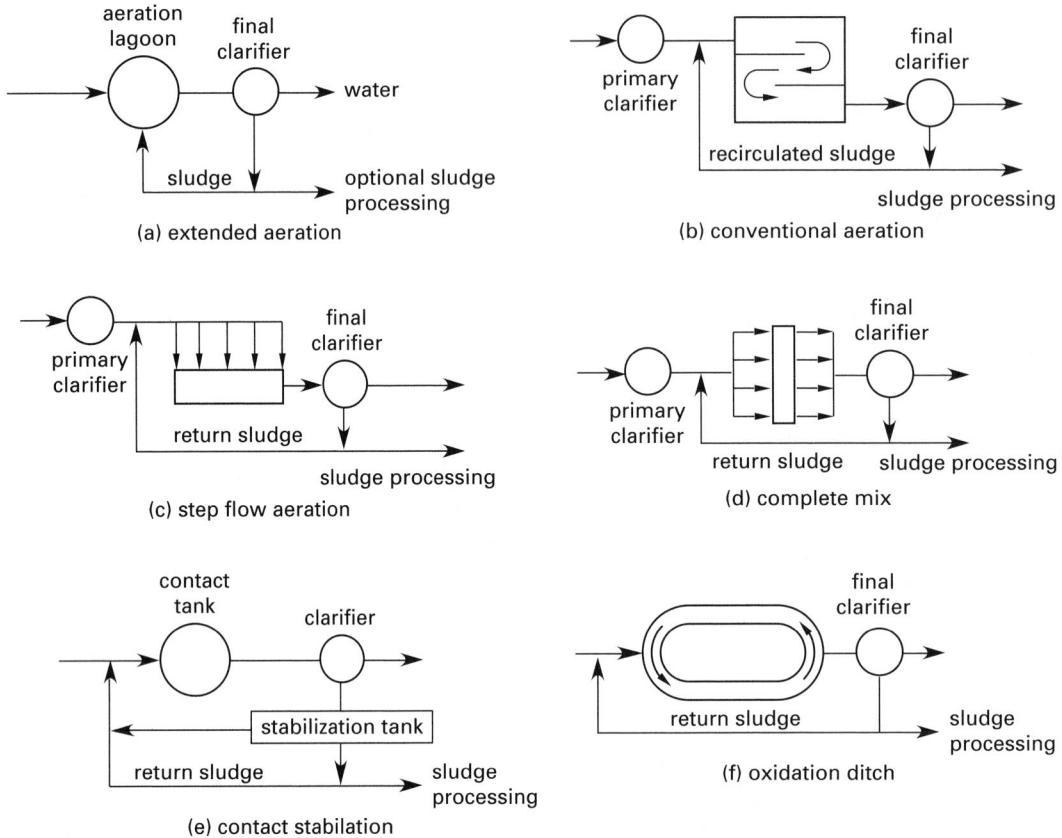

(a) extended aeration

(b) conventional aeration

(c) step flow aeration

(d) complete mix

(e) contact stabilation

(f) oxidation ditch

Table 30.2 *Representative Operating Conditions for Aeration*[a]

type of aeration	plant flow rate (MGD)	mean cell residence time, θ_c (d)	oxygen required (lbm/lbm BOD removed)	waste sludge (lbm/lbm BOD removed)	total plant BOD load (lbm/day)	aerator BOD load, L_{BOD} (lbm/day-1000 ft³)	F:M (lbm/ lbm-day)	MLSS (mg/L)	R (%)	η_{BOD} (%)
conventional	0–0.5	7.5	0.8–1.1	0.4–0.6	0–1000	30	0.2–0.5	1500–3000	30	90–95
	0.5–1.5	7.5–6.0			1000–3000	30–40				
	1.5 up	6.0			3000 up	40				
contact stabilization	0–0.5	3.0[b]	0.8–1.1	0.4–0.6	0–1000	30	0.2–0.5	1000–3000[b]	100	85–90
	0.5–1.5	3.0–2.0[b]	0.4–0.6		1000–3000	30–50				
	1.5 up	1.5–2.0[b]	0.4–0.6		3000 up	50				
extended	0–0.5	24	1.4–1.6	0.15–0.3	all	10.0	0.05–0.1	3000–6000	100	85–95
	0.5–1.5	20				12.5				
	1.5 up	16				15.0				
high rate	0–0.5	4.0	0.7–0.9	0.5–0.7	2000 up	100	1.0 or less	4000–10,000	100	80–85
	0.5–1.5	3.0								
	1.5 up	2.0								
step aeration	0–0.5	7.5			0–1000	30	0.2–0.5	2000–3500	50	85–95
	0.5–1.5	7.5–5.0			1000–3000	30–50				
	1.5 up	5.0			3000 up	50				
high purity oxygen		1.0–3.0				100–200	0.6–1.5	6000–8000	50	90–95
oxidation ditch		12–36				5–30				

[a]compiled from a variety of sources
[b]in contact unit only

Table 30.3 *Ten States' Standards for Activated Sludge Processes*

process	aeration tank organic loading—lbm BOD$_5$/day per 1000 ft^3 (kg/day-m^3)	F:M lbm BOD$_5$/day per lbm MLVSS	MLSS^a mg/L
conventional step aeration complete mix	40 (0.64)	0.2–0.5	1000–3000
contact stabilization	50^b (0.80)	0.2–0.6	1000–3000
extended aeration oxidation ditch	15 (0.24)	0.05–0.1	3000–5000

(Multiply lbm/day-1000 ft^3 by 0.016 to obtain kg/day·m^3.)
(Multiply lbm/day-lbm by 1.00 to obtain kg/day·kg.)
^{a}MLSS values are dependent upon the surface area provided for sedimentation and the rate of sludge return as well as the aeration process.
^{b}Total aeration capacity; includes both contact and reaeration capacities. Normally the contact zone equals 30 to 35% of the total aeration capacity.
Reprinted from *Recommended Standards for Sewage Works*, 1997, Sec. 92.31, Great Lakes-Upper Mississippi River Board of State Sanitary Engineers, published by Health Education Service, Albany, NY.

4. FINAL CLARIFIERS

Table 30.4 lists typical operating characteristics for clarifiers in activated sludge processes. Sludge should be removed rapidly from the entire bottom of the clarifier.

Table 30.4 *Characteristics of Final Clarifiers for Activated Sludge Processes*

type of aeration	design flow (MGD)	minimum detention time (hr)	maximum overflow rate (gpd/ft^2)
conventional, high rate and step	< 0.5	3.0	600
	0.5–1.5	2.5	700
	> 1.5	2.0	800
contact stabilization	< 0.5	3.6	500
	0.5–1.5	3.0	600
	> 1.5	2.5	700
extended aeration	< 0.05	4.0	300
	0.05–0.15	3.6	300
	> 0.15	3.0	600

(Multiply MGD by 0.0438 to obtain m^3/s.)
(Multiply gal/day-ft^2 by 0.0407 m^3/day·m^2.)

5. SLUDGE PARAMETERS

The bacteria and other suspended material in the mixed liquor is known as *mixed liquor suspended solids* (MLSS) and is measured in mg/L. Suspended solids are further divided into fixed solids and volatile solids. *Fixed solids* (also referred to as *nonvolatile solids*) are those inert solids that are left behind after being fired in a furnace. *Volatile solids* are essentially those carbonaceous solids that are consumed in the furnace. The volatile solids are considered to be the measure of solids capable of being digested. Approximately 60 to 75% of sludge solids are volatile. The volatile material in the mixed liquor is known as the *mixed liquor volatile suspended solids* (MLVSS).

The organic material in the incoming wastewater constitutes "food" for the activated organisms. The food arrival rate is given by Eq. 30.1. S_o is usually taken as the incoming BOD$_5$, although COD is used in rare situations.

$$F = S_o Q_o \qquad 30.1$$

The mass of microorganisms is determined from the *volatile suspended solids concentration*, X, in the aeration tank.

$$M = V_a X \qquad 30.2$$

The *food-to-microorganism* (F:M) *ratio* is given by Eq. 30.3. For conventional aeration, typical values are 0.20 to 0.50 lbm/lbm-day (0.20 to 0.50 kg/kg·day), though values between 0.05 and 1.0 have been reported. θ in Eq. 30.3 is the *hydraulic detention time*.

$$\begin{aligned} \text{F:M} &= \frac{S_{o,\text{mg/L}} Q_{o,\text{MGD}}}{V_{a,\text{MG}} X_{\text{mg/L}}} \\ &= \frac{S_{o,\text{mg/L}}}{\theta_{\text{days}} X_{\text{mg/L}}} \end{aligned} \qquad 30.3$$

Some authorities include the entire suspended solids content (MLSS), not just the volatile portion (X), when calculating the food-to-microorganism ratio, although this interpretation is less common.

$$\text{F:M} = \frac{S_{o,\text{mg/L}} Q_{o,\text{MGD}}}{V_{a,\text{MG}} \text{MLSS}_{\text{mg/L}}} \qquad 30.4$$

The liquid fraction of the wastewater passes through an activated sludge process in a matter of hours. However, the sludge solids are recycled continuously and have an average stay much longer in duration. There are two measures of *sludge age*: the *mean cell residence time* (also known as the *age of the suspended solids* and *solids residence time*), essentially the age of the microorganisms, and *age of the BOD*, essentially the age of the food. For conventional aeration, typical values of the mean cell residence time are 6 to 15 days for high-quality effluent and sludge.

$$\theta_c = \frac{V_a X}{Q_e X_e + Q_w X_w} \qquad 30.5$$

The *BOD sludge age* (*age of the BOD*) is the reciprocal of the food-to-microorganism ratio.

$$\theta_{\text{BOD}} = \frac{1}{\text{F:M}}$$
$$= \frac{V_{a,\text{MG}} X_{\text{mg/L}}}{S_{o,\text{mg/L}} Q_{o,\text{MGD}}} \qquad 30.6$$

The *sludge volume index* (SVI) is a measure of the sludge's settleability. SVI can be used to determine the tendency toward *sludge bulking*. (See Sec. 13.) SVI is determined by taking 1 L of mixed liquor and measuring the volume of settled solids after 30 min. SVI is the volume in mL occupied by 1 g of settled volatile and nonvolatile suspended solids.

$$\text{SVI} = \frac{\left(1000 \, \frac{\text{mg}}{\text{g}}\right) V_{\text{settled,mL/L}}}{\text{MLSS}_{\text{mg/L}}} \qquad 30.7$$

The total suspended solids concentration in the recirculated sludge can be found from Eq. 30.8. This equation is useful in calculating the solids concentration for any sludge that is wasted from the return sludge line. (Suspended sludge solids wasted include both fixed and volatile portions.)

$$\text{TSS}_{\text{mg/L}} = \frac{\left(1000 \, \frac{\text{mg}}{\text{g}}\right)\left(1000 \, \frac{\text{mL}}{\text{L}}\right)}{\text{SVI}_{\text{mL/g}}} \qquad 30.8$$

6. SOLUBLE BOD ESCAPING TREATMENT

S is a variable used to indicate the growth-limiting substrate in solution. Specifically, S (without a subscript) is defined as the soluble BOD_5 escaping treatment (i.e.,

the effluent BOD_5 leaving the activated sludge process). This may need to be calculated from the effluent suspended solids and the fraction, G, of the suspended solids that is ultimately biodegradable.

$$\text{BOD}_e = \text{BOD}_{\text{escaping treatment}}$$
$$+ \text{BOD}_{\text{effluent suspended solids}}$$
$$= S + S_e$$
$$= S + 1.42 fG X_e \qquad 30.9$$

Care must be taken to distinguish between the standard five-day BOD and the ultimate BOD. The ratio of these two parameters, typically approximately 70%, is

$$f = \frac{\text{BOD}_5}{\text{BOD}_u} \qquad 30.10$$

In some cases, the terms in the product $1.42 fG$ in Eq. 30.9 are combined, and S_e is known directly as a percentage of X_e. However, this format is less common.

7. PROCESS EFFICIENCY

The *treatment efficiency* (*BOD removal efficiency, removal fraction*, etc.), typically 90 to 95% for conventional aeration, is given by Eq. 30.11. S_o is the BOD as received from the primary settling tank. If the overall plant efficiency is wanted, the effluent S is taken as the total BOD_5 of the effluent. Depending on convention, efficiency can also be based on only the soluble BOD escaping treatment (i.e., S from Eq. 30.9).

$$\eta_{\text{BOD}} = \frac{S_o - S}{S_o} \qquad 30.11$$

8. PLUG FLOW AND STIRRED TANK MODELS

Conventional aeration and complete-mix aeration represent two extremes for continuous-flow reactions. Ideal conventional aeration is described by a *plug-flow reaction* (PFR, also known as *tubular flow*) kinetic model. (The term *reactor* is commonly used to describe the aeration tank.) In a plug-flow reactor, material progresses along the reaction path, changing in composition as it goes. Mixing is considered to be ideal in the lateral plane but totally absent in the longitudinal direction. That is, the properties (in this case, BOD and suspended solids) change along the path, as though only a "plug" of material was progressing forward.

A complete-mix aerator, however, is a type of *continuous-flow stirred tank reactor* (CSTR or CFSTR). In a CSTR model, the material is assumed to be homogeneous in properties throughout as a result of continuous perfect mixing. It is assumed that incoming material is immediately diluted to the tank concentration, and the outflow properties are the same as the tank properties.

Modifications of the basic kinetic models are made to account for return/recycle and wastage. Many, but not all, of the same equations can be used to describe the performance of both the PFR and CSTR models.

Figure 30.3 *Kinetic Model Process Variables*

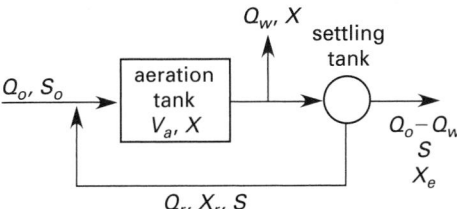

(a) plug flow reactor with recycle

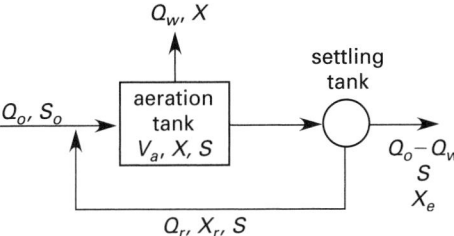

(b) continuously stirred tank reactor
with recycle

Until the late 1950s, nearly all activated sludge aeration tanks were long and narrow with length-to-width ratios greater than 5, and performance was assumed to be predicted by PFR equations. However, extensive testing has shown that substantial longitudinal mixing actually occurs, and many conventional systems are actually CSTRs. In fact, the long basin configuration gives an oxidation ditch the attributes of a PFR while maintaining CSTR characteristics.

The kinetic model for the PFR is complex, but it can be simplified under certain conditions. If the ratio of $\theta_c/\theta > 5$, the concentration of microorganisms in the influent to the reactor (aeration tank) will be approximately the same as in the effluent from the reactor.

$$\frac{1}{\theta_c} = \frac{\mu_m(S_o - S)}{S_o - S + (1+R)K_s \ln\left(\frac{S_i}{S}\right)} - k_d \quad \text{[PFR only]}$$

30.12

The *maximum specific growth rate*, μ_m, is calculated from the first-order reaction constant, k.

$$\mu_m = kY \qquad 30.13$$

The influent concentration after dilution with recycle flow is

$$S_i = \frac{S_o + RS}{1 + R} \quad \text{[PFR only]} \qquad 30.14$$

The *rate of substrate utilization* as the waste passes through the aerating reactor is

$$r_{su} = \frac{-\mu_m SX}{Y(K_s + S)} \quad \text{[PFR only]} \qquad 30.15$$

For the CSTR, the *specific substrate utilization* (*specific utilization*), U, is the food-to-microorganism ratio multiplied by the fractional process efficiency.

$$U = \eta(\text{F:M}) = \frac{-r_{su}}{X}$$
$$= \frac{S_o - S}{\theta X} \qquad 30.16$$

For the CSTR, the equivalent relationship between mean cell residence time, food-to-microorganism ratio, and specific utilization rate, U, is given by Eq. 30.17.

$$\frac{1}{\theta_c} = Y(\text{F:M})\eta - k_d$$
$$= -Y\left(\frac{r_{su}}{X}\right) - k_d$$
$$= YU - k_d \quad \text{[CSTR only]} \qquad 30.17$$

Y and k_d are the kinetic coefficients describing the bacterial growth process. Y is the *yield coefficient*, the mass of cells formed to the mass of substrate (food) consumed, in mg/mg. k_d is the *endogenous decay coefficient* (rate) of the microorganisms in day^{-1}. For a specified waste, biological community, and set of environmental conditions, Y and k_d are fixed. Equation 30.17 is a linear equation of the general form $y = mx + b$. Figure 30.4 illustrates how these coefficients can be determined graphically from bench-scale testing.

Figure 30.4 *Kinetic Growth Constants*[a]

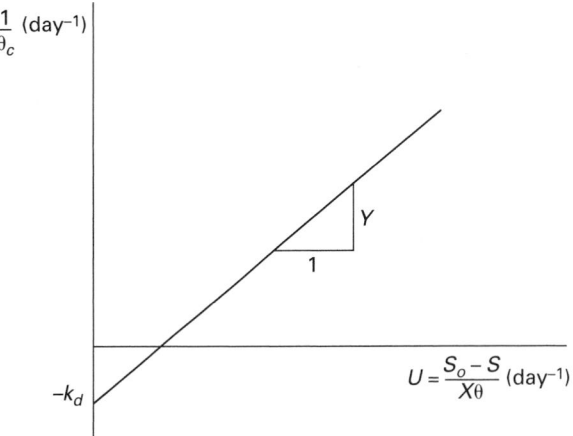

[a]As determined from bench-scale-activated sludge studies using a continuous flow stirred tank reactor without recycle

For a continuously stirred tank reactor, the kinetic model predicts the effluent substrate concentration.

$$S = \frac{K_s(1 + k_d\theta_c)}{\theta_c(\mu_m - k_d) - 1} \quad \text{[CSTR only]} \qquad 30.18$$

The reactor volume, V_a, is

$$V_a = \theta Q_o$$
$$= \frac{\theta_c Q_o Y(S_o - S)}{X(1 + k_d\theta_c)} \quad \text{[PFR and CSTR]} \qquad 30.19$$

The *hydraulic retention time*, θ, for the reactor is

$$\theta = \frac{V_a}{Q_o} \quad \text{[PFR and CSTR]} \qquad 30.20$$

The *hydraulic retention time*, θ, for the system is

$$\theta_s = \frac{V_a + V_s}{Q_o} \quad \text{[PFR and CSTR]} \qquad 30.21$$

The average concentration of microorganisms, X, in the reactor is

$$X = \frac{\left(\dfrac{\theta_c}{\theta}\right) Y (S_o - S)}{1 + k_d \theta_c} \quad \text{[PFR and CSTR]} \qquad 30.22$$

The *observed yield* is

$$Y_{\text{obs}} = \frac{Y}{1 + k_d \theta_c} \quad \text{[PFR and CSTR]} \qquad 30.23$$

The volatile portion of the mass of activated sludge that must be wasted each day is given by Eq. 30.24. The total (fixed and volatile) mass of sludge wasted each day must be calculated from the ratio of MLVSS and MLSS.

$$\begin{aligned} P_{x,\text{kg/day}} &= \frac{Q_{w,\text{m}^3/\text{day}} X_{r,\text{mg/L}}}{1000 \, \dfrac{\text{g}}{\text{kg}}} \\ &= \frac{Y_{\text{obs,mg/mg}} Q_{o,\text{m}^3/\text{day}} (S_o - S)_{\text{mg/L}}}{1000 \, \dfrac{\text{g}}{\text{kg}}} \\ &\qquad\qquad\qquad \text{[PFR and CSTR]} \qquad 30.24 \end{aligned}$$

The sludge that must be wasted is reduced by the sludge that escapes in the clarifier effluent, which is ideally zero. The actual sludge solids escaping can be calculated as

$$m_{e,\text{solids}} = Q_e X_e \quad \text{[PFR and CSTR]} \qquad 30.25$$

The *mean cell residence time* (see Eq. 30.5) can be calculated from Eq. 30.26. The sludge wasting rate, Q_w, can also be found from this equation. In most activated-sludge processes, waste sludge is drawn from the sludge recycle line. Therefore, Q_w is the *cell wastage rate* from the recycle line, and X_r is the microorganism concentration in the return sludge line.

$$\begin{aligned} \theta_c &= \frac{V_a X}{Q_w X_r + Q_e X_e} \\ &= \frac{V_a X}{Q_w X_r + (Q_o - Q_w) X_e} \quad \text{[PFR and CSTR]} \\ &\qquad\qquad\qquad\qquad\qquad 30.26 \end{aligned}$$

Since the average mass of cells in the effluent, X_e, is very small, Eq. 30.27 gives the mean cell residence time for sludge wastage from the return line. Centrifuge testing

can be used to determine the ratio of X/X_r. (There is another reason for omitting X_e from the calculation: It may be too small to measure conveniently.)

$$\theta_c \approx \frac{V_a X}{Q_w X_r} \quad \text{[PFR and CSTR]} \qquad 30.27$$

For wastage directly from the aeration basin, $X_r = X$. In this case, Eq. 30.28 shows that the process can be controlled by wasting a flow equal to the volume of the aeration tank divided by the mean cell residence time.

$$\theta_c \approx \frac{V_a}{Q_w} \quad \text{[CSTR]} \qquad 30.28$$

9. ATMOSPHERIC AIR

Atmospheric air is a mixture of oxygen, nitrogen, and small amounts of carbon dioxide, water vapor, argon, and other inert gases. If all constituents except oxygen are grouped with the nitrogen, the air composition is as given in Table 30.5. It is necessary to supply by weight $1/0.2315 = 4.32$ masses of air to obtain one mass of oxygen. The average molecular weight of air is 28.9.

Table 30.5 Composition of Air[a]

component	fraction by mass	fraction by volume
oxygen	0.2315	0.209
nitrogen	0.7685	0.791
ratio of nitrogen to oxygen	3.320	3.773[b]
ratio of air to oxygen	4.320	4.773

[a]Inert gases and CO_2 are included in N_2.
[b]This value is also reported by various sources as 3.76, 3.78, and 3.784.

10. AERATION TANKS

The *aeration period* is the same as the *hydraulic detention time* and is calculated without considering recirculation.

$$\theta = \frac{V_a}{Q_o} \qquad 30.29$$

The organic (BOD) *volumetric loading* rate (in kg BOD_5/day·m^3) for the aeration tank is

$$L_{\text{BOD}} = \frac{S_o Q_o}{V_a \left(1000 \, \dfrac{\text{mg·m}^3}{\text{kg·L}}\right)} \qquad 30.30$$

In the United States, volumetric loading is often specified in lbm/day-1000 ft^3.

$$L_{\text{BOD}} = \frac{S_{o,\text{mg/L}} Q_{o,\text{MGD}} \left(8.345 \, \dfrac{\text{lbm-L}}{\text{MG-mg}}\right) (1000)}{V_{a,\text{ft}^3}} \qquad 30.31$$

The *rate of oxygen transfer* from the air to the mixed liquor during aeration is given by Eq. 30.32. K_t is a macroscopic transfer coefficient that depends on the equipment and characteristics of the mixed liquor and that has typical dimensions of 1/time. The dissolved oxygen deficit, D in mg/L, is given by Eq. 30.33. β is the mixed liquor's *oxygen saturation coefficient*, approximately 0.8 to 0.9. β corrects for the fact that the mixed liquor does not absorb as much oxygen as pure water does. The minimum dissolved oxygen content is approximately 0.5 mg/L, below which the processing would be limited by oxygen. However, a lower limit of 2 mg/L is typically specified.

$$\dot{m}_{\text{oxygen}} = K_t D \qquad \text{30.32}$$

$$D = \beta \text{DO}_{\text{saturated water}} - \text{DO}_{\text{mixed liquor}} \qquad \text{30.33}$$

The *oxygen demand* is given by Eq. 30.34. The factor 1.42 is the theoretical gravimetric ratio of oxygen required for carbonaceous organic material based on an ideal stoichiometric reaction. Equation 30.34 neglects nitrogenous demand, which can also be significant.

$$\dot{m}_{\text{oxygen,kg/day}} = \left(\frac{Q_{o,\text{m}^3/\text{day}}(S_o - S)_{\text{mg/L}}}{f \left(1000 \ \frac{\text{mg·m}^3}{\text{kg·L}} \right)} \right)$$
$$- 1.42 P_{x,\text{kg/day}} \qquad \text{[PFR and CSTR]}$$
$$\text{30.34}$$

Air is approximately 23.2% oxygen by mass. Considering a *transfer efficiency* of η_{transfer} (typically less than 10%), the air requirement is calculated from the oxygen demand.

$$\dot{m}_{\text{air}} = \frac{\dot{m}_{\text{oxygen}}}{0.232 \eta_{\text{transfer}}} \qquad \text{30.35}$$

The volume of air required is given by Eq. 30.36. The density of air is approximately 0.075 lbm/ft^3 (1.2 kg/m^3). Typical volumes for conventional aeration are 500 to 900 ft^3/lbm BOD$_5$ (30 to 55 m^3/kg BOD$_5$) for F:M ratios greater than 0.3 and 1200 to 1800 ft^3/lbm BOD$_5$ (75 to 115 m^3/kg BOD$_5$) for F:M ratios less than 0.3 day^{-1}. Air flows in SCFM (*standard cubic feet per minute*) are based on a temperature of 70°F (21°C) and pressure of 14.7 psia (101 kPa).

$$\dot{V}_{\text{air}} = \frac{\dot{m}_{\text{air}}}{\rho_{\text{air}}} \qquad \text{[PFR and CSTR]} \qquad \text{30.36}$$

Aeration equipment should be designed with an excess-capacity safety factor of 1.5 to 2.0. *Ten States' Standards* requires 200% of calculated capacity for air diffusion systems, which is assumed to be 1500 ft^3/lbm BOD$_5$ (94 m^3/kg BOD$_5$) [TSS-1997, 92.332].

The air requirement per unit volume (m^3/m^3, calculated using consistent units) is V_{air}/Q_o. The air requirement in m^3 per kilogram of soluble BOD$_5$ removed is

$$\frac{\dot{V}_{\text{air}} \left(1000 \ \frac{\text{mg·m}^3}{\text{kg·L}} \right)}{Q_o(S_o - S)} \qquad \text{30.37}$$

Example 30.1

Mixed liquor ($\beta = 0.9$, DO = 3 mg/L) is aerated at 20°C. The transfer coefficient is 2.7 h^{-1}. What is the rate of oxygen transfer?

Solution

From App. 22.D, at 20°C, the saturated oxygen content of pure water is 9.2 mg/L. The oxygen deficit is found from Eq. 30.33.

$$D = \beta \text{DO}_{\text{saturated water}} - \text{DO}_{\text{mixed liquor}}$$
$$= (0.9) \left(9.17 \ \frac{\text{mg}}{\text{L}} \right) - 3 \ \frac{\text{mg}}{\text{L}}$$
$$= 5.25 \ \text{mg/L}$$

The rate of oxygen transfer is found from Eq. 30.32.

$$\dot{m} = K_t D = (2.7 \ \text{h}^{-1}) \left(5.25 \ \frac{\text{mg}}{\text{L}} \right)$$
$$= 14.18 \ \text{mg/L·h}$$

11. AERATION POWER AND COST

The work required to compress a unit mass of air from atmospheric pressure to the discharge pressure depends on the nature of the compression. Various assumptions can be made, but generally an *isentropic compression* is assumed. Inefficiencies (thermodynamic and mechanical) are combined into a single *compressor efficiency*, η_c, which is essentially an *isentropic efficiency* for the compression process.

The ideal power to compress a volumetric flow rate, $\dot{V}$, or mass flow rate, $\dot{m}$, from pressure p_1 to pressure p_2 in an isentropic steady-flow compression process is given by Eq. 30.38. k is the ratio of specific heats, which has an appropriate value of 1.4. c_p is the specific heat at constant pressure, which has a value of 0.24 Btu/lbm-°R (1005 J/kg·K). R_{air} is the specific gas constant, 53.3 ft-lbf/lbm-°R (287 J/kg·K). In Eq. 30.38, temperatures must be absolute. Volumetric and mass flow rates are per second.

$$P_{\text{ideal,kW}} = -\left(\frac{k p_1 \dot{V}_1}{(k-1) \left(1000 \ \frac{\text{W}}{\text{kW}} \right)} \right) \left(1 - \left(\frac{p_2}{p_1} \right)^{(k-1)/k} \right)$$
$$= -\left(\frac{k \dot{m} R_{\text{air}} T_1}{(k-1) \left(1000 \ \frac{\text{W}}{\text{kW}} \right)} \right) \left(1 - \left(\frac{p_2}{p_1} \right)^{(k-1)/k} \right)$$
$$= -\left(\frac{c_p \dot{m} T_1}{1000 \ \frac{\text{W}}{\text{kW}}} \right) \left(1 - \left(\frac{p_2}{p_1} \right)^{(k-1)/1} \right)$$
$$\text{[SI]} \ \textit{30.38(a)}$$

$$P_{ideal,hp} = -\left(\frac{kp_1\dot{V}_1}{(k-1)\left(550\,\frac{\text{ft-lbf}}{\text{hp-sec}}\right)}\right)\left(1-\left(\frac{p_2}{p_1}\right)^{(k-1)/k}\right)$$

$$= -\left(\frac{k\dot{m}R_{air}T_1}{(k-1)\left(550\,\frac{\text{ft-lbf}}{\text{hp-sec}}\right)}\right)\left(1-\left(\frac{p_2}{p_1}\right)^{(k-1)/k}\right)$$

$$= -\left(\frac{c_p J\dot{m}T_1}{550\,\frac{\text{ft-lbf}}{\text{hp-sec}}}\right)\left(1-\left(\frac{p_2}{p_1}\right)^{(k-1)/k}\right)$$

[U.S.]   *30.38(b)*

The actual compression power is

$$P_{actual} = \frac{P_{ideal}}{\eta_c} \qquad 30.39$$

The cost of running the compressor for some duration, t, is

$$\text{total cost} = C_{kW\text{-}hr}P_{actual}t \qquad 30.40$$

12. RETURN RATE/RECYCLE RATIO

For any given wastewater and treatment facility, the influent flow rate, BOD, and tank volume cannot be changed. Of all the variables appearing in Eq. 30.3, only the suspended solids can be controlled. Therefore, the primary process control variable is the amount of organic material in the wastewater. This is controlled by the sludge return rate. Equation 30.41 gives the sludge *return ratio* (*recycle ratio*), which is typically 0.20 to 0.30. (Some authorities use the symbol α to represent the recycle ratio.)

$$R = \frac{Q_r}{Q_o} \qquad 30.41$$

The actual recirculation rate can be found by writing a suspended solids mass balance around the inlet to the reactor and solving for the recirculation rate, Q_r, noting $X_o = 0$.

$$X_oQ_o + X_rQ_r = X(Q_o + Q_r) \qquad 30.42$$

$$\frac{Q_r}{Q_o + Q_r} = \frac{X}{X_r} \qquad 30.43$$

The theoretical required return rate for the current MLSS content can be calculated from the *sludge volume index* (SVI; see Sec. 5). Equation 30.44 assumes that the settling tank responds identically to the graduated 1000 mL cylinder used in the SVI test. Equations 30.43 and 30.44 are analogous.

$$\frac{Q_r}{Q_o + Q_r} = \frac{V_{settled,mL/L}}{1000\,\frac{\text{mL}}{\text{L}}} \qquad 30.44$$

$$R = \frac{V_{settled,mL/L}}{1000\,\frac{\text{mL}}{\text{L}} - V_{settled,mL/L}}$$

$$= \frac{1}{\dfrac{10^6}{(SVI_{mL/g})(MLSS_{mg/L})} - 1} \qquad 30.45$$

Equation 30.45 is based on settling and represents a theoretical value, but it does not necessarily correspond to the actual recirculation rate. Greater-than-necessary recirculation returns clarified liquid unnecessarily, and less recirculation leaves settled solids in the clarifier.

Figure 30.5 *Theoretical Relationship Between SVI Test and Return Rate*

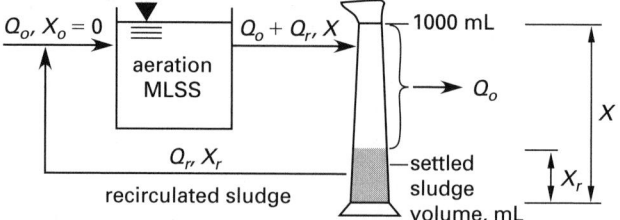

Example 30.2

Two 1 L samples of mixed liquor are taken from an aerating lagoon. After settling for 30 min in a graduated cylinder, 250 mL of solids have settled out in the first sample. The total suspended solids concentration in the second sample is found to be 2300 mg/L. (a) What is the sludge volume index? (b) What is the theoretical required sludge recycle rate?

Solution

(a) Use Eq. 30.7.

$$SVI = \frac{1000V_{settled,mL/L}}{X_{mg/L}}$$

$$= \frac{\left(1000\,\frac{\text{mg}}{\text{g}}\right)\left(250\,\frac{\text{mL}}{\text{L}}\right)}{2300\,\frac{\text{mg}}{\text{L}}} = 109\text{ mL/g}$$

(b) The theoretical required recycle rate is given by Eq. 30.45.

$$R = \frac{V_{settled,mL/L}}{1000 - V_{settled,mL/L}}$$

$$= \frac{250\text{ mL}}{1000\text{ mL} - 250\text{ mL}} = 0.33$$

Example 30.3

An activated sludge plant receives 4.0 MGD of wastewater with a BOD of 200 mg/L. The primary clarifier removes 30% of the BOD. The sludge is aerated for 6 hr. The oxygen transfer efficiency is 6%, and one pound of oxygen is required for each pound of BOD oxidized. The food-to-microorganism ratio based on MLSS is 0.33. The SVI is 100. 30% of the volatile solids (MLVSS) are wasted each day from the aeration tank. The MLVSS/MLSS ratio is 0.7. The surface settling rate of the secondary clarifier is 800 gal/day-ft^2. The final effluent has a BOD of 10 mg/L.

What are the (a) secondary BOD removal fraction, (b) aeration tank volume, (c) MLSS, (d) total suspended solids content in the recirculated sludge, (e) theoretical recirculation rate, (f) mass of MLSS wasted each day, (g) clarifier surface area, and (h) volumetric air requirements?

Solution

(a) Wastewater leaves primary settling with a BOD of $(1 - 0.3)(200 \text{ mg/L}) = 140 \text{ mg/L}$. The efficiency of the activated sludge processing is

$$\eta_{\text{BOD}} = \frac{140 \, \frac{\text{mg}}{\text{L}} - 10 \, \frac{\text{mg}}{\text{L}}}{140 \, \frac{\text{mg}}{\text{L}}} = 0.929 \quad (92.9\%)$$

(b) The aeration tank volume is

$$V_a = Q_o t = \frac{(4 \text{ MGD}) \left(10^6 \, \frac{\text{gal}}{\text{MG}} \right) (6 \text{ hr})}{\left(24 \, \frac{\text{hr}}{\text{day}} \right) \left(7.48 \, \frac{\text{gal}}{\text{ft}^3} \right)}$$

$$= 1.34 \times 10^5 \text{ ft}^3$$

(c) The total MLSS can be determined from the food-to-microorganism ratio given. Use Eq. 30.4.

$$\text{MLSS} = \frac{S_o Q_o}{(\text{F:M}) V_a}$$

$$= \frac{\left(140 \, \frac{\text{mg}}{\text{L}} \right) (4 \text{ MGD}) \left(10^6 \, \frac{\text{gal}}{\text{MG}} \right)}{\left(0.33 \, \frac{\text{lbm}}{\text{lbm-day}} \right) (1.34 \times 10^5 \text{ ft}^3) \left(7.481 \, \frac{\text{gal}}{\text{ft}^3} \right)}$$

$$= 1693 \text{ mg/L}$$

(d) The suspended solids content in the recirculated sludge is given by Eq. 30.8.

$$\text{TSS}_{\text{mg/L}} = \frac{\left(1000 \, \frac{\text{mg}}{\text{g}} \right) \left(1000 \, \frac{\text{mL}}{\text{L}} \right)}{\text{SVI}_{\text{mL/g}}}$$

$$= \frac{\left(1000 \, \frac{\text{mg}}{\text{g}} \right) \left(1000 \, \frac{\text{mL}}{\text{L}} \right)}{100 \, \frac{\text{mL}}{\text{g}}}$$

$$= 10{,}000 \text{ mg/L}$$

(e) The recirculation ratio can be calculated from the SVI test. However, the settled volume must first be calculated.

Use Eq. 30.7.

$$V_{\text{settled,mL/L}} = \frac{\text{MLSS}_{\text{mg/L}} \text{SVI}_{\text{mL/g}}}{1000 \, \frac{\text{mg}}{\text{g}}}$$

$$= \frac{\left(1693 \, \frac{\text{mg}}{\text{L}} \right) \left(100 \, \frac{\text{mL}}{\text{g}} \right)}{1000 \, \frac{\text{mg}}{\text{g}}}$$

$$= 169.3 \text{ mL/L}$$

Use Eq. 30.45.

$$R = \frac{V_{\text{settled,mL/L}}}{1000 - V_{\text{settled,mL/L}}}$$

$$= \frac{169.3 \, \frac{\text{mL}}{\text{L}}}{1000 \, \frac{\text{mL}}{\text{L}} - 169.3 \, \frac{\text{mL}}{\text{L}}}$$

$$= 0.204$$

The recirculation is given by Eq. 30.41.

$$Q_r = RQ_o = (0.204)(4 \text{ MGD}) = 0.816 \text{ MGD}$$

(f) Since 30% of the MLVSS is wasted each day and wasting is from the aeration tank, 30% of the MLSS is wasted. That is, 30% of everything is wasted because everything is uniformly mixed.

The sludge wasted is

$$(0.30)(4 \text{ MGD}) \left(1693 \, \frac{\text{mg}}{\text{L}} \right) \left(8.345 \, \frac{\text{lbm-L}}{\text{mg-MG}} \right)$$

$$= 16{,}954 \text{ lbm/day}$$

(g) The clarifier surface area is

$$A = \frac{Q_t}{\text{v}^*} = \frac{Q_r + Q_o}{\text{v}^*}$$

$$= \frac{(0.81 \text{ MGD} + 4 \text{ MGD}) \left(10^6 \, \frac{\text{gal}}{\text{MG}} \right)}{800 \, \frac{\text{gal}}{\text{day-ft}^2}}$$

$$= 6010 \text{ ft}^2$$

(h) From part (a), the fraction of BOD reduction in the activated sludge process aeration (exclusive of the wasted sludge) is 0.929.

Since one pound of oxygen is required for each pound of BOD removed, the required oxygen mass is

$$m_{\text{oxygen}} = (0.929)(4 \text{ MGD}) \left(140 \, \frac{\text{mg}}{\text{L}} \right) \left(8.345 \, \frac{\text{lbm-L}}{\text{mg-MG}} \right)$$

$$= 4341 \text{ lbm/day}$$

The density of air is approximately 0.075 lbm/ft^3, and air is 20.9% oxygen by volume. Considering the efficiency of the oxygenation process, the air required is

$$\dot{V}_{\text{air}} = \frac{4341 \, \frac{\text{lbm}}{\text{day}}}{(0.209)(0.06)\left(0.075 \, \frac{\text{lbm}}{\text{ft}^3}\right)}$$

$$= 4.62 \times 10^6 \, \text{ft}^3/\text{day}$$

13. OPERATIONAL DIFFICULTIES

Sludge bulking refers to a condition in which the sludge does not settle out. Since the solids do not settle, they leave the sedimentation tank and cause problems in subsequent processes. The sludge volume index can often (but not always) be used as a measure of settling characteristics. If the SVI is less than 100, the settling process is probably operating satisfactorily. If SVI > 150, the sludge is bulking. Remedies include addition of lime, chlorination, additional aeration, and a reduction in MLSS.

Some sludge may float back to the surface after settling, a condition known as *rising sludge*. Rising sludge occurs when nitrogen gas is produced from denitrification of the nitrates and nitrites. Remedies include increasing the return sludge rate and decreasing the mean cell residence time. Increasing the speed of the sludge scraper mechanism to dislodge nitrogen bubbles may also help.

It is generally held that after 30 min of settling, the sludge should settle to between 20 and 70% of its original volume. If it occupies more than 70%, there are too many solids in the aeration basin, and more sludge should be wasted. If the sludge occupies less than 20%, less sludge should be wasted.

Sludge washout (solids washout) can occur during a period of peak flow or even with excess recirculation. Washout is the loss of solids from the sludge blanket in the settling tank. This often happens with insufficient wasting, such that solids build up (to more than 70% in a settling test), hinder settling in the clarifier, and cannot be contained.

14. QUANTITIES OF SLUDGE

The procedure for determining the volume and mass of sludge from activated sludge processing is essentially the same as for sludge from sedimentation and any other processes. The primary variables, other than mass and volume, are density and/or specific gravity.

$$m_{\text{wet}} = V\rho_{\text{sludge}} = V(\text{SG}_{\text{sludge}})\rho_{\text{water}} \qquad 30.46$$

Raw sludge is approximately 95 to 99% water, and the specific gravity of sludge is only slightly greater than 1.0. The actual specific gravity of sludge can be calculated from the fractional solids content, s, and specific

gravity of the sludge solids, which is approximately 2.5. (If the sludge specific gravity is known, Eq. 30.47 can be solved for the specific gravity of the solids.) In Eq. 30.47, $1 - s$ is the fraction of the sludge that is water.

$$\frac{1}{\text{SG}_{\text{sludge}}} = \frac{1-s}{1} + \frac{s}{\text{SG}_{\text{solids}}}$$

$$= \frac{1 - s_{\text{fixed}} - s_{\text{volatile}}}{1} + \frac{s_{\text{fixed}}}{\text{SG}_{\text{fixed solids}}}$$

$$+ \frac{s_{\text{volatile}}}{\text{SG}_{\text{volatile solids}}} \qquad 30.47$$

The volume of sludge can be estimated from its dried mass. Since the sludge specific gravity is approximately 1.0, the sludge volume is

$$V_{\text{sludge,wet}} = \frac{m_{\text{dried}}}{s\rho_{\text{sludge}}} \approx \frac{m_{\text{dried}}}{s\rho_{\text{water}}} \qquad 30.48$$

The dried mass of sludge solids from primary settling basins is easily determined from the decrease in solids. The decrease in suspended solids, ΔSS, due to primary settling is approximately 50% of the total incoming suspended solids.

$$m_{\text{dried}} = (\Delta\text{SS})_{\text{mg/L}} Q_{o,\text{MGD}} \left(8.345 \, \frac{\text{lbm-L}}{\text{MG-mg}}\right) \qquad 30.49$$

The dried mass of solids for biological filters and secondary aeration (e.g., activated sludge) can be estimated on a macroscopic basis as a fraction of the change in BOD. In Eq. 30.50, K is the *cell yield*, also known as the *removal efficiency*, the fraction of the total influent BOD that ultimately appears as excess (i.e., settled) biological solids. The cell yield can be estimated from the food-to-microorganism ratio from Fig. 30.6. (Note that the cell yield, K, and yield coefficient, Y_{obs}, are not the same since K is based on total incoming BOD and Y_{obs} is based on consumed BOD. The term "cell yield" is often used for both Y_{obs} and K.)

$$m_{\text{dried}} = KS_{o,\text{mg/L}} Q_{o,\text{MGD}} \left(8.345 \, \frac{\text{lbm-L}}{\text{MG-mg}}\right)$$

$$= Y_{\text{obs}}(S_{o,\text{mg/L}} - S) Q_{o,\text{MGD}} \left(8.345 \, \frac{\text{lbm-L}}{\text{MG-mg}}\right)$$

$$30.50$$

Environmental

Figure 30.6 *Approximate Food:Microorganism Ratio Versus Cell Yield,[a] K*

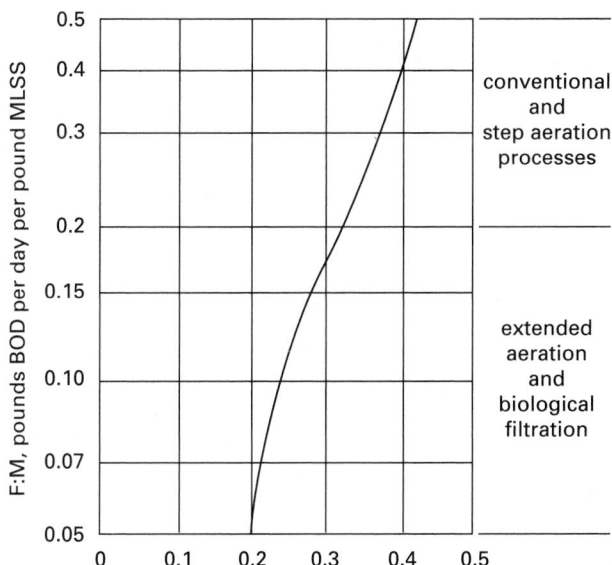

fraction of BOD converted to excess solids, *K*

[a]calculated assuming an effluent BOD of approximately 30 mg/L

Water and Wastewater Technology 3/E. by Hammer/Hammer, © 1996. Reprinted by permission of Prentice-Hall, Inc., Upper Saddle River, NJ.

Table 30.6 *Characteristics of Sludge*

origin of sludge	fraction solids, s	dry mass (lbm/day-person)
primary settling tank	0.06–0.08	0.12
trickling filter	0.04–0.06	0.04
mixed primary settling and trickling filter	0.05	0.16
conventional activated sludge	0.005–0.01	0.07
mixed primary and conventional activated sludge	0.02–0.03	0.19
high-rate activated sludge	0.025–0.05	0.06
mixed primary settling and high-rate activated sludge	0.05	0.18
extended aeration activated sludge	0.02	0.02
filter backwashing water	0.01–0.1	–
softening sludge	0.03–0.15	–

(Multiply lbm/day-person by 0.45 to obtain kg/day·person.)

Example 30.4

A trickling filter plant processes 4 MGD of domestic wastewater with 190 mg/L BOD and 230 mg/L suspended solids. The primary sedimentation tank removes 50% of the suspended solids and 30% of the BOD. The cell yield for the trickling filter is 0.25. What is the total sludge volume from the primary sedimentation tank and trickling filter if the combined solids content of the sludge is 5%?

Solution

The mass of solids removed in primary settling is given by Eq. 30.49.

$$
\begin{aligned}
m_{\text{dried}} &= (\Delta\text{SS})_{\text{mg/L}} Q_{o,\text{MGD}} \left(8.345 \ \frac{\text{lbm-L}}{\text{MG-mg}}\right) \\
&= (0.5)\left(230 \ \frac{\text{mg}}{\text{L}}\right)(4 \ \text{MGD})\left(8.345 \ \frac{\text{lbm-L}}{\text{MG-mg}}\right) \\
&= 3839 \ \text{lbm/day}
\end{aligned}
$$

The mass of solids removed from the trickling filter's clarifier is calculated from Eq. 30.50.

$$
\begin{aligned}
m_{\text{dried}} &= K\Delta S_{o,\text{mg/L}} Q_{o,\text{MGD}} \left(8.345 \ \frac{\text{lbm-L}}{\text{MG-mg}}\right) \\
&= (0.25)(1-0.3)\left(190 \ \frac{\text{mg}}{\text{L}}\right)(4 \ \text{MGD}) \\
&\quad \times \left(8.345 \ \frac{\text{lbm-L}}{\text{MG-mg}}\right) \\
&= 1110 \ \text{lbm/day}
\end{aligned}
$$

The volume is calculated from Eq. 30.48.

$$
\begin{aligned}
V &= \frac{m_{\text{dried}}}{s\rho_{\text{water}}} \\
&= \frac{3839 \ \frac{\text{lbm}}{\text{day}} + 1110 \ \frac{\text{lbm}}{\text{day}}}{(0.05)\left(62.4 \ \frac{\text{lbm}}{\text{ft}^3}\right)} \\
&= 1586 \ \text{ft}^3/\text{day}
\end{aligned}
$$

15. SLUDGE THICKENING

Waste-activated sludge (WAS) has a typical solids content of 0.5 to 1.0%. *Thickening* of sludge is used to reduce the volume of sludge prior to digestion or dewatering. Thickening is accomplished by decreasing the liquid fraction $(1-s)$, thereby increasing the solids fraction, s. Equation 30.48 shows that the volume of wet sludge is inversely proportional to its solids content.

For dewatering, thickening to at least 4% solids (i.e., 96% moisture) is required; for digestion, thickening to at least 5% solids is required. Depending on the nature of the sludge, polymers may be used with all of the thickening methods. Other chemicals (e.g., lime for stabilization through pH control and potassium permanganate to react with sulfides) can also be used. However, because of cost and increased disposal problems, the trend is toward the use of fewer inorganic chemicals with new thickening or dewatering applications unless special conditions apply.

Gravity thickening occurs in circular sedimentation tanks similar to primary and secondary clarifiers. The settling process is categorized into four zones: the *clarification zone*, containing relatively clear *supernatant*; the *hindered settling zone*, where the solids move downward at essentially a constant rate; the *transition zone*, characterized by a decrease in solids settling rate; and the *compression zone*, where motion is essentially zero.

In *batch gravity thickening*, the tank is filled with thin sludge and allowed to stand. Supernatant is decanted, and the tank is topped off with more thin sludge. The operation is repeated continually or a number of times before the underflow sludge is removed. A heavy-duty (deep-truss) scraper mechanism pushes the settled solids into a hopper in the tank bottom, and the clarified effluent is removed in a peripheral weir. A doubling of solids content is usually possible with gravity thickening. Table 30.7 contains typical characteristics of gravity thickening tanks.

With *dissolved air flotation* (DAF) *thickening* (DAFT), fine air bubbles are released into the sludge as it enters the DAF tank. The solids particles adhere to the air bubbles and float to the surface where they are skimmed away as scum. The scum has a solids content of approximately 4%. Up to 85% of the total solids may be recovered in this manner, and chemical flocculants (e.g., polymers) can increase the recovery to 97 to 99%.

Table 30.7 *Typical Characteristics of Gravity Thickening Tanks*

shape	circular
minimum number of tanks[a]	2
overflow rate	600–800 gal/day-ft²
	(24–33 m³/day·m²)
maximum dry solids loading	
primary sludge	22 lbm/day-ft²
	(107 kg/day·m²)
primary and trickling filter	
sludge	15 lbm/day-ft²
	(73 kg/day·m²)
primary and modified	
aeration activated sludge	12 lbm/day-ft²
	(59 kg/day·m²)
primary and conventional	
aeration activated sludge	8 lbm/day-ft²
	(39 mg/day·m²)
waste-activated sludge	4 lbm/day-ft²
	(20 kg/day·ft²)
minimum detention time	6 hr
minimum sidewater depth	10 ft (3 m)
minimum freeboard	1.5 ft (0.45 m)

[a]unless alternative methods of thickening are available

With *gravity belt thickening* (GBT), sludge is spread over a drainage belt. Multiple stationary plows in the path of the moving belt may be used to split, turn, and recombine the sludge so that water does not pool on top of the sludge. After thickening, the sludge cake is removed from the belt by a doctor blade.

Gravity thickening is usually best for sludges from primary and secondary settling tanks, while dissolved air flotation and centrifugal thickening are better suited for activated sludge. The current trend is toward using gravity thickening for primary sludge and flotation thickening for activated sludge, then blending the thickened sludge for further processing.

Not all of the solids in sludge are captured in thickening processes. Some sludge solids escape with the liquid portion. Table 30.8 lists typical solids capture fractions.

16. SLUDGE STABILIZATION

Raw sludge is too bulky, odorous, and putrescible to be dewatered easily or disposed of by land spreading. Such sludge can be "stabilized" through chemical addition or by digestion prior to dewatering and landfilling. (Sludge that is incinerated does not need to be stabilized.) Stabilization converts the sludge to a stable, inert form that is essentially free from odors and pathogens.

Sludge can be stabilized chemically by adding lime dust to raise the pH. A rule of thumb is that adding enough lime to raise the pH to 12 for at least 2 hr will inhibit or kill bacteria and pathogens in the sludge. Kiln and cement dust is also used in this manner.

Table 30.8 *Typical Thickening Solids Fraction and Capture Efficiencies*

operation	solids capture range (%)	solids content range (%)
gravity thickeners		
primary sludge only	85–92	4–10
primary and waste-activated	80–90	2–6
flotation thickeners		
with chemicals	90–98	3–6
without chemicals	80–95	3–6
centrifuge thickeners		
with chemicals	90–98	4–8
without chemicals	80–90	3–6
vacuum filtration		
with chemicals	90–98	15–30
belt filter press		
with chemicals	85–98	15–30
filter press		
with chemicals	90–98	20–50
centrifuge dewatering		
with chemicals	85–98	10–35
without chemicals	55–90	10–30

Alternatively, ammonia compounds can be added to the sludge prior to land spreading. This has the additional advantage of producing a sludge with high plant nutrient value.

17. AEROBIC DIGESTION

Aerobic digestion occurs in an open holding tank digester and is preferable with stabilized primary and combined primary-secondary sludges. Up to 70% (typically 40 to 50%) of the volatile solids can be removed in an aerobic digester. Mechanical aerators are used in a manner similar to aerated lagoons. Construction details of aerobic digesters are similar to those of aerated lagoons.

18. ANAEROBIC DIGESTION

Anaerobic digestion occurs in the absence of oxygen. Anaerobic digestion is more complex and more easily upset than aerobic digestion. However, it has a lower operating cost.

Three types of anaerobic bacteria are involved. The first type converts organic compounds into simple fatty or amino acids. The second group of *acid-formed bacteria* converts these compounds into simple organic acids, such as acetic acid. The third group of *acid-splitting bacteria* converts the organic acids to methane, carbon dioxide, and some hydrogen sulfide. The third phase takes the longest time and sets the rate and loadings. The pH should be 6.7 to 7.8, and temperature should be in the mesophilic range of 85 to 100°F (29 to 38°C) for the methane-producing bacteria to be effective. Sufficient alkalinity must be added to buffer acid production.

Table 30.9 *Typical Characteristics of Aerobic Digesters*

minimum number of units	2
size	2–3 ft^3/person
	(0.05–0.08 m^3/person)
aeration period	
activated sludge only	10–15 days
activated sludge and	
primary settling sludge	12–18 days
mixture of primary sludge	
and activated or	
trickling filter sludge	15–20 days
volatile solids loading	0.1–0.3 lbm/day-ft^3
	(1.6–4.8 kg/day·m^3)
volatile solids reduction	40–50%
sludge age	
primary sludge	25–30 days
activated sludge	15–20 days
oxygen required (primary	
sludge)	1.6–1.9 lbm O$_2$/lbm
	BOD removed
	(1.6–1.9 kg O$_2$/kg
	BOD removed)
minimum dissolved oxygen	
level	1–2 mg/L
mixing energy required by	
mechanical aerators	0.75–1.5 hp/1000 ft^3
	(0.02–0.04 kW/m^3)
maximum depth	15 ft (4.5 m)

A simple single-stage sludge digestion tank consists of an inlet pipe, outlet pipes for removing both the digested sludge and the clear supernatant liquid, a dome, an outlet pipe for collecting and removing the digester gas, and a series of heating coils for circulating hot water.

In a single-stage, floating-cover digester, sludge is brought into the tank at the top. The contents of the digester stratify into four layers: scum on top, clear supernatant, a layer of actively digesting sludge, and a bottom layer of concentrated sludge. Some of the contents may be withdrawn, heated, and returned in order to maintain a proper digestion temperature.

Figure 30.7 *Simple Anaerobic Digester*

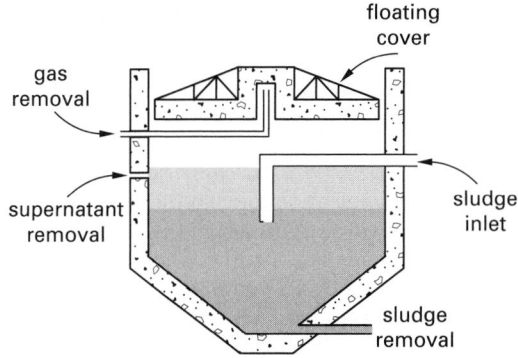

Supernatant is removed along the periphery of the digester and returned to the input of the processing plant. Digested sludge is removed from the bottom and dewatered prior to disposal. Digester gas is removed from the gas dome. Heat from burning the methane can be used to warm the sludge that is withdrawn or to warm raw sludge prior to entry.

A reasonable loading and well-mixed digester can produce a well-stabilized sludge in 15 days. With poorer mixing, 30 to 60 days may be required.

A single-stage digester performs the functions of digestion, gravity thickening, and storage in one tank. In a two-stage process, two digesters in series are used. Heating and mechanical mixing occur in the first digester. Since the sludge is continually mixed, it will not settle. Settling and further digestion occur in the unheated second tank.

Germany began experimenting with large spherical and egg-shaped digesters in the 1950s. These and similar designs are experiencing popularity in the United States. Volumes are on the order of 100,000 to 400,000 ft^3 (3000 to 12,000 m^3), and digesters can be constructed from either prestressed concrete or steel. Compared with conventional cylindrical digesters, the egg-shaped units minimize problems associated with the accumulation of solids in the lower corners and collection of foam and scum on the upper surfaces. The shape also promotes better mixing that enhances digestion. However, variations in the shape greatly affect grit collection and required mixing power. Mixing can be accomplished with uncombined gas mixing, mechanical mixing using an impeller and draft-tube, and pumped recirculation. Typical volatile suspended solids loading rates are 0.062 to 0.17 lbm VSS/ft^3-day (1 to 2.8 kg VSS/m^3·day) with a 15 to 25 day retention time. Operational problems include grit accumulation, gas leakage, and foaming.

19. METHANE PRODUCTION

Wastes that contain biologically degradable organics and are free of substances that are toxic to microorganisms (e.g., acids) will be biodegraded. Methane gas is a product of anaerobic degradation. Under anaerobic conditions, degradation of organic wastes yields methane. Inorganic wastes are not candidates for biological treatment and do not produce methane.

The theoretical volume of methane produced from sludge digestion is 5.61 ft^3 (0.35 m^3) per pound (per kilogram) of BOD converted. Wasted sludge does not contribute to methane production, and the theoretical amount is reduced by various inefficiencies. Therefore, the actual volume of methane produced is predicted by Eq. 30.51. The total volume of digester gas is larger than the methane volume alone since methane constitutes about $2/3$ of the total volume. E is the efficiency of waste utilization, typically 0.6 to 0.9. S_o is the ultimate BOD of the influent. ES_o can be replaced with $S_o - S$, as the two terms are equivalent.

$$V_{methane, m^3/day} = \left(0.35\,\frac{m^3}{kg}\right)\left(\frac{ES_{o,mg/L}Q_{m^3/day}}{1000\,\frac{g}{kg}} - 1.42P_{x,kg/day}\right)$$

$$\text{[SI]} \quad 30.51(a)$$

$$V_{methane, ft^3/day} = \left(5.61\,\frac{ft^3}{lbm}\right)\left(\begin{array}{c}(ES_{o,mg/L}Q_{MGD}) \\ \times \left(8.345\,\frac{lbm\text{-}L}{MG\text{-}mg}\right) \\ - 1.42P_{x,lbm/day}\end{array}\right)$$

$$\text{[U.S.]} \quad 30.51(b)$$

The heating value of digester gas listed in Table 30.10 is approximately 600 Btu/ft^3 (22 MJ/m^3). This value is appropriate for digester gas with the composition given: 65% methane and 35% carbon dioxide by volume. Carbon dioxide is not combustible and does not contribute to the heating effect. The actual heating value will depend on the volumetric fraction of methane. The temperature and pressure also affect the gas volume. At 60°F (15.6°C) and 1 atm, the pure methane has a lower heating value of 913 Btu/ft^3 (34.0 MJ/m^3). The higher heating value of 1013 Btu/ft^3 (37.7 MJ/m^3) should not normally be used, since the water vapor produced during combustion is not allowed to condense out.

The total heating energy is

$$q = (LHV)V_{fuel} \qquad 30.52$$

Table 30.10 *Typical Characteristics of Single-Stage Heated Anaerobic Digester*

minimum number of units	2
size	
trickling filter sludge	3–4 ft^3/person (0.08–0.11 m^3/person)
primary and secondary sludge	6 ft^3 (0.16 m^3/person)
volatile solids loading	0.13–0.2 lbm/day-ft^3 (2.2–3.3 kg/day·m^3)
temperature	85–100°F; 95–98°F optimum (29–38°C; 35–36.7°C optimum)
pH	6.7–7.8; 6.9–7.2 optimum
gas production	7–10 ft^3/lbm volatile solids (0.42–0.60 m^3/kg volatile solids)
	0.5–1.0 ft^3/day-person (0.014–0.027 m^3/day·person)
gas composition	65% methane; 35% carbon dioxide
gas heating value	500–700; 600 Btu/ft^3 typical (19–26 MJ/m^3; 22 MJ/m^3)
sludge final moisture content	90–95%
sludge detention time	30–90 days (conventional) 15–25 days (high rate)
depth	20–45 ft (6–14 m)
minimum freeboard	2 ft (0.6 m)

20. HEAT TRANSFER AND LOSS

In an adiabatic reactor, energy is required only to heat the influent sludge to the reactor temperature. In a nonadiabatic reactor, energy will also be required to replace energy lost to the environment through heat transfer from the exposed surfaces.

The energy required to increase the temperature of sludge from T_1 to T_2 is given by Eq. 30.53. The specific heat, c_p, of sludge is assumed to be the same as for water, 1.0 Btu/lbm-°F (4190 J/kg·°C).

$$q = m_{sludge}c_p(T_2 - T_1)$$
$$= V_{sludge}\rho_{sludge}c_p(T_2 - T_1) \qquad 30.53$$

The energy lost from a heated digester can be calculated in several ways. If the surface temperature is known, the convective heat lost from a heated surface to the surrounding air depends on the convective outside *film coefficient*, h. The film coefficient depends on the surface temperature, air temperature, orientation, and prevailing wind. Heat losses from digester sides, top, and other surfaces must be calculated separately and combined.

$$q = hA_{surface}(T_{surface} - T_{air}) \qquad 30.54$$

If the temperature of the digester contents is known, it can be used with the *overall coefficient of heat transfer,*

U (a combination of inside and outside film coefficients), as well as the conductive resistance of the digester walls to calculate the heat loss.

$$q = UA_{\text{surface}}(T_{\text{contents}} - T_{\text{air}}) \qquad 30.55$$

If the temperatures of the inner and outer surfaces of the digester walls are known, the energy loss can be calculated as a conductive loss. k is the average thermal conductivity of the digester wall, and L is the wall thickness.

$$q = \frac{kA(T_{\text{inner}} - T_{\text{outer}})}{L} \qquad 30.56$$

21. SLUDGE DEWATERING

Once sludge has been thickened, it can be digested or dewatered prior to disposal. At water contents of 75%, sludge can be handled with shovel and spade, and is known as *sludge cake*. A 50% moisture content represents the general lower limit for conventional drying methods. Several methods of dewatering are available, including vacuum filtration, pressure filtration, centrifugation, sand and gravel drying beds, and lagooning.

Generally, vacuum filtration and centrifugation are used with undigested sludge. A common form of vacuum filtration occurs in a *vacuum drum filter*. A hollow drum covered with filter cloth is partially immersed in a vat of sludge. The drum turns at about 1 rpm. Suction is applied from within the drum to attract sludge to the filter surface and to extract moisture. The sludge cake is dewatered to about 75 to 80% moisture content (solids content of 20 to 25%) and is then scraped off the belt by a blade or delaminated by a sharp bend in the belt material. This degree of dewatering is sufficient for sanitary landfill. A higher solids content of 30 to 33% is needed, however, for direct incineration unless external fuel is provided. Chemical flocculants (e.g., polymers or ferric chloride, with or without lime) may be used to collect finer particles on the filter drum. Performance is measured in terms of the amount of dry solids removed per unit area (kg/m²·hr). Typical performance is 3.5 lbm/hr-ft² (17 kg/hr·m²).

Belt filter presses accept a continuous feed of conditioned sludge and use a combination of gravity drainage and mechanically applied pressure. Sludge is applied to a continuous-loop woven belt (cotton, wool, nylon, polyester, woven stainless steel, etc.) where the sludge begins to drain by gravity. As the belt continues on, the sludge is compressed under the action of one or more rollers. Belt presses typically have belt widths of 80 to 140 in (2.0 to 3.5 m) and are loaded at the rate of 60 to 450 lbm/ft-hr (90 to 680 kg/m·h). Moisture content of the cake is typically 70 to 80%.

Pressure filtration is usually chosen only for sludges with poor dewaterability (such as waste-activated sludge) and where it is desired to dewater to a solids content higher than 30%. *Recessed plate filter presses*

operate by forcing sludge at pressures of up to 120 psi (830 kPa) into cavities of porous filter cloth. The filtrate passes through the cloth, leaving the filter cake behind. When the filter becomes filled with cake, it is taken offline and opened. The filter cake is either manually or automatically removed.

Filter press sizing is based on a continuity of mass. In Eq. 30.57, s is the solids content as a decimal fraction in the filter cake.

$$\begin{aligned}
V_{\text{press}}&\rho_{\text{filter cake}}s_{\text{filter cake}} \\
&= V_{\text{sludge,per cycle}}s_{\text{sludge}}\rho_{\text{sludge}} \\
&= V_{\text{sludge,per cycle}}s_{\text{sludge}}\rho_{\text{water}}\text{SG}_{\text{sludge}} \qquad 30.57
\end{aligned}$$

Centrifuges can operate continuously and reduce water content to about 70%, but the effluent has a high percentage of suspended solids. The types of centrifuges used include solid bowl (both cocurrent and countercurrent varieties) and imperforate basket decanter centrifuges. With *solid bowl centrifuges*, sludge is continuously fed into a rotating bowl where it is separated into a dilute stream (the *centrate*) and a dense cake consisting of approximately 70 to 80% water. Since the centrate contains fine, low-density solids, it is returned to the wastewater treatment system. Equation 30.58 gives the number of gravities acting on a rotating particle. Rotational speed is typically 1500 to 2000 rpm. A polymer may be added to increase the capture efficiency (same as reducing the solids content of the centrate) and, accordingly, increase the feed rate.

$$G = \frac{\omega^2 r}{g} \qquad 30.58$$

Sand drying beds are preferable when digested sludge is to be disposed of in a landfill since sand beds produce sludge cake with a low moisture content. Drying beds are enclosed by low concrete walls. A broken stone/gravel base 8 to 20 in (20 to 50 cm) thick is covered by 4 to 8 in (10 to 20 cm) of sand. Drying occurs primarily through drainage, not evaporation. Water that seeps through the sand is removed by a system of drainage pipes. Typically, sludge is added to a depth of about 8 to 12 in (20 to 30 cm) and allowed to dry for 10 to 30 days.

There is no theory-based design method, and drying beds are sized by rules of thumb. The drying area required is approximately 1 to 1.5 ft² (0.09 to 1.4 m²) per person for primary digested sludge, 1.25 to 1.75 ft² (0.11 to 0.16 m²) per person for primary and trickling filter digested sludge, 1.75 to 2.5 ft² (0.16 to 0.23 m²) per person for primary and waste-activated digested sludge, and 2 to 2.5 ft² (0.18 to 0.23 m²) per person for primary and chemically precipitated digested sludge. This area can be reduced by approximately 50% if transparent covers are placed over the beds, creating a "greenhouse effect" and preventing the spread of odors.

Other types of drying beds include paved beds (both drainage and decanting types), vacuum-assisted beds, and artificial media (e.g., stainless steel wire and high-density polyurethane panels) beds.

Sludge dewatering by freezing and thawing in exposed beds is effective during some months in parts of the country.

Example 30.5

A belt filter press handles an average flow of 19,000 gal/day of thickened sludge containing 2.5% solids by weight. Operation is 8 hr/day and 5 days/wk. The belt filter press loading is specified as 500 lbm/m-hr (lbm/hr per meter of belt width). The dewatered sludge is to have 25% solids by weight; the suspended solids concentration in the filtrate is 900 mg/L. The belt is continuously washed on its return path at the rate of 24 gal/m-min. The specific gravities of the sludge feed, dewatered cake, and filtrate are 1.02, 1.07, and 1.01 respectively. (a) What size belt is required? (b) What is the volume of filtrate rejected? (c) What is the solids capture fraction?

Solution

(a) The daily masses of wet and dry influent sludge are

$$m_{\text{wet}} = V\rho_{\text{sludge}} = V\rho_{\text{water}}\text{SG}_{\text{sludge}}$$

$$= \left(19{,}000 \ \frac{\text{gal}}{\text{day}}\right)\left(\frac{\left(62.4 \ \frac{\text{lbm}}{\text{ft}^3}\right)(1.02)}{7.48 \ \frac{\text{gal}}{\text{ft}^3}}\right)$$

$$= 1.617 \times 10^5 \ \text{lbm/day}$$

$$m_{\text{dry}} = (\text{fraction solids})m_{\text{wet}}$$

$$= (0.025)\left(1.617 \times 10^5 \ \frac{\text{lbm}}{\text{day}}\right)$$

$$= 4042 \ \text{lbm/day}$$

Since the influent is received continuously but the belt filter press operates only 5 days a week for 8 hours per day, the average hourly processing rate is

$$\frac{\left(4042 \ \frac{\text{lbm}}{\text{day}}\right)\left(\frac{7 \ \text{days}}{5 \ \text{days}}\right)}{8 \ \frac{\text{hr}}{\text{day}}} = 707 \ \text{lbm/hr}$$

The belt size required is

$$w = \frac{707 \ \frac{\text{lbm}}{\text{hr}}}{500 \ \frac{\text{lbm}}{\text{m-hr}}} = 1.41 \ \text{m}$$

Use one 1.5 m belt filter press. A second standby unit may also be specified.

(b) The mass of solids received each day is

$$\left(707 \ \frac{\text{lbm}}{\text{hr}}\right)\left(8 \ \frac{\text{hr}}{\text{day}}\right) = 5656 \ \text{lbm/day}$$

The fractional solids in the filtrate is

$$s_{\text{filtrate}} = \frac{900 \ \frac{\text{mg}}{\text{L}}}{\left(1000 \ \frac{\text{mg}}{\text{g}}\right)\left(1000 \ \frac{\text{g}}{\text{L}}\right)} = 0.0009$$

The solids balance is

$$\text{influent solids} = \text{sludge cake solids} + \text{filtrate solids}$$

$$5656 \ \frac{\text{lbm}}{\text{day}} = \left(\frac{Q_{\text{cake,gal/day}}}{7.48 \ \frac{\text{gal}}{\text{ft}^3}}\right)\left(62.4 \ \frac{\text{lbm}}{\text{ft}^3}\right)$$

$$\times \ (1.07)(0.25)$$

$$+ \left(\frac{Q_{\text{filtrate,gal/day}}}{7.48 \ \frac{\text{gal}}{\text{ft}^3}}\right)\left(62.4 \ \frac{\text{lbm}}{\text{ft}^3}\right)$$

$$\times \ (1.01)(0.0009)$$

The total liquid flow rate is the sum of the sludge flow rate and the washwater flow rate.

$$Q_{\text{total}} = Q_{\text{cake}} + Q_{\text{filtrate}}$$

$$= \left(19{,}000 \ \frac{\text{gal}}{\text{day}}\right)\left(\frac{7 \ \text{days}}{5 \ \text{days}}\right)$$

$$+ \left(24 \ \frac{\text{gal}}{\text{m-min}}\right)(1.5 \ \text{m})\left(60 \ \frac{\text{min}}{\text{hr}}\right)\left(8 \ \frac{\text{hr}}{\text{day}}\right)$$

$$= 43{,}880 \ \text{gal/day}$$

Solving the solids and liquid flow rate equations simultaneously,

$$Q_{\text{cake}} = 2396 \ \text{gal/day}$$

$$Q_{\text{filtrate}} = 41{,}484 \ \text{gal/day}$$

(c) The solids in the filtrate are

$$m_{\text{filtrate}} = \left(\frac{41{,}484 \ \frac{\text{gal}}{\text{day}}}{7.48 \ \frac{\text{gal}}{\text{ft}^3}}\right)$$

$$\times \left(62.4 \ \frac{\text{lbm}}{\text{ft}^3}\right)(1.01)(0.0009)$$

$$= 315 \ \text{lbm/day}$$

The solids capture fraction is

$$\frac{\text{influent solids} - \text{filtrate solids}}{\text{influent solids}}$$

$$= \frac{5656 \ \frac{\text{lbm}}{\text{day}} - 315 \ \frac{\text{lbm}}{\text{day}}}{5656 \ \frac{\text{lbm}}{\text{day}}}$$

$$= 0.944$$

22. SLUDGE DISPOSAL

Liquid sludge and sludge cake disposal options are limited to landfills; incineration; nonagricultural land application such as strip-mine reclamation; agricultural land application; commercial applications (known as *distribution and marketing* and *beneficial reuse*) such as composting, special monofills, and surface impoundments; and various state-of-the-art methods. Different environmental contaminants are monitored with each disposal method, making the options difficult to evaluate.

If a satisfactory site is available, sludge, sludge cake, screenings, and scum can be discarded in municipal *landfills*. Approximately 40% of the sludge produced in the United States is disposed of in this manner. This has traditionally been the cheapest disposal method, even when transportation costs are included.

The sludge composition and its potential effects on landfill leachate quality must be considered. Sludge with minute concentrations of monitored substances (e.g., heavy metals such as cadmium, chromium, copper, lead, nickel, and zinc) may be classified as hazardous waste, requiring disposal in hazardous waste landfills.

Sludge-only *monofills* and other surface impoundments may be selected in place of municipal landfills. Monofills are typically open pits protected by liners, slurry cutoff walls, and levees (as protection against floods). Monofills may include a pug mill to combine sludge with soil from the site.

Approximately 25% of the sludge produced in the United States is disposed of in a *beneficial reuse program* such as surface spreading (i.e., "land application") and/or composting. Isolated cropland, pasture, and forest land (i.e., "silvaculture") is perfect for surface spreading. However, sludge applied to the surface may need to be harrowed into the soil soon after spreading to prevent water pollution from sludge runoff and possible exposure to or spread of bacteria. Alternatively, the sludge can be injected 12 to 18 in (300 to 460 mm) below grade directly from distribution equipment. The actual application rate depends on sludge strength (primarily nitrogen content), the condition of the receiving soil, the crop on plants using the applied nutrients, and the spreading technology.

By itself, filter cake is not highly useful as a fertilizer since its nutrient content is low. However, it is a good filler and soil conditioner for more nutrient-rich fertilizers. Other than as filler, filter cake has few practical applications.

Where sludges have low metals contents, *composting* (in static piles or vessels) can be used. *Compost* is a dry, odor-free soil conditioner with applications in turf grass, landscaping, land reclamation, and other horticultural industries. These markets can help to defray the costs of the program. Control of odors, cold weather operations, and concern about airborne pathogens are problems with *static pile composting*. Expensive air-quality scrubbers add to the cost of processing. With *in-vessel composting*, dewatered raw sludge is mixed with a bulking material, placed in an enclosed tank, and mixed with air in the tank.

Incineration is the most expensive treatment, though it results in almost total destruction of volatile solids, is odor free, and is independent of the weather. Approximately 20% of sludge is disposed of in this manner. A solids content of 30 to 33% is needed for direct incineration without the need to provide external fuel (i.e., so that the sludge burns itself). Air quality controls, equipment, and the regulatory permitting needed to meet strict standards may make incineration prohibitively expensive in some areas. The incineration ash produced represents a disposal problem of its own.

State-of-the-art options include using sludge solids in bricks, tiles, and other building materials; below-grade injection; subsurface injection wells; multiple-effect (Carver-Greenfield) evaporation; combined incineration with solid waste in "mass burn" units; and fluidized-bed gasification. The first significant installation attempting to use sludge solids as powdered fuel for power generation was the 25 MW Hyperion Energy Recovery System at the Hyperion treatment plant in Los Angeles. This combined-cycle cogeneration plant unsuccessfully attempted to operate using the Carver-Greenfield process and a fluidized-bed gasifier to convert sludge solids to fuel.

Ocean dumping, once a popular option for coastal areas, has been regulated out of existence and is no longer an option for new or existing facilities.

Environmental

31 Municipal Solid Waste

Nomenclature

a	distance	ft	m
A	area	ft^2	m^2
b	distance	ft	m
B	volumetric fraction	–	–
C	concentration	mg/L	mg/L
CF	compaction factor	–	–
D	depth of flow	ft	m
g	acceleration of gravity	ft/sec^2	m/s^2
g_c	gravitational constant	ft-lbm/ lbf-sec^2	–
G	MSW generation rate, per capita	lbm/day	kg/day
H	total hydraulic head	ft	m
HRR	heat release rate	Btu/hr-ft^2	W/m^2
HV	heating value	Btu/lbm	J/kg
i	hydraulic gradient	ft/ft	m/m
K	hydraulic conductivity	ft/hr	m/s
K_s	solubility	–	L/L
L	spacing or distance	ft	m
LF	loading factor	–	–
m	mass	lbm	kg
MW	molecular weight	lbm/lbmole	g/mol
N	population size	–	–
O	overlap	–	–
p	pressure	lbf/ft^2	Pa
Q	flow rate	ft^3/sec	m^3/s
Q	recharge rate	ft^3/ft^2-sec	m^3/m^2·s
R	radius of influence	ft	m
R^*	universal gas constant	ft-lbf/ lbmole-°R	atm·L/ mol·K
SFC	specific feed characteristics	Btu/lbm	J/kg
t	time	sec	s
T	absolute temperature	°R	K
V	volume	ft^3	m^3
x	mass fraction	–	–

Symbols

γ	specific weight	lbf/ft^3	–
ρ	density	lbm/ft^3	kg/m^3

Subscripts

c	compacted
i	gas component i
MSW	municipal solid waste
o	original
s	solubility
t	total

1. MUNICIPAL SOLID WASTE

Municipal solid waste (MSW, known for many years as *"garbage"*) consists of the solid material discarded by a community, including excess food, containers and packaging, residential garden wastes, other household discards, and light industrial debris. Hazardous wastes, including paints, insecticides, lead car batteries, used crankcase oil, dead animals, raw animal wastes, and radioactive substances present special disposal problems and are not included in MSW.

MSW is generated at the average rate of approximately 5 to 8 lbm/capita-day (2.3 to 3.6 kg/capita·day). Five lbm/capita-day (2.3 kg/capita·day) is commonly used in design studies.

Although each community's MSW has its own characteristics, Table 31.1 gives average composition in the United States. The moisture content of MSW is approximately 20%.

Environmental

Table 31.1 *Typical Characteristics of MSW[a,b]*

component	percentage by weight
paper, cardboard	35.6
yard waste	20.1
food waste	8.9
metal	8.9
glass	8.4
plastics	7.3
wood	4.1
rubber, leather	2.8
cloth	2.0
other (ceramic, stone, dirt)	1.9

[a]United States national average
[b]Dry composition

2. LANDFILLS

Municipal solid waste has traditionally been disposed of in *municipal solid waste landfills* (MSWLs, for many years referred to as "*dumps*"). Other wastes, including incineration ash and water treatment sludge, may be included with MSW in some landfills.

The fees charged to deposit waste are known as *tipping fees*. Tipping fees are generally higher in the eastern United States (where landfill sites are becoming scarcer). Tipping fees for resource recovery plants are approximately double those of traditional landfills.

Many early landfills were not designed with liners, not even simple compacted-soil bottom layers. Design was often by rules of thumb, such as the bottom of the landfill had to be at least 5 ft (1.5 m) above maximum elevation of groundwater and 5 ft (1.5 m) above bedrock. Percolation through the soil was believed sufficient to make the leachate bacteriologically safe. However, inorganic pollutants can be conveyed great distances, and the underlying soil itself can no longer be used as a barrier to pollution and to perform filtering and cleansing functions.

Landfills without bottom liners are known as *natural attenuation* (NA) *landfills*, since the native soil is used to reduce the concentration of leachate components. Landfills lined with clay (pure or bentonite-amended soil) or synthetic liners are known as *containment landfills*.

Landfills can be incrementally filled in a number of ways. With the *area method*, solid waste is spread and compacted on flat ground before being covered. With the *trench method*, solid waste is placed in a trench and covered with the trench soil. In the *progressive method* (also known as the *slope/ramp method*), cover soil is taken from the front (toe) of the working face.

Landvaults are essentially above-ground piles of MSW placed on flat terrain. This design, with appropriate covers, liners, and instrumentation, has been successfully used with a variety of industrial wastes, including incineration ash and wastewater sludge. However, except when piggybacking a new landfill on top of a closed landfill, it is not widely used with MSW.

Waste is placed in layers, typically 2 to 3 ft thick (0.6 to 0.9 m), and is compacted before soil is added as a cover. Two to five passes by a tracked bulldozer are sufficient to compact the MSW to 800 to 1500 lbm/yd^3 (470 to 890 kg/m^3). (A density of 1000 lbm/yd^3 or 590 kg/m^3 is used in design studies.) The *lift* is the height of the covered layer, as shown in Fig. 31.2. When the landfill layer has reached full height, the ratio of solid waste volume to soil cover volume will be approximately between 4:1 and 3:1.

Figure 31.1 *Landfill Creation Methods*

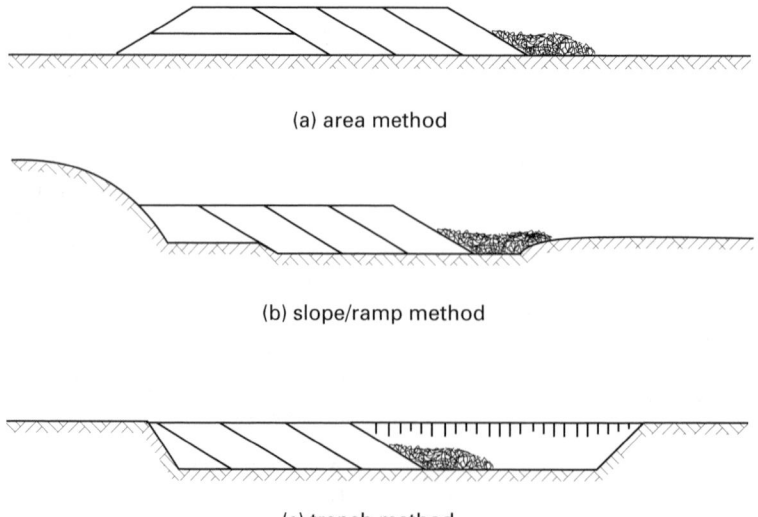

(a) area method

(b) slope/ramp method

(c) trench method

The *cell height* is typically taken as 8 ft (2.4 m) for design studies, although it can actually be much higher. The height should be chosen to minimize the cover material requirement consistent with the regulatory requirements. Cell slopes will be less than 40°, and typically 20° to 30°.

Figure 31.2 *Landfill Cells*

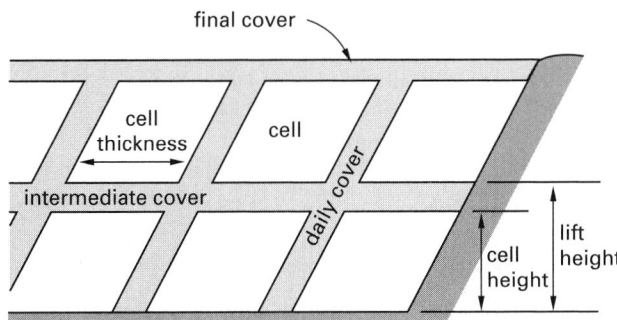

Each *cell* is covered by a soil layer 0.5 to 1.0 ft (0.15 to 0.3 m) in thickness. The final top and side covers should be at least 2 ft (0.6 m) thick. Daily, intermediate, and final soil covers are essential to proper landfill operation. The daily and intermediate covers prevent fly emergence, discourage rodents, reduce odors, and reduce wind-blown litter. In the event of ignition, the soil layers form fire stops within the landfill.

The final soil cover (cap) is intended to reduce or eliminate entering moisture. The soil cover also contains and channels landfill gas and provides a pleasing appearance and location for growing vegetation.

3. LANDFILL CAPACITY

The *compaction factor*, CF, is a multiplicative factor used to calculate the compacted volume for a particular type of waste and compaction method. The compacted volume of MSW or components within the waste is

$$V_c = (CF)V_o \qquad 31.1$$

The daily increase in landfill volume is predicted by Eq. 31.2. N is the population size, G is the per-capita MSW generation rate, and γ is the landfill overall specific weight calculated as a weighted average of soil and compacted MSW specific weights. Soils have specific weights between 70 and 130 lbf/ft³ (densities between 1100 and 2100 kg/m³). A value of 100 lbf/ft³ (1600 kg/m³) is commonly used in design studies.

$$\Delta V_{\text{day}} = \frac{NG(\text{LF})}{\gamma} = \frac{NG(\text{LF})}{\rho g} \qquad [\text{SI}] \qquad 31.2(a)$$

$$\Delta V_{\text{day}} = \frac{NG(\text{LF})}{\gamma} = \frac{NG(\text{LF})g_c}{\rho g} \qquad [\text{U.S.}] \qquad 31.2(b)$$

The *loading factor*, LF, is 1.25 for a 4:1 volumetric ratio and is calculated from Eq. 31.3 for other ratios.

$$\text{LF} = \frac{V_{\text{MSW}} + V_{\text{cover soil}}}{V_{\text{MSW}}} \qquad 31.3$$

Most large-scale sanitary landfills do not apply daily cover to deposited waste. Time, cost, and reduced capacity are typically cited as the reasons that the landfill is not covered with soil. In the absence of such cover, the loading factor has a value of 1.0.

4. SUBTITLE D LANDFILLS

Since 1992, new and expanded municipal landfills in the United States must satisfy strict design regulations and are designated "*Subtitle D landfills*," referring to Subtitle D of the Resource Conservation and Recovery Act (RCRA). The regulations apply to any landfill designed to hold municipal solid waste, biosolids, and ash from MSW incineration.

Construction of landfills is prohibited near sensitive areas such as airports, floodplains, wetlands, earthquake zones, and geologically unstable terrain. Air quality control methods are required to control emission of dust, odors (from hydrogen sulfide and volatile organic vapors), and landfill gas. Runoff from storms must be controlled.

Subtitle D landfills must have *double-liner systems*, defined as "two or more liners and a leachate collection system above and between such liners." While states can specify greater protection, the minimum (basic) bottom layer requirements are a 30-mil flexible PVC membrane liner (FML) and at least 2 ft of compacted soil with a maximum hydraulic conductivity of 1.2×10^{-5} ft/hr (1×10^{-7} cm/s). If the membrane is high-density polyethylene (HDPE), the minimum thickness is 60 mils. 30-mil PVC costs less than 60-mil HDPE, but the 60-mil product offers superior protection.

The preferred double liner consists of two FMLs separated by a drainage layer (approximately 0.5 m thick) of sand, gravel, or *drainage netting* and placed on low-permeability soil. This is the standard design for *Subtitle C* hazardous waste landfills. The advantage of selecting the preferred design for municipal landfills is realized in the permitting process.

The minimum thickness is 20 mils for the top FML, and the maximum hydraulic conductivity of the cover soil is 1.8×10^{-5} ft/hr (1.5×10^{-7} cm/s).

A series of wells and other instrumentation are required to detect high hydraulic heads and the accumulation of heavy metals and volatile organic compounds (VOCs) in the leachate.

Environmental

Figure 31.3 *Subtitle D Liner and Cover Detail*

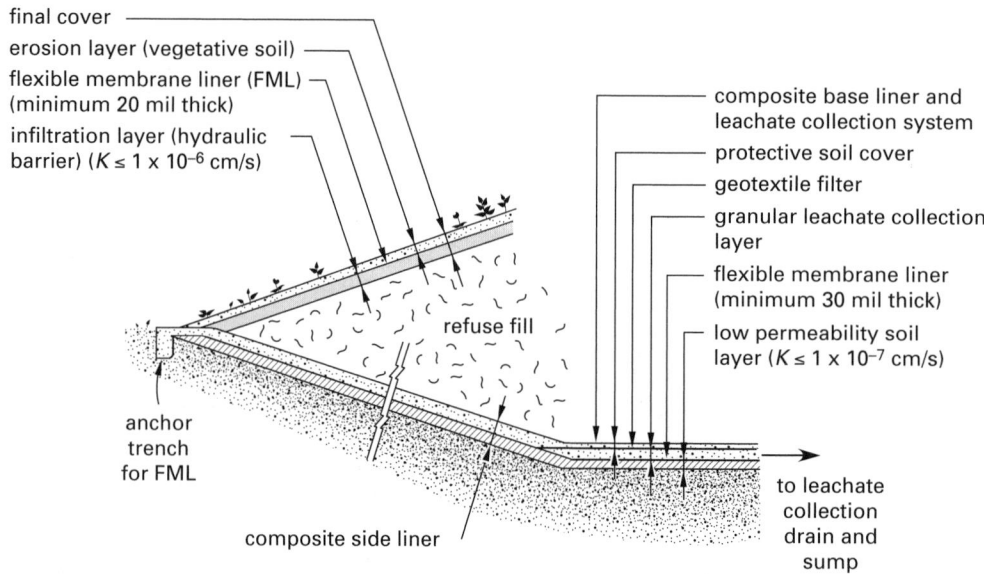

5. CLAY LINERS

Liners constructed of well-compacted bentonite clays fail by fissuring (as during freeze-thaw cycles), dessication (i.e., drying out), chemical interaction, and general disruption. Permeability increases dramatically with organic fluids (hydrocarbons) and acidic or caustic leachates. Clays need to be saturated and swollen to retain their high impermeabilities. Permeabilities measured in the lab as 1.2×10^{-5} ft/hr (10^{-7} cm/s) can actually be 1.2×10^{-3} ft/hr (10^{-5} cm/s) in the field.

Although a 0.5 to 1.0 ft (0.15 to 0.30 m) thickness could theoretically contain the leachate, the probability of success is not high when typical exposure issues are considered. Clay liners with thicknesses of 4 to 5 ft (1.2 to 1.5 m) will still provide adequate protection, even when the top half has degraded.

6. FLEXIBLE MEMBRANE LINERS

Synthetic *flexible membrane liners* (FMLs, the term used by the United States EPA), also known as *synthetic membranes, synthetic membrane liners* (SMLs), *geosynthetics, geomembranes,* and liners are highly attractive because of their negligible permeabilities, and good chemical resistance. The term *geocomposite liner* is used when referring to a combination of FML and one or more compacted clay layers.

FML materials include polyvinyl chloride (PVC); chlorinated polyethylene (CPE); low-density polyethylene (LDPE); linear, low-density polyethylene (LLDPE); very low density polyethylene (VLDPE); high-density polyethylene (HDPE); chlorosulfonated polyethylene, also known as Hypalon® (CSM or CSPE); ethylene

propylene dienemonomer (EP or EPDM); polychloroprene, also known as neoprene (CR); isobutylene isoprene, also known as butyl (IIR); and oil resistant polyvinyl chloride (ORPVC). LDPE covers polyethylene with densities in the range of about 0.900 to 0.935 g/cm³, while HDPE covers the range from 0.935 to 0.970 g/cm³.

Currently, HDPE is preferred for municipal landfills and hazardous waste sites because of its high tensile strength, high toughness (resistance to tear and puncture), and ease of seaming. HDPE has captured the majority of the FML market.

Some FMLs can be manufactured with an internal nylon or dacron mesh, known as a *scrim*, or a spread coating on one side. The resulting membrane is referred to as a *reinforced* or *supported membrane*. An "R" is added to the name to indicate the reinforcement (e.g., "HDPE-R"). Membranes without internal scrims are *unreinforced* or *unsupported membranes*.

FMLs can be embossed or textured to increase the friction between the FML and any soil layer adjacent to it.

FMLs are sealed to the original grade and restrained against movement by deep *anchor trenches* around the periphery of the landfill.

FMLs fail primarily by puncturing, tearing, and long-term exposure to ultraviolet (UV) rays. Freezing temperatures may also weaken them. FMLs are protected from UV radiation by a soil backfill with a minimum thickness of approximately 6 in (15 cm). The cover soil should be placed promptly.

7. LANDFILL CAPS

The landfill *cap* (*top cap*) is the *final cover* placed over the landfill. Under Subtitle D regulations, caps cannot be any more permeable than the bottom liners. Therefore, only preexisting landfills can be capped with clay.

Caps fail by desiccation, penetration by roots, rodent and other animal activity, freezing and thawing cycles, erosion, settling, and breakdown of synthetic liners from sunlight exposure. Failure can also occur when the cover collapses into the disposal site due to voids opening below or excess loading (rainwater ponding) above.

Synthetic FML caps offer several advantages over clay caps—primarily lower permeability, better gas containment, faster installation, and easier testing and inspection.

Preferred landfill design incorporates maximum slopes of 1:1 or 1:1.5 (horizontal:vertical), with horizontal tiers (benches) every so often. However, many landfills have slopes as steep as 1:3 (horizontal:vertical). This is steep enough to cause concern about cap slope instability. Although clay caps can be constructed this steep, synthetic membranes have lower *interfacial friction angles* and are too smooth for use on steep slopes. Slippage between different layers in the cap (e.g., between the FML and the soil cover) is known as *veneer failure*. Some relief is possible with *textured synthetics*.

A common cap system is a textured synthetic with a nonwoven geotextile. Vertical cap collapse can be prevented by incorporating a high-strength plastic *geogrid*.

Table 31.2 Typical Interfacial Friction Angles

interface	typical friction angle
nonwoven geotextile/smooth FML	10°–13°
compacted sand/smooth FML	15°–25°
compacted clay/smooth FML	5°–20°
nonwoven geotextile/textured FML	25°–35°
compacted sand/textured FML	20°–35°
compacted clay/textured FML	15°–32°

8. LANDFILL SITING

Sanitary landfills should be designed for a 5 yr minimum life. Sanitary landfill sites are selected on the basis of many relevant factors, including (a) economics and the availability of inexpensive land; (b) location, including ease of access, transport distance, acceptance by the population, and aesthetics; (c) availability of cover soil, if required; (d) wind direction and speed, including odor, dust, and erosion considerations; (e) flat topography; (f) dry climate and low infiltration rates; (g) location (elevation) of the water table; (h) low risk of aquifer contamination; (i) types and permeabilities of underlying strata; (j) avoidance of winter freezing; (k) future growth and capacity requirements; and (l) ultimate use.

Suitable landfill sites are becoming difficult to find, and once identified, they are subjected to a rigorous permitting process. They are also objected to by residents near the site and MSW transport corridor. This is referred to as the *NIMBY syndrome* (not in my backyard). People agree that landfills are necessary, but they do not want to live near them.

9. VECTORS

In landfill parlance, a *vector* is an insect or arthropod, rodent, or other animal of public health significance capable of causing human discomfort or injury, or capable of harboring or transmitting the causative agent of human disease. Vectors include flies, domestic rats, field rodents, mosquitoes, wasps, and cockroaches. Standards for vector control should state the vector thresholds (e.g., "six or more flies per square yard").

10. ULTIMATE LANDFILL DISPOSITION

When filled to final capacity and closed, the covered landfill can be used for grassed green areas, shallow-rooted agricultural areas, and recreational areas (soccer fields, golf, etc.). Light construction uses are also possible, although there may be problems with gas accumulation, corrosion of pipes and foundation piles, and settlement.

Settlement will generally be uneven. Settlement in areas with high rainfall can be up to 20% of the overall landfill height. In dry areas, the settlement may be much less, 2 to 3%. Little information on bearing capacity of landfills is available. Some studies have suggested capacities of 500 to 800 lbf/ft^2 (24 to 38 kPa).

11. LANDFILL GAS

Once covered, the organic material within a landfill cell provides an ideal site for decomposition. Aerobic conditions prevail for approximately the first month. Gaseous products of the initial aerobic decomposition include CO_2 and H_2O. The following are typical reactions involving carbohydrates and stearic acid.

$$C_6H_{12}O_6 + 6O_2 \rightarrow 6CO_2 + 6H_2O \qquad 31.4$$
$$C_{18}H_{36}O_2 + 26O_2 \rightarrow 18CO_2 + 18H_2O \qquad 31.5$$

After the first month or so, the decomposing waste will have exhausted the oxygen from the cell. Digestion continues anaerobically, producing a low-heating value *landfill gas* (LFG) consisting of approximately 50% methane (CH_4), 50% carbon dioxide, and trace amounts of other gases (e.g., CO, N_2, and H_2S). Decomposition occurs at temperatures of 100 to 120°F (40 to 50°C), but may increase to up to 160°F (70°C).

$$C_6H_{12}O_6 \rightarrow 3CO_2 + 3CH_4 \qquad 31.6$$
$$C_{18}H_{36}O_2 + 8H_2O \rightarrow 5CO_2 + 13CH_4 \qquad 31.7$$

LFG is essentially saturated with water vapor when formed. However, if the gas collection pipes are vertical or sloped, some of the water vapor will condense on the pipes and drain back into the landfill. Most of the remaining moisture is removed in condensate traps located along the gas collection system line.

From Dalton's law, the total gas pressure within a landfill is the sum of the partial pressures of the component gases. Partial pressure is volumetrically weighted and can be found from the volumetric fraction, B, of the gas.

$$p_t = p_{CH_4} + p_{CO_2} + p_{H_2O} + p_{N_2} + p_{other} \quad \text{31.8}$$
$$p_i = B_i p_t \quad \text{31.9}$$

If uncontrolled, LFG will migrate to the surface. If the LFG accumulates, an explosion hazard results, since methane is highly explosive in concentrations of 5 to 15% by volume. Thus, methane will pass through the explosive concentration as it is being diluted. Other environmental problems, including objectionable odors, can also occur. Therefore, various methods are used to prevent gas from escaping or spreading laterally.

Landfill gases can be collected by either passive or active methods. *Passive collection* uses the pressure of the gas itself as the driving force. Passive collection is applicable for sites that generate low volumes of gas and where offsite migration of gas is not expected. This generally applies to small municipal landfills—those with volumes less than approximately 50,000 yd^3 (40 000 m^3). Common passive control methods include (a) isolated gas (pressure relief) vents in the landfill cover, with or without a common flare; (b) perimeter interceptor gravel trenches; (c) perimeter barrier trenches (e.g., slurry walls); (d) impermeable barriers within the landfill; and (e) sorptive barriers in the landfill.

Active gas collection draws the landfill gas out with a vacuum from *extraction wells*. Various types of horizontal and vertical wells can be used through the landfill to extract gas in vertical movement, while perimeter facilities are used to extract gases in lateral movement. Vertical wells are usually perforated PVC pipes packed in sleeves of gravel and bentonite clay. PVC pipe is resistant to the chlorine and sulfur compounds present in the landfill.

Determining the location of isolated vents is essentially heuristic—one vent per 10,000 yd^3 (7500 m^3) of landfill is probably sufficient. Regardless, extraction wells should be spaced with overlapping zones of influence. If the *radius of influence*, R, of each well and the desired fractional overlap, O, are known, the spacing between wells is given by Eq. 31.10. For example, if the extraction wells are placed on a square grid with spacing of $L = 1.4R$, the overlap will be 60%. For a 100% overlap, the spacing would equal the radius of influence.

$$\frac{L}{R} = 2 - O \quad \text{31.10}$$

In many locations, the LFG is incinerated in flares. However, emissions from flaring are problematic. Alternatives to flaring include using the LFG to produce hot water or steam for heating or electricity generation. During the 1980s, reciprocating engines and combustion turbines powered by LFG were tried. However, such engines generated relatively high emissions of their own due to impurities and composition variations in the fuel. True Rankine-cycle power plants (generally without reheat) avoid this problem, since boilers are less sensitive to impurities.

One problem with using LFG commercially is that LFG is withdrawn from landfills at less than atmospheric pressure. Conventional furnace burners need approximately 5 psig at the boiler front. Low-pressure burners that require 2 psig are available, but they are expensive. Therefore, some of the plant power must be used to pressurize the LFG in blowers.

Although production is limited, LFG is produced for a long period after a landfill site is closed. Production slowly drops 3 to 5% annually to approximately 30% of its original value after about 20 to 25 yr, which is considered to be the economic life of a gas-reclamation system. The theoretical ultimate production of LFG has been estimated by other researchers as 15,000 ft^3 per ton (0.45 m^3/kg) of solid waste, with an estimated volumetric gas composition of 54% methane and 46% carbon dioxide. However, unfavorable and nonideal conditions in the landfill often reduce this yield to approximately 1000 to 3000 ft^3/ton (0.03 to 0.09 m^3/kg) or even lower.

Table 31.3 *Properties of Methane and Carbon Dioxide*

	CH$_4$	CO$_2$
color	none	none
odor	none	none
density, at STP		
(g/L)	0.717	1.977
(lbm/ft^3)	0.0447	0.123
specific gravity,		
at STP, ref. air	0.554	1.529
solubility (760 mm Hg,		
20°C), volumes in one		
volume of water	0.33	0.88
solubility, qualitative	slight	moderate

12. LANDFILL LEACHATE

Leachates are liquid wastes containing dissolved and finely suspended solid matter and microbial waste produced in landfills. Leachate becomes more concentrated as the landfill ages. Leachate forms from liquids brought into the landfill, water run-on, and precipitation. Leachate in a natural attenuation landfill will contaminate the surrounding soil and groundwater. In a lined containment landfill, leachate will percolate downward through the refuse and collect at the first landfill liner.

Leachate must be removed to reduce hydraulic head on the liner and to reduce unacceptable concentrations of hazardous substances.

When a layer of liquid sludge is disposed of in a landfill, the consolidation of the sludge by higher layers will cause the water to be released. This released water is known as *pore-squeeze liquid*.

Some water will be absorbed by the MSW and will not percolate down. The quantity of water that can be held against the pull of gravity is referred to as the *field capacity*, FC. The potential quantity of leachate is the amount of moisture within the landfill in excess of the FC.

In general, the amount of leachate produced is directly related to the amount of external water entering the landfill. Theoretically, the leachate generation rate can be determined by writing a water mass balance on the landfill. This can be done on a preclosure and a postclosure basis. Typical units for all the terms are units of length (e.g., "1.2 in of rain") or mass per unit volume (e.g., "a field capacity of 4 lbm/yd^3").

The preclosure leachate generation rate is

$$
\begin{array}{l}
\text{preclosure} \\
\text{leachate} \\
\text{generation}
\end{array}
=
\begin{array}{l}
\text{moisture released by incoming waste,} \\
\quad \text{including pore-squeeze liquid} \\
\\
+ \text{ precipitation} \\
- \text{ moisture lost due to evaporation} \\
- \text{ field capacity}
\end{array}
\qquad \textit{31.11}
$$

The postclosure water balance is

$$
\begin{array}{l}
\text{postclosure} \\
\text{leachate} \\
\text{generation}
\end{array}
=
\begin{array}{l}
\text{precipitation} \\
- \text{ surface runoff} \\
- \text{ evapotranspiration} \\
- \text{ moisture lost in formation} \\
\quad \text{of landfill gas and other} \\
\quad \text{chemical compounds} \\
- \text{ water vapor removed along} \\
\quad \text{with landfill gas} \\
- \text{ change in soil moisture storage}
\end{array}
\qquad \textit{31.12}
$$

13. LEACHATE MIGRATION FROM LANDFILLS

From Darcy's law (Ch. 21), migration of leachate contaminants that have passed through liners into aquifers or the groundwater table is proportional to the hydraulic conductivity, K, and the hydraulic gradient, i. Hydraulic conductivities of clay liners are 1.2×10^{-7} to 1.2×10^{-5} ft/hr (10^{-9} to 10^{-7} cm/s). However, the properties of clay liners can change considerably over

time due to interactions with materials in the landfill. If the clay dries out (desiccates), it will be much more permeable. For synthetic FMLs, hydraulic conductivities are 1.2×10^{-10} to 1.2×10^{-7} ft/hr (10^{-12} to 10^{-9} cm/s). The average permeability of high-density polyethylene is approximately 1.2×10^{-11} ft/hr (1×10^{-13} cm/s).

$$ Q = KiA \qquad \textit{31.13} $$

$$ i = \frac{dH}{dL} \qquad \textit{31.14} $$

14. GROUNDWATER DEWATERING

It may be possible to prevent or reduce contaminant migration by reducing the elevation of the groundwater table (GWT). This is accomplished by dewatering the soil with relief-type *extraction drains (relief drains)*. The *ellipse equation*, also known as the *Donnan formula*, used for calculating pipe spacing, L, in draining agricultural fields, can be used to determine the spacing of groundwater dewatering systems. In Eq. 31.15, K is the hydraulic conductivity with units of length/time, a is the distance between the pipe and the impervious layer barrier (a is zero if the pipe is installed on the barrier), b is the maximum allowable table height above the barrier, and Q is the *recharge rate*, also known as the *drainage coefficient*, with dimensions of length/time. The units of K and Q must be on the same time basis.

$$ L = 2\sqrt{\left(\frac{K}{Q}\right)(b^2 - a^2)} = 2\sqrt{\frac{b^2 - a^2}{i}} \qquad \textit{31.15} $$

Equation 31.15 is often used because of its simplicity, but the accuracy is only approximately $\pm 20\%$. Therefore, the calculated spacing should be decreased by 10 to 20%.

For pipes above the impervious stratum, the total discharge per unit length of each pipe (in ft^3/sec-ft or m^3/s·m) is given by Eq. 31.16. H is the maximum height of the water table above the pipe invert elevation. D is the average depth of flow.

$$ Q_{\text{unit length}} = \frac{2\pi KHD}{L} \qquad \textit{31.16} $$

$$ D = a + \frac{H}{2} \qquad \textit{31.17} $$

For pipes on the impervious stratum, the total discharge per unit length from the ends of each pipe (in ft^3/sec-ft or m^3/s·m) is

$$ Q_{\text{unit length}} = \frac{4KH^2}{L} \qquad \textit{31.18} $$

Equation 31.19 gives the total discharge per pipe. If the pipe drains from both ends, the discharge per end would be half of that amount.

$$ Q_{\text{pipe}} = LQ_{\text{unit length}} \qquad \textit{31.19} $$

Figure 31.4 *Geometry for Groundwater Dewatering Systems*

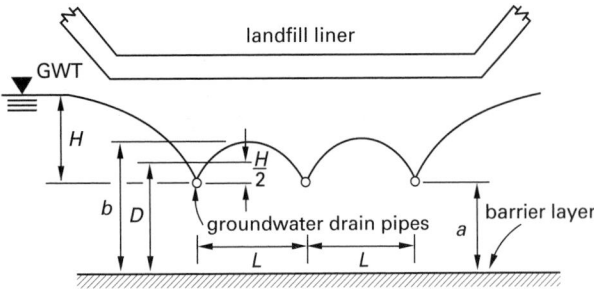

(a) pipes above impervious stratum

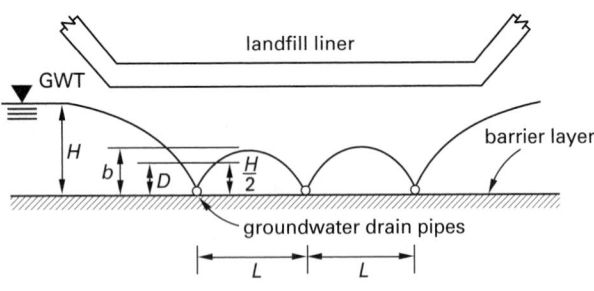

(b) pipes on impervious stratum

15. LEACHATE RECOVERY SYSTEMS

At least two distinct leachate recovery systems are required in landfills: one within the landfill to limit the hydraulic head of leachate that has reached the top liner, and another to catch the leachate that has passed through the top liner and drainage layer and has reached the bottom liner.

By removing leachate at the first liner, the hydrostatic pressure on the liner is reduced, minimizing the pressure gradient and hydraulic movement through the liner. A pump is used to raise the collected leachate to the surface once a predetermined level has been reached. Tracer compounds (e.g., lithium compounds or radioactive hydrogen) can be buried with the wastes to signal migration and leakage.

Leachate collection and recovery systems fail because of clogged drainage layers and pipe, crushed collection pipes due to waste load, pump failures, and faulty design.

16. LEACHATE TREATMENT

Leachate is essentially a very strong municipal wastewater, and it tends to become more concentrated as the landfill ages. Leachate from landfills contains extremely high concentrations of compounds and cannot be discharged directly into rivers or other water sources.

Leachate is treated with biological (i.e., trickling filter and activated sludge) and physical/chemical processes very similar to those used in wastewater treatment plants. For large landfills, these treatment facilities are located on the landfill site. A typical large landfill treatment facility would include an equalization tank, a primary clarifier, a first-stage activated sludge aerator and clarifier, a second-stage activated sludge aerator and clarifier, and a rapid sand filter. Additional equipment for sludge dewatering and digestion would also be required. Liquid effluent would be discharged to the municipal wastewater treatment plant.

17. LANDFILL MONITORING

Monitoring is conducted at sanitary landfills to ensure that no contaminants are released from the landfill. Monitoring is conducted in the vadose zone for gases and liquid, in the groundwater for leachate movements, and in the air for air quality monitoring.

Landfills use *monitoring wells* located outside of the covered cells and extending past the bottom of the disposal site into the aquifer. In general, individual monitoring wells should extend into each stratum upstream and downstream (based on the hydraulic gradient) of the landfill.

18. CARBON DIOXIDE IN LEACHATE

Carbon dioxide is formed during both aerobic and anaerobic decomposition. Carbon dioxide combines with water to produce *carbonic acid*, H_2CO_3. In the absence of other mitigating compounds, this will produce a slightly acidic leachate.

The concentration of carbon dioxide (in mg/L) in the leachate (assumed to be water) can be calculated from the *absorption coefficient (solubility)*, K_s. The absorption coefficient for carbon dioxide at 32°F (0°C) and 1 atm is approximately $K_s = 0.88$ L/L. This can be converted to mg/L with the ideal gas law.

$$\frac{m_g}{V_L} = \frac{p(\text{MW})}{R^* T}$$

$$= \frac{B_{CO_2} p_t \left(44 \ \frac{\text{g}}{\text{mol}}\right)}{\left(0.08206 \ \frac{\text{atm-L}}{\text{mol-K}}\right) T_K} \qquad 31.20$$

$$T_K = T_{°C} + 273 \qquad 31.21$$

At a typical internal landfill temperature of 100°F (38°C), this equation reduces to

$$\frac{m_g}{V_L} = 1.72 B_{CO_2} \qquad 31.22$$

The concentration of carbon dioxide in the leachate is

$$C_{CO_2,\text{mg/L}} = \left(\frac{m_g}{V_L}\right) K_{s,\text{L/L}} \left(1000 \ \frac{\text{mg}}{\text{g}}\right) \qquad 31.23$$

Table 31.4 *Typical Absorption Coefficients*
(L/L at 0° C and 1 atm)

hydrogen	0.017
nitrogen	0.015
oxygen	0.028
carbon monoxide	0.025
methane	0.33
carbon dioxide	0.88

19. INCINERATION OF MUNICIPAL SOLID WASTE

Incineration of MSW results in a 90% reduction in waste disposal volume and a mass reduction of 75%.

In *no-boiler incinerators*, MSW is efficiently burned without steam generation. However, incineration at large installations is often accompanied by recycling and steam and/or electrical power generation. Incinerator/generator facilities with separation capability are known as *resource-recovery plants*. Facilities with boilers are referred to as *waste-to-steam plants*. Facilities with boilers and electrical generators are referred to as *waste-to-energy (WTE) facilities*. In WTE plants, the combustion heat is used to generate steam and electrical power.

Mass burning is the incineration of unprocessed MSW, typically on stoker grates or in rotary and waterwall combustors to generate steam. With mass burning, MSW is unloaded into a pit and then moved by crane to the furnace conveying mechanism. Approximately 27% of the MSW remains as ash, which consists of glass, sand, stones, aluminum, and other noncombustible materials. It and the flyash are usually collected and disposed of in municipal landfills. (The U.S. Environmental Protection Agency (EPA) has ruled that ash from the incineration of MSW is not a hazardous waste. However, this is hotly contested and is subject to ongoing evaluation.) Capacities of typical mass burn units vary from less than 400 tons/day (360 Mg/d) to a high of 3000 tons/day (2700 Mg/d), with the majority of units processing 1000 to 2000 tons/day (910 to 1800 Mg/d).

MSW, as collected, has a heating value of approximately 4500 Btu/lbm (10 MJ/kg), though higher values have been reported. For 1000 tons/day (1000 Mg/d) of MSW incinerated, the yields are approximately 150,000 to 200,000 lbm/hr (75 to 100 Mg/h) of 650 to 750 psig (4.5 to 5.2 MPa) steam at 700 to 750°F (370 to 400°C) and 25 to 30 MW (27.6 to 33.1 MW) of gross electrical power. Approximately 10% of the electrical power is used internally, and units generating much less than approximately 10 MW (gross) may use all of their generated electrical power internally.

The *burning rate* varies from approximately 40 to 60 lbm/hr-ft^2 (200 to 300 kg/h·m^2) and is the fueling rate divided by the total effective grate area. Maximum heat release rates are approximately 300,000 Btu/hr-ft^2 (940 kW/m^2). The *heat release rate*, HRR, is defined as

$$\text{HRR} = \frac{(\text{fueling rate})(\text{HV})}{\text{total effective grate area}} \qquad 31.24$$

Figure 31.5 *Typical Mass-Burn Waste-to-Energy Plant*

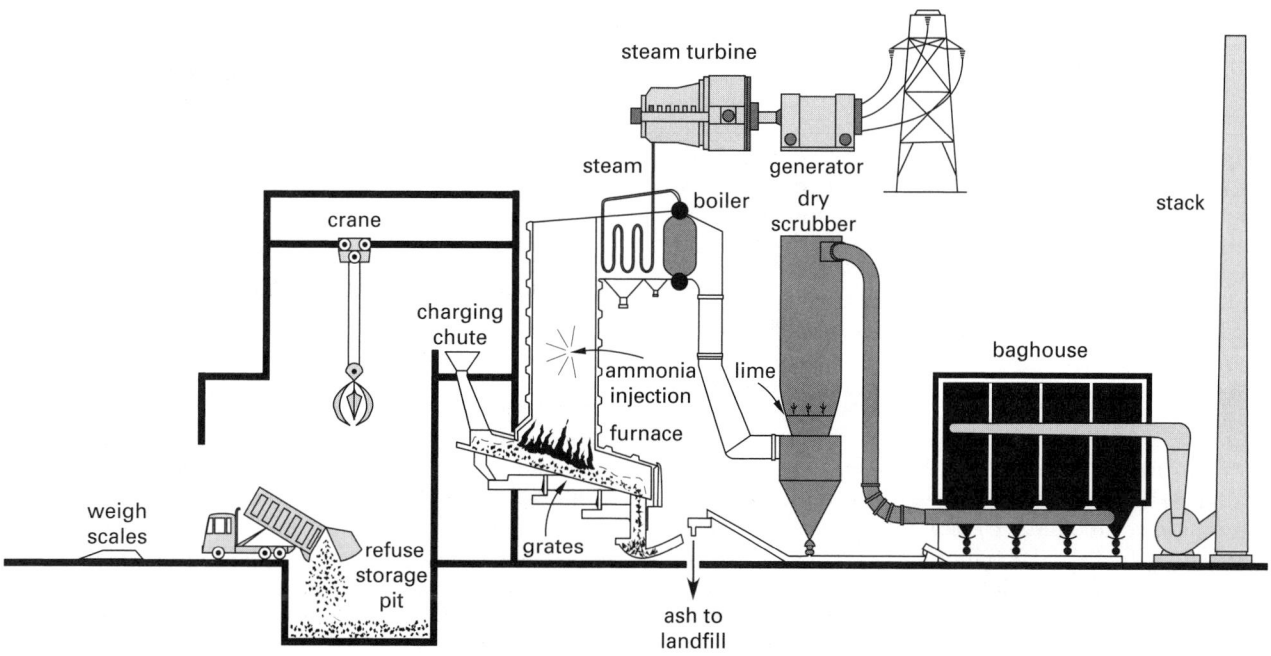

20. REFUSE-DERIVED FUEL

Refuse-derived fuel (RDF) is derived from MSW. First-generation RDF plants use "crunch and burn" technology. In these plants, the MSW is shredded after ferrous metals are removed magnetically. First-generation RDF plants suffer from the same problems that have plagued early mass burn units, including ash with excessive quantities of noncombustible materials such as glass, grit and sand, and aluminum.

Second-generation plants incorporate screens and air classifiers to reduce noncombustible materials and to increase the recovery of some materials. *Material recovery facilities* (MRFs) specialize in sorting out recyclables from MSW. The ash content is reduced, and the energy content of the RDF is increased. The MSW is converted into RDF pellets $2^1/_2$ to 6 in (6.4 to 15 cm) in size and is introduced through feed ports above a traveling grate. Some of fuel is burned in suspension, with the rest burned on the grate. Grate speed is varied so that the fuel is completely incinerated by the time it reaches the ash rejection ports at the front of the burner.

Table 31.5 *Typical Ultimate Analyses of MSW and RDF*

| | percentage by weight | |
element	MSW	RDF
carbon	26.65	31.00
water	25.30	27.14
ash	23.65	13.63
oxygen	19.61	22.72
hydrogen	3.61	4.17
chlorine	0.55	0.66
nitrogen	0.46	0.49
sulfur	0.17	0.19
TOTAL	100.00	100.00

In large (2000 to 3000 tons/day (1800 to 2700 Mg/d)), third-generation RDF plants, illustrated in Fig. 31.6, more than 95% of the original MSW combustibles are retained while reducing *mass yield* (i.e., the ratio of RDF mass to MSW mass) to below 85%.

RDF has a heating value of approximately 5500 to 5900 Btu/lbm (12 to 14 MJ/kg). The moisture and ash contents of RDF are approximately 24% and 12%, respectively.

The performance of a typical third-generation facility burning RDF is similar to a mass-burn unit: For each 1000 tons/day (1000 Mg/d) of MSW collected, approximately 150,000 to 250,000 lbm/hr (75 to 125 Mg/h) of 750 to 850 psig (5.2 to 5.9 Mpa) steam at 750 to 825°F (400 to 440°C) can be generated, resulting in approximately 30 to 40 MW (33.1 to 44.1 MW) of electrical power.

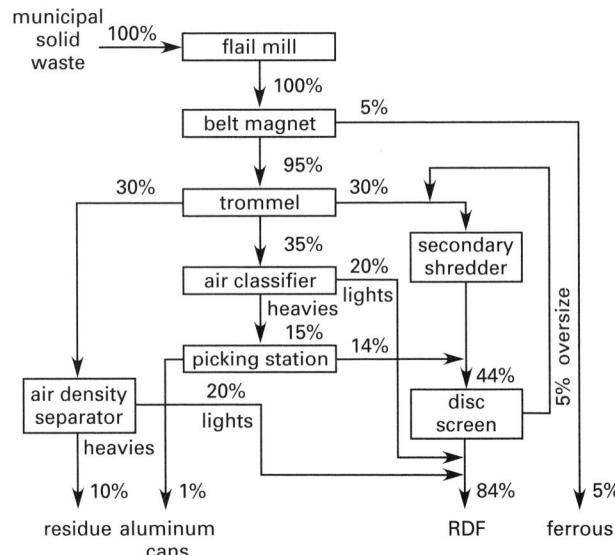

Figure 31.6 *Typical RDF Processing*

Natural gas is introduced at startup to bring the furnace up to the required 1800°F (1000°C) operating temperature. Natural gas can also be used upon demand, as when a load of particularly wet RDF lowers the furnace temperature.

Coal, oil, or natural gas can be used as a back-up fuel if RDF is unavailable or as an intended part of a *co-firing* design. Co-firing installations are relatively rare, and many have discontinued burning RDF due to poor economic performance. Typical problems with co-firing are (a) upper furnace wall slagging, (b) decreased efficiencies in electrostatic precipitators, (c) increased boiler tube corrosion, (d) excessive amounts of bottoms ash, and (e) difficulties in receiving and handling RDF.

RDF is a low-sulfur fuel, but like MSW, RDF is high in ash and chlorine. Relative to coal, RDF produces less SO_2 but more hydrogen chloride. Bottom ash can also contain trace organics and lead, cadmium, and other heavy metals.

21. FLUIDIZED-BED COMBUSTION

In *fluidized-bed combustion* (FBC), combustion is efficient and excess air required is low (e.g., 25 to 50%). Destruction is essentially complete due to the long residence time (5 to 8 sec for gases, and even longer for solids), high turbulence, and exposure to oxygen. Combustion control is simple, and combustion is often maintained within a 15°F (8°C) band. Both overautogeneous and subautogenous feeds can be handled. Due to the thermal mass of the bed, the FBC temperature drops slowly—at the rate of about 10°F/hr (5°C/h) after shut-down. Start-up is fast and operation can be intermittent.

Relative to other hazardous waste disposal methods, NOx and metal emissions are low, and organic content of the ash is very low (e.g., below 1%). Because of the long residence time, FBC systems do not usually require afterburners or *secondary combustion chambers* (SCCs). Due to the turbulent combustion process, the combustion temperature can be 200 to 300°F (110 to 170°C) lower than in rotary kilns. These factors translate into fuel savings, fewer or no NOx emissions, and lower metal emissions.

Most limitations of FBC systems relate to the feed. The feed must be small in size and roughly regular in shape so that it can be fluidized. This makes FBCs ideal for non-atomizable liquids, slurries, sludges, tars, and granular solids. It is more appropriate (and economical) to destroy atomizable liquid wastes in a boiler or special injection furnace and large bulky wastes in a rotary kiln or incinerator.

FBCs cannot usually be used when the feed material or its ash melts below the bed temperature. If the feed melts, it will agglomerate and defluidize the bed. When it is necessary to burn a feed with a low melting temperature, two methods can be used. The bed can be operated as a *chemically active bed*, where operation is close to but below the ash melting point of 1350°F to 1450°F (730°C to 790°C). A small amount of controlled melting is permitted to occur, and the agglomerated ash is removed.

When a higher temperature is desired in order to destroy hazardous materials, the melting point of feed can be increased by injecting low-cost additives (e.g., calcium hydroxide or kaolin clay). These combine with alkali metals salts to form refractory-like materials with melting points in the 1950°F to 2350°F (1070°C to 1290°C) range.

The design of a fluidizing-bed incinerator starts with a heat balance. Energy enters the FBC during combustion of the feed and auxiliary fuel and from sensible heat contributions of the air and fuel. Some of the heat is recovered and reused. The remainder of the heat is lost through sensible heating of the combustion products, water vapor, excess air, and ash, and through radiation from the combustor vessel and associated equipment. Since these items may not all be known in advance, the following assumptions are reasonable.

- Radiation losses are approximately 5%.
- Sensible heat losses from the ash are minimal, particularly if the feed contains significant amounts of moisture.
- Excess air will be approximately 40%.
- The combustion temperature will be approximately 1400°F (760°C).
- If air is preheated, the preheat temperature will be approximately 1000°F (540°C).

The fuel's *specific feed characteristic* (SFC) is the ratio of the higher (gross) heating value to the moisture content. Typical values range from a low of approximately 1000 Btu/lbm (2300 kJ/kg) for wastewater treatment plant biosolids to over 60,000 Btu/lbm (140 MJ/kg) for barks, sawdust, and RDF. (Although the units are the same, the specific feed characteristic is not the same as the heating value.) Making the previous assumptions and using the SFC as an indicator, the following generalizations can be made.

- Fuels with an SFC less than 2600 Btu/lbm (6060 kJ/kg) require a 1000°F (540°C) hot windbox, and are autogeneous at that value and subautogeneous below that value. The SFC drops to 2400 Btu/lbm (5600 kJ/kg) with air preheated to 1200°F (650°C). Below the subautogeneous SFC, water evaporation is the controlling design factor and auxiliary fuel is required.

- Fuels with an SFC of 4000 Btu/lbm (9300 kJ/kg) are autogeneous with a cold windbox and are overautogeneous above that value. Combustion is the controlling design factor for overautogenous SFCs.

Example 31.1

A wastewater treatment sludge is dewatered to 75% water by weight. The solids have a heating value of 6500 Btu/lbm (15 000 kJ/kg). The dewatered sludge enters a fluidized-bed combustor with a 1000°F (540°C) windbox at the rate of 15,000 lbm/hr (1.9 kg/s). (a) What is the specific feed characteristic? (b) Approximately what energy (in Btu/hr) must the auxiliary fuel supply?

SI Solution

(a) The dewatered sludge consists of 25% combustible solids and 75% moisture. The total mass of sludge required to contain 1.0 kg water is

$$m_{\text{sludge}} = \frac{m_w}{x_w}$$
$$= \frac{1 \text{ kg}}{0.75} = 1.333 \text{ kg}$$

The mass of combustible solids per kilogram of water is

$$m_{\text{solids}} = 0.25 m_{\text{sludge}} = (0.25)(1.333 \text{ kg})$$
$$= 0.3333 \text{ kg}$$

The specific feed characteristic is

$$\text{SFC} = \frac{\text{total heat of combustion}}{m_w} = \frac{m_{\text{solid}} \text{HV}}{m_w}$$
$$= \frac{(0.3333 \text{ kg})\left(15\,000 \, \frac{\text{kJ}}{\text{kg}}\right)\left(1000 \, \frac{\text{J}}{\text{kJ}}\right)}{1 \text{ kg}}$$
$$= 5.0 \times 10^6 \text{ J/kg}$$

(b) Making the listed assumptions, operation with a 540°C hot windbox requires an SFC of approximately 6060 kJ/kg. Therefore, the energy per kilogram of water in the fuel that an auxiliary fuel must provide is

$$(6060 \text{ kJ}) \left(1000 \; \frac{\text{J}}{\text{kJ}}\right) - 5 \times 10^6 \text{ J} = 1.06 \times 10^6 \text{ J}$$

The energy supplied by the auxiliary fuel is

$$\dot{m}_{\text{fuel}} x_w \times \text{SFC deficit}$$
$$= \left(1.9 \; \frac{\text{kg}}{\text{s}}\right)(0.75)\left(1.06 \times 10^6 \; \frac{\text{J}}{\text{kg}}\right)$$
$$= 1.51 \times 10^6 \text{ J/s} \quad (1.5 \text{ MW})$$

Customary U.S. Solution

(a) The dewatered sludge consists of 25% combustible solids and 75% moisture. The total mass of sludge required to contain 1.0 lbm water is

$$m_{\text{sludge}} = \frac{m_w}{x_w}$$
$$= \frac{1 \text{ lbm}}{0.75} = 1.333 \text{ lbm}$$

The mass of combustible solids per pound of water is

$$m_{\text{solids}} = 0.25 m_{\text{sludge}} = (0.25)(1.333 \text{ lbm})$$
$$= 0.3333 \text{ lbm}$$

The specific feed characteristic is

$$\text{SFC} = \frac{\text{total heat of combustion}}{m_w} = \frac{m_{\text{solid}} \text{HV}}{m_w}$$
$$= \frac{(0.3333 \text{ lbm}) \left(6500 \; \frac{\text{Btu}}{\text{lbm}}\right)}{1 \text{ lbm}}$$
$$= 2166 \text{ Btu/lbm}$$

(b) Making the listed assumptions, operation with a 1000°F hot windbox requires an autogenous SFC of approximately 2600 Btu/lbm. Therefore, the energy per pound of water in the fuel that an auxiliary fuel must provide is

$$2600 \text{ Btu} - 2166 \text{ Btu} = 434 \text{ Btu}$$

The energy supplied by the auxiliary fuel is

$$\dot{m}_{\text{fuel}} x_w \times \text{SFC deficit}$$
$$= \left(15{,}000 \; \frac{\text{lbm}}{\text{hr}}\right)(0.75)\left(434 \; \frac{\text{Btu}}{\text{lbm}}\right)$$
$$= 4.88 \times 10^6 \text{ Btu/hr}$$

32 Pollutants in the Environment

Environmental

Part 1: General Concepts

Pollution prevention (sometimes referred to as "P2") can be achieved in a number of ways. The most desirable method is reduction at the source. Reduction is accomplished through process modifications, raw material quantity reduction, substitution of materials, improvements in material quality, and increased efficiencies. However, traditional *end-of-pipe* treatment and disposal processes after the pollution is generated remain the main focus. *Pollution control* is the limiting of pollutants in a planned and systematic manner.

Table 32.1 *Pollution Prevention Hierarchy (from most to least desirable)*

source reduction
recycling
waste separation and concentration
waste exchange
energy/material recovery
waste treatment
disposal

Pollution control, hazardous waste, and other environmental regulations vary from nation to nation and are constantly changing. Therefore, regulation-specific issues, including timetables, permit application processes, enforcement, and penalties for violations, are either omitted from this chapter or discussed in general terms. For a similar reason, few limits on pollutant emission rates and concentrations are given in this chapter.[1] Those that are given should be considered merely representative and typical of the general range of values.

Part 2: Types and Sources of Pollution

2. THE ENVIRONMENT

Specific regulations often deal with parts of the *environment*, such as the atmosphere (i.e., the "air"), oceans and other surface water, subsurface water, and the soil. *Nonattainment areas* are geographical areas identified

---

[1]Another factor complicating the publication of specific regulations is that the maximum-permitted concentrations and emissions depend on the size and nature of the source.

by regulation that already do not currently meet *national ambient air quality standards* (NAAQS). Nonattainment is usually the result of geography, concentrations of industrial facilities, and excessive vehicular travel, and can apply to any of the regulated substances (e.g., ozone, oxides of sulfur and nitrogen, and heavy metals).

3. POLLUTANTS

A *pollutant* is a material or substance that is accidentally or intentionally introduced to the environment in a quantity that exceeds what occurs naturally. Not all pollutants are toxic or hazardous, but the issue is moot when regulations limiting permissible concentrations are specific. As defined by regulations in the United States, *hazardous air pollutants* (HAPs), also known as *air toxics*, consist of trace metals (e.g., lead, beryllium, mercury, cadmium, nickel, and arsenic) and other substances for a total of approximately 200 "listed" substances.[2]

Another categorization defined by regulation in the United States separates pollutants into *criteria pollutants* (e.g., sulfur dioxide, nitrogen oxides, carbon monoxide, volatile organic compounds, particulate matter, and lead) and *noncriteria pollutants* (e.g., fluorides, sulfuric acid mist, reduced sulfur compounds, vinyl chloride, asbestos, beryllium, mercury, and other heavy metals).

4. WASTES

Process wastes are generated during manufacturing. *Intrinsic wastes* are part of the design of the product and manufacturing process. Examples of intrinsic wastes are impurities in the reactants and raw materials, byproducts, residues, and spent materials. Reduction of intrinsic wastes usually means redesigning the product or manufacturing process. *Extrinsic wastes*, on the other hand, are usually reduced by administrative controls, maintenance, training, or recycling. Examples of extrinsic wastes are fugitive leaks and discharges during material handling, testing, or process shutdown.

Solid wastes are garbage, refuse, biosolids, and containerized solid, liquid, and gaseous wastes.[3] *Hazardous waste* is defined as solid waste, alone or in combination with other solids, that because of its quantity, concentration, or physical, chemical, or infectious characteristics may either (a) cause an increase in mortality or

serious (irreversible or incapacitating) illness, or (b) pose a present or future hazard to health or the environment when improperly treated, stored, transported, disposed of, or managed.

A substance is *reactive* if it reacts violently with water; if the pH is less than 2 or greater than 12.5, it is *corrosive*; if its flash point is less than $140°F$ ($60°C$), it is *ignitable*. The *toxicity characteristic leaching procedure* (TCLP)[4] test is used to determine if the substance is *toxic*. The TCLP tests for the presence of 8 metals (e.g., chromium, lead, and mercury) and 25 organic compounds (e.g., benzene and chlorinated hydrocarbons).

Hazardous wastes can be categorized by the nature of their sources. *F-wastes* (e.g., spent solvents and distillation residues) originate from nonspecific sources; *K-wastes* (e.g., separator sludge from petroleum refining) are generated from industry-specific sources; *P-wastes* (e.g., hydrogen cyanide) are acutely hazardous discarded commercial products, off-specification products, and spill residues; *U-wastes* (e.g., benzene and hydrogen sulfide) are other discarded commercial products, off-specification products, and spill residues.

Two notable "rules" pertain to hazardous wastes. The *mixture rule* states that any solid waste mixed with hazardous waste becomes hazardous. The *derived from rule* states that any waste derived from the treatment of a hazardous waste (e.g., ash from the incineration of hazardous waste) remains a hazardous waste.[5]

5. POLLUTION SOURCES

A *pollution source* is any facility that produces pollution. A *generator* is any facility that generates hazardous waste. The term "major" (i.e., a *major source*) is defined differently for each class of nonattainment areas.[6]

The combustion of fossil fuels (e.g., coal, fuel oil, and natural gas) to produce steam in electrical generating plants is the most significant source of air-borne pollution. As such, this industry is among the most highly regulated.

Specific regulations pertain to generators of hazardous waste. Generators of more than certain amounts (e.g., 100 kg/month) must be registered (i.e., with the Environmental Protection Agency, EPA). Restrictions are placed on generators in the areas of storage, personnel training, shipping, treatment, and disposal.

---

[2]The most dangerous air toxics include asbestos, benzene, cadmium, carbon tetrachloride, chlorinated dioxins and dibenzofurans, chromium, ethylene dibromide, ethylene dichloride, ethylene oxide and methylene chloride, radionuclides, vinyl chloride, and emissions from coke ovens. Most of these substances are carcinogenic.

[3]The term *biosolids* is replacing the term *sludge* as it refers to organic waste produced from biological wastewater treatment processes. Sludge from industrial processes and flue gas cleanup (FGC) devices retains its name.

[4]Probably no subject in this book has more acronyms than environmental engineering. All of the acronyms used in this chapter are in actual use; none were invented for the benefit of the chapter.

[5]Hazardous waste should not be incinerated with nonhazardous waste, as all of the ash would be considered hazardous by these rules.

[6]A nonattainment area is classified as marginal, moderate, serious, severe, or extreme based on the average pollution (e.g., ozone) level measured in the area.

6. ENVIRONMENTAL IMPACT REPORTS

New installations and large-scale construction projects must be assessed for potential environmental damage before being approved by state and local building officials. The assessment is documented in an *environmental impact report* (EIR). The EIR evaluates all of the potential ways that a project could affect the ecological balance and environment. A report that alleges the absence of any environmental impacts is known as a *negative declaration* ("negative dec").

The following questions and issues are typically addressed in an EIR.

1. the nature of the proposed project

2. the nature of the project area, including distinguishing natural and man-made characteristics

3. current and proposed percentage uses by zoning: residential, commercial, industrial, public, and planned development

4. current and proposed percentage uses by application: built-up, landscaped, agricultural, paved streets and highways, paved parking, surface and aerial utilities, railroad, and vacant and unimproved

5. the nature and degree that the earth will be altered: (a) changes in topology from excavation and earth movement to and from the site, (b) changes in slope, (c) changes in chemical composition of the soil, and (d) changes in structural capacity of the soil as a result of compaction, tilling, shoring, or moisture content

6. the nature of known geologic hazards such as earthquake faults and soil instability (subsidence, landslides, and severe erosion)

7. increases in dust, smoke, or air pollutants generated by hauling during construction

8. changes in path, direction, and capacity of natural draining or tendency to flood

9. changes in erosion rates on and off the site

10. the extent to which the project will affect the potential use, extraction, or conservation of the earth's resources, such as crops, minerals, and groundwater

11. after construction is complete, the nature and extent to which the completed project's sewage, drainage, airborne particulate matter, and solid waste will affect the quality and characteristics of the soil, water, and air in the immediate project area and in the surrounding community

12. the quantity and source of fresh water consumed as a result of the project

13. effects on the plant and animal life currently onsite

14. effects on any unique (i.e., not found anywhere else in the city, county, state, or nation) natural or man-made features

15. effect on any historically significant or archeological site

16. changes to the view of the site from points around the project

17. changes affecting wilderness use, open space, landscaping, recreational use, and other aesthetic considerations

18. effects on the health and safety of the people in the project area

19. changes to existing noise levels

20. changes in the number of people who will (a) live or (b) work in the project area

21. changes in the burden placed on roads, highways, intersections, railroads, mass transit, or other elements of the transportation system

22. changes in the burden placed on other municipal services, including sewage treatment; health, fire, and police services; utility companies; etc.

23. the extent to which hazardous materials will be involved, generated, or disposed of during or after construction

Part 3: Environmental Issues

7. INTRODUCTION

Engineers face many environmental issues. This part of the chapter discusses (in alphabetical order) some of them. In some cases, "listed substances" (e.g., those that are specifically regulated) are discussed. In other cases, environmental issues are discussed in general terms.

Environmental engineering covers an immense subject area, and this chapter is merely an introduction to some of the topics. Some subjects "belong" to other engineering disciplines. For example, coal-fired plants are typically designed by mechanical engineers. Other subjects, such as the storage and destruction of nuclear wastes, are specialized topics subject to changing politics, complex legislation, and sometimes-untested technologies. Finally, some wastes are considered nonhazardous industrial wastes and are virtually unregulated. They are not discussed in this chapter either. Processing of medical wastes, personal safety, and the physiological effects of exposure, as important as they are, are beyond the scope of this chapter.

Other philosphical and political issues, such as nuclear fuel versus fossil fuel, plastic bags versus paper bags, and disposable diapers versus cloth diapers, are similarly not covered.

8. ACID GAS

Acid gas generally refers to *sulfur trioxide*, SO_3, in flue gas.[7,8] *Sulfuric acid*, H_2SO_4, formed when sulfur trioxide combines with water, has a low vapor pressure and, consequently, a high boiling point. Hydrochloric (HCl), hydrofluoric (HF), and nitric (HNO_3) acids also form in smaller quantities. However, unlike sulfuric acid, they do not lower the vapor pressure of water significantly. Therefore, any sulfuric acid present will control the dew point.

Sulfur trioxide has a large affinity for water, forming a strong acid even at very low concentrations. At the elevated temperatures in a stack, sulfuric acid attacks steel, almost all plastics, cement, and mortar. Sulfuric acid can be prevented from forming by keeping the temperature of the flue gas above the dew-point temperature. This may require preheating equipment prior to start-up and postheating during shutdown.

Hydrochloric acid does not normally occur unless the fuel has a high chlorine content, as do chlorinated solvents, municipal solid wastes (MSW), and refuse-derived fuels (RDF). Hydrochloric acid formed during the combustion of MSW and RDF can be removed by semidry scrubbing. HCl removal efficiencies of 90 to 99% are common. (See also Acid Rain and Sulfur Oxides.)

9. ACID RAIN

Acid rain consists of weak solutions of sulfuric, hydrochloric, and to a lesser extent, nitric acids. These acids are formed when emissions of sulfur oxides (SOx), hydrogen chloride (HCl), and nitrogen oxides (NOx) return to the ground in rain, fog, or snow, or as dry particles and gases. Acid rain affects lakes and streams, damages buildings and monuments, contributes to reduced visibility, and affects certain forest tree species. Acid rain may also represent a health hazard. (See also Acid Gas and Sulfur Oxides.)

10. ALLERGENS AND MICROORGANISMS

Allergens such as molds, viruses, bacteria, animal droppings, mites, cockroaches, and pollen can cause allergic reactions in humans. One form of bacteria, *legionella*, causes the potentially fatal *Legionnaire's disease*. Inside

buildings, allergens and microorganisms become particularly concentrated in standing water, carpets, HVAC filters and humidifier pads, and in locations where birds and rodents have taken up residence. Legionella bacteria can also be spread by aerosol mists generated by cooling towers and evaporative condensers if the bacteria are present in the recirculating water.

Some buildings cause large numbers of people to simultaneously become sick, particularly after a major renovation or change has been made. This is known as *sick building syndrome* or *building-related illness*. This phenomenon can be averted by using building materials that do not release vapor over time (e.g., as does plywood impregnated with formaldehyde) or harbor other irritants. Carpets can accumulate dusts. Repainting, wallpapering, and installing of new flooring can release new airborne chemicals. Areas must be flushed with fresh air until all noticeable effects have been eliminated.

Care must also be taken to ensure that filters in the HVAC system do not harbor microorganisms and are not contaminated by bird or rodent droppings. Air intakes must not be located near areas of chemical storage or parking garages.

11. ASBESTOS

Asbestos is a fibrous silicate mineral material that is inert, strong, and incombustible. Once released into the air, its fibers are light enough to stay airborne for a long time.

Asbestos has typically been used in woven and compressed forms in furnace insulation, gaskets, pipe coverings, boards, roofing felt, and shingles, and has been used as a filler and reinforcement in paint, asphalt, cement, and plastic. Asbestos is no longer banned outright in industrial products. However, regulations, well-publicized health risks associated with cancer and *asbestosis*, and potential liabilities have driven producers to investigate alternatives.

No single product has emerged as a suitable replacement for all asbestos applications. (Almost all replacements are more costly.) *Fiberglass* is an insulator with superior tensile strength but low heat resistance. Fiberglass has a melting temperature of approximately $1000°F$ ($538°C$). However, fiberglass treated with hydrochloric acid to leach out most of the silica (SiO_2) can withstand 2000 to $3500°F$ (1090 to $1930°C$).

In typical static sealing applications, *aramid fibers* (known by the trade names Kevlar™ and Twaron™) are particularly useful up to approximately $800°F$ ($427°C$). However, aramid fibers cannot withstand the caustic, high-temperature environment encountered in curing concrete.

Table 32.2 lists asbestos substitutes by application.

---

[7]The term *stack gas* is used interchangeably with *flue gas*.
[8]Sulfur dioxide normally is not a source of acidity in the flue gas.

Table 32.2 Asbestos Substitutes

application	substitute material
insulation boards	mineral wool, fiberglass, foams (polyurethane, cellulose, styrene, and polyimide)
braided packing seals	polytetrafluoroethylene (PTFE), carbon, graphite
flange and furnace door gaskets	*reinforcing fibers*: cellulose, carbon, glass, polyvinyl alcohol, polyamide, polyester, polyacrylonitrile, aramid, and polyolefin; *binders*: nitrile butadiene rubber (NBR)
concrete filler; flame retardant	fiberglass, graphite, polypropylene, acrylics
thixotropic^a agent in paints, coatings, adhesives, and sealants	polyethylene
additive to plastics and fiberglass to improve heat resistance	vermiculite

^{a}A *thixotropic substance* is thick and gel-like when stationary but becomes free-flowing when stirred or agitated.

12. BOTTOM ASH

Ash is the residue left after combustion of a fossil fuel. *Bottom ash* (*bottoms ash* or *bottoms*) is the ash that is removed from the combustor after a fuel is burned. (The rest of the ash is flyash.) The ash falls through the combustion grates into quenching troughs below. It may be continuously removed by *submerged scraper conveyors* (SSCs), screw-type devices, or ram dischargers. The ash can be dewatered to approximately 15% moisture content by compression or by being drawn up a dewatering slope. Bottom ash is combined with conditioned flyash on the way to the ash storage bunker.[9] Most combined ash is eventually landfilled.

13. DISPOSAL OF ASH

Combined ash (bottom ash and flyash) is usually landfilled. Other occasional uses for high-quality combustion ash (not ash from the incineration of municipal solid waste) include roadbed subgrades, road surfaces ("ashphalt"), and building blocks.

[9]Approximately half of the electrical generating plants in the United States use wet flyash handling.

14. CARBON DIOXIDE

Carbon dioxide, though an environmental issue, is not a hazardous material and is not regulated as a pollutant.[10] Carbon dioxide is not an environmental or human toxin, is not flammable, and is not explosive. Skin contact with solid or liquid carbon dioxide presents a freezing hazard. Other than the remote potential for causing frostbite, carbon dioxide has no long-term health effects. Its major hazard is that of asphyxiation, by excluding oxygen from the lungs.

15. CARBON MONOXIDE

Carbon monoxide, CO, is formed during incomplete combustion of carbon in fuels. This is usually the result of an oxygen deficiency at lower temperatures. Carbon monoxide displaces oxygen in the bloodstream, so it represents an asphyxiation hazard. Carbon monoxide does not contribute to smog.

Generation of carbon monoxide can be minimized by furnace monitoring and control. For industrial sources, the American Boiler Manufacturers Association (ABMA) recommends limiting carbon monoxide to 400 ppm (corrected to 3% O_2) in oil- and gas-fired industrial boilers. This value can usually be met with reasonable ease. Local ordinances may be more limiting, however.

Most carbon monoxide released in highly populated areas comes from vehicles. Vehicular traffic may cause the CO concentration to exceed regulatory limits. For this reason, *oxygenated fuels* are required to be sold in those areas during certain parts of the year. Oxygenated gasoline has a minimum oxygen content of approximately 2.0%. Oxygen is increased in gasoline with additives such as ethanol (ethyl alcohol) or methyl tertiary butyl ether (MTBE).

Minimization of carbon monoxide is compromised by efforts to minimize nitrogen oxides. Control of these pollutants is inversely related.

16. CHLOROFLUOROCARBONS

Most atmospheric oxygen is in the form of two-atom molecules, O_2. However, there is a thin layer in the stratosphere about 12 miles up where *ozone* molecules, O_3, are found in large quantities. Ozone filters out ultraviolet radiation that damages crops and causes skin cancer.

Chlorofluorocarbons (i.e., chlorinated fluorocarbons, such as Freon™) contribute to the deterioration of the Earth's ozone layer. Ozone in the atmosphere is depleted in a complex process involving pollutants, wind patterns, and atmospheric ice. As chlorofluorocarbon molecules rise through the atmosphere, solar energy breaks the chlorine free. The chlorine molecules attach

[10]Industrial exposure is regulated by the U.S. Occupational Safety and Health Administration (OSHA).

themselves to ozone molecules, and the new structure eventually decomposes into chlorine oxide and normal oxygen, O_2. The depletion process is particularly pronounced in the Antarctic because that continent's dry, cold air is filled with ice crystals on whose surfaces the chlorine and ozone can combine. Also, the prevailing winter wind isolates and concentrates the chlorofluorocarbons.

Table 32.3 Typical Replacement Compounds for Chlorofluorocarbons

designation	applications
HCFC 22	low- and medium-temperature refrigerant; blowing agent; propellant
HCFC 123	replacement for CFC-11; industrial chillers and applications where potential for exposure is low; somewhat toxic; blowing agent; replacement for perchloroethylene (dry cleaning fluid)
HCFC 124	industrial chillers; blowing agent
HFC 134a	replacement for CFC-12; medium-temperature refrigeration systems; centrifugal and reciprocating chillers; propellant
HCFC 141b	replacement for CFC-11 as a blowing agent; solvent
HCFC 142b	replacement for CFC-12 as a blowing agent; propellant
IPC (isopropyl chloride)	replacement for CFC-11 as a blowing agent

The depletion is not limited to the Antarctic, but occurs throughout the northern hemisphere, including virtually all of the United States. At various points during the year in the Northern Hemisphere, the ozone deficit is as much as 6%.

The 1987 Montreal Protocol (conference) resulted in an international agreement to phase out world-wide production of chemical compounds that have ozone-depletion characteristics.

The Clean Air Act (Title VI) follows the Montreal Protocol and prohibits production of chlorofluorocarbons (and Halon) in the United States. The 1990 Clean Air Act required that class I chemicals, including chlorofluorocarbons (CFCs), Halons, and carbon tetrachloride, be phased out by the year 2000. The fumigant methyl chloride, also a class I chemical, was given a phase-out deadline of 2002. The 1990 Clean Air Act required that class II chemicals, including hydrochlorofluorocarbons (HCFCs), be phased out by the year 2030.

Special allowances are made for aviation safety, national security, and fire suppression and explosion prevention if safe or effective substitutes are not available for those purposes. Excise taxes are used as interim disincentives for those who produce the compounds. Large reserves and recycling, however, probably ensure that chlorofluorocarbons and Halons will be in use for many years after the deadlines have passes.

Possible replacements for chlorofluorocarbons (CFCs) include hydrochlorofluorocarbons (HCFCs) and hydrofluorocarbons (HFCs), both of which are environmentally more benign than CFCs, and blends of HCFCs and HFCs. The additional hydrogen atoms in the molecules make them less stable, allowing nearly all chlorine to dissipate in the lower atmosphere before reaching the ozone layer. The lifetime of HCFC molecules is 2 to 25 years, compared with 100 years or longer for CFCs. The net result is that HCFCs have only 2 to 10% of the ozone-depletion ability of CFCs. HFCs have no chlorine and thus cannot deplete the ozone layer.

Most chemicals intended to replace CFCs still have chlorine, but at reduced levels. Additional studies are determining if HCFCs and HFCs accumulate in the atmosphere, how they decompose, and whether any byproducts could damage the environment.

17. COOLING TOWER BLOWDOWN

State-of-the art reuse programs in *cooling towers* (CTs) may recirculate water 15 to 20 times before it is removed through blowdown. Pollutants such as metals, herbicides, and pesticides originally in the makeup water are concentrated to five or six times their incoming concentrations. Most CTs are constructed with copper alloy condenser tubes, so the recirculated water becomes contaminated with copper ions as well. CT water also usually contains chlorine compounds or other biocides added to inhibit biofouling. Ideally, no water should leave the plant (i.e., a *zero-discharge facility*). If discharged, CT blowdown must be treated prior to disposal.

18. CONDENSER COOLING WATER

Approximately half of the electrical generation plants in the United States use *once-through (OT) cooling water*. The discharged cooling water may be a chemical or thermal pollutant. Since fouling in the main steam condenser significantly reduces performance, water can be treated by the intermittent addition of chlorine, chlorine dioxide, bromine, or ozone, and these chemicals may be present in residual form. *Total residual chlorine* (TRC) is regulated. Methods of chlorine control include *targeted chlorination* (the frequent application of small amounts of chlorine where needed) and *dechlorination*.

19. DIOXINS

Dioxins are a family of chlorinated dibenzo-*p*-dioxins (CDDs). The term *dioxin*, however, is commonly used to refer to the specific congener 2,3,7,8-tetrachlorodibenzo-*p*-dioxin (TCDD). Primary sources of dioxin include herbicides containing 2,4-T, 2,4,5-trichlorophenol, and hexachlorophene. Other potential sources include incinerated municipal and industrial waste, leaded gasoline exhaust, chlorinated chemical wastes, incinerated polychlorinated biphenyls (PCBs), and any combustion in the presence of chlorine.

The exact mechanism of dioxin formation during incineration is complex but probably requires free chlorine (in the form of HCl vapor), heavy metal concentrations (often found in the ash), and a critical temperature window of 570 to 840°F (300 to 450°C). Dioxins in incinerators probably form near waste heat boilers, which operate in this temperature range.

Dioxin destruction is difficult because it is a large organic molecule with a high boiling point. Most destruction methods rely on high temperature since temperatures of 1550°F (850°C) denature the dioxins. Other methods include physical immobilization (i.e., vitrification), dehalogenation, oxidation, and catalytic cracking using catalysts such as platinum.

Dioxins liberated during the combustion of municipal solid waste (MSW) and refuse-derived fuel (RDF) can be controlled by the proper design and operation of the furnace combustion system. Once formed, they can be removed by end-of-pipe processes, including activated charcoal (AC) injection. Success has also been reported using the vanadium oxide catalyst used for NOx removal.

20. DUST, GENERAL

Dust or *fugitive dust* is any solid particulate matter (PM) that becomes airborne, with the exception of PM emitted from the exhaust stack of a combustion process. Nonhazardous fugitive dusts are commonly generated when a material (e.g., coal) is unloaded from trucks and railcars. Dusts are also generated by manufacturing, construction, earth-moving, sand blasting, demolition, and vehicle movement.

Dusts pose three types of hazards. (a) Inhalation of airborne dust or vapors, particularly those that carry hazardous compounds, is the major concern. Even without toxic compounds, odors can be objectionable. Dusts are easily observed and can cover cars and other objects left outside. (b) Dusts can transport hazardous materials, contaminating the environment far from the original source. (c) In closed environments, even nontoxic dusts can represent an explosion hazard.

Dust emission reduction from *spot sources* (e.g., manufacturing processes such as grinders) is accomplished by *inertial separators* such as cyclone separators. Potential dust sources (e.g., truck loads and loose piles) can be covered, and dust generation can be reduced by spraying water mixed with other compounds.

There are three mechanisms of dust control by spraying. (a) In *particle capture* (as occurs in a spray curtain at a railcar unloading station), suspended particles are knocked down, wetted, and captured by liquid droplets. (b) In *bulk agglomeration* (as when a material being carried on a screw conveyor is sprayed), the moisture keeps the dust with the material being transported. (c) Spraying roads and coal piles to inhibit wind-blown dust is an example of *surface stabilization*.

Wetting agents are *surfactant* formulations added to water to improve water's ability to wet and agglomerate fine particles.[11] The resulting solution can be applied as liquid spray or as a foam.[12] *Humectant binders* (e.g., magnesium chloride and calcium chloride) and adhesive binders (e.g., waste oil) may also be added to the mixture to make the dust adhere to the contact surface if other water-based methods are ineffective.[13]

Surface stabilization of materials stored outside and exposed to wind, rain, freeze-thaw cycles, and ultraviolet radiation is enhanced by the addition of *crusting agents*.

21. DUST, COAL

Clean coals, western low-sulfur coal, eastern low-sulfur coal, eastern high-sulfur coal, low-rank lignite coal, and varieties in between have their own peculiar handling characteristics.[14] *Dry ultra-fine coal* (DUC) and coal slurries have their own special needs.

Western coals have a lower sulfur content, but because they are easily fractured, they generate more dust. Western coals also pose higher fire and explosion hazards than eastern coals. Water misting or foam must be applied to coal cars, storage piles, and conveyer transfer points. Adequate ventilation in storage silos and bunkers is also required, with the added benefit of reducing methane accumulation.

DUC is as fine as talcum (10 μm with less than 10% moisture).[15] It must be containerized for transport, because traditional railcars are not sufficiently airtight. Also, since the minimum oxygen content for combustion is 14 to 15%, DUC should be maintained in a pressurized, oxygen-depleted environment until used. Pneumatic flow systems are used to transport DUC through the combustion plant.

---

[11]A *surface-acting agent (surfactant)* is a soluble compound that reduces a liquid's surface tension or reduces the interfacial tension between a liquid and a solid.

[12]Foam is an increasingly popular means of reducing the potential for explosions in secondary coal crushers.

[13]A *humectant* is a substance that absorbs or retains moisture.

[14]A valuable resource for this subject is NFPA 850, *Fire Protection for Fossil Fueled Steam and Combustion Turbine Electrical Generating Plants*, National Fire Protection Association, Quincy, MA.

[15]A μm is a *micrometer*, 10^{-6} m, and is commonly referred to as a *micron*.

Environmental

22. FUGITIVE EMISSIONS

Equipment leaks from plant equipment are known as *fugitive emissions* (FEs). Leaks from pump and valve seals are common sources, though compressors, pressure-relief devices, connectors, sampling systems, closed vents, and storage vessels are also potential sources. FEs are reduced administratively by *leak detection and repair programs* (LDARs).

Common causes of fugitive emissions are (a) equipment in poor condition, (b) off-design pump operation, (c) inadequate seal characteristics, and (d) inadequate boiling-point margin in the seal chamber. Other pump/shaft/seal problems that can increase emissions include improper seal-face material, excessive seal-face loading and seal-face deflections, and improper pressure balance ratio.

Inadequate *boiling-point margin* (BPM) results in poor seal performance and face damage. BPM is the difference between the seal chamber temperature and the initial boiling point temperature of the pumped product at the seal chamber pressure. When seals operate close to the BPM, seal-generated heat can cause the pumped product between the seal faces to flash and the seal to run dry. A minimum BPM of 15°F (9°C) or a 25 psig (172 kPa) pressure margin is recommended to avoid flashing. Even greater margins will result in longer seal life and reduced emissions.

23. GASOLINE-RELATED POLLUTION

Gasoline-related pollution is primarily in the form of unburned hydrocarbons, nitrogen oxides, lead, and carbon monoxide. (The requirement for lead-free gasoline has severely curtailed gasoline-related lead pollution.) Though a small reduction in gasoline-related pollution can be achieved by blending detergents into gasoline, reformulation is required for significant improvements. Table 32.4 lists typical characteristics for traditional and reformulated gasolines.

Table 32.4 *Typical Gasoline Characteristics*

fuel parameter	traditional	reformulated
sulfur	150 ppmw[a]	40 ppmw
benzene	2% by vol	1% by vol
olefins	9.9% by vol	4% by vol
aromatic hydrocarbons	32% by vol	25% by vol
oxygen	0	1.8 to 2.2% by wt
T90	330°F (166°C)	300°F (149°C)
T50	220°F (104°C)	210°F (99°C)
RVP	8.5 psig (59 kPa)	7 psig (48 kPa)

[a]ppmw = parts per million by weight

The potential for smog formation can be reduced by reformulating gasoline and/or reducing the summer *Reid vapor pressure* (RVP) of the blend. Regulatory requirements to reduce the RVP can be met, in the short run, by using less butane in the gasoline. However, reformulating to remove pentanes is required when the RVP is required to achieve 7 psig (48 kPa) or below. Sulfur content can be reduced by pretreating the refinery feed in hydrodesulfurization (HDS) units. Heavier gasoline components must be removed to lower the 50% and 90% *distillation temperatures* (T50 and T90).

24. GLOBAL WARMING

The *global warming* hypothesis is that increased levels of atmospheric carbon dioxide, CO_2, from the combustion of carbon-rich fossil fuels and other *greenhouse gases* (e.g., water vapor, methane, nitrous oxide, and chlorofluorocarbons) trap an increasing amount of solar radiation in a *greenhouse effect*, gradually increasing the Earth's temperature. It is claimed that the most recent cycle of increases in carbon dioxide began with the industrial revolution.

Recent studies have shown that atmospheric carbon dioxide is increasing at the rate of about 1% per year. (For example, in one year carbon dioxide might increase from 350 ppmv to 354 ppmv. By comparison, oxygen is approximately 209,500 ppmv.) There has also been a 100% increase in atmospheric methane since the beginning of the industrial revolution. According to some researchers, there has been a corresponding global temperature increase of approximately 0.9°F (0.5°C) since the year 1890.[16]

In addition to a temperature increase, other evidence cited by supporters of the global warming hypothesis are several record-breaking hot summers, widespread aberrations in the traditional seasonal weather patterns (e.g., hurricane-like storms in England), and a 4 to 12 in (10 to 30 cm) rise in sea level over the last century.[17]

Although global warming is generally accepted, its human-made causes are not. The global warming hypothesis is disputed by some scientists and has not been proved to be an absolute truth. Arguments against the hypothesis center around the fact that manufactured carbon dioxide is a small fraction of what is naturally released (e.g., by wetlands, in rain forest fires, and during volcanic eruptions). It is argued that, in the face of such massive contributors, and since the Earth's temperature has remained essentially constant for millenia, the Earth already possesses some built-in mechanism, currently not perfectly understood, that reduces the Earth's temperature swings.

The environmentalists' claim that there will be a temperature increase of 3 to 9°F (1.7 to 5°C) by the year 2100 and that this increase will be catastrophic is also

---

[16]Other researchers can detect no discernible upward trend, and some offer a counter-argument. Based on the retreat of the northern-most lines capable of growing oranges since 1850, some believe that the weather is generally becoming colder, not hotter.
[17]The rise in the level of the oceans is disputed.

disputed by many scientists. On the contrary, most scientists agree that the deprivation and human suffering that would result from a reduction or cessation in the burning of fossil fuels will be significantly higher than the effects of a slightly increased temperature.

Regardless of the validity of the global warming hypothesis, some major power generation industries have adopted goals of reducing carbon dioxide emissions. However, efforts to reduce carbon dioxide emissions by converting fossil fuel from one form to another are questioned by many engineers. Natural gas produces the least amount of carbon dioxide of any fossil fuel. Therefore, conversion of coal to a gas or liquid fuel in order to lower the carbon-to-hydrogen ratio would appear to lessen carbon dioxide emissions at the point of final use. However, the conversion processes consume energy derived from carbon-containing fuel. This additional consumption, taken over all sites, results in a net increase in carbon dioxide emission of 10 to 200%, depending on the process.

The use of ethanol as an alternative for gasoline is also problematic. Manufacturing processes that produce ethanol give off (at least) twice as much carbon dioxide as the gasoline being replaced produces during combustion.

Most synthetic fuels are intrinsically less efficient (based on their actual heating values compared with those theoretically obtainable from the fuels' components in elemental form). This results in an increase in the amount of fuel consumed. Thus, fossil fuels should be used primarily in their raw forms until cleaner sources of energy are available.

25. LEAD

Lead, even in low concentrations, is toxic. Inhaled lead accumulates in the blood, bones, and vital organs. It can produce stomach cramps, fatigue, aches, and nausea. It causes irreparable damage to the brain, particularly in young and unborn children, and high blood pressure in adults. At high concentrations, lead damages the nervous system and can be fatal.

Lead was outlawed in paint in the late 1970s.[18] Lead has also been removed from most gasoline blends. Lead continues to be used in large quantities in automobile batteries and plating and metal-finishing operations. However, these manufacturing operations are tightly regulated. Lead enters the atmosphere during the combustion of fossil fuels and the smelting of sulfide ore. Lead enters lakes and streams primarily from acid mine drainage.

For lead in industrial and municipal wastewater, current remediation methods include pH adjustment with lime or alkali hydroxides, coagulation-sedimentation, reverse osmosis, and zeolite ion exchange.

26. NITROGEN OXIDES

Nitrogen oxides (NOx) are one of the primary causes of smog formation.[19] NOx from the combustion of coal is primarily *nitric oxide* (NO) with small quantities of *nitrogen dioxide* (NO_2).[20] NO_2 can be a primary or a secondary pollutant. Although some NO_2 is emitted directly from combustion sources, it is also produced from the oxidation of nitric oxide, NO, in the atmosphere.

NOx is produced in two ways: (a) *thermal NOx* produced at high temperatures from free nitrogen and oxygen, and (b) *fuel NOx* (or *fuel-bound NOx*) formed from the decomposition/combustion of fuel-bound nitrogen.[21] When natural gas and light distillate oil are burned, almost all of the NOx produced is thermal. Residual fuel oil, however, can have a nitrogen content as high as 0.3%, and 50 to 60% of this can be converted to NOx. Coal has an even higher nitrogen content.[22]

Thermal NOx is usually produced in small (but significant) quantities when excess oxygen is present at the highest temperature point in a furnace, such that nitrogen (N_2) can dissociate.[23] Dissociation of N_2 and O_2 is negligible, and little or no thermal NOx is produced below approximately 3000°F (1650°C).

Formation of thermal NOx is approximately exponential with temperature. For this reason, many NOx-reduction techniques attempt to reduce the *peak flame temperature* (PFT). NOx formation can be reduced also by injecting urea or ammonia reagents directly into the furnace.[24] The relationship between NOx production and excess oxygen is somewhat inconclusive, but NOx production appears to vary directly with the square root of the oxygen concentration.

In existing plants, retrofit NOx-reduction techniques include fuel-rich combustion (i.e., staged air burners), flue gas recirculation, changing to a low-nitrogen fuel, reduced air preheat, installing low-NOx burners, and

---

[18]Workers can be exposed to lead if they strip away lead-based paint, demolish old buildings, or weld or cut metals that have been coated with lead-based paint.

[19]In Los Angeles County, approximately 75% of the NOx is produced by vehicles, with the remainder being produced by combustion operations, such as boilers and heaters.

[20]Other oxides are produced in insignificant amounts. These include *nitrous oxide* (N_2O), N_2O_4, N_2O_5, and NO_3, all of which are eventually oxidized to (and reported as) NO_2.

[21]Some engineers further divide the production of fuel-bound NOx into low- and high-temperature processes and declare a third fuel-related NOx-production method known as *prompt-NOx*. Prompt-NOx is the generation of the first 15 to 20 ppm of NOx from partial combustion of the fuel at lower temperatures.

[22]In order of increasing fuel-bound NOx production potential, common boiler fuels are: methanol, ethanol, natural gas, propane, butane, ultralow-nitrogen fuel oil, fuel oil No. 2, fuel oil No. 6, and coal.

[23]The *mean residence time* at the high temperature points of the combustion gases is also an important parameter. At temperatures below 2500°F (1370°C), several minutes of exposure may be required to generate any significant quantities of NOx. At temperatures above 3000°F (1650°C), dissociation can occur in less than a second. At the highest temperatures—3600°F (1980°C) and above—dissociation takes less than a tenth of a second.

[24]*Urea* (NH_2CONH_2), also known as *carbamide urea*, is a water-soluble organic compound prepared from ammonia. Urea has significant biological and industrial usefulness.

using overfire air.[25] Low-NOx burners using controlled flow/split flame or internal fuel staging technology are essentially direct replacement (i.e., "plug in") units, differing from the original burners primarily in nozzle configuration. However, some fuel supply and air modifications may also be needed. Use of overfire air requires windbox modifications and separate ducts.

Lime spray-dryer scrubbers of the type found in electrical generating plants do not remove all of the NOx. Further reduction requires that the remaining NOx be destroyed. Reburn, selective catalytic reduction (SCR), and selective noncatalytic methods are required.

Scrubbing, incineration, and other end-of-pipe methods typically have been used to reduce NOx emissions from stationary sources such as gas turbine-generators, although these methods are unwieldy for the smallest units. Water/steam injection, catalytic combustion, and *selective catalytic reduction* (SCR) are particularly suited for gas-turbine and combined-cycle installations. SCR is also effective in NOx reduction in all heater and boiler applications.

27. ODORS

Odors of unregulated substances can be eliminated at their source, contained by sealing and covering, diluted to unnoticeable levels with clean air, or removed by simple water washing, chemical scrubbing (using acid, alkali, or sodium hypochlorite), bioremediation, and activated carbon adsorption.

28. OIL SPILLS IN NAVIGABLE WATERS

Intentional and accidental releases of oil in navigable waters are prohibited. Deleterious effects of such spills include large-scale biological (i.e., sea life and wildlife) damage and destruction of scenic and recreational sites. Long-term toxicity can be harmful for microorganisms that normally live in the sediment.[26]

29. OZONE

Ground-level ozone is a secondary pollutant. Ozone is not usually emitted directly, but is formed from hydrocarbons and nitrogen oxides (NO and NO_2) in the presence of sunlight. *Oxidants* are byproducts of reactions between combustion products. Nitrogen oxides react with other organic substances (e.g., hydrocarbons) to form the oxidants ozone and peroxyacyl nitrates (PAN) in complex *photochemical reactions*. Ozone and PAN are usually considered to be the major components of smog.[27] (See also Smog.)

30. PARTICULATE MATTER

Particulate matter (PM), also known as *aerosols*, is defined as all particles that are emitted by a combustion source. Particulate matter with aerodynamic diameters of less than or equal to a nominal 10 μm is known as *PM-10*. Particulate matter is generally inorganic in nature. It can be categorized into metals (or heavy metals), acids, bases, salts, and nonmetallic inorganics.

Metallic inorganic PM from incinerators is controlled with baghouses or electrostatic precipitators (ESPs), while nonmetallic inorganics are removed by scrubbing (wet absorption). Flue gas PM, such as flyash and lime particles from desulfurization processes, can be removed by fabric baghouses and electrostatic precipitators. These processes must be used with other processes that remove NOx and SO_2.

High temperatures cause the average flue gas particle to decrease in size toward or below 10 μm. With incineration, emission of trace metals into the atmosphere increases significantly. Because of this, incinerators should not operate above 1650°F (900°C.)[28]

31. PCBs

Polychlorinated biphenyls (PCBs) are organic compounds (i.e., *chlorinated organics*) manufactured in oily liquid and solid forms through the late 1970s, and subsequently prohibited. PCBs are carcinogenic and can cause skin lesions and reproductive problems. PCBs build up, rather than dissipate, in the food chain, accumulating in fatty tissues. Most PCBs were used as dielectric and insulating liquids in large electrical transformers and capacitors, and in ballasts for fluorescent lights (which contain capacitors). PCBs were also used as heat transfer and hydraulic fluids, as dye carriers in carbonless copy paper, and as plasticizers in paints, adhesives, and caulking compounds.

Incineration of PCB liquids and PCB-contaminated materials (usually soil) has long been used as an effective mediation technique. However, PCB-contaminated soil apparently can be effectively treated in situ by adding high-calcium flyash or (more expensive) aluminum powder. The reduction mechanism is unclear, but catalysis with an unidentified catalyst is most likely. The destruction of PCB is total. (The addition of quicklime alone has not been supported as a treatment method.)

Specialized PCB processes targeted at cleaning PCB from spent oil are also available. Final PCB concentrations are below detectable levels, and the cleaned oil can be recycled or used as fuel.

32. PESTICIDES

The term *traditional organochlorine pesticide* refers to a narrow group of persistent pesticides, including DDT,

---

[25]Air is injected into the furnace at high velocity over the combustion bed to create turbulence and to provide oxygen.
[26]This has given rise to a new *sediment testing technology*.
[27]The term *oxidant* sometimes is used to mean the original emission products of NOx and hydrocarbons.

---

[28]The upper temperature limit for biosolids (sewage sludge) incineration may be regulated.

the '*drins*' (aldrin, endrin, and dieldrin), chlordane, endosulfan, hexachlorobenzene, lindane, mirex, and toxaphene. Traditional organochlorine insecticides used extensively between the 1950s and early 1970s have been widely banned because of their environmental persistence. There are, however, notable exceptions, and environmental levels of traditional organochlorine pesticides (especially DDT) are not necessarily declining throughout the world, especially in developing countries and countries with malaria. For example, DDT is still used for vector control of malaria, among others, particularly in India. DDT, with its half-life of up to 60 years, does not always remain in the country where it is used. This helps to explain why levels of one of DDT's metabolites, DDE, are actually increasing in fish in the Great Lakes, despite DDT's being phased out in the U.S. and Canada during the 1970s. The semi-volatile nature of the chemicals means that at high temperatures they will tend to evaporate from the land, only to condense in cooler air. This *global distillation* is thought to be responsible for levels of organochlorines increasing in the Arctic.

The term *chlorinated pesticides* refers to a much wider group of insecticides, fungicides, and herbicides that contain organically bound chlorine. A major difference between traditional organochlorine and other chlorinated pesticides is that the former have been perceived to have high persistence and build up in the food chain, and the latter do not. However, even some chlorinated pesticides are persistent in the environment. There is little information available concerning the overall environmental impact of chlorinated pesticides. Far more studies exist concerning the effects of traditional organochlorines in the public domain than on chlorinated pesticides in general. Pesticides that have, for example, active organophosphate (OP), carbamate, or triazine parts of their molecules are chlorinated pesticides and may pose long-lasting environmental dangers.

In the U.S., about 30 to 40% of pesticides are chlorinated. All the top five pesticides are chlorinated (the sixth is methyl bromide, which is a serious organohalogen ozone depleter, and the seventh is trifluralin, which contains the halogen fluorine). Worldwide, half of the ten top-selling herbicides are chlorinated (alachlor, metolachlor, 2,4-D, cyanazine, and atrazine). Four of the top ten insecticides are chlorinated (chlorpyrifos, fenvalerate, endosulfan, and cypermethrin). Four (propiconazole, chlorothalonil, prochloraz, and triadimenol) of the ten most popular fungicides are chlorinated.

33. PLASTICS

Plastics, of which there are six main chemical polymers, generally do not degrade once disposed of, and are therefore considered a disposal issue. Disposal is not a problem per se, however, since plastics are lightweight, inert, and do not harm the environment when discarded.

A distinction is made between *biodegrading* and *recycling*. Most plastics, such as the polyethylene bags used to protect pressed shirts from the dry cleaners and to mail some magazines, are not biodegradable but are recyclable. Also, all plastics can be burned for their fuel value.

The collection and sorting problems often render low-volume plastic recycling efforts uneconomical. Complicating the drive toward recycling is the fact that many of the six different types cannot be distinguished visually, and they cannot be recycled successfully when intermixed. Also, some plastic products consist of layers of different polymers that cannot be separated mechanically.

Table 32.5 Polymers

polymer	common use
low-density polyethylene (LDPE)	grocery bags; food wrap
high-density polyethylene (HDPE)	detergent; milk bottles; oil containers; toys
polyethylene terephthalate (PET)	clear beverage containers
polypropylene (PP)	labels; bottles; housewares;
polyvinyl chloride (PVC)	clear bottles
polystyrene (PS)	styrofoam cups; "clam shell" food containers

Sorting in low-volume applications is performed visually and manually. Commercial high-volume methods include hydrocycloning, flotation with flocculation (for all polymers), x-ray fluorescence (primarily for PVC detection), and near-infrared spectroscopy (primarily for separating PVC, PET, PP, PE, and PS). Mass spectroscopy is also promising, but has yet to be commercialized.

Unsorted plastics can be melted and reformed into some low-value products. This operation is known as *downcycling*, since each successful cycle further degrades the material. This method is suitable only for a small fraction of the overall recyclable plastic.

Other operations that can reuse the compounds found in plastic products are hydrogenation, pyrolysis, and gasification. *Hydrogenation* is the conversion of mixed plastic scrap to "syncrude," synthetic crude oil, in a high-temperature (i.e., 750 to 880°F (400 to 470°C)), high-pressure (i.e., 2200 to 4400 psig (15 to 30 MPa)), hydrogen-rich atmosphere. Since the end product is a crude oil substitute, hydrogenation operations must be integrated into refinery or petrochemical operations.

Gasification and pyrolysis are stand-alone operations that do not require integration with a refinery. *Pyrolysis* takes place in a fluidized bed between 750°F and 1475°F (400°C and 800°C). Cracked polymer gas or other inert gas fluidizes the sand bed, which promotes good mixing and heat transfer, resulting in liquid and gaseous petroleum products.

Gasification operates at higher temperatures, 1650 to 3600°F (900 to 2000°C), and lower pressures, around 870 psig (6 MPa). The waste stream is pyrolyzed at lower temperatures before being processed by the gasifier. The gas can be used on-site to generate steam. Gasification has the added advantage of being able to treat the entire municipal solid waste stream, avoiding the need for sorting plastics.

The development and use of a biodegradable plastic is an alternative to recycling. Original research in degradable plastics focused on nondegradable polymers mixed with small amounts of materials that reacted with sunlight and soil microbes. Such products, however, were criticized by environmentalists as being waste that broke down into smaller nondegradable pieces. Polymers that are derived from agricultural sources, rather than from petroleum, are now being investigated in the search for totally biodegradable plastics.

Success has been reported with polymers based on blackstrap molasses (a sugar production waste residue) and potato or corn starches. Another promising degradable polymer is polyhydroxybutyrate-valerate (PBHV), derived from the bacterium *alcaligenes eutrophus*. The bacteria, when deprived of phosphorus, feed on organic acids and sugars, converting more than 80% of their body mass into PBHV.

Some engineers point out that biodegrading is not even a desirable characteristic for plastics and that being nonbiodegradable is not harmful. Biodegrading converts materials (such as the paper bags often preferred over plastic bags) to water and carbon dioxide, contributing to the greenhouse effect without even receiving the energy benefit of incineration. Biodegrading of most substances also results in gases and leachates that can be more harmful to the environment than the original substance. In a landfill, biodegrading serves no useful purpose, since the space occupied by the degraded plastic does not create additional useful space (volume).

34. RADON

Radon gas is a radioactive gas produced from the natural decay of radium within the rocks beneath a building.[29] Radon accumulates in unventilated areas (e.g., basements), in stagnant water, and in air pockets formed when the ground settles beneath building slabs. Radon also can be brought into the home by radon-saturated well water used in baths and showers. The EPA's action level of 4 pCi/L for radon in air is contested by many as being too high.

Radon mitigation methods include (a) pressurizing to prevent the infiltration of radon, (b) installing of

---

[29] Radon also can be generated in concrete made from slag derived from uranium mines. Such mines are primarily located in Colorado, Florida, Oregon, and South Dakota.

depressurization systems to intercept radon below grade and vent it safely, (c) removing radon-producing soil, and (d) abandoning radon-producing sites.

35. RAINWATER RUNOFF

Rainwater percolating through piles of coal, flyash, mine tailings, and other stored substances can absorb toxic compounds and eventually make its way into the earth, possibly contaminating soil and underground aquifers.

36. RAINWATER RUNOFF FROM HIGHWAYS

The *first flush* of a storm is generally considered to be the first half-inch of storm runoff or the runoff from the first 15 min of the storm. Along highways and other paved transportation corridors, the first flush contains potent pollutants such as petroleum products, asbestos fibers from brake pads, tire rubber, and fine metal dust from wearing parts. Under the National Pollutant Discharge Elimination System (NPDES), stormwater runoff in newly developed watersheds must be cleaned before it reaches existing drainage facilities, and runoff must be maintained at or below the present undeveloped runoff rate.

A good stormwater system design generally contains two separate basins or a single basin with two discrete compartments. The function of one compartment is water quality control, and the function of the other is peak runoff control.

The *water quality compartment* (WQC) should normally have sufficient volume and discharge rates to provide a minimum of 1 hour of detention time for 90 to 100% of the first flush volume. "Treatment" in the WQC consists of sedimentation of suspended solids and evaporation of volatiles. A removal goal of 75% of the suspended solids is reasonable in all but the most environmentally sensitive areas.

In environmentally sensitive areas, a filter berm of sand, a sand chamber, or a sand filter bed can further clarify the discharge from the WQC.

After the WQC becomes full from the first flush, subsequent runoff will be diverted to the *peak-discharge compartment* (PDC). This is done by designing a junction structure with an inlet for the incoming runoff and separate outlets from each compartment. If the elevation of the WQC outlet is lower than the inlet to the PDC, the first flush will be retained in the WQC.

Sediment from the WQC chamber and any filters should be removed every 1 to 3 years, or as required. The sediment must be properly handled, as it may be considered to be hazardous waste under the EPA's "mixture" and "derived-from" rules and its "contained-in" policy.

When designing chambers and filters, sizing should accommodate the first flush of a 100-year storm. The top of the berm between the WQC and PDC should include a minimum of 1 ft (0.3 m) of freeboard. In all cases, an emergency overflow weir should allow a storm greater than the design storm to discharge into the PDC or receiving water course.

Minimum chamber and berm width is approximately 8 ft (2.4 m). Optimum water depth in each chamber is approximately 2 to 5 ft (0.6 to 1.5 m). Each compartment can be sized by calculating the divided flows and staging each compartment. The outlet from the WQC should be sufficient to empty the compartment in approximately 24 to 28 hours after a 25-year storm.

A *filter berm* is essentially a sand layer between the WQC and the receiving chamber that filters water as it flows between the two compartments. The filter should be constructed as a layer of sand placed on geosynthetic fabric, protected with another sheet of geosynthetic fabric, and covered with coarse gravel, another geosynthetic cover, and finally a layer of medium stone. The sand, gravel, and stone layers should all have a minimum thickness of 1 ft (0.3 m). The rate of permeability is controlled by the sand size and front-of-fill material. Permeability calculations should assume that 50% of the filter fabric is clogged.

A *filter chamber* consists of a concrete structure with a removable filter pack. The filter pack consists of geosynthetic fabric wrapped around a plastic frame (core) that can be removed for backwashing and maintenance. The filter is supported on a metal screen mounted in the concrete chamber. The outlet of the chamber should be located at least 1 ft (0.3 m) behind the filter pack. The opening's size will determine the discharge rate through the filter, which should be designed as less than 2 ft/sec (60 cm/s) assuming that 50% of the filter area is clogged.

A *sand filter bed* is similar to the filter beds used for tertiary sewage treatment. The sand filter consists of a series of 4 in (100 mm) perforated PVC pipes in a gravel bed. The gravel bed is covered by geotextile fabric and 8 in (200 mm) of fine-to-medium sand. The perforated underdrains lead to the outlet channel or chamber.

37. SMOG

Photochemical smog (usually, just *smog*) consists of ground-level ozone and peroxyacyl nitrates (PAN). Smog is produced by the sunlight-induced reaction of ozone *precursors*, primarily nitrogen dioxide (NOx), hydrocarbons, and volatile organic compounds (VOCs). NOx and hydrocarbons are emitted by combustion sources such as automobiles, refineries, and industrial boilers. VOCs are emitted by manufacturing processes, dry cleaners, gasoline stations, print shops, painting operations, and municipal wastewater treatment plants. (See also Ozone.)

38. SMOKE

Smoke results from incomplete combustion and indicates unacceptable combustion conditions. In addition to being a nuisance problem, smoke contributes to air pollution and reduced visibility. Smoke generation can be minimized by proper furnace monitoring and control.

Opacity can be measured by a variety of informal and formal methods, including transmissometers mounted on the stack. The sum of the *opacity* (the fraction of light blocked) and the *transmittance* (the fraction of light transmitted) is 1.0.

Optical density is calculated from Eq. 32.1. The *smoke spot number* (SSN) can also be used to quantify smoke levels.

$$\text{optical density} = \log_{10}\left(\frac{1}{1-\text{opacity}}\right) \qquad 32.1$$

Visible moisture plumes with opacities of 40% are common at large steam-generators even when there are no unburned hydrocarbons emitted. High-sulfur fuels and the presence of ammonium chloride (a byproduct of some ammonia-injection processes) seem to increase formation of visible plumes. Moisture plumes from saturated gas streams can be avoided by reheating prior to discharge to the atmosphere.

39. SPILLS, HAZARDOUS

Contamination by a hazardous material can occur accidentally (e.g., a spill) or intentionally (e.g., a previously used chemical-holding lagoon). Soil that has been contaminated with hazardous materials from spills or leaks from *underground storage tanks* (commonly known as *UST wastes*) is itself a hazardous waste.

The type of waste determines what laws are applicable, what permits are required, and what remediation methods are used.[30] With contaminated soil, spilled substances can be (a) solid and nonhazardous or (b) nonhazardous liquid petroleum products (e.g., "UST nonhazardous"), and Resource Conservation and Recovery Act- (RCRA-) listed (c) hazardous substances, and (d) toxic substances.

Cleaning up a hazardous waste requires removing the waste from whatever air, soil, and water (lakes, rivers, and oceans) that have been contaminated. The term *remediation* is sometimes used to mean the corrective steps taken to return the environment to its original condition. *Stabilization* refers to the act of reducing the waste concentrations to lower levels so that the waste can be transported, stored, or landfilled.

Remediation methods are classified as available or innovative. *Available methods* can be implemented immediately without being further tested. *Innovative methods* are new, unproven methods in various stages of study.

The remediation method used depends on the waste type. The two most common available methods are

---

[30]Permits must be obtained from the U.S. Environmental Protection Agency (EPA) whenever hazardous wastes (including hazardous-waste contaminated soil) are incinerated.

Environmental

incineration and landfilling after stabilization.[31] Incineration can occur in rotary kilns, injection incinerators, infrared incineration, and fluidized-bed combustors. Landfilling requires the contaminated soil to be stabilized chemically or by other means prior to disposal. Innovative technologies for general VOC-contaminated soil include vacuum extraction, bioremediation, thermal desorption, and soil washing. Innovative methods for dioxin removal include oxygen-enriched incineration, plasma incineration, chemical dehalogenation, and ultraviolet radiation.

40. SULFUR OXIDES

Sulfur oxides (SOx), consisting of *sulfur dioxide* (SO_2) and *sulfur trioxide* (SO_3), are the primary cause of acid rain. *Sulfurous acid* (H_2SO_3) and sulfuric acid (H_2SO_4) are produced when oxides of sulfur react with moisture in the flue gas. Both of these acids are corrosive.

$$SO_2 + H_2O \longrightarrow H_2SO_3 \qquad 32.2$$
$$SO_3 + H_2O \longrightarrow H_2SO_4 \qquad 32.3$$

Fuel switching (*coal substitution/blending* (CS/B)) is the burning of low-sulfur fuel. (Low-sulfur fuels are approximately 0.25 to 0.65% sulfur by weight, compared to high-sulfur coals with 2.4 to 3.5% sulfur.) However, unlike nitrogen oxides, which can be prevented during combustion, formation of sulfur oxides cannot be avoided when low-cost, high-sulfur fuels are burned.

Some air quality regulations regarding SOx production may be met by a combination of options. These options include fuel switching, flue gas scrubbing, derating, and allowance trading. The most economical blend of these options will vary from location to location.

In addition to fuel switching, available technology options for retrofitting existing coal-fired plants include wet scrubbing, dry scrubbing, sorbent injection, repowering with clean coal technology (CCT), and co-firing with natural gas.

41. TIRES

Discarded tires are more than a disposal problem. At 15,000 Btu/lbm (35 MJ/kg), their energy content is 80% of crude oil. Discarded tires are wasted energy resources. While tires can be incinerated, other processes can be used to gasify them to produce clean synthetic gas (i.e., *syngas*). The low-sulfur, hydrogen-rich syngas can be subsequently burned in combined-cycle plants or used as a feedstock for ammonia and methanol production, or the hydrogen can be recovered for separate use. Tires can also be converted in a nonchemical process into a strong asphalt-rubber pavement. In one process, asphalt binder is mixed with ground tire rubber and held at 375°F (190°C) for 30 to 60 minutes. Pavement is then produced in the normal fashion.

42. VOLATILE INORGANIC COMPOUNDS

Volatile inorganic compounds (VICs) include H_2S, NOx (except N_2O), SO_2, HCl, NH_3, and many other less common compounds.

43. VOLATILE ORGANIC COMPOUNDS

Volatile organic compounds (VOCs) (e.g., benzene, chloroform, formaldehyde, methylene chloride, napthalene, phenol, toluene, and trichloroethylene) are highly soluble in water. (VOCs are listed in Tables 34.2 and 34.7.) VOCs that have leaked from storage tanks or have been discharged often end up in groundwater and drinking supplies.

There are a large number of methods for removing VOCs, including incineration, chemical scrubbing with oxidants, water washing, air or steam stripping, activated charcoal adsorption processes, SCR (selective catalytic reduction), and bioremediation. Incineration of VOCs is fast and 99%+ effective, but incineration requires large amounts of fuel and produces NOx. Using SCR with heat recovery after incineration reduces the energy input but adds to the expense.

44. WATER VAPOR

Water vapor emitted from sources is not generally considered to be a pollutant.

---

[31]Other innovative technologies include in situ and ex situ bioremediation, chemical treatment, in situ flushing, in situ vitrification, soil vapor extraction, soil washing, solvent extraction, and thermal desorption.

33 Disposition of Hazardous Materials

1. GENERAL STORAGE

Storage of *hazardous materials (hazmats)* is often governed by local building codes in addition to state and federal regulations. Types of construction, maximum floor areas, and building layout may all be restricted.[1]

Good engineering judgment is called for in areas not specifically governed by the building code. Engineering consideration will need to be given to the following aspects of storage facility design: (a) spill containment provisions, (b) chemical resistance of construction and storage materials, (c) likelihood of and resistance to explosions, (d) exiting, (e) ventilation, (f) electrical design, (g) storage method, (h) personnel emergency equipment, (i) security, and (j) spill cleanup provisions.

2. STORAGE TANKS

Underground storage tanks (USTs) have traditionally been used to store bulk chemicals and petroleum products. Fire and explosion risks are low with USTs, but subsurface pollution is common since inspection is limited. Since 1988, the U.S. Environmental Protection Agency (EPA) has required USTs to have secondary containment, corrosion protection, and leak detection. UST operators also must carry insurance in an amount sufficient to clean up a tank failure.

Because of the cost of complying with UST legislation, above-ground storage tanks (ASTs) are becoming more popular. AST strengths and weaknesses are the reverse of USTs: ASTs reduce pollution caused by leaks, but the expected damage due to fire and explosion is greatly increased. Because of this, some local ordinances prohibit all ASTs for petroleum products.[2]

---

[1]For example, flammable materials stored in rack systems are typically limited to heights of 25 ft (8.3 m).

[2]The American Society of Petroleum Operations Engineers (AS-POE) policy statement states, "Above-ground storage of liquid hydrocarbon motor fuels is inherently less safe than underground storage. Above-ground storage of Class 1 liquids (gasoline) should be prohibited at facilities open to the public."

The following factors should be considered when deciding between USTs and ASTs: (a) space available, (b) zoning ordinances, (c) secondary containment, (d) leak detection equipment, (e) operating limitations, and (f) economics.

Most ASTs are constructed of carbon or stainless steel. These provide better structural integrity and fire resistance than fiberglass-reinforced plastic and other composite tanks. Tanks can be either field-erected or factory-fabricated (capacities greater than approximately 50,000 gal (190 kL). Factory-fabricated ASTs are usually designed according to UL-142 (Underwriters Laboratories *Standard for Safety*), which dictates steel type, wall thickness, and characteristics of compartments, bulkheads, and fittings. Most ASTs are not pressurized, but those that are must be designed in accordance with the ASME *Boiler and Pressure Vessel Code*, Section VIII.

NFPA 30 ("Flammable and Combustible Liquids," National Fire Protection Association, Quincy, MA) specifies the minimum separation distances between ASTs, other tanks, structures, and public right-of-ways. The separation is a function of tank type, size, and contents. NFPA 30 also specifies installation, spill control, venting, and testing.

ASTs must be double-walled, concrete-encased, or contained in a dike or vault to prevent leaks and spills, and they must meet fire codes. Dikes should have a capacity in excess (e.g., 110 to 125%) of the tank volume. ASTs (as do USTs) must be equipped with overfill prevention systems. Piping should be above-ground wherever possible. Reasonable protection against vandalism and hunters' bullets is also necessary.[3]

Though it's a good idea, ASTs are not typically required to have leak-detection systems. Methodology for leak detection is evolving, but currently includes vacuum or pressure monitoring, electronic gauging, and optical and sniffing sensors. Double-walled tanks may also be fitted with sensors within the interstitial space.

Operationally, ASTs present special problems. In hot weather, volatile substances vaporize and represent an additional leak hazard. In cold weather, viscous contents may need to be heated (often by steam tracing).

---

[3]Approximately 20% of all spills from ASTs result from vandalism.

Environmental

ASTs are not necessarily less expensive than USTs, but they are generally thought to be so. Additional hidden costs of regulatory compliance, secondary containment, fire protection, and land acquisition must also be considered.

3. DISPOSITION OF HAZARDOUS WASTES

When a hazardous waste is disposed of, it must be taken to a registered *treatment, storage, or disposal facility* (TSDF). The EPA's *land ban* specifically prohibits the disposal of hazardous wastes on land prior to treatment. Incineration at sea is also prohibited. Wastes must be treated to specific maximum concentration limits by specific technology prior to disposal in landfills.

Once treated to specific regulated concentrations, hazardous waste residues can be disposed of by incineration or landfilling, or less frequently, by deep-well injection. All disposal facilities must meet detailed design and operational standards.

34 Environmental Remediation

Nomenclature

Symbol	Description	US	SI
A	area	ft^2	m^2
B	cyclone inlet width	ft	m
C	concentration	various	various
C	ESP constant	–	–
D	diameter	ft	m
g	gravitational acceleration	ft/sec^2	m/s^2
G	gas loading rate	ft^3/min-ft^2	m^3/s·m^2
h	head	ft	m
H	cyclone inlet height	ft	m
H	Henry's law constant	atm	kPa
HHV	higher heating value	MMBtu/lbm	kJ/kg
HTU	height of a transfer unit	ft	m
k	reaction rate constant	1/sec	1/s
K	constant	–	–
K_G	coefficient of gas mass transfer	ft/sec	m/s
$K_G a$	gas mass transfer coefficient	1/sec	1/s
$K_L a$	liquid mass transfer coefficient	1/sec	1/s
L	length	ft	m
L	liquid loading rate	ft^3/min-ft^2	m^3/s·m^2
m	mass	lbm	kg
NTU	number of transfer units	–	–
p	partial pressure	atm	atm
Q	volumetric flow rate	ft^3/sec	m^3/s
r	cyclone radius	ft	m
R	stripping factor	–	–
S	separation factor	–	–
SFC	specific feed characteristics	Btu/lbm	J/kg
t	time	sec	s
T	temperature	°R	K
v	velocity	ft/sec	m/s
w	drift velocity	ft/sec	m/s
x	fraction by weight	–	–
z	packing height	ft	m

Symbols

Symbol	Description	US	SI
α	coefficient of linear thermal expansion	ft/ft-°F	m/m·°C
η	efficiency	%	%

Environmental

Superscripts

*	equilibrium

Subscripts

a	air
G	gas
L	liquid
O	overall
p	plate
t	total
w	water

1. INTRODUCTION

This chapter discusses (in alphabetical order) the methods and equipment that can be used to reduce or eliminate pollution. Legislation often requires use of the *best available control technology* (BACT—also known as the *best available technology*—BAT) and the *maximum achievable control technology* (MACT), *lowest achievable emission rate* (LAER), and *reasonably available control technology* (RACT) in the design of pollution-prevention systems.

2. ABSORPTION, GAS (GENERAL)

Gas *absorption processes* remove a gaseous substance (the *target substance*) from a gas stream by dissolving it in a liquid solvent.[1] Absorption can be used for flue gas cleanup (FGC) to remove sulfur dioxide, hydrogen sulfide, hydrogen chloride, chlorine, ammonia, nitrogen oxides, and light hydrocarbons. Gas absorption equipment includes packed towers, spray towers and chambers, and venturi absorbers.[2]

3. ABSORPTION, GAS (SPRAY TOWERS)

In a general spray tower or spray chamber (i.e., a *wet scrubber*), liquid and gas flow countercurrently or crosscurrently. The gas moves through a liquid spray that is carried downward by gravity. A mist eliminator removes entrained liquid from the gas flow. The liquid can be recirculated. The removal efficiency is moderate.

---

[1]Scrubbing, gas absorption, and stripping are distinguished by their target substances, the carrier flow phase, and the directions of flow. *Scrubbing* is the removal of particulate matter from a gas flow by exposing the flow to a liquid or slurry spray. *Gas absorption* is a countercurrent operation for the removal of a target gas from a gas mixture by exposing the mixture to a liquid bath or spray. *Stripping*, also known as *gas desorption* (also a countercurrent operation), is the removal of a dissolved gas or other volatile component from liquid by exposing the liquid to air or steam. Stripping is the reverse operation of gas absorption.

Packed towers can be used for both gas absorption and stripping processes, and the processes look similar. The fundamental difference is that in gas absorption processes, the target substance (i.e., the substance to be removed) is in the gas flow moving up the tower, and in stripping processes, the target substance is in the liquid flow moving down the tower.

[2]Spray and packed towers, though they are capable, are not generally used for desulfurization of furnace or incineration combustion gases, as scrubbers are better suited for this task.

Spray towers are characterized by their low pressure drops (typically 1 to 2 iwg[3] (0.25 to 0.5 kPa)) and their liquid-to-gas ratios (typically 20 to 100 gal/1000 ft^3 (3 to 14 L/m^3)). For self-contained units, fan power is also low—approximately 3×10^{-4} kW/ft^3 (0.01 kW/m^3) of gas moved.

Flooding of spray towers is an operational difficulty where the liquid spray is carried up the column by the gas stream. Flooding occurs when the gas stream velocity approaches the *flooding velocity*. To prevent this, the tower diameter is chosen such that the gas velocity is approximately 50 to 75% of the flooding velocity.

Flue gas desulfurization (FGD) wet scrubbers can remove approximately 90 to 95% of the sulfur, with efficiencies of 98% claimed by some installations. Stack effluent leaves at approximately 150°F (65°C) and is saturated (or nearly saturated) with moisture. Dense steam plumes may be present unless the scrubbed stream is reheated.

Collected sludge waste and flyash are removed within the scrubber by purely inertial means. After being dewatered, the sludge is landfilled.[4]

Figure 34.1 *Spray Tower Absorber*

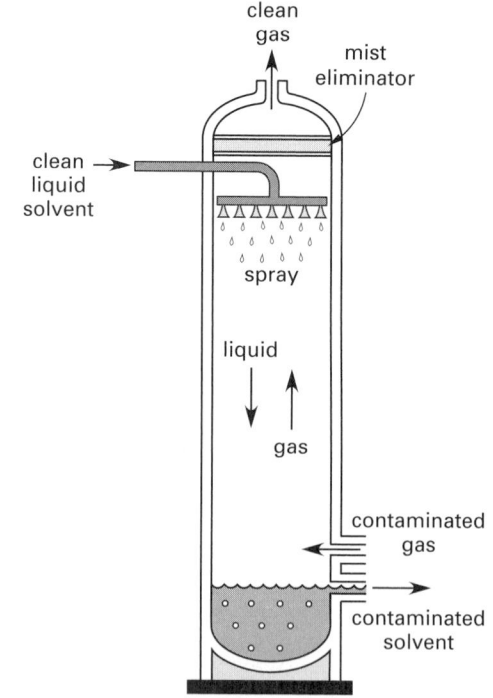

Spray towers, though simple in concept and requiring little energy to operate, are not simple to operate and maintain. Other disadvantages of spray scrubbers include high water usage, generation of wastewater or

---

[3] "iwg" is the abbreviation for "inches water gage," also referred to as iwc for "inches water column" and "inches of water."

[4]A typical 500 MW plant burning high-sulfur fuel can produce as much as 10^7 ft^3 (300,000 m^3) of dewatered sludge per year.

sludge, and low efficiency at removing particles smaller than 5 μm. Relative to dry scrubbing, the production of wet sludge and the requirement for a sludge-handling system is the major disadvantage of wet scrubbing.

4. ABSORPTION, GAS (IN PACKED TOWERS)

In a packed tower, clean liquid flows from top to bottom over the tower packing media, usually consisting of synthetic engineered shapes designed to maximize liquid-surface contact area. (See Fig. 34.11.) The contaminated gas flows countercurrently, from bottom to top, although crosscurrent designs also exist. As in spray towers, the liquid can be recirculated, and a mist-eliminator is used.

Pressure drops are in the 1 to 8 iwg (0.25 to 2.0 kPa) range. Typical liquid-to-gas ratios are 10 to 20 gal/1000 ft^3 (1 to 3 L/m^3). Although the pressure drop is greater than in spray towers, the removal efficiency is much higher.

Figure 34.2 *Packed Bed Spray Tower*

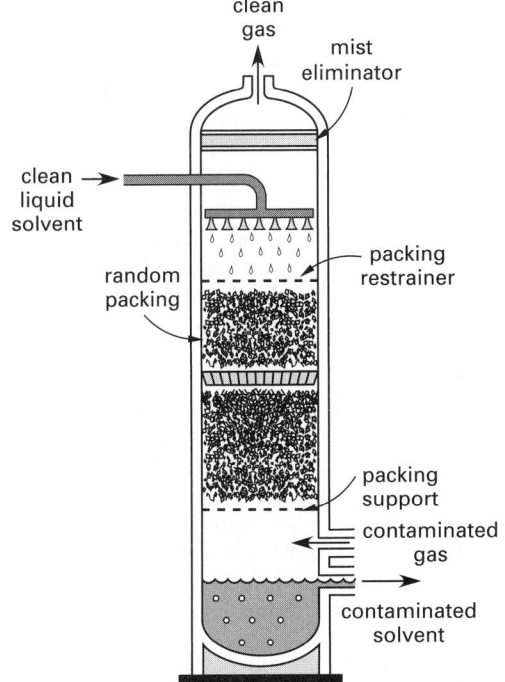

5. ADSORPTION (ACTIVATED CARBON)

Granular activated carbon (GAC), also known as *activated carbon* and *activated charcoal* (AC), processes are effective in removing a number of compounds, including volatile organic compounds (VOCs), heavy metals (e.g., lead and mercury), and dioxins.[5] AC is an effective

adsorbent[6] for both air and water streams, and a VOC removal efficiency of 99% can be achieved. AC can be manufactured from almost any carbonaceous raw material, but wood, coal, and coconut shell are widely used.

Pollution control processes use AC in both solvent capture-recovery (i.e., recycling) and capture-destruction processes. AC is available in powder and granular form. Granules are preferred for use in recovery systems since the AC can be regenerated when *breakthrough*, also known as *breakpoint*, occurs. This is when the AC has become saturated with the solvent and traces of the solvent begin to appear in the exit air. Until breakthrough, removal efficiency is essentially constant. The *retentivity* of the AC is the ratio of adsorbed solvent mass to carbon mass.

6. ADSORPTION (SOLVENT RECOVERY)

Activated carbon (AC) for solvent recovery is used in a cyclic process where it is alternately exposed to the target substance and then regenerated by the removal of the target substance. For solvent recovery to be effective, the VOC inlet concentration should be at least 700 ppm. Regeneration of the AC is accomplished by heating, usually by passing low-pressure (e.g., 5 psig) steam over the AC to raise the temperature above the solvent-capture temperature.

There are three main processes for solvent recovery. In a traditional *open-loop recovery system*, a countercurrent air stripper separates VOC from the incoming stream. The resulting VOC air/vapor stream passes through an AC bed. Periodically, the air/vapor stream is switched to an alternate bed, and the first bed is regenerated by passing steam through it. The VOC is recovered from the steam. The VOC-free air stream is freely discharged. Operation of a *closed-loop recovery system* is the same as an open loop system, except that the VOC-free air stream is returned to the air stripper.

For VOC-laden gas flows from 5000 to 100,000 SCFM (2.3 to 47 kL/s), a third type of solvent recovery system involves a *rotor concentrator*. VOCs are continuously adsorbed onto a multilayer, corrugated wheel whose honeycomb structure has been coated with powdered AC. The wheel area is divided into three zones: one for adsorption, one for desorption (i.e., regeneration), and one for cooling. Each of the zones is isolated from the other by tight sealing. The wheel rotates at low speed, continuously exposing new portions of the streams to the appropriate zones. Though the equipment is expensive, operational efficiencies are high— 95 to 98%.

---

[5]The term *activated* refers to the high-temperature removal of tarry substances from the interior of the carbon granule, leaving a highly porous structure.

[6]An *adsorbent* is a substance with high surface area per unit weight, an intricate pore structure, and a hydrophobic surface. An *adsorbent material* traps substances in fluid (liquid and gaseous) form on its exposed surfaces. In addition to activated carbon, other common industrial adsorbents include alumina, bauxite, bone char, Fuller's earth, magnesia, and silica gel.

Figure 34.3 *Stripping-AC Process with Recovery*

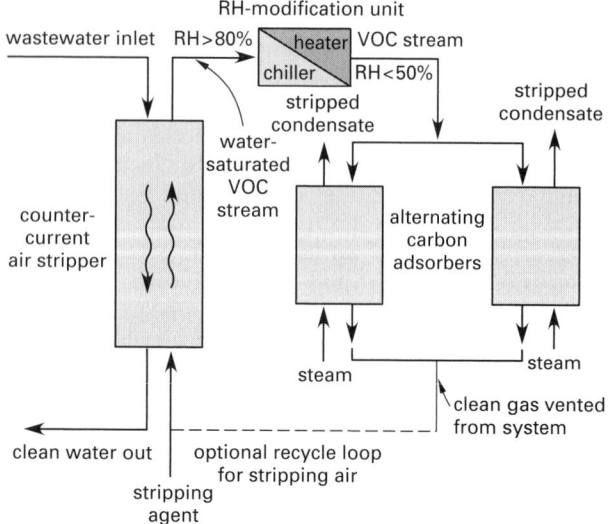

9. ADVANCED OXIDATION

The term *advanced oxidation* refers to the use of ozone, hydrogen peroxide, ultraviolet radiation, and other exotic methods that produce free hydroxyl radicals (OH^-). (See also Sec. 38: Scrubbing, Chemical.)

Table 34.1 *Relative Oxidation Powers of Common Oxidants*

oxidant	oxidation power (relative to Cl)
fluorine	2.25
hydroxyl radical (OH)	2.05
ozone (O_3)	1.52
permanganate radical (MnO_4)	1.23
chlorine dioxide (ClO_2)	1.10
hypochlorous acid	1.10
chlorine	1.00
bromine	0.80
iodine	0.40
oxygen	0.29

7. ADSORPTION, HAZARDOUS WASTE

AC is particularly attractive for flue gas cleanup (FGC) at installations that burn spent oil, electrical cable, biosolids (i.e., sewage sludge), waste solvents, or tires as supplemental fuels. As with liquids, flue gases can be cleaned by passing them through fixed AC. Heavy metal dioxins are quickly adsorbed by the AC—in the first 8 in (20 cm) or so. AC injection (direct or spray) can remove 60 to 90% of the target substance present.

The spent AC creates its own waste disposal problem. For some substances (e.g., dioxins), AC can be incinerated. Heavy metals in AC must be removed by a wash process. Another process in use is vitrifying the AC and flyash into a glassy, unleachable substance. *Vitrification* is a high-temperature process that turns incinerator ash into a safe, glass-like material. In some processes, heavy-metal salts are recovered separately. No gases and no hazardous wastes are formed.

8. ADVANCED FLUE GAS CLEANUP

Most air pollution control systems reduce NOx in the burner and remove SOx in the stack. *Advanced flue gas cleanup* (AFGC) methods combine processes to remove both NOx and SOx in the stack. Particulate matter is removed by an electrostatic precipitator or baghouse as is typical. Promising AFGC methods include wet scrubbing with metal chelates such as ferrous ethylenediaminetetraacetate (Fe(II)-EDTA) (NOx/SOx removal efficiencies of 60%/90%), adding sodium hydroxide injection to dry scrubbing operations (35% NOx removal), in-duct sorbent injection of urea (80%/ NOx removal), in-duct sorbent injection of sodium bicarbonate (35% NOx removal), and the NOXSO process using a fluidized-bed of sodium-impregnated alumina sorbent at approximately 250°F (120°C) (70 to 90%/90% NOx/SOx removal).

10. BAGHOUSES

Baghouses have a reputation for excellent particulate removal, down to 0.005 grains/ft³ (0.18 grains/m³) (dry), with particulate emissions of 0.01 grains/ft³ (0.35 grains/m³) being routine. Removal efficiencies are in excess of 99% and are often as high as 99.99% (weight basis). Fabric filters have a high efficiency for removing particular matter less than 10 μm in size. Because of this, baghouses are effective at collecting air toxics, which preferentially condense on these particles at the baghouse operating temperature—less than 300°F (150°C).

Baghouse filter fabric has micro-sized holes but may be felted or woven, with the material depending on the nature of the flue gas and particulates. Woven fabrics are better suited for lower filtering velocities in the 1 to 2 ft/min (0.3 to 0.7 m/min) range, while felted fabrics are better at 5 ft/min (1.7 m/min). The fabric is often used in tube configuration, but envelopes (i.e., flat bags) and pleated cartridges are also available.

The particulate matter collects on the outside of the bag, forming a cake-like coating. If lime has been introduced into the stream in a previous step, some of the cake will consist of unreacted lime. As the gases pass through this cake, additional neutralizing takes place.

When the pressure drop across the filter reaches a preset limit (usually 6 to 8 iwg (1.5 to 2 kPa)), the cake is dislodged by mechanical shaking, reverse-air cleaning, or pulse-jetting. The dislodged flyash cake falls into collection hoppers below the bags. Flyash is transported by pressure- or vacuum-conveying systems to the conditioning system (consisting of surge bins, rotary feeders, and pug mills).

Environmental

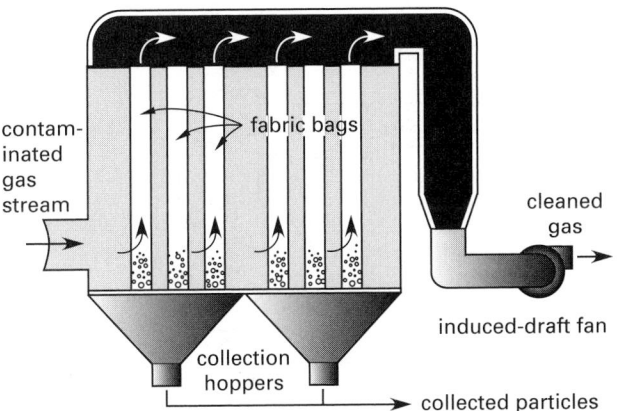

Figure 34.4 Typical Baghouse

Baghouses are characterized by their air-to-cloth ratios and pressure drop. The *air-to-cloth ratio*, also known as *filter ratio, superficial face velocity*, and *filtering velocity*, is the ratio of the air volumetric flow rate in ft^3/min (m^3/s) to the exposed surface area in ft^2 (m^2). After canceling, the units are ft/min (m/s), hence the name *filtering velocity*. The higher the ratio, the smaller the baghouse and the higher the pressure drop. Shaker- and reverse-air baghouses with woven fabrics have typical air-to-cloth ratios ranging from 2.0 to 2.5 ft/min (0.01 to 0.0125 m/s). Pulse-jet collectors with felted fabrics have higher ratios, ranging from 3.5 to 5.0 ft/min (0.0175 to 0.025 m/s).

Advantages of baghouses include high efficiency and performance that is essentially independent of flow rate, particle size, and particle (electrical) resistivity. Also, baghouses produce the lowest opacity (generally less than 10, which is virtually invisible). Disadvantages include clogging, difficult cleaning, and bag breakage.

11. BIOREMEDIATION

Bioremediation encompasses the methods of biofiltration, bioreaction, bioreclamation, activated sludge, trickle filtration, fixed-film biological treatment, landfilling, and injection wells (for in situ treatment of soils and groundwater). Bioremediation relies on microorganisms in a moist, oxygen-rich environment to oxidize solid, liquid, or gaseous organic compounds, producing carbon dioxide and water. Bioremediation is effective for removing volatile organic compounds (VOCs) and easy-to-degrade organic compounds such as BTEXs (benzene, toluene, ethylbenzene, and xylene).[7] Wood-preserving wastes, such as creosote and other polynuclear aromatic hydrocarbons (PAHs), can also be treated.

Bioremediation can be carried out in open tanks, packed columns, beds of porous synthetic materials, composting piles, or soil. The effectiveness of bioremediation depends on the nature of the process, the time and

---

[7]BTEXs are common ingredients in gasoline.

physical space available, and the degradability of the substance. Table 34.2 categorizes gases according to their general degradabilities.

Table 34.2 Degradability of Volatile Organic and Inorganic Gases

rapidly degradable VOCs	rapidly reactive VICs	slowly degradable VOCs	very slowly degradable VOCs
alcohols	H_2S	hydrocarbons[a]	halogenated
aldehydes	NOx (but	phenols	hydrocarbons[b]
ketones	not N_2O)	methylene	polyaromatic
ethers	SO_2	chloride	hydrocarbons
esters	HCl		CS_2
organic acids	NH_3		
amines	PH_3		
thiols	SiH_4		
other	HF		
molecules			
containing O,			
N, or S			
functional groups			

[a]Aliphatics degrade faster than aromatics, such as xylene, toluene, benzene, and styrene.
[b]such as trichloroethylene, trichloroethane, carbon tetrachloride, and pentachlorophenol

Though slow, limited by microorganisms with the specific affinity for the chemicals present, and susceptible to compounds that are toxic to the microorganisms, bioremediation has the advantage of destroying substances rather than merely concentrating them. Bioremediation is less effective when a variety of different compounds are present simultaneously.

12. BIOFILTRATION

The term *biofiltration* refers to the use of composting and soil beds. A *biofilter* is a bed of soil or compost through which runs a distribution system of perforated pipe. Contaminated air or liquid flows through the pipes and into the bed. Volatile organic compounds (VOCs) are oxidized to CO_2 by microorganisms. Volatile inorganic compounds (VICs) are oxidized to acids (e.g., HNO_3 and H_2SO_4) and salts.

Biofilters require no fuel or chemicals when processing VOCs, and the operational lifetime is essentially infinite. For VICs, the lifetime depends on the soil's capacity to neutralize acids. Though reaction times are long and absorption capacities are low (hence the large areas required), the oxidation continually regenerates (rather than depletes) the treatment capacity. Once operational, biofiltration is probably the least expensive method of eliminating VOCs and VICs.

The *removal efficiency* of a biofilter is given by Eq. 34.1. k is an empirical *reaction rate constant*, and t is the *bed residence time* of the carrier fluid (water or air) in the bed. The reaction rate constant depends on the temperature but is primarily a function of the biodegradability of the target substance. Biofiltration typically removes 80 to 99% of volatile organic and inorganic compounds (VOCs and VICs).

$$\eta = 1 - \frac{C_{\text{out}}}{C_{\text{in}}} = 1 - e^{-kt} \qquad \textit{34.1}$$

13. BIOREACTION

Bioreactors (reactor tanks) are open or closed tanks containing dozens or hundreds of slowly rotating disks covered with a biological film of microorganisms (i.e., colonies). Closed tanks can be used to maintain anaerobic or other design atmosphere conditions. For example, methanotrophic bacteria are useful in breaking down chlorinated hydrocarbons (e.g., trichlorethylene (TCE), dichloroethylene (DCE), and vinyl chloride (VC)) that would otherwise be considered nonbiodegradable. Methanotrophic bacteria derive their food from methane gas that is added to the bioreactor. An enzyme, known as MMO, secreted by the bacteria breaks down the chlorinated hydrocarbons.

14. BIOVENTING

Bioventing is the treatment of contaminated soil in a large plastic-covered tank. Clean air, water, and nutrients are continuously supplied to the tank while off-gases are suctioned off. The off-gas is cleaned with activated carbon (AC) adsorption or with thermal or catalytic oxidation prior to discharge. Bioventing has been used successfully to remove volatile hydrocarbon compounds (e.g., gasoline and BTEX compounds) from soil.

15. COAL CONDITIONING

Coal intended for electrical generating plants can be modified into "self-scrubbing" coal by conditioning prior to combustion. Conditioning consists of physical separation by size, by cleaning (e.g., cycloning to remove noncombustible material, including up to 90% of the pyritic sulfur), and by the optional addition of limestone and other additives to capture SO_2 during combustion. Small-sized material is pelletized to reduce loss of fines and particulate emissions (dust).

16. CYCLONE SEPARATORS

Inertial separators of the *double-vortex, single cyclone* variety are suitable for collecting medium- and large-sized (i.e., greater than 15 μm) particles from spot sources. During operation, particulate matter in the incoming gas stream spirals downward at the outside and upward at the inside.[8] The particles, because of their greater mass, move toward the outside wall, where they drop into a collection bin.

---

[8]The number of gas revolutions varies approximately between 0.5 and 10.0, with averages of 1.5 revolutions for simple cyclones and 5 revolutions for high-efficiency cyclones.

Figure 34.5 *Double-Vortex, Single Cyclone*

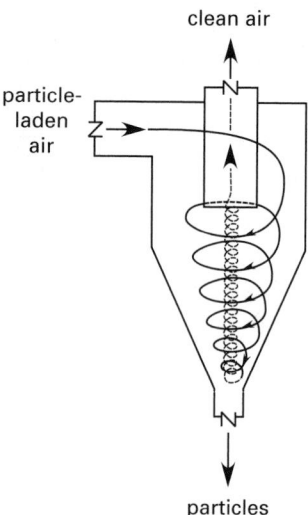

The *cut size* is the diameter of particles collected with a 50% efficiency. Separation efficiency varies directly with (a) particle diameter, (b) particle density, (c) inlet velocity, (d) cyclone body length (ratio of cyclone body diameter to outlet diameter), (e) smoothness of inner wall, (f) number of gas revolutions, and (g) amount of particles in the flow. Collection efficiency decreases with increases in (a) gas viscosity, (b) gas density, (c) cyclone diameter, (d) gas outlet diameter, (e) gas inlet duct width, and (f) inlet area.

Collection efficiencies are not particularly high, and dusts (5 to 10 μm) are too fine for most cyclones. For geometrically similar cyclones, the collection efficiency varies directly with the dimensionless *separation factor, S*.

$$S = \frac{v_{\text{inlet}}^2}{rg} \qquad \textit{34.2}$$

Pressure drop, h, in feet of air across a cyclone can be roughly (i.e., with an accuracy of only approximately $\pm 30\%$) estimated by Eq. 34.3. H and B are the height and width of the rectangular cyclone inlet duct, respectively; D is the gas exit duct diameter; and K is an empirical constant that varies from approximately 7.5 to 18.4, with $K = 13$ being a common design value.

$$h = \frac{KBHv_{\text{inlet}}^2}{2gD^2} \qquad \textit{34.3}$$

17. DECHLORINATION

Industrial and municipal wastewaters containing excessive amounts of *total residual chlorine* (TRC) must be dechlorinated prior to discharge. Sulfur dioxide, sodium metabisulfate, and sulfite salts are effective dechlorinating agents. For sulfur dioxide, the dose is approximately 0.9 lbm per pound (kg per kg) of chlorine to be removed. Reaction is essentially instantaneous, being completed within ten pipe diameters at turbulent flow. For open channels, a submerged weir may be necessary to obtain the necessary turbulence. TRC is reduced to less than detectable levels.

18. ELECTROSTATIC PRECIPITATORS

Electrostatic precipitators (ESPs) are used to remove particulate matter from gas streams. Collection efficiency for particulate matter is usually in the 95 to 99% range. ESPs are preferred over scrubbers because they are more economical to operate, dependable, and predictable, and because they don't produce a moisture plume.

ESPs used on steam generator/electrical utility units treat gas that is approximately 280 to 300°F (140 to 150°C) with moisture being superheated.[9] This high temperature enhances buoyancy and plume dissipation. However, ESPs generally cannot be used with moist flows, mists, or sticky or hygroscopic particles. Scrubbers should be used in those cases. Relatively humid flows can be treated, although entrained water droplets can insulate particles, lowering their resistivities.

In operation, the gas passes over negatively charged tungsten *corona wires* or grids. Particles are attracted to positively charged collection plates.[10] The speed at which the particles move toward the plate is known as the *drift velocity*, w. Drift velocity is approximately 0.20 to 0.30 ft/sec (0.06 to 0.09 m/s), with 0.25 ft/sec (0.075 m/s) being a reasonable design value. Periodically, *rappers* vibrate the collection plates and dislodge the particles, which drop into collection hoppers.

One of the most important factors affecting the collection efficiency is the *specific collection area* (SCA), the area (in ft^2) of the collection plates divided by the volumetric flow rate (in actual ft^3/min) of the air. The higher the SCA, the higher will be the collection efficiency, with 900 ft^2/1000 ft^3/min (3000 m^2/1000 m^3/min) being at the high end of the scale. Pressure drop across ESPs is low (e.g., less than approximately 0.5 iwg (0.13 kPa)).

Use of ESPs is most efficient when particles have resistivities in the 10^{10} to 10^{14} Ω·cm range. If the particle resistivity is much lower, particles may give up their charge upon contacting the collection plate and become reentrained in (i.e., reintroduced into) the gas. If the resistivity is too high, the particles may be difficult to dislodge and may insulate the collection plates. Very small particles (0.1 to 1.0 μm) reduce the corona efficiency, and particles less than 2 to 3 μm are more easily reentrained.

The theoretical efficiency of a plate precipitator is calculated from Eq. 34.4.[11] K is a measure of the ease with which the particles can be captured, and C depends on the precipitator size, gas velocity and volume, and voltage. K and C are constants for a particular installation. t is the time the particle remains in the electrical field. w is the drift velocity, A_p is the collection plate area, and Q is the volumetric air flow rate.

$$\eta = 1 - K^{Ct} = 1 - e^{-wA_\mathrm{p}/Q} \qquad \textit{34.4}$$

Advantages of ESPs include high reliability, low maintenance, low power requirements, and low pressure drop. Disadvantages are sensitivity to particle size and resistivity, and the need to heat ESPs during start-up and shut-down to avoid corrosion from acid gas condensation.

Table 34.3 *Typical Electrostatic Precipitator Design Parameters*

parameter	typical range U.S.	typical range SI
efficiency	90 to 98%	
gas velocity	2 to 4 ft/sec	0.6 to 1.2 m/s
gas temperature		
standard	≤700°F	≤370°C
high-temperature	≤1000°F	≤540°C
special	≤1300°F	≤700°C
drift velocity	0.1 to 0.7 ft/sec	0.03 to 0.21 m/s
treatment/ residence time	2 to 10 sec	
draft pressure loss	0.1 to 0.5 iwg	0.025 to 0.125 kPa
plate spacing	12 to 16 in	30 to 41 cm
plate height	30 to 50 ft	9 to 15 m
plate length/ height ratio	1.0 to 2.0	
applied voltage	30 to 75 kV	

19. FLUE GAS RECIRCULATION

NOx emissions can be reduced when thermal dissociation is the primary NOx source (as it is when low-nitrogen fuels such as natural gas are burned) by recirculating a portion (15 to 25%) of the flue gas back into the furnace. This process is known as *flue gas recirculation* (FGR). The recirculated gas absorbs heat energy from the flame and lowers the peak temperature. Thermal NOx formation can be reduced by up to 50%. The recirculated gas should not be more than 600°F (315°C).

20. FLUIDIZED-BED COMBUSTORS

Fluidized-bed combustors (FBCs) are increasingly being used in steam/electric generation systems and for destruction of hazardous wastes. A *bubbling bed FBC,*

---

[9]The flue gas, at approximately 1400°F (760°C), is cooled to this temperature range in a conditioning tower. Injected water increases the moisture content of the flue gas to approximately 25% by volume.

[10]The *tubular ESP*, used for collecting moist or sticky particles, is a variation on this design.

---

[11]Design of ESPs is almost entirely empirical, being based on previous experience with similar processes or pilot studies.

as shown in Fig. 34.6, consists of four major components: (a) a windbox (plenum) that receives the fluidizing/combustion air, (b) an air distribution plate that transmits the air at 10 to 30 ft/sec (3 to 9 m/s) from the windbox to the bed and prevents the bed material from sifting into the windbox, (c) the fluid bed of inert material (usually sand or product ash), and (d) the freeboard area above the bed.

During the operation of a bubbling bed fluidized-bed combustor, the inert bed is levitated by the upcoming air, taking on many characteristics of a fluid (hence, the name FBC). The bed "boils" vigorously, offering an extremely large heat-transfer area, resulting in the thorough mixing of feed combustibles and air. Combustible material usually represents less than 1% of the bed mass, so the rest of the bed acts as a large thermal flywheel.

Figure 34.6 *Fluidized-Bed Combustor*

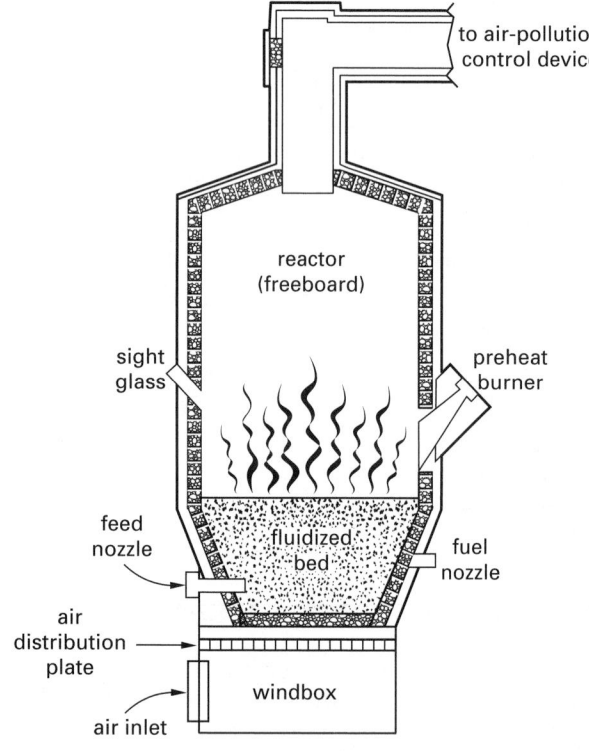

Combustion of volatile materials is completed in the freeboard area. Ash is generally reduced to small size so that it exits with the flue gas. In some cases (e.g., deliberate pelletization or wastes with high ash content), ash can accumulate in the bed. The fluid nature of the bed allows the ash to float on its surface, where it is removed through an overflow drain.

Most FBC systems use forced-air with a single blower at the front end. If there are significant losses due to heat recovery or pollution control systems, an exhaust fan may also be used. In that case, the *null (balanced draft) point* should be in the freeboard area.

Large variations in the composition of the flue gases (known as *puffing*) is minimized by the long residence time and the large heat reservoir of the bed. Air pollution control equipment common to most boilers and incinerators is used with fluidized-bed combustors. Either wet scrubbers or baghouses can be used.

The temperature of the bed can be as high as approximately 1900°F (1040°C), though in most applications, temperatures this high are neither required nor desirable.[12] Most systems operate in the 1400 to 1650°F (760 to 900°C) range.

Three main options exist for reducing temperatures with overautogeneous fuels:[13] (a) Water can be injected into the bed. This has the disadvantage of reducing downstream heat recovery. (b) Excess air can be injected. This requires the entire system to be sized for the excess air, increasing its cost. (c) Heat-exchange coils can be placed within the bed itself, in which case, the name *fluidized-bed boiler* is applicable.

Air can be preheated to approximately 1000 to 1250°F (540 to 680°C) for use with subautogenous fuels, or auxiliary fuel can be used.

Contempory FBC boilers for steam/electricity generation are typically of the *circulating fluidized-bed boiler design*. An important aspect of FBC boiler operation is the in-bed gas desulfurization and dechlorination that occurs when limestone and other solid reagents are injected into the combustion area. In addition, circulating fluidized-bed boilers have very low NOx emissions (i.e., less than 200 ppm).

21. INJECTION WELLS

Properly treated and stabilized liquid and low-viscosity wastes can be injected under high pressure into appropriate strata 2000 to 6000 ft (600 to 1800 m) below the surface. The wastes displace natural fluids, and the injection well is capped to maintain the pressure. Injection wells fail primarily by waste plumes through fractures, cracks, fault slips, and seepage around the well casing.

22. INCINERATORS, FLUIDIZED-BED COMBUSTION

In an FBC, combustion is efficient and excess air required is low (e.g., 25 to 50%). Destruction is essentially complete due to the long residence time (5 to 8 sec for gases, and even longer for solids), high turbulence, and exposure to oxygen. Combustion control is simple, and combustion is often maintained within a 15°F (8°C) band. Both overautogeneous and subautogenous feeds can be handled. Due to the thermal mass

---

[12] It is a common misconception that extremely high temperatures are required to destroy hazardous waste.

[13] A *subautogenous waste* has a heating value too low to sustain combustion and requires a supplemental fuel. Conversely, an *overautogeneous waste* has a heating value in excess of what is required to sustain combustion and requires temperature control.

of the bed, the FBC temperature drops slowly—at the rate of about 10°F/hr (5°C/h) after shutdown. Start-up is fast and operation can be intermittent.

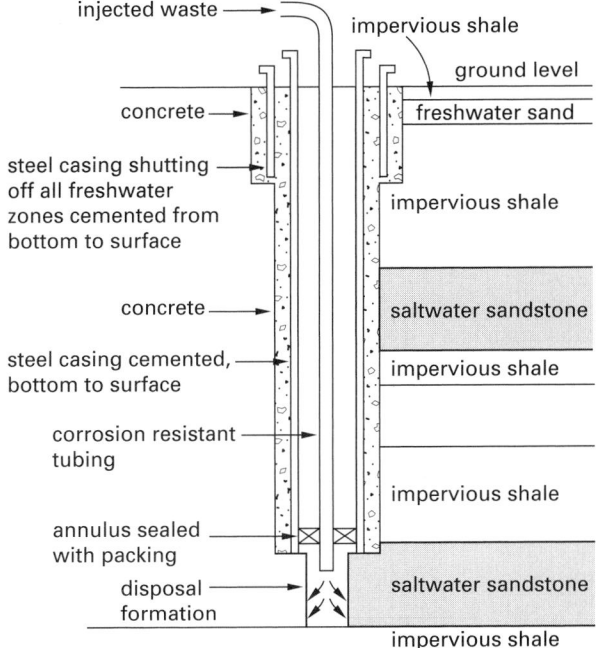

Figure 34.7 *Typical Injection Well Installation*

Relative to other hazardous waste disposal methods, NOx and metal emissions are low, and the organic content of the ash is very low (e.g., below 1%). Because of the long residence time, FBC systems do not usually require afterburners or secondary combustion chambers (SCCs). Due to the turbulent combustion process, the combustion temperature can be 200 to 300°F (110 to 170°C) lower than in rotary kilns. These factors translate into fuel savings, fewer or no NOx emissions, and lower metal emissions.

Most limitations of FBC systems relate to the feed. The feed must be of a form (small size and roughly regular in shape) that can be fluidized. This makes FBCs ideal for non-atomizable liquids, slurries, sludges, tars, and granular solids. It is more appropriate (and economical) to destroy atomizable liquid wastes in a boiler or special injection furnace and large bulky wastes in a rotary kiln or incinerator.

FBCs are usually inapplicable when the feed material or its ash melts below the bed temperature.[14] If the feed melts, it will agglomerate and defluidize the bed. When it is necessary to burn a feed with low melting temperature, two methods can be used. The bed can be operated as a *chemically active bed*, where operation is close to but below the ash melting point of 1350 to 1450°F (730 to 790°C). A small amount of controlled

melting is permitted to occur, and the agglomerated ash is removed.

When a higher temperature is desirable in order to destroy hazardous materials, the melting point of feed can be increased by injecting low-cost additives (e.g., calcium hydroxide or kaolin clay). These combine with salts of alkali metals to form refractory-like materials with melting points in the 1950 to 2350°F (1070 to 1290°C) range.

The design of a fluidizing-bed incinerator starts with a heat balance. Energy enters the FBC during combustion of the feed and auxiliary fuel and from sensible heat contributions of the air and fuel. Some of the heat is recovered and reused. The remainder of the heat is lost through sensible heating of the combustion products, water vapor, excess air, and ash, and through radiation from the combustor vessel and associated equipment. Since these items may not all be known in advance, the following assumptions are reasonable.

- Radiation losses are approximately 5%.

- Sensible heat losses from the ash are minimal, particularly if the feed contains significant amounts of moisture.

- Excess air will be approximately 40%.

- Combustion temperature will be approximately 1400°F (760°C).

- If air is preheated, the preheat temperature will be approximately 1000°F (540°C).

The fuel's *specific feed characteristic* (SFC) is the ratio of the higher (gross) heating value to the moisture content. Typical values range from a low of approximately 1000 Btu/lbm (2300 kJ/kg) for wastewater treatment plant biosolids to more than 60,000 Btu/lbm (140 MJ/kg) for barks, sawdust, and RDF.[15] Making the previous assumptions and using the SFC as an indicator, the following generalizations can be made:

- Fuels with SFCs less than 2600 Btu/lbm (6060 kJ/kg) require a 1000°F (540°C) hot windbox, and are autogenous at that value and subautogenous below that value. The SFC drops to 2400 Btu/lbm (5600 kJ/kg) with air preheated to 1200°F (650° C). Below the subautogenous SFC, water evaporation is the controlling design factor, and auxiliary fuel is required.

- Fuels with an SFC of 4000 Btu/lbm (9300 kJ/kg) are autogenous with a cold windbox and are overautogenous above that value. Combustion is the controlling design factor for overautogenous SFCs.

---

[14]The melting temperature of a eutectic mixture of two components may be lower than the individual melting temperatures.

[15]Although the units are the same, the specific feed characteristic is not the same as the heating value.

Example 34.1

A biosolid sludge is dewatered to 75% water by weight. The solids have a higher heating value of 6500 Btu/lbm. The dewatered sludge enters a fluidized-bed combustor with a 1000°F windbox at the rate of 15,000 lbm/hr. (a) What is the specific feed characteristic? (b) Approximately what energy (in Btu/hr) must the auxiliary fuel supply?

Solution

(a) The dewatered sludge consists of 25% combustible solids and 75% moisture. The total mass of sludge required to contain 1.0 lbm of water is

$$m_{\text{sludge}} = \frac{m_w}{x_w}$$

$$= \frac{1 \text{ lbm}}{0.75} = 1.333 \text{ lbm}$$

The mass of combustible solids is

$$m_{\text{solids}} = 0.25 m_{\text{sludge}} = (0.25)(1.333 \text{ lbm})$$

$$= 0.3333 \text{ lbm}$$

The specific feed characteristic is

$$\text{SFC} = \frac{\text{total heat of combustion}}{m_w} = \frac{m_{\text{solid}} \text{HHV}}{m_w}$$

$$= \frac{(0.3333 \text{ lbm})\left(6500 \dfrac{\text{Btu}}{\text{lbm}}\right)}{1 \text{ lbm}}$$

$$= 2166 \text{ Btu/lbm}$$

(b) Making the assumptions discussed, operation with a 1000°F hot windbox requires a SFC of approximately 2600 Btu/lbm. Therefore, the energy per pound of water in the fuel that an auxiliary fuel must provide is

$$2600 \text{ Btu} - 2166 \text{ Btu} = 434 \text{ Btu}$$

The energy supplied by the auxiliary fuel is

$$\dot{m}_{\text{fuel}} x_w (\text{SFC deficit}) = \left(15,000 \dfrac{\text{lbm}}{\text{hr}}\right)(0.75)$$

$$\times \left(434 \dfrac{\text{Btu}}{\text{lbm}}\right)$$

$$= 4.88 \times 10^6 \text{ Btu/hr}$$

23. INCINERATION, GENERAL

Most rotary kiln and liquid injection incinerators have primary and *secondary combustion chambers* (SCCs). Kiln temperatures are approximately 1200 to 1400°F (650 to 760°C) for soil incineration and up to 1700°F

(930°C) for other waste types.[16] SCC temperatures are higher—1800°F (980°C) for most hazardous wastes and 2200°F (1200°C) for liquid PCBs. The waste heat may be recovered in a boiler, but the combustion gas must be cooled prior to further processing. *Thermal ballast* can be accomplished by injecting large amounts of excess air (typical when liquid fuels are burned) or by quenching with a water-spray (typical in rotary kilns).

SCCs are necessary to destroy toxics in the off-gases. SCCs are vertical units with high-swirl vortex-type burners. These produce high *destruction removal efficiencies* (DREs) with low retention times (e.g., 0.5 sec) and moderate-to-high temperatures, even for chlorinated compounds. When soil with fine clay is incinerated, fines can build up in the SCC, causing slagging and other problems. A refractory-lined cyclone located after the primary combustion chamber can be used to reduce particle carryover to the SCC.

Prior to full operation, incinerators must be tested in a trial burn with a *principal organic hazardous constituent* (POHC) that is in or has been added to the waste. The POHC must be destroyed with a DRE of at least 99.99% by weight.

Table 34.4 *Representative Incinerator Performance*

	type of incinerator/use			
			soil incinerators	
	rotary kiln	liquid injection	hazard-ous	nonhaz-ardous
waste heating value,				
(Btu/lbm)	15,000	20,000	0	0
(MJ/kg)	35	46	0	0
kiln temp				
(°F)	1700	–	1650	850
(°C)	925	–	900	450
SCC temp				
(°F)	1800–2200	2000–2200	1800	1400
(°C)	980–1200	1100–1200	980	750
SCC mean residence time (sec)	2	2	2	1
O_2 in stack gas (%)	10%	12%	9%	6%

For nontoxic organics, a DRE of 95% is a common requirement. Hazardous wastes require a DRE of 99.99%. Certain hazardous wastes, including PCBs and dioxins, require a 99.9999% DRE. This is known as the *six nines rule.*

---

[16]Temperatures higher than 1400°F (760°C) may cause incinerated soil to vitrify and clog the incinerator.

Emission limitations of some pollutants depend on the incoming concentration and the height of the stack.[17] Thus, in certain circumstances, simply raising the stack is the most effective method of being in compliance. This is considered justified on the basis that ground-level concentrations will be lower with higher stacks.

Rules of thumb regarding incinerator performance are:

- Stoichiometric combustion requires approximately 725 lbm (330 kg) of air for each million Btu of fuel or waste burned.

- 100% excess air is required.

- Stack gas dew point is approximately 180°F (80°C).

- Water-spray-quenched flue gas is approximately 40% moisture by weight.

Example 34.2

What is the mass (in lbm/hr) of water required to spray-quench combustion gases from the incineration of hazardous waste with a heating rate of 50 MMBtu/hr? No additional fuel is added to the incinerator.

Solution

Use the rules of thumb. Assume 100% excess air is required. MM is a standard power generation abbreviation for "million." The total dry air required is

$$m_a = (2)\left(50\ \frac{\text{MMBtu}}{\text{hr}}\right)\left(725\ \frac{\text{lbm}}{\text{MMBtu}}\right)$$
$$= 72{,}500\ \text{lbm/hr}\quad[\text{dry}]$$

Since the quenched combustion gas is 40% water by weight, it is 60% dry air by weight. The total mass of wet combustion gas produced per hour is

$$m_t = \frac{m_a}{x_a} = \frac{m_a}{0.6} = \frac{72{,}500\ \dfrac{\text{lbm}}{\text{hr}}}{0.6}$$
$$= 1.21 \times 10^5\ \text{lbm/hr}$$

The required mass of quenching water is

$$m_w = x_w m_t = 0.4 m_t = (0.4)\left(1.21 \times 10^5\ \frac{\text{lbm}}{\text{hr}}\right)$$
$$= 4.84 \times 10^4\ \text{lbm/hr}$$

24. INCINERATION, HAZARDOUS WASTES

Most hazardous waste incinerators use rotary kilns. Waste in solid and paste form enters a rotating drum where it is burned at 1850 to 2200°F (1000 to 1200°C).

---

[17]Some of the metallic pollutants treated this way include antimony, arsenic, barium, beryllium, cadmium, chromium, lead, mercury, silver, and thallium.

(See Table 34.4 for other representative performance characteristics.) Slag is removed at the bottom, and toxic gases exit to a tall, vortex secondary combustion chamber (SCC). Gases remain in the SCC for 2 to 4 sec where they are completely burned at approximately 1850°F (1000°C). Liquid wastes are introduced into and destroyed by the SCC as well. Heat from off-gases may be recovered in a boiler. Typical flue gas cleaning processes include electrostatic precipitation, two-stage scrubbing, and NOx removal.

Common problems with hazardous waste incinerators include (a) inadequate combustion efficiency (easily caused by air leakage in the drum and uneven fuel loading) resulting in incomplete combustion of the primary organic hazardous component (POHC), emission of CO, NOx, and *products of incomplete combustion* (also known as *partially incinerated compounds* or PICS) and metals; (b) meeting low dioxin limits; and (c) minimizing the toxicity of slag and flyash.

These problems are addressed by (a) reducing air leaks in the drum and (b) introducing air to the SCC through multiple sets of ports at specific levels. Gas is burned in the SCC in substoichiometric conditions, with the vortex ensuring adequate mixing to obtain complete combustion. Dioxin formation can be reduced by eliminating the waste-heat recovery process, since the lower temperatures present near waste-heat boilers are ideal for dioxin formation. Once formed, dioxin is removed by traditional end-of-pipe methods.

25. INCINERATION, INFRARED

Infrared incineration (II) is effective for reducing dioxins to undetectable levels. The basic II system consists of a waste feed conveyor, an electrical-heated primary chamber, a gas-fired afterburner, and a typical flue gas cleanup (FGC) system (i.e., scrubber, electrostatic precipitator, and/or baghouse). Electrical heating elements heat organic wastes to their combustion temperatures. Off-gas is burned in a secondary combustion chamber (SCC).

26. INCINERATION, LIQUIDS

Liquid-injection incinerators can be used for atomizable liquids. Such incinerators have burners that fire directly into a refractory-lined chamber. If the liquid waste contains salts or metals, a downfired liquid-injection incinerator is used with a submerged quench to capture the molten material. A typical flue gas cleanup (FGC) system (i.e., scrubber, electrostatic precipitator, and/or baghouse) completes the system.

Incineration of organic liquid wastes usually requires little external fuel, since the wastes are overautogenous and have good heating values. The heating value is approximately 20,000 Btu/lbm (47 MJ/kg) for solvents and approximately 8000 to 18,000 Btu/lbm (19 to 42 MJ/kg) for chlorinated compounds.

27. INCINERATION, OXYGEN-ENRICHED

Oxygen-enriched incineration is intended primarily for dioxin removal and is operationally similar to that of a rotary kiln. However, the burner includes oxidant jets. The jets aspirate furnace gases to provide more oxygen for combustion. Apparent advantages are low NOx production and increased incinerator feed rates.

28. INCINERATION, PLASMA

A wide variety of solid, liquid, and gaseous wastes can be treated in a *plasma incinerator*. Wastes are heated by an electric arc to higher than 5000°F (2760°C), dissociating them into component atoms. Upon cooling, atoms recombine into hydrogen gas, nitrogen gas, carbon monoxide, hydrochloric acid, and particulate carbon. The ash cools to a nonleachable, vitrified matrix. Off-gases pass through a normal train of cyclone, baghouse, and scrubbing operations. The process has a very high DRE. However, energy requirements are high.

29. INCINERATION, SOIL

Incineration can completely decontaminate soil. Incinerators are often thought of as being fixed facilities, as are cement kilns and special-use (e.g., Superfund) incinerators. However, mobile incinerators can be brought to sites when the soil quantities are large (e.g., 2000 to 100,000 tons (1800 to 90,000 Mg)) and enough setup space is available.[18] (See also Incineration, Solids and Sludges.)

30. INCINERATION, SOLIDS AND SLUDGES

Rotary kilns and fluidized-bed incinerators are commonly used to incinerate solids and sludges. The feed system depends on the waste's physical characteristics. Ram feeders are used for boxed or drummed solids. Bulk solids are fed via chutes or screw feeders. Sludges are fed via lances or by premixing with solids.

The constant rotation of the shell moves wastes through rotary kilns and promotes incineration. External fuel is required in rotary kilns if the heating value of the waste is below 1200 Btu/lbm (2800 kJ/mg) (i.e., is subautogenous). Additional fuel is required in the secondary chamber, as well.

Fluidized-bed incinerators work best when the waste is consistent in size and texture. An important benefit is the ability to introduce limestone and other solid reagents to the bed in order to remove HCl and SO_2.

31. INCINERATION, VAPORS

Vapor incinerators, also known as *afterburners* and *flares*, convert combustible materials (gases, vapors, and particulate matter) in the stack gas to carbon dioxide and water. Afterburners can be either direct-flame or catalytic in operation.

32. LOW EXCESS-AIR BURNERS

NOx formation in gas-fired boilers can be reduced by maintaining excess air below 5%.[19] *Low excess-air burners* use a forced-draft and self-recirculating combustion chamber configuration to approximate multistaged combustion.

33. LOW-NOx BURNERS

Low (or *ultralow*) *NOx burners* (LNB) in gas-fired applications use a combination of staged-fuel burning and internal flue gas recirculation (FGR). Recirculation within a burner is induced by either the pressure of the fuel gas or other agents (e.g., medium-pressure steam or compressed air).

34. MECHANICAL SEALS

Fugitive emissions from pumps and other rotating equipment can be reduced or eliminated using current technology by the proper selection and installation of mechanical seals. The three major classes of mechanical seals are single seals, tandem seals (dual seals placed next to each other), and double seals (dual seals mounted back-to-back or face-to-face).

The most economical and reliable sealing device is a *single seal*. It has a minimum number of parts and requires no support devices. However, since the pumped product is usually the lubricant for the seal face, small amounts of the product escape into the environment. Emissions are generally below 1000 ppmv, and are often below 100 ppmv.

Tandem seals consist of two seal assemblies separated by a buffer fluid at a pressure lower than the seal-chamber liquid. The primary inner seal operates at full pump pressure. The outer seal operates in the buffer fluid. Tandem seals can achieve zero emissions when used with a vapor recovery system, provided that the pumped product's specific gravity is less than that of the buffer fluid and the product is immiscible with the buffer fluid.

If the buffer fluid used in a tandem dual seal is a controlled substance, emissions from the outer seal must also meet the emission limits. Seals using glycol as the buffer fluid typically achieve zero emissions. Seals using diesel oil or kerosene have emissions in the 25 to 100 ppmv range.

---

[18]Modified asphalt batch processing plants can be used to incinerate soils contaminated with low-heating value, nonchlorinated hydrocarbons.

[19]Reducing excess air from 30% to 10%, for example, can reduce NOx emissions by 30%.

Double seals are recommended for hazardous fluids with specific gravities less than 0.4 and are a good choice when a vapor recovery system is not available. (Low specific-gravity liquids do not lubricate the seal well.) Double seals consist of two seal assemblies connected by a common collar; they operate with a barrier fluid kept at a pressure higher than that of the pumped product. Double seals can reduce the emission rate to zero.

Figure 34.8 shows the relationship between typical emissions and specific gravity for the different seal types.

Figure 34.8 *Representative Emissions to Atmosphere for Mechanical Seals*[a] *(1 cm from source)*

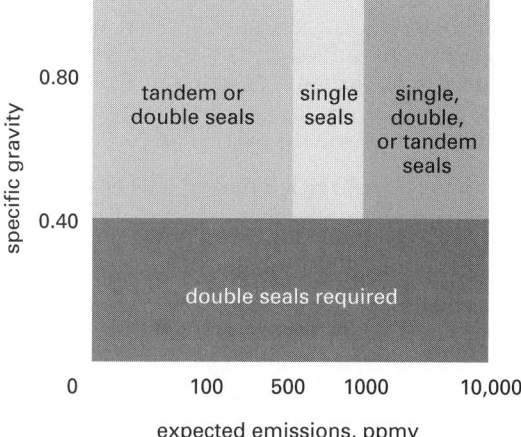

[a]maximum seal size, 6 in (153 mm); maximum pressure, 600 psig (40 bar); maximum speed, 3600 rpm

35. MULTIPLE PIPE CONTAINMENT

The U.S. Environmental Protection Agency (EPA) requires *multiple pipe containment* (MPC) in the storage and transmission of hazardous fluids.[20] MPC systems consist of a carrier pipe or bundle of pipes, a common casing, and a leak detector consisting of redundant instrumentation with automatic shutdown features. Casings are generally not pressurized. If one of the pipes within the bundle develops a leak, the fluid will remain in the casing and the detector will signal an alarm and automatically shut off the liquid flow.

Some of the many factors that go into the design of an MPC system are (a) the number of pipes to be grouped together; (b) the weight and strength of the pipes, carrier, and supports; (c) the compatibility of pipe materials and fluids; (d) differential expansion of the container and carrier pipes; and (e) a method of leak detection.

Carrier pipes within the casing should be supported and separated from each other by internal supports (i.e., perforated baffles within the casing). Carrier pipes

---

[20]In addition to multiple pipe containment, other factors that contribute to reduction of fugitive emissions are (a) the use of tongue-in-groove flanges, and (b) the reduction of as many nozzles as possible from storage tanks and reactor vessels.

pass through oval-shaped holes in the internal supports. Holes are oval and oversized to allow flexing of the carrier pipes due to expansion. Other holes through the internal supports allow venting and draining of the casing in the event of a leak. If multiple detectors are used, isolation baffles can be included to separate the casing into smaller sections.

When a pipe bundle changes elevation as well as direction, the carrier pipes may change their positions relative to one another. A pipe on the outside (i.e., at 3 o'clock) may end up being the lower pipe (i.e., at 6 o'clock).[21] This change in relative positioning is known as *rotation*. Rotation complicates the accommodation of pipes that must enter and exit the bundle at specific locations.

Rotation occurs when a change in direction is located at the top or bottom of an elevation change. Therefore, unless rotation can be tolerated, changes in direction should not be combined with changes in elevation. In Fig. 34.9(a), pipe A remains to the left of pipe B. In Fig. 34.9(b), pipe A starts out to the left of pipe B and ends up above pipe B.

Figure 34.9 *Rotation in Pipe Bundles*

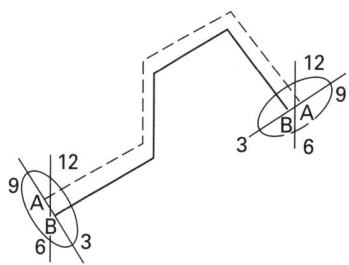

(a) no rotation

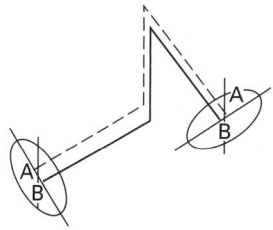

(b) rotation

Another problem associated with MPC systems is expansion of the containment casing and carrier pipes caused by differences in ambient and fluid temperatures and the use of different pipe materials. For pipes containing direct changes, there must be space at each elbow to permit the pipe expansion. If both ends of a pipe are fixed, expansion and contraction will cause the pipes to flex unless Z-bends or loops are included.

Differential expansion must be calculated from the largest expected temperature difference using Eq. 34.5.

$$\Delta L = \alpha L \Delta T \qquad \textit{34.5}$$

---

[21]It is important to consistently "face" the same way (e.g., in the direction of flow).

Table 34.5 *Approximate Coefficients of Thermal Expansion of Piping Materials (at 70° F (21° C))*

pipe material	U.S. (ft/ft-°F)	SI (m/m·°C)
carbon steel	6.1×10^{-6}	1.1×10^{-5}
chlorinated polyvinyl chloride (CPVC)	3.8×10^{-5}	6.8×10^{-5}
copper	9.5×10^{-6}	1.7×10^{-5}
fiberglass-reinforced polyethylene (FRP)	8.5×10^{-6}	1.5×10^{-5}
polyethylene (PE)	8.3×10^{-5}	1.5×10^{-4}
polyvinyl chloride (PVC)	3.0×10^{-5}	5.4×10^{-5}
stainless steel	9.1×10^{-6}	1.6×10^{-5}

(Multiply ft/ft-°F by 1.8 to get m/m·°C.)

Computer programs are commonly used to design pipe networks. These programs are helpful in locating guides and anchors and in performing a stress analysis on the final design. Drains, purge points, baffles, and joints are generally chosen by experience.

36. OZONATION

Ozonation is one of several advanced oxidation methods capable of reducing pollutants from water. Ozone is a powerful oxidant produced by electric discharge through liquid oxygen. Ozone is routinely used in the water treatment industry for disinfection.

Table 34.6 *Oxidation of Industrial Wastewater Pollutants*[a]

	Cl_2	ClO_2	$KMnO_4$	O_3	H_2O_2	OH^-
amines	C	P	P	P	P	C
ammonia	C	N	P	N	N	N
bacteria	C	C	C	C	P	C
carbohydrates	P	P	P	P	N	C
chlorinated solvents	P	P	P	P	N	C
phenols	P	C	C	C	P	C
sulfides	C	C	C	C	C	C

[a]C = complete reaction; P = partially effective; N = not effective

37. SCRUBBING, GENERAL

Scrubbing is the act of removing particulate matter from a gaseous stream, although substances in gaseous and liquid forms may also be removed.[22]

38. SCRUBBING, CHEMICAL

Chemical scrubbing using oxidizing compounds such as chlorine, ozone, hypochlorite, or permanganate rapidly

destroys volatile organic compounds (VOCs). Efficiencies are typically 95% for highly reactive substances, but are lower for hydrocarbons and substances with low reactivities.

39. SCRUBBING, DRY

Dry scrubbing, also known as *dry absorption* and *semi-dry scrubbing*, is one form of sorbent injection. It is commonly used to remove SO_2 from flue gas. In operation, flue gases pass through a scrubbing chamber where a slurry of lime and water is sprayed through them.[23] The slurry is produced by high-speed rotary atomizers. The sorbent (lime, limestone, hydrated lime, sodium bicarbonate, or sodium sesquicarbonate–trona–reagent) is injected into the flue gas in either dry or slurry form. The flue gas heat drives off the water. In either case, a dry powder is carried through the flue gas system. An electrostatic precipitator or baghouse captures the flyash and calcium sulfate particulates.

Dry scrubbing has an SO_2 removal efficiency of approximately 50 to 75%, with some installations reporting 90% efficiencies. NOx removal is not usually intended and is essentially zero. Though the removal efficiency is lower than with wet scrubbing, an advantage of dry scrubbing is that the waste is dry, requiring no sludge-handling equipment. The lower efficiency may be sufficient in older power plants with less stringent regulations. (See also Sorbent Injection.)

Particulate removal efficiencies of 90% can be achieved. Since wet scrubbers normally have a lower particulate removal efficiency than baghouses, they are usually combined with electrostatic precipitators.

40. SCRUBBING, WET

See Absorption, Gas (Spray Towers).

41. SELECTIVE CATALYTIC REDUCTION

Selective catalytic reduction (SCR) uses ammonia in combination with a catalyst to reduce nitrogen oxides (NOx) to nitrogen gas and water.[24] Vaporized ammonia is injected into the flue gas; the mixture passes through a catalytic reaction bed approximately 0.5 to 1.0 sec later. The reaction bed can be constructed from honeycomb plates or parallel-ridged plates. Alternatively, the reaction bed may consist of a packed bed of rings, pellets, or other shapes. The catalyst lowers the NOx decomposition activation energy. NOx and NH_3 combine on the catalyst's surface, producing nitrogen and water.

---

[22]As it relates to air pollution control, the term *scrubbing* is loosely used. The term may be used in reference to any process that removes any substance from flue gas. In particular, any process that removes SO_2 by passing flue gas through lime or limestone is referred to as *scrubbing*.

[23]In addition to lime, limestone, sodium carbonate, and magnesium oxide can be used. Because they cost less, lime and limestone are the most common.

[24]Vanadium oxide, titanium, and platinum are metallic catalysts; zeolites and ceramic catalysts are also used.

One mole of ammonia is required for every mole of NOx removed. However, in order to maximize the reduction efficiency, approximately 5 to 10 ppm of unreacted ammonia is left behind.[25] The chemical reaction is

$$O_2 + 4NO + 4NH_3 \longrightarrow 4N_2 + 6H_2O$$

The optimal temperature for SCR is 600 to 700°F (315 to 370°C). Gas velocities are typically around 20 ft/sec (6 m/s). The pressure drop is 3 to 4 iwg (0.75 to 1 kPa). NOx removal efficiency is typically 90% but can range from 70 to 95% depending on the application.[26] SCR produces no liquid waste.

Several conditions can cause deactivation of the catalyst. *Poisoning* occurs when trace quantities of specific materials (e.g., arsenic, lead, phosphorous, other heavy metals, silicon, and sulfur from SO_2) react with the catalyst and lower its activity. Poisoning by SO_2 can be reduced by keeping the flue gas temperature above 608°F (320°C) and/or using poison-resisting compounds. *Masking* (or *plugging*) occurs when the catalytic surface becomes covered with fine particle dust, unburned solids, or ammonium salts. Internal cleaning devices remove surface contaminants such as ash deposits and are used to increase the activity of the equipment.

42. SELECTIVE NONCATALYTIC REDUCTION

Selective noncatalytic reduction (SNCR) involves injecting ammonia or urea into the upper parts of the combustion chamber (or into a thermally favorable location downstream of the combustion chamber) to reduce NOx.[27] SNCR is effective when the oxygen content is low (e.g., 1%) and the combustion temperature is controlled. If the temperature is too high, the NH_3 will react more with oxygen than with NOx, forming even more NOx. If the temperature is too low, the reactions slow and unreacted ammonia enters the flue gas. The optimal temperature range is 1600 to 1750°F (870 to 950°C) for NH_3 injection and up to 1900°F (1040°C) for urea. (In general, the temperature ranges for NH_3 and urea are similar. However, various hydrocarbon additives can be used with urea to lower the temperature range.)

---

[25]Leftover ammonia in the flue gas is referred to as *ammonia slip* and is usually measured in ppm. 50 to 100 ppm would be considered an excessive ammonia slip. Since NOx and NH_3 react on a 1:1 molar basis, slip is easily calculated. In the following equations, the units can be either molar or volumetric (ppmvd, lbm/hr, mol/hr, scfm, etc.), but they must be consistent.

$$NH_{3,slip} = NH_{3,feed} - NH_{3,reacted}$$
$$= NH_{3,feed} - (NOx_{in} - NOx_{out})$$
$$= NH_{3,feed} - (NOx_{in})(\text{removal efficiency})$$

[26]Removal efficiencies of 95 to 99% are possible when SCR is used for VOC removal.
[27]*Urea*, which decomposes to NH_3 and carbon dioxide inside the combustion chamber, is safer and easier to handle.

The reactions are

$$6NO + 4NH_3 \longrightarrow 5N_2 + 6H_2O$$
$$6NO_2 + 8NH_3 \longrightarrow 7N_2 + 12H_2O$$

The NOx reduction efficiency is approximately 20 to 50% with an ammonia slip of 20 to 30 ppm. The actual efficiency is highly dependent on the injection geometries and interrelations between ammonia slip, ash, and sulfur.

Since the NOx reduction efficiency is relatively low, the SNCR process cannot usually satisfy NOx regulations by itself. The use of urea must be balanced against the potential for ammonia slip and the conversion of NO to nitrous oxide. These and other problems relating to formation of ammonium salts have kept SNCR from gaining widespread popularity in small installations.

43. SOIL WASHING

Soil washing is effective in removing heavy metals, wood preserving wastes (PAHs), and BTEX compounds from contaminated soil. Soil washing is a two-step process. In the first step, soil is mixed with water to dissolve the contaminants. Additives are used as required to improve solubility. In the second step, additional water is used to flush the soil and to separate the fine soil from coarser particles. (Semivolatile materials concentrate in the fines.) Metals are extracted by adding chelating agents to the wash water.[28] The contaminated wash water is subsequently treated.

44. SORBENT INJECTION

Sorbent injection (FSI) involves injecting a limestone slurry directly into the upper parts of the combustion chamber (or into a thermally favorable location downstream) to reduce SOx. Heat calcines the limestone into reactive lime. Fast drying prevents wet particles from building up in the duct. Lime particles are captured in a scrubber with or without an electrostatic precipitator (ESP). With only an ESP, the SOx removal efficiency is approximately 50%.

45. SPARGING

Sparging is the process of using air injection wells to bubble air through groundwater. The air pushes volatile contaminants into the overlying soil above the aquifer where they can be captured by vacuum extraction.

---

[28]A *chelate* is a ring-like molecular structure formed by unshared electrons of neighboring atoms. A *chelating agent (chelant)* is an organic compound in which atoms form bonds with metals in solution. By combining with metal ions, chelates control the damaging effects of trace metal contamination. Ethylenediaminetetraacetic acid (EDTA) types are the leading industrial chelants.

46. STAGED COMBUSTION

Staged combustion methods are primarily used with gas-fired burners to reduce formation of nitrogen oxides (NOx). Both staged-air burner and staged-fuel burner systems are used. *Staged-air burner systems* reduce NOx production 20 to 35% by admitting the combustion air through primary and secondary paths around each fuel nozzle. The fuel burns partially in the fuel-rich zone. Fuel-borne nitrogen is converted to compounds that are subsequently oxidized to nitrogen gas. Secondary air completes combustion and controls the flame size and shape. Combustion temperature is lowered by recirculation of combustion products within the burner. Staged burners have few disadvantages, the main one being longer flames.

In *staged-fuel burner systems*, a portion of the fuel gas is initially burned in a fuel-lean (air-rich) combustion. The peak flame temperature is reduced, with a corresponding reduction in thermal NOx production. The remainder of the fuel is injected through secondary nozzles. Combustion gases from the first stage dilute the combustion in the second stage, reducing peak temperature and oxygen content. Reductions in NOx formation of 50 to 60% are possible. Flame length is less than with staged-air burners, and the required excess air is lower.

47. STRIPPING, AIR

Air strippers are primarily used to remove volatile organic compounds (VOCs) or other target substances from water. In operation, contaminated water enters a stripping tower at the top, and fresh air enters at the bottom. The effectiveness of the process depends on the volatility of the compound, its temperature and concentration, and the liquid-air contact area. However, removal efficiencies of 80 to 90% (and above) are common for VOCs.

There are three types of stripping towers—*packed towers* filled with synthetic *packing media* (polypropylene balls, rings, saddles, etc.), *spray towers*, and (less frequently for VOC removal) *tray towers* with horizontal trays arranged in vertical layers. *Redistribution rings (wall wipers)* prevent channeling down the inside of the tower. (*Channeling* is the flow of liquid through a few narrow paths rather than an over-the-bed packing.) The stripping air is generated by a small blower at the column base. A mist eliminator at the top eliminates entrained water from the air.

As the contaminated water passes over packing media in a packed tower, the target substance leaves the liquid and enters the air where the concentration is lower. The mole fraction of the target substance in the water, x, decreases; the mole fraction of the target substance in the air, y, increases.

The discharged air, known as *off-gas*, is discharged to a process that destroys or recovers the target substance. (Since the quantities are small, recovery is rarer.)

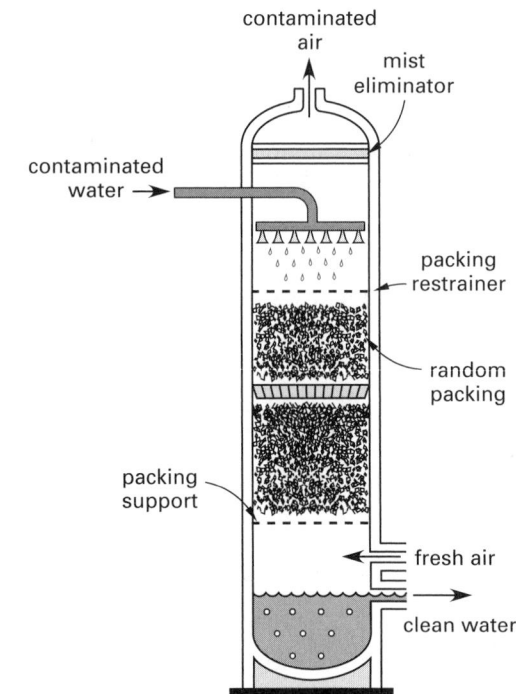

Figure 34.10 Schematic of Air Stripping Operation

Destruction of the target substance can be accomplished by flaring, carbon absorption, and incineration. Since flaring is dangerous and carbon absorption creates a secondary waste if the AC is not regenerated, incineration is often preferred.

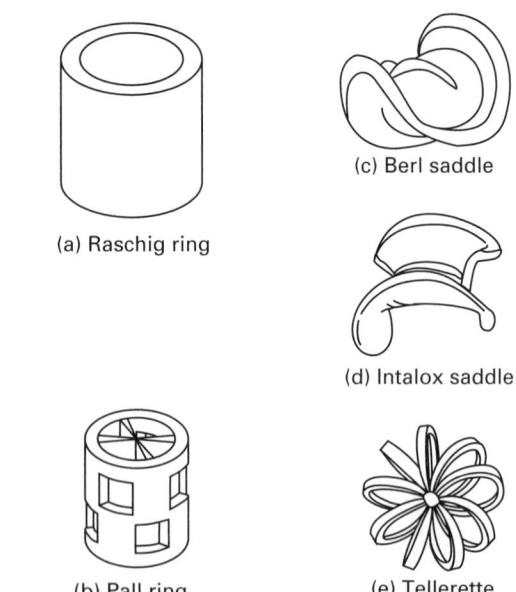

Figure 34.11 Packing Media Types

Henry's law, as it applies to water treatment, states that at equilibrium, the vapor pressure of target substance A is directly proportional to the target substance's mole fraction, x_A. (The maximum mole fraction is the substance's *solubility*.) Table 34.7 lists representative values of Henry's law constant, H.[29] When multiple compounds are present in the water, the *key component* is the one that is the most difficult to remove (i.e., the component with the highest concentration and lowest Henry's constant).

$$p_A = H_A x_A \qquad 34.6$$

The liquid *mass-transfer coefficient*, $K_L a$, is the product of the *coefficient of liquid mass transfer*, K, and the *interfacial area* per volume of packing, a. The mass transfer coefficient is a measure of the efficiency of the air stripper as a whole. The higher the mass transfer coefficient, the higher the efficiency. The $K_L a$ value is largely a function of the size and shape of the packing material, and it is usually obtained from the packing manufacturer or theoretical correlations.

Table 34.7 *Typical Henry's Law Constants for Selected Volatile Organic Compounds (at low pressures and 25° C)*

VOC	Henry's law constant, H (atm)[a]
1,1,2,2-tetrachloroethane	24.02
1,1,2-trichloroethane	47.0
propylene dichloride	156.8
methylene chloride	177.4
chloroform	188.5
1,1,1-trichloroethane	273.56
1,2-dichloroethene	295.8
1,1-dichloroethane	303.0
hexachloroethane	547.7
hexachlorobutadiene	572.7
trichloroethylene	651.0
1,1-dichloroethene	834.03
perchloroethane	1596.0
carbon tetrachloride	1679.17

(Multiply atm by 101.3 to obtain kPa.)
[a]Henry's law indicates that the units of H are atmospheres per mole fraction. Mole fraction is dimensionless and does not appear in the units for H.

For some types of packing, the gas-phase resistance is greater than the liquid-phase resistance, and the gas mass-transfer coefficient, $K_G a$, may be given.

The *stripping factor*, R (the reciprocal of the *absorption factor*), is given by Eq. 34.7. Units for the gas and liquid flow rates, G and L, depend on the correlations used to solve the stripping equations. The air flow rate is limited by the acceptable pressure drop through the tower.

$$R = \frac{HG}{L} \qquad 34.7$$

[29]Henry's law is fairly accurate for most gases when the partial pressure is less than 1 atm. If the partial pressure is greater than 1 atm, Henry's law constants will vary with partial pressure.

The *transfer unit method* is a convenient way of designing and analyzing the performance of stripping towers. A *transfer unit* is a measure of the difficulty of the mass-transfer operation and depends on the solubility and concentrations. The overall number of transfer units is expressed as NTU_{OG} or NTU_{OL} (alternatively, N_{OG} or N_{OL}), depending on whether the gas or liquid resistance dominates.[30] The NTU value depends on the incoming and desired concentrations and the material flow rates.[31]

The *height of the packing*, z, is the effective height of the tower. It is calculated from the NTUs and the *height of a transfer unit*, HTU, using Eq. 34.8. The HTU value depends on the packing media.

$$z = (\text{HTU}_{OG})(\text{NTU}_{OG}) = (\text{HTU}_{OL})(\text{NTU}_{OL}) \qquad 34.8$$

Ten percent additional height should be added to the theoretical value as a safety factor. Packing heights greater than 30 to 40 ft (9.1 to 12.2 m) are not recommended since the packing might be crushed, and greater heights produce little or no increase in removal efficiency.

The blower power depends on the packing shape and size and the air-to-water ratio. Typical pressure drops are 0.5 to 1.0 iwg per vertical foot (0.38 to 0.75 kPa/m) of packing. To minimize the blower power requirement, the ratio of tower diameter to gross packing dimension should be between 8 and 15. Blower fan power increases with increasing air-to-water ratios.

A significant problem for air strippers removing contaminants from water is fouling of the packing media through biological growth and solids (e.g., iron complexes) deposition. Biological growth can be inhibited by continuous or batch use of a *biocide* that does not interfere with the off-gas system.[32] Another problem is *flooding*, which occurs when excess liquid flow rates impede the flow of the air.

48. STRIPPING, STEAM

Steam stripping is more effective than air stripping for removing semi- and non-volatile compounds, such as

[30]The gas film resistance controls when the solubility of the target substance in the liquid is high. Conversely, when the solubility is low, the liquid film resistance controls. In flue gas cleanup (FGC) work, the gas film resistance usually controls.
[31]Stripping is one of many mass-transfer operations studied by chemical engineers. Determining the number of transfer units required and the height of a single transfer unit are typical chemical engineering calculations.
[32]A *biocide* is a chemical that kills living things, particularly microorganisms. Biocide categories include chlorinated isocyanurates (used in swimming pool disinfectants and dishwashing detergents), sodium bromide, inorganics (used in wood treatments), quaternaries ("quats") used in hard surface cleaners and sanitizers, and iodophors (used in human skin disinfectants). By comparison, *biostats* do not kill microorganisms already present, but they retard further growth of microorganisms from the moment they are incorporated. Biostats are organic acids and salts (e.g., sodium and potassium benzoate, sorbic acid, and potassium sorbate used "to preserve freshness" in foods).

Environmental

diesel fuel, oil, and other organic compounds with boiling points up to approximately 400°F (200°C). Operation of a steam stripper is similar to that of an air stripper, except that steam is used in place of the air. Steam strippers can be operated at or below atmospheric pressure. Higher vacuums will remove greater amounts of the compound.

49. THERMAL DESORPTION

Thermal desorption is primarily used to remove volatile organic compounds (VOCs) from contaminated soil. The soil is heated directly or indirectly to approximately 1000°F (540°C) to evaporate the volatiles. This method differs from incineration in that the released gases are not burned but are captured in a subsequent process step (e.g., activated carbon filtration).

50. VACUUM EXTRACTION

Vacuum extraction is used to remove many types of volatile organic compounds (VOCs) from soil. The VOCs are pulled from the soil through a well dug in the contaminated area. Air is withdrawn from the well, vacuuming volatile substances with it.

51. VAPOR CONDENSING

Some vapors can be removed simply by cooling them to below their dew points. Traditional contact- (open) and surface- (closed) condensers can be used.

52. VITRIFICATION

Vitrification melts and forms slag and ash wastes into glass-like pellets. Heavy metals and toxic compounds cannot leach out, and the pellets can be disposed of in hazardous waste landfills. Vitrification can occur in the incineration furnace or in a stand-alone process. Stand-alone vitrification occurs in an electrically heated vessel where the temperature is maintained at 2200 to 2370°F (1200 to 1300°C) for up to 20 hr or so. Since the electric heating is nonturbulent, flue gas cleaning systems are not needed.

53. INDUSTRIAL WASTEWATER TREATMENT PROCESSES

Many industrial processes use water for cooling, rinsing, or mixing. Such water must be treated prior to being discharged to *publicly owned treatment works* (POTWs).[33] Table 34.8 lists some of the polluting characteristics of industrial wastewaters.

---

[33]Rainwater runoff from some industrial plants can also be a hazardous waste.

Table 34.8 *Types of Pollution from Industrial Wastewater*

ammonia
biochemical oxygen demand (BOD)
carbon, total organic (TOC)
chemical oxygen demand (COD)
chloride
flow rate
metals, soluble
metals, nonsoluble
nitrate
nitrite
organic compounds, acid-extractable
organic compounds, base/neutral extractable
organic compounds, volatile (VOC)
pH
phosphorous
sodium
solids, total suspended (TSS)
sulfate
surfactants
temperature
whole-effluent toxicity (LC_{50})

Most large-volume industrial wastewaters go through the following processes: (a) flow equalization, (b) neutralization, (c) oil and grease removal, (d) suspended solids removal, (e) metals removal, and (f) VOC removal. These processes are similar, in many cases, to processes with the same names used to treat municipal wastewater in a POTW.

Flow equalization reduces the chance of under- and overloading a treatment process. Equalization of both hydraulic and chemical loading is required. *Hydraulic loading* is usually equalized by storing wastewater during high flow periods and discharging it during periods of low flow. *Chemical loading* is equalized by use of an equalization basin with mechanical mixing or air agitation. Mixing with air has the added advantages of oxidizing reducing compounds, stripping away volatiles, and eliminating odors.

There are two reasons for water *neutralization*. First, the pH of water is regulated and must be between 6.0 and 9.0 when discharged. Second, water that is too acidic or alkaline may not be properly processed by subsequent, particularly biological, processes. The optimum pH for biological processes is between 6.5 and 7.5. The effectiveness of the processes is greatly reduced with pHs below 4.0 and above 9.5.

Acidic water is neutralized by adding lime (oxides and hydroxides of calcium and magnesium), limestone, or some other caustic solution. Alkaline waters are treated with sulfuric or hydrochloric acid or carbon dioxide gas.[34]

Large volumes of oil and grease float to the surface and are skimmed off. Smaller volumes may require a dissolved-air process. Removing emulsified oil is more complex and may require use of chemical coagulants.

---

[34]In water, carbon dioxide gas forms *carbonic acid*.

The method used to remove suspended solids depends on the solid size. Most industrial plants do not require strainers, bar screens, or fine screens to remove large solids (i.e., those larger than 1 in (25 mm)). *Grit* (i.e., sand and gravel) is removed in grit chambers by simple sedimentation. Fine solids are categorized into *settleable solids* (diameters more than 1 μm) and *colloids* (diameters between 0.001 μm and 1 μm). Settleable solids and colloids are removed in a sedimentation tank, with chemical coagulants or dissolved-air flotation being used to assist colloidal particles in settling out.

Most metal (e.g., lead) removal occurs by precipitating its hydroxide, although precipitation as carbonates or sulfides is also used.[35] A caustic substance (e.g., lime) is added to the water to raise the pH below the solubility limit of the metal ion. When they come out of solution, the metallic compounds are *flocculated* into larger flakes that ultimately settle out. Floc is mechanically removed as inorganic heavy-metal sludge, which has a 96 to 99% water content by weight. The sludge is dewatered in a drying bed, vacuum filter, or filter press to 65 to 85% water. Depending on its composition, the sludge can be landfilled or treated as a hazardous waste by incineration or other methods.

Volatile organic compounds (VOCs) such as benzene and toluene are removed by air stripping or adsorption in activated charcoal (AC) towers. Nonvolatile organic compounds (NVOCs) are removed by biological processes such as lagooning, trickle filtration, and activated sludge.

Although destruction of biological health hazards is not always needed for industrial wastewater, chemical oxidants (see Table 34.1) are still needed to destroy odors, control bacterial growth downstream, and eliminate sulfur compounds.[36] Chemical oxidants can also reduce heavy metals (e.g., iron, manganese, silver, and lead) that were not removed in previous operations.

---

[35]Ion exchange, activated charcoal, and reverse osmosis methods can be used but may be more expensive.

[36]Meat-packing and dairy (e.g., cheese) plants are examples of industrial processes that require chlorination to destroy bacteria in wastewater.

Topic IV: Geotechnical

Geotechnical

35 Soil Properties and Testing

Nomenclature

A	area	ft^2	m^2
c	cohesion	lbf/ft^2	Pa
C_c	compression index	–	–
C_u	uniformity coefficient	–	–
C_z	coefficient of curvature	–	–
CBR	California bearing ratio	–	–
D	diameter	ft	m
D_r	relative density	–	–
e	void ratio	–	–
f_R	friction ratio	%	%
F	shape factor	various	various
h	head	ft	m
i	hydraulic gradient	–	–
I_g	group index	–	–
K	coefficient of permeability	ft/sec	m/s
L	flow path length	ft	m
LI	liquidity index	–	–
LL	liquid limit	–	–
m	mass	lbm	kg
n	porosity	–	–
N	number of blows	–	–
OCR	overconsolidation ratio	–	–
p	pressure	lbf/ft^2	Pa
P	load	lbf	N
PI	plasticity index	–	–
PL	plastic limit	–	–
PPS	percent pore space	–	–
q_c	tip resistance	lbf/ft^2	Pa
q_s	sleeve friction resistance	lbf/ft^2	Pa
Q	flow quantity	ft^3/sec	m^3/s
r	radius	ft	m
R	Hveem's resistance	–	–
RC	relative compaction	%	%
s	shear strength	lbf/ft^2	Pa
S	degree of saturation	%	%
S	strength	lbf/ft^2	Pa
S_t	sensitivity	–	–
SG[1]	specific gravity	–	–
SPT	Standard Penetration Resistance (test)	blows/ft	blows/m
t	time	sec	s
u	pore pressure	lbf/ft^2	Pa
v	velocity	ft/sec	m/s
V	volume	ft^3	m^3
w	water content	%	%
W	weight	lbf	–
z	depth	ft	m

Symbols

α	angle of failure plane	deg	deg
γ	unit weight	lbf/ft^3	–
δ	displacement	ft	m
ϵ	strain	–	–
θ	angle of stress plane	deg	deg
ρ	mass density	lbm/ft^3	kg/m^3
σ	normal stress	lbf/ft^2	Pa
τ	shear stress	lbf/ft^2	Pa
ϕ	angle of internal friction	deg	deg

Subscripts

A	axial
b	buoyant
B	borrow
c	cone tip
d	dry
D	deviator
f	final or failure
F	fill
g	air (gas) or group
h	horizontal

[1] As a peculiarity of soils engineering, the specific gravity is usually given the symbol G, as opposed to this book which uses SG throughout.

Geotechnical

i	initial
l	liquid
max	maximum
min	minimum
n	normal
p	plastic
R	radial
s	shrinkage, sleeve, or solids
sat	saturated
t	total
u	undrained
uc	unconfined compression
v	void (gas and water)
w	water
z	zero air voids

1. SOIL PARTICLE SIZE DISTRIBUTION

Soil is an aggregate of loose mineral and organic particles. *Rock*, on the other hand, exhibits strong and permanent cohesive forces between the mineral particles. From an engineering perspective, the distinction between soil and rock relates to the workability of materials. For example, one practical definition is that soil can be excavated with a backhoe while rock needs to be blasted. The distinction between soil and rock can also be made based on strength, density, and other quantifiable parameters. A geologist or soil scientist might be more interested in how a material has been formed than its workability, and thus might distinguish between soil and rock differently.

The primary mineral components of any soil are gravel, sand, silt, and clay. Organic material can also be present in surface samples. Gravel and sand are classified as coarse-grained soils, while inorganic silt and clay are classified as fine-grained soils. Particle size limits for defining gravel, sand, silt, and clay used in different classification schemes are given in Table 35.1.

The particle size distribution for a coarse soil is found from a sieve test. In a *sieve test*, the soil is passed through a set of sieves of consecutively smaller openings. The particle size distribution for the finer particles of a soil is determined from a *hydrometer test*. This test is based on Stokes' Law, which relates the speed of a particle falling out of suspension to its diameter and solid density. The results of both the sieve and hydrometer tests are graphed as a *particle size distribution*.

Table 35.2 *Sieve Sizes*

sieve size	opening size (mm)
4 in	100
3 in	75
2 in	50
$1\frac{1}{2}$ in	37.5
1 in	25
$\frac{3}{4}$ in	19
$\frac{1}{2}$ in	12.5
$\frac{3}{8}$ in	9.5
no. 4	4.75
no. 8	2.36
no. 10	2.00
no. 16	1.18
no. 20	0.850
no. 30	0.600
no. 40	0.425
no. 50	0.300
no. 60	0.250
no. 70	0.212
no. 100	0.150
no. 140	0.106
no. 200	0.075

specified in ASTM E-11, Table 1

Figure 35.1 *Typical Particle Size Distribution*

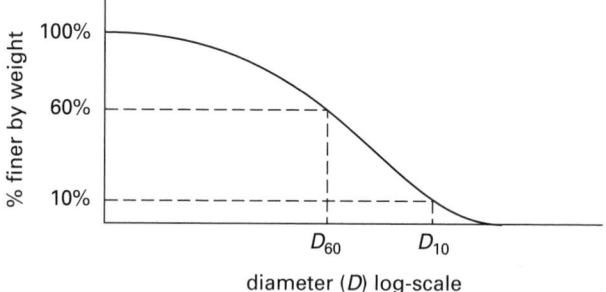

Table 35.1 *Classification of Soil Particle Sizes*

		sizes (mm)			
system	date	gravel	sand	silt	clay
Bureau of Soils	1890	1–100	0.05–1	0.005–0.05	< 0.005
Atterberg	1905	2–100	0.2–2	0.002–0.2	< 0.002
MIT	1931	2–100	0.06–2	0.002–0.06	< 0.002
USDA	1938	2–100	0.05–2	0.002–0.05	< 0.002
Unified (or USCS)	1953	4.75–75	0.075–4.75	< 0.075	
ASTM	1967	4.75–75	0.075–4.75	< 0.075	
AASHTO	1970	2–75	0.075–2	0.002–0.075	0.001–0.002

Figure 35.2 *Particle Size Distribution Chart (blank)*

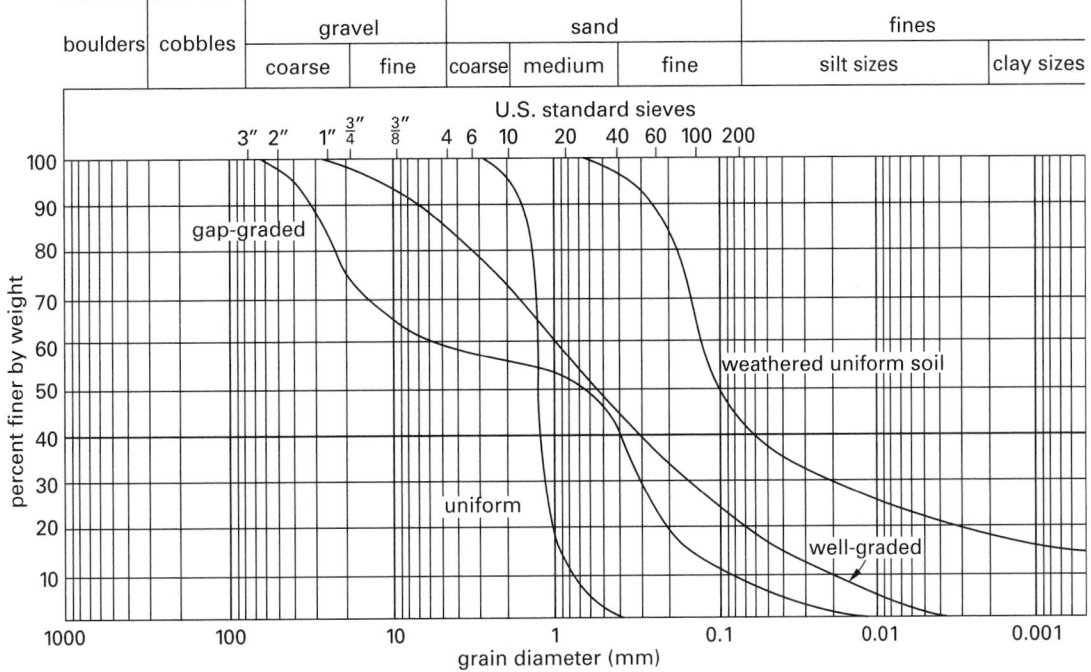

The *Hazen uniformity coefficient*, C_u given by Eq. 35.1, indicates the general shape of the particle size distribution. The diameter, for which only 10% of the particles are finer, is designated as D_{10} and is known as the *effective grain size*.

$$C_u = \frac{D_{60}}{D_{10}} \qquad 35.1$$

If the uniformity coefficient is less than 4 or 5, the soil is considered uniform in particle size, which means that all particle sizes fall within a narrow range. *Well-graded soils* have uniformity coefficients greater than 10 and have a continuous, wide range of particle sizes. *Gap-graded soils* are missing one or more ranges of particle sizes.

Another index of distribution shape is the *coefficient of curvature* (also known as the *coefficient of gradation*), given by Eq. 35.2. The diameters for which 30% and

Figure 35.3 *Uniformity in Particle Size Distributions*

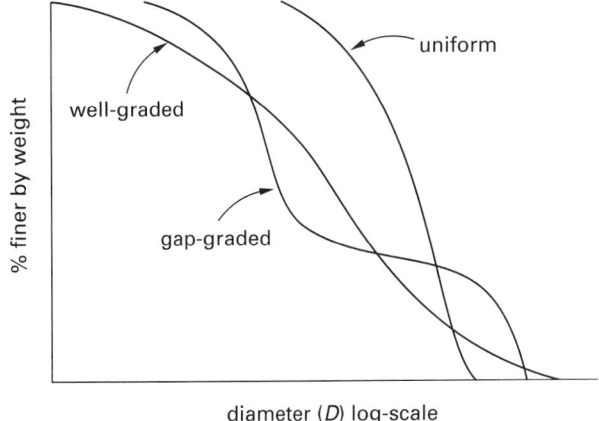

60% of the particles are finer are designated as D_{30} and D_{60}, respectively. Typical shape parameters are listed in Table 35.3.

$$C_z = \frac{(D_{30})^2}{D_{10}D_{60}} \qquad 35.2$$

Table 35.3 *Typical Uniformity Coefficients*

soil	C_u	C_z
gravel	> 4	1–3
fine sand	5–10	1–3
coarse sand	4–6	
mixture of silty sand and gravel	15–300	
mixture of clay, sand, silt, and gravel	25–1000	

Equations 35.1 and 35.2, as well as various soil classification schemes, require knowing percentages of soil passing through specific sieve sizes. When sieve data are incomplete, the needed values can be interpolated by plotting known data on a *particle size distribution chart*.

2. SOIL CLASSIFICATION

Formal soil classification schemes have been established by a number of organizations. These schemes standardize the way that soils are described and group similar soils by the characteristics that are important in determining behavior. The classification of a soil depends mostly on the percentages of gravel, sand, silt, and clay. Usually it will also depend on special characteristics of the silt and clay fractions.

A distinction is made between silt and clay. What we call *clays* for classification purposes depends on the existence of clay minerals in the *fines fraction* (silt and clay sizes) of the soil, as well as size of the particles. These clay minerals have a different composition and behavior than silts and coarse-grained soils.

Clays display the characteristic of *plasticity*. For the most part, particles that are clay size are also clay minerals. The plasticity characteristics of the fines fraction of a soil is measured in the laboratory by the *Atterberg limit tests*. (See Sec. 14.) The Atterberg limits used most in classification are the liquid limit and plasticity index. In the field, where laboratory tests are not available, clays can be informally distinguished from silts by using the following simple "visual identification" tests.

Dry strength test: A small brick of soil is molded and allowed to air dry. The brick is broken and a small ($1/8$ in or 3 mm) fragment is taken between thumb and finger. A silt fragment will break easily, whereas clay will not.

Dilatancy test: A small sample is mixed with water to form a thick slurry. When the sample is squeezed, water will flow back into a silty sample quickly. The return rate will be much lower for clay.

Plasticity test: A moist soil sample is rolled into a thin ($1/8$ in or 3 mm) thread. As the thread dries, silt will be weak and friable, but clay will be tough.

Table 35.4 *AASHTO Soil Classification System*

	granular materials (35% or less passing no. 200 sieve)							silt-clay materials (more than 35% passing no. 200 sieve)				
	A-1		A-3	A-2				A-4	A-5	A-6	A-7	A-8
	A-1-a	A-1-b		A-2-4	A-2-5	A-2-6	A-2-7				A-7-5 or A-7-6	
sieve analysis: % passing no. 10 no. 40 no. 200	50 max 30 max 15 max	50 max 25 max	51 min 10 max	35 max	35 max	35 max	35 max	36 min	36 min	36 min	36 min	
characteristics of fraction passing no. 40: LL: liquid limit PI: plasticity index	6 max		NP	40 max 10 max	41 min 10 max	40 max 11 min	41 min 11 min	40 max 10 max	41 min 10 max	40 max 11 min	41 min 11 min	
usual types of significant constituents	stone fragments gravel and sand		fine sand	silty or clayey gravel and sand				silty soils		clayey soils		peat, highly organic soils
general subgrade rating	excellent to good							fair to poor				unsatisfactory

Classification procedure: Using the test data, proceed from left to right in the chart. The correct group will be found by process of elimination. The first group from the left consistent with the test data is the correct classification. The A-7 group is subdivided into A-7-5 or A-7-6, depending on the plastic limit. For plastic limit PL = LL − PI less than 30, the classification is A-7-6. For plastic limit PL = LL − PI greater than or equal to 30, it is A-7-5. NP means non-plastic.

Dispersion test: A sample of soil is dispersed in water. The time for the particles to settle is measured. Sand settles in 30 to 60 sec. Silt settles in 15 to 60 min, and clay remains in suspension for a long time.

Organic matter can also be present in soil. Generally, the greater the organic content, the darker the soil color. Organic matter can have a significant effect (usually negative) on the mechanical properties of the soil.

3. AASHTO SOIL CLASSIFICATION

The American Association of State Highway Transportation Officials' (AASHTO) classification system is based on the sieve analysis, liquid limit, and plasticity index. The best soils suitable for use as roadway subgrades are classified as A-1. Highly organic soils not suitable for roadway subgrades are classified as A-8. Soils can also be classified into subgroups. The AASHTO classification methodology is given in Table 35.4.

The *group index*, given by Eq. 35.3, may be added to the group classification. The group index is a means of comparing soils within a group, not between groups. A soil with a group index of zero is a good subgrade material within its particular group. Group indexes of 20 or higher represent poor subgrade materials. The group index, I_g, is reported to the nearest whole number but is reported as zero if it is calculated to be negative. (The liquid limit, LL, and plasticity index, PI, are discussed in Sec. 14.)

$$I_g = (F_{200} - 35)(0.2 + 0.005(LL - 40))$$
$$+ 0.01(F_{200} - 15)(PI - 10) \qquad 35.3$$

F_{200} in Eq. 35.3 is the percentage of soil that passes through a no. 200 sieve. The quantities $LL - 40$ and $PI - 10$ may be negative. For the A-2-6 and A-2-7 subgroups, only the second term in Eq. 35.3 is used in calculating the group index.

4. UNIFIED SOIL CLASSIFICATION

The *Unified Soil Classification System* (USCS) is described in Table 35.5. Like the AASHTO system, it is based on the grain size distribution, liquid limit, and plasticity index of the soil.

Soils are classified into USCS groups that are designated by a group symbol and a corresponding group name. The symbols each contain two letters: The first represents the most significant particle size fraction, and the second is a descriptive modifier. Some categories require dual symbols.

Coarse-grained soils are divided into two categories: gravel soils (symbol G) and sand soils (symbol S). Sands and gravels are further subdivided into four subcategories as follows.

symbol W: well-graded, fairly clean
symbol C: significant amounts of clay
symbol P: poorly graded, fairly clean
symbol M: significant amounts of silt

Fine-grained soils are divided into three categories: inorganic silts (symbol M), inorganic clays (symbol C), and organic silts and clays (symbol O).[2] These three are subdivided into two subcategories as follows.

symbol L: low compressibilities (LL less than 50)
symbol H: high compressibilities (LL 50 or greater)

Example 35.1

Determine the classification of an inorganic soil with the characteristics listed using (a) the AASHTO and (b) the USCS classification systems.

soil size (mm)	fraction retained on sieve	
< 0.002	0.19	LL = 53
0.002–0.005	0.12	PL = 22
0.005–0.05	0.36	
0.05–0.075	0.04	$F_{200} = 0.04 + 0.36 + 0.12$
		$+ 0.19$
		$= 0.71$
0.075–2.0	0.29	
> 2.0	0	

Solution

The plasticity index is given by Eq. 35.23.

$$PI = LL - PL = 53 - 22 = 31$$

(a) From Table 35.2, the no. 200 sieve has an opening of 0.075 mm. F_{200} is 0.71, so from Table 35.4, the soil is first classified as a silt-clay material (more than 35% passing a no. 200 sieve). For a liquid limit of 53 and a plasticity index of 31, the classification is A-7-5 or A-7-6. Since the plastic limit is 22 (less than 30), the classification is A-7-6. The group index is

$$I_g = (71 - 35)(0.2 + 0.005(53 - 40))$$
$$+ 0.01(71 - 15)(31 - 10)$$
$$= 21.3$$

Round 21.3 to the nearest whole number. The soil classification is A-7-6 (21).

(b) From Table 35.5, the soil is first classified as a fine-grained soil (more than 50% passing the no. 200 sieve). The liquid limit is greater than 50 (LL = 53).

Since the soil plots above the A-line, it is classified as CH: a highly plastic clay.

---

[2]The symbol M comes from the Swedish *mjala* (meaning silt) and *mo* (meaning very fine sand).

Geotechnical

Table 35.5 *Unified Soil Classification System*

major division		group symbol	laboratory classification criteria: finer than 200 sieve (%)	laboratory classification criteria: supplementary requirements	soil description
coarse-grained (over 50% by weight coarser than no. 200 sieve)	gravelly soils (over half of coarse fraction larger than no. 4)	GW	0–5[a]	D_{60}/D_{10} greater than 4; $D_{30}^2/(D_{60} \times D_{10})$ between 1 and 3	well-graded gravels, sandy gravels
		GP	0–5[a]	not meeting above gradation for GW	gap-graded or uniform gravels, sandy gravels
		GM	12 or more[a]	PI less than 4 or below A-line	silty gravels, silty sandy gravels
		GC	12 or more[a]	PI over 7 and above A-line	clayey gravels, clayey sandy gravels
	sandy soils (over half of coarse fraction finer than no. 4)	SW	0–5[a]	D_{60}/D_{10} greater than 6; $D_{30}^2/(D_{60} \times D_{10})$ between 1 and 3	well-graded, gravelly sands
		SP	0–5[a]	not meeting above gradation requirements	gap-graded or uniform sands, gravelly sands
		SM	12 or more[a]	PI less than 4 or below A-line	clayey sands, clayey gravelly sands
		SC	12 or more[a]	PI over 7 and above A-line	
fine-grained (over 50% by weight finer than no. 200 sieve)	low compressibility (liquid limit less than 50)	ML	plasticity chart		silts, very fine sands, silty or clayey fine sands, micaceous silts
		CL	plasticity chart		low plasticity clays, sandy or silty clays
		OL	plasticity chart, organic odor or color		organic silts and clays of high plasticity
	high compressibility (liquid limit more than 50)	MH	plasticity chart		micaceous silts, diatomaceous silts, volcanic ash
		CH	plasticity chart		highly plastic clays and sandy clays
		OH	plasticity chart, organic odor or color		organic silts and clays of high plasticity
soils with fibrous organic matter		Pt	fibrous organic matter; will char, burn, or glow		peat, sandy peats, and clayey peat

[a]For soils having 5 to 12% passing the no. 200 sieve, use a dual symbol such as GW-GC.

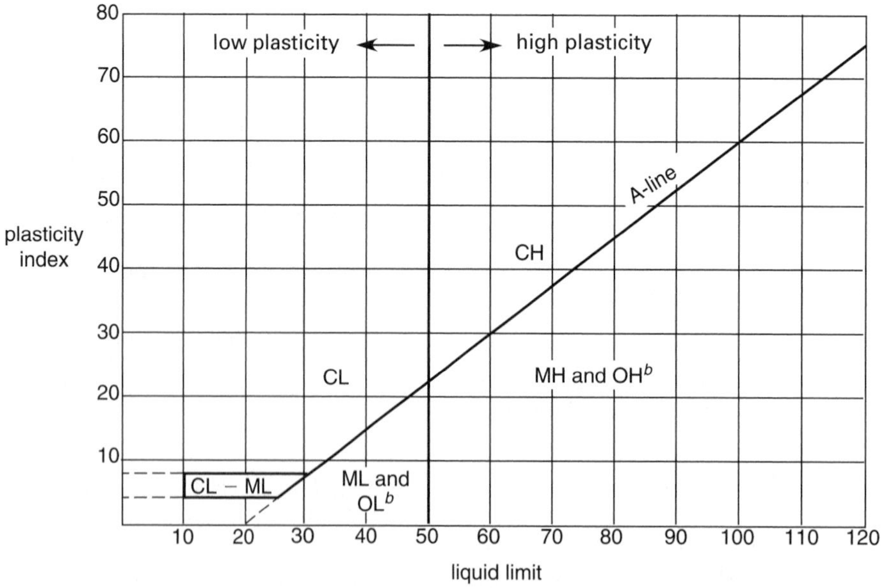

Plasticity chart for the classification of fine-grained soils. Tests made on fraction finer than no. 40 sieve, 0.425 mm.

[b] Distinguishing between M and O classifications requires identifying organic components by observation, odor, or other testing.

5. MASS-VOLUME RELATIONSHIPS

Soil consists of solid soil particles separated by voids. The voids can be filled with either air or water. Thus, soil is a three-phase material (solids, water, and air). The percentages by volume, mass, and weight of these three constituents are used to calculate the aggregate properties.

Figure 35.4 *Soil Phases*

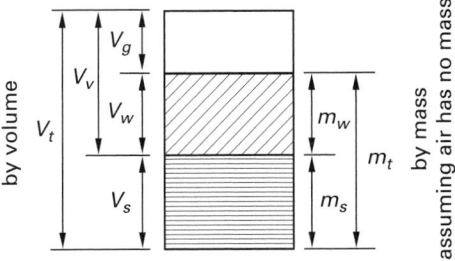

The *porosity*, n, is the ratio of the volume of voids to the total volume.

$$n = \frac{V_v}{V_t} = \frac{V_g + V_w}{V_g + V_w + V_s} \qquad 35.4$$

The *void ratio*, e, is the ratio of the volume of voids to the volume of solids.

$$e = \frac{V_v}{V_s} = \frac{V_g + V_w}{V_s} \qquad 35.5$$

The *moisture content (water content)*, w, is the ratio of the mass of water to the mass of solids, usually expressed in percent[3].

$$w = \frac{m_w}{m_s} \times 100\% \qquad 35.6$$

The *degree of saturation*, S, is the percentage of the volume of water to the total volume of voids.[3] This indicates how much of the void space is filled with water. If all the voids are filled with water, then the volume of air is zero and the sample's degree of saturation is 100%. The sample is said to be fully saturated.

$$S = \frac{V_w}{V_v} \times 100\%$$
$$= \frac{V_w}{V_g + V_w} \times 100\% \qquad 35.7$$

The *density* (also called the *total density*), ρ, is the ratio of the total mass to the total volume. The total density may be referred to as the *moist density* (*wet density*) above the water table and as the *saturated density* below the water table.

$$\rho = \frac{m_t}{V_t} = \frac{m_w + m_s}{V_g + V_w + V_s} \qquad 35.8$$

[3]Both water content and saturation are reported as percentages; however, in working problems it is easiest to use the decimal form.

In the United States, weight is commonly used in soil calculations instead of mass. The corresponding density is called *unit weight*, γ. Weight is calculated $W = mg/g_c$, where g_c is the gravitational constant, 32.2 ft-lbm/lbf-sec^2. In Eqs. 35.8 and 35.10 through 35.14, γ can be substituted for ρ whenever convenient.[4]

$$\gamma = \rho g \qquad \text{[SI]} \qquad 35.9(a)$$
$$\gamma = \frac{\rho g}{g_c} \qquad \text{[U.S.]} \qquad 35.9(b)$$

The *dry density*, ρ_d, is the ratio of the solid mass to the total volume.

$$\rho_d = \frac{m_s}{V_t} = \frac{m_s}{V_g + V_w + V_s} \qquad 35.10$$

If the water content is known, the dry density of a moist sample can be found from Eq. 35.11.

$$\rho_d = \frac{m_t}{(1+w)V_t} = \frac{\rho}{1+w} \qquad 35.11$$

The *buoyant density (submerged density)*, ρ_b, is the difference between the total density and the density of water.

$$\rho_b = \rho - \rho_w \qquad 35.12$$

The density of the solid constituents, ρ_s, is the ratio of the mass of the solids to the volume of the solids. This would also be the density of soil if there were no voids. For that reason, ρ_s is also known as the *solid density* and *zero-voids density*.

$$\rho_s = \frac{m_s}{V_s} \qquad 35.13$$

The *percent pore space*, PPS, is the ratio of the volume of the voids to the total volume, in percent.

$$\text{PPS} = \frac{V_v}{V_t} \times 100\%$$
$$= \left(1 - \frac{V_s}{V_t}\right) \times 100\%$$
$$= \left(1 - \frac{\rho_d}{\rho_s}\right) \times 100\% \qquad 35.14$$

The *specific gravity*, SG, of the solid constituents is given by Eq. 35.15. For practical purposes, the density of water, ρ_w, is 62.4 lbm/ft^3 (1000 kg/m^3). The specific gravity of most soils is within the range of 2.65 to 2.70, with clays as high as 2.9 and organic soils as low as 2.5.

$$\text{SG} = \frac{\rho_s}{\rho_w} \qquad 35.15$$

The *relative density*, D_r (also denoted as the *density index*, I_d), is the density of a granular soil relative to

[4]There is no concept of weight in SI units. The choice to work in weight or mass in U.S. units is somewhat arbitrary, because as long as the local gravitational field is essentially standard, the mass in lbm is numerically equal to the weight in lbf.

the minimum and maximum densities achieved for that soil in a laboratory test. This index is not applicable to clays, because clays do not densify in this particular laboratory test. The relative density is equal to 1 for a very dense soil and 0 for a very loose soil.

$$D_r = \frac{e_{\max} - e}{e_{\max} - e_{\min}} \qquad 35.16$$

Typical values of some of these soil parameters are given in Table 35.6. Useful relationships for computing the above soil indexes are summarized in Table 35.7.

Table 35.6 Typical Soil Characteristics

description	n	e	w_{sat}	ρ_d $\dfrac{\text{lbm}}{\text{ft}^3}$ $\left(\dfrac{\text{kg}}{\text{m}^3}\right)$		ρ_{sat} $\dfrac{\text{lbm}}{\text{ft}^3}$ $\left(\dfrac{\text{kg}}{\text{m}^3}\right)$	
sand, loose and uniform	0.46	0.85	0.32	90	(1440)	118	(1890)
sand, dense and uniform	0.34	0.51	0.19	109	(1750)	130	(2080)
sand, loose and mixed	0.40	0.67	0.25	99	(1590)	124	(1990)
sand, dense and mixed	0.30	0.43	0.16	116	(1860)	135	(2160)
glacial clay, soft	0.55	1.20	0.45	76	(1200)	110	(1760)
glacial clay, stiff	0.37	0.60	0.22	106	(1700)	125	(2000)

(Multiply lbm/ft^3 by 16.02 to obtain kg/m^3.)

Example 35.2

What is the degree of saturation for a sand sample with SG = 2.65, ρ = 115 lbm/ft^3 (1840 kg/m^3), and w = 17%?

SI Solution

Draw the various mass and volume phases on a phase diagram. Since an actual mass or volume of soil is not specified, arbitrarily assume any convenient volume, usually 1.0 m^3.

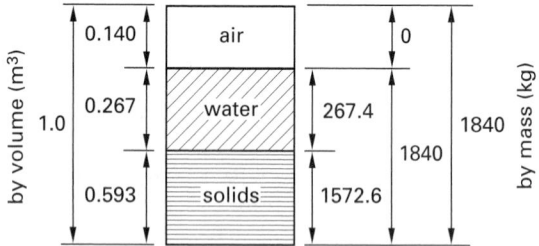

From Eq. 35.6,

$$m_w = w m_s = 0.17 m_s$$

But, $m_t = m_w + m_s = 1840$ kg. (The mass of any air voids is negligible.)

$$0.17 m_s + m_s = 1840 \text{ kg}$$
$$m_s = 1572.6 \text{ kg}$$
$$m_w = 267.4 \text{ kg}$$

The solids volume is given by Eq. 35.13.

$$V_s = \frac{m_s}{\rho_s}$$

From Eq. 35.15, $\rho_s = (\text{SG})\rho_w$.

$$V_s = \frac{m_s}{(\text{SG})\rho_w}$$
$$= \frac{1572.6 \text{ kg}}{(2.65)(1000 \text{ kg}/\text{m}^3)}$$
$$= 0.593 \text{ m}^3$$

Similarly, the water volume is

$$V_w = \frac{m_w}{\rho_w}$$
$$= \frac{267.4 \text{ kg}}{1000 \dfrac{\text{kg}}{\text{m}^3}}$$
$$= 0.267 \text{ m}^3$$

The air volume is

$$V_g = V_t - V_s - V_w$$
$$= 1 \text{ m}^3 - 0.593 \text{ m}^3 - 0.267 \text{ m}^3 = 0.140 \text{ m}^3$$

The degree of saturation is given by Eq. 35.7.

$$S = \frac{V_w}{V_g + V_w} \times 100\%$$
$$= \frac{0.267 \text{ m}^3}{0.140 \text{ m}^3 + 0.267 \text{ m}^3} \times 100\%$$
$$= 66\%$$

Customary U.S. Solution

Draw the various mass and volume phases on a phase diagram. Since an actual mass or volume of soil is not specified, arbitrarily assume any convenient volume, usually 1.0 ft^3.

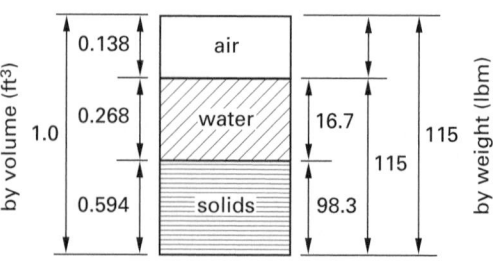

Table 35.7 *Soil Indexing Formulas*

property		saturated sample (m_s, m_w, SG are known)	unsaturated sample (m_s, m_w, SG, V_t are known)	supplementary formulas relating measured and computed factors			
volume components							
V_s	volume of solids		$\dfrac{m_s}{(\text{SG})\rho_w}$	$V_t - (V_g + V_w)$	$V_t(1-n)$	$\dfrac{V_t}{1+e}$	$\dfrac{V_v}{e}$
V_w	volume of water		$\dfrac{m_w}{\rho_w^*}$	$V_v - V_g$	SV_v	$\dfrac{SV_t e}{1+e}$	$SV_s e$
V_g	volume of gas or air	zero	$V_t - (V_s + V_w)$	$V_v - V_w$	$(1-S)V_v$	$\dfrac{(1-S)V_t e}{1+e}$	$(1-S)V_s e$
V_v	volume of voids	$\dfrac{m_w}{\rho_w^*}$	$V_t - \dfrac{m_s}{(\text{SG})\rho_w}$	$V_t - V_s$	$\dfrac{V_s n}{1-n}$	$\dfrac{V_t e}{1+e}$	$V_s e$
V_t	total volume of sample	$V_s + V_w$	measured $(V_g + V_w + V_s)$	$V_s + V_g + V_w$	$\dfrac{V_s}{1-n}$	$V_s(1+e)$	$\dfrac{V_v(1+e)}{e}$
n	porosity		$\dfrac{V_v}{V_t}$	$1 - \dfrac{V_s}{V_t}$	$1 - \dfrac{m_s}{(\text{SG})V_t \rho_w}$	$\dfrac{e}{1+e}$	
e	void ratio		$\dfrac{V_v}{V_s}$	$\dfrac{V_t}{V_s} - 1$	$\dfrac{(\text{SG})V_t \rho_w}{m_s} - 1$	$\dfrac{m_w(\text{SG})}{m_s S}$	$\dfrac{n}{1-n} \left| w\left(\dfrac{\text{SG}}{S}\right)\right.$
mass for specific sample							
m_s	mass of solids		measured	$\dfrac{m_t}{1+w}$	$(\text{SG})V_t \rho_w (1-n)$	$\dfrac{m_w(\text{SG})}{eS}$	$V_s(\text{SG})\rho_w$
m_w	mass of water		measured	wm_s	$S\rho_w V v$	$\dfrac{em_s S}{\text{SG}}$	$V_t \rho_d w$
m_t	total mass of sample		$m_s + m_w$	$m_s(1+w)$			
mass for sample of unit volume (density)							
ρ_d	dry density	$\dfrac{m_s}{V_s + V_w}$	$\dfrac{m_s}{V_g + V_w + V_s}$	$\dfrac{m_t}{V_t(1+w)}$	$\dfrac{(\text{SG})\rho_w}{1+e}$	$\dfrac{(\text{SG})\rho_w}{1+\dfrac{w(\text{SG})}{S}}$	$\dfrac{\rho}{1+w}$
ρ	wet density (density with moisture)	$\dfrac{m_s + m_w}{V_s + V_w}$	$\dfrac{m_s + m_w}{V_t}$	$\dfrac{m_t}{V_t}$	$\dfrac{(SG + Se)\rho_w}{1+e}$	$\dfrac{(1+w)\rho_w}{\dfrac{w}{S}+\dfrac{1}{\text{SG}}}$	$\rho_d(1+w)$
ρ_{sat}	saturated density	$\dfrac{m_s + m_w}{V_s + V_w}$	$\dfrac{m_s + V_v \rho_w}{V_t}$	$\dfrac{m_s}{V_t} + \left(\dfrac{e}{1+e}\right)\rho_w$	$\dfrac{(SG + e)\rho_w}{1+e}$	$\dfrac{(1+w)\rho_w}{w+\dfrac{1}{\text{SG}}}$	
ρ_b	buoyant (submerged) density		$\rho_{\text{sat}} - \rho_w^*$	$\dfrac{m_s}{V_t} - \left(\dfrac{1}{1+e}\right)\rho_w^*$	$\left(\dfrac{SG+e}{1+e}-1\right)\rho_w^*$	$\left(\dfrac{1-\dfrac{1}{\text{SG}}}{w+\dfrac{1}{\text{SG}}}\right)\rho_w^*$	

(continued)

Table 35.7 Soil Indexing Formulas (continued)

property		saturated sample (m_s, m_w, SG are known)	unsaturated sample (m_s, m_w, SG, V_t are known)	supplementary formulas relating measured and computed factors		
combined relations						
w	water content		$\dfrac{m_w}{m_s}$	$\dfrac{m_t}{m_s} - 1$	$\dfrac{Se}{SG}$	$S\left(\dfrac{\rho_w^*}{\rho_d} - \dfrac{1}{SG}\right)$
S	degree of saturation	100%	$\dfrac{V_w}{V_v}$	$\dfrac{m_w}{V_v \rho_w^*}$	$\dfrac{w(SG)}{e}$	$\dfrac{w}{\dfrac{\rho_w^*}{\rho_d} - \dfrac{1}{SG}}$
SG	specific gravity of solids		$\dfrac{m_s}{V_s \rho_w}$	$\dfrac{Se}{w}$		

ρ_w is the density of water. Where noted with an asterisk (*), use the actual density of water at the recorded temperature. In other cases, use 62.4 lbm/ft³ or 1000 kg/m³.

From Eq. 35.6,

$$m_w = wm_s = 0.17m_s$$

However, $m_t = m_w + m_s = 115$ lbm. (The mass of any air voids is negligible.)

$$0.17m_s + m_s = 115 \text{ lbm}$$
$$m_s = 98.3 \text{ lbm}$$
$$m_w = 16.7 \text{ lbm}$$

The solids volume is given by Eq. 35.13.

$$V_s = \frac{m_s}{\rho_s} = \frac{m_s}{(SG)\rho_w}$$
$$= \frac{98.3 \text{ lbm}}{(2.65)\left(62.4 \, \dfrac{\text{lbm}}{\text{ft}^3}\right)}$$
$$= 0.594 \text{ ft}^3$$

Similarly, the water volume is

$$V_w = \frac{m_w}{\rho_w}$$
$$= \frac{16.7 \text{ lbm}}{62.4 \, \dfrac{\text{lbm}}{\text{ft}^3}}$$
$$= 0.268 \text{ ft}^3$$

The air volume is

$$V_g = V_t - V_s - V_w$$
$$= 1 \text{ ft}^3 - 0.594 \text{ ft}^3 - 0.268 \text{ ft}^3$$
$$= 0.138 \text{ ft}^3$$

The degree of saturation is given by Eq. 35.7.

$$S = \frac{V_w}{V_g + V_w} \times 100\%$$
$$= \frac{0.268 \text{ ft}^3}{0.138 \text{ ft}^3 + 0.268 \text{ ft}^3} \times 100\%$$
$$= 66\%$$

Example 35.3

Borrow soil is used to fill a 100,000 yd³ (75,000 m³) depression. The borrow soil has the following characteristics: density = 96.0 lbm/ft³ (1540 kg/m³), water content = 8%, and specific gravity of the solids = 2.66. The final in-place dry density should be 112.0 lbm/ft³ (1790 kg/m³), and the final water content should be 13%.

(a) How many cubic yards (cubic meters) of borrow soil are needed? (b) Assuming no evaporation loss, what water mass is needed to achieve 13% moisture? (c) What will be the density of the in-place fill after a long rain?

SI Solution

Draw the phase diagrams for both the borrow and compacted fill soils. Use subscript B for borrow soil and F for fill soil, and work with 1 m³ of fill material.

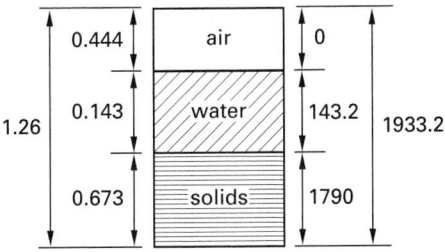

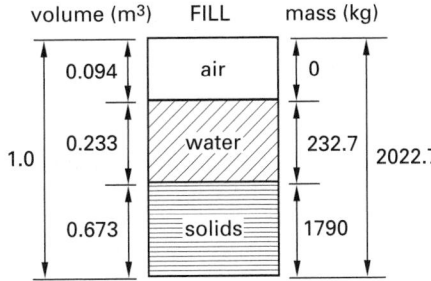

step 1: The air has no mass. The soil content (mass) is the same at both locations. (That is, getting 1790 kg of soil solids in the fill requires taking 1790 kg of borrow soil solids.) Per cubic meter of fill, the mass of borrow soil solids is

$$m_{s,B} = m_{s,F} = 1790 \text{ kg}$$

step 2: The mass of the water in 1 m³ of fill is found from Eq. 35.6.

$$m_{w,F} = w m_{s,F} = (0.13)(1790 \text{ kg}) = 232.7 \text{ kg}$$

step 3: The total mass and density of the fill are

$$m_{t,F} = m_{s,F} + m_{w,F}$$
$$= 1790 \text{ kg} + 232.7 \text{ kg} = 2022.7 \text{ kg}$$
$$\rho_F = \frac{m_{t,F}}{V_t} = \frac{2022.7 \text{ kg}}{1 \text{ m}^3}$$
$$= 2022.7 \text{ kg/m}^3 \text{ fill}$$

step 4: Since $\rho_s = m_s/V_s$, the volume of the solids in the fill is

$$V_{s,F} = \frac{m_{s,F}}{\rho_{s,F}}$$
$$= \frac{1790 \text{ kg}}{(2.66)\left(1000 \dfrac{\text{kg}}{\text{m}^3}\right)}$$
$$= 0.673 \text{ m}^3$$

step 5: Similarly, the volume of the water in the fill is

$$V_{w,F} = \frac{m_{w,F}}{\rho_{w,F}}$$
$$= \frac{232.7 \text{ kg}}{1000 \dfrac{\text{kg}}{\text{m}^3}}$$
$$= 0.233 \text{ m}^3$$

step 6: The air volume in the fill is

$$V_g = V_t - V_s - V_w$$
$$= 1 \text{ m}^3 - 0.673 \text{ m}^3 - 0.233 \text{ m}^3$$
$$= 0.094 \text{ m}^3$$

step 7: The mass of the water per cubic meter of fill in the borrow soil is given by Eq. 35.6.

$$m_{w,B} = w m_{s,B}$$
$$= (0.08)(1790 \text{ kg}) = 143.2 \text{ kg}$$

step 8: The total mass of the borrow soil per cubic meter of fill is

$$m_{t,B} = m_{s,B} + m_{w,B}$$
$$= 1790 \text{ kg} + 143.2 \text{ kg} = 1933.2 \text{ kg}$$

step 9: The total volume of the borrow soil per cubic meter of fill is

$$V_{t,B} = \frac{m_{t,B}}{\rho_B}$$
$$= \frac{1933.2 \text{ kg}}{1540 \dfrac{\text{kg}}{\text{m}^3}}$$
$$= 1.26 \text{ m}^3$$

step 10: The mass of solids in 1 m³ of fill is the same as the mass of solids in 1.26 m³ of borrow soil. Since the mass is the same, the volume of solids is also the same, even though the total soil density is different. The volume of solids was calculated in step 4.

$$V_{s,B} = V_{s,F} = 0.673 \text{ m}^3$$

step 11: The volume of water in the borrow soil per cubic meter of fill is

$$V_{w,B} = \frac{m_{w,B}}{\rho_w}$$
$$= \frac{143.2 \text{ kg}}{1000 \dfrac{\text{kg}}{\text{m}^3}}$$
$$= 0.143 \text{ m}^3$$

Geotechnical

step 12: The air volume in the borrow soil per cubic foot of fill is

$$V_{g,B} = V_{t,B} - V_{s,B} - V_{w,B}$$
$$= 1.26 \text{ m}^3 - 0.143 \text{ m}^3 - 0.673 \text{ m}^3$$
$$= 0.444 \text{ m}^3$$

step 13a: Since 1.26 m^3 of borrow soil is required for every 1 m^3 of fill, the volume of borrow soil required to fill the depression is

$$V_{\text{required},B} = \left(\frac{1.26 \text{ m}^3}{1 \text{ m}^3}\right)(75\,000 \text{ m}^3)$$
$$= 94\,500 \text{ m}^3$$

step 13b: The amount of water needed to achieve 13% moisture is the difference in mass of water between the fill and borrow soil. The actual moisture in the compacted borrow soil is

$$m_{B,\text{total}} = V_{B,\text{total}}\left(\frac{m_{w,B}}{V_{t,B}}\right)$$
$$= \frac{(94,500 \text{ m}^3)(143.2 \text{ kg})}{1.26 \text{ m}^3}$$
$$= 1.074 \times 10^7 \text{ kg}$$

The required moisture in the fill soil is

$$m_{F,\text{total}} = V_{F,\text{total}}\left(\frac{m_{w,F}}{V_{t,F}}\right)$$
$$= \frac{(75,000 \text{ m}^3)(232.7 \text{ kg})}{1 \text{ m}^3}$$
$$= 1.745 \times 10^7 \text{ kg}$$

The required additional moisture is

$$m_{w,\text{required}} = m_{F,\text{total}} - m_{B,\text{total}}$$
$$= 1.745 \times 10^7 \text{ kg} - 1.074 \times 10^7 \text{ kg}$$
$$= 6.71 \times 10^6 \text{ kg}$$

step 13c: After a rain, all the void spaces will be filled with water and the density will be the saturated density.

$$m_w = (V_w + V_g)\rho_w$$
$$= (0.233 \text{ m}^3 + 0.094 \text{ m}^3)\left(1000 \frac{\text{kg}}{\text{m}^3}\right)$$
$$= 327.0 \text{ kg}$$
$$\rho = \frac{m_t}{V_t} = \frac{m_s + m_w}{V_t}$$
$$= \frac{1790 \text{ kg} + 327.0 \text{ kg}}{1 \text{ m}^3}$$
$$= 2117 \text{ kg/m}^3$$

Customary U.S. Solution

Draw the phase diagrams for both the borrow and compacted fill soils. Use subscript B for borrow soil and F for fill soil, and work with 1 ft^3 of fill material.

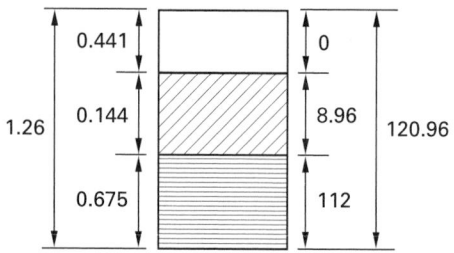

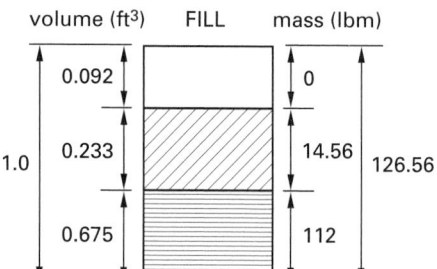

step 1: The air has no mass. The soil content (mass) is the same at both locations. (That is, getting 112 lbm of soil solids in the fill requires taking 112 lbm of borrow soil solids.) Per cubic foot of fill, the mass of borrow soil solids is

$$m_{s,B} = m_{s,F} = 112 \text{ lbm}$$

step 2: The mass of the water in 1 ft^3 of fill is given by Eq. 35.6.

$$m_{w,F} = w m_{s,F} = (0.13)(112 \text{ lbm}) = 14.56 \text{ lbm}$$

step 3: The total mass and density of the fill are

$$m_{t,F} = m_{s,F} + m_{w,F}$$
$$= 112 \text{ lbm} + 14.56 \text{ lbm} = 126.56 \text{ lbm}$$
$$\rho_F = \frac{126.56 \text{ lbm}}{1 \text{ ft}^3}$$
$$= 126.56 \text{ lbm/ft}^3 \text{ fill}$$

step 4: Since $\rho_s = m_s/V_s$, the volume of the solids in the fill is

$$V_{s,F} = \frac{m_{s,F}}{\rho_{s,F}}$$
$$= \frac{112 \text{ lbm}}{(2.66)\left(62.4 \frac{\text{lbm}}{\text{ft}^3}\right)}$$
$$= 0.675 \text{ ft}^3$$

step 5: Similarly, the volume of the water in the fill is

$$V_{w,F} = \frac{m_{w,F}}{\rho_{w,F}}$$

$$= \frac{14.56 \text{ lbm}}{62.4 \dfrac{\text{lbm}}{\text{ft}^3}}$$

$$= 0.233 \text{ ft}^3$$

step 6: The air volume in the fill is

$$V_g = V_t - V_s - V_w$$

$$= 1 \text{ ft}^3 - 0.675 \text{ ft}^3 - 0.233 \text{ ft}^3$$

$$= 0.092 \text{ ft}^3$$

step 7: The mass of the water per cubic foot of fill in the borrow soil is given by Eq. 35.6.

$$m_{w,B} = w m_{s,B}$$

$$= (0.08)(112 \text{ lbm}) = 8.96 \text{ lbm}$$

step 8: The total mass of the borrow soil per cubic foot of fill is

$$m_{t,B} = m_{s,B} + m_{w,B}$$

$$= 112 \text{ lbm} + 8.96 \text{ lbm} = 120.96 \text{ lbm}$$

step 9: The total volume of the borrow soil per cubic foot of fill is

$$V_{t,B} = \frac{m_{t,B}}{\rho_B}$$

$$= \frac{120.96 \text{ lbm}}{96 \dfrac{\text{lbm}}{\text{ft}^3}}$$

$$= 1.26 \text{ ft}^3$$

step 10: The mass of solids in 1 ft^3 of fill is the same as the mass of solids in 1.26 ft^3 of borrow soil. Since the mass is the same, the volume of solids is also the same, even though the total soil density is different. The volume of solids was calculated in step 4.

$$V_{s,B} = V_{s,F} = 0.675 \text{ ft}^3$$

step 11: The volume of water in the borrow soil per cubic foot of fill is

$$V_{w,B} = \frac{m_{w,B}}{\rho_w}$$

$$= \frac{8.96 \text{ lbm}}{62.4 \dfrac{\text{lbm}}{\text{ft}^3}}$$

$$= 0.144 \text{ ft}^3$$

step 12: The air volume in the borrow soil per cubic foot of fill is

$$V_{g,B} = V_{t,B} - V_{s,B} - V_{w,B}$$

$$= 1.26 \text{ ft}^3 - 0.675 \text{ ft}^3 - 0.144 \text{ ft}^3$$

$$= 0.441 \text{ ft}^3$$

step 13a: Since 1.26 ft^3 of borrow soil are required for every 1 ft^3 of fill, the volume of borrow soil required to fill the depression is

$$V_{\text{required},B} = \left(\frac{1.26 \text{ ft}^3}{1 \text{ ft}^3} \right)(100{,}000 \text{ yd}^3)$$

$$= 126{,}000 \text{ yd}^3$$

step 13b: The amount of water needed to achieve 13% moisture is the difference in mass of water between the fill and borrow soil. The actual moisture in the compacted borrow soil is

$$m_{B,\text{total}} = V_{B,\text{total}} \left(\frac{m_{w,B}}{V_{t,B}} \right)$$

$$= \frac{(126{,}000 \text{ yd}^3) \left(27 \dfrac{\text{ft}^3}{\text{yd}^3} \right)(8.96 \text{ lbm})}{1.26 \text{ ft}^3}$$

$$= 2.42 \times 10^7 \text{ lbm}$$

The required moisture in the fill soil is

$$m_{F,\text{total}} = V_{F,\text{total}} \left(\frac{m_{w,F}}{V_{t,F}} \right)$$

$$= (100{,}000 \text{ yd}^3) \left(27 \dfrac{\text{ft}^3}{\text{yd}^3} \right) \left(14.56 \dfrac{\text{lbm}}{\text{ft}^3} \right)$$

$$= 3.93 \times 10^7 \text{ lbm}$$

The required additional moisture is

$$m_{w,\text{required}} = m_{F,\text{total}} - m_{B,\text{total}}$$

$$= 3.93 \times 10^7 \text{ lbm} - 2.42 \times 10^7 \text{ lbm}$$

$$= 1.51 \times 10^7 \text{ lbm}$$

step 13c: After a rain, all the void spaces will be filled with water and the density will be the saturated density.

$$m_w = (V_w + V_g)\rho_w$$

$$= (0.233 \text{ ft}^3 + 0.092 \text{ ft}^3) \left(62.4 \dfrac{\text{lbm}}{\text{ft}^3} \right)$$

$$= 20.28 \text{ lbm}$$

$$\rho = \frac{m_t}{V_t}$$

$$= \frac{m_s + m_w}{V_t}$$

$$= \frac{112.0 \text{ lbm} + 20.28 \text{ lbm}}{1.0 \text{ ft}^3}$$

$$= 132.3 \text{ lbm/ft}^3$$

Geotechnical

6. SWELL

Swell occurs when clayey soils are used at lower loadings and/or higher moisture contents than existed prior to excavation. A small percentage may need to be added to calculated borrow-soil volumes to account for swell. Actual swell percentages are difficult to predict, but are typically less than 5%. Soils with large swell percentages are probably not suitable for foundation fill.

7. EFFECTIVE STRESS

Each of the three phases (solids, water, and air) in soil have different characteristics. Unlike other construction materials that essentially perform as homogeneous media, the three phases in soil interact with each other: Water and air can enter or leave the soil, and the soil grains can be pushed closer together. The water and air in the soil voids cannot support shear stress, so shear is supported entirely through grain-to-grain contact.

The *effective stress*, σ', is the portion of the total stress that is supported through grain contact. It is the difference between the *total stress*, σ, and the *pore water pressure*, u. (The pore water pressure is sometimes called the *neutral stress* because it is equal in all directions; that is, it has no shear stress component.) The effective stress is the average stress on a plane through the soil, not the actual contact stress between two soil particles (which can be much higher).

$$\sigma' = \sigma - u \qquad\qquad 35.17$$

The stresses in a soil element at depth z below the ground surface include those due to the weight above that element (called the *overburden stress*) and any buoyant forces that the water in the soil voids exerts. The stress in a soil column can be divided into total stress, effective stress, and pore pressure components. Above the water table, total and effective stresses are equal. Below the water table, the pore water exerts a buoyant force, and this hydrostatic stress is subtracted from the total stress to obtain the effective stress.

Example 35.4

The sand in the profile illustrated is saturated above the water table by virtue of capillary action. Given the soil properties illustrated, draw the total stress, effective stress, and pore pressure diagrams.

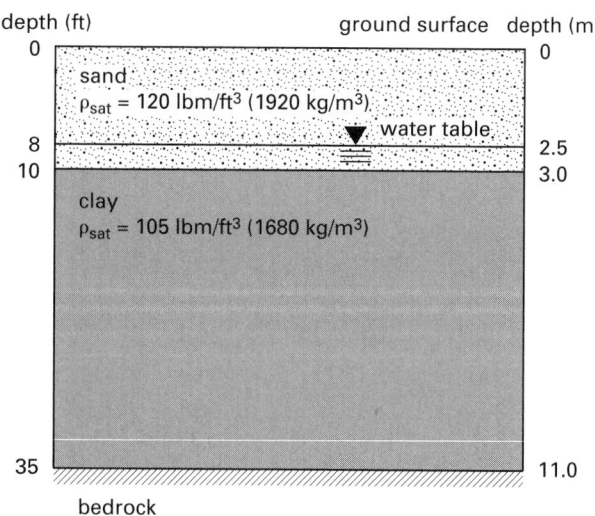

SI Solution

The total stress profile is a profile of the total weight of the soil column with depth. The total stress at a depth of 3 m is

$$\sigma_{3\text{ m}} = \rho g z$$
$$= \left(1920 \ \frac{\text{kg}}{\text{m}^3}\right)\left(9.81 \ \frac{\text{m}}{\text{s}^2}\right)(3 \text{ m})$$
$$= 56\,505 \text{ Pa} \quad (56.5 \text{ kPa})$$

At a depth of 11 m, the total stress is

$$\sigma_{11\text{ m}} = \sigma_{3\text{ m}} + \rho g z$$
$$= 56\,505 \text{ Pa} + \left(1680 \ \frac{\text{kg}}{\text{m}^3}\right)\left(9.81 \ \frac{\text{m}}{\text{s}^2}\right)$$
$$\times (11 \text{ m} - 3 \text{ m})$$
$$= 188\,351 \text{ Pa} \quad (188.4 \text{ kPa})$$

The pore water pressure above the water table is zero. Below the water table, the pore water pressure is equal to the weight of the water column. At a depth of 11 m, the pore water pressure is

$$u_{11\text{ m}} = \rho_w g z$$
$$= \left(1000 \ \frac{\text{kg}}{\text{m}^3}\right)\left(9.81 \ \frac{\text{m}}{\text{s}^2}\right)(8.5 \text{ m})$$
$$= 83\,355 \text{ Pa} \quad (83.4 \text{ kPa})$$

The effective stress profile is determined by subtracting the pore water pressure from the total stress, or by using buoyant weights below the water table. Above the water table, the total and effective stresses are equal.

At a depth of 2.5 m, the effective stress is

$$\sigma'_{2.5\text{ m}} = \rho g z$$
$$\left(1920 \ \frac{\text{kg}}{\text{m}^3}\right)\left(9.81 \ \frac{\text{m}}{\text{s}^2}\right)(2.5 \text{ m})$$
$$= 47\,088 \text{ Pa} \quad (47.1 \text{ kPa})$$

At a depth of 3 m, the effective stress is

$$\sigma'_{3\text{ m}} = \sigma_{2.5\text{ m}} + \rho g z$$

$$= 47\,088 \text{ Pa} + \left(1920\ \frac{\text{kg}}{\text{m}^3} - 1000\ \frac{\text{kg}}{\text{m}^3}\right)$$

$$\times (0.5 \text{ m})\left(9.81\ \frac{\text{m}}{\text{s}^2}\right)$$

$$= 51\,601 \text{ Pa} \quad (51.6 \text{ kPa})$$

At a depth of 11 m, the effective stress is

$$\sigma'_{11\text{ m}} = 51\,601 \text{ Pa} + \left(1680\ \frac{\text{kg}}{\text{m}^3}\right.$$

$$\left. - 1000\ \frac{\text{kg}}{\text{m}^3}\right)\left(9.81\ \frac{\text{m}}{\text{s}^2}\right)(8 \text{ m})$$

$$= 104\,967 \text{ Pa} \quad (105.0 \text{ kPa})$$

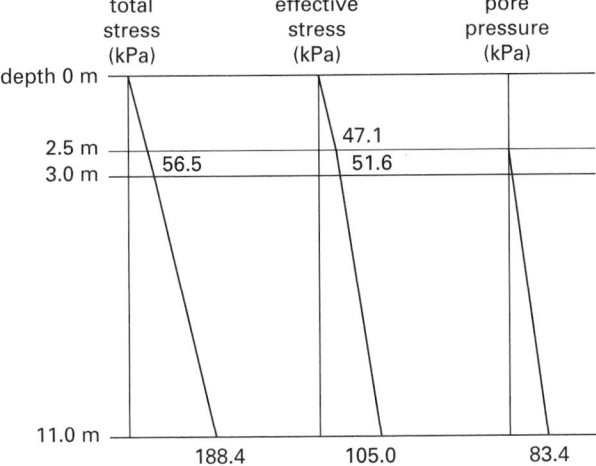

Customary U.S. Solution

The total stress profile is a profile of the total weight of the soil column with depth. The total stress at a depth of 10 ft is

$$\sigma_{10\text{ ft}} = \gamma z$$

$$= \left(120\ \frac{\text{lbf}}{\text{ft}^3}\right)(10 \text{ ft})$$

$$= 1200 \text{ lbf/ft}^2$$

At a depth of 35 ft, the total stress is

$$\sigma_{35\text{ ft}} = \sigma_{10\text{ ft}} + \gamma z$$

$$= 1200\ \frac{\text{lbf}}{\text{ft}^2} + \left(105\ \frac{\text{lbf}}{\text{ft}^3}\right)(25 \text{ ft})$$

$$= 3825 \text{ lbf/ft}^2$$

The pore water pressure above the water table is zero. Below the water table, the pore water pressure is equal to the weight of the water column. At a depth of 35 ft, the pore water pressure is

$$u_{35\text{ ft}} = \left(62.4\ \frac{\text{lbf}}{\text{ft}^3}\right)(35 \text{ ft} - 8 \text{ ft})$$

$$= 1684.8 \text{ lbf/ft}^2$$

The effective stress profile is determined by subtracting the pore water pressure from the total stress, or by using buoyant densities below the water table. Above the water table, the total and effective stresses are equal.

At a depth of 8 ft, the effective stress is

$$\sigma'_{8\text{ ft}} = \gamma z$$

$$= \left(120\ \frac{\text{lbf}}{\text{ft}^3}\right)(8 \text{ ft}) = 960 \text{ lbf/ft}^2$$

At a depth of 10 ft, the effective stress is

$$\sigma'_{10\text{ ft}} = 960\ \frac{\text{lbf}}{\text{ft}^2} + \left(120\ \frac{\text{lbf}}{\text{ft}^3} - 62.4\ \frac{\text{lbf}}{\text{ft}^3}\right)(2 \text{ ft})$$

$$= 1075.2 \text{ lbf/ft}^2$$

At a depth of 35 ft, the effective stress is

$$\sigma'_{35\text{ ft}} = 1075.2\ \frac{\text{lbf}}{\text{ft}^2} + \left(105\ \frac{\text{lbf}}{\text{ft}^3} - 62.4\ \frac{\text{lbf}}{\text{ft}^3}\right)(25 \text{ ft})$$

$$= 2140.2 \text{ lbf/ft}^2$$

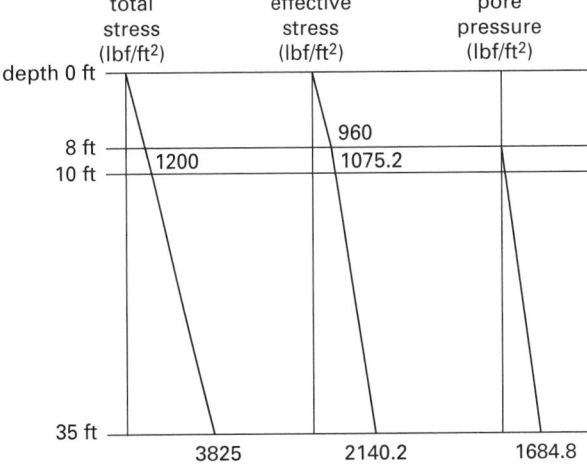

8. STANDARDIZED SOIL TESTING PROCEDURES

Soil testing procedures and equipment vary from laboratory to laboratory and depend on specific project needs. To compare results, consistent methodologies are necessary. For this reason, government and professional agencies have developed standards for commonly performed tests. Table 35.8 lists some of the soils tests that have been standardized by the American Society for Testing and Materials (ASTM). Most of the procedures referred to in this chapter have been chosen in accordance with ASTM standards.

Table 35.8 *ASTM Standard Soils Tests*

test	ASTM designation*	title (abbreviated)
AASHTO Soil Classification	D3282	Classification of Soils and Soil-Aggregate Mixtures for Highway Construction Purposes
USCS Soil Classification	D2487	Classification of Soils for Engineering Purposes
Standard Penetration Test	D1586	Penetration Test and Split-Barrel Sampling of Soils
Cone Penetrometer Test	D3441	Deep, Quasi-Static, Cone and Friction Cone Penetration Tests of Soil
Proctor/Modified Proctor Test	D698 D1557	Laboratory Compaction Characteristics —Using Standard Effort —Using Modified Effort
In-Place Density Test	D2922 D2167 D1556	Nuclear Methods (Shallow Depth) Rubber-Balloon Method Sand-Cone Method
Atterberg Limit Tests	D4318	Liquid Limit, Plastic Limit, and Plasticity Index of Soils
Permeability Tests	D2434	Permeability of Granular Soils (Constant Head)
Consolidation Tests	D2435	One-Dimensional Consolidation Properties of Soils
Direct Shear Tests	D3080	Direct Shear Test of Soils Under Consolidated Drained Conditions
Triaxial Stress Tests	D2850 D4767	UU Tests, Cohesive Soils CU Tests, Cohesive Soils
Unconfined Compressive Strength Test	D2166	Unconfined Compressive Strength of Cohesive Soil
CBR Test	D1883	CBR (California Bearing Ratio) of Laboratory Compacted Soils
Plate Bearing Value Test	D1195	Repetitive Static Plate Load Tests of Soils and Flexible Pavement Components
Hveem's Resistance Value Test	D2844	Resistance R-Value and Expansion Pressure of Compacted Soils

*Adapted from the 1997 *Annual Book of ASTM Standards*

9. STANDARD PENETRATION TEST

A common soils test is the *standard penetration test* (SPT), which is performed in situ as part of the drilling and sampling operation. (The term "in situ" is synonymous with "in place.") The SPT measures resistance to the penetration of a standard split-spoon sampler that is driven by a 140 lbm (63.5 kg) hammer dropped from a height of 30 in (0.76 m). The number of blows required to drive the sampler a distance of 12 in (0.305 m) after an initial penetration of 6 in (0.15 m) is referred to as the *N-value* or *standard penetration resistance* in blows per foot.

The measured value of N is inconsistent from operator to operator because different drill rig systems deliver energy input that deviates from the theoretical value. Therefore, the N-value obtained in the field is converted to a standardized N-value, N'. In addition, because the N-value is sensitive to overburden pressure, corrections are applied to reference the value to a standard overburden stress, usually 2000 lbf/ft^2 (95.76 kPa).

The N-value has been correlated with many other mechanical properties, including shear modulus, unconfined compressive strength, angle of internal friction, and relative density. The correlations work best with cohesionless soils. Table 35.9 relates N to the relative density and friction angle.

Table 35.9 *Relationship between SPT (N), Relative Density (D_r), and Angle of Internal Friction (ϕ) (cohesionless soils)*

type of soil	SPT, N	relative density, D_r	angle of internal friction, ϕ Peck, et al. (1974)	Meyerhof (1956)
very loose sand	< 4	< 0.02	< 29	< 30
loose sand	4–10	0.2–0.4	29–30	30–35
medium sand	10–30	0.4–0.6	30–36	35–40
dense sand	30–50	0.6–0.8	36–41	40–45
very dense sand	> 50	> 0.8	> 41	> 45

Reprinted with permission from *Foundation Engineering Handbook*, by Hsai-Yang Fang, copyright © 1991 by Van Nostrand Reinhold.

10. CONE PENETROMETER TEST

The *cone penetrometer test* (CPT) has gained popularity as an alternative or adjunct to the standard penetration test. The standard cone penetrometer is a cylinder with an area of 1.55 in^2 (10 cm^2), tipped with a cone that has a 60° point. The cone is pushed into the soil at a continuous rate of 2 to 4 ft/min (10 to 20 mm/s). Resistance is measured separately at the tip of the cone (tip resistance, q_c, in lbf/ft^2 or kPa) and along the sides (sleeve friction resistance, q_s, in lbf/ft^2 or kPa). A data acquisition system collects nearly continuous data with depth.

One advantage of the cone penetration test is that a continuous record is obtained with depth, in contrast with the standard penetration test that is performed usually once every few feet. The cone penetration test is quite good for classifying both sands and clays. The cone can also be altered to measure pore water pressure or shear wave velocity.

The disadvantages to the cone test are: (a) No sample is retrieved, so some additional borings are usually required to confirm classification results or to obtain samples for further tests; (b) the cone and its accompanying equipment require a high initial investment to acquire; and (c) the cone is easily damaged or stuck and cannot be used in very stiff or gravelly soils.

For sands, the *friction ratio*, f_R, is less than 1%; for clays, f_R is larger, up to 5% or more.

$$f_R = \frac{q_s}{q_c} \times 100\% \qquad \qquad 35.18$$

11. PROCTOR TEST

Soils are compacted to increase stability and strength, enhance resistance to erosion, decrease permeability, and decrease compressibility. This is usually accomplished by placing the soil in *lifts* (i.e., layers) of a few inches to a few feet thick, and then mechanically compacting the lifts. Compaction equipment can densify the soil by static loading, impact, vibration, and/or kneading actions. *Grid rollers* can be used to break up and compact rocky soil. Cohesive soils respond best to the kneading action that is provided by a *sheepsfoot roller*. *Rubber-tired* or *smooth-wheeled rollers* are especially useful for final finishes. Roller compactors with vibration capabilities are well-suited for cohesionless soils. These types of compaction equipment are shown in Fig. 35.5.

The specification given to the grading contractor sets forth the minimum acceptable density as well as a range of acceptable water content values. The minimum density is specified as a *relative compaction* (RC), the percentage of the maximum value determined in the laboratory, ρ_d^*.

$$\mathrm{RC} = \frac{\rho_d}{\rho_d^*} \times 100\% \qquad \qquad 35.19$$

The basic laboratory test used to determine the maximum dry density of compacted soils is the *Proctor test*. In this procedure, a soil sample is compacted into a $\frac{1}{30}$ ft^3 (944 cm^3) mold in 3 layers by 25 hammer blows on each layer. The hammer has a mass of 5.5 lbm (2.5 kg) and is dropped 12 in (305 mm). The dry density of the sample can be found from Eq. 35.10. This procedure is repeated for various water contents, and a graph similar to Fig. 35.6 is obtained. ρ_d^* is known as the *maximum dry density*, or density at 100% compaction. w^* is known as the *optimum water content*.

Figure 35.5 Compaction Equipment

(a) smooth-wheel roller

(b) rubber-tired roller

(c) sheepsfoot roller

(d) grid roller

Figures reproduced with permission of Compaction America, Inc.

Figure 35.6 Proctor Test Curve

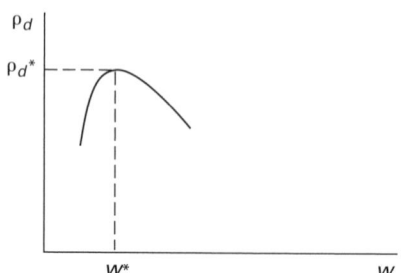

Soils close to the optimum water content require less compactive effort to achieve the required relative compaction. Usually a range of acceptable water contents, bracketing the optimum water content, is specified in addition to the relative compaction.

12. MODIFIED PROCTOR TEST

The *modified Proctor test* is similar to the Proctor test except that the soil is compacted in five layers with a 10 lbm (4.5 kg) hammer falling 18 in (457 mm). The result is a denser soil that is more representative of the compaction densities that can be achieved with modern equipment. The modified Proctor curve for a soil always lies above and to the left of the standard Proctor curve. Table 35.10 can be used by adding 10 to 20 lbm/ft^3 to the densities and subtracting 3 to 10% from the moisture contents.

For a given water content, perfect compaction will result in saturation, since all air will be removed. The dry densities corresponding to saturation at each water content can be plotted versus water content, and the result is known as a *zero air voids curve*. The zero air voids curve always lies above the Proctor test curve, since that test cannot expel all air.

Table 35.10 *Typical Values of Optimum Moisture Content and Suggested Relative Compactions (based on standard Proctor test)*

class group symbol	description	range of maximum dry densities (lbm/ft³)	range of optimum moisture content (%)	recommended percentage of Proctor maximum (%) class[a] 1	2	3
GW	well-graded, clean gravels, gravel-sand mixtures	125–135	11–8	97	94	90
GP	poorly graded clean gravels, gravel-sand mixtures	115–125	14–11	97	94	90
GM	silty gravels, poorly graded gravel-sand silt	120–135	12–8	98	94	90
GC	clayey gravels, poorly graded gravel-sand-clay	115–130	14–9	98	94	90
SW	well-graded clean sands, gravelly sands	110–130	16–9	97	95	91
SP	poorly graded clean sands, sand-gravel mix	100–120	21–12	98	95	91
SM	silty sands, poorly graded sand-silt mix	110–125	16–11	98	95	91
SM-SC	sand-silt-clay mix with slightly plastic fines	110–130	15–11	99	96	92
SC	clayey sands, poorly graded sand-clay mix	105–125	19–11	99	96	92
ML	inorganic silts and clayey silts	95–120	24–12	100	96	92
ML-CL	mixture of organic silt and clay	100–120	22–12	100	96	92
CL	inorganic clays of low-to-medium plasticity	95–120	24–12	100	96	92
OL	organic silts and silt-clays, low plasticity	80–100	33–21	–	96	93
MH	inorganic clayey silts, elastic silts	70–95	40–24	–	97	93
CH	inorganic clays of high plasticity	75–105	36–19	–	–	93
OH	organic and silty clays	65–100	45–21	–	97	93

(Multiply lbm/ft³ by 16.02 to obtain kg/m³.)
[a]Class 1 uses include the upper 9 ft (2.7 m) of fills supporting one- and two-story buildings, the upper 3 ft (0.9 m) of subgrade under pavements, and the upper 1 ft (0.3 m) of subgrade under floors. Class 2 uses include deeper parts of fills under buildings and pavements as well as earth dams. All other fills requiring some degree of strength or incompressibility are classified as class 3.

Figure 35.7 *Typical Zero Air Voids Curve*

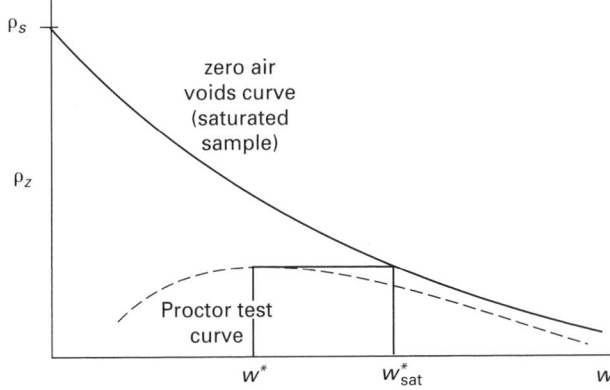

The *zero air voids density* (i.e., *dry unit weight at zero air voids*) is the ratio of the mass of the solids to the total volume if the sample is fully compacted (i.e., there is no air in the voids).

$$\rho_z = \frac{m_s}{V_w + V_s} \qquad 35.20$$

The theoretical dry density of the zero air voids curve is calculated from Eq. 35.21.

$$\rho_z = \frac{\rho_w}{w + \dfrac{1}{SG}} \qquad 35.21$$

The maximum value of the zero air voids density occurs at $w = 0$. At that point, the maximum zero air voids density is equal to the density of the solid itself (as calculated from the solid specific gravity).

$$\rho_s = (SG)\rho_w \qquad 35.22$$

ρ_s and ρ_d^* are not the same, however, since air voids exist in the ρ_d^* case and ρ_d^* occurs at w^*.

Example 35.5

A Proctor test using a $\frac{1}{30}$ ft³ (0.9443 L) mold is performed on a sample of soil.

test no.	sample net mass (lbm)	(kg)	water content (%)
1	4.28	1.941	7.3
2	4.52	2.050	9.7
3	4.60	2.087	11.0
4	4.55	2.064	12.8
5	4.50	2.041	14.4

If 0.032 ft³ (0.00091 m³) of compacted soil tested at a construction site had a mass of 3.87 lbm (1.755 kg) wet and 3.74 lbm (1.696 kg) dry, what is the percentage of compaction?

SI Solution

From Eq. 35.11, the dry density of sample 1 is

$$\rho_d = \frac{m_t}{(1+w)V_t}$$

$$= \frac{1.941 \text{ kg}}{(1+0.073)(0.9443L)\left(0.001\frac{\text{m}^3}{\text{L}}\right)}$$

$$= 1915.7 \text{ kg/m}^3$$

The following table is constructed from the results of all five tests.

test no.	dry density (kg/m³)
1	1915.7
2	1979.0
3	1991.1
4	1937.7
5	1889.4

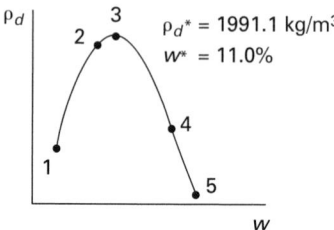

Since the peak is near point 3, take the maximum dry density to be 1991.1 kg/m³. The sample dry density is

$$\rho_d = \frac{m_d}{V} = \frac{1.696 \text{ kg}}{0.00091 \text{ m}^3}$$

$$= 1863.7 \text{ kg/m}^3$$

The percentage of compaction is

$$\frac{1863.7 \frac{\text{kg}}{\text{m}^3}}{1991.1 \frac{\text{kg}}{\text{m}^3}} \times 100\% = 94\%$$

Customary U.S. Solution

From Eq. 35.11, the dry density of sample 1 is

$$\rho_d = \frac{m_t}{(1+w)V_t}$$

$$= \frac{4.28 \text{ lbm}}{(1+0.073)\left(\frac{1}{30} \text{ ft}^3\right)}$$

$$= 119.7 \text{ lbm/ft}^3$$

The following table is constructed from the results of all five tests.

test no.	dry density (lbm/ft³)
1	119.7
2	123.6
3	124.3
4	121.0
5	118.0

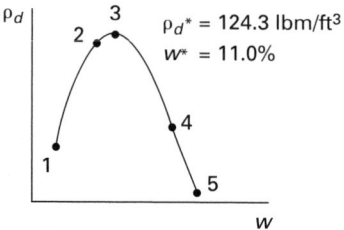

Since the peak is near point 3, take the maximum dry density to be 124.3 lbm/ft³. The sample dry density is

$$\rho_d = \frac{m_d}{V} = \frac{3.74 \text{ lbm}}{0.032 \text{ ft}^3}$$

$$= 116.9 \text{ lbm/ft}^3$$

The percentage of compaction is

$$\frac{116.9 \frac{\text{lbm}}{\text{ft}^3}}{124.3 \frac{\text{lbm}}{\text{ft}^3}} \times 100\% = 94\%$$

13. IN-PLACE DENSITY TESTS

The *in-place density test*, also known as the *field density test*, determines the density of a compacted fill to see if it meets specifications. A 3 to 5 in (8 to 13 cm) deep hole with smooth sides is dug into the compacted soil. All soil taken from the hole is saved and weighed before the water content can change. The hole is filled with sand or a water-filled rubber balloon. The volume of the hole is calculated from the known mass and density of the sand or water. The required densities of the compacted soil are given by Eqs. 35.8 and 35.11.

Another way to determine the density of in-place soil is by using a *nuclear gauge* (*nuclear moisture/density gauge*). This device has a probe containing radioactive material that is inserted into a hole punched into the compacted soil. The rate of radiation penetration through the soil is detected and used to determine both the wet density and the water content of the soil. The dry density can be calculated with Eq. 35.11.

14. ATTERBERG LIMIT TESTS

A clay soil can behave like a solid, semi-solid, plastic solid, or liquid, depending on the water content. The water contents corresponding to the transitions between these states are known as the *Atterberg limits*.[5] Each of the Atterberg limits varies with the clay content, type of clay mineral, and ions (cations) contained in the clay. Tests that determine two of these Atterberg limits—the plastic limit and the liquid limit—are frequently used to classify clay soils.

The *plastic limit* (PL or w_p) is the water content corresponding to the transition between the semi-solid and plastic state. The *liquid limit* (LL or w_l) is the water content corresponding to the transition between the plastic and liquid state. A third limit that is occasionally used in soils engineering is the *shrinkage limit* (SL or w_s), which is the water content corresponding to the transition between a brittle solid and a semi-solid.[6]

The liquid limit test is performed with a special apparatus. A soil sample is placed in a shallow container and the sample is parted in half with a grooving tool. The container is dropped a distance of 0.4 in (10 mm) repeatedly until the sample has rejoined for a length of $1/2$ in (13 mm). The liquid limit is defined as the water content at which the soil rejoins at exactly 25 blows. The test is repeated at different water contents, and the water content corresponding to the liquid limit is found by interpolation using a *flow curve* (a plot of water content versus logarithm of number of blows, N).

When a soil has a liquid limit of 100, the weight of water equals the weight of the dry soil (i.e., $w = 100\%$). A liquid limit of 50 means that the soil at the liquid limit is two-thirds soil and one-third water.

Sandy soils have low liquid limits—on the order of 20. In such soils, the test is of little significance in judging load-carrying capacities. Silts and clays can have significant liquid limits—most clays have liquid limits less than 100, but they can be as high as 1000.

The plastic limit test consists of rolling a soil sample into a $1/8$ in (3 mm) thread. The sample will crumble at that diameter when it is at the plastic limit. The sample is remolded to remove moisture and rolled into a thread repeatedly until the plastic limit is reached. The water content is determined for three such samples, and the average value is the plastic limit.

Sands and most silts have no plastic limit at all. They are known as *nonplastic soils*. The test is of little significance in judging the relative load-carrying capacities of such soils. The plastic limit of clays and plastic silts can be from 0 to 100 or more, but it is usually less than 40.

The difference between the liquid and plastic limits is known as the *plasticity index*, PI. The plasticity index indicates the range in moisture content over which the soil is in a plastic condition. In this condition it can be deformed and still hold together without crumbling. A large plasticity index (i.e., greater than 20) shows that considerable water can be added before the soil becomes liquid. The plasticity index correlates with strength, deformation properties, and insensitivity.

$$PI = LL - PL \qquad 35.23$$

The difference between the plastic and shrinkage limits is known as the *shrinkage index*, SI. The shrinkage index indicates the range in moisture content over which the soil is in a semi-solid condition. The shrinkage limit is the water content at which further drying out of the soil does not decrease the volume of the soil. Below the shrinkage limit, air enters the voids and water content decreases are not accompanied by decreases in volume. The test consists of drying a brick of soil to remove all interstitial (removable) water, and then adding water until all the voids are filled.

$$SI = PL - SL \qquad 35.24$$

The *consistency* of a clay means the water content relative to the Atterberg limits. This is represented by the *liquidity index*, LI.[7] A liquidity index between 0 and 1 indicates that the water content is between the plastic limit and the liquid limit; a liquidity index greater than one indicates that the water content is above the liquid limit.

$$LI = \frac{w - PL}{PI} \qquad 35.25$$

Example 35.6

Atterberg limit tests were performed on a silty clay. The water contents obtained in the plastic limit test were 31.6%, 33.5%, and 30.9%. The following data were obtained in the liquid limit test. What are the (a) liquid limit, (b) plastic limit, and (c) plasticity index?

no. of blows	water content
12	54.8%
18	54.2%
23	52.9%
35	51.9%

Solution

(a) For the liquid limit, plot the logarithm of the number of blows versus the water content. The liquid limit corresponding to the water content at 25 blows is 53%.

---

[5] Although standard practice varies, it is suggested that Atterberg limits be reported as whole numbers rather than as percentages, to emphasize that they are an index of behavior, not a soil characteristic.

[6] The Atterberg limits are sometimes given the symbols w_L (liquid limit), w_P (plastic limit), and w_S (shrinkage limit).

[7] Alternative symbols are IL for liquidity index and IP for plasticity index.

Geotechnical

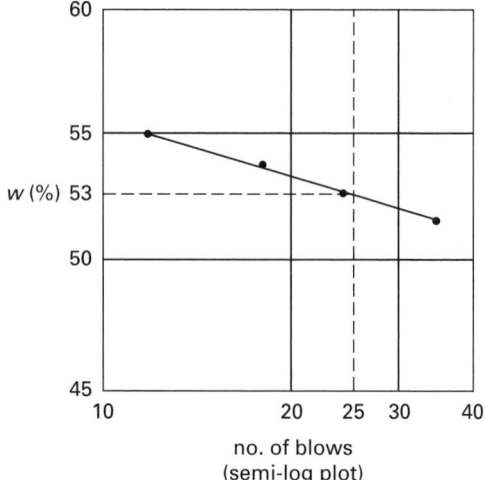

no. of blows
(semi-log plot)

(b) The plastic limit is the average of the three test results.

$$PL = \frac{31.6 + 33.5 + 30.9}{3} = 32$$

(c) The plasticity index is given by Eq. 35.23.

$$PI = LL - PL = 53 - 32 = 21$$

15. PERMEABILITY TESTS

Permeability of a soil is a measure of continuous voids. It is not enough for a soil to have large voids. The voids must also be connected for water to flow through them. A permeable material permits a significant flow of water. The flow of water through a permeable aquifer or soil is given by Eq. 35.26, known as *Darcy's law*.

$$Q = \text{v}A_{\text{gross}} \qquad 35.26$$
$$\text{v} = Ki \qquad 35.27$$

The area in Eq. 35.26 is the total cross-sectional area of the aquifer, not the actual area in flow. Water can only flow through the open spaces between the solids. This open space area is nA.

Typical values of the coefficient of permeability, K, are given in Table 35.11. Clays are considered relatively impervious, while sands and gravels are pervious. For comparison, the permeability of concrete is approximately 10^{-10} cm/s.

For uniform sands with 0.1 mm $\leq D_{10} \leq$ 3 mm and $C_u \leq 5$, permeability is approximately given by *Hazen's formula*, Eq. 35.28. D_{10} is the *effective grain size*—the diameter for which only 10% of the particles are finer. The coefficient C is 0.4 to 0.8 for very fine sand (poorly sorted) or fine sand with appreciable fines; 0.8 to 1.2 for medium sand (well sorted) or coarse sand (poorly sorted); and 1.2 to 1.5 for coarse sand (well sorted and clean).

$$K_{\text{cm/s}} \approx C(D_{10,\text{mm}})^2 \qquad 35.28$$

Table 35.11 *Typical Permeabilities*

group symbol	typical coefficient of permeability (cm/s)
GW	2.5×10^{-2}
GP	5×10^{-2}
GM	$> 5 \times 10^{-7}$
GC	$> 5 \times 10^{-8}$
SW	$> 5 \times 10^{-4}$
SP	$> 5 \times 10^{-4}$
SM	$> 2.5 \times 10^{-5}$
SM-SC	$> 10^{-6}$
SC	$> 2.5 \times 10^{-7}$
ML	$> 5 \times 10^{-6}$
ML-CL	$> 2.5 \times 10^{-7}$
CL	$> 5 \times 10^{-8}$
OL	$-$
MH	$> 2.5 \times 10^{-7}$
CH	$> 5 \times 10^{-8}$
OH	$-$

Hazen's formula provides only a crude estimate, and actual values can vary by two orders of magnitude. Accurate values are calculated from controlled permeability tests using constant- or falling-head *permeameters*. In a *constant-head test*, the volume of water, V, percolating through the soil over time is measured. The test is applicable to coarse-grained soils with $K > 10^{-3}$ cm/s.

$$K = \frac{VL}{hAt} \qquad \text{[constant head]} \qquad 35.29$$

In *falling-head tests*, the change in head over time is measured as the water percolates through the soil. The test is applicable to fine-grained soils with $K < 10^{-3}$ cm/s.

$$K = \left(\frac{A'L}{At}\right) \ln\left(\frac{h_i}{h_f}\right) \qquad \text{[falling head]} \qquad 35.30$$

Figure 35.8 *Permeameters*

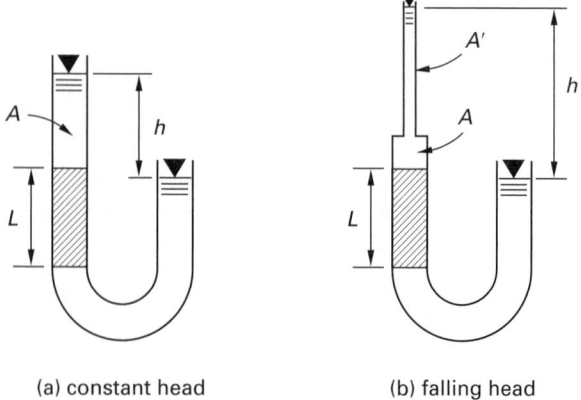

(a) constant head   (b) falling head

The permeability can also be determined with an in situ test in the field. For the *auger-hole method* (a falling-head test), the combination of area and length variables may be known as the *shape factor* or *conductivity coefficient*, F. For example, for a cased hole below the water table of length L and radius r whose impervious casing extends all the way to the hole bottom and whose liquid level rises from h_i to h_f in time t (approaching the water table level), the shape factor and permeability are

$$F = \frac{11r}{2} \qquad\qquad 35.31$$

$$K = \left(\frac{\pi r^2}{Ft}\right)\ln\left(\frac{h_i}{h_f}\right)$$

$$= \left(\frac{2\pi r}{11t}\right)\ln\left(\frac{h_i}{h_f}\right) \qquad \text{[auger hole]} \qquad 35.32$$

Other in-field tests can use cased holes with constant head or uncased holes with constant and variable head. The shape factors for these tests are not the same as that in Eq. 35.31.

Example 35.7

The permeability of a semi-impervious soil was evaluated in a falling-head permeameter whose head decreased from 40 to 15 in (100 to 38 cm) in 5 min. The body diameter was 5 in (13 cm), the standpipe diameter was 0.1 in (0.3 cm), and the sample length was 2 in (5 cm). What was the permeability of the soil?

SI Solution

Use Eq. 35.30.

$$
\begin{aligned}
K &= \left(\frac{A'L}{At}\right)\ln\left(\frac{h_i}{h_f}\right) \\
&= \left(\frac{\frac{1}{4}\pi(0.3\ \text{cm})^2(5\ \text{cm})}{\frac{1}{4}\pi(13\ \text{cm})^2(5\ \text{min})\left(60\ \frac{\text{s}}{\text{min}}\right)}\right) \\
&\quad\times \ln\left(\frac{100\ \text{cm}}{38\ \text{cm}}\right) \\
&= 8.6\times 10^{-6}\ \text{cm/s}
\end{aligned}
$$

Customary U.S. Solution

Use Eq. 35.30.

$$
\begin{aligned}
K &= \left(\frac{A'L}{At}\right)\ln\left(\frac{h_i}{h_f}\right) \\
&= \left(\frac{\frac{1}{4}\pi(0.1\ \text{in})^2(2\ \text{in})}{\frac{1}{4}\pi(5\ \text{in})^2(5\ \text{min})\left(60\ \frac{\text{sec}}{\text{min}}\right)}\right) \\
&\quad\times \ln\left(\frac{40\ \text{in}}{15\ \text{in}}\right) \\
&= 2.6\times 10^{-6}\ \text{in/sec}
\end{aligned}
$$

16. CONSOLIDATION TESTS

Consolidation tests (also known as *confined compression tests* and *oedometer tests*) start with a disc of soil (usually clay) confined by a metal ring. The faces of the disc are covered with porous plates. The disc sandwich is loaded and submerged in water. Static loads are applied in increments, and the vertical displacement is measured with time for each load increment. The testing time is very long, usually 24 hr per load increment, since seepage through clay soils is very slow. When the displacement rate levels off, the final void ratio is determined for that increment. The load versus the void ratio for all increments is plotted together as an *e-log p curve*.[8]

Figure 35.9 *Consolidation Test Apparatus*

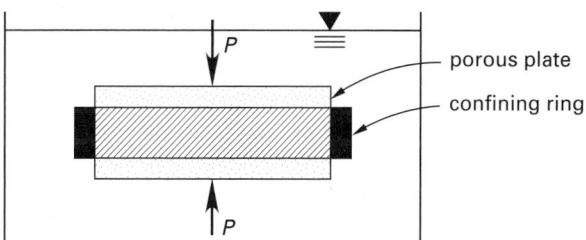

Figure 35.10 shows an *e-log p* curve for a soil sample from which the load has been removed at point m, allowing the clay to recover.

Figure 35.10 *e-log p Curve*

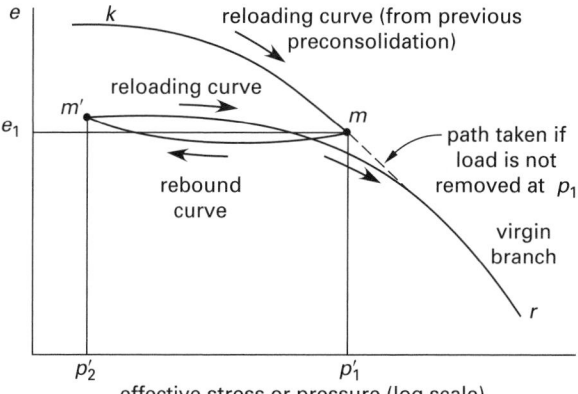

The line segment m-r is known as the *virgin compression line* or *virgin consolidation line*. In this region, the clay is considered to be *normally consolidated*, which means that the present load has never been exceeded.

Line m-m' is a *rebound curve*. Line m'-r is known as a *reloading curve*. Such curves result when a normally consolidated clay is unloaded and then reloaded. Since

---

[8]The use of pressure (p) instead of stress (σ) for the discussion on consolidation follows convention. Pressure refers to the external forces on the soil. Some soils texts substitute stress for pressure to emphasize that the soil response is dependent on internal forces in the soil, but otherwise the methodology is exactly the same.

it has carried a higher load in the past, the soil is considered to be *overconsolidated* or *preloaded* in the rebound-reload region.

Notice that point m' can only be reached by loading the soil to a pressure of p'_1 and then removing the pressure. Although the pressure of the clay at p'_2 is essentially the same as at the start of the test, the void ratio has been reduced.

The *overconsolidation ratio*, OCR, is defined by Eq. 35.33. p'_o is the present or in situ overburden pressure; p'_{max} is the maximum past pressure or *preconsolidation pressure*. An overconsolidation ratio of 1 means that the soil is normally consolidated; an overconsolidation ratio greater than 1 means that the soil is overconsolidated. In rare circumstances, such as during construction or rapid underwater deposition, a soil may be "underconsolidated"; that is, it has not yet come to equilibrium with its present load.

$$\text{OCR} = \frac{p'_{max}}{p'_o} \qquad 35.33$$

The shape of the e-log p curve will depend on the degree of remolding or disturbance, as shown in Fig. 35.11. A highly disturbed soil will show a gradual transition between overconsolidated and normally consolidated behavior.

Figure 35.11 *Consolidation Curves*

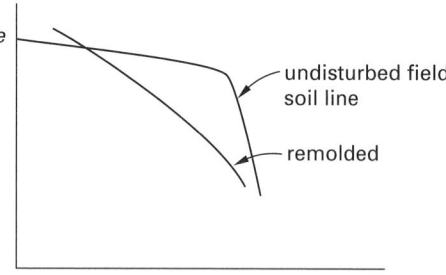

it has carried a higher load in the past... Laboratory results can be used to find the preconsolidation pressure (point m in Fig. 35.10). One procedure is the *Casagrande Method*, as illustrated in Fig. 35.12. This procedure starts by drawing two lines—a tangent line and a horizontal line—through the point of maximum curvature, which is determined by eye. The resulting angle is bisected. Then a tangent extension (line k) is drawn to the virgin compression line. The intersection of this tangent and the bisection line defines the preconsolidation pressure, p'_{max}.

Figure 35.12 *Casagrande Method*

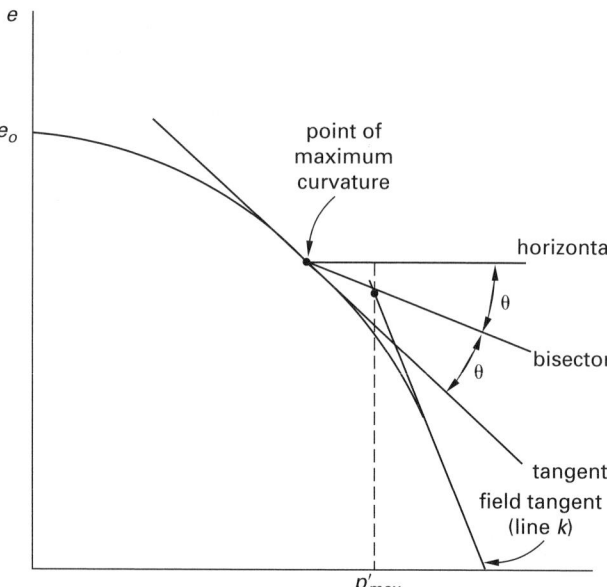

effective stress or pressure (log scale)

The field virgin compression line, line k in Fig. 35.12, can be used to predict consolidation of the soil under any loading greater than p'_{max}. The *compression index*, C_c, or *compressibility index*, is the negative of the logarithmic slope of line k and is given by Eq. 35.34, where points 1 and 2 correspond to any two points on line k.[9]

$$C_c = -\frac{e_1 - e_2}{\log_{10}\left(\dfrac{p'_1}{p'_2}\right)} \qquad 35.34$$

It is common practice to plot strain, ϵ, versus log p for consolidation test data, as well as void ratio versus log p as shown in Fig. 35.10. The slope of the ϵ-log p line is $C_{\epsilon c}$ and is related to the compression index, C_c.

$$C_{\epsilon c} = \frac{C_c}{1 + e_0} \qquad 35.35$$

If the clay is soft and near its liquid limit, the compression index can be approximated by Eq. 35.36.

$$C_c \approx (0.009)(\text{LL} - 10) \qquad 35.36$$

When the loading is removed from a soil sample, it will have a tendency to swell. The *swelling index* and *reconsolidation index* can be found from the (negative of the) logarithmic slopes of the rebound and reloading curves, respectively. For practical purposes, these indices are equal to each other and found as the slope of the line that bisects the rebound and reload curves.

17. DIRECT SHEAR TEST

The *direct shear test* is a relatively simple test used to determine the relationship of shear strength to consolidation stress. In this test, a disc of soil is inserted

---

[9]Although the compression index is, strictly speaking, the slope of the virgin compression line and therefore negative, it is typically reported and used as a positive number. Equation 35.34 is a convenient form for calculating C_c as a positive number.

into the direct shear box. The box has a top half and a bottom half that can slide laterally with respect to each other. A normal stress, σ_n, is applied vertically, and then one half of the box is moved laterally relative to the other at a constant rate. Measurements of vertical and horizontal displacement, δ, and horizontal shear load, P_h, are taken. The test is usually repeated at three different vertical normal stresses.

Because of the box configuration, failure is forced to occur on a horizontal plane. Results from each test are plotted as horizontal displacement versus horizontal stress, τ_h (horizontal force divided by the nominal area). *Failure* is determined as the maximum value of horizontal stress achieved. The vertical normal stress and failure stress from each test are then plotted in Mohr's circle space of normal stress versus shear stress.

Figure 35.13 Graphing Direct-Shear Test Results

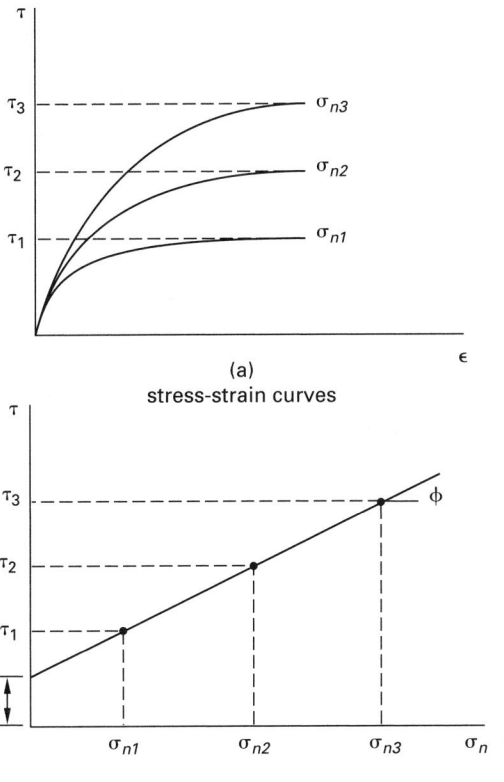

(a)
stress-strain curves

(b)
Mohr's failure envelope

A line drawn through all of the test values is called the *failure envelope* (*failure line* or *rupture line*). The equation for the failure envelope is given by *Coulomb's equation*, which relates the strength of the soil, S, to the normal stress on the failure plane.[10,11,12]

$$S = \tau = c + \sigma \tan \phi \qquad 35.37$$

[10]Equation 35.37 is also known as the Mohr-Coulomb equation.
[11]The ultimate shear strength may be given the symbol S in some soils books.
[12]τ and σ in Coulomb's equation are the shear stress and normal stress, respectively, on the failure plane at failure.

ϕ is known as the *angle of internal friction*.[13] c is the *cohesion intercept*, a characteristic of cohesive soils. Representative values of ϕ and c are given in Table 35.12.

Table 35.12 Typical Strength Characteristics
(above the water table)

group symbol	cohesion (as compacted) c lbf/ft^2 (kPa)	cohesion (saturated) c_{sat} lbf/ft^2 (kPa)	effective stress friction angle ϕ
GW	0	0	> 38°
GP	0	0	> 37°
GM	–	–	> 34°
GC	–	–	> 31°
SW	0	0	38°
SP	0	0	37°
SM	1050 (50)	420 (20)	34°
SM-SC	1050 (50)	300 (14)	33°
SC	1550 (74)	230 (11)	31°
ML	1400 (67)	190 (9)	32°
ML-CL	1350 (65)	460 (22)	32°
CL	1800 (86)	270 (13)	28°
OL	–	–	–
MH	1500 (72)	420 (20)	25°
CH	2150 (100)	230 (11)	19°
OH	–	–	–

(Multiply lbf/ft^2 by 0.04788 to obtain kPa.)

18. TRIAXIAL STRESS TEST

The *triaxial test* is a more sophisticated method than the direct shear test for determining the strength of soils. In the triaxial test apparatus, a cylindrical sample is stressed completely around its peripheral surface by pressurizing the sample chamber. This pressure is referred to as the *confining stress*. Then, the soil is loaded vertically to failure through a top piston. The confining stress is kept constant while the axial stress is varied. The radial component of the confining stress is called the *radial stress*, σ_R, and represents the minor principal stress, σ_3. The combined stresses at the ends of the sample (confining stress plus applied vertical stress) are called the *axial stress*, σ_A, and represent the major principal stress, σ_1.[14]

Results of a triaxial test at a given chamber pressure are plotted as a stress-strain curve. Two such examples are illustrated in Fig. 35.14. The axial component of

[13]In a physical sense, the angle of internal friction for cohesionless soils is the angle from the horizontal naturally formed by a pile. For example, a uniform fine sand makes a pile with a slope of approximately 30°. For most soils, the natural angle of repose will not be the same as the angle of internal friction, due to the effects of cohesion.
[14]In reality, the triaxial test apparatus is a "biaxial" device because it controls stresses in only two directions: radial and axial.

strain is measured and presented. The stress is the difference between the axial and radial stresses, also called the *deviator stress*, σ_D. The point of failure is usually chosen as the maximum deviator stress in the test (Fig. 35.14(a)). It can alternatively be chosen as the stress difference for which the strain is 20%, if the test does not reach a maximum before then (Fig. 35.14(b)).

Figure 35.14 *Triaxial Test Stress-Strain Curves*

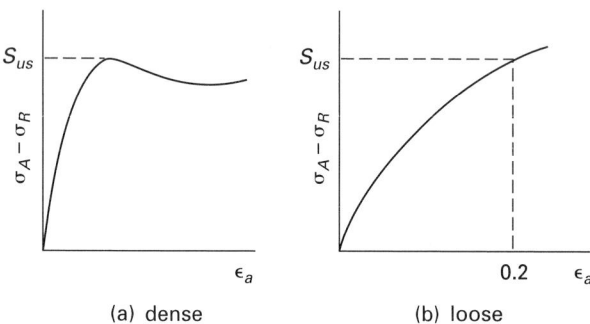

(a) dense   (b) loose

The normal and shear stresses on any plane can be found from the combined stress equations. The angle α is measured counterclockwise from the horizontal (i.e., counterclockwise from the plane that σ_1 acts on). Compression is considered positive.

$$\sigma_\alpha = \tfrac{1}{2}(\sigma_A + \sigma_R)$$
$$+ \tfrac{1}{2}(\sigma_A - \sigma_R)\cos 2\alpha \qquad 35.38$$
$$\tau_\alpha = \tfrac{1}{2}(\sigma_A - \sigma_R)\sin 2\alpha \qquad 35.39$$

These equations generate points on Mohr's circle, which can easily be constructed once σ_A and σ_R are known. Representative test results are shown in Fig. 35.15 for two different samples of the same soil that were both tested to failure. The failure envelope is the tangent to all of the Mohr's circles at failure.[15] The ultimate shear strength, S_u, can be read directly from the y-axis or can be calculated from Eq. 35.37.

For drained sands and gravels and for normally consolidated clays, the cohesion, c, is zero. Therefore, it is theoretically possible to draw the rupture line with only one test.

The angle of internal friction for cohesionless soils can be calculated from Eq. 35.40.

$$\frac{\sigma_1}{\sigma_3} = \frac{1 + \sin\phi}{1 - \sin\phi} \quad [c = 0] \qquad 35.40$$

---

[15]In the direct shear test, the stresses on the failure plane are measured directly, because the horizontal plane is forced to be the failure plane. In the triaxial test, the measured stresses are principal stresses, and therefore a Mohr's construction must be made to determine the stresses on the failure plane.

Figure 35.15 *Mohr's Circle of Stress*

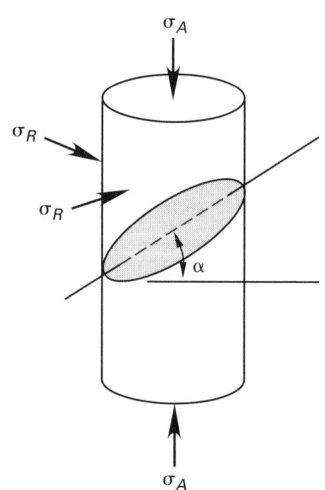

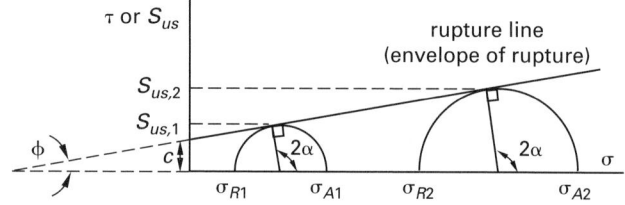

The plane of failure is inclined at the following angle.

$$\alpha = 45° + \tfrac{1}{2}\phi \qquad 35.41$$

For saturated clays in quick (undrained) shear, it is commonly assumed that $\phi = 0$. This would be represented as a horizontal rupture line. For this condition only, the undrained strength, S_{us}, is

$$S_{us} = c = \frac{\sigma_D}{2} \qquad 35.42$$

Figure 35.16 *Rupture Lines*

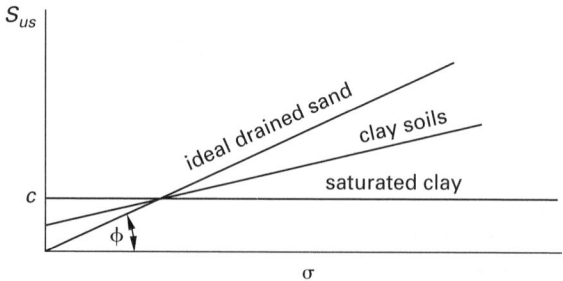

If a soil is saturated, an increase in load may be supported by either the soil grains at their contact points or by an increase in the pore water pressure. How much each mode contributes to the support depends on the rate of loading and the permeability of the soil. Since water cannot carry a shear stress, the effective stress provides the shear strength of soils. The effective stress is the stress that is carried by the soil grains. The strength can be calculated from Eq. 35.43, where c' and ϕ' are the *effective stress parameters*.

$$s = c' + \sigma' \tan \phi' \qquad 35.43$$

The various categories of triaxial tests are shown in Table 35.13. Triaxial tests can be performed on consolidated or unconsolidated specimens, and they can be performed drained or undrained.

Table 35.13 *Categories of Triaxial Test*

	drained	undrained
consolidated	S-test CD test	R-test CU test
unconsolidated	(not performed)	Q-test UU test

The term "consolidated" refers to the state of the sample just prior to the shearing phase (when the axial stress, σ_A, is varied). In most triaxial tests, the soil is consolidated, which means that a stress simulating the in situ overburden stress is applied and the soil is allowed to drain freely until it is at equilibrium with this stress. The consolidation stress is applied as the pressure in the surrounding chamber.

The terms "drained" and "undrained" refer to conditions during the shearing phase. The distinction is only important for specimens that are fully saturated; for dry or partially saturated specimens, the behavior in an undrained test is similar to that in a drained test. In an undrained test, no pore water is allowed to move in or out of the sample, and the test is usually performed rapidly. In these tests, the shearing phase results in pore water pressure development. In a drained test, water is allowed to move in or out freely, and the test is performed slowly enough that the water drains before pore water pressure develops. The speed of testing is determined by the soil permeability. Clay soils must be tested quite slowly, and a drained test may take several weeks.

A slow, drained test following consolidation is known as an *S-test* (*slow test*) or a *consolidated-drained test* (*CD test*). A rapid, undrained test following consolidation is known as an *R-test* or *consolidated-undrained test* (*CU test*). The *Q-test* (*quick test*) is also known

as an *unconsolidated-undrained test* (*UU test*).[16] In the Q-test, a chamber pressure is applied to the soil prior to the shearing phase but the soil is not allowed to drain or consolidate. Such a test is justified only with high-permeability (e.g., 10^{-3} cm/s or more) soils.[17]

R-tests are used to determine the effective stress parameters, c' and ϕ'. In the absence of pore pressure measurements, though, R-tests can only determine the total stress parameters c and ϕ.

Example 35.8

A sample of dry sand was taken and a triaxial test was performed. The added axial stress causing failure was 5.43 tons/ft^2 (520 kPa) when the radial stress was 1.5 tons/ft^2 (145 kPa). (a) What is the angle of internal friction? (b) What is the angle of the failure plane?

SI Solution

Draw a Mohr's circle to plot the stresses in two dimensions. The axial stress at failure is the combination of axial and radial loads.

$$\sigma_{A,f} = p_R + p_A$$
$$= 145 \text{ kPa} + 520 \text{ kPa}$$
$$= 665 \text{ kPa}$$

The center of the Mohr's circle is

$$\frac{\sigma_R + \sigma_{A,f}}{2} = \frac{145 \text{ kPa} + 665 \text{ kPa}}{2}$$
$$= 405 \text{ kPa}$$

The radius of the circle is the distance from the center of the circle to the radial stress.

$$405 \text{ kPa} - 145 \text{ kPa} = 260 \text{ kPa}$$

(a) The failure envelope is tangent to the Mohr's circle. (It is assumed that $c = 0$ for dry sand.) From trigonometry,

$$\phi = \arcsin\left(\frac{260 \text{ kPa}}{405 \text{ kPa}}\right) = 40°$$

(b) The angle of the failure plane is given by Eq. 35.41.

$$\alpha = 45° + \tfrac{1}{2}(40°) = 65°$$

---

[16]The S-test, R-test, and Q-test designations are by the U.S. Army Corps of Engineers. The letter "R" for the R-test was chosen because R is between Q and S in the alphabet.
[17]There is no category for unconsolidated, drained testing. The theoretical justifications for performing Q-tests do not apply to drained testing of unconsolidated specimens.

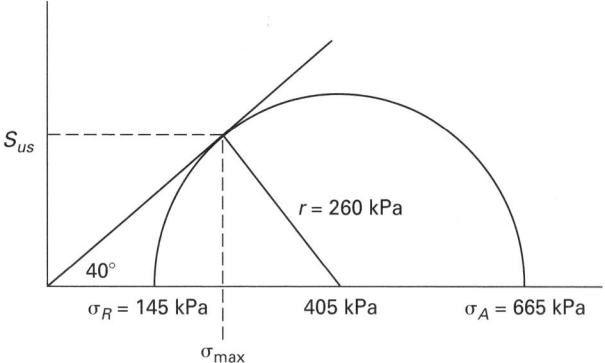

19. VANE-SHEAR TEST

The shear strength of a low-strength, homogeneous cohesive soil (e.g., clay) can be measured in situ by use of a *vane-shear apparatus*, consisting of a four-bladed vane on a vertical shaft. The blades are pushed into the soil and a torque is applied to the shaft until the apparatus rotates. When the soil is stressed to its shear strength, the vanes will rotate in the soil. Since the soil fails along a cylindrical surface, the shearing resistance can be calculated from the vane dimensions and the applied torque.

The sensitivity of the soil can be determined if the test is repeated after turning the sample several times and allowing the soil to remold.

20. UNCONFINED COMPRESSIVE STRENGTH TEST

In an *unconfined compressive strength test*, a cylinder of cohesive soil is loaded axially to compressive failure. Unlike in the triaxial test, there is no radial stress applied before the shearing phase of testing. This test can only be performed on soils that can stand without confinement, usually clays. The unconfined compressive strength test is equivalent to a UU-test (Q-test) in which the chamber pressure is zero.

The unconfined compressive strength is given by Eq. 35.44. The undrained shear strength is calculated as one half of the unconfined compressive strength.

$$S_{\mathrm{uc}} = \frac{P}{A} \qquad \textit{35.44}$$

$$s_u = \frac{S_{\mathrm{uc}}}{2} \qquad \textit{35.45}$$

Customary U.S. Solution

Draw a Mohr's circle to plot the stresses in two dimensions. The axial stress at failure is the combination of axial and radial loads.

$$\begin{aligned}
\sigma_{A,f} &= p_R + p_A \\
&= 1.5 \ \frac{\mathrm{tons}}{\mathrm{ft}^2} + 5.43 \ \frac{\mathrm{tons}}{\mathrm{ft}^2} \\
&= 6.93 \ \mathrm{tons/ft}^2
\end{aligned}$$

The center of the Mohr's circle is

$$\frac{\sigma_R + \sigma_{A,f}}{2} = \frac{1.5 \ \dfrac{\mathrm{tons}}{\mathrm{ft}^2} + 6.93 \ \dfrac{\mathrm{tons}}{\mathrm{ft}^2}}{2}$$
$$= 4.215 \ \mathrm{tons/ft}^2$$

The radius of the circle is the distance from the center of the circle to the radial stress.

$$4.215 \ \frac{\mathrm{tons}}{\mathrm{ft}^2} - 1.5 \ \frac{\mathrm{tons}}{\mathrm{ft}^2} = 2.715 \ \frac{\mathrm{tons}}{\mathrm{ft}^2}$$

(a) The failure envelope is tangent to the Mohr's circle. (It is assumed that $c = 0$ for dry sand.) From trigonometry,

$$\phi = \arcsin\left(\frac{2.715 \ \dfrac{\mathrm{tons}}{\mathrm{ft}^2}}{4.215 \ \dfrac{\mathrm{tons}}{\mathrm{ft}^2}}\right) = 40°$$

(b) The angle of the failure plane is given by Eq. 35.41.

$$\alpha = 45° + \tfrac{1}{2}(40°) = 65°$$

21. SENSITIVITY

Clay will become softer as it is worked, and clay soils can turn into viscous liquids during construction. This tendency, known as *sensitivity*, is determined by measuring the ultimate strength of two samples, one that is undisturbed before testing and the other that has been fully disturbed, usually by mechanically remolding the soil. Although sensitivity is normally determined as the ratio of unconfined compression strengths, the two strength values compared could be from any two identical tests.

$$S_t = \frac{S_{\mathrm{undisturbed}}}{S_{\mathrm{remolded}}} \qquad \textit{35.46}$$

Table 35.14 *Sensitivity Classifications*

sensitivity	class
1–4	insensitive clays
4–8	sensitive
8–16	extra sensitive
> 16	quick

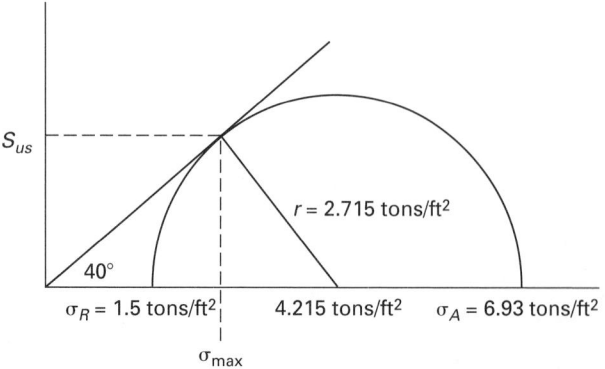

22. CALIFORNIA BEARING RATIO TEST

The *California bearing ratio* (CBR) test is used to determine the suitability of a soil for use as a subbase in pavement sections.[18] It is most reliable when used to test cohesive soils. The test measures the relative load required to cause a standard 3 in^2 (19.3 cm^2) plunger to penetrate a water-saturated soil specimen at a specific rate to a specific depth. The word "relative" is used because the actual load is compared to a standard load derived from a sample of crushed stone. The ratio is multiplied by 100, and the result is reported without a percentage symbol.

The resulting data will be in the form of inches of penetration versus load. This data can be plotted as shown in Fig. 35.17. If the plot is concave upward (curve B), the steepest slope is extended downward to the x-axis. This point is taken as the zero penetration point, and all penetration values are adjusted accordingly.

Figure 35.17 *Plotting CBR Test Data*

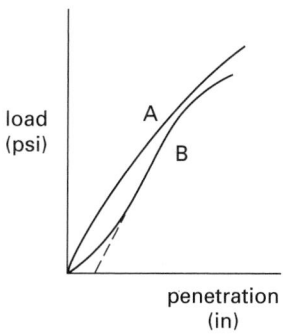

Standard loads for crushed stone are given in Table 35.15. For a plunger of 3 in^2 (19.3 cm^2), the CBR is the ratio of the load for a 0.1 in (2.54 mm) penetration divided by 1000 lbf/in^2 (6900 kPa). The CBR for 0.2 in (5.08 mm) should also be calculated. The test should be repeated if $\text{CBR}_{0.2} > \text{CBR}_{0.1}$. If the results are similar, $\text{CBR}_{0.2}$ should be used.

$$\text{CBR} = \frac{\text{actual load}}{\text{standard load}} \times 100 \qquad 35.47$$

Table 35.15 *Standard CBR Loads*

depth of penetration (in)	standard load (lbf/in^2 (MPa))
0.1	1000 (7)
0.2	1500 (10)
0.3	1900 (13)
0.4	2300 (16)
0.5	2600 (18)

(Multiply in by 25.4 to obtain mm.)
(Multiply lbf/in^2 by 0.00689 to obtain MPa.)

[18]California's Department of Transportation (CALTRANS) was the first to make use of the CBR test. However, other states and the Army Corps of Engineers have adopted CBR testing techniques. These states have, generally, retained the California bearing ratio test name for the test.

Table 35.16 *Typical CBR Values*

group symbol	range of CBR values
GW	40–80
GP	30–60
GM	20–60
GC	20–40
SW	20–40
SP	10–40
SM	10–40
SM-SC	5–30
SC	5–20
ML	≤ 15
ML-CL	–
CL	≤ 15
OL	≤ 5
MH	≤ 10
CH	≤ 15
OH	≤ 5

Example 35.9

The following load data are collected for a 3 in^2 plunger test. What is the California bearing ratio?

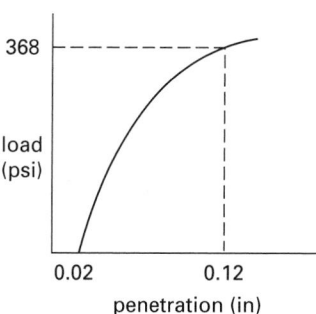

penetration (in)	load (lbf/in^2)
0.020	0
0.025	20
0.050	130
0.075	230
0.100	320
0.125	380
0.150	470
0.175	530
0.200	600
0.250	700
0.300	830

Solution

After graphing the data, it is apparent that a 0.02 in correction is required. Therefore, the 0.1 in load is read from the graph as a 0.12 in load.

(Since an 0.02 deflection was recorded for a zero load, the first 0.02 of soil must consist of uncompacted fluff, which would not occur in a roadbed.)

From Eq. 35.47,

$$\text{CBR}_{0.1} = \frac{368\ \frac{\text{lbf}}{\text{in}^2}}{1000\ \frac{\text{lbf}}{\text{in}^2}} \times 100$$

$$= 36.8$$

$$\text{CBR}_{0.2} = \frac{645\ \frac{\text{lbf}}{\text{in}^2}}{1500\ \frac{\text{lbf}}{\text{in}^2}} \times 100$$

$$= 43.0$$

Since $\text{CBR}_{0.2}$ is greater than $\text{CBR}_{0.1}$, the test should be repeated.

23. PLATE BEARING VALUE TEST

The *plate bearing value test* is performed on compacted soil in the field and provides an indication of the shear strength of pavement components. A standard diameter round steel plate is set over the soil on a bedding of fine sand or plaster of paris. Smaller diameter plates are placed on top of the bottom plate to ensure rigidity. After the plate is seated by a quick but temporary load, it is loaded to a deflection of about 0.04 in (1 mm). This load is maintained until the deflection rate decreases to 0.001 in/min (0.03 mm/min). The load is then released. The deflection prior to loading, the final deflection, and the deflection each minute are recorded.

The test is repeated 10 times. For each repetition of each load, the endpoint deflection is found for which the deflection rate is exactly 0.001 in/min (0.03 mm/min). The loads are then corrected for dead weights of jacks, plates, and so on.

The corrected load versus the corrected deflection is graphed for the 10th repetition. The bearing value is the interpolated load that would produce a deflection of 0.5 in (12 mm). The *subgrade modulus* (*modulus of subgrade reaction*), k, is the slope of the line (in psi per inch) in the loading range encountered by the soil.

Figure 35.18 *10th Repetition Bearing Load*

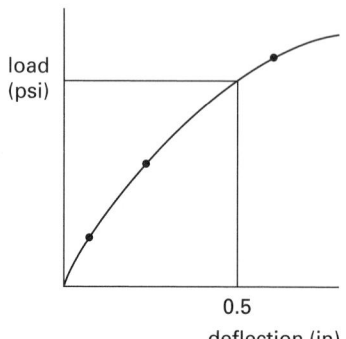

Table 35.17 *Typical Values of the Subgrade Modulus*

group symbol	range of subgrade modulus, k (psi/in (kPa/mm))
GW	300–500 (80–140)
GP	250–400 (68–110)
GM	100–400 (27–110)
GC	100–300 (27–80)
SW	200–300 (54–80)
SP	200–300 (54–80)
SM	100–300 (27–80)
SM-SC	100–300 (27–80)
SC	100–300 (27–80)
ML	100–200 (27–54)
ML-CL	–
CL	50–200 (14–54)
OL	50–100 (14–27)
MH	50–100 (14–27)
CH	50–150 (14–41)
OH	25–100 (6.8–27)

(Multiply psi/in by 0.2714 to obtain kPa/mm.)

24. HVEEM'S RESISTANCE VALUE TEST

Hveem's resistance value test is performed to evaluate the suitability of a soil for use in pavement sections. The term "resistance" refers to the ability of a soil to resist lateral deformation when a vertical load acts upon it. When displacement does occur, the soil moves out and away from the applied load.

The *R-value* (*resistance value*) of a soil is measured in a *stabilometer test*. The *R*-value will range from zero (the resistance of water) to 100 (the approximate resistance of steel). Typical values are clay, 5–15; sandy clay, 10–20; clayey silts, 20–35; sandy silts, 20–55; silty sands, 25–70; sand, 40–75; gravel, 20–80; and good crushed rock, 75–90.

The *R*-value is determined using soil samples that are compacted as they would be during normal construction. They are tested as near to saturation as possible to give the lowest expected *R*-value. Thus, the *R*-value represents the worst possible state the soil might attain during use.

When a compacted soil expands due to the absorption of water, damage such as pavement distortion may result. A determination of expansion pressures is usually performed along with the *R*-value test.

25. CLASSIFICATION OF ROCKS

Geologists classify rocks into three groups, according to the major Earth processes that formed them. The three rock groups are igneous, sedimentary, and metamorphic rocks.

Igneous rock is formed from melted rock that has cooled and solidified. When rock is buried deep within the Earth, it melts due to the high pressure and temperature. When magma cools slowly, usually at depths of thousands of feet, crystals grow from the molten liquid, forming a coarse-grained rock. When magma cools rapidly, usually at or near the Earth's surface, the crystals are extremely small, and a fine-grained rock results. Rocks formed from cooling magma are known as *intrusive igneous rocks*; rocks formed from cooling lava are known as *extrusive igneous rocks*. A wide variety of rocks are formed by different cooling rates and different chemical compositions of the original magma. Obsidian (volcanic glass), granite, basalt, and andesite porphyry are four of the many types of igneous rock. *Granite* is composed of three main minerals quartz (grey), mica (black), and feldspar (white).

Sedimentary rocks such as chalk, sandstone, and clay are formed at the surface of the Earth. They are layered accumulations of sediments—fragments of rocks, minerals, or animal or plant material. Temperatures and pressures are low at the Earth's surface, and sedimentary rocks are easily identified by their appearance and by the diversity of minerals they contain. Most sedimentary rocks become cemented together by minerals and chemicals or are held together by electrical attraction. Some, however, remain loose and virtually unconsolidated. The layers are originally parallel to the Earth's surface. Sand and gravel on beaches or in river bars consolidate into *sandstone*. Compacted and dried mud flats harden into *shale*. Sediments of mud and shells settling on the floors of lagoons form sedimentary *chalk*.

There are several ways to classify sedimentary rocks, but a common one is according to the process which led to their deposition. Rocks formed from particles of older eroded rocks are known as *clastic sedimentary rocks*. These include sandstones and clays. Rocks formed from plant and animal remains are known as *organic sedimentary rocks*. Examples include limestone, chalk, and coal. Rocks formed from chemical action are known as *chemical sedimentary rocks*. These include sedimentary iron ores, evaporites such as rock salt (halite), and to some extent, flint, limestone, and chert.

Sometimes sedimentary and igneous rocks buried in the Earth's crust are subjected to pressures so intense or heat so high that the rocks are completely changed. They then become *metamorphic rocks*. Common metamorphic rocks include slate, schist, gneiss, and marble. The process of metamorphism does not melt the rocks, but it does transform them into denser, more compact rocks. New minerals are created, either by rearrangement of mineral components or by reactions with fluids that enter the rocks. Pressure or temperature can even change previously metamorphosed rocks into new types. In this way, limestone can become marble and shale can be converted into slate.

Geotechnical

36 Shallow Foundations

Nomenclature

A	area	ft^2	m^2
B	width or diameter	ft	m
c	cohesion	lbf/ft^2	Pa
C	correction factor	–	–
d	correction factor	–	–
D	depth	ft	m
DL	dead load	lbf	N
D_R	relative density	%	%
F	factor of safety	–	–
K	constant	–	–
L	length	ft	m
LL	live load	lbf	N
M	moment	ft-lbf	N·m
N	bearing capacity factor	–	–
N	number of blows	–	–
p	pressure	lbf/ft^2	Pa
P	load	lbf	N
q	bearing capacity	lbf/ft^2	Pa
q	uniform surcharge	lbf/ft	N/m
RC	relative compaction	%	%
S	strength	lbf/ft^2	Pa
SPT	standard penetration resistance	blows/ft	blows/m

Symbols

ϵ	eccentricity	ft	m
γ	specific weight	lbf/ft^3	–
ρ	mass density	lbm/ft^3	kg/m^3
ϕ	angle of internal friction	deg	deg

Subscripts

a	allowable
b	buoyant (submerged)
B	width
c	cohesive
c	correction
d	dry
f	footing
L	length
n	blow count
q	surcharge
sat	saturated
u	undrained
uc	unconfined compression
ult	ultimate
w	water
γ	density (as a subscript)

1. SHALLOW FOUNDATIONS

A *foundation* is the part of an engineered structure that transmits the structure's forces into the soil or rock that supports it. The shape, depth, and materials of the foundation design depend on many factors including the structural loads, the existing ground conditions, and local material availability.

The term *shallow foundation* refers to a foundation system in which the depth of the foundation is shallow relative to its width, usually $D_f/B \leq 1$. This category of foundations includes spread footings, continuous (or wall) footings, and mats.

The main considerations in designing shallow foundations are ensuring against bearing capacity failures and excessive settlements. *Bearing capacity* is the ability of the soil to support the foundation loads without shear failure. *Settlement* is the tendency of soils to deform (densify) under applied loads. Since structures can tolerate only a limited amount of settlement, foundation design will often be controlled by settlement criteria, because soil usually deforms significantly before it fails in shear. Methods of calculating foundation settlements are reviewed in Ch. 40. The most damaging settlements are *differential settlements*—those that are not uniform across the supported area. Excessive settlements may only lead to minor damage such as cracked floors and walls, windows and doors that do not operate correctly, and so on. However, bearing capacity failures have the potential to cause major damage or collapse.

2. SAND VERSUS CLAY

Ordinarily, sand makes a good foundation material. Sand is usually strong, and it drains quickly. It may have an initial, "immediate" settlement upon first loading, but this usually is small and is complete before most architectural elements (drywall, brick, glass), which are rigid and more susceptible to settlement damage, are constructed. However, it behaves poorly in excavations because it lacks cohesion. When sand is loose and saturated, it can become "quick" (i.e., liquefy), and a major loss in supporting strength occurs.

Clay is generally good in excavations but poor in foundations. Clay strength is usually lower than that of sand. Clays retain water, are relatively impermeable, and do not drain freely. Settlement in clays continues beyond the end of construction, and significant settlement can continue for years or even indefinitely. Large volume changes result from *consolidation*, which is the squeezing out of water from the pores as the soil comes to equilibrium with the applied loads.

A slightly different approach to calculating bearing capacity is used for sands and clays. Clays are not free-draining, so an undrained approach is used as the critical condition. But soils do not always fall into a pure category of "sand" or "clay." Silt, for example, is an intermediate soil type that will behave like a sand or a clay depending on the fraction of clay minerals it possesses. Many soils are a mixture of sand, silt, and clay. It is necessary to use engineering judgment when applying the bearing capacity equations to intermediate soils.

3. GENERAL CONSIDERATIONS FOR FOOTINGS

A *footing* is an enlargement at the base of a load-supporting column that is designed to transmit forces to the soil. The area of the footing will depend on the load and the soil characteristics. Several types of footings are used. A *spread footing* is a footing used to support a single column. This is also known as an *individual column footing* or an *isolated footing*. A *continuous footing*, also known as a *wall footing*, is a long footing supporting a continuous wall. A *combined footing* is a footing carrying more than one column. A *cantilever footing* is a combined footing that supports a column and an exterior wall or column.

Footings should be designed according to the following general considerations.

- The footing should be located below the frost line and below the level affected by seasonal moisture content changes.

- Footings do not need to be any lower than the bottom of the highest inadequate stratum (layer). Inadequate strata include disturbed or compressible soils, uncompacted fills, and soils that are susceptible to erosion or scour.

- The foundation should be safe against overturning, sliding, and uplift. The resultant of the applied load should coincide with the middle third of the footing.

- The allowable soil pressure should not be exceeded.

- Loose sand should be densified prior to putting footings on it. "Loose" is roughly defined as $D_R <$

50% for natural sands and RC $<$ 90% (modified Proctor) for compacted sands or fill.

- Footings should be sized to the nearest 3 in (0.075 m) greater than or equal to the theoretical size.

Figure 36.1 *Types of Footings*

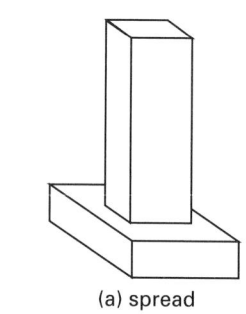

(a) spread

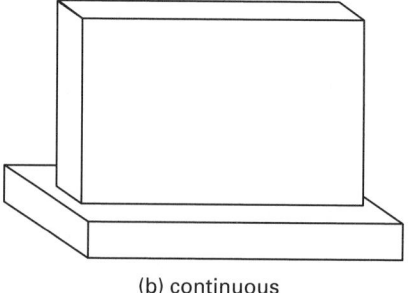

(b) continuous

4. ALLOWABLE BEARING CAPACITY

The *allowable bearing capacity* (also known as the *net allowable bearing pressure* or *safe bearing pressure*) is the net pressure in excess of the overburden stress that will not cause shear failure or excessive settlements. This is the soil pressure that is used to design the foundation. (The term "allowable" means that a factor of safety has already been applied.) When data from soil tests are unavailable, Table 36.1 can be used for preliminary calculations.

Table 36.1 *Typical Allowable Soil Bearing Capacities[a]*

type of soil	allowable pressure	
	(tons/ft^2)	(kPa)
massive crystalline bedrock	2	200
sedimentary and foliated rock	1	100
sandy gravel and/or gravel (GW and GP)	1	100
sand, silty sand, clayey sand, silty gravel, and clayey gravel (SW, SP, SM, SC, GM, and GC)	0.75	75
clay, sandy clay, silty clay, and clayey silt (CL, ML, MH, and CH)	0.5	50

(Multiply tons/ft^2 by 95.8 to obtain kPa.)
[a]Inclusive of a factor of safety

5. GENERAL BEARING CAPACITY EQUATION

With *general shear failures*, the soil resists an increased load until a sudden failure occurs. *Local shear failure* occurs in looser, more compressible soils and at high bearing pressures. The boundaries between these types of behavior are not distinct, and the methods of calculating general shear failure are commonly used for most soil conditions.

The *ultimate* (or *gross*) *bearing capacity* for a shallow wall footing is given by Eq. 36.1, which is known as the *Terzaghi-Meyerhof equation*. The equation is valid for both sandy and clayey soils. p_q is an additional surface surcharge, if any.

$$q_{\text{ult}} = \tfrac{1}{2}\rho g B N_\gamma + c N_c + (p_q + \rho g D_f)N_q$$

$$\text{[SI]} \qquad 36.1(a)$$

$$q_{\text{ult}} = \tfrac{1}{2}\gamma B N_\gamma + c N_c + (p_q + \gamma D_f)N_q$$

$$\text{[U.S.]} \qquad 36.1(b)$$

Various researchers have made improvements on the theory supporting this equation, leading to somewhat different terms and sophistication in evaluating N_γ, N_c, and N_q. The approaches differ in the assumptions made of the shape of the failure zone beneath the footing. However, the general form of the equation is the same in most cases.

Figure 36.2 and Table 36.2 can be used to evaluate the *capacity factors* N_γ, N_c, and N_q in Eq. 36.1. Alternatively, Table 36.3 can be used. The bearing capacity factors in Table 36.2 are based on Terzaghi's 1943 studies. The values in Table 36.3 are based on Meyerhof's 1955 studies and others, and have been widely used. Other values are also in use.

Figure 36.2 Terzaghi Bearing Capacity Factors

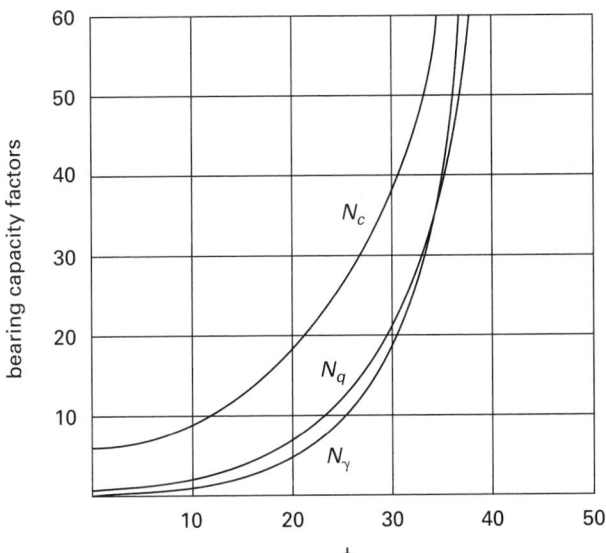

Table 36.2 Terzaghi Bearing Capacity Factors for General Shear[a]

ϕ	N_c	N_q	N_γ
0	5.7	1.0	0.0
5	7.3	1.6	0.5
10	9.6	2.7	1.2
15	12.9	4.4	2.5
20	17.7	7.4	5.0
25	25.1	12.7	9.7
30	37.2	22.5	19.7
34	52.6	36.5	35.0
35	57.8	41.4	42.4
40	95.7	81.3	100.4
45	172.3	173.3	297.5
48	258.3	287.9	780.1
50	347.5	415.1	1153.2

[a]Curvilinear interpolation may be used. Do not use linear interpolation.

Table 36.3 Meyerhof and Vesic Bearing Capacity Factors for General Shear[a]

ϕ	N_c	N_q	N_γ	$N_\gamma{}^b$
0	5.14	1.0	0.0	0.0
5	6.5	1.6	0.07	0.5
10	8.3	2.5	0.37	1.2
15	11.0	3.9	1.1	2.6
20	14.8	6.4	2.9	5.4
25	20.7	10.7	6.8	10.8
30	30.1	18.4	15.7	22.4
32	35.5	23.2	22.0	30.2
34	42.2	29.4	31.2	41.1
36	50.6	37.7	44.4	56.3
38	61.4	48.9	64.1	78.0
40	75.3	64.2	93.7	109.4
42	93.7	85.4	139.3	155.6
44	118.4	115.3	211.4	224.6
46	152.1	158.5	328.7	330.4
48	199.3	222.3	526.5	496.0
50	266.9	319.1	873.9	762.9

[a]Curvilinear interpolation (graphical or methods from Ch. 12) may be used. Do not use linear interpolation.
[b]As predicted by the Vesic equation, $N_\gamma = 2(N_q + 1)\tan\phi$.

Equation 36.1 is appropriate for a foundation in a continuous wall footing. Corrections for various footing geometries, called *shape factors*, are presented in Tables 36.4 and 36.5 using the parameters identified in Fig. 36.3. The bearing capacity factors N_c and N_γ are multiplied by the appropriate shape factors when they are used in Eq. 36.1.

Figure 36.3 Spread Footing Dimensions

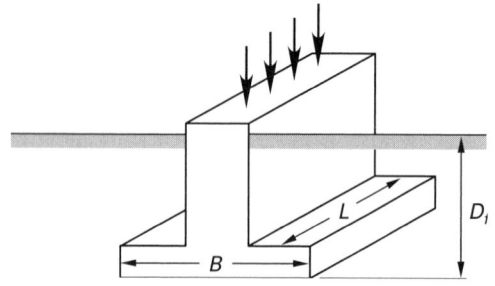

Table 36.4 N_c Bearing Capacity Factor Multipliers
for Various Values of B/L

B/L	multiplier
1 (square)	1.25
0.5	1.12
0.2	1.05
0.0	1.00
1 (circular)	1.20

Table 36.5 N_γ Multipliers for Various Values of B/L

B/L	multiplier
1 (square)	0.85
0.5	0.90
0.2	0.95
0.0	1.00
1 (circular)	0.70

Several researchers have recommended corrections to N_c to account for footing depth. (Corrections to N_q for footing depth have also been suggested. No corrections to N_γ for footing depth have been suggested.) There is considerable variation in the method of calculating this correction, if it is used at all. A multiplicative correction factor, d_c, which is used most often, has the form

$$d_c = 1 + \frac{KD_f}{B} \qquad 36.2$$

K is a constant for which values of 0.2 and 0.4 have been proposed. The depth factor correction is applied to N_c along with the shape factor correction in Eq. 36.1.

Once the ultimate bearing capacity is determined, it is corrected by the *overburden*, giving the *net bearing capacity*. This is the net pressure the soil can support beyond the pressure applied by the existing overburden.

$$q_{\text{net}} = q_{\text{ult}} - \rho g D_f \qquad [\text{SI}] \qquad 36.3(a)$$
$$q_{\text{net}} = q_{\text{ult}} - \gamma D_f \qquad [\text{U.S.}] \qquad 36.3(b)$$

The *allowable bearing capacity* is determined by dividing the net capacity by a factor of safety. The safety factor accounts for the uncertainties in evaluating soil properties and anticipated loads, and also on the

amount of risk involved in building the structure. A safety factor between 2 and 3 (based on q_{net}) is common for average conditions. Smaller safety factors are sometimes used for transient load conditions such as from wind and seismic forces.

$$q_a = \frac{q_{\text{net}}}{F} \qquad 36.4$$

6. BEARING CAPACITY OF CLAY

Clay is often soft and fairly impermeable. When loads are first applied to saturated clay, the pore pressure increases. For a short time, this pore pressure does not dissipate, and the angle of internal friction should be taken as $\phi = 0°$. This is known as the $\phi = 0°$ case or the *undrained case*, which is the critical condition for saturated clays. As discussed in Chap. 35, the undrained shear strength of clays is equal to the cohesion, which is one-half of the unconfined compressive strength.

$$S_u = c = \frac{S_{\text{uc}}}{2} \qquad 36.5$$

If $\phi = 0°$, then $N_\gamma = 0$ and $N_q = 1$. If there is no surface surcharge (i.e., $p_q = 0$), the ultimate bearing capacity is given by Eq. 36.6.

$$q_{\text{ult}} = cN_c + \rho g D_f \qquad [\text{SI}] \qquad 36.6(a)$$
$$q_{\text{ult}} = cN_c + \gamma D_f \qquad [\text{U.S.}] \qquad 36.6(b)$$

$$q_{\text{net}} = q_{\text{ult}} - \rho g D_f = cN_c \qquad [\text{SI}] \qquad 36.7(a)$$
$$q_{\text{net}} = q_{\text{ult}} - \gamma D_f = cN_c \qquad [\text{U.S.}] \qquad 36.7(b)$$

The allowable clay loading is based on a factor of safety, which is typically taken as 3 for clay.

$$q_a = \frac{q_{\text{net}}}{F} \qquad 36.8$$

From Eqs. 36.6 and 36.7, it is evident that the cohesion term dominates the bearing capacity in cohesive soil.

Example 36.1

An individual square column footing carries an 83,800 lbf (370 kN) dead load and a 75,400 lbf (335 kN) live load. The unconfined compressive strength of the supporting clay is 0.84 tons/ft^2 (80 kPa), and its specific weight is $\gamma = 115$ lbf/ft^3 ($\rho = 1840$ kg/m^3). $\phi = 0$. The footing is covered by a 6.0 in (0.15 m) thick basement slab whose upper surface is flush with the original grade. The footing thickness is initially unknown. Neglect depth correction factors. Do not design the structural steel. Specify the footing size and thickness.

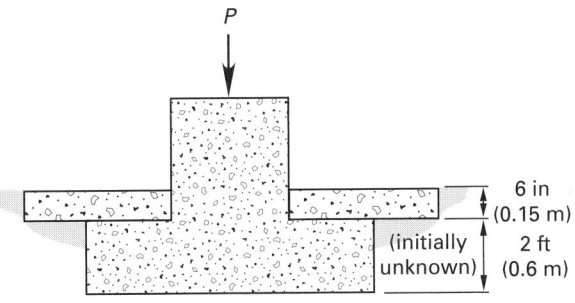

SI Solution

The strategy for determining the depth and width of the footing is to calculate the allowable bearing pressure, assume a width and depth that will support the dead load and live load, and then compare the actual pressures on the footing with the allowable pressures, considering the effect of soil displacement by the concrete footing and slab.

From Eqs. 36.4 and 36.7, the allowable bearing capacity is

$$q_a = \frac{q_{\text{net}}}{F} = \frac{cN_c}{F}$$

From Table 36.2, the bearing capacity factor is $N_c = 5.7$. From Table 36.4, for square footings, the shape factor is 1.25. The cohesion is estimated from the unconfined compressive strength and Eq. 36.5.

$$c = \frac{S_{\text{uc}}}{2} = \frac{80 \text{ kPa}}{2} = 40 \text{ kPa}$$

Using a factor of safety of 3, the allowable pressure is

$$\begin{aligned} q_a &= \frac{cN_c}{F} \\ &= \frac{(40 \text{ kPa})(5.7)(1.25)}{3} = 95 \text{ kPa} \end{aligned}$$

The total load on the column is

$$P = DL + LL = 370 \text{ kN} + 335 \text{ kN} = 705 \text{ kN}$$

The approximate area required is

$$A = \frac{705 \text{ kN}}{95 \text{ kPa}} = 7.42 \text{ m}^2$$

Try a 2.8 m square footing (area = 7.84 m²). At this point, a footing thickness would be determined based on concrete design considerations, and it is reasonable to assume a 0.6 m footing thickness.

The actual pressure under the footing due to the applied column load is

$$p_{\text{actual}} = \frac{705 \text{ kN}}{7.84 \text{ m}^2} = 89.9 \text{ kPa} < q_a \quad [\text{OK}]$$

This first iteration did not consider the concrete weight. The concrete density is approximately 2400 kg/m³.

Therefore, the pressure surcharge due to 1 m² of concrete floor is

$$\begin{aligned} p_{\text{floor slab}} &= (1 \text{ m})(1 \text{ m})(0.15 \text{ m}) \left(2400 \ \frac{\text{kg}}{\text{m}^3} \right) \left(9.81 \ \frac{\text{m}}{\text{s}^2} \right) \\ &= 3.53 \text{ kPa} \end{aligned}$$

Similarly, the footing itself has weight. The footing extends 0.6 m down.

$$\begin{aligned} p_{\text{footing}} &= (1 \text{ m})(1 \text{ m})(0.6 \text{ m}) \left(2400 \ \frac{\text{kg}}{\text{m}^3} \right) \left(9.81 \ \frac{\text{m}}{\text{s}^2} \right) \\ &= 14.13 \text{ kPa} \end{aligned}$$

Equation 36.3 gives the allowable pressure in excess of the soil surcharge. The footing bottom is 0.75 m below the original grade, so the soil surcharge is

$$\begin{aligned} p_{\text{soil}} &= (0.75 \text{ m}) \left(1840 \ \frac{\text{kg}}{\text{m}^3} \right) \left(9.81 \ \frac{\text{m}}{\text{s}^2} \right) \\ &= 13.54 \text{ kPa} \end{aligned}$$

Therefore, the total pressure under the footing is

$$\begin{aligned} p_{\text{total}} &= 89.9 \text{ kPa} + 3.53 \text{ kPa} + 14.13 \text{ kPa} - 13.54 \text{ kPa} \\ &= 94.02 \text{ kPa} \end{aligned}$$

This is the net actual pressure to be compared to the allowable pressure.

$$p_{\text{net,actual}} = 94.02 \text{ kPa}$$

This is essentially the same as q_a (94.7 kPa).

Customary U.S. Solution

The strategy for determining the depth and width of the footing is to calculate the allowable bearing pressure, assume a width and depth that will support the dead load and live load, and then compare the actual pressures on the footing with the allowable pressures, considering the effect of soil displacement by the concrete footing and slab.

From Eqs. 36.4 and 36.7, the allowable bearing capacity is

$$q_a = \frac{q_{\text{net}}}{F} = \frac{cN_c}{F}$$

From Table 36.2, the bearing capacity factor is $N_c = 5.7$. From Table 36.4, for square footings, the shape factor is 1.25. The cohesion is estimated from the unconfined compressive strength and Eq. 36.5.

$$\begin{aligned} c = \frac{S_{\text{uc}}}{2} &= \frac{0.84 \ \frac{\text{tons}}{\text{ft}^2}}{2} \\ &= 0.42 \text{ tons/ft}^2 \end{aligned}$$

Using a factor of safety of 3, the allowable pressure is

$$q_a = \frac{cN_c}{F}$$

$$= \frac{\left(0.42 \; \frac{\text{tons}}{\text{ft}^2}\right)(5.7)(1.25)}{3} = 0.99 \; \text{tons/ft}^2$$

The total load on the column is

$$P = \text{DL} + \text{LL}$$

$$= \frac{83,800 \; \text{lbf} + 75,400 \; \text{lbf}}{2000 \; \frac{\text{lbf}}{\text{ton}}}$$

$$= 79.6 \; \text{tons}$$

The approximate area required is

$$A = \frac{79.6 \; \text{tons}}{0.99 \; \frac{\text{tons}}{\text{ft}^2}} = 80.4 \; \text{ft}^2$$

Try a 9 ft 3 in square footing (area = 85.6 ft^2).

At this point, a footing thickness would be determined based on concrete design considerations, and it is reasonable to assume a 2 ft footing thickness.

The actual pressure under the footing due to the applied column load is

$$p_{\text{actual}} = \frac{79.6 \; \text{tons}}{85.6 \; \text{ft}^2}$$

$$= 0.93 \; \frac{\text{tons}}{\text{ft}^2} < q_a \quad [\text{OK}]$$

This first iteration did not consider the concrete weight. The concrete specific weight is approximately 150 lbf/ft^3. Therefore, the pressure surcharge due to 1 ft^2 of concrete floor slab is

$$p_{\text{floor slab}} = \frac{(1 \; \text{ft})(1 \; \text{ft})(6 \; \text{in})\left(150 \; \frac{\text{lbf}}{\text{ft}^3}\right)}{\left(2000 \; \frac{\text{lbf}}{\text{ton}}\right)\left(12 \; \frac{\text{in}}{\text{ft}}\right)}$$

$$= 0.04 \; \text{tons/ft}^2$$

Similarly, the footing itself has weight. The footing extends 2 ft down.

$$p_{\text{footing}} = \frac{(1 \; \text{ft})(1 \; \text{ft})(2 \; \text{ft})\left(150 \; \frac{\text{lbf}}{\text{ft}^3}\right)}{2000 \; \frac{\text{lbf}}{\text{ton}}}$$

$$= 0.15 \; \text{tons/ft}^2$$

Equation 36.3 gives the allowable pressure in excess of the soil surcharge. The footing bottom is 2.5 ft below the original grade, so the soil surcharge is

$$p_{\text{soil}} = \frac{(2.5 \; \text{ft})\left(115 \; \frac{\text{lbf}}{\text{ft}^3}\right)}{2000 \; \frac{\text{lbf}}{\text{ton}}}$$

$$= 0.14 \; \text{tons/ft}^2$$

The total pressure under the footing is

$$p_{\text{total}} = 0.93 \; \frac{\text{tons}}{\text{ft}^2} + 0.04 \; \frac{\text{tons}}{\text{ft}^2}$$

$$+ 0.15 \; \frac{\text{tons}}{\text{ft}^2} - 0.14 \; \frac{\text{tons}}{\text{ft}^2}$$

$$= 0.98 \; \text{tons/ft}^2$$

This is the net actual pressure to be compared to the allowable pressure.

$$p_{\text{net,actual}} = 0.98 \; \text{tons/ft}^2$$

This is essentially the same as q_a (0.99 tons/ft^2).

7. BEARING CAPACITY OF SAND

The *cohesion*, c, of ideal sand is zero. The ultimate bearing capacity can be derived from Eq. 36.1 by setting $c = 0$.

$$q_{\text{ult}} = \tfrac{1}{2}B\rho g N_\gamma + (p_q + \rho g D_f)N_q \quad \text{[SI]} \quad 36.9(a)$$

$$q_{\text{ult}} = \tfrac{1}{2}B\gamma N_\gamma + (p_q + \gamma D_f)N_q \quad \text{[U.S.]} \quad 36.9(b)$$

The net bearing capacity when there is no surface surcharge (i.e., $p_q = 0$) is

$$q_{\text{net}} = q_{\text{ult}} - \rho g D_f$$
$$= \tfrac{1}{2}B\rho g N_\gamma + \rho g D_f(N_q - 1)$$
$$\text{[SI]} \quad 36.10(a)$$

$$q_{\text{net}} = q_{\text{ult}} - \gamma D_f$$
$$= \tfrac{1}{2}B\gamma N_\gamma + \gamma D_f(N_q - 1)$$
$$\text{[U.S.]} \quad 36.10(b)$$

From Eqs. 36.9 and 36.10, it is evident that the depth term $\rho g D_f N_q$ dominates the bearing capacity in cohesionless soil. A small increase in depth increases the bearing capacity substantially.

The allowable bearing capacity is based on a factor of safety, which is typically taken as 2 for sand.

$$q_a = \frac{q_{\text{net}}}{F}$$

$$= \left(\frac{B}{F}\right)\left(\tfrac{1}{2}\rho g N_\gamma + \rho g(N_q - 1)\left(\frac{D_f}{B}\right)\right)$$
$$\text{[SI]} \quad 36.11(a)$$

$$q_a = \frac{q_{\text{net}}}{F}$$

$$= \left(\frac{B}{F}\right)\left(\tfrac{1}{2}\gamma N_\gamma + \gamma(N_q - 1)\left(\frac{D_f}{B}\right)\right)$$
$$\text{[U.S.]} \quad 36.11(b)$$

For common applications (cohesionless soil, settlement governing, footing width greater than 2–4 ft, and no groundwater within B of bottom of footing), Eq. 36.11

can be simplified. The quantity in brackets in Eq. 36.11 is constant for specific D_f/B ratios, and the bearing capacity factors depend on ϕ, which can be correlated to the *standard penetration test* (SPT) N-value in blows per foot. Assuming $F = 2$, $\gamma = 100$ lbf/ft^3, and $D_f < B$, Eq. 36.12 can be derived for use with spread footings. Equation 36.12 is illustrated by the rightmost part of Fig. 36.4.

$$q_a = 0.11 C_n N \quad [\text{in tons/ft}^2, \ B > 2\text{–}4 \text{ ft}, \ N \leq 50] \quad \textbf{36.12}$$

Corrections are required for shallow water tables. See Sec. 36.14. No correction is usually made to Eq. 36.12 if the density is different from 100 lbf/ft^3. However, the equation assumes that the overburden load (γD_f) is approximately 1 ton/ft^2. This means that the N-values will correspond to data from a depth of 10 to 15 ft below the original surface, not the basement surface. If the footing is to be installed close to the original surface, then a correction factor is required. (The correction is actually a correction for N. If corrected N-values are known, C_n may be neglected.) The potential failure zone extends into the soil below the footing, and it is common practice to evaluate N from the bottom of the footing to a depth B below the footing. The lowest average value of N from this zone is used to calculate the bearing capacity.

Table 36.6 Overburden Corrections

overburden (tons/ft^2 (kPa))	C_n
0 (0)	2
0.25 (24)	1.45
0.5 (48)	1.21
1.0 (96)	1.00
1.5 (144)	0.87
2.0 (192)	0.77
2.5 (240)	0.70
3.0 (287)	0.63
3.5 (335)	0.58
4.0 (383)	0.54
4.5 (431)	0.50
5.0 (479)	0.46

(Multiply tons/ft^2 by 95.8 to obtain kPa.)

For a given sand settlement, the soil pressure will be greatest in intermediate-width ($B = 2$ to 4 ft) footings. This is illustrated in Fig. 36.4. Equation 36.12 should not be used for small-width footings, since bearing pressure governs. For wide footings (i.e., $B > 2$ to 4 ft), settlement governs. Equation 36.12 was derived to ensure total settlement on sand would be 1 in or less.

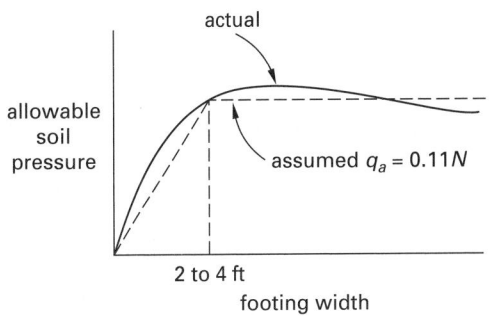

Figure 36.4 Soil Pressure on Sand (Constant Settlement)

8. BEARING CAPACITY OF ROCK

If bedrock can be reached by excavation, the allowable pressure is likely to be determined by local codes. A minimum safety factor of 3 based on the unconfined compressive strength is typical. For most rock beds, the design will be based on settlement characteristics, not strength.

9. EFFECTS OF WATER TABLE ON FOOTING DESIGN

The presence of a water table within the soil support zone may or may not affect bearing capacity, according to the following principles. (Also, see Sec. 36.14.)

General Principle 1: For cohesive soils (e.g., clay), the location of the water table does not affect the bearing capacity, and the effect of the water table is disregarded. (Strictly speaking, the water table has some effect, but the theory predicts none. If $\phi = 0$, then $N_\gamma = 0$. Also, $N_q = 1$, which is essentially zero. These two terms "zero out" the density terms in Eq. 36.1.)

General Principle 2: For sand, use the *submerged density* $\gamma_b = \gamma_{\text{sat}} - 62.4$ lbf/ft^3 (or, $\rho_b = \rho_{\text{sat}} - 1000$ kg/m^3) in the equation for bearing capacity, Eq. 36.1. Since the submerged density is approximately half of the dry (drained) density, it is commonly stated that the bearing capacity of a footing with the water table at the ground surface is half of the dry bearing capacity, varying linearly to full strength at a distance B below the footing base. However, a more accurate estimate can be obtained from the following cases.

(a) When the water table is at the base of the footing, use the submerged density in the first term of Eq. 36.1 only. Since $c = 0$ for sand, the bearing capacity is

$$q_{\text{ult}} = \tfrac{1}{2}\rho_b g B N_\gamma + \rho_d g D_f N_q \quad [\text{SI}] \quad \textbf{36.13(a)}$$
$$q_{\text{ult}} = \tfrac{1}{2}\gamma_b B N_\gamma + \gamma_d D_f N_q \quad [\text{U.S.}] \quad \textbf{36.13(b)}$$

(b) When the water table is at the surface, use the submerged density in both the first and third terms of Eq. 36.1.

$$q_{\text{ult}} = \tfrac{1}{2}\rho_b g B N_\gamma + \rho_b g D_f N_q \quad \text{[SI]} \quad \textit{36.14(a)}$$

$$q_{\text{ult}} = \tfrac{1}{2}\gamma_b B N_\gamma + \gamma_b D_f N_q \quad \text{[U.S.]} \quad \textit{36.14(b)}$$

(c) When the water table is between the base of the footing and the surface, linear interpolation is used between cases (a) and (b). This is equivalent to using the submerged density in the first term as in (a) above, and calculating the third term in Eq. 36.1 as

$$\left(\rho_d g D_w + \left(\rho_d - 1000\ \frac{\text{kg}}{\text{m}^3}\right)g(D_f - D_w)\right)N_q$$

$$= \left(\rho_d g D_f + \left(1000\ \frac{\text{kg}}{\text{m}^3}\right)g(D_w - D_f)\right)N_q$$

$$\text{[SI]} \quad \textit{36.15(a)}$$

$$\left(\gamma_d D_w + \left(\gamma_d - 62.4\ \frac{\text{lbf}}{\text{ft}^3}\right)(D_f - D_w)\right)N_q$$

$$= \left(\gamma_d D_f + \left(62.4\ \frac{\text{lbf}}{\text{ft}^3}\right)(D_w - D_f)\right)N_q$$

$$\text{[U.S.]} \quad \textit{36.15(b)}$$

General Principle 3: For sand, if the water table depth, D_w, is greater than $D_f + B$ (i.e., more than a distance B below the base of the footing), the bearing capacity is not affected. Calculate the bearing capacity from Eq. 36.1 as if there was no water table.

$$q_{\text{ult}} = \tfrac{1}{2}\rho_d g B N_\gamma + \rho_d g D_f N_q \quad \text{[SI]} \quad \textit{36.16(a)}$$

$$q_{\text{ult}} = \tfrac{1}{2}\gamma_d B N_\gamma + \gamma_d D_f N_q \quad \text{[U.S.]} \quad \textit{36.16(b)}$$

General Principle 4: For sand, if the water table depth, D_w, is between D_f and $D_f + B$, the bearing capacity is calculated using the drained density in the $D_f N_q$ term, and a weighted average of the drained and submerged densities in the BN_γ term.

Figure 36.5 *Water Table Beneath a Footing*

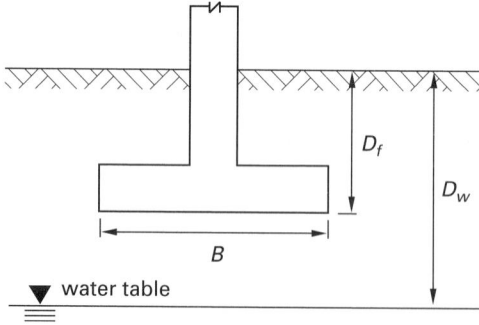

10. ECCENTRIC LOADS ON RECTANGULAR FOOTINGS

If a rectangular footing carries a moment in addition to its vertical load, an eccentric loading situation is created. Under these conditions, the footing bearing capacity should be analyzed assuming an area in which the size is reduced by twice the eccentricity.

$$\epsilon_B = \frac{M_B}{P}; \quad \epsilon_L = \frac{M_L}{P} \qquad \textit{36.17}$$

$$L' = L - 2\epsilon_L; \quad B' = B - 2\epsilon_B \qquad \textit{36.18}$$

$$A' = L'B' \qquad \textit{36.19}$$

This area reduction places the equivalent force at the centroid of the reduced area. The equivalent width B' should be used in Eq. 36.1, and both B' and L' should be used in the ratio B'/L' used to determine shape factors. The footing bearing capacity is reduced in two ways. First, a smaller B in Eq. 36.1 results in a smaller q_{ult} and q_a. Second, a smaller q_a results in a smaller allowable load (i.e., $P = q_a B'L'$).

Figure 36.6 *Footing with Overturning Moment*

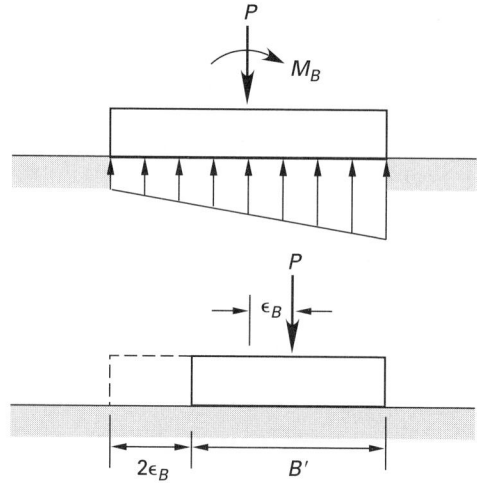

Although the eccentricity is independent of the footing dimensions, a trial-and-error solution may be necessary when designing footings. Trial and error is not required when analyzing a rectangular footing of known dimensions.

Assuming $M_L = 0$ and $\epsilon_B = \epsilon$ and disregarding the concrete and overburden weights, the actual soil pressure distribution is given by Eq. 36.20. B' and L' should not be used in Eq. 36.20 because these variables place the load at the centroid of the reduced area, assuming a uniform pressure distribution.

$$p_{\text{max}},\ p_{\text{min}} = \left(\frac{P}{BL}\right)\left(1 \pm \frac{6\epsilon}{B}\right) \qquad \textit{36.20}$$

If the eccentricity, ϵ, is sufficiently large, a negative soil pressure will result. Since soil cannot carry a tensile stress, such stresses are neglected. This results in a reduced area to carry the load.

If the resultant force is within the middle third of the footing, the contact pressures will be distributed over the entire footing. That is, the maximum eccentricity without incurring a reduction in footing contact area will be $B/6$.

11. GENERAL CONSIDERATIONS FOR RAFTS

A *raft* or *mat* foundation is a combined footing slab that usually covers the entire area beneath a building and supports all walls and columns. (The term *pad* is also used.) A raft foundation should be used (at least for economic reasons) any time the individual footings would constitute half or more of the area beneath a building. Rafts are also used to combine foundations with basement floor slabs, to minimize differential settlements over compressible soils, or to provide resistance to uplift.

12. RAFTS ON CLAY

The net or ultimate bearing capacity for rafts on clay is found in the same manner as for shallow footings. Since the size of the raft is usually fixed by the building size (within a few feet), the only method available to increase the allowable loading is to lower the elevation (increase D_f) of the raft.

The factor of safety produced by a raft construction is given by Eq. 36.21, which can also be solved to give the required values of D_f if the factor of safety is known. The factor of safety should be at least 3 for normal loadings, but it may be reduced to 2 during temporary extreme loading.

$$F = \frac{cN_c}{\dfrac{\text{total load}}{\text{raft area}} - \rho g D_f} \qquad \text{[SI]} \quad \textbf{\textit{36.21(a)}}$$

$$F = \frac{cN_c}{\dfrac{\text{total load}}{\text{raft area}} - \gamma D_f} \qquad \text{[U.S.]} \quad \textbf{\textit{36.21(b)}}$$

If the denominator in Eq. 36.21 is small, the factor of safety is very large. If the denominator is zero, the raft is said to be a *fully compensated foundation*. In a fully compensated foundation, the loads imposed on the foundation by the structure are exactly equal to the weight of the soil that is excavated. For D_f less than the fully compensated depth, the raft is said to be a *partially compensated foundation*.

Example 36.2

A raft foundation is to be designed for a 120 ft × 200 ft (36 m × 60 m) building with a total loading of 5.66×10^7 lbf (2.5×10^5 kN). The clay specific weight is $\gamma = 115$ lbf/ft³ ($\rho = 1840$ kg/m³), and the clay has an average unconfined compressive strength of 0.3 tons/ft² (28.7 kPa). Neglect depth correction factors. (a) What should be the raft depth, D_f, for full compensation? (b) What should be the raft depth for a factor of safety of 3?

SI Solution

(a) For full compensation, by Eq. 36.21,

$$\frac{\text{total load}}{\text{raft area}} - \rho g D_f = 0$$

$$
\begin{aligned}
D_f &= \frac{\dfrac{\text{total load}}{\text{raft area}}}{\rho g} \\[2ex]
&= \frac{\dfrac{(2.5 \times 10^5 \text{ kN})\left(1000 \dfrac{\text{N}}{\text{kN}}\right)}{(36 \text{ m})(60 \text{ m})}}{\left(1840 \dfrac{\text{kg}}{\text{m}^3}\right)\left(9.81 \dfrac{\text{m}}{\text{s}^2}\right)} \\[2ex]
&= 6.4 \text{ m}
\end{aligned}
$$

(b) From Eq. 36.21,

$$F = \frac{cN_c}{\dfrac{\text{total load}}{\text{raft area}} - \rho g D_f}$$

From Table 36.2, $N_c = 5.7$. Since $B/L = 36$ m/60 m $= 0.6$, use an N_c multiplier of 1.15 from Table 36.4. The cohesion is calculated from Eq. 36.5.

$$c = \frac{S_{uc}}{2} = \frac{28.7 \text{ kPa}}{2} = 14.35 \text{ kPa}$$

$$D_f = \frac{\dfrac{\text{total load}}{\text{raft area}} - \dfrac{cN_c}{F}}{\rho g}$$

$$
= \frac{\dfrac{(2.5 \times 10^5 \text{ kN})\left(1000 \dfrac{\text{N}}{\text{kN}}\right)}{(36 \text{ m})(60 \text{ m})} - \dfrac{(14.35 \text{ kPa})\left(1000 \dfrac{\text{N}}{\text{kN}}\right)(5.7)(1.15)}{3}}{\left(1840 \dfrac{\text{kg}}{\text{m}^3}\right)\left(9.81 \dfrac{\text{m}}{\text{s}^2}\right)}
$$

$$= 4.7 \text{ m}$$

Customary U.S. Solution

(a) For full compensation, use Eq. 36.21.

$$\frac{\text{total load}}{\text{raft area}} - \gamma D_f = 0$$

$$
\begin{aligned}
D_f &= \frac{\dfrac{\text{total load}}{\text{raft area}}}{\gamma} \\[2ex]
&= \frac{\dfrac{5.66 \times 10^7 \text{ lbf}}{(120 \text{ ft})(200 \text{ ft})}}{115 \dfrac{\text{lbf}}{\text{ft}^3}} \\[2ex]
&= 20.5 \text{ ft}
\end{aligned}
$$

(b) From Eq. 36.21,

$$F = \frac{cN_c}{\dfrac{\text{total load}}{\text{raft area}} - \gamma D_f}$$

From Table 36.2, $N_c = 5.7$. Since $B/L = 120$ ft$/200$ ft $= 0.6$, use an N_c multiplier of 1.15 from Table 36.4. The cohesion is calculated from Eq. 36.5.

$$c = \frac{S_{uc}}{2} = \frac{\left(0.3 \dfrac{\text{ton}}{\text{ft}^2}\right)\left(2000 \dfrac{\text{lbf}}{\text{ton}}\right)}{2}$$

$$= 300 \text{ lbf/ft}^2$$

$$D_f = \frac{\dfrac{\text{total load}}{\text{raft area}} - \dfrac{cN_c}{F}}{\gamma}$$

$$= \frac{\dfrac{5.66 \times 10^7 \text{ lbf}}{(120 \text{ ft})(200 \text{ ft})} - \dfrac{\left(300 \dfrac{\text{lbf}}{\text{ft}^2}\right)(5.7)(1.15)}{3}}{115 \dfrac{\text{lbf}}{\text{ft}^3}}$$

$$= 14.8 \text{ ft}$$

13. RAFTS ON SAND

Rafts on sand are well protected against bearing capacity failure. Therefore, settlement will govern the design. Since differential settlement will be much smaller for various locations on the raft (due to the raft's rigidity), differential settling is not a factor, and the allowable soil pressure may be doubled (compared to Eq. 36.12).

$$q_a = 0.22 C_n N \quad \text{[in tons/ft}^2\text{]} \qquad \textbf{36.22}$$

N should always be at least 5 after correcting for overburden. (Note that the depth of the potential failure zone, $D_f + B$, is very large for a raft foundation.) Otherwise, the sand should be compacted or a pier or pile foundation should be used.

To calculate the factor of safety in bearing, the net bearing capacity from Eq. 36.3 should be compared with the actual (net) bearing pressure. The actual (net) bearing pressure is

$$p = \frac{\text{total load}}{\text{raft area}} - \rho g D_f \qquad \text{[SI]} \quad \textbf{36.23(a)}$$

$$p = \frac{\text{total load}}{\text{raft area}} - \gamma D_f \qquad \text{[U.S.]} \quad \textbf{36.23(b)}$$

14. SHALLOW WATER TABLE CORRECTION

If the simplified analyses in Eqs. 36.12 and 36.22 (as proposed by Peck, Hanson, and Thornburn) are used, a multiplicative correction is required when the water table is at the surface or below, down to a distance B below the footing depth, D_f. The values of allowable bearing pressure from Eqs. 36.12 and 36.22 are multiplied by C_w from Eq. 36.24.

$$C_w = 0.5 + 0.5 \left(\frac{D_w}{D_f + B}\right) \qquad \textbf{36.24}$$

37 Rigid Retaining Walls

Nomenclature

A	area	ft^2	m^2
B	base width	ft	m
c	cohesion	lbf/ft^2	Pa
c_A	adhesion	lbf/ft^2	Pa
F	factor of safety	–	–
g	acceleration of gravity	ft/sec^2	m/s^2
h	height of the water table	ft	m
H	height of soil	ft	m
k	earth pressure coefficient	–	–
K	earth pressure constant	lbf/ft^3	N/m^3
L	heel length	ft	m
L	line surcharge	lbf/ft	N/m
m	the fraction x/H	–	–
M	moment	ft-lbf	N·m
n	the fraction y/H	–	–
p	pressure	lbf/ft^2	Pa
q	uniform surcharge	lbf/ft^2	Pa
R	force	lbf	N
S	strip surcharge	lbf/ft^2	Pa
t	thickness	ft	m
V	vertical force surcharge	lbf	N
W	vertical force (weight)	lbf	N
x	distance in the x-direction	ft	m
y	distance in the y-direction	ft	m
y	moment arm	ft	m

Symbols

α	angle of failure plane	deg	deg
β	slope of backfill	deg	deg
γ	specific weight	lbf/ft^3	–
δ	angle of external friction	deg	deg
ϵ	eccentricity	ft	m
θ	angle of the resultant force	deg	deg
λ	rake angle of retaining wall face	deg	deg
μ	pore pressure	lbf/ft^2	Pa
ρ	density	lbm/ft^3	kg/m^3
ϕ	angle of internal friction	deg	deg

Subscripts

a	active
A	adhesion
eff	effective
eq	equivalent
h	horizontal
o	at rest
OT	overturning
p	passive
q	surcharge
R	resultant
sat	saturated
SL	sliding
v	vertical
w	water

1. TYPES OF RETAINING WALL STRUCTURES

A *gravity wall* is a high-bulk structure that relies on self-weight and the weight of the earth over the heel to resist overturning. *Semi-gravity walls* are similar though less massive. A *buttress wall* depends on compression ribs (*buttresses*) between the stem and the toe to resist flexure and overturning. *Counterfort walls* depend on tension ribs between the stem and the heel to resist flexure and overturning. A *cantilever wall* resists overturning through a combination of the soil weight over the heel and the resisting pressure under the base.

A *cantilever retaining wall* consists of a base, a stem, and an optional key.[1] The stem may have a constant thickness, or it may be tapered. The taper is known as *batter*. The *batter decrement* is the change in stem thickness per unit of vertical distance. Batter is used to "disguise" bending (deflection) that would otherwise make it appear as if the wall were failing. It also reduces the quantity of material needed at the top of the wall where less strength is needed.

Cantilever retaining walls are generally intended to be permanent and are made of cast-in-place poured concrete. However, retaining walls may also be constructed

---

[1]The use of a key may increase the cost of installing the retaining wall by more than just the cost of extending the thickness of the base by the depth of the key. To install a key, the contractor will have to hand-shovel the keyway or change buckets on the backhoe. Arranging any vertical key steel is also time consuming.

from reinforced masonry block, stacked elements of various types, closely spaced driven members, railroad ties, and heavy lumber.

Gabion walls, consisting of stacked layers of rock-filled wire cages (baskets), may be economical if adequate rock is available nearby. Gabions are applicable to waterway lining, embankment control, spillways, and stilling basins as well as retaining walls. Other stacked-wall structures use interlocking concrete blocks, with and without tension cabling. Geosynthetic fabrics may be needed with gabion and stacked-block walls to prevent the migration of fines.

Figure 37.1 *Types of Retaining Walls*

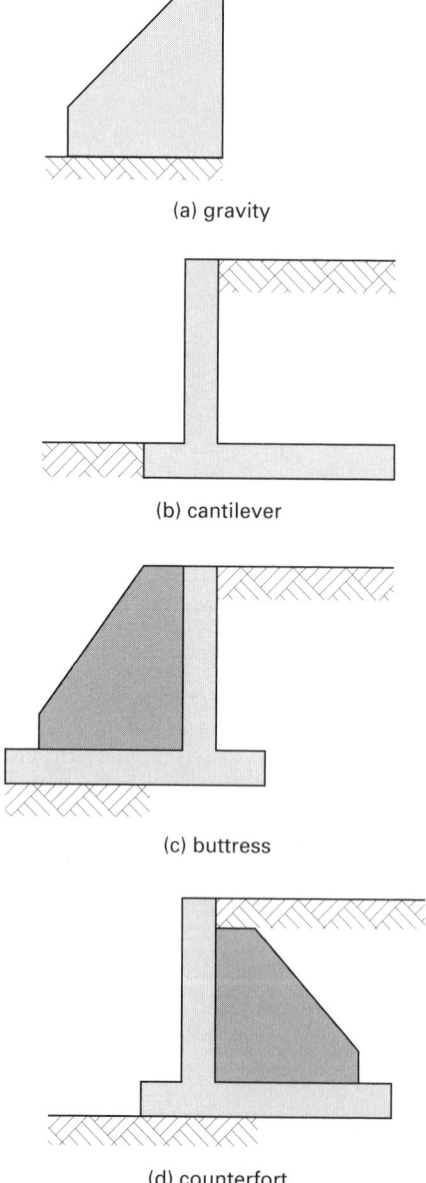

(a) gravity

(b) cantilever

(c) buttress

(d) counterfort

2. COHESIVE AND GRANULAR SOILS

The nature of the backfilled or retained soil greatly affects the design of retaining walls. The two main soil classifications are granular and cohesive soils. *Cohesive soils* are clay-type soils with angles of internal friction, ϕ, of close to zero. *Granular soils* (also referred to as *noncohesive soils* and *cohesionless soils*) are sand- and gravel-type soils with values of cohesion, c, of close to zero. It is understood that granular soils also encompass "moist sands" and "drained sands."

3. EARTH PRESSURE

Earth pressure is the force per unit area exerted by soil on the retaining wall. Generally, the term is understood to mean "horizontal earth pressure." *Active earth pressure* (also known as *tensioned soil pressure* and *forward soil pressure*) is present behind a retaining wall that moves away from and tensions the remaining soil. *Passive earth pressure* (also known as *backward soil pressure* and *compressed soil pressure*) is present in front of a retaining wall that moves toward and compresses the soil.

Figure 37.2 *Active and Passive Earth Pressure*

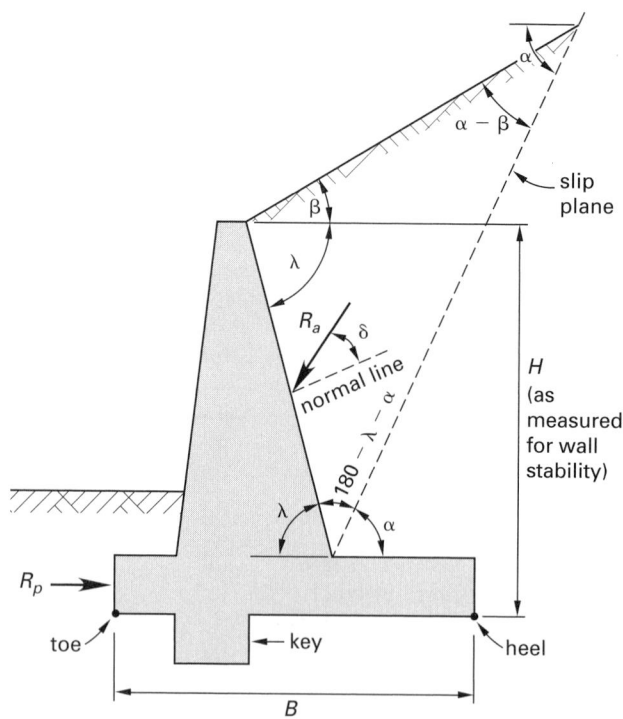

There are three common earth pressure theories. The *Rankine earth pressure theory* assumes that failure occurs along a flat plane behind the wall inclined at an angle α from the horizontal (counterclockwise being positive). The area above the failure plane is referred to as the *active zone*. The Rankine theory disregards friction between the wall and the soil.

$$\alpha = 45° + \frac{\phi}{2} \quad \text{[Rankine]} \qquad 37.1$$

The *Coulomb earth pressure theory* also assumes a flat failure plane, but the effect of wall friction is included. Where friction is significant, the Coulomb theory can predict a lower active pressure than the Rankine theory. The angle of the failure plane depends on both ϕ and δ, but not on H or γ. α is measured from the horizontal, with counterclockwise being positive.

$$\alpha = \phi + \arctan$$
$$\times \left(\frac{-\tan\phi + \sqrt{\tan\phi(\tan\phi + \cot\phi)(1 + \tan\delta\cot\phi)}}{1 + \tan\delta(\tan\phi + \cot\phi)} \right)$$

$$\text{[Coulomb]} \quad 37.2$$

The *log-spiral theory* assumes that the failure surface will be curved. The sophistication of a log-spiral solution is probably warranted only for the passive case, and even then, only when δ is large (i.e., larger than approximately 15°).

4. VERTICAL SOIL PRESSURE

Vertical soil pressure, p_v, is caused by the soil's own weight and is calculated in the same manner as for a fluid column. H is measured from the bottom of the base.

$$p_v = \rho g H \qquad \text{[SI]} \qquad 37.3(a)$$
$$p_v = \gamma H \qquad \text{[U.S.]} \qquad 37.3(b)$$

5. ACTIVE EARTH PRESSURE

Equation 37.4 is the equation for horizontal active earth pressure with level backfill for all soil types (i.e., sandy soils and clayey soils). k_a is the *coefficient of active earth pressure* (*active earth pressure coefficient*). Depending on the value of the cohesion, c, p_a can become negative (i.e., develop tension cracks) above the wall base.

$$p_a = p_v k_a - 2c\sqrt{k_a} \qquad 37.4$$

The method of calculating k_a depends on the assumptions made. Equation 37.5 is the most general form of the active Coulomb equation. This form allows for sloping backfill (angle β), inclined active-side wall face (angle λ), and friction between the soil and wall face (angle δ).

$$k_a = \frac{\sin^2(\lambda + \phi)}{\sin^2\lambda \sin(\lambda - \delta)\left(1 + \sqrt{\dfrac{\sin(\phi + \delta)\sin(\phi - \beta)}{\sin(\lambda - \delta)\sin(\lambda + \beta)}}\right)^2}$$

$$\text{[Coulomb]} \quad 37.5$$

Table 37.1 *External Friction Angles*[a]

interface materials[b]	friction angle, δ
concrete or masonry against the following foundation materials	
clean, sound rock	35
clean gravel, gravel-sand mixtures, and coarse sand	29–31
clean fine-to-medium sand, silty medium-to-coarse sand, and silty or clayey gravel	24–29
clean fine sand and silty or clayey fine-to-medium sand	19–24
fine sandy silt and non-plastic silt	17–19
very stiff clay and hard residual or preconsolidated clay	22–26
medium-stiff clay, stiff clay, and silty clay	17–19
steel sheet piles against the following soils	
clean gravel, gravel-sand mixtures, and well-graded rock fill with spall	22
clean sand, silty sand-gravel mixtures, and single-size hard rock fill	17
silty sand and gravel or sand mixed with silt or clay	14
fine sandy silt and nonplastic silt	11
formed concrete or concrete sheet piles against the following soils	
clean gravel, gravel-sand mixtures, and well-graded rock fill with spalls	22–26
clean sand, silty-sand-gravel mixtures, and single-size hard rock fill	17–22
silty sand and gravel or sand mixed with silt or clay	17
fine sandy silt and nonplastic silt	14
miscellaneous combinations of structural materials	
masonry on masonry, igneous, and metamorphic rocks	
dressed soft rock on dressed soft rock	35
dressed hard rock on dressed soft rock	33
dressed hard rock on dressed hard rock	29
masonry on wood (cross-grain)	26
steel on steel at sheet-steel interlocks	17

[a]For material not listed, use $\delta = {}^2\!/_3\phi$.
[b]Angles given are ultimate values that require significant movement before failure occurs.
Source: *Foundations and Earth Structures*, NAVFAC DM-7.2 (1982), p. 7.2-63, Table 1

The Rankine theory disregards wall friction. The most general Rankine equation is derived by setting $\delta = 0$ in Eq. 37.5.

$$k_a = \cos\beta \left(\frac{\cos\beta - \sqrt{\cos^2\beta - \cos^2\phi}}{\cos\beta + \sqrt{\cos^2\beta - \cos^2\phi}} \right) \quad \text{[Rankine]}$$

$$37.6$$

If the backfill is horizontal ($\beta = 0$) and the wall face is vertical ($\lambda = 90°$), then

$$k_a = \frac{1}{k_p} = \tan^2 \left(45° - \frac{\phi}{2} \right)$$

$$= \frac{1 - \sin\phi}{1 + \sin\phi} \quad \left[\begin{array}{l} \text{Rankine: horizontal} \\ \text{backfill; vertical face} \end{array} \right] \quad 37.7$$

For saturated clays, the angle of internal friction, ϕ, is zero. As long as tension cracks do not develop near the top of the retaining wall, $k_a = 1$ and Eq. 37.4 resolves to

$$p_a = p_v - 2c \quad [\phi = 0] \qquad 37.8$$

For granular soils, $c = 0$. In that case, Eq. 37.4 resolves to

$$p_a = k_a p_v \quad [c = 0] \qquad 37.9$$

Since Eq. 37.9 describes a triangular pressure distribution, the *total active resultant* acts $H/3$ above the base for both the Rankine and Coulomb cases and is, per unit width of wall,

$$R_a = \tfrac{1}{2} p_a H = \tfrac{1}{2} k_a \rho g H^2 \quad \text{[SI]} \qquad 37.10(a)$$

$$R_a = \tfrac{1}{2} p_a H = \tfrac{1}{2} k_a \gamma H^2 \quad \text{[U.S.]} \qquad 37.10(b)$$

Since wall friction is disregarded with the Rankine case, the earth pressure resultant is normal to the wall. For the Coulomb case, there is a downward force component, and the earth pressure resultant is directed at the angle δ above the normal, or $90° - \delta$ from the wall.

$$\theta_R = 90° \text{ from the wall} \quad \text{[Rankine]} \qquad 37.11$$

$$\theta_R = 90° - \delta \text{ from the wall} \quad \text{[Coulomb]} \qquad 37.12$$

6. PASSIVE EARTH PRESSURE

Equation 37.13 is the equation for horizontal passive earth pressure with level backfill for all soil types (i.e., sandy soils and clayey soils). k_p is the *coefficient of passive earth pressure (passive earth pressure coefficient)*.

$$p_p = p_v k_p + 2c\sqrt{k_p} \qquad 37.13$$

The method of calculating k_p depends on the assumptions made. Equation 37.14 is the most general form of the passive Coulomb equation. This form allows for sloping backfill (angle β), inclined active-side wall face

(angle λ), and friction between the soil and wall face (angle δ).

$$k_p = \frac{\sin^2(\lambda - \phi)}{\sin^2\lambda \sin(\lambda + \delta) \left(1 - \sqrt{\dfrac{\sin(\phi + \delta)\sin(\phi + \beta)}{\sin(\lambda + \delta)\sin(\lambda + \beta)}} \right)^2}$$

$$\text{[Coulomb]} \quad 37.14$$

The Rankine theory disregards wall friction. The most general Rankine equation is derived by setting $\delta = 0$ in Eq. 37.14.

$$k_p = \cos\beta \left(\frac{\cos\beta + \sqrt{\cos^2\beta - \cos^2\phi}}{\cos\beta - \sqrt{\cos^2\beta - \cos^2\phi}} \right) \quad \text{[Rankine]}$$

$$37.15$$

If the backfill is horizontal ($\beta = 0$) and the wall face is vertical ($\lambda = 90°$), then

$$k_p = \frac{1}{k_a} = \tan^2 \left(45° + \frac{\phi}{2} \right)$$

$$= \frac{1 + \sin\phi}{1 - \sin\phi} \quad \left[\begin{array}{l} \text{Rankine: horizontal} \\ \text{backfill; vertical face} \end{array} \right] \quad 37.16$$

For saturated clays, the angle of internal friction, ϕ, is zero and $k_p = 1$. Eq. 37.13 resolves to

$$p_p = p_v + 2c \quad [\phi = 0] \qquad 37.17$$

For granular soils, $c = 0$. In that case, Eq. 37.13 resolves to

$$p_p = k_p p_v \quad [c = 0] \qquad 37.18$$

$$R_p = \tfrac{1}{2} p_p H = \tfrac{1}{2} k_p \rho g H^2 \quad \text{[SI]} \qquad 37.19(a)$$

$$R_p = \tfrac{1}{2} p_p H = \tfrac{1}{2} k_p \gamma H^2 \quad \text{[U.S.]} \qquad 37.19(b)$$

Passive earth pressure may not always be present during the life of the wall. When a retaining wall is first backfilled, or when the toe of the wall is excavated for repair or other work, the passive pressure may be absent. Because of this, the restraint from passive pressure is usually disregarded in factor-of-safety calculations.

7. AT-REST SOIL PRESSURE

The active and passive pressures predicted by the Rankine and Coulomb theories assume the wall moves slightly. Lateral movement as little as $H/200$ is sufficient for the active/passive distributions to develop. In some situations, however, the soil is completely confined and cannot move. This "at rest" case is appropriate for soil next to bridge abutments, basement walls restrained at their tops, walls bearing on rock, and walls with soft-clay backfill, as well as for sand deposits of infinite depth and extent.

The horizontal pressure at rest depends on the *coefficient of earth pressure at rest*, k_o, which varies from 0.4 to 0.5 for untamped sand, from 0.5 to 0.7 for normally consolidated clays, and from 1.0 and up for overconsolidated clays.

$$p_o = k_o p_v \qquad 37.20$$

$$k_o \approx 1 - \sin \phi \qquad 37.21$$

$$R_o = \tfrac{1}{2} k_o \rho g H^2 \qquad \text{[SI]} \qquad 37.22(a)$$

$$R_o = \tfrac{1}{2} k_o \gamma H^2 \qquad \text{[U.S.]} \qquad 37.22(b)$$

Table 37.2 *Typical Range of Earth Pressure Coefficients*

condition	granular soil	cohesive soil
active	0.20–0.33	0.25–0.5
passive	3–5	2–4
at rest	0.4–0.6	0.4–0.8

8. GRAPHICAL SOLUTIONS

Appendix 37.A provides a convenient graphical method of evaluating the horizontal and vertical earth pressures for various soil types. Since average values of soil density have been incorporated, K_h and K_v have units of lbf/ft^3 (lbf/ft^2 per lineal foot of wall). H is defined as shown: the height of the soil directly above the heel.

$$R_{a,h} = \tfrac{1}{2} K_h H^2 \qquad 37.23$$

$$R_{a,v} = \tfrac{1}{2} K_v H^2 \qquad 37.24$$

$$R_a = \sqrt{R_{a,h}^2 + R_{a,v}^2} \qquad 37.25$$

$$\theta = \arctan\left(\frac{R_v}{R_h}\right) \qquad 37.26$$

Most complications involving irregular and stratified backfill, irregular surcharges, and sloping water tables require more complex calculations. However, App. 37.B provides a way of evaluating retaining walls with *broken slope backfill*.

9. SURCHARGE LOADING

A *surcharge* is an additional force applied at the exposed upper surface of the restrained soil. A surcharge can result from a uniform load, point load, line load, or strip load. (Line and strip loads are parallel to the wall.)

With a *uniform load surcharge* of q (in lbf/ft^2 (N/m^2)) at the surface, there will be an additional active force, R_q. R_q acts horizontally at $H/2$ above the base. This surcharge resultant is in addition to the backfill active force that acts at $H/3$ above the base.

$$p_q = k_a q \qquad 37.27$$

$$R_q = k_a q H \times (\text{wall width}) \qquad 37.28$$

If a vertical *point load surcharge* (e.g., a truck wheel), V_q, in pounds (newtons) is applied a distance x back from the wall face, the approximate distribution of pressure behind the wall can be found from Eq. 37.29 or 37.30. These equations assume elastic soil performance and a Poisson's ratio of 0.5. The coefficients have been adjusted to bring the theoretical results into agreement with observed values.

$$p_q = \frac{1.77 V_q m^2 n^2}{H^2 (m^2 + n^2)^3} \qquad [m > 0.4] \qquad 37.29$$

$$p_q = \frac{0.28 V_q n^2}{H^2 (0.16 + n^2)^3} \qquad [m \leq 0.4] \qquad 37.30$$

$$R_q \approx \frac{0.78 V_q}{H} \qquad [m = 0.4] \qquad 37.31$$

$$R_q \approx \frac{0.60 V_q}{H} \qquad [m = 0.5] \qquad 37.32$$

$$R_q \approx \frac{0.46 V_q}{H} \qquad [m = 0.6] \qquad 37.33$$

$$m = \frac{x}{H} \qquad 37.34$$

$$n = \frac{y}{H} \qquad 37.35$$

For a *line load surcharge*, L_q (in lbf/ft (N/m)), the distribution of pressure behind the wall is given by Eqs. 37.36 and 37.38. m and n are as defined in Eqs. 37.34 and 37.35.

$$p_q = \frac{4 L_q m^2 n}{\pi H (m^2 + n^2)^2} \qquad [m > 0.4] \qquad 37.36$$

$$R_q = \frac{0.64 L_q}{m^2 + 1} \qquad 37.37$$

$$p_q = \frac{0.203 L_q n}{H (0.16 + n^2)^2} \qquad [m \leq 0.4] \qquad 37.38$$

$$R_q = 0.55 L_q \qquad 37.39$$

Sidewalks, railways, and roadways parallel to the retaining wall are examples of *strip surcharges*. The effect of a strip surcharge, S_q (in lbf/ft^2 (N/m^2)), is covered in most soils textbooks.

Figure 37.3 *Surcharges*

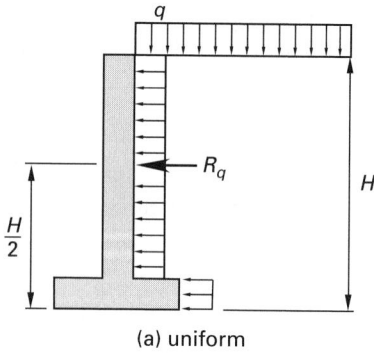

(a) uniform

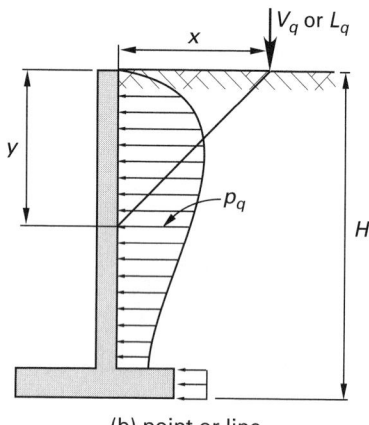

(b) point or line

10. EFFECTIVE STRESS

Rarely in a properly designed retaining wall is the *groundwater table* (GWT) above the base. However, with submerged construction or when drains become plugged, a significant water table can exist behind the wall. The *pore pressure* (i.e., the *hydrostatic pressure*), μ, behind the wall at a point h below the water table is

$$\mu = \rho_w g h \quad \text{[SI]} \quad \textit{37.40(a)}$$

$$\mu = \gamma_w h \quad \text{[U.S.]} \quad \textit{37.40(b)}$$

The soil pressure behind the wall depends on its saturated density. The saturated soil density can be calculated from the porosity, n, or void ratio, e.

$$\rho_{\text{sat}} = \rho_{\text{dry}} + n\rho_w$$

$$= \rho_{\text{dry}} + \left(\frac{e}{1+e}\right)\rho_w \quad \text{[SI]} \quad \textit{37.41(a)}$$

$$\gamma_{\text{sat}} = \gamma_{\text{dry}} + n\gamma_w$$

$$= \gamma_{\text{dry}} + \left(\frac{e}{1+e}\right)\gamma_w \quad \text{[U.S.]} \quad \textit{37.41(b)}$$

The water provides a buoyant effect, since each sand particle is submerged. The *effective pressure* (generally referred to as the *effective stress*) is the difference between the *total pressure* and the pore pressure.

$$p_v = g(\rho_{\text{sat}}H - \rho_w h) \quad \text{[SI]} \quad \textit{37.42(a)}$$

$$p_v = \gamma_{\text{sat}}H - \gamma_w h \quad \text{[U.S.]} \quad \textit{37.42(b)}$$

The total horizontal pressure from the submerged sand is the sum of the hydrostatic pressure and the lateral earth pressure considering the buoyant effect.

$$p_h = g\big(\rho_w h + k_a(\rho_{\text{sat}}H - \rho_w h)\big)$$

$$= g\big(k_a\rho_{\text{sat}}H + (1-k_a)\rho_w h\big) \quad \text{[SI]} \quad \textit{37.43(a)}$$

$$p_h = \gamma_w h + k_a(\gamma_{\text{sat}}H - \gamma_w h)$$

$$= k_a\gamma_{\text{sat}}H + (1-k_a)\gamma_w h \quad \text{[U.S.]} \quad \textit{37.43(b)}$$

The increase in horizontal pressure above the saturated condition is the *equivalent hydrostatic pressure* (*equivalent fluid pressure*) caused by the *equivalent fluid weight* (*equivalent fluid density, equivalent fluid specific weight,* etc.), γ_{eq}. Considering typical values of k_a, the equivalent specific weight of water behind a retaining wall is typically taken as 45 lbf/ft^3 (720 kg/m^3).

$$\rho_{\text{eq}} = (1-k_a)\rho_w \quad \text{[SI]} \quad \textit{37.44(a)}$$

$$\gamma_{\text{eq}} = (1-k_a)\gamma_w \quad \text{[U.S.]} \quad \textit{37.44(b)}$$

The effective pressure can be quite low due to submergence and tension cracking in cohesive clays. However, all retaining walls should be designed for a minimum effective density of 30 lbf/ft^3 (480 kg/m^3).

11. CANTILEVER RETAINING WALLS: ANALYSIS

Retaining walls must have sufficient resistance against overturning and sliding, and they must possess adequate structural strength against bending outward. The maximum soil pressure under the base must be less than the allowable soil pressure. The following procedure can be used to analyze a retaining wall whose dimensions are known.

step 1: Determine the horizontal active earth pressure resultants from the backfill and all surcharges, $R_{a,h,i}$. Determine the points of application and moment arms, $y_{a,h,i}$, above the base.

step 2: Unless it is reasonable to assume that they will always be present, restraint from passive distributions is disregarded. If it is to be considered, however, determine the passive earth pressure.

step 3: Find all of the vertical forces acting at the base. These include the weights of the retaining wall itself, the soil directly above the heel, and the soil directly above the toe. Each of these weights can be evaluated individually by dividing the concrete and soil into areas with simple geometric shapes. The specific weight of concrete is almost always taken as 150 lbf/ft^3 (2400 kg/m^3). Find the location of the centroid of each shape and the moment arm, x_i. Either the base of the heel or the base of the toe can be chosen as the pivot point. However, it is more common to use distances from the toe.

$$W_i = g\rho_i A_i \quad \text{[SI]} \quad \textit{37.45(a)}$$

$$W_i = \gamma_i A_i \quad \text{[U.S.]} \quad \textit{37.45(b)}$$

step 4: Find the net moment about the toe from the forces found in steps 1 through 3.

$$M_{\text{toe}} = \sum W_i x_i - R_{a,h} y_a + R_{a,v} x_a \quad \textit{37.46}$$

step 5: Determine the location, x_R, and eccentricity, ϵ, of the vertical force component. The eccentricity is the distance from the center of the base to the vertical force resultant. Eccentricity should be less than $B/6$ for the entire base to be in compression.

$$x_R = \frac{M_{\text{toe}}}{R_{a,v} + \sum W_i} \quad \textit{37.47}$$

$$\epsilon = \left| \frac{B}{2} - x_R \right| \quad \textit{37.48}$$

step 6: Check the factor of safety, F, against overturning. The factor of safety generally should be greater than 1.5 for granular soils and 2.0 for cohesive soils.

$$F_{\text{OT}} = \frac{M_{\text{resisting}}}{M_{\text{overturning}}}$$
$$= \frac{\sum W_i x_i + R_{a,v} x_{a,v}}{R_{a,h} y_{a,h}} \quad \textit{37.49}$$

step 7: Find the maximum pressure (at the toe) and the minimum pressure (at the heel) on the base. The maximum pressure should not exceed the allowable pressure.

$$p_{v,\text{max}}, \ p_{v,\text{min}} = \left(\frac{\sum W_i + R_{a,v}}{B} \right)$$
$$\times \left(1 \pm \left(\frac{6\epsilon}{B} \right) \right) \quad \textit{37.50}$$

step 8: Calculate the resistance against sliding. The active pressure is resisted by friction and adhesion between the base and the soil, and in the case of a keyed base, also by the shear strength of the soil. Equation 37.51 is for use when the wall has a key, and then only for the compressed soil in front of the key. Equation 37.52 is for use with a keyless base and for tensioned soil behind the key.

$$R_{\text{SL}} = (\Sigma W_i + R_{a,v}) \tan \phi + c_A B \quad \textit{37.51}$$

$$R_{\text{SL}} = (\Sigma W_i + R_{a,v}) \tan \delta + c_A B \quad \textit{37.52}$$

The adhesion, c_A, is zero for granular soil. $\tan \delta$ is referred to as a coefficient of friction and is given the symbol k or f by some authorities. In the absence of more sophisticated information, $\tan \delta$ is approximately 0.60 for rock, 0.55 for sand without silt, 0.45 for sand with silt, 0.35 for silt, and 0.30 for clay.

step 9: Calculate the factor of safety against sliding. A lower factor of safety (which may be referred to as the "minimum" factor of safety), 1.5, is permitted when the passive resultant is disregarded. If the passive resultant is included, the factor of safety (referred to as the "maximum" factor of safety) should be higher (e.g., 2). If the factor of safety is too low, the base length (B) can be increased, or a vertical key can be used.

$$F_{\text{SL}} = \frac{R_{\text{SL}}}{R_{a,h}} \quad \textit{37.53}$$

Figure 37.4 *Elements Contributing to Vertical Force^a (step 3)*

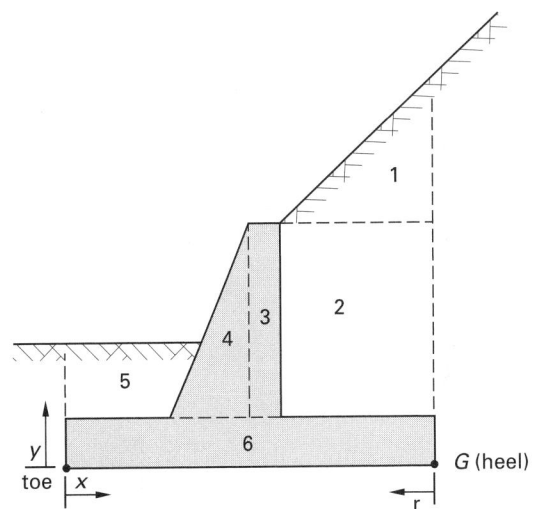

^{a}Some retaining walls may not have all elements.

Figure 37.5 *Resultant Distribution on the Base (step 7)*

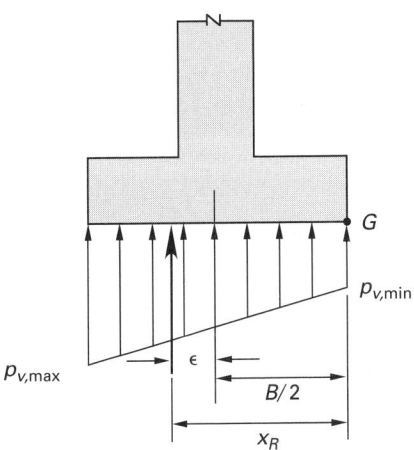

Example 37.1

The unkeyed retaining wall shown has been designed for a backfill of coarse-grained sand with silt having a weight of 125 lbf/ft³. The angle of internal friction, ϕ, is 30°, and the angle of external friction, δ, is 17°. The backfill is sloped as shown. The maximum allowable soil pressure is 3000 lbf/ft². The adhesion is 950 lbf/ft². Use a graphical solution to check the factor of safety against sliding.

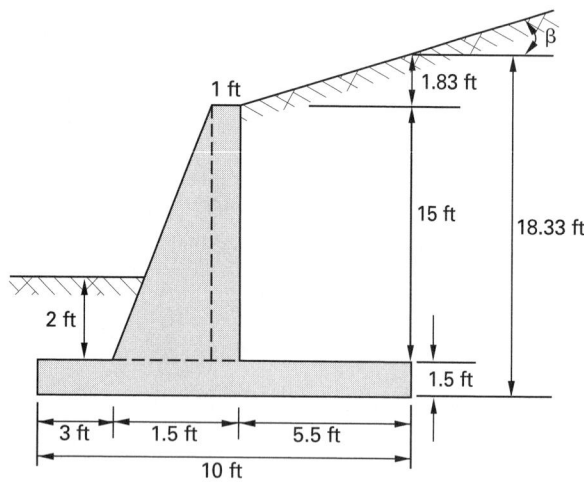

Solution

step 1: $\beta = \arctan\left(\dfrac{1.83 \text{ ft}}{5.5 \text{ ft}}\right) = \arctan(\frac{1}{3}) = 18.4°$

(H is defined as shown in App. 37.A.) From App. 37.A with a type-2 fill, $k_v = 10$ lbf/ft² and $k_h = 40$ lbf/ft². (Eq. 37.6 could also be used and R then broken into components.)

$$R_{a,v} = \tfrac{1}{2}k_v H^2 = \left(\tfrac{1}{2}\right)\left(10\ \frac{\text{lbf}}{\text{ft}^2}\right)(18.33 \text{ ft})^2$$
$$= 1680 \text{ lbf}$$

$$R_{a,h} = \tfrac{1}{2}k_h H^2 = \left(\tfrac{1}{2}\right)\left(40\ \frac{\text{lbf}}{\text{ft}^2}\right)(18.33 \text{ ft})^2$$
$$= 6720 \text{ lbf}$$

$R_{a,h}$ is located 18.33 ft/3 = 6.11 ft above the bottom of the base.

step 2: Disregard the passive earth pressure.

step 3: Calculate the weights of soil and concrete. Refer to Fig. 37.4.

(per foot of wall)

i	area (ft²)	γ (lbf/ft³)	W_i (lbf)	x_i (ft)	M_i (ft-lbf)
1	$(0.5)(5.5)(1.83) = 5.03$	125	629	8.17	5139
2	$(5.5)(15) = 82.5$	125	10,313	7.25	74,769
3	$(1)(15) = 15$	150	2250	4.0	9000
4	$(0.5)(0.5)(15) = 3.75$	150	563	3.33	1875
5	$(2)(3) = 6$	125	750	1.5	1125
6	$(1.5)(10) = 15$	150	2250	5.0	11,250
		totals	16,755		103,158

step 4: Find the moment about the toe. Notice that $R_{a,v}$ goes through the heel and has a moment arm equal to the base length. Use Eq. 37.46.

$$M_{\text{toe}} = \sum W_i x_i - R_{a,h} y_a + R_{a,v} x_a$$
$$= 103{,}158 \text{ ft-lbf} - (6720 \text{ lbf})(6.11 \text{ ft})$$
$$\quad + (1680 \text{ lbf})(10 \text{ ft})$$
$$= 78{,}899 \text{ ft-lbf}$$

step 5: Use Eq. 37.47.

$$x_R = \frac{M_{\text{toe}}}{R_{a,v} + \sum W_i}$$
$$= \frac{78{,}899 \text{ ft-lbf}}{1680 \text{ lbf} + 16{,}755 \text{ lbf}}$$
$$= 4.28 \text{ ft}$$

Use Eq. 37.48.

$$\epsilon = \frac{B}{2} - x_R = \frac{10 \text{ ft}}{2} - 4.28 \text{ ft} = 0.72 \text{ ft}$$

Since 0.72 ft < 10 ft/6, the base is in compression everywhere.

step 6: (This step skipped.)

step 7: $p_{v,\text{max}} = \left(\dfrac{\sum W_i + R_{a,v}}{B}\right)\left(1 + \dfrac{6\epsilon}{B}\right)$

$$= \left(\frac{16{,}755 \text{ lbf} + 1680 \text{ lbf}}{10 \text{ ft}}\right)$$
$$\times \left(1 + (6)\left(\frac{0.72 \text{ ft}}{10 \text{ ft}}\right)\right)$$
$$= 2640 \text{ lbf/ft}^2 \quad [< 3000 \text{ lbf/ft}^2, \text{ so OK}]$$

step 8: Use Eq. 37.52.

$$R_{\text{SL}} = \left(\sum W_i + R_{a,v}\right)\tan\delta + c_A B$$
$$= (16{,}755 \text{ lbf} + 1680 \text{ lbf})(\tan 17°)$$
$$\quad + \left(950\ \frac{\text{lbf}}{\text{ft}^2}\right)(10 \text{ ft})$$
$$= 15{,}136 \text{ lbf}$$

step 9: $F_{\text{SL}} = \dfrac{R_{\text{SL}}}{R_{a,h}} = \dfrac{15{,}136 \text{ lbf}}{6720 \text{ lbf}}$
$$= 2.25 \quad [> 1.5, \text{ so OK}]$$

12. RETAINING WALLS: DESIGN

A retaining wall is most likely to fail structurally at its base due to the applied moment. In this regard, a retaining wall is similar to a cantilever beam with a nonuniform load. The following characteristics are typical of retaining walls and can be used as starting points for more detailed designs.

- To prevent frost heaving, the top of the base should be below the frost line. This establishes a minimum wall height, H.

- The heel length, L, can be determined analytically by setting the active soil moment equal to the "resisting moment" due to the weight of the soil above the heel.

$$R_{a,h}\left(\frac{H}{3}\right) \approx (\text{soil weight})\left(\frac{L}{2}\right) \qquad 37.54$$

- The toe should project approximately $B/3$ beyond the stem face.

$$B \approx \tfrac{3}{2}L \qquad 37.55$$

- If an analytical method is not used to determine its size, the base should be proportioned such that

$B = 0.4H$ [granular backfills and nominal surcharges]

$B = 0.5H$ [granular backfills and heavier surcharges]

$B = 0.6H$ [cohesive backfills and nominal surcharges]

$B = 0.7H$ [cohesive backfills and heavier surcharges]

- The stem thickness, t_{stem}, should be proportioned such that

$$8 < \frac{H}{t_{\text{stem}}} < 12 \qquad 37.56$$

- The base thickness should be approximately equal to the thickness of the stem at the base, with a minimum of 12 in (300 mm). Alternatively, the base should be proportioned such that

$$10 < \frac{H}{t_{\text{base}}} < 14 \qquad 37.57$$

- The *batter decrement* should be approximately $^1/_4$ in to $^1/_2$ in per vertical foot.

- The minimum stem thickness at the top should be approximately 12 in (300 mm).

Geotechnical

38 Piles and Deep Foundations

Nomenclature

A	area	ft^2	m^2
b	group width	ft	m
B	diameter or width	ft	m
c	undrained shear strength (cohesion)	lbf/ft^2	Pa
c_A	adhesion	lbf/ft^2	Pa
D	depth	ft	m
E	energy per blow	ft-lbf	J
f	friction coefficient	–	–
F	factor of safety	–	–
g	acceleration of gravity	ft/sec^2	m/s^2
h	depth below the water table	ft	m
H	height	ft	m
k	coefficient of lateral earth pressure	–	–
L	length	ft	m
N	capacity factor	–	–
p	perimeter	ft	m
S	average penetration per blow	ft	m
w	group length	ft	m
W	weight	lbf	–

Symbols

α	adhesion factor	–	–
β	effective stress factor	–	–
γ	specific weight	lbf/ft^3	–
δ	coefficient of external friction	–	–
η	efficiency	–	–
μ	pore pressure	lbf/ft^2	Pa
ρ	density	lbm/ft^3	kg/m^3
σ	earth pressure	–	–
ϕ	coefficient of internal friction	–	–

Subscripts

a	allowable
c	critical
e	effective
f	friction or footing
G	group
h	horizontal
p	pile tip
s	skin
ult	ultimate
v	vertical
w	water

1. INTRODUCTION

Piles are slender members that are hammered, drilled, or jetted into the ground. They provide strength in soils that would otherwise be too weak to support a foundation. Piles may be constructed of timber, steel, or prestressed reinforced concrete. *Composite piles* include any combination of timber, steel, concrete, and fiberglass (e.g., a concrete-filled steel or concrete-filled fiberglass casing). Other deep foundations include micropiles, auger-cast piles, I-piles, and H-piles (i.e., steel piles with H-shaped cross sections).

Friction piles derive the majority of their loadbearing ability from the skin friction between the soil and the pile. *Point-bearing piles* derive their loadbearing ability from the support of the layer at the tip. (Point-bearing piles are used to transfer loads to rock or firm layers below.) Skin friction capacity and point-bearing capacity are simultaneously present to some degree in all piles. However, one mode is usually dominant.

The capacity of a pile to support static loadings is easily calculated. The capacity of a pile exposed to dynamic loads from earthquakes, explosions, and other vibrations is relevant in many cases, but dynamic capacity is more difficult to determine. All calculations in this chapter are for static capacity.

The *ultimate static bearing capacity*, Q_{ult}, of a single pile is the sum of its point-bearing and skin-friction capacities. For piles supporting a compressive load, the pile weight is balanced by the overburden and is not considered. Other than its effect on the undrained shear strength, the position of the groundwater table has no effect on the ultimate bearing capacity.

$$Q_{\text{ult}} = Q_p + Q_f \qquad 38.1$$

The allowable capacity of a pile depends on the factor of safety, which is typically 2 to 3 for both compression and tension piles, the lower value of 2 being used when the capacity can be verified by pile loading tests. Even lower values can be specified in unusual and extreme cases.

$$Q_a = \frac{Q_{ult}}{F} \qquad 38.2$$

Pile capacities do not consider settlement, which might actually be the controlling factor.

2. PILE CAPACITY FROM DRIVING DATA

The *safe load* (*safe bearing value*) can be calculated empirically from installation data using the *Engineer News Record* (*ENR* or *Engineering News*) equations. For a pile driven in by a drop hammer of weight W_{hammer}, falling a distance H, driving a weight of W_{driven} (including the pile weight), and penetrating an average distance S per blow during the last five blows, the maximum allowable vertical load per pile, Q_a, referred to as *design capacity* and *pile resistance*, is calculated as $1/6$ of the ultimate strength and, incorporating the conversion from feet to inches, is

$$Q_{a,lbf} = \frac{Q_{ult}}{FS} = \frac{2W_{hammer,lbf}H_{fall,ft}}{S_{in} + 1} \quad \left[\begin{array}{c}\text{drop} \\ \text{hammer}\end{array}\right] \quad 38.3$$

$$= \frac{2W_{hammer,lbf}H_{fall,ft}}{S_{in} + 0.1}$$

$$\left[\begin{array}{c}\text{single-acting steam hammer;} \\ \text{driven weight} < \text{striking weight}\end{array}\right] \quad 38.4$$

$$Q_{a,lbf} = \frac{2W_{hammer,lbf}H_{fall,ft}}{S_{in} + 0.1\left(\dfrac{W_{driven}}{W_{hammer}}\right)}$$

$$\left[\begin{array}{c}\text{single-acting steam hammer;} \\ \text{driven weight} > \text{striking weight}\end{array}\right] \quad 38.5$$

For double-acting (i.e., powered) hammers, the energy, E_{ft-lbf}, transferred to the pile on each stroke replaces the WH potential energy term. Typical values of energy per blow for timber piles are 7500 to 14,000 ft-lbf (10 to 19 kJ) for single-acting hammers and 14,000 ft-lbf (19 kJ) for double-acting hammers. For concrete and steel piles, the typical energy per blow is 15,000 to 20,000 ft-lbf (20 to 27 kJ).

$$Q_{a,lbf} = \frac{2E_{ft-lbf}}{S_{in} + 0.1}$$

$$\left[\begin{array}{c}\text{double-acting steam hammer;} \\ \text{driven weight} < \text{striking weight}\end{array}\right] \quad 38.6$$

$$Q_{a,lbf} = \frac{2E_{ft-lbf}}{S_{in} + 0.1\left(\dfrac{W_{driven}}{W_{hammer}}\right)}$$

$$\left[\begin{array}{c}\text{double-acting steam hammer;} \\ \text{driven weight} > \text{striking weight}\end{array}\right] \quad 38.7$$

Correlations with standard penetration tests (i.e., those that derive the N-value) have also been established and can be used with adequate testing.

3. THEORETICAL POINT-BEARING CAPACITY

The theoretical point-bearing capacity (also known as the *tip resistance* and *point capacity*), Q_p, of a single pile can be calculated in much the same manner as for a footing. Since the pile size, B, is small, the $1/2\gamma BN\gamma$ term is generally omitted.

$$Q_p = A_p\left(\tfrac{1}{2}\rho gBN_\gamma + cN_c + \rho gD_fN_q\right) \quad \text{[SI]} \quad 38.8(a)$$

$$Q_p = A_p\left(\tfrac{1}{2}\gamma BN_\gamma + cN_c + \gamma D_fN_q\right) \quad \text{[U.S.]} \quad 38.8(b)$$

For cohesionless (granular) soil, $c = 0$. For sands and silts, the tip point-bearing capacity increases down to a *critical depth*, D_c, after which it is essentially constant. For loose sands (i.e., with relative densities less than 30%), the critical depth is taken as $10B$. For dense sands (relative densities above 70%), the critical depth is $20B$. Between relative densities of 30% and 70%, the critical depth can be interpolated between $10B$ and $20B$.

$$Q_p = A_p\rho gDN_q \quad \text{[cohesionless; } D \le D_c] \quad \text{[SI]} \quad 38.9(a)$$

$$Q_p = A_p\gamma DN_q \quad \text{[cohesionless; } D \le D_c] \quad \text{[U.S.]} \quad 38.9(b)$$

For cohesive soils with $\phi = 0°$, $N_q = 1$ and the γD_f term is approximately cancelled by the pile weight. $N_c = 9$ for driven piles of virtually all conventional dimensions.

$$Q_p = A_pcN_c \approx 9A_pc \quad \text{[cohesive]} \quad 38.10$$

Table 38.1 *Meyerhof Values of N_q For Piles*[a]

ϕ	20°	25°	28°	30°	32°	34°	36°	38°	40°	42°	45°
driven	8	12	20	25	35	45	60	80	120	160	230
drilled	4	5	8	12	17	22	30	40	60	80	115

[a]Significant variation in N_q has been reported by various researchers. The actual value is highly dependent on ϕ and the installation method.

Table 38.2 *Values of N_c for Driven Piles*[a]
(square and cylindrical perimeters)

$\dfrac{D_f}{B}$	N_c
0	6.3
1	7.8
2	8.5
≥ 4	9

[a]Not for use with drilled piles.

Derived from *Foundations and Earth Structures Design Manual*, NAVFAC, DM 7.2, 1982, Fig. 2, p. 7.2-196.

4. THEORETICAL SKIN-FRICTION CAPACITY

The skin-friction capacity, Q_f (also known as *side resistance, skin resistance,* and *shaft capacity*), is given by Eq. 38.11. L_e is the *effective pile length,* which can be estimated as the pile length less the depth of the *seasonal variation,* if any.

$$Q_f = A_s f_s = p f_s L_e = p f_s (L - \text{seasonal variation})$$
$$\textbf{38.11}$$

For piles passing through two or more layers, the skin friction capacity of each layer is calculated separately, considering only the thickness, L_i, of one layer at a time. Individual layer capacities are summed. Alternatively, a weighted average value of the skin friction coefficient can be calculated by weighting the individual coefficients by the layer thicknesses.

$$Q_f = p \sum f_{s,i} L_{e,i} \qquad \textbf{38.12}$$

The *skin friction coefficient* (*unit shaft friction* or *side friction factor*), f_s, includes both cohesive and adhesive components. The external friction angle, δ, is zero (as is ϕ) for saturated clay, and is generally taken as $^2/_3\phi$ in the absence of specific information. The *adhesion,* c_A, should be obtained from testing, or it can be estimated from the undrained shear strength (cohesion), c. For rough concrete, rusty steel, and corrugated metal, $c_A = c$. For smooth concrete, $0.8c \le c_A \le c$. For clean steel, $0.5c \le c_A \le 0.9c$.

$$f_s = c_A + \sigma_h \tan \delta \qquad \textbf{38.13}$$

The α-*method* determines the *adhesion factor,* α, as the ratio of the skin friction factor, f_s, to the undrained shear strength (cohesion), c. As in Table 38.4, the values of α are correlated with c.

$$f_s = \alpha c \qquad \textbf{38.14}$$

For sand, the lateral earth pressure, σ_h, depends on the depth, down to a critical depth, D_c, after which it is essentially constant. Below a depth of D_c, skin friction is directly proportional to the length. For loose sands (i.e., relative densities less than 30%), the critical depth is taken as $10B$. For dense sands (relative densities above 70%), the critical depth is $20B$. Between relative densities of 30% and 70%, the critical depth can be interpolated between $10B$ and $20B$.

$$\sigma_h = k_s \sigma'_v = k_s (\rho g D - \mu) \qquad \text{[SI]} \qquad \textbf{38.15(a)}$$
$$\sigma_h = k_s \sigma'_v = k_s (\gamma D - \mu) \qquad \text{[U.S.]} \qquad \textbf{38.15(b)}$$

For compression piles driven in clay and silt, the *coefficient of lateral earth pressure,* k_s, is 1.0. For piles driven in sand, the coefficient varies from 1.0 to 2.0, with the higher value applicable to the most dense sands. For drilled piles, the coefficients of lateral earth pressure are approximately half of the values for driven piles.

For jetted piles, the coefficients are approximately 25% of the driven values.

The *pore pressure,* μ, a distance h below the water table is the hydrostatic pressure at that depth.

$$\mu = \rho_w g h \qquad \text{[SI]} \qquad \textbf{38.16(a)}$$
$$\mu = \gamma_w h \qquad \text{[U.S.]} \qquad \textbf{38.16(b)}$$

For piles through layers of cohesionless sand, $c = c_A = 0$.

$$Q_f = p k_s \tan \delta \sum L_i \sigma' \quad \text{[cohesionless]} \qquad \textbf{38.17}$$

For piles in clay, $\tan \delta = 0$, $f_s = c_A$, and $k_s = 1$.

$$Q_f = p \sum c_A L_i \quad \text{[cohesive]} \qquad \textbf{38.18}$$

With the β-*method* for cohesive clay, the friction capacity is estimated as a fraction of the average effective vertical stress (as evaluated halfway down the pile). Values of β are correlated with pile length (see Table 38.5). For a pile passing through one layer, the friction capacity is

$$Q_f = p \beta \sigma' L \quad \text{[cohesive]} \qquad \textbf{38.19}$$

Figure 38.1 *Lateral Earth Pressure Distribution on a Pile in Sand*

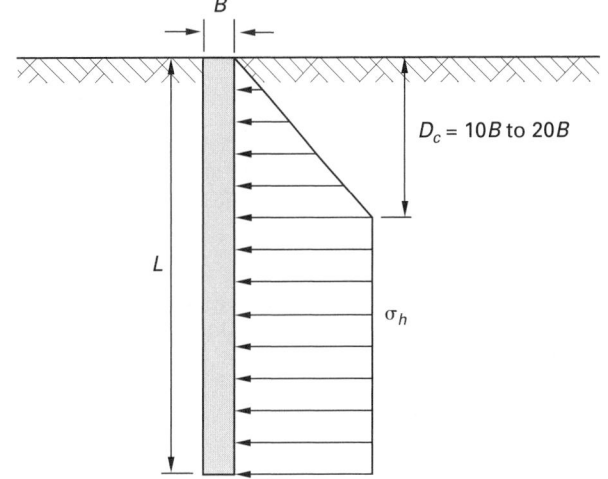

5. H-PILES

Analysis of steel H-pile capacity is similar to that of other piles except for the perimeter calculation and friction angle. The area between the flanges is assumed to fill with soil that moves with the pile. In calculating the skin area, A_s, the perimeter should be taken as the block perimeter of the pile. The friction angle should be the average of the soil-on-steel and soil-on-soil cases.

Similarly, the tip capacity should be calculated using the area enclosed by the block perimeter.

Example 38.1

A round smooth concrete pile with a diameter of 11 in is driven 60 ft into wetlands clay. The clay's undrained shear strength and specific weight are 1400 lbf/ft² and

Table 38.3 *Recommended Values of Adhesion*
(piles in clay)

pile type	consistency of clay	cohesion, c lbf/ft^2 (kPa)	adhesion, c_A lbf/ft^2 (kPa)
timber or concrete	very soft	0–250 (0–12)	0–250 (0–12)
	soft	250–500 (12–24)	250–480 (12–23)
	medium stiff	500–1000 (24–48)	480–750 (23–36)
	stiff	1000–2000 (48–96)	750–950 (36–45)
	very stiff	2000–4000 (96–192)	950–1300 (45–62)
steel	very soft	0–250 (0–12)	0–250 (0–12)
	soft	250–500 (12–24)	250–460 (12–22)
	medium stiff	500–1000 (24–48)	460–700 (22–34)
	stiff	1000–2000 (48–96)	700–720 (34–34.5)
	very stiff	2000–4000 (96–192)	720–750 (34.5–36)

(Multiply lbf/ft^2 by 0.04788 to obtain kPa.)
Source: *Foundations and Earth Structures Design Manual*, NAVFAC, DM 7.2, 1982, Fig. 2, p. 7.2-196.

Table 38.4 *Typical Values[a] of the Adhesion Factor, α*
(for use with Eq. 38.14)

cohesion, c lbf/ft^2 (kPa)	α range of values	average
500 (24)	–	1.0
1000 (48)	0.56–0.96	0.83
2000 (96)	0.34–0.83	0.56
3000 (144)	0.26–0.78	0.43

(Multiply lbf/ft^2 by 0.04788 to obtain kPa.)
[a]Reported values vary widely.

Table 38.5 *Typical Values of β [a]*
(for use with Eq. 38.19)

pile length, L ft (m)	skin friction factor, β
0 (0)	0.3
25 (7.5)	0.3
50 (15)	0.3
75 (23)	0.27
100 (30)	0.23
125 (38)	0.20
150 (45)	0.18
175 (53)	0.17
200 (60)	0.16

(Multiply ft by 0.3048 to obtain m.)
[a]For driven piles in soft and medium clays with $c < 2000$ lbf/ft^2 (96 GPa). After Meyerhof.

120 lbf/ft^3, respectively. At high tide, the water table extends to the ground surface. What is the allowable bearing capacity of the pile?

Solution

$$B = \frac{11 \text{ in}}{12 \ \dfrac{\text{in}}{\text{ft}}} = 0.917 \text{ ft}$$

The pile's end and surface areas are

$$A_p = \tfrac{\pi}{4}B^2 = \left(\frac{\pi}{4}\right)(0.917 \text{ ft})^2 = 0.66 \text{ ft}^2$$

$$A_s = \pi BL = \pi(0.917 \text{ ft})(60 \text{ ft}) = 172.9 \text{ ft}^2$$

The tip capacity is given by Eq. 38.10.

$$Q_p = A_p c N_c = 9 A_p c$$

$$= \frac{(9)(0.66 \text{ ft}^2)\left(1400 \ \dfrac{\text{lbf}}{\text{ft}^2}\right)}{1000 \ \dfrac{\text{lbf}}{\text{kip}}}$$

$$= 8.3 \text{ kips}$$

Estimate $c_A = 0.8c$ for smooth concrete. The friction capacity is given by Eq. 38.11.

$$Q_f = A_s f_s = A_s c_A$$

$$= \frac{(172.9 \text{ ft}^2)(0.8)\left(1400 \ \dfrac{\text{lbf}}{\text{ft}^2}\right)}{1000 \ \dfrac{\text{lbf}}{\text{kip}}}$$

$$= 193.6 \text{ kips}$$

The total capacity is

$$Q_{\text{ult}} = Q_p + Q_f = 8.3 \text{ kips} + 193.6 \text{ kips} = 201.9 \text{ kips}$$

With a factor of safety of 3, the allowable load is

$$Q_a = \frac{Q_{\text{ult}}}{F} = \frac{201.9 \text{ kips}}{3} = 67.3 \text{ kips}$$

6. TENSILE CAPACITY

Tension piles are intended to resist upward forces. Basements and buried tanks below the water level may require tension piles to prevent "floating away." However, tall buildings subjected to overturning moments also need to resist pile pull-out.

Unlike piles loaded in compression, the *pull-out capacity* of piles does not include the tip capacity. The pullout capacity includes the weight of the pile and the shaft resistance (skin friction). Tensile shaft resistance is calculated similarly to, but not necessarily the same as, compressive shaft resistance. For example, the lateral earth pressure coefficient, k_s, is approximately 50 to 75% of the equivalent values in compression.

7. CAPACITY OF PILE GROUPS

Some piles are installed in groups, spaced approximately 3 to 3.5 times the pile diameter apart. The piles function as a group due to the use of a concrete load-transfer cap encasing all of the pile heads. The weight of the cap subtracts from the gross group capacity. The capacity due to the pile cap resting on the ground (as a spread footing) is disregarded.

For cohesionless (granular) soils, the capacity of a pile group is taken as the sum of the individual capacities, although the actual capacity will be greater. In-situ tests should be used to justify any increases.

For cohesive soils, the group capacity is taken as the smaller of (a) the sum of the individual capacities and (b) the capacity assuming block action. The block action capacity is calculated assuming that the piles form a large pier whose dimensions are the group's perimeter. The block depth, L, is the distance from the surface to the depth of the pile points. The width, w, and length, b, of the pier are the length and width of the pile group as measured from the outsides (not centers) of the outermost piles.

The average undrained shear strength, c_1, along the depth of the piles is used to calculate the skin friction capacity. The undrained shear strength at the pile tips, c_2, is used to calculate the end-bearing capacity. A factor of safety of 3 is used to determine the allowable capacity. Any increase in friction capacity due to lateral soil pressure is disregarded.

$$Q_s = 2(b+w)L_e c_1 \qquad 38.20$$
$$Q_p = 9c_2 bw \qquad 38.21$$
$$Q_{\text{ult}} = Q_s + Q_p \qquad 38.22$$
$$Q_a = \frac{Q_{\text{ult}}}{F} \qquad 38.23$$

The group capacity can be more or less than the sum of the individual pile capacities. The *pile group efficiency*, η_G, is

$$\eta_G = \frac{\text{group capacity}}{\sum \text{individual capacities}} \qquad 38.24$$

For pile groups in clay, the efficiency is typically less than 100% for spacings up to approximately $8B$, above which an efficiency of 100% is maintained for all reasonable pile spacings. For pile groups in sand, efficiencies are similar, although some researchers report efficiencies higher than 100% at very close (e.g., $2B$) spacings. Efficiencies higher than 100% are rarely counted on in designs.

Figure 38.2 Pile Group

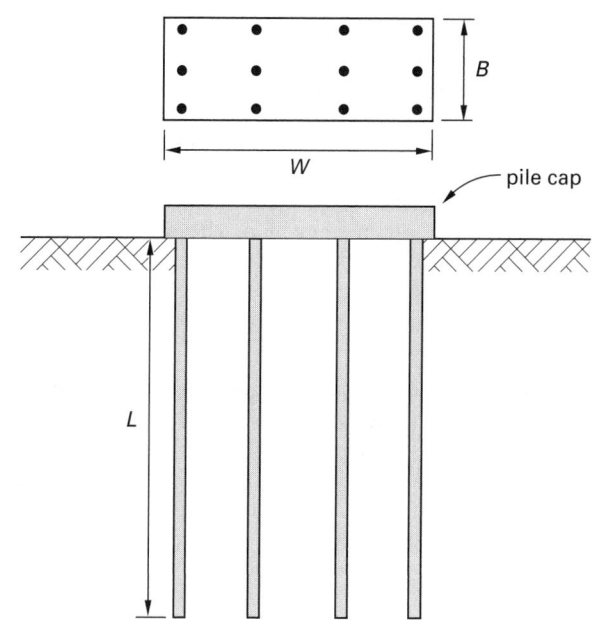

8. SETTLEMENT OF PILES AND PILE GROUPS

Piles bearing on rock essentially do not settle. Piles in sand experience minimal settlement. There are few theoretical or empirical methods of evaluating settlement of piles in sand.

Piles in clay may experience significant settling. The settlement of a pile group can be estimated by assuming that the support block (used to calculate the group capacity) extends to a depth of only two thirds of the pile length. Settlement above $^2/_3 L$ is assumed to be negligible. Below the $^2/_3 L$ depth, the pressure distribution spreads out at a vertical:horizontal rate of 2:1 (i.e., at an angle of approximately 60° from the horizontal). The consolidation of the layers encountering the pressure distribution is the pile group settlement. The presence of the lower $L/3$ pile length is disregarded.

Geotechnical

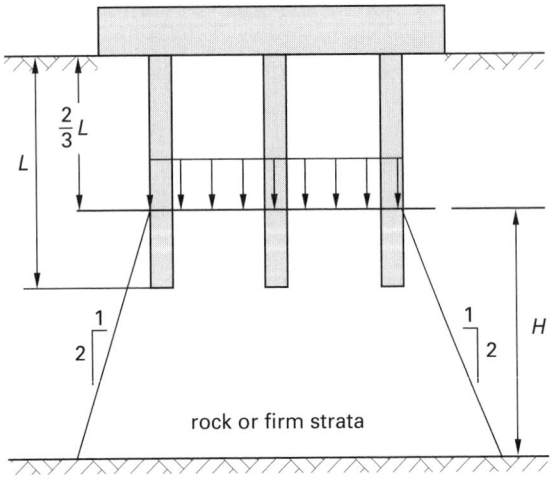

Figure 38.3 Settlement of Pile Groups

9. DOWNDRAG AND ADFREEZE FORCES

Downdrag occurs when the skin friction force is in the same direction as the axial load. *Negative skin friction* occurs when the settlement of the surrounding soil exceeds the downward movement of the pile shaft. This can occur when a pile passes through an underconsolidated layer of fill that is consolidating under its own weight, but can also be due to a lowering of the water table or placement of surcharges.

Pile support is also affected by the freeze-thaw process. *Frost jacking* occurs when the frozen ground moves upward (*frost heave*) and takes a pile with it. The forces from this process are known as *adfreeze forces*. When the soil subsequently thaws in the summer, a downdrag force will be experienced. Thus, a pile may experience upward adfreeze forces each winter and downdrag forces each summer. The effects of such forces are particularly pronounced in piles penetrating permafrost.

10. MICROPILES

Machine-drilled micropiles (*small-diameter grouted piles, minipiles, pin piles, root piles,* and *needle piles*) are piles with diameters 4 to 10 in (100 to 250 mm) in diameter, 65 to 100 ft (20 to 30 m) long. Significant column strength is obtained from thick wall sections, usually occupying 50% of the cross-sectional area.

Micropiles are installed using rotary drilling techniques. (The micropile can serve as its own drill bit.) After installation, grout is used to fill the interior of the pile, as well as the tip and any annular voiding. Micropiles are used when traditional pile driving is prevented by restricted access. (This is usually the case in urban areas and in seismic retrofits.)

11. PIERS

A *pier* is a deep foundation with a significant cross-sectional area. A pier differs from a pile in its diameter (larger), load-carrying capacity (much larger), and installation method (usually cased within an excavation). In other respects, the analysis of pier foundations is similar to that of piles.

39

Excavations

Nomenclature

B	width of cut	ft	m
c	undrained shear strength (cohesion)	lbf/ft^2	Pa
D	depth of embedment	ft	m
f	actual stress	lbf/ft^2	Pa
F	factor of safety	–	–
F	force	lbf	N
F_b	allowable bending stress	lbf/ft^2	Pa
g	acceleration of gravity	ft/sec^2	m/s^2
H	height	ft	m
k	coefficient of earth pressure	–	–
L	distance	ft	m
M	moment	ft-lbf	N·m
N	capacity factor	–	–
N_o	stability number	–	–
p	pressure	lbf/ft^2	Pa
q	uniform surcharge	lbf/ft^2	Pa
S	section modulus	ft^3	m^3
w	uniform loading	lbf/ft	N/m
y	vertical distance	ft	m

Symbols

γ	specific weight	lbf/ft^3	–
ρ	density	lbm/ft^3	kg/m^3
ϕ	angle of internal friction	deg	deg

Subscripts

a	allowable or active
b	bending
c	critical
p	passive
v	vertical

1. EXCAVATION[1]

Excavation is the removal of soil to allow construction of foundations and other permanent features below the finished level of the grade. It is usually done with machinery, although small areas may be excavated by hand. When a relatively narrow, long excavation is done for piping or for narrow footings and foundation walls, it is called *trenching*.

For large excavations, excess soil has to be removed from the site. However, to minimize cost, it is best to use the soil elsewhere on the site for backfill or in contour modification.

Because excavations can pose a hazard to workers, unshored sides of soil should be no steeper than their natural angle of repose or not greater than a slope of $1^1/_2$ horizontal to 1 vertical.

For shallow excavations in open areas, the sides of the excavation can be sloped without the need for some supporting structure. However, if the depth increases or the excavation walls need to be vertical in confined locations, temporary support is required. Shoring and bracing are used to temporarily support adjacent buildings and other construction with posts, timbers, and beams when excavation is proceeding and to temporarily support the sides of an excavation.

2. BRACED CUTS

Bracing is used when temporary trenches for water, sanitary, and other lines are opened in soil. A *braced cut* is an excavation in which the active earth pressure from one bulkhead is used to support the facing bulkhead. The bulkhead members in contact with the soil are known as sheeting, sheathing, sheet piling, lagging, and, rarely, poling. The *box-shoring* and *close-sheeting* methods of support are shown in Fig. 39.1. In box shoring, each of the upright-strut units is known as a *set*. (A partial support system known as *skeleton shoring* is not shown.) Deeper braced cuts may be constructed with vertical *soldier piles* (typically steel H-sections) with horizontal timber sheeting members between the soldiers.

The load is transferred to the struts at various points, so the triangular active pressure distribution does not develop. Since struts are installed as the excavation goes

---

[1]Chapter 39 Secs. 1 and 14 are reprinted with permission from *Standard Handbook for Civil Engineers*, 4th edition, by Frederick Merritt, © 1996 by The McGraw-Hill Companies.

Figure 39.1 *Shoring of Braced Cuts*

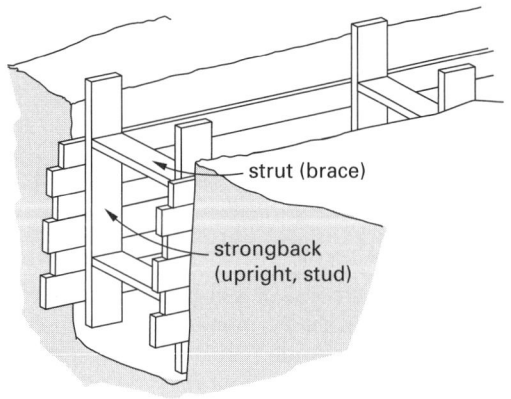

(a) box shoring

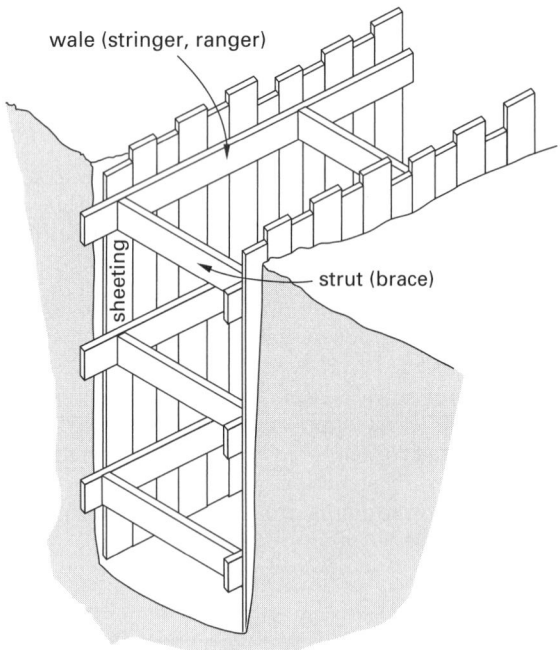

(b) close sheeting

down, the upper part of the wall deflects very little due to the strut restraint. The pressure on the upper part of the wall is considerably higher than is predicted by the active earth pressure equations. Failure in soils above the water table generally occurs by wale (stringer) crippling followed by strut buckling.

The soil removed from the excavation is known as the *spoils*. Spoils should be placed far enough from the edge of the cut so that they do not produce a surcharge lateral loading.

The bottom of the excavation is referred to as the *base of the cut*, *mud line* (or *mudline*), *dredge line*, and *toe of the excavation*.

Excavations below the water table should be dewatered prior to cutting.

3. BRACED CUTS IN SAND

The analysis of braced cuts is approximate due to the extensive bending of the sheeting. For drained sand, the pressure distribution is approximately uniform with depth. (The *Tschebotarioff trapezoidal pressure distribution*, Fig. 39.2(b), has also been proposed for sand.) The maximum lateral pressure is given by Eq. 39.1. (Chapter 37 presents the coefficient of earth pressure, k_a.)

$$p_{max} = 0.65 k_a \rho g H \qquad \text{[SI]} \qquad 39.1(a)$$

$$p_{max} = 0.65 k_a \gamma H \qquad \text{[U.S.]} \qquad 39.1(b)$$

Figure 39.2 *Cuts in Sand*

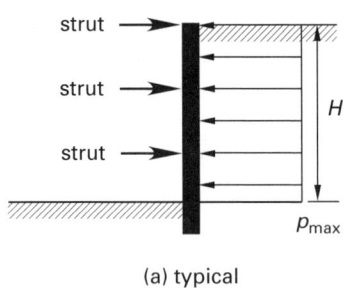

(a) typical

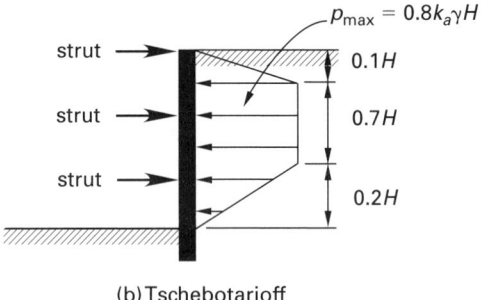

(b) Tschebotarioff

4. BRACED CUTS IN STIFF CLAY

For undrained clay (typical of cuts made rapidly in comparison to drainage times), $\phi = 0°$. In this case, the lateral pressure distribution depends on the average undrained shear strength (cohesion) of the clay. If $\gamma H/c \leq 4$, the clay is stiff and the pressure distribution is as given in Fig. 39.3. (The quantity $\gamma H/c$ is known as the *stability number* and is sometimes given the symbol N_o. More information on this is available in Ch. 40 Sec. 10.) The value of p_{max} is affected by many variables. Use the lower range of values of p_{max} in Eq. 39.2 when movement is minimal or when the construction period is short.

$$0.2 \rho g H \leq p_{max} \leq 0.4 \rho g H \qquad \text{[SI]} \qquad 39.2(a)$$

$$0.2 \gamma H \leq p_{max} \leq 0.4 \gamma H \qquad \text{[U.S.]} \qquad 39.2(b)$$

Except when the cut is underlain by deep, soft, normally consolidated clay, the maximum pressure can be estimated as

$$p_{\max} = k_a \rho g H \qquad [\text{SI}] \qquad \textit{39.3(a)}$$

$$p_{\max} = k_a \gamma H \qquad [\text{U.S.}] \qquad \textit{39.3(b)}$$

$$k_a = 1 - \frac{4c}{\rho g H} \qquad [\text{SI}] \qquad \textit{39.4(a)}$$

$$k_a = 1 - \frac{4c}{\gamma H} \qquad [\text{U.S.}] \qquad \textit{39.4(b)}$$

Figure 39.3 *Cuts in Stiff Clay*

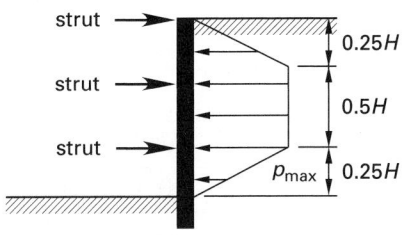

5. BRACED CUTS IN SOFT CLAY

If $\gamma H/c \geq 6$, the clay is soft and the lateral pressure distribution will be as shown in Fig. 39.4. Except for cuts underlain by deep, soft, normally consolidated clays, the maximum pressure is

$$p_{\max} = \rho g H - 4c \qquad [\text{SI}] \qquad \textit{39.5(a)}$$

$$p_{\max} = \gamma H - 4c \qquad [\text{U.S.}] \qquad \textit{39.5(b)}$$

If $6 \leq \gamma H/c \leq 8$, the bearing capacity of the soil is probably sufficient to prevent shearing and upward heave. Simple braced cuts should not be attempted if $\gamma H/c > 8$.

Figure 39.4 *Cuts in Soft Clay*

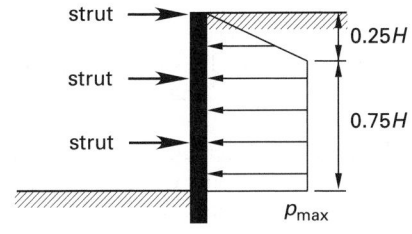

6. BRACED CUTS IN MEDIUM CLAY

If $4 < \gamma H/c < 6$, the soft and stiff clay cases should both be evaluated. The case that results in the greater pressure should be used when designing the bracing.

7. ANALYSIS/DESIGN OF BRACED EXCAVATIONS

Since braced excavations with more than one strut are statically indeterminate, strut forces and sheet piling moments may be evaluated by assuming continuous

beam action or hinged beam action (i.e., with hinges at the strut points). In the absence of a true indeterminate analysis, the axial strut compression can be determined by making several strategic assumptions.

A hinge (point of zero moment) is assumed to exist at the second (from the top) strut support point. If moments above that point are taken, a force in the first strut will be derived. Next, a hinge is assumed at the third (from the top) strut support point and moments are taken about that point (including moments from the top strut, now known), giving the force in the second strut. The process is repeated for all lower struts, finishing by assuming the existence of a fictitious support at the base of the cut in order to determine the force in the last real strut. The pressure distribution on any embedment below the base of the cut is disregarded.

The required section modulus of the sheet piling is given by Eq. 39.6. Typical values of F_b are given in Table 39.2.

$$S = \frac{M_{\max,\text{sheet piling}}}{F_b} \qquad \textit{39.6}$$

Struts are selected as column members in order to support the axial loads with acceptable slenderness ratios. Wales are selected as uniform-loaded beams, although cofferdam wales are simultaneously loaded in compression and are designed as beam-columns.

8. STABILITY OF BRACED EXCAVATIONS IN CLAY

Heave may occur at the bottom of cuts in soft clay. The depth of excavation at which a heaving failure can be expected to occur is referred to as the *critical height of excavation*, H_c. B is the width of the excavation. In Eqs. 39.7 and 39.8, it is assumed that there is sufficient distance between the base of the cut and any hard stratum below to allow the heave failure surface to fully form. Friction and adhesion on the back of the sheeting are disregarded.

$$H_c = \frac{5.7c}{\rho g - \sqrt{2}\left(\dfrac{c}{B}\right)} \qquad [H < B] \quad [\text{SI}] \qquad \textit{39.7(a)}$$

$$H_c = \frac{5.7c}{\gamma - \sqrt{2}\left(\dfrac{c}{B}\right)} \qquad [H < B] \quad [\text{U.S.}] \qquad \textit{39.7(b)}$$

$$H_c = \frac{N_c c}{\rho g} \qquad [H > B] \qquad [\text{SI}] \qquad \textit{39.8(a)}$$

$$H_c = \frac{N_c c}{\gamma} \qquad [H > B] \qquad [\text{U.S.}] \qquad \textit{39.8(b)}$$

Approximate values (originally compiled by Alec Westley Skempton) of the bearing capacity factor N_c for infinitely long trenches in clay are given in Table 39.1. Different values are necessary for circular, square, and rectangular cofferdam-like excavations that are sheeted on all sides.

Geotechnical

Equations 39.7(a) and 39.7(b) are for analysis. For design, it is common to apply a factor of safety of 1.5 to the undrained shear strength (cohesion), c. (If the factor of safety is less than 1.25–1.5, the sheeting must be carried down below the base of the cut until a factor of safety of 1.25–1.5 is achieved.) For the common case where $H > B$, where sheeting terminates at the base of the cut, and where there is a surface surcharge loading, q, the factor of safety is

$$F = \frac{N_c c}{\rho g H + q} \qquad \text{[SI]} \qquad 39.9(a)$$

$$F = \frac{N_c c}{\gamma H + q} \qquad \text{[U.S.]} \qquad 39.9(b)$$

If the factor of safety is less than 1.25–1.5, the sheeting must be carried to a depth approximately $B/2$ below the base of the cut.

Table 39.1 *Approximate Skempton Values of N_c for Infinitely Long Trenches in Clay (for use in Eqs. 39.8 and 39.9)*

H/B	N_c
0	5.1
1	6.3
2	7.0
3	7.3
≥ 4	7.5

9. STABILITY OF BRACED EXCAVATIONS IN SAND

For braced excavations in sand, the stability is independent of H and B. However, stability does depend on density, friction angle, and seepage conditions. When the water table is at least a distance of B below the base of the cut and the drained (dry, moist, etc.) sand density below the cut is the same as behind the cut, the factor of safety is

$$F = 2N_\gamma k_a \tan\phi \qquad 39.10$$

If the water table is stationary (i.e., not rising or falling) at the base of the cut,

$$F = 2N_\gamma \left(\frac{\rho_{\text{submerged}}}{\rho_{\text{drained}}}\right) k_a \tan\phi \qquad \text{[SI]} \qquad 39.11(a)$$

$$F = 2N_\gamma \left(\frac{\gamma_{\text{submerged}}}{\gamma_{\text{drained}}}\right) k_a \tan\phi \qquad \text{[U.S.]} \qquad 39.11(b)$$

If there is continual upward seepage through the base of the cut, the uplift pressure must also be evaluated.

10. SHEET PILING

Steel *sheet piles* (also known as *trench sheets*) are blade-like rolled steel sections that are driven into the ground to provide lateral support. Sheet piles are placed side by side and generally have interlocking edges ("clutches")

that enable the individual sheet piles to act as a continuous wall. Sheet piles are driven into the unexcavated ground ahead of the trenching operation.

Sheet piling is selected according to its geometry (i.e., flat, waffle, Z-, etc.) and the required section modulus. ASTM A572 is a typical steel, with the grade indicating the minimum yield point in ksi. The allowable design stress, F_b, is usually taken as 65% of the minimum yield stress, with some increases for temporary overstresses typically being allowed.

Table 39.2 *Suggested Allowable Design Stress in Sheet Piling*

	minimum yield point		allowable design stress[a]	
steel grade	ksi	MPa	ksi	MPa
A572 grade 55	55	380	35	240
A572 grade 50	50	345	32	220
A572 grade 45	45	310	29	200
A328 regular carbon	38.5	265	25	170

(Multiply ksi by 6.9 to obtain MPa.)
[a] 65% of the minimum yield stress

Adapted from *U.S. Steel Sheet Piling Design Manual*, U.S. Steel, 1974

11. ANALYSIS/DESIGN OF FLEXIBLE BULKHEADS

There are two types of flexible bulkheads: untied and tied (i.e., anchored). If the bulkhead is untied, it is designed as a cantilever beam. This is known as a *cantilever wall* or *cantilever bulkhead*.

The first step in the design of flexible bulkheads is determining the lateral pressure distribution due to the restrained soil and surcharges. Below the base of the cut, the passive pressure distribution opposes the active pressure distribution. The point at which the net pressure distribution (i.e., the shear) is zero is taken as the *pivot point* (or *hinge point, inflection point,* or *point of counterflexure*). This corresponds to the point of maximum moment in the bulkhead.

The theoretical depth of embedment, D, is chosen such that the sum of moments (taken about any convenient point) is zero. Thus, the pivot point must simultaneously satisfy both force and moment equilibrium conditions. Figure 39.5 illustrates the pressure distributions for cohesionless sand, defining the cross-over point located a distance y above the bottom edge of the bulkhead. Once y is calculated, the pressure distributions can be defined and the sum of moments calculated about the pivot point. The solution procedure for cohesive clay is more complex since the pressure distribution depends on the backfill material and changes with time.

$$y = \frac{k_p D^2 - k_a (H+D)^2}{(k_p - k_a)(H + 2D)} \qquad 39.12$$

Figure 39.5 *Shear on a Cantilever Wall in Uniform Granular Soil*

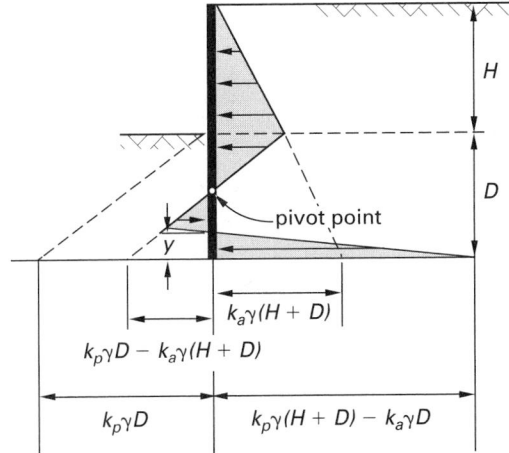

12. ANCHORED BULKHEADS

An *anchored bulkhead* is supported at its base by its embedment. The bulkhead is anchored near the top by *tie rods* (*tendons, ties, tie backs,* etc.) projecting back into the soil. These rods can terminate at deadmen, vertical piles, walls, beams, or various other types of anchors. To be effective, the anchors must be located outside of the failure zone (i.e., behind the slip plane).

Figure 39.6 *Anchored Bulkhead*

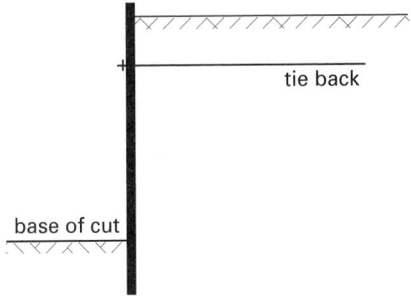

tie back

base of cut

Anchored bulkheads can fail in several ways. (a) The soil's base layer can heave (i.e., fail due to inadequate bearing capacity). The soil will shear along a circular arc passing under the bulkhead. (b) The anchorage can fail. (c) The toe embedment at the base of the cut can fail. This is referred to as *toe kick-out* and *toe wash-out* (d) Rarely, the sheeting can fail.

13. ANALYSIS/DESIGN OF TIED BULKHEADS

The *anchor pull* (tension in the tie) is found by setting to zero the sum of all horizontal loads on the bulkhead, including the passive loads on the embedded portion.

If the dimensions are known, the factor of safety against toe failure is found by taking moments about the anchor point on the bulkhead. (Alternatively, if the dimensions

are not known, the depth of embedment can be found by taking moments about the anchor point.)

The maximum bending moment in the bulkhead sheeting is found by taking moments about the hinge point, where the shear is zero. If the dimensions are not known, the location of this point must be initially assumed. For firm and dense embedment soils, the hinge point is assumed to be at the base of the cut. For loose and weak soils, it is 1 to 2 ft below the base of the cut. For a soft layer over a hard layer, it is at the depth of the hard layer. The stress in the bulkhead is

$$f = \frac{M}{S} \qquad 39.13$$

The required section modulus is

$$S = \frac{M}{F_b} \qquad 39.14$$

With multiple ties, the strut forces are statically indeterminate. Although exact methods (e.g., moment distribution) could be used, it is more expedient to assign portions of the active distribution to the struts based on the tributary areas. The area tributary to a strut is bounded by imaginary lines running midway between it and an adjacent strut, in all four directions.

14. COFFERDAMS

Temporary walls or enclosures for protecting an excavation are called *cofferdams*. Generally, one of the most important functions of a cofferdam is to permit work to be carried out on a nearly dry site.

Cofferdams should be planned so that they can be easily dismantled for reuse. Since they are temporary, safety factors can be small, 1.25 to 1.5, when all probable loads are accounted for in the design. But design stresses should be kept low when stresses, unit pressure, and bracing reactions are uncertain. Design should allow for construction loads and the possibility of damage from construction equipment. For cofferdams in water, the design should provide for dynamic effect of flowing water and impact of waves. The height of the cofferdam should be adequate to keep out floods that occur frequently.

Double-Wall Cofferdams

Double-wall cofferdams consist of two lines of sheetpiles tied to each other; the space between is filled with sand (Fig. 39.7). These may be erected in water to enclose large areas. For sheetpiles driven to irregular rock or gravel, or onto boulders, the bottom of the space between walls may be plugged with a thick layer of tremie concrete to seal gaps below the tips of the sheeting. Double-wall cofferdams are likely to be more watertight than single-wall ones and can be used to greater depths.

Figure 39.7 *Double-Wall Cofferdam*

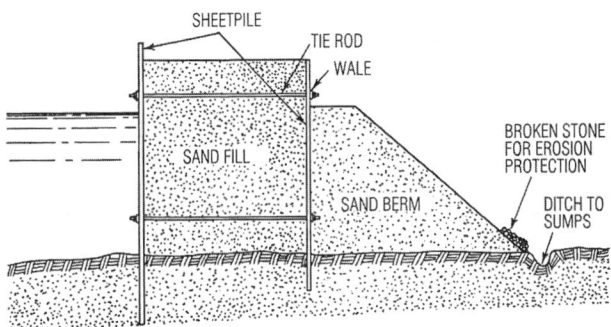

Figure 39.8 *Cellular Sheetpile Cofferdam*

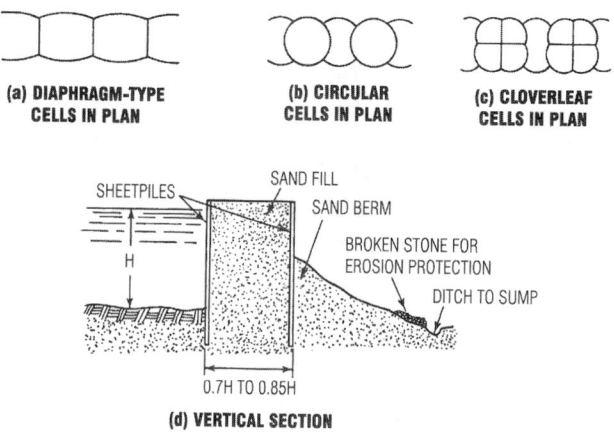

A berm may be placed against the outside face of a cofferdam for stability. If so, it should be protected against erosion. For this purpose, riprap, woven mattresses, streamline fins or jetties, or groins may be used. If the cofferdam rests on rock, a berm needs to be placed on the inside only if required to resist sliding, overturning, or shearing.

On sand, a substantial berm must be provided so that water has a long path to travel to enter the cofferdam (Fig. 39.7). (The amount of percolation is proportional to the length of path and the head.) Otherwise, the inside face of the cofferdam may settle, and the cofferdam may overturn as water percolates under the cofferdam and causes a quick, or boiling, excavation bottom. An alternative to the wide berm is wider spacing of the cofferdam walls. This is more expensive but has the added advantage that the top of the fill can be used by construction equipment.

Cellular Cofferdams

Used in construction of dams, locks, wharves, and bridge piers, cellular cofferdams are suitable for enclosing large areas in deep water. These enclosures are composed of relatively wide units. Average width of a cellular cofferdam on rock should be 0.70 to 0.85 times the head of water against the outside. When constructed on sand, a cellular cofferdam should have an ample berm on the inside to prevent the excavation bottom from becoming quick (Fig. 39.8(d)).

Steel sheetpiles interlocked form the cells. One type of cell consists of circular arcs connected by straight diaphragms (Fig. 39.8(a)). Another type comprises circular cells connected by circular arcs (Fig. 39.8(b)). Still another type is the coverleaf, composed of large circular cells subdivided by straight diaphragms (Fig. 39.8(c)). The cells are filled with sand. The internal shearing resistance of the sand contributes substantially to the strength of the cofferdam. For this reason, it is unwise to fill a cofferdam with clay or silt. Weepholes on the inside sheetpiles drain the fill, thus relieving the hydrostatic pressure on those sheets and increasing the shear strength of the fill.

In circular cells, lateral pressure of the fill causes only ring tension in the sheetpiles. The maximum stress in the pile interlocks usually is limited to 8000 lbf/in

(1400 kN/m). This in turn limits the maximum diameter of the circular cells. Because of numerous uncertainties, this maximum generally is set at 60 ft (18 m). When larger-size cells are needed, the cloverleaf type may be used.

Circular cells are preferred to the diaphragm type because each circular cell is a self-supporting unit. It may be filled completely to the top before construction of the next cell starts. (Unbalanced fills in a cell may distort straight diaphragms.) When a circular cell has been filled, the top may be used as a platform for construction of the next cell. Also, circular cells require less steel per linear foot of cofferdam. The diaphragm type, however, may be made as wide as desired.

When the sheetpiles are being driven, care must be taken to avoid breaking the interlocks. The sheetpiles should be accurately set and plumbed against a structurally sound template. They should be driven in short increments, so that when uneven bedrock or boulders are encountered, driving can be stopped before the cells or interlocks are damaged. Also, all the piles in a cell should be started until the cell is ringed. This can reduce jamming troubles with the last piles to be installed for the cell.

Single-Wall Cofferdams

Single-wall cofferdams form an enclosure with only one line of sheeting. If there will be no water pressure on the sheeting, they may be built with *soldier beams* (piles extended to the top of the enclosure) and horizontal wood lagging (Fig. 39.9). If there will be water pressure, the cofferdam may be constructed of sheetpiles. Although they require less wall material than double-wall or cellular cofferdams, single-wall cofferdams generally require bracing on the inside. Also, unless the bottom is driven into a thick, impervious layer, they may leak excessively at the bottom. There may also be leakage at interlocks. Furthermore, there is danger of flooding and collapse due to hydrostatic forces when these cofferdams are unwatered.

Figure 39.9 *Soldier Beams and Wood Lagging Retaining the Sides of an Excavation*

Figure 39.10 *Types of Cofferdams Bracing Include Compression Rings; Bracing, Diagonal (Rakers) or Cross Lot; Wales, and Tiebacks*

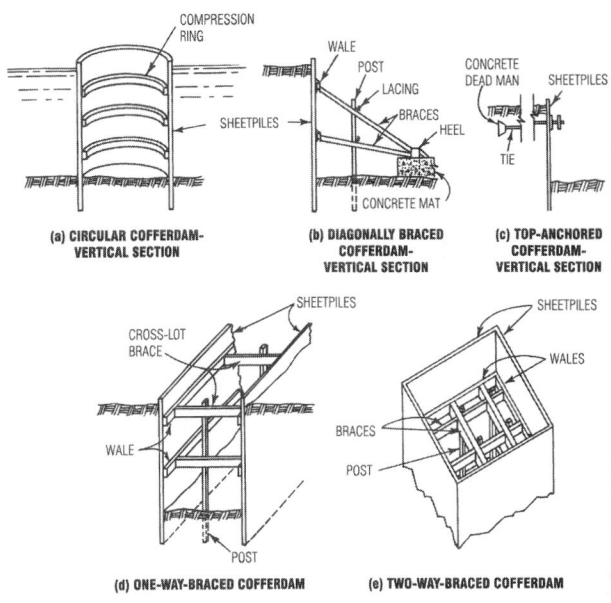

For marine applications, therefore, it is advantageous to excavate, drive piles, and place a seal of tremie concrete without unwatering single-wall sheetpile cofferdams. Often, it is advisable to predredge the area before the cofferdam is constructed, to facilitate placing of bracing and to remove obstructions to pile driving. Also, if blasting is necessary, it would severely stress the sheeting and bracing if done after they were installed.

For buildings, single-wall cofferdams must be carefully installed. Small movements and consequent loss of ground usually must be prevented to avoid damaging neighboring structures, streets, and utilities. Therefore, the cofferdams must be amply braced. Sheeting close to an existing structure should not be a substitute for underpinning.

Bracing

Cantilevered sheetpiles may be used for shallow single-wall cofferdams in water or on land where small lateral movement will not be troublesome. Embedment of the piles in the bottom must be deep enough to insure stability. Design usually is based on the assumptions that lateral passive resistance varies linearly with depth and the point of inflection is about two-thirds the embedded length below the surface. In general, however, cofferdams require bracing.

Cofferdams may be braced in many ways. Figure 39.10 shows some commonly used methods. Circular cofferdams may be braced with horizontal rings (see Fig. 39.10(a)). For small rectangular cofferdams, horizontal braces, or wales, along sidewalls and end walls may be connected to serve only as struts. For larger cofferdams, diagonal bracing (Fig. 39.10(b)) or cross-lot bracing (Fig. 39.10(d) and (e)) is necessary. When space is available at the top of an excavation, pile tops can be anchored with concrete dead men anchored in grouted sockets in the rock (Fig. 39.11).

Figure 39.11 *Vertical Section Showing Prestressed Tiebacks for Soldier Beams*

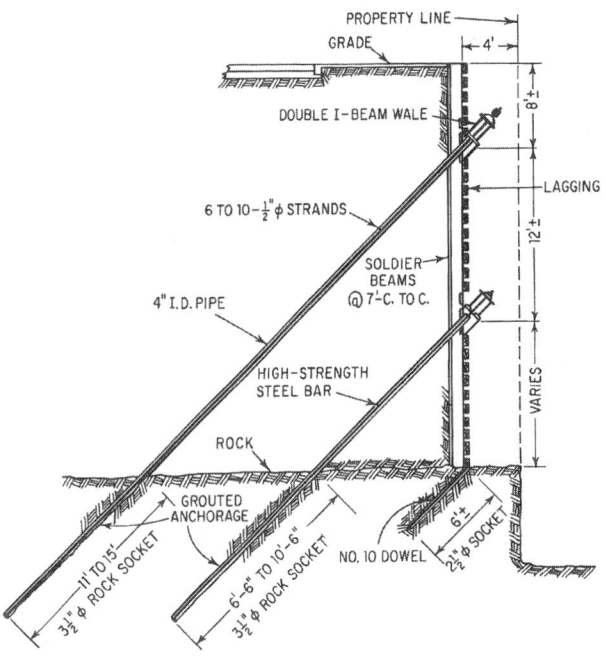

Horizontal cross braces should be spaced to minimize interference with excavation, form construction, concreting, and pile driving. Spacing of 12 and 18 ft (3.6 and 5.4 m) is common. Piles and wales selected should be strong enough as beams to permit such spacing. In marine applications, divers often have to install the wales and braces underwater. To reduce the amount of such work, tiers of bracing may be prefabricated and lowered into the cofferdam from falsework or from the top set of wales and braces, which is installed above the water surface. In some cases, it may be advanageous to prefabricate and erect the whole cage of bracing before the sheetpiles are driven. Then, the cage, suported on piles, can serve also as a template for driving the sheetpiles.

All wales and braces should be forced into bearing with the sheeting by wedges and jacks.

When pumping cannot control leakage into a cofferdam, excavation may have to be carried out in compressed air. This requires a sealed working chamber, access shafts, and air locks, as for pneumatic caissons. Other techniques, such as use of a tremie concrete seal or chemical solidification or freezing of the soil, if practicable, however, will be more economical.

Braced sheetpiles may be designed as continuous beams subjected to uniform loading for earth and to loading varying linearly with depth for water. (Actually, earth pressure depends on the flexibility of the sheeting and relative stiffness of supports.) Wales may be designed for uniform loading. Allowable unit stresses in the wales, struts, and ties may be taken at half the elastic limit for the materials because the construction is temporary and the members are exposed to view. Distress in a member can easily be detected and remedial steps taken quickly.

Soldier beams and horizontal wood sheeting are a variation of single-wall cofferdams often used where impermeability is not required. The soldier beams, or piles, are driven vertically into the ground to below the bottom of the proposed excavation. Spacing usually ranges from 5 to 10 ft (1.5 to 3 m) as shown in Table 39.3. (The wood lagging can be used in the thicknesses shown in Table 39.3 because of arching of the earth between successive soldier beams.)

As excavation proceeds, the wood boards are placed horizontally between the soldiers (Fig. 39.9). Louvers or packing spaces, 1 to 2 in (2.5 to 5 cm) high, are left between the boards so that earth can be tamped behind them to hold them in place. Hay, geotechnical fabric, or plastic sheet may also be stuffed behind the boards to keep the ground from running through the gaps. The louvers permit drainage of water to relieve hydrostatic pressure on the sheeting and thus allow use of a lighter bracing system. The soldiers may be braced directly with horizontal or inclined struts; or wales and braces may be used.

Advantages of soldier-beam construction include fewer piles; the sheeting does not have to extend below the excavation bottom, as do sheetpiles; and the soldiers can be driven more easily in hard ground than can sheetpiles. Varying the spacing of the soldiers permits avoidance of underground utilities. Use of heavy sections for the piles allows wide spacing of wales and braces. But the soldiers and lagging, as well as sheetpiles, are no substitute for underpinning; it is necessary to support and underpin even light adjoining structures.

Liner-plate cofferdams may be used for excavating circular shafts. The plates are placed in horizontal rings as excavation proceeds. Stamped from a steel plate usually about 16 in (40 cm) high and 3 ft (0.9 m) long, and light enough to be carried by one person, liner plates have inward-turned flanges along all edges. Top and bottom flanges provide a seat for successive rings. End flanges permit easy bolting of adjoining plates in a ring. The plates also are corrugated for added stiffness. Large-diameter cofferdams may be constructed by bracing the liner plates with steel beam rings.

Vertical-lagging cofferdams, with horizontal-ring bracing, also may be used for excavating circular shafts. The method is similar to that used for Chicago caissons. It is similarly restricted to soils that can stand without support in depths of 3 to 5 ft (0.9 to 1.5 m) for a short time.

Table 39.3 *Usual Maximum Spans of Horizontal Sheeting with Soldier Piles*

nominal thickness of sheeting (in)	in well-drained soils (ft)	in cohesive soils with low shear resistance (ft)
2	5	4.5
3	8.5	6
4	10	8

40 Special Soil Topics

Nomenclature

a	acceleration	ft/sec^2	m/s^2
a_v	coefficient of compressibility	ft^2/lbf	m^2/N
A	area	ft^2	m^2
B	footing width	ft	m
B	trench width	ft	m
c	undrained shear strength (cohesion)	lbf/ft^2	Pa
C	Marston's constant	–	–
C_α	coefficient of compression	–	–
C_c	compression index	–	–
C_r	recompression index	–	–
C_v	coefficient of consolidation	–	–
CR	compression ratio	–	–
d	depth factor	–	–
D	depth to firm base below	ft	m
D	diameter of pipe	ft	m
e	void ratio	–	–
F	factor of safety	–	–
g	acceleration of gravity	ft/sec^2	m/s^2
h	distance below water table	ft	m
H	depth or cut height	ft	m
H	layer thickness	ft	m
I	influence value	–	–
K	coefficient of permeability	ft/sec	m/s
L	distance	ft	m
LF	load factor	–	–
LL	liquid limit	percent	percent
N_o	stability number	–	–

p	pressure	lbf/ft^2	Pa
p'	effective pressure	lbf/ft^2	Pa
P	load	lbf	N
r	distance	ft	m
r	radius of round footing	ft	m
r_d	stress reduction factor	–	–
RR	recompression ratio	–	–
S	settlement	ft	m
SG	specific gravity	–	–
t	time	sec	s
T_v	time factor	–	–
U_z	degree of consolidation	–	–
V	volume	ft^3	m^3
w	load per unit length	lbf/ft	N/m
w	moisture content	–	–
x	distance	ft	m
z	distance (depth)	ft	m

Symbols

α	secondary consolidation index	–	–
β	slope angle	deg	deg
γ	specific weight	lbf/ft^3	–
μ	pore pressure	lbf/ft^2	Pa
ρ	density	lbm/ft^3	kg/m^3
σ	stress	lbf/ft^2	Pa
τ	shear stress	lbf/ft^2	Pa
ϕ	angle of internal friction	deg	deg

Subscripts

c	compression
d	drainage
eff	effective
h	horizontal
i	inside
n	natural
o	original
r	recompression
v	vertical
z	at depth z

1. PRESSURE FROM APPLIED LOADS: BOUSSINESQ'S EQUATION

The increase in vertical pressure (stress), Δp_v, caused by an application of a point load, P, at the surface can be found from *Boussinesq's equation*. This equation assumes that the footing width, B, is small compared to the depth, h, and that the soil is semi-infinite, elastic, isotropic, and homogeneous. It is sometimes convenient to solve Eq. 40.1 by using specially prepared *Boussinesq contour charts* included as Apps. 40.A and 40.B.

$$\Delta p_v = \frac{3h^3 P}{2\pi z^5}$$

$$= \left(\frac{3P}{2\pi h^2}\right)\left(\frac{1}{1 + \left(\frac{r}{h}\right)^2}\right)^{5/2} \quad [h > 2B] \quad 40.1$$

Figure 40.1 *Pressure at a Point*

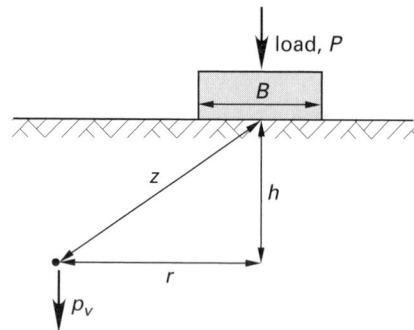

Not all situations are adequately described by the Boussinesq theory. The *Westergaard case* applies to layered materials as they exist in multilayered highway pavement sections. The effect of such layering is to reduce the stresses substantially below those predicted by the Boussinesq theory.

2. PRESSURE FROM APPLIED LOADS: ZONE OF INFLUENCE

The approximate stress at a depth can be determined by assuming the geometry of the affected area (i.e., the *zone of influence*). Approximate methods can be reasonably accurate in nonlayered homogeneous soils when $1.5 < h/B < 5$.

The zone of influence is defined by the angle of the *influence cone*. This angle is typically assumed to be 51° for point loads and 60° from the horizontal for uniformly loaded circular and rectangular footings. (Accordingly, this method is sometimes referred to as the *60° method*.) Alternatively, since 60° is an approximation anyway, for ease of computation, the angle is taken as 63.4°, corresponding to a 2:1 (vertical:horizontal) influence cone angle.

For any depth, the area of the cone of influence is the area of the horizontal plane enclosed by the influence boundaries, as determined geometrically. For a rectangular $B \times L$ footing with the 60° method, the area of influence at depth h is

$$A = \left(B + 2(h\cot 60°)\right)\left(L + 2(h\cot 60°)\right) \quad 40.2$$

For a rectangular $B \times L$ footing assuming a 2:1 influence, the area of influence is

$$A = (B + h)(L + h) \quad 40.3$$

The increase in pressure at that location is

$$\Delta p_v = \frac{P}{A} \quad 40.4$$

Figure 40.2 *Zone of Influence*

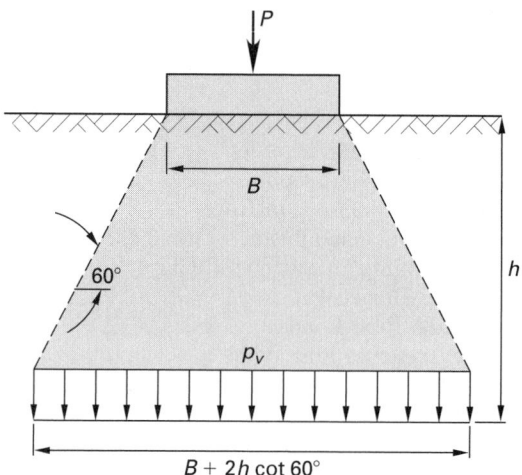

3. PRESSURE FROM APPLIED LOADS: INFLUENCE CHART

If a footing or mat foundation is oddly shaped or large compared to the depth where the unknown pressure is wanted, the vertical pressure at a point can be estimated by using an *influence chart (Newmark chart)*. A Newmark influence chart is a graphical solution to the Boussinesq case. Each influence chart has an indicated scale (the "key") and an influence value that is equal to the reciprocal of the number of squares in the chart.

To use the influence chart, the implied scale is determined by making the indicated distance A-B equal to the depth at which the pressure is wanted. Using this scale, a plan view of the footing or mat foundation is drawn on a piece of tracing paper. The tracing paper is placed over the influence chart such that the center of the chart coincides with the location under the footing where the pressure is wanted. Then, the number of squares under the footing drawing is counted. Partial squares are counted as fractions. The pie-shaped areas in the center circle are counted as complete squares. The *influence value* is read from the influence chart, and Eq. 40.5 is used to calculate the pressure.

$$\Delta p_v = (\text{influence value})(\text{no. of squares})$$
$$\times (\text{applied pressure}) \quad 40.5$$

Figure 40.3 *Influence Chart*

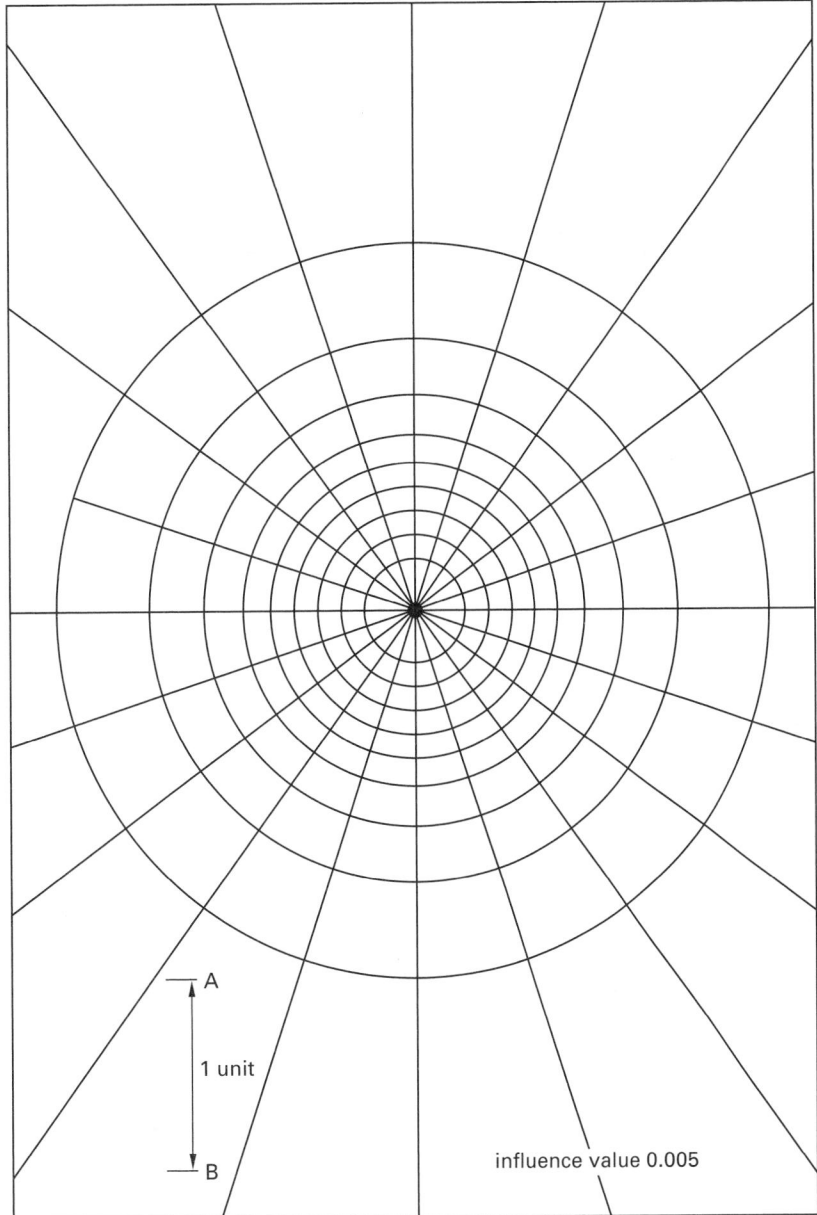

A

1 unit

B

influence value 0.005

Geotechnical

4. SETTLING

Settling is generally due to *consolidation* (i.e., a decrease in void fraction) of the supporting soil. There are three distinct periods of consolidation. (a) *Immediate settling*, also known as *elastic settling*, occurs immediately after the structure is constructed. Immediate settling is the major settling component in sandy soils. (b) In clayey soils, *primary consolidation* occurs gradually due to the extrusion of water from the void spaces. (c) *Secondary consolidation* occurs in clayey soils at a much slower rate after the primary consolidation has finished. Since plastic readjustment of the soil grains (including progressive fracture of the grains themselves)

is the primary mechanism, the magnitude of secondary consolidation is considerably less than primary consolidation.

Since settling is greater for higher foundation pressures, specific settlement limits (e.g., 1 in) are directly related to the maximum allowable pressures. This, in turn, can be used to size footings and other foundations.

5. CLAY CONDITION

For clays, the method used to calculate consolidation depends on the clay condition. Figure 40.4 shows a typical *consolidation curve* showing void ratio, *e*, versus applied effective pressure, p' (or effective stress, σ'), for a previously consolidated clayey soil, graphed on an $e\text{-}\log p$ curve. The curve shows a *recompression segment* (*preconsolidated* or *overconsolidated segment*) and the *virgin compression branch* (*normally consolidated segment*).

If the pressure is increased from p_1 to p_2, the consolidation will be in the recompression zone, which has the lower slope, C_r. C_r is typically one fifth to one tenth of C_c. If the pressure is increased above p_2, consolidation is more dramatic as the slope C_c is higher. If the pressure is increased from p_1 to p_3, the two consolidations are added together to obtain the total settlement.

Figure 40.4 *Consolidation Curve for Clay*

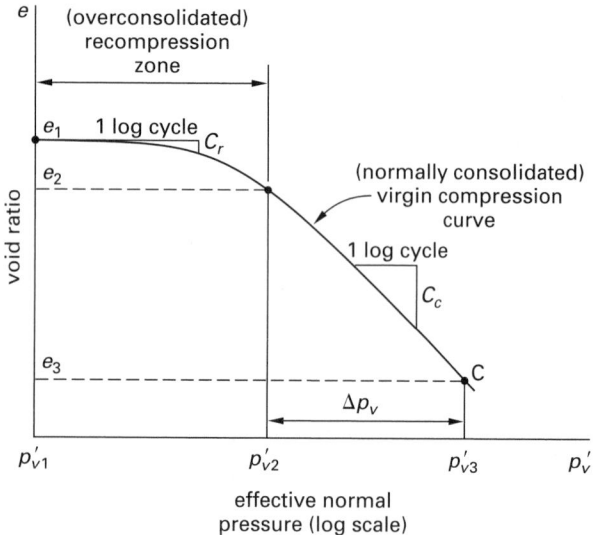

6. CONSOLIDATION PARAMETERS

Along the recompression line, the *recompression index*, C_r, is the logarithmic slope of the recompression segment. C_c and C_r are reported as positive numbers even though the slopes of the lines they represent are negative.

$$C_r = \frac{-(e_1 - e_2)}{\log_{10}\left(\dfrac{p_1}{p_2}\right)} = \frac{e_2 - e_1}{\log_{10}\left(\dfrac{p_1}{p_2}\right)} \qquad 40.6$$

The *compression index* is the logarithmic slope of the primary consolidation curve. $p_2 + \Delta p_v$ is the total pressure after the load has been applied (or removed).

$$C_c = \frac{-(e_2 - e_3)}{\log_{10}\left(\dfrac{p_2}{p_3}\right)} = \frac{e_3 - e_2}{\log_{10}\left(\dfrac{p_2}{p_3}\right)}$$

$$= \frac{\Delta e}{\log_{10}\left(\dfrac{p_2}{p_2 + \Delta p_v}\right)} \qquad 40.7$$

If necessary, the compression index can be estimated from other soil parameters, although these correlations are not highly accurate (i.e., have an error of 30% or more). For clays, a general expression is given by Eq. 40.8. The term e_o represents the "original" void ratio of the clay, often used as an estimate of the average void ratio over the range of interest. Referring to Fig. 40.4, e_o will either be e_1 or e_2, depending on whether the clay is originally overconsolidated or normally consolidated, respectively.

$$C_c \approx 1.15(e_o - 0.35) \quad \text{[clays]} \qquad 40.8$$

For normally consolidated inorganic soils (clays) with sensitivities less than 4,

$$C_c \approx 0.009(\text{LL} - 10) \qquad 40.9$$

For organic soils (e.g., peat), the compression index depends on the natural moisture content.

$$C_c \approx 0.0155w \qquad 40.10$$

For *varved clays* (i.e., clays that are layered with fine and coarse varieties),

$$C_c \approx (1 + e_o)\big(0.1 + 0.006(w_n - 25)\big) \qquad 40.11$$

Some authorities define the *compression ratio*, CR (not to be confused with C_r), and the *recompression ratio*, RR (also known as the *swell* or *swelling index*), as follows.

$$\text{CR} = \frac{C_c}{1 + e_o} \qquad 40.12$$

$$\text{RR} = \frac{C_r}{1 + e_o} \qquad 40.13$$

For saturated soils, the void ratio, e_o, can be calculated from the water content.

$$e_o = w_o(\text{SG}) \quad \text{[saturated]} \qquad 40.14$$

7. PRIMARY CONSOLIDATION

When clay layers are loaded to a higher pressure, water is "squeezed" from the voids. Water is lost over a long period of time, though the rate of loss decreases with time. The loss of water results in a consolidation of the clay layer and a settlement of the soil surface. The long-term consolidation due to water loss is the *primary consolidation*, S_{primary}. For an overconsolidated clay layer of thickness H, the consolidation is

$$S_{\text{primary}} = \frac{H \Delta e}{1 + e_o}$$

$$= \frac{H C_r \log_{10}\left(\dfrac{p'_o + \Delta p'_v}{p'_o}\right)}{1 + e_o}$$

$$= H(\text{RR}) \log_{10}\left(\frac{p'_o + \Delta p'_v}{p'_o}\right) \quad \text{[overconsolidated]}$$

$$40.15$$

The primary consolidation in a normally consolidated clay layer of thickness H is

$$
\begin{aligned}
S_{\text{primary}} &= \frac{H\Delta e}{1 + e_o} \\
&= \frac{HC_c \log_{10}\left(\dfrac{p'_o + \Delta p'_v}{p'_o}\right)}{1 + e_o} \\
&= H(\text{CR})\log_{10}\left(\frac{p'_o + \Delta p'_v}{p'_o}\right) \quad \begin{bmatrix} \text{normally} \\ \text{consolidated} \end{bmatrix}
\end{aligned}
$$

$$\text{40.16}$$

$\Delta p'_v$ is the increase in effective vertical pressure. Boussinesq's equation, stress contour charts, or influence charts can be used to determine Δp_v.

p'_o is the original *effective pressure* (*effective stress*) at the midpoint of the clay layer and directly below the foundation. The average effective pressure is the sum of the following items, depending on whether the layer is above or below the *ground water table* (GWT). H is the layer thickness. Notice that in Eq. 40.16, p' is the effective pressure, exclusive of the pore pressure. Only the effective pressure causes consolidation.

$$p_v = \gamma_{\text{layer}} H \quad [\text{above GWT}] \qquad \text{40.17}$$

$$
\begin{aligned}
p'_v &= \gamma_{\text{layer}} H - \gamma_{\text{water}} h \\
&= p_v - \mu \quad [\text{below GWT}]
\end{aligned}
$$

$$\text{40.18}$$

$$\mu = \gamma_{\text{water}} h \qquad \text{40.19}$$

The *pore pressure*, μ, can be determined in a number of ways, including piezometer readings (i.e., the height of water in a standpipe).

8. PRIMARY CONSOLIDATION RATE

Consolidation of clay is a continuous process, though the rate decreases with time. The time for a single layer to reach a specific consolidation is given by Eq. 40.20. H_d is the full consolidated layer thickness if drainage is from one surface only (i.e., *single drainage* or *one-way drainage*). If drainage is through both the top and bottom surfaces (i.e., *double drainage* or *two-way drainage*), H_d is half the consolidated layer's thickness. The units of t will depend on the units of C_v.

$$t = \frac{T_v H_d^2}{C_v} \qquad \text{40.20}$$

The *coefficient of consolidation*, C_v, with typical units of ft^2/day (m^2/day), is assumed to remain constant over small variations in the void ratio, e. C_v is evaluated in the laboratory by solving Eq. 40.20 for it, and using observed time and thickness from an oedometer test.

$$C_v = \frac{K(1 + e_o)}{a_v \gamma_{\text{water}}} \qquad \text{40.21}$$

The *coefficient of compressibility*, a_v, can be found from the void ratio and the effective pressure (stress) for any two different loadings. The coefficient of compressibility is not really constant—it decreases with increasing stress. However, it is assumed constant over small variations.

$$a_v = \frac{-(e_2 - e_1)}{p'_2 - p'_1} \qquad \text{40.22}$$

T_v is a dimensionless *time factor* that depends on the *degree of consolidation*, U_z. U_z is the fraction of the total consolidation that is expected. For values of U_z larger than 0.60, Table 40.1 can be used.

$$T_v = \tfrac{1}{4}\pi U_z^2 \quad [U_z < 0.60] \qquad \text{40.23}$$

Table 40.1 Approximate Time Factors[a]

U_z	T_v
0.10	0.008
0.20	0.031
0.30	0.071
0.40	0.126
0.50	0.197
0.55	0.238
0.60	0.287
0.65	0.340
0.70	0.403
0.75	0.477
0.80	0.567
0.85	0.684
0.90	0.848
0.95	1.129
0.99	1.781
1.0	∞

[a] Calculated from
$$T_v = 1.781 - 0.933 \log\left(100\,(1 - U_z)\right) \quad [U_z > 0.60]$$

Figure 40.5 Consolidation Parameters

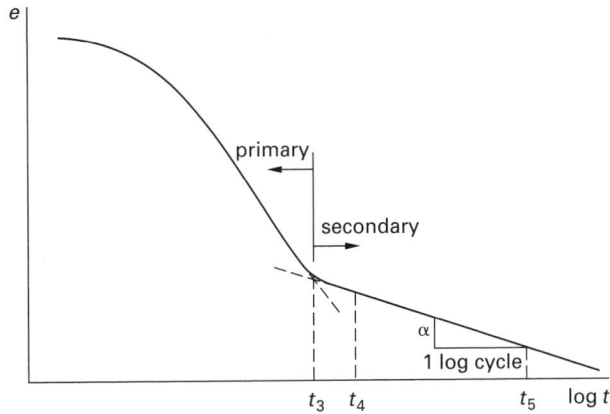

9. SECONDARY CONSOLIDATION

Following the completion of primary consolidation (at $T_v = 1$), the rate of consolidation decreases markedly. The continued consolidation is known as *secondary consolidation.* Secondary consolidation is a gradual continuation of consolidation that continues long after the cessation of primary consolidation. Secondary consolidation may not occur at all, as is the case in granular soils. Secondary consolidation may be a major factor for inorganic clays and silts, as well as for highly organic soils.

Secondary consolidation can be identified on a chart of void ratio (or settlement) versus logarithmic time. The region of secondary consolidation is characterized by a distinct slope reduction. The plot can be used to obtain important parameters necessary to calculate the magnitude and rate of secondary consolidation.

The final void ratio, e_3, at the end of the primary consolidation period is read directly from the chart. The logarithmic slope of the secondary compression line is the *secondary compression index,* α, which generally varies from 0 to 0.03, seldom exceeding 0.04.

$$\alpha = \frac{-(e_5 - e_4)}{\log\left(\dfrac{t_5}{t_4}\right)} \qquad 40.24$$

The *coefficient of secondary consolidation,* C_α, is given by Eq. 40.25. e_o is the "original" void ratio, often used as an approximation of the average void ratio over the range of interest.

$$C_\alpha = \frac{\alpha}{1 + e_o} \qquad 40.25$$

The secondary consolidation during the period t_4 to t_5 is

$$S_{\text{secondary}} = C_\alpha H \log_{10}\left(\frac{t_5}{t_4}\right) \qquad 40.26$$

Example 40.1

A 40 ft × 60 ft raft foundation on sand is constructed and loaded as shown. The sand has already settled. The clay layer is normally consolidated. (a) What long-term primary settlement can be expected at the center of the raft? (b) What long-term primary settlement can be expected at the corner of the raft?

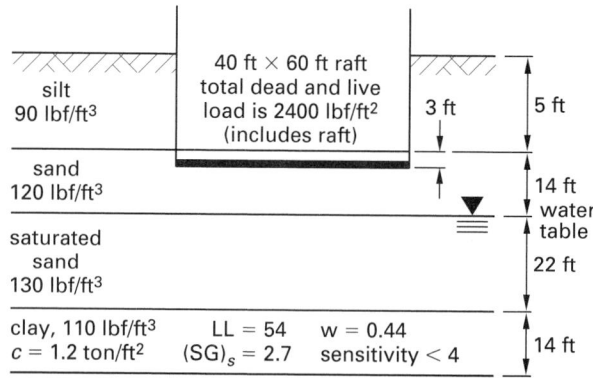

Solution

step 1: Only the clay will contribute to long-term settlement. Calculate the original pressure at the center of the clay layer because this represents the average conditions in the clay. The pressure consists of components from the various layers above it.

silt layer: $(5 \text{ ft})\left(90 \dfrac{\text{lbf}}{\text{ft}^3}\right) = 450 \text{ lbf/ft}^2$

drained sand layer: $(14 \text{ ft})\left(120 \dfrac{\text{lbf}}{\text{ft}^3}\right) = 1680 \text{ lbf/ft}^2$

submerged sand layer: $(22 \text{ ft})\left(130 \dfrac{\text{lbf}}{\text{ft}^3} - 62.4 \dfrac{\text{lbf}}{\text{ft}^3}\right)$
$= 1487 \text{ lbf/ft}^2$

clay layer: $\left(\frac{1}{2}\right)(14 \text{ ft})\left(110 \dfrac{\text{lbf}}{\text{ft}^3} - 62.4 \dfrac{\text{lbf}}{\text{ft}^3}\right)$
$= 333 \text{ lbf/ft}^2$

total: $p_o = 3950 \text{ lbf/ft}^2$

step 2: The applied pressure at the base of the raft is the dead and live loading less the overburden due to the excavated sand and silt.

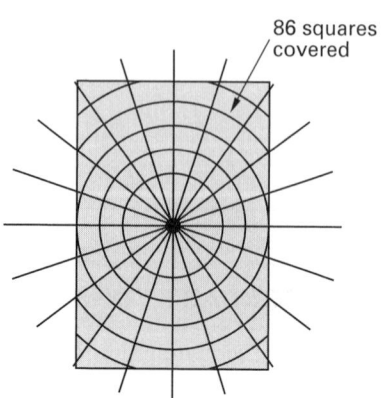

86 squares covered

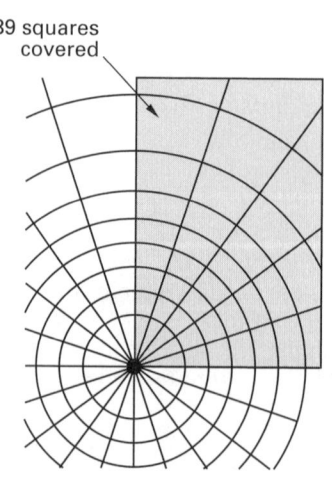

39 squares covered

$$p_{\text{net,center}} = 2400\ \frac{\text{lbf}}{\text{ft}^2} - (5\text{ ft})\left(90\ \frac{\text{lbf}}{\text{ft}^3}\right)$$
$$- (3\text{ ft})\left(120\ \frac{\text{lbf}}{\text{ft}^3}\right)$$
$$= 1590\ \text{lbf/ft}^2$$

Use Fig. 40.3 to calculate the pressure increase due to the loading. The distance from the bottom of the raft to the midpoint of the clay layer is $36\text{ ft} - 3\text{ ft} + 7\text{ ft} = 40\text{ ft}$. Using a scale of 1 in:40 ft, the raft is $1\text{ in} \times 1\frac{1}{2}$ in. When the raft is centered over the influence chart, 86 squares are covered. When the raft's corner is placed over the center of the chart, 39 squares are covered.

The midlayer pressure increases are given by Eq. 40.5.

$$p_{v,\text{center}} = (\text{influence value})$$
$$\times (\text{no. of squares})$$
$$\times (\text{applied pressure})$$
$$= (0.005)(86)\left(1590\ \frac{\text{lbf}}{\text{ft}^2}\right)$$
$$= 680\ \text{lbf/ft}^2$$
$$p_{v,\text{corner}} = (0.005)(39)\left(1590\ \frac{\text{lbf}}{\text{ft}^2}\right)$$
$$= 310\ \text{lbf/ft}^2$$

step 3: Estimate the original void content of the clay. Use Eq. 40.14.
$$e_o \approx w(\text{SG}) = (0.44)(2.7) = 1.188$$

step 4: Estimate the compression index. Use Eq. 40.9.
$$C_c \approx 0.009(\text{LL} - 10)$$
$$= (0.009)(54 - 10) = 0.396$$

step 5: The sensitivity is low, so the clay is considered to be normally consolidated. The settlements are given by Eq. 40.16.

$$S_{\text{center}} = \frac{HC_c\left(\log_{10}\left(\dfrac{p_o + p_v}{p_o}\right)\right)}{1 + e_o}$$

$$= \frac{(14\text{ ft})(0.396)\left(\log_{10}\left(\dfrac{3950\ \frac{\text{lbf}}{\text{ft}^2} + 680\ \frac{\text{lbf}}{\text{ft}^2}}{3950\ \frac{\text{lbf}}{\text{ft}^2}}\right)\right)}{1 + 1.188}$$

$$= 0.175\text{ ft}$$

$$S_{\text{corner}} = \frac{(14\text{ ft})(0.396)\left(\log_{10}\left(\dfrac{3950\ \frac{\text{lbf}}{\text{ft}^2} + 310\ \frac{\text{lbf}}{\text{ft}^2}}{3950\ \frac{\text{lbf}}{\text{ft}^2}}\right)\right)}{1 + 1.188}$$

$$= 0.083\text{ ft}$$

10. SLOPE STABILITY IN SATURATED CLAY

The maximum slope for cuts in cohesionless (i.e., drained) sand is the *angle of internal friction*, ϕ, also known as the *angle of repose*. However, the maximum slope for cuts in cohesive soils is more difficult to determine. For saturated clay with $\phi = 0°$, the *Taylor slope stability chart* can be used to determine the factor of safety against slope failure. The Taylor chart makes the following assumptions: (a) There is no open water outside of the slope. (b) There are no surcharges or tension cracks. (c) Shear strength is derived from cohesion only and is constant with depth. (d) Failure takes place as rotation on a circular arc.

The Taylor chart relates a dimensionless depth factor, d, to a dimensionless stability number, N_o. The depth factor, d, is the quotient of the vertical distance from the toe of the slope to the firm base below the clay layer, D, and the slope height, H (the depth of the cut).

$$d = \frac{D}{H} \qquad \textit{40.27}$$

Equation 40.28 gives the relationship between the *stability number* and the cohesive factor of safety. The effective specific weight is used when the clay is submerged. The minimum acceptable factor of safety is approximately 1.3 to 1.5.

$$F_{\text{cohesive}} = \frac{N_o c}{\gamma_{\text{eff}} H} \qquad \textit{40.28}$$

$$\gamma_{\text{eff}} = \gamma_{\text{saturated}} - \gamma_{\text{water}} \qquad \textit{40.29}$$

The Taylor chart shows that toe circle failures occur in slopes steeper than 53°. For slopes of less than 53°, *slope circle failure, toe circle failure,* or *base circle failure* may occur.

Figure 40.6 *Taylor Slope Stability*
($\phi = 0°$)

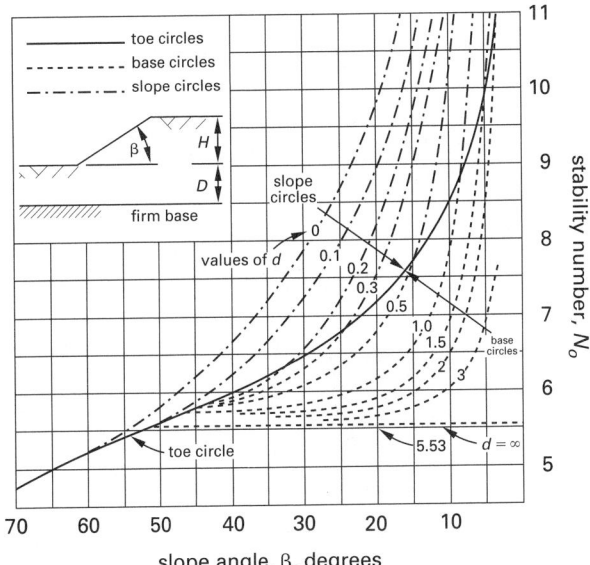

Source: *Soil Mechanics*, NAVFAC Design Manual DM-7.1, May 1982

Geotechnical

Figure 40.7 *Types of Slope Failures*

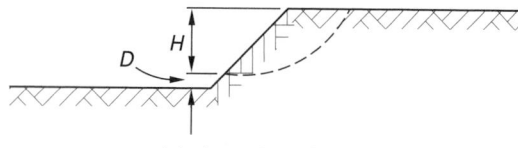

(a) slope circle failure

(b) toe circle failure

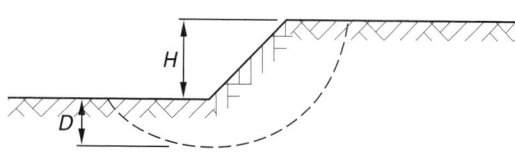

(c) base circle failure

Example 40.2

A submerged trench is excavated in a layer of soft mud. The trench walls are sloped at a vertical:horizontal ratio of 1.5:1. The mud has a saturated density of 100 lbf/ft³ and a cohesion of 400 lbf/ft². A factor of safety against cohesive slope failure of 1.5 is required. (a) What type of failure will occur? (b) What depth of cut can be made?

Solution

(a) The cut angle is

$$\beta = \arctan\left(\frac{1.5}{1}\right) = 56°$$

Since this is greater than 53°, the failure will be toe circle.

(b) From Fig. 40.6, the stability number for $\beta = 56°$ is approximately 5.4. Since the clay is submerged, use the effective unit weight.

$$\gamma_{eff} = \gamma_{saturated} - \gamma_{water}$$
$$= 100\ \frac{lbf}{ft^3} - 62.4\ \frac{lbf}{ft^3} = 37.6\ lbf/ft^3$$

Solve Eq. 40.28 for depth of cut.

$$H = \frac{N_o c}{F\gamma_{eff}}$$
$$= \frac{(5.4)\left(400\ \frac{lbf}{ft^2}\right)}{(1.5)\left(37.6\ \frac{lbf}{ft^3}\right)}$$
$$= 38.3\ ft$$

11. LOADS ON BURIED PIPES

If a pipe is placed in an excavated trench and subsequently backfilled, the pipe must be strong enough to support the vertical soil load in addition to any loads from surface surcharges. The magnitude of the load supported depends on the amount of backfill, type of soil, and pipe stiffness. For rigid pipes (e.g., concrete, cast iron, ductile iron, and older clay types) that cannot deform and that are placed in narrow trenches (e.g., 2 to 3 pipe diameters), the approximate dead load per unit length can be calculated from *Marston's formula*, Eq. 40.30. Typical values of C and γ are given in Table 40.2. B is the trench width at the top of the pipe.

$$w = C\rho g B^2 \qquad \text{[SI]} \qquad 40.30(a)$$
$$w = C\gamma B^2 \qquad \text{[U.S.]} \qquad 40.30(b)$$

The *dead load pressure* is the dead load divided by the trench width at the pipe's top.

$$p = \frac{w}{B} \qquad\qquad 40.31$$

Flexible pipes (e.g., steel, plastic, and copper) are able to develop a horizontal restraining pressure equal to the vertical pressure if the backfill is well compacted. There is evidence that flexible pipes shrink away from the soil under high compressive loads, thus reducing the soil pressure significantly. D is the pipe diameter.

$$w = C\rho g B D \qquad \text{[SI]} \qquad 40.32(a)$$
$$w = C\gamma B D \qquad \text{[U.S.]} \qquad 40.32(b)$$

Figure 40.8 *Pipes in Backfilled Trenches*

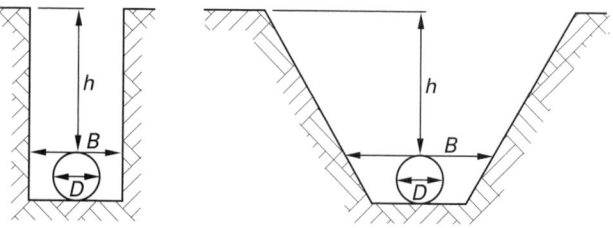

If a pipe is placed on undisturbed ground and covered with fill (*broad fill* or *embankment fill*), the load is given by Eq. 40.33. Values of C_p are given in Table 40.3.

$$w = C_p \rho g D^2 \qquad \text{[SI]} \qquad 40.33(a)$$
$$w = C_p \gamma D^2 \qquad \text{[U.S.]} \qquad 40.33(b)$$

Equation 40.30 shows that the trench width is an important factor in determining whether or not the pipe has sufficient capacity. For a given depth, as the trench width increases, so does the load on the pipe. There is a width, however, beyond which no additional load is carried by a pipe. The limiting trench width, known as the *transition trench width*, is found by setting Eqs. 40.30 and 40.33 equal to each other.

$$B_{transition} = D\sqrt{\frac{C_p}{C}} \qquad 40.34$$

Boussinesq's equation can be used to calculate the live load on a pipe due to an external loading at the surface. For traffic loads, or if the surface loading is sudden, and if there is less than 3 ft of cover on the pipe, the calculated load should be increased by a multiplicative *impact factor*. (See Table 40.4.)

Table 40.2 *Approximate Buried Pipe Loading Coefficients*[a,b]

	backfill material				
specific weight (lbf/ft^3)	cohesionless granular material 100	sand and gravel 100	saturated topsoil 100	clay 120	saturated clay 130
h/B^c	values of C				
1	0.82	0.84	0.86	0.88	0.90
2	1.40	1.45	1.50	1.55	1.62
3	1.80	1.90	2.00	2.10	2.20
4	2.05	2.22	2.33	2.49	2.65
5	2.20	2.45	2.60	2.80	3.03
6	2.35	2.60	2.78	3.04	3.33
7	2.45	2.75	2.95	3.23	3.57
8	2.50	2.80	3.03	3.37	3.76
10	2.55	2.92	3.17	3.56	4.04
12	2.60	2.97	3.24	3.68	4.22
∞	2.60	3.00	3.25	3.80	4.60

(Multiply lbf/ft^3 by 16.018 to obtain kg/m^3.)
[a]Typical values are given. Other values may apply.
[b]Not for use with jacked conduit.
[c]Ratio of depth of backfill (measured from top of pipe) to trench width.

Table 40.3 *Typical Values of C_p*

h/D	rigid pipe, rigid surface, noncohesive backfill	flexible pipe, average conditions
1	1.2	1.1
2	2.8	2.6
3	4.7	4.0
4	6.7	5.4
6	11.0	8.2
8	16.0	11.0

Table 40.4 *Impact Factors for Buried Pipes*

by depth, ft (m)	(general use)
< 1 (0.3)	1.3
1–2 (0.3–0.6)	1.2
2–3 (0.6–0.9)	1.1
> 3 (> 0.9)	1.0
by use ($h < 3$ ft)	
highway	1.5
railway	1.75
airfield runway	1.00
airfield taxiway, apron	1.50

12. ALLOWABLE PIPE LOADS

The loadbearing ability of concrete pipes is determined in three-edge bearing tests. The *D-load strength*, reported in pounds per foot of pipe length per foot of inside diameter, is the crushing load that causes a 0.01 in (0.25 mm) wide crack. The *crushing strength (laboratory strength, cracking strength,* or *ultimate strength*) per unit length of pipe is calculated by multiplying the crushing load by the pipe's inside diameter.

$$\text{crushing strength} = (\text{D-load strength})D_i \quad \textbf{40.35}$$

The allowable load for a given pipe is obtained by dividing the known crushing strength by a factor of safety and multiplying by a bedding load factor. The factor of safety, F, varies from 1.25 for flexible pipe to 1.50 for rigid (including reinforced concrete) pipe.

$$w_{\text{allowable}} = \left(\begin{array}{c} \text{known pipe} \\ \text{crushing strength} \end{array} \right) \left(\frac{\text{LF}}{F} \right) \quad \text{[analysis]}$$
$$\textbf{40.36}$$

The *bedding load factor*, LF, depends on the bedding method. Class A bedding uses a concrete cradle. Class B bedding is compacted granular fill to half of the pipe's diameter. Class C is compacted granular fill or densely compacted backfill less than half the pipe's diameter. With class D, there is no bedding. The pipe is placed on a flat subgrade and backfilled.

Table 40.5 *Bedding Load Factors*

class A	class B	class C	class D
4.8[a]	1.9	1.5	1.1
3.4[b]			
2.8[c]			

[a]pipe with 1.0% reinforcing steel
[b]pipe with 0.4% reinforcing steel
[c]plain concrete

Source: *Soil Mechanics,* NAVFAC Design Manual DM-7.1, May 1982, Fig. 18, p. 7.1-186.

13. SLURRY TRENCHES AND WALLS

Slurry trenches are nonstructural barriers created by chemically solidifying soils and are used to dewater construction sites, contain hazardous groundwater contaminants, and hydraulically isolate holding ponds and lagoons. *Slurry walls* are reinforced semistructural walls used where more seepage control is needed than can be provided by H-pile or sheet-pile construction. The terms *cut-off wall* and *containment wall* are ambiguous and could refer to either slurry trenches or slurry walls.

Large trenches are opened with a backhoe or other conventional equipment. Trenches 4 to 12 in (10 to 39 cm) wide can be opened hydraulically. A slurry trenching fluid containing 1 to 6% *bentonite* (a naturally occurring clay) by weight fills the trench during excavation. The trench is backfilled, often with a mixture of bentonite and previously excavated soil. If the backfill is

PROFESSIONAL PUBLICATIONS, INC.

Geotechnical

soil, the term *soil bentonite (SB) slurry trench* is used. The backfill displaces the trenching fluid, and the trench is extended in a continuous operation. If the trench is filled with a bentonite and cement slurry, the term *cement bentonite slurry trench* is used. Trenches are "keyed" at the bottom by trenching some distance into an impermeable layer (i.e., an *aquiclude*) below.

Bentonite is not very chemically resistant, particularly when it is fully hydrated. Furthermore, its permeability is on the order of 10^{-7} to 10^{-8} cm/s, so a synthetic liner is required to contain hazardous waste.

Slurry walls are constructed by excavating a trench, supporting the open trench with bentonite slurry to prevent ground collapse, inserting a prefabricated rebar cage, and displacing the slurry with cast-in-place *tremie concrete*.

14. COFFERDAMS AND CAISSONS

A *cofferdam* is a temporary structure built to enclose a construction site. Cofferdams may be built to exclude both soil and water from the worksite. Single-walled types are used inland to restrain soil, and double-walled types are used on rivers.

A *caisson* is a permanent structure built in place for the purpose of supporting a bridge or building. A caisson becomes an integral part of the foundation. Some caissons are prebuilt and sunk into place, while others are built in place underwater.

15. GEOTEXTILES

Geotextiles (also known as *filter cloth, reinforcing fabric,* and *support membrane*) are fabrics used to stabilize and retain soil. (*Geosynthetics* is a term more appropriately applied to the synthetic sheets used as impermeable barriers.) Geotextiles are synthetic fabrics made from wood pulp (rayon and acetate), silica (fiberglass),

and petroleum (nylon-polyamide, polyolefin, polyester, and polystyrene). Polyolefin includes polyethylene and polypropylene. There are both woven and nonwoven varieties. Modern geotextiles are not subject to biological and chemical degradation. Commercial geotextiles have weights of 4 to 22 oz/yd^2 (140 to 750 g/m^2).

Geotextiles are used to prevent the mixing of dissimilar materials (e.g., an aggregate base and soil), which would reduce the support strength. Geotextiles provide a filtering function, allowing free passage of water while restraining soil movement. Other uses include pavement support, subgrade reinforcement, drainage, erosion control, and silt containment.

A significant fraction of all geotextiles is used in highway repair. A layer of geotextile spread at the base of a new roadway may reduce the amount of sand and gravel required while helping to control drainage. Geotextiles can also be used to prevent the infiltration of fine clays into underdrains. In another application, geotextiles strengthen flexible pavements when placed directly under the surface layer. An innovative use of geotextiles is to retard the onset of reflective cracking in flexible pavements.

Nonwoven fabrics appear to hold asphalt better than woven varieties. The mechanical properties of nonwoven fabrics are uniform in all directions. Nonwovens also have higher permeabilities. (Fabrics should have minimum permeabilities equal to the adjacent soil, and preferably at least ten times the soil's permeability.) However, *woven fabrics* are lower in cost and are often used for subgrade separation. Wovens have higher tensile strength, higher moduli, and lower elongation than nonwovens.

There is significant debate on the merits of woven versus nonwoven varieties. Most departments of transportation sidestep the woven-nonwoven issue by specifying minimum performance specifications and allowing contractors to make the selection.

Table 40.6 Typical DOT Minimum Geotextile Specifications (woven and nonwoven)

property	application			
	subsurface drainage	stabilization	rip rap filter	erosion control
grab tensile strength, lbf	100	200	250	120–200
elongation (minimum/maximum), %	20/100	20/70	20/70	15/30
Mullen burst time, min	150–200	250	500	200–400
puncture time, min	40–60	75	115	40–80
trapezoidal tear time, min	35–50	50	50	50–100
seam strength, lbf	100	200	250	120
permeability, cm/s	0.01–0.02	0.001	0.02	—

(Multiply lbf by 4.45 to obtain N.)
(Multiply cm/s by 0.0328 to obtain ft/sec.)

16. SOIL NAILING

Soil nailing is a slope-stabilization method that involves installing closely spaced "nails" in the soil/rock face to increase its overall shear strength. Nails are actually straight lengths of 60 ksi or 75 ksi rebar, typically no. 7 to no. 10, with a length of 75 to 100% of the cut height. The nails are passive in that they are not pretensioned when they are installed. The nails become tensioned only when the soil deflects laterally. Nails are typically inclined downward from the insertion point at 15° from the horizontal, but may be inclined anywhere from 10° to 30° to avoid obstructions or utilities.

In practice, a structural concrete facing (e.g., *shotcrete*) is first placed on the face over a wire mesh support grid. This facing connects the nails and supports the soil between them. Then holes are drilled through the concrete facing and soil/rock behind on a 4 to 6 ft (1.2 to 1.8 m) grid. Nails are inserted and grouted using a portland cement grout to bond the nails to the surrounding soil.

17. TRENCHLESS METHODS

Open-cut methods of pipeline installation and replacement may be the least expensive methods, particularly when lines are close to the surface. Trenchless methods can be used to create and extend existing pipelines, primarily sewer lines, in other situations and where the disruption of trenching is not acceptable. Trenchless installation options include pipe jacking, microtunneling, auger boring, and impact ramming. Renovation options include sliplining, inversion lining, and bursting methods. One problem common to all forms of piping system renovation is that, in providing new lines, connections get sealed off and must be reopened with robotic cutters. Another problem with relining is a reduction in capacity, though some improvement may result from improved hydraulic properties of the lining.

18. LIQUEFACTION

Liquefaction is a sudden drop in shear strength that can occur in soils of saturated cohesionless particles such as sand. The lower shear strength is manifested as a drop in bearing capacity. Continued cycles of reversed shear in a saturated sand layer can cause pore pressures to increase, which in turn decreases the effective stress and shear strength. When the shear strength drops to zero, the sand liquefies. In effect, the soil turns into a liquid, allowing everything it previously supported to sink.

Conditions most likely to contribute to or indicate a potential for liquefaction include (a) a lightly loaded sand layer within 50 to 65 ft (15 to 20 m) of the surface, (b) uniform particles of medium size, (c) saturated conditions below the water table, and (d) a low-penetration test value (i.e., a low N-value).

The *cyclic stress ratio* is a numerical rating of the potential for liquefaction in sands with depths up to 40 ft (12 m). This is the ratio of the average cyclic shear stress, τ_h, developed on the horizontal surfaces of the sand as a result of the earthquake loading to the initial vertical effective stress, σ_o', acting on the sand layer before the earthquake forces were applied. a_{max} is the maximum (effective peak) acceleration at the ground surface. σ_o is the total overburden pressure on the sand under consolidation. σ_o' is the initial effective overburden pressure. r_d is a stress reduction factor that varies from 1 at the ground surface to about 0.9 at a depth of 30 ft (9 m).

$$\frac{\tau_{h,\text{ave}}}{\sigma_o'} \approx 0.65 \left(\frac{a_{max}}{g} \right) \left(\frac{\sigma_o}{\sigma_o'} \right) r_d \qquad 40.37$$

The critical value of $\tau_{h,\text{ave}}/\sigma_o'$ that causes liquefaction must be determined from field or laboratory testing. A factor of safety of 1.3 to 1.5 is typical.

Geotechnical

Topic V: Structural

Chapter

(continued)

Structural

Chapter

41

Determinate Statics

Nomenclature

a	distance to lowest cable point	ft	m
c	parameter of the catenary	ft	m
d	distance or diameter	ft	m
D	diameter	ft	m
F	force	lbf	N
H	horizontal cable force	lbf	N
L	length	ft	m
M	moment	ft-lbf	N·m
n	number of sheaves	–	
r	position vector or radius	ft	m
R	reaction force	lbf	N
s	distance along cable	ft	m
S	sag	ft	m
T	tension	lbf	N
w	load per unit length	lbf/ft	N/m
W	weight	lbf	–
x	horizontal distance or position	ft	m
y	vertical distance or position	ft	m
z	distance or position along z-axis	ft	m

Symbols

ϵ	pulley loss factor	–	–
η	pulley efficiency	–	–
θ	angle	deg	deg
ϕ	angle	deg	deg

Subscripts

O	origin
P	point P
R	resultant

1. INTRODUCTION TO STATICS

Statics is a part of the subject known as *engineering mechanics*.[1] It is the study of rigid bodies that are stationary. To be stationary, a rigid body must be in static equilibrium. In the language of statics, a stationary rigid body has no *unbalanced forces* acting on it.

---

[1]Engineering mechanics also includes the subject of dynamics. Interestingly, the subject of mechanics of materials (i.e., strength of materials) is not part of engineering mechanics.

2. INTERNAL AND EXTERNAL FORCES

An *external force* is a force on a rigid body caused by other bodies. The applied force can be due to physical contact (i.e., pushing) or close proximity (e.g., gravitational, magnetic, or electrostatic forces). If unbalanced, an external force will cause motion of the body.

An *internal force* is one that holds parts of the rigid body together. Internal forces are the tensile and compressive forces within parts of the body as found from the product of stress and area. Although internal forces can cause deformation of a body, motion is never caused by internal forces.

3. UNIT VECTORS

A *unit vector* is a vector of unit length directed along a coordinate axis.[2] In the rectangular coordinate system, there are three unit vectors, **i**, **j**, and **k**, corresponding to the three coordinate axes, x, y, and z, respectively.[3] Unit vectors are used in vector equations to indicate direction without affecting magnitude. For example, the vector representation of a 97 N force in the negative x-direction would be written as $\mathbf{F} = -97\mathbf{i}$.

4. CONCENTRATED FORCES

A *force* is a push or pull that one body exerts on another. A *concentrated force*, also known as a *point force*, is a vector having magnitude, direction, and location (i.e., point of application) in three-dimensional space. In this chapter, the symbols **F** and F will be used to represent the vector and its magnitude, respectively.[4]

The vector representation of a three-dimensional force is given by Eq. 41.1. Of course, vector addition is required.

$$\mathbf{F} = F_x\mathbf{i} + F_y\mathbf{j} + F_z\mathbf{k} \qquad 41.1$$

Figure 41.1 *Components and Direction Angles of a Force*

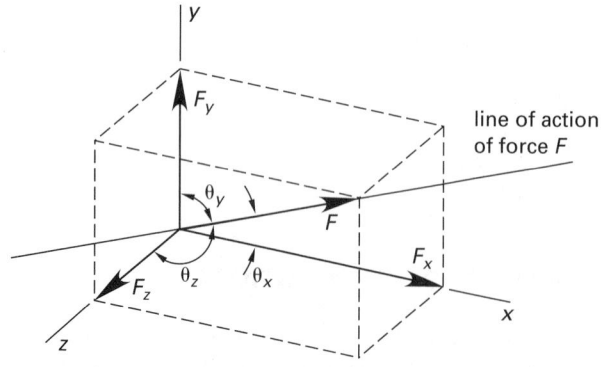

---

[2]Although polar, cylindrical, and spherical coordinate systems can have unit vectors also, this chapter is concerned only with the rectangular coordinate system.

[3]There are other methods of representing vectors, in addition to bold letters. For example, the unit vector **i** is represented in other sources as $\bar{i}$ and $\hat{i}$.

[4]As with the unit vectors, the symbols **F**, $\overline{F}$, and $\hat{F}$ are used in other sources interchangeably to represent the same vector.

If **u** is a *unit vector* in the direction of the force, the force can be represented as

$$\mathbf{F} = F\mathbf{u} \qquad 41.2$$

The components of the force can be found from the *direction cosines*, the cosines of the true angles made by the force vector with the x-, y-, and z-axes.

$$F_x = F\cos\theta_x \qquad 41.3$$

$$F_y = F\cos\theta_y \qquad 41.4$$

$$F_z = F\cos\theta_z \qquad 41.5$$

$$F = \sqrt{F_x^2 + F_y^2 + F_z^2} \qquad 41.6$$

The *line of action* of a force is the line in the direction of the force extended forward and backward. The force, **F**, and its unit vector, **u**, are along the line of action.

5. MOMENTS

Moment is the name given to the tendency of a force to rotate, turn, or twist a rigid body about an actual or assumed pivot point. (Another name for moment is *torque*, although torque is used mainly with shafts and other power-transmitting machines.) When acted upon by a moment, unrestrained bodies rotate. However, rotation is not required for the moment to exist. When a restrained body is acted upon by a moment, there is no rotation.

An object experiences a moment whenever a force is applied to it.[5] Only when the line of action of the force passes through the center of rotation (i.e., the actual or assumed pivot point) will the moment be zero.

Moments have primary dimensions of length $\times$ force. Typical units are foot-pounds, inch-pounds, and newton-meters.[6]

6. MOMENT OF A FORCE ABOUT A POINT

Moments are vectors. The moment vector, $\mathbf{M_O}$, for a force about point O is the *cross product* of the force, **F**, and the vector from point O to the point of application of the force, known as the *position vector*, **r**. The scalar product $|\mathbf{r}|\sin\phi$ is known as the *moment arm*, d.

$$\mathbf{M_O} = \mathbf{r} \times \mathbf{F} \qquad 41.7$$

$$M_O = |\mathbf{M_O}| = |\mathbf{r}|\,|\mathbf{F}|\sin\theta = d|\mathbf{F}| \quad [\theta \leq 180°] \qquad 41.8$$

---

[5]The moment may be zero, as when the moment arm length is zero, but there is a (trivial) moment nevertheless.

[6]Units of kilogram-force-meter have also been used in metric countries.

Foot-pounds and newton-meters are also the units of energy. To distinguish between moment and energy, some authors reverse the order of the units. Therefore, pound-feet and meter-newtons become the units of moment. This convention is not universal and is unnecessary since the context is adequate to distinguish between the two.

The line of action of the moment vector is normal to the plane containing the force vector and the position vector. The sense (i.e., the direction) of the moment is determined from the *right-hand rule*.

> *Right-hand rule:* Place the position and force vectors tail to tail. Close your right hand and position it over the pivot point. Rotate the position vector into the force vector, and position your hand such that your fingers curl in the same direction as the position vector rotates. Your extended thumb will coincide with the direction of the moment.[7]

Figure 41.2 *Right-Hand Rule*

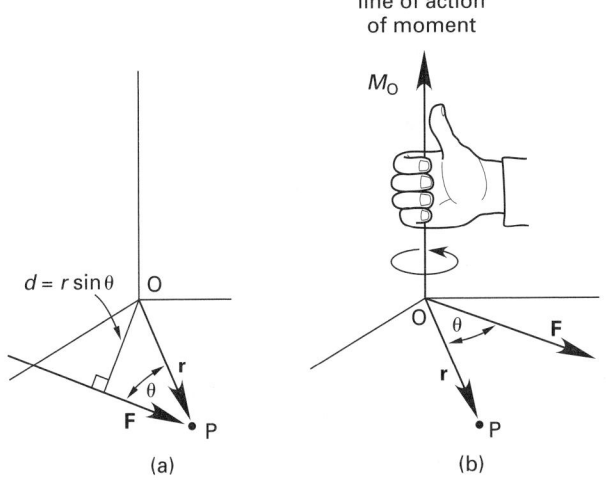

Figure 41.3 *Moment of a Force About a Line*

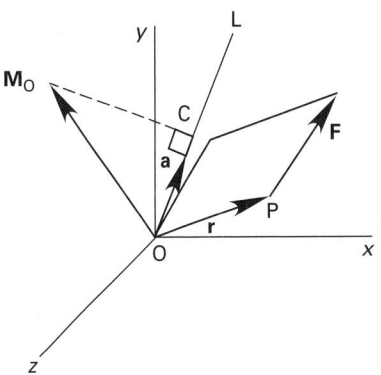

The moment M_{OL} of a force **F** about a line OL is the projection OC of the moment $\mathbf{M}_O$ onto the line. Equation 41.10 gives the moment of a force about a line. **a** is the unit vector directed along the line, and a_x, a_y, and a_z are the direction cosines of the axis OL. Notice that Eq. 41.10 is a dot product (i.e., a scalar).

$$M_{OL} = \mathbf{a} \cdot \mathbf{M}_O = \mathbf{a} \cdot (\mathbf{r} \times \mathbf{F})$$

$$= \begin{vmatrix} a_x & a_y & a_z \\ x_P - x_O & y_P - y_O & z_P - z_O \\ F_x & F_y & F_z \end{vmatrix} \quad \textit{41.10}$$

If point O is the origin, then Eq. 41.10 reduces to Eq. 41.11.

$$M_{OL} = \begin{vmatrix} a_x & a_y & a_z \\ x & y & z \\ F_x & F_y & F_z \end{vmatrix} \quad \textit{41.11}$$

7. VARIGNON'S THEOREM

Varignon's theorem is a statement of how the total moment is derived from a number of forces acting simultaneously at a point.

> *Varignon's theorem:* The sum of individual moments about a point caused by multiple concurrent forces is equal to the moment of the resultant force about the same point.

$$(\mathbf{r} \times \mathbf{F}_1) + (\mathbf{r} \times \mathbf{F}_2) + \cdots = \mathbf{r} \times (\mathbf{F}_1 + \mathbf{F}_2 + \cdots) \quad \textit{41.9}$$

8. MOMENT OF A FORCE ABOUT A LINE

Most rotating machines (motors, pumps, flywheels, etc.) have a fixed rotational axis. That is, the machines turn around a line, not around a point. The moment of a force about the rotational axis is not the same as the moment of the force about a point. In particular, the moment about a line is a scalar.[8]

9. COMPONENTS OF A MOMENT

The direction cosines of a force (vector) can be used to determine the components of the moment about the coordinate axes.

$$M_x = M \cos \theta_x \quad \textit{41.12}$$

$$M_y = M \cos \theta_y \quad \textit{41.13}$$

$$M_z = M \cos \theta_z \quad \textit{41.14}$$

Alternatively, the following three equations can be used to determine the components of the moment from a force applied at point (x, y, z) referenced to an origin at $(0, 0, 0)$.

$$M_x = yF_z - zF_y \quad \textit{41.15}$$

$$M_y = zF_x - xF_z \quad \textit{41.16}$$

$$M_z = xF_y - yF_x \quad \textit{41.17}$$

The resultant moment magnitude can be reconstituted from its components.

$$M = \sqrt{M_x^2 + M_y^2 + M_z^2} \quad \textit{41.18}$$

---

[7]The direction of a moment also corresponds to the direction a right-hand screw would progress if it was turned in the direction that rotates **r** into **F**.

[8]Some sources say that the moment of a force about a line can be interpreted as a moment directed along the line. However, this interpretation does not follow from vector operations.

10. COUPLES

Any pair of equal, opposite, and parallel forces constitutes a *couple*. A couple is equivalent to a single moment vector. Since the two forces are opposite in sign, the x-, y-, and z-components of the forces cancel out. Therefore, a body is induced to rotate without translation. A couple can be counteracted only by another couple. A couple can be moved to any location within the plane without affecting the equilibrium requirements.

Figure 41.4 *Couple*

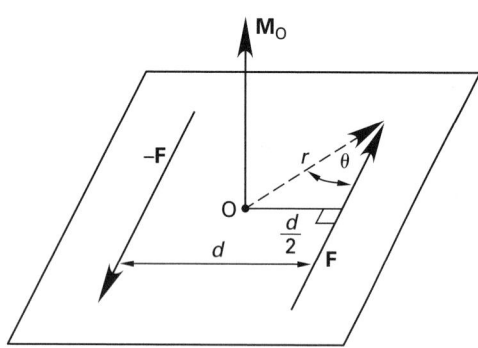

In Fig. 41.4, the equal but opposite forces produce a moment vector $\mathbf{M}_O$ of magnitude Fd. The two forces can be replaced by this moment vector, which can be moved to any location on a body. (Such a moment is known as a *free moment, moment of a couple,* or *coupling moment.*)

$$M_O = 2rF\sin\theta = Fd \qquad 41.19$$

11. EQUIVALENCE OF FORCES AND FORCE-COUPLE SYSTEMS

If a force, F, is moved a distance d from the original point of application, a couple, M, equal to Fd must be added to counteract the induced couple. The combination of the moved force and the couple is known as a *force-couple system*. Alternatively, a force-couple system can be replaced by a single force located a distance $d = M/F$ away.

12. RESULTANT FORCE-COUPLE SYSTEMS

The equivalence described in the previous section can be extended to three dimensions and multiple forces. Any collection of forces and moments in three-dimensional space is statically equivalent to a single resultant force vector plus a single resultant moment vector. (Either or both of these resultants can be zero.)

The x-, y-, and z-components of the resultant force are the sums of the x-, y-, and z-components of the individual forces, respectively.

$$F_{R,x} = \sum_i (F\cos\theta_x)_i \qquad 41.20$$

$$F_{R,y} = \sum_i (F\cos\theta_y)_i \qquad 41.21$$

$$F_{R,z} = \sum_i (F\cos\theta_z)_i \qquad 41.22$$

The resultant moment vector is more complex. It includes the moments of all system forces around the reference axes plus the components of all system moments.

$$M_{R,x} = \sum_i (yF_z - zF_y)_i + \sum_i (M\cos\theta_x)_i \quad 41.23$$

$$M_{R,y} = \sum_i (zF_x - xF_z)_i + \sum_i (M\cos\theta_y)_i \quad 41.24$$

$$M_{R,z} = \sum_i (xF_y - yF_x)_i + \sum_i (M\cos\theta_z)_i \quad 41.25$$

13. LINEAR FORCE SYSTEMS

A *linear force system* is one in which all forces are parallel and applied along a straight line. A straight beam loaded by several concentrated forces is an example of a linear force system.

Figure 41.5 *Linear Force System*

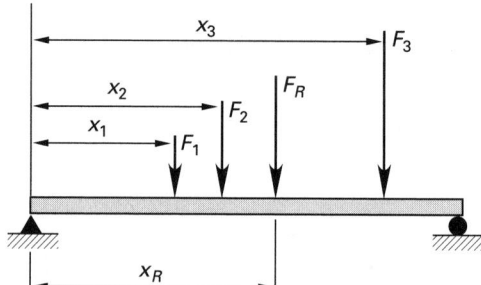

For the purposes of statics, all of the forces in a linear force system can be replaced by an *equivalent resultant force*, F_R, equal to the sum of the individual forces. The location of the equivalent force coincides with the location of the centroid of the force group.

$$F_R = \sum_i F_i \qquad 41.26$$

$$x_R = \frac{\sum_i F_i x_i}{\sum_i F_i} \qquad 41.27$$

14. DISTRIBUTED LOADS

If an object is continuously loaded over a portion of its length, it is subject to a *distributed load*. Distributed loads result from *dead load* (i.e., self-weight), hydrostatic pressure, and materials distributed over the object.

If the load per unit length at some point x is $w(x)$, the statically equivalent concentrated load, F_R, can be found from Eq. 41.28. The equivalent load is the area under the loading curve.

$$F_R = \int_{x=0}^{x=L} w(x)\, dx \qquad 41.28$$

Figure 41.6 *Distributed Loads on a Beam*

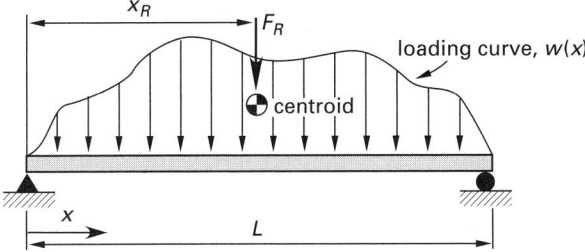

The location, x_R, of the equivalent load is calculated from Eq. 41.29. The location coincides with the centroid of the area under the loading curve and is referred to in some problems as the *center of pressure*.

$$x_R = \frac{\int_{x=0}^{x=L} x\, w(x)\, dx}{F_R} \qquad 41.29$$

For a straight beam of length L under a uniform transverse loading of w pounds per foot (newtons per meter),

$$F_R = wL \qquad 41.30$$

$$x_R = \frac{L}{2} \qquad 41.31$$

For a straight beam of length L under a triangular distribution that increases from zero (at $x = 0$) to w (at $x = L$),

$$F_R = \frac{wL}{2} \qquad 41.32$$

$$x_R = \frac{2L}{3} \qquad 41.33$$

Figure 41.7 *Special Cases of Distributed Loading*

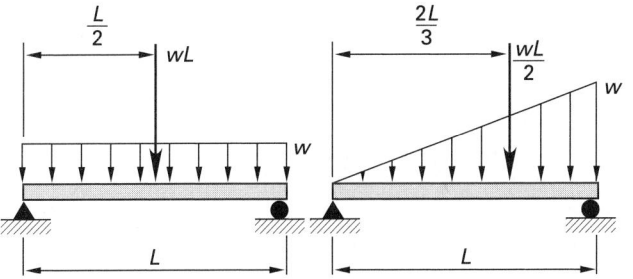

Example 41.1

Find the magnitude and location of the two equivalent forces on the two spans of the beam.

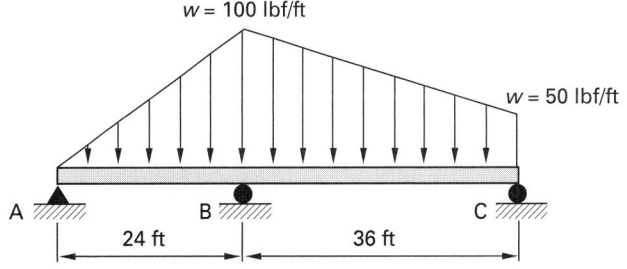

Solution

Span A–B
The area under the triangular loading curve is

$$A = \tfrac{1}{2}bh = \left(\tfrac{1}{2}\right)(24\text{ ft})\left(100\ \frac{\text{lbf}}{\text{ft}}\right)$$
$$= 1200\text{ lbf}$$

The centroid of the loading triangle is located at

$$x_{R,\text{A–B}} = \tfrac{2}{3}b = \left(\tfrac{2}{3}\right)(24\text{ ft})$$
$$= 16\text{ ft}$$

Span B–C
The area under the loading curve consists of a uniform load of 50 lbf/ft over the entire span B–C, plus a triangular load that starts at zero at point C and increases to 50 lbf/ft at point B. The area under the loading curve is

$$A = wL + \tfrac{1}{2}bh$$
$$= \left(50\ \frac{\text{lbf}}{\text{ft}}\right)(36\text{ ft}) + \left(\tfrac{1}{2}\right)(36\text{ ft})\left(50\ \frac{\text{lbf}}{\text{ft}}\right)$$
$$= 2700\text{ lbf}$$

The centroid of the trapezoidal loading curve[9] is located at

$$x_{R,\text{B–C}} = \left(\frac{h}{3}\right)\left(\frac{b+2t}{b+t}\right)$$
$$= \left(\frac{36\text{ ft}}{3}\right)\left(\frac{50\text{ ft} + (2)(100\text{ ft})}{50\text{ ft} + 100\text{ ft}}\right) = 20\text{ ft}$$

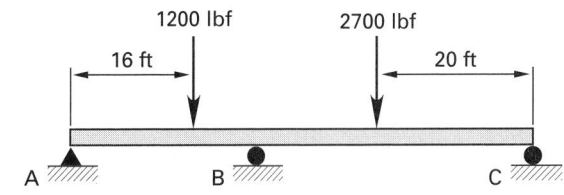

---

[9]The formula for the location of a centroid in a trapezoidal area is found in App. 42.A.

15. MOMENT FROM A DISTRIBUTED LOAD

The total force from a uniformly distributed load w over a distance x is wx. For the purposes of statics, the uniform load can be replaced by a concentrated force of wx located at the centroid of the distributed load, that is, at the midpoint, $x/2$, of the load. Therefore, the moment taken about one end of the distributed load is

$$M_{\text{distributed load}} = \text{force} \times \text{distance}$$
$$= wx\left(\frac{x}{2}\right) = \tfrac{1}{2}\,wx^2 \qquad 41.34$$

In general, the moment of a distributed load, uniform or otherwise, is the product of the total force and the distance to the centroid of the distributed load.

16. TYPES OF FORCE SYSTEMS

The complexity of methods used to analyze a statics problem depends on the configuration and orientation of the forces. Force systems can be divided into the following categories.

- *concurrent force system:* All of the forces act at the same point.

- *collinear force system:* All of the forces share the same line of action.

- *parallel force system:* All of the forces are parallel (though not necessarily in the same direction).

- *coplanar force system:* All of the forces are in a plane.

- *general three-dimensional system:* This category includes all other combinations of nonconcurrent, nonparallel, and noncoplanar forces.

17. CONDITIONS OF EQUILIBRIUM

An object is static when it is stationary. To be stationary, all of the forces on the object must be in equilibrium.[10] For an object to be in equilibrium, the resultant force and moment vectors must both be zero.

$$\mathbf{F}_R = \sum \mathbf{F} = 0 \qquad 41.35$$

$$F_R = \sqrt{F_{R,x}^2 + F_{R,y}^2 + F_{R,z}^2} = 0 \qquad 41.36$$

$$\mathbf{M}_R = \sum \mathbf{M} = 0 \qquad 41.37$$

$$M_R = \sqrt{M_{R,x}^2 + M_{R,y}^2 + M_{R,z}^2} = 0 \qquad 41.38$$

Since the square of any nonzero quantity is positive, Eqs. 41.39 through 41.44 follow directly from Eqs. 41.36 and 41.38.

$$F_{R,x} = 0 \qquad 41.39$$

$$F_{R,y} = 0 \qquad 41.40$$

---

[10]Thus the term *static equilibrium*, though widely used, is redundant.

$$F_{R,z} = 0 \qquad 41.41$$

$$M_{R,x} = 0 \qquad 41.42$$

$$M_{R,y} = 0 \qquad 41.43$$

$$M_{R,z} = 0 \qquad 41.44$$

Equations 41.39 through 41.44 seem to imply that six simultaneous equations must be solved in order to determine whether a system is in equilibrium. While this is true for general three-dimensional systems, fewer equations are necessary with most problems. Table 41.1 can be used as a guide to determine which equations are most helpful in solving different categories of problems.

Table 41.1 Number of Equilibrium Conditions
Required to Solve Different Force Systems

type of force system	two-dimensional	three-dimensional
general	3	6
coplanar	3	3
concurrent	2	3
parallel	2	3
coplanar, parallel	2	2
coplanar, concurrent	2	2
collinear	1	1

18. TWO- AND THREE-FORCE MEMBERS

Members limited to loading by two or three forces are special cases of equilibrium. A *two-force member* can be in equilibrium only if the two forces have the same line of action (i.e., are collinear) and are equal but opposite. In most cases, two-force members are loaded axially, and the line of action coincides with the member's longitudinal axis. By choosing the coordinate system so that one axis coincides with the line of action, only one equilibrium equation is needed.

A *three-force member* can be in equilibrium only if the three forces are concurrent or parallel. Stated another way, the force polygon of a three-force member in equilibrium must close on itself.

19. REACTIONS

The first step in solving most statics problems is to determine the reaction forces (i.e., the *reactions*) supporting the body. The manner in which a body is supported determines the type, location, and direction of the reactions. Conventional symbols are often used to define the type of support (such as pinned, roller, etc.). Examples of the symbols are shown in Table 41.2.

For beams, the two most common types of supports are the roller support and the pinned support. The *roller support*, shown as a cylinder supporting the beam, supports vertical forces only. Rather than support a horizontal force, a roller support simply rolls into a new equilibrium position. Only one equilibrium equation

Table 41.2 *Types of Two-Dimensional Supports*

type of support	reactions and moments	number of unknowns[a]
simple, roller, rocker, ball, or frictionless surface	reaction normal to surface, no moment	1
cable in tension, or link	reaction in line with cable or link, no moment	1
frictionless guide or collar	reaction normal to rail, no moment	1
built-in, fixed support	two reaction components, one moment	3
frictionless hinge, pin connection, or rough surface	reaction in any direction, no moment	2

[a]The number of unknowns is valid for two-dimensional problems only.

(i.e., the sum of vertical forces) is needed at a roller support. Generally, the terms *simple support* and *simply supported* refer to a roller support.

The *pinned support*, shown as a pin and clevis, supports both vertical and horizontal forces. Two equilibrium equations are needed.

Generally, there will be vertical and horizontal components of a reaction when one body touches another. However, when a body is in contact with a *frictionless surface*, there is no frictional force component parallel to the surface. Therefore, the reaction is normal to the contact surfaces. The assumption of frictionless contact is particularly useful when dealing with systems of spheres and cylinders in contact with rigid supports. Frictionless contact is also assumed for roller and rocker supports.[11]

20. DETERMINACY

When the equations of equilibrium are independent, a rigid body force system is said to be *statically determinate*. A statically determinate system can be solved for all unknowns, which are usually reactions supporting the body.

When the body has more supports than are necessary for equilibrium, the force system is said to be *statically indeterminate*. In a statically indeterminate system, one or more of the supports or members can be removed or reduced in restraint without affecting the equilibrium position.[12] Those supports and members are known as *redundant members*. The number of redundant members is known as the *degree of indeterminacy*. Figure 41.8 illustrates several common indeterminate structures.

A statically indeterminate body requires additional equations to supplement the equilibrium equations. The additional equations typically involve deflections and depend on mechanical properties of the body.

21. TYPES OF DETERMINATE BEAMS

Figure 41.9 illustrates the terms used to describe determinate beam types.

22. FREE-BODY DIAGRAMS

A *free-body diagram* is a representation of a body in equilibrium. It shows all applied forces, moments, and reactions. Free-body diagrams do not consider the internal structure or construction of the body, as Fig. 41.10 illustrates.

[11]Frictionless surface contact, which requires only one equilibrium equation, should not be confused with a frictionless pin connection, which requires two equilibrium equations. A pin connection with friction introduces a moment at the connection, increasing the number of required equilibrium equations to three.

[12]An example of a support reduced in restraint is a pinned joint replaced by a roller joint. The pinned joint restrains the body vertically and horizontally, requiring two equations of equilibrium. The roller joint restrains the body vertically only and requires one equilibrium equation.

Figure 41.8 *Examples of Indeterminate Systems*

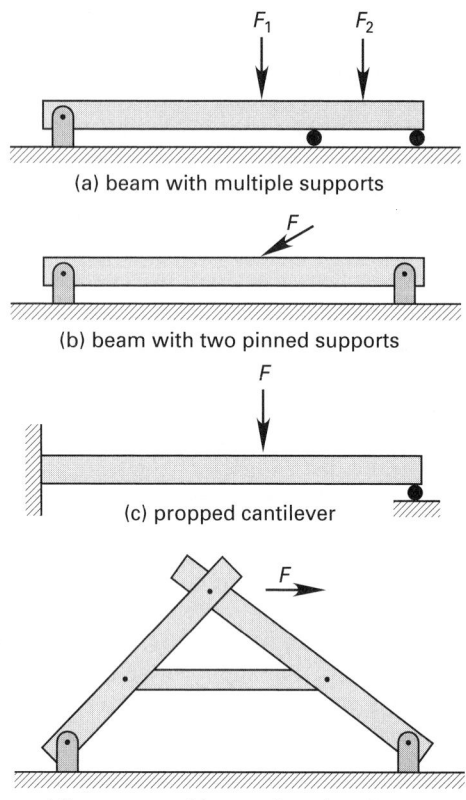

(a) beam with multiple supports

(b) beam with two pinned supports

(c) propped cantilever

(d) structure with two pinned supports

Figure 41.9 *Types of Determinate Beams*

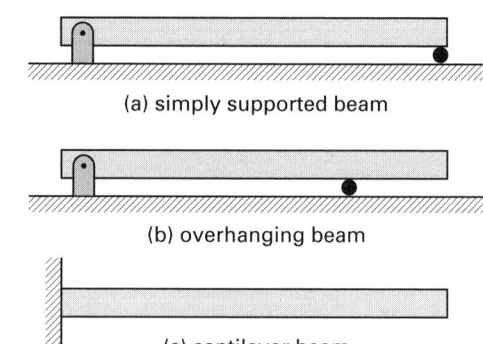

(a) simply supported beam

(b) overhanging beam

(c) cantilever beam

Figure 41.10 *Bodies and Free Bodies*

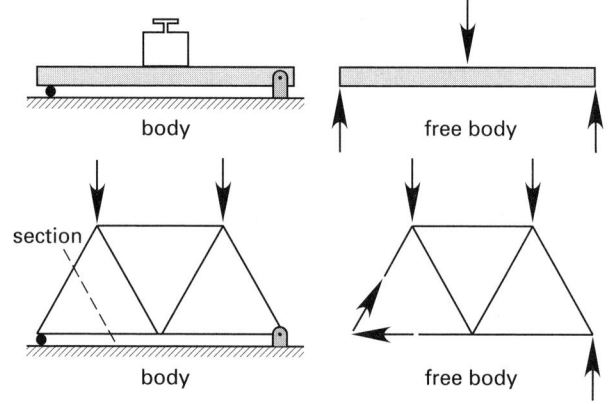

Since the body is in equilibrium, the resultants of all forces and moments on the free body are zero. In order to maintain equilibrium, any portions of the body that are removed must be replaced by the forces and moments those portions impart to the body. Typically, the body is isolated from its physical supports in order to help evaluate the reaction forces. In other cases, the body may be sectioned (i.e., cut) in order to determine the forces at the section.

23. FINDING REACTIONS IN TWO DIMENSIONS

The procedure for finding determinate reactions in two-dimensional problems is straightforward. Determinate structures will have either a roller support and a pinned support or two roller supports.

step 1: Establish a convenient set of coordinate axes. (To simplify the analysis, one of the coordinate directions should coincide with the direction of the forces and reactions.)

step 2: Draw the free-body diagram.

step 3: Resolve the reaction at the pinned support (if any) into components normal and parallel to the coordinate axes.

step 4: Establish a positive direction of rotation (e.g., clockwise) for purposes of taking moments.

step 5: Write the equilibrium equation for moments about the pinned connection. (By choosing the pinned connection as the point about which to take moments, the pinned connection reactions do not enter into the equation.) This will usually determine the vertical reaction at the roller support.

step 6: Write the equilibrium equation for the forces in the vertical direction. Usually, this equation will have two unknown vertical reactions.

step 7: Substitute the known vertical reaction from step 5 into the equilibrium equation from step 6. This will determine the second vertical reaction.

step 8: Write the equilibrium equation for the forces in the horizontal direction. Since there is a maximum of one unknown reaction component in the horizontal direction, this step will determine that component.

step 9: If necessary, combine the vertical and horizontal force components at the pinned connection into a resultant reaction.

Example 41.2

Determine the reactions, R_1 and R_2, on the following beam.

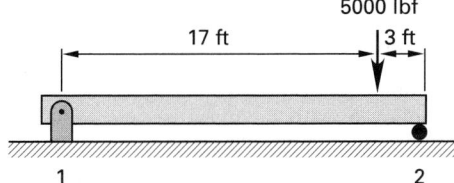

Solution

step 1: The x- and y-axes are established parallel and perpendicular to the beam.

step 2: The free-body diagram is

step 3: R_1 is a pinned support. Therefore, it has two components, $R_{1,x}$ and $R_{1,y}$.

step 4: Assume clockwise moments are positive.

step 5: Take moments about the left end and set them equal to zero. Use Eq. 41.37.

$$\sum M_{\text{left end}} = (5000 \text{ lbf})(17 \text{ ft}) - R_2(20 \text{ ft})$$
$$= 0$$
$$R_2 = 4250 \text{ lbf}$$

step 6: The equilibrium equation for the vertical direction is given by Eq. 41.43.

$$\sum F_y = R_{1,y} + R_2 - 5000 \text{ lbf}$$
$$= 0$$

step 7: Substituting R_2 into the vertical equilibrium equation,

$$R_{1,y} + 4250 \text{ lbf} - 5000 \text{ lbf} = 0$$
$$R_{1,y} = 750 \text{ lbf}$$

step 8: There are no applied forces in the horizontal direction. Therefore, the equilibrium equation is given by Eq. 41.39.

$$\sum F_x = R_{1,x} + 0$$
$$= 0$$
$$R_{1,x} = 0$$

24. COUPLES AND FREE MOMENTS

Once a couple on a body is known, the derivation and source of the couple are irrelevant. When the moment on a body is 80 N·m, it does not make any difference whether the force is 40 N with a lever arm of 2 m, or 20 N with a lever arm of 4 m, and so on. Therefore, the point of application of a couple is disregarded when writing the moment equilibrium equation. For this reason, the term *free moment* is used synonymously with *couple*.

Figure 41.11 illustrates two diagrammatic methods of indicating the application of a free moment.

Figure 41.11 *Free Moments*

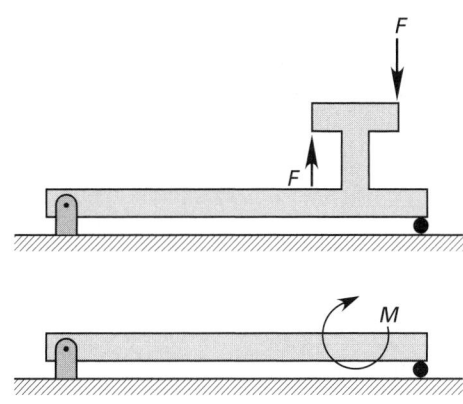

Example 41.3

What is the reaction R_2 for the beam shown?

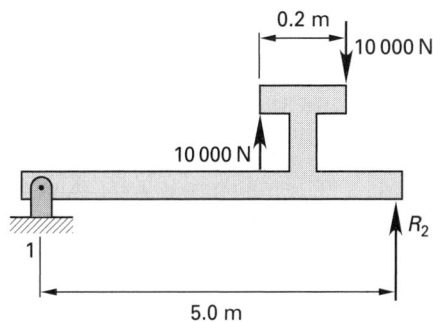

Solution

The two couple forces are equal and cancel each other as they come down the stem of the tee bracket. Therefore, there are no applied vertical forces.

The couple has a value given by Eq. 41.7.

$$M = (10\,000 \text{ N})(0.2 \text{ m}) = 2000 \text{ N·m} \quad \text{[clockwise]}$$

Choose clockwise as the direction for positive moments. Taking moments about the pinned connection and using

Eq. 41.37,

$$\sum M = 2000 \text{ N·m} - R_2(5 \text{ m})$$
$$= 0$$
$$R_2 = 400 \text{ N}$$

25. INFLUENCE LINES FOR REACTIONS

An *influence line* (also known as an *influence graph* and *influence diagram*) is a graph of the magnitude of a reaction as a function of the load placement.[13] The x-axis of the graph corresponds to the location on the body (along the length of a beam). The y-axis corresponds to the magnitude of the reaction.

By convention (and to generalize the graph for use with any load), the load is taken as one force unit. Therefore, for an actual load of F units, the actual reaction, R, is the product of the actual load and the influence line ordinate.

$$R = F \times \text{influence line ordinate} \qquad 41.45$$

Example 41.4

Draw the influence line for the left reaction for the beam shown.

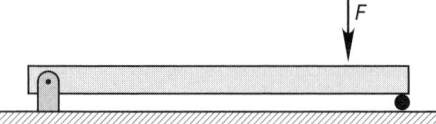

Solution

If the unit load is at the left end, the left reaction will be 1.0. If the unit load is at the right end, it will be supported entirely by the right reaction, so the left reaction will be zero. The influence line for the left reaction varies linearly for intermediate load placement.

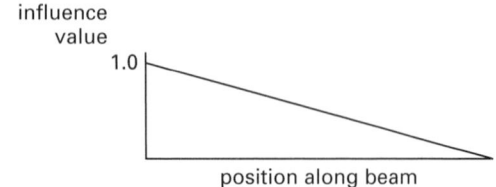

26. HINGES

Hinges are added to structures to prevent translation while permitting rotation. A frictionless hinge can support a force, but it cannot transmit a moment. Since

the moment is zero at a hinge, a structure can be sectioned at the hinge and the remainder of the structure can be replaced by only a force.

Example 41.5

Calculate the reaction R_3 and the hinge force on the two-span beam shown.

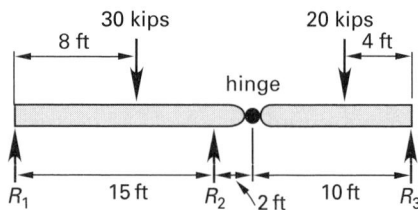

Solution

At first, this beam may appear to be statically indeterminate since it has three supports. However, the moment is known to be zero at the hinge. Therefore, the hinged portion of the span can be isolated.

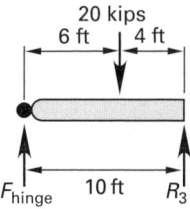

Reaction R_3 is found by taking moments about the hinge. Assume clockwise moments are positive.

$$\sum M_{\text{hinge}} = (20{,}000 \text{ lbf})(6 \text{ ft}) - R_3(10 \text{ ft}) = 0$$
$$R_3 = 12{,}000 \text{ lbf}$$

The hinge force is found by summing vertical forces on the isolated section.

$$\sum F_y = F_{\text{hinge}} + 12{,}000 \text{ lbf} - 20{,}000 \text{ lbf} = 0$$
$$F_{\text{hinge}} = 8000 \text{ lbf}$$

27. LEVERS

A *lever* is a simple mechanical machine with the ability to increase an applied force. The ratio of the load-bearing force to applied force (i.e., the *effort*) is known as the *mechanical advantage* or *force amplification*. As Fig. 41.12 shows, the mechanical advantage is equal to the ratio of lever arms.

$$\begin{aligned}\text{mechanical} \\ \text{advantage}\end{aligned} = \frac{F_{\text{load}}}{F_{\text{applied}}} = \frac{\text{applied force lever arm}}{\text{load lever arm}}$$
$$= \frac{\text{distance moved by applied force}}{\text{distance moved by load}} \qquad 41.46$$

---

[13]Influence diagrams can also be drawn for moments, shears, and deflections.

Figure 41.12 Lever

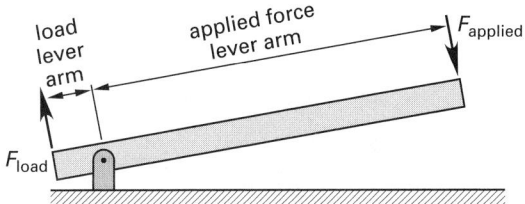

28. PULLEYS

A *pulley* (also known as a *sheave*) is used to change the direction of an applied tensile force. A series of pulleys working together (known as a *block and tackle*) can also provide *pulley advantage* (i.e., mechanical advantage). A *hoist* is any device used to raise or lower an object. A hoist may contain one or more pulleys.

If the pulley is attached by a bracket or cable to a fixed location, it is said to be a *fixed pulley*. If the pulley is attached to a load, or if the pulley is free to move, it is known as a *free pulley*.

Most simple problems disregard friction and assume that all ropes are parallel.[14] In such cases, the pulley advantage is equal to the number of ropes coming to and going from the load-carrying pulley. The diameters of the pulleys are not factors in calculating the pulley advantage.

In other cases, a *loss factor*, ϵ, is used to account for rope rigidity. For most wire ropes and chains with 180° contact, the loss factor at low speeds varies between 1.03 and 1.06. The loss factor is the reciprocal of the *pulley efficiency*, η.

$$\epsilon = \frac{\text{applied force}}{\text{load}} = \frac{1}{\eta} \qquad 41.47$$

29. AXIAL MEMBERS

An *axial member* is capable of supporting axial forces only and is loaded only at its joints (i.e., ends). This type of performance can be achieved through the use of frictionless bearings or smooth pins at the ends. Since the ends are assumed to be pinned (i.e., rotation-free), an axial member cannot support moments. The weight of the member is disregarded or is included in the joint loading.

An axial member can be in either tension or compression. It is common practice to label forces in axial members as (T) or (C) for tension or compression, respectively. Alternatively, tensile forces can be written as positive numbers, while compressive forces are written as negative numbers.

[14]Although the term *rope* is used here, the principles apply equally well to wire rope, cables, chains, belts, etc.

The members in simple trusses are assumed to be axial members. Each member is identified by its endpoints, and the force in a member is designated by the symbol for the two endpoints. For example, the axial force in a member connecting points C and D will be written as **CD**. Similarly, $\mathbf{EF}_y$ is the y-component of the force in the member connecting points E and F.

For equilibrium, the resultant forces at the two joints must be equal, opposite, and collinear. This applies to the total (resultant) force as well as to the x- and y-components at those joints.

Table 41.3 *Mechanical Advantages of Rope-Operated Machines*

	fixed sheave	free sheave	ordinary pulley block (n sheaves)	differential pulley block
F_{ideal}	W	$\dfrac{W}{2}$	$\dfrac{W}{n}$	$\left(\dfrac{W}{2}\right)\left(1-\dfrac{d}{D}\right)$
F to raise load	ϵW	$\dfrac{\epsilon W}{1+\epsilon}$	$\dfrac{\epsilon^n(\epsilon-1)W}{\epsilon^n-1}$	$\dfrac{\left(\epsilon^2-\dfrac{d}{D}\right)W}{1+\epsilon}$
F to lower load	$\dfrac{W}{\epsilon}$	$\dfrac{W}{1+\epsilon}$	$\left(\dfrac{\frac{1}{\epsilon}-1}{1-\epsilon^n}\right)W$	$\left(\dfrac{\epsilon W}{1+\epsilon}\right)\left(\dfrac{1}{\epsilon^2}-\dfrac{d}{D}\right)$
ratio of distance of force to distance of load	1	2	n	$\dfrac{2D}{D-d}$

30. FORCES IN AXIAL MEMBERS

The line of action of a force in an axial member coincides with the longitudinal axis of the member. Depending on the orientation of the coordinate axis system, the direction of the longitudinal axis will have both x- and y-components. Therefore, the force in an axial member will generally have both x- and y-components.

The following four general principles are helpful in determining the force in an axial member.

- A horizontal member carries only horizontal loads. It cannot carry vertical loads.

- A vertical member carries only vertical loads. It cannot carry horizontal loads.

- The vertical component of an axial member's force is equal to the vertical component of the load applied to the member.

- The total and component forces in an inclined member are proportional to the sides of the triangle outlined by the member and the coordinate axes.[15]

Example 41.6

Member BC is an inclined axial member pinned at both ends and oriented as shown. A vertical 1000 N force is applied to the top end. What are the x- and y-components of the force in member BC? What is the total force in the member?

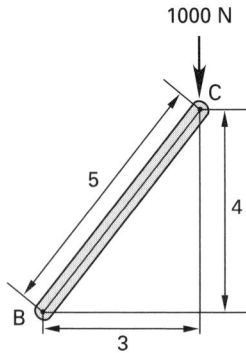

Solution

From the third principle,

$$\mathbf{BC}_y = 1000 \text{ N}$$

From the fourth principle,

$$\begin{aligned}
\mathbf{BC}_x &= \left(\frac{3}{4}\right) \mathbf{BC}_y \\
&= \left(\frac{3}{4}\right) (1000 \text{ N}) \\
&= 750 \text{ N}
\end{aligned}$$

The resultant force in member BC can be calculated from the Pythagorean theorem. However, it is easier to use the fourth principle.

$$\begin{aligned}
\mathbf{BC} &= \left(\frac{5}{4}\right) \mathbf{BC}_y \\
&= \left(\frac{5}{4}\right) (1000 \text{ N}) \\
&= 1250 \text{ N}
\end{aligned}$$

Example 41.7

The l2 ft long axial member FG supports an axial force of 180 lbf. What are the x- and y-components of the applied force?

---

[15]This is an application of the principle of similar triangles.

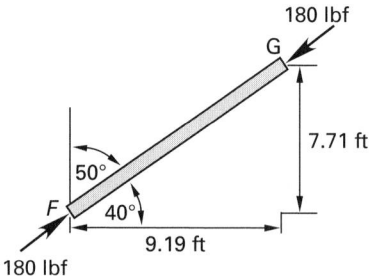

Solution

method 1: direction cosines

The x- and y-direction angles are 40° and 90° − 40° = 50°, respectively.

$$\begin{aligned}
\mathbf{FG}_x &= (180 \text{ lbf}) \cos 40° = 137.9 \text{ lbf} \\
\mathbf{FG}_y &= (180 \text{ lbf}) \cos 50° = 115.7 \text{ lbf}
\end{aligned}$$

method 2: similar triangles

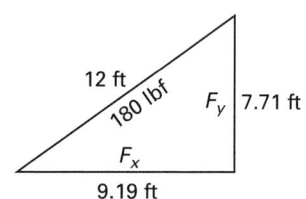

$$\begin{aligned}
\mathbf{FG}_x &= \left(\frac{9.19 \text{ ft}}{12 \text{ ft}}\right) (180 \text{ lbf}) = 137.9 \text{ lbf} \\
\mathbf{FG}_y &= \left(\frac{7.71 \text{ ft}}{12 \text{ ft}}\right) (180 \text{ lbf}) = 115.7 \text{ lbf}
\end{aligned}$$

31. TRUSSES

A *truss* or *frame* is a set of pin-connected axial *members* (i.e., *two-force members*). The connection points are known as *joints*. Member weights are disregarded, and truss loads are applied only at joints. A *structural cell* consists of all members in a closed loop of members. For the truss to be stable (i.e., to be a *rigid truss*), all of the structural cells must be triangles. Figure 41.13 identifies *chords*, *end posts*, *panels*, and other elements of a typical *bridge truss*.

A *trestle* is a braced structure spanning a ravine, gorge, or other land depression in order to support a road or rail line. Trestles are usually indeterminate, have multiple earth contact points, and are more difficult to evaluate than simple trusses.

Figure 41.13 *Parts of a Bridge Truss*

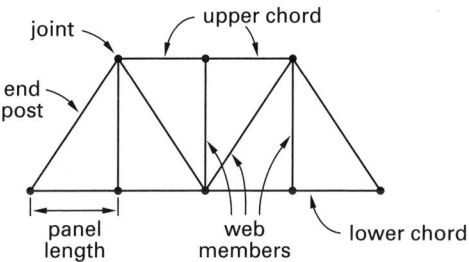

Several types of trusses have been given specific names. Some of the more common types of named trusses are shown in Fig. 41.14.

Figure 41.14 *Special Types of Trusses*

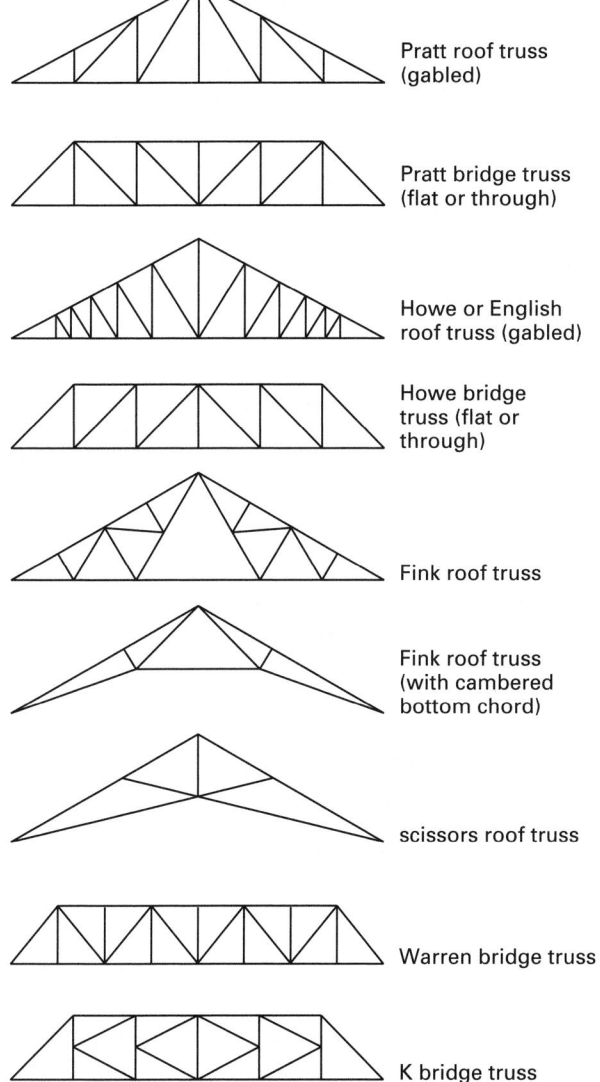

Pratt roof truss (gabled)

Pratt bridge truss (flat or through)

Howe or English roof truss (gabled)

Howe bridge truss (flat or through)

Fink roof truss

Fink roof truss (with cambered bottom chord)

scissors roof truss

Warren bridge truss

K bridge truss

Truss loads are considered to act only in the plane of a truss. Therefore, trusses are analyzed as two-dimensional structures. Forces in truss members hold the various truss parts together and are known as *internal forces*. The internal forces are found by applying equations of equilibrium to appropriate free-body diagrams.

Although free-body diagrams of truss members can be drawn, this is not usually done. Instead, free-body diagrams of the pins (i.e., the joints) are drawn. A pin in compression will be shown with force arrows pointing toward the pin, away from the member. (Similarly, a pin in tension will be shown with force arrows pointing away from the pin, toward the member.)[16]

With typical bridge trusses supported at the ends and loaded downward at the joints, the upper chords are almost always in compression, and the end panels and lower chords are almost always in tension.

32. DETERMINATE TRUSSES

A truss will be statically determinate if Eq. 41.48 holds.

$$\text{no. of members} = 2(\text{no. of joints}) - 3 \qquad 41.48$$

If the left-hand side is greater than the right-hand side (i.e., there are *redundant members*), the truss is statically indeterminate. If the left-hand side is less than the right-hand side, the truss is unstable and will collapse under certain types of loading.

Equation 41.48 is a special case of the following general criterion.

$$
\begin{aligned}
&\text{no. of members} \\
&\quad + \text{no. of reactions} \\
&\qquad - 2(\text{no. of joints}) = 0 \quad \text{[determinate]} \\
&\qquad\qquad\qquad\qquad\quad\; > 0 \quad \text{[indeterminate]} \\
&\qquad\qquad\qquad\qquad\quad\; < 0 \quad \text{[unstable]} \qquad 41.49
\end{aligned}
$$

Furthermore, Eq. 41.48 is a necessary, but not sufficient, condition for truss stability. It is possible to arrange the members in such a manner as to not contribute to truss stability. This will seldom be the case in actual practice, however.

33. ZERO-FORCE MEMBERS

Forces in truss members can sometimes be determined by inspection. One of these cases is where there are *zero-force members*. A third member framing into a joint already connecting two collinear members carries no internal force unless there is a load applied at that joint. Similarly, both members forming an apex of the truss are zero-force members unless there is a load applied at the apex.

---

[16]The method of showing tension and compression on a truss drawing may appear incorrect. This is because the arrows show the forces on the pins, not on the members.

Figure 41.15 *Zero-Force Members*

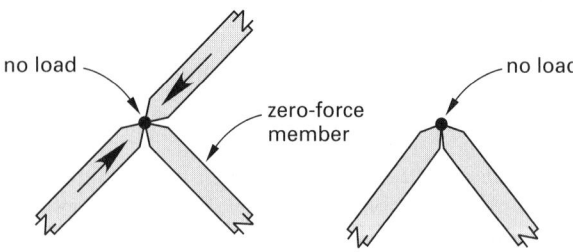

34. METHOD OF JOINTS

The *method of joints* is one of three methods that can be used to find the internal forces in each truss member. This method is useful when most or all of the truss member forces are to be calculated. Because this method advances from joint to adjacent joint, it is inconvenient when a single isolated member force is to be calculated.

The method of joints is a direct application of the equations of equilibrium in the x- and y-directions. Traditionally, the method starts by finding the reactions supporting the truss. Next, the joint at one of the reactions is evaluated, which determines all the member forces framing into the joint. Then, knowing one or more of the member forces from the previous step, an adjacent joint is analyzed. The process is repeated until all the unknown quantities are determined.

At a joint, there may be up to two unknown member forces, each of which can have dependent x- and y-components.[17] Since there are two equilibrium equations, the two unknown forces can be determined. Even though determinate, however, the sense of a force will often be unknown. If the sense cannot be determined by logic, an arbitrary decision can be made. If the incorrect direction is chosen, the force will be negative.

Example 41.8

Use the method of joints to calculate the force **BD** in the truss shown.

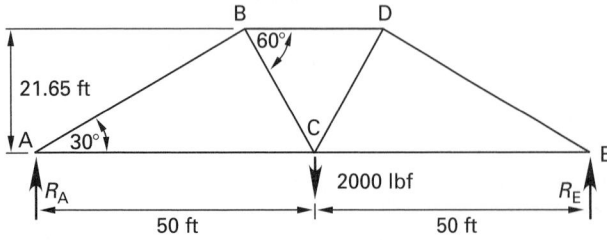

---

[17]Occasionally, there will be three unknown member forces. In that case, an additional equation must be derived from an adjacent joint.

Solution

First, find the reactions. Assume clockwise is positive and take moments about point A.

$$\sum M_A = (2000 \text{ lbf})(50 \text{ ft}) - R_E(50 \text{ ft} + 50 \text{ ft}) = 0$$
$$R_E = 1000 \text{ lbf}$$

Since the sum of forces in the y-direction is also zero,

$$\sum F_y = R_A + 1000 \text{ lbf} - 2000 \text{ lbf} = 0$$
$$R_A = 1000 \text{ lbf}$$

There are three unknowns at joint B (and also at D). Therefore, the analysis must start at joint A (or E) where there are only two unknowns (forces **AB** and **AC**).

The free-body diagram of pin A is shown. The direction of R_A is known to be upward. The directions of forces **AB** and **AC** can be assumed, but logic can be used to determine them. Only the vertical component of **AB** can oppose R_A. Therefore, **AB** is directed downward. (This means that member AB is in compression.) Similarly, **AC** must oppose the horizontal component of **AB**. Therefore, **AC** is directed to the right. (This means that member AC is in tension.)

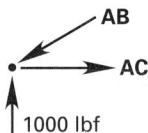

Resolve force **AB** into horizontal and vertical components using trigonometry, direction cosines, or similar triangles. (R_A and **AC** are already parallel to an axis.) Then, use the equilibrium equations to determine the forces.

By inspection, $AB_y = 1000$ lbf.

$$\mathbf{AB}_y = \mathbf{AB}\sin 30°$$
$$1000 \text{ lbf} = \mathbf{AB}(0.5)$$
$$\mathbf{AB} = 2000 \text{ lbf} \quad (C)$$
$$\mathbf{AB}_x = \mathbf{AB}\cos 30°$$
$$= (2000 \text{ lbf})(0.866)$$
$$= 1732 \text{ lbf}$$

Now, draw the free-body diagram of pin B. (Notice that the direction of force **AB** is toward the pin, just as it was for pin A.) Although the true directions of the forces are unknown, they can be determined logically. The direction of force **BC** is chosen to counteract the vertical component of force **AB**. The direction of force **BD** is chosen to counteract the horizontal components of forces **AB** and **BC**.

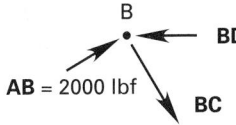

$\mathbf{AB}_x$ and $\mathbf{AB}_y$ are already known. Resolve the force $\mathbf{BC}$ into horizontal and vertical components.

$$\mathbf{BC}_x = \mathbf{BC}\sin 30° = \mathbf{BC}(0.5)$$
$$\mathbf{BC}_y = \mathbf{BC}\cos 30° = \mathbf{BC}(0.866)$$

Now, write the equations of equilibrium for point B.

$$\sum F_x = 1732 + 0.5\mathbf{BC} - \mathbf{BD} = 0$$
$$\sum F_y = 1000 \text{ lbf} - 0.866\mathbf{BC} = 0$$

From the second equation, $\mathbf{BC} = 1155$ lbf. Substituting this into the first equation,

$$1732 \text{ lbf} + (0.5)(1155 \text{ lbf}) - \mathbf{BD} = 0$$
$$\mathbf{BD} = 2310 \text{ lbf} \quad (C)$$

Since $\mathbf{BD}$ turned out to be positive, its direction was chosen correctly.

The direction of the arrow indicates that the member is compressing the pin. Consequently, the pin is compressing the member. Member BD is in compression.

If the process is continued, all forces can be determined. However, the truss is symmetrical, and it is not necessary to evaluate every joint to calculate all forces.

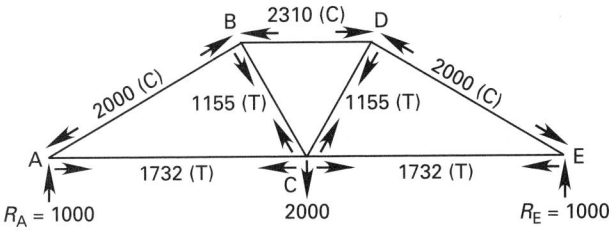

35. CUT-AND-SUM METHOD

The *cut-and-sum method* can be used to find forces in inclined members. This method is strictly an application of the vertical equilibrium condition ($\Sigma F_y = 0$).

The method starts by finding all of the support reactions on a truss. Then a cut is made through the truss in such a way as to pass through one inclined or vertical member only. (At this point, it should be clear that the vertical component of the inclined member must balance all of the external vertical forces.) The equation for vertical equilibrium is written for the free body of the remaining truss portion.

Example 41.9

Find the force in member BC for the truss in Ex. 41.8.

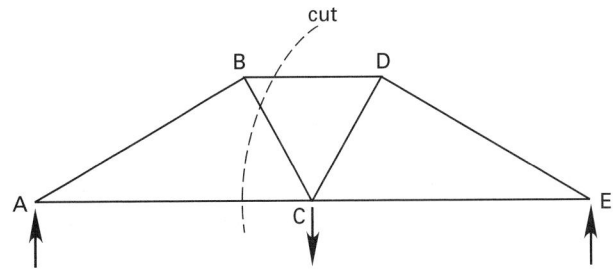

Solution

The reactions were determined in Ex. 41.8. The truss is cut in such a way as to pass through member BC but through no other inclined member. The free body of the remaining portion of the truss is

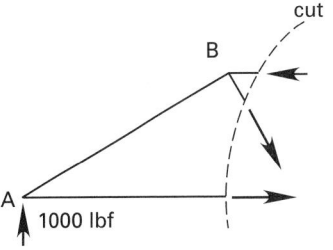

The vertical equilibrium equation is

$$\sum F_y = R_A - \mathbf{BC}_y$$
$$= 0$$
$$= 1000 \text{ lbf} - 0.866\mathbf{BC}$$
$$= 0$$
$$\mathbf{BC} = 1155 \text{ lbf} \quad (T)$$

36. METHOD OF SECTIONS

The *method of sections* is a direct approach to finding forces in any truss member. This method is convenient when only a few truss member forces are unknown.

As with the previous two methods, the first step is to find the support reactions. Then a cut is made through the truss, passing through the unknown member.[18] Finally, all three conditions of equilibrium are applied as needed to the remaining truss portion. (Since there are three equilibrium equations, the cut cannot pass through more than three members in which the forces are unknown.)

---

[18] Knowing where to cut the truss is the key part of this method. Such knowledge is developed only by practice.

Example 41.10

Find the forces in members CD and CE. The support reactions have already been determined.

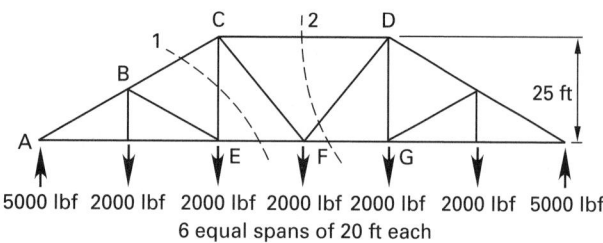

5000 lbf 2000 lbf 2000 lbf 2000 lbf 2000 lbf 2000 lbf 5000 lbf

6 equal spans of 20 ft each

Solution

To find the force **CE**, the truss is cut at section 1.

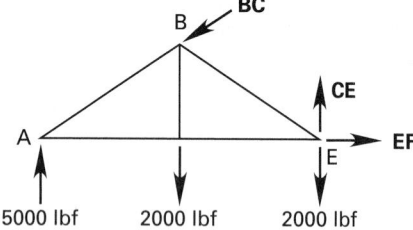

5000 lbf   2000 lbf   2000 lbf

Taking moments about point A will eliminate all of the unknown forces except **CE**. Assume clockwise moments are positive.

$$\sum M_\mathrm{A} = (2000 \text{ lbf})(20 \text{ ft}) + (2000 \text{ lbf})(40 \text{ ft})$$
$$- (40 \text{ ft})\mathbf{CE} = 0$$
$$\mathbf{CE} = 3000 \text{ lbf} \quad (T)$$

To find the force **CD**, the truss is cut at section 2.

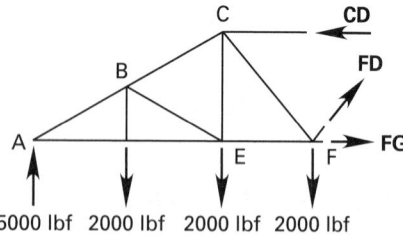

5000 lbf   2000 lbf   2000 lbf   2000 lbf

Taking moments about point F will eliminate all unknowns except **CD**. Assume clockwise moments are positive.

$$\sum M_\mathrm{F} = (5000 \text{ lbf})(60 \text{ ft}) - (25)\mathbf{CD}$$
$$- (2000 \text{ lbf})(20 \text{ ft}) - (2000 \text{ lbf})(40 \text{ ft}) = 0$$
$$\mathbf{CD} = 7200 \text{ lbf} \quad (C)$$

37. SUPERPOSITION OF LOADS

Superposition is a term used to describe the process of determining member forces by considering loads one at a time. Suppose, for example, that the force in member FG is unknown and that the truss carries three loads. If the method of superposition is used, the force in member FG (call it $\mathbf{FG}_1$) is determined with only the first load acting on the truss. $\mathbf{FG}_2$ and $\mathbf{FG}_3$ are similarly found. The true member force **FG** is found by adding $\mathbf{FG}_1$, $\mathbf{FG}_2$, and $\mathbf{FG}_3$.

Superposition should be used with discretion since trusses can change shape under load. If a truss deflects such that the load application points are significantly different from those in the undeflected truss, superposition cannot be used for that truss.

In simple truss analysis, change of shape under load is neglected. Superposition, therefore, can be assumed to apply.

38. TRANSVERSE TRUSS MEMBER LOADS

Truss members are usually designed as axial members, not as beams. Trusses are traditionally considered to be loaded at joints only. Figure 41.16, however, illustrates cases of nontraditional *transverse loading* that can actually occur. For example, a truss member's own weight would contribute to a uniform load, as would a severe ice buildup.

Figure 41.16 Transverse Truss Member Loads

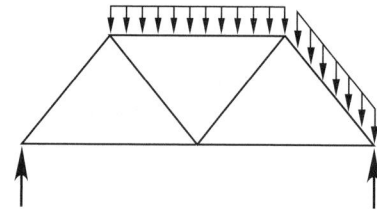

Transverse loads add two solution steps to a truss problem. First, the truss member must be individually considered as a beam simply supported at its pinned connections, and the reactions needed to support the transverse loading must be found. These reactions become additional loads applied to the truss joints, and the truss can then be evaluated in the normal manner.

The second step is to check the structural adequacy (deflection, bending stress, shear stress, buckling, etc.) of the truss member under transverse loading.

39. CABLES CARRYING CONCENTRATED LOADS

An *ideal* cable is assumed to be completely flexible, massless, and incapable of elongation. It therefore acts as an axial two-force tension member between points of concentrated loading. In fact, the term *tension* or *tensile force* is commonly used in place of "member force" when dealing with cables.

Figure 41.17 Cable with Concentrated Load

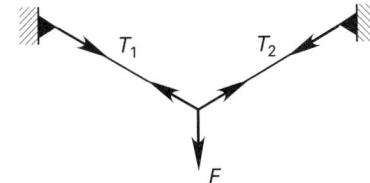

The methods of joints and sections used in truss analysis can be used to determine the tensions in cables carrying concentrated loads. After separating the reactions into x- and y-components, it is particularly useful to sum moments about one of the reaction points. All cables will be found to be in tension, and (with vertical loads only) the horizontal tension component will be the same in all cable segments. Unlike the case of a rope passing over a series of pulleys, however, the total tension in the cable will not be the same in every cable segment.

Example 41.11

What are the tensions **AB**, **BC**, and **CD**?

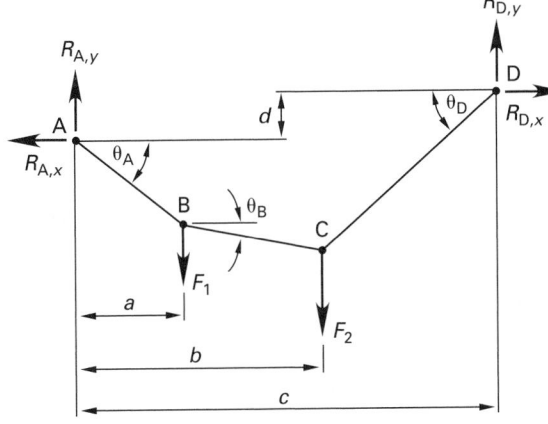

Solution

Separate the two reactions into x- and y-components. (The total reactions R_A and R_D are also the tensions **AB** and **CD**, respectively.)

$$R_{A,x} = -\mathbf{AB}\cos\theta_A$$
$$R_{A,y} = -\mathbf{AB}\sin\theta_A$$
$$R_{D,x} = -\mathbf{CD}\cos\theta_D$$
$$R_{D,y} = -\mathbf{CD}\sin\theta_D$$

Next, take moments about point A to find **CD**. Assume clockwise to be positive.

$$\sum M_A = aF_1 + bF_2 - d\mathbf{CD}_x + c\mathbf{CD}_y = 0$$

None of the applied loads are in the x-direction. Therefore, the only horizontal loads are the x-components of the reactions. To find tension **AB**, take the entire cable as a free body. Then, sum the external forces in the x-direction.

$$\sum F_x = R_{D,x} - R_{A,x} = 0$$
$$\mathbf{CD}\cos\theta_D = \mathbf{AB}\cos\theta_A$$

The x-component of force is the same in all cable segments. To find **BC**, sum the x-direction forces at point B.

$$\sum F_x = \mathbf{BC}_x - \mathbf{AB}_x = 0$$
$$\mathbf{BC}\cos\theta_B = \mathbf{AB}\cos\theta_A$$

40. PARABOLIC CABLES

If the distributed load per unit length, w, on a cable is constant with respect to the horizontal axis (as is the load from a bridge floor), the cable will be parabolic in shape.[19] This is illustrated in Fig. 41.18.

Figure 41.18 Parabolic Cable

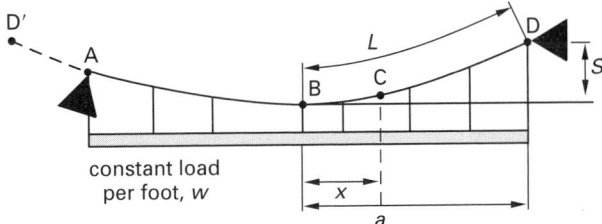

If the location of the maximum sag (i.e., the lowest cable point) is known, the horizontal component of tension, H, can be found by taking moments about a reaction point. If the cable is cut at the maximum sag point, B, the cable tension on the free body will be horizontal since there is no vertical component to the cable. Cutting the cable in Fig. 41.18 at point B and taking moments about point D will determine the minimum cable tension, H.

$$\sum M_D = wa\left(\frac{a}{2}\right) - HS = 0 \qquad \textit{41.50}$$

$$H = \frac{wa^2}{2S} \qquad \textit{41.51}$$

Since the load is vertical everywhere, the horizontal component of tension is constant everywhere in the cable. The tension, T_C, at any point C can be found by applying the equilibrium conditions to the cable segment BC.

$$T_{C,x} = H = \frac{wa^2}{2S} \qquad \textit{41.52}$$

$$T_{C,y} = wx \qquad \textit{41.53}$$

$$T_C = \sqrt{(T_{C,x})^2 + (T_{C,y})^2}$$
$$= w\sqrt{\left(\frac{a^2}{2S}\right)^2 + x^2} \qquad \textit{41.54}$$

---

[19]The parabolic case can also be assumed with cables loaded only by their own weight (e.g., telephone and trolley wires), if both ends are at the same elevations and if the sag is no more than 10% of the distance between supports.

The angle of the cable at any point is

$$\tan \theta = \frac{wx}{H} \qquad 41.55$$

The tension and angle are maximum at the supports.

If the lowest sag point, point B, is used as the origin, the shape of the cable is

$$y(x) = \frac{wx^2}{2H} \qquad 41.56$$

The approximate length of the cable from the lowest point to the support (i.e., length BD) is

$$L \approx a \left(1 + \left(\tfrac{2}{3}\right) \left(\frac{S}{a}\right)^2 - \left(\tfrac{2}{5}\right) \left(\frac{S}{a}\right)^4 \right) \qquad 41.57$$

Example 41.12

A pedestrian foot bridge has two suspension cables and a flexible floor weighing 28 lbf/ft. The span of the bridge is 100 ft. When the bridge is empty, the tension at point C is 1500 lbf. Assuming a parabolic shape, what is the maximum cable sag, S?

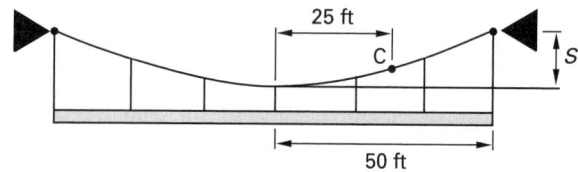

Solution

Since there are two cables, the floor weight per suspension cable is

$$w = \frac{28 \ \dfrac{\text{lbf}}{\text{ft}}}{2}$$

$$= 14 \ \text{lbf/ft}$$

From Eq. 41.54,

$$T_C = w \sqrt{\left(\frac{a^2}{2S}\right)^2 + x^2}$$

$$1500 \ \text{lbf} = 14 \ \frac{\text{lbf}}{\text{ft}} \sqrt{\left(\frac{(50 \ \text{ft})^2}{2S}\right)^2 + (25 \ \text{ft})^2}$$

$$S = 12 \ \text{ft}$$

41. CABLES CARRYING DISTRIBUTED LOADS

An idealized tension cable with a distributed load is similar to a linkage made up of a very large number of axial members. The cable is an axial member in the sense that the internal tension acts tangentially to the cable everywhere.

Since the load is vertical everywhere, the horizontal component of cable tension is constant along the cable. The cable is horizontal at the point of lowest sag. There is no vertical tension component, and the cable tension is minimum. By similar reasoning, the cable tension is maximum at the supports.

Figure 41.19 illustrates a general cable with a distributed load. The shape of the cable will depend on the relative distribution of the load. A free-body diagram of segment BC is also shown.

Figure 41.19 *Cable with Distributed Load*

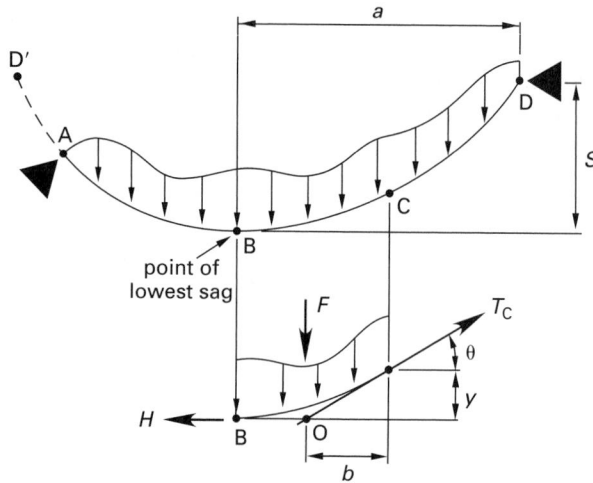

F is the resultant of the distributed load on segment BC, T is the cable tension at point C, and H is the tension at the point of lowest sag (i.e., the point of minimum tension). Since segment BC taken as a free body is a three-force member, the three forces (H, F, and T) must be concurrent to be in equilibrium. The horizontal component of tension can be found by taking moments about point C.

$$\sum M_C = Fb - Hy = 0 \qquad 41.58$$

$$H = \frac{Fb}{y} \qquad 41.59$$

Also, $\tan \theta = y/b$. Therefore,

$$H = \frac{F}{\tan \theta} \qquad 41.60$$

The basic equilibrium conditions can be applied to the free-body cable segment BC to determine the tension in the cable at point C.

$$\sum F_x = T_C \cos \theta - H = 0 \qquad 41.61$$

$$\sum F_y = T_C \sin \theta - F = 0 \qquad 41.62$$

The resultant tension at point C is

$$T_C = \sqrt{H^2 + F^2} \qquad 41.63$$

42. CATENARY CABLES

If the distributed load is constant along the length of the cable, as it is with a loose cable loaded by its own weight, the cable will have the shape of a *catenary*. A vertical axis catenary has a shape determined by Eq. 41.64, where c is a constant and cosh is the *hyperbolic cosine*. The quantity x/c is in radians.[20]

$$y(x) = c \, \cosh\left(\frac{x}{c}\right) \qquad 41.64$$

Referring to Fig. 41.20, the vertical distance, y, to any point C on the catenary is measured from a reference plane located a distance c below the point of greatest sag, point B. The distance c is known as the *parameter of the catenary*. Although the value of c establishes the location of the x-axis, the value of c does not correspond to any physical distance, nor is the reference plane the ground level.

Figure 41.20 *Catenary Cable*

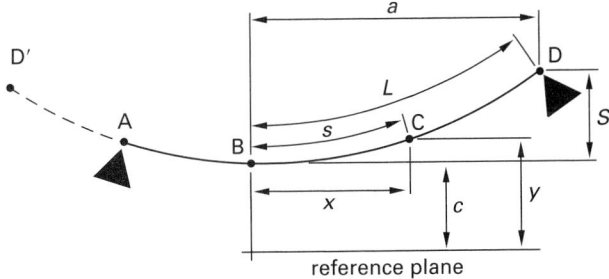

In order to define the cable shape and determine cable tensions, it is necessary to have enough information to calculate c. For example, if a and S are known, Eq. 41.67 can be solved by trial and error for c.[21] Once c is known, the cable geometry and forces are determined by the remaining equations.

For any point C, the equations most useful in determining the shape of the catenary are

$$y = \sqrt{s^2 + c^2} = c \left(\cosh\left(\frac{x}{c}\right)\right) \qquad 41.65$$

$$s = c \left(\sinh\left(\frac{x}{c}\right)\right) \qquad 41.66$$

$$\text{sag} = S = y_{\text{D}} - c = c \left(\cosh\left(\frac{a}{c}\right) - 1\right) \qquad 41.67$$

$$\tan \theta = \frac{s}{c} \qquad 41.68$$

The equations most useful in determining the cable tensions are

$$H = wc \qquad 41.69$$

$$F = ws \qquad 41.70$$

---

[20]In order to use Eqs. 41.64 through 41.67, you must reset your calculator from degrees to radians.

[21]Because obtaining the solution may require trial and error, it will be advantageous to assume a parabolic shape if the cable is taut. (See Ftn. 19.) The error will generally be small.

$$T = wy \qquad 41.71$$

$$\tan \theta = \frac{ws}{H} \qquad 41.72$$

$$\cos \theta = \frac{H}{T} \qquad 41.73$$

Example 41.13

A cable 100 m long is loaded by its own weight. The maximum sag is 25 m, and the supports are on the same level. What is the distance between the supports?

Solution

Since the two supports are on the same level, the cable length, L, between the point of maximum sag and support D is half of the total length.

$$L = \frac{100 \text{ m}}{2} = 50 \text{ m}$$

Writing Eqs. 41.65 and 41.67 for point D (with $S = 25$ m),

$$y_{\text{D}} = c + S = \sqrt{L^2 + c^2}$$

$$c + 25 \text{ m} = \sqrt{(50 \text{ m})^2 + c^2}$$

$$c = 37.5 \text{ m}$$

Substituting a for x and $L = 50$ for s in Eq. 41.66,

$$s = c \left(\sinh\left(\frac{x}{c}\right)\right)$$

$$50 \text{ m} = (37.5 \text{ m}) \left(\sinh\left(\frac{a}{37.5}\right)\right)$$

$$a = 41.2 \text{ m}$$

The distance between supports is

$$2a = (2)(41.2 \text{ m}) = 82.4 \text{ m}$$

43. CABLES WITH ENDS AT DIFFERENT ELEVATIONS

A cable will be asymmetrical if its ends are at different elevations. In some cases, as shown in Fig. 41.21, the cable segment will not include the lowest point B. However, if the location of the theoretical lowest point can be derived, the positions and elevations of the cable supports will not affect the analysis. The same procedure is used in proceeding from the theoretical point B to either support. In fact, once the theoretical shape of a cable has been determined, the supports can be relocated anywhere along the cable line without affecting the equilibrium of the supported segment.

Figure 41.21 *Asymmetrical Segment of Symmetrical Cable*

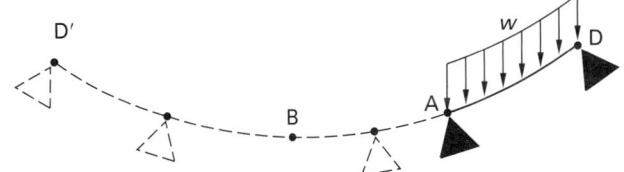

44. TWO-DIMENSIONAL MECHANISMS

A two-dimensional *mechanism* (*machine*) is a nonrigid structure. Although parts of the mechanism move, the relationships between forces in the mechanism can be determined by statics. In order to determine an unknown force, one or more of the mechanism components must be considered as a free body. All input forces and reactions must be included on this free body. In general, the resultant force on such a free body will not be in the direction of the member.

Several free bodies may be needed for complicated mechanisms. Sign conventions of acting and reacting forces must be strictly adhered to when determining the effect of one component on another.

Example 41.14

A 70 N·m couple is applied to the mechanism shown. All connections are frictionless hinges. What are the x- and y-components of the reactions at B?

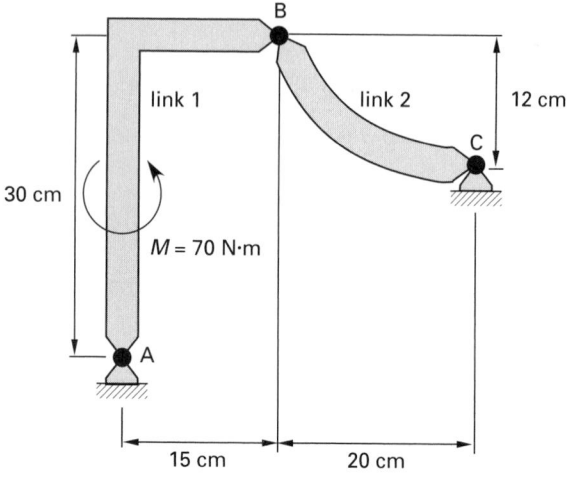

Solution

Isolate links 1 and 2 and draw their free bodies.

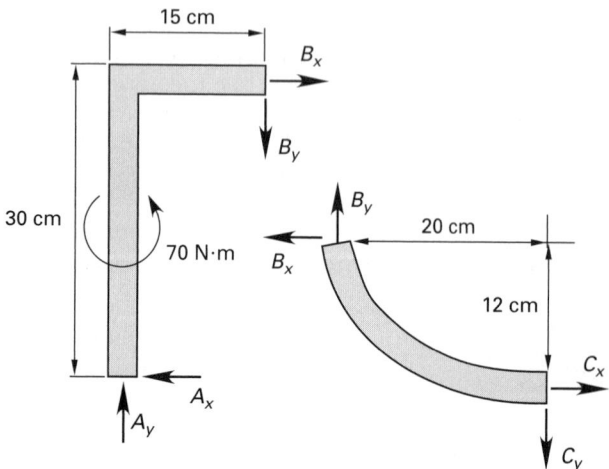

Assume clockwise moments are positive. Take moments about point A on link 1.

$$\sum M_A = B_x(0.3 \text{ m}) + B_y(0.15 \text{ m}) - 70 \text{ N·m} = 0$$

Assume clockwise moments are positive. Take moments about point C on link 2.

$$\sum M_C = B_y(0.20 \text{ m}) - B_x(0.12 \text{ m}) = 0$$

Solving these two equations simultaneously determines the force at joint B.

$$B_x = 179 \text{ N}$$
$$B_y = 108 \text{ N}$$

45. EQUILIBRIUM IN THREE DIMENSIONS

The basic equilibrium equations can be used with vector algebra to solve a three-dimensional statics problem. When a manual calculation is required, however, it is often more convenient to write the equilibrium equations for one orthogonal direction at a time, thereby avoiding the use of vector notation and reducing the problem to two dimensions. The following method can be used to analyze a three-dimensional structure.

step 1: Establish the $(0, 0, 0)$ origin for the structure.

step 2: Determine the (x, y, z) coordinates of all load and reaction points.

step 3: Determine the x-, y-, and z-components of all loads and reactions. This is accomplished by using direction cosines calculated from the (x, y, z) coordinates.

step 4: Draw a *coordinate free body* of the structure for each of the three coordinate axes. Include only forces, reactions, and moments that affect the coordinate free body.

step 5: Apply the basic two-dimensional equilibrium equations.

Example 41.15

Beam AC is supported at point A by a frictionless ball joint and at points B and C by cables. A 100 lbf load is applied vertically to point C, and a 180 lbf load is applied horizontally to point B. What are the cable tensions T_1 and T_2?

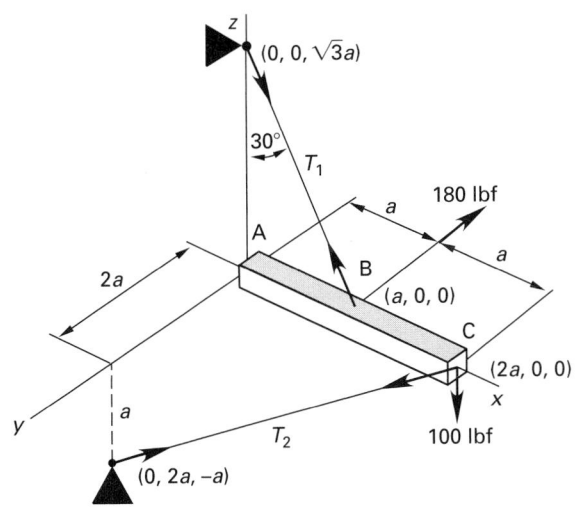

Solution

steps 1 and 2: Point A has already been established as the origin. The locations of all support and load points are shown on the illustration.

step 3: By inspection, for the 180 lbf horizontal load at point B,

$$F_x = 0$$
$$F_y = -180 \text{ lbf}$$
$$F_z = 0$$

By inspection, for the 100 lbf vertical load at point C,

$$F_x = 0$$
$$F_y = 0$$
$$F_z = -100 \text{ lbf}$$

The length of cable 1 is

$$L_1 = \sqrt{(a-0)^2 + (0-0)^2 + (0-\sqrt{3}a)^2}$$
$$= 2a$$

The direction cosines of the force from cable 1 at point B are

$$\cos\theta_x = \frac{d_x}{L_1} = \frac{0-a}{2a} = -0.5$$

$$\cos\theta_y = \frac{d_y}{L_1} = \frac{0-0}{2a} = 0$$

$$\cos\theta_z = \frac{d_z}{L_1} = \frac{\sqrt{3}a - 0}{2a} = 0.866$$

Therefore, the components of the tension in cable 1 are

$$T_{1,x} = -0.5\,T_1$$
$$T_{1,y} = 0$$
$$T_{1,z} = 0.866\,T_1$$

Similarly, for cable 2,

$$L_2 = \sqrt{(2a-0)^2 + (0-2a)^2 + \left(0-(-a)\right)^2}$$
$$= 3a$$

The direction cosines for the force from cable 2 at point C are

$$\cos\theta_x = \frac{d_x}{L_2} = \frac{0-2a}{3a} = -0.667$$

$$\cos\theta_y = \frac{d_y}{L_2} = \frac{2a-0}{3a} = 0.667$$

$$\cos\theta_z = \frac{d_z}{L_2} = \frac{-a-0}{3a} = -0.333$$

Therefore, the components of the tension in cable 2 are

$$T_{2,x} = -0.667\,T_2$$
$$T_{2,y} = 0.667\,T_2$$
$$T_{2,z} = -0.333\,T_2$$

step 4: The three coordinate free-body diagrams are

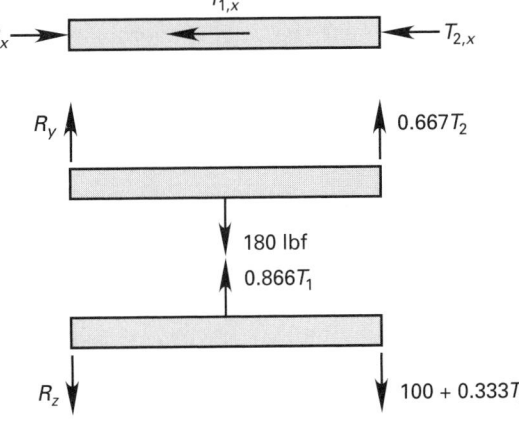

step 5: Tension T_2 can be found by taking moments about point A on the y-coordinate free body.

$$\sum M_A = (0.667T_2)(2a) - (180 \text{ lbf})(a) = 0$$
$$T_2 = 135 \text{ lbf}$$

Tension T_1 can be found by taking moments about point A on the z-coordinate free body.

$$\sum M_A = (0.866\,T_1)(a) - (0.333)(135\text{ lbf})(2a)$$
$$- (100\text{ lbf})(2a) = 0$$
$$T_1 = 335\text{ lbf}$$

46. TRIPODS

A *tripod* is a simple three-dimensional truss (frame) that consists of three axial members. One end of each member is connected at the *apex* of the tripod, while the other ends are attached to the supports. All connections are assumed to allow free rotation in all directions.

Figure 41.22 *Tripod*

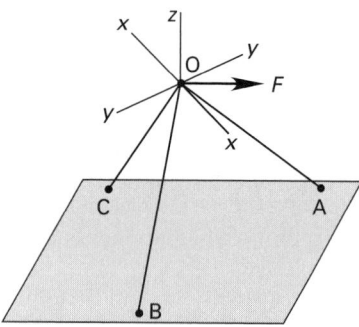

The general solution procedure given in the preceding section can be made more specific for tripods.

step 1: Establish the apex as the origin.

step 2: Determine the x-, y-, and z-components of the force applied to the apex.

step 3: Determine the (x, y, z) coordinates of points A, B, and C—the three support points.

step 4: Determine the length of each tripod leg from the coordinates of the support points.

$$L = \sqrt{x^2 + y^2 + z^2} \qquad 41.74$$

step 5: Determine the direction cosines for the leg forces at the apex. For leg A, for example,

$$\cos\theta_{A,x} = \frac{x_A}{L} \qquad 41.75$$

$$\cos\theta_{A,y} = \frac{y_A}{L} \qquad 41.76$$

$$\cos\theta_{A,z} = \frac{z_A}{L} \qquad 41.77$$

step 6: Write the x-, y-, and z-components of each leg force in terms of the direction cosines. For leg A, for example,

$$F_{A,x} = F_A \cos\theta_{A,x} \qquad 41.78$$
$$F_{A,y} = F_A \cos\theta_{A,y} \qquad 41.79$$
$$F_{A,z} = F_A \cos\theta_{A,z} \qquad 41.80$$

step 7: Write the three sum-of-forces equilibrium equations for the apex.

$$F_{A,x} + F_{B,x} + F_{C,x} + F_x = 0 \qquad 41.81$$
$$F_{A,y} + F_{B,y} + F_{C,y} + F_y = 0 \qquad 41.82$$
$$F_{A,z} + F_{B,z} + F_{C,z} + F_z = 0 \qquad 41.83$$

Example 41.16

Determine the force in each leg of the tripod.

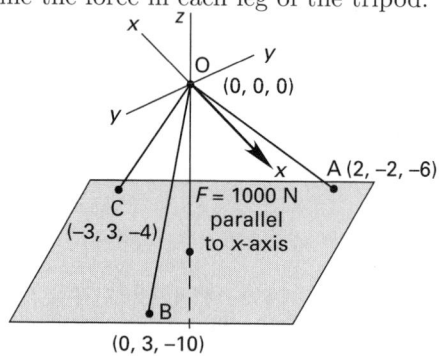

Solution

step 1: The origin is at the apex.

step 2: By inspection, $F_x = +1000$ N. All other components are zero.

steps 3 and 4: The direction cosines for the tripod legs have been calculated and are presented in the following table.

member	x^2	y^2	z^2	L^2	L	$\cos\theta_x$	$\cos\theta_y$	$\cos\theta_z$
OA	4	4	36	44	6.63	0.3015	−0.3015	−0.9046
OB	0	9	100	109	10.44	0.0	0.2874	−0.9579
OC	9	9	16	34	5.83	−0.5146	0.5146	−0.6861

steps 5 and 6: The equilibrium equations are

$$0.3015\,F_A + \qquad 0\,F_B - 0.5146\,F_C + 1000 = 0$$
$$-0.3015\,F_A + 0.2874\,F_B + 0.5146\,F_C \qquad = 0$$
$$-0.9046\,F_A - 0.9579\,F_B - 0.6861\,F_C \qquad = 0$$

The solution to these simultaneous equations is

$$F_A = +1531\text{ N} \quad (T)$$

$$F_B = -3480\text{ N} \quad (C)$$

$$F_C = +2841\text{ N} \quad (T)$$

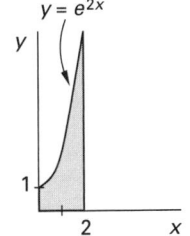

42 Properties of Areas

Nomenclature

A	area	units^2
b	base distance	units
c	distance to extreme fiber	units
d	separation distance	units
h	height distance	units
I	moment of inertia	units^4
J	polar moment of inertia	units^4
L	length	units
P	product of inertia	units^4
Q	first moment of the area	units^3
r	radius	units
r	radius of gyration	units
S	section modulus	units^3
V	volume	units^3
x	distance in the x-direction	units
y	distance in the y-direction	units

Symbols

θ	angle	rad

Subscripts

o	with respect to the origin
c	centroidal

1. CENTROID OF AN AREA

The *centroid* of an area is analogous to the center of gravity of a homogeneous body.[1] The centroid is often described as the point at which a thin homogeneous plate would balance. This definition, however, combines the definitions of centroid and center of gravity and implies that gravity is required to identify the centroid, which is not true.

---

[1] The analogy has been simplified. A three-dimensional body also has a centroid. The centroid and center of gravity will coincide when the body is homogeneous.

The location of the centroid of an area bounded by the x- and y-axes and the mathematical function $y = f(x)$ can be found by the *integration method* by using Eqs. 42.1 through 42.4. The centroidal location depends only on the geometry of the area and is identified by the coordinates (x_c, y_c). Some references place a bar over the coordinates of the centroid to indicate an average point, such as $(\overline{x}, \overline{y})$.

$$x_c = \frac{\int x \, dA}{A} \qquad 42.1$$

$$y_c = \frac{\int y \, dA}{A} \qquad 42.2$$

$$A = \int f(x) \, dx \qquad 42.3$$

$$dA = f(x) \, dx = g(y) \, dy \qquad 42.4$$

The locations of the centroids of *basic shapes*, such as triangles and rectangles, are well known. The most common basic shapes have been included in App. 42.A. There should be no need to derive centroidal locations for these shapes by the integration method.

The centroid of a complex area can be found from Eqs. 42.5 and 42.6 if the area can be divided into the basic shapes in App. 42.A. This process is simplified when all or most of the subareas adjoin the reference axis. Example 42.1 illustrates this method.

$$x_c = \frac{\sum_i A_i x_{ci}}{\sum_i A_i} \qquad 42.5$$

$$y_c = \frac{\sum_i A_i y_{ci}}{\sum_i A_i} \qquad 42.6$$

Example 42.1

An area is bounded by the x- and y-axes, the line $x = 2$, and the function $y = e^{2x}$. Find the x-component of the centroid.

Structural

Solution

First, use Eq. 42.3 to find the area.

$$A = \int f(x)\, dx = \int_{x=0}^{x=2} e^{2x}\, dx$$

$$= \left[\tfrac{1}{2} e^{2x} \right]_0^2 = 27.3 - 0.5 = 26.8 \text{ units}^2$$

Since y is a function of x, dA must be expressed in terms of x. From Eq. 42.4,

$$dA = f(x)\, dx = e^{2x}\, dx$$

Finally, use Eq. 42.1 to find x_c.

$$x_c = \frac{\int x\, dA}{A} = \frac{1}{26.8} \int_{x=0}^{x=2} x e^{2x}\, dx$$

$$= \left(\frac{1}{26.8} \right) \left[\tfrac{1}{2} x e^{2x} - \tfrac{1}{4} e^{2x} \right]_0^2 = 1.54 \text{ units}$$

Example 42.2

Find the y-coordinate of the centroid of the area shown.

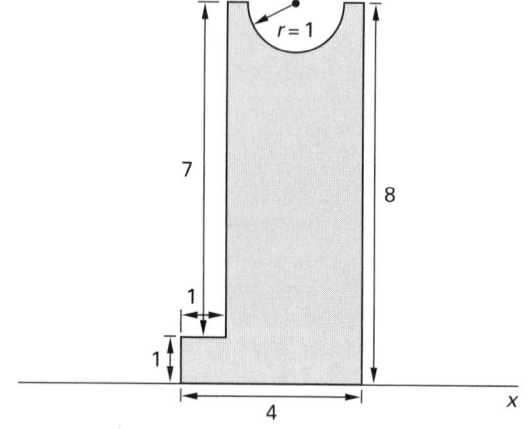

Solution

The x-axis is the reference axis. The area is divided into basic shapes of a 1×1 square, a 3×8 rectangle, and a half-circle of radius 1. (The area could also have been divided into 1×4 and 3×7 rectangles and the half-circle, but the 3×7 rectangle would not then adjoin the x-axis.)

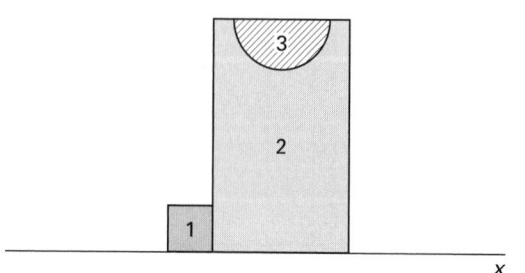

First, calculate the areas of the basic shapes. Notice that the half-circle area is negative since it represents a cutout.

$$A_1 = (1.0)(1.0) = 1.0 \text{ units}^2$$

$$A_2 = (3.0)(8.0) = 24.0 \text{ units}^2$$

$$A_3 = -\tfrac{1}{2}\pi r^2 = -\tfrac{1}{2}\pi(1.0)^2 = -1.57 \text{ units}^2$$

Next, find the y-components of the centroids of the basic shapes. Most are found by inspection, but App. 42.A can be used for the half-circle. Notice that the centroidal location for the half-circle is positive.

$$y_{c1} = 0.5 \text{ units}$$

$$y_{c2} = 4.0 \text{ units}$$

$$y_{c3} = 8.0 - 0.424 = 7.576 \text{ units}$$

Finally, use Eq. 42.6.

$$y_c = \frac{\sum A_i y_{ci}}{\sum A_i}$$

$$= \frac{(1.0)(0.5) + (24.0)(4.0) + (-1.57)(7.576)}{1.0 + 24.0 - 1.57}$$

$$= \frac{0.5 + 96.0 - 11.9}{23.43} = 3.61 \text{ units}$$

2. FIRST MOMENT OF THE AREA

The quantity $\int x\, dA$ is known as the *first moment of the area* or *first area moment* with respect to the y-axis. Similarly, $\int y\, dA$ is known as the first moment of the area with respect to the x-axis. By rearranging Eqs. 42.1 and 42.2, it is obvious that the first moment of the area can be calculated from the area and centroidal distance.

$$Q_y = \int x\, dA = x_c A \qquad\qquad 42.7$$

$$Q_x = \int y\, dA = y_c A \qquad\qquad 42.8$$

In basic engineering, the two primary applications of the first moment concept are to determine centroidal locations and shear stress distributions. In the latter application, the first moment of the area is known as the *statical moment*.

3. CENTROID OF A LINE

The location of the *centroid of a line* is defined by Eqs. 42.9 and 42.10, which are analogous to the equations used for centroids of areas.

$$x_c = \frac{\int x\, dL}{L} \qquad\qquad 42.9$$

$$y_c = \frac{\int y\, dL}{L} \qquad\qquad 42.10$$

Since equations of lines are typically in the form $y = f(x)$, dL must be expressed in terms of x or y.

$$dL = \left(\sqrt{\left(\frac{dy}{dx}\right)^2 + 1} \right) dx \qquad 42.11$$

$$dL = \left(\sqrt{\left(\frac{dx}{dy}\right)^2 + 1} \right) dy \qquad 42.12$$

4. THEOREMS OF PAPPUS-GULDINUS

The *Theorems of Pappus-Guldinus* define the surface and volume of revolution (i.e., the surface area and volume generated by revolving a curve around a fixed axis).

- *Theorem I:* The area of a surface of revolution is equal to the product of the length of the generating curve and the distance traveled by the centroid of the curve while the surface is being generated.

Figure 42.1 Surface of Revolution

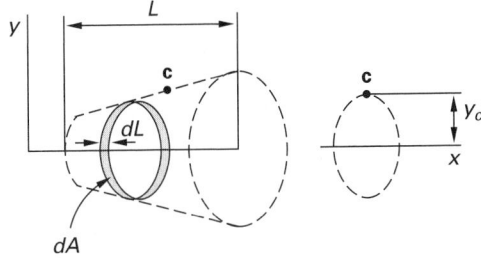

When a part, dL, of a line L is revolved about the x-axis, a differential ring having surface area dA is generated.

$$dA = 2\pi y \, dL \qquad 42.13$$

$$A = \int dA = 2\pi \int y \, dL = 2\pi y_c L \qquad 42.14$$

- *Theorem II:* The volume of a surface of revolution is equal to the generating area times the distance traveled by the centroid of the area in generating the volume.

Figure 42.2 Volume of Revolution

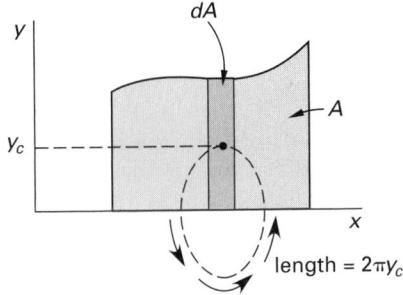

When a differential plane area, dA, is revolved about the x-axis and does not intersect the y-axis, it generates a ring of volume dV.

$$dV = \pi y^2 dx = \pi y \, dA \qquad 42.15$$

$$V = 2\pi y_c A \qquad 42.16$$

5. MOMENT OF INERTIA OF AN AREA

The *moment of inertia*, I, of an area is needed in mechanics of materials problems. It is convenient to think of the moment of inertia of a beam's cross-sectional area as a measure of the beam's ability to resist bending. Thus, given equal loads, a beam with a small moment of inertia will bend more than a beam with a large moment of inertia.

Since the moment of inertia represents a resistance to bending, it is always positive. Since a beam can be unsymmetrical (e.g., a rectangular beam) and can be stronger in one direction than another, the moment of inertia depends on orientation. Therefore, a reference axis or direction must be specified.

The moment of inertia taken with respect to one of the axes in the rectangular coordinate system is sometimes referred to as the *rectangular moment of inertia*.

The symbol I_x is used to represent a moment of inertia with respect to the x-axis. Similarly, I_y is the moment of inertia with respect to the y-axis. I_x and I_y do not normally combine and are not components of some resultant moment of inertia.

Any axis can be chosen as the reference axis, and the value of the moment of inertia will depend on the reference selected. The moment of inertia taken with respect to an axis passing through the area's centroid is known as the *centroidal moment of inertia*, I_{cx} or I_{cy}. The centroidal moment of inertia is the smallest possible moment of inertia for the shape.

The *integration method* can be used to calculate the moment of inertia of a function that is bounded by the x- and y-axes and a curve $y = f(x)$. From Eqs. 42.17 and 42.18, it is apparent why the moment of inertia is also known as the *second moment of the area* or *second area moment*.

$$I_x = \int y^2 \, dA \qquad 42.17$$

$$I_y = \int x^2 \, dA \qquad 42.18$$

$$dA = f(x) \, dx = g(y) \, dy \qquad 42.19$$

The moments of inertia of the *basic shapes* are well known and are listed in App. 42.A.

Example 42.3

What is the centroidal moment of inertia with respect to the x-axis of a rectangle 5.0 units wide and 8.0 units tall?

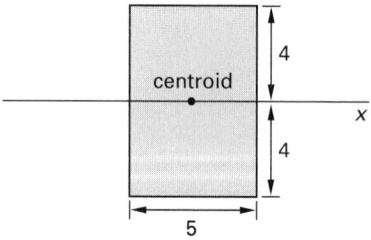

Solution

Since the centroidal moment of inertia is needed, the reference line passes through the centroid. From App. 42.A, the centroidal moment of inertia is

$$I_{cx} = \frac{bh^3}{12} = \frac{(5)(8)^3}{12} = 213.3 \text{ units}^4$$

Example 42.4

What is the moment of inertia with respect to the y-axis of the area bounded by the y-axis, the line $y = 8.0$, and the parabola $y^2 = 8x$?

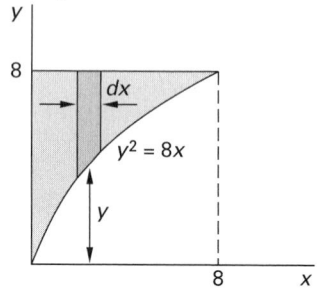

Solution

This problem is more complex than it first appears, since the area is above the curve, bounded not by $y = 0$ but by $y = 8$. In particular, dA must be determined correctly.

$$y = \sqrt{8x}$$

Use Eq. 42.4.

$$dA = (8 - f(x))\, dx = (8 - y)\, dx$$
$$= (8 - \sqrt{8x})\, dx$$

Equation 42.18 is used to calculate the moment of inertia with respect to the y-axis.

$$I_y = \int x^2\, dA = \int\limits_0^8 x^2 \left(8 - \sqrt{8x}\right) dx$$

$$= \left[\left(\frac{8}{3}\right) x^3 - \left(\frac{4\sqrt{2}}{7}\right) x^{7/2}\right]_0^8 = 195.0 \text{ units}^4$$

6. PARALLEL AXIS THEOREM

If the moment of inertia is known with respect to one axis, and the moment of inertia with respect to another, the parallel axis can be calculated from the *parallel axis theorem*, also known as the *transfer axis theorem*. This theorem is used to evaluate the moment of inertia of areas that are composed of two or more basic shapes. In Eq. 42.20, d is the distance between the centroidal axis and the second, parallel axis.

$$I_{\text{parallel axis}} = I_c + Ad^2 \qquad \textit{42.20}$$

The second term in Eq. 42.20 is often much larger than the first term. Areas close to the centroidal axis do not affect the moment of inertia considerably. This principle is exploited by structural steel shapes that derive bending resistance from *flanges* located away from the centroidal axis. The *web* does not contribute significantly to the moment of inertia.

Figure 42.3 *Structural Steel W-Shape*

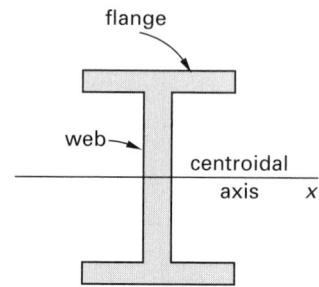

Example 42.5

Find the moment of inertia about the x-axis for the inverted-T area shown.

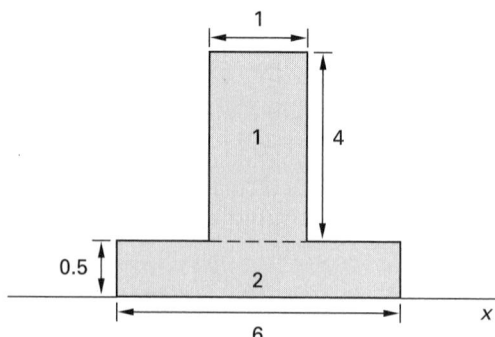

Solution

The area is divided into two basic shapes: 1 and 2. From App. 42.A, the moment of inertia of basic shape 2 with respect to the x-axis is

$$I_{x2} = \frac{bh^3}{3} = \frac{(6.0)(0.5)^3}{3} = 0.25 \text{ units}^4$$

The moment of inertia of basic shape 1 about its own centroid is

$$I_{cx1} = \frac{bh^3}{12} = \frac{(1)(4)^3}{12} = 5.33 \text{ units}^4$$

The x-axis is located 2.5 units from the centroid of basic shape 1. Therefore, from the parallel axis theorem, Eq. 42.20, the moment of inertia of basic shape 1 about the x-axis is

$$I_{x1} = 5.33 + (4)(2.5)^2 = 30.33 \text{ units}^4$$

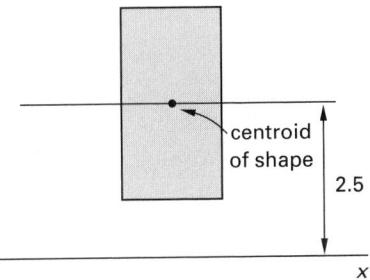

centroid of shape

2.5

x

The total moment of inertia of the T-area is

$$I_x = I_{x1} + I_{x2} = 30.33 \text{ units}^4 + 0.25 \text{ units}^4$$
$$= 30.58 \text{ units}^4$$

Example 42.6

Find the moment of inertia about the horizontal centroidal axis for the inverted-T area shown in Ex. 42.5.

Solution

The first step is to find the location of the centroid. The areas and centroidal locations (with respect to the x-axis) of the two basic shapes are

$$A_1 = (4.0)(1.0) = 4.0 \text{ units}^2$$
$$A_2 = (0.5)(6.0) = 3.0 \text{ units}^2$$
$$y_{c1} = 2.5 \text{ units}$$
$$y_{c2} = 0.25 \text{ units}$$

From Eq. 42.6, the composite centroid is located at

$$y_c = \frac{A_1 y_{c1} + A_2 y_{c2}}{A_1 + A_2} = \frac{(4.0)(2.5) + (3.0)(0.25)}{4.0 + 3.0}$$
$$= 1.536 \text{ units}$$

The distances between the centroids of the basic shapes and the composite shape are

$$d_1 = 2.5 - 1.536 = 0.964 \text{ units}$$
$$d_2 = 1.536 - 0.25 = 1.286 \text{ units}$$

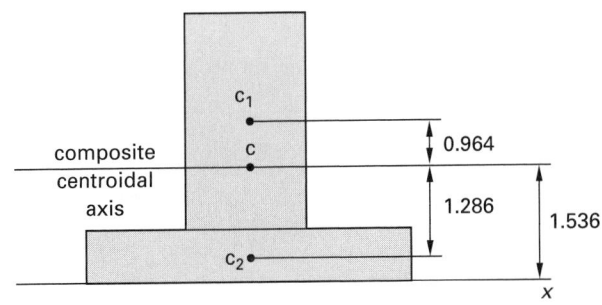

composite centroidal axis

c_1

c

c_2

0.964

1.286

1.536

x

The centroidal moments of inertia of the basic shapes with respect to an axis parallel to the x-axis are

$$I_{cx1} = \frac{bh^3}{12} = \frac{(1)(4)^3}{12} = 5.33 \text{ units}^4$$
$$I_{cx2} = \frac{bh^3}{12} = \frac{(6)(0.5)^3}{12} = 0.0625 \text{ units}^4$$

Using Eq. 42.20, the centroidal moment of inertia of the inverted-T area is

$$I_{cx} = I_{cx1} + A_1 d_1^2 + I_{cx2} + A_2 d_2^2$$
$$= 5.33 + (4.0)(0.964)^2 + 0.0625 + (3.0)(1.286)^2$$
$$= 14.07 \text{ units}^4$$

7. POLAR MOMENT OF INERTIA

The *polar moment of inertia*, J, is required in torsional shear stress calculations.[2] It can be thought of as a measure of an area's resistance to torsion (twisting). The definition of a polar moment of inertia of a two-dimensional area requires three dimensions because the reference axis for a polar moment of inertia of a plane area is perpendicular to the plane area.

The polar moment of inertia can be derived from Eq. 42.21.

$$J = \int (x^2 + y^2)\, dA \qquad 42.21$$

It is often easier to use the *perpendicular axis theorem* to quickly calculate the polar moment of inertia.

- *perpendicular axis theorem:* The polar moment of inertia of a plane area about an axis normal to the plane is equal to the sum of the moments of inertia about any two mutually perpendicular axes lying in the plane and passing through the given axis.

$$J = I_x + I_y \qquad 42.22$$

Since the two perpendicular axes can be chosen arbitrarily, it is most convenient to use the centroidal moments of inertia.

$$J = I_{cx} + I_{cy} \qquad 42.23$$

[2]The symbols I_z and I_{xy} are also encountered and are more consistent with the nomenclature. However, the symbol J is more common.

Structural

Example 42.7

What is the centroidal polar moment of inertia of a circular area of radius r?

Solution

From App. 42.A, the centroidal moment of inertia of a circle with respect to the x-axis is

$$I_{cx} = \frac{\pi r^4}{4}$$

Since the area is symmetrical, I_{cy} and I_{cx} are the same. From Eq. 42.23,

$$J_c = I_{cx} + I_{cy} = \frac{\pi r^4}{4} + \frac{\pi r^4}{4} = \frac{\pi r^4}{2}$$

8. RADIUS OF GYRATION

Every nontrivial area has a centroidal moment of inertia. Usually, some portions of the area are close to the centroidal axis and other portions are farther away. The *transverse radius of gyration*, or just *radius of gyration*, r, is an imaginary distance from the centroidal axis at which the entire area can be assumed to exist without affecting the moment of inertia. Despite the name "radius," the radius of gyration is not limited to circular shapes or polar axes. This concept is illustrated in Fig. 42.4.

Figure 42.4 *Radius of Gyration*

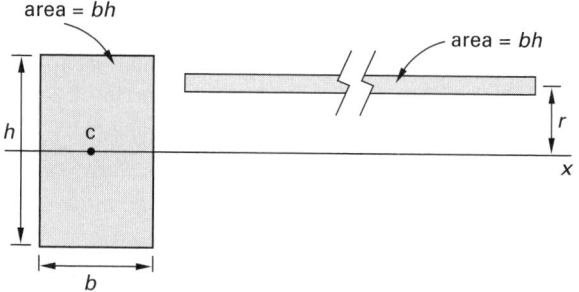

The method of calculating the radius of gyration is based on the parallel axis theorem. If all of the area is located a distance r from the original centroidal axis, there will be no I_c term in Eq. 42.20. Only the Ad^2 term will contribute to the moment of inertia.

$$I = r^2 A \qquad 42.24$$

$$r = \sqrt{\frac{I}{A}} \qquad 42.25$$

The concept of *least radius of gyration* comes up frequently in column design problems. (The column will tend to buckle about an axis that produces the smallest radius of gyration.) Usually, finding the least radius of gyration for symmetrical sections means solving Eq. 42.25 twice: once with I_x to find r_x and once with I_y to find r_y. The smallest value of r is the least radius of gyration.

The analogous quantity in the polar system is

$$r = \sqrt{\frac{J}{A}} \qquad 42.26$$

Just as the polar moment of inertia, J, can be calculated from the two rectangular moments of inertia, the polar radius of gyration can be calculated from the two rectangular radii of gyration.

$$r^2 = r_x^2 + r_y^2 \qquad 42.27$$

Example 42.8

What is the radius of gyration of the rectangular shape in Ex. 42.3?

Solution

The area of the rectangle is

$$A = bh = (5)(8) = 40 \text{ units}^2$$

From Eq. 42.25, the radius of gyration is

$$r_x = \sqrt{\frac{I_x}{A}} = \sqrt{\frac{213.3}{40}} = 2.31 \text{ units}$$

2.31 units is the distance from the centroidal x-axis that an infinitely long strip with an area of 40 square units would have to be located in order to have a moment of inertia of 213.3 units^4.

9. PRODUCT OF INERTIA

The *product of inertia*, P_{xy}, of a two-dimensional area is found by multiplying each differential element of area by its x- and y-coordinate and then summing over the entire area.

$$P_{xy} = \int xy \, dA \qquad 42.28$$

The product of inertia is zero when either axis is an axis of symmetry. Since the axes can be chosen arbitrarily, the area may be in one of the negative quadrants, and the product of inertia may be negative.

The parallel axis theorem for products of inertia is given by Eq. 42.29. (Both axes are allowed to move to new positions.) x_c' and y_c' are the coordinates of the centroid in the new coordinate system.

$$P_{x'y'} = P_{c,xy} + x_c' y_c' A \qquad 42.29$$

Figure 42.5 *Calculating the Product of Inertia*

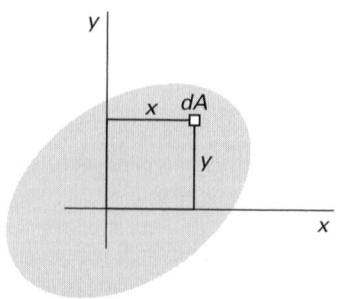

Figure 42.6 *Parallel Axis Theorem for Products of Inertia*

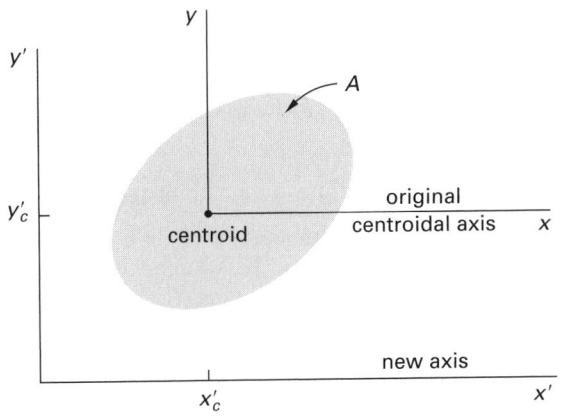

$$P_{x'y'} = P_{c,xy} + x'_c y'_c A$$

10. SECTION MODULUS

In the analysis of beams, the outer compressive (or tensile) surface is known as the *extreme fiber*. The distance, c, from the centroidal axis of the beam cross section to the extreme fiber is the "distance to the extreme fiber." The *section modulus*, S, combines the centroidal moment of inertia and the distance to the extreme fiber.

$$S = \frac{I_c}{c} \qquad 42.30$$

11. ROTATION OF AXES

Figure 42.7 shows rotation of the x-y axes through an angle, θ, into a new set of u-v axes, without rotating the area. If the moments and product of inertia of the area are known with respect to the old x-y axes, the new properties can be calculated from Eqs. 42.31 through 42.33.

$$\begin{aligned} I_u &= I_x \cos^2\theta - 2P_{xy}\sin\theta\cos\theta + I_y \sin^2\theta \\ &= \tfrac{1}{2}(I_x + I_y) + \tfrac{1}{2}(I_x - I_y)\cos 2\theta \\ &\quad - P_{xy}\sin 2\theta \qquad 42.31 \end{aligned}$$

$$\begin{aligned} I_v &= I_x \sin^2\theta + 2P_{xy}\sin\theta\cos\theta + I_y \cos^2\theta \\ &= \tfrac{1}{2}(I_x + I_y) - \tfrac{1}{2}(I_x - I_y)\cos 2\theta \\ &\quad + P_{xy}\sin 2\theta \qquad 42.32 \end{aligned}$$

$$\begin{aligned} P_{uv} &= I_x \sin\theta\cos\theta + P_{xy}(\cos^2\theta - \sin^2\theta) \\ &\quad - I_y \sin\theta\cos\theta \\ &= \tfrac{1}{2}(I_x - I_y)\sin 2\theta + P_{xy}\cos 2\theta \qquad 42.33 \end{aligned}$$

Since the polar moment of inertia about a fixed axis perpendicular to any two orthogonal axes in the plane is constant, the polar moment of inertia is unchanged by the rotation.

$$J_{xy} = I_x + I_y = I_u + I_v = J_{uv} \qquad 42.34$$

Figure 42.7 *Rotation of Axes*

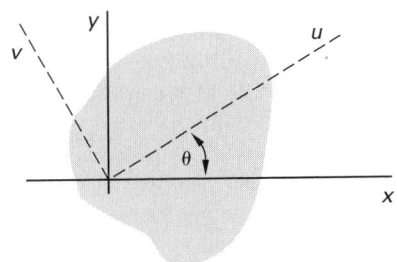

Example 42.9

What is the centroidal area moment of inertia of a 6×6 square that is rotated $45°$ from its "flat" orientation?

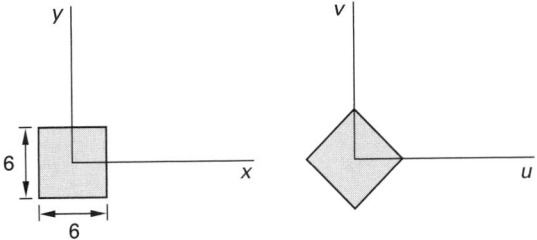

Solution

The centroidal moments of inertia with respect to the x- and y-axes are

$$I_x = I_y = \frac{s^4}{12} = \frac{(6)^4}{12} = 108 \text{ units}^4$$

Since the centroidal x- and y-axes are axes of symmetry, the product of inertia is zero.

Use Eq. 42.31.

$$\begin{aligned} I_u &= I_x \cos^2\theta - 2P_{xy}\sin\theta\cos\theta + I_y \sin^2\theta \\ &= (108)\cos^2 45° - 0 + (108)\sin^2 45° \\ &= 108 \text{ units}^4 \end{aligned}$$

The centroidal moment of inertia of a square is the same regardless of rotation angle.

12. PRINCIPAL AXES

Referring to Fig. 42.7, there is one angle, θ, that will maximize the moment of inertia, I_u. This angle can be found from calculus by setting $dI_u/d\theta = 0$. The resulting equation defines two angles, one that maximizes I_u and one that minimizes I_u.

$$\tan 2\theta = \frac{-2P_{xy}}{I_x - I_y} \qquad 42.35$$

Structural

The two angles that satisfy Eq. 42.35 are 90° apart. The set of u-v axes defined by Eq. 42.35 are known as *principal axes*. The moments of inertia about the principal axes are defined by Eq. 42.36 and are known as the *principal moments of inertia*.

$$I_{max,min} = \tfrac{1}{2}(I_x + I_y) \pm \sqrt{\tfrac{1}{4}(I_x - I_y)^2 + P_{xy}^2} \quad \textit{42.36}$$

13. MOHR'S CIRCLE

Once I_x, I_y, and P_{xy} are known, *Mohr's circle* can be drawn to graphically determine the moments of inertia about the principal axes. The procedure for drawing Mohr's circle is given as follows.

step 1: Determine I_x, I_y, and P_{xy} for the existing set of axes.

step 2: Draw a set of I-P_{xy} axes.

step 3: Plot the center of the circle, point **c**, by calculating distance c along the I-axis.

$$c = \tfrac{1}{2}(I_x + I_y) \quad \textit{42.37}$$

Figure 42.8 *Mohr's Circle*

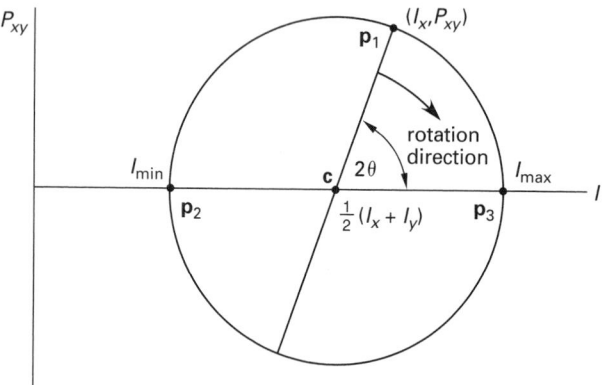

step 4: Plot the point $\mathbf{p}_1 = (I_x, P_{xy})$.

step 5: Draw a line from point $\mathbf{p}_1$ through center **c** and extend it an equal distance below the I-axis. This is the diameter of the circle.

step 6: Using the center **c** and point $\mathbf{p}_1$, draw the circle. An alternate method of constructing the circle is to draw a circle of radius r.

$$r = \sqrt{\tfrac{1}{4}(I_x - I_y)^2 + P_{xy}^2} \quad \textit{42.38}$$

step 7: Point $\mathbf{p}_2$ defines I_{min}. Point $\mathbf{p}_3$ defines I_{max}.

step 8: Determine the angle θ as half of the angle 2θ on the circle. This angle corresponds to I_{max}. (The axis giving the minimum moment of inertia is perpendicular to the maximum axis.) The sense of this angle and the sense of the rotation are the same. That is, the direction that the diameter would have to be turned in order to coincide with the I_{max}-axis has the same sense as the rotation of the x-y axes needed to form the principal u-v axes.

Figure 42.9 *Principal Axes from Mohr's Circle*

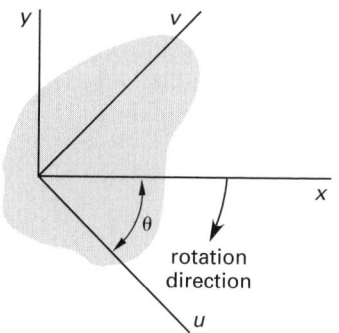

43 Material Properties and Testing

Nomenclature

A	area	in^2	m^2
b	width	in	m
B	bulk modulus	lbf/in^2	MPa
C	constant	–	–
C_V	impact energy	in-lbf	J
d	diameter of impression	in	mm
D	diameter	in	m
e	engineering strain	in/in	m/m
E	modulus of elasticity	lbf/in^2	MPa
F	force	lbf	N
G	shear modulus	lbf/in^2	MPa
J	polar moment of inertia	in^4	m^4
k	exponent	1/hr	1/h
k	strength derating factor	–	–
K	stress concentration factor	–	–
K	strength coefficient	lbf/in^2	MPa
L	length	in	m
LYS	lower yield strength	lbf/in^2	MPa
n	exponent	–	–
N	number of cycles	–	–
P	force of impression	lbf	N
q	fatigue notch sensitivity factor	–	–
q	reduction in area	–	–
r	radius	in	m

s	stress	lbf/in^2	MPa
S	strength	lbf/in^2	MPa
t	depth (thickness)	in	mm
t	time	hr	h
T	torque	in-lbf	N·m
U_R	modulus of resilience	lbf/in^2	MPa
U_T	modulus of toughness	lbf/in^2	MPa
UYS	upper yield strength	lbf/in^2	MPa

Symbols

β	Andrade's beta	$hr^{-1/3}$	$h^{-1/3}$
γ	angle of twist	rad	rad
δ	elongation	in	m
ϵ	true strain or creep	in/in	m/m
θ	angle of rupture	deg	deg
θ	shear strain	rad	rad
ν	Poisson's ratio	–	–
σ	true stress	lbf/in^2	MPa
τ	shear stress	lbf/in^2	MPa
ϕ	angle of internal friction	deg	deg

Subscripts

c	compressive
e	endurance
f	fatigue or fracture
o	original
p	particular
s	shear
t	tensile
u	ultimate
y	yield

1. TENSILE TEST

Many useful material properties are derived from the results of a standard *tensile test*. In this test, a prepared material sample (i.e., a *specimen*) is axially loaded in tension, and the resulting elongation, δ, is measured as the load, F, increases. A *load-elongation curve* of tensile test data for a ductile ferrous material (e.g., low-carbon steel or other BCC transition metal) is shown in Fig. 43.1.

When elongation is plotted against the applied load, the graph is applicable only to an object with the same length and area as the test specimen. To generalize the test results, the data are converted to stresses and strains by the use of Eqs. 43.1 and 43.2.[1]

---

[1] The most common *test specimen* in the United States has a length of 2.00 in and a diameter of 0.505 in. Since the cross-sectional area of this *0.505 bar* is 0.2 in^2, the stress in lbf/in^2 is calculated by multiplying the force in pounds by five.

Figure 43.1 *Typical Tensile Test Results for a Ductile Material*

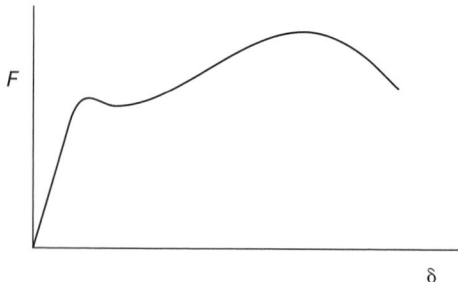

Figure 43.2 *Typical Stress-Strain Curve for Steel*

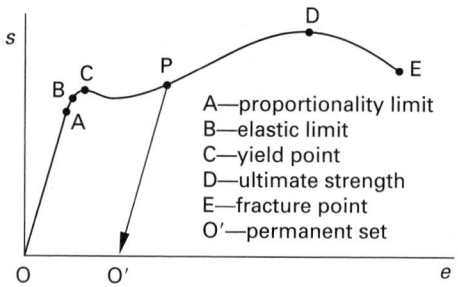

A—proportionality limit
B—elastic limit
C—yield point
D—ultimate strength
E—fracture point
O'—permanent set

Engineering stress, s (usually called *stress*), is the load per unit original area. Typical engineering stress units are lbf/in^2 and MPa. *Engineering strain*, e (usually called *strain*), is the elongation of the test specimen expressed as a percentage or decimal fraction of the original length. The units in/in and m/m are also used for strain.

$$s = \frac{F}{A_o} \qquad 43.1$$

$$e = \frac{\delta}{L_o} \qquad 43.2$$

If the stress-strain data are plotted, the shape of the resulting line will be essentially the same as the force-elongation curve, although the scales will change.

Segment O-A in Fig. 43.2 is a straight line. The relationship between the stress and strain in this linear region is given by *Hooke's law*, Eq. 43.3. The slope of line segment O-A is the *modulus of elasticity*, E, also known as *Young's modulus*. Table 43.1 lists approximate values of the modulus of elasticity for materials at room temperature. The modulus of elasticity will be lower at higher temperatures.[2]

$$s = Ee \qquad 43.3$$

The stress at point A in Fig. 43.2 is known as the *proportionality limit* (i.e, the maximum stress for which the linear relationship is valid). Strain in the *proportional region* is called *proportional strain*.

The *elastic limit*, point B in Fig. 43.2, is slightly higher than the proportionality limit. As long as the stress is kept below the elastic limit, there will be no *permanent set* (permanent deformation) when the stress is removed. Strain that disappears when the stress is

---

[2]For steel at higher temperatures, the modulus of elasticity is reduced approximately as follows.

temperature		
°F	°C	% of original value
70	20	100%
400	200	90%
800	425	75%
1000	540	65%
1200	650	60%

Table 43.1 *Approximate Modulus of Elasticity of Representative Materials at Room Temperature*

material	lbf/in^2	GPa
aluminum alloys	$10–11\times10^6$	70–80
brass	$15–16\times10^6$	100–110
cast iron	$15–22\times10^6$	100–150
cast iron, ductile	$22–25\times10^6$	150–170
cast iron, malleable	$26–27\times10^6$	180–190
copper alloys	$17–18\times10^6$	110–112
glass	$7–12\times10^6$	50–80
magnesium alloys	6.5×10^6	45
molybdenum	47×10^6	320
nickel alloys	$26–30\times10^6$	180–210
steel, hard[a]	30×10^6	210
steel, soft[a]	29×10^6	200
steel, stainless	$28–30\times10^6$	190–210
titanium	$15–17\times10^6$	100–110

(Multiply lbf/in^2 by 6.89×10^{-6} to obtain GPa.)
[a]Common values given.

removed is known as *elastic strain*, and the stress is said to be in the *elastic region*. When the applied stress is removed, the *recovery* is 100%, and the material follows the original curve back to the origin.

If the applied stress exceeds the elastic limit, the recovery will be along a line parallel to the straight line portion of the curve, as shown in line segment P-O'. The strain that results (line O-O') is *permanent set* (i.e, a permanent deformation). The terms *plastic strain* and *inelastic strain* are used to distinguish this behavior from elastic strain.

For steel, the *yield point*, point C, is very close to the elastic limit. For all practical purposes, the *yield strength* or *yield stress*, S_y (or S_{yt} to indicate yield in tension), can be taken as the stress that accompanies the beginning of plastic strain. Yield strengths are reported in lbf/in^2, ksi, and MPa.[3]

Figure 43.2 does not show the full complexity of the stress-strain curve near the yield point. Rather than being smooth, the curve is ragged near the yield point. At the upper yield strength, there is a pronounced drop

---

[3]A *kip* is a thousand pounds. *ksi* is the abbreviation for kips per square inch (thousands of lbf/in^2).

(i.e., "drop of beam") in load-carrying ability to a plateau yield strength after the initial yielding occurs. The plateau value is known as the *lower yield strength* and is commonly reported as the yield strength.

The *ultimate strength* or *tensile strength*, S_u (or S_{ut} to indicate an ultimate tensile strength), point D in Fig. 43.2, is the maximum stress the material can support without failure.

The *breaking strength* or *fracture strength*, S_f, is the stress at which the material actually fails (point E in Fig. 43.2). For ductile materials, the breaking strength is less than the ultimate strength due to the necking down in the cross-sectional area that accompanies high plastic strains.

Figure 43.3 *Upper and Lower Yield Strengths*

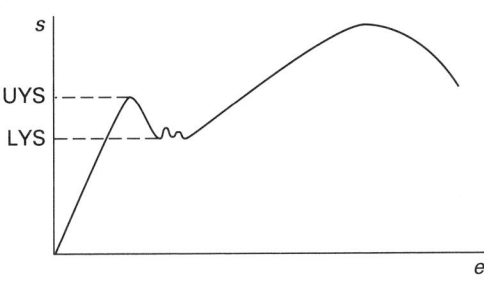

Table 43.2 *Approximate Yield Strengths of Representative Materials*

	yield strength	
material	lbf/in^2	MPa
iron and steel		
1020	43,000	300
A36	36,000	250
stainless (304)	43,000	300
pure	24,000	160
copper		
beryllium	130,000	900
brass	11,000	75
pure	10,000	70
aluminum		
2024	50,000	345
6061	21,000	145
pure	5000	35
titanium		
alloy 6% Al, 4% V	160,000	1100
pure	20,000	140
nickel		
hastelloy	55,000	380
inconel	40,000	280
monel	35,000	240
pure	20,000	140

(Multiply lbf/in^2 by 6.89×10^{-3} to obtain MPa.)

2. STRESS-STRAIN CHARACTERISTICS OF NONFERROUS METALS

Most nonferrous materials, such as aluminum, magnesium, copper, and other FCC and HCP metals, do not have well-defined yield points. The stress-strain curve starts to bend at low stresses, as illustrated by Fig. 43.4. In such cases, the yield strength is commonly defined as the stress that will cause a 0.2% *parallel offset* (i.e., a plastic strain of 0.002).[4] However, the yield strength can also be defined by other offset values (e.g., 0.1% for metals and 1.0% for plastics).

The yield strength is found by extending a line from the offset strain value parallel to the linear portion of the curve until it intersects the curve.

Figure 43.4 *Typical Stress-Strain Curve for a Nonferrous Metal*

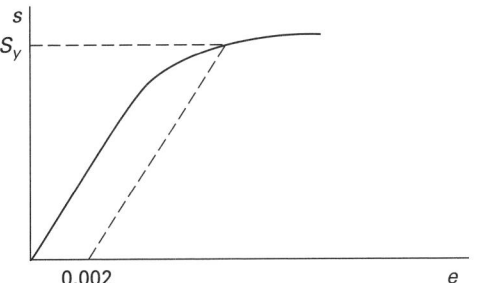

With nonferrous metals, the difference between parallel offset and total strain characteristics is important. Sometimes, the yield point will be determined as the stress accompanying *0.5% total strain* (i.e., a strain of 0.005).

3. STRESS-STRAIN CHARACTERISTICS OF BRITTLE MATERIALS

Brittle materials, such as glass, cast iron, and ceramics, can support only small strains before they fail catastrophically (i.e., without warning). As the stress is increased, the elongation is linear and Hooke's law (Eq. 43.3) can be used to predict the strain. Failure occurs within the linear region, and there is very little, if any, necking down. Since the failure occurs at a low strain, brittle materials are not ductile. (The words "brittle" and "ductile" are antonyms.) Figure 43.5 is typical of the stress-strain curve of a brittle material.

---

[4]In Great Britain, the 0.2% offset strength is known as the *proof stress*.

Figure 43.5 *Stress-Strain Curve of a Brittle Material*

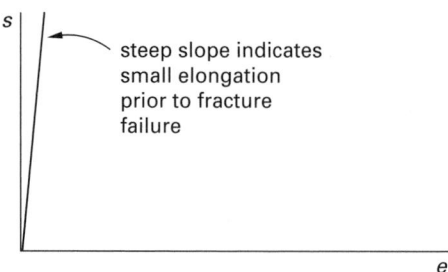

4. SECANT MODULUS

The modulus of elasticity, E, is usually determined from the steepest portion of the stress-strain curve. (This avoids the difficulty of locating the starting part of the curve.) For materials operating in the nonlinear region, the *secant modulus* gives the average ratio of stress to strain. The secant modulus is the slope of the straight line connecting the origin and the point of operation. Some designs using elastomers or concrete may be based on the secant modulus.

Figure 43.6 *Secant Modulus*

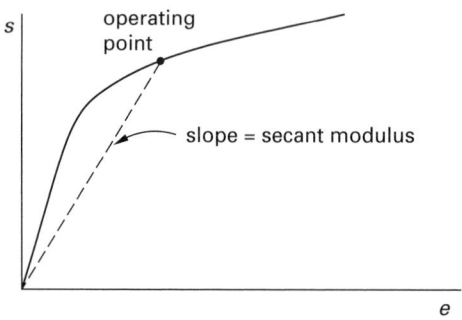

5. POISSON'S RATIO

As a specimen elongates axially during a tensile test, it will also decrease slightly in diameter or breadth. For any specific material, the percentage decrease in diameter, known as the *lateral strain*, will be a fraction of the *axial strain*. The ratio of the lateral strain to the axial strain is known as *Poisson's ratio*, ν, which is taken as approximately 0.3 for most metals.

$$\nu = \frac{e_{\text{lateral}}}{e_{\text{axial}}} = \frac{\dfrac{\Delta D}{D_o}}{\dfrac{\delta}{L_o}} \qquad 43.4$$

Poisson's ratio applies only to elastic strain. When the stress is removed, the lateral strain disappears along with the axial strain.

6. STRAIN HARDENING AND NECKING DOWN

When the applied stress exceeds the yield strength, the specimen will experience plastic deformation and will strain harden. (Plastic deformation is primarily due to the shear stress-induced movement of dislocations.) Since the specimen volume is constant (i.e., $A_o L_o = AL$), the cross-sectional area decreases. Initially, the strain hardening more than compensates for the decrease in area, so the material's strength increases and the engineering stress increases with larger strains.

Table 43.3 *Approximate Values of Poisson's Ratio*

material	Poisson's ratio
liquids	0.50^a
rubber	0.49
thermosetting plastics	0.40–0.45
aluminum	0.32–0.34 $(0.33)^b$
magnesium	0.35
copper	0.33–0.36 $(0.33)^b$
titanium	0.34
brass	0.33–0.36
stainless steel	0.30
steel	0.26–0.30 $(0.30)^b$
nickel	0.30
beryllium	0.27
cast iron	0.21–0.33 $(0.27)^b$
glass (SiO_2)	0.21–0.27 $(0.23)^b$
diamond	0.20

[a] Limiting value.
[b] Commonly used for design.

Eventually, a point is reached when the available strain hardening and increase in strength cannot keep up with the decrease in cross-sectional area. The specimen then begins to neck down (at some local weak point), and all subsequent plastic deformation is concentrated at the neck. The cross-sectional area decreases even more rapidly thereafter since only a small portion of the specimen volume is strain hardening. The engineering stress decreases to failure.

Figure 43.7 shows necking down in two different specimens tested to failure. Very ductile materials pull out to a point, while most moderately ductile materials exhibit a *cup-and-cone failure*. (Failed brittle materials, not shown, do not exhibit any significant reduction in area.)

Figure 43.7 *Types of Tensile Ductile Failure*

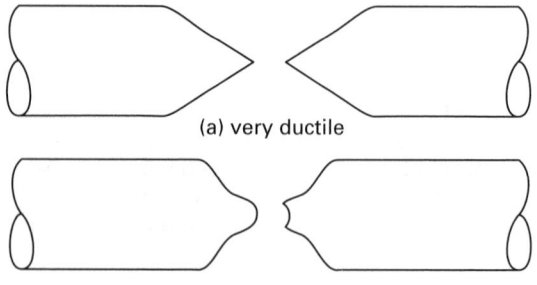

(a) very ductile

(b) ductile, cup and cone

7. TRUE STRESS AND STRAIN

Engineering stress, given by Eq. 43.1, is calculated for all stress levels from the original cross-sectional area. However, during a tensile test, the area of a specimen decreases as the stress increases. The decrease is only slight in the elastic region but is much more significant after plastic deformation begins.

If the stress is calculated from the instantaneous area, it is known as *true stress* or *physical stress*, σ. Equation 43.5 assumes a homogenous strain distribution along the gage length and that there is no change in total volume with strain, which are valid up to the point of necking. (Fractional reduction in area, $q = (A_o - A)/A_o$, used in Eq. 43.5, is expressed as a number between 0 and 1.)

$$\sigma = \frac{F}{A} = \frac{F}{\left(1 - \dfrac{A_o - A}{A_o}\right) A_o} = \frac{F}{(1 - q) A_o}$$

$$= s(1 + e) = \frac{s}{(1 - \nu e)^2} \quad \begin{bmatrix} \text{circular} \\ \text{specimens} \end{bmatrix} \quad \textit{43.5}$$

Engineering strain, given by Eq. 43.2, is calculated from the original length, although the actual length increases during the tensile test. The *true strain*, *physical strain*, or *log strain*, ϵ, is found from Eq. 43.6.

$$\epsilon = \int_{L_0}^{L} \frac{dL}{L} = \ln\left(\frac{L}{L_o}\right)$$

$$= \ln(1 + e) \quad \text{[prior to necking]} \quad \textit{43.6}$$

Since the plastic deformation occurs through a shearing process, there is essentially no volume decrease during elongation.

$$A_o L_o = AL \qquad \textit{43.7}$$

Therefore, true strain can be calculated from the cross-sectional areas and, for a circular specimen, from diameters. If necking down has occurred, true strain must be calculated from the areas or diameters, not the lengths.

$$\epsilon = \ln\left(\frac{A_o}{A}\right) = \ln\left(\frac{D_o}{D}\right)^2 = 2\ln\left(\frac{D_o}{D}\right) \qquad \textit{43.8}$$

Figure 43.8 compares engineering and true stresses and strains for a ferrous alloy. A graph of true stress and true strain is known as a *flow curve*. Log σ can also be plotted against log ϵ, resulting in a straight-line relationship.

Figure 43.8 True and Engineering Stresses and Strains for a Ferrous Alloy

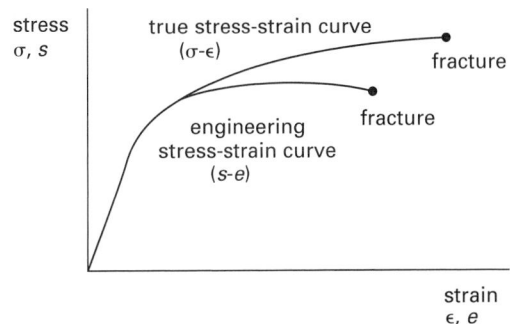

The flow curve of many metals in the plastic region can be expressed by the relationship of Eq. 43.9, known as a *power curve*. K is known as the *strength coefficient*, and n is the *strain-hardening exponent*. Values of both vary greatly with material, composition, and heat treatment. n can vary from 0 (for a perfectly inelastic solid) to 1.0 (for an elastic solid). Typical values are between 0.1 and 0.5. For annealed steel with 0.05% carbon, for example, $K \approx 77{,}000$ psi and $n = 0.26$.

$$\sigma = K\epsilon^n \qquad \textit{43.9}$$

Although true stress and strain are more accurate, almost all engineering work is based on engineering stress and strain, which is justifiable for two reasons: (a) design using ductile materials is limited to the elastic region where engineering and true values differ little, and (b) the reduction in area of most parts at their service stresses is not known; only the original area is known.

Example 43.1

The engineering stress in a solid tension member was 47,000 lbf/in² at failure. The reduction in area was 80%. What were the true stress and strain at failure?

Solution

Since engineering stress, s, is F/A_o, from Eq. 43.5 the true stress is

$$\sigma = \frac{s}{1 - \text{reduction in area}}$$

$$= \frac{47{,}000 \; \dfrac{\text{lbf}}{\text{in}^2}}{1 - 0.80} = 235{,}000 \; \text{lbf/in}^2$$

From Eq. 43.8, the true strain is

$$\epsilon = \ln\left(\frac{1}{1 - 0.80}\right) = 1.61 \quad (161\%)$$

8. DUCTILITY

A material that deforms and elongates a great deal before failure is said to be a *ductile material*. (Steel, for example, is a ductile material.) The *percent elongation*, short for *percent elongation at failure*, is the total plastic strain at failure. (Percent elongation does not include the elastic strain, because even at ultimate failure the material snaps back an amount equal to the elastic strain.)

$$\begin{aligned} \frac{\text{percent}}{\text{elongation}} &= \frac{L_f - L_o}{L_o} \times 100\% \\ &= e_f \times 100\% \end{aligned} \qquad 43.10$$

The value of the final strain to be used in Eq. 43.10 is found by extending a line from the failure point downward to the strain axis, parallel to the linear portion of the curve. This is equivalent to putting the two broken specimen pieces together and measuring the total length.

Figure 43.9 *Percent Elongation*

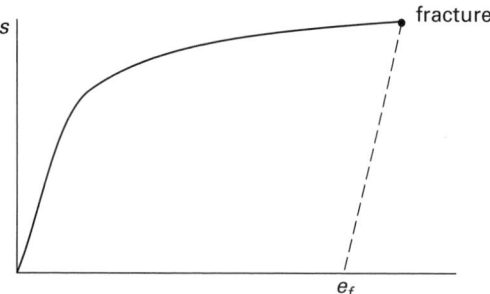

Highly ductile materials exhibit large percent elongations. However, percent elongation is not the same as *ductility*.

$$\text{ductility} = \frac{\text{ultimate failure strain}}{\text{yielding strain}} \qquad 43.11$$

The *reduction in area* (at the point of failure), expressed as a percentage or decimal fraction, is a third measure of a material's ductility. The reduction in area due to necking down will be 50% or greater for ductile materials and less than 10% for brittle materials.[5]

$$\frac{\text{reduction}}{\text{in area}} = \frac{A_o - A_f}{A_o} \times 100\% \qquad 43.12$$

9. STRAIN ENERGY

Strain energy, also known as *internal work*, is the energy per unit volume stored in a deformed material. The strain energy is equivalent to the work done by the applied tensile force. Simple work is calculated as the product of a force moving through a distance.

$$\text{work} = \text{force} \times \text{distance} = \int F \, dL \qquad 43.13$$

$$\frac{\text{work per}}{\text{unit volume}} = \int \frac{F \, dL}{AL} = \int_0^{\epsilon_{\text{final}}} \sigma \, d\epsilon \qquad 43.14$$

---

[5] *Notch-brittle materials* have reductions in area that are moderate (e.g., 25 to 35%) when tested in the usual manner but close to zero when the test specimen is given a small notch or crack.

This work per unit volume corresponds to the area under the true stress-strain curve. Units are in-lbf/in³ (i.e., inch-pounds (a unit of energy) per cubic inch (a unit of volume)), usually shortened to lbf/in² (MPa). (Equation 43.14 cannot be simplified further because stress is not proportional to strain for the entire curve.)

10. RESILIENCE

A *resilient material* is able to absorb and release *strain energy* without permanent deformation. *Resilience* is measured by the *modulus of resilience*, also known as the *elastic toughness*, which is the strain energy per unit volume required to reach the yield point. This is represented by the area under the stress-strain curve up to the yield point. Since the stress-strain curve is essentially a straight line up to that point, the area is triangular.

$$U_R = \int_0^{\epsilon_y} \sigma \, d\epsilon = E \int_0^{\epsilon_y} \epsilon \, d\epsilon = \frac{E\epsilon_y^2}{2} = \frac{S_y \epsilon_y}{2} \qquad 43.15$$

The modulus of resilience varies greatly for steel. It can be more than ten times higher for high-carbon spring steel ($U_R = 320$ lbf/in², 2.2 MPa) than for low-carbon steel ($U_R = 20$ lbf/in², 0.14 MPa).

Figure 43.10 *Modulus of Resilience*

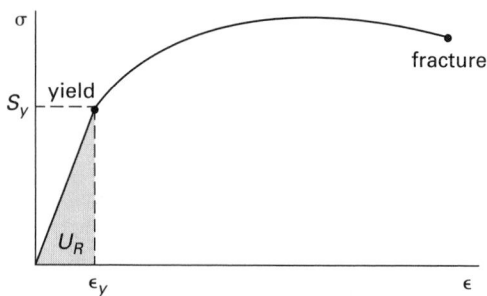

11. TOUGHNESS

A *tough material* will be able to withstand occasional high stresses without fracturing. Products subjected to sudden loading, such as chains, crane hooks, railroad couplings, and so on, should be tough. One measure of a material's *toughness* is the *modulus of toughness* (i.e, the strain energy or work per unit volume required to cause fracture). This is the total area under the stress-strain curve, and is given the symbol U_T. Since the area is irregular the modulus of toughness cannot be exactly calculated by a simple formula. However, the modulus of toughness of ductile materials (with large strains at failure) can be approximately calculated from either Eq. 43.16 or Eq. 43.17.

$$U_T \approx S_u \epsilon_u \quad [\text{ductile}] \qquad 43.16$$

$$U_T \approx \left(\frac{S_y + S_u}{2} \right) \epsilon_u \quad [\text{ductile}] \qquad 43.17$$

For brittle materials, the stress-strain curve may be either linear or parabolic. If the curve is parabolic, Eq. 43.18 approximates the modulus of toughness.

$$U_T \approx \tfrac{2}{3} S_u \epsilon_u \quad \text{[brittle]} \qquad 43.18$$

Figure 43.11 *Modulus of Toughness*

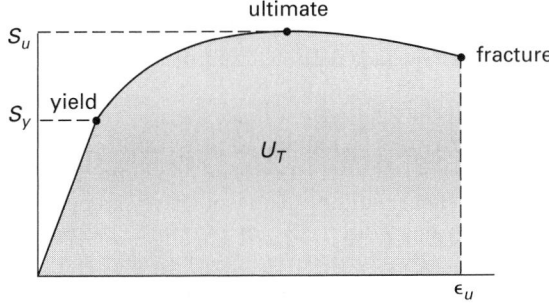

12. UNLOADING AND RELOADING

If the load is removed after a specimen is stressed elastically, the material will return to its original state. If the load is removed after a specimen is stressed into the plastic region, the *unloading curve* will follow a sloped path back to zero stress. The slope of the unloading curve will be equal to the original modulus of elasticity, E, illustrated by Fig. 43.12.

If this same material is subsequently reloaded, the *reloading curve* will follow the previous unloading curve up to the continuation of the original stress-strain curve. Therefore, the *apparent yield stress* of the reloaded specimen will be higher. This extra strength is the result of the strain hardening that has occurred.[6] Although the material will have a higher strength, its ductility will have been reduced.

Figure 43.12 *Unloading and Reloading Curves*

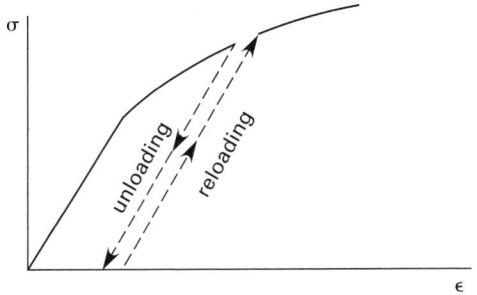

13. COMPRESSIVE STRENGTH

Compressive strength, S_{uc} (i.e, ultimate strength in compression), is an important property for brittle materials such as concrete and cast iron that are primarily loaded in compression only.[7] The compressive strengths of these materials are much greater than their tensile

strengths, whereas the compressive strengths for ductile materials, such as steel, are the same as their tensile yield strengths.

Within the linear (elastic) region, Hooke's law is valid for compression of both brittle and ductile materials.

The failure mechanism for ductile materials is plastic deformation alone. Such materials do not rupture in compression. Thus, a ductile material can support a load long after the material is distorted beyond a useful shape.

Figure 43.13 *Compressive Failures*

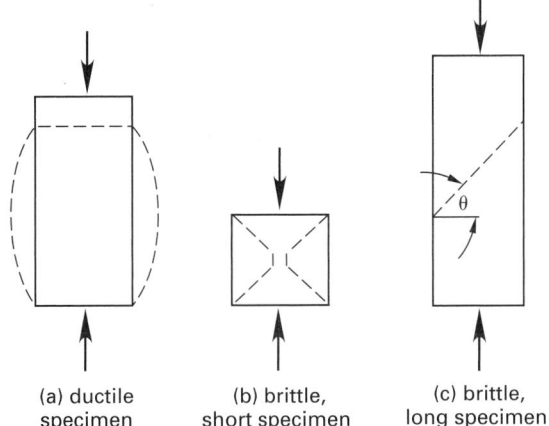

(a) ductile specimen (b) brittle, short specimen (c) brittle, long specimen

The failure mechanism for brittle materials is shear along an inclined plane. The characteristic plane and hourglass failures for brittle materials are shown in Fig. 43.13.[8] Theoretically, only *cohesion* contributes to compressive strength, and the *angle of rupture* (i.e., the incline angle), θ, should be 45°. In real materials, however, internal friction also contributes strength. The angles of rupture for cast iron, concrete, brick, and so on vary roughly between 50° and 60∘. If the *angle of internal friction*, ϕ, is known for the material, the angle of rupture can be calculated exactly from *Mohr's theory of rupture*.

$$\theta = 45° + \frac{\phi}{2} \qquad 43.19$$

14. TORSION TEST

Figure 43.14 illustrates a simple cube loaded by a shear stress, τ. The volume of the cube does not decrease when loaded, but the shape changes. The *shear strain* is the angle, θ, expressed in radians. The shear strain is proportional to the shear stress, analogous to Hooke's law for tensile loading. G is the *shear modulus*, also known as the *modulus of shear, modulus of elasticity in shear*, and *modulus of rigidity*.

$$\tau = G\theta \qquad 43.20$$

---

[6]The additional strength is lost if the material is subsequently annealed.

[7]f_c' is commonly used as the symbol for the compressive strength of concrete.

[8]The *hourglass failure* appears when the material is too short for a complete failure surface to develop.

Figure 43.14 *Cube Loaded in Shear*

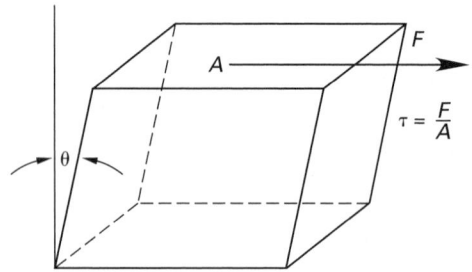

$$\tau = \frac{F}{A}$$

Table 43.4 *Approximate Values of Shear Modulus*

material	lbf/in²	GPa
aluminum	3.8×10^6	26
brass	5.5×10^6	38
copper	6.2×10^6	43
cast iron	8.0×10^6	55
magnesium	2.4×10^6	17
steel	11.5×10^6	79
stainless steel	10.6×10^6	73
titanium	6.0×10^6	41
glass	4.2×10^6	29

(Multiply lbf/in² by 6.89×10^{-6} to obtain GPa.)

The shear modulus can be calculated from the modulus of elasticity and Poisson's ratio and, therefore, can be derived from the results of a tensile test.

$$G = \frac{E}{2(1 + \nu)} \qquad 43.21$$

The shear stress can also be calculated from a torsion test, as illustrated by Fig. 43.15. Equation 43.22 relates the angle of twist (in radians) to the shear modulus.

$$\gamma = \frac{TL}{JG} = \frac{\tau L}{rG} \quad \text{[radians]} \qquad 43.22$$

Figure 43.15 *Uniform Bar in Torsion*

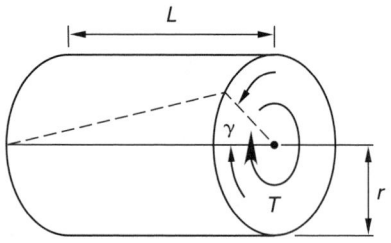

The *shear strength*, S_s or S_{ys}, of a material is the maximum shear stress that the material can support without yielding in shear. (The ultimate shear strength property is rarely encountered.) For ductile materials, *maximum shear stress theory* predicts the shear strength as one-half of the tensile yield strength. A more accurate relationship is derived from the *distortion energy theory* (also known as *von Mises theory*).

$$S_{ys} = \frac{S_{yt}}{\sqrt{3}} = 0.577 S_{yt} \qquad 43.23$$

15. RELATIONSHIP BETWEEN THE ELASTIC CONSTANTS

The elastic constants (modulus of elasticity, shear modulus, bulk modulus, and Poisson's ratio) are related in elastic materials. Table 43.5 lists the common relationships.

16. FATIGUE TESTING

A material can fail after repeated stress loadings even if the stress level never exceeds the ultimate strength, a condition known as *fatigue failure*.

Table 43.5 *Relationships Between Elastic Constants*

elastic constants	in terms of				
	E, ν	E, G	B, ν	B, G	E, B
E	–	–	$3(1 - 2\nu)B$	$\dfrac{9BG}{3B + G}$	–
ν	–	$\dfrac{E}{2G} - 1$	–	$\dfrac{3B - 2G}{2(3B + G)}$	$\dfrac{3B - E}{6B}$
G	$\dfrac{E}{2(1 + \nu)}$	–	$\dfrac{3(1 - 2\nu)B}{2(1 + \nu)}$	–	$\dfrac{3EB}{9B - E}$
B	$\dfrac{E}{3(1 - 2\nu)}$	$\dfrac{GE}{3(3G - E)}$	–	–	–

The behavior of a material under repeated loadings is evaluated by a fatigue test. A specimen is loaded repeatedly to a specific stress amplitude, s, and the number of applications of that stress required to cause failure, N, is counted. Rotating beam tests that load the specimen in bending are more common than alternating deflection and push-pull tests but are limited to round specimens.[9]

Figure 43.16 *Rotating Beam Test*

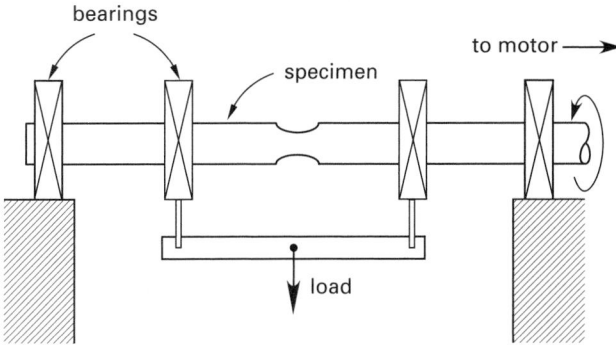

This procedure is repeated for different stresses, using eight to fifteen specimens. The results of these tests are graphed, resulting in an *S-N curve* (i.e., stress-number of cycles) shown in Fig. 43.17.

For an alternating stress test, the stress plotted on the *S-N* curve can be the maximum, minimum, or mean value. The choice depends on the method of testing as well as the intended application. The maximum stress should be used in rotating beam tests, since the mean stress is zero. For cyclic, one-dimensional bending, the maximum and mean stresses are commonly used.

Figure 43.17 *Typical S-N Curve for Steel*

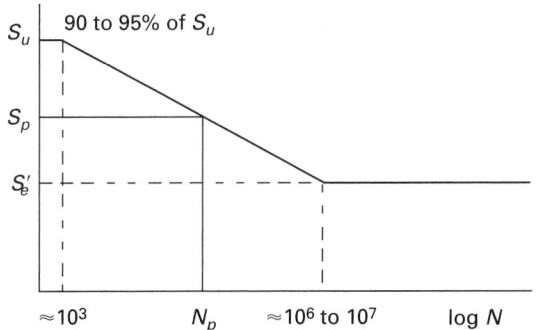

For a specific stress level, say S_p in Fig. 43.17, the number of cycles required to cause failure, N_p, is the *fatigue life*. S_p is the *fatigue strength* corresponding to N_p.

---

[9]In the design of ductile steel buildings, the static case is assumed up to 20,000 cycles. However, in critical applications (such as nuclear steam vessels, turbines, and so on) that experience temperature swings, fatigue failure can occur with a smaller number of cycles due to *cyclic strain*, not due to *cyclic stress*.

For steel subjected to fewer than approximately 10^3 loadings, the fatigue strength starts at the ultimate strength and drops to 90 to 95% of the ultimate strength at 10^3 cycles. (Although *low-cycle fatigue* theory has its own peculiarities, a part experiencing a small number of cycles can usually be designed or analyzed for static loading.) The curve is linear between 10^3 and 10^6 cycles if a logarithmic N-scale is used. Beyond 10^6 to 10^7 cycles, there is no further decrease in strength.

Therefore, below a certain stress level, called the *endurance limit* or *endurance strength*, S'_e, the material will withstand an almost infinite number of loadings without experiencing failure.[10] This is characteristic of steel and titanium. Therefore, if a dynamically loaded part is to have an infinite life, the stress must be kept below the endurance limit. The ratio S'_e/S_u is known as the *endurance ratio* or *fatigue ratio*. For carbon steel, the endurance ratio is approximately 0.4 for pearlitic, 0.60 for ferritic, and 0.25 for martensitic microstructures. For martensitic alloy steels, it is approximately 0.35.

For steel whose microstructure is unknown, the endurance strength is given approximately by Eq. 43.24.[11]

$$S'_{e,\text{steel}} \begin{cases} = 0.5 S_u & [S_u < 200{,}000 \text{ lbf/in}^2] \\ & [S_u < 1.4 \text{ GPa}] \\ = 100{,}000 \text{ lbm/in}^2 & [S_u > 200{,}000 \text{ lbf/in}^2] \\ (700 \text{ MPa}) & [S_u > 1.4 \text{ GPa}] \end{cases} \quad 43.24$$

For cast iron, the endurance ratio is lower.

$$S'_{e,\text{cast iron}} = 0.4 S_u \quad\quad 43.25$$

Steel and titanium are the most important engineering materials that have well-defined endurance limits. Many nonferrous metals and alloys, such as aluminum, magnesium, and copper alloys, do not have well-defined endurance limits. The strength continues to decrease with cyclic loading and never levels off. In such cases, the endurance limit is taken as the stress that causes failure at 10^8 or 5×10^8 loadings. Alternatively, the endurance strength is approximated by Eq. 43.26.

$$S'_{e,\text{aluminum}} = \begin{cases} 0.3 S_u & [\text{cast}] \\ 0.4 S_u & [\text{wrought}] \end{cases} \quad 43.26$$

---

[10]Most endurance tests use some form of sinusoidal loading. However, the fatigue and endurance strengths do not depend much on the shape of the loading curve. Only the maximum amplitude of the stress is relevant. Therefore, the endurance limit can be used with other types of loading (sawtooth, square wave, random, etc.).
[11]The coefficient in Eq. 43.24 actually varies between 0.25 and 0.6. However, 0.5 is commonly quoted.

Figure 43.18 *Typical S-N Curve for Aluminum*

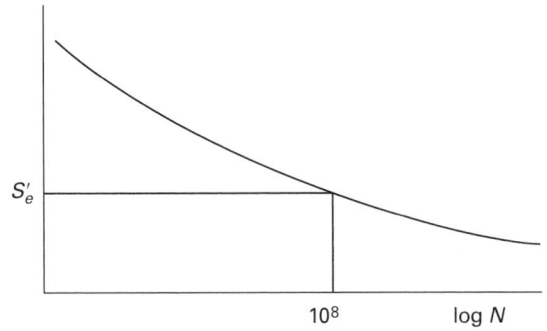

Figure 43.19 *Surface Finish Reduction Factors for Endurance Strength*

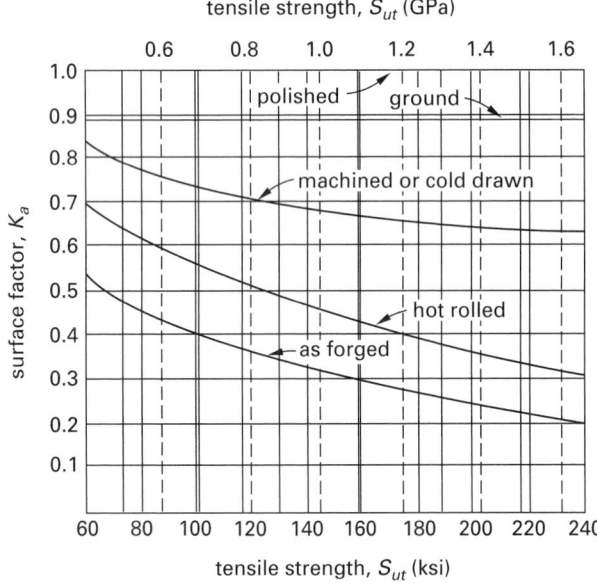

Reprinted with permission from *Mechanical Engineering Design*, 3rd ed., by Joseph Edward Shigley, copyright © 1977, The McGraw-Hill Companies.

The yield strength is an irrelevant factor in cyclic loading. Fatigue failures are fracture failures; they are not yielding failures. They start with microscopic cracks at the material surface. Some of the cracks are present initially; others form when repeated cold working reduces the ductility in strain-hardened areas. These cracks grow minutely with each loading. Since cracks start at the location of surface defects, the endurance limit is increased by proper treatment of the surface. Such treatments include polishing, surface hardening, shot peening, and filleting joints.

The endurance limit is not a true property of the material since the other significant influences, particularly surface finish, are never eliminated. However, representative values of S'_e obtained from ground and polished specimens provide a baseline to which other factors can be applied to account for the effects of surface finish,

temperature, stress concentration, notch sensitivity, size, environment, and desired reliability. These other influences are accounted for by fatigue strength reduction (derating) factors, k_i, which are used to calculate a working endurance strength, S_e, for the material.

$$S_e = \prod k_i S'_e \qquad 43.27$$

Since a rough surface significantly decreases the endurance strength of a specimen, it is not surprising that notches (and other features that produce stress concentration) do so as well. In some cases, the theoretical tensile *stress concentration factor*, K_t, due to notches and other features can be determined theoretically or experimentally. The ratio of the fatigue strength of a polished specimen to the fatigue strength of a notched specimen at the same number of cycles is known as the *fatigue notch factor*, K_f, also known as the *fatigue stress concentration factor*. The *fatigue notch sensitivity*, q, is a measure of the degree of agreement between the stress concentration factor and the fatigue notch factor.

$$q = \frac{K_f - 1}{K_t - 1} \quad [K_f > 1] \qquad 43.28$$

17. TESTING OF PLASTICS

With reasonable variations, mechanical properties of plastics are evaluated using the same methods as for metals.[12] Although temperature is an important factor in the testing of plastics, tests for tensile strength, endurance, hardness, toughness, and creep rate are similar or the same as for metals. Figure 43.20 illustrates typical tensile test results.[13]

Unlike metals, which follow Hooke's law, plastics are non-Hookean. (They may be Hookean for a short-duration loading.) The modulus of elasticity, for example, changes with stress level, temperature, time, and chemical environment. A plastic that appears to be satisfactory under one set of conditions can fail quickly under slightly different conditions. Therefore, properties of plastics determined from testing (and from tables) should be used only to compare similar materials, not to predict long-term behavior. Plastic tests are used to determine material specifications, not performance specifications.

---

[12]For example, plastic specimens for tensile testing can be produced by injection molding as well as by machining from compression-molded plaques, rather than machining from bar stock.

[13]Plastics are sensitive to the rate of loading. Figure 43.20 illustrates tensile performance based on a 2 in/min loading rate. However, for a fast loading rate (e.g., 2 in/sec), most plastics would exhibit brittle performance. On the other hand, given a slow loading rate (e.g., 2 in/month), most would behave as a soft and flexible plastic. Therefore, with different rates of loading, all three types of stress-strain performance shown in Fig. 43.20 can be obtained from the same plastic.

Figure 43.20 *Typical Tensile Test Performance for Plastics (loaded at 2 in/min)*

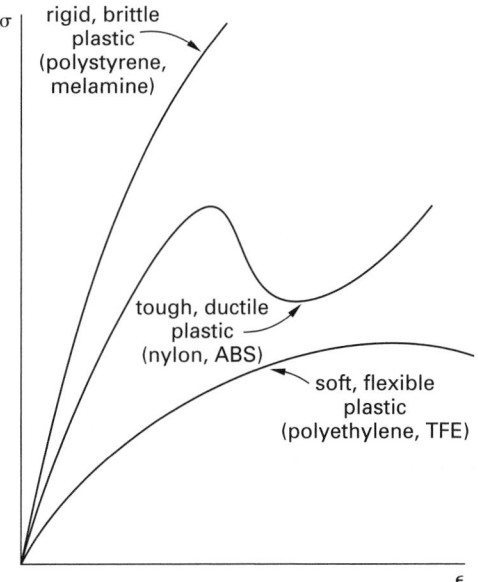

18. NONDESTRUCTIVE TESTING

Nondestructive testing (NDT) or *nondestructive evaluation* (NDE) is used when it is impractical or uneconomical to perform destructive sampling on manufactured products and their parts. Typical applications of NDT are inspection of helicopter blades, cast aluminum wheels, and welds in nuclear pressure vessels. Some procedures are particularly useful in providing quality monitoring on a continuous, real-time basis. In addition to visual processes, the main types of nondestructive testing are magnetic particle, eddy current, liquid penetrant, ultrasonic imaging, acoustic emission, and infrared testing, as well as radiography.

The *visual-optical* process differs from normal visual inspection in the use of optical scanning systems, borescopes, magnifiers, and holographic equipment. Flaws are identified as changes in light intensity (reflected, transmitted, or refracted), color changes, polarization changes, or phase changes. This method is limited to the identification of surface flaws or interior flaws in transparent materials.

Magnetic particle testing takes advantage of the attraction of ferromagnetic powders (e.g., the *Magnaflux*™ *process*) and fluorescent particles (e.g., the *Magnaglow*™ *process*) to leakage flux at surface flaws in magnetic materials. The particles accumulate and become visible at such flaws when an intense magnetic field is set up in a workpiece.

This method can locate most surface flaws (such as cracks, laps, and seams) and, in some special cases, subsurface flaws. The procedure is fast and simple to interpret. However, workpieces must be ferromagnetic and clean. Following the test, demagnetization may be required. A high-current power source is required.

Eddy current testing uses alternating current from a test coil to induce eddy currents in electrically conducting, metallic objects. Flaws and other material properties affect the current flow. The change in current is monitored by a detection circuit or on a meter or screen. This method can be used to locate defects of many types, including cracks, voids, inclusions, and weld defects, as well as to find changes in composition, structure, hardness, and porosity.

Intimate contact between the material and the test coil is not required. Operation can be continuous, automatic, and monitored electronically. Sensitivity is easily adjusted. Therefore, this method is ideal for unattended continuous processing. Many variables, however, can affect the current flow, and only electrically conducting materials can be tested with this method.

With *infrared testing*, infrared radiation emitted from objects can be detected and correlated with quality. Any discontinuities that interrupt heat flow, such as flaws, voids, and inclusions, can be detected.

Infrared testing requires access to only one side and is highly sensitive. It is applicable to complex shapes and

On a short-term basis, plastics behave elastically. They distort when loaded and spring back when unloaded. Under prolonged loading, however, creep (cold flow) becomes significant. When loading is removed, there is some instantaneous recovery, some delayed recovery, and some permanent deformation. The recovery might be complete if the load is removed within 10 hours, but if the loading is longer (e.g., 100 hours), recovery may be only partial. Because of this behavior, plastics are subjected to various other tests.

Additional tests used to determine the mechanical properties of plastics include deflection temperature, long-term (e.g., 3000 hours) tensile creep, creep rupture, and *stress-relaxation* (long-duration, constant-strain tensile testing at elevated temperatures).[14] Because some plastics deteriorate when exposed to light, plasma, or chemicals, performance under these conditions can be evaluated, as can the insulating and dielectric properties.

The *creep modulus* (also known as *apparent modulus*), determined from tensile creep testing, is the instantaneous ratio of stress to creep strain. The creep modulus decreases with time. The *deflection temperature* test indicates the dimensional stability of a plastic at high temperatures. A plastic bar is loaded laterally (as a beam) to a known stress level, and the temperature of the bar is gradually increased. The temperature at which the deflection reaches 0.010 in (0.254 mm) is taken as the *thermal deflection temperature* (TDT). The *Vicat softening point*, primarily used with polyethylenes, is the temperature at which a loaded standard needle penetrates 1 mm when the temperature is uniformly increased at a standard rate.

---

[14]Plastic pipes have their own special tests (ASTM D1598).

assemblies of dissimilar components but is relatively slow. The detection can be performed electronically. Results are affected by variations in material size, coatings, and colors, and hot spots can be hidden by cool surface layers.

Liquid penetrant testing is based on a fluorescent dye being drawn by capillary action into surface defects. A developer substance is commonly used to aid in visual inspection. This method can be used with any nonporous material, including metals, plastics, and glazed ceramics. It is capable of finding cracks, porosities, pits, seams, and laps.

Liquid penetrant tests are simple, can be used with complex shapes, and can be performed on site. Workpieces must be clean and nonporous. However, only small surface defects are detectable.

In *ultrasound imaging testing (ultrasonics)*, mechanical vibrations in the 0.1 to 25 MHz range are induced by pressing a piezoelectric transducer against a workpiece. The transmitted waves are normally reflected back, but the waves are scattered by interior defects. The results are interpreted by reading a screen or meter. The method can be used for metals, plastics, glass, rubber, graphite, and concrete. It is excellent for detecting internal defects such as inclusions, cracks, porosities, laminations, and changes in material structure.

Ultrasound testing is extremely flexible. It can be automated and is very fast. Results can be recorded or interpreted electronically. Penetration through thick steel layers is possible. Direct contact (or immersion in a fluid) is required, but only one surface needs to be accessible. Rough surfaces and complex shapes may cause difficulties, however. A related method, *acoustic emission monitoring*, is used to test pressurized systems.

There are two types of *holographic NDT methods*. *Acoustic holography* is a form of ultrasonic testing that passes an ultrasonic beam through the workpiece (or through a medium such as water surrounding the workpiece) and measures the displacement of the workpiece (or medium). With suitable processing, a three-dimensional hologram is formed that can be visually inspected.

In one form of *optical holography*, a hologram of the unloaded workpiece is imposed on the actual workpiece. If the workpiece is then loaded (stressed), the observed changes (e.g., deflections) from the holographic image will be non-uniform when discontinuities and defects are present.

Radiography (i.e., *nuclear sensing*) uses neutron, X-ray, gamma-ray (e.g., Ce-137), and isotope (e.g., Co-60) sources. (When neutrons are used, the method is known as *neutron radiography* or *neutron gaging*.) The intensity of emitted radiation is changed when the rays pass through defects, and the intensity changes are monitored on a fluoroscope or recorded on film. This method

can be used to detect internal defects, changes in material structure, thickness, and the absence of internal workpieces. It is also used to check liquid levels in filled containers.

Up to 30 in (0.75 m) of steel can be penetrated by X-ray sources. Gamma sources, which are more portable and lower in cost than X-ray sources, can be used with steel up to 10 in (0.25 m).

Radiography requires access to both sides of the workpiece. Radiography involves some health risk, and there may be government standards associated with its use. Electrical power and cooling water may be required in large installations. Shielding and film processing are also required, making this the most expensive form of nondestructive testing.

19. HARDNESS TESTING

Hardness tests measure the capacity of a surface to resist deformation. The main use of hardness testing is to verify heat treatments, an important factor in product service life. Through empirical correlations, it is also possible to predict the ultimate strength and toughness of some materials.

The *Brinell hardness test* is used primarily with iron and steel castings. The *Brinell hardness number*, BHN, is determined by pressing a hardened steel ball into the surface of a specimen. The diameter of the resulting depression is correlated to the hardness. The standard ball is 10 mm in diameter and loads are 500 kg and 3000 kg for soft and hard materials, respectively.

The Brinell hardness number is the load per unit contact area. If a load, P (in kilograms), is applied through a steel ball of diameter D (in millimeters) and produces a depression of diameter d (in millimeters) and depth t (in millimeters), the Brinell hardness number is calculated from Eq. 43.29.

$$\begin{aligned} \text{BHN} &= \frac{P}{A_{\text{contact}}} = \frac{P}{\pi D t} \\ &= \frac{2P}{\pi D \left(D - \sqrt{D^2 - d^2} \right)} \end{aligned} \qquad 43.29$$

For heat-treated plain-carbon and medium-alloy steels, the ultimate tensile strength in lbf/in^2 can be approximately calculated from the steel's Brinell hardness number.

$$S_u \approx 500(\text{BHN}) \qquad 43.30$$

The *Rockwell hardness test* is similar to the Brinell test. A steel ball or diamond spheroconical penetrator (known as a *brale indentor*) is pressed into the material. The machine applies an initial load (10 kg) that sets the penetrator below surface imperfections. Then a significant load is applied. The Rockwell hardness is determined from the depth of penetration and is read directly from a dial.

Table 43.6 *Hardness Penetration Tests*

test	penetrator	diagram	measured dimension	hardness
Brinell	sphere	(a)	diameter, d	$\text{BHN} = \dfrac{2P}{\pi D \left(D - \sqrt{D^2 - d^2}\right)}$
Rockwell[a]	sphere or penetrator	(b)	depth, t	$R = C_1 - C_2 t$
Vickers	square pyramid	(b)	mean diagonal, d_1	$\text{VHN} = \dfrac{1.854P}{d_1^2}$
Meyer	sphere	(a)	diameter, d	$\text{MHN} = \dfrac{4P}{\pi d^2}$
Meyer-Vickers	square pyramid	(b)	mean diagonal, d_1	$M_V = \dfrac{2P}{d_1^2}$
Knoop	asymmetrical pyramid	(c)	long diagonal, L	$K = \dfrac{14.2P}{L^2}$

[a]C_1 and C_2 are constants that depend on the scale.

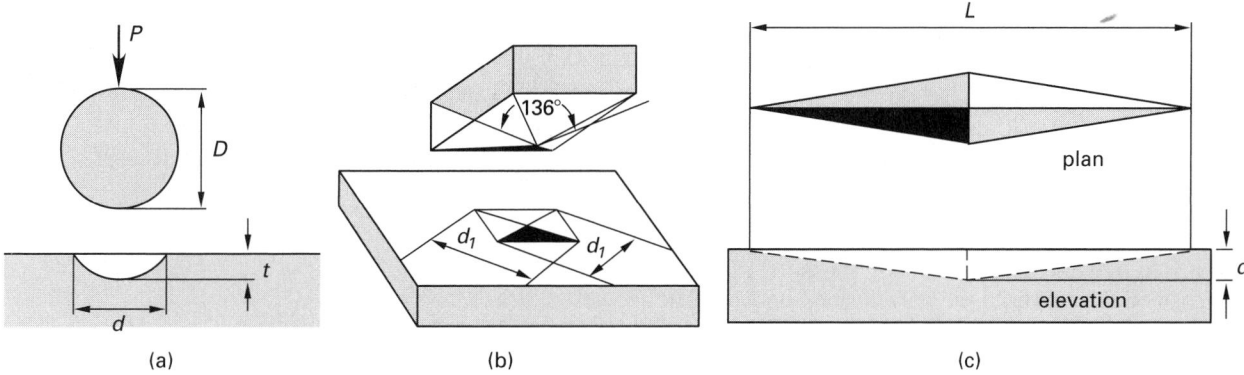

(a)   (b)   (c)

Although a number of Rockwell scales (A through G) exist, the B and C scales are commonly used for steel. The *Rockwell B scale* is used with a steel ball for mild steel and high-strength aluminum. The *Rockwell C scale* is used with the brale indentor for hard steels having ultimate tensile strengths up to 300 ksi (2 GPa). The *Rockwell A scale* has a wide range and can be used with both soft materials (such as annealed brass) and hard materials (such as cemented carbides).

Other penetration hardness tests include the *Meyer*, *Vickers*, *Meyer-Vickers*, and *Knoop* tests, as described in Table 43.6.

Cutting hardness is a measure of the force per unit area to cut a chip at low speed.

$$\text{cutting hardness} = \frac{F}{bt} \qquad 43.31$$

Figure 43.21 *Cutting Hardness*

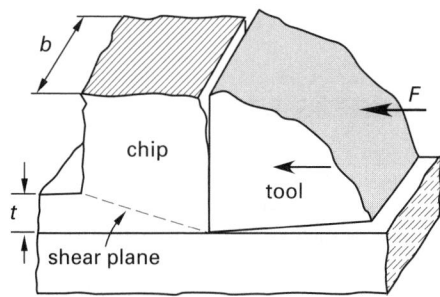

The *scratch hardness test*, also known as the *Mohs test*, compares the hardness of the material to that of minerals. Minerals of increasing hardness are used to scratch the sample. The resulting *Mohs scale* hardness can be used or correlated to other hardness scales, as in Fig. 43.22.

Structural

Figure 43.22 *Mohs Hardness Scale*

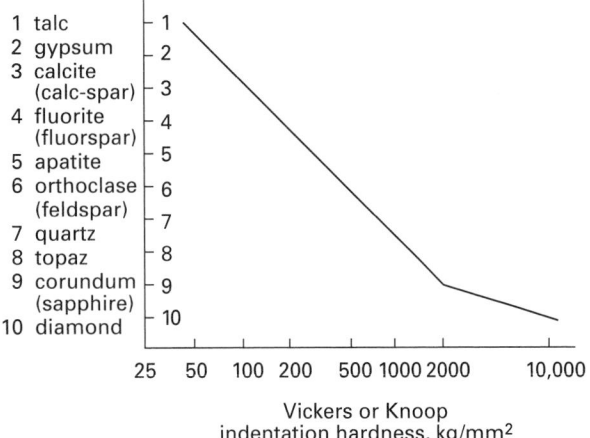

1 talc
2 gypsum
3 calcite (calc-spar)
4 fluorite (fluorspar)
5 apatite
6 orthoclase (feldspar)
7 quartz
8 topaz
9 corundum (sapphire)
10 diamond

Vickers or Knoop
indentation hardness, kg/mm²

Table 43.7 *Correlations Between Hardness Scales for Steel*

Brinell number	Vickers number	Rockwell numbers		scleroscope number
		C	B	
780	1150	70	. . .	106
712	960	66	. . .	95
653	820	62	. . .	87
601	717	58	. . .	81
555	633	55	120	75
514	567	52	119	70
477	515	49	117	65
429	454	45	115	59
401	420	42	113	55
363	375	38	110	51
321	327	34	108	45
293	296	31	106	42
277	279	29	104	39
248	248	24	102	36
235	235	22	99	34
223	223	20	97	32
207	207	16	95	30
197	197	13	93	29
183	183	9	90	27
166	166	4	86	25
153	153	. . .	82	23
140	140	. . .	78	21
131	131	. . .	74	20
121	121	. . .	70	. . .
112	112	. . .	66	. . .
105	105	. . .	62	. . .
99	99	. . .	59	. . .
95	95	. . .	56	. . .

The *file hardness* test is a combination of the cutting and scratch tests. Files of known hardness are drawn across the sample. The file ceases to cut the material when the material and file hardnesses are the same.

All of the preceding hardness tests are *destructive tests* because they mar the material surface. However, *ultrasonic tests* and various *rebound tests* (e.g., the *Shore hardness test* and the *scleroscopic hardness test*) are *nondestructive tests*. In a rebound test, a standard object, usually a diamond-tipped hammer, is dropped from a standard height onto the sample. The height of the rebound is measured and correlated to other hardness scales.

The various hardness tests do not measure identical properties of the material, so correlations between the various scales are not exact. For steel, the Brinell and Vickers hardness numbers are approximately the same below values of 320 Brinell. Also, the Brinell hardness is approximately ten times the Rockwell C hardness (R_c) for $R_c > 20$. Table 43.7 is an accepted correlation between several of the scales for steel. The table should not be used for other materials.

20. TOUGHNESS TESTING

During World War II, the United States experienced spectacular failures in approximately 25% of its Liberty ships and T-2 tankers. The mild steel plates of these ships were connected by welds that lost their ductility and became brittle at winter temperatures. Some of the ships actually broke into two sections. Such *brittle failures* are most likely to occur when three conditions are met: (a) triaxial stress, (b) low temperature, and (c) rapid loading.

Toughness is a measure of the material's ability to yield and absorb highly localized and rapidly applied stresses.

Notch toughness is evaluated by measuring the *impact energy* that causes a notched sample to fail.[15]

In the *Charpy test*, popular in the United States, a standardized beam specimen is given a 45° notch. The specimen is then centered on simple supports with the notch down. A falling pendulum striker hits the center of the specimen. This test is performed several times with different heights and different specimens until a sample fractures.

The kinetic energy expended at impact, equal to the initial potential energy less the rebound or follow-through height of the pendulum striker, is calculated from measured heights. It is designated C_V and is expressed in either foot-pounds (ft-lbf) or joules (J).[16] The energy required to cause failure is a measure of toughness.

---

[15]Without a notch, the specimen would experience uniaxial stress (tension and compression) at impact. The notch allows triaxial stresses to develop. Most materials become more brittle under triaxial stresses than under uniaxial stresses.

[16]In Europe, the energy is often expressed per unit cross section of specimen area.

Figure 43.23 *Charpy Test*

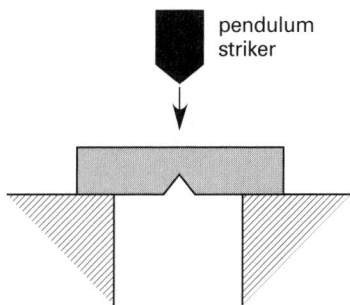

At 70°F (21°C), the energy required to cause failure ranges from 45 ft-lbf (60 J) for carbon steels to approximately 110 ft-lbf (150 J) for chromium-manganese steels. As temperature is reduced, however, the toughness decreases. In BCC metals, such as steel, at a low enough temperature the toughness decreases sharply. The transition from high-energy ductile failures to low-energy brittle failures begins at the *fracture transition plastic (FTP) temperature*.

Since the transition occurs over a wide temperature range, the *transition temperature* (also known as the *ductile transition temperature*) is taken as the temperature at which an impact of 15 ft-lbf (20 J) will cause failure. (15 ft-lbf is used for low-carbon ship steels. Other values may be used with other materials.) This occurs at approximately 30°F (−1°C) for low-carbon steel.

The appearance of the fractured surface is also used to evaluate the transition temperature. The fracture can be fibrous (from shear fracture) or granular (from cleavage fracture), or a mixture of both. The fracture planes are studied and the percentages of ductile failure plotted against temperature. The temperature at which the failure is 50% fibrous and 50% granular is known as the *fracture appearance transition temperature*, FATT.

Table 43.8 *Approximate Ductile Transition Temperatures*

type of steel	transition ductile temperature, °F
carbon steel	30°
high-strength, low-alloy steel	0°, to 30°
heat-treated, high-strength, carbon steel	−25°
heat-treated, construction alloy steel	−40° to −80°

Not all materials have a ductile-brittle transition. Aluminum, copper, other FCC metals, and most HCP metals do not lose their toughness abruptly. Figure 43.24 illustrates the failure energy curves for several different types of materials.

Figure 43.24 *Failure Energy versus Temperature*

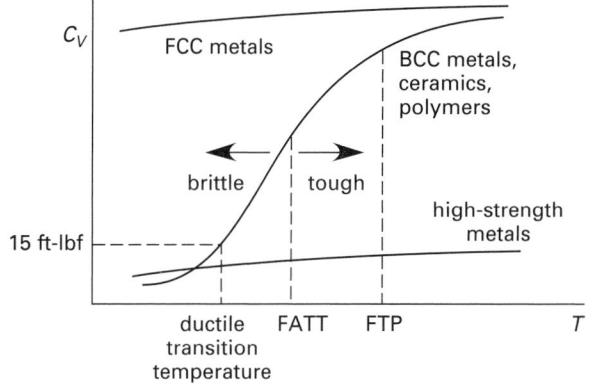

Another toughness test is the *Izod test*. This is illustrated in Fig. 43.25 and is similar to the Charpy test in its use of a notched specimen. The height to which a swinging pendulum follows through after causing the specimen to fail determines the energy of failure.

Figure 43.25 *Izod Test*

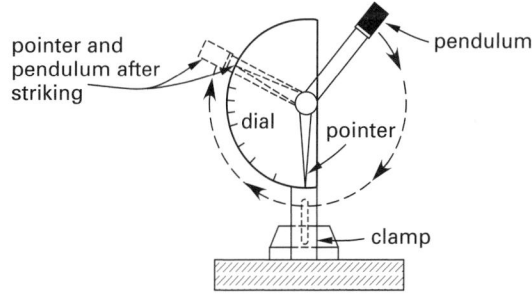

21. CREEP TEST

Creep or *creep strain* is the continuous yielding of a material under constant stress. For metals, creep is negligible at low temperatures (i.e., less than half of the absolute melting temperature), although the usefulness of nonreinforced plastics as structural materials is seriously limited by creep at room temperature.

During a *creep test*, a low tensile load of constant magnitude is applied to a specimen, and the strain is measured as a function of time. The *creep strength* is the stress that results in a specific creep rate, usually 0.001% or 0.0001% per hour. The *rupture strength*, determined from a *stress-rupture test*, is the stress that results in a failure after a given amount of time, usually 100, 1000, or 10,000 hours.

If strain is plotted as a function of time, three different curvatures will be apparent following the initial elastic extension.[17] During the first stage, the *creep*

---

[17]In Great Britain, the initial elastic elongation is considered the first stage. Therefore, creep has four stages in British nomenclature.

rate ($d\epsilon/dt$) decreases since strain hardening (dislocation generation and interaction with grain boundaries and other barriers) is occurring at a greater rate than annealing (annihilation of dislocations, climb, cross-slip, and some recrystallization). This is known as *primary creep.*

During the second stage, the creep rate is constant, with strain hardening and annealing occurring at the same rate. This is known as *secondary creep* or *cold flow.* During the third stage, the specimen begins to neck down, and rupture eventually occurs. This region is known as *tertiary creep.*

The secondary creep rate is lower than the primary and tertiary creep rates. The secondary creep rate, represented by the slope (on a log-log scale) of the line during the second stage, is temperature and stress dependent. This slope increases at higher temperatures and stresses. The creep rate curve can be represented by the following empirical equation, known as *Andrade's equation.*

$$\epsilon = \epsilon_o \left(1 + \beta t^{1/3}\right) e^{kt} \qquad 43.32$$

Figure 43.26 *Stages of Creep*

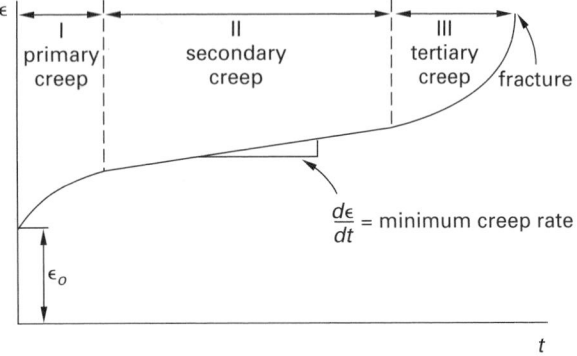

Dislocation climb (glide and creep) is the primary creep mechanism, although diffusion creep and grain boundary sliding also contribute to creep on a microscopic level. On a larger scale, the mechanisms of creep involve slip, subgrain formation, and grain-boundary sliding.

Figure 43.27 *Effect of Stress on Creep Rates*

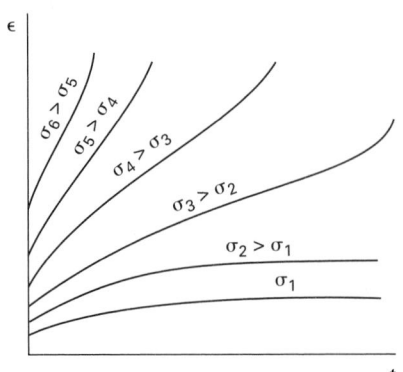

22. EFFECTS OF IMPURITIES AND STRAIN ON MECHANICAL PROPERTIES

Anything that restricts the movement of dislocations will increase the strength of metals and reduce ductility. Alloying materials, impurity atoms, imperfections, and other dislocations produce stronger materials. This is illustrated in Fig. 43.28.

Figure 43.28 *Effect of Impurities on Mechanical Properties*

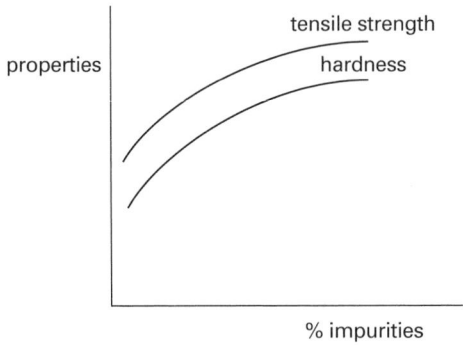

Additional dislocations are generated by the plastic deformation (i.e., cold working) of metals, and these dislocations can strain-harden the metal. Figure 43.29 shows the effect of strain-hardening on mechanical properties.

Figure 43.29 *Effect of Strain-Hardening on Mechanical Properties*

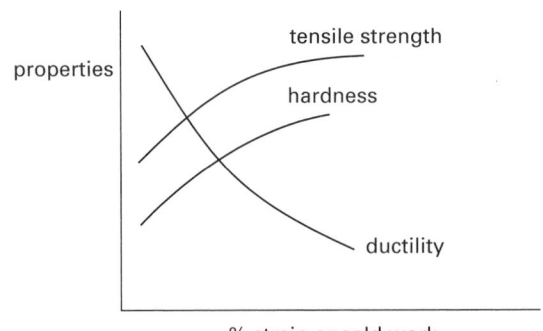

23. CLASSIFICATION OF MATERIALS

When used to describe engineering materials, the terms "strong" and "tough" are not synonymous. Similarly, "weak," "soft," and "brittle" have different engineering meanings. A *strong material* has a high ultimate strength, whereas a *weak material* has a low ultimate strength. A *tough material* will yield greatly before breaking, whereas a *brittle material* will not. (A brittle material is one whose strain at fracture is less than approximately 0.5%.) A *hard material* has a high modulus of elasticity, whereas a *soft material* does not. Figure 43.30 illustrates some of the possible combinations of these classifications.

Figure 43.30 *Types of Engineering Materials*

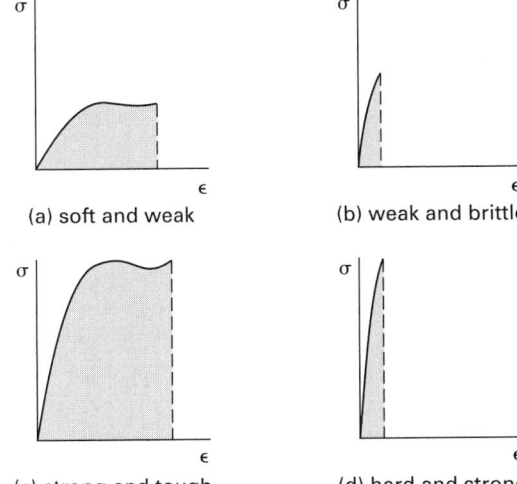

(a) soft and weak

(b) weak and brittle

(c) strong and tough

(d) hard and strong

44 Strength of Materials

Nomenclature

a	width	in	m
A	area	in^2	m^2
b	width	in	m
c	distance to extreme fiber	in	m
C	constant	–	–
C	couple	in-lbf	N·m
d	distance, depth, or diameter	in	m
e	eccentricity	in	m
E	modulus of elasticity	lbf/in^2	MPa
F	force	lbf	N
G	shear modulus	lbf/in^2	MPa
h	height	in	m
I	moment of inertia	in^4	m^4
J	polar moment of inertia	in^4	m^4
k	spring constant	lbf/in	N/m
K	stress concentration factor	–	–
L	length	in	m
M	moment	in-lbf	N·m
P	force	lbf	N
Q	statical moment	in^3	m^3
r	radius	in	m
R	reaction	lbf	N
R	rigidity	–	–
S	force	lbf	N
S	section modulus	in^3	m^3
t	thickness	in	m
T	temperature	°F	°C
T	torque	in-lbf	N·m
u	unit force	lbf	N
U	energy	in-lbf	N·m
V	vertical shear force	lbf	N
V	volume	in^3	m^3
w	load per unit length	lbf/in	N/m
x	location	in	m
y	location	in	m

Symbols

α	coefficient of linear thermal expansion	1/°F	1/°C
β	coefficient of volumetric thermal expansion	1/°F	1/°C
γ	coefficient of area thermal expansion	1/°F	1/°C
δ	deformation	in	m
ϵ	strain	–	–
θ	angle	deg	deg
ν	Poisson's ratio	–	–
ρ	radius of curvature	in	m
σ	normal stress	lbf/in^2	MPa
τ	shear stress	lbf/in^2	MPa
ϕ	angle	rad	rad

Subscripts

0	nominal
a	alternating
b	bending
c	centroidal
i	inside
j	jth member
l	left
m	mean
o	original
r	range or right
R	right
th	thermal
w	web

Structural

1. BASIC CONCEPTS

Strength of materials (known also as *mechanics of materials*) deals with the elastic behavior of loaded engineering materials.[1] This subject draws heavily on the topics in Chaps. 42 and 43.

Stress is force per unit area, F/A. Typical units of stress are lbf/in^2, ksi (thousands of pounds per square inch), and MPa. Although there are many names given to stress, there are only two primary types, differing in the orientation of the loaded area. With *normal stress*, σ, the area is normal to the force carried. With *shear stress*, τ, the area is parallel to the force.

Figure 44.1 *Normal and Shear Stress*

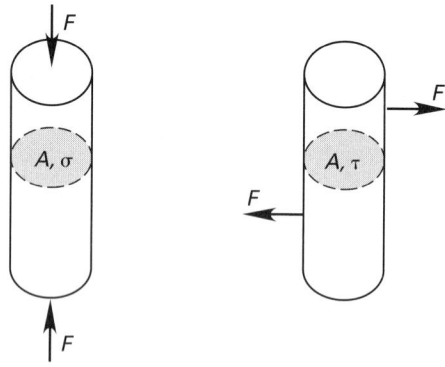

Strain, ϵ, is elongation expressed on a fractional or percentage basis. It may be listed as having units of in/in, mm/mm, and percent, or no units at all. A strain in one direction will be accompanied by strains in orthogonal directions in accordance with Poisson's ratio. *Dilation* is the sum of the strains in the three coordinate directions.

$$\text{dilation} = \epsilon_x + \epsilon_y + \epsilon_z \qquad 44.1$$

2. HOOKE'S LAW

Hooke's law is a simple mathematical statement of the relationship between elastic stress and strain: Stress is proportional to strain. For normal stress, the constant of proportionality is the *modulus of elasticity* (*Young's modulus*), E.

$$\sigma = E\epsilon \qquad 44.2$$

For shear stress, the constant of proportionality is the *shear modulus*, G.

$$\tau = G\phi \qquad 44.3$$

---

[1]Plastic behavior and ultimate strength design are not covered in this chapter.

Figure 44.2 *Application of Hooke's Law*

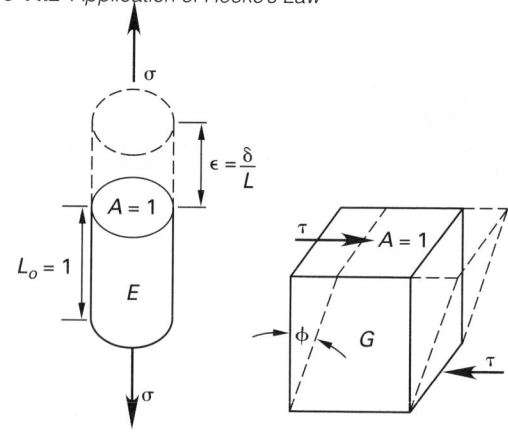

(a) normal stress on a unit cylinder

(b) shear stress on a unit cube

3. ELASTIC DEFORMATION

Since stress is F/A and strain is δ/L_o, Hooke's law can be rearranged in form to give the elongation of an axially loaded member with a uniform cross section experiencing normal stress. Tension loading is considered positive; compressive loading is negative.

$$\delta = L_o\epsilon = \frac{L_o\sigma}{E} = \frac{L_oF}{EA} \qquad 44.4$$

The actual length of a member under loading is given by Eq. 44.5. The algebraic sign of the deformation must be observed.

$$L = L_o + \delta \qquad 44.5$$

4. TOTAL STRAIN ENERGY

The energy stored in a loaded member is equal to the work required to deform the member. Below the proportionality limit, the total *strain energy* for a member loaded in tension or compression is given by Eq. 44.6.

$$U = \tfrac{1}{2}F\delta = \frac{F^2 L_o}{2AE} = \frac{\sigma^2 L_o A}{2E} \qquad 44.6$$

5. STIFFNESS AND RIGIDITY

Stiffness is the amount of force required to cause a unit of deformation (displacement) and is often referred to as a *spring constant*. Typical units are pounds per inch and newtons per meter. The stiffness of a spring or other structure can be calculated from the deformation equation by solving for F/δ. Equation 44.7 is valid for tensile and compressive normal stresses. For torsion and bending, the stiffness equation will depend on how the deflection is calculated.

$$k = \frac{F}{\delta} \quad \text{[general form]} \qquad 44.7(a)$$

$$k = \frac{AE}{L_o} \quad \text{[normal stress form]} \qquad 44.7(b)$$

When more than one spring or resisting member share the load, the relative stiffnesses are known as *rigidities*. Rigidities have no units, and the individual rigidity values have no significance. A ratio of two rigidities, however, indicates how much stiffer one member is compared to another. Equation 44.8 is one method of calculating rigidity in a multi-member structure. (Since rigidities are relative numbers, they can be multiplied by the least common denominator to obtain integer values.)

$$R_j = \frac{k_j}{\sum\limits_i k_i} \qquad 44.8$$

Figure 44.3 *Stiffness and Rigidity*

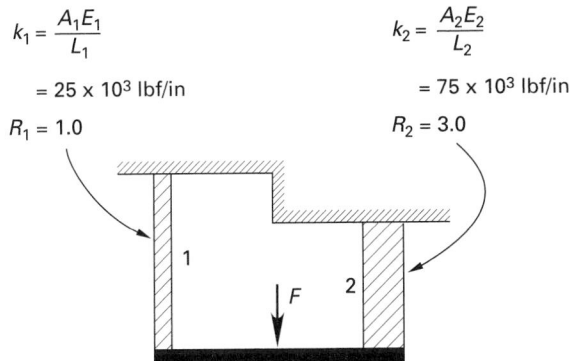

$$k_1 = \frac{A_1 E_1}{L_1}$$
$$= 25 \times 10^3 \text{ lbf/in}$$
$$R_1 = 1.0$$

$$k_2 = \frac{A_2 E_2}{L_2}$$
$$= 75 \times 10^3 \text{ lbf/in}$$
$$R_2 = 3.0$$

Rigidity is proportional to the reciprocal of deflection. *Flexural rigidity* is the reciprocal of deflection in members that are acted upon by a moment (i.e., are in bending), although that term may also be used to refer to the product, EI, of the modulus of elasticity and the moment of inertia.

6. THERMAL DEFORMATION

If the temperature of an object is changed, the object will experience length, area, and volume changes. The magnitude of these changes will depend on the *coefficient of linear expansion*, α, which is widely tabulated for solids. The *coefficient of volumetric expansion*, β, is encountered less often for solids but is used extensively with liquids and gases.

$$\Delta L = \alpha L_o (T_2 - T_1) \qquad 44.9$$
$$\Delta A = \gamma A_o (T_2 - T_1) \qquad 44.10$$
$$\gamma \approx 2\alpha \qquad 44.11$$
$$\Delta V = \beta V_o (T_2 - T_1) \qquad 44.12$$
$$\beta \approx 3\alpha \qquad 44.13$$

It is a common misconception that a hole in a plate will decrease in size when the plate is heated (because the surrounding material "squeezes in" on the hole). However, changes in temperature affect all dimensions the same way. In this case, the circumference of the hole is a linear dimension that follows Eq. 44.9. As the circumference increases, the hole area also increases.

Table 44.1 *Deflection and Stiffness for Various Systems (due to bending moment alone)*

system	maximum deflection (x)	stiffness (k)
	$\dfrac{Fh}{AE}$	$\dfrac{AE}{h}$
	$\dfrac{Fh^3}{3EI}$	$\dfrac{3EI}{h^3}$
	$\dfrac{Fh^3}{12EI}$	$\dfrac{12EI}{h^3}$
	$\dfrac{wL^4}{8EI}$	$\dfrac{8EI}{L^3}$
	$\dfrac{Fh^3}{12E(I_1 + I_2)}$	$\dfrac{12E(I_1 + I_2)}{h^3}$
	$\dfrac{FL^3}{48EI}$	$\dfrac{48EI}{L^3}$
(w is load per unit length)	$\dfrac{5wL^4}{384EI}$	$\dfrac{384EI}{5L^3}$
	$\dfrac{FL^3}{192EI}$	$\dfrac{192EI}{L^3}$
(w is load per unit length)	$\dfrac{wL^4}{384EI}$	$\dfrac{384EI}{L^3}$

Figure 44.4 *Thermal Expansion of an Area*

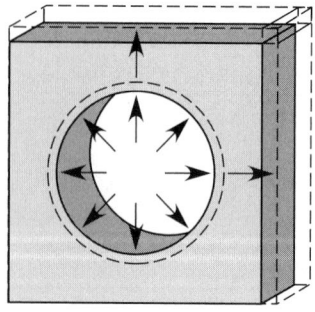

If Eq. 44.9 is rearranged, an expression for the *thermal strain* is obtained.

$$\epsilon_{\text{th}} = \frac{\Delta L}{L_o} = \alpha(T_2 - T_1) \qquad 44.14$$

Thermal strain is handled in the same manner as strain due to an applied load. For example, if a bar is heated but is not allowed to expand, the stress can be calculated from the thermal strain and Hooke's law.

$$\sigma_{\text{th}} = E\epsilon_{\text{th}} \qquad 44.15$$

Low values of the coefficient of expansion, such as with Pyrex™ glassware, result in low thermally induced stresses and high insensitivity to temperature extremes. Differences in the coefficients of expansion of two materials are used in *bimetallic elements*, such as thermostatic springs and strips.

Table 44.2 *Average Coefficients of Linear Thermal Expansion (multiply all values by 10^{-6})*

substance	1/°F	1/°C
aluminum alloy	12.8	23.0
brass	10.0	18.0
cast iron	5.6	10.1
chromium	3.8	6.8
concrete	6.7	12.0
copper	8.9	16.0
glass (plate)	4.9	8.9
glass (Pyrex™)	1.8	3.2
invar	0.39	0.7
lead	15.6	28.0
magnesium alloy	14.5	26.1
marble	6.5	11.7
platinum	5.0	9.0
quartz, fused	0.2	0.4
steel	6.5	11.7
tin	14.9	26.9
titanium alloy	4.9	8.8
tungsten	2.4	4.4
zinc	14.6	26.3

(Multiply 1/°F by 9/5 to obtain 1/°C.)
(Multiply 1/°C by 5/9 to obtain 1/°F.)

Example 44.1

A replacement steel rail ($L = 20.0$ m, $A = 60 \times 10^{-4}$ m^2) was installed when its temperature was 5°C. The rail was installed tightly in the line, without an allowance for expansion. If the rail ends are constrained by adjacent rails and if the spikes prevent buckling, what is the compressive force in the rail at 25°C?

Solution

From Table 44.2, the coefficient of linear expansion for steel is 11.7×10^{-6} 1°C. From Eq. 44.14, the thermal strain is

$$\epsilon_{\text{th}} = \alpha(T_2 - T_1) = \left(11.7 \times 10^{-6}\ \frac{1}{°C}\right)(25°C - 5°C)$$
$$= 2.34 \times 10^{-4}\ \text{m/m}$$

The modulus of elasticity of steel is 20×10^{10} N/m^2 (20×10^4 MPa). The compressive stress is given by Hooke's law. Use Eq. 44.15.

$$\sigma_{\text{th}} = E\epsilon_{\text{th}} = \left(20 \times 10^{10}\ \frac{\text{N}}{\text{m}^2}\right)\left(2.34 \times 10^{-4}\ \frac{\text{m}}{\text{m}}\right)$$
$$= 4.68 \times 10^7\ \text{N/m}^2$$

The compressive force is

$$F = \sigma_{\text{th}}A = \left(4.68 \times 10^7\ \frac{\text{N}}{\text{m}^2}\right)(60 \times 10^{-4}\ \text{m}^2)$$
$$= 281\,000\ \text{N}$$

7. STRESS CONCENTRATIONS

A *geometric stress concentration* occurs whenever there is a discontinuity or non-uniformity in an object. Examples of non-uniform shapes are stepped shafts, plates with holes and notches, and shafts with keyways. It is convenient to think of stress as streamlines within an object. There will be a stress concentration wherever local geometry forces the streamlines closer together.

Figure 44.5 *Streamline Analogy to Stress Concentrations*

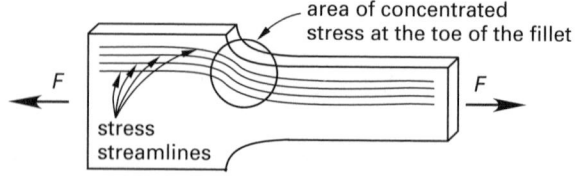

Stress values determined by simplistic F/A, Mc/I, or Tr/J calculations will be greatly understated. *Stress concentration factors (stress risers)* are correction factors used to account for the non-uniform stress distributions. The symbol K is often used, but this is not universal. The actual stress is determined as the product of the stress concentration factor, K, and the *nominal stress*, σ_0. Values of the stress concentration factor are almost always greater than 1.0 and can run as high as 3.0 and above. The exact value for a given application must be determined from extensive experimentation or from published tabulations of standard configurations.

$$\sigma' = K\sigma_0 \qquad 44.16$$

Stress concentration factors are normally not applied to members with multiple redundancy, for static loading of ductile materials, or where local yielding around the discontinuity reduces the stress. For example, there will be many locations of stress concentration in a lap rivet connection. However, the stresses are kept low by design, and stress concentration is disregarded.

Stress concentration factors are not applicable to every point on an object; they apply only to the point of maximum stress. For example, with filleted shafts, the maximum stress occurs at the toe of the fillet. Therefore, the stress concentration factor should be applied to the stress calculated from the smaller section's properties. For objects with holes or notches, it is important to know if the nominal stress to which the factor is applied is calculated from an area that includes or excludes the holes or notches.

In addition to geometric stress concentrations, there are also *fatigue stress concentrations*. The *fatigue stress concentration* factor is the ratio of the fatigue strength without a stress concentration to the fatigue stress with a stress concentration. Fatigue stress concentration factors depend on the material, material strength, and geometry of the stress concentration (i.e., radius of the notch). Fatigue stress concentration factors can be less than the geometric factors from which they are computed.

8. COMBINED STRESSES (BIAXIAL LOADING)

Loading is rarely confined to a single direction. Many practical cases have different normal and shear stresses on two or more perpendicular planes. Sometimes, one of the stresses may be small enough to be disregarded, reducing the analysis to one dimension. In other cases, however, the shear and normal stresses must be combined to determine the maximum stresses acting on the material.

For any point in a loaded specimen, a plane can be found where the shear stress is zero. The normal stresses associated with this plane are known as the *principal stresses*, which are the maximum and minimum stresses acting at that point in any direction.

For two-dimensional (biaxial) loading (i.e., two normal stresses combined with a shearing stress), the normal and shear stresses on a plane whose normal line is inclined an angle θ from the horizontal can be found from Eqs. 44.17 and 44.18. Proper sign convention must be adhered to when using the combined stress equations. The positive senses of shear and normal stresses are

shown in Fig. 44.6. As is usually the case, tensile normal stresses are positive; compressive normal stresses are negative. In two dimensions, shear stresses are designated as clockwise (positive) or counterclockwise (negative).[2]

$$\sigma_\theta = \tfrac{1}{2}(\sigma_x + \sigma_y) + \tfrac{1}{2}(\sigma_x - \sigma_y)\cos 2\theta + \tau \sin 2\theta$$
$$44.17$$

$$\tau_\theta = -\tfrac{1}{2}(\sigma_x - \sigma_y)\sin 2\theta + \tau \cos 2\theta \qquad 44.18$$

Figure 44.6 *Sign Convention for Combined Stress*

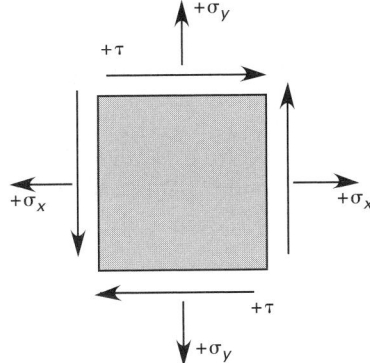

At first glance, the orientation of the shear stresses may seem confusing. However, the arrangement of stresses shown produces equilibrium in the x- and y-directions without causing rotation. Other than a mirror image or a trivial rotation of the arrangement shown in Fig. 44.6, no arrangement of shear stresses will produce equilibrium.

The maximum and minimum values (as θ is varied) of the normal stress, σ_θ, are the *principal stresses*, which can be found by differentiating Eq. 44.17 with respect to θ, setting the derivative equal to zero, and substituting θ back into Eq. 44.17. Equation 44.19 is derived in this manner. A similar procedure is used to derive the *extreme shear stresses* (i.e., maximum and minimum shear stresses) in Eq. 44.20 from Eq. 44.18. (The term *principal stress* implies a normal stress, never a shear stress.)

$$\sigma_1, \sigma_2 = \tfrac{1}{2}(\sigma_x + \sigma_y) \pm \tau_1 \qquad 44.19$$

$$\tau_1, \tau_2 = \pm\tfrac{1}{2}\sqrt{(\sigma_x - \sigma_y)^2 + (2\tau)^2} \qquad 44.20$$

Figure 44.7 *Stresses on Inclined Plane*

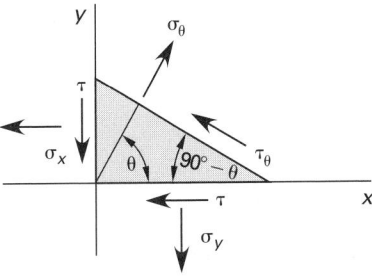

[2]Some sources refer to the shear stress as τ_{xy}; others use the symbol τ_z. When working in two dimensions only, the subscripts xy and z are unnecessary and confusing conventions.

The angles of the planes on which the normal stresses are minimum and maximum are given by Eq. 44.21. θ is measured from the x-axis, clockwise if negative and counterclockwise if positive. Equation 44.21 will yield two angles, 90° apart. These angles can be substituted back into Eqs. 44.17 and 44.18 to determine which angle corresponds to the minimum normal stress and which angle corresponds to the maximum normal stress.[3]

$$\theta_{\sigma_1,\sigma_2} = \tfrac{1}{2}\arctan\left(\frac{2\tau}{\sigma_x - \sigma_y}\right) \qquad 44.21$$

The angles of the planes on which the shear stress is minimum and maximum are given by Eq. 44.22. These planes will be 90° apart and will be rotated 45° from the planes of principal normal stresses. As with Eq. 44.21, θ is measured from the x-axis, clockwise if negative and counterclockwise if positive. Generally, the sign of a shear stress on an inclined plane will be unimportant.

$$\theta_{\tau_1,\tau_2} = \tfrac{1}{2}\arctan\left(\frac{\sigma_x - \sigma_y}{-2\tau}\right) \qquad 44.22$$

Example 44.2

Find the maximum normal and shear stresses on the object shown. Determine the angle of the plane of principal normal stresses.

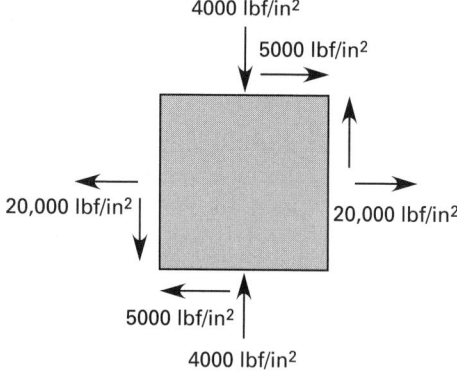

Solution

Find the principal shear stresses first. The applied 4000 lbf/in² compressive stress is negative. Equation 44.20 can be used directly.

$$\tau_1 = \tfrac{1}{2}\sqrt{(\sigma_x - \sigma_y)^2 + (2\tau)^2}$$

$$= \tfrac{1}{2}\sqrt{\left(20{,}000\ \frac{\text{lbf}}{\text{in}^2} - \left(-4000\ \frac{\text{lbf}}{\text{in}^2}\right)\right)^2 + \left((2)\left(5000\ \frac{\text{lbf}}{\text{in}^2}\right)\right)^2}$$

$$= 13{,}000\ \text{lbf/in}^2$$

---

[3]Alternatively, the following procedure can be used to determine the direction of the principal planes. Let σ_x be the algebraically larger of the two given normal stresses. The angle between the direction of σ_x and the direction of σ_1, the algebraically larger principal stress, will always be less than 45°.

From Eq. 44.19, the maximum normal stress is

$$\sigma_1 = \tfrac{1}{2}(\sigma_x + \sigma_y) + \tau_1$$

$$= \left(\tfrac{1}{2}\right)\left(20{,}000\ \frac{\text{lbf}}{\text{in}^2} + \left(-4000\ \frac{\text{lbf}}{\text{in}^2}\right)\right) + 13{,}000\ \frac{\text{lbf}}{\text{in}^2}$$

$$= 21{,}000\ \text{lbf/in}^2 \quad [\text{tension}]$$

The angle of the principal normal stresses is given by Eq. 44.21.

$$\theta = \tfrac{1}{2}\arctan\left(\frac{2\tau}{\sigma_x - \sigma_y}\right)$$

$$= \tfrac{1}{2}\arctan\left(\frac{(2)\left(5000\ \frac{\text{lbf}}{\text{in}^2}\right)}{20{,}000\ \frac{\text{lbf}}{\text{in}^2} - \left(-4000\ \frac{\text{lbf}}{\text{in}^2}\right)}\right)$$

$$= \left(\tfrac{1}{2}\right)(22.6°, 202.6°)$$

$$= 11.3°, 101.3°$$

It is not obvious which angle produces which normal stress. One of the angles can be substituted back into the general equation (Eq. 44.17) for σ_θ.

$$\sigma_{11.3°} = \left(\tfrac{1}{2}\right)\left(20{,}000\ \frac{\text{lbf}}{\text{in}^2} + \left(-4000\ \frac{\text{lbf}}{\text{in}^2}\right)\right)$$

$$+ \left(\tfrac{1}{2}\right)\left(20{,}000\ \frac{\text{lbf}}{\text{in}^2} - \left(-4000\ \frac{\text{lbf}}{\text{in}^2}\right)\right)$$

$$\times \cos\big((2)(11.3°)\big) + \left(5000\ \frac{\text{lbf}}{\text{in}^2}\right)\sin\big((2)(11.3°)\big)$$

$$= 21{,}000\ lbf/in^2$$

Thus, the 11.3° angle corresponds to the maximum normal stress of 21,000 lbf/in².

9. MOHR'S CIRCLE FOR STRESS

Mohr's circle can be constructed to graphically determine the principal stresses.

Figure 44.8 *Mohr's Circle for Stress*

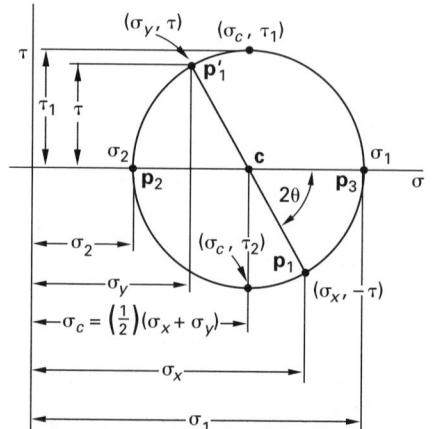

step 1: Determine the applied stresses: σ_x, σ_y, and τ. (Tensile normal stresses are positive; compressive normal stresses are negative. Clockwise shear stresses are positive; counterclockwise shear stresses are negative.)

step 2: Draw a set of σ-τ axes.

step 3: Plot the center of the circle, point **c**, by calculating $\sigma_c = \frac{1}{2}(\sigma_x + \sigma_y)$.

step 4: Plot the point $\mathbf{p}_1 = (\sigma_x, -\tau)$. (Alternatively, plot $\mathbf{p}_1'$ at $(\sigma_y, +\tau)$.)

step 5: Draw a line from point $\mathbf{p}_1$ through center **c** and extend it an equal distance beyond the σ-axis. This is the diameter of the circle.

step 6: Using the center **c** and point $\mathbf{p}_1$, draw the circle. An alternative method is to draw a circle of radius r about point **c**.

$$r = \sqrt{\frac{1}{4}(\sigma_x - \sigma_y)^2 + \tau^2} \qquad 44.23$$

step 7: Point $\mathbf{p}_2$ defines the smaller principal stress, σ_2. Point $\mathbf{p}_3$ defines the larger principal stress, σ_1.

step 8: Determine the angle θ as half of the angle 2θ on the circle. This angle corresponds to the larger principal stress, σ_1. On Mohr's circle, angle 2θ is measured counterclockwise from the $\mathbf{p}_1$-$\mathbf{p}_1'$ line to the horizontal axis.

Example 44.3

Construct Mohr's circle for Ex. 44.2.

Solution

$$\sigma_c = \frac{1}{2}(\sigma_x + \sigma_y)$$
$$= \left(\frac{1}{2}\right)\left(20{,}000 \ \frac{\text{lbf}}{\text{in}^2} + \left(-4000 \ \frac{\text{lbf}}{\text{in}^2}\right)\right)$$
$$= 8000 \ \text{lbf/in}^2$$
$$r = \sqrt{\frac{1}{4}(\sigma_x - \sigma_y)^2 + \tau^2}$$
$$= \sqrt{\begin{array}{c} \left(\frac{1}{4}\right)\left(20{,}000 \ \frac{\text{lbf}}{\text{in}^2} - \left(-4000 \ \frac{\text{lbf}}{\text{in}^2}\right)\right)^2 \\ + \left(5000 \ \frac{\text{lbf}}{\text{in}^2}\right)^2 \end{array}}$$
$$= 13{,}000 \ \text{lbf/in}^2$$

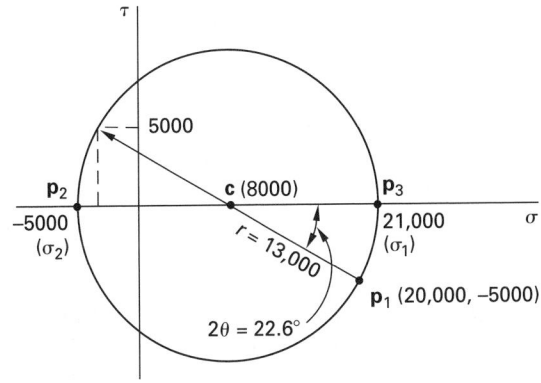

10. IMPACT LOADING

If a load is applied to a structure suddenly, the structure's response will be composed of two parts: a transient response (which decays to zero) and a steady-state response. (These two parts are also known as the *dynamic* and *static responses*.) It is not unusual for the transient loading to be larger than the steady-state response.

Although a *dynamic analysis* of the structure is preferred, the procedure is lengthy and complex. Therefore, arbitrary multiplicative factors may be applied to the steady-state stress to determine the maximum transient. For example, if a load is applied quickly as compared to the natural period of vibration of the structure (e.g., the classic definition of an *impact load*), a dynamic factor of 2.0 might be used. Actual dynamic factors should be determined or validated by testing.

The energy-conservation method (i.e., the work-energy principle) can be used to determine the maximum stress due to a falling mass. The total change in potential energy of the mass from the change in elevation and the deflection δ) is equated to the appropriate expression for total strain energy (see Sec. 4).

11. SHEAR AND MOMENT

Shear at a point is the sum of all vertical forces acting on an object. It has units of pounds, kips, tons, newtons, and so on. Shear is not the same as shear stress, since the area of the object is not considered.

A typical application is shear at a point on a beam, V, defined as the sum of all vertical forces between the point and one of the ends.[4] The direction (i.e., to the left or right of the point) in which the summation proceeds is not important. Since the values of shear will differ only in sign for summations to the left and right ends, the direction that results in the fewest calculations should be selected.

---

[4]The conditions of equilibrium require that the sum of all vertical forces on a beam be zero. However, the *shear* can be nonzero because only a portion of the beam is included in the analysis. Since that portion extends to the beam end in one direction only, shear is sometimes called *resisting shear* or *one-way shear*.

$$V = \sum_{\substack{\text{point to} \\ \text{one end}}} F_i \qquad\qquad 44.24$$

Shear is taken as positive when there is a net upward force to the left of a point and negative when there is a net downward force between the point and the left end.

Moment at a point is the total bending moment acting on an object. In the case of a beam, the moment, M, will be the algebraic sum of all moments and couples located between the investigation point and one of the beam ends. As with shear, the number of calculations required to calculate the moment can be minimized by careful choice of the beam end.[5]

$$M = \sum_{\substack{\text{point to} \\ \text{one end}}} F_i d_i + \sum_{\substack{\text{point to} \\ \text{one end}}} C_i \qquad 44.25$$

Moment is taken as positive when the upper surface of the beam is in compression and the lower surface is in tension (see Fig. 44.12). Since the beam ends will usually be higher than the midpoint, it is commonly said that "a positive moment will make the beam smile."

12. SHEAR AND BENDING MOMENT DIAGRAMS

The value of the shear and moment, V and M, will depend on location along the beam. Both shear and moment can be described mathematically for simple loadings, but the formulas are likely to become discontinuous as the loadings become more complex. It is much more convenient to describe the shear and moment functions graphically. Graphs of shear and moment as functions of position along the beam are known as *shear* and *moment diagrams*. Drawing these diagrams does not require knowing the shape or area of the beam.

The following guidelines and conventions should be observed when constructing a *shear diagram*.

- The shear at any point is equal to the sum of the loads and reactions from the point to the left end.

- The magnitude of the shear at any point is equal to the slope of the moment line at that point.

$$V = \frac{dM}{dx} \qquad\qquad 44.26$$

- Loads and reactions acting upward are positive.

- The shear diagram is straight and sloping over uniformly distributed loads.

- The shear diagram is straight and horizontal between concentrated loads.

- The shear is a vertical line and is undefined at points of concentrated loads.

The following guidelines and conventions should be observed when constructing a *bending moment diagram*. By convention, the moment diagram is drawn on the compression side of the beam.

- The moment at any point is equal to the sum of the moments and couples from the point to the left end.[6]

- Clockwise moments about the point are positive.

- The magnitude of the moment at any point is equal to the area under the shear line up to that point. This is equivalent to the integral of the shear function.

$$M = \int V \, dx \qquad\qquad 44.27$$

- The *maximum moment* occurs where the shear is zero.

- The moment diagram is straight and sloping between concentrated loads.

- The moment diagram is curved (parabolic upward) over uniformly distributed loads.

These principles are illustrated in Fig. 44.9.

Figure 44.9 *Drawing Shear and Moment Diagrams*

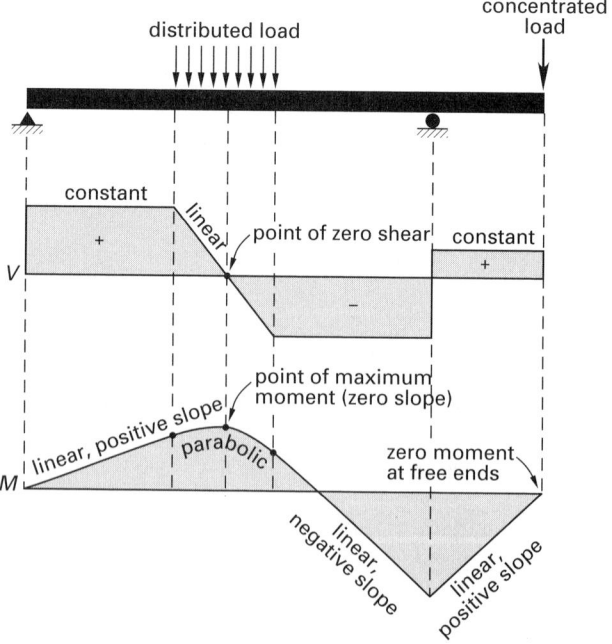

---

[5] The conditions of equilibrium require that the sum of all moments on a beam be zero. However, the *moment* can be nonzero because only a portion of the beam is included in the analysis. Since that portion extends to the beam end in one direction only, moment is sometimes called *bending moment, resisting moment,* or *one-way moment.*

[6] If the beam is cantilevered with its built-in end at the left, the fixed-end moment will be unknown. In that case, the moment must be calculated to the right end of the beam.

Example 44.4

Draw the shear and bending moment diagrams for the following beam.

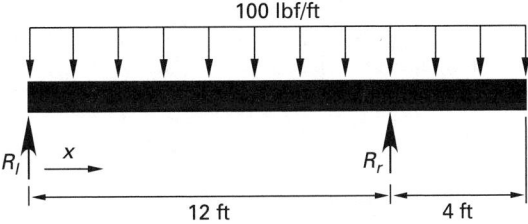

Solution

First, determine the reactions. The uniform load of $100x$ can be assumed to be concentrated at $x/2$.

$$R_r = \frac{\left(\frac{1}{2}\right)(16 \text{ ft})(16 \text{ ft})\left(100 \frac{\text{lbf}}{\text{ft}}\right)}{12 \text{ ft}} = 1066.7 \text{ lbf}$$

$$R_l = (16 \text{ ft})\left(100 \frac{\text{lbf}}{\text{ft}}\right) - R_r = 533.3 \text{ lbf}$$

The shear diagram starts at +533.3 at the left reaction but decreases linearly at the rate of 100 lbf/ft between the two reactions. Measuring x from the left end, the shear line goes through zero at

$$x = \frac{533.3 \text{ lbf}}{100 \frac{\text{lbf}}{\text{ft}}} = 5.333 \text{ ft}$$

The shear just to the left of the right reaction is

$$533.3 \text{ lbf} - (12 \text{ ft})\left(100 \frac{\text{lbf}}{\text{ft}}\right) = -666.7 \text{ lbf}$$

The shear just to the right of the right reaction is

$$-666.7 \text{ lbf} + R_r = +400 \text{ lbf}$$

To the right of the right reaction, the shear diagram decreases to zero at the same constant rate: 100 lbf/ft. This is sufficient information to draw the shear diagram.

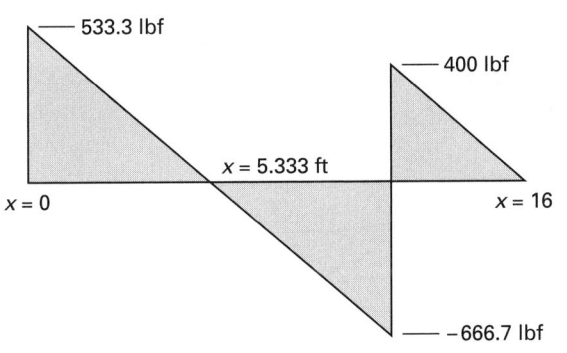

The bending moment at a distance x to the right of the left end has two parts. The left reaction of 533.3 lbf acts with moment arm x. The moment between the two reactions is

$$M_x = 533.3x - 100x\left(\frac{x}{2}\right)$$

This equation describes a parabolic section (curved upward) with a peak at $x = 5.333$ ft, where the shear is zero. The maximum moment is

$$M_{x=5.333 \text{ ft}} = (533.3 \text{ lbf})(5.333 \text{ ft})$$
$$- \left(50 \frac{\text{lbf}}{\text{ft}}\right)(5.333 \text{ ft})^2$$
$$= 1422.0 \text{ ft-lbf}$$

The moment at the right reaction (where $x = 12$ ft) is

$$M_{x=12 \text{ ft}} = (533.3 \text{ lbf})(12 \text{ ft}) - \left(50 \frac{\text{lbf}}{\text{ft}}\right)(12 \text{ ft})^2$$
$$= -800 \text{ ft-lbf}$$

The right end is a free end, so the moment is zero. The moment between the right reaction and the right end could be calculated by summing moments to the left end, but it is more convenient to sum moments to the right end. Measuring x from the right end, the moment is derived only from the uniform load.

$$M = 100x\left(\frac{x}{2}\right) = 50x^2$$

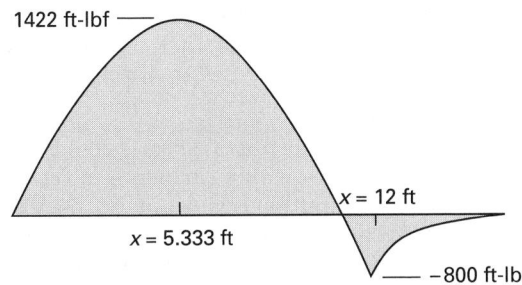

This is sufficient information to draw the moment diagram. Once the maximum moment is located, no attempt is made to determine the exact curvature. The point where $M = 0$ is of limited interest, and no attempt is made to determine the exact location.

Notice that the cross-sectional area of the beam was not needed in this example.

13. SHEAR STRESS IN BEAMS

Shear stress is generally not the limiting factor in most designs. However, it can control (or be limited by code) in wood and concrete beams and in thin tubes.

The average shear stress experienced at a point along the length of a beam depends on the shear, V, at that point and the area, A, of the beam. The shear can be found from the shear diagram.

$$\tau = \frac{V}{A} \qquad 44.28$$

In most cases, the entire area, A, of the beam is used in calculating the average shear stress. However, in flanged beam calculations it is assumed that only the web carries the average shear stress.[7] The flanges are not included in shear stress calculations.

$$\tau = \frac{V}{t_w d} \qquad 44.29$$

Figure 44.10 *Web of a Flanged Beam*

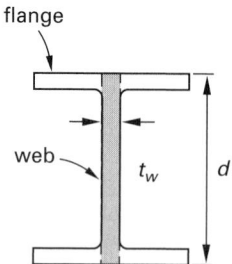

Shear stress is also induced in a beam due to flexure (i.e., bending). Figure 44.6 shows that for biaxial loading, identical shear stresses exist simultaneously in all four directions. One set of parallel shears (a couple) counteracts the rotational moment from the other set of parallel shears. The horizontal shear exists even when the loading is vertical (e.g., when a horizontal beam is loaded by a vertical force). For that reason, the term *horizontal shear* is sometimes used to distinguish it from the applied shear load.

The exact value of the horizontal shear stress is dependent on the location, y_1, within the depth of the beam. The shear stress distribution is given by Eq. 44.30. The shear stress is zero at the top and bottom surfaces of the beam and is usually maximum at the neutral axis (i.e., the center).

$$\tau_{y_1} = \frac{QV}{Ib} \qquad 44.30$$

In Eq. 44.30, V is the vertical shear at the point along the length of the beam where the shear stress is wanted. I is the beam's centroidal moment of inertia, and b is the width of the beam at the depth y_1 within the beam where the shear stress is wanted. Q is the *statical moment* of the area, as defined by Eq. 44.31.

$$Q = \int_{y_1}^{c} y \, dA \qquad 44.31$$

---

[7]This is more than an assumption; it is a fact. There are several reasons the flanges do not contribute to shear resistance, including a non-uniform shear stress distribution in the flanges. This non-uniformity is too complex to be analyzed by elementary methods.

Figure 44.11 *Shear Stress Distribution Within a Rectangular Beam*

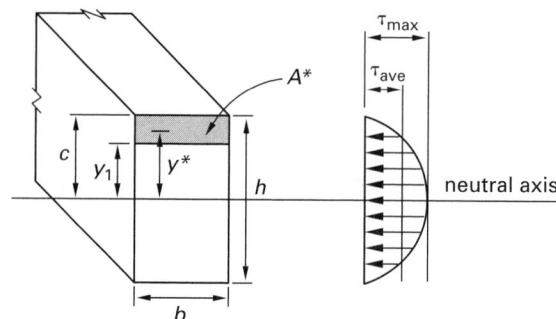

For rectangular beams, $dA = b \, dy$. Then, the statical moment of the area A^* above layer y_1 is equal to the product of the area and the distance from the centroidal axis to the centroid of the area.

$$Q = y^* A^* \qquad 44.32$$

Equation 44.33 calculates the maximum shear stress in a rectangular beam. It is 50% higher than the average shear stress.

$$\tau_{\text{max,rectangular}} = \frac{3V}{2A} = \frac{3V}{2bh} \qquad 44.33$$

For a beam with a circular cross section, the maximum shear stress is

$$\tau_{\text{max,circular}} = \frac{4V}{3A} = \frac{4V}{3\pi r^2} \qquad 44.34$$

For a hollow cylinder used as a beam, the maximum shear stress occurs at the plane of the neutral axis and is

$$\tau_{\text{max,hollow cylinder}} = \frac{2V}{A} \qquad 44.35$$

14. BENDING STRESS IN BEAMS

Normal stress occurs in a bending beam, as shown in Fig. 44.12, where the beam is acted upon by a *transverse force*. Although it is a normal stress, the term *bending stress* or *flexural stress* is used to indicate the source of the stress. The lower surface of the beam experiences tensile stress (which causes lengthening). The upper surface of the beam experiences compressive stress (which causes shortening). There is no normal stress along a horizontal plane passing through the centroid of the cross section, a plane known as the *neutral plane* or the *neutral axis*.

Figure 44.12 *Normal Stress Due to Bending*

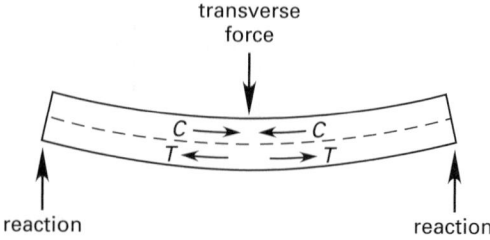

Bending stress varies with location (depth) within the beam. It is zero at the neutral axis and increases linearly with distance from the neutral axis as predicted by Eq. 44.36.

$$\sigma_b = \frac{-My}{I_c} \qquad 44.36$$

Figure 44.13 *Bending Stress Distribution in a Beam*

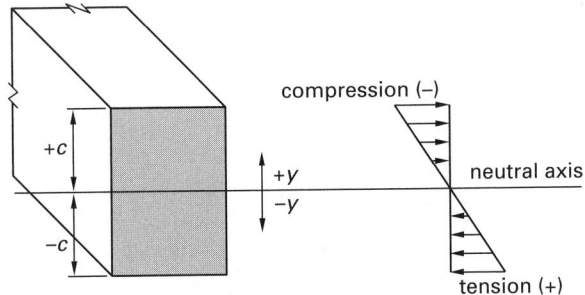

The bending moment, M, is used in Eq. 44.36. I_c is the centroidal moment of inertia of the beam's cross section. The negative sign in Eq. 44.36, required by the convention that compression is negative, is commonly omitted.

Since the maximum stress will govern the design, y can be set equal to c to obtain the *extreme fiber stress*. c is the distance from the neutral axis to the *extreme fiber* (i.e., the top or bottom surface most distant from the neutral axis).

$$\sigma_{b,\text{max}} = \frac{Mc}{I_c} \qquad 44.37$$

Equation 44.37 shows that the maximum bending stress will occur where the moment is maximum. The region immediately adjacent to the point of maximum bending moment is called the *dangerous section* of the beam. The dangerous section can be found from a bending moment or shear diagram.

For any given beam cross section, I_c and c are fixed. Therefore, these two terms can be combined into the *section modulus, S*.[8]

$$\sigma_{b,\text{max}} = \frac{M}{S} \qquad 44.38$$

$$S = \frac{I_c}{c} \qquad 44.39$$

Since $c = h/2$, the section modulus of a rectangular $b \times h$ section ($I_c = bh^3/12$) is

$$S_{\text{rectangular}} = \frac{bh^2}{6} \qquad 44.40$$

Example 44.5

The beam in Ex. 44.4 has a 6 in $\times$ 8 in cross section. What are the maximum shear and bending stresses in the beam?

---

[8]The symbol Z is also commonly used for the section modulus.

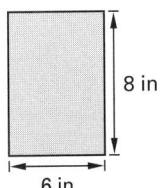

Solution

The maximum shear (from the shear diagram) is 666.7 lbf. (The negative sign is disregarded.) From Eq. 44.33, the maximum shear stress in a rectangular beam is

$$\tau_{\text{max}} = \frac{3V}{2A} = \frac{(3)(666.7 \text{ lbf})}{(2)(6 \text{ in})(8 \text{ in})}$$
$$= 20.8 \text{ lbf/in}^2$$

The centroidal moment of inertia is

$$I_c = \frac{bh^3}{12} = \frac{(6 \text{ in})(8 \text{ in})^3}{12} = 256 \text{ in}^4$$

The maximum bending moment (from the bending moment diagram) is 1422 ft-lbf. From Eq. 44.37, the maximum bending stress is

$$\sigma_{b,\text{max}} = \frac{Mc}{I_c} = \frac{(1422 \text{ ft-lbf})\left(12 \frac{\text{in}}{\text{ft}}\right)(4 \text{ in})}{256 \text{ in}^4}$$
$$= 266.6 \text{ lbf/in}^2$$

15. STRAIN ENERGY DUE TO BENDING MOMENT

The elastic strain energy due to a bending moment stored in a beam is

$$U = \frac{1}{2EI} \int M^2(x)\, dx \qquad 44.41$$

The use of Eq. 44.41 is illustrated by Ex. 44.10.

16. ECCENTRIC LOADING OF AXIAL MEMBERS

If a load is applied through the centroid of a tension or compression member's cross section, the loading is said to be *axial loading* or *concentric loading*. *Eccentric loading* occurs when the load is not applied through the centroid.

If an axial member is loaded eccentrically, it will bend and experience bending stress in the same manner as a beam. Since the member experiences both axial stress and bending stress, it is known as a *beam-column*. In Fig. 44.14, e is known as the *eccentricity*.

Figure 44.14 *Eccentric Loading of an Axial Member*

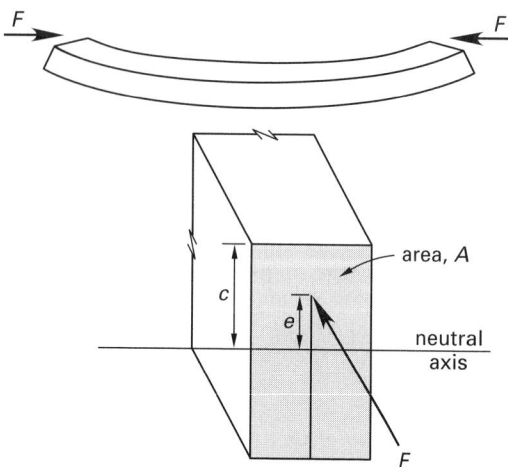

Figure 44.16 *Kerns of Common Cross Sections*

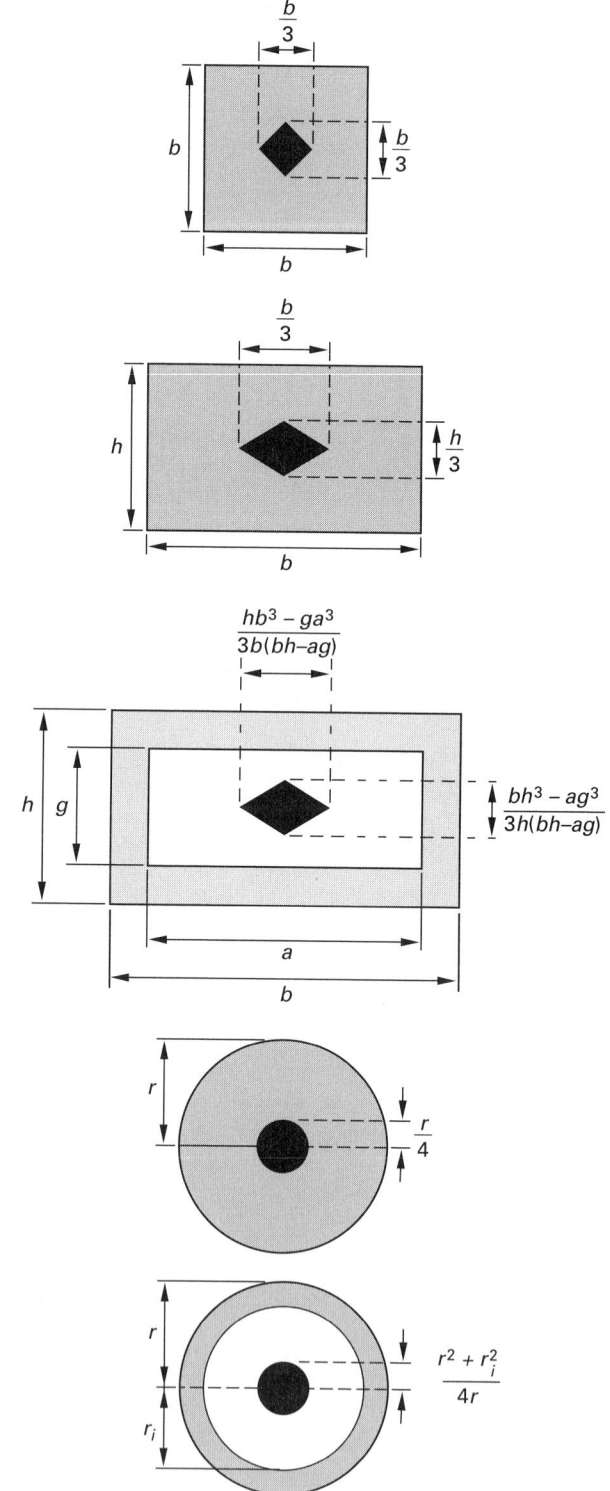

Both the axial stress and bending stress are normal stresses oriented in the same direction; therefore, simple addition can be used to combine them. Combined stress theory is not applicable. By convention, F is negative if the force compresses the member (as shown in Fig. 44.14).

$$\sigma_{\text{max,min}} = \frac{F}{A} \pm \frac{Mc}{I_c} \qquad 44.42$$

$$= \frac{F}{A} \pm \frac{Fec}{I_c} \qquad 44.43$$

If a pier or column (primarily designed as a compression member) is loaded with an eccentric compressive load, part of the section can still be placed in tension. Tension will exist when the Mc/I_c term in Eq. 44.42 is larger than the F/A term. It is particularly important to eliminate or severely limit tensile stresses in concrete and masonry piers, since these materials cannot support tension.

Figure 44.15 *Tension in a Pier*

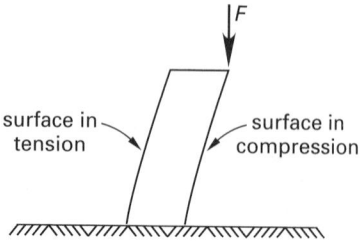

Regardless of the size of the load, there will be no tension as long as the eccentricity is low. In a rectangular member, the load must be kept within a rhombus-shaped area formed from the middle thirds of the centroidal axes. This area is known as the *core*, *kern*, or *kernel*. Figure 44.16 illustrates the kernel for other cross sections.

Example 44.6

A built-in hook with a cross section of 1 in × 1 in carries a load of 500 lbf, but the load is not in line with the centroidal axis of the hook's neck. What are the minimum stresses in the neck? Is the neck in tension everywhere?

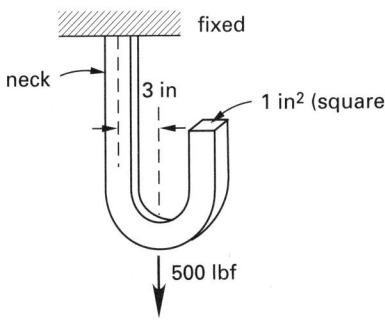

fixed
neck
3 in
1 in² (square)
500 lbf

Solution

The centroidal moment of inertia of a 1 in × 1 in section is

$$I_c = \frac{bh^3}{12} = \frac{(1 \text{ in})(1 \text{ in})^3}{12} = 0.0833 \text{ in}^4$$

The hook is eccentrically loaded with an eccentricity of 3 in. From Eq. 44.43, the total stress is the sum of the direct axial tension and the bending stress. As the hook bends to reduce the eccentricity, the inner face of the neck will receive a tensile bending stress. The outer face of the neck will receive a compressive bending stress.

$$\begin{aligned} \sigma_{\text{max,min}} &= \frac{F}{A} \pm \frac{Fec}{I} \\ &= \frac{500 \text{ lbf}}{1 \text{ in}^2} \pm \frac{(500 \text{ lbf})(3 \text{ in})(0.5 \text{ in})}{0.0833 \text{ in}^4} \\ &= 500 \, \frac{\text{lbf}}{\text{in}^2} \pm 9000 \, \frac{\text{lbf}}{\text{in}^2} \\ &= +9500 \text{ lbf/in}^2, \ -8500 \text{ lbf/in}^2 \end{aligned}$$

The 500 lbf/in² direct stress is tensile, and the inner face experiences a total tensile stress of 9500 lbf/in². However, the compressive bending stress of 9000 lbf/in² counteracts the direct tensile stress, resulting in an 8500 lbf/in² compressive stress at the outer face of the neck.

17. BEAM DEFLECTION: DOUBLE INTEGRATION METHOD

The deflection and the slope of a loaded beam are related to the moment and shear by Eqs. 44.44 through 44.48.

$$y = \text{deflection} \qquad \qquad 44.44$$

$$y' = \frac{dy}{dx} = \text{slope} \qquad \qquad 44.45$$

$$y'' = \frac{d^2y}{dx^2} = \frac{M(x)}{EI} \qquad \qquad 44.46$$

$$y''' = \frac{d^3y}{dx^3} = \frac{V(x)}{EI} \qquad \qquad 44.47$$

If the *moment function*, $M(x)$, is known for a section of the beam, the deflection at any point can be found from Eq. 44.48.

$$y = \frac{1}{EI} \int \left(\int M(x) \, dx \right) dx \qquad \qquad 44.48$$

In order to find the deflection, constants must be introduced during the integration process. These constants can be found from Table 44.3.

Table 44.3 Beam Boundary Conditions

end condition	y	y'	y''	V	M
simple support	0				0
built-in support	0	0			
free end			0	0	0
hinge					0

Example 44.7

Find the tip deflection of the beam shown. EI is 5×10^{10} lbf-in² everywhere.

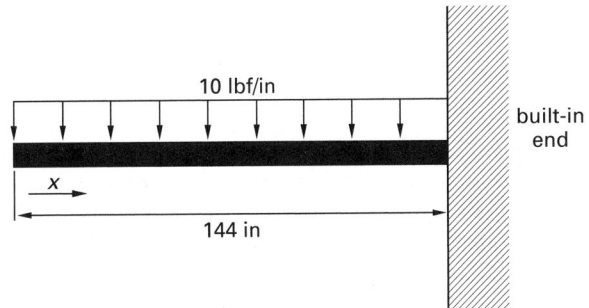

10 lbf/in
built-in end
x
144 in

Solution

The moment at any point x from the left end of the beam is

$$M(x) = -10x \left(\tfrac{1}{2} x \right) = -5x^2$$

This is negative by the left-hand rule convention. From Eq. 44.46,

$$y'' = \frac{M(x)}{EI}$$

$$EIy'' = M(x) = -5x^2$$

$$EIy' = \int -5x^2 \, dx = -\tfrac{5}{3} x^3 + C_1$$

Since $y' = 0$ at a built-in support (Table 44.3) and $x = 144$ in at the built-in support,

$$0 = \left(-\tfrac{5}{3} \right) (144)^3 + C_1$$

$$C_1 = 4.98 \times 10^6$$

$$EIy = \int \left(-\tfrac{5}{3} x^3 + 4.98 \times 10^6 \right) dx$$

$$= -\tfrac{5}{12} x^4 + (4.98 \times 10^6)x + C_2$$

Again, $y = 0$ at $x = 144$ in, so $C_2 = -5.38 \times 10^8$ lbf-in^3. Therefore, the deflection as a function of x is

$$y = \left(\frac{1}{EI}\right)\left(\left(-\tfrac{5}{12}\right)x^4 + (4.98 \times 10^6)x - (5.38 \times 10^8)\right)$$

At the tip $x = 0$, so the deflection is

$$y_{\text{tip}} = \frac{-5.38 \times 10^8 \text{ lbf-in}^3}{5 \times 10^{10} \text{ lbf-in}^2} = -0.0108 \text{ in}$$

18. BEAM DEFLECTION: MOMENT AREA METHOD

The moment area method is a semigraphical technique that is applicable whenever slopes of deflection beams are not too great. This method is based on the following two theorems.

- *Theorem I:* The angle between tangents at any two points on the *elastic line* of a beam is equal to the area of the moment diagram between the two points divided by EI. That is,

$$\phi = \int \frac{M(x)\,dx}{EI} \qquad \text{44.49}$$

- *Theorem II:* One point's deflection away from the tangent of another point is equal to the *statical moment* of the bending moment between those two points divided by EI. That is,

$$y = \int \frac{xM(x)\,dx}{EI} \qquad \text{44.50}$$

If EI is constant, the statical moment $\int xM(x)\,dx$ can be calculated as the product of the total moment diagram area times the horizontal distance from the point whose deflection is wanted to the centroid of the moment diagram.

If the moment diagram has positive and negative parts (areas above and below the zero line), the statical moment should be taken as the sum of two products, one for each part of the moment diagram.

Example 44.8

Find the deflection, y, and the angle, ϕ, at the free end of the cantilever beam shown. Neglect the beam weight.

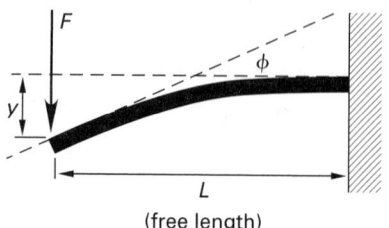

(free length)

Solution

The deflection angle, ϕ, is the angle between the tangents at the free and built-in ends (Theorem I). The moment diagram is

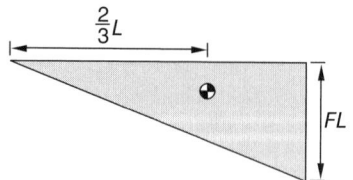

The area of the moment diagram is

$$\tfrac{1}{2}(FL)(L) = \tfrac{1}{2}FL^2$$

From Eq. 44.49,

$$\phi = \frac{FL^2}{2EI}$$

From Eq. 44.50,

$$y = \left(\frac{FL^2}{2EI}\right)\left(\tfrac{2}{3}L\right) = \frac{FL^3}{3EI}$$

Example 44.9

Find the deflection of the free end of the cantilever beam shown. Neglect the beam weight.

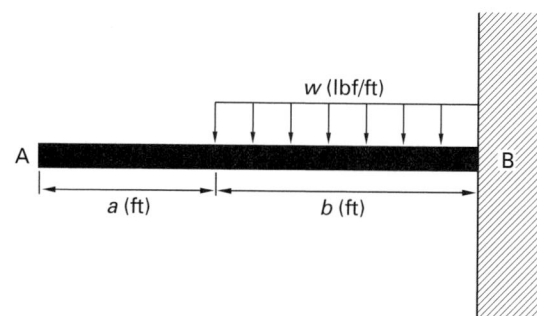

Solution

The distance from point A (where the deflection is wanted) to the centroid is $a + 0.75b$. The area of the moment diagram is $wb^3/6$. From Theorem II,

$$y = \left(\frac{wb^3}{6EI}\right)(a + 0.75b)$$

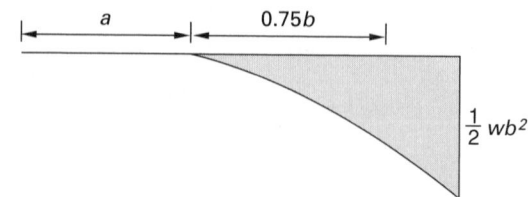

19. BEAM DEFLECTION: STRAIN ENERGY METHOD

The deflection at a point of load application can be found by the strain energy method. This method equates the external work to the total internal strain energy. Since work is a force moving through a distance (which in this case is the deflection), Eq. 44.51 holds true.

$$\tfrac{1}{2}Fy = \sum U \qquad 44.51$$

Example 44.10

Find the deflection at the tip of the stepped beam shown.

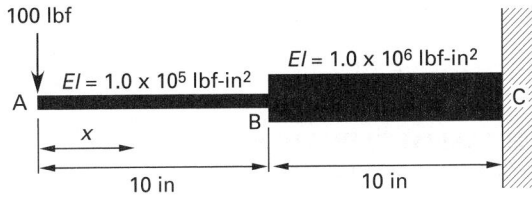

Solution

In section A–B, $M(x) = 100x$ in-lbf.

From Eq. 44.41,

$$U = \frac{1}{2EI} \int M^2(x)\,dx$$

$$= \frac{1}{(2)(1 \times 10^5 \text{ lbf-in}^2)} \int_0^{10 \text{ in}} (100x)^2 = 16.67 \text{ in-lbf}$$

In section B–C, $M = 100x$.

$$U = \frac{1}{(2)(1 \times 10^6 \text{ lbf-in}^2)} \int_{10 \text{ in}}^{20 \text{ in}} (100x)^2 = 11.67 \text{ in-lbf}$$

Equating the internal work (U) and the external work,

$$\sum U = W$$

$$16.67 \text{ in-lbf} + 11.67 \text{ in-lbf} = \left(\tfrac{1}{2}\right)(100 \text{ lbf})y$$

$$y = 0.567 \text{ in}$$

20. BEAM DEFLECTION: CONJUGATE BEAM METHOD

The *conjugate beam method* changes a deflection problem into one of drawing moment diagrams. The method has the advantage of being able to handle beams of varying cross sections (e.g., stepped beams) and materials. It has the disadvantage of not easily being able to handle beams with two built-in ends. The following steps constitute the conjugate beam method.

step 1: Draw the moment diagram for the beam as it is actually loaded.

step 2: Construct the M/EI diagram by dividing the value of M at every point along the beam by EI at that point. If the beam is of constant cross section, EI will be constant, and the M/EI diagram will have the same shape as the moment diagram. However, if the beam cross section varies with x, I will change. In that case, the M/EI diagram will not look the same as the moment diagram.

step 3: Draw a conjugate beam of the same length as the original beam. The material and the cross-sectional area of this conjugate beam are not relevant.

   (a) If the actual beam is simply supported at its ends, the conjugate beam will be simply supported at its ends.

   (b) If the actual beam is simply supported away from its ends, the conjugate beam has hinges at the support points.

   (c) If the actual beam has free ends, the conjugate beam has built-in ends.

   (d) If the actual beam has built-in ends, the conjugate beam has free ends.

step 4: Load the conjugate beam with the M/EI diagram. Find the conjugate reactions by methods of statics. Use the superscript * to indicate conjugate parameters.

step 5: Find the conjugate moment at the point where the deflection is wanted. The deflection is numerically equal to the moment as calculated from the conjugate beam forces.

Example 44.11

Find the deflections at the two load points. EI has a constant value of 2.356×10^7 lbf-in^2.

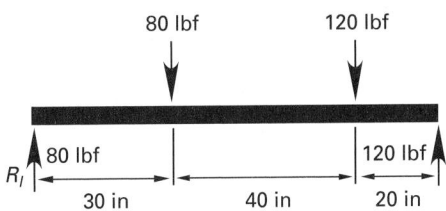

Solution

   step 1: The moment diagram for the actual beam is

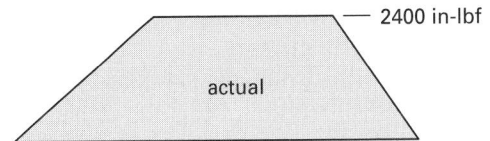

steps 2, 3, and 4: Since the beam cross section is constant, the conjugate load has the same shape as the original moment diagram. The peak load on the conjugate beam is

$$\frac{2400 \text{ in-lbf}}{2.356 \times 10^7 \text{ lbf-in}^2} = 1.019 \times 10^{-4} \text{ 1/in}$$

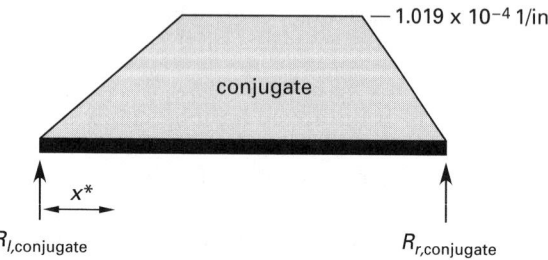

The conjugate reaction, $R_{l,\text{conjugate}}$, is found by the following method. The loading diagram is assumed to be made up of a rectangular load and two negative triangular loads. The area of the rectangular load (which has a centroid at $x_{\text{conjugate}} = 45$ in) is $(90 \text{ in})(1.019 \times 10^{-4} \text{ 1/in}) = 9.171 \times 10^{-3}$.

Similarly, the area of the left triangle (which has a centroid at $x_{\text{conjugate}} = 10$ in) is $(\frac{1}{2})(30 \text{ in})(1.019 \times 10^{-4} \text{ 1/in}) = 1.529 \times 10^{-3}$. The area of the right triangle (which has a centroid at $x_{\text{conjugate}} = 83.33$ in) is $(\frac{1}{2})(20 \text{ in})(1.019 \times 10^{-4} \text{ 1/in}) = 1.019 \times 10^{-3}$.

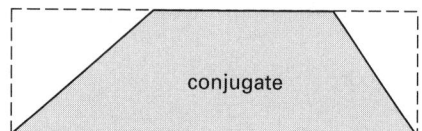

$$\sum M_L^* = (90 \text{ in})R_{r,\text{conjugate}} + (1.019 \times 10^{-3})$$
$$\times (83.3 \text{ in})$$
$$+ (1.529 \times 10^{-3})(10 \text{ in})$$
$$- (9.171 \times 10^{-3})(45 \text{ in})$$
$$= 0$$
$$R_{r,\text{conjugate}} = 3.472 \times 10^{-3}$$

Then,

$$R_{l,\text{conjugate}} = (9.171 - 1.019$$
$$- 1.529 - 3.472) \times 10^{-3}$$
$$= 3.151 \times 10^{-3}$$

step 5: The conjugate moment at $x_{\text{conjugate}} = 30$ is the deflection of the actual beam at that point.

$$M_{\text{conjugate}} = (3.151 \times 10^{-3})(30 \text{ in})$$
$$+ (1.529 \times 10^{-3})(30 \text{ in} - 10 \text{ in})$$
$$- (9.171 \times 10^{-3})\left(\frac{30 \text{ in}}{90 \text{ in}}\right)(15 \text{ in})$$
$$= 7.926 \times 10^{-2} \text{ in}$$

The conjugate moment (the deflection) at the rightmost load is

$$M_{\text{conjugate}} = (3.472 \times 10^{-3})(20 \text{ in})$$
$$+ (1.019 \times 10^{-3})(13.3 \text{ in})$$
$$- (9.171 \times 10^{-3})\left(\frac{20 \text{ in}}{90 \text{ in}}\right)(10 \text{ in})$$
$$= 6.261 \times 10^{-2} \text{ in}$$

21. BEAM DEFLECTION: TABLE LOOK-UP METHOD

Appendix 44.A is a compilation of the most commonly used beam deflection formulas. These formulas should never need to be derived and should be used whenever possible. They are particularly useful in calculating deflections due to multiple loads using the principle of superposition.

The actual deflection of very *wide beams* (i.e., those whose widths are larger than 8 or 10 times the thickness) is less than that predicted by the equations in App. 44.A for elastic behavior. (This is particularly true for leaf springs.) The large width prevents lateral expansion and contraction of the beam material, reducing the deflection. For wide beams, the calculated deflection should be reduced by multiplying by $(1-\nu^2)$.

22. BEAM DEFLECTION: SUPERPOSITION

When multiple loads act simultaneously on a beam, all of the loads contribute to deflection. The principle of *superposition* permits the deflections at a point to be calculated as the sum of the deflections from each individual load acting singly.[9] This principle is valid as long as none of the deflections is excessive and all stresses are kept less than the yield point of the beam material.

23. INFLECTION POINTS

The *inflection point* (also known as a *point of contraflexure*) on a horizontal beam in elastic bending occurs where the curvature changes from concave up to concave down, or vice versa. There are three ways of determining the inflection point.

1. If the elastic deflection equation, $y(x)$, is known, the inflection point can be found consistent with normal calculus methods (i.e., by determining the value of x for which $y''(x) = M(x) = 0$).

---

[9]The principle of superposition is not limited to deflections. It can also be used to calculate the shear and moment at a point and to draw the shear and moment diagrams.

2. From Eq. 44.46, $y''(x) = M(x)/EI$. y'' is also the reciprocal of the *radius of curvature*, ρ, of the beam. Therefore,

$$y''(x) = \frac{1}{\rho(x)} = \frac{M(x)}{EI} \qquad 44.52$$

Since the flexural rigidity, EI, is always positive, the radius of curvature, $\rho(x)$, changes sign when the moment equation, $M(x)$, changes sign.

3. If a shear diagram is known, the inflection point can sometimes be found by noting the point at which the positive and negative shear areas on either side of the point balance.

24. TRUSS DEFLECTION: STRAIN ENERGY METHOD

The deflection of a truss at the point of a single load application can be found by the *strain energy method* if all member forces are known. This method is illustrated by Ex. 44.12.

Example 44.12

Find the vertical deflection of point A under the external load of 707 lbf. $AE = 10^6$ lbf for all members. The internal forces have already been determined.

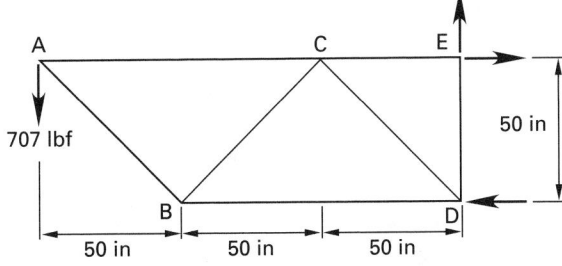

Solution

The length of member AB is $\sqrt{(50 \text{ in})^2 + (50 \text{ in})^2} = 70.7$ in. From Eq. 44.6, the internal strain energy in member AB is

$$U = \frac{F^2 L_o}{2AE} = \frac{(-1000 \text{ lbf})^2 (70.7 \text{ in})}{(2)(10^6 \text{ lbf})} = 35.4 \text{ in-lbf}$$

Similarly, the energy in all members can be determined.

member	L (in)	F (lbf)	U (in-lbf)
AB	70.7	−1000	+35.4
BC	70.7	+1000	+35.4
AC	100	+707	+25.0
BD	100	−1414	+100.0
CD	70.7	−1000	+35.4
CE	50	+2121	+112.5
DE	50	+707	+12.5
			$\overline{356.2}$

The work done by a constant force F moving through a distance y is Fy. In this case, the force increases with y. The average force is $\frac{1}{2}F$. The external work is $W_{\text{ext}} = \left(\frac{1}{2}\right)(707 \text{ lbf})y$, so

$$\left(\tfrac{1}{2}\right)(707 \text{ lbf})y = 356.2 \text{ in-lbf}$$
$$y = 1 \text{ in}$$

25. TRUSS DEFLECTION: VIRTUAL WORK METHOD

The *virtual work method* (also known as the *unit load method*) is an extension of the strain energy method. It can be used to determine the deflection of any point on a truss.

step 1: Draw the truss twice.

step 2: On the first truss, place all the actual loads.

step 3: Find the forces, S, due to the actual applied loads in all the members.

step 4: On the second truss, place a dummy one-unit load in the direction of the desired displacement.

step 5: Find the forces, u, due to the one-unit dummy load in all members.

step 6: Find the desired displacement from Eq. 44.53. The summation is over all truss members that have nonzero forces in *both* trusses.

$$\delta = \sum \frac{SuL}{AE} \qquad 44.53$$

Example 44.13

What is the horizontal deflection of joint F on the truss shown? Use $E = 3 \times 10^7$ lbf/in^2. Joint A is restrained horizontally. Member lengths and areas are listed in the accompanying table.

Solution

steps 1 and 2: Use the truss as drawn.

step 3: The forces in all the truss members are summarized in step 5.

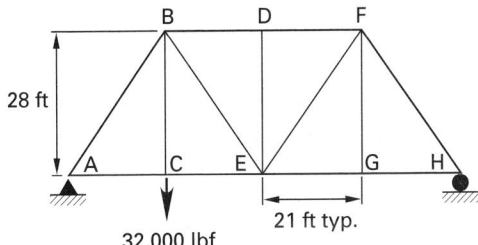

Structural

step 4: Draw the truss and load it with a unit horizontal force at point F.

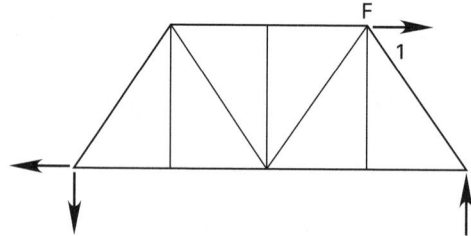

step 5: Find the forces, *u*, in all members of the second truss. These are summarized in the following table. Notice the sign convention: + for tension and − for compression.

member	S (lbf)	u (lbf)	L (ft)	A (in²)	$\dfrac{SuL}{AE}$ (ft)
AB	−30,000	5/12	35	17.5	-8.33×10^{-4}
CB	32,000	0	28	14	0
EB	−10,000	−5/12	35	17.5	2.75×10^{-4}
ED	0	0	28	14	0
EF	10,000	5/12	35	17.5	2.78×10^{-4}
GF	0	0	28	14	0
HF	−10,000	−5/12	35	17.5	2.78×10^{-4}
BD	−12,000	1/2	21	10.5	-4.00×10^{-4}
DF	−12,000	1/2	21	10.5	-4.00×10^{-4}
AC	18,000	3/4	21	10.5	9.00×10^{-4}
CE	18,000	3/4	21	10.5	9.00×10^{-4}
EG	6000	1/4	21	10.5	1.00×10^{-4}
GH	6000	1/4	21	10.5	1.00×10^{-4}
					12.01×10^{-4}

Since 12.01×10^{-4} is positive, the deflection is in the direction of the dummy unit load. In this case, the deflection is to the right.

26. MODES OF BEAM FAILURE

Beams can fail in different ways, including excessive deflection, local buckling, lateral buckling, and rotation.

Excessive deflection occurs when a beam bends more than a permitted amount.[10] The deflection is elastic and no yielding occurs. For this reason, the failure mechanism is sometimes called *elastic failure.* Although the beam does not yield, the excessive deflection may cause cracks in plaster and sheetrock, misalignment of doors and windows, and occupant concern and lack of confidence in the structure.

Local buckling is an overload condition that occurs near large concentrated loads. Such locations include where a column frames into a supporting girder or a reaction point. *Vertical buckling* and *web crippling*, two types

---

[10]The Uniform Building Code, as well as the steel, concrete, and timber codes, specifies maximum permitted deflections in terms of beam length.

of local buckling, can be eliminated by use of *stiffeners.* Such stiffeners can be referred to as *intermediate stiffeners*, *bearing stiffeners*, *web stiffeners*, and *flange stiffeners*, depending on the location and technique of stiffening.

Figure 44.17 *Local Buckling and Stiffeners*

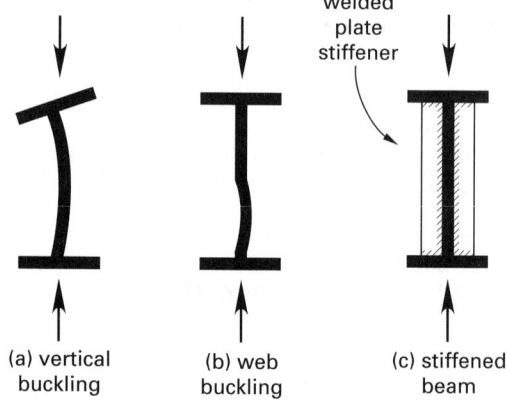

Lateral buckling, such as illustrated in Fig. 44.18, occurs when a long, unsupported member rolls out of its normal plane. To prevent lateral buckling, either the beam's compression flange must be supported continuously or at frequent intervals along its length, or the beam must be restrained against twisting about its longitudinal axis.

Figure 44.18 *Lateral Buckling and Flange Support*

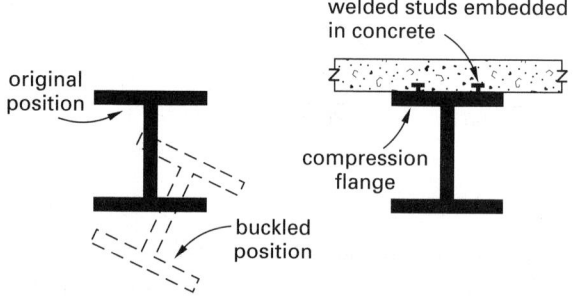

Rotation is an inelastic (plastic) failure of the beam. When the bending stress at a point exceeds the strength of the beam material, the material yields. As the beam yields, its slope changes. Since the beam appears to be rotating at a hinge at the yield point, the term *plastic hinge* is used to describe the failure mechanism.

Figure 44.19 *Beam Failure by Rotation*

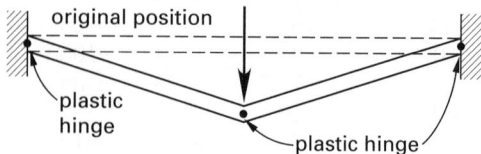

27. CURVED BEAMS

Many members (e.g., hooks, chain links, clamps, and machine frames) have curved main axes. The distribution of bending stress in a curved beam is nonlinear. Compared to a straight beam, the stress at the inner radius is higher because the inner radius fibers are shorter. Conversely, the stress at the outer radius is lower because the outer radius fibers are longer. Also, the neutral axis is shifted from the center inward toward the center of curvature.

Since the process of finding the neutral axis and calculating the stress amplification is complex, tables and graphs are used for quick estimates and manual computations. The forms of these computational aids vary, but the straight-beam stress is generally multiplied by factors, K, to obtain the stresses at the extreme faces. The factor values depend on the beam cross section and radius of curvature.

$$\sigma_{\text{curved}} = K\sigma_{\text{straight}}$$
$$= \frac{KMc}{I}$$

Table 44.4 is typical of compilations for round and rectangular beams. Factors K_A and K_B are the multipliers for the inner (high stress) and outer (low stress) faces. The ratio h/r is the fractional distance that the neutral axis shifts inward toward the radius of curvature.

Table 44.4 *Curved Beam Correction Factors*

solid rectangular section	r/c	K_A	K_B	h/r
	1.2	2.89	0.57	0.305
	1.4	2.13	0.63	0.204
	1.6	1.79	0.67	0.149
	1.8	1.63	0.70	0.112
	2.0	1.52	0.73	0.090
	3.0	1.30	0.81	0.041
	4.0	1.20	0.85	0.021
	6.0	1.12	0.90	0.0093
	8.0	1.09	0.92	0.0052
	10.0	1.07	0.94	0.0033

solid circular section	r/c	K_A	K_B	h/r
	1.2	3.41	0.54	0.224
	1.4	2.40	0.60	0.151
	1.6	1.96	0.65	0.108
	1.8	1.75	0.68	0.084
	2.0	1.62	0.71	0.069
	3.0	1.33	0.79	0.03
	4.0	1.23	0.84	0.016
	6.0	1.14	0.89	0.007
	8.0	1.10	0.91	0.0039
	10.0	1.08	0.93	0.0025

28. COMPOSITE STRUCTURES

A *composite structure* is one in which two or more different materials are used. Each material carries part of an applied load. Examples of composite structures include steel-reinforced concrete and steel-plated timber beams.

Most simple composite structures can be analyzed using the *method of consistent deformations*, also known as the *area transformation method*. This method assumes that the strains are the same in both materials at the interface between them. Although the strains are the same, the stresses in the two adjacent materials are not equal, since stresses are proportional to the moduli of elasticity.

The following steps comprise an analysis method based on area transformation.

step 1: Determine the modulus of elasticity for each of the materials used in the structure.

step 2: For each of the materials used, calculate the *modular ratio, n.*

$$n = \frac{E}{E_{\text{weakest}}} \qquad 44.54$$

E_{weakest} is the smallest modulus of elasticity of any of the materials used in the composite structure. For two materials that experience the same strains (i.e., are perfectly bonded), n is also the ratio of stresses.

step 3: For all of the materials except the weakest, multiply the actual material stress area by n. Consider this expanded *(transformed)* area to have the same composition as the weakest material.

step 4: If the structure is a tension or compression member, the distribution or placement of the transformed area is not important. Just assume that the transformed areas carry the axial load. For beams in bending, the transformed area can add to the width of the beam, but it cannot change the depth of the beam or the thickness of the reinforcement.

step 5: For compression or tension numbers, calculate the stresses in the weakest and stronger materials.

$$\sigma_{\text{weakest}} = \frac{F}{A_t} \qquad 44.55$$

$$\sigma_{\text{stronger}} = \frac{nF}{A_t} \qquad 44.56$$

step 6: For beams in bending, proceed through step 9. Find the centroid of the transformed beam.

step 7: Find the centroidal moment of inertia of the transformed beam $I_{c,t}$.

step 8: Find V_{max} and M_{max} by inspection or from the shear and moment diagrams.

step 9: Calculate the stresses in the weakest and stronger materials.

$$\sigma_{\text{weakest}} = \frac{Mc_{\text{weakest}}}{I_{c,t}} \qquad 44.57$$

$$\sigma_{\text{stronger}} = \frac{nMc_{\text{stronger}}}{I_{c,t}} \qquad 44.58$$

Example 44.14

A short circular steel core is surrounded by a copper tube. The assemblage supports an axial compressive load of 100,000 lbf. The core and tube are well bonded, and the load is applied uniformly. Find the compressive stress in the inner steel core and the outer copper tube.

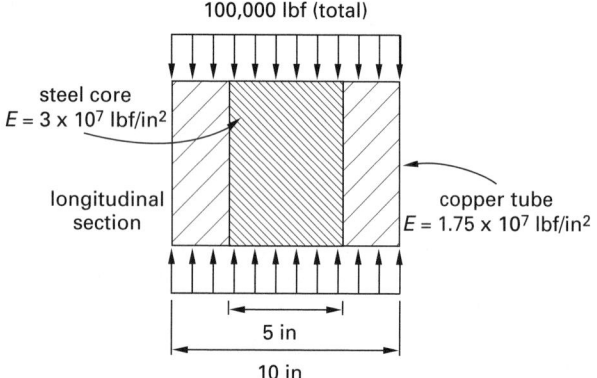

Solution

The moduli of elasticity are given in the illustration. From step 2, the modular ratio is

$$n = \frac{E_{\text{steel}}}{E_{\text{copper}}} = \frac{3 \times 10^7 \, \frac{\text{lbf}}{\text{in}^2}}{1.75 \times 10^7 \, \frac{\text{lbf}}{\text{in}^2}} = 1.714$$

The actual cross-sectional area of the steel is

$$A_{\text{steel}} = \frac{\pi}{4}d^2 = \left(\frac{\pi}{4}\right)(5 \text{ in})^2$$
$$= 19.63 \text{ in}^2$$

The actual cross-sectional area of the copper is

$$A_{\text{copper}} = \frac{\pi}{4}\left(d_o^2 - d_i^2\right) = \left(\frac{\pi}{4}\right)\left((10 \text{ in})^2 - (5 \text{ in})^2\right)$$
$$= 58.90 \text{ in}^2$$

The steel is the stronger material. Its area must be expanded to an equivalent area of copper. From step 3, the total transformed area is

$$A_t = A_{\text{copper}} + nA_{\text{steel}}$$
$$= 58.90 \text{ in}^2 + (1.714)(19.63 \text{ in}^2) = 92.55 \text{ in}^2$$

Since the two pieces are well bonded and the load is applied uniformly, both pieces experience identical strains. From step 5, the compressive stresses are

$$\sigma_{\text{copper}} = \frac{F}{A_t} = \frac{-100,000 \text{ lbf}}{92.55 \text{ in}^2} = -1080 \text{ lbf/in}^2$$
$$[\text{compression}]$$

$$\sigma_{\text{steel}} = \frac{nF}{A_t} = n\sigma_{\text{copper}} = (1.714)\left(-1080 \, \frac{\text{lbf}}{\text{in}^2}\right)$$
$$= -1851 \text{ lbf/in}^2$$

Example 44.15

At a particular point along the length of a steel-reinforced wood beam, the moment is 40,000 ft-lbf. Assume the steel reinforcement is lag-bolted to the wood at regular intervals along the beam. What is the maximum bending stress in the wood and steel?

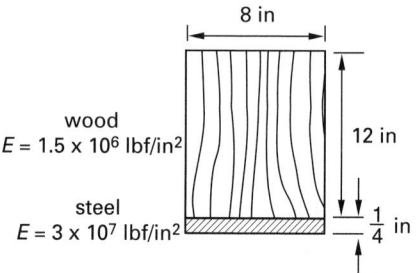

Solution

The moduli of elasticity are given in the illustration. From step 2, the modular ratio is

$$n = \frac{E_{\text{steel}}}{E_{\text{wood}}} = \frac{3 \times 10^7 \, \frac{\text{lbf}}{\text{in}^2}}{1.5 \times 10^6 \, \frac{\text{lbf}}{\text{in}^2}} = 20$$

The actual cross-sectional area of the steel is

$$A_{\text{steel}} = (0.25 \text{ in})(8 \text{ in}) = 2 \text{ in}^2$$

The steel is the stronger material. Its area must be expanded to an equivalent area of wood. Since the depth of the beam and reinforcement cannot be increased (step 4), the width must increase. The width of the transformed steel plate is

$$b' = nb = (20)(8 \text{ in}) = 160 \text{ in}$$

The centroid of the transformed section is located 4.45 in from the horizontal axis. The centroidal moment of inertia of the transformed section is $I_{c,t} = 2211.5 \text{ in}^4$. (The calculations for centroidal location and moment of inertia are not presented here.)

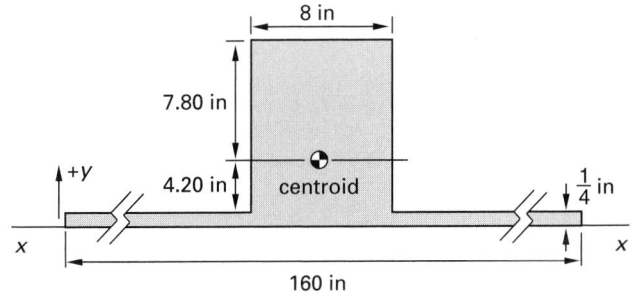

Since the steel plate is bolted to the wood at regular intervals, both pieces experience the same strain. From step 9, the stresses in the wood and steel are

$$\sigma_{\text{max,wood}} = \frac{Mc_{\text{wood}}}{I}$$

$$= \frac{(40{,}000 \text{ ft-lbf})\left(12\ \dfrac{\text{in}}{\text{ft}}\right)(7.8 \text{ in})}{2211.5 \text{ in}^4}$$

$$= 1693 \text{ lbf/in}^2$$

$$\sigma_{\text{max,steel}} = \frac{nMc_{\text{steel}}}{I}$$

$$= \frac{(20)(40{,}000 \text{ ft-lbf})\left(12\ \dfrac{\text{in}}{\text{ft}}\right)(4.45 \text{ in})}{2211.5 \text{ in}^4}$$

$$= 19{,}317 \text{ lbf/in}^2$$

45 Basic Elements of Design

Nomenclature

a	dimension	ft	m
A	area	ft^2	m^2
b	dimension	ft	m
b	width	ft	m
c	distance from neutral axis to extreme fiber	ft	m
C	circumference	ft	m
d	diameter	ft	m
e	eccentricity	ft	m
E	energy	ft-lbf	J
E	modulus of elasticity	lbf/ft^2	Pa
f	coefficient of friction	–	–
F	force	lbf	N
FS	factor of safety	–	–
g	acceleration of gravity	ft/sec^2	m/s^2
g	gravitational constant	ft-lbm/lbf-sec^2	–
G	shear modulus	lbf/ft^2	Pa
h	height	ft	m
I	interference	ft	ft
I	moment of inertia	ft^4	m^4
J	polar moment of inertia	ft^4	m^4
k	stiffness	lbf/ft	N/m
K	end-restraint coefficient	–	–
L	length	ft	m
m	mass	lbm	kg
M	moment	ft-lbf	N·m
n	modular ratio	–	–
n	number of connectors	–	–
N	normal force	lbf	N
p	perimeter	ft	m
p	pressure	lbf/ft^2	Pa
q	shear flow	lbf/ft	N/m
r	radius	ft	m
r	radius of gyration	ft	m
S	strength	lbf/ft^2	Pa
SR	slenderness ratio	–	–
t	thickness	ft	m
T	torque	ft-lbf	N·m
U	energy	ft-lbf	J
y	weld size	ft	m

Symbols

α	thread half-angle	deg	deg
δ	deflection	in	m
γ	angle of twist	rad	rad
ϵ	strain	ft/ft	m/m
θ	lead angle	deg	deg
θ	shear strain	rad	rad
ν	Poisson's ratio	–	–
σ	normal stress	lbf/ft^2	Pa
τ	shear stress	lbf/ft^2	Pa
ϕ	angle	rad	rad

Subscripts

a	allowable
c	centroidal, circumferential, or collar
cr	critical
e	Euler or effective
eq	equivalent
h	hoop
i	inside or initial
l	longitudinal
m	mean
o	outside
p	bearing or potential
r	radial or rope
sh	sheave
t	tension, thread, or transformed
T	torque
ut	ultimate tensile
v	vertical
w	wire
y	yield

Structural

PROFESSIONAL PUBLICATIONS, INC.

1. SLENDER COLUMNS

Very short compression members are known as *piers*. Long compression members are known as *columns*. Failure in piers occurs when the applied stress exceeds the yield strength of the material. However, very long columns fail by sideways *buckling* long before the compressive stress reaches the yield strength. Buckling failure is sudden, often without significant initial sideways bending. The load at which a column fails is known as the *critical load* or *Euler load*.

The *Euler load* is the theoretical maximum load that an initially straight column can support without buckling. For columns with frictionless or pinned ends, this load is given by Eq. 45.1. r is the *radius of gyration*.

$$F_e = \frac{\pi^2 EI}{L^2} = \frac{\pi^2 EA}{\left(\dfrac{L}{r}\right)^2} \qquad 45.1$$

The corresponding column stress is given by Eq. 45.2. In order to use Euler's theory, this stress cannot exceed half of the compressive yield strength of the column material.

$$\sigma_e = \frac{F_e}{A} = \frac{\pi^2 E}{\left(\dfrac{L}{r}\right)^2} \qquad 45.2$$

The quantity L/r is known as the *slenderness ratio*. Long columns have high slenderness ratios. The smallest slenderness ratio for which Eq. 45.2 is valid is the *critical slenderness ratio*. Typical critical slenderness ratios range from 80 to 120. The critical slenderness ratio becomes smaller as the compressive yield strength increases.

L is the longest unbraced column length. If a column is braced against buckling at some point between its two ends, the column is known as a *braced column*, and L will be less than the full column height. Columns with rectangular cross sections have two radii of gyration, r_x and r_y, and therefore, will have two slenderness ratios. The largest slenderness ratio will govern the design.

Columns do not always have frictionless or pinned ends. Often, a column will be fixed ("clamped," "built in," etc.) at its top and base. In such cases, the *effective length*, L', must be used in place of L in Eqs. 45.1 and 45.2.

$$L' = KL \qquad 45.3$$

$$\sigma_e = \frac{F_e}{A} = \frac{\pi^2 E}{\left(\dfrac{L'}{r}\right)^2} \qquad 45.4$$

K is the *end restraint coefficient*, which varies from 0.5 to 2.0 according to Table 45.1. For most real columns, the design values of K should be used since infinite stiffness of the supporting structure is not achievable.

Table 45.1 *Theoretical End Restraint Coefficients*

illus.	end conditions	K ideal	recommended for design
(a)	both ends pinned	1	1.0*
(b)	both ends built in	0.5	0.65*–0.90
(c)	one end pinned, one end built in	0.707	0.80*–0.90
(d)	one end built in, one end free	2	2.0–2.1*
(e)	one end built in, one end fixed against rotation but free	1	1.2*
(f)	one end pinned, one end fixed against rotation but free	2	2.0*

*AISC values

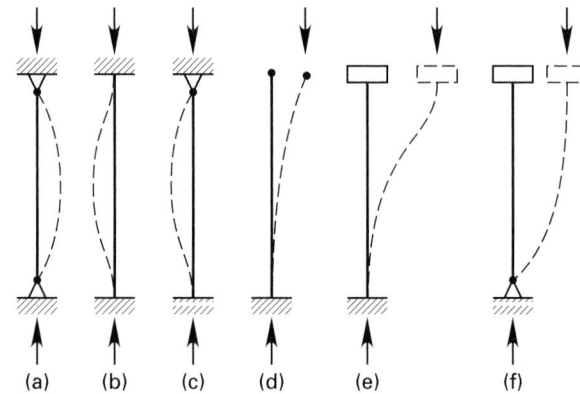

Euler's curve for columns, line BCD in Fig. 45.1, is generated by plotting the *Euler stress* (Eq. 45.2) versus the slenderness ratio. Since the material's compressive yield strength cannot be exceeded, a horizontal line AC is added to limit applications to the region below. Theoretically, members with slenderness ratios less than $(SR)_C$ could be treated as pure compression members. However, this is not done in practice.

Figure 45.1 *Euler's Curve*

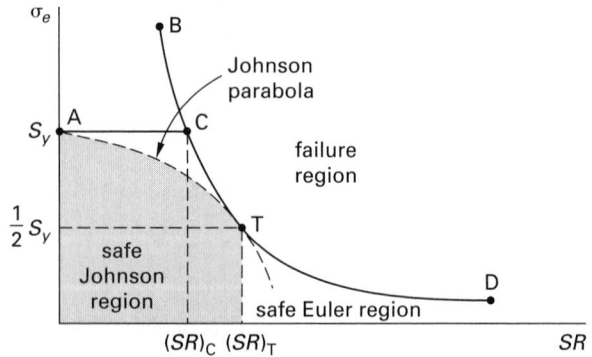

Defects in materials, errors in manufacturing, inabilities to achieve theoretical end conditions, and eccentricities frequently combine to cause column failures in the region around point C. Therefore, this region is excluded by designers.

The empirical Johnson procedure used to exclude the failure area is to draw a parabolic curve from point A through a tangent point T on the Euler curve at a stress of $^{1}/_{2}S_{y}$. The corresponding value of the slenderness ratio of any end restraint coefficient, K, is

$$(SR)_{T} = \frac{1}{K}\sqrt{\frac{2\pi^{2}E}{S_{y}}} \qquad 45.5$$

Example 45.1

A steel member is used as an 8.5 ft long column. The ends are pinned. What is the maximum allowable compressive stress in order to have a factor of safety of 3.0? Use the following data for the column.

$$E = 2.9 \times 10^{7} \text{ lbf/in}^{2}$$

$$S_{yt} = 36,000 \text{ lbf/in}^{2}$$

$$r = 0.569 \text{ in}$$

$$(SR)_{T} = \frac{1}{1.0}\sqrt{\frac{2\pi^{2}\left(29 \times 10^{6}\ \dfrac{\text{lbf}}{\text{in}^{2}}\right)}{36,000\ \dfrac{\text{lbf}}{\text{in}^{2}}}} = 126$$

Solution

First, check the slenderness ratio to see if this is a long column.

$$\frac{L}{r} = \frac{(8.5 \text{ ft})\left(12\ \dfrac{\text{in}}{\text{ft}}\right)}{0.569 \text{ in}} = 179.3 \quad [> 126, \text{ so OK}]$$

From Eq. 45.2, the Euler stress is

$$\sigma_{e} = \frac{\pi^{2}E}{\left(\dfrac{L}{r}\right)^{2}} = \frac{\pi^{2}\left(2.9 \times 10^{7}\ \dfrac{\text{lbf}}{\text{in}^{2}}\right)}{(179.3)^{2}}$$

$$= 8903 \text{ lbf/in}^{2}$$

Since 8903 lbf/in^{2} is less than half of the yield strength of 36,000 lbf/in^{2}, the Euler formula is valid. The allowable working stress is

$$\sigma_{a} = \frac{\sigma_{e}}{\text{FS}} = \frac{8903\ \dfrac{\text{lbf}}{\text{in}^{2}}}{3} = 2968 \text{ lbf/in}^{2}$$

2. INTERMEDIATE COLUMNS

Columns with slenderness ratios less than the critical slenderness ratio but that are too long to be short piers are known as *intermediate columns*. The *parabolic formula* (also known as the *J. B. Johnson formula*) is used to describe the parabolic line between points A and T on Fig. 45.1. The critical stress is given by Eq. 45.6, where a and b are curve-fit constants.

$$\sigma_{cr} = \frac{P_{cr}}{A} = a - b\left(\frac{KL}{r}\right)^{2} \qquad 45.6$$

It is commonly assumed that the stress at point A is S_{y} and the stress at point T is $S_{y}/2$. In that case, the parabolic formula becomes

$$\sigma_{cr} = S_{y} - \left(\frac{1}{E}\right)\left(\frac{S_{y}}{2\pi}\right)^{2}\left(\frac{KL}{r}\right)^{2} \qquad 45.7$$

3. ECCENTRICALLY LOADED COLUMNS

Accidental eccentricities are introduced during the course of normal manufacturing, so the load on real columns is rarely axial. The *secant formula* is one of the methods available for determining the critical column stress and critical load with eccentric loading.[1]

$$\sigma_{max} = \sigma_{ave}(1 + \text{amplification factor})$$

$$= \left(\frac{F}{A}\right)\left(1 + \left(\frac{ec}{r^{2}}\right)\sec\left(\frac{\pi}{2}\sqrt{\frac{F}{F_{e}}}\right)\right)$$

$$= \left(\frac{F}{A}\right)\left(1 + \left(\frac{ec}{r^{2}}\right)\sec\left(\frac{L}{2r}\sqrt{\frac{F}{AE}}\right)\right)$$

$$= \left(\frac{F}{A}\right)\left(1 + \left(\frac{ec}{r^{2}}\right)\sec\phi\right) \qquad 45.8$$

$$\phi = \tfrac{1}{2}\left(\frac{L}{r}\right)\sqrt{\frac{F}{AE}} \qquad 45.9$$

For a given *eccentricity*, e, or eccentricity ratio, ec/r^{2}, and an assumed value of the buckling load, F, Eq. 45.9 is solved by trial and error for the slenderness ratio, L/r. Equations 45.8 and 45.9 converge quickly to the known L/r ratio when assumed values of F are substituted. (L/r is smaller when F is larger.)

4. THIN-WALLED CYLINDRICAL TANKS

In general, tanks under internal pressure experience circumferential, longitudinal, and radial stresses. If the wall thickness is small, the radial stress component is negligible and can be disregarded. A cylindrical tank is a *thin-walled tank* if its wall thickness-to-internal radius ratio is less than approximately 0.1.[2]

$$\frac{t}{d_{i}} = \frac{t}{2r_{i}} < 0.1 \qquad [\text{thin-walled}] \qquad 45.10$$

The *hoop stress*, σ_{h}, also known as *circumferential stress* and *tangential stress*, for a cylindrical thin-walled tank

---

[1]The design of timber, steel, and reinforced concrete building columns is very code-intensive. None of the theoretical methods presented in this section is acceptable for building design.
[2]There is overlap in the thin-wall/thick-wall criterion. The limiting ratios between thin- and thick-walled cylinders are matters of the accuracy desired. Lamé's solution can always be used, for both thick- and thin-walled cylinders.

under internal pressure is derived from the free-body diagram of a cylinder half.[3] Since the cylinder is assumed to be thin-walled, it is not important which radius (e.g., inner, mean, or outer) is used in Eq. 45.11. However, the inner radius is used by common convention.

$$\sigma_h = \frac{pr}{t} \qquad 45.11$$

Figure 45.2 *Stresses in a Thin-Walled Tank*

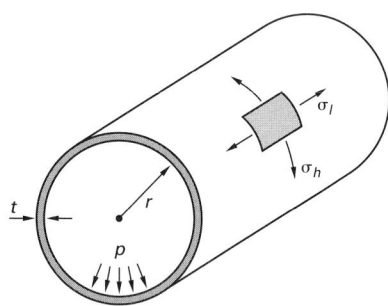

The axial forces on the ends of the cylindrical tank produce a stress, known as the *longitudinal stress* or *long stress*, σ_l, directed along the tank's longitudinal axis.

$$\sigma_l = \frac{pr}{2t} \qquad 45.12$$

Unless the tank is subject to torsion, there is no shear stress. Accordingly, the hoop and long stresses are the principal stresses. They do not combine into larger stresses. Their combined effect should be evaluated according to the appropriate failure theory.

The increase in length due to pressurization is easily determined from the longitudinal strain.

$$\Delta L = L\epsilon_l$$
$$= L\left(\frac{\sigma_l - \nu\sigma_h}{E}\right) \qquad 45.13$$

The increase in circumference (from which the radial increase can also be determined) due to pressurization is

$$\Delta C = C\epsilon_h$$
$$= \pi d_o \left(\frac{\sigma_h - \nu\sigma_l}{E}\right) \qquad 45.14$$

Example 45.2

A thin-walled pressurized tank is supported at both ends, as shown. Points A and B are located midway

---

[3]There is no simple way, including the more exact Lamé solutions, of evaluating theoretical stresses in thin-walled cylinders under external pressure, since failure is by collapse, not yielding. However, empirical equations exist for predicting the *collapsing pressure*.

---

between the supports and at the upper and lower surfaces, respectively. Evaluate the maximum stresses on the tank.

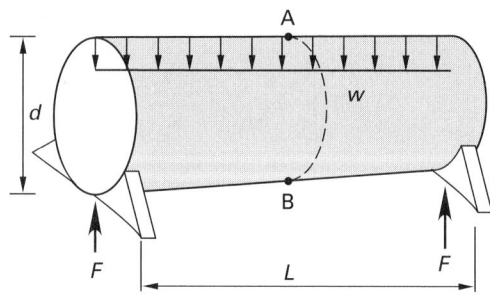

Solution

Since the tank is thin-walled, the radial stress is essentially zero. The hoop and longitudinal stresses are given by Eqs. 45.11 and 45.12, respectively. In addition, point A experiences a bending stress. The bending stress is

$$\sigma_b = \frac{Mc}{I}$$
$$c = \frac{d}{2}$$
$$M = \frac{FL}{2} = \frac{wL^2}{8}$$
$$I = \left(\frac{d}{2}\right)^3 \pi t$$

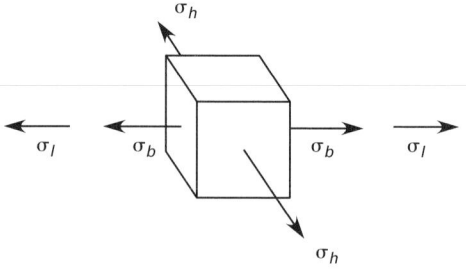

The bending stress is compressive at point A and tensile at point B. At point B, the bending stress has the same sign (tensile) as the longitudinal stress, and these two stresses add to each other. There is no torsional stress, so the resultant normal stresses are the principal stresses.

$$\sigma_1 = \sigma_h$$
$$\sigma_2 = \sigma_l + \sigma_b$$

5. THICK-WALLED CYLINDERS

A thick-walled cylinder has a wall thickness-to-radius ratio greater than 0.1 (i.e., a wall thickness-to-diameter ratio greater than 0.05). Figure 45.3 illustrates a thick-walled tank under either internal or external pressures.

In thick-walled tanks, radial stress is significant and cannot be disregarded. In *Lamé's solution*, a thick-walled cylinder is assumed to be made up of thin laminar rings. This method shows that the radial and circumferential stresses vary with location within the tank wall. (The term *circumferential stress* is preferred over *hoop stress* when dealing with thick-walled cylinders.) Compressive stresses are negative.

$$\sigma_c = \dfrac{r_i^2 p_i - r_o^2 p_o + \dfrac{(p_i - p_o)\, r_i^2 r_o^2}{r^2}}{r_o^2 - r_i^2} \qquad 45.15$$

Figure 45.3 *Thick-Walled Cylinder*

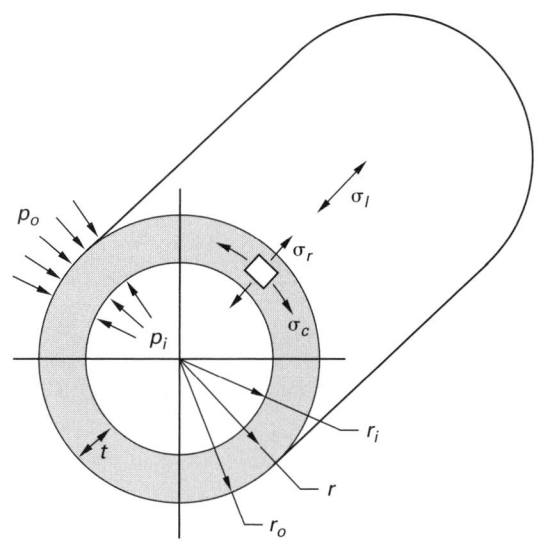

$$\sigma_r = \dfrac{r_i^2 p_i - r_o^2 p_o - \dfrac{(p_i - p_o)\, r_i^2 r_o^2}{r^2}}{r_o^2 - r_i^2} \qquad 45.16$$

$$\sigma_l = \dfrac{p_i r_i^2}{r_o^2 - r_i^2} \quad \left[\begin{array}{l} p_o \text{ does not act}\\ \text{longitudinally on the ends}\end{array}\right] \; 45.17$$

At every point in the cylinder, the circumferential, radial, and long stresses are the principal stresses. Unless an external torsional shear stress is added, it is not necessary to use the combined stress equations. Failure theories can be applied directly.

The cases of main interest are those of internal or external pressure only. The stress equations for these cases are summarized in Table 45.2. The maximum shear and normal stresses occur at the inner surface for both internal and external pressure.

Table 45.2 *Stresses in Thick-Walled Cylinders*[a]

stress	external pressure, p	internal pressure, p
$\sigma_{c,o}$	$\dfrac{-\left(r_o^2 + r_i^2\right) p_o}{r_o^2 - r_i^2}$	$\dfrac{2 r_i^2 p_i}{r_o^2 - r_i^2}$
$\sigma_{r,o}$	$-p_o$	0
$\sigma_{c,i}$	$\dfrac{-2 r_o^2 p_o}{r_o^2 - r_i^2}$	$\dfrac{\left(r_o^2 + r_i^2\right) p_i}{r_o^2 - r_i^2}$
$\sigma_{r,i}$	0	$-p_i$
τ_{max}	$\frac{1}{2}\sigma_{c,i}$	$\frac{1}{2}\left(\sigma_{c,i} + p_i\right)$

[a]Table 45.2 can be used with thin-walled cylinders. However, in most cases it will not be necessary to do so.

The *diametral strain* (which is the same as the *circumferential* and *radial strains*) is given by Eq. 45.18. Radial stresses are always compressive (hence they are negative), and algebraic signs must be observed with Eq. 45.18. Since the circumferential and radial stresses depend on location within the wall thickness, the strain can be evaluated at inner, outer, and any intermediate locations within the wall.

$$\begin{aligned} \epsilon &= \frac{\Delta d}{d} = \frac{\Delta C}{C} = \frac{\Delta r}{r} \\ &= \frac{\sigma_c - \nu(\sigma_r + \sigma_l)}{E} \end{aligned} \qquad 45.18$$

6. THIN-WALLED SPHERICAL TANKS

There is no unique axis in a spherical tank or in the spherical ends of a cylindrical tank. Therefore, the hoop and long stresses are identical.

$$\sigma = \frac{pr}{2t} \qquad 45.19$$

7. INTERFERENCE FITS

When assembling two pieces, interference fitting is often more economical than pinning, keying, or splining. The assembly operation can be performed in a hydraulic press, either with both pieces at room temperature or after heating the outer piece and cooling the inner piece. The former case is known as a *press fit* or *interference fit*; the latter as a *shrink fit*.

If two cylinders are pressed together, the pressure acting between them will expand the outer cylinder (placing it into tension) and will compress the inner cylinder. The *interference*, I, is the difference in dimensions between

Structural

the two cylinders. *Diametral interference* and *radial interference* are both used.[4]

$$
\begin{aligned}
I_{\text{diametral}} &= 2I_{\text{radial}} \\
&= d_{o,\text{inner}} - d_{i,\text{outer}} \\
&= |\Delta d_{o,\text{inner}}| + |\Delta d_{i,\text{outer}}|
\end{aligned} \qquad 45.20
$$

If the two cylinders have the same length, the thick-wall cylinder equations can be used. The materials used for the two cylinders do not need to be the same. Since there is no longitudinal stress from an interference fit and since the radial stress is negative, the strain from Eq. 45.18 is

$$
\begin{aligned}
\epsilon &= \frac{\Delta d}{d} = \frac{\Delta C}{C} = \frac{\Delta r}{r} \\
&= \frac{\sigma_c - \nu \sigma_r}{E}
\end{aligned} \qquad 45.21
$$

Equation 45.22 applies to the general case where both cylinders are hollow and have different moduli of elasticity and Poisson's ratios. The outer cylinder is designated as the *hub*; the inner cylinder is designated as the *shaft*. If the shaft is solid, use $r_{i,\text{shaft}} = 0$ in Eq. 45.22.

$$
\begin{aligned}
I_{\text{diametral}} &= 2I_{\text{radial}} \\
&= \left(\frac{2pr_{o,\text{shaft}}}{E_{\text{hub}}} \right) \left(\frac{r_{o,\text{hub}}^2 + r_{o,\text{shaft}}^2}{r_{o,\text{hub}}^2 - r_{o,\text{shaft}}^2} + \nu_{\text{hub}} \right) \\
&\quad + \left(\frac{2pr_{o,\text{shaft}}}{E_{\text{shaft}}} \right) \left(\frac{r_{o,\text{shaft}}^2 + r_{i,\text{shaft}}^2}{r_{o,\text{shaft}}^2 - r_{i,\text{shaft}}^2} - \nu_{\text{shaft}} \right)
\end{aligned}
$$

$$45.22$$

In the special case where the shaft is solid and is made from the same material as the hub, the diametral interference is given by Eq. 45.23.

$$
\begin{aligned}
I_{\text{diametral}} &= 2I_{\text{radial}} \\
&= \left(\frac{4pr_{\text{shaft}}}{E} \right) \left(\frac{1}{1 - \left(\dfrac{r_{\text{shaft}}}{r_{o,\text{hub}}} \right)^2} \right)
\end{aligned} \qquad 45.23
$$

The maximum assembly force required to overcome friction during a press-fitting operation is given by Eq. 45.24. The coefficient of friction is highly variable. Values in the range of 0.03 to 0.33 have been reported. This relationship is approximate because the coefficient of friction is not known with certainty and the assembly force affects the pressure, p, through Poisson's ratio.

$$
F_{\text{max}} = fN = 2\pi f p r_{o,\text{shaft}} L_{\text{interface}} \qquad 45.24
$$

---

[4]Theoretically, the interference can be given to either the inner or outer cylinder, or it can be shared by both cylinders. However, in the case of a surface-hardened shaft with a standard diameter, the interference is usually all given to the disk. Otherwise, it may be necessary to machine the shaft and remove some of the hardened surface.

The maximum torque that the press-fitted hub can withstand or transmit is given by Eq. 45.25. This can be greater or less than the shaft's torsional shear capacity. Both values should be calculated.

$$
T_{\text{max}} = 2\pi f p r_{o,\text{shaft}}^2 L_{\text{interface}} \qquad 45.25
$$

Most interference fits are designed to keep the contact pressure or the stress below a given value. Designs of interference fits limited by strength generally use the distortion energy failure criterion. That is, the maximum shear stress is compared with the shear strength determined from the failure theory.

Example 45.3

A steel cylinder has inner and outer diameters of 1.0 in and 2.0 in, respectively. The cylinder is pressurized internally to 10,000 lbf/in^2. The modulus of elasticity is 2.9×10^7 lbf/in^2, and Poisson's ratio is 0.3. What is the radial strain at the inside face?

Solution

The longitudinal stress is

$$
\begin{aligned}
\sigma_l &= \frac{F}{A} = \frac{p_i \pi r_i^2}{\pi(r_o^2 - r_i^2)} = \frac{p_i r_i^2}{r_o^2 - r_i^2} \\
&= \frac{\left(10{,}000 \ \dfrac{\text{lbf}}{\text{in}^2} \right) (0.5 \ \text{in})^2}{(1.0 \ \text{in})^2 - (0.5 \ \text{in})^2} \\
&= 3333 \ \text{lbf/in}^2
\end{aligned}
$$

The stresses at the inner face are found from Table 45.2.

$$
\begin{aligned}
\sigma_{c,i} &= \frac{(r_o^2 + r_i^2)p}{r_o^2 - r_i^2} \\
&= \frac{\left((1.0 \ \text{in})^2 + (0.5 \ \text{in})^2 \right) \left(10{,}000 \ \dfrac{\text{lbf}}{\text{in}^2} \right)}{(1.0 \ \text{in})^2 - (0.5 \ \text{in})^2} \\
&= 16{,}667 \ \text{lbf/in}^2 \\
\sigma_{r,i} &= -p = -10{,}000 \ \text{lbf/in}^2
\end{aligned}
$$

The circumferential and radial stresses increase the radial strain; the longitudinal stress decreases the radial strain. The radial strain is

$$
\begin{aligned}
\frac{\Delta r}{r} &= \frac{\sigma_{c,i} - \nu(\sigma_{r,i} + \sigma_l)}{E} \\
&= \frac{16{,}667 \ \dfrac{\text{lbf}}{\text{in}^2} - (0.3)\left(-10{,}000 \ \dfrac{\text{lbf}}{\text{in}^2} + 3333 \ \dfrac{\text{lbf}}{\text{in}^2} \right)}{2.9 \times 10^7 \ \dfrac{\text{lbf}}{\text{in}^2}} \\
&= 6.44 \times 10^{-4}
\end{aligned}
$$

Example 45.4

A hollow aluminum cylinder is pressed over a hollow brass cylinder as shown. Both cylinders are 2 in long. The interference is 0.004 in. The average coefficient of friction during assembly is 0.25. (a) What is the maximum shear stress in the brass? (b) What initial disassembly force is required to separate the two cylinders?

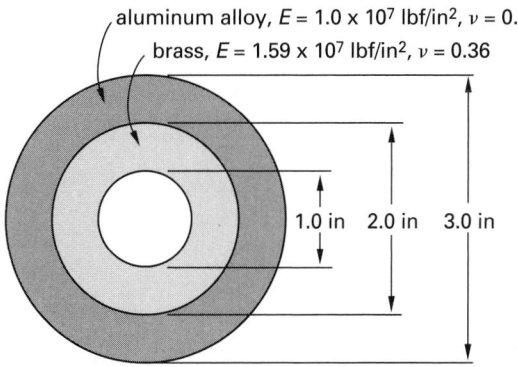

aluminum alloy, $E = 1.0 \times 10^7$ lbf/in², $\nu = 0.33$

brass, $E = 1.59 \times 10^7$ lbf/in², $\nu = 0.36$

1.0 in   2.0 in   3.0 in

Solution

(a) Work with the aluminum outer cylinder, which is under internal pressure.

$$\sigma_{c,i} = \frac{(r_o^2 + r_i^2)p}{r_o^2 - r_i^2}$$
$$= \frac{\left((1.5 \text{ in})^2 + (1.0 \text{ in})^2\right)p}{(1.5 \text{ in})^2 - (1.0 \text{ in})^2}$$
$$= 2.6p$$
$$\sigma_{r,i} = -p$$

From Eq. 45.18, the diametral strain is

$$\epsilon = \frac{\sigma_{c,i} - \nu(\sigma_{r,i} + \sigma_l)}{E}$$
$$= \frac{2.6p - (0.33)(-p)}{1.0 \times 10^7 \ \frac{\text{lbf}}{\text{in}^2}}$$
$$= 2.93 \times 10^{-7}p$$
$$\Delta d = \epsilon d = (2.93 \times 10^{-7}p)(2.0 \text{ in})$$
$$= 5.86 \times 10^{-7}p$$

Now work with the brass inner cylinder, which is under external pressure. Use Table 45.2.

$$\sigma_{c,o} = \frac{-(r_o^2 + r_i^2)p}{r_o^2 - r_i^2}$$
$$= \frac{-\left((1.0 \text{ in})^2 + (0.5 \text{ in})^2\right)p}{(1.0 \text{ in})^2 - (0.5 \text{ in})^2}$$
$$= -1.667p$$
$$\sigma_{r,o} = -p$$

From Eq. 45.18, the diametral strain is

$$\epsilon = \frac{\sigma_{c,o} - \nu(\sigma_{r,o} + \sigma_l)}{E}$$
$$= \frac{-1.667p - (0.36)(-p)}{1.59 \times 10^7 \ \frac{\text{lbf}}{\text{in}^2}}$$
$$= -0.822 \times 10^{-7}p$$
$$\Delta d = \epsilon d = (-0.822 \times 10^{-7}p)(2.0 \text{ in})$$
$$= -1.644 \times 10^{-7}p$$

The diametral interference is known to be 0.004 in. From Eq. 45.20,

$$I_{\text{diametral}} = |\Delta d_{o,\text{inner}}| + |\Delta d_{i,\text{outer}}|$$
$$0.004 \text{ in} = |5.86 \times 10^{-7}p| + |-1.644 \times 10^{-7}p|$$
$$p = 5330 \text{ lbf/in}^2$$

From Table 45.2, the circumferential stress at the inner face of the brass (under external pressure) is

$$\sigma_{c,i} = \frac{-2r_o^2 p}{r_o^2 - r_i^2}$$
$$= \frac{(-2)(1.0 \text{ in})^2 \left(5330 \ \frac{\text{lbf}}{\text{in}^2}\right)}{(1.0 \text{ in})^2 - (0.5 \text{ in})^2}$$
$$= -14{,}213 \text{ lbf/in}^2$$

Also from Table 45.2, the maximum shear stress is

$$\tau_{\text{max}} = (0.5)\sigma_{c,i} = (0.5)\left(-14{,}213 \ \frac{\text{lbf}}{\text{in}^2}\right)$$
$$= -7107 \text{ lbf/in}^2$$

(b) The initial force necessary to disassemble the two cylinders is the same as the maximum assembly force. Use Eq. 45.24.

$$F_{\text{max}} = 2\pi f p r_{\text{shaft}} L_{\text{interface}}$$
$$= (2\pi)(0.25)\left(5330 \ \frac{\text{lbf}}{\text{in}^2}\right)(1 \text{ in})(2 \text{ in})$$
$$= 16{,}745 \text{ lbf}$$

8. STRESS CONCENTRATIONS FOR PRESS-FITTED SHAFTS IN FLEXURE

When a shaft carrying a press-fitted hub (whose thickness is less than the shaft length) is loaded in flexure, there will be an increase in shaft bending stress in the vicinity of the inner hub edge. The fatigue life of the shaft can be seriously affected by this stress increase. The extent of the increase depends on the magnitude of the bending stress, σ_b, and the contact pressure, p, and can be as high as 2.0 or more.

Structural

Some designs attempt to reduce the increase in shaft stress by grooving the disk (to allow the disk to flex). Other designs rely on various treatments to increase the fatigue strength of the shaft. For an unmodified simple press-fit, the multiplicative stress concentration factor (to be applied to the bending stress calculated from $\sigma_b = Mc/I$) is given by Fig. 45.4.

Figure 45.4 *Stress Concentration Factor for Press Fit*

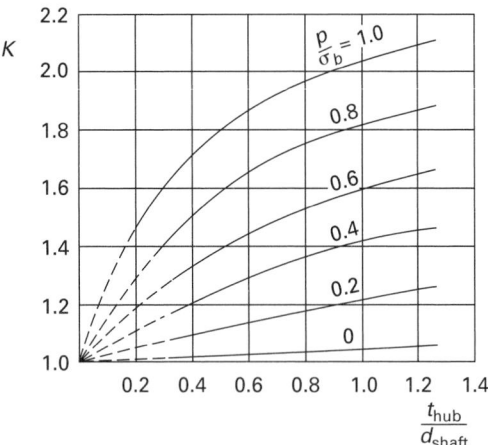

"Fatigue of Shafts at Fitted Members, with a Related Photoelastic Analysis," reproduced from *Transactions of the ASME*, Vol. 57, © 1935, and Vol. 58, © 1936, with permission of the American Society of Mechanical Engineers.

9. BOLTS

There are three leading specifications for bolt thread families: ANSI, ISO metric, and DIN metric.[5] ANSI (essentially identical to SAE, ASTM, ISO-inch standard) is widely used in the United States. DIN (Deutches Institute für Normung) fasteners are widely available and broadly accepted.[6] ISO metric (the International Organization for Standardization, which first met in 1961) fasteners are used in large volume by U.S. car manufacturers. The CEN (European Committee for Standardization) standards promulgated by the European Community (EC) have essentially adopted the ISO standards.

An American National (Unified) thread is specified by the sequence of parameters S($\times$L)-N-F-A-(H-E), where S is the thread outside diameter (nominal size), L is the optional shank length, N is the number of threads per inch, F is the thread pitch family, A is the class (allowance), and H and E are the optional hand and engagement length designations. The letter R can be added to the thread pitch family to indicate that the thread roots are radiused (for better fatigue resistance).

[5]Other fastener families include the Italian UNI, Swiss VSM, Japanese JIS, and United Kingdom's BS series.
[6]To add to the confusion, many DIN standards are identical to ISO standards, with only slight differences in the tolerance ranges. However, the standards are not interchangeable in every case.

For example, a $^3/_8 \times$ 1-16UNC-2A bolt is $^3/_8$ in in diameter, 1 in in length, and has 16 Unified Coarse threads per inch rolled with a class 2A accuracy.[7] A UNRC bolt would be identical except for radiused roots. Table 45.3 lists some (but not all) values for these parameters.

Table 45.3 *Representative American National (Unified) Bolt Thread Designations[a]*

S: Size
  1 through 12
  $^1/_4''$ through $^9/_{16}''$ in $^1/_{16}''$ increments
  $^5/_8''$ through $1^1/_2''$ in $^1/_8''$ increments
  $1^3/_4''$ through $4''$ in $^1/_4''$ increments

F: Thread Family
  UNC and NC—Unified Coarse[b]
  UNF and NF—Unified Fine[b]
  UNEF and NEF—Unified Extra Fine[c]
  8N—8 threads per inch
  12UN and 12N—12 threads per inch
  16UN and 16N—16 threads per inch
  UN, UNS, and NS—special series

A: Allowance (A—external threads, B—internal threads)[d]
  1A and 1B—liberal allowance for ease of assembly with dirty or damaged threads
  2A and 2B—normal production allowance (sufficient for plating)
  3A and 3B—close tolerance work with no allowance

H: Hand
  blank—right-hand thread
  LH—left-hand thread

[a]In addition to fastener thread families, there are other special-use threads such as Acme, stub, square, buttress, and worm series.
[b]Previously known as United States Standard or American Standard.
[c]The UNEF series is the same as the SAE (Society of Automotive Engineers) fine series.
[d]Allowance classes 2 and 3 (without the A and B designation) were used prior to industry transition to the Unified classes.

The *grade* of a bolt indicates the fastener material and is marked on the bolt cap.[8] In this regard, the marking depends on whether an SAE grade or ASTM designation is used. The minimum *proof load* (i.e., the maximum stress the bolt can support without acquiring a permanent set) increases with the grade. (The term *proof strength* is less common.) Table 45.4 lists how the caps of the bolts are marked to distinguish among the major grades.[9] If a bolt is manufactured in the United States, its cap must also show the logo or mark of the manufacturer.

A metric thread is specified by an M or MJ and a diameter and a pitch in millimeters, in that order. Thus, M10 $\times$ 1.5 is a thread having a nominal major diameter of 10 mm and a pitch of 1.5 mm. The MJ series have rounded root fillets and larger minor diameters.

[7]Threads are generally rolled, not cut, into a bolt.
[8]The *type* of a structural bolt should not be confused with the *grade* of a structural rivet.
[9]Optional markings can also be used.

Head markings on metric bolts indicate their *property class* and correspond to the approximate tensile strength in MPa/100. For example, a bolt marked 8.8 would correspond to a medium carbon, quenched and tempered bolt with an approximate tensile strength of 880 MPa. (The minimum of the tensile strength range for property class 8.8 is 830 MPa.)

Table 45.4 *Selected Steel Bolt Grades and Designations*

LC = low-carbon; MC = medium-carbon; Q&T = quenched and tempered; CD = cold-drawn

(Subject fo change and requires validation for design use.)

(Not for use with stainless steel.)

standard	head marking[i]	material type	proof load (ksi)		minimum tensile strength (ksi)		minimum yield strength (ksi)	
SAE grades								
grade 1	none	LC or MC	33^h		55	60^n		36^n
grade 2	none	LC or MC	55^a	33^b	74^a	60^b	57^a	36^b
grade 4	none	CD MC	–		115			
grade 5	3 tics, 360°	Q&T MC	85^c	74^d	120^c	105^d	92^c	81^d
grade 5.1	3 tics, 180°		85^p		120^p			
grade 5.2	3 tics, 120°	Q&T LC martensite	85^q		120^q		92^q	
grade 7	5 tics, 360°	Q&T MC alloy	105^h		133^h		115^h	
grade 8	6 tics, 360°	Q&T MC alloy	120^h		150^h		130^h	
grade 8.1	none		120^h		150^h		130^h	
grade 8.2	6 tics, 180°	Q&T LC martensite	120^s		150^s			
ISO designations								
class 4.6	none	LC or MC	225 MPa		400 MPa			
class 4.8			310 MPa		420 MPa			
class 5.8	none	LC or MC	380° MPa		520° MPa			
class 8.8	8.8 or 88	Q&T MC	580^j	600^k MPa	800^j	830^k MPa	640^j	660^k MPa
class 9.8	9.8		650^r MPa		900^r MPa			
class 10.9	10.9 or 109	Q&T alloy steel	830^l MPa		1040^l MPa		940^l	MPa
class 12.9	12.9		970^m MPa		1220^m MPa		1100^m	MPa
ASTM designations								
A307 grades A, B	none	LC	–		60			
A325 type 1	3 tics, 360°, A325	Q&T MC	85^e	74^f	120^e	105^f	92^e	81^f
A325 type 2	3 tics, 120°, A325	Q&T LC martensite	85^e	74^f	120^e	105^f	92^e	
A325 type 3	<u>A325</u>	Q&T weathering steel	85^e	74^f	120^e	105^f	92^e	81^f
A354 grade BC	BC	Q&T alloy steel	105^g		125^g		109^g	
A354 grade BB	BB	Q&T alloy steel	80^g		105^g		83^g	
A354 grade BD	6 tics, 360°	Q&T alloy steel	120^h		150^h		130^h	
A449	3 tics, 360°	Q&T MC	85^c	74^d	120^c	105^d	92^c	81^d
A490 type 1	A490	Q&T alloy steel	120^t		$150–170^t$		130^t	
A490 type 3	<u>A490</u>	Q&T weathering steel	–		–			

(Multiply ksi by 6894.8 to obtain kPa.)

[a] $1/4$–$3/4$ in [j] 5–15 mm [q] $1/4$–1 in
[b] $3/4$–$1 1/2$ in [k] 16–72 mm [r] 1.6–16 mm
[c] $1/4$–1 in [l] 5–100 mm [s] $1/4$–1 in
[d] 1–$1 1/2$ in [m] 1.6–100 mm [t] $1/2$–$1 1/2$ in
[e] $1/2$–1 in [n] $1/4$–$1 1/4$ in
[f] $1 1/8$ in–$1 1/2$ in [o] 5–24 mm
[g] $1/4$–$2 1/2$ in [p] No. 6–$3/8$ in
[h] $1/4$–$1 1/2$ in
[i] Tics are spread over the arc indicated.

Table 45.5 *Dimensions of American Unified Standard Threaded Bolts*[a]

nominal size	threads per inch	major diameter (in)	minor area (in^2)	tensile stress area (in^2)
coarse series				
1/4	20	0.2500	0.0269	0.0318
5/16	18	0.3125	0.0454	0.0524
3/8	16	0.3750	0.0678	0.0775
7/16	14	0.4375	0.0933	0.1063
1/2	13	0.5000	0.1257	0.1419
9/16	12	0.5625	0.162	0.182
5/8	11	0.6250	0.202	0.226
3/4	10	0.7500	0.302	0.334
7/8	9	0.8750	0.419	0.462
1	8	1.0000	0.551	0.606
fine series				
1/4	28	0.2500	0.0326	0.0364
5/16	24	0.3125	0.0524	0.0580
3/8	24	0.3750	0.0809	0.0878
7/16	20	0.4375	0.1090	0.1187
1/2	20	0.5000	0.1486	0.1599
9/16	18	0.5625	0.189	0.203
5/8	18	0.6250	0.240	0.256
3/4	16	0.7500	0.351	0.373
7/8	14	0.8750	0.480	0.509
1	12	1.0000	0.625	0.663

(Multiply in by 25.4 to obtain mm.)
(Multiply in^2 by 645 to obtain mm^2.)
[a]Based on ANSI B1.1-1974.

10. RIVET AND BOLT CONNECTIONS

Figure 45.5 illustrates a tension *lap joint* connection using rivet or bolt connectors.[10] Unless the plate material is very thick, the effects of eccentricity are disregarded. A connection of this type can fail in shear, tension, or bearing. A common design procedure is to determine the number of connectors based on shear stress and then to check the bearing and tensile stresses.

Figure 45.5 *Tension Lap Joint*

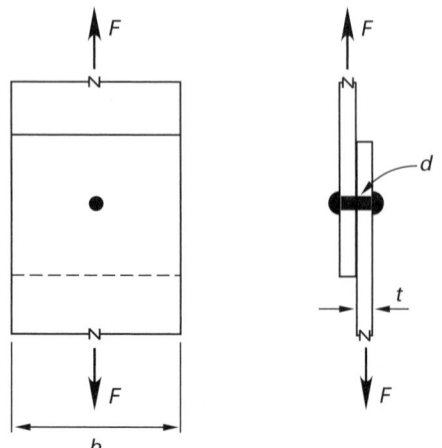

---

[10]Rivets are no longer used in building construction, but they are still extensively used in manufacturing.

One of the failure modes is shearing of the connectors. In the case of *single shear*, each connector supports its proportionate share of the load. In *double shear*, each connector has two shear planes, and the stress per connector is halved.[11] The shear stress in a cylindrical connector is

$$\tau = \frac{F}{\frac{\pi}{4}d^2} \qquad 45.26$$

The number of required connectors, as determined by shear, is

$$n = \frac{\tau}{\text{allowable shear stress}} \qquad 45.27$$

Figure 45.6 *Single and Double Shear*

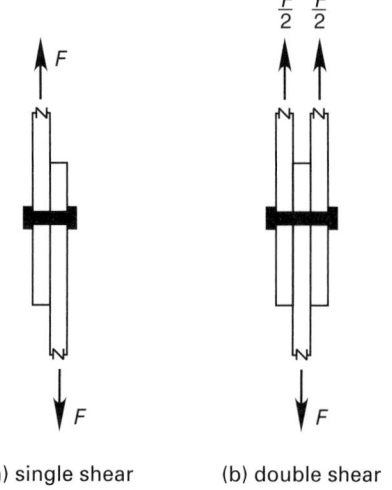

(a) single shear   (b) double shear

The plate can fail in tension. If there are n connector holes of diameter d in a line across the width, b, of the plate, the cross-sectional area in the plate remaining to resist the tension is

$$A_t = t(b - nd) \qquad 45.28$$

The number of connectors across the plate width must be chosen to keep the tensile stress less than the allowable stress. The maximum tensile stress in the plate will be

$$\sigma_t = \frac{F}{A_t} \qquad 45.29$$

The plate can also fail by *bearing* (i.e., crushing). The number of connectors must be chosen to keep the actual *bearing stress* below the allowable bearing stress. For one connector, the bearing stress in the plate is

$$\sigma_p = \frac{F}{dt} \qquad 45.30$$

$$n = \frac{\sigma_p}{\text{allowable bearing stress}} \qquad 45.31$$

---

[11]"Double shear" is not the same as "double rivet" or "double butt." *Double shear* means that there are two shear planes in one rivet. *Double rivet* means that there are two rivets along the force path. *Double butt* refers to the use of two backing plates (i.e., "scabs") used on either side to make a tension connection between two plates. Similarly, *single butt* refers to the use of a single backing plate to make a tension connection between two plates.

The plate can also fail by shear tear-out, as illustrated in Fig. 45.7. The shear stress is

$$\tau = \frac{F}{2A} = \frac{F}{2t\left(L - \dfrac{d}{2}\right)} \qquad 45.32$$

Figure 45.7 *Shear Tear-Out*

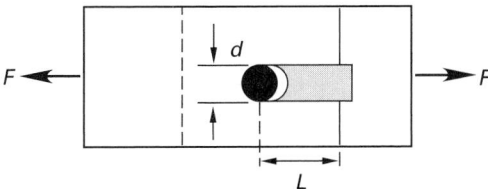

The *joint efficiency* is the ratio of the strength of the joint divided by the strength of a solid (i.e., unpunched or undrilled) plate.

11. BOLT PRELOAD

Consider the ungasketed connection shown in Fig. 45.8. The load varies from $F_{\min}$ to $F_{\max}$. If the bolt is initially snug but without initial tension, the force in the bolt also will vary from $F_{\min}$ to $F_{\max}$. If the bolt is tightened so that there is an initial *preload force*, F_i, greater than $F_{\max}$ in addition to the applied load, the bolt will be placed in tension and the parts held together will be in compression.[12] When a load is applied, the bolt tension will increase even more, but the compression in the parts will decrease.

Figure 45.8 *Bolted Tension Joint with Varying Load*

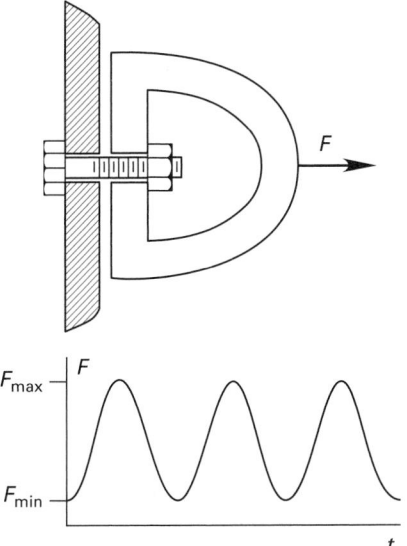

---

[12]If the initial preload force is less than $F_{\max}$, the bolt may still carry a portion of the applied load. Equation 45.37 can be solved for the value of F that will result in a loss of compression ($F_{\text{parts}} = 0$) and cause the bolt to carry the entire applied load.

The amount of compression in the parts will vary as the applied load varies. Thus, the clamped members will carry some of the applied load, since this varying load has to "uncompress" the clamped part as well as lengthen the bolt. The net result is the reduction of the variation of the force in the bolt. The initial tension produces a larger mean stress, but the overall result is the reduction of the alternating stress. Thus, preloading is an effective method of reducing the alternating stress in bolted tension connections.

It is convenient to define the *spring constant, k*, of the bolt. The *grip, L*, is the thickness of the parts being connected by the bolt (not the bolt length). It is common to use the nominal diameter of the bolt, disregarding the reduction due to threading.

$$k_{\text{bolt}} = \frac{F}{\Delta L} = \frac{A_{\text{bolt}} E_{\text{bolt}}}{L} \qquad 45.33$$

The actual spring constant for a bolted part, k_{part}, is difficult to determine if the clamped area is not small and well defined. The only accurate way to determine the stiffness of a part in a bolted joint is through experimentation. If the clamped parts are flat plates, various theories can be used to calculate the effective load-bearing areas of the flanges, but doing so is a laborious process.

One simple rule of thumb is that the bolt force spreads out to three times the bolt-hole diameter. Of course, the hole diameter needs to be considered (i.e., needs to be subtracted) in calculating the effective force area. If the modulus of elasticity is the same for the bolt and the clamped parts, using this rule of thumb, the larger area results in the parts being eight times stiffer than the bolts.

$$k_{\text{parts}} = \frac{A_{e,\text{parts}} E_{\text{parts}}}{L} \qquad 45.34$$

If the clamped parts have different moduli of elasticity, including if a gasket constitutes one of the layers compressed by the bolt, the composite spring constant can be found from Eq. 45.35.[13]

$$\frac{1}{k_{\text{parts,composite}}} = \frac{1}{k_1} + \frac{1}{k_2} + \frac{1}{k_3} + \cdots \qquad 45.35$$

The bolt and the clamped parts all carry parts of the applied load, F_{applied}. F_i is the initial preload force.

$$F_{\text{bolt}} = F_i + \frac{k_{\text{bolt}} F_{\text{applied}}}{k_{\text{bolt}} + k_{\text{parts}}} \qquad 45.36$$

$$F_{\text{parts}} = \frac{k_{\text{parts}} F_{\text{applied}}}{k_{\text{bolt}} + k_{\text{parts}}} - F_i \qquad 45.37$$

---

[13]If a soft washer or gasket is used, its spring constant can control Eq. 45.35.

O-ring (metal and elastomeric) seals permit metal-to-metal contact and affect the effective spring constant of the parts very little. However, the seal force tends to separate bolted parts and must be added to the applied force. The seal force can be obtained from the seal deflection and seal stiffness or from manufacturer's literature.

For static loading, recommended amounts of preloading often are specified as a percentage of the *proof strength* (or *proof load*) in psi.[14] For bolts, the proof load is slightly less than the yield strength. Traditionally, preload has been specified conservatively as 75% of proof for reusable connectors and 90% of proof for one-use connectors.[15] Connectors with some ductility can safely be used beyond the yield point, and 100% is now in widespread use.[16] When understood, advantages of preloading to 100% of proof load often outweigh the disadvantages.[17]

If the applied load varies, the forces in the bolt and parts will also vary. In that case, the preload must be determined from an analysis of the Goodman line.

Tightening of a tension bolt will induce a torsional stress in the bolt.[18] Where the bolt is to be locked in place, the torsional stress can be removed without greatly affecting the preload by slightly backing off the bolt. If the bolt is subject to cyclic loading, the bolt will probably slip back by itself, and it is reasonable to neglect the effects of torsion in the bolt altogether. (This is the reason that well-designed connections allow for a loss of 5 to 10% of the initial preload during routine use.)

Stress concentrations at the beginning of the threaded section are significant in cyclic loading.[19] To avoid a reduction in fatigue life, the alternating stress used in the Goodman line should be multiplied by an appropriate stress concentration factor, K. For fasteners with rolled threads, an average factor of 2.2 for SAE grades 0 to 2 (metric grades 3.6 to 5.8) is appropriate. For SAE grades 4 to 8 (metric grades 6.6 to 10.9), an average

factor of 3.0 is appropriate. Stress concentration factors for the fillet under the bolt head are different, but lower than these values. Stress concentration factors for cut threads are much higher.

The stress in a bolt depends on its load-carrying area. This area is typically obtained from a table of bolt properties. In practice, except for loading near the bolt's failure load, working stresses are low, and the effects of threads usually are ignored, so the area is based on the major (nominal) diameter.

$$\sigma_{\text{bolt}} = \frac{KF}{A} \qquad 45.38$$

12. BOLT TORQUE TO OBTAIN PRELOAD

During assembly, the preload tension is not monitored directly. Rather, the torque required to tighten the bolt is used to determine when the proper preload has been reached. Methods of obtaining the required preload include the standard torque wrench, the *run-of-the-nut method* (e.g., turning the bolt some specific angle past snugging torque), *direct-tension indicating* (DII) washers, and computerized automatic assembly.

The standard manual torque wrench does not provide precise, reliable preloads, since the fraction of the torque going into bolt tension is variable.[20] Torque-, angle-, and time-monitoring equipment, usually part of an automated assembly operation, is essential to obtaining precise preloads on a consistent basis. It automatically applies the snugging torque and specified rotation, then checks the results with torque and rotation sensors. The computer warns of out-of-spec conditions.

The *Maney formula* is a simple relationship between the initial bolt tension, F_i, and the installation torque, T. The *torque coefficient*, K_T (also known as the *bolt torque factor* and the *nut factor*) used in Eq. 45.40 depends mainly on the coefficient of friction, f. The torque coefficient for lubricated bolts generally varies from 0.15 to 0.20, and a value of 0.2 is commonly used.[21] With antiseize lubrication, it can drop as low as 0.12. (The torque coefficient is not the same as the coefficient of friction.)

$$T = K_T d_{\text{bolt}} F_i \qquad 45.39$$

$$K_T = \frac{f_c r_c}{d_{\text{bolt}}} + \left(\frac{r_t}{d_{\text{bolt}}}\right)\left(\frac{\tan\theta + f_t \sec\alpha}{1 - f_t \tan\theta \sec\alpha}\right) \qquad 45.40$$

$$\tan\theta = \frac{\text{lead per revolution}}{2\pi r_t} \qquad 45.41$$

f_c is the coefficient of friction at the collar (fastener bearing face). r_c is the mean collar radius (i.e., the

---

[14]This is referred to as a "rule of thumb" specification, because a mathematical analysis is not performed to determine the best preload.

[15]Some U.S. military specifications call for 80% of proof load in tension fasteners and only 30% for shear fasteners. The object of keeping the stresses below yielding is to be able to reuse the bolts.

[16]Even under normal elastic loading of a bolt, local plastic deformation occurs in the bolt-head fillet and thread roots. Since the stress-strain curve is nearly flat at the yield point, a small amount of elongation into the plastic region does not increase the stress or tension in the bolt.

[17]The disadvantages are: (a) Field maintenance probably won't be possible, as manually running up bolts to 100% proof will result in many broken bolts. (b) Bolts should not be reused, as some will have yielded. (c) The highest-strength bolts do not exhibit much plastic elongation and ordinarily should not be run up to 100% proof load.

[18]An argument for the conservative 75% of proof load preload limit is that the residual torsional stress will increase the bolt stress to 90% or higher anyway, and the additional 10% needed to bring the preload up to 100% probably won't improve economic performance much.

[19]Stress concentrations are frequently neglected for static loading.

---

[20]Even with good lubrication, about 50% of the torque goes into overcoming friction between the head and collar/flange, another 40% is lost in thread friction, and only the remaining 10% goes into tensioning the connector.

[21]With a coefficient of friction of 0.15, the torque coefficient is approximately 0.20 for most bolt sizes, regardless of whether the threads are coarse or fine.

effective radius of action of the friction forces on the bearing face). r_t is the effective radius of action of the frictional forces on the thread surfaces. Similarly, f_t is the coefficient of friction between the thread contact surfaces. θ is the *lead angle*, also known as the *helix thread angle*. α is the thread half angle (30° for UNF threads), and $d_{t,m}$ is the mean thread diameter.

13. FILLET WELDS

(Also, see the discussion of welds in Ch. 66.)

The common *fillet weld* is shown in Fig. 45.9. Such welds are used to connect one plate to another. The applied load, F, is assumed to be carried in shear by the *effective weld throat*. The *effective throat size*, t_e, is related to the weld size, y, by Eq. 45.42.

$$t_e = 0.707y \qquad 45.42$$

Neglecting any increased stresses due to eccentricity, the shear stress in a fillet lap weld is

$$\tau = \frac{F}{bt_e} \qquad 45.43$$

Figure 45.9 Fillet Lap Weld and Symbol

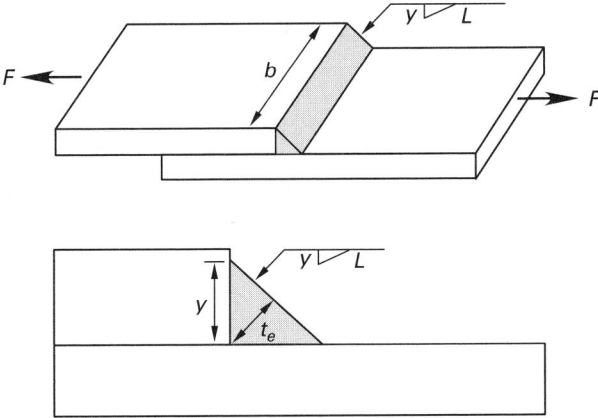

Weld (filler) metal should have a strength equal to or greater than the base material. Properties of filler metals are readily available from their manufacturers and, for standard rated welding rod, from engineering handbooks.

14. CIRCULAR SHAFT DESIGN

Shear stress occurs when a shaft is placed in torsion. The shear stress at the outer surface of a bar of radius r, which is torsionally loaded by a torque, T, is

$$\tau = G\theta = \frac{Tr}{J} \qquad 45.44$$

The total strain energy due to torsion is

$$U = \frac{T^2 L}{2GJ} \qquad 45.45$$

J is the shaft's polar moment of inertia. For a solid round shaft,

$$J = \frac{\pi r^4}{2} = \frac{\pi d^4}{32} \qquad 45.46$$

For a hollow round shaft,

$$J = \frac{\pi}{2}\left(r_o^4 - r_i^4\right) \qquad 45.47$$

If a shaft of length L carries a torque T, the angle of twist (in radians) will be

$$\gamma = \frac{L\theta}{r} = \frac{TL}{GJ} \qquad 45.48$$

Figure 45.10 Torsional Deflection of a Circular Shaft

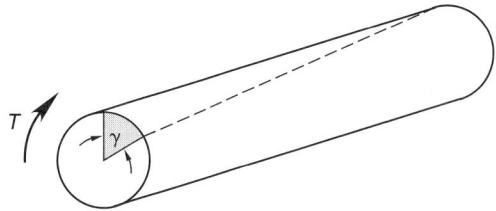

G is the *shear modulus*. For steel, it is approximately equal to 11.5×10^6 lbf/in² (8.0×10^4 MPa). The shear modulus also can be calculated from the modulus of elasticity.

$$G = \frac{E}{2(1+\nu)} \qquad 45.49$$

The torque, T, carried by a shaft spinning at n revolutions per minute is related to the transmitted horsepower.

$$T_{\text{in-lbf}} = \frac{(63{,}025)(\text{horsepower})}{n_{\text{rpm}}} \qquad 45.50$$

If a statically loaded shaft without axial loading experiences a bending stress, $\sigma_x = Mc/I$ (i.e., is loaded in flexure), in addition to torsional shear stress, $\tau = Tr/J$, the maximum shear stress from the combined stress theory is

$$\tau_{\text{max}} = \sqrt{\left(\frac{\sigma_x}{2}\right)^2 + \tau^2} \qquad 45.51$$

$$\tau_{\text{max}} = \frac{16}{\pi d^3}\sqrt{M^2 + T^2} \qquad 45.52$$

The equivalent normal stress from the distortion energy theory is

$$\sigma' = \frac{16}{\pi d^3}\sqrt{4M^2 + 3T^2} \qquad 45.53$$

The diameter can be determined by setting the shear and normal stresses equal to the maximum allowable shear (as calculated from the maximum shear stress theory, $S_y/2(\text{FS})$, or from the distortion energy theory, $\sqrt{3}S_y/2(\text{FS})$, and normal stresses, respectively).

Equations 45.51 and 45.52 should not be used with dynamically loaded shafts (i.e., those that are turning). Fatigue design of shafts should be designed according to a specific code (e.g., ANSI or ASME) or should use a fatigue analysis (e.g., Goodman, Soderberg, or Gerber).

Example 45.5

The press-fitted aluminum alloy-brass cylinder described in Ex. 45.4 is used as a shaft. The press fit is adequate to maintain nonslipping contact between the two materials. The shaft carries a steady torque of 24,000 in-lbf. There is no bending stress. What is the maximum torsional shear stress in the (a) aluminum and (b) brass?

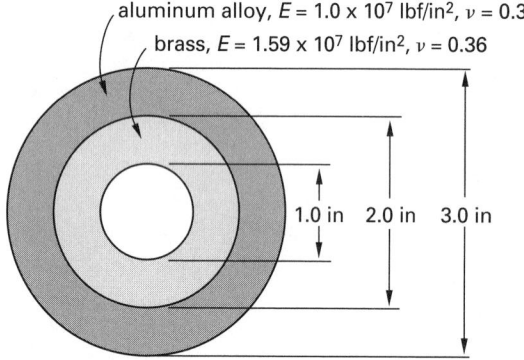

aluminum alloy, E = 1.0 x 10^7 lbf/in^2, ν = 0.33
brass, E = 1.59 x 10^7 lbf/in^2, ν = 0.36

1.0 in   2.0 in   3.0 in

Solution

The stronger material (as determined from the shear modulus, G) should be converted to an equivalent area of the weaker material.

For the aluminum, from Eq. 45.49,

$$G_{\text{aluminum}} = \frac{E}{2(1+\nu)}$$
$$= \frac{1.0 \times 10^7 \; \frac{\text{lbf}}{\text{in}^2}}{(2)(1+0.33)}$$
$$= 3.76 \times 10^6 \; \text{lbf/in}^2$$

For the brass, from Eq. 45.49,

$$G_{\text{brass}} = \frac{E}{2(1+\nu)}$$
$$= \frac{1.59 \times 10^7 \; \frac{\text{lbf}}{\text{in}^2}}{(2)(1+0.36)}$$
$$= 5.85 \times 10^6 \; \text{lbf/in}^2$$

The brass is the stronger material. The modular shear ratio is

$$n = \frac{G_{\text{brass}}}{G_{\text{aluminum}}}$$
$$= \frac{5.85 \times 10^6 \; \frac{\text{lbf}}{\text{in}^2}}{3.76 \times 10^6 \; \frac{\text{lbf}}{\text{in}^2}}$$
$$= 1.56$$

The polar moment of inertia of the aluminum is

$$J_{\text{aluminum}} = \frac{\pi}{2}(r_o^4 - r_i^4)$$
$$= \left(\frac{\pi}{2}\right)\left((1.5 \text{ in})^4 - (1.0 \text{ in})^4\right)$$
$$= 6.38 \text{ in}^4$$

The equivalent polar moment of inertia of the brass is

$$J_{\text{brass}} = n\left(\frac{\pi}{2}\right)(r_o^4 - r_i^4)$$
$$= (1.56)\left(\frac{\pi}{2}\right)\left((1.0 \text{ in})^4 - (0.5 \text{ in})^4\right)$$
$$= 2.30 \text{ in}^4$$

The total equivalent polar moment of inertia is

$$J_{\text{total}} = J_{\text{aluminum}} + J_{\text{brass}}$$
$$= 6.38 \text{ in}^4 + 2.30 \text{ in}^4 = 8.68 \text{ in}^4$$

(a) The maximum torsional shear stress in the aluminum occurs at the outer edge. Use Eq. 45.44.

$$\tau = \frac{Tr}{J} = \frac{(24{,}000 \text{ in-lbf})(1.5 \text{ in})}{8.68 \text{ in}^4}$$
$$= 4147 \text{ lbf/in}^2$$

(b) Using the composite structures analysis methodology, the maximum torsional shear stress in the brass is

$$\tau = \frac{nTr}{J} = \frac{(1.56)(24{,}000 \text{ in-lbf})(1.0 \text{ in})}{8.68 \text{ in}^4}$$
$$= 4313 \text{ lbf/in}^2$$

15. TORSION IN THIN-WALLED, NONCIRCULAR SHELLS

Shear stress due to torsion in a thin-walled, noncircular shell (also known as a *closed box*) acts around the perimeter of the shell, as shown in Fig. 45.11. The shear stress, τ, is given by Eq. 45.54. A is the area enclosed by the centerline of the shell.

$$\tau = \frac{T}{2At} \qquad \textit{45.54}$$

Figure 45.11 *Torsion in Thin-Walled Shells*

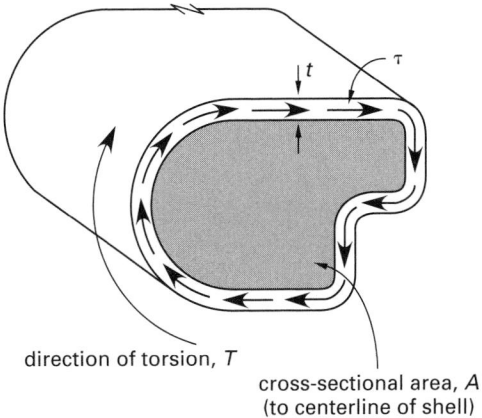

direction of torsion, T

cross-sectional area, A
(to centerline of shell)

The shear stress at any point is not proportional to the distance from the centroid of the cross section. Rather, the *shear flow*, q, around the shell is constant, regardless of whether the wall thickness is constant or variable.[22] The shear flow is the shear per-unit length of the centerline path.[23] At any point where the shell thickness is t,

$$q = \tau t = \frac{T}{2A} \quad \text{[constant]} \qquad 45.55$$

When the wall thickness, t, is constant, the angular twist depends on the perimeter, p, of the shell as measured along the centerline of the shell wall.

$$\gamma = \frac{TLp}{4A^2 tG} \qquad 45.56$$

16. TORSION IN SOLID, NONCIRCULAR MEMBERS

When a noncircular solid member is placed in torsion, the shear stress is not proportional to the distance from the centroid of the cross section. The maximum shear usually occurs close to the point on the surface that is nearest the centroid.

Shear stress, τ, and angular deflection, γ, due to torsion are functions of the cross-sectional shape. They cannot be specified by simple formulas that apply to all sections. Table 45.6 lists the governing equations for several basic cross sections. These formulas have been derived by dividing the member into several concentric thin-walled closed shells and summing the torsional strength, T, provided by each shell.

---

[22]The concept of shear flow can also be applied to a regular beam in bending, although there is little to be gained by doing so. Removing the dimension b in the general beam shear stress equation, $q = VQ/I$.
[23]Shear flow is not analogous to magnetic flux or other similar quantities because the shear flow path does not need to be complete (i.e., does not need to return to its starting point).

Table 45.6 *Torsion in Solid, Noncircular Shapes*

cross section	K in formula $\gamma = TL/KG$	τ (max)
ellipse (2a wide, 2b tall)	$\dfrac{\pi a^3 b^3}{a^2 + b^2}$	$\dfrac{2T}{\pi ab^2}$ (maximum at ends of minor axis)
square (side a)	$0.1406a^4$	$\dfrac{T}{0.208a^3}$ (maximum at midpoint of each side)
rectangle (2a wide, 2b tall)	*	$\dfrac{T(3a + 1.8b)}{8a^2 b^2}$ (maximum at midpoint of each longer side)
equilateral triangle (side a)	$\dfrac{a^4 \sqrt{3}}{80}$	$\dfrac{20T}{a^3}$ (maximum at midpoint of each side)
slotted tube (radius r, thickness t)	$\dfrac{2\pi r t^3}{3}$	$\dfrac{T(6\pi r + 1.8t)}{4\pi^2 r^2 t^2}$ (maximum along both edges remote from ends)
I-beam	$\dfrac{2bt_f^3 + ht_w^3}{3}$	$\dfrac{3Tt_f}{2bt_f^3 + ht_w^3}$ $[t_w < t_f]$

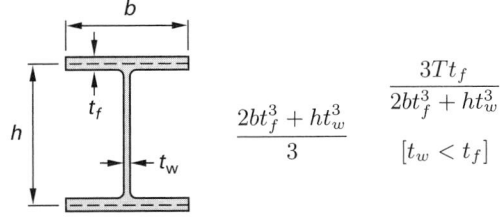

$$*ab^3 \left(\frac{16}{3} - \left(\frac{3.36b}{a} \right) \left(1 - \frac{b^4}{12a^4} \right) \right)$$

17. SHEAR CENTER FOR BEAMS

A beam with a symmetrical cross section supporting a transverse force that is offset from the longitudinal centroidal axis will be acted upon by a torsional moment, and the beam will tend to "roll" about a longitudinal axis known as the *bending axis* or *torsional axis*. For solid, symmetrical cross sections, this bending axis passes through the centroid of the cross section. However, for an asymmetrical beam (e.g., a channel beam on its side), the bending axis passes through the *shear center* (*torsional center* or *center of twist*), not the centroid. The shear center is a point that does not experience rotation (i.e., is a point about which all other points rotate) when the beam is in torsion.

Figure 45.12 *Channel Beam in Pure Bending*
(shear resultant, V, directed through shear center, O)

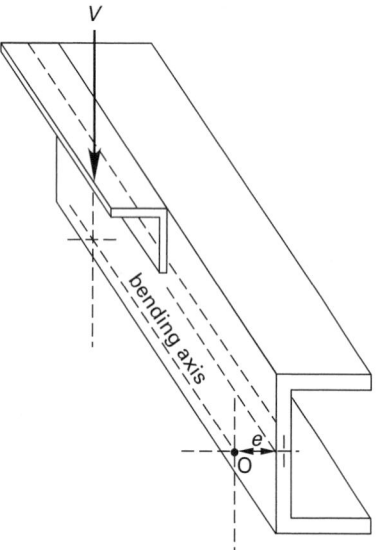

For beams with transverse loading, simple bending without torsion can only occur if the transverse load (*shear resultant* or *shear force of action*) is directed through the shear center. Otherwise, a torsional moment calculated as the product of the shear resultant and the torsional eccentricity will cause the beam to twist. The *torsional eccentricity* is the distance between the line of action of the shear resultant and the shear center.

The location of the shear center for any particular beam geometry is determined by setting the torsional moment equal to the shear resisting moment, as calculated from the total shear flow and the appropriate moment arm. In practice, however, shear centers for common shapes are located in tables similar to Fig. 45.13.

18. ECCENTRICALLY LOADED BOLTED CONNECTIONS

An eccentrically loaded connection is illustrated in Fig. 45.14. The bracket's natural tendency is to rotate about the centroid of the connector group. The shear

stress in the connectors includes both the direct vertical shear and the torsional shear stress. The sum of these shear stresses is limited by the shear strength of the critical connector, which in turn determines the capacity of the connection as limited by bolt shear strength.[24]

Figure 45.13 *Shear Centers of Selected Thin-Walled Open Sections*[a]

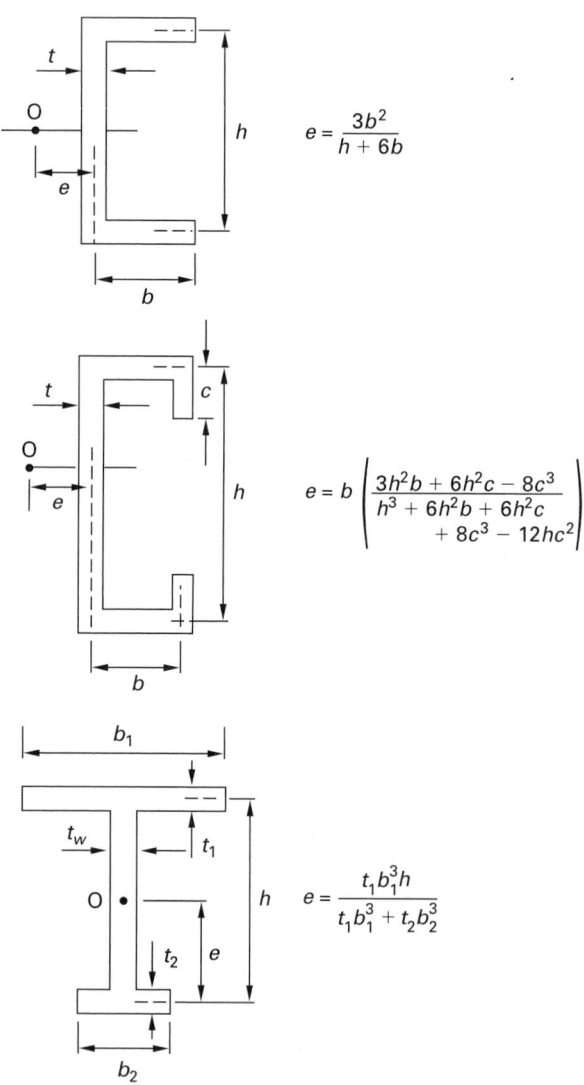

[a] Distance e from shear center, O, measured from the reference point shown. Distance e is not the torsional eccentricity.

---

[24] This type of analysis is known as an *elastic analysis* of the connection. Although it is traditional, it tends to greatly understate the capacity of the connection.

Figure 45.14 *Eccentrically Loaded Connection*

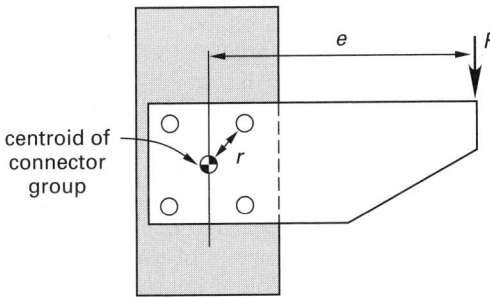

Analysis of an eccentric connection is similar to the analysis of a shaft under torsion. The shaft torque, T, is analogous to the moment, Fe, on the connection. The shaft's radius corresponds to the distance from the centroid of the fastener group to the *critical fastener*. The critical fastener is the one for which the vector sum of the vertical and torsional shear stresses is the greatest.

$$\tau = \frac{Tr}{J} = \frac{Fer}{J} \qquad \qquad 45.57$$

The polar moment of inertia, J, is calculated from the parallel axis theorem. Since bolts and rivets have little resistance to twisting in their holes, their individual polar moments of inertia are omitted.[25] Only r^2A terms in the parallel axis theorem are used. r_i is the distance from the fastener group centroid to the centroid (i.e., center) of the ith fastener, which has an area of A_i.

$$J = \sum_i r_i^2 A_i \qquad \qquad 45.58$$

The torsional shear stress is directed perpendicularly to a line between each fastener and the connector group centroid. The direction of the shear stress is the same as the rotation of the connection.

Figure 45.15 *Direction of Torsional Shear Stress*

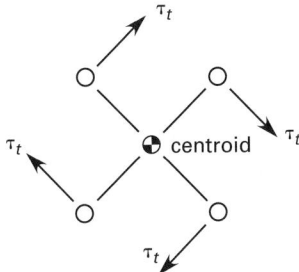

Once the torsional shear stress has been determined in the critical fastener, it is added in a vector sum to the direct vertical shear stress. The direction of the vertical shear stress is the same as that of the applied force.

$$\tau_v = \frac{F}{nA} \qquad \qquad 45.59$$

---

[25] In spot-welded and welded stud connections, the torsional resistance of each connector can be considered.

Typical connections gain great strength from the frictional slip resistance between the two surfaces. By preloading the connection bolts, the normal force between the plates is greatly increased. The connection strength from friction will rival or exceed the strength from bolt shear in connections designed to take advantage of preload.

Example 45.6

All fasteners used in the bracket shown have a nominal $1/2$ in diameter. What is the stress in the most critical fastener?

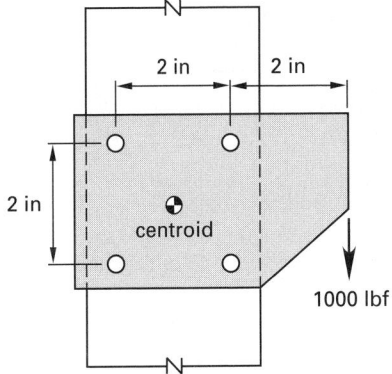

Solution

Since the fastener group is symmetrical, the centroid is centered within the four fasteners. This makes the eccentricity of the load equal to 3 in. Each fastener is located a distance r from the centroid, where

$$r = \sqrt{x^2 + y^2} = \sqrt{(1 \text{ in})^2 + (1 \text{ in})^2} = 1.414 \text{ in}$$

The area of each fastener is

$$A_i = \frac{\pi}{4}d^2 = \left(\frac{\pi}{4}\right)(0.5 \text{ in})^2 = 0.1963 \text{ in}^2$$

Using the parallel axis theorem for polar moments of inertia and disregarding the individual torsional resistances of the fasteners,

$$J = \sum_i r_i^2 A_i = (4)\big((1.414 \text{ in})^2(0.1963 \text{ in}^2)\big)$$
$$= 1.570 \text{ in}^4$$

The torsional shear stress on each fastener is

$$\tau_t = \frac{Fer}{J} = \frac{(1000 \text{ lbf})(3 \text{ in})(1.414 \text{ in})}{1.570 \text{ in}^4} = 2702 \text{ lbf/in}^2$$

Each torsional shear stress can be resolved into a horizontal shear stress, τ_{tx}, and a vertical shear stress, τ_{ty}. Both of these components are equal to

$$\tau_{tx} = \tau_{ty} = \frac{\sqrt{2}\left(2702 \frac{\text{lbf}}{\text{in}^2}\right)}{2} = 1911 \text{ lbf/in}^2$$

The direct vertical shear downward is

$$\tau_v = \frac{F}{nA} = \frac{1000 \text{ lbf}}{(4)(0.1963 \text{ in}^2)}$$
$$= 1274 \text{ lbf/in}^2$$

The two right fasteners have vertical downward components of torsional shear stress. The direct vertical shear is also downward. These downward components add, making both right fasteners critical.

The total stress in each of these fasteners is

$$\tau = \sqrt{\tau_{tx}^2 + (\tau_{ty} + \tau_v)^2}$$
$$= \sqrt{\left(1911 \frac{\text{lbf}}{\text{in}^2}\right)^2 + \left(1911 \frac{\text{lbf}}{\text{in}^2} + 1274 \frac{\text{lbf}}{\text{in}^2}\right)^2}$$
$$= 3714 \text{ lbf/in}^2$$

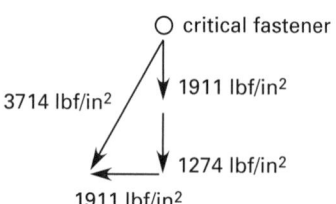

19. ECCENTRICALLY LOADED WELDED CONNECTIONS

The traditional elastic analysis of an eccentrically loaded welded connection is virtually the same as for a bolted connection, with the additional complication of having to determine the polar moment of inertia of the welds.[26] This can be done either by taking the welds as lines or by assuming each weld has an arbitrary thickness, t. After finding the centroid of the weld group, the rectangular moments of inertia of the individual welds are taken about that centroid using the parallel axis theorem. These rectangular moments of inertia are added to determine the polar moment of inertia. This laborious process can be shortened by use of App. 45.A.

The torsional shear stress, calculated from Mr/J (where r is the distance from the centroid of the weld group to the most distant weld point), is added vectorially to the direct shear to determine the maximum shear stress at the critical weld point.

20. FLAT PLATES

Flat plates under uniform pressure are separated into two edge-support conditions: simply supported and

---

[26]Steel building design does not use an elastic analysis to design eccentric brackets, either bolted or welded. The design methodology is highly proceduralized and codified.

Table 45.7 *Flat Plates Under Uniform Pressure*

shape	edge condition	maximum stress	deflection at center
circular	simply supported	$\dfrac{\frac{3}{8}pr^2(3+\nu)}{t^2}$ (at center)	$\dfrac{\frac{3}{16}pr^4(1-\nu)(5+\nu)}{Et^3}$
	built-in	$\dfrac{\frac{3}{4}pr^2}{t^2}$ (at edge)	$\dfrac{\frac{3}{16}pr^4(1-\nu^2)}{Et^3}$
rectangular	simply supported	$\dfrac{C_1pb^2}{t^2}$ (at center)	$\dfrac{C_2pb^4}{Et^3}$
	built-in	$\dfrac{C_3pb^2}{t^2}$ (at centers of long edges)	$\dfrac{C_4pb^4}{Et^3}$

$\frac{a}{b}$	1.0	1.2	1.4	1.6	1.8	2	3	4	5	∞
C_1	0.287	0.376	0.453	0.517	0.569	0.610	0.713	0.741	0.748	0.750
C_2	0.044	0.062	0.077	0.091	0.102	0.111	0.134	0.140	0.142	0.142
C_3	0.308	0.383	0.436	0.487	0.497	0.500	0.500	0.500	0.500	0.500
C_4	0.0138	0.0188	0.023	0.025	0.027	0.028	0.028	0.028	0.028	0.028

built-in edges.[27] Commonly accepted working equations are summarized in Table 45.7. It is assumed that (a) the plates are of "medium" thickness (meaning that the thickness is equal to or less than one-fourth of the minimum dimension of the plate), (b) the pressure is no more than will produce a maximum deflection less than or equal to one-half of the thickness, (c) the plates are constructed of isotropic, elastic material, and (d) the stress does not exceed the yield strength.

Example 45.7

A steel pipe with an inside diameter of 10 in (254 mm) is capped by welding round mild steel plates on its ends. The allowable stress is 11,100 lbf/in² (77 MPa). The internal gage pressure in the pipe is maintained at 500 lbf/in² (3.5 MPa). What plate thickness is required?

SI Solution

A fixed edge approximates the welded edges of the plate. From Table 45.7, the maximum bending stress is

$$\sigma_{\max} = \frac{3pr^2}{4t^2}$$

$$t = \sqrt{\frac{3pr^2}{4\sigma_{\max}}}$$

$$= \sqrt{\frac{(3)(3.5 \text{ MPa})\left(\dfrac{254 \text{ mm}}{2}\right)^2}{(4)(77 \text{ MPa})}}$$

$$= 23.4 \text{ mm}$$

Customary U.S. Solution

A fixed edge approximates the welded edges of the plate. From Table 45.7, the maximum bending stress is

$$\sigma_{\max} = \frac{3pr^2}{4t^2}$$

$$t = \sqrt{\frac{3pr^2}{4\sigma_{\max}}}$$

$$= \sqrt{\frac{(3)\left(500 \dfrac{\text{lbf}}{\text{in}^2}\right)\left(\dfrac{10 \text{ in}}{2}\right)^2}{(4)\left(11,100 \dfrac{\text{lbf}}{\text{in}^2}\right)}}$$

$$= 0.919 \text{ in}$$

21. SPRINGS

An *ideal spring* is assumed to be perfectly elastic within its working range. The deflection is assumed to follow

Hooke's law.[28] The *spring constant*, k, is also known as the *stiffness, spring rate, scale,* and *k-value*.[29]

$$F = k\delta \qquad\qquad 45.60$$

$$k = \frac{F_1 - F_2}{\delta_1 - \delta_2} \qquad\qquad 45.61$$

A spring stores energy when it is compressed or extended. By the *work-energy principle*, the energy storage is equal to the work required to displace the spring. The potential energy of a spring whose ends have been displaced a total distance δ is

$$\Delta E_p = \tfrac{1}{2}k\delta^2 \qquad\qquad 45.62$$

If a mass, m, is dropped from height h onto a spring, the compression, δ, can be found by equating the change in potential energy to the energy storage.

$$mg(h + \delta) = \tfrac{1}{2}k\delta^2 \qquad \text{[SI]} \quad 45.63(a)$$

$$m\left(\frac{g}{g_c}\right)(h + \delta) = \tfrac{1}{2}k\delta^2 \qquad \text{[U.S.]} \; 45.63(b)$$

Within the elastic region, this energy can be recovered by restoring the spring to its original unstressed condition. It is assumed that there is no permanent set, and no energy is lost through external friction or *hysteresis* (internal friction) when the spring returns to its original length.[30]

The entire applied load is felt by each spring in a series of springs linked end-to-end. The *equivalent (composite) spring constant* for springs in series is

$$\frac{1}{k_{\text{eq}}} = \frac{1}{k_1} + \frac{1}{k_2} + \frac{1}{k_3} + \cdots \qquad 45.64$$

Springs in parallel (e.g., concentric springs) share the applied load. The equivalent spring constant for springs in parallel is

$$k_{\text{eq}} = k_1 + k_2 + k_3 + \cdots \qquad 45.65$$

22. WIRE ROPE

Wire rope is constructed by first winding individual *wires* into *strands* and then winding the strands into rope. Wire rope is specified by its diameter and numbers of strands and wires. The most common *hoisting cable* is 6×19, consisting of six strands of 19 wires each,

---

[27]Fixed-edge conditions are theoretical and are seldom achieved in practice. Considering this fact and other simplifying assumptions that are made to justify the use of Table 45.7, a value of $\nu = 0.3$ can be used without loss of generality.

[28]A spring can be perfectly elastic even though it does not follow Hooke's law. The deviation from proportionality, if any, occurs at very high loads. The difference in theoretical and actual spring forces is known as the *straight-line error*.

[29]Another unfortunate name for the spring constant, k, that is occasionally encountered is the *spring index*. This is not the same as the spring index, C, used in helical coil spring design. The units will determine which meaning is intended.

[30]There is essentially no hysteresis in properly formed compression, extension, or open-wound helical torsion springs.

wound around a core. This configuration is sometimes referred to as "standard wire rope." Other common configurations are 6×7 (stiff *transmission* or *haulage rope*), 8×19 (extra-flexible hoisting rope), and the abrasion-resistant 6×37. The diameter and area of a wire rope are based on the circle that just encloses the rope.

Figure 45.16 *Wire Rope Cross Sections*

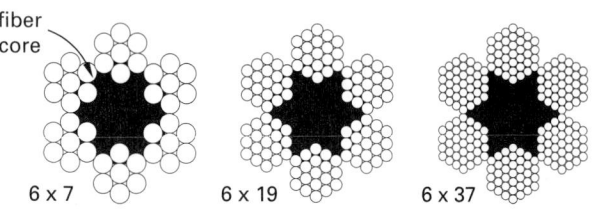

Wire rope can be obtained in a variety of materials and cross sections. In the past, wire ropes were available in iron, cast steel, traction steel (TS), mild plow steel (MPS), and plow steel (PS) grades. Modern wire ropes are generally available only in improved plow steel (IPS) and extra-improved plow steel (EIP) grades.[31] *Monitor* and *blue center steels* are essentially the same as improved plow steel.

Table 45.8 *Minimum Strengths of Wire Materials*

material	ultimate strength, $S_{ut,w}$ (ksi)
iron	65
cast steel	140
extra-strong cast steel	160
plow steel	175–210
improved plow steel	200–240
extra-improved plow steel	240–280

(Multiply ksi by 6.8947 to obtain MPa.)

In Table 45.9, the ultimate strength, $S_{ut,r}$, is the ultimate tensile load that the rope can carry without breaking. This is different from the ultimate strength, $S_{ut,w}$, of each wire given in Table 45.8. The rope's tensile strength will be only 80 to 95% of the combined tensile strengths of the individual wires. The modulus of elasticity for steel ropes is more a function of how the rope is constructed than the type of steel used. (See Table 45.10.)

While manufacturer's data should be relied on whenever possible, general properties of 6×19 wire rope are given in Table 45.9.

Table 45.9 *Properties of 6 × 19 Steel Wire Rope (improved plow steel, fiber core)*

diameter		mass		tensile strength[a,b], $S_{ut,r}$	
in	(mm)	lbm/ft	(kg/m)	tons[c]	(tonnes)
$1/4$	(6.4)	0.11	(0.16)	2.74	(2.49)
$3/8$	(9.5)	0.24	(0.35)	6.10	(5.53)
$1/2$	(13)	0.42	(0.63)	10.7	(9.71)
$5/8$	(16)	0.66	(0.98)	16.7	(15.1)
$7/8$	(22)	1.29	(1.92)	32.2	(29.2)
$1\,1/8$	(29)	2.13	(3.17)	52.6	(47.7)
$1\,3/8$	(35)	3.18	(4.73)	77.7	(70.5)
$1\,5/8$	(42)	4.44	(6.61)	107	(97.1)
$1\,7/8$	(48)	5.91	(8.80)	141	(128)
$2\,1/8$	(54)	7.59	(11.3)	179	(162)
$2\,3/8$	(60)	9.48	(14.1)	222	(201)
$2\,5/8$	(67)	11.6	(17.3)	268	(243)

[a]Add $7\,1/2$% for wire ropes with steel cores.
[b]Deduct 10% for galvanized wire ropes.
[c]tons of 2000 pounds

The central core can be of natural (e.g., hemp) or synthetic fibers or, for higher-temperature use, steel strands or cable. Core designations are FC for *fiber core*, IWRC for *independent wire rope core*, and WSC for *wire-strand core*. Wire rope is protected against corrosion by lubrication carried in the saturated fiber core. Steel-cored ropes are approximately 7.5% stronger than fiber-cored ropes.

Structural rope, structural strand, and *aircraft cabling* are similar in design to wire rope but are intended for permanent installation in bridges and aircraft, respectively. Structural rope and strand are galvanized to prevent corrosion, while aircraft cable is usually manufactured from corrosion-resistant steel.[32,33] Galvanized ropes should not be used for hoisting, as the galvanized coating will be worn off. Structural rope and strand have a nominal tensile strength of 220 ksi (1.5 GPa) and a modulus of elasticity of approximately 20,000 ksi (140 GPa) for diameters between $3/8$ and 4 in (0.95 and 10.2 cm).

The most common winding is *regular lay* in which the wires are wound in one direction and the strands are wound around the core in the opposite direction. Regular lay ropes do not readily kink or unwind. Wires and strands in *lang lay* ropes are wound in the same direction, resulting in a wear-resistant rope that is more prone to unwinding. Lang lay ropes should not be used to support loads that are held in free suspension.

In addition to considering the primary tensile dead load, the significant effects of bending and sheave-bearing pressure must be considered when selecting wire rope. Self-weight may also be a factor for long cables. Appropriate dynamic factors should be applied to allow for

---

[31]The term "plow steel" is somewhat traditional, as hard-drawn AISI 1070 or AISI 1080 might actually be used.

[32]Manufacture of structural rope and strand in the United States are in accordance with ASTM A603 and ASTM A586, respectively.
[33]Galvanizing usually reduces the strength of wire rope by approximately 10%.

acceleration, deceleration, and impacts, or the acceleration forces. In general for hoisting and hauling, the working load should not exceed 20% of the breaking strength (i.e., a minimum factor of safety of 5 should be used).[34]

If d_w is the nominal wire diameter in inches,[35] d_r is the nominal wire rope diameter in inches, and d_{sh} is the sheave diameter in inches, the stress from bending around a drum or sheave is given by Eq. 45.66. E_w is the modulus of elasticity of the wire material (approximately 3×10^7 psi (207 GPa for steel), not the rope's modulus of elasticity, though the latter is widely used in this calculation.[36]

$$\sigma_{\text{bending}} = \frac{d_w E_w}{d_{\text{sh}}} \qquad \textit{45.66}$$

To reduce stress and eliminate permanent set in wire ropes, the diameter of the sheave should be kept as large as is practical, ideally 45 to 90 times the rope diameter. Alternatively, the minimum diameter of the sheave or drum may be stated as 400 times the diameter of the individual outer wires in the rope.[37] Table 45.10 lists minimum diameters for specific rope types.

Table 45.10 *Typical Characteristics of Steel Wire Rope*

configuration	mass (lbm/ft)	area (in²)	minimum sheave diameter	modulus of elasticity (psi)
6×7	$1.50d_r^2$	$0.380d_r^2$	42–$72d_r$	14×10^6
6×19	$1.60d_r^2$	$0.404d_r^2$	30–$45d_r$	12×10^6
6×37	$1.55d_r^2$	$0.404d_r^2$	18–$27d_r$	11×10^6
8×19	$1.45d_r^2$	$0.352d_r^2$	21–$31d_r$	10×10^6

(Multiply lbm/ft by 1.49 to obtain kg/m.)
(Multiply in² by 6.45 to obtain cm².)
(Multiply psi by 6.9×10^{-6} to obtain GPa.)

To prevent wear and fatigue of the sheave or drum, the radial bearing pressure should be kept as low as possible. Actual maximum bearing pressures are highly dependent on the sheave material, type of rope, and application. For 6×19 wire ropes, the acceptable bearing pressure can be as low as 500 psi (3.5 MPa) for cast-iron sheaves and as high as 2500 psi (17 MPa) for alloy steel sheaves. The approximate bearing pressure of the wire rope on the sheave or drum depends on the tensile force in the rope and is given by Eq. 45.67.

$$p_{\text{bearing}} = \frac{2F_t}{d_r d_{\text{sh}}} \qquad \textit{45.67}$$

Fatigue failure in wire rope can be avoided by keeping the ratio $p_{\text{bearing}}/S_{ut,w}$ below approximately 0.014 for 6×19 wire rope.[38] ($S_{ut,w}$ is the ultimate tensile strength of the wire material, not of the rope.)

---

[34] Factors of safety are much higher and may be as high as 8 to 12 for elevators and hoists carrying passengers.

[35] For 6×19 standard wire rope, the outer wires are typically 1/13 to 1/16 of the rope diameter. For 6×7 haulage rope, the ratio is approximately 1/9.

[36] Although E in Eq. 45.66 is often referred to as "the modulus of elasticity of the wire rope," it is understood that E is actually the modulus of elasticity of the wire rope material.

[37] For elevators and mine hoists, the sheave-to-wire diameter ratio may be as high as 1000.

[38] A maximum ratio of 0.001 is often quoted for wire rope regardless of configuration.

46

Structural Analysis I

Nomenclature

A	area	ft^2	m^2
d	distance	ft	m
E	modulus of elasticity	lbf/ft^2	Pa
F	force	lbf	N
I	area moment of inertia	ft^4	m^4
L	length	ft	m
M	moment	ft-lbf	N·m
R	reaction	lbf	N
S	force	lbf	N
T	temperature	°F	°C
u	force	lbf	N
w	distributed load	lbf/ft	N/m
y'	slope	ft/ft	m/m

Symbols

α	coefficient of thermal expansion	1/°F	1/°C
δ	deformation	ft	m
θ	angle	deg	rad

Subscripts

c	concrete
o	original
st	steel

1. INTRODUCTION TO INDETERMINATE STATICS

A structure that is *statically indeterminate* is one for which the equations of statics are not sufficient to determine all reactions, moments, and internal forces. Additional formulas involving deflection are required to completely determine these variables.

Although there are many configurations of statically indeterminate structures, this chapter is primarily concerned with beams on more than two supports, trusses with more members than are required for rigidity, and miscellaneous composite structures.

2. DEGREE OF INDETERMINACY

The *degree of indeterminacy* (*degree of redundancy*) is equal to the number of reactions or members that would have to be removed in order to make the structure statically determinate. For example, a two-span beam on three simple supports is indeterminate (redundant) to the first degree. The degree of indeterminacy of a pin-connected truss is given by Eq. 46.1.

$$\begin{matrix} \text{degree of} \\ \text{indeterminacy} \end{matrix} = 3 + \begin{matrix} \text{no. of} \\ \text{members} \end{matrix} - \left(2 \times \begin{matrix} \text{no. of} \\ \text{joints} \end{matrix} \right) \quad 46.1$$

3. INDETERMINATE BEAMS

Three common configurations of beams can easily be recognized as being statically indeterminate. These are the *continuous beam*, *propped cantilever beam*, and *fixed-end beam* illustrated in Fig. 46.1.

Figure 46.1 *Types of Indeterminate Beams*

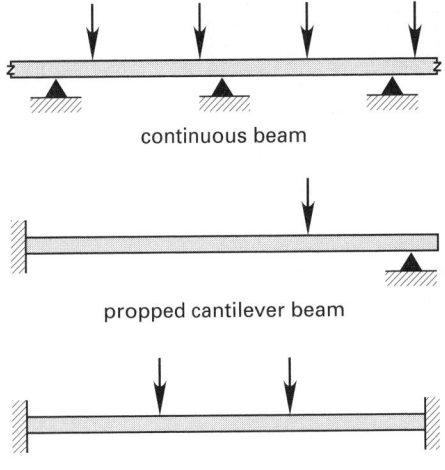

continuous beam

propped cantilever beam

fixed-end beam

4. REVIEW OF ELASTIC DEFORMATION[1]

When an axial force, F, acts on an object with length L, cross-sectional area A, and modulus of elasticity E, the deformation[2] is

$$\delta = \frac{FL}{AE} \quad 46.2$$

---

[1]This subject is covered in greater detail in Ch. 44.
[2]The terms *deformation* and *elongation* are often used interchangeably in this context.

When an object with initial length L_o and coefficient of thermal expansion α experiences a temperature change of ΔT degrees, the deformation is

$$\delta = \alpha L_o \Delta T \qquad 46.3$$

5. CONSISTENT DEFORMATION METHOD

The *consistent deformation method*, also known as the *compatibility method*, is one of the methods of solving indeterminate problems. This method is simple to learn and to apply. First, geometry is used to develop a relationship between the deflections of two different members (or for one member at two locations) in the structure. Then, the deflection equations for the two different members at a common point are written and equated, since the deformations must be the same at a common point. This method is illustrated by the following examples.

Example 46.1

A pile is constructed of concrete with a steel jacket. What are the forces in the steel and concrete if a load F is applied? Assume the end caps are rigid and the steel-concrete bond is perfect.

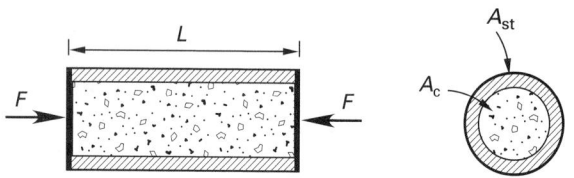

Solution

Let F_c and F_{st} be the loads carried by the concrete and steel, respectively. Then,

$$F_c + F_{st} = F$$

The deformation of the steel is given by Eq. 46.2.

$$\delta_{st} = \frac{F_{st} L}{A_{st} E_{st}}$$

Similarly, the deflection of the concrete is

$$\delta_c = \frac{F_c L}{A_c E_c}$$

But, $\delta_c = \delta_{st}$ since the bonding is perfect. Therefore,

$$\frac{F_c L}{A_c E_c} - \frac{F_{st} L}{A_{st} E_{st}} = 0$$

The first and last equations are solved simultaneously for F_c and F_{st}.

$$F_c = \frac{F}{1 + \dfrac{A_{st} E_{st}}{A_c E_c}}$$

$$F_{st} = \frac{F}{1 + \dfrac{A_c E_c}{A_{st} E_{st}}}$$

Example 46.2

A uniform bar is clamped at both ends and the axial load applied near one of the supports. What are the reactions?

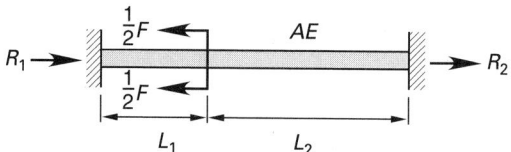

Solution

The first required equation is

$$R_1 + R_2 = F$$

The shortening of section 1 due to the reaction R_1 is

$$\delta_1 = \frac{-R_1 L_1}{AE}$$

The elongation of section 2 due to the reaction R_2 is

$$\delta_2 = \frac{R_2 L_2}{AE}$$

However, the bar is continuous, so $\delta_1 = -\delta_2$. Therefore,

$$R_1 L_1 = R_2 L_2$$

The first and last equations are solved simultaneously to find R_1 and R_2.

$$R_1 = \frac{F}{1 + \dfrac{L_1}{L_2}}$$

$$R_2 = \frac{F}{1 + \dfrac{L_2}{L_1}}$$

Example 46.3

The non-uniform bar shown is clamped at both ends. What are the reactions if a temperature change of ΔT is experienced?

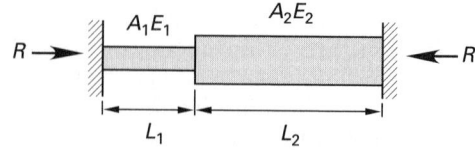

Solution

The thermal deformations of sections 1 and 2 can be calculated directly. Use Eq. 46.3.

$$\delta_1 = \alpha_1 L_1 \Delta T$$
$$\delta_2 = \alpha_2 L_2 \Delta T$$

The total deformation is $\delta = \delta_1 + \delta_2$. However, the deformation can also be calculated from the principles of mechanics.

$$\delta = \frac{RL_1}{A_1 E_1} + \frac{RL_2}{A_2 E_2}$$

These equations can be solved directly for R.

$$R = \frac{(\alpha_1 L_1 + \alpha_2 L_2)\Delta T}{\dfrac{L_1}{A_1 E_1} + \dfrac{L_2}{A_2 E_2}}$$

Example 46.4

The beam shown is supported by dissimilar members. What are the forces in the members? Assume the bar is rigid and remains horizontal.[3] The beam's mass is insignificant.

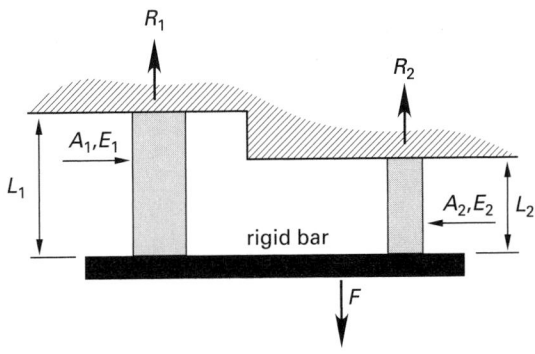

Solution

The required equilibrium condition is

$$R_1 + R_2 = F$$

The elongations of the two tension members are

$$\delta_1 = \frac{R_1 L_1}{A_1 E_1}$$
$$\delta_2 = \frac{R_2 L_2}{A_2 E_2}$$

Since the horizontal bar remains horizontal, $\delta_1 = \delta_2$.

$$\frac{R_1 L_1}{A_1 E_1} = \frac{R_2 L_2}{A_2 E_2}$$

---

[3]This example is easily solved by summing moments about a point on the horizontal beam.

The first and last equations are solved simultaneously to find R_1 and R_2.

$$R_1 = \frac{F}{1 + \dfrac{L_1 A_2 E_2}{L_2 A_1 E_1}}$$

$$R_2 = \frac{F}{1 + \dfrac{L_2 A_1 E_1}{L_1 A_2 E_2}}$$

Example 46.5

The beam shown is supported by dissimilar members. The bar is rigid but is not constrained to remain horizontal. The beam's mass is insignificant. Develop the simultaneous equations needed to determine the reactions in the vertical members.

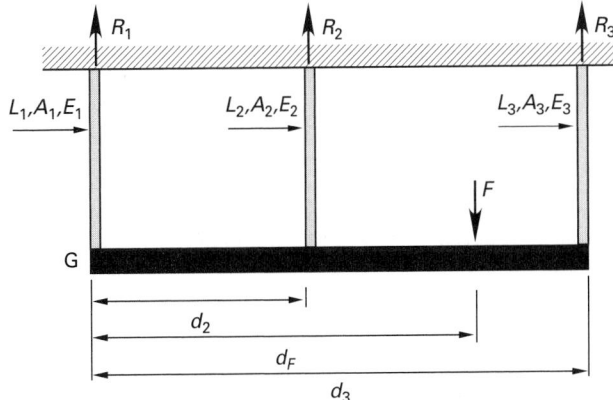

Solution

The forces in the supports are R_1, R_2, and R_3. Any of these may be tensile (positive) or compressive (negative).

$$R_1 + R_2 + R_3 = F$$

The changes in length are given by Eq. 46.2.

$$\delta_1 = \frac{R_1 L_1}{A_1 E_1}$$
$$\delta_2 = \frac{R_2 L_2}{A_2 E_2}$$
$$\delta_3 = \frac{R_3 L_3}{A_3 E_3}$$

Since the bar is rigid, the deflections will be proportional to the distance from point G.

$$\delta_2 = \delta_1 + \left(\frac{d_2}{d_3}\right)(\delta_3 - \delta_1)$$

Moments can be summed about point G to give a third equation.

$$M_G = R_3 d_3 + R_2 d_2 - F d_F = 0$$

Example 46.6

A load is supported by three tension members. Develop the simultaneous equations needed to find the forces in the three members.

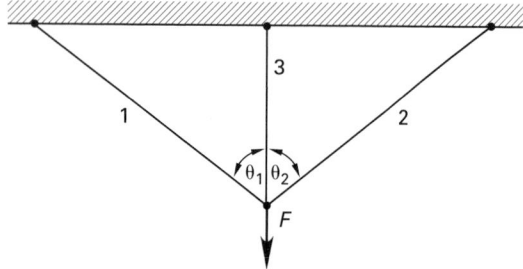

Solution

The equilibrium requirement is

$$F_{1y} + F_3 + F_{2y} = F$$
$$F_1 \cos \theta_1 + F_3 + F_2 \cos \theta_2 = F$$

Assuming the elongations are small compared to the member lengths, the angles θ_1 and θ_2 are unchanged. Then, the vertical deflections are the same for all three members.

$$\frac{F_1 L_1 \cos \theta_1}{A_1 E_1} = \frac{F_3 L_3}{A_3 E_3} = \frac{F_2 L_2 \cos \theta_2}{A_2 E_2}$$

These equations can be solved simultaneously to find F_1, F_2, and F_3. (It may be necessary to work with the x-components of the deflections in order to find a third equation.)

6. SUPERPOSITION METHOD

Two-span (three-support) beams and propped cantilevers are indeterminate to the first degree. Their reactions can be determined from a variation of the consistent deformation procedure known as the *superposition method*.[4] This method requires finding the deflection with one or more supports removed and then satisfying the known conditions.

step 1: Remove enough redundant supports to reduce the structure to a statically determinate condition.

step 2: Calculate the deflections at the previous locations of redundant supports. Use consistent sign conventions.

---

[4]Superposition can also be used with higher-order indeterminate problems. However, the simultaneous equations that must be solved may make superposition unattractive for manual calculations.

step 3: Apply each redundant support as a load, and find the deflections at the redundant support points as functions of the redundant support forces.

step 4: Use superposition to combine (i.e., add) the deflections due to the actual loads and the redundant support loads. The total deflections must agree with the known deflections (usually zero) at the redundant support points.

Example 46.7

A propped cantilever is loaded by a concentrated force at midspan. Determine the reaction, S, at the prop.

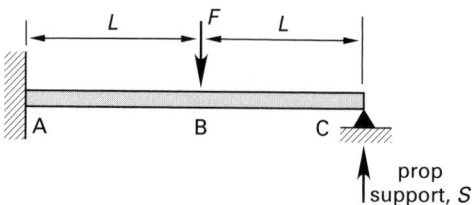

Solution

Start by removing the unknown prop reaction at point C. The cantilever beam is then statically determinate. The deflection and slope at point B can be found or derived from the beam equations. Use App. 44.A, Case 1.

$$\text{deflection:} \quad \delta_B = \frac{-FL^3}{3EI}$$

$$\text{slope:} \quad y'_B = \frac{-FL^2}{2EI}$$

The slope remains constant to the right of point B. Therefore, the deflection at point C due to the load at point B is

$$\delta_{C,F} = \delta_B + y'_B L$$
$$= \frac{-5FL^3}{6EI}$$

The upward deflection at the cantilever tip due to the prop support, S, alone is given by App. 44.A, Case 1.

$$\delta_{C,S} = \frac{S(2L)^3}{3EI} = \frac{8SL^3}{3EI}$$

Now, it is known that the actual deflection at point C is zero (the boundary condition). Therefore, the prop support, S, can be determined as a function of the applied load.

$$\delta_{C,S} + \delta_{C,F} = 0$$
$$\frac{8SL^3}{3EI} - \frac{5FL^3}{6EI} = 0$$
$$S = \frac{5F}{16}$$

7. THREE-MOMENT EQUATION

A *continuous beam* has two or more spans (i.e., three or more supports) and is statically indeterminate. The *three-moment equation* is a method of determining the reactions on continuous beams. It relates the moments at any three adjacent supports. The three-moment method can be used with a two-span beam to directly find all three reactions.

When a beam has more than two spans, the equation must be used with three adjacent supports at a time, starting with a support whose moment is known. (The moment is known to be zero at a simply supported end. For a cantilever end, the moment depends only on the loads on the cantilever portion.)

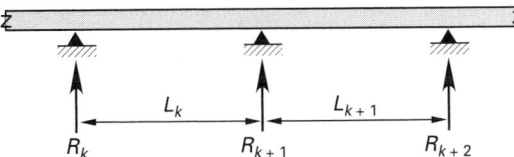

Figure 46.2 Portion of a Continuous Beam

In its most general form, the three-moment equation is applicable to beams with non-uniform cross sections. In Eq. 46.4, I_k is the moment of inertia of span k.

$$\frac{M_k L_k}{I_k} + (2M_{k+1})\left(\frac{L_k}{I_k} + \frac{L_{k+1}}{I_{k+1}}\right) + \frac{M_{k+2} L_{k+1}}{I_{k+1}}$$
$$= -6\left(\frac{A_k a}{I_k L_k} + \frac{A_{k+1} b}{I_{k+1} L_{k+1}}\right) \quad 46.4$$

Equation 46.4 uses the following special nomenclature.

a distance from the left support to the centroid of the moment diagram on the left span

b distance from the right support to the centroid of the moment diagram on the right span

I_k the moment of inertia of the open span between supports k and $k+1$

L_k length of the span between supports k and $k+1$

M_k bending moment at support k

A_k area of moment diagram between supports k and $k+1$, assuming that the span is simply and independently supported

The products Aa and Ab are known as *first moments of the areas*. It is convenient to derive simplified expressions for Aa and Ab for commonly encountered configurations. Several are presented in Fig. 46.3.

Figure 46.3 Simplified Three-Moment Equation Terms

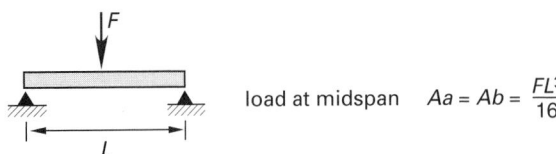

load at midspan $Aa = Ab = \dfrac{FL^3}{16}$

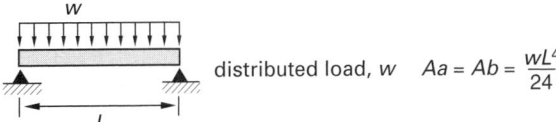

distributed load, w $Aa = Ab = \dfrac{wL^4}{24}$

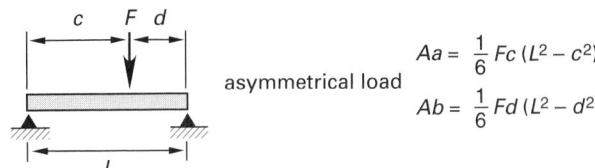

asymmetrical load

$$Aa = \frac{1}{6}Fc\,(L^2 - c^2)$$
$$Ab = \frac{1}{6}Fd\,(L^2 - d^2)$$

For beams with uniform cross sections, the moment of inertia terms can be eliminated.

$$M_k L_k + (2M_{k+1})(L_k + L_{k+1}) + M_{k+2} L_{k+1}$$
$$= -6\left(\frac{A_k a}{L_k} + \frac{A_{k+1} b}{L_{k+1}}\right) \quad 46.5$$

Example 46.8

Find the four reactions supporting the beam. EI is constant.

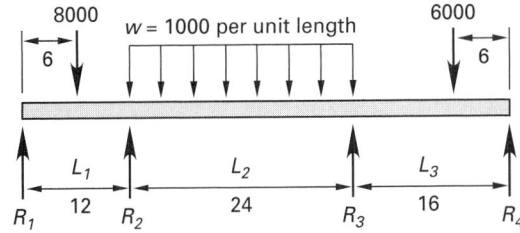

Solution

Spans 1 and 2:

Since the three-moment method can be applied to only two spans at a time, work first with the left and middle spans (spans 1 and 2).

From Fig. 46.3, the quantities $A_1 a$ and $A_2 b$ are

$$A_1 a = \frac{FL^3}{16} = \frac{(8000)(12)^3}{16} = 864{,}000$$

$$A_2 b = \frac{wL^4}{24} = \frac{(1000)(24)^4}{24} = 13{,}824{,}000$$

Since the left end of the beam is simply supported, M_1 is zero. Therefore, the three-moment equation (Eq. 46.5) becomes

$$(2M_2)(L_1 + L_2) + M_3 L_2 = -6\left(\frac{A_1 a}{L_1} + \frac{A_2 b}{L_2}\right)$$

$$(2M_2)(12 + 24) + M_3(24) = (-6)\left(\frac{864{,}000}{12} + \frac{13{,}824{,}000}{24}\right)$$

After simplification,

$$3M_2 + M_3 = -162{,}000$$

Spans 2 and 3:

From the previous calculations,

$$A_2 a = A_2 b = 13{,}824{,}000$$

From Fig. 46.3 for the third span,

$$\begin{aligned}A_3 b &= \tfrac{1}{6}Fd(L^2 - d^2)\\ &= \left(\tfrac{1}{6}\right)(6000)(6)\big((16)^2 - (6)^2\big)\\ &= 1{,}320{,}000\end{aligned}$$

Since the right end is simply supported, $M_4 = 0$ and the three-moment equation is

$$M_2 L_2 + (2M_3)(L_2 + L_3) = -6\left(\frac{A_2 a}{L_2} + \frac{A_3 b}{L_3}\right)$$

$$M_2(24) + (2M_3)(24 + 16) = (-6)\left(\frac{13{,}824{,}000}{24} + \frac{1{,}320{,}000}{16}\right)$$

After simplifying,

$$0.3M_2 + M_3 = -49{,}388$$

There are two equations in two unknowns (M_2 and M_3). A simultaneous solution yields

$$M_2 = -41{,}708$$
$$M_3 = -36{,}875$$

Finding reactions:

M_2 can be written in terms of the loads and reactions to the left of support 2. Assuming clockwise moments are positive,

$$M_2 = 12R_1 - (6)(8000) = -41{,}708$$
$$R_1 = 524.3$$

Now that R_1 is known, moments can be taken from support 3 to the left.

$$M_3 = (36)(524.3) + 24R_2 - (30)(8000) - (12)(24{,}000)$$
$$= -36{,}875$$
$$R_2 = 19{,}677.1$$

Similarly, R_4 and R_3 can be determined by working from the right end to the left. Assuming counterclockwise moments are positive,

$$M_3 = 16R_4 - (10)(6000) = -36{,}875$$
$$R_4 = 1445.3$$
$$M_2 = (40)(1445) + 24R_3 - (34)(6000) - (12)(24{,}000)$$
$$= -41{,}708$$
$$R_3 = 16{,}353.3$$

Check:

It is a good idea to check for equilibrium in the vertical direction.

$$\sum \text{loads} = 8000 + 24{,}000 + 6000$$
$$= 38{,}000$$
$$\sum \text{reactions} = 524.3 + 19{,}677.1 + 1445.3 + 16{,}353.3$$
$$= 38{,}000$$

8. FIXED-END MOMENTS

When the end of a beam is constrained against rotation, it is said to be a *fixed end* (also known as a *built-in end*). The ends of fixed-end beams are constrained to remain horizontal. Cantilever beams have a single fixed end. Some beams, as illustrated in Fig. 46.1, have two fixed ends and are known as *fixed-end beams*.[5]

Fixed-end beams are inherently indeterminate. To reduce the work required to find end moments and reactions, tables and books of fixed-end moments are often used. (See App. 47.A.)

9. INDETERMINATE TRUSSES

It is possible to manually calculate the forces in all members of an indeterminate truss. However, due to the time required, it is preferable to limit such manual calculations to trusses that are indeterminate to the first degree. The following *dummy unit load method* can be used to solve trusses with a single redundant member.

step 1: Draw the truss twice. Omit the redundant member on both trusses. (There may be a choice of redundant members.)

step 2: Load the first truss (which is now determinate) with the actual loads.

step 3: Calculate the forces, S, in all of the members. Assign a positive sign to tensile forces.

---

[5]The definition is loose. The term *fixed-end beam* can also be used to mean any indeterminate beam with at least one built-in end (e.g., a propped cantilever).

step 4: Load the second truss with two unit forces acting collinearly toward each other along the line of the redundant member.

step 5: Calculate the force, u, in each of the members.

step 6: Calculate the force in the redundant member from Eq. 46.6.

$$S_{\text{redundant}} = \frac{-\sum \left(\dfrac{SuL}{AE}\right)}{\sum \left(\dfrac{u^2L}{AE}\right)} \qquad 46.6$$

If AE is the same for all members,

$$S_{\text{redundant}} = \frac{-\sum SuL}{\sum u^2L} \qquad 46.7$$

The true force in member j of the truss is

$$F_{j,\text{true}} = S_j + S_{\text{redundant}}\,u_j \qquad 46.8$$

Example 46.9

Find the force in members BC and BD. $AE = 1$ for all members except for CB, which is 2, and AD, which is 1.5.

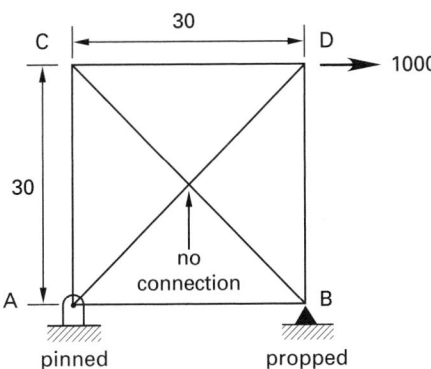

Solution

The two trusses are shown appropriately loaded.

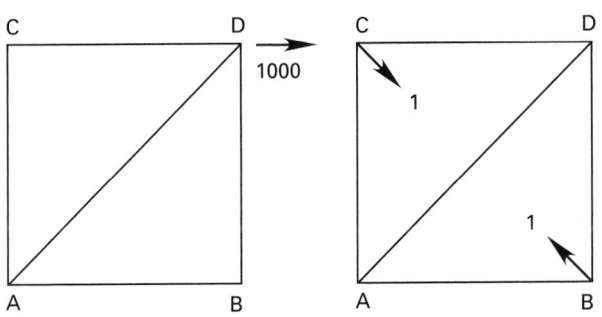

member	L	AE	S	u	$\dfrac{SuL}{AE}$	$\dfrac{u^2L}{AE}$
AB	30	1	0	−0.707 (C)	0	15
BD	30	1	−1000 (C)	−0.707	21,210	15
DC	30	1	0	−0.707	0	15
CA	30	1	0	−0.707	0	15
CB	42.43	2	0	1.0	0	21.22
AD	42.43	1.5	1414 (T)	1.0	39,997	28.29
					61,207	109.51

From Eq. 46.6,

$$S_{\text{BC}} = \frac{-61{,}207}{109.51}$$
$$= -558.9 \quad (\text{C})$$

From Eq. 46.8,

$$F_{\text{BD,true}} = -1000 + (-558.9)(-0.707)$$
$$= -604.9 \quad (\text{C})$$

10. INFLUENCE DIAGRAMS

Shear, moment, and reaction influence diagrams (influence lines) can be drawn for any point on a beam or truss. This is a necessary first step in the evaluation of stresses induced by moving loads. It is important to realize, however, that the influence diagram applies only to one point on the beam or truss.

Influence Diagrams for Beam Reactions

In a typical problem, the load is fixed in position and the reactions do not change. If a load is allowed to move across a beam, the reactions will vary. An influence diagram can be used to investigate the value of a chosen reaction as the load position varies.

To make the influence diagram as general in application as possible, the load is taken as 1 lbf. As an example, consider a 20 ft, simply supported beam and determine the effect on the left reaction of moving a 1 lbf load across the beam.

If the load is directly over the right reaction ($x = 0$), the left reaction will not carry any load. Therefore, the ordinate of the influence diagram is zero at that point. (Even though the right reaction supports 1 lbf, this influence diagram is being drawn for one point only—the left reaction.) Similarly, if the load is directly over the left reaction ($x = L$), the ordinate of the influence diagram will be 1. Basic statics can be used to complete the rest of the diagram, as shown in Fig. 46.4.

Figure 46.4 *Influence Diagram for Reaction of Simple Beam*

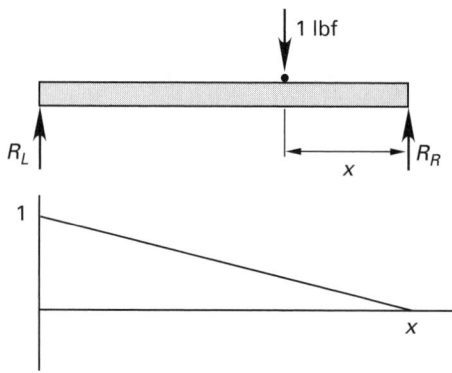

Use this rudimentary example of an influence diagram to calculate the left reaction for any placement of any load (not just 1 lbf loads) by multiplying the actual load by the ordinate of the influence diagram.

$$R_L = P \times \text{ordinate} \qquad 46.9$$

Even though the influence diagram was drawn for a point load, it can still be used when the beam carries a uniformly distributed load. In the case of a uniform load of w distributed over the beam from x_1 to x_2, the left reaction can be calculated from Eq. 46.10.

$$R_L = \int_{x_1}^{x_2} (w \times \text{ordinate})dx$$
$$= w \times \text{area under curve} \qquad 46.10$$

Example 46.10

A 500 lbf load is placed 15 ft from the right end of a 20 ft, simply supported beam. Use the influence diagram to determine the left reaction.

Solution

Since the influence line increases linearly from 0 to 1, the ordinate is the ratio of position to length. That is, the ordinate is $15/20 = 0.75$. The left reaction is

$$R_L = (0.75)(500 \text{ lbf}) = 375 \text{ lbf}$$

Example 46.11

A uniform load of 15 lbf/ft is distributed between $x = 4$ ft and $x = 10$ ft along a 20 ft, simply supported beam. What is the left reaction?

Solution

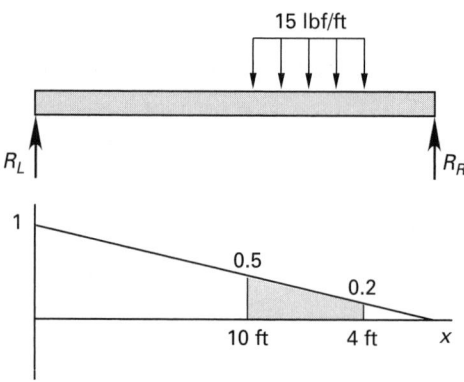

From Eq. 46.10, the left reaction can be calculated from the area under the influence diagram between the limits of loading.

$$\text{area} = \left(\tfrac{1}{2}\right)(10 \text{ ft})(0.5) - \left(\tfrac{1}{2}\right)(4 \text{ ft})(0.2) = 2.1 \text{ ft}$$

The left reaction is

$$R_L = \left(15 \ \frac{\text{lbf}}{\text{ft}}\right)(2.1 \text{ ft}) = 31.5 \text{ lbf}$$

Finding Reaction Influence Diagrams Graphically

Since the reaction will always have a value of 1 when the unit load is directly over the reaction and since the reaction is always directly proportional to the distance x, the reaction influence diagram can be easily determined from the following steps.

step 1: Remove the support being investigated.

step 2: Displace (lift) the beam upward a distance of one unit at the support point. The resulting beam shape will be the shape of the reaction influence diagram.

Example 46.12

What is the approximate shape of the reaction influence diagram for reaction 2?

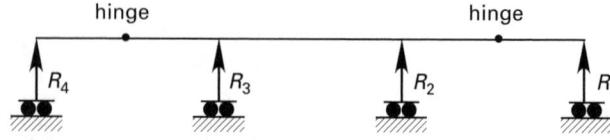

Solution

Pushing up at reaction 2 such that the deflection is one unit results in the shown shape.

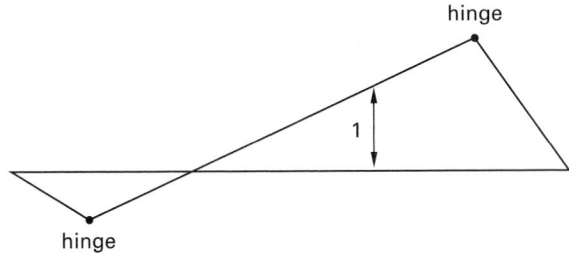

Influence Diagrams for Beam Shears

A shear influence diagram (not the same as a shear diagram) illustrates the effect on the shear at a particular point in the beam of moving a load along the beam's length. As an illustration, consider point A along the simply supported beam of length 20.

In all cases, principles of statics can be used to calculate the shear at point A as the sum of loads and reactions on the beam from point A to the left end. (With the appropriate sign convention, summation to the right end could be used as well.) If the unit load is placed between the right end ($x = 0$) and point A, the shear at point A will consist only of the left reaction, since there are no other loads between point A and the left end. From the reaction influence diagram, the left reaction varies linearly. At $x = 12$ ft, the location of point A, the shear is $V = R_L = 12/20 = 0.6$.

Figure 46.5 *Shear Influence Diagram for Simple Beam*

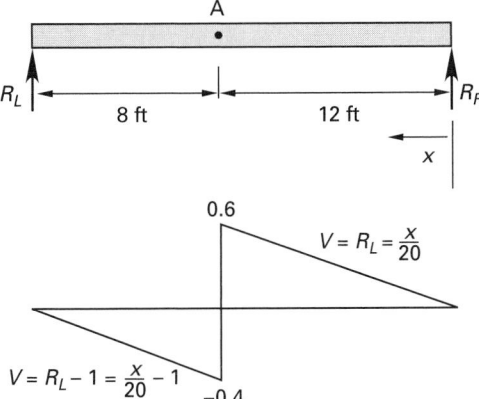

When the unit load is between point A and the left end, the shear at point A is the sum of the left reaction (upward and positive) and the unit load itself (downward and negative). Therefore, $V = R_L - 1$. At $x = 12$ ft, the shear is $V = 0.6$ lbf $- 1$ lbf $= -0.4$ lbf.

Figure 46.5 is the shear influence diagram. In the diagram, notice that the shear goes through a reversal of 1. It is also helpful to note that the slopes of the two inclined sections are the same.

Shear influence diagrams are used in the same manner as reaction influence diagrams. The shear at point A for any position of the load can be calculated by multiplying the ordinate of the diagram by the actual load. Distributed loads are found by multiplying the uniform load by the area under the diagram between the limits of loading. If the loading extends over positive and negative parts of the curve, the sign of the area is considered when performing the final summation.

If it is necessary to determine the distribution of loading that will produce the maximum shear at a point whose influence diagram is available, the load should be positioned in order to maximize the area under the diagram.[6] This can be done by "covering" either all of the positive area or all of the negative area.[7]

Shear Influence Diagrams by Virtual Displacement

A difficulty in drawing shear influence diagrams for continuous beams on more than two supports is finding the reactions. The method of *virtual displacement* or *virtual work* can be used to find the influence diagram without going through that step.

step 1: Replace the point being investigated (i.e., point A) with an imaginary link with unit length. (It may be necessary to think of the link as having a length of 1 ft, but the link does not add to or subtract from any length of the beam.) If the point being investigated is a reaction, place a hinge at that point and lift the hinge upward a unit distance.

step 2: Push the two ends of the beam (with the link somewhere in between) toward each other a very small amount until the linkage is vertical. The distance between supports does not change, but the linkage allows the beam sections to assume a slope. The sections to the left and right of the linkage displace δ_1 and δ_2, respectively, from their equilibrium positions. The slope of both sections is the same. Points of support remain in contact with the beam.

step 3: Determine the ratio of δ_1 and δ_2. Since the slope on the two sections is the same, the longer section will have the larger deflection. If $L = a + b$ is the length of the beam, the relationships between the deflections can be determined from Eqs. 46.11 through 46.13.

$$\delta_1 + \delta_2 = 1$$
$$\frac{\delta_1}{\delta_2} = \frac{a}{b} \qquad 46.11$$
$$\delta_1 = \left(\frac{a}{L}\right)\delta \qquad 46.12$$

---

[6]If the *minimum shear* is requested, the maximum negative shear is implied. The minimum shear is not zero in most cases.
[7]Usually, the dead load is assumed to extend over the entire length of the beam. The uniform live loads are distributed in any way that will cause the maximum shear.

Figure 46.6 *Virtual Beam Displacements*

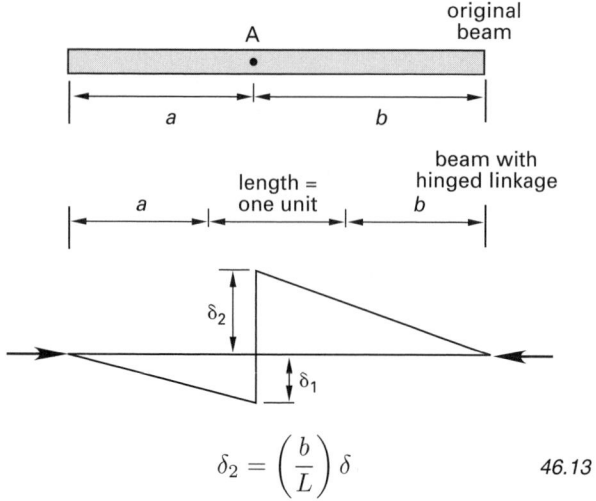

$$\delta_2 = \left(\frac{b}{L}\right)\delta \qquad\qquad 46.13$$

Since $\delta = \delta_1 + \delta_2$ was chosen as one, Eqs. 46.12 and 46.13 really give the relative proportions of the unit link that extend below and above the reference line in Fig. 46.6.

Knowing that the total shear reversal through point A is one unit and that the slopes are the same, the relative proportions of the reversal below and above the line will determine the shape of the displaced beam. The shape of the influence diagram is the shape taken on by the beam.

step 4: As required, use equations of straight lines to obtain the shear influence ordinate as a function of position along the beam.

Example 46.13

For the simply supported beam shown, draw the shear influence diagram for a point 10 ft from the right end.

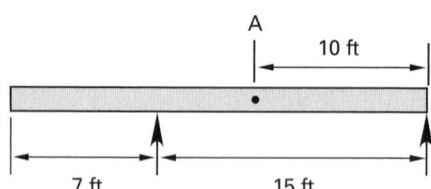

Solution

If a unit link is placed at point A and the beam ends are pushed together, the following shape will result. Notice that the beam must remain in contact with the points of support, and that the two slopes are the same.

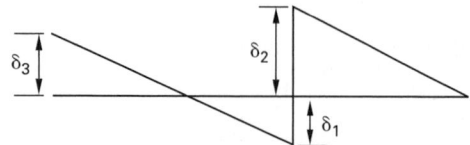

The overhanging 7 ft of beam do not change the shape of the shear influence diagram between the supports. The deflections can be evaluated assuming a 15 ft long beam.

$$\delta_1 = \frac{5 \text{ ft}}{15 \text{ ft}} = 0.33$$

$$\delta_2 = \frac{10 \text{ ft}}{15 \text{ ft}} = 0.67$$

The slope in both sections of the beam is the same. This slope can be used to calculate δ_3.

$$m = \frac{\delta_1}{a} = \frac{0.33}{5 \text{ ft}} = 0.066 \; 1/\text{ft}$$

$$\delta_3 = (7 \text{ ft})\left(0.066 \; \frac{1}{\text{ft}}\right) = 0.46$$

Example 46.14

Where should a uniformly distributed load be placed on the following beam to maximize the shear at section A?

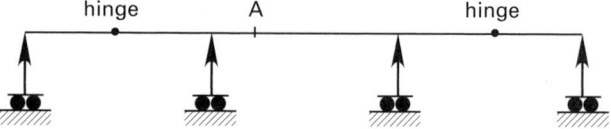

Solution

Using the principle of virtual displacement, the following shear influence diagram results by inspection. (It is not necessary to calculate the relative displacements to answer this question. It is only necessary to identify the positive and negative parts of the influence diagram.)

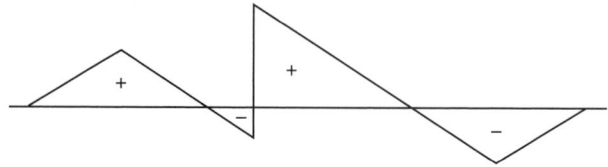

To maximize the shear, the uniform load should be distributed either over all positive or all negative sections of the influence diagram.

Moment Influence Diagrams by Virtual Displacement

A moment influence diagram (not the same as a moment diagram) gives the moment at a particular point for any location of a unit load. The method of virtual displacement can be used in this situation to simplify finding the moment influence diagram.

step 1: Replace the point being investigated (i.e., point A) with an imaginary hinge.

step 2: Rotate the beam one unit rotation by applying equal but opposite moments to each of the two beam sections. Except where the point being investigated is at a support, this unit rotation can be achieved simply by "pushing up" on the beam at the hinge point.

Figure 46.7 *Moment Influence Diagram by Virtual Displacement*

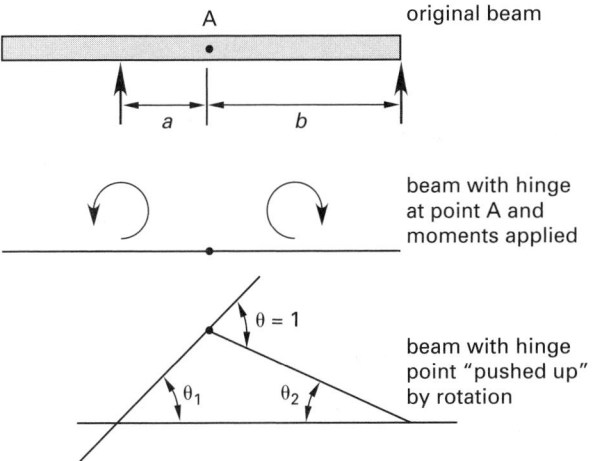

step 3: The angles made by the sections on either side of the hinge will be proportional to the lengths of the opposite sections. (Since the angle is small for a virtual displacement, the angle and its tangent, or slope, are the same.)

$$\theta_1 = \frac{b}{L} \qquad 46.14$$

$$\theta_2 = \frac{a}{L} \qquad 46.15$$

$$L = a + b \qquad 46.16$$

Example 46.15

What are the approximate shapes of the moment influence diagrams for points A and B on the beam shown?

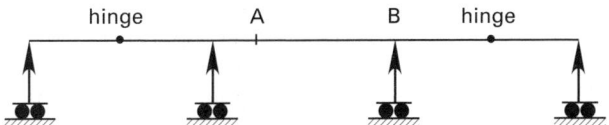

Solution

By placing an imaginary hinge at point A and rotating the two adjacent sections of the beam, the following shape results.

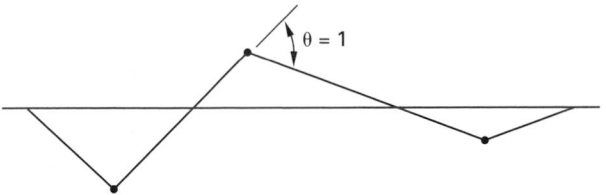

The moment influence diagram for point B is found by placing an imaginary hinge at point B and applying a rotating moment. Since the beam must remain in contact with all supports, and since there is no hinge between the two middle supports, the moment influence diagram must be horizontal in that region.

Shear Influence Diagrams on Cross-Beam Decks

When girder-type construction is used to construct a road or bridge deck, the loads will not be applied directly to the girder. Rather, the loads will be transmitted to the girder at panel points from cross beams. Figure 46.8 shows a typical construction detail involving girders and cross beams.

Figure 46.8 *Cross-Beam Decking*

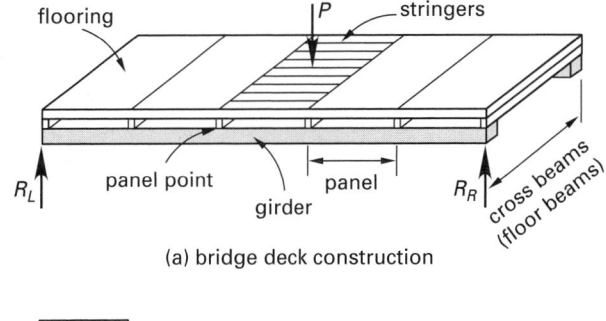

(a) bridge deck construction

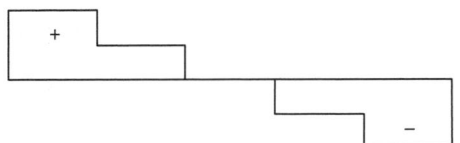

(b) shear diagram for girder

A load applied to the deck stringers will be transmitted to the girder only at the panel points. Because the girder experiences a series of concentrated loads, the shear between panel points is horizontal. Since the shear is always constant between panel points, we speak of *panel shear* rather than shear at a point. Accordingly, shear influence diagrams are drawn for a panel, not for a point. Moment influence diagrams are similarly drawn for a panel.

Influence Diagrams on Cross-Beam Decks

Shear and moment influence diagrams for girders with cross beams are identical to simple beams, except for the panel being investigated. Once the influence diagram has been drawn for the simple beam, the influence diagram ordinates at the ends of the panel being investigated are connected to obtain the influence diagram for the girder. This is illustrated in Fig. 46.9.

Figure 46.9 *Comparison of Influence Diagrams for Simple Beams and Girders (Panel bc)*

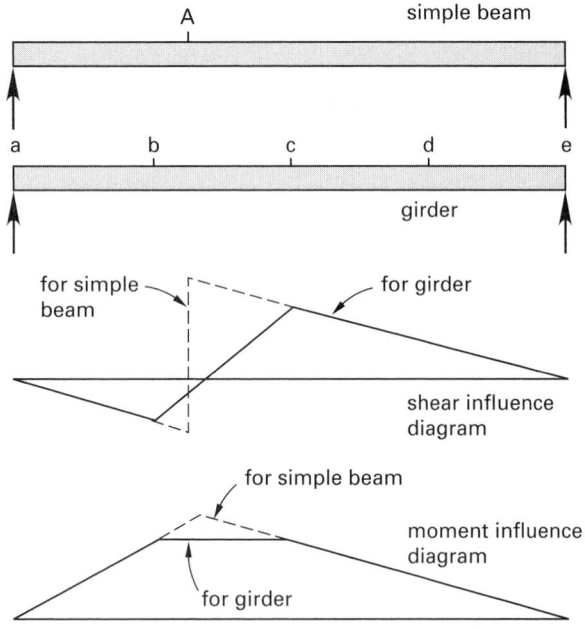

Influence Diagrams for Truss Members

Since members in trusses are assumed to be axial members, they cannot carry shears or moments. Therefore, shear and moment influence diagrams do not exist for truss members. However, it is possible to obtain an influence diagram showing the variation in axial force in a given truss member as the load varies in position.

There are two general cases for finding forces in truss members. The force in a horizontal truss member is proportional to the moment across the member's panel. The force in an inclined truss member is proportional to the shear across that member's panel.

So, even though we may only want the axis load in a truss member, it is still necessary to construct the shear and moment influence diagrams for the entire truss in order to determine the applications of loading on the truss that produce the maximum shear and moment across the member's panel.

Example 46.16

Draw the influence diagram for vertical shear in panel DF of the through truss shown. What is the maximum force in member DG if a 1000 lbf load moves across the truss?

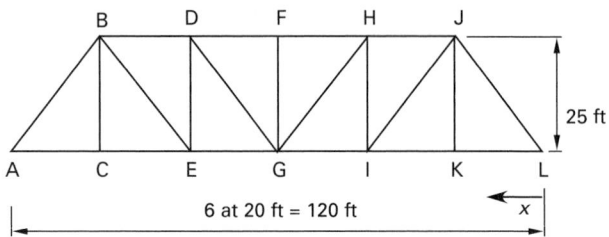

Solution

Allow a unit load to move from joint L to joint G along the lower chords. If the unit vertical load is at a distance x from point L, the right reaction will be $+\left(1-(x/120)\right)$. The unit load itself has a value of -1, so the shear at distance x is just $-x/120$.

Allow a unit load to move from joint A to joint E along the lower chords. If the unit load is a distance x from point L, the left reaction will be $x/120$, and the shear at distance x will be $(x/120) - 1$.

These two lines can be graphed.

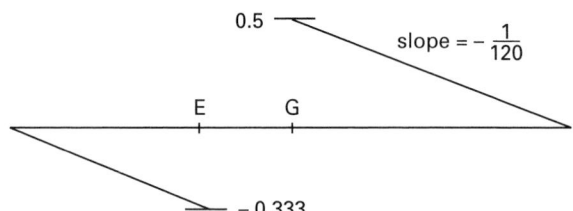

The influence line is completed by connecting the two lines as shown. Therefore, the maximum shear in panel DF will occur when a load is at point G on the truss.

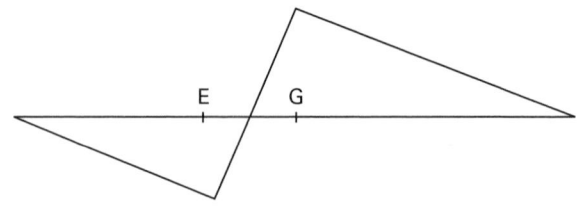

If the 1000 lbf load is at point G, the two reactions at points A and L will each be 500 lbf. The cut-and-sum method can be used to calculate the force in member DG simply by evaluating the vertical forces on the free-body to the left of point G.

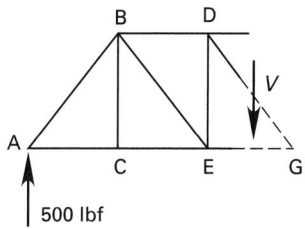

500 lbf

For equilibrium to occur, V must be 500. This vertical shear is entirely carried by member DG. The length of member DG is

$$\sqrt{(20 \text{ ft})^2 + (25 \text{ ft})^2} = 32 \text{ ft}$$

The force in member DG is

$$\left(\frac{32 \text{ ft}}{25 \text{ ft}}\right) 500 \text{ lbf} = 640 \text{ lbf}$$

Example 46.17

Draw the moment influence diagram for panel DF on the truss shown in Ex. 46.16. What is the maximum force in member DF if a 1000 lbf load moves across the truss?

Solution

The left reaction is $x/120$ where x is the distance from the unit load to the right end. If the unit load is to the right of point G, the moment can be found by summing moments from point G to the left. The moment is $(x/120)(60) = 0.5x$.

If the unit load is to the left of point E, the moment will again be found by summing moments about point G. The distance between the unit load and point G is $x - 60$.

$$\left(\frac{x}{120}\right)(60) - (1)(x - 60) = 60 - 0.5x$$

These two lines can be graphed. The moment for a unit load between points E and G is obtained by connecting the two end points of the lines derived above. Therefore, the maximum moment in panel DF will occur when the load is at point G on the truss.

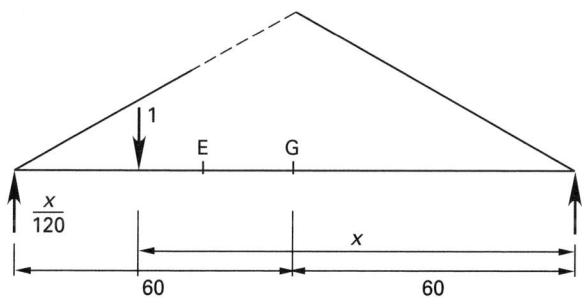

If the 1000 lbf load is at point G, the two reactions at points A and L will each be 500 lbf. The method of sections can be used to calculate the force in member DF by taking moments about joint G.

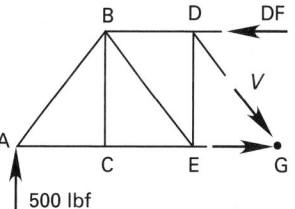

500 lbf

$$\sum M_\text{G}: (500 \text{ lbf})(60 \text{ ft}) - (\text{DF})(25 \text{ ft}) = 0$$
$$\text{DF} = 1200 \text{ lbf}$$

11. MOVING LOADS ON BEAMS

Global Maximum Moment Anywhere on Beam

If a beam supports a single moving load, the maximum bending and shearing stresses at any point can be found by drawing the moment and shear influence diagrams for that point. Once the positions of maximum moment and maximum shear are known, the stress at the point in question can be found from Mc/I.

If a simply supported beam carries a set of moving loads (which remain equidistant as they travel across the beam), the following procedure can be used to find the *dominant load*. (The dominant load is the one that occurs directly over the point of maximum moment.)

step 1: Calculate and locate the resultant of the load group.

step 2: Assume that one of the loads is dominant. Place the group on the beam such that the distance from one support to the assumed dominant load is equal to the distance from the other support to the resultant of the load group.

step 3: Check to see that all loads are on the span and that the shear changes sign under the assumed dominant load. If the shear does not change sign under the assumed dominant load, the maximum moment may occur when only some of the load group is on the beam. If it does change sign, calculate the bending moment under the assumed dominant load.

step 4: Repeat steps 2 and 3, assuming that the other loads are dominant.

step 5: Find the maximum shear by placing the load group such that the resultant is a minimum distance from a support.

Placement of Load Group to Maximize Local Moment

In the design of specific members or connections, it is necessary to place the load group in a position that will maximize the load on those members or connections. The procedure for finding these positions of local maximum loadings is different from the global maximum procedures.

The solution to the problem of local maximization is somewhat trial-and-error oriented. It is aided by use of the influence diagram. In general, the variable being evaluated (reaction, shear, or moment) is maximum when one of the wheels is at the location or section of interest.

When there are only two or three wheels in the load group, the various alternatives can be simply evaluated by using the influence diagram for the variable being evaluated. When there are many loads in the load group (e.g., a train loading), it may be advantageous to use heuristic rules for predicting the dominant wheel.

47 Structural Analysis II

Nomenclature

A	cross-sectional area	ft^2	m^2
C	coefficient	–	–
DF	distribution factor	–	–
E	modulus of elasticity	lbf/ft^2	Pa
f	flexibility coefficient	ft	m
F	force	lbf	N
FEM	fixed-end moment	ft-lbf	N·m
G	shear modulus	lbf/ft^2	Pa
h	height	ft	m
I	degree of indeterminacy	–	–
I	moment of inertia	ft^4	m^4
J	polar moment of inertia	ft^4	m^4
k	stiffness	lbf/ft	N/m
K	relative stiffness	–	–
L	length	ft	m
m	moment	ft-lbf	N·m
M	total applied moment	ft-lbf	N·m

MD	displaced moment	ft-lbf	N·m
N	axial force	lbf	N
P	force	lbf	N
Q	restraining force	lbf	N
R	reaction	lbf	N
S	settlement	ft	m
T	temperature	°F	°C
T	torsional moments (torque)	ft-lbf	N·m
U	energy	ft-lbf	J
V	shear	lbf	N
V	volume	ft^3	m^3
w	unit loading	lbf/ft	N/m
W	work	ft-lbf	J
z	distance	ft	m

Symbols

α	coefficient of thermal expansion	1/°F	1/°C
δ	displacement	ft	m
Δ	displacement	ft	m
ϵ	strain	ft/ft	m/m
θ	rotation	rad	rad
σ	internal stress	lbf/ft^2	Pa

Subscripts

a	due to axial loading
f	flexure
m	due to moment
p	from applied loads, or plastic
Q	from unit load
t	due to torsion
th	thermal
u	ultimate
v	shear, or due to shear

1. INTRODUCTION TO STRUCTURAL ANALYSIS

Structural analysis determines the reactions, internal forces, stresses, and deformations induced in a structure by loading. While loading from external forces is the most common situation, stress and deformation can also be caused by internal temperature changes and settling of foundations.

Prior to widespread use of computers in engineering design and analysis, various exact and approximate methods of manual analysis were developed for specific types of structures. While these traditional manual methods continue to be useful, the need for a variety of specialized techniques has been reduced significantly. This chapter concentrates on the manual techniques that remain most relevant in present-day structural analysis.

Structural analysis is categorized as either *static* or *dynamic*, depending on the need to evaluate inertial forces caused by acceleration. A *static analysis* does not consider inertial forces. In a *dynamic analysis*, the various forces and deformations are determined as functions of time. While computer analysis has enhanced the application of dynamic analysis in certain areas (e.g., earthquake loading), most structures continue to be designed using static analyses based on the maximum expected dynamic loads. This chapter is limited to a review of several common methods of structural analysis used for static loading.

Structural analysis can also be categorized on the basis of the level of refinement that is incorporated. *Linear elastic first-order analysis* is used to evaluate the majority of structures built today. The basic assumptions are that (a) the structure is perfectly linear and elastic in behavior, and (b) deformations can be neglected. The principle of *proportionality* is assumed to hold, so changing all of the loads by a certain percentage results in an equal percentage change in all of the internal forces, reactions, and deflections.

Figure 47.1 *Linear Elastic Behavior*

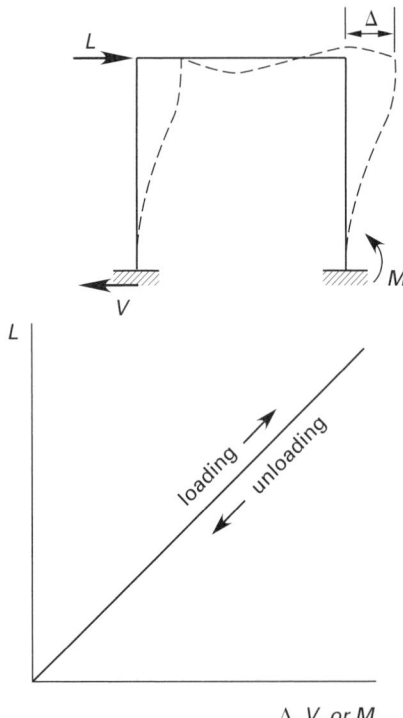

With a *linear elastic second-order analysis*, the assumption of elastic linear behavior is maintained, but the deflections are considered in the equilibrium equations. In a second-order analysis, equilibrium between internal and external forces is achieved in the final deformed configuration, not in the initial undeformed geometry.

Since the deformations of the structure cannot be computed precisely until the internal forces are obtained, a second-order solution is iterative. Due to computational difficulties in manual solutions, second-order

effects have been traditionally included as corrections applied during the design process. The moment magnifier method used to design columns in the ACI code and LRFD Manual are typical examples. There is an increasing trend, however, to introduce second-order effects in computerized analyses. The P-Δ method described in Sec. 18 is a simplified second-order analysis technique.

With an *inelastic first-order analysis*, the potential for inelastic, disproportional behavior is considered, but equilibrium is assumed to occur in the undeformed configuration. *Plastic analysis* is a popular version of this method. The objective of plastic analysis is to obtain the *ultimate load* (maximum load) that the structure can carry. When supporting an ultimate load, the structure becomes a mechanism. A *mechanism* is a structure that can deform by rotation around a number of points. In highly stressed structures, these points are referred to as *plastic hinges* since the material at those points is loaded in the plastic region. Plastic analysis is often used in the design of continuous beams, where it is particularly easy to apply.

Figure 47.2 *Failure Mechanisms*

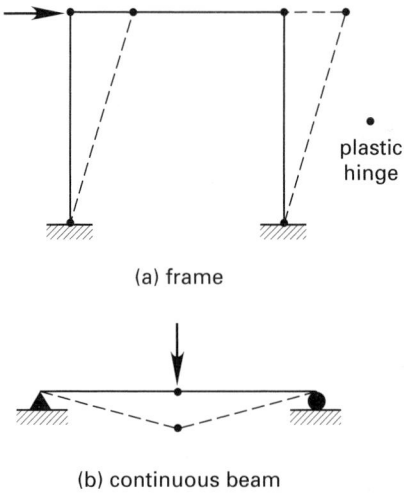

With an *inelastic second-order analysis*, inelastic response in the material is permitted, and the equilibrium equations are satisfied in the deformed configuration. Inelastic second-order analysis is a well-developed tool, but it is too complex to be used in most designs.

2. TRADITIONAL METHODS

Statically determinate structures can easily be solved by applying the equations of equilibrium, and advanced methods are seldom needed. Therefore, specialized methods of structural analysis generally address indeterminate structures. Indeterminate structural analysis procedures can be classified as either force methods or displacement (deformation, deflection, etc.) methods, depending on whether forces or displacements are used as basic unknowns. The classical *moment distribution* and *slope deflection methods* are displacement-based solutions. The *stiffness method*, typically used as the basis of modern computerized structural analysis, also is a displacement method. The *flexibility method* is a force-based approach.

3. REVIEW OF WORK AND ENERGY

For a simple structure deflection a distance Δ in the direction of the applied force P, a simple mathematical definition of *work* is the product of force times displacement. Analogously, the work of a torque T (or moment) causing a rotation of ϕ radians is torque times rotation.

$$W = P\Delta \quad \text{[linear displacement]} \qquad 47.1$$

$$W = T\phi \quad \text{[rotation]} \qquad 47.2$$

The change in energy of an ideal conservative system to which work has been added is equal to the work.

$$W = U_2 - U_1 \qquad 47.3$$

4. REVIEW OF LINEAR DEFORMATION

The deformation of a tension or compression member due to axial force P is

$$\Delta = \frac{PL}{AE} \qquad 47.4$$

5. THERMAL LOADING

Structures subjected to temperature changes can be evaluated using the methods in this chapter by adding the appropriate thermally induced forces and moments to the externally applied loads. For example, the thermally induced axial load in a constrained member with a uniform temperature change is

$$P_{\text{th}} = \frac{\Delta_{\text{constrained}} AE}{L} = \alpha(T_2 - T_1)\left(\frac{LAE}{L}\right)$$
$$= \alpha(T_2 - T_1)AE \qquad 47.5$$

Bending moments will be induced in a structural member that experiences a temperature variation across its thickness (depth). If the temperature varies linearly with depth, and if the temperature difference is zero at the neutral axis (depth $h/2$), the thermally induced moment will be given by Eq. 47.6.

$$M = \alpha(T_{\text{extreme fiber}} - T_{\text{neutral axis}})\left(\frac{EI}{\frac{h}{2}}\right) \qquad 47.6$$

Any arbitrary linear variation can be treated as the sum of a uniform temperature change plus a linear variation with a value of zero at the section's neutral axis. In Eqs. 47.5 and 47.6, tension is considered to be positive,

and compression is negative. For an axially constrained beam experiencing an increase in temperature at its upper surface, the beam will bow upwards (to lengthen the upper surface), and the moment will be negative.

6. DUMMY UNIT LOAD METHOD

The *energy method*, typically referred to as the *dummy load method*, is based on the *virtual work principle*.

> If a structure that is in equilibrium with a set of loads is given an arbitrary (virtual) deformation, the external work calculated from the applied loads and the virtual displacements is equal to the internal work of the internal stresses and the virtual strains.

Figure 47.3 *Virtual Work Concepts*

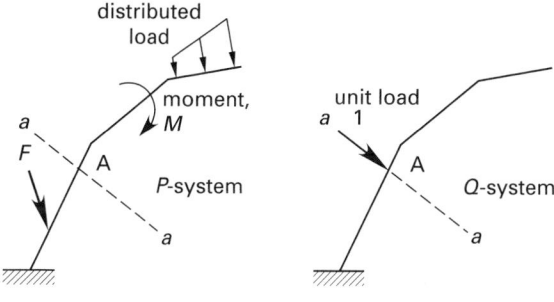

(a) external loading (b) dummy unit loading

Figure 47.3 illustrates a generalized structure subjected to a general load distribution and whose component of deformation at point A in the direction of line aa is desired. The analysis begins by considering the structure to be loaded by a unit load at point A along line aa. For convenience, the system of actual loads is designated as the "P-system," and the unit loading is designated as the "Q-system." The work of unit load can be calculated from the Q-system stresses and the P-system strains.

$$W_Q = 1 \times \Delta = \int \sigma_Q \epsilon_P \, dV \qquad 47.7$$

The integral in Eq. 47.7 is calculated as a discrete summation; the product is evaluated separately for stresses caused by moments, shears, and axial forces. Some of the terms in Eq. 47.8 may be zero or may be significantly negligible to be neglected.

$$W_Q = W_m + W_v + W_a + W_t \qquad 47.8$$

$$W_m = \int \left(\frac{m_Q m_P}{EI}\right) ds \qquad 47.9$$

$$W_v = \int \left(\frac{V_Q V_P}{GA}\right) ds \qquad 47.10$$

$$W_a = \int \left(\frac{N_Q N_P}{EA}\right) ds \qquad 47.11$$

$$W_t = \int \left(\frac{T_Q T_P}{GJ}\right) ds \qquad 47.12$$

The sign convention used with Eqs. 47.8 through 47.12 is arbitrary, but it must be used consistently. If a certain direction is assumed positive in the P-system, the same direction must be taken as positive in the Q-system. A positive value of Δ indicates that the deflection is in the direction of the unit load.

To compute rotations rather than deflections, Eqs. 47.8 through 47.12 can still be used. The Q-system is loaded by a unit moment at the point where the rotation is wanted. m_Q, V_Q, N_Q, and T_Q are internal forces due to the unit moment rather than a unit load.

7. BEAM DEFLECTIONS BY THE DUMMY UNIT LOAD METHOD

A beam is a member that reacts to applied loads by bending. For most applications, axial stresses and torsion are both zero or negligible. And, except for deep beams (spans $\leq 3 \times$ depth), it is acceptable to assume that the work due to shear loading is zero. In this case, Eq. 47.8 reduces to

$$W_Q = W_m = \int \left(\frac{m_Q m_P}{EI} \right) ds \qquad 47.13$$

Example 47.1

The simply supported cantilever beam shown is loaded uniformly. Shear deformation is negligible. What is the expression for the deflection at the cantilever end?

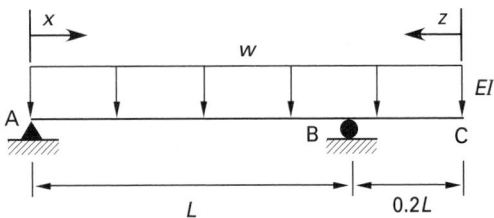

Solution

First, determine the moments, m_P, due to the applied loads. The moment diagram and the defining equations are determined from principles of statics. (The location x has its origin at A and increases from left to right, while z starts at point C and increases from right to left.)

$$m_P = 0.48wLx - 0.5wx^2 \quad \text{[segment AB]}$$

$$m_P = -0.5wz^2 \quad \text{[segment BC]}$$

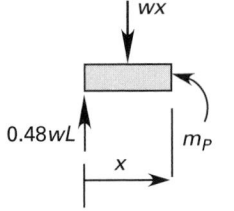 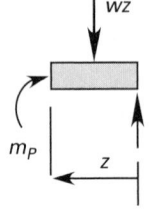

Next, determine the moments due to the unit load.

$$m_Q = -0.2x \quad \text{[segment AB]}$$

$$m_Q = -z \quad \text{[segment BC]}$$

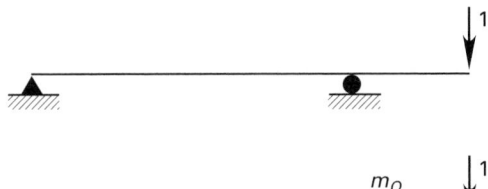

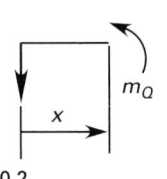 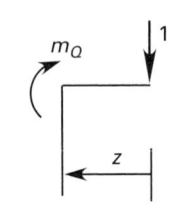

The deflection is the work done by a unit load. Substituting these expressions for m_P and m_Q into Eq. 47.13,

$$\Delta = \int_0^L \frac{(0.48wLx - 0.5wx^2)(-0.2x)}{EI} dx$$

$$+ \int_0^{0.2L} \frac{(-0.5wz^2)(-z)}{EI} dz$$

$$= \frac{-0.0068wL^4}{EI} \quad \text{[upward deflation]}$$

8. TRUSS DEFLECTIONS BY THE DUMMY UNIT LOAD METHOD

The applied loads on a truss are carried by axial forces solely, so Eq. 47.8 reduces to

$$W_Q = W_a = \int \left(\frac{N_Q N_P}{EA} \right) ds \qquad 47.14$$

Since the axial forces are constant in the truss members, the integral is evaluated as a summation over all of the members.

$$W_Q = W_a = \sum_i \left(\frac{N_{Q,i} N_{P,i} L_i}{E_i A_i} \right) \qquad 47.15$$

Example 47.2

Use the dummy load method to determine the expression for the deflection at point B in the truss shown.

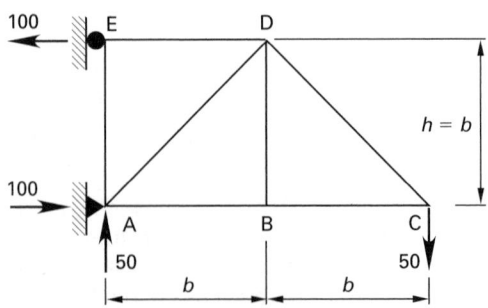

Solution

The unit-loaded truss is

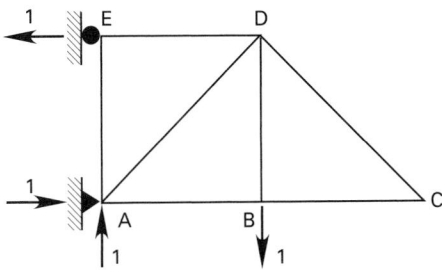

The forces in the bars due to the applied load (on the *P*-system) as well as due to the unit load (on the *Q*-system) are determined by traditional methods of statics.

bar	N_P	N_Q	$N_P N_Q L_i$
AB	-50.0	0.0	0.0
BC	-50.0	0.0	0.0
CD	70.71	0.0	0.0
DE	100.0	1.0	$100.0b$
EA	0.0	0.0	0.0
AD	-70.71	-1.41	$100.0b$
BD	0.0	1.0	0.0
			$\sum 200.0b$

The deflection is the work done by a unit load. Use Eq. 47.11.

$$\Delta = W_a = \sum_i \left(\frac{N_{Q,i} N_{P,i} L_i}{E_i A_i} \right)$$
$$= 200b/AE$$

9. FRAME DEFLECTIONS BY THE DUMMY UNIT LOAD METHOD

For typically proportioned frames, a good approximation of the deflection can usually be obtained by considering only the flexural contribution. There are cases, however, where axial and shear deformations are significant enough to be included. In particular, shear deformations can be significant in structures with deep members, and axial deformations can have an important effect on the lateral displacements of tall frames subjected to wind or earthquake forces.

In rigid frames that resist loading primarily by flexure, using the unit dummy load method to calculate deflection is based on Eq. 47.9.

Example 47.3

Use the dummy load method to determine an expression for the vertical deflection at point C in the frame shown. Neglect shear and axial deformations.

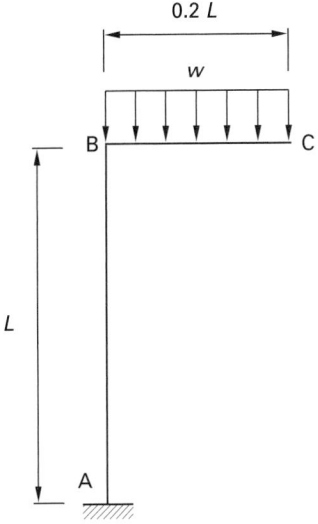

Solution

The moments on the frame as loaded are

$$m_P = 0.02wL^2 \quad \text{[bar AB]}$$
$$m_P = 0.5wz^2 \quad \text{[bar BC]}$$

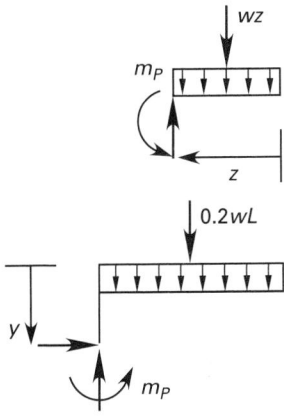

For the unit-loaded frame, the corresponding moments are

$$m_Q = 0.2L \quad \text{[bar AB]}$$
$$m_Q = z \quad \text{[bar BC]}$$

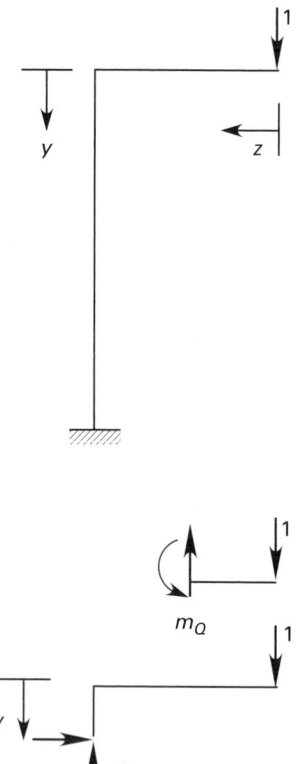

Substituting these expressions into Eq. 47.9,

$$\Delta = \int_0^L \frac{(0.02wL^2)(0.2L)}{EI}dy + \int_0^{0.2L} \frac{(0.5wz^2)z}{EI}dz$$

$$= \frac{0.0042wL^4}{EI}$$

10. CONJUGATE BEAM METHOD

The conjugate beam method is a practical procedure for computing deflections in beams. In the conjugate beam method, a "conjugate beam" is loaded with the moment diagram for the real beam divided by the flexural stiffness product, *EI*. The conjugate beam has the same length as the real beam, but the support types are changed according to Table 47.1. When this is done, (a) the shear in the conjugate beam will be the same (numerically) as the slope on the real beam, and (b) the moment on the conjugate beam will be the same (numerically) as the deflection in the real beam.

Table 47.1 Supports for Conjugate Beams

real beam	conjugate beam
exterior simple support	exterior simple support
interior support	hinge
hinge	interior support
free end	fixed end
fixed end	free end

Figure 47.4 *Examples of Conjugate Beams*

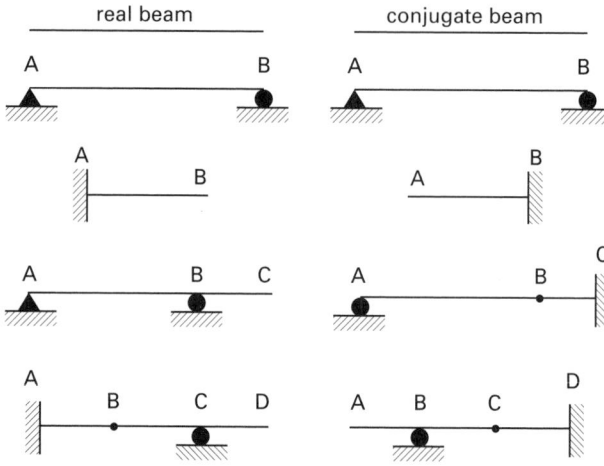

If the sign convention is such that downward loads on the real beam produce positive moments (the usual assumption), a positive moment on the conjugate beam corresponds to deflection downward. Likewise, slopes in the first and third quadrant are associated with shears that are up on the right side of a free-body diagram.

Example 47.4

Use the conjugate beam method to determine an expression for the deflection at point C on the cantilever beam shown. The cross section and modulus of elasticity are constant along the length of the beam.

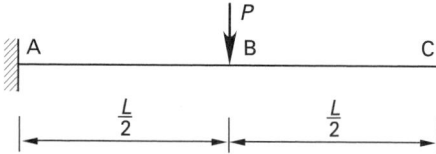

Solution

The moment diagram for the real beam is shown. The conjugate beam has a free end at the left (because the real beam is fixed at that point) and a fixed end at the right (because the real beam is free at that point). The loading of the conjugate beam is M/EI, but since EI is constant, the loading distribution on the conjugate beam is the same as the moment diagram on the real beam.

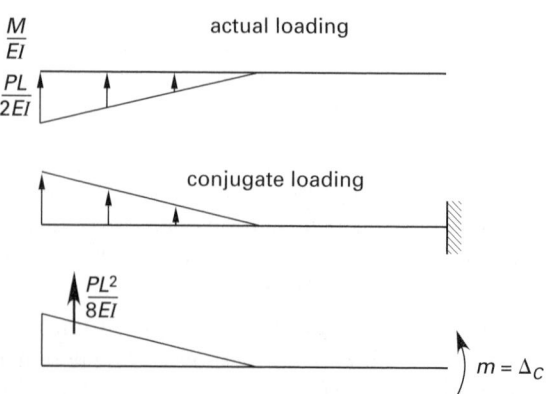

The moment on the conjugate beam at point C is equal to the deflection on the real beam.

$$\Delta_C = \left(\frac{PL^2}{8EI}\right)\left(\frac{L}{2} + \frac{2}{3}\left(\frac{L}{2}\right)\right)$$

$$= \frac{5PL^3}{48EI}$$

11. INTRODUCTION TO THE FLEXIBILITY METHOD

The *flexibility method* (also known as the *method of consistent deformations*) is one of the fundamental methods of indeterminate structural analysis. In practice, a manual application of this method is only convenient with structures having degrees of indeterminacy (I) of 1, 2, and (with effort) 3, since the number of deflection terms that must be calculated is $0.5I^2 + 1.5I$. The basis of the flexibility method is explained with the aid of Fig. 47.5(a), which shows a three-span continuous beam with an arbitrary loading.

Figure 47.5 *Flexibility Method*

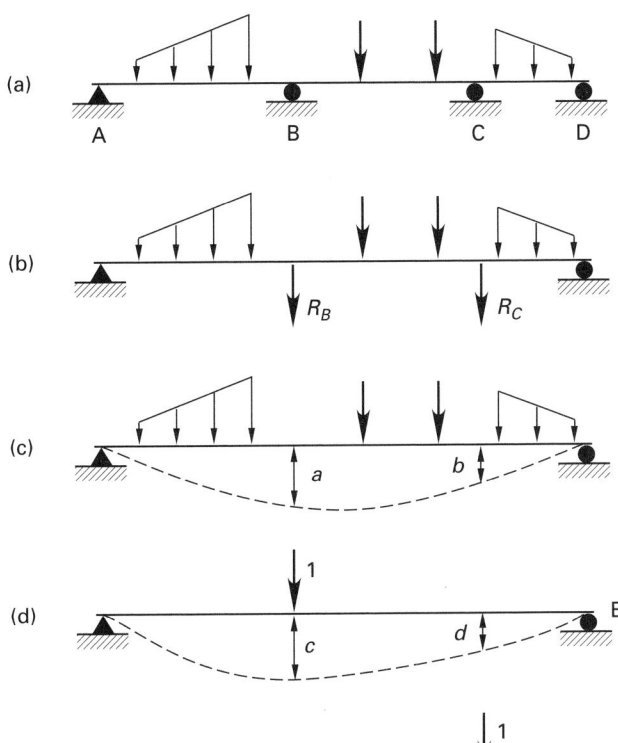

The beam in Fig. 47.5(a) is indeterminate to the second degree because there are five reactions and only three equations of statics to determine the reactions. However, we may consider any two of the four vertical reactions as being applied loads of unknown magnitudes.

(The forces selected are known as *redundants*.) The number of redundants that must be removed to render the structure statically determinate is known as the *degree of static indeterminacy, I*. The structure with the redundants removed is known as the *primary structure*.

In Fig. 47.5(a), the vertical reactions at B and C may be considered applied loads on a simply supported statically determinate beam. Since the beam must remain in contact with the supports at reaction points, the values of the deflections at points B and C must be zero. Figures 47.5(a) and 47.5(b) are equivalent.

To illustrate how the equations of compatibility are set up, assume that the deflections at B and C are to be computed using superposition. Referring to Fig. 47.5(c), the first components are the deflections from the applied loading, which are designated as a and b. The deflections due to the reaction at B, R_B, are proportional to the magnitude of the reaction, so it is convenient to compute the deflections for a unit reaction and then to scale them by R_B. Designate the deflections from $R_B = 1$ as c and d (Fig. 47.5(d)) and those from $R_C = 1$ as e and f (Fig. 47.5(e)). With this notation, the conditions of zero deflection at support points B and C can be mathematically expressed as Eqs. 47.16 and 47.17. These two equations can be solved simultaneously to obtain R_B and R_C. Once the reactions are known, any quantity of interest can be readily computed using statics.

$$a + cR_B + eR_C = 0 \qquad \textit{47.16}$$
$$b + dR_B + fR_C = 0 \qquad \textit{47.17}$$

This method can be modified to account for settling of the supports, as well. If there was a settlement of 0.5 in at B, then the equations would be

$$a + cR_B + eR_C = 0.5 \text{ in} \qquad \textit{47.18}$$
$$b + dR_B + fR_C = 0 \qquad \textit{47.19}$$

12. BASIC FLEXIBILITY METHOD PROCEDURE

The following steps constitute the procedure for using the flexibility method.

step 1: Transform the structure into a statically determinate structure by eliminating redundant reactions, moments, or internal forces. A set of redundants for a given indeterminate structure is not unique. Although some choices may lead to less numerical effort than others, all selections are equally valid. It is important to keep in mind, however, that the statically determinate primary system that remains must be stable.

For trusses, the degree of static indeterminacy is a combination of external indeterminacy (from redundant reactions) and internal indeterminacy (from redundant members). The total degree of indeterminacy in trusses is

$$I = \text{no. of reactions}$$
$$+ \text{no. of members (bars)}$$
$$- 2(\text{no. of joints}) \qquad 47.20$$

If Eq. 47.20 is zero, the truss is statically determinate; if it is negative, the truss is unstable. In cases where $I \geq 0$, the truss must also be checked for stability.

Figure 47.6 *Frame and Its Primary Structures*

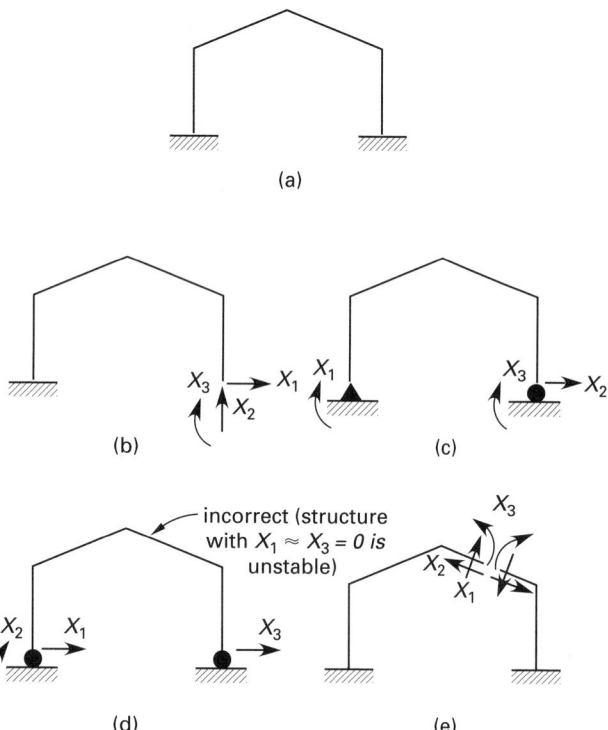

Figure 47.6(a) illustrates a frame that is indeterminate to the third degree ($I = 3$). Three possible primary structures obtained by considering some of the reactions as redundants are shown as Fig. 47.6(b), (c), (d), and (e). However, Fig. 47.6(d) is inappropriate because the primary structure cannot resist horizontal forces and, therefore, is unstable. The redundants also need not be reactions at all, but can be taken as internal forces. For example, the axial, shear, and moment at an arbitrarily located cut can be selected as redundants as in Fig. 47.6(e). When internal forces are taken as redundants, the equilibrium forces act on both sides of the cut with opposite directions. For convenience, the redundants are designated as $X_1, X_2, \dots X_n$, regardless of whether they are forces or moments. These X_i are collected in a vector $\mathbf{X}$.

$$\mathbf{X} = \begin{bmatrix} X_1 \\ X_2 \\ \vdots \\ X_n \end{bmatrix} \qquad 47.21$$

step 2: Compute the deflections (or rotations) due to the applied loads at each of the locations (and in the directions) of the redundants. These deflections are designated as $\delta_1, \delta_2, \dots, \delta_n$. These δ_i terms are collected in vector $\boldsymbol{\delta}$.

$$\boldsymbol{\delta} = \begin{bmatrix} \delta_1 \\ \delta_2 \\ \vdots \\ \delta_n \end{bmatrix} \qquad 47.22$$

For beams, it is generally appropriate to compute the deflections accounting only for flexural effects. For trusses, the deflections are due to axial elongations only. In most practical cases, frames can be solved accounting only for flexural deformations.

step 3: Compute the deflections for unit values of each of the redundants. These deflections are known as *flexibility coefficients* and are designated as $f_{i,j}$, where $f_{i,j}$ is the deflection at location i due to a unit value of the redundant at location j. The flexibility coefficients are collected in a flexibility matrix $\mathbf{F}$. Matrix $\mathbf{F}$ is symmetrical since $f_{i,j} = f_{j,i}$.

$$\mathbf{F} = \begin{bmatrix} f_{1,1} & f_{1,2} & \cdots & f_{1,n} \\ & f_{2,2} & \cdots & f_{2,n} \\ & & \cdots & f_{3,n} \\ & & & \vdots \\ & & & f_{n,n} \end{bmatrix} \qquad 47.23$$

step 4: Determine the settlements, if any, at the redundant supports. Collect these settlements in the matrix $\mathbf{S}$. Where there is no settling, all entries in $\mathbf{S}$ are zero (i.e., $\mathbf{S}$ is the null matrix).

step 5: Write the set of simultaneous equations that impose the geometrical requirements of compatibility. Using the format previously introduced, these equations are written in vector format.

$$\mathbf{FX} = \mathbf{S} - \boldsymbol{\delta} \qquad 47.24$$

13. SYSTEMATIC FLEXIBILITY METHOD PROCEDURE

Most of the work in applying the flexibility method is associated with finding all of the deflections needed to set up the compatibility equations. While these deflections can be obtained using any applicable technique, the dummy load method is most frequently used in practice. The following systematic approach can be used to integrate these two methods.

step 1: Select the redundants and their positive directions.

step 2: Determine equations that describe how internal forces vary throughout the structure. (For visualization purposes, it is often convenient to write these expressions on a sketch of the structure.) Designate the effects of the loads on the primary structure as "$X = 0$".

step 3: Derive equations for the relevant internal forces due to the unit values of each redundant. Designate the effects of the first redundant as "$X_1 = 1$", for the second as "$X_2 = 1$," and so on.

step 4: Compute the deflection terms. (Refer to Eqs. 47.7 through 47.11.) For the terms in F, obtain $f_{j,n}$ by integrating the product of the internal forces for $X_j = 1$ times $X_n = 1$. For the terms in $\boldsymbol{\delta}$, obtain δ_n by integrating the product of the internal forces for $X = 0$ times $X_n = 1$.

Example 47.5

Use the flexibility method to draw the moment diagram for the frame shown. Neglect axial and shear deformations in the computations. Consider EI to be constant for both members.

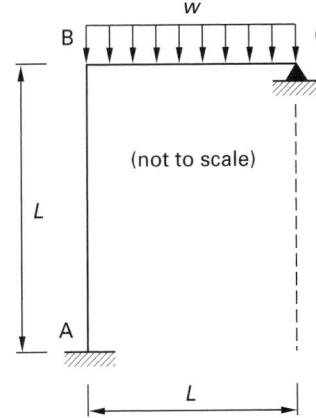

Solution

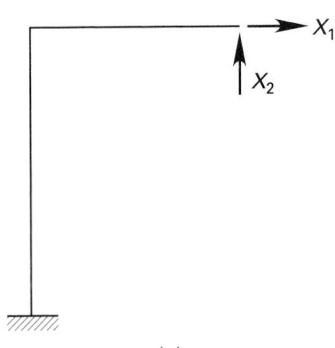

(a)

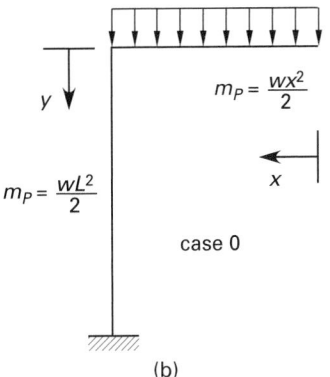

(b)

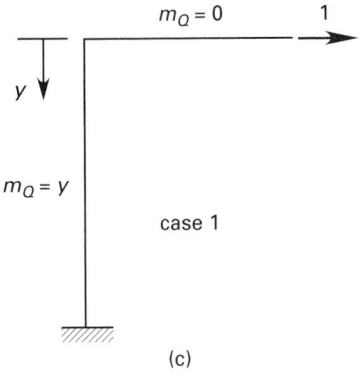

(c)

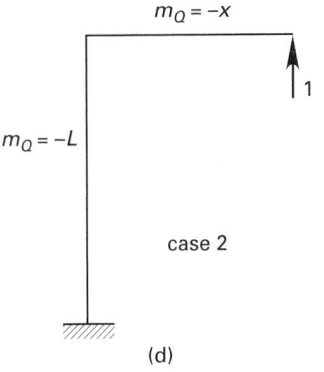

(d)

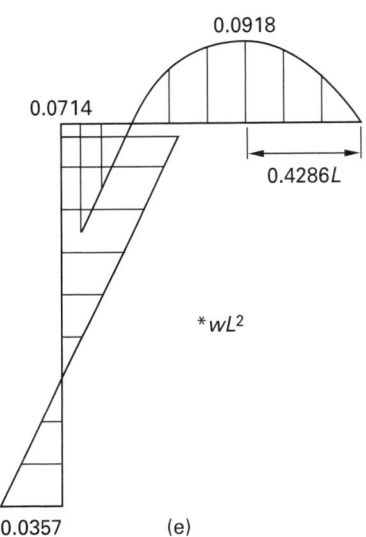

(e)

Refer to the illustration. From (a), the frame is indeterminate to the second degree. Select the two reactions at point C as the redundants. Expressions for the moments in the primary structure due to the loading as well as from the unit values of the redundants are shown in (b). From (c) and (d), the required deflection and flexibility coefficients are determined in the following expressions. Once the reactions are known, the moment diagram can be easily computed as shown in (e).

$$\delta_1 = \int_0^L \frac{wL^2 y}{2EI} dy = \frac{wL^4}{4EI}$$

$$\delta_2 = \int_0^L \frac{-wx^3}{2EI} dx + \int_0^L \frac{-wL^3}{2EI} dy = \frac{-5wL^4}{8EI}$$

$$f_{1,1} = \int_0^L \frac{y^2}{EI} dy = \frac{L^3}{3EI}$$

$$f_{2,2} = \int_0^L \frac{x^2}{EI} dx + \int_0^L \frac{(-L)^2}{EI} dy = \frac{4L^3}{3EI}$$

$$f_{1,2} = f_{2,1} = \int_0^L \frac{y(-L)}{EI} dy = \frac{-L^3}{2EI}$$

Substitute these values into Eq. 47.24 and solve.

$$\frac{L^3}{EI} \begin{bmatrix} \frac{1}{3} & -\frac{1}{2} \\ -\frac{1}{2} & \frac{4}{3} \end{bmatrix} \begin{bmatrix} X_1 \\ X_2 \end{bmatrix} = \frac{wL^4}{EI} \begin{bmatrix} -\frac{1}{4} \\ \frac{5}{8} \end{bmatrix}$$

$$\begin{bmatrix} X_1 \\ X_2 \end{bmatrix} = wL \begin{bmatrix} -0.107 \\ 0.429 \end{bmatrix}$$

14. STIFFNESS METHOD

The *stiffness method* is based on the fact that the behavior of any structure can be viewed as the superposition of the effect of loads on a structure with joints that do not displace and the effect of the joint displacements. (The term "displacement" is used in a generic sense to mean both translation and rotation.)

With the stiffness method, the traditional degree of static indeterminacy is not considered when obtaining a solution. Instead, the number of independent joint displacements, known as the *degree of kinematic indeterminacy*, determines the number of simultaneous equations. While the equations in the flexibility method describe conditions of compatibility, equations in the stiffness method are statements of equilibrium at the joints.

Figure 47.7 *Stiffness Method*

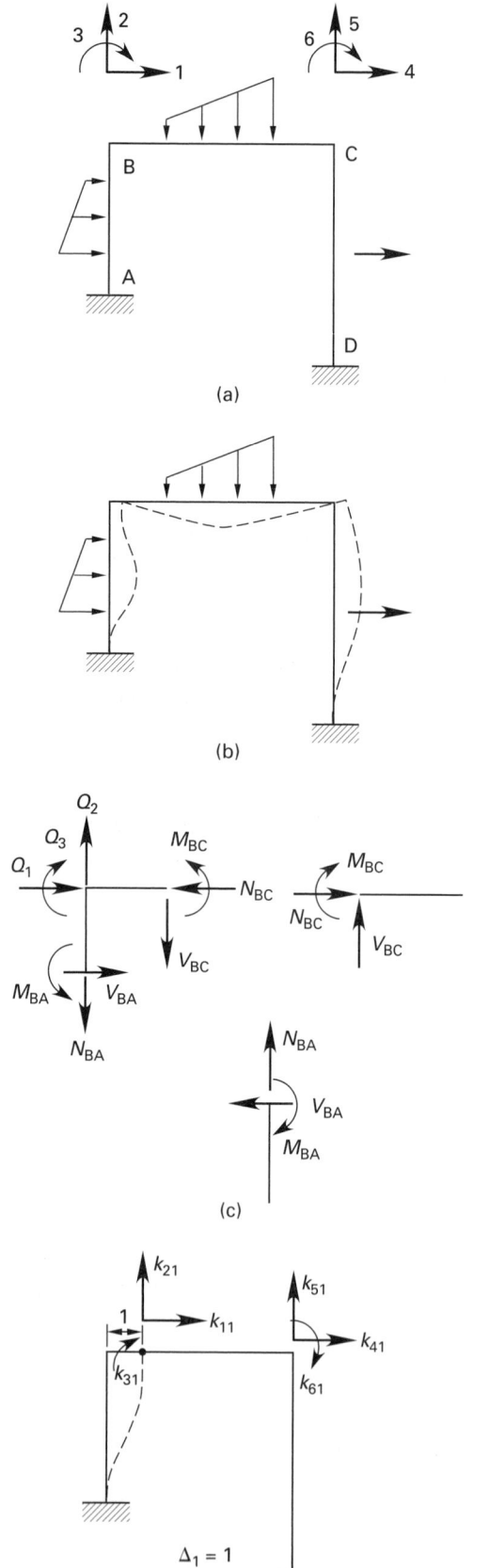

(a)

(b)

(c)

(d)

(Continued)

Figure 47.7 *Stiffness Method—(Continued)*

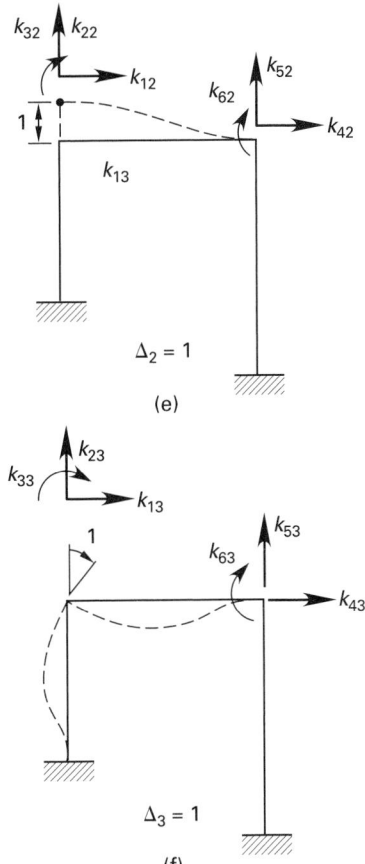

$\Delta_2 = 1$

(e)

$\Delta_3 = 1$

(f)

Consider the frame shown in Fig. 47.7(a). As shown in Fig. 47.7(b), the members are first assumed to behave as doubly fixed-ended beams, which is equivalent to stating that the joints do not translate or rotate. The magnitudes of the external joint forces needed to "lock" the joints are shown in Fig. 47.7(c) and can be readily obtained using simple statics. For example, restraining forces, Q, at joint B can be found from the equilibrium conditions at joint B.

$$V_{\mathrm{BA}} - N_{\mathrm{BC}} + Q_1 = 0 \quad [x\text{-direction}] \quad 47.25$$

$$-N_{\mathrm{BA}} - V_{\mathrm{BC}} + Q_2 = 0 \quad [y\text{-direction}] \quad 47.26$$

$$-M_{\mathrm{BA}} - M_{\mathrm{BC}} + Q_3 = 0 \quad [\text{moments}] \quad 47.27$$

In many cases, the resulting locked joint will result in a fixed-end beam. The fixed-end moments (FEMs) can be read directly from App. 47.A for most common loading conditions.

Since there are actually no restraining forces acting, the effect that the negative of the restraining Q forces have on the structure must be determined. A convenient way to do this is to compute the joint forces needed to impose a unit value at each of the possible joint movements (also known as *degrees of freedom*, DOF) while keeping the others restrained. These forces are easy to compute. Any joint load distribution can be expressed as a linear combination of the forces. The deformed structure with unit displacements imposed on the first three DOFs is illustrated in Figs. 47.7(d) through 47.7(f).

With beams, the stiffness method is simplified by the fact that the DOFs are only joint rotations. Truss bars are pinned at the joints, so for trusses, only the two translational DOFs need to be considered at each joint. Since the loads in trusses are also applied at the joints, the analysis of locked joints leads to zero bar forces. For frames, there are typically three DOFs at each joint. However, solutions assuming the bar elongations to be negligible are often quite accurate. This simplification can be used to reduce the number of DOFs for manual solutions.

The external forces required to impose the unit displacements are designated using the letter k and two subscripts. The first subscript designates the DOF where the force is applied; the second subscript designates the DOF where the unit displacement was imposed. Thus, $k_{i,j}$ is the force at DOF i due to a unit displacement at DOF j. (The terms "displacement" and "force" are used in a generalized sense.) Some computational effort is saved by recognizing that the stiffness matrix, **k**, is symmetrical, and $k_{i,j} = k_{j,i}$. The coefficients constituting **k** are the forces required to impose unit displacements at the end of an individual bar. The typical case of a prismatic member is depicted in Fig. 47.8.

Figure 47.8 *Forces in the Prismatic Bar Due to Unit Displacements*

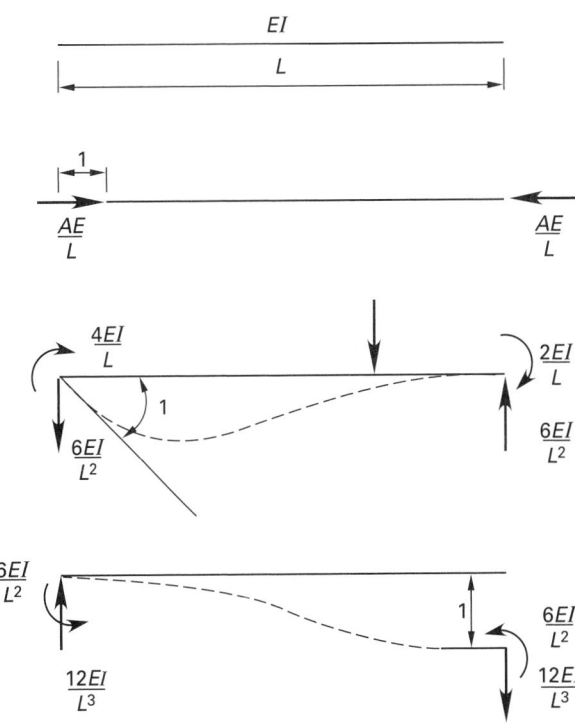

Equation 47.28 indicates that the Q forces must vanish (i.e., be zero). (The vector Δ collects the displacements at all of the DOFs.)

$$\mathbf{k}\Delta + \mathbf{Q} = 0 \quad 47.28$$

It is customary to write Eq. 47.28 in terms of a *load vector*, **L**, that is the negative of the restraining force vector.

$$k\Delta + L \qquad 47.29$$

The actual conditions are obtained by summing the conditions that exist in the locked condition and the conditions from the unit displacements scaled by the appropriate **Δ** values.

With the flexibility method, the number of simultaneous equations is equal to the degree of static indeterminacy. With the stiffness method, the number of simultaneous equations equals the number of degrees of freedom. When the number of equations is the same, the stiffness method is typically faster than the flexibility. This is because the terms in the equations are obtained from tabulated fixed-end moments and stiffness coefficients using basic principles of statics. In contrast, in the flexibility method, one must go through the time-consuming process of computing the deflections from basic principles. Closed form solutions are typically unavailable for the loading conditions encountered.

Example 47.6

Use the stiffness method to draw the moment diagram for the three-span beam shown. The product of the modulus of elasticity (E) and the centroidal moment of inertia (I) of the cross section is $EI = 28 \times 10^6$ in²-kips.

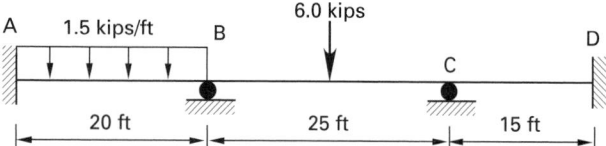

Solution

Since the supports at A and D are fixed, the only DOFs are the rotations at B and C. If these DOFs are locked, the structure will be made up of three fixed-ended beams. From App. 47.A, the restraining moments are

$$\text{FEM}_{AB} = \text{FEM}_{BA} = \frac{wL^2}{12} = \frac{\left(1.5 \; \frac{\text{kips}}{\text{ft}}\right)(20 \text{ ft})^2}{12}$$
$$= 50 \text{ ft-kips}$$

$$\text{FEM}_{BC} = \text{FEM}_{CB} = \frac{PL}{8} = \frac{(6 \text{ kips})(25 \text{ ft})}{8}$$
$$= 18.75 \text{ ft-kips}$$

For equilibrium at joints B and C, the restraining moment vector, **Q**, is

$$\mathbf{Q} = [31.25 \text{ ft-kips}, \; 18.75 \text{ ft-kips}]$$

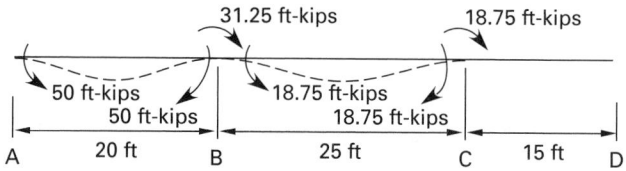

The deformed configuration corresponding to a unit rotation at point B (DOF no. 1) is shown. Let clockwise be the positive direction. Calculate the nonzero moments at the ends of the bars for this configuration. (All other moments are zero.)

$$M_{AB} = \frac{2EI}{L} = \frac{(2)(28 \times 10^6 \text{ in}^2\text{-kips})}{(20 \text{ ft})\left(12 \; \frac{\text{in}}{\text{ft}}\right)^2}$$
$$= 19{,}444 \text{ ft-kips}$$

$$M_{BA} = \frac{4EI}{L} = \frac{(4)(28 \times 10^6 \text{ in}^2\text{-kips})}{(20 \text{ ft})\left(12 \; \frac{\text{in}}{\text{ft}}\right)^2}$$
$$= 38{,}889 \text{ ft-kips}$$

$$M_{BC} = \frac{4EI}{L} = \frac{(4)(28 \times 10^6 \text{ in}^2\text{-kips})}{(25 \text{ ft})\left(12 \; \frac{\text{in}}{\text{ft}}\right)^2}$$
$$= 31{,}111 \text{ ft-kips}$$

$$M_{CB} = \frac{2EI}{L} = \frac{(2)(28 \times 10^6 \text{ in}^2\text{-kips})}{(25 \text{ ft})\left(12 \; \frac{\text{in}}{\text{ft}}\right)^2}$$
$$= 15{,}556 \text{ ft-kips}$$

Consider the rotational equilibrium requirement at joints B and C.

$$k_{1,1} = M_{BA} + M_{BC} = 38{,}889 \text{ ft-kips} + 31{,}111 \text{ ft-kips}$$
$$= 70{,}000 \text{ ft-kips}$$

$$k_{2,1} = M_{CB} + M_{CD} = 15{,}556 \text{ ft-kips} + 0$$
$$= 15{,}556 \text{ ft-kips}$$

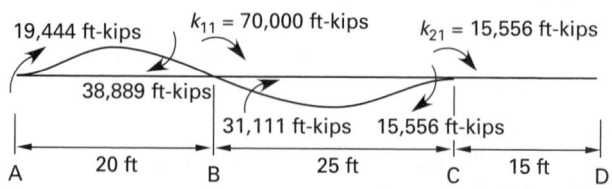

Use the same procedure to determine the moments induced by the deformed structure corresponding to a unit rotation at DOF no. 2. (Note that the coefficients

$k_{2,1}$ and $k_{1,2}$ are equal since the stiffness matrix is symmetrical.)

$$k_{2,2} = 82{,}963 \text{ ft-kips}$$

$$k_{1,2} = 15{,}556 \text{ ft-kips}$$

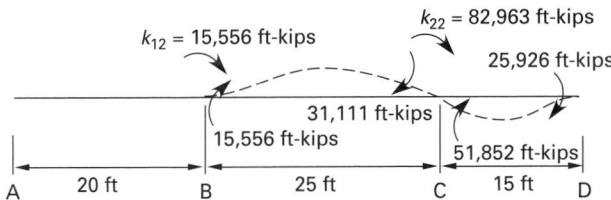

The stiffness matrix is

$$\mathbf{k} = \begin{bmatrix} 70{,}000 & 15{,}556 \\ 15{,}556 & 82{,}963 \end{bmatrix}$$

Use Eq. 47.29. The "deflections" are actually rotations, so use θ to represent the joint rotations.

$$\mathbf{k}\boldsymbol{\theta} = -\mathbf{Q}$$

$$\begin{bmatrix} 70{,}000 & 15{,}556 \\ 15{,}556 & 82{,}963 \end{bmatrix} \begin{bmatrix} \theta_B \\ \theta_C \end{bmatrix} = \begin{bmatrix} -31.25 \\ -18.75 \end{bmatrix}$$

The solution to this system of equations is $\theta_B = -0.000413$ rad and $\theta_C = -0.000149$ rad.

The actual moments at the ends of the members are obtained by superposition of the locked cases and the scaled contributions of the moments calculated from the unit values. For example, the moment at end M_{BA} is

$$M_{BA} = 50 \text{ ft-kips} + (-0.000413 \text{ rad})(38{,}889 \text{ ft-kips})$$
$$= 33.92 \text{ ft-kips}$$

Moments that are not at the ends of the bars can be obtained from statics once the end moments have been computed.

15. MOMENT DISTRIBUTION METHOD

The *moment distribution method* was developed by Prof. Hardy Cross in the early 1930s. The procedure is based on a locked-joint situation similar to stiffness. The name "moment distribution" derives from the fact that the external moments needed to obtain the locked condition are "distributed" using an iterative approach.

Although it is possible in principle to deal with axial deformations, practical moment distribution always assumes that the members do not change in length as a result of the applied loads. Moment distribution is particularly simple in cases where the fixed-length assumption is sufficient to ensure that the joints of the structure are fixed in space: This case is typically referred to as "structures without sideway." Moment

distribution can also be used when the joints in the structure displace (i.e., where there is sideway), although the procedure is not as efficient.

The moment required to induce a unit rotation at the end of a bar when the far end is assumed to be fixed is known as the *stiffness*, K. Referring to Fig. 47.8, the flexural stiffness is $K = 4EI/L$. The *carryover factor*, COF, is defined as the ratio of the moment that appears at the far end to the applied moment. Referring to Fig. 47.8, the carryover factor is COF = $1/2$.

Consider a situation where a number of bars meet at a common rigid joint while all of the far ends are fixed. Assume that the joint is subjected to an external applied moment, M. Each of the bars will support some portion of the applied moment. With m_i as the moment supported by bar i, the *distribution factor* for bar i (at the joint in question) is the fraction of the total applied moment distributed to another joint.

$$\mathrm{DF}_i = \frac{m_i}{M} \qquad \textit{47.30}$$

All of the bars that meet at the common joint rotate by an equal amount, θ. The moment that is induced by a rotation is the product of the flexural stiffness times the rotation.

$$m_i = K_i \theta \qquad \textit{47.31}$$

The sum of all the moments in the bars equal the applied moment.

$$M = \sum_i m_i$$
$$= \sum_i K_i \theta \qquad \textit{47.32}$$

Combining Eqs. 47.30 and 47.32,

$$\mathrm{DF}_i = \frac{K_i}{\displaystyle\sum_i K_i} \qquad \textit{47.33}$$

The sum of all of the distribution factors at a rigid joint is 1.

$$\sum_i \mathrm{DF}_i = 1 \qquad \textit{47.34}$$

16. MOMENT DISTRIBUTION PROCEDURE: NO SIDESWAY

step 1: Draw a line diagram representation of the structure to be analyzed. This diagram will be used to record intermediate results.

step 2: Except for cantilevered sections that have $K = 0$, compute the flexural stiffness for each bar using Eq. 47.35 or Eq. 47.36.

$$K = \frac{4EI}{L} \quad \text{[opposite end fixed]} \qquad \textit{47.35}$$

$$K = \frac{3EI}{L} \quad \begin{bmatrix} \text{opposite end pinned} \\ \text{or cantilevered} \end{bmatrix} \qquad \textit{47.36}$$

step 3: Use Eq. 47.33 to compute the distribution factor at each end of the bars. Record the values on the drawing of the structure. Since the distribution factor is a ratio of flexural stiffnesses, one can always take a common factor (e.g., the product EI) from the numerator and the denominator without affecting the results. It is convenient to remember that DF = 0 at fixed ends (representing infinite stiffness) and DF = 1 at simple supports.

step 4: Obtain the fixed-end moments from App. 47.A. (Adopt a consistent sign convention, such as "clockwise moments at the ends are positive.") Record the fixed-end moments on the drawing at the ends of each bar.

step 5: Go to any joint and compute the "unbalanced moment" (or "out of balance," OOB) by adding all the moments at the ends of the bars that meet there. Distribute the unbalanced moment to each bar. The distributed moments are equal to the product of the unbalanced moment times the distribution factor with the reversed sign. (Although the order used for releasing and locking the joints is arbitrary, convergence is fastest if the joint with the largest unbalance is handled first.)

step 6: Carry the distributed moment to the far end of the corresponding bar by multiplying the distributed moment by the *carryover factor*. The carryover factor is 0.5 for joints with opposite ends that are fixed; it is 0 for joints with opposite ends that are pinned or cantilevered.

step 7: Repeat steps 5 and 6 until the unbalanced moment at the most unbalanced joint is insignificant (typically, less than 1% of the largest fixed-end moment).

step 8: Calculate the true moments at each bar end as the sum of moments that have been recorded at that end.

Example 47.7

Use the moment distribution method to determine the end moments for the continuous beam shown.

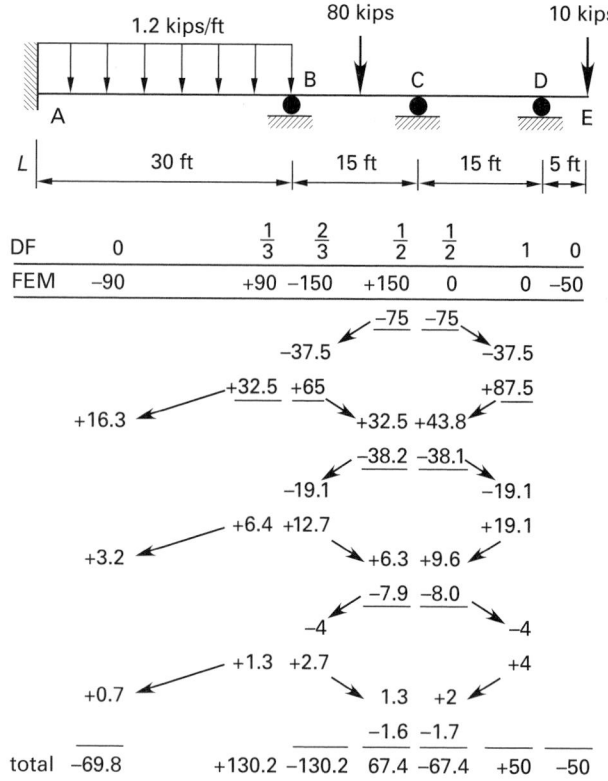

Solution

Use Eq. 47.35 to calculate the flexural stiffness of each bar. Since EI is constant and only the relative values of K affect the results, take $4EI$ as the longest span (30 ft in this case). Then,

$$K_{AB} = \frac{4EI}{L} = \frac{30 \text{ ft}}{30 \text{ ft}} = 1$$

$$K_{BC} = \frac{30 \text{ ft}}{15 \text{ ft}} = 2$$

$$K_{CD} = \frac{30 \text{ ft}}{15 \text{ ft}} = 2$$

$$K_{DE} = 0 \quad [\text{cantilever}]$$

Determine the distribution factors.

$$\text{DF}_{AB} = 0 \quad [\text{fixed end}]$$

$$\text{DF}_{BA} = \frac{K_{BA}}{\sum_i K_i}$$

$$= \frac{1}{1+2} = 1/3$$

$$\text{DF}_{BC} = \frac{2}{1+2} = 2/3$$

$$\text{DF}_{CB} = \frac{2}{2+2} = 1/2$$

$$\text{DF}_{CD} = \frac{2}{2+2} = 1/2$$

$$\text{DF}_{DC} = \frac{2}{2+0} = 1 \quad [\text{free end}]$$

The fixed-end moments are determined from App. 47.A.

$$\text{FEM}_{AB} = \text{FEM}_{BA} = \frac{wL^2}{12}$$

$$= \frac{\left(1.2 \, \frac{\text{kips}}{\text{ft}}\right)(30 \text{ ft})^2}{12} = 90 \text{ ft-kips}$$

$$\text{FEM}_{BC} = \text{FEM}_{CB} = PL/8$$

$$= \frac{(80 \text{ kips})(15 \text{ ft})}{8} = 150 \text{ ft-kips}$$

$$\text{FEM}_{DE} = (10 \text{ kips})(5 \text{ ft}) = 50 \text{ ft-kips}$$

$$\text{FEM}_{ED} = 0 \quad [\text{free end}]$$

Start the moment distribution at joint C. The unbalanced moment of 150 ft-kips is reversed in sign and multiplied by the distribution factors, both of which are one half. This puts a balancing correction of −75 ft-kips on either side of joint C, for a total balancing correction of −150 ft-kips. (The horizontal lines under the moments of −75 ft-kips indicate that joint C has now been balanced.) Multiply the balancing moments by the carryover factors (0.5) and carry the products over to the far ends of BC and CD.

Continue by moving to joint B. The unbalanced moment is 90 ft-kips − 150 ft-kips − 37.5 ft-kips = −97.5 ft-kips, requiring a total balancing moment of 97.5 ft-kips. The balance is distributed according to the distribution factors of one third and two thirds, resulting in 32.5 ft-kips and 65 ft-kips on the left and right sides of the joint, respectively. When the balance is multiplied by the COF, half of the distributed moments are carried over to the far side. (The horizontal lines indicate that the joint is balanced.)

The procedure continues in this manner as the carryover values decrease in magnitude. Finish at joint C by distributing the unbalanced moment of 2 ft-kips + 1.3 ft-kips, but do not perform the carryover step. (The process should finish with a distribution so that the joint will be balanced.)

The end moments are obtained by summing all the entries at each location.

17. STRUCTURES WITH SIDESWAY

Analyzing structures with sidesway with the moment distribution procedure uses superposition of elementary joint-restrained cases. Consider the two-story frame shown in Fig. 47.9(a). This frame has two sway DOFs: the horizontal translations at each of the two floors. The first step in the analysis is to obtain the solution for the structure when the translations are restricted. This step is schematically depicted in Fig. 47.9(b). The forces, R_1 and R_2, against the fictitious restraints that are keeping the frame from swaying are obtained after the moment distribution is completed.

Figure 47.9 *Frames with Sidesway*

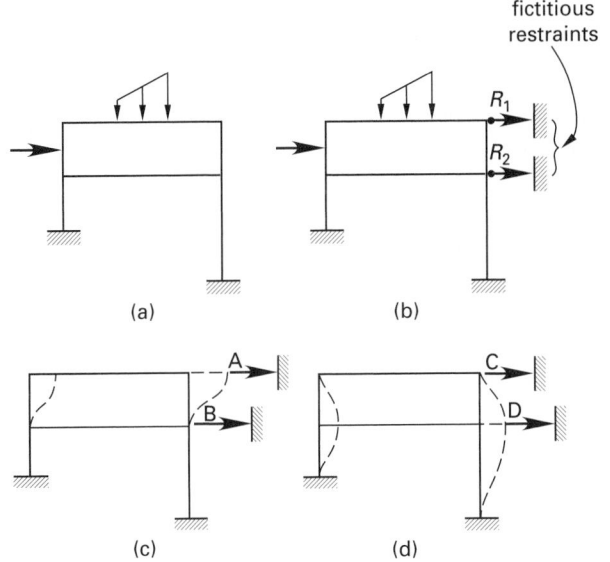

The next step consists of obtaining moment diagrams associated with arbitrary translations at each of the locations where the restraints were previously introduced. In these analyses, the translations are initially imposed with the joints held against rotation. The unbalanced moments generated are distributed using the standard moment distribution approach. The situation when the first restraint is "dragged" (while the second is held in place) is depicted prior to the moment distribution in Fig. 47.9(c). The deformed configuration when the second restraint is dragged an arbitrary amount is shown in Fig. 47.9(d). It is helpful to recall that the fixed-end moment generated by a relative displacement of magnitude Δ is

$$\text{FEM} = \frac{6EI\Delta}{L^2} \qquad 47.37$$

Once the moment distribution is complete, the forces at the fictitious restraints are computed using statics. For example, the force in the restraint at the second story is equal to the sum of the shear forces in the two columns.

The final solution is obtained as a linear combination of the no-sidesway case and scaled versions of the case with sidesway. The scaling is obtained by recognizing that the actual forces in the restraints are zero. For example, the requirements for the two-story frame of Fig. 47.9 are given by Eqs. 47.38 and 47.39, in which A, B, C, and D are the forces in the restraints for the unit displacement condition, and α and β are the scaling factors.

$$R_1 + \alpha\text{A} + \beta\text{C} = 0 \qquad 47.38$$
$$R_2 + \alpha\text{B} + \beta\text{D} = 0 \qquad 47.39$$

Designating the displaced moments from the sidesway evaluation at the first and second floors as MD_1 and MD_2, respectively, the complete solution for the moment at any particular joint is

$$M = M_{\text{sidesway restrained}} + \alpha\text{MD}_1 + \beta\text{MD}_2 \qquad 47.40$$

18. SECOND-ORDER (P-Δ) ANALYSIS

In a first-order analysis, the loads are supported and equilibrium is achieved on the structural geometry that exists prior to the application of the loads. However, this is an approximation since equilibrium is actually achieved in a deformed configuration. In a second-order analysis, the effects of deformations on the equilibrium requirements are considered.

Figure 47.10 Second-Order Effect on a Cantilever

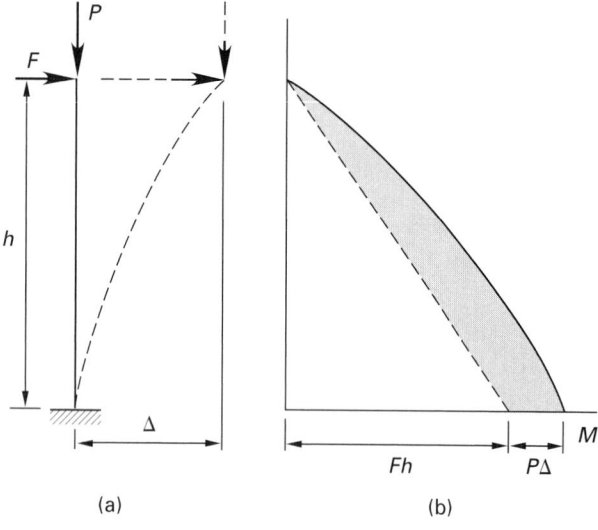

(a)   (b)

The approximation implicit in a first-order analysis is readily apparent in the cantilever column in Fig. 47.10(a). For the loading shown, the moment at the base in a first-order analysis is $M = Fh$. However, in the deformed configuration, the total moment at the base is actually $M = Fh + P\Delta$, as shown in Fig. 47.10(b). The total deflection Δ results from load F and the moments induced by both F and P when acting on the deformed geometry.

Second-order analyses of complex structures are best performed on computers, and these methods are increasingly being used. (ACI 318 and LRFD encourage second-order analysis as the preferred approach for incorporating P-Δ effects in the design of slender columns.)

19. SIMPLIFIED SECOND-ORDER ANALYSIS

Second-order effects include bending members with respect to their chord, usually referred to as the P-δ effect, and the effects of joint translation, generally designated as the P-Δ effect. Structures where second-order effects are important are typically flexible and unbraced. In such structures, virtually all the difference between the correct solution and the results predicted by a first-order analysis derives from the P-Δ effect.

A simple iterative approach that incorporates the P-Δ effect in the analysis of building frames is schematically illustrated in Fig. 47.11. The steps are as follows.

step 1: Compute the deflections of the structure using a first-order analysis.

step 2: Consider that all the vertical loads are carried by an auxiliary structure consisting of a vertical stack of truss bars that is stabilized by horizontal links to the structure. This auxiliary structure experiences the displacements of the actual building.

step 3: Calculate the forces in the links (forces A, B, and C in Fig. 47.11(b)) from the free-body diagram of the auxiliary structure.

step 4: Repeat step 1, adding the forces in the link bars to the applied lateral loads. The solution converges when the forces in the links do not change significantly from one iteration to the next.

Figure 47.11 P-Δ Analysis

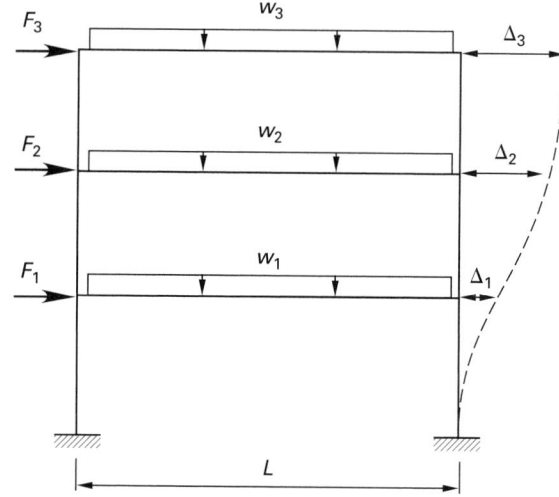

(a) loaded structure

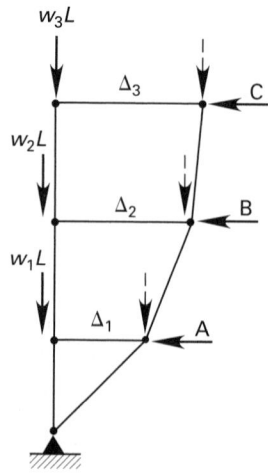

(b) auxiliary structure

20. PLASTIC ANALYSIS

A *plastic analysis* determines the maximum load that a structure can sustain. Alternatively, plastic analysis determines the plastic moment capacity required to support a certain factored load. Although plastic analysis is applicable to all aspects of structural design, it is primarily used in the design of continuous steel beams.

In practical applications, plastic analysis assumes that a hinge is created when the moment at any section reaches the plastic moment. These hinges are known as *plastic hinges* because they are not points of zero moment, but locations where the moment is constant and equal to the plastic capacity M_p. The maximum load-carrying capacity of the structure is attained when there are sufficient hinges to turn the structure into a mechanism.

21. PLASTIC ANALYSIS OF BEAMS

Either the *equilibrium method* or the *virtual work method* can be used in a plastic analysis of a beam. The virtual work method is used in the following analysis, which determines an unknown plastic moment capacity for a given factored load. (This is the usual case when designing steel beams.)

step 1: Postulate a number of plastic hinges that turn the beam into a mechanism. (If more than one mechanism is possible, all must be considered. The mechanism requiring the largest moment capacity governs.)

step 2: Impose a unit displacement on the point of the structure that deflects the most after the mechanism forms.

step 3: Calculate the total internal work at hinges 1, 2, ..., i. (The values of the rotations, $\theta_{p,i}$, at all of the hinges are known once the unit deflection is imposed.)

$$W_{\text{internal}} = \sum_i M_{p,i}\theta_{p,i} \qquad \textbf{47.41}$$

step 4: Calculate the total external work performed by loads 1, 2, ..., j from the loads, P_j, and the corresponding displacements, Δ_j. The replacement of distributed loads by their resultants must be done on each of the segments of the mechanism independently. The loads P_i will typically include an appropriate load factor.

$$W_{\text{external}} = \sum_j P_j\Delta_j \qquad \textbf{47.42}$$

step 5: Equate the internal work to the external work and solve for M_p.

Example 47.8

Use plastic analysis to calculate the maximum plastic moment from the loads acting on the two-span beam shown.

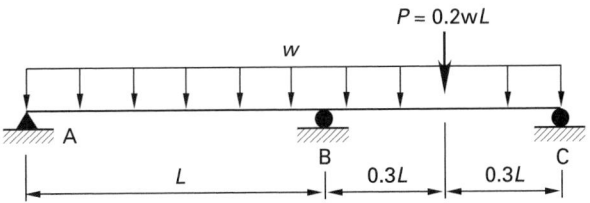

Solution

Two distinct mechanisms are possible.

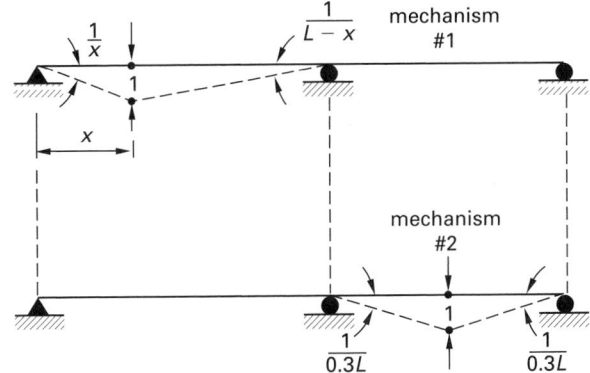

Consider the first mechanism involving failure in span AB. Due to a lack of symmetry, the location of the positive plastic hinge is not known in advance. Designating the distance between the support at A and the positive-moment plastic hinge as x, the external work is given by Eq. 47.42.

$$W_{\text{external}} = 0.5wx + 0.5w(L - x) = 0.5wL$$

The internal work is given by Eq. 47.41.

$$W_{\text{internal}} = U_{\text{internal}} = M_P\left(\frac{1}{x} + \frac{1}{L - x}\right)$$
$$+ M_P\left(\frac{1}{L - x}\right)$$

Equate the external and internal work.

$$M_p = \frac{wLx(L - x)}{2(L + x)}$$

Determine the location of the maximum moment by taking the derivative with respect to x and equate the result to zero.

$$x_{\text{maximum } M_p} = L\left(\sqrt{2} - 1\right)$$

The maximum moment is found by substituting x into the expression for M_p.

$$M_{p,\text{max}} = 0.08579wL^2$$

Now, consider the mechanism involving failure in span BC. The external work for a unit deflection is

$$W_{\text{external}} = (0.3wL + 0.2wL)(1) = 0.5wL$$

The internal work is

$$W_{\text{internal}} = U_{\text{internal}} = M_p \left(\frac{2}{0.3L} \right) + M_p \left(\frac{1}{0.3L} \right)$$

The plastic moment is found by setting the external and internal works equal to each other.

$$M_p = 0.05wL^2$$

The plastic moment is larger for the mechanism in span AB and thus controls.

22. APPROXIMATE METHOD: ASSUMED INFLECTION POINTS

Approximate methods are useful when an exact solution is not needed, when time is short, or when it is desired to quickly check an exact solution. If the location of a point of inflection on a statically indeterminate structure is known or assumed, that knowledge can be used to generate moments for the remainder of the structure.

The curvature on a structure changes between positive and negative at a point of inflection. Accordingly, the moment is zero at that point. If the point of inflection is assumed, then the moment at that point is also known. The applicability of this method depends on being able to predict the locations of the inflection points. Figure 47.12 provides reasonable predictions of these locations.

Example 47.9

Determine the fixed-end moment at joint A.

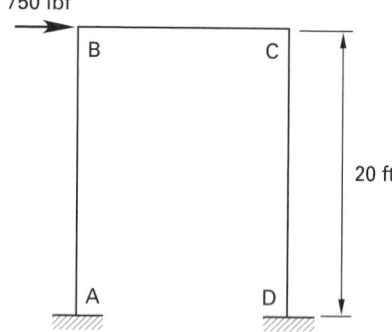

Solution

Assume that a point of inflection occurs on each vertical member at $(0.55)(20 \text{ ft}) = 11 \text{ ft}$ above the supports. By symmetry and equilibrium, the shear at the inflection point is

$$V = \frac{750 \text{ lbf}}{2} = 375 \text{ lbf}$$

The moment at the base is

$$M = (11 \text{ ft})(375 \text{ lbf}) = 4125 \text{ ft-lbf}$$

Figure 47.12 *Approximate Locations of Inflection Points*

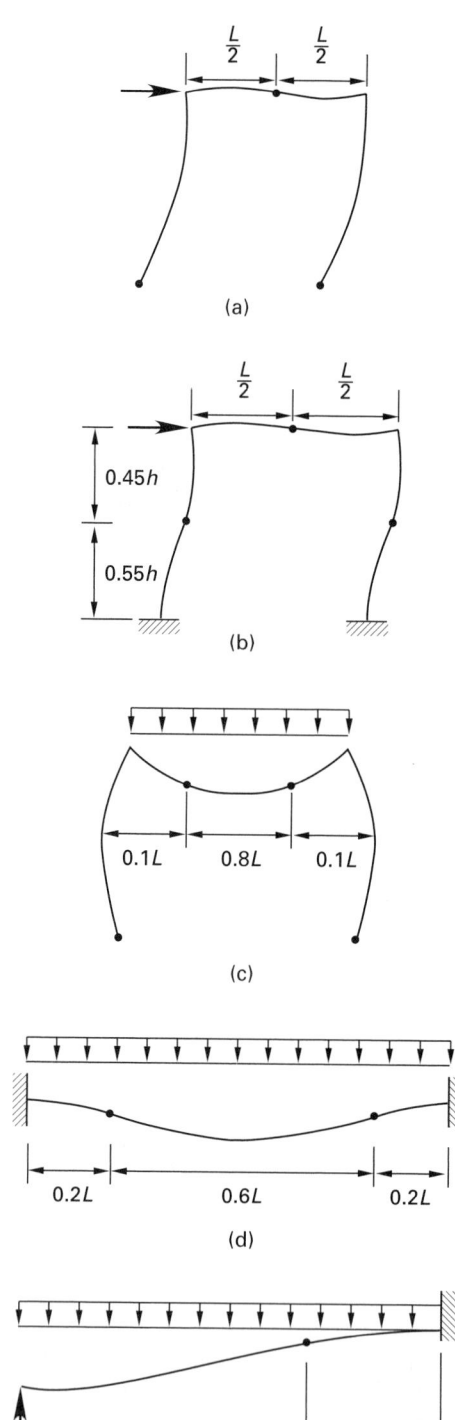

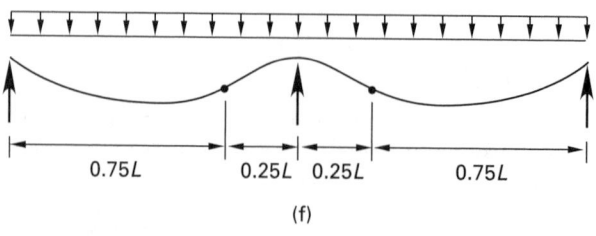

The fixed-end moment is counterclockwise. Therefore, $M_{AB} = -4125$ ft-lbf.

23. APPROXIMATE METHOD: MOMENT COEFFICIENTS

With certain restrictions, ACI 318 permits continuous beams and slabs constructed of reinforced concrete to be designed using tabulated moment coefficients. This section deals specifically with beams and one-way slabs (slabs that support moments in one direction only, ACI 318 8.3). Approximate design of two-way slabs [ACI 318 13.6], known as the *direct design method*, follows different rules and is not covered here.

The maximum moments at the ends and midpoints of continuously loaded spans are taken as fractions of the distributed load function, wL^2, where L is the span length between supports when computing shears and moments within the span. L is the average of adjacent spans when computing moment over a support. w is the ultimate factored load. The moment at some point along the beam is obtained from the *moment coefficient*, C_1. Table 47.2 contains the moment coefficients allowed by ACI 318.

$$M = C_1 w L^2 \qquad 47.43$$

The method of moment coefficients can be used when the following conditions are met: (a) The load is continuously distributed, (b) construction is not prestressed, (c) there are two or more spans, (d) the longest span length is less than 20% longer than the shortest, and (e) the beams are prismatic, having the same cross section along their lengths.

Table 47.2 *ACI Moment Coefficients*

condition	C_1
positive moments near midspan	
end spans	
simply supported	$\frac{1}{11}$
built-in support	$\frac{1}{14}$
interior spans	$\frac{1}{16}$
negative moments at exterior face of first interior support	
2 spans	$\frac{1}{9}$
3 or more spans	$\frac{1}{10}$
negative moments at other faces or interior supports	
all cases	$\frac{1}{11}$
negative moments at face of all supports	
slabs with spans not exceeding 10 ft, and beams with ratio of sum of column stiffnesses to beam stiffness exceed 8 at each end of the span	$\frac{1}{12}$
negative moments at exterior built-in support	
support is a cross beam or girder (spandrel beam)	$\frac{1}{24}$
support is a column	$\frac{1}{16}$

24. APPROXIMATE METHOD: SHEAR COEFFICIENTS

When the conditions in Sec. 23 are met, ACI 318 8.3 also permits approximating the critical shear using *shear coefficients*. For shear, the design coefficient is used with the average span loading, $\frac{1}{2}wL$. In Eq. 47.44, C_2 has a value of 1.15 for end members at the first interior support. For shear at the face of all other supports, $C_2 = 1$.

$$V = C_2 \left(\frac{wL}{2} \right) \qquad 47.44$$

Example 47.10

Determine the critical moment and draw the moment diagram for the uniformly loaded, continuous beam shown. Support A is simple; supports BC and D are built-in column supports. The ultimate factored loading is 500 lbf/ft. All of the conditions necessary to use ACI moment coefficients are satisfied.

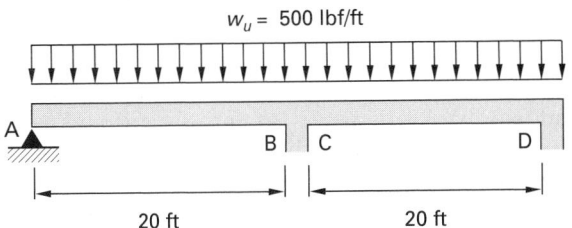

Solution

The moment at point A is zero since it is a simple support. Point B is an exterior face of the first interior support over a two-span beam. The moment there is

$$M_B = \left(-\frac{1}{9} \right) \left(500 \; \frac{\text{lbf}}{\text{ft}} \right) (20 \text{ ft})^2 = -22{,}222 \text{ ft-lbf}$$

Point C is also an exterior face of a first interior support (counting from the opposite end).

$$M_C = M_B = -22{,}222 \text{ ft-lbf}$$

Point D is an exterior column support.

$$M_D = \left(-\frac{1}{16} \right) \left(500 \; \frac{\text{lbf}}{\text{ft}} \right) (20 \text{ ft})^2 = -12{,}500 \text{ ft-lbf}$$

The left span is a simply supported end span. The maximum positive moment is

$$M_L = \left(\frac{1}{11} \right) \left(500 \; \frac{\text{lbf}}{\text{ft}} \right) (20 \text{ ft})^2 = 18{,}182 \text{ ft-lbf}$$

The right span is an end-span with a built-in support. The maximum positive moment is

$$M_R = \left(\frac{1}{14} \right) \left(500 \; \frac{\text{lbf}}{\text{ft}} \right) (20 \text{ ft})^2 = 14{,}286 \text{ ft-lbf}$$

The critical moment diagram is

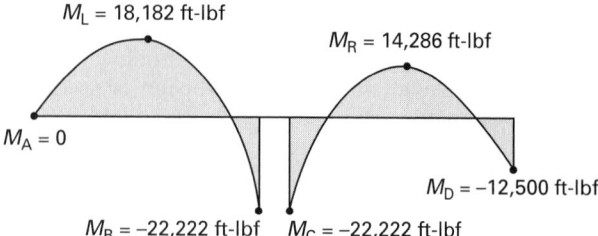

$M_L = 18,182$ ft-lbf

$M_R = 14,286$ ft-lbf

$M_A = 0$

$M_D = -12,500$ ft-lbf

$M_B = -22,222$ ft-lbf $M_C = -22,222$ ft-lbf

25. APPROXIMATE METHOD: ENVELOPE OF MAXIMUM SHEAR

Since a live load can be placed or distributed anywhere along the beam length, it is convenient to draw the applied shear loading as a function of location. Such a drawing is known as a *shear envelope*. Since beams (particularly concrete beams) must remain ductile in failure, the shear envelope should be drawn accurately, if not conservatively. A conservative *envelope of maximum shear* for simple spans can be drawn by assuming a maximum shear condition for the ends (usually uniform dead and live loading over the entire span) and assuming a different maximum shear condition for the beam's midspan (usually, uniform live loading over the first half of the span), even though these two conditions cannot occur simultaneously. The shear envelope is assumed to vary linearly in between, which makes interpolation of shear at positions along the beam relatively simple.

48 Properties of Concrete and Reinforcing Steel

Nomenclature

A	area	in^2	mm^2
c	distance from neutral axis to extreme fiber	in	mm
D	diameter	in	mm
e	normal strain	in/in	mm/mm
E	modulus of elasticity	lbf/in^2	MPa
f	strength	lbf/in^2	MPa
I	moment of inertia	in^4	mm^4
L	length	in	mm
M	moment	in-lbf	N·mm
P	force	lbf	N
s	normal stress	lbf/in^2	MPa
w	specific weight	lbf/ft^3	kg/m^3

Subscripts

c	compressive
ct	compressive tensile
r	rupture

1. INTRODUCTION

Concrete (*portland cement concrete*) is a mixture of cement, aggregates, water, and air. The cement paste consists of a mixture of portland cement and water. The paste binds the coarse and fine aggregates into a rock-like mass as the paste hardens during the chemical reaction (*hydration*). Table 48.1 lists the approximate volumetric percentage of each ingredient.

Table 48.1 *Typical Volumetric Proportions of Concrete Ingredients*

component	air-entrained	non-air-entrained
coarse aggregate	31%	31%
fine aggregate	28%	30%
water	18%	21%
cement	15%	15%
air	8%	3%

2. CEMENT

Portland cement is produced by burning a mixture of lime and clay in a rotary kiln and grinding the resulting mass. Cement has a specific weight (density) of approximately 195 lbf/ft^3 (3120 kg/m^3) and is packaged in standard sacks ("bags") weighing 94 lbf (40 kg).

ASTM C-150 describes the five classifications of portland cement.

Type I—Normal portland cement: This is a general-purpose cement used whenever sulfate hazards are absent and when the heat of hydration will not produce a significant rise in the temperature of the cement. Typical uses are sidewalks, pavement, beams, columns, and culverts.

Type II—Modified portland cement: This cement has a moderate sulfate resistance, but is generally used in hot weather for the construction of large structures. Its heat rate and total heat generation are lower than those of normal portland cement.

Type III—High-early strength portland cement: This type develops its strength quickly. It is suitable for use when a structure must be put into early use or when long-term protection against cold temperatures is not feasible. Its shrinkage rate, however, is higher than those of types I and II, and extensive cracking may result.

Type IV—Low-heat portland cement: For massive concrete structures such as gravity dams, low-heat cement is required to maintain a low temperature during curing. The ultimate strength also develops more slowly than for the other types.

Type V—Sulfate-resistant portland cement: This type of cement is appropriate when exposure to sulfate concentration is expected. This typically occurs in regions having highly alkaline soils.

Types I, II, and III are available in two varieties: normal and air-entraining (designated by an "A" suffix). The compositions of the three types of air-entraining portland cement (types IA, IIA, and IIIA) are similar to types I, II and III, respectively, with the exception that an air-entraining admixture is added.

Many states use modified concrete mixes in critical locations in order to reduce *concrete-disintegration cracking* ("D-cracking") caused by the freeze-thaw cycle. Coarse aggregates are the primary cause of D-cracking, so the maximum coarse aggregate size is reduced. However, a higher cement paste content causes shrinkage cracking during setting, leading to increased water penetration and corrosion of reinforcing steel. The cracking can be reduced or eliminated by using *shrinkage-compensating cement,* known as "type-K cement" (named after ASTM C-846 type E-1(K)).

Type-K cement (often used in bridge decks) contains an aluminate that expands during setting, offsetting the shrinkage. The net volume change is near zero. The resulting concrete is referred to as *shrinkage-compensating concrete.*

Special cement formulations are needed to reduce *alkali-aggregate reactivity* (AAR)—the reaction of the alkalis in the cement with compounds in the sand and gravel aggregate. AAR produces long-term distress in the forms of network cracking and spalling (popouts) in otherwise well-designed structures. AAR takes on two forms: the more common *alkali-silica reaction* (ASR) and the less-common *alkali-carbonate reaction* (ACR). ASR is countered by using low-alkali cement (ASTM C-150) with an equivalent alkali content of less than 0.60% (as sodium oxide), using lithium-based admixtures, or "sweetening" the mixture by replacing approximately 30% of the aggregate with crushed limestone. ACR is not effectively controlled by using low-alkali cements. Careful selection, blending, and sizing of the aggregate are needed to minimize ACR.

3. AGGREGATE

Because aggregate makes up 60–75% of the total concrete volume, its properties influence the behavior of freshly mixed concrete and the properties of hardened concrete. Aggregates should consist of particles with sufficient strength and resistance to exposure conditions such as freezing and thawing cycles. Also, they should not contain materials that will cause the concrete to deteriorate.

Most sand and rock aggregate has a specific weight of approximately 165 lbf/ft^3 (2640 kg/m^3) corresponding to a specific gravity of 2.64.

Fine aggregate consists of natural sand or crushed stone up to $1/4$ in (6 mm), with most particles being smaller than 0.2 in (5 mm). Aggregates, whether fine or coarse, must conform to certain standards to achieve the best engineering properties. They must be strong, clean, hard, and free of absorbed chemicals. Fine aggregates must meet the particle-size distribution (grading).

The seven standard ASTM C-33 sieves for fine aggregates have openings ranging from 0.150 mm (no. 100 sieve) to $3/8$ in (9.5 mm). The fine aggregate should have not more than 45% passing any sieve and retained on the next consecutive sieve, and its fineness modulus should be not less than 2.3 or more than 3.1. The *fineness modulus* is an empirical factor obtained by adding the cumulative weight percentages retained on each of a specific series (usually no. 4, no. 8, no. 16, no. 30, no. 50, and no. 100 for the fine aggregate) of sieves and dividing the sum by 100. (The dust or pan percentage is not included in calculating the cumulative percentage retained.) The higher the fineness modulus, the coarser will be the gradation.

Coarse aggregates consist of natural gravel or crushed rock, with pieces large enough to be retained on a no. 4 sieve (openings of 0.2 in or 4.75 mm). In practice, coarse aggregate is generally between $3/8$ in and $1\frac{1}{2}$ in (9.5 to 38 mm) in size. Also, coarse aggregates should meet the gradation requirements of ASTM C-33, which specifies 13 standard sieve sizes for coarse aggregate.

Coarse aggregate has three main functions in a concrete mix: (a) to act as relatively inexpensive filler, (b) to provide a mass of particles that are capable of resisting the applied loads, and (c) to reduce the volume changes that occur during the setting of the cement-water mixture.

4. WATER

Water in concrete has three functions. (a) Water reacts chemically with the cement. This chemical reaction is known as *hydration.* (b) Water wets the aggregate. (c) The water and cement mixture, which is known as *cement paste,* lubricates the concrete mixture and allows it to flow.

Water has a standard density of 62.4 lbf/ft^3 (1000 kg/m^3). 7.48 gal occupy 1 ft^3 (1000 L occupy 1 m^3). One ton (2000 lbf) of water has a volume of 240 gal.

Any potable water that has no pronounced odor or taste can be used for producing concrete. (With some quality restrictions, the ACI code also allows nonpotable water to be used in concrete mixing.) Impurities in water may affect the setting time, strength, and corrosion resistance. Water used in mixing concrete should be clean and free from injurious amounts of oils, acids, alkalis, salt, organic materials, and other substances that could damage the concrete or reinforcing steel.

5. ADMIXTURES

Admixtures are routinely used to modify the performance of concrete. Advantages include higher strength, durability, chemical resistance, and workability; controlled rate of hydration; and reduced shrinkage and cracking. Accelerating and retarding admixtures fall into several different categories, as classified by ASTM C-494.

Type A: water-reducing

Type B: set-retarding

Type C: set-accelerating

Type D: water-reducing and set-retarding

Type E: water-reducing and set-accelerating

Type F: high-range water-reducing

Type G: high-range water-reducing and set-retarding

ASTM C-260 covers air-entraining admixtures, which enhance freeze-thaw durability. ASTM C-1017 deals exclusively with plasticizers to produce flowing concrete. ACI 212 recognizes additional categories, including corrosion inhibitors and dampproofing. Finally, microsilica, fly ash, and synthetic fibers are routinely used in concrete.

Water-reducing admixtures disperse the cement particles throughout the plastic concrete, reducing water requirements by 5–10%. Water that would otherwise be trapped within the cement floc remains available to fluidize the concrete. Although water is necessary to produce concrete, using lesser amounts increases strength and durability and decreases permeability and shrinkage. The same slump can be obtained with less water.

High-range water reducers, also known as *superplasticizers*, function via the same mechanisms as regular water reducers. However, the possible water reduction is greater (e.g., 12–30%). Dramatic increases in slump, workability, and strength are achieved. *High-slump concrete* is suitable for use in sections that are heavily reinforced and in areas where consolidation cannot otherwise be attained. Also, concrete can be pumped at lower pump pressures, so the lift and pumping distance can be increased. Overall, superplasticizers reduce the cost of mixing, pumping, and finishing concrete.

Set accelerators increase the rate of cement hydration, shortening the setting time and increasing the rate of strength development. They are useful in cold weather (below 35–40°F or 2–4°C) or when urgent repairs are needed. While calcium chloride ($CaCl_2$) is a very effective accelerator, nonchloride, noncorrosive accelerators can provide comparable performance. (The ACI code does not allow chloride to be added to concrete used in prestressed construction, in concrete containing aluminum embedments, or in concrete cast against galvanized stay-in-place steel forms.)

Set retarders are used in hot environments and where the concrete must remain workable for an extended period of time, allowing extended haul and finishing times.

A higher ultimate strength will also result. Most retarders also have water-reducing properties.

Air entraining mixtures create microscopic air bubbles in the concrete. This improves the durability of hardened concrete subject to freezing and thawing cycles. The wet workability is improved, while bleeding and segregation are reduced.

A waste product of coal-burning power-generation stations, *fly ash* is the most common *pozzolanic additive*. As cement sets, calcium silicate hydrate and calcium hydroxide are formed. While the former is a binder that holds concrete together, calcium hydroxide does not contribute to binding. However, fly ash reacts with some of the calcium hydroxide to increase the binding. Also, since fly ash acts as a microfiller between cement particles, strength and durability are increased while permeability is reduced. When used as a replacement for less than 45% of the portland cement, fly ash meeting ASTM C-618 contributes to resistance to scaling from road deicing chemicals.

Microsilica (silica fume) is an extremely fine particulate material, approximately 1/100th the size of cement particles. It is a waste product of electric arc furnaces. It acts as a "super pozzolan." Adding 5–15% microsilica will increase the pozzolanic reaction as well as provide a microfiller to reduce permeability.

Microsilica reacts with calcium hydroxide in the same manner as fly ash. It is customarily used to achieve strengths in the 8000–9000 psi (55–62 MPa) range.

Corrosion-resisting compounds are intended to inhibit rusting of the reinforcing steel and prestressing strands. Calcium nitrate is commonly used to inhibit the corrosive action of chlorides. It acts by forming a passivating protective layer on the steel. Calcium nitrate has essentially no effect on the mechanical and plastic properties of concrete.

In the *Devlo admixture system*, the cement particles are coated with a stabilizer, halting the hydration process indefinitely. Setting can be reinitiated at will hours or days later. The manufacturer claims that the treatment has no effect on the concrete when it hardens.

6. SLUMP

The four basic concrete components (cement, sand, coarse aggregate, and water) are mixed together to produce a homogeneous concrete mixture. The *consistency* and *workability* of the mixture affect the concrete's ability to be placed, consolidated, and finished without segregation or bleeding. The slump test is commonly used to determine consistency and workability.

The *slump test* consists of completely filling a slump cone mold in three layers of about one third of the mold volume. Each layer is rodded 25 times with a round, spherical-nosed steel rod of $5/8$ in (16 mm) diameter. When rodding the subsequent layers, the previous layers beneath are not penetrated by the rod. After rodding,

Structural

the mold is removed by raising it carefully in the vertical direction. The slump is the difference in the mold height and the resulting concrete pile height. Typical values are 1–4 in (25–100 mm).

Concrete mixtures that do not slump appreciably are known as *stiff mixtures*. Stiff mixtures are inexpensive because of the large amounts of coarse aggregate. However, placing time and workability are impaired. Mixtures with large slumps are known as *wet mixtures* (*watery mixtures*) and are needed for thin castings and structures with extensive reinforcing. Slumps for concrete that is machine-vibrated during placement can be approximately one third less than for concrete that is consolidated manually.

7. DENSITY

The density, also known as *weight density, unit weight*, and *specific weight*, of normal-weight concrete varies from about 140 lbf/ft^3 to about 160 lbf/ft^3 (2240 to 2560 kg/m^3), depending on the specific gravities of the constituents. For most calculations involving normal-weight concrete, the density may be taken as 145 to 150 lbf/ft^3 (2320 to 2400 kg/m^3). Lightweight concrete can have a density as low as 90 lbf/ft^3 (1450 kg/m^3).

8. COMPRESSIVE STRENGTH

The concrete's *compressive strength*, f'_c, is the maximum stress a concrete specimen can sustain in compressive axial loading. It is also the primary parameter used in ordering concrete. When one speaks of "6000 psi (41 MPa) concrete," the compressive strength is being referred to. Compressive strength is expressed in psi or MPa. (MPa is equivalent to N/mm^2, which is also commonly quoted.) SI compressive strength may be written as "Cxx" (e.g., "C20"), where xx is the compressive strength in MPa.

Typical compressive strengths range from 4000 psi to 6000 psi (27–41 MPa) for traditional structural concrete, though concrete for residential slabs-on-grade and foundations will be lower in strength (e.g., 3000 psi). 6000 psi (41 MPa) concrete is used in the manufacture of some concrete pipes, particularly those that are jacked in.

Cost is approximately proportional to concrete's compressive strength—a rule that applies to high-performance concrete as well as traditional concrete. For example, if 5000 psi (34 MPa) concrete costs $50 per cubic yard, then 14,000 psi concrete will cost approximately $140 per cubic yard.

Compressive strength is controlled by selective proportioning of the cement, coarse and fine aggregates, water, and various admixtures. However, the compressive strength of traditional concrete is primarily dependent on the mixture's water/cement ratio (see Fig. 48.1). Provided that the mix is of a workable consistency,

strength varies directly with the cement/water ratio. (This is *Abram's strength law*, named after Dr. Duff Abrams, who formulated the law in 1918.)

The standard ASTM compressive test specimen mold is a cylinder with a 6 in (150 mm) diameter and a 12 in (300 mm) height. Steel molds are more expensive than plastic molds, but they provide greater rigidity. (Some experts say specimens from steel molds test 3 to 15% higher.) The specimen is axially loaded to failure at a specific rate. The concrete is cured for a specific amount of time (three days, a week, 28 days, or more) at a specific temperature. Plain or lime-saturated heated water baths, as well as dry "hot boxes" heated by incandescent lights, are used for this purpose at some testing labs. To ensure uniform loading, the ends are smoothed by grinding or are capped in sulfur. For testing of very high-strength concrete, the ends may need to be ground glass-smooth in lapidary machines.

Since the ultimate load for 15,000 psi (103 MPa) and higher concrete exceeds the capacity (typically 300,000 lbf (1.34 MN)) of most testing machines, testing firms are switching to smaller cylinders with diameters of 4 in (100 mm) and heights of 8 in (200 mm) rather than purchasing 400,000–600,000 lbf (1.8–2.7 MN) machines.

The compressive strength is calculated as the maximum axial load, P, divided by the cross-sectional area, A, of the cylinder. Since as little as 0.1 in difference in the diameter can affect the test results by 5%, the diameter must be measured precisely.

$$f'_c = \frac{P}{A} \qquad 48.1$$

Compressive strength is normally measured on the 28th day after the specimens are cast. Since the strength of concrete increases with time, all values of f'_c must be stated with respect to a known age. If no age is given, a strength at a "standard" 28-day age is assumed.

9. STRESS-STRAIN RELATIONSHIP

The stress-strain relationship for concrete is dependent on its strength, age at testing, rate of loading, nature of the aggregates, cement properties, and type and size of specimens. Typical stress-strain curves for concrete specimens loaded in compression at 28 days of age under a normal rate of loading are shown in Fig. 48.2.

10. MODULUS OF ELASTICITY

The *modulus of elasticity* (also known as *Young's modulus*) is defined as the ratio of stress to strain in the elastic region. Unlike steel, the modulus of elasticity of concrete varies with compressive strength. Since the slope of the stress-strain curve varies with the applied stress, there are several ways of calculating the modulus of elasticity. Figure 48.3 shows a typical stress-strain

Figure 48.1 *Typical Concrete Compressive Strength Characteristics*

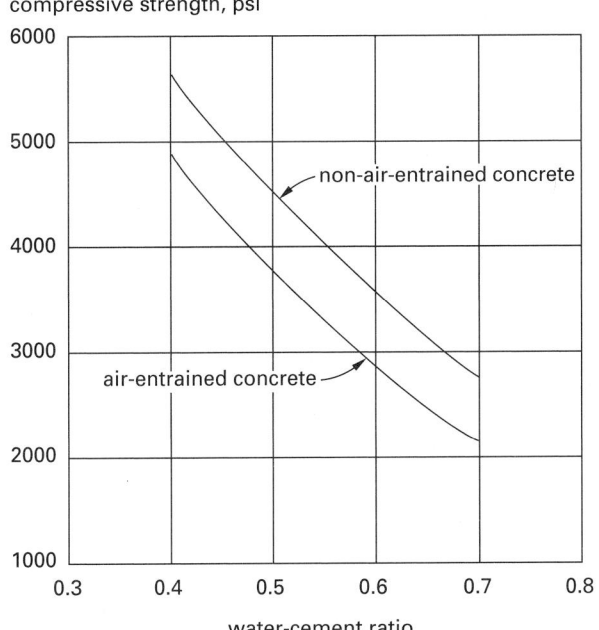

curve for concrete with the *initial modulus*, the *tangent modulus*, and the *secant modulus* indicated.

The *secant modulus of elasticity* is specified by the American Concrete Institute Code for use with specific weights between 90 and 155 lbf/ft^3 (1440 and 2480 kg/m^3). Equation 48.2 is used for both instantaneous and long-term deflection calculations. w_c is in lbf/ft^3 (kg/m^3), and E_c and f'_c are in lbf/in^2 (MPa) [ACI 318 8.5.1].

$$E_c = w_c^{1.5} 0.043 \sqrt{f'_c} \qquad \text{[SI]} \qquad \textit{48.2(a)}$$

$$E_c = w_c^{1.5} 33 \sqrt{f'_c} \qquad \text{[U.S.]} \qquad \textit{48.2(b)}$$

For normal-weight concrete, the ACI code suggests Eq. 48.3, corresponding to a specific weight of approximately 145 lbf/ft^3 (2320 kg/m^3) [ACI 318 8.5.1].

$$E_c = 5000 \sqrt{f'_c} \qquad \text{[SI]} \qquad \textit{48.3(a)}$$

$$E_c = 57{,}000 \sqrt{f'_c} \qquad \text{[U.S.]} \qquad \textit{48.3(b)}$$

Figure 48.2 *Typical Concrete Stress-Strain Curves*

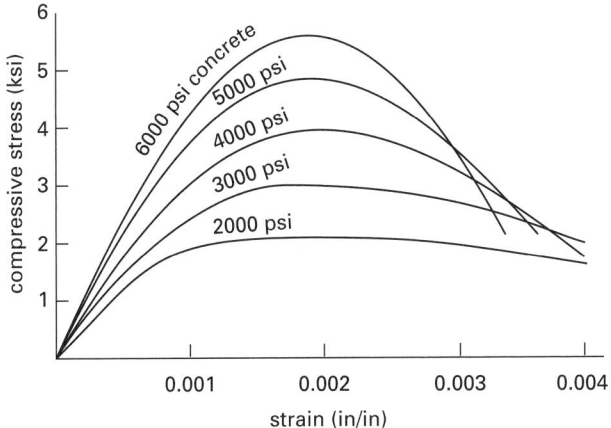

11. SPLITTING TENSILE STRENGTH

The extent and size of cracking in concrete structures are affected to a great extent by the tensile strength of the concrete. The ASTM C-496 *split cylinder testing procedure* is the standard test to determine the tensile strength of concrete. A 6 in × 12 in (150 mm × 300 mm) cast or drill-core cylinder is placed on its side as in Fig. 48.4, and the minimum load, P, that causes the cylinder to split in half is used to calculate the splitting tensile strength.

$$f_{ct} = \frac{2P}{\pi DL} \qquad \textit{48.4}$$

ACI 318 5.1.4 suggests that the splitting tensile strength f_{ct} can be calculated from correlations with compressive strength, and ACI 318 11.2 indirectly gives such correlations.

Figure 48.3 *Concrete Moduli of Elasticity*

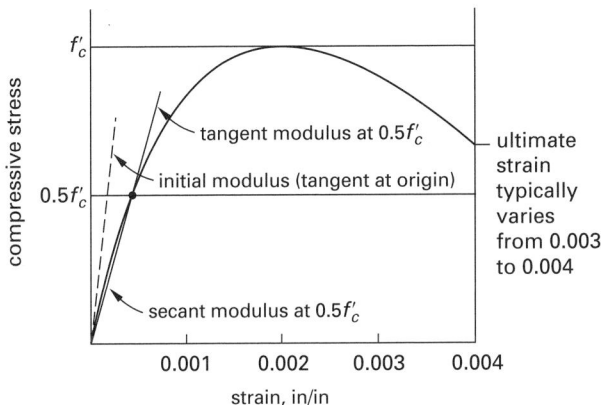

$$f_{ct} = 6.7 \sqrt{f'_c} \qquad \text{[normal weight]} \qquad \textit{48.5}$$

$$f_{ct} = 5.7 \sqrt{f'_c} \qquad \text{[sand lightweight]} \qquad \textit{48.6}$$

$$f_{ct} = 5 \sqrt{f'_c} \qquad \text{[all-lightweight]} \qquad \textit{48.7}$$

Figure 48.4 Splitting Tensile Strength Test

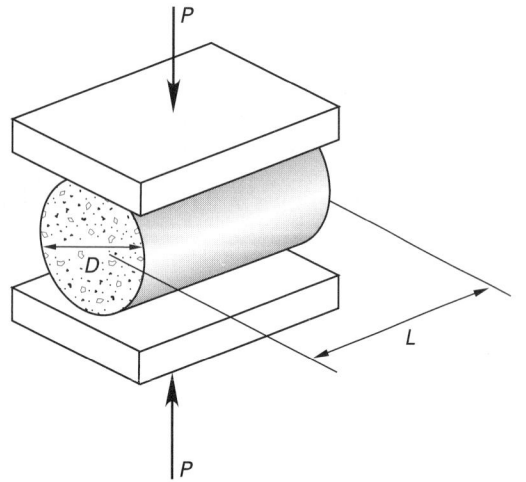

12. MODULUS OF RUPTURE

The tensile strength of concrete in flexure is known as the *modulus of rupture*, f_r, and is an important parameter for evaluating cracking and deflection in beams. The tensile strength of concrete is relatively low, about 10 to 15% (and occasionally up to 20%) of the compressive strength. ASTM C-78 gives the details of beam testing using *third point loading*. The modulus of rupture is calculated from Eq. 48.8.

$$f_r = \frac{Mc}{I} \quad \text{[tension]} \qquad \textbf{48.8}$$

Equation 48.8 gives higher values for tensile strength than the *splitting tensile strength test* because the stress distribution in concrete is not linear as is assumed in Eq. 48.4. For normal-weight concrete, the ACI code prescribes that Eq. 48.9 should be used for modulus of rupture calculations. For all-lightweight concrete, the modulus of rupture is taken as 75% of the calculated values. Other special rules for lightweight concrete may apply [ACI 318 9.5.2.3].

$$f_r = 0.62\sqrt{f'_c} \qquad \text{[SI]} \qquad \textbf{48.9(a)}$$

$$f_r = 7.5\sqrt{f'_c} \qquad \text{[U.S.]} \qquad \textbf{48.9(b)}$$

13. SHEAR STRENGTH

Concrete's true *shear strength* is difficult to determine in the laboratory because shear failure is seldom pure and is typically affected by other stresses in addition to the shear stress. Reported values of shear strength vary greatly with the test method used, but they are a small percentage (e.g., 25% or less) of the ultimate compressive strength.

14. POISSON'S RATIO

Poisson's ratio is the ratio of the lateral strain to the axial strain. It varies in concrete from 0.11 to 0.23, with typical values being 0.17 to 0.21.

15. REINFORCING STEEL

Steel is an alloy consisting almost entirely of iron. It also contains small quantities of carbon, silicon, manganese, sulfur, phosphorus, and other elements. Carbon has the greatest effect on the steel's properties. The carbon content is normally less than 0.5% by weight, with 0.2 to 0.3% being common percentages.

The density of steel is essentially unaffected by its composition, and a value of 0.283 lbf/in^3 (7820 kg/m^3) can be used.

Reinforcing steel may be formed from billet steel, axle steel, or rail steel. Most modern reinforcing bars are made from new billet steel. (Special bars of titanium, stainless steel, corrosion-resistant alloys, and glass fiber composites may see extremely limited use in corrosion-sensitive applications.) The following ASTM designations are used for steel reinforcing bars.

ASTM A-615: carbon steel, grades 40, 60, and 75 (symbol "S")

ASTM A-996: rail steel, grades 50 and 60 (symbols "R" and "⊥"; only "R" is permitted to be used by ACI 318), and axle steel, grades 40 and 60 (symbol "A")

ASTM A-706: low-alloy steel, grade 60 (symbol "W")

Reinforcing steel used for concrete structures comes in the form of bars (known as "rebar"), welded wire reinforcement, or wires. Reinforcing bars can be plain or deformed; however, most bars are manufactured deformed to increase the bond between concrete and steel. Figure 48.5 shows how the surface is deformed by rolling a pattern on the bar surface. The patterns used vary with the manufacturer.

Plain round reinforcing bars are designated by their nominal diameters in fractions of an inch or in millimeters (e.g., $1/2$ in, $5/8$ in, etc.) Deformed bars are also round, with sizes designed in numbers of eighths of an inch or in millimeters. Standard deformed bars are manufactured in sizes no. 3 to no. 11, with two special large sizes, no. 14 and no. 18, also available on special order. Metric (SI) bar designations are based on a "soft" conversion of the bar diameter to millimeters. For example, the traditional no. 3 bar has a nominal diameter of $3/8$ in, and this bar has a metric designation of no. 10 because it has a 9.5 mm diameter (approximately 10 mm). Hard metric conversions, where bars with slightly different diameters would be used in metric projects, were once considered but ultimately rejected. Hard conversions are used in Canadian bar sizes.

Figure 48.5 Deformed Bars

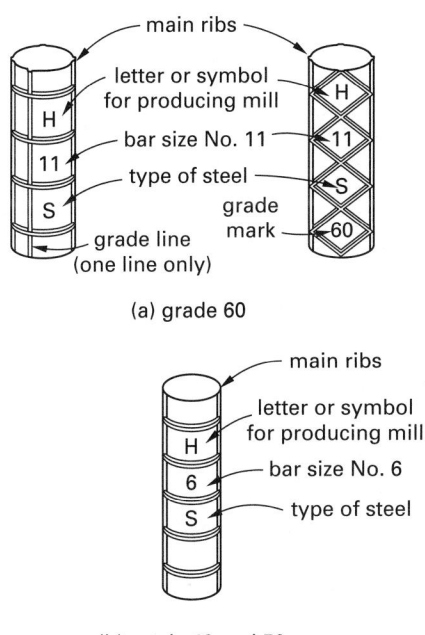

(a) grade 60

(b) grade 40 and 50

Figure 48.6 Typical Stress-Strain Curve for Ductile Steel

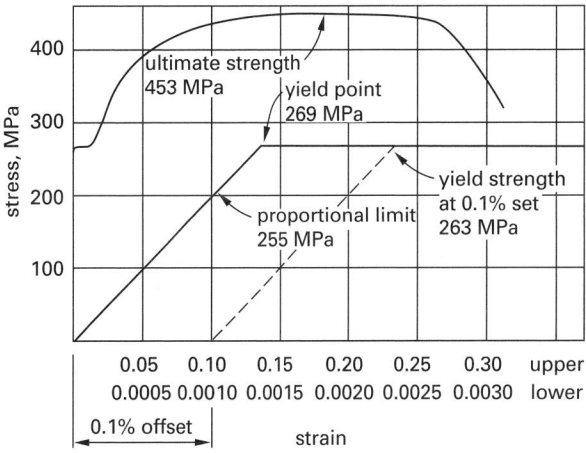

16. MECHANICAL PROPERTIES OF STEEL

A typical stress-strain curve for ductile structural steel is shown in Fig. 48.6. The curve consists of elastic, plastic, and strain-hardening regions.

The *elastic region* is the portion of the stress-strain curve where the steel will recover its size and shape upon release of load. Within the *plastic region*, the material shows substantial deformation without noticeable change of the stress. Plastic deformation, or *permanent set*, is any deformation that remains in the material after the load has been removed. In the *strain-hardening region*, additional stress is necessary to produce additional strain. This portion of the stress-strain diagram is not important from the design point of view, because so much strain occurs that the functionality of the material is affected.

The slope of the linear portion within the elastic range is the *modulus of elasticity*. The modulus of elasticity for all types of ductile reinforcing steels is taken to be 29×10^6 psi (200 GPa) [ACI 318 8.5.2]. (The modulus of elasticity is sometimes quoted as 30×10^6 psi (207 GPa). This value may apply to hard steels used for other purposes, but not to steel used for reinforcing concrete.)

The steel *grade* corresponds to its nominal yield tensile strength in ksi (thousands of pounds per square inch). Grade 60 steel with a yield strength of 60 ksi (413 MPa) is the most common, though grades 40 and 50 are also available upon request. SI values of steel strength are obtained using "soft" conversions. That is, steel properties are the same in customary U.S. and SI values. Only the designations are different.

17. CHLORIDE CORROSION

Chloride has been used as a concrete additive for a long time. In some fresh concrete, calcium chloride is deliberately added to the mix as a low-cost means of increasing early strength. In cold weather, chloride speeds up the initial set before the cement paste freezes.

However, corrosion from chlorides is a major problem for steel-reinforced concrete. On roads and bridges, de-icing chemicals and environmental salt corrode from the outside in. In buildings, chloride from concrete accelerators works from the inside out. In both, chloride ions migrate through cracks to attack steel reinforcing bars.

Once corrosion has begun, the steel is transformed into expanding rust, putting pressure on surrounding concrete and causing the protective concrete layer to spall.

The ACI code limits chloride concentrations in four categories: prestressed concrete, reinforced concrete exposed to chloride, reinforced concrete protected from moisture in service, and other reinforced concrete construction. At the lowest end of these limits, the code permits 0.06% chloride in prestressed concrete. At the high end, the code permits 1.00% chloride in reinforced concrete that is kept dry. (These percentages are on a weight basis, as ion. A 1% chloride concentration is approximately equivalent to 2% calcium chloride, since $CaCl_2$ is approximately 63% chloride by weight.)

18. ELECTRICAL PROTECTION OF REBAR

There are two major methods of using electricity to prevent chloride corrosion. *Cathodic protection* lowers the active corrosion potential of the reinforcing steel to immune or passive levels. Current is supplied from an external direct-current source. The current flows from an anode embedded in the concrete or from a surface-mounted anode mesh covered with 1.5–2 in (38–51 mm) of concrete. From there, the current passes through the electrolyte (the water and salt in the pores of the concrete) to the steel. The steel acts as the cathode, which is protected.

Installation of cathodic protection is simple. Power consumption is very low. Maintenance is essentially zero. Large areas should be divided into "zones," each with its own power supply and monitoring equipment. This protection scheme has initial and ongoing expenses, as the system and power supply must remain in place throughout the life of the structure.

An alternative method, developed in Europe, also attaches steel mesh electrodes to the surface and uses a current. However, the treatment is maintained only for a limited amount of time (1 to 2 months). An electrolytic cellulose paste is sprayed on and the current is applied. Chloride ions migrate from the rebar and are replaced by alkali ions from the paste. This raises the pH and forms a passivating oxide layer around the rebar. The mesh and cellulose are removed and discarded when the treatment is complete. Protection lasts for years. This electrolytic treatment may not be effective with epoxy-coated bars, prestressed structures (where chemical reactions can cause embrittlement), and bridge decks (due to the duration of shutdown required for treatment).

19. COATED REBAR

Corrosion-resistant *epoxy-coated rebar* ("green bar" or "purple bar") has been in use in the United States since the 1970s, most of it in bridge decks and other structures open to traffic. After proper cleaning, standard rebar is given a 0.005–0.012 in (0.13–0.30 mm) electrostatic spray coating of epoxy. The epoxy is intended to be flexible enough to permit subsequent bending and cold working. Epoxy-coated bars must comply with ASTM standards A-775 or A-934.

Epoxy's effectiveness, once described as "maintenance-free in corrosive environments," is now questioned by some. Corrosion protection is apparently less than expected when poor manufacturing quality and installation practices remove or damage portions of the coating. Corrosion of coated bars can be reduced by coating after bending and by using alternative "pipeline coatings," which are thicker but less flexible than epoxy.

Epoxy-coated rebar is now seen as one of several fabrication methods needed to prevent or slow chloride corrosion in *bridge decks*, including cathodic protection, lateral and longitudinal prestressing, less-porous and low-slump concrete, thicker (3 in (75 mm) or more) topping layers, interlayer membranes and asphalt concrete, latex-modified or silica fume concrete overlays, corrosion-resistant additives, surface sealers (with or without overlays), galvanizing steel rebar, polymer impregnation, and polymer concrete.

20. HIGH-PERFORMANCE CONCRETE

ACI defines *high-performance concrete* (HPC) as "concrete meeting special combinations of performance and uniformity requirements that cannot always be achieved routinely using conventional constituents and normal mixing, placing, and curing practices" [ACI 116R-2000]. High-strength concrete (HSC), a special case of high-performance concrete, is defined by ACI Committee 363 on high-strength concrete as concrete having a specified compressive strength for design of 8000 psi (55 MPa). However, the strength threshold at which concrete is considered high-strength depends on regional factors, such as characteristics and availability of raw materials, production capabilities, and the experience of local ready-mix producers in making high-strength concrete. Specifying concrete with strengths of 6000–10,000 psi (41–70 MPa) for high-rise construction is now routine. Ready-mix plants can deliver 9000–15,000 psi (62–103 MPa) concrete, and 20,000 psi (138 MPa) concrete has seen limited use.

High strength is achieved in a variety of ways. Superplasticizers and other mineral admixtures (e.g., pozzolans such as fly ash, silica fume, and precipitated silica) are the main components affecting strength.

HSC allows engineers to design structures with smaller structural members (i.e., columns and beams), reducing dead load and increasing usable space. Though HSC is more costly than standard concrete, the long-term benefits to the building owner can be substantial. The benefits of HSC use, however, may not be realized in areas where concrete member dimensions are governed by building code requirements concerning fire safety and so forth.

21. FIBER-REINFORCED CONCRETE

Fiber-reinforced concrete (FRC) contains small steel, polymer (polyolefin or polypropylene), carbon, or glass fibers approximately 0.5–2 in (12–50 mm) long dispersed randomly throughout the mix. (The term *carbon-fiber-reinforced concrete*, CFRC, is used with carbon reinforcement.) Synthetic fibers are specified by ASTM C-1116, steel fibers by ASTM A-820.

As setting concrete loses water, it shrinks. Fibers are added to control surface cracking during the setting process. These fibers intersect cracks that form during shrinkage. For control of surface cracking, fibers are added in a proportion of approximately 0.1% by volume corresponding to 1.5 lbm/ft^3 (24 kg/m^3) of concrete.

Fibers can also be used to carry a portion of the service load, though the volumetric fraction must be much higher: 4–20%. Steel fibers are commonly used in airport pavements, industrial floors, bridge decks, and tunnel lining. Glass fibers are used in precast products. The fibers carry the load once microcracking from flexural loading has occurred, at approximately 3500 psi (24 MPa). Microreinforced concretes with compressive strengths of 30,000 psi (207 MPa) and higher have been reported.

22. POLYMER CONCRETE

Polymer concrete (polymer-portland cement concrete, PPCC) is a material primarily used for rapid repair (sealing or overlaying) of concrete, usually bridge decks. Sometimes dubbed "MMA concrete" after one of its components, it is delivered as two components that are mixed together prior to use: (a) a premixed powder of fine aggregates coated with polymers, initiators, and pigments; and (b) a liquid methyl-methacrylate (MMA) monomer. Setting is rapid, though complete curing may take several weeks. Final performance is similar or superior to traditional concrete.

Polymer concrete can be precast into products with compressive strengths up to 15,000 psi (103 MPa) and flexural strengths up to 300 psi (2.1 MPa). It is water- and corrosion-resistant, thin, lightweight, and colorfast, making it ideal for some bridge components.

23. LIGHTWEIGHT CONCRETE

Aggregate of rotary kiln-expanded shales or clays having a specific weight of 70 lbf/ft³ (1120 kg/m³) or less is known as *lightweight aggregate*, also known as *ASTM C330 aggregate*. It is used in the production of *lightweight concrete*. Lightweight concrete in which only the coarse aggregate is lightweight is known as *sand-lightweight concrete*. If both the coarse and fine aggregates are lightweight, the concrete is known as *all-lightweight concrete*. Unless noted otherwise, concrete in this book is assumed to be *normal-weight concrete*.

24. SELF-PLACING CONCRETE

Self-placing concrete (also known as *self-compacting concrete*, *self-consolidating concrete*, *flowable concrete*, and *nonvibration concrete*) is used to reduce labor needs and construction time, particularly where congestion of steel reinforcing bars make the consolidation of concrete difficult. Such concrete does not require vibration and thus reduces site noise and placement defects. Self-placing concrete relies on viscosity agents to produce liquid-like flow characteristics.

25. ROLLER-COMPACTED CONCRETE

Roller-compacted concrete (RCC), also known as *roll-crete*, is a very lean, no-slump, almost-dry concrete, similar to damp gravel in consistency. It is primarily used in dams and other water-control structures, though use has been extended to heavy-duty pavements (e.g., logging roads, freight yards, and truck stopping areas) where a perfectly smooth surface is not required. RCC is attractive because of its low cost. Requiring less time and labor to complete, an RCC dam costs about half the price of a conventional concrete gravity dam and about one third less than an earth or a rockfill dam.

Scrapers and bulldozers spread and compact RCC in broad 1–2 ft (30–71 cm) thick lifts, and vibratory rollers or dozers consolidate it to the specified density with multiple (4–10) passes. A compressive strength of approximately 1000 to 4500 psi (7–30 MPa) is achieved, identical to the strength of a conventional mixture with the same water-cement ratio.

Seepage in RCC dams occurs primarily through the interfaces between the lifts, but this seepage can be controlled or eliminated with a variety of practices. The abutments and one or both faces should be constructed from normal concrete. Each new lift of RCC should be bedded in a ¹/₄–¹/₂ in (6–12 mm) thin layer of high-slump cement-paste mortar. The time between lifts can be reduced, and slightly wetter mixes can be used. Other methods of reducing seepage include using more cement, fly ash, and water in the core of the dam; incorporating a PVC membrane between the cast-in-place face and the RCC; and building up the lifts in small 6 in (150 mm) layers. Since RCC tends to clump in temperatures greater than 90°F (32°C), in extreme cases placement can be limited to winter months or cooler nights.

26. SOIL CEMENT

Soil cement is a combination of soil, water, and a small amount of cement (about 10%). Soil cement is low in price, strong, and easy to work with, can take repeated wetting and heavy wave action, and can stand up to freeze-thaw cycles. Placed soil cement looks like concrete or rock. It has proven to be an ideal material for pavement base courses (where it is referred to as *cement-treated base*) but can also be used for river bank protection, reservoir and channel lining, facing for earthfill dams, seepage control, pipe bedding, and foundation stabilization.

27. CONTROLLED LOW-STRENGTH MATERIAL

Controlled low-strength material (CLSM) is a mixture of fly ash, fine aggregates, cement (about 5%), and water. Delivered in a semi-fluid state, it flows readily into place, needs no tamping or vibration, and achieves a compressive strength of 100 psi (700 kPa) within 24 hr. It has proven to be an efficient and economical backfilling material for culverts, bridge abutments, and trenches.

49 Concrete Proportioning, Mixing, and Placing

Nomenclature

f	fraction moisture	–	–
p	pressure	lbf/ft^2	Pa
R	rate of forming	ft/hr	m/s
SG	specific gravity	–	–
T	temperature	$°F$	$°C$
V	volume	ft^3	m^3
W	weight	lbf	–

Symbols

γ	specific weight	lbf/ft^3	–
ρ	density	lbm/ft^3	kg/m^3

1. DESIGN CONSIDERATIONS

Concrete can be designed for compressive strength or durability in its hardened state. In its wet state, concrete should have good workability. *Workability* relates to the effort required to transport, place, and finish wet concrete without segregation or bleeding. Workability is often closely correlated with slump. All other design requirements being met, the most economical mix should be selected.

Durability is defined as the ability of concrete to resist environmental exposure or service loadings. One of the most destructive environmental factors is the freeze/thaw cycle. ASTM C-666, "Standard Test Method for Resistance of Concrete to Rapid Freezing and Thawing," is the standard laboratory procedure for determining the freeze-thaw durability of hardened concrete. This test determines a *durability factor*, the number of freeze-thaw cycles required to produce a certain amount of deterioration.

ACI 318 4.2.2 places maximum limits on the water-cement ratio and minimum limits on the strength for concrete with special exposures, including concrete exposed to freeze-thaw cycles, deicing chemicals, and chloride, and installations requiring low permeability.

Specifying an air-entrained concrete will improve the durability of concrete subject to freeze-thaw cycles or deicing chemicals. The amount of entrained air needed will depend on the exposure conditions and the size of coarse aggregate, as prescribed in ACI 318 4.2.1.

Table 49.1 *Typical Slumps by Application*

	slump, in (mm)	
application	maximum	minimum
reinforced footings and foundations	3 (76)	1 (25)
plain footings and substructure walls	3 (76)	1 (25)
slabs, beams, and reinforced walls	4 (102)	1 (25)
reinforced columns	4 (102)	1 (25)
pavements and slabs	3 (76)	1 (25)
heavy mass construction	2 (51)	1 (25)
roller-compacted concrete	0	0

(Multiply in by 25.4 to obtain mm.)

2. STRENGTH ACCEPTANCE TESTING

When extensive statistical data are not available, acceptance testing of laboratory-cured specimens can be evaluated per ACI 318 5.6.3.3. This section states that the compressive strength of concrete is considered satisfactory if both of the following criteria are met: (a) No single test (average of two cylinders) falls below the specified compressive strength, f'_c, by more than 500 psi (3.45 MPa) when f'_c is 5000 psi (34.5 MPa) or less, or by more than $0.10f'_c$ when f'_c is more than 5000 psi (34.5 MPa); (b) the average of any three consecutive test strengths equals or exceeds the specified compressive strength.

3. BATCHING

All concrete ingredients are weighed or volumetrically measured before being mixed, a process known as *batching*. Weighing is more common because it is simple and accurate. However, water and liquid admixtures can be added by either volume or weight. The following accuracies are commonly specified or assumed in concrete batching: cement, 1%; water, 1%; aggregates, 2%; and admixtures, 3%. ACI 318 4.2.1 specifies the accuracy of air entrainment as $1^1/_2\%$.

4. WATER-CEMENT RATIO AND CEMENT CONTENT

Concrete strength is inversely proportional to the *water-cement ratio* (ACI 318 uses the more general term *water-cementious materials ratio*), the ratio of the amount of water to the amount of cement in a mixture, usually stated as a decimal by weight. Typical values are approximately 0.45–0.60 by weight. (Alternatively, the water-cement ratio may be stated as the number of gallons of water per 94 lbf sack of cement, in which case typical values are 5–7 gal.)

A mix is often described by the number of sacks of cement needed to produce 1 yd^3 of concrete. For example, a mix using 6 sacks of cement per cubic yard would be described as a "6-sack mix." Another method of describing the cement content is the *cement factor*, which is the number of cubic feet of cement per cubic yard of concrete.

Example 49.1

A concrete mixture using 340 lbf of water per cubic yard has a water-cement ratio of 0.60 by weight. (a) Determine the ideal weight and volume of cement needed to produce 1.0 yd^3 of concrete. (b) How many sacks of cement are needed to produce a 4 in thick concrete slab 9 ft × 4 ft with this mix?

Solution

(a) The cement requirement is found from the water-cement ratio.

$$W_{\text{cement}} = \frac{W_{\text{water}}}{0.6} = \frac{340 \, \dfrac{\text{lbf}}{\text{yd}^3}}{0.6}$$
$$= 567 \, \text{lbf/yd}^3$$

From Table 49.2, the specific weight of cement is 195 lbf/ft^3. The volume of the cement is

$$V_{\text{cement}} = \frac{W_{\text{cement}}}{\gamma_{\text{cement}}}$$
$$= \frac{567 \, \dfrac{\text{lbf}}{\text{yd}^3}}{195 \, \dfrac{\text{lbf}}{\text{ft}^3}}$$
$$= 2.91 \, \text{ft}^3/\text{yd}^3$$

(b) The slab volume is

$$V_{\text{slab}} = \frac{(4 \text{ in})(9 \text{ ft})(4 \text{ ft})}{\left(12 \, \dfrac{\text{in}}{\text{ft}}\right)\left(27 \, \dfrac{\text{ft}^3}{\text{yd}^3}\right)}$$
$$= 0.444 \, \text{yd}^3$$

The cement weight needed is

$$W_{\text{cement}} = (0.444 \, \text{yd}^3)\left(567 \, \frac{\text{lbf}}{\text{yd}^3}\right) = 252 \, \text{lbf}$$

From Table 49.2, each sack weighs 94 lbf. The number of sacks is

$$\frac{252 \, \text{lbf}}{94 \, \dfrac{\text{lbf}}{\text{sack}}} = 2.7 \, \text{sacks} \quad [3 \text{ sacks}]$$

5. PROPORTIONING MIXES

The oldest method of proportioning concrete is the *arbitrary proportions method* (*arbitrary volume method* or *arbitrary weight method*). Ingredients of average properties are assumed, and various proportions are logically selected. Tabular or historical knowledge of mixes and compressive strengths may be referred to. Proportions of cement, fine aggregate, and coarse aggregate are designated (in that sequence). For example, 1:2:3 means that one part of cement, two parts of fine aggregate, and three parts of coarse aggregate are combined. The proportions are generally in terms of weight; volumetric ratios are rarely used. (In the rare instances where the mix proportions are volumetric, the ratio values must be multiplied by the bulk densities to get the weights of the constituents. Then weight ratios may be calculated and the absolute volume method applied directly.)

For more critical applications, the actual ingredients can be tested in various logical proportions. This is known as the *trial batch method* or the *trial mix method*. This method is more time-consuming initially, since cured specimens must be obtained for testing.

Although some "extra" concrete ends up being brought to the job site, once determined, mix quantities should be rounded up, not down. Otherwise, the concrete volume delivered may be short.

6. ABSOLUTE VOLUME METHOD

The *yield* is the volume of wet concrete produced in a batch. Typical units are cubic yards (referred to merely as "yards") but may also be cubic feet or cubic meters. The yield that results from mixing known quantities of ingredients can be found from the *absolute volume method*, also known as the *solid volume method* and *consolidated volume method*. This method uses the specific gravities or densities for all the ingredients to calculate the absolute volume each will occupy in a unit volume of concrete. The absolute volume (solid volume, consolidated volume, etc.) is

$$V_{absolute} = \frac{m}{(SG)\rho_{water}} \qquad [SI] \qquad \textbf{49.1(a)}$$

$$V_{absolute} = \frac{W}{(SG)\gamma_{water}} \qquad [U.S.] \qquad \textbf{49.1(b)}$$

The absolute volume method assumes that, for granular materials such as cement and aggregates, there will be no voids between particles. Therefore, the amount of concrete is the sum of the solid volumes of cement, sand, coarse aggregate, and water.

To use the absolute volume method, it is necessary to know the solid densities of the constituents. In the absence of other information, Table 49.2 can be used.

Table 49.2 *Summary of Approximate Properties of Concrete Components*

cement
    specific weight   195 lbf/ft^3 (3120 kg/m^3)
    specific gravity   3.13–3.15
    weight of one sack   94 lbf (42 kg)
fine aggregate
    specific weight   165 lbf/ft^3 (2640 kg/m^3)
    specific gravity   2.64
coarse aggregate
    specific weight   165 lbf/ft^3 (2640 kg/m^3)
    specific gravity   2.64
water
    specific weight   62.4 lbf/ft^3 (1000 kg/m^3)
                                    7.48 gal/ft^3 (1000 L/m^3)
                                    8.34 lbf/gal (1 kg/L)
                                    239.7 gal/ton (1 L/kg)
    specific gravity   1.00

(Multiply lbf/ft^3 by 16 to obtain kg/m^3.)
(Multiply lbf by 0.45 to obtain kg.)

7. ADJUSTMENTS FOR WATER AND AIR

In most problems, the weights and volumes of the components must be adjusted for air entrainment and aggregate water content. This determines the *adjusted weights* of the components.

The *saturated, surface-dry* (SSD) condition occurs when the aggregate holds as much water as it can without trapping any free water between the aggregate particles. Calculations of yield should be based on the SSD densities, and the water content should be adjusted (increased or decreased) to account for any deviation from the SSD condition. Any water in the aggregate above the SSD water content must be subtracted from the water requirements. Any moisture content deficit below the SSD water content must be added to the water requirements.

The phrase "5% excess water" (or similar) is ambiguous and can result in confusion when used to specify batch quantities. There are two methods in field use, differing in whether the percentage is calculated on a wet or dry basis. Table 49.3 summarizes the batching relationships for both bases. (Though Table 49.3 is written for sand, it can also be used for coarse aggregate.)

There is also a similar confusion about air entrainment. The volumetric basis can be calculated either with or without entrained air. If "5% air" means that the concrete volume is 5% air, then the solid volume is only 95% of the final volume. The solid volume should be divided by 0.95 to obtain the final volume. However, if "5% air" means that the concrete volume is increased by 5% when the air is added, the solid volume should be multiplied by 1.05 to obtain the final volume.

The bases of these percentages must be known. If they are not, then either definition could apply. Regardless, the final volume is not affected significantly by either interpretation.

Table 49.3 *Dry and Wet Basis Calculations*

	dry basis	wet basis
fraction moisture, f	$\dfrac{W_{excess\ water}}{W_{SSD\ sand}}$	$\dfrac{W_{excess\ water}}{W_{SSD\ sand} + W_{excess\ water}}$
weight of sand, $W_{wet\ sand}$	$W_{SSD\ sand} + W_{excess\ water}$ $(1+f)W_{SSD\ sand}$	$W_{SSD\ sand} + W_{excess\ water}$ $\left(\dfrac{1}{1-f}\right)W_{SSD\ sand}$
weight of SSD sand, $W_{SSD\ sand}$	$\dfrac{W_{wet\ sand}}{1+f}$	$(1-f)W_{wet\ sand}$
weight of excess water, $W_{excess\ water}$	$fW_{SSD\ sand}$ $\dfrac{fW_{wet\ sand}}{1+f}$	$fW_{wet\ sand}$ $\dfrac{fW_{SSD\ sand}}{1-f}$

Structural

Example 49.2

A mix is designed as 1:1.9:2.8 by weight. The water-cement ratio is 7 gal per sack. (a) What is the concrete yield in cubic feet? (b) How much sand, coarse aggregate, and water is needed to make 45 yd³ of concrete?

Solution

(a) The solution can be tabulated. Refer to Table 49.2.

material	ratio	weight per sack of cement (lbf)	solid density (lbf/ft³)	absolute volume (ft³/sack)
cement	1.0	$1 \times 94 = 94$	195	$\dfrac{94}{195} = 0.48$
sand	1.9	$1.9 \times 94 = 179$	165	$\dfrac{179}{165} = 1.08$
coarse	2.8	$2.8 \times 94 = 263$	165	$\dfrac{263}{165} = 1.60$
water				$\dfrac{7}{7.48} = 0.94$
				$\overline{4.10}$

The solid yield is 4.10 ft³ of concrete per sack of cement.

(b) The number of one-sack batches is

$$\frac{(45 \text{ yd}^3)\left(27 \dfrac{\text{ft}^3}{\text{yd}^3}\right)}{4.10 \dfrac{\text{ft}^3}{\text{sack}}} = 296.3 \text{ sacks}$$

Order a minimum of 297 sacks of cement. Calculate the remaining order quantities from the cement weight and the mix ratios.

$$W_{\text{sand}} = 1.9 W_{\text{cement}}$$
$$= (1.9)\left(94 \dfrac{\text{lbf}}{\text{sack}}\right)(297 \text{ sacks})$$
$$= 53{,}044 \text{ lbf}$$

$$W_{\text{coarse aggregate}} = (2.8)\left(94 \dfrac{\text{lbf}}{\text{sack}}\right)(297 \text{ sacks})$$
$$= 78{,}170 \text{ lbf}$$

$$V_{\text{water}} = \left(7 \dfrac{\text{gal}}{\text{sack}}\right)(297 \text{ sacks}) = 2079 \text{ gal}$$

Example 49.3

50 ft³ of 1:2.5:4 (by weight) concrete are to be produced. The ingredients have the following properties.

ingredient	SSD density (lbf/ft³)	moisture (dry basis from SSD)
cement	197	—
fine aggregate	164	5% excess (free moisture)
coarse aggregate	168	2% deficit (absorption)

5.5 gal of water are to be used per sack, and the mixture is to have 6% entrained air. What are the ideal order quantities expressed in tons?

Solution

Proceed as in Ex. 49.2.

material	ratio	weight per sack of cement (lbf)	solid density (lbf/ft³)	absolute volume (ft³/sack)
cement	1.0	$1 \times 94 = 94$	197	$\dfrac{94}{197} = 0.477$
sand	2.5	$2.5 \times 94 = 235$	164	$\dfrac{235}{164} = 1.433$
coarse	4.0	$4.0 \times 94 = 376$	168	$\dfrac{376}{168} = 2.238$
water				$\dfrac{5.5}{7.48} = 0.735$
				$\overline{4.883}$

The solid yield is 4.883 ft³ of concrete per sack of cement. The yield with 6% air is

$$\frac{4.883 \dfrac{\text{ft}^3}{\text{sack}}}{1 - 0.06} = 5.19 \text{ ft}^3/\text{sack}$$

The ideal number of one-sack batches required is

$$\frac{50 \text{ ft}^3}{5.19 \dfrac{\text{ft}^3}{\text{sack}}} = 9.63 \text{ sacks}$$

The required sand weight as ordered (not SSD) is

$$\frac{(9.63 \text{ sacks})(1 + 0.05)\left(94 \dfrac{\text{lbf}}{\text{sack}}\right)(2.5)}{2000 \dfrac{\text{lbf}}{\text{ton}}} = 1.19 \text{ tons}$$

The required coarse aggregate weight as ordered (not SSD) is

$$\frac{(9.63 \text{ sacks})(1 - 0.02)\left(94 \dfrac{\text{lbf}}{\text{sack}}\right)(4.0)}{2000 \dfrac{\text{lbf}}{\text{ton}}} = 1.77 \text{ tons}$$

From Table 49.2, the excess water contained in the sand is

$$(1.19 \text{ tons})\left(\frac{0.05}{1 + 0.05}\right)\left(239.7 \dfrac{\text{gal}}{\text{ton}}\right) = 13.58 \text{ gal}$$

The water needed to bring the coarse aggregate to SSD conditions is

$$(1.77 \text{ tons})\left(\frac{0.02}{1 - 0.02}\right)\left(239.7 \dfrac{\text{gal}}{\text{ton}}\right) = 8.66 \text{ gal}$$

The total water needed is

$$\left(5.5\ \frac{\text{gal}}{\text{sack}}\right)(9.63\ \text{sacks}) + 8.66\ \text{gal} - 13.58\ \text{gal} = 48.0\ \text{gal}$$

8. MIXING

Mixing coats the surface of all aggregate particles with cement paste and blends all the ingredients into a uniform, homogeneous mass. The mixing method depends on many factors, including the quantity of concrete required for the job, the location of the job, the availability of the ingredients, and transportation costs.

Stationary mixers are mixing units that are to be moved from one job site to another. Large stationary mixers can also be found in ready-mix plants, usually located off the job site. They are available in sizes from 2 to 12 yd^3 (0.6 to 9.1 m^3).

Mobile batcher mixers are special trucks that proportion each batch by volume, feeding and continuously mixing aggregates, water, and cement in the mixer.

9. TRANSPORTING AND HANDLING

The homogeneity of the concrete mix must not change during transport and handling at the job site. To achieve the predetermined desired wet and hardened properties, concrete should be transported to the job site without delay. With delays, the concrete may begin to set, making consolidation difficult and increasing the possibility of voids (also known as *honeycombs*) in the finished product. Delays also increase the possibility of *segregation*, which occurs when the coarse aggregate separates from the sand-cement mortar.

10. PLACING

Before placing concrete, forms must be clean and free from any foreign substances such as oil, ice, and snow. All surfaces must be moistened, especially in hot weather and when concrete is to be placed on subgrades. All reinforcing steel bars and other embedded items should be secured in their place and be free from rust, scale, and oil.

Concrete should be placed in a manner that eliminates or minimizes segregation. It should not be dumped in piles and then leveled. Concrete should be placed continuously in horizontal layers.

11. CONSOLIDATION

Freshly placed concrete must be compacted to eliminate entrapped air, rock pockets, and voids. Consolidation can be done manually or mechanically, depending on the workability of concrete, the shape and size of formwork, and the spacing between reinforcing bars. Workable concrete can be consolidated easily by hand-rodding, while stiff mixtures with low water-cement ratios must be consolidated using mechanical methods such as vibration.

Vibration is the most common technique for consolidating concrete. Vibration can be applied internally or externally. Both types of vibration are characterized by the amplitude and frequency of vibration (expressed in vibrations per minute (vpm)). Internal vibration is normally used to consolidate concrete in columns, beams, and slabs. External vibration is accomplished by using vibrating tables, surface vibrators such as vibratory screeds, plate vibrators, or vibratory hand floats or trowels. External vibration is used in cases of thin and heavily reinforced concrete members. Also it is used for stiff mixes where internal vibrators cannot be used.

12. CURING

The strength that concrete achieves depends on the *curing process*. During curing, the temperature and humidity must be controlled. Concrete strength increases with age as long as moisture and favorable temperatures are present for hydration of the cement. Therefore, concrete should be kept saturated or nearly saturated until the chemical reaction between water and cement is completed. Loss of water will reduce the hydration process and cause the concrete to shrink and crack.

There are three primary ways of keeping concrete moist and at a favorable temperature: (a) Water can be copiously applied to make it available to meet the hydration demands. Flat surfaces can be ponded or immersed. Vertical and horizontal surfaces can be sprayed or fogged. Saturated fabric coverings such as cotton mats, burlap, and rugs can be placed over the concrete. (b) The mixing water can be prevented from evaporating by sealing the surface. Impervious paper, plastic sheets, and membrane-forming curing compounds are commonly used. (c) The rate of strength development can be accelerated by providing heat. Live steam, heating coils, and electrically heated forms or pads are used.

13. HOT-WEATHER CONCRETING

Concrete should be mixed, transported, placed, and finished at a temperature that will not affect its properties in the fresh or hardened state. Hot weather will increase evaporation and water demand, decrease the slump, and increase the rate of setting. Counteracting these effects by adding more water to concrete will decrease strength and durability of concrete in the hardened state.

Previous studies have shown that increasing the concrete temperature from 50–100°F (10–38°C) would require an additional 33 lbf of water per cubic yard (20 kg per cubic meter) just to maintain a 3 in (75 mm) slump. This additional water would decrease the compressive strength by 12 to 15%. Such a loss of strength could disqualify the concrete mix.

Structural

Rather than increase the water, an effort is made to limit the temperatures at which the concrete is placed. Concrete having a temperature between 50°F and 60°F (10°C and 16°C) is generally considered ideal, but placement in this temperature range is not always possible. However, many specifications require that concrete be placed when its temperature is less than 85 to 90°F (29 to 32°C).

14. COLD-WEATHER CONCRETING

Concrete temperature has a direct effect on the strength and workability of concrete. Therefore, in cold weather, steps should be taken to ensure a safe and desirable temperature for concrete during mixing, placing, finishing, and curing.

Publication ACI 306R defines cold weather as existing when the mean daily temperature is less than 40°F (4°C) for more than three consecutive days. At such temperatures, the rate of cement hydration decreases, delaying the strength development. Not only must concrete be delivered at the proper temperature, but the temperature of the forms, reinforcing steel, and ground must be considered. Concrete should not be placed on frozen concrete or on frozen ground.

The mix temperature at the time of pouring can be raised by heating the ingredients. Water is easily heated, but it is not advisable to exceed a temperature of 140–180°F (60–82°C), as the concrete may *flash set* (i.e., harden suddenly). Aggregate can also be steam-heated prior to use.

15. FORMWORK

Formwork refers to the system of boards, ties, and bracing required to construct the mold in which wet concrete is placed. Formwork must be strong enough to withstand the weight and pressure created by the wet concrete and must be easy to erect and remove.

Types of Forms

Forms are constructed out of a variety of materials. Unless the concrete is finished in some way, the shape and pattern of the formwork will affect the appearance of the final product. Wood grain, knotholes, joints, and other imperfections in the form will show in their negative image when the form is removed. Plywood is the most common forming material. It is usually $^3/_4$ in (19 mm) thick and is coated on one side with oil, a water-resistant glue, or plastic to prevent water from penetrating the wood and to increase the reusability of the form. Oil on forms also prevents adhesion of the concrete so the forms are easier to remove. The plywood is supported with solid wood framing, which is braced or shored as required. Figure 49.1 shows two typical wood-framed forms.

Prefabricated steel forms are often used because of their strength and reusability. They are often employed for

Figure 49.1 *Concrete Framework*

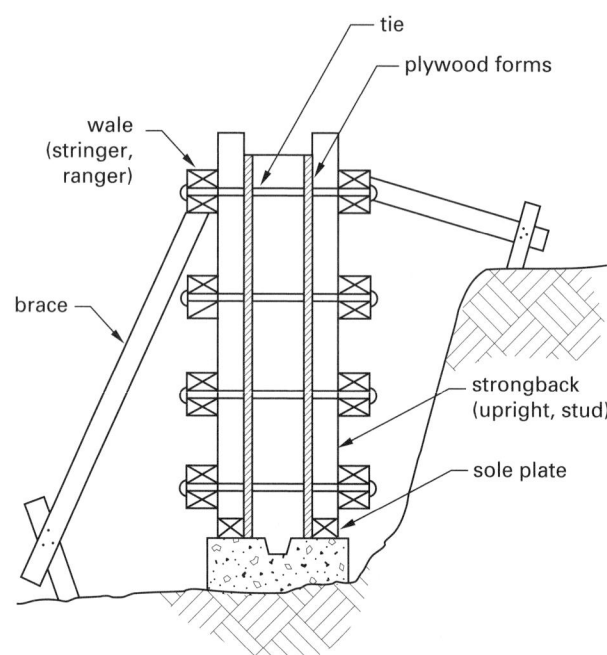

(a) wall formwork

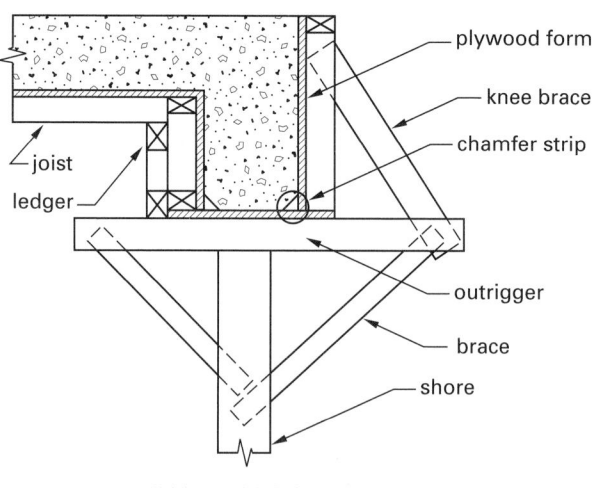

(b) beam/slab formwork

forming one-way joist systems, waffle slabs, round columns, and other special shapes.

Other types of forms include glass-fiber reinforced plastic, hardboard, and various kinds of proprietary systems. Plastic forms are manufactured with a variety of patterns embedded in them. These patterns are transferred to the concrete and constitute the final surface. Special form liners can also be used to impart a deeply embossed pattern.

For exposed architectural surfaces, a great deal of consideration must be given to the method and design of the formwork because the pattern of joints and form ties will be visible. Joints are often emphasized with rustication strips, continuous pieces of neoprene, wood, or other material that when removed shows a deep reveal in the concrete.

Form ties are metal wires or rods used to hold opposite sides of the form together and also to prevent their collapse. When the forms are removed, the wire remains in the concrete, and the excess is twisted or cut off. Some form ties are threaded rods that can be unscrewed and reused. Tie holes are made with cone-shaped heads placed against the concrete form. When these are removed, a deep, round hole is left that allows the tie to be cut off below the surface of the concrete. These holes can remain exposed as a design feature or can be patched with grout.

Special Forms

Most formwork is designed and constructed to remain in place until the concrete cures sufficiently to stand on its own. However, with *slip forming*, the formwork moves as the concrete cures. Slip forming is used to form continuous surfaces such as curbs and gutters, open channels, tunnels, and high-rise building cores. The entire form is constructed along with working platforms and supports for the jacking assembly. In vertical slip forming, the form moves continuously at about 6 in to 12 in (150 to 300 mm) per hour. Various types of jacking systems are used to support the form as it moves upward. With horizontal slip forming, as would be used in roadway construction, speeds of 200–300 ft/hr (60–90 m/h) are possible.

Flying forms are large fabricated sections of framework that are removed, once the concrete has cured, to be reused in forming an identical adjacent section. They are often used in buildings with highly repetitive units, such as hotels and apartments. After forming the floor for a hotel, for example, the form assembly is slid outside the edge of the building where it is lifted by crane to the story above. Once that floor is poured, the process is repeated.

Economy in Formwork

Because one of the biggest expenses for cast-in-place concrete is formwork, the overall cost can be minimized by following some basic guidelines. To begin with, forms should be reusable as much as possible. This implies uniform bay sizes, beam depths, column widths, opening sizes, and other major elements. Slab thicknesses should be kept constant without offsets, as should walls. Of course, structural requirements will necessitate variations in many elements, but it is often less expensive to use a little more concrete to maintain a uniform dimension than to form offsets.

Accuracy Standards

Because of the nature of the material and forming methods, concrete construction cannot be perfect; there are certain tolerances that are industry standards. Construction attached to concrete must be capable of accommodating these tolerances. For columns, piers, and walls, the maximum variation in plumb will be plus or minus $1/4$ in (6 mm) in any 10 ft (3050 mm) length. The same tolerance applies for horizontal elements such as ceilings, beam soffits, and slab soffits.

The maximum variation out of plumb for the total height of the structure is 1 in (25 mm) for interior columns and $1/2$ in (13 mm) for corner columns for buildings up to 100 ft (30 m) tall, while the maximum variation for the total length of the building is plus or minus 1 in (25 mm). Elevation control points for slabs on grade can vary up to $1/2$ in (13 mm) in any 10 ft (3050 mm) bay and plus or minus $3/4$ in (19 mm) for the total length of the structure.

For elevated, formed slabs the tolerance is plus or minus $3/4$ in (19 mm). Finished concrete floors can be specified anywhere from plus or minus $1/8$ in in 10 ft (3 mm in 3 m) for very flat slabs to plus or minus $1/2$ in (13 mm) for bullfloated slabs.

16. LATERAL PRESSURE ON FORMWORK

Formwork must be strong enough to withstand hydraulic loading from the concrete during curing. The hydraulic load is greatest immediately after pouring. As the concrete sets up, it begins to support itself, and the lateral force against the formwork is reduced. Publication ACI 347 predicts the maximum lateral pressure for regular (type I) concrete with a 4 in (100 mm) slump (or less), ordinary work, and internal vibration. The maximum pressure depends on the temperature, T, and the vertical rate of pour, R. (Equations 49.2 and 49.3 are valid only with customary U.S. units.)

$$p_{\text{max,psf}} = 150 + 9000 \left(\frac{R_{\text{ft/hr}}}{T_{\circ \text{F}}} \right) \quad [R \le 7 \text{ ft/hr}] \quad 49.2$$

$$p_{\text{max,psf}} = 150 + \frac{43{,}400}{T_{\circ \text{F}}} + 2800 \left(\frac{R_{\text{ft/hr}}}{T_{\circ \text{F}}} \right)$$
$$[R > 7 \text{ ft/hr}] \quad 49.3$$

50 Reinforced Concrete: Beams

Nomenclature

a	deep beam shear span (face of support to load)	in	mm
a	depth of rectangular stress block	in	mm
A	area	in^2	mm^2
b	width (of compression member)	in	mm
c	distance from neutral axis to extreme compression fiber	in	mm
c_c	clear cover from the nearest surface in tension to the surface of the flexural tension reinforcement	in	mm
C	compressive force	lbf	N
d	beam effective depth (to tension steel centroid)	in	mm
d	diameter	in	mm
D	dead load	lbf	N
E	modulus of elasticity	lbf/in^2	MPa
f	stress	lbf/in^2	MPa
f_c'	compressive strength	lbf/in^2	MPa
f_r	modulus of rupture	lbf/in^2	MPa
f_s	stress in tension steel	lbf/in^2	MPa
f_y	tension steel yield strength	lbf/in^2	MPa
F	force	lbf	N
h	overall height (thickness)	in	mm
I	moment of inertia	in^4	mm^4
l_n	clear span length	in	mm
L	live load	lbf	N
L	total span length	in	mm
L_r	roof live load	lbf	N
M	moment	in-lbf	N·mm
M_a	maximum moment	in-lbf	N·mm
n	modular ratio	–	–
s	spacing of longitudinal reinforcement nearest tension zone	in	mm
s	stirrup spacing	in	mm
T	tensile force	lbf	N
U	required strength or strength required to resist factored loads	lbf	N
V	shear strength	lbf	N
w	load per unit length	lbf/in	N/mm
w	unit weight	lbf/ft^3	kN/m^3
W	wind load	lbf	N
y	distance	in	mm

Symbols

β_1	ratio of depth of equivalent stress block to depth of actual neutral axis	–	–
Δ	deflection	in	mm
ε	strain	–	–
θ	angle	deg	deg
λ	distance from centroid of compressed area to extreme compression fiber	in	mm
λ	long-term deflection factor	–	–
ε	long-term deflection factor	–	–
ρ	reinforcement ratio	–	–
ϕ	strength reduction factor	–	–

Subscripts

a	long term
add	additional

Structural

b	balanced condition or bar
c	concrete, critical, or cover
cb	concrete balanced
cr	cracked (cracking)
d	dead
e	effective
g	gross
h	horizontal
i	immediate
l	live
max	maximum
min	minimum
n	nominal (strength)
req	required
s	steel or reinforcement
sb	steel balanced
u	ultimate (factored)
v	shear
w	web
y	yield (in steel)

1. INTRODUCTION

The design of reinforced concrete beams is governed by the provisions of the American Concrete Institute code ACI 318. Many of the code equations are not homogeneous—the units of the result cannot be derived from those of the input. Whenever nonhomogeneous equations appear in this chapter, the units listed in the nomenclature should be used.

2. STEEL REINFORCING

Figure 50.1 illustrates typical reinforcing used in a simply supported reinforced concrete beam. The straight longitudinal bars resist the tension induced by bending, and the *stirrups* resist the diagonal tension resulting from shear stresses. As shown in Fig. 50.1(b), it is also possible to bend up the longitudinal bars to resist diagonal tension near the supports. This alternative, however, is rarely used because the saving in stirrups tends to be offset by the added cost associated with bending the longitudinal bars. In either case, the stirrups are usually passed underneath the bottom steel for anchoring. Prior to pouring the concrete, the horizontal steel is supported on *bolsters* (*chairs*), of which there are a variety of designs.

3. TYPES OF BEAM CROSS SECTIONS

The design of reinforced concrete beams is based on the assumption that concrete does not resist any tensile stress. A consequence of this assumption is that the effective shape of a cross section is determined by the part of the cross section that is in compression. Consider the design of monolithic slab-and-girder systems. As Fig. 50.2 shows, moments are negative (i.e., there is tension on the top fibers) near the columns. The effective section in the region of negative moments is rectangular (Fig. 50.2(b)). Elsewhere, moments are positive,

and the effective section may be either rectangular or T-shaped, depending on the depth of the compressed region (Fig. 50.2(c)).

Figure 50.1 *Typical Reinforcement in Beams*

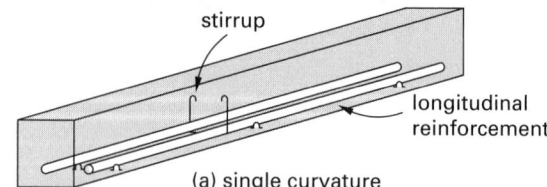

(a) single curvature

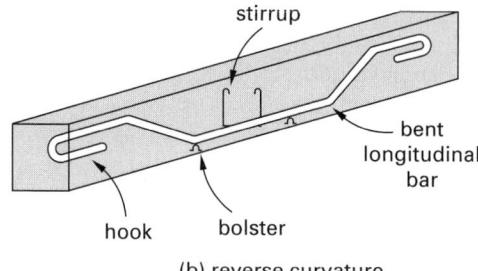

(b) reverse curvature

Figure 50.2 *Slab-Beam Floor System*

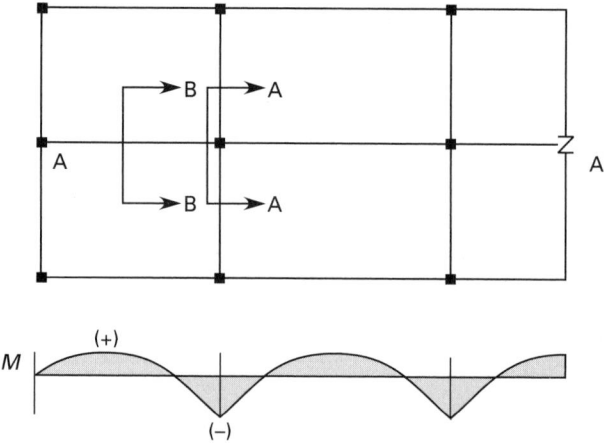

(a) monolithic slab-beam floor system

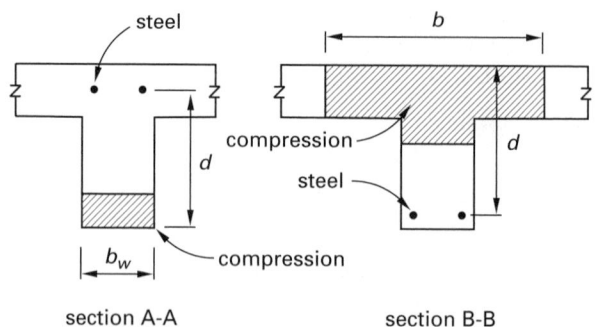

section A-A

section B-B

(b) effective beam cross section in region of negative moments

(c) effective beam cross section in region of positive moments

Reinforced concrete sections can be singly or doubly reinforced. Doubly reinforced sections are used when concrete in the compression zone cannot develop the required moment strength. Doubly reinforced sections are also used to reduce long-term deformations resulting from creep and shrinkage.

Depending on the ratio of beam clear span to the depth of the cross section, beams may be considered "regular" or deep. Deep beams are members loaded on one face and supported on the opposite face so that compression struts can develop between the loads and the support. Deep beems have either: (a) clear spans, l_n, equal to or less than four times the overall member depth, h, or (b) regions loaded with concentrated loads within twice the member depth from the face of the support. The same definition applies for flexural design [ACI 318 10.7.1] as well as for shear design [ACI 318 11.8.1].

4. ALLOWABLE STRESS VS. STRENGTH DESIGN

Structural members can be sized using two alternative design procedures. With the *allowable stress method*, stresses induced in the concrete and the steel are estimated assuming that the behavior is linearly elastic. The member is sized so that the computed stresses do not exceed certain predetermined values. With the *strength design method*, the actual stresses are not the major concern. Rather, a strength is provided that will support factored (i.e., amplified) loads. Although both methods produce adequate designs, strength design is considered to be more rational. This chapter only covers strength design provisions. ACI 318 no longer contains provisions for *allowable stress design*.

It is important to recognize that even though beam cross sections are sized on the basis of strength, the effects (i.e., moments, shears, deflections, etc.) of the factored loads are computed using elastic analysis.

5. SERVICE LOADS, FACTORED LOADS, AND LOAD COMBINATIONS

The objective in all designs is to ensure that the safety margin against any possible collapse under the service loads is adequate. In ACI 318, this margin is achieved by designing beam strength to be equal to or greater than the effect of the service loads amplified by appropriate load factors.

The term *service loads* designates the loads (forces or moments) that are expected to be actually imposed on a structure during its service life, and for design purposes are taken from building codes. The term *factored loads* designates the service loads increased by various amplifying *load factors*. The load factors depend both on the uncertainty of the various loads as well as on the load combination being considered. Load factors and load combinations are specified in ACI 318 9.2 and C.2. The factored loads are designated using the subscript

u (ultimate). The moment and shear due to factored loads are M_u and V_u, respectively. These are also the *required strengths* in moment and shear, respectively.

According to ACI 318 9.2, required strength U must be at least equal to the factored loads in Eqs. 50.1 through 50.7.

When dead and live loads only are considered,

$$U = 1.4D \qquad 50.1$$
$$U = 1.2D + 1.6L \qquad 50.2$$

When dead and live loads and wind are considered,

$$U = 1.2D + 1.6L + 0.8W \qquad 50.3$$
$$U = 1.2D + L + 1.6W \qquad 50.4$$
$$U = 0.9D + 1.6W \qquad 50.5$$

When earthquake forces are considered, the earthquake loading has a factor of 1.0, as it is derived from factored loads in the seismic code. Members must also satisfy the requirements of Eqs. 50.1 and 50.2.

$$U = 1.2D + L + E \qquad 50.6$$
$$U = 0.9D + E \qquad 50.7$$

The effect of one or more transient loads not acting simultaneously (i.e., the effects of live loads being zero in Eq. 50.2, live or wind loads being zero in Eqs. 50.3, 50.4, and 50.5, and live or earthquake loads being zero in Eqs. 50.6 and 50.7) needs to be investigated. The load factor on L in Eqs. 50.3, 50.4, and 50.6 can be reduced to 0.5 for all structures except garages, areas occupied as places of public assembly, and all areas where the live load is greater than 100 lbf/ft^2.

According to ACI 318 App. C, when only dead and live loads act, the load combination that is used for design is

$$U = 1.4D + 1.7L \qquad 50.8$$

With wind or earthquake loading, the load combinations are

$$U = 0.75(1.4D + 1.7L) + (1.6W \text{ or } E) \qquad 50.9$$
$$U = 0.9D + (1.6W \text{ or } E) \qquad 50.10$$

A load factor of 1.2 applies to the maximum tendon jacking force in a post-tensioned anchorage zone [ACI 318 9.2.5 and C.2.7].

Example 50.1

The simply supported beam shown is subjected to a dead load of 1.2 kips/ft and a live load of 0.8 kip/ft in addition to its own dead weight. What moment should be used to determine the steel reinforcement at the center of the beam?

Structural

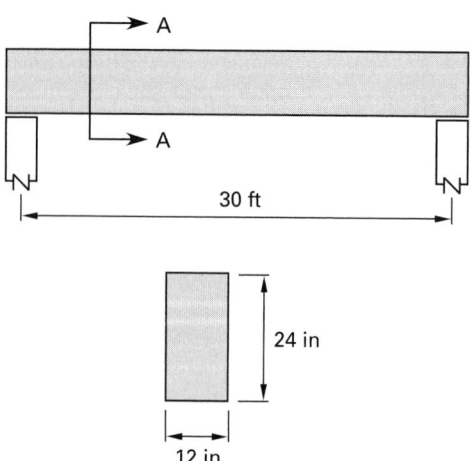

section A-A

Solution Using ACI 318 App. C

The specific weight of steel-reinforced concrete is approximately 150 lbf/ft^3. The weight of the beam is

$$\frac{(12 \text{ in})(24 \text{ in}) \left(150 \frac{\text{lbf}}{\text{ft}^3}\right)}{\left(12 \frac{\text{in}}{\text{ft}}\right)^2 \left(1000 \frac{\text{lbf}}{\text{kip}}\right)} = 0.3 \text{ kip/ft}$$

The total dead load is

$$D = 1.2 \frac{\text{kips}}{\text{ft}} + 0.3 \frac{\text{kip}}{\text{ft}}$$
$$= 1.5 \text{ kips/ft}$$

The ACI 318 App. C factored load is

$$w_u = 1.4D + 1.7L$$
$$= (1.4) \left(1.5 \frac{\text{kips}}{\text{ft}}\right) + (1.7) \left(0.8 \frac{\text{kip}}{\text{ft}}\right)$$
$$= 3.46 \text{ kips/ft}$$

For a uniformly loaded beam, the factored moment is maximum at the center of the beam and is

$$M_u = \frac{w_u L^2}{8} = \frac{\left(3.46 \frac{\text{kips}}{\text{ft}}\right) (30 \text{ ft})^2}{8}$$
$$= 389 \text{ ft-kips}$$

Solution Using ACI 318 Ch. 9

The specific weight of steel-reinforced concrete is approximately 150 lbf/ft^3. The weight of the beam is

$$\frac{(12 \text{ in})(24 \text{ in}) \left(150 \frac{\text{lbf}}{\text{ft}^3}\right)}{\left(12 \frac{\text{in}}{\text{ft}}\right)^2 \left(1000 \frac{\text{lbf}}{\text{kip}}\right)} = 0.3 \text{ kip/ft}$$

The total dead load is

$$D = 1.2 \frac{\text{kips}}{\text{ft}} + 0.3 \frac{\text{kip}}{\text{ft}}$$
$$= 1.5 \text{ kips/ft}$$

The ACI 318 Ch. 9 factored load is

$$w_u = 1.2D + 1.6L$$
$$= (1.2) \left(1.5 \frac{\text{kips}}{\text{ft}}\right) + (1.6) \left(0.8 \frac{\text{kip}}{\text{ft}}\right)$$
$$= 3.08 \text{ kips/ft}$$

For a uniformly loaded beam, the factored moment is maximum at the center of the beam and is

$$M_u = \frac{w_u L^2}{8} = \frac{\left(3.08 \frac{\text{kips}}{\text{ft}}\right) (30 \text{ ft})^2}{8}$$
$$= 346 \text{ ft-kips}$$

6. DESIGN STRENGTH AND DESIGN CRITERIA

ACI 318 uses the subscript n to indicate a "nominal" quantity. A *nominal value* can be interpreted as being in accordance with theory for the specified dimensions and material properties. The nominal moment and shear strengths are designated M_n and V_n, respectively.

The *design strength* is the product of the *nominal strength* times a *strength reduction factor* (also known as a *capacity reduction factor*), ϕ.

$$\text{design strength} = (\text{nominal strength})\phi \qquad 50.11$$

The design criteria for all sections in a beam are

$$\phi M_n \geq M_u \qquad 50.12(a)$$
$$\phi V_n \geq V_u \qquad 50.12(b)$$

ACI 318 has two sets of design criteria for reinforced and prestressed concrete structural members: those for a so-called "unified design procedure" in the main body of the code, and those for an alternative (traditional ACI) design procedure in App. B.

According to the unified design provisions in the body of the ACI code, strength reduction factors are determined on the basis of the strain conditions in the reinforcement farthest from the extreme compression face. Prior to the adoption of these provisions by ACI in 2002, strength reduction factors depended only on the type of loading (axial load, flexure, or both) on the section. The unified design provisions are intended to bring uniformity to the design bases for nonprestressed and prestressed concrete members subjected to flexure and axial loads, and eliminate many of the inconsistencies in the previous design requirements.

The following definitions are relevant to the unified design provisions. These definitions are used when determining strength reduction factors and design strengths. They can be found in Ch. 2 of the ACI code.

Net tensile strain, ε_t, is the tensile strain in the extreme tension steel at nominal strength, exclusive of strains due to effective prestress, creep, shrinkage, and temperature. The net tensile strain is caused by external axial loads and/or bending moments at a section when the concrete strain at the extreme compression fiber reaches its assumed limit of 0.003. Generally speaking, the net tensile strain can be used as a measure of excessive cracking or excessive deflection.

Extreme tension steel is the reinforcement (prestressed or nonprestressed) that is the farthest from the extreme compression fiber.

Fig. 50.3 depicts the location of the extreme tension steel for two sections with different reinforcement arrangements, both of which assume the top fiber of the section is the extreme compression fiber. The distance from the extreme compression fiber to the centroid of the extreme tension steel is denoted in the illustration as d_t. The net tensile strain in the extreme tension steel due to the external loads can be determined from a strain compatibility analysis for sections with multiple layers of reinforcement. For sections with one layer of reinforcement, the net tensile strain can be determined from the strain diagram by using similar triangles.

Figure 50.3 *Location of Extreme Tension Steel and Net Tensile Strain at Nominal Strength*

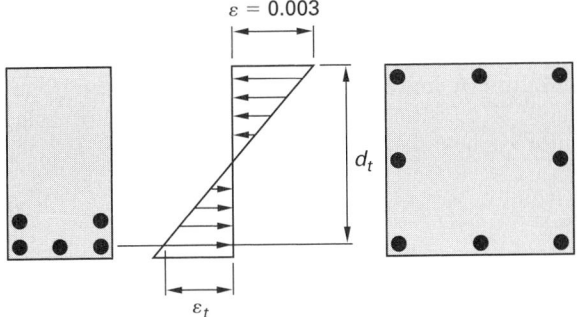

The compression-controlled strain limit is the net tensile strain corresponding to a balanced condition. The definition of a balanced strain condition (also see Sec. 8), given in ACI Sec. 10.3.2, is unchanged from previous editions of the code. A balanced strain condition exists in a cross section when tension reinforcement reaches the strain corresponding to its specified yield strength just as the concrete strain in the extreme compression fiber reaches its assumed limit of 0.003.

For grade 60 reinforcement and all prestressed reinforcement, ACI Sec. 10.3.3 permits the compression-controlled strain limit to be taken equal to 0.002. For grade 60 bars, this limit is actually equal to $f_y/E_s =$

$60,000 \text{ lbf/in}^2 / 29,000,000 \text{ lbf/in}^2 = 0.00207$. f_y and E_s are the specified yield strength and modulus of elasticity of the nonprestressed reinforcement, respectively. For other grades of nonprestressed steel, the limit must be calculated from the ratio f_y/E_s.

A compression-controlled section is a cross section in which the net tensile strain is less than or equal to the compression-controlled strain limit. When the net tensile strain is small, a brittle failure condition is expected. In such cases, there is little warning of impending failure. Cross sections of compression members such as columns, subject to significant axial compression, are usually compression controlled.

A tension-controlled section is a cross section in which the net tensile strain is greater than or equal to 0.005. The net tensile strain limit of 0.005 applies to both nonprestressed and prestressed reinforcement and provides ductile behavior for most designs. When the net tensile strain is greater than or equal to 0.005, the section is expected to have sufficient ductility so that ample warning of failure in the form of visible cracking and deflection should occur. Cross sections of flexural members such as beams, if not heavily reinforced, are usually tension controlled.

Some sections have a net tensile strain between the limits for compression-controlled and tension-controlled sections. An example of this is a section subjected to a small axial load and a large bending moment. These sections are called *transition sections*.

Strength Reduction Factors, ϕ

In editions of the code prior to 2002, the appropriate strength reduction factor used in design depended on the type of loading to which members would be subjected. For example, for members that would be subjected to flexure without axial load, ϕ was equal to 0.90.

According to the unified design provisions, strength reduction factors are a function of the net tensile strain. ACI 9.3.2 and C.3.2 contain strength reduction factors for tension-controlled sections, compression-controlled sections, and sections in which the net tensile strain is between the limits for tension-controlled and compression-controlled sections. The variation in ϕ with respect to ε_t according to ACI 9.3.2 is depicted Fig. 50.4. Variation of ϕ with respect to ε_t in ACI C.3.2 is depicted in ACI Fig. RC.3.2.

Figure 50.4 *Variation of Strength Reduction Factors, ϕ, with Net Tensile Strain, ε_t [ACI 318 9.3.2]*

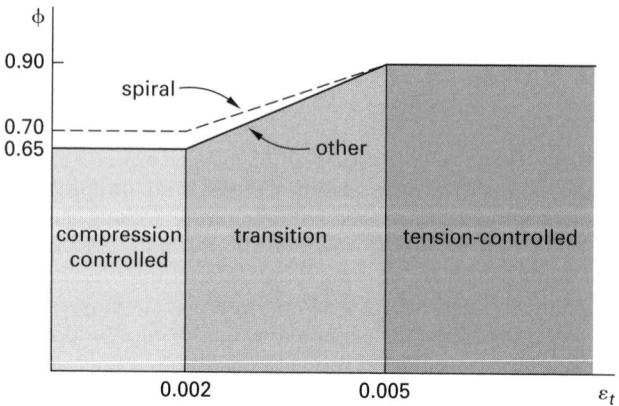

For compression-controlled sections, ϕ is equal to 0.70 for members with spiral reinforcement conforming to ACI 10.9.3 and is equal to 0.65 for other members in ACI 9.3.2. For compression-controlled sections designed according to ACI 318 App. C, ϕ is equal to 0.75 for members with spiral reinforcement conforming to ACI 10.9.3 and is equal to 0.70 for other members in ACI C.3.2. For tension-controlled sections, ϕ is equal to 0.90 in both ACI 9.3.2 and ACI C.3.2. For sections that fall between these two limits, the code permits the strength reduction factor to be determined considering linear variation.

The following equations can be used to determine ϕ in the transition region. For sections with spiral reinforcement,

$$\phi = 0.57 + 67\varepsilon_t \quad \text{[ACI 9.3.2]} \qquad \textit{50.13(a)}$$

$$\phi = 0.65 + 50\varepsilon_t \quad \text{[ACI C.3.2]} \qquad \textit{50.13(b)}$$

For other sections,

$$\phi = 0.48 + 83\varepsilon_t \quad \text{[ACI 9.3.2]} \qquad \textit{50.14(a)}$$

$$\phi = 0.57 + 67\varepsilon_t \quad \text{[ACI C.3.2]} \qquad \textit{50.14(b)}$$

ACI Figs. R9.3.2 and RC.3.2 also contain equations to determine ϕ as a function of the ratio c/d_t. c is the distance from the extreme compression fiber to the neutral axis at nominal strength.

The nominal shear strength must be reduced by

$$\phi = 0.75 \quad \text{[ACI 9.3.2]} \qquad \textit{50.15(a)}$$

$$\phi = 0.85 \quad \text{[ACI C.3.2]} \qquad \textit{50.15(b)}$$

7. MINIMUM STEEL AREA

ACI 318 sets limits on the minimum and maximum areas of tension steel. The minimum area of tension steel is

$$A_{s,\min} = \frac{\sqrt{f_c'}\,b_w d}{4 f_y} \geq \frac{1.4 b_w d}{f_y} \qquad \text{[SI]} \qquad \textit{50.16(a)}$$

$$A_{s,\min} = \frac{3\sqrt{f_c'}\,b_w d}{f_y} \geq \frac{200 b_w d}{f_y} \qquad \text{[U.S.]} \qquad \textit{50.16(b)}$$

When the bending moment produces tension in the flange of a statically determinate T-beam, the value of $A_{s,\min}$ calculated from Eq. 50.16 must be increased. For this condition, ACI 318 10.5.2 requires that the minimum steel be taken as the smaller of what is calculated from either Eq. 50.16 (with b_w set equal to the width of the flange) or Eq. 50.17.

$$A_{s,\min} = \frac{\sqrt{f_c'}\,b_w d}{2 f_y} \qquad \text{[SI]} \qquad \textit{50.17(a)}$$

$$A_{s,\min} = \frac{6\sqrt{f_c'}\,b_w d}{f_y} \qquad \text{[U.S.]} \qquad \textit{50.17(b)}$$

Minimum steel provisions of Eqs. 50.16 and 50.17 do not have to be followed if the area of steel provided exceeds the calculated amount required by at least 33% (ACI 318 10.5.3).

8. MAXIMUM STEEL AREA

In a *ductile failure mode*, the steel yields in tension before the concrete crushes in compression (the reverse would be a *brittle failure mode*). The area of tension steel for which steel yielding occurs simultaneously with concrete crushing in compression is known as the *balanced steel area*, A_{sb}. An *under-reinforced beam* is defined as one having a steel area less than A_{sb}. An *over-reinforced beam* has more steel than A_{sb}.

ACI 318 B.10.3.3 (alternative provisions) ensures ductile failure by limiting the maximum amount of tension steel to 75% of the balanced reinforcement. For a singly reinforced beam, the maximum steel area is

$$A_{s,\max} = 0.75 A_{\text{sb}} \qquad \textit{50.18}$$

The balanced steel area is

$$A_{\text{sb}} = \frac{0.85 f_c' A_{\text{cb}}}{f_y} \qquad \textit{50.19}$$

Figure 50.5 illustrates that the compression area for balanced conditions, A_{cb}, extends from the most compressed fiber to a depth equal to a_b.

$$a_b = \beta_1 \left(\frac{600}{600 + f_y} \right) d \qquad \text{[SI]} \qquad \textit{50.20(a)}$$

$$a_b = \beta_1 \left(\frac{87{,}000}{87{,}000 + f_y} \right) d \quad \text{[U.S.]} \qquad \textit{50.20(b)}$$

The factor β_1 is defined as

$$\beta_1 = 0.85 \quad [f_c' \leq 27.6 \text{ MPa}] \qquad \text{[SI]} \qquad \textit{50.21(a)}$$

$$\beta_1 = 0.85 \quad [f'_c \le 4000 \text{ lbf/in}^2] \qquad \text{[U.S.]} \quad \textit{50.21(b)}$$

$$\beta_1 = 0.85 - (0.05)\left(\frac{f'_c - 27.6}{6.9}\right) \ge 0.65$$
$$\text{[SI]} \quad \textit{50.21(c)}$$

$$\beta_1 = 0.85 - (0.05)\left(\frac{f'_c - 4000}{1000}\right) \ge 0.65$$
$$\text{[U.S.]} \quad \textit{50.21(d)}$$

Combining Eqs. 50.19 and 50.20 with $A_{cb} = 0.85\beta_1 a_b b$, for rectangular sections, the balanced steel area is

$$A_{\text{sb}} = \left(\left(\frac{0.85\beta_1 f'_c}{f_y}\right)\left(\frac{600}{600 + f_y}\right)\right) bd$$
$$\text{[SI]} \quad \textit{50.22(a)}$$

$$A_{\text{sb}} = \left(\left(\frac{0.85\beta_1 f'_c}{f_y}\right)\left(\frac{87,000}{87,000 + f_y}\right)\right) bd$$
$$\text{[U.S.]} \quad \textit{50.22(b)}$$

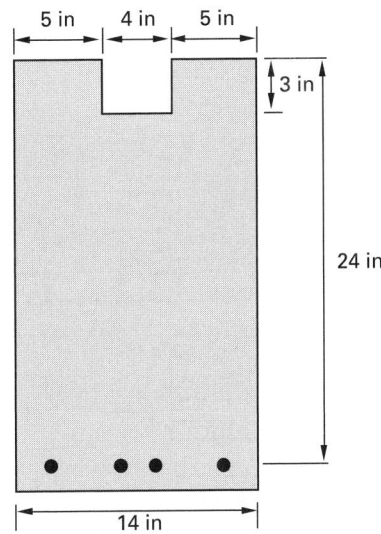

Figure 50.5 Beam at Balanced Condition

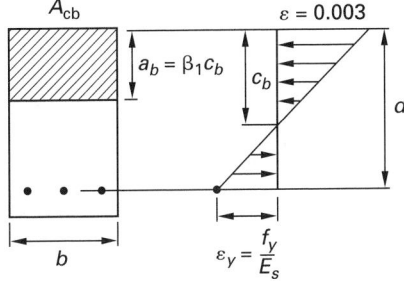

The unified (ACI 318 Ch. 10) design provisions limit the maximum reinforcement in a flexural member (with factored axial load less than $0.1f'_c A_g$) to that which would result in a net tensile strain, ε_t, not less than 0.004 [ACI 318 10.3.5]. For a singly reinforced section with grade 60 reinforcement, this is equivalent to a maximum reinforcement ratio of $0.714\rho_b$. The ACI App. B limit of $\rho_{\text{max}} = 0.75\rho_b$ results in a net tensile strain of 0.00376. At the net tensile strain limit of 0.004, the strength reduction factor of ACI 9.3.2, ϕ, is reduced to 0.812.

It is almost always advantageous to limit the net tensile strain in flexural members to a minimum of 0.005, which is equivalent to a maximum reinforcement ratio of $0.63\rho_b$ [ACI 318 RC3.2.2], even though the code permits higher amounts of reinforcement that produce lower net tensile strains. Where member size is limited and extra strength is needed, it is best to use compression reinforcement to limit the net tensile strain so that the section is tension controlled.

Example 50.2

For the beam cross section shown, assume the centroid of the tension steel is at a depth of 24 in. $f'_c = 4000$ lbf/in^2, and $f_y = 60,000$ lbf/in^2. Calculate the maximum area of reinforcing steel that can be used.

Solution Using ACI 318 App. B

This is not a rectangular beam, so the balanced steel area cannot be calculated from Eq. 50.22.

The depth of the equivalent rectangular stress block at the balanced condition is obtained from Eq. 50.20(b) with $\beta_1 = 0.85$.

$$\begin{aligned}
a_b &= \beta_1 \left(\frac{87,000}{87,000 + f_y}\right) d \\
&= (0.85)\left(\frac{87,000}{87,000 + 60,000}\right)(24 \text{ in}) \\
&= 12.07 \text{ in}
\end{aligned}$$

The area A_{cb} corresponding to the depth a_b is

$$\begin{aligned}
A_{\text{cb}} &= (10 \text{ in})(3 \text{ in}) + (14 \text{ in})(12.07 \text{ in} - 3 \text{ in}) \\
&= 156.98 \text{ in}^2
\end{aligned}$$

Substitute into Eq. 50.19.

$$\begin{aligned}
A_{\text{sb}} &= \frac{0.85 f'_c A_{\text{cb}}}{f_y} \\
&= \frac{(0.85)\left(4000 \,\dfrac{\text{lbf}}{\text{in}^2}\right)(156.98 \text{ in}^2)}{60,000 \,\dfrac{\text{lbf}}{\text{in}^2}} \\
&= 8.89 \text{ in}^2
\end{aligned}$$

The maximum steel permitted without the use of compression steel is given by Eq. 50.18.

$$\begin{aligned}
A_{s,\text{max}} &= 0.75 A_{\text{sb}} \\
&= (0.75)(8.89 \text{ in}^2) \\
&= 6.67 \text{ in}^2
\end{aligned}$$

Solution Using ACI 318 Ch. 10

Use the unified design provisions in ACI 318 Ch. 10.

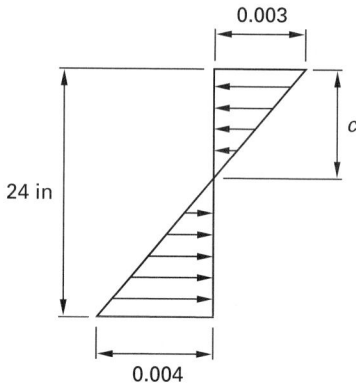

Neutral axis depth, c, corresponding to minimum permissible $\varepsilon_t = 0.004$, is

$$\frac{0.003}{0.003 + 0.004} = \frac{c}{24 \text{ in}}$$

$$c = \left(\frac{3}{7}\right)(24 \text{ in})$$

$$= 10.29 \text{ in}$$

The corresponding steel area is $A_{s,\max}$.

From force equilibrium (see Sec. 10),

$$\left(\begin{array}{c} (10 \text{ in})(0.85)(10.29 \text{ in}) \\ + (4 \text{ in})(0.85) \\ \times (10.29 \text{ in} - 3 \text{ in}) \end{array} \right)(0.85)\left(4000 \ \frac{\text{lbf}}{\text{in}^2}\right)$$

$$= A_{s,\max}\left(60,000 \ \frac{\text{lbf}}{\text{in}^2}\right)$$

$$A_{s,\max} = 6.36 \text{ in}^2$$

9. STEEL COVER AND BEAM WIDTH

The following details must be considered in the design of steel-reinforced concrete beams. The beam widths in Table 50.2 are derived from them.

- All steel, including shear reinforcement, must be adequately covered by concrete. A minimum of 1.5 in of cover is generally required. Table 50.1 provides more specific details.

- The minimum horizontal *clear distance* between bars is the larger of one bar diameter or 1 in [ACI 318 7.6].

- The minimum vertical clear distance between steel layers is 1 in.

- Clear distance beween bars must be $^4/_3$ of the maximum aggregate size [ACI 318 3.3.2(c)]. For typical $^3/_4$ in aggregate, this is equivalent to a minimum clear distance of 1 in.

- Although it is not an ACI requirement, it is common practice to use a ratio of beam depth to width that satisfies Eq. 50.23.

$$1.5 \leq d/b \leq 2.5 \qquad \textit{50.23}$$

Table 50.1 *Minimum Cover on Nonprestressed Steel*

cast against and permanently exposed to earth	3 in
exposed to earth or weather	
no. 5 bars or smaller and W31 or D31 wire	$1\frac{1}{2}$ in
no. 6 bars or larger	2 in
not exposed to weather or earth	
beams, girders, or columns	$1\frac{1}{2}$ in
slabs, walls, or joists	
no. 11 bars or smaller	$\frac{3}{4}$ in
no. 14 and no. 18 bars	$1\frac{1}{2}$ in

(Multiply in by 2.54 to obtain cm.)
Source: ACI 318 7.7.1

Table 50.2 *Minimum Beam Widths*[a,b] *for Beams with 1½ in Cover (inches)*

size of bar	number of bars in a single layer of reinforcement							add for each additional bar[c]
	2	3	4	5	6	7	8	
no. 4	6.8	8.3	9.8	11.3	12.8	14.3	15.8	1.50
no. 5	6.9	8.5	10.2	11.8	13.4	15.0	16.7	1.63
no. 6	7.0	8.8	10.5	12.3	14.0	15.8	17.5	1.75
no. 7	7.2	9.0	10.9	12.8	14.7	16.5	18.4	1.88
no. 8	7.3	9.3	11.3	13.3	15.3	17.3	19.3	2.00
no. 9	7.6	9.8	12.2	14.3	16.6	18.8	21.1	2.26
no. 10	7.8	10.4	12.9	15.5	18.0	20.5	23.1	2.54
no. 11	8.1	10.9	13.8	16.6	19.4	22.2	25.0	2.82
no. 14	8.9	12.3	15.7	19.1	22.5	25.9	29.3	3.40
no. 18	10.6	15.1	19.6	24.1	28.6	33.1	37.6	4.51

(Multiply in by 25.4 to obtain mm.)
[a]Using no. 3 stirrups. If stirrups are not used, deduct 0.75 in.
[b]The minimum inside radius of a 90° stirrup bend is 2 times the stirrup diameter. An allowance has been included in the beam widths to achieve a full bend radius.
[c]For additional horizontal bars, the beam width is increased by adding the value in the last column.

10. NOMINAL MOMENT STRENGTH OF SINGLY REINFORCED SECTIONS

The moment capacity of a reinforced concrete cross section derives from the couple composed of the tensile force in the steel and the compressive force in the concrete. These forces are equal. The steel force is limited to the product of the steel area times the steel yield strength. Therefore, the nominal moment capacity is

$$M_n = f_y A_s \times \text{lever arm} \qquad \textit{50.24}$$

The maximum nominal moment capacity will occur when the steel yields and the lever arm reaches a maximum value. However, since the true distribution of compressive forces at ultimate capacity is nonlinear, it is difficult to determine the lever arm length. To simplify the analysis, ACI 318 permits replacing the true nonlinear distribution of stresses with a simple uniform distribution where the stress equals $0.85f_c'$ [ACI 318 10.2.7]. This is known as the *Whitney assumption*. As shown in Fig. 50.6(c), β_1 is the ratio of the depth of the equivalent rectangular stress block to the true depth of the neutral axis.

The nominal moment capacity can be computed for any singly reinforced section with the following procedure.

step 1: Compute the tension force.

$$T = f_y A_s \qquad 50.25$$

step 2: Calculate the area of concrete that, at a stress of $0.85f_c'$, gives a force equal to T.

$$A_c = \frac{T}{0.85f_c'} = \frac{f_y A_s}{0.85f_c'} \qquad 50.26$$

step 3: Locate the centroid of the area A_c. Designating the distance from the centroid of the compression region to the most compressed fiber as λ,

$$M_n = A_s f_y (d - \lambda) \qquad 50.27$$

Figure 50.6 *Conditions at Maximum Moment*

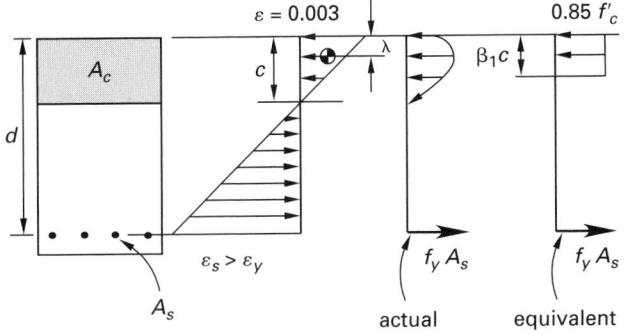

(a) strain distribution

(b) compressive stress distribution

(c) equivalent rectangular compressive stress block

Example 50.3

The cross section of the beam shown is reinforced with 3.0 in² of steel. (The cross section and material properties of this example are the same as those used in

Ex. 50.2.) $f_c' = 4000$ lbf/in², and $f_y = 60,000$ lbf/in². Calculate the design moment capacity, ϕM_n.

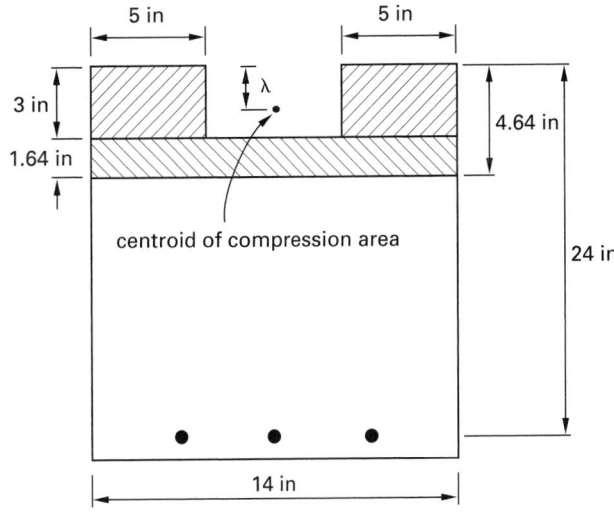

Solution

Assume that the tension steel yields at maximum moment. The area of concrete required to balance the steel force at yield is given by Eq. 50.26.

$$A_c = \frac{f_y A_s}{0.85f_c'}$$

$$= \frac{\left(60,000 \ \frac{\text{lbf}}{\text{in}^2}\right)(3 \ \text{in}^2)}{(0.85)\left(4000 \ \frac{\text{lbf}}{\text{in}^2}\right)}$$

$$= 52.94 \ \text{in}^2$$

The thickness of the 14 in wide compression zone is

$$52.94 \ \text{in}^2 = (2)(5 \ \text{in})(3 \ \text{in}) + (14 \ \text{in})t$$

$$t = 1.64 \ \text{in}$$

Sum moments of the areas about the top of the section and equate the sum to $A_c \lambda$.

$$(2)(5 \ \text{in})(3 \ \text{in})(1.5 \ \text{in})$$
$$+ (1.64 \ \text{in})(14 \ \text{in})\left(3 \ \text{in} + \frac{1.64 \ \text{in}}{2}\right)$$
$$= (52.94 \ \text{in}^2)\lambda$$
$$\lambda = 2.51 \ \text{in}$$

The nominal moment capacity is obtained from Eq. 50.27.

$$M_n = A_s f_y (d - \lambda)$$

$$= \frac{(3 \ \text{in}^2)\left(60,000 \ \frac{\text{lbf}}{\text{in}^2}\right)(24 \ \text{in} - 2.51 \ \text{in})}{\left(12 \ \frac{\text{in}}{\text{ft}}\right)\left(1000 \ \frac{\text{lbf}}{\text{kip}}\right)}$$

$$= 322.4 \ \text{ft-kips}$$

Structural

The design moment capacity is

$$\phi M_n = (0.9)(322.4 \text{ ft-kips})$$
$$= 290 \text{ ft-kips}$$

Verify that the tensile steel yields at ultimate loading. Use the ACI alternative provisions. In Ex. 50.2, the balanced steel area was found to be 8.89 in². 3.0 in² is less than the balanced area—and less than the maximum area—so the steel yields at ultimate loading.

Using unified design,

$$a = 4.64 \text{ in}$$
$$c = \frac{a}{\beta_1} = \frac{4.64 \text{ in}}{0.85}$$
$$= 5.46 \text{ in}$$
$$\varepsilon_t = \left(\frac{d-c}{c}\right) 0.003$$
$$= \left(\frac{24 \text{ in} - 5.46 \text{ in}}{5.46 \text{ in}}\right)(0.003)$$
$$= 0.010$$

Since 0.010 is greater than 0.005, the section is tension controlled, and the tension reinforcement yields at ultimate loading.

11. BEAM DESIGN: SIZE KNOWN, REINFORCEMENT UNKNOWN

If the required strength, M_u, due to the factored loads is known or can be computed, beam design reduces to determining the amount of reinforcing steel required. Although formulas for determining reinforcing steel in rectangular cross sections can be derived, an iterative approach is sufficiently simple.

Equation 50.28 is derived by changing the $\geq$ sign in Eq. 50.12(a) to an equality and substituting Eq. 50.27 for M_n.

$$A_s = \frac{M_u}{\phi f_y(d - \lambda)} \qquad 50.28$$

Equation 50.28 is easily solved if a value for λ is first assumed. (A value of $0.1d$ is typically a good estimate.) The design procedure consists of estimating λ, solving for A_s from Eq. 50.28, and checking the moment capacity for the selected steel using the actual associated λ. If the values of the capacity, ϕM_n, and the demand, M_u, are not sufficiently close, the area of steel is adjusted (i.e., extrapolated) by multiplying it by $(M_u/\phi M_n)$. Even though the relationship between steel area and moment is linear only over small variations, this procedure seldom requires additional iterations because the adjusted steel area will be very close to the exact solution.

It should be noted that, in actual practice, it is customary to provide the required steel area using at most two different bar sizes. For this reason, and because only a limited number of bar sizes is available, some excess steel is typically unavoidable.

Example 50.4

The beam cross section in Exs. 50.2 and 50.3 is used in a simply supported beam having a span of 25 ft. The beam is subjected to a service dead load (which includes its own weight) of 1.2 kips/ft and a live load of 2.0 kips/ft. $f_c' = 5000 \text{ lbf/in}^2$, and $f_y = 60,000 \text{ lbf/in}^2$. Determine the required steel area for the maximum moment.

Solution Using ACI 318 App. C

The ACI 318 App. C factored load is

$$w_u = 1.4D + 1.7L$$
$$= (1.4)\left(1.2 \frac{\text{kips}}{\text{ft}}\right) + (1.7)\left(2.0 \frac{\text{kips}}{\text{ft}}\right)$$
$$= 5.08 \text{ kips/ft}$$

The maximum moment is

$$M_u = \frac{w_u L^2}{8} = \frac{\left(5.08 \frac{\text{kips}}{\text{ft}}\right)(25 \text{ ft})^2}{8}$$
$$= 396.9 \text{ ft-kips}$$

Estimate λ.

$$\lambda = 0.1d = (0.1)(24 \text{ in})$$
$$= 2.4 \text{ in}$$

From Eq. 50.28,

$$A_s = \frac{M_u}{\phi f_y(d - \lambda)}$$
$$= \frac{(396.9 \text{ ft-kips})\left(12 \frac{\text{in}}{\text{ft}}\right)}{(0.9)\left(60 \frac{\text{kips}}{\text{in}^2}\right)(24 \text{ in} - 2.4 \text{ in})}$$
$$= 4.08 \text{ in}^2$$

Compute M_n for this steel area. From Eq. 50.26,

$$A_c = \frac{f_y A_s}{0.85 f_c'} = \frac{(4.08 \text{ in}^2)\left(60 \frac{\text{kips}}{\text{in}^2}\right)}{(0.85)\left(5 \frac{\text{kips}}{\text{in}^2}\right)}$$
$$= 57.60 \text{ in}^2$$

The thickness of the 14 in wide compression zone is

$$57.60 \text{ in}^2 = (2)(5 \text{ in})(3 \text{ in}) + (14 \text{ in})t$$
$$t = 1.97 \text{ in}$$

Locate the centroid of the compression zone.

$$(10 \text{ in})(3 \text{ in})(1.5 \text{ in}) + (1.97 \text{ in})(14 \text{ in})\left(3 \text{ in} + \frac{1.97 \text{ in}}{2}\right)$$

$$= (57.60 \text{ in}^2)\lambda$$

$$\lambda = 2.69 \text{ in}$$

The nominal moment capacity obtained from Eq. 50.27 is

$$M_n = A_s f_y(d - \lambda)$$

$$= \frac{(4.08 \text{ in}^2)\left(60 \dfrac{\text{kips}}{\text{in}^2}\right)(24 \text{ in} - 2.69 \text{ in})}{12 \dfrac{\text{in}}{\text{ft}}}$$

$$= 434.7 \text{ ft-kips}$$

The design moment capacity is

$$\phi M_n = (0.9)(434.7 \text{ ft-kips})$$

$$= 391.2 \text{ ft-kips}$$

Since the design moment capacity (391.2 ft-kips) is slightly less than the factored moment (396.9 ft-kips), the "exact" steel area required is nominally larger than the estimated value of 4.08 in². The steel area is extrapolated as

$$A_{s,\text{new}} = A_s\left(\frac{M_u}{\phi M_n}\right)$$

$$= \left(\frac{396.9 \text{ ft-kips}}{391.2 \text{ ft-kips}}\right)(4.08 \text{ in}^2)$$

$$= 4.14 \text{ in}^2$$

The value of ϕM_n for this adjusted steel area is 396.4 ft-kips, essentially the same as M_u.

Solution Using ACI 318 Ch. 9

The ACI 318 Ch. 9 factored load is

$$w_u = 1.2D + 1.6L$$

$$= (1.2)\left(1.2 \frac{\text{kips}}{\text{ft}}\right) + (1.6)\left(2.0 \frac{\text{kips}}{\text{ft}}\right)$$

$$= 4.64 \text{ kips/ft}$$

The maximum moment is

$$M_u = \frac{w_u L^2}{8} = \frac{\left(4.64 \dfrac{\text{kips}}{\text{ft}}\right)(25 \text{ ft})^2}{8}$$

$$= 362.5 \text{ ft-kips}$$

Estimate λ.

$$\lambda = 0.1d = (0.1)(24 \text{ in})$$

$$= 2.4 \text{ in}$$

From Eq. 50.28, assuming a tension-controlled section,

$$A_s = \frac{M_u}{\phi f_y(d - \lambda)}$$

$$= \frac{(362.5 \text{ ft-kips})\left(12 \dfrac{\text{in}}{\text{ft}}\right)}{(0.9)\left(60 \dfrac{\text{kips}}{\text{in}^2}\right)(24 \text{ in} - 2.4 \text{ in})}$$

$$= 3.73 \text{ in}^2$$

Compute M_n for this steel area. From Eq. 50.26,

$$A_c = \frac{f_y A_s}{0.85 f_c'} = \frac{(3.73 \text{ in}^2)\left(60 \dfrac{\text{kips}}{\text{in}^2}\right)}{(0.85)\left(5 \dfrac{\text{kips}}{\text{in}^2}\right)}$$

$$= 52.66 \text{ in}^2$$

The thickness of the 14 in wide compression zone is

$$52.66 \text{ in}^2 = (2)(5 \text{ in})(3 \text{ in}) + (14 \text{ in})t$$

$$t = 1.62 \text{ in}$$

Locate the centroid of the compression zone.

$$(10 \text{ in})(3 \text{ in})(1.5 \text{ in}) + (1.62 \text{ in})(14 \text{ in})\left(3 \text{ in} + \frac{1.62 \text{ in}}{2}\right)$$

$$= (52.66 \text{ in}^2)\lambda$$

$$\lambda = 2.50 \text{ in}$$

The nominal moment capacity obtained from Eq. 50.27 is

$$M_n = A_s f_y(d - \lambda)$$

$$= \frac{(3.73 \text{ in}^2)\left(60 \dfrac{\text{kips}}{\text{in}^2}\right)(24 \text{ in} - 2.50 \text{ in})}{12 \dfrac{\text{in}}{\text{ft}}}$$

$$= 401 \text{ ft-kips}$$

The design moment capacity is

$$\phi M_n = (0.9)(401 \text{ ft-kips})$$

$$= 360.9 \text{ ft-kips}$$

Since the design moment capacity (360.9 ft-kips) is slightly less than the factored moment (362.5 ft-kips), the "exact" steel area required is nominally larger than the estimated value of 3.73 in². The steel area is extrapolated as

$$A_{s,\text{new}} = A_s\left(\frac{M_u}{\phi M_n}\right) = \left(\frac{362.5 \text{ ft-kips}}{360.9 \text{ ft-kips}}\right)(3.73 \text{ in}^2)$$

$$= 3.75 \text{ in}^2$$

The value of ϕM_n for this adjusted steel area is 362.6 ft-kips, essentially the same as M_u.

Structural

Check that the section is tension controlled. For $A_s = 3.75$ in^2,

$$A_c = 52.94 \text{ in}^2$$

$$t = 1.64 \text{ in}$$

The stress block depth is

$$a = 3 \text{ in} + 1.64 \text{ in} = 4.64 \text{ in}$$

From Eq. 50.21(d), with 5000 psi concrete, $\beta_1 = 0.80$. The corresponding neutral axis depth is

$$c = \frac{a}{\beta_1} = \frac{4.64 \text{ in}}{0.80}$$
$$= 5.8 \text{ in}$$

The net tensile strain is

$$\varepsilon_t = \left(\frac{24 \text{ in} - 5.8 \text{ in}}{5.8 \text{ in}}\right)(0.003) = 0.0094$$

Since 0.0094 is greater than 0.005, the section is tension controlled.

12. BEAM DESIGN: SIZE AND REINFORCEMENT UNKNOWN

If beam size (i.e., cross section) and area of reinforcement are both unknown, there will not be a unique solution. In principle, the procedure involves first choosing a beam cross section and then sizing the steel in the usual manner. The beam is sized by (a) selecting a depth to ensure that deflections are not a problem (see Sec. 17) and (b) selecting a width so that the ratio of depth to width is reasonable. Then, the steel is checked to see that the corresponding net tensile strain is not less than 0.004, or that the steel area does not exceed the limit of $0.75A_{sb}$ and that the steel area is larger than or equal to the minimum value prescribed in ACI 318.

Portions of the design procedure can be streamlined for rectangular cross sections by deriving an explicit formula relating the nominal moment capacity to the relevant design parameters. For a rectangular cross section,

$$\lambda = \frac{A_c}{2b} \qquad 50.29$$

Combining Eqs. 50.27, 50.28, and 50.29,

$$M_n = A_s f_y d \left(1 - \frac{A_s f_y}{1.7 f'_c bd}\right) \qquad 50.30$$

A convenient form of Eq. 50.30 is obtained by expressing the steel area in terms of the *steel ratio (reinforcement ratio)*, ρ, which is defined as

$$\rho = \frac{A_s}{bd} \qquad 50.31$$

Combining Eqs. 50.30 and 50.31,

$$M_n = \rho bd^2 f_y \left(1 - \frac{\rho f_y}{1.7 f'_c}\right)$$
$$= A_s f_y \left(d - \frac{A_s f_y}{1.7 f'_c b}\right) \qquad 50.32$$

The balanced steel ratio is

$$\rho_{\text{sb}} = \left(\frac{0.85\beta_1 f'_c}{f_y}\right)\left(\frac{600}{600 + f_y}\right) \qquad \text{[SI]} \quad 50.33(a)$$

$$\rho_{\text{sb}} = \left(\frac{0.85\beta_1 f'_c}{f_y}\right)\left(\frac{87{,}000}{87{,}000 + f_y}\right) \qquad \text{[U.S.]} \quad 50.33(b)$$

The minimum steel is derived from Eq. 50.16.

$$\rho_{\min} = \frac{\sqrt{f'_c}}{4 f_y} \geq \frac{1.4}{f_y} \qquad \text{[SI]} \quad 50.34(a)$$

$$\rho_{\min} = \frac{3\sqrt{f'_c}}{f_y} \geq \frac{200}{f_y} \qquad \text{[U.S.]} \quad 50.34(b)$$

The following steps constitute a procedure for designing a beam with a rectangular cross section.

step 1: Select a value of ρ between the minimum and maximum values. The maximum value corresponds to $\varepsilon_t = 0.004$ in unified design and is equal to $0.75\rho_{\text{sb}}$ by the alternative provisions of App. B. It is widely held that values of $\rho f_y / f'_c$ near 0.18 lead to economical beams. (Larger values will lead to smaller concrete cross sections.)

step 2: Calculate the required value of bd^2 from Eq. 50.32.

step 3: Choose appropriate values of b and d. A good choice is to keep d/b between 1.50 and 2.5. For beams with a single layer of tension steel, the value of d is approximately equal to $h - 2.5$ in. Furthermore, explicit computation of deflections can be obviated if a depth is chosen that is larger than a certain fraction of the span. (See Table 50.4.) The dimensions b and h are usually rounded to the nearest whole inch.

step 4: If the actual bd^2 quantity is notably different from the bd^2 calculated in step 2, use Eq. 50.32 to recalculate the associated ρ value.

step 5: Calculate the required steel area as

$$A_s = \rho bd \qquad 50.35$$

step 6: Select reinforcing steel bars to satisfy the distribution and placement requirements of ACI 318. (Refer to Tables 50.2 and 50.3.)

Table 50.3 *Total Areas for Various Numbers of Bars*

bar size	nominal diameter (in)	weight (lbf/ft)	number of bars									
			1	2	3	4	5	6	7	8	9	10
no. 3	0.375	0.376	0.11	0.22	0.33	0.44	0.55	0.66	0.77	0.88	0.99	1.10
no. 4	0.500	0.668	0.20	0.40	0.60	0.80	1.00	1.20	1.40	1.60	1.80	2.00
no. 5	0.625	1.043	0.31	0.62	0.93	1.24	1.55	1.86	2.17	2.48	2.79	3.10
no. 6	0.750	1.502	0.44	0.88	1.32	1.76	2.20	2.64	3.08	3.52	3.96	4.40
no. 7	0.875	2.044	0.60	1.20	1.80	2.40	3.00	3.60	4.20	4.80	5.40	6.00
no. 8	1.000	2.670	0.79	1.58	2.37	3.16	3.95	4.74	5.53	6.32	7.11	7.90
no. 9	1.128	3.400	1.00	2.00	3.00	4.00	5.00	6.00	7.00	8.00	9.00	10.0
no. 10	1.270	4.303	1.27	2.54	3.81	5.08	6.35	7.62	8.89	10.16	11.43	12.70
no. 11	1.410	5.313	1.56	3.12	4.68	6.24	7.80	9.36	10.92	12.48	14.04	15.60
no. 14[a]	1.693	7.650	2.25	4.50	6.75	9.00	11.25	13.5	15.75	18.00	20.25	22.50
no. 18[a]	2.257	13.60	4.00	8.00	12.0	16.0	20.00	24.0	28.00	32.00	36.00	40.00

(Multiply in^2 by 645 to obtain mm^2.)
[a]No. 14 and no. 18 bars are typically used in columns only.

step 7: Design the shear reinforcement. (See Sec. 20.)
step 8: Check cracking requirements. (See Sec. 13.)
step 9: Check deflection. (See Sec. 15.)

Example 50.5

The span of a beam with the cross section shown is 10 ft, and the ends are built in. $f_c' = 3500$ lbf/in², and $f_y = 40{,}000$ lbf/in². The maximum load is controlled by the capacity in the negative moment region. Based solely on flexural requirements, what is the maximum uniform live load the beam can carry?

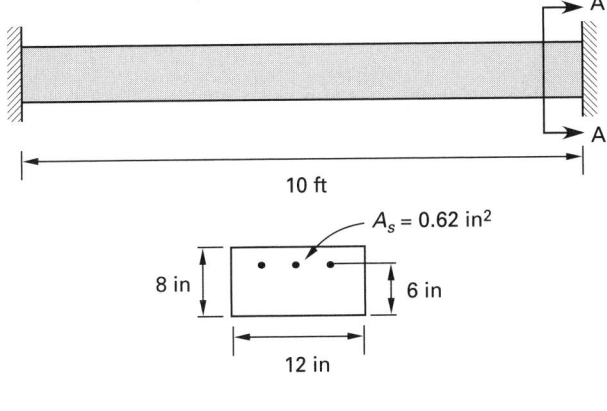

section A-A

Solution Using ACI 318 App. C

The beam is rectangular, so Eqs. 50.31 and 50.32 can be used to calculate M_n.

$$\rho = \frac{A_s}{bd} = \frac{0.62 \text{ in}^2}{(12 \text{ in})(6 \text{ in})}$$

$$= 0.00861$$

From Eq. 50.32,

$$M_n = \rho b d^2 f_y \left(1 - \frac{\rho f_y}{1.7 f_c'}\right)$$

$$= (0.00861)(12 \text{ in})(6 \text{ in})^2 \left(40 \frac{\text{kips}}{\text{in}^2}\right)$$

$$\times \left(1 - \frac{(0.00861)\left(40 \frac{\text{kips}}{\text{in}^2}\right)}{(1.7)\left(3.5 \frac{\text{kips}}{\text{in}^2}\right)}\right)$$

$$= 140.2 \text{ in-kips}$$

For a beam fixed at both ends, the maximum negative moment occurs at the ends and is

$$M_u = \phi M_n = \frac{w_u L^2}{12}$$

Solve for the factored uniform load.

$$w_u = \frac{12\phi M_n}{L^2} = \frac{(12)(0.9)(140.2 \text{ in-kips})}{(10 \text{ ft})^2 \left(12 \frac{\text{in}}{\text{ft}}\right)}$$

$$= 1.26 \text{ kips/ft}$$

Using a unit weight of 150 lbf/ft³ for the concrete, the weight of the beam is

$$w_d = \frac{\left(150 \frac{\text{lbf}}{\text{ft}^3}\right)(12 \text{ in})(8 \text{ in})}{\left(1000 \frac{\text{lbf}}{\text{kip}}\right)\left(12 \frac{\text{in}}{\text{ft}}\right)^2}$$

$$= 0.10 \text{ kips/ft}$$

Assume no additional dead load acts on the beam. Using ACI 318 App. C,

$$w_u = 1.26 \frac{\text{kips}}{\text{ft}} = 1.4D + 1.7L$$

$$= (1.4)\left(0.10 \frac{\text{kip}}{\text{ft}}\right) + 1.7w_l$$

Solve for the maximum live load.

$$w_l = 0.66 \text{ kips/ft}$$

Solution Using ACI 318 Ch. 9

The beam is rectangular, so Eqs. 50.31 and 50.32 can be used to calculate M_n.

$$\rho = \frac{A_s}{bd} = \frac{0.62 \text{ in}^2}{(12 \text{ in})(6 \text{ in})}$$

$$= 0.00861$$

From Eq. 50.32,

$$M_n = \rho bd^2 f_y \left(1 - \frac{\rho f_y}{1.7 f_c'}\right)$$

$$= (0.00861)(12 \text{ in})(6 \text{ in})^2 \left(40 \frac{\text{kips}}{\text{in}^2}\right)$$

$$\times \left(1 - \frac{(0.00861)\left(40 \frac{\text{kips}}{\text{in}^2}\right)}{(1.7)\left(3.5 \frac{\text{kips}}{\text{in}^2}\right)}\right)$$

$$= 140.2 \text{ in-kips}$$

For a beam fixed at both ends, the maximum negative moment occurs at the ends and is

$$M_u = \phi M_n = \frac{w_u L^2}{12}$$

Solve for the factored uniform load.

$$w_u = \frac{12 \phi M_n}{L^2} = \frac{(12)(0.9)(140.2 \text{ in-kips})}{(10 \text{ ft})^2 \left(12 \frac{\text{in}}{\text{ft}}\right)}$$

$$= 1.26 \text{ kips/ft}$$

Using a unit weight of 150 lbf/ft³ for the concrete, the weight of the beam is

$$w_d = \frac{\left(150 \frac{\text{lbf}}{\text{ft}^3}\right)(12 \text{ in})(8 \text{ in})}{\left(1000 \frac{\text{lbf}}{\text{kip}}\right)\left(12 \frac{\text{in}}{\text{ft}}\right)^2}$$

$$= 0.10 \text{ kips/ft}$$

Assume no additional dead load acts on the beam. Using ACI 318 Ch. 9,

$$w_u = 1.26 \frac{\text{kips}}{\text{ft}} = 1.2D + 1.6L$$

$$= (1.2)\left(0.10 \frac{\text{kip}}{\text{ft}}\right) + 1.6w_l$$

Solve for the maximum live load.

$$w_l = 0.71 \text{ kips/ft}$$

Example 50.6

Design a rectangular beam with only tension reinforcement to carry service moments of 34.3 ft-kips (dead) and 30 ft-kips (live). $f_c' = 3500$ lbf/in², and $f_y = 40{,}000$ lbf/in². No. 3 bars are used for shear reinforcing. (Do not design stirrup placement.)

Solution Using ACI 318 App. C

The ACI 318 App. C factored moment is

$$M_u = 1.4M_d + 1.7M_l$$

$$= (1.4)(34.3 \text{ ft-kips}) + (1.7)(30 \text{ ft-kips})$$

$$= 99.02 \text{ ft-kips}$$

Assuming a tension-controlled section,

$$M_n = \frac{M_u}{\phi}$$

$$= \frac{99.02 \text{ ft-kips}}{0.9}$$

$$= 110.02 \text{ ft-kips}$$

Select a reinforcement ratio $\rho = 0.18 f_c'/f_y$ to achieve a reasonably economical design.

$$\rho = \frac{(0.18)\left(3500 \frac{\text{lbf}}{\text{in}^2}\right)}{40{,}000 \frac{\text{lbf}}{\text{in}^2}}$$

$$= 0.0158$$

Substitute ρ and M_n into Eq. 50.32 and solve for bd^2.

$$bd^2 = \frac{M_n}{\rho f_y \left(1 - \frac{\rho f_y}{1.7 f_c'}\right)}$$

$$= \frac{(110.02 \text{ ft-kips})\left(12 \frac{\text{in}}{\text{ft}}\right)\left(1000 \frac{\text{lbf}}{\text{kip}}\right)}{(0.0158)\left(40{,}000 \frac{\text{lbf}}{\text{in}^2}\right)}$$

$$\times \left(1 - \frac{(0.0158)\left(40{,}000 \frac{\text{lbf}}{\text{in}^2}\right)}{(1.7)\left(3500 \frac{\text{lbf}}{\text{in}^2}\right)}\right)$$

$$= 2337 \text{ in}^3$$

Choose $d/b = 1.8$.

$$d^3 = (1.8)(2337 \text{ in}^3) = 4207 \text{ in}^3$$
$$d = 16.14 \text{ in}$$
$$b = \frac{16.14 \text{ in}}{1.8} = 8.97 \text{ in}$$

The total depth includes the effective depth (16.14 in), the diameter of the stirrup ($^3/_8$ in), the cover (1.5 in), and half the diameter of the longitudinal bars. Assuming no. 8 bars, the theoretical depth is

$$h = 16.14 \text{ in} + 0.375 \text{ in} + 1.5 \text{ in} + 0.5 \text{ in}$$
$$= 18.515 \text{ in}$$

The overall dimensions are typically rounded to the nearest inch. Select $b = 9$ in and $h = 19$ in. The actual d is $h - 2.375 \text{ in} = 16.625 \text{ in}$.

It is appropriate to solve for the reinforcement ratio, ρ, corresponding to the final dimensions. This can be easily done with Eq. 50.32 and the actual value of bd^2.

$$bd^2 = (9 \text{ in})(16.625 \text{ in})^2$$
$$= 2487.5 \text{ in}^3$$

The resulting value, $\rho = 0.0147$, is smaller than the initial value of 0.0158 because the dimensions were rounded up. For this low reinforcement ratio, the section is going to be tension controlled. A formal check is unnecessary. The required steel area is

$$A_s = \rho bd = (0.0147)(9 \text{ in})(16.625 \text{ in})$$
$$= 2.2 \text{ in}^2$$

Select three no. 8 bars, which gives a total area of 2.37 in^2. Although this completes the first cycle of beam design, a check of required beam width from Table 50.2 shows that the reinforcement will not fit. The beam should be redesigned with either larger b or d, or both.

Solution Using ACI 318 Ch. 9

The ACI 318 Ch. 9 factored moment is

$$M_u = 1.2M_d + 1.6M_l$$
$$= (1.2)(34.3 \text{ ft-kips}) + (1.6)(30 \text{ ft-kips})$$
$$= 89.16 \text{ ft-kips}$$

Assuming a tension-controlled section,

$$M_n = \frac{M_u}{\phi} = \frac{89.16 \text{ ft-kips}}{0.9}$$
$$= 99.07 \text{ ft-kips}$$

Select a reinforcement ratio $\rho = 0.18 f_c'/f_y$ to achieve a reasonably economical design.

$$\rho = \frac{(0.18)\left(3500 \, \dfrac{\text{lbf}}{\text{in}^2}\right)}{40,000 \, \dfrac{\text{lbf}}{\text{in}^2}}$$
$$= 0.0158$$

Substitute ρ and M_n into Eq. 50.32 and solve for bd^2.

$$bd^2 = \frac{M_n}{\rho f_y \left(1 - \dfrac{\rho f_y}{1.7 f_c'}\right)}$$
$$= \frac{(99.07 \text{ ft-kips})\left(12 \, \dfrac{\text{in}}{\text{ft}}\right)\left(1000 \, \dfrac{\text{lbf}}{\text{kip}}\right)}{(0.0158)\left(40,000 \, \dfrac{\text{lbf}}{\text{in}^2}\right)}$$
$$\times \left(1 - \frac{(0.0158)\left(40,000 \, \dfrac{\text{lbf}}{\text{in}^2}\right)}{(1.7)\left(3500 \, \dfrac{\text{lbf}}{\text{in}^2}\right)}\right)$$
$$= 2104 \text{ in}^3$$

Choose $d/b = 1.8$.

$$d^3 = (1.8)(2104 \text{ in}^3) = 3787 \text{ in}^3$$
$$d = 15.59 \text{ in}$$
$$b = \frac{15.59 \text{ in}}{1.8} = 8.66 \text{ in}$$

The total depth includes the effective depth (15.58 in), the diameter of the stirrup ($^3/_8$ in), the cover (1.5 in), and half the diameter of the longitudinal bars. Assuming no. 8 bars, the theoretical depth is

$$h = 15.58 \text{ in} + 0.375 \text{ in} + 1.5 \text{ in} + 0.5 \text{ in}$$
$$= 17.955 \text{ in}$$

The overall dimensions are typically rounded to the nearest inch. Select $b = 9$ in and $h = 18$ in. The actual d is $h - 2.375 \text{ in} = 15.625 \text{ in}$.

It is appropriate to solve for the reinforcement ratio, ρ, corresponding to the final dimensions. This can be done with Eq. 50.32 (by solving a quadratic) and the actual value of bd^2.

$$bd^2 = (9 \text{ in})(15.625 \text{ in})^2$$
$$= 2197.3 \text{ in}^3$$

The resulting value, $\rho = 0.0151$, is smaller than the initial value of 0.0158 because the dimensions were rounded up. For this low reinforcement ratio, the section is going to be tension controlled. A formal check is unnecessary. The required steel area is

$$A_s = \rho b d = (0.0151)(9 \text{ in})(15.625 \text{ in})$$
$$= 2.12 \text{ in}^2$$

Select three no. 8 bars, which gives a total area of 2.37 in². Although this completes the first cycle of beam design, a check of required beam width from Table 50.2 shows that the reinforcement will not fit. The beam should be redesigned with either larger b or d, or both.

13. SERVICEABILITY: CRACKING

Flexural design procedures for reinforced concrete beams are based on strength. *Serviceability* considerations involve checking that deflections and crack widths are not excessive at service loads.

The low tensile strength of concrete makes cracking of concrete beams inevitable. It is possible, however, to keep the width of the cracks sufficiently small so that the tension steel remains protected by the cover concrete. Acceptable crack widths range from 0.006–0.016 in (0.15–0.41 mm), clearly having no effect on the appearance of the member. While cracking reduces the stiffness of the beam, it has no significant effect on strength.

The basic goal in the control of cracking is to ensure that the elongation of the tension side of the beam is distributed between a large number of closely spaced small cracks instead of being taken up in a few widely spaced large cracks. If it is concluded that unacceptable cracking will occur, then smaller reinforcing bars must be used to redistribute the steel around a greater area.

Since the elongation of the tension steel is proportional to the stress acting on it, the potential importance of cracking is larger when high-strength steels are used. Experience has shown that cracking is usually not a problem for beams reinforced with steel bars having a yield strength equal to or less than 40,000 lbf/in² (276 MPa).

ACI 318 Ch. 10 provides a simplified approach for evaluating cracking in beams with reinforcement. The method does not deal directly with crack widths, but instead controls the spacing of the reinforcement layer closest to the tension face.

$$s_{\max} = 15\left(\frac{40{,}000}{f_s}\right) - 2.5 c_c \le 12\left(\frac{40{,}000}{f_s}\right) \qquad 50.36$$

f_s is the computed stress in the tension steel under service loads in psi. ACI 318 10.6.4 permits this value to be calculated using linear elastic theory, or in lieu of an elastic analysis, f_s may be approximated as $2/3 f_y$. The parameters s, c_c, and others are illustrated in Fig. 50.7.

For the usual case of beams with grade 60 reinforcement, 2 in of clear cover to the main reinforcement, and with $f_s = 40$ ksi (the maximum steel stress permitted) the maximum bar spacing is 10 in [ACI 318 R10.6.4].

14. CRACKED MOMENT OF INERTIA

The *cracked moment of inertia*, also referred to as the *transformed moment of inertia*, is the moment of inertia computed for the section under the assumption that the concrete carries no tension and behaves linearly in compression. Referring to Fig. 50.8, the procedure for computing I_{cr} is as follows.

step 1: Calculate the *modular ratio*, n, the ratio of the moduli of elasticity of the steel and the concrete. The modulus of elasticity for steel is almost always 29,000 kips/in² (200 GPa).

Figure 50.7 *Parameters for Cracking Calculation*

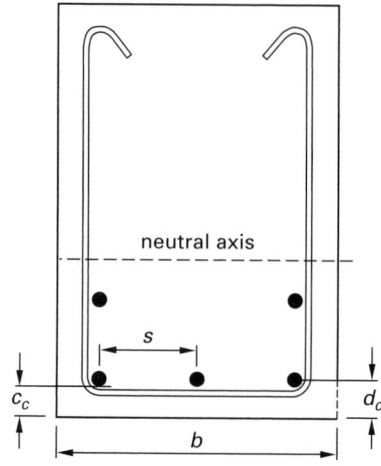

$$n = \frac{E_s}{E_c} \qquad\qquad 50.37$$

$$E_c = 4700\sqrt{f_c'} \qquad \text{[SI]} \qquad 50.38(a)$$

$$E_c = 57{,}000\sqrt{f_c'} \qquad \text{[U.S.]} \qquad 50.38(b)$$

step 2: Multiply the tension steel areas by n and the compression steel areas by $n - 1$. This converts the steel areas to equivalent concrete areas. (The "-1" term for the compression steel accounts for the fact that the steel is replacing a concrete area.)

step 3: Determine the distance, c_s, to the neutral axis. All concrete below the neutral axis is assumed to be ineffective.

step 4: Calculate I_{cr} using the parallel axis theorem. The moments of inertia of the steel areas with respect to their local centroid are assumed to be negligible.

There are many cases where the cross section is rectangular and singly reinforced. Closed-form expressions can be derived to simplify the computation of c_s and I_{cr}. Referring to Fig. 50.8,

$$\frac{bc_s^2}{2} = n A_s(d - c_s) \qquad\qquad 50.39$$

This is a quadratic equation with the solution

$$c_s = \left(\frac{nA_s}{b}\right)\left(\sqrt{1+\frac{2bd}{nA_s}}-1\right) \qquad 50.40$$

Equation 50.40 can also be written in terms of the steel ratio.

$$c_s = n\rho d\left(\sqrt{1+\frac{2}{n\rho}}-1\right) \qquad 50.41$$

The cracked moment of inertia for a rectangular, singly reinforced beam is

$$I_{\mathrm{cr}} = \frac{bc_s^3}{3}+nA_s(d-c_s)^2 \qquad 50.42$$

Figure 50.8 *Parameters for Cracked Moment of Inertia*

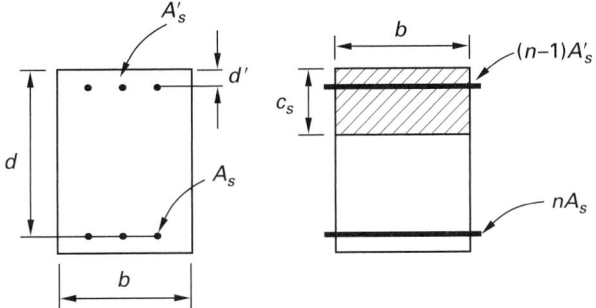

(a) reinforced section   (b) transformed section

15. SERVICEABILITY: DEFLECTIONS

Reinforced concrete beam deflections have two components: (a) an immediate deflection that occurs as a result of the strains induced by the applied loads, and (b) a long-term deflection that develops due to creep and shrinkage of the concrete. Immediate (instantaneous) deflections, Δ_i, are computed using any available general method based on elastic linear response. However, the effect of cracking on the flexural stiffness of the beam must be considered.

Cracking takes place whenever the bending moment at a cross section exceeds the *cracking moment*, M_{cr}. Since moments from the applied loads vary along the beam length, certain portions of the beam will be cracked while others will be uncracked. This makes a rigorous analysis that takes into account the actual variation of the flexural stiffness along the beam length essentially impractical. Therefore, ACI 318 provides a simplified method for obtaining an effective moment of inertia that is assumed to be constant along the entire beam length. The *effective moment of inertia*, I_e, is given by ACI 318 9.5.2.3 as

$$I_e = \left(\frac{M_{\mathrm{cr}}}{M_a}\right)^3 I_g + \left(1-\left(\frac{M_{\mathrm{cr}}}{M_a}\right)^3\right)I_{\mathrm{cr}} \le I_g \qquad 50.43$$

The *cracking moment* is

$$M_{\mathrm{cr}} = \frac{f_r I_g}{y_t} \qquad 50.44$$

In Eq. 50.44, I_g is the *gross moment of inertia*, $I_g = bh^3/12$, f_r is the *modulus of rupture* (i.e., the stress at which the concrete is assumed to crack), and y_t is the distance from the center of gravity of the gross section (neglecting reinforcement) to the extreme tension fiber. For a rectangular cross section, $y_t = h/2$. For stone or gravel concrete (normal-weight concrete), the average modulus of rupture used for deflection calculations is

$$f_r = 0.7\sqrt{f_c'} \qquad \text{[SI]} \qquad 50.45(a)$$
$$f_r = 7.5\sqrt{f_c'} \qquad \text{[U.S.]} \qquad 50.45(b)$$

For all-lightweight concrete, f_r as calculated from Eq. 50.45 is reduced 25% (i.e., is multiplied by a factor of 0.75). If the fine aggregate is sand but the coarse aggregate is lightweight, f_r is reduced by 15% (i.e., is multiplied by a factor of 0.85).

Since cracking is not reversible, M_a should be taken as the largest bending moment due to service loads. In particular, if an immediate deflection due to the live load is desired, the deflection should be based on the live load, but I_e should be computed from Eq. 50.43 using the value of M_a corresponding to dead plus live load.

When a beam is continuous, the reinforcement will vary along the beam length. Then, the cracked inertia, I_{cr}, will also vary along the length. ACI 318 permits using the value of I_e that corresponds to the positive moment region (with M_a taken also as the maximum positive service moment), or I_e can be taken as the average of the values for the regions of maximum negative and maximum positive moment [ACI 318 9.5.2.4].

Example 50.7

The service moment on the beam shown is 100 ft-kips. The concrete compressive strength is 4000 lbf/in², and the modular ratio is 8. Determine the (a) gross, (b) cracked, and (c) effective moments of inertia.

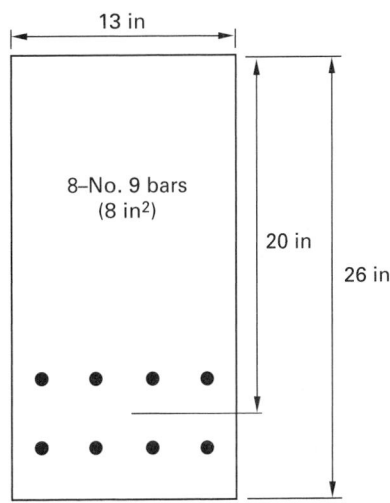

Solution

(a) The gross moment of inertia is

$$I_g = \frac{bh^3}{12} = \frac{(13\text{ in})(26\text{ in})^3}{12}$$

$$= 19{,}041\text{ in}^4$$

(b) Since the section is rectangular, Eqs. 50.41 and 50.42 can be used.

$$\rho = \frac{A_s}{bd} = \frac{8\text{ in}^2}{(13\text{ in})(20\text{ in})}$$

$$= 0.0308$$

$$c_s = n\rho d \left(\sqrt{1 + \frac{2}{n\rho}} - 1 \right)$$

$$= (8)(0.0308)(20\text{ in}) \left(\sqrt{1 + \frac{2}{(8)(0.0308)}} - 1 \right)$$

$$= 9.95\text{ in}$$

$$I_{cr} = \frac{bc_s^3}{3} + nA_s(d - c_s)^2$$

$$= \frac{(13\text{ in})(9.95\text{ in})^3}{3} + (8)(8\text{ in}^2)(20\text{ in} - 9.95\text{ in})^2$$

$$= 10{,}733\text{ in}^4$$

(c) The modulus of rupture is given by Eq. 50.45.

$$f_r = 7.5\sqrt{f_c'} = 7.5\sqrt{4000\ \frac{\text{lbf}}{\text{in}^2}}$$

$$= 474\text{ lbf/in}^2$$

The cracking moment is given by Eq. 50.44.

$$M_{cr} = \frac{f_r I_g}{y_t}$$

$$= \frac{\left(0.474\ \frac{\text{kips}}{\text{in}^2} \right) (19{,}041\text{ in}^4)}{(13\text{ in}) \left(12\ \frac{\text{in}}{\text{ft}} \right)}$$

$$= 57.86\text{ ft-kips}$$

$$\left(\frac{M_{cr}}{M_a} \right)^3 = \left(\frac{57.86\text{ ft-kips}}{100\text{ ft-kips}} \right)^3 = 0.194$$

The effective moment of inertia is given by Eq. 50.43.

$$I_e = \left(\frac{M_{cr}}{M_a} \right)^3 I_g + \left(1 - \left(\frac{M_{cr}}{M_a} \right)^3 \right) I_{cr}$$

$$= (0.194)(19{,}041\text{ in}^4) + (1 - 0.194)(10{,}733\text{ in}^4)$$

$$= 12{,}345\text{ in}^4$$

16. LONG-TERM DEFLECTIONS

Long-term deflection is caused by creep and shrinkage of the concrete. The long-term deformation develops rapidly at first and then slows down, being largely complete after approximately five years. In the procedure specified in ACI 318, the *long-term deformation*, Δ_a, is obtained by multiplying by an amplification factor the immediate deflection produced by the portion of the load that is sustained. In computing the sustained part of the immediate deflection, judgment is required to decide what fraction of the prescribed service live load can be assumed to be acting continuously. The long-term deflection is taken as

$$\Delta_a = \lambda \Delta_i \qquad \textit{50.46}$$

$$\lambda = \frac{i}{1 + 50\rho'} \qquad \textit{50.47}$$

ρ' is the compression steel ratio (A_s'/bd) at midspan for simply supported and continuous beams, and at supports for cantilevers. i is an empirical factor that accounts for the rate of the additional deflection. i is 1.0 at three months, 1.2 at six months, 1.4 at one year, and 2 at five years or more.

The total deflection is the sum of the instantaneous plus the long-term deflections.

17. MINIMUM BEAM DEPTHS TO AVOID EXPLICIT DEFLECTION CALCULATIONS

For beams not supporting or attached to nonstructural elements sensitive to deflections, explicit deflection calculations are not required if the beam has been designed with a certain minimum thickness. The minimum thickness, h, is given in Table 50.4.

Table 50.4 *Minimum Beam Thickness Unless Deflections are Computed* ($f_y = 60{,}000$ lbf/in^2)[a]

construction	minimum h (fraction of span length)
simply supported	$\frac{1}{16}$
one end continuous	$\frac{1}{18.5}$
both ends continuous	$\frac{1}{21}$
cantilever	$\frac{1}{8}$

[a]Corrections are required for other steel strengths. See Eqs. 50.48 and 50.49.

Source: ACI 318 Table 9.5(a)

The values in Table 50.4 apply to normal-weight concrete reinforced with $f_y = 60{,}000$ lbf/in^2 (400 MPa) steel. The table values are multiplied by the adjustment factor in Eq. 50.48 for other steel strengths.

$$0.4 + \frac{f_y}{700} \qquad \text{[SI]} \qquad \textit{50.48(a)}$$

$$0.4 + \frac{f_y}{100{,}000} \qquad \text{[U.S.]} \qquad \textit{50.48(b)}$$

For structural lightweight concrete with a unit weight, w, between 90 and 120 lbf/ft^3 (14.13 and 18.83 kN/m^3), multiply the table values by

$$1.65 - 0.0318w \geq 1.09 \qquad \text{[SI]} \qquad \textit{50.49(a)}$$

$$1.65 - 0.005w \geq 1.09 \qquad \text{[U.S.]} \qquad \textit{50.49(b)}$$

18. MAXIMUM ALLOWABLE DEFLECTIONS

When the minimum thicknesses of Table 50.4 are not used, or if the beam supports nonstructural elements that are sensitive to deformations, the deflections must be explicitly checked by computation. The limits prescribed by ACI 318 for the computed deflections are presented in Table 50.5.

The deflection limits in the first row of Table 50.5 are not intended to safeguard against ponding. Ponding should be checked by deflection calculations, taking into consideration the added deflections due to accumulated ponded water and considering long-term effects of all sustained loads, camber, construction tolerances, and reliability of provisions for drainage.

19. DESIGN OF T-BEAMS

Although a beam can be specifically cast with a T-shaped cross section, that is rarely done. T-beam behavior usually occurs in monolithic beam-slab (one-way) systems. The general beam design procedure presented earlier in this chapter applies to the design of T-beams. However, determining the width of the flange (i.e., the slab) that is effective in resisting compressive loads is code-sensitive. ACI 318 also specifies requirements pertaining to the distribution of the reinforcement.

The compressive stresses in the flange of a T-beam decrease with distance from the centerline of the web. ACI 318 specifies using an *effective width*, which is a width over which the compressive stress is assumed uniform.

It is assumed that a stress of $0.85f'_c$ acting over the effective width provides approximately the same compressive force as that realized from the actual variable stress distribution over the total flange width. The effective width of the flange for T- and L-shaped beams is illustrated in Fig. 50.9. The effective width in each case is the smallest value from the options listed [ACI 318 8.10].

Figure 50.9 *Effective Flange Width (one-way reinforced slab systems) (ACI 318 8.10)*

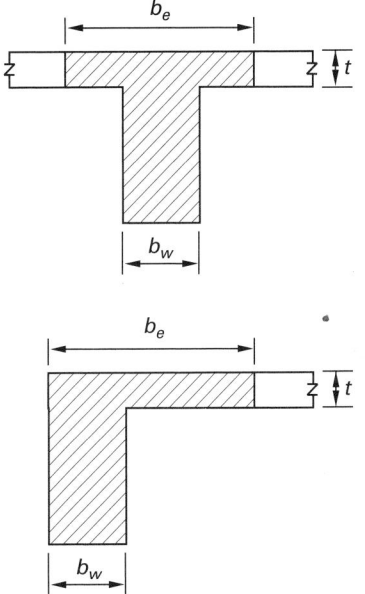

Table 50.5 *Maximum Allowable Computed Deflections*

type of member	deflection to be considered	deflection limitation
flat roofs not supporting or attached to nonstructural elements likely to be damaged by large deflections	immediate deflection due to live load	$\frac{1}{180}$
floors not supporting or attached to nonstructural elements likely to be damaged by large deflections	immediate deflection due to live load	$\frac{1}{360}$
roof or floor construction supporting or attached to nonstructural elements likely to be damaged by large deflections	that part of the total deflection that occurs after attachment of the nonstructural elements	$\frac{1}{480}$
roof or floor construction supporting or attached to nonstructural elements not likely to be damaged by large deflections	that part of the total deflection that occurs after attachment of the nonstructural elements	$\frac{1}{240}$

Source: ACI 318 Table 9.5(b)

Case 1: Beams with flanges on each side of the web [ACI 318 8.10.2]

Effective width (including the compression area of the stem) for a T-beam is the minimum of

one fourth of the beam's span length, or

the stem width plus 16 times the thickness of the slab, or

the beam spacing

Case 2: Beams with an L-shaped flange [ACI 318 8.10.3]

Effective width (including the compression area of the stem) is the minimum of

the stem width plus one twelfth of the beam's span length, or

the stem width plus 6 times the thickness of the slab, or

the stem width plus one half of the clear distance between beam webs

Case 3: Isolated T-Beams [ACI 318 8.10.4]

If a T-beam is not part of a floor system but the T-shape is used to provide additional compression area, the flange thickness shall not be less than one half of the width of the web. The effective width of the flange shall not be more than 4 times the width of the web.

When a T-beam is subjected to a negative moment, the tension steel must be placed in the flange. ACI 318 10.6.6 requires that this tension steel not all be placed inside the region of the web but that it be spread out into the flange over a distance whose width is the smaller of (a) the effective flange width or (b) one tenth of the span. This provision ensures that cracking on the top surface will be distributed.

If the T-beam is an isolated member, loading will produce bending of the flange in the direction perpendicular to the beam span. Adequate transverse steel must be provided to prevent a bending failure of the flange.

Example 50.8

A monolithic slab-beam floor system is supported on a column grid of 18 ft on centers as shown. The dimensions of the cross section for the beams running in the north-south direction have been determined. The positive moments in the north-south beams are $M_d = 200$ ft-kips, and $M_l = 340$ ft-kips. $f_c' = 3000$ lbf/in^2, and $f_y = 50,000$ lbf/in^2. Design the positive steel reinforcement.

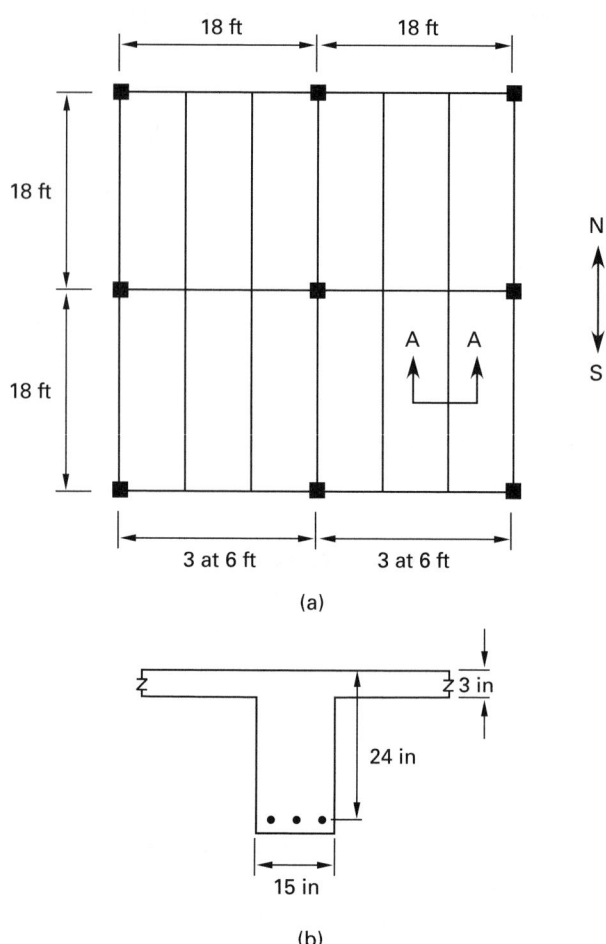

(a)

(b)

Solution Using ACI 318 App. C

The effective width of the flange is determined as the smallest of

$$\frac{L}{4} = \frac{(18 \text{ ft})\left(12 \, \frac{\text{in}}{\text{ft}}\right)}{4} = 54 \text{ in} \quad \text{[controls]}$$

$$b_w + 16t = 15 \text{ in} + (16)(3 \text{ in}) = 63 \text{ in}$$

$$\text{beam spacing} = (6 \text{ ft})\left(12 \, \frac{\text{in}}{\text{ft}}\right) = 72 \text{ in}$$

The effective flange area is

$$(3 \text{ in})(54 \text{ in}) = 162 \text{ in}^2$$

The ACI 318 App. C factored moment is

$$\begin{aligned} M_u &= 1.4M_d + 1.7M_l \\ &= (1.4)(200 \text{ ft-kips}) + (1.7)(340 \text{ ft-kips}) \\ &= 858 \text{ ft-kips} \end{aligned}$$

Estimate λ as $(0.1)(24 \text{ in}) = 2.4$ in. The steel area required is given by Eq. 50.28. Assuming a tension-controlled section,

$$A_s = \frac{M_u}{\phi f_y(d - \lambda)}$$

$$= \frac{(858 \text{ ft-kips})\left(12 \dfrac{\text{in}}{\text{ft}}\right)}{(0.9)\left(50 \dfrac{\text{kips}}{\text{in}^2}\right)(24 \text{ in} - 2.4 \text{ in})}$$

$$= 10.59 \text{ in}^2$$

The area of concrete in compression is given by Eq. 50.17.

$$A_c = \frac{f_y A_s}{0.85 f'_c}$$

$$= \frac{\left(50 \dfrac{\text{kips}}{\text{in}^2}\right)(10.59 \text{ in}^2)}{(0.85)\left(3 \dfrac{\text{kips}}{\text{in}^2}\right)}$$

$$= 207.65 \text{ in}^2$$

This area includes the flange plus a depth into the web of

$$\frac{207.65 \text{ in}^2 - 162 \text{ in}^2}{15 \text{ in}} = 3.04 \text{ in}$$

The centroid of the compressed area measured from the top is determined as follows.

$$(162 \text{ in}^2)(1.5 \text{ in}) + (207.65 \text{ in}^2 - 162 \text{ in}^2)$$
$$\times \left(3 \text{ in} + \frac{3.04 \text{ in}}{2}\right) = (207.65 \text{ in}^2)\lambda$$

$$\lambda = 2.16 \text{ in}$$

The nominal moment capacity is given by Eq. 50.27.

$$M_n = A_s f_y(d - \lambda)$$

$$= \frac{(10.59 \text{ in}^2)\left(50 \dfrac{\text{kips}}{\text{in}^2}\right)(24 \text{ in} - 2.16 \text{ in})}{12 \dfrac{\text{in}}{\text{ft}}}$$

$$= 963.7 \text{ ft-kips}$$

The design moment capacity is

$$\phi M_n = (0.9)(963.7 \text{ ft-kips})$$
$$= 867.3 \text{ ft-kips}$$

Since the design moment capacity (867.32 ft-kips) is slightly larger than the factored moment (858 ft-kips), the steel area required is nominally smaller than the estimated value of 10.59 in². An incrementally better estimate of the steel area is

$$A_s = \left(\frac{858 \text{ ft-kips}}{867.3 \text{ ft-kips}}\right)(10.59 \text{ in}^2)$$

$$= 10.48 \text{ in}^2$$

Check that the steel does not exceed the maximum allowed in ACI 318 App. B. The depth of the rectangular stress block at balanced condition is given by Eq. 50.20.

$$a_b = \beta_1\left(\frac{87{,}000}{87{,}000 + f_y}\right)d$$

$$= (0.85)\left(\frac{87{,}000}{87{,}000 + 50{,}000}\right)(24 \text{ in})$$

$$= 12.95 \text{ in}$$

The corresponding concrete area is

$$A_{cb} = (54 \text{ in})(3 \text{ in}) + (12.95 \text{ in} - 3 \text{ in})(15 \text{ in})$$
$$= 311.25 \text{ in}^2$$

The balanced steel is obtained from Eq. 50.19.

$$A_{sb} = \frac{0.85 f'_c A_{cb}}{f_y}$$

$$= \frac{(0.85)\left(3 \dfrac{\text{kips}}{\text{in}^2}\right)(311.25 \text{ in}^2)}{50 \dfrac{\text{kips}}{\text{in}^2}}$$

$$= 15.87 \text{ in}^2$$

The maximum steel is given by Eq. 50.18.

$$A_{s,\max} = 0.75 A_{sb} = (0.75)(15.87 \text{ in}^2)$$
$$= 11.90 \text{ in}^2$$

The maximum steel is larger than the required steel area, so the design is adequate. Since the two values are relatively close, the minimum steel requirement is not checked explicitly. Since the required steel is high, two layers may be required.

Whether the section is tension controlled needs to be determined to justify a strength reduction factor of 0.9. The stress block depth is

$$a = 3 \text{ in} + 3.04 \text{ in}$$
$$= 6.04 \text{ in}$$

The corresponding neutral axis depth is

$$c = \frac{a}{\beta_1} = \frac{6.04 \text{ in}}{0.85}$$
$$= 7.11 \text{ in}$$

The net tensile strain is

$$\varepsilon_t = \left(\frac{24 \text{ in} - 7.11 \text{ in}}{7.11 \text{ in}}\right)(0.003)$$
$$= 0.0071 > 0.005$$

Therefore, the section is tension controlled.

Solution Using ACI 318 Ch. 9

The effective width of the flange is determined as the smallest of

$$\frac{L}{4} = \frac{(18 \text{ ft})\left(12 \frac{\text{in}}{\text{ft}}\right)}{4} = 54 \text{ in} \quad \text{[controls]}$$

$$b_w + 16t = 15 \text{ in} + (16)(3 \text{ in}) = 63 \text{ in}$$

$$\text{beam spacing} = (6 \text{ ft})\left(12 \frac{\text{in}}{\text{ft}}\right) = 72 \text{ in}$$

The effective flange area is

$$(3 \text{ in})(54 \text{ in}) = 162 \text{ in}^2$$

The ACI 318 Ch. 9 factored moment is

$$\begin{aligned} M_u &= 1.2 M_d + 1.6 M_l \\ &= (1.2)(200 \text{ ft-kips}) + (1.6)(340 \text{ ft-kips}) \\ &= 784 \text{ ft-kips} \end{aligned}$$

Estimate λ as $(0.1)(24 \text{ in}) = 2.4 \text{ in}$. The steel area required is given by Eq. 50.28. Assuming a tension-controlled section,

$$\begin{aligned} A_s &= \frac{M_u}{\phi f_y (d - \lambda)} \\ &= \frac{(784 \text{ ft-kips})\left(12 \frac{\text{in}}{\text{ft}}\right)}{(0.9)\left(50 \frac{\text{kips}}{\text{in}^2}\right)(24 \text{ in} - 2.4 \text{ in})} \\ &= 9.68 \text{ in}^2 \end{aligned}$$

The area of concrete in compression is given by Eq. 50.26.

$$\begin{aligned} A_c &= \frac{f_y A_s}{0.85 f'_c} \\ &= \frac{\left(50 \frac{\text{kips}}{\text{in}^2}\right)(9.68 \text{ in}^2)}{(0.85)\left(3 \frac{\text{kips}}{\text{in}^2}\right)} \\ &= 189.80 \text{ in}^2 \end{aligned}$$

This area includes the flange plus a depth into the web of

$$\frac{189.80 \text{ in}^2 - 162 \text{ in}^2}{15 \text{ in}} = 1.85 \text{ in}$$

The centroid of the compressed area measured from the top is determined as follows.

$$(162 \text{ in}^2)(1.5 \text{ in}) + (189.80 \text{ in}^2 - 162 \text{ in}^2)$$
$$\times \left(3 \text{ in} + \frac{1.85 \text{ in}}{2}\right) = (189.80 \text{ in}^2)\lambda$$

$$\lambda = 1.86 \text{ in}$$

The nominal moment capacity is given by Eq. 50.27.

$$\begin{aligned} M_n &= A_s f_y (d - \lambda) \\ &= \frac{(9.68 \text{ in}^2)\left(50 \frac{\text{kips}}{\text{in}^2}\right)(24 \text{ in} - 1.86 \text{ in})}{12 \frac{\text{in}}{\text{ft}}} \\ &= 893 \text{ ft-kips} \end{aligned}$$

The design moment capacity is

$$\begin{aligned} \phi M_n &= (0.9)(893 \text{ ft-kips}) \\ &= 804 \text{ ft-kips} \end{aligned}$$

Since the design moment capacity (804 ft-kips) is slightly larger than the factored moment (784 ft-kips), the steel area required is nominally smaller than the estimated value of 9.68 in². An incrementally better estimate of the steel area is

$$\begin{aligned} A_s &= \left(\frac{784 \text{ ft-kips}}{804 \text{ ft-kips}}\right)(9.68 \text{ in}^2) \\ &= 9.44 \text{ in}^2 \end{aligned}$$

Check that the section is tension controlled.

The stress block depth is

$$\begin{aligned} a &= 3 \text{ in} + 1.85 \text{ in} \\ &= 4.85 \text{ in} \end{aligned}$$

The corresponding neutral axis depth is

$$\begin{aligned} c &= \frac{a}{\beta_1} = \frac{4.85 \text{ in}}{0.85} \\ &= 5.71 \text{ in} \end{aligned}$$

The net tensile strain is

$$\begin{aligned} \varepsilon_t &= \left(\frac{24 \text{ in} - 5.71 \text{ in}}{5.71 \text{ in}}\right)(0.003) \\ &= 0.0096 > 0.005 \end{aligned}$$

Therefore, the section is tension controlled. The tension reinforcement ratio by definition is below the maximum allowed. Since the tension reinforcement ratio is relatively high, the minimum steel requirement is not checked explicitly. The design is adequate. Since the required steel is high, two layers may be required.

20. SHEAR STRESS

In addition to producing bending moments, beam loads also produce shear forces. Shear forces induce diagonal tension stresses that lead to diagonal cracking. The typical pattern of cracks induced by shear forces is depicted in Fig. 50.10. If a diagonal crack forms and the beam does not contain shear reinforcement, failure will occur abruptly. In order to have ductile beams (where the controlling mechanism is the ductile yielding of the tension steel), shear reinforcement is designed conservatively.

Figure 50.10 *Typical Pattern of Shear Cracks*

The maximum shear at a given location does not necessarily result from placing the live load over the complete beam. For a simple beam, for example, the shear at the centerline is zero if the live load is placed over the full span. If only half the span is loaded, the shear from live load at the center will equal $w_l L/8$.

A *shear envelope* is a plot of the maximum shear that can occur at any given section allowing for a variable position of the live load. In practice, the construction of the exact shear envelope is usually unnecessary since an approximate envelope obtained by connecting the maximum possible shear at the supports with the maximum possible value at the center of the spans is sufficiently accurate. Of course, the dead load shear must be added to the live load shear envelope.

In the design of beams, it is customary to call out the span length as the distance from the centerline of supports. Although the shear computed is largest immediately adjacent to a reaction, shears corresponding to locations that are within the support width are not physically meaningful. Furthermore, for the typical case where the beam is supported from underneath, the compression induced by the reaction increases the shear strength in the vicinity of the support.

ACI 318 11.1.3 takes this enhanced strength into account by specifying that the region of the beam from a distance d to the face of the support can be designed for the same shear force that exists at a distance d from the support face. For this provision to apply, the reaction must induce compression in the member (i.e., the beam must not be hanging from a support), and there must be no concentrated loads acting within the distance d. If concentrated loads exist within this distance, or if the reaction does not induce compression, it is necessary to provide for the actual shear forces occurring all the way to the face of the support. (See Fig. 50.11.)

21. SHEAR REINFORCEMENT

Shear reinforcement can take several forms. Vertical stirrups are used most often, but inclined stirrups are occasionally seen, as is welded wire fabric. Longitudinal steel that is bent up at 30° or more and enters the compression zone also contributes to shear reinforcement.

In seismically active regions, in order to increase the ductility of beams that are part of lateral load-resisting frames, the concrete in the compression zone must be adequately confined. This confinement is typically attained by providing closely spaced transverse reinforcement in the form of closed ties. These closed ties, which resist shear forces in the same way that stirrups do, are known as *hoops*.

The basic equation governing design for shear forces is Eq. 50.50. For shear, $\phi = 0.85$ when ACI 318 App. C load combinations are used, and $\phi = 0.75$ when ACI 318 Ch. 9 load combinations are used.

$$\phi V_n = V_u \qquad \qquad 50.50$$

Figure 50.11 *Locations of Critical Section*

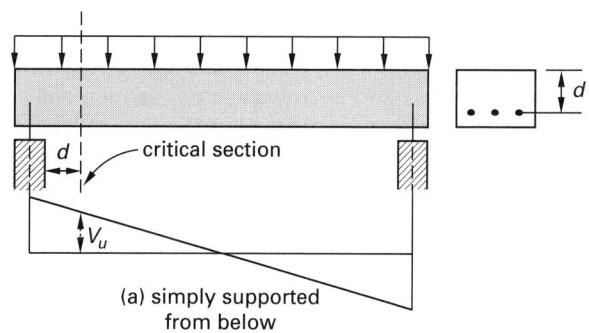

(a) simply supported from below

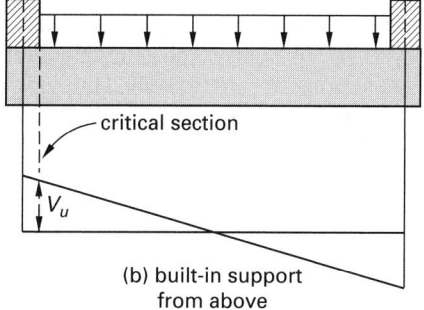

(b) built-in support from above

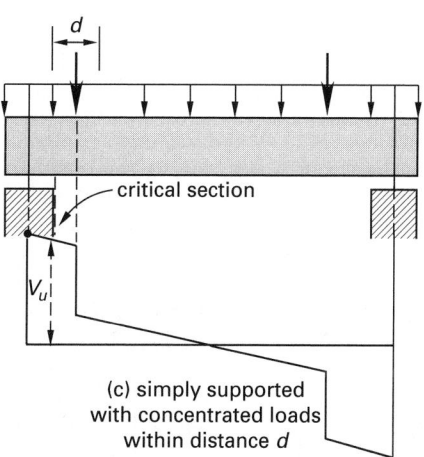

(c) simply supported with concentrated loads within distance d

The beam's shear capacity, V_n, is determined semi-empirically as the sum of the concrete shear capacity, V_c, and the shear capacity that derives from the presence of shear reinforcement, V_s.

$$V_n = V_c + V_s \qquad 50.51$$

(The term *nominal shear strength* is ambiguous in concrete design. The concrete shear strength, V_c, is often referred to as the *nominal concrete shear strength*. The sum of $V_c + V_s$ is referred to as the *nominal beam shear strength*. Common usage attributes the term "nominal" to both.)

22. SHEAR STRENGTH PROVIDED BY CONCRETE

The shear capacity provided by the concrete, V_c, results from the strength in shear of the uncracked compression zone, a contribution from aggregate interlock across the cracks, and dowel action from the tension steel that crosses the diagonal crack. ACI 318 11.3 provides two equations to calculate V_c. One equation is a refined expression that accounts for the effect of the moment and the amount of longitudinal steel on the shear strength.

$$V_c = \left(\frac{\sqrt{f'_c}}{7} + \left(\frac{120}{7}\right) \rho_w \left(\frac{V_u d}{M_u}\right) \right) b_w d \le 0.29 \sqrt{f'_c} b_w d$$
$$\text{[SI]} \qquad 50.52(a)$$

$$V_c = \left(1.9\sqrt{f'_c} + 2500 \rho_w \left(\frac{V_u d}{M_u}\right) \right) b_w d \le 3.5 \sqrt{f'_c} b_w d$$
$$\text{[U.S.]} \qquad 50.52(b)$$

The other equation is a simple expression computed solely from the strength of the concrete and the size of the cross section. In practice, the simpler expression is used. If the shear capacity is inadequate, the refined equation can be used.

$$V_c = \tfrac{1}{6}\sqrt{f'_c} b_w d \qquad \text{[SI]} \qquad 50.53(a)$$

$$V_c = 2\sqrt{f'_c} b_w d \qquad \text{[U.S.]} \qquad 50.53(b)$$

ACI 318 8.11.8 permits an increase of 10% in the shear force that can be assigned to the concrete in the ribs of floor joist construction. (The term *joist construction* refers to closely spaced T-beams with tapered webs.) To qualify for the 10% increase in V_c, the ribs must be at least 4 in (10 cm) wide, have a depth of no more than 3.5 times the minimum width of the rib, and have a clear spacing not exceeding 30 in (76.2 cm).

For circular members, the area used to calculate V_c can be considered a rectangle with width equal to the circular diameter and effective depth, d, of 0.8 times the circular diameter [ACI 318 11.3.3].

23. SHEAR STRENGTH PROVIDED BY SHEAR REINFORCEMENT

The shear capacity provided by the steel reinforcement, V_s, is derived by considering a freebody diagram of the beam with an idealized diagonal crack at 45° and computing the vertical component of the force developed by the reinforcement intersecting the crack. It is assumed that all the steel is yielding when the strength is attained. For the typical case of vertical stirrups, as shown in Fig. 50.12, the shear capacity is

$$V_s = \frac{A_v f_{yt} d}{s} \qquad 50.54$$

The area A_v includes all of the legs in a single stirrup. For example, for the typical U-shaped stirrup, A_v is twice the area of the stirrup bar. For stirrups inclined at an angle θ from the horizontal, ACI 318 11.5.7.4 gives the shear strength contribution as

$$V_s = \frac{A_v f_{yt}(\sin\theta + \cos\theta)d}{s} \qquad 50.55$$

If the shear reinforcement is provided by a single bar or a single group of parallel bars all bent up at the same distance from the support, ACI 318 11.5.7.5 gives the shear strength contribution as Eq. 50.56(b), where θ is the angle between the member's bent-up reinforcement and its longitudinal axis.

$$V_s = A_v f_y \sin\theta \le 0.25\sqrt{f'_c} b_w d \qquad \text{[SI]} \quad 50.56(a)$$

$$V_s = A_v f_y \sin\theta \le 3\sqrt{f'_c} b_w d \qquad \text{[U.S.]} \quad 50.56(b)$$

Figure 50.12 *Contribution of Vertical Stirrups to Shear Capacity*

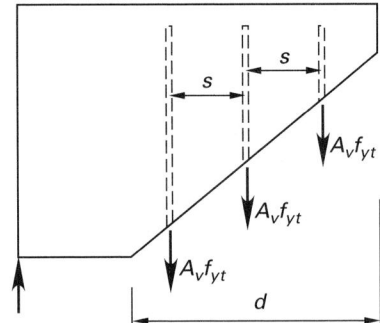

24. SHEAR REINFORCEMENT LIMITATIONS

In order to control the diagonal compressive stresses in the concrete resulting from shear, ACI 318 limits the maximum shear that can be assigned to the shear reinforcement [ACI 318 11.5.7.9].

$$V_s \le \tfrac{2}{3}\sqrt{f'_c} b_w d \qquad \text{[SI]} \qquad 50.57(a)$$

$$V_s \le 8\sqrt{f'_c} b_w d \qquad \text{[U.S.]} \qquad 50.57(b)$$

The maximum total shear force permitted depends on whether Eq. 50.52 or Eq. 50.53 is used to compute the concrete contribution. For the common case where V_c is computed with the simple expression, the maximum shear force V_u is

$$V_{u,max} = \tfrac{5}{6}\phi\sqrt{f'_c} b_w d \qquad \text{[SI]} \qquad 50.58(a)$$

$$V_{u,max} = 10\phi\sqrt{f'_c} b_w d \qquad \text{[U.S.]} \qquad 50.58(b)$$

25. STIRRUP SPACING

Whenever stirrups are required, ACI 318 11.5.4 and 11.5.5 specify that the spacing, s, shall not be larger than calculated in Eqs. 50.59 and 50.60. This maximum spacing ensures that at least one stirrup crosses each potential diagonal crack. The second limit in Eq. 50.60, which applies when the shear is high, ensures that at least two stirrups cross each potential shear crack.

$$s_{\max} = \min\{60 \text{ cm or } d/2\} \quad [V_s \le \tfrac{2}{3}\sqrt{f_c'}b_wd]$$
$$[\text{SI}] \quad 50.59(a)$$

$$s_{\max} = \min\{24 \text{ in or } d/2\} \quad [V_s \le 4\sqrt{f_c'}b_wd]$$
$$[\text{U.S.}] \quad 50.59(b)$$

$$s_{\max} = \min\{30 \text{ cm or } d/4\} \quad [V_s > \tfrac{2}{3}\sqrt{f_c'}b_wd]$$
$$[\text{SI}] \quad 50.60(a)$$

$$s_{\max} = \min\{12 \text{ in or } d/4\} \quad [V_s > 4\sqrt{f_c'}b_wd]$$
$$[\text{U.S.}] \quad 50.60(b)$$

Though not specified by the code, the minimum practical spacing is about 3 in (76 mm). If stirrups need to be closer than 3 in (76 mm), a larger bar size should be used. It is common to place the first stirrup at a distance of $s/2$ from the face of the support, but not closer than 2 in (51 mm). However, the first stirrup can be placed a full space from the face.

When a diagonal crack forms, the stress transfers from the concrete to the stirrups. To ensure that the stirrups have sufficient strength (i.e., a minimum strength of at least $0.75\sqrt{f_c'}$, but no less than 50 lbf/in^2 (340 kPa)), ACI 318 11.5.6.3 requires all stirrups to have a minimum area of

$$A_{v,\min} = \tfrac{1}{8}\sqrt{f_c'}\frac{b_w s}{f_{yt}} \ge \frac{b_w s}{3f_{yt}} \quad [\text{SI}] \quad 50.61(a)$$

$$A_{v,\min} = 0.75\sqrt{f_c'}\frac{b_w s}{f_{yt}} \ge \frac{50b_w s}{f_{yt}} \quad [\text{U.S.}] \quad 50.61(b)$$

26. NO-STIRRUP CONDITIONS

If the shear force is low enough, the concrete in the compression zone will be able to resist it without the need for shear reinforcement. Logically, shear reinforcement is not needed if $\phi V_c \ge V_u$. However, per ACI 318 11.5.6.1, this criterion can be used only if (a) the depth of the beam is less than or equal to 10 in (254 mm), (b) the depth is not larger than 2.5 times the flange thickness, and (c) the depth is not greater than one-half the width of the web. In all other cases, ACI 318 is more conservative by a factor of two. Shear reinforcement can be eliminated only in regions of the beam where

$$\frac{\phi V_c}{2} \ge V_u \qquad 50.62$$

27. SHEAR REINFORCEMENT DESIGN PROCEDURE

The following procedure will design shear reinforcement for an entire beam. Some steps can be omitted when designing for a limited portion of the beam.

step 1: Calculate the shear envelope for the factored loads. If Eq. 50.52 is to be used to compute V_c, the moment diagram is also needed. The shear at any section is V_u.

step 2: Calculate V_c, generally using the simple option.

step 3: Determine the portions of the beam where the factored shear is less than or equal to $\phi V_c/2$. A graphical representation of the shear envelope drawn to scale can be helpful in locating these regions. Stirrups are not required in these regions.

step 4: Use Eq. 50.54 to compute the capacity, $V_{s,\min}$, provided by minimum shear reinforcement. Minimum reinforcement is obtained by setting the spacing of the stirrups to the largest amount allowed by Eq. 50.59 and A_v to the smallest bar choice that will satisfy Eq. 50.61.

step 5: Add V_c from step 2 to $V_{s,\min}$ from step 4 to obtain the shear capacity of the beam with minimum shear reinforcement. Use this minimum reinforcement in regions where $\phi V_c/2 < V_u < \phi(V_c + V_{s,\min})$.

step 6: Calculate the spacing of stirrups for the regions of the beam where $V_u > \phi(V_c + V_{s,\min})$. At any location, the shear reinforcement needs to supply a strength given by

$$V_{s,\text{req}} = \frac{V_u}{\phi} - V_c \qquad 50.63$$

The maximum value permitted for $V_{s,\text{req}}$ is given by Eq. 50.57. If the value computed in Eq. 50.63 exceeds this limit, adequate shear strength cannot be obtained with shear reinforcement. Either the section size or the concrete strength (or both) must be increased.

Calculate the spacing derived from Eq. 50.64. The spacing from Eq. 50.64 must not exceed the limits presented in Eqs. 50.59 and 50.60.

$$s = \frac{A_v f_{yt} d}{V_{s,\text{req}}} \qquad 50.64$$

28. ANCHORAGE OF SHEAR REINFORCEMENT

ACI 318 requires that the shear reinforcement extend into the compression and tension regions as far as cover requirements and proximity to other bars permit. In order to develop its full capacity, shear reinforcement must be anchored at both ends by one of two means. In

Structural

a U-shaped stirrup, one end of the steel is bent around the longitudinal steel, anchoring that end. For no. 5 or smaller stirrups and for stirrups constructed of no. 6, no. 7, or no. 8 bars with $f_{yt} \leq 40{,}000$ lbf/in^2 (276 MPa), the other end may be considered to be developed if it is wrapped with a standard hook around longitudinal reinforcement [ACI 318 12.13.2.1].

For no. 6, no. 7, or no. 8 stirrups with $f_{yt} > 40{,}000$ psi (276 MPa), the other end may be considered to be developed if it has a standard hook around longitudinal reinforcement plus an embedment distance between midheight of the member and the outside of the hook given by Eq. 50.65 [ACI 318 12.13.2.2].

$$\text{embedment length} \geq \frac{0.17 d_b f_{yt}}{\sqrt{f_c'}} \quad \text{[SI]} \qquad 50.65(a)$$

$$\text{embedment length} \geq \frac{0.014 d_b f_{yt}}{\sqrt{f_c'}} \quad \text{[U.S.]} \qquad 50.65(b)$$

Figure 50.13 *Anchorage Requirements for Shear Reinforcement (ACI 318 12.13.2)*

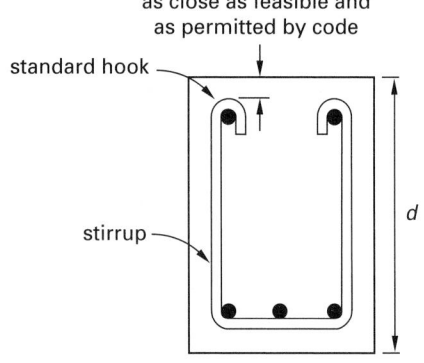

(a)
No. 5 and smaller, all f_{yt} values; No. 6, No. 7, No. 8 stirrups with $f_{yt} \leq 40{,}000$ psi

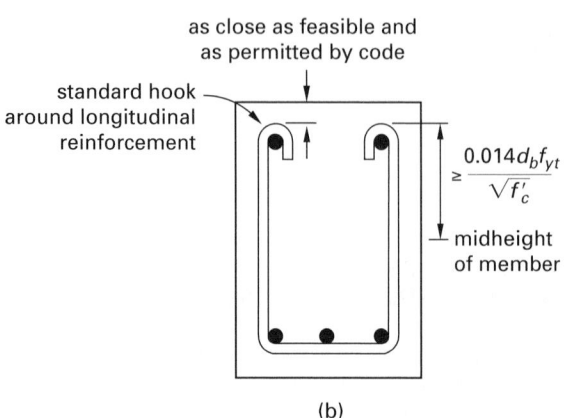

(b)
No. 6, No. 7, No. 8 stirrups with $f_{yt} > 40{,}000$ psi

29. DOUBLY REINFORCED CONCRETE BEAMS

In singly reinforced beams, steel is used to withstand tensile-only forces. There are instances, however, where it is appropriate to use steel reinforcement to resist compressive forces. The most common situation where a doubly reinforced beam is needed is found when the moment capacity of the singly reinforced section, with the maximum area of steel permitted, is not sufficient to carry the applied moment. In such cases, steel can be provided in excess of the maximum allowed for a singly reinforced section if the added tensile force is balanced by steel in compression. Another less common reason for using compression steel is to control long-term deformations. Indeed, the factor λ, which determines the increase in deflection due to creep and shrinkage (see Eq. 50.47), decreases as the percentage of compression steel in the section increases. Sections that have steel in the tension and the compression regions are referred to as *doubly reinforced beams*.

There are also situations where cross sections are doubly reinforced as a result of detailing, not as a result of needing compression steel for strength. For example, in the negative moment region of a typical monolithic beam-slab system the tension steel is at the top of the section, but compression steel is always present regardless of whether it is needed. This is because ACI 318 requires that a certain fraction of the steel used for the positive moment be extended into the support. (ACI 318 12.11.1 requires that at least one third of the reinforcement from a section of maximum moment be extended at least 6 in (15 cm) into a simple support. If the member is continuous, at least one fourth of the reinforcement from the section of maximum positive moment must extend at least 6 in (15 cm) into the support.) ACI 318 7.13 contains requirements for structural integrity that often mandate positive reinforcement at the supports. In moderate to high seismic applications, top as well as bottom reinforcement is mandated at every section.

30. STRENGTH ANALYSIS OF DOUBLY REINFORCED SECTIONS

Figure 50.14 illustrates a doubly reinforced concrete section and the profile of strains and stresses when the flexural capacity is attained. The following iterative procedure calculates the nominal moment capacity of a doubly reinforced beam section with a known shape and reinforcement.

step 1: Compute the tension force in the steel as $T_s = f_y A_s$.

step 2: Estimate the stress in the compression steel. (Since the analysis process is iterative, $f_s' = 0$ is a valid first estimate.)

step 3: Calculate the area of concrete, A_c, which at a stress of $0.85 f_c'$ carries a compressive force equal to $T_s - A_s' f_s'$.

$$A_c = \frac{f_y A_s - f'_s A'_s}{0.85 f'_c} \qquad 50.66$$

The depth of the equivalent compressed area is a. The actual neutral axis is at $c = a/\beta_1$, where β_1 is given by Eq. 50.21.

step 4: Compute the strain in the compression steel as

$$\varepsilon'_s = \left(\frac{0.003}{c}\right)(c - d') \qquad 50.67$$

step 5: Compute the stress in the compression steel as

$$f'_s = E_s \varepsilon'_s \le f_y \qquad 50.68$$

step 6: Compare the stress assumed in step 2 with the value computed in step 5. If they differ significantly (say, by more than 10%), replace the assumed value with the result from Eq. 50.68 and repeat steps 3 through 5. If they are reasonably close (say, within 10%), go to step 7.

step 7: Locate the centroid of the area A_c.

step 8: The nominal moment capacity is the sum of the concrete-steel couple and the steel-steel couple. (Equation 50.69 is the general form of Eq. 50.27 for singly reinforced sections with $f'_s = 0$ and $A'_s = 0$.)

$$M_n = (A_s - A'_s)f_y(d - \lambda) + A'_s f'_s(d - d') \qquad 50.69$$

Figure 50.14 *Parameters for Doubly Reinforced Beam*

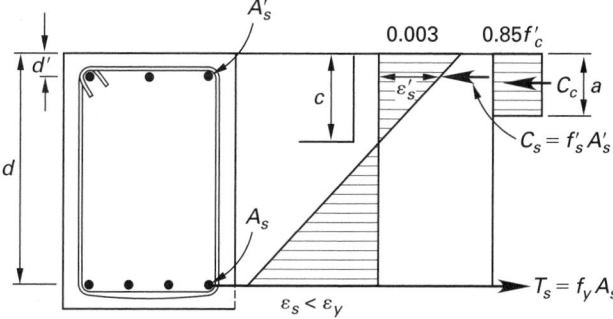

31. DESIGN OF DOUBLY REINFORCED SECTIONS

The design of a doubly reinforced beam begins by determining the difference between the applied moment and the maximum that can be carried with tension steel. This moment deficit must be provided by a couple formed by a force from additional tension steel ($A_{s,add}f_y$) balanced with a force from compression steel ($A'_s f'_s$). It should be noted that, at ultimate, while the steel added in tension will always yield, the compression steel may or may not yield.

The following procedure can be used to design a doubly reinforced beam.

step 1: Calculate the maximum area of tension steel permitted if the section is singly reinforced. (See Eq. 50.18, or use the lowest permissible value of net tensile strain, $\varepsilon_t = 0.004$.)

step 2: Calculate the moment capacity for the maximum steel area as singly reinforced. Designate this moment as M_{nc}. For a rectangular cross section, M_{nc} can be conveniently computed using Eq. 50.32, with the steel ratio ρ taken as 75% of that given by Eq. 50.33 when using the alternative provisions of App. B.

step 3: The nominal moment deficit is

$$M'_n = \frac{M_u}{\phi} - M_{nc} \qquad 50.70$$

step 4: Compute the area of tension steel $A_{s,add}$ that must be added to the area computed in step 1.

$$A_{s,add} = \frac{M'_n}{f_y(d - d')} \qquad 50.71$$

The total tension steel area is

$$A_s = A_{s,max} + A_{s,add} \qquad 50.72$$

step 5: Compute the stress in the compression steel from Eqs. 50.67 and 50.68. In these formulas, the depth of the neutral axis, c, is that for the singly reinforced section with the area of steel in step 1.

step 6: Calculate the area of compression steel from Eq. 50.73.

$$A'_s = \left(\frac{f_y}{f'_s}\right) A_{s,add} \qquad 50.73$$

step 7: Complete the design as for a singly reinforced section. (ACI 318 7.11.1 requires that the compression steel be enclosed by ties or stirrups satisfying size and spacing limitations that are the same as for ties in columns.)

Example 50.9

The factored loads acting on a reinforced concrete beam produce a maximum design moment $M_u = 275$ ft-kips. $f'_c = 3000$ lbf/in², and $f_y = 60,000$ lbf/in². The width of the beam is limited to 12 in, and the effective depth must not exceed 18 in. Design the reinforcing steel.

Solution Using ACI 318 App. B

Determine if the moment that can be resisted with maximum steel as a singly reinforced section is larger than the applied moment M_u.

The balanced steel area is obtained from Eq. 50.22.

$$\beta_1 = 0.85$$

$$A_{sb} = \left(\frac{0.85\beta_1 f_c'}{f_y}\right)\left(\frac{87{,}000}{87{,}000 + f_y}\right)bd$$

$$= \left(\frac{(0.85)(0.85)\left(3000\ \dfrac{\text{lbf}}{\text{in}^2}\right)}{60{,}000\ \dfrac{\text{lbf}}{\text{in}^2}}\right)$$

$$\times \left(\frac{87{,}000}{87{,}000 + 60{,}000}\right)(12\ \text{in})(18\ \text{in})$$

$$= 4.62\ \text{in}^2$$

The maximum amount of steel as a singly reinforced section is

$$A_s = (0.75)(4.62\ \text{in}^2) = 3.47\ \text{in}^2$$

The concrete area is given by Eq. 50.26.

$$A_c = \frac{f_y A_s}{0.85 f_c'} = \frac{\left(60{,}000\ \dfrac{\text{lbf}}{\text{in}^2}\right)(3.47\ \text{in}^2)}{(0.85)\left(3000\ \dfrac{\text{lbf}}{\text{in}^2}\right)}$$

$$= 81.6\ \text{in}^2$$

The depth of the equivalent rectangular stress block is

$$a = \frac{81.6\ \text{in}^2}{12\ \text{in}} = 6.8\ \text{in}$$

The centroid of the compressed area is $\lambda = (6.8\ \text{in})/2 = 3.40\ \text{in}$ from the top. The nominal moment capacity with maximum reinforcement is given by Eq. 50.27.

$$M_{nc} = A_s f_y(d - \lambda)$$

$$= \left(\frac{(3.47\ \text{in}^2)\left(60\ \dfrac{\text{kips}}{\text{in}^2}\right)(18\ \text{in} - 3.40\ \text{in})}{12\ \dfrac{\text{in}}{\text{ft}}}\right)$$

$$= 253.3\ \text{ft-kips}$$

Since $\phi M_{nc} < M_u$, compression reinforcement is needed. The nominal moment deficit is

$$M_n' = \frac{M_u}{\phi} - M_{nc}$$

$$= \frac{275\ \text{ft-kips}}{0.9} - 253.3\ \text{ft-kips}$$

$$= 52.2\ \text{ft-kips}$$

The area of tension steel, $A_{s,\text{add}}$, that must be added to the maximum permitted as a singly reinforced beam is

$$A_{s,\text{add}} = \frac{M_n'}{f_y(d - d')}$$

$$= \frac{(52.2\ \text{ft-kips})\left(12\ \dfrac{\text{in}}{\text{ft}}\right)}{\left(60\ \dfrac{\text{kips}}{\text{in}^2}\right)(18\ \text{in} - 2.5\ \text{in})}$$

$$= 0.67\ \text{in}^2$$

The total tension steel area is

$$A_s = 3.47\ \text{in}^2 + 0.67\ \text{in}^2 = 4.14\ \text{in}^2$$

In order to find the required compression steel area, it is necessary to determine the strain at the level where the compression steel is to be placed when the beam is reinforced with $A_{s,\text{max}}$. In this problem, assume that the centroid of the compression steel is 2.5 in from the top of the section to account for 1.5 in of cover, the stirrup bar diameter, and half of a compression bar diameter.

The neutral axis depth is $c = a/\beta_1 = 6.8\ \text{in}/0.85 = 8.0\ \text{in}$. Using Eqs. 50.67 and 50.68,

$$\varepsilon_s' = \frac{0.003(c - d')}{c}$$

$$= \frac{(0.003)(8.0\ \text{in} - 2.5\ \text{in})}{8.0\ \text{in}}$$

$$= 0.00206$$

$$f_s' = E_s \varepsilon_s' = \left(29{,}000\ \frac{\text{kips}}{\text{in}^2}\right)(0.00206)$$

$$= 59.7\ \text{kips/in}^2 \leq f_y$$

The compression steel area is

$$A_s' = \left(\frac{f_y}{f_s'}\right)A_{s,\text{add}} = \left(\frac{60{,}000\ \dfrac{\text{lbf}}{\text{in}^2}}{59{,}700\ \dfrac{\text{lbf}}{\text{in}^2}}\right)(0.66\ \text{in}^2)$$

$$= 0.66\ \text{in}^2$$

Put three no. 11 bars in the tension region, giving 4.68 in^2 of reinforcement. Put two no. 6 bars in the compression region, giving 0.88 in^2 of area.

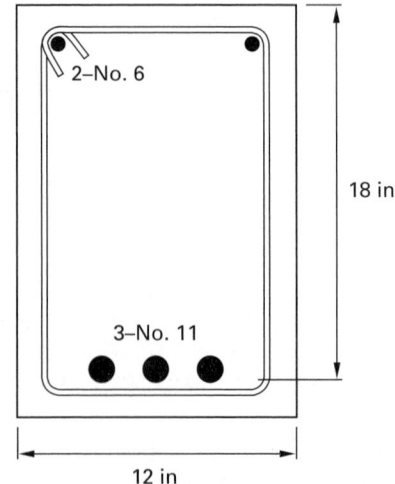

Solution Using ACI 318 Ch. 10

Determine if the moment that can be resisted with maximum steel as a singly reinforced section is larger than the applied moment M_u.

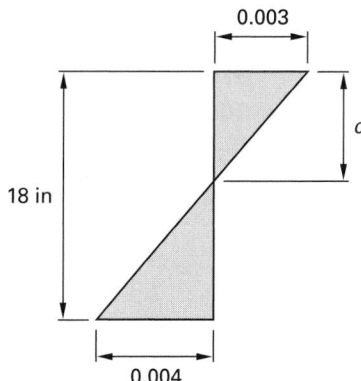

Neutral axis depth, c, corresponding to minimum permissible $\varepsilon_t = 0.004$, is

$$\frac{0.003}{0.003 + 0.004} = \frac{c}{18 \text{ in}}$$

$$c = \left(\frac{3}{7}\right)(18 \text{ in})$$

$$= 7.71 \text{ in}$$

The corresponding steel area is $A_{s,\max}$.

From a force equilibrium (see Sec. 10),

$$(12 \text{ in})(0.85)(7.71 \text{ in})(0.85)\left(3000 \, \frac{\text{lbf}}{\text{in}^2}\right)$$

$$= A_{s,\max}(60{,}000 \text{ lbf/in}^2)$$

$$A_{s,\max} = 3.34 \text{ in}^2$$

The depth of the equivalent rectangular stress block is

$$a = 0.85c = (0.85)(7.71 \text{ in})$$

$$= 6.55 \text{ in}$$

The centroid of the compressed area is $\lambda = (6.55 \text{ in})/2 = 3.28$ in from the top. The nominal moment capacity with maximum reinforcement is given by Eq. 50.27.

$$M_{nc} = A_s f_y (d - \lambda)$$

$$= \left(\frac{(3.34 \text{ in}^2)\left(60 \, \frac{\text{kips}}{\text{in}^2}\right)(18 \text{ in} - 3.28 \text{ in})}{12 \, \frac{\text{in}}{\text{ft}}}\right)$$

$$= 245.8 \text{ ft-kips}$$

Since $\phi M_{nc} < M_u$, compression reinforcement is needed. The nominal moment deficit is

$$M'_n = \frac{M_u}{\phi} - M_{nc}$$

$$= \frac{275 \text{ ft-kips}}{0.9} - 245.8 \text{ ft-kips}$$

$$= 59.8 \text{ ft-kips}$$

The area of tension steel, $A_{s,\text{add}}$, that must be added to the maximum permitted as a singly reinforced beam is

$$A_{s,\text{add}} = \frac{M'_n}{f_y(d - d')}$$

$$= \frac{(59.8 \text{ ft-kips})\left(12 \, \frac{\text{in}}{\text{ft}}\right)}{\left(60 \, \frac{\text{kips}}{\text{in}^2}\right)(18 \text{ in} - 2.5 \text{ in})}$$

$$= 0.77 \text{ in}^2$$

The total tension steel area is

$$A_s = 3.34 \text{ in}^2 + 0.77 \text{ in}^2$$

$$= 4.11 \text{ in}^2$$

In order to find the required compression steel area, it is necessary to determine the strain at the level where the compression steel is to be placed when the beam is reinforced with $A_{s,\max}$. In this problem, assume that the centroid of the compression steel is 2.5 in from the top of the section to account for 1.5 in of cover, the stirrup bar diameter, and half of a compression bar diameter.

Using Eqs. 50.67 and 50.68,

$$\varepsilon'_s = \frac{0.003(c - d')}{c}$$

$$= \frac{(0.003)(7.71 \text{ in} - 2.5 \text{ in})}{7.71 \text{ in}}$$

$$= 0.00203$$

$$f'_s = E_s \varepsilon'_s = \left(29{,}000 \, \frac{\text{kips}}{\text{in}^2}\right)(0.00203)$$

$$= 58.9 \text{ kips/in}^2 \leq f_y$$

The compression steel area is

$$A'_s = \left(\frac{f_y}{f'_s}\right) A_{s,\text{add}}$$

$$= \left(\frac{60 \, \frac{\text{kips}}{\text{in}^2}}{58.9 \, \frac{\text{kips}}{\text{in}^2}}\right)(0.77 \text{ in}^2)$$

$$= 0.78 \text{ in}^2$$

Put three no. 11 bars in the tension region, giving 4.68 in^2 of reinforcement, which is marginally greater than the 4.11 in^2 required. Put two no. 6 bars in the compression region, giving 0.88 in^2 of area.

32. DEEP BEAMS

Deep beams are members loaded on one face and supported on the opposite face so that compression struts can develop between the loads and the supports. They have either (a) clear spans, l_n, equal to or less than four times the overall member depth, h; or (b) regions loaded with concentrated loads within twice the member depth from the face of the support. The same definition applies for flexural design [ACI 318 10.7.1] and for shear design [ACI 318 11.8.1].

The detailed provisions for shear design of deep flexural members in editions of ACI 318 prior to 2002 have been removed. ACI Sec. 11.8 now states only that deep beams must be designed using nonlinear analysis as permitted in either ACI Sec. 10.7.1 or App. A (strut and tie models).

Since the expected orientation of shear cracks in deep beams is more nearly vertical than in shallow beams, horizontal reinforcement as well as vertical reinforcement is required. ACI 318 requires that the area of shear reinforcement perpendicular to the span, A_v, be not less than $0.0025 b_w s$, and that s may not exceed $d/5$ and 12 in. The area of shear reinforcement parallel to the span, A_{vh}, may not be less than $0.0015 b_w s_2$, and s_2 may not exceed $d/5$ nor 12 in. ACI 318 limits the shear strength, V_n, of deep beams to $10\sqrt{f_c'} b_w d$.

51 Reinforced Concrete: Slabs

Nomenclature

A	area	in^2	m^2
b	width of compressive member	in	m
C	torsional stiffness constant	in^4	m^4
d	effective depth	in	m
E	modulus of elasticity	lbf/in^2	Pa
f'_c	compressive strength	lbf/in^2	Pa
f_y	yield strength	lbf/in^2	Pa
I	gross moment of inertia	in^4	m^4
l	span length for one-way slab (centerline-to-centerline of supports)	in	m
l_1	span length in the direction moments are being computed	in	m
l_2	length of panel in the direction normal to that for which moments are being computed	in	m
l_n	clear span length in the direction moments are being computed (face-to-face of supports)	in	m
M_o	total factored static moment	in-lbf	N·m
q_u	factored uniform load per unit length	lbf/ft	N/m
s	spacing	in	m
t	thickness of slab	in	m
V	shear strength	lbf/in^2	m
w_u	factored uniform load per unit area of beam or one-way slab	lbf/ft^2	N/m^2
x	shorter dimension	in	m
y	longer dimension	in	m

Symbols

α	ratio of flexural stiffness of beams in comparison to slab	–	–
β	ratio of clear spans in long-to-short directions of a two-way slab	–	–
β_t	edge beam torsional stiffness constant	–	–
λ	distance from centroid of compressed area to extreme compression fiber	in	m
ρ	reinforcement ratio for tension	–	–
ϕ	strength reduction factor	–	–

Subscripts

b	bar or beam
c	concrete
m	mean
s	slab or steel
sr	steel required
t	temperature
u	uniform or ultimate
w	web

1. INTRODUCTION

Slabs are structural elements whose lengths and widths are large in comparison to their thicknesses. Shear is generally carried by the concrete without the aid of shear reinforcement (which is difficult to place and anchor in shallow slabs). Longitudinal reinforcement is used to resist bending moments. Slab thickness is typically governed by deflection criteria or fire rating requirements.

A primary issue associated with the design of slabs is the computation of the moments induced by the applied loads. Since slabs are highly indeterminate two-dimensional structures, an exact analysis to obtain the distribution of moments is impractical. Fortunately, moments can be obtained using simplified techniques. In fact, in many instances, slabs can be designed assuming that all the load is carried by moments in one direction only. These slabs are known as one-way slabs and are discussed in detail starting in Sec. 2. A practical approach for obtaining the distribution of moments in the more general two-dimensional case is presented starting in Sec. 8.

2. ONE-WAY SLABS

Floor slabs are typically supported on all four sides. If the slab length is more than twice the slab width, a uniform load will produce a deformed surface that has little curvature in the direction parallel to the long dimension. Given that moments are proportional to curvature, bending moments will be significant only in the short direction. Slabs designed under the assumption that bending takes place in only one direction are known as *one-way slabs*.

In one-way slabs, the internal forces are computed by taking a strip of unit width and treating it like a beam. The torsional restraint introduced by the supporting beams is typically neglected. For example, the moments per unit width in the short direction for the floor system shown in Fig. 51.1(a) would be obtained by analyzing the four-span continuous beam depicted in Fig. 51.1(b).

Figure 51.1 *One-Way Slab*

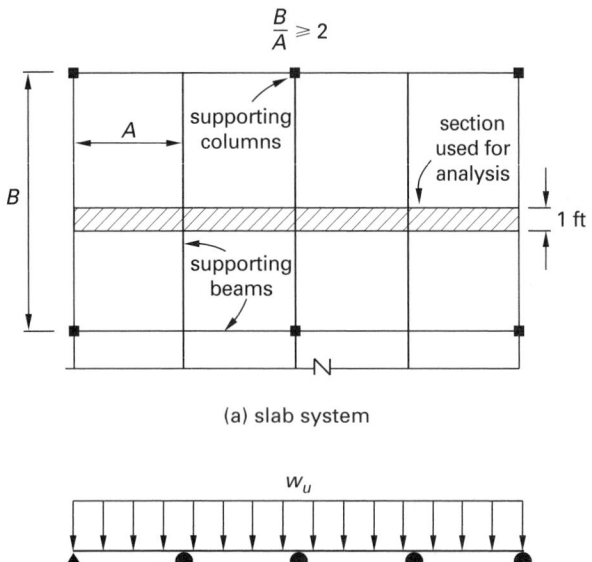

(a) slab system

(b) model used to obtain
moments and shears

3. TEMPERATURE STEEL

Although one-way slabs are designed to be capable of carrying the full applied load by spanning in a single direction, ACI 318 7.12 requires that reinforcement for shrinkage and temperature stresses be provided normal to the main flexural steel. This is known as *temperature steel*. Temperature steel is required only in structural slabs, not slabs cast against the earth and retaining wall footings. The minimum reinforcement ratio for shrinkage and temperature control (based on the gross area) is 0.0020 (grade-40 and grade-50 steel) and 0.0018 (grade-60 steel). For steels with yield stresses exceeding 60 kips/in² (413.7 MPa), the minimum reinforcement ratio is

$$\rho_t = 0.0018 \left(\frac{413.7}{f_y} \right) \quad \text{[SI]} \qquad 51.1(a)$$

$$\rho_t = 0.0018 \left(\frac{60{,}000}{f_y} \right) \quad \text{[U.S.]} \qquad 51.1(b)$$

The minimum reinforcement ratio cannot be less than 0.0014.

Although slabs are typically designed for uniform loads, slabs also experience significant concentrated loads. Temperature steel makes one-way slabs less vulnerable

to excessive cracking from moments parallel to the long dimension that are induced by such concentrated loads. Temperature steel must be spaced no farther apart than five times the slab thickness nor 18 in (46.0 cm) [ACI 318 7.12.2.2].

Although no. 3 bars may be used, it is common practice to use no. 4 bars, which are stiffer and, therefore, bend less during construction handling and installation.

4. MINIMUM THICKNESS FOR DEFLECTION CONTROL

Deflections of one-way slabs can be computed using the procedures described for beams in Ch. 50. If the slab is not supporting or attached to partitions that are likely to be damaged by large deflections, the computation of deflections can be obviated if the thickness, t, equals or exceeds the values in Table 51.1.

Table 51.1 *Minimum One-Way Slab Thickness[a,b]*
(unless deflections are computed)

construction	minimum thickness, t (fraction of span length[c])
simply supported	$\frac{1}{20}$
one end continuous	$\frac{1}{24}$
both ends continuous	$\frac{1}{28}$
cantilever	$\frac{1}{10}$

[a] ACI 318 Table 9.5(a)
[b] For normal-weight concrete reinforced with $f_y = 60{,}000$ lbf/in² (413.7 MPa) steel
[c] For slabs built integrally with supports, the minimum depth can be based on the clear span [ACI 318 8.7.3]. For slabs that are not built integrally with supports, the span length (for the purposes of this table) equals the clear span plus the thickness of the slab but need not exceed the centerline-to-centerline distance [ACI 318 8.7.1].

5. ANALYSIS USING ACI COEFFICIENTS

Although the shear and moments in one-way slabs can be computed using standard indeterminate structural analysis, ACI 318 8.3.3 specifies a simplified method that can be used when the following conditions are satisfied: (a) There are two or more spans. (b) Spans are approximately equal, with the longer of two adjacent spans not longer than the shorter by more than 20%. (c) The loads are uniformly distributed. (d) The ratio of live to dead loads is no more than 3. (e) The slab has a uniform thickness. The formulas for calculating moments and shears when these conditions apply were presented in Ch. 47.

6. SLAB DESIGN FOR FLEXURE

The procedure for selecting steel reinforcement is identical to that for beams, except that moments per unit width are used. Designating the required area of steel

Table 51.2 *Average Steel Area per Foot of Width*

bar size number	nominal diameter (in)	spacing of bars in inches												
		2	$2\frac{1}{2}$	3	$3\frac{1}{2}$	4	$4\frac{1}{2}$	5	$5\frac{1}{2}$	6	7	8	9	10	12
3	0.375	0.66	0.53	0.44	0.38	0.33	0.29	0.26	0.24	0.22	0.19	0.17	0.15	0.13	0.11
4	0.500	1.18	0.94	0.78	0.67	0.59	0.52	0.47	0.43	0.39	0.34	0.29	0.26	0.24	0.20
5	0.625	1.84	1.47	1.23	1.05	0.92	0.82	0.74	0.67	0.61	0.53	0.46	0.41	0.37	0.31
6	0.750	2.65	2.12	1.77	1.51	1.32	1.18	1.06	0.96	0.88	0.76	0.66	0.59	0.53	0.44
7	0.875	3.61	2.88	2.40	2.06	1.80	1.60	1.44	1.31	1.20	1.03	0.90	0.80	0.72	0.60
8	1.000		3.77	3.14	2.69	2.36	2.09	1.88	1.71	1.57	1.35	1.18	1.05	0.94	0.78
9	1.128		4.80	4.00	3.43	3.00	2.67	2.40	2.18	2.00	1.71	1.50	1.33	1.20	1.00
10	1.270			5.06	4.34	3.80	3.37	3.04	2.76	2.53	2.17	1.89	1.69	1.52	1.27
11	1.410			6.25	5.36	4.69	4.17	3.75	3.41	3.12	2.68	2.34	2.08	1.87	1.56

(Multiply in by 25.4 to obtain mm.)
(Multiply in² by 645 to obtain mm².)

per unit (1 ft) width as A_{sr} and the area of one bar as A_b, the spacing of the reinforcement, s, is given by Eq. 51.2. The maximum spacing for the main steel is three times the slab thickness but no more than 18 in (46 cm) (ACI 318 7.6.5). The smallest bar size typically used for flexural resistance is no. 4. Table 51.2 can be used to select the steel reinforcement if the steel area per foot of width is known.

$$s = \frac{A_b}{A_{sr}}(12 \text{ in}) \qquad 51.2$$

Although the design procedures for beams and slabs are similar, there are some differences. (a) The minimum cover for the steel is $^3/_4$ in (19 mm). (b) The minimum steel ratio is equal to that used for shrinkage and temperature. The equations for minimum steel used in beams do not apply.

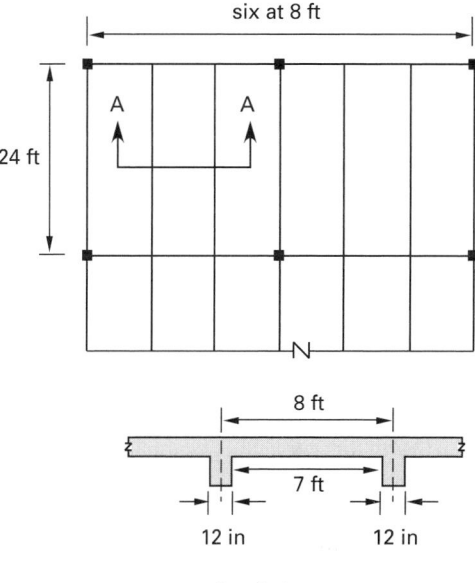

section A-A

7. SLAB DESIGN FOR SHEAR

Due to the small thickness, shear reinforcement is difficult to anchor and is seldom used in one-way slabs. Because one-way slabs are wide members, the requirement for no shear reinforcement is $\phi V_c \geq V_u$. As in beams, the critical section for shear is at a distance d from the face of the support if the support induces compression in the vertical direction.

Example 51.1

A floor system with columns on a 24 ft grid is shown. Beams in the north-south direction spaced at 8 ft on centers are used to create one-way slab action. The floor is subjected, in addition to its own weight, to dead and live loads of 20 lbf/ft² and 50 lbf/ft², respectively. $f'_c = 3000$ lbf/in², and $f_y = 60,000$ lbf/in². No. 4 bars are used for flexural reinforcement, and no. 3 bars are to be used for temperature steel. Select the thickness of the slab and design the flexural reinforcement.

Solution Using ACI 318 App. C

Begin by determining the required thickness. The clear span is 7 ft. From Table 51.1, for an edge panel,

$$t \geq \frac{l}{24}$$

$$= \frac{(7 \text{ ft})\left(12 \frac{\text{in}}{\text{ft}}\right)}{24}$$

$$= 3.5 \text{ in} \quad [\text{say 4 in}]$$

(The interior panels could be made thinner than the edge panels.)

The slab weight is

$$\frac{\left(150 \frac{\text{lbf}}{\text{ft}^3}\right)(4 \text{ in})}{12 \frac{\text{in}}{\text{ft}}} = 50 \text{ lbf/ft}^2$$

The ACI 318 App. C factored load is

$$w_u = 1.4D + 1.7L$$
$$= (1.4)\left(50\ \frac{\text{lbf}}{\text{ft}^2} + 20\ \frac{\text{lbf}}{\text{ft}^2}\right) + (1.7)\left(50\ \frac{\text{lbf}}{\text{ft}^2}\right)$$
$$= 183\ \text{lbf/ft}^2 \quad (0.183\ \text{kips/ft}^2)$$

Next, obtain the factored moments using moment coefficients with Eq. 47.43 and shear coefficients with Eq. 47.44. Consider a 1 ft wide strip.

The exterior negative moment is

$$\frac{M}{l_2} = \frac{w_u l_n^2}{24} = \frac{\left(0.183\ \frac{\text{kip}}{\text{ft}^2}\right)(7\ \text{ft})^2}{24}$$
$$= 0.374\ \text{ft-kip/ft}$$

The interior negative moment is

$$\frac{M}{l_2} = \frac{w_u l_n^2}{12} = \frac{\left(0.183\ \frac{\text{kip}}{\text{ft}^2}\right)(7\ \text{ft})^2}{12}$$
$$= 0.747\ \text{ft-kip/ft}$$

The positive moment in the exterior span is

$$\frac{M}{l_2} = \frac{w_u l_n^2}{14} = \frac{\left(0.183\ \frac{\text{kip}}{\text{ft}^2}\right)(7\ \text{ft})^2}{14}$$
$$= 0.641\ \text{ft-kip/ft}$$

The positive moment in the interior spans is

$$\frac{M}{l_2} = \frac{w_u l_n^2}{16} = \frac{\left(0.183\ \frac{\text{kip}}{\text{ft}^2}\right)(7\ \text{ft})^2}{16}$$
$$= 0.560\ \text{ft-kip/ft}$$

The maximum shear per unit width is

$$\frac{V}{l_2} = \frac{1.15 w_u l_n}{2} = \frac{(1.15)\left(0.183\ \frac{\text{kip}}{\text{ft}^2}\right)(7\ \text{ft})}{2}$$
$$= 0.737\ \text{kip/ft}$$

Compute the effective depth. Using no. 4 bars,

$$d = t - \text{cover} - \frac{d_b}{2}$$
$$= 4\ \text{in} - 0.75\ \text{in} - \frac{0.5\ \text{in}}{2}$$
$$= 3\ \text{in}$$

Check the shear capacity.

$$V_c = 2\sqrt{f_c'}b_w d$$
$$= \frac{2\sqrt{3000\ \frac{\text{lbf}}{\text{in}^2}}\left(12\ \frac{\text{in}}{\text{ft}}\right)(3\ \text{in})}{1000\ \frac{\text{lbf}}{\text{kip}}}$$
$$= 3.94\ \text{kips/ft}$$
$$\phi V_c = (0.85)(3.94\ \text{kips})$$
$$= 3.35\ \text{kips/ft} > 0.737\ \text{kip/ft} \quad [\text{OK}]$$

Design the flexural reinforcement. The interior negative moment is the largest. This will require the steel to be placed nearer to the top surface than the bottom surface in this region (if not on both surfaces). As with beams, initially estimate λ.

$$\lambda = 0.1d = (0.1)(3\ \text{in})$$
$$= 0.3\ \text{in}$$

From Ch. 50, Eq. 50.28, the steel required to resist the maximum moment is

$$A_s = \frac{M_u}{\phi f_y (d - \lambda)}$$
$$= \frac{\left(0.747\ \frac{\text{ft-kip}}{\text{ft}}\right)\left(12\ \frac{\text{in}}{\text{ft}}\right)}{(0.9)\left(60\ \frac{\text{kips}}{\text{in}^2}\right)(3\ \text{in} - 0.3\ \text{in})}$$
$$= 0.0615\ \text{in}^2/\text{ft}$$

The minimum steel is $(0.0018)(12\ \text{in/ft})(3\ \text{in}) = 0.0648$ in^2/ft, which is larger than the steel required to resist the maximum moment. Therefore, the minimum steel controls at all locations. Note that $\phi = 0.9$ for tension-controlled sections is fully justified when the reinforcement is so light.

The maximum bar spacing per ACI 318 is $3h = (3)(4\ \text{in}) = 12\ \text{in}$. Since the area of a no. 4 bar is 0.2 in^2, the maximum spacing controls. Use no. 4 bars on 12 in centers for positive and negative moments.

Temperature steel is required parallel to the long direction. Using no. 3 bars, the required spacing is

$$s = \frac{A_b}{A_{\text{sr}}}(12\ \text{in}) = \frac{(0.11\ \text{in}^2)\left(12\ \frac{\text{in}}{\text{ft}}\right)}{0.0648\ \frac{\text{in}^2}{\text{ft}}}$$
$$= 20.37\ \text{in}$$

The maximum spacing for temperature steel is the minimum of five times the slab thickness: $(5)(4 \text{ in}) = 20 \text{ in}$ or 18 in. Use no. 3 bars on 18 in centers in the long direction as temperature reinforcement.

Solution Using ACI 318 Ch. 9

Begin by determining the required thickness. The clear span is 7 ft. From Table 51.1, for an edge panel,

$$t \geq \frac{l}{24}$$

$$= \frac{(7 \text{ ft}) \left(12 \, \frac{\text{in}}{\text{ft}}\right)}{24}$$

$$= 3.5 \text{ in} \quad [\text{say 4 in}]$$

(The interior panels could be made thinner than the edge panels.)

The slab weight is

$$\frac{\left(150 \, \frac{\text{lbf}}{\text{ft}^3}\right)(4 \text{ in})}{12 \, \dfrac{\text{in}}{\text{ft}}} = 50 \text{ lbf/ft}^2$$

The ACI 318 Ch. 9 factored load is

$$w_u = 1.2D + 1.6L$$

$$= (1.2)\left(50 \, \frac{\text{lbf}}{\text{ft}^2} + 20 \, \frac{\text{lbf}}{\text{ft}^2}\right) + (1.6)\left(50 \, \frac{\text{lbf}}{\text{ft}^2}\right)$$

$$= 164 \text{ lbf/ft}^2 \quad (0.164 \text{ kip/ft}^2)$$

Next, obtain the factored moments using moment coefficients with Eq. 47.43 and shear coefficients with Eq. 47.44. Consider a 1 ft wide strip.

The exterior negative moment is

$$\frac{M}{l_2} = \frac{w_u l_n^2}{24} = \frac{\left(0.164 \, \frac{\text{kip}}{\text{ft}^2}\right)(7 \text{ ft})^2}{24}$$

$$= 0.335 \text{ ft-kip/ft}$$

The interior negative moment is

$$\frac{M}{l_2} = \frac{w_u l_n^2}{12} = \frac{\left(0.164 \, \frac{\text{kip}}{\text{ft}^2}\right)(7 \text{ ft})^2}{12}$$

$$= 0.670 \text{ ft-kip/ft}$$

The positive moment in the exterior span is

$$\frac{M}{l_2} = \frac{w_u l_n^2}{14} = \frac{\left(0.164 \, \frac{\text{kip}}{\text{ft}^2}\right)(7 \text{ ft})^2}{14}$$

$$= 0.574 \text{ ft-kip/ft}$$

The positive moment in the interior spans is

$$\frac{M}{l_2} = \frac{w_u l_n^2}{16} = \frac{\left(0.164 \, \frac{\text{kip}}{\text{ft}^2}\right)(7 \text{ ft})^2}{16}$$

$$= 0.502 \text{ ft-kip/ft}$$

The maximum shear is

$$\frac{V}{l_2} = \frac{1.15 w_u l_n}{2} = \frac{(1.15)\left(0.164 \, \frac{\text{kip}}{\text{ft}^2}\right)(7 \text{ ft})}{2}$$

$$= 0.660 \text{ kip/ft}$$

The design using the ACI 318 App. C load combinations will be the same as the ACI 318 Ch. 9 design, because the design is governed by minimum steel requirements throughout.

8. TWO-WAY SLABS

Slabs are classified as *two-way slabs* when the ratio of long-to-short sides is no greater than 2. A two-way slab supported on a column grid without the use of beams (Fig. 51.2(a)) is known as a *flat plate*. A modified version of a flat plate, where the shear capacity around the columns is increased by thickening the slab in those regions, is known as a *flat slab* (Fig. 51.2(b)). The thickened part of the flat plate is known as a *drop panel*. Another two-way slab often used without beams between column lines is a *waffle slab* (Fig. 51.2(c)). In a waffle slab, the forms used to create the voids are omitted around the columns to increase resistance to punching shear. A two-way slab system supported on beams is illustrated in Fig. 51.2(d).

Figure 51.2 *Two-Way Slabs*

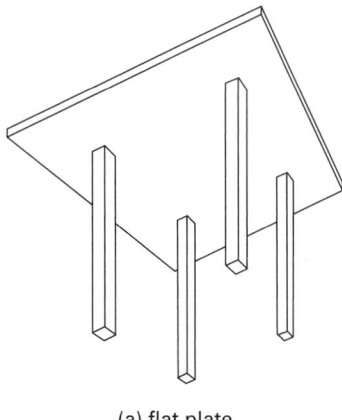

(a) flat plate

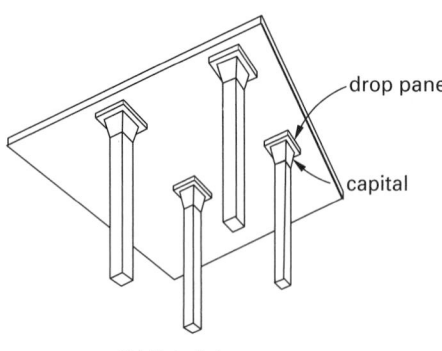

(b) flat slab

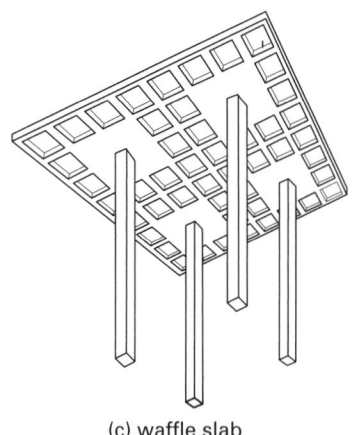

(c) waffle slab

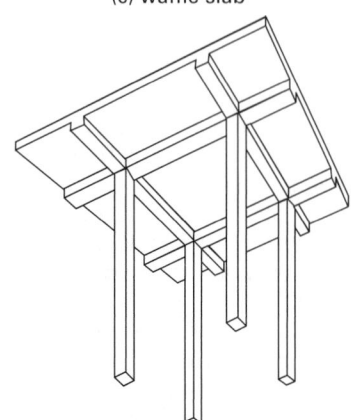

(d) two-way slab with beams

The moments used to design the reinforcement in two-way slabs, whether with or without beams, are obtained in the same manner as for beams and one-way slabs: The slab system and supporting columns are reduced to a series of one-dimensional frames running in both directions. As Fig. 51.3 illustrates, the beams in the frames are wide elements whose edges are defined by cuts midway between the columns. In ACI 318, the dimensions associated with the direction in which moments are being computed are identified by the subscript 1 and those in the transverse direction by the subscript 2. Note from Fig. 51.3 that the width of the wide beam is the average of the transverse panel dimensions on either side of the column line. For the edge frames, the width is half the width (centerline-to-centerline) of the first panel.

Figure 51.3 *Wide Beam Frame*
(used in the analysis of two-way slabs)

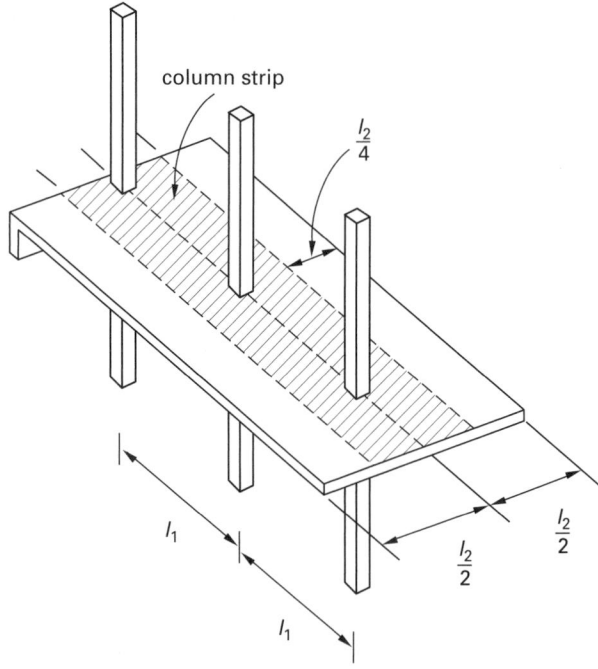

If the small effect of edge panels is neglected, and if the slab system is assumed to be uniformly loaded (from symmetry), the vertical shears and torques in the edges of the wide beams will be zero. There are two steps in the analysis of two-way slabs using the procedure in ACI 318: (a) calculating the longitudinal distribution of the moments (variation in the direction l_1), and (b) distributing the moment at any cross section across the width of the wide beams.

ACI 318 provides two alternative methods for computing the longitudinal distribution of moments. In the *equivalent frame method* (EFM), the moments are obtained from a structural analysis of the wide beam frame. In this analysis, it is customary to treat the columns as fixed one level above and below the level

Table 51.3 *Distribution of Moments in Exterior Spans (fraction of M_o)*

	exterior edge unrestrained	slab with beams between all supports	slabs without beams between interior supports		exterior edge fully restrained
			without edge beam	with edge beam	
interior negative factored moment	0.75	0.70	0.70	0.70	0.65
positive factored moment	0.63	0.57	0.52	0.50	0.35
exterior negative factored moment	0	0.16	0.26	0.30	0.65

being considered. (Special provisions that apply to the definition of the stiffnesses of the elements in the equivalent frame, and to other related aspects, are presented in ACI 318 13.7.) In the second alternative, known as the *direct design method* (DDM), the moments are obtained using a simplified procedure conceptually similar to that introduced for one-way slabs [ACI 318 13.6].

The procedure to distribute the moment computed at any section of the wide beam across the width of the wide beam is not dependent on whether the EFM or the DDM is used.

9. DIRECT DESIGN METHOD

The DDM can be used when the following conditions are met: (a) There are a minimum of three spans. (b) Panels are rectangular with a ratio of long-to-short side (center-to-center of supports) of no more than 2. (c) Successive span lengths do not differ by more than one third of the longest span. (d) Columns are not offset by more than 10% of the span in the direction of the offset.[1] (e) The loading consists of uniformly distributed gravity loads. (f) The service live load does not exceed two times the dead load. (g) If beams are present, the relative stiffness in two perpendicular directions, $\alpha_1 l_2^2 / \alpha_2 l_1^2$, is not less than 0.2 nor greater than 5.0.

The following steps constitute the direct design method.

step 1: Divide the floor system in each direction into wide beams as shown in Fig. 51.3.

step 2: Calculate the total statical moment in each span from Eq. 51.3 [ACI 318 13.6.2.2].

$$M_o = \frac{w_u l_2 l_n^2}{8} \qquad \textit{51.3}$$

This moment is the maximum moment in a simple beam of span l_n that carries the total load ($w_u l_2$). In Eq. 51.3, the span l_2 refers to the width of the wide beam being considered. l_n is measured face-to-face of columns or other supports. However, $l_n \geq 0.65 l_1$. For the computation of minimum thickness in two-way slabs, l_n is taken as the distance face-to-face of supports in slabs without beams and face-to-face of beams or other supports in other cases.

step 3: Divide the total moment, M_o, in each span into positive and negative moments. For interior spans, the negative factored moment is calculated as $0.65 M_o$, and the positive factored moment is $0.35 M_o$. The total moment in end (exterior) spans is distributed according to the coefficients in Table 51.3.

step 4: Divide the width of the wide beam into column-strip and middle-strip regions. A *column strip* is a design strip with a width on each side of the column centerline equal to $0.25 l_2$ or $0.25 l_1$, whichever is less. A *middle strip* is a design strip bounded by two column strips.

step 5: Design the column strip for the fractions of the moment at each section according to Table 51.4.

step 6: Design the middle strip for the fractions of the moment at each section not assigned to the column strip.

10. FACTORED MOMENTS IN SLAB BEAMS

A "beam," for the purpose of designing two-way slabs, is illustrated in Fig. 51.4. A beam includes the portion of the slab on each side extending a distance equal to the projection of the beam above or below the slab, whichever is largest, but not greater than four times the slab thickness.

---

[1]For design purposes, the moments are computed neglecting the column offsets. This procedure will lead to an adequate design if the column offset does not exceed the limit specified. For larger offsets, neither the DDM nor the EFM apply. It is customary in these cases to calculate the moments in the slab using a finite element model.

The distribution of moments in two-way slabs depends on the relative stiffness of the beams (with respect to the slab without beams). ACI 318 designates this relative stiffness as α, which is the ratio of the flexural stiffness of the beam (as defined in Fig. 51.4) to the flexural stiffness of a slab of width equal to that of the wide beam (i.e., the width of a slab bounded laterally by the centerlines of adjacent panels).

$$\alpha = \frac{E_{cb}I_b}{E_{cs}I_s} \qquad 51.4$$

When beams are part of the column strip, they are proportioned to resist 85% of the column strip moment if $\alpha_1 l_2/l_1 \geq 1$. For values of $\alpha_1 l_2/l_1$ between 0 and 1, linear interpolation is used to select the moment to be assigned to the beam, with zero percent assigned when $\alpha_1 l_2/l_1 = 0$. The value of l_2 in the previous expressions refers to the average of the values for the panels on either side of the column line (i.e., to the width of the wide beam).

Table 51.4 Distribution of Moments in Column Strips[a,b,c] (fraction of moment at section, M_o)

(a) interior negative moment

$\dfrac{l_2}{l_1}$		0.5	1.0	2.0
$\alpha_1 \dfrac{l_2}{l_1} = 0$ (no beams)		0.75	0.75	0.75
$\alpha_1 \dfrac{l_2}{l_1} \geq 1$		0.90	0.75	0.45

(b) exterior negative moment

$\dfrac{l_2}{l_1}$			0.5	1.0	2.0
$\alpha_1 \dfrac{l_2}{l_1} = 0$	$\beta_t = 0$		1.00	1.00	1.00
$\alpha_1 \dfrac{l_2}{l_1} = 0$	$\beta_t \geq 2.5$		0.75	0.75	0.75
$\alpha_1 \dfrac{l_2}{l_1} \geq 1$	$\beta_t = 0$		1.00	1.00	1.00
$\alpha_1 \dfrac{l_2}{l_1} \geq 1$	$\beta_t \geq 2.5$		0.90	0.75	0.45

(c) positive factored moment

$\dfrac{l_2}{l_1}$		0.5	1.0	2.0
$\alpha_1 \dfrac{l_2}{l_1} = 0$ (no beams)		0.60	0.60	0.60
$\alpha_1 \dfrac{l_2}{l_1} \geq 1$		0.90	0.75	0.45

[a]Linear interpolation is used between the values shown.
[b]Middle strips are designed for the fraction of the moment not assigned to the column strip.
[c]Refer to ACI 318 Secs. 13.6.4.1, 13.6.4.2, and 13.6.4.3.

Figure 51.4 Two-Way Slab Beams (monolithic or fully composite construction) (ACI 318 13.2.4)

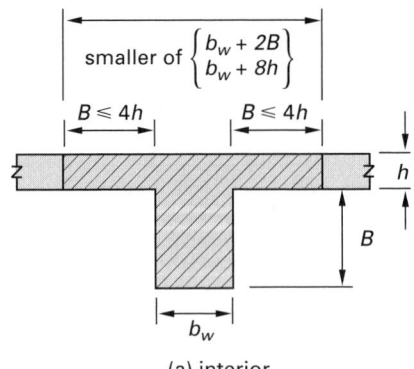

smaller of $\begin{cases} b_w + 2B \\ b_w + 8h \end{cases}$

$B \leq 4h$ $B \leq 4h$

(a) interior

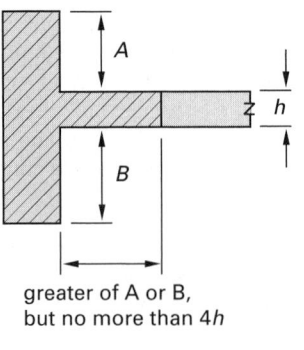

greater of A or B, but no more than 4h

(b) exterior

11. COMPUTATION OF β_t

The distribution of the negative moment across the width of a slab at the exterior edge depends not only on the relative beam stiffness and the ratio l_2/l_1, but also on the stiffness in torsion of edge beams. ACI 318 uses the parameter β_t to quantify the relative torsional stiffness of the edge beam.

$$\beta_t = \frac{E_{cb}C}{2E_{cs}I_s} \qquad 51.5$$

$$C = \sum \left(1 - 0.63\left(\frac{x}{y}\right)\right)\left(\frac{x^3 y}{3}\right) \qquad 51.6$$

The summation in Eq. 51.6 is taken over all of the separate rectangles that make up the edge beam (which is typically L-shaped as illustrated in Fig. 51.4). The division into separate rectangles that leads to the largest value of C should be used.

12. DEFLECTIONS IN TWO-WAY SLABS

Accurate estimates of deflections in two-way slabs are difficult to obtain. In practice, rather than attempting to calculate deflections accurately, two-way slabs should be sized to satisfy the minimum thickness values given in ACI 318 9.5.3.

Table 51.5 *Minimum Thickness for Slabs Without Interior Beams*
(longest clear span divided by value given)

	without drop panels			with drop panels		
	exterior panels (in)		interior panels (in)	exterior panels (in)		interior panels (in)
yield strength of reinforcement (lbf/in^2)	without edge beams	with edge beams		without edge beams	with edge beams	
40,000	33	36	36	36	40	40
60,000	30	33	33	33	36	36
75,000	28	31	31	31	34	34

For slabs without beams or with beams that are only placed between exterior columns (i.e., slabs without beams between interior supports), minimum thickness is specified as the largest clear span (face-to-face of supports) divided by the values listed in Table 51.5. For the values in Table 51.5 for slabs with drop panels to be applicable, drop panels must project below the slab at least one-quarter of the slab thickness beyond the drop and must extend in each direction at least one-sixth the length of the corresponding span. The thickness of a slab without drop panels may not be less than 5 in. Slabs with drop panels may not be less than 4 in thick.

For the purpose of calculating minimum thickness in slabs with beams between interior supports, there are three possibilities. For $\alpha_m \leq 0.2$, the minimum thickness is computed neglecting the beams. For $0.2 < \alpha_m \leq 2.0$, the minimum thickness is given by Eq. 51.7 where β is the ratio of clear spans in the long-to-short directions.

$$t = \frac{l_n \left(0.8 + \dfrac{f_y}{200,000} \right)}{36 + 5\beta(\alpha_m - 0.2)}$$

$$\geq 5 \text{ in} \qquad\qquad 51.7$$

For $\alpha_m > 2.0$, the minimum thickness is

$$t = \frac{l_n \left(0.8 + \dfrac{f_y}{200,000} \right)}{36 + 9\beta}$$

$$\geq 3.5 \text{ in} \qquad\qquad 51.8$$

When the stiffness ratio α of the edge beam is less than 0.8, the minimum thickness of the edge panel shall be at least 10% larger than the value obtained from Eqs. 51.7 and 51.8.

Structural

52 Reinforced Concrete: Short Columns

Nomenclature

A	area	in^2	m^2
c	clear distance between bars	in	m
c	neutral axis depth	in	m
d	wire diameter	in	m
D	circular column diameter	in	m
d'	distance from extreme compression fiber to the centroid of the compression reinforcement	in	m
d_s	distance from the extreme tension fiber to the centroid of the tension reinforcement	in	m
f'_c	compressive strength	lbf/in^2	Pa
f_y	yield strength	lbf/in^2	Pa
h	plan dimension perpendicular to the axis of bending	in	m
k	effective length factor	–	–
l_u	clear height of column	in	m
M	end moment	ft-kips	N·m
M_b	moment that together with P_b leads to a balanced strain diagram	ft-kips	N·m
P_b	axial force that together with M_b leads to a balanced strain diagram	kips	N
P_o	maximum nominal axial load capacity	kips	N
P_u	factored axial load	kips	N
r	radius of gyration	in	m
s	spiral pitch	in	m

Symbols

β	column strength factor		
Δ	deflection	in	m
ε	strain	in/in	m/m
ε_t	net tensile strain	in/in	m/m
ρ_g	longitudinal reinforcement ratio		
ρ_s	spiral reinforcement ratio		
ϕ	strength reduction factor		

Subscripts

b	balanced or braced
c	core or compressive
g	gross
n	nominal
s	spiral or steel
sp	spiral wire
st	longitudinal steel
u	unbraced

1. INTRODUCTION[1]

Columns are vertical members whose primary purpose is to transfer axial compression to lower members. In many practical situations, columns are subjected not only to axial compression but also to significant bending moments. Another factor sometimes increasing column loads is the P-Δ effect. The P-Δ effect refers to the increase in moment that results when the column sways, adding eccentricity. Columns where the P-Δ effect is significant are known as *long columns*. In this chapter we focus on the design of *short columns* only, that is, columns where the P-Δ effect can be neglected. (Long columns are covered in Ch. 53.)

Columns that are part of braced structures are considered to be short columns if Eq. 52.1 holds. The quantity $k_b l_u/r$ is the column's *slenderness ratio*. M_1 is the smaller (in absolute value) of the two end moments acting on a column. M_2 is the larger (in absolute value) of the two end moments acting on a column. The ratio M_1/M_2 is positive if the member is bent in single curvature, and it is negative if the member is bent in double curvature. The ratio M_1/M_2 cannot be less than -0.5. In Eq. 52.1, it is always conservative to take $k_b = 1$. [ACI 318 10.12.2]

$$\frac{k_b l_u}{r} \le 34 - 12 \left(\frac{M_1}{M_2} \right) \le 40 \qquad \textit{52.1}$$

Columns that are part of unbraced structures are considered to be short columns if Eq. 52.2 holds.

$$\frac{k_u l_u}{r} \le 22 \quad [k_u > 1] \qquad \textit{52.2}$$

The effective length factors, k_b and k_u, used in Eqs. 52.1 and 52.2, are discussed in Ch. 53. (ACI 318 uses the notation k to represent both k_b and k_u.)

2. TIED COLUMNS

Reinforced concrete columns with transverse reinforcement in the form of closed ties or hoops are known as

---

[1]Not specifically covered in this chapter are *pedestals*—upright compression members with ratios of unsupported height to average least lateral dimension not exceeding 3.

tied columns. Figure 52.2 illustrates a number of typical configurations. The following construction details are specified by ACI 318 for tied columns.

Figure 52.1 *Tied and Spiral Columns*

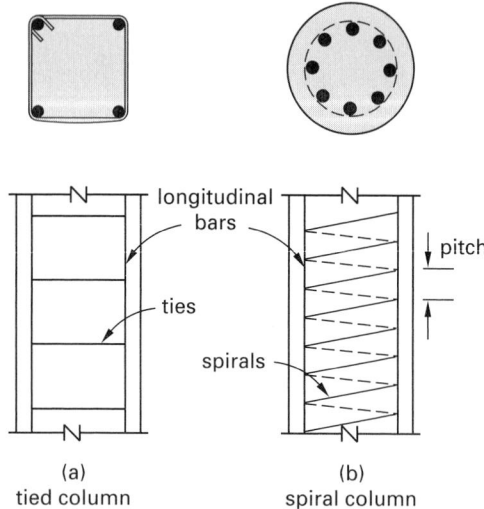

(a)
tied column

(b)
spiral column

- Longitudinal bars must have a clear distance between bars of at least 1.5 times the bar diameter, but not less than 1.5 in [ACI 318 7.6.3].

- Ties must be at least no. 3 if the longitudinal reinforcement consists of bars no. 10 in size or smaller. For bars larger than no. 10, or when bundles are used in the longitudinal reinforcement, the minimum tie size is no. 4 [ACI 318 7.10.5.1]. (The maximum practical tie size is normally no. 5.)

- The concrete cover must be at least 1.5 in over the outermost surface of the tie steel [ACI 318 7.7.1].

- At least four longitudinal bars are needed for columns with square or circular ties [ACI 318 10.9.2].

- The ratio of longitudinal steel area to the gross column area (see Eq. 52.12) must be between 0.01 and 0.08 [ACI 318 10.9.1]. (The lower limit keeps the column from behaving like a plain concrete member. The upper limit keeps the column from being too congested.)

$$0.01 \leq \rho_g \leq 0.08 \qquad \textbf{52.3}$$

- Center-to-center spacing of ties must not exceed the smallest of 16 longitudinal bar diameters, 48 diameters of the tie, or the least column dimension [ACI 318 7.10.5.2].

- Every corner and alternating longitudinal bar must be supported by a tie corner. The included angle of the tie cannot be more than 135° [ACI 318 7.10.5.3]. (Tie corners are not relevant to tied columns with longitudinal bars placed in a circular pattern.)

- No longitudinal bar can be more than 6 in away from a bar that is properly restrained by a corner [ACI 318 7.10.5.3].

Figure 52.2 *Types of Tied Columns*

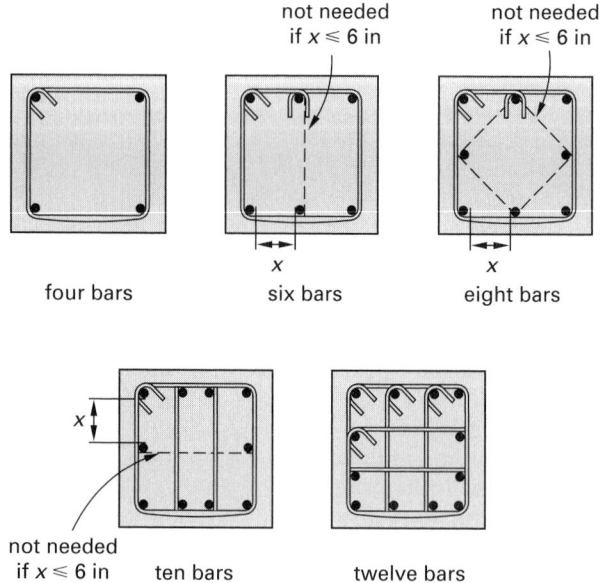

four bars

not needed
if $x \leq 6$ in

six bars

not needed
if $x \leq 6$ in

eight bars

not needed
if $x \leq 6$ in   ten bars

twelve bars

3. SPIRAL COLUMNS

Reinforced concrete columns with transverse reinforcement in the form of a continuous spiral are known as *spiral columns.* The following construction details are specified by ACI 318 for spiral columns.

- Longitudinal bars must have a clear distance between bars of at least 1.5 times the bar diameter but not less than 1.5 in [ACI 318 7.6.3].

- The minimum spiral wire diameter is ³/₈ in [ACI 318 7.10.4.2]. (The maximum practical spiral diameter is ⁵/₈ in.)

- The clear distance between spirals cannot exceed 3 in or be less than 1 in [ACI 318 7.10.4.3].

- The concrete cover should be at least 1.5 in from the outermost surface of the spiral steel [ACI 318 7.7.1].

- At least six longitudinal bars are to be used for spiral columns [ACI 318 10.9.2].

- The ratio of the longitudinal steel area to the gross column area must be between 0.01 and 0.08 [ACI 318 10.9.1].

- Splicing of spiral reinforcement is covered by ACI 318 7.10.4.5.

ACI 318 also requires that the ratio of the volume of spiral reinforcement to the volume of the column core (concrete placed inside the spiral) be no less than the value in Eq. 52.4. A_c is the area of the core, measured to the outside surface of the spiral reinforcement, and A_g is the gross section area.

$$\rho_s = 0.45 \left(\frac{A_g}{A_c} - 1\right)\left(\frac{f'_c}{f_{yt}}\right) \qquad 52.4$$

Once the spiral wire diameter is selected, the spiral pitch needed to satisfy Eq. 52.4 can be easily computed from Eq. 52.5, where $A_{\rm sp}$ is the area of the spiral wire and D_c is the diameter of the core measured to the outside surface of the spiral.

$$s \approx \frac{4A_{\rm sp}}{\rho_s D_c} \qquad 52.5$$

The clear distance between the spirals is related to the spiral pitch, s, by Eq. 52.6 where $d_{\rm sp}$ is the diameter of the spiral wire.

$$\text{clear distance} = s - d_{\rm sp} \qquad 52.6$$

The design of spiral reinforcing consists of the following steps: (a) Select a wire size. (A starting estimate for a spiral wire size can be obtained from Table 52.1.) (b) Evaluate Eqs. 52.4 and 52.5 to determine the required pitch. (c) Compute the clear spacing from Eq. 52.6. (d) Make sure the limit on clear spacing is satisfied.

Table 52.1 *Typical Spiral Wire Sizes*[a]

column diameter	spiral wire size (in)
up to 15 in, using no. 10 bars or smaller	$\frac{3}{8}$
up to 15 in, using bars larger than no. 10	$\frac{1}{2}$
16 to 22 in	$\frac{1}{2}$
23 in and larger	$\frac{5}{8}$

(Multiply in by 25.4 to obtain mm.)
[a]The ACI code does not specify spiral wire size.

4. DESIGN FOR SMALL ECCENTRICITY

The relative importance of bending and axial compression in the design of a column is measured by the *normalized eccentricity*. This parameter is defined as the ratio of the maximum moment acting on the column to the product of the axial load and the dimension of the column perpendicular to the axis of bending, $M_u/P_u h$.

Limits on normalized eccentricity are no longer part of ACI 318, but they can still be used to obtain approximate estimates of the maximum moment that will not affect the design for a given axial load. In an exact solution, the eccentricity limit will be a function of the percent of steel used and the type of reinforcement pattern. As described in Sec. 6, when the normalized eccentricity is less than approximately 0.1 for tied columns or 0.05 for spiral columns, the column can be designed without explicit consideration of flexural effects.

The design of a column for small eccentricity is based on Eq. 52.7. For tied columns, ACI 318 App. C specifies a *strength reduction factor* (or *capacity reduction factor*) of $\phi = 0.70$, while ACI 318 Ch. 9 specifies $\phi = 0.65$; the *strength factor*, β, is 0.80. For spiral columns, ACI 318 App. C specifies $\phi = 0.75$, while ACI 318 Ch. 9 specifies $\phi = 0.70$; $\beta = 0.85$. (ACI 318 does not actually use the symbol β as a parameter for columns. ACI 318 10.3.6.1 prescribes the values without specifying a symbol.)

$$\phi\beta P_o \geq P_u \qquad 52.7$$

In Eq. 52.7, P_o is the nominal axial load strength at zero eccentricity. The product $\phi\beta P_o$ is known as the *design strength* for a column with small eccentricity. This strength is computed by assuming that, at the attainment of maximum capacity, the concrete is stressed to $0.85f'_c$ and the steel is stressed to f_y.

$$P_o = 0.85f'_c(A_g - A_{\rm st}) + f_y A_{\rm st} \qquad 52.8$$

Although the nominal capacity, P_o, is the same for tied and spiral columns, their behavior is different once the peak capacity is attained. In particular, while tied columns fail in a relatively brittle manner, spiral columns, having better confinement of the core, undergo significant deformation before significant deterioration of the load-carrying capacity ensues. ACI 318 takes into consideration the increased reliability of spiral columns by providing higher values for ϕ and β. The ultimate load, P_u, is determined by structural analysis.

Using ACI 318 App. C,

$$\begin{aligned} P_u = &\ 1.4(\text{dead load axial force}) \\ &+ 1.7(\text{live load axial force}) \qquad 52.9(a) \end{aligned}$$

Using ACI 318 Ch. 9,

$$\begin{aligned} P_u = &\ 1.2(\text{dead load axial force}) \\ &+ 1.6(\text{live load axial force}) \qquad 52.9(b) \end{aligned}$$

The basic design expression for columns with small eccentricity is obtained by substituting Eq. 52.8 into Eq. 52.7. Passing ϕ and β to the right-hand side,

$$0.85f'_c(A_g - A_{\rm st}) + A_{\rm st}f_y = \frac{P_u}{\phi\beta} \qquad 52.10$$

A convenient form of Eq. 52.10 is obtained by expressing the steel area as a function of the reinforcement ratio, ρ_g. If the column size is not known, a value of ρ_g between the limits of 0.01 and 0.08 is selected, and

Structural

Eq. 52.11 is solved for the required gross column area, A_g. (Larger values of ρ_g decrease the gross column area. Typical starting values of ρ_g are approximately 0.02 to 0.025.) If the section size is known and only the steel reinforcement needs to be obtained, then the equation is solved for ρ_g, and the area of steel is obtained from Eq. 52.12.

$$A_g\left(0.85f_c'(1-\rho_g)+\rho_g f_y\right)=\frac{P_u}{\phi\beta} \qquad \textbf{52.11}$$

$$\rho_g=\frac{A_{\text{st}}}{A_g} \qquad \textbf{52.12}$$

Example 52.1

Calculate the design strength of the short spiral column shown. Assume the loading has a low eccentricity. $f_c' = 3500$ lbf/in^2, and $f_y = 40{,}000$ lbf/in^2.

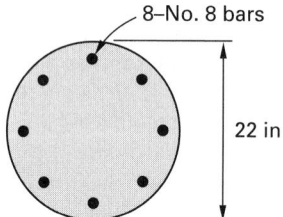

8–No. 8 bars

22 in

Solution

The gross area of the column is

$$A_g=\frac{\pi D_g^2}{4}=\frac{\pi(22\ \text{in})^2}{4}=380.1\ \text{in}^2$$

Since there are eight no. 8 bars, the area of steel is

$$A_{\text{st}}=(8)(0.79\ \text{in}^2)=6.32\ \text{in}^2$$

Substitute into Eq. 52.8.

$$
\begin{aligned}
P_o &= 0.85f_c'(A_g-A_{\text{st}})+f_y A_{\text{st}}\\
&= \left((0.85)\left(3500\ \frac{\text{lbf}}{\text{in}^2}\right)(380.1\ \text{in}^2-6.32\ \text{in}^2)\right.\\
&\quad \left.+\left(40{,}000\ \frac{\text{lbf}}{\text{in}^2}\right)(6.32\ \text{in}^2)\right)\left(\frac{1}{1000\ \frac{\text{lbf}}{\text{kip}}}\right)\\
&= 1364.8\ \text{kips}
\end{aligned}
$$

The ACI 318 App. C design strength is

$$\phi\beta P_o=(0.75)(0.85)(1364.8\ \text{kips})=870\ \text{kips}$$

The ACI 318 Ch. 9 design strength is

$$\phi\beta P_o=(0.70)(0.85)(1364.8\ \text{kips})=812\ \text{kips}$$

Example 52.2

Design a spiral column to carry a factored load of 375 kips. Use $f_c' = 3000$ lbf/in^2 and $f_y = 40{,}000$ lbf/in^2.

Solution Using ACI 318 App. C

Assume $\rho_g = 0.02$, which is within the allowable range.

Use Eq. 52.11 with the ACI 318 App. C ϕ factor of 0.75.

$$
\begin{aligned}
A_g &= \frac{P_u}{\phi\beta\left(0.85f_c'(1-\rho_g)+\rho_g f_y\right)}\\
&= \frac{375\ \text{kips}}{(0.75)(0.85)\left((0.85)\left(3\ \dfrac{\text{kips}}{\text{in}^2}\right)(1-0.02)\right.}\\
&\qquad \overline{\left.+\,(0.02)\left(40\ \dfrac{\text{kips}}{\text{in}^2}\right)\right)}\\
&= 178.3\ \text{in}^2
\end{aligned}
$$

The outside diameter is

$$D_g=\sqrt{\frac{4A_g}{\pi}}=\sqrt{\frac{(4)(178.3\ \text{in}^2)}{\pi}}$$
$$=15.07\ \text{in}\quad[\text{say }15.25\ \text{in}]$$

With 1.5 in of cover, the diameter of the core is

$$D_c=15.25\ \text{in}-3\ \text{in}=12.25\ \text{in}$$

From Eq. 52.12, the area of steel is

$$A_{\text{st}}=(0.02)(178.3\ \text{in}^2)=3.57\ \text{in}^2$$

ACI 318 requires at least six bars to be used. Several possibilities exist. For example, six no. 7 bars, nine no. 6 bars, or twelve no. 5 bars would all give the required area. Choose nine no. 6 bars after checking the clear spacing (not shown here).

The required spiral reinforcement ratio is obtained from Eq. 52.4. The area ratio is calculated from the ratio of the squared diameters.

$$
\begin{aligned}
\rho_s &= 0.45\left(\frac{A_g}{A_c}-1\right)\left(\frac{f_c'}{f_y}\right)\\
&= (0.45)\left(\left(\frac{15.25\ \text{in}}{12.25\ \text{in}}\right)^2-1\right)\left(\frac{3000\ \dfrac{\text{lbf}}{\text{in}^2}}{40{,}000\ \dfrac{\text{lbf}}{\text{in}^2}}\right)\\
&= 0.0186
\end{aligned}
$$

Assume a $^3/_8$ in spiral wire. The maximum pitch is obtained from Eq. 52.5.

$$s=\frac{4A_{\text{sp}}}{\rho_s D_c}=\frac{(4)(0.11\ \text{in}^2)}{(0.0186)(12.25\ \text{in})}$$
$$=1.93\ \text{in}$$

Use a spiral pitch equal to 1.75 in.

The clear space between spirals is 1.75 in − 0.375 in = 1.375 in. This is between 1 and 3 in and satisfies ACI 318 7.10.4.3.

Solution Using ACI 318 Ch. 9

Assume $\rho_g = 0.02$, which is within the allowable range.

Use Eq. 52.11 with the ACI 318 Ch. 9 ϕ factor of 0.70.

$$A_g = \frac{P_u}{\phi\beta\left(0.85f'_c(1-\rho_g)+\rho_g f_y\right)}$$

$$= \frac{375 \text{ kips}}{(0.70)(0.85)\left((0.85)\left(3\,\dfrac{\text{kips}}{\text{in}^2}\right)(1-0.02)\right.}$$
$$\left.+\,(0.02)\left(40\,\dfrac{\text{kips}}{\text{in}^2}\right)\right)$$

$$= 191.0 \text{ in}^2$$

The outside diameter is

$$D_g = \sqrt{\frac{4A_g}{\pi}} = \sqrt{\frac{(4)(191.0 \text{ in}^2)}{\pi}}$$

$$= 15.59 \text{ in} \quad [\text{say } 15.75 \text{ in}]$$

With 1.5 in of cover, the diameter of the core is

$$D_c = 15.75 \text{ in} - 3 \text{ in} = 12.75 \text{ in}$$

From Eq. 52.12, the area of steel is

$$A_{\text{st}} = (0.02)(191.0 \text{ in}^2) = 3.82 \text{ in}^2$$

ACI 318 requires at least six bars to be used. Several possibilities exist. For example, nine no. 6 bars or twelve no. 5 bars would both give the required area. Choose nine no. 6 bars after checking the clear spacing (not shown here).

The required spiral reinforcement ratio is obtained from Eq. 52.4. The area ratio is calculated from the ratio of the squared diameters.

$$\rho_s = 0.45\left(\frac{A_g}{A_c}-1\right)\left(\frac{f'_c}{f_{yt}}\right)$$

$$= (0.45)\left(\left(\frac{15.75 \text{ in}}{12.75 \text{ in}}\right)^2-1\right)\left(\frac{3000\,\dfrac{\text{lbf}}{\text{in}^2}}{40{,}000\,\dfrac{\text{lbf}}{\text{in}^2}}\right)$$

$$= 0.0178$$

Assume a $^3/_8$ in spiral wire. The maximum pitch is obtained from Eq. 52.5.

$$s = \frac{4A_{\text{sp}}}{\rho_s D_c} = \frac{(4)(0.11 \text{ in}^2)}{(0.0178)(12.75 \text{ in})}$$

$$= 1.94 \text{ in}$$

Use a spiral pitch equal to 1.75 in.

The clear space between spirals is 1.75 in − 0.375 in = 1.375 in. This is between 1 and 3 in and satisfies ACI 318 7.10.4.3.

5. INTERACTION DIAGRAMS

Based on approximate eccentricity limits, when the maximum moment acting on a column is larger than approximately $0.1P_u h$ for tied columns or $0.05P_u h$ for spiral columns, the design must account explicitly for the effect of bending. In practice, design of columns subjected to significant bending moments is based on interaction diagrams. An *interaction diagram* is a plot of the values of bending moments and axial forces corresponding to the strength (nominal or design) of a reinforced concrete section.

Figure 52.3 illustrates that, for small compressive loads, the moment capacity initially increases, then at a certain axial load it reaches a maximum and subsequently decreases, becoming zero when the axial load equals P_o from Eq. 52.8. The point in the interaction diagram where the moment is maximum is referred to as the *balanced point* because at that point, the yielding of the extreme tension reinforcement coincides with the attainment of a strain of 0.003 at the extreme compression fiber of concrete.

Figure 52.3 *Nominal Interaction Diagram*

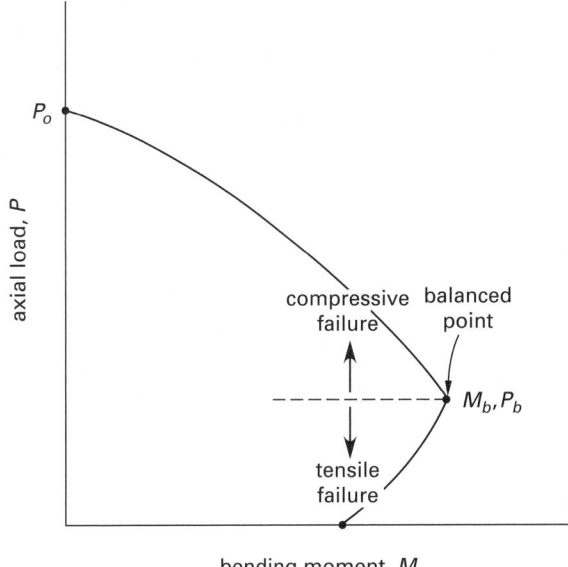

Points on a *nominal interaction diagram* are obtained from the following procedure. (Refer to Figs. 52.4 and 52.5.)

step 1: Assume an arbitrary neutral axis depth, c. (For $c > c_b$, points above the balanced point will be obtained.)

step 2: Calculate $a = \beta_1 c$ and the associated area of concrete in compression, A_c. (β_1 is specified in Eq. 50.21.) The force in the concrete is $0.85f'_c A_c$.

step 3: Compute the forces in the steel bars (or layers). This is done by computing the strain from the geometry of the strain diagram, obtaining the associated stress, and multiplying by the appropriate steel area.

step 4: Compute the nominal axial load, P_n, as the sum of all the forces in the cross section. (For a sufficiently small neutral axis depth, the axial force resultant will be tensile. Although the portion of the interaction diagram corresponding to tension is not usually of interest, points on the negative side need not be discarded since they can help in drawing the interaction diagram.)

step 5: Compute the nominal moment, M_n, by summing moments of all the forces about the plastic centroid. The *plastic centroid* is the location of the resultant force when all of the concrete is compressed at $0.85 f_c'$ and all of the steel is yielding. This corresponds to the point where the line of action of P_o intersects the section. For the typical case of symmetrically reinforced rectangular or circular column sections, the plastic centroid coincides with the centroid of the section.

An approximate and conservative nominal interaction diagram can be sketched by drawing straight lines between the balanced point at (M_b, P_b) and (O, P_o), and between (M_b, P_b) and the nominal moment capacity for pure bending.

Figure 52.4 *Calculation of Points on a Nominal Interaction Diagram*

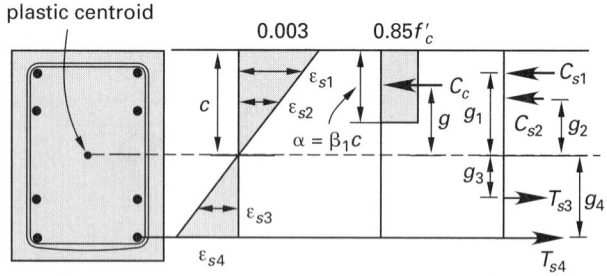

$$P_n = C_c + C_{s1} + C_{s2} - T_{s3} - T_{s4}$$

$$M_n = C_c g + C_{s1} g_1 + C_{s2} g_2 + T_{s3} g_3 + T_{s4} g_4$$

A *design interaction diagram* can be constructed from the nominal interaction diagram in two steps: (a) Multiply by the appropriate values of ϕ. (b) Limit the maximum allowable compression to $\beta \phi P_o$. Although ϕ is constant (either 0.70/0.75 per ACI 318 App. C or 0.65/0.70 per ACI 318 Ch. 9 for most of the diagram), consistency with the design of beams requires that a transition to 0.9 be made as the axial load decreases to very low values.

Figure 52.5 *Relation Between Nominal and Design Interaction Diagrams (per ACI 318 App. C)*

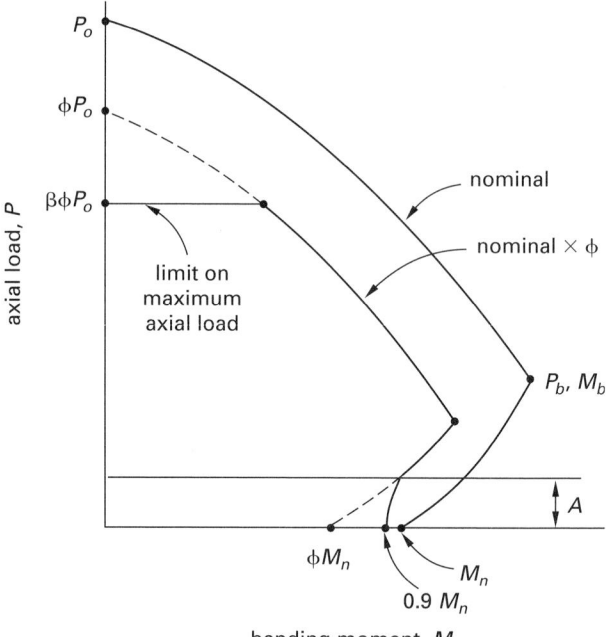

$$A = 0.1 f_c' A_g \qquad \left[F_y \le 60{,}000\ \frac{\text{lbf}}{\text{in}^2}\ \text{and}\ \frac{h - d' - d_s}{d} \ge 0.7 \right]$$

$$A = \text{smaller of} \begin{cases} 0.1 f_c' A_g & \text{[otherwise]} \\ \phi P_b \end{cases}$$

In the typical case where $f_y \le 60{,}000$ lbf/in^2 (457 GPa) and $(h - d' - d_s)/h \ge 0.7$, ACI 318 C.3.2.2 indicates that the transition (which is a linear function of P_u) begins when $P_u = 0.1 f_c' A_g$. For other conditions, the transition starts at $P_u = 0.1 f_c' A_g$ or $P_u = \phi P_b$, whichever is smaller.

Figure 52.5 illustrates the conversion of a nominal interaction diagram to the design interaction diagram per ACI 318 App. C. Note that the ϕ transition is easily implemented (with sufficient accuracy) by joining the last value where ϕ is constant to 0.9 times the nominal pure bending strength.

Alternatively, a transition is made in accordance with ACI 318 9.3.2.2 between compression-controlled and tension-controlled sections. For sections in which the net tensile strain is between the limits for compression-controlled and tension-controlled sections, the strength reduction factor is permitted to be linearly increased from that for compression-controlled sections (either 0.65 or 0.70) to 0.90 as the net tensile strain increases from the compression-controlled strain limit (which is equal to 0.002 in the typical case of $f_y = 60{,}000$ lbf/in^2) to 0.005.

Figure 52.6 illustrates the conversion of a nominal interaction diagram to the design interaction diagram per ACI 318 Ch. 9. Note that the balanced point and the point corresponding to the beginning of compression-controlled sections (those with ε_t equal to 0.002) are the same. The subscript "tc" in the figure refers to the beginning of tension-controlled sections (those with ε_t equal to 0.005).

Figure 52.6 *Relation Between Nominal and Design Interaction Diagrams (per ACI 318 Ch. 9)*

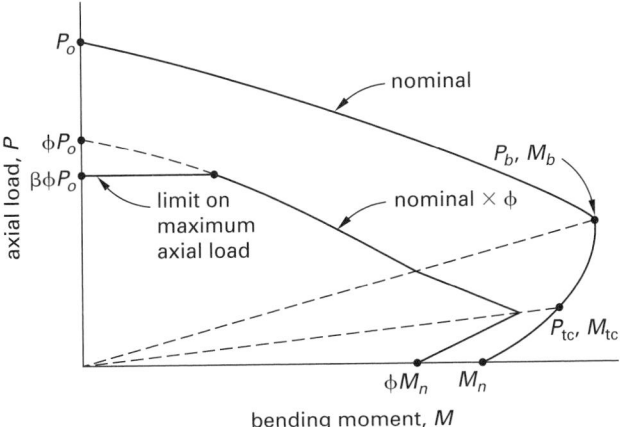

6. DESIGN FOR LARGE ECCENTRICITY

Although interaction diagrams for columns with unusual shapes need to be drawn on a case-by-case basis, routine design of rectangular or circular columns is carried out using dimensionless interaction diagrams available from several sources. Typical dimensionless interaction diagrams are presented in App. 52.A.

The use of dimensionless interaction diagrams is straightforward. In App. 52.A, each diagram includes envelope lines (most of the diagrams have eight lines). Each line corresponds to a percent of steel from the minimum of 1% to the maximum of 8%. Each diagram corresponds to a particular set of material properties and to a particular value of γ that is the ratio of the parallel distance between the outer steel layers to the gross section dimension perpendicular to the neutral axis. For practical purposes, γ can be estimated as $(h-5)/h$, where h is in inches. For rectangular cross sections, there is also a need to determine if the steel is to be placed on the two faces parallel to the neutral axis or if a uniform distribution around the column is to be used. In this regard, it is worth noting that the circles represent only the reinforcing pattern, not the exact positioning.

The following procedure can be used to design columns for large eccentricity using standard interaction diagrams.

step 1: If the column size is not known, an initial estimate can be obtained from Eq. 52.11 using an arbitrarily selected value of ρ_g between the limits allowed. Since the moment will increase the steel demand above that for no moment, a value of ρ_g somewhat smaller than the desired target should be selected.

step 2: Select the type of reinforcement pattern desired and estimate the value of γ. Choose the appropriate interaction diagram. Using a diagram for a γ value smaller than the actual leads to conservative results.

step 3: Evaluate the coordinates $P_u/\phi f'_c A_g$ and $P_u e/\phi f'_c A_g h$ (where $P_u e = M_u$) and plot this point on the diagram, assuming a value of ϕ. The dashed line corresponding to $\varepsilon_t = 0.002$ (compression-controlled section) and to $\varepsilon_t = 0.005$ (tension-controlled section) on the diagrams can be used as guides when assuming a value for the strength reduction factor. If the point is outside the last curve for $\rho_g = 0.08$, the section size is too small. In that case, the section size should be increased and this step repeated. If the point is inside the curve for $\rho_g = 0.01$, the section is oversized. In that case, the dimensions should be reduced and this step should be repeated.

step 4: Verify the assumed ϕ value. In most cases, the section will turn out to be clearly compression controlled or tension controlled, and verification will involve no effort. In rare situations where this is not the case, calculate the strain in the extreme tension steel from a strain compatibility analysis for the selected size and arrangement of reinforcement. Determine the strength reduction factor on the basis of the magnitude of this strain and compare it to the value assumed in step 3. If the calculated and assumed values vary by more than a few percentage points, repeat the preceding steps based on the calculated strength reduction factor from this step.

Once the size and steel reinforcement are selected, the design is completed by selecting the transverse reinforcement. This step is carried out exactly the same way as in the case of columns with small eccentricity. Although seldom an issue in the design of columns, the adequacy of the cross section in shear should also be checked when significant bending moments exist.

53 Reinforced Concrete: Long Columns

Nomenclature

A	area	in^2
d	diameter	in
D	dead load	kips
E	modulus of elasticity	lbf/in^2
f'_c	compressive strength of concrete	lbf/in^2
h	dimension perpendicular to the axis of bending	in
I	moment of inertia	in^4
I_{se}	moment of inertia of steel in a column	in^4
k	effective length factor	–
L	live load	kips
l	span length of flexural member, center-to-center of joints	in
l_c	height of column, center-to-center of joints	in
l_u	clear height of column	in
M_c	amplified moment for braced conditions	ft-kips
M_1	smaller (in absolute value) of the two end moments acting on a column	ft-kips
M_2	larger (in absolute value) of the two end moments acting on a column	ft-kips
P	load	kips
Q	stability index	–
r	radius of gyration	in
V	shear force	kips
W	wind load	kips

Symbols

β_d	stiffness reduction factor reflecting long-term effects	–
δ	amplification factor	–
Δ_o	interstory deformation	in
Ψ	relative stiffness parameter	–

Subscripts

b	beam or braced
c	column or concrete
d	dead
g	gross
l	live
ns	no sway
s	sway or story
u	ultimate (factored) or unbraced

1. INTRODUCTION

A typical assumption used in the calculation of internal forces in structures is that the change in geometry resulting from deformations is sufficiently small to be neglected. This type of analysis, known as a *first-order analysis*, is an approximation that is often sufficiently accurate. There are other instances, however, where the changes in geometry resulting from the structural deformations have a notable effect and must be considered.

Figure 53.1 Effect of Deformation on Internal Forces

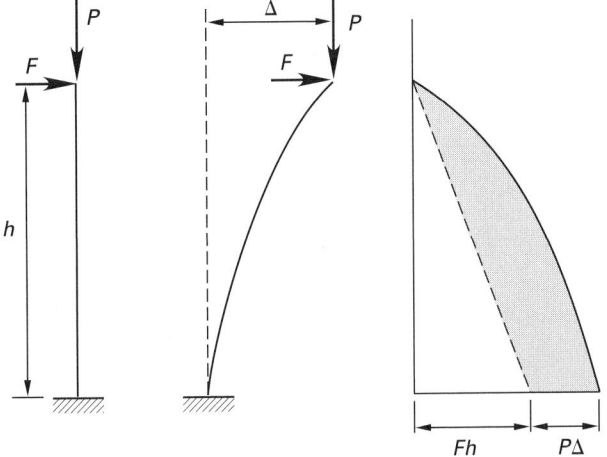

(a) applied load   (b) deflected shape   (c) story movement

To illustrate how the deformations of a structure affect the internal forces, consider the cantilever column subjected to compressive and lateral loads shown in Fig. 53.1. The moment at the base of the column in a first-order solution is the product of the lateral load F times the height h. (The actual moment at the base, however, is $Fh + P\Delta$, where Δ is the deflection at the tip of the cantilever due to the combined action of both F and P.) An unconservative design results if the moment $P\Delta$ is significant (in comparison to Fh) and is neglected in the design. Although the changes in geometry have an influence on the axial forces also, these effects are typically small and are neglected in design.

The rigorous evaluation of second-order moments in large structures is generally considered too involved for routine design. A practical alternative, however, is to estimate the true moments by multiplying the first-order solution by appropriately defined amplification factors—this is the approach taken by the ACI Code in its treatment of long columns.

ACI 318 contains simplified criteria for determining when slenderness amplification factors do not have to be calculated. These criteria were presented in Ch. 52.

2. BRACED AND UNBRACED COLUMNS

A fundamental determination that has to be made in the process of computing the amplification factors for columns is whether the column is part of a braced or an unbraced structure. The magnitude of the vertical load required to induce buckling when sway is possible is typically much smaller than when sway is restrained.

Conceptually, a column is considered braced if its buckling mode shape does not involve translation of the end points. It is important to recognize that a structure may be braced even though *diagonal bracing* is not provided. For example, if the frame in Fig. 53.2(a) is assumed elastic and the vertical loading is progressively increased, columns AB and DE will eventually buckle without joint translation. Wall CF provides the necessary bracing in this case.

Figure 53.2 *Braced and Unbraced Frames*

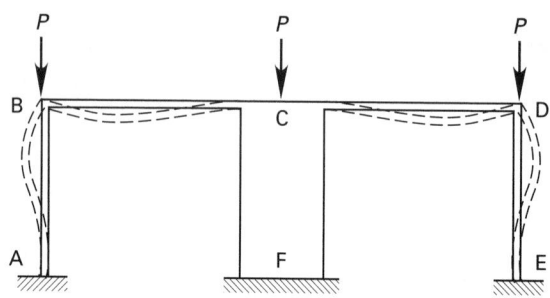

(a) frame braced by wall

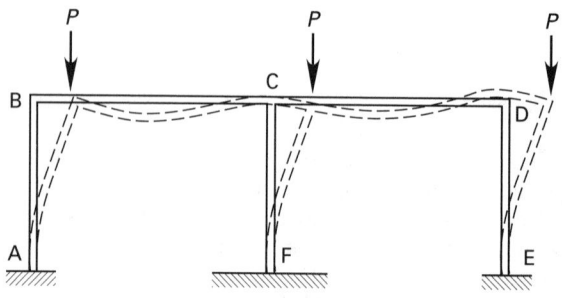

(b) unbraced frame

When the wall is replaced by a column, however, buckling takes place with lateral translation of the joints so the structure must be considered unbraced, as in Fig. 53.2(b). Note that because floor systems are essentially rigid against distortions in their own plane, it is not even necessary for a wall to be located in the

same plane of a column to provide adequate bracing. The only requirement is that the arrangement of walls be such that significant torsional stiffness is provided so that the slab cannot rotate about a wall while a certain frame buckles.

There are instances where inspection is not sufficient to establish the distinction between braced and unbraced structures. For example, as the size of the wall CF in Fig. 53.2 is reduced, the condition passes from braced to unbraced. It is difficult to determine when the transition occurs. ACI 318 provides the following two quantitative methods to distinguish between braced and unbraced conditions. In particular, columns in any given level of a building can be treated as braced when either one of the following two criteria is satisfied.

1. The ratio of the moment from a second-order analysis to the first-order moment is less than or equal to 1.05.

2. The *stability coefficient, Q,* is less than or equal to 0.05, where Q is given by [ACI 318 10.11.4.2].

$$Q = \frac{\sum P_u \Delta_o}{V_{us} l_c} \qquad 53.1$$

$\sum P_u$ is the sum of the axial forces in the columns at the level in question, Δ_o is the *story drift* (the relative displacement between the floor and the roof of the story), V_{us} is the shear force in the story, and l_c is the column height measured from centerline-to-centerline of the joints (essentially the same as the story height).

From a practical standpoint, Eq. 53.1 is the simplest criterion since it can be checked without the need to perform a second-order analysis of the structure.

3. EFFECTIVE LENGTH

The buckling load of a column is dependent on how the ends are supported. The end-restraint conditions affect the column's *effective length*. The effective length is calculated as the product of an effective length factor, k, and the actual length of the column.

$$\text{effective length} = kl \qquad 53.2$$

The *effective length factor* depends on the relative stiffness of the columns and beams at the ends of the column and also on whether the column is part of a braced or an unbraced frame. ACI 318 uses the symbol k to identify the effective length factor for both braced and unbraced conditions. However, it is clearer to use k_b for the braced condition and k_u for unbraced condition.

A convenient way to obtain the effective length factor is by using a *Jackson and Moreland alignment chart,* Fig. 53.3. The effective length is obtained by connecting the values of the relative stiffness parameter, Ψ, for each end of the column.

$$\Psi = \frac{\sum_{\text{columns}} \dfrac{EI}{l_c}}{\sum_{\text{beams}} \dfrac{EI}{l_b}} \qquad 53.3$$

Figure 53.3 *Effective Length Factors (Jackson and Moreland chart)*

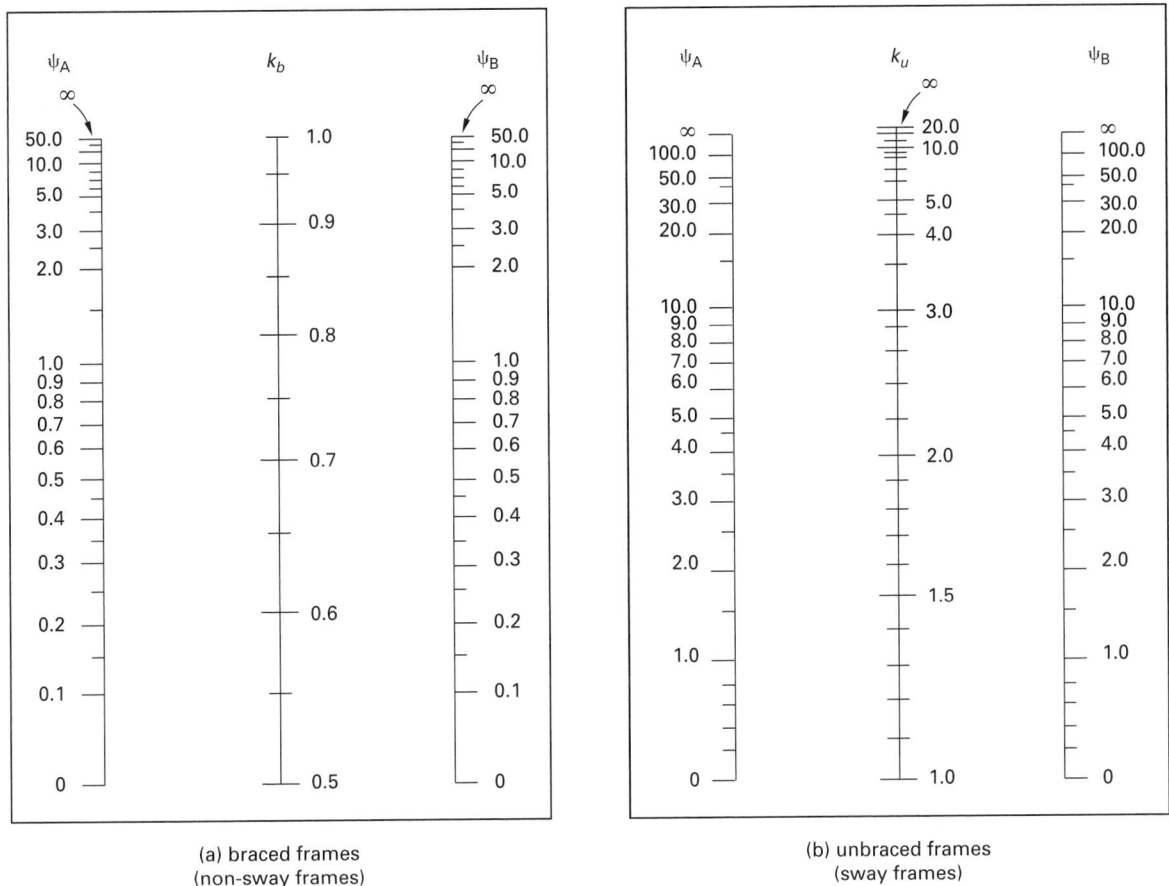

(a) braced frames
(non-sway frames)

(b) unbraced frames
(sway frames)

Reprinted by permission of the American Concrete Institute from *ACI Publication 318-95*, Fig. R10.12.1, copyright © 1995. Unchanged in *ACI Publication 318-05*.

4. SECTION PROPERTIES (SECOND ORDER ANALYSIS)

Cracks and other factors will reduce member strength and increase deformations. Since second-order moments depend on the actual deformations, it is important to use section properties that reflect the influence of cracking and the effect of the duration of the loads on the stiffness of the structure. ACI 318 10.11.1 requires an analysis of such properties, but it permits use of the following values in the absence of a more detailed analysis. (Values for walls are also provided in this section of the ACI.)

modulus of elasticity of concrete	$57{,}000\sqrt{f_c'}$
area	A_g
moments of inertia of columns	$0.70I_g$
moments of inertia of beams	$0.35I_g$
moments of inertia of plates and slabs	$0.25I_g$

I_g for T-beams and one-way beam-slab systems is computed using the effective flange width shown in Fig. 50.9 [ACI 318 8.10]. I_g for two-way beam-slab systems is computed using the effective flange width shown in Fig. 53.4 [ACI 318 13.2.4].

For the unusual situation where a building is subjected to sustained lateral loads (e.g., from earth pressure) the moments of inertia are divided by the factor $1+\beta_d$. β_d is defined as the ratio of the maximum factored sustained shear within a story to the total factored shear in that story.

5. EFFECTIVE FLEXURAL STIFFNESS: BRACED FRAMES

To evaluate the buckling load of a column, the effective length (kl_u) and the *flexural stiffness EI* are needed. This last term must include possible reductions due to cracking, nonlinear effects in the stress-strain curve and, in the case of sustained loads, the increase in curvatures that occurs as a result of creep deformations. ACI 318 10.12.3 provides two formulas that provide conservative estimates of the flexural stiffness. Equation 53.5 is typically used since it can be evaluated without knowing the steel reinforcement.

$$EI = \frac{0.2E_cI_g + E_sI_{se}}{1 + \beta_d} \quad \text{[braced frames]} \qquad 53.4$$

$$EI = \frac{0.4E_cI_g}{1 + \beta_d} \quad \text{[braced frames]} \qquad 53.5$$

In Eq. 53.4, I_{se} is the moment of inertia of the steel bars in a column cross section, calculated as the sum of the products of areas times the square of the distances to the centroidal axis.

Figure 53.4 *Two-Way Slab Sections Considered as Parts of Beams (ACI 318 13.2.4)*

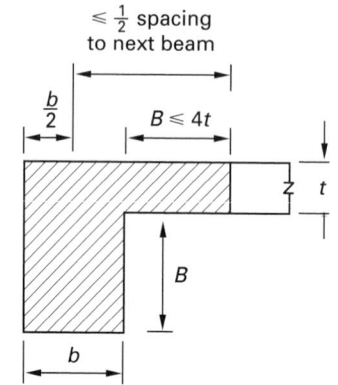

(a) exterior section

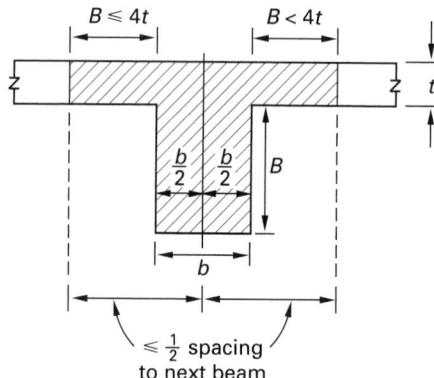

(b) interior section

There are two cases for evaluating β_d: (a) When the internal forces are from gravity loads only, β_d is the ratio of the permanent fraction of factored axial load to the total factored axial load. (b) When the internal forces are derived from gravity and lateral loads, β_d is the ratio of the permanent fraction of factored lateral load to the total factored lateral load. Thus, for load combinations that include wind or earthquake loads, $\beta_d = 0$.

6. BUCKLING LOAD

The column *buckling load* for a slender column is given by the Euler formula. The values of flexural stiffness and effective length previously determined in Secs. 3 and 5 must be used.

$$P_c = \frac{\pi^2 EI}{(kl_u)^2} \qquad 53.6$$

7. COLUMNS IN BRACED STRUCTURES (NON-SWAY FRAMES)

When the columns in a story are braced, the design is carried out using a factored axial load obtained from a first-order analysis and for a moment M_c given by Eq. 53.7. M_2 is the larger factored end moment from a first-order solution.

$$M_c = \delta_{ns} M_2 \qquad 53.7$$

The amplification factor given by ACI 318 10.12.3 is

$$\delta_{ns} = \frac{C_m}{1 - \dfrac{P_u}{0.75 P_c}} \geq 1.0 \qquad 53.8$$

The buckling load, P_c, is obtained from Eq. 53.6 with $k = k_b$ and EI given by either Eq. 53.4 or 53.5.

The parameter C_m is used to account for the shape of the first-order moment diagram. For columns with transverse loads between the endpoints, $C_m = 1$. For columns without transverse loads between the endpoints, Eq. 53.9 can be used. The quantity M_1/M_2 is negative when the column is bent in reverse curvature. (When $M_1 = M_2 = 0$, use $M_1/M_2 = 1$ [ACI 318 10.12.3.1].)

$$C_m = 0.6 + 0.4\left(\frac{M_1}{M_2}\right) \geq 0.4 \qquad 53.9$$

ACI 318 10.12.3.2 provides a minimum value for the moment M_2, as given by Eq. 53.10.

$$M_{2,\text{min,in-lbf}} = P_{u,\text{lbf}}(0.6 + 0.03 h_{\text{in}}) \qquad 53.10$$

ACI 318 10.12.2 specifies the conditions necessary to disregard slenderness effects in non-sway frames.

8. COLUMNS IN UNBRACED STRUCTURES (SWAY FRAMES)

Columns in unbraced structures should be designed for the axial load computed from a first-order analysis and for the end moments M_1 and M_2, which are computed as

$$M_1 = M_{1,ns} + \delta_s M_{1s} \qquad 53.11$$
$$M_2 = M_{2,ns} + \delta_s M_{2s} \qquad 53.12$$

The amplification factor, δ_s, is given by

$$\delta_s = \frac{1}{1 - \dfrac{\sum P_u}{0.75 \sum P_c}} \geq 1.0 \qquad 53.13$$

Alternatively, if $Q \leq {}^1/_3$,

$$\delta_s = \frac{1}{1 - Q} \geq 1 \qquad 53.14$$

The amplification factor, δ_s, is not computed for a single given column but rather for all the columns in a given story. This is because all of the columns in a story must buckle with sway if any one does.

The value of each one of the buckling loads, P_c in Eq. 53.13, is obtained using Eq. 53.6 with the effective length factor taken as k_u.

While Eqs. 53.11 and 53.12 give the amplified moments at the column ends in an unbraced structure, the moment may be even larger at some point along the height of the column. In fact, the moments within the column height cannot be obtained from the moments at the ends alone. The deflection of the column with respect to its chord affects the moments within the length.

ACI 318 10.13.5 indicates when further moment checking is necessary. When the condition of Eq. 53.15 is satisfied, an explicit check on other moments is necessary. In that case, the moment within the length is computed by assuming that the column is braced and has the end moments given by Eqs. 53.11 and 53.12.

$$\frac{l_u}{r} \geq \frac{35}{\sqrt{\dfrac{P_u}{f_c' A_g}}} \qquad 53.15$$

For circular columns, $r = d/4$. For rectangular columns, $r = 0.289h$. ACI 318 10.11.2 permits use of the approximation $r = 0.3h$ for rectangular columns.

To ensure that the safety margin against global instability is adequate in unbraced frames, ACI 318 requires that the maximum amplification factor, δ_s (computed for the factored dead and live loads corresponding to $U = 1.2D + 1.6L$) be less than or equal to 2.5 [ACI 318 10.13.6]. In checking stability under a maximum vertical load, the value of β_d is taken as the ratio of the axial force for the factored, sustained axial load (usually dead load) to the total factored axial load.

Example 53.1

A nine-floor (nine-story) reinforced concrete building is supported on 20 in by 20 in columns located on a 20 ft by 20 ft rectangular grid. The stories are 12 ft in height with the exception of the first story, which is 18 ft high. The first story column height can be taken as 17 ft. The effective length factor, k_u, for column B2 is 1.5. The slab thickness is 6 in. The actions in column B2 for dead load, live load, and wind loading in the N-S direction are as shown. The moments from the live load shown are obtained by positioning the live load on the panels located to the north of column B2 only. Moments in the E-W direction (about the y-y axis) are negligible.

To account for the low probability of all stories above being fully loaded, the live load over all the stories above is assumed to be 40% of the nominal value for the purpose of computing the amplification δ_s at level 1. To account for the fact that their effective lengths are longer as a result of having only one beam framing into their joints in the direction of the analysis, the critical load of the columns along axes 1-1 and 3-3 is assumed to be

0.75 times the critical load of the columns along axis 2-2. The average dead load per floor is 150 lbf/ft², and the basic (unreduced) floor live load is 50 lbf/ft².

Compute the (a) design axial load and (b) moment in the first level for column B2. Consider the load combination $U = 0.75(1.4D + 1.7L + 1.7W)$ for ACI 318 App. C solutions, and $U = 1.2D + 0.5L + 1.6W$ for ACI 318 Ch. 9 solutions.

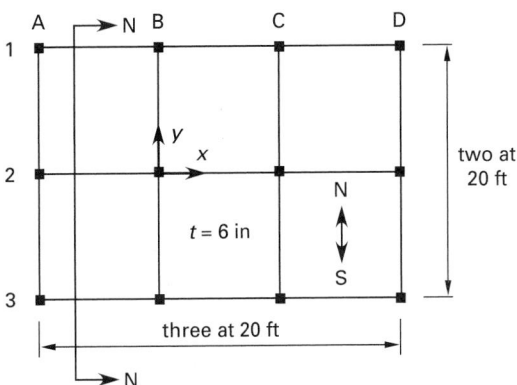

(a) column plan view

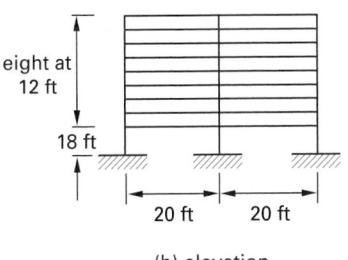

(b) elevation

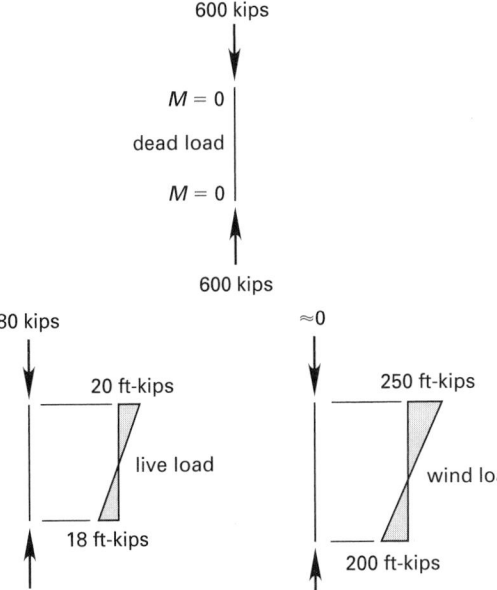

(c) dead, live, and wind load
(reductions considered)

Determine the dead loading.

The dead load of each 20 in × 20 in column is approximately

$$D_{\text{column}} = \frac{(20 \text{ in})(20 \text{ in})\left(150 \dfrac{\text{lbf}}{\text{ft}^3}\right)(8 \text{ stories}) \times \left(12 \dfrac{\text{ft}}{\text{story}}\right)}{\left(144 \dfrac{\text{in}^2}{\text{ft}^2}\right)\left(1000 \dfrac{\text{lbf}}{\text{kip}}\right)}$$

$$= 40 \text{ kips}$$

Including the roof, the floor dead load is

$$D_{\text{floor}} = \frac{(20 \text{ ft})(20 \text{ ft})\left(150 \dfrac{\text{lbf}}{\text{ft}^2}\right)(9 \text{ floors})}{1000 \dfrac{\text{lbf}}{\text{kip}}}$$

$$= 540 \text{ kips}$$

The total dead load per column is

$$D = D_{\text{column}} + D_{\text{floor}} = 40 \text{ kips} + 540 \text{ kips}$$

$$= 580 \text{ kips} \quad (600 \text{ kips})$$

Determine the live loading.

$$L = \frac{(60 \text{ ft})(40 \text{ ft})\left(50 \dfrac{\text{lbf}}{\text{ft}^2}\right)(8 \text{ stories})}{(12 \text{ columns})\left(1000 \dfrac{\text{lbf}}{\text{kip}}\right)}$$

$$= 80 \text{ kips}$$

Solution Using ACI 318 App. C

(a) The ACI 318 App. C design axial force is

$$P_u = (0.75)\big((1.4)(600 \text{ kips}) + (1.7)(80 \text{ kips})\big)$$

$$= 732.0 \text{ kips}$$

(b) The amplification factor δ_s is given by Eq. 53.13. $\sum P_u$ is the total factored load above the level in question. Taking into consideration the reduction of the live load,

$$\sum P_u = (0.75)\left((1.4)\left(0.15 \dfrac{\text{kip}}{\text{ft}^2}\right)\right.$$

$$+ (1.7)(0.4)\left(0.05 \dfrac{\text{kip}}{\text{ft}^2}\right)\bigg)$$

$$\times \big((9)(60 \text{ ft})(40 \text{ ft})\big)$$

$$= 3952.8 \text{ kips}$$

$\sum P_c$ is the sum of the buckling loads of all the columns in the level. Compute this for the columns along axis 2-2.

$$E_c = \frac{57{,}000\sqrt{4000 \dfrac{\text{lbf}}{\text{in}^2}}}{1000 \dfrac{\text{lbf}}{\text{kip}}}$$

$$= 3605 \text{ kips/in}^2$$

$$I_g = \frac{bh^3}{12} = \frac{(20 \text{ in})(20 \text{ in})^3}{12}$$

$$= 13{,}333 \text{ in}^4$$

$$\beta_d = 0 \quad [\text{sway frame, no sustained lateral load}]$$

$$EI = \frac{0.4E_c I_g}{1 + \beta_d}$$

$$= (0.4)\left(3605 \dfrac{\text{kips}}{\text{in}^2}\right)(13{,}333 \text{ in}^4)$$

$$= 19.23 \times 10^6 \text{ kip-in}^2$$

Substitute into Eq. 53.6.

$$P_c = \frac{\pi^2 EI}{(kl_u)^2}$$

$$= \frac{\pi^2(19.23 \times 10^6 \text{ kip-in}^2)}{\left((1.5)(17 \text{ ft})\left(12 \dfrac{\text{in}}{\text{ft}}\right)\right)^2}$$

$$= 2024 \text{ kips}$$

Taking the reduction in axial loading into account,

$$\sum P_c = \big((0.75)(8) + 4\big)(2024 \text{ kips}) = 20{,}240 \text{ kips}$$

The amplification factor is computed from Eq. 53.13.

$$\delta_s = \frac{1}{1 - \dfrac{\sum P_u}{0.75 \sum P_c}}$$

$$= \frac{1}{1 - \dfrac{3952.8 \text{ kips}}{(0.75)(20{,}240 \text{ kips})}}$$

$$= 1.35$$

$$M_{1,\text{ns}} = (0.75)(1.7)(18 \text{ ft-kips}) = 23.0 \text{ ft-kips}$$

$$M_{2,\text{ns}} = (0.75)(1.7)(20 \text{ ft-kips}) = 25.5 \text{ ft-kips}$$

$$M_{1,s} = (0.75)(1.7)(200 \text{ ft-kips}) = 255.0 \text{ ft-kips}$$

$$M_{2,s} = (0.75)(1.7)(250 \text{ ft-kips}) = 318.8 \text{ ft-kips}$$

Using Eqs. 53.11 and 53.12, the amplified moments at the column ends are

$$M_1 = M_{1,\text{ns}} + \delta_s M_{1,s}$$

$$= 23 \text{ ft-kips} + (1.35)(255.0 \text{ ft-kips})$$

$$= 367 \text{ ft-kips}$$

$$M_2 = M_{2,\text{ns}} + \delta_s M_{2,s}$$

$$= 25.5 \text{ ft-kips} + (1.35)(318.8 \text{ ft-kips})$$

$$= 456 \text{ ft-kips}$$

Solution Using ACI 318 Ch. 9

(a) The ACI 318 Ch. 9 design axial force is

$$P_u = (1.2)(600 \text{ kips}) + (0.5)(80 \text{ kips})$$
$$= 760.0 \text{ kips}$$

(b) The amplification factor δ_s is given by Eq. 53.13. $\sum P_u$ is the total factored load above the level in question. There are nine areas that contribute to loading. Taking into consideration the reduction of the live load,

$$\sum P_u = \left((1.2)\left(0.15 \frac{\text{kip}}{\text{ft}^2} \right) + (0.5)(0.4)\left(0.05 \frac{\text{kip}}{\text{ft}^2} \right) \right)$$
$$\times (9)(60 \text{ ft})(40 \text{ ft})$$
$$= 4104.0 \text{ kips}$$

$\sum P_c$ is the sum of the buckling loads of all the columns in the level. Compute this for the columns along axis 2-2.

$$E_c = \frac{57,000\sqrt{4000 \dfrac{\text{lbf}}{\text{in}^2}}}{1000 \dfrac{\text{lbf}}{\text{kip}}}$$
$$= 3605 \text{ kips/in}^2$$

$$I_g = \frac{bh^3}{12} = \frac{(20 \text{ in})(20 \text{ in})^3}{12}$$
$$= 13,333 \text{ in}^4$$

$\beta_d = 0$ [sway frame, no sustained lateral load]

$$EI = \frac{0.4 E_c I_g}{1 + \beta_d}$$
$$= (0.4)\left(3605 \frac{\text{kips}}{\text{in}^2} \right)(13,333 \text{ in}^4)$$
$$= 19.23 \times 10^6 \text{ kip-in}^2$$

Substitute into Eq. 53.6.

$$P_c = \frac{\pi^2 EI}{(kl_u)^2}$$
$$= \frac{\pi^2(19.23 \times 10^6 \text{ kip-in}^2)}{\left((1.5)(17 \text{ ft})\left(12 \frac{\text{in}}{\text{ft}} \right) \right)^2}$$
$$= 2024 \text{ kips}$$

Taking the reduction in axial loading into account,

$$\sum P_c = \big((0.75)(8) + 4 \big)(2024 \text{ kips})$$
$$= 20,240 \text{ kips}$$

The amplification factor is computed from Eq. 53.13.

$$\delta_s = \frac{1}{1 - \dfrac{\sum P_u}{0.75 \sum P_c}}$$
$$= \frac{1}{1 - \dfrac{4104.0 \text{ kips}}{(0.75)(20,240 \text{ kips})}}$$
$$= 1.37$$

$$M_{1,ns} = (0.5)(18 \text{ ft-kips}) = 9.0 \text{ ft-kips}$$
$$M_{2,ns} = (0.5)(20 \text{ ft-kips}) = 10.0 \text{ ft-kips}$$
$$M_{1,s} = (1.6)(200 \text{ ft-kips}) = 320.0 \text{ ft-kips}$$
$$M_{2,s} = (1.6)(250 \text{ ft-kips}) = 400.0 \text{ ft-kips}$$

Using Eqs. 53.11 and 53.12, the amplified moments at the column ends are

$$M_1 = M_{1,ns} + \delta_s M_{1,s}$$
$$= 9 \text{ ft-kips} + (1.37)(320.0 \text{ ft-kips})$$
$$= 447 \text{ ft-kips}$$
$$M_2 = M_{2,ns} + \delta_s M_{2,s}$$
$$= 10.0 \text{ ft-kips} + (1.37)(400.0 \text{ ft-kips})$$
$$= 558 \text{ ft-kips}$$

Structural

54 Reinforced Concrete: Walls and Retaining Walls

Nomenclature

A	area	ft^2	m^2
b	batter decrement	ft/ft	m/m
B	base length	ft	m
d	depth (to reinforcement)	ft	m
d	diameter of bar	ft	m
e	eccentricity	ft	m
f'_c	compressive strength	lbf/ft^2	Pa
f_y	yield strength	lbf/ft^2	Pa
F	force	lbf	N
FS	factor of safety	–	–
h	overall thickness of wall	ft	m
H	wall height	ft	m
k	effective length factor	–	–
l_c	vertical distance between sections	ft	m
l_u	unsupported clear height	ft	m
M	moment	ft-lbf	N·m
P_u	factored axial load at given eccentricity	lbf	N
r	radius of gyration	ft	m
R	coefficient of resistance	lbf/ft^2	Pa
R	earth pressure resultant	lbf	N
t	thickness	ft	m
V	shear	lbf	N
w	uniform distributed load	lbf/ft	N/m
w	width	ft	m
y	vertical distance	ft	m

Symbols

γ	specific weight	lbf/ft^3	–
λ	distance from centroid of compressed area to extreme compression fiber	in	m
ρ	reinforcement ratio	–	–
ϕ	strength reduction factor	–	–

Subscripts

a	active
b	base or bar
c	compressive or concrete
f	friction
g	gross
h	horizontal
n	nominal
OT	overturning
p	passive
s	steel
SL	sliding
u	ultimate or unsupported
w	wall
y	yield

1. NONBEARING WALLS

Nonbearing walls support their own weight and, occasionally, lateral wind and seismic loads. The following minimum details are specified by ACI 318 for designing nonbearing walls.

Minimum vertical reinforcement [ACI 318 14.3.2]

1. 0.0012 times the gross concrete area for deformed no. 5 bars or smaller and $f_y \geq 60{,}000$ psi

2. 0.0015 times the gross concrete area for other deformed bars

3. 0.0012 times gross concrete area for smooth or deformed welded wire reinforcement not larger than W31 or D31

Minimum horizontal reinforcement [ACI 318 14.3.3]

1. 0.0020 times the gross concrete area for deformed no. 5 bars or smaller and $f_y \geq 60{,}000$ psi

2. 0.0025 times the gross concrete area for other deformed bars

3. 0.0020 times the gross concrete area for smooth or deformed welded wire reinforcement not larger than W31 or D31

Number of reinforcing layers [ACI 318 14.3.4]

1. Wall thickness > 10 in: two layers; one layer containing from $1/2$ to $2/3$ of the total steel placed not less than 2 in and not more than $1/3h$ from the exterior surface; the other layer placed at a distance not less than $3/4$ in and no more than $1/3h$ from the interior surface.

2. Wall thickness $\leq$ 10 in: ACI 318 does not specify two layers of reinforcement.

Spacing of vertical and horizontal reinforcement [ACI 318 14.3.5]

The spacing of vertical and horizontal reinforcement may not exceed three times the wall thickness or 18 in.

Need for ties [ACI 318 14.3.6]

The vertical reinforcement does not have to be enclosed by ties unless (a) the vertical reinforcement is greater than 0.01 times the gross concrete area, or (b) where the vertical reinforcement is not required as compression reinforcing.

Thickness [ACI 318 14.6.1]

Thickness cannot be less than 4 in or $^1/_{30}$ times the least distance between members that provide lateral support.

2. BEARING WALLS: EMPIRICAL METHOD

The majority of concrete walls in buildings are *bearing walls* (*load-carrying walls*). In addition to their own weights, bearing walls support vertical and lateral loads. There are two methods for designing bearing walls. They may be designed (a) by the ACI empirical design method, (b) by the ACI alternate design method, or (c) as compression members using the strength design provisions for flexure and axial loads. The minimum design requirements for nonbearing walls must also be satisfied for bearing walls.

The empirical method may be used only if the resultant of all factored loads falls within the middle third of the wall thickness. This is the case with short vertical walls with approximately concentric loads. ACI 318 14.5.2 gives the empirical design equation as

$$P_u \leq \phi P_{n,w} \leq 0.55 \phi f'_c A_g \left(1 - \left(\frac{k l_c}{32 h} \right)^2 \right) \quad \textbf{54.1}$$

The strength reduction factor, ϕ, is that for compression-controlled sections. The *effective length factor*, k, depends on the end conditions of the wall. According to ACI 318 14.5.2, $k = 0.8$ for walls braced at the top and bottom against lateral translation and restrained against rotation at the top and/or bottom. $k = 1.0$ for walls braced at the top and bottom against lateral translation and not restrained against rotation at either end. $k = 2.0$ for walls not braced against lateral translation.

Use of the ACI empirical equation is further limited to the following conditions. The wall thickness, h, must not be less than $^1/_{25}$ times the supported length or height, whichever is shorter, nor may it be less than 4 in. Exterior basement walls and foundation walls must be at least 7.5 in thick. The following additional provisions apply to all walls designed by ACI 318 Ch. 14. (a) To be considered effective for beam reaction or other concentrated load, the length of the wall must not exceed the center-to-center distance between reactions or the width of bearing plus four times the wall thickness [ACI 318 14.2.4]. (b) The wall must be anchored to the floors or to columns and other structural elements of the building [ACI 318 14.2.6].

3. BEARING WALLS: STRENGTH DESIGN METHOD

If the wall has a nonrectangular cross section (as in ribbed wall panels) or if the eccentricity of the force resultant is greater than one sixth of the wall thickness, the wall must be designed as a column subject to axial loading and bending.

4. SHEAR WALLS

Shear walls are designed to resist lateral wind and seismic loads. Reinforced concrete walls have high in-plane stiffness, making them suitable for resisting lateral forces. The ductility provided by the reinforcing steel ensures a ductile-type failure. Ductile failure is the desirable failure mode because it gives warning and enough time for the occupants of the building to escape prior to total collapse.

5. DESIGN OF RETAINING WALLS

Retaining walls are essentially vertical cantilever beams, with the additional complexities of nonuniform loading and soil-to-concrete contact. The wall is designed to resist moments from the lateral earth pressure. The following procedure can be used to design the structural reinforcement.

step 1: Gather information needed to design the wall.

H	wall height
B	base length
t_b	thickness of base
b	batter decrement (change in stem width per unit height)
$R_{a,h}$	horizontal active earth pressure resultant per unit wall width
y_a	height of the active earth pressure resultant
$R_{a,v}$	vertical active earth pressure resultant per unit wall width

Figure 54.1 *Retaining Wall Dimensions (step 1)*

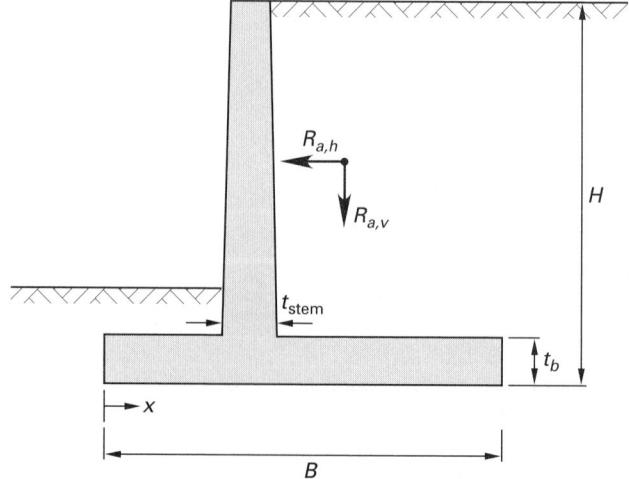

step 2: Select a reinforcement ratio to control deflections.

$$\rho \approx 0.02 \qquad 54.2$$

step 3: Calculate a trial coefficient of resistance for the stem.

$$R_{u,\text{trial}} = \rho f_y \left(1 - \left(\frac{\rho f_y}{(2)(0.85)f_c'} \right) \right) \qquad 54.3$$

step 4: Calculate the factored moment, $M_{u,\text{stem}}$, at the base of the stem per unit width, w, from the active earth pressure resultant. Include forces from surcharge loads, but disregard reductions from passive distributions.

Per ACI 318 App. C,

$$M_{u,\text{stem}} = 1.7 R_{a,h} y_a \qquad 54.4(a)$$

Per ACI 318 Ch. 9,

$$M_{u,\text{stem}} = 1.6 R_{a,h} y_a \qquad 54.4(b)$$

Figure 54.2 *Base of Stem Details (step 4)*

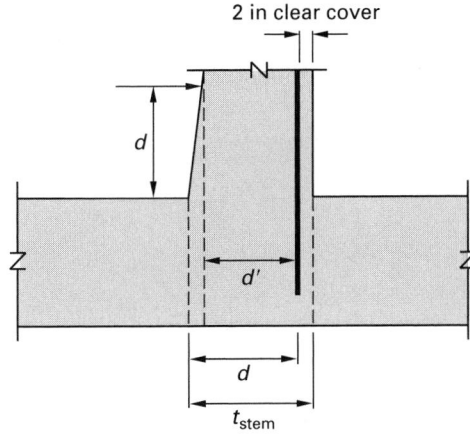

2 in clear cover

step 5: Calculate the required stem thickness at the base. Since the ultimate moment is per unit wall width, w will be the width of that unit (e.g., 12 in for moments per foot). $\phi = 0.90$ for flexure. If ρ has been selected arbitrarily, the stem thickness will not be unique.

$$d = \sqrt{\frac{M_{u,\text{stem}}}{\phi R_u w}} \qquad 54.5$$

2 in of cover is required for no. 6 bars or larger when exposed to earth or weather. The total thickness of the stem at the base is

$$t_{\text{stem}} = d + \text{cover} + \tfrac{1}{2} d_b \qquad 54.6$$

step 6: Calculate the factored shear, $V_{u,\text{stem}}$, on the stem per unit width at the critical section.

Include all of the active and surcharge pressure loading from the top of the retaining wall down to the critical section, a distance d from the base.

Per ACI 318 App. C,

$$V_{u,\text{stem}} = 1.7 V_{\text{active}} + 1.7 V_{\text{surcharge}} \qquad 54.7(a)$$

Per ACI 318 Ch. 9,

$$V_{u,\text{stem}} = 1.6 V_{\text{active}} + 1.6 V_{\text{surcharge}} \qquad 54.7(b)$$

step 7: Calculate the unreinforced concrete shear strength for a unit wall width, w. $\phi = 0.85$ for shear at the critical section a distance d from the bottom of the stem per ACI 318 App. C, or $\phi = 0.75$ per ACI 318 Ch. 9. d' is the thickness of the compression portion of the stem at a distance d from the bottom of the stem. (One-way shear is assumed.) $d' = d$ when the batter decrement is zero.

$$\phi V_n = 2 \phi w d' \sqrt{f_c'} \qquad 54.8$$

step 8: If $\phi V_n \geq V_{u,\text{stem}}$, the stem thickness is adequate. Otherwise, either the stem thickness must be increased or higher-strength concrete specified.

Figure 54.3 *Heel Details (step 9)*

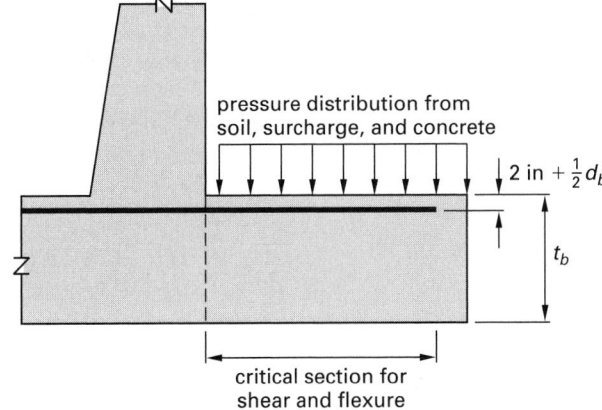

pressure distribution from soil, surcharge, and concrete

2 in + $\frac{1}{2} d_b$

t_b

critical section for shear and flexure

step 9: If the base design is not known, design the base heel as a cantilever beam. The procedure is similar to the previous steps. The critical section for shear checking is located at the face of the stem. (Since tension occurs in the heel, the critical section cannot be assumed to occur a distance d from the face of the stem. The entire heel contributes to V_u [ACI 318 11.1.3]). Include shear for a unit width of base from the vertical active pressure, the surcharge, and the heel's self-weight. Disregard the upward pressure distribution as well as any effect the key has on the pressure distributions.

Per ACI 318 App. C,

$$V_{u,\text{base}} = 1.4V_{\text{soil}} + 1.4V_{\text{heel weight}} + 1.7V_{\text{surcharge}} \quad \text{54.9(a)}$$

Per ACI 318 Ch. 9,

$$V_{u,\text{base}} = 1.2V_{\text{soil}} + 1.2V_{\text{heel weight}} + 1.6V_{\text{surcharge}} \quad \text{54.9(b)}$$

Per ACI 318 App. C,

$$M_{u,\text{base}} = 1.4M_{\text{soil}} + 1.4M_{\text{heel weight}} + 1.7M_{\text{surcharge}} \quad \text{54.10(a)}$$

Per ACI 318 Ch. 9,

$$M_{u,\text{base}} = 1.2M_{\text{soil}} + 1.2M_{\text{heel weight}} + 1.6M_{\text{surcharge}} \quad \text{54.10(b)}$$

Since the base is cast against the earth, a cover of 3 in is required for reinforcing steel. If the base thickness, t_b, is significantly different from what was used to calculate M_u, a second iteration may be necessary.

Complete the heel reinforcement design by calculating the development length. The distance from the face of the stem to the end of the heel reinforcement must equal or exceed the development length. The heel reinforcement must also extend a distance equal to the development length past the stem reinforcement into the toe.

Figure 54.4 *Toe Details (step 10)*

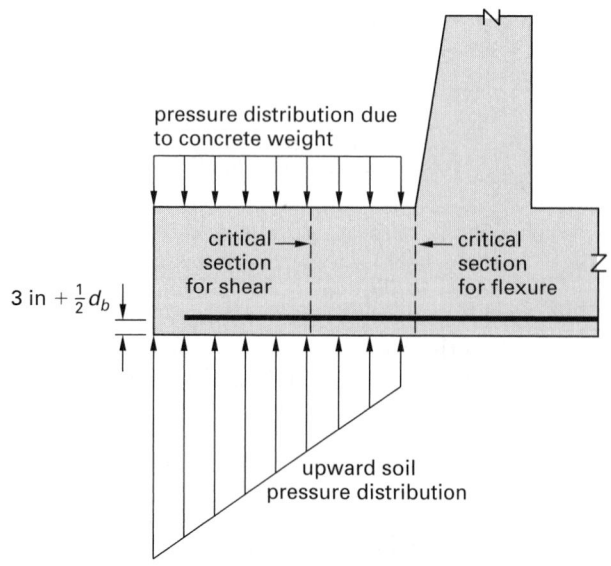

pressure distribution due to concrete weight

critical section for shear

critical section for flexure

$3 \text{ in} + \frac{1}{2}d_b$

upward soil pressure distribution

step 10: The procedure for designing the toe is similar to that for the heel. The toe loading is assumed to be caused by the toe self-weight and the upward soil pressure distribution. The passive soil loading is disregarded. For one-way shear, the critical section is located a distance d from the outer face of the stem. For flexure, the critical section is at the outer face of the stem. (ACI 318 C.2.3 specifies a factor of 1.4 for concrete and soil dead loads, and the factor is 1.7 for earth pressure. ACI 318 9.2.1 specifies a factor of 1.2 for concrete and soil dead loads, and the factor is 1.6 for earth pressure. The toe soil pressure distribution is the result of the horizontal active earth pressure, so a factor of 1.7 or 1.6 is used.) It is not appropriate to use a 0.9 factor for dead loads that counteract live loads, as the toe dead load will never be absent. It is likely that shear will determine the toe thickness.

Per ACI 318 App. C,

$$V_{u,\text{toe}} = 1.7V_{\text{toe pressure}} - 1.4V_{\text{toe weight}} \quad \text{54.11(a)}$$

Per ACI 318 Ch. 9,

$$V_{u,\text{toe}} = 1.6V_{\text{toe pressure}} - 1.2V_{\text{toe weight}} \quad \text{54.11(b)}$$

Per ACI 318 App. C,

$$M_{u,\text{toe}} = 1.7M_{\text{toe pressure}} - 1.4M_{\text{toe weight}} \quad \text{54.12(a)}$$

Per ACI 318 Ch. 9,

$$M_{u,\text{toe}} = 1.6M_{\text{toe pressure}} - 1.2M_{\text{toe weight}} \quad \text{54.12(b)}$$

An alternate interpretation, less conservative, is that the upward soil distribution is the result of two live loads, the active earth pressure, and the counteracting toe weight. This would result in forms of Eqs. 54.11 and 54.12 that multiplied both terms by 1.7 or 1.6.

step 11: Calculate the reinforcement ratio, ρ, and the steel area, A_s. Select reinforcement to meet distribution requirements.

$$M_n = \frac{M_u}{\phi} = \rho f_y \left(1 - \frac{\rho f_y}{(2)(0.85)f'_c}\right) bd^2 \quad \text{54.13}$$

step 12: As the stem goes up, less reinforcement is required. Similar calculations should be performed at one or two other heights above the base to obtain a curve of M_u versus height. Select reinforcement to meet the moment requirements.

step 13: Check the development length of the stem reinforcement. The vertical stem height should be adequate without checking. However, it may be difficult to achieve full development in the base. The reinforcement can be extended into the key, if present. Otherwise hooks, bends, or smaller bars should be used.

step 14: Calculate the horizontal temperature and shrinkage reinforcement per unit width from the average stem thickness and the reinforcement ratios in ACI 318 14.3.3. ACI 318 14.3.4 requires that walls greater than 10 in thick have at least $^1/_2$ and not more than $^2/_3$ of this reinforcement placed in a layer on the exposed side. Allocate two thirds of this reinforcement to the outside stem face, since that surface is alternately exposed to day and night temperature extremes. The remaining one third should be located on the soil side of the stem, which is maintained at a more constant temperature by the insulating soil. Refer to ACI 318 14.3.4 for specific face cover distances.

step 15: If the design assumed a drained backfill, provide for drainage behind the wall. Weep holes should be placed regularly.

Example 54.1

The cantilever retaining wall shown supports 17 ft of earth above the toe of a cut. The backfill is level and has a unit weight of 100 lbf/ft³. The angle of internal friction is 35°. The allowable soil pressure under service loading is 5000 lbf/ft². The coefficient of friction between the concrete and soil is 0.60. 3500 psi concrete and 60,000 psi steel are used. (a) What is the equivalent fluid density behind the wall? (b) What is the active earth pressure resultant? (c) What is the overturning moment? (d) What is the resisting moment? (e) What is the factor of safety against overturning? (f) What is the factor of safety against sliding? (g) Is any part of the heel soil in tension? (h) What is the maximum pressure below the toe? (i) Specify the stem reinforcement. (j) Investigate the development length of the stem bars in the base. (k) Design the temperature steel for the stem. (l) Design the bending steel for the heel.

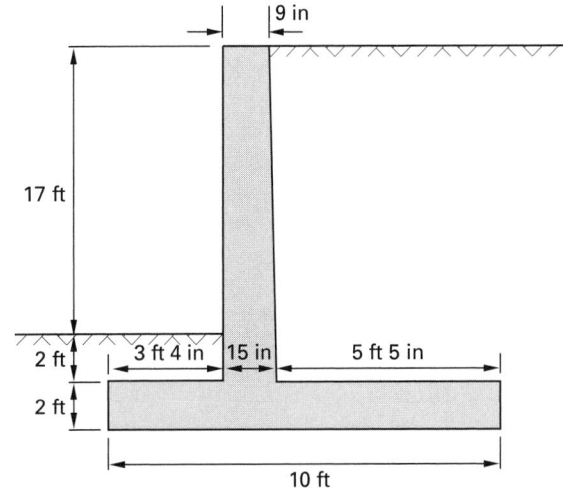

Solution

(a) The active earth pressure coefficient is

$$k_a = \frac{1 - \sin\phi}{1 + \sin\phi} = \frac{1 - \sin 35°}{1 + \sin 35°}$$
$$= 0.271$$
$$\gamma_{\text{equivalent}} = k_a \gamma_{\text{soil}}$$
$$= (0.271)\left(100 \; \frac{\text{lbf}}{\text{ft}^3}\right)$$
$$= 27.1 \; \text{lbf/ft}^3$$

(b) The active height is

$$H = 17 \text{ ft} + 2 \text{ ft} + 2 \text{ ft} = 21 \text{ ft}$$
$$R_a = \tfrac{1}{2} k_a \gamma H^2$$
$$= \frac{(0.5)(0.271)\left(100 \; \dfrac{\text{lbf}}{\text{ft}^3}\right)(21 \text{ ft})^2}{1000 \; \dfrac{\text{lbf}}{\text{kip}}}$$
$$= 5.976 \text{ kips/ft}$$

(c) The overturning moment from the active resultant is

$$M = R_a y_a = R_a \left(\frac{H}{3}\right)$$
$$= \left(5.976 \; \frac{\text{kips}}{\text{ft}}\right)\left(\frac{21 \text{ ft}}{3}\right)$$
$$= 41.83 \text{ ft-kips/ft}$$

(d) Determine the weights per unit width (1 ft) and their moment arms (from the toe). Area 1 is the soil above the heel.

$$W_1 = \frac{\left(5\frac{5}{12} \text{ ft}\right)(19 \text{ ft})\left(100 \frac{\text{lbf}}{\text{ft}^3}\right)}{1000 \frac{\text{lbf}}{\text{kip}}}$$

$$= 10.29 \text{ kips/ft}$$

$$x_1 = 10 \text{ ft} - \left(\frac{1}{2}\right)\left(5\frac{5}{12}\right) \text{ ft} = 7.29 \text{ ft}$$

$$M_1 = W_1 x_1 = \left(10.29 \frac{\text{kips}}{\text{ft}}\right)(7.29 \text{ ft})$$

$$= 75.02 \text{ ft-kips/ft}$$

The other weights are found similarly.

area	weight (kips/ft)	moment arm (ft)	moment (ft-kips/ft)
1: heel soil	10.29	7.29	75.02
2: batter soil	0.48	4.42	2.12
3: batter stem	0.72	4.25	3.06
4: 9 in stem	2.14	3.71	7.92
5: base	3.0	5.00	15.0
totals	16.63 $\frac{\text{kips}}{\text{ft}}$		103.12 $\frac{\text{ft-kips}}{\text{ft}}$

(e) The factor of safety against overturning is

$$\text{FS}_{\text{OT}} = \frac{103.12 \dfrac{\text{ft-kips}}{\text{ft}}}{41.83 \dfrac{\text{ft-kips}}{\text{ft}}}$$

$$= 2.46 \quad [> 1.5, \text{ so OK}]$$

(f) The friction force resisting sliding is

$$R_{\text{SL}} = f\sum W_i = (0.60)\left(16.63 \frac{\text{kips}}{\text{ft}}\right)$$

$$= 9.98 \text{ kips/ft}$$

$$\text{FS}_{\text{SL}} = \frac{9.98 \dfrac{\text{kips}}{\text{ft}}}{5.976 \dfrac{\text{kips}}{\text{ft}}}$$

$$= 1.67 \quad [> 1.5, \text{ so OK}]$$

(g) Locate the resultant from the toe.

$$x_R = \frac{M_R - M_{\text{OT}}}{\sum W_i}$$

$$= \frac{103.12 \dfrac{\text{ft-kips}}{\text{ft}} - 41.83 \dfrac{\text{ft-kips}}{\text{ft}}}{16.63 \dfrac{\text{kips}}{\text{ft}}}$$

$$= 3.68 \text{ ft}$$

Since the resultant acts in the middle third, there is no uplift on the footing. The heel soil is not in tension.

(h) The eccentricity is

$$\epsilon = \frac{10 \text{ ft}}{2} - 3.68 \text{ ft} = 1.32 \text{ ft}$$

Perform all calculations per foot of base width.

$$P = \sum W_i = 16.63 \text{ kips/ft}$$

$$A = 10 \text{ ft}^2/\text{ft}$$

$$M = (1.32 \text{ ft})(16.63 \text{ kips}) = 21.95 \text{ ft-kips/ft}$$

$$S = \frac{bh^2}{6} = \frac{(1 \text{ ft})(10 \text{ ft})^2}{6} = 16.67 \text{ ft}^3/\text{ft}$$

$$q = \frac{P}{A} \pm \frac{M}{S}$$

$$= \frac{16.63 \dfrac{\text{kips}}{\text{ft}}}{10 \dfrac{\text{ft}^2}{\text{ft}}} \pm \frac{\left(21.95 \dfrac{\text{ft-kips}}{\text{ft}}\right)\left(1 \dfrac{\text{ft}}{\text{ft}}\right)}{16.67 \dfrac{\text{ft}^3}{\text{ft}}}$$

$$= 1.663 \frac{\text{kips}}{\text{ft}^2} \pm 1.32 \frac{\text{kips}}{\text{ft}^2}$$

$$q_{\text{toe}} = 1.66 \frac{\text{kips}}{\text{ft}^2} + 1.32 \frac{\text{kips}}{\text{ft}^2} = 2.98 \frac{\text{kips}}{\text{ft}^2}$$

$$q_{\text{heel}} = 1.66 \frac{\text{kips}}{\text{ft}^2} - 1.32 \frac{\text{kips}}{\text{ft}^2} = 0.35 \frac{\text{kip}}{\text{ft}^2}$$

Solution Using ACI 318 App. C

(i) Assume a 1 in diameter stem bar. Subtracting 2 in of cover for a formed surface, the effective depth of the wall is approximately

$$d \approx 15 \text{ in} - 2 \text{ in} - \frac{1 \text{ in}}{2} = 12.5 \text{ in}$$

The distance from the base to the critical stem section is also 12.5 in.

Using ACI 318 App. C, calculate the factored shear of the critical section.

$$V_u = (1.7)\left(\tfrac{1}{2}\right)\gamma_{\text{equivalent}}L^2$$

$$= (1.7)\left(\tfrac{1}{2}\right)\left(27.1 \frac{\text{lbf}}{\text{ft}^3}\right)\left(21 \text{ ft} - 2 \text{ ft} - \frac{12.5 \text{ in}}{12 \dfrac{\text{in}}{\text{ft}}}\right)^2$$

$$= 7429 \text{ lbf/ft}$$

Assuming a 1 in diameter bar and considering the taper,

$$d' = 12.5 \text{ in} - \frac{\left(\dfrac{12.5 \text{ in}}{12 \dfrac{\text{in}}{\text{ft}}}\right)(15 \text{ in} - 9 \text{ in})}{17 \text{ ft} + 2 \text{ ft}}$$

$$= 12.2 \text{ in}$$

Use Eq. 54.8.

$$\phi V_n = 2\phi wd'\sqrt{f'_c}$$

$$= (2)(0.85)(12 \text{ in})(12.2 \text{ in})\sqrt{3500 \frac{\text{lbf}}{\text{in}^2}}$$

$$= 14{,}724 \text{ lbf}$$

Since $\phi V_n > V_u$, the stem thickness is acceptable. The moment at the base of the stem is

$$M_u = (1.7)\left(\tfrac{1}{2}\right)\gamma_{\text{equivalent}}(H - t_b)^2\left(\frac{H - t_b}{3}\right)$$

$$= (1.7)\left(\tfrac{1}{2}\right)\left(27.1 \frac{\text{lbf}}{\text{ft}^3}\right)\left(\frac{(21 \text{ ft} - 2 \text{ ft})^3}{3}\right)$$

$$= 52{,}666 \text{ ft-lbf}$$

Estimate λ.

$$\lambda = 0.1d = (0.1)(12.5 \text{ in}) = 1.25 \text{ in}$$

$$\phi = 0.9 \text{ for beams in flexure}$$

Use Eq. 50.28.

$$A_s = \frac{M_u}{\phi f_y(d - \lambda)}$$

$$= \frac{(52{,}666 \text{ ft-lbf})\left(12 \frac{\text{in}}{\text{ft}}\right)}{(0.90)\left(60{,}000 \frac{\text{lbf}}{\text{in}^2}\right)(12.5 \text{ in} - 1.25 \text{ in})}$$

$$= 1.04 \text{ in}^2 \quad [\text{per foot of wall}]$$

This area can be provided with no. 9 bars at 12 in spacing.

(j) Check the development length. Refer to ACI 318 12.2.2. $\psi_t = 1$ (vertical bars), $\psi_e = 1$ (uncoated), and $\lambda = 1$ (normal weight concrete).

$$l_d = \frac{d_b f_y \psi_t \psi_e \lambda}{20\sqrt{f'_c}} \quad [\text{using no. 9 bars}]$$

$$= \frac{(1.125 \text{ in})\left(60{,}000 \frac{\text{lbf}}{\text{in}^2}\right)(1)(1)(1)}{20\sqrt{3500 \frac{\text{lbf}}{\text{in}^2}}}$$

$$= 57 \text{ in}$$

This is greater than the footing thickness. Either the footing thickness should be increased, hooks should be used on the ends of the stem bars, or a smaller-diameter bar should be used. A more sophisticated analysis can also be performed.

(k) Using 15 in as the stem thickness, conservatively the temperature steel required in the stem is

$$A_{s,\text{temperature}} = 0.0020bt_{\text{stem}}$$

$$= (0.0020)(12 \text{ in})(15 \text{ in}) = 0.36 \text{ in}^2/\text{ft}$$

Give two thirds of this to the front face and one third to the back face.

$$\left(\tfrac{2}{3}\right)\left(0.36 \frac{\text{in}^2}{\text{ft}}\right) = 0.24 \text{ in}^2/\text{ft}$$

From Table 51.2, use no. 5 bars spaced at 12 in, or no. 4 bars spaced at 10 in.

$$\left(\tfrac{1}{3}\right)\left(0.36 \frac{\text{in}^2}{\text{ft}}\right) = 0.12 \text{ in}^2/\text{ft}$$

Use no. 4 bars spaced at 12 in, or no. 3 bars spaced at 10 in.

Place the temperature steel no closer than 2 in from the front face [ACI 318 14.3.4].

(l) Disregard the soil pressure under the heel. Using ACI 318 App. C, the factored uniform load above the heel is

$$w_u = \frac{(1.4)\left((19 \text{ ft})\left(100 \frac{\text{lbf}}{\text{ft}^3}\right) + (2 \text{ ft})\left(150 \frac{\text{lbf}}{\text{ft}^3}\right)\right)}{1000 \frac{\text{lbf}}{\text{kip}}}$$

$$= 3.08 \text{ kips/ft}$$

Since the base is cast against earth, the bottom cover required is 3 in. The distance from the vertical stem steel to heel edge is

$$L = (5 \text{ ft } 5 \text{ in}) + 3 \text{ in} = 5 \text{ ft } 8 \text{ in} \ (5.67 \text{ ft})$$

$$M_u = \tfrac{1}{2}w_u L^2 = \left(\tfrac{1}{2}\right)\left(3.08 \frac{\text{kips}}{\text{ft}}\right)(5.67 \text{ ft})^2$$

$$= 49 \text{ ft-kips}$$

The heel extension (from the stem's soil face) is

$$L = 5 \text{ ft } 5 \text{ in} \ (5.42 \text{ ft})$$

$$V_u = w_u L = \left(3.08 \frac{\text{kips}}{\text{ft}}\right)(5.42 \text{ ft})$$

$$= 16.7 \text{ kips}$$

The heel thickness is 2 ft. Assume a 1 in diameter bar will be used. The heel depth is

$$d = t_{\text{heel}} - \text{top cover} - \frac{d_b}{2}$$

$$= (2 \text{ ft})\left(12 \frac{\text{in}}{\text{ft}}\right) - 2 \text{ in} - \frac{1 \text{ in}}{2}$$

$$= 21.5 \text{ in}$$

$$\phi V_n = 2\phi bd\sqrt{f'_c}$$

$$= \frac{(2)(0.85)(12 \text{ in})(21.5 \text{ in})\sqrt{3500 \frac{\text{lbf}}{\text{in}^2}}}{1000 \frac{\text{lbf}}{\text{kip}}}$$

$$= 26.0 \text{ kips} \quad \begin{bmatrix} > V_u = 16.7 \text{ kips,} \\ \text{so OK} \end{bmatrix}$$

Solve for the required flexural steel using Eq. 54.13.

$$M_n = \frac{M_u}{\phi} = \rho b d^2 f_y \left(1 - \frac{\rho f_y}{1.7 f_c'}\right)$$

$$\left(\frac{49 \text{ ft-kips}}{0.9}\right)\left(12 \frac{\text{in}}{\text{ft}}\right)$$

$$= \rho(12 \text{ in})(21.5 \text{ in})^2 \left(60 \frac{\text{kips}}{\text{in}^2}\right)$$

$$\times \left(1 - \frac{\rho\left(60 \frac{\text{kips}}{\text{in}^2}\right)}{(1.7)\left(3.5 \frac{\text{kips}}{\text{in}^2}\right)}\right)$$

$$\rho = 0.002$$

The minimum reinforcement ratio is

$$\rho_{\min} = \frac{200}{f_y} = \frac{200}{60{,}000 \frac{\text{lbf}}{\text{in}^2}} = 0.0033$$

The required steel area per foot of wall width is

$$A_s = \rho b d = (0.0033)(12 \text{ in})(21.5 \text{ in}) = 0.85 \text{ in}^2$$

This can be provided with no. 9 bars at 12 in spacing. Note that $\rho = \left(\frac{4}{3}\right)(0.002) = 0.0027$ could also have been used.

Solution Using ACI 318 Ch. 9

(i) Assume a 1 in diameter stem bar. Subtracting 2 in of cover for a formed surface, the effective depth of the wall is approximately

$$d \approx 15 \text{ in} - 2 \text{ in} - \frac{1 \text{ in}}{2} = 12.5 \text{ in}$$

The distance from the base to the critical stem section is also 12.5 in.

Using ACI 318 Ch. 9, calculate the factored shear at the critical section.

$$V_u = (1.6)\left(\tfrac{1}{2}\right)\gamma_{\text{equivalent}} L^2$$

$$= (1.6)\left(\tfrac{1}{2}\right)\left(27.1 \frac{\text{lbf}}{\text{ft}^3}\right)\left(21 \text{ ft} - 2 \text{ ft} - \frac{12.5 \text{ in}}{12 \frac{\text{in}}{\text{ft}}}\right)^2$$

$$= 6992 \text{ lbf/ft}$$

Assuming a 1 in diameter bar and considering the taper,

$$d' = 12.5 \text{ in} - \frac{\left(\dfrac{12.5 \text{ in}}{12 \frac{\text{in}}{\text{ft}}}\right)(15 \text{ in} - 9 \text{ in})}{17 \text{ ft} + 2 \text{ ft}}$$

$$= 12.2 \text{ in}$$

Use Eq. 54.8.

$$\phi V_n = 2\phi w d' \sqrt{f_c'}$$

$$= (2)(0.75)(12 \text{ in})(12.2 \text{ in})\sqrt{3500 \frac{\text{lbf}}{\text{in}^2}}$$

$$= 12{,}992 \text{ lbf}$$

Since $\phi V_n > V_u$, the stem thickness is acceptable. The moment at the base of the stem is

$$M_u = (1.6)\left(\tfrac{1}{2}\right)\gamma_{\text{equivalent}}(H - t_b)^2 \left(\frac{H - t_b}{3}\right)$$

$$= (1.6)\left(\tfrac{1}{2}\right)\left(27.1 \frac{\text{lbf}}{\text{ft}^3}\right)\left(\frac{(21 \text{ ft} - 2 \text{ ft})^3}{3}\right)$$

$$= 49{,}568 \text{ ft-lbf}$$

Estimate λ.

$$\lambda = 0.1d = (0.1)(12.5 \text{ in}) = 1.25 \text{ in}$$

$$\phi = 0.9 \text{ for beams in flexure}$$

Use Eq. 50.28.

$$A_s = \frac{M_u}{\phi f_y(d - \lambda)} = \frac{(49{,}568 \text{ ft-lbf})\left(12 \frac{\text{in}}{\text{ft}}\right)}{(0.90)\left(60{,}000 \frac{\text{lbf}}{\text{in}^2}\right)(12.5 \text{ in} - 1.25 \text{ in})}$$

$$= 0.98 \text{ in}^2 \quad [\text{per foot of wall}]$$

This area can be provided with no. 9 bars at 12 in spacing.

(j) Check the development length. Refer to ACI 318 12.2.2. $\psi_t = 1$ (vertical bars), $\psi_e = 1$ (uncoated), and $\lambda = 1$ (normal weight concrete).

$$l_d = \frac{d_b f_y \psi_t \psi_e \lambda}{20\sqrt{f_c'}} \quad [\text{using no. 9 bars}]$$

$$= \frac{(1.125 \text{ in})\left(60{,}000 \frac{\text{lbf}}{\text{in}^2}\right)(1)(1)(1)}{20\sqrt{3500 \frac{\text{lbf}}{\text{in}^2}}}$$

$$= 57 \text{ in}$$

This is greater than the footing thickness. Either the footing thickness should be increased, hooks should be used on the ends of the stem bars, or a smaller-diameter bar should be used. A more sophisticated analysis can also be performed.

(k) Using 15 in as the stem thickness, conservatively the temperature steel required in the stem is

$$A_{s,\text{temperature}} = 0.0020 b t_{\text{stem}}$$

$$= (0.0020)(12 \text{ in})(15 \text{ in}) = 0.36 \text{ in}^2/\text{ft}$$

Give two thirds of this to the front face and one third to the back face.

$$\left(\tfrac{2}{3}\right)\left(0.36\ \frac{\text{in}^2}{\text{ft}}\right) = 0.24\ \text{in}^2/\text{ft}$$

From Table 51.2, use no. 5 bars spaced at 12 in, or no. 4 bars spaced at 10 in.

$$\left(\tfrac{1}{3}\right)\left(0.36\ \frac{\text{in}^2}{\text{ft}}\right) = 0.12\ \text{in}^2/\text{ft}$$

Use no. 4 bars spaced at 12 in, or no. 3 bars spaced at 10 in.

Place the temperature steel no closer than 2 in from the front face [ACI 318 14.3.4].

(1) Disregard the soil pressure under the heel. Using ACI 318 Ch. 9, the factored uniform load above the heel is

$$w_u = \frac{(1.2)\left((19\ \text{ft})\left(100\ \frac{\text{lbf}}{\text{ft}^3}\right) + (2\ \text{ft})\left(150\ \frac{\text{lbf}}{\text{ft}^3}\right)\right)}{1000\ \dfrac{\text{lbf}}{\text{kip}}}$$

$$= 2.64\ \text{kips/ft}$$

Since the base is cast against earth, the bottom cover required is 3 in. The distance from the vertical stem steel to heel edge is

$$L = (5\ \text{ft}\ 5\ \text{in}) + 3\ \text{in} = 5\ \text{ft}\ 8\ \text{in}\ (5.67\ \text{ft})$$

$$M_u = \tfrac{1}{2} w_u L^2 = \left(\tfrac{1}{2}\right)\left(2.64\ \frac{\text{kips}}{\text{ft}}\right)(5.67\ \text{ft})^2$$

$$= 42.4\ \text{ft-kips}$$

The heel extension (from the stem's soil face) is

$$L = 5\ \text{ft}\ 5\ \text{in}\ (5.42\ \text{ft})$$

$$V_u = w_u L = \left(2.64\ \frac{\text{kips}}{\text{ft}}\right)(5.42\ \text{ft})$$

$$= 14.3\ \text{kips}$$

The heel thickness is 2 ft. Assume a 1 in diameter bar will be used. The heel depth is

$$d = t_{\text{heel}} - \text{top cover} - \frac{d_b}{2}$$

$$= (2\ \text{ft})\left(12\ \frac{\text{in}}{\text{ft}}\right) - 2\ \text{in} - \frac{1\ \text{in}}{2}$$

$$= 21.5\ \text{in}$$

$$\phi V_n = 2\phi b d \sqrt{f_c'}$$

$$= \frac{(2)(0.75)(12\ \text{in})(21.5\ \text{in})\sqrt{3500\ \dfrac{\text{lbf}}{\text{in}^2}}}{1000\ \dfrac{\text{lbf}}{\text{kip}}}$$

$$= 22.9\ \text{kips} \quad \begin{bmatrix} > V_u = 14.3\ \text{kips,} \\ \text{so OK} \end{bmatrix}$$

Solve for the required flexural steel using Eq. 54.13.

$$M_n = \frac{M_u}{\phi} = \rho b d^2 f_y \left(1 - \frac{\rho f_y}{1.7 f_c'}\right)$$

$$\left(\frac{42.4\ \text{ft-kips}}{0.9}\right)\left(12\ \frac{\text{in}}{\text{ft}}\right)$$

$$= \rho(12\ \text{in})(21.5\ \text{in})^2 \left(60\ \frac{\text{kips}}{\text{in}^2}\right)$$

$$\times \left(1 - \frac{\rho\left(60\ \dfrac{\text{kips}}{\text{in}^2}\right)}{(1.7)\left(3.5\ \dfrac{\text{kips}}{\text{in}^2}\right)}\right)$$

$$\rho = 0.0017$$

The minimum reinforcement ratio is

$$\rho_{\text{min}} = \frac{200}{f_y} = \frac{200}{60,000\ \dfrac{\text{lbf}}{\text{in}^2}} = 0.0033$$

The required steel area per foot of wall width is

$$A_s = \rho b d = (0.0033)(12\ \text{in})(21.5\ \text{in}) = 0.85\ \text{in}^2$$

This can be provided with no. 9 bars at 12 in spacing.

Note that $\rho = \left(\tfrac{4}{3}\right)(0.0017) = 0.0023$ could also have been used.

55 Reinforced Concrete: Footings

Nomenclature

A	area	in^2
A_f	base area of footing per ACI 318 15.2.2	in^2
A_{ff}	largest footing area geometrically similar to the column	in^2
A_{sd}	area of steel parallel to the short direction in a rectangular footing	in^2
b	dimension	in
b_o	punching shear area perimeter	in
B	one of the plan dimensions of a rectangular footing	in
d	effective footing depth (to steel layer)	in
d_b	diameter of a bar	in
e	distance from the critical section to the edge	in
f'_c	compressive strength of concrete	lbf/in^2
f_y	yield strength of steel	lbf/in^2
h	overall height (thickness) of footing	in
H	distance from soil surface to footing base	in
I	moment of inertia	in^4
J	polar moment of inertia of punching shear surface	in^4
L	one of the plan dimensions of a rectangular footing	in
M_u	factored moment in a column at the juncture with the footing	ft-kips
P_u	column load or wall load per unit length (factored)	kips
q	pressure under footing	lbf/ft^2
R	resultant of soil pressure distribution	lbf
t	wall thickness	in
v_c	nominal concrete shear stress	lbf/in^2
v_u	shear stress due to factored loads	lbf/in^2
V_c	nominal concrete shear strength	lbf

Symbols

α	strength enhancement factor	–
β	ratio of long side to short side of a footing	–
β_c	ratio of long side to short side of a column	–
ψ_e	epoxy coating factor	–
ψ_t	reinforcement location factor	–
γ	specific weight	lbf/ft^3
γ_v	fraction of moment that affects the punching shear stress	–
ϵ	load eccentricity	ft
λ	distance from centroid of compression area to extreme compression fiber	in
λ	lightweight aggregate factor	–
ϕ	capacity reduction factor	–

Subscripts

a	allowable
b	bar
c	concrete
d	dead
db	dowel bar
f	footing
g	gross
h	hook
l	live
n	net
p	punching shear or bearing
s	soil, steel, or service
u	ultimate or factored
v	shear
y	yield

1. INTRODUCTION

Footings are designed in two steps: (a) selecting the footing area, and (b) selecting the footing thickness and reinforcement. The footing area is chosen so that the soil contact pressure is within limits. The footing thickness and the reinforcement are chosen to keep the shear and bending stresses in the footing within permissible limits.

Although the footing area is obtained from the unfactored service loads, the footing thickness and reinforcement are calculated from factored loads. This is because the weight of the footing and the overburden contribute to the contact stress, but these forces do not induce shears or bending moments in the footing.

For the general case of a footing with dimensions L and B (see Fig. 55.1), carrying a vertical (downward) axial service load, P_s, and a service moment, M_s, the service soil pressure is

$$q_s = \frac{P_s}{A_f} + \gamma_c h + \gamma_s (H - h) \pm \frac{M_s \left(\frac{B}{2}\right)}{I_f} \leq q_a \quad 55.1$$

$$I_f = \frac{1}{12} L B^3 \quad\quad\quad\quad\quad\quad 55.2$$

Figure 55.1 *General Footing Loading*

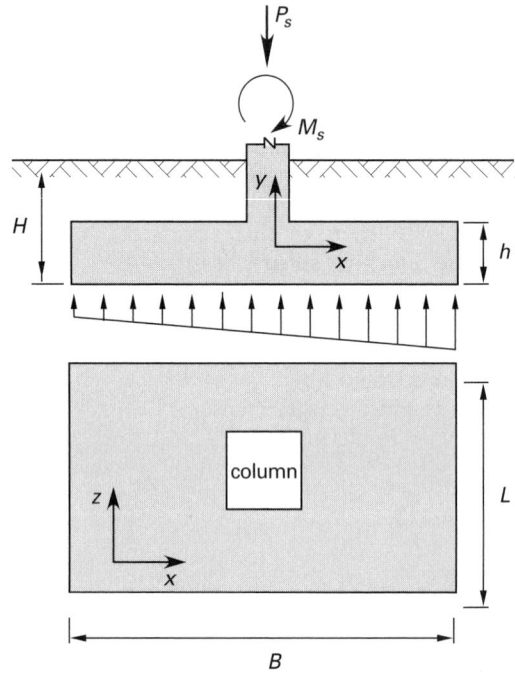

2. WALL FOOTINGS

For a wall footing, the factored load per unit length is

$$q_u = \frac{P_u}{B} \quad\quad\quad 55.3$$

Per ACI 318 App. C,
$$P_u = 1.4 P_d + 1.7 P_l \quad\quad 55.4(a)$$
Per ACI 318 Ch. 9,
$$P_u = 1.2 P_d + 1.6 P_l \quad\quad 55.4(b)$$

The critical plane for shear is assumed to be located a distance d (the effective footing depth) from the wall face, where d is measured to the center of the flexural steel layer. The thickness of wall footings is selected to satisfy the shear stress requirement of Eq. 55.5. v_u is the shear stress at a distance d from the face of the wall, v_c is the shear stress of the concrete. $\phi = 0.85$ per ACI 318 App. C, or $\phi = 0.75$ per ACI 318 Ch. 9.

$$\phi v_c \geq v_u \quad\quad\quad 55.5$$

$$v_c = 2\sqrt{f_c'} \quad\quad\quad 55.6$$

Referring to Fig. 55.2, the shear stress at the critical section is

$$v_u = \left(\frac{q_u}{d}\right)\left(\frac{B - t}{2} - d\right) \quad\quad 55.7$$

Combining Eqs. 55.5 and 55.7,

$$d = \frac{q_u (B - t)}{2(q_u + \phi v_c)} \quad\quad\quad 55.8$$

In a wall footing, the main steel (x) resists flexure and is perpendicular to the wall face. Orthogonal steel (z) runs parallel to the wall and is placed in contact with the main steel. The total thickness of the wall footing is

$$h = d + \tfrac{1}{2}(\text{diameter of } x \text{ bars})$$
$$+ \text{diameter of } z \text{ bars} + \text{cover}$$
$$\left[\begin{array}{c}\text{orthogonal reinforcement} \\ \text{under main steel}\end{array}\right] \quad 55.9(a)$$

$$h = d + \tfrac{1}{2}(\text{diameter of } x \text{ bars}) + \text{cover}$$
$$\left[\begin{array}{c}\text{orthogonal reinforcement} \\ \text{over main steel}\end{array}\right] \quad 55.9(b)$$

The minimum cover (below the bars) for cast-in-place concrete cast against and permanently exposed to earth is 3 in [ACI 318 7.7.1]. The minimum depth above the bottom reinforcement is 6 in [ACI 318 15.7]. Therefore, the minimum total thickness of a footing is

$$h \geq 6 \text{ in} + \text{diameter of } x \text{ bars}$$
$$+ \text{diameter of } z \text{ bars} + 3 \text{ in} \quad 55.10$$

Figure 55.2 *Critical Section for Shear in Wall Footing*

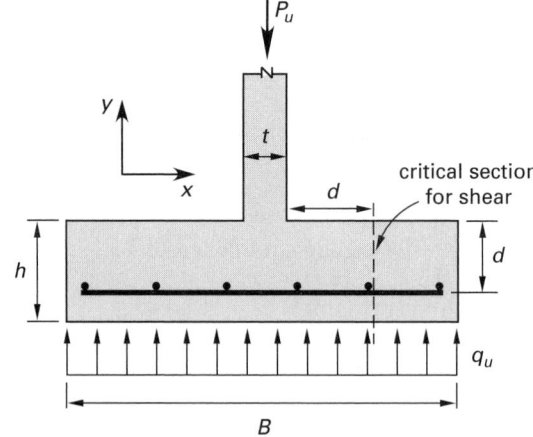

3. COLUMN FOOTINGS

The effective depth, d (and thickness, h), of column footings is controlled by shear strength. Two shear failure mechanisms are considered: one-way shear and two-way shear. The footing depth is taken as the larger of the two values calculated. In the majority of cases, the required depth is controlled by two-way shear.

For *one-way shear* (also known as *single-action shear* and *wide-beam shear*), the critical sections are a distance d from the face of the column. Equation 55.11 applies for the case of uniform pressure distribution (i.e., where there is no moment). The section with the largest values of e controls.

$$v_u = \frac{q_u e}{d} \quad\quad\quad 55.11$$

Figure 55.3 *Critical Section for One-Way Shear*

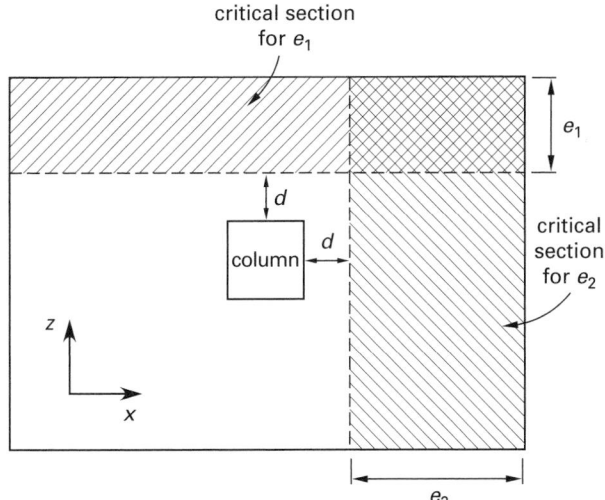

In addition to the footing failing in shear as a wide beam, failure can also occur in *two-way shear* (also known as *double-action shear* and *punching shear*). In this failure mode, the column and an attached concrete piece punch through the footing. Although the failure plane is actually inclined outward, the failure surface is assumed for simplicity to consist of vertical planes located a distance $d/2$ from the column sides.

The area in punching is given by Eq. 55.12. b_1 is the length of the critical area parallel to the axis of the applied column moment. Similarly, b_2 is the length of the critical area normal to the axis of the applied column moment.

$$A_p = 2(b_1 + b_2)d \qquad 55.12$$

Figure 55.4 *Critical Section for Two-Way Shear*

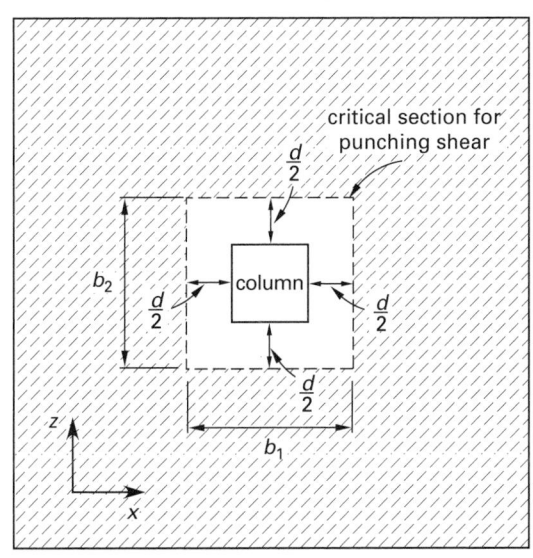

Assume for generality that the column supports a moment. The maximum shear stress on the critical shear plane is computed as the sum of the shear stresses resulting from the axial load and the moment. The punching shear stress is given by Eq. 55.13.

$$v_u = \frac{P_u - R}{A_p} + \frac{\gamma_v M_u(0.5b_1)}{J} \qquad 55.13$$

In Eq. 55.13, R is the resultant of the factored soil pressure acting over the area $b_1 b_2$. The stresses from the moment cancel over this area, so R is

$$R = \frac{P_u b_1 b_2}{A_f} \qquad 55.14$$

γ_v is the fraction of the moment M_u that is assumed to contribute to shear stress [ACI 318 11.12.6.1 and ACI 318 13.5.3.2].

$$\gamma_v = 1 - \frac{1}{1 + \frac{2}{3}\sqrt{\frac{b_1}{b_2}}} \qquad 55.15$$

When the column is located at the center of the footing, the constant J is given by Eq. 55.16 [ACI 318 R11.12.6.2].

$$J = \left(\frac{db_1^3}{6}\right)\left(1 + \left(\frac{d}{b_1}\right)^2 + 3\left(\frac{b_2}{b_1}\right)\right) \qquad 55.16$$

The nominal concrete shear stress for punching shear is given by Eq. 55.17.

$$v_c = (2 + y)\sqrt{f_c'} \qquad 55.17$$
$$y = \min\{2,\ 4/\beta_c,\ 40d/b_o\} \qquad 55.18$$
$$\beta_c = \frac{\text{column long side}}{\text{column short side}} \qquad 55.19$$
$$b_o = \frac{A_p}{d} = 2(b_1 + b_2) \qquad 55.20$$

Determining the footing thickness is accomplished through trial and error. First, obtain the required depth for punching shear using the following procedure.

step 1: Assume a value of d.
step 2: Compute b_1 and b_2.
step 3: Evaluate A_p, J, and γ_v.
step 4: Compute R.
step 5: Compute v_u.
step 6: Calculate v_c.
step 7: If $v_u \geq \phi v_c$, increase d and repeat from step 2.

Structural

If $v_u < \phi v_c$, the punching shear strength is adequate. If the excess in capacity is substantial, reduce d and repeat from step (2). Second, check the adequacy of the footing depth for one-way shear.

The effective depth for shear is typically assumed to be the distance from the top of the footing to the average depth of the two perpendicular steel reinforcement layers. (The two steel layers may use different bar diameters, so the average depth may not correspond to the contact point.) The total thickness of a column footing is

$$h = d + \tfrac{1}{2}(\text{diameter of } x \text{ bars} \\ + \text{diameter of } z \text{ bars}) + \text{cover} \qquad 55.21$$

As with wall footings, the minimum cover (below the bars) for cast-in-place concrete cast against and permanently exposed to earth is 3 in [ACI 318 7.7.1]. The minimum depth above the bottom reinforcement is 6 in [ACI 318 15.7]. Therefore, the minimum total thickness of a footing is

$$h \geq 6 \text{ in} + \text{diameter of } x \text{ bars} \\ + \text{diameter of } z \text{ bars} + 3 \text{ in} \qquad 55.22$$

4. SELECTION OF FLEXURAL REINFORCEMENT

For the purpose of designing flexural reinforcement, footings are treated as cantilever beams carrying an upward-acting, nonuniform distributed load. The critical section for flexure is located at the face of the column or wall. (Masonry walls are exceptions. The critical section is halfway between the center and the edge of the wall [ACI 318 15.4.2].) The moment acting at the critical section from a uniformly distributed soil pressure distribution (i.e., as when there is no column moment) is

$$M_u = \frac{q_u L l^2}{2} \quad \text{[no column moment]} \qquad 55.23$$

The required steel area is given by Eq. 55.24. As with beams, a good starting point is $\lambda = 0.1d$.

$$A_s = \frac{M_u}{\phi f_y (d - \lambda)} \qquad 55.24$$

The effective depth of the flexural reinforcement will differ in each perpendicular direction. However, it is customary to design both layers using a single effective depth. While the conservative approach is to use the smaller effective depth for both layers, it is not unusual in practice to design the steel in the x and z directions based on the average effective depth.

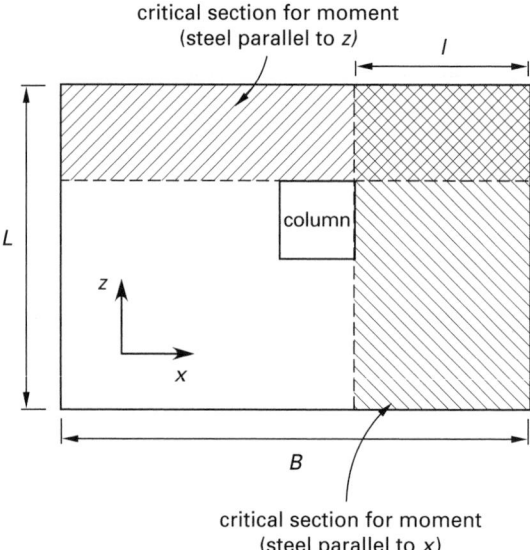

Figure 55.5 *Critical Section for Moment*

The following provisions apply.

- According to ACI 318 10.5.4, for footings of uniform thickness, the minimum area of tension reinforcement in the direction of the span is the same as that required by 7.12. The maximum spacing of this reinforcement is not to exceed three times the thickness nor 18 in. The minimum steel ratio (based on the gross cross-sectional area) in any direction is 0.002 for $f_y = 40$ kips/in^2 steel and 0.0018 for $f_y = 60$ kips/in^2 steel [ACI 7.12.2.1].

- In one-way wall footings and square two-way footings, the steel is distributed uniformly [ACI 318 15.4.3].

- In rectangular two-way footings, the steel parallel to the long dimension, A_s, is distributed uniformly. However, the total steel parallel to the short dimension, A_{sd}, is divided into two parts [ACI 318 15.4.4.2].

$$A_1 = A_{sd} \left(\frac{2}{\beta + 1} \right) \qquad 55.25$$
$$A_2 = A_{sd} - A_1 \qquad 55.26$$

β is the ratio of the long footing side to the short footing side. The steel area A_1 is distributed uniformly in a band centered on the column and having a width equal to the short dimension of the footing. The area A_2 is distributed uniformly in the remaining part of the footing.

- The maximum spacing center-to-center of bars cannot exceed the lesser of $3h$ (three times the overall footing thickness) or 18 in [ACI 318 10.5.4]. (The 18 in maximum spacing usually controls in footings.)

- The steel bars must be properly anchored.

Figure 55.6 *Distribution of Steel*
(parallel to the short dimension)

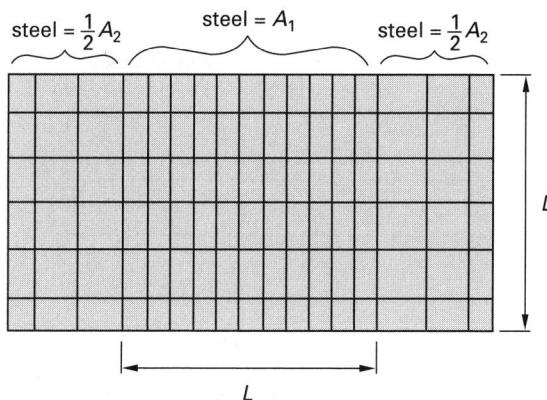

Figure 55.7 *Standard Hooks*

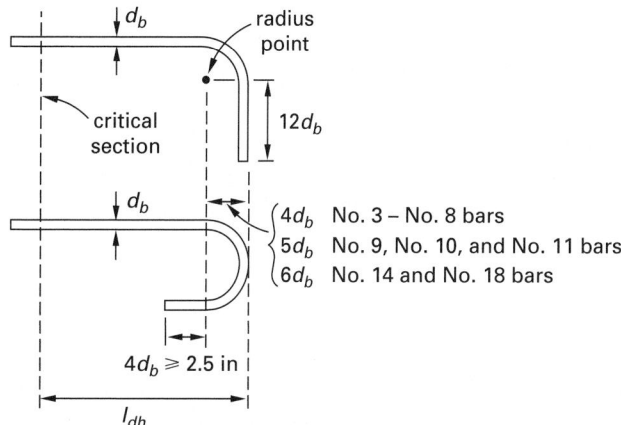

Used with permission of the American Concrete Institute.

5. DEVELOPMENT LENGTH OF FLEXURAL REINFORCEMENT

Either ACI 318 12.2.2 or ACI 318 12.2.3 can be used to calculate the development length in normal-weight concrete of straight deformed bars of diameter d_b in tension. However, both sections offer a number of complex alternatives and require knowledge that may not be available. Using the simpler section [ACI 318 12.2.2], the development length for straight bars in tension in normal-weight concrete can be conservatively estimated as

$$l_d = \frac{3d_b f_y \psi_t \psi_e \lambda}{50\sqrt{f'_c}} \geq 12 \text{ in} \quad \text{[no. 6 bars and smaller]} \quad 55.27$$

$$l_d = \frac{3d_b f_y \psi_t \psi_e \lambda}{40\sqrt{f'_c}} \geq 12 \text{ in} \quad \text{[no. 7 bars and larger]} \quad 55.28$$

The *epoxy coating factor*, ψ_e, the reinforcement location factor, ψ_t, and the *lightweight aggregate factor*, λ, are given in ACI 318 12.2.4.

$\psi_e = 1.5$ epoxy-coated, clear spacing $\leq 6d_b$
$\psi_e = 1.2$ for all other epoxy-coated bars
$\psi_e = 1.0$ uncoated bars

$\psi_t = 1.3$ horizontal reinforcement > 12 in below development length or splice
$\psi_t = 1.0$ for all other situations

$\lambda = 1.3$ lightweight concrete
$\lambda = 1.0$ normal-weight concrete

Equation 55.29 can be used for bars in tension with standard hooks [ACI 318 12.5]. However, ψ_e can only take on the values of 1.0 and 1.2.

$$l_{dh} = \frac{0.02\psi_e d_b f_y}{\sqrt{f'_c}} \geq 8d_b \geq 6 \text{ in} \quad 55.29$$

6. TRANSFER OF FORCE AT COLUMN BASE

Column and wall forces are transferred to footings by direct bearing and by forces in steel bars. The steel bars used to transfer forces from one member to the other are known as *dowels* or *dowel bars*. Dowel bars are left to extend above the footing after the footing concrete has hardened, and they are spliced into the column or wall reinforcement prior to pouring the column or wall concrete.

It is common practice to use at least four dowel bars in a column. Where a significant moment must be transferred from a column to a footing, the dowel steel provided is typically chosen to match the column steel. Generally, the number of dowels and longitudinal bars are equal. The minimum dowel area is [ACI 318 15.8.2.1]

$$A_{db,\min} = 0.005A_g \quad 55.30$$

The dowels must extend into the footing no less than the compressive *development length*, l_{dc}, of the dowels [ACI 318 12.3.2]. For straight bars in compression,

$$l_{dc} = 0.02d_b \left(\frac{f_y}{\sqrt{f'_c}}\right) \geq 0.0003d_b f_y \geq 8 \text{ in} \quad 55.31$$

If the dowel bar is bent in an L-shape, the development length must be achieved over the vertical portion. (While hooks and bends can be used to decrease the required development length for dowels acting in tension, hooks and bends are ineffective for compressive forces.) Since lap splices are considered to be class-B splices [ACI 12.15.2], the dowel bar must extend into the column or wall 1.3 times the development length of the dowel or 1.3 times the development length of the bar in the column or wall, whichever is larger [ACI 318 12.15.1].

The footing thickness sets a limit on the length of dowel that is effective. If the footing depth is less than the development length, smaller bars must be used. The maximum effective dowel length is

$$l_d \leq h - \text{cover} - \text{diameter of } x \text{ steel}$$
$$- \text{diameter of } z \text{ steel} \qquad 55.32$$

The maximum bearing strength of column concrete is $f_{\text{bearing}} = 0.85 f_c'$ [ACI 318 10.17.1]. The load carried by bearing on the column is given by Eq. 55.33. Note that in ACI 318, A_c is A_1.

$$P_{\text{bearing,column}} = 0.85 \phi f_{c,\text{column}}' A_c \qquad 55.33$$

However, the effective bearing capacity of the footing is actually greater than $f_{\text{bearing}} A_c$, since a redistribution of stressed area over the larger footing area will increase the effective allowable bearing load. The load carried by bearing on the footing concrete is given by Eq. 55.34 [ACI 318 10.17.1]. For bearing on concrete, as in Eqs. 55.33 and 55.34, $\phi = 0.70$ [ACI 318 C.3.2.4], or $\phi = 0.65$ [ACI 318 9.3.2.4].

$$P_{\text{bearing,footing}} = 0.85 \alpha \phi f_{c,\text{footing}}' A_c \qquad 55.34$$

α accounts for the enhancement in the capacity of the concrete that results because the loaded area is only a fraction of the total area of the footing.

$$\alpha = \sqrt{\frac{A_{ff}}{A_c}} \leq 2 \qquad 55.35$$

If the footing thickness and geometry are such that the frustrum requirements of ACI 318 10.17.1 are satisfied, A_{ff} [A_2 in ACI 318] is the largest footing area that is geometrically similar to the column and that does not extend beyond any of the edges of the footing [ACI 318 10.17.1]. For a square column and square footing, A_{ff} is not permitted to exceed 4 times the bearing area (A_1 in ACI 318). ACI 318 Fig. R10.17 depicts this limitation.

For the case of pure axial loading (no moment), the required area of the dowel bars is given by Eq. 55.36, in which ϕ has a value of 0.70 per ACI 318 App. C or 0.65 per ACI 318 Ch. 9.

$$A_{db} = \frac{P_u - \text{smaller of} \left\{ \begin{array}{c} P_{\text{bearing,column}} \\ P_{\text{bearing,footing}} \end{array} \right\}}{\phi f_y}$$
$$\geq A_{db,\text{min}} \qquad 55.36$$

Example 55.1

A 111 in square footing supports a 12 in square column. The service loads are 82.56 kips dead load and 75.4 kips live load. Normal weight concrete is used. Material properties are $f_c' = 4000$ lbf/in^2, and $f_y = 60{,}000$ lbf/in^2. Design the footing depth and reinforcement.

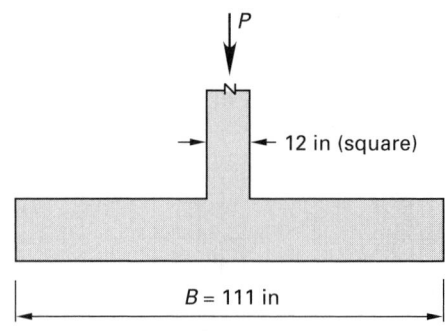

Solution Using ACI 318 App. C

step 1: Assume punching shear controls the depth.

$$b_1 = b_2 = 12 \text{ in} + (2)\left(\frac{d}{2}\right) = 12 \text{ in} + d$$

The area resisting the punching shear is given by Eq. 55.12.

$$A_p = 2(b_1 + b_2)d$$
$$= 2(12 \text{ in} + d + 12 \text{ in} + d)d = 48d + 4d^2$$

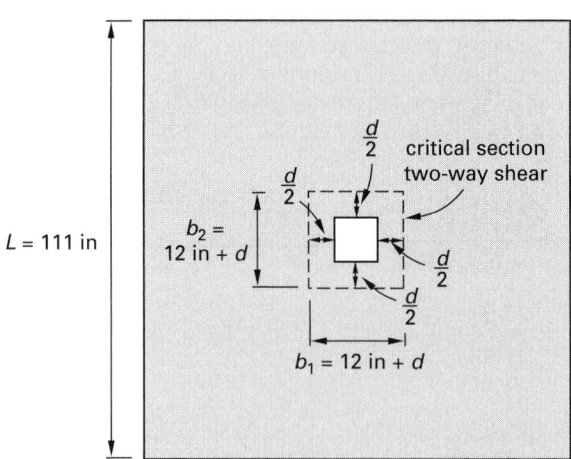

The ACI 318 App. C factored load is

$$P_u = 1.4 P_d + 1.7 P_l$$
$$= (1.4)(82.56 \text{ kips}) + (1.7)(75.4 \text{ kips})$$
$$= 243.76 \text{ kips}$$

The upward force from the soil pressure that acts to reduce the total punching shear force is given by Eq. 55.14.

$$R = \frac{P_u b_1 b_2}{A_f}$$
$$= \frac{(243.76 \text{ kips})(12 \text{ in} + d)(12 \text{ in} + d)}{(111 \text{ in})^2}$$
$$= \left(0.0198 \, \frac{\text{kip}}{\text{in}^2}\right)(12 \text{ in} + d)^2$$

There is no moment. The punching shear stress is given by Eq. 55.13.

$$v_u = \frac{P_u - R}{A_p}$$

$$= \frac{243.76 \text{ kips} - \left(0.0198 \dfrac{\text{kips}}{\text{in}^2}\right)(12 \text{ in} + d)^2}{48d + 4d^2}$$

Using $y = 2$, the allowable stress for punching shear is given by Eq. 55.17.

$$v_c = (2 + y)\sqrt{f_c'}$$

$$= (2 + 2)\sqrt{4000 \dfrac{\text{lbf}}{\text{in}^2}} = 252.98 \text{ lbf/in}^2$$

Equate the capacity to the shear stress. From Eq. 55.5,

$$\phi v_c = v_u$$

$$\frac{(0.85)\left(252.98 \dfrac{\text{lbf}}{\text{in}^2}\right)}{1000 \dfrac{\text{lbf}}{\text{kip}}} = \frac{\left(\begin{array}{c}243.76 \text{ kips} \\ -\left(0.0198 \dfrac{\text{kips}}{\text{in}^2}\right)(12 \text{ in} + d)^2\end{array}\right)}{48d + 4d^2}$$

$$d = 11.5 \text{ in}$$

The $40d/b_o$ limitation on y can now be checked.

$$b_o = 2(b_1 + b_2)$$

$$= (2)(2)(12 \text{ in} + d)$$

$$= (2)(2)(12 \text{ in} + 11.5 \text{ in})$$

$$= 9 \text{ in}$$

$$\frac{40d}{b_o} = \frac{(40)(11.5 \text{ in})}{94 \text{ in}} = 4.89 > 2$$

step 2: Check the capacity of the footing in one-way shear.

The critical section for one-way shear is at a distance d from the face of the column.

$$e = (0.5)(111 \text{ in} - 12 \text{ in}) - d$$

$$= (0.5)(111 \text{ in} - 12 \text{ in}) - 11.5 \text{ in}$$

$$= 38 \text{ in}$$

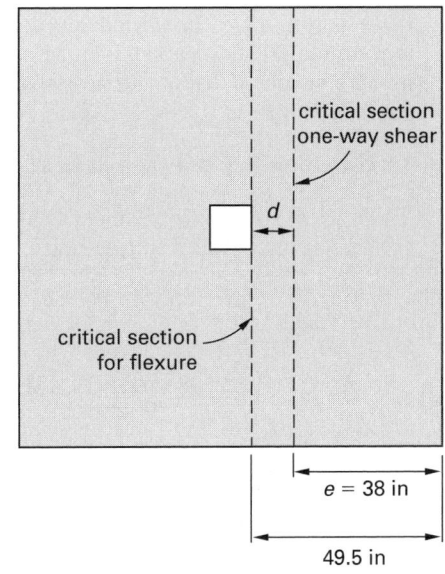

The contact pressure is

$$q_u = \frac{P_u}{A_f} = \frac{243.76 \text{ kips}}{(111 \text{ in})^2}$$

$$= 0.0198 \text{ kip/in}^2$$

The shear stress at the critical section is given by Eq. 55.11.

$$v_u = \frac{q_u e}{d}$$

$$= \frac{\left(0.0198 \dfrac{\text{kip}}{\text{in}^2}\right)(38 \text{ in})\left(1000 \dfrac{\text{lbf}}{\text{kip}}\right)}{11.5 \text{ in}}$$

$$= 65.43 \text{ lbf/in}^2$$

The nominal shear stress capacity for one-way shear is given by Eq. 55.6.

$$v_c = 2\sqrt{f_c'} = 2\sqrt{4000 \dfrac{\text{lbf}}{\text{in}^2}}$$

$$= 126.49 \text{ lbf/in}^2$$

Compare v_u and ϕv_c to determine if the value of d is adequate.

$$65.43 \dfrac{\text{lbf}}{\text{in}^2} < (0.85)\left(126.49 \dfrac{\text{lbf}}{\text{in}^2}\right) \quad [\text{OK}]$$

step 3: Design the flexural reinforcement.

Assuming no. 6 bars and taking the average for both layers, the overall depth of the footing is

$$h = 11.5 \text{ in} + 0.75 \text{ in} + 3.0 \text{ in} = 15.25 \text{ in}$$

Structural

It is customary to use a footing depth that is rounded to the nearest inch. Use $h = 16$ in. Using no. 6 bars, the actual (average) effective depth is

$$d = 16 \text{ in} - 3 \text{ in} - 0.75 \text{ in} = 12.25 \text{ in}$$

The critical section for moment is at the face of the column. From the free-body diagram,

$$M_u = \frac{q_u L l^2}{2}$$

$$= \frac{\left(0.0198 \, \dfrac{\text{kip}}{\text{in}^2}\right)(111 \text{ in})}{2}$$

$$\times (49.5 \text{ in})(49.5 \text{ in})$$

$$= 2692.6 \text{ in-kips}$$

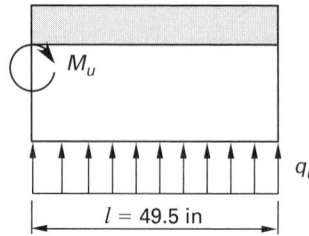

As with beams, start by assuming λ.

$$\lambda = 0.1d = (0.1)(12.25 \text{ in}) = 1.22 \text{ in}$$

The steel area is given by Eq. 55.24.

$$A_s = \frac{M_u}{\phi f_y (d - \lambda)}$$

$$= \frac{2692.6 \text{ in-kips}}{(0.9)\left(60 \, \dfrac{\text{kips}}{\text{in}^2}\right)(12.25 \text{ in} - 1.22 \text{ in})}$$

$$= 4.52 \text{ in}^2$$

Check the assumed value of λ. Use Eq. 50.17.

$$A_c = \frac{f_y A_s}{0.85 f_c'}$$

$$= \frac{\left(60 \, \dfrac{\text{kips}}{\text{in}^2}\right)(4.52 \text{ in}^2)}{(0.85)\left(4 \, \dfrac{\text{kips}}{\text{in}^2}\right)}$$

$$= 79.76 \text{ in}^2$$

Since the compressed zone is rectangular with a width of 111 in, the value of λ is given by Eq. 50.29.

$$\lambda = \frac{A_c}{2b} = \frac{79.76 \text{ in}^2}{(2)(111 \text{ in})} = 0.36 \text{ in}$$

Using $\lambda = 0.36$ in, the revised steel area is given by Eq. 55.24.

$$A_s = \frac{M_u}{\phi f_y (d - \lambda)}$$

$$= \frac{2692.6 \text{ in-kips}}{(0.9)\left(60 \, \dfrac{\text{kips}}{\text{in}^2}\right)(12.25 \text{ in} - 0.36 \text{ in})}$$

$$= 4.19 \text{ in}^2$$

The minimum steel is

$$A_{s,\text{min}} = (0.0018)(111 \text{ in})(16 \text{ in})$$

$$= 3.20 \text{ in}^2 \quad [\text{does not control}]$$

step 4: Check the development length of the flexural reinforcement. For uncoated no. 6 bars, using Eq. 55.27, where $\psi_t = 1.0$, $\psi_e = 1.0$, and $\lambda = 1.0$ per ACI 318 12.2.4,

$$l_d = \frac{3 d_b f_y \psi_t \psi_e \lambda}{50 \sqrt{f_c'}}$$

$$= \frac{(3)(0.75 \text{ in})\left(60{,}000 \, \dfrac{\text{lbf}}{\text{in}^2}\right)(1.0)(1.0)(1.0)}{50 \sqrt{4000 \, \dfrac{\text{lbf}}{\text{in}^2}}}$$

$$= 42.7 \text{ in}$$

The available distance for development (with a 2 in cover at the tip of the bar) is 49.5 in − 2 in = 47.5 in. Since the available distance exceeds l_d, the anchorage is adequate. Use 10 no. 6 bars spaced evenly in each direction. Using 2 in end cover on each bar end, the spacing is

$$\frac{111 \text{ in} - (2)(2 \text{ in}) - (10 \text{ bars})(0.75 \text{ in})}{9 \text{ gaps}}$$

$$= 11 \text{ in} > 2 d_b$$

The spacing does not exceed the maximum permissible value of 18 in.

Solution Using ACI 318 Ch. 9

step 1: Assume punching shear controls the depth.

$$b_1 = b_2 = 12 \text{ in} + (2)\left(\frac{d}{2}\right) = 12 \text{ in} + d$$

The area resisting the punching shear is given by Eq. 55.12.

$$A_p = 2(b_1 + b_2)d$$

$$= 2(12 \text{ in} + d + 12 \text{ in} + d)d = 48d + 4d^2$$

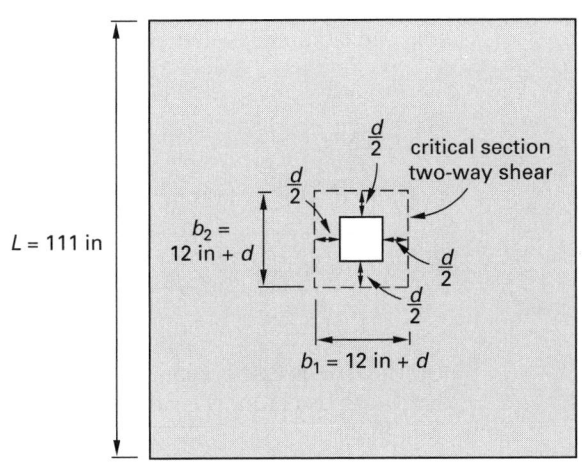

The ACI 318 Ch. 9 factored load is

$$P_u = 1.2P_d + 1.6P_l$$
$$= (1.2)(82.56 \text{ kips}) + (1.6)(75.4 \text{ kips})$$
$$= 219.71 \text{ kips}$$

The upward force from the soil pressure that acts to reduce the total punching shear force is given by Eq. 55.14.

$$R = \frac{P_u b_1 b_2}{A_f}$$
$$= \frac{(219.71 \text{ kips})(12 \text{ in} + d)(12 \text{ in} + d)}{(111 \text{ in})^2}$$
$$= \left(0.0178 \frac{\text{kip}}{\text{in}^2}\right)(12 \text{ in} + d)^2$$

There is no moment. The punching shear stress is given by Eq. 55.13.

$$v_u = \frac{P_u - R}{A_p}$$
$$= \frac{219.71 \text{ kips} - \left(0.0178 \frac{\text{kips}}{\text{in}^2}\right)(12 \text{ in} + d)^2}{48d + 4d^2}$$

Using $y = 2$, the allowable stress for punching shear is given by Eq. 55.17.

$$v_c = (2 + y)\sqrt{f_c'}$$
$$= (2 + 2)\sqrt{4000 \frac{\text{lbf}}{\text{in}^2}} = 252.98 \text{ lbf/in}^2$$

Equate the capacity to the shear stress. From Eq. 55.5,

$$\phi v_c = v_u$$

$$\frac{(0.75)\left(252.98 \frac{\text{lbf}}{\text{in}^2}\right)}{1000 \frac{\text{lbf}}{\text{kip}}} = \frac{\left(\begin{array}{c} 219.71 \text{ kips} \\ - \left(0.0178 \frac{\text{kips}}{\text{in}^2}\right)(12 \text{ in} + d)^2 \end{array}\right)}{48d + 4d^2}$$

$$d = 11.7 \text{ in}$$

The $40d/b_o$ limitation on y can now be checked.

$$b_o = 2(b_1 + b_2)$$
$$= (2)(2)(12 \text{ in} + d)$$
$$= (2)(2)(12 \text{ in} + 11.7 \text{ in})$$
$$= 9 \text{ in}$$
$$\frac{40d}{b_o} = \frac{(40)(11.5 \text{ in})}{94 \text{ in}} = 4.98 > 2$$

step 2: Check the capacity of the footing in one-way shear.

The critical section for one-way shear is at a distance d from the face of the column.

$$e = (0.5)(111 \text{ in} - 12 \text{ in}) - d$$
$$= (0.5)(111 \text{ in} - 12 \text{ in}) - 11.7 \text{ in}$$
$$= 37.8 \text{ in}$$

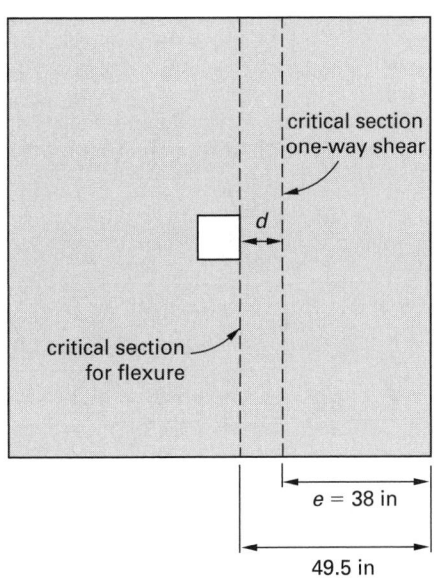

The contact pressure is

$$q_u = \frac{P_u}{A_f} = \frac{219.71 \text{ kips}}{(111 \text{ in})^2}$$
$$= 0.0178 \text{ kip/in}^2$$

The shear stress at the critical section is given by Eq. 55.11.

$$v_u = \frac{q_u e}{d}$$
$$= \frac{\left(0.0178 \frac{\text{kip}}{\text{in}^2}\right)(37.8 \text{ in})\left(1000 \frac{\text{lbf}}{\text{kip}}\right)}{11.7 \text{ in}}$$
$$= 57.51 \text{ lbf/in}^2$$

The nominal shear stress capacity for one-way shear is given by Eq. 55.6.

$$v_c = 2\sqrt{f'_c} = 2\sqrt{4000 \; \frac{\text{lbf}}{\text{in}^2}}$$
$$= 126.49 \; \text{lbf/in}^2$$

Compare v_u and ϕv_c to determine if the value of d is adequate.

$$57.51 \; \frac{\text{lbf}}{\text{in}^2} < (0.75)\left(126.49 \; \frac{\text{lbf}}{\text{in}^2}\right) \quad \text{[OK]}$$

step 3: Design the flexural reinforcement.

Assuming no. 6 bars and taking the average for both layers, the overall depth of the footing is

$$h = 11.7 \; \text{in} + 0.75 \; \text{in} + 3.0 \; \text{in} = 15.45 \; \text{in}$$

It is customary to use a footing depth that is rounded to the nearest inch. Use $h = 16$ in. Using no. 6 bars, the actual (average) effective depth is

$$d = 16 \; \text{in} - 3 \; \text{in} - 0.75 \; \text{in} = 12.25 \; \text{in}$$

The critical section for moment is at the face of the column. From the free-body diagram,

$$M_u = \frac{q_u L l^2}{2}$$
$$= \frac{\left(0.0178 \; \frac{\text{kip}}{\text{in}^2}\right)(111 \; \text{in})}{2}$$
$$\frac{\times (49.5 \; \text{in})(49.5 \; \text{in})}{2}$$
$$= 2420.6 \; \text{in-kips}$$

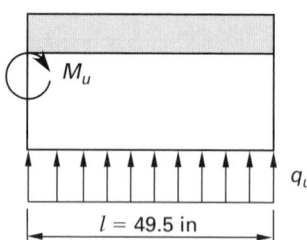

M_u

q_u

$l = 49.5$ in

As with beams, start by assuming λ.

$$\lambda = 0.1d = (0.1)(12.25 \; \text{in}) = 1.22 \; \text{in}$$

The steel area is given by Eq. 55.24.

$$A_s = \frac{M_u}{\phi f_y (d - \lambda)}$$
$$= \frac{2420.6 \; \text{in-kips}}{(0.9)\left(60 \; \frac{\text{kips}}{\text{in}^2}\right)(12.25 \; \text{in} - 1.22 \; \text{in})}$$
$$= 4.06 \; \text{in}^2$$

Check the assumed value of λ. Use Eq. 50.17.

$$A_c = \frac{f_y A_s}{0.85 f'_c}$$
$$= \frac{\left(60 \; \frac{\text{kips}}{\text{in}^2}\right)(4.06 \; \text{in}^2)}{(0.85)\left(4 \; \frac{\text{kips}}{\text{in}^2}\right)}$$
$$= 71.64 \; \text{in}^2$$

Since the compressed zone is rectangular with a width of 111 in, the value of λ is given by Eq. 50.20.

$$\lambda = \frac{A_c}{2b} = \frac{71.64 \; \text{in}^2}{(2)(111 \; \text{in})} = 0.32 \; \text{in}$$

Using $\lambda = 0.32$ in, the revised steel area is given by Eq. 55.24.

$$A_s = \frac{M_u}{\phi f_y (d - \lambda)}$$
$$= \frac{2420.6 \; \text{in-kips}}{(0.9)\left(60 \; \frac{\text{kips}}{\text{in}^2}\right)(12.25 \; \text{in} - 0.32 \; \text{in})}$$
$$= 3.76 \; \text{in}^2$$

The minimum steel is

$$A_{s,\min} = (0.0018)(111 \; \text{in})(16 \; \text{in})$$
$$= 3.20 \; \text{in}^2 \quad \text{[does not control]}$$

step 4: Check the development length of the flexural reinforcement. For uncoated no. 6 bars, using Eq. 55.27, where $\psi_t = 1.0$, $\psi_e = 1.0$, and $\lambda = 1.0$ per ACI 318 12.2.4,

$$l_d = \frac{3 d_b f_y \psi_t \psi_e \lambda}{50 \sqrt{f'_c}}$$
$$= \frac{(3)(0.75 \; \text{in})\left(60{,}000 \; \frac{\text{lbf}}{\text{in}^2}\right)(1.0)(1.0)}{50\sqrt{4000 \; \frac{\text{lbf}}{\text{in}^2}}}$$
$$= 42.7 \; \text{in}$$

The available distance for development (with a 2 in cover at the tip of the bar) is 49.5 in − 2 in = 47.5 in. Since the available distance exceeds l_d, the anchorage is adequate. Use 10 no. 6 bars spaced evenly in each direction. Using 2 in end cover on each bar end, the spacing is

$$\frac{111 \; \text{in} - (2)(2 \; \text{in}) - (10 \; \text{bars})(0.75 \; \text{in})}{9 \; \text{gaps}}$$
$$= 11 \; \text{in} > 2 d_b$$

The spacing does not exceed the maximum permissible value of 18 in.

56 Pretensioned Concrete

Nomenclature

A	area	in^2	mm^2
A'	area of compression reinforcement	in^2	mm^2
A_{ps}	area of prestressed reinforcement in tension zone	in^2	mm^2
A_s	area of nonprestressed tension reinforcement	in^2	mm^2
b	width of compression face of member	in	mm
d	distance from extreme compression fiber to centroid of nonprestressed tension reinforcement ($\geq 0.8h$ (ACI 318 11.0))	in	mm
d'	distance from extreme compression fiber to centroid of compression reinforcement	in	mm
d_p	distance from extreme compression fiber to centroid of prestressed reinforcement	in	mm
D	dead load	lbf	N
e	eccentricity	in	mm
E_c	modulus of elasticity of concrete	psi	Pa
f'_c	specified compressive strength of concrete	psi	Pa
f'_{ci}	compressive strength of concrete at time of initial prestress	psi	Pa
f_{pu}	specified tensile strength of prestressing tendons	psi	Pa
f_{py}	specified yield strength of prestressing tendons	psi	Pa
f_{se}	effective stress in prestressed reinforcement after all prestress losses	psi	Pa
f_y	specified yield strength of nonprestressed reinforcement	psi	Pa
h	overall thickness	in	mm
I	moment of inertia	in^4	mm^4
L	live load	lbf	N
M	moment	in-lbf	N·m
P	normal load	lbf	N
w	unit weight of concrete	lbf/ft^3	–
y	distance	in	mm

Symbols

β_1	equivalent stress block ratio	–	–
Δ	deflection	in	mm
γ_p	tendon prestressing factor	–	–
λ	long-term deflection factor	–	–
ρ	nonprestressed reinforcement ratio	–	–
ρ_p	prestressed reinforcement ratio	–	–
ϕ	strength reduction factor (load factor)	–	–
ω	$\rho f_y / f'_c$	–	–
ω'	$\rho' f_y / f'_c$	–	–

Subscripts

c	concrete
ci	concrete, at the time of initial prestress
cr	cracking
d	dead load
e	effective
l	live load
p	prestressing tendons
ps	prestressed reinforcement in tension zone
s	nonprestressed tension reinforcement
sp	steel prestressing
u	ultimate (factored) tensile
y	yield

1. INTRODUCTION

Concrete is strong under compressive stresses but weak under tensile stresses. The tensile strength of concrete varies from 8 to 14% of its compressive strength. Because of the low tensile strength of concrete, flexural cracks develop at early stages of loading.

In traditional reinforced concrete structures, steel provides the tensile strength that the concrete is incapable of developing. However, it is also possible to impose an initial state of stress such that, when the stresses from the applied loads are added, no tensile stresses occur. The initial state of stress is provided by high-strength tendons compressing the concrete. The prestressing force keeps all or most of the member's cross section in compression.

2. PRETENSIONING AND POST-TENSIONING

Prestressing can be achieved by two different procedures: pretensioning and post-tensioning. In *pretensioned construction*, the tendons are pulled to the specified tensile force before placing the concrete. Then, concrete is poured around the tendons. After the concrete reaches the specified strength, f'_{ci}, the tendons are cut at the anchorages. The compressive stress caused by the shortening of the wires transfers to the concrete through steel-concrete bond.

In *post-tensioned construction*, the tendons are free to slide within the member inside tubes, sheaths, or conduits. (Early tendons were paper-wrapped bundles of greased wires.) After the concrete hardens, the tendons are tensioned by anchoring them to one end and pulling ("jacking") from the other. Tubes may or may not be subsequently grouted ("bonded") to prevent corrosion. The tubes can also be filled with corrosion-resistant grease.

3. BONDED AND UNBONDED TENDONS

Unbonded tendons can be used in relatively shallow construction. Unbonded construction has made the post-tensioning of longer-span flat plates, shallow beams, and slabs practical and economical.

Grouting bonds tendons directly to the concrete. If a bonded tendon breaks at any point, it is still capable of providing local prestress for part of the structure. A failed unbonded tendon, on the other hand, stops contributing when it fails. Unbonded failure is sudden; bonded failure is gradual. Because of this, bonded tendons may be required in areas subject to seismic loading. (However, unbonded tendons were originally used in nuclear power plants where they could be replaced as needed.)

The ultimate strength of a simple beam with unbonded tendons is about 30% less than the same beam with bonded tendons. The ultimate strength of a continuous unbonded beam or slab versus a bonded beam or slab is about 20% less. Unbonded tendons do not help in controlling crack sizes and growth.

4. BENEFITS AND DISADVANTAGES

Prestressed members are shallower in depth than conventional reinforced concrete members for the same spans and loading conditions. In general, the depth of a prestressed concrete member is usually about 65 to 80% of the depth of the equivalent conventionally reinforced concrete member. This can lead to savings due to reductions in concrete and steel. However, the materials needed for prestressed construction are of higher quality and are therefore more expensive, so the cost savings may be minimal.

Prestressing significantly reduces the amount of cracking in a beam at service load levels. Since uncracked sections have a larger moment of inertia than cracked sections, the deflection of a prestressed beam for a given applied load is less than that of a conventionally reinforced beam. The increased stiffness of prestressed construction permits smaller, lighter sections to be used,

thereby reducing dead weight and some components of the total cost. Other benefits are increased protection of the steel from the environment and the fact that the prestressing can be used to induce camber.

The dead weight of nonprestressed concrete members in excess of 70–90 ft (21–27 m) long can become excessive, resulting in greater long-term deflection and cracking. Therefore, the best construction technique for long-span concrete members is prestressed concrete. In fact, very long concrete spans, such as those found in cable-stayed bridges, can only be constructed with prestressing.

Prestressed concrete members have high toughness (i.e., the ability to absorb energy under impact loads). The fatigue resistance of prestressed concrete members is also high due to the low steel stress variation.

The disadvantages of prestressed construction are the costs associated with the prestressing itself and the increased costs that derive from higher quality materials, inspection, testing, and quality control.

Figure 56.1 *Prestressing Effect on a Simple Beam*

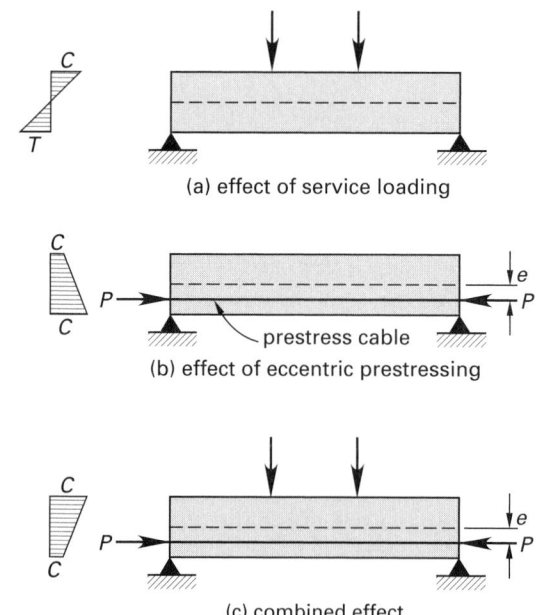

(a) effect of service loading

(b) effect of eccentric prestressing

(c) combined effect

5. PRESTRESSING LIMITATIONS

When a live load acting on a beam is a large fraction of the total load but acts intermittently, it may not be possible to design the beam so that it has no tensile stresses under full live load and still behaves appropriately when the live load is not present. High prestressing required for a full live load may be such that, when the live load is not present, the beam deforms upward excessively. This upward camber can be exacerbated by long-term effects from creep and shrinkage. Very large prestressing forces may also produce enough axial shortening in the beam to damage other framing members.

In such cases, it is better to allow some tensile stresses to develop when the beam is fully loaded. Members in which tensile stresses are allowed under service loads are termed *partially prestressed members*. When no tensile stresses are permitted, the term *fully prestressed members* is utilized.

6. CONSTRUCTION METHODS AND MATERIALS

The performance of prestressed concrete structures is related directly to the quality of the materials used. Strict production methods and quality assurance are needed at the various stages of production, construction, and maintenance.

High-quality concrete and steel tendons are used. Minimum center-to-center spacing of strands is covered in ACI 318 7.6.7.1. The two major qualities of concrete are strength and long-term endurance. Concrete strengths of 5000 and 6000 psi (34.5 and 41.3 MPa) are common. The specified concrete strength f'_{cc} must be achieved by the time of prestressing.

Accelerating agents containing chlorides are excluded because they contribute to strand corrosion.

7. CREEP AND SHRINKAGE

Long-term deflections (deformations) related to creep and shrinkage can reduce the prestressing forces and can lead to unexpected failure. Both creep and shrinkage are time-dependent phenomena. *Creep* is the deformation that occurs under a constant load over a period of time. Creep depends primarily on loading and time, but is also influenced by the composition of concrete, the environmental conditions, and the size of the member.

Shrinkage is a volume change that is unrelated to load application. Shrinkage is closely related to creep. As a general rule, concrete that has a low volume change also has a low creep tendency.

Creep and shrinkage contribute to a loss of prestressing force as well as increased deflection. For prestressed members without compression reinforcement, the long-term deflection can be estimated from the immediate deflection by use of a *long-term deflection factor*, λ. For members without compression reinforcement, λ has values of 1.0 for three months, 1.2 for six months, 1.4 for twelve months, and 2.0 for five years or more [ACI 318 9.5.2.5].

$$\Delta_{\text{long term}} = \lambda \Delta_{\text{immediate}} \qquad 56.1$$

8. PRESTRESS LOSSES

Several effects tend to reduce the prestress, including tendon seating, creep and shrinkage in the concrete,

elastic shortening of the member from the applied compressive stress, movements at the anchorage sitting, relaxation of the steel tendons, and friction between the tendons and the tubes (in the case of ungrouted post-tensioned construction). The strains associated with these losses are close to the strain associated with yielding of conventional 40,000 psi (270 MPa) reinforcement, so regular-grade steel cannot be used for prestressing. The tendons used for prestressing are manufactured from steel with an ultimate tensile strength (f_{pu}) of 250 ksi or 270 ksi (1.70 GPa or 1.90 GPa), and the typical losses average 25 to 35 ksi (170 to 240 MPa).

High early strength concrete, such as that made with type III cement, with a compressive strength, f'_c, of 4000–10,000 psi (27.6–69.0 MPa) is commonly used. This concrete sets up quickly and suffers smaller elastic compression losses. Due to the higher quality of concrete, the *modular ratio*, n, is typically lower than in conventional construction.

$$n = \frac{E_s}{E_c} \qquad 56.2$$

Losses of prestressing force occur in the prestressing steel due to relaxation of steel and creep and shrinkage of the concrete. This type of loss is known as *time-dependent loss*. On the other hand, there are immediate elastic losses due to fabrication or construction techniques. Those include elastic shortening of concrete, anchorage losses, and frictional losses. The exact amounts of these losses are difficult to estimate analytically, though methods exist. There are also several empirical methods for estimating prestress losses. Table 56.1 shows the lump-sum losses permitted by AASHTO. (A *lump-sum loss* is a rule-of-thumb loss that is applied to the design without more analytical methods of determination.)

Table 56.1 AASHTO Lump-Sum Losses[a,b] in psi (MPa)

type of prestressing steel	total loss	
	4000 psi concrete (27.6 MPa concrete)	5000 psi concrete (34.5 MPa concrete)
pretensioning strand		45,000 (310)
post-tensioning wire or strand	32,000 (221)	33,000 (228)
bars	22,000 (152)	23,000 (159)

(Multiply psi by 0.00689 to obtain MPa.)
[a] For normal-weight concrete, normal prestress levels, and average exposure conditions. Use exact methods for exceptionally long spans and unusual designs.
[b] Losses due to friction are excluded. Such losses should be calculated according to AASHTO specifications.

Adapted from *Standard Specifications for Highway Bridges*, 17th ed., 2002, AASHTO, Div. 1, Table 9.16.2.2, p. 236.

9. DEFLECTIONS

Prestressed concrete members are generally thinner than conventional reinforced concrete members. This makes them subject to increased deflection. Deflection consists of the short-term (instantaneous) and long-term effects.

Several assumptions are made when calculating deflection. (a) The modulus of elasticity of concrete is given by the ACI empirical equation $E_c = 33w^{1.5}\sqrt{f'_c}$ [ACI 318 8.5.1]. (b) The moment of inertia is calculated for the concrete cross-sectional area. (c) Superposition can be used to calculate the combined deflection due to applied loads and the opposing *camber*. (d) All of the strands act as a single tendon with a combined steel area. (e) Deflection computations can be based on the center of gravity of the prestressing strands (CGS). (f) Deflections resulting from shear deformations can be ignored.

Figure 56.2 *Midspan Deflections from Prestressing*[a]

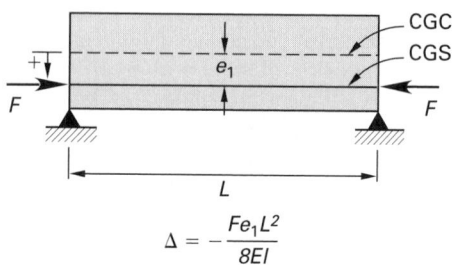

$$\Delta = -\frac{Fe_1 L^2}{8EI}$$

(a) straight tendons

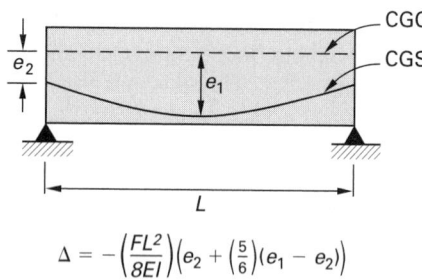

$$\Delta = -\left(\frac{FL^2}{8EI}\right)\left(e_2 + \left(\frac{5}{6}\right)(e_1 - e_2)\right)$$

(b) parabolically draped tendons

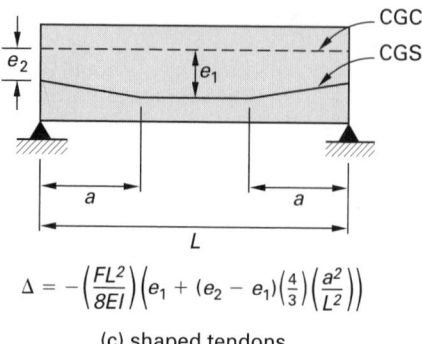

$$\Delta = -\left(\frac{FL^2}{8EI}\right)\left(e_1 + (e_2 - e_1)\left(\frac{4}{3}\right)\left(\frac{a^2}{L^2}\right)\right)$$

(c) shaped tendons

[a]CGC—centroid of concrete; CGS—centroid of prestressing tendons

10. STRANDS

Prestressing tendons can be single wires, strands composed of several wires twisted together, and high-strength bars. Seven-wire strands, manufactured by twisting six wires around a larger, straight central wire, are the most widely used. Strands come in two grades, 250 ksi and 270 ksi (1.70 GPa and 1.90 GPa), representing the ultimate tensile strength. Different diameters are available.

The high-strength steel strands are stress-relieved to reduce residual stresses caused by cold-working during winding. Stress-relieving consists of heating the steel to about 930°F (500°C). The stress-relieving process improves the ductility of steel and reduces the stress relaxation. *Stress relaxation* is the loss of prestress when wires or strands are subject to constant strain. It is similar to creep in concrete, except that creep is a change in strain whereas relaxation in steel is a loss in steel stress.

Aramid tendons are manufactured from aramid fibers. Aramid tendons are the focus of ongoing research and experimental use. Tensile rigidities are approximately one-fifth of those from prestressing steel, so uses are limited to special applications.

Figure 56.3 *Typical Cross Sections for Seven-Wire Prestressing Tendons*

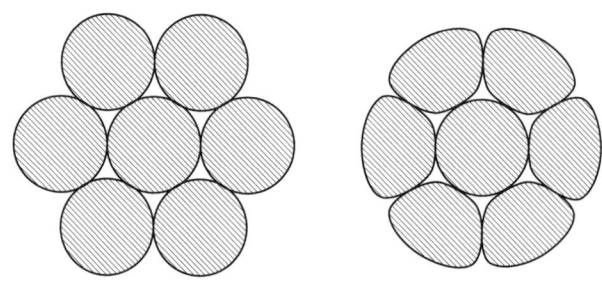

11. CORROSION PROTECTION

Any reduction in the cross-sectional area of prestressing tendons reduces the nominal moment strength of the prestressed section. Therefore, prestressing tendons should be protected against corrosion. In pretensioned members protection against corrosion is provided by the concrete surrounding the tendon, while in post-tensioned members grouting is commonly used to prevent moisture from entering the ducts. However, concrete with grouted tendons is susceptible to freeze-induced cracking if grouting occurs during the winter. Therefore, other options, including heat-sealed and extruded coatings, can be specified.

Table 56.2 ASTM Standard Prestressing Tendons

type[a]		nominal diameter (in)	nominal area (in²)	nominal weight (lbf/ft)
grade 250[b]	$\frac{1}{4}$	(0.250)	0.036	0.122
seven-wire	$\frac{5}{16}$	(0.313)	0.058	0.197
strand	$\frac{3}{8}$	(0.375)	0.080	0.272
	$\frac{7}{16}$	(0.438)	0.108	0.367
	$\frac{1}{2}$	(0.500)	0.144	0.490
	—	(0.600)	0.216	0.737
grade 270[b]	$\frac{3}{8}$	(0.375)	0.085	0.290
seven-wire	$\frac{7}{16}$	(0.438)	0.115	0.390
strand	$\frac{1}{2}$	(0.500)	0.153	0.520
		(0.600)	0.217	0.740
prestressing	0.192		0.029	0.098
wire	0.196		0.030	0.100
	0.250		0.049	0.170
	0.276		0.060	0.200
smooth	$\frac{3}{4}$		0.44	1.50
prestressing	$\frac{7}{8}$		0.60	2.04
bars	1		0.78	2.67
	$1\frac{1}{8}$		0.99	3.38
	$1\frac{1}{4}$		1.23	4.17
	$1\frac{3}{8}$		1.48	5.05
deformed	$\frac{5}{8}$		0.28	0.98
prestressing	$\frac{3}{4}$		0.42	1.49
bars	1		0.85	3.01
	$1\frac{1}{4}$		1.25	4.39
	$1\frac{3}{8}$		1.58	5.56

(Multiply in by 25.4 to obtain mm.)
(Multiply in² by 645 to obtain mm².)
(Multiply lbf/ft by 1.51 to obtain kg/m.)
[a]All sizes in all types may not be readily available.
[b]Strand grade is f_{pu} in kips/in².

12. MAXIMUM STRESSES

The maximum stresses in concrete and reinforcement permitted by ACI are presented in Table 56.3. Prestressed flexural members are classified in ACI 318 18.3.3 as Class U, Class T, or Class C based on the computed extreme fiber stress, f_t, at service loads in the precompressed tensile zone, as follows.

(a) Class U: $f_t \le 7.5\sqrt{f'_c}$

(b) Class T: $7.5\sqrt{f'_c} < f_t \le 12\sqrt{f'_c}$

(c) Class C: $f_t > 12\sqrt{f'_c}$

Prestressed two-way slab systems must be designed as Class U, with $f_t \le 6\sqrt{f'_c}$.

For Class U and Class T flexural members, stresses at service loads are permitted to be calculated using the uncracked section. For Class C flexural members, stresses at service loads must be calculated using the cracked transformed section. Class C flexural members are subject to the deflection and crack control requirements of ACI 318.

The maximum permissible stresses permitted by AASHTO are presented in Table 56.4. The ACI classification of prestressed flexural members has not yet been adopted by AASHTO.

AASHTO also specifies the following strength reduction factors and load factors for prestressed concrete construction: for post-tensioned anchorage zones exposed to maximum tendon jacking force, load factor $= 1.2$; for factory-produced precast prestressed members, $\phi = 1$; for post-tensioned cast-in-place members, $\phi = 0.95$; for shear, $\phi = 0.90$; for anchorage zones with normal-weight concrete, $\phi = 0.85$; and for anchorage zones with lightweight concrete, $\phi = 0.70$ [AASHTO *Bridges*, Div. 1, Sec. 9.14].

13. ACI CODE PROVISIONS FOR STRENGTH

The ACI code contains the following strength-related provisions. (AASHTO specifications are similar but not identical to those in ACI 318 App. B.)

(1) The moment due to factored loads, M_u, cannot exceed the design moment strength, ϕM_n.

(2) The design moment strength, ϕM_n, must be at least 1.2 times the cracking moment, M_{cr}, calculated from the modulus of rupture [ACI 318 18.8.2].

(3) The area of prestressed and non-prestressed reinforcement used to compute the design moment strength, ϕM_n, of prestressed flexural members is limited in ACI 318 App. B to ensure that failure is initiated by yielding of the steel and not by crushing of the concrete in compression. (Exact values are specified in ACI 318 B.18.8.1.) In ACI 318 Ch. 18, prestressed concrete sections, like reinforced concrete sections, are classified as either tension controlled, transition, or compression controlled, in accordance with ACI 318 Secs. 10.3.3 and 10.3.4. The appropriate ϕ factor from ACI 318 Sec. 9.3.2 applies.

The design moment strength, ϕM_n, is calculated in the same manner as for a conventional reinforced section. The only aspect that is specific to prestressed construction is the fact that the tension in the steel at failure is computed from formulas that apply for specific circumstances. For example, in ACI 318 18.7.2, the common case of members with bonded tendons, the ultimate stress in the tendons, f_{ps}, is

$$f_{ps} = f_{pu}\left(1 - \left(\frac{\gamma_p}{\beta_1}\right)\left(\rho_p\left(\frac{f_{pu}}{f'_c}\right) + \left(\frac{d}{d_p}\right)(\omega - \omega')\right)\right)$$

[bonded tendons] *56.3*

Table 56.3 *ACI Maximum Stresses in Prestressed Concrete*

Concrete Stresses in Flexure

Stresses in concrete immediately after prestress transfer (before time-dependent prestress losses) shall not exceed the following:

(1)	extreme fiber stress in compression		$0.60f'_{ci}$
(2)	extreme fiber stress in tension except as permitted in (3)	[U.S.]	$3\sqrt{f'_{ci}}$
		[SI]	$\frac{1}{4}\sqrt{f'_{ci}}$
(3)	extreme fiber stress in tension at ends of simply supported members	[U.S.]	$6\sqrt{f'_{ci}}$
		[SI]	$\frac{1}{2}\sqrt{f'_{ci}}$

For Class U and Class T prestressed flexural members, stresses in concrete at service loads (based on uncracked section properties and measured after all prestress losses) shall not exceed the following:

(1)	extreme fiber stress in compression due to prestress plus sustained load	$0.45f'_c$
(2)	extreme fiber stress in compression due to prestress plus total load	$0.60f'_c$

Prestressing Steel Stresses

(1)	due to tendon jacking force (but not greater than the lesser of $0.80f_{pu}$ or the maximum value recommended by the manufacturer of prestressing tendons or anchorage devices)	$0.94f_{py}$
(2)	immediately after prestress transfer (but not more than $0.74f_{pu}$)	$0.82f_{py}$
(3)	post-tensioning tendons at anchorage devices and couplers, immediately after force transfer	$0.70f_{pu}$

Source: ACI 318 18.4 and 18.5

γ_p is the tendon prestressing factor, equal to 0.55 when $0.80 \leq f_{py}/f_{pu} < 0.85$; 0.40 when $0.85 \leq f_{py}/f_{pu} < 0.90$; and 0.28 when $f_{py}/f_{pu} \geq 0.90$.

ρ_p is the *prestressing steel ratio.*

$$\rho_p = \frac{A_{ps}}{bd_p} \qquad \qquad 56.4$$

If any compression reinforcement is taken into account, the negative part of Eq. 56.3 cannot be less than 0.17, and the distance from the extreme compression fiber to the centroid of the compression reinforcement, d', cannot be greater than $0.15d_p$, where d_p is the distance from the extreme compression fiber to the centroid of the prestressed reinforcement.

$$\rho_p\left(\frac{f_{pu}}{f'_c}\right) + \left(\frac{d}{d_p}\right)(\omega - \omega') \geq 0.17 \qquad 56.5$$

For Eq. 56.3 to apply, the stress in the tendons after all losses have been subtracted shall not be less than $0.5f_{pu}$ [ACI 318 18.7.2]. Other formulas for f_{ps} can be found in ACI 318 18.7.2.

The *cracking moment*, M_{cr}, is computed as the moment required to induce a tensile stress equal to the modulus or rupture. In this calculation, the force in the tendons after all losses must be used. Long-term losses are

frequently assumed to be 35,000 psi (240 MPa) for pretensioned construction and 25,000 psi (170 MPa) for post-tensioned construction.

14. ANALYSIS OF PRESTRESSED BEAMS

If a beam's tendon layout is known, the flexural strength to support a particular load distribution can be determined.

Service Load Review

step 1: Compute the moment of inertia for the cross section. If the beam is post-tensioned and the tendons are not grouted, the area of steel of the tendons should not be considered.

step 2: Calculate the eccentricity, e, of the tendons as the distance from the centroid of the tendons to the centroid of the gross concrete section.

step 3: Compute the stress distribution in the concrete at transfer from Eq. 56.6. P is the net prestress force acting on the beam before losses, M_s is the moment due to the loads on the beam at transfer (typically the self-weight only), and y is the distance from

Table 56.4 *AASHTO Maximum Stresses in Prestressed Concrete*

Concrete Stresses in Flexure

Stresses in concrete immediately after prestress transfer (before time-dependent prestress losses) shall not exceed the following:

(1) extreme fiber stress in compression

- pretensioned members $0.60f'_{ci}$
- post-tensioned members $0.55f'_{ci}$

(2) extreme fiber stress in tension

- precompressed tensile zone[a] —
- other areas without bonded reinforcement, lesser of

  [U.S.] 200 psi and $3\sqrt{f'_c}$

  [SI] 1.38 MPa and $\frac{1}{4}\sqrt{f'_c}$

- other areas with bonded reinforcement

  [U.S.] $7.5\sqrt{f'_c}$

  [SI] $\frac{5}{8}\sqrt{f'_c}$

Stresses in concrete at service loads (after all prestress losses) shall not exceed the following:

(1) compressive stresses under all load combinations except as stated in (2) and (3) below $0.60f'_c$

(2) extreme fiber stress in compression $0.40f'_c$

(3) extreme fiber stress in tension in precompressed tensile zone

- with bonded reinforcement, normal exposure

  [U.S.] $6\sqrt{f'_c}$

  [SI] $\frac{1}{2}\sqrt{f'_c}$

- with bonded reinforcement, in coastal areas and other severe corrosive exposures

  [U.S.] $3\sqrt{f'_c}$

  [SI] $\frac{1}{2}\sqrt{f'_c}$

- without bonded reinforcement 0

Cracking stress

(1) normal-weight concrete

  [U.S.] $7.5\sqrt{f'_c}$

  [SI] $\frac{5}{8}\sqrt{f'_c}$

(2) sand-lightweight concrete

  [U.S.] $6.3\sqrt{f'_c}$

  [SI] $0.525\sqrt{f'_c}$

(3) all other lightweight concrete

  [U.S.] $5.5\sqrt{f'_c}$

  [SI] $0.458\sqrt{f'_c}$

Anchorage bearing stress (post-tensioned anchored at service loads), lesser of

  [U.S.] $0.9f'_{ci}$ and 3000 psi

  [SI] $0.9f'_{ci}$ and 20.7 MPa

Prestressing Steel Stresses

(1) after seating

- pretensioned members

  stress-relieved strands $0.70f_{pu}$

  low-relaxation strands $0.75f_{pu}$

- post-tensioned members $0.70f_{pu}$

  at the end of seating loss zone[b] $0.83f_{py}$

(2) at service load after losses[c] $0.80f_{py}$

[a]AASHTO does not specify tension limits for the precompressed tensile zone. Other specifications may apply.
[b]Overstressing up to $0.90f_{py}$ for short periods of time to offset seating and friction losses as long as the final anchorage stress does not exceed specified values. Includes bonded prestressed strands. Other limitations apply.
[c]Includes bonded prestressed strands.

Source: *Standard Specifications for Highway Bridges*, 17th ed., 2002, AASHTO, Div. 1, Sec. 9.15.

the centroidal axis to the fiber in question. y is positive upward. Negative stresses indicate compression.

$$f_c = \frac{-P}{A} + \frac{Pey}{I} - \frac{M_s y}{I} \qquad 56.6$$

step 4: Compute the stresses in the concrete when the full service load is acting. These stresses are also computed with Eq. 56.6 with the following changes: P is the force in the tendons after all losses, and M_s is the moment due to the full service load.

Strength Review

step 5: Compute the ultimate stress in the tendons from Eq. 56.3. (Use the appropriate ACI 318 code equation if the tendon is not bonded.)

step 6: Use the stress from step 5 to calculate the design moment strength, ϕM_n. Compare this to the maximum moment due to factored loads, M_u. If ϕ is based on a tension-controlled section, it must be verified that the section is in fact tension controlled.

step 7: Use Eq. 56.6 to compute the cracking moment. The cracking moment is the value of M_s required to induce a tension stress equal to the modulus of rupture. Check to see if the requirement $\phi M_n \geq 1.2 M_{cr}$ is satisfied.

15. SHEAR IN PRESTRESSED SECTIONS

The shear design requirement for prestressed concrete members is the same as for conventionally reinforced concrete members.

$$V_u < \phi V_n \qquad 56.7$$

For prestressed and reinforced concrete members, the nominal shear strength, V_n, is the sum of the shear strengths provided by the concrete and the steel.

$$V_n = V_c + V_s \qquad 56.8$$

For prestressed concrete members, ACI 318 11.1.3.2 states that the critical section for computing the maximum factored shear V_u is located at a distance of $h/2$ from the face of the support. This is different from the provisions for conventional reinforced concrete members where the critical section is located at a distance d from the face of the support.

57
Composite Concrete and Steel Bridge Girders

Nomenclature

b_e	effective slab width	in	mm
b_f	flange width	in	mm
b_o	center-to-center spacing of beams (girders)	in	mm
E_c	modulus of elasticity of concrete	lbf/in²	MPa
E_s	modulus of elasticity of steel	lbf/in²	MPa
f'_c	compressive strength	lbf/in²	MPa
L	beam (girder) span	ft	m
n	modular ratio	–	–
t	slab thickness	in	mm

Subscripts

b	bottom
c	compressive or concrete
e	effective
f	flange
o	center-to-center
s	steel
t	top

1. INTRODUCTION

Any conventionally reinforced concrete structure could be considered a composite structure since it contains two dissimilar materials. However, the term "composite member" normally refers to the combination of concrete with structural steel in a member as shown in Fig. 57.1.

The composite construction shown in Fig. 57.1(a) is widely used for highway bridges. A cast-in-place reinforced concrete slab is bonded to a steel beam, producing a stronger and stiffer structure. Fig. 57.2 shows a typical highway bridge utilizing composite construction.

There are two authoritative documents governing composite design: (a) AASHTO's *Standard Specifications for Highway Bridges* (*Bridges* or AASHTO), and (b) the *AISC Manual*. The requirements imposed by these two documents are similar, though not always exactly the same. In addition, ACI 318 applies to the concrete parts of composite construction.

Figure 57.1 *Composite Members*

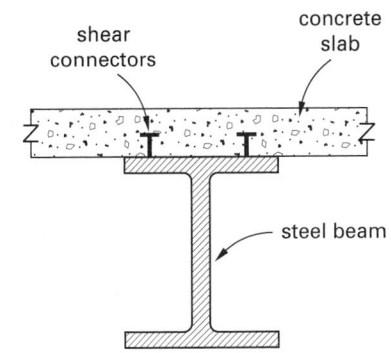

(a) concrete deck slab with steel beam (side view)

(b) concrete-encased steel column (end view)

Figure 57.2 *Highway Bridge Deck*

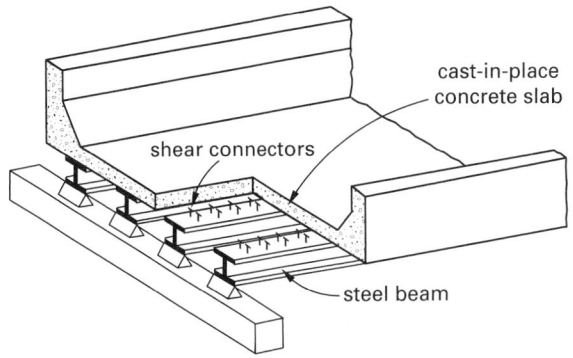

2. COMPOSITE ACTION

Composite construction depends on *composite action* between the two materials. Composite action means that the steel and the concrete components act together to resist loading. There is a tendency for slip to occur at the interface between the concrete slab and the steel beam. Slippage nullifies composite action. The tendency to slip is maximum near the supports, where the vertical shear (and horizontal) forces are the greatest. Tendency to slip is zero at the point of maximum bending moment.

To prevent slippage from occurring, *shear connectors* (straight studs, L-connectors, or steel channels) are welded to the top flange of the steel beam as shown in Fig. 57.3. When the concrete is poured, the encased studs bond the beam to the slab.

According to the AASHTO specifications, the following criteria for shear connectors should be met. (a) The clear depth of concrete cover over the tops of all shear connectors must be not less than 2 in. (b) The connectors must penetrate at least 2 in above the bottom of the slab. (c) The clear distance between the edge of the beam flange and the edge of the connector must be not less than 1 in. (d) The welds for channel shear connectors must be at least $^3/_{16}$ in fillets [*Bridges*, Div. 1, Sec. 10.38.2]. Adjacent stud shear connectors shall not be closer than 4 diameters center to center.

AASHTO specifications also contain details of requirements that the shear connectors be designed for fatigue and checked for strength [*Bridges*, Div. 1, Sec. 10.38.5].

Figure 57.3 *Types of Shear Connectors*

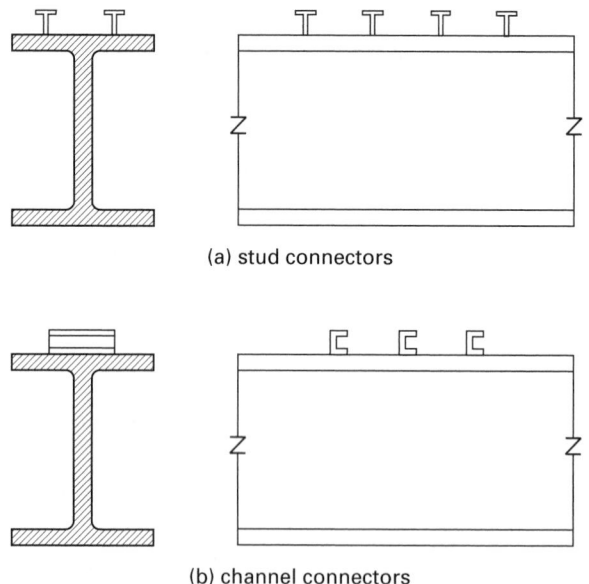

(a) stud connectors

(b) channel connectors

3. ADVANTAGES

The primary advantages of composite construction are as follows. (a) The load-carrying capacity of a composite system is greater than that of the same system in the absence of composite action. (b) A composite system is stronger and more rigid than a noncomposite system. (c) Smaller and shallower steel beams can be used. This will provide economic savings in the steel to

be used for a given project. The savings in weight can reach about 20–30%. (d) Spans can be longer without exceeding the allowable deflection.

4. DISADVANTAGES

The disadvantages of composite construction are as follows. (a) There are additional materials and labor costs related to the shear connectors. However, the savings from using smaller sections generally offset the additional costs of the connectors. (b) The construction of composite members requires additional supervision and quality control.

Despite these disadvantages of composite construction, a large number of medium-span highway bridges in the United States are constructed using the slab-beam composite system.

5. EFFECTIVE SLAB WIDTH

In the slab-beam composite system used for bridge decking, the steel beams (girders) are placed parallel to each other in the direction parallel to the traffic. The steel beams can be standard W shapes with cover plates to increase the resistance of the beam in the regions of high moment, or they can be built-up plate girders.

The concrete slab is designed for bending in the transverse direction (i.e., is designed as a one-way slab). However, all of the slab may not be considered effective in resisting compressive stresses. The portion of the slab that resists the compressive stresses is the *effective width*, b_e. The *AISC Manual* [Sec. I3] states that the effective width of an interior beam with slab extending on both sides is the smaller of the following.

$$b_e = \text{smaller of } \{L/4; \ b_o\} \qquad 57.1$$

According to the AASHTO specifications [*Bridges*, Div. 1, Sec. 10.38.3.1], the effective width of the compression flange for an interior girder may not exceed the following.

$$b_e = \text{smallest of } \{L/4; \ 12t; \ b_o\} \qquad 57.2$$

For an exterior girder, the AASHTO specifications [*Bridges*, Div. 1, Sec. 10.38.3.2] state that the effective width may not exceed the following.

$$b_e = \text{smallest of } \{L/12; \ 6t; \ \tfrac{1}{2}b_o\} \qquad 57.3$$

Figure 57.4 *Effective Width*

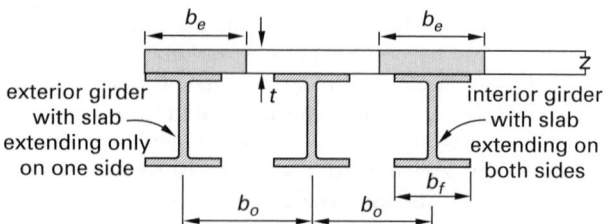

6. SECTION PROPERTIES

In order to compute the stresses for a nonhomogeneous cross section, the section properties (e.g., centroidal axis location, moment of inertia, and section modulus) are computed using transformed cross sections.

The concrete modulus of elasticity may be calculated from the ACI formula. The minimum *modular ratio*, *n*, permitted by AASHTO is 6 [*Bridges*, Div. 1, Sec. 10.38.1.3].

$$n = \frac{E_s}{E_c} \qquad 57.4$$

The transformed slab width is calculated by dividing b_e by n. This "transforms" the concrete slab to steel. The neutral axis and moment of inertia are calculated for the transformed section.

Table 57.1 Modular Ratio[a]

compressive strength, f_c'		modular ratio[b]
(psi)	(MPa)	n
3000	21	9
3500	24	8.5
4000	28	8
4500	31	7.5
5000	35	7
6000	42	6.5
AASHTO recommendations[c]		
2000–2300	14–16	11
2400–2800	17–19	10
2900–3500	20–24	9
3600–4500	25–31	8
4600–5900	32–41	7
> 6000	> 41	6

(Multiply psi by 0.00689 to obtain MPa.)
[a]Based on a concrete unit (specific) weight of 145 lbf/ft³ and a steel modulus of elasticity of 29,000,000 psi
[b]The modular ratio has been rounded to the nearest whole number. ACI 318 used to contain a specific provision permitting this.
[c]*Bridges*, Div. 1, Sec. 10.38.1.3

7. STRENGTH OF COMPOSITE SECTIONS

The ultimate strength of the composite section depends on the (a) section properties of the steel beam (girder), (b) yield strength of the steel beam, (c) compressive strength of concrete used for the slab, and (d) capacity of the shear connectors.

In analyzing a composite section, it is assumed that the shear connectors transfer the total shear at the steel-concrete interface. It is also assumed that concrete carries only compressive stress, and the small tensile stress capacity of concrete is ignored.

When analyzing or designing a composite section, the effects of noncomposite and composite loads should be considered. Also, the method of construction of composite members (with or without shoring) will have an effect on the actual stresses resulting from loads.

Noncomposite loads are those loads acting on the steel beam before the concrete slab hardens and can contribute to the strength. These loads include the weight of the steel beam, the weight of the fresh concrete slab, and the weight of formwork and bracing members. In unshored construction, the flexural stresses caused by these loads are resisted by the steel beam alone. *Composite loads* are those loads that act on the section after concrete has hardened. Composite loads include live loads from traffic, snow, and ice, and any additional dead load such as that of future roadway surfacing, sidewalks, railing, and permanent utilities fixtures.

Erection of composite members can be done either without shoring (unshored) or with shoring (shored). *Unshored construction* is the simplest construction method. The steel beams are placed first and used to support the formwork, fresh concrete, and live loads of construction crew and equipment. The steel beam acts noncompositely. It supports all noncomposite loads by itself.

With *shored construction*, the steel beams are supported by temporary falsework until the concrete cures. All noncomposite loads, including the steel beam itself, are supported by the falsework. After the concrete cures, the falsework is removed and the section acts compositely to resist all noncomposite and composite loads. Shored construction results in a reduction in the service load stresses.

The ultimate moment capacity of the composite section depends on the location of the neutral axis of the composite section. There are two possible locations of the neutral axis. (a) If the neutral axis is within the concrete slab, the slab is capable of resisting the total compressive force. In this case, the slab is "adequate." (b) If the neutral axis is within the steel beam below the concrete slab, the slab resists only a portion of the compressive force. The remainder is carried by the steel beam. The slab is said to be "inadequate." The calculation of the ultimate moment capacities in both cases is similar to calculations for T-beam sections.

Structural

58 Structural Steel: Introduction

Nomenclature

b	width	in
d	depth	in
E	modulus of elasticity	ksi
f	computed stress	ksi
F	strength or allowable stress	ksi
G	shear modulus	ksi
h	clear distance between flanges	in
I	area moment of inertia	in^4
t	thickness	in
V	shear	kips
w	width	in
y	distance	in

Symbols

α	coefficient of thermal expansion	1/°F
ν	Poisson's ratio	–
ρ	density	lbm/ft^3
ϕ	resistance factor	–

Subscripts

c	centroidal
f	flange
n	nominal
t	tensile
u	ultimate (maximum) tensile
y	yield

1. STEEL NOMENCLATURE AND UNITS

In contrast to concrete nomenclature, it is traditional in steel design to use the uppercase letter F to indicate strength or allowable stress. Furthermore, such strengths or maximum stresses are specified in ksi. For example, $F_y = 36$ ksi is a steel with a yield stress of 36 ksi. Similarly, V_n is the nominal shear strength and ϕM_n is the flexural design strength, both in ksi. Actual or computed stresses are given the symbol of lowercase f. Computed stresses are also specified in ksi. For example, f_t is a computed tensile stress in ksi.

In the United States, steel design is carried out exclusively in customary U.S. (inch-pound) units.

2. TYPES OF STRUCTURAL STEEL

The term *structural steel* refers to a number of steels that, because of their economy and desirable mechanical properties, are suitable for load-carrying members in structures. In the United States, the customary way to specify a structural steel is to use an ASTM (American Society for Testing and Materials) designation. For ferrous metals, the designation has the prefix letter "A" followed by two or three numerical digits (e.g., ASTM A36, ASTM A992). The general requirements for such steels are covered under ASTM A6 specifications. Basically, three groups of hot-rolled structural steels are available for use in buildings: carbon steels, high-strength low-alloy steels, and quenched and tempered alloy steels.

Carbon steels use carbon as the chief strengthening element. These are divided into four categories based on the percentages of carbon: *low-carbon* (less than 0.15%), *mild-carbon* (0.15% to 0.29%), *medium-carbon* (0.30% to 0.59%), and *high-carbon* (0.60% to 1.70%). The most widely used all-purpose structural steel, ASTM A36, belongs to the mild-carbon category, and it has a maximum carbon content varying from 0.25% to 0.29%, depending on thickness. Carbon steels used in structures have minimum yield stresses ranging from 36 to 55 ksi. Table 58.1 gives the typical properties of A36 steel. An increase in carbon content raises the yield stress but reduces ductility, making welding more difficult. The maximum percentages of other elements of carbon steels are: 1.65% manganese, 0.60% silicon, and 0.60% copper.

High-strength low-alloy steels (HSLA) having yield stresses from 40 to 70 ksi are available under several ASTM designations. In addition to carbon and manganese, these steels contain one or more alloying elements (e.g., columbium, vanadium, chromium, silicon, copper, and nickel) that improve strength and other mechanical properties. The term "low-alloy" is arbitrarily used to indicate that the total of all alloying elements is limited to 5%. No heat treatment is used in the manufacture of HSLA steels. These steels generally have greater atmospheric corrosion resistance than the

Structural

carbon steels. ASTM designations A242, A441, A572, A588, and A992, among others, belong to this group.

Quenched and tempered alloy steels have yield stresses of 70 to 100 ksi. These steels of higher strengths are obtained by heat-treating low-alloy steels. The heat treatment consists of quenching (rapid cooling) and tempering (reheating). ASTM designations A514, A852, and A709 belong to this category.

3. STEEL PROPERTIES

Each structural steel is produced to specified minimum mechanical properties as required by the specific ASTM designation by which it is identified. Some properties of steel (such as the modulus of elasticity and density) are essentially independent of the type of steel. Other properties (such as the tensile strength and the yield stress) depend not only on the type of steel but also on the size or thickness of the piece. The mechanical properties of structural steel are generally determined from tension tests on small specimens in accordance with standard ASTM procedures. The results of a tension test are displayed on a stress-strain diagram. Figure 58.1 shows typical diagrams of three different types of steels.

Figure 58.1 *Typical Stress-Strain Curves for Structural Steels*

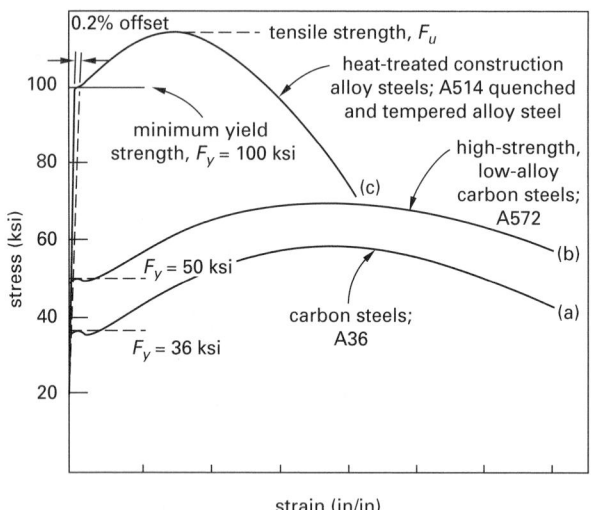

Yield stress, F_y, is that unit tensile stress at which the stress-strain curve exhibits a well-defined increase in strain without an increase in stress. For carbon and HSLA steels that show sharply defined yield points, F_y represents the stress at the obvious yield point. For the heat-treated steels that do not show a definite *yield point*, F_y represents the yield strength. The *yield strength* is defined as the stress corresponding to a specified deviation (usually 0.2%) from perfectly elastic behavior. The elastic behavior is characterized by the initial, nearly vertical straight-line portion of the stress-strain diagram. The yield stress property is extremely important in structural design because it serves as a limiting value of a member's usefulness.

Tensile strength, F_u, is the largest unit stress that the material achieves in a tension test. The *modulus of elasticity*, E, is the slope of the initial straight-line portion of the stress-strain diagram. For all structural steels, it is usually taken as 29,000 ksi for design calculations. *Ductility* is the ability of the material to undergo large inelastic deformations without fracture. In a tension test, it is generally measured by percent elongation for a specified gage length (usually 2 or 8 in). This property allows redistribution of stresses in continuous members and at points of high local stresses, such as those at holes or other discontinuities.

Toughness is the ability of a specimen to absorb energy and is characterized by the area under a stress-strain curve. *Weldability* is the ability of steel to be welded without changing its basic mechanical properties. Generally, weldability decreases with increases in carbon and manganese. *Poisson's ratio* is the ratio of transverse strain to longitudinal strain. Poisson's ratio is essentially the same for all structural steels and has a value of 0.3 in the elastic range. *Shear modulus* is the ratio of shearing stress to shearing strain during the initial elastic behavior.

Table 58.1 *Typical Properties of Structural Steel*

	A992*/A572, grade 50	A36
modulus of elasticity, E	29,000 ksi (as designated by AISC)	
tensile yield strength, F_y	50 ksi	36 ksi (up to 8 in thickness)
tensile strength, F_u	65 ksi (min)	58 ksi (min)
endurance strength	30 ksi (approximate)	
density, ρ	490 lbm/ft^3	
Poisson's ratio, ν	0.30 (average)	
shear modulus, G	11,200 ksi (as designated by AISC)	
coefficient of thermal expansion, α	6.5×10^{-6} 1/°F (average)	
specific heat (32°F–212°F)	0.107 Btu/lbm-°F	

*A992 steel is the *de facto* material for rolled W-shapes, having replaced A36 and A572 in most designs for new structures.

(Multiply ksi by 6.9 to obtain MPa.)
(Multiply in by 25.4 to obtain mm.)
(Multiply lbm/ft^3 by 16 to obtain kg/m^3.)
(Multiply °F^{-1} by 9/5 to obtain °C^{-1}.)

Appendix 58.A lists common structural steels along with their minimum yield stresses, tensile strengths, and uses. Under normal conditions, characterized by a fairly narrow temperature range (usually taken as −30°F to 120°F), the yield stress, tensile strength, and modulus of elasticity of a structural steel remain virtually constant. But when steel members are subjected to the elevated temperatures of a fire, a significant reduction in strength and rigidity occurs over time. (See App. 58.B.) For that reason, structural steels are given a spray-applied fire-resistant coating.

4. STRUCTURAL SHAPES

Many different structural shapes are available. The dimension and weight is added to the designation to uniquely identify the shape. For example, W30 × 132 refers to a W-shape with an overall depth of approximately 30 in that weighs 132 lbf/ft. The term *hollow structural sections* (HSS) is used to describe round and rectangular tubular members, which are often used as struts in trusses and space frames. Table 58.2 lists structural shape designations, and Figure 58.2 shows common structural shapes.

Table 58.2 Structural Shape Designations

shape	designation
wide flange beam	W
American standard beam	S
bearing piles	HP
miscellaneous (those that cannot be classified as W, S or HP)	M
American standard channel	C
miscellaneous channel	MC
angle	L
structural tee (cut from W or S or M)	WT or ST
structural tubing	TS
round and rectangular tubing	HSS
pipe	pipe
plate	PL
bar	bar

Figure 58.2 Structural Shapes

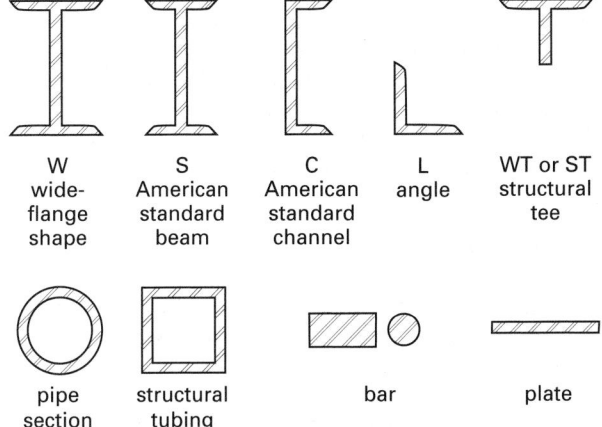

| W wide-flange shape | S American standard beam | C American standard channel | L angle | WT or ST structural tee |

pipe section | structural tubing | bar | plate

Figure 58.3 illustrates several combinations of shapes that are used in construction. The double-angle combination is particularly useful for carrying axial loads. Combinations of W-shapes and channels, channels with channels, or channels with angles are used for a variety of special applications, including struts and light crane rails. Properties for certain combinations have been tabulated in the *AISC Manual*.

Figure 58.3 Typical Combined Sections

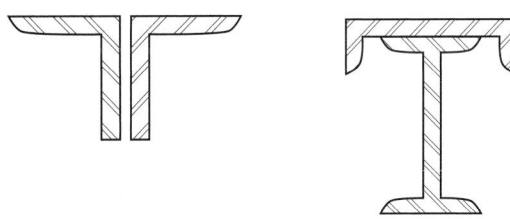

Occasionally, it will be desirable to provide additional bending or compressive strength to a shape by adding plates. It is generally easy to calculate the properties of the built-up section from the properties of the shape and plate. The following characteristics can be used when it is necessary to specify plate reinforcement.

- Plate widths should not be the same as the flange width (b_f), due to difficulty in welding. Widths should be somewhat larger or smaller. It is better to keep the plate width as close to b_f as possible, as width-thickness ratios specified in the *AISC Manual* may govern.

- Width and length tolerances smaller than $1/8$ in are not practical. Table 58.3 (from the *AISC Manual*, Part 1, Standard Mill Practice) should be used when specifying the nominal plate width.

- Not every plate exists in the larger thicknesses. Unless special plates are called for, the following thickness guidelines should be used.

  $1/32$ in increments up to $1/2$ in

  $1/16$ in increments from $9/16$ in to 1 in

  $1/8$ in increments from $1 1/8$ in to 3 in

  $1/4$ in increments for $3 1/4$ in and above

Table 58.3 Width Tolerance for Universal Mill Plates

| thickness (in) | width (in) | | |
	8 to 20 exclusive	20 to 36 exclusive	36 and above
0 to $\frac{3}{8}$, exclusive	$\frac{1}{8}$	$\frac{3}{16}$	$\frac{5}{16}$
$\frac{3}{8}$ to $\frac{5}{8}$, exclusive	$\frac{1}{8}$	$\frac{1}{4}$	$\frac{3}{8}$
$\frac{5}{8}$ to 1, exclusive	$\frac{3}{16}$	$\frac{5}{16}$	$\frac{7}{16}$
1 to 2, inclusive	$\frac{1}{4}$	$\frac{3}{8}$	$\frac{1}{2}$
over 2 to 10, inclusive	$\frac{3}{8}$	$\frac{7}{16}$	$\frac{9}{16}$
over 10 to 15, inclusive	$\frac{1}{2}$	$\frac{9}{16}$	$\frac{5}{8}$

(Multiply in by 25.4 to obtain mm.)

Example 58.1

A W30 × 124 shape must be reinforced to achieve the strong-axis bending strength of a W30 × 173 shape by welding plates to both flanges. All steel is A36. The plates are welded continuously to the flanges. What size plate is required if all plate sizes are available?

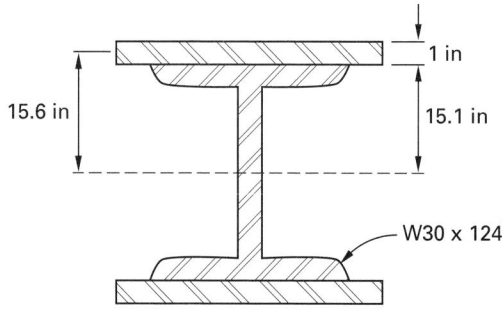

Solution

The moments of inertia for the W30×124 and W30×173 beams are 5360 in^4 and 8230 in^4, respectively. The difference in moments of inertia to be provided by the supplemental plates is

$$I_{\text{plates}} = 8230 \text{ in}^4 - 5360 \text{ in}^4 = 2870 \text{ in}^4$$

For ease of welding, assume the plate thickness, t, will be approximately the same as the flange thickness. For the W30 × 124 beam, the flange thickness is 0.930 in, so choose a plate thickness of 1.0 in. If w is the required plate width, the centroidal moment of inertia of the two plates acting together is

$$I_{c,\text{plates}} = \frac{(\text{number of plates})wh^3}{12}$$
$$= \frac{2w(1 \text{ in})^3}{12} = w/6$$

The depth of the W30 × 124 beam is 30.2 in. Therefore, the distance from the neutral axis to the plate centroid is

$$y_c = \frac{d}{2} + \frac{t}{2}$$
$$= \frac{30.2 \text{ in}}{2} + \frac{1 \text{ in}}{2} = 15.6 \text{ in}$$

From the parallel axis theorem, the moment of inertia of the two plates about the neutral axis is

$$I_{\text{plates}} = \frac{w}{6} + 2w(1)(15.6)^2 = 486.9w$$

The required moment of inertia is 2870 in^4.

$$w = \frac{2870 \text{ in}^4}{486.9 \text{ in}^3} = 5.89 \text{ in} \quad [\text{use 6 in}]$$

5. SPECIFICATIONS AND BUILDING CODES

Although the terms "specification" and "building code" are often used interchangeably in the context of steel design, there is a difference between the two. A *specification* is a set of guidelines or recommendations put forth by a group of experts in the field of steel research and design with the intent of ensuring safety. A specification is not legally enforceable unless it is a part of a building code.

The design of steel buildings in the United States is principally based on the specifications of the American Institute of Steel Construction (AISC), a nonprofit trade association representing and serving the fabricated structural steel industry in the United States. AISC's *Specification for Structural Steel Buildings* (referred to as the *AISC Specification* in this book) covers the two permitted methods of design—the allowable stress design (ASD) and the load and resistance factor design (LRFD). The ASD and LRFD specifications are contained in AISC's *Steel Construction Manual, 13 ed.*, (referred to as the *AISC Manual* in this book), and both are equally valid methods to use with structures when the *AISC Manual* applies. In addition to the ASD and LRFD specifications, the *AISC Manual* contains a wealth of information on products available, design aids, examples, erection guidelines, and other applicable specifications.

A *building code* is a broad-based document covering all facets of safety, such as design loadings, occupancy limits, plumbing, electrical requirements, and fire protection. Building codes are adopted by states, cities, or other government bodies as a legally enforceable means of protecting public safety and welfare. There are three common model codes that are integrated, fully or partially, into many state or local building codes. These are: Building Officials and Codes Administrators International Incorporated (BOCA), *National Building Code*; International Conference of Building Officials (ICBO), *Uniform Building Code* (UBC); and Southern Building Code Congress International (SBCCI), *Standard Building Code*. The best aspects of the three model codes have been combined by the International Code Council (ICC) into the *International Building Code* (IBC), intended to be adopted in other countries as well as throughout the United States.

6. PHILOSOPHIES OF DESIGN

There are two philosophies of design in current use in the United States: *allowable strength design* (ASD) and *load and resistance factor design* (LRFD). For steel design, the allowable strength design (previously called *allowable stress design* before publication of the 13th edition of the *AISC Manual*) has been the primary one used for over 100 years.

The ASD design philosophy is based on the premise that structural members remain elastic when subject to applied loads. According to this philosophy, a structural member is designed so that its computed strength under service or working loads does not exceed available strength. The available strengths are prescribed by the building codes or specifications to provide a factor of safety against attaining some limiting strength such as that defined by yielding or buckling.

The LRFD design philosophy, also referred to as *limit states design*, is the predominant design philosophy for concrete structures. The philosophy was first introduced for steel structures in 1986 with the publication of the first edition of AISC's *LRFD Manual of Steel Construction*. "Limit state" is a general term meaning a condition at which a structure or some part of it ceases to fulfill its intended function. There are two categories of limit states: strength and serviceability. The *strength limit states* that are the primary concerns of designers include plastic strength, fracture, buckling, fatigue, and so on. These affect the safety or load-carrying capacity of a structure. The *serviceability limit states* refer to the performance under normal service loads and pertain to uses and/or occupancy of structures, including excessive deflection, drift, vibration, and cracking. Subsequent chapters in this book use the LRFD method in examples unless otherwise noted.

7. LOADS

Structures are designed to resist many types of loads including dead loads, live loads, snow loads, wind loads, and earthquake loads. The complete design must take into account all effects of these loads, including all applicable load combinations. Building codes (state, municipal, or other) usually provide minimum loads for a designer's use in a particular area. In the absence of such provisions, *ASCE Standard 7*, "Minimum Design Loads for Buildings and Other Structures" (ASCE7), may be referred to. For LRFD design, the load combinations in ASCE7 Sec. 2.3 apply and for ASD design, the load combinations in Sec. 2.4 apply.

8. FATIGUE LOADING

The effects of fatigue loading are generally not considered except in the cases of bridge and connection design. If a load is to be applied and removed less than 20,000 times (roughly equivalent to twice daily for 25 years), as would be the case in a conventional building, no provision for repeated loading is necessary. However, some designs, such as for crane runway girders and supports, must consider the effects of fatigue. Design for fatigue loading is covered in App. 3 of the *AISC Specification*.

9. MOST ECONOMICAL SHAPE

Since a major part of the cost of using a rolled shape in construction is the cost of raw materials, the lightest shape possible that will satisfy the structural requirements should usually be used. Thus, generally speaking, the most *economical shape* is the structural shape that has the lightest weight per foot and that has the required strength. The beam selection table and chart are designed to make choosing economical shapes possible.

Prior to 1995, there was a significant cost differential between ASTM A36 steel and the stronger ASTM grade 50 steel. However, that cost differential vanished when steel producers began to make a *dual-certified* steel that met the more restrictive criteria from the specifications for both ASTM A36 and A572 grade 50. Consequently, A992 steel with a minimum specified yield strength of 50 ksi is now standard for new construction in the United States.

10. DESIGNING WITH WEATHERING STEEL

Weathering steel is a term used to describe A588 steel when it is used unpainted, particularly as part of bridge substructures. The natural layer of oxidation (rust) that forms provides all the protection needed for most applications. The higher cost of the material is offset by eliminating the need to paint and repaint the structure. (A588 steel usually costs somewhat less than painted A572 and A992 grade 50 steels, which also have a 50 ksi yield strength.)

If used without a careful evaluation of environmental location and detailing, however, weathering steel can experience significant corrosion and cracking. Successful use depends on considering several factors in the design. (a) The use of A588 steel in locations that interfere with the development of the protective oxide coating should be avoided. For example, contact between the steel and vegetation, masonry, wood, or other materials should be avoided so that weathering can be maintained on a natural basis. (b) Retention of water and debris on steel surfaces should be avoided. In bridge decks, rely on camber rather than deck drains that can be easily clogged by roadside debris. (c) "Jointless," continuous construction should be used, or the steel beneath joints should be protected and sealed by galvanizing. (d) Surface runoff from the deck to the steel below should be prevented. Similarly, the "tunnel effect," whereby road salts are sucked onto the substructure due to limited space between it and surrounding terrain or features, should be eliminated. (e) Areas where debris has collected should be flushed and cleaned regularly to prevent the buildup of corrosive salts.

To prevent unsightly staining of concrete abutments and piers in urban areas, the first 5 ft to 6 ft (1.5 m to 1.8 m) of the steel structure can be painted.

11. DESIGNING WITH HIGH-STRENGTH STEEL

100 ksi (690 MPa) high-strength structural steel is readily available, and rods and prestressing steel can have strengths near or over 200 ksi (1380 MPa). These steels

Structural

are frequently used in structural steel connectors (A514 quenched and tempered steel with a 100 ksi (690 MPa) yield, for example). However, high-strength steels have particular characteristics that must be considered.

Higher strengths are obtained at the expense of decreased ductility and increased brittleness. Very high-strength steels have very little or almost no margin of strength above the yield strength. The design stress that controls is often half of the tensile strength, and it may be lower than the fraction of yield used in lower-strength steels. In the same vein, the transition temperature, below which the steel loses its ductility, must be considered.

High-strength steels used in structural shapes are sensitive to welding-induced hydrogen cracking. Welding may change the grain structure significantly, introducing discontinuities (flaws) in the steel structure that can serve as sites for fractures. Also, welding may introduce residual stresses, particularly with thick plates and complex joints, unless preheating and postheating are used. Some complex joints may need to be fabricated offsite, heated in an annealing furnace, and allowed to cool slowly in order to reduce the high levels of residual stresses introduced by welding.

High-strength steels are sensitive to fatigue (dynamic loading) and the rate of stress application. Bridges, for example, are subjected to thousands or millions of vehicle loads over their lifetimes. The design stress in such cases is limited to the endurance strength, which might be only a fraction of the design stress in the absence of fatigue.

High-strength steels are also more sensitive to corrosion, leading to *stress-corrosion cracking*. Even environments where the strongest corrosive is rainwater may cause high-strength bolts to fail.

For applications where the use of high-strength steel is problematic, several options exist. (a) Use a lower-strength steel or a steel with a higher fracture toughness. Some state highway departments limit the strength of connectors used in thick members for this reason.

(b) Closely monitor pretensioning of bolts. (c) Use accepted procedures for welding and weld inspection. (d) Design for redundancy. For example, in bridge design, a larger number of smaller stringers is preferred over ever-larger rolled sections or large sections built up by welding. (e) Predict fatigue-crack rates in order to determine realistic service lives of structures subjected to repeated loadings, such as bridges and offshore structures. (f) Design for inspection. (g) Use proper inspection techniques. For example, bolt tension cannot be adequately field-checked by using a torque wrench.

12. HIGH-PERFORMANCE STEEL

The term *high-performance steel* (HPS), as used in bridge design, generally refers to HPS 70W grade steel plate. This 70 ksi (485 MPa) steel meets ASTM A709 specifications and is tougher than traditional grades. Available primarily in plate form, it is finding increasing use in bridge steel plate girders where the cost savings can range from 10 to 20% since less material is needed.

Most HPS 70W plate is quenched and tempered, a manufacturing operation that involves additional heating and cooling steps, and that limits plate lengths to approximately 50 ft (15 m) and thicknesses to approximately 2 in (51 mm). Advanced compositions may allow HPS plate to be manufactured in longer and thicker sections without quenching and tempering. True structural sections may also then be possible.

As with all high-strength steel, welding protocol must be strictly controlled to avoid hydrogen cracking. Specific techniques and different consumables (e.g., rod, flux, and shield gas) are required for each combination of steels to be joined.

59

Structural Steel: Beams

Nomenclature

A	area	in^2
A_1	area of steel bearing concentrically on a concrete support	in^2
A_2	maximum area of the portion of the supporting surface that is geometrically similar and concentric with loaded area	in^2
b	width	in
B	width of base plate	in
c	torsional constant	–
C	coefficient	–
C_w	warping constant	in^6
d	depth	in
E	modulus of elasticity	ksi
f	computed stress	ksi
f'_c	concrete compressive strength	ksi
F	strength or allowable stress	ksi
h	clear distance between flanges	in
h_o	distance between flange centroids	in
I	moment of inertia	in^4
J	torsional constant	in^4
k	distance from bottom of beam to web toe of fillet	in
l	distance between lateral supports	in
L	span length	in
L_b	length between laterally braced points	in
L_p	limiting laterally unbraced length for yielding	in
L_r	limiting laterally unbraced length for buckling	in
M	moment	in-kips
n	cantilever dimension of bearing plate	in
N	length of bearing of applied load	in
P	load	kips
r	radius of gyration	in
r_{ts}	effective radius of gyration of the compression flange	in
R	reaction or concentrated load	kips
R_m	cross-section monosymmetric parameter	–
R_n	nominal strength	–
S	elastic section modulus	in^3
t	thickness	in
V	shear	kips
w	load per unit length	kips/in
x	distance	in
Z	plastic section modulus	in^3

Symbols

Δ	deflection	in
θ	angle	deg
ϕ	resistance factor	–
Ω	safety factor	–

Subscripts

b	bending or braced
c	centroidal
cr	critical
D	dead load
f	flange
L	live load
n	net or nominal
p	bearing or plastic
u	ultimate
v	shear
w	web or warping
x	strong axis
y	yield or weak axis

1. TYPES OF BEAMS

Beams primarily support transverse loads (i.e., loads that are applied at right angles to the longitudinal axis of the member). They are subjected primarily to flexure (bending). Although some axial loading is practically unavoidable in any structural member, the effect of axial loads is generally negligible, and the member can be treated strictly as a beam. If an axial compressive load of substantial magnitude is also present with transverse loads, the member is called a *beam-column*.

Beams are often designated by names that are representative of some specialized functions: A *girder* is a major beam that often provides supports for other beams. A *stringer* is a main longitudinal beam, usually supporting

Structural

bridge decks. A *floor beam* is a transverse beam in bridge decks. A *joist* is a light beam that supports a floor. A *lintel* is a beam spanning an opening (a door or window), usually in masonry construction. A *spandrel* is a beam on the outside perimeter of a building that supports, among other loads, the exterior wall. A *purlin* is a beam that supports a roof and frames between or over supports, such as roof trusses or rigid frames. A *girt* is a light beam that supports only the lightweight exterior sides of a building, as is typical in pre-engineered metal buildings.

Commonly used beam cross sections are standard hot-rolled shapes including the W, S, M, C, T, and L shapes. Doubly symmetrical shapes, such as W, S, and M sections, are the most efficient. They have excellent flexural strength and relatively good lateral strength for their weight. Channels have reasonably good flexural strength but poor lateral strength, and they require horizontal bracing or some other lateral support. Tees and angles are suitable only for light loads.

The flexural strength of a rolled section can be improved by adding flange plates. But if the loadings are too heavy or the spans are too long for a standard rolled section, a plate girder may be necessary. *Plate girders* are built up from plates in I (most common), H, or box shapes of any depth. A *box shape* is used if depth is restricted or if lateral stability is a problem.

2. BEAM BENDING PLANES

The property tables for W shapes contain two moments of inertia, I_x and I_y, for each beam. There are several ways of referring to the *plane of bending*. Figure 59.1(a) shows a W-shape beam as it is used typically. The value of I_x should be used to calculate bending stress. This bending mode is referred to as "bending about the major (or strong) axis." However, it is also referred to as "loading in the plane of the web." Figure 59.1(b) shows a W shape "bending about the minor (or weak) axis" This mode is also referred to as "loading perpendicular to the plane of the web."

Figure 59.1 *Beam Bending Planes*

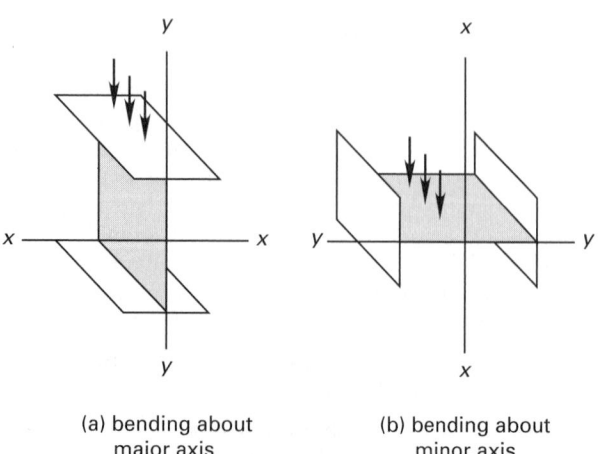

(a) bending about   (b) bending about
major axis   minor axis

3. BENDING STRENGTH IN STEEL BEAMS

Design of simple beams is based on comparing the moment due to the loads to a nominal flexural strength modified by a safety factor. For compact sections with adequate lateral support, the nominal flexural strength is given by

$$M_n = F_y Z_x \quad \text{[AISC Eq. F2-1]} \qquad 59.1$$

The nominal flexural strength is modified by the safety factor, Ω_b, for bending; therefore, the design equation becomes

$$M \leq \frac{M_n}{\Omega_b} \qquad 59.2$$

M is the moment due to the loads, and $\Omega_b = 1.67$ per *AISC Specification* F1.

4. COMPACT SECTIONS

Compact sections have width-thickness ratios of their compression elements that satisfy the limits of Table B4.1 of the *AISC Specification*. Therefore, compact sections are permitted to achieve higher strengths in many instances. To be compact, the flanges of a beam must be continuously connected to the web. Therefore, a built-up section or plate girder constructed with intermittent welds does not qualify. In addition, Eqs. 59.3 and 59.4 must be satisfied by standard rolled shapes without flange stiffeners. Equation 59.3 applies only to flanges in flexural compression. Equation 59.4 applies only to webs in flexural compression [*AISC Specification*, Table B4.1 with $b = b_f/2$].

$$\frac{b_f}{2t_f} \leq 0.38\sqrt{\frac{E}{F_y}} \quad \begin{bmatrix} \text{flanges in flexural} \\ \text{compression only} \end{bmatrix} \qquad 59.3$$

$$\frac{h}{t_w} \leq 3.76\sqrt{\frac{E}{F_y}} \quad \begin{bmatrix} \text{webs in flexural} \\ \text{compression only} \end{bmatrix} \qquad 59.4$$

Compactness, as Eqs. 59.3 and 59.4 show, depends on the steel strength. Most rolled W-shapes are compact at lower values of F_y. However, a 36 ksi beam may be compact, while the same beam in 50 ksi steel may not be. Tables 3-10 and 3-11 in the *AISC Manual* indicate the design moment of W and channel shapes based on unbraced lengths.

5. LATERAL BRACING

To prevent *lateral torsional buckling* (illustrated in Fig. 59.2), a beam's compression flange must be supported at frequent intervals. Complete support is achieved when a beam is fully encased in concrete or has its flange welded or bolted along its full length (see Fig. 59.3). In many designs, however, lateral support is provided only at regularly spaced intervals. The actual spacing between points of lateral bracing is designated as L_b.

Figure 59.2 *Lateral Buckling in a Beam*

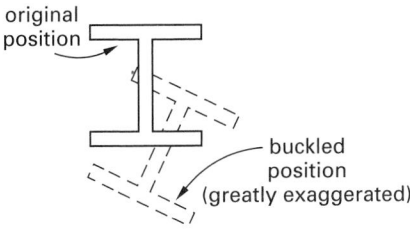

Figure 59.3 *Compression Flange Bracing Using Headed Studs*

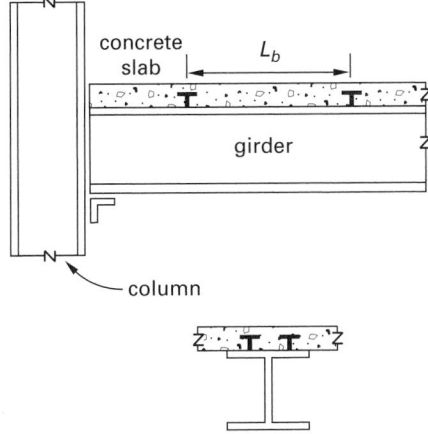

For the purpose of determining flexural design strength (M_n/Ω), two limits are placed on spacing: $L_p < L_b \leq L_r$ and $L_b > L_r$. (See Fig. 59.11.) For most shapes, L_p is calculated from Eq. 59.5 and L_r is calculated from Eq. 59.6.

$$L_p = 1.76 r_y \sqrt{\frac{E}{F_y}} \quad \text{[AISC Eq. F2-5]} \qquad 59.5$$

$$L_r = 1.95 r_{ts} \frac{E}{0.7 F_y}$$
$$\times \sqrt{\frac{Jc}{S_x h_o}}$$
$$\times \sqrt{1 + \sqrt{1 + 6.76 \left(\left(\frac{0.7 F_y}{E}\right) \left(\frac{S_x h_o}{Jc}\right) \right)^2}}$$
$$\text{[AISC Eq. F2-6]} \quad 59.6$$

r_{ts} is the *effective radius of gyration* used to determine L_r for the lateral torsional buckling state. For doubly symmetrical, compact shapes bending about the major axis, it is accurately (and, conservatively) estimated as the radius of gyration of the compression flange plus one-sixth of the radius of gyration of the web. It is tabulated in the *AISC Manual*, Part 1, Shape Tables.

$$r_{ts} = \frac{b_f}{\sqrt{12 \left(1 + \frac{h t_w}{6 b_f t_f}\right)}} \qquad 59.7$$

r_{ts} can also be calculated from Eq. 59.8 [AISC Eq. F2-7] in which both the elastic section modulus, S_x, and the *warping constant*, C_w, are tabulated in the Shape Tables.

$$r_{ts}^2 = \frac{\sqrt{I_y C_w}}{S_x} \qquad 59.8$$

The warping constant, C_w, is a measure of a shape's resistance to failure by lateral buckling. For an assembly of standard shapes, each shape's warping constant can be summed to give the assembly's warping coefficient. An approximation of the warping constant for I-shapes is given by Eq. 59.9.[1] Obtaining exact values for more complex built-up sections requires substantial effort.

$$C_w = \frac{h^2 I_y}{4} \approx \frac{(d - t_f)^2 t_f b_f^3}{24} \qquad 59.9$$

The symbol J denotes the *torsional constant*.[2] The torsional constant is a measure of the shape's resistance to failure by twisting. Values of J are tabulated in the *AISC Manual's*, Part 1, Shape Tables for rolled shapes and angles.[3] For any other I-beam with element thickness, t, the torsional constant is

$$J = \frac{1}{3} \int_0^b t^3 \, ds \approx \frac{1}{3} \sum b t^3 \quad [b > t; \ b/t > 10] \qquad 59.10$$

The second form of Eq. 59.10 shows that J can be closely approximated for I-shapes and angles by dividing the cross-sectional shape into mutually exclusive rectangular pieces (as though the shape was being constructed of plate girders) and the quantities $bt^3/3$ calculated for each, and then summed. The small contributions of the fillet radii are disregarded with this method. (This approximation cannot be used for closed shapes.)

For doubly symmetrical I-shapes, $c = 1$ [AISC Eq. F2-8a]. For a channel, c is calculated from AISC Eq. F2-8b.

$$c = \frac{h_o}{2} \sqrt{\frac{I_y}{C_w}} \qquad 59.11$$

6. LATERAL TORSIONAL BUCKLING

The laterally unsupported length of a beam can fail in lateral torsional buckling due to the applied moment.

---

[1]Equation 59.9 assumes the entire beam is solid, without bolt holes. However, hole geometry greatly affects the value. As such, it is not a conservative estimate.
[2]This J is the *St. Venant torsional constant*, not the polar moment of inertia (i.e., the sum of the moments of inertia about the x- and y-axes), even though (1) the same variable is used, (2) the units are the same, and (3) the values are identical for circular members. For I beams and other shapes, the values are significantly different. For example, for a W8 × 24 shape, the polar moment of inertia is approximately 101 in⁴, whereas the torsional constant is only 0.35 in⁴. Using the polar moment of inertia for the torsional constant will result in grossly under-calculating stress.
[3]For double angles and other assemblies of shapes, the combined torsional constant is simply the sum of the torsional constants of each angle.

Lateral torsional buckling is fundamentally similar to the flexural buckling of a beam and flexural torsional buckling of a column subjected to axial loading. If the laterally unbraced length, L_b, is less than or equal to a plastic length, L_p (see Eq. 59.5), then lateral torsional buckling will not be a problem, and the beam will develop its full plastic strength, M_p. However, if L_b is greater than L_p, then lateral torsional buckling will occur and the moment capacity of the beam will be reduced below the plastic strength, M_p.

Usually, the bending moment varies along the unbraced length of a beam. The worst case scenario is for beams subjected to uniform bending moments along their unbraced lengths. For this situation, the *AISC Manual* specifies a value $C_b = 1$ for the *lateral torsional buckling modification factor*, also known as the *beam bending coefficient, moment modification factor*, and in the past, as the *moment gradient multiplier*. For non-uniform moment loadings, C_b is greater than 1.0, effectively increasing the strength of the beam. C_b is equal to 1.0 for uniform bending moments, and is always greater than 1.0 for non-uniform bending moments. C_b may not exceed a value of 3.0. Of course, the increased moment capacity for the non-uniform moment case cannot be more than M_p. If the calculated strength value, M_n, is greater than M_p, it must be reduced to M_p.

$$M_{n,\text{non-uniform moment}} = C_b M_{n,\text{uniform moment}} \qquad 59.12$$

C_b can always be conservatively assumed as 1.0. In fact, for cantilever beams where the free ends are unbraced, $C_b = 1.0$. However, with C_b values being as high as 3, substantial material savings are possible by not being conservative.

Various theoretical and experimental methods exist for calculating C_b. The most convenient method is to read C_b directly from App. 59.A. *AISC Specification* Eq. F1-1 provides a method for calculating C_b when both ends of the unsupported segment are braced and the beam is bent in single or double curvature. In Eq. 59.13, M_{max} is the magnitude of maximum bending moment in L_b. M_A, M_B, and M_C are the magnitudes of the bending moments at the quarter point, midpoint, and three-quarter point, respectively, along the unbraced segment of length L_b. R_m is the cross-section monosymmetric parameter having a value of 1.0 for doubly symmetric members as well as singly symmetric members bent in single curvature. Refer to *AISC Specification* Sec. F1 for other cases.

$$C_b = \frac{12.5 M_{\text{max}}}{2.5 M_{\text{max}} + 3 M_A + 4 M_B + 3 M_C} \times R_m \leq 3.0$$
$$59.13$$

7. FLEXURAL DESIGN STRENGTH: I-SHAPES BENDING ABOUT MAJOR AXIS

For I-shapes loaded in the plane of their webs and bending about the major axis, the nominal flexural strength will be $M_p = F_y Z_x$ or less. If $L_b \leq L_p$, or if the bracing

is continuous and the beam is compact, the nominal flexural strength, M_n, is M_p. If $L_r \geq L_b > L_p$, then

$$M_n = C_b \left(M_p - (M_p - 0.7 F_y S_x) \left(\frac{L_b - L_p}{L_r - L_p} \right) \right)$$
$$\leq M_p \quad \text{[AISC Eq. F2-2]} \qquad 59.14$$

If $L_b > L_r$, or if there is no bracing at all between support points, then

$$M_n = F_{\text{cr}} S_x \leq M_p \quad \text{[AISC Eq. F2-3]} \qquad 59.15$$

$$F_{\text{cr}} = \frac{C_b \pi^2 E}{\left(\frac{L_b}{r_{ts}} \right)^2} \sqrt{1 + 0.078 \frac{Jc}{S_x h_o} \left(\frac{L_b}{r_{ts}} \right)^2}$$
$$\text{[AISC Eq. F2-4]} \qquad 59.16$$

8. FLEXURAL DESIGN STRENGTH: WEAK-AXIS BENDING

If a doubly symmetrical rolled shape is placed such that bending will occur about its weak axis, the nominal flexural strength is $M_p = F_y Z_y \leq 1.6 F_y S_y$. The shape must be compact, and other conditions may also apply. However, this is not an efficient use of the beam, so this configuration is seldom used. Nevertheless, weak-axis bending may occur, particularly in beam-columns.

9. SHEAR STRENGTH IN STEEL BEAMS

It is assumed that only the web carries shear in W-shapes. The ultimate shear strength is compared against the design shear strength. The nominal shear strength, V_n, in the web of a beam is $0.6 F_y A_w C_v$. (Different limitations apply to shear stresses in bolts, rivets, and plate girders.)

Figure 59.4 *Nomenclature and Terminology for Steel W-Shape Beam*

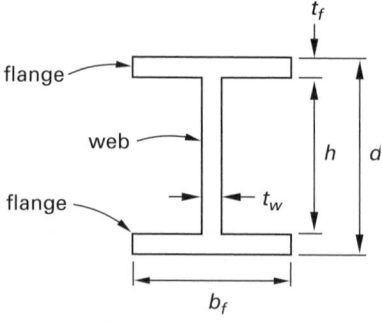

10. CONCENTRATED WEB FORCES

Local buckling is a factor in the vicinity of large concentrated loads. Such loads may occur at a reaction point or where a column frames into a supporting girder. *Web yielding* and *web crippling*, two types of local buckling shown in Fig. 59.5, can be reduced or eliminated by use of *stiffeners*. If the load is applied uniformly over a large enough area (say, along N inches of beam flange or more), no stiffeners will be required.

Figure 59.5 *Local Buckling*

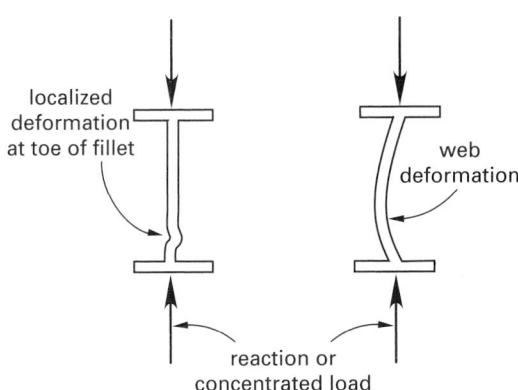

(a) web yielding   (b) web crippling

Regarding web yielding, the limit state is determined using $\phi = 1.00$ (LRFD) or $\Omega = 1.50$ (ASD). The nominal strength, R_n, is specified by Eq. 59.17 [AISC Eq. J10-2] for interior loads and Eq. 59.18 [AISC Eq. J10-3] for loads at beam ends (see Fig. 59.6).

Figure 59.6 *Nomenclature for Web Yielding Calculations*

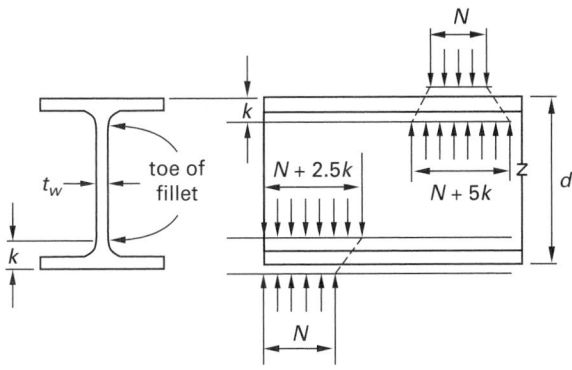

$$R_n = (5k + N)F_{yw}t_w \qquad 59.17$$

$$R_n = (2.5k + N)F_{yw}t_w \qquad 59.18$$

Regarding web crippling, the limit state is determined using $\phi = 0.75$ (LRFD) or $\Omega = 2.00$ (ASD). The nominal strength, R_n, is specified by Eq. 59.19 [AISC Eq. J10-4] for interior loads and by Eq. 59.20 [AISC Eq. J10-5a] and Eq. 59.21 [AISC Eq. J10-5b] for reactions at beam ends.

$$R_n = 0.80t_w^2\left(1 + 3\left(\frac{N}{d}\right)\left(\frac{t_w}{t_f}\right)^{1.5}\right)\sqrt{\frac{EF_{yw}t_f}{t_w}}$$

[AISC Eq. J10-4]   *59.19*

For $N/d \leq 0.2$,

$$R_n = 0.40t_w^2\left(1 + 3\left(\frac{N}{d}\right)\left(\frac{t_w}{t_f}\right)^{1.5}\right)\sqrt{\frac{EF_{yw}t_f}{t_w}}$$

[AISC Eq. J10-5a]   *59.20*

For $N/d > 0.2$,

$$R_n = 0.40t_w^2\left(1 + \left(\frac{4N}{d} - 0.2\right)\left(\frac{t_w}{t_f}\right)^{1.5}\right)\sqrt{\frac{EF_{yw}t_f}{t_w}}$$

[AISC Eq. J10-5b]   *59.21*

Intermediate stiffeners (i.e., web stiffeners spaced throughout a stock rolled shape) are never needed with rolled shapes but are typically used in plate girders. (For built-up beams, diagonal buckling requirements should also be checked.) *Bearing stiffeners* (see Fig. 59.7) are typically web stiffening plates welded to the webs and flanges of rolled sections.

Figure 59.7 *Bearing Stiffeners*

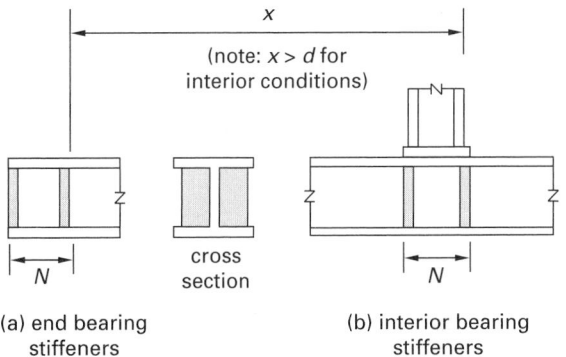

(a) end bearing stiffeners   (b) interior bearing stiffeners

Flange stiffeners (see Fig. 59.8) are typically angles placed at the web-flange corner used to keep the flange perpendicular to the web. Flange stiffeners cannot be used in place of bearing stiffeners.

Figure 59.8 *Flange Stiffeners*

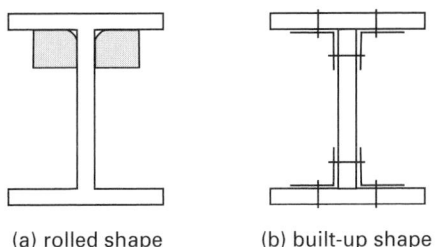

(a) rolled shape   (b) built-up shape

11. SERVICEABILITY AND BEAM DEFLECTIONS

In addition to being safe, a structure with all its components must be serviceable. While stresses in a beam due to moment and shear must be within allowable limits to ensure safety, the beam cannot be too flexible. Deflection of the beam usually limits the flexibility. Among the many reasons for avoiding excessive deflections are the effects on nonstructural elements (e.g., doors, windows, and partitions), undesirable vibrations, and the proper functioning of the roof drainage systems.

Steel beam deflections are calculated using traditional beam equations based on the principles of strength of materials. For common beams and loadings the *AISC Manual*, Beam section, Beam Diagrams and Formulas, Tables 3-22 and 3-23, contains deflection formulas. Deflection limitations are typically unique to each design situation and are generally expressed in terms of some fraction of the span length. The *AISC Manual* suggests, but does not require, general guidelines to maintain appearance and occupant confidence in a structure (Ch. L of the *AISC Manual Commentary*. Special provisions in the *AISC Manual*, App. 2 K2, are used to check for ponding.)

12. ANALYSIS OF STEEL BEAMS

Although bending strength is the primary criterion for beams, a complete analysis of a steel beam of known cross section also includes checking for shear strength and deflection. A flowchart for beam analysis (moment capacity criterion only) is shown in Fig. 59.9. The elastic deflection limit is $L/360$ for beams and girders supporting plastered ceilings [*AISC Manual Commentary*, Ch. L].

Example 59.1

Based on moments only, determine the maximum allowable superimposed uniformly distributed load that a W21 × 55 beam of A992 steel can carry on a simple span of 36 ft. Consider the compression flange to be braced at (a) 6 ft intervals, (b) 12 ft intervals, and (c) 18 ft intervals based on flexural strength only.

Solution

For A992 steel, $F_y = 50$ ksi. Obtain the properties of a W21 × 55 beam from the *AISC Manual*, Shape Table 1-1.

Check for compactness.

$$S_x = 110 \text{ in}^3$$

$$r_{ts} = 2.11 \text{ in}$$

$$\frac{b_f}{2t_f} = 7.87 < 0.38\sqrt{\frac{E}{F_y}} = 9.15 \quad \text{[Eq. 59.3]}$$

$$\frac{h}{t_w} = 50.0 < 3.76\sqrt{\frac{E}{F_y}} = 90.55 \quad \text{[Eq. 59.4]}$$

The shape is compact.

From the *AISC Manual*, Table 3-2, for W21 × 55, $L_p = 6.11$ ft and $L_r = 17.4$ ft.

(a) The unbraced length is $L_b = 6$ ft.

Since $L_b < L_p$, the design flexural strength is

$$\phi_b M_n = \phi_b M_p = \phi_b F_y Z_x$$
$$= (0.9)\left(50\ \frac{\text{kips}}{\text{in}^2}\right)(126 \text{ in}^3)\left(\frac{1 \text{ ft}}{12 \text{ in}}\right)$$
$$= 473 \text{ ft-kips}$$

The actual bending moment is

$$M_u = \frac{w_u L^2}{8} = \frac{w_u (36 \text{ ft})^2}{8} = (162 \text{ ft}^2)w_u$$

Equate the flexural design strength to the actual bending moment.

$$473 \text{ ft-kips} = (162 \text{ ft}^2)w_u$$
$$w_u = 2.92 \text{ kips/ft}$$

Let w equal the superimposed load and w_D equal the weight of the beam. As given by ASCE7 Sec. 2.3, the factored load, w_u, is

$$w_u = 1.2w_D + 1.6w$$
$$2.92\ \frac{\text{kips}}{\text{ft}} = (1.2)\left(0.055\ \frac{\text{kips}}{\text{ft}}\right) + 1.6w$$
$$w = 1.784 \text{ kips/ft}$$

(b) The unbraced length is $L_b = 12$ ft.

Since $L_p < L_b < L_r$, the nominal bending moment is given by Eq. 59.12. From App. 59.A, $C_b = 1.45$ and $C_b = 1.01$. Use the more conservative 1.01 value.

$$M_n = C_b\left(M_p - (M_p - 0.7F_y S_x)\left(\frac{L_b - L_p}{L_r - L_p}\right)\right) \leq M_p$$

$$= 1.01\left(\begin{array}{l}525 \text{ ft-kips} \\[2pt] 525 \text{ ft-kips} - \left(\begin{array}{l}525 \text{ ft-kips} \\[2pt] -\dfrac{(0.7)\left(50\ \frac{\text{kips}}{\text{in}^2}\right)(110 \text{ in}^3)}{12\ \frac{\text{in}}{\text{ft}}}\end{array}\right) \\[10pt] \times\left(\dfrac{12 \text{ ft} - 6.11 \text{ ft}}{17.4 \text{ ft} - 6.11 \text{ ft}}\right)\end{array}\right)$$

$$= 423 \text{ ft-kips} \quad (< 525 \text{ ft-kips})$$

$$\phi M_n = (0.9)(423 \text{ ft-kips})$$
$$= 381 \text{ ft-kips}$$
$$381 \text{ ft-kips} = (162 \text{ ft}^2)w_u$$
$$w_u = 2.35 \text{ kips/ft}$$
$$w_u = 1.2w_D + 1.6w$$
$$2.35\ \frac{\text{kips}}{\text{ft}} = (1.2)\left(0.055\ \frac{\text{kips}}{\text{ft}}\right) + 1.6w$$
$$w = 1.43 \text{ kips/ft}$$

(c) The unbraced length is $L_b = 18$ ft.

Figure 59.9 *Flowchart for LRFD Analysis of Beams (W-shape, strong-axis bending, moment criteria only)*
(SPEC = Specifications; part 16 of the AISC Manual)

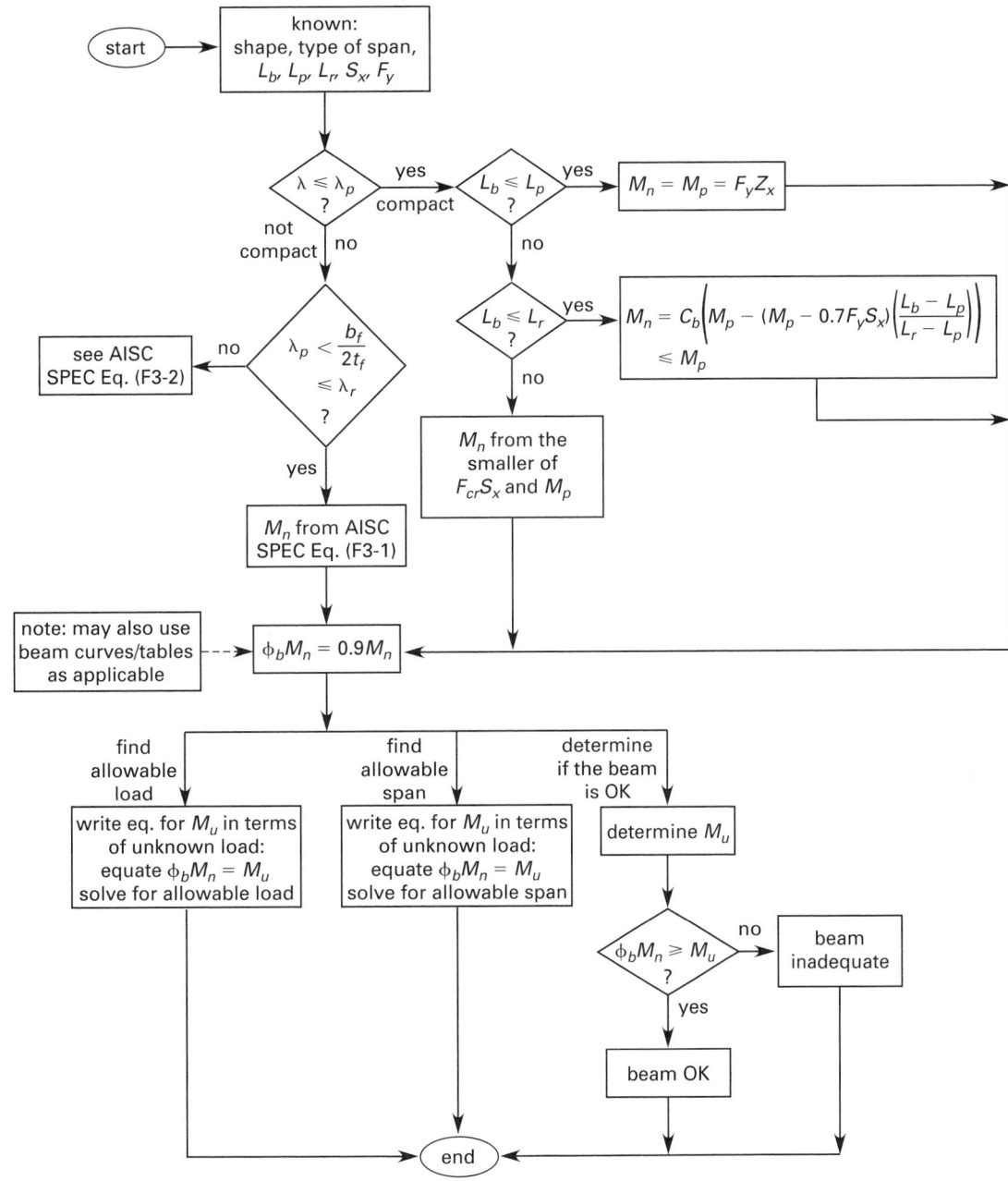

Adapted from *Applied Structural Steel Design*, 3rd ed., by Spiegel/Limbrunner, copyright © 1997. Reprinted by permission of Prentice-Hall, Inc., Upper Saddle River, NJ.

$L_b > L_r$, so the nominal bending moment is $M_n = F_{cr}S_x \le M_p$. Compute F_{cr}. From App. 59.A, $C_b = 1.3$.

$$\frac{L_b}{r_{ts}} = \frac{(18 \text{ ft}) \left(12 \frac{\text{in}}{\text{ft}}\right)}{2.11 \text{ in}} = 102.4$$

$$F_{cr} = \frac{C_b \pi^2 E}{\left(\frac{L_b}{r_{ts}}\right)^2} \sqrt{1 + 0.078 \frac{Jc}{S_x h_o} \left(\frac{L_b}{r_{ts}}\right)^2}$$

$$= \frac{1.3\pi^2 \left(29\,000 \frac{\text{kips}}{\text{in}^2}\right)}{(102.4)^2}$$

$$\times \sqrt{\begin{array}{l} 1 + 0.078 \\ \times \left(\frac{(1.24 \text{ in}^4)(1)}{(110 \text{ in}^3)(20.3 \text{ in})}\right) \\ \times (102.4)^2 \end{array}}$$

$$= 42.8 \frac{\text{kips}}{\text{in}^2}$$

$$M_n = F_{cr} S_x \le M_p$$

$$= \left(42.8 \frac{\text{kips}}{\text{in}^2}\right) (110 \text{ in}^3) \left(\frac{1 \text{ ft}}{12 \text{ in}}\right)$$

$$= 392.3 \text{ ft-kips} \le 525 \text{ ft-kips}$$

$$\phi_b M_n = (0.9)(392.3 \text{ ft-kips}) = 353.1 \text{ ft-kips}$$

$$w_u = \frac{8\phi_b M_n}{L^2}$$

$$= \frac{(8)(353.1 \text{ ft-kips})}{(36 \text{ ft})^2} = 2.180 \text{ kips/ft}$$

The superimposed load is

$$2.180 \frac{\text{kips}}{\text{ft}} = (1.2) \left(0.055 \frac{\text{kips}}{\text{ft}}\right) + 1.6w$$

$$w = 1.32 \text{ kips/ft}$$

13. DESIGN OF STEEL BEAMS

Steel beams can be designed with either the *allowable strength method* (also known as the *elastic design method*) or the *plastic design method* (also known as the *load factor method* and the *ultimate strength design method*). Simple one-span beams are usually designed by the allowable strength method since there is no advantage to using plastic design.

A detailed flowchart for design of steel beams is given in Fig. 59.10. The *AISC Manual* also provides three design aids for choosing beams: the Design Selection Tables [AISC Tables 3-2 through 3-5], the Maximum Total Uniform Load Tables [AISC Tables 3-6 through 3-9], and the Plots of Available Flexural Strength versus Unbraced Length Tables [AISC Tables 3-10 through 3-11].

Design Selection Tables

The Z_x beam selection tables (AISC Tables 3-2 to 3-5) are easy to use, and they provide a method for quickly selecting economical beams. Their use assumes that either the required plastic section modulus, Z, or the required design plastic moment, $M_u = \phi M_p$, is known, and one of these two criteria is used to select the beam. The tabulated values of $\phi_v V_n$ and I_x may be used to determine shear design strength or deflection, and the designer must ensure that $L_b \le L_p$. Most tables for W shapes assume that $F_y = 50$ ksi.

Beams are grouped in the beam selection table. Within a group, the most economical shape is listed at the top in bold print. This is the beam that should generally be used, even if the moment-resisting capacity and section modulus are greater than necessary. The weight of the most economical beam will be less than the beam in the group that most closely meets the structural requirements.

Maximum Total Uniform Load Table

In the Maximum Total Uniform Load tables [AISC Tables 3-6 to 3-9], loads are tabulated for $F_y = 50$ ksi steels for W shapes, and for 36 ksi for S, C, and MC shapes. Loads can be read directly from the tables when $L_b \le L_p$ for compact and noncompact W, S, C, and MC shapes. When $L_r \ge L_b > L_p$ or $L_b > L_r$, the tables are not applicable.

The tables are convenient for selecting laterally supported simple beams with equal concentrated loads spaced as shown in the *AISC Manual* table of Concentrated Load Equivalents [AISC Table 3-22a]. Except for short spans where shear controls the design, the beam load table may be entered with an equivalent uniform load.

Available Moment vs. Unbraced Length Tables

When the unbraced length L_b is greater than L_p, the Available Moment vs. Unbraced Length tables (AISC Table 3-10 for W shapes and Table 3-11 for channels) should be used. Table 3-10 can be used for 50 ksi steel with unbraced lengths up to 100 ft. (Table 3-11 covers channels with 36 ksi steel.) Each beam that is plotted on the chart is shown with a profile similar to that in Fig. 59.11.

The table is entered knowing L_b and $M_u = \phi M_n$. (The plots allow for beam self-weight, which should be deducted when calculating the maximum uniform load. When the unbraced length varies along the beam span, use the longest unbraced length.) The nearest solid line above the intersection of the L_b and M_u values is the most economical beam. Dotted line sections mean that a lighter beam exists that has the same capacity. As with the beam selection table, shear stress and deflection must still be checked.

The Available Moment vs. Unbraced Length tables implicitly assume the beam bending moment coefficient, C_b, is 1, and this remains a conservative assumption. However, when a non-uniform moment gradient is desired, C_b can be calculated using AISC Eq. F1-1.

Figure 59.10 *Flowchart for LRFD Beam Design (W and M shapes—moment, shear, and deflection criteria)*
(SPEC = Specifications; part 16 of the AISC *Manual)*

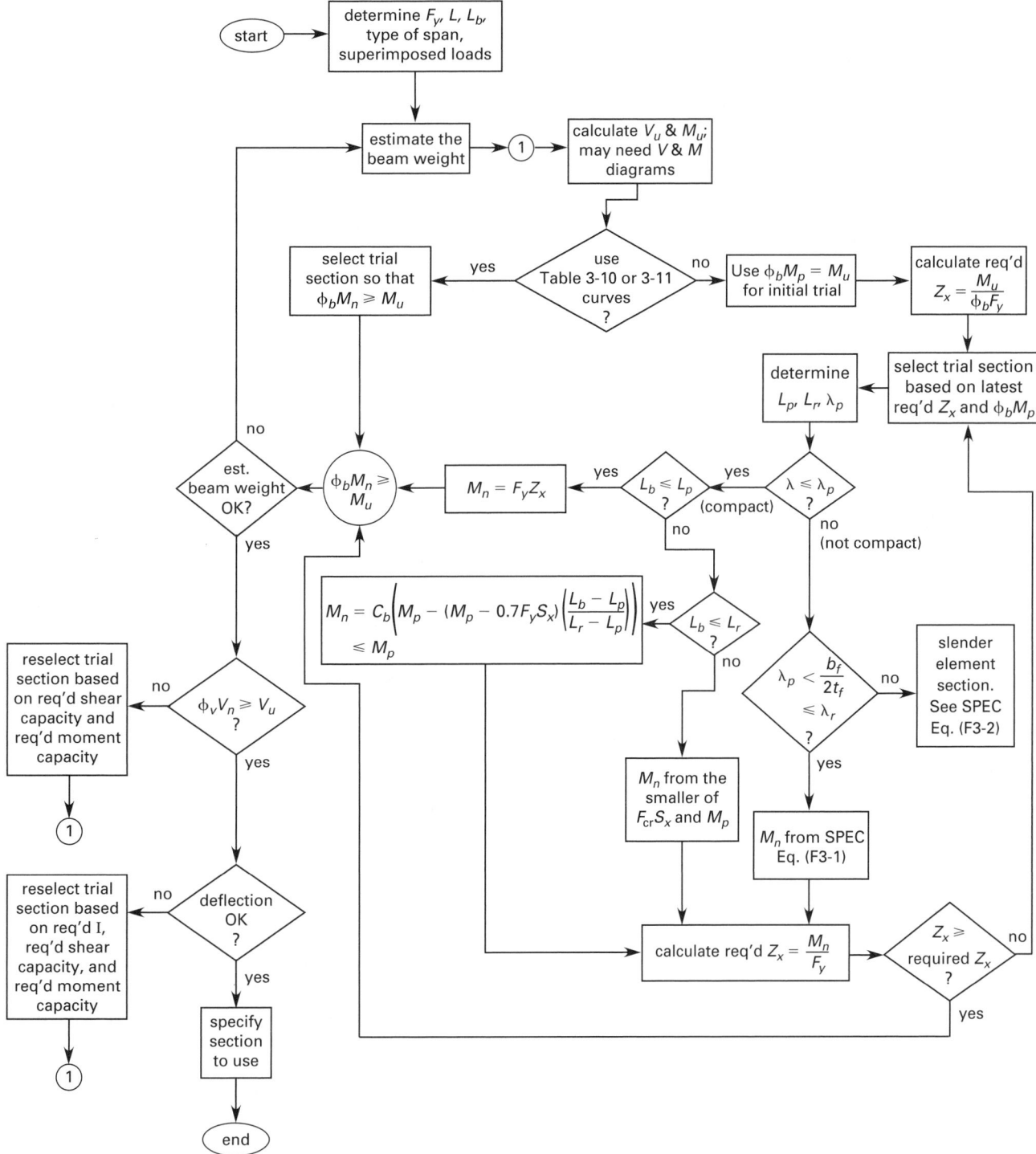

Adapted from *Applied Structural Steel Design*, 3rd ed., by Spiegel/Limbrunner, copyright © 1997. Reprinted by permission of Prentice-Hall, Inc., Upper Saddle River, NJ.

Figure 59.11 *Available Moment vs. Unbraced Length*

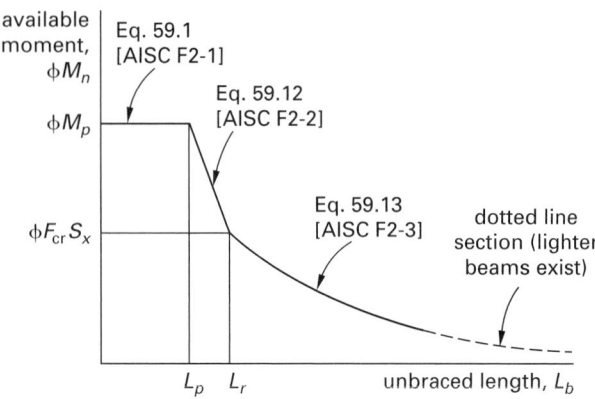

Example 59.2

Using the LRFD method and inelastic analysis, select the lightest W-shape of A992 steel for the beam and service loading shown. The load does not include the weight of the beam. Assume the compression flange is braced at the ends and at intervals of 8 ft. Consider moment, shear, and deflection. The maximum allowable deflection for the live load is $L/360$.

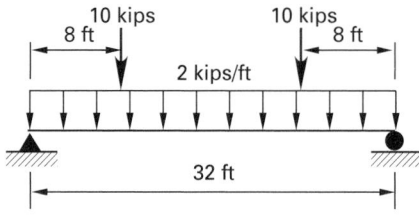

Solution

Design for moment.

$$M_u = \phi\left(\frac{wL^2}{8} + Px\right)$$

$$= (1.6)\left(\frac{\left(2\ \frac{\text{kips}}{\text{ft}}\right)(32\ \text{ft})^2}{8} + (10\ \text{kips})(8\ \text{ft})\right)$$

$$= 537.6\ \text{ft-kips}$$

The required plastic section modulus is

$$Z_x = \frac{M_u}{\phi F_y} = \frac{(537.6\ \text{ft-kips})\left(12\ \frac{\text{in}}{\text{ft}}\right)}{(0.9)(50\ \text{ksi})} = 143.4\ \text{in}^3$$

From Table 3-2, try W21 × 68 (in boldface type)

$$Z_x = 160\ \text{in}^3$$
$$\phi M_p = 600\ \text{ft-kips}$$
$$L_p = 6.36\ \text{ft}$$
$$L_r = 18.7\ \text{ft}$$

Since $L_p < L_b = 8\ \text{ft} \leq L_r$, use Eq. 59.12.

$$\phi M_n = \phi C_b\left(M_p - (M_p - 0.7F_yS_x)\left(\frac{L_b - L_p}{L_r - L_p}\right)\right) \leq \phi M_p$$

$$= (0.9)(1.0)$$

$$\times \left(\begin{array}{c} 667\ \text{ft-kips} \\ -\left(\begin{array}{c}667\ \text{ft-kips} \\ -(0.7)\left(50\ \frac{\text{kips}}{\text{in}^2}\right)(140\ \text{in}^3)\left(\frac{1\ \text{ft}}{12\ \text{in}}\right)\end{array}\right) \\ \times\left(\frac{8\ \text{ft} - 6.36\ \text{ft}}{18.7 - 6.36\ \text{ft}}\right) \end{array}\right)$$

$$= 569\ \text{ft-kips} \leq 600\ \text{ft-kips} \quad [\text{OK}]$$

The total moment including moment due to self-weight is

$$M_u = 538\ \text{ft-kips} + (1.2)\left(\frac{\left(0.068\ \frac{\text{kips}}{\text{ft}}\right)(32\ \text{ft})^2}{8}\right)$$

$$= 548\ \text{ft-kips}$$

$$\phi M_n = 569\ \text{ft-kips} \geq 548\ \text{ft-kips} \quad [\text{OK}]$$

Check the shear.

For this beam, the maximum shear occurs at the support and is equal to the reaction.

$$V_u = \phi\left(\frac{wL + 2P}{2}\right) + 1.2\left(\frac{w_D L}{2}\right)$$

$$= (1.6)\left(\frac{\left(2.068\ \frac{\text{kips}}{\text{ft}}\right)(32\ \text{ft}) + (2)(10\ \text{kips})}{2}\right)$$

$$+ (1.2)\left(\frac{\left(0.068\ \frac{\text{kips}}{\text{ft}}\right)(32\ \text{ft})}{2}\right)$$

$$= 70.2\ \text{kips}$$

From Table 3-2, the maximum permissible web shear for the W21 × 68 is 273 kips.

$$70.2\ \text{kips} < 273\ \text{kips} \quad [\text{OK}]$$

Check the live load deflection.

From Table 3-2, I_x for W21 × 68 is 1480 in⁴. The maximum allowable deflection is

$$\Delta = \frac{L}{360} = \frac{(32\ \text{ft})\left(12\ \frac{\text{in}}{\text{ft}}\right)}{360} = 1.07\ \text{in}$$

From the *AISC Manual*, Part 3, Beam Diagrams and Formulas, or from App. 44.A, the actual deflection is

$$\Delta = \frac{5wL^4}{384EI} + \frac{Pa(3L^2 - 4a^2)}{24EI}$$

$$= \frac{(5)\left(2\ \dfrac{\text{kips}}{\text{ft}}\right)(32\ \text{ft})^4\left(12\ \dfrac{\text{in}}{\text{ft}}\right)^3}{(384)\left(29{,}000\ \dfrac{\text{kips}}{\text{in}^2}\right)(1480\ \text{in}^4)}$$

$$+ \frac{(10\ \text{kips})(8\ \text{ft})\left(12\ \dfrac{\text{in}}{\text{ft}}\right)^3\left((3)(32\ \text{ft})^2 - (4)(8\ \text{ft})^2\right)}{(24)\left(29{,}000\ \dfrac{\text{kips}}{\text{in}^2}\right)(1480\ \text{in}^4)}$$

$$= 1.48\ \text{in}\quad [\text{NG}]$$

Calculate the required moment of inertia. Since deflection is inversely proportional to moment of inertia,

$$I_{x,\text{req}} = \left(\frac{1.48\ \text{in}}{1.07\ \text{in}}\right)(1480\ \text{in}^2)$$

$$= 2047\ \text{in}^4$$

Therefore, from Table 3-3 in the *AISC Manual* select W24 × 76.

14. PLASTIC DESIGN OF CONTINUOUS BEAMS

Plastic design and *inelastic design* are other names for *ultimate strength design*. The plastic design procedure is explicitly incorporated into the LRFD methods in the *AISC Manual*. When the design must be in accordance with App. 1 of the *AISC Manual*, the following method can be used to design continuous beams and frames. It must not be used with noncompact shapes, crane runway rails, A514 steel, or steels with yield strengths in excess of 65 ksi. The basic procedure consists of the following steps.

step 1: Obtain the factored loading. (Additional terms are needed when wind and earthquake loads are present.)

$$\text{factored load} = 1.2 \times \text{dead load}$$
$$+ 1.6 \times \text{live load} \qquad 59.22$$

step 2: Based on the factored loading, calculate the maximum (plastic) moment, M_p. (This moment cannot be calculated from the elastic moment diagrams in Tables 3-22 and 3-23 in Part 3 of the *AISC Manual.*) The distances to the maximum moment are taken from Table 59.1. Assume the left end span is a beam

with a fixed support at one end and a simple support at the other. The other spans are considered fixed-fixed (see Fig. 59.12).

Figure 59.12 *Location of Plastic Moments on a Uniformly Loaded Continuous Beam (plastic hinges shown as •)*

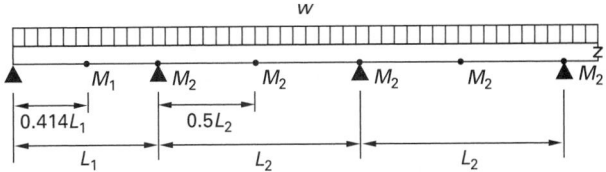

step 3: If beam selection is to be made according to the required *plastic section modulus*, calculate Z_x.

$$Z_x = \frac{M_p}{F_y} \qquad 59.23$$

step 4: Use the Flexural Design Tables in Part 3 of the *AISC Manual* to select an economical beam.

step 5: Check that the nominal shear strength V_n based on plastic failure does not exceed the design shear strength (calculated from AISC Eq. G2-1).

$$V_n = 0.6F_y A_w C_v \qquad 59.24$$

step 6: Specify web (intermediate) stiffeners at loading points where plastic hinges are expected.

step 7: Determine lateral bracing requirements. Compression flange support (lateral bracing) is required at points where plastic hinges will form. In addition, the distance between points of lateral support must be less than or equal to the distance given by Eq. 59.25 [*AISC Specification* Eq. A1-7].

$$L_{pd} = \left(0.12 + 0.076\left(\frac{M_1}{M_2}\right)\right)\left(\frac{E}{F_y}\right)r_y$$

$$59.25$$

M_1 is the smaller (absolute value) end moment, and M_2 is the larger end moment of the unbraced segment. The ratio M_1/M_2 is positive when moments cause reverse curvature and negative for single curvature.

Some easing of the l_{cr} distance is allowed for the last hinge to form in the failure mechanism, since the last hinge does not have to rotate as much as do the previous hinges. However, normal bracing lengths for elastic design must still be met. The term *mechanism* (i.e., a mechanical device with its own joints and elbows) is used to describe each of the different ways that the beam can fail plastically. Each of the spans may experience hinges at different locations and/or loadings and thus represents a distinctly unique mechanism.

This procedure assumes that the Plastic Design Selection Table will be used and, therefore, compactness is assured. The procedure also omits thickness checks that must be performed when the beam carries axial loads in addition to transverse loads.

15. ULTIMATE PLASTIC MOMENTS

Step 2 requires calculating the maximum moment based on plastic theory. This moment is not the same as the moment determined from elastic theory. Table 59.1 can be used in a simple case to determine the ultimate moment, M_p. Locations of plastic hinges are also indicated in the table.

A uniformly loaded, continuous beam is a case that occurs frequently. The ultimate moments can be derived

Table 59.1 *Maximum Plastic Moments (plastic hinges shown as •)*

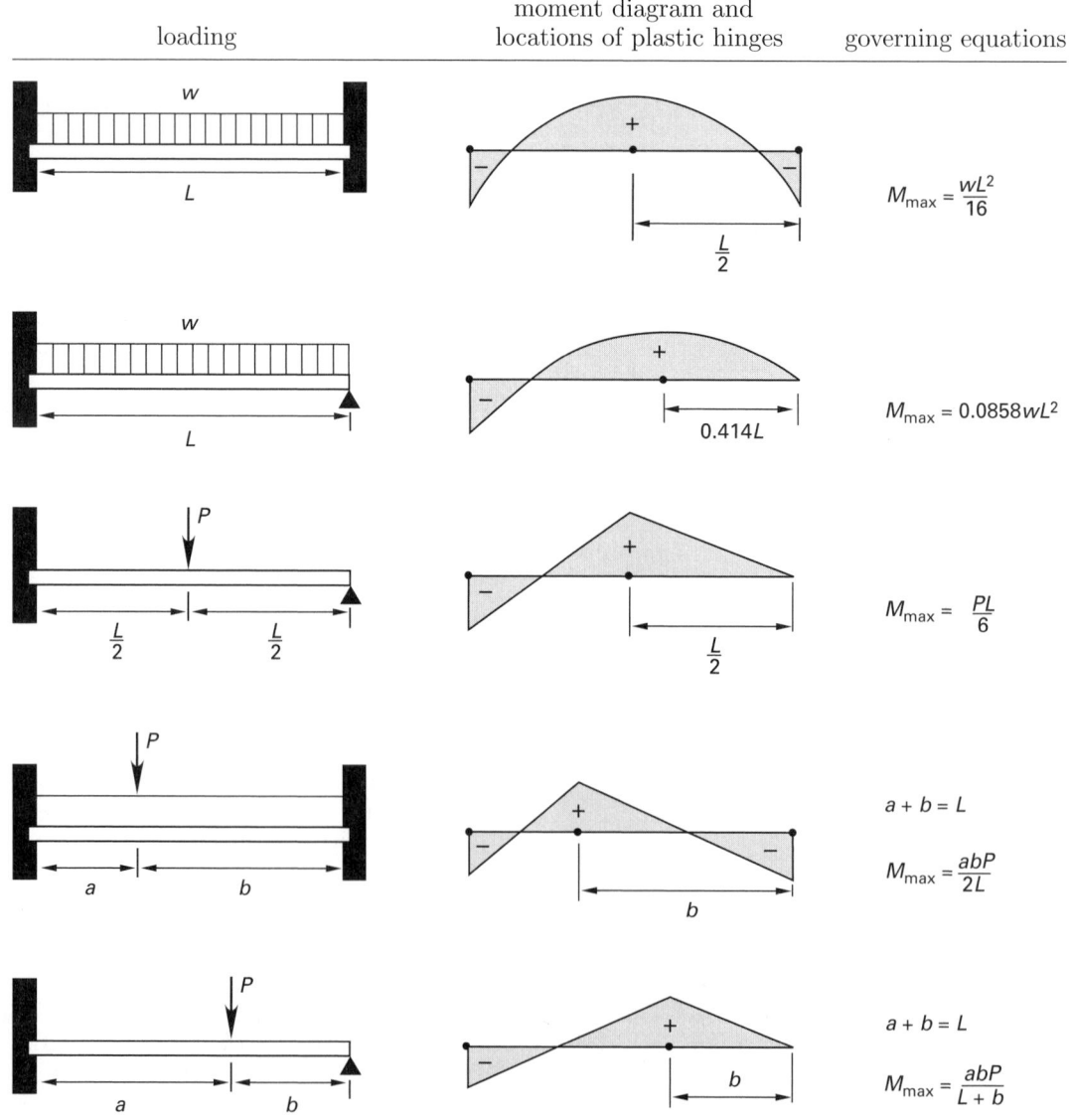

from the cases in Fig. 59.12. Both ultimate moments should be checked, since the interior hinges will form first if the span lengths are short. (The end span hinges will form first if all spans are the same length.)

$$M_1 = \left(\tfrac{3}{2} - \sqrt{2}\right) wL_1^2$$

$$\approx 0.0858 wL_1^2 \qquad \textit{59.26}$$

$$M_2 = \frac{wL^2}{16} \qquad \textit{59.27}$$

16. ULTIMATE SHEARS

The ultimate shear in step 5 may or may not be the factored shear on the beam. The shear, by definition, is the slope of the moment diagram. In the case of a uniformly loaded beam with built-in ends, the effect of plastic failure is merely to move the baseline without changing the overall shape of the moment diagram. Since there is no change in the slope, the ultimate shear is merely the factored shear.

In the case of continuous beams, however, the effect of plastic failure will change the slope of the lines in the moment diagram. Graphical or analytical means can be used to obtain the maximum slope. For a uniformly loaded continuous beam as shown in Fig. 59.12, the maximum shear induced will be

$$V_{\max} = \frac{wL}{2} + \frac{M_{\max}}{L} = 0.5858 wL \qquad \textit{59.28}$$

Example 59.3

Use plastic design to select a W-shape beam (A992 steel) to support a dead load (including the beam's own weight) of 1000 lbf/ft and a live load of 3500 lbf/ft over the beam shown. The beam is simply supported at all three points. The flange is supported continuously.

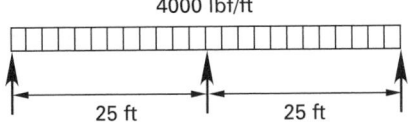

4000 lbf/ft

25 ft   25 ft

Solution

step 1: The factored load is

$$w_u = (1.2)\left(1000\ \frac{\text{lbf}}{\text{ft}}\right) + (1.6)\left(3500\ \frac{\text{lbf}}{\text{ft}}\right)$$

$$= 6800\ \text{lbf/ft}$$

step 2: The maximum moment on an end-span of a uniformly loaded continuous beam is given in Table 59.1. From Eq. 59.26 (i.e., the required M),

$$M_{\max} = 0.0858 wL^2$$

$$= (0.0858)\left(6800\ \frac{\text{lbf}}{\text{ft}}\right)(25\ \text{ft})^2$$

$$= 364{,}650\ \text{ft-lbf} \quad (365\ \text{ft-kips})$$

step 3: The selection can be made on the basis of required plastic moment. This step is skipped.

step 4: Entering the Flexural Design Tables (*AISC Manual* Part 3, Table 3-10), a W21 × 48 beam is chosen. The available strength is 398 ft-kips, which is more than required. However, this is the economical beam with a capacity exceeding the requirements.

step 5: The maximum (required) shear on an end span of a uniformly loaded, continuous beam is given by Eq. 59.28.

$$V_{\text{req'd}} = 0.5858 wL$$

$$= (0.5858)\left(6800\ \frac{\text{lbf}}{\text{ft}}\right)(25\ \text{ft})$$

$$= 99{,}586\ \text{lbf}$$

From the AISC Shape Table 1-1, $d = 20.60$ in and $t_w = 0.350$ in. The available shear is

$$V_{\text{available}} = \phi F_v A = \phi 0.6 F_y A$$

$$= (1.0)(0.6)\left(50{,}000\ \frac{\text{lbf}}{\text{in}^2}\right)$$

$$\times (0.350\ \text{in})(20.60\ \text{in})$$

$$= 217{,}000\ \text{lbf}$$

Alternatively, from *AISC Manual* Table 3-2, $\phi V_n = 217$ kips.

Since $V_{\text{req'd}} < V_{\text{available}}$, shear is not a problem.

17. CONSTRUCTING PLASTIC MOMENT DIAGRAMS

There are several methods of determining the shape of the plastic moment diagram. The method presented here is a semigraphical approach that can be used to quickly draw moment diagrams and locate points where plastic hinges will form.

step 1: Consider the beam as a series of simply supported spans.

step 2: Draw the elastic bending moment on each span.

step 3: Construct the modified baseline. This is a jointed set of straight lines that meets the following conditions: (a) The baseline meets the horizontal axis at simply supported exterior ends. This is consistent with the requirement that the moment be zero at free and simply supported ends. (It is not necessary for the baseline to reach the horizontal axis at built-in ends. In fact, the moment is usually nonzero at such points.) (b) The slope of the baseline changes only at points of support. (c) The baselines for all spans connect at points of support. The baseline is located to minimize the maximum ordinate along the entire length of the beam (along all spans). Therefore, one span will control. (Do not consider the distance between the baseline and the horizontal axis when minimizing the maximum ordinate.)

step 4: The maximum ordinate determines M_p.

step 5: Hinges form at points of maximum ordinates and wherever else required to support full rotation. The moments at hinges that form simultaneously are identical.

Example 59.4

Draw the plastic moment diagram for the uniformly loaded, two-span beam shown.

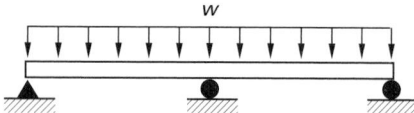

Solution

steps 1 and 2: The moment diagram of a simply supported, uniformly loaded single span is drawn twice.

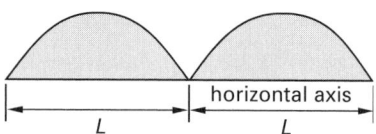

step 3: Referring to Table 59.1, the modified baseline is chosen to minimize the distance between the curved line and the baseline.

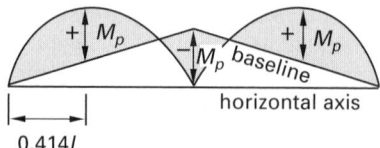

step 4: The maximum ordinate is determined visually to be near the middle of the end spans.

step 5: If hinges form near the middle of the end spans, a hinge must also form over the center support. Otherwise, the beam could not rotate in failure.

The final moment diagram is drawn by "straightening out" the modified baseline.

$$M_{\max} = 0.0858wL^2$$

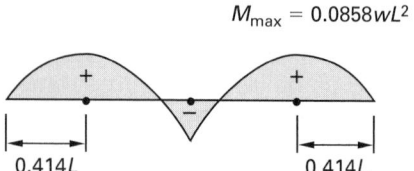

Example 59.5

Draw the plastic moment diagram for the beam of constant cross section shown.

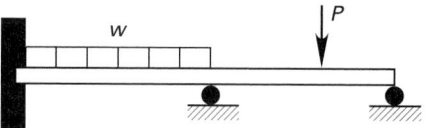

Solution

Since the actual numerical loads are unknown, it is not possible to determine which of the two spans controls. If the left span controls, then three hinges must form simultaneously. This forces the baseline to be horizontal along the left span. The baseline along the right span is fixed by its end points.

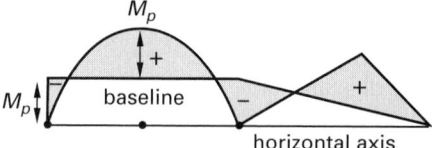

If the right span controls, then two hinges must form simultaneously. The moments at these two hinge points are equal. The baseline along the left span is chosen to minimize the positive and negative ordinates.

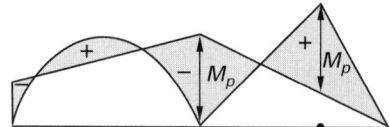

18. UNSYMMETRICAL BENDING

Typically, a wide-flange beam is oriented with its web in the plane of the loading (i.e., the plane of the weak axis), and the bending is said to occur about the strong axis or the x-x axis of beam cross section. The x-x and y-y axes of a beam cross section are known as the *principal axes* for which the product of inertia is zero. A beam is said to be subjected to *unsymmetrical bending* when bending occurs about an axis other than one of the principal axes. When the applied loads are not in a plane with either of the principal axes or when loads are simultaneously applied from more than one direction, unsymmetrical bending (also called *biaxial bending*) is the result.

There are two cases of unsymmetrical bending: (1) loads applied through the shear center and (2) loads not applied through the shear center. The *shear center* of a cross section is that point through which the loads must act if there is to be no twisting (torsion) of the beam. The location of the shear center is determined from principles of mechanics of materials by equating the internal resisting torsional moment to the external torque. For a wide-flange section, however, the shear center and the centroid are the same point. In either case, an appropriate interaction equation is used for design or analysis purpose.

In case (1), the load applied through the shear center may be resolved into two mutually perpendicular components along the principal (x-x and y-y) axes. Thus, referring to Fig. 59.13(a) where θ is the angle between the applied load F and the y-y axis, the components of F are $F_x = F \cos \theta$ and $F_y = F \sin \theta$. The moments M_x and M_y about the principal axes are computed from F_x and F_y, respectively. (Note: The force F_x creates a moment about the x-x axis and F_y creates a moment about the y-y axis.) The stresses due to moments M_x and M_y are computed separately for bending about each axis as: $f_{bx} = M_x/S_x$ and $f_{by} = M_y/S_y$.

Finally, to check the adequacy of a section, Eq. 59.29 (AISC *interaction equation* H2-1), which serves as the design criterion since the allowable bending stresses with respect to the x-x and y-y axes are different, is used. (Note that AISC uses the subscript "w" to relate to the major principal axis bending and "z" for the minor principal axis bending, while this book uses x and y.)

$$\left| \frac{f_a}{F_a} + \frac{f_{by}}{F_{by}} + \frac{f_{bx}}{F_{bx}} \right| \le 1.0 \qquad \textbf{59.29}$$

Since the axial stress for a beam in pure bending is zero, the first term of the equation drops out and Eq. 59.29 becomes

$$\frac{f_{bx}}{F_{bx}} + \frac{f_{by}}{F_{by}} \le 1.0 \qquad \textbf{59.30}$$

Since loads are frequently applied from different directions on the top flange, case (2) (where load does not pass through the shear center) is a common occurrence. The top flange alone is assumed to resist the lateral force component. For a typical wide-flange shape, the section modulus of the top flange is approximately equal to $S_y/2$, and the stress due to bending about the y-y axis becomes $f_{by} = 2M_y/S_y$.

Example 59.6

A W8 × 21 beam of A992 steel is used as a roof purlin on a simple span of 15 ft. The roof slope is 4:12. The beam carries a uniformly distributed gravity live load of 0.52 kip/ft that passes through the centroid of the section. Assuming that the beam has full lateral support provided by the roofing above, check the adequacy of the section.

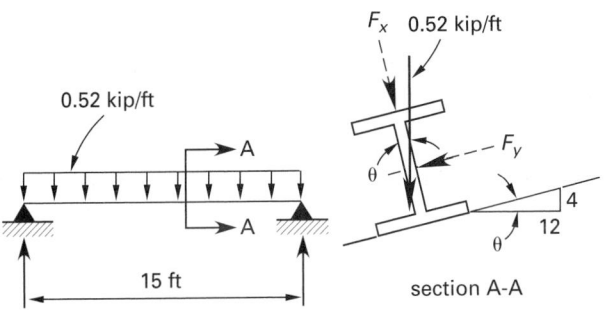

Figure 59.13 *Cases of Unsymmetrical Bending*

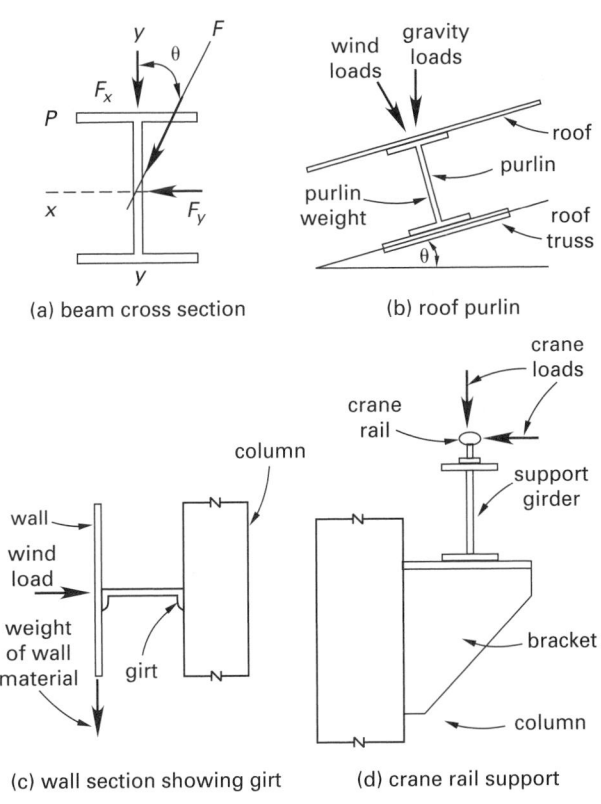

(a) beam cross section

(b) roof purlin

(c) wall section showing girt

(d) crane rail support

Source: *Applied Structural Steel Design*, 3rd ed., by Spiegel/ Limbrunner, copyright © 1997. Reprinted by permission of Prentice-Hall, Inc., Upper Saddle River, NJ.

Solution

Properties of W8 × 21 are $S_x = 18.2$ in^3 and $S_y = 3.71$ in^3. With a weight per foot of 21 lbf/ft, the uniform gravity load is $w = 0.52$ kip/ft.

Resolve the load into its components.

$$\theta = \tan^{-1}\left(\frac{4}{12}\right) = 18.4°$$

$$w_x = w \cos \theta = \left(0.52 \ \frac{\text{kip}}{\text{ft}}\right) \cos 18.4° = 0.49 \ \text{kip/ft}$$

$$w_{ux} = (1.6)\left(0.49 \ \frac{\text{kip}}{\text{ft}}\right) = 0.784 \ \text{kip/ft}$$

$$w_y = w \sin \theta = \left(0.52 \ \frac{\text{kip}}{\text{ft}}\right) \sin 18.4° = 0.16 \ \text{kip/ft}$$

$$w_{uy} = (1.6)\left(0.16 \ \frac{\text{kip}}{\text{ft}}\right) = 0.256 \ \text{kip/ft}$$

Compute the moments.

$$M_{ux} = \frac{w_x l^2}{8} = \frac{\left(0.784 \ \frac{\text{kip}}{\text{ft}}\right)(15 \ \text{ft})^2}{8}$$
$$= 22.0 \ \text{ft-kips}$$

$$M_{uy} = \frac{w_y l^2}{8} = \frac{\left(0.256 \ \frac{\text{kip}}{\text{ft}}\right)(15 \ \text{ft})^2}{8}$$
$$= 7.2 \ \text{ft-kips}$$

Compute the required bending stresses.

$$f_{bx} = \frac{M_{ux}}{S_x} = \frac{(22.0 \ \text{ft-kips})\left(12 \ \frac{\text{in}}{\text{ft}}\right)}{18.2 \ \text{in}^3}$$
$$= 14.5 \ \text{ksi}$$

$$f_{by} = \frac{M_{uy}}{S_y} = \frac{(7.2 \ \text{ft-kips})\left(12 \ \frac{\text{in}}{\text{ft}}\right)}{3.71 \ \text{in}^3}$$
$$= 23.3 \ \text{ksi}$$

Determine the available stresses from the available moments.

For bending about the x-x axis, since the beam has full lateral support, from AISC Table 3-2 or Table 3-10, for W8 × 21, $\phi M_n = 76.5$ ft-kips.

$$F_{bx} = \frac{\phi_b M_{nx}}{S_x} = \frac{(76.5 \ \text{ft-kips})\left(12 \ \frac{\text{in}}{\text{ft}}\right)}{18.2 \ \text{in}^3} = 50.4 \ \text{ksi}$$

For bending about the y-y axis, from AISC Table 3-4, $\phi M_n = 21.3$ ft-kips.

$$F_{by} = \frac{\phi_b M_{ny}}{S_y} = \frac{(21.3 \ \text{ft-kips})\left(12 \ \frac{\text{in}}{\text{ft}}\right)}{3.71 \ \text{in}^3} = 68.9 \ \text{ksi}$$

Check the interaction equation, Eq. 59.30.

$$\frac{f_{bx}}{F_{bx}} + \frac{f_{by}}{F_{by}} \le 1$$
$$\frac{14.5 \ \text{ksi}}{50.4 \ \text{ksi}} + \frac{23.3 \ \text{ksi}}{68.9 \ \text{ksi}} = 0.63 < 1.0 \quad [\text{OK}]$$

Therefore, the W8 × 21 section is adequate.

19. BRIDGE CRANE MEMBERS

A bridge crane member is a beam that carries a set of moving loads that remain equidistant as they travel across the beam. For a simply supported beam, (a) the absolute maximum moment occurs under one of the concentrated loads when that load and the resultant of the set of loads are equidistant from the centerline of the beam, and (b) the absolute maximum shear will occur just next to one of the supports when the load group is placed such that the resultant is at a minimum distance from the support.

Two other factors that are usually taken into consideration in the design of bridge crane members are fatigue and impact loading.

20. HOLES IN BEAMS

A beam is usually connected to other structural members. If such connections are made with bolts, holes are punched or drilled in the web or flanges of the beam. In addition, to provide space for utilities such as electrical conduits and ventilation ducts, sometimes relatively large holes are cut in the beam webs. Holes in a beam result in capacity reduction. Two such reductions are recognized: (a) holes in beam webs reduce the shear capacity, and (b) holes in beam flanges reduce the moment capacity. In general, web holes should be located at sections of low shear and centered on the neutral axis to avoid high bending stresses, and the flange holes should be located at sections of low bending moment.

The effect of flange holes is to reduce the moment of inertia, and hence, reduce moment capacity. The reduced moment of inertia is based upon the effective tension flange area given by *AISC Specification* C, Sec. F13.

21. BEAM BEARING PLATES

Not all beams are supported by connections to other structural steel members (e.g., beams or columns). Some beams are supported by bearing on concrete or masonry members, such as walls or pilasters. Since masonry and concrete are weaker than steel, the beam reaction must be distributed over a large enough area to keep the average bearing pressure on concrete or masonry within allowable limits. A *bearing plate* is used for this purpose.

The design of a bearing plate is based on *AISC Manual*, Part 14, Secs. J7 and J8 of the *AISC Specification*, and on the following criteria. (a) It must be long enough so that local web yielding or web crippling of the beam does not occur. (b) It must be large enough to ensure the required bearing strength, R_u or R_a, under the plate does not exceed the available bearing strength, ϕR_n or R_n/Ω. (c) It must be thick enough so that the moment in the plate at the critical section is less than the available moment.

A procedure for the design of beam bearing plates on concrete supports is given in *AISC Manual*, Part 14. *AISC Specification* Sec. J7 gives the nominal bearing strength R_n for various types of bearing. For column bases and bearing on concrete, *AISC Specification* Sec. J8 gives provisions so that column loads and moments are properly transferred to the footings and foundations. For a plate covering the full area of a concrete support, where P_p is the nominal bearing strength and

Figure 59.14 *Nomenclature for Beam Bearing Plate*

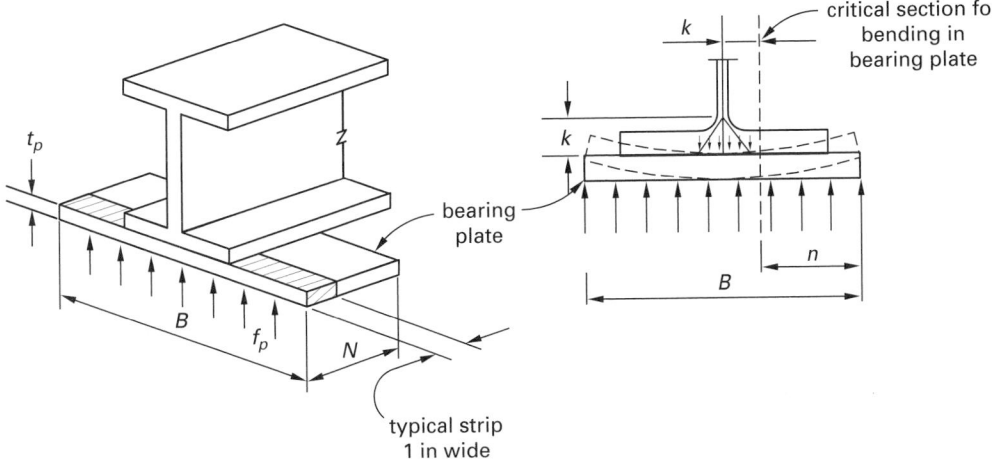

A_1 is the area of steel concentrically bearing on a concrete support,

$$P_p = 0.85 f'_c A_1 \quad \text{[AISC Eq. J8-1]} \qquad 59.31$$

For a plate covering less than the full area of a concrete support, the area of steel concentrically bearing on a concrete support is given by Eq. 59.32. Area A_2 is the maximum area of the portion of the supporting surface that is geometrically similar to and concentric with the loaded area.

$$P_p = 0.85 f'_c A_1 \sqrt{\frac{A_2}{A_1}} \leq 1.7 f'_c A_1 \quad \text{[AISC Eq. J8-2]} \quad 59.32$$

The minimum bearing length, N (in inches), is obtained as the larger of the two values, using Eqs. 59.33 and 59.34 (corresponding to *AISC Manual* values found in Table 9-4), based on local web yielding ϕR_n and web crippling criteria, respectively, as described in *AISC Specification* Sec. J10.2.

$$N = \frac{\phi R_n - \phi R_1}{\phi R_2} \qquad 59.33$$

$$N = \frac{\phi R_n - \phi R_3}{\phi R_4} \qquad 59.34$$

After establishing the bearing length, N, the width, B, is obtained from Eq. 59.35. Round up to an integral number of inches to provide a shelf for welding the bearing plate to the beam.

$$B = \frac{A_1}{N} \qquad 59.35$$

The required moment, M_u, is limited by the pressure developed under the plate, f_p, from the factored vertical reaction, R_u, which is assumed to be uniform and is calculated as

$$f_p = \frac{R_a}{A_1} = \frac{R_a}{BN} \quad \text{[ASD]} \qquad 59.36(a)$$

$$f_p = \frac{R_u}{A_1} = \frac{R_u}{BN} \quad \text{[LRFD]} \qquad 59.36(b)$$

The cantilever dimension, n, is determined from

$$n = \frac{B}{2} - k \qquad 59.37$$

The thickness, t, is specified in $^1/_8$ in increments. The minimum thickness is

$$t = \sqrt{\frac{2R_a n^2 \Omega}{A_1 F_y}} = \sqrt{\frac{2 f_p n^2 \Omega}{F_y}} \quad \text{[ASD]} \qquad 59.38(a)$$

$$t = \sqrt{\frac{2R_u n^2}{\phi A_1 F_y}} = \sqrt{\frac{2 f_p n^2}{\phi F_b}} \quad \text{[LRFD]} \qquad 59.38(b)$$

With LRFD, the available moment, ϕM_n, is limited by F_b, which, for rectangular plates, is the same as for weak-axis bending: $F_b = 0.9 F_y$. For ASD, $\Omega = 1.67$. For flexure, $phi = 0.90$.

Example 59.7

A W21 × 68 beam is supported on a 12 in thick concrete wall ($f'_c = 3500$ lbf/in²). The beam reaction, ϕR_n, is 68 kips, and the length of bearing is limited to 8 in, considering a minimum edge distance of 2 in. The beam is A992, and the plate is A36 steel. Design the bearing plate using LRFD.

Solution

From the *AISC Manual*, Table 9-4, Beam Bearing Constants, $\phi R_1 = 63.7$ kips, $\phi R_2 = 21.5$ kips/in, $\phi R_3 = 84.3$ kips, and $\phi R_4 = 5.95$ kips/in. Using Eqs. 59.33 and 59.34,

$$N = \frac{\phi R_n - \phi R_1}{\phi R_2}$$
$$= \frac{68 \text{ kips} - 63.7 \text{ kips}}{21.5 \dfrac{\text{kips}}{\text{in}}}$$
$$= 0.2 \text{ in} < 8 \text{ in} \quad [\text{OK}]$$
$$N = \frac{\phi R_n - \phi R_3}{\phi R_4}$$
$$= \frac{68 \text{ kips} - 84.3 \text{ kips}}{5.95 \dfrac{\text{kips}}{\text{in}}}$$
$$= -2.74 \text{ in} < 8 \text{ in} \quad [\text{OK}]$$

Conservatively estimate the required area, A_1, by rearranging Eq. 59.31.

$$A_1 = \frac{P_p}{0.85\phi_c f_c'}$$
$$A_{1,\text{required}} = \frac{68 \text{ kips}}{(0.85)(0.6)\left(3.5 \dfrac{\text{kips}}{\text{in}^2}\right)} = 38.1 \text{ in}^2$$

This requires a minimum width of

$$B_{\min} = \frac{A_1}{N} = \frac{38.1 \text{ in}^2}{8 \text{ in}} = 4.76 \text{ in}$$

For a W21 × 68, the flange width is $b_f = 8.27$ in, so use $B = 10$ in.

$$A_{\text{provided}} = (8 \text{ in})(10 \text{ in}) = 80 \text{ in}^2$$

The area of concrete support is

$$A_2 = \big(8 \text{ in} + (2)(2 \text{ in})\big)\big(10 \text{ in} + (2)(2 \text{ in})\big) = 168 \text{ in}^2$$

Recalculate to reduce plate area.

Try $N = 6$ in. The area of concrete support, rearranging Eq. 59.32, is

$$A_2 = \frac{P_p^2}{(0.85)^2 f_c'^2 A_1}$$
$$= \frac{(68 \text{ kips})^2}{(0.85)^2 \left(3.5 \dfrac{\text{kips}}{\text{in}^2}\right)^2 (38.1 \text{ in}^2)}$$
$$= 13.7 \text{ in}^2 < 60 \text{ in}^2$$
$$A_{\text{provided}} = (6 \text{ in})(10 \text{ in}) = 60 \text{ in}^2$$

Therefore, 6 in bearing length is adequate.

For W21 × 68, $k = 1.19$ in. The plate cantilever dimension is

$$n = \frac{B}{2} - k = \frac{10 \text{ in}}{2} - 1.19 \text{ in} = 3.81 \text{ in}$$

The available bending moment, ϕM_n, is limited by

$$F_b = 0.9 F_y = (0.9)\left(36 \frac{\text{kips}}{\text{in}^2}\right) = 32.4 \text{ kips/in}^2$$

The required moment, M_u, is limited by

$$f_p = \frac{R_u}{BN} = \frac{68 \text{ kips}}{60 \text{ in}^2} = 1.133 \text{ kips/in}^2$$

From Eq. 59.38, the plate thickness is

$$t = \sqrt{\frac{2 f_p n^2}{\phi F_b}}$$
$$= \sqrt{\frac{(2)\left(1.133 \dfrac{\text{kips}}{\text{in}^2}\right)(3.81 \text{ in})^2}{(0.90)\left(32.4 \dfrac{\text{kips}}{\text{in}^2}\right)}}$$
$$= 1.06 \text{ in} \quad [\text{use } 1^{1}/_4 \text{ in thick plate}]$$

The bearing plate is $1^{1}/_4 \times 6 \times 10$.

60 Structural Steel: Tension Members

Nomenclature

A	cross-sectional area	in^2
b	width	in
d	depth or diameter	in
F	strength or allowable stress	ksi
g	lateral gage spacing of adjacent holes	in
l	weld length	in
L	member length	in
P	applied load or load capacity	kips
r	radius of gyration	in
s	longitudinal spacing adjacent holes	in
SR	slenderness ratio	–
t	thickness	in
U	reduction coefficient	–
w	plate width	in

Subscripts

e	net effective
g	gross
h	hole
n	net or nominal
t	tensile
u	ultimate
v	net shear
y	yield or y-axis

Symbols

ϕ	resistance factor	–

1. INTRODUCTION

Tension members occur in a variety of structures including bridge and roof trusses, transmission towers, wind bracing systems in multistoried buildings, and suspension and cable-stayed bridges. Wire cables, rods, eyebars, structural shapes, and built-up members are typically used as tension members.

Figure 60.1 *Common Tension Members*

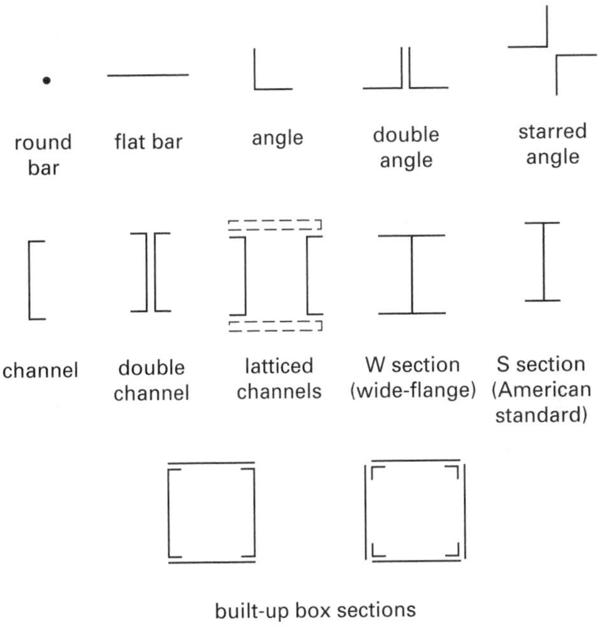

round bar, flat bar, angle, double angle, starred angle

channel, double channel, latticed channels, W section (wide-flange), S section (American standard)

built-up box sections

2. AXIAL TENSILE STRENGTH

Tension members are designed such that the required axial strength is less than the available strength.

$$P_u \le \phi_t P_n \qquad 60.1$$

3. GROSS AREA

The gross area, A_g, is the original unaltered cross-sectional area of the member. The gross area for the plate shown in Fig. 60.2 is

$$A_g = bt \qquad 60.2$$

4. NET AREA

Plates and shapes are connected to other plates and shapes by means of welds, rivets, and bolts. If rivets or bolts are used, holes must be punched or drilled in the member. As a result, the member's cross-sectional area at the connection is reduced. The load-carrying capacity of the member may also be reduced, depending on the size and location of the holes. The *net area* actually available to resist tension is illustrated in Fig. 60.2(b).

Figure 60.2 *Tension Member with Unstaggered Rows of Holes*

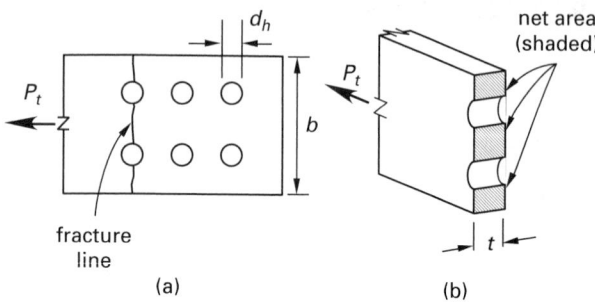

(a)   (b)

In a multiconnector connection, the *gage spacing, g*, is the lateral center-to-center spacing of two adjacent holes perpendicular to the direction of the applied load as shown in Fig. 60.4. The *pitch spacing, s*, is the longitudinal spacing of any two adjacent holes in the direction for the applied load. Gage and pitch spacing are shown in Fig. 60.3.

When space limitations (such as a limit on dimension x in Fig. 60.3(a) or odd connection geometry as in Fig. 60.3(b)) make it necessary to use more than one row of fasteners, the reduction in cross-sectional area can be minimized by using a staggered arrangement of fasteners. For multiple rows of fasteners (staggered or not) at a connection, more than one potential fracture line may exist. For each possible fracture line, the net width, b_n, is calculated by subtracting the hole diameters, d_h, in the potential failure path from the gross width, b, and then adding a correction factor $s^2/4g$ for each diagonal leg in the failure path. (There is one diagonal leg, BC, in the failure path ABCD in Fig. 60.3(c). Thus, the term $s^2/4g$ would be added once.)

$$b_n = b - \sum d_h + \sum \frac{s^2}{4g} \qquad 60.3$$

Punched bolt holes should be $^1/_{16}$ in larger than nominal fastener dimensions. (This is sometimes referred to as a *standard hole* [*AISC Specification* B2].) However, due to difficulties in producing uniform punched holes in field-produced assemblies, another $^1/_{16}$ in should be added to the hole diameter. (Note the additional $^1/_{16}$ is advised, but not specified, in the *AISC Manual*.) Thus, the hole diameter, d_h, in Eq. 60.3 is taken as the fastener diameter, d, plus $^1/_8$ in.

For members of uniform thickness (e.g., plates), the net area, A_n, is calculated as the net width, b_n, times the thickness, t.

$$A_n = b_n t \qquad 60.4$$

Figure 60.3 *Tension Members with Staggered Holes*

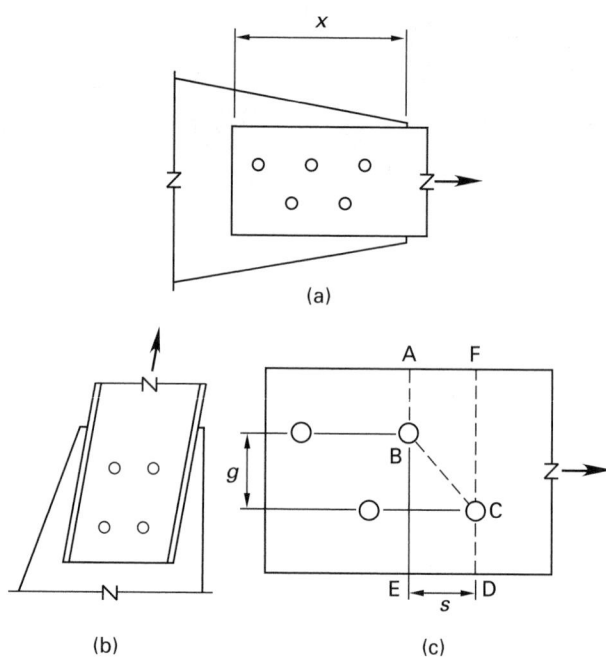

(a)

(b)   (c)

For members of nonuniform thickness (e.g., channels), a more useful formula for net area, A_n, is Eq. 60.5.

$$A_n = A_g - \sum d_h t + \left(\sum \frac{s^2}{4g} \right) t \qquad 60.5$$

The *critical net area* is the net area having the least value. It is obtained by checking all possible failure paths. For example, in Fig. 60.3(c), paths ABCD and DCF must both be checked. In Fig. 60.4, where the number of bolts is not the same in all rows, both paths ABDE and ABCDE must be checked.

Figure 60.4 *Tension Member with Uniform Thickness and Unstaggered Holes*

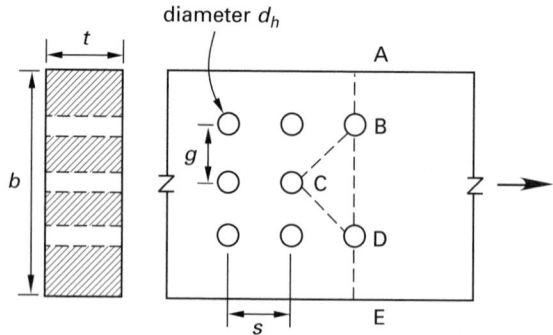

If connectors in a tension lap splice are arranged in two or more unstaggered rows and if the rows have unequal numbers of fasteners, each fracture line should be checked for tension capacity assuming that the previous fracture lines have absorbed a proportionate share of the load. For example, if three bolts are arranged in two rows—one row with two bolts and the next row with one bolt—the net section of the first fracture line (the width less two bolt diameters) should be checked for tension capacity with two-thirds of the total applied load.

5. EFFECTIVE NET AREA

When a tension member frames into a supporting member, some of the load-carrying ability will be lost unless all connectors are in the same plane and all elements of the tension member are connected to the support. (An angle connected to its support only by one of its legs is a case in which load-carrying ability would be lost.) A *shear lag* reduction coefficient, U, is used to calculate the effective net area. U can be taken as 1.0 if all cross-sectional elements are connected to the support to transmit the tensile force. When the load is transmitted by bolts through only some of the cross-sectional elements of the tension member, the *effective net area*, A_e, is computed from Eq. 60.6 [AISC Eq. D3-1]. U takes on values of 0.90, 0.85, or 0.60 as given in Table 60.1.

$$A_e = UA_n \qquad 60.6$$

Table 60.1 Shear Lag Reduction Coefficient (U) Values for Bolted Connections

(a) For W, M, S, or HP shapes with flange widths not less than two thirds of the depth, and for structural tees cut from these shapes (provided the connection is to the flanges) with three or more fasteners per line in the direction of stress): $U = 0.90$.

(b) For W, M, S, or HP structural shapes not meeting the conditions of paragraph (a), structural tees cut from these shapes, and all other shapes, including built-up cross sections, with three or more fasteners per line in the direction of stress: $U = 0.85$.

(c) For all members with only two fasteners per line in the direction of stress: $U = 0.60$.

Source: *AISC Specification* Table D3.1

In addition to the reduction in the net area by U, the effective net area for splice and gusset plates must not exceed 85% of the gross area, regardless of the number of holes. Tests have shown that as few as one hole in a plate will reduce the strength of a plate by at least 15%. It is a good idea to limit the effective net area to 85% of the gross area for all connections (not just splices and gusset plates) with holes in plates.

When the load is transmitted by welds through only some of the cross-sectional elements of the member, the effective net area, A_e, shall be computed from Eq. 60.7 [*AISC Specification* Table D3.1, case 3]. U takes on the values in Table 60.1, except that paragraph (c) is not applicable.

AISC Specification Table D3.1 furnishes two special cases (cases 3 and 4) of welded connections.

(a) When a load is transmitted by *transverse* welds to some but not all of the cross-sectional elements of W, M, or S shapes and structural tees cut from these shapes, the effective net area, A_e, shall be taken as the area of the directly connected elements, and U is taken as 1.0.

(b) When a load is transmitted to a plate by only *longitudinal* welds along both edges at the end of the plate, the length of each weld shall be not less than the width of the plate. The effective net area, A_e, shall be computed by Eq. 60.7, where A_g is the gross area, and $U = 1.0$, 0.87, or 0.75 as given in Table 60.2.

$$A_e = UA_g \qquad 60.7$$

Table 60.2 Shear Lag Reduction Coefficient (U) Values for Welded Longitudinal Connections[a]

$l \geq 2w$	$U = 1.0$
$2w > l \geq 1.5w$	$U = 0.87$
$1.5w > l \geq w$	$U = 0.75$

[a]l = weld length, in; w = plate width (distance between welds), in

Source: *AISC Specification* Table D3.1

6. DESIGN TENSILE STRENGTH

The design (available) strength, $\phi_t P_n$, and the allowable strength, P_n/Ω_t, of tension members shall be the lower value obtained according to the limit states of yielding in the gross section and the limit states of fracture (rupture) in the net section according to AISC Sec. D.2.

$$P_n = F_y A_g \quad \text{[yielding criterion; AISC Eq. D2-1]} \qquad 60.8$$
$$P_n = F_u A_e \quad \text{[fracture criterion; AISC Eq. D2-2]} \qquad 60.9$$

7. BLOCK SHEAR STRENGTH

For tension members, another possible failure mode exists. In this mode, a segment or "block" of material at the end of the member tears out as shown in Fig. 60.5. Depending on the connection at the end of a tension member, such *block shear* failure can occur in either the tension member itself or in the member to which it is attached (e.g., a gusset plate). Block shear failure represents a combination of two failures—a shear failure along a plane through the bolt holes or welds and a tension failure along a perpendicular plane—occurring simultaneously.

Block shear strength, R_n, is computed from Eq. 60.10 [AISC Eq. J4-5]. A_{gv} is the gross area subject to shear, A_{nt} is the net area subject to tension, and A_{nv} is the net area subject to shear. Where the tension stress is uniform, $U_{bs} = 1$, and where the tension stress is nonuniform, $U_{bs} = 0.5$.

$$R_n = 0.6 F_u A_{nv} + U_{bs} F_u A_{nt}$$

$$\leq 0.6 F_y A_{gv} + U_{bs} F_u A_{nt} \qquad \left[\begin{array}{c}\text{block shear criterion;} \\ \text{AISC Eq. J4-5}\end{array}\right]$$

$$60.10$$

Figure 60.5 *Block Shear Failures*

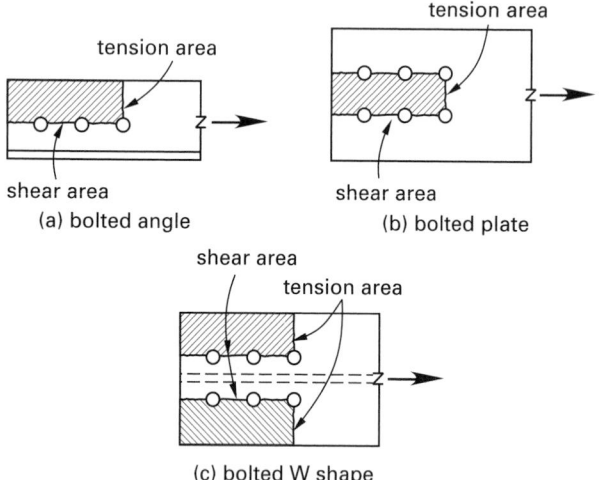

(a) bolted angle

(b) bolted plate

(c) bolted W shape

Figure 60.6 *Flowchart for Analysis of Tension Members Using LRFD*

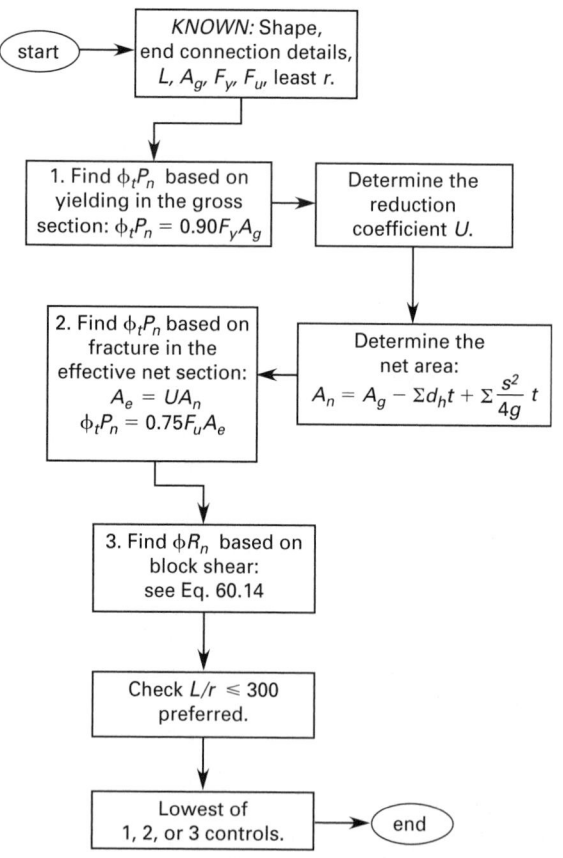

Adapted from *Applied Structural Steel Design*, 3rd. ed., by Leonard Spiegel and George F. Limbrunner, copyright © 1997. Reprinted by permission of Prentice-Hall, Inc., Upper Saddle River, NJ.

8. SLENDERNESS RATIO

Where structural shapes are used in tension, *AISC Specification* D1 lists a preferred (but not required) maximum slenderness ratio (SR) of 300. Rods and wires are excluded from this recommendation.

$$\text{SR} = \frac{L}{r_y} \qquad 60.11$$

9. ANALYSIS OF TENSION MEMBERS

The analysis of a tension member is essentially the determination of its tensile strength, P_n, as the smallest of the three values obtained from Eqs. 60.8 (AISC Eq. D2-1), 60.9 (AISC Eq. D2-2), and 60.10 (AISC Eq. J4-5). A flowchart for tension member analysis is shown in Fig. 60.6.

Example 60.1

A long tensile member is constructed by connecting two plates as shown. Each plate is $1/2$ in $\times$ 9 in, A572 grade 50 steel with an ultimate strength of 65 ksi. The fasteners are $1/2$ in diameter bolts. Using LRFD, determine the maximum tensile load the connection can support. Disregard the shear strength of the bolts and disregard eccentricity.

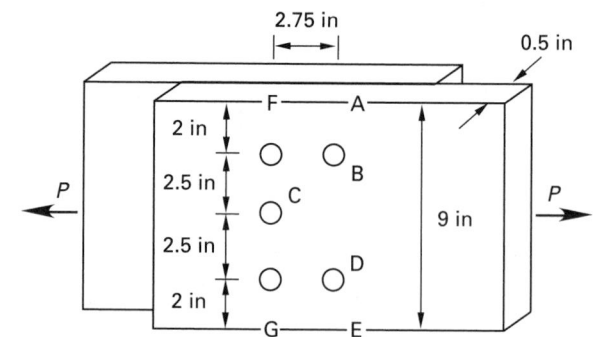

Solution

The gross area of the plate is

$$A_g = (0.5 \text{ in})(9 \text{ in}) = 4.5 \text{ in}^2$$

The effective hole diameter includes $1/8$ in allowance for clearance and manufacturing tolerances.

$$d = 0.5 \text{ in} + 0.125 \text{ in} = 0.625 \text{ in}$$

The net area of the connection must be evaluated in three ways: paths ABDE, ABCDE, and FCG. Path ABDE does not have any diagonal runs. The net area is given by Eq. 60.5.

$$A_{n,\text{ABDE}} = A_g - \sum d_h t + \left(\sum \frac{s^2}{4g}\right) t$$

$$= (0.5 \text{ in})\left(9 \text{ in} - (2 \text{ holes})\left(0.625 \frac{\text{in}}{\text{hole}}\right)\right)$$

$$= 3.875 \text{ in}^2$$

To determine the net area of path ABCDE, the quantity $s^2/4g$ must be calculated. The *pitch*, s, is shown as 2.75 in, and the *gage*, g, is shown as 2.5 in.

$$\frac{s^2}{4g} = \frac{(2.75 \text{ in})^2}{(4)(2.5 \text{ in})} = 0.756 \text{ in}$$

There are two diagonals in path ABCDE. The net area is

$$A_{n,\text{ABCDE}} = (0.5 \text{ in})\big((9 \text{ in} - (3)(0.625 \text{ in})$$
$$+ (2)(0.756 \text{ in})\big)$$
$$= 4.319 \text{ in}^2$$

The net area of path FCG is

$$A_{n,\text{FCG}} = (0.5 \text{ in})\big((9 \text{ in} - (3)(0.625 \text{ in})\big)$$
$$= 3.5625 \text{ in}^2$$

The smallest area is $A_{n,\text{FCG}}$, which is less than 85% of A_g. The design tensile strength $\phi_t P_n$ of the connection based on the gross section is given by Eq. 60.8. ϕ_t is 0.90.

$$\phi_t P_n = \phi_t F_y A_g = (0.90)\left(50 \text{ } \frac{\text{kips}}{\text{in}^2}\right)(4.5 \text{ in}^2)$$
$$= 202.5 \text{ kips}$$

Since all of the connections are in the same plane, $U = 1$. Based on the net section, the design tensile strength is determined from Eqs. 60.6 and 60.9. ϕ_t is 0.75.

$$A_e = U A_n = (1)(3.5625 \text{ in}^2) = 3.5625 \text{ in}^2$$

$$\phi_t P_n = \phi_t F_u A_e = (0.75)\left(65 \text{ } \frac{\text{kips}}{\text{in}^2}\right)(3.5625 \text{ in}^2)$$
$$= 173.7 \text{ kips}$$

The design tensile strength is the smaller value of 173.7 kips.

10. DESIGN OF TENSION MEMBERS

The design of a tension member involves selection of a cross-sectional shape that is suitable for simple connections and has adequate gross area, net area, block shear strength, and radius of gyration (for a preferred limit on slenderness ratio). A flowchart for tension member design is shown in Fig. 60.7.

Example 60.2

Select a 25 ft long W shape of A992 steel to carry factored dead and live tensile loads of 468 kips. The shape will be used as a main member with loads transmitted to framing members through the bolted flanges only.

Solution

Assume the design strength is limited by the net area remaining after deductions for bolt holes. The design tensile strength in the effective (net) area of the member is determined from Eq. 60.9. ϕ_t is 0.75.

$$\phi_t P_n = \phi_t F_u A_e = (0.75)\left(65 \text{ } \frac{\text{kips}}{\text{in}^2}\right) A_e$$
$$= (48.8 \text{ kips/in}^2) A_e$$

Since the web does not transmit the tensile strength, $U = 0.90$ is used, and $A_e = U A_n$. The required area is

$$A_n = \frac{P_u}{0.75 F_u U} = \frac{468 \text{ kips}}{(0.75)\left(65 \text{ } \frac{\text{kips}}{\text{in}^2}\right)(0.90)}$$
$$= 10.67 \text{ in}^2$$

From AISC Table 5-1, a W21 × 44 member has a gross area of 13.0 in². The minimum radius of gyration is $r_y = 1.26$ in. The maximum slenderness ratio is

$$\text{SR} = \frac{L}{r_y} = \frac{(25 \text{ ft})\left(12 \text{ } \frac{\text{in}}{\text{ft}}\right)}{1.26 \text{ in}}$$
$$= 238.1$$

The slenderness ratio is less than the suggested limit of 300. The yield limit state should be checked when the bolt size and arrangement are designed.

11. THREADED MEMBERS IN TENSION

Nominal tensile stress on threaded parts made from approved steels may not exceed $0.75 F_u$ [*AISC Specification* Table J3.2]. This limit applies to static loading, regardless of whether or not threads are present in the shear plane. The area to be used when calculating the maximum tensile load is the gross or nominal area, as determined from the outer extremity of the threaded section. The nominal area of bolts is calculated from the nominal bolt diameter [*AISC Specification* Table J3.1].

Figure 60.7 *Flowchart for the Design of Tension Members Using LRFD*

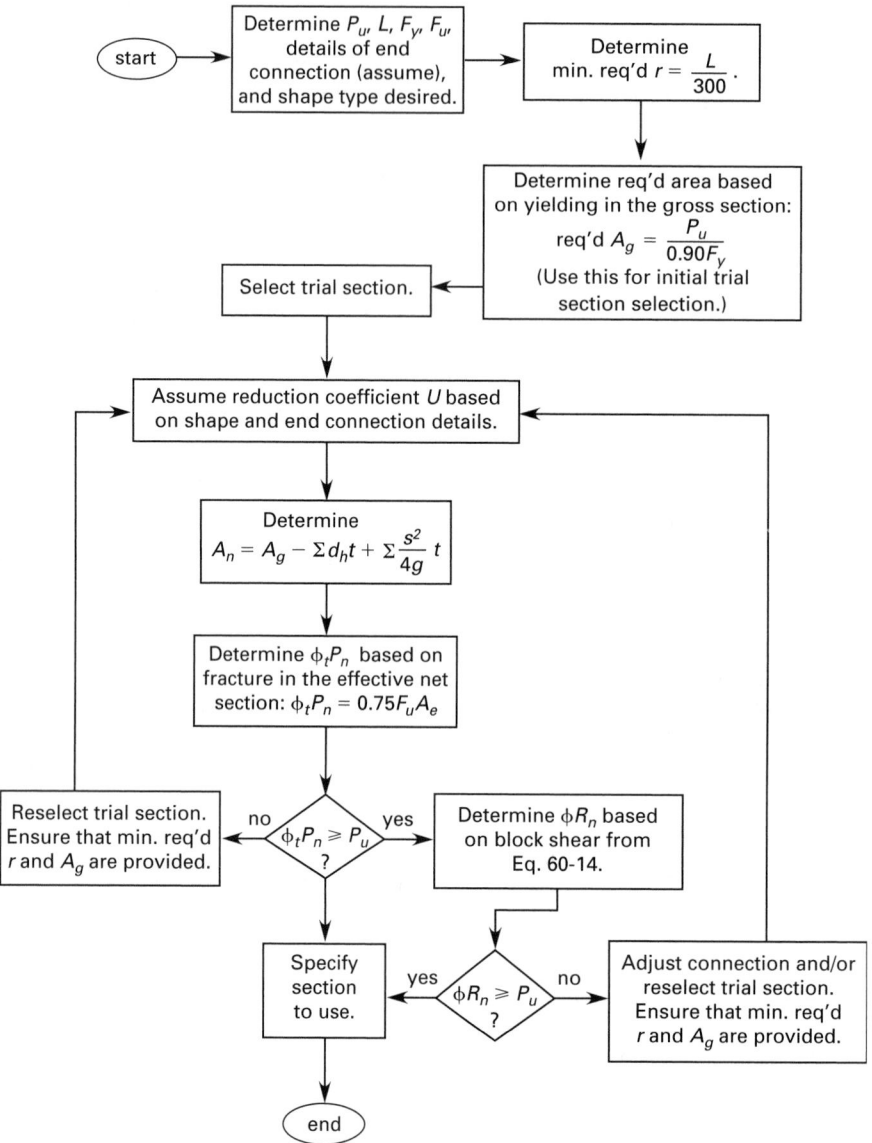

Adapted from *Applied Structural Steel Design*, 3rd. ed., by Leonard Spiegel and George F. Limbrunner, copyright © 1997. Reprinted by permission of Prentice-Hall, Inc., Upper Saddle River, NJ.

61 Structural Steel: Compression Members

Nomenclature

A	cross-sectional area	in^2
b	width	in
B	base plate dimension	in
d	depth	in
D	outside diameter	in
E	modulus of elasticity	kips/in^2
f	stress	in^2
F	strength or allowable stress	kips/in^2
G	end condition coefficient	–
H	constant in width-thickness calculation	–
I	moment of inertia	in^4
k_c	restraint coefficient	–
K	effective length factor	–
L	column length or dimension in base plate calculations	in
m	dimension	in
n	dimension	in
N	base plate dimension	in
P	axial load	kips
Q	strength reduction factor	–
r	radius of gyration	in
SR	slenderness ratio	–
t	thickness	in

Subscripts

a	axial, allowable, available, or area
c	compression
cr	critical
e	Euler or effective
f	flange
g	gross
n	nominal
p	bearing
s	strength
u	ultimate
x	strong axis
y	yield or weak axis

Symbols

λ	baseplate slenderness factor	–
ϕ	resistance factor	–
Ω	safety factor	–

1. INTRODUCTION

Structural members subjected to axial compressive loads are often called by names identifying their functions. Of these, the best-known are *columns*, the main vertical compression members in a building frame. Other common compression members include *chords* in trusses and *bracing members* in frames.

The selection of a particular shape for use as a compression member depends on the type of structure, the availability, and the connection methods. Load-carrying capacity varies approximately inversely with the slenderness ratio, so stiff members are generally required. Rods, bars, and plates, commonly used as tension members, are too slender to be used as compression members unless they are very short or lightly loaded.

For building columns, *W* shapes having nominal depths of 14 in or less are commonly used. These sections, being rather square in shape, are more efficient than others for carrying compressive loads. (Deeper sections are more efficient as beams.) *Pipe sections* are satisfactory for small or medium loads. Pipes are often used as columns in long series of windows, in warehouses, and in basements and garages. In the past, square and *rectangular tubing* saw limited use, primarily due to the difficulty in making bolted or riveted connections at the ends. Modern welding techniques have essentially eliminated this problem.

Built-up sections are needed in large structures for very heavy loads that cannot be supported by individual rolled shapes. For bracing and compression members in light trusses, *single-angle members* are suitable. However, *equal-leg angles* may be more economical than *unequal-leg angles* because their least radius of gyration values are greater for the same area of steel. For *top chord* members of bolted roof trusses, a pair of angles (usually unequal legs, with long legs back-to-back to give a better balance between the radius of gyration values about the x- and y- axes) are used with or without gusset plates. In welded roof trusses, where gusset plates are unnecessary, *structural tees* are used as top chord members.

Box sections have large radii of gyration. The section shown in Fig. 61.1(i) is often seen in towers and crane booms. In Fig. 61.1(i) through 61.1(m), solid lines represent the main elements of *box sections* that are continuous for the full length of the compression member. The dashed lines represent discontinuous elements—typically plates, *lacing bars*, or *perforated cover plates*. The discontinuous elements do not usually contribute to load-carrying capacity; they serve only to space the continuous elements. Depending on the design, *cover plates* may or may not contribute load-carrying capacity.

Single *channels* are not satisfactory for the average compression member because the radii of gyration about their weak axes are almost negligible unless additional lateral support in the weak direction is provided.

There are three general modes by which axially loaded compression members can fail. These are flexural buckling, local buckling, and torsional buckling. *Flexural buckling* (also called *Euler buckling*) is the primary type of buckling. *Local buckling* is where one or more parts of a member's cross section fail in a small region, before the other modes of failure can occur. This type of failure is a function of the width-thickness ratios of the parts of the cross-section. *Torsional buckling*, not covered in this chapter, occurs in some members (e.g., single angles and tees) but is not significant in a majority of cases.

2. EULER'S COLUMN BUCKLING THEORY

Column design and analysis are based on the Euler buckling load theory. However, specific factors of safety and slenderness ratio limitations distinguish design and analysis procedures from purely theoretical concepts.

An *ideal column* is initially perfectly straight, isotropic, and free of residual stresses. When loaded to the *buckling* (or *Euler*) *load*, a column will fail by sudden buckling (bending). The Euler column buckling load for an ideal, pin-ended, concentrically loaded column is given by Eq. 61.1. The modulus of elasticity term, E, implies that Eq. 61.1 is valid as long as the loading remains in the elastic region.

$$P_e = \frac{\pi^2 E I}{L^2} \qquad 61.1$$

The *buckling stress*, F_e, is derived by dividing both sides of Eq. 61.1 by the area, A, and using the relationship $I = A r^2$, where r is the radius of gyration.

$$F_e = \frac{P_e}{A} = \frac{\pi^2 E}{\left(\dfrac{L}{r}\right)^2} \qquad 61.2$$

As Eq. 61.2 shows, the buckling stress is not a function of the material strength. Rather, it is a function of the ratio L/r, the *slenderness ratio*, SR. As the slenderness ratio increases, the buckling stress decreases, meaning that as a column becomes longer and more slender, the load that causes buckling becomes smaller. Equation 61.2 is convenient for design and is valid for all types of cross sections and all grades of steel.

3. EFFECTIVE LENGTH

Real columns usually do not have pin-connected ends. The restraints placed on a column's ends greatly affect its stability. Therefore, an *effective length factor*, K

Figure 61.1 *Typical Compression Members' Cross Sections*

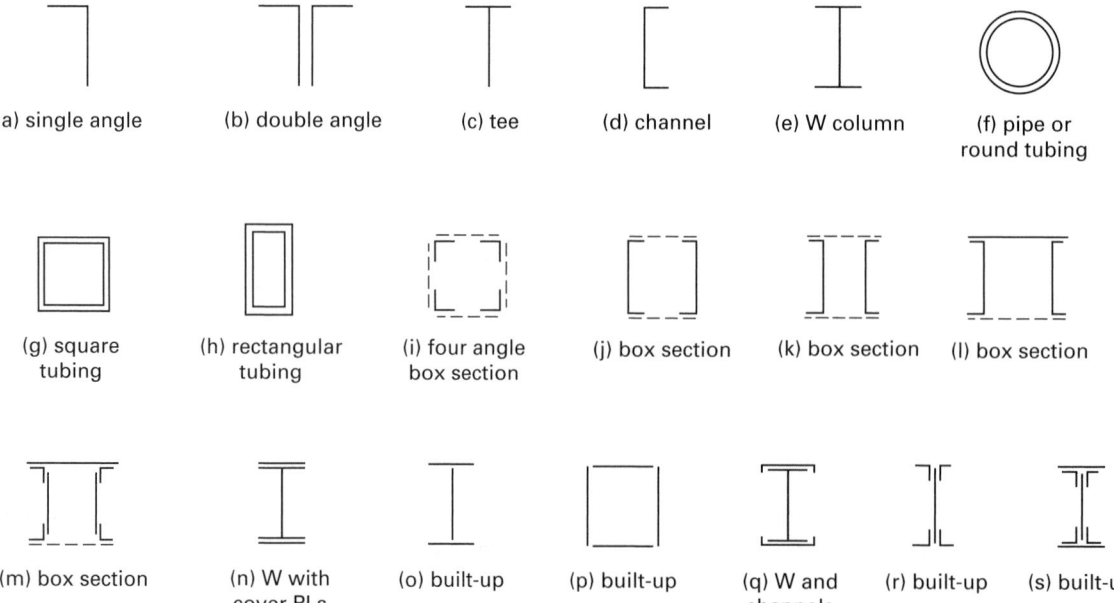

(a) single angle   (b) double angle   (c) tee   (d) channel   (e) W column   (f) pipe or round tubing

(g) square tubing   (h) rectangular tubing   (i) four angle box section   (j) box section   (k) box section   (l) box section

(m) box section   (n) W with cover PLs   (o) built-up   (p) built-up   (q) W and channels   (r) built-up   (s) built-up (also known as an *end-restraint factor*), is used to modify the unbraced length. The product, KL, of the effective length factor and the unbraced length is known as the *effective length* of the column. The effective length approximates the length over which a column actually buckles. The effective length can be longer or shorter than the actual unbraced length.

$$F_e = \frac{\pi^2 E}{\left(\dfrac{KL}{r}\right)^2} \quad \text{[AISC Eq. E3-4]} \qquad 61.3$$

Values of the effective length factor depend upon the rotational restraint at the column ends as well as on the resistance provided against any lateral movement along the column length (i.e., whether or not the column is braced against sidesway). Theoretically, restraint at each end can range from complete *fixity* (though this is impossible to achieve in practice) to zero fixity (e.g., as with a freestanding signpost or flagpole end). Table 61.1 lists recommended values of K for use with steel columns. For braced columns (sidesway inhibited), $K \leq 1$, whereas for unbraced columns (sidesway uninhibited), $K > 1$.

Table 61.1 *Effective Length Factors*[a]

end no. 1	end no. 2	K
built-in	built-in	0.65
built-in	pinned	0.80
built-in	rotation fixed, translation free	1.2
built-in	free	2.1
pinned	pinned	1.0
pinned	rotation fixed, translation free	2.0

[a]These are slightly different from the theoretical values often quoted for use with Euler's equation.

Source: *AISC Commentary*, Table C-C2.2

The values of K in Table 61.1 do not require prior knowledge of the column size or shape. However, if the columns and beams of an existing design are known, the alignment charts in Fig. 61.2 can be used to obtain a more accurate effective length factor.

To use Fig. 61.2 [AISC Figs. C-C2.3 and C-C2.4], the *end condition coefficients*, G_A and G_B, are first calculated from Eq. 61.4 for the two column ends. (The alignment charts are symmetrical. Either end can be designated A or B.)

Figure 61.2 *Effective Length Factor Alignment Charts*

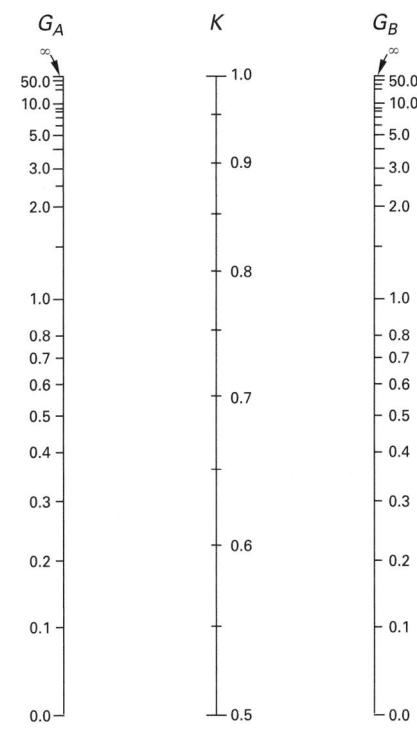

(a) sidesway inhibited

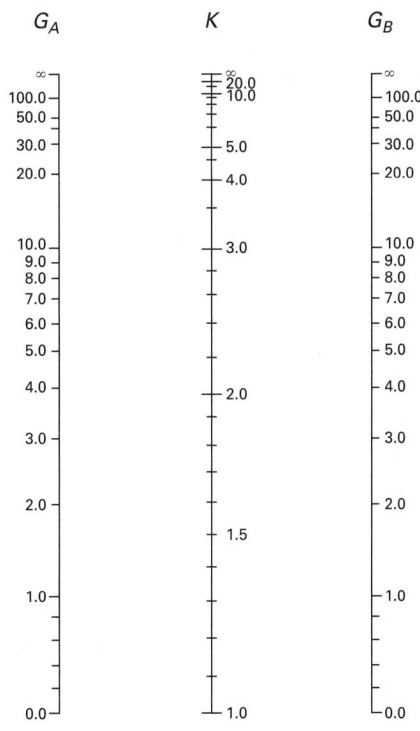

(b) sidesway uninhibited

$$G = \frac{\sum \left(\dfrac{I}{L}\right)_{\text{columns}}}{\sum \left(\dfrac{I}{L}\right)_{\text{beams}}} \qquad 61.4$$

In calculating G, only beams and columns in the plane of bending (i.e., that resist the tendency to buckle) are included in the summation. Also, only rigidly attached beams and columns are included, since pinned connections do not resist moments. For ground-level columns, one of the column ends will not be framed to beams or other columns. In that case, $G = 10$ (theoretically, $G = \infty$) is used for pinned ends and $G = 1$ (theoretically, $G = 0$) is used for rigid footing connections.

4. GEOMETRIC TERMINOLOGY

Figure 61.3 shows a W shape used as a column. The I_y moment of inertia is smaller than I_x. In this case, the column is said to "buckle about the minor axis" if the failure (buckling) mode is as shown.

Since buckling about the minor axis is the expected buckling mode, bracing for the minor axis is usually provided at intermediate locations to reduce the unbraced length. Associated with each column are two unbraced lengths, L_x and L_y. In Fig. 61.3(b), L_x is the full column height. However, L_y is half the column height, assuming that the brace is placed at midheight.

Another important geometric characteristic is the *radius of gyration*. Since there are two moments of inertia, there are also two radii of gyration. Since I_y is smaller than I_x, r_y will be smaller than r_x. r_y is known as the *least radius of gyration*.

Figure 61.3 *Minor Axis Buckling and Bracing*

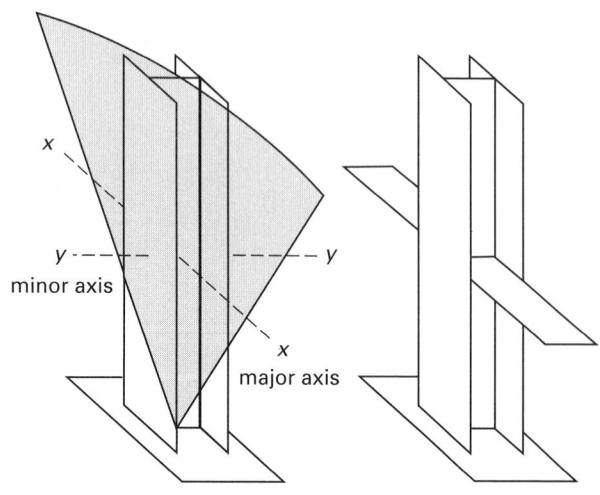

(a) minor axis buckling   (b) minor axis bracing

5. SLENDERNESS RATIO

Steel columns are categorized as long columns and intermediate columns, depending on their slenderness ratios, SR. Since there are two values of the radius of gyration, r (and two values of K and L), corresponding to the x- and y-directions, there will be two slenderness ratios.

$$\text{SR} = \frac{KL}{r} \qquad 61.5$$

The SR of compressive members preferably should not exceed 200 [AISC Sec. E2].

6. BUCKLING OF REAL COLUMNS

Euler's buckling load formula assumes concentrically loaded, perfect columns—conditions that are difficult to achieve in the real world. Real columns always have some flaws—curvature, variations in material properties, and residual stresses (of 10 to 15 ksi, mainly due to uneven cooling after hot-rolling). Furthermore, concentric loading is not possible.

Residual stresses are of particular importance with slenderness ratios of 40 to 120—a range that includes a very large percentage of columns. Because of the residual stresses, early localized yielding occurs at points in the column cross section, and buckling strength is appreciably reduced.

7. DESIGN COMPRESSIVE STRENGTH

The available column strength varies with the slenderness ratio. Figure 61.4 shows two different regions, representing *inelastic buckling* and *elastic buckling*, separated by the limit $4.71\sqrt{E/F_y}$. (Very short compression members—those with effective lengths less than about 2 ft—are governed by different requirements.)

Figure 61.4 *Available Compressive Stress Versus Slenderness Ratio*

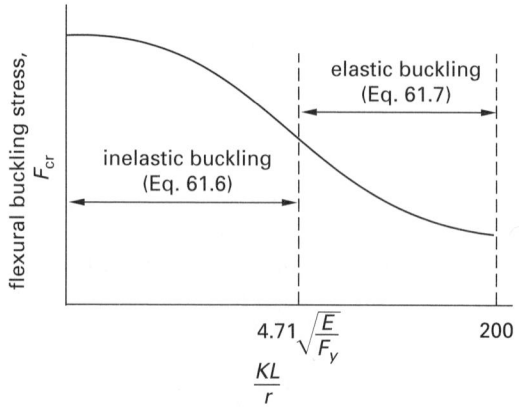

For intermediate columns ($KL/r \leq 4.71\sqrt{E/F_y}$) that fail by inelastic buckling, Eq. 61.6 gives the design stress [AISC Eq. E3-2].

$$F_{\text{cr}} = \left(0.658^{F_y/F_e}\right) F_y \qquad \textit{61.6}$$

The parabolic curve of Eq. 61.6 accounts for the inelastic behavior due to residual stresses and initial crookedness in a real column.

For long columns ($KL/r > 4.71\sqrt{E/F_y}$) that fail by elastic buckling, Eq. 61.7 [AISC Eq. E3-3] must be used. Although Eq. 61.6 is a function of F_y, Eq. 61.7 is independent of F_y. F_e is the elastic critical buckling stress (Eq. 61.3).

$$F_{\text{cr}} = 0.877 F_e \qquad \textit{61.7}$$

8. ANALYSIS OF COLUMNS

Columns are analyzed by verifying that the required strength, P_u or P_a, does not exceed the design strength, $\phi_c P_n$, or the allowable strength, P_n/Ω_c. The nominal axial strength, P_n, is

$$P_n = F_{\text{cr}} A_g \quad \text{[AISC Eq. E3-1]} \qquad \textit{61.8}$$

step 1: Determine the shape properties A, r_x, and r_y from the *AISC Manual*. Determine the unbraced lengths, L_x and L_y.

step 2: Obtain K_x and K_y from Table 61.1 or from the alignment chart, Fig. 61.2.

step 3: Calculate the maximum slenderness ratio as

$$\text{SR} = \text{larger of} \left\{ \begin{array}{c} \dfrac{K_x L_x}{r_x} \\[2mm] \dfrac{K_y L_y}{r_y} \end{array} \right\} \qquad \textit{61.9}$$

step 4: Calculate the limit, $4.71\sqrt{E/F_y}$.

step 5: Using SR, determine the design stress, F_{cr}, from either AISC Table 4-1 or Table 4-22. Alternatively, calculate the design stress, F_{cr}, from either Eq. 61.6 or 61.7, depending on the value of SR.

step 6: Compute the allowable strength or the design compressive strength and compare against the required strength.

$$P_a = \frac{P_n}{\Omega_c} = \frac{F_{\text{cr}} A_g}{1.67} \quad \text{[ASD]} \quad \textit{61.10(a)}$$

$$\phi_c P_n = 0.90 F_{\text{cr}} A_g \quad \text{[LRFD]} \quad \textit{61.10(b)}$$

9. DESIGN OF COLUMNS

Trial-and-error column selection is difficult. Accordingly, Part 4 of the *AISC Manual* contains column selection tables [AISC Table 4-2] that make it fairly easy

to select a column based on the required column capacity. These tables assume that buckling will occur first about the minor axis. Steps 4 through 7 in the following procedure accommodate the case where buckling occurs first about the major axis.

step 1: Determine the load to be carried. Include an allowance for the column weight.

step 2: Determine the effective length factors, K_y and K_x, for the column. (If Eq. 61.4 is to be used, an initial column may need to be assumed.) Calculate the effective length assuming that buckling will be about the minor axis.

$$\text{effective length} = K_y L_y \qquad \textit{61.11}$$

step 3: Enter the table and locate a column that will support the required load with an effective length of $K_y L_y$.

step 4: Check for buckling in the strong direction. Calculate $K_x L'_x$ from Eq. 61.12. (The ratio r_x/r_y is tabulated in the column tables.)

$$K_x L'_x = \frac{K_x L_x}{\dfrac{r_x}{r_y}} \qquad \textit{61.12}$$

step 5: If $K_x L'_x < K_y L_y$, the column is adequate and the procedure is complete. Go to step 8.

step 6: If $K_x L'_x > K_y L_y$ but the column chosen can support the load at a length of $K_x L'_x$, the column is adequate and the procedure is complete. Go to step 8.

step 7: If $K_x L'_x > K_y L_y$ but the column chosen cannot support the load at a length of $K_x L'_x$, choose a larger member that will support the load at a length of $K_x L'_x$. (The ratio r_x/r_y is essentially constant.)

step 8: If sufficient information on other members framing into the column is available, use the alignment charts (Fig. 61.2) to check the values of K.

Example 61.1

Select an A992 steel W14 shape to support a 2000 kip concentric live load. The unbraced length is 11 ft in both directions. Use $K_y = 1.2$ and $K_x = 0.80$. Use the LRFD method.

Solution

step 1: Assume the weight of the column is approximately 500 lbf/ft. The load to be carried is

$$P_u = (1.6)(2000 \text{ kips})$$

$$+ (1.2) \left(\frac{\left(500 \, \frac{\text{lbf}}{\text{ft}} \right)(11 \text{ ft})}{1000 \, \frac{\text{lbf}}{\text{kip}}} \right)$$

$$= 3207 \text{ kips}$$

step 2: From Eq. 61.11, the effective length for minor axis bending is

$$K_y L_y = (1.2)(11 \text{ ft}) = 13.2 \text{ ft} \quad [\text{say } 13 \text{ ft}]$$

step 3: From the AISC column Table 4-1 for 50 ksi steel, select a W14×283 shape with an available strength, $\phi_c P_n$, of 3380 kips.

step 4: From the column table, $r_x/r_y = 1.63$. From Eq. 61.12,

$$K_x L'_x = \frac{K_x L_x}{\dfrac{r_x}{r_y}}$$

$$= \frac{(0.80)(11 \text{ ft})}{1.63} = 5.40 \text{ ft}$$

step 5: Since 5.40 ft < 13 ft, the column selected is adequate.

Example 61.2

Select a 25 ft A992 W shape column to support a live load of 375 kips. The base is rigidly framed in both directions. The top is rigidly framed in the weak direction and fixed against rotation in the strong direction, but translation in the strong direction is possible.

Solution

step 1: Assume the column weight is approximately 80 lbf/ft. The required axial strength is

$$P_u = (1.6)(375 \text{ kips}) + (1.2) \left(\frac{\left(80 \, \frac{\text{lbf}}{\text{ft}} \right)(25 \text{ ft})}{1000 \, \frac{\text{lbf}}{\text{kip}}} \right)$$

$$= 602 \text{ kips}$$

step 2: From Table 61.1, the end restraint coefficients are $K_y = 0.65$ and $K_x = 1.2$. The effective lengths are

$$K_y L_y = (0.65)(25 \text{ ft}) = 16.25 \text{ ft}$$
$$K_x L_x = (1.2)(25 \text{ ft}) = 30 \text{ ft}$$

step 3: From the column table for 50 ksi steel, find a column capable of supporting 602 kips with an effective length of 16 ft. Try a W12×65 shape with a capacity of 639 kips.

step 4: From the column table, $r_x/r_y = 1.75$. Using Eq. 61.12,

$$K_x L'_x = \frac{K_x L_x}{\dfrac{r_x}{r_y}}$$

$$= \frac{30 \text{ ft}}{1.75} = 17.1 \text{ ft}$$

step 5: Since 17.1 ft > 16.25 ft, the strong-axis buckling controls. Use the column table again to find an effective length of 17.1 ft. The column capacity is found by interpolation to be 613 kips. Since 613 kips > 602 kips, this column is adequate.

10. LOCAL BUCKLING

Local buckling of a plate element in a rolled shape or built-up compression member may occur before Euler buckling. The ability of plate sections to carry compressive loads without buckling is determined by the *width-thickness ratio*, b/t. If a shape is selected from the AISC column-selection tables, the width-thickness ratios do not generally need to be evaluated. However, compression members constructed from structural tees, structural tubing, and plates must be checked.

For the purpose of specifying limiting width-thickness ratios, compression elements are divided into stiffened elements and unstiffened elements. *Stiffened elements* are supported along two parallel edges. Examples are webs of W shapes and the sides of box beams. *Unstiffened elements* are supported along one edge only. Flanges of W shapes and legs of angles are unstiffened elements.

Figure 61.5 *Stiffened and Unstiffened Compressive Elements*

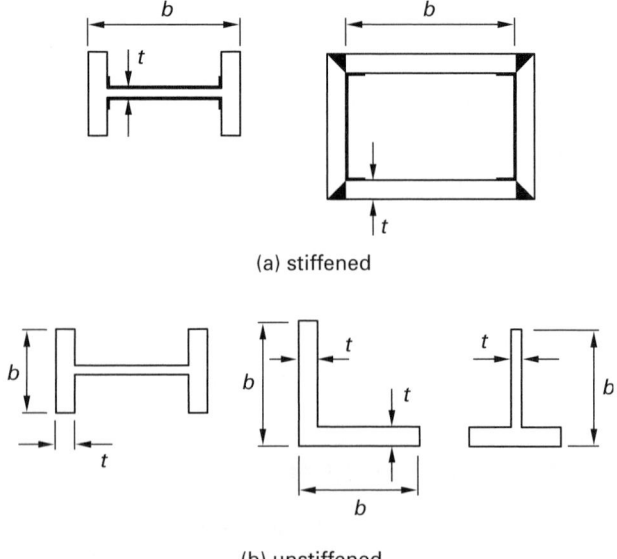

(a) stiffened

(b) unstiffened

To prevent local buckling, *AISC Specification* Sec. B4 requires Eq. 61.13 to be met if the plate element is to be fully effective. Values of H are listed in Table 61.2.

$$\frac{b}{t} \leq \frac{H}{\sqrt{F_y}} \qquad \textit{61.13}$$

Circular tubular sections whose ratios of outside diameter to wall thickness satisfy Eq. 61.14 are considered to be fully effective [AISC Table B.4.1, case 15].

$$\frac{D}{t} \leq 0.11 \left(\frac{E}{F_y} \right) \qquad \textit{61.14}$$

Table 61.2 *H Values for Width-Thickness Ratios*

element	H
unstiffened elements	
stems of tees	127
double angles in contact	95
compression flanges of beams	95
angles or plates projecting from girders,	
columns, or other compression members,	
and compression flanges of plate girders[a,b]	$109\sqrt{k_c}$
stiffeners on plate girders	95
flanges of tees and I-beams (use $b_f/2$)	95
single-angle struts or separated	
double-angle struts	76
stiffened elements	
square and rectangular box sections	190
other uniformly compressed elements	253

[a] $k_c = 4/\sqrt{h/t}$; $0.35 < k_c < 0.76$
[b] Other provisions govern the width-thickness ratios of plate girder flanges, as well.
Source: *AISC Specification* Table B4.1

For compression members with slender elements that exceed width-thickness requirements, the boundary between elastic and inelastic buckling and the inelastic buckling stress is modified by a reduction factor, Q, as defined in *AISC Specification* Sec. E7. For slender unstiffened compression elements, a reduction factor, Q_s, is computed, and for slender stiffened compression elements, a reduction factor, Q_a, is computed based on the effective area of the cross section. The reduction factor, Q, is the product of Q_s and Q_a. The critical buckling stress is reduced by the Q factor.

Example 61.3

Two A36 L9 × 4 × $^1/_2$ angles are used with a $^3/_8$ in gusset plate to create a truss compression member. The short legs are back to back, making the long legs unstiffened elements. Can the combination fully develop compressive stresses?

Solution

From AISC Table B4.1, the limitation on the unstiffened separated double angles is given by

$$0.45\sqrt{\frac{E}{F_y}} = 0.45\sqrt{\frac{29{,}000 \text{ ksi}}{36 \text{ ksi}}} = 12.77$$

The actual width-thickness ratio is

$$\frac{b}{t} = \frac{9 \text{ in}}{0.5 \text{ in}} = 18$$

Since $18 > 12.77$, local buckling will control. A reduced stress factor must be used when calculating the allowable compressive load on the truss member.

11. MISCELLANEOUS COMBINATIONS IN COMPRESSION

Miscellaneous shapes, including round and rectangular tubing, single and double angles, and built-up sections, can be used as compression members, as shown in Fig. 61.6.

Figure 61.6 *Miscellaneous Compression Members*

Generally, sections with distinct strength advantages in one plane (e.g., double angles or tees) should be used when bending is confined to that plane. For example, a double-angle member could be used as a compression strut in a roof truss or as a spreader bar for hoisting large loads.

Design and analysis of these miscellaneous compression members is similar to the design and analysis of W-shape columns. The same equations are used for calculating the design stress, F_{cr}. It is essential to check the width-thickness ratios for all elements in the compression members, including both flanges and stems of tees.

Where spot welding or stitch riveting is used to combine two shapes into one (as is done with double angles), the spacing of the connections must be sufficient to prevent premature buckling of one of the shapes. The KL/r value affected by the spacing of intermediate connectors must be modified by the equations given in Sec. E6 of the *AISC Specification*.

Column load tables have been prepared for many combinations of double angles, tees, round pipe, and structural tubing. Single-angle compression members are

2 L9 × 4 × $\frac{1}{2}$

$\frac{3}{8}$ in gusset plate

difficult to load concentrically, and special attention is required when designing the connections.

Example 61.4

Design a 10 ft long A36 double-angle strut to support an axial compressive load of 40 kips. Assume $K = 1$ in both directions. Gusset plates at the ends are $^3/_8$ in thick.

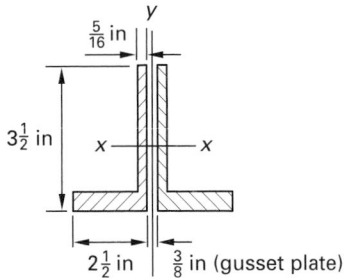

Solution

(Slightly different answers will be obtained if *AISC Manual* Table 4-22 is used.)

ASD Method Solution

Self weight is disregarded since the weight of the strut will be insignificant compared to the axial load.

From the Double-Angle Column Table 4-9 of the *AISC Manual*, try two L$3^1/_2 \times 2^1/_2 \times {}^3/_8$. This angle has an available strength, P_n/Ω, of 48.6 kips in the x-direction and 43.8 kips in the y-direction. The available strength is the lesser of these two values. Since 43.8 kips > 40 kips, the angle is OK.

The available strength about the x-axis can be computed from Eqs. 61.6 or 61.7 as follows. From the table, the properties are

$$A = 4.22 \text{ in}^2$$
$$r_x = 1.10 \text{ in}$$

The slenderness ratio, KL/r, is

$$\frac{K_x L}{r_x} = \frac{(1)(10 \text{ ft})\left(12 \frac{\text{in}}{\text{ft}}\right)}{1.10 \text{ in}} = 109$$

The limiting slenderness ratio between inelastic and elastic buckling is

$$4.71\sqrt{\frac{E}{F_y}} = 4.71\sqrt{\frac{29{,}000 \frac{\text{kips}}{\text{in}^2}}{36 \frac{\text{kips}}{\text{in}^2}}} = 134$$

Since $109 < 134$, Eq. 61.6 controls. The Euler stress is calculated from Eq. 61.12.

$$F_e = \frac{\pi^2 E}{\left(\dfrac{KL}{r}\right)^2} = \frac{\pi^2 \left(29{,}000 \dfrac{\text{kips}}{\text{in}^2}\right)}{(109)^2}$$

$$= 24.09 \text{ kips/in}^2$$

The flexural buckling stress, F_{cr}, is

$$F_{\text{cr}} = \left(0.658^{F_y/F_e}\right) F_y = \left(0.658^{\frac{36 \text{ kips/in}^2}{24.09 \text{ kips/in}^2}}\right)\left(36 \frac{\text{kips}}{\text{in}^2}\right)$$

$$= 19.26 \text{ kips/in}^2$$

The available strength is

$$\frac{P_n}{\Omega} = \frac{A_g F_{\text{cr}}}{\Omega} = \frac{(4.22 \text{ in}^2)\left(19.26 \dfrac{\text{kips}}{\text{in}^2}\right)}{1.67}$$

$$= 48.7 \text{ kips}$$

The value given for the available strength in the y-direction is a function of connector spacing and is computed from a modified KL/r determined from equations in *AISC Specification* Sec. E6.

LRFD Method Solution

Since the dead load is neglected, the required strength P_u is

$$P_u = (1.6)(40 \text{ kips}) = 64 \text{ kips}$$

From the Double-Angle Column tables in Part 4 of the *AISC Manual*, try two L$3 {}^1/_2 \times 2^1/_2 \times {}^3/_8$. This combination has an available strength, ϕP_n, equal to 73.0 kips in the x-direction and 65.8 kips in the y-direction. The available strength is the lesser of these two values. Since 65.8 kips > 64 kips the angle is OK.

The flexural buckling stress with respect to the x-axis is computed as before. The available strength is given by

$$\phi P_n = \phi A_g F_{\text{cr}} = (0.9)(4.22 \text{ in}^2)\left(19.26 \frac{\text{kips}}{\text{in}^2}\right)$$

$$= 73.1 \text{ kips}$$

12. COLUMN BASE PLATES

Column loads transmitted to masonry and concrete foundations must not exceed the available bearing strengths of the base material. If the area of the foundation, A_2, exceeds the area of the column base plate, A_1, the available bearing stress may be increased by $\sqrt{A_2/A_1}$, but not more than a factor of 2. The available bearing stress, F_p, is given by

$$F_p = 0.85 f'_c \sqrt{\frac{A_2}{A_1}} \qquad 61.15$$

The area of the base plate, A_1, is given by

$$A_{\text{plate}} = \frac{\text{column load}}{F_p} \qquad 61.16$$

Base plate dimensions are typically specified in whole inches. Therefore, the actual plate area will be somewhat larger than the required plate area. The actual bearing pressure is

$$f_p = \frac{\text{column load}}{\text{actual plate area}} \qquad 61.17$$

It is generally desirable to have $m = n$ in Fig. 61.7. This approximately occurs when

$$A_1 = \frac{\text{column load}}{F_p} \qquad 61.18$$

$$N = 2m + 0.95d \qquad 61.19$$

$$B = \frac{A_1}{N} \qquad 61.20$$

$$X = \left(\frac{4db_f}{(d + b_f)^2} \right) \left(\frac{P_u}{\phi_c P_p} \right) \quad \text{[LRFD]} \quad 61.21(a)$$

$$X = \left(\frac{4db_f}{(d + b_f)^2} \right) \left(\frac{\Omega_c P_a}{P_p} \right) \quad \text{[ASD]} \quad 61.21(b)$$

It is assumed that part of the base plate outboard from a $0.95d \times 0.8b_f$ rectangle acts as a uniformly loaded cantilever, resulting in a bending stress, F_b, in the cantilever portion. The critical base cantilever dimension, l, in Eq. 61.23 is limited to the larger of m, n, or $\lambda n'$, where $n' = \frac{1}{4}\sqrt{db_f}$. Once the column and base plate sizes have been determined, the flange slenderness ratio, λ, is given by Eq. 61.22. λ can always be conservatively taken as 1.0.

$$\lambda = \frac{2\sqrt{X}}{1 + \sqrt{1 - X}} \leq 1 \qquad 61.22$$

Eq. 61.23 can be used to calculate the minimum plate thickness. Thicknesses should be specified in $\frac{1}{8}$ in increments up to $1\frac{1}{4}$ in, and in $\frac{1}{4}$ in increments thereafter.

$$t_{\min} = l\sqrt{\frac{2P_u}{0.9 F_y BN}} \quad \text{[LRFD]} \qquad 61.23(a)$$

$$t_{\min} = l\sqrt{\frac{3.33 P_a}{F_y BN}} \quad \text{[ASD]} \qquad 61.23(b)$$

Figure 61.7 *Column Base Plate*

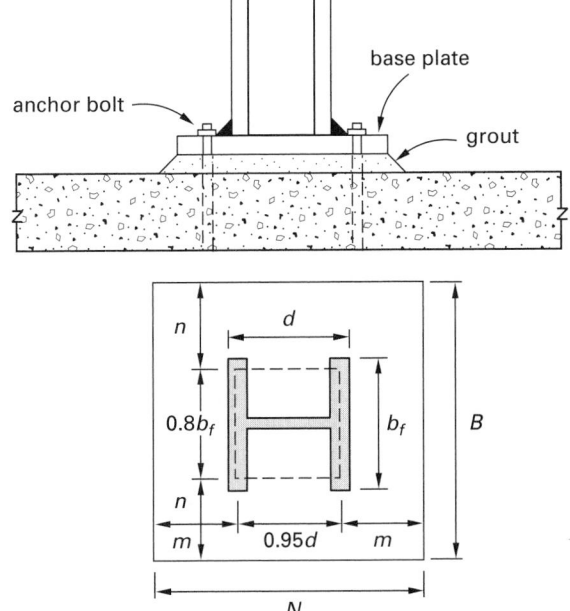

62 Structural Steel: Beam-Columns

Nomenclature

A	cross-sectional area	in^2
AF	amplification factor	–
E	modulus of elasticity	$kips/in^2$
f	computed stress	$kips/in^2$
F	strength or allowable stress	$kips/in^2$
I	moment of inertia	in^4
K	effective length factor	–
L_b	unbraced length	in
L	live load	kips
M	moment	in-kips
M_A	absolute value of moment at quarter point of the unbraced segment	in-kip
M_B	absolute value of moment at centerline of the unbraced segment	in-kip
M_c	available flexural strength	in-kip
M_C	absolute value of moment at three-quarter point of the unbraced segment	in-kip
M_{max}	absolute value of maximum moment in the unbraced segment	in-kip
M_r	required flexural strength	in-kip
P	axial load	kips
P_c	available axial strength	kips
P_r	required axial strength	kips
r	radius of gyration	in
R_m	cross-section mono-symmetry parameter	–
S	elastic section modulus	in^3
SR	slenderness ratio	–
Z	plastic section modulus	in^3

Symbols

Δ	beam deflection	in
ϕ	resistance factor	–
Ω	safety factor	–

Subscripts

a	available, allowable, or axial
b	bending or unbraced
c	available
e	Euler
n	nominal
o	out of plane
r	required
u	ultimate
x	strong axis
y	weak axis or yield

1. INTRODUCTION

A member that is acted upon by a compressive force and a bending moment is known as a *beam-column*. The bending moment can be due to an eccentric load or a true lateral load.

For pure beams and columns, second-order effects are neglected. Only first-order effects based on undeformed geometry need to be considered (i.e., for a member subjected to a moment only, or when the same member carries an axial compressive load alone).

When a member is acted upon by both bending moment and an axial load, the two stresses cannot be added directly to obtain the so-called combined stresses. Additional stresses resulting from a secondary moment must be taken into account, especially when the axial compressive load is large. The secondary moment is caused by the P-Δ *effect* and has a magnitude of $P\Delta$. The deflection, Δ, results from the initial lateral deflection due to the bending moment, but causes further bending of its own and produces secondary stresses.

The secondary moment and the associated stress are known as *second-order effects*, since they are dependent on the deformed geometry of the member. *Moment amplification factors* are used to account for the second-order effects. An expression for maximum combined stress can be written as Eq. 62.1. f_{bx} and f_{by} are the maximum bending stresses caused by bending moments about x- and y-axes, respectively. $(AF)_x$ and $(AF)_y$ are the corresponding amplification factors.

$$f_{max} = f_a + (AF)_x f_{bx} + (AF)_y f_{by} \qquad 62.1$$

Unfortunately, even with the inclusion of secondary moment effects, computed values of combined stresses are not particularly useful, since there are no code-specified allowable combined stresses for beam-columns. Design and analysis of beam-columns generally attempt to transform moments into equivalent axial loads (i.e., the *equivalent axial compression method*) or make use of interaction equations. Interaction-type equations are best suited for beam-column analysis and validation, since so much (e.g., area, moment of inertia) needs to be known about a shape. Equivalent axial compression methods are better suited for design.

Equation 62.1 is converted to an *interaction equation* by dividing both sides by f_{max} and then by substituting applicable allowable stresses for f_{max}. In Eq. 62.2, if one or more applied stresses is zero, the member acts as an axially loaded column, a beam bending about one or both axes, or a beam-column bending about one axis only.

$$\frac{f_a}{F_a} + (\text{AF})_x \left(\frac{f_{bx}}{F_{bx}}\right) + (\text{AF})_y \left(\frac{f_{by}}{F_{by}}\right) = 1.0 \qquad 62.2$$

2. FLEXURAL/AXIAL COMPRESSION

The flexural/compressive force relationships, as presented in the *AISC Specification* H1.1, are as follows.

For $P_r/P_c \geq 0.2$, AISC Eq. H1-1a gives

$$\frac{P_r}{P_c} + \left(\frac{8}{9}\right)\left(\frac{M_{rx}}{M_{cx}} + \frac{M_{ry}}{M_{cy}}\right) \leq 1.0 \qquad 62.3$$

For $P_r/P_c < 0.2$, AISC Eq. H1-1b gives

$$\frac{P_r}{2P_c} + \frac{M_{rx}}{M_{cx}} + \frac{M_{ry}}{M_{cy}} \leq 1.0 \qquad 62.4$$

P_r is the required axial compressive strength, P_c is the available axial compressive strength, M_r is the required flexural strength, M_c is the available flexural strength, x relates to the strong axis bending, and y relates to the weak axis bending.

For doubly symmetric members in flexural/compressive forces mainly in one plane, the two independent limit states (i.e., in-plane stability and out-of-plane buckling or flexural-torsional buckling) can be considered separately.

According to *AISC Specification* H1.3, analysis of in-plane instability uses AISC Eq. H1-1 with P_c, M_r, and M_c determined in the plane of bending. For out-of-plane buckling, AISC Eq. H1-2 applies, where P_{co} is the available compressive strength out of the plane of bending, and M_{cx} is the available flexural-torsional strength for strong axis flexure determined from *AISC Specification* Ch. F.

$$\frac{P_r}{P_{co}} + \left(\frac{M_r}{M_{cx}}\right)^2 \leq 1.0 \quad [\text{AISC Eq. H1-2}]$$

If bending occurs only in the weak axis, the moment ratio may be neglected.

The lateral-torsional buckling coefficient, C_b, accounts for beams and girders being restrained against rotation about their longitudinal axis when both ends of the unsupported segment are braced. From AISC Eq. F1-1,

$$C_b = \left(\frac{12.5 M_{max}}{2.5 M_{max} + 3M_A + 4M_B + 3M_C}\right) R_m \leq 3.0$$

$$62.4$$

3. SECOND ORDER EFFECTS

In frames and structures subject to sway, the effect of deflection (i.e., second order effects) on the required strengths of members subjected to both axial and bending loads is accounted for by a multiplicative amplification factor, B_1, applied to the required elastic nominal moment strength. B_1 is not applied to the axial strength. *AISC Specification* Sec. C, specifies two such factors, B_1 for use between points of bracing, and B_2 for use at points of bracing. This accounts for the nickname, "B_1-B_2 procedure" given to this method. B_2 is used to account for frame deflections such as drift, while B_1 is used to account for member deflections. In Eq. 62.6 [AISC Eq. C2-2], C_m can be conservatively taken as 1.0 for beam-columns subjected to transverse loading between the supports, or evaluated based on end moments. α is 1.0 for LRFD, and is 1.6 for ASD. P_{e1} is the Euler buckling load calculated with $K = 1$ for the end restraint factor in the plane of buckling and assuming zero sidesway. (See *AISC Specification* Sec. C2.)

$$M_r = B_1 M_n \qquad 62.5$$

$$B_1 = \frac{C_m}{1 - \dfrac{\alpha P_r}{P_{e1}}} \geq 1 \qquad 62.6$$

4. ANALYSIS OF BEAM-COLUMNS

The following procedure can be used for beam-column analysis.

step 1: Calculate the required axial compression strength, P_r, using appropriate load combinations.

step 2: Calculate the slenderness ratios for both bending modes.

step 3: Based on the larger slenderness ratio, determine the design axial compressive strength, P_c, using the procedure described in Ch. 61.

step 4: Calculate the ratio P_r/P_c.

step 5: Calculate the moment magnifiers, B_1 and B_2.

step 6: Calculate the required flexural strengths M_{rx} and M_{ry}.

step 7: Calculate the design flexural strength, M_{cx} and M_{cy}, using the procedure described in Ch. 59.

step 8: Determine the adequacy of the design using the applicable interactions equation. If $P_r/P_c \geq 0.2$, use Eq. 62.3. If $P_r/P_c < 0.2$, use Eq. 62.4.

Example 62.1

A W14×120 A992 shape has been chosen to carry an unfactored axial compressive load of 200 kips and a 250 ft-kips moment about its strong axis. The unsupported length is 20 ft. Sidesway is permitted in the direction of bending. Use $K = 1$. Determine if the column is adequate.

Solution

Assume the dead load is neglible.

LRFD Method

$$P_u = (1.6)(200 \text{ kips}) = 320 \text{ kips}$$

$$M_u = (1.6)(250 \text{ kips}) = 400 \text{ ft-kips}$$

$$KL = 20 \text{ ft}$$

$$P_c = \phi_c P_n = 1180 \text{ kips}$$

$$L_b = 20 \text{ ft}$$

From AISC Table 3-10,

$$M_{cx} = \phi M_{nx} = 743 \text{ ft-kips}$$

$$\frac{P_u}{\phi_c P_n} = \frac{320 \text{ kips}}{1180 \text{ kips}} = 0.271$$

Since $0.27 > 0.20$, use

$$\frac{P_r}{P_c} + \left(\frac{8}{9}\right)\left(\frac{M_{rx}}{M_{cx}} + \frac{M_{ry}}{M_{cy}}\right) \le 1$$

$$= \frac{320 \text{ kips}}{1180 \text{ kips}} + \left(\frac{8}{9}\right)\left(\frac{400 \text{ ft-kips}}{743 \text{ ft-kips}} + 0\right) \le 1$$

$$0.750 < 1 \quad \text{[OK]}$$

ASD Method

$$P_r = 200 \text{ kips}$$

$$M_{rx} = 250 \text{ ft-kips}$$

$$KL = 20 \text{ ft}$$

$$P_c = \frac{P_n}{\Omega} = 783 \text{ kips}$$

From AISC Table 3-10,

$$M_{cx} = \frac{M_{nx}}{\Omega} = 495 \text{ ft-kips}$$

$$\frac{P_r}{P_c} = \frac{200 \text{ kips}}{783 \text{ kips}} = 0.26$$

Since $0.26 > 0.20$, use

$$\frac{P_r}{P_c} + \left(\frac{8}{9}\right)\left(\frac{M_{rx}}{M_{cx}} + \frac{M_{ry}}{M_{cy}}\right) \le 1$$

$$= \frac{200 \text{ kips}}{783 \text{ kips}} + \left(\frac{8}{9}\right)\left(\frac{250 \text{ ft-kips}}{495 \text{ ft-kips}} + 0\right) \le 1$$

$$0.70 \le 1 \quad \text{[OK]}$$

4. DESIGN OF BEAM-COLUMNS

In order to minimize the trial-and-error process of selecting a shape for combined axial and bending loads, Eqs. 62.3 and 62.4 can be modified to allow the use of the tables in Combined Axial and Bending tables in Part 6 of the *AISC Manual*. From the Combined Axial and Bending tables, the values of p, b_x, and b_y can be used to solve the modified forms of Eqs. 62.2 and 62.3 (AISC Eqs. H1-1a and H1-1b).

For $P_r/P_c \ge 0.2$,

$$pP_r + b_x M_{rx} + b_y M_{ry} \le 1.0 \quad \text{[large axial loads]}$$

$$62.7$$

For $P_r/P_c < 0.2$,

$$\frac{pP_r}{2} + \left(\frac{9}{8}\right)(b_x M_{rx} + b_y M_{ry}) \le 1.0 \quad \text{[small axial loads]}$$

$$62.8$$

The following procedure can be used to determine an equivalent axial load.

step 1: Determine the effective length, KL, based on weak axis bending and bracing.

step 2: Use Table 6-1 in Part 6 of the *AISC Manual* and select a trial shape.

step 3: From Table 6-1, select p, b_x, and b_y corresponding to KL for that shape.

step 4: An estimate of the adequacy of the shape is given by Eqs. 62.7 and 62.8.

step 5: Repeat steps 2 through 4 until a reasonable shape is found.

step 6: Determine the adequacy of the design using the applicable interactions equation. If $P_r/P_c \ge 0.2$, use Eq. 62.2. If $P_r/P_c < 0.2$, use Eq. 62.3.

This procedure tends to oversize beam-columns. More economical members can be found if the member initially selected is used as a starting point for a subsequent trial-and-error solution.

Example 62.2

Select a 50 ksi shape to carry an axial load of 200 kips and a maximum moment of 250 ft-kips about the strong axis. The unsupported length is 20 ft. Assume $K = 1$.

Solution

LRFD Method

$$P_u = (1.6)(200 \text{ kips}) = 320 \text{ kips}$$
$$M_u = (1.6)(250 \text{ ft-kips}) = 400 \text{ ft-kips}$$

Try a W14 × 132. From Table 6-1 in the *AISC Manual*,

$$p = 0.772 \times 10^{-3} \text{ 1/kips}$$
$$b_x = 1.08 \times 10^{-3} \text{ 1/ft-kips}$$

Assuming a large axial load, an estimate of the adequacy of the shape is given by Eq. 62.7.

$$pP_r + b_x M_{rx} + b_y M_{ry} \leq 1.0$$
$$\left(0.772 \times 10^{-3} \frac{1}{\text{kips}}\right)(320 \text{ kips})$$
$$+ \left(1.08 \times 10^{-3} \frac{1}{\text{ft-kips}}\right)(400 \text{ ft-kips})$$
$$= 0.679 < 1.0 \quad [\text{OK}]$$

From Table 6-1, a W14 × 132 has properties of

$$A = 38.8 \text{ in}^2$$
$$S_x = 209 \text{ in}^3$$
$$I_x = 1530 \text{ in}^4$$
$$Z_x = 234 \text{ in}^3$$
$$L_p = 13.3 \text{ ft}$$
$$L_r = 56.0 \text{ ft}$$

$$P_c = \phi_c P_n = 1300 \text{ kips}$$

The Euler buckling load is

$$P_e = \frac{\pi^2 E I_x}{(KL_x)^2} = \frac{\pi^2 \left(29,000 \frac{\text{kips}}{\text{in}^2}\right)(1530 \text{ in}^4)}{\left((1)(20 \text{ ft})\left(12 \frac{\text{in}}{\text{ft}}\right)\right)^2}$$
$$= 7603 \text{ kips}$$

The *x-x* axis flexural magnifier is

$$B_1 = \frac{1.0}{1 - (1.0)\left(\dfrac{320 \text{ kips}}{7603 \text{ kips}}\right)} = 1.044$$

$$M_{ux} = (1.044)(400 \text{ ft-kips}) = 417.6 \text{ ft-kips}$$

The yielding limit state is

$$M_{nx} = M_p = F_y Z_x = \frac{\left(50 \frac{\text{kips}}{\text{in}^2}\right)(234 \text{ in}^3)}{12 \frac{\text{in}}{\text{ft}}}$$
$$= 975 \text{ ft-kips}$$

The lateral-torsional buckling limit state is

$$L_p < L_b < L_r = 13.3 \text{ ft} < 20 \text{ ft} < 56 \text{ ft}$$

From App. 59.A or the *AISC Manual* Table 3-1, C_b is 1.14. Use Eq. 59.14.

$$M_{nx} = C_b \left(M_p - (M_p - 0.7 F_y S_x)\left(\frac{L_b - L_p}{L_r - L_p}\right)\right) \leq M_p$$

$$= 1.14 \left(\begin{array}{c} 975 \text{ ft-kips} \\ - \left(\begin{array}{c} 975 \text{ ft-kips} \\ - \dfrac{(0.7)\left(36 \frac{\text{kips}}{\text{in}^2}\right)(209 \text{ in}^3)}{12 \frac{\text{in}}{\text{ft}}} \end{array}\right) \\ \times \left(\dfrac{20 \text{ ft} - 13.3 \text{ ft}}{56 \text{ ft} - 13.3 \text{ ft}}\right) \end{array}\right)$$

$$= 1016 \text{ ft-kips} > 975 \text{ ft-kips} \quad [M_p \text{ controls}]$$

$$\phi_b = 0.90$$

$$M_{cx} = \phi_b M_{nx} = (0.9)(975 \text{ ft-kips}) = 877.5 \text{ ft-kips}$$

$$\frac{P_u}{\phi_c P_n} = \frac{320 \text{ kips}}{1300 \text{ kips}} = 0.25$$

Since $0.25 > 0.2$ use Eq. 62.3,

$$\frac{P_r}{P_c} + \left(\frac{8}{9}\right)\left(\frac{M_{rx}}{M_{cx}} + \frac{M_{ry}}{M_{cy}}\right) \leq 1.0$$

$$\frac{320 \text{ kips}}{1300 \text{ kips}} + \left(\frac{8}{9}\right)\left(\frac{417.6 \text{ ft-kips}}{877.5 \text{ ft-kips}} + 0\right) \leq 1.0$$

$$0.67 \leq 1.0 \quad [\text{OK}]$$

ASD Method

$$P_a = 200 \text{ kips}$$
$$M_{ax} = 250 \text{ kips}$$

Try a W14 × 132. From Table 6-1 in the *AISC Manual*,

$$p = 1.16 \times 10^{-3} \text{ 1/kips}$$
$$b_x = 1.62 \times 10^{-3} \text{ 1/ft-kips}$$

Assuming a large axial load, an estimate of the adequacy of the shape is given by

$$\left(1.16 \times 10^{-3} \frac{1}{\text{kips}}\right)(200 \text{ kips})$$
$$+ \left(1.62 \times 10^{-3} \frac{1}{\text{ft-kips}}\right)(250 \text{ ft-kips})$$
$$= 0.637 < 1.0 \quad [\text{OK}]$$

From Table 6-1, a W14 × 132 has properties of

$$A = 38.8 \text{ in}^2$$
$$S_x = 209 \text{ in}^3$$
$$I_x = 1530 \text{ in}^4$$
$$Z_x = 234 \text{ in}^3$$
$$L_p = 13.3 \text{ ft}$$
$$L_r = 56.0 \text{ ft}$$

$$P_c = \frac{P_n}{\Omega_c} = 862 \text{ kips}$$

The Euler buckling load is

$$P_e = \frac{\pi^2 E I_x}{(KL_x)^2} = \frac{\pi^2 \left(29{,}000 \ \dfrac{\text{kips}}{\text{in}^2}\right)(1530 \text{ in}^4)}{\left((1)(20 \text{ ft})\left(12 \ \dfrac{\text{in}}{\text{ft}}\right)\right)^2}$$

$$= 7603 \text{ kips}$$

The x-x axis flexural magnifier is

$$B_1 = \frac{1.0}{1 - (1.6)\left(\dfrac{200 \text{ kips}}{7603 \text{ kips}}\right)} = 1.044$$

$$M_{ax} = (1.044)(250 \text{ ft-kips}) = 261.0 \text{ ft-kips}$$

The yielding limit state is

$$M_{nx} = M_p = F_y Z_x = \frac{\left(50 \ \dfrac{\text{kips}}{\text{in}^2}\right)(234 \text{ in}^3)}{12 \ \dfrac{\text{in}}{\text{ft}}}$$

$$= 975 \text{ ft-kips}$$

The lateral-torsional buckling limit state is

$$L_p < L_b < L_r = 13.3 \text{ ft} < 20 \text{ ft} < 56 \text{ ft}$$

From App. 59.A or the *AISC Manual* Table 3-1, C_b is 1.14. Use Eq. 59.14.

$$M_{nx} = C_b\left(M_p - (M_p - 0.7F_y S_x)\left(\frac{L_b - L_p}{L_r - L_p}\right)\right) \le M_p$$

$$= 1.14\left(\begin{array}{l} 975 \text{ ft-kips} \\ \qquad -\left(\begin{array}{l} 975 \text{ ft-kips} \\ \qquad -\dfrac{(0.7)\left(36 \ \dfrac{\text{kips}}{\text{in}^2}\right)(209 \text{ in}^3)}{12 \ \dfrac{\text{in}}{\text{ft}}} \end{array}\right) \\ \qquad \times \left(\dfrac{20 \text{ ft} - 13.3 \text{ ft}}{56 \text{ ft} - 13.3 \text{ ft}}\right) \end{array}\right)$$

$$= 1016 \text{ ft-kips} > 975 \text{ ft-kips} \quad [M_p \text{ controls}]$$

$$\Omega_b = 1.67$$

$$M_{cx} = \frac{M_{nx}}{\Omega_b} = \frac{975 \text{ kips}}{1.67} = 584 \text{ kips}$$

$$\frac{P_a}{\dfrac{P_n}{\Omega_c}} = \frac{200 \text{ kips}}{862 \text{ kips}} = 0.232$$

Since 0.237 > 0.2, use Eq. 62.3.

$$\frac{P_r}{P_c} + \left(\frac{8}{9}\right)\left(\frac{M_{rx}}{M_{cx}} + \frac{M_{ry}}{M_{cy}}\right) \le 1.0$$

$$\frac{200 \text{ kips}}{862 \text{ kips}} + \left(\frac{8}{9}\right)\left(\frac{261 \text{ ft-kips}}{589 \text{ ft-kips}} + 0\right) \le 1.0$$

$$0.63 \le 1.0 \quad [\text{OK}]$$

Structural

63 Structural Steel: Built-Up Sections

Nomenclature

a	clear distance between transverse stiffeners	in
A	cross-sectional area	in^2
b	width	in
C_v	ratio of critical web stress to shear web stress	–
d	depth of girder	in
D	factor depending upon transverse stiffeners	–
E	modulus of elasticity	ksi
f	computed stress	ksi
F	strength or allowable stress	ksi
F	stress or strength	ksi
h	clear distance between flanges (girders without corner radii fillets)	in
I	area moment of inertia	in^4
k_c	factor in width-thickness calculation	–
k_v	shear buckling coefficient for girder webs	–
l	unbraced length	in
L	girder length	in
r	radius of gyration	in
t	thickness	in
V	shear	lbf
w	uniform load	kips/in

Subscripts

a	axial, allowable, or available
c	available
cr	critical
f	flange
n	nominal
p	plastic
r	available
r	required
st	stiffener

T	section comprising the compression flange plus one third of the compression web area
u	required or ultimate
v	shear
w	web
y	yield

Symbols

Δ	deflection	in
λ	width thickness ratio	–
ϕ	resistance factor	–
Ω	safety factor	–

1. INTRODUCTION

Built-up sections (previously and commonly referred to in the *AISC Specification* as *plate girders*) are used when beams with moments of inertia larger than standard mill shapes are required. Figure 63.1 illustrates two different methods of constructing plate girders. One is built up by bolting or riveting. The other is by welding. A *hybrid girder* is a plate girder that uses stronger steel for the flanges than for the web. Hybrid girders can be less expensive than traditional plate girders.

Two different definitions of "depth" are encountered with plate girders: the *web depth*, h, and the *girder depth*, d. Unlike for rolled shapes, only the web depth, h, is used for shear stress calculations. For depth thickness (h/t) ratios, the *AISC Specification* requires h to be the clear distance between flanges. For bolted or riveted flanges, this distance is the separation between the last row of fasteners. For welded flanges, the distinction is not made.

The girder depth, d, is generally chosen to be approximately one tenth of the span, although the ratio can vary from 1:5 to 1:15.

2. DEPTH-THICKNESS RATIOS

The required web thickness, t_w, depends on the spacing of intermediate stiffeners. If intermediate stiffeners are spaced no more than $1.5h$ (i.e., 150% of the girder depth), then Eq. 63.1 gives the maximum permissible depth-thickness ratio. If h is known, the required thickness, t_w, can be obtained from Eq. 63.1 [AISC Eq. F13-3].

$$\left(\frac{h}{t_w}\right)_{\max} \leq 11.7\sqrt{\frac{E}{F_y}} \qquad 63.1$$

Figure 63.1 *Elements of a Plate Girder*

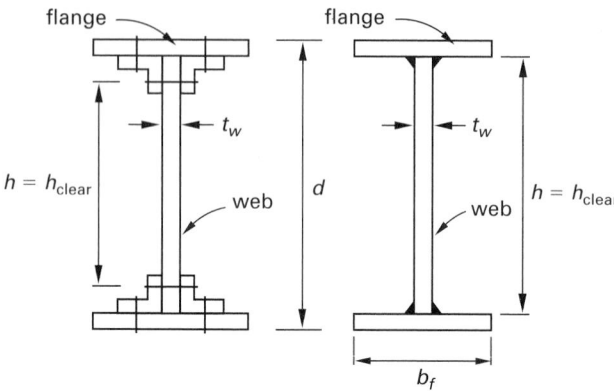

If intermediate stiffeners are spaced more than $1.5h$, then Eq. 63.2 governs the depth-thickness ratio [AISC Eq. F13-4].

$$\left(\frac{h}{t_w}\right)_{\max} \leq \frac{0.42E}{F_y} \qquad 63.2$$

If h/t_w satisfies Eq. 63.2, is less than 260, with a ratio of web area to flange area less than 10, and if the shear stress is less than the allowable value, no intermediate stiffeners are required. (Bearing stiffeners are still required at reaction and loading points, however.) In other cases, actual stiffener spacing will depend on the shear stress [*AISC Specification* Sec. E6].

3. SHEAR STRENGTH

The *AISC Specification* gives two methods for calculating shear strength. The first method, which is presented in this book, does not utilize the post buckling strength of the member (*tension field action*). Tension field action is presented in Sec. G3 of the *AISC Specification*.

The available shear strength, $\phi_v V_n$ or V_n/Ω, must equal or exceed the required shear strength, V_u or V_a. *AISC Specification* Ch. G covers the design of members for shear. The nominal shear strength, V_n, for stiffened or unstiffened webs, in accordance with the limit states of shear yielding and buckling is

$$V_n = 0.6F_y A_w C_v \quad \text{[AISC Eq. G2-1]} \qquad 63.3$$

A_w is the web area (the overall depth multiplied by the web thickness, dt_w), and C_v is the web shear coefficient. For webs of I-shaped members, where $h/t_w \leq 2.24\sqrt{E/F_y}$, $C_v = 1.0$ [AISC Eq. G2-2]. For webs of all other singly or doubly symmetric members (except round HSS), C_v is determined as follows. For $h/t_w \leq 1.10\sqrt{k_v E/F_y}$,

$$C_v = 1 \quad \text{[AISC Eq. G2-3]} \qquad 63.4$$

For $1.10\sqrt{k_v E/F_y} < h/t_w \leq 1.37\sqrt{k_v E/F_y}$,

$$C_v = \frac{1.10\sqrt{\dfrac{k_v E}{F_y}}}{\dfrac{h}{t_w}} \quad \text{[AISC Eq. G2-4]} \qquad 63.5$$

For $h/t_w > 1.37\sqrt{k_v E/F_y}$,

$$C_v = \frac{1.51 E k_v}{\left(\dfrac{h}{t_w}\right)^2 F_y} \quad \text{[AISC Eq. G2-5]} \qquad 63.6$$

a is the clear distance between intermediate stiffeners. When a is very large (i.e., $a/h > 3.0$), or when there are no stiffeners, the web plate buckling coefficient, k_v, is 5. Equation 63.12 gives the formula for k_v. (Refer also to *AISC Specification*, Sec. G2, for more cases.)

4. DESIGN OF GIRDER WEBS AND FLANGES

Generally, plate girder webs are designed based on shear stress using AISC Tables 3-16 and 3-17. The available shear stress is a function of a/h and h/t_w. The design of girder flanges depends on an assumption regarding flexural strength. Flexural strength is a function of the web thickness. If $h/t < 640/\sqrt{F_y}$, then the flexural strength equals $F_y Z_x$; otherwise, a smaller moment must be assumed.

Equation 63.7, derived from basic mechanics principles, gives an initial estimate of the required flange area. Since the moment varies along the beam length, girder flanges do not have to be the same thickness along the entire plate girder length. It is possible to substitute thinner flanges near the ends of beams or to add cover plates where additional flexural capacity is needed.

$$A_f = \frac{M_p}{F_y h} - \frac{ht}{4} \qquad 63.7$$

Once the flange area is known, trial flange widths and thicknesses can be evaluated.

$$A_f = b_f t_f \qquad 63.8$$

The limitations in available plate thicknesses should be considered when choosing t_f. Alternatively, b_f can be chosen based on b_f/d ratios, which typically vary between 0.2 and 0.3. Flange plate widths are typically rounded up to a convenient value, often to the nearest 2 in.

5. WIDTH-THICKNESS RATIOS

The width-thickness ratios specified in Chap. 61 apply to plate girders as well. Specifically, for unstiffened plates such as plate girder flanges, cases 2, 4, and 14, *AISC Specification* Table B.4.1 should be used.

For compact sections,

$$\frac{b_f}{2t_f} \leq \frac{64.7}{\sqrt{F_{yf}}} \qquad 63.9$$

For non-compact sections,

$$\frac{b_f}{2t_f} \leq 162\sqrt{\frac{k_c}{F_L}} \qquad 63.10$$

$$0.35 \leq \frac{4}{\sqrt{\frac{h}{t_w}}} \leq 0.76 \qquad 63.11$$

F_L may be taken to equal $0.7F_y$ for symmetrical members.

For stiffened plates, such as plate girder webs, Eqs. 63.1 and 63.2 [AISC Eqs. G1-1 and G1-2] must be satisfied also.

6. REDUCTION IN FLANGE STRENGTH

Once a trial design of the plate girder has been determined, the flexural strength can be computed. There are several possible limit states for girders, including yielding, inelastic lateral-torsional buckling, elastic lateral-torsional buckling, and local buckling. For members with compact webs and flanges and with adequate lateral support, the flexural strength, M_n, is given by $F_y Z_x$. As the distance between lateral supports increases, flexural strength is give by *AISC Specification* Eqs. F2-2 and F2-4.

For members with compact webs and non-compact flanges, flexural strength is given by the equations listed in Sec. F3 of the *AISC Specification*; for members with non-compact webs and flanges, flexural strength is given by the equations in Sec. F4.

7. LOCATION OF FIRST (OUTBOARD) STIFFENERS

The first intermediate stiffeners can be located where the shear stress exceeds Eq. 63.3 [AISC Eq. G2-1]. In practice, a trial distance a ($a/h < 1.0$) is selected as the separation between the end panel and the first intermediate stiffener. The design shear strength, $\phi_v V_n$, and the allowable shear strength, V_n/Ω_v, are calculated and compared to the required shear strength. If ϕV_n is greater than V_u, or if V_n/Ω_v is greater than V_a, the location is adequate. Otherwise, a smaller value of the spacing, a, should be tried.

8. LOCATION OF INTERIOR STIFFENERS

If $h/t_w \leq 2.46\sqrt{E/F_y}$, or where required shear strength is less than or equal to available shear strength, then intermediate (i.e., transverse) stiffeners are not required.

[*AISC Specification* Sec. G2.2]. Otherwise, the horizontal spacing, a, of interior intermediate stiffeners should not exceed the value determined from Eq. 63.13 [AISC Sec. G2.1]. For unstiffened webs of doubly symmetric shapes, with $h/t_w < 260$, $k_v = 5$. For stiffened webs,

$$k_v = 5 + \frac{5}{\left(\frac{a}{h}\right)^2} \qquad 63.12$$

$k_v = 5$ when $a/h > 3.0$, or

$$\frac{a}{h} > \left(\frac{260}{\frac{h}{t_w}}\right)^2 \qquad 63.13$$

With the value of k_v calculated from Eq. 63.12, the coefficient C_v can be calculated from Eqs. 63.4, 63.5, and 63.6, and the shear strength can be computed from Eq. 63.3. In beams where the maximum shear occurs near the ends, the spacing chosen for the first interior stiffener will be adequate for the entire beam length.

If Eq. 63.13 is met and if $C_v \leq 1.0$, then Eq. 63.14 can be used in place of Eq. 63.4 [AISC Eq. G2-1]. This is the *tension field action equation*. Use of Eq. 63.14 places an additional restraint on the allowable bending stress in the girder web.

For $h/t_w > 1.10\sqrt{k_v E/F_y}$,

$$V_n = 0.6F_y A_w \left(C_v + \frac{1 - C_v}{1.15\sqrt{1 + \left(\frac{a}{h}\right)^2}} \right) \quad \text{[AISC G3-2]}$$

$$63.14$$

For $h/t_w \leq 1.10\sqrt{k_v E/F_y}$,

$$V_n = 0.6F_y A_w \qquad 63.15$$

In lieu of equations, Tables 3-17a and 3-17b of the *AISC Manual* Part 3 can be used.

9. DESIGN OF INTERMEDIATE STIFFENERS

Intermediate stiffeners, as shown in Fig. 63.2, are used to support the compression flange and prevent buckling. They may be constructed from plates or angles and may be used either singly or in pairs. Unlike stiffeners that transmit reactions and loads, intermediate stiffeners do not need to extend completely from the top to bottom flanges, but they must be fastened to the compression flange to resist uplift. The *AISC Specification* contains limitations on weld and rivet spacing.

Intermediate stiffeners are sized by their gross steel area, as calculated from the steel area in contact with the compression flange. Thus, the width and thickness are used to calculate the stiffener area, not the width and depth. Equation 63.16 gives the steel area at a particular location [AISC Eq. G3-3]. This area can be divided between two stiffeners or given to a single stiffener. In Eq. 63.16, D is 1.0 for stiffeners furnished in pairs and is 2.4 for single plate stiffeners. D is 1.8 for single-angle stiffeners. In addition to meeting *AISC Specification* Sec. G2.2, the following two limitations must also be met when designing intermediate stiffeners.

$$A_{\text{st}} > \frac{F_y}{F_{y,\text{st}}} \left(0.15 D_s h t_w (1 - C_v) \frac{V_r}{V_c} - 18 t_w^2 \right) \geq 0$$

63.16

$$\left(\frac{b}{t} \right)_{\text{st}} \leq 0.56 \sqrt{\frac{E}{F_{y,\text{st}}}}$$

63.17

$(b/t)_{\text{st}}$ is the stiffener width-thickness ratio, $F_{y,\text{st}}$ is the specified minimum yield stress of the stiffener, C_v is the coefficient as defined by AISC Sec. G2.1, V_r is the required strength at the location of the stiffener, and V_c is the available shear strength as defined in AISC Sec. G3.2.

Figure 63.2 *Intermediate Stiffeners*

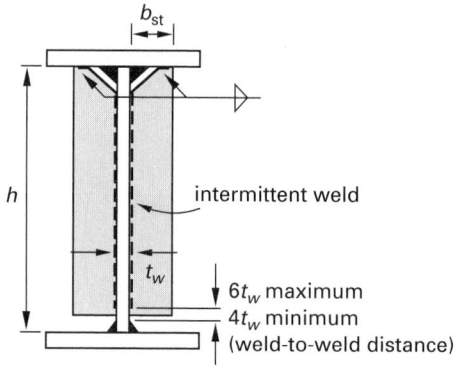

10. DESIGN OF BEARING STIFFENERS

Bearing stiffeners are used to transmit loads and reactions from one flange to the other. Bearing stiffeners that transmit loads and reactions may extend from flange to flange. There is no single formula for calculating the thickness of bearing stiffeners. Stiffener thickness is the maximum thickness determined from evaluations of the bearing stress, width-thickness ratio, column stress, and compression yield criteria.

The bearing pressure on bearing stiffeners is limited to $0.90F_y$ [*AISC Specification* Sec. J10.8]. Such stiffeners must essentially extend from the web to the edge of the flanges. Therefore, this criterion establishes one method of determining the bearing stiffener thickness. However, other criteria must also be met. (See the following section.) The width is essentially fixed by the flange dimension. So only the thickness needs to be determined. Also, bearing stiffeners should have close contact with the flanges.

Since the stiffener is loaded as a column, it must satisfy the width-thickness ratio for an unstiffened element.

$$\frac{b_{\text{st}}}{t_{\text{st}}} \leq \frac{95}{\sqrt{F_y}}$$

63.18

The stiffener should be designed as a column with an effective length equal to $0.75h$, with a cross section composed of a pair of stiffeners, and a strip of the web having a width of either (a) $25t_w$ at interior concentrated loads or (b) $12t_w$ at the ends of members [*AISC Specification* Sec. J10.8].

For the purpose of determining the slenderness ratio, the radius of gyration, r, can be determined exactly or it can be approximated as 0.25 times the stiffener edge-to-edge distance.

$$\frac{l}{r} = \frac{0.75h}{0.25(2b_{\text{st}} + t_w)}$$

63.19

Figure 63.3 *Bearing Stiffener (top view)*

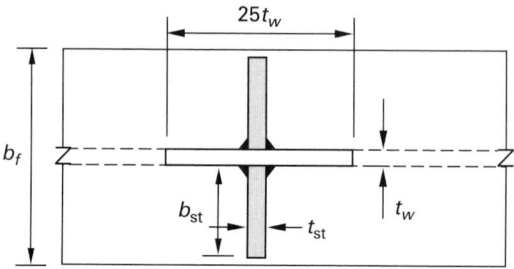

Once the l/r ratio is known, it can be used (as Kl/r) to determine the flexural buckling stress, F_{cr}. However, some of the web supports the load. Specifically, 25 times the web thickness is the contributing area at an interior concentrated load. Therefore, the required stiffener thickness based on column stress is

$$t_{\text{st}} = \frac{\dfrac{\text{load}}{\phi F_{cr}} - 25t_w^2}{2b_{\text{st}}}$$

63.20

Compression yielding of stiffeners should also be considered when determining stiffener thickness. Available compressive strength is limited to $0.9F_y$. Therefore, the thickness is

$$t_{\text{st}} = \frac{\dfrac{\text{load}}{0.9F_y}}{2b_{\text{st}}}$$

63.21

If column stability is not the factor controlling stiffener thickness, the larger thickness could conceivably increase the slenderness ratio and reduce the allowable compressive stress even further. This should be checked, but it is not likely to be a factor.

11. SPECIAL DESIGN CONSIDERATIONS FOR CONCENTRATED FORCES

The *AISC Specification* Sec. J10 contains provisions to determine if the web is capable of supporting the concentrated forces (loads or reactions) without experiencing local web yielding, web crippling, or sidesway web buckling. The need for bearing stiffeners to preclude local web yielding is determined from AISC Eqs. J10-2 and J10-3. AISC Eqs. J10-4, J10-5a, and J10-5b determine the maximum load that can be applied before stiffeners are required from web-crippling criterion. If stiffeners are provided and extend at least one-half the web depth, Eqs. J10-4, J10-5a, and J10-5b need not be checked. Equations J10-6 and J10-7 are used to check if bearing stiffeners are needed to prevent sidesway web buckling. If the term $(d_c/t_w)/(l/b_f)$ in Eqs. J10-6 and J10-7 exceeds 2.3 or 1.7, respectively, then these two equations need not be checked.

12. DESIGN PROCESS SUMMARY

The following steps summarize plate girder design. Using Tables 3-16 and 3-17 in Part 3 of the *AISC Manual* will help design the web.

step 1: Select the overall depth.

step 2: Select a trial web size using the required shear strength.

step 3: Select a trial flange size using the required flexural strength.

step 4: Check the available flexural strength of the trial section.

step 5: Check for tension field action.

step 6: Design the intermediate stiffeners (if any).

step 7: Check for web resistance to applied concentrated loads and/or support reactions and design bearing stiffeners if needed.

step 8: Design all required welds.

Example 63.1

Design a symmetrical welded plate girder of A992 steel with no intermediate stiffeners to carry a load of 2.5 kips/ft on a simple span of 60 ft. The depth of the girder is restricted to 50 in. The compression flange is laterally supported throughout the girder span.

Solution

LRFD Method

Assume the dead load is 0.5 kips/ft.

The required shear is

$$V_u = (1.2)(15 \text{ kips}) + (1.6)(75 \text{ kips}) = 138 \text{ kips}$$

$$w_u = (1.2)\left(0.5 \frac{\text{kips}}{\text{ft}}\right) + (1.6)\left(2.5 \frac{\text{kips}}{\text{ft}}\right) = 4.6 \text{ kips/ft}$$

The required flexural strength is

$$M_u = \frac{\left(4.6 \frac{\text{kips}}{\text{ft}}\right)(60 \text{ ft})^2}{8} = 2070 \text{ ft-kips}$$

Assume a $44 \times {}^3/_8$ in plate for a web, and two $16 \times {}^3/_4$ in plates for the flanges.

The material and geometric properties for ASTM A992 are

$$F_y = 50 \text{ ksi}$$
$$F_u = 65 \text{ ksi}$$
$$t_w = 0.375 \text{ in}$$
$$t_f = 0.75 \text{ in}$$
$$b_f = 16 \text{ in}$$
$$h = 44 \text{ in}$$
$$d = 44 \text{ in} + (2)(0.75 \text{ in}) = 45.5 \text{ in}$$

The required shear strength is

$$R_u = \left((1.2)\left(0.5 \frac{\text{kips}}{\text{ft}}\right) + (1.6)\left(2.5 \frac{\text{kips}}{\text{ft}}\right)\right)(30 \text{ ft})$$
$$= 138 \text{ kips}$$

Determine if stiffeners are required.

$$A_w = dt_w = (45.5 \text{ in})(0.375 \text{ in}) = 17.06 \text{ in}^2$$
$$\frac{h}{t_w} = \frac{44 \text{ in}}{0.375 \text{ in}} = 117 < 260 \quad [\text{per AISC Sec. G2.1(i)}]$$
$$k_v = 5$$

Then, from G2.1(b)(iii),

$$1.37\sqrt{\frac{k_v E}{F_y}} = 1.37\sqrt{\frac{(5)\left(29{,}000 \frac{\text{kips}}{\text{in}^2}\right)}{50 \frac{\text{kips}}{\text{in}^2}}} = 73.8$$

Since $h/t_w = 117 > 73.8$, use AISC Eq. G2-5 to calculate the web shear coefficient, C_v.

$$C_v = \frac{1.51 E k_v}{\left(\frac{h}{t_w}\right)^2 F_y} = \frac{(1.51)\left(29{,}000 \frac{\text{kips}}{\text{in}^2}\right)(5)}{(117)^2\left(50 \frac{\text{kips}}{\text{in}^2}\right)} = 0.320$$

Calculate the nominal shear strength, V_n.

$$V_n = 0.6 F_y A_w C_v = (0.6)\left(50 \frac{\text{kips}}{\text{in}^2}\right)(17.06 \text{ in}^2)(0.320)$$
$$= 163.8 \text{ kips}$$

Check the available (design) shear strength without stiffeners.

$$\phi_v = 0.90$$
$$\phi_v V_n = (0.90)(163.8 \text{ kips}) = 147 \text{ kips}$$
$$147 \text{ kips} > 138 \text{ kips}$$

Since the design shear strength is greater than the required shear strength, stiffeners are not required.

Check the bending moment and the required flexural strength.

$$w_u = (1.2)\left(0.5 \ \frac{\text{kips}}{\text{ft}}\right) + (1.6)\left(2.5 \ \frac{\text{kips}}{\text{ft}}\right) = 4.6 \ \text{kips/ft}$$

$$M_u = \frac{w_u L^2}{8} = \frac{\left(4.6 \ \dfrac{\text{kips}}{\text{ft}}\right)(60 \ \text{ft})^2}{8} = 2070 \ \text{ft-kips}$$

Calculate I_{gr} using the following table.

section	dimension	A (in^2)	y (in)	A_y^2 (in^4)	I_o (in^4)	I_{gr} (in^4)
web	$^3/_8 \times 44$	16.5	0	0	2662	2662
flange	$^3/_4 \times 16$	12.0	22.375	6007.7	0.5625	6008.8
flange	$^3/_4 \times 16$	12.0	22.375	6007.7	0.5625	6008.8
					total	14,680

$$c = \frac{d}{2} = \frac{45.5 \ \text{in}}{2} = 22.75 \ \text{in}$$

Calculate S_x.

$$S_x = \frac{I_{gr}}{c} = \frac{14{,}680 \ \text{in}^4}{22.75 \ \text{in}} = 645.3 \ \text{in}^3$$

$$\frac{M_u}{S_x} = \frac{(2070 \ \text{ft-kips})\left(12 \ \dfrac{\text{in}}{\text{ft}}\right)}{645.3 \ \text{in}^3} = 38.5 \ \text{kips/in}^2$$

$$M_{yc} = M_{yt} = M_y = F_y S_x = \frac{\left(50 \ \dfrac{\text{kips}}{\text{in}^2}\right)(645 \ \text{in}^3)}{12 \ \dfrac{\text{in}}{\text{ft}}}$$

Compute Z_x.

$$Z_x = A_f(h + t_f) + \frac{t_w h^2}{4}$$
$$= (16 \ \text{in})(0.75 \ \text{in})(44 \ \text{in} + 0.75 \ \text{in})$$
$$+ \frac{(0.375 \ \text{in})(44 \ \text{in})^2}{4}$$
$$= 718.5 \ \text{in}^3$$

$$M_p = F_y Z_x = \frac{\left(50 \ \dfrac{\text{kips}}{\text{in}^2}\right)(718.5 \ \text{in}^3)}{12 \ \dfrac{\text{in}}{\text{ft}}}$$
$$= 2994 \ \text{ft-kips}$$

Check for compactness.

For the flange,

$$\lambda_f = \frac{b_f}{2t_f} = \frac{16 \ \text{in}}{(2)(0.75 \ \text{in})} = 10.67$$

Since $\lambda_p < \lambda_f < \lambda_r$ as given in Sec. B4 of the *AISC Specification*, the flange is non-compact.

For the web,

$$\lambda_w = \frac{h}{t_w} = \frac{44 \ \text{in}}{0.375 \ \text{in}} = 117$$

Since $\lambda_p < \lambda_f < \lambda_r$ as given in Sec. B4 of the *AISC Specification*, the web is non-compact.

Since both the web and flange are non-compact, the flexural strength is given by Sec. F4 in the *AISC Specification*.

$$h_c = h \quad \text{[welded built-up shapes]}$$
$$\lambda = \frac{h_c}{t_w} = \frac{44 \ \text{in}}{0.375 \ \text{in}} = 117.3$$

The cross-section is symmetric, so case 9 applies.

$$\lambda_{pw} = \left[\begin{array}{c}\text{Table B4.1}\\\text{case 9}\end{array}\right] = 3.76\sqrt{\frac{E}{F_y}}$$
$$= 3.76\sqrt{\frac{29{,}000 \ \dfrac{\text{kips}}{\text{in}^2}}{50 \ \dfrac{\text{kips}}{\text{in}^2}}} = 90.55$$

$$\lambda_{rw} = \left[\begin{array}{c}\text{Table B4.1}\\\text{case 9}\end{array}\right] = 5.70\sqrt{\frac{E}{F_y}} = 5.70\sqrt{\frac{29{,}000 \ \dfrac{\text{kips}}{\text{in}^2}}{50 \ \dfrac{\text{kips}}{\text{in}^2}}}$$
$$= 137.27$$

The *web plastification factor*, R_{pc}, is found from AISC Eq. F4-9b.

$$R_{pc} = \frac{M_p}{M_{yc}} - \left(\frac{M_p}{M_{yc}} - 1\right)\left(\frac{\lambda - \lambda_{pw}}{\lambda_{rw} - \lambda_{pw}}\right) \leq \frac{M_p}{M_{yc}}$$
$$= \frac{2994 \ \text{ft-kips}}{2688 \ \text{ft-kips}} - \left(\frac{2994 \ \text{ft-kips}}{2688 \ \text{ft-kips}} - 1\right)$$
$$\times \left(\frac{117.3 - 90.55}{137.27 - 90.55}\right) \leq \frac{2994 \ \text{ft-kips}}{2688 \ \text{ft-kips}}$$
$$= 1.049$$

The flexural strength, M_n, as limited by compression flange yielding is given by Eq. F4-1 in the *AISC Specification*. For compression flange yielding,

$$M_n = R_{pc}M_{yc} = (1.049)(2688 \ \text{ft-kips}) = 2820 \ \text{ft-kips}$$

Lateral-torsional buckling is not a limit state since the compression flange is fully supported laterally.

The flexural strength, M_n, as limited by compression flange local buckling is given by Eq. F4-12 in the *AISC*

Specification. Use the following process to evaluate AISC Eq. F4-12.

Determine F_L. The plate girder is symmetrical, so

$$\frac{S_{xt}}{S_{xc}} = 1$$

$$F_L = 0.7F_y = (0.7)\left(50\ \frac{\text{kips}}{\text{in}^2}\right) = 35\ \text{kips/in}^2$$

Determine k_c as defined in AISC Table B4.1, footnote (a).

$$k_c = 0.35 \le \frac{4}{\sqrt{\dfrac{h}{t_w}}} \le 0.76$$

$$0.35 \le \frac{4}{\sqrt{\dfrac{44\ \text{in}}{0.375\ \text{in}}}} = 0.369 \le 0.76$$

Since $0.369 \le 0.76$, 0.369 controls.

From AISC Table B4.1, case 2,

$$\lambda_{pf} = 0.38\sqrt{\frac{E}{F_y}} = 0.38\sqrt{\frac{29{,}000\ \dfrac{\text{kips}}{\text{in}^2}}{50\ \dfrac{\text{kips}}{\text{in}^2}}}$$

$$= 9.2$$

$$\lambda_{rf} = 0.95\sqrt{\frac{k_c E}{F_L}} = 0.95\sqrt{\frac{(0.369)\left(29{,}000\ \dfrac{\text{kips}}{\text{in}^2}\right)}{35\ \dfrac{\text{kips}}{\text{in}^2}}}$$

$$= 16.6$$

From AISC Eq. F4-12,

$$M_n = \left(R_{pc}M_{yc} - (R_{pc}M_{yc} - F_L S_{xc})\left(\frac{\lambda - \lambda_{pf}}{\lambda_{rf} - \lambda_{pf}}\right)\right)$$

$$= \left(\begin{array}{l}(1.049)(2688\ \text{ft-kips} \\[4pt] - \left(\begin{array}{l}(1.049)(2688\ \text{ft-kips}) \\[4pt] - \dfrac{\left(35\ \dfrac{\text{kips}}{\text{in}^2}\right)(645.3\ \text{in}^3)}{12\ \dfrac{\text{in}}{\text{ft}}}\end{array}\right) \\[4pt] \times \left(\dfrac{10.67 - 9.2}{16.6 - 9.2}\right)\end{array}\right)$$

$$= 2633\ \text{ft-kips}$$

Since 2633 ft-kips > 2820 ft-kips, the lower value controls.

$$M_n = 2633\ \text{ft-kips} \quad [\text{controls}]$$

The available flexural strength, ϕM_n, is given by

$$\phi M_n = (0.9)(2633\ \text{ft-kips}) = 2370\ \text{ft-kips}$$

Since $M_u < \phi M_n$ (2070 ft-kips < 2370 ft-kips), flexural strength is OK.

Check live load deflection.

$$\Delta_{\max} = \frac{L}{360} = \frac{(60\ \text{ft})\left(12\ \dfrac{\text{in}}{\text{ft}}\right)}{360} = 2.0\ \text{in}$$

$$I_{s,\text{req'd}} = \frac{5wl^4}{384E\Delta_{\max}}$$

$$= \frac{(5)\left(2.5\ \dfrac{\text{kips}}{\text{ft}}\right)(60\ \text{ft})^4\left(12\ \dfrac{\text{in}}{\text{ft}}\right)^3}{(384)\left(29{,}000\ \dfrac{\text{kips}}{\text{in}^2}\right)(2.0\ \text{in})}$$

$$= 12{,}569\ \text{in}^4$$

Check the minimum section modulus required.

$$S_{\text{req'd}} = \frac{I}{c} = \frac{12{,}569\ \text{in}^4}{22.75\ \text{in}} = 552\ \text{in}^3$$

Since the actual section modulus is greater than the required section modulus, a $44 \times {}^3/_8$ web PL with two 16 in $\times$ ${}^3/_4$ in flanges is OK.

ASD Method

Assume the dead load is 0.5 kips/ft.

The required shear is

$$V_a = 15\ \text{kips} + 75\ \text{kips} = 90\ \text{kips}$$

The required flexural strength is

$$w_a = 0.5\ \frac{\text{kips}}{\text{ft}} + 2.5\ \frac{\text{kips}}{\text{ft}} = 3.0\ \text{kips/ft}$$

$$M_a = \frac{w_a L^2}{8} = \frac{\left(3\ \dfrac{\text{kips}}{\text{ft}}\right)(60\ \text{ft})^2}{8} = 1350\ \text{ft-kips}$$

Assume a $44 \times {}^3/_8$ in plate for a web, and two $16 \times {}^3/_4$ in plates for the flanges.

The material and geometric properties for ASTM A992 are

$$F_y = 50\ \text{ksi}$$
$$F_u = 65\ \text{ksi}$$
$$t_w = 0.375\ \text{in}$$
$$t_f = 0.75\ \text{in}$$
$$b_f = 16\ \text{in}$$
$$h = 44\ \text{in}$$
$$d = 44\ \text{in} + (2)(0.75\ \text{in}) = 45.5\ \text{in}$$

The required shear strength is

$$R_a = (0.5 + 2.5)(30 \text{ kips}) = 90 \text{ kips}$$

Determine if stiffeners are required.

$$A_w = dt_w = (45.5 \text{ in})(0.375 \text{ in}) = 17.06 \text{ in}^2$$

$$\frac{h}{t_w} = \frac{44 \text{ in}}{0.375 \text{ in}} = 117 < 260 \quad [\text{per AISC Sec. G2.1(i)}]$$

$$k_v = 5$$

Then, from G2.1(b)(iii),

$$1.37\sqrt{\frac{k_v E}{F_y}} = 1.37\sqrt{\frac{(5)\left(29{,}000 \, \dfrac{\text{kips}}{\text{in}^2}\right)}{50 \, \dfrac{\text{kips}}{\text{in}^2}}} = 73.8$$

Since $h/t_w = 117 > 73.8$, use AISC Eq. G2-5 to calculate the web shear coefficient, C_v.

$$C_v = \frac{1.51 E k_v}{\left(\dfrac{h}{t_w}\right)^2 F_y} = \frac{(1.51)\left(29{,}000 \, \dfrac{\text{kips}}{\text{in}^2}\right)(5)}{(117)^2 \left(50 \, \dfrac{\text{kips}}{\text{in}^2}\right)} = 0.320$$

Calculate the nominal shear strength, V_n.

$$V_n = 0.6 F_y A_w C_v = (0.6)\left(50 \, \frac{\text{kips}}{\text{in}^2}\right)(17.06 \text{ in}^2)(0.320)$$

$$= 163.8 \text{ kips}$$

Check the available shear strength without stiffeners.

$$\Omega_v = 1.67$$

$$\frac{V_n}{\Omega_v} = \frac{170 \text{ kips}}{1.67} = 102 \text{ kips}$$

$$102 \text{ kips} > 90 \text{ kips}$$

Since the allowable shear strength is greater than the required shear strength, stiffeners are not required.

Check the bending moment and the required flexural strength.

$$w_a = 0.5 \, \frac{\text{kips}}{\text{ft}} + 2.5 \, \frac{\text{kips}}{\text{ft}} = 3 \, \frac{\text{kips}}{\text{ft}}$$

$$M_a = \frac{w_a L^2}{8} = \frac{\left(3 \, \dfrac{\text{kips}}{\text{ft}}\right)(60 \text{ ft})^2}{8} = 1350 \text{ ft-kips}$$

As was shown with the LRFD method, the flange and web are non-compact and the flexural strength, M_n, is

$$M_n = 2633 \text{ ft-kips}$$

The available flexural strength, M_n/Ω, is given by

$$\frac{M_n}{\Omega} = \frac{2633 \text{ ft-kips}}{1.67} = 1577 \text{ ft-kips}$$

Since $M_a < M_n/\Omega$, flexural strength is OK.

Check live load deflection.

$$\Delta_{\max} = \frac{L}{360} = \frac{(60 \text{ ft})\left(12 \, \dfrac{\text{in}}{\text{ft}}\right)}{360} = 2.0 \text{ in}$$

$$I_{s,\text{req'd}} = \frac{5 w l^4}{384 E \Delta_{\max}}$$

$$= \frac{(5)\left(2.5 \, \dfrac{\text{kips}}{\text{ft}}\right)(60 \text{ ft})^4 \left(12 \, \dfrac{\text{in}}{\text{ft}}\right)^3}{(384)\left(29{,}000 \, \dfrac{\text{kips}}{\text{in}^2}\right)(2.0 \text{ in})}$$

$$= 12{,}569 \text{ in}^4$$

Check the minimum section modulus required.

$$S_{\text{req'd}} = \frac{I}{c} = \frac{12{,}569 \text{ in}^4}{22.75 \text{ in}} = 552 \text{ in}^3$$

Since the actual section modulus is greater than the required section modulus, a $44 \times {}^3/_8$ web PL with two $16 \text{ in} \times {}^3/_4 \text{ in}$ flanges is OK.

64 Structural Steel: Composite Beams

Nomenclature

a	slab thickness	in
A	area	in^2
b	effective concrete slab width	in
C	compressive force	–
d	depth	in
E	steel modulus of elasticity	kips/in^2
f'_c	compressive strength of concrete	kips/in^2
F	strength or allowable stress	kips/in^2
h	total depth from bottom of steel beam to top of concrete	in
I	moment of inertia	in^4
k	distance	in
K	distance	in
L	beam span	ft
M	moment	ft-kips
N_1	number of shear connectors required between the point of maximum moment and the point of zero moment	–
Q	allowable horizontal shear per shear connector	kips
t	thickness of concrete in compression	in
T	tensile force	kips
w	uniform load	kips/in
$Y1$	distance from top of steel beam to plastic neutral axis	in
$Y2$	distance from top of steel beam to concrete flange force	in

Subscripts

a	available or allowable
b	lower bound
c	concrete or effective concrete flange
D	dead load
DL	construction dead load
eff	effective
f	flange
g	gross
L	live load
n	nominal
s	steel
st	steel
t	top of concrete slab
u	ultimate
y	yield

Symbols

Δ	deflection
ϕ	resistance factor
Ω	safety factor

1. INTRODUCTION

Composite construction usually refers to a construction method in which a cast-in-place concrete slab is bonded to steel beams, girders, or decking below, such that the two materials act as a single unit. Although there are varying degrees of composite action, the concrete usually becomes the compression flange of the composite beam, and the steel carries the tension.

There are two types of composite beams: (a) steel beams fully encased in concrete so that the natural bond between the two materials holds them together, and (b) unencased steel beams attached to the concrete slab (solid or on formed steel deck) by mechanical anchorage (i.e., shear connectors). In this chapter, only the more common second type is discussed.

Figure 64.1 *Typical Composite Beam Cross Section*

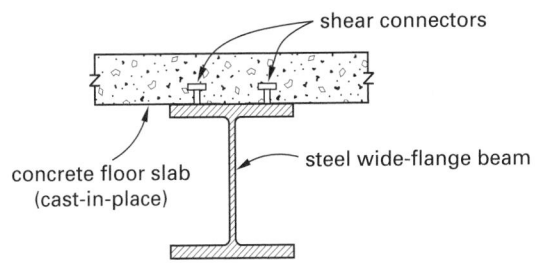

Composite construction is often used for bridge decks and building floors. In general, a composite floor system is stronger and stiffer than its noncomposite counterpart using the same size beams. For a given steel member cross section, the advantages of composite construction are increased load capacity and span length. Although there are no major disadvantages in composite construction, two limitations—the effect of continuity and long-term deflection (creep)—should be recognized.

Composite construction is carried out with or without temporary *shoring*. If the steel is erected and concrete is poured without temporary shoring (i.e., construction is unshored), the combination will act compositely to carry only the loads that are applied after the concrete cures. Conversely, in fully shored construction, a temporary support carries both the steel and concrete weights until the concrete has cured. When the shore is removed, the beam acts compositely to carry the steel and concrete weights in addition to live loads applied later.

The design of steel beams/girders can be based on the assumption of composite action in accordance with the *AISC Specification* Ch. I, provided that (a) the concrete slab supported by the steel beams is adequate to effectively serve as the flange of a composite T-beam, and (b) the concrete and steel are adequately tied together with mechanical anchorage (shear connectors) to act as one unit with no relative slippage. Development of full or partial composite action depends on the quantity of shear connectors. However, usually it is not necessary, and sometimes it is not feasible, to provide full composite action.

2. EFFECTIVE WIDTH OF CONCRETE SLAB

In a composite beam, the bending stresses in the concrete slab depend on the spacing of the steel beams. The slab bending stresses are uniformly distributed across the compression zone for closely spaced beams. However, the stresses vary considerably and nonlinearly across the compression flange when beams are placed far apart. The farther away a part of the slab is from the steel beam, the smaller is the stress in it.

AISC Specification Sec. I3.1a simplifies this issue of varying stresses by replacing the actual slab with an "equivalent" narrower or "effective" slab that has a constant stress. This equivalent slab is considered to carry the same total compression as that supported by the actual slab. For an interior beam with slab extending on both sides, the effective width, b, of the concrete slab is given by Eq. 64.1, where L is the beam span and s is the spacing between beam centerlines. (For an exterior beam, s is replaced by $e + s/2$, where e is the distance from the beam centerline to the edge of the slab.)

$$b = \text{smaller of} \left\{ \begin{array}{c} \dfrac{L}{4} \\ s \end{array} \right\} \quad \text{[interior beams]} \quad \textbf{64.1}$$

3. SECTION PROPERTIES

For serviceability limit states, the estimated moment of inertia, I, must be calculated. It is difficult to calculate an accurate value, so a lower bound value is listed in Table 3-20 in Part 3 of the *AISC Manual*. Section I3 of the *AISC Specification Commentary* describes the process used to compute the lower bound value. For W shapes used as composite beams, the lower bound moment of inertia, I_{lb}, is given by

$$I_{nb} = I_s + A_s \left(Y_{\text{ENA}} - \frac{d}{2} \right)^2$$
$$+ \left(\frac{\sum Q_n}{F_y} \right) (d + Y_2 - Y_{\text{ENA}})^2 \quad \textbf{64.2}$$

The nomenclature is illustrated in Fig. 64.2.

I_s	moment of inertia of the steel shape
A_s	area of steel shape

$$Y_{\text{ENA}} \qquad \dfrac{\dfrac{A_s d}{2} + \left(\dfrac{\sum Q_n}{F_y} \right) (d + Y_2)}{A_s + \dfrac{\sum Q_n}{F_y}}$$

d	depth of steel beam
$\sum Q_n$	sum of the shear strength of the studs
$\sum Q_n / F_y$	equivalent concrete area
Y_2	distance from the top of the compression flange to the concrete force

Figure 64.2 *Deflection Design Model for Composite Beams*

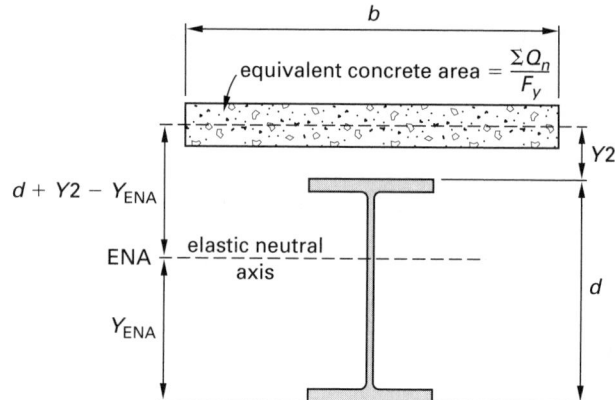

4. AVAILABLE FLEXURAL STRENGTH

When shoring is provided, the composite section composed of the slab and beam acts as a single unit and carries the entire dead and live load. If shoring is not provided, then the dead load of the steel and concrete is resisted by the steel beam alone and the composite beam resists only the live load.

The maximum available flexural strength, M_n, for a composite beam is based on plastic analysis and depends on the location of the *plastic neutral axis*, PNA. The maximum available flexural strength, M_n, for a composite beam occurs when the PNA is at the top of the flange; therefore, the steel shape is all in tension and the concrete is all in compression. The value for M_n is given by Eq. 64.3. A_g is the gross steel area and d is the total beam depth.

$$M_n = A_g F_y \left(\frac{d}{2} \right) \quad \textbf{64.3}$$

Equation 64.3 requires that there be sufficient shear capacity in the studs to transfer the tension force in the steel, $A_g F_y$, to the concrete. This is called *full composite action*. If sufficient shear capacity is not provided, then the PNA drops into the steel flange or web as shown in Fig. 64.3. Y_1 is the distance from the top of the flange to the PNA. Then, only a portion of the steel section is in tension and a portion shares the compressive force with the concrete, and the available flexural strength is reduced. This is called *partial composite action*. This may be useful when the full capacity of the composite beam is not necessary to resist the required moment. Cost savings may be made in the reduction of the number of required shear studs.

AISC Table 3-19 lists available flexural strengths for various values of Y_1.

5. SHEAR CONNECTORS

Although a natural bond is created between a steel beam and a concrete slab, this bond cannot be relied upon to ensure composite action. Mechanical *shear connectors* (also called *stud connectors*) attached to the top of the steel beam must be provided as a means of positive connection between the steel beam and the concrete slab. *AISC Specification* Sec. I gives the material, placing, and spacing requirements, and specifies that the entire horizontal shear at the junction of the steel beam and the concrete slab is assumed to be transferred by shear connectors. Steel stud connectors must conform to AWS D1.1.

The connectors required on each side of the point of maximum moment in an area of positive moment may be uniformly distributed between that point and adjacent points of zero moment. The required number of such connectors, N_1, can be calculated from Eq. 64.4 or 64.5. The nominal shear strength, Q_n, of one of the connectors is given in *AISC Specification* Sec. I3.2d; AISC Table 3-21 also gives the tabulated values for Q_n.

$$N_1 = \frac{A_g F_y}{Q_n} \quad \text{[full composite action]} \qquad 64.4$$

$$N_1 = \frac{C_c}{Q_n} \quad \text{[partial composite action]} \qquad 64.5$$

AISC Specification Sec. I3.2d(5) requires that the number of shear connectors between the maximum positive or negative bending moment and the adjacent zero moment section be equal to the horizontal shear force, $\sum Q_n$, for shear connectors [AISC Eq. I3-1c], divided by the nominal strength of one shear connector.

$$N = \frac{\sum Q_n}{Q_n} \qquad 64.6$$

6. DESIGN OF COMPOSITE BEAMS

Design of composite beams is based on the criteria of bending stress, shear stress, and deflection. Tables 3-19 and 3-20 in Part 3 of the *AISC Manual* are useful in design. Table 3-19 lists available flexural strength, and Table 3-20 lists lower bounds on the moment of inertia.

Figure 64.3 *Strength Design Models for Composite Beams*

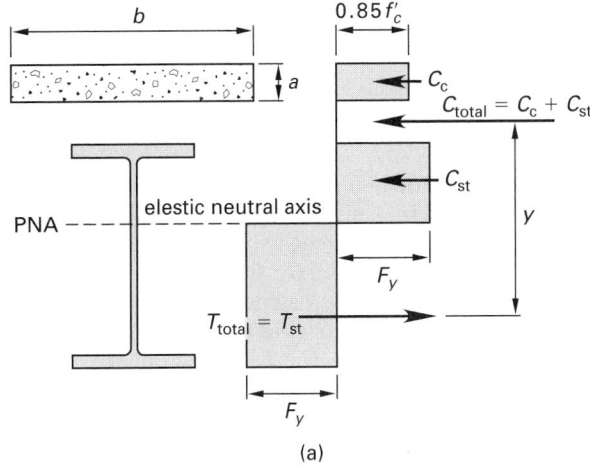

(a)

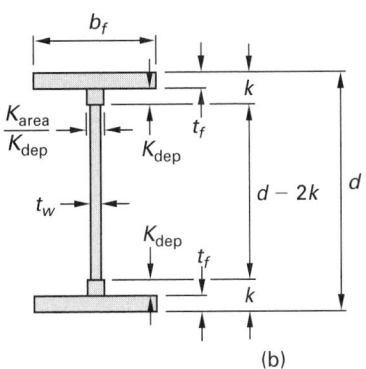

(b)

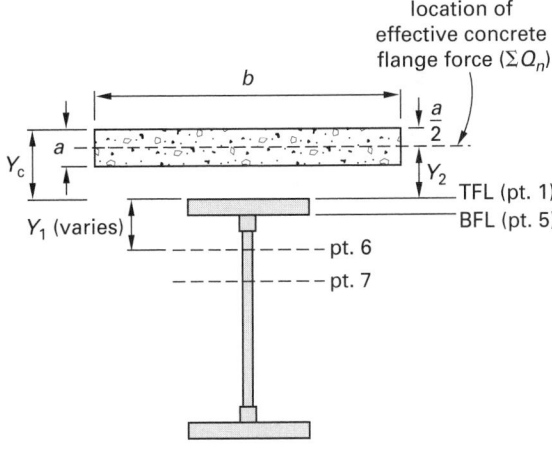

Y_1 = distance from top of steel flange to any of the seven tabulated PNA locations

$$\sum Q_n (\text{@ pt. 6}) = \frac{\sum Q_n (\text{@ pt. 5}) + \sum Q_n (\text{@ pt. 7})}{2}$$

$$\sum Q_n (\text{@ pt. 7}) = 0.25 F_y A_s$$

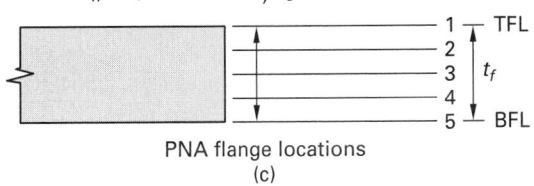

PNA flange locations
(c)

These tables can be used for full and partial composite action with a solid slab or a slab on formed-steel-deck without restriction on concrete strength, modular ratio, or effective width.

Example 64.1

Check the adequacy of a simply supported interior floor beam of composite construction, assuming full composite action and shored construction. The deflection checked during construction considers the weight of concrete as contributing to construction dead load (DL), and limits deflection to a $2^1/_2$ in maximum to facilitate concrete placement. Use the LRFD method and the following data.

ASTM A992 steel strength	$F_y = 50$ ksi
beam	W21 × 62
beam span	$L = 40$ ft
beam spacing	$s = 8$ ft
concrete strength	$f'_c = 3$ ksi
unit weight	150 lbf/ft^3
slab thickness	$t = 4$ in
floor live load	100 psf
partition load	20 psf

Solution

From the *AISC Manual*, for W21 × 62, $A = 18.3$ in^2, $d = 21.0$ in, and $t_w = 0.4$ in.

The loading from the 4 in slab is

$$\left(\frac{4 \text{ in}}{12 \frac{\text{in}}{\text{ft}}} \right) (8 \text{ ft}) \left(150 \frac{\text{lbf}}{\text{ft}^3} \right) = 400 \text{ lbf/ft}$$

The loading from the W21 × 62 steel beam is 62 lbf/ft.

The total dead load is

$$w_D = \frac{400 \frac{\text{lbf}}{\text{ft}} + 62 \frac{\text{lbf}}{\text{ft}}}{1000 \frac{\text{lbf}}{\text{kip}}}$$

$$= 0.462 \text{ kip/ft}$$

The dead load moment is

$$M_D = \frac{w_D L^2}{8} = \frac{\left(0.462 \frac{\text{kip}}{\text{ft}} \right) (40 \text{ ft})^2}{8}$$

$$= 92.4 \text{ ft-kips}$$

After the concrete has hardened, the loading from the live load and partition will be 100 psf + 20 psf = 120 psf.

$$w_L = \frac{\left(120 \frac{\text{lbf}}{\text{ft}^2} \right) (8 \text{ ft})}{1000 \frac{\text{lbf}}{\text{kip}}}$$

$$= 0.96 \text{ kip/ft}$$

$$M_L = \frac{w_L L^2}{8} = \frac{\left(0.96 \frac{\text{kip}}{\text{ft}} \right) (40 \text{ ft})^2}{8}$$

$$= 192 \text{ ft-kips}$$

The maximum moment is

$$M_{\max} = M_D + M_L = 92.4 \text{ ft-kips} + 192 \text{ ft-kips}$$

$$= 284.4 \text{ ft-kips}$$

The maximum shear is

$$V_{\max} = \frac{(w_D + w_L)L}{2}$$

$$= \frac{\left(0.462 \frac{\text{kip}}{\text{ft}} + 0.96 \frac{\text{kip}}{\text{ft}} \right) (40 \text{ ft})}{2} = 28.44 \text{ kips}$$

The effective width of the concrete slab is the smaller of

$$b = \frac{L}{4} = \frac{(40 \text{ ft}) \left(12 \frac{\text{in}}{\text{ft}} \right)}{4} = 120 \text{ in}$$

$$b = s = (8 \text{ ft}) \left(12 \frac{\text{in}}{\text{ft}} \right) = 96 \text{ in} \quad \text{[governs]}$$

LRFD Method

$$w_u = (1.2) \left(0.462 \frac{\text{kips}}{\text{ft}} \right) + (1.6) \left(0.96 \frac{\text{kips}}{\text{ft}} \right)$$

$$= 2.09 \frac{\text{kips}}{\text{ft}}$$

$$M_u = \frac{w_u L^2}{8} = \frac{\left(2.09 \frac{\text{kips}}{\text{ft}} \right) (40 \text{ ft})^2}{8} = 418 \text{ ft-kips}$$

Use AISC Tables 3-19, 3-20, and 3-21.

Determine $b_{\text{eff}} = 96$ in and assuming $a = 1$,

$$Y2 = \left(t_{\text{slab}} - \frac{a}{2} \right) = \left(4 \text{ in} - \frac{1 \text{ in}}{2} \right) = 3.5 \text{ in}$$

With a PNA location 5 (i.e., at the bottom of the flange, BFL),

$$\phi M_n = 813 \text{ kips} > 418 \text{ kips}$$

Check the beam deflection and the available shear strength. The construction dead load is

$$w_{DL} = (8\ \text{ft})\left(\dfrac{4\ \text{in}}{12\ \frac{\text{ft}}{\text{in}}}\right)\left(150\ \dfrac{\text{lbf}}{\text{ft}^3}\right)\left(\dfrac{1\ \text{kip}}{1000\ \text{lbf}}\right)$$

$$= 0.4\ \text{kips/ft}$$

The required moment of inertia to limit deflection during construction is

$$I_{\text{req'd}} = \dfrac{5w_{DL}L^4}{384E\Delta_{\max}}$$

$$= \dfrac{(5)\left(0.4\ \dfrac{\text{kips}}{\text{ft}}\right)(40\ \text{ft})^4\left(12\ \dfrac{\text{in}}{\text{ft}}\right)^3}{(384)\left(29{,}000\ \dfrac{\text{kips}}{\text{in}^2}\right)(2.5\ \text{in})}$$

$$= 318\ \text{in}^4$$

From AISC Table 3-20, a W21 × 62 has

$$I_x = 1330\ \text{in}^4 \gg 318\ \text{in}^4 \quad [\text{OK}]$$

From AISC Table 3-19,

$$\sum Q_n = 406\ \text{kips}$$

Check a.

$$a = \dfrac{\sum Q_n}{0.85 f'_c b} = \dfrac{406\ \text{kips}}{(0.85)(3\ \text{ksi})(8\ \text{ft})\left(12\ \dfrac{\text{in}}{\text{ft}}\right)}$$

$$= 1.66\ \text{in} > 1.0\ \text{assumed}\quad [\text{NG}]$$

Recalculate $Y2 = 4\ \text{in} - 1.66\ \text{in}/2 = 3.17\ \text{in}$.

For a $Y2 \approx 3.0$, and a PNA location of 5,

$$\phi M_n = 798\ \text{ft-kips} > 418\ \text{ft-kips}\quad [\text{OK}]$$

Check the live load deflection.

$$\Delta_{L} = \dfrac{L}{360} = \dfrac{(40\ \text{ft})\left(12\ \dfrac{\text{in}}{\text{ft}}\right)}{360} = 1.33\ \text{in}$$

A lower bound moment of inertia for composite beams is found in AISC Table 3-20.

For a W21 × 62, with $Y2 = 3.0$, and a PNA at location 7, $I_{LB} = 1990\ \text{in}^4$.

Then,

$$\Delta_{L} = \dfrac{5w_{L}L^4}{384EI_{LB}}$$

$$= \dfrac{(5)\left(0.96\ \dfrac{\text{kips}}{\text{ft}}\right)(40\ \text{ft})^4\left(12\ \dfrac{\text{in}}{\text{ft}}\right)^3}{(384)\left(29{,}000\ \dfrac{\text{kips}}{\text{in}^2}\right)(1990\ \text{in}^4)}$$

$$= 0.96\ \text{in} < 1.33\ \text{in}\quad [\text{OK}]$$

Determine if the beam has sufficient available shear strength.

$$V_u = \left(\dfrac{40\ \text{ft}}{2}\right)\left(\begin{array}{l}(1.2)\left(0.462\ \dfrac{\text{kip}}{\text{ft}}\right)\\[4pt] + (1.6)\left(0.96\ \dfrac{\text{kip}}{\text{ft}}\right)\end{array}\right)$$

$$= 41.8\ \text{kips}$$

$$V_n = 0.6F_y A_w C_v\quad [\text{AISC Eq. G2-1}]$$

$$= (0.6)\left(50\ \dfrac{\text{kips}}{\text{in}^2}\right)\big((0.40)(20\ \text{in})\big)(1) = 240\ \text{kips}$$

$$\phi V_n = (0.9)(240\ \text{kips}) = 216\ \text{kips} > 41.8\ \text{kips}$$

Since the available shear strength is greater than the required shear strength, it is OK.

Determine the required number of shear stud connectors.

From AISC Table 3-21, with 3 ksi concrete, conservatively assume a perpendicular deck, one $^3/_4$ in diameter weak stud per rib.

$$Q_n = 17.2\ \text{kips/stud}$$

Per AISC Sec. I3.2d(5),

$$\dfrac{\sum Q_n}{Q_n} = \dfrac{406\ \text{kips}}{17.2\ \dfrac{\text{kips}}{\text{stud}}} = 23.6$$

[24 on each side of the beam centerline]

The total number of studs per beam is 48.

ASD Method

Check for adequacy using the following information and determine the flexural strength.

$$w_a = 0.462\ \dfrac{\text{kips}}{\text{ft}} + 0.96\ \dfrac{\text{kips}}{\text{ft}} = 1.42\ \dfrac{\text{kips}}{\text{ft}}$$

$$M_a = \dfrac{w_a L^2}{8} = \dfrac{\left(1.42\ \dfrac{\text{kips}}{\text{ft}}\right)\left(40\ \dfrac{\text{kips}}{\text{ft}}\right)^2}{8}$$

$$= 284\ \text{ft-kips}$$

Use AISC Tables 3-19, 3-20, and 3-21.

Determine $b_{eff} = 96$ in and assuming $a = 1$,

$$Y2 = \left(t_{slab} - \frac{a}{2}\right) = \left(4 \text{ in} - \frac{1 \text{ in}}{2}\right) = 3.5 \text{ in}$$

With a PNA location 5 (i.e., at the bottom of the flange, BFL),

$$\frac{M_n}{\Omega_b} = 541 \text{ kips} \geq 284 \text{ kips}$$

Check the beam deflection and the available shear strength. The construction dead load is

$$w_{DL} = (8 \text{ ft})\left(\frac{4 \text{ in}}{12 \frac{\text{ft}}{\text{in}}}\right)\left(150 \frac{\text{lbf}}{\text{ft}^2}\right)\left(\frac{1 \text{ kip}}{1000 \text{ lbf}}\right)$$
$$= 0.4 \frac{\text{kips}}{\text{ft}}$$

The moment of inertia required to limit deflection during construction is

$$I_{req'd} = \frac{5 w_{DL} L^4}{384 E \Delta_{max}}$$
$$= \frac{(5)\left(0.4 \frac{\text{kips}}{\text{ft}}\right)(40 \text{ ft})^4 \left(12 \frac{\text{in}}{\text{ft}}\right)^3}{(384)\left(29{,}000 \frac{\text{kips}}{\text{in}^2}\right)(2.5 \text{ in})}$$
$$= 318 \text{ in}^4$$

From Table 3-20, a W21 × 62 has

$$I_x = 1330 \text{ in}^4 \gg 318 \text{ in}^4 \quad [\text{OK}]$$

From Table 3-19,

$$\sum Q_n = 406 \text{ kips}$$

Check a.

$$a = \frac{\sum Q_n}{0.85 f'_c b} = \frac{406 \text{ kips}}{(0.85)(3 \text{ ksi})(8 \text{ ft})\left(12 \frac{\text{in}}{\text{ft}}\right)}$$
$$= 1.66 \text{ in} > 1.0 \text{ assumed} \quad [\text{NG}]$$

Recalculate $Y2 = 4 \text{ in} - 1.66 \text{ in}/2 = 3.17 \text{ in}$.

For a $Y2 \approx 3.0$, and a PNA of 5,

$$\frac{M_n}{\Omega_b} = 531 \text{ ft-kips} > 284 \text{ ft-kips} \quad [\text{OK}]$$

Check the live load deflection.

$$\Delta_L = \frac{L}{360} = \frac{(40 \text{ ft})\left(12 \frac{\text{in}}{\text{ft}}\right)}{360} = 1.33 \text{ in}$$

A lower bound moment of inertia for composite beams is found in AISC Table 3-20.

For a W21 × 62, with $Y2 = 3.0$, and a PNA at location 7, $I_{LB} = 1990 \text{ in}^4$.

Then,

$$\Delta_L = \frac{5 w_L L^4}{384 E I_{LB}}$$
$$= \frac{(5)\left(0.96 \frac{\text{kips}}{\text{ft}}\right)(40 \text{ ft})^4 \left(12 \frac{\text{in}}{\text{ft}}\right)^3}{(384)\left(29{,}000 \frac{\text{kips}}{\text{in}^2}\right)(1990 \text{ in}^4)}$$
$$= 0.96 \text{ in} < 1.33 \text{ in} \quad [\text{OK}]$$

Determine if the beam has sufficient available shear strength.

$$V_a = \left(\frac{40 \text{ ft}}{2}\right)\left(1.422 \frac{\text{kips}}{\text{ft}}\right) = 28.4 \text{ kips}$$
$$\frac{V_n}{\Omega} = \frac{240 \text{ kips}}{1.67} = 143.7 \text{ kips}$$
$$143.7 \text{ kips} > 28.4 \text{ kips}$$

Since the available shear strength is greater than the required shear strength, it is OK.

Determine the required number of shear stud connectors.

From Table 3-21, with 3 ksi concrete, conservatively assume a perpendicular deck, one $^3/_4$ in diameter weak stud per rib.

$$Q_n = 17.2 \text{ kips/stud}$$

Per AISC, Sec. I3.2d(5),

$$\frac{\sum Q_n}{Q_n} = \frac{406 \text{ kips}}{17.2 \frac{\text{kips}}{\text{stud}}} = 23.6$$

[24 on each side of the beam centerline]

The total number of studs per beam is 48.

65 Structural Steel: Connectors

Nomenclature

A_b	nominal body area of a bolt	in^2
c	distance from neutral axis to extreme fiber	in
C	constant	–
d	bolt diameter	in
D_u	ratio of mean installed bolt pretension to specified minimum bolt pretension	–
e	eccentricity	in
f	computed stress	in
F	strength or allowable stress	kips/in^2
h	couple separation distance	in
H	horizontal force	kips
I	moment of inertia	in^4
h_{sc}	hole factor	–
M	moment	in-kips
N_s	number of slip planes	–
P	force transmitted by one bolt	kips
R_n	slip resistance	kips
t	thickness of the critical connected part	in
T	tensile force	kips
T_b	specified pretension	kips
V	shear	kips

Subscripts

b	bolt
n	nominal
t	tension
th	threaded
u	ultimate
v	shear

Symbols

μ	mean slip coefficient
ϕ	resistance coefficient
Ω	safety factor

1. INTRODUCTION

There are two types of structural *connectors* (also referred to as *fasteners*): rivets and bolts. *Rivets*, however, have become virtually obsolete because of their low strength, high cost of installation, installed variability, and other disadvantages. The two types of *bolts* generally used in the connections of steel structures are common bolts and high-strength bolts. *Common bolts* are classified by ASTM as A307 bolts. The two basic types of *high-strength bolts* have ASTM designations of A325 and A490.

A *concentric connection* is one for which the applied load passes through the centroid of the fastener group. If the load is not directed through the fastener group centroid, the connection is said to be an *eccentric connection*.

At low loading, the distribution of forces among the fasteners is very nonuniform, since friction carries some of the load. However, at higher stresses (i.e., those near yielding), the load is carried equally by all fasteners in the group. Stress concentration factors are not normally applied to connections with multiple redundancy.

In many connection designs, materials with different strengths will be used. The item manufactured from the material with the minimum strength is known as the *critical part*, and the critical part controls the design.

Connections using bolts and rivets are analyzed and designed similarly. Such connections can place fasteners in direct shear, torsional shear, tension, or any combination of shear and tension. In accordance with *AISC Specification* Sec. B3.1, theoretical design by elastic, inelastic, or plastic analysis is permitted.

The analysis presented in this chapter for connection design and analysis assumes static loading. Provisions for fatigue loading are contained in the *AISC Specification* App. 3.

2. HOLE SPACING AND EDGE DISTANCES

The minimum distance between centers of standard, oversized, or slotted holes is $2^2/_3$ times the nominal fastener diameter; a distance of $3d$ is preferred [*AISC Specification* Sec. J3.3]. The longitudinal spacing between

Figure 65.1 *Typical Bolted Connections*

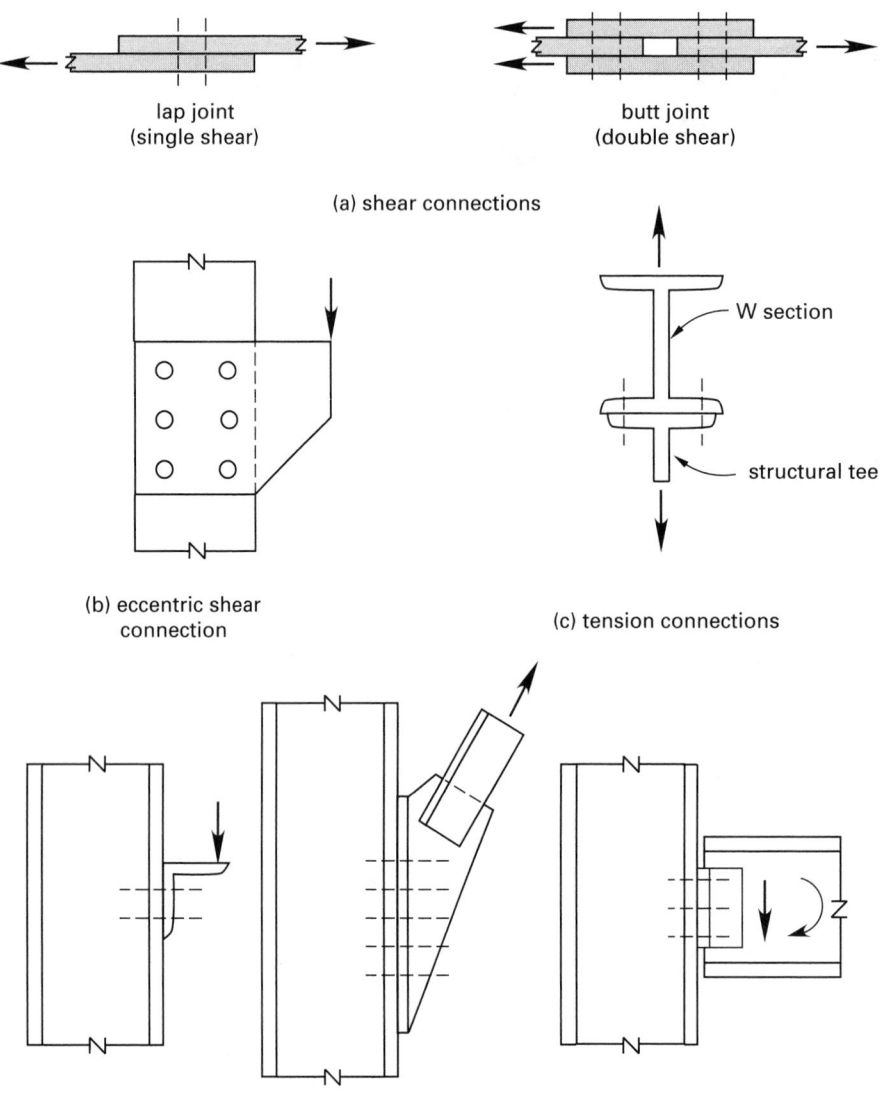

(a) shear connections

(b) eccentric shear connection

(c) tension connections

(d) combined shear and tension connections

bolts along the line of action of the force is specified in Section J3.5 of the *AISC Specification* and is a function of painted surfaces and material.

The minimum distance from the hole center to the edge of a member is approximately 1.75 times the nominal diameter for sheared edges, and approximately 1.25 times the nominal diameter for rolled or gas-cut edges, both rounded to the nearest $1/8$ in [*AISC Specification* J3-4]. (For exact values, refer to Table J3.4 in the *AISC Specification.*)

For parts in contact, the maximum edge distance from the center of a fastener to the edge in contact is 12 times the plate thickness or 6 in, whichever is less [*AISC Specification* J3.5].

3. BEARING AND SLIP-CRITICAL CONNECTIONS

A distinction is made between bearing and slip-critical connections. A *bearing connection* relies on the shearing resistance of the fasteners to resist loading. In effect, it is assumed that the fasteners are loose enough to allow the plates to slide slightly, bringing the fastener shanks into contact with the holes. The area surrounding the hole goes into bearing, hence the name. Connections using rivets, welded studs, and A307 bolts are always considered to be bearing connections. However, high-strength bolts can also be used in bearing connections.

If the fasteners are constructed from high-strength steel, a high preload can be placed on the bolts. This preload will clamp the plates together, and friction alone will keep the plates from sliding. Such connections are known as *slip-critical connections*. The fastener shanks never come into contact with the plate holes. Bolts constructed from A325 and A490 steels are suitable for slip-critical connections.

4. AVAILABLE LOADS FOR FASTENERS

The *AISC Manual* contains tables of available loads for common connector types and materials. The design strength of a fastener can also be determined by multiplying its area by a nominal stress. Except for rods with upset ends, the area to be used is calculated simply from the fastener's nominal (unthreaded and undriven) dimension.

Table 65.1 lists nominal stresses for common connector types for tension and shear. (Available bearing stress is covered in the following section.) Reductions are required for use with oversized holes and connector patterns longer than 50 in. Available stresses in fasteners can be increased by one third for temporary exposure to wind and seismic loading.

Table 65.1 Nominal Fastener Stresses for Static Loading

type of connector	F_{nt} (ksi)	F_{nv} (ksi) bearing
A307 common bolts	45	24
A325 high-strength bolts		
no threads in shear plane	90	60
threads in shear plane	90	48
A490 high-strength bolts		
no threads in shear plane	113	75
threads in shear plane	113	60

Source: Based on *AISC Specification* Table J3.2.

For fasteners made from approved steels, including A449, A572, and A588 alloys, the nominal tensile stress is $F_{nt} = 0.75F_u$, whether or not threads are in the shear plane. For bearing connections using the same approved steels, the nominal shear stress is $F_{nv} = 0.40F_u$ if threads are present in the shear plane and $F_{nv} = 0.50F_u$ if threads are excluded from the shear plane [*AISC Specification* Table J3.2].

If fasteners are exposed to both tension and shear, *AISC Specification* J3.7 for bearing-type and J3.9 for slip-critical connections should be used to determine the maximum tensile stress.

For A325 and A490 bolts used in slip-critical connections with tension, the available shear stresses given in Table 65.1 must be determined in accordance with *AISC Specification* Sec. J3.8]

The design slip resistance, ϕR_n, and the allowable slip resistance, R_n/Ω, are determined from Eq. 65.1 [AISC Eq. J3-4].

$$R_n = \mu D_u h_{sc} T_b N_s \qquad \text{65.1}$$

The LRFD serviceability limit state is $\phi = 1.00$, and for a required strength level it is 0.85. The ASD serviceability limit state is 1.50, and for a required strength level it is 1.76. The *mean slip coefficient*, μ, is 0.35 for a C1.A surface (class A surfaces include unpainted, clean mill scale surfaces and wire brushed, sand blasted, phosphated, zinc silicate painted, and roughened and galvanized surfaces) and 0.50 for a C1.B surface (class B surfaces include unpainted blast-cleaned steel surfaces and blast-cleaned surfaces with class B coatings). The ratio of mean installed bolt pretension to a specified minimum bolt pretension, D_u, is 1.13. The *hole factor*, h_{sc}, is 1.00 for standard size holes, 0.85 for oversized and short-slotted holes, and 0.70 for long slotted holes. N_s is the number of slip planes, and the minimum fastener tension, T_b, is per Table 65.2.

Table 65.2 Minimum Bolt Pretension (kips)[a]

bolt size (in)	A325 bolts	A490 bolts
$\frac{1}{2}$	12	15
$\frac{5}{8}$	19	24
$\frac{3}{4}$	28	35
$\frac{7}{8}$	39	49
1	51	64
$1\frac{1}{8}$	56	80
$1\frac{1}{4}$	71	102
$1\frac{3}{8}$	85	121
$1\frac{1}{2}$	103	148

[a] A325 and A490 bolts are required to be tightened to 70% of their minimum tensile strength.

Source: Based on *AISC Specification* Table J3.1.

Table 65.2 gives the minimum pretension for specific bolt sizes and types. For bolts of other sizes, or for bolts manufactured from other steels, the minimum pretension is given by Eq. 65.2, where $A_{b,th}$ is the tensile stress area of the bolt.

$$T_{b,\min} = 0.70A_{b,th}F_u \quad \text{[rounded]} \qquad \text{65.2}$$

5. AVAILABLE BEARING STRENGTH

Bearing strength should be evaluated in bearing connections. Theoretically, bearing should not be a problem in friction-type connections because the bolts never bear on the pieces assembled. However, in the event there is slippage due to insufficient tension in the connectors, bearing should be checked anyway.

The available bearing strength on projected areas of connectors in shear connections is given by Eq. 65.3.

[*AISC Specification* Eq. J3-1]. F_n is the nominal tensile stress F_{nt}, or shear stress, F_{nv}, from AISC Table J3.2.

$$R_n = F_n A_b \qquad \text{65.3}$$

If the connectors, plates, or shapes have different strengths, then the bearing capacity of each component must be checked.

6. CONCENTRIC SHEAR CONNECTIONS

In a concentric shear connection, fasteners are subject to direct shear only. The number of fasteners required in the connection is determined by considering the fasteners in shear, the plate in bearing, and the effective net area of the plate in tension.

Example 65.1

Two $^1/_4 \times 8$ A36 steel plates are joined with a lap joint using $^3/_4$ in A325-N bolts. Design a bearing connection with threads in the shear plane to carry a concentric live load of 25 kips.

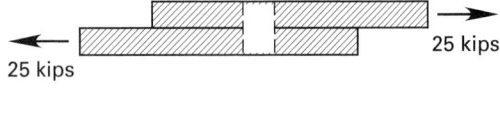

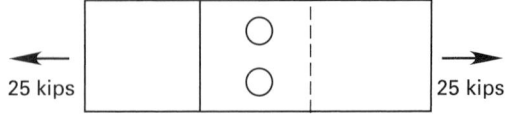

Solution

The area of the $^3/_4$ in bolt is

$$A_v = \frac{\pi d^2}{4} = \frac{\pi (0.75 \text{ in})^2}{4} = 0.442 \text{ in}^2$$

LRFD Method

Find the required strength.

$$\text{tension} = (1.6)(25 \text{ kips}) = 40 \text{ kips}$$
$$\text{shear} = (1.6)(25 \text{ kips}) = 40 \text{ kips}$$

Calculate f_v.

$$f_v = \frac{V_u}{A_v} = \frac{40 \text{ kips}}{0.442 \text{ in}^2} = 90.5 \text{ ksi} \quad [\text{required}]$$

$$\phi F_{nv} = (0.75)\left(48 \frac{\text{kips}}{\text{in}^2}\right) = 36 \text{ ksi} \quad [\text{available}]$$

The number of required bolts is

$$\frac{90.5 \frac{\text{kips}}{\text{in}^2}}{36 \frac{\text{kips}}{\text{in}^2}} = 2.5$$

Use four rivets for the symmetry of the pattern.

Check for combined tension and shear using AISC Eq. (J3-3a), where $\phi = 0.75$.

$$F'_{nt} = 1.3 F_{nt} - \left(\frac{F_{nt}}{\phi F_{nv}}\right) f_v \leq F_{nt}$$

$$= (1.3)\left(90 \frac{\text{kips}}{\text{in}^2}\right)$$

$$- \left(\frac{90 \frac{\text{kips}}{\text{in}^2}}{(0.75)\left(48 \frac{\text{kips}}{\text{in}^2}\right)}\right)\left(\frac{90.5 \frac{\text{kips}}{\text{in}^2}}{4 \text{ bolts}}\right)$$

$$= 60.4 \frac{\text{kips}}{\text{in}^2} < 90 \frac{\text{kips}}{\text{in}^2}$$

$$R_n = F'_{nt} A_b = \left(60.4 \frac{\text{kips}}{\text{in}^2}\right)(0.442 \text{ in}^2)$$

$$= 26.7 \text{ kips} \quad [\text{AISC Eq. J3-1}]$$

The design tensile strength is ϕR_n.

$$\phi R_n = (0.75)(26.7 \text{ kips}) = 20 \text{ kips}$$

Multiply the design tensile strength by the number of required bolts.

$$\left(20 \frac{\text{kips}}{\text{bolt}}\right)(4 \text{ bolts}) = 80 \text{ kips} > 40 \text{ kips}$$

80 kips is greater than the required tension strength, so it is OK.

ASD Method

Find the required strength.

$$\text{tension} = 25 \text{ kips}$$
$$\text{shear} = 25 \text{ kips}$$

Calculate f_v.

$$f_v = \frac{V}{A_v} = \frac{25 \text{ kips}}{0.442 \text{ in}^2} = 56.6 \text{ ksi} \quad [\text{required}]$$

$$\frac{F_{nv}}{\Omega} = \frac{48}{2} = 24 \text{ ksi} \quad [\text{available}]$$

The number of required bolts is

$$\frac{56.6 \frac{\text{kips}}{\text{in}^2}}{24 \frac{\text{kips}}{\text{in}^2}} = 2.4$$

Use four bolts for the symmetry of the pattern.

Check for combined tension and shear using AISC Eq. (J3-3a), where $\Omega = 2.00$.

$$F'_{nt} = 1.3F_{nt} - \left(\frac{\Omega F_{nt}}{F_{nv}}\right)f_v \le F_{nt}$$

$$= (1.3)\left(90\ \frac{\text{kips}}{\text{in}^2}\right)$$

$$-\left(\frac{(2.00)\left(90\ \frac{\text{kips}}{\text{in}^2}\right)}{48}\right)\left(\frac{56.6\ \frac{\text{kips}}{\text{in}^2}}{4\ \text{bolts}}\right)$$

$$= 63.9\ \frac{\text{kips}}{\text{in}^2} < 90\ \frac{\text{kips}}{\text{in}^2}$$

From Eq. 65.3,

$$R_n = F'_{nt}A_b = \left(63.9\ \frac{\text{kips}}{\text{in}^2}\right)(0.442\ \text{in}^2)$$

$$= 28.2\ \text{kips}\quad[\text{AISC Eq. J3-1}]$$

The allowable tensile strength is

$$\frac{R_n}{\Omega} = \frac{28.2\ \text{kips}}{2.00} = 14.1\ \text{kips}$$

Multiply the allowable tensile strength by the number of required bolts.

$$\left(14.1\ \frac{\text{kips}}{\text{bolt}}\right)(4\ \text{bolts}) = 56.4\ \text{kips} > 25\ \text{kips}$$

56.4 kips is greater than the required tension strength, so it is OK.

To determine bolt-hole locations, AISC J3.3 specifies that the minimum distance between holes shall not be less than $2\frac{2}{3}$ times the nominal diameter, but a distance of $3d$ is preferred.

Assuming sheared edges for a $^3/_4$ bolt diameter, AISC Table J3.4 specifies the minimum edge spacing to be $1\frac{1}{4}$ in.

Based on these requirements, a trial layout is made.

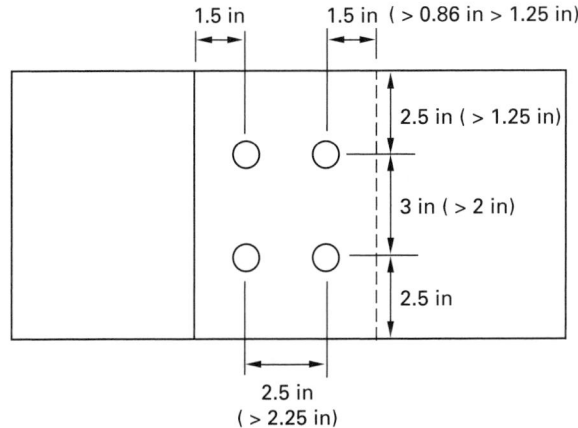

7. ECCENTRIC SHEAR CONNECTIONS

An *eccentric shear connection* is illustrated in Fig. 65.2. The connection's tendency to rotate around the centroid of the fastener group is the torsion (or torque) on the connection, equal to the product Pe. This torsion is resisted by the shear stress in the fasteners. Friction is not assumed to contribute to the rotational resistance of the connection.

Basically, only one analytic method is available: the Uniform Force Method in *AISC Manual*, Part 13. The traditional elastic approach has been covered in Ch. 45. This method, although providing a simplified solution, generally underestimates the capacity of an eccentric shear connection. In some cases, the actual capacity may be as much as twice the capacity calculated from the elastic model.

Figure 65.2 *Eccentric Shear Connection*

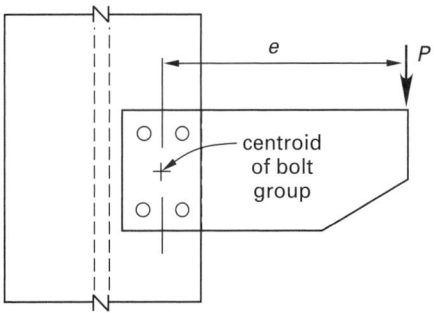

8. ULTIMATE STRENGTH OF ECCENTRIC SHEAR CONNECTIONS

Design and analysis using one of several available ultimate strength procedures is preferred over the use of elastic methods. In practice, these procedures involve more table look-up than theory. The design strength of a fastener group is calculated from the product of tabulated coefficients found in Tables 7-7 through 7-14 in Part 7 of the *AISC Manual*. For example, in Eq. 65.4, R_n is the least nominal strength of one bolt determined from the limit states of bolt shear strength, bearing strength, and slip resistance. Coefficient C is found from Tables 7-7 through 7-14, and is a function of bolt geometry and eccentricity.

$$P = CR_n \qquad 65.4$$

9. TENSION CONNECTIONS

Figure 65.3 illustrates a basic *tension connection*. As the name implies, fasteners in this hanger-type connection are subjected to tensile stresses. The nominal strength of bolts in tension is equal to the bolt area, A_b, multiplied by the nominal tension stress, F_{nt}. Nominal tension stresses in bolts are listed in Table J3.2 in the *AISC Specification*.

$$P = A_bF_{nt} \qquad 65.5$$

Figure 65.3 Fasteners in Tension Connection

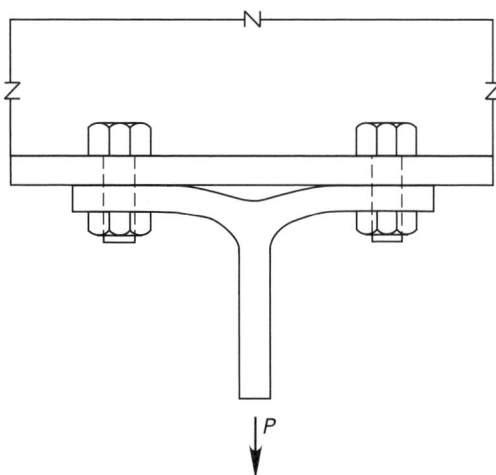

10. COMBINED SHEAR AND TENSION CONNECTIONS

The tendency of the flanges to act as cantilever beams when a load is applied to them is called *prying action*. This must be considered in the design of tension connections. The effect of prying action is to increase the tension in the bolts.

Consider the simple framing connection shown in Fig. 65.4. If the vertical shear is assumed to act along line A, the bolts in the column connection will be put into tension by prying action. The most highly stressed (in tension) will be the topmost fasteners.

Figure 65.4 Combined Shear and Tension Connection

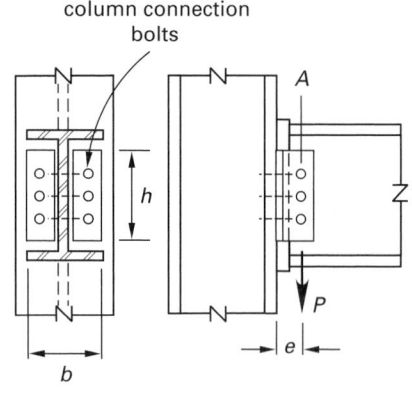

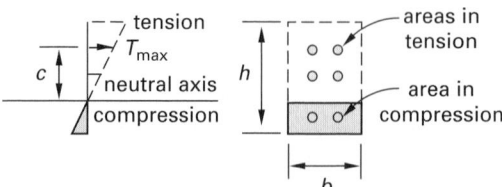

The moment on the connection is easily calculated, and the uppermost line of connectors is most likely to experience the greatest effect from prying. However, the performance of bolted connections with prying action is too complex to be evaluated without knowing certain factors and assuming others. The type of connectors, the type of connection (friction or bearing), and the amount of pretension all affect the methodology used. In friction-type connection with pretensioned bolts, the prying will separate the pieces, elongating the bolts and reducing the stress in the bolts. Another complexity is the yielding that occurs to reduce the applied moment.

Part 7 of the *AISC Manual* outlines two methods for designing connections with combined shear and tension depending on assumptions regarding the neutral axis of the bolt group. The more direct and more conservative method is to assume the neutral axis at the center of gravity of the bolt group. The bolts in the tension region are subject to tension, shear, and prying action if any exists. Bolts in the compressive region are subject to shear alone. The required shear strength per bolt is merely the required shear force divided by the number of bolts. The required tension strength is the eccentric moment divided by the number of bots in the tension region and the moment arm. Prying action must also be considered.

11. FRAMING CONNECTIONS

All connections are assumed to have some rigidity and are divided into two categories—FR, or rigid connections (previously referred to as Type 1, rigid connections), and PR, or semi-rigid connections (previously referred to as Type 3, semi-rigid connections). *Rigid connections* are assumed to be sufficiently rigid to maintain the angle of rotation between two connected parts like a beam and a column. *Semi-rigid connections* are assumed to lack the rigidity to maintain the rotation between two connect parts.

A simple connection (previously referred to as a Type 2, simple framing connection) is a PR connection without any rigidity. It is generally assumed that a FR connection will be able to develop at least 90% of a moment developed by a perfectly rigid connection. Other PR connections fall somewhere between the simple connection and the FR connection.

12. SIMPLE FRAMING CONNECTIONS

Simple framing connections are designed to be as flexible as possible. Design is predetermined by use of the standard tables of framed beam connections in the *AISC Manual*. Construction methods include use of beam seats and clips to beam webs.

Connections to beams and columns can be by bolting and/or welding, and such fastening methods can be used on either the beams or the columns. Welded connections, stiffened connections, and the use of top seats do not necessarily imply a moment-resisting connection.

Direct shear determines the number of bolts required in the column connection. It is common to neglect the effect of eccentricity in determining shear stresses in riveted and bolted beam connections to webs. It is also common to neglect the effect of eccentricity in determining shear stresses in riveted and bolted column connections. However, the eccentricity can be considered, which will add tension stress to the shear stress on column fasteners.

Angle thickness must be checked for allowable bearing stress. Angles should be checked for direct shear, as well. Since the connection is designed to rotate, the angle (for seated beams) should be checked for bending stress as its free lip bends.

Figure 65.5 Bolted Simple Framing Connections

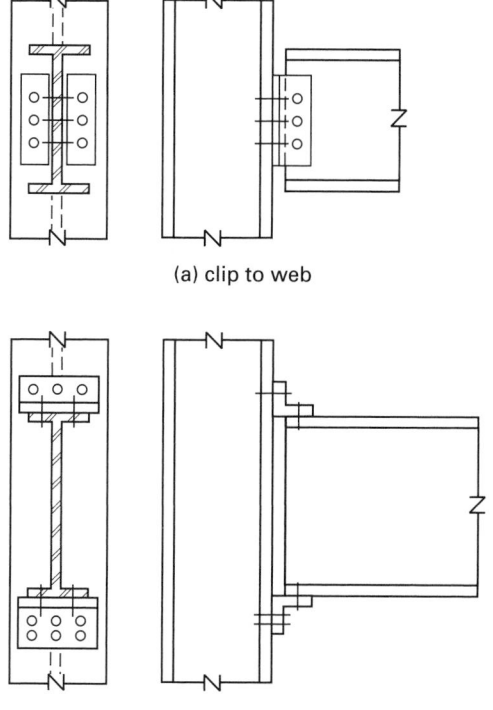

(a) clip to web

(b) seated beam

13. MOMENT-RESISTING FRAMING CONNECTIONS

Moment-resisting connections transmit their vertical (shearing) load through the same types of connections as simple connections, typically through connections at the beam web. However, the flanges of the beam are also rigidly connected to the column. These top and bottom flange connections are in tension and compression respectively, and serve to transmit the moment.

Figure 65.6 Bolted Moment-Resisting Framing Connections

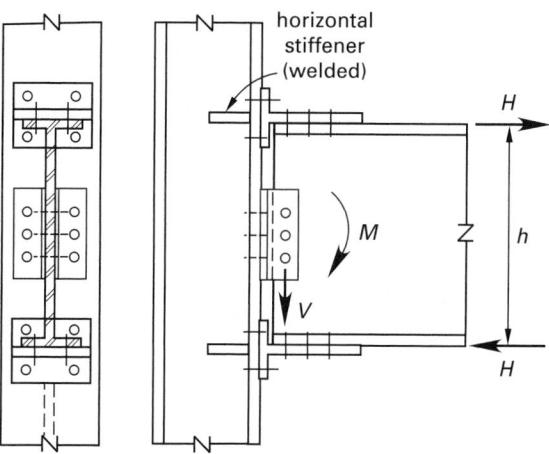

Since moment transfer is through the flanges by tension and compression connections, the design of such connections involves ensuring adequate strength in tension and compression. Design of the shear transfer mechanism (i.e., the web connections) is essentially the same as for simple connections. Also, in order to prevent localized failure of the column flanges, horizontal stiffeners (between the column flanges) may be needed.

The moment-resisting ability of a simple connection can be increased to essentially any desired value by increasing the distance between the tension and compression connections. Figure 65.7 illustrates how this could be accomplished by using an intermediate plate between the beam flanges and the column. The horizontal tensile and compressive forces, H, can be calculated from Eq. 65.6.

$$H = \pm \frac{M}{h} \qquad 65.6$$

Figure 65.7 Bolted Simple Connection with Increased Moment Resistance

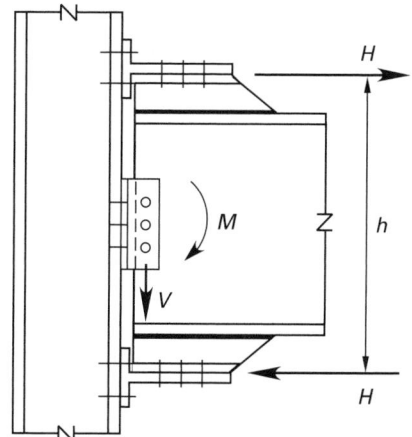

Special attention to moment-resisting connections is critical (and required) in regions of seismicity. Designs must specifically compensate for brittle fracture. (See Fig. 65.8.) This is a specialized area of structural engineering requiring knowledge of seismic design concepts.

Structural

Figure 65.8 *Welded Moment-Resisting Connections[a]*

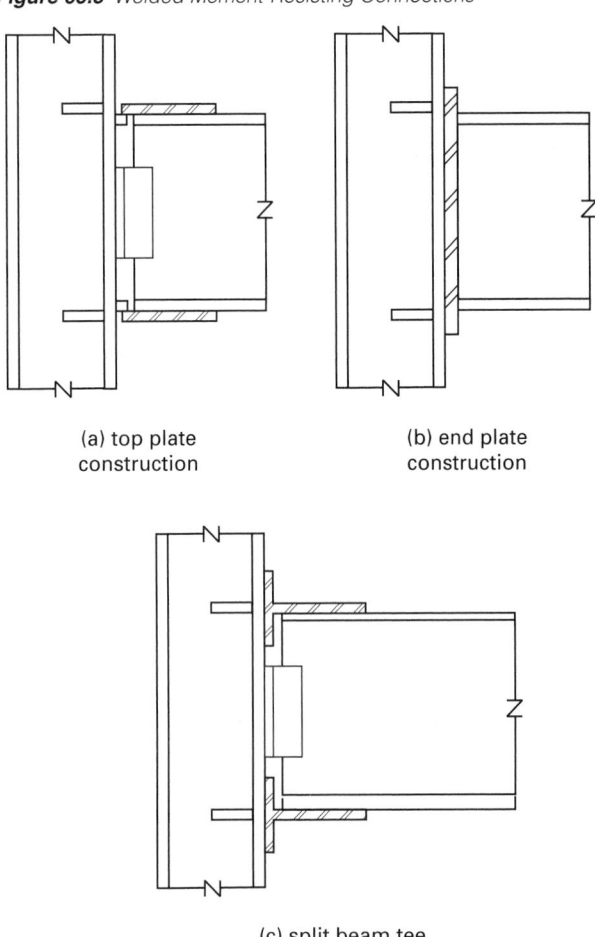

(a) top plate
construction

(b) end plate
construction

(c) split beam tee
construction

[a]Not intended to transmit seismically induced loading without special detailing.

66 Structural Steel: Welding

Nomenclature

A	area	in^2
c	radial distance	in
d	distance	in
D	weld size in multiples of $^1/_{16}$ in	–
e	eccentricity	in
f	computed stress	kips/in^2
F	strength or allowable stress	kips/in^2
I	moment of inertia	in^4
I_p	polar moment of inertia	in^4
l	weld length	in
L	length	in
M	moment	in-kips
P	force	kips
r_a	resultant shear stress	kips/in^2
r_m	torsional shear stress	kips/in^2
r_p	direct shear stress	kips/in^2
R_w	weld strength (resistance)	kips/in
S	section modulus	in^3
t	thickness	in
T	tensile force	kips
w	weld size	in
y	distance	in

Subscripts

a	required
e	effective
n	nominal
t	total
u	ultimate
v	shear
w	weld
y	yield

Symbols

ϕ	resistance coefficient
Ω	safety factor

1. INTRODUCTION

Welding is a joining process in which the surfaces of two pieces of structural steel (the *base metal*) are heated to a fluid state, usually with the addition of other molten steel (called *weld metal* or *filler metal*) from the *electrodes*. The properties of the base metal and those of the deposited weld metal should be similar, but the weld metal should be stronger than the metals it connects.

The American Welding Society *Structural Welding Code—Steel* (AWS D1.1) is the governing standard for structural welding and it is incorporated by reference, with a few differences, into the *AISC Specification*, Part 8.

Two procedures are typically used for structural welding. The *shielded metal arc welding* (SMAW) process is used both in the shop and in the field. Electrodes available for SMAW welding are designated as E60XX, E70XX, E80XX, E90XX, E100XX, and E110XX. The letter "E" denotes an electrode. The first two digits indicate the electrode's ultimate tensile strength (F_u) in ksi; thus, the tensile strength of the weld metal ranges from 60 ksi to 110 ksi. The XX designations indicate how the electrodes are used.

The *submerged arc welding* (SAW) process is primarily used in the shop. The electrodes for the SAW process are specified somewhat differently. Each designation is a combination of flux (that shields the weld) and electrode classifications. For example, in the designation F7X-E7XX, F indicates a granular flux, the first digit following F represents the tensile strength requirement of the resulting weld (e.g., 7 means 70 ksi), E represents an electrode, the first digit following the E represents the minimum tensile strength of the weld metal (e.g., 7 means 70 ksi), and the XX represents numbers related to the use. The most commonly used electrode for structural work is E70 because it is compatible with all grades of steel having yield stresses (F_y) up to 60 ksi.

2. TYPES OF WELDS AND JOINTS

Figure 66.1 shows four common types of welds: fillet, groove, plug, and slot. Each type of weld has certain advantages that determine the extent of its use. The *fillet weld* is the most commonly used type in structural connections due to its overall economy and its ease of fabrication. The *groove weld* is primarily used to connect structural members aligned in the same plane. It is more costly as it often requires extensive edge preparation and precise fabrication. The use of *slot welds* and *plug welds* is rather limited—they are used principally in combination with fillet welds, where the size of the connection limits the length available for fillet welds.

Figure 66.1 *Types of Welds*

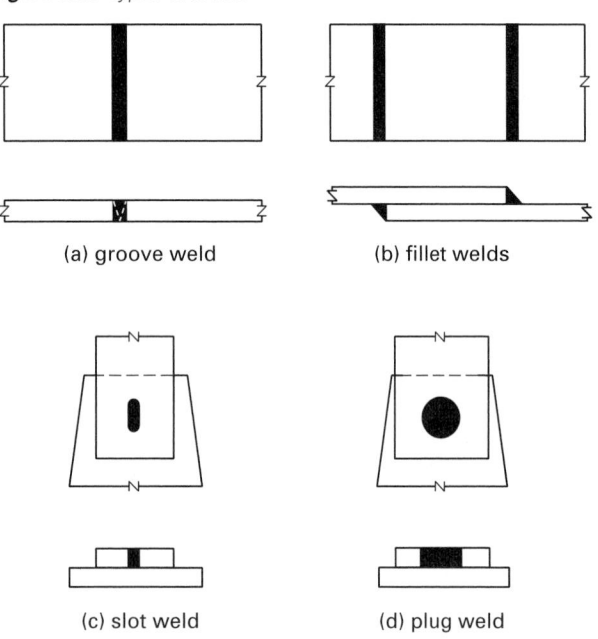

(a) groove weld   (b) fillet welds

(c) slot weld   (d) plug weld

There are five basic types of joints: butt, lap, tee, corner, and edge, as shown in Fig. 66.2. The type of joint depends on factors such as type of loading, shape and size of members connected, joint area available, and relative costs of various types of welds. The *lap joint* is the most common type because of ease of fitting (i.e., there is no need for great precision in fabrication) and ease of joining (i.e., there is no need for special preparation of edges being joined).

Figure 66.2 *Types of Welded Joints*

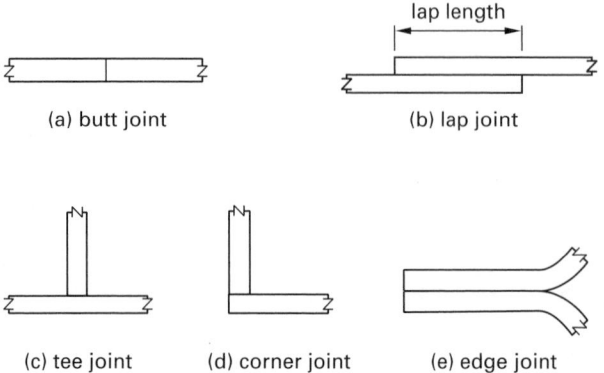

(a) butt joint   (b) lap joint

(c) tee joint   (d) corner joint   (e) edge joint

Figure 66.3 *Standard Weld Symbols*[a]

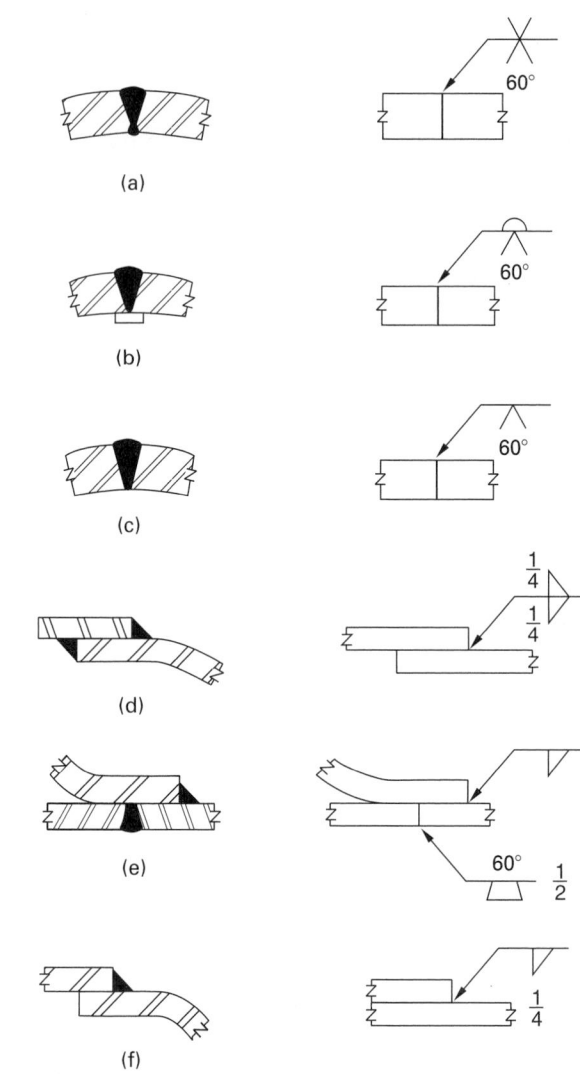

(a)

(b)

(c)

(d)

(e)

(f)

[a]Designations: (a) double-weld butt joint, (b) single-weld butt joint with integral backing strip, (c) single-weld butt joint without backing strip, (d) double-full fillet lap joint, (e) single-full fillet lap joint with plug welds, (f) single-full fillet lap joint without plug welds.

3. FILLET WELDS

An enlarged view of a fillet weld is shown in Fig. 66.4. The applied load is assumed to be carried by the weld throat, which has an effective dimension of t_e.

Figure 66.4 *Fillet Weld*

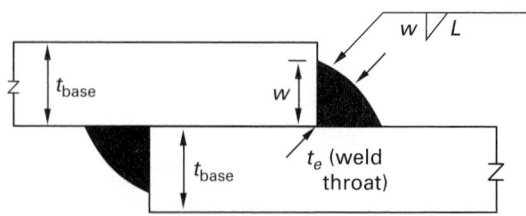

The effective weld throat thickness, t_e, depends on the type of welding used. For hand-held, shielded metal arc welding (SMAW) processes,

$$t_e = 0.707w \qquad 66.1$$

The effective throat of a fillet weld may be larger than that given by Eq. 66.1 if consistent penetration beyond the root of the weld is provided. Appropriate tests are required to justify an increase in the effective throat [*AISC Specification* Sec. J2.2a].

Weld sizes, w, of $3/16$ in, $1/4$ in, and $5/16$ in are desirable because they can be made in a single pass. ($5/16$ in is an appropriate limit for SMAW, especially when using hand-held rods. If SAW is used, up to $1/2$ in welds can be made in one pass.) However, fillet welds from $3/16$ in to $1/2$ in can be made in $1/16$ in increments. For welds larger than $1/2$ in, every $1/8$ in weld size can be made.

In almost all instances, loads carried by welds result in shear stresses in the weld material, regardless of the weld group orientation. The available shear strength on the weld throat is [AISC Table J2.5]

$$F_v = 0.60 F_{u,\text{rod}} \qquad 66.2$$

Shear and tensile strengths of the weld may not be greater than the nominal strength per unit area of the base metal times the cross-sectional area of the base metal, or the nominal strength of the weld metal per unit area times the area of the weld [*AISC Specification* Sec. J2.4]. For tensile or compressive loads, the allowable strength parallel to the weld axis is the same as for the base metal.

To simplify the analysis and design of certain types of welded connections, it is convenient to define the shear resistance per unit length of weld as Eq. 66.3. (When the base material thickness is small, the weld strength may be limited by the shear strength of the base metal. This should also be checked.)

$$R_w = 0.30 t_e F_{u,\text{rod}} \qquad 66.3$$

Several special restrictions apply to fillet welds [*AISC Specification* Sec. J2.2b].

- Minimum weld size depends on the thickness of the thicker of the two parts joined, except that the weld size need not exceed the thickness of the thinner part. (When weld size is increased to satisfy these minimums, the capacity is not increased. Weld strength is based on the theoretical weld size calculated.)

- For materials less than $1/4$ in thick, the maximum weld size along edges of connecting parts is equal to the material thickness. For materials thicker than $1/4$ in, the maximum size must be $1/16$ in less than the material thickness.

- The minimum-length weld for full strength analysis is four times the weld size. (If this criterion is not met, the weld size is reduced to one fourth of the weld length.)

- If the strength required of a welded connection is less than would be obtained from a full-length weld of the smallest size, then an intermittent weld can be used. The minimum length of an intermittent weld is four times the weld size or $1\frac{1}{2}$ in, whichever is greater.

- The minimum amount of lap length for lap joints (as illustrated in Fig. 66.2) is five times the thinner plate's thickness, but not less than 1 in.

Table 66.1 *Minimum Fillet Weld Size*

thickness of thinnest part joined (in)	minimum w (in)
to $\frac{1}{4}$ inclusive	$\frac{1}{8}$
over $\frac{1}{4}$ to $\frac{1}{2}$ inclusive	$\frac{3}{16}$
over $\frac{1}{2}$ to $\frac{3}{4}$ inclusive	$\frac{1}{4}$
over $\frac{3}{4}$	$\frac{5}{16}$

From *AISC Specification* Table J2.4.

4. CONCENTRIC TENSION CONNECTIONS

If the weld group centroid is in line with the applied load, the loading is concentric. Equation 66.4 can be used to design or evaluate the connection.

$$f_v = \frac{P}{A_{\text{weld}}} = \frac{P}{l_{\text{weld}} t_e} \qquad 66.4$$

Example 66.1

Two $1/2 \times 8$ A36 steel plates are lap-welded using E70 electrodes with a shielded metal arc process. Size the weld to carry a concentric tensile loading of 50 kips.

Solution

Good design will weld both joining ends to the base plate. The total weld length will be

$$l_{\text{weld}} = (2)(8 \text{ in}) = 16 \text{ in}$$

This length meets the minimum-length specification of five times the thinner plate's thickness. From Table 66.1, the minimum weld thickness for a $1/2$ in plate is $3/16$ in. Try $w = 3/16$ in.

LRFD Method

The required strength is

$$R_n = P_u = (1.6)(50 \text{ kips}) = 80 \text{ kips}$$

The required strength per inch is

$$\frac{80 \text{ kips}}{16 \text{ in}} = 5 \frac{\text{kips}}{\text{in}}$$

The available strength as given in AISC Ch. 8, where D is the weld size in sixteenths of an inch, and l is the length, is

$$\begin{aligned}
\frac{\phi R_n}{l} &= 1.392D \\
&= (1.392)(3) \\
&= 4.17 \text{ kips/in} \quad [\text{AISC p. 8-8}] \\
&< 5 \text{ kips/in} \quad [\text{no good}]
\end{aligned}$$

Since the available strength is less than the required strength, try a $1/4$ in weld.

$$\begin{aligned}
\frac{\phi R_n}{l} &= 1.392D \\
&= (1.392)(4) \\
&= 5.568 \text{ kips/in} \quad [\text{AISC p. 8-8}] \\
&> 5 \text{ kips/in} \quad [\text{OK}]
\end{aligned}$$

Since $\phi R_n > R_n = P_u$, use a $1/4$ in weld.

ASD Method

The required strength is

$$P_a = 50 \text{ kips}$$

The required strength per inch is

$$\frac{50 \text{ kips}}{16 \text{ in}} = 3.125 \text{ kips/in}$$

The available strength as given in the *AISC Manual*, Part 8, where D is the weld size in sixteenths of an inch, and l is the length, is

$$\begin{aligned}
\frac{R_n}{\Omega l} &= 0.928D \\
&= (0.928)(3) \\
&= 2.78 \text{ kips/in} \quad [\text{AISC p. 8-8}] \\
&< 3.125 \text{ kips/in} \quad [\text{NG}]
\end{aligned}$$

Since the available strength is less than the required strength, try a $1/4$ in weld.

$$\begin{aligned}
\frac{R_n}{\Omega l} &= 0.928D \\
&= (0.928)(4) \\
&= 3.712 \text{ kips/in} \quad [\text{AISC p. 8-8}] \\
&> 3.125 \text{ kips/in} \quad [\text{OK}]
\end{aligned}$$

Since $R_n/\Omega > P_a$, use a $1/4$ in weld.

The maximum weld size allowed is $1/2$ in $- 1/16$ in $= 7/16$ in. The $1/4$ in weld can be used.

5. BALANCED TENSION WELDS

When tension is applied along an unsymmetrical member, the tensile force will act along a line passing through the centroid of the member. In that instance, it may be desirable or necessary to design a *balanced weld group* in order to have the force also pass through the centroid of the weld group. (*AISC Specification* Sec. J1.7 exempts single and double angles from the need to balance welds when loading is static. All other static loading configurations, and angles subjected to fatigue loading, must have balanced welds.)

Figure 66.5 Balanced Weld Group

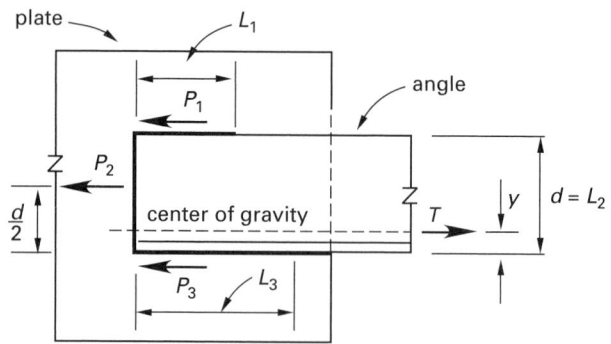

The procedure for designing the unequal weld lengths assumes that the end weld (if present) acts at full shear stress with a resultant passing through its mid-height (e.g., $d/2$ in Fig. 66.5). The forces in the other welds are assumed to act along the edges of the angle. Moments are taken about a point on the line of action of either longitudinal weld, resulting in Eq. 66.5.

$$P_3 = T\left(1 - \frac{y}{d}\right) - \frac{P_2}{2} \qquad 66.5$$

$$P_2 = R_w L_2 = R_w d \qquad 66.6$$

$$P_1 = T - P_2 - P_3 \qquad 66.7$$

$$L_1 = \frac{P_1}{R_w} \qquad 66.8$$

$$L_3 = \frac{P_3}{R_w} \qquad 66.9$$

6. ECCENTRICALLY LOADED WELDED CONNECTIONS

There are two possible loading conditions that will result in an eccentrically loaded weld group: (a) loading in the plane of the weld group (as shown in Fig. 66.6) causing combined shear and torsion, and (b) loading out of the plane of the weld group causing combined shear and bending (as shown in Fig. 66.7).

The complexity of the stress distribution in shear/torsion connections makes accurate analysis and design based on pure mechanics principles impossible. Therefore, simplifications are made and the assumed shear stress in the most critical location is calculated. This type of problem can be dealt with in two ways, as described in the *AISC Manual*, Part 8: the elastic method and the instantaneous center of rotation method.

Figure 66.6 *Welded Connection in Combined Shear and Torsion*

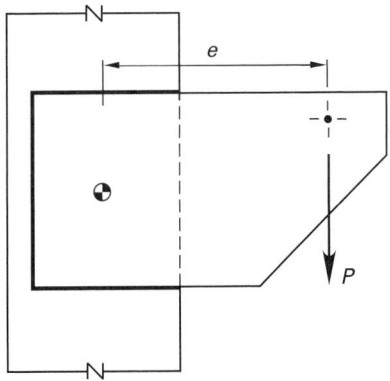

7. ELASTIC METHOD FOR ECCENTRIC SHEAR/TORSION CONNECTIONS

The elastic method (as discussed in the *AISC Manual*, Part 8) starts by finding the centroid of the weld group. The centroid can be located by weighting the weld areas (calculated from the weld throat size and length) by their distances from an assumed axis, or the welds can be treated as lines and their lengths used in place of their areas. Appendix 45.A is useful in this latter case. (In an analysis of a weld group, the throat size will be known and welds can be treated as areas. In design, the throat size will be unknown, and it is common to treat the welds as lines. Alternatively, a variable weld size can be used and carried along in a design problem. However, the results will be essentially identical.)

Once the centroid is located, the polar area moment of inertia I_p is calculated. AISC Fig. 8-6, Part 8, provides detailed information for this parameter. This is usually accomplished by calculating and adding the moments of inertia with respect to the horizontal and vertical axes. For the purpose of calculating the moment of inertia, the welds may also be treated as either areas or lines.

$$I_p = I_x + I_y \qquad 66.10$$

The distance from the weld group centroid to the critical location in the weld group is next determined, usually by inspection. This distance is used to calculate the torsional shear stress.

$$r_m = \frac{Mc}{I_p} = \frac{Pec}{I_p} \qquad 66.11$$

The direct shear stress is easily calculated from the total weld area (or length, if the throat size is not known).

$$r_p = \frac{P}{A} \qquad 66.12$$

Vector addition is used to combine the torsional and direct shear stresses (i.e., the resultant shear stress).

$$r_a = \sqrt{(r_{m,y} + r_p)^2 + (r_{m,x})^2} \qquad 66.13$$

Although the elastic method is simple, it does not result in a consistent factor of safety and may result in an excessively conservative design of connections.

Example 66.2

A 50 ksi plate bracket is attached using shielded metal arc welding with E70 electrodes to the face of a 50 ksi column as shown. Buckling and bending of the plate and column can be neglected. Using ASD, what size fillet weld is required?

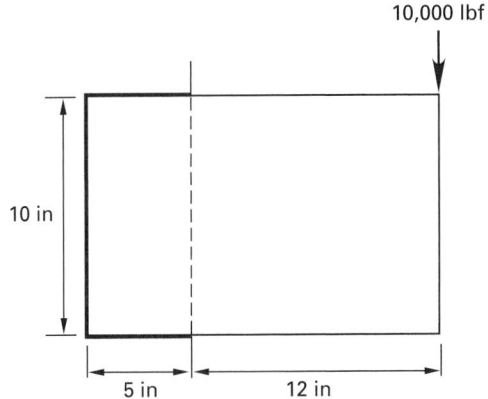

Solution

step 1: Assume the weld has an effective throat thickness, t.

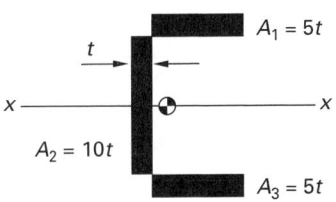

step 2: Find the centroid of the weld group. By inspection, $y_c = 0$. For the three welds,

$$A_1 = 5t$$
$$\overline{x}_1 = 2.5 \text{ in}$$
$$A_2 = 10t$$
$$\overline{x}_2 = 0$$
$$A_3 = 5t$$
$$\overline{x}_3 = 2.5 \text{ in}$$
$$\overline{x}_c = \frac{(5t)(2.5 \text{ in}) + (10t)(0) + (5t)(2.5 \text{ in})}{5t + 10t + 5t}$$
$$= 1.25 \text{ in}$$

step 3: Determine the centroidal moment of inertia of the weld group about the x-axis. Use the parallel axis theorem for areas 1 and 3.

$$I_x = \frac{t(10 \text{ in})^3}{12}$$
$$+ (2)\left(\frac{(5 \text{ in})t^3}{12} + (5 \text{ in})t(5 \text{ in})^2\right)$$
$$= 333.33t + 0.833t^3$$

Since t will be small (probably less than 0.5 in), the t^3 term can be neglected. So, $I_x = 333.33t$.

step 4: Determine the centroidal moment of inertia of the weld group about the y-axis.

$$I_y = \frac{(10 \text{ in})t^3}{12} + (10 \text{ in})t(1.25 \text{ in})^2$$
$$+ (2)\left(\frac{t(5 \text{ in})^3}{12} + (5 \text{ in})t(1.25 \text{ in})^2\right)$$
$$= 52.08t + 0.833t^3 \approx 52.08t$$

step 5: The polar moment of inertia is

$$I_p = I_x + I_y = 333.33t + 52.08t = 385.4t$$

step 6: By inspection, the maximum shear stress will occur at point A.

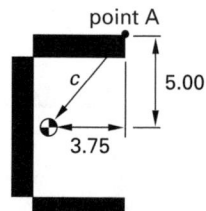

point A

5.00

c

3.75

$$c = \sqrt{(3.75 \text{ in})^2 + (5 \text{ in})^2} = 6.25 \text{ in}$$

step 7: The applied moment is

$$M = Pe = (10,000 \text{ lbf})(12 \text{ in} + 3.75 \text{ in})$$
$$= 157,500 \text{ in-lbf}$$

step 8: The torsional shear stress is given by Eq. 66.11.

$$r_{m,t} = \frac{Mc}{I_p} = \frac{(157,000 \text{ in-lbf})(6.25 \text{ in})}{385.4t}$$
$$= 2554.2/t$$

This shear stress is directed at right angles to the line r. The x- and y-components of the stress can be determined from geometry.

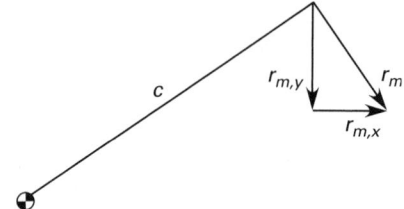

$$r_{m,y} = \left(\frac{3.75 \text{ in}}{6.25 \text{ in}}\right)\left(\frac{2554.2}{t}\right) = 1532.5/t$$
$$r_{m,x} = \left(\frac{5.00 \text{ in}}{6.25 \text{ in}}\right)\left(\frac{2554.2}{t}\right) = 2043.4/t$$

step 9: The direct shear stress from Eq. 66.12 is

$$r_{p,y} = \frac{P}{A} = \frac{10,000 \text{ lbf}}{t(5 \text{ in} + 10 \text{ in} + 5 \text{ in})}$$
$$= 500/t$$

step 10: Using Eq. 66.13, the resultant shear stress at point A is

$$r_a = \sqrt{\left(\frac{1532.5 + 500}{t}\right)^2 + \left(\frac{2043.4}{t}\right)^2}$$
$$= 2882.1/t$$

step 11: For a 50 ksi base metal with a 70 ksi rod, the rod strength controls the available strength. The design strength is given by Eq. 66.2 divided by the strength reduction factor, which is equal to 2.0.

$$\frac{F_w}{\Omega} = \frac{0.6F_u}{2.0} = \frac{(0.6)\left(70 \dfrac{\text{kips}}{\text{in}^2}\right)}{2.0}$$
$$= 21 \text{ ksi}$$

step 12: Equate the allowable stress to the shear stress at point A.

$$\frac{2882.1}{t} = \left(21 \frac{\text{kips}}{\text{in}^2}\right)\left(1000 \frac{\text{lbf}}{\text{kip}}\right)$$
$$t = 0.137 \text{ in}$$

Use a $^3/_8$ in (0.1875 in) weld.

Example 66.3

Use App. 45.A to calculate the polar moment of inertia of the weld group in Ex. 66.2.

Solution

From App. 45.A with $b = 5$ in and $d = 10$ in,

$$I_p = \left(\frac{(8)(5 \text{ in})^3 + (6)(5 \text{ in})(10 \text{ in})^2 + (10 \text{ in})^3}{12} \right.$$
$$\left. - \frac{(5 \text{ in})^4}{(2)(5 \text{ in}) + (10 \text{ in})} \right) t$$
$$= 385.42t$$

(This is the same result as was calculated in Ex. 66.2.)

8. AISC INSTANTANEOUS CENTER OF ROTATION METHOD FOR ECCENTRIC SHEAR/TORSION CONNECTIONS

An eccentrically loaded weld group has a tendency to rotate about a point called the *instantaneous center of rotation* (IC). The location of this point depends on the eccentricity, the geometry of the weld group, and the deformation of weld elements. The ultimate shear strength of a weld group is obtained by linking the load-deformation relationship of individual weld elements to the correct location of the instantaneous center. The AISC instantaneous center of rotation method calculates the capacity of an eccentrically loaded welded connection as Eq. 66.14.

$$R_n = CC_1 Dl \qquad 66.14$$

C and C_1 are coefficients from the Eccentrically Loaded Weld Groups Tables 8-3 through 8-11 in Part 8 of the *AISC Manual*. These coefficients depend on the load configuration and electrode type. D is the number of multiples of $^1/_{16}$ in in the weld size. l is the weld length.

Example 66.4

Use the *AISC Manual* to calculate the capacity of the weld group designed in Ex. 66.2.

Solution

Referring to AISC Table 8-8, p. 8-90,

$$l = 10 \text{ in}$$
$$kl = 5 \text{ in}$$
$$al = 12 \text{ in} + 3.75 \text{ in} = 15.75 \text{ in}$$

For a $^1/_4$ in weld and E70 electrodes, the following co-efficients are needed.

$$k = \frac{5 \text{ in}}{10 \text{ in}} = 0.5$$
$$a = \frac{15.75 \text{ in}}{10 \text{ in}} = 1.575$$
$$C = 1.10 \quad \text{[interpolated]}$$
$$C_1 = 1 \quad \text{[for E70 electrodes from Table 8-3]}$$
$$D = \frac{\frac{1}{4} \text{ in}}{\frac{1}{16} \text{ in}} = 4$$

From Eq. 66.14, the capacity of the connection is

$$R_n = CC_1 Dl = (1.10)(1)(4)(10) = 44 \text{ kips}$$

9. ELASTIC METHOD FOR OUT OF PLANE LOADING

Figure 66.7 illustrates a welded connection that must support both direct shear and a bending moment. Such a connection may be designed or analyzed by either an elastic method or AISC Table 8-4. Even though the maximum shear and maximum bending moment do not actually occur at the same place, the elastic method makes simplifying assumptions to combine the nominal stresses as vectors.

Figure 66.7 Welded Connection in Combined Shear and Bending

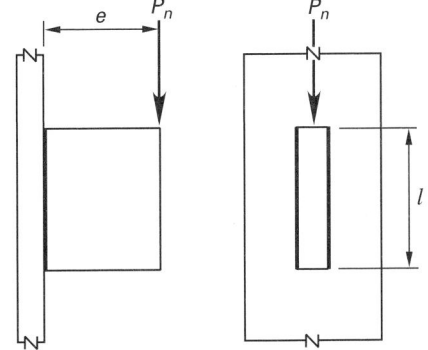

The nominal shear stress is

$$f_v = \frac{P_n}{A_{\text{weld}}} = \frac{P_n}{2Lt_e} \qquad 66.15$$

The nominal bending stress is easily calculated if the section modulus, S, of the weld group can be obtained from App. 45.A.

$$f_b = \frac{Mc}{I} = \frac{M}{S} = \frac{P_n e}{S} \qquad 66.16$$

The resultant stress is

$$f = \sqrt{f_v^2 + f_b^2} \qquad 66.17$$

The resultant stress must be less than the allowable stress, as calculated from Eq. 66.2.

10. WELDED CONNECTIONS IN BUILDING FRAMES

Part 10 of the *AISC Manual* contains tables listing capacities of various welded (E70XX electrodes) shear connections: framed beam connections, seated beam connections, stiffened seated beam connections, and end plate shear connections. The *AISC Manual* also contains several procedures for designing moment connections (both FR and PR). These procedures specifically assume top plate or end plate construction (as in Fig. 65.8) in order to transmit moments. Construction with split beam tees and the accompanying *prying action* are not addressed. See Ch. 65 for more information.

Welded beam-column connections intended to transfer moment in seismically active areas require special consideration to ensure that the connection behaves in a ductile manner.

67

Properties of Masonry

Nomenclature

A	area	in^2	mm^2
d_b	nominal diameter of reinforcement	in	mm
E_m	modulus of elasticity of masonry in compression	lbf/in^2	MPa
f'_g	specified compressive strength of grout	lbf/in^2	MPa
f'_m	specified compressive strength of masonry	lbf/in^2	MPa
F_s	allowable tensile stress in steel reinforcement	lbf/in^2	MPa
I	moment of inertia of masonry	in^4	mm^4
K	least distance among masonry cover, clear spacing between adjacent reinforcement, or $5d_b$	in	mm
l_d	embedment length of straight reinforcement	in	mm
l_e	equivalent embedment length provided by a standard hook	in	mm
r	radius of gyration	in	mm
S	section modulus per unit length	in^3/ft	mm^3/m

Symbols

γ	reinforcement size factor	–	–
ϕ	strength reduction factor	–	–

Subscripts

b	bar
g	grout
m	masonry
n	net
s	steel

1. INTRODUCTION

Masonry units used in the United States include concrete, clay (brick as well as structural clay tile), glass block, and stone. Concrete and clay masonry are used for the majority of structural masonry construction.

The relevant structural properties of clay and concrete masonry are similar, although test methods and requirements can vary significantly between the two materials. Requirements listed in this chapter are those of ACI 530/ASCE 5/TMS 402, *Building Code Requirements for Masonry Structures*, referred to as the *MSJC* (Masonry Standards Joint Committee) *Code*, and ACI 530.1/ASCE 6/TMS 602, *Specification for Masonry Structures*, referred to as the *MSJC Specification*.

2. UNIT SIZES AND SHAPES OF CONCRETE MASONRY

Concrete masonry units (CMUs) are manufactured in different sizes, shapes, colors, and textures to achieve a number of finishes and functions. The most common shapes are shown in Fig. 67.1. Typical concrete masonry unit nominal face dimensions are 8 in × 16 in (203 mm × 406 mm), with nominal thicknesses of 4, 6, 8, 10, and 12 in (102, 152, 203, 254, and 305 mm).

The shapes shown in Fig. 67.2 have been developed specifically for reinforced construction. Open-ended units can be placed around reinforcing bars.

In addition to the units shown, special unit shapes have been developed to maximize certain performance characteristics such as insulation, sound absorption, and resistance to water penetration.

3. UNIT SIZES AND SHAPES OF CLAY MASONRY

Brick (i.e., clay masonry) is manufactured by either extrusion or molding; both processes make it easy to produce unique features. (Very large shapes can be difficult to manufacture because of problems with proper drying and firing.) Clay brick is available in many sizes, most of which are based on a 4 in × 4 in (102 mm × 102 mm) module. Typical clay brick sizes are shown in Fig. 67.3. Actual dimensions are typically $^3/_8$ in (9.5 mm) less than the nominal dimensions. For example, the thickness of a 4 in face shell brick would be $3^5/_8$ in.

Structural

Figure 67.1 Typical Concrete Masonry Units

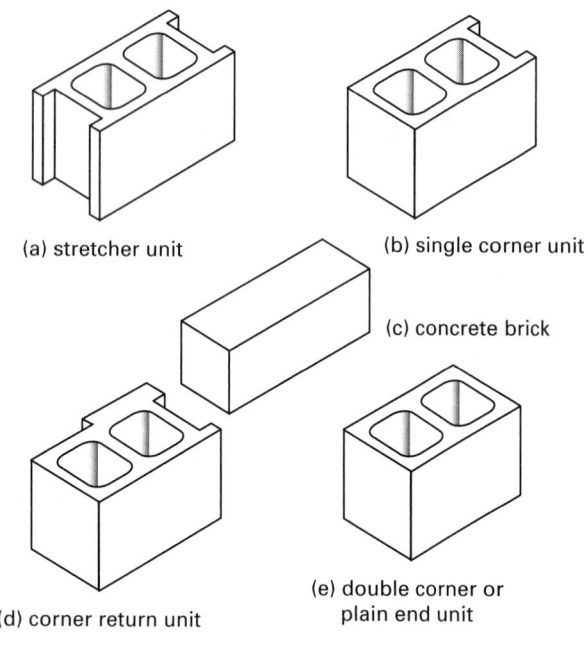

(a) stretcher unit

(b) single corner unit

(c) concrete brick

(d) corner return unit

(e) double corner or plain end unit

Figure 67.2 Concrete Masonry Units Accommodating Reinforcement

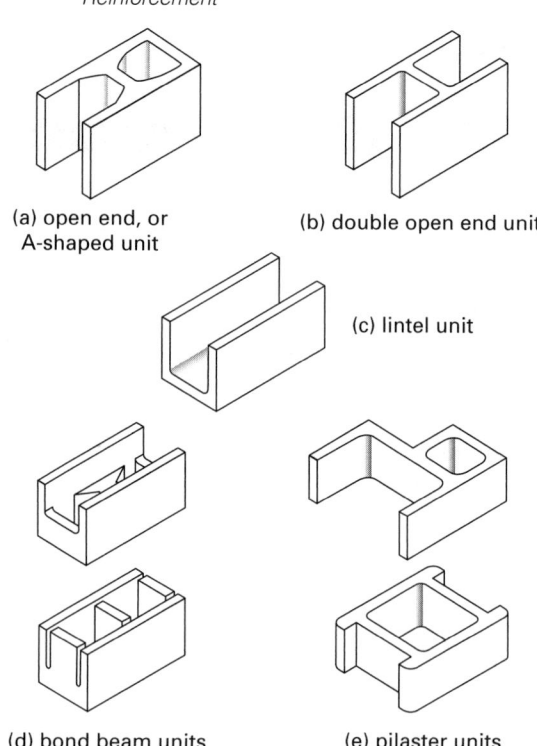

(a) open end, or A-shaped unit

(b) double open end unit

(c) lintel unit

(d) bond beam units

(e) pilaster units

Figure 67.3 Typical Brick Sizes (Nominal Dimensions[a])

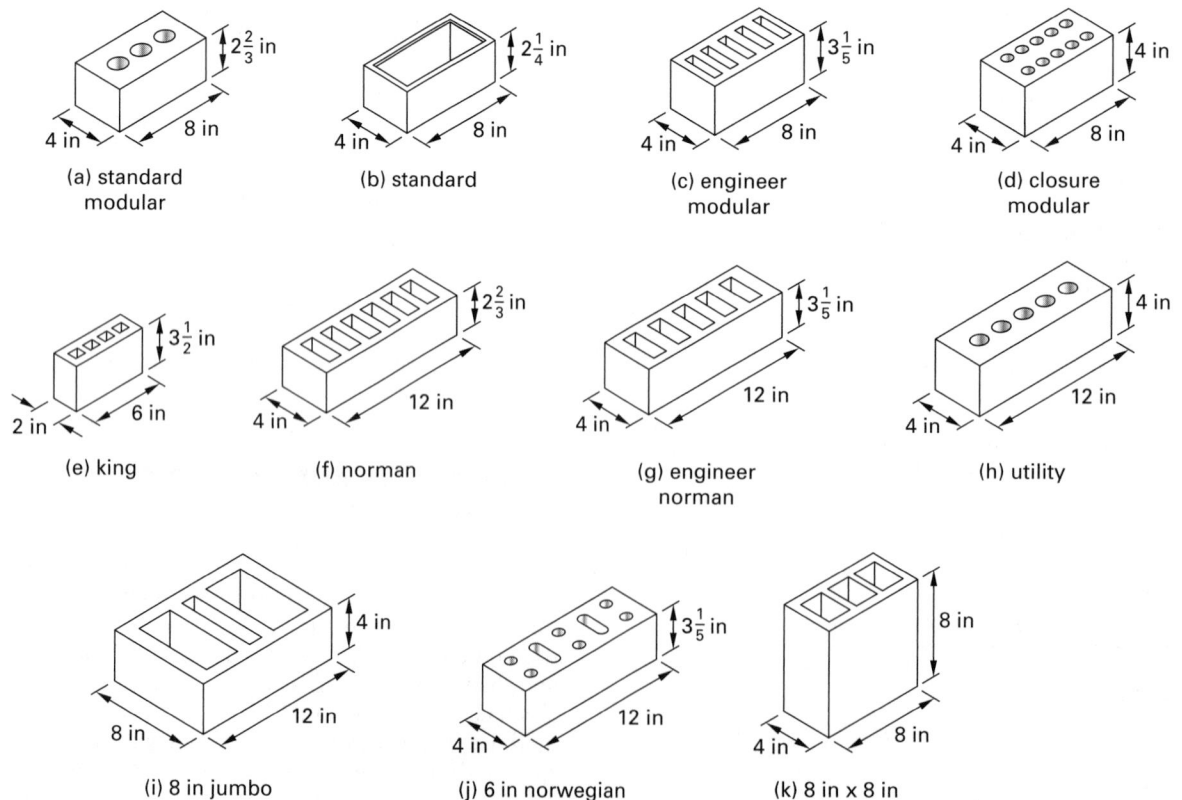

(a) standard modular — $2\frac{2}{3}$ in, 4 in, 8 in

(b) standard — $2\frac{1}{4}$ in, 4 in, 8 in

(c) engineer modular — $3\frac{1}{5}$ in, 4 in, 8 in

(d) closure modular — 4 in, 4 in, 8 in

(e) king — $3\frac{1}{2}$ in, 2 in, 6 in

(f) norman — $2\frac{2}{3}$ in, 4 in, 12 in

(g) engineer norman — $3\frac{1}{5}$ in, 4 in, 12 in

(h) utility — 4 in, 4 in, 12 in

(i) 8 in jumbo — 4 in, 8 in, 12 in

(j) 6 in norwegian — $3\frac{1}{5}$ in, 4 in, 12 in

(k) 8 in x 8 in — 8 in, 4 in, 8 in

[a]Actual dimensions are typically $^3/_8$ in (9.5 mm) less than nominal dimensions.

Table 67.1 *Nominal and Actual Brick Dimensions*[a]

	nominal size[b,g] (in)	specified size[c,d,e] (in)	vertical coursing
modular series[f]			
modular (standard modular)	$2^2/_3 \times 8 \times 4$	$3^5/_8 \times 2^1/_4 \times 7^5/_8$	three courses to 8 in
engineer modular	$3^1/_5 \times 8 \times 4$	$3^5/_8 \times 2^3/_4 \times 7^5/_8$	five courses to 16 in
closure modular (utility brick)	$4 \times 8 \times 4$	$3^5/_8 \times 3^5/_8 \times 7^5/_8$	two courses to 8 in
roman	$2 \times 12 \times 4$	$3^5/_8 \times 1^5/_8 \times 11^5/_8$	four courses to 8 in
norman	$2^2/_3 \times 12 \times 4$	$3^5/_8 \times 2^1/_4 \times 11^5/_8$	three courses to 8 in
utility (utility norman)	$4 \times 12 \times 4$	$3^5/_8 \times 3^5/_8 \times 11^5/_8$	two courses to 8 in
engineer norman (jumbo norman)	$3^1/_5 \times 4 \times 12$	$3^5/_8 \times 2^3/_4 \times 11^5/_8$	five courses to 16 in
meridian	$4 \times 4 \times 16$	$3^5/_8 \times 3^5/_8 \times 15^5/_8$	two courses to 8 in
6 in through-wall meridian	$4 \times 6 \times 16$	$5^5/_8 \times 3^5/_8 \times 15^5/_8$	two courses to 8 in
8 in through-wall meridian	$4 \times 8 \times 16$	$7^5/_8 \times 3^5/_8 \times 15^5/_8$	two courses to 8 in
double meridian	$8 \times 4 \times 16$	$3^5/_8 \times 7^5/_8 \times 15^5/_8$	two courses to 16 in
double through-wall meridian	$8 \times 8 \times 16$	$7^5/_8 \times 7^5/_8 \times 15^5/_8$	two courses to 16 in
nonmodular series[f]			
standard (standard brick)		$3^5/_8 \times 2^1/_4 \times 8$	three courses to 8 in
standard engineer		$3^5/_8 \times 2^3/_4 \times 8$	five courses to 16 in
closure standard		$3^5/_8 \times 3^5/_8 \times 8$	two courses to 8 in
king		$2^3/_4$ (up to 3) $\times 2^5/_8$ (up to $2^3/_4$) $\times 9^5/_8$ (up to $9^3/_4$)	five courses to 16 in
queen		$2^3/_4$ (up to 3) $\times 2^3/_4$ $\times 7^5/_8$ (up to 8)	five courses to 16 in

(Multiply in by 25.4 to obtain mm.)

[a]Other nonstandard units exist. For additional unit types, see Fig. 67.3.

[b]Used with $^3/_8$ in mortar joint thickness.

[c]Specified size is the anticipated manufactured size. Actual size is the size actually manufactured. Slight differences in actual sizes between manufacturers may exist.

[d]Standard specification of dimensions is width $\times$ height $\times$ length.

[e]Height and length are the face dimensions, as they are showing when the brick is laid as a stretcher.

[f]Standardized in 1992 by the Brick Industry Association and the National Association of Brick Distributors.

[g]Nominal size is the installed size including mortar joint thickness.

Brick types used in structural applications include building brick, face brick, and hollow brick. *Building brick* is typically used as a backing material or in other applications where appearance is not an issue. *Face brick* is used in both structural and nonstructural masonry where an appealing appearance is a requirement. *Hollow brick* is identical to face brick but has a larger core area, which allows hollow brick to be reinforced. (See Table 67.1.)

4. DIMENSIONS

Masonry unit dimensions are listed in the following order: width $\times$ height $\times$ length. Nominal dimensions are used for planning bond patterns and modular layout with respect to door and window openings. Actual dimensions of masonry units are typically $^3/_8$ in (9.5 mm) less than nominal dimensions, so that the 4 or 8 in (102 or 203 mm) module size is maintained with $^3/_8$ in (9.5 mm) mortar joints.

5. COMPRESSIVE STRENGTH

Compressive strength of a masonry assembly varies with time, unit properties, mortar type, and grout, if used. Strength is typically tested at 28 days. The specified compressive strength of masonry, f'_m, is noted in the project documents and is used to establish allowable stresses for masonry elements.

Compliance with specified compressive strength is verified by either the unit strength method or the prism test method. The *unit strength method* uses the net area compressive strength of the units and the mortar type to establish the compressive strength of the masonry assembly. Criteria from the *MSJC Specification* are listed in Tables 67.2 and 67.3.

Table 67.2 *Compressive Strength of Clay Masonry*

net area compressive strength of clay masonry units, lbf/in² (MPa)		net area compressive strength of masonry, lbf/in² (MPa)
type M or S mortar	type N mortar	
1700 (11.7)	2100 (14.5)	1000 (6.9)
3350 (23.1)	4150 (28.6)	1500 (10.3)
4950 (34.1)	6200 (42.7)	2000 (13.8)
6600 (45.5)	8250 (56.9)	2500 (17.2)
8250 (56.9)	10,300 (71.0)	3000 (20.7)
9900 (68.3)	–	3500 (24.1)
13,200 (91.0)	–	4000 (27.6)

Source: *MSJC Specification* 1.4 B(2)a. Based on the compressive strength of clay masonry units and the type of mortar used in construction.

Table 67.3 *Compressive Strength of Concrete Masonry*

net area compressive strength of concrete masonry units, lbf/in² (MPa)		net area compressive strength of masonry[a] lbf/in² (MPa)
type M or S mortar	type N mortar	
1250 (8.6)	1300 (9.0)	1000 (6.9)
1900 (13.1)	2150 (14.8)	1500 (10.3)
2800 (19.3)	3050 (21.0)	2000 (13.8)
3750 (25.8)	4050 (27.9)	2500 (17.2)
4800 (33.1)	5250 (36.2)	3000 (20.7)

[a]For units of less than 4 in (102 mm) height, use 85% of the values listed.

Source: *MSJC Specification* 1.4 B(2)b. Based on the compressive strength of concrete masonry units and the type of mortar used in construction.

Concrete masonry units are required by ASTM C90 to have an average net area compressive strength of 1900 lbf/in² (13.1 MPa). This corresponds to f'_m values of 1500 lbf/in² and 1350 lbf/in² (10.3 and 9.3 MPa) for mortar types M or S and type N, respectively. f'_m values up to 4000 lbf/in² (27.6 MPa) are attainable using high-strength concrete masonry units.

Clay units typically have strengths ranging from 3000 lbf/in² (20.7 MPa) to over 20,000 lbf/in² (138 MPa), averaging between 6000 lbf/in² and 10,000 lbf/in² (41.3 and 68.9 MPa) in North America. The required minimum average compressive strength of clay brick is 3000 lbf/in² (20.7 MPa) for brick grade SW.

The *prism test method* determines the compressive strength of masonry at 28 days or at the designated test age. Prisms are constructed from materials representative of those to be used in the construction. Typical concrete masonry prisms are two units high, where brick prisms are five units high. A set of three prisms is tested in compression until failure. The individual prism strengths are then corrected to account for the aspect ratio of the prism and averaged together. The corrected value represents the net area compressive strength of masonry, f'_m. Checking that proper proportions of certified materials were used in the test provides verification of the net area compressive strength. The *MSJC Specification* does not require field testing of masonry units, mortar, grout, or completed assemblies.

6. DENSITY

Density, or unit weight, of a masonry unit is described in terms of dry weight per cubic foot. Larger densities result in increased wall and building weight, heat capacity, sound transmission loss, and thermal conductivity (resulting in lower R-values and decreased fire resistance).

7. MODULUS OF ELASTICITY

The *MSJC Code* defines the modulus of elasticity (elastic modulus) in compression as the *chord modulus* (defined as the slope of a line intersecting the stress-strain curve at two points, neither of which is the origin of the curve) from a stress value of 5 to 33% of the compressive strength of masonry. Moduli of elasticity are determined from testing or tables, or estimated from *MSJC* Sec. 1.8.2.2.1.

$$E_m = 700 f'_m \quad \text{[clay masonry]} \qquad 67.1$$

$$E_m = 900 f'_m \quad \text{[concrete masonry]} \qquad 67.2$$

8. BOND STRENGTH

Bond describes both the amount of unit/mortar contact and the strength of adhesion. Bond varies with mortar properties (air and water content, water retention), unit surface characteristics (texture and suction), workmanship (pressure applied to joint during tooling, time between spreading mortar and placing units), and curing. In unreinforced masonry design, relative bond strengths are reflected in allowable flexural tension values.

9. ABSORPTION

Absorption describes the amount of water a masonry unit can hold when it is saturated. For concrete masonry, absorption is determined either in terms of weight of water per cubic foot of concrete, or as a percent by dry weight of concrete. For a given aggregate type and gradation, absorption is an indication of how well the concrete was compacted during manufacture. This, in turn, influences compressive strength, tensile strength, and durability.

The maximum permissible absorption per ASTM C90 for concrete masonry ranges from 13 to 18 lbm/ft^3 (208 to 288 kg/m^3), depending on the concrete density.

For brick, there are two categories of absorption: water absorption and *initial rate of absorption* (IRA). *Water absorption* tests are used to calculate a *saturation coefficient* for brick, which is an indicator of durability.

The IRA indicates how much water a brick draws in during the first minute of contact with water. Typical IRA values range between 5 and 40 g/min-30 in^2 (0.2 and 2.05 kg/min·m^2). IRA affects mortar and grout bond. If the IRA is too high, the brick absorption can impair the strength and extent of the bond with mortar and grout. High-suction brick should be wetted prior to laying to reduce suction (although the brick should be surface dry when laid). Very low-suction brick should be covered and kept dry on the jobsite.

In the laboratory, IRA is measured using oven-dried brick, which results in a higher measured IRA than if a moist brick were tested. The field test for IRA, on the other hand, is performed without drying the units. The laboratory test will indicate the order of magnitude of the IRA, where the field test indicates if additional wetting is necessary to lay the units.

10. DURABILITY

Clay brick *durability* may be predicted from either (a) the compressive strength, absorption, and saturation coefficient or (b) the compressive strength and ability to pass 50 freeze-thaw cycles.

Brick durability is reflected in *weathering grades* (severe, moderate, or negligible weathering), defined in ASTM standards. These standards also indicate weather conditions in the United States and define where each grade is required. Brick that meets the severe weathering (SW) grade is commonly available.

11. MORTAR

Mortar structurally bonds units together, seals joints against air and moisture penetration, accommodates small wall movements, and bonds to joint reinforcement, ties, and anchors.

Mortar types are defined in ASTM C270. Four mortar types—M, S, N, and O—are included and vary by the proportions of ingredients. Types M, S, and N are the most commonly used.

Type N mortar is a medium-strength mortar suitable for exposed masonry above grade. It is typically recommended for exterior walls subject to severe exposure, for chimneys, and for parapet walls.

Type S mortar has a relatively high compressive strength and will typically develop the highest flexural bond strength (all other variables being equal). Type M mortar has high compressive strength and excellent durability. Both type M and type S mortar are recommended for below-grade construction and reinforced masonry.

Type O mortar is a relatively low compressive strength mortar suitable for limited exterior use and interior use. It should not be used in freezing or moist environments.

Mortar type is one factor used to determine f'_m, if the unit strength method is used. Mortar type is also used to determine allowable compressive strengths for empirically designed masonry and allowable flexural tension for unreinforced masonry.

The *MSJC Code* restricts the use of some mortars for particular applications. In seismic design categories D, E, and F, for example, type N mortar and masonry cement mortar are prohibited in lateral force-resisting systems [*MSJC* Sec. 1.14.6.6]. Mortar type M or S must be used for empirical design of foundation walls [*MSJC* Sec. 5.6.3.1]. Glass block requires use of type S or N mortar [*MSJC* Sec. 7.6].

12. MORTAR COMPRESSIVE STRENGTH

Laboratory tests are for preconstruction evaluation of mortars and use low water content mortars. Results establish the general strength characteristics of the mortar mix. Field tests, on the other hand, establish quality control of mortar production and use mortars with high water contents, typical of those used in the work. Field test results are not required to meet the minimum compressive strength values contained in the property specification of ASTM C270.

Field test results should not be compared to lab test results, since the higher water content adversely affects the compressive strength of mortar, but also enhances the bond strength.

13. GROUT

Grout is a fluid cementitious mixture used to fill masonry cores or cavities to increase structural performance. Grout is most commonly used in reinforced construction to bond steel reinforcement to the masonry. The collar joint of a double wythe masonry wall is sometimes grouted to allow composite action.

Grout modulus of elasticity is calculated as $500f'_g$ [*MSJC* Sec. 1.8.2.4].

Structural

14. GROUT SLUMP

Grout slump is determined using the same procedure that is used to measure concrete slump. Grout slump should be between 8 and 11 in (203 and 279 mm) to facilitate complete filling of the grout space and proper performance [*MSJC Specification* 2.6 B2]. When relatively quick water loss is expected due to high temperatures and/or highly absorptive masonry units, a higher slump should be maintained. The high initial water content of grout compensates for water absorption by the masonry units after placement. Thus, grout gains high strength despite the high initial water-to-cement ratios.

15. GROUT COMPRESSIVE STRENGTH

When grout compressive strength testing is required, ASTM Test Method C1019 is used. Grout specimens are formed in a mold made up of masonry units with the same absorption and moisture content characteristics as those being used on the job. The masonry units remove excess water from the grout to more closely replicate grout strength in the wall. Concrete test methods should not be used for grout, as grout cubes or cylinders formed in nonabsorptive molds will produce unreliable results.

After curing, the grout specimens are capped and tested. Although a lower aggregate-to-cement ratio generally produces higher grout strengths, there is no direct relationship between the two.

Grout compressive strength is required to equal the specified compressive strength of masonry, but cannot be less than 2000 lbf/in² (13.8 MPa) [*MSJC Specification* 1.4 B(2)b(3)b].

16. STEEL REINFORCING BARS

Reinforcement increases masonry strength and ductility, providing increased resistance to applied loads and (in the case of horizontal reinforcement) to shrinkage cracking. The two principal types of reinforcement used in masonry are reinforcing bars and cold-drawn wire products.

Reinforcing bars are placed vertically and/or horizontally in the masonry and are grouted into position.

MSJC Specification 3.4B covers reinforcement installation requirements to ensure elements are placed as assumed in the design. *MSJC Sec.* 1.13.4 provides for a minimum masonry and grout cover around reinforcing bars to reduce corrosion and ensure sufficient clearance for grout and mortar to surround reinforcement and accessories, so that stresses can be properly transferred. Bar positioners are often used to hold reinforcement in place during grouting.

Reinforcing bars for masonry are the same as those used in concrete construction, although for masonry, bars larger than no. 11 are not permitted. Bars larger than no. 9 are not permitted when using strength design [*MSJC Sec.* 3.3.3.1] and should generally be avoided otherwise. Core size and masonry cover requirements may further restrict the maximum bar size that can be used. For bar reinforcement in tension, $F_s = 24,000$ lbf/in² (165 MPa) for grade-60 steel [*MSJC Sec.* 2.3.2.1].

17. DEVELOPMENT LENGTH

Development length (*anchorage* or *embedment length*) ensures that forces can be transferred. Reinforcing bars can be anchored by embedment length, hooks, or mechanical devices. Reinforcing bars anchored by embedment length rely on interlock of the bar deformations with grout and on the masonry cover being sufficient to prevent splitting from the reinforcing bar to the free surface.

The current edition of the *MSJC Code* has harmonized the development length, l_d, provisions for allowable stress design, ASD (Ch. 2), and strength design, SD (Ch. 3). The required development length for wires in tension is given by Eq. 67.3, but it cannot be less than 6 in (152 mm) for wire [*MSJC Sec.* 2.1.10.2].

$$l_d = 0.22d_bF_s \quad \text{[SI]} \quad 67.3(a)$$
$$l_d = 0.0015d_bF_s \quad \text{[U.S.]} \quad 67.3(b)$$

When *epoxy-coated wire* is used, the development length determined by Eq. 67.3 must be increased by 50% [*MSJC Sec.* 2.1.10.2].

The required development length of reinforcing bars is determined using Eq. 67.4 [*MSJC Secs.* 2.1.10.3 and 3.3.3.3].

$$l_d = \frac{1.57d_b^2 f_y \gamma}{K\sqrt{f_m'}} \quad \text{[SI]} \quad 67.4(a)$$
$$l_d = \frac{0.13d_b^2 f_y \gamma}{K\sqrt{f_m'}} \quad \text{[U.S.]} \quad 67.4(b)$$

The distance K is the least value among the masonry cover, clear bar spacing, or $5d_b$. γ is 1.0 for nos. 3–5 bars (M#10–16), 1.3 for nos. 6–7 bars (M#19–22), and 1.5 for nos. 8–9 bars (M#25–29). (In ASD, γ is 1.5 for nos. 8–11 bars (M#25–36).)

The required embedment may not be less than 12 in (305 mm). When epoxy-coated reinforcing bars are used, the development length determined by Eq. 67.4 must be increased by 50%.

Reinforcement splices must have a minimum lap equal to the development length determined by Eq. 67.4. Mechanical or welded splices must be capable of developing 125% of the reinforcement yield strength, f_y.

The equivalent embedment length for standard hooks, l_e, in tension is slightly different in the ASD and SD chapters. For ASD, $l_e = 11.25d_b$ [*MSJC Sec.* 2.1.10.5.1], while for SD $l_e = 13d_b$ [*MSJC Sec.* 3.3.3.2].

18. HORIZONTAL JOINT REINFORCEMENT

Joint reinforcement is a welded wire assembly consisting of two or more longitudinal wires connected with cross wires. It is placed in horizontal mortar bed joints, most commonly to control wall cracking associated with thermal or moisture shrinkage or expansion. Secondary functions include (a) metal tie system for bonding adjacent masonry wythes and (b) structural steel reinforcement.

For wire joint reinforcement in tension, $F_s = 30,000$ lbf/in^2 (207 MPa) [*MSJC* Sec. 2.3.2.1].

19. BOND PATTERNS

Allowable design stresses, lateral support criteria, and minimum thickness requirements for masonry are based primarily on structural testing and research on wall panels laid in running bond. *Running bond* is defined as construction where the head joints in successive courses are horizontally offset at least one-quarter the unit length [*MSJC* Sec. 1.6]. The typical offset is half the unit length. When a different bond pattern is used, its influence on the compressive and flexural strength of the wall must be considered. Allowable flexural tensile stresses and masonry modulus of rupture [*MSJC* Secs. 2.2.3.2 and 3.1.8.2, respectively] assign different design values based on bond pattern.

Structural

68 Masonry Walls

Nomenclature

a	depth of an equivalent compression zone at nominal strength	in	mm
A	area	in^2	mm^2
A_n	net cross-sectional area of masonry	in^2	mm^2
A_s	cross-sectional area of steel per unit length of wall	in^2/ft	mm^2/m
A_{st}	total area of laterally tied longitudinal reinforced steel in reinforced masonry	in^2	mm^2
A_v	cross-sectional area of shear reinforcement	in^2	mm^2
b	width	in	mm
b_w	for partially grouted walls, width of grouted area (i.e., width of grout cell plus width of the two adjacent webs)	in	mm
c	distance from neutral axis to extreme fiber in bending	in	mm
d	distance from extreme compression fiber to centroid of tension reinforcement	in	mm
d	effective depth	in	mm
e	eccentricity	in	mm
e_u	eccentricity of P_{uf}	in	mm
E_m	modulus of elasticity of masonry in compression	lbf/in^2	MPa
E_s	modulus of elasticity of steel	lbf/in^2	MPa
f_a	compressive stress in masonry due to axial load alone	lbf/in^2	MPa
f_b	stress in masonry due to flexure alone	lbf/in^2	MPa
f'_g	specified compressive strength of grout	lbf/in^2	MPa
f'_m	specified compressive strength of masonry	lbf/in^2	MPa
f_r	modulus of rupture for masonry	lbf/in^2	MPa
f_s	stress in reinforcement	lbf/in^2	MPa
f_v	shear stress in masonry	lbf/in^2	MPa
f_y	yield strength of steel for reinforcement and anchors	lbf/in^2	MPa
F_a	allowable compressive stress due to axial load alone	lbf/in^2	MPa
F_b	allowable stress due to flexure only	lbf/in^2	MPa
F_s	allowable reinforcement tensile stress	lbf/in^2	MPa
F_t	allowable tensile stress	lbf/in^2	MPa
F_v	allowable shear stress in masonry	lbf/in^2	MPa
g	ratio of distance between compression steel and tension steel to overall wall depth	–	–
h	effective height	in	mm
I	moment of inertia	in^4	mm^4
I_{cr}	moment of inertia of the cracked cross section of the member	in^4	mm^4
I_g	moment of inertia of the gross cross section of the member	in^4	mm^4
I_n	moment of inertia of the net cross section of the member	in^4	mm^4
j	ratio of distance between centroid of flexural compressive forces and centroid of tensile forces to depth	–	–
k	ratio of the distance between compression face of wall and neutral axis to the effective depth	–	–
l	clear span between supports	in	mm
M	moment	in-lbf	N·m
M_{cr}	nominal masonry cracking moment	in-lbf	N·m

Structural

M_m resisting moment assuming
  masonry governs   in-lbf   N·m
M_n nominal moment strength   in-lbf   N-m
M_R resisting moment of wall
  (the lesser of M_m and M_s)   in-lbf   N·m
M_s resisting moment assuming
  steel governs   in-lbf   N·m
M_{ser} service moment at midheight
  of member, including P-Δ effects   in-lbf   N-m
M_u factored moment   in-lbf   N-m
n modular ratio (E_1/E_2 or f_1/f_2)   –   –
N_u factored compressive force acting
  normal to shear surface   lbf   N
N_v compressive force acting normal to
  shear surface   lbf   N
P axial load   lbf   N
P_a allowable compressive force in
  reinforced masonry due to
  axial load   lbf   N
P_e Euler buckling load   lbf   N
P_n nominal axial strength   lbf   N
P_u factored axial load   lbf   N
P_{uf} factored load from tributary
  floor or roof areas   lbf   N
Q statical moment   in^3 mm^3
r radius of gyration   in   mm
R seismic response modification factor   –   –
s spacing of reinforcement   in   mm
S section modulus   in^3 mm^3
t_{fs} face shell thickness of masonry
  unit   in   mm
v shear stress   lbf/in^2 MPa
V design shear force   lbf   N
V_m shear strength provided by masonry   lbf   N
V_n nominal shear strength   lbf   N
V_R resisting shear of wall   lbf   N
V_s shear strength provided by
  shear reinforcement   lbf   N
V_u factored shear force   lbf   N
w uniformly distributed load   lbf/ft^2 MPa
x_c distance from centroid of element
  to neutral axis of section   in   mm

Symbols

δ_s horizontal deflection at midheight
  due to service loads   in   mm
ρ ratio of tensile steel area to
  compressive area, bd –   –
ρ_t ratio of steel area to gross
  area of masonry, bt –   –
ϕ strength reduction factor   –   –

Subscripts

a allowable
bal   balanced
br   brick
c centroidal
CMU   concrete masonry unit
e Euler
fs   face shell
g grout
m masonry
n net
R resisting

s steel
tr   transformed
u factored load effects
v shear
y yield

1. METHODS OF DESIGN

Building Code Requirements for Masonry Structures (the *MSJC Code*) defines a wall as a vertical element with a horizontal length-to-thickness ratio greater than three, and used to enclose space [*MSJC* Sec. 1.6]. This differentiates walls from columns, which are subject to additional design limitations. Masonry walls can serve as veneer facings subject to out-of-plane flexure; load-bearing walls subject to axial compression, flexure, or both; and shear walls. While walls are typically designed to span vertically (between floors) or horizontally (between intersecting walls or pilasters), they can also be designed for two-way bending to limit deflection and increase stability.

Masonry structures are currently designed using one of several methods, including empirical procedures, allowable stress design, and strength design. Allowable stress design (ASD) is currently the most widely used of the three methods, although strength design (SD) is gaining popularity.

2. EMPIRICAL WALL DESIGN

Empirical design is a procedure of proportioning and sizing masonry elements. The criteria are based on historical experience rather than analytical methods. This method is conservative for most masonry construction. It is an expedient method for typical loadbearing structures, exterior curtain walls, and interior partitions.

Prescriptive criteria in Ch. 5 of the *MSJC Code* govern vertical and lateral load resistance of walls. Certain limitations apply to empirical designs, helping to ensure the construction and design loads are consistent with the experience used to establish the empirical criteria. The code prohibits its use in *seismic design categories* (SDC) D, E, and F; in any seismic-resisting system in SDC B or C; and for lateral force-resisting elements when the basic wind speed exceeds 110 mph (145 km/hr), or the building height exceeds 35 ft (10.67 m) in height. Additional restrictions are included for non-lateral force-resisting elements based on combinations of buiding height and basic wind speed. This *MSJC Code* chapter includes requirements for wall lateral support, allowable stresses, anchorage for lateral support, and shear wall spacing.

3. ALLOWABLE STRESS WALL DESIGN

The allowable stress design (ASD) method compares expected or calculated design stresses due to expected loads to allowable design stresses. Loads are unfactored;

that is, the service loads are such that they may be expected to occur during the structure's life. When calculated design stresses are less than allowable stresses, the design is acceptable.

Allowable stresses are determined by reducing expected material strengths by appropriate factors of safety. In addition, Sec. 2.1.2.3 of the *MSJC Code* permits allowable stresses to be increased by one third for load combinations including wind or seismic loads. (Governing local or state building codes may not permit this one-third increase in allowable stress). Masonry allowable stresses may vary with unit type, mortar type, compressive strength, and other factors. Stronger masonry will have higher allowable stresses.

The following assumptions form the basis of ASD for masonry. (a) Stress is linearly proportional to strain for stresses within the allowable stress (elastic) range. (b) Masonry materials combine to form a homogeneous, isotropic material. Therefore, sections that are plane before bending remain plane after bending. (c) Reinforcement, when used, is perfectly bonded to masonry.

ASD of *unreinforced masonry* is covered in Sec. 2.2 of the *MSJC Code*. When unreinforced, masonry is designed to resist tensile and compressive loads; the tensile strength of any steel in the wall is neglected. Tension due to axial loads, such as wind uplift, is not permitted in unreinforced masonry. Compressive stress from dead loads can be used to offset axial tension, although if axial tension develops, the wall must be reinforced to resist the tension. However, flexural tension due to bending may be resisted by the unreinforced masonry.

MSJC Sec. 2.3 gives requirements for reinforced masonry subjected to axial compression and tension, flexure, and shear. The tensile strength of masonry units, mortar, and grout is neglected, as is the compressive resistance of steel reinforcement, unless confined by lateral ties to prevent buckling in accordance with provisions in the *MSJC Code*. (These confinement provisions are virtually impossible to meet for walls.) Axial tension and flexural tension are resisted entirely by steel reinforcement.

4. STRENGTH DESIGN

The strength design (SD) method compares factored loads to the design strength of the material. Expected loads are multiplied by load factors to provide a factor of safety. The ultimate strength of the wall section is reduced by a strength reduction factor, ϕ, that accounts for variations in material properties and workmanship. Although this same method is typically used to design reinforced concrete elements, different design parameters apply to masonry. Chapter 3 of the *MSJC Code* covers SD of masonry.

In SD, a compressive masonry stress of $0.80f'_m$ is assumed to be rectangular and uniformly distributed over an equivalent compression zone, bounded by the compression face of the masonry, with a depth, $a = 0.80c$.

The maximum usable strain at the extreme compression fiber of the masonry is limited to 0.0035 for clay masonry and 0.0025 for concrete masonry [*MSJC* Sec. 3.3.2].

The code defines the strength reduction factors, ϕ, as 0.90 for reinforced masonry subjected to flexure or axial load, 0.60 for unreinforced masonry subjected to flexure or axial loads, and 0.80 for shear design of masonry [*MSJC* Sec. 3.1.4].

The maximum amount of reinforcement permitted in a masonry wall is limited by code to ensure that ductility is achieved. The specific limitation is based on the ductility demand as indicated by the *seismic response modification factor*, R, used in design as well as the type of force (in plane or out of plane). Reinforcement is limited so that the steel will reach a prescribed strain level before the masonry crushes. The maximum reinforcing limits are based on less restrictive assumptions. Depending on the design conditions, the steel must reach 1.5, 3, or 4 times its yield strain [*MSJC* Sec. 3.3.3.5].

5. PROPERTIES OF BUILT-UP SECTIONS

Wall design is greatly assisted by using tabulated properties of various built-up sections. Appendices 68.A through D are convenient tabulations of useful properties.

In Apps. 68.A through D, the net cross-sectional area of masonry, A_n, and the section modulus, S, are based on the minimum cross-sectional area (i.e., the mortared and grouted areas) of the wall [*MSJC* Sec. 1.9.1]. Moment of inertia, I, and radius of gyration, r, are based on the average cross-sectional area, which is taken through the masonry unit [*MSJC* Sec. 1.9.3].

On horizontal bed joints, masonry units are typically mortared on the face shells only. Webs are not typically mortared unless they are adjacent to a grouted core, in which case, they are mortared to confine the grout. Section properties listed reflect these practices.

6. ASD WALL DESIGN: FLEXURE— UNREINFORCED

Examples of walls subjected to out-of-plane flexure are exterior nonloadbearing walls subjected to wind or seismic loads, along with retaining walls. The self-weight of the wall can be neglected where the wall design will be governed by tension, such as for a one-story wall. In this case, the wall weight will offset tension in the wall, so the assumption is conservative.

In ASD, the actual wall stress is calculated as $f_b = M/S$ and compared to the allowable stress, F_b, obtained from Table 68.1. Out-of-plane flexure produces tension stress in the mortar normal to the bed joints.

Table 68.1 *Allowable Flexural Tensile Stress for Clay and Concrete Masonry in psi (MPa)*

	mortar types			
	portland cement/lime or mortar cement		masonry cement or air entrained portland cement/lime	
masonry type	M or S	N	M or S	N
normal to bed joints				
solid units	40 (0.28)	30 (0.21)	24 (0.17)	15 (0.10)
hollow units[a]				
ungrouted	25 (0.17)	19 (0.13)	15 (0.10)	9 (0.06)
fully grouted	65 (0.45)	63 (0.43)	61 (0.42)	58 (0.40)
parallel to bed joints in running bond				
solid units	80 (0.55)	60 (0.41)	48 (0.33)	30 (0.21)
hollow units				
ungrouted and partially grouted	50 (0.35)	38 (0.26)	30 (0.21)	19 (0.13)
fully grouted	80 (0.55)	60 (0.41)	48 (0.33)	30 (0.21)

[a]For partially grouted masonry, allowable stresses shall be determined on the basis of linear interpolation between hollow units that are fully grouted or ungrouted and hollow units based on the amount of grouting.

Source: *MSJC* Table 2.2.3.2

Example 68.1

Evaluate the tensile stresses in a one-story masonry wall for a wind load of 20 lbf/ft². The wall is 12 ft high and simply supported at the top and bottom. 12 in hollow concrete masonry and portland cement-lime type S mortar are used.

Solution

From Table 68.1, $F_b = 25$ lbf/in², which can be increased by one third for wind loads. Therefore,

$$F_b = \left(1 + \tfrac{1}{3}\right)\left(25.0 \ \frac{\text{lbf}}{\text{in}^2}\right) = 33.3 \ \text{lbf/in}^2$$

From App. 68.B, $S = 159.9$ in³/ft.

Determine the maximum moment.

$$M = \frac{wl^2}{8} = \frac{\left(20 \ \dfrac{\text{lbf}}{\text{ft}^2}\right)(12 \ \text{ft})^2 \left(12 \ \dfrac{\text{in}}{\text{ft}}\right)}{8}$$
$$= 4320 \ \text{in-lbf/ft}$$

Calculate the stress and compare to the allowable stress.

$$f_b = \frac{M}{S} = \frac{4320 \ \dfrac{\text{in-lbf}}{\text{ft}}}{159.9 \ \dfrac{\text{in}^3}{\text{ft}}}$$
$$= 27.0 \ \text{lbf/in}^2 < 33.3 \ \text{lbf/in}^2 \quad [\text{OK}]$$

Note that the wall is overstressed if the one-third increase in allowable stress is not used. However, if the compressive stress due to the masonry dead load is included in the analysis, the net flexural tension stress is less than 25 lbf/in².

7. SD WALL DESIGN: FLEXURE—UNREINFORCED

SD of unreinforced walls subjected to flexure is similar to ASD. All contributions from reinforcement are ignored, and principles of engineering mechanics are applied. The nominal compressive strength of the masonry is taken as $0.80f'_m$, and the nominal flexural tension strength is taken as the modulus of rupture, f_r. Values for f_r are given in Table 68.2 [*MSJC* Table 3.1.8.2.1]. The maximum factored tension and compression forces must not exceed the calculated respective strengths as decreased by the strength reduction factor of 0.60 for unreinforced masonry.

8. ASD WALL DESIGN: FLEXURE—REINFORCED

The structural model for ASD of fully grouted flexural members is shown in Fig. 68.1. The allowable stress for grade 60 steel in tension is $F_s = 24,000$ lbf/in² (165 MPa) [*MSJC* Sec. 2.3.2.1].

Figure 68.1 Stress Distribution for Fully Grouted Masonry

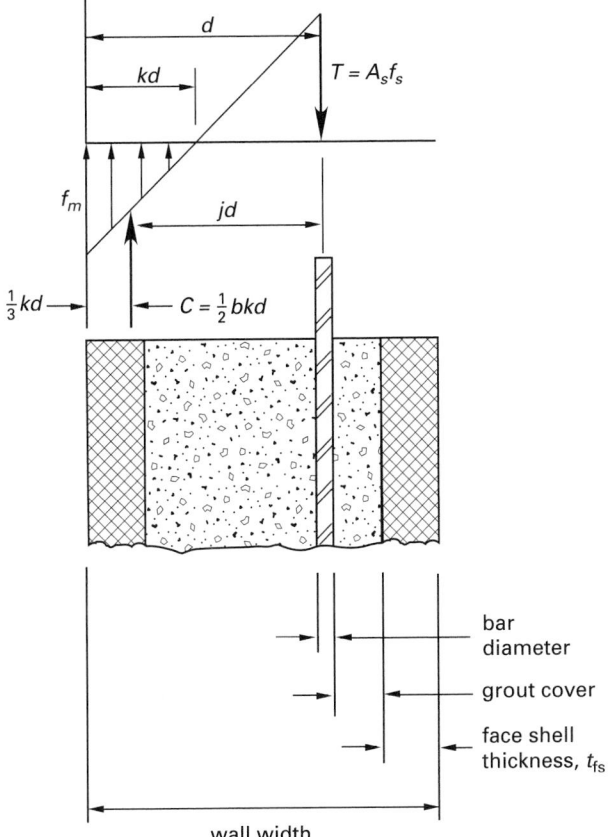

From *MSJC* Sec. 2.3.3.2.2, the allowable bending stress for masonry flexural members in compression is

$$F_b = \tfrac{1}{3} f'_m \qquad \textit{68.1}$$

From *MSJC* Sec. 2.3.5.2.2, the allowable shear stress for masonry flexural members is given by Eq. 68.2, although shear due to out-of-plane bending seldom governs the design.

$$F_v = \sqrt{f'_m} \qquad \textit{68.2}$$

Fully grouted masonry walls respond in bending similar to reinforced concrete beams. Analysis of these walls is based on the straight-line theory for beams subjected to bending. The equations governing the analysis follow the traditional ASD equations for reinforced concrete.

$$\rho = \frac{A_s}{bd} \qquad \textit{68.3}$$

$$k = \sqrt{2\rho n + (\rho n)^2} - \rho n \qquad \textit{68.4}$$

$$j = 1 - \frac{k}{3} \qquad \textit{68.5}$$

$$M_m = F_b bd^2 \left(\frac{jk}{2}\right) \qquad \textit{68.6}$$

$$M_s = A_s F_s jd \qquad \textit{68.7}$$

$$M_R = \text{the lesser of } M_m \text{ and } M_s \qquad \textit{68.8}$$

$$V_R = F_v bd \qquad \textit{68.9}$$

Section 2.3.3.3.1 of the *MSJC Code* limits the width of the compression area, b, used in stress calculations to the smallest of: the center-to-center bar spacing, six times the wall thickness, or 72 in (1829 mm) for running bond masonry.

Table 68.2 Modulus of Rupture for Clay and Concrete Masonry in psi (kPa)

direction of flexural tensile stress and masonry type	mortar types			
	portland cement/lime or mortar cement		masonry cement or air entrained portland cement/lime	
	M or S	N	M or S	N
normal to bed joints in running or stack bond				
solid units	100 (689)	75 (517)	60 (413)	38 (262)
hollow units*				
ungrouted	63 (431)	48 (331)	38 (262)	23 (158)
fully grouted	163 (1124)	158 (1089)	153 (1055)	145 (1000)
parallel to bed joints in running bond				
solid units	200 (1379)	150 (1033)	120 (827)	75 (517)
hollow units				
ungrouted and partially grouted	125 (862)	95 (655)	75 (517)	48 (331)
fully grouted	200 (1379)	150 (1033)	120 (827)	75 (517)
parallel to bed joints in stack bond	0 (0)	0 (0)	0 (0)	0 (0)

*For partially grouted masonry, modulus of rupture values shall be determined on the basis of linear interpolation between hollow units that are fully grouted and ungrouted based on amount (percentage) of grouting.

Source: *MSJC* Table 3.1.8.2.1

Determining the balanced condition allows the designer to know whether reinforcement stress or masonry stress governs the design. A *balanced condition* occurs when both the steel and masonry reach their respective allowable stresses at the same time (i.e., $f_m = F_b$ and $f_s = F_s$), although the balanced design is not necessarily the most cost-effective. If there is less reinforcement than that required for a balanced condition, moment capacity is governed by F_s. The balanced condition is determined by setting masonry compressive force, C, equal to reinforcement tensile force, T.

$$\rho_{\text{bal}} = \frac{nF_b}{2F_s \left(n + \dfrac{F_s}{F_b} \right)} \quad \text{[balanced]} \qquad 68.10$$

When analyzing *partially grouted masonry* (i.e., where some core voids are filled and others are not), there are two cases to consider: (a) neutral axis falling within the face shell of the masonry, and (b) neutral axis falling within the core area. When the neutral axis falls within the face shell (i.e., when kd is less than the face shell thickness), the analysis is identical to the analysis of the fully grouted masonry wall.

When the neutral axis falls within the core area of the masonry, the masonry wall responds to bending similar to a reinforced concrete T-beam. The grouted area and the face shell react to forces as a single unit, rather than as two separate beams. (T-beam performance seldom occurs in walls less than 10 in thick.) The analysis of partially grouted walls is summarized in Eqs. 68.11 through 68.16. (b is the actual distance between corresponding points in a wall segment containing one grouted cell.)

$$k = \frac{-A_s n - t_{\text{fs}}(b - b_w)}{db_w}$$

$$+ \frac{\sqrt{\begin{array}{c}(A_s n + t_{\text{fs}}(b - b_w))^2 \\ + t_{\text{fs}}^2 b_w(b - b_w) + 2db_w A_s n\end{array}}}{db_w}$$

$$\qquad 68.11$$

$$j = \left(\frac{1}{kdb_w + t_{\text{fs}}(b - b_w)\left(2 - \dfrac{t_{\text{fs}}}{kd}\right)} \right)$$

$$\times \left(kb_w \left(d - \frac{kd}{3} \right) + \left(\frac{2t_{\text{fs}}(b - b_w)}{kd^2} \right) \right.$$

$$\times \left. \left((kd - t_{\text{fs}}) \left(d - \frac{t_{\text{fs}}}{2} \right) + \left(\frac{t_{\text{fs}}}{2} \right) \left(d - \frac{t_{\text{fs}}}{3} \right) \right) \right)$$

$$\qquad 68.12$$

$$M_m = \tfrac{1}{2} F_b k d b_w \left(d - \frac{kd}{3} \right) + F_b t_{\text{fs}}(b - b_w)$$

$$\times \left(\left(1 - \frac{t_{\text{fs}}}{kd} \right) \left(d - \frac{t_{\text{fs}}}{2} \right) + \left(\frac{t_{\text{fs}}}{2kd} \right) \left(d - \frac{t_{\text{fs}}}{3} \right) \right)$$

$$\qquad 68.13$$

$$M_s = A_s F_s j d \qquad 68.14$$

$$M_R = \text{the lesser of } M_m \text{ and } M_s \qquad 68.15$$

$$V_R = F_v \left(bt_{\text{fs}} + b_w(d - t_{\text{fs}}) \right) \qquad 68.16$$

Table 68.3 *Reinforcing Steel Areas, A_s, in in^2/ft*

bar spacing (in)	reinforcing bar size						
	no. 3	no. 4	no. 5	no. 6	no. 7	no. 8	no. 9
8	0.166	0.295	0.460	0.663	0.902	1.178	1.491
16	0.083	0.147	0.230	0.331	0.451	0.589	0.746
24	0.055	0.098	0.153	0.221	0.301	0.393	0.497
32	0.041	0.074	0.115	0.166	0.225	0.295	0.373
40	0.033	0.059	0.092	0.133	0.180	0.236	0.298
48	0.028	0.049	0.077	0.110	0.150	0.196	0.249
56	0.024	0.042	0.066	0.095	0.129	0.168	0.213
64	0.021	0.037	0.058	0.083	0.113	0.147	0.186
72	0.018	0.033	0.051	0.074	0.100	0.131	0.166

Example 68.2

Design a fully grouted cantilever concrete masonry retaining wall for a maximum design moment of 12,960 in-lbf/ft. The wall is 6 ft high. Use 8 in units with $f'_m = 1500 \text{ lbf/in}^2$.

Solution

Work on a per unit width basis.

Place steel on the tension (soil) side of the wall. Choose $d = 4.75$ in. Use fully grouted masonry.

For an initial steel estimate, assume $j = 0.9$.

$$E_m = 900 f'_m = (900) \left(1500 \; \frac{\text{lbf}}{\text{in}^2} \right)$$

$$= 1,350,000 \text{ lbf/in}^2$$

$$n = \frac{E_s}{E_m} = \frac{29,000,000 \; \dfrac{\text{lbf}}{\text{in}^2}}{1,350,000 \; \dfrac{\text{lbf}}{\text{in}^2}}$$

$$= 21.5$$

From Eq. 68.1,

$$F_b = \tfrac{1}{3} f'_m = \left(\tfrac{1}{3} \right) \left(1500 \; \frac{\text{lbf}}{\text{in}^2} \right)$$

$$= 500 \text{ lbf/in}^2$$

Obtain an initial estimate of the steel required. From Eq. 68.7,

$$A_s = \frac{M}{F_s j d} = \frac{12,960 \; \dfrac{\text{in-lbf}}{\text{ft}}}{\left(24,000 \; \dfrac{\text{lbf}}{\text{in}^2} \right)(0.9)(4.75 \text{ in})}$$

$$= 0.126 \text{ in}^2/\text{ft}$$

From Table 68.3, try no. 6 bars at 32 in on center, giving $A_s = 0.166$ in^2/ft.

$$\rho = \frac{A_s}{bd} \quad \text{[Eq. 68.3]}$$

$$= \frac{0.166 \, \dfrac{\text{in}^2}{\text{ft}}}{\left(12 \, \dfrac{\text{in}}{\text{ft}}\right)(4.75 \text{ in})}$$

$$= 0.0029$$

$$k = \sqrt{2\rho n + (\rho n)^2} - \rho n \quad \text{[Eq. 68.4]}$$

$$= \sqrt{(2)(0.0029)(21.5) + (0.0029)^2(21.5)^2} \\ - (0.0029)(21.5)$$

$$= 0.296$$

$$j = 1 - \frac{k}{3} \quad \text{[Eq. 68.5]}$$

$$= 1 - \frac{0.296}{3} = 0.901$$

$$M_m = F_b bd^2 \left(\frac{jk}{2}\right) \quad \text{[Eq. 68.6]}$$

$$= \left(500 \, \frac{\text{lbf}}{\text{in}^2}\right)\left(12 \, \frac{\text{in}}{\text{ft}}\right)(4.75 \text{ in})^2\left(\frac{(0.901)(0.296)}{2}\right)$$

$$= 18,052 \text{ in-lbf/ft}$$

$$M_s = A_s F_s jd \quad \text{[Eq. 68.7]}$$

$$= \left(0.166 \, \frac{\text{in}^2}{\text{ft}}\right)\left(24,000 \, \frac{\text{lbf}}{\text{in}^2}\right)(0.901)(4.75 \text{ in})$$

$$= 17,051 \text{ in-lbf/ft}$$

$$M_R = 17,051 \text{ in-lbf/ft} > 12,960 \text{ in-lbf/ft} \quad \text{[OK]}$$

9. ASD WALL DESIGN: AXIAL COMPRESSION AND FLEXURE— UNREINFORCED

Many masonry walls are designed for both axial compression and flexure. Examples include: load-bearing exterior walls (subject to wind and/or seismic loads), tall nonloadbearing exterior walls (where the self-weight of the wall decreases wall capacity), and walls with an eccentric vertical load. Figure 69.2 shows the bending moment and transverse loading distributions assumed in the design of such walls.

Axial capacity decreases with increasing wall height. For very tall slender walls, elastic buckling is the limiting factor. For most masonry walls, however, the main effect of slenderness is the development of additional bending due to deflection (known as the P-Δ effect). The *MSJC Code* uses h/r to evaluate the performance of hollow and partially grouted masonry.

Figure 68.2 *Unreinforced Walls Subject to Axial Compression and Flexure*

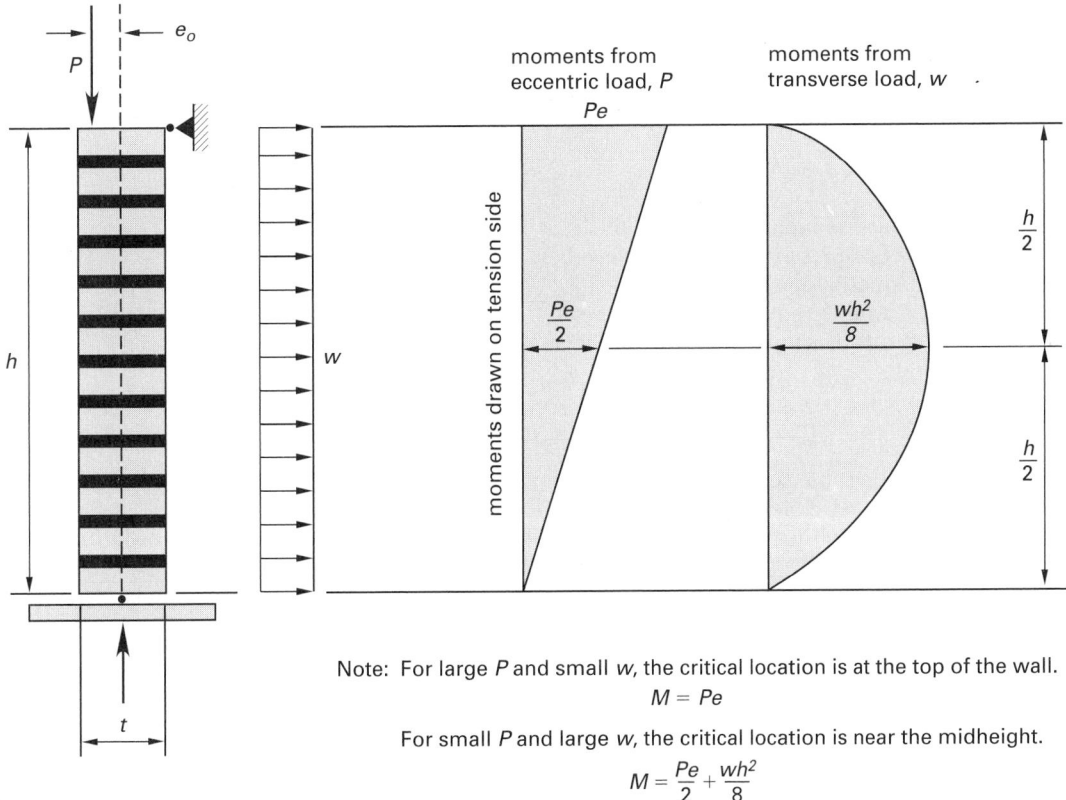

Note: For large P and small w, the critical location is at the top of the wall.

$$M = Pe$$

For small P and large w, the critical location is near the midheight.

$$M = \frac{Pe}{2} + \frac{wh^2}{8}$$

Walls subject to compressive loads only may be in pure compression or may be subject to flexure. For unreinforced walls where the compressive load is within a distance S/A_n (i.e., $t/6$ for solid sections) from the centroid of the wall, there will be no net tensile stress due to load eccentricity.

The critical section for walls subject to eccentric load and out-of-plane loads (such as wind or seismic) may be either at the top of the wall or at midheight, depending on the magnitudes of the loads. Both locations should be checked. Unreinforced walls subjected to axial compression, flexure, or both must satisfy Eqs. 68.17 through 68.22 [*MSJC* Sec. 2.2.3]. In Eq. 68.19, e should be the actual or estimated eccentricity. It should not include virtual eccentricity due to lateral load.

$$\frac{f_a}{F_a} + \frac{f_b}{F_b} \leq 1 \qquad \text{68.17}$$

$$P \leq \tfrac{1}{4} P_e \qquad \text{68.18}$$

$$P_e = \left(\frac{\pi^2 E_m I_n}{h^2}\right)\left(1 - 0.577\left(\frac{e}{r}\right)\right)^3 \qquad \text{68.19}$$

$$F_a = \tfrac{1}{4} f'_m \left(1 - \left(\frac{h}{140r}\right)^2\right) \quad [h/r \leq 99] \quad \text{68.20}$$

$$F_a = \tfrac{1}{4} f'_m \left(\frac{70r}{h}\right)^2 \quad [\text{with } h/r > 99] \quad \text{68.21}$$

$$F_b = \tfrac{1}{3} f'_m \qquad \text{68.22}$$

Equation 68.17 recognizes that compressive stresses due to both axial load and bending may occur simultaneously, and it assumes a straight-line interaction between axial and flexural compression stresses. This is often referred to as the *unity equation*. Equation 68.18 is intended to safeguard against buckling due to an eccentric axial load. This replaces arbitrary slenderness limits used in previous codes.

The effect of wall slenderness is included in Eqs. 69.20 and 69.21. The allowable compressive stress is reduced by the final term in each equation based on its h/r ratio. The axial capacity of the wall is reduced by 50% when its slenderness reaches the transition value ($h/r = 99$) from Eq. 69.20 to Eq. 69.21.

Example 68.3

Determine the maximum allowable concentric axial load on an 8 in hollow unreinforced concrete masonry wall rising vertically 12 ft. $f'_m = 2000$ psi.

Solution

From App. 68.A,

$$r = 2.84 \text{ in}$$
$$A_n = 30 \text{ in}^2/\text{ft}$$
$$I_n = 334 \text{ in}^4/\text{ft}$$
$$S = 81.0 \text{ in}^3/\text{ft}$$

$$\frac{h}{r} = \frac{(12 \text{ ft})\left(12 \frac{\text{in}}{\text{ft}}\right)}{2.84 \text{ in}}$$
$$= 50.7 \quad [< 99]$$

From Eq. 68.20,

$$F_a = \tfrac{1}{4} f'_m \left(1 - \left(\frac{h}{140r}\right)^2\right)$$
$$= \left(\tfrac{1}{4}\right)\left(2000 \frac{\text{lbf}}{\text{in}^2}\right)\left(1 - \left(\frac{50.7}{140}\right)^2\right)$$
$$= 434 \text{ lbf/in}^2$$

For concentric loading,

$$P = A_n F_a = \left(30 \frac{\text{in}^2}{\text{ft}}\right)\left(434 \frac{\text{lbf}}{\text{in}^2}\right)$$
$$= 13{,}020 \text{ lbf/ft}$$

Check buckling. From Eq. 67.2,

$$E_m = 900 f'_m = (900)\left(2000 \frac{\text{lbf}}{\text{in}^2}\right)$$
$$= 1{,}800{,}000 \text{ lbf/in}^2$$
$$e = 0$$
$$P_e = \left(\frac{\pi^2 E_m I_n}{h^2}\right)\left(1 - 0.577\left(\frac{e}{r}\right)\right)^3$$
$$= \left(\frac{\pi^2 \left(1.8 \times 10^6 \frac{\text{lbf}}{\text{in}^2}\right)\left(334 \frac{\text{in}^4}{\text{ft}}\right)}{(144 \text{ in})^2}\right)(1)^3$$
$$= 286{,}150 \text{ lbf/ft}$$
$$P \leq \tfrac{1}{4} P_e$$
$$\leq \left(\tfrac{1}{4}\right)\left(286{,}150 \frac{\text{lbf}}{\text{ft}}\right)$$
$$\leq 71{,}540 \text{ lbf/ft} \quad [\text{OK}]$$

Example 68.4

Repeat Ex. 68.3 assuming an eccentricity of 1.5 in.

$$F_a = 434 \text{ lbf/in}^2$$
$$f_a = \frac{P}{A_n}$$
$$f_b = \frac{M}{S} = \frac{Pe}{S}$$
$$F_b = \tfrac{1}{3} f'_m \quad [\text{Eq. 69.22}]$$
$$= \left(\tfrac{1}{3}\right)\left(2000 \frac{\text{lbf}}{\text{in}^2}\right)$$
$$= 667 \text{ lbf/in}^2$$

$$\frac{\dfrac{P}{A}}{F_a} + \frac{\dfrac{Pe}{S}}{F_b} = 1 \quad \text{[Eq. 69.17]}$$

$$\frac{\dfrac{P}{30\,\dfrac{\text{in}^2}{\text{ft}}}}{434\,\dfrac{\text{lbf}}{\text{in}^2}} + \frac{\dfrac{P(1.5\ \text{in})}{81\,\dfrac{\text{in}^3}{\text{ft}}}}{667\,\dfrac{\text{lbf}}{\text{in}^2}} = 1$$

$$P = 9563\ \text{lbf/ft}$$

Check buckling.

$$P_e = \left(\frac{\pi^2 E_m I_n}{h^2}\right)\left(1 - 0.577\left(\frac{e}{r}\right)\right)^3$$

$$= \left(\frac{\pi^2\left(1.8\times10^6\,\dfrac{\text{lbf}}{\text{in}^2}\right)\left(334\,\dfrac{\text{in}^4}{\text{ft}}\right)}{(144\ \text{in})^2}\right)$$

$$\times\left(1 - (0.577)\left(\frac{1.5\ \text{in}}{2.84\ \text{in}}\right)\right)^3$$

$$= 96{,}160\ \text{lbf/ft}$$

$$P \le \tfrac{1}{4}P_e = \left(\tfrac{1}{4}\right)\left(96{,}160\,\frac{\text{lbf}}{\text{ft}}\right) = 24{,}041\ \text{lbf/ft} \quad \text{[OK]}$$

Check tension.

$$-f_a + f_b < F_t$$

$$-\frac{P}{A_n} + \frac{Pe}{S} < F_t = 19\,\frac{\text{lbf}}{\text{in}^2} \quad \text{[for type N mortar]}$$

$$-\frac{9563\,\dfrac{\text{lbf}}{\text{ft}}}{30\,\dfrac{\text{in}^2}{\text{ft}}} + \frac{\left(9563\,\dfrac{\text{lbf}}{\text{ft}}\right)(1.5\ \text{in})}{81\,\dfrac{\text{in}^3}{\text{ft}}} = -142\ \text{lbf/in}^2$$

The section is in compression, so there is no net tension.

10. SD WALL DESIGN: AXIAL COMPRESSION AND FLEXURE —UNREINFORCED

SD of unreinforced masonry walls is performed in accordance with principles of engineering mechanics with the masonry compressive stresses limited to $0.80f'_m$ and the tension stresses limited in accordance with values in Table 68.2 [*MSJC* Table 3.1.8.2.1]. The strength reduction factor is 0.60 for unreinforced masonry. The nominal axial capacity is modified on the basis of the wall slenderness and computed using *MSJC* Sec. 3.2.3 as

$$P_n = 0.80\left(0.80A_n f'_m\left(1 - \left(\frac{h}{140r}\right)^2\right)\right)$$

$$\text{[for } h/r \le 99]$$

$$P_n = 0.80\left(0.80A_n f'_m\left(\frac{70r}{h}\right)^2\right)$$

$$\text{[for } h/r > 99]$$

From *MSJC* Sec. 3.1.3,

$$P_u \le \phi P_n$$

P_u is calculated using the applicable load factors from the code. The *MSJC Code* does not require checking the unity equation (Eq. 69.17) or any other interaction formula for SD of unreinforced walls.

11. ASD WALL DESIGN: AXIAL COMPRESSION AND FLEXURE —REINFORCED

Reinforced walls subjected to axial compression and/or flexure are designed according to the following requirements. From *MSJC* Sec. 2.3.3.2.2,

$$F_b = \tfrac{1}{3}f'_m \qquad 68.23$$

The compressive force due to axial loading is limited by *MSJC* Sec. 2.3.3.2.1 to

$$P_a = (0.25f'_m A_n + 0.65A_{\text{st}}F_s)\left(1 - \left(\frac{h}{140r}\right)^2\right)$$

$$[h/r \le 99] \qquad 68.24$$

$$P_a = (0.25f'_m A_n + 0.65A_{\text{st}}F_s)\left(\frac{70r}{h}\right)^2$$

$$[h/r > 99] \qquad 68.25$$

Many design approaches are used to design reinforced masonry walls for combined axial compression and flexure, typically using iterative procedures (usually by means of computer programs) or graphic methods (i.e., interaction diagrams).

Figures 68.3 and 68.4 show interaction diagrams for cases where compression controls and where tension controls, respectively. The controlling condition is determined by comparing k to k_b, where

$$k_b = \frac{F_b}{F_b + \dfrac{F_s}{n}} \qquad 68.26$$

These diagrams are applicable to fully grouted masonry and to partially grouted masonry where the neutral axis falls within the face shell. Therefore, the designer must verify that $kd < t_{\text{fs}}$ when using these diagrams.

Example 68.5

Determine the amount of steel needed to adequately reinforce an 8 in fully grouted concrete masonry wall subjected to a compressive load of 1000 lbf/ft with an eccentricity at the midheight of the wall of 6 in. Use the following values.

$$f'_m = 1500\ \text{lbf/in}^2$$
$$F_b = 500\ \text{lbf/in}^2$$
$$n = 21.5$$
$$\frac{h}{r} = 68$$
$$A_n = 91.5\ \text{in}^2/\text{ft}$$
$$F_s = 24{,}000\ \text{lbf/in}^2$$

Solution

Calculate the relevant parameters for the interaction diagrams.

$$\frac{P}{F_b bt} = \frac{1000 \ \frac{\text{lbf}}{\text{ft}}}{\left(500 \ \frac{\text{lbf}}{\text{in}^2}\right)(12 \ \text{in})(7.625 \ \text{in})}$$

$$= 0.022$$

$$\frac{Pe}{F_b bt^2} = \frac{\left(1000 \ \frac{\text{lbf}}{\text{ft}}\right)(6 \ \text{in})}{\left(500 \ \frac{\text{lbf}}{\text{in}^2}\right)(12 \ \text{in})(7.625 \ \text{in})^2}$$

$$= 0.017$$

$$k_b = \frac{F_b}{F_b + \dfrac{F_s}{n}} \quad \text{[Eq. 68.26]}$$

$$= \frac{500 \ \frac{\text{lbf}}{\text{in}^2}}{500 \ \frac{\text{lbf}}{\text{in}^2} + \dfrac{24{,}000 \ \frac{\text{lbf}}{\text{in}^2}}{21.5}}$$

$$= 0.309$$

From Fig. 68.3, $k = 0.14$. Since $k < k_b$, tension governs and Fig. 68.4 should be used. From Fig. 68.4, $n\rho_t = 0.014 = \rho_t F_s / F_b$. For bonded materials experiencing the same strain, $n = f_1/f_2$.

$$\rho_t = \frac{0.014 F_b}{F_s} = \frac{(0.014)\left(500 \ \frac{\text{lbf}}{\text{in}^2}\right)}{24{,}000 \ \frac{\text{lbf}}{\text{in}^2}}$$

$$= 0.00029$$

$$A_s = \rho_t bt = (0.00029)(12 \ \text{in})(7.625 \ \text{in})$$

$$= 0.027 \ \text{in}^2/\text{ft}$$

From Table 68.3, choose no. 3 bars at 48 in on center for a steel area of $A_s = 0.028 \ \text{in}^2/\text{ft}$.

Check the allowable compressive force. For $h/r < 99$, use Eq. 68.26.

$$P_a = (0.25 f'_m A_n + 0.65 A_{\text{st}} F_s)\left(1 - \left(\frac{h}{140r}\right)^2\right)$$

$$= \left((0.25)\left(1500 \ \frac{\text{lbf}}{\text{in}^2}\right)\left(91.5 \ \frac{\text{in}^2}{\text{ft}}\right)\right.$$

$$+ (0.65)\left(0.028 \ \frac{\text{in}^2}{\text{ft}}\right)\left(24{,}000 \ \frac{\text{lbf}}{\text{in}^2}\right)\bigg)$$

$$\times \left(1 - \left(\frac{68}{140}\right)^2\right)$$

$$= 26{,}550 \ \text{lbf/ft} > 1000 \ \text{lbf/ft} \quad \text{[OK]}$$

12. SD WALL DESIGN: AXIAL COMPRESSION AND FLEXURE —REINFORCED

SD of reinforced walls for flexure and axial compression is similar to reinforced concrete design, although the values for many of the design parameters are different. The equivalent rectangular stress block has a depth of $0.80c$, and the maximum usable compressive masonry stress is assumed to be $0.80 f'_m$. The maximum usable compression fiber strain is assumed to be 0.0035 for clay masonry and 0.0025 for concrete masonry [*MSJC* Sec. 3.3.2].

Flexure of walls is generally due to out-of-plane loads such as wind, soil, earthquake, or eccentric axial loads. *MSJC* Sec. 3.3.5 presents the design requirements. Note that the effects of floor load eccentricity and wall deflection are included. The factored design moment and strength are computed from *MSJC* Sec. 3.3.5.4.

$$M_u = \frac{w_u h^2}{8} + P_{uf}\frac{e_u}{2} + P_u \delta_u \qquad \text{68.27}$$

$$M_u \leq \phi M_n \qquad \text{68.28}$$

$$M_n = (A_s f_y + P_u)\left(d - \frac{a}{2}\right) \qquad \text{68.29}$$

$$a = \frac{A_s f_y + P_u}{0.80 f'_m b} \qquad \text{68.30}$$

Determination of wall moment and axial load capacities are easily calculated using computer solutions or a spreadsheet to calculate various combinations of permitted values based on assumed strain distributions.

The design solution, however, must limit the reinforcement to ensure ductility as defined in *MSJC* Sec. 3.3.3.5. Walls designed using a seismic response modification factor, $R > 1.5$, must be capable of achieving a steel strain of 1.5 times yield prior to reaching the maximum permitted masonry compression strain. All walls where $M_u/V_{ud} \geq 1$ must be capable of achieving this steel strain.

The *MSJC Code* does not have any specific slenderness limitations for walls that have a relatively low axial stress ($< 0.05 f'_m$). Walls with a higher axial stress are limited to a slenderness ratio no greater than 30; however, a factored axial stress exceeding $0.20 f'_m$ is not permitted [*MSJC* Sec. 3.3.5.4]. Slenderness is addressed by placing limits on wall deflection. *MSJC* Sec. 3.3.5.5 requires that the horizontal midheight deflection of walls due to service loads, δ_s, must not exceed $0.007h$. The P-Δ effect and effects of masonry cracking must be included. The cracking moment is calculated from the modulus of rupture (Table 68.2) and the section modulus,

$$M_{cr} = S f_r \qquad \text{68.31}$$

When the service loads moment does not exceed the cracking moment ($M_{\text{ser}} < M_{cr}$), *MSJC* Eq. 3-30 applies.

$$\delta_s = \frac{5 M_{\text{ser}} h^2}{48 E_m I_g} \qquad \text{68.32}$$

Figure 68.3 *Interaction Diagram for Reinforced Wall (compression controls)*

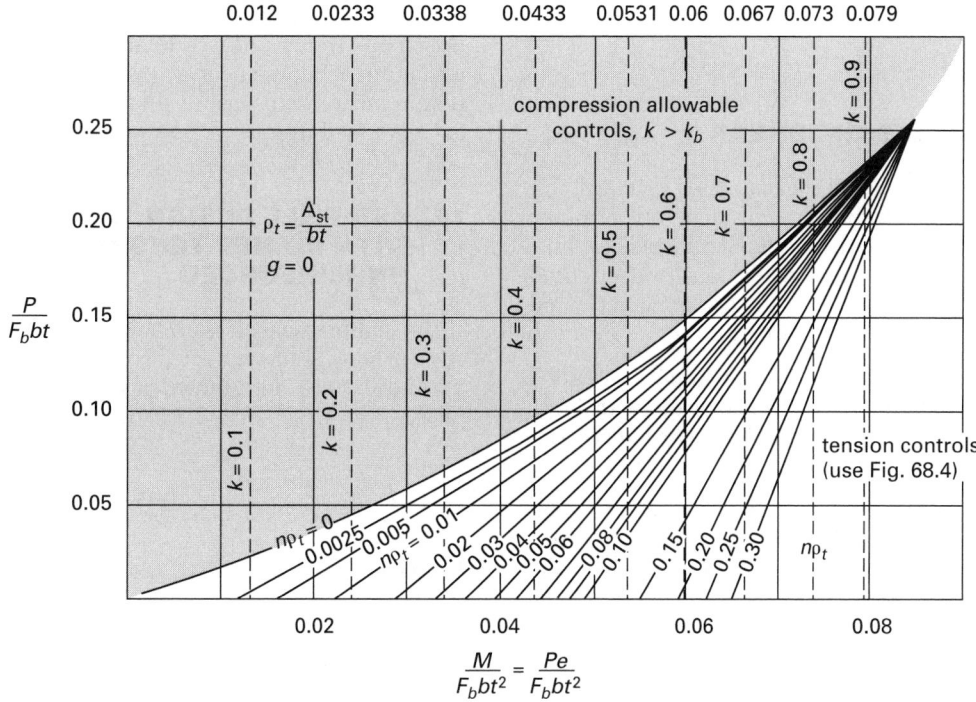

Reprinted by permission of the American Concrete Institute from *Masonry Designer's Guide*, copyright © 1993.

Figure 68.4 *Interaction Diagram for Reinforced Wall (tension controls)*

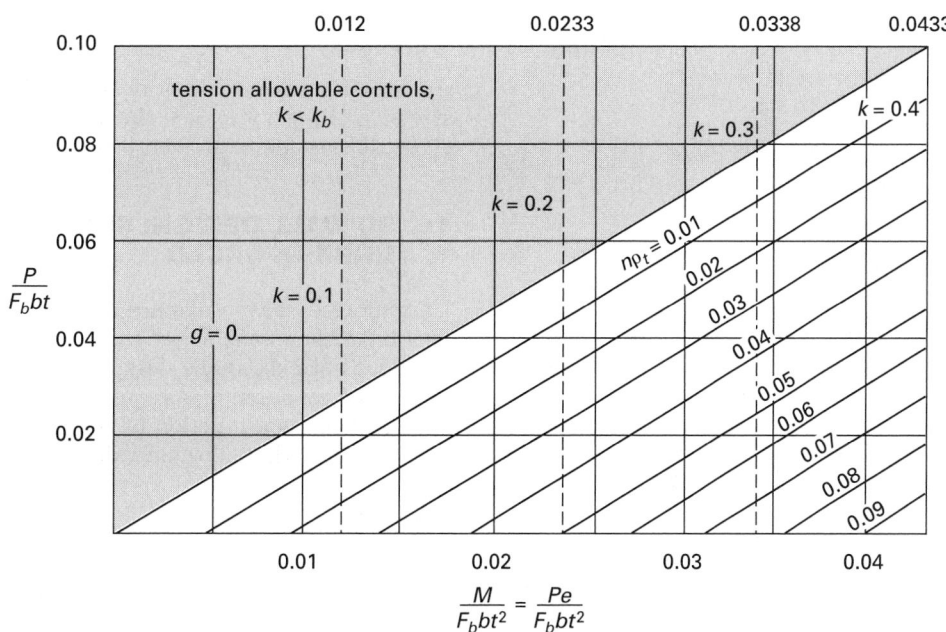

Reprinted by permission of the American Concrete Institute from *Masonry Designer's Guide*, copyright © 1993.

When $M_{cr} < M_{ser} < M_n$, *MSJC* Eq. 3-31 applies.

$$\delta_s = \frac{5M_{cr}h^2}{48E_mI_g} + \frac{5(M_{ser} - M_{cr})h^2}{48E_mI_{cr}} \qquad \textbf{68.33}$$

13. ASD WALL DESIGN: SHEAR

Design of masonry walls for shear focuses primarily on shear walls, since flexural shear due to out-of-plane bending seldom governs the design (whereas in-plane shear often controls). Shear walls must be oriented along both axes of the building, since seismic and wind lateral loads can occur in any direction.

Small openings in shear walls do not impact intended performance to a great degree. Shear walls with large openings can be thought of as smaller shear panels connected by panels subject to shear and flexure.

Loads distributed to shear walls depend on the relative rigidities of the floor diaphragms and shear walls. *Flexible diaphragms* transfer loads in proportion to the distance between the load and shear wall, whereas *rigid diaphragms* transfer loads in proportion to the relative stiffness of the walls. To design walls as rigid diaphragms, the proportions must meet the ratios in Table 68.4.

Shear wall design depends on the wall axial load for two reasons. First, walls in compression have greater shear resistance. Second, compression offsets flexural tension from lateral loads, which may eliminate the need for shear reinforcement.

Table 68.4 *Maximum Span-to-Width Ratios for Rigid Floor Diaphragms*

floor construction	span-to-width
cast-in-place solid concrete slab	5:1
precast concrete interconnected	4:1
metal deck with concrete fill	3:1
metal deck with no fill	2:1
cast-in-place gypsum deck (roof)	3:1

If there is no net flexural tension, the shear wall is designed as an unreinforced wall [*MSJC* Sec. 2.3.5.1]. Shear reinforcement is required only if $f_v > F_v$. If there is flexural tension in the wall, the wall must be reinforced to resist that tension. The design is checked to determine if that reinforcement is sufficient to resist shear as well. f_v is calculated according to *MSJC* Sec. 2.3.5.2.1.

If $f_v > F_v$, shear reinforcement must be provided. F_v is recalculated according to *MSJC* Sec. 2.3.5.2.3, assuming that all of the design shear is resisted by the shear reinforcement steel. If f_v still exceeds F_v, a larger wall must be used to resist shear.

MSJC Sec. 1.14 gives restrictions and requirements for structures based on the applicable seismic design category (SDC). Unreinforced shear walls are not permitted in SDC C, D, E, or F [*MSJC* Secs. 1.14.5, 1.14.6, 1.14.7]. Based on the SDC, the code may require the addition of minimum reinforcement at prescribed locations. These requirements help ensure ductility in the walls during extreme loadings.

14. ASD WALL DESIGN: SHEAR WALLS WITH NO NET TENSION— UNREINFORCED

If the compressive stress, P/A_n, exceeds flexural stress, M/S, the wall has no net flexural tension. In this case, f_v and F_v are determined according to the following procedure [*MSJC* Sec. 2.2.5.1].

$$f_v = \frac{VQ}{I_nb} = \frac{3V}{2A_n} \quad \text{[rectangular sections]} \qquad \textbf{68.34}$$

Under *MSJC* Sec. 2.2.5.2, allowable in-plane shear stress, F_v, is the least of

(a) $1.5\sqrt{f'_m}$ **68.35**

(b) 120 lbf/in^2 (0.83 MPa)   **68.36**

(c) $v + 0.45N_v/A_n$ **68.37**

In (c), v has values of 37 lbf/in² (0.26 MPa) for masonry in running bond that is not grouted solid, 37 lbf/in² (0.26 MPa) for masonry in other than running bond with open end units that are grouted solid, 60 lbf/in² (0.41 MPa) for masonry in running bond that is grouted solid, and 15 lbf/in² (0.10 MPa) for masonry in other than running bond with other than open end units that are grouted solid.

As for other stress conditions in ASD, the calculated stress, f_v, may not exceed the allowable F_v. If $f_v > F_v$, the wall must be reinforced or increased in size.

15. SD WALL DESIGN: SHEAR WALLS— UNREINFORCED

Unreinforced shear walls are addressed in *MSJC Code* Sec. 3.2.4. For walls where the factored shear strength, V_u, does not exceed the design shear strength, no special shear reinforcing is required. The nominal shear strength, V_n, is a function of the net cross section and is the least of the following values.

(a) $3.8A_n\sqrt{f'_m}$ **68.38**

(b) $300A_n$ **68.39**

(c) $56A_n + 0.45N_u$ $\left[\begin{array}{c}\text{for running bond masonry} \\ \text{not grouted solid}\end{array}\right]$ **68.40**

(d) $90A_n + 0.45N_u$ $\left[\begin{array}{c}\text{for running bond masonry} \\ \text{grouted solid}\end{array}\right]$ **68.41**

Values of V_n are reduced for patterns other than running bond. If $V_u > \phi V_n$, then the wall must be reinforced for shear or increased in size. The strength reduction factor, ϕ, is 0.80 for masonry subjected to shear [*MSJC* Sec. 3.1.4.3].

16. ASD WALL DESIGN: SHEAR WALLS WITH NET TENSION—REINFORCED

Walls subjected to flexural tension are reinforced to resist the tension. In this case, calculated shear stress in the masonry is determined by Eq. 68.44 [*MSJC* Sec. 2.3.5.2].

$$f_v = \frac{V}{bd} \qquad\qquad 68.42$$

Where reinforcement is not provided to resist all of the calculated shear, F_v is determined from *MSJC* Sec. 2.3.5.2.2. The quantity M/Vd is always taken as a positive value.

$$F_v = \tfrac{1}{3}\left(4 - \frac{M}{Vd}\right)\sqrt{f'_m}$$
$$< 80 - 45\left(\frac{M}{Vd}\right) \quad [M/Vd < 1] \qquad 68.43$$
$$F_v = \sqrt{f'_m}$$
$$< 35 \text{ lbf/in}^2 \quad (0.24 \text{ MPa}) \quad [M/Vd \geq 1] \quad 68.44$$

If $f_v < F_v$, the section is satisfactory. If $f_v > F_v$, shear reinforcement is required, and F_v is recalculated with reinforcement provided to resist all of the calculated shear. *MSJC* Sec. 2.3.5.2.3 specifies the allowable shear stress as shown in Eq. 68.45.

$$F_v = \tfrac{1}{2}\left(4 - \frac{M}{Vd}\right)\sqrt{f'_m}$$
$$< 120 - 45\left(\frac{M}{Vd}\right) \quad [M/Vd < 1] \qquad 68.45$$
$$F_v = 1.5\sqrt{f'_m}$$
$$< 75 \text{ lbf/in}^2 \quad (0.52 \text{ MPa}) \quad [M/Vd \geq 1] \quad 68.46$$

If f_v still exceeds F_v, the wall size must be increased.

When shear reinforcement is required to resist all of the calculated shear, the *MSJC Code* contains the following requirements. The minimum area of shear reinforcement placed parallel to the direction of applied shear force according to *MSJC* Sec. 2.3.5.3 is

$$A_v = \frac{Vs}{F_s d} \qquad\qquad 68.47$$

Maximum spacing of shear reinforcement is the lesser of $d/2$ or 48 in (1219 mm). Reinforcement must also be provided perpendicular to the shear reinforcement. This reinforcement must be uniformly distributed, have a maximum spacing of 8 ft (2.44 m), and have a minimum reinforcement area of at least $^1/_3 A_v$ [*MSJC* Secs. 2.3.5.3.1 and 2.3.5.3.2].

17. SD WALL DESIGN: SHEAR WALLS— REINFORCED

Walls subjected to in-plane loads (shear walls) are addressed in *MSJC* Secs. 3.3.4.1.2 and 3.3.6. Shear forces may be shared by the masonry and the shear reinforcement as indicated by the following equation.

$$V_n = V_m + V_s \quad [3.2.4.1.2] \qquad\qquad 68.48$$
$$\leq 6A_n\sqrt{f'_m} \quad [\text{when } M_u/V_u d_v < 0.25] \qquad 68.49$$
$$\leq 4A_n\sqrt{f'_m} \quad [\text{when } M_u/V_u d_v > 1.0] \qquad 68.50$$

When $0.25 < M_u/V_u d_v < 1.0$, the value of V_n may be interpolated.

The nominal shear strength provided by the masonry, V_m, is computed from *MSJC* Sec. 3.3.4.1.2.1 as

$$V_m = \left(4 - 1.75\frac{M_u}{V_u d_v}\right)A_n\sqrt{f'_m} + 0.25P_u \quad 68.51$$

If $M_u/V_u d_v$ is conservatively taken as its maximum value of 1.0 and the axial forcee is minimal, this equation is

$$V_m = 2.25A_n\sqrt{f'_m} \qquad\qquad 68.52$$

The nominal shear strength provided by shear reinforcement is calculated from *MSJC* Sec. 3.3.4.1.2.3 as

$$V_s = 0.5\frac{A_v}{s}f_y d_v \qquad\qquad 68.53$$

The *MSJC Code* limits the maximum reinforcement permitted in SD shear wall design, similar to the limits for SD wall design related to axial compression and flexure. However, *MSJC* Sec. 3.3.3.5.1 requires that shear walls that are part of structures designed for a seismic response modification factor, R, greater than 1.5 must be capable of achieving a steel strain of five times yield prior to reaching the maximum permitted masonry strain. This requirement is severe and may control the design.

The flexural and axial strength of shear walls is calculated using the criteria for nomial strength of beams, piers, and columns [*MSJC* Sec. 3.3.4.1.1]. The following equations are used to compute the nominal axial capacity.

$$P_n = 0.80(0.80f'_m(A_n - A_s) + f_y A_s)\left(1 - \left(\frac{h}{140r}\right)^2\right)$$
$$[\text{when } h/r \leq 99] \quad 68.54$$

$$P_n = 0.80(0.80f'_m(A_n - A_s) + f_y A_s)\left(\frac{70r}{h}\right)^2$$
$$[\text{when } h/r > 99] \quad 68.55$$

The nominal axial strength, P_n, when reduced by the strength reduction factor, must not exceed the factored axial loads. The strength reduction factor, ϕ, is 0.80 for masonry subjected to shear [*MSJC* Sec. 3.1.4.3].

$$P_u \le \phi P_n$$

18. MULTI-WYTHE CONSTRUCTION

Multiple-wythe masonry walls are analyzed using the same procedures outlined in previous sections. In some cases, however, determination of wall thickness or other parameters may vary, depending on the type of wall and how it is loaded.

Multi-wythe walls can be designed for *composite action* (i.e., where the two wythes act as one unit to resist loads) or for *noncomposite action*. Masonry veneer is a nonstructural facing that transfers the out-of-plane load to the backing (structural wythe). Chapter 6 of the *MSJC Code* contains requirements for both anchored and adhered veneer. SD of multi-wythe construction is not addressed by the code.

19. MULTI-WYTHE CONSTRUCTION—COMPOSITE ACTION USING ASD

Multi-wythe walls designed for composite action are tied together by either wall ties or masonry headers. (Wall ties are more common and provide more ductility than headers.) The ties or headers ensure adequate shear transfer across the collar joint. When wall ties are used, the space between the wythes (collar joint) must be filled with mortar or grout. To help ensure shear transfer, specific requirements govern the size and spacing of the headers and ties, and shear stresses are limited.

For composite action, stresses are calculated using section properties based on the minimum transformed net cross-sectional area of the composite member. Areas of dissimilar materials are transformed in accordance with relative elastic moduli ratios.

Shear stresses at the interfaces between wythes and collar joints or within headers are limited to the following: mortared collar joints, 5 lbf/in² (0.034 MPa); grouted collar joints, 10 lbf/in² (0.069 MPa); headers, $\sqrt{}$ (unit compressive strength of header), in lbf/in² (MPa), over net area of header [*MSJC* Sec. 2.1.5.2.2].

When bonded by headers, the headers must be uniformly distributed. The sum of their cross-sectional areas must be at least 4% of the wall surface area. Headers must be embedded at least 3 in (76 mm) into each wythe [*MSJC* Sec. 2.1.5.2.3].

Wythes bonded using wall ties are subject to the following restrictions [*MSJC* Sec. 2.1.5.2.4].

- Maximum spacing between ties is 36 in (914 mm) horizontally and 24 in (610 mm) vertically.

- With a W1.7 wire, one wall tie is needed for every $2^{2}/_{3}$ ft² (0.25 m²) of wall. With W2.8 wire, one wall tie is needed for every $4^{1}/_{2}$ ft² (0.42 m²) of wall.

Cross wires of joint reinforcement or rectangular wall ties may be used as wall ties. "Z" wall ties can be used only on walls constructed of solid masonry units.

Example 68.6

Determine the maximum lateral wind load that can be sustained by a nonloadbearing multi-wythe composite wall constructed of 8 in concrete masonry units (CMUs), a 3 in grouted collar joint, and 4 in meridian face brick. Both wythes are constructed with type-S portland cement-lime mortar. The wall spans vertically 20 ft and can be assumed to be simply supported laterally at the top and bottom. Neglect the self-weight of the wall and utilize the one-third increase in allowable stress for wind loads. The following properties are to be used.

	CMU	brick	grout
f'_m or f'_g (in lbf/in²)	1500	2500	2000
E_m (in lbf/in²)	1,350,000	1,750,000	1,000,000

Solution

$$n_{\mathrm{br}} = \frac{E_{m,\mathrm{br}}}{E_{m,\mathrm{CMU}}} = \frac{1{,}750{,}000 \ \dfrac{\mathrm{lbf}}{\mathrm{in}^2}}{1{,}350{,}000 \ \dfrac{\mathrm{lbf}}{\mathrm{in}^2}} = 1.3$$

$$n_g = \frac{E_{m,\mathrm{g}}}{E_{m,\mathrm{CMU}}} = \frac{1{,}000{,}000 \ \dfrac{\mathrm{lbf}}{\mathrm{in}^2}}{1{,}350{,}000 \ \dfrac{\mathrm{lbf}}{\mathrm{in}^2}} = 0.74$$

The widths of the transformed areas (relative to a 12 in long CMU) are

brick: $(1.3)(12 \text{ in}) = 15.6 \text{ in}$

grout: $(0.74)(12 \text{ in}) = 8.9 \text{ in}$

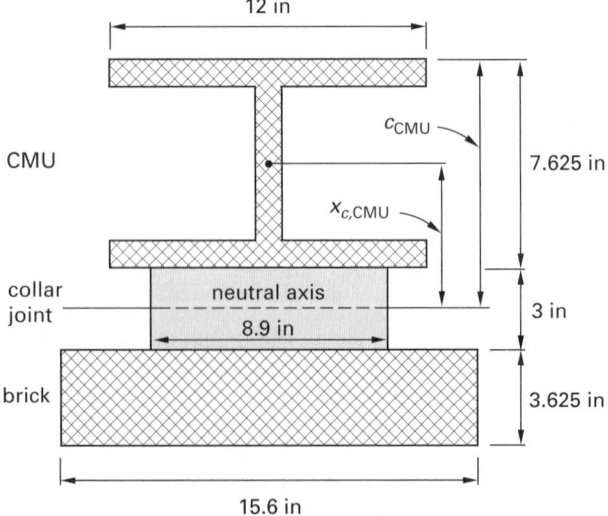

Determine the distance from the centroid of the transformed area to the centroid of the CMU.

The width of the 4 in face brick is $3^5/_8$ in (3.625 in). From App. 68.A, $A_{n,\text{CMU}} = 30$ in²/ft.

$$x_{c,\text{CMU}} = \frac{\sum Ax_i}{\sum A}$$

$$= \frac{\begin{array}{l}(30\,\text{in}^2)(0\,\text{in}) \\[4pt] + (15.6\,\text{in})(3.625\,\text{in})\left(\dfrac{7.625\,\text{in}}{2} + 3\,\text{in} + \dfrac{3.625\,\text{in}}{2}\right) \\[8pt] + (8.9\,\text{in})(3\,\text{in})\left(\dfrac{7.625\,\text{in}}{2} + \dfrac{3.0\,\text{in}}{2}\right)\end{array}}{30\,\text{in}^2 + (15.6\,\text{in})(3.625\,\text{in}) + (8.9\,\text{in})(3\,\text{in})}$$

$$= 5.56\,\text{in} \quad [\text{below the centroid of the CMU}]$$

Determine the transformed moment of inertia.

From App. 68.B, $I_{\text{CMU}} = 308.7$ in⁴/ft.

From App. 68.C, $I_{\text{br}} = 42.9$ in⁴/ft.

$$\begin{aligned}
I_{\text{tr}} &= \sum (I + Ax_c^2) \\
&= 308.7\,\text{in}^4 + (30\,\text{in}^2)(5.56\,\text{in})^2 \\
&\quad + \frac{(8.9\,\text{in})(3\,\text{in})^3}{12} \\
&\quad + (8.9\,\text{in})(3\,\text{in})\left(5.56\,\text{in} - \frac{7.625\,\text{in}}{2} - \frac{3\,\text{in}}{2}\right)^2 \\
&\quad + (42.9\,\text{in}^4)\left(\frac{15.6\,\text{in}}{12\,\text{in}}\right) \\
&\quad + (15.6\,\text{in})(3.625\,\text{in}) \\
&\quad \times \left(\frac{7.625\,\text{in}}{2} + 3\,\text{in} + \frac{3.625\,\text{in}}{2} - 5.56\,\text{in}\right)^2 \\
&= 1847\,\text{in}^4 \quad [\text{per foot of wall}]
\end{aligned}$$

Determine the allowable flexural tension from Table 68.1.

$$F_{b,\text{CMU}} = \left(25\,\frac{\text{lbf}}{\text{in}^2}\right)(1.33) = 33.3\,\text{lbf/in}^2$$

$$F_{b,\text{br}} = \left(40\,\frac{\text{lbf}}{\text{in}^2}\right)(1.33) = 53.2\,\text{lbf/in}^2$$

Determine the maximum allowable moment for each wythe.

$$c_{\text{CMU}} = x_{c,\text{CMU}} + \frac{d_{\text{CMU}}}{2} = 5.56\,\text{in} + \frac{7.625\,\text{in}}{2}$$

$$= 9.37\,\text{in}$$

$$\text{CMU: } M = \frac{F_b I_{\text{tr}}}{c}$$

$$= \frac{\left(33.3\,\dfrac{\text{lbf}}{\text{in}^2}\right)(1847\,\text{in}^4)}{9.37\,\text{in}}$$

$$= 6564\,\text{in-lbf}$$

$$c_{\text{brick}} = d_{\text{total}} - c_{\text{CMU}}$$

$$= 7.625\,\text{in} + 3\,\text{in} + 3.625\,\text{in} - 9.37\,\text{in}$$

$$= 4.88\,\text{in}$$

$$\text{brick: } M = \frac{F_b I_{\text{tr}}}{c n_{\text{br}}}$$

$$= \frac{\left(53.2\,\dfrac{\text{lbf}}{\text{in}^2}\right)(1847\,\text{in}^4)}{(4.88\,\text{in})(1.3)}$$

$$= 15{,}489\,\text{in-lbf}$$

This design is limited by the allowable flexural tension stress in the CMU wythe: $M_{\text{max}} = 6564$ in-lbf.

Since the wall is simply supported laterally at its top and bottom, $M_{\text{max}} = wl^2/8$.

$$w = \frac{8M_{\text{max}}}{l^2} = \frac{(8)\left(6564\,\dfrac{\text{in-lbf}}{\text{ft}}\right)}{(20\,\text{ft})^2\left(12\,\dfrac{\text{in}}{\text{ft}}\right)}$$

$$= 10.9\,\text{lbf/ft}^2$$

Check the shear at the collar joint. For simply supported walls,

$$V = \frac{wl}{2} = \frac{\left(10.9\,\dfrac{\text{lbf}}{\text{ft}^2}\right)(20\,\text{ft})}{2}$$

$$= 109\,\text{lbf/ft}$$

The statical moment at the CMU/collar joint interface is

$$Q = (15.6\,\text{in})(3.625\,\text{in})(3.07\,\text{in}) + (8.9\,\text{in})(3\,\text{in})(0.25\,\text{in})$$

$$= 180.3\,\text{in}^3$$

Use Eq. 68.34.

$$f_v = \frac{VQ}{I_{\text{tr}} b} = \frac{(109\,\text{lbf})(180.3\,\text{in}^3)}{(1847\,\text{in}^4)(8.9\,\text{in})}$$

$$= 1.2\,\frac{\text{lbf}}{\text{in}^2} < 10\,\frac{\text{lbf}}{\text{in}^2} \quad [\text{OK}]$$

20. MULTI-WYTHE CONSTRUCTION—NONCOMPOSITE ACTION USING ASD

For noncomposite design, each wythe is designed to individually resist the applied loads. Unless a more detailed analysis is performed, the following guidelines and restrictions apply [*MSJC* Sec. 2.1.5.3.1].

- Collar joints may not contain headers, grout, or mortar.
- Gravity loads from supported horizontal members are resisted by the wythe nearest to the center of

the span of the supported member. Any resulting bending moment about the weak axis of the wall is distributed to each wythe in proportion to its relative stiffness.

- In-plane loads are carried only by the wythe subjected to the load.

- Out-of-plane loads are resisted by all wythes in proportion to their relative flexural stiffness.

- Stresses are determined using the net cross-sectional area of the member or part of member under consideration and, where applicable, using the transformed area concept described previously.

- Collar joint width is limited to 4.5 in (114 mm) unless a detailed wall tie analysis is performed.

Wythes can be connected using wall ties or adjustable ties. If wall ties are used, the spacings described previously under composite action are applicable. Adjustable ties are subject to the following restrictions [*MSJC* Sec. 2.1.5.3.2].

- One tie must be provided for each 1.77 ft^2 (0.16 m^2) of wall area.

- The maximum horizontal and vertical spacing is 16 in (406 mm).

- Adjustable ties may not be used when the misalignment of bed joints from one wythe to the other exceeds $1^1/_4$ in (32 mm).

- Maximum clearance between connecting parts of the tie is $^1/_{16}$ in (1.6 mm).

- Pintle ties shall have at least two pintle legs of wire size W2.8.

69 Masonry Columns

Symbols

ρ_t	ratio of steel area to gross area of masonry, bt	–	–
ϕ	strength reduction factor	–	–

Subscripts

a	allowable
b	bending
k	kern
m	masonry
n	net
s	steel

Nomenclature

A	area	in^2	mm^2
A_n	net cross-sectional area of masonry	in^2	mm^2
A_s	cross-sectional area of reinforcing steel	in^2	mm^2
A_{st}	area of laterally tied longitudinal steel in a column or pilaster	in^2	mm^2
b	column dimension	in	mm
e	eccentricity	in	mm
e_k	kern eccentricity	in	mm
f'_m	compressive strength of masonry	lbf/in^2	MPa
f_y	specified steel yield strength	lbf/in^2	MPa
F_b	allowable compressive stress due to flexure alone	lbf/in^2	MPa
F_s	allowable stress in reinforcement	lbf/in^2	MPa
g	ratio of distance between tension steel and compression steel to overall column depth	–	–
h	effective height	in	mm
M	maximum moment occurring simultaneously with design shear force V	in-lbf	N·m
n	modular ratio	–	–
P	axial load	lbf	N
P_a	allowable load		
$P_{biaxial}$	maximum allowable design load for the specified biaxial load	lbf	N
P_n	nominal axial strength	lbf	N
P_o	maximum allowable design load for zero eccentricity	lbf	N
P_u	factored axial load	lbf	N
P_x	maximum allowable design load for the specified eccentricity in the x direction	lbf	N
P_y	maximum allowable design load for the specified eccentricity in the y direction	lbf	N
r	radius of gyration	in	mm
R	seismic response modification factor	–	–
S	section modulus	in^3	mm^3
t	thickness of column	in	mm

1. MASONRY COLUMNS

Columns are isolated structural elements subject to axial loads and often to flexure. To distinguish columns from short walls, *Building Code Requirements for Masonry Structures* (the *MSJC Code*) defines a column as an isolated vertical member whose length-to-thickness ratio does not exceed three and whose height-to-thickness ratio is greater than four [*MSJC Sec. 1.6*].

Pilasters are sometimes confused with columns. Whereas columns are isolated elements, pilasters are integral parts of masonry walls. A pilaster is a thickened wall section that typically projects beyond one or both faces. It provides lateral stability to the masonry wall and may carry axial loads as well.

Concrete masonry columns are most often constructed of hollow masonry units, with grout and reinforcement in the masonry cores. Brick columns are typically constructed of solid units laid with an open center, as shown in Fig. 69.1. In both cases, units in successive courses should be laid in a running bond (i.e., overlapping) pattern to avoid a continuous vertical mortar joint up the side of the column.

Columns are designed for axial compression, flexure, and shear as needed. Design methods and assumptions are essentially the same as those presented for masonry walls. However, three significant differences may exist. (a) Due to the structural importance of columns, the *MSJC Code* imposes additional criteria for minimum size, slenderness, and reinforcement. (b) Biaxial bending may be significant in columns. (c) Compression reinforcement may be necessary. Since all columns must be reinforced, unreinforced masonry column design is not permitted.

MSJC Code Ch. 2 presents the requirements for allowable stress design (ASD) while Ch. 3 presents strength design (SD) requirements. Columns may not be designed using *MSJC Code* Ch. 5, which describes empirical design of masonry.

2. SLENDERNESS

Whereas slenderness effects are not a major consideration with most walls, they can be significant for columns. Load-carrying capacity can be reduced due to either buckling or to additional bending moments caused by deflection (P-Δ effects). In the *MSJC Code*, slenderness effects are included in the calculation of permitted compressive stress for reinforced masonry. Columns are required by the *MSJC Code* to have nominal side dimensions of at least 8 in (203 mm) [*MSJC* Secs. 2.1.6.1 and 3.3.4.4.4]. ASD limits the ratio of the effective height to its least dimension to 25 [*MSJC* Sec. 2.1.6.2], while SD limits the distance between supports to 30 times its least dimension [*MSJC* Sec. 3.3.4.4.4].

Figure 69.1 *Column Reinforcement*

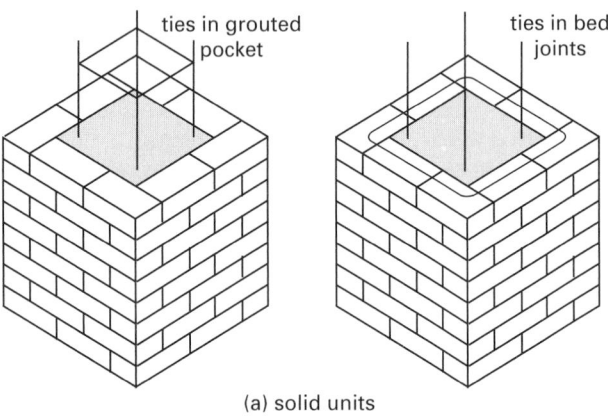

(a) solid units

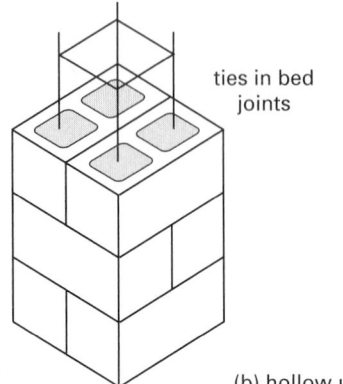

(b) hollow units

The effective height of a column varies with end conditions. *Effective height* is equal to the distance between points of inflection of the member's buckled configuration. If the restraint conditions are unknown, the clear height between supports should be used.

3. REINFORCEMENT

The *MSJC Code* requires a minimum amount of reinforcement to prevent brittle collapse of columns. The general design provisions for reinforced masonry described in Ch. 69 are used to analyze masonry columns, as well.

The area of vertical column reinforcement must be between $0.0025A_n$ and $0.04A_n$. A minimum of four reinforcement bars is required, which allows ties to provide a confined core of masonry [*MSJC* Secs. 2.1.6.4 and 3.3.4.4.1]. Vertical bars must have a clear distance between bars of at least $1^1/_2$ times the nominal bar diameter, but not less than $1^1/_2$ in (38 mm) [*MSJC* Sec. 1.13.3.2].

Lateral ties in columns enclose the vertical bars and provide support to prevent buckling of column reinforcement acting in compression and resistance to diagonal tension for columns subjected to shear. Requirements are based on those for concrete columns and are summarized as follows [*MSJC* Secs. 2.1.6.5 and 3.3.4.4.2].

- Lateral ties must be at least $^1/_4$ in (6.4 mm) in diameter, with maximum vertical spacing of 16 longitudinal bar diameters, 48 lateral tie bar or wire diameters, or the least cross-sectional dimension of the member.

- Every corner and alternate longitudinal bar must be supported by a tie corner. Tie corners may not have an included angle of more than 135°. No vertical bar can be more than 6 in (152 mm) away from such a laterally supported bar.

- Lateral ties may be placed in either a mortar joint or in grout, although placement in grout is preferable. Ties in contact with the vertical reinforcing bars are more effective in preventing buckling and result in more ductile behavior. In seismic design categories (SDC) C, D, E, and F, the *MSJC Code* requires ties to be embedded in grout, be spaced no more than 8 in (230 mm) on center, and be at least $^3/_8$ in (9.5 mm) diameter.

- Where longitudinal bars are located around the perimeter of a circle, a complete circular lateral tie is permitted. (The lap length for circular ties is 48 tie diameters.)

- Lateral ties must be located within one-half the lateral tie spacing at the top and bottom of the column.

- Where beams or brackets frame into a column from four directions, lateral ties may be terminated within 3 in (76 mm) below the lowest reinforcement.

When designing devices used to transfer lateral support to columns, a minimum force of 1000 lbf (4448 N) must be used [*MSJC* Sec. 2.1.8.3]. In SDC C, D, E, and F, the *MSJC Code* mandates further prescriptive requirements, including ties around anchor bolts and size and spacing of column ties.

4. ASD DESIGN FOR AXIAL LOAD AND BENDING

The eccentricity of column loads is an important consideration because of slenderness effects. Eccentricity can be due to eccentric axial loads, lateral loads, or a column that is out of plumb. For column design, *MSJC* Sec. 2.1.6.3 requires a minimum eccentricity equal to 0.1 times each side dimension. This requirement is intended to account for construction imperfections that may introduce eccentricities. If the actual eccentricity exceeds the code minimum, the actual eccentricity is used in the design. Stresses about each principal axis must be checked independently.

The method used to design the column depends on the magnitude of the axial load relative to the bending moment. There are three possibilities: (a) the column is in pure compression and the allowable axial load is governed by the allowable axial compressive force, P_a; (b) the allowable moment and axial force are governed by the allowable flexural compressive stress, F_b; (c) the allowable moment and axial force are governed by the allowable tensile stress in the reinforcement, F_s.

Figure 69.2 *Minimum Design Eccentricity*

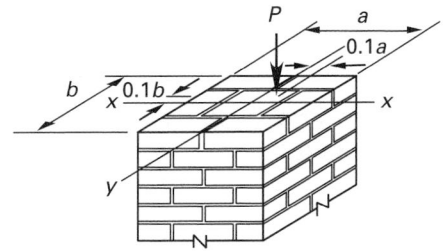

5. ASD DESIGN FOR PURE COMPRESSION

When the eccentricity falls within the *kern* of the section (the middle third), the entire section will be in compression. Transformed longitudinal column steel can be included using a transformation factor of n, although neglecting the steel is conservative. Equations for walls in compression are used to determine capacity [*MSJC* Sec. 2.3.3.2.1].

$$P_a = (0.25f'_m A_n + 0.65A_{st}F_s)\left(1 - \left(\frac{h}{140r}\right)^2\right)$$

$$[h/r \le 99] \qquad 69.1$$

$$P_a = (0.25f'_m A_n + 0.65A_{st}F_s)\left(\frac{70r}{h}\right)^2$$

$$[h/r > 99] \qquad 69.2$$

Example 69.1

Determine the maximum column height for the clay brick column shown. The column carries a concentric axial load of 100,000 lbf. $f'_m = 2500$ lbf/in^2, $A_n = 244$ in^2, $S = 636$ in^3, and $r = 4.51$ in.

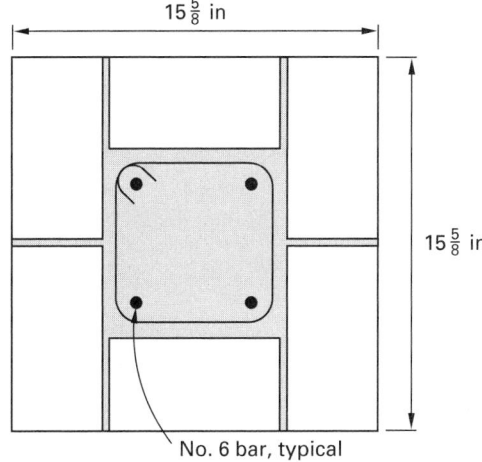

15$\frac{5}{8}$ in

15$\frac{5}{8}$ in

No. 6 bar, typical

Solution

The *MSJC Code* allows a maximum height of

$$(25)(15.625 \text{ in}) = 390.6 \text{ in}$$

Check the load capacity. The minimum required eccentricity is

$$e = (0.1)(15.625 \text{ in}) = 1.56 \text{ in}$$

$$e_k = \frac{t}{6} = \frac{15.625 \text{ in}}{6} = 2.6 \text{ in}$$

Since $e < e_k$, the section is in pure compression.

Neglecting the steel and assuming $h/r < 99$, Eq. 69.1 becomes

$$P_a = 0.25f'_m A_n \left(1 - \left(\frac{h}{140r}\right)^2\right)$$

$$100,000 \text{ lbf} = (0.25)\left(2500 \frac{\text{lbf}}{\text{in}^2}\right)(244 \text{ in}^2)$$

$$\times \left(1 - \left(\frac{h}{(140)(4.51 \text{ in})}\right)^2\right)$$

$$h = 370 \text{ in} < 390.6 \text{ in}$$

Check the h/r ratio.

$$h/r = \frac{370 \text{ in}}{4.51 \text{ in}} = 82 < 99 \quad [\text{OK}]$$

The maximum column height is 370 in.

6. SD DESIGN FOR PURE COMPRESSION

SD for pure compression is similar to ASD, with the nominal axial strength reduced based on the effects of slenderness. Using *MSJC* Sec. 3, the following equations

Structural

are used to compute the nominal axial capacity [*MSJC* Sec. 3.3.4.1.1].

$$P_n = 0.80 \left(0.80 f'_m (A_n - A_s) + f_y A_s \right)$$
$$\times \left(1 - \left(\frac{h}{140r} \right) \right) \quad \text{[when } h/r \leq 99] \quad \textbf{69.3}$$

$$P_n = 0.80 \left(0.80 f'_m (A_n - A_s) + f_y A_s \right)$$
$$\times \left(\frac{70r}{h} \right)^2 \quad \text{[when } h/r > 99] \qquad \textbf{69.4}$$

The factored axial loads must not exceed the nominal axial strength, P_n, reduced by the strength reduction factor.

$$P_u \leq \phi P_n \qquad \textbf{69.5}$$

$\phi = 0.90$ for reinforced masonry subjected to flexure, axial load, or combinations thereof [*MSJC* Sec. 3.1.4.1].

7. ASD DESIGN FOR COMPRESSION AND BENDING

For large eccentricities, the section is subject to axial compression and flexure, and the interaction of vertical load and bending moment must be checked as for reinforced masonry walls, typically using interaction diagrams or computer solutions.

The interaction diagrams shown in Apps. 69.A through F were developed for rectangular columns with equal areas of top and bottom steel, symmetrically placed and adequately tied. Three cases are presented representing three values of the steel spacing ratio, g. Appendices 69.A through C assume the allowable masonry compressive stress controls the design. Appendices 69.D through F assume allowable steel tensile stress governs. If it is unknown whether masonry or steel stress governs, the required value of ρ_t should be calculated from both interaction diagrams, and the larger steel area should be used.

Example 69.2

For the column in Ex. 69.1, assume a compressive load occurs at an eccentricity of 4.7 in. Determine the maximum load if the effective column height is 18 ft. The modular ratio, n, is 12. Use interaction curves for $g = 0.6$.

Solution

Use Eq. 69.1 as in Ex. 69.1.

$$P_a = 0.25 f'_m A_n \left(1 - \left(\frac{h}{140r} \right)^2 \right)$$

$$= (0.25) \left(2500 \ \frac{\text{lbf}}{\text{in}^2} \right) (244 \ \text{in}^2)$$

$$\times \left(1 - \left(\frac{(18 \ \text{ft}) \left(12 \ \frac{\text{in}}{\text{ft}} \right)}{(140)(4.51 \ \text{in})} \right)^2 \right)$$

$$= 134{,}650 \ \text{lbf}$$

$$\frac{e}{t} = \frac{4.7 \ \text{in}}{15.625 \ \text{in}} = 0.30$$

$$n\rho_t = \frac{n A_s}{bt} = \frac{(12)(1.76 \ \text{in}^2)}{(15.625 \ \text{in})(15.625 \ \text{in})} = 0.087$$

From App. 69.B, $P/F_b bt = 0.31$. The allowable bending stress is limited to one-third of the compressive strength.

$$P = 0.31 F_b bt$$

$$= (0.31) \left(\tfrac{1}{3} \right) \left(2500 \ \frac{\text{lbf}}{\text{in}^2} \right) (15.625 \ \text{in})(15.625 \ \text{in})$$

$$= 63{,}070 \ \text{lbf} < P_a \quad \text{[OK]}$$

8. SD DESIGN FOR COMPRESSION AND BENDING

Columns subjected to axial compression and flexure are affected by the interaction of forces for both tension and compression stresses. SD of reinforced masonry columns is similar to reinforced concrete design, although the values for many of the design parameters are different. The design assumptions are given in *MSJC* Sec. 3.3.2. The equivalent rectangular stress block has a depth, $a = 0.80c$, and the maximum usable compressive masonry stress is assumed to be $0.80f'_m$. The maximum usable compression fiber strain is assumed to be 0.0035 for clay masonry and 0.0025 for concrete masonry. Generally, an interaction diagram or computer solution is utilized to evaluate the capacity of a masonry column section. The stress and strain relationships are entered into a spreadsheet to quickly generate a custom interaction diagram for a specific column configuration.

The nominal axial capacity of a column is reduced based on its slenderness. Equations 70.3 and 70.4 are also applicable for columns subjected to axial compression and flexure. In addition to the minimum reinforcement criteria, $A_s \geq 0.0025A_n$, Ch. 3 of the *MSJC Code* places a limit on the maximum amount of reinforcement that is permitted [*MSJC* Sec. 3.3.3.5]. This limit ensures that the tension steel has strained sufficiently to achieve ductile behavior. For columns in structures designed with a seismic response modification factor, $R > 1.5$, the reinforcing steel must reach a strain of 1.5 times yield before the masonry reaches its maximum compressive strain. All columns where $M_u/V_ud \geq 1$ must achieve this strain in the tension steel.

9. BIAXIAL BENDING

Biaxial bending is due to the eccentricity of the vertical load about both principal axes of the column. Corner columns are particularly subject to biaxial bending. In this case, the neutral axis is not parallel to either principal axis. This requires an iterative process to determine the depth and angle of the neutral axis that satisfies the equilibrium of the axial force and bending moments about both principal axes.

The following interaction relationship may be used to determine an approximate allowable axial load under biaxial bending of columns. This equation applies to columns with significant axial load (i.e., $P > 0.1P_o$).

$$\frac{1}{P_{\text{biaxial}}} = \frac{1}{P_x} + \frac{1}{P_y} - \frac{1}{P_o} \qquad 69.6$$

Example 69.3

Assume the compressive load on the column in Ex. 69.2 has an additional eccentricity of 7.8 in along the y-axis. Determine the maximum compressive load.

Solution

From Ex. 69.2, $P_x = 63{,}070$ lbf (281 kN) and $P_o = 134{,}650$ lbf (599 kN).

Considering eccentricity about the y-axis,

$$\frac{e}{t} = \frac{7.8 \text{ in}}{15.625 \text{ in}} = 0.5$$
$$n\rho_t = 0.087$$

From App. 69.B, $P/F_b bt = 0.19$.

$$P_y = (0.19)\left(\tfrac{1}{3}\right)\left(2500 \, \frac{\text{lbf}}{\text{in}^2}\right)(15.625 \text{ in})(15.625 \text{ in})$$
$$= 38{,}656 \text{ lbf}$$

Use Eq. 69.6.

$$\frac{1}{P_{\text{biaxial}}} = \frac{1}{P_x} + \frac{1}{P_y} - \frac{1}{P_o}$$
$$= \frac{1}{63{,}070 \text{ lbf}} + \frac{1}{38{,}656 \text{ lbf}} - \frac{1}{134{,}650 \text{ lbf}}$$
$$P_{\text{biaxial}} = 29{,}156 \text{ lbf}$$

Topic VI: Transportation

Transportation

70 Properties of Solid Bodies

Nomenclature

a	acceleration	ft/sec^2	m/s^2
d	distance	ft	m
g	gravitational acceleration	ft/sec^2	m/s^2
g_c	gravitational constant	lbm-ft/lbf-sec^2	n.a.
h	height	ft	m
I	mass moment of inertia	lbm-ft^2	kg·m^2
k	radius of gyration	ft	m
L	length	ft	m
m	mass	lbm	kg
r	radius	ft	m
V	volume	ft^3	m^3
w	weight	lbf	N

Symbols

ρ	density	lbm/ft^3	kg/m^3

Subscripts

c	centroidal
i	inner
o	outer

1. CENTER OF GRAVITY

A solid body will have both a center of gravity and a centroid, but the locations of these two points will not necessarily coincide. The earth's attractive force, called *weight*, can be assumed to act through the *center of gravity* (also known as the *center of mass*). Only when the body is homogeneous will the *centroid of the volume* coincide with the center of gravity.[1]

---

[1] The study of nonhomogeneous bodies is beyond the scope of this book. Homogeneity is assumed for all solid objects.

For simple objects and regular polyhedrons, the location of the center of gravity can be determined by inspection. It will always be located on an axis of symmetry. The location of the center of gravity can also be determined mathematically if the object can be described mathematically.

$$x_c = \frac{\int x \, dm}{m} \qquad 70.1$$

$$y_c = \frac{\int y \, dm}{m} \qquad 70.2$$

$$z_c = \frac{\int z \, dm}{m} \qquad 70.3$$

If the object can be divided into several smaller constituent objects, the location of the composite center of gravity can be calculated from the centers of gravity of each of the constituent objects.

$$x_c = \frac{\sum m_i x_{ci}}{\sum m_i} \qquad 70.4$$

$$y_c = \frac{\sum m_i y_{ci}}{\sum m_i} \qquad 70.5$$

$$z_c = \frac{\sum m_i z_{ci}}{\sum m_i} \qquad 70.6$$

2. MASS AND WEIGHT

The mass, m, of a homogeneous solid object is calculated from its mass density and volume. Mass is independent of the strength of the gravitational field.

$$m = \rho V \qquad 70.7$$

The weight, w, of an object depends on the strength of the gravitational field, g.

$$w = mg \qquad \text{[SI]} \qquad 70.8(a)$$

$$w = \frac{mg}{g_c} \qquad \text{[U.S.]} \qquad 70.8(b)$$

3. INERTIA

Inertia (the *inertial force* or *inertia vector*), $m\mathbf{a}$, is the resistance the object offers to attempts to accelerate it (i.e., change its velocity) in a linear direction. Although the mass, m, is a scalar quantity, the acceleration, $\mathbf{a}$, is a vector.

4. MASS MOMENT OF INERTIA

The *mass moment of inertia* measures a solid object's resistance to changes in rotational speed about a specific axis. I_x, I_y, and I_z are the mass moments of inertia with respect to the x-, y-, and z-axes. They are not components of a resultant value.[2]

The *centroidal mass moment of inertia*, I_c, is obtained when the origin of the axes coincides with the object's center of gravity. Although it can be found mathematically from Eqs. 70.9 through 70.11, it is easier to use App. 70.A for simple objects.

$$I_x = \int (y^2 + z^2)\, dm \qquad 70.9$$

$$I_y = \int (x^2 + z^2)\, dm \qquad 70.10$$

$$I_z = \int (x^2 + y^2)\, dm \qquad 70.11$$

5. PARALLEL AXIS THEOREM

Once the centroidal mass moment of inertia is known, the *parallel axis theorem* is used to find the mass moment of inertia about any parallel axis.

$$I_{\text{any parallel axis}} = I_c + md^2 \qquad 70.12$$

For a composite object, the parallel axis theorem must be applied for each of the constituent objects.

$$I = I_{c,1} + m_1 d_1^2 + I_{c,2} + m_2 d_2^2 + \cdots \qquad 70.13$$

6. RADIUS OF GYRATION

The *radius of gyration*, k, of a solid object represents the distance from the rotational axis at which the object's entire mass could be located without changing the mass moment of inertia.

$$k = \sqrt{\frac{I}{m}} \qquad 70.14$$

$$I = k^2 m \qquad 70.15$$

7. PRINCIPAL AXES

An object's mass moment of inertia depends on the orientation of axes chosen. The *principal axes* are the axes for which the *products of inertia* are zero. Equipment rotating about a principal axis will draw minimum power during speed changes.

Finding the principal axes through calculation is too difficult and time consuming to be used with most rotating equipment. Furthermore, the rotating axis is generally fixed. Therefore, *balancing operations* are used to change the distribution of mass about the rotational axis. A device, such as a rotating shaft, flywheel, or crank, is said to be *statically balanced* if its center of mass lies on the axis of rotation. It is said to be *dynamically balanced* if the center of mass lies on the axis of rotation and the products of inertia are zero.

---

[2]At first, it may be confusing to use the same symbol, I, for area and mass moments of inertia. However, the problem types are distinctly dissimilar, and both moments of inertia are seldom used simultaneously.

71 Kinematics

Nomenclature

a	acceleration	ft/sec^2	m/s^2
d	distance	ft	m
g	gravitational acceleration	ft/sec^2	m/s^2
H	height	ft	m
l	length	ft	m
n	rotating speed	rpm	rpm
r	radius	ft	m
R	earth's radius	ft	m
R	range	ft	m
s	distance	ft	m
t	time	sec	s
T	flight time	sec	s
v	velocity	ft/sec	m/s
z	elevation	ft	m

Symbols

α	angular acceleration	rad/sec^2	rad/s^2
β	angle	deg	deg
γ	angle	deg	deg
θ	angular position	rad	rad
ϕ	angle or latitude	deg	deg
ω	angular velocity	rad/sec	rad/s

Subscripts

0	initial
ϕ	transverse
a	acceleration
c	Coriolis
H	to maximum altitude
n	normal
O	center
r	radial
t	tangential

1. INTRODUCTION TO KINEMATICS

Dynamics is the study of moving objects. The subject is divided into kinematics and kinetics. *Kinematics* is the study of a body's motion independent of the forces on the body. It is a study of the geometry of motion without consideration of the causes of motion. Kinematics deals only with relationships among position, velocity, acceleration, and time. (Kinetics is covered in Ch. 73.)

2. PARTICLES AND RIGID BODIES

Bodies in motion can be considered *particles* if rotation is absent or insignificant. Particles do not possess rotational kinetic energy. All parts of a particle have the same instantaneous displacement, velocity, and acceleration.

A *rigid body* does not deform when loaded and can be considered a combination of two or more particles that remain at a fixed, finite distance from each other. At any given instant, the parts (particles) of a rigid body can have different displacements, velocities, and accelerations.

3. COORDINATE SYSTEMS

The position of a particle is specified with reference to a *coordinate system*. The description takes the form of an ordered sequence $(q_1, q_2, q_3, \ldots)$ of numbers called *coordinates*. A coordinate can represent a position along an axis, as in the rectangular coordinate system, or it can represent an angle, as in the polar, cylindrical, and spherical coordinate systems.

Transportation

In general, the number of *degrees of freedom* is equal to the number of coordinates required to completely specify the state of an object. If each of the coordinates is independent of the others, the coordinates are known as *holonomic coordinates*.

The state of a particle is completely determined by the particle's location. In three-dimensional space, the locations of particles in a system of m particles must be specified by $3m$ coordinates. However, the number of required coordinates can be reduced in certain cases. The position of each particle constrained to motion on a surface (i.e., on a two-dimensional system) can be specified by only two coordinates. A particle constrained to moving on a curved path requires only one coordinate.[1]

The state of a rigid body is a function of orientation as well as position. Six coordinates are required to specify the state: three for orientation and three for location.

4. CONVENTIONS OF REPRESENTATION

Consider the particle shown in Fig. 71.1. Its position (as well as its velocity and acceleration) can be specified in three primary forms: vector form, rectangular coordinate form, and unit vector form.

Figure 71.1 *Position of a Particle*

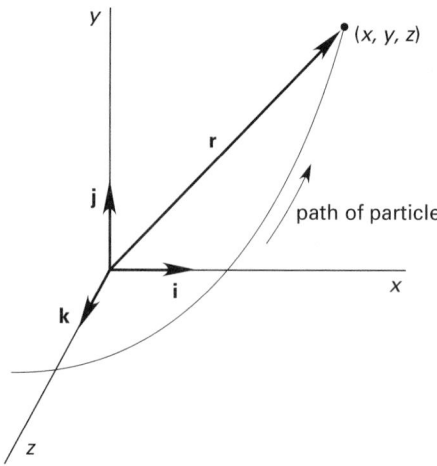

The vector form of the particle's position is $\mathbf{r}$, where the vector $\mathbf{r}$ has both magnitude and direction. The rectangular coordinate form is (x, y, z). The unit vector form is

$$\mathbf{r} = x\mathbf{i} + y\mathbf{j} + z\mathbf{k} \qquad 71.1$$

5. LINEAR PARTICLE MOTION

A *linear system* is one in which particles move only in straight lines. (Another name is *rectilinear system*.) The relationships among position, velocity, and acceleration for a linear system are given by Eqs. 71.2 through

---

71.4. When values of t are substituted into these equations, the position, velocity, and acceleration are known as *instantaneous values*.

$$s(t) = \int v(t)\, dt = \int \left(\int a(t)\, dt \right) dt \qquad 71.2$$

$$v(t) = \frac{ds(t)}{dt} = \int a(t)\, dt \qquad 71.3$$

$$a(t) = \frac{dv(t)}{dt} = \frac{d^2 s(t)}{dt^2} \qquad 71.4$$

The average velocity and acceleration over a period from t_1 to t_2 are

$$v_{ave} = \frac{\int_1^2 v(t)\, dt}{t_2 - t_1} = \frac{s_2 - s_1}{t_2 - t_1} \qquad 71.5$$

$$a_{ave} = \frac{\int_1^2 a(t)\, dt}{t_2 - t_1} = \frac{v_2 - v_1}{t_2 - t_1} \qquad 71.6$$

Example 71.1

A particle is constrained to move along a straight line. The velocity and location are both zero at $t = 0$. The particle's velocity as a function of time is

$$v(t) = 8t - 6t^2$$

(a) What are the acceleration and position functions?
(b) What is the instantaneous velocity at $t = 5$?

Solution

(a)
$$a(t) = \frac{d\,v(t)}{dt} = \frac{d(8t - 6t^2)}{dt}$$
$$= 8 - 12t$$

$$s(t) = \int v(t)\, dt = \int (8t - 6t^2)\, dt$$
$$= 4t^2 - 2t^3 \text{ when } s(t = 0) = 0$$

(b) Substituting $t = 5$ into the $v(t)$ function,

$$v(5) = (8)(5) - (6)(5)^2$$
$$= -110 \quad \text{[backward]}$$

6. DISTANCE AND SPEED

The terms "displacement" and "distance" have different meanings in kinematics. *Displacement* (or *linear displacement*) is the net change in a particle's position as determined from the position function, $s(t)$. *Distance traveled* is the accumulated length of the path traveled during all direction reversals, and it can be found by adding the path lengths covered during periods in which the velocity sign does not change. Thus, distance is always greater than or equal to displacement.

$$\text{displacement} = s(t_2) - s(t_1) \qquad 71.7$$

---

[1]The curve can be a straight line, as in the case of a mass hanging on a spring and oscillating up and down. In this case, the coordinate will be a linear coordinate.

Similarly, "velocity" and "speed" have different meanings: *velocity* is a vector, having both magnitude and direction; *speed* is a scalar quantity, equal to the magnitude of velocity. When specifying speed, direction is not considered.

Example 71.2

What distance is traveled during the period $t = 0$ to $t = 6$ by the particle described in Ex. 71.1?

Solution

Start by determining when, if ever, the velocity becomes negative. (This can be done by inspection, graphically, or algebraically.) Solving for the roots of the velocity equation, the velocity changes from positive to negative at

$$t = \tfrac{4}{3}$$

The initial displacement is zero. From the position function, the position at $t = {}^4/_3$ is

$$s\left(\tfrac{4}{3}\right) = (4)\left(\tfrac{4}{3}\right)^2 - (2)\left(\tfrac{4}{3}\right)^3 = 2.37$$

The displacement while the velocity is positive is

$$\Delta s = s\left(\tfrac{4}{3}\right) - s(0) = 2.37 - 0$$
$$= 2.37$$

The position at $t = 6$ is

$$s(6) = (4)(6)^2 - (2)(6)^3 = -288$$

The displacement while the velocity is negative is

$$\Delta s = s(6) - s\left(\tfrac{4}{3}\right) = -288 - 2.37$$
$$= -290.37$$

The total distance traveled is

$$2.37 + 290.37 = 292.74$$

7. UNIFORM MOTION

The term *uniform motion* means uniform velocity. The velocity is constant and the acceleration is zero. For a constant velocity system, the position function varies linearly with time.

$$s(t) = s_0 + \mathrm{v}t \qquad \text{71.8}$$
$$\mathrm{v}(t) = \mathrm{v} \qquad \text{71.9}$$
$$a(t) = 0 \qquad \text{71.10}$$

Figure 71.2 Constant Velocity System

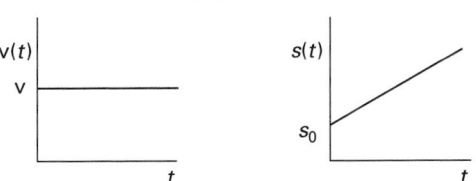

8. UNIFORM ACCELERATION

The acceleration is constant in many cases. (Gravitational acceleration, where $a = g$, is a notable example.) If the acceleration is constant, the a term can be taken out of the integrals in Eqs. 71.2 and 71.3.

$$a(t) = a \qquad \text{71.11}$$
$$\mathrm{v}(t) = a \int dt = \mathrm{v}_0 + at \qquad \text{71.12}$$
$$s(t) = a \iint dt^2 = s_0 + \mathrm{v}_0 t + \tfrac{1}{2}at^2 \qquad \text{71.13}$$

Figure 71.3 Uniform Acceleration

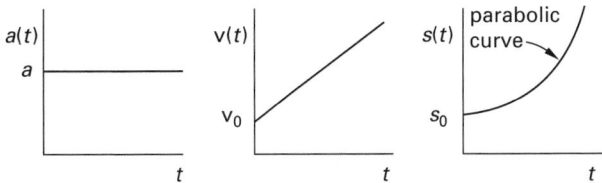

Table 71.1 summarizes the equations required to solve most uniform acceleration problems.

Table 71.1 Uniform Acceleration Formulas[a]

to find	given these	use this equation
a	$t, \mathrm{v}_0, \mathrm{v}$	$a = \dfrac{\mathrm{v} - \mathrm{v}_0}{t}$
a	t, v_0, s	$a = \dfrac{2s - 2\mathrm{v}_0 t}{t^2}$
a	$\mathrm{v}_0, \mathrm{v}, s$	$a = \dfrac{\mathrm{v}^2 - \mathrm{v}_0^2}{2s}$
s	t, a, v_0	$s = \mathrm{v}_0 t + \tfrac{1}{2}at^2$
s	$a, \mathrm{v}_0, \mathrm{v}$	$s = \dfrac{\mathrm{v}^2 - \mathrm{v}_0^2}{2a}$
s	$t, \mathrm{v}_0, \mathrm{v}$	$s = \tfrac{1}{2}t(\mathrm{v}_0 + \mathrm{v})$
t	$a, \mathrm{v}_0, \mathrm{v}$	$t = \dfrac{\mathrm{v} - \mathrm{v}_0}{a}$
t	a, v_0, s	$t = \dfrac{\sqrt{\mathrm{v}_0^2 + 2as} - \mathrm{v}_0}{a}$
t	$\mathrm{v}_0, \mathrm{v}, s$	$t = \dfrac{2s}{\mathrm{v}_0 + \mathrm{v}}$
v_0	t, a, v	$\mathrm{v}_0 = \mathrm{v} - at$
v_0	t, a, s	$\mathrm{v}_0 = \dfrac{s}{t} - \tfrac{1}{2}at$
v_0	a, v, s	$\mathrm{v}_0 = \sqrt{\mathrm{v}^2 - 2as}$
v	t, a, v_0	$\mathrm{v} = \mathrm{v}_0 + at$
v	a, v_0, s	$\mathrm{v} = \sqrt{\mathrm{v}_0^2 + 2as}$

[a]The table can be used for rotational problems by substituting α, ω, and θ for a, v, and s, respectively.

Transportation

Example 71.3

A locomotive traveling at 80 kph locks its wheels and skids 95 m before coming to a complete stop. If the deceleration is constant, how many seconds will it take for the locomotive to come to a standstill?

Solution

First, convert the 80 kph to meters per second.

$$v_0 = \frac{\left(80 \ \frac{km}{h}\right)\left(1000 \ \frac{m}{km}\right)}{3600 \ \frac{s}{h}} = 22.22 \ m/s$$

In this problem, $v_0 = 22.2$ m/s, $v = 0$, and $s = 95$ m are known. t is the unknown. From Table 71.1,

$$t = \frac{2s}{v_0 + v} = \frac{(2)(95 \ m)}{22.22 + 0}$$
$$= 8.55 \ s$$

9. LINEAR ACCELERATION

Linear acceleration means that the acceleration increases uniformly with time. Figure 71.4 shows how the velocity and position vary with time.[2]

Figure 71.4 *Linear Acceleration*

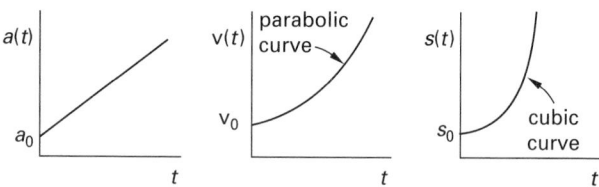

10. PROJECTILE MOTION

A *projectile* is placed into motion by an initial impulse. (Kinematics deals only with dynamics during the flight. The force acting on the projectile during the launch phase is covered in kinetics.) Neglecting air drag, once the projectile is in motion it is acted upon only by the downward gravitational acceleration (i.e., its own weight). Thus, projectile motion is a special case of motion under constant acceleration.

Consider a general projectile set into motion at an angle of ϕ (from the horizontal plane) and initial velocity v_0. Its range is R, the maximum altitude attained is H, and the total flight time is T. In the absence of air drag, the following rules apply to the case of a level target.[3]

---

[2]Because of the successive integrations, if the acceleration function is a polynomial of degree n, the velocity function will be a polynomial of degree $n + 1$. Similarly, the position function will be a polynomial of degree $n + 2$.

[3]The case of projectile motion with air friction cannot be handled in kinematics, since a retarding force acts continuously on the projectile. In kinetics, various assumptions (e.g., friction varies linearly with the velocity or with the square of the velocity) can be made to include the effect of air friction.

- The trajectory is parabolic.

- The impact velocity is equal to initial velocity, v_0.

- The impact angle is equal to the initial launch angle, ϕ.

- The range is maximum when $\phi = 45°$.

- The time for the projectile to travel from the launch point to the apex is equal to the time to travel from apex to impact point.

- The time for the projectile to travel from the apex of its flight path to impact is the same time an initially stationary object would take to fall a distance H.

Table 71.3 contains the solutions to most common projectile problems. These equations are derived from the laws of uniform acceleration and conservation of energy.

Example 71.4

A projectile is launched at 600 ft/sec (180 m/s) with a 30° inclination from the horizontal. The launch point is on a plateau 500 ft (150 m) above the plane of impact. Neglecting friction, find the maximum altitude, H, above the plane of impact, the total flight time, T, and the range, R.

SI Solution

The maximum altitude above the impact plane includes the height of the plateau and the elevation achieved by the projectile.

$$H = z + \frac{v_0^2 \sin^2 \phi}{2g}$$
$$= 150 \ m + \frac{\left(180 \ \frac{m}{s}\right)^2 (\sin^2 30°)}{(2)\left(9.81 \ \frac{m}{s^2}\right)} = 563 \ m$$

The total flight time includes the time to reach the maximum altitude and the time to fall from the maximum altitude to the impact plane below.

$$T = t_H + t_{fall}$$
$$= \frac{v_0 \sin \phi}{g} + \sqrt{\frac{2H}{g}}$$
$$= \frac{\left(180 \ \frac{m}{s}\right)(\sin 30°)}{9.81 \ \frac{m}{s^2}} + \sqrt{\frac{(2)(563 \ m)}{9.81 \ \frac{m}{s^2}}}$$
$$= 19.9 \ s$$

Table 71.2 Projectile Motion Equations
(φ may be negative for projection downward)

	level target	target above	target below	horizontal projection
				$v_0 = v_{x'}$ $\phi = 0°$
$x(t)$	$v_0 \cos \phi \, t$			$v_0 t$
$y(t)$	$v_0 \sin \phi \, t - \frac{1}{2} g t^2$			$H - \frac{1}{2} g t^2$
$v_x(t)^a$	$v_0 \cos \phi$			v_0
$v_y(t)^b$	$v_0 \sin \phi - gt$			$-gt$
$v(t)^c$	$\sqrt{v_0^2 - 2gy} = \sqrt{v_0^2 - 2gt v_0 \sin \phi + g^2 t^2}$			$\sqrt{v_0^2 + g^2 t^2}$
$v(y)^d$	$\sqrt{v_0^2 - 2gy}$			$\sqrt{v_0^2 + 2g(H - y)}$
H	$\dfrac{v_0^2 \sin^2 \phi}{2g}$	$\dfrac{v_0^2 \sin^2 \phi}{2g}$	$z + \dfrac{v_0^2 \sin^2 \phi}{2g}$	$\frac{1}{2} g T^2$
R	$v_0 T \cos \phi$			$v_0 T$
	$\dfrac{v_0^2 \sin 2\phi}{g}$	$\left(\dfrac{v_0 \cos \phi}{g}\right)\left(v_0 \sin \phi + \sqrt{v_0^2 \sin^2 \phi - 2gz}\right)$	$\left(\dfrac{v_0 \cos \phi}{g}\right)\left(v_0 \sin \phi + \sqrt{2gz + v_0^2 \sin^2 \phi}\right)$	
T	$\dfrac{R}{v_0 \cos \phi}$			$\sqrt{\dfrac{2H}{g}}$
	$\dfrac{2 v_0 \sin \phi}{g}$	$\dfrac{v_0 \sin \phi}{g} + \sqrt{\dfrac{(2)(H - z)}{g}}$	$\dfrac{v_0 \sin \phi}{g} + \sqrt{\dfrac{2H}{g}}$	
t_H	$\dfrac{v_0 \sin \phi}{g} = \dfrac{T}{2}$	$\dfrac{v_0 \sin \phi}{g}$		

[a] horizontal velocity component
[b] vertical velocity component
[c] resultant velocity as a function of time
[d] resultant velocity as a function of vertical elevation above the launch point

The x-component of velocity is

$$v_x = v_0 \cos\phi = \left(180 \ \frac{\text{m}}{\text{s}}\right)(\cos 30°)$$

$$= 156 \ \text{m/s}$$

The range is

$$R = v_x T = \left(156 \ \frac{\text{m}}{\text{s}}\right)(19.9 \ \text{s})$$

$$= 3100 \ \text{m}$$

Customary U.S. Solution

The maximum altitude above the impact plane is given by Table 71.2.

$$H = 500 \ \text{ft} + \frac{\left(600 \ \frac{\text{ft}}{\text{sec}}\right)^2 (\sin^2 30°)}{(2)\left(32.2 \ \frac{\text{ft}}{\text{sec}^2}\right)}$$

$$= 1898 \ \text{ft}$$

The total flight time is

$$T = \frac{\left(600 \ \frac{\text{ft}}{\text{sec}}\right)(\sin 30°)}{32.2 \ \frac{\text{ft}}{\text{sec}^2}} + \sqrt{\frac{(2)(1898 \ \text{ft})}{32.2 \ \frac{\text{ft}}{\text{sec}^2}}}$$

$$= 20.2 \ \text{sec}$$

The maximum range is

$$R = v_0 T \cos\phi = \left(600 \ \frac{\text{ft}}{\text{sec}}\right)(20.2 \ \text{sec})(\cos 30°)$$

$$= 10{,}500 \ \text{ft} \quad (1.99 \ \text{mi})$$

Example 71.5

A bomber flies horizontally at 275 mph at an altitude of 9000 ft. At what viewing angle, ϕ, from the bomber to the target should the bombs be dropped?

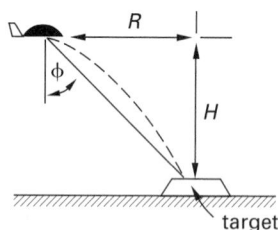

Solution

This is a case of horizontal projection. The falling time depends only on the altitude of the bomber. From Table 71.2,

$$T = \sqrt{\frac{2H}{g}} = \sqrt{\frac{(2)(9000 \ \text{ft})}{32.2 \ \frac{\text{ft}}{\text{sec}^2}}}$$

$$= 23.64 \ \text{sec}$$

If air friction is neglected, the bomb has the same horizontal velocity as the bomber. Since the time of flight, T, is known, the distance traveled during that time can be calculated.

$$R = v_0 T = \frac{\left(275 \ \frac{\text{mi}}{\text{hr}}\right)\left(5280 \ \frac{\text{ft}}{\text{mi}}\right)(23.64 \ \text{sec})}{3600 \ \frac{\text{sec}}{\text{hr}}}$$

$$= 9535 \ \text{ft}$$

The viewing angle is found from trigonometry.

$$\phi = \arctan\left(\frac{R}{H}\right) = \arctan\left(\frac{9535 \ \text{ft}}{9000 \ \text{ft}}\right)$$

$$= 46.7°$$

11. ROTATIONAL PARTICLE MOTION

Rotational particle motion (also known as *angular motion* and *circular motion*) is motion of a particle around a circular path. There are 2π radians per complete revolution.

Figure 71.5 *Rotational Particle Motion*

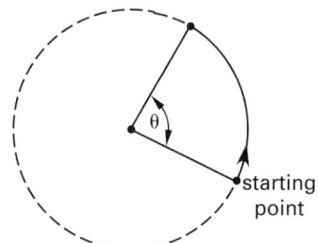

The behavior of a rotating particle is defined by its *angular position*, θ, *angular velocity*, ω, and *angular acceleration*, α, functions. These variables are analogous to the $s(t)$, $v(t)$, and $a(t)$ functions for linear systems. Angular variables can be substituted one-for-one in place of linear variables in most equations.

The relationships among angular position, velocity, and acceleration for a rotational system are given by Eqs. 71.14 through 71.16. When values of t are substituted into these equations, the position, velocity, and acceleration are known as *instantaneous values*.

$$\theta(t) = \int \omega(t)\,dt = \iint \alpha(t)\,dt^2 \qquad \text{71.14}$$

$$\omega(t) = \frac{d\theta(t)}{dt} = \int \alpha(t)\,dt \qquad \text{71.15}$$

$$\alpha(t) = \frac{d\omega(t)}{dt} = \frac{d^2\theta(t)}{dt^2} \qquad \text{71.16}$$

The average velocity and acceleration are

$$\omega_{\text{ave}} = \frac{\int_1^2 \omega(t)\,dt}{t_2 - t_1} = \frac{\theta_2 - \theta_1}{t_2 - t_1} \qquad \text{71.17}$$

$$\alpha_{\text{ave}} = \frac{\int_1^2 \alpha(t)\,dt}{t_2 - t_1} = \frac{\omega_2 - \omega_1}{t_2 - t_1} \qquad \text{71.18}$$

Example 71.6

A turntable starts from rest and accelerates uniformly at 1.5 rad/sec². How many revolutions will it take before a rotational speed of 33⅓ rpm is attained?

Solution

First, convert 33⅓ rpm into radians per second. Since there are 2π radians per complete revolution,

$$\omega = \frac{\left(33\frac{1}{3}\ \frac{\text{rev}}{\text{min}}\right)\left(2\pi\ \frac{\text{rad}}{\text{rev}}\right)}{60\ \frac{\text{sec}}{\text{min}}} = 3.49\ \text{rad/sec}$$

α, ω_0, and ω are known. θ is unknown. (This is analogous to knowing a, v_0, and v, and not knowing s.) From Table 71.1,

$$\theta = \frac{\omega^2 - \omega_0^2}{2\alpha} = \frac{\left(3.49\ \frac{\text{rad}}{\text{sec}}\right)^2 - \left(0\ \frac{\text{rad}}{\text{sec}}\right)^2}{(2)\left(1.5\ \frac{\text{rad}}{\text{sec}^2}\right)}$$

$$= 4.06\ \text{rad}$$

Converting from radians to revolutions,

$$n = \frac{4.06\ \text{rad}}{2\pi\ \frac{\text{rad}}{\text{rev}}} = 0.65\ \text{rev}$$

Example 71.7

A flywheel is brought to a standstill from 400 rpm in 8 sec. (a) What was its average angular acceleration in rad/sec during that period? (b) How far (in radians) did the flywheel travel?

Solution

(a) The initial rotational speed must be expressed in radians per second. Since there are 2π radians per revolution,

$$\omega_0 = \frac{\left(400\ \frac{\text{rev}}{\text{min}}\right)\left(2\pi\ \frac{\text{rad}}{\text{rev}}\right)}{60\ \frac{\text{sec}}{\text{min}}}$$

$$= 41.9\ \text{rad/sec}$$

t, ω_0, and ω are known, and α is unknown. These variables are analogous to t, v_0, v, and a in Table 71.1.

$$\alpha = \frac{\omega - \omega_0}{t} = \frac{0 - 41.9\ \frac{\text{rad}}{\text{sec}}}{8\ \text{sec}}$$

$$= -5.24\ \text{rad/sec}^2$$

(b) t, ω_0, and ω are known, and θ is unknown. Again, from Table 71.1,

$$\theta = \tfrac{1}{2}t(\omega + \omega_0) = \left(\tfrac{1}{2}\right)(8\ \text{sec})\left(41.9\ \frac{\text{rad}}{\text{sec}} + 0\right)$$

$$= 168\ \text{rad} \quad (26.7\ \text{rev})$$

12. RELATIONSHIP BETWEEN LINEAR AND ROTATIONAL VARIABLES

A particle moving in a curvilinear path will also have instantaneous linear velocity and linear acceleration. These linear variables will be directed tangentially to the path and, therefore, are known as *tangential velocity* and *tangential acceleration*, respectively. In general, the linear variables can be obtained by multiplying the rotational variables by the path radius, r.

$$v_t = \omega r \qquad\qquad\qquad 71.19$$

$$v_{t,x} = v_t \cos\phi = \omega r \cos\phi \qquad 71.20$$

$$v_{t,y} = v_t \sin\phi = \omega r \sin\phi \qquad 71.21$$

$$a_t = \frac{dv_t}{dt} = \alpha r \qquad\qquad 71.22$$

If the path radius is constant, as it would be in rotational motion, the linear distance (i.e., the *arc length*) traveled is

$$s = \theta r \qquad\qquad\qquad 71.23$$

13. NORMAL ACCELERATION

A moving particle will continue tangentially to its path unless constrained otherwise. For example, a rock twirled on a string will move in a circular path only as long as there is tension in the string. When the string is released, the rock will move off tangentially.

Figure 71.6 *Tangential Variables*

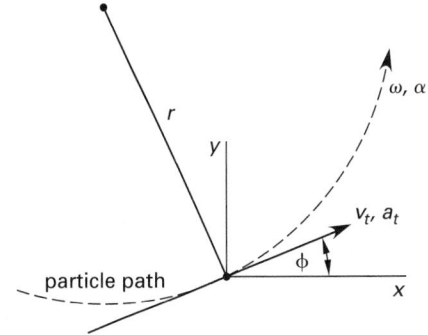

The twirled rock is acted upon by the tension in the string. In general, a restraining force will be directed toward the center of rotation. Whenever a mass experiences a force, an acceleration is acting.[4] The acceleration has the same sense as the applied force (i.e., is directed toward the center of rotation). Since the inward acceleration is perpendicular to the tangential velocity and acceleration, it is known as *normal acceleration*, a_n.

$$a_n = \frac{v_t^2}{r} = r\omega^2 = v_t\omega \qquad 71.24$$

The *resultant acceleration*, a, is the vector sum of the tangential and normal accelerations. The magnitude of the resultant acceleration is

$$a = \sqrt{a_t^2 + a_n^2} \qquad 71.25$$

The x- and y-components of the resultant acceleration are

$$a_x = a_n \sin\phi \pm a_t \cos\phi \qquad 71.26$$
$$a_y = a_n \cos\phi \mp a_t \sin\phi \qquad 71.27$$

The normal and tangential accelerations can be expressed in terms of the x- and y-components of the resultant acceleration (not shown in Fig. 71.7).

$$a_n = a_x \sin\phi \pm a_y \cos\phi \qquad 71.28$$
$$a_t = a_x \cos\phi \mp a_y \sin\phi \qquad 71.29$$

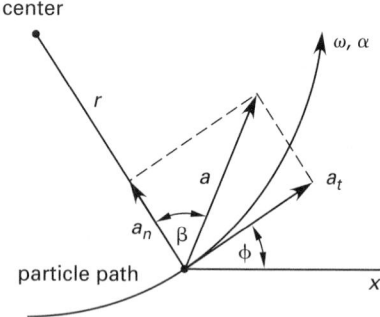

Figure 71.7 *Normal Acceleration*

14. CORIOLIS ACCELERATION

Consider a particle moving with linear radial velocity v_r away from the center of a flat disk rotating with constant velocity ω. Since $v_t = \omega r$, the particle's tangential velocity will increase as it moves away from the center of rotation. This increase is said to be produced by the tangential *Coriolis acceleration*, a_c.

$$a_c = 2v_r\omega \qquad 71.30$$

---

[4]This is a direct result of Newton's second law of motion, covered in Ch. 73.

Figure 71.8 *Coriolis Acceleration on a Rotating Disk*

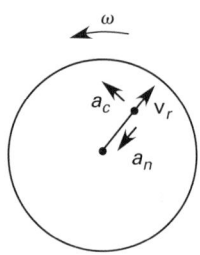

Coriolis acceleration also acts on particles moving on rotating spheres. Consider an aircraft flying with constant air speed v from the equator to the north pole while the earth rotates below it. Three accelerations act on the aircraft: normal, radial, and Coriolis accelerations, shown in Fig. 71.9. The Coriolis acceleration depends on the latitude, ϕ, because the earth's tangential velocity is less near the poles than at the equator.

$$a_n = r\omega^2 = R\omega^2 \cos\phi \qquad 71.31$$
$$a_r = \frac{v^2}{R} \qquad 71.32$$
$$a_c = 2\omega v_x = 2\omega v \sin\phi \qquad 71.33$$

Figure 71.9 *Coriolis Acceleration on a Rotating Sphere*

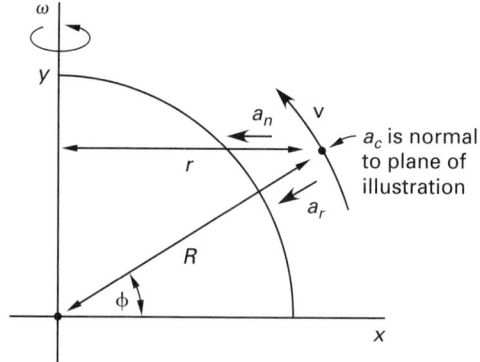

Example 71.8

A slider moves with a constant velocity of 20 ft/sec along a rod rotating at 5 rad/sec. What is the magnitude of the slider's total acceleration when the slider is 4 ft from the center of rotation?

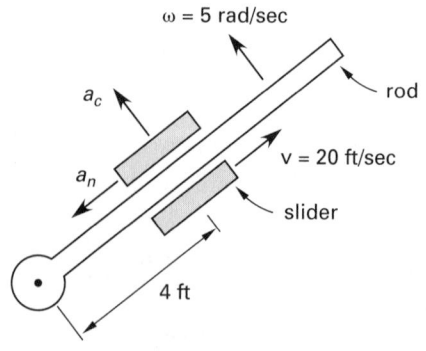

Solution

The normal acceleration is given by Eq. 71.31.

$$a_n = r\omega^2 = (4 \text{ ft}) \left(5 \frac{\text{rad}}{\text{sec}}\right)^2$$

$$= 100 \text{ ft/sec}^2$$

The Coriolis acceleration is given by Eq. 71.33.

$$a_c = 2\text{v}\omega = (2) \left(20 \frac{\text{ft}}{\text{sec}}\right) \left(5 \frac{\text{rad}}{\text{sec}}\right)$$

$$= 200 \text{ ft/sec}^2$$

The total acceleration is given by 71.25.

$$a = \sqrt{a_n^2 + a_c^2} = \sqrt{\left(100 \frac{\text{ft}}{\text{sec}^2}\right)^2 + \left(200 \frac{\text{ft}}{\text{sec}^2}\right)^2}$$

$$= 223.6 \text{ ft/sec}^2$$

15. PARTICLE MOTION IN POLAR COORDINATES

In polar coordinates, the path of a particle is described by a radius vector, $\mathbf{r}$, and an angle, ϕ. Since the velocity of a particle is not usually directed radially out from the center of the coordinate system, it can be divided into two perpendicular components. The terms *normal* and *tangential* are not used with polar coordinates. Rather, the terms *radial* and *transverse* are used. Figure 71.10 illustrates the *radial* and *transverse components* of velocity in a polar coordinate system.

Figure 71.10 also illustrates the unit radial and unit transverse vectors, $\mathbf{e}_r$ and $\mathbf{e}_\phi$, used in the vector forms of the motion equations.

position: $\mathbf{r} = r\mathbf{e}_r$ *71.34*

velocity: $\mathbf{v} = \text{v}_r\mathbf{e}_r + \text{v}_\phi\mathbf{e}_\phi = \dfrac{dr}{dt}\mathbf{e}_r + r\dfrac{d\phi}{dt}\mathbf{e}_\phi$ *71.35*

acceleration: $\mathbf{a} = a_r\mathbf{e}_r + a_\phi\mathbf{e}_\phi$

$$= \left(\frac{d^2r}{dt^2} - r\left(\frac{d\phi}{dt}\right)^2\right)\mathbf{e}_r$$

$$+ \left(r\frac{d^2\phi}{dt^2} + 2\frac{dr}{dt}\frac{d\phi}{dt}\right)\mathbf{e}_\phi \qquad \textit{71.36}$$

Figure 71.10 *Radial and Transverse Components*

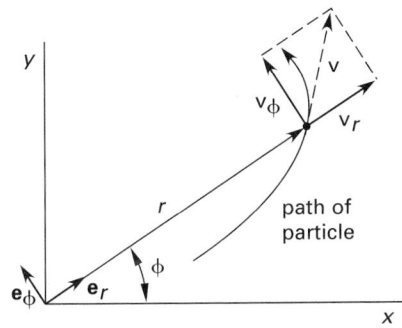

The magnitudes of the radial and transverse components of velocity and acceleration are given by Eqs. 71.37 through 71.40.

$$\text{v}_r = \frac{dr}{dt} \qquad \textit{71.37}$$

$$\text{v}_\phi = r\frac{d\phi}{dt} \qquad \textit{71.38}$$

$$a_r = \frac{d^2r}{dt^2} - r\left(\frac{d\phi}{dt}\right)^2 \qquad \textit{71.39}$$

$$a_\phi = r\frac{d^2\phi}{dt^2} + 2\frac{dr}{dt}\frac{d\phi}{dt} \qquad \textit{71.40}$$

If the radial and transverse components of acceleration and velocity are known, they can be used to calculate the tangential and normal accelerations in a rectangular coordinate system.

$$a_t = \frac{a_r\text{v}_r + a_\phi\text{v}_\phi}{\text{v}_t} \qquad \textit{71.41}$$

$$a_n = \frac{a_\phi\text{v}_r - a_r\text{v}_\phi}{\text{v}_t} \qquad \textit{71.42}$$

16. RELATIVE MOTION

The term *relative motion* is used when motion of a particle is described with respect to something else in motion. The particle's position, velocity, and acceleration may be specified with respect to another moving particle or with respect to a moving frame of reference, known as a *Newtonian* or *inertial frame of reference*.

In Fig. 71.11, two particles, A and B, are moving with different velocities along a straight line. The separation between the two particles at any specific instant is the *relative position*, $s_{B/A}$, of B with respect to A, calculated as the difference between their two *absolute positions*.

$$s_{B/A} = s_B - s_A \qquad \textit{71.43}$$

Figure 71.11 *Relative Motion of Two Particles*

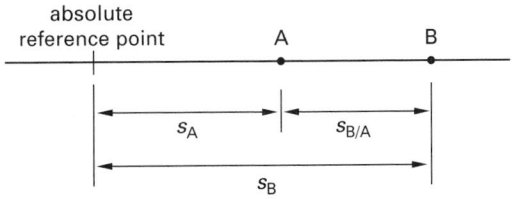

Similarly, the *relative velocity* and *relative acceleration* of B with respect to A are the differences between the two *absolute velocities* and *absolute accelerations*, respectively.

$$\text{v}_{B/A} = \text{v}_B - \text{v}_A \qquad \textit{71.44}$$

$$a_{B/A} = a_B - a_A \qquad \textit{71.45}$$

Particles A and B are not constrained to move along a straight line. However, the subtraction must be done in vector or graphical form in all but the simplest cases.

$$\mathbf{s}_{B/A} = \mathbf{s}_B - \mathbf{s}_A \qquad \textit{71.46}$$

$$\mathbf{v}_{B/A} = \mathbf{v}_B - \mathbf{v}_A \qquad \textit{71.47}$$

$$\mathbf{a}_{B/A} = \mathbf{a}_B - \mathbf{a}_A \qquad \textit{71.48}$$

Since vector subtraction and addition operations can be performed graphically, many relative motion problems can be solved by a simplified graphical process.

Example 71.9

A stream flows at 5 kph. At what upstream angle, ϕ, should a 10 kph boat be piloted in order to reach the shore directly opposite the initial point?

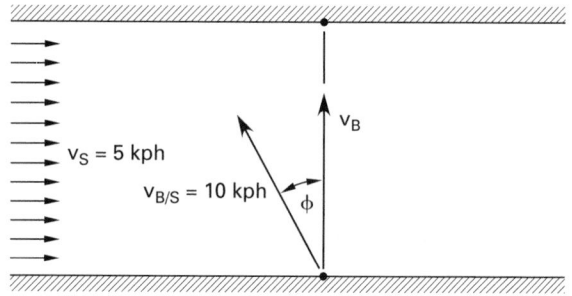

Solution

From Eq. 71.47, the absolute velocity of the boat, v_B, with respect to the shore is equal to the vector sum of the absolute velocity of the stream, v_S, and the relative velocity of the boat with respect to the stream, $v_{B/S}$. The magnitudes of these two velocities are known.

$$v_B = v_S + v_{B/S}$$

Since vector addition is accomplished graphically by placing the two vectors head to tail, the angle can be determined from trigonometry.

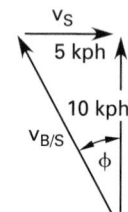

$$\sin\phi = \frac{v_S}{v_{B/S}} = \frac{5 \text{ kph}}{10 \text{ kph}} = 0.5$$

$$\phi = \arcsin 0.5 = 30°$$

Example 71.10

A stationary member of a marching band tosses a 2.0 ft long balanced baton straight up into the air and then begins walking forward at 4 mph. At a particular moment, the baton is 20 ft in the air and is falling back toward the earth with a velocity of 30 ft/sec. The tip of the baton is rotating at 140 rpm in the orientation shown.

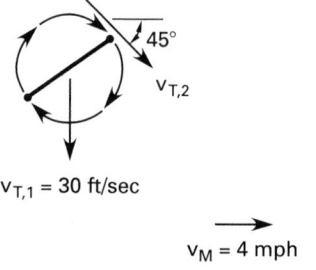

(a) What is the absolute speed of the baton tip with respect to the ground? (b) What is the relative speed of the baton tip with respect to the band member?

Solution

(a) The baton tip has two absolute velocity components. The first, with a magnitude of $v_{T,1} = 30$ ft/sec, is directed vertically downward. The second, with a magnitude of $v_{T,2}$, is directed as shown in the figure. The baton's radius, r, is 1 ft. From Eq. 71.19,

$$v_{T,2} = r\omega = \frac{(1 \text{ ft})\left(140 \ \frac{\text{rev}}{\text{min}}\right)\left(2\pi \ \frac{\text{rad}}{\text{rev}}\right)}{60 \ \frac{\text{sec}}{\text{min}}}$$

$$= 14.7 \text{ ft/sec}$$

The vector sum of these two absolute velocities is the velocity of the tip, $\mathbf{v}_T$, with respect to the earth.

$$\mathbf{v}_T = \mathbf{v}_{T,1} + \mathbf{v}_{T,2}$$

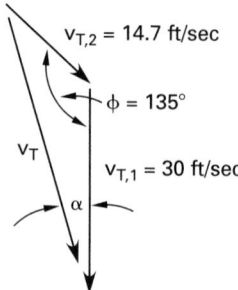

The velocity of the tip is found from the law of cosines.

$$v_T = \sqrt{v_{T,1}^2 + v_{T,2}^2 - 2v_{T,1}v_{T,2}\cos\phi}$$

$$= \sqrt{(30)^2 + (14.7)^2 - (2)(30)(14.7)(\cos 135°)}$$

$$= 41.7 \text{ ft/sec}$$

Angle α is found from the law of sines.

$$\frac{\sin \alpha}{14.7 \ \dfrac{\text{ft}}{\text{sec}}} = \frac{\sin 135°}{41.7 \ \dfrac{\text{ft}}{\text{sec}}}$$

$$\alpha = 14.4°$$

The band member's absolute velocity, v_M, is

$$v_M = \frac{\left(4 \ \dfrac{\text{mi}}{\text{hr}}\right)\left(5280 \ \dfrac{\text{ft}}{\text{mi}}\right)}{3600 \ \dfrac{\text{sec}}{\text{hr}}} = 5.87 \ \text{ft/sec}$$

(b) From Eq. 71.44, the velocity of the tip with respect to the band member is

$$v_{T/M} = v_T - v_M$$

Subtracting a vector is equivalent to adding its negative. The velocity triangle is as shown. The law of cosines is used again to determine the relative velocity.

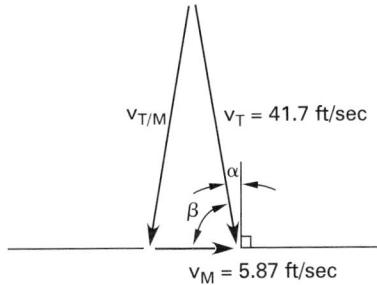

$$\beta = 90° - \alpha = 90° - 14.4° = 75.6°$$

$$\begin{aligned}
v_{T/M} &= \sqrt{v_T^2 + v_M^2 - 2v_T v_M \cos \beta} \\
&= \sqrt{(41.7)^2 + (5.87)^2 - (2)(41.7)(5.87)(\cos 75.6°)} \\
&= 40.6 \ \text{ft/sec} \quad (27.7 \ \text{mph})
\end{aligned}$$

17. DEPENDENT MOTION

When the position of one particle in a multiple-particle system depends on the position of one or more other particles, the motions are said to be "dependent." A block-and-pulley system with one fixed rope end, as illustrated by Fig. 71.12, is a *dependent system*.

Figure 71.12 *Dependent System*

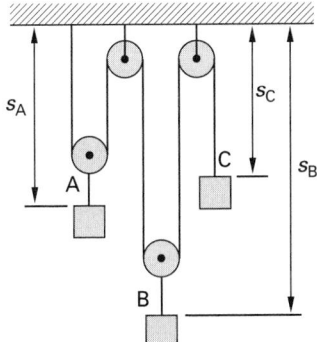

The following statements define the behavior of a dependent block-and-pulley system.

- Since the length of the rope is constant, the sum of the rope segments representing distances between the blocks and pulleys is constant. By convention, the distances are measured from the top of the block to the support point.[5] Since in Fig. 71.12 there are two ropes supporting block A, two ropes supporting block B, and one rope supporting block C,

$$2s_A + 2s_B + s_C = \text{constant} \qquad \textit{71.49}$$

- Since the position of the nth block in an n-block system is determined when the remaining $n - 1$ positions are known, the number of *degrees of freedom* is one less than the number of blocks.

- The movement, velocity, and acceleration of a block supported by two ropes are half the same quantities of a block supported by one rope.

- The relative relationships between the blocks' velocities or accelerations are the same as the relationships between the blocks' positions. For Fig. 71.12,

$$2v_A + 2v_B + v_C = 0 \qquad \textit{71.50}$$
$$2a_A + 2a_B + a_C = 0 \qquad \textit{71.51}$$

18. GENERAL PLANE MOTION

Rigid body *plane motion* can be described in two dimensions. Examples include rolling wheels, gear sets, and linkages. Plane motion can be considered as the sum of a translational component and a rotation about a fixed axis, as illustrated by Fig. 71.13.

Figure 71.13 *Components of Plane Motion*

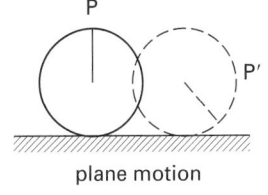

plane motion

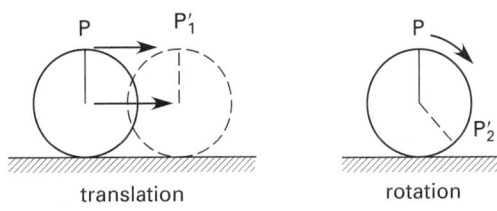

translation   rotation

---

[5]In measuring distances, the finite diameters of the pulleys and the lengths of rope wrapped around the pulleys are disregarded.

19. ROTATION ABOUT A FIXED AXIS

Analysis of the rotational component of a rigid body's plane motion can sometimes be simplified if the location of the body's instantaneous center is known. Using the instantaneous center reduces many relative motion problems to simple geometry. The *instantaneous center* (also known as the *instant center* and IC) is a point at which the body could be fixed (pinned) without changing the instantaneous angular velocities of any point on the body. Thus, with the angular velocities, the body seems to rotate about a fixed instantaneous center.

The instantaneous center is located by finding two points for which the absolute velocity directions are known. Lines drawn perpendicular to these two velocities will intersect at the instantaneous center. (This graphical procedure is slightly different if the two velocities are parallel, as Fig. 71.14 shows. In that case, use is made of the fact that the tangential velocity is proportional to the distance from the instantaneous center.) For a rolling wheel, the instantaneous center is the point of contact with the supporting surface.

Figure 71.14 *Graphical Method of Finding the Instantaneous Center*

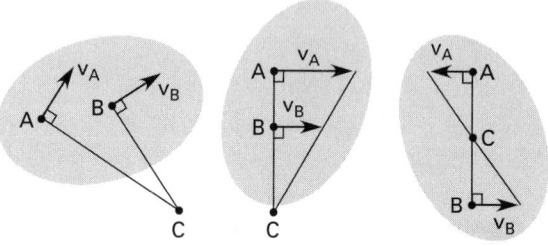

The absolute velocity of any point P on a wheel rolling (Fig. 71.15) with translational velocity, v_O, can be found by geometry. Assume that the wheel is pinned at C and rotates with its actual angular velocity, $\omega = v_O/r$. The direction of the point's velocity will be perpendicular to the line of length l between the instantaneous center and the point.

$$v = l\omega = \frac{lv_O}{r} \qquad \textit{71.52}$$

Figure 71.15 *Instantaneous Center of a Rolling Wheel*

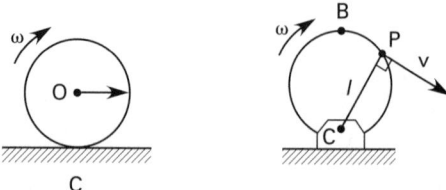

Equation 71.52 is valid only for a velocity referenced to the instantaneous center, point C. Table 71.3 can be used to find the velocities with respect to other points.

Table 71.3 *Relative Velocities of a Rolling Wheel*

	reference point		
point	O	C	B
v_O	0	$v_O \rightarrow$	$\leftarrow v_O$
v_C	$\leftarrow v_O$	0	$\leftarrow 2v_O$
v_B	$v_O \rightarrow$	$2v_O \rightarrow$	0

Example 71.11

A truck with 35 in diameter tires travels at a constant 35 mph. What is the absolute velocity of point P on the circumference of the tire?

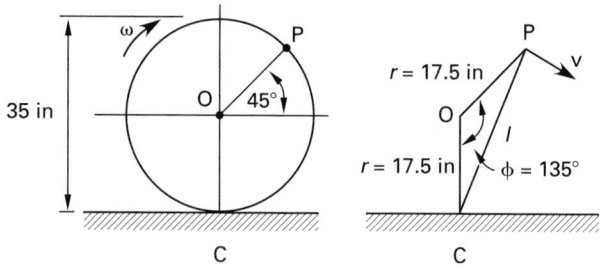

Solution

The translational velocity of the center of the wheel is

$$v_O = \frac{\left(35\ \frac{\text{mi}}{\text{hr}}\right)\left(5280\ \frac{\text{ft}}{\text{mi}}\right)}{3600\ \frac{\text{sec}}{\text{hr}}} = 51.33\ \text{ft/sec}$$

The wheel radius is

$$r = \frac{35\ \text{in}}{(2)\left(12\ \frac{\text{in}}{\text{ft}}\right)} = 1.458\ \text{ft}$$

The angular velocity of the wheel is

$$\omega = \frac{v_O}{r} = \frac{51.33\ \frac{\text{ft}}{\text{sec}}}{1.458\ \text{ft}} = 35.21\ \text{rad/sec}$$

The instantaneous center is the contact point, C. The law of cosines is used to find the distance l.

$$l^2 = r^2 + r^2 - 2r^2 \cos\phi = 2r^2(1 - \cos\phi)$$
$$l = \sqrt{(2)(1.458\ \text{ft})^2(1 - \cos 135°)} = 2.694\ \text{ft}$$

From Eq. 71.52, the absolute velocity of point P is

$$v_P = l\omega = (2.694\ \text{ft})\left(35.21\ \frac{\text{rad}}{\text{sec}}\right)$$
$$= 94.9\ \text{ft/sec}$$

20. INSTANTANEOUS CENTER OF ACCELERATION

The *instantaneous center of acceleration* is used to compute the absolute acceleration of a point as if a body were in pure rotation about that point. It is the same as the instantaneous center of rotation only for a body starting from rest and accelerating uniformly with angular acceleration, α.

$$a = l\alpha = \frac{la_O}{r} \qquad 71.53$$

In general, the instantaneous center of acceleration, C_a, will be deflected an angle β from the absolute acceleration vectors, as shown in Fig. 71.16. The relationship among the angle β, the instantaneous acceleration, α, and the instantaneous velocity, ω, is

$$\tan\beta = \frac{\alpha}{\omega^2} \qquad 71.54$$

Figure 71.16 *Instantaneous Center of Acceleration*

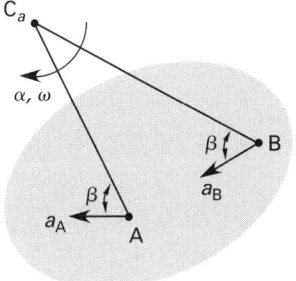

The absolute acceleration, a, determined from Eq. 71.53 is the same as the *resultant acceleration* in Fig. 71.7.

21. SLIDER RODS

The absolute velocity of any point, P, on a slider rod assembly can be found from the instantaneous center concept. The instantaneous center, C, is located by extending perpendiculars from the velocity vectors.

Figure 71.17 *Instantaneous Center of Slider Rod Assembly*

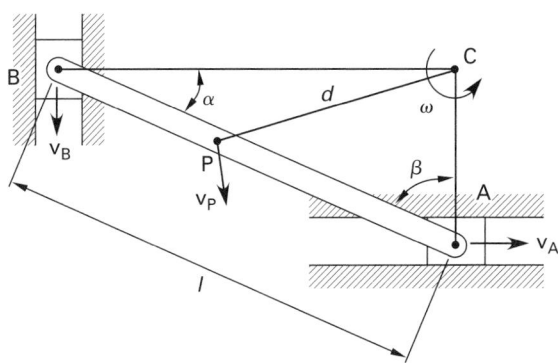

If the velocity with respect to point C of one end of the slider is known, say v_A, then v_B can be found from geometry. Since the slider can be assumed to rotate about point C with angular velocity ω,

$$\omega = \frac{v_A}{AC} = \frac{v_A}{l\cos\beta} = \frac{v_B}{BC} = \frac{v_B}{l\cos\alpha} \qquad 71.55$$

Since $\cos\alpha = \sin\beta$,

$$v_B = v_A\tan\beta \qquad 71.56$$

If the velocity with respect to point C of any other point P is required, it can be found from

$$v_P = d\omega \qquad 71.57$$

22. SLIDER-CRANK ASSEMBLIES

Figure 71.18 illustrates a slider-crank assembly for which points A and D are in the same plane and at the same elevation. The instantaneous velocity of any point, P, on the rod can be found if the distance to the instantaneous center is known.

$$v_P = d\omega_1 \qquad 71.58$$

Figure 71.18 *Slider-Crank Assembly*

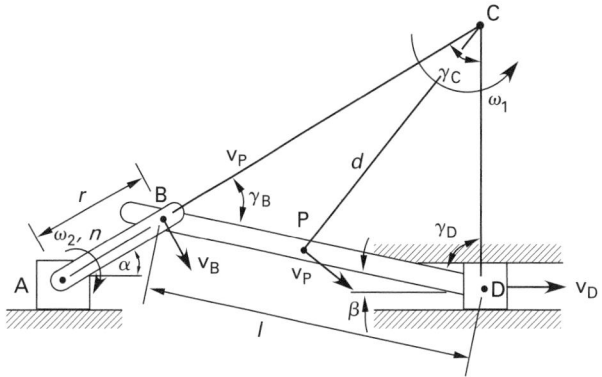

The tangential velocity of point B on the crank with respect to point A is perpendicular to the end of the crank. Slider D moves with a horizontal velocity. The intersection of lines drawn perpendicular to these velocity vectors locates the instantaneous center, point C. At any given instant, the connecting rod seems to rotate about point C with instantaneous angular velocity ω_1. The velocity of point D, v_D, is

$$v_D = (CD)\omega_1 = v_B\left(\frac{CD}{BC}\right) = \omega_2\left(\frac{AB \times CD}{BC}\right) \qquad 71.59$$

Similarly, the velocity of point B, v_B, is

$$v_B = (AB)\omega_2 = (BC)\omega_2 = v_D\left(\frac{BC}{CD}\right) \qquad 71.60$$

$$\omega_2 = \frac{2\pi n}{60} \qquad 71.61$$

Transportation

The following geometric relationships exist between the various angles.

$$\frac{\omega_1}{\omega_2} = \frac{AB}{BC} \qquad 71.62$$

$$\frac{\sin \alpha}{l} = \frac{\sin \beta}{r} \qquad 71.63$$

$$\frac{\sin \gamma_D}{BC} = \frac{\sin \gamma_B}{CD} = \frac{\sin \gamma_C}{l} \qquad 71.64$$

$$\gamma_B = \alpha + \beta \qquad 71.65$$

$$\gamma_D = 90 - \beta \qquad 71.66$$

$$\gamma_C = 90 - \alpha \qquad 71.67$$

72 Kinetics

C	constant used in space mechanics	$1/\text{ft}$	$1/\text{m}$
d	diameter	ft	m
e	coefficient of restitution	–	–
e	superelevation	ft/ft	m/m
E	energy	ft-lbf	J
f	coefficient of friction	–	–
F	force	lbf	N
g	acceleration due to gravity	ft/sec^2	m/s^2
g_c	gravitational constant	lbm-ft/ lbf-sec^2	n.a.
G	grade	ft/ft	m/m
G	universal gravitational constant	$\text{lbf-ft}^2/\text{lbm}^2$	$\text{N·m}^2/\text{kg}^2$
h	angular momentum	$\text{ft}^2\text{-lbm/sec}$	$\text{m}^2\text{·kg/s}$
h	height	ft	m
i	slippage fraction	–	–
I	mass moment of inertia	lbm-ft^2	kg·m^2
Imp	angular impulse	lbf-ft-sec	N·m·s
Imp	linear impulse	lbf-sec	N·s
k	spring constant	lbf/ft	N/m
m	mass	lbm	kg
m	slope	–	–
$\dot{m}$	mass flow rate	lbm/sec	kg/s
M	mass of the earth	lbm	kg
n	rotational speed	rev/sec	rev/s
N	normal force	lbf	N
p	momentum	lbf-sec	N·s
P	power	ft-lbf/sec	W
Q	flow rate	gal/hr	L/h
r	radius	ft	m
R	gear ratio	–	–
s	distance	ft	m
s'	fuel economy	mi/gal	km/L
t	time	sec	s
T	torque	ft-lbf	N·m
v	velocity	ft/sec	m/s
w	weight	lbf	–
W	work	ft-lbf	J
y	superelevation	ft	m

Nomenclature

a	acceleration	ft/sec^2	m/s^2
a	coefficient of rolling resistance	ft	m
a	semimajor axis length	ft	m
A	area	ft^2	m^2
b	semiminor axis length	ft	m
BSFC	brake specific fuel consumption	lbm/hp-hr	kg/kW·h
C	coefficient	–	–
C	coefficient of viscous damping (linear)	lbf-sec/ft	N·s/m
C	coefficient of viscous damping (quadratic)	$\text{lbf-sec}^2/\text{ft}^2$	$\text{N·s}^2/\text{m}^2$

Symbols

α	angular acceleration	rad/sec^2	rad/s^2
δ	deflection	ft	m
ϵ	eccentricity	–	–
η	efficiency	–	–
θ	angular position	rad	rad
ρ	density	lbm/ft^3	kg/m^3
ϕ	angle	deg or rad	deg or rad
ω	angular velocity	rad/sec	rad/s

Subscripts

0	initial
b	braking
c	centripetal

C	instant center
D	drag
f	final or frictional
g	gravitational
i	inertial
k	kinetic (dynamic)
m	mechanical
n	normal
O	center or centroidal
p	periodic
pend	pendulum
proj	projectile
r	rolling
s	static
t	tangential or terminal
w	wedge

1. INTRODUCTION TO KINETICS

Kinetics is the study of motion and the forces that cause motion. Kinetics includes an analysis of the relationship between the force and mass for translational motion and between torque and moment of inertia for rotational motion. Newton's laws form the basis of the governing theory in the subject of kinetics.

2. RIGID BODY MOTION

The most general type of motion is *rigid body motion*. There are five types.

- *pure translation*: The orientation of the object is unchanged as its position changes. (Motion can be in straight or curved paths.)

- *rotation about a fixed axis*: All particles within the body move in concentric circles about the *axis of rotation*.

- *general plane motion*: The motion can be represented in two dimensions (i.e., the *plane of motion*).

- *motion about a fixed point*: This describes any three-dimensional motion with one fixed point, such as a spinning top or a truck-mounted crane. The distance from a fixed point to any particle in the body is constant.

- *general motion*: This is any motion not falling into one of the other four categories.

Figure 72.1 illustrates the terms yaw, pitch, and roll as they relate to general motion. *Yaw* is a left or right swinging motion of the leading edge. *Pitch* is an up or down swinging motion of the leading edge. *Roll* is rotation about the leading edge's longitudinal axis.

Figure 72.1 *Yaw, Pitch, and Roll*

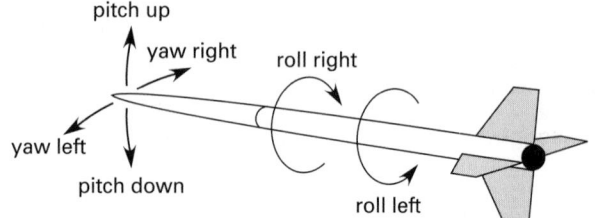

3. STABILITY OF EQUILIBRIUM POSITIONS

Stability is defined in terms of a body's relationship with an equilibrium position. *Neutral equilibrium* exists if a body, when displaced from its equilibrium position, remains in its displaced state. *Stable equilibrium* exists if the body returns to the original equilibrium position after experiencing a displacement. *Unstable equilibrium* exists if the body moves away from the equilibrium position. These terms are illustrated by Fig. 72.2.

Figure 72.2 *Types of Equilibrium Positions*

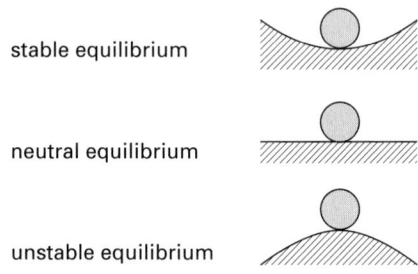

4. CONSTANT FORCES

Force is a push or a pull that one body exerts on another, including gravitational, electrostatic, magnetic, and contact influences. Forces that do not vary with time are *constant forces*.

Strictly speaking, actions of other bodies on a rigid body are known as *external forces*. External forces are responsible for external motion of a body. *Internal forces* hold together parts of a rigid body.

5. LINEAR MOMENTUM

The vector *linear momentum* (usually just *momentum*) is defined by Eq. 72.1.[1] It has the same direction as the velocity vector. Momentum has units of force $\times$ time (e.g., lbf-sec or N·s).

$$\mathbf{p} = m\mathbf{v} \qquad \text{[SI]} \qquad 72.1(a)$$

$$\mathbf{p} = \frac{m\mathbf{v}}{g_c} \qquad \text{[U.S.]} \qquad 72.1(b)$$

Momentum is conserved when no external forces act on a particle. If no forces act on the particle, the velocity and direction of the particle are unchanged. The *law of conservation of momentum* states that the linear momentum is unchanged if no unbalanced forces act on the particle. This does not prohibit the mass and velocity from changing, however. Only the product of mass and

---

[1]The symbols $\mathbf{P}$, **mom**, mv, and others are also used for momentum. Some authorities assign no symbol and just use the word momentum.

velocity is constant. Depending on the nature of the problem, momentum can be conserved in any or all of the three coordinate directions.

$$\sum m_0 \mathbf{v}_0 = \sum m_f \mathbf{v}_f \qquad 72.2$$

6. BALLISTIC PENDULUM

Figure 72.3 illustrates a *ballistic pendulum*. A projectile of known mass but unknown velocity is fired into a hanging target (the *pendulum*). The projectile is captured by the pendulum, which moves forward and upward. Kinetic energy is not conserved during impact because some of the projectile's kinetic energy is transformed into heat. However, momentum is conserved during impact, and the movement of the pendulum can be used to calculate the impact velocity of the projectile.[2]

Figure 72.3 Ballistic Pendulum

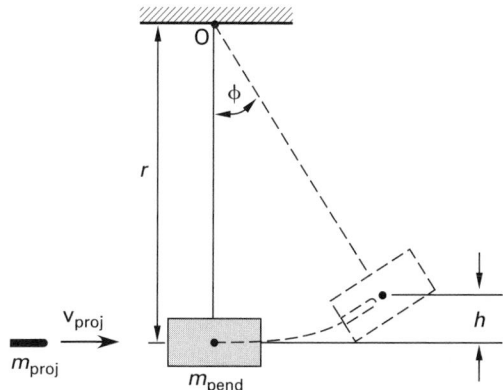

Since no external forces act on the block during impact, the momentum of the system is conserved.

$$\mathbf{p}_{\text{before impact}} = \mathbf{p}_{\text{after impact}} \qquad 72.3$$

$$m_{\text{proj}} \mathbf{v}_{\text{proj}} = (m_{\text{proj}} + m_{\text{pend}}) \mathbf{v}_{\text{pend}} \qquad 72.4$$

Although kinetic energy before impact is not conserved, the total remaining energy after impact is conserved. That is, once the projectile has been captured by the pendulum, the kinetic energy of the pendulum-projectile combination is converted totally to potential energy as the pendulum swings upward.

$$\left(\tfrac{1}{2}\right) (m_{\text{proj}} + m_{\text{pend}}) \mathbf{v}_{\text{pend}}^2 = (m_{\text{proj}} + m_{\text{pend}}) gh \qquad 72.5$$

$$\mathbf{v}_{\text{pend}} = \sqrt{2gh} \qquad 72.6$$

The relationship between the rise of the pendulum, h, and the swing angle, ϕ, is

$$h = r(1 - \cos\phi) \qquad 72.7$$

---

[2]In this type of problem, it is important to be specific about when the energy and momentum are evaluated. During impact, kinetic energy is not conserved, but momentum is conserved. After impact, as the pendulum swings, energy is conserved but momentum is not conserved because gravity (an external force) acts on the pendulum during its swing.

Since the time during which the force acts is not well defined, there is no single equivalent force that can be assumed to initiate the motion. Any force that produces the same impulse over a given contact time will be applicable.

7. ANGULAR MOMENTUM

The vector *angular momentum* (also known as *moment of momentum*) taken about a point O is the moment of the linear momentum vector. Angular momentum has units of distance × force × time (e.g., ft-lbf-sec or N·m·s). It has the same direction as the rotation vector and can be determined by use of the right-hand rule. (That is, it acts in a direction perpendicular to the plane containing the position and linear momentum vectors.)

$$\mathbf{h}_{\text{O}} = \mathbf{r} \times m\mathbf{v} \qquad \text{[SI]} \qquad 72.8(a)$$

$$\mathbf{h}_{\text{O}} = \frac{\mathbf{r} \times m\mathbf{v}}{g_c} \qquad \text{[U.S.]} \qquad 72.8(b)$$

Any of the methods normally used to evaluate cross-products can be used with angular momentum. The scalar form of Eq. 72.8 is

$$h_{\text{O}} = rmv \sin\phi \qquad \text{[SI]} \qquad 72.9(a)$$

$$h_{\text{O}} = \frac{rmv \sin\phi}{g_c} \qquad \text{[U.S.]} \qquad 72.9(b)$$

Figure 72.4 Angular Momentum

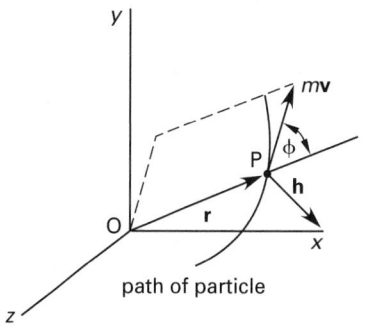

path of particle

For a rigid body rotating about an axis passing through its center of gravity located at point O, the scalar value of angular momentum is given by Eq. 72.10.

$$h_{\text{O}} = I\omega \qquad \text{[SI]} \qquad 72.10(a)$$

$$h_{\text{O}} = \frac{I\omega}{g_c} \qquad \text{[U.S.]} \qquad 72.10(b)$$

8. NEWTON'S FIRST LAW OF MOTION

Much of this chapter is based on Newton's laws of motion. *Newton's first law of motion* can be stated in several forms.

common form: A particle will remain in a state of rest or will continue to move with constant velocity unless an unbalanced external force acts on it.

law of conservation of momentum form: If the resultant external force acting on a particle is zero, then the linear momentum of the particle is constant.

9. NEWTON'S SECOND LAW OF MOTION

Newton's second law of motion is stated as follows.

second law: The acceleration of a particle is directly proportional to the force acting on it and inversely proportional to the particle mass. The direction of acceleration is the same as the force of direction.

This law can be stated in terms of the force vector required to cause a change in momentum. The resultant force is equal to the rate of change of linear momentum.

$$\mathbf{F} = \frac{d\mathbf{p}}{dt} \qquad \text{72.11}$$

If the mass is constant with respect to time, the scalar form of Eq. 72.11 is[3]

$$F = m\left(\frac{dv}{dt}\right) = ma \qquad \text{[SI]} \qquad \text{72.12(a)}$$

$$F = \left(\frac{m}{g_c}\right)\left(\frac{dv}{dt}\right) = \frac{ma}{g_c} \qquad \text{[U.S.]} \qquad \text{72.12(b)}$$

Equation 72.12 can be written in rectangular coordinates form (i.e., in terms of x- and y- component forces), in polar coordinates form (i.e, tangential and normal components), and in cylindrical coordinates form (i.e., radial and transverse components).

Although Newton's laws do not specifically deal with rotation, there is an analogous relationship between torque and change in angular momentum. For a rotating body, the torque, $\mathbf{T}$, required to change the angular momentum is

$$\mathbf{T} = \frac{d\mathbf{h}_0}{dt} \qquad \text{72.13}$$

If the moment of inertia is constant, the scalar form of Eq. 72.13 is

$$T = I\left(\frac{d\omega}{dt}\right) = I\alpha \qquad \text{[SI]} \qquad \text{72.14(a)}$$

$$T = \left(\frac{I}{g_c}\right)\left(\frac{d\omega}{dt}\right) = \frac{I\alpha}{g_c} \qquad \text{[U.S.]} \qquad \text{72.14(b)}$$

---

[3]Equation 72.12 shows that force is a scalar multiple of acceleration. Any consistent set of units can be used. For example, if both sides are divided by the acceleration of gravity (i.e., so that acceleration in Eq. 72.12 is in gravities), the force will have units of *g-forces* or *gees* (i.e., multiples of the gravitational force).

Example 72.1

The acceleration in m/s^2 of a 40 kg body is specified by the equation

$$a(t) = 8 - 12t$$

What is the instantaneous force acting on the body at $t = 6$ s?

Solution

The acceleration is

$$a(6) = 8\,\frac{m}{s^2} - \left(12\,\frac{m}{s^3}\right)(6\text{ s}) = -64\text{ m/s}^2$$

From Newton's second law, the instantaneous force is

$$F = ma = (40\text{ kg})\left(-64\,\frac{m}{s^2}\right)$$
$$= -2560\text{ N}$$

Example 72.2

During start-up, a 4.0 ft diameter pulley with centroidal moment of inertia of 1610 lbm-ft^2 is subjected to tight-side and loose-side belt tensions of 200 lbf and 100 lbf, respectively. A frictional torque of 15 ft-lbf is acting to resist pulley rotation. (a) What is the angular acceleration? (b) How long will it take the pulley to reach a speed of 120 rpm?

Solution

(a) From Eq. 72.24, the net torque is

$$T = rF_{\text{net}} = (2\text{ ft})(200\text{ lbf} - 100\text{ lbf}) - 15\text{ ft-lbf}$$
$$= 185\text{ ft-lbf}$$

From Eq. 72.14, the angular acceleration is

$$\alpha = \frac{g_c T}{I} = \frac{\left(32.2\,\dfrac{\text{lbm-ft}}{\text{lbf-sec}^2}\right)(185\text{ ft-lbf})}{1610\text{ lbm-ft}^2}$$
$$= 3.7\text{ rad/sec}^2$$

(b) The rotational speed is

$$\omega = \frac{\left(120\,\dfrac{\text{rev}}{\text{min}}\right)\left(2\pi\,\dfrac{\text{rad}}{\text{rev}}\right)}{60\,\dfrac{\text{sec}}{\text{min}}}$$
$$= 12.6\text{ rad/sec}$$

This is a case of constant angular acceleration starting from rest.

$$t = \frac{\omega}{\alpha} = \frac{12.6\,\dfrac{\text{rad}}{\text{sec}}}{3.7\,\dfrac{\text{rad}}{\text{sec}^2}}$$
$$= 3.4\text{ sec}$$

10. CENTRIPETAL FORCE

Newton's second law says there is a force for every acceleration a body experiences. For a body moving around a curved path, the total acceleration can be separated into tangential and normal components. By Newton's second law, there are corresponding forces in the tangential and normal directions. The force associated with the normal acceleration is known as the *centripetal force*.[4]

$$F_c = ma_n = \frac{mv_t^2}{r} \qquad \text{[SI]} \qquad \textbf{72.15(a)}$$

$$F_c = \frac{mv_t^2}{g_c r} \qquad \text{[U.S.]} \qquad \textbf{72.15(b)}$$

Figure 72.5 *Centripetal Force*

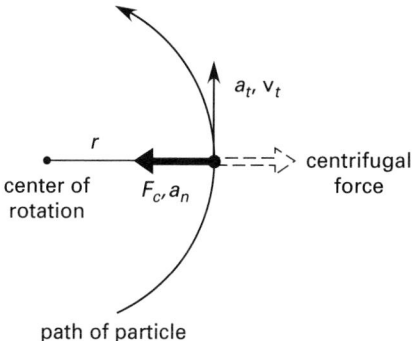

The centripetal force is a real force on the body toward the center of rotation. The so-called *centrifugal force* is an apparent force on the body directed away from the center of rotation. The centripetal and centrifugal forces are equal in magnitude but opposite in sign.

Example 72.3

A 4500 lbm (2000 kg) car travels at 40 mph (65 kph) around a curve with a radius of 200 ft (60 m). What is the centripetal force?

SI Solution

The tangential velocity is

$$v_t = \frac{\left(65 \, \frac{km}{h}\right)\left(1000 \, \frac{m}{km}\right)}{3600 \, \frac{s}{h}} = 18.06 \text{ m/s}$$

From Eq. 72.15(a), the centripetal force is

$$F_c = \frac{mv_t^2}{r} = \frac{(2000 \text{ kg})\left(18.06 \, \frac{m}{s}\right)^2}{60 \text{ m}}$$
$$= 10\,900 \text{ N}$$

[4]The term *normal force* is reserved for the plane reaction in friction calculations.

Customary U.S. Solution

The tangential velocity is

$$v_t = \frac{\left(40 \, \frac{mi}{hr}\right)\left(5280 \, \frac{ft}{mi}\right)}{3600 \, \frac{sec}{hr}} = 58.7 \text{ ft/sec}$$

From Eq. 72.15(b), the centripetal force is

$$F_c = \frac{mv_t^2}{g_c r}$$
$$= \frac{(4500 \text{ lbm})\left(58.7 \, \frac{ft}{sec}\right)^2}{\left(32.2 \, \frac{\text{lbm-ft}}{\text{lbf-sec}^2}\right)(200 \text{ ft})}$$
$$= 2400 \text{ lbf}$$

An unbalanced rotating body (vehicle wheel, clutch disk, rotor of an electrical motor, etc.) will experience a dynamic *unbalanced force*. Though the force is essentially centripetal in nature and is given by Eq. 72.15, it is generally difficult to assign a value to the radius. For that reason, the force is often determined directly on the rotating body or from the deflection of its supports.

Since the body is rotating, the force will be experienced in all directions perpendicular to the axis of rotation. If the supports are flexible, the force will cause the body to vibrate, and the frequency of vibration will essentially be the rotational speed. If the supports are rigid, the bearings will carry the unbalanced force and transmit it to other parts of the frame.

11. NEWTON'S THIRD LAW OF MOTION

Newton's third law of motion is as follows.

> *third law:* For every acting force between two bodies, there is an equal but opposite reacting force on the same line of action.

$$\mathbf{F}_{\text{reacting}} = -\mathbf{F}_{\text{acting}} \qquad \textbf{72.16}$$

12. DYNAMIC EQUILIBRIUM

An accelerating body is not in static equilibrium. Accordingly, the familiar equations of statics ($\sum F = 0$ and $\sum M = 0$) do not apply. However, if the *inertial force*, $m\mathbf{a}$, is included in the static equilibrium equation, the body is said to be in *dynamic equilibrium*.[5,6]

[5]Other names for the inertial force are *inertia vector* (when written as $m\mathbf{a}$), *dynamic reaction*, and *reversed effective force*. The term $\sum \mathbf{F}$ is known as the *effective force*.
[6]*Dynamic* and *equilibrium* are contradictory terms. A better term is *simulated equilibrium*, but this form has not caught on.

This is known as *D'Alembert's principle*. Since the inertial force acts to oppose changes in motion, it is negative in the summation.

$$\sum \mathbf{F} - m\mathbf{a} = 0 \qquad \text{[SI]} \qquad 72.17(a)$$

$$\sum \mathbf{F} - \frac{m\mathbf{a}}{g_c} = 0 \qquad \text{[U.S.]} \qquad 72.17(b)$$

It should be clear that D'Alembert's principle is just a different form of Newton's second law, with the ma term transposed to the left-hand side.

The analogous rotational form of the dynamic equilibrium principle is

$$\sum \mathbf{T} - I\boldsymbol{\alpha} = 0 \qquad \text{[SI]} \qquad 72.18(a)$$

$$\sum \mathbf{T} - \frac{I\boldsymbol{\alpha}}{g_c} = 0 \qquad \text{[U.S.]} \qquad 72.18(b)$$

13. FLAT FRICTION

Friction is a force that always resists motion or impending motion. It always acts parallel to the contacting surfaces. The frictional force, F_f, exerted on a stationary body is known as *static friction*, *Coulomb friction*, and *fluid friction*. If the body is moving, the friction is known as *dynamic friction* and is less than the static friction.

The actual magnitude of the frictional force depends on the *normal force*, N, and the *coefficient of friction*, f, between the body and the surface.[7] For a body resting on a horizontal surface, the normal force is the weight of the body.

$$N = mg \qquad \text{[SI]} \qquad 72.19(a)$$

$$N = \frac{mg}{g_c} \qquad \text{[U.S.]} \qquad 72.19(b)$$

Figure 72.6 *Frictional and Normal Forces*

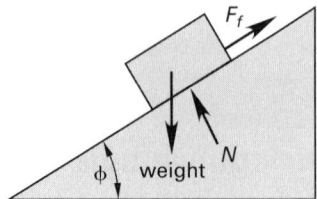

If the body rests on an inclined surface, the normal force is calculated from the weight.

$$N = mg \cos \phi \qquad \text{[SI]} \qquad 72.20(a)$$

$$N = \frac{mg \cos \phi}{g_c} \qquad \text{[U.S.]} \qquad 72.20(b)$$

---

[7]The symbol μ is also widely used by engineers to represent the coefficient of friction.

The maximum static frictional force, F_f, is the product of the coefficient of friction, f, and the normal force, N. (The subscripts s and k are used to distinguish between the static and dynamic (kinetic) coefficients of friction.)

$$F_{f,\max} = f_s N \qquad 72.21$$

The frictional force acts only in response to a disturbing force. If a small disturbing force (i.e., a force less than $F_{f,\max}$) acts on a body, then the frictional force will equal the disturbing force, and the maximum frictional force will not develop. This is known as the *equilibrium phase*. The *motion impending phase* is where the disturbing force equals the maximum frictional force, $F_{f,\max}$. Once motion begins, however, the coefficient of friction drops slightly, and a lower frictional force opposes movement. These cases are illustrated in Fig. 72.7.

Figure 72.7 *Frictional Force versus Disturbing Force*

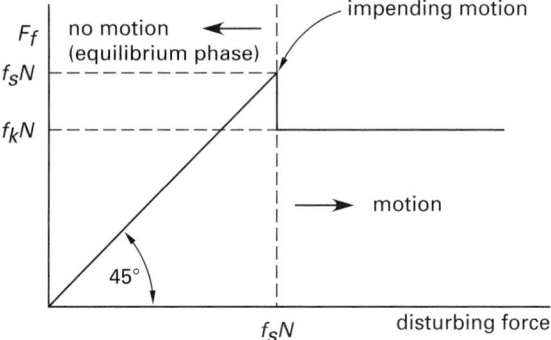

A body on an inclined plane will not begin to slip down the plane until the component of weight parallel to the plane exceeds the frictional force. If the plane's inclination angle can be varied, the body will not slip until the angle reaches a critical angle known as the *angle of repose* or *angle of static friction*, ϕ. Equation 72.22 relates this angle to the coefficient of static friction.

$$\tan \phi = f_s \qquad 72.22$$

Tabulations of coefficients of friction distinguish between types of surfaces and between static and dynamic cases. They might also list values for dry conditions and oiled conditions. The term *dry* is synonymous with *nonlubricated*. The ambiguous term *wet*, although a natural antonym for *dry*, is sometimes used to mean *oily*. However, it usually means wet with water, as in tires on a wet roadway after a rain. Typical values of the coefficient of friction are given in Table 72.1.[8]

---

[8]Experimental and reported values of the coefficient of friction vary greatly from researcher to researcher and experiment to experiment. The values in Table 72.1 are more for use in solving practice problems than serving as the last word in available data.

Table 72.1 *Typical Coefficients of Friction*

material	condition	dynamic	static
cast iron on cast iron	dry	0.15	1.00
plastic on steel	dry	0.35	0.45
grooved rubber on pavement	dry	0.40	0.55
bronze on steel	oiled	0.07	0.09
steel on graphite	dry	0.16	0.21
steel on steel	dry	0.42	0.78
steel on steel	oiled	0.08	0.10
steel on asbestos-faced steel	dry	0.11	0.15
steel on asbestos-faced steel	oiled	0.09	0.12
press fits (shaft in hole)	oiled	–	0.10–0.15

A special case of the angle of repose is the *angle of internal friction*, ϕ, of soil, grain, or other granular material. The angle made by a pile of granular material depends on how much friction there is between the granular particles. Liquids have angles of internal friction of zero, because they do not form piles.

Figure 72.8 *Angle of Internal Friction*

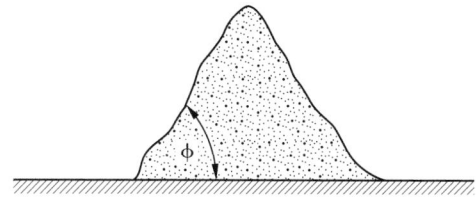

14. WEDGES

Wedges are machines that are able to raise heavy loads. The wedge angles are chosen so that friction will keep the wedge in place once it is driven between the load and support. As with any situation where friction is present, the frictional force is parallel to the contacting surfaces.

Figure 72.9 *Typical Wedge Problem*

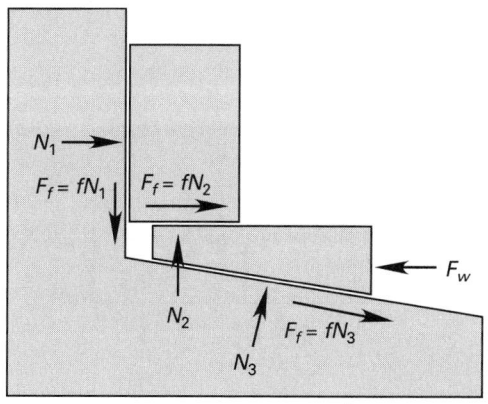

15. BELT FRICTION

Friction from a belt, rope, or band wrapped around a pulley or sheave is responsible for the transfer of torque. Except at start-up, one side of the belt (the tight side) will have a higher tension than the other (the slack side). The basic relationship between these belt tensions and the coefficient of friction neglects centrifugal effects and is given by Eq. 72.23.[9] (The angle of wrap, ϕ, must be expressed in radians.)

$$\frac{F_{\max}}{F_{\min}} = e^{f\phi} \qquad 72.23$$

The net transmitted torque is

$$T = (F_{\max} - F_{\min})\, r \qquad 72.24$$

Figure 72.10 *Belt Friction*

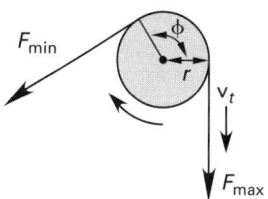

The power transmitted by the belt running at tangential velocity v_t is given by Eq. 72.25.[10]

$$P = (F_{\max} - F_{\min})\, v_t \qquad 72.25$$

The change in belt tension caused by the centrifugal force should be considered when the velocity or belt mass is very large. Equation 72.26 can be used, where m is the mass per unit length of belt.

$$\frac{F_{\max} - mv_t^2}{F_{\min} - mv_t^2} = e^{f\phi} \qquad \text{[SI]} \qquad 72.26(a)$$

$$\frac{F_{\max} - \dfrac{mv_t^2}{g_c}}{F_{\min} - \dfrac{mv_t^2}{g_c}} = e^{f\phi} \qquad \text{[U.S.]} \qquad 72.26(b)$$

16. ROLLING RESISTANCE

Rolling resistance is a force that opposes motion, but it is not friction. Rather, it is caused by the deformation of the rolling body and the supporting surface. Rolling resistance is characterized by a *coefficient of*

---

[9]This equation does not apply to V-belts. V-belt design and analysis is dependent on the cross-sectional geometry of the belt.
[10]When designing a belt system, the horsepower to be transmitted should be multiplied by a *service factor* to obtain the *design power*. Service factors range from 1.0 to 1.5 and depend on the nature of the power source, the load, and the starting characteristics.

rolling resistance, a, which has units of length.[11] Since this deformation is very small, the rolling resistance in the direction of motion is

$$F_r = \frac{mga}{r} \qquad \text{[SI]} \qquad 72.27(a)$$

$$F_r = \frac{mga}{rg_c} = \frac{wa}{r} \qquad \text{[U.S.]} \qquad 72.27(b)$$

Figure 72.11 *Rolling Resistance*

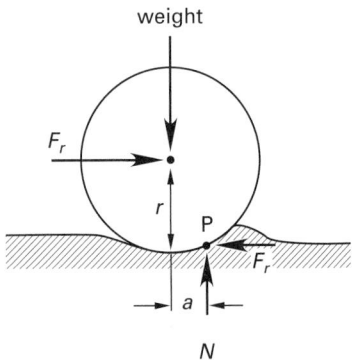

The term *coefficient of rolling friction, f_r,* is occasionally encountered, although friction is not the cause of rolling resistance.

$$f_r = \frac{F_r}{w} = \frac{a}{r} \qquad\qquad 72.28$$

17. MOTION OF RIGID BODIES

When a rigid body experiences pure translation, its position changes without any change in orientation. At any instant, all points on the body have the same displacement, velocity, and acceleration. The behavior of a rigid body in translation is given by Eqs. 72.29 and 72.30. All equations are written for the center of mass. (These equations represent Newton's second law written in component form.)

$$\sum F_x = ma_x \quad \text{[consistent units]} \qquad 72.29$$

$$\sum F_y = ma_y \quad \text{[consistent units]} \qquad 72.30$$

When a torque acts on a rigid body, the rotation will be about the center of gravity unless the body is constrained otherwise. In the case of rotation, the torque and angular acceleration are related by Eq. 72.31.

$$T = I\alpha \qquad \text{[SI]} \qquad 72.31(a)$$

$$T = \frac{I\alpha}{g_c} \qquad \text{[U.S.]} \qquad 72.31(b)$$

---

[11]Rolling resistance is traditionally derived by assuming the roller encounters a small step in its path a distance a in front of the center of gravity. The forces acting on the roller are the weight and driving force acting through the centroid and the normal force and rolling resistance acting at the contact point. Equation 72.27 is derived by taking moments about the contact point, P.

Euler's equations of motion are used to analyze the motion of a rigid body about a fixed point, O. This class of problem is particularly difficult because the mass moments of products of inertia change with time if a fixed set of axes is used. Therefore, it is more convenient to define the x-, y-, and z-axes with respect to the body. Such an action is acceptable because the angular momentum about the origin, $\mathbf{h}_O$, corresponding to a given angular velocity, ω, is independent of the choice of coordinate axes.

An infinite number of axes can be chosen. (A general relationship between moments and angular momentum is given in most dynamics textbooks.) However, if the origin is at the mass center and the x-, y-, and z-axes coincide with the principal axes of inertia of the body (such that the product of inertia is zero), the angular momentum of the body about the origin (i.e., point O at (0,0,0)) is given by the simplified relationship

$$\mathbf{h}_O = I_x\omega_x\mathbf{i} + I_y\omega_y\mathbf{j} + I_z\omega_z\mathbf{k} \qquad 72.32$$

The three scalar Euler equations of motion can be derived from this simplified relationship.

$$\sum M_x = I_x\alpha_x - (I_y - I_z)\omega_y\omega_z \qquad 72.33$$

$$\sum M_y = I_y\alpha_y - (I_z - I_x)\omega_z\omega_x \qquad 72.34$$

$$\sum M_z = I_z\alpha_z - (I_x - I_y)\omega_x\omega_y \qquad 72.35$$

Example 72.4

A 5000 lbm truck skids with a deceleration of 15 ft/sec^2. (a) What is the coefficient of sliding friction? (b) What are the frictional forces and normal reactions (per axle) at the tires?

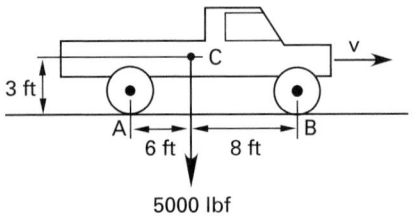

Solution

(a) The free-body diagram of the truck in equilibrium with the inertial force is shown.

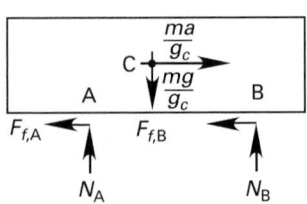

The equation of dynamic equilibrium in the horizontal direction is

$$\sum F_x = \frac{ma}{g_c} - F_{f,\text{A}} - F_{f,\text{B}} = 0$$

$$= \frac{ma}{g_c} - Nf = 0$$

$$\frac{(5000 \text{ lbm})\left(15 \dfrac{\text{ft}}{\text{sec}^2}\right)}{32.2 \dfrac{\text{lbm-ft}}{\text{lbf-sec}^2}} - (5000 \text{ lbf})f = 0$$

The coefficient of friction is

$$f = 0.466$$

(b) The vertical reactions at the tires can be found by taking moments about one of the contact points.

$$\sum M_{\text{A}}: \quad 14 N_{\text{B}} - (6 \text{ ft})(5000 \text{ lbf})$$

$$- (3 \text{ ft})\left(\frac{5000 \text{ lbm}}{32.2 \dfrac{\text{lbm-ft}}{\text{lbf-sec}^2}}\right)\left(15 \frac{\text{ft}}{\text{sec}^2}\right) = 0$$

$$N_{\text{B}} = 2642 \text{ lbf}$$

The remaining vertical reaction is found by summing vertical forces.

$$\sum F_y: \quad N_{\text{A}} + N_{\text{B}} - \frac{mg}{g_c} = 0$$

$$N_{\text{A}} + 2642 \text{ lbf} - 5000 \text{ lbf} = 0$$

$$N_{\text{A}} = 2358 \text{ lbf}$$

The horizontal frictional forces at the front and rear axles are

$$F_{f,\text{A}} = (0.466)(2358 \text{ lbf}) = 1099 \text{ lbf}$$

$$F_{f,\text{B}} = (0.466)(2642 \text{ lbf}) = 1231 \text{ lbf}$$

18. CONSTRAINED MOTION

Figure 72.12 shows a cylinder (or sphere) on an inclined plane. If there is no friction, there will be no torque to start the cylinder rolling. Regardless of the angle, the cylinder will slide down the incline in *unconstrained motion*.

Figure 72.12 Unconstrained Motion

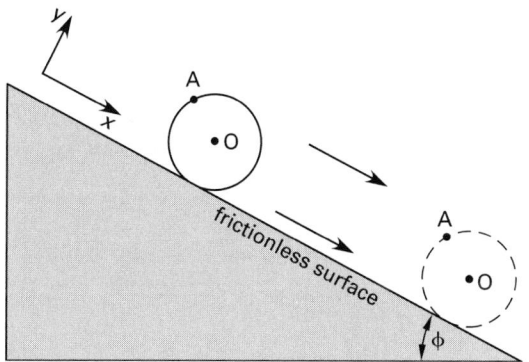

The acceleration sliding down the incline can be calculated by writing Newton's second law for an axis parallel to the plane.[12] Once the acceleration is known, the velocity can be found from the constant-acceleration equations.

$$ma_{\text{O},x} = mg \sin \phi \qquad \textit{72.36}$$

If friction is sufficiently large, or if the inclination is sufficiently small, there will be no slipping. This condition occurs if

$$\phi < \arctan f_s \qquad \textit{72.37}$$

The frictional force acting at the cylinder's radius r supplies a torque that starts and keeps the cylinder rolling. The frictional force is

$$F_f = fN \qquad \textit{72.38}$$

$$F_f = fmg \cos \phi \qquad [\text{SI}] \qquad \textit{72.39(a)}$$

$$F_f = \frac{fmg \cos \phi}{g_c} \qquad [\text{U.S.}] \qquad \textit{72.39(b)}$$

Figure 72.13 Constrained Motion

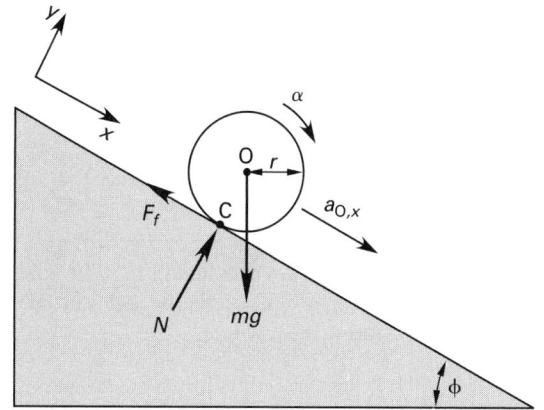

Thus, with no slipping, the cylinder has two degrees of freedom (the x-directional and rotation), and motion of the center of mass must simultaneously satisfy (i.e., is constrained by) two equations. (This excludes motion perpendicular to the plane.) This is called *constrained motion*.

$$mg \sin \phi - F_f = ma_{\text{O},x} \quad [\text{consistent units}] \qquad \textit{72.40}$$

$$F_f r = I_{\text{O}} \alpha \quad [\text{consistent units}] \qquad \textit{72.41}$$

The mass moment of inertia used in calculating angular acceleration can be either the centroidal moment of inertia, I_{O}, or the moment of inertia taken about the contact point, I_{C}, depending on whether torques (moments) are evaluated with respect to point O or point C, respectively.

---

[12]Most inclined plane problems are conveniently solved by resolving all forces into components parallel and perpendicular to the plane.

If moments are evaluated with respect to point O, the coefficient of friction, f, must be known, and the centroidal moment of inertia (found from a table) can be used. If moments are evaluated with respect to the contact point, the frictional and normal forces drop out of the torque summation. The cylinder instantaneously rotates as though it were pinned at point C. If the centroidal moment of inertia, I_O, is known, the parallel axis theorem can be used to find the required moment of inertia.

$$I_C = I_O + mr^2 \qquad 72.42$$

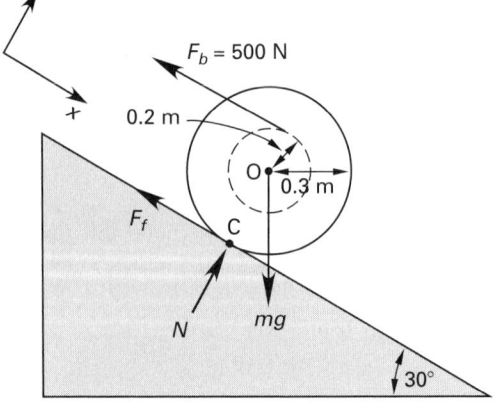

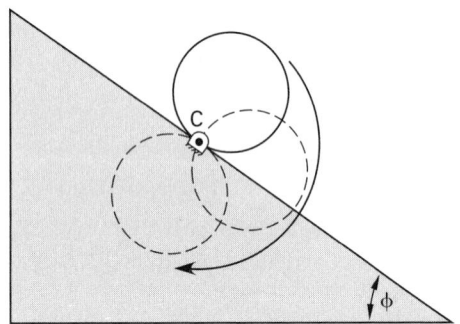

Figure 72.14 *Instantaneous Center of a Constrained Cylinder*

When there is no slipping, the cylinder will roll with constant linear and angular acceleration. The distance traveled by the center of mass can be calculated from the angle of rotation.

$$s_O = r\theta \qquad 72.43$$

If $\phi \geq \arctan f_s$, the cylinder will simultaneously roll and slide down the incline. The analysis is similar to the no-sliding case, except that the coefficient of sliding friction is used. Once sliding has started, the inclination angle can be reduced to $\arctan f_k$, and rolling with sliding will continue.

Example 72.5

A 150 kg cylinder with radius 0.3 m is pulled up a plane inclined at 30° as fast as possible without the cylinder slipping. The coefficient of friction is 0.236. There is a groove in the cylinder at radius = 0.2 m. A rope in the groove applies a force of 500 N up the ramp. What is the linear acceleration of the cylinder?

Solution

To solve by summing forces in the x-direction:

The normal force is

$$N = mg\cos\phi = (150 \text{ kg})\left(9.81 \ \frac{\text{m}}{\text{s}^2}\right)(\cos 30°)$$
$$= 1274.4 \text{ N}$$

The frictional maximum (friction impending) force is

$$F_f = fN = (0.236)(1274.4 \text{ N}) = 300.8 \text{ N}$$

The summation of forces in the x-direction is

$$ma_{O,x} = mg\sin\phi - F_f - F_b$$

$$a_{O,x} = \frac{(150 \text{ kg})\left(9.81 \ \frac{\text{m}}{\text{s}^2}\right)(\sin 30°)}{150 \text{ kg}}$$
$$\phantom{a_{O,x}} = \frac{\quad - 300.8 \text{ N} - 500 \text{ N}}{150 \text{ kg}}$$
$$= -0.433 \text{ m/s}^2 \quad [\text{up the incline}]$$

To solve by taking moment about the contact point:[13]

From App. 71.A, the centroidal mass moment of inertia of the cylinder is

$$I_O = \tfrac{1}{2}mr^2 = (0.5)(150 \text{ kg})(0.3 \text{ m})^2$$
$$= 6.75 \text{ kg·m}^2$$

---

[13]This example can also be solved by summing moments about the center. If this is done, the governing equations are

$$I_O = \tfrac{1}{2}mr^2$$
$$M_O = I_O\alpha = F_b r' - F_f r$$
$$\tfrac{1}{2}mr^2\alpha = F_b r' - fmg\cos\phi r$$
$$a_{O,x} = r\alpha$$

The mass moment of inertia with respect to the contact point, C, is given by the parallel axis theorem.

$$I_C = I_O + mr^2 = \tfrac{1}{2}mr^2 + mr^2 = \tfrac{3}{2}mr^2$$
$$= \left(\tfrac{3}{2}\right)(150 \text{ kg})(0.3 \text{ m})^2 = 20.25 \text{ kg·m}^2$$

The x-component of the weight acts through the center of gravity. (This term dropped out when moments were taken with respect to the center of gravity.)

$$(mg)_x = (150 \text{ kg})\left(9.81 \ \frac{\text{m}}{\text{s}^2}\right)(\sin 30^\circ)$$
$$= 735.8 \text{ N}$$

The summation of torques about point C gives the angular acceleration with respect to point C.

$$(735.8 \text{ N})(0.3 \text{ m})$$
$$- (500 \text{ N})(0.3 \text{ m} + 0.2 \text{ m}) = (20.25 \text{ kg·m}^2)\alpha$$
$$\alpha = -1.445 \text{ rad/s}^2$$

The linear acceleration can be calculated from the angular acceleration and the distance between points C and O.

$$a_{O,x} = r\alpha = (0.3 \text{ m})\left(-1.445 \ \frac{\text{rad}}{\text{s}^2}\right)$$
$$= -0.433 \text{ m/s}^2 \quad [\text{up the incline}]$$

19. CABLE TENSION FROM AN ACCELERATING SUSPENDED MASS

When a mass hangs motionless from a cable, or when the mass is moving with a uniform velocity, the cable tension will equal the weight of the mass. However, when the mass is accelerating, the weight must be reduced by the inertial force. If the mass experiences a downward acceleration equal to the gravitational acceleration, there is no tension in the cable. Thus, the two cases shown in Fig. 72.15 are not the same.

Figure 72.15 Cable Tension from a Suspended Mass

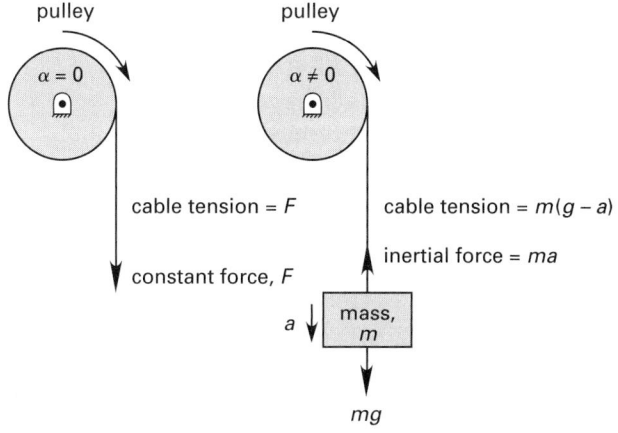

Example 72.6

A 10.0 lbm (4.6 kg) mass hangs from a rope wrapped around a 2.0 ft (0.6 m) diameter pulley (centroidal moment of inertia of 70 lbm-ft^2 (2.9 kg·m^2)). (a) What is the angular acceleration of the pulley? (b) What is the linear acceleration of the mass?

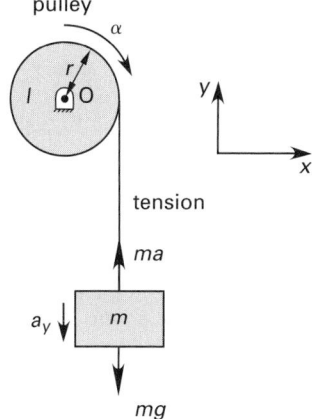

SI Solution

(a) The two equations of motion are

$$\sum F_{y,\text{mass: tension}} + ma - mg = 0$$
$$\text{tension} - (4.6 \text{ kg})\left(9.81 \ \frac{\text{m}}{\text{s}^2}\right) = -(4.6 \text{ kg})a_y$$
$$\sum M_{O,\text{pulley: (tension)}}\, r = I\alpha$$
$$(\text{tension})(0.3 \text{ m}) = (2.9 \text{ kg·m}^2)\alpha$$

Both a_y and α are unknown but are related by $a_y = r\alpha$. Substituting into the $\sum F$ equation and eliminating the tension,

$$\frac{(2.9 \text{ kg·m}^2)\alpha}{0.3 \text{ m}} - 45.1 \text{ N} = -(4.6 \text{ kg})(0.3 \text{ m})\alpha$$
$$9.67\alpha - 45.1 = -1.38\alpha$$
$$\alpha = 4.08 \text{ rad/s}^2$$

(b) The linear acceleration of the mass is

$$a_y = r\alpha = (0.3 \text{ m})\left(4.08 \ \frac{\text{rad}}{\text{s}^2}\right)$$
$$= 1.22 \text{ m/s}^2$$

Customary U.S. Solution

(a) The two equations of motion are

$$\sum F_{y,\text{mass: tension}} + \frac{ma_y}{g_c} - \frac{mg}{g_c} = 0$$
$$\text{tension} - \frac{(10 \text{ lbm})\left(32.2 \ \frac{\text{ft}}{\text{sec}^2}\right)}{32.2 \ \frac{\text{lbm-ft}}{\text{lbf-sec}^2}} = -\frac{(10 \text{ lbm})a_y}{32.2 \ \frac{\text{lbm-ft}}{\text{lbf-sec}^2}}$$
$$\sum M_{O,\text{pulley: (tension)}}\, r = I\alpha$$
$$(\text{tension})(1.0 \text{ ft}) = (70 \text{ lbm-ft}^2)\alpha$$
$$a_y = r\alpha$$

(b) Substituting into the $\sum F$ equation and eliminating the tension,

$$\frac{\left(70 \text{ lbm-ft}^2\right)\alpha}{\left(32.2 \dfrac{\text{lbm-ft}}{\text{lbf-sec}^2}\right)(1.0 \text{ ft})} - 10 \text{ lbf} = -\frac{(10 \text{ lbm})(1.0 \text{ ft})\alpha}{32.2 \dfrac{\text{lbm-ft}}{\text{lbf-sec}^2}}$$

$$2.174\,\alpha - 10 = -0.311\,\alpha$$

$$\alpha = 4.02 \text{ rad/sec}^2$$

$$a_y = r\alpha = (1.0 \text{ ft})\left(4.02 \frac{\text{rad}}{\text{sec}^2}\right)$$

$$= 4.02 \text{ ft/sec}^2$$

Example 72.7

A 300 lbm cylinder ($I_O = 710$ lbm-ft^2) has a narrow groove cut in it as shown. One end of the cable is wrapped around the cylinder in the groove, while the other end supports a 200 lbm mass. The pulley is massless and frictionless, and there is no slipping. Starting from a standstill, what are the linear accelerations of the 200 lbm mass and the cylinder?

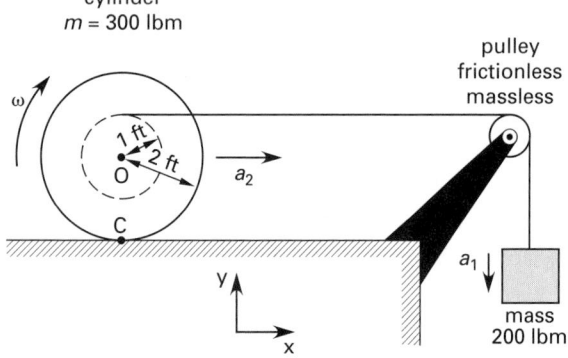

Solution

Since there is no slipping, there is friction between the cylinder and the plane. However, the coefficient of friction is not given. Therefore, moments must be taken about the contact point (the instantaneous center). The moment of inertia about the contact point is

$$I_C = I_O + mr^2$$

$$= 710 \text{ lbm-ft}^2 + (300 \text{ lbm})(2 \text{ ft})^2 = 1910 \text{ lbm-ft}^2$$

The first equation is a summation of forces on the mass.

$$\sum F_y: \text{tension} + \frac{ma_1}{g_c} - \frac{mg}{g_c} = 0$$

The second equation is a summation of moments about the instantaneous center. The frictional force passes through the instantaneous center and is disregarded.

$$\sum M_C: (\text{tension})(2.0 + 1.0 \text{ ft}) = \frac{I_C\alpha}{g_c}$$

Since there are two unknowns, a third equation is needed. This is the relationship between the linear and angular accelerations. a_2 is the acceleration of point O, located 2 ft from point C. a_1 is the acceleration of the cable, whose groove is located 3 ft from point C.

$$\alpha = \frac{a}{r} = \frac{a_2}{2.0 \text{ ft}} = \frac{a_1}{3.0 \text{ ft}}$$

$$a_1 = \tfrac{3}{2}a_2$$

Solving the three equations simultaneously yields

$$\text{cylinder: } a_2 = 10.4 \text{ ft/sec}^2$$

$$\text{mass: } a_1 = 15.6 \text{ ft/sec}^2$$

$$\text{tension} = 103 \text{ lbf}$$

$$\alpha = 5.2 \text{ rad/sec}^2$$

20. IMPULSE

Impulse, **Imp**, is a vector quantity equal to the change in momentum.[14] Units of linear impulse are the same as for linear momentum: lbf-sec and N·s. Units of lbf-ft-sec and N·m·s are used for angular impulse. Equations 72.44 and 72.45 define the scalar magnitudes of *linear impulse* and *angular impulse*. Figure 72.16 illustrates that impulse is represented by the area under the F-t (or T-t) curve.

$$\text{Imp} = \int_{t_1}^{t_2} F\,dt \quad [\text{linear}] \qquad 72.44$$

$$\text{Imp} = \int_{t_1}^{t_2} T\,dt \quad [\text{angular}] \qquad 72.45$$

Figure 72.16 Impulse

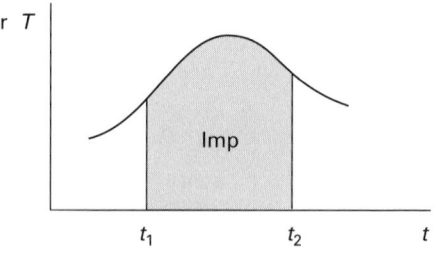

If the applied force or torque is constant, impulse is easily calculated. A large force acting for a very short period of time is known as an *impulsive force*.

$$\text{Imp} = F(t_2 - t_1) \quad [\text{linear}] \qquad 72.46$$

$$\text{Imp} = T(t_2 - t_1) \quad [\text{angular}] \qquad 72.47$$

---

[14] Although **Imp** is the most common notation, engineers have no universal symbol for impulse. Some authors use I, **I**, and i, but these symbols can be mistaken for moment of inertia. Other authors merely use the word *impulse* in their equations.

If the impulse is known, the average force acting over the duration of the impulse is

$$F_{\text{ave}} = \frac{\text{Imp}}{\Delta t} \qquad 72.48$$

21. IMPULSE-MOMENTUM PRINCIPLE

The change in momentum is equal to the applied *impulse*. This is known as the *impulse-momentum* principle. For a linear system with constant force and mass, the scalar magnitude form of this principle is

$$\text{Imp} = \Delta \text{p} \qquad 72.49$$

$$F(t_2 - t_1) = m(\text{v}_2 - \text{v}_1) \quad [\text{SI}] \qquad 72.50(a)$$

$$F(t_2 - t_1) = \frac{m(\text{v}_2 - \text{v}_1)}{g_c} \quad [\text{U.S.}] \qquad 72.50(b)$$

For an angular system with constant torque and moment of inertia, the analogous equations are

$$T(t_2 - t_1) = I(\omega_2 - \omega_1) \quad [\text{SI}] \qquad 72.51(a)$$

$$T(t_2 - t_1) = \frac{I(\omega_2 - \omega_1)}{g_c} \quad [\text{U.S.}] \qquad 72.51(b)$$

Example 72.8

A 1.62 oz (0.046 kg) marble attains a velocity of 170 mph (76 m/s) in a hunting slingshot. Contact with the sling is 1/25th of a second. What is the average force on the marble during contact?

SI Solution

From Eq. 72.50(a), the average force is

$$F = \frac{m\Delta\text{v}}{\Delta t} = \frac{(0.046 \text{ kg})\left(76 \dfrac{\text{m}}{\text{s}}\right)}{\dfrac{1}{25} \text{ s}}$$

$$= 87.4 \text{ N}$$

Customary U.S. Solution

The mass of the marble is

$$m = \frac{1.62 \text{ oz}}{16 \dfrac{\text{oz}}{\text{lbm}}} = 0.101 \text{ lbm}$$

The velocity of the marble is

$$\text{v} = \frac{\left(170 \dfrac{\text{mi}}{\text{hr}}\right)\left(5280 \dfrac{\text{ft}}{\text{mi}}\right)}{3600 \dfrac{\text{sec}}{\text{hr}}} = 249.3 \text{ ft/sec}$$

From Eq. 72.50(b), the average force is

$$F = \frac{m\Delta\text{v}}{g_c \Delta t} = \frac{(0.101 \text{ lbm})\left(249.3 \dfrac{\text{ft}}{\text{sec}}\right)}{\left(32.2 \dfrac{\text{lbm-ft}}{\text{lbf-sec}^2}\right)\left(\dfrac{1}{25} \text{ sec}\right)}$$

$$= 19.5 \text{ lbf}$$

Example 72.9

A 2000 kg cannon fires a 10 kg projectile horizontally at 600 m/s. It takes 0.007 s for the projectile to pass through the barrel and 0.01 s for the cannon to recoil. The cannon has a spring mechanism to absorb the recoil. (a) What is the cannon's initial recoil velocity? (b) What force is exerted on the recoil spring?

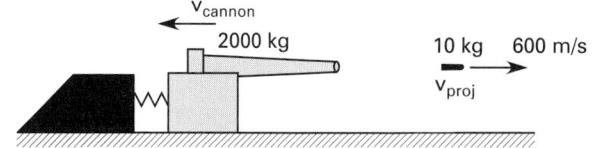

Solution

(a) The accelerating force is applied to the projectile quickly, and external forces such as gravity and friction are not a significant factor. Therefore, momentum is conserved.

$$\sum p: \quad m_{\text{proj}}\Delta\text{v}_{\text{proj}} = m_{\text{cannon}}\Delta\text{v}_{\text{cannon}}$$

$$(10 \text{ kg})\left(600 \dfrac{\text{m}}{\text{s}}\right) = (2000 \text{ kg})(\text{v}_{\text{cannon}})$$

$$\text{v}_{\text{cannon}} = 3 \text{ m/s}$$

(b) From Eq. 72.50(a), the recoil force is

$$F = \frac{m\Delta\text{v}}{\Delta t} = \frac{(2000 \text{ kg})\left(3 \dfrac{\text{m}}{\text{s}}\right)}{0.01 \text{ s}}$$

$$= 6 \times 10^5 \text{ N}$$

22. IMPULSE-MOMENTUM PRINCIPLE IN OPEN SYSTEMS

The impulse-momentum principle can be used to determine the forces acting on flowing fluids (i.e., in open systems). This is the method used to calculate forces in jet engines and on pipe bends, and forces due to other changes in flow geometry. Equation 72.52 is rearranged in terms of a mass flow rate.

$$F = \frac{m\Delta\text{v}}{\Delta t} = \dot{m}\Delta\text{v} \quad [\text{SI}] \qquad 72.52(a)$$

$$F = \frac{m\Delta\text{v}}{g_c\Delta t} = \frac{\dot{m}\Delta\text{v}}{g_c} \quad [\text{U.S.}] \qquad 72.52(b)$$

Example 72.10

Air enters a jet engine at 1500 ft/sec (450 m/s) and leaves at 3000 ft/sec (900 m/s). The thrust produced is 10,000 lbf (44 500 N). Disregarding the small amount of fuel added during combustion, what is the mass flow rate?

SI Solution

From Eq. 72.52(a),

$$\dot{m} = \frac{F}{\Delta v} = \frac{44\,500 \text{ N}}{900 \, \dfrac{\text{m}}{\text{s}} - 450 \, \dfrac{\text{m}}{\text{s}}}$$

$$= 98.9 \text{ kg/s}$$

Customary U.S. Solution

From Eq. 72.52(b),

$$\dot{m} = \frac{Fg_c}{\Delta v} = \frac{(10{,}000 \text{ lbf})\left(32.2 \, \dfrac{\text{lbm-ft}}{\text{lbf-sec}^2}\right)}{3000 \, \dfrac{\text{ft}}{\text{sec}} - 1500 \, \dfrac{\text{ft}}{\text{sec}}}$$

$$= 215 \text{ lbm/sec}$$

Example 72.11

20 kg of sand fall continuously each second on a conveyor belt moving horizontally at 0.6 m/s. What power is required to keep the belt moving?

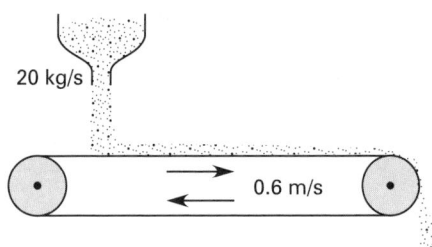

Solution

From Eq. 72.52(a), the force on the sand is

$$F = \dot{m}\Delta v = \left(20 \, \frac{\text{kg}}{\text{s}}\right)\left(0.6 \, \frac{\text{m}}{\text{s}}\right)$$

$$= 12 \text{ N}$$

The power required is

$$P = Fv = (12 \text{ N})\left(0.6 \, \frac{\text{m}}{\text{s}}\right)$$

$$= 7.2 \text{ W}$$

Example 72.12

A 6×9, $^5/_8$ in diameter hoisting cable (area of 0.158 in^2, modulus of elasticity of 12×10^6 lbf/in^2) carries a 1000 lbm load at its end. The load is being lowered vertically at the rate of 4 ft/sec. When 200 ft of cable have been reeled out, the take-up reel suddenly locks. Neglect the cable mass. What are the (a) cable stretch,

(b) maximum dynamic force in the cable, (c) maximum dynamic stress in the cable, and (d) approximate time for the load to come to a stop vertically?

Solution

(a) The stiffness of the cable is

$$k = \frac{F}{x} = \frac{AE}{L}$$

$$= \frac{(0.158 \text{ in}^2)\left(12 \times 10^6 \, \dfrac{\text{lbf}}{\text{in}^2}\right)}{(200 \text{ ft})\left(12 \, \dfrac{\text{in}}{\text{ft}}\right)}$$

$$= 790 \text{ lbf/in}$$

Neglecting the cable mass, the kinetic energy of the moving load is

$$\text{KE} = \frac{m v^2}{2g_c}$$

$$= \frac{(1000 \text{ lbm})\left(4 \, \dfrac{\text{ft}}{\text{sec}}\right)^2 \left(12 \, \dfrac{\text{in}}{\text{ft}}\right)}{(2)\left(32.2 \, \dfrac{\text{lbm-ft}}{\text{lbf-sec}^2}\right)}$$

$$= 2981 \text{ in-lbf}$$

By the work-energy principle, the decrease in kinetic energy is equal to the work of lengthening the cable (i.e., the energy stored in the spring).

$$\Delta\text{KE} = \tfrac{1}{2}k\delta^2$$

$$2981 \text{ in-lbf} = \left(\tfrac{1}{2}\right)\left(790 \, \frac{\text{lbf}}{\text{in}}\right)\delta^2$$

$$\delta = 2.75 \text{ in}$$

(b) The maximum dynamic force in the cable is

$$F = k\delta = \left(790 \, \frac{\text{lbf}}{\text{in}}\right)(2.75 \text{ in})$$

$$= 2173 \text{ lbf}$$

(c) The maximum dynamic tensile stress in the cable is

$$\sigma = \frac{F}{A} = \frac{2173 \text{ lbf}}{0.158 \text{ in}^2}$$

$$= 13{,}750 \text{ lbf/in}^2$$

(d) Since the tensile force in the cable increases from zero to the maximum while the load decelerates, the average decelerating force is half of the maximum force. From the impulse momentum principle, Eq. 70.50,

$$F\Delta t = \frac{m\Delta v}{g_c}$$

$$\left(\tfrac{1}{2}\right)(2173 \text{ lbf})\Delta t = \frac{(1000 \text{ lbm})\left(4 \, \dfrac{\text{ft}}{\text{sec}}\right)}{\left(32.2 \, \dfrac{\text{lbm-ft}}{\text{lbf-sec}^2}\right)}$$

$$\Delta t = 0.114 \text{ sec}$$

23. IMPACTS

According to Newton's second law, momentum is conserved unless a body is acted upon by an external force such as gravity or friction from another object. In an *impact* or *collision*, contact is very brief and the effect of external forces is insignificant. Therefore, momentum is conserved, even though energy may be lost through heat generation and deformation of the bodies.

Consider two particles, initially moving with velocities v_1 and v_2 on a collision path, as shown in Fig. 72.17. The conservation of momentum equation can be used to find the velocities after impact, v_1' and v_2'. (Observe algebraic signs with velocities.)

$$m_1 v_1 + m_2 v_2 = m_1 v_1' + m_2 v_2' \qquad 72.53$$

Figure 72.17 *Direct Central Impact*

The impact is said to be an *inelastic impact* if kinetic energy is lost. (Other names for an inelastic impact are *plastic impact* and *endoergic impact*.[15]) The impact is said to be *perfectly inelastic* or *perfectly plastic* if the two particles stick together and move on with the same final velocity.[16] The impact is said to be *elastic* only if kinetic energy is conserved.

$$m_1 v_1^2 + m_2 v_2^2 = m_1 v_1'^2 + m_2 v_2'^2 \Big|_{\text{elastic impact}} \qquad 72.54$$

24. COEFFICIENT OF RESTITUTION

A simple way to determine whether the impact is elastic or inelastic is by calculating the *coefficient of restitution*, e. The collision is inelastic if $e < 1.0$, perfectly inelastic if $e = 0$, and elastic if $e = 1.0$. The coefficient of restitution is the ratio of relative velocity differences along a mutual straight line. (When both impact velocities are not directed along the same straight line, the coefficient of restitution should be calculated separately for each velocity component.)

$$
\begin{aligned}
e &= \frac{\text{relative separation velocity}}{\text{relative approach velocity}} \\
&= \frac{v_1' - v_2'}{v_2 - v_1} \qquad 72.55
\end{aligned}
$$

---

[15]Theoretically, there is also an *exoergic impact* (i.e., one in which kinetic energy is gained during the impact). However, this can occur only in special cases, such as in nuclear reactions.
[16]In traditional textbook problems, clay balls should be considered perfectly inelastic.

25. REBOUND FROM STATIONARY PLANES

Figure 72.18 illustrates the case of an object rebounding from a massive, stationary plane.[17] This is an impact where $m_2 = \infty$ and $v_2 = 0$. The impact force acts perpendicular to the plane, regardless of whether the impact is elastic or inelastic. Therefore, the x-component of velocity is unchanged. Only the y-component of velocity is affected, and then only if the impact is inelastic.

$$v_x = v_x' \qquad 72.56$$

Figure 72.18 *Rebound from a Stationary Plane*

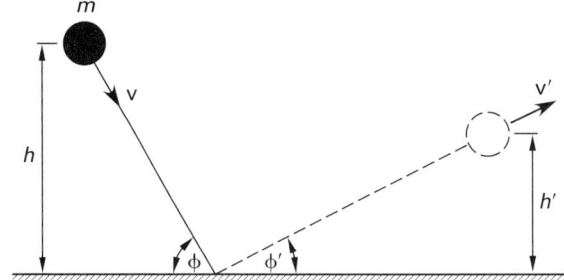

The coefficient of restitution can be used to calculate the *rebound angle*, *rebound height*, and *rebound velocity*.

$$e = \frac{\sin \phi'}{\sin \phi} = \sqrt{\frac{h'}{h}} = \frac{-v_y'}{v_y} \qquad 72.57$$

Example 72.13

A golf ball dropped vertically from a height of 8.0 ft (2.4 m) onto a hard surface rebounds to a height of 6.0 ft (1.8 m). What are the (a) impact velocity, (b) rebound velocity, and (c) coefficient of restitution?

SI Solution

The impact velocity of the golf ball is found by equating the initial potential energy to the incident kinetic energy.

$$\tfrac{1}{2} m v^2 = mgh$$

$$v = \sqrt{2gh} = \sqrt{(2)\left(9.81 \ \frac{m}{s^2}\right)(2.4 \ m)}$$

$$= -6.86 \ m/s \quad [\text{negative because down}]$$

The rebound velocity can be found from the rebound height.

$$v' = \sqrt{2gh'} = \sqrt{(2)\left(9.81 \ \frac{m}{s^2}\right)(1.8 \ m)}$$

$$= 5.94 \ m/s$$

---

[17]The particle path is shown as a straight line in Fig. 72.18 for convenience. The path will be a straight line only when the particle is dropped straight down. Otherwise, the path will be parabolic.

From Eq. 72.55 with $v_2 = v_2' = 0$ (or Eq. 72.57),

$$e = \frac{5.94 \, \frac{m}{s} - 0}{0 - \left(-6.86 \, \frac{m}{s}\right)} = 0.87$$

Customary U.S. Solution

The impact velocity is

$$v = \sqrt{2gh} = \sqrt{(2) \left(32.2 \, \frac{ft}{sec^2}\right) (8.0 \, ft)}$$

$$= -22.7 \, ft/sec \quad \text{[negative because down]}$$

Similarly, the rebound velocity is

$$v' = \sqrt{(2) \left(32.2 \, \frac{ft}{sec^2}\right) (6.0 \, ft)} = 19.7 \, ft/sec$$

From Eq. 72.55 with $v_2 = v_2' = 0$ (or from Eq. 72.57),

$$e = \frac{19.7 \, \frac{ft}{sec} - 0}{0 - \left(-22.7 \, \frac{ft}{sec}\right)} = 0.87$$

26. COMPLEX IMPACTS

The simplest type of impact problem is the direct central impact, shown in Fig. 72.17. An impact is said to be a *direct impact* when the velocities of the two bodies are perpendicular to the contacting surfaces. *Central impact* occurs when the force of the impact is along the line of connecting centers of gravity. Round bodies (i.e., spheres) always experience central impact, whether or not the impact is direct.

When the velocities of the bodies are not along the same line, the impact is said to be an *oblique impact*, as illustrated in Fig. 72.19. The coefficient of restitution can be used to find the x-components of the resultant velocities. Since impact is central, the y-components of velocities will be unaffected by the collision.

$$v_{1y} = v'_{1y} \qquad \textit{72.58}$$

$$v_{2y} = v'_{2y} \qquad \textit{72.59}$$

$$e = \frac{v'_{1x} - v'_{2x}}{v_{2x} - v_{1x}} \qquad \textit{72.60}$$

$$m_1 v_{1x} + m_2 v_{2x} = m_1 v'_{1x} + m_2 v'_{2x} \qquad \textit{72.61}$$

Figure 72.19 *Central Oblique Impact*

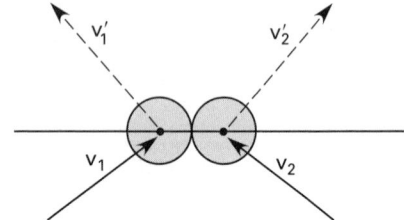

Eccentric impacts are neither direct nor central. The coefficient of restitution can be used to calculate the linear velocities immediately after impact along a line normal to the contact surfaces. Since the impact is not central, the bodies will rotate. Other methods must be used to calculate the rate of rotation.

$$e = \frac{v'_{1n} - v'_{2n}}{v_{2n} - v_{1n}} \qquad \textit{72.62}$$

Figure 72.20 *Eccentric Impact*

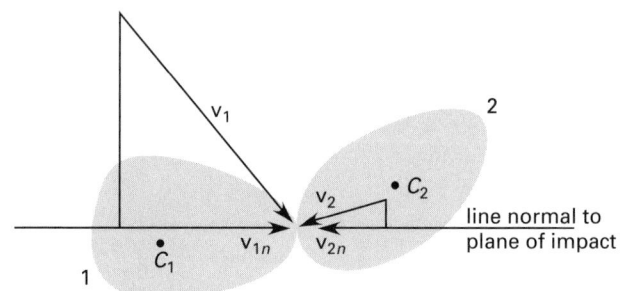

27. VELOCITY-DEPENDENT FORCE

A force that is a function of velocity is known as a *velocity-dependent force*. A common example of a velocity-dependent force is the *viscous drag* a particle experiences when falling through a fluid. There are two main cases of viscous drag: linear and quadratic.

A *linear velocity-dependent force* is proportional to the first power of the velocity. A linear relationship is typical of a particle falling slowly through a fluid (i.e., viscous drag in laminar flow). In Eq. 72.63, C is a constant of proportionality known as the *viscous coefficient* or *coefficient of viscous damping*.

$$F_b = Cv \qquad \textit{72.63}$$

In the case of a particle falling slowly through a viscous liquid, the differential equation of motion and its solution are derived from Newton's second law.

$$mg - Cv = ma \qquad \text{[SI]} \qquad \textit{72.64(a)}$$

$$\frac{mg}{g_c} - Cv = \frac{ma}{g_c} \qquad \text{[U.S.]} \qquad \textit{72.64(b)}$$

$$v(t) = v_t \left(1 - e^{-(Ct/m)}\right) \qquad \text{[SI]} \qquad \textit{72.65(a)}$$

$$v(t) = v_t \left(1 - e^{-(Cg_c t/m)}\right) \quad \text{[U.S.]} \qquad \textit{72.65(b)}$$

Equation 72.65 shows that the velocity asymptotically approaches a final value known as the *terminal velocity*, v_t. For laminar flow, the terminal velocity is

$$v_t = \frac{mg}{C} \qquad \text{[SI]} \qquad \textit{72.66(a)}$$

$$v_t = \frac{mg}{Cg_c} \qquad \text{[U.S.]} \qquad \textit{72.66(b)}$$

A *quadratic velocity-dependent force* is proportional to the second power of the velocity. A quadratic relationship is typical of a particle falling quickly through a fluid (i.e., turbulent flow).

$$F_b = Cv^2 \qquad \textbf{72.67}$$

In the case of a particle falling quickly through a liquid under the influence of gravity, the differential equation of motion is

$$mg - Cv^2 = ma \qquad \text{[SI]} \qquad \textbf{72.68(a)}$$

$$\frac{mg}{g_c} - Cv^2 = \frac{ma}{g_c} \qquad \text{[U.S.]} \qquad \textbf{72.68(b)}$$

For turbulent flow, the terminal velocity is

$$v_t = \sqrt{\frac{mg}{C}} \qquad \text{[SI]} \qquad \textbf{72.69(a)}$$

$$v_t = \sqrt{\frac{mg}{Cg_c}} \qquad \text{[U.S.]} \qquad \textbf{72.69(b)}$$

If a skydiver falls far enough, a turbulent terminal velocity of approximately 125 mph (200 kph) will be achieved. With a parachute (but still in turbulent flow), the terminal velocity is reduced to approximately 25 mph (40 kph).

28. VARYING MASS

Integral momentum equations must be used when the mass of an object varies with time. Most varying mass problems are complex, but the simplified case of an ideal rocket can be evaluated. This discussion assumes constant gravitational force, constant fuel usage, and constant exhaust velocity. (For brevity of presentation, all of the following equations are presented in consistent form.)

The forces acting on the rocket are its thrust, F, and gravity. Newton's second law is

$$F(t) - F_{\text{gravity}} = \frac{d}{dt}(m(t)v(t)) \qquad \textbf{72.70}$$

If $\dot{m}$ is the constant fuel usage, the thrust and gravitational forces are

$$F(t) = \dot{m}v_{\text{exhaust,absolute}} = \dot{m}(v_{\text{exhaust}} - v(t)) \qquad \textbf{72.71}$$

$$F_{\text{gravity}} = m(t)g \qquad \textbf{72.72}$$

The velocity as a function of time is found by solving the following differential equation.

$$\dot{m}v_{\text{exhaust,absolute}} - m(t)g = \dot{m}v(t) + m(t)\left(\frac{dv(t)}{dt}\right) \qquad \textbf{72.73}$$

$$v(t) = v_0 - gt + v_{\text{exhaust}} \ln\left(\frac{m_0}{m_0 - \dot{m}t}\right) \qquad \textbf{72.74}$$

The final *burnout velocity*, v_f, depends on the initial mass, m_0, and the final mass, m_f.

$$v_f = v_0 - g\left(\frac{m_0}{\dot{m}}\right)\left(1 - \frac{m_f}{m_0}\right) + v_{\text{exhaust}} \ln\left(\frac{m_0}{m_f}\right) \qquad \textbf{72.75}$$

A simple relationship exists for a rocket starting from standstill in a gravity-free environment (i.e., $v_0 = 0$ and $g = 0$).

$$\frac{m_f}{m_0} = e^{-v_f/v_{\text{exhaust}}} \qquad \textbf{72.76}$$

29. CENTRAL FORCE FIELDS

When the force on a particle is always directed toward or away from a fixed point, the particle is moving in a *central force field*. Examples of central force fields are gravitational fields (*inverse-square attractive fields*) and electrostatic fields (*inverse-square repulsive fields*). Particles traveling in inverse-square attractive fields can have circular, elliptical, parabolic, or hyperbolic paths. Particles traveling in inverse-square repulsive fields always travel in hyperbolic paths.

The fixed point, O in Fig. 72.21, is known as the *center of force*. The magnitude of the force depends on the distance between the particle and the center of force.

Figure 72.21 Motion in a Central Force Field

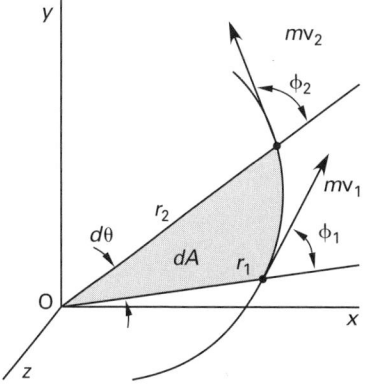

The angular momentum of a particle moving in a central force field is constant.

$$\mathbf{h}_O = \mathbf{r} \times m\mathbf{v} = \text{constant} \qquad \textbf{72.77}$$

Equation 72.77 can be written in scalar form.

$$r_1 m v_1 \sin \phi_1 = r_2 m v_2 \sin \phi_2 \qquad \textbf{72.78}$$

For a particle moving in a central force field, the *areal velocity* is constant.

$$\text{areal velocity} = \frac{dA}{dt} = \tfrac{1}{2}r^2 \frac{d\theta}{dt} = \frac{h_O}{2m} \qquad \textbf{72.79}$$

30. NEWTON'S LAW OF GRAVITATION

For a particle far enough away from a large body, gravity can be considered to be a central force field. *Newton's law of gravitation*, also known as *Newton's law of universal gravitation*, describes the force of attraction between the two masses. The law states that the attractive gravitational force between the two masses is directly proportional to the product of masses, is inversely proportional to the square of the distance between their centers of mass, and is directed along a line passing through the centers of gravity of both masses.

$$F = \frac{Gm_1m_2}{r^2} \qquad \text{72.80}$$

G is *Newton's gravitational constant* (*Newton's universal constant*). Approximate values of G for the earth are given in Table 72.2 for different sets of units. For an earth-particle combination, the product Gm_{earth} has the value of 4.39×10^{14} lbf-ft^2/lbm (4.00×10^{14} N·m^2/kg).

Table 72.2 *Approximate Values of the Universal Constant, G*

6.673×10^{-11}	N·m^2/kg^2
6.673×10^{-8}	cm^3/g·s^2
3.436×30^{-8}	lbf-ft^2/slug^2
3.320×10^{-11}	lbf-ft^2/lbm^2
3.436×10^{-8}	ft^4/lbf-sec^4

31. KEPLER'S LAWS OF PLANETARY MOTION

Kepler's three *laws of planetary motion* are as follows.

- *the law of orbits*: The path of each planet is an ellipse with the sun at one focus.

- *the law of areas*: The radius vector drawn from the sun to a planet sweeps equal areas in equal times. (The areal velocity is constant. This is equivalent to the statement, "The angular velocity is constant.")

$$\frac{dA}{dt} = \text{constant} \qquad \text{72.81}$$

- *the law of periods*: The square of a planet's periodic time is proportional to the cube of the semi-major axis of its orbit.

$$t_p^2 \propto a^3 \qquad \text{72.82}$$

The *periodic time* referenced in Kepler's third law is the time required for a satellite to travel once around the parent body. If the parent body rotates in the same plane and with the same periodic time as the satellite, the satellite will always be above the same point on the parent body. This condition defines a *geostationary orbit*. Referring to Fig. 72.22, the periodic time is

$$t_p = \frac{2\pi ab}{h_0} = \frac{\pi ab}{\dfrac{dA}{dt}} = \frac{\pi ab}{\text{areal velocity}} \qquad \text{72.83}$$

Figure 72.22 *Planetary Motion*

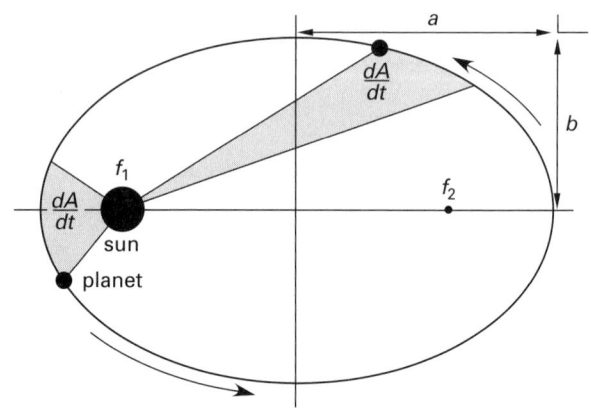

32. SPACE MECHANICS

Figure 72.23 illustrates the motion of a satellite that is released in a path parallel to the earth's surface. (The dotted line represents the launch phase and is not relevant to the analysis.)

Figure 72.23 *Space Mechanics Nomenclature*

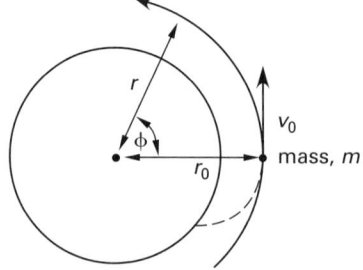

At the instant of release, the magnitude of the angular momentum, **h**, is

$$\mathbf{h} = h_0 = r_0 m v_0 \qquad \text{72.84}$$

The force exerted on the satellite by the earth is given by Newton's law of gravitation. For any angle, ϕ, swept out by the satellite, the separation distance can be found from Eq. 72.85. C is a constant.

$$\frac{1}{r} = \frac{GM}{h^2} + C\cos\phi \qquad \text{72.85}$$

$$C = \frac{1}{r_0} - \frac{GM}{h^2} \qquad \text{72.86}$$

The *orbit eccentricity*, ϵ, can be calculated and used to determine the type of orbit, as listed in Table 72.3.

$$\epsilon = \frac{Ch^2}{GM} \qquad 72.87$$

Table 72.3 Orbit Eccentricities

value of ϵ	type of orbit
> 1.0	nonreturning hyperbola
$= 1.0$	nonreturning parabola
< 1.0	ellipse
$= 0.0$	circle

The limiting value of the initial release velocity, $v_{0,max}$, to prevent a nonreturning orbit ($\epsilon = 1.0$) is known as the *escape velocity*. For the earth, the escape velocity is approximately 7.0 mi/sec (25,000 mph, or 11.2 km/s).

$$v_{escape} = v_{0,max} = \sqrt{\frac{2GM}{r_0}} \qquad 72.88$$

The release velocity, $v_{0,circular}$, that results in a circular orbit is

$$v_{0,circular} = \sqrt{\frac{GM}{r_0}} \qquad 72.89$$

For an elliptical orbit, the minimum separation is known as the *perigee distance*. The maximum separation distance is known as the *apogee*. The terms perigee and apogee are traditionally used for earth satellites. The terms *perihelion* (closest) and *aphelion* (farthest) are used to describe distances between the sun and earth.

33. ROADWAY BANKING

If a vehicle travels in a circular path with instantaneous radius r and tangential velocity v_t, it will experience an apparent centrifugal force. The centrifugal force is resisted by a combination of roadway banking (*superelevation*) and *sideways friction*.[18] If the roadway is banked so that friction is not required to resist the centrifugal force, the superelevation angle, ϕ, can be calculated from Eq. 72.90.[19]

$$\tan\phi = \frac{v_t^2}{gr} \qquad 72.90$$

Equation 72.90 can be solved for the *normal speed* corresponding to the geometry of a curve.

$$v_t = \sqrt{gr\tan\phi} \qquad 72.91$$

---

[18]The *superelevation* is the slope (in ft/ft or m/m) in the transverse direction (i.e., across the roadway).
[19]Generally it is not desirable to rely on roadway banking alone, since a particular superelevation angle would correspond to only a single speed.

When friction is used to counteract some of the centrifugal force, the *side friction factor*, f, between the tires and roadway is incorporated into the calculation of the superelevation angle.

$$e = \tan\phi = \frac{v_t^2 - fgr}{gr + fv_t^2} \qquad 72.92$$

If the banking angle, ϕ, is set to zero, Eq. 72.92 can be used to calculate the maximum velocity of a vehicle making a turn when there is no banking.

For highway design, *superelevation rate*, e, is the amount of rise or fall of the cross slope per unit amount of horizontal width (i.e., the tangent of the slope angle above or below horizontal). Customary U.S. units are expressed in feet per foot, such as 0.06 ft/ft, or inch fractions per foot, such as $3/4$ in/ft. SI units are millimeters per meter, such as 60 mm/m. The slope can also be expressed as a percent cross slope, such as 6% cross slope; or as a ratio, 1:17. The *total superelevation*, y, is the difference in heights of the inside and outside edges of the curve.

When the speed, superelevation, and radius are such that no friction is required to resist sliding, the curve is said to be "balanced." There is no tendency for a vehicle to slide up or down the slope at the *balanced speed*. At any speed other than the balanced speed, some friction is needed to hold the vehicle on the slope. Given a friction factor, f, the required value of the superelevation rate, e, can be calculated.

Figure 72.24 Roadway Banking

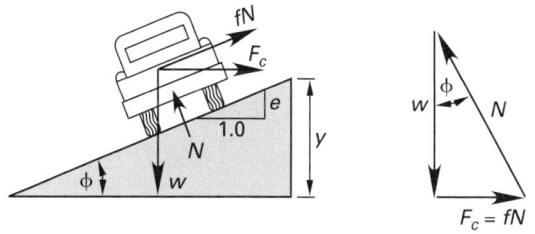

Equation 72.92 can be solved for the curve radius. For small banking angles (i.e., $\phi \le 8°$), this simplifies to

$$r \approx \frac{v_t^2}{g(e+f)} \qquad 72.93$$

$$r_m = \frac{v_{km/h}^2}{127(e_{m/m} + f)} \qquad [SI] \quad 72.93(a)$$

$$r_{ft} = \frac{v_{mph}^2}{15(e_{ft/ft} + f)} \qquad [U.S.] \quad 72.93(b)$$

Example 72.14

A 4000 lbm (1800 kg) car travels at 40 mph (65 km/h) around a banked curve with a radius of 500 ft (150 m).

What should the superelevation be so that the tire friction is not needed to prevent the car from sliding?

SI Solution

From Eq. 72.93,

$$e + f = \frac{v^2}{127r}$$

$$e + 0 = \frac{\left(65 \; \dfrac{km}{hr}\right)^2}{(127)(150 \text{ m})} = 0.222 \text{ m/m}$$

Customary U.S. Solution

From Eq. 72.93,

$$e + f = \frac{v^2}{15r}$$

$$e + 0 = \frac{\left(40 \; \dfrac{mi}{hr}\right)^2}{(15)(500 \text{ ft})} = 0.213 \text{ ft/ft}$$

Example 72.15

A 4000 lbm car travels at 40 mph around a banked curve with a radius of 500 ft. What should be the superelevation angle so that tire friction is not needed to prevent the car from sliding?

Solution

The tangential velocity of the car is

$$v_t = \frac{\left(40 \; \dfrac{mi}{hr}\right)\left(5280 \; \dfrac{ft}{mi}\right)}{3600 \; \dfrac{sec}{hr}} = 58.7 \text{ ft/sec}$$

From Eq. 72.90,

$$\phi = \arctan\left(\frac{v_t^2}{gr}\right) = \arctan\left(\frac{\left(58.7 \; \dfrac{ft}{sec}\right)^2}{\left(32.2 \; \dfrac{ft}{sec^2}\right)(500 \text{ ft})}\right)$$

$$= 12.1°$$

Example 72.16

A vehicle is traveling at 70 mph when it enters a circular curve of a test track. The curve radius is 240 ft. The sideways sliding coefficient of friction between the tires and the roadway is 0.57. (a) At what minimum angle from the horizontal must the curve be banked in order to prevent the vehicle from sliding off the top of the curve? (b) If the roadway is banked at 20° from the horizontal, what is the maximum vehicle speed such that no sliding occurs?

Solution

(a) The speed of the vehicle is

$$v_t = \frac{\left(70 \; \dfrac{mi}{hr}\right)\left(5280 \; \dfrac{ft}{mi}\right)}{3600 \; \dfrac{sec}{hr}}$$

$$= 102.7 \text{ ft/sec}$$

From Eq. 72.92, the required banking angle is

$$\phi = \arctan\left(\frac{v_t^2 - fgr}{gr + fv_t^2}\right)$$

$$= \frac{\left(102.7 \; \dfrac{ft}{sec}\right)^2 - (0.57)\left(32.2 \; \dfrac{ft}{sec^2}\right)(240 \text{ ft})}{\left(32.2 \; \dfrac{ft}{sec^2}\right)(240 \text{ ft}) + (0.57)\left(102.7 \; \dfrac{ft}{sec}\right)^2}$$

$$= 24.1°$$

(b) Solve Eq. 72.92 for the velocity.

$$v_t = \sqrt{\frac{rg(\tan\phi + f)}{1 - f\tan\phi}}$$

$$= \sqrt{\frac{(240 \text{ ft})\left(32.2 \; \dfrac{ft}{sec^2}\right)(\tan 20° + 0.57)}{1 - (0.57)\tan 20°}}$$

$$= 95.4 \text{ ft/sec} \quad (65.1 \text{ mi/hr})$$

73 Roads and Highways: Capacity Analysis

Nomenclature

ADT	average daily traffic	vpd	vpd
c	capacity	vph/lane	vph/lane
C	total cycle length	sec	s
d	separation distance	ft	m
D	density	vpm/lane	vpk/lane
D	directional factor	–	–
DDHV	directional design hourly volume	vph	vph
DHV	design hourly volume	vph	vph
E	passenger car equivalent	–	–
f	adjustment factor	–	–
F	adjustment to free-flow speed	mi/hr	km/h
FFS	free-flow speed	mi/hr	km/h
g	effective green time	sec	s
h	headway	sec	s
K	ratio of DHV to AADT	–	–
L	expected system length (includes service)	–	–
L	lost time	sec	s
L_q	expected queue length	–	–
M	pedestrian space	ft^2/ped	m^2/ped
N	number of lanes	–	–
$p\{n\}$	probability of n customers in the system	–	–
P	fraction	–	–
PHF	peak hour factor	–	–
R	ratio	–	–
s	number of parallel servers	–	–
s	saturation flow rate	vph/green	vph/green
S	speed	mi/hr	km/h
t	time	sec	s
v	rate of flow during peak 15 min period	vph/lane	vph/lane
V	volume	vph	vph
W	expected time in the system (includes service)	sec	s
W	width	ft	m
W_q	expected time in the queue	sec	s
X	v/c ratio of lane group	–	–

Symbols

λ	mean arrival rate	1/sec	1/s
μ	mean service rate per server	1/sec	1/s
ρ	utilization factor (λ/μ)	–	–

Subscripts

a	area
A	access
bb	bus blocking
B	bus
c	critical
E	effective
f	free-flow
g	grade
HV	heavy vehicle
I	ideal
j	jam
LC	lateral clearance
LT	left turn

Transportation

LW   lane width
m maximum
M median
np   no passing
o ideal or critical
p population, peak, or platoon
R recreational vehicles
RT   right turn
T trucks and buses
w width

1. STANDARD TRAFFIC REFERENCES

The standard references for the subject matter in this chapter include the *Highway Capacity Manual*, referred to as the "HCM" (Transportation Research Board, National Research Council); *A Policy on Geometric Design of Highways and Streets*, known as the "Green Book," "PGDHS," or just "GDHS" (AASHTO); and the *Manual on Uniform Traffic Control Devices*, referred to as the "MUTCD" (Federal Highway Administration). Except for trivial exercises, traffic/transportation designs and analyses cannot be performed without these references. Calculations using the HCM and Green Book may be performed in either customary U.S. or SI units.

Many of the methods used to design and analyze specific facilities are highly proceduralized. These methods often rely on empirical correlations and graphical/nomographical solutions rather than on pure engineering fundamentals. In that sense, they are not difficult to implement. However, each feature uses a unique procedure, different nomenclature, and specific criteria. Due to the variety and complexity of these procedures, it is not possible to present them without making significant reference to the major supporting documents, particularly the HCM. Even with the supporting documents, it soon becomes apparent why the majority of traffic design is computer assisted. Analysis of existing facilities is somewhat more tractable.

In recent years and in response to mounting liability issues, these references (the Green Book, in particular) have backed away from appearing to be mandatory standards by softening their unequivocal language. These documents are currently considered to be "suggested policies" by their originating organizations. State departments of transportation are expected to accept responsibility for proper evaluation and implementation of these "nonstandards."

2. ABBREVIATIONS AND UNITS

Following standard practice, the units and abbreviations shown in Table 73.1 are used in this chapter. (Explicit units are shown in numerical calculations to facilitate cancellation.) Some HCM abbreviations differ from this book and from commonly used SI abbreviations, so some inconsistencies may be noted.

- The letter "p" signifying "per" is not used in SI units. While the HCM still uses "pcph" for calculations in English units, it has adopted "pc/h" for calculations in SI units.

- "ln" is used for "lane" in SI calculations.

- The word "seat" is spelled out to avoid confusion with "s" for "seconds."

- "Minute" is abbreviated "min"; "minimum" is abbreviated "Min."

Table 73.1 Abbreviations

abbreviation	meaning
ln	lane
km/h	kilometers per hour
mph	miles per hour
p	people, person, or pedestrian
pc	passenger car
pcph	passenger cars per hour
pcphg	passenger cars per hour of green signal
pcphpl	passenger cars per hour per lane
pcphgpl	passenger cars per hour of green signal per lane
pc/km/ln	passenger cars per kilometer per lane
pcpmpl	passenger cars per mile per lane
ped	pedestrian
pers	person
veh	vehicle
vph	vehicles per hour
veh/km	vehicles per kilometer
vph/lane	vehicles per hour per lane
vpm	vehicles per mile

3. FACILITIES TERMINOLOGY

Although common usage does not always distinguish between highways, freeways, and other types of roadways, the major traffic/transportation references are more specific. For example, "vehicle" and "car" have different meanings. "Vehicle" encompasses trucks, buses, and recreational vehicles as well as passenger cars.

A *freeway* is a divided corridor with at least two lanes in each direction that operates in an *uninterrupted flow* mode (i.e., without *fixed elements* such as signals, stop signs, and at-grade intersections). *Access points* are limited to ramp locations. Since grades, curves, and other features can change along a freeway, performance measures (e.g., capacity) are evaluated along shorter *freeway segments*.

Multilane and two-lane *highways*, on the other hand, contain some fixed elements and access points from at-grade intersections, though relatively uninterrupted flow can occur if signal spacing is greater than 2 mi (3.2 km). Where signal spacing is less than 2 mi (3.2 km), the roadway is classified as an *urban* street (or *arterial*), and flow is considered to be interrupted. An urban street has a significant amount of driveway access, while an arterial does not. *Divided highways* have separate roadbeds for the opposing directions, whereas *undivided highways* do not.

Smaller roadways are classified as *local roads* and *streets*. All roadways can be classified as *urban*, *suburban*, or *rural*, depending on the surrounding population density. *Urban areas* have populations greater than 5000. *Rural areas* are outside the boundaries of urban areas.

Table 73.2 *General Functional Classifications of Roadways*[a]

road designation	ADT (vpd)
local road	2000 or less
collector road	2000–12,000
arterial/urban road	12,000–40,000
freeway	30,000 and above

[a]Classifications can also be established based on percentages of total length and travel volume.

4. DESIGN VEHICLES

Standardized *design vehicles* have been established to ensure that geometric features will accommodate all commonly sized vehicles. Standardized design vehicles are given specific designations in the Green Book, as listed in Table 73.3. The Green Book also contains information on *wheelbase* and *turning radius*. (See Green Book Exhibits 2-1 through 2-23.)

5. LEVELS OF SERVICE

A user's quality of service through or over a specific facility (e.g., over a highway, through an intersection, across a crosswalk) is classified by a *level of service* (LOS). Levels of service are designated A through F. Level A represents unimpeded flow, which is ideal but only possible when the volume of traffic is small. Level F represents a highly impeded, packed condition. Generally, level E will have the maximum flow rate (i.e., capacity).

The desired design condition is generally between levels A and F. Economic considerations favor higher volumes and more obstructed levels of service. However, political considerations favor less obstructed levels of service. The parameter used to define the level of service varies with the type of facility, as listed in Table 73.4.

Since levels of service can vary considerably during an hour, capacity and LOS evaluations focus on the peak 15 min of flow.

6. SPEED PARAMETERS

Several measures of vehicle speed are used in highway design and capacity calculations. Most measures will not be needed in every capacity calculation. The *design speed* is the maximum safe speed that can be maintained over a specified section of roadway when conditions are so favorable that the design features of the roadway govern. Most elements of roadway design depend on the design speed. The *legal speed* on a roadway section

Table 73.3 *Standard Design Vehicles*

designation	symbol[a]	dimensions, ft (m) height		width		length	
passenger car	P	4.25	(1.3)	7	(2.1)	19	(5.8)
single-unit truck	SU	11–13.5	(3.4–4.1)	8.0	(2.4)	30	(9.2)
single-unit bus	BUS-40						
	(BUS-12)	12.0	(3.7)	8.5	(2.6)	40	(12.2)
articulated bus	A-BUS	11.0	(3.4)	8.5	(2.6)	60	(18.3)
combination trucks							
intermediate							
semi-trailer	WB-40						
	(WB-12)	13.5	(4.1)	8.0	(2.4)	45.5	(13.9)
large semi-trailer	WB-50						
	(WB-15)	13.5	(4.1)	8.5	(2.6)	55	(16.8)
double-bottom							
semi-trailer							
full-trailer	WB-67D						
	(WB-20D)	13.5	(4.1)	8.5	(2.6)	73.3	(22.4)
recreational vehicles							
motor home	MH	12.0	(3.7)	8	(2.4)	30	(9.2)
car and camper trailer	P/T	10.0	(3.1)	8	(2.4)	48.7	(14.8)
car and boat trailer	P/B	–		8	(2.4)	42	(12.8)
motor home and boat trailer	MH/B	12.0	(3.7)	8	(2.4)	53	(16.2)

(Multiply ft by 0.3048 to obtain m.)
[a]Symbols in parentheses represent the SI designations.

Adapted from Green Book (2004) Exhibit 2-1, which has additional categories.

Table 73.4 *Primary Measures of Level of Service*

type of facility	level of service parameter
freeways	
basic freeway segment	density (pc/mi/ln or pc/km/ln)
weaving areas	speed (mph or km/h)
ramp junctions	density (pc/mi/ln or pc/km/ln)
multilane highways	density (pc/mi/ln or pc/km/ln)
two-lane rural highways	speed or percent time spent following
signalized intersections	average control delay (sec/veh or s/veh)
unsignalized intersections	average control delay (sec/veh or s/veh)
urban streets	average travel speed (mph or km/h)
mass transit	various (pers/seat, veh/hr, people/hr, p/seat, veh/h, p/h)
pedestrians	space per pedestrian or delay (ft^2/ped or m^2/p)
bicycles	event, delay (sec/veh or s/veh)

Adapted from *Highway Capacity Manual* Exh. 3-1.

is often set at approximately the *85th percentile speed*, determined by observation of a sizable sample of vehicles. In suburban and urban areas, the legal speed limit is often influenced by additional considerations such as visibility at intersections, the presence of driveways, parking and pedestrian activity, population density, and other local factors. Typical design speeds are given in Table 73.5.

The *average highway speed*, AHS, is the weighted average of the observed speeds within a highway section based on each subsection's proportional contribution to total mileage. The *operating speed* is the highest overall speed at which a driver can safely travel on a given highway under favorable weather conditions and prevailing traffic conditions. The *running speed* is the speed over a specified section of roadway equal to the distance divided by the running time, or the total time required to travel over the roadway section, disregarding any stationary time. The *average running speed* is the running speed for all traffic equal to the distance summation for all cars divided by the time summation for all cars.

The *average spot speed*, also known as the *time mean speed*, is the arithmetic mean of the instantaneous speeds of all cars at a particular point. The *average travel speed* is the speed over a specified section of highway, including operational delays such as stops for traffic signals. The *free-flow speed* is measured using the mean speed of passenger cars under low to moderate flow conditions (up to 1300 pcphpl). (The term

operating speed, as used by AASHTO and in previous HCMs, is similar to free-flow speed when evaluated at low-volume conditions.) *Space mean speed* in a specific time period is calculated by taking the total distance traveled by all vehicles and dividing by the total of the travel times of all vehicles. *Crawl speed* is the maximum sustained speed that heavy vehicles can maintain on a given extended upgrade.

The HCM uses average travel speed as the primary defining parameter. Except for LOS F, the average travel and average running speeds are identical.

Table 73.5 *Minimum Design Speeds (mph (km/h))* *(local and rural roads)*

design volumes, ADT	terrain		
	level	rolling	mountainous
≥ 2000	50 (80)	40 (60)	30 (50)
1500–2000	50 (80)	40 (60)	30 (50)
400–1500	50 (80)	40 (60)	30 (50)
250–400	40 (60)	30 (50)	20 (30)
50–250	30 (50)	30 (50)	20 (30)
< 50	30 (50)	20 (30)	20 (30)

(Multiply mph by 1.609 to obtain km/h.)

From *A Policy on Geometric Design of Highways and Streets*, Exhibit 5-1, copyright © 2004 by the American Association of State Highway and Transportation Officials, Washington, D.C. Used by permission.

7. SPOT SPEED STUDIES

The analysis of *spot speed data* draws on traditional statistical methods. The *mean*, *mode*, and *median speeds* are commonly determined. Certain percentile rank speeds may also be needed. For example, the *50th percentile speed* (i.e., the median speed) and the 85th percentile speed (i.e., the *design speed*) can be found from the frequency distribution. The *pace interval* (also known as the *pace range*, *group pace*, or just *pace*) is the 10 mph speed range containing the most observations.

8. VOLUME PARAMETERS

There are several volume parameters, and not all parameters will be needed in every capacity investigation. It is important to note if volumes are for both directions combined, for all lanes of one direction, or just for one lane.

The *average daily traffic*, ADT, and *average annual daily traffic*, AADT, may be one- or two-way. The *design hour volume*, DHV, is evaluated for the design year. DHV is usually the 30th highest hourly expected volume in the design year, hence its nickname "30th hour volume." It is not an average or a maximum. DHV is two-way unless noted otherwise. The ratio of DHV to AADT is designated as the "*K-factor*." Default values of 0.09 (urban) and 0.10 (rural) are used in the HCM.

$$K = \frac{\text{DHV}}{\text{AADT}} \qquad \textit{73.1}$$

The *directional factor*, D, is the percentage of the dominant *peak* flow direction. It can range from up to 80% for rural roadways at peak hours to 50% for central business district traffic, though values between 55 and 65% are more common. A default value of 0.60 is used in the HCM. The *directional design hour volume*, DDHV, is calculated as the product of the directional factor, D (also known as the "*D-factor*"), and DHV.

$$DDHV = D(DHV) = DK(AADT) \qquad 73.2$$

The *rate of flow*, v, is the equivalent hourly rate at which vehicles pass a given point during a given time interval of less than 1 hr, usually 15 min. Thus, the rate of flow changes every 15 min.

The *design capacity* is the maximum volume of traffic that the roadway can handle. The *ideal capacity*, c, for freeways is considered to be 2400 passenger cars per hour per lane (pcphpl). Different values apply for other levels of service, speeds, and types of facilities. The actual flow rate, v, can be used to calculate the *volume-capacity ratio*, v/c, for a particular level of service, i.

$$\text{volume-capacity ratio}_i = (v/c)_i \qquad 73.3$$

The *maximum service flow rate* is the capacity in passenger cars per hour per lane under ideal conditions for a particular level of service, i.

$$v_{m,i} = c\,(v/c)_{m,i} \qquad 73.4$$

The *peak hour factor*, PHF, is the ratio of the total actual hourly volume to the peak rate of flow within the hour.

$$PHF = \frac{\text{actual hourly volume}_{\text{vph}}}{\text{peak rate of flow}_{\text{vph}}}$$

$$= \frac{V_{\text{vph}}}{v_p} = \frac{V_{\text{vph}}}{4V_{15\text{ min,peak}}} \qquad 73.5$$

The *service flow rate* at level of service i is the actual capacity in passenger cars per hour under nonideal conditions. Service flow rate is calculated from the maximum service flow rate, the number of lanes, N, and other factors that correct for nonideal conditions.

$$V_i = (v_{m,i})N(\text{adjustment factors}) \qquad 73.6$$

Example 73.1

Using the following traffic counts, determine the (a) peak hourly traffic volume and (b) peak hour factor.

interval	volume (veh)
4:15–4:30	520
4:30–4:45	580
4:45–5:00	670
5:00–5:15	790
5:15–5:30	700
5:30–5:45	630
5:45–6:00	570
6:00–6:15	510

Solution

(a) For the 1 hr time period of 4:15 through 5:15, the total hourly traffic volume was

520 veh + 580 veh + 670 veh + 790 veh = 2560 vph

The hourly traffic volumes for the subsequent 1 hr intervals are found similarly.

4:15–5:15	2560 vph
4:30–5:30	2740 vph
4:45–5:45	2790 vph
5:00–6:00	2690 vph
5:15–6:15	2410 vph

The maximum hourly volume is 2790 vph.

(b) The peak 15 min volume occurs between 5:00 and 5:15 and is 790. From Eq. 73.5,

$$PHF = \frac{V_{\text{vph}}}{4V_{15\text{ min,peak}}}$$

$$= \frac{2790\ \dfrac{\text{veh}}{\text{hr}}}{\left(4\ \dfrac{\text{periods}}{\text{hr}}\right)\left(790\ \dfrac{\text{veh}}{\text{period}}\right)}$$

$$= 0.883$$

9. TRIP GENERATION

Trip generation requires estimating the number of trips that will result from a particular population or occupancy. Estimates can be obtained from general or area-specific tables or correlations. There are many general trip generation correlations that can be used if specific, targeted data are not available. Since predictive formulas are highly dependent on the characteristics of the local area and population as well as on time of day and year, numerous assumptions must be made and verified before such correlations are relied upon.

Although linear equations with constant coefficients are often used, in the form of Eq. 73.7, more sophisticated correlations are easily obtained. Many take on the forms of either Eq. 73.8(a) or the equivalent 73.8(b). The calling population parameter in these equations may be population quantity, number of homes, square footage (as in the case of retail shops), or any other convenient characteristic that can be quantified.

$$\text{no. of trips} = a + b\left(\frac{\text{calling population}}{\text{parameter}}\right) \qquad 73.7$$

$$\log(\text{no. of trips}) = A + B\log\left(\frac{\text{calling population}}{\text{parameter}}\right)$$
$$73.8(a)$$

$$\text{no. of trips} = \frac{C}{\left(\dfrac{\text{calling population}}{\text{parameter}}\right)^D} \qquad 73.8(b)$$

The best trip data are actual counts obtained from *automatic traffic recorders* (ATRs). These devices can be used where roads are already in existence. Since ATRs count the number of vehicles (i.e., sets of axles) that pass over their detectors, counts must be corrected for trailer truck movement.

Logical methods, including straight-line extrapolation or exponential growth, should be used in the estimation of future traffic counts. Expansion factors can be determined for each axle classification.

Example 73.2

An ATR record indicates that 966 axles passed over its detector. It is known that 10% of the vehicles were five-axle trucks. How many two-axle passenger cars and five-axle trucks actually passed over the detector?

Solution

The number of axles passing over the detector was

$$(2)(\text{no. of cars}) + (5)(\text{no. trucks}) = 966$$

The fraction of trucks is known to be 10%.

$$\frac{\text{no. of trucks}}{\text{no. of cars} + \text{no. of trucks}} = 0.10$$

Solving these two equations simultaneously,

$$\text{no. of cars} = 378$$

$$\text{no. of trucks} = 42$$

10. SPEED, FLOW, AND DENSITY RELATIONSHIPS

With uninterrupted flow, the speed of travel decreases as the number of cars occupying a freeway or multilane highway increases. The *free-flow speed*, S_f or FFS, is the maximum speed for which the density does not affect travel speed. Free-flow speed can be determined from actual measurements or estimated from the ideal value. Free-flow speed is measured in the field as the average speed of passenger cars when flow rates are less than 1300 pcphpl.

Density, D, is defined as the number of vehicles per mile per lane (vpm/lane or vpk/lane) or, for pedestrians, in pedestrians per unit area. The *critical density*, D_o (also referred to as the *optimum density*), is the density at which maximum capacity occurs. The *jam density*, D_j, is the density when the vehicles (or pedestrians) are all at a standstill. The average travel speed, S, for any given density can be related to the free-flow speed by Eq. 73.9.

$$S = S_f \left(1 - \frac{D}{D_j}\right) \quad \text{73.9}$$

The *flow* (*rate of flow*), v (number of vehicles or volume), crossing a point per hour per lane (vph/lane) is given in Eq. 73.10.

$$v = SD = \frac{3600 \; \frac{\text{sec}}{\text{hr}}}{\text{headway} \; \frac{\text{sec}}{\text{veh}}} \quad \text{73.10}$$

Spacing is the distance between common points (e.g., the front bumper) on successive vehicles. The *headway* is the time between successive vehicles.

$$\text{spacing}_{\text{m/veh}} = \frac{1000 \; \frac{\text{m}}{\text{km}}}{D_{\text{vpk/lane}}} \quad \text{[SI]} \quad \text{73.11(a)}$$

$$\text{spacing}_{\text{ft/veh}} = \frac{5280 \; \frac{\text{ft}}{\text{mi}}}{D_{\text{vpm/lane}}} \quad \text{[U.S.]} \quad \text{73.11(b)}$$

$$\text{headway}_{\text{s/veh}} = \frac{\text{spacing}_{\text{m/veh}}}{\text{space mean speed}_{\text{m/s}}} \quad \text{[SI]} \quad \text{73.12(a)}$$

$$\text{headway}_{\text{sec/veh}} = \frac{\text{spacing}_{\text{ft/veh}}}{\text{space mean speed}_{\text{ft/sec}}} \quad \text{[U.S.]} \quad \text{73.12(b)}$$

$$v_{\text{vph}} = \frac{3600 \; \frac{\text{sec}}{\text{hr}}}{\text{headway}_{\text{sec/veh}}} \quad \text{73.13}$$

Gap is an important concept, particularly in unsignalized intersections where pedestrians or cross-street traffic are waiting for opportunities to cross. A *gap* is the time interval between vehicles in a major traffic stream. The *critical gap* is the minimum time interval between successive vehicles that will permit pedestrians or cross-street traffic to enter the intersection.

As Figs. 73.1 through 73.3 indicate, the *optimum density* (*critical density*), *optimum speed* (*critical speed*), and *maximum flow* (*maximum capacity*) pass through the same point.

11. LANE DISTRIBUTION

It may be necessary to estimate the distribution of truck traffic on the various lanes of a multilane facility. The distribution of lane use varies widely, being dependent on traffic regulations, traffic composition, speed, volume, number and location of access points, origin-destination patterns of drivers, development environment, and local driver habits. Because there are so many factors, there are no typical values available for lane distribution.

Lane distribution values based on various vehicle types for selected freeways are provided in the HCM, but they are not intended to represent typical values.

12. VEHICLE EQUIVALENTS

It is convenient to designate a standard unit of measure for vehicular flow. Passenger cars make up the majority of highway traffic, so it is natural to use passenger cars as the unit. Other types of vehicles are converted to *passenger car equivalents*. Since trucks, buses, and recreational vehicles (known as RVs) take up more space

Figure 73.1 *Space Mean Speed Versus Density*

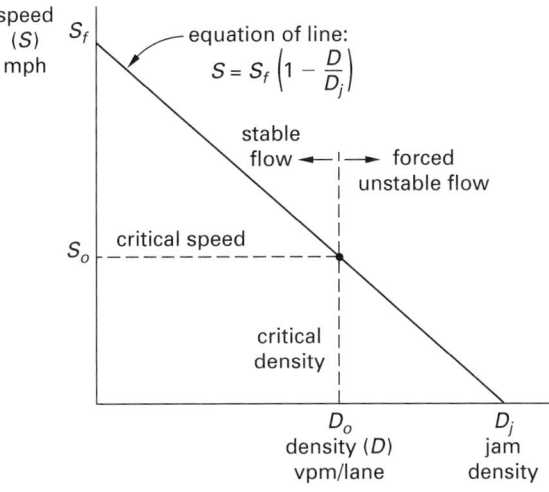

on a roadway than cars and since they tend to travel more slowly up grades, they degrade the quality of travel more than the same number of cars would. Therefore, their traffic volumes are converted to equivalent passenger car volumes, E, when computing flow.

Passenger car equivalents for trucks and RVs on grades depend on the percent grade, the length of the grade, the number of lanes, and the percentage of trucks and buses. Table 73.6 lists passenger car equivalents for general conditions. However these values are applicable only to long sections of highways. For operation on specific grades of specific lengths, the HCM must be used.

High-occupancy vehicles (HOVs) include taxis, buses, and carpool vehicles. Special HCM procedures apply to HOV lanes.

Table 73.6 *Passenger Car Equivalents on Extended General Freeway Segments[a]*

terrain	E_T (trucks and buses)	E_R (RVs)
level	1.5	1.2
rolling	2.5	2.0
mountainous	4.5	4.0

[a]Primarily for use in determining approximate capacity during planning stages when specific alignments are not known.

Reproduced with permission by the Transportation Research Board. In *Highway Capacity Manual 2000*, Exhibit 21-8, copyright © 2000 by the Transportation Research Board, National Research Council, Washington, D.C.

Figure 73.2 *Space Mean Speed Versus Flow*

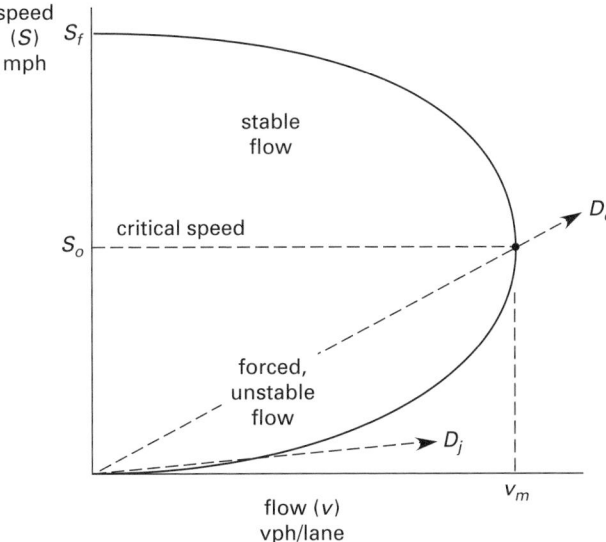

13. FREEWAYS

(Capacity analysis of basic freeway segments is covered in the HCM, Ch. 23.)

The process of determining *level of service* of a freeway section generally involves determining the vehicular density, D. Freeway conditions are classified into levels of service A through F. Level A represents conditions where there are no physical restrictions on operating speeds. Since there are only a few vehicles on the freeway, operation at highest speeds is possible. However, the traffic volume is small. Level F represents stop-and-go, low-speed conditions with poor safety and maneuverability. The desired design condition is between levels A and F. Typically, levels B and C are chosen for initial design in rural areas, and levels C and D are used for initial design in suburban and urban areas.

$$D = \frac{v_p}{S} \qquad 73.14$$

The actual level of service is determined by comparing the actual density (in pcpmpl) with the density limits given in Table 73.7. Other criteria may also be used, such as speed, service flow rate, and the v/c ratio for a given design speed.

The actual service flow rate, v, per lane for a particular level of service is calculated from the ideal capacity, c,

Figure 73.3 *Flow Versus Density*

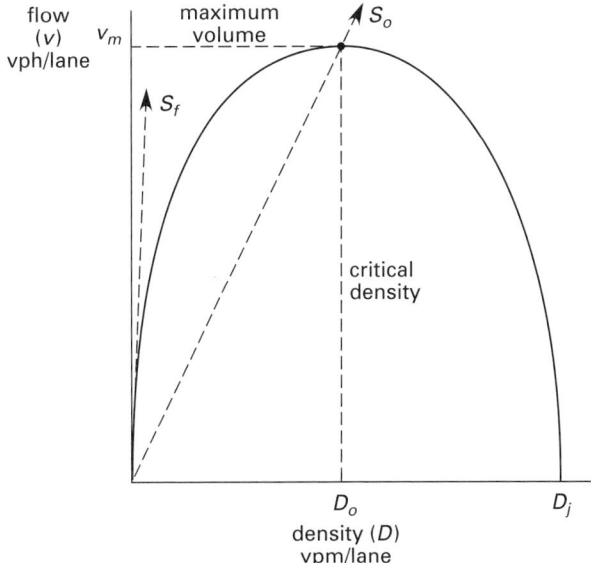

for ideal conditions and the actual v/c ratio. The maximum *service flow rate*, v_m, per lane for a particular level of service is calculated from the ideal capacity, c, for ideal conditions and the maximum v/c ratio. Table 73.8 contains maximum v/c ratios for given design speeds and levels of service.

$$v = c(v/c) \qquad 73.15$$

$$v_{m,i} = c(v/c)_{m,i} \qquad 73.16$$

The *flow rate*, V, can be calculated from the number of lanes in the analysis, N; a factor to adjust for the presence of heavy vehicles such as buses, trucks, and recreational vehicles, f_{HV}; and a factor to adjust for the effect of the driver population, f_p. The passenger car equivalent flow rate for the peak 15 min is shown as v_p (pcphpl).

$$V = v_p(\text{PHF})Nf_{HV}f_p \qquad 73.17$$

Unless other information is available, the HCM suggests using estimated PHF values of 0.88 and 0.92 for rural and urban-suburban freeways, respectively.

The *heavy vehicle factor*, f_{HV}, is a function of the truck, bus and recreational vehicle fractions and is given by Eq. 73.18. E_T and E_R are the passenger car equivalents of trucks/buses and recreational vehicles, respectively, as given in the HCM as a function of specific grades. The fraction of trucks and buses, P_T, is sometimes referred to as the *T-factor*. Trucks do not include light delivery vans.

$$f_{HV} = \frac{1}{1 + P_T(E_T - 1) + P_R(E_R - 1)} \qquad 73.18$$

The *driver population adjustment*, f_p, is 1.0 for weekday or commuter traffic (drivers familiar with the route) and 0.75 to 0.90 for weekend, recreational, or other types of traffic that do not use the available space as efficiently. Engineering judgment is required in selecting the population's adjustment factor in those instances.

For freeway analysis, *free-flow speed*, FFS, is determined from the *basic free-flow speed* (BFFS) with adjustments for lane width (f_{LW}), right shoulder clearance (f_{LC}), number of lanes (f_N), and interchange density (f_{ID}). These factors are given in Ch. 23 of the HCM.

$$\text{FFS} = \text{BFFS} - f_{LW} - f_{LC} - f_N - f_{ID} \qquad 73.19$$

Table 73.7 *Levels of Service for Basic Freeway Sections*

level	density range (pc/mi/ln)	description
A	0–11	free flow with low volumes and high speeds
B	>11–18	stable flow, but speeds are beginning to be restricted by traffic conditions
C	>18–26	stable flow, but most drivers cannot select their own speed
D	>26–35	approaching unstable flow, and maneuvering room is noticeably limited
E	>35–45	unstable flow with short stoppages, and maneuvering room is severely limited
F	> 45	forced flow at crawl speeds; localized lines of vehicles

(Multiply pc/mi/ln by 0.621 to obtain pc/km/ln.)
Adapted from *Highway Capacity Manual 2000*, p. 23-3.

Figure 73.4 *Speed-Flow Relationships*

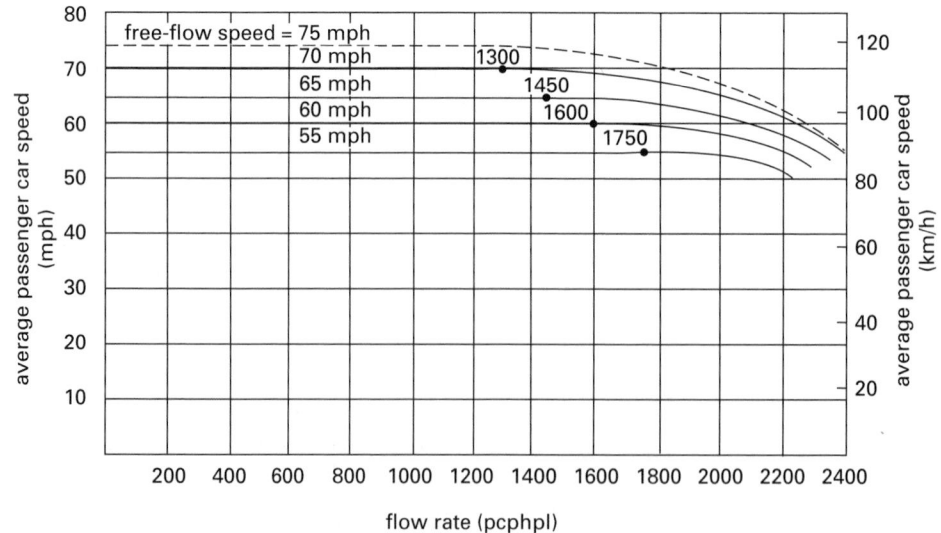

Table 73.8 Level of Service Criteria for Basic Freeway Sections

criteria	A	B	C	D	E
			LOS		
FFS = 75 mi/h					
maximum density (pc/mi/ln)	11	18	26	35	45
minimum speed (mi/h)	75.0	74.8	70.6	62.2	53.3
maximum v/c	0.34	0.56	0.76	0.90	1.00
maximum service flow rate (pc/h/ln)	820	1350	1830	2170	2400
FFS = 70 mi/h					
maximum density (pc/mi/ln)	11	18	26	35	45
minimum speed (mi/h)	70.0	70.0	68.2	61.5	53.3
maximum v/c	0.32	0.53	0.74	0.90	1.00
maximum service flow rate (pc/h/ln)	770	1260	1770	2150	2400
FFS = 65 mi/h					
maximum density (pc/mi/ln)	11	18	26	35	45
minimum speed (mi/h)	65.0	65.0	64.6	59.7	52.2
maximum v/c	0.30	0.50	0.71	0.89	1.00
maximum service flow rate (pc/h/ln)	710	1170	1680	2090	2350
FFS = 60 mi/h					
maximum density (pc/mi/ln)	11	18	26	35	45
minimum speed (mi/h)	60.0	60.0	60.0	57.6	51.1
maximum v/c	0.29	0.47	0.68	0.88	1.00
maximum service flow rate (pc/h/ln)	660	1080	1560	2020	2300
FFS = 55 mi/h					
maximum density (pc/mi/ln)	11	18	26	35	45
minimum speed (mi/h)	55.0	55.0	55.0	54.7	50.0
maximum v/c	0.27	0.44	0.64	0.85	1.00
maximum service flow rate (pc/h/ln)	600	990	1430	1910	2250

Note: The exact mathematical relationship between density and v/c has not always been maintained at LOS boundaries because of the use of rounded values. Density is the primary determinant of LOS. The speed criterion is the speed at maximum density for a given LOS.

(Multiply mph by 1.609 to obtain km/h.)
(Multiply pcpmpl by 0.621 to obtain pc/km/ln.)

Reproduced with permission by the Transportation Research Board. In *Highway Capacity Manual 2000*, Exhibit 23-2, copyright © 2000 by the Transportation Research Board, National Research Council, Washington, D.C.

Example 73.3

A four-lane (two lanes in each direction) freeway passes through rolling terrain in an urban area. The freeway is constructed with 11 ft lanes and abutment walls 2 ft from the outer pavement edges of both slow lanes. The one-direction peak hourly volume during the weekday commute is 1800 vph. Traffic includes 3% buses and 5%

trucks. There is one interchange per mile. The peak hour factor is 0.90. The posted speed limit is 65 mph. The base free-flow speed is 70 mph. (a) What is the passenger car equivalent flow rate per lane? (b) What is the speed during peak-hour travel? (c) What is the density? (d) What is the weekday peak-hour level of service?

Solution

(a) Trucks and buses have the same vehicle equivalents, so the "truck" fraction is 8%. There are no RVs. From Table 73.6, $E_T = 2.5$. From Eq. 73.18,

$$f_{HV} = \frac{1}{1 + P_T(E_T - 1)}$$
$$= \frac{1}{1 + (0.08)(2.5 - 1)} = 0.893$$

Convert the volume to flow rate per lane. The commute population knows the route, so $f_p = 1.0$. From Eq. 73.17,

$$v_p = \frac{V}{(\text{PHF})N f_{HV} f_P}$$
$$= \frac{1800 \text{ vph}}{(0.90)(2)(0.893)(1)} = 1119 \text{ pcphpl}$$

(b) The base free-flow speed was given as 70 mph. The actual free-flow speed is calculated from Eq. 73.19.

$$f_{LW} = 1.9 \text{ mph} \quad [\text{HCM Exh. 23-4}]$$
$$f_{LC} = 2.4 \text{ mph} \quad [\text{HCM Exh. 23-5}]$$
$$f_N = 4.5 \text{ mph} \quad [\text{HCM Exh. 23-6}]$$
$$f_{ID} = 2.5 \text{ mph} \quad [\text{HCM Exh. 23-7}]$$

$$\text{FFS} = \text{BFFS} - f_{LW} - f_{LC} - f_N - f_{ID}$$
$$= 70 \text{ mph} - 1.9 \text{ mph} - 2.4 \text{ mph} - 4.5 \text{ mph}$$
$$\quad - 2.5 \text{ mph}$$
$$= 58.7 \text{ mph} \quad [\text{say 59 mph}]$$

(c) The density is

$$D = \frac{v_p}{S} = \frac{1119 \text{ pcphpl}}{59 \frac{\text{mi}}{\text{hr}}}$$
$$= 19.0 \text{ pc/mi-ln (pcpmpl)}$$

(d) From Table 73.8 with a free-flow speed of 60 mph, the level of service (based on density) is C.

Example 73.4

An urban freeway segment is being designed for an AADT of 60,000 weekday commuters traveling at 60 mph in 12 ft lanes. Heavy trucks constitute 5% of the total traffic. The directionality factor is 75%. The K-factor is 9.8%. The freeway segment consists of 1 mi of

4% upward grade. The lateral clearances are 10 ft on the right and 6 ft on the left. Assume reasonable values for PHF and FFS. How many lanes are needed for LOS D?

Solution

Use Eq. 73.2 to convert AADT to design hourly volume.

$$\text{DDHV} = DK(\text{AADT}) = (0.75)(0.098)\left(60{,}000 \; \frac{\text{veh}}{\text{day}}\right)$$

$$= 4410 \text{ vph}$$

From HCM Exh. 23-9, for a 4% upgrade 1 mi in length with 5% trucks, the passenger car equivalent for trucks is 2.5. From Eq. 73.18,

$$f_{\text{HV}} = \frac{1}{1 + P_T(E_T - 1)}$$

$$= \frac{1}{1 + (0.05)(2.5 - 1)} = 0.930$$

Assume a free-flow speed of 65 mph and peak hour factor of 0.92.

From Table 73.8 with FFS = 65 and LOS D, the maximum service flow rate is 2090 pcphpl.

The adjustment factor for a driver population of weekday commuters is $f_p = 1.0$.

$$V_{2 \text{ lanes}} = v_p(\text{PHF})N f_{\text{HV}} f_p$$

$$= \left(2090 \; \frac{\text{pc}}{\text{hr-ln}}\right)(0.92)(2 \text{ ln})(0.930)(1)$$

$$= 3576 \text{ vph } [< 4410]$$

$$V_{3 \text{ lanes}} = v_p(\text{PHF})N f_{\text{HV}} f_p$$

$$= \left(2090 \; \frac{\text{pc}}{\text{hr-ln}}\right)(0.92)(3 \text{ ln})(0.930)(1)$$

$$= 5364 \text{ vph } [> 4410]$$

Three lanes are required.

14. MULTILANE HIGHWAYS

(Capacity analysis of multilane highways is covered in the HCM, Ch. 21.)

Multilane highways are not completely access controlled. The *free-flow speed*, FFS, is the theoretical speed of a vehicle under all actual conditions except interference by other vehicles. Free-flow speed is the theoretical speed when density is zero, but is essentially unchanged for densities up to 1400 pcphpl. Free-flow speed is determined from the base free-flow speed (BFFS) with adjustments for median type (f_M), lane width (f_{LW}), total lateral clearance (f_{LC}), and density of access points (f_A). The base free-flow speed is covered in HCM Ch. 12, but is approximately 5 mph greater than the speed limit for 50 and 55 mph speed limits. Values of

Table 73.9 *Level of Service Criteria for Multilane Highways*

				LOS		
free-flow speed	criteria	A	B	C	D	E
60 mi/h	maximum density (pc/mi/ln)	11	18	26	35	40
	average speed (mi/h)	60.0	60.0	59.4	56.7	55.0
	maximum volume to capacity ratio (v/c)	0.30	0.49	0.70	0.90	1.00
	maximum service flow rate (pc/h/ln)	660	1080	1550	1980	2200
55 mi/h	maximum density (pc/mi/ln)	11	18	26	35	41
	average speed (mi/h)	55.0	55.0	54.9	52.9	51.2
	maximum v/c	0.29	0.47	0.68	0.88	1.00
	maximum service flow rate (pc/h/ln)	600	990	1430	1850	2100
50 mi/h	maximum density (pc/mi/ln)	11	18	26	35	43
	average speed (mi/h)	50.0	50.0	50.0	48.9	47.5
	maximum v/c	0.28	0.45	0.65	0.86	1.00
	maximum service flow rate (pc/h/ln)	550	900	1300	1710	2000
45 mi/h	maximum density (pc/mi/ln)	11	18	26	35	45
	average speed (mi/h)	45.0	45.0	45.0	44.4	42.2
	maximum v/c	0.26	0.43	0.62	0.82	1.00
	maximum service flow rate (pc/h/ln)	490	810	1170	1550	1900

Note: The exact mathematical relationship between density and volume to capacity ratio (v/c) has not always been maintained at LOS boundaries because of the use of rounded values. Density is the primary determinant of LOS. LOS F is characterized by highly unstable and variable traffic flow. Prediction of accurate flow rate, density, and speed at LOS F is difficult.

Reproduced with permission by the Transportation Research Board. In *Highway Capacity Manual 2000*, Exhibit 21-2, copyright © 2000 by the Transportation Research Board, National Research Council, Washington, D.C.

the adjustments are obtained from Table 73.9 and the appropriate tables in the HCM, Ch. 21. (Though similar in concept to freeway adjustments, some highway values are different. Accordingly, different symbols are used.)

$$\text{FFS} = \text{BFFS} - f_M - f_{\text{LW}} - f_{\text{LC}} - f_A \qquad 73.20$$

The *15 min passenger car equivalent flow rate*, v_p, in pcphpl is given by Eq. 73.21. V is the volume of vehicles passing a point each hour, f_{HV} is the same heavy-vehicle factor used in freeway analysis, and PHF is the peak hour factor. Where specific local data are not available, the HCM recommends reasonable estimates of PHF of 0.88 for rural multilane highways and 0.92 for urban multilane highways. For congested conditions, PHF = 0.95 is a reasonable assumption. The driver population adjustment factor, f_p, is 1.0 for familiar drivers (i.e., weekday commuters) and down to 0.85 otherwise.

$$v_p = D_m(\text{FFS}) = \frac{V}{N(\text{PHF})f_{\text{HV}}f_p} \qquad 73.21$$

Example 73.5

An undivided suburban multilane highway segment with four 11 ft lanes (two in each direction) is used by weekday commuters. There is no median, and the two lanes in each direction are separated by a striped centerline. The estimated free-flow speed under ideal conditions is 60 mph. The segment contains 4% up- and downgrades 0.8 mi long. The fraction of recreational vehicles on this segment is essentially zero. However, the fraction of trucks is 10%. The lateral clearance on the right-hand side of the slow lane is 2 ft from the pavement edge. There are no points of entry to the highway segment.

(a) What is the hourly volume in the upgrade direction for LOS C? (b) What is the capacity of the downgrade section? (c) What is the LOS in the downgrade section during the morning peak if the peak-hour traffic volume is 1300 vph?

Solution

(a) The base free-flow speed is given as BFFS = 60 mph.

The median type adjustment factor for undivided highways is found from the HCM, Exh. 21-6, as $f_M = 1.6$.

The lane width adjustment factor for 11 ft lanes is found from the HCM, Exh. 21-4, as $f_{\text{LW}} = 1.9$.

For undivided highways with only a striped centerline, the left side clearance is zero. However, the median factor accounts for the proximity of the two opposing lanes. The left-edge lateral clearance is taken as 6 ft per the HCM. The total lateral clearance is 6 ft + 2 ft = 8 ft. The adjustment factor for lateral clearance for four-lane highways is found in HCM, Exh. 21-5, as $f_{\text{LC}} = 0.9$.

If there are no access points, Exh. 21-7 of the HCM gives the access-point density adjustment factor as $f_A = 0$.

Use Eq. 73.20.

$$\begin{aligned}
\text{FFS} &= \text{BFFS} - f_M - f_{\text{LW}} - f_{\text{LC}} - f_A \\
&= 60\ \frac{\text{mi}}{\text{hr}} - 1.6\ \frac{\text{mi}}{\text{hr}} - 1.9\ \frac{\text{mi}}{\text{hr}} - 0.9\ \frac{\text{mi}}{\text{hr}} - 0 \\
&= 55.6\ \text{mi/hr}
\end{aligned}$$

Round to 55 mi/hr in order to use data from HCM Exh. 21-2. The maximum density at LOS C is 26 pcpmpl.

Use Eq. 73.21.

$$\begin{aligned}
v_p = D_m(\text{FFS}) &= \left(26\ \frac{\text{pc}}{\text{mi-lane}}\right)\left(55.6\ \frac{\text{mi}}{\text{hr}}\right) \\
&= 1445\ \text{pc/hr-lane (pcphpl)}
\end{aligned}$$

From Exh. 21-9 of the HCM, the passenger car equivalent of trucks on a 4% grade 0.8 mi long with 10% trucks is $E_T = 2.5$. From Eq. 73.18, the heavy-vehicle adjustment factor is

$$\begin{aligned}
f_{\text{HV}} &= \frac{1}{1 + P_T(E_T - 1)} \\
&= \frac{1}{1 + (0.10)(2.5 - 1)} \\
&= 0.870
\end{aligned}$$

Assume a value of 0.92 for the peak hour factor as recommended by the HCM. Solve Eq. 73.21 for the volume.

$$\begin{aligned}
V &= v_p N(\text{PHF})f_{\text{HV}}f_p \\
&= \left(1445\ \frac{\text{pc}}{\text{hr-lane}}\right)(2\ \text{lanes})(0.92)(0.870)(1.0) \\
&= 2313\ \text{vph}
\end{aligned}$$

(b) From Exh. 21-2 of the HCM, at 55 mph, a maximum capacity of 2100 pcphpl will be reached at LOS E.

From Exh. 21-11 of the HCM, the passenger car equivalent of trucks on a 4% downgrade 0.8 mi long is $E_T = 1.5$. From Eq. 73.18, the heavy-vehicle adjustment factor is

$$\begin{aligned}
f_{\text{HV}} &= \frac{1}{1 + P_T(E_T - 1)} \\
&= \frac{1}{1 + (0.10)(1.5 - 1)} \\
&= 0.952
\end{aligned}$$

Use Eq. 73.21 to calculate the hourly volume.

$$\begin{aligned}
V &= v_p N(\text{PHF})f_{\text{HV}}f_p \\
&= \left(2100\ \frac{\text{pc}}{\text{hr-lane}}\right)(2\ \text{lanes})(0.92)(0.952)(1.0) \\
&= 3678\ \text{vph}
\end{aligned}$$

Transportation

(c) Use Eq. 73.21 to calculate the service flow rate.

$$
\begin{aligned}
v_p &= \frac{V}{N(\text{PHF}) f_{\text{HV}} f_p} \\
&= \frac{1300 \, \dfrac{\text{pc}}{\text{hr}}}{(2)(0.92)(0.952)(1.0)} \\
&= 742 \text{ pc/hr-lane (pcphpl)}
\end{aligned}
$$

From part (a), the free-flow speed is 55.6 mph. From Eq. 73.21, the peak density is

$$
\begin{aligned}
D_m &= \frac{v_p}{\text{FFS}} = \frac{742 \, \dfrac{\text{pc}}{\text{hr-lane}}}{55.6 \, \dfrac{\text{mi}}{\text{hr}}} \\
&= 13.3 \text{ pc/mi-lane (pcpmpl)}
\end{aligned}
$$

Use Exh. 21-2 of the HCM for 55 mi/hr. Since 11 pcpmpl < 13.3 pcpmpl < 18 pcpmpl, the level of service is B.

15. SIGNALIZED INTERSECTIONS

(Capacity analysis of signalized intersections is covered in the HCM, Ch. 16.)

Signalized intersections are controlled by signals operating in two or more phases. Each phase consists of three intervals: green, amber (i.e., "yellow"), and red. For a typical intersection with two streets crossing at 90° to each other, a *two-phase signal* is one that has one phase for each axis of travel (e.g., one phase of north-south movements and one phase of east-west movements). A *three-phase signal* provides one of the roads with a left-turn phase. In a *four-phase signal*, both roads have left-turn phases.

Level of service for signalized intersections is defined in terms of control delay in the intersection. *Control delay* consists of only the portion of delay attributable to the control facility (e.g., initial deceleration delay, queue move-up time, stopped delay, and final acceleration delay), but not geometric delay or incident delay. Delay can be measured in the field, or it can be estimated using procedures as outlined in the HCM. LOS criteria are stated in terms of control delay per vehicle for a 15 min analysis period, as given in Table 73.10. Because of this criterion, an intersection may be operating below its maximum capacity but have an unacceptable delay, and therefore it would be classified as LOS F (failure). This is a different meaning for "LOS F," which generally indicates that demand exceeds capacity.

The capacity of a signalized intersection is calculated for each lane group. *Capacity* is the maximum rate of flow for the subject lane group that may pass through the intersection under prevailing traffic, roadway, and signalization conditions. The rate of flow is generally measured or projected for a 15 min period. Capacity is given in terms of vph and is dependent on many factors, including the width of approach, parking conditions, traffic direction (one- or two-way traffic), environment, bus and truck traffic, and percentage of turning vehicles.

Rather than calculate the capacity of the entire intersection, a total volume-capacity (v/c) flow ratio is computed for all the critical lane groups within the intersection as a measure of the overall intersection performance. The *critical lane groups* are those lane groups that have the maximum *flow ratio*, v/s, also known as *saturation flow ratio*, for each signal phase. The *critical movements* consume the maximum amount of time during each signal phase. For example, in a two-phase signal at a typical cross-street intersection, opposing lane groups move during the same green time (phase). Generally, one of the lane groups, or *approaches*, will have a higher flow ratio (v/s) and will require more green time than the other approach. This would be the critical lane group, X_{ci}, for that phase.

In capacity analysis, volume-capacity ratios (v/c) are computed for each intersection movement and for all of the critical movements together. The v/c ratios are determined by dividing the peak 15 min rate of flow on an approach or lane group by the capacity of the approach or lane group. Both the traffic flow and the geometric characteristics are taken into consideration. The objective is to provide the minimum number of lane groups to adequately characterize the intersection operation.

The capacity of an approach or lane group is given by Eq. 73.22. s_i is the *saturation flow rate* in vphg for lane group i (i.e., in vphgpl), which is the flow rate per lane at which vehicles can pass through a signalized intersection in a stable moving queue. g_i/C is the *effective green ratio* for the lane group.

$$
c_i = s_i \left(\frac{g_i}{C} \right) \tag{73.22}
$$

$$
s_{\text{vphgpl}} = \frac{3600 \, \dfrac{\text{sec}}{\text{hr}}}{\text{saturation headway}_{\text{sec/veh}}} \tag{73.23}
$$

The ratio of flow to capacity (v/c) is known as the *degree of saturation* and *volume-capacity ratio*. For convenience and consistency with the literature, the degree of saturation for lane group or approach i is designated as X_i instead of $(v/c)_i$, which might be confused with level of service i. When the flow rate equals the capacity, X_i equals 1.00; when flow rate is zero, X_i is zero.

$$
X_i = \left(\frac{v}{c} \right)_i = \frac{v_i}{s_i \left(\dfrac{g_i}{C} \right)} = \frac{v_i C}{s_i g_i} \tag{73.24}
$$

When the overall intersection is to be evaluated with respect to its geometry and the total cycle time, the concept of the critical v/c ratio, X_c, is used. The *critical v/c ratio* is usually obtained for the overall intersection considering only the critical lane groups or approaches. For a typical cross intersection with a two-, three-, or four-phase signal, once the total cycle length is selected, the time for each phase is proportioned according to the critical ratios, X_{ci}, of each phase. Other factors can affect the detail adjustment of the phase length such as lost time, pedestrian crossings, and approach conditions. In Eq. 73.25, C is the total cycle length and L is the *lost time* per cycle. Time is lost when the intersection is not used effectively, as during start-up through the intersection. Lost time includes start-up time and some of the amber signal time. In conservative studies, lost time includes all of the amber signal time.

$$X_c = \sum_i \left(\frac{v}{s}\right)_{ci} \left(\frac{C}{C - L}\right) \qquad 73.25$$

Equation 73.25 can be used to estimate the signal timing for the intersection if a critical v/c ratio is specified for the intersection. This equation can also be used to estimate the overall sufficiency of the intersection by substituting the specified maximum permitted cycle length and determining the resultant critical v/c ratio for the intersection.

When the critical v/c ratio is less than 1.00, the cycle length provided is adequate for all critical movements to pass through the intersection if the green time is proportionately distributed to the different phases. If the total green time is not proportionately distributed to the different phases, it is possible to have a critical v/c ratio of less than 1.00, but one or more individual oversaturated movements may occur within a cycle.

Saturation flow rate, s, is defined as the total maximum flow rate on the approach or group of lanes that can pass through the intersection under prevailing traffic and roadway conditions when 100% of the effective green time is available. Saturation flow rate is expressed in units of vehicles per hour of effective green time (vphg) for a given lane group.

Saturation flow rate is calculated from the ideal saturation flow rate, s_o, which is assumed to be 1900 passenger cars per hour of green per lane (abbreviated "pcphgpl"). Adjustments are made for lane width (f_w), heavy vehicles (f_{HV}), grade (f_g), existence of parking lanes (f_p), stopped bus blocking (f_{bb}), type of area (f_a), lane utilization (f_{LU}), right-hand turns (f_{RT}), left-hand turns (f_{LT}), and pedestrians and bicycles turning left and right (f_{Lpb} and f_{Rpb}). All of the adjustments are tabulated in the HCM.

$$s = s_o N f_w f_{HV} f_g f_p f_{bb} f_a f_{LU} f_{RT} f_{LT} f_{Lpb} f_{Rpb} \qquad 73.26$$

Other important concepts in signalized intersection analysis are arrival type and platoon ratio. The *arrival type* (AT) is a categorization of the quality of progression through the intersection. There are six categories—arrival types 1 through 6. Arrival type 1 represents a "very poor progression" where 80% or more of the lane group arrives at the start of the red signal phase. Arrival type 6 represents near-ideal "exceptional progression" with the percentage of arrivals at the start

of the green signal phase approaching 100%. The *platoon ratio*, R_p, is the ratio of the fraction of all vehicles in movement arriving during the green phase and the green signal time fraction. The default value is 1.00 for arrival type 3, which corresponds to random arrivals. HCM Exh. 16-4 quantifies arrival type by platoon ratio.

$$R_p = \frac{P_{\text{green}}}{\dfrac{g}{C}} \qquad 73.27$$

Table 73.10 Level of Service Criteria for Signalized Intersections

level of service	control delay per vehicle (sec)
A	≤ 10.0
B	> 10.0 and ≤ 20.0
C	> 20.0 and ≤ 35.0
D	> 35.0 and ≤ 55.0
E	> 55.0 and ≤ 80.0
F	> 80.0

Reproduced with permission by the Transportation Research Board. In *Highway Capacity Manual 2000*, Exhibit 16-2, copyright © 2000 by the Transportation Research Board, National Research Council, Washington, D.C.

Example 73.6

A signalized intersection, without pedestrian access and located in a central business district, has an approach with two 11 ft lanes on a 2% downgrade. 10% of the traffic consists of heavy trucks, but there are no buses or RVs. Both lanes are through lanes; no turns are permitted.

Figure 73.5 Elements of an Intersection

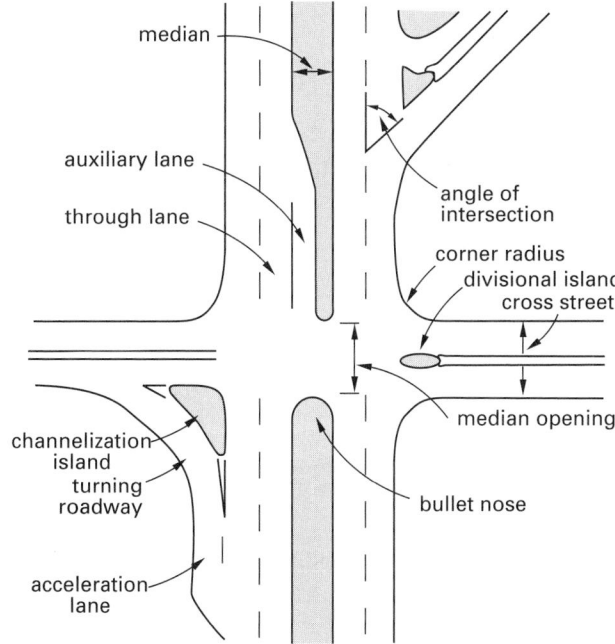

There are no parking lanes. Arrivals to the intersection are random. What is the saturation flow rate of the approach?

Solution

Refer to the HCM Exh. 16-7 to obtain the required adjustment factors.

$$s_o = 1900 \text{ pcphgpl}$$

$$N = 2 \text{ lanes}$$

$$f_w = 1 + \frac{W - 12}{30} = 1 + \frac{11 - 12}{30} = 0.967$$

$$E_T = 2.0 \text{ per HCM p. 16-10}$$

$$f_{HV} = \frac{100}{100 + \%HV(E_T - 1)} = \frac{100}{100 + (10)(2.0 - 1)}$$
$$= 0.909$$

$$f_g = 1 - \frac{\%G}{200} = 1 - \frac{-2}{200} = 1.010$$

$$f_p = 1.000 \text{ for no parking}$$

$$f_{bb} = 1.000 \text{ for no bus blocking}$$

$$f_a = 0.900 \text{ for a central business district (CBD)}$$

$$f_{LU} = 0.952 \text{ from HCM Exh. 10-23}$$

$$f_{RT} = 1.000$$

$$f_{LT} = 1.000$$

$$f_{Lpb} = 1.000$$

$$f_{Rpb} = 1.000$$

$$s = s_o N f_w f_{HV} f_g f_p f_{bb} f_a f_{LU} f_{RT} f_{LT} f_{Lpb} f_{Rpb}$$

$$= \left(1900 \ \frac{\text{pc}}{\text{hrg-lane}} \right) (2 \text{ lanes})(0.967)(0.909)(1.010)$$

$$\times (1.000)(1.000)(0.90)(0.952)(1.000)(1.000)$$

$$\times (1.000)(1.000)$$

$$= 2890 \text{ pcphg}$$

16. WARRANTS FOR INTERSECTION SIGNALING

The MUTCD (Sec. 4C) contains methodology for determining when four-way traffic signals should be considered for an intersection. Specifically, charts and tables help determine if one or more of the conditions (known as *warrants*) that might justify traffic signals is present. The traffic signal warrants have become progressively more quantitative with each MUTCD edition, and it is no longer possible to perform the analysis without consulting the MUTCD charts and tables.

Identifying a warrant is necessary but not sufficient to require signalization. Conditions that might exclude signalization even when a warrant is satisfied include an overall decrease in safety or effectiveness of the intersection and disruption of progressive traffic flow.

An engineering study of traffic volumes and destinations (i.e., turns) is required in order to use the MUTCD warrants. Counts of vehicles by type (heavy trucks,

passenger cars and light trucks, public-transit vehicles, and in some locations, bicycles) and of pedestrians are required. Also needed is knowledge of speed limits, accident history, local needs of the elderly and disabled, and geometric design.

In some situations, traffic control signals (as compared to stop sign control) might undesirably increase vehicular delay and change the frequency of certain types of crashes. Alternatives to traffic control signals include installing signs along the major street to warn road users of an approaching stop-sign controlled intersection, increasing the intersection sight distances by relocating stop lines and making other geometric changes, installing measures to reduce speeds on the approaches, installing flashing warning beacons, installing or increasing roadway lighting to improve nighttime performance, restricting turning movements at particular times or throughout the day, and reducing the number of vehicles per approach lane by adding lanes on minor approach street.

Intersection delays might also be reduced by increasing the capacity of the roadway in one or more directions. Widening a minor approach road will decrease that road's required green time. Such widening can easily be achieved by eliminating parking on that road in the vicinity of the intersection. In all cases of proposed roadway widening, a comparison should be made of the increased pedestrian crossing time to the green time saved due to improved vehicular flow.

- *Warrant 1, Eight-Hour Vehicular Volume:* The MUTCD specifies two alternative conditions, A and B, that if satisfied will result in Warrant 1. MUTCD Table 4C-1 establishes minimum vehicular volumes in both directions, justifying signalization based on pure volume (condition A) and based on potential for interruption of continuous traffic (condition B). Satisfaction of either condition for any eight hours during a standard day is sufficient for Warrant 1. A combination of condition A and condition B uses 80% of the base warrant volumes. When the major street speed exceeds 40 mph (70 km/h), or in isolated communities with a population of less than 10,000, volumes are 70% of the base warrant volumes for conditions A or B, or 56% of the base volume for a combination of A and B.

- *Warrant 2, Four-Hour (Average Hourly) Vehicular Volume:* The MUTCD provides two figures (Figs. 4C-1 and 4C-2) for determining whether the traffic approaching the intersection from one direction is the principal reason for evaluating signalization. MUTCD Fig. 4C-1 applies generally; Fig. 4C-2 applies to intersections with 85th percentile approach speeds in excess of 40 mph (70 kph) or in isolated communities with populations of less than 10,000. Satisfaction of either condition for any four hours during an average day is sufficient for Warrant 2.

- *Warrant 3, Peak Hour:* This warrant is designed to identify intersections whose minor-street traffic experiences undue delays for a minimum of one hour per day. It is expected that this warrant will be applied in cases where high-occupancy facilities (e.g., office and industrial complexes) attract or discharge large numbers of vehicles over a short period of time. Both graphical (MUTCD Figs. 4C-3 and 4C-4) and comparative minimum cut-offs for vehicle volume are given. Satisfaction of either condition for any single hour during an average day is sufficient for Warrant 3.

- *Warrant 4, Pedestrian Volume:* This warrant is designed to provide relief for pedestrians crossing a high-volume main street at intersections and in the middle of the block. In order to be applicable, the nearest existing signal must be at least 300 ft (90 m) away. Volume requirements include 100 pedestrians during each of any four hours or 190 pedestrians during any single hour during an average day and fewer than 60 gaps in the traffic stream per hour that would be large enough to allow pedestrians to safely cross.

- *Warrant 5, School Crossing:* A large number of student pedestrians crossing a major street may justify signalization. The warrant requires a minimum of 20 students during the highest crossing hour and the determination that there are fewer adequate crossing gaps in the traffic stream than there are minutes in the period when children are crossing. There cannot be another traffic control signal closer than 300 ft (90 m). All other methodologies for providing crossing opportunities should be considered before applying Warrant 5.

- *Warrant 6, Coordinated Signal System:* Signals can be justified if they will induce desirable platooning and progressive movement of vehicles. This warrant can be applied to streets on which the traffic is predominantly one-way or two-way, as long as resulting signals are at least 1000 ft (300 m) apart, and the existing signals are so far apart that vehicular platooning is not achieved.

- *Warrant 7, Crash Experience:* This warrant is based on crash frequency. Three conditions are specified, all of which must be met: (A) Alternative remediation and enforcement efforts have been unsuccessful in reducing crash frequency. (B) Five or more crashes involving reportable personal injury or property damage and that would be been prevented by signalization have occurred within a 12-month period. (C) The intersection experiences a minimum level of traffic (see the MUTCD) during each of any 8 hours of an average day.

- *Warrant 8, Roadway Network:* Signalization may be considered to encourage concentration and organization of traffic flow on a roadway network.

Routes that can be considered for this warrant must be part of a current or future highway system that serves as the principle roadway network for through-traffic flow or includes highways entering or serving a city. This warrant requires either an intersection volume of 1000 vehicles per hour during the peak hour of a typical weekday and is expected to meet Warrants 1, 2, and 3 during an average weekday in the future, or the intersection has or soon will have a volume of at least 1000 vehicles per hour during each of any five hours of the weekend (e.g., Saturday or Sunday).

17. FIXED-TIME CYCLES

Fixed-time controllers are the least expensive and simplest to use. They are most efficient only where traffic can be accurately predicted. Fixed-time controllers are necessary if sequential intersections or intersections spaced less than 1/2 mi (0.8 km) apart are to be coordinated.

In general, the fixed signal cycle length should be between 35 and 120 sec. Green cycle lengths with fixed-time controllers should be chosen to clear all waiting traffic in 95% of the cycles. Usually, the 85th percentile speed is used in preliminary studies. Since the green cycle must handle peak loads, level of service is sacrificed during the rest of the day, unless the cycle length is changed during the day with *multi-dial controllers.*

Amber time is usually 3 to 6 sec. Six sec may be used for higher speed roadways. A short all-red clearance interval may be provided after the amber signal to clear the intersection. It is also necessary to check that pedestrians can cross the intersection in the available walk time. (See Sec. 21.) A short all-red clearance interval is suggested after the green walk signal terminates.

Determining cycle lengths of signalized intersections is covered in the HCM, Ch. 16. Queuing models and simulation can also be used to determine or check cycle lengths in complex situations.

18. TIME-SPACE DIAGRAMS

To minimize the frustration of drivers who might otherwise have to stop at every traffic signal encountered while traveling in a corridor, it is desirable to coordinate adjacent fixed-time signals. With *alternate mode operation,* every other signal will be green at the same time. By the time a vehicle moving at a specific speed has traveled the distance between two adjoining signals, the signal being approached will turn green. For *double-alternate mode operation,* two adjacent pairs of signals will have the same color (i.e., red or green), while the following two adjacent signals will have the opposite color. Alternate mode is preferred. Unless all of the signals are equidistant, it is unlikely that all inconvenience will be eliminated from the traffic stream.

Transportation

Signal coordination is essentially achieved by setting the controller's *offset*. *Offset* is the time from the end of one controller's green cycle to the end of the next controller's green cycle. The following graphical procedure can be used to establish the offset by drawing *time-space diagrams* (*space-time diagrams*). The horizontal axis represents distance (typical scales are 1 in:100 ft or 1 in:200 ft), and the vertical axis represents time (usually in seconds).

step 1: Draw the main and intersecting streets to scale along the horizontal scale. When signals are separated by short distances, the cross-street widths should be drawn accurately.

step 2: Assume or obtain the actual average travel speed along the main street. (This is rarely the posted speed limit.) Starting at the lower left corner, draw a diagonal line representing the average speed.

step 3: Make an initial assumption for the cycle length. For a two-way street, the cycle length should be either two times (alternate mode) or four times (double-alternate mode) the travel time at the average speed between intersections of average separation. The *effective green time* includes that portion (e.g., half) of the amber time that vehicles will continue to move through the intersection. Similarly, the *effective red time* includes the remainder of the amber time. The general timing guidelines of 35 sec minimum and 120 sec maximum apply.

Figure 73.6 *Time-Space Diagram Construction*

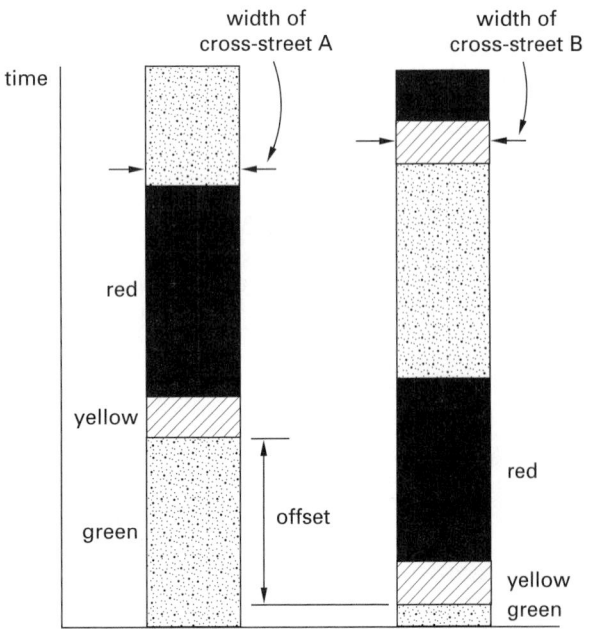

distance along Main Street

Figure 73.7 *Time-Space Diagram Bandwidth*

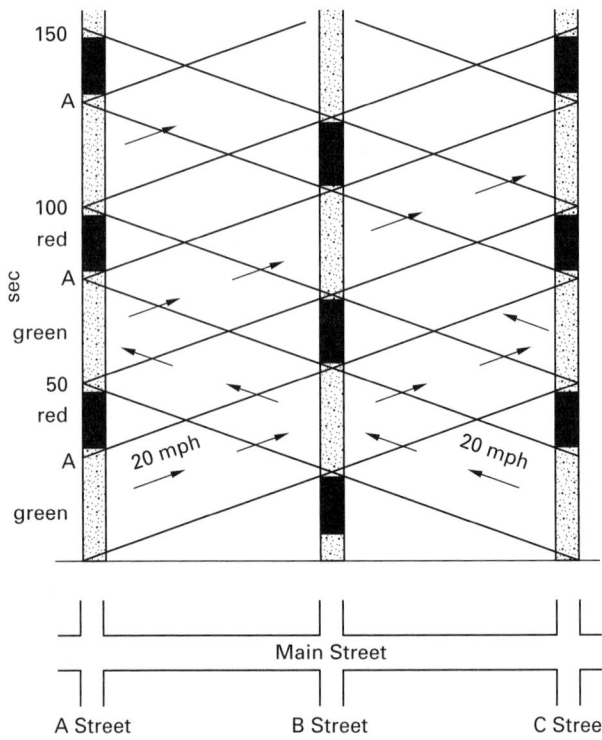

step 4: Minimize the conflict between effective green and effective red periods at adjacent intersections. This can be done heuristically by cutting strips of paper for each intersection and marking off green and red periods according to the time scale. The strips are placed on the time-space diagram with offsets determined by the average speed and average intersection separation.

step 5: Check to see if the assumed cycle can accommodate the heaviest traveled intersection.

Figure 73.7 shows a time-space diagram with diagonal lines drawn between green cycle limits. (The two sets of diagonal lines represent the traffic flows in the two opposing directions.) The vertical distance between the diagonal lines indicates the green "window" of travel, known as the *bandwidth*. The *platoon length* is the horizontal distance between diagonal lines. The signal offset is the time difference (measured in the vertical direction) between the start of successive signals' green periods.

19. TRAFFIC-ACTIVATED TIMING

Traffic-activated controllers (or *activated controllers*) vary the green time periods in relation to the approach volume of traffic. Fully activated controllers manage all approaches based on traffic volume. Semi-activated controllers use a traffic-activated signal for at least one approach and a fixed-time signal for at least one of the other approaches. Both types are initially more expensive than fixed-time controllers. However, activated controllers do a better job of controlling flow, and they are better accepted by drivers.

Since cycle changes with activated controllers adjust to the arriving traffic volume, less detailed traffic counts are needed. Four parameters must be specified to completely define the timing sequence: initial period, vehicle time period, maximum time period, and clearance allowance.

- *Initial period:* The initial period must allow enough time for traffic stopped between the stop line and the detector to begin moving. The number of cars between the stop line and the detector is calculated from Eq. 73.28.

$$\text{no. of cars in initial period} = \frac{\text{distance between line and detector}}{\text{car length}} \quad \textit{73.28}$$

The first car can be assumed to cross the stop line approximately 5 sec after the green signal appears. The next car will require 3 sec more. Subsequent cars between the detector and stop line require $2\frac{1}{4}$ sec. Studies have shown the average start-up lot time and average arrival headway to be approximately half of these values. However, these values accommodate the slowest of drivers.

- *Vehicle period:* The vehicle period must be long enough to allow a car crossing the detector (moving at the slowest reasonable speed) to get to the intersection before the amber signal appears. It is not necessary to have the vehicle get entirely through the intersection during the green, since the amber period will provide additional time. In a 30 mph (50 km/h) zone, a speed of approximately 20 mph (30 km/h) into the intersection is reasonable.
- *Maximum period:* The maximum period is the maximum delay that the opposing traffic can tolerate. For a main street 60 sec is typical, and 30 to 40 sec is appropriate for a side street. The period should never exceed 120 sec.
- *Amber period:* The amber clearance period can be determined from the time required to perceive the light, brake, and stop the vehicle, plus the assumed average speed into the intersection. Amber periods of 3 to 6 sec are typical.
- *Green period:* The green period is the smaller of the sum of initial and vehicle periods and the maximum period.

20. WALKWAYS

(Capacity analysis of pedestrians in walkways and crosswalks is covered in the HCM, Ch. 18.)

The HCM recommends using an average walking speed of 4.0 ft/sec (1.22 m/s) and a 3 sec starting delay if elderly users constitute less than 20% of the walkway

users. However, some people can walk as fast as 6 ft/sec (1.8 m/s), and 30 to 40% walk slower than 4 ft/sec (1.2 m/s). If more than 20% of walkway users are elderly, it may be appropriate to use a design speed of 3.3 ft/sec (1.0 m/s) and a 4 to 5 sec starting delay. Physically challenged individuals require additional consideration. An upgrade of 10% or greater reduces walking speed by 0.3 ft/sec (0.1 m/s).

The level of service (LOS) for pedestrians in walkways, sidewalks, and queuing areas is categorized in much the same way as for freeway and highway vehicles. Table 73.11 relates important parameters to the level of service. The primary criterion for determining pedestrian level of service is *space* (the inverse of density) per pedestrian. This, in turn, affects the speed at which pedestrians can walk. Mean speed and flow rate are supplementary criteria.

The *peak pedestrian flow rate*, $v_{p,15}$, is the number of pedestrians passing a particular point during the 15 min peak period (ped/15 min). The *effective walkway width*, W_E, is determined by subtracting any unusable width, including perceived "buffer zones" from adjacent features, from the total walkway width (ft or m). (Refer to HCM Exh. 18-1.) The average *pedestrian unit flow rate* (also known as *unit width flow*), v, is the number of pedestrians passing a particular point per unit width of walkway (ped/min-ft).

$$v = \frac{v_{p,15}}{15W_E} \quad \textit{73.29}$$

Pedestrian speed, S, is the average pedestrian walking speed (ft/sec). *Pedestrian density* is the average number of pedestrians per unit area (ped/ft^2 or ped/m^2). The reciprocal of pedestrian density is *pedestrian space*, M (ft^2/ped or m^2/ped).

$$v = SD = \frac{S}{M} \quad \textit{73.30}$$

Maximum pedestrian *capacity* in walkways is 23 ped/min-ft, pedestrians per minute per foot of walkway width (75 ped/min-m). This occurs when the space is approximately 5 to 9 ft^2/ped (0.47 to 0.84 m^2/ped). Capacity drops significantly as space per pedestrian decreases, and movement effectively stops when space is reduced to 2 to 4 ft^2/ped (0.19 to 0.37 m^2/ped).

The HCM reports that impeded flow starts at 530 ft^2/ped (49 m^2/ped), which is equivalent to 0.5 ped/min-ft (1.6 ped/min·m). These values are taken as the limits for LOS A. Also reported is that jammed flow in platoons starts at 11 ft^2/ped (1 m^2/ped), corresponding to 18 ped/min-ft (59 ped/min·m), which are used as the thresholds for LOS F. A *platoon* is a group of pedestrians walking together in a group.

21. CROSSWALKS

Analysis of pedestrians in *crosswalks* is slightly different than in pure walkways. Walking is affected by signaling, turning vehicles, platooning, and interception of the platoon of pedestrians coming from the opposite side.

Table 73.11 *Pedestrian Level of Service on Walkways and Sidewalks*

LOS	pedestrian space in ft²/ped	average speed in ft/sec	flow rate in ped/min-ft[a]	volume-capacity (v/c) ratio
A	> 60	> 4.25	≤ 5	≤ 0.21
B	> 40–60	> 4.17–4.25	> 5–7	> 0.21–0.31
C	> 24–40	> 4.00–4.17	> 7–10	> 0.31–0.44
D	> 15–24	> 3.75–4.0	> 10–15	> 0.44–0.65
E	> 8–15	> 2.50–3.75	> 15–23	> 0.65–1.0
F	≤ 8	≤ 2.50	variable	variable

(Multiply ft²/ped by 0.0929 to obtain m²/ped.)
(Multiply ft/sec by 0.3048 to obtain m/s.)
(Multiply ped/min-ft by 3.28 to obtain ped/min·m.)
[a]pedestrians per minute per foot width of walkway

Reproduced with permission by the Transportation Research Board. In *Highway Capacity Manual 2000*, Exhibit 18-3, copyright © 2000 by the Transportation Research Board, National Research Council, Washington, D.C.

22. PARKING

The minimum parallel street parking *stall width* is commonly taken as 7 ft (2.1 m) in a residential area and 8 to 11 ft (2.4 to 3.3 m) in a commercial or industrial area. This width accommodates the vehicle and its separation from the curb. Stalls 7 ft (2.1 m) wide are substandard and should be limited to residential areas and attendant-parked lots. Widths larger than 9 ft (2.7 m) are appropriate in shopping areas where package-loading is expected. If the width is specified as 10 to 12 ft (3.0 to 3.7 m), the street parking corridor can be used for delivery trucks or subsequently converted to an extra traffic lane or bicycle path. The minimum length of a parallel street parking stall is 18 ft (5.4 m). In order to accommodate most cars, longer lengths, 20 to 26 ft (6.0 to 7.8 m), may be used.

Figure 73.8 illustrates parallel street parking near an intersection as recommended by the Green Book. The 20 to 28 ft (6 to 8.4 m) clearance from the last stall to the intersection is required to prevent vehicles from using the parking lane for right-turn movements.

Figure 73.8 *Green Book Parallel Parking Design*

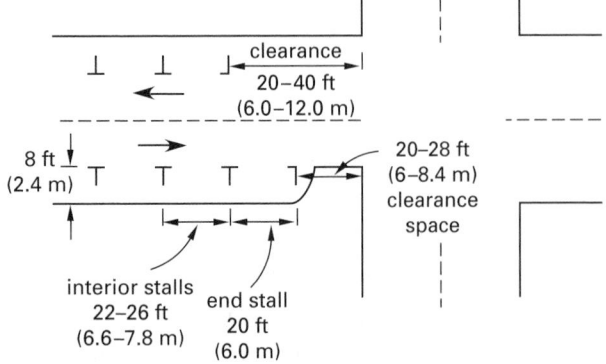

Adapted from Green Book (2004), Exh. 4-31.

Tandem parking (also known as *travers tandem parking* and *double-alternate parking*) provides two shortened stalls placed end to end with a single maneuver zone available for each of the two stalls. A parking vehicle pulls into the stall space and maneuver zone. The vehicle is removed from the traffic stream in about 4 sec. (With the traditional parallel parking procedure, traffic may halt for as much as 30 sec or more while a vehicle attempts to back into the parking space.) The original (1970) tandem design required 56 ft (17.1 m): two 20 ft (6.1 m) stalls and a 16 ft (4.9 m) maneuver zone. With modern smaller cars, the required space can be reduced to 45 to 50 ft (13.7 to 15.3 m): two 18 ft (5.5 m) stalls and a 9 to 14 ft (2.7 to 4.3 m) maneuver zone. Other dimensions may work. Typical tandem parking geometry is given in Fig. 73.9.

Diagonal parking (*angle parking*) street parking can be specified with angles to the curb of 45°, 60°, 75°, and 90°. The effects of diagonal parking on lane width can be determined from trigonometry. The significant disadvantage of impaired vision when backing up should be considered when designing angle parking.

Figure 73.9 *Typical Tandem Curb Parking Geometry*

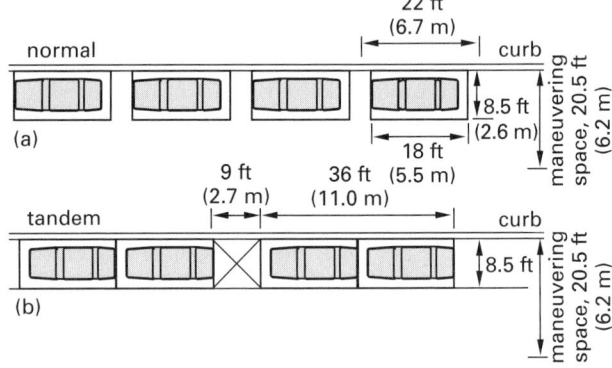

In designing *parking lots*, the maximum capacity of the lot can be calculated by dividing the gross lot area by the minimum area per car (e.g., 280 to 320 ft² (26 to 30 m²) depending on the types of cars in the community). Figure 73.10 and accompanying Table 73.12 show a typical module with two 90° spaces and an aisle between that can be used as a template for quick layout of spaces. Even though most textbooks show the need for a 26 ft aisle when using 9 ft wide by 18.5 ft long stalls set at 90°, a slightly smaller 60 ft module using 18 ft long stalls and a 24 ft wide two-way aisle has proven more than adequate for general traffic in much of the United States. The practical limits needed for door opening space between cars and driver or passenger access to vehicles suggest a stall width of no less than 8.5 ft (2.6 m), unless vehicles are segregated by general size. Widths in excess of 9 ft (2.7 m) and lengths in excess of 18.5 ft (5.6 m) are rarely necessary. For selected spaces with restricted access, such as the end space against a wall or the last space at the end of a blind aisle, additional width should be added. In many jurisdictions, local zoning requirements

Table 73.12 Parking Lot Dimensions (9 × 18.5 ft stall)

dimension (ft)	on diagram[a]	angle			
		45°	60°	75°	90°
stall width, parallel to aisle	A	12.7	10.4	9.3	9.0
stall length of line	B	25.0	22.0	20.0	18.5
stall depth to wall	C	17.5	19.0	19.5	18.5
aisle width between stall lines	D	12.0	16.0	23.0	26.0
stall depth, interlock	E	15.3	17.5	18.8	18.5
module, wall to interlock	F	44.8	52.5	61.3	63.0
module, interlocking	G	42.6	51.0	61.0	63.0
module, interlock to curb face	H	42.8	50.2	58.8	60.5
bumper overhang (typical)	I	2.0	2.3	2.5	2.5
offset	J	6.3	2.7	0.5	0.0
setback	K	11.0	8.3	5.0	0.0
cross aisle, one-way	L	14.0	14.0	14.0	14.0
cross aisle, two-way	–	24.0	24.0	24.0	24.0

(Multiply ft by 0.3048 to obtain m.) [a]See Fig. 73.10.

Figure 73.10 Parking Lot Layout

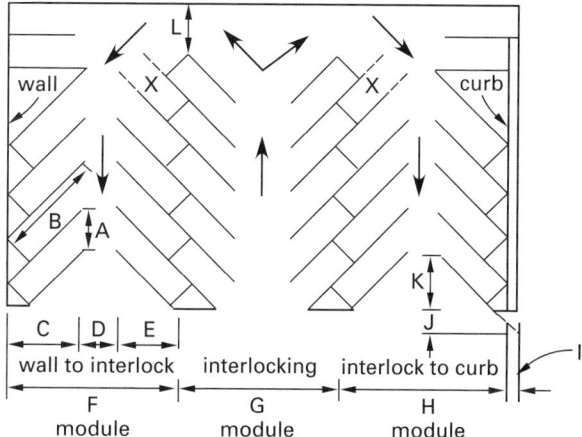

X = stall not accessible in certain layouts

set the minimum stall size and other parameters (such as fire access lanes) that may exceed standard published references.

Space efficiency and access convenience are optimized in parking lots having a minimum of odd-shaped boundaries and that are more or less rectangular in shape. Some handbooks suggest 60° stall angles, one-way aisles, and stall overlaps to obtain space efficiency. These types of layout can show good parking densities but sometimes result in odd space configurations and many unusable spaces, leading to losses in efficiency. For instance, one-way aisles are susceptible to blockage due to careless parking or maneuvering delays. Layout and line stripe maintenance of skew angles with overlapping spaces are more difficult, especially where detail adjustments are necessary to accommodate lot perimeter irregularities. Diagonal layouts quite often have odd-shaped leftover spaces here or there that become subject to improper parking when the lot is crowded.

More recent experience has shown that 90° stalls and two-way aisles provide high customer acceptance for random-arrival situations. Breaking up long aisles with a cross passage every 20 spaces or so allows users to move freely among the aisles while hunting for an available space and provides additional exit paths should blockage occur. Shorter aisles placed 90° from the access lanes allow vehicles to quickly enter and leave available spaces, thereby improving the level of service of the lot. Many cars today are equipped with power steering and have more compact outlines, making them easily maneuverable into smaller spaces. This can reduce driver frustration and the potential for vehicle damage. As with any design question, several layout arrangements should be considered before the final one is selected.

ADA (Americans with Disabilities Act) *accessible parking spaces* (i.e., "handicapped parking") are required in all parking lots for visitors, customers, and employees. The total number of accessible spaces is given in Table 73.13. The number of designated van-accessible spaces is obtained by dividing the total number of accessible spaces by eight and rounding up, with no less than one van-accessible space. All remaining spaces can be car accessible.

Table 73.13 Minimum Number of Accessible Parking Spaces[a]

total number of parking spaces in lot	total minimum number of car- and van-accessible[b] parking spaces
1–25	1
26–50	2
51–75	3
76–100	4
101–150	5
151–200	6
201–300	7
301–400	8
401–500	9
501–1000	2% of total
> 1000	20 plus 1 for each 100 over 1000

[a]Subject to changes in ongoing legislation
[b]One-eighth of total minimum accessible spaces must be van accessible.
Source: ADA Standards for Accessible Design 4.1.2(5)

Accessible spaces may be dispersed among multiple lots with accessible entrances. Spaces should be located closest to accessible building entrances. An accessible curb cut is required from the parking lot to the general walkway. The accessible route must be at least 3 ft (0.9 m) wide. The slope along the accessible route should not be greater than 1:12 in the direction of travel.

Car-accessible spaces should be 8 ft (2.4 m) wide, with a 5 ft (1.5 m) accessible aisle alongside. Two adjacent accessible spaces can share one aisle. Van-accessible spaces should be 8 ft (2.4 m) wide, with an 8 ft (2.4 m) access aisle. Vans with lifts generally exit on the passenger side.) Van-accessible spaces also require a 98 in (2490 mm) minimum height clearance. All surfaces must be stable and slip-resistant, and have a maximum

Transportation

cross slope of 2%. Signage with the accessibility symbol must be provided. Special van-accessible signage is required for appropriate spaces.

Other facts that must be considered when designing a parking lot include the location of driveways, lighting, landscaping, sidewalks, and entry/exit systems (i.e., gates, doors, ticketing, etc.). Only after these factors have been specified can the layout be designed to maximize the number of parking places.

23. HIGHWAY INTERCHANGES

Highway interchanges allow traffic to enter or leave a highway. Interchange locations are affected by the volume of traffic expected on the interchange, convenience, and required land area. The frequency of interchanges along a route should be sufficient to allow weaving between the interchanging traffic. Factors affecting the type of interchange chosen include cost, available land, total flow volume, volume of left turns, and volume of weaving movement. Except in extreme cases, several different designs may be capable of satisfying the design requirements.

Diamond interchanges are suitable for major-road–minor-road intersections. Diamond interchanges can handle fairly large volume roadways and can accommodate some left turns at grade. Left turns must be made directly on the minor highway. They lend themselves to staged construction. The frontage roads and/or ramps can be constructed, leaving the freeway lanes to be built at a later date. The right-of-way costs are low, since little additional area around the freeway is required.

Diamond interchanges force traffic using a ramp to substantially reduce its speed. Since the capacity per ramp is limited to approximately 1000 vph, diamond interchanges cannot be used for freeway-to-highway intersections.

Since diamond interchanges (except for the single-point interchange) place two signalized intersections in close proximity, it is possible for *demand starvation* to occur. Demand starvation occurs when a portion of the green signal of the downstream intersection is unused due to delays or blockage at the upstream intersection. Demand starvation can be caused by suboptimal signal

Figure 73.11 Interchange Types

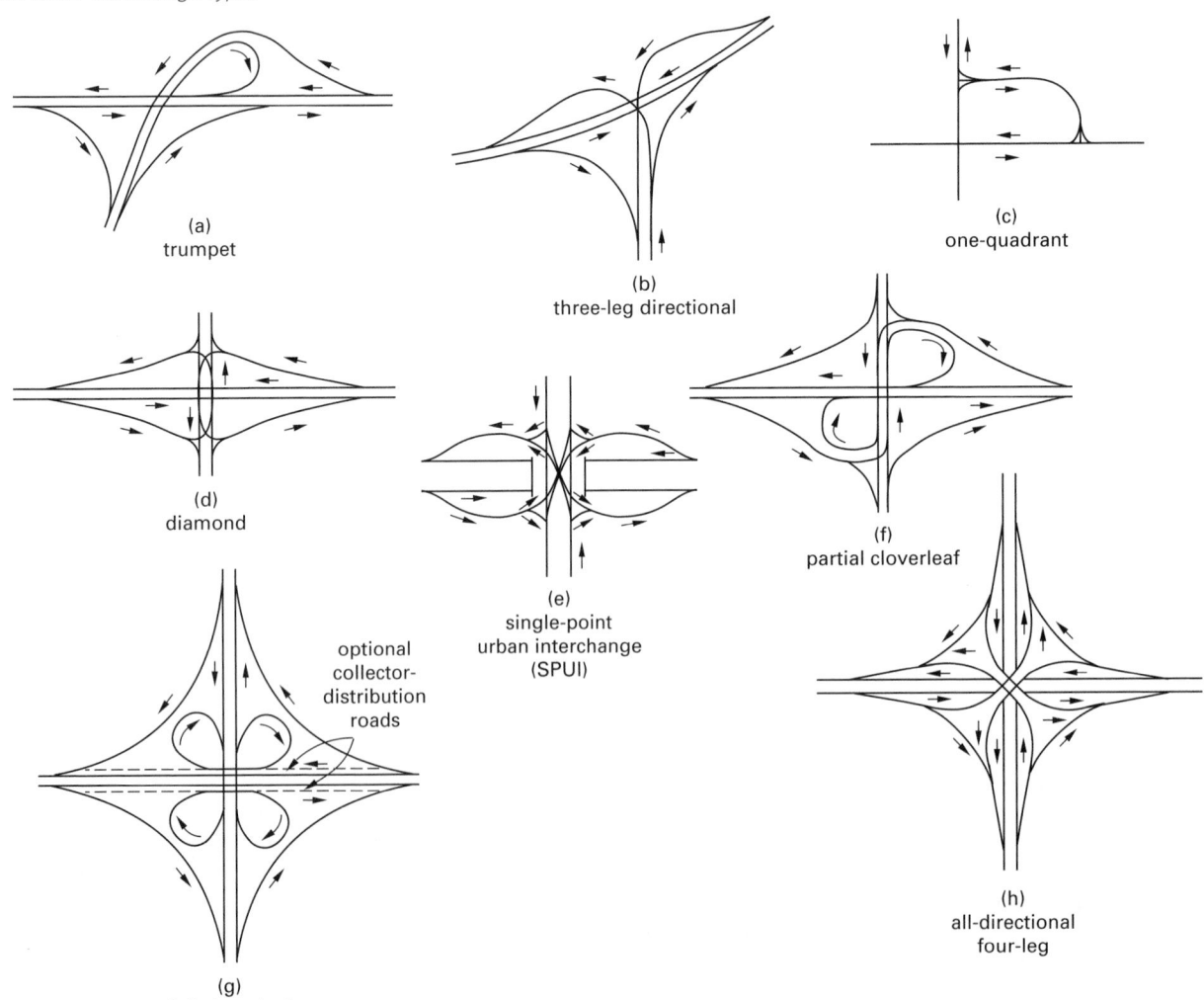

From *A Policy on Geometric Design of Highways and Streets*, Exh. 10-1, copyright © 2004, by the American Association of State Highway and Transportation Officials, Washington, D.C. Used by permission.

Figure 73.12 *Adaptability of Interchanges on Freeways as Related to Types of Intersecting Facilities*

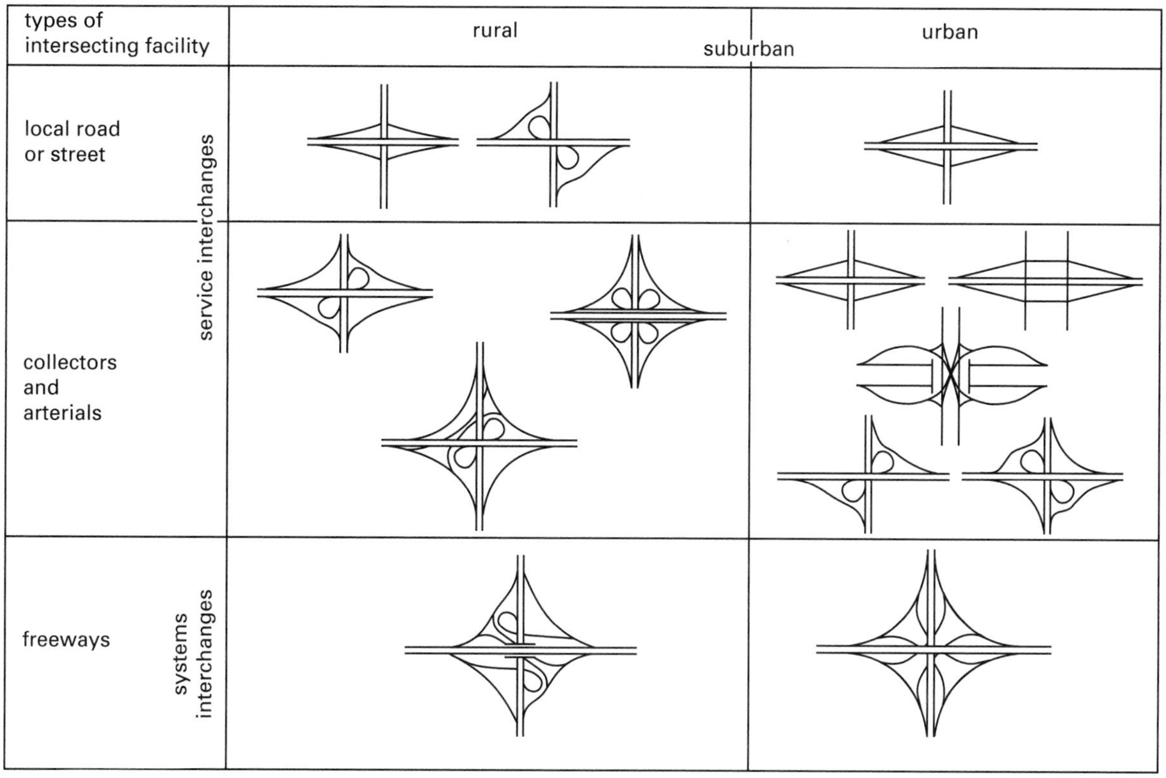

From *A Policy on Geometric Design of Highways and Streets*, Exh. 10-43, copyright © 2004, by the American Association of State Highway and Transportation Officials, Washington, D.C. Used by permission.

coordination, as well as by queues from the downstream intersection effectively blocking departures from the upstream intersection.

Cloverleaf interchanges incorporate loop connections and allow nonstop left turn movement. *Partial cloverleafs* (known as *parclos*) provide for nonstop left turns along selected routes only, whereas *full cloverleafs* provide nonstop turning for all four traffic directions. (A *loop* or *loop ramp* is a 270° turn in the direction opposite the final direction of travel.) Cloverleafs provide for free flow by separating the traffic in both directions. However, they require large rights-of-way. Cloverleaf intersections slow traffic from the design speed and require short weaving distances. Turning traffic follows a circuitous route. The practical limit of capacity of loop ramps is approximately 800 to 1200 vph. Cloverleaf interchanges should not be used to connect two freeways with large volumes of turning traffic.

Both diamond and cloverleaf intersections can be improved by various means, including by the use of a third level and collector-distributor roads to increase the speed and volume of weaving sections. *Directional interchanges* (*all-directional interchanges*) allow direct or semidirect connections for left turn movements. The design speeds of connections normally are near the design speed of the through lanes, and large traffic volumes can be handled without significant weaving. However, large rights-of-way are required. Directional

interchanges are expensive because of the structures provided. Directional interchanges may contain three or four layers or be of *rotary bridge* design.

When there are only three approach legs, T-, Y-, or *trumpet interchanges* can be used.

The *single-point urban interchange* (SPUI), also known as the *urban interchange* and *single-point diamond interchange*, is a type of diamond interchange with a single signalized intersection controlling all left turns. All right turns are free flow, which increases the capacity above that of traditional diamond interchanges.

The main advantage of SPUIs is that they require only a narrow right-of-way, thereby reducing land acquisition cost. The main disadvantage, as with all traffic bridges, is a high construction cost. There are additional geometric design features that require careful consideration, such as the elliptical left-hand turning path, pedestrian accommodations, and the difficulty in accommodating freeways approaching with high skew angles (e.g., more than 30°).

24. WEAVING AREAS

(Analysis and design procedures for weaving areas are covered in the HCM, Ch. 24.)

Weaving is the crossing of at least two traffic streams traveling in the same general direction along a length of highway without traffic control. Weaving is an issue that must be considered in interchange selection, and interchanges without weaving are favored over interchanges with weaving. Weaving areas require increased lane-change maneuvers and result in increased traffic turbulence. The length of weaving area is measured from the *merge gore area* at a point where the right edge of the freeway shoulder lane and the left edge of the merging lane(s) are 2 ft (0.6 m) apart to a point at the *diverge gore area* where the two edges are 12 ft (3.7 m) apart.

Weaving configuration refers to the relative placement and number of entry lanes and exit lanes for a roadway section. The configuration can have a major impact on how much lane changing is required. Configuration is based on the number of required lane changes that must be performed by the two weaving flows in the section. The HCM illustrates three configuration types, and criteria for each are listed in Table 73.14. Figures 73.13, 73.14, and 73.15 illustrate type-A, B, and C weaving areas.

Table 73.14 Configuration Type Versus Minimum Number of Required Lane Changes

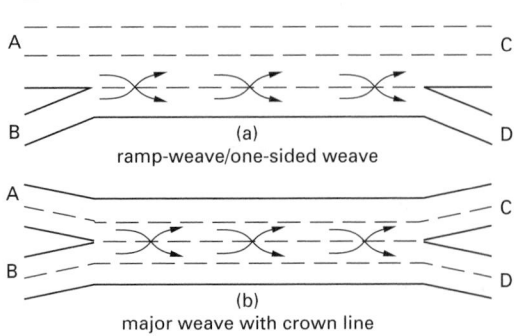

minimum number of required lane changes for merging into mainline traffic (movement a)	minimum number of required lane changes for weaving out of mainline traffic (movement b)		
	0	1	≥ 2
0	type-B	type-B	type-C
1	type-B	type-A	–
≥ 2	type-C	–	–

Reproduced with permission by the Transportation Research Board. In *Highway Capacity Manual 2000*, Exhibit 24-5, copyright © 2000 by the Transportation Research Board, National Research Council, Washington, D.C.

Figure 73.13 Type-A Weaving Areas

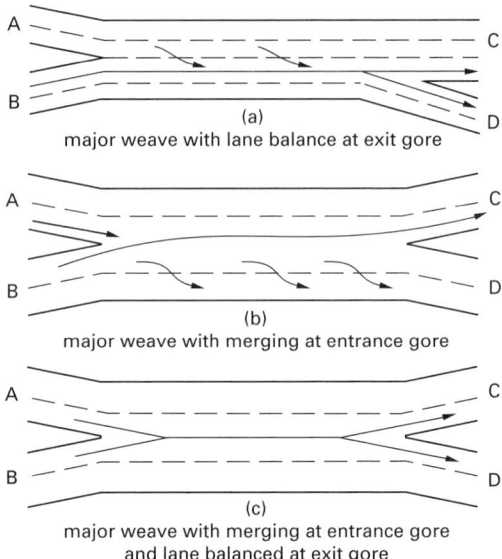

(a)
ramp-weave/one-sided weave

(b)
major weave with crown line

Reproduced with permission by the Transportation Research Board. In *Highway Capacity Manual 2000*, Exhibit 13-8, copyright © 2000 by the Transportation Research Board, National Research Council, Washington, D.C.

Figure 73.14 Type-B Weaving Areas

(a)
major weave with lane balance at exit gore

(b)
major weave with merging at entrance gore

(c)
major weave with merging at entrance gore and lane balanced at exit gore

Reproduced with permission by the Transportation Research Board. In *Highway Capacity Manual 2000*, Exhibit 13-9, copyright © 2000 by the Transportation Research Board, National Research Council, Washington, D.C.

Figure 73.15 Type-C Weaving Areas

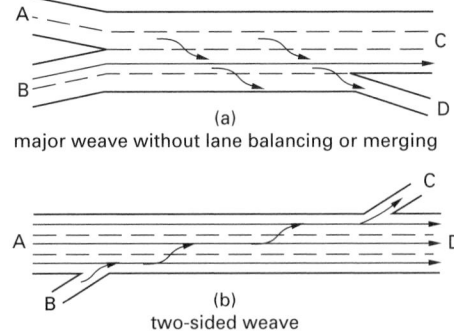

(a)
major weave without lane balancing or merging

(b)
two-sided weave

Reproduced with permission by the Transportation Research Board. In *Highway Capacity Manual 2000*, Exhibit 13-10, copyright © 2000 by the Transportation Research Board, National Research Council, Washington, D.C.

25. TRAFFIC CALMING

Although traditional designs based on the HCM and the Green Book have intended to maximize the flow of traffic, *traffic calming* features are introduced to slow the flow. Such features include traffic circles, narrow streets, curb extensions, and textured crosswalks. These features, which generally are contrary to traditional design guidelines in the Green Book, are in response to a new urbanism that encourages consideration of historical, community, and aesthetic factors.

26. ECONOMIC EVALUATION

Road user costs include fuel, tires, oil, repairs, time, and accidents. Such costs affect design decisions where there are delays due to congestion, stops, turning, and so on. A reduction in road-user costs can be used to economically justify (i.e., as an economic warrant) interchanges and other features. Reductions in road-user costs are generally far greater than the increased cost of travel times that interchanges cause (compared to at-grade intersections).

Generally, public projects (such as interchanges) are justified using a benefit-cost ratio of annual benefits to annual capital costs. *Annual benefits* is the difference in road-user costs between the existing and improved conditions. *Annual capital costs* is the sum of interest and the amortization of the cost of the improvement. When staged construction is anticipated, incremental costs must be used to justify the construction of future subsequent stages.

Much can be done to improve the safety of some roadway sections. Features such as breakaway poles, cushioned barriers, barriers separating two directions of traffic, and direction channeling away from abutments are common. These features must be economically justified, particularly where they are to be retrofitted to existing highways. The justification is usually the value of personal injury and property damage avoided by the installation of such features.

There are three general classifications of accidents: those with property damage only, those with injury combined with property damage, and those involving death with property damage. The cost of each element (i.e., death, injury, and property damage) can be evaluated from insurance records, court awards, state disability records, and police records. Actuaries may be able to prepare estimates of remaining potential earnings. Also, federal agencies such as the U.S. National Safety Council and OSHA are active in monitoring and maintaining similar statistics.

27. CONGESTION PRICING

Congestion pricing (CP) is the charging of higher tolls (on toll roads and bridges) during peak traffic hours. CP limits the use of roadways during peak periods by providing an economic disincentive. The disincentive can be a toll (also called a *cordon charge* or *road-access charge*) on an otherwise free road. It can also be a fuel tax, or even a car purchase and ownership taxation. On toll facilities, tolls can be raised only during peak hours. Alternatively, additional lanes can be constructed to collect tolls from single-occupancy vehicles.

Congestion pricing is a politically sensitive concept. It did not meet with significant success before the 1990s. However, political and environmental issues are making it easier for decision makers to embrace the concept.

28. QUEUING MODELS

A *queue* is a waiting line. Time spent in a system (i.e., the *system time*) includes the *waiting time* (time spent waiting to be served) and the *service time*. *Queuing theory* can be used to predict the system time, W, the time spent in the queue, W_q, the average queue length, L_q, and the probability that a given number of customers will be in the queue, $p\{n\}$.

Over the years, many different queuing models have been developed to accommodate different service policies and populations. For example, some multiserver processes (e.g., tellers drawing from a line of banking customers) draw from a single queue. Other processes (e.g., bridge tolltakers) have their own queues. Most models predict performance only for steady-state operation, which means that the service facility has been open and in operation for some time. Start-up performance often must be evaluated by simulation.

The following basic relationships are valid for all queuing models predicting steady-state performance. $1/\lambda$ is the average time between arrivals, and $1/\mu$ is the average service time per server. For a system to be viable, λ must be less than μs, where s is the number of servers. The *utilization factor*, ρ, is defined as the ratio λ/μ.

$$L = \lambda W \qquad \text{73.31}$$

$$L_q = \lambda W_q \qquad \text{73.32}$$

$$W = W_q + \frac{1}{\mu} \qquad \text{73.33}$$

Most queuing models are mathematically complex and fairly specialized. However, two models are important because they adequately predict the performance of simple queuing processes drawing from typical populations. These are the M/M/1 single-server model and the M/M/s multi-server model.

29. M/M/1 SINGLE-SERVER MODEL

In the M/M/1 single-server model, a single server ($s = 1$) draws from an infinite calling population. The service times are exponentially distributed with mean μ. The specific service time distribution is defined by Eq. 73.34.

$$f(t) = \mu e^{-\mu t} \qquad \text{73.34}$$

As a consequence of using the exponential distribution, the probability of a customer's remaining service time exceeding h (after already spending time with the server) is given by Eq. 73.35. The probability is not affected by the time a customer has already spent with the server.

$$P\{t > h\} = e^{-\mu h} \qquad \text{73.35}$$

Arrival rates are described by a Poisson distribution with mean λ. The probability of x customers arriving in any period is

$$p\{x\} = \frac{e^{-\lambda}\lambda^x}{x!} \qquad \text{73.36}$$

Transportation

The following relationships describe an M/M/1 model.

$$p\{0\} = 1 - \rho \qquad\qquad 73.37$$

$$p\{n\} = p\{0\}\rho^n \qquad\qquad 73.38$$

$$W = \frac{1}{\mu - \lambda}$$

$$= W_q + \frac{1}{\mu} = \frac{L}{\lambda} \qquad\qquad 73.39$$

$$W_q = \frac{\rho}{\mu - \lambda} = \frac{L_q}{\lambda} \qquad\qquad 73.40$$

$$L = \frac{\lambda}{\mu - \lambda} = L_q + \rho \qquad\qquad 73.41$$

$$L_q = \frac{\rho\lambda}{\mu - \lambda} \qquad\qquad 73.42$$

Example 73.7

A state's truck weigh station has the ability to handle an average of 20 trucks per hour. Trucks arrive at the average rate of 12 per hour. Performance is described by an M/M/1 model. Find the steady-state values of (a) the time spent waiting to be weighed, (b) the time spent in the queue, (c) the number of trucks waiting to be weighed, (d) the number of trucks in the queue, and (e) the probability that there will be five trucks waiting to be weighed at any time.

Solution

$$\rho = \frac{\lambda}{\mu} = \frac{12 \ \frac{\text{trucks}}{\text{hr}}}{20 \ \frac{\text{trucks}}{\text{hr}}} = 0.6$$

(a) $$W = \frac{1}{\mu - \lambda} = \frac{1}{20 \ \frac{\text{trucks}}{\text{hr}} - 12 \ \frac{\text{trucks}}{\text{hr}}}$$

$$= 0.125 \ \text{hr/truck}$$

(b) $$W_q = \frac{\rho}{\mu - \lambda}$$

$$= \frac{0.6}{20 \ \frac{\text{trucks}}{\text{hr}} - 12 \ \frac{\text{trucks}}{\text{hr}}}$$

$$= 0.075 \ \text{hr/truck}$$

(c) $$L = \frac{\lambda}{\mu - \lambda}$$

$$= \frac{12 \ \frac{\text{trucks}}{\text{hr}}}{20 \ \frac{\text{trucks}}{\text{hr}} - 12 \ \frac{\text{trucks}}{\text{hr}}} = 1.5$$

(d) $$L_q = \frac{\rho\lambda}{\mu - \lambda}$$

$$= \frac{(0.6)\left(12 \ \frac{\text{trucks}}{\text{hr}}\right)}{20 \ \frac{\text{trucks}}{\text{hr}} - 12 \ \frac{\text{trucks}}{\text{hr}}} = 0.9$$

(e) $$p\{0\} = 1 - \rho = 1 - 0.6 = 0.4$$

$$p\{5\} = p\{0\}\rho^5 = (0.4)(0.6)^5 = 0.031$$

30. M/M/s MULTI-SERVER MODEL

Similar assumptions are made for the M/M/s model as for the M/M/1 model, except that there are s servers instead of 1. Each server has an average service rate of μ and draws from a single common calling line. Therefore, the first person in line goes to the first server that is available. Each server does not have its own line. (This model can be used to predict the performance of a multiple-server system where each server has its own line if customers are allowed to change lines so that they go to any available server.) The following equations describe the steady-state performance of an M/M/s system with $\rho = \lambda/\mu s$.

$$W = W_q + \frac{1}{\mu} \qquad\qquad 73.43$$

$$W_q = \frac{L_q}{\lambda} \qquad\qquad 73.44$$

$$L_q = \frac{p\{0\}\rho\left(\frac{\lambda}{\mu}\right)^s}{s!(1-\rho)^2} \qquad\qquad 73.45$$

$$L = L_q + \frac{\lambda}{\mu} \qquad\qquad 73.46$$

$$p\{0\} = \frac{1}{\dfrac{\left(\frac{\lambda}{\mu}\right)^s}{s!\left(1 - \frac{\lambda}{s\mu}\right)} + \displaystyle\sum_{j=0}^{s-1}\dfrac{\left(\frac{\lambda}{\mu}\right)^j}{j!}} \qquad\qquad 73.47$$

$$p\{n\} = \frac{p\{0\}\left(\frac{\lambda}{\mu}\right)^n}{n!} \quad [n \le s] \qquad\qquad 73.48$$

$$p\{n\} = \frac{p\{0\}\left(\frac{\lambda}{\mu}\right)^n}{s!s^{n-s}} \quad [n > s] \qquad\qquad 73.49$$

Figure 73.16 is a graphical solution to the $M/M/s$ multiple server model.

Figure 73.16 *Mean Number in System (L) for M/M/s System*

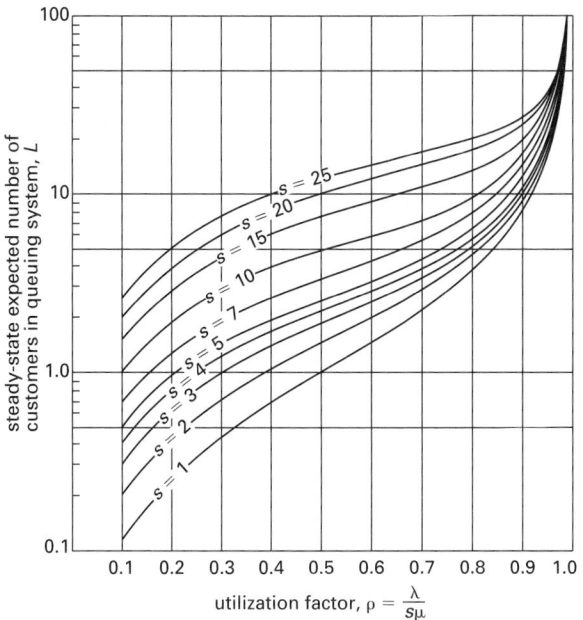

steady-state expected number of customers in queuing system, L

utilization factor, $\rho = \dfrac{\lambda}{s\mu}$

Reprinted from *Operations Research*, 6th Ed., by Frederick S. Hillier and Gerald J. Lieberman, Holden-Day, Inc., with permission of McGraw-Hill, Inc., © 1974.

Example 73.8

A company has several identical machines operating in parallel. The average breakdown rate is 0.7 machines per week. There is one repair station for the entire company. It takes a maintenance worker one entire week to repair a machine, although the time is reduced in proportion to the number of maintenance workers assigned to the repair. Each maintenance worker is paid $400 per week. Machine downtime is valued at $800 per week. Other costs (additional tools, etc.) are to be disregarded. What is the optimum number of maintenance workers at the repair station?

Solution

Assume the number of breakdowns per week is a Poisson distribution. Then, this example can be solved with queuing theory. Using two or more maintenance workers only decreases the repair time, so this is a single-server model, even with multiple maintenance workers.

The average number of machines breaking down each week is the mean arrival rate, $\lambda = 0.7$ per week. With one worker, the repair rate, μ, is 1.0 per week.

The average time a machine is out of service (waiting for its turn to be repaired and during the repair) is W, the "time in the system."

$$W = \frac{1}{\mu - \lambda} = \frac{1}{1 - 0.7}$$
$$= 3.33 \text{ wk}$$

With one maintenance worker, the average downtime cost in a week is

$$\left(\frac{\text{downtime cost}}{\text{machine-wk}} \right) (\text{no. of machines})(\text{no. of wk})$$

$$= (\text{downtime cost})\lambda W$$

$$= \left(800 \ \frac{\$}{\text{machine-wk}} \right) (0.7 \text{ machines})(3.33 \text{ wk})$$

$$= (\$800)(2.33) = \$1864$$

However, the product of λ and W is the same as the average number of machines in the system, L. The average number of machines in the system, L (i.e., being repaired or waiting for repair), is

$$L_1 = \frac{\lambda}{\mu - \lambda} = \frac{0.7}{1 - 0.7}$$
$$= 2.33$$

The total average weekly cost with one worker is the sum of the costs of the worker and the downtime.

$$C_{t,1} = (1 \text{ worker}) \left(400 \ \frac{\$}{\text{wk}} \right) + (2.33)(\$800) = \$2264$$

With two workers, the values are

$$\mu = (2 \text{ workers}) \left(\frac{1}{\text{worker-wk}} \right) = 2/\text{wk}$$

$$L_2 = \frac{0.7}{2 - 0.7} = 0.538$$

$$C_{t,2} = (2) \left(400 \ \frac{\$}{\text{wk}} \right) + (0.538)(\$800) = \$1230$$

With three workers, the values are

$$\mu = (3 \text{ workers}) \left(\frac{1}{\text{worker-wk}} \right) = 3/\text{wk}$$

$$L_3 = \frac{0.7}{3 - 0.7} = 0.304$$

$$C_{t,3} = (3) \left(400 \ \frac{\$}{\text{wk}} \right) + (0.304)(\$800) = \$1443$$

Adding workers will increase the costs above C_3. Two maintenance workers should staff the maintenance station.

Transportation

Example 73.9

Twenty identical machines are in operation. The hourly reliability for any one machine is 90%. (That is, the probability of a machine breaking down in any given hour is 10%.) The cost of downtime is $5 per hour. Each broken machine requires one technician for repair, and the average repair time is one hour. If all technicians are busy, broken machines wait idle. Each technician costs $2.5 per hour. How many separate technicians should be used?

Solution

This is a multiple-server model. It is assumed that the $M/M/s$ assumptions are satisfied. The mean arrival rate, λ, is $(0.10)(20) = 2$ per hour. The repair rate, μ, is 1 per hour. Clearly, one technician cannot handle the workload, nor can two technicians. The average number of machines in the system, L (i.e., being repaired or waiting for repair), is calculated for two, three, four, and five servers. The total cost per hour is calculated as

$$C = 2.5s + 5L$$

s	p_0	L_q	L	cost per hour
2	–	–	–	infinite
3	0.11	0.91	2.91	22.0
4	0.13	0.17	2.17	20.9
5	0.13	0.04	2.04	22.7

The minimum hourly cost is achieved with four technicians.

31. AIRPORT RUNWAY DESIGNATION

Analysis of airport capacity is based on traffic counts and determination of a "design aircraft." The Federal Aviation Administration (FAA) has prepared Federal Advisory Circulars outlining the requirements for each airplane design group and airport or runway designation.

Runways are usually designated based on the *magnetic azimuth* of the runway centerline. A designation will consist of a number and, in the case of parallel runways, a letter. (Single-digit runway designation numbers are not preceded by zeros.) On single runways, dual parallel runways, and triple parallel runways, the designation number will be the whole number nearest one-tenth of the magnetic azimuth when viewed from the direction of the approach. For example, if the magnetic azimuth was 192°, the runway designation marking would be 19. For a magnetic azimuth of 57°, the runway designation would be 6. For a magnetic azimuth ending with the number 5, such as 135°, the runway designation can be either 13 or 14. In the case of parallel runways, each runway designation number is supplemented by a letter (i.e., L for left and R for right) to indicate their relative positions as viewed from an approaching airplane (e.g., 18L or 18R).

32. DETOURS

A *detour* is a temporary rerouting of road users onto an existing highway in order to avoid a temporary traffic control zone. A *diversion* is a temporary rerouting of road users onto a temporary roadway around the work zone. If they are used, warning and taper lengths are generally not included in detour lengths.

33. TEMPORARY TRAFFIC CONTROL ZONES

Temporary traffic control zones are areas of a roadway where the normal conditions (width, speed, direction, route, etc.) are changed by police or other authorized officials, signs, or *temporary traffic control (channelization) devices* (e.g., cones, tubular markers, drums, and barricades). Temporary traffic control zones may be used in areas of construction and maintenance, or in response to an incident (accident, natural disaster, or other emergency). The control zone extends from the first notification point to the last temporary traffic control device.

Properly implemented temporary traffic control zones are composed of four sections (*areas*): (1) advance warning area, (2) transition area, (3) activity area, and (4) termination area. These areas are illustrated in Fig. 73.17.

Except in areas where the advance warning area may be eliminated because the activity area does not interfere with normal traffic, road users will learn of the traffic control zone in the *advance warning area*. The warning can be accomplished by signage, flashing light trailers, or by rotating lights and strobes on parked vehicles. Three consecutive points of warning are required. Table 73.15 summarizes the MUTCD's suggested distances between the three consecutive warning signs.

In the transition area, vehicles are redirected out of their normal paths by the use of *tapers*. A taper is created by using channelization devices and/or other pavement markings. Channelization design is specified by the distance between devices and the overall length of the transition area. The MUTCD specifies that, except for downstream and one-lane, two-way tapers, the maximum distance between channelization devices is 1 ft for every mph of speed (0.2 m for every kph). For downstream and one-lane, two-way tapers, the maximum distance between channelization devices is approximately 20 ft (6.1 m).

The minimum taper length depends on the type of transition. Table 73.16 summarizes the MUTCD's suggested *taper lengths*, L. In Eqs. 73.50 and 73.51, S is the posted speed limit, the off-peak 85th percentile speed prior to work starting, or the anticipated operating speed.

$$L_{\mathrm{m}} = \frac{W_{\mathrm{m}} S_{\mathrm{km/h}}^2}{155} \quad [S \leq 60 \text{ km/h}]$$

$$[\text{SI}] \quad 73.50(a)$$

$$L_{\text{ft}} = \frac{W_{\text{ft}}S_{\text{mph}}^2}{60} \quad [S \le 40 \text{ mph}]$$

$$[\text{U.S.}] \quad 73.50(b)$$

$$L_{\text{m}} = \frac{W_{\text{m}}S_{\text{km/h}}}{1.6} \quad [S > 70 \text{ km/h}]$$

$$[\text{SI}] \quad 73.51(a)$$

$$L_{\text{ft}} = W_{\text{ft}}S_{\text{mph}} \quad [S > 45 \text{ mph}]$$

$$[\text{U.S.}] \quad 73.51(b)$$

Figure 73.17 *Sections of a Temporary Traffic Control Zone*

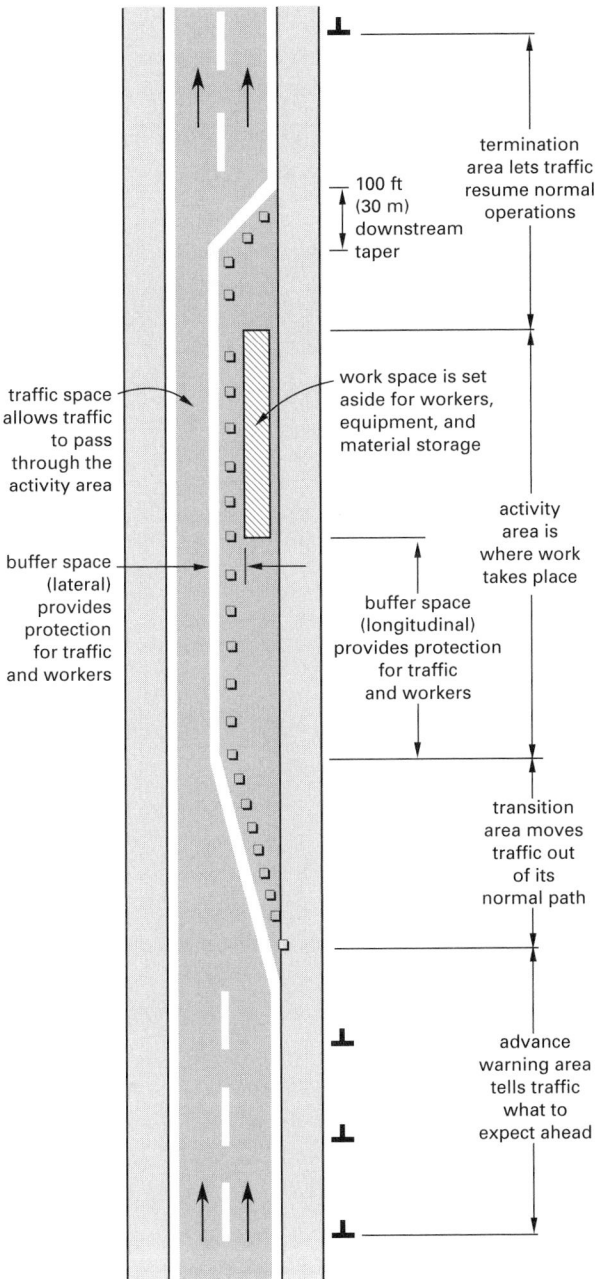

Reprinted with permission from Fig. 6C-1 of the *Manual on Uniform Traffic Control Devices*, 2003 ed., U.S. Department of Transportation, Federal Highway Administration, 2003.

Merging tapers (where two lanes merge) are longer than *nonmerging tapers* and require the full taper length since vehicles are required to merge into a common road space. When more than one lane must be closed, merged traffic should be permitted to travel approximately $2L$ before the second lane is merged. *Shifting tapers* (where traffic moves to an entirely different travel path) require a minimum length of $L/2$ but can benefit from longer lengths. *Shoulder tapers* preceding shoulder construction on an untraveled improved shoulder with a minimum width of 8 ft (2.4 m) that might be mistaken as a driving lane require a minimum length of $L/3$. When two-directional traffic on a two-way road is slowed, stopped, and alternately channeled into a single lane by a *flagger*, *two-way traffic tapers* with a maximum length of 100 ft (30 m) are used, one at each end. *Downstream tapers* are useful in providing a visual cue to drivers but are optional. When used, they have lengths of 100 ft per lane.

As shown in Fig. 73.17, the *activity area* consists of the actual work space, traffic space, and buffer space. The *work space* is the part of the highway that is closed to road users and is set aside for workers, equipment, and materials. It also includes a shadow vehicle if one is used upstream. Work spaces are separated from passing vehicles by channelization devices or temporary barriers.

The *traffic space* is the part of the highway in which road users are routed through the activity area. The *buffer space* is a lateral and/or longitudinal (with respect to the work space) area that separates road user flow from the work space or an unsafe area. It might provide some recovery space for an errant vehicle. The buffer space should not be used for storage of equipment, vehicles, or supplies. If a shadow vehicle is used, the buffer space ends at the bumper first encountered. If a lateral buffer area is provided, its width is chosen using engineering judgment.

Transportation

Table 73.15 *Suggested Advance Warning Sign Spacing*

	distance between signs		
road type	distance between the third sign encountered and the beginning of the transition	distance between the second sign encountered and the third sign	distance between the first sign and the second sign
urban (low speed)[a]	100 ft (30 m)	100 ft (30 m)	100 ft (30 m)
urban (high speed)[a]	350 ft (100 m)	350 ft (100 m)	350 ft (100 m)
rural	500 ft (150 m)	500 ft (150 m)	500 ft (150 m)
expressway/freeway	1000 ft (300 m)	1500 ft (450 m)	2640 ft (300 m)

[a]The speed category is to be determined by the highway agency.

Source: Table 6C-1 of the *Manual on Uniform Traffic Control Devices*, 2003 ed., U.S. Department of Transportation, Federal Highway Administration, 2003.

Table 73.16 *Taper Length Criteria for Temporary Traffic Control Zones*

type of taper	taper length, L
merging taper	L minimum
shifting taper	$0.5L$ minimum
shoulder taper	$0.33L$ minimum
one-lane, two-way traffic taper	100 ft (30 m) maximum
downstream taper	100 ft (30 m) minimum per lane

Source: Table 6C-3 of the *Manual on Uniform Traffic Control Devices*, 2003 ed., U.S. Department of Transportation, Federal Highway Administration, 2003.

74 Vehicle Dynamics and Accident Analysis

Nomenclature

a	acceleration	ft/sec^2	m/s^2
A	area	ft^2	m^2
ADT	average daily traffic	veh/day	veh/day
BSFC	brake-specific fuel consumption	lbm/hp-hr	kg/kW·h
C	coefficient	–	–
d	diameter	ft	m
DBP	drawbar pull	lbf	N
f	coefficient of friction	–	–
F	force	lbf	N
g	acceleration due to gravity	ft/sec^2	m/s^2
g_c	gravitational constant	ft-lbm/ lbf-sec^2	–
G	grade	decimal	decimal
i	loss fraction	–	–
K	air resistance coefficient	various	various
m	mass	lbm	kg
n	number of axles	–	–
n	rotational speed	rev/sec	rev/s
N	normal force	lbf	N
P	power	hp	kW
Q	volumetric quantity	gal	L
r	radius	ft	m
R	accident rate	various	various
R	gear ratio	–	–
R	resistance	lbf	N
s	distance	ft	m
s'	fuel economy	mi/gal	km/L
t	time	sec	s
T	torque	ft-lbf	N·m
v	velocity	ft/sec	m/s
w	average axle loading	lbf	–
w	weight	lbf	–

Symbols

η	efficiency	–	–
θ	angle	deg	deg
ρ	density	lbm/ft^3	kg/m^3
ϕ	angle	deg	deg

Subscripts

b	braking
c	curvature
D	drag
f	frictional
g	grade
i	inertial
m	mechanical
o	initial
p	perception-reaction
r	rolling
t	tangential

1. VEHICLE DYNAMICS

A moving surface vehicle (e.g., car, truck, or bus) is retarded by five major forms of resistance: inertia, grade, rolling, curve, and air resistance. The *inertia resistance* (*inertial resisting force*) is calculated from the vehicle mass and acceleration.

$$F_i = ma \qquad \text{[SI]} \qquad 74.1(a)$$

$$F_i = \frac{ma}{g_c} = \frac{wa}{g} \qquad \text{[U.S.]} \qquad 74.1(b)$$

For a vehicle ascending a constant incline ϕ, the *grade resistance* is the component of the vehicle weight, w, acting parallel and down a frictionless inclined surface. The *grade* or *gradient*, G, is the slope in the direction of roadway. Grade is usually specified in percent. For example, a roadway with a 6% grade would increase in elevation 6 ft for every 100 ft of travel. For grades commonly encountered in road and highway design, there is essentially no difference in the road length and the horizontal measurement.

$$F_g = w \sin \phi$$

$$\approx w \tan \phi = \frac{wG\%}{100} \qquad 74.2$$

The *rolling resistance* can be calculated theoretically as presented in Ch. 73. For analysis of vehicle dynamics, the coefficient of rolling friction, f_r, used in Eq. 74.3 is

Transportation

generally considered to be constant.[1] For modern vehicles, it is reported as approximately 0.010 to 0.015 and is approximately constant at 13.5 lbf per 1000 lbm of mass (130 N per kg) for all speeds up to 60 mph (97 kph). For higher speeds, it should be increased 10% for every 10% increase in speed above 60 mph (97 kph). Rolling resistance is even higher for gravel roads and roads in poor condition.

$$F_r = f_r w \cos \phi$$
$$\approx f_r w \qquad\qquad 74.3$$

Curve resistance is the force exerted on the vehicle by its front wheels when the direction of travel is changing. Components of the curve resistance also impede the vehicle's forward motion, decreasing the fuel economy. As with centrifugal force, it is a function of the curve radius and vehicle speed. Curve resistance is generally obtained experimentally or from a table of typical values. When the vehicle is traveling in a straight direction, the curve resistance is zero.

Air resistance is the drag force given by Eq. 74.4, where A is the frontal cross-sectional area.

$$F_D = \frac{C_D A \rho v^2}{2} \qquad \text{[SI]} \qquad 74.4(a)$$

$$F_D = \frac{C_D A \rho v^2}{2 g_c} \qquad \text{[U.S.]} \qquad 74.4(b)$$

In the absence of any information about the drag coefficient and/or air density, the average air resistance for modern cars can be approximated from Eq. 74.5, where all of the constants have been accumulated into an *air resistance coefficient*.

$$F_D = K A v^2 \approx 0.0011 A_{\mathrm{m}^2} v_{\mathrm{km/h}}^2$$
$$\text{[SI]} \qquad 74.5(a)$$

$$F_D = K A v^2 \approx 0.0006 A_{\mathrm{ft}^2} v_{\mathrm{mi/hr}}^2$$
$$\text{[U.S.]} \qquad 74.5(b)$$

The *propulsion power* (excluding the power used by power steering, air conditioner, and other accessories) is given by Eq. 74.6, where the sum of all resistances may be referred to as the *tractive force* or *tractive effort*. (Appropriate conversions can be made to horsepower and kilowatts.) For a typical vehicle, only about 50% of the nominal engine horsepower (brake power) rating is available at the flywheel for propulsion (i.e., after accessories). If the power is known, Eq. 74.6 can be solved for the maximum acceleration or velocity. Since the velocity also appears in the expression for drag force, a trial and error solution may be needed.

$$P = (F_i + F_g + F_r + F_c + F_D) v \qquad 74.6$$

---

[1]An approximate relationship between speed and coefficient of rolling resistance for vehicles operating on typical paved roads was reported in 1957 by Taborek as the following (customary U.S. units).

$$f_r = 0.01 \left(1 + \frac{v_{\mathrm{ft/sec}}}{147} \right)$$

Regardless of how much power a vehicle's engine may be capable of developing, there is a point beyond which the use of additional power will merely result in spinning the driving tires. (The maximum tractive force is $f_s w$.) Beyond that point, no additional tractive effort can be generated to overcome resistance. The maximum *tractive force* as limited by wheel spinning is found by summing moments about some point on the vehicle, typically the tire-roadway contact point for the non-driven wheels. The moment from the unknown tractive force balances the moments from all of the resistance forces.[2]

Analysis of performance under acceleration and deceleration is the same as for any body under uniform acceleration. A vehicle's instantaneous forward velocity and tangential velocity of the tires are the same.

$$v_{\mathrm{vehicle}} = v_{t,\mathrm{tire}} = \pi d_{\mathrm{tire}} n_{\mathrm{tire}} \qquad 74.7$$

The engine-wheel speed ratio depends on the transmission and differential gear ratios. The *transmission gear ratio* is the ratio of the engine speed to the driveshaft speed. Unless there is an *overdrive gear*, the typical transmission gear ratio used for highest speeds is approximately 1:1. The differential gear ratio (typically 2.5:1 to 4:1) is the ratio of driveshaft to tire speeds.

$$n_{\mathrm{wheel}} = \frac{n_{\mathrm{engine}}}{R_{\mathrm{transmission}} R_{\mathrm{differential}}} \qquad 74.8$$

$$R_{\mathrm{transmission}} = \frac{n_{\mathrm{engine}}}{n_{\mathrm{driveshaft}}} \qquad 74.9$$

$$R_{\mathrm{differential}} = \frac{n_{\mathrm{driveshaft}}}{n_{\mathrm{wheel}}} \qquad 74.10$$

The relationship between brake power and engine-generated torque is

$$P_{\mathrm{kW}} = \frac{T_{\mathrm{N \cdot m}} n_{\mathrm{rpm}}}{9549} \qquad \text{[SI]} \qquad 74.11(a)$$

$$P_{\mathrm{hp}} = \frac{T_{\mathrm{in\text{-}lbf}} n_{\mathrm{rpm}}}{63{,}025} \qquad \text{[U.S.]} \qquad 74.11(b)$$

Only 75 to 90% of the engine power or torque reaches the rear wheels. This represents the mechanical efficiency, η_m. The remainder is lost in the transmission, differential, and other gear-reduction devices. The engine-generated tractive effort reaching the driving wheels is

$$F_{\mathrm{tractive}} = \frac{\eta_m T R_{\mathrm{transmission}} R_{\mathrm{differential}}}{r_{\mathrm{tire}}} \qquad 74.12$$

Equation 74.13 gives the relationship between vehicle speed and engine speed. The loss fraction, i, is typically

---

[2]Unless specific information is known about the location of the vehicle's center of gravity and other properties, assumptions may need to be made about the point of application of the resistance forces.

0.02 to 0.05 and primarily accounts for slippage in the clutch and automatic transmission.

$$v = \frac{2\pi n_{\text{rev/sec}}(1-i)}{R_{\text{transmission}} R_{\text{differential}}} \qquad 74.13$$

The mass and volumetric rates of fuel consumption can be found from the horsepower and brake specific fuel consumption.

$$\dot{m}_{\text{fuel,kg/h}} = P_{\text{brake,kW}}(\text{BSFC}_{\text{kg/kW·h}})$$
$$\qquad\qquad\qquad\qquad\qquad \text{[SI]} \qquad 74.14(a)$$

$$\dot{m}_{\text{fuel,lbm/hr}} = P_{\text{brake,hp}}(\text{BSFC}_{\text{lbm/hp-hr}})$$
$$\qquad\qquad\qquad\qquad\qquad \text{[U.S.]} \qquad 74.14(b)$$

$$\dot{Q}_{\text{fuel,L/h}} = \frac{\dot{m}_{\text{fuel,kg/h}}\left(1000\,\dfrac{\text{L}}{\text{m}^3}\right)}{\rho_{\text{kg/m}^3}} \qquad \text{[SI]} \qquad 74.15(a)$$

$$\dot{Q}_{\text{fuel,gal/hr}} = \frac{\dot{m}_{\text{fuel,lbm/hr}}\left(7.48\,\dfrac{\text{gal}}{\text{ft}^3}\right)}{\rho_{\text{lbm/ft}^3}} \qquad \text{[U.S.]} \qquad 74.15(b)$$

The distance traveled per volume of fuel (also known as *fuel economy*), s', is

$$s'_{\text{fuel,km/L}} = \frac{v_{\text{km/h}}}{\dot{Q}_{\text{fuel,L/h}}} \qquad \text{[SI]} \qquad 74.16(a)$$

$$s'_{\text{fuel,mi/gal}} = \frac{v_{\text{mi/hr}}}{\dot{Q}_{\text{fuel,gal/hr}}} \qquad \text{[U.S.]} \qquad 74.16(b)$$

Example 74.1

A 2500 lbm (1100 kg) offroad utility vehicle has a frontal area of 22 ft^2 (2.0 m^2). At 30 mph (50 km/h), the vehicle has a maximum tractive force of 320 lbf (1400 N). While traveling offroad over gravel in still air, the rolling resistance is 50 lbf per ton of vehicle weight (0.25 N/kg). The air resistance coefficient, K, is 0.002 lbf-hr^2/ft^2-mi^2 (0.04 N·hr^2/m^2·km^2). What straight gravel grade can the vehicle climb while maintaining a constant forward velocity of 30 mph (50 km/h)?

SI Solution

Since the grade is straight, there is no curve resistance. Since the speed is constant, there is no inertial resistance. Since the air resistance coefficient has units, it is not a coefficient of drag. The air resistance is

$$F_D = KAv^2 = \left(\frac{0.04\ \text{N·h}^2}{\text{m}^2\text{·km}^2}\right)(2.0\ \text{m}^2)\left(50\ \frac{\text{km}}{\text{h}}\right)^2$$
$$= 200\ \text{N}$$

The rolling resistance is

$$F_r = \left(0.25\ \frac{\text{N}}{\text{kg}}\right)(1100\ \text{kg})$$
$$= 275\ \text{N}$$

The grade resistance is

$$F_g = F_{\text{tractive}} - F_D - F_r = 1400\ \text{N} - 200\ \text{N} - 275\ \text{N}$$
$$= 925\ \text{N}$$

The maximum grade is given by Eq. 74.2.

$$G = \frac{F_g}{w} = \frac{F_g}{mg} = \frac{925\ \text{N}}{(1100\ \text{kg})\left(9.81\ \dfrac{\text{m}}{\text{s}^2}\right)}$$
$$= 0.0857\ \ (8.6\%)$$

Customary U.S. Solution

Since the grade is straight, there is no curve resistance. Since the speed is constant, there is no inertial resistance. Since the air resistance coefficient has units, it is not a coefficient of drag. The air resistance is

$$F_D = KAv^2 = \left(0.002\ \frac{\text{lbf-hr}^2}{\text{ft}^2\text{-mi}^2}\right)(22\ \text{ft}^2)\left(30\ \frac{\text{mi}}{\text{hr}}\right)^2$$
$$= 39.6\ \text{lbf}$$

A vehicle with a mass of 2500 lbm has a weight of 2500 lbf in standard gravity. The rolling resistance is

$$F_r = \frac{\left(50\ \dfrac{\text{lbf}}{\text{ton}}\right)(2500\ \text{lbf})}{2000\ \dfrac{\text{lbf}}{\text{ton}}}$$
$$= 62.5\ \text{lbf}$$

The grade resistance is

$$F_g = F_{\text{tractive}} - F_D - F_r = 320\ \text{lbf} - 39.6\ \text{lbf} - 62.5\ \text{lbf}$$
$$= 217.9\ \text{lbf}$$

The maximum grade is given by Eq. 74.2.

$$G = \frac{F_g}{w} = \frac{217.9\ \text{lbf}}{2500\ \text{lbf}}$$
$$= 0.08716\ \ (8.7\%)$$

2. DYNAMICS OF STEEL-WHEELED RAILROAD ROLLING STOCK

The kinematic analysis of rail rolling stock is similar to that of vehicles, though different variables and terminology are used. The *drawbar pull*, DBP, is the net tractive force (behind the locomotive) and is the force available for hauling cars. Drawbar pull is calculated from the *car resistance*, CR; *accelerating force*, AF; *tractive force at the driving axles*, TFDA; and *locomotive resistance*, LR. The *tonnage rating* is the DBP divided by the total

Transportation

resistance per ton for an average car. The *ruling gradient* is the maximum grade that occurs along the route and that requires the full DBP for an extended period of time. Ruling grades limit the weight (length) of a train that can be accommodated with a single locomotive.

$$\text{DBP} = \text{TFDA} - \text{LR}$$
$$= \text{CR} + \text{AF} \qquad \textit{74.17}$$

The approximate tractive effort at the driving axles of a *diesel-electric locomotive unit* is given by Eq. 74.18. Typical efficiencies, η_{drive}, of diesel-electric drive systems vary from 0.82 to 0.93, with 0.82 being typically used. The efficiency of diesel-electric locomotives is essentially constant throughout the speed range.

$$F_{\text{tractive},N} = \frac{175 P_{\text{kW,rated}}\eta_{\text{drive}}}{\text{v}_{\text{km/h}}} \qquad [\text{SI}] \qquad \textit{74.18(a)}$$

$$F_{\text{tractive,lbf}} = \frac{375 P_{\text{hp,rated}}\eta_{\text{drive}}}{\text{v}_{\text{mph}}} \qquad [\text{U.S.}] \qquad \textit{74.18(b)}$$

The approximate frontal area of a locomotive or passenger rail car is 100 to 120 ft² (9.0 to 11 m²). Freight cars have 85 to 90 ft² (7.7 to 8.1 m²) of frontal area, though this value varies greatly with the nature of the load. Incorporating the frontal area term, the air resistance coefficient, K, has representative values of 0.094 lbf-hr²-ton/mi² for containers on flat cars, 0.16 lbf-hr²-ton/mi² for trucks and trailers on flat cars, and 0.07 lbf-hr²-ton/mi² for all other standard rail units.

The *modified Davis equation*, applicable to modern, large, streamlined rolling stock moving at fast speeds, gives the *level tangent resistance*, R. w is the average loading per axle, and wn is the total weight calculated as the average axle load times the number of axles.

$$R_{\text{lbf/ton}} = 0.6 + \frac{20}{w_{\text{tons}}} + 0.01\text{v}_{\text{mph}} + \frac{K\text{v}_{\text{mph}}^2}{w_{\text{tons}}n} \qquad [\text{U.S. only}]$$
$$\textit{74.19}$$

The *incidental resistances* include effects due to grade, curvature, and wind. The *grade resistance* is 20 lbf per ton for each percentage grade (i.e., essentially the weight times the grade). The *curve resistance* is approximately 0.8 lbf/ton (0.004 N/kg) per degree of curvature. Railroad curve resistance is considered equivalent to a grade resistance of 0.04% per degree of curvature. When the grade is "compensated," the compensated grade resistance includes the curve resistance, and curve resistance can be disregarded when calculating power requirements.

3. COEFFICIENT OF FRICTION

In most cases, a vehicle's braking system is able to provide more braking force than can be transmitted to the pavement. Therefore, the maximum deceleration is limited by the coefficient of friction between the tires and pavement.

The *coefficient of friction*, f, between a vehicle and the supporting roadway is the frictional force, F_f, divided by the normal force, N. The *normal force* is essentially the total weight of the vehicle, w, on all but the most extreme grades. The coefficient of friction is dependent on the condition of the vehicle's tires, the type and condition of the pavement, and the weather conditions.

$$f = \frac{F_f}{N} \qquad \textit{74.20}$$

There are two coefficients of friction: static and dynamic (kinetic). (The coefficients of friction may also be referred to as *coefficients of road adhesion*.) The *coefficient of static friction* is larger than the coefficient of dynamic (kinetic) friction. While a vehicle's tires are rotating, the relative velocity between a point of contact on the tire and roadway is zero and the coefficient of static friction controls.

Once a vehicle enters a skid, however, the *coefficient of dynamic friction* controls. Therefore, a vehicle held to its maximum braking deceleration without entering a skid (i.e., the skid is impending) will take less distance to come to a complete stop than if the vehicle locks up its tires and skids to a stop.

The coefficient of friction is not constant throughout the braking maneuver, but varies inversely with speed. This level of sophistication is not normally considered, and an average value that is representative of the speed and conditions is used.

Table 74.1 *Typical Coefficients of Skidding Friction*

speed in mph (km/h)	type of pavement[a]				
	AC	SA	RA	PCC	wet[b]
vehicles with new tires					
11 (18)	0.74	0.75	0.78	0.76	–
20 (30)	0.76	0.75	0.76	0.73	0.40
30 (50)	0.79	0.79	0.74	0.78	0.36
40 (65)	0.75	0.75	0.74	0.76	0.33
50 (80)					0.31
60 (100)					0.30
70 (115)					0.29
vehicles with badly worn tires					
11 (18)	0.61	0.66	0.73	0.68	–
20 (30)	0.60	0.57	0.65	0.50	0.40
30 (50)	0.57	0.48	0.59	0.47	0.36
40 (65)	0.48	0.39	0.50	0.33	0.33
50 (80)					0.31
60 (100)					0.30
70 (115)					0.29

(Multiply mph by 1.609 to obtain km/h.)
[a] AC—asphalt concrete, dry; SA—sand asphalt, dry; RA—rock asphalt, dry; PCC—portland cement concrete, dry; wet—all wet pavements.

[b] Most design problems are based on wet pavement.

Table compiled from a variety of sources; values vary widely.

4. ANTILOCK BRAKES

When a vehicle begins to skid, the driver becomes unable to steer. An *antilock (antiskid) braking system* (ABS), also known as a *brake assist system* (BAS), detects an impending skid and alternately removes and applies hydraulic pressure from the hydraulic brake system. This is equivalent to the driver rapidly "pumping" the brakes, a common antiskid strategy. The application of the hydraulic pressure pulses is approximately 18 times per second. The effect is to allow the tires to rotate slightly, preventing the skid and allowing the driver to steer. There are no skid marks with antilock brakes.

On wet pavement and ice, ABS will generally bring the vehicle to a stop in a shorter distance than if the vehicle were to go into a skid. Since the static coefficient of friction in a "skid-impending" braking maneuver is greater than the dynamic coefficient, the ABS braking distance on dry pavement should also be less. However, inasmuch as the average driver is unable to apply "just the right amount" of braking to achieve the maximum deceleration, the braking distance is essentially unchanged, and the primary benefit is being able to steer. On gravel and snow-covered roads, ABS braking distances can be greater than distances without ABS, since the tires of a skidding vehicle will dig through the upper layers of loose gravel or snow to find a solid layer with a greater coefficient of friction.

While ABS is active, a loud grating sound is heard and the brake pedal rapidly pulsates. This alarms some drivers and causes them to pump their brakes or remove pressure entirely from the brake pedal, eliminating the benefits of the ABS system. Also, some drivers tend to jerk the steering wheel during an ABS-assisted stop, causing the vehicle to go off course, often into an opposing lane. Because of these problems, many car insurance companies no longer offer discounts for ABS-equipped vehicles.

5. STOPPING DISTANCE

Stopping distance includes the distance traveled before the brakes are applied as well as the distance during the braking maneuver. The *braking perception-reaction time*, t_p, is also referred to as the *PIEV time*, using an acronym for perception, identification, emotion, and volition. PIEV time varies widely from person to person. Though the median value is approximately 0.90 sec for unexpected (unanticipated) events in complex roadway situations, individuals with slow reaction times may require up to 3.5 sec. AASHTO's Green Book Ch. 3 lists 2.5 sec as the value used to determine the minimum *stopping sight distances*, and is appropriate for approximately 90% of the population.

$$s_{\text{stopping}} = \text{v}t_p + s_b \qquad \text{74.21}$$

6. BRAKING AND DECELERATION RATE

The maximum deceleration that can be developed in dry weather by a vehicle with tires and brakes in good condition is about 25 ft/sec^2 (7.5 m/s^2), and in wet conditions, the maximum deceleration provided by modern vehicles, tires, and roads easily exceeds 11.2 ft/sec^2 (3.4 m/s^2). (Acceleration and deceleration are sometimes specified in miles per hour per second, mphps, or kilometers per hour per second, kphps. Multiply mphps by 1.467 to obtain ft/sec^2, and multiply kphps by 0.278 to obtain m/s^2.) However, decelerations of 14 ft/sec^2 (4.2 m/s^2) are experienced by occupants as uncomfortable and alarming panic stops. 11.2 ft/sec^2 (3.4 m/s^2), corresponding to the 90th percentile of commonly experienced deceleration rates, is the approximate upper limit of desirable decelerations, and 9 ft/sec^2 (2.7 m/s^2) is the approximate maximum comfortable deceleration from high-speed travel.

If a vehicle locks its brakes and begins to skid on level ground, the deceleration rate will be $f_{\text{dynamic}}g$.

$$a = fg = f\left(9.81 \ \frac{\text{m}}{\text{s}^2}\right) \qquad \text{[SI]} \qquad \textit{74.22(a)}$$

$$a = fg = f\left(32.2 \ \frac{\text{ft}}{\text{sec}^2}\right) \qquad \text{[U.S.]} \qquad \textit{74.22(b)}$$

7. BRAKING AND SKIDDING DISTANCE

The braking/skidding distance required to come to a complete stop depends on friction and grade and can be calculated from Eq. 74.23. This distance is exclusive of the time traveled during PIEV time.

$$s_b = \frac{\text{v}^2}{2g(f\cos\theta + \sin\theta)} \qquad \text{74.23}$$

The incline angle, θ, is small, so $\cos\theta \approx 1.0$, and $\sin\theta \approx \tan\theta = G$. The decimal grade, G, is positive when the skid is uphill and negative when the skid is downhill.

$$s_b = \frac{\text{v}^2}{2g(f + G)}$$
$$\approx \frac{\text{v}_{\text{km/h}}^2}{254(f + G)} \qquad \text{[SI]} \qquad \textit{74.24(a)}$$

$$s_b = \frac{\text{v}^2}{2g(f + G)}$$
$$\approx \frac{\text{v}_{\text{mph}}^2}{30(f + G)} \qquad \text{[U.S.]} \qquad \textit{74.24(b)}$$

Equations 74.23 and 74.24 do not apply directly to situations where the vehicle is brought to an abrupt stop by a collision or when the driver stops braking during the skid and continues on at a slower speed. In such cases, the actual skid length can be calculated as the difference in skidding distance to come to a complete stop from the initial speed and the skidding distance to come to complete stop from the final speed (i.e., the impact speed in the case of collisions).

Example 74.2

A 4000 lbm (1800 kg) car traveling at 80 mph (130 km/h) locks up its wheels, decelerates at a constant rate, and slides 580 ft (175 m) along a level road before stopping. (a) Disregarding perception-reaction time, how much time does it take to stop? (b) What is the acceleration? (c) What is the retarding force? (d) What is the coefficient of friction between the tires and the road?

SI Solution

(a) This is a case of uniform acceleration. The initial velocity is

$$v_o = \frac{\left(130 \ \frac{km}{h}\right)\left(1000 \ \frac{m}{km}\right)}{3600 \ \frac{s}{h}} = 36.1 \text{ m/s}$$

The stopping distance, s, is 175 m. The final velocity, v, is 0.

$$t = \frac{2s}{v_o + v} = \frac{(2)(175 \text{ m})}{36.1 \ \frac{m}{s} + 0} = 9.69 \text{ s}$$

(b) Use the uniform acceleration formulas.

$$a = \frac{v - v_o}{t} = \frac{0 - 36.1 \ \frac{m}{s}}{9.69 \text{ s}}$$
$$= -3.7 \text{ m/s}^2$$

(c) $\quad F_f = ma = (1800 \text{ kg})\left(3.7 \ \frac{m}{s^2}\right)$
$$= 6660 \text{ N}$$

(d) $\quad f = \dfrac{F_f}{N} = \dfrac{F_f}{mg} = \dfrac{6660 \text{ N}}{(1800 \text{ kg})\left(9.81 \ \frac{m}{s^2}\right)} = 0.38$

Customary U.S. Solution

(a) This is a case of uniform acceleration. The initial velocity is

$$v_o = \frac{\left(80 \ \frac{mi}{hr}\right)\left(5280 \ \frac{ft}{mi}\right)}{3600 \ \frac{sec}{hr}}$$
$$= 117.3 \text{ ft/sec}$$

The stopping distance, s, is 580 ft. The final velocity, v, is 0.

$$t = \frac{2s}{v_o + v} = \frac{(2)(580 \text{ ft})}{117.3 \ \frac{ft}{sec} + 0}$$
$$= 9.89 \text{ sec}$$

(b) Use the uniform acceleration formulas.

$$a = \frac{v - v_o}{t} = \frac{0 - 117.3 \ \frac{ft}{sec}}{9.89 \text{ sec}}$$
$$= -11.9 \text{ ft/sec}^2$$

(c) Use Eq. 74.1.

$$F_f = \frac{ma}{g_c}$$
$$= \left(\frac{4000 \text{ lbm}}{32.2 \ \frac{ft\text{-}lbm}{lbf\text{-}sec^2}}\right)\left(11.9 \ \frac{ft}{sec^2}\right)$$
$$= 1480 \text{ lbf}$$

(d) $\quad f = \dfrac{F_f}{N} = \dfrac{1480 \text{ lbf}}{4000 \text{ lbf}} = 0.37$

Example 74.3

A car traveling at 80 mph (130 km/h) locks up its wheels and skids up a 3% incline before crashing into a stationary massive concrete pedestal and coming to a complete stop. The skid marks leading up to the pedestal are 300 ft (90 m) long. The coefficient of friction between the tires and road is 0.35. (a) How far would the car have skidded if it had not hit the pedestal? (b) What was the speed of the car at impact?

SI Solution

(a) Use Eq. 74.24. The grade is positive since the car skids uphill.

$$s_b = \frac{v_{km/h}^2}{254(f + G)} = \frac{\left(130 \ \frac{km}{h}\right)^2}{(254)(0.35 + 0.03)}$$
$$= 175.1 \text{ m}$$

(b) Since the skid marks were only 90 m long, the vehicle would have skidded 175.1 m − 90 m = 85.1 m further. Solve Eq. 74.24 for the speed.

$$85.1 \text{ m} = \frac{v_{km/h}^2}{254(f + G)} = \frac{v^2}{(254)(0.35 + 0.03)}$$
$$v = 90.6 \text{ km/h}$$

Customary U.S. Solution

(a) Use Eq. 74.24. The grade is positive since the car skids uphill.

$$s_b = \frac{v_{mph}^2}{30(f+G)} = \frac{\left(80 \; \frac{mi}{hr}\right)^2}{(30)(0.35+0.03)}$$
$$= 561.4 \; ft$$

(b) Since the skid marks were only 300 ft long, the vehicle would have skidded 561.4 ft − 300 ft = 261.4 ft further. Solve Eq. 74.24 for the speed.

$$261.4 \; ft = \frac{v_{mph}^2}{30(f+G)} = \frac{v^2}{(30)(0.35+0.03)}$$
$$v = 54.6 \; mph$$

8. SPEED DEGRADATION ON UPHILL GRADES

Most modern passenger cars traveling on highways are capable of negotiating uphill grades of 4 to 5% without speed decreases below their initial level-highway speeds. (Older cars with high mass-to-power ratios and some smaller-sized "economy" vehicles may experience speed decreases.)

Heavy trucks experience greater speed degradations than passenger cars. The primary variables affecting actual speed decreases are the grade steepness, the grade length, and the truck's mass-to-power ratio. *Mass-to-power ratios* are commonly stated in pounds per horsepower (lbm/hp) and kilograms per kilowatt (kg/kW). (Multiply lbm/hp by 0.6083 to obtain kg/kW.) Chapter 3 of AASHTO's Green Book contains simple graphs of speed decreases for "heavy trucks" with mass-to-power ratios of 200 lbm/hp (120 kg/kW) and recreational vehicles entering ascending grades at 55 mph (88.5 km/h).

9. ANALYSIS OF ACCIDENT DATA

Accident data are compiled and evaluated to identify hazardous features and locations, set priorities for safety improvements, support economic analyses, and identify patterns, causes (i.e., driver, highway, or vehicle), and possible countermeasures.

Accidents are classified into three *severity categories*, depending on whether there is (a) property damage only (referred to as *PDO accidents*), (b) personal injury, or (c) fatalities. The *severity ratio* is defined as the ratio of the number of injury and fatal accidents divided by the total number of all accidents (including PDO accidents.)

It is common to prioritize intersections according to the *accident rate, R*. The accident rate may be determined for PDO, personal injury, and fatal accidents, or the total thereof. The accident ratio is the ratio of the number of accidents per year to the average daily traffic, ADT. The rate is reported as an RMEV (i.e., rate per million entering vehicles) taking into consideration vehicles entering an intersection from all directions.

$$R = \frac{(\text{no. of accidents})(10^6)}{(\text{ADT})(\text{no. of years})\left(365 \; \frac{days}{yr}\right)} \qquad 74.25$$

Routes (segments or links) between points are prioritized according to the accident rate per mile (per kilometer), calculated as the ratio of the number of accidents per year to the ADT per mile (kilometer) of length, counting traffic from all directions in the intersection. For convenience, the rate may be calculated per 100 million vehicle miles (HMVM).

$$R = \frac{(\text{no. of accidents})(10^8)}{(\text{ADT})(\text{no. of years})\left(365 \; \frac{days}{yr}\right)L_{mi}} \qquad 74.26$$

To identify the most dangerous features, it is necessary to determine what the accidents at an intersection or along a route had in common. This can be done in a number of ways, one of which is by drawing a collision diagram. A *collision diagram* is essentially a true-scale cumulative drawing of the intersection showing details of all of the accidents in the study. Dates, times, weather conditions, types of maneuvers, directions, severity, types of accidents, and approximate locations are recorded on the diagram.

Common *accident-type* categories include: rear-end, right-angle, left-turn, fixed-object, sideswipe, parked-vehicle, run off road, head-on, bicycle-related, and pedestrian-related.

Hypothesis testing can be used to determine if installed countermeasures are being effective. The *null hypothesis* is that the highway improvements have brought no significant decrease in accident rate. The counterhypothesis is that the highway improvements have brought significant decreases. A chi-squared test with a 5% confidence level is used to evaluate the two hypotheses.

10. ROAD SAFETY FEATURES

Road safety features are installed to protect public life and property and to reduce traffic-related lawsuits against highway and transportation departments. The most common actions include the installation of illumination, guardrails, and impact attenuators, as well as the relocation of dangerous facilities.

Guardrails are used on roadways where there is a severe slope or vertical dropoff to the side of the road,

Transportation

ditches, permanent bodies of water, embankments, and roadside obstacles (e.g., boulders, retaining walls, and sign and signal supports). Guardrails with turned-down ends are now prohibited in new installations and should be upgraded. Such guardrails, rather than protecting motorists from a fixed impact or spearing, often cause *vaulting* (also known as *launching* or *ramping*) and subsequent rollover.

Impact attenuators are used to provide crash protection at bridge pillars, center piers, gore areas, butterfly signs, light posts, and ends of concrete median barriers. These are mainly of the *crash cushion* variety (sand barrel, water barrel, and Hi-Dro® water-tube arrays), sandwich-type units, and crushable cartridges (including the G-R-E-A-T® "Guard Rail Energy Absorbing Terminal" manufactured by Energy Absorption Systems, Chicago, IL, which uses crushable Hex-Foam® units surrounded by a framework of triple-corrugated steel guardrail).

Slip- and breakaway- (frangible coupling) bases are used on poles of roadside luminaires and signs to prevent vehicles from "wrapping around" the poles. Hinged and weakened bases for high-voltage power poles, however, may violate other codes (e.g., the *National Electric Code*) designed to prevent high-voltage wires from dropping on the public below. Other solutions, such as placing utility lines underground, increasing lateral offset or longitudinal pole spacing, or dividing the load over multiple poles, should be considered.

75 Flexible Pavement Design

Nomenclature

a	layer strength coefficients	1/in	–
AADT	average annual daily traffic	vpd	vpd
CBR	California bearing ratio	–	–
D	depth (thickness)	ft	m
D_D	directional distribution factor	–	–
D_L	lane distribution factor	–	–
E	modulus of elasticity	lbf/in^2	Pa
ESAL	equivalent single-axle loads	–	–
f_c'	compressive strength	lbf/in^2	Pa
FF	fatigue fraction	–	–
g	growth rate	percent decimal	percent decimal
G	grade	–	–
G	specific gravity	–	–
GF	growth rate factor	–	–
k	modulus of subgrade reaction	lbf/in^3	N/m^3
L	linear yield	ft/ton	m/tonne
LEF	load equivalency factor	–	–
m	drainage coefficient	–	–
M_R	resilient modulus	lbf/in^2	Pa
p_o	initial serviceability	–	–
p_t	terminal serviceability	–	–
P	percentage	percent	percent
PSI	present serviceability index	–	–
r	spreading rate	lbf/yd^2	kg/m^2
R	soil resistance value	–	–
R_p	production rate	ton/hr	tonne/h
S	standard deviation	various	various
SN	structural number	–	–
t	thickness	ft	m
TF	truck factor	–	–
u_f	relative fatigue damage	–	–
v	velocity	ft/min	m/min
V	volume	ft^3	m^3
VFA	voids filled with asphalt	percent	percent
VMA	voids in mineral aggregate	percent	percent
VTM	total voids	percent	percent
w	width	ft	m
w_{18}	design lane traffic	–	–
$\hat{w}_{18}$	all-lane, two-directional traffic	–	–
W	weight	lbf	N

Symbols

γ	specific weight	lbf/ft^3	–
ρ	density	lbm/ft^3	kg/m^3

Subscripts

a	aggregate, air, or apparent
b	base, bulk, or asphalt
BS	base
ca	coarse aggregate
e	effective
f	fatigue
fa	fine aggregate
mm	maximum
o	overall or original
p	production
s	specific
SB	subbase
t	terminal

1. ASPHALT CONCRETE PAVEMENT

Hot mix asphalt (HMA), commonly referred to as *asphalt concrete* (AC), *bituminous mix* (BM), and sometimes *hot mix asphaltic concrete* (HMAC), is a mixture

Transportation

of *asphalt cement* (*asphalt binder*) and well-graded, high-quality aggregate. The mixture is heated and compacted by a paving machine into a uniform dense mass.

AC pavement is used in the construction of traffic lanes, auxiliary lanes, ramps, parking areas, frontage roads, and shoulders. AC pavement adjusts to limited amounts of differential settlement. It is easily repaired, and additional thicknesses can be placed at any time to withstand increased usage and loading. Over time, its nonskid properties do not significantly deteriorate. However, as the asphalt oxidizes, it loses some of its flexibility and cohesion, often requiring resurfacing sooner than would be needed with portland cement concrete. Asphalt concrete is not normally chosen where water is expected to permeate the surface layer.

Most AC pavement is usually applied over several other structural layers, not all of which are necessarily AC layers. A *full-depth asphalt pavement* consists of asphalt mixtures in all courses above the subgrade. Since asphalt and asphalt-treated bases are stronger than untreated granular bases, the surface pavement can be thinner. Other advantages of full-depth pavements include a potential decrease in trapped water within the pavement, a decrease in the moisture content of the subgrade, and little or no reduction in subgrade strength.

If the asphalt layer is thicker than approximately 4 in (102 mm) and is placed all in one lift, or if lift layers are thicker than 4 in (102 mm), the construction is said to be *deep strength asphalt pavement*. Using deep lifts to place layers of hot-mix asphalt concrete is advantageous for several reasons. (a) Thicker layers hold heat longer, making it easier to roll the layer to the required density. (b) Lifts can be placed in cooler weather. (c) One lift of a given thickness is more economical to place than multiple lifts equaling the same thickness. (d) Placing one lift is faster than placing several lifts. (e) Less distortion of the asphalt course will result than if thin lifts are rolled.

Modern advances in asphalt concrete paving include *Superpave*™ (Sec. 24) and *stone matrix asphalt* (Sec. 25). Both of these are outgrowths of the U.S. Strategic Highway Research Program (SHRP).

2. OTHER ASPHALT APPLICATIONS

Other types of asphalt products used in paving include *emulsified asphalts*, both anionic and cationic (depending on the charge of the emulsifying agent), and *cutback asphalts*, which are graded based on the speed at which the volatile substance used to liquefy the asphalt cement evaporates. Cutback asphalts may be referred to as RC, MC, and SC, referring to *rapid cure*, *medium cure*, and *slow cure* agents, respectively. Other applications of asphalt cement in construction include cold-mix asphalt mixtures, recycled road mixtures (both hot and cold), and surface treatments, which include seal coats, chip seals, slurry seals, and fog-coats.

3. ASPHALT GRADES

Asphalt varies significantly in its properties. In general, softer grades are used in colder climates to resist the expansion and contraction of the asphalt concrete caused by thermal changes. Harder grades of asphalt are specified in warmer climates to protect against rutting.

In the past, asphalt cement has been graded by *viscosity grading* and *penetration grading* methods. *Penetration grading* is based upon the penetration of a standard-sized needle loaded with a mass of 100 g in 5 sec at 77°F (25°C). A penetration of 40 to 50 (in units of mm) is graded as hard, and a penetration of 200 to 300 is soft. Intermediate ranges of 60 to 70, 85 to 100, and 120 to 150 have also been established. A penetration-graded asphalt is identified by its range and the word "pen" (e.g., "120 to 150 pen").

Viscosity grading is based on a measure of the absolute viscosity (tendency to flow) at 140°F (60°C). A viscosity measurement of AC-40 is graded as hard, and a measurement of AC-2.5 is soft. Intermediate ranges of AC-5, AC-10, and AC-20 have been established.

To meet the minimum specified viscosity requirements, oil suppliers have typically added more "light ends" (liquid petroleum distillates) to soften asphalt, regardless of the effects that such additions might have on structural characteristics. Concerns about the effects of such additions include oily buildup in mixing plant baghouses, light-end evaporation during pavement life, and possible reduced roadway performance.

Most state transportation and highway departments have stopped using penetration grading and have switched to performance grading. The *performance grading* (PG) system, which is used in Superpave and other designs, eliminates the structural concerns by specifying other characteristics. PG grading is specified by two numbers, such as PG 64-22. These two numbers represent the high and low (seven-day average) pavement temperature in degrees Celsius that the project location will likely experience in its lifetime. Both high and low temperature ratings have been established in 6° increments.

PG ratings can be achieved with or without polymers, at the supplier's preference. However, the "rule of 90" says that if the sum of the absolute values of the two PG numbers is 90 or greater, then a polymer-modified asphalt is needed. Asphalts without polymers can be manufactured "straight-run."

The use of *recycled asphalt pavement* (RAP) in the mixture has an effect on the asphalt grade specified. However, specifying less than 15 to 25% RAP generally does not require a different asphalt cement grade to be used.

Table 75.1 Superpave Binder Grades[a] (°C)

high-temperature grades	low-temperature grades
PG 46	−34, −40, −46
PG 52	−10, −16, −22, −28, −34, −40, −46
PG 58	−16, −22, −28, −34, −40
PG 64	−10, −16, −22, −28, −34, −40
PG 70	−10, −16, −22, −28, −34, −40
PG 76	−10, −16, −22, −28, −34
PG 82	−10, −16, −22, −28, −34

[a] For example, PG 52-16 would be applicable in the range of 52°C down to −16°C.

4. AGGREGATE

The *mineral aggregate* component of the asphalt mixture comprises 90 to 95% of the weight and 75 to 85% of the mix volume. Mineral aggregate consists of sand, gravel, or crushed stone. The size and grading of the aggregate is important, as the minimum lift thickness depends on the maximum aggregate size. Generally, the minimum lift thickness should be at least three times the nominal maximum aggegate size. However, compaction is not an issue with open-graded mixes, since it is intended that the final result be very open. Therefore, the maximum-size aggregate can be as much as 80% of the lift thickness.

Generally, *coarse aggregate* is material retained on a no. 8 sieve (2.36 mm openings), *fine aggregate* is material passing through a no. 8 sieve (2.36 mm), and *mineral filler* is fine aggregate for which at least 70% passes through a no. 200 sieve (0.075 mm openings). The fine aggregate should not contain organic materials.

Aggregate size grading is done by sieving. Results may be expressed as either the percent passing through the sieve or the percent retained on the sieve, where the percent passing and the percent retained add up to 100%.

The maximum size of the aggregate is determined from the smallest sieve through which 100% of the aggregate passes. The *nominal maximum size* is designated as the largest sieve that retains some (but not more than 10%) of the aggregate.

An *asphalt mix* is specified by its nominal size and a range of acceptable passing percentages for each relevant sieve size, such as is presented in Table 75.2. The properties of the final mixture are greatly affected by the grading of aggregate. Aggregate grading can be affected by stockpile handling, cold-feed proportioning, degradation at impact points (more of an issue with batch plants), dust collection, and the addition of baghouse fines to the mix. Aggregates not only can become segregated; they also can be contaminated with other materials in adjacent stockpile.

Dense-graded mixtures contain enough fine, small, and medium particles to fill the majority of void space between the largest particles without preventing direct contact between all of the largest particles. Aggregate for most HMA pavement is *open graded*, which means that insufficient fines and sand are available to fill all of the volume between the aggregate. Asphalt cement is expected to fill such voids. *Gap-graded mixtures* (as used in stone matrix asphalt and Superpave) are mixtures where two sizes (very large and very small) predominate.

Open-graded friction courses (OGFCs) have experienced problems with asphalt stripping, drain down, and raveling. Properly formulated, mixes perform reasonably well, though they are susceptible to problems when the voids become plugged, retain moisture, or become oxidized.

5. PAVEMENT PROPERTIES

Several properties are desirable in an asphalt mixture, including stability, durability, flexibility, fatigue resistance, skid resistance, impermeability, and workability. *Stability* refers to the ability to resist permanent deformation, usually at high temperatures over a long time period. It depends on the amount of internal friction present in the mixture, which is in turn dependent on aggregate shape and surface texture, mix density, and asphalt viscosity. Lack of stability results in rutting in wheel paths, shoving at intersections, bleeding or flushing, and difficulty in compaction.

Durability refers to the ability of a mixture to resist disintegration by weathering. A mixture's durability is enhanced by a high asphalt content, dense aggregate gradation, and high density. Lack of durability results in raveling, early aging of the asphalt cement, and stripping.

A mixture's *flexibility* refers to its ability to conform to the gradual movement due to temperature changes or settlement of the underlying pavement layers. Flexibility is enhanced by a high asphalt content, open gradation, and low asphalt viscosity. A lack of flexibility can result in transverse or block cracking and shear failure cracking.

Fatigue resistance refers to the ability to withstand repeated wheel loads. A mixture's fatigue resistance can be enhanced with the presence of dense aggregate gradation and high asphalt content. Lack of fatigue resistance will result in fatigue or alligator cracking.

Skid resistance refers to the ability to resist tire slipping or skidding. An asphalt mixture with the optimum asphalt content, angular surface aggregates, and hard, durable aggregates will provide adequate skid resistance. The *polishing characteristics* of the aggregate need to be determined before use, either by lab polishing, insoluble residue tests, or previous service records. ASTM procedure C131, "Resistance to Degradation of Small-Size Coarse Aggregate by Abrasion and Impact in

Table 75.2 *Typical Composition of Asphalt Concrete*

sieve size		mix designation and nominal maximum size of aggregate				
		$1^1/_2$ in (37.5 mm)	1 in (25.0 mm)	$^3/_4$ in (19.0 mm)	$^1/_2$ in (12.5 mm)	$^3/_8$ in (9.5 mm)
		total percent passing (by weight)				
2 in	(50 mm)	100	–	–	–	–
$1^1/_2$ in	(37.5 mm)	90 to 100	100	–	–	–
1 in	(25.0 mm)	–	90 to 100	100	–	–
$^3/_4$ in	(19.0 mm)	56 to 80	–	90 to 100	100	–
$^1/_2$ in	(12.5 mm)	–	56 to 80	–	90 to 100	100
$^3/_8$ in	(9.5 mm)	–	–	56 to 80	–	90 to 100
no. 4	(4.75 mm)	23 to 53	29 to 59	35 to 65	44 to 74	55 to 85
no. 8[a]	(2.36 mm)	15 to 41	19 to 45	23 to 49	28 to 58	32 to 67
no. 16	(1.18 mm)	–	–	–	–	–
no. 30	(0.60 mm)	–	–	–	–	–
no. 50	(0.30 mm)	4 to 16	5 to 17	5 to 19	5 to 21	7 to 23
no. 100	(0.15 mm)	–	–	–	–	–
no. 200[b]	(0.075 mm)	0 to 5	1 to 7	2 to 8	2 to 10	2 to 10
asphalt cement weight percent of total mixture[c]		3 to 8	3 to 9	4 to 10	4 to 11	5 to 12
		suggested coarse aggregate size numbers				
		4 and 67 or 4 and 68	5 and 7 or 57	67 or 68 or 6 and 8	7 or 78	8

[a]In considering the total grading characteristics of an asphalt paving mixture, the amount passing the no. 8 (2.36 mm) sieve is a significant and convenient field control point between fine and coarse aggregate. Gradings approaching the maximum amount permitted to pass the no. 8 (2.36 mm) sieve will result in pavement surfaces having a comparatively fine texture, while gradings approaching the minimum amount passing the no. 8 (2.36 mm) sieve will result in surfaces with a comparatively coarse texture.

[b]The material passing the no. 200 (0.075 mm) sieve may consist of fine particles of the aggregates or mineral filler, or both. It shall be free from organic matter and clay particles and have a plasticity index not greater than 4 when tested in accordance with ASTM Method D423 and Method D424.

[c]The quantity of asphalt cement is given in terms of weight percent of the total mixture. The wide difference in the specific gravity of various aggregates, as well as a considerable difference in absorption, results in a comparatively wide range in the limiting amount of asphalt cement specified. The amount of asphalt required for a given mixture should be determined by appropriate laboratory testing or on the basis of past experience with similar mixtures, or by a combination of both.

Reprinted with permission of the Asphalt Institute from *The Asphalt Handbook, Manual Series No. 4 (MS-4)*, 1989 ed., Fig. 3.6, copyright © 1989.

Los Angeles Machine," known as the *Los Angeles Abrasion Test* and the *Los Angeles Rattler Test*, is used to determine the durability of the aggregate and quality of long-term skid resistance. Prepared aggregate is placed in a drum with steel balls. The drum is rotated a specific number of times (e.g., 500), and the mass loss is measured.

A mixture's *impermeability* refers to its resistance to the passage of water and air. Impermeability is enhanced with a dense aggregate gradation, high asphalt content, and increased compaction. Permeable mixtures may result in stripping, raveling, and early hardening.

Workability refers to the ease of placement and compaction. A mixture's workability is enhanced by a proper asphalt content, smaller-size coarse aggregates, and proper mixing and compacting temperatures.

6. PROBLEMS AND DEFECTS

Various terms are used to describe the problems and defects associated with the mixing, placing, and functionality of HMA pavements.

- *alligator cracks:* interconnected cracks forming a series of small blocks resembling the marking on alligator skin or chicken wire

- *bleeding:* forming a thin layer of asphalt that has migrated upward to the surface (same as "flushing")

- *blowing:* see "pumping"

- *blow-up:* a disintegration of the pavement in a limited area

- *channeling:* see "rutting"

- *cold-cracking:* separation of pavement caused by temperature extremes, usually observed as cracks running perpendicular to the road's centerline

- *corrugations:* plastic deformation characterized by ripples across the pavement

- *cracking:* separation of pavement caused by loading, temperature extremes, and fatigue

- *disintegration:* breakup of pavement into small, loose pieces

- *drain-down:* liquid asphalt draining through the aggregate in a molten stage

- *faulting:* a difference in the elevations of the edges of two adjacent slabs

- *flushing:* see "bleeding"

- *lateral spreading cracking:* longitudinal cracking that occurs when the two edges of a road embankment supporting a pavement move away from the center

- *placement problems:* tender mixes

- *plastic instability:* excessive displacement under traffic

- *polishing:* aggregate surfaces becoming smooth and rounded (and subsequently, slippery) under the action of traffic

- *pothole:* a bowl-shaped hole in the pavement

- *pumping:* a bellows-like movement of the pavement, causing trapped water to be forced or ejected through cracks and joints

- *raveling:* a gradual, progressive loss of surface material caused by the loss of fine and increasingly larger aggregate from the surface, leaving a pock-marked surface

- *reflective cracking (crack reflection):* cracking in asphalt overlays that follow the crack or joint pattern of layers underneath

- *rutting:* channelized depressions that occur in the normal paths of wheel travel

- *scaling:* the peeling away of an upper layer

- *shoving:* pushing the pavement around during heavy loading, resulting in bulging where the pavement abuts an immobile edge

- *slippage cracking:* cracks in the pavement surface, sometimes crescent-shaped, that point in the direction of the wheel forces

- *spalling:* breaking or chipping of the pavement at joints, cracks, and edges

- *streaking:* alternating areas (longitudinal or transverse) of asphalt caused by uneven spraying

- *stripping:* asphalt not sticking to aggregate, water susceptibility

- *tender mix:* a slow-setting pavement that is difficult to roll or compact, and that shoves (slips or scuffs) under normal loading

- *upheaving:* a local upward displacement of the pavement due to swelling of layers below

- *washboarding:* see "corrugations"

7. ASPHALT MODIFIERS

By itself, asphalt performance cannot be predicted with a good degree of confidence, particularly in the modern mixtures (e.g., Superpave) that place greater demands on asphalt. *Asphalt modifiers* are added to improve the characteristics and reliability of the asphalt binder, improve the performance of the HMA, and reduce cost. Table 75.3 lists some of the types of modifiers.

Table 75.3 *Types of Asphalt Modifiers*

category	examples
mineral fillers	dust, lime, portland cement, and carbon black
extenders	sulfur and lignin
rubbers	natural latex, synthetic latex (styrene-butadiene), block copolymer (styrene-butadiene-styrene), and reclaimed rubber
plastics	polyethylene, polypropylene, ethylene-vinyl-acetate, and polyvinyl chloride
fibers	asbestos[a], rock wool, polypropylene, and polyester
oxidants	manganese and other mineral salts
antioxidants	lead compounds, carbon, and calcium salts
hydrocarbons	recycling oils and rejuvenating oils
antistrip materials	amines and lime

[a]Asbestos may be prohibited in current formulations.

One major type of modifier is *polymers*, which come in two varieties: rubber types and plastic types. Rubber-type "elastomers" (e.g., latexes, block copolymers, and reclaimed rubber) toughen asphalt, whereas plastic-type "plastomers" (polybutadiene and polyisoprene) stiffen it. *Polymer-modified asphalt* (PMA) is more resistant to rutting, cold-cracking, and other durability problems.

In hot-mix, dense-graded mixtures, *synthetic latex* is the most common additive, providing improved resistance to rutting, shoving, and thermal cracking; improved load-carrying ability; improved fracture resistance; improved durability; and water resistance. Synthetic latex is also the most common additive in chip seals, used to improve chip retention and resistance to bleeding and chip embedment.

Lime and other anti-strip commercial additives in HMA reduce pavement raveling and strength degradation in the presence of water. Antioxidant additives increase asphalt's tensile strength and stiffness.

8. ASPHALT MIXERS AND PLANTS

The earliest asphalt mixers were of the *batch mixer* variety. Aggregate was heated and dried in a rotating drum exposed to a burner flame. Mixing with asphalt cement took place in a separate pugmill.

The highest-quality asphalt concrete mixtures are produced in stationary mixing plants, and these are referred to as *plant mixes*.

In modern *drum mixers*, graded aggregate is dried and heated by a large burner flame at 2500 to 2800°F (1370 to 1540°C) at one end of a rotating drum. Aggregate and asphalt cement are added, and at some point, *recycled asphalt pavement* (RAP) can be added. In *parallel-flow mixers* (also known as *center-entry mixers*), virgin aggregate moves in the same direction as the burner flame. RAP is added farther into the drum, away from the main flame, so as not to burn off the RAP's asphalt cement. RAP must be heated essentially entirely by conduction. RAP cannot be added until the mixture

temperature is down to about 800°F (430°C), otherwise the RAP asphalt cement will be burned off as "blue smoke."

Parallel flow mixers can make up to 50% of their mix from RAP, but not much more, since smaller quantities of virgin aggregate cannot transfer enough heat by conduction to the RAP to soften it. Also, if the virgin aggregate has a high moisture content, the RAP fraction must be decreased. Inasmuch as most paving specifications limit RAP to maximums of 25 to 40%, parallel-flow mixers can still be used effectively with recycled materials.

In modern *counterflow mixers*, graded virgin aggregate enters from the end and moves by rotation and gravity toward the burner flame. Counterflow uses fuel more efficiently, and it is easier to use with RAP. Emissions and air pollution rules are easier to satisfy.

Tough air pollution standards apply to asphalt plants. Applicable standards affect the release of carbon monoxide, total hydrocarbons, NOx gases, and particulates. Current technology, however, make "zero opacity" operations possible, without any visible emissions.

9. WEIGHT-VOLUME RELATIONSHIPS

The objectives of an asphalt mixture design are to provide enough asphalt cement to adequately coat, waterproof, and bind the aggregate; provide adequate stability for traffic demands; provide enough air voids to avoid bleeding and loss of stability; and produce a mixture with adequate workability to permit efficient placement and compaction.

Figure 75.1 illustrates the weight-volume relationships for asphalt concrete. The total weight of an asphalt

Figure 75.1 *Weight-Volume Relationships for Asphalt Mixtures*

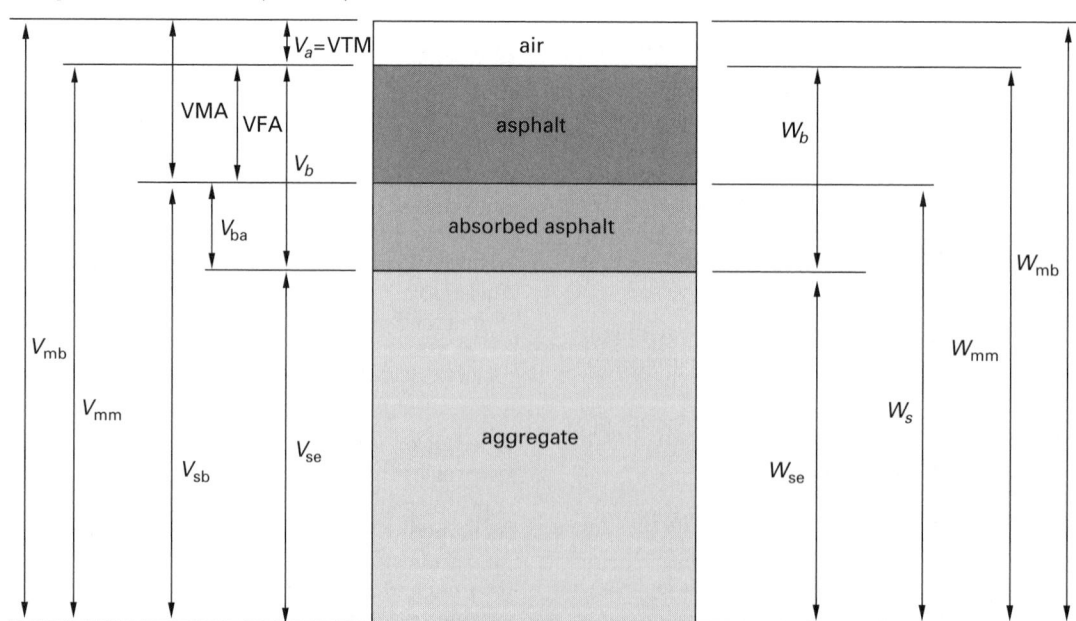

mixture is the sum of the weight of the asphalt and the aggregate. The total volume is the sum of the volume of aggregate and the asphalt not absorbed by the aggregate, plus the air voids.

Although the specific gravity of the asphalt cement can vary widely, the percentage of the total mixture weight contributed by asphalt cement varies from 3 to 8% (by weight) for large nominal maximum sizes (e.g., $1\frac{1}{2}$ in or 37.5 mm) and from 5 to 12% for small nominal maximum sizes (e.g., $\frac{3}{8}$ in or 9.5 mm).

The *surface area method* can be used as a starting point in mix design for determining the percent of asphalt needed. The percentage of asphalt, P_b, is

$$\begin{aligned}
P_b = {}& (100\%)(\text{aggregate surface area}) \\
& \times (\text{asphalt thickness}) \\
& \times (\text{specific weight of asphalt}) \qquad \textit{75.1}
\end{aligned}$$

(The aggregate surface area is measured in units of ft^2/lbf or m^2/kg, the asphalt thickness is in in or mm, and the specific weight of asphalt is in lbf/ft^3 or kg/m^3.)

The aggregate surface area is obtained by multiplying the weight (mass) of the aggregate by a *surface area factor* (ft^2/lbf or m^2/kg). The factor must be known, as it is different for each type of aggregate and for each sieve size. In practice, each surface area factor is multiplied by the percent (converted to a decimal) passing each associated sieve, not by the actual aggregate weight, and the products are summed. The total is the surface area in ft^2/lbf (m^2/kg) for the aggregate mixture.

10. PLACEMENT AND PAVING EQUIPMENT

Paving machines consist of a loading hopper, distribution equipment, and an adjustable *screed*. Hot paving mix is brought from the plant in trucks and loaded directly from the trucks into the paving machine. *Low-lift loaders* are able to pick asphalt paving mix that has been dumped in windrows in front of the paver by trucks. Normally, a mixture *paving temperature* in the range of 250 to 285°F (120 to 140°C) is satisfactory. (Compaction must be completed before the mix cools below 185°F (85°C).)

Hydrostatic drive systems are the norm, eliminating mechanical shifting and keeping the operation smooth. One-pass paving is not generally possible, and roller compaction is required. However, *high-density screeds* are used to achieve densities as high as 92% of specification, about 10% more than with conventional screeds. A tamper bar before the screed also improves final density. High-density screeds are larger (i.e., more expensive) and require more power, or they must run at a slower speed. High-density screeds are preferred with cold-recycling since the rejuvenated mixtures tend to be stiffer than virgin materials.

Self-widening machines (*extending screeds*) enable a paving contractor to change paving widths "on the fly,"

without having to stop and add or remove parts. Extending screeds provide the same compaction as the main screen.

Paving machines can run on rubber tires or crawlers. Crawlers provide better traction on soft subgrades, but they must be loaded on lowboys to be moved any great distance. Rubber-tired pavers have greater mobility but less traction.

Large asphalt *paving machines* can place a layer, or *lift*, of asphalt concrete with a thickness of 1 to 10 in (25 to 250 mm) over a width of 6 to 32 ft (1.8 to 9.8 m) at a forward speed of 10 to 70 ft/min (3 to 21 m/min).

The asphalt pavement *linear yield*, also known as the *length of spread*, L, is calculated from the volumetric relationship of the layer being placed.

$$L_{\text{m/tonne}} = \frac{1000 \, \dfrac{\text{kg}}{\text{tonne}}}{w_{\text{m}} t_{\text{m}} \rho_{\text{compacted,kg/m}^3}} \qquad \text{[SI]} \qquad \textit{75.2(a)}$$

$$L_{\text{ft/ton}} = \frac{2000 \, \dfrac{\text{lbf}}{\text{ton}}}{w_{\text{ft}} t_{\text{ft}} \gamma_{\text{compacted,lbf/ft}^3}} \qquad \text{[U.S.]} \qquad \textit{75.2(b)}$$

Some paving is placed on the basis of weight per unit area and depends on the spreading rate, r.

$$L_{\text{m/tonne}} = \frac{1000 \, \dfrac{\text{kg}}{\text{tonne}}}{r_{\text{kg/m}^2} w_{\text{m}}} \qquad \text{[SI]} \qquad \textit{75.3(a)}$$

$$L_{\text{ft/ton}} = \frac{18{,}000 \, \dfrac{\text{lbf-ft}^2}{\text{ton-yd}^2}}{r_{\text{lbf/yd}^2} w_{\text{ft}}} \qquad \text{[U.S.]} \qquad \textit{75.3(b)}$$

The actual speed, v, will depend on many factors and should not exceed the value for which a quality pavement is produced. The forward speed should also coincide with the plant production rate, R_p, of asphalt concrete.

$$\begin{aligned}
\text{v}_{\text{m/min}} &= \frac{R_{p,\text{tonne/h}} L_{\text{m/tonne}}}{60 \, \dfrac{\text{min}}{\text{h}}} \\[2ex]
&= \frac{R_{p,\text{tonne/h}} \left(1000 \, \dfrac{\text{kg}}{\text{tonne}}\right)\left(1000 \, \dfrac{\text{mm}}{\text{m}}\right)}{w_{\text{m}} t_{\text{mm}} \rho_{\text{kg/m}^3} \left(60 \, \dfrac{\text{min}}{\text{hr}}\right)}
\end{aligned}$$
$$\text{[SI]} \qquad \textit{75.4(a)}$$

$$\begin{aligned}
\text{v}_{\text{ft/min}} &= \frac{R_{p,\text{ton/hr}} L_{\text{ft/ton}}}{60 \, \dfrac{\text{min}}{\text{hr}}} \\[2ex]
&= \frac{R_{p,\text{ton/hr}} \left(2000 \, \dfrac{\text{lbf}}{\text{ton}}\right)\left(12 \, \dfrac{\text{in}}{\text{ft}}\right)}{w_{\text{ft}} t_{\text{in}} \gamma_{\text{lbf/ft}^3} \left(60 \, \dfrac{\text{min}}{\text{hr}}\right)}
\end{aligned}$$
$$\text{[U.S.]} \qquad \textit{75.4(b)}$$

Transportation

11. ROLLING EQUIPMENT

Most specifications now require 92 to 97% density, which means rolling is necessary even with high-density screeds. Roller forward speeds are usually between 1 and $2^1/_2$ mph. Compaction must be completed before the mix cools below 185°F (85°C).

Production efficiency is higher with *vibratory rollers* than with *static rollers*, but consistency, noise, and control are sometimes issues. Because of this, many specifications still require a compaction train that includes static rollers. Modern technology is moving toward *density-on-the-run meters* attached to vibratory rollers. These gauges can determine density on the spot, allowing the operator to adjust vibration frequency and amplitude to maintain a consistent density.

Rubber-tired *pneumatic rollers* provide a kneading action in the finish roll. This is particularly important with chip-seal and other thin-surface treatments.

Lift thickness is dependent on the type of compaction equipment used. When static steel-wheeled rollers are used, the maximum lift thickness that can be compacted is about 3 in (75 mm). When pneumatic or vibratory rollers are used, the maximum thickness that can be compacted is almost unlimited. Generally, lift thickness is limited to 6 to 8 in (150 to 200 mm). Proper placement is problematic with greater thicknesses.

12. CHARACTERISTICS OF ASPHALT CONCRETE

(The following material is consistent with the methods presented in *The Asphalt Handbook, MS-4*, 1989 ed., Asphalt Institute.)

There are three specific gravity terms used to describe aggregate. The *apparent specific gravity*, G_{sa}, of an aggregate (coarse or fine) is

$$G_{sa} = \frac{m}{V_{aggregate}\rho_{water}} \quad \text{[SI]} \qquad 75.5(a)$$

$$G_{sa} = \frac{W}{V_{aggregate}\gamma_{water}} \quad \text{[U.S.]} \qquad 75.5(b)$$

Each of the components, including the coarse aggregate, fine aggregate, and mineral filler, has its own specific gravity (G) and proportion (P) by weight in the total mixture. The *bulk specific gravity*, G_{sb}, of the combined aggregate is

$$G_{sb} = \frac{P_1 + P_2 + \dots P_n}{\dfrac{P_1}{G_1} + \dfrac{P_2}{G_2} + \dots + \dfrac{P_n}{G_n}} \qquad 75.6$$

ASTM procedure C127, "Specific Gravity and Absorption of Coarse Aggregate," is used to measure the specific gravity of coarse aggregate without having to measure its volume. In Eqs. 75.7 through 75.9, A represents the oven-dried weight, B is the weight at saturated-surface dry conditions, and C is the submerged weight of the aggregate in water.

$$G_{sa} = \frac{A}{A - C} \qquad 75.7$$

$$G_{sb} = \frac{A}{B - C} \qquad 75.8$$

$$\text{absorption} = \frac{(100\%)(B - A)}{A} \qquad 75.9$$

ASTM procedure C128, "Specific Gravity and Absorption of Fine Aggregate," is used for the fine aggregate. In Eqs. 75.10 through 75.12, A is the oven-dried weight, B represents the weight of a *pycnometer* filled with water, S is the weight of the soil sample (standardized as 500 g), and C is the weight of a pycnometer filled with the soil sample and water added to the calibration mark.

$$G_{sa} = \frac{A}{B + A - C} \qquad 75.10$$

$$G_{sb} = \frac{A}{B + S - C} \qquad 75.11$$

$$\text{absorption} = \frac{(100\%)(S - A)}{A} \qquad 75.12$$

The *bulk specific gravity*, G_{sb}, of the aggregate mixture is found from Eq. 75.6. The *apparent specific gravity* of the aggregate mixture, G_{sa}, is also calculated from Eq. 75.6, except that the apparent specific gravities are used in place of the bulk specific gravities.

The *effective specific gravity* of the aggregate, G_{se}, is calculated from the maximum specific gravity, the specific gravity of the asphalt, G_b, and the proportion of the asphalt in the total mixture, P_b, expressed as a percentage. The effective specific gravity is always between the bulk and apparent specific gravities.

$$G_{se} = \frac{100\% - P_b}{\dfrac{100\%}{G_{mm}} - \dfrac{P_b}{G_b}} \qquad 75.13$$

ASTM procedure D2041, "Theoretical Maximum Density (Rice Method)," is used to measure the *maximum specific gravity*, G_{mm}, of the paving mixture, which is the specific gravity if the mix had no air voids. However, in practice, some air voids always remain and this maximum specific gravity cannot be achieved.

The maximum specific gravity is calculated from the proportions of aggregate and asphalt, P_s and P_b, respectively, in percent, and the specific gravity of the asphalt, G_b.

$$G_{mm} = \frac{100\%}{\dfrac{P_s}{G_{se}} + \dfrac{P_b}{G_b}} \qquad 75.14$$

If ASTM D2041 is used to determine the maximum specific gravity, G_{mm}, Eq. 75.15 is used. A is the mass of an oven-dried sample, D is the mass of a container filled with water at 77°F (25°C), and E is the mass of the container filled with the sample and water at 77°F (25°C).

$$G_{mm} = \frac{A}{A + D - E} \qquad 75.15$$

By convention, the *absorbed asphalt*, P_{ba}, is expressed as a percentage by weight of the aggregate, not as a percentage of the total mixture.

$$P_{ba} = \frac{(100\%)G_b(G_{se} - G_{sb})}{G_{sb}G_{se}} \qquad 75.16$$

The *effective asphalt content* of the paving mixture, P_{be}, is the total percentage of asphalt in the mixture less the percentage of asphalt lost by absorption into the aggregate.

$$P_{be} = P_b - \frac{P_{ba}P_s}{100\%} \qquad 75.17$$

The *bulk specific gravity of the compacted mixture*, G_{mb}, is determined through testing according to ASTM procedure D2726, "Bulk Density of Cores (SSD)."

The *percent VMA* (*voids in mineral aggregate*) in the compacted paving mixture is the total of the air voids volume and the volume of aggregate available for binding.

$$VMA = 100\% - \frac{G_{mb}P_s}{G_{sb}} \qquad 75.18$$

The percent *total air voids*, P_a, in the compacted mixture represents the air spaces between coated aggregate particles. The total air voids in the compacted mixture should be 3 to 5%. With higher void percentages, voids can interconnect and allow air and moisture to permeature the pavement. If air voids are less than 3%, there will be inadequate room for expansion of the asphalt binder in hot weather. Below 2%, the asphalt becomes plastic and unstable.

$$P_a = VTM = \frac{(100\%)(G_{mm} - G_{mb})}{G_{mm}} \qquad 75.19$$

The *voids filled with asphalt*, VFA, indicates how much of the VMA contains asphalt and is found using Eq. 75.20.

$$VFA = \frac{(100\%)(VMA - VTM)}{VMA} \qquad 75.20$$

Nondestructive *nuclear gauge testing* (with the gauge in the surface or backscatter position) can be used to determine actual relative density or percent air voids of new pavement. *Extraction methods* (which use dangerous and expensive chemicals) of determining density and air voids are now essentially outdated.

13. MARSHALL MIX TEST PROCEDURE

The *Marshall test method* is a density voids analysis and a stability-flow test of the compacted test specimen. The size of a test specimen for a Marshall test is 2.5 in by 4 in (64 mm by 102 mm) in diameter. Specimens are prepared using a specified procedure for heating, mixing, and compacting the asphalt aggregate mixtures. When preparing data for a Marshall mix

design, the stability of the samples that are not 2.5 in (63.5 mm) high must be multiplied by the correlation ratios in Table 75.4.

Figure 75.2 *Volumes in a Compacted Asphalt Specimen (see also Fig. 75.1)*

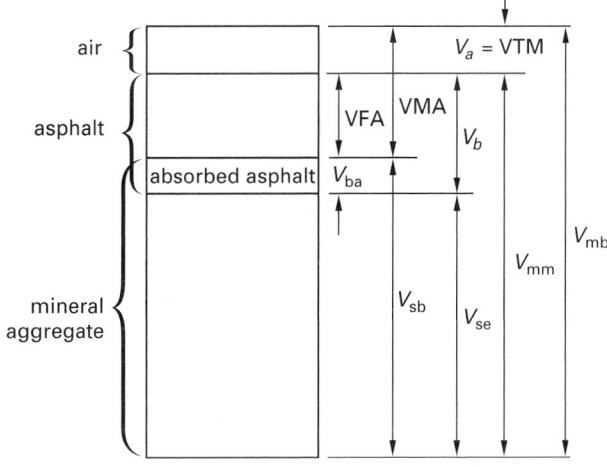

VMA = volume of voids in mineral aggregate
V_{mb} = bulk volume of compacted mix
V_{mm} = voidless volume of paving mix
V_a = volume of air voids
V_b = volume of asphalt
V_{ba} = volume of absorbed asphalt
V_{sb} = volume of mineral aggregate (by bulk specific gravity)
V_{se} = volume of mineral aggregate (by effective specific gravity)

Table 75.4 *Stability Correlation Factors*

specimen volume, cm³ (in³)	approximate specimen thickness, mm (in)	correlation ratio
406–420 (25.0–25.6)	50.8 (2.0)	1.47
421–431 (25.7–26.3)	52.4 (2.06)	1.39
432–443 (26.4–27.0)	54.0 (2.13)	1.32
444–456 (27.1–27.8)	55.6 (2.25)	1.25
457–470 (27.9–28.7)	57.2 (2.25)	1.19
471–482 (28.7–29.4)	58.7 (2.31)	1.14
483–495 (29.5–30.2)	60.3 (2.37)	1.09
496–508 (30.2–31.0)	61.9 (2.44)	1.04
509–522 (31.1–31.8)	63.5 (2.5)	1.00
523–535 (31.9–32.6)	64.0 (2.52)	0.96
536–546 (32.7–33.3)	65.1 (2.56)	0.93
547–559 (33.4–34.1)	66.7 (2.63)	0.89
560–573 (34.2–34.9)	68.3 (2.69)	0.86
574–585 (35.0–35.7)	71.4 (2.81)	0.83
586–598 (35.8–36.5)	73.0 (2.87)	0.81
599–610 (36.6–37.2)	74.6 (2.94)	0.78
611–625 (37.3–38.1)	76.2 (3.0)	0.76

(Multiply cm³ by 0.061 to obtain in³.)
(Multiply mm by 0.0394 to obtain in.)

Reprinted with permission of the Asphalt Institute from *The Asphalt Handbook, Manual Series No. 4 (MS-4)*, 1989 ed., Table 4.3, copyright © 1989.

The objective of a Marshall test is to find the optimum asphalt content for the blend or gradation of aggregates. To accomplish this, a series of test specimens is prepared for a range of different asphalt contents so that when plotted, the test data will show a well-defined optimum value. Tests are performed in $1/2\%$ increments of asphalt contents, with at least two asphalt content samples above the optimum and at least two below.

After each sample is prepared, it is placed in a mold and compacted with 35, 50, or 75 hammer blows, as specified by the design traffic category. The compaction hammer is dropped from a height of 18 in (457 mm), and after compaction the sample is removed and subjected to the bulk specific gravity test, stability and flow test, and density and voids analysis. Results of these tests are plotted, a smooth curve giving the best fit is drawn for each set of data, and an optimum asphalt content determined that meets the criteria of (a) maximum stability and unit weight and (b) median of limits for percent air voids from Table 75.5.

The *stability* of the test specimen is the maximum load resistance in pounds that the test specimen will develop at 140°F (60°C). The *flow value* is the total movement (strain) in units of $1/100$ in occurring in the specimen between the points of no load and maximum load during the stability test.

The *optimum asphalt content* for the mix is the numerical average of the values for the asphalt content. This value represents the most economical asphalt content that will satisfactorily meet all of the established criteria.

Figure 75.3 *Marshall Testing*

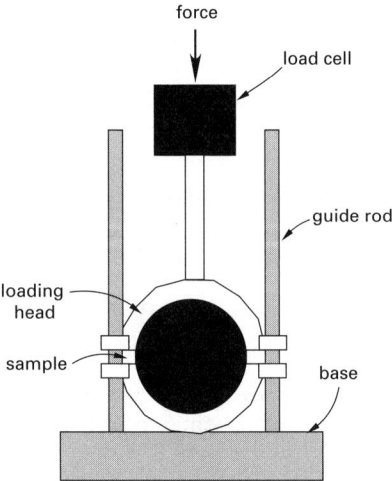

(a) device

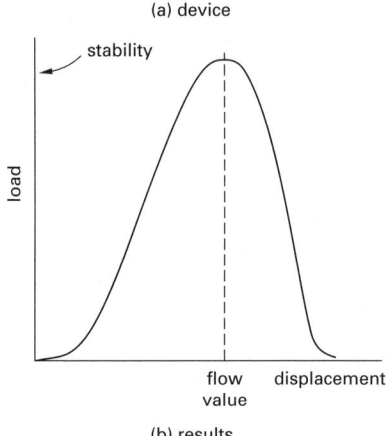

(b) results

Table 75.5 *Marshall Mix Design Criteria*

mix criteria	light traffic ESAL< 10^4 surface and base min–max	medium traffic 10^4 <ESAL< 10^6 surface and base min–max	heavy traffic ESAL> 10^6 surface and base min–max
compactive effort, no. of blows/face	35	50	75
stability, N (lbf)	3336 (750)–NA	5338 (1200)–NA	8006 (1800)–NA
flow, 0.25 mm (0.01 in)	8–18	8–16	8–14
air voids, %	3–5	3–5	3–5
VMA, %	(See Fig. 75.5.)		

(Multiply N by 0.225 to obtain lbf.)

Reprinted with permission of the Asphalt Institute from *The Asphalt Handbook, Manual Series No. 4 (MS-4)*, 1989 ed., Table 4.4, copyright © 1989.

Figure 75.4 *Typical Marshall Mix Design Test Results*

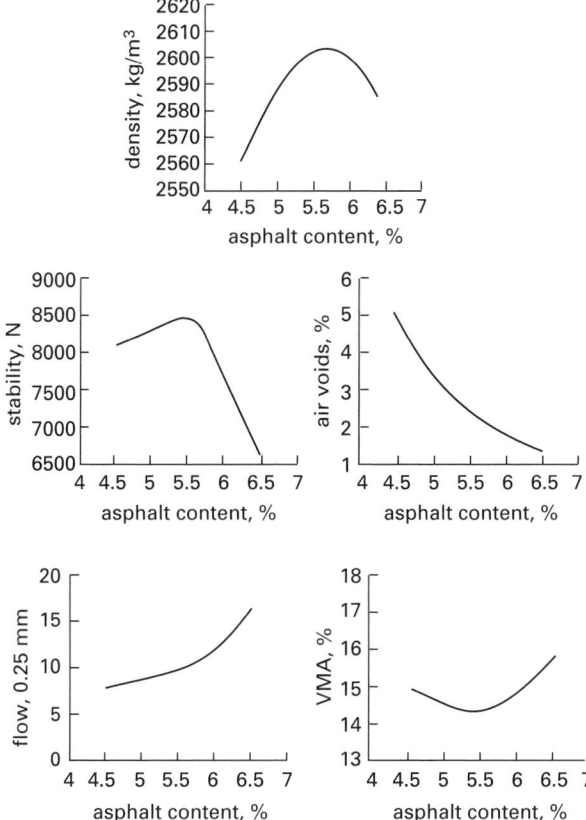

Figure 75.5 *VMA Criterion for Mix Design*

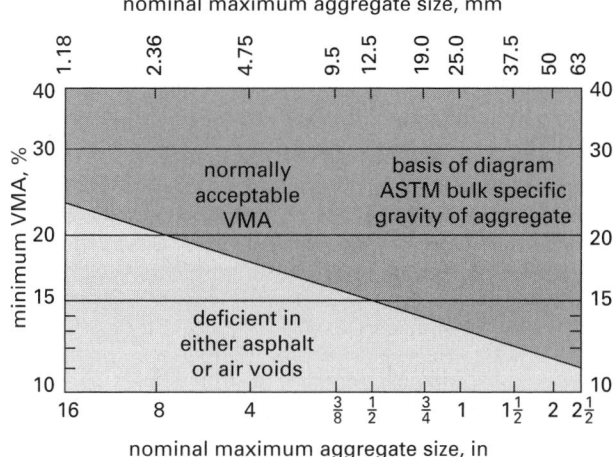

Example 75.1

An asphalt mix uses an aggregate blend of 56% coarse aggregate (specific gravity of 2.72) and 44% fine aggregate (specific gravity of 2.60). The maximum aggregate size for the mixture is $^3/_4$ in (19 mm). The resulting asphalt mixture is to have a specific gravity of 2.30. The asphalt content is to be selected on the basis of light traffic and the following Marshall test data.

maximum content, % by weight of mix	Marshall stability, N	flow, 0.25 mm	theoretical density, kg/m³	mixture specific gravity
4	4260	10.0	2300	2.42
5	4840	12.3	2330	2.44
6	5380	14.4	2340	2.44
7	5060	16.0	2314	2.40
8	3590	19.0	2250	2.30

Solution

step 1: Calculate air voids for each asphalt content and graph these values versus asphalt content. For example, for the 4% asphalt mixture, using Eq. 75.19,

$$\text{VTM}_{4\%} = \frac{(100\%)(G_{\text{mm}} - G_{\text{mb}})}{G_{\text{mm}}}$$

$$= \frac{(100\%)(2.42 - 2.30)}{2.42} = 4.96\%$$

step 2: Calculate the voids in mineral aggregate for each asphalt content. Use Eq. 75.6.

$$G_{\text{sb}} = \frac{100\%}{\dfrac{P_{\text{ca}}}{G_{\text{ca}}} + \dfrac{P_{\text{fa}}}{G_{\text{fa}}}}$$

$$= \frac{100\%}{\dfrac{56\%}{2.72} + \dfrac{44\%}{2.60}} = 2.67$$

Use Eq. 75.18.

$$\text{VMA}_{4\%} = 100\% - \frac{G_{\text{mb}}P_s}{G_{\text{sb}}}$$

$$= 100\% - \frac{(2.30)(96\%)}{2.67} = 17.3\%$$

step 3: Graph the results.

Transportation

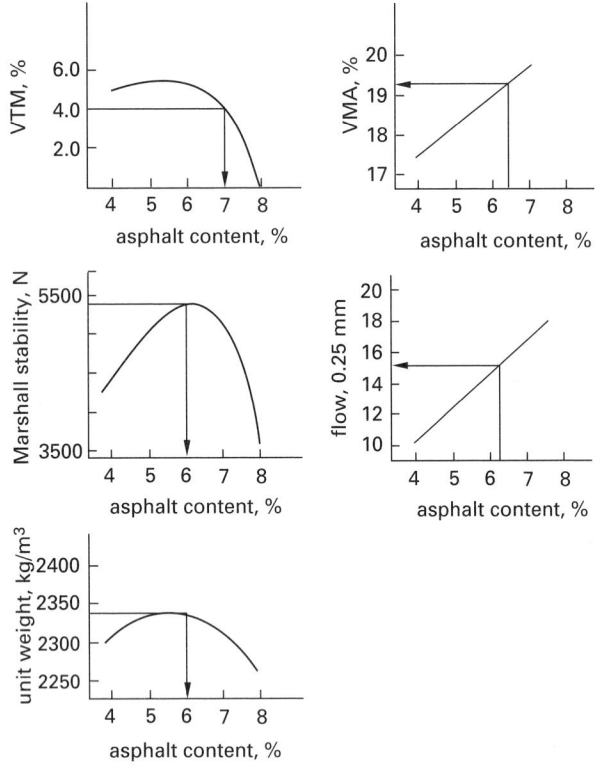

step 4: Obtain asphalt contents for maximum stability, maximum unit weight and median air voids (4% from Table 75.5) from the graphs.

step 5: Determine the optimum AC content.

$$\text{optimum AC} = \frac{\begin{array}{c}\text{AC}_{\text{stability}} + \text{AC}_{\text{unit weight}} \\ + \text{AC}_{4\% \text{ air voids}}\end{array}}{3}$$

$$= \frac{6.0\% + 6.0\% + 7.0\%}{3}$$

$$= 6.33\%$$

step 6: Check stability against the criteria for light traffic. Stability $\approx 5280 > 3336$ [OK]. Check VMA = 19.25 against Fig. 75.5 [OK]. Check flow = 14.8 against Table 75.5 [OK].

14. TRAFFIC

(The information included in this section is consistent with the *AASHTO Guide for the Design of Pavement Structures (AASHTO Guide)*, 1993.)

The AASHTO pavement design method requires that all traffic be converted into *equivalent single-axle loads* (ESALs). This is the number of 18,000 lbf single axles (with dual tires) on pavements of specified strength that would produce the same amount of traffic damage over the design life of the pavement.

Appendix D of the *AASHTO Guide* gives *load equivalency factors* (LEFs) for flexible pavements for various *terminal serviceability indices*, p_t. A p_t value of 2.5 is

assumed unless other information is available. Appendices 75.A, 75.B, and 75.C in this book can be used for $p_t = 2.5$.

$$\text{ESALs} = (\text{no. of axles})(\text{LEF}) \qquad 75.21$$

Since the structural number, SN, is not known until the design is complete, initial values of 3 and 5 are assumed for the structural number of low-volume roads and high-volume roads, respectively. Once the design is complete, the ESALs can be recomputed and the design verified.

15. TRUCK FACTORS

Truck factors, TFs, are the average LEF for a given class of vehicle and are computed using *loadometer data*. The truck factor is calculated as the total ESALs for all axles divided by the number of trucks.

$$\text{TF} = \frac{\text{ESALs}}{\text{no. of trucks}} \qquad 75.22$$

Example 75.2

Using the axle loading data given, calculate the truck factor for 165 total trucks with five or more axles, assuming SN = 5 and $p_t = 2.5$.

Solution

ESALs are obtained by multiplying the number of axles by the LEF for each entry. Obtain LEF values from Apps. 75.A through 75.C.

axle load, lbf	no. of axles	×	LEF	=	ESALs
	single axle				
<3000	0		0.0002		0.000
4000	1		0.002		0.002
8000	6		0.034		0.204
10,000	144		0.088		12.672
16,000	16		0.623		9.968
28,000	1		5.39		5.390
	tandem axles				
<6000	0		0.0003		0.000
10,000	14		0.007		0.098
16,000	21		0.047		0.987
22,000	44		0.180		7.920
27,000	42		0.430		18.039
30,000	44		0.658		28.952
32,000	21		0.857		17.997
34,000	101		1.09		110.090
36,000	43		1.38		59.340
			total ESALs for all axles:		271.659

The truck factor is

$$\text{TF} = \frac{\text{ESALs}}{\text{no. of trucks}} = \frac{271.659 \text{ ESALs}}{165 \text{ trucks}}$$

$$= 1.65 \text{ ESAL/truck}$$

16. DESIGN TRAFFIC

Once the truck factors are computed, the design ESALs can be computed from the distribution of vehicle classes in the AADT and the expected growth rate. If the 20 yr ESAL (i.e., the total number of vehicles over 20 years) is to be predicted from the current (first) year ESAL and if a constant growth rate of $g\%$ per year is assumed, traffic *growth factors*, GFs (also known as *projection factors*), for growth rates, g, can be read as the $(F/A, g\%, 20)$ factors from economic analysis tables. $ESAL_{20}$ is the total number of vehicles over 20 years, not the 20th year traffic.

$$ESAL_{20} = (ESAL_{\text{first year}})(GF) \qquad 75.23$$

The *design traffic* is calculated as the product of the AADT, the fraction of AADT that represents truck traffic, the days in one year, and the growth factor over the design life. The ESAL is obtained by multiplying by the truck factor.

The *directional distribution factor*, D_D, is used to account for the differences in loading according to road direction. It is usually assumed to be 50%. Table 75.6 gives recommended values for *lane distribution factors*, D_L, on multilane facilities. The ESALs for the design lane are computed from Eq. 75.24.

$$w_{18} = D_D D_L \hat{w}_{18} \qquad 75.24$$

Table 75.6 Lane Distribution Factors, D_L

no. of lanes in each direction	fraction of ESALs in design lane
1	1.00
2	0.80–1.00
3	0.60–0.80
4	0.50–0.75

From *Guide for Design of Pavement Structures*, p. II-9, copyright © 1993, by the American Association of State Highway and Transportation Officials, Washington, D.C. Used by permission.

Example 75.3

Given the following traffic data in columns 1 through 3, as well as the truck factors, compute the design ESALs for a 20 yr period and a growth rate of 2%. Disregard contributions of passenger cars to ESALs.

Solution

Example 75.4

The traffic described in Ex. 75.3 has a directional traffic split of 60%/40%, and there are two lanes in each direction. What are the ESALs for the design lane?

Solution

From Table 75.6, $D_L \approx 0.9$ (the midpoint of the 0.8–1.0 range).

From Eq. 75.24,

$$\begin{aligned} w_{18} &= D_D D_L \hat{w}_{18} \\ &= (0.6)(0.9)(43{,}009{,}022 \text{ ESALs}) \\ &= 23{,}224{,}872 \text{ ESALs} \end{aligned}$$

17. STANDARD VEHICLE CLASSIFICATIONS AND DESIGNATIONS

For designing road geometry, AASHTO defines design vehicles according to the following categories: passenger car (P), single-unit truck (SU), single-unit bus (BUS), intermediate-length semitrailer (WB-40), large semitrailer (WB-50), and semitrailer/full trailer combination (WB-60). In the case of WB vehicles, the number represents the approximate wheelbase distance between the front (cab) axle and the last trailer axle.

Classification into axle types is also common. An axle may be defined as "single" or "tandem," and each axle may have single or dual tires. *Spread-tandem axles*, where two axles are separated by more than 96 in (2.2 m), generally are classified as two single axles. (All axles in a truck are included in the calculation of ESALs. For example, a typical tractor/trailer would contribute three quantities to the ESALs—one for the tractor front axle, one for the tractor tandem axles, and one for the trailer tandem axles.)

Except in theoretical stress studies, no attempt is made to account for the number of tires per axle. Although stresses at shallow depths are caused principally by individual wheels acting singly, stresses at greater depths are highest approximately midway between adjacent

(vehicle type) [given]	(AADT) [given]	$\times$	(truck fraction of AADT) [given]	$\times$	$\left(\dfrac{\text{days}}{\text{yr}}\right)$	$\times$	(GF) $=$	(design traffic)	$\times$	(design TF) [given]	$=$	ESAL
single units												
2-axle, 4-tire	12,000		0.11		365		24.30	11,707,740		0.0122		142,834
2-axle, 6-tire	12,000		0.06		365		24.30	6,386,040		0.1890		1,206,962
3+ axles	12,000		0.04		365		24.30	4,257,360		0.1303		554,734
tractor semi-trailers												
3-axle	12,000		0.02		365		24.30	2,128,680		0.8646		1,840,457
4-axle	12,000		0.02		365		24.30	2,128,680		0.6560		1,396,414
5-axle	12,000		0.15		365		24.30	15,965,100		2.3719		37,867,621

total design ESALs: 43,009,022

Transportation

Table 75.7 *Standard Truck Loadings*
(All loads are axle loads in lbf (kg).)[a]

load designation	F_1	F_2	F_3	d_1^b, ft (m)	d_2^b, ft (m)
H20-44	8000 (3636)	32,000(14 545)	0	14(4)	
H15-44	6000 (2727)	24,000 (10 909)	0	14 (4)	
H10-44	4000 (1818)	16,000 (7273)	0	14 (4)	
HS20-44	8000 (3636)	32,000 (14 545)	32,000 (14 545)	14 (4)	14-30 (4-8.3)
HS15-44	6000 (2727)	24,000 (10 909)	24,000 (10 909)	14 (4)	14-30 (4-8.3)
P5	26,000 (11 793)	48,000 (21 772)	48,000 (21 772)	18 (5.4)	18 (5.4)
3	16,000 (7273)	17,000 (7273)	17,000 (7273)	15 (4.2)	4 (1.1)
3S2	(See Fig. 75.6.)				
3-3	(See Fig. 75.6.)				

[a]The mass, in lbm, supported per axle is numerically the same as the force, in lbf, experienced by the pavement.
[b]If the separation between axles is variable, the distance that produces the maximum stress in the section should be used.

Figure 75.6 *Standard Truck Loadings*
(all loads in kips)

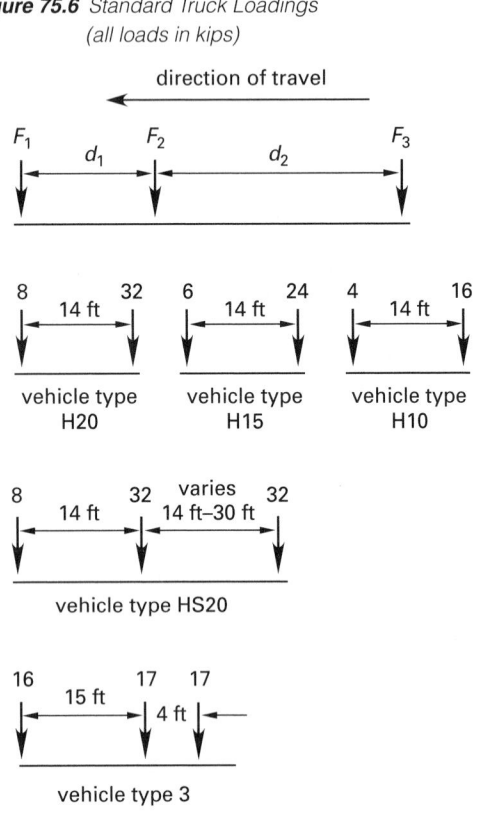

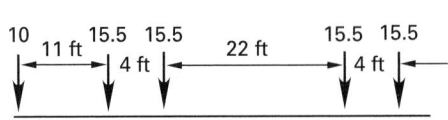

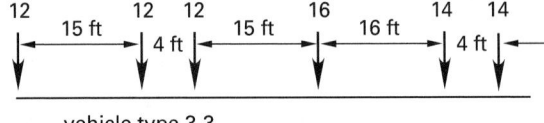

Figure 75.7 *Types of Axles and Axle Sets*

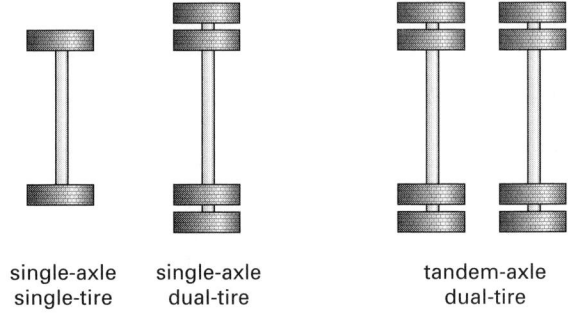

single-axle single-tire   single-axle dual-tire   tandem-axle dual-tire

tires. Deep stresses due to dual-wheeled axles are approximately the same as for single-wheeled axles. Therefore, required pavement thickness is determined by the total axle load, not the number of tires.

Table 75.7 and Fig. 75.6 illustrate standard truck loads commonly used for design.

18. AASHTO METHOD OF FLEXIBLE PAVEMENT DESIGN

The 1993 *AASHTO Guide for the Design of Pavement Structures* (*AASHTO Guide*) is the basis of the conventional flexible pavement design method presented in this chapter. The *AASHTO Guide* is a conservative methodology, so average values can be used for all design variables. The design is based on four main design variables: time, traffic, reliability, and environment. Performance criteria, material properties, and pavement structural characteristics are also considered.

Time: The *analysis period* is the length of time that a given design strategy covers (the *design life* or *design period*). The *performance period* is the time that the initial pavement structure is expected to perform adequately before needing rehabilitation. For example, on a high-volume urban roadway, the analysis period should be 30 to 50 yr, which may include the initial performance period and several *rehabilitation periods* following overlays or maintenance operations.

Traffic: The traffic counts are converted into standard 18 kip ESALs.

Reliability: Reliability considerations ensure that the structure will last for the designated design period. They take into consideration variations in traffic and performance predictions. Facilities that are considered to be more critical are designed using higher reliability factors. It is necessary to select an *overall standard deviation, S_o,* for reliability to account for traffic and pavement performance that is representative of local conditions. Based on historical information obtained during the AASHTO Road Test, appropriate standard deviations for flexible pavements are 0.4 to 0.5.

Table 75.8 *Suggested Levels of Reliability for Various Functional Classifications*

functional classification	recommended level of reliability (percent)	
	urban	rural
interstates and freeways	85–99.9	80–99.9
principal arterials	80–99	75–95
collectors	80–95	75–95
local	50–80	50–80

From *Guide for Design of Pavement Structures*, Sec. II, p. II-9, Table 2.2, copyright © 1993 by the American Association of State Highway and Transportation Officials, Washington, D.C. Used by permission.

19. PERFORMANCE CRITERIA

The *terminal pavement serviceability index, p_t,* represents the lowest pavement serviceability index that can be experienced before rehabilitation, resurfacing, or reconstruction is required. Suggested levels are between 2 and 3, with 2.5 recommended for major highways and 2.0 recommended for less important roads. If costs are to be kept low, the design traffic volume or design period should be reduced. Terminal serviceability should not be reduced, as small changes in it will result in large differences in pavement design.

The *actual initial pavement serviceability, p_o,* represents the actual ride quality of the new roadway immediately after it is installed. This value is not usually known during the initial design, but it may be assumed to be 4.2 for flexible pavement and 4.5 for rigid pavement.

The *change in pavement serviceability index, ΔPSI,* is calculated as the difference between the initial pavement serviceability index and the terminal pavement serviceability index.

$$\Delta\text{PSI} = p_o - p_t \qquad \text{75.25}$$

20. LAYER STRENGTHS

Prior to designing a flexible pavement, the strengths of the pavement layers and the underlying soil must be determined or assumed.

There may be a considerable range in the strength values of the underlying soil. If the range is small for a given area, the lowest value should be selected for design. If there are a few exceptionally low values that come from one area, it may be possible to specify replacing that area's soil with borrow soil to increase the localized weak spots. If there are changing geological formations along the route that modify the value, it may be necessary to design different pavement sections as appropriate.

The *effective roadbed soil resilient modulus, M_R,* must be determined. The resilient modulus can either be measured in the laboratory using the AASHTO T274 test procedure, or it can be predicted from correlations with nondestructive deflection measurements. For roadbed materials, laboratory resilient modulus tests should be performed on representative soils under different representative seasonal moisture conditions. The effective roadbed modulus represents the combined effect of all the seasonal modulus values. The *relative damage value, u_f,* is estimated using the vertical scale on the right side of Fig. 75.8 or by Eq. 75.26. All u_f values are then summed, and the total is divided by the number of seasons, giving the average u_f value. The corresponding effective subgrade modulus value is read from the vertical scale.

$$u_f = (1.18 \times 10^8)M_R^{-2.32} \qquad \text{75.26}$$

The resilient modulus, M_R, is the same as the modulus of elasticity, E, of the soil. It is not the same as the *modulus of subgrade reaction, k,* used in rigid pavement design, although the two are related. For positive values of the resilient modulus, $M_R \approx 19.4k$.

Different state departments of transportation use different methods of classifying materials. The *California bearing ratio* (CBR) and *soil resistance value* (R-value) are widely used. It may be necessary to convert one known strength parameter to another for use with a particular design procedure. Equations 75.27 and 75.28 are used to make such conversions. However, correlations of the resilient modulus with other measured parameters, such as the R-value, are usually poor.

$$M_R = 1500(\text{CBR}) \quad [\textit{AASHTO Guide Eq. 1.5.1}] \qquad \text{75.27}$$

To convert from the R-value to the resilient modulus for fine-grained soils, use Eq. 75.28.

$$M_R = 1000 + 555R \qquad \left[\begin{array}{c} \textit{AASHTO Guide Eq. 1.5.3} \\ R \leq 20 \end{array}\right]$$

$$\text{75.28}$$

Figure 75.8 *Effective Roadbed Soil Resilient Modulus Estimation Chart (serviceability criteria)*

month	roadbed soil modulus, (psi) M_R	relative damage, u_f
January		
February		
March		
April		
May		
June		
July		
August		
September		
October		
November		
December		
	summation: Σu_f =	

average: $\bar{u}_f = \frac{\Sigma u_f}{n}$ = ______

effective roadbed soil resilient modulus, M_R (psi) = ______ (corresponds to $\bar{u}_f$)

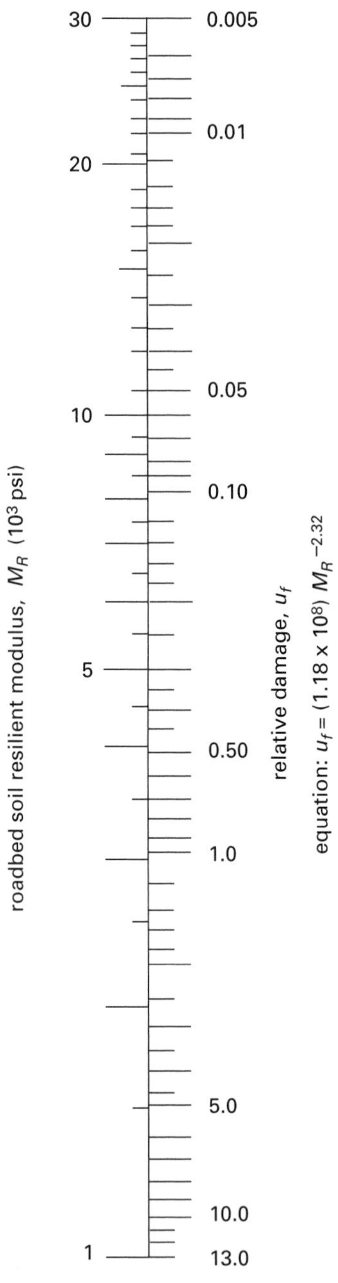

From *Guide for Design of Pavement Structures*, Sec. II, p. II-14, Fig. 2.3, copyright © 1993 by the American Association of State Highway and Transportation Officials, Washington, D.C. Used by permission.

Figure 75.9 is a nomograph provided by AASHTO for determining the design structural number for specific design conditions: estimated future traffic, w_{18}, over the performance period; the reliability, R, which assumes all input variables are average values; the overall standard deviation, S_o; the effective resilient modulus of the roadbed material, M_R; and the design serviceability loss, ΔPSI.

Figure 75.9 *AASHTO Nomograph for Flexible Pavement Design*

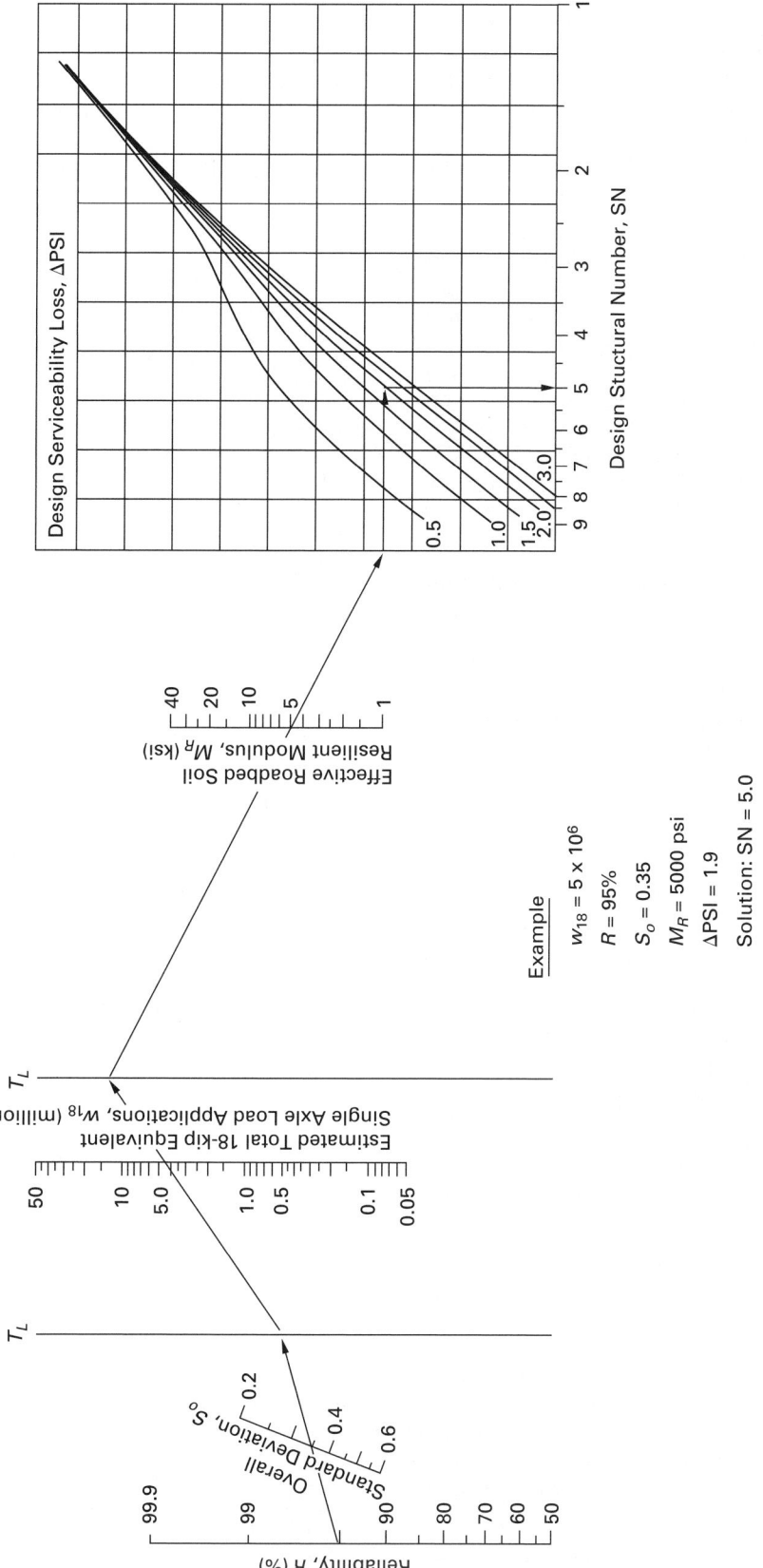

Example

$w_{18} = 5 \times 10^6$
$R = 95\%$
$S_o = 0.35$
$M_R = 5000$ psi
ΔPSI = 1.9
Solution: SN = 5.0

From *Guide for Design of Pavement Structures*, Sec. II, p. II-32, Fig. 3.1, copyright © 1993 by the American Association of State Highway and Transportation Officials, Washington, D.C. Used by permission.

21. PAVEMENT STRUCTURAL NUMBER

AASHTO combines pavement layer properties and thicknesses into one variable called the design *structural number*, SN. Once the structural number for a pavement is determined, a set of pavement layer thicknesses are chosen. When combined, these layer thicknesses must provide the load-carrying capacity corresponding to the design structural number.

Equation 75.29 is the AASHTO *layer-thickness equation*. D_1, D_2, and D_3 represent actual thicknesses (in inches) of surface, base, and subbase courses, respectively. (If a subbase layer is not used, the third term is omitted.) The a_i are the *layer coefficients*, also known as *strength coefficients*. m_2 and m_3 represent *drainage coefficients* for base and subbase layers, respectively. Typical values of drainage coefficients vary from 0.4 to 1.40, as recommended in Table 2.4 of the *AASHTO Guide*. Values greater than 1.0 are assigned to bases and subbases with good/excellent drainage and are seldom saturated [*AASHTO Guide* Table 2.4].

Theoretically, any combinations of thicknesses that satisfy Eq. 75.29 will work. However, minimum layer thicknesses result from construction techniques and strength requirements. The thickness of the flexible pavement layers should be rounded to the nearest $1/2$ in (12 mm). When selecting layer thicknesses, cost effectiveness as well as placement and compaction issues must be considered to avoid impractical design.

$$SN = D_1 a_1 + D_2 a_2 m_2 + D_3 a_3 m_3 \qquad 75.29$$

The layer coefficients, a_i, vary from material to material, but the values given in Table 75.10 can be used for general calculations. The *AASHTO Guide* contains correlations for all three layer coefficients as functions of the materials' elastic (resilient) moduli. Although the a_1 correlation is purely graphical (see *AASHTO Guide* Fig. 2.5, p. II-18), mathematical correlations are given for a_2 and a_3 for granular materials.

$$a_2 = 0.249(\log_{10} E_{BS}) - 0.977$$
$$[\textit{AASHTO Guide} \text{ p. II-20}] \qquad 75.30$$
$$a_3 = 0.227(\log_{10} E_{SB}) - 0.839$$
$$[\textit{AASHTO Guide} \text{ p. II-22}] \qquad 75.31$$

Flexible pavements are layered systems and are designed accordingly. First, the required structural number over the native soil is determined, followed by the structural number over the subbase and base layers, using applicable strength values for each. The maximum allowable thickness of any layer can be computed from the differences between the computed structural numbers.

This method should not be used to determine the structural number required above subbase or base materials having moduli of resilience greater than 40,000 psi (275 MPa). Layer thicknesses for materials above high-strength subbases and bases should be based on cost effectiveness and minimum practical thickness considerations.

Table 75.9 *Minimum Thickness*[a]

traffic (ESAL)	asphalt concrete in (mm)	aggregate base[b] in (mm)
< 50,000	1.0 (25) (or, surface treatment)	4 (100)
50,001–150,000	2 (50)	4 (100)
150,001–500,000	2.5 (63)	4 (100)
500,001–2,000,000	3 (75)	6 (150)
2,000,001–7,000,000	3.5 (88)	6 (150)
> 7,000,000	4 (100)	6 (150)

(Multiply in by 25.4 to obtain mm.)
[a] Minimum thicknesses may also be specified by local agencies and by contract.
[b] Includes cement-, lime-, and asphalt-treated bases and subbases.

From *Guide for Design of Pavement Structures*, Sec. II, p. II-35, copyright © 1993 by the American Association of State Highway and Transportation Officials, Washington, D.C. Used by permission.

Figure 75.10 *Procedure for Determining Thicknesses of Layers Using a Layered Analysis Approach*[a]

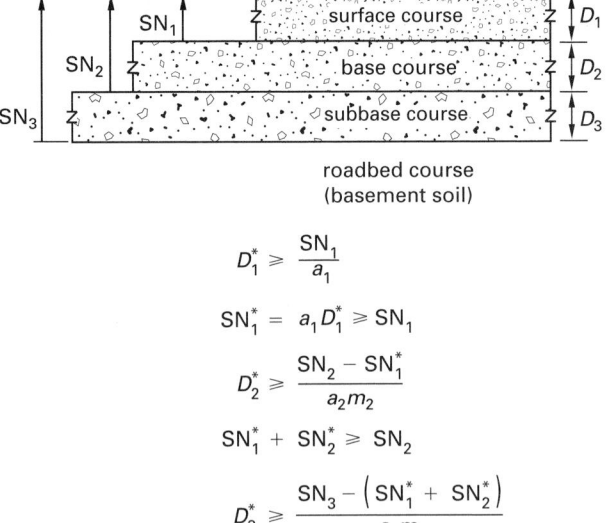

$$D_1^* \geq \frac{SN_1}{a_1}$$

$$SN_1^* = a_1 D_1^* \geq SN_1$$

$$D_2^* \geq \frac{SN_2 - SN_1^*}{a_2 m_2}$$

$$SN_1^* + SN_2^* \geq SN_2$$

$$D_3^* \geq \frac{SN_3 - \left(SN_1^* + SN_2^*\right)}{a_3 m_3}$$

[a] Asterisks indicate values actually used, which must be equal to or greater than the required values.

Example 75.5

Determine the pavement thicknesses for two pavement layers over subgrade that have a required structural number of 5. Layer 1 consists of asphalt concrete with a strength coefficient, a_1, of 0.44 and a minimum thickness of 3.00 in. Layer 2 is a granular base with a strength coefficient, a_2, of 0.13; a drainage coefficient, m_2, of 1.00; a minimum thickness of 6.00 in; and a maximum thickness of 25 in.

Table 75.10 Typical Layer Strength Coefficients[a] (1/in)

	value	range
subbase coefficient, a_3		
sandy gravel[b]	0.11	
sand, sandy clay		0.05–0.10
lime-treated soil	0.11	
lime-treated clay, gravel		0.14–0.18
base coefficient, a_2		
sandy gravel[b]	0.07	
crushed stone	0.14	0.08–0.14
cement treated base (CTB)	0.27	0.15–0.29
seven-day $f'_c > 650$ psi		
(4.5 MPa)	0.23	
400–650 psi (2.8–4.5 MPa)	0.20	
< 400 psi (2.8 MPa)	0.15	
bituminous treated base (BTB)		
coarse	0.34	
sand	0.30	
lime-treated base		0.15–0.30
soil cement	0.20	
lime/fly ash base		0.25–0.30
surface course coefficient, a_1		
plant mix[b]	0.44	
recycled AC,		
3 in or less	0.40	0.40–0.44
4 in or more	0.42	0.40–0.44
road mix	0.20	
sand-asphalt	0.40	

(Multiply 1/in by 0.0394 to obtain 1/mm.)
(Multiply psi by 6.89 to obtain kPa.)
[a] The AASHTO method correlates layer coefficients with resilient modulus.
[b] The average value for materials used in the original AASHTO road tests were:

asphaltic concrete surface course	0.44
crushed stone base course	0.14
sandy gravel subbase	0.11

Compiled from a variety of sources.

Solution

Trial 1: Try a 3 in thick bituminous pavement. Use Eq. 75.29.

$$SN_1 = \left(0.44 \, \frac{1}{in}\right)(3.00 \text{ in}) = 1.32$$

Determine the thickness of aggregate base required.

$$D_2 = \frac{SN_2 - SN_1}{a_2 m_2}$$

$$= \frac{5 - 1.32}{\left(0.13 \, \dfrac{1}{in}\right)(1.00)}$$

$$= 28.3 \text{ in} \quad [> 25 \text{ in}]$$

Trial 2: Try a 5 in thick bituminous pavement.

$$SN_1 = \left(0.44 \, \frac{1}{in}\right)(5.00 \text{ in}) = 2.20$$

$$D_2 = \frac{SN_2 - SN_1}{a_2 m_2}$$

$$= \frac{5 - 2.20}{\left(0.13 \, \dfrac{1}{in}\right)(1.00)}$$

$$= 21.5 \text{ in} \quad [OK]$$

22. ASPHALT INSTITUTE METHOD OF FULL-DEPTH FLEXIBLE PAVEMENT DESIGN

(The Asphalt Handbook, MS-4, 1989 edition, is the source for the information in this section. The Asphalt Institute pavement design method was developed in 1981. It makes several assumptions regarding material properties, enabling the design to be read from a series of charts. Like the AASHTO method, it uses 18 kip ESALs as a measure of traffic loading.)

Two designs are considered: full-depth asphalt (asphalt concrete over asphalt subgrade) and asphalt concrete over untreated granular base. Environmental conditions are considered for the effects of temperature on asphalt stiffness and the effects of freezing and thawing on subgrade and untreated base materials.

The following steps constitute the design procedure.

step 1: Determine the 20 yr, 18 kip equivalent single-axle loading (ESAL) for the pavement.

step 2: The resilient modulus of the subgrade is the design parameter, and it is a function of traffic. Select a design subgrade percentile based on Table 75.11 and design ESALs.

Table 75.11 Design Resilient Modulus Percentiles (highways)

traffic, ESALs	design subgrade percentile
≤ 10,000	60
10,000–1,000,000	75
≥ 1,000,000	87.5

Reprinted with permission of the Asphalt Institute from *The Asphalt Handbook, Manual Series No. 4 (MS-4)*, 1989 ed., Table 11.1, copyright © 1989.

Table 75.12 Minimum Asphalt Concrete Surface Thickness

traffic, ESALs	traffic conditions	min. asphalt concrete, in (mm)
≤ 10,000	light rural roads	1 (25)
10,000–1,000,000	medium truck traffic	1.5 (40)
≥ 1,000,000	heavy truck traffic	2.0 (50)

(Multiply in by 25.4 to obtain mm.)

Reprinted with permission of the Asphalt Institute from *The Asphalt Handbook, Manual Series No. 4 (MS-4)*, 1989 ed., Table 11.6, copyright © 1989.

Transportation

step 3: For full-depth bituminous asphalt pavement, use Fig. 75.11 or 75.12 to determine minimum pavement thickness. For asphalt concrete over granular base, determine the minimum asphalt concrete surface thickness according to Fig. 75.13 or Fig. 75.14.

step 4: Select the aggregate base thickness as 6 or 12 in.

step 5: From Figs. 75.13 and 75.14, find the asphalt concrete thickness to be used over the corresponding base thickness.

Figure 75.11 *Full-Depth Asphalt Concrete (SI units)*

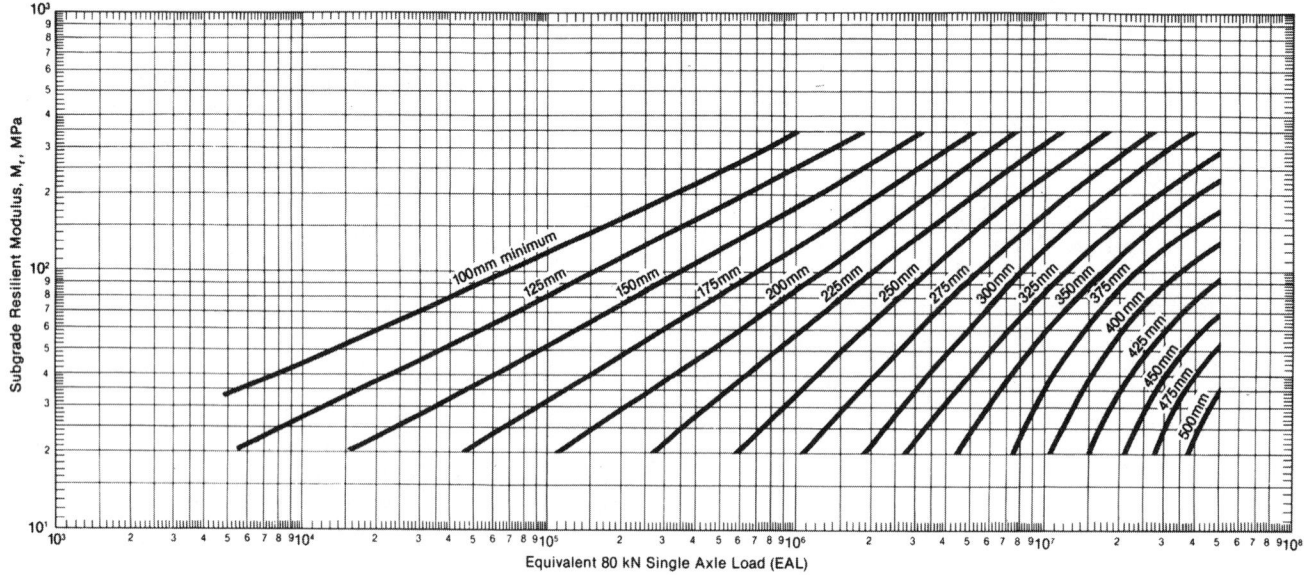

Reprinted with permission of the Asphalt Institute from *The Asphalt Handbook, Manual Series No. 4 (MS-4)*, 1989 ed., Fig. 11.5, copyright © 1989.

Figure 75.12 *Full-Depth Asphalt Concrete (customary U.S. units)*

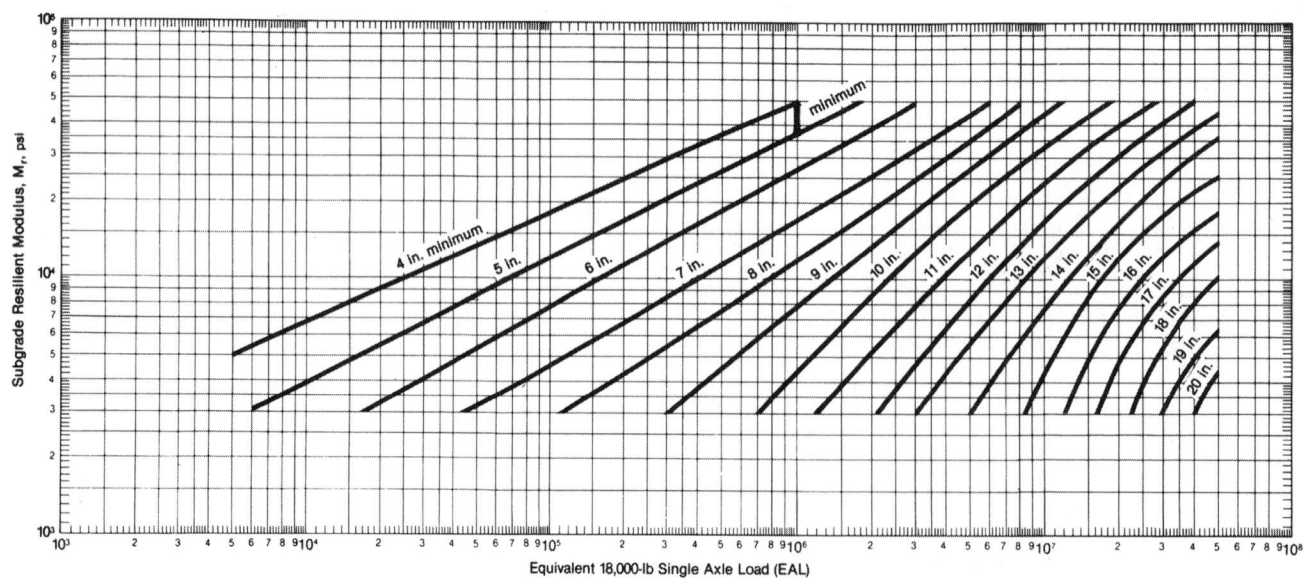

Reprinted with permission of the Asphalt Institute from *The Asphalt Handbook, Manual Series No. 4 (MS-4)*, 1989 ed., Fig. 11.8, copyright © 1989.

Figure 75.13 Asphalt Concrete on Untreated Aggregate Base
(SI units)

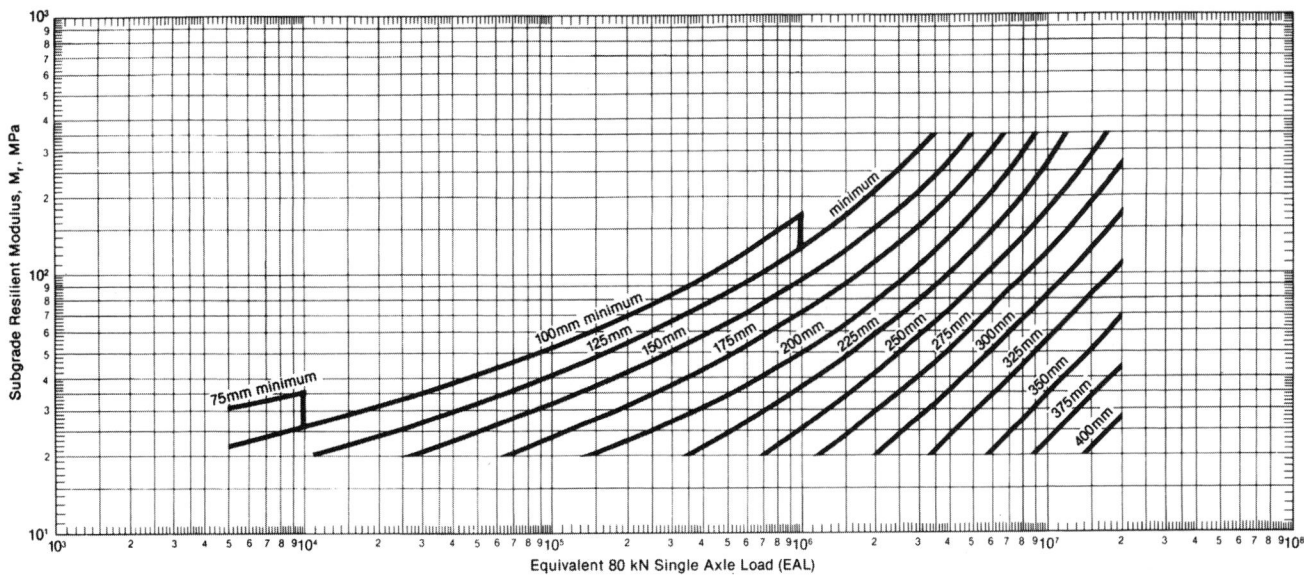

Reprinted with permission of the Asphalt Institute from *The Asphalt Handbook, Manual Series No. 4 (MS-4)*, 1989 ed., Fig. 11.7, copyright © 1989.

Figure 75.14 Asphalt Concrete on Untreated Aggregate Base
(customary U.S. units)

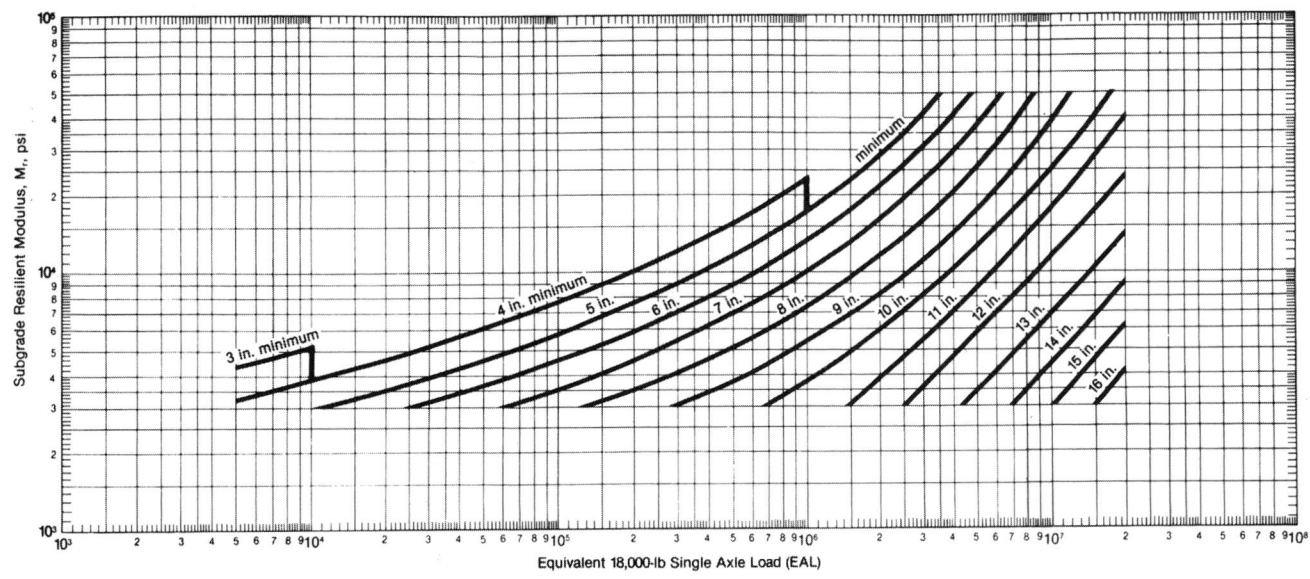

Reprinted with permission of the Asphalt Institute from *The Asphalt Handbook, Manual Series No. 4 (MS-4)*, 1989 ed., Fig. 11.10, copyright © 1989.

Example 75.6

The following California bearing ratio (CBR) values were obtained from multiple tests on the base soil over a proposed roadway: 7, 6, 8, 4, 6, 9, 5, and 10. Determine the design resilient modulus, M_R, for a traffic level of 100,000.

Solution

Use Eq. 75.27 for each sample.

CBR	M_R, psi	number greater	quantity percentage
10	15,000	1	$\left(\frac{1}{8}\right)(100\%) = 12.5\%$
9	13,500	2	25%
8	12,000	3	37.5%
7	10,500	4	50%
6	9000	6	75%
6	9000	6	75%
5	7500	7	87.5%
4	6000	8	100%

Plot the values. From the plot and Table 75.11, the required resilient modulus at 100,000 ESALs is the 75th percentile, and $M_R = 9000$ psi.

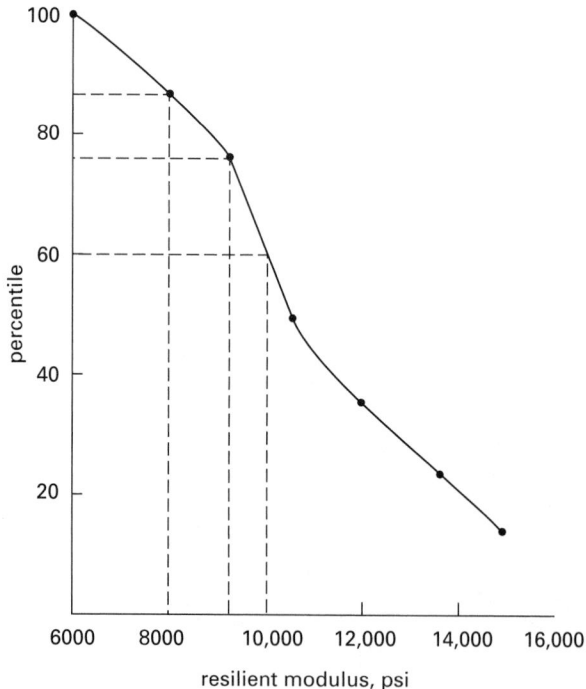

Example 75.7

Design a full-depth asphalt pavement for a subgrade modulus of 6000 psi and 1,000,000 ESALs.

Solution

Use Fig. 75.12. At 1,000,000 ESALs, the full-depth pavement design is 9.5 in.

Example 75.8

Design an asphalt concrete pavement over a granular base for a subgrade modulus of 6000 psi and 1,000,000 ESALs.

Solution

Use Fig. 75.14. At 1,000,000 ESALs and a 6000 psi granular base, the required asphalt pavement thickness is 8.5 in.

23. ASPHALT PAVEMENT RECYCLING

There are a variety of processes that collectively comprise *recycled asphalt pavement* (RAP) technology. The primary benefit of recycling is cost effectiveness, although strength and reliability do not appear to be affected. *Full-depth reclamation* turns a roadway into the base material for a new surface. The original roadway pavement and some of the underlying material are removed, pulverized, and reused.

Profiling, also known as *surface recycling*, is a modification of the visible surface of the pavement. *Cold planing* is the removal of asphalt pavement by special pavement milling and planing machines (i.e., *profiling equipment*). The resulting pavement with long striations provides a good skid-resistant surface. Cold planing is used to restore a road to an even surface. The material removed is 95% aggregate and 5% asphalt cement. Conventional milling machines run at 5 to 10 ft/min (1.5 to 3 m/min) with a 1 to 2 in (25 to 50 mm) cut. High-capacity machines can run at up to 80 ft/min (24 m/min) with a 4 to 6 in (100 to 150 mm) cut (though half this speed is more typical).

Cold in-place recycling employs a *cold train* of equipment to mill 1 to 6 in (25 to 150 mm) of pavement, crush, size, add asphalt rejuvenating agents and/or light asphalt oil in a pug mill, and finally redeposit and recompact the mixture in a conventional manner. A cold-recycled mix will have at least 6% voids, requiring some type of seal or overlay to keep air and water from entering. The overlay can be a sand, slurry, or chip seal. To achieve the required density, roller compaction is required.

Hot in-place recycling, also known as *hot in-place remixing* and *surface-recycling*, begins by heating the pavement to above 250°F (120°C) with high-intensity indirect propane heaters. Then, a scarifier loosens and removes a layer of up to 2 in (50 mm) of softened asphalt concrete. The removed material is mixed with a rejuvenator in a traveling pugmill, augered out laterally, redistributed, and compressed by a screed and rollers. A petroleum-based agent may be subsequently applied to restore the asphalt cement's adhesive qualities, or up to 2 in (50 mm) of additional new hot-mix asphalt cement can be applied over the replaced surface. Alternatively, a heavier petroleum product can be applied to help the recycled surface withstand direct traffic. The entire paving train runs at approximately 2 ft/min. The pavement can be used almost immediately.

Hot mix recycling is used in the majority of pavement rehabilitation projects. This process removes the existing asphalt pavement, hauls it to an offsite plant, and blends it with asphalt cement and virgin rock aggregate. Batch plants can blend 20 to 40% RAP with virgin material, though 10 to 20% is more typical. Drum mixers can blend as much as 50% RAP with virgin material.

Microwave asphalt recycling produces hot mix with 100% RAP. Warm air is first used to dry and pre-heat the reclaimed pavement. Microwave radiation is used to heat the stones in the mixture (since asphalt is not easily heated by microwaves) to approximately 300°F (150°C) without coking any hydrocarbons. Rejuvenating agents restore the pavement's original characteristics, or new aggregate can be used to improve performance.

24. SUPERPAVE™

Superior Performing Asphalt Pavement (Superpave) is an outgrowth of the U.S. *Strategic Highway Research Program* completed in 1993. Superpave is one of several methods of greatly increasing the stone-on-stone contact in the pavement, thus improving its loadbearing capacity. Toward that end, Superpave mixtures pay close attention to voids in the mineral aggregate (VMA), air voids in the mix, and voids filled with asphalt (VFA). Specifications also are set for coarse aggregate angularity and flat-and-elongated particles. The use of sand with rounded edges is reduced.

Three elements make up the Superpave concept. Level 1 mix design includes specifying asphalts by a set of performance-based *binder specifications*. Specifications include stiffness, dynamic shear (stiffness at high and medium temperatures), bending (stiffness at low temperatures), and in some cases direct tension. The binder specifications relate laboratory tests to actual field performance. Level 1 mix design is currently used by most highway agencies and is replacing older methods of asphalt call-outs.

Level 2 mix design includes performance-based tests to measure primary mixture performance factors: fatigue cracking, permanent deformation, low-temperature (thermal) cracking, aging, and water sensitivity.

Level 3 mix design is a computer-aided volumetric mix design and analysis incorporating test results, geographical location, and climatological data as well as new mix-testing technology such as the Superpave *gyratory compactor*, SGC. The laboratory gyratory compactor replaces the Marshall hammer for compacting mixture specimens.

Use of RAP in Superpave pavements is more difficult to quantify since the source and characteristics of RAP are not known in advance.

25. STONE MATRIX ASPHALT

Stone matrix asphalt (SMA), also known as *split mastic, stone-filled asphalt* and *stone mastic asphalt*, is a design method that has been imported from Europe. SMA has withstood years of punishment under Europe's heavier axle loads and studded tires without undue deterioration or rutting.

SMA uses single-sized cubical stones, typically 100% crushed rock or stone, in close contact. In the United States, typically 100% of an SMA aggregate will pass through a $^3/_4$ in sieve and 80 to 90% will pass through a $^1/_2$ in sieve, but only 25 to 30% will pass through a no. 4 sieve. (By contrast, a conventional mix will pass about 55% through a no. 4 sieve.) A rich mortar of asphalt cement, fibers, and fine aggregate (i.e., sand) is used to fill the voids and prevent asphalt drain-down. The sand fraction contains a large amount of manufactured crushed sand that may be washed out if it contains too many fine particles. The mortar provides durability.

SMA requires different aggregates, different mixing technology, and different paving methods. Because of this, it is a specialty tool for highways that carry the heaviest traffic.

SMA contains a higher fraction (6 to 8%) of hard, low-penetration grade asphalt cement. The mixture often incorporates cellulose fibers, mineral (rock wool) fibers, or, less frequently, polymers to stiffen the mix and improve toughness.

When placed, SMA compacts less than $^1/_8$ in (3 mm) per inch of thickness, and very little at all under 240°F (115°C). Paving rates are lower because the mix is stiffer. Since SMA compacts less and sets up faster, rollers have to follow closely. Most contractors roll SMA with a team of three 10- or 12-ton rollers, one working right behind the other, all closely following the paver.

SMA is currently more expensive than conventional mixes, but the general consensus is that lower life-cycle costs and extended life more than offset the initial cost premium.

Although SMA can be (and has been) designed with the traditional 50- and 75-blow Marshall mix methods, cutting-edge efforts include using the PG grading system for specifying asphalt cements. Like Superpave, SMA's strength comes from its stone-on-stone contact. SMA differs from Superpave, however, in the methods used to specify the asphalt and to design and test the mixture. However, the differences are becoming less distinct.

26. ADVANCED, ALTERNATIVE, AND EXPERIMENTAL FLEXIBLE PAVEMENTS

Asphalt rubber (AR) is a general term referring to a wide range of paving products that combine virgin or reclaimed rubber with asphalt cement. *Rubber-modified asphalt*, also known as *crumb-rubber modified (CRM) asphalt*, is made from scrap rubber from tires.

Transportation

In the "wet process" (*reacted asphalt process* or *asphalt hot rubber mix* (AHRM)), the rubber is blended into hot asphalt, creating an asphalt-rubber binder with 10 to 25% rubber by weight. The mixture is held at 375 to 400°F (190 to 205°C) and agitated for 20 to 60 min. In the "dry process" (*rubberized asphalt process*), *crumb rubber*, also known as *crumb rubber modifier* (CRM), in the form of pea-sized beads replaces about 3% of the stone aggregate.

Asphalt rubber has attracted considerable attention in recent years. Federal requirements mandating the use of crumb rubber have been eliminated, mainly out of concerns over the health effects on paving workers. But some states continue to require it. Concerns have been raised about recyclability and combustible roads catching fire. Experience in California seems to indicate that asphalt rubber may result in pavement sections half the thickness of conventional asphalt concrete pavement. Asphalt rubber also holds heat longer than conventional mixes, allowing night placement at temperatures as low as 50°F (10°C). Other preliminary tests are mixed, indicating higher costs without significant structural or durability benefits.

Fast-track pavement is a fast-setting (using type-III portland cement) concrete. Fast-track pavement is used for asphalt overlays, where quick restoration without significant traffic interruption is needed. 4 in (100 mm) layers typically set up in 12 hr and are ready for traffic in 18 hr.

White-topping (also known as *ultrathin white-topping*, UTW) is a rehabilitation method used for rutted asphalt. It creates an economic surface for high-use areas. The technique involves placing a thin (e.g., 75 to 100 mm) layer of fiber-reinforced, high-performance concrete over rutted asphalt. The process forms a highly durable bond (lasting 3 to 4 times longer than asphalt), and it does not need the regular milling or maintenance that asphalt does. The white-topping has the flexibility of asphalt and the loadbearing capacity of concrete. Although asphalt can be placed less expensively per volume than concrete, the life-cycle costs of white-topping repair appear to be lower.

Roller-compacted concrete (RCC) is a zero-slump mixture that is placed and compacted. RCC is good for pothole repair.

Trashphalt is asphalt paving using a wide variety of scrap and waste materials including glass, rubber tire scrap, fly ash, steel refining waste, porcelain from crushed plumbing fixtures, roofing shingles, concrete, and brick. Concern has been raised that such paving materials will probably preclude pavement recycling efforts in later years.

Coal-derived *synthetic asphalt* is a technology that has become less important than when it was first developed. As long as oil prices remain low, it is unlikely that synthetic asphalt will be used much.

Sulfur concrete (*sulfur-asphalt concrete, sulfur-extended concrete, sulphex,* etc.) is manufactured by heating sulfur (20 to 40% by weight) to approximately 280 to 290°F (138 to 143°C) until it is molten, combining with asphalt, and then adding mineral filler and aggregate. (Below 300°F (150°C), elemental sulfur is the primary vapor component. Above that, sulfur dioxide and hydrogen sulfide become the dominant pollutants. Therefore, molten sulfur must be kept below 300°F (150°C).) First-generation sulfur concrete using pure sulfur experienced early rutting. Second-generation sulfur concrete used sulfur that had been chemically modified to reduce brittleness, and it performs better. There is currently little incentive to perfect sulfur concrete.

27. SUBGRADE DRAINAGE

Subgrade drains are applicable whenever the following conditions exist: (a) high groundwater levels that reduce subgrade stability and provide a source of water for frost action, (b) subgrade soils of silts and very fine sands that become quick or spongy when saturated, (c) water seepage from underlying water-bearing strata, (d) drainage path of higher elevations intercepts sag curves with low-permeability subgrade soils below.

Figure 75.15 illustrates typical subgrade drain placement. In general, drains should not be located too close to the pavement (to prevent damage to one while the other is being worked on), and some provision should be made to prevent the infiltration of silt and fines into the drain. Roofing felt or geotextile sleeves can be placed around the drains for this purpose.

Figure 75.15 Typical Subgrade Drain Details

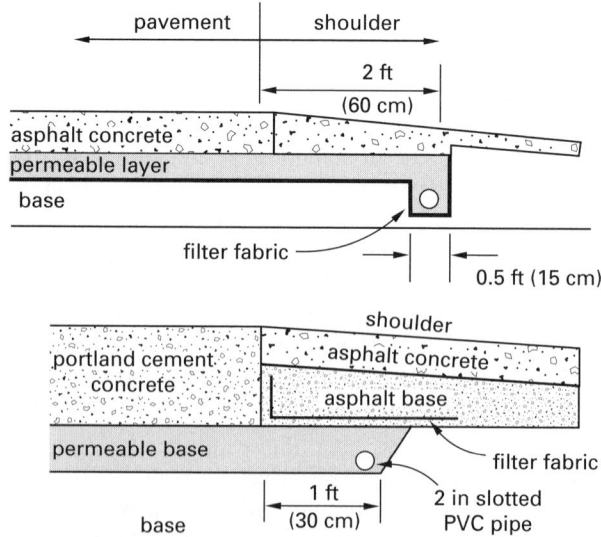

28. DAMAGE FROM FROST AND FREEZING

Frost heaving and reduced subgrade strength, along with the accompanying pumping during spring thaw, can quickly destroy a pavement. The following techniques can be used to reduce damage to a flexible pavement in areas susceptible to frost: (a) constructing stronger and thicker pavement sections, (b) lowering the water table by use of subdrains and drainage ditches, (c) using layers of coarse sands or waterproof sheets beneath the pavement surface to reduce capillary action, (d) removing and replacing frost-susceptible materials to a level beneath the zone of frost penetration, and (e) using rigid foam sheets to insulate and reduce the depth of frost penetration.

76

Rigid Pavement Design

Nomenclature

A	area	ft^2	m^2
CBR	California bearing ratio	–	–
C_d	drainage coefficient	–	–
D	depth (thickness)	ft	m
f_c'	concrete compressive strength	lbf/in^2	Pa
f_s	steel stress	lbf/in^2	Pa
f_t	allowable working stress	lbf/in^2	Pa
E_c	modulus of elasticity of concrete	lbf/in^2	Pa
F	friction factor	–	–
G	specific gravity	–	–
J	load transfer coefficient	–	–
k	modulus of subgrade reaction	lbf/in^3	N/m^3
L	slab length	ft	m
LSF	load safety factor	–	–
M_R	resilient modulus	lbf/in^2	Pa
p_t	terminal serviceability	–	–
P	percentage	percent	percent
PSI	present serviceability index	–	–
R	reliability	–	–
S	soil support value	–	–
S_c'	modulus of rupture	lbf/in^2	Pa
S_o	overall standard deviation	–	–
SN	structural number	–	–
t	thickness	ft	m
V	volume	ft^3	m^3
w	slab width	ft	m
W	weight	lbf	–
Y	spacing	ft	m

Symbols

γ	specific weight	lbf/ft^3	–
ρ	density	lbm/ft^3	kg/m^3

Subscripts

s	steel
t	transverse

1. RIGID PAVEMENT

Information for this chapter section is primarily derived from the *AASHTO Guide for the Design of Pavement Structures* (*"AASHTO Guide"*), 1993 edition. The Portland Cement Association (PCA) has also published a methodology for designing rigid pavements in its *Thickness Design for Concrete Highway and Street Pavements* (*"PCA Pavement Manual"*), 1984 edition.

The Federal Highway Administration's (FHWA) Long-Term Pavement Performance (LTPP) program is a long-term study designed to provide the data required to better understand pavement performance. Based on data from the LTPP studies, guidelines have been prepared to aid in the design of better-performing asphalt and Portland cement concrete (PCC) pavements, more accurately predict pavement performance, and reduce dependence on empirical design procedures in favor of more advanced mechanistic design procedures have been prepared. Alternative guidelines for PCC have been published as a supplement to the 1993 *AASHTO Guide for Design of Pavement Structures*. The guidelines are meant to help reduce the likelihood of cracking or faulting in new or reconstructed pavements by providing tools to tailor pavement designs to the base course and underlying soil layers at the project site. These design procedures can be used in place of or in conjunction with the 1993 procedures for rigid pavement design.

Portland cement concrete (PCC) pavement is the most common form of *rigid pavement* because of its excellent durability and long service life. Its raw materials are readily available and reasonably inexpensive, it is easily formed, it withstands exposure to water without deterioration, and it is recyclable. The primary disadvantages of PCC pavement are that it can lose its original nonskid surface over time, it must be used with an even subgrade and only where uniform settling is expected, and it may rise and fall (i.e., fault) at transverse joints.

Transportation

PCC pavement is placed with *slip-form construction* methods. A stiff concrete is placed in front of a paving train that distributes, vibrates, screeds, and finishes the layer. Shoulders are usually constructed of asphalt pavement later, since shoulder traffic is much lower.

PCC pavement may be reinforced or nonreinforced, and joints may or may not be included. The four main types of concrete pavement are *plain jointed concrete pavement* (JCP), *jointed reinforced concrete pavement* (JRCP) *continuously reinforced concrete pavement* (CRCP), and *prestressed concrete pavement* (PCP). *Construction joints* are required in both reinforced and nonreinforced pavement. (See Sec. 9.)

Use of *prestressed concrete pavement* (PCP) is rare in the United States. The major advantage of prestressing is a reduction (up to 50%) in thickness, making it more applicable to airfield pavements, which are thicker. Initial testing seems to indicate a superior, stronger, longer lasting pavement, but with added cost and complexity. Although the *AASHTO Guide* provides some design guidance, few state transportation departments, vendors, or contractors are available to teach the technology.

2. MIXTURE PROPORTIONING

The following steps constitute the Portland Cement Association's *absolute volume method* for proportioning a PCC mixture.

step 1: Select an appropriate slump from Table 76.1.

step 2: Select the maximum aggregate size to be smaller than one fifth of the narrowest form dimension, one third of the slab depth, and three fourths of the clear space between the rebar.

step 3: Select a trial water-cement ratio as the lower of the two values required for durability from Table 76.2, and for strength from Table 76.3.

step 4: Estimate the mixing water and air void contents from Table 76.4.

step 5: Calculate cement content by dividing the mixing water content by the water-cement ratio.

step 6: Estimate the coarse aggregate content from Table 76.5. This gives a ratio of the volume of coarse aggregate to the volume of concrete.

step 7: Estimate the fine aggregate content. Subtract the volumes of all other ingredients from the unit volume of the concrete to determine the unit volume of fine aggregate.

Table 76.1 *Recommended Slumps*

	slump (in)	
concrete construction	maximum[a]	minimum
pavements, slabs, reinforced foundations, walls, footings, caissons, and substructure walls	3	1
beams, reinforced walls, and building columns	4	1
mass concrete	3	1

[a]May be increased 1 in consolidation by hand methods such as rodding or spading.

Reprinted with permission of the Portland Cement Association from the *Design and Control of Concrete Mixtures*, 14th ed., Table 9-6, copyright © 2002.

Table 76.2 *Maximum Water-Cement Ratios*

exposure condition	maximum water-cement ratio by weight for normal-weight concrete
concrete protected from exposure to freezing and thawing or application of deicing chemicals	select water-cement ratio on basis of strength, workability, and finishing needs
concrete intended to have low permeability when exposed to water	0.50
concrete exposed to deicers or freezing and thawing in a moist condition	0.45
for corrosion protection for reinforced concrete exposed to deicing salts, brackish water, seawater, or spray from these sources	0.40

Reprinted with permission of the Portland Cement Association from the *Design and Control of Concrete Mixtures*, 14th ed., Table 9-1, copyright © 2002.

Table 76.3 *Typical Relationship Between Water-Cement Ratio and Compressive Strength*

	water-cement ratio by weight	
compressive strength at 28 days (lbf/in²)	non-air-entrained concrete	air-entrained concrete
7000	0.33	—
6000	0.41	0.32
5000	0.48	0.40
4000	0.57	0.48
3000	0.68	0.59
2000	0.82	0.74

(Multiply lbf/in² by 6.89 to obtain kPa.)

Reprinted with permission of the Portland Cement Association from the *Design and Control of Concrete Mixtures*, 14th ed., Table 9-3, copyright © 2002.

Table 76.4 *Approximate Mixing Water and Target Air Content Requirements for Different Slumps and Maximum Sizes of Aggregate*

slump (in)	lbf/yd^3 of concrete for indicated maximum sizes of aggregate							
	$\frac{3}{8}$ in	$\frac{1}{2}$ in	$\frac{3}{4}$ in	1 in	$1\frac{1}{2}$ in	2 in	3 in	6 in
non-air-entrained concrete								
1 to 2	350	335	315	300	275	260	220	190
3 to 4	385	365	340	325	300	285	245	210
6 to 7	410	385	360	340	315	300	270	–
approximate amount of entrapped air in non-air-entrained concrete, percent	3	2.5	2	1.5	1	0.5	0.3	0.2
air-entrained concrete								
1 to 2	305	295	280	270	250	240	205	180
3 to 4	340	325	305	295	275	265	225	200
6 to 7	365	345	325	310	290	280	260	–
recommended average total air content, percent, for level of exposure:								
mild exposure	4.5	4.0	3.5	3.0	2.5	2.0	1.5	1.0
moderate exposure	6.0	5.5	5.0	4.5	4.5	3.5	3.5	3.0
severe exposure	7.5	7.0	6.0	6.0	5.5	5.0	4.5	4.0

Reprinted with permission of the Portland Cement Association from the *Design and Control of Concrete Mixtures*, 14th ed., Table 9-5, copyright © 2002.

Table 76.5 *Bulk Volume of Coarse Aggregate per Unit Volume*

nominal maximum size of aggregate (in)	bulk volume of dry-rodded coarse aggregate per unit volume of concrete for different fineness moduli of fine aggregate			
	2.40	2.60	2.80	3.00
$\frac{3}{8}$	0.50	0.48	0.46	0.44
$\frac{1}{2}$	0.59	0.57	0.55	0.53
$\frac{3}{4}$	0.66	0.64	0.62	0.60
1	0.71	0.69	0.67	0.65
$1\frac{1}{2}$	0.75	0.73	0.71	0.69
2	0.78	0.76	0.74	0.72
3	0.82	0.80	0.78	0.76
6	0.87	0.85	0.83	0.81

Reprinted with permission of the Portland Cement Association from the *Design and Control of Concrete Mixtures*, 14th ed., Table 9-4, copyright © 2002.

Example 76.1

Design a concrete mixture for use in an exposed bridge pier. The bridge is located where exposure is severe, so air entrainment will be used. Specifications call for a 28 day compressive strength of 4000 lbf/in^2. The rebar has a clear spacing of 1 in. The fine aggregate has a fineness modulus of 2.60, and the coarse aggregate has a dry-rodded unit weight of 97 lbf/ft^3. The coarse aggregate bulk specific gravity is 2.65, and the fine aggregate bulk specific gravity is 2.63. Cement has a specific gravity of 3.15.

Solution

step 1: From Table 76.1, the recommended slump is 3 in.

step 2: The maximum aggregate size is $^3/_4$ of the clear spacing between rebar.

$$\left(\tfrac{3}{4}\right)(1 \text{ in}) = 3/4 \text{ in}$$

step 3: Determine a trial water/cement ratio. From Table 76.2, $w/c = 0.40$. From Table 76.3, $w/c = 0.48$. Select the lower value of 0.40.

step 4: Select mixing water and air content. From Table 76.4, mixing water is 305 lbf/yd^3 of concrete. The volume of mixing water is

$$V_{\text{water}} = \frac{305 \ \dfrac{\text{lbf}}{\text{yd}^3}}{62.4 \ \dfrac{\text{lbf}}{\text{ft}^3}} = 4.89 \text{ ft}^3/\text{yd}^3$$

From Table 76.4, the air content is 6.0%.

Therefore, the volume of air in a cubic yard of concrete is

$$V_{\text{air}} = \left(27 \ \frac{\text{ft}^3}{\text{yd}^3}\right)(0.06) = 1.62 \text{ ft}^3/\text{yd}^3$$

step 5: The cement per cubic yard of concrete is

$$W_{cement} = \frac{W_{water}}{\frac{w}{c}} = \frac{305 \ \frac{lbf}{yd^3}}{0.40}$$

$$= 762.5 \ lbf/yd^3$$

The volume of cement required is

$$V_{cement} = \frac{W_{cement}}{(SG)\gamma_w}$$

$$= \frac{762.5 \ \frac{lbf}{yd^3}}{(3.15)\left(62.4 \ \frac{lbf}{ft^3}\right)} = 3.88 \ ft^3/yd^3$$

step 6: Determine the coarse aggregate content. The volume fraction of dry-rodded coarse aggregate from Table 76.5 is 0.64. The dry-rodded volume of coarse aggregate per cubic yard is

$$V_{coarse} = (0.64)\left(27 \ \frac{ft^3}{yd^3}\right) = 17.28 \ ft^3/yd^3$$

The weight of coarse aggregate is

$$W_{coarse} = V\gamma_{coarse} = \left(17.28 \ \frac{ft^3}{yd^3}\right)\left(97 \ \frac{lbf}{ft^3}\right)$$

$$= 1676.2 \ lbf/yd^3$$

The volume of coarse aggregate is

$$V_{coarse} = \frac{W_{coarse}}{(SG)\gamma_w}$$

$$= \frac{1676.2 \ \frac{lbf}{yd^3}}{(2.65)\left(62.4 \ \frac{lbf}{ft^3}\right)}$$

$$= 10.14 \ ft^3/yd^3$$

step 7: Determine the fine aggregate content. The fine aggregate takes up all space remaining after allowances have been made for the water, air, cement, and coarse aggregate.

$$V_{fine} = 27 \ \frac{ft^3}{yd^3} - 4.89 \ \frac{ft^3}{yd^3} - 1.62 \ \frac{ft^3}{yd^3}$$

$$- 3.88 \ \frac{ft^3}{yd^3} - 10.14 \ \frac{ft^3}{yd^3}$$

$$= 6.47 \ ft^3/yd^3$$

The weight of the fine aggregate is

$$W_{fine} = V_{fine}(SG)\gamma_w$$

$$= \left(6.47 \ \frac{ft^3}{yd^3}\right)(2.63)\left(62.4 \ \frac{lbf}{ft^3}\right)$$

$$= 1061.8 \ lbf/yd^3$$

3. AASHTO METHOD OF RIGID PAVEMENT DESIGN

All four types (see Sec. 1) of PCC rigid pavement may be designed using the AASHTO method. The AASHTO rigid pavement design concepts, terminology, and procedures are similar to those used by AASTHO for flexible pavement design. Similar and parallel concepts include the principal design variables (time, traffic, reliability and environment), serviceability, load equivalency factors, truck factors, growth factors, layer coefficients, structural number concept, lane and directional distribution factors, and ESAL loading.

The primary differences include using the soil modulus of subgrade reaction, k; concrete modulus of rupture, S_c'; drainage coefficient, C_D; joint J-value; frictional factor, F; and standard deviation, S_o.

Based on information obtained during the AASHTO Road Test, an appropriate overall standard deviation, S_o, for rigid pavements is 0.35. This is slightly different than for flexible pavements.

4. LAYER MATERIAL STRENGTHS

The material used for each pavement layer must be characterized. In the AASHTO method, the *modulus of subgrade reaction*, k (also known as the k-value, and essentially the modulus of elasticity), is the primary performance factor for the soil. Its units are psi/in (i.e., pressure per unit length) or lbf/in^3. The resilient modulus concept can also be used, and the two moduli are linearly related. Correlations with other parameters are not generally exact, as Fig. 76.1 indicates.

$$k = \frac{M_R}{19.4} \tag{76.1}$$

The *effective slab support k-value* incorporates the support abilities of all the layers. AASHTO and PCA (see Tables 76.6 and 76.7) have published methods for incorporating the strength of all the layers.

Concrete's average *modulus of rupture* (flexural strength), S_c', is found using a *third-point loading flexure test* at 28 days, as specified by AASHTO T97 or ASTM C78. The construction specification modulus of rupture should not be used, as it is too conservative.

Concrete's *modulus of elasticity* is found using the ACI-318 relationship.

$$E_c = 57,000\sqrt{f_c'} \quad [E_c \ and \ f_c' \ in \ lbf/in^2] \tag{76.2}$$

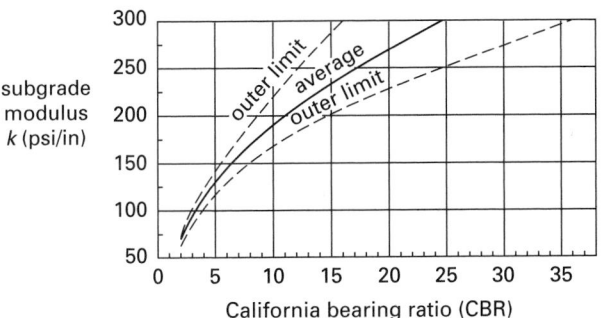

Figure 76.1 *Approximate Correlation Between CBR and Subgrade Modulus*

Table 76.6 *Effective Slab Support (Subbase) Modulus with Untreated Subbase (lbf/in^3)*

subgrade k-value (lbf/in^3)	subbase thickness (in)			
	4 in	6 in	9 in	12 in
50	65	75	85	110
100	130	140	160	190
200	220	230	270	320
300	320	330	370	430

Reprinted with permission from the American Concrete Pavement Association from the *Thickness Design for Concrete Highway and Street Pavements*, Table 1, copyright © 1984.

Table 76.7 *Effective Slab Support (Subbase) Modulus with Cement-Treated Subbase (lbf/in^3)*

subgrade k-value (lbf/in^3)	subbase thickness (in)			
	4 in	6 in	8 in	10 in
50	170	230	310	390
100	280	400	520	640
200	470	640	830	–

Reprinted with permission from the American Concrete Pavement Association from the *Thickness Design for Concrete Highway and Street Pavements*, Table 2, copyright © 1984.

Example 76.2

A subgrade has a California bearing ratio value of 10. A 6 in cement-treated subbase is to be used. What is the approximate effective subbase k-value?

Solution

From Fig. 76.1, the subgrade k-value is 200 lbf/in^3. From Table 76.7, the effective subbase k-value is 640 lbf/in^3.

5. PAVEMENT DRAINAGE

The *drainage coefficient*, C_d, is a function of the drainage quality and presence of moisture. If the ground is saturated much of the time or the drainage is poor, C_d may be as low as 0.70. If the pavement is well-drained and conditions are dry, C_d may be as high as 1.25. In cases where the conditions are not specific, $C_d = 1.00$ is assumed.

6. LOAD TRANSFER AND DOWELS

A *load transfer coefficient*, J (also known as the *J-value*), accounts for load distribution across joints and cracks. The J-value is affected by the use and placement of dowels, aggregate interlock, and tied shoulders. Approximate J-values for various conditions are given as

asphalt shoulders with dowels	$J = 3.2$
asphalt shoulders without dowels	$J = 4.0$
concrete shoulders with dowels	$J = 2.8$
concrete shoulders without dowels	$J = 3.8$

A rule of thumb for specifying the dowel diameter is the slab thickness divided by 8 and then rounded up to the nearest standard bar size. Typical dowel spacing is 12 in (305 mm), and typical dowel length is 18 in (457 mm).

7. PAVEMENT DESIGN METHODOLOGY

Slab thickness is determined from Fig. 76.2 for each k-value, and then rounded to the nearest 0.5 in (13 mm). To use Fig. 76.2 to determine the slab thickness, the following variables are required: (a) effective k-value; (b) estimated future traffic in ESALs using the rigid pavement equivalency factors; (c) reliability level, R; (d) overall standard deviation, S_o; (e) design serviceability loss, ΔPSI; (f) concrete modulus of elasticity, E_c; (g) concrete modulus of rupture, S'_c; (h) load transfer coefficient, J; and (i) drainage coefficient, C_d.

8. STEEL REINFORCING

Reinforcing steel for rigid pavements may consist of deformed reinforcing bar, smooth wire mesh, or deformed wire fabric. The purpose of the steel is not to add structural strength, but to hold cracks tightly together and restrain their growth. As the slab contracts, the contraction is resisted by friction with underlying material. Restraint of contraction causes tensile stresses to be at a maximum at mid-slab. When tensile stresses exceed the tensile strength of the concrete, cracks develop. Because transverse cracking is not expected in slabs less than 15 ft (4.2 m) long, reinforcement is not used in short slabs.

Although AASHTO provides a nomograph ("Reinforcement Design Chart for Jointed Reinforced Concrete Pavement, *AASHTO Guide* Fig. 3.8, p. II-52), the percentage of reinforcing steel is easily calculated.

$$P_s = \frac{L_{\text{ft}} F(100\%)}{2f_{s,\text{lbf/in}^2}} \qquad 76.3$$

Figure 76.2a *AASHTO Design Chart for Rigid Pavement*[a]

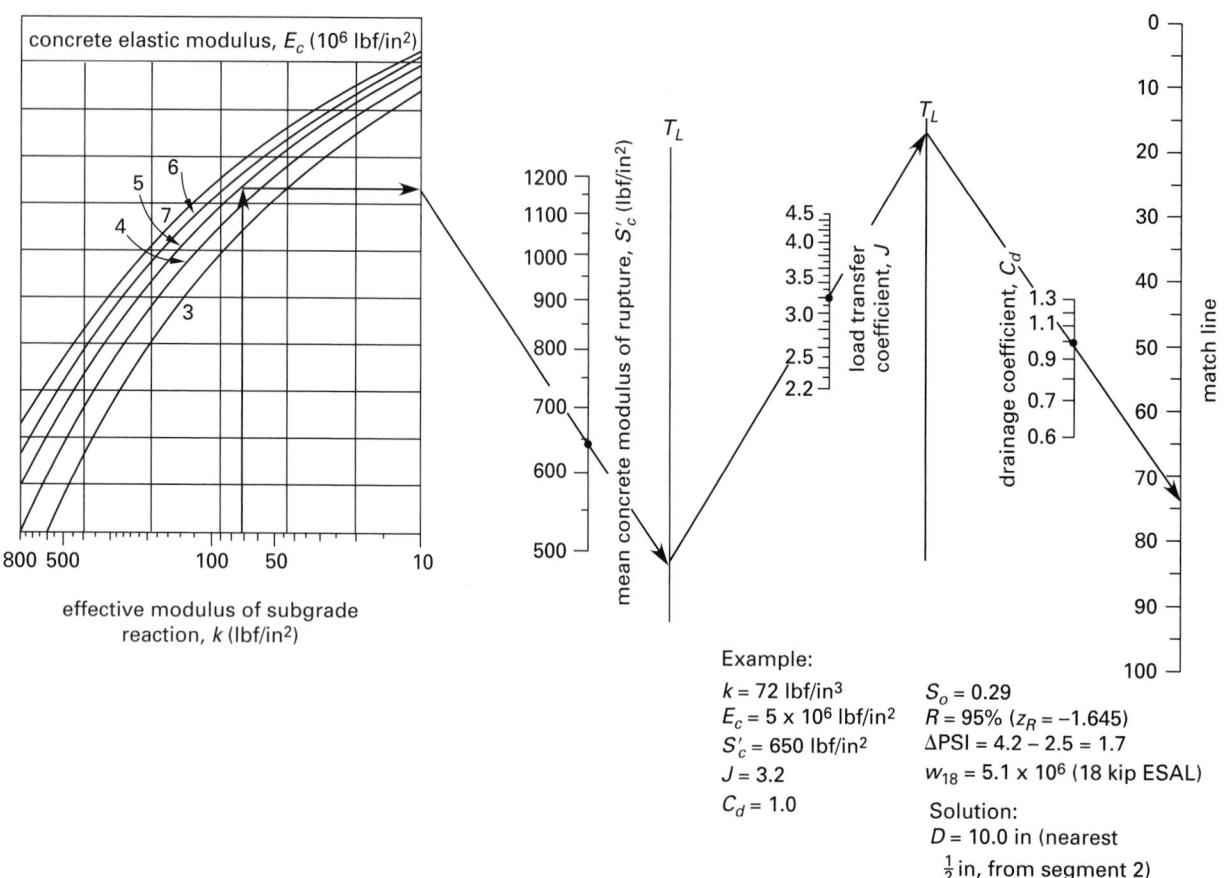

Example:

$k = 72$ lbf/in^3
$E_c = 5 \times 10^6$ lbf/in^2
$S'_c = 650$ lbf/in^2
$J = 3.2$
$C_d = 1.0$

$S_o = 0.29$
$R = 95\%$ ($z_R = -1.645$)
ΔPSI $= 4.2 - 2.5 = 1.7$
$w_{18} = 5.1 \times 10^6$ (18 kip ESAL)

Solution:
$D = 10.0$ in (nearest
$\frac{1}{2}$ in, from segment 2)

[a]Based on mean value of each variable.

From *Guide for Design of Pavement Structures*, Sec. II, p. II-45, Fig. 3.7, copyright © 1993 by the American Association of State Highway and Transportation Officials, Washington, D.C. Used by permission.

Figure 76.2b *AASHTO Design Chart for Rigid Pavement*[a]

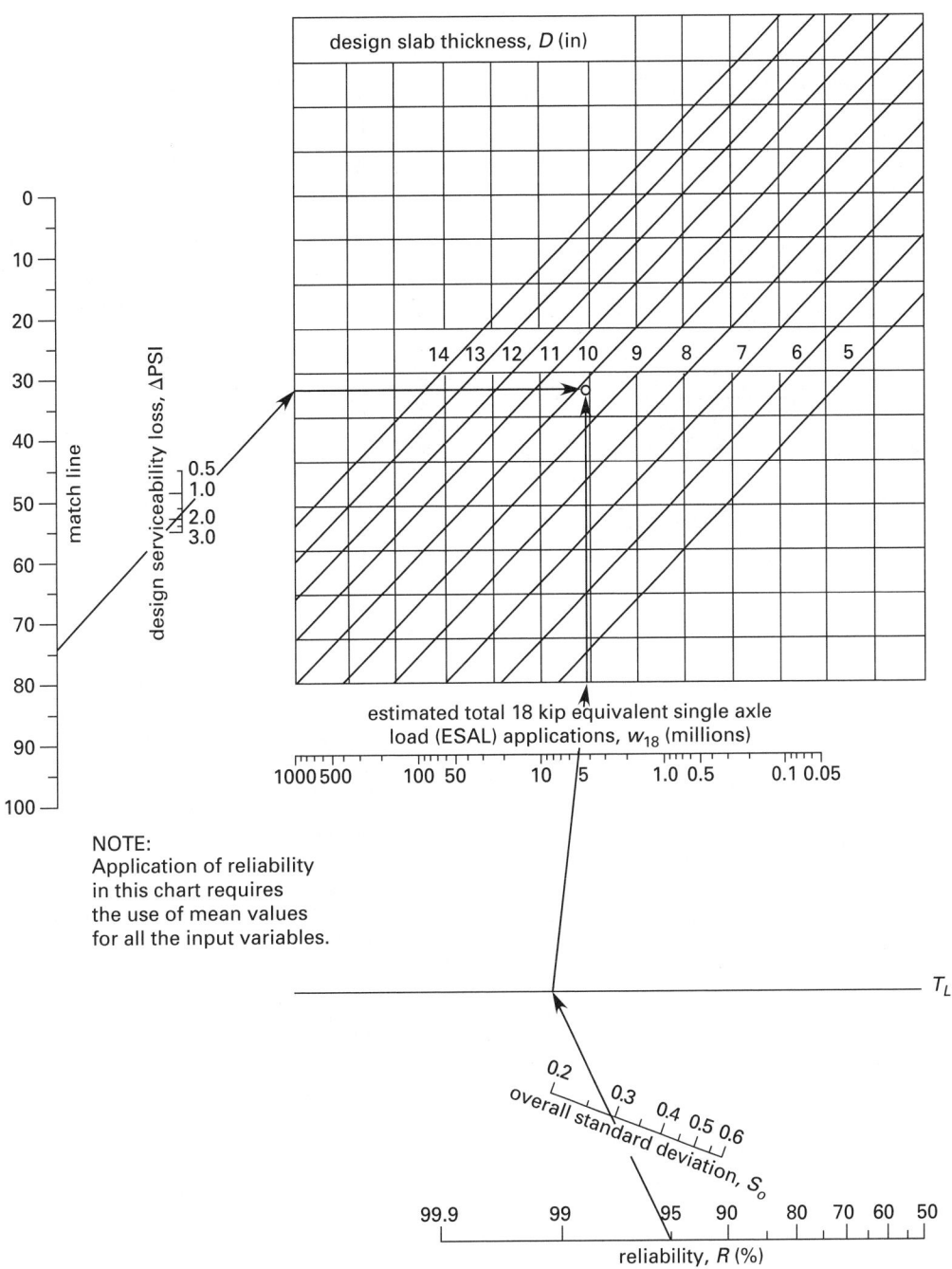

NOTE:
Application of reliability
in this chart requires
the use of mean values
for all the input variables.

[a]Based on mean value of each variable.

From *Guide for Design of Pavement Structures*, Sec. II, p. II-45, Fig. 3.7, copyright © 1993 by the American Association of State
Highway and Transportation Officials, Washington, D.C. Used by permission.

Figure 76.3 *Recommended Maximum Tie Bar Spacing for PCC Pavement (1/2 in diameter tie bars, grade-40 steel, and subgrade friction factor of 1.5)*

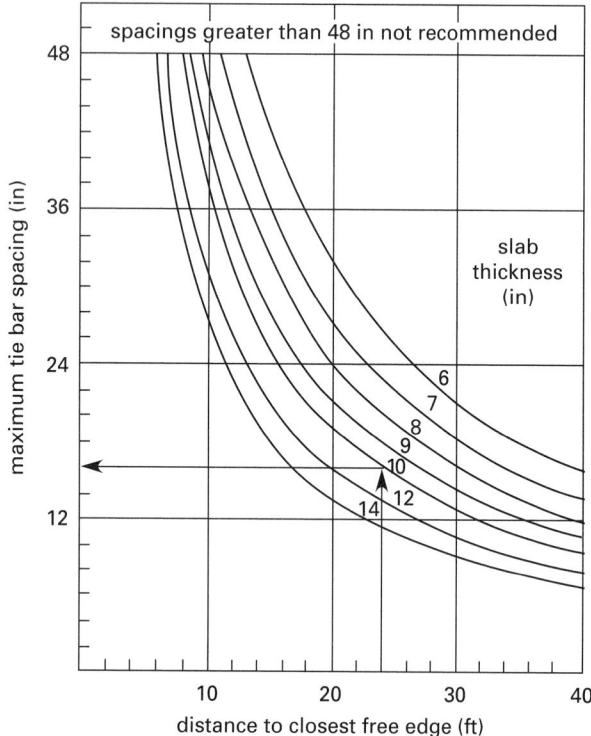

Example: distance from free edge: 24 ft
D = 10 in
Answer: spacing = 16 in

From *Guide for Design of Pavement Structures*, Part II, Fig. 3.13, copyright © 1993 by the American Association of State Highway and Transportation Officials, Washington, D.C. Used by permission.

Figure 76.4 *Recommended Maximum Tie Bar Spacing for PCC Pavement (5/8 in diameter tie bars, grade-40 steel, and subgrade friction factor of 1.5)*

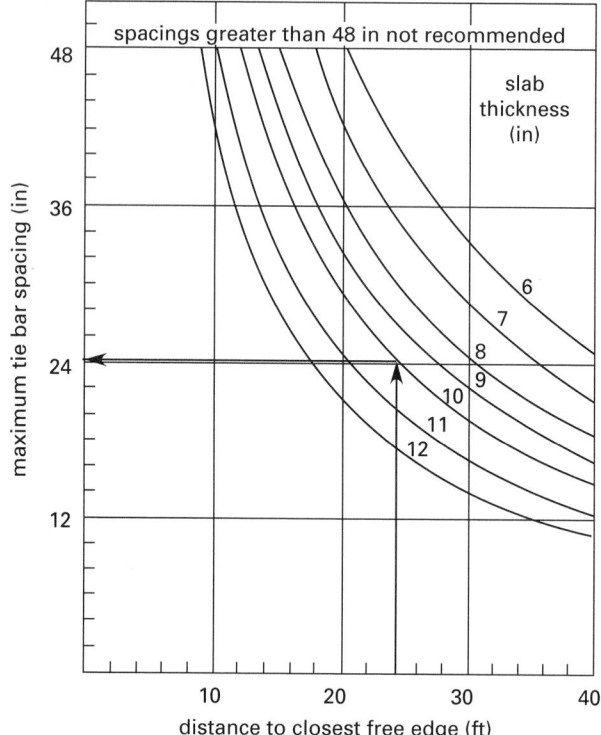

Example: distance from free edge: 24 ft
D = 10 in
Answer: spacing = 24 in

From *Guide for Design of Pavement Structures*, Part II, Fig. 3.14, copyright © 1993 by the American Association of State Highway and Transportation Officials, Washington, D.C. Used by permission.

The required reinforcement is expressed as a percentage, P_s, of the cross-sectional area. The *slab length*, L, represents the distance between the free, or untied, joints and has a substantial effect on the maximum concrete tensile stresses and the amount of steel required. *Steel working stress*, f_s, is assumed to be 75% of the steel yield strength. For example, grade-60 steel would be designed for 45 kips/in².

The *friction factor*, F, represents the frictional resistance between the bottom of the slab and the top of the subbase or subgrade. Typical values of F are 0.9 for natural subgrade, 1.5 for crushed stone or gravel, 1.8 for stabilized material, and 2.2 for placement on another surface treatment [*AASHTO Guide* Table 2.8, p. II-28].

To determine the amount of transverse reinforcement, Eq. 76.3 may still be used if L is taken as the side-to-side width of the slab. The transverse spacing between rebar, Y, is calculated from the cross-sectional area of the rebar, A_s; the percent of transverse steel, P_t; and the slab thickness, D.

$$P_t = \frac{A_s(100\%)}{YD} \qquad 76.4$$

Tie bars are dowels, usually deformed steel bars 24 to 40 in long, used to connect two abutting slabs (i.e., between two lanes poured at different times), to create a longitudinal joint. Figures 76.3 and 76.4 give tie bar spacing for $^1/_2$ in and $^5/_8$ in diameter deformed bars. To use these figures, the closest distance to a free edge, defined as any joint not having tie bars or as the distance between the edges of pavement, must be known. Tie bar spacing increases with decreases in the distance to a free edge since the steel tensile stress is maximum at the center of the slab. Spacing also increases with a decrease in slab thickness.

Example 76.3

Design a jointed reinforced concrete pavement with a 60 ft slab length, two 12 ft lanes tied together, doweled joints, and doweled asphalt shoulders. Assume the pavement is placed in a wet location with poor drainage. The subgrade is very weak clay with a California bearing ratio of 3. The friction factor, F, is 1.8. The design reliability is 90%. The concrete has a compressive strength of 4000 lbf/in² and a modulus of rupture of 650 lbf/in², both at 28 days. The traffic on the design lane is projected to be 20×10^6 ESALs. An 8 in cement-treated subbase will be used. Grade-40 steel reinforcement and $^1/_2$ in diameter tie bars will be used.

Determine the (a) slab thickness, (b) length and spacing of dowel bars, (c) longitudinal and transverse reinforcement, and (d) tie bars for use in the longitudinal joints.

Solution

(a) From Fig. 76.1, the modulus of subgrade reaction is $k = 100$ lbf/in^3. From Table 76.7 with the subbase, the effective subgrade modulus is $k = 520$ lbf/in^3.

Use Eq. 76.2 to estimate E_c.

$$E_c = 57,000\sqrt{f'_c} = 57,000\sqrt{4000 \text{ lbf/in}^2}$$
$$= 3.6 \times 10^6 \text{ lbf/in}^2$$

From Sec. 6, the load transfer coefficient for asphalt shoulders with dowels is $J = 3.2$.

For poor draining, the drainage coefficient is $C_d = 0.7$.

Reliability, R, is given as 90%.

For PCC pavement, use an overall standard deviation, S_o, of 0.35.

Assume "normal" serviceability values of $p_o = 4.5$ and $p_t = 2.5$.

$$\Delta \text{PSI} = 4.5 - 2.5 = 2.0$$

From Fig. 76.2, the slab thickness is 12.5 in.

(b) Use the rule of thumb for dowel bars.

$$\text{diameter} = \frac{\text{thickness}}{8} = \frac{12.5 \text{ in}}{8}$$
$$= 1.56 \text{ in} \quad [\text{use } 1^5/_8 \text{ in}]$$

Use an average spacing of 12 in and a length of 18 in.

(c) Design the longitudinal and transverse reinforcement.

With grade 40 steel, the working stress is

$$0.75f_y = (0.75)\left(40 \frac{\text{kips}}{\text{in}^2}\right) = 30 \text{ kips/in}^2$$

The friction factor is given as $F = 1.8$.

The percentage of longitudinal steel is given by Eq. 76.3.

$$P_s = \frac{LF(100\%)}{2f_s}$$
$$= \frac{(60 \text{ ft})(1.8)(100\%)}{(2)\left(30,000 \frac{\text{lbf}}{\text{in}^2}\right)} = 0.18\%$$

Since the two lanes are tied together, both must be considered for horizontal steel. The width is 24 ft.

From Eq. 76.3, the area of transverse steel is

$$P_t = \frac{LF(100\%)}{2f_s}$$
$$= \frac{(24 \text{ ft})(1.8)(100\%)}{(2)\left(30,000 \frac{\text{lbf}}{\text{in}^2}\right)} = 0.072\%$$

(d) Since $^1/_2$ in diameter tie bars are used, Fig. 76.3 is applicable. The slab thickness is 12.5 in. The distance from the joint to a free edge is 12 ft. The tie bar spacing is read as approximately 26 in.

9. PAVEMENT JOINTS

Figure 76.5 shows the three standard forms of joints used on a concrete pavement. *Contraction joints (control joints, weakened plane joints, dummy joints)* are usually sawn into the pavement to a depth of one fourth (transverse joints) or one third (longitudinal joints) of the slab thickness, but they can also be hand-formed while the concrete is wet or formed by pouring around inserts. The purpose of the joint is to create a thinner pavement section in one location (a weak line to encourage shrinkage cracking along that line). Contraction joints also relieve tensile stresses in the pavement. The uneven crack joint provides for load transfer between the slab sections by aggregate interlock. To prevent deicing sand and water from infiltrating the joints, a hot pour or preformed joint sealer can be used in the joint.

With nonreinforced construction, contraction joints should be placed at regular intervals to allow for the shrinkage of the concrete during curing, drying, and thermal changes. Twenty-four times the thickness of the slab and a ratio of slab width to slab length ≤ 1.25 are typical rules of thumb used to select the distance between contraction joints for nonreinforced pavement.

Contraction joints can be placed at regular intervals and at 90° angles to the centerline. However, skewing the transverse joints and/or placing them at random intervals minimizes the amount of faulting as well as noise and impact from vehicular traffic. State highway departments have standard joint patterns for spacing and standard skew angles.

Hinge joints, also known as *warping joints*, are similar to contraction joints. However, hinge joints are longitudinal joints, generally placed along the centerline of a highway.

Construction joints are used transversely between construction periods and longitudinally between pavement lanes. Longitudinal joints are usually spaced to coincide with lane markings at 8 and 12 ft (SI) intervals, but spacing depends primarily on paver width. They are usually keyed, formed by attaching a metal or wooden keyway form at middepth. With slip-form pavers, the keyway is formed by the slipform as it advances. Dowels are not generally used with longitudinal joints.

Isolation joints (expansion joints) are used to relieve compressive (i.e., buckling) stresses in the pavement where it adjoins another structure. Expansion joints are placed where the pavement abuts bridges, at intersections, where concrete and asphalt pavements abut each other, and at end-of-day pours. Transverse joints are placed across the direction of roadway travel.

Use of *dowel bars* in transverse joints depends on the service conditions. They are not needed in residential or other low-traffic streets, but they may be needed with heavy truck traffic. If one end of a dowel is lubricated, the slab will slip over the dowel as the slab expands and contracts, providing a good transfer of load. However, most dowels are inserted dry. Dowel insertion can be done automatically with modern paving train equipment, or they can be inserted manually.

Figure 76.5 *Types of Concrete Pavement Joints*

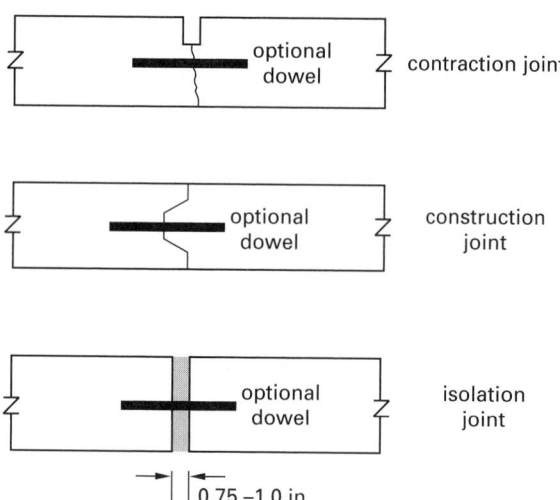

10. SURFACE SEALING

Sealing of PCC surfaces is typically used to prevent chloride ions (from deicing salts) from penetrating the concrete and corroding reinforcing and supporting steel. Sealing is predominantly used to protect steel bridge decks. It is not commonly done on open roads.

In the past, natural linseed oil has traditionally been used. However, modern commercial formulations include silane/siloxane-type sealers and alkali silicate-type sealers. Silane/siloxane sealers are classified as "penetrating sealers," as they retain their properties after the upper surface layer of sealer has been worn away.

11. RECYCLING

Recycled PCC pavement material may be used as aggregate base, subbase, or aggregate for new pavement. Recycling is not only economically favorable; it also solves the problem of disposing of large amounts of PCC pavement. Recycled aggregate may actually be more durable than in the original state because it has already gone through multiple freeze-thaw cycles. Recycling can be used with lean concrete (*econocrete*) and slipform pavers.

In typical operation, diesel pile hammers, resonant pavement breakers, or drop hammers are used to break up the pre-existing concrete. A backhoe separates broken pavement and steel, and the steel is also recovered. The concrete is removed, crushed, screened, and reused. Most recycling is done on-site.

12. PAVEMENT GROOVING

Grooving is a method of increasing skid resistance and reducing hydroplaning on all types of pavements. Grooves permit the water to escape under tires and prevent buildup of water on the surface. The recommended groove is $1/4$ in wide and $1/4$ in deep (6 mm by 6 mm), with a center-to-center spacing of $1\frac{1}{4}$ in (32 mm). So that the grooving does not weaken the pavement surface, the minimum spacing is approximately $1\frac{1}{8}$ in (29 mm). To be effective in removing water, the maximum spacing is approximately 2 in (50 mm).

Grooving should be used only with structurally adequate pavements. If the pavement is in poor condition, it should be rehabilitated before grooving. Grooves should be continuous. In the case of special surfaces such as airfields where pavement widths exceed standard roadway widths, transverse rather than longitudinal grooves may be used.

13. ADVANCED, ALTERNATIVE, AND EXPERIMENTAL PAVEMENTS

Dual-layer pavements are claimed to yield quieter rides and reduced road noise. The paving machine lays down two layers of different concretes simultaneously—a lower, thicker layer containing large aggregate for economy, and an upper thinner layer made up of smaller stones. The upper stones are subsequently exposed by washing away cement paste before hardening.

Until recently, few *thin-overlay* options were available. Delamination problems have been experienced with thin, *grout-bonded concrete* placed over existing concrete surfaces. Epoxy-bonded overlays have been more successful. Thin $3/4$ to 1 in (19 to 25 mm) *polymer concrete* overlays are successful and are becoming more common. Polyester-styrene polymer, a catalyst, aggregate smaller than $1/2$ in, and sand are used. Good bonding is obtained by abrasive blasting (e.g., shotblasting) followed by vacuuming to clean and roughen the original surface. The overlay sets up in less than half an hour.

Flyash (class-C) modified concrete appears to be more susceptible to sulfate attack than normal concrete. When flyash concrete contains a high calcium content, the concrete is also more prone to alkali silica reactions. Adding flyash appears to decrease durability with no significant increase in strength.

Rubcrete, using ground-up rubber in place of some of the aggregate, is generally not applicable to structural applications, including upper pavement layers.

Various other innovations, including *latex-modified concrete* (LMC) and *steel fiber-reinforced* concrete, are available for special applications and adventurous contractors.

Transportation

77 **Plane Surveying**

Nomenclature

A	area	ft^2	m^2
A	azimuth	deg	deg
BM	benchmark or monument	ft	m
BS	backsight	ft	m
C	correction	various	various
d	distance	ft	m
D	distance between two points	ft	m
DMD	double meridian distance	ft	m
E	error	ft	m
E	modulus of elasticity	lbf/ft^2	Pa
FS	foresight	ft	m
h	height or correction	ft	m
HI	height of instrument	ft	m
k	number of observations	–	–
K	stadia interval factor	–	–
L	length	ft	m
P	tension	lbf	N
R	rod reading	ft	m
s	sample standard deviation	various	various
T	temperature	°F	°C
W	relative weight	–	–
x	distance or location	ft	m
y	difference in elevation	ft	m
y	distance or location	ft	m

Symbols

α	angle	deg	deg
α	coefficient of linear thermal expansion	ft/ft-°F	m/m·°C
β	angle	deg	deg
θ	angle	deg	deg
μ	mean	various	various

Subscripts

c	curvature
i	observation number
p	probable
P	pull
r	refraction
s	sag or standarized
T	temperature

1. ERROR ANALYSIS: MEASUREMENTS OF EQUAL WEIGHT

There are still many opportunities for errors in surveying, although calculators and modern equipment have significantly reduced the magnitudes of most errors. The purpose of error analysis is not to eliminate errors but rather to estimate their magnitudes. In some cases, it is also appropriate to assign the errors to the appropriate measurements.

The *expected value* of a measurement, also known as the *most likely value* or *probable value*, is the value that has the highest probability of being correct. If a series of k measurements is taken, the most probable value, x_p, is the average (mean) of those measurements, as shown in

Eq 77.1. For related measurements whose sum should equal some known quantity, the most probable values are the observed values corrected by an equal part of the total error.

$$x_p = \frac{x_1 + x_2 + \cdots + x_k}{k} \qquad 77.1$$

Measurements of a given quantity are assumed to be normally distributed. If a quantity has a mean μ and a sample standard deviation s, the probability is 50% that a measurement of that quantity will fall within the range of $\mu \pm 0.6745s$. The quantity $0.6745s$ is known as the *probable error*. The *probable ratio of precision* is $\mu/0.6745s$. The interval between the extremes is known as the 50% *confidence interval*. Other confidence limits are easily obtained from the normal tables. For example, for a 95% interval, replace 0.6745 with 1.96, and for a 99% interval, replace 0.6745 with 2.57.

The *probable error of the mean*, E_{mean}, of k observations is given by Eq. 77.2.

$$\begin{aligned} E_{\text{mean}} &= \frac{0.6745s}{\sqrt{k}} \\ &= \frac{E_{\text{total},k \text{ measurements}}}{\sqrt{k}} \qquad 77.2 \end{aligned}$$

Example 77.1

The interior angles of a triangular traverse were measured as $63°$, $77°$, and $41°$. Each measurement was made once, and all angles were measured with the same precision. What are the most probable interior angles?

Solution

The sum of angles in a triangle should equal $180°$. The error in measurement is

$$(63° + 77° + 41°) - 180° = +1°$$

The correction required to make the sum total $180°$ is $-1°$, which is proportioned equally among the three angles. The most probable values are then

$$63° - \frac{+1°}{3} = 62.67°$$

$$77° - \frac{+1°}{3} = 76.67°$$

$$41° - \frac{+1°}{3} = 40.67°$$

Example 77.2

A critical distance was measured twelve times. The mean value was 423.7 ft with a sample standard deviation, s, of 0.31 ft. What are the 50% confidence limits for the distance?

Solution

From Eq. 77.2, the standard error of the mean value is

$$E_{\text{mean}} = \frac{0.6745s}{\sqrt{k}} = \frac{(0.6745)(0.31 \text{ ft})}{\sqrt{12}}$$

$$= 0.06 \text{ ft}$$

The probability is 50% that the true distance is within the limits of 423.7 ± 0.06 ft.

Example 77.3

The true length of a tape is 100 ft. The most probable error of a measurement with this tape is 0.01 ft. What is the expected error if the tape is used to measure out a distance of 1 mi?

Solution

1 mi consists of 5280 ft. The number of individual measurements will be 5280 ft/100 ft = 52.8, or 53 measurements. The most probable error is given by Eq. 77.2.

$$E_{\text{total},k \text{ measurements}} = E_{\text{mean}}\sqrt{k}$$

$$= (0.01 \text{ ft})\sqrt{53} = 0.073 \text{ ft}$$

2. ERROR ANALYSIS: MEASUREMENTS OF UNEQUAL WEIGHT

Some measurements may be more reliable than others. It is reasonable to weight each measurement with its relative reliability. Such weights can be determined subjectively. More frequently, however, they are determined from relative frequencies of occurrence or from the relative inverse squares of the probable errors.

The probable error and 50% confidence interval for weighted observations can be found from Eq. 77.3. x_i represents the ith observation, and w_i represents its relative weight.

$$E_{p,\text{weighted}} = 0.6745\sqrt{\frac{\sum \left(w_i(\overline{x} - x_i)^2\right)}{(k-1)\sum w_i}} \qquad 77.3$$

For related weighted measurements whose sum should equal some known quantity, the most probable weighted values are corrected inversely to the relative frequency of observation. Weights can also be calculated when the probable errors are known. The weights are the relative squares of the probable errors.

Example 77.4

An angle was measured five times by five equally competent crews on similar days. Two of the crews obtained a value of $39.77°$, and the remaining three crews obtained a value of $39.74°$. What is the probable value of the angle?

Solution

Weight the values by the relative frequency of observation.

$$\theta = \frac{(2)(39.77°) + (3)(39.74°)}{5}$$

$$= 39.75°$$

Example 77.5

A distance has been measured by three different crews. The measurements' probable errors are given. What is the most probable value?

crew 1: 1206.40 ± 0.03 ft

crew 2: 1206.42 ± 0.05 ft

crew 3: 1206.37 ± 0.07 ft

Solution

The sum of squared probable errors is

$$(0.03 \text{ ft})^2 + (0.05 \text{ ft})^2 + (0.07 \text{ ft})^2 = 0.0083 \text{ ft}^2$$

The weights to be applied to the three measurements are

$$\frac{0.0083 \text{ ft}^2}{(0.03 \text{ ft})^2} = 9.22 \text{ ft}$$

$$\frac{0.0083 \text{ ft}^2}{(0.05 \text{ ft})^2} = 3.32 \text{ ft}$$

$$\frac{0.0083 \text{ ft}^2}{(0.07 \text{ ft})^2} = 1.69 \text{ ft}$$

The most probable length is

$$\frac{\begin{array}{l}(1206.40 \text{ ft})(9.22 \text{ ft}) \\ + (1206.42 \text{ ft})(3.32 \text{ ft}) \\ + (1206.37 \text{ ft})(1.689 \text{ ft})\end{array}}{9.22 \text{ ft} + 3.32 \text{ ft} + 1.69 \text{ ft}} = 1206.40 \text{ ft}$$

Example 77.6

What is the 50% confidence interval for the measured distance in Ex. 77.5?

Solution

Work with the decimal portion of the answer.

$$\bar{x} = \frac{0.40 + 0.42 + 0.37}{3}$$

$$\approx 0.40$$

i	x_i	$\bar{x} - x_i$	$(\bar{x} - x_i)^2$	w_i	$w_i(\bar{x} - x_i)^2$
1	0.40	0	0	9.22	0
2	0.42	−0.02	0.0004	3.32	0.0013
3	0.37	0.03	0.0009	1.69	0.0015
			total	14.23	0.0028

Use Eq. 77.3.

$$E_{p,\text{weighted}} = 0.6745\sqrt{\frac{\sum\left(w_i(\bar{x} - x_i)^2\right)}{(k-1)\sum w_i}}$$

$$= 0.6745\sqrt{\frac{0.0028}{(3-1)(14.23)}}$$

$$= 0.0067$$

The 50% confidence interval is 1206.40 ± 0.0067 ft.

Example 77.7

The interior angles of a triangular traverse were repeatedly measured. What is the probable value for angle 1?

angle	value	number of measurements
1	63°	2
2	77°	6
3	41°	5

Solution

The total of the angles is

$$63° + 77° + 41° = 181°$$

So, $-1°$ must be divided among the three angles. These corrections are inversely proportional to the number of measurements. The sum of the measurement inverses is

$$\tfrac{1}{2} + \tfrac{1}{6} + \tfrac{1}{5} = 0.5 + 0.167 + 0.2$$

$$= 0.867$$

The most probable value of angle 1 is then

$$63° + \frac{\left(\tfrac{1}{2}\right)(-1°)}{0.867} = 62.42°$$

Example 77.8

The interior angles of a triangular traverse were measured. What is the most probable value of angle 1?

angle	value
1	$63° \pm 0.01°$
2	$77° \pm 0.03°$
3	$41° \pm 0.02°$

Solution

The total of the angles is

$$63° + 77° + 41° = 181°$$

Transportation

So, $-1°$ must be divided among the three angles. The corrections are proportional to the square of the probable errors.

$$(0.01)^2 + (0.03)^2 + (0.02)^2 = 0.0014$$

The most probable value of angle 1 is

$$63° + \frac{(0.01)^2(-1°)}{0.0014} = 62.93°$$

3. ERRORS IN COMPUTED QUANTITIES

When independent quantities with known errors are added or subtracted, the error of the result is given by Eq. 77.4. The squared terms under the radical are added regardless of whether the calculation is addition or subtraction.

$$E_{sum} = \sqrt{E_1^2 + E_2^2 + E_3^2 + \cdots} \qquad 77.4$$

The error in the product of two quantities, x_1 and x_2, which have known errors, E_1 and E_2, is given by Eq. 77.5.

$$E_{product} = \sqrt{x_1^2 E_2^2 + x_2^2 E_1^2} \qquad 77.5$$

Example 77.9

An EDM instrument manufacturer has indicated that the measurement error with a particular instrument is ± 0.04 ft with another error of ± 10 ppm. The instrument is used to measure a distance of 3000 ft. What is the expected error of measurement?

Solution

There are two independent errors here. Since both parts can be positive or negative, they may have opposite signs. The variable error is

$$E = (3000 \text{ ft})\left(\frac{10}{1{,}000{,}000}\right) = 0.03 \text{ ft}$$

Use Eq. 77.4.

$$E = \sqrt{E_1^2 + E_2^2}$$
$$= \sqrt{(0.04 \text{ ft})^2 + (0.03 \text{ ft})^2} = 0.05 \text{ ft}$$

Example 77.10

The sides of a rectangular section were determined to be 1204.77 ± 0.09 ft and 765.31 ± 0.04 ft, respectively. What is the probable error in area?

Solution

Use Eq. 77.5.

$$E_{area} = \sqrt{x_1^2 E_2^2 + x_2^2 E_1^2}$$
$$= \sqrt{\begin{array}{c}(1204.77 \text{ ft})^2 (0.04 \text{ ft})^2 \\ + (765.31 \text{ ft})^2 (0.09 \text{ ft})^2\end{array}}$$
$$= 84.06 \text{ ft}^2$$

4. ORDERS OF ACCURACY

The significance of a known measurement error can be quantified by a ratio of parts (e.g., 1 ft in 10,000 ft, or 1:10,000). A distance measurement error is divided by the distance measured. A traverse closure error is divided by the sum of all traverse leg lengths to determine the fractional error.

Survey accuracy is designated by its *order*, with *first order* being the most accurate. In the United States, standard accuracies have been established for horizontal and vertical control, as well as for closure and triangulation errors. Theoretically, any organization or agency could establish its own standards, but the phrase "order of accuracy" generally refers to standards specified by the Federal Geodetic Control Committee. The American Land Title Association (ALTA) and the American Congress on Surveying and Mapping (ASCM) have also issued standards for land title surveys of small parcels.

There are three different national *orders* (i.e., *levels*) of *accuracy*: first, second, and third order. Some states include a fourth order of accuracy (less than 1:5000). There are two different classes of accuracy in second- and third-order surveys: Classes I and II. These are described in Table 77.1.

Increased use of very precise GPS techniques has created a demand for greater control. The National Geodetic Survey (NGS) has established both order A (accuracy of 1:10,000,000) and order B (accuracy of 1:1,000,000) points.

For land title surveys of parcels using ALTA/ACSM specifications, the *degree of accuracy* should be based on the intended use of the land parcel, without regard to its present use, provided the intended use is known. Four general survey classes are defined using various state regulations and accepted practices.

Class A, Urban Surveys: Surveys of land lying within or adjoining a city or town. This also includes the surveys of commercial and industrial properties, condominiums, townhouses, apartments, and other multiunit developments, regardless of geographic location.

Class B, Suburban Surveys: Surveys of land lying outside urban areas. This land is used almost exclusively for single-family residences or residential subdivisions.

Class C, Rural Surveys: Surveys of land such as farms and other undeveloped land outside the suburban areas that may have a potential for future development.

Class D, Mountain and Marshland Surveys: Surveys of land that normally lie in remote areas with difficult terrain and that usually have limited potential for development.

Table 77.1 *Federal Standards of Traverse Closure Errors*

order of accuracy	maximum error	application
first	1:100,000	primary control nets; precise scientific studies
second		support for primary control; control for large engineering projects
class I	1:50,000	
class II	1:20,000	
third		small-scale engineering projects; large-scale mapping projects
class I	1:10,000	
class II	1:5000	

5. TYPES OF SURVEYS

Plane surveys disregard the curvature of the earth. A plane survey is appropriate if the area is small. This is true when the area is not more than 12 mi (19 km) in any one direction. *Geodetic surveys* consider the curvature of the earth. *Zoned surveys,* as used in various *State Plane Coordinate Systems* and in the *Universal Transverse Mercator* (UTM) system, allow computations to be performed as if on a plane while accomodating larger areas.

6. SURVEYING METHODS

A *stadia survey* requires the use of a transit, theodolite, or engineer's level as well as a rod for reading elevation differences and a tape for measuring horizontal distances. Stadia surveys are limited by the sighting capabilities of the instrument as well as by the terrain ruggedness.

In a *plane table survey*, a plane table is used in conjunction with a *telescopic instrument*. The plane table is a drawing board mounted on a tripod in such a way that the board can be leveled and rotated without disturbing the azimuth. The primary use of the combination of the plane table and telescope is in field compilation of maps, for which the plane table is much more versatile than the transit.

Total station surveys integrate theodolites, electronic distance measurement (EDM), and data recorders, collecting vertical and horizontal data in a single operation. *Manual total stations* use conventional optical-reading theodolites. *Automatic total stations* use electronic theodolites. *Data collectors* work in conjunction with total stations to store data electronically. Previously determined data can be downloaded to the data recorders for use in the field to stake out or field-locate construction control points and boundaries.

Triangulation is a method of surveying in which the positions of survey points are determined by measuring the angles of triangles defined by the points. Each survey point or monument is at a corner of one or more triangles. The survey lines form a network of triangles. The three angles of each triangle are measured. Lengths of triangle sides are calculated from trigonometry. The positions of the points are established from the measured angles and the computed sides.

Triangulation is used primarily for geodetic surveys, such as those performed by the National Geodetic Survey. Most first- and second-order control points in the national control network have been established by triangulation procedures. The use of triangulation for transportation surveys is minimal. Generally, triangulation is limited to strengthening traverses for control surveys.

Trilateration is similar to triangulation in that the survey lines form triangles. In trilateration, however, the lengths of the triangles' sides are measured. The angles are calculated from the side lengths. Orientation of the survey is established by selected sides whose directions are known or measured. The positions of trilaterated points are determined from the measured distances and the computed angles.

Photogrammetric surveys are conducted using aerial photographs. The advantages of using photogrammetry are speed of compilation, reduction in the amount of surveying required to control the mapping, high accuracy, faithful reproduction of the configuration of the ground by continuously traced contour lines, and freedom from interference by adverse weather and inaccessible terrain. Disadvantages include difficulty in areas containing heavy ground cover, high cost of mapping small areas, difficulty of locating contour lines in flat terrain, and the necessity for field editing and field completion. Photogrammetry works with ground control panels that are 3 to 4 times farther apart than with conventional surveys. But a conventional survey is still required in order to establish control initially.

Airborne LIDAR (Light Detection and Ranging) units are aircraft-mounted laser systems designed to measure the 3D coordinates of a passive target. This is achieved by combining a laser with positioning and orientation measurements. The laser measures the distance to the ground surface or target, and when combined with the position and orientation of the sensors, yields the 3D position of the target. Unlike EDM, which uses phase shifts of a continuous laser beam, LIDAR measures the time of flight for a single laser pulse to make the round trip from the source to receiver. LIDAR systems typically use a single wavelength, either 1064 nm or 532 nm,

Transportation

corresponding to the infrared and green areas of the electronic spectrum, respectively. Time can be measured to approximately $1/3$ ns, corresponding to a distance resolution of approximately 5 cm.

7. GLOBAL POSITIONING SYSTEM

The NAVSTAR (Navigation Satellite Timing and Ranging) *Global Positioning System* (GPS) is a one-way (satellite to receiver) ranging system. The advantages of GPS are speed and access. The main disadvantages are cost of equipment and dependence on satellite availability/visibility. Four satellites must be visible for 3D work, and a clear view of the sky is needed.

GPS determines positions without reference to any other point. It has the advantage of allowing work to proceed day or night, rain or shine, in fair weather or foul. It is not necessary to have clear lines of sight between stations, or to provide mountaintop stations for *inversability*.

GPS uses precisely synchronized clocks on each end—atomic clocks in the satellites and quartz clocks in the receivers. Frequency shift (as predicted by the Doppler effect) data is used to determine positions. Accurate positions anywhere on the earth's surface can easily be determined.

GPS can be used in two modes. *Standalone navigational mode* yields positional accuracies of ± 2 to 10 m; *differential* or *relative positioning* yields centimeter accuracies.

In relative positioning, one GPS receiver is placed on a point whose coordinates have been established in the National Geodetic Reference System, and one or more other receivers are placed at other points whose coordinates are to be determined. All receivers observe the same satellites at the same time for about an hour, more or less, depending on the accuracy required and the conditions of observation. A computer analysis (using proprietary software from the GPS equipment manufacturer) of the observed and recorded data determines the coordinates of the unknown points, the distance and elevation between the known point and each unknown point, and the direction (azimuth) from the known point to each unknown point. With relative positioning, the distance between points can be measured to 1 to 2 cm, and better than 1 part in 100,000.

Terms used when discussing GPS accuracy are *GDOP*, *HDOP*, and *VDOP*, standing for *geodetic, horizontal, and vertical dilution of position.* These numbers are inversely proportional to the volume of a pyramid that is formed by the position of the receiver and the four satellites being observed. In effect, GDOP is a measure of the geometry of the solution. A low GDOP number provides the best geometry and the most accurate solution.

GPS receivers are generally used in a fixed (stationary) position, hence the term *static GPS surveying*. With *rapid static mode* using a more sophisticated receiver,

occupation times can be reduced from 45 to 60 min to 10 to 15 min for short baselines.

Using a modified observation procedure known as *kinematic GPS surveying*, the coordinates of any point can be determined in 45 to 60 sec. Kinematic GPS uses one stationary receiver and one or more roving receivers. However, this method is procedurally demanding and very unforgiving. For example, if the "lock" on a satellite is lost, a previously observed point must be reoccupied to continue the survey.

Pseudokinematic GPS surveying is similar to kinematic GPS in field procedure but similar to static GPS in processing. Rover units occupy each point for approximately 10 min, then reoccupy the positions later during a reverse run of the route.

GPS is less suited to elevation measurements. It is used in conjunction with traditional surveying methods to determine accurate elevations. Inaccuracies in elevation derive from the difference between the *spheroid* (the mathematical model selected by the software to best fit the surface of the entire earth) and the *geoid* (the actual shape of the earth at sea level). Depending on location, this difference in surface elevation can be 2 to 35 m.

8. INERTIAL SURVEY SYSTEMS

Inertial survey systems (ISSs) determine a position on the earth by analyzing the movement of a transport vehicle, usually a light-duty truck or helicopter in which the equipment is installed. An ISS system consists of a precise clock, a computer, a recording device, sensitive accelerometers, and gyroscopes. The equipment measures the acceleration of the vehicle in all three axes and converts the acceleration into distance. The production rate is 8 to 10 mph when mounted in tired vehicles. Helicopter ISSs can proceed at 50 to 70 mph.

To reduce errors and achieve horizontal accuracies of 1 part in 10,000 or even 20,000, the circuit is "double run" in opposite directions. This cancels some instrumentation errors caused by the earth's rotation and magnetic field. Vertical accuracies are less.

9. GEOGRAPHIC INFORMATION SYSTEMS

A *Geographic Information System* (GIS) is a computerized database management system used to capture, store, retrieve, analyze, and display spatial data in map and overlay form. GIS replaces rolls and piles of time-bleached paper prints stored in drawers.

The integrated database contains spatial information, literal information (e.g., descriptions), and characteristics (e.g., land cover and soil type). Users can pose questions involving both position and characteristics.

GIS can produce "intelligent supermaps" at the touch of a button. Details (e.g., lines, features, text) on maps that are *vector images* can be individually addressed, edited, colored, moved, and so on. Maps that consist of *raster images*, on the other hand, can be edited on a pixel-by-pixel basis, but the individual details cannot be modified.

10. UNITS

In most of the United States, survey work is conducted using the foot and decimal parts for distance. The SI system, using meters to measure distance, is widely used throughout the world. In the United States, the SI system has been adopted by only some state highway departments.

In the United States, angles are measured in the *sexagesimal system* with degrees, minutes, and seconds. However, some countries have divided the circle into 400 units called *gons* (or *grads*). Thus, a right angle would have a measure of 100.0000 gon.

11. POSITIONS

Modern equipment permits surveyors to obtain positioning information directly from the theodolite.

It is not necessary to make a series of distance and angle measurements in the field and then return to the office to perform calculations to determine an unknown position.

Two methods are used to specify positions: (a) by latitude and longitude, and (b) by rectangular (Cartesian) coordinates measured from a reference point. The *state plane coordinate systems* are rectangular systems using a partial latitude/longitude system as baseline references.

12. BENCHMARKS

"Benchmark" is the common name given to permanent monuments of known vertical positions. Monuments with known horizontal positions are referred to as *control stations* or *triangulation stations*. A nail or hub in the pavement, the flange bolts of a fire hydrant, or the top of a concrete feature (e.g, curb) can be used as a *temporary benchmark*. The elevations of temporary benchmarks are generally found in field notes and local official filings.

Official benchmarks (monuments) installed by surveyors and engineers generally consist of a bronze (or other inert material) disk set in the top of a concrete post/pillar. (Bronze caps in iron pipes were once used, but these have limited lives due to corrosion.) Stainless steel pins and invar rods are also used. Bronze disks and identification plates are stamped, at the minimum, with the name of the agency or the state registration number of the individual setting the benchmark. Official USGS benchmarks are stamped with a unique identification code. Official benchmarks may or may not have the elevations indicated. If not, an official directory can be consulted for the information.

First-order monuments are widely spaced, but they are more accurately located for use in large-area control. Within the first-order network are a number of more closely spaced second-order monuments, followed by third-order and fourth-order networks of ever smaller scopes. Each lower-order monument is referenced to a higher-order monument.

Vertical positions (i.e., elevations) are measured above a reference surface or *datum*, often taken as *mean sea level*. For small projects, a local datum on a temporary permanent benchmark can be used.

13. DISTANCE MEASUREMENT: TAPING

Gunter's chain (developed around 1620), also known as a *surveyor's chain*, consisted of 100 links totaling 66 ft. It was superseded by steel tape in the 1900s. However, steel tapes (i.e., "engineers' chains") are still sometimes referred to as "chains," and measuring distance with a steel tape is still called "chaining." The earliest tapes were marked at each whole foot, with only the last foot being subdivided into hundredths, similar to a Gunter's chain. Tapes exactly 100 ft long were called *cut chains*. Other tapes were 101 ft long and were called *add chains*. Tapes are now almost universally subdivided into hundreds along their entire lengths. In practice, this enables the rear chainman to hold the zero mark on the point being measured from, while the head chainman reads the distance directly off the face of the tape.

Tapes are relatively inexpensive and easy to use, but they are mainly used only when measuring or staking short distances. Low-coefficient tapes made from nickel-steel alloy known as *invar* are fairly insensitive to temperature. The coefficient of thermal expansion, α, of invar is 2 to 5.5×10^{-7} ft/ft-°F (3.6 to 9.9×10^{-7} m/m·°C), and its modulus of elasticity, E, is 2.1×10^{7} lbf/in² (145 GPa). For steel tapes, α is 6.45×10^{-6} ft/ft-°F (1.16×10^{-5} m/m·C) and the modulus of elasticity, E, is 2.9×10^{7} lbf/in² (200 GPa). Some woven tapes (e.g., linen, dacron) may have embedded copper fibers to provide strength. Other tapes are manufactured from fiberglass, which neither stretch excessively nor conduct electricity.

Tapes can be marked with customary U.S. units, SI units, or both (on either side). They come in a variety of lengths, up to 100 m, though 100 ft and 30 m tapes are the most common.

Tape readings are affected by temperature, tension, and sag. Correction equations are given in Eqs. 77.6, 77.7, and 77.8 and are applied according to Table 77.2. L is the length of tape, α is the coefficient of thermal expansion, T and P are the temperature and tension at which the measurement was made, T_s and P_s are the temperature and tension at which the tape was standardized, A is the cross-sectional area of the tape, and E is the modulus of elasticity of the tape.

$$C_T = L\alpha(T - T_s) \quad \text{[temperature]} \qquad 77.6$$

$$C_P = \frac{(P - P_s)L}{AE} \quad \text{[tension]} \qquad 77.7$$

$$C_s = \pm\left(\frac{W^2 L^3}{24P^2}\right) \quad \text{[sag]} \qquad 77.8$$

Table 77.2 Correction for Surveyor's Tapes

	measured flat	measured suspended
standardized flat	none	subtract C_s
standardized suspended	add C_s	none

14. DISTANCE MEASUREMENT: TACHYOMETRY

Tachyometric distance measurement involves sighting through a small angle at a distant scale. The angle may be fixed and the length measured (*stadia method*), or the length may be fixed and the angle measured (*European method*). Stadia measurement consists of observing the apparent locations of the horizontal crosshairs on a distant stadia rod. The interval between the two rod readings is called the *stadia interval* or the *stadia reading*.

The distance is directly related to the distance between the telescope and the rod. For rod readings, R_1 and R_2 (both in ft), the distance from the instrument to the rod is found from Eq. 77.9. The *instrument factor*, C, is the sum of the focal length and the distance from the plumb bob (or center of the instrument) to the forward lens. It varies from 0.6 to 1.4 ft but is typically set by the manufacturer at 1 ft. C is zero for internal focusing telescopes. The *interval factor*, K, often has a value of 100.

$$x = K(R_2 - R_1) + C \qquad 77.9$$

If the sighting is inclined, as it is in Fig. 77.2, it is necessary to find both horizontal and vertical distances. These can be determined from Eqs. 77.10 and 77.11, in which y is measured from the telescope to the sighting rod center. The height of the instrument above the ground must be known to calculate the elevation of the object above the ground.

$$x = K(R_2 - R_1)\cos^2\theta + C\cos\theta \qquad 77.10$$
$$y = \tfrac{1}{2}K(R_2 - R_1)\sin 2\theta + C\sin\theta \qquad 77.11$$

Figure 77.1 Horizontal Stadia Measurement

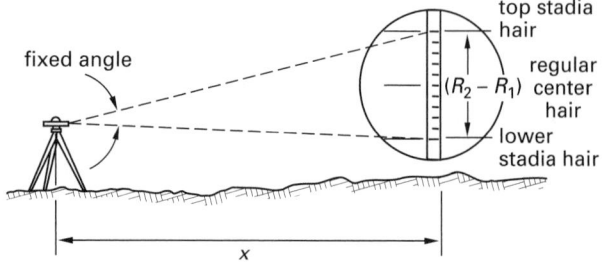

Figure 77.2 Inclined Stadia Measurement

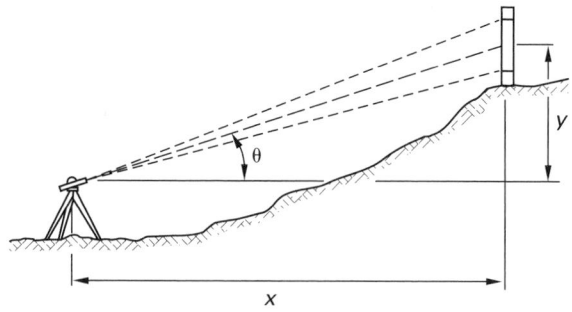

15. DISTANCE MEASUREMENT: EDM

Long-range laser (infrared and microwave in older units) *electronic distance measurement* (EDM) equipment is capable of measuring lines up to 10 mi long in less than 1 sec with an accuracy of better than $^1/_3$ ft (100 mm). Shorter ranges are even more precise. EDM can measure lines 1 mi long in less than 5 sec with an accuracy of about 0.01 ft. Light beams are reflected back by glass *retroreflectors* (precisely ground trihedral prisms). Common EDM accuracy ranges up to 2 mi (3.2 km) are $\pm$ 3 to 7 mm and $\pm$ 2 to 7 ppm. Shorter distances enjoy even better accuracies.

An *optical plummet* allows the instrument or retroprism to be placed precisely over a point on the ground, in spite of winds that would make a string *plumb bob* swing.

16. STATIONING

In route surveying, lengths are divided into 100 ft or 100 m sections called *stations*. The word "station" can mean both a location and a distance. A station is a length of 100 ft or 100 m, and the unit of measure is frequently abbreviated "sta." (30 m and 1000 m stationing are also used.) When a length is the intended meaning, the unit of measure will come after the numerical value; for example, "the length of curve is 4 sta." When a location is the intended meaning, the unit of measure will come before the number. For example, "the point of intersection is at sta 4." The location is actually a distance from the starting point. Therefore, the two meanings are similar and related.

Interval *stakes* along an established route are ordinarily laid down at *full station* intervals. If a marker stake is placed anywhere else along the line, it is called a *plus station* and labeled accordingly. A stake placed 825 ft (251 m) from station 0+00 is labeled "8+25" (2+51). Similarly, a stake placed 2896 ft (883 m) from a reference point at station 10+00 (10+00) is labeled "38+96" (18+83).

17. LEVELS

Automatic and self-leveling laser *levels* have all but replaced the optical level for leveling and horizontal alignment. A tripod-mounted laser can rotate and project a laser plane 300 to 2000 ft in diameter. A receiver mounted on a rod with visual and audible signals will alert the rodman to below-, at-, and above-grade conditions.

18. ELEVATION MEASUREMENT

Leveling is the act of using an engineer's level (or other leveling instrument) and rod to measure a vertical distance (*elevation*) from an arbitrary level surface.

If a level sighting is taken on an object with actual height h_a, the *curvature of the earth* will cause the object to appear taller by an amount h_c. In Eq. 77.12, x is measured in feet along the curved surface of the earth.

$$h_c = \left(2.4 \times 10^{-8} \, \frac{1}{\text{ft}}\right) x_{\text{ft}}^2 \quad \text{[U.S.]} \qquad 77.12$$

Atmospheric refraction will make the object appear shorter by an amount h_r.

$$h_r = \left(3.0 \times 10^{-9} \, \frac{1}{\text{ft}}\right) x_{\text{ft}}^2 \quad \text{[U.S.]} \qquad 77.13$$

The corrected rod reading (actual height) is

$$\begin{aligned} h_a &= R_{\text{observed}} + h_r - h_c \\ &= R_{\text{observed}} - \left(2.1 \times 10^{-8} \, \frac{1}{\text{ft}}\right) x_{\text{ft}}^2 \quad \text{[U.S.]} \end{aligned} \qquad 77.14$$

Figure 77.3 *Curvature and Refraction Effects*

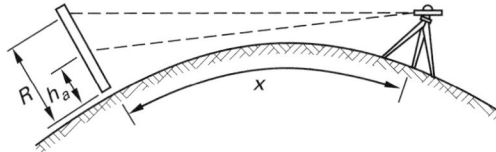

19. ELEVATION MEASUREMENT: DIRECT LEVELING

A *telescope* is part of the sighting instrument. A *transit* or *theodolite* is often referred to as a "telescope" even though the telescope is only a part of the instrument. (Early telescoping theodolites were so long that it was not possible to invert, or "transit," them. Hence the name "transit" is used for inverting telescopes.) An *engineer's level (dumpy level)* is a sighting device with a telescope rigidly attached to the level bar. A sensitive level vial is used to ensure level operation. The engineer's level can be rotated, but in its basic form it cannot be elevated. (An *alidade* consists of the base and telescopic part of the transit but not the tripod or leveling equipment.)

In a *semi-precise level*, also known as a *prism level*, the level vial is visible from the eyepiece end. In other respects, it is similar to the engineer's level. The *precise*

or *geodetic level* has even better control of horizontal angles. The bubble vial is magnified for greater accuracy.

Transits and levels are used with rods. *Leveling rods* are used to measure the vertical distance between the line of sight and the point being observed. The *standard rod* is typically made of fiberglass and is extendable. It is sometimes referred to as a *Philadelphia rod*. *Precise rods*, made of wood-mounted invar, are typically constructed in one piece and are spring-loaded in tension to avoid sagging.

With *direct leveling*, a level is set up at a point approximately midway between the two points whose difference in elevation is desired. The vertical *backsight (plus sight)* and *foresight (minus sight)* are read directly from the rod. HI is the height of the instrument above the ground, and h_{rc} is the correction for refraction and curvature.

Referring to Fig. 77.4,

$$y_{\text{A-L}} = R_{\text{A}} - h_{\text{rc,A-L}} - \text{HI} \qquad 77.15$$
$$y_{\text{L-B}} = \text{HI} + h_{\text{rc,L-B}} - R_{\text{B}} \qquad 77.16$$

The difference in elevations between points A and B is

$$\begin{aligned} y_{\text{A-B}} &= y_{\text{A-L}} + y_{\text{L-B}} \\ &= R_{\text{A}} - R_{\text{B}} + h_{\text{rc,L-B}} - h_{\text{rc,A-L}} \end{aligned} \qquad 77.17$$

If the backsight and foresight distances are equal (or approximately so), then the effects of refraction and curvature cancel.

$$y_{\text{A-B}} = R_{\text{A}} - R_{\text{B}} \qquad 77.18$$

Figure 77.4 *Direct Leveling*

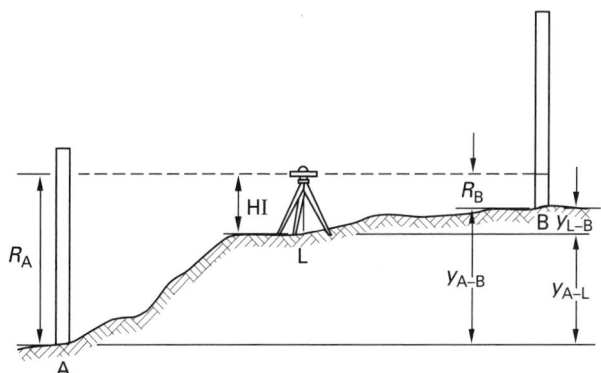

20. ELEVATION MEASUREMENT: DIFFERENTIAL LEVELING

Differential leveling is the consecutive application of direct leveling to the measurement of large differences in elevation. There is usually no attempt to exactly balance the foresights and backsights. Therefore, there is no record made of the exact locations of the level positions. Furthermore, the path taken between points need not be along a straight line connecting them, as only the elevation differences are relevant. If greater accuracy is desired without having to accurately balance the foresight and backsight distances, it is possible to eliminate most of the curvature and refraction error by balancing the sum of the foresights against the sum of the backsights.

The following abbreviations are used with differential leveling.

BM	benchmark or monument
TP	turning point
FS	foresight (also known as a minus sight)
BS	backsight (also known as a plus sight)
HI	height of the instrument
L	level position

HI (or H.I.) is the distance between the instrument axis and the datum. For differential leveling, the datum established for all elevations is used to define the HI. However, in stadia measurements, HI may be used (and recorded in the notes) to represent the height of the instrument axis above the ground.

Example 77.11

The following readings were taken during a differential leveling survey between benchmarks 1 and 2. All values given are in feet. What is the difference in elevations between these two benchmarks?

station	BS	FS	elevation
BM1	7.11		721.05
TP1	8.83	1.24	
TP2	11.72	1.11	
sta 1+00		2.92	
sta 2+00		6.43	
BM2		10.21	

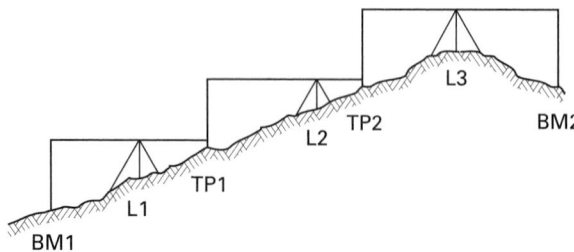

Solution

The first measurement is shown in larger scale. The height of the instrument is

$$HI1 = \text{elev}_{BM} + BS$$
$$= 721.05 \text{ ft} + 7.11 \text{ ft} = 728.16 \text{ ft}$$

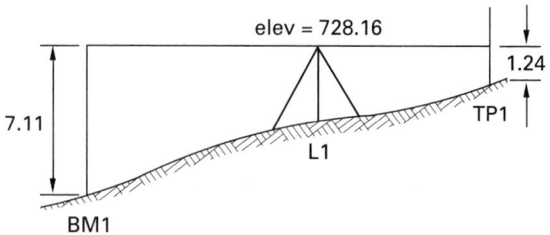

The height of the instrument at the second level position is

$$HI2 = HI1 + BS - FS$$
$$= 728.16 \text{ ft} + 8.83 \text{ ft} - 1.24 \text{ ft}$$
$$= 735.75 \text{ ft}$$

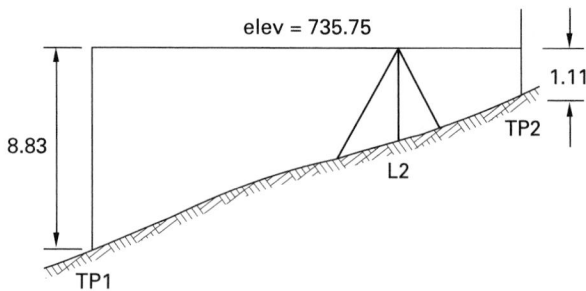

The height of the instrument at the third level position is

$$HI3 = HI2 + BS - FS$$
$$= 735.75 \text{ ft} + 11.72 \text{ ft} - 1.11 \text{ ft}$$
$$= 746.36 \text{ ft}$$

The elevation of sta 1+00 is

$$\text{elev}_{\text{sta } 1+00} = HI3 - FS$$
$$= 746.36 \text{ ft} - 2.92 \text{ ft} = 743.44 \text{ ft}$$

The elevation of sta 2+00 is

$$\text{elev}_{\text{sta } 2+00} = HI3 - FS$$
$$= 746.36 \text{ ft} - 6.43 \text{ ft}$$
$$= 739.93 \text{ ft}$$

The elevation of BM2 is

$$\text{elev}_{BM2} = HI3 - FS$$
$$= 746.36 \text{ ft} - 10.21 \text{ ft}$$
$$= 736.15 \text{ ft}$$

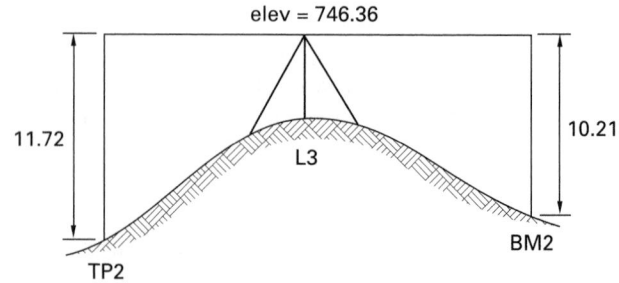

The difference in elevation is

$$\text{elev}_{\text{BM2}} - \text{elev}_{\text{BM1}} = 736.15 \text{ ft} - 721.05 \text{ ft} = 15.1 \text{ ft}$$

Check using the backsight and foresight sums.

The backsight sum is

$$7.11 \text{ ft} + 8.83 \text{ ft} + 11.72 \text{ ft} = 27.66 \text{ ft}$$

The foresight sum is

$$1.24 \text{ ft} + 1.11 \text{ ft} + 10.21 \text{ ft} = 12.56 \text{ ft}$$

The difference in elevation is

$$27.66 \text{ ft} - 12.56 \text{ ft} = 15.1 \text{ ft} \quad [\text{OK}]$$

21. ELEVATION MEASUREMENT: INDIRECT LEVELING

Indirect leveling, illustrated in Fig. 77.5, does not require a backsight. (A backsight reading can still be taken to eliminate the effects of curvature and refraction.) In Fig. 77.5, distance AC has been determined. Within the limits of ordinary practice, angle ACB is 90°. Including the effects of curvature and refraction, the elevation difference between points A and B is

$$y_{\text{A-B}} = \text{AC}\tan\alpha + 2.1 \times 10^{-8}(\text{AC})^2 \quad [\text{U.S.}] \quad \textit{77.19}$$

If a backsight is taken from B to A and angle β is measured, then

$$y_{\text{A-B}} = \text{AC}\tan\beta - 2.1 \times 10^{-8}(\text{AC})^2 \quad [\text{U.S.}] \quad \textit{77.20}$$

Adding Eqs. 77.19 and 77.20 and dividing by 2,

$$y_{\text{A-B}} = \tfrac{1}{2}\text{AC}(\tan\alpha + \tan\beta) \quad \textit{77.21}$$

Figure 77.5 *Indirect Leveling*

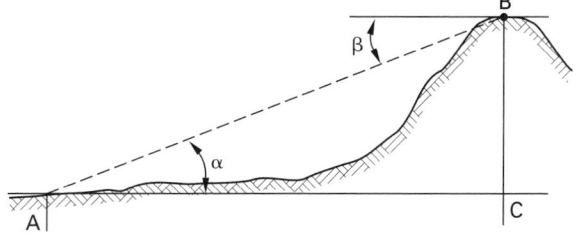

22. EQUIPMENT AND METHODS USED TO MEASURE ANGLES

A *transit* is a telescope that measures vertical angles as well as horizontal angles. The conventional *engineer's transit* (*surveyor's transit*) used in the past had a $6^1/_4$ in

diameter metal circle and vernier. It could be clamped vertically and used as a level. Few instruments of this nature are still available.

With a *theodolite*, horizontal and vertical angles are measured by looking into viewpieces. There may be up to four viewpieces, one each for the sighting-in telescope, the compass, the horizontal and vertical angles, and the optical plummet. Theodolites may be scale-reading optical, micrometer-reading optical, or electronic. Angles read electronically interface with EDM equipment and data recorders.

23. DIRECTION SPECIFICATION

The direction of any line can be specified by an angle between it and some other reference line, known as a *meridian*. If the meridian is arbitrarily chosen, it is called an *assumed meridian*. If the meridian is a true north-to-south line passing through the true north pole, it is called a *true meridian*. If the meridian is parallel to the earth's magnetic field, it is known as a *magnetic meridian*. A rectangular grid may be drawn over a map with any arbitrary orientation. If so, the vertical lines are referred to as *grid meridians*.

A true meridian differs from a magnetic meridian by the *declination* (*magnetic declination* or *variation*). If the north end of a compass points to the west of true meridian, the declination is referred to as a *west declination* or *minus declination*. Otherwise, it is referred to as an *east declination* or *plus declination*. Plus declinations are added to the magnetic compass azimuth to obtain the true azimuth. Minus declinations are subtracted from the magnetic compass azimuth.

A direction (i.e., the variation of a line from its meridian) may be specified in several ways. Directions are normally specified as either bearings or azimuths referenced to either north or south. In the United States, most control work is performed using directions stated as azimuths. Most construction projects and property surveys specify directions as bearings.

Azimuth: An azimuth is given as a clockwise angle from the reference direction, either "from the north" or "from the south" (e.g., "NAz 320"). Azimuths may not exceed 360°.

Deflection angle: The angle between a line and the prolongation of a preceding line is a deflection angle. Such measurements must be labeled as "right" for clockwise angles and "left" for counterclockwise angles.

Angle to the right: An angle to the right is a clockwise deflection angle measured from the preceding to the following line.

Azimuths from the back line: Same as angle to the right.

Figure 77.6 *Calculation of Bearing Angle*

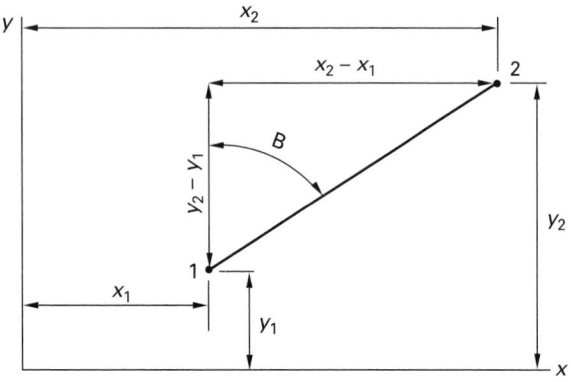

Figure 77.7 *Equivalent Angle Measurements*

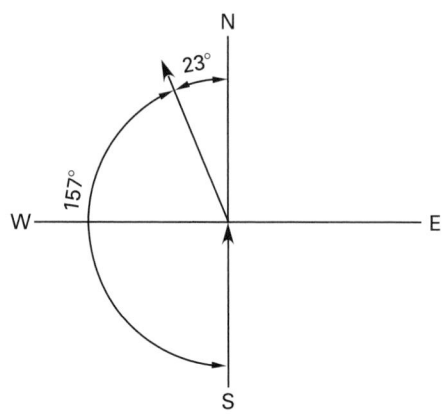

azimuth from the south: 157°
azimuth from the north: 337°
deflection angle: 23° L
angle to the right: 157°
bearing: N23° W

Bearing: The bearing of a line is referenced to the quadrant in which the line falls and the angle that the line makes with the meridian in that quadrant. It is necessary to specify the two cardinal directions that define the quadrant in which the line is found. The north and south directions are always specified first. A bearing contains an angle and a direction from a reference line, either north or south (e.g., "N45° E"). A bearing may not have an angular component exceeding 90°.

Bearings and azimuths are calculated from trigonometric and geometric relationships. From Fig. 77.6, the tangent of the bearing angle B of the line between the two points is given by Eq. 77.22.

$$\tan B = \frac{x_2 - x_1}{y_2 - y_1} \qquad 77.22$$

The distance D between the two points is given by Eq. 77.23.

$$D = \sqrt{(y_2 - y_1)^2 + (x_2 - x_1)^2} \qquad 77.23$$

If the numerator in Eq. 77.22 is positive, the line from point 1 to 2 bears east. If the denominator is positive, the line bears north. Determining the bearing and length of a line from the coordinates of the two points is called *inversing the line*.

The *azimuth* A_N of a line from point 1 to 2 measured from the north meridian is given by Eq. 77.24.

$$\tan A_N = \frac{x_2 - x_1}{y_2 - y_1} \qquad 77.24$$

The azimuth A_S of the same line from point 1 to 2 measured from the south meridian is given by Eq. 77.25.

$$\tan A_S = \frac{x_1 - x_2}{y_1 - y_2} \qquad 77.25$$

Example 77.12

A group of hikers on the Pacific Crest Trail (magnetic declination of +17°) uses a magnetic compass to sight in on a distant feature. What is the true azimuth if the observed angle is 42°?

Solution

A plus declination is added to the observed azimuth. Therefore, the true azimuth is $42° + 17° = 59°$.

24. LATITUDES AND DEPARTURES

The *latitude* of a line is the distance that the line extends in a north or south direction. A line that runs toward the north has a positive latitude; a line that runs toward the south has a negative latitude. The *departure* of a line is the distance that the line extends in an east or west direction. A line that runs toward the east has a positive departure; a line that runs toward the west has a negative departure.

If the x- and y-coordinates of two points are known, the latitude and departure of the line between the two points are determined by Eqs. 77.26 and 77.27.

latitude of line from point 1 to 2: $y_2 - y_1$ 77.26
departure of line from point 1 to 2: $x_2 - x_1$ 77.27

Figure 77.8 *Latitudes and Departures*

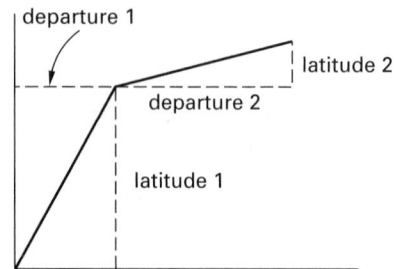

25. TRAVERSES

A *traverse* is a series of straight lines whose lengths and directions are known. A traverse that does not come back to its starting point is an *open traverse*. A traverse that comes back to its starting point is a *closed traverse*. The polygon that results from closing a traverse is governed by the following two geometric requirements.

1. The sum of the deflection angles is 360°.
2. The sum of the interior angles of a polygon with n sides is $(n-2)(180°)$.

There are three ways to specify angles in traverses: *deflection angles*, *interior angles* (also known as *explement angles*), and *station angles*. These are illustrated in Fig. 77.9.

Figure 77.9 *Angles Used in Defining Traverses*

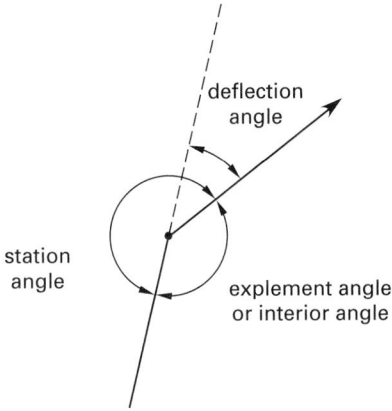

26. BALANCING

Refined field measurements are not the goal of measurement *balancing* (also known as *adjusting*). Balancing is used simply to accommodate the mathematical necessity of having a balanced column of figures. Balancing is considered inappropriate by some surveyors, since errors are not eliminated, only distributed to all of the measurements. Balancing does not make inaccurate measurements more accurate—it only makes inaccurate measurements more inaccurate. It is particularly inappropriate when a "blunder" (a measurement grossly in error that should be remeasured) is fixed by balancing.

27. BALANCING CLOSED TRAVERSE ANGLES

Due to measurement errors, variations in magnetic declination, and local magnetic attractions, it is likely that the sum of angles making up the interior angles will not exactly equal $(n-2)(180°)$. The following adjusting procedure can be used to distribute the angle error of closure among the angles.

step 1: Calculate the interior angle at each vertex from the observed bearings.

step 2: Subtract $(n-2)(180°)$ from the sum of the interior angles.

step 3: Unless additional information in the form of numbers of observations or probable errors is available, assume the angle error of closure can be divided equally among all angles. Divide the error by the number of angles.

step 4: Find a line whose bearing is assumed correct; that is, find a line whose bearing appears unaffected by errors, variations in magnetic declination, and local attractions. Such a line may be chosen as one for which the forward and back bearings are the same. If there is no such line, take the line whose difference in forward and back bearings is the smallest.

step 5: Start with the assumed correct line and add (or subtract) the prorated error to each interior angle.

step 6: Correct all bearings except the one for the assumed correct line.

Example 77.13

Adjust the angles on the four-sided closed traverse whose magnetic foresights and backsights are as follows.

line	bearing
AB	N 25° E
BA	S 25° W
BC	S 84° E
CB	N 84.1° W
CD	S 13.1° E
DC	N 12.9° W
DA	S 83.7° W
AD	N 84° E

Solution

step 1: The interior angles are calculated from the bearings.

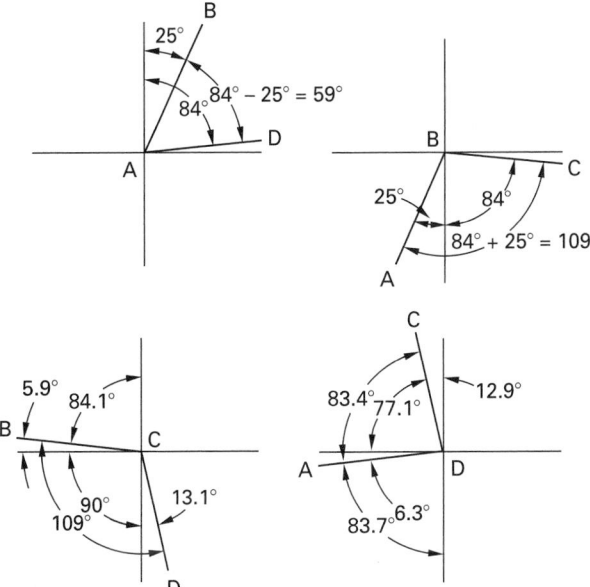

step 2: The sum of the angles is

$$59° + 109° + 109° + 83.4° = 360.4°$$

For a four-sided traverse, the sum of interior angles should be $(4 - 2)(180°) = 360°$, so a correction of $-0.4°$ must be divided evenly among the four angles. Since the backsight and foresight bearings of line AB are the same, it is assumed that the 25° bearings are the most accurate. The remaining bearings are then adjusted to use the corrected angles.

line	bearing
AB	N 25° E
BA	S 25° W
BC	S 83.9° E
CB	N 83.9° W
CD	S 12.8° E
DC	N 12.8° W
DA	S 83.9° W
AD	N 83.9° E

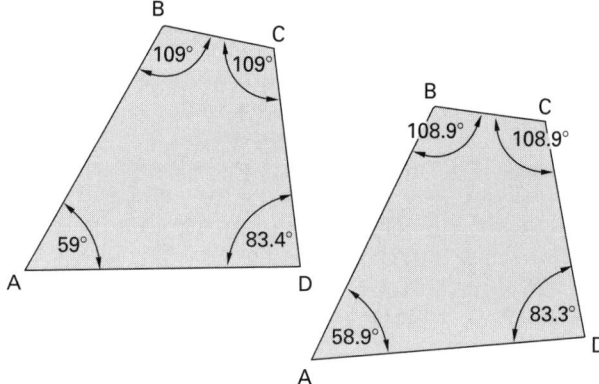

28. BALANCING CLOSED TRAVERSE DISTANCES

In a closed traverse, the algebraic sum of the latitudes should be zero. The algebraic sum of the departures should also be zero. The actual non-zero sums are called *closure in latitude* and *closure in departure*, respectively.

The *traverse closure* is the line that will exactly close the traverse. Since latitudes and departures are orthogonal, the closure in latitude and closure in departure can be considered as the rectangular coordinates and the traverse closure length calculated from the Pythagorean theorem. The coordinates will have signs opposite the closures in departure and latitude. That is, if the closure in departure is positive, point A will lie to the left of point A′, as shown in Fig. 77.10.

The length of a traverse closure is calculated from the Pythagorean theorem.

$$L = \sqrt{\begin{array}{l}(\text{closure in departure})^2 \\ + (\text{closure in latitude})^2\end{array}}$$ *77.28*

To balance a closed traverse, the traverse closure must be divided among the various legs of the traverse. (Of course, if it is known that one leg was poorly measured due to difficult terrain, all of the error may be given to that leg.) This correction requires that the latitudes and departures be known for each leg of the traverse.

Computer-assisted traverse balancing systems offer at least two balancing methods: compass rule, transit rule, *Crandall method*, and *least squares*. The use of one rule or the other is often arbitrary or a matter of convention.

The most common balancing method is the *compass rule*, also known as the *Bowditch method*: the ratio of a leg's correction to the total traverse correction is equal to the ratio of leg length to the total traverse length, with the signs reversed. The compass rule is used when the angles and distances in the traverse are considered equally precise.

$$\frac{\text{leg departure correction}}{\text{closure in departure}} = \frac{-\text{leg length}}{\text{total traverse length}}$$ *77.29*

$$\frac{\text{leg latitude correction}}{\text{closure in latitude}} = \frac{-\text{leg length}}{\text{total traverse length}}$$ *77.30*

If the angles are precise but the distances are less precise (such as when the distances have been determined by taping through rugged terrain), the *transit rule* is a preferred method of distributing the correction to the traverse legs. This rule distributes the closure error in proportion to the absolute values of the latitudes and departures.

$$\frac{\text{leg departure correction}}{\text{closure in departure}}$$
$$= -\left(\frac{\text{leg departure}}{\text{sum of departure absolute values}}\right)$$ *77.31*

$$\frac{\text{leg latitude correction}}{\text{closure in latitude}}$$
$$= -\left(\frac{\text{leg latitude}}{\text{sum of latitude absolute values}}\right)$$ *77.32*

Figure 77.10 *Traverse Closure*

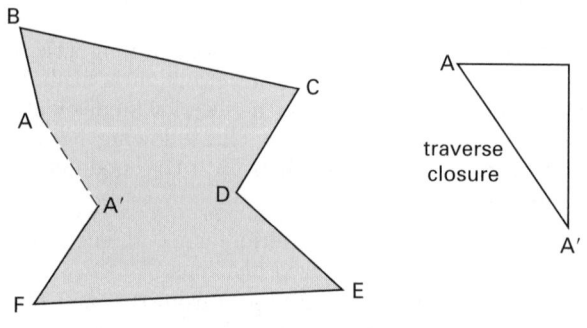

Example 77.14

A closed traverse consists of seven legs, the total of whose lengths is 2705.13 ft. Leg CD has a departure of 443.56 ft and a latitude of 219.87 ft. The total closure in departure for the traverse was +0.41 ft; the total closure in latitude was −0.29 ft. Use the compass rule to determine the corrected latitude and departure for leg CD.

Solution

The length of leg CD is

$$L_{CD} = \sqrt{(443.56 \text{ ft})^2 + (219.87 \text{ ft})^2}$$
$$= 495.06 \text{ ft}$$

Use the compass rule, Eq. 77.30.

$$\frac{\text{latitude correction}}{-0.29 \text{ ft}} = -\frac{495.06 \text{ ft}}{2705.13 \text{ ft}}$$

$$\text{latitude correction} = 0.05 \text{ ft}$$

The corrected latitude is

$$219.87 \text{ ft} + 0.05 \text{ ft} = 219.92 \text{ ft}$$

$$\frac{\text{departure correction}}{+0.41 \text{ ft}} = -\frac{495.06 \text{ ft}}{2705.13 \text{ ft}}$$

$$\text{departure correction} = -0.08 \text{ ft}$$

The corrected departure is

$$443.56 \text{ ft} - 0.08 \text{ ft} = 443.48 \text{ ft}$$

29. RECONSTRUCTING MISSING SIDES AND ANGLES

If one or more sides or angles of a traverse are missing or cannot be determined by measurement, they can sometimes be reconstructed from geometric and trigonometric principles.

The traverse shown in Fig. 77.11(a) has one leg missing. However, line EA can be reconstructed from its components EE′ and E′A. These components are equal to the sum of the departures and the sum of the latitudes, respectively, with the signs changed. The angle can be determined from the ratio of the sides. The length EA can be found from Eq. 77.28.

Figure 77.11(b) shows a traverse that has two adjacent legs missing. The traverse can be closed as long as some length/angle information is available. The technique is to close the traverse by using the same method for when one leg is missing. This will give the line DA. Then, the triangle EAD can be completed from other information.

Figure 77.11(c) shows a traverse with two nonadjacent legs missing. Since the latitudes and departures of two parallel lines are equal, this can be solved by closing the

traverse and shifting one missing leg as a parallel leg, to be adjacent to the other missing leg. This reduces the problem to the case of two adjacent legs missing.

Figure 77.11 *Completing Partial Traverses*

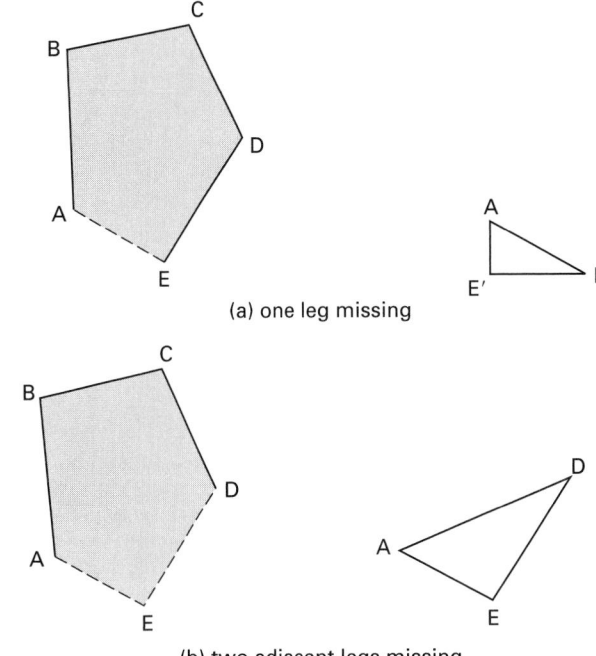

(a) one leg missing

(b) two adjacent legs missing

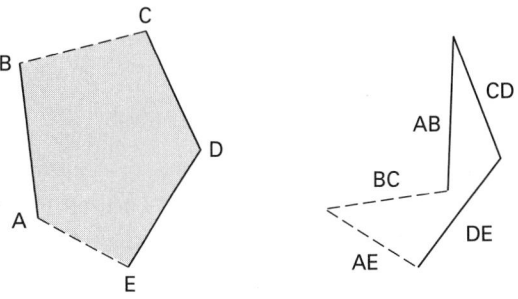

(c) two non-adjacent legs missing

30. TRAVERSE AREA: METHOD OF COORDINATES

The area of a simple traverse can be found by dividing the traverse into a number of geometric shapes and summing their areas.

If the coordinates of the traverse leg end points are known, the *method of coordinates* can be used. The coordinates can be x-y coordinates referenced to some arbitrary set of axes, or they can be sets of departure and latitude. In Eq. 77.33, x_n is substituted for x_0. Similarly, x_1 is substituted for x_{n+1}.

$$A = \frac{1}{2} \left| \left(\sum_{i=1}^{n} y_i (x_{i-1} - x_{i+1}) \right) \right| \qquad 77.33$$

The area calculation is simplified if the coordinates are written in the following form.

$$\frac{x_1}{y_1} \times \frac{x_2}{y_2} \times \frac{x_3}{y_3} \times \frac{x_4}{y_4} \times \frac{x_1}{y_1} \quad \text{etc.}$$

The area is

$$A = \tfrac{1}{2} \left| \sum \text{ of full line products} \right.$$
$$\left. - \sum \text{ of broken line products} \right| \quad \textit{77.34}$$

Example 77.15

Calculate the area of a triangle with x-y coordinates of its corners given as (3,1), (5,1), and (5,7).

$$\frac{3}{1} \times \frac{5}{1} \times \frac{5}{7} \times \frac{3}{1}$$

Solution

Use Eq. 77.33.

$$A = \left(\tfrac{1}{2}\right)\big((3)(1) + (5)(7) + (5)(1) - (1)(5)$$
$$- (1)(5) - (7)(3)\big)$$
$$= \left(\tfrac{1}{2}\right)(43 - 31)$$
$$= 6$$

31. TRAVERSE AREA: DOUBLE MERIDIAN DISTANCE

If the latitudes and departures are known, the *double meridian distance* (DMD) method can be used to calculate the area of a traverse. Equation 77.35 defines a double meridian distance for a leg.

$$\text{DMD}_{\text{leg } i} = \text{DMD}_{\text{leg } i-1} + \text{departure}_{\text{leg } i-1}$$
$$+ \text{departure}_{\text{leg } i}$$
$$\textit{77.35}$$

Special rules are required to handle the first and last legs. The DMD of the first course is defined as the departure of that course. The DMD of the last course is the negative of its own departure. The traverse area is calculated from Eq. 77.36. A tabular approach is the preferred manual method of using the DMD method.

$$A = \tfrac{1}{2} \left| \sum (\text{latitude}_{\text{leg } i} \times \text{DMD}_{\text{leg } i}) \right| \quad \textit{77.36}$$

Example 77.16

The latitudes and departures (in ft) of a six-leg traverse have been determined as given. Use the DMD method to calculate the traverse area.

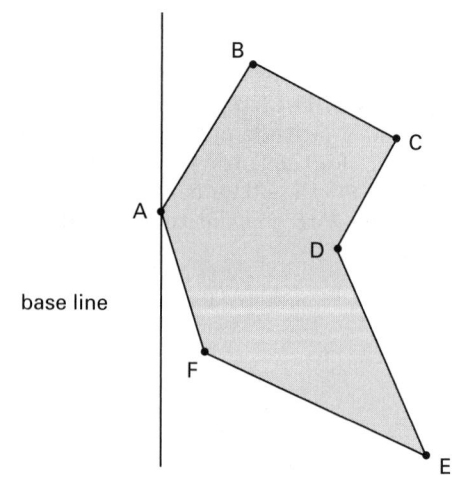

leg	latitude (ft)	departure (ft)
AB	200	200
BC	−100	200
CD	−200	−100
DE	−300	200
EF	200	−400
FA	200	−100

Solution

By the special starting rule, the DMD of the first leg, AB, is the departure of leg AB. The DMD of leg BC is given by Eq. 77.35.

$$\text{DMD}_{\text{BC}} = \text{DMD}_{\text{AB}} + \text{departure}_{\text{AB}} + \text{departure}_{\text{BC}}$$
$$= 200 \text{ ft} + 200 \text{ ft} + 200 \text{ ft}$$
$$= 600 \text{ ft}$$

The other DMDs are calculated similarly.

leg	latitude (ft)	departure (ft)	DMD (ft)	lat × DMD (ft^2)
AB	200	200	200	40,000
BC	−100	200	600	−60,000
CD	−200	−100	700	−140,000
DE	−300	200	800	−240,000
EF	200	−400	600	120,000
FA	200	−100	100	20,000
			total	−260,000

The area is given by Eq. 77.36.

$$A = \tfrac{1}{2} \left| \sum (\text{latitude}_{\text{leg } i} \times \text{DMD}_{\text{leg } i}) \right|$$
$$= \left(\tfrac{1}{2}\right)(260{,}000 \text{ ft}^2) = 130{,}000 \text{ ft}^2$$

32. AREAS BOUNDED BY IRREGULAR BOUNDARIES

Areas of sections with irregular boundaries, such as creek banks, cannot be determined precisely, and approximation methods must be used. If the irregular side can be divided into a series of cells of width d, either the trapezoidal rule or Simpson's rule can be used. (If the coordinates of all legs in the irregular boundary are known, the DMD method can be used.)

If the irregular side of each cell is fairly straight, the *trapezoidal rule* as given in Eq. 77.37 can be used.

$$A = d \left(\frac{h_1 + h_n}{2} + \sum_{i=2}^{n-1} h_i \right) \qquad 77.37$$

If the irregular side of each cell is curved or parabolic, then *Simpson's rule* (sometimes referred to as *Simpson's 1/3 rule*) as given in Eq. 77.38 can be used.

$$A = \left(\frac{d}{3} \right) h_1 - h_n + 2 \underbrace{\sum_{i=3}^{n-1} h_i}_{\substack{\text{for all odd numbers} \\ \text{starting at } i=3}} + \underbrace{4 \sum_{i=2}^{n-1} h_i}_{\substack{\text{for all even numbers} \\ \text{starting at } i=2}} \qquad 77.38$$

Figure 77.12 Area Bounded by Irregular Areas

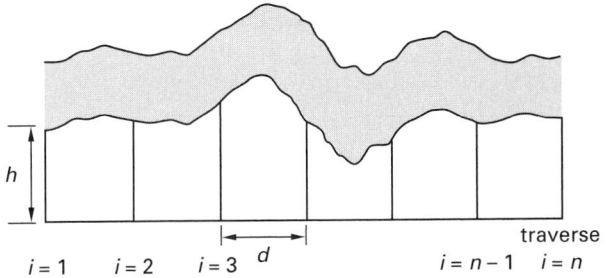

33. PHOTOGRAMMETRY

Photogrammetry (aerial mapping) uses photography to obtain distance measurements. Either vertical or oblique (inclined) photographs can be used, as shown in Fig. 77.13. The controlling dimensions are taken between large paint marks or flags set out on the ground that are visible in the aerial photographs. Adjacent photographs overlap by 60% or more, forming a "stereo pair." Stereo pair photographs enable the operator to see the ground in three dimensions so that elevations of the features can also be determined.

The *scale* of a photograph is the ratio of the dimension on the photo and the dimension on the ground. It is also the ratio of the focal length and the flight altitude. Precise mappings use scales of 1 in = 50 to 500 ft, whereas more general mappings might use 1 in = 600 to 1300 ft. Scale is constantly changing, since the elevation above the ground level depends on the surface terrain. Also, the distance from the camera to the ground directly below the camera will be less than the distance from

the camera to points on the outer fringes of the photograph. Therefore, an average distance from camera to ground level is used as the flight altitude. The number of photographs and the number of flight paths required depend on the film size and lap percentages.

$$\begin{aligned} \text{scale} &= \frac{\text{focal length}}{\text{flight altitude}} \\ &= \frac{\text{length in photograph}}{\text{true length}} \end{aligned} \qquad 77.39$$

Figure 77.13 Photogrammetric Images

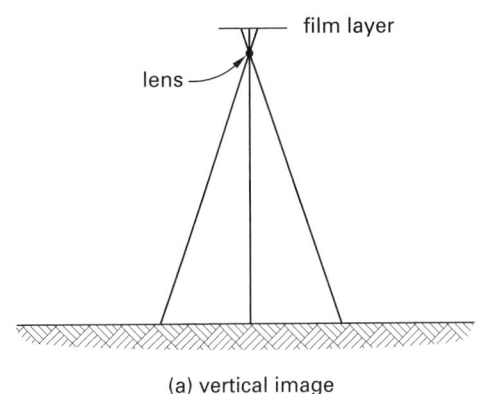

(a) vertical image

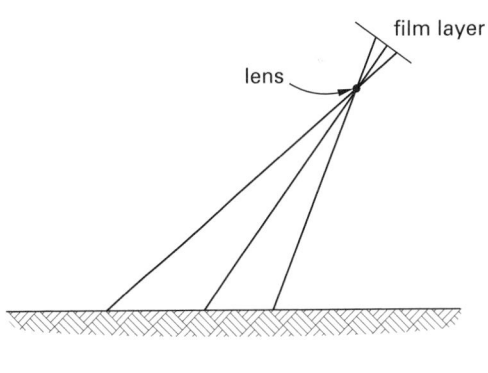

(b) oblique image

Example 77.17

Aerial mapping is being used with a scale of 1 in:1000 ft. The film holder uses 8 in by 8 in film. The focal length of the camera is 6 in. The area to be mapped is square, 6 mi on each side. A side-lap from photo to photo of 30% is desired, as well as an end-lap from photo to photo of 60%. The plane travels at 150 mph.

(a) What is the area covered by each photograph? (b) At what altitude should the plane fly while taking photographs? (c) How far apart will the flight paths be? (d) How many flight paths are required? (e) How many photographs are required per flight path? (f) How many photographs will be taken altogether? (g) How frequently should the photographs be taken?

Solution

(a) The photograph area is

$$(8 \text{ in})(8 \text{ in}) = 64 \text{ in}^2$$

The area covered by each photograph is

$$(64 \text{ in}^2)\left(1000 \frac{\text{ft}}{\text{in}}\right)^2 = 64{,}000{,}000 \text{ ft}^2$$

(b) The flight altitude is calculated using Eq. 77.39.

$$\begin{aligned}
\text{altitude} &= \frac{\text{focal length}}{\text{scale}} \\
&= \frac{6 \text{ in}}{\dfrac{1 \text{ in}}{1000 \text{ ft}}} = 6000 \text{ ft}
\end{aligned}$$

(c) Each photograph covers 64,000,000 ft^2, or an area 8000 ft on each side. With a 30% overlap, the distance between flight paths is calculated as

$$(1 - 0.30)(8000 \text{ ft}) = 5600 \text{ ft}$$

(d) The number of flight paths is

$$\frac{(6 \text{ mi})\left(5280 \dfrac{\text{ft}}{\text{mi}}\right)}{5600 \text{ ft}} = 5.7 \quad [\text{use } 6]$$

(e) With an end lap of 60%, the distance between photos along the flight path is

$$(8000 \text{ ft})(1 - 0.60) = 3200 \text{ ft}$$

The number of photographs per flight path is

$$\frac{(6 \text{ mi})\left(5280 \dfrac{\text{ft}}{\text{mi}}\right)}{3200 \text{ ft}} = 9.9 \quad [\text{use } 10]$$

(f) The total number of photos will be (6)(10) = 60.

(g) At 150 mph, the frequency of camera shots will be

$$\frac{3200 \text{ ft}}{\dfrac{\left(150 \dfrac{\text{mi}}{\text{hr}}\right)\left(5280 \dfrac{\text{ft}}{\text{mi}}\right)}{3600 \dfrac{\text{sec}}{\text{hr}}}}$$

$$= 14.5 \text{ sec/photo}$$

34. PUBLIC LAND SYSTEM

The *U.S. Rectangular Surveying System* was devised with the objective of locating, marking, and fixing subdivisions "for all time." The land was divided into *tracts* approximately 24 mi by 24 mi by means of meridians and parallels of latitude. These tracts were then divided into 16 *townships*, which were approximately 6 mi on a side. The last step was to divide the township into 36 *sections*, each approximately 1 mi by 1 mi. Each township has an area of approximately 36 mi^2, containing 36 sections, each 1 mi^2 in area. Sections were further divided into quarter sections.

Townships were labeled using a range and township number. The *township number* assigned to the parcel of land was determined according to its location relative to the standard parallel of latitude defining the northern and southern boundaries. The *range number* assigned to that parcel was determined according to its location relative to the principal meridian defining the eastern and western boundaries. For example, the parcel of land that is 2 rows south of the standard parallel and 3 rows east of the principal meridian bordering the township would be numbered "T2S, R3E." Figure 77.14 illustrates the numbering of townships and ranges for a 24 mi^2 tract of land. The township's 36 sections are numbered as shown in Fig. 77.15, beginning in the northeasternmost section and ending in the southeasternmost section.

Figure 77.14 *Standard Parallels and Guide Meridians*

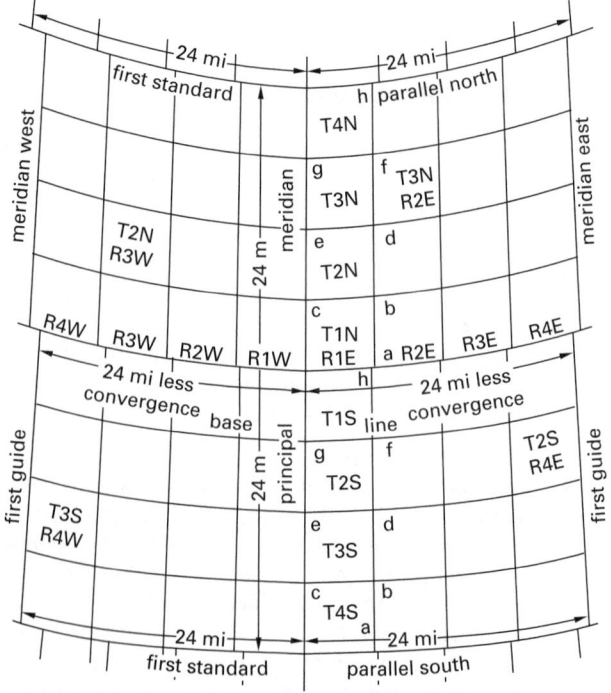

Figure 77.15 Subdivision of Township

6	5	4	3	2	1
7	8	9	10	11	12
18	17	16	15	14	13
19	20	21	22	23	24
30	29	28	27	26	25
31	32	33	34	35	36

Example 77.18

Give a legal description of the parcel of land shown, located in Section 7 of T12S, R9W, of the 10 Principal Meridian, located in Williams Township, Pennsylvania.

Solution

Beginning with the smallest identifiable quarter section of land, the legal identification would be as follows.

The $NE^{1}/_4$ of the $NW^{1}/_4$ of the $NE^{1}/_4$ of the $SW^{1}/_4$ of Section 7, T12S, R9W, of the 10 Principal Meridian, located in Williams Township, Pennsylvania.

35. TOPOGRAPHIC MAPS

Topographic maps are graphic representations of the surface of the earth. They contain a plan view of the land, scales to measure distance, bearings to indicate direction, coordinate systems to locate features, symbols of natural and artificial features, and contour lines to show elevation, slope, and relief of the landscape. The standard USGS mapping symbols are shown in Fig. 77.16.

The *map scale* can be specified in several ways: a fractional scale (e.g., "1:50,000") indicating that one unit of distance on the map corresponds to a number of the same units on the ground, an explicit verbal scale (e.g., "1 cm to 1 km"), and a graphic scale consisting of a calibrated bar or line.

The *vertical exaggeration* is the distortion of the vertical scale of a topographic profile used to emphasize relief

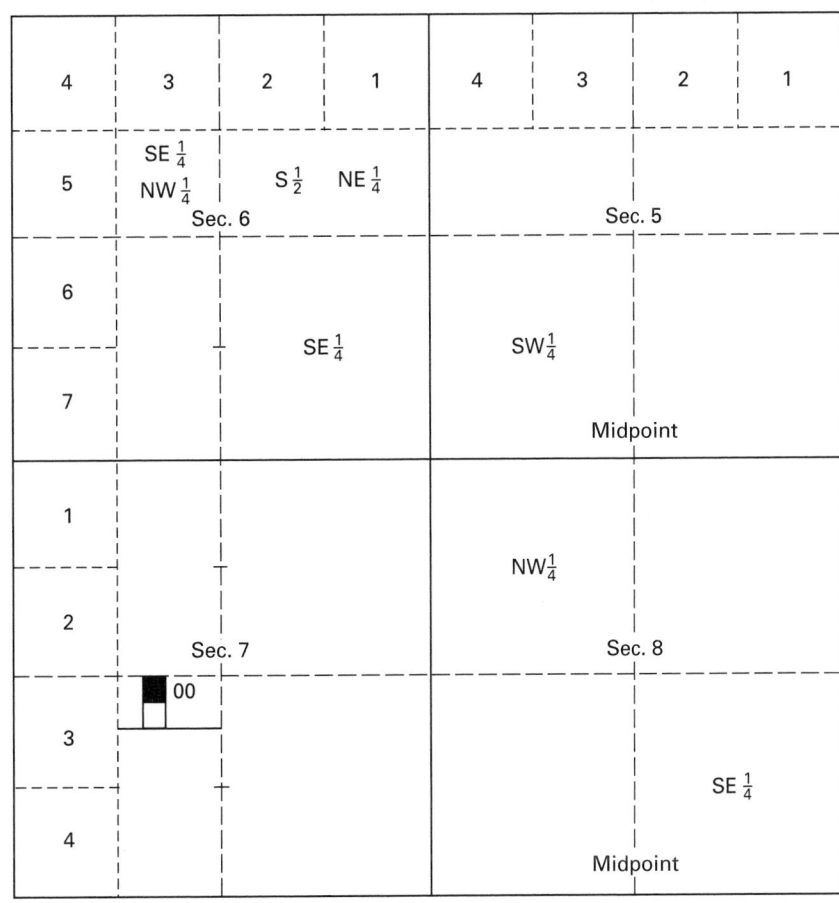

Transportation

and slope. The vertical exaggeration is calculated as the vertical scale divided by the horizontal scale.

Official USGS topographic maps can be selected by scale or coverage angle. Available scales include 1:24,000 or 1:25,000, 1:62,500, 1:100,000, and 1:250,000. Some 1:500,000 maps are also available. The two most common coverage angles are (a) 7.5′ (1/8°) quadrangle maps, which extend 7.5′ of latitude from north to south and 7.5′ of longitude from east to west; and (b) 15′ (1/4°) quadrangle maps, which extend 15′ of latitude from north to south and 15′ of longitude from east to west. North is always at the top of the map.

Contour lines are imaginary lines connecting points of equal elevation. They can be used to show elevation, *relief* (the difference in elevations between two points), and slope. The *contour interval* (e.g., 10 ft or 20 ft) is constant on any given map. Every fifth line is an *index contour*, printed thicker and marked with the elevation for reference. When ground is level, no contours are shown. Closely spaced contours indicate a steep slope. Merged contours indicate a cliff.

Two types of coordinate systems are given on most topographic maps: (a) latitude-longitude and (b) township-range. The township-range system is only used in the western and southern United States, whereas the latitude-longitude system is used throughout the world.

The *Universal Transverse Mercator* (UTM) grid generally appears on topographic maps. A UTM coordinate is essentially the x-y coordinates on the map. The first half of the UTM digits are the x-coordinates; the second half are the y-coordinates. 4-, 6-, 8-, and 10-digit coordinates can be specified, with some interpolation on the map being required. For complete identification, the map number, zone, or other regional information must also be specified.

Figure 77.16 *Standard USGS Topographic Map Symbols*

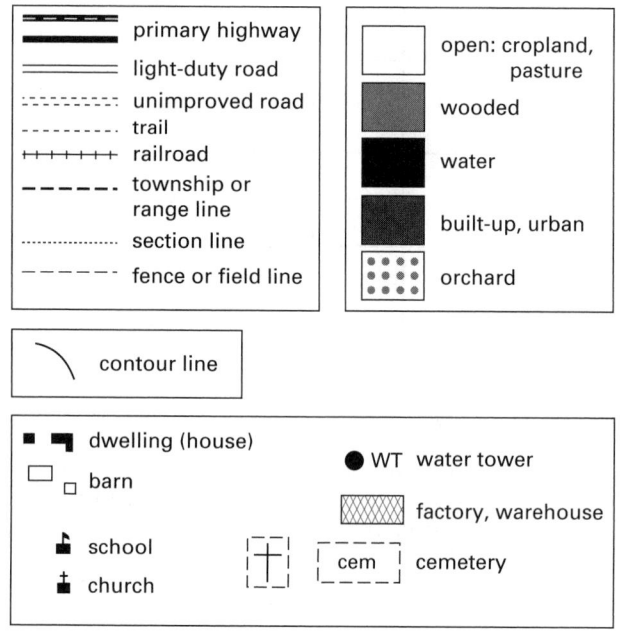

78 Horizontal, Compound, Vertical, and Spiral Curves

Nomenclature

A	absolute value of the algebraic grade difference	percent	percent
C	chord length	ft	m
C	clearance	ft	m
D	degree of curve	deg	deg
e	superelevation rate	ft/ft	m/m
E	equilibrium elevation	ft	m
E	external distance	ft	m
f	coefficient of friction	–	–
F	force	lbf	N
g	acceleration due to gravity	ft/sec^2	m/s^2
G	gauge	ft	m
G	grade	decimal	decimal
h	height	ft	m
HSO	horizontal sightline offset	ft	m
I	intersection angle	deg	deg
K	length of vertical curve per percent grade difference	ft/%	m/%
l	length	ft	m
L	length of curve	ft	m
L	superelevation runoff	ft	m
m	mass	lbm	kg
M	middle ordinate	ft	m
p	cross slope (rate)	ft/ft	m/m
R	curve radius	ft	m
S	sight distance	ft	m
SRR	superelevation runout rate	ft/ft	m/m
t	time	sec	s
T	tangent length	ft	m
v	velocity	ft/sec	m/s
W	lane width	ft	m
W	offset (maximum)	ft	m
x	distance from BVC	sta	sta
x	tangent distance	ft	m

Symbols

α	angle	deg	deg
γ	angle	deg	deg
θ	angle	deg	deg
ϕ	angle	deg	deg

Subscripts

c	centrifugal or circular curve
eff	effective
f	frictional
p	perception-reaction
R	runout
s	side friction or spiral
t	tangential

1. HORIZONTAL CURVES

A *horizontal circular curve* is a circular arc between two straight lines known as *tangents*. When traveling in a particular direction, the first tangent encountered is the *back tangent (approach tangent)*, and the second tangent encountered is the *forward tangent (departure tangent or ahead tangent)*.

The geometric elements of a horizontal circular curve are shown in Fig. 78.1. Table 78.1 lists the standard terms and abbreviations used to describe the elements. Equations 78.1 through 78.7 describe the basic relationships between the elements. The *intersection angle (deflection angle)*, I, has units of degrees unless indicated otherwise. (Equations that contain the *degree of curve* term, D, are only to be used with customary U.S. units.)

Transportation

$$R = \frac{5729.578}{D} \quad \text{[U.S.—arc definition]} \qquad 78.1$$

$$R = \frac{50}{\sin \frac{D}{2}} \quad \text{[U.S.—chord definition]} \qquad 78.2$$

$$L = \frac{2\pi R I}{360°} = R I_{\text{radians}} = \frac{100 I}{D} \quad \text{[U.S.]} \qquad 78.3$$

$$T = R \tan \frac{I}{2} \qquad 78.4$$

$$E = R \left(\sec \frac{I}{2} - 1 \right) = R \tan \frac{I}{2} \tan \frac{I}{4} \qquad 78.5$$

$$\text{HSO} = R \left(1 - \cos \frac{I}{2} \right) = \frac{C}{2} \tan \frac{I}{4} \qquad 78.6$$

$$C = 2R \sin \frac{I}{2} = 2T \cos \frac{I}{2} \qquad 78.7$$

Table 78.1 *Horizontal Curves: Abbreviations and Terms*

preferred

c	short chord (any straight distance from one point on the curve to another)
C	long chord (chord PC to PT); same as LC
D	degree of curve
E	external distance (the distance from the vertex to the midpoint of the curve)
I	intersection angle; central angle of curve; deflection angle between back and forward tangents
L	length of the curve (the length of the curve from the PC to the PT)
LC	long chord (chord PC to PT); same as C
HSO	horizontal sightline offset (the distance from the centerline of the inside lane to the midpoint of the long chord) known prior to 2004 as the middle ordinate, M
MPC	midpoint of curve
PC	point of curvature (the point where the back tangent ends and the curve begins)
PCC	point of continuing curve (the transition point between curves in a compound curve)
PI	point of intersection of back and forward tangents
POC	(any) point on the curve
POCT	point of curve tangent
POST	(any) point on the semitangent
POT	(any) point on the tangent
PT	point of tangency (the point where the curve ends and the forward tangent begins)
R	radius of the curve
RP	radius point (center of curve)
T	(semi-) tangent distance from V to PC or from V to PT
V	vertex of the tangent intersection point

alternate

BC	beginning of curve (same as PC and TC)
CT	change from curve to tangent (same as PT and EC)
Δ	interior angle (same as I)
EC	end of curve (same as PT and CT)
O	center of circle (same as RP)
PVI	point of vertical intersection (same as PI)
TC	a change from a tangent to a curve (same as BC and PC)

Figure 78.1 *Horizontal Curve Elements*

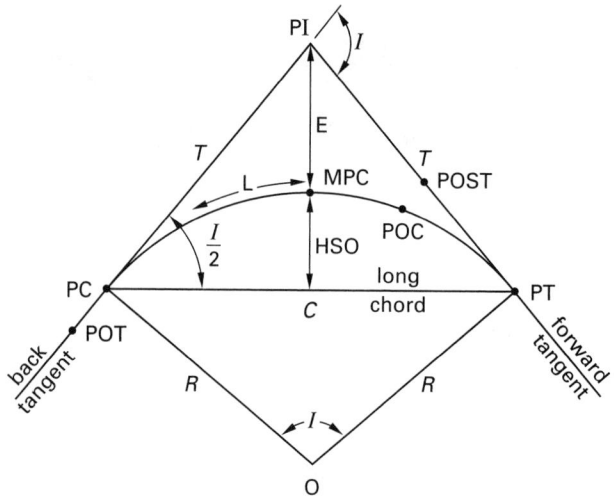

Example 78.1

Two tangents—the first entering a horizontal curve and the second leaving the horizontal curve—have bearings of N10° W and N12° E, respectively. The curve radius is 1300 ft. Determine the intersection angle.

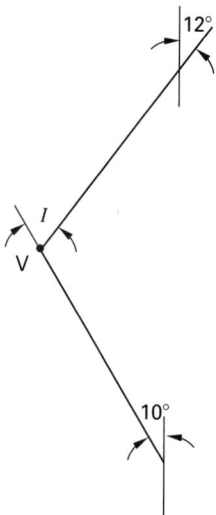

Solution

Since the two tangents do not originate from a single point, the intersection angle is the deflection angle between the two tangents.

$$I = 10° + 12° = 22°$$

2. DEGREE OF CURVE

In the United States, curvature of city streets, highways, and railway curves can be specified by either the radius, R, or the *degree of curve*, D. There is no parallel concept in metric highway design. Therefore, curves are only denoted by radius when using metric units.

In most highway work, the *length of the curve* is understood to be the actual curved arc length, and the degree of the curve is the angle subtended by an arc of 100 ft. When the degree of curve is related to an arc of 100 ft, it is said to be calculated on an *arc basis*.

$$D = \frac{(360°)(100)}{2\pi R} = \frac{5729.578}{R} \quad \text{[arc basis]} \qquad \textbf{78.8}$$

Railway curves have very large radii. The radii are so large that short portions of the curves are essentially straight. In railroad surveys, the *chord basis* is used, and the degree of curve has a different definition. The railroad degree of curve is the angle subtended by a chord of 100 ft. In that case, the degree of curve and radius are related by Eq. 78.9. Once the radius is calculated, other arc-basis equations (Eqs. 78.1 through 78.7) can be used.

$$\sin \frac{D}{2} = \frac{50}{R} \quad \text{[chord basis]} \qquad \textbf{78.9}$$

Where the radius is large (4° curves or smaller), the difference between the arc length and the chord length is insignificant. Therefore, the length of the curve in railroad practice is equal to the number of 100 ft chords.

$$L \approx \left(\frac{I}{D}\right)(100 \text{ ft}) \qquad \textbf{78.10}$$

Figure 78.2 *Horizontal Railroad Curve (chord basis)*

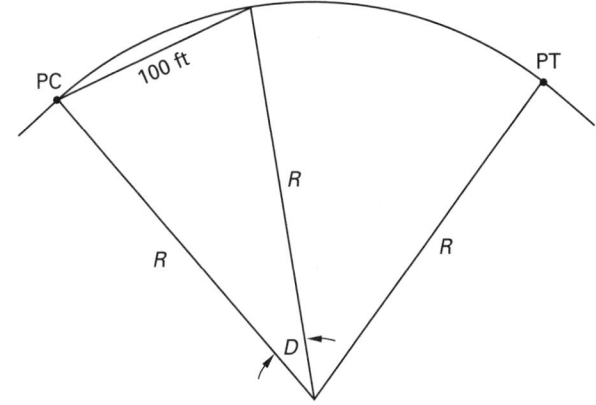

3. STATIONING ON A HORIZONTAL CURVE

When the route is initially laid out between PIs, the curve is undefined. The "route" distance is measured from PI to PI. The route distance changes, though, when the curve is laid out. Stationing along the curve is continuous, as a vehicle's odometer would record the distance. The PT station is equal to the PC station plus the curve length.

$$\text{sta PT} = \text{sta PC} + L \qquad \textbf{78.11}$$
$$\text{sta PC} = \text{sta PI} - T \qquad \textbf{78.12}$$

In route surveying, stationing is carried ahead continuously from a starting point or hub designated as station 0+00, and called *station zero plus zero zero*. The term *station* is applied to each subsequent 100 ft (or 100 m) length, where a stake is normally set. Also, the term *station* is applied to any point whose position is given by its total distance from the beginning hub. Thus, station 8+33.2 is a unique point 833.2 ft (or 833.2 m) from the starting point, measuring along the survey line. The partial length beyond the full station 8 is 33.2 ft (or 33.2 m) and is termed a *plus*. Moving or looking toward increasing stations is called *ahead stationing*. Moving or looking toward decreasing stations is called *back stationing*. Offsets from the centerline are either left or right looking ahead on stationing.

Normal curve layout follows the convention of showing increased stationing (ahead stationing) from left to right on the plan sheet, or from the bottom to the top of the sheet. Moving ahead on stationing, the first point of the curve is the *point of curvature* (PC). This point is alternatively called the point of *beginning of curve* (BC). Stationing is carried ahead along the arc of the curve to the end point of the curve, called the *point of tangency* (PT). This end point is alternatively called the point of *end of curve* (EC). The PI is stationed ahead from the PC. Therefore, the difference in stationing along the back curve tangent is the curve tangent length. The ahead tangent is rarely stationed to avoid the confusion of creating two stations for the PC.

Example 78.2

An interior angle of 8.4° is specified for a 2° (arc basis) horizontal curve. The forward PI station is 64+27.46. Locate the PC and PT stations.

Solution

Use Eq. 78.8.

$$R = \frac{(360°)(100 \text{ ft})}{2\pi D} = \frac{(360°)(100 \text{ ft})}{(2\pi)(2°)}$$
$$= 2864.79 \text{ ft}$$

Use Eqs. 78.4 and 78.3.

$$T = R \tan \frac{I}{2} = (2864.79 \text{ ft}) \tan \frac{8.4°}{2}$$
$$= 210.38 \text{ ft}$$
$$L = RI \frac{2\pi}{360°} = (2864.79 \text{ ft})(8.4°)\frac{2\pi}{360°}$$
$$= 420.00 \text{ ft}$$

The PC and PT points are located at

$$\text{sta PC} = \text{sta PI} - T = (64+27.46) - 210.38 \text{ ft}$$
$$= 62+17.08$$
$$\text{sta PT} = \text{sta PC} + L = (62+17.08) + 420.00 \text{ ft}$$
$$= 66+37.08$$

4. CURVE LAYOUT BY DEFLECTION ANGLE

Construction survey stakes should be placed at the PC, at the PT, and at all full stations. (In the United States, 100 ft stations are the rule. Metric stationing may be 100 or 1000 m, depending on the organization.) Stakes may also be required at quarter or half stations and at all other critical locations.

The *deflection angle method* is a common method used for staking out the curve. A *deflection angle* is the angle between the tangent and a chord. Deflection angles are related to corresponding arcs by the following principles.

1. The deflection angle between a tangent and a chord (Fig. 78.3(a)) is half of the arc's subtended angle.

2. The angle between two chords (Fig. 78.3(b)) is half of the arc's subtended angle.

In Fig. 78.3(a), angle V-PC-A is a deflection angle between a tangent and a chord. Using Principle 1,

$$\alpha = \angle\text{V-PC-A} = \frac{\beta}{2} \qquad \textit{78.13}$$

Angle β can be found from the following relationships.

$$\frac{\beta}{360°} = \frac{\text{arc length PC-A}}{2\pi R} \qquad \textit{78.14}$$

$$\frac{\beta}{I} = \frac{\text{arc length PC-A}}{L} \qquad \textit{78.15}$$

The chord distance PC-A is given by Eq. 78.16. The entire curve can be laid out from the PC by sighting the deflection angle V-PC-A and taping the chord distance PC-A.

$$C_{\text{PC-A}} = 2R\sin\alpha = 2R\sin\frac{\beta}{2} \qquad \textit{78.16}$$

Figure 78.3 *Circular Curve Deflection Angle*

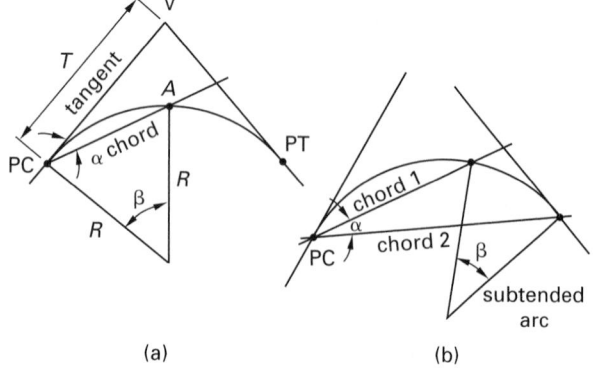

(a)   (b)

Example 78.3

A circular curve is to be constructed with a 225 ft radius and an intersection angle of 55°. The separation between the stakes along the arc is 50 ft. (a) Determine the chord lengths between stakes. (b) Assuming the curve is laid out from point to point, specify the first and last deflection angles. (c) Determine the length of the final (partial) chord.

Customary U.S. Solution

(a) The central angle for an arc of 50 ft is given by Eq. 78.14.

$$\beta = \frac{(360°)(\text{arc length})}{2\pi R}$$

$$= \frac{(360°)(50\text{ ft})}{2\pi(225\text{ ft})} = 12.732°$$

From Eq. 78.16, the required chord length for full 50 ft arcs is

$$C = 2R\sin\frac{\beta}{2} = (2)(225\text{ ft})\sin\frac{12.732°}{2}$$

$$= 49.90\text{ ft}$$

(b) The first central angle is 12.732°. From Principle 1, the first deflection angle is half of this or 6.366°.

12.732° goes into 55° four times with a remainder of 4.072°. The last deflection angle (sighting to the PT) is 2.036°. Use Eq. 78.16.

(c)

$$C = (2)(225\text{ ft})\sin\frac{4.072°}{2}$$

$$= 15.98\text{ ft}$$

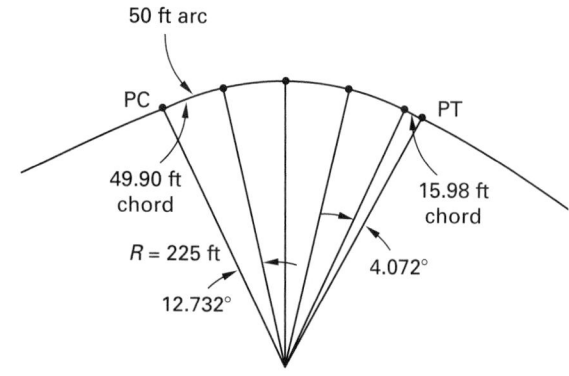

5. TANGENT OFFSETS

A *tangent offset*, y, is the perpendicular distance from an extended tangent line to the curve. A *tangent distance*, x, is the distance along the tangent to a perpendicular point. Tangent offsets for circular curves can be calculated from Eq. 78.17.

$$y = R(1 - \cos\beta)$$

$$= R - \sqrt{R^2 - x^2} \qquad \textit{78.17}$$

$$\beta = \arcsin\frac{x}{R} = \arccos\left(\frac{R-y}{R}\right) \qquad \textit{78.18}$$

$$x = R\sin\beta = \sqrt{2Ry - y^2} \qquad \textit{78.19}$$

Offsets from *parabolic curves* (*parabolic flares, parabolic tapers*, or *curb flares*) are calculated from Eq. 78.20. W is the maximum offset, L is the length of the *flare* (*taper*), x is the distance along the baseline, and y is the offset.

$$\frac{y}{W} = \frac{x^2}{L^2} \qquad 78.20$$

Figure 78.4 Tangent Offset

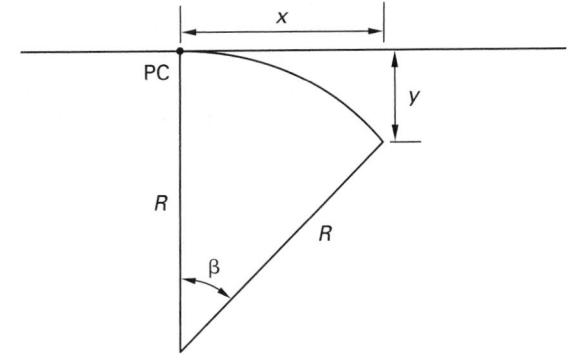

and α is the deflection angle between the tangent and NQ.

$$NP = \text{tangent distance} = NQ \cos \alpha \qquad 78.21$$
$$PQ = \text{tangent offset} = NQ \sin \alpha \qquad 78.22$$

The short chord distance is

$$NQ = C = 2R \sin \alpha \qquad 78.23$$
$$NP = (2R \sin \alpha) \cos \alpha$$
$$= C \cos \alpha \qquad 78.24$$
$$PQ = (2R \sin \alpha) \sin \alpha$$
$$= 2R \sin^2 \alpha \qquad 78.25$$

7. CURVE LAYOUT BY CHORD OFFSET

The *chord offset method* is a third method for laying out horizontal curves. This method is also suitable for short curves. The method is named for the way in which the measurements are made, which is by measuring distances along the main chord from the instrument location at PC.

$$NR = \text{chord distance} = NQ \cos \left(\frac{I}{2} - \alpha \right)$$
$$= (2R \sin \alpha) \cos \left(\frac{I}{2} - \alpha \right)$$
$$= C \cos \left(\frac{I}{2} - \alpha \right) \qquad 78.26$$
$$RQ = \text{chord offset} = NQ \sin \left(\frac{I}{2} - \alpha \right)$$
$$= (2R \sin \alpha) \sin \left(\frac{I}{2} - \alpha \right)$$
$$= C \sin \left(\frac{I}{2} - \alpha \right) \qquad 78.27$$

6. CURVE LAYOUT BY TANGENT OFFSETS

The *tangent offset method* (station offset method) can be used to lay out horizontal curves. This method is typically used on short curves. The method is named for the way in which the measurements are made, which is by measuring offsets from the tangent line.

Figure 78.5 Tangent and Chord Offset Geometry

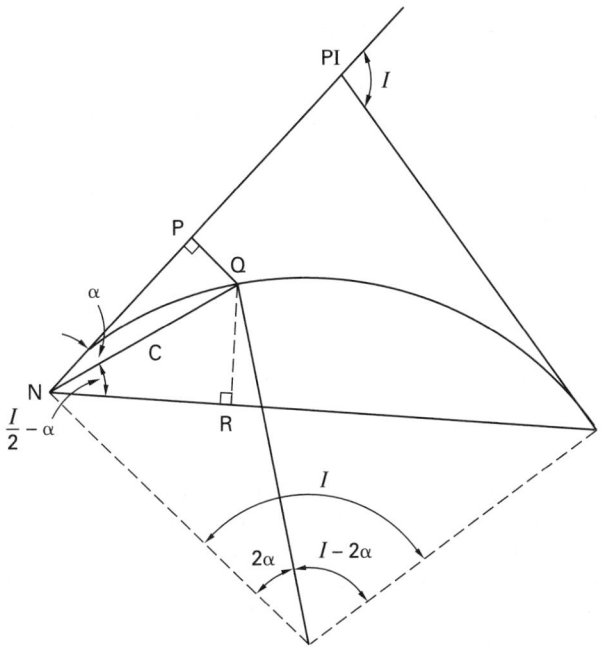

Consider the two right triangles NPQ and NRQ in Fig. 78.5. Point Q is any point along the circular curve,

8. HORIZONTAL CURVES THROUGH POINTS

Occasionally, it is necessary to design a horizontal curve to pass through a specific point. The following procedure can be used. (Refer to Fig. 78.6.)

step 1: Calculate α and m from x and y. (If α and m are known, skip this step.)

$$\alpha = \arctan \frac{y}{x} \qquad 78.28$$
$$m = \sqrt{x^2 + y^2} \qquad 78.29$$

step 2: Calculate γ. Since $90° + I/2 + \alpha + \gamma = 180°$,

$$\gamma = 90° - \frac{I}{2} - \alpha \qquad 78.30$$

Transportation

step 3: Calculate ϕ.

$$\phi = 180° - \arcsin\left(\frac{\sin\gamma}{\cos\dfrac{I}{2}}\right) \qquad 78.31$$

step 4: Calculate θ.

$$\theta = 180° - \gamma - \phi \qquad 78.32$$

step 5: Calculate the curve radius, R, from the law of sines.

$$\frac{\sin\theta}{m} = \frac{\sin\phi\cos\dfrac{I}{2}}{R} \qquad 78.33$$

Figure 78.6 *Horizontal Curve Through a Point (I known)*

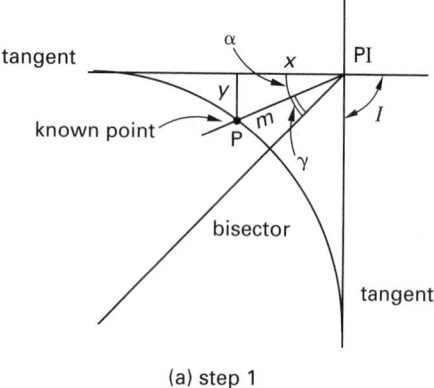

(a) step 1

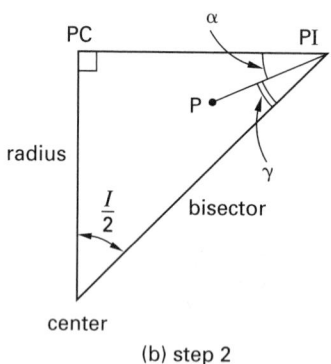

(b) step 2

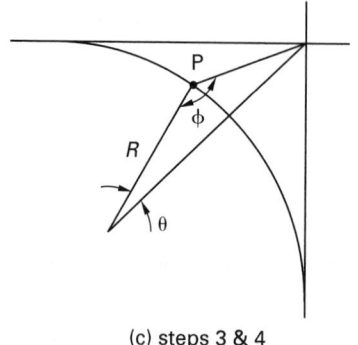

(c) steps 3 & 4

9. COMPOUND HORIZONTAL CURVES

A *compound curve* comprises two or more curves of different radii that share a common tangent point, with their centers on the same side of the common tangent. The PT for the first curve and the PC for the second curve coincide. This point is the *point of continuing curve* (PCC).

Compound curves are often used on low-speed roadways, such as on entrance or exit ramps. Their use is reserved for applications where design constraints, such as topography or high land cost, preclude the use of a circular or spiral curve. For safety, the radius of the larger curve should be less than or equal to four thirds times the radius of the smaller curve (three halves on interchanges).

Figure 78.7 *Compound Circular Curve*

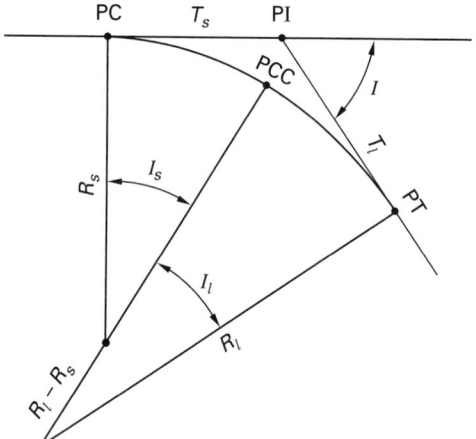

10. SUPERELEVATION

[In this book, superelevation rate, e, is a decimal number. In the AASHTO Green Book, it is a percentage.]

If a vehicle travels in a circular path with instantaneous radius R and tangential velocity (i.e., speed) v_t, it will experience an apparent centrifugal force, F_c.

$$F_c = \frac{mv_t^2}{R} \quad \text{[consistent units]} \qquad 78.34$$

The centrifugal force can be resisted by banking, friction, or a combination of the two. If it is intended that banking (no friction) alone will resist the centrifugal force, the banking angle, ϕ, is given by Eq. 78.35. The elevation difference between the inside and outside edges of the curve is the *superelevation*. The transverse slope of the roadway has units of ft/ft (m/m), hence the synonymous name of *superelevation rate*, e, for slope. (However, runoff and transition "rates" are not the same.)

$$e = \tan\phi = \frac{v^2}{gR} \quad \text{[consistent units]} \qquad 78.35$$

Some of the centrifugal force is usually resisted by *sideways friction*. (One-half of the theoretical superelevation rate is usually adequate.) The quantity $1 - f\tan\phi$, implicit in Eq. 72.92, is very nearly 1 in roadway work, leading to a simplified expression for roadway banking. Equation 78.36 is AASHTO's *simplified curve formula 3-9*, used for determining the superelevation rate when friction is relied upon to counteract some of the centrifugal force. The sum, $e + f_s$, may be referred to as the *centrifugal factor*.

$$e = \tan\phi = \frac{v^2}{gR} - f_s \quad \text{[consistent units]} \quad 78.36$$

If the velocity, v, is expressed in common units, Eq. 78.36 becomes

$$e = \tan\phi = \frac{v_{km/h}^2}{127R} - f_s \quad \text{[SI]} \quad 78.37(a)$$

$$e = \tan\phi = \frac{v_{mph}^2}{15R} - f_s \quad \text{[U.S.]} \quad 78.37(b)$$

For very large values of R, Eqs. 78.36 and 78.37 become negative. However, curves with very large radii do not need to be superelevated. The normal slope of the terrain and crown limit the lower value of e.

In general, a lower *banking angle* is used in urban areas than in rural areas. (For low-speed urban streets, use of superelevation is optional [AASHTO Green Book].) For arterial streets in downtown areas, the maximum superelevation rate is approximately 0.04 to 0.06. For arterial streets in suburban areas and on freeways where there is no snow or ice, the maximum superelevation rate is approximately 0.10 to 0.12. For arterial streets and freeways that experience snow and ice, the maximum should be 0.06 to 0.08.

Since the maximum superelevation rate is approximately 0.08 or 0.10, Eq. 78.35 can be used to calculate the minimum curve radius if the speed is known.

Many studies have been performed to determine values of the *side friction factor*, f_s, and most departments of transportation have their own standards. One methodology is to assume the side friction factor to be 0.16 for speeds less than 30 mph (50 km/h). Equations 78.38 or 78.39 can be used for higher speeds.

$$f_s = 0.16 - \frac{0.01(v_{mph} - 30)}{10} \quad \text{[< 50 mph]} \quad 78.38$$

$$f_s = 0.14 - \frac{0.02(v_{mph} - 50)}{10} \quad \text{[50 to 70 mph]} \quad 78.39$$

Although the sophistication may be unwarranted, the friction factor for sideways slipping, f_s, can be differentiated from the straight-ahead friction factor. For sideways slipping, the friction factor may be referred to as the *side friction factor, lateral ratio, cornering ratio,* or *unbalanced centrifugal ratio*.

Example 78.4

A low speed rural street has a design speed of 25 mph, a curve radius of 500 ft, and a maximum superelevation rate of 8%. What is the superelevation rate according to the AASHTO Green Book?

Solution

Superelevation rate problems are almost always solved by use of the Green Book superelevation tables. Use Exh. 3-27. Enter from the top with a design speed of 25 mph, proceed down the 25 mph column to the curve radius of 500 ft (499 ft in the table), and read the superelevation of 5.0% in the left column.

11. TRANSITIONS TO SUPERELEVATION

Transitions from crowned sections to superelevated sections should be gradual. When a curve is to be superelevated, it is first necessary to establish a rotational axis (often referred to as a "point") on the cross section. This is the axis around which the pavement will be rotated (longitudinally) to gradually change to the specified superelevated cross slope. The rotational axis ("point") is a longitudinal axis parallel to the instantaneous direction of travel. The location of the rotational point varies with the basic characteristics of the typical section. The following guidelines can be used.

1. For two-lane and undivided highways, the axis of rotation is generally at (along) the original crown of the roadway. However, it may also be the edge of the outside or inside lane.

2. On divided highways with relatively wide depressed medians, the axis of rotation can be at the crown of each roadway or at the edge of the lane or shoulder nearest the median of each roadway. Placing the axis at the crown results in median edges at different elevations, but it reduces the elevation differential between extreme pavement edges. If it is likely that the highway will be widened in the future, it will be desirable to rotate the pavement cross-slope about the inside lane or shoulder.

3. On divided highways with narrow raised medians and moderate superelevation rates, the axis of rotation should be at the center of the median. If the combination of pavement width and superelevation rate results in substantial differences between pavement edge elevations, the axis of rotation should be at the edge of the lane or shoulder nearest the median of each roadway. If an at-grade crossing is located on the superelevated curve, the impact of intersecting traffic should be considered in selecting the axis of rotation.

4. On divided highways with concrete median barriers, the median elevation must be the same for both directions of travel. Rotation should occur at the barrier gutter.

For maximum comfort and safety, superelevation should be introduced and removed uniformly over a length adequate for the likely travel speeds. The total length of

Transportation

the *superelevation transition distance* is the sum of the tangent (crown) runout and the superelevation runoff. The following design factors should be considered in designing the superelevation transition distance.

1. *Tangent runout*, T_R, also known as *tangent runoff* and *crown runoff*, is a gradual change from a normal crowned section to a point where the *adverse cross slope* on the outside of the curve has been removed. When the adverse cross slope has been removed, the elevation of the outside pavement edge will be equal to the centerline elevation. The inside pavement edge will be unchanged. The rate of removal is usually the same as the superelevation runoff rate (SRR).

2. *Superelevation runoff*, L, is a gradual change from the end of the tangent runout to a cross section that is fully superelevated. The *superelevation runoff rate* (also known as the *transition rate*), SRR, is the rate at which the normal cross-slope of the roadway is transitioned to the superelevated cross-slope. The superelevation runoff rate is expressed in units of cross-slope elevation per unit width per unit length of traveled roadway.

   For single lanes, a common superelevation runoff rate is 1 ft per foot of width for every 200 ft (60 m) length, expressed as 1:200. For speeds less than 50 mph, or if conditions are restrictive and if sufficient room is not available for a 1:200 transition rate, more abrupt rates may be used.

3. Tangent runout and superelevation runoff distances may be calculated using Eqs. 78.40 and 78.41. w is the lane width, and p is the rate of cross slope.

$$T_R = \frac{wp}{\text{SRR}} \qquad 78.40$$

$$L = \frac{we}{\text{SRR}} \qquad 78.41$$

4. Superelevation runoff rate on a curve may be determined by a *time rule* (e.g., "runoff shall be completed within 4 sec at the design speed") or by a *speed rule* (e.g., "3 ft of runoff for every mph of design speed"), regardless of the initial superelevation. Time and speed rules depend on the specifying agency.

5. On circular curves, the superelevation runoff should be developed 60 to 90% on the tangent and 40 to 10% on the curve, with a large majority of state highway agencies using a rule of two thirds on the tangent and one third on the curve. This results in two thirds of the full superelevation at the beginning and ending of the curve. This is a compromise between placing the entire transition on the tangent section, where superelevation is not needed, and placing the transition on the curve, where full superelevation is needed. AASHTO Green Book Exhibit 3-33 gives specific runoff recommendations based on speed and number of rotated lanes.

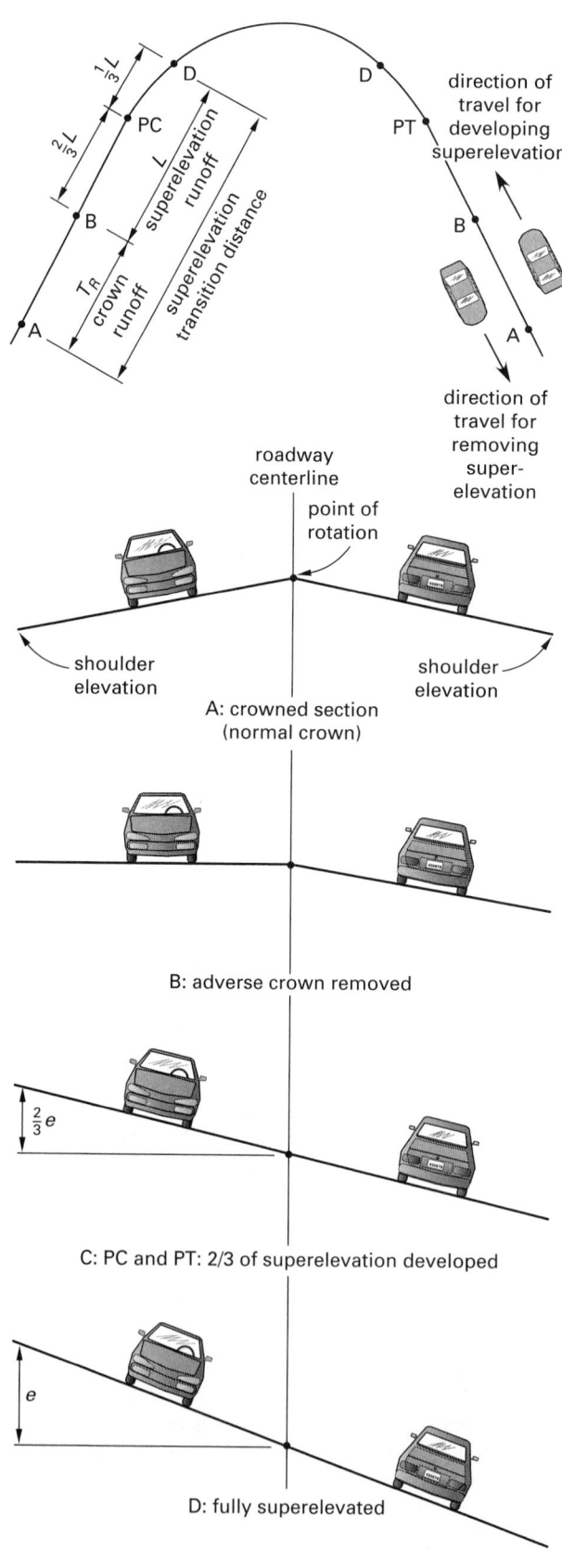

Figure 78.8 *Roadway Profile Around Circular Curve*

6. On spiral curves, the superelevation is developed entirely within the length of the spiral.

7. All shoulders should slope away from the traveled lanes. The angular breaks at the pavement edges of the superelevated roadways should be rounded by the insertion of vertical curves. The minimum curve length in feet should be approximately numerically equal to the design speed in mph.

Example 78.5

One end of a horizontal curve on a two-lane highway has 12 ft lanes and a crown cross-slope of 0.02 ft/ft. The PC is at sta 10+00. The transition rate is 1:400, and the required superelevation is 0.04. The $^2/_3$-$^1/_3$ rule is used to apportion the runoff to the curve. Calculate the stationing where (a) the superelevation runoff begins, (b) the tangent runout begins, and (c) the curve is fully superelevated.

Solution

(a) Use Eqs. 78.40 and 78.41.

$$T_R = \frac{wp}{\text{SRR}} = \frac{(12\text{ ft})\left(0.02\ \dfrac{\text{ft}}{\text{ft}}\right)}{\dfrac{1}{400}\ \dfrac{\text{ft}}{\text{ft}}}$$

$$= 96\text{ ft}$$

$$L = \frac{we}{\text{SRR}} = \frac{(12\text{ ft})\left(0.04\ \dfrac{\text{ft}}{\text{ft}}\right)}{\dfrac{1}{400}\ \dfrac{\text{ft}}{\text{ft}}}$$

$$= 192\text{ ft}$$

The superelevation should be developed two thirds before the curve and one third after. Therefore, the superelevation begins at $2L/3$ before the PC.

$$\text{sta }10{+}00 - \frac{2L}{3} = \text{sta }10{+}00 - \left(\tfrac{2}{3}\right)(192\text{ ft})$$

$$= \text{sta }10{+}00 - 128\text{ ft} = \text{sta }8{+}72$$

(b) The tangent runout begins at

$$\text{sta }8{+}72 - 96\text{ ft} = \text{sta }7{+}76$$

(c) The pavement should be fully superelevated at $L/3$ after the PC, at station

$$\text{sta }10{+}00 + \frac{L}{3} = \text{sta }10{+}00 + \frac{192\text{ ft}}{3}$$

$$= \text{sta }10{+}00 + 64\text{ ft} = \text{sta }10{+}64$$

12. SUPERELEVATION OF RAILROAD LINES

The method of specifying superelevation for railroad lines is somewhat different than for roadways. The *equilibrium elevation*, E, of the outer rail relative to the inner rail is calculated from Eq. 78.42. The effective gauge, G_{eff}, is the center-to-center rail spacing.

$$E = \frac{G_{\text{eff}}\text{v}^2}{gR} \quad \text{[railroads]} \qquad 78.42$$

13. STOPPING SIGHT DISTANCE

The *stopping sight distance* is the distance required by a vehicle traveling at the design speed to stop before reaching a stationary object that has suddenly appeared in its path. Calculated stopping sight distances are the minimums that should be provided at any point on any roadway. Greater distances should be provided wherever possible.

Stopping sight distance is the sum of two distances: the distance traveled during driver perception and reaction time, and the distance traveled during brake application. The equations used to calculate sight distance assume that the driver's eyes are 3.5 ft (1080 mm) above the surface of the roadway. For stopping sight distances, it is assumed that the object being observed has a height of 2.0 ft (600 mm).

Equation 78.43 can be used to calculate the stopping sight distance, S, for straight-line travel on a constant grade. In Eq. 78.43, the grade, G, is in decimal form and is negative if the roadway is downhill. The coefficient of friction, f, is evaluated for a wet pavement. The perception-reaction time, t_p, often is taken as 2.5 sec for all but the most complex conditions. If the deceleration rate, a, is known, the friction factor, f, can be replaced with a/g, where g is the acceleration of gravity (9.81 m/s^2 for SI calculations, and 32.2 ft/sec^2 for U.S. calculations).

$$S = \left(0.278\ \frac{\dfrac{\text{m}}{\text{s}}}{\dfrac{\text{km}}{\text{h}}}\right)t_p\text{v}_{\text{km/h}} + \frac{\text{v}^2_{\text{km/h}}}{254(f+G)}$$

$$\text{[SI]} \qquad 78.43(a)$$

$$S = \left(1.47\ \frac{\dfrac{\text{ft}}{\text{sec}}}{\dfrac{\text{mi}}{\text{hr}}}\right)t_p\text{v}_{\text{mph}} + \frac{\text{v}^2_{\text{mph}}}{30(f+G)}$$

$$\text{[U.S.]} \qquad 78.43(b)$$

"Desirable values" of stopping sight distance, S, are listed in Table 78.2 for various design speeds.

14. PASSING SIGHT DISTANCE

The *passing sight distance* is the length of open roadway ahead necessary to pass without meeting an oncoming vehicle. Passing sight distance is applicable only to two-lane, two-way highways. Passing sight distance is not relevant on multilane highways.

Transportation

Table 78.2 *AASHTO Minimum Stopping Sight Distances (Level Roadways)[a,b]*

design speed ((mph)(km/h))	brake reaction distance ((ft) (m))	braking distance on level ((ft) (m))	stopping sight distance calculated ((ft) (m))	stopping sight distance design ((ft) (m))
15 (20)	55.1 (13.9)	21.6 (4.6)	76.7 (18.5)	80 (20)
20 (30)	73.5 (20.9)	38.4 (10.3)	111.9 (31.2)	115 (35)
25 (40)	91.9 (27.8)	60.0 (18.4)	151.9 (46.2)	155 (50)
30 (50)	110.3 (34.8)	86.4 (28.7)	196.7 (63.5)	200 (65)
35 (60)	128.6 (41.7)	117.6 (41.3)	246.2 (83.0)	250 (85)
40 (70)	147.0 (48.7)	153.6 (56.2)	300.6 (104.9)	305 (105)
45 (80)	165.4 (55.6)	194.4 (73.4)	359.8 (129.0)	360 (130)
50 (90)	183.8 (62.6)	240.0 (92.9)	423.8 (155.5)	425 (160)
55 (100)	202.1 (69.5)	290.3 (114.7)	492.4 (184.2)	495 (185)
60 (110)	220.5 (76.5)	345.5 (138.8)	566.0 (215.3)	570 (220)
65 (120)	238.9 (83.4)	405.5 (165.2)	644.4 (248.6)	645 (250)
70 (130)	257.3 (90.4)	470.3 (193.8)	727.6 (284.2)	730 (285)
75	275.6	539.9	815.5	820
80	294.0	614.3	908.3	910

[a]Brake reaction distance predicated on a time of 2.5 sec; deceleration rate of 11.2 ft/sec^2 (3.4 m/s^2) used to determine calculated sight distance.
[b]Use AASHTO Green Book Exh. 3-2 for roadways on grades.
Adapted from *A Policy on Geometric Design of Highways and Streets*, Exh. 3-1, copyright © 2004 by the American Association of Highway and Transportation Officials, Washington, D.C. Used by permission.

Passing sight distances assume that the driver's eyes are 3.5 ft (1080 mm) above the surface of the roadway. The object being viewed is assumed to be at a height of 3.5 ft (1080 mm). Exhibit 3-7 in the AASHTO Green Book tabulates passing sight distances for two-lane highways.

15. MINIMUM HORIZONTAL CURVE LENGTH FOR STOPPING DISTANCE

A horizontal circular curve is shown in Fig. 78.9. Obstructions along the inside of curves, such as retaining walls, cut slopes, trees, buildings, and bridge piers, can limit the available (chord) sight distance. Often, a curve must be designed that will simultaneously provide the required stopping sight distance while maintaining a clearance from a roadside obstruction.

The equations for calculating the geometry of a horizontal curve to see around an obstruction assume that the stopping sight distance is less than the curve length (i.e., $S \leq L$). The stopping sight distance and length along the centerline of the inside lane of the curve are the same. Given a stopping sight distance and a curve radius, the clear area *horizontal sightline offset*, HSO, can be calculated. (Prior to 2004, HSO was known as the *middle ordinate* and given the symbol M.) The angles are given in degrees, not radians.

$$S = \left(\frac{R}{28.65}\right)\left(\arccos\frac{R - \text{HSO}}{R}\right) \quad \text{78.44}$$

$$\text{HSO} = R(1 - \cos\theta) = R\left(1 - \cos\frac{DS}{200}\right)$$

$$= R\left(1 - \cos\frac{28.65S}{R}\right) \quad \text{78.45}$$

Decision sight distance should be added on a horizontal curve whenever a driver is confronted with additional information that may unduly complicate the highway information the driver must process. Ten seconds of decision time is considered to be the minimum for all but the most simple avoidance maneuvers.

The curve length methods presented in this section are based on horizontal tangent grades only. If a vertical grade occurs in conjunction with a circular curve, these methods cannot be used.

Figure 78.9 *Horizontal Curve with Obstructions*

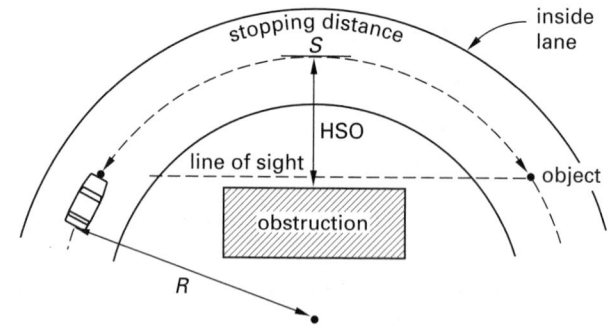

16. VERTICAL CURVES

Vertical curves are used to change the grade of a highway. Most vertical curves take the shape of an *equal-tangent parabola*. Such curves are symmetrical about the vertex. Since the grades are very small, the actual arc length of the curve is approximately equal to the chord length BVC-EVC. Vertical curves are measured as the horizontal distance between grade points, regardless of the slope of the grade. Table 78.3 lists the standard abbreviations used to describe geometric elements of vertical curves.

A vertical parabolic curve is completely specified by the two grades and the curve length. Alternatively, the *rate of grade change per station*, R, can be used in place of curve length. The rate of grade change per station is given by Eq. 78.46. Units of %/sta are the same as $\text{ft}/(\text{sta})^2$.

$$R = \frac{G_2 - G_1}{L} \quad \text{[may be negative]} \qquad 78.46$$

Equation 78.47 defines an equal-tangent parabolic curve. x is the distance to any point on the curve, measured in stations beyond the BVC, and elev is measured in ft. The same reference point is used to measure all elevations.

$$\text{elev}_x = \frac{R}{2}x^2 + G_1 x + \text{elev}_{\text{BVC}} \qquad 78.47$$

The maximum or minimum elevation will occur when the slope is equal to zero (the high point or low point, respectively), which is located as determined by Eq. 78.48. This point is known as the *turning point*. In sag vertical curves, the turning point (low point) is the location at which catch basins should be installed.

$$x_{\text{turning point}} = \frac{-G_1}{R} \quad \text{[in stations]} \qquad 78.48$$

The *middle ordinate distance* is found from Eq. 78.49.

$$M_{\text{ft}} = \frac{AL_{\text{sta}}}{8} \qquad 78.49$$

Figure 78.10 Symmetrical Parabolic Vertical Curve

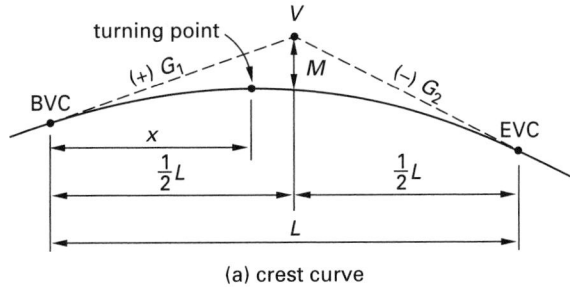

(a) crest curve

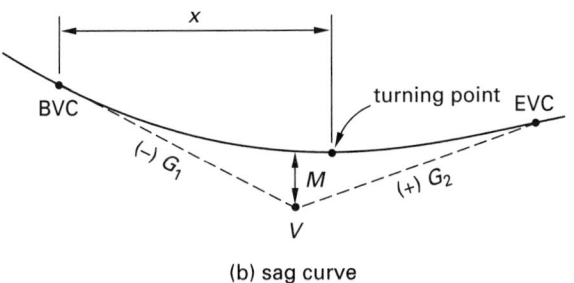

(b) sag curve

Table 78.3 Vertical Curves: Abbreviations and Terms

preferred

A	change in gradient, $\lvert G_2 - G_1 \rvert$ [always positive]
BVC	beginning of the vertical curve
EVC	end of the vertical curve
G_1	grade from which the stationing starts, in percent
G_2	grade toward which the stationing heads, in percent
L	length of curve
M	middle ordinate
R	rate of change in grade per station
V	vertex (the intersection of the two tangents)

alternate

E	tangent offset at V (same as M)
EVT	end of vertical tangency (same as EVC and PVT)
PVC	same as BVC
PVI	same as V
PVT	same as EVC
VPI	vertical point of intersection (same as V)

Example 78.6

A crest vertical curve with a length of 400 ft connects grades of +1.0% and −1.75%. The vertex is located at station 35+00 and elevation 549.20 ft. What are the elevations of the (a) BVC, (b) EVC, and (c) all full stations on the curve?

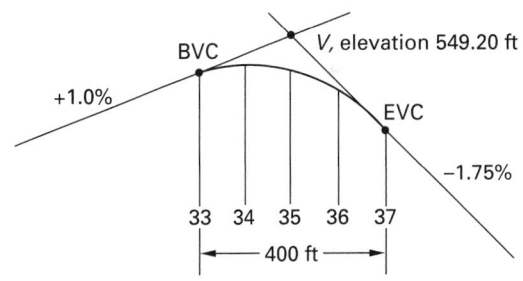

Solution

The curve length is 4 stations (400 ft).

(a)
$$\text{elev}_{\text{BVC}} = \text{elev}_V - G_1 \frac{L}{2}$$
$$= 549.20 \text{ ft} - \left(1 \frac{\text{ft}}{\text{sta}}\right)\left(\frac{4 \text{ sta}}{2}\right)$$
$$= 547.20 \text{ ft}$$

(b)
$$\text{elev}_{\text{EVC}} = \text{elev}_V - G_2 \frac{L}{2}$$
$$= 549.20 \text{ ft} - \left(1.75 \frac{\text{ft}}{\text{sta}}\right)\left(\frac{4 \text{ sta}}{2}\right)$$
$$= 545.70 \text{ ft}$$

Transportation

(c) Use Eq. 78.46.

$$R = \frac{G_2 - G_1}{L}$$

$$= \frac{-1.75\% - 1\%}{4 \text{ sta}}$$

$$= -0.6875 \text{ \%/sta}$$

$$\frac{R}{2} = \frac{-0.6875 \dfrac{\%}{\text{sta}}}{2}$$

$$= -0.3438 \text{ \%/sta} \quad [\text{same as } -0.3438 \text{ ft/sta}^2]$$

The equation of the curve is

$$\text{elev}_x = \frac{R}{2}x^2 + G_1 x + \text{elev}_{\text{BVC}}$$

$$= -0.3438 x^2 + \left(1 \ \frac{\text{ft}}{\text{sta}}\right) x + 547.20 \text{ ft}$$

At sta 34+00, $x = 34 - 33 = 1$ sta.

$$\text{elev}_{34+00} = \left(-0.3438 \ \frac{\text{ft}}{\text{sta}^2}\right)(1 \text{ sta})^2 + 1 \text{ ft} + 547.20 \text{ ft}$$

$$= 547.86 \text{ ft}$$

Similarly,

$$\text{elev}_{35+00} = \left(-0.3438 \ \frac{\text{ft}}{\text{sta}^2}\right)(2 \text{ sta})^2 + 2 \text{ ft} + 547.20 \text{ ft}$$

$$= 547.82 \text{ ft}$$

$$\text{elev}_{36+00} = \left(-0.3438 \ \frac{\text{ft}}{\text{sta}^2}\right)(3 \text{ sta})^2 + 3 \text{ ft} + 547.20 \text{ ft}$$

$$= 547.11 \text{ ft}$$

Example 78.7

A vertical sag curve with vertex located at sta 67+15 has a low point at sta 66+89. The grade into the curve is -2%, and the grade out of the curve is $+3\%$. What is the length of curve?

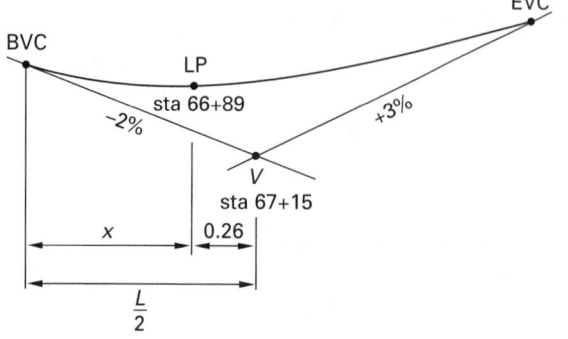

Solution

The location of the low point is defined by Eqs. 78.46 and 78.48.

$$R = \frac{G_2 - G_1}{L} = \frac{3 \ \dfrac{\%}{\text{sta}} - \left(-2 \ \dfrac{\%}{\text{sta}}\right)}{L} = \frac{5 \ \dfrac{\%}{\text{sta}}}{L}$$

$$x = \frac{-G_1}{R} = \frac{-\left(-2 \ \dfrac{\%}{\text{sta}}\right)}{\dfrac{5 \ \dfrac{\%}{\text{sta}}}{L}}$$

$$= 0.4L$$

The distance between the low point and the vertex is $(67 + 15) - (66 + 89) = 0.26$ sta. The distance between BVC and the vertex is

$$\frac{L}{2} = x + 0.26 \text{ sta}$$

Substituting $x = 0.4L$,

$$\frac{L}{2} = 0.4L + 0.26 \text{ sta}$$

$$L = 2.6 \text{ sta} \quad (260 \text{ ft})$$

17. VERTICAL CURVES THROUGH POINTS

If a curve is to have some minimum clearance from an obstruction as shown in Fig. 78.11, the curve length generally will not be known in advance. If the station and elevation of the point, P, the station and elevation of the BVC or the vertex, and the gradient values G_1 and G_2 are known, the curve length can be determined explicitly.

step 1: Find the elevation of points E, F, and G.

step 2: Calculate the constant (no physical significance) s.

$$s = \sqrt{\frac{\text{elev}_E - \text{elev}_G}{\text{elev}_E - \text{elev}_F}} \qquad 78.50$$

step 3: Solve for L directly.

$$L = \frac{2d(s + 1)}{s - 1} \qquad 78.51$$

Figure 78.11 *Vertical Curve with an Obstruction*

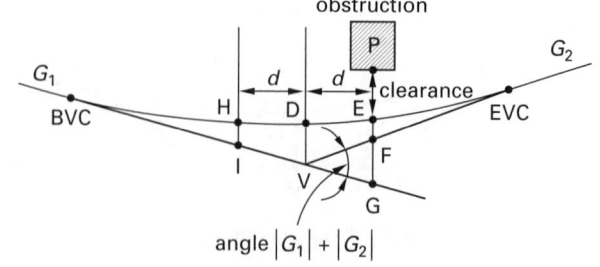

Example 78.8

Determine the length of a sag vertical curve that passes through a point at elevation 614.00 ft and sta 17+00. The grades are $G_1 = -4.2\%$ and $G_2 = +1.6\%$. The BVC is at sta 13+00, elevation = 624.53 ft.

Solution

The distance from the BVC to the point is

$$x = 17 \text{ sta} - 13 \text{ sta}$$

$$= 4 \text{ sta}$$

At the given point for this curve,

elev = 614.00 ft

$$R = \frac{G_2 - G_1}{L} = \frac{1.6 \frac{\%}{\text{sta}} - \left(-4.2 \frac{\%}{\text{sta}}\right)}{L} = \frac{5.8 \frac{\%}{\text{sta}}}{L}$$

The equation of the curve is

$$\text{elev} = \frac{Rx^2}{2} + G_1 x + \text{elev}_{\text{BVC}}$$

$$614.00 \text{ ft} = \frac{\left(\frac{5.8 \frac{\%}{\text{sta}}}{L}\right)(4 \text{ sta})^2}{2}$$

$$+ \left(-4.2 \frac{\text{ft}}{\text{sta}}\right)(4 \text{ sta}) + 624.53 \text{ ft}$$

$$-10.53 \text{ ft} = \frac{46.4}{L} - 16.8 \text{ ft}$$

$$L = 7.4 \text{ sta} \quad (740 \text{ ft})$$

18. VERTICAL CURVE TO PASS THROUGH TURNING POINT

Another common situation involving equal-tangent vertical curves is calculating the length of a sag or crest curve needed to pass through a turning point, TP, at a particular elevation. Equation 78.52 can be used to determine the curve length directly. Grades G_1 and G_2 and the elevations of the BVC or the PVI (point V) and the turning point must be known.

$$L = \frac{2(G_2 - G_1)(\text{elev}_V - \text{elev}_{\text{TP}})}{G_1 G_2}$$

$$= \frac{2(G_2 - G_1)(\text{elev}_{\text{BVC}} - \text{elev}_{\text{TP}})}{G_1^2} \quad 78.52$$

19. MINIMUM VERTICAL CURVE LENGTH FOR SIGHT DISTANCES (CREST CURVES)

Crest curve lengths are generally determined based on stopping sight distances. (Headlight sight distance and rider comfort control the design of sag vertical curves.) The passing sight distance could also be used, except that the required length of the curve would be much greater than that based only on the stopping sight distance. Since the curve length determines the extent of the earthwork required, and it is easier (i.e., less expensive) to prohibit passing on crest curves than to perform the earthwork required to achieve the required passing sight distance, only the stopping sight distance is usually considered in designing the curve length.

Two factors affect the sight distance: the algebraic difference, A, between gradients of the intersecting tangents, and the length of the vertical curve, L. With a small algebraic difference in grades, the length of the vertical curve may be short. However, to obtain the same sight distance with a large algebraic difference in grades, a much longer vertical curve is needed.

Table 78.4 implies that the choice of a curve length is a simple selection of sight distance based on design speed. This is actually the case in simple curve length design problems where the design speed, v, and grades, G_1 and G_2, are known. The required stopping distance is determined from Table 78.2 or calculated from Eq. 78.43 or 78.44. The required curve length, L, is calculated from the formulas in Table 78.4. Since it is initially unknown, curve length is calculated for both the $S < L$ and $S > L$ cases. The calculated curve length that is inconsistent with its assumption is discarded.

The constants in Table 78.4 are based on specific heights of objects and driver's eyes above the road surface. In general, the sight distance over the crest of a vertical curve is given by Eqs. 78.53 and 78.54, where h_1 is the height of the eyes of the driver and h_2 is the height of the object sighted, both in feet.

$$L = \frac{AS^2}{200\left(\sqrt{h_1} + \sqrt{h_2}\right)^2} \quad [S < L] \qquad 78.53$$

$$L = 2S - \frac{200\left(\sqrt{h_1} + \sqrt{h_2}\right)^2}{A} \quad [S > L] \qquad 78.54$$

The sight distance under an overhead structure to see an object beyond a sag vertical curve is given by Eqs. 78.55 and 78.56. Various assumptions, including eye and tail light (object) heights and beam divergence, were made by AASHTO in developing the following equations.

$$L = \frac{AS^2}{800(C - 1.5)} \quad [S < L] \text{ [SI]} \qquad 78.55(a)$$

$$L = \frac{AS^2}{800(C - 5)} \quad [S < L] \quad [\text{U.S.}] \qquad 78.55(b)$$

$$L = 2S - \frac{800(C - 1.5)}{A} \quad [S > L]$$
$$[\text{SI}] \qquad 78.56(a)$$

$$L = 2S - \frac{800(C - 5)}{A} \quad [S > L]$$
$$[\text{U.S.}] \qquad 78.56(b)$$

Transportation

Table 78.4 AASHTO Required Lengths of Curves on Grades[a]

	stopping sight distance[b] (crest curves)	passing sight distance[c] (crest curves)	stopping sight distance (sag curves)
SI units			
$S < L$	$L = \dfrac{AS^2}{658}$	$L = \dfrac{AS^2}{864}$	$L = \dfrac{AS^2}{120 + 3.5S}$
$S > L$	$L = 2S - \dfrac{658}{A}$	$L = 2S - \dfrac{864}{A}$	$L = 2S - \dfrac{120 + 3.5S}{A}$
U.S. units			
$S < L$	$L = \dfrac{AS^2}{2158}$	$L = \dfrac{AS^2}{2800}$	$L = \dfrac{AS^2}{400 + 3.5S}$
$S > L$	$L = 2S - \dfrac{2158}{A}$	$L = 2S - \dfrac{2800}{A}$	$L = 2S - \dfrac{400 + 3.5S}{A}$

[a]$A = |G_2 - G_1|$, absolute value of the algebraic difference in grades, in percent.
[b]The driver's eye is 3.5 ft (1080 mm) above road surface, viewing an object 2.0 ft (600 mm) high.
[c]The driver's eye is 3.5 ft (1080 mm) above road surface, viewing an object 3.5 ft (1080 mm) high.
Compiled from *A Policy on Geometric Design of Highways and Streets*, Ch. 3, copyright © 2004 by the American Association of State Highway and Transportation Officials, Washington, D.C.

20. DESIGN OF CREST CURVES USING *K*-VALUE

The *K-value method* of analysis used in AASHTO's Green Book is a simplified and more conservative method of choosing a stopping sight distance for a crest vertical curve. The length of vertical curve per percent grade difference, K, is the ratio of the curve length, L, to grade difference, A.

$$K = \frac{L}{A} = \frac{L}{|G_2 - G_1|} \quad \text{[always positive]} \qquad \textbf{78.57}$$

The $L = KA$ relationship is conveniently linear. In order to facilitate rapid calculation of curve lengths, AASHTO has prepared several graphs. Figures 78.12 and 78.13 give minimum curve lengths for crest vertical curves. Since for a fixed grade difference the speed determines the stopping distance, every value of speed has a corresponding value of K. Thus, the curves in the figures are identified concurrently with the speed and the K-value. It is not necessary to specify both A and L in a design problem, as knowing K is sufficient.

The simplified procedure is to select one of the curves based on the speed or the K-value, and then read the curve length that corresponds to the grade difference, A. K-values shown on the graphs have been rounded for design.

Figure 78.12 Design Controls for Crest Vertical Curves (SI Units)

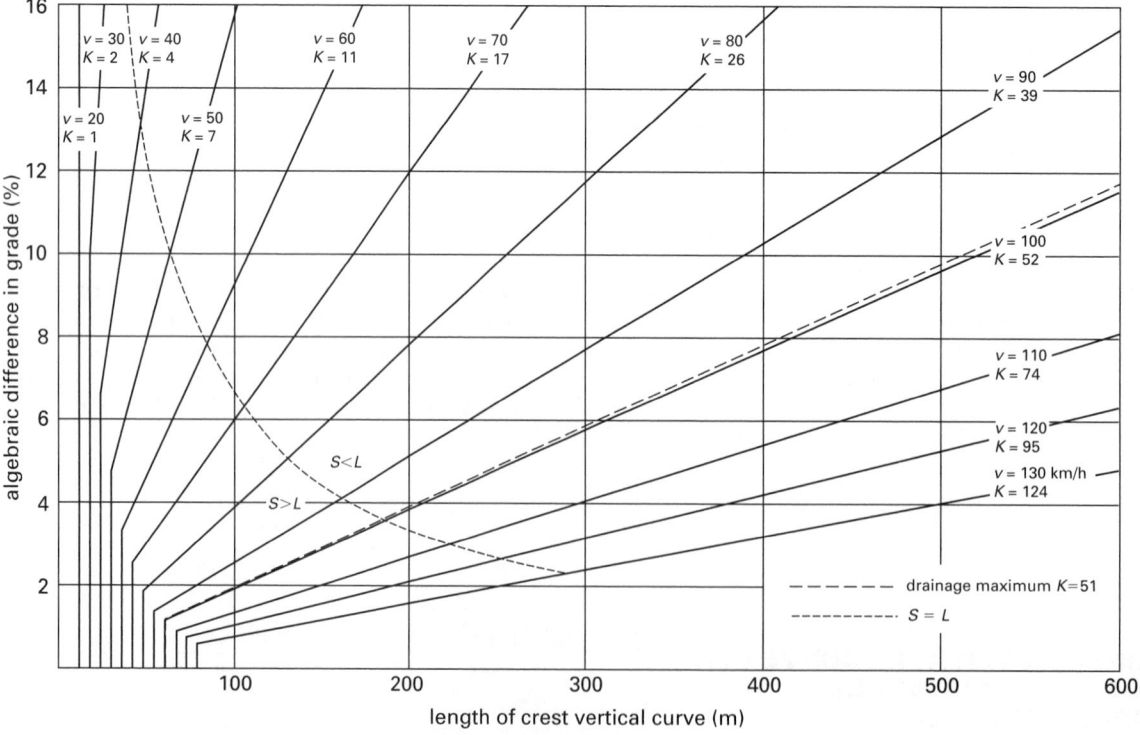

From *A Policy on Geometric Design of Highways and Streets*, Exh. 3-71, copyright © 2004 by the American Association of State Highway and Transportation Officials, Washington, D.C. Used by permission.

Figure 78.13 *Design Controls for Crest Vertical Curves (Customary U.S. Units)*

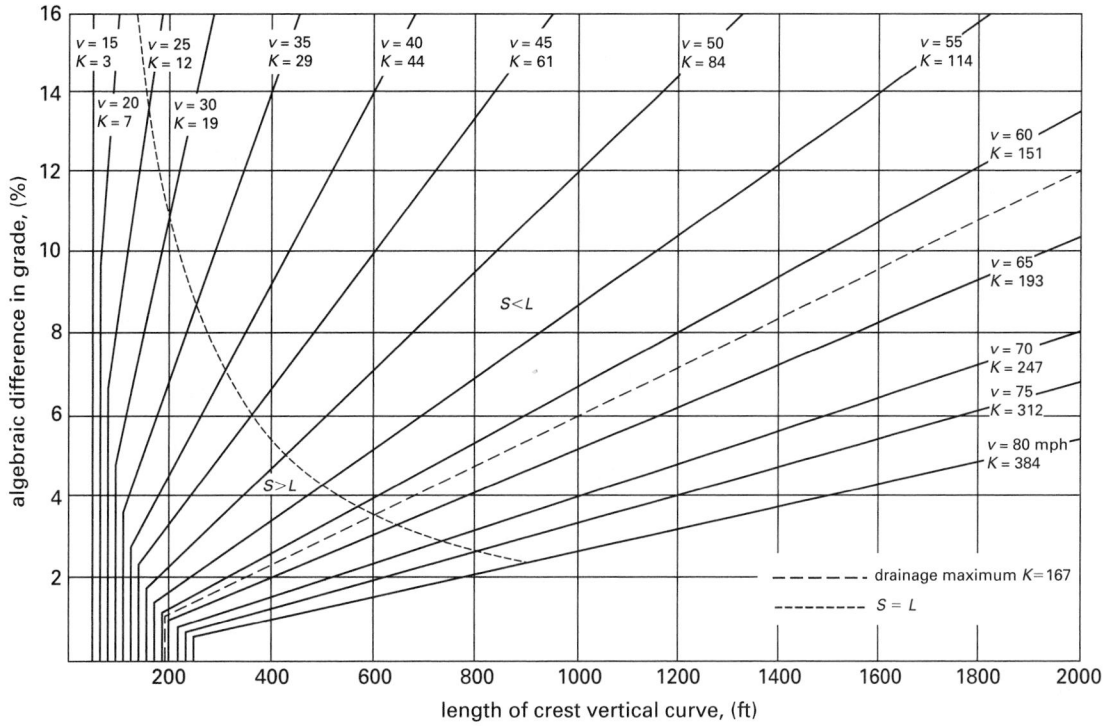

The minimum curve lengths in the AASHTO figure have been determined by other overriding factors, the most important of which are experience and state requirements. Curve lengths calculated from Table 78.4 for $S > L$ often do not represent desirable design practice and are replaced by estimated values of three times the design speed (0.6 times the design speed in km/h). This is consistent with the minimum curve lengths of 100 to 300 ft (30 to 90 m) prescribed by most states. The estimated solutions in the AASHTO figures are also justified on the basis that the longer curve lengths are obtained inexpensively when the difference in grades is small.

Example 78.9

A car is traveling up a 1.25% grade of a crest curve with a design speed of 40 mph. The descending grade is −2.75%. What is the required length of curve for minimum proper stopping sight distance?

$$A = |G_2 - G_1| = |-2.75\% - 1.25\%| = 4.0\%$$

Solution

Table 78.2 gives the minimum stopping sight distance at 40 mph as $S = 305$ ft. Refer to Table 78.4. Assume $S > L$.

$$L = 2S - \frac{2158}{A}$$
$$= (2)(305 \text{ ft}) - \frac{2158}{4.0} = 70.5 \text{ ft} \quad [S > L \text{ verified}]$$

Using Table 78.4 and assuming $S < L$,

$$L = \frac{AS^2}{2158}$$
$$= \frac{(4.0)(305 \text{ ft})^2}{2158} = 172.4 \text{ ft} \quad [S < L \text{ not verified}]$$

Since 305 ft is greater than 172.4 ft ($S > L$), the second assumption is not valid. The required curve length by formula is 70.5.

Since $L < S$, the curve length specifications will be affected by common usage and/or various state minimums. From Fig. 78.13, using the 40 mph line and $A = 4$, it is apparent that the operating condition is still in the sloped region to the left of the $S = L$ line, so an arbitrary minimum value does not apply. However, the straight $L = KA$ line will deviate from the formula significantly. Since $K = 44$, the required curve length is

$$L = KA = \left(44 \, \frac{\text{ft}}{\%}\right)(4\%) = 176 \text{ ft}$$

21. MINIMUM VERTICAL CURVE LENGTH FOR HEADLIGHT SIGHT DISTANCE: SAG CURVES

A *sag curve* should be designed so that a vehicle's headlights will illuminate a minimum distance of road ahead equal to the stopping sight distance. When full roadway lighting is available and anticipated to be available for the foreseeable future, designing for the *headlight sight distance* (also known as *light distance*) may not be necessary.

22. MINIMUM VERTICAL CURVE LENGTH FOR COMFORT: SAG CURVES

In a sag curve, gravitational and centrifugal forces combine to simultaneously act on the driver and passengers and comfort becomes the design control. A formula for calculating the minimum length of curve to keep the added acceleration on passengers below 1 ft/sec² (0.3 m/s²) is given in Eq. 78.58. The curve length for comfort may be shorter or longer than the safe passing or stopping sight distances.

$$L_m = \frac{A v_{km/h}^2}{395} \quad \text{[SI]} \quad \textit{78.58(a)}$$

$$L_{ft} = \frac{A v_{mph}^2}{46.5} \quad \text{[U.S.]} \quad \textit{78.58(b)}$$

23. DESIGN OF SAG CURVES USING K-VALUE

The AASHTO Green Book contains a graphical method for determining the minimum lengths of sag vertical curves also. Factors taken into consideration are the headlight sight distance, rider comfort, drainage control, and rules of thumb. Figures 78.14 and 78.15 provide graphical methods of determining the lengths using the K-value concept. As with crest curves, the procedure is to select one of the curves based on the speed or the K-value (from Eq. 78.57), and read the curve length that corresponds to the grade difference, A.

24. UNEQUAL TANGENT (UNSYMMETRICAL) VERTICAL CURVES

Not all vertical curves are symmetrical about their vertices. Figure 78.16 illustrates a curve in which the distance from the BVC to the vertex, l_1, is not the same as the distance from the vertex to the EVC, l_2. To evaluate the curve, a line v_1v_2 is drawn parallel with AB such that $Av_1 = v_1V$ and $Vv_2 = v_2B$. This line divides the curve into two halves of equal-tangent parabolic vertical curves: the first from A to K, and the second from K to B. The elevation of v_1 is the average of that of A and V; the elevation of v_2 is the average of that of V and B. Therefore, CK = KV. The vertical distance, CV, is defined by Eq. 78.59.

$$CV = \left(\frac{l_1 l_2}{L} \right) A \quad \textit{78.59}$$

In Eq. 78.59, l_1, l_2, and L are expressed in stations, and A is expressed in percent. Eq. 78.60 can be used to solve for the distance CK.

$$CK = KV = \left(\frac{l_1 l_2}{2L} \right) A \quad \textit{78.60}$$

Figure 78.14 *Design Controls for Sag Vertical Curves (SI units)*

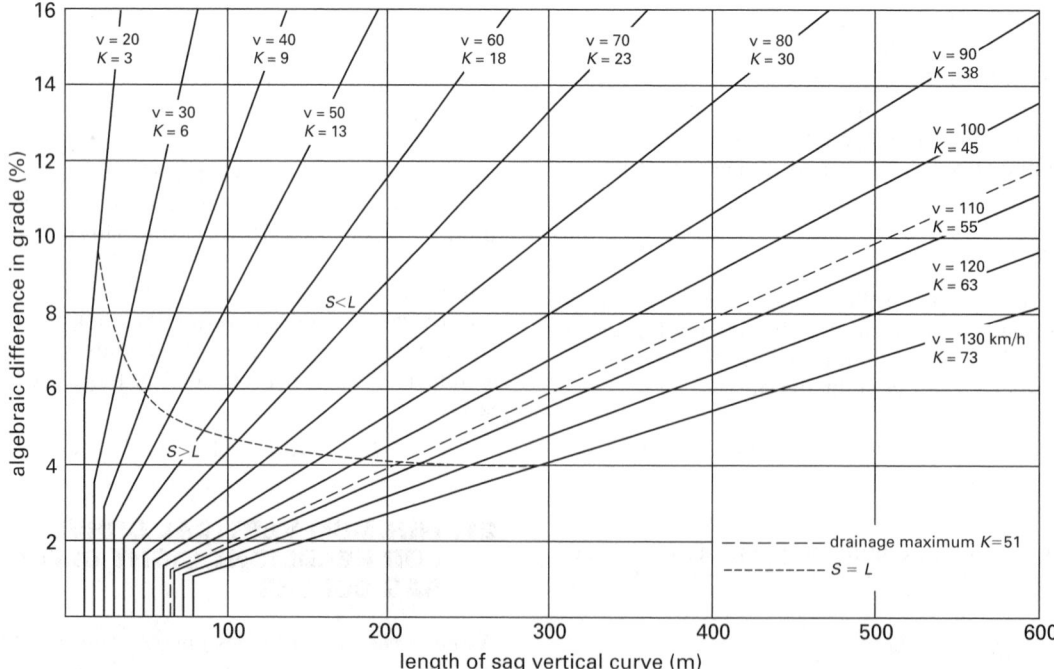

Figure 78.15 *Design Controls for Sag Vertical Curves (customary U.S. units)*

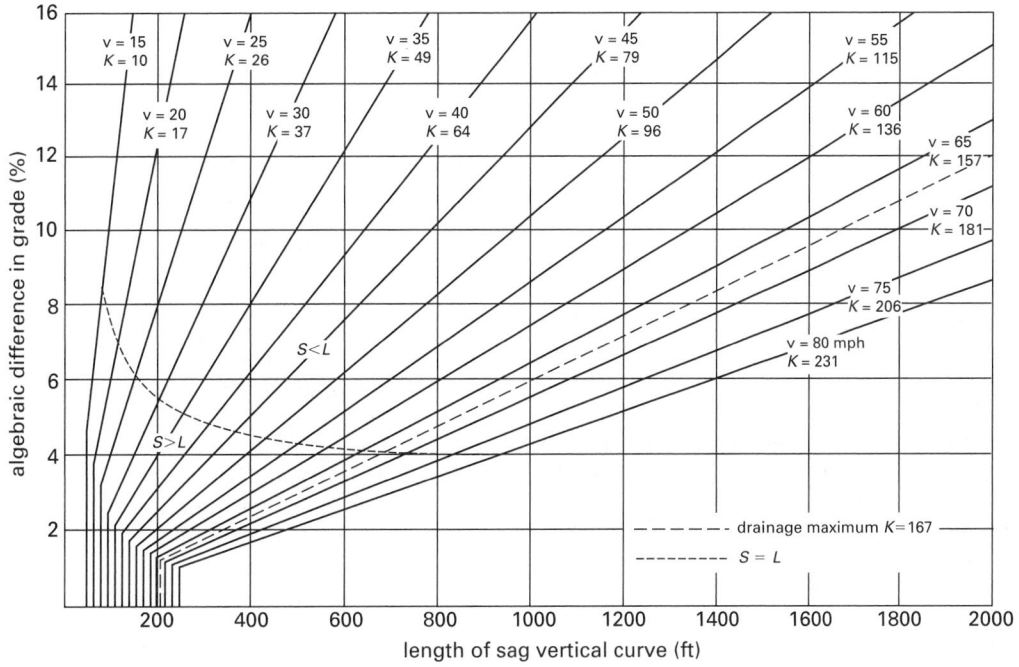

From *A Policy on Geometric Design of Highways and Streets*, Exh. 3-74, copyright © 2004 by the American Association of State Highway and Transportation Officials, Washington, D.C. Used by permission.

The grade, G', of the line $v_1 v_2$ is the same as that of AB, found by Eq. 78.61.

$$G' = \frac{\text{elev}_B - \text{elev}_A}{L} \qquad 78.61$$

Figure 78.16 *Unequal Tangent Vertical Curve*

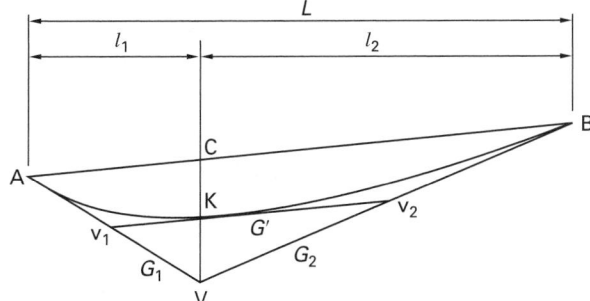

25. SPIRAL CURVES

Spiral curves (also known as *transition curves* and *easement curves*) are used to produce a gradual transition from tangents to circular curves. A spiral curve is a curve of gradually changing radius and gradually changing degree of curvature. Figure 78.17 illustrates the geometry of spiral curves connecting tangents with a circular curve of radius R_c and degree of curvature D. The *entrance spiral* begins at the left at the TS (tangent to spiral) and ends at the SC (spiral to curve). The circular curve begins at the SC and ends at the CS (curve to spiral). The *exit spiral* begins at the CS and ends at the ST (spiral to tangent).

Figure 78.17 *Spiral Curve Geometry*

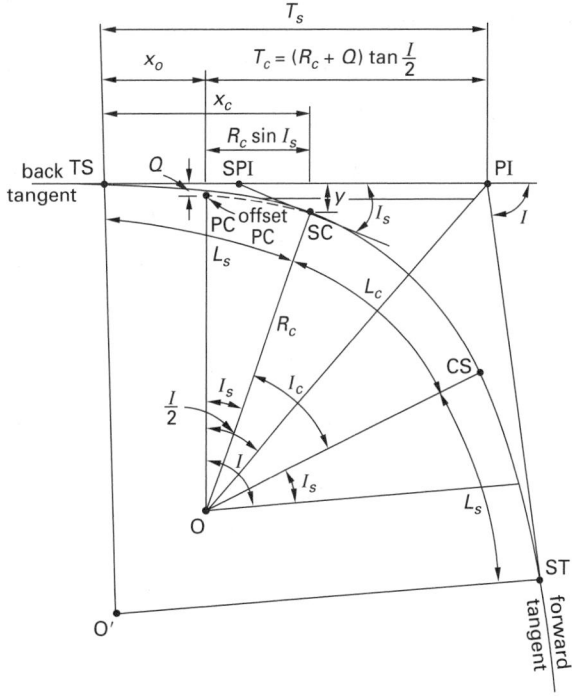

Entrance and exit spirals are geometrically identical. Their length, L_s, is the arc distance from the TS to the SC (or CS to ST). This length is selected to provide sufficient distance for introducing the curve's superelevation. The length of a spiral curve can be adjusted to between 75% and 200% of the theoretical value to meet other design criteria.

There are various ways of selecting the spiral length, L_s, including rules of thumb, tables, formulas, and codes. Equation 78.62 is a modification of the 1909 *Shortt equation*.

$$L_{s,\text{m}} \approx \frac{0.035 \text{v}_{\text{mph}}^3}{R_{\text{m}}} \qquad \text{[SI]} \qquad 78.62(a)$$

$$L_{s,\text{ft}} \approx \frac{1.6 \text{v}_{\text{mph}}^3}{R_{\text{ft}}} \qquad \text{[U.S.]} \qquad 78.62(b)$$

A tangent to the entrance spiral at the SC projected to the back tangent locates the SPI (spiral point of intersection). The angle at the SPI between the two tangents is the *spiral angle*, I_s. (As with circular curves, the symbol I and Δ are both used for spiral angle.) The spiral's degree of curvature changes uniformly from $0°$ at TS to D (the circular curve's degree) at SC. Since the change is uniform, the average degree of curve over the spiral's length is $D/2$. The spiral angle is given by Eq. 78.63. L_s is the length of spiral curve in ft, and I_s, I_c, and D are in degrees. Since an oblique line intersects two parallel lines at the same skew angle, the angle between lines O-SC and O'-TS and the angle between O-SC and O-PC are the same.

$$I_s = \left(\frac{L_s}{100}\right)\left(\frac{D}{2}\right) = \frac{L_s D}{200} \qquad 78.63$$

The *total deflection angle* is equal to the intersection angle, I, of the two intersecting tangents.

$$I = I_c + 2I_s \qquad 78.64$$

The *total length of the curve system* is

$$L = L_c + 2L_s \qquad 78.65$$

As with circular curves, spiral curves can be laid out using deflection angles or tangent offsets. The total deflection angle, α_s, between lines TS-PI and TS-SC is

$$\alpha_s = \tan^{-1}\frac{y}{x} \approx \frac{y}{x} \approx \frac{I_s}{3} \qquad [I_s < 20°] \qquad 78.66$$

At any other point, P, along the spiral curve, spiral angles, α_P, are proportional to the square of the distance from the TS to the point. I_P is the central spiral angle at any point P whose distance from TS is L_P.

$$\frac{\alpha_\text{P}}{\alpha_s} = \frac{I_\text{P}}{I_s} = \left(\frac{L_\text{P}}{L_s}\right)^2 \qquad 78.67$$

For any spiral curve length L up to L_s, the tangent offset is

$$y = x\tan\alpha_\text{p} \approx x\alpha_{\text{p,radians}}$$

$$= \frac{xI_{s,\text{radians}}}{3}$$

$$= \left(\frac{xL_s D_{\text{degrees}}}{(200)(3)}\right)\left(\frac{\pi}{180°}\right)$$

$$= \frac{xL_s}{6R_c} \qquad [I_s < 20°] \qquad 78.68$$

$$T_c = (R_c + Q)\tan\frac{I}{2} \qquad 78.69$$

$$Q = y - R_c(1 - \cos I_s)$$

$$= \frac{L_s^2}{6R_c} - R_c(1 - \cos I_s) \qquad 78.70$$

Table 78.5 *Spiral Curve Abbreviations*

CS	curve to spiral point
I_s	spiral angle
L_c	curve length
L_P	distance to any point, P
L_s	length of spiral
Q	offset of ghost PC to new tangent (tangent shift)
R_c	radius of circular curve
SC	spiral to curve point
SPI	spiral to point of intersection
ST	spiral to tangent point
TS	tangent to spiral point

Example 78.10

A 300 ft long spiral curve is used as a transition to a $6°$ circular curve. The intersection angle of the tangents is $85°$. The station of the PI is 70+00. Determine the stations of the (a) TS, (b) SC, (c) CS, and (d) ST.

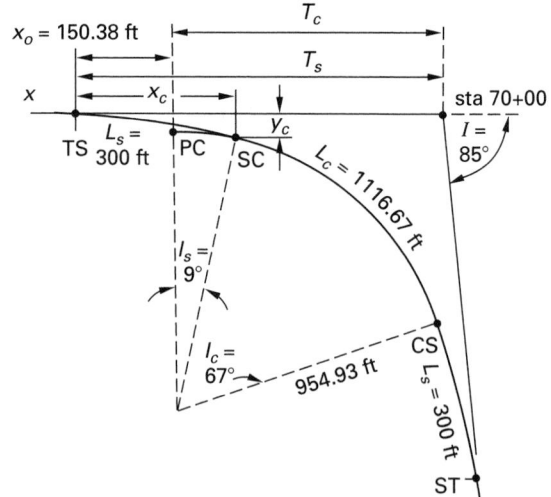

Solution

Use Eq. 78.63.

$$I_s = \frac{L_s D}{200} = \frac{(300 \text{ ft})(6°)}{200} = 9°$$

The circular curve's subtended angle between the SC and CS is

$$I_c = I - 2I_s = 85° - (2)(9°) = 67°$$

Use Eq. 78.1 to calculate the circular curve's radius.

$$R_c = \frac{5729.578}{D_c} = \frac{5729.578}{6°} = 954.93 \text{ ft}$$

Use Eq. 78.10 to calculate the length of the circular curve.

$$L_c = 100 \left(\frac{I_c}{D_c} \right) = (100 \text{ ft}) \left(\frac{67°}{6°} \right)$$
$$= 1116.67 \text{ ft}$$

Use Eq. 78.68 to calculate the offset from the tangent at the SC. At that point, $x \approx L_s$.

$$y = \frac{L_s^2}{6R_c} = \frac{(300 \text{ ft})^2}{(6)(954.93 \text{ ft})}$$
$$= 15.71 \text{ ft}$$

Since $9° < 20°$, the tangent distance can be approximated as

$$x_c = \frac{y_c}{\tan \dfrac{I_s}{3}}$$
$$= \frac{15.71 \text{ ft}}{\tan \dfrac{9°}{3}} = 299.76 \text{ ft}$$

From Fig. 78.17,

$$x_o = x_c - R_c \sin I_s = 299.76 \text{ ft} - (954.93 \text{ ft})(\sin 9°)$$
$$= 150.38 \text{ ft}$$

Use Eq. 78.70.

$$Q = \frac{L_s^2}{6R_c} - R_c(1 - \cos I_s)$$
$$= \frac{(300 \text{ ft})^2}{(6)(954.93 \text{ ft})} - (954.93 \text{ ft})(1 - \cos 9°)$$
$$= 3.95 \text{ ft}$$

Use Eq. 78.69.

$$T_c = (R_c + Q) \tan \frac{I}{2}$$
$$= (954.93 \text{ ft} + 3.95 \text{ ft}) \tan \frac{85°}{2}$$
$$= 878.65 \text{ ft}$$
$$T_s = x_o + T_c = 150.38 \text{ ft} + 878.65 \text{ ft} = 1029.03 \text{ ft}$$

(a) sta TS = sta PI $- T_s$
$$= (70+00) - 1029.03 \text{ ft} = 59+70.97$$

(b) sta SC = sta TS $+ L_s$
$$= (59+70.97) + 300 \text{ ft} = 62+70.97$$

(c) sta CS = sta SC $+ L_c$
$$= (62+70.97) + 1116.67 \text{ ft} = 73+87.64$$

(d) sta ST = sta CS $+ L_s = (73+87.64) + 300 \text{ ft}$
$$= 76+87.64$$

26. AIRPORT PAVEMENT GRADES

Longitudinal grades of airport runways should be limited to 1.5% for transport airports and 2.0% for utility airports. Similarly, the maximum grade change from one runway grade to another grade should be limited to 1.5% for transport airports and 2.0% for utility airports. The maximum grade on the first and last quarter of the runway distance should be 0.8% for transport airports. The minimum lengths of vertical curves (PC to PT) should be 1000 ft/% grade change for transport airports and 300 ft/% grade change for utility airports. The minimum separation between points of intersection (PIs) of runway vertical curves should be $(1000 \text{ ft})(|G_1| + |G_2|)$ for transport airports and $(250 \text{ ft})(|G_1|+|G_2|)$ for utility airports.

27. RAILROAD GRADES

Table 78.6 gives general recommendations for maximum rates of change of gradients for railroad curves.

Table 78.6 *Maximum Rate of Change of Gradient on Railroad Lines (percent per 100 ft station)*

track line rating	curve type	
	sag	crest
high-speed main line	0.05	0.10
secondary or branch line	0.10	0.20

Transportation

Topic VII: Construction

Construction

79 Construction Earthwork

Nomenclature

a	dimension	ft	m
A	area	ft^2	m^2
b	dimension	ft	m
d	distance	ft	m
h	height	ft	m
HI	height of the instrument	ft	m
L	length	ft	m
L	load factor	–	–
r	radius	ft	m
s	side-slope	–	–
V	volume	ft^3	m^3
w	width	ft	m

Symbols

ϕ	angle	deg	deg

Subscripts

b	bank measure
c	compacted
l	loose measure
m	mean

1. DEFINITION

Earthwork is the excavation, hauling, and placing of soil, rock, gravel, or other material found below the surface of the earth. This definition also includes the measurement of such material in the field, the computation in the office of the volume of such material, and the determination of the most economical method of performing such work. Specific information about excavation analysis, bracing, bulkheads, and cofferdams is in Ch. 39.

2. UNIT OF MEASURE

In the United States, the *cubic yard* (i.e., the "yard") is the unit of measure for earthwork. However, the volume and density of earth changes under natural conditions and during the operations of excavation, hauling, and placing.

3. SWELL AND SHRINKAGE

The volume of a loose pile of excavated earth will be greater than the original, in-place natural volume. If the earth is compacted after it is placed, the volume may be less than its original volume.

The volume of the earth in its natural state is known as *bank-measure*. The volume during transport is known as *loose-measure*. The volume after compaction is known as *compacted-measure*.

When earth is excavated, it increases in volume because of an increase in voids. The change in volume of earth from its natural to loose state is known as *swell*. Swell is expressed as a percentage of the natural volume. A soils *load factor*, L, in a particular excavation environment is the inverse of the *swell factor*, the sum of 1 and the decimal swell.

$$V_l = \left(\frac{100\% + \% \text{ swell}}{100\%} \right) V_b = \frac{V_b}{L} \qquad 79.1$$

The decrease in volume of earth from its natural state to its compacted state is known as *shrinkage*. Shrinkage also is expressed as a percent decrease from the natural state.

$$V_c = \left(\frac{100\% - \% \text{ shrinkage}}{100\%} \right) V_b \qquad 79.2$$

As an example, 1 yd^3 in the ground may become 1.2 yd^3 loose-measure and 0.85 yd^3 after compaction. The swell would be 20%, and the shrinkage would be 15%. Swell and shrinkage vary with soil types.

Refer to Table 79.1 for load factors and swell percentages for many different soil types.

Construction

Table 79.1 *Typical Swell and Load Factors of Materials*

material	swell, %	load factor
clay		
dry	40	0.72
wet	40	0.72
clay and gravel		
dry	40	0.72
wet	40	0.72
coal, anthracite	35	0.74
coal, bituminous	35	0.74
earth, loam		
dry	25	0.80
wet	25	0.80
gravel		
dry	12	0.89
wet	12	0.89
gypsum	74	0.57
hardpan	50	0.67
limestone	67	0.60
rock, well blasted	65	0.60
sand		
dry	12	0.89
wet	12	0.89
sandstone	54	0.65
shale and soft rock	65	0.60
slate	65	0.60
traprock	65	0.61

Adapted from *Standard Handbook for Civil Engineers*.

4. CLASSIFICATION OF MATERIALS

Excavated material is usually classified as *common excavation* or *rock excavation*. Common excavation is soil.

In highway construction, common road excavation is soil found in the roadway. *Common borrow* is soil found outside the roadway and brought in to the roadway. Borrow is necessary where there is not enough material in the roadway excavation to provide for the embankment.

5. CUT AND FILL

Earthwork that is to be excavated is known as *cut*. Excavation that is placed in embankment is known as *fill*.

Payment for earthwork is normally either for cut and not for fill, or for fill and not for cut. In highway work, payment is usually for cut; in dam work, payment is usually for fill. To pay for both would require measuring two different volumes and paying for moving the same earth twice.

6. FIELD MEASUREMENT

Cut and fill volumes can be computed from slope-stake notes, from plan cross sections, or by photogrammetric methods.

7. CROSS SECTIONS

Cross sections are profiles of the earth taken at right angles to the centerline of an engineering project (such as a highway, canal, dam, or railroad). A cross section for a highway is shown in Fig. 79.1.

Figure 79.1 *Typical Highway Cross Section*

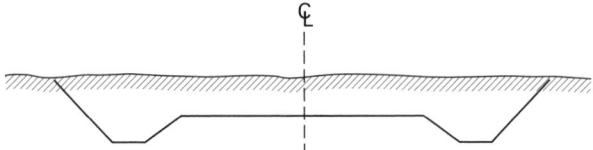

8. ORIGINAL AND FINAL CROSS SECTIONS

To obtain volume measurement, cross sections are taken before construction begins and after it is completed. By plotting the cross section at a particular station both before and after construction, a sectional view of the change in the profile of the earth along a certain line is obtained. The change along this line appears on the plan as an area. By using these areas at various intervals along the centerline, and by using distance between the areas, volume can be computed.

9. TYPICAL SECTIONS

Typical sections show the cross section view of the project as it will look on completion, including all dimensions (Fig. 79.2). Highway projects usually show several typical sections including cut sections, fill sections, and sections showing both cut and fill. Interstate highway plans also show access-road sections and sections at ramps.

10. DISTANCE BETWEEN CROSS SECTIONS

Cross sections are usually taken at each full station and at breaks in the ground along the centerline. In taking cross sections, it must be assumed that the change in the earth's surface from one cross section to the next is uniform, and that a section halfway between the cross sections is an average of the two. If the ground breaks appreciably between any two full-stations, one or more cross sections between full-stations must be taken. This is referred to as *taking sections at pluses*. Figure 79.3 shows eleven stations at which cross sections should be taken.

In rock excavation, or any other expensive operation, cross sections should be taken at intervals of 50 ft (15 m)

Figure 79.2 *Typical Completed Section*

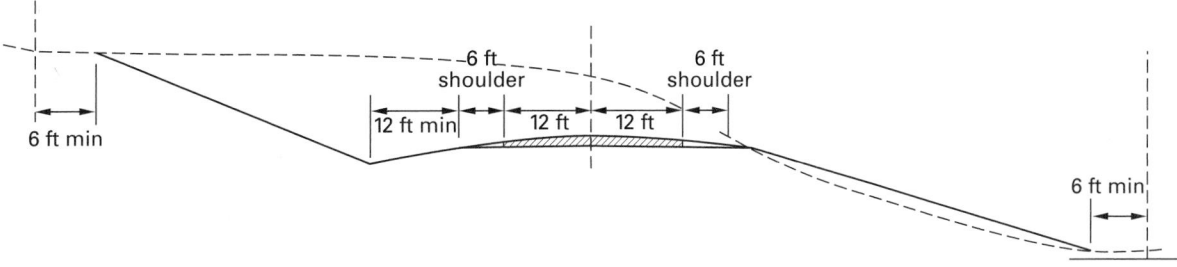

or less. Cross sections should always be taken at the PC and PT of a curve. Plans should also show a section on each end of a project (where no construction is to take place) so that changes caused by construction will not be abrupt.

Figure 79.3 *Cross-Section Locations*

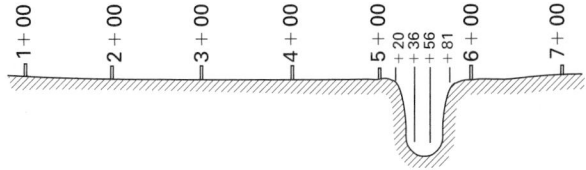

Where a cut section of a highway is to change to a fill section, several additional cross sections are needed. Such sections are shown in Fig. 79.4.

Figure 79.4 *Cut Changing to Fill*

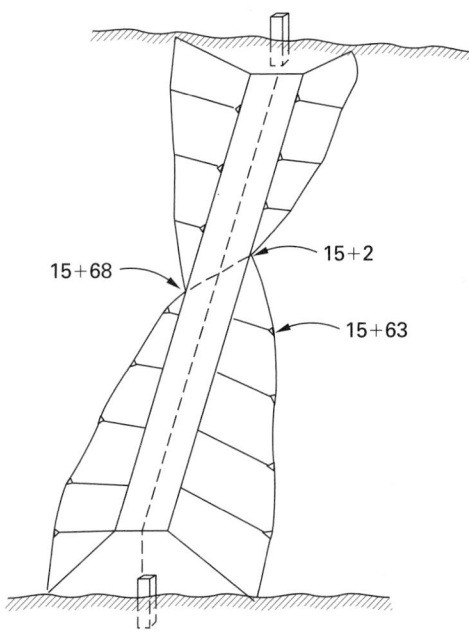

11. GRADE POINT

The point where a fill section meets the natural ground (where a cut section begins) is known as the *grade point*.

12. VOLUMES OF PILES

The volumes of piles of soil, recycled pavement, and paving materials can be calculated if the pile shapes are assumed. The *angle of repose*, ϕ, depends on the material, but is approximately 30° for smooth gravel, 40° for sharp gravel, 25 to 35° for dry sand, 30 to 45° for moist sand, 20 to 40° for wet sand, and 37° for cement.

$$V = \left(\frac{h}{3}\right)\pi r^2 \quad \text{[cone]} \tag{79.3}$$

$$V = \left(\frac{h}{6}\right)b(2a + a_1) = \tfrac{1}{6}hb\left(3a - \frac{2h}{\tan\phi}\right)$$
$$\text{[wedge]} \tag{79.4}$$

$$
\begin{aligned}
V &= \left(\frac{h}{6}\right)\left(ab + (a + a_1)(b + b_1) + a_1 b_1\right)\\
&= \left(\frac{h}{6}\right)\left(ab + 4\left(a - \frac{h}{\tan\phi_1}\right)\left(b - \frac{h}{\tan\phi_2}\right)\right.\\
&\quad \left. + \left(a - \frac{2h}{\tan\phi_1}\right)\left(b - \frac{2h}{\tan\phi_2}\right)\right)
\end{aligned}
$$
$$\begin{bmatrix} \text{frustrum of a} \\ \text{rectangular pyramid} \end{bmatrix} \tag{79.5}$$

$$a_1 = a - \frac{2h}{\tan\phi_1} \tag{79.6}$$

$$b_1 = b - \frac{2h}{\tan\phi_2} \tag{79.7}$$

Construction

Figure 79.5 *Pile Shapes*

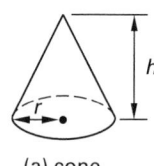

(a) cone

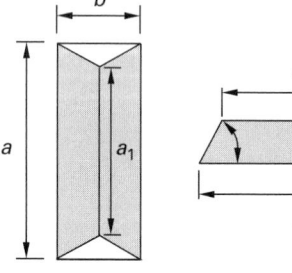

(b) wedge

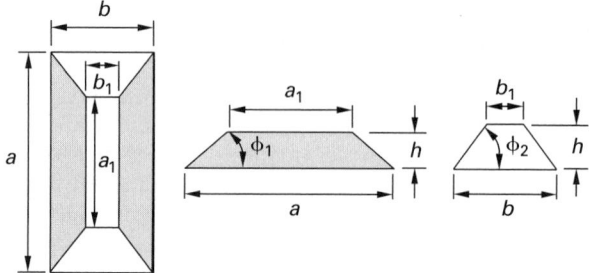

(c) frustrum of a rectangular pyramid

13. EARTHWORK VOLUMES

A three-dimensional soil volume between two points is known as a soil *prismoid* or *prism*. The prismoid (prismatic) volume must be calculated in order to estimate hauling requirements. Such volume is generally expressed in units of cubic yards ("yards") or cubic meters. There are two methods of calculating the prismoid volume: the average end area method and the prismoidal formula method.

Figure 79.6 *Soil Prismoid*

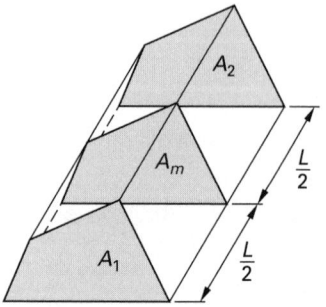

14. AVERAGE END AREA METHOD

With the *average end area method*, the volume is calculated by averaging the two end areas and multiplying by the prism length. This disregards the slopes and orientations of the ends and sides, but is sufficiently accurate for most earthwork calculations. When the end area is complex, it may be necessary to use a planimeter or to plot the area on fine grid paper and simply count the squares. The average end area method usually overestimates the actual soils volume, favoring the contractor in earthwork cost estimates.

$$V = \frac{L(A_1 + A_2)}{2} \qquad 79.8$$

The precision obtained from the average end area method is generally sufficient unless one of the end areas is very small or zero. In that case, the volume should be computed as a pyramid or truncated pyramid.

$$V_{\text{pyramid}} = \frac{LA_{\text{base}}}{3} \qquad 79.9$$

15. PRISMOIDAL FORMULA METHOD

The *prismoidal formula* is preferred when the two end areas differ greatly or when the ground surface is irregular. It generally produces a smaller volume than the average end area method and thus favors the owner-developer in earthwork cost estimating.

The prismoidal formula uses the mean area, A_m, midway between the two end sections. In the absence of actual observed measurements, the dimensions of the middle area can be found by averaging the similar dimensions of the two end areas. The middle area is not found by averaging the two end areas.

$$V = \left(\frac{L}{6}\right)(A_1 + 4A_m + A_2) \qquad 79.10$$

When using the prismoidal formula, the volume is not found as LA_m, although that quantity is usually sufficiently accurate for estimating purposes.

16. BORROW PIT

It is often necessary to borrow earth from an adjacent area to construct embankments. Normally, the *borrow pit* area is laid out in a rectangular grid with 10 ft (3 m), 50 ft (15 m), or even 100 ft (30 m) squares. Elevations are determined at the corners of each square by leveling before and after excavation so that the cut at each corner can be computed.

Points outside the cut area are established on the grid lines so that the lines can be reestablished after excavation is completed.

Example 79.1

Excavation of a borrow pit has been surveyed and recorded in Fig. 79.7. What are the volumes of excavation from prisms A0-B0-B1-A1 and E2-F2-E3?

Figure 79.7 *Depth of Excavation (Cut) in a Borrow Pit Area*

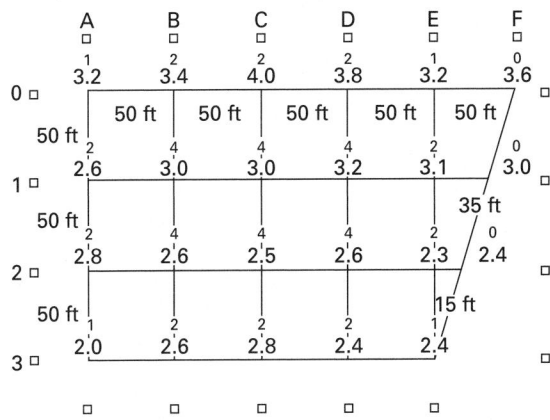

Solution

Volumes are computed by multiplying the average cut by the area of the figure. The volume of the prism A0-B0-B1-A1 is

$$V = \left(\frac{(50 \text{ ft})(50 \text{ ft})}{27 \ \dfrac{\text{ft}^3}{\text{yd}^3}} \right)$$
$$\times \left(\frac{3.2 \text{ ft} + 3.4 \text{ ft} + 3.0 \text{ ft} + 2.6 \text{ ft}}{4} \right)$$
$$= 282.4 \text{ yd}^3$$

The volume of the prism E2-F2-E3 is

$$V = \left(\frac{(50 \text{ ft})(15 \text{ ft})}{(2)\left(27 \ \dfrac{\text{ft}^3}{\text{yd}^3}\right)} \right) \left(\frac{2.3 \text{ ft} + 2.4 \text{ ft} + 2.4 \text{ ft}}{3} \right)$$
$$= 33 \text{ yd}^3$$

Instead of computing volumes of prisms represented by squares separately, all square-based prisms can be computed collectively by multiplying the area of one square by the sum of the cut at each corner times the number of times that cut appears in any square, divided by 4. For instance, on the second line from the top in Fig. 79.7, which is line 1, 2.6 appears in two squares, 3.0 appears in four squares, 3.0 appears in four squares, 3.2 appears in four squares, and 3.1 appears in two squares. In the figure, the small number above the cut indicates the number of times the cut is used in averaging the cuts for the square prisms.

Example 79.2

Calculate the volume of earth excavated from the borrow pit shown in Fig. 79.7.

Solution

The volume of the squares is

$$V = \left(\frac{50 \text{ ft}}{\left(27 \ \dfrac{\text{ft}^3}{\text{yd}^3}\right)(4)} \right)(50 \text{ ft})$$
$$\times \left(\begin{array}{l} 3.2 \text{ ft} + (2)(3.4 \text{ ft}) + (2)(4.0 \text{ ft}) \\ \quad + (2)(3.8 \text{ ft}) + 3.2 \text{ ft} + (2)(2.6 \text{ ft}) \\ \quad + (4)(3.0 \text{ ft}) + (4)(3.0 \text{ ft}) + (4)(3.2 \text{ ft}) \\ \quad + (2)(3.1 \text{ ft}) + (2)(2.8 \text{ ft}) + (4)(2.6 \text{ ft}) \\ \quad + (4)(2.5 \text{ ft}) + (4)(2.6 \text{ ft}) + (2)(2.3 \text{ ft}) \\ \quad + 2.0 + (2)(2.6 \text{ ft}) + (2)(2.8 \text{ ft}) \\ \quad + (2)(2.4 \text{ ft}) + 2.4 \text{ ft} \end{array} \right)$$
$$= 3194 \text{ yd}^3$$

The volume of the trapezoids is

$$V = \left(\frac{50 \text{ ft} + 35 \text{ ft}}{(2)\left(27 \ \dfrac{\text{ft}^3}{\text{yd}^3}\right)} \right)(50 \text{ ft})$$
$$\times \left(\frac{3.2 \text{ ft} + 3.6 \text{ ft} + 3.1 \text{ ft} + 3.0 \text{ ft}}{4} \right)$$
$$= 254 \text{ yd}^3$$

$$V = \left(\frac{35 \text{ ft} + 15 \text{ ft}}{(2)\left(27 \ \dfrac{\text{ft}^3}{\text{yd}^3}\right)} \right)(50 \text{ ft})$$
$$\times \left(\frac{3.1 \text{ ft} + 3.0 \text{ ft} + 2.3 \text{ ft} + 2.4 \text{ ft}}{4} \right)$$
$$= 125 \text{ yd}^3$$

The volume of the triangle is

$$V = \left(\frac{(15 \text{ ft})}{(2)\left(27 \ \dfrac{\text{ft}^3}{\text{yd}^3}\right)} \right)(50 \text{ ft})$$
$$\times \left(\frac{2.3 \text{ ft} + 2.4 \text{ ft} + 2.4 \text{ ft}}{3} \right)$$
$$= 33 \text{ yd}^3$$

The total volume of earth excavated is

$$V = 3194 \text{ yd}^3 + 254 \text{ yd}^3 + 125 \text{ yd}^3 + 33 \text{ yd}^3 = 3606 \text{ yd}^3$$

Construction

17. MASS DIAGRAMS

A *profile diagram* is a cross section of the existing ground elevation along a route alignment. The elevation of the route is superimposed to identify the *grade points*, the points where the final grade coincides with the natural elevation.

A *mass diagram* is a record of the cumulative earthwork volume moved along an alignment, usually plotted below profile sections of the original ground and finished grade. The mass diagram can be used to establish a finished grade that balances cut-and-fill volumes and minimizes long hauls.

After volumes of cut and fill between stations have been computed, they are tabulated as shown in Table 79.2. The cuts and fills are then added, and the cumulative yardage at each station is recorded in the table. It is this cumulative yardage that is plotted as an ordinate. In Fig. 79.8, the baseline serves as the x-axis and cumulative yardage that has a plus sign is plotted above the baseline. Cumulative yardage that has a minus sign is plotted below the baseline. The scale is chosen for convenience. 1 in = 5 ft is typical, although 1 in = 10 ft or 20 ft can be used in steep terrain.

Figure 79.8 *Baseline and Centerline Profile Mass Diagram*

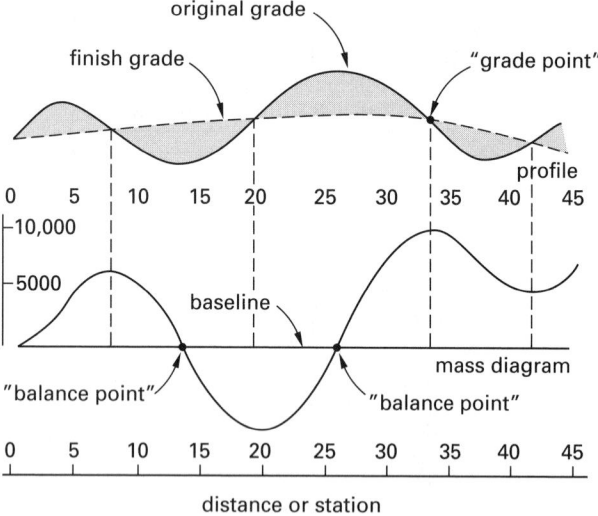

In Fig. 79.8, the mass diagram is plotted on the lower half of the sheet, and the centerline profile of the project is plotted on the upper half.

A rising line on the mass diagram represents areas of excavation; a falling line represents areas of embankment. The local minima and maxima on the mass diagram identify the *grade points*. Vertical distances on a mass diagram represent volumes of material (areas on the profile diagram). Areas in a mass diagram represent the product of volume and distance.

A *balance line* is a horizontal line drawn between two adjacent *balance points* in a crest or sag area. The volumes of excavation and embankment between the

Table 79.2 Typical Cut and Fill Calculations
(all volumes are in cubic yards)

sta	cut +	fill −	cum sum	sta	cut +	fill −	cum sum
0			0	23			−4710
	184				1676		
1			+184	24			−3034
	622				1676		
2			+806	25			−1358
	1035				1860		
3			+1841	26			+502
	1268				1917		
4			+3109	27			+2419
	1231				1839		
5			+4340	28			+4258
	919				1611		
6			+5259	29			+5869
	503				1338		
7			+5762	30			+7207
	164	21			1029		
8			+5905	31			+8236
	12	190			652		
9			+5727	32			+8888
		616			357		
10			+5111	33			+9245
		942			150	39	
11			+4169	34			+9356
		1150			52	236	
12			+3019	35			+9172
		1500				465	
13			+1519	36			+8707
		1773				712	
14			−254	37			+7995
		1755				904	
15			−2009	38			+7091
		1540				904	
16			−3549	39			+6187
		1262				757	
17			−4811	40			+5430
		932				516	
18			−5743	41			+4914
		546				280	
19			−6289	42			+4634
		172				127	
20			−6461	43			+4507
	178					98	
21			−6283	44			+4409
	568					20	
22			−5715	45			+4389
	1005						

points are equal. Thus, a contractor can plan on using the earth volume from the excavation on one side of the grade point for the embankment on the other side of the grade point. In Fig. 79.9, a balance line has been drawn that intersects the mass diagram at two points.

These points represent the inclusive stations for which the cut-and-fill volumes are equal. That is, the cut soil (area B) can be used for the fill soil (area A).

Figure 79.9 *Balance Line Between Two Points*

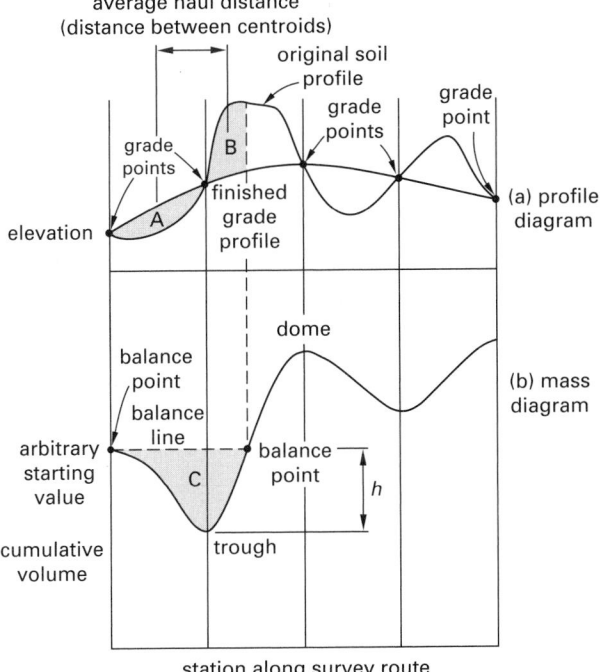

It is important in planning and construction to know the points along the centerline a particular section of cut that will balance a particular section of fill. For example, assume that a cut section extends from sta 12+25 to sta 18+65, and a fill section extends from sta 18+65 to sta 26+80. Also, assume that the excavated material will exactly provide the material needed to make the embankment. Then, cut balances fill, and sta 12+25 and 26+80 are balance points.

A balance point occurs where the mass curve crosses the baseline. In Fig. 79.10, a balance point falls between sta 13 and 14. The ordinate of 13 is +1519; the ordinate of 14 is −254. Therefore, the curve fell 1519 yd³ + 254 yd³ = 1773 yd³ in 100 ft or 17.73 yd³/ft. The curve crosses the baseline at a distance of 1519/17.73 = 86 ft from sta 13 (13+86).

The average *haul distance* is the separation of the centroids of the excavation and embankment areas on the profile diagram. This distance is usually determined for each section of earthwork rather than for the entire mass of earthwork in a project.

The average haul can also be calculated from the mass diagram. For customary U.S. usage, the curve area under a balance line represents the total haul in *yard-stations*, which is the number of cubic yards moved in 100 ft (30 m). This figure can be used by the contractor to determine hauling costs. The height of the curve, or the distance between the balance line and the minimum or maximum, represents the *solidity* or the volume of the total haul. The average haul distance is found by dividing the curve area (area C) by the curve height, *h*.

The *freehaul distance* is the maximum distance, as specified in the construction contract, that the contractor is expected to transport earth without receiving additional payment. Typically, the freehaul distance is between 500 and 1000 ft (150 to 300 m). Any soil transported more than the freehaul distance is known as *overhaul*.

Figure 79.10 *Mass Diagram Showing Sub-Bases*

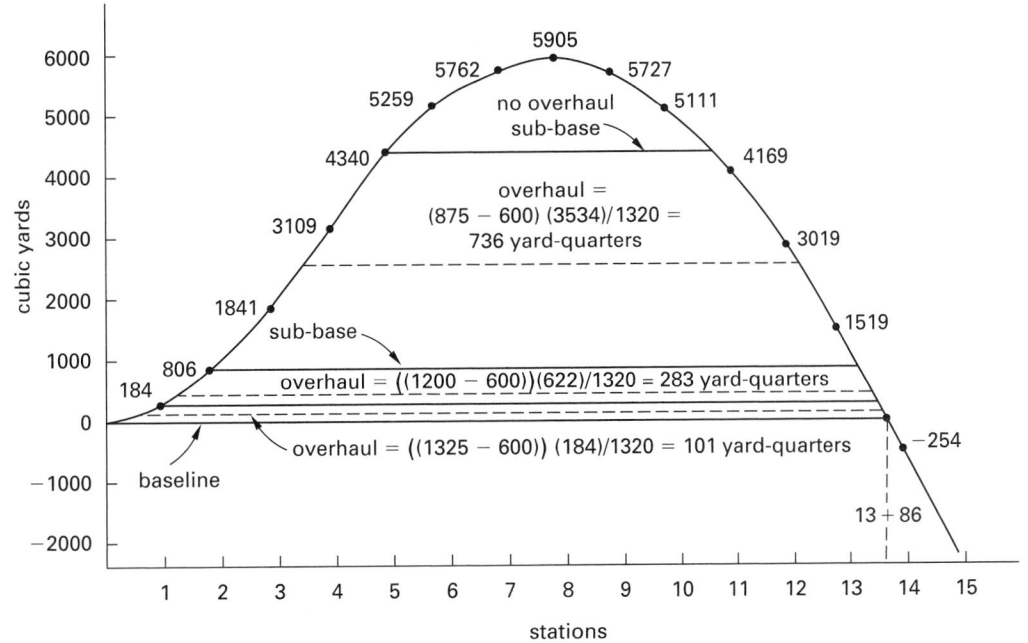

Freehaul and overhaul can be determined from the mass diagram. The procedure begins by drawing a balance line equal in length to the freehaul distance. The enclosed area on the mass diagram represents material that will be hauled with no extra cost. The actual volume moved (the solidity) is the vertical distance between the balance line and the maximum or minimum.

The overhaul is found directly from the overhaul area on the mass diagram. It can also be calculated indirectly from the overhaul volume and overhaul distance. The overhaul volume is determined from the maximum height of the overhaul area on the mass diagram, or from the overhaul area on the profile diagram. The overhaul distance is found as the separation in overhaul centroids on the profile diagram.

Figure 79.11 *Freehaul and Overhaul*

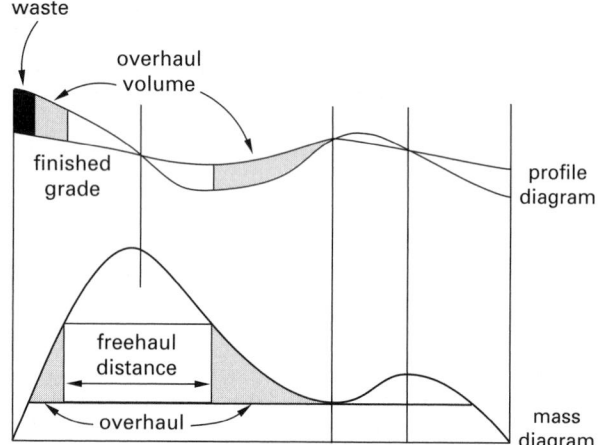

Sub-bases are horizontal balance lines that divide an area of the mass diagram between two balance points into trapezoids for the purpose of more accurately computing overhaul. In Fig. 79.10, the top sub-base is less than 600 ft in length, the freehaul distance for this project. Therefore, the volume of earth represented by the area above this line will be hauled a distance less than the free-haul distance, and no payment will be made for overhaul. All the volume represented by the area below this top sub-base will receive payment for overhaul.

The area between two sub-bases is nearly trapezoidal. The average length of the bases of a trapezoid can be measured in feet, and the altitude of the trapezoid can be measured in cubic yards of earth. The product of these two quantities can be expressed in *yard-quarters*. A "yard-quarter" is a cubic yard of earthwork transported one quarter mile. If free haul is subtracted from the length, the quantity can be expressed as overhaul.

Sub-bases are drawn at distinct breaks in the mass curve. Distinct breaks in Fig. 79.10 can be seen at sta 1+00(184), 2+00(806), and 5+00(4340). After sub-bases are drawn, a horizontal line is drawn midway between the sub-bases. This line represents the average haul for the volume of earth between the two sub-bases. If a horizontal scale is used, the length of the average haul can be determined by scaling. This line, shown as a dashed line in Fig. 79.10, scales 875 ft for the area between the top sub-bases. The free haul is subtracted from this in Table 79.2.

The volume of earth between the two sub-bases is found by subtracting the ordinate of the lower sub-base from the ordinate of the upper sub-base. These ordinates are found in Table 79.2.

Multiplying average haul minus free haul in feet by volume of earth in cubic yards gives overhaul in yard-feet. Dividing by 1320 ft gives yard-quarters.

When factors such as *shrinkage, swell, loss during transport*, and *subsidence* can be estimated, they are included in determining embankment volumes. A 5 to 15% excess is usually included in the figure. This is achieved by increasing all fill volumes by the necessary percentage.

80 Construction Staking and Layout

1. STAKING

Surveying markers are referred to as *construction stakes*, *alignment stakes*, *offset stakes*, *grade stakes*, or *slope stakes*, depending on their purpose. Stakes come in different sizes, including $1 \times 2 \times 18$ markers, $5/16 \times 1^1/_2 \times 24$ *half-lath stakes*, and various lengths (usually 12 or 18 in) of 2×2 *hubs* and 1×1 *guineas*. (All measurements are in inches.) Fill stakes can be ripped diagonally lengthwise to give them a characteristic shape. Pin flags, full-length lagging, or colored paint may be used to provide further identification and visibility.

Distances are measured in feet or meters, with precisions depending on the nature of the feature being documented. A precision of 0.1 ft (30 mm) vertically and 0.5 ft (150 mm) horizontally is considered standard with earthwork (e.g., setting slope stakes). A precision of 0.01 ft (3 mm) is considered standard for locating curb and alignment stake positions and elevations so that the feature can be built to within 0.1 ft (30 mm). Bridges and buildings are located with higher precision.

2. STAKE MARKINGS

Stake markings vary greatly from agency to agency. The abbreviations shown in Table 80.1 and the procedures listed in this chapter are typical but not universal.

The front (the side facing the construction) and back are marked permanently using pencil, carpenter's crayon, or permanent ink markers. Stakes are read from top to bottom. The front of the stake is marked with *header information* (e.g., RPSS, offset distance) and *cluster information* (e.g., horizontal and vertical measurements, slope ratio). The header is separated from the first cluster by a double horizontal line. Multiple clusters are separated by single horizontal lines. All cluster information is measured in the same direction from the same point.

Words like "from" and "to" are understood and are seldom written on a stake. For example, "2.0 FC" and "4.3 FG" mean "2.0 feet from the face of the curb" and "4.3 feet above the finished grade," respectively.

The back of a stake is used to record the station and other information, including literal descriptions (e.g.,

"at ramp"). Other information identifying the survey may also be included on the back. Actual elevations, when included, are marked on the thin edge of the stake. Figure 80.1 illustrates how a construction stake would be marked to identify a trench for a storm drain.

Figure 80.1 Construction Stake for Storm Drain

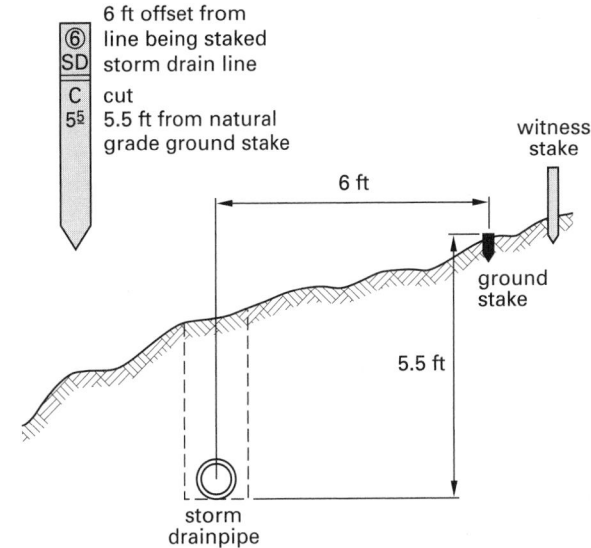

A stake both locates and identifies a specific point. A distance referenced on a stake is measured from the natural ground surface at the point where the stake is driven. Another system involves measuring from a tack or a reference mark on the stake itself. A *crow's foot*, which consists of a horizontal line drawn on the stake with a vertical arrow pointing to it, may be used. A crow's foot should be drawn on the edge of the stake.

Many times, measurements are taken from a *hub stake*, also referred to as a *ground stake* or *reference point stake*, which is driven flush with the ground. No markings are made on hub stakes, since such stakes are entirely below grade.

Hub stakes are located, identified, and protected by witness stakes or guard stakes. A *witness stake* calls attention to a hub stake but does not itself locate a specific point. A *guard stake* may be driven at an angle with its top over the flush-driven hub stake.

Alignment stakes marked with the station indicate centerline alignment along a proposed roadway. They are placed every half, whole, or even-numbered station

Construction

Table 80.1 Common Stake Marking Abbreviations

@	from the		INV	invert
¼SR	quarter point of slope rounding		ISS	intermediate slope stake
½SR	midpoint of slope rounding		JT	joint trench
ABUT	abutment		L	length or left
BC	begin curve		L/2	midpoint
BCH	bench		L/4	quarter point
BCR	begin curb return		LIP	lip
BEG	begin or beginning		L/O	line only
BK	back		LOL	lay-out line
BL	baseline		LT	left
BM	benchmark		MC	middle of curve
BR	bridge		MH	manhole
BSR	begin slope rounding		MP	midpoint
BSW	back of sidewalk		MSR	midpoint of slope rounding
BVC	begin vertical curve		OG	original ground
C	cut		O/S	offset
CL	centerline		PC	point of curvature
CB	catch basin		PCC	point of compound curve
CF	curb face		PG	pavement grade or profile grade
CGS	contour grading stake		PI	point of intersection
CONT	contour		POC	point on curve
CP	control point or catch point		POL	point on line
CR	curb return		POT	point on tangent
CS	curb stake		PP	power pole
CURB	curb		PPP	pavement plane projected
DAY	daylight		PRC	point of reverse curvature
DI	drop inlet or drainage inlet		PT	point of tangency
DIT	ditch		PVC	point of vertical curvature
DL	daylight		PVT	point of vertical tangency
DMH	drop manhole		QSR	quarterpoint of slope rounding
DS	drainage stake		R	radius
E	flow line		RGS	rough grade stake
EC	end of curve		RT	right
ECR	end of curb return		RP	radius point or reference point
EL	elevation		RPSS	reference point for slope stake
ELECT	electrical		ROW	right of way
ELEV	elevation		R/W	right of way
END	end of ending		SD	storm drain
EOP	edge of pavement		SE	superelevation
EP	edge of pavement		SHLD	shoulder
ES	edge of shoulder		SHO	shoulder
ESR	end slope rounding		SL	stationing line
ETW	edge of traveled way		SR	slope rounding
EVC	end of vertical curve		SS	sanitary sewer or slope stake
EW	end wall		STA	station
F	fill		STR	structure
FC	face of curb		SW	sidewalk
FDN	foundation		TBC	top back of curb
FE	fence		TBM	temporary benchmark
FG	finish grade		TC	toe of curb or top of curve
FGS	final grade stake		TOE	toe
FH	fire hydrant		TOP	top
FL	flow line		TP	turning point
FTG	footing		TW	traveled way
G	grade		VP	vent pipe
GRT	grate		WALL	wall
GTR	gutter		WL	water line
GUT	gutter		WM	water meter
HP	hinge point		WV	water valve
INL	inlet		WW	wingwall
INT	intersection			

Compiled from various sources. Not intended to be exhaustive. Differences may exist from agency to agency.

along tangents with uniform grade, and at every quarter or half station along horizontal and vertical curve alignments.

Offset stakes are used to mark excavations. Offset stakes are offset from the actual edge to protect the stakes from earthmoving and other construction equipment. The offset distance is circled on the stake and is separated from the subsequent data by a double line. For close-in work, the offset distance may be standardized, such as 2 ft (61 cm). For highway work, offset stakes can be set at 25, 50, or 100 ft (8, 15, or 30 m) from the alignment centerline. Unless a separate ground stake or hub is used, distances (e.g. cuts, fills, and distances to centerline) marked on an offset stake refer to the point of insertion, not to the imaginary point located the offset distance away.

Slope stakes indicate *grade points*—points where the cuts and fills begin and the planned side slopes intersect the natural ground surface. Slope stakes are marked "SS" to indicate their purpose. Typically, they are placed with a 10 ft (3 m) offset. Figure 80.2 illustrates the use of slope stakes along three adjacent sections of a proposed highway.

In addition to indicating the grade point, the front of slope stakes are marked to indicate the nature of the earthwork (i.e., "C" for "cut" and "F" for "fill"), the offset distance, the type of line being staked, the distance from the centerline or control line, the slope to finished grade, and the elevation difference between the grade point and the finished grade. Distances from the centerline can be marked "L" or "R" to indicate whether the stake points are to the left or the right of centerline when looking up-station (i.e., ahead on stationing). The station is marked on the stake back.

For example, a stake's front face markings "C 4.2 FG @ 38.4L CL 2:1" would be interpreted as "cut with 2:1 slope is required; stake set point is 4.2 ft above the finished grade; stake is 38.4 ft to the left of the centerline of the roadway." The use of FG is optional.

Slope stakes can be driven vertically or at an angle, depending on convention. When driven at an angle, slope stakes slant outward, pointing away from the earthwork, when fill is required; and they slope inward, toward the earthwork, when a cut is required. Stakes are set with the front face toward the construction.

Hubs and other ground stakes without offsets are not used with fill stakes, since they would be covered during construction. For shallow and moderate fills, a fill stake alone is used. A crow's foot may be marked on the stake to indicate the approximate finished grade, or the top of the stake may be set to coincide with the approximate finished grade. For deeper fills, offset stakes are required.

Not all earthwork produces a straight slope. Some earthwork results in a smooth-curved ground surface between two lateral points. The points where the grading begins and ends are marked "BSR" (*begin slope rounding*) and

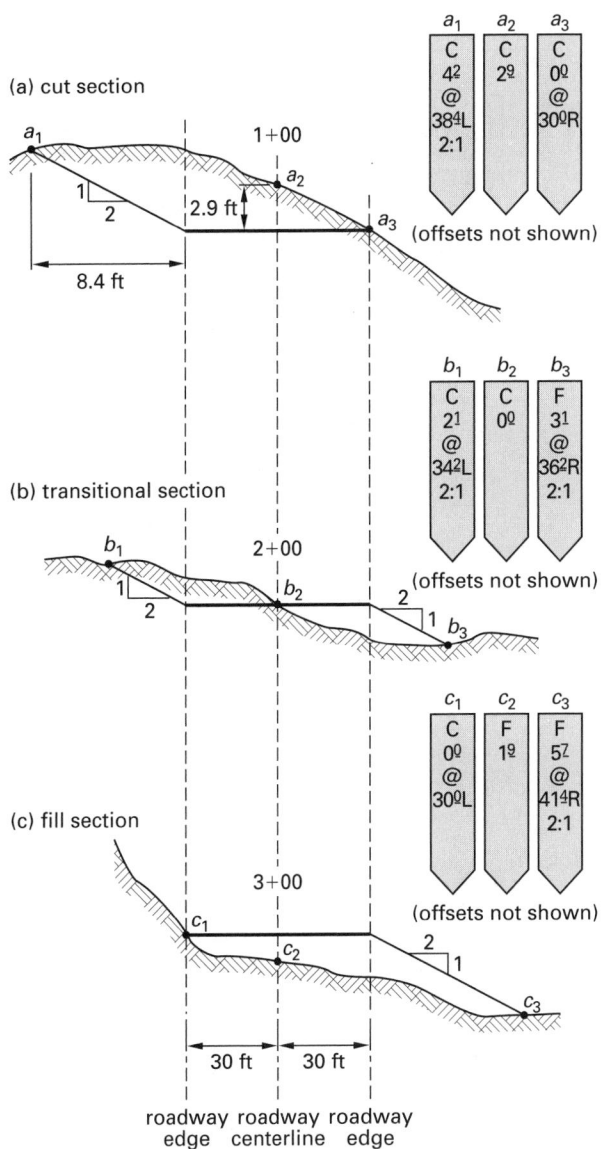

Figure 80.2 *Slope Stakes Along a Highway*

"ESR" (*end slope rounding*), rather than the analogous grade point.

A line of stakes at adjacent grade or ESR points is known as a *daylight line*. Therefore, the term *daylight stake* can be used when referring to the stakes marking the grade or ESR points.

Example 80.1

While checking the work of a surveying crew along a highway, you encounter the ground stake and its accompanying witness stake as shown. What is the elevation at the toe of the slope documented by the witness stake?

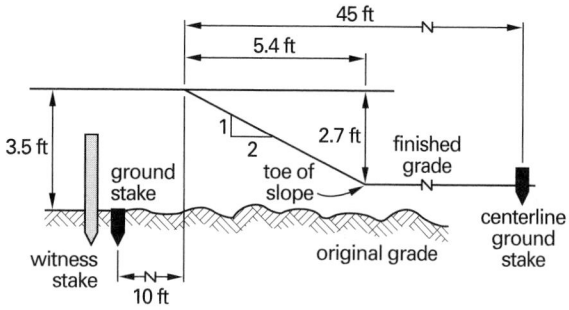

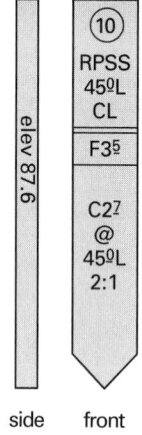

side   front

Solution

The witness stake indicates that the ground stake is a reference point slope stake located 45.0 ft to the left of centerline when looking up-station. The elevation of the ground stake, marked on the side of the witness stake, is 87.6 ft. The earthwork starts out with a 3.5 ft fill. Then, the earth is cut away at a 2:1 slope until the finished grade is reached, 2.7 ft below.

The elevation at the toe of the slope is the same as the elevation of the finished grade.

$$\text{elev}_{\text{slope toe}} = \text{elev}_{\text{ground stake}} + \text{fill} - \text{cut}$$

$$= 87.6 \text{ ft} + 3.5 \text{ ft} - 2.7 \text{ ft}$$

$$= 88.4 \text{ ft}$$

3. ESTABLISHING SLOPE STAKE MARKINGS

The markings on a construction stake are determined from a survey of the natural ground surface. This survey requires two individuals: one, the *leveler* or *instrument person*, to work the instrument, and the other, the *rod person*, to hold the leveling rod. The elevation of the instrument (or the elevation of the ground at the instrument location) and the height of the instrument above the ground must be known if actual elevations are

to be marked on a stake. The term "height of the instrument" (HI) is the elevation of the instrument above the reference datum.

$$\text{HI} = \text{elev}_{\text{ground}} + \begin{array}{c}\text{instrument height}\\ \text{above the ground}\end{array}$$

$$= \text{elev}_{\text{ground}} + \text{ground rod} \qquad 80.1$$

The *rod reading*, or *ground rod*, is the sighting made through a leveling instrument at the rod held vertically at the grade point. The *grade rod*, short for *grade rod reading* and *rod reading for grade*, is the reading that would be observed on an imaginary rod held on the finished grade elevation. The grade rod is calculated from the planned grade elevation and the height of the instrument.

$$\text{grade rod} = \text{HI} - \text{elev}_{\text{grade}} \qquad 80.2$$

The distance, h, marked on a construction stake, or the cut or fill, is the distance between natural and finished grade elevations and is easily calculated from the ground and grade rods. The cut or fill stake marking is the difference between grade and ground rod elevations. The actual steps taken to calculate the stake marking depend on whether the earthwork is a cut or a fill and whether the instrument is above or below the finished grade. Drawing a diagram will clarify the algebraic steps and prevent sign errors.

$$h = \text{grade rod} - \text{ground rod} \qquad 80.3$$

The distance from grade point to centerline or other control is also written on the stake. If w is the width of the finished surface, s is the side slope ratio (as horizontal:vertical), and h is the cut or fill at the grade point, then the horizontal distance, d, from the grade point to the centerline stake of the finished surface is calculated using Eq. 80.4. In the field, the stake location is found by trial and error.

$$d = \frac{w}{2} + hs \qquad 80.4$$

Figure 80.3 *Determining Stake Location*

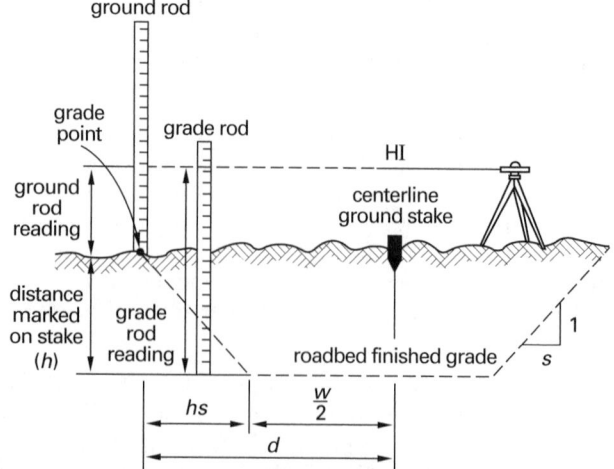

81 Building Codes and Materials Testing

1. INTRODUCTION

Local jurisdictions (including states) may write their own building codes, but in most cases a model code is adopted into law by reference. A *model code* is one that has been written by a group comprised of experts knowledgeable in the field, without reference to any particular geographical area. Adopting a model code allows a city, county, or district to have a complete, workable building code without the difficulty and expense of writing its own. If certain provisions need to be added or changed to suit the particular requirements of a municipality, the model code is enacted with modification. Even when a city or state writes its own code, that code is usually based on a model code. Exceptions include some large cities, such as New York and Chicago, and a few states that have adopted the Life Safety Code.

The primary model code is the International Building Code (IBC), first published in 2000 by the International Code Council (ICC) and updated every three years. The IBC is an amalgam of the work of the three code-writing groups that previously published the three model codes in the United States. The IBC combines provisions of all three of the previous model codes and is organized in the same format that the three code-writing groups used in the most recent editions of their codes. Jurisdictions in all states have adopted one or more of the family of international codes and some states have adopted the International Building Codes on a statewide level or have used it as a basis for writing their own codes.

The three model codes previously published and still used by some jurisdictions in the United States include the following.

- the Uniform Building Code (UBC), used in the western and central portions of the United States and published by the International Conference of Building Officials (ICBO)

- the BOCA National Building Code, used in the northeastern part of the country and published by the Building Officials and Code Administrators International (BOCA)

- the Standard Building Code (SBC), used in much of the southeastern United States and published by the Southern Building Code Congress International (SBCCI)

The Life Safety Code, published by the National Fire Protection Association, is also used by some jurisdictions.

The eventual use of one model code throughout the United States will bring consistency and make it easier for designers and architects to work across the country. However, to complicate matters, the National Fire Protection Association (NFPA) has written its own code. First publication of the NFPA 5000 Building Code was in 2002. At the time of this writing, NFPA 5000 has not been well received and very few jurisdictions in the United States use it.

The material in this chapter and in this book is based on the IBC, which is expected to become the most commonly used model code in the United States. Of course, building design and construction must conform to whatever code is in force in the locale where the structure is erected.

2. BUILDING REGULATIONS

Building codes are only one type of regulation affecting the design and construction of buildings. Additional requirements that may be applicable include legal and administrative regulations at the federal, state, and local levels. For example, a state may enforce environmental protection rules, while the building codes used in that state may not regulate environmental impact at all.

State and Federal Regulations

Most states have agencies that regulate building in some way. In addition to a state building code, a state government may enforce energy codes, environmental regulations, fabric flammability standards, and specific rules relating to state government buildings, institutions, hospital buildings, housing, emergency services, and other facilities.

At the national level, several federal agencies may regulate a construction project, ranging from military construction to the building of federal prisons. Certain federal agencies may also regulate or issue rules covering a specific aspect of construction, such as the safety-glazing requirement issued by the Consumer Product Safety Commission (CPSC).

Local Regulations

Local codes may include amendments to the model building code in use. These amendments usually pertain to specific concerns or needs of a geographical region or are provisions designed to alleviate local problems that are not addressed in the model codes. For example, a local amendment in a mountainous area might require a higher snow-load factor for roof design based on the local climate.

Local regulations may also affect construction and operation of hospitals, nursing homes, restaurants, schools, and similar institutions, as well as police, fire, and emergency response agencies.

3. BUILDING CODES

The IBC and each of the three previously published model codes are prescriptive as opposed to performance based. This means that the code describes specific materials and methods of construction, or how a building component or design must be designed, as opposed to how it is supposed to function. In most cases, the codes refer to nationally recognized standards of materials and testing so that, if a building component meets the test standard, it can be used. New or untested materials and construction methods can be used if they can pass the performance-based testing or if they can otherwise be shown to meet the requirements of the code.

Building codes are written to protect the health, safety, and welfare of the public. As such, all three model codes are based on the concept of the "least acceptable risk." This is the minimum level required for building and occupant safety (even though just "meeting the code" is not always the best construction for a given circumstance).

Building codes differ from zoning ordinances, easements, deed restrictions, and other regulations affecting the use and planning of land. Zoning ordinances, for example, deal with the use of a piece of property, the density of buildings within a district and on a lot, the locations of buildings on a property, floor-area ratio, building height, and parking.

Legal Basis of Codes

In the United States, the authority for adopting and enforcing building codes is one of the police powers given to the states by the Tenth Amendment to the United States Constitution. Each state, in turn, may retain those powers or delegate some of them to lower levels of government, such as counties or cities. Because of this division of power, the authority for adopting and enforcing building codes varies among the states.

Building codes are usually adopted and enforced by local governments, either by a municipality or, in the case of sparsely populated areas, a county or district. A few states write their own codes or adopt a model code statewide.

Codes are enacted as laws, just as any other local regulation would be. Before construction, a building code is enforced through the permit process, which requires that builders submit plans and specifications to the authority having jurisdiction (AHJ) for checking and approval before a building permit is issued. During construction, the AHJ conducts inspections to verify that building is proceeding according to the approved plans.

Even though code enforcement is the responsibility of the local building department or the AHJ, the architect and engineer are responsible for designing a building in conformance with all applicable codes and regulations. This is because an architect or engineer is required by registration laws to practice lawfully in order to protect the health, safety, and welfare of the public.

Adjuncts to Building Codes

In addition to building codes, there are companion codes that govern other aspects of construction. The same groups that publish the model building codes publish these companion codes. For example, the ICC also publishes the International Residential Code, the International Fire Code, the International Mechanical Code, the International Plumbing Code, and the International Zoning Code, among others.

The electrical code used by all jurisdictions is the National Electrical Code (NEC), published by the National Fire Protection Association (NFPA). In order to maintain greater uniformity in building regulations, the ICC does not publish an electrical code, but relies on the NEC.

Model codes also make extensive use of industry standards that are developed by trade associations, such as the Gypsum Association; government agencies; standards-writing organizations, such as the American Society for Testing and Materials (ASTM) and the National Fire Protection Association (NFPA); and standards-approving groups, such as the American National Standards Institute (ANSI). Standards are adopted into a building code by reference name, number, and date of latest revision. For example, most codes adopt by reference the American National Standard ICC/ANSI A117.1-1998, *Accessible and Usable Buildings and Facilities*. This standard was developed by the ICC based on previous ANSI accessibility standards and is approved by ANSI.

4. TESTING AND MATERIAL STANDARDS

All approved materials and construction assemblies referred to in building codes are required to be manufactured according to accepted methods or tested by approved agencies according to standardized testing procedures, or both. There are hundreds of standardized tests and product standards for building materials and constructions. Some of the more common ones are listed in this section. Information about additional soils tests is included in Ch. 35, and concrete testing is included in Chs. 48 and 49.

As previously stated, standards are developed by trade associations, standards-writing organizations, and government agencies. By themselves, standards have no legal standing. Only when they are referred to in a building code and that code is adopted by a governmental jurisdiction do standards become enforceable.

Standards-Writing Organizations

The American Society for Testing and Materials (ASTM) is one organization that publishes thousands of standards and test procedures that prescribe, in detail, such things as how the test apparatus must be set up, how materials must be prepared for the test, the length of the test, and other requirements. If a product manufacturer has one of its materials successfully tested, it will indicate in its product literature what tests the material has passed. Standards are developed through the work of committees of experts in a particular field. Although ASTM does not actually perform tests, its procedures and standards are used by testing agencies.

The National Fire Protection Association (NFPA) is another private, voluntary organization that develops standards related to the causes and prevention of destructive fires. NFPA publishes hundreds of codes and standards in a multivolume set that covers the entire scope of fire prevention including sprinkler systems, fire extinguishers, hazardous materials, fire fighting, and much more. As mentioned earlier in this chapter, NFPA has published its own building code, NFPA 5000™.

Other standards-writing organizations are typically industry trade groups that have an interest in a particular material, product, or field of expertise. Examples of such trade groups include the American Society of Heating, Refrigerating and Air-Conditioning Engineers (ASHRAE), the Illuminating Engineering Society (IES), the Gypsum Association (GA), the American Concrete Institute (ACI), the American Iron and Steel Institute (AISI), and the American Institute of Timber Construction (AITC), among others. There are hundreds of these construction trade organizations.

The American National Standards Institute (ANSI) is a well-know organization in the field, but unlike the other standards groups, ANSI does not develop or write standards. Instead, it approves standards developed by other organizations and works to avoid duplications between different standards. The ANSI approval process ensures industry consensus for a standard and avoids duplication of standards. For example, ANSI 108, *Specifications for Installation of Ceramic Tile*, was developed by the Tile Council of America and reviewed by a large committee of widely varying industry representatives. Although the ANSI approval processes does not necessarily represent unanimity among committee members, it requires much more than a simple majority and requires that all views and objections be considered, and that a concerted effort be made toward their resolution.

Testing Laboratories

When a standard describes a test procedure or requires one or more tests in its description of a material or product, a testing laboratory must perform the test. A standards- writing organization may also provide testing, but in most cases a Nationally Recognized Testing Laboratory (NRTL) must perform the test. An NRTL is an independent laboratory recognized by the Occupational Safety and Health Administration (OSHA) to test products to the specifications of applicable product safety standards.

One of the most well known NTRLs is Underwriters Laboratories (UL). Among other activities, UL develops standards and tests products for safety. When a product successfully passes the prescribed test, it is give a UL label. There are several types of UL labels, and each means something different.

When a complete and total product is successfully tested, it is *listed*. This means that the product passed the safety test and is manufactured under the UL follow-up services program. Such a product receives a *listed label*.

Another type of label is the *classified label*. This means that samples of the product were tested for certain types of uses only. In addition to the classified label, the product must also carry a statement specifying the conditions that were tested for. This allows field inspectors and others to determine if the product is being used correctly.

One of the most common uses of UL testing procedures is for doors and other opening protections. For example, fire doors are required to be tested in accordance with UL 10B, *Fire Tests of Door Assemblies*, and to carry a UL label. The results of UL tests and products that are listed are published in UL's *Building Materials Directory*.

Types of Tests and Standards

There are hundreds of types of tests and standards for building materials and assemblies that examine a wide range of properties from fire resistance to structural integrity to durability to stain resistance. Building codes indicate what tests or standards a particular type of material must satisfy in order to be considered acceptable for use.

Construction

ASTM has established a number of standard soils tests, which are summarized in Ch. 38, Secs. 11 through 25. Other materials testing procedures are summarized in Ch. 43.

Destructive vs. Nondestructive Tests

Construction materials testing, a form of field quality control, helps to determine conformance to project specifications during construction, thus preventing costly damages resulting from substandard materials and/or unsatisfactory materials installation.

Nondestructive testing (NDT), also called nondestructive evaluation (NDE) and nondestructive inspection (NDI), is testing that does not destroy the test object. NDE is vital for constructing and maintaining all types of components and structures. To detect different defects such as cracking and corrosion, there are different methods of testing available.

- *Magnetic particle testing* (MT) identifies surface and near-surface discontinuities in ferromagnetic materials.

- *Dye penetrant testing* (PT) identifies surface discontinuities on alloy steel, aluminum, magnesium, brass, bronze, plastic, and glass objects using color contrast and fluorescence.

- *Radiographic inspection* (RT) identifies internal defects of a variety of materials using x-rays.

- *Ultrasonic testing* (UT) is a highly specialized field that is performed in both field and laboratory settings.

While destructive testing usually provides a more reliable assessment of the state of the test object, destruction of the test object usually makes this type of test more costly to the test object's owner than nondestructive testing. Destructive testing is also inappropriate in many circumstances, such as forensic investigation. That there is a tradeoff between the cost of the test and its reliability favors a strategy in which most test objects are inspected nondestructively; destructive testing is performed on a sampling of test objects that are drawn randomly for the purpose of characterizing the testing reliability of the nondestructive test.

5. FIRE RESISTIVITY

The most important types of tests for building components are those that rate the ability of a construction assembly to prevent the passage of fire and smoke from one space to another, and tests that rate the degree of flammability of a finish material. The following summaries include common construction materials tests and fire testing for building products.

Three tests are commonly used for fire-resistive assembly ratings. These include ASTM E-119, NFPA 252, and NFPA 257.

ASTM E-119

One of the most commonly used tests for fire resistance of construction assemblies is ASTM E-119, *Standard Methods of Fire Tests of Building Construction and Materials*. This test involves building a sample of the wall or floor/ceiling assembly in the laboratory and applying a standard fire on one side of it using controlled gas burners. Monitoring devices measure temperature and other aspects of the test as it proceeds.

There are two parts to the E-119 test, the first of which measures heat transfer through the assembly. The goal of this test is to determine the temperature at which the surface or adjacent materials on the side of the assembly not exposed to the heat source will combust. The second is the "hose stream" test, which uses a high-pressure hose stream to simulate how well the assembly stands up to impacts from falling debris and the cooling and eroding effects of water. Overall, the test evaluates an assembly's ability to prevent the passage of fire, heat, and hot gases for a given amount of time. A similar test for doors is NFPA 252, *Fire Tests of Door Assemblies*.

Construction assemblies tested according to ASTM E-119 are given a rating according to time. In general terms, this rating indicates the amount of time an assembly can resist a standard test fire without failing. The ratings are 1 hr, 2 hr, 3 hr, or 4 hr. Doors and other opening assemblies can also be given 20 min, 30 min, and 45 min ratings.

NFPA 252

NFPA 252, *Fire Tests of Door Assemblies*, evaluates the ability of a door assembly to resist the passage of flame, heat, and gases. It establishes a time-based fire-endurance rating for the door assembly, and the hose stream part of the test determines if the door will stay within its frame when subjected to a standard blast from a fire hose after the door has been subjected to the fire-endurance part of the test. Similar tests include UL10B and UL10C.

NFPA 257

NFPA 257, *Standard on Fire Test for Window and Glass Block Assemblies*, prescribes specific fire and hose stream test procedures to establish a degree of fire protection in units of time for window openings in fire-resistive walls. It determines the degree of protection to the spread of fire, including flame, heat, and hot gasses.

Flammability tests for building and finish materials determine the following.

- whether or not a material is flammable, and if so, whether it simply burns with applied heat or supports combustion (adds fuel to the fire)

- the degree of flammability (how fast fire spreads across it)

- how much smoke and toxic gas it produces when ignited

6. FIRE-RESISTANCE STANDARDS

Building codes recognize that there is no such thing as a fireproof building; there are only degrees of fire resistance. Because of this, building codes specify requirements for two broad classifications of fire resistance as mentioned in the previous section: resistance of materials and assemblies, and surface burning characteristics of finish materials. While architects would consider both in the design process, engineers are concerned primarily with the former.

Construction Materials and Assemblies

In the first type of classification, the amount of fire resistance that a material or construction assembly must have is specified in terms of an hourly rating as determined by ASTM E-119 for walls, ceiling/floor assemblies, columns, beam enclosures, and similar building elements. Codes also specify what time rating doors and glazing must have as determined by NFPA 252 or NFPA 257, respectively. For example, exit-access corridors are often required to have at least a 1 hr rating, and the door assemblies in such a corridor may be required to have a 20 min rating.

Building codes typically include tables indicating what kinds of construction meet various hourly ratings. Other sources of information for acceptable construction assemblies include Underwriters Laboratories' *Building Materials Directory*, manufacturers' proprietary product literature, and other reference sources.

Various building elements must be protected with the types of construction specified in IBC Table 601 (see Table 81.1) and elsewhere in the code. When a fire-resistive barrier is built, any penetrations in the barrier must also be fire rated. This includes doors, windows, and ducts. Duct penetrations are protected with fire dampers placed in line with the wall. If a fire occurs, a fusible link in the damper closes a louver that maintains the rating of the wall. The fire resistance ratings of

Table 81.1 *Fire-Resistive Rating Requirements for Building Elements (hours) [IBC Table 601]*

building elements	type I A	type I B	type II A^e	type II B	type III A^e	type III B	type IV HT	type V A^e	type V B
structural frame^a including columns, girders, trusses	3^b	2^b	1	0	1	0	HT	1	0
bearing walls exterior^g	3	2	1	0	2	2	2	1	0
bearing walls interior	3^b	2^b	1	0	1	0	1/HT	1	0
nonbearing walls and partitions exterior	see IBC Table 602								
nonbearing walls and partitions interior^f	0	0	0	0	0	0	see IBC Sec. 602.4.6	0	0
floor construction including supporting beams and joists	2	2	1	0	1	0	HT	1	0
roof construction including supporting beams and joists	$1^1/_2{}^c$	$1^{c,d}$	$1^{c,d}$	$0^{c,d}$	$1^{c,d}$	$0^{c,d}$	HT	$1^{c,d}$	0

For SI: 1 ft = 304.8 mm.

^{a}The structural frame shall be considered to be the columns and the girders, beams, trusses and spandrels having direct connections to the columns and bracing members designed to carry gravity loads. The members of floor or roof panels which have no connection to the columns shall be considered secondary members and not a part of the structural frame.

^{b}Roof supports: Fire-resistance ratings of structural frame and bearing walls are permitted to be reduced by 1 hr where supporting a roof only.

^{c}Except in Factory-Industrial (F-I), Hazardous (H), Mercantile (M) and Moderate Hazard Storage (S-1) occupancies, fire protection of structural members shall not be required, including protection of roof framing and decking where every part of the roof construction is 20 ft or more above any floor immediately below. Fire-retardant-treated wood members shall be allowed to be used for such unprotected members.

^{d}In all occupancies, heavy timber shall be allowed where a 1 hr or less fire-resistance rating is required.

^{e}An approved automatic sprinkler system in accordance with IBC Sec. 903.3.1.1 shall be allowed to be substituted for 1 hr fire-resistance-rated construction, provided such system is not otherwise required by other provisions of the code or used for an allowable area increase in accordance with IBC Sec. 506.3 or an allowable height increase in accordance with IBC Sec. 504.2. The 1 hr substitution for the fire resistance of exterior walls shall not be permitted.

^{f}Not less than the fire-resistance rating required by other sections of this code.

^{g}Not less than the fire-resistance rating based on fire separation distance (see IBC Table 602).

International Building Code 2006. Copyright 2006. Washington, DC: International Code Council, Inc. Reproduced with permission. All rights reserved. www.iccsafe.org.

Construction

existing building components are important in determining the construction type of the building. This is discussed in more detail in a later section on classification based on construction type.

It is important to note that many materials by themselves do not create a fire-rated barrier. It is the construction assembly of which they are a part that is fire resistant. A 1 hr rated suspended ceiling, for example, must use rated ceiling tile, but it is the assembly of tile, the suspension system, and the structural floor above that carries the 1 hr rating. In a similar way, a 1 hr rated partition may consist of a layer of $^5/_8$ in (15.9 mm) Type X gypsum board attached to both sides of a wood or metal stud according to certain conditions. A single piece of gypsum board cannot have a fire-resistance rating by itself, except under special circumstances defined by the IBC.

7. ADMINISTRATIVE REQUIREMENTS OF BUILDING CODES

All building codes include a chapter dealing with the administration of the code itself. Provisions that are normally part of the administrative chapter include what codes apply, the duties and powers of the building official, the permit process, what information is required on construction documents, fees for services, how inspections are handled, and what kinds of inspections are required. Also included are the requirements for issuing a certificate of occupancy, instruction on how violations are handled, and the provisions for appealing the decisions of the building official concerning the application and interpretation of the code.

8. REQUIREMENTS BASED ON OCCUPANCY

Occupancy refers to the type of use of a building or interior space, such as an office, a restaurant, a private residence, or a school. Uses are grouped by occupancy based on similar life-safety characteristics, fire hazards, and combustible contents.

The idea behind occupancy classification is that some uses are more hazardous than others. For example, a building where flammable liquids are used is more dangerous than a single-family residence. Also, residents of a nursing home will have more trouble exiting than will young school children who have participated in fire drills. In order to achieve equivalent safety in building design, each occupancy group varies by fire protection requirements, area and height limitations, type of construction restrictions (as described in Sec. 9), and means of egress.

There are additional requirements for special occupancy types that include covered mall buildings, high-rise buildings, atriums, underground buildings, motor-vehicle-related occupancies, hazardous occupancies, and institutional occupancies, among others.

Occupancy Groups

Every building or portion of a building is classified according to its use and is assigned an occupancy group. This is true of the IBC as well as Canadian model codes and the three former U.S. model codes still used in some jurisdictions. The IBC classifies occupancies into 10 major groups.

A	assembly
B	business
E	educational
F	factory and industrial
H	hazardous
I	institutional
M	mercantile
R	residential
S	storage
U	utility

Six of these groups are further divided into categories to distinguish subgroups that define the relative hazard of the occupancy. For example, in the assembly group, an A-1 occupancy includes assembly places, usually with fixed seats, used to view performing arts or motion pictures, while an A-2 occupancy includes places designed for food and/or drink consumption. Table 81.2 shows a brief summary of the occupancy groups and subgroups and gives some examples of each. This table is not complete, and the IBC should be consulted for specific requirements.

Knowing the occupancy classification is important in determining other building requirements, such as the maximum area, the number of floors allowed, and how the building must be separated from other structures. The occupancy classification also affects the following.

- calculation of occupant load
- egress design
- interior finish requirements
- use of fire partitions and fire barriers
- fire detection and suppression systems
- ventilation and sanitation requirements
- other special restrictions particular to any given classification

Mixed Occupancy and Occupancy Separation

When a building or area of a building contains two or more occupancies, it is considered to be of *mixed occupancy* or *mixed use*. This is quite common. For instance, the design of a large office space can include office occupancy (a B occupancy) adjacent to an auditorium used for training, which would be an assembly occupancy (A occupancy). Each occupancy must be separated from other occupancies with a fire barrier of the hourly rating (shown in Table 81.3) as defined

Table 81.2 Occupancy Groups Summary

occupancy	description	examples
A-1	assembly with fixed seats for viewing of performances or movies	movie theaters, live performance theaters
A-2	assembly for food and drink consumption	bars, restaurants, clubs
A-3	assembly for worship, recreation, etc., not classified elsewhere	libraries, art museums, conference rooms with more than 50 occupants
A-4	assembly for viewing of indoor sports	arenas
A-5	assembly for outdoor sports	stadiums
B	business for office or service transactions	offices, banks, educational above the 12th grade, post office
E	educational by > 5 people through 12th grade	grade schools, high schools, day cares with more than 5 children ages greater than 2.5 yr
F-1	factory moderate hazard	see code
F-2	factory low hazard	see code
H	hazardous—see code	see code
I-1	> 16 ambulatory people on 24 hr basis	assisted living, group home, convalescent facilities
I-2	medical care on 24 hr basis	hospitals, skilled care nursing
I-3	> 5 people restrained	jails, prisons, reformatories
I-4	daycare for > 5 adults or infants (< 2.5 yr)	daycare for infants
M	mercantile	department stores, markets, retail stores, drug stores, sales rooms
R-1	residential for transient lodging	hotels and motels
R-2	residential with 3 or more units	apartments, dormitories, condominiums, convents
R-3	1 or 2 dwelling units with attached uses or child care < 6, less than 24 hr care	bed and breakfast, small child care
R-4	residential assisted living where number of occupants > 5 but < 16	small assisted living
S	storage—see code	see code
U	utility—see code	see code
dwellings	must use International Residential Code	

Note: This is just a brief summary of the groups and examples of occupancy groups. Refer to the IBC for a complete list or check with local building officials when a use is not clearly stated or described in the code.

by the particular code that applies. The idea is to increase the fire protection between occupancies as the relative hazard increases.

9. CLASSIFICATION BASED ON CONSTRUCTION TYPE

Every building is classified into one of five major types of construction based on the fire-resistance rating (protection) of its major construction components. The purpose of this classification is to protect the structural elements of a building from fire and collapse, and to divide the building into compartments so that a fire in one area will be contained long enough to allow people to evacuate the building and firefighters to arrive.

The major construction components, under the IBC, include the structural frame, bearing walls, nonbearing walls, exterior walls, floor construction, and roof construction. The five types of construction are Type I, II, III, IV, and V. Type I buildings are the most fire resistive, while Type V are the least fire resistive. Type I and II buildings are noncombustible, while Types III, IV, and V are considered combustible. The building type categories and fire-resistance rating requirements for each building element are shown in Table 81.1. The fire-resistance requirements for exterior, nonbearing walls are based on the distance from the building to the property line, the type of construction, and the occupancy group as shown in Table 81.4. Detailed requirements for the various construction types are contained in Ch. 6 of the IBC.

Construction

Table 81.3 Incidental Use Areas [IBC Table 302.1.1]

room or area	separation and/or protection
furnace room where any piece of equipment is over 400,000 Btu/hr input	1 hr or provide automatic fire extinguishing system
rooms with any boiler over 15 psi and 10 horsepower	1 hr or provide automatic fire extinguishing system
refrigerant machinery rooms	1 hr or provide automatic sprinkler system
parking garage (IBC Sec. 406.2)	2 hr; or 1 hr and provide automatic fire extinguishing system
hydrogen cutoff rooms, not classified as Group H	1 hr in Group B, F, M, S, and U occupancies. 2 hr in Group A, E, I, and R occupancies
incinerator rooms	2 hr and automatic sprinkler system
paint shops, not classified as Group H, located in occupancies other than Group F	2 hr; or 1 hr and provide automatic fire extinguishing system
laboratories and vocational shops, not classified as Group H, located in Group E or I-2 occupancies	1 hr or provide automatic fire extinguishing system
laundry rooms over 100 ft^2	1 hr or provide automatic fire extinguishing system
storage rooms over 100 ft^2	1 hr or provide automatic fire extinguishing system
Group I-3 cells equipped with padded surfaces	1 hr
Group I-2 waste and linen collection rooms	1 hr
waste and linen collection rooms over 100 ft^2	1 hr or provide automatic fire extinguishing system
stationary lead-acid battery systems having a liquid capacity of more than 100 gal used for facility standby power, emergency power, or uninterrupted power supplies	1 hr in Group B, F, M, S, and U occupancies. 2 hr in Group A, E, I, and R occupancies

For SI: 1 ft^2 = 0.0929 m^2, 1 lbf/in^2 = 6.9 kPa, 1 British thermal unit = 0.293 watts, 1 horsepower = 746 W, 1 gal = 3.785 L.

International Building Code 2006. Copyright 2006. Washington, DC: International Code Council, Inc. Reproduced with permission. All rights reserved. www.iccsafe.org.

In combination with occupancy groups, building type sets limits on the area and height of buildings. For example, Type I buildings of any occupancy (except certain hazardous occupancies) can be of unlimited area and height, while Type V buildings are limited to only a few thousand square feet in area and one to three stories in height, depending on their occupancy. The type and amount of combustibles due to the building's use and construction affect its safety. Limiting height and area based on construction type and occupancy recognizes that it becomes more difficult to fight fires, provide time for egress, and rescue people as buildings get larger and higher.

Table 81.4 Fire-Resistance Rating Requirements for Exterior Walls Based on Fire Separation Distance^{a,e} [IBC Table 602]

fire separation distance, x (ft)	type of construc-tion	group H	group F-1, M, S-1	group A, B, E, F-2, I, R^b, S-2, U
$x < 5^c$	all	3	2	1
$5 \leq x < 10$	IA	3	2	1
	others	2	1	1
$10 \leq x < 30$	IA, IB	2	1	1^d
	IIB, VB	1	0	0
	others	1	1	1^d
$x \geq 30$	all	0	0	0

For SI: 1 ft = 304.8 mm

^{a}Load-bearing exterior walls must also comply with the fire-resistance rating requirements of IBC Table 601.

^{b}Group R-3 and Group U when used as accessory to Group R-3, as applicable in IBC Sec. 101.2, are not required to have a fire-resistance rating where the fire separation distance is 3 ft or more.

^{c}See IBC Sec. 503.2 for party walls.

^{d}Open parking garages complying with Sec. 406 are not required to have a fire-resistance rating.

^{e}The fire-resistance rating of an exterior wall is determined based on the fire separation distance of the exterior wall and the story in which the wall is located.

10. ALLOWABLE FLOOR AREA AND HEIGHTS OF BUILDINGS

Chapter 5 of the IBC sets forth the requirements for determining maximum height (in stories as well as feet or meters) and area of a building based on construction type. It also gives the allowed occupancy, and then presents conditions under which the height and area may be increased. The concept is that the more hazardous a building is the smaller it should be, making it easier to fight a fire and easier for occupants to exit in an emergency.

Table 503 of the IBC and similar tables in the other model codes give the maximum allowable area, per floor, of a building. A portion of IBC Table 503 is reproduced here in Table 81.5. This basic area can be multiplied by the number of stories up to a maximum of three stories under the IBC and up to two stories under the UBC.

If the building is equipped throughout with an approved automatic sprinkler system, the area and height can be increased. For one-story buildings, the area can be tripled, and for multistory buildings, the area can be doubled. The maximum height can be increased by 20 ft (6 m), and the number of stories can be increased by one. Both area and height increases are allowed in combination.

If more than 25% of the building's perimeter is located on a public way or open space, the basic allowable area may be increased according to various formulas. Except for Group H, Divisions 1, 2, and 5 (hazardous occupancies), a Type I building may be of unlimited floor area and unlimited height, while other construction types are limited.

The basic allowable height and building area table (Table 81.5) can be used in one of two ways. If the occupancy and construction type are known, simply find the intersection of the row designating "occupancy" and the column designating "type," read the permitted area or height, and then increase the areas according to the percentages allowed for sprinklers and perimeter open space. More often, the occupancy and required floor area are known from the building program and a determination must be made on the required construction type that will allow construction of a building that meets the client's size needs. This is typically part of the pre-design work of a project.

Table 81.5 Allowable Height and Building Areas[a] [as per IBC Table 503]
(Height limitations shown as stories and feet above grading plane. Area limitations as determined by the definition of "area, building" per floor.)

| | | type of construction[a] | | | | | | | | |
| | | type I | | type II | | type III | | type IV | type V | |
group	height (ft or stories, S) area (ft^2, A)	A	B	A	B	A	B	HT	A	B
		UL	160	65	55	65	55	65	50	40
A-1	S	UL	5	3	2	3	2	3	2	1
	A	UL	UL	15,500	8,500	14,000	8,500	15,000	11,500	5,500
A-2	S	UL	11	3	2	3	2	3	2	1
	A	UL	UL	15,500	9,500	14,000	9,500	15,000	11,500	6,000
A-3	S	UL	11	3	2	3	2	3	2	1
	A	UL	UL	15,500	9,500	14,000	9,500	15,000	11,500	6,000
A-4	S	UL	11	3	2	3	2	3	2	1
	A	UL	UL	15,500	9,500	14,000	9,500	15,000	11,500	6,000
A-5	S	UL	UL	UL	UL	UL	UL	UL	UL	UL
	A	UL	UL	UL	UL	UL	UL	UL	UL	UL
B	S	UL	11	5	4	5	4	5	3	2
	A	UL	UL	37,500	23,000	28,500	19,000	36,000	18,000	9,000
E	S	UL	5	3	2	3	2	3	1	1
	A	UL	UL	26,500	14,500	23,500	14,500	25,500	18,500	9,500
F-1	S	UL	11	4	2	3	2	4	2	1
	A	UL	UL	25,000	15,500	19,000	12,000	33,500	14,000	8,500
F-2	S	UL	11	5	3	4	3	5	3	2
	A	UL	UL	37,500	23,000	28,500	18,000	50,500	21,000	13,000
H-1	S	1	1	1	1	1	1	1	1	NP
	A	21,000	16,500	11,000	7,000	9,500	7,000	10,500	7,500	NP
H-2[d]	S	UL	3	2	1	2	1	2	1	1
	A	21,000	16,500	11,000	7,000	9,500	7,000	10,500	7,500	3,000
H-3[d]	S	UL	6	4	2	4	2	4	2	1
	A	UL	60,000	26,500	14,000	17,500	13,000	25,500	10,000	5,000
H-4	S	UL	7	5	3	5	3	5	3	2
	A	UL	UL	37,500	17,500	28,500	17,500	36,000	18,000	6,500
H-5	S	4	4	3	3	3	3	3	3	2
	A	UL	UL	37,500	23,000	28,500	19,000	36,000	18,000	9,000
I-1	S	UL	9	4	3	4	3	4	3	2
	A	UL	55,000	19,000	10,000	16,500	10,000	18,000	10,500	4,500
I-2	S	UL	4	2	1	1	NP	1	1	NP
	A	UL	UL	15,000	11,000	12,000	NP	12,000	9,500	NP
I-3	S	UL	4	2	1	2	1	2	2	1
	A	UL	UL	15,000	10,000	10,500	7,500	12,000	7,500	5,000
I-4	S	UL	5	3	2	3	2	3	1	1
	A	UL	60,500	26,500	13,000	23,500	13,000	25,500	18,500	9,000
M	S	UL	11	4	4	4	4	4	3	1
	A	UL	UL	21,500	12,500	18,500	12,500	20,500	14,000	9,000
R-1	S	UL	11	4	4	4	4	4	3	2
	A	UL	UL	24,000	16,000	24,000	16,000	20,500	12,000	7,000
R-2	S	UL	11	4	4	4	4	4	3	2
	A	UL	UL	24,000	16,000	24,000	16,000	20,500	12,000	7,000
R-3	S	UL	11	4	4	4	4	4	3	3
	A	UL	UL	UL	UL	UL	UL	UL	UL	UL
R-4	S	UL	11	4	4	4	4	4	3	2
	A	UL	UL	24,000	16,000	24,000	16,000	20,500	12,000	7,000
S-1	S	UL	11	4	3	3	3	4	3	1
	A	UL	48,000	26,000	17,500	26,000	17,500	25,500	14,000	9,000
S-2[b,c]	S	UL	11	5	4	4	4	5	4	2
	A	UL	79,000	39,000	26,000	39,000	26,000	38,500	21,000	13,500
U[c]	S	UL	5	4	2	3	2	4	2	1
	A	UL	35,500	19,000	8,500	14,000	8,500	18,000	9,000	5,500

For SI: 1 ft = 304.8 mm, 1 ft^2 = 0.0929 m^2. UL = Unlimited, NP = Not permitted.
a. See the following sections for general exceptions to IBC Table 503:
  1. IBC Section 504.2, Allowable height increase due to automatic sprinkler system installation.
  2. IBC Section 506.2, Allowable area increase due to street frontage.
  3. IBC Section 506.3, Allowable area increase due to automatic sprinkler system installation.
b. For open parking structures, see IBC Sec. 406.3.
c. For private garages, see IBC Sec. 406.1.
d. See IBC Sec. 415.5 for limitations.

82 Construction and Jobsite Safety

1. INTRODUCTION

Common sense can go a long way in preventing many jobsite accidents. However, scheduling, economics, lack of concern, carelessness, and laxity are often more likely to guide actions than common sense. For that reason, the "science" of accident prevention is highly regulated. Workers' safety is regulated by the federal Occupational Safety and Health Act (OSHA). State divisions (e.g., Cal-OSHA) are charged with enforcing federal and state safety regulations in the United States. Surface and underground mines, which share many hazards with the construction industry and which have others specific to them, are regulated by the federal Mine Safety and Health Act (MSHA). Other countries may have more restrictive standards.

All of the federal OSHA regulations are published in the Congressional Federal Register (CFR). The two main categories of standards are the "1910 standards," which apply to general industry, and the "1926 standards," which apply to the construction industry.

This chapter covers only a minuscule portion of the standards and safety issues facing civil engineers and construction workers. Only the briefest descriptions of these complex issues and the regulations that govern them are provided.

2. TRENCHING AND EXCAVATION

Soils are classified as stable rock and types A, B, or C, with type C being the most unstable. Type A soils are cohesive soils with unconfined compressive strengths of 1.5 tons per square foot (tsf) (144 kPa) or greater. Type B soils are cohesive soils with compressive strengths of 0.5 to 1.5 tsf (48 to 144 kPa) and some granular soils. Type C soils are cohesive soils with compressive strengths less than 0.5 tsf (48 to 144 kPa), and some granular soils. Soils classified as types A and B may need to be reclassified as type C following rain and flooding [OSHA 1926.652, App. A to Subpart P].

Except for excavations entirely in stable rock, excavations deeper than 5 ft (1.5 m) in all types of earth must be protected from cave-in and collapse [OSHA 1926.652]. Excavations less than 5 ft (1.5 m) deep are usually exempt but may also need to be protected when inspection indicates that hazardous ground movement is possible.

Timber and aluminum shoring (hydraulic, pneumatic, and screw-jacked) and trench shields that meet the requirements of OSHA 1926.652 App. C to Subpart P may be used in excavations up to 20 ft (6 m) deep.

Sloping and benching the trench walls may be substituted for shoring. For long-term use, the maximum allowable slope (H:V) for type A soils is $^3/_4$:1 (53° from the horizontal). For type B soils, it is 1:1 (45°). For type C soils, it is $1^1/_2$:1 (34°). These slopes shall be reduced 50% if the soil shows signs of distress. Sloped walls in excavations deeper than 20 ft (6 m) must be designed by a professional engineer. Greater slopes are permitted for short-term usage in excavations less than 12 ft (3.67 m) deep.

In trenches 4 ft (1.2 m) deep or more, ladders, stairways, or ramps are required with a maximum lateral spacing of 25 ft (7.5 m) [OSHA 1926.651(b)(2)].

Spoils and other equipment that could fall into a trench or excavation must be kept at least 2 ft (0.6 m) from the edge of a trench unless secured in some other fashion [OSHA 1926.651(l)].

3. CHEMICAL HAZARDS

OSHA's Hazard Communication Standard [OSHA 1910.1200] requires that the dangers of all chemicals purchased, used, or manufactured be known to employees. The hazards are communicated in a variety of ways, including labeling containers, training employees, and providing ready access to *material safety data sheets* (MSDSs).

OSHA has suggested a nonmandatory standard form for the MSDS, but other forms are acceptable as long as they contain the same (or more) information. The information contained on an MSDS consists of following categories.

Chemical identity: the identity of the substance as it appears on the label.

Construction

Section I: Manufacturer's name and contact information: manufacturer's name, address, telephone number, and emergency phone number; date the MSDS was prepared; and an optional signature of the preparer.

Section II: Hazardous ingredients/identity information: list of the hazardous components by chemical identity and other common names; OSHA *permissible exposure limit* (PEL), ACGIH *threshold level value* (TLV), and other recommended exposure limits; percentage listings of the hazardous components is optional.

Section III: Physical/chemical characteristics: boiling point, vapor pressure, vapor density, specific gravity, melting point, evaporation rate, solubility in water, physical appearance, and odor.

Section IV: Fire and explosion hazard data: flash point (and method used to determine it), flammability limits, extinguishing media, special fire-fighting procedures, and unusual fire and explosion hazards.

Section V: Reactivity data: stability, conditions to avoid, incompatibility (materials to avoid), hazardous decomposition or by-products, and hazardous polymerization (and conditions to avoid).

Section VI: Health hazard data: routes of entry into the body (inhalation, skin, ingestion), health hazards (acute = immediate, or chronic = builds up over time), carcinogenicity (NTP, IARC monographs, OSHA regulated), signs and symptoms of exposure, medical conditions generally aggravated by exposure, and emergency and first-aid procedures.

Section VII: Precautions for safe handling and use: steps to be taken in case the material is released or spilled, waste disposal method, precautions to be taken in handling or storage, and other precautions.

Section VIII: Control measures: respiratory protection (type to be specified), ventilation (local, mechanical exhaust, special, or other), protective gloves, eye protection, other protective clothing or equipment, and work/hygienic practices.

Section IX: Special precautions and comments: safe storage and handling, types of labels or markings for containers, and Department of Transportation (DOT) policies for handling the material.

4. CONFINED SPACES AND HAZARDOUS ATMOSPHERES

Employees entering confined spaces (e.g., excavations, sewers, tanks) must be properly trained, supervised, and equipped. Atmospheres in confined spaces must be monitored for oxygen content and other harmful contaminants. Oxygen content must be maintained at 19.5% or higher unless a breathing apparatus is provided. Employees entering deep confined excavations must wear harnesses with lifelines [OSHA 1926.651].

5. POWER LINE HAZARDS

Employees operating cranes or other overhead material-handling equipment (e.g., concrete boom trucks, backhoe arms, and raised dumptruck boxes) must be aware of the possibility of inadvertent power line contact. Prior to operation, the site must be thoroughly inspected for the danger of power line contact. OSHA provides specific minimum requirements for safe operating distances. For example, for lines of 50 kV or less, all parts of the equipment must be kept at least 10 ft (3 m) from the power line [OSHA 1926.550(a)(15)]. A good rule of thumb for voltages greater than 50 kV is a clearance of 35 ft (10.5 m). However, the exact OSHA clearance requirement can be calculated from Eq. 82.1.

$$\text{line clearance} = 3 \text{ m} + (10.2 \text{ mm})(V_{kV} - 50 \text{ kV})$$
$$\text{[SI]} \quad 82.1(a)$$

$$\text{line clearance} = 10 \text{ ft} + (0.4 \text{ in})(V_{kV} - 50 \text{ kV})$$
$$\text{[U.S.]} \quad 82.1(b)$$

6. FALL AND IMPACT PROTECTION

Fall protection can take the form of barricades, walkways, bridges (with guardrails), nets, and fall-arrest systems. Personal fall-arrest systems include lifelines, lanyards, and deceleration devices. Such equipment is attached to an anchorage at one end and to the body-belt or body hardness at the other. All equipment is to be properly used, certified, and maintained. Employees are to be properly trained in the equipment's use and operation.

Employees must be protected from impalement hazards from exposed rebar [OSHA 1926.701(b)]. A widely used method for covering rebar ends has been plastic *mushroom caps*, often orange or yellow in color. However, OSHA no longer considers plastic caps adequate for anything more than scratch protection. Commercially available steel-reinforced caps and wooden troughs capable of withstanding a 250 lbm/10 ft drop test without breakthrough can still be used.

Head protection, usually in the form of a helmet (*hardhat*), is part of a worker's *personal protective equipment* (PPE). Head protection is required where there is a danger of head injuries from impact, flying or falling objects, electrical shock, or burns [OSHA 1910.132(a) and (c)]. Head protection should be nonconductive when there is electrical or thermal danger [OSHA 1910.335(a)(1)(iv)].

7. NOISE

OSHA sets maximum limits on daily sound exposure. The "all-day" 8 hr noise level limit is 90 dBA. This is higher than the maximum level permitted in other countries (e.g., 85 dBA in Germany and Japan). In the

United States, employees may not be exposed to steady sound levels above 115 dBA, regardless of the duration. Impact sound levels are limited to 140 dBA.

Hearing protection, educational programs, periodic examinations, and other actions are required for workers whose 8 hr exposure is more than 85 dBA or whose noise dose exceeds 50% of the *action levels.*

Table 82.1 *Typical Permissible Noise Exposure Levels[a]*

sound level (dBA)	exposure (hr/day)
90	8
92	6
95	4
97	3
100	2
102	$1\frac{1}{2}$
105	1
110	$\frac{1}{2}$
115	$\frac{1}{4}$ or less

[a]without hearing protection
Source: OSHA, Sec. 1910.95, Table G-16

8. SCAFFOLDS

Scaffolds are any temporary elevated platform (supported or suspended) and its supporting structure (including points of anchorage), used for supporting employees, materials, or both. Construction and use of scaffolds is regulated in detail by OSHA Std. 1926.451. A few of the regulations are summarized in the following paragraphs.

Each employee who performs work on a scaffold must be trained by a person qualified to recognize the hazards associated with the type of scaffold used and to understand the procedures to control or minimize those hazards. The training shall include such topics as the nature of any electrical hazards, fall hazards, falling object hazards, the maintenance and disassembly of the fall protection systems, the use of the scaffold, handling of materials, the capacity, and the maximum intended load.

Fall protection (guardrail systems and personal fall arrest systems) must be provided for each employee on a scaffold more than 10 ft (3.1 m) above a lower level.

Each scaffold and scaffold component must have the capacity to support without failure its own weight and at least 4 times the maximum intended load applied or transmitted to it. Suspension ropes and connecting hardware must support 6 times the intended load. Scaffolds and scaffold components may not be loaded in excess of their maximum intended loads or rated capacities, whichever is less.

The scaffold platform must be planked or decked as fully as possible. It cannot deflect more than $1/_{60}$ of the span when loaded.

The work area for each scaffold platform and walkway must be at least 18 in (46 cm) wide. When the work area must be less than 18 in (46 cm) wide, guardrails and/or personal fall arrest systems must still be used.

Access must be provided when the scaffold platforms are more than 2 ft (0.6 m) above or below a point of access. Direct access is acceptable when the scaffold is not more than 14 in (36 cm) horizontally and not more than 24 in (61 cm) vertically from the other surfaces. Cross braces cannot be used as a means of access.

A competent person with the authority to require prompt corrective action is required to inspect the scaffold, scaffold components, and ropes on suspended scaffolds before each work shift and after any occurrence which could affect the structural integrity.

Construction

Topic VIII: Systems, Management, and Professional

Systems, Mgmt,
and Professional

83 Electrical Systems and Equipment

Nomenclature

a	turns ratio	–	–
A	area	ft^2	m^2
B	susceptance	S	S
C	capacitance	F	F
d	diameter	ft	m
E	energy usage (demand)	kW-hr	kW·h
f	linear frequency	Hz	Hz
G	conductance	S	S
i	varying current	A	A
I	current	A	A
l	length	ft	m
L	inductance	H	H
n	speed	rpm	rpm
N	number of turns	–	–
p	number of poles	–	–
pf	power factor	–	–
P	power	W	W
Q	charge	C	C
Q	reactive power	VAR	VAR
R	resistance	Ω	Ω
s	slip	–	–
sf	service factor	–	–
S	apparent power	VA	VA
SR	speed regulation	–	–
t	time	sec	s
T	period	sec	s
T	torque	ft-lbf	N·m
v	varying voltage	V	V
V	voltage	V	V
VR	voltage regulation	–	–
X	reactance	Ω	Ω
Y	admittance	S	S
Z	impedance	Ω	Ω

Symbols

η	efficiency	–	–
θ	phase angle	rad	rad
ρ	resistivity	$\Omega\text{-ft}$	$\Omega\cdot\text{m}$
σ	conductivity	$1/\Omega\text{-ft}$	$1/\Omega\cdot\text{m}$
ϕ	angle	rad	rad
ω	angular frequency	rad/sec	rad/s

Subscripts

a	armature
ave	average
C	capacitive
Cu	copper
e	equivalent
eff	effective

f field
l line
L inductive
m maximum
p phase or primary
R resistive
s secondary
t total

1. ELECTRIC CHARGE

The charge on an electron is one *electrostatic unit* (esu). Since an esu is very small, electrostatic charges are more conveniently measured in *coulombs* (C). One coulomb is approximately 6.24×10^{18} esu. Another unit of charge, the *faraday*, is sometimes encountered in the description of ionic and plating processes. One faraday is equal to one mole of electrons, approximately 96,500 C.

2. CURRENT

Current, I, is the movement of electrons. By historical convention, current moves in a direction opposite to the flow of electrons (i.e., current flows from the positive terminal to the negative terminal). Current is measured in *amperes* (A) and is the time rate change in charge.

$$I = \frac{dQ}{dt} \qquad \text{83.1}$$

3. VOLTAGE SOURCES

A net *voltage*, V, causes electrons to move, hence the common synonym *electromotive force* (emf).[1] With *direct-current* (DC) voltage sources, the voltage may vary in amplitude, but not in polarity. In simple problems where a battery serves as the voltage source, the magnitude is also constant.

With *alternating-current* (AC) voltage sources, the magnitude and polarity both vary with time. Due to the method of generating electrical energy, AC voltages are typically sinusoidal.

4. RESISTIVITY AND RESISTANCE

Resistance is the property of a *resistor* or resistive circuit to impede current flow.[2] A circuit with zero resistance is a *short circuit*, whereas an *open circuit* has infinite resistance. Adjustable resistors are known as *potentiometers* and *rheostats*.

---

[1] The symbol E (derived from the name electromotive force) has been commonly used to represent voltage induced by electromagnetic induction.
[2] *Resistance* is not the same as *inductance* (Sec. 20), which is the property of a device to impede a *change* in current flow.

Resistance of a circuit element depends on the *resistivity*, ρ (in Ω-in or $\Omega \cdot$cm), of the material and the geometry of the current path through the element. The area, A, of circular conductors is often measured in *circular mils*, abbreviated *cmils*, the area of a 0.001 in diameter circle.

$$R = \frac{\rho l}{A} \qquad \text{83.2}$$

$$A_{\text{cmils}} = \left(\frac{d_{\text{in}}}{0.001} \right)^2 \qquad \text{83.3}$$

$$A_{\text{in}^2} = (7.854 \times 10^{-7})(A_{\text{cmils}}) \qquad \text{83.4}$$

Resistivity depends on temperature. For most conductors, resistivity increases linearly with temperature. The resistivity of standard *IACS (International Annealed Copper Standard)* copper wire at $20°$C is approximately

$$
\begin{aligned}
\rho_{\text{IACS Cu},20°\text{C}} &= 1.7241 \times 10^{-6} \ \Omega \cdot \text{cm} \\
&= 0.67879 \times 10^{-6} \ \Omega\text{-in} \\
&= 10.371 \ \Omega\text{-cmil/ft} \qquad \text{83.5}
\end{aligned}
$$

Example 83.1

What is the resistance of a parallelepiped (1 cm $\times$ 1 cm $\times$ 1 m long) of IACS copper if current flows between the two smaller faces?

Solution

From Eqs. 83.2 and 83.5, the resistance is

$$
\begin{aligned}
R = \frac{\rho l}{A} &= \frac{(1.7241 \times 10^{-6} \ \Omega \cdot \text{cm})(1 \ \text{m}) \left(100 \ \frac{\text{cm}}{\text{m}} \right)}{(1 \ \text{cm})(1 \ \text{cm})} \\
&= 1.724 \times 10^{-4} \ \Omega
\end{aligned}
$$

5. CONDUCTIVITY AND CONDUCTANCE

The reciprocals of resistivity and resistance are *conductivity* (σ) and *conductance* (G), respectively. The unit of conductance is the *siemens* (S).[3]

$$\sigma = \frac{1}{\rho} \qquad \text{83.6}$$

$$G = \frac{1}{R} \qquad \text{83.7}$$

Percent conductivity is the ratio of a substance's conductivity to the conductivity of standard IACS copper. (See Sec. 4.)

$$
\begin{aligned}
\% \ \text{conductivity} &= \frac{\sigma}{\sigma_{\text{Cu}}} \times 100\% \\
&= \frac{\rho_{\text{Cu}}}{\rho} \times 100\% \qquad \text{83.8}
\end{aligned}
$$

---

[3] The siemens is the inverse of an ohm and is the same as the obsolete unit, the *mho*.

6. OHM'S LAW

The *voltage drop*, also known as the *IR drop*, across a circuit or circuit element with resistance R is given by *Ohm's law*.[4]

$$V = IR \quad \text{[DC circuits]} \qquad 83.9$$

$$v(t) = i(t)R \quad \text{[AC circuits]} \qquad 83.10$$

7. POWER IN DC CIRCUITS

The *power*, P (in watts), dissipated across two terminals with resistance R and voltage drop V is given by *Joule's law*, Eq. 83.11.

$$P = IV = I^2 R = \frac{V^2}{R} = V^2 G \quad \text{[DC circuits]} \qquad 83.11$$

$$P(t) = i(t)v(t) = i(t)^2 R = \frac{v(t)^2}{R} = v(t)^2 G$$
$$\text{[AC circuits]} \qquad 83.12$$

8. ELECTRICAL CIRCUIT SYMBOLS

Figure 83.1 illustrates symbols typically used to diagram electrical circuits in this book.

Figure 83.1 *Symbols for Electrical Circuit Elements*

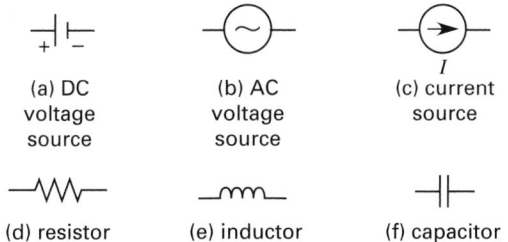

(a) DC voltage source

(b) AC voltage source

(c) current source

(d) resistor

(e) inductor

(f) capacitor

(g) transformer

9. RESISTORS IN COMBINATION

Resistors in series are added to obtain the total (equivalent) resistance of a circuit.

$$R_e = R_1 + R_2 + R_3 + \cdots \quad \text{[series]} \qquad 83.13$$

Resistors in parallel are combined by adding their reciprocals. This is a direct result of the fact that conductances in parallel add.

$$G_e = G_1 + G_2 + G_3 + \cdots \quad \text{[parallel]} \qquad 83.14$$

$$\frac{1}{R_e} = \frac{1}{R_1} + \frac{1}{R_2} + \frac{1}{R_3} + \cdots \quad \text{[parallel]} \qquad 83.15$$

---

[4]This book uses the convention that uppercase letters represent fixed, maximum, or effective values, and lowercase letters represent values that change with time.

For two resistors in parallel, the equivalent resistance is

$$R_e = \frac{R_1 R_2}{R_1 + R_2} \quad \text{[two parallel resistors]} \qquad 83.16$$

10. SIMPLE SERIES CIRCUITS

Figure 83.2 illustrates a simple series DC circuit and its equivalent circuit.

- The current is the same through all circuit elements.

$$I = I_{R_1} = I_{R_2} = I_{R_3} \qquad 83.17$$

- The equivalent resistance is the sum of the individual resistances.

$$R_e = R_1 + R_2 + R_3 \qquad 83.18$$

- The equivalent applied voltage is the sum of all voltage sources (polarities considered).

$$V_e = \pm V_1 \pm V_2 \qquad 83.19$$

- The sum of all of the voltage drops across the components in the circuit (a *loop*) is equal to the equivalent applied voltage. This fact is known as *Kirchhoff's voltage law*.

$$V_e = \Sigma I R_j = I \Sigma R_j = I R_e \qquad 83.20$$

Figure 83.2 *Simple Series DC Circuit and Its Equivalent*

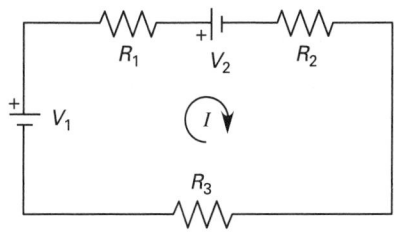

(a) original series circuit

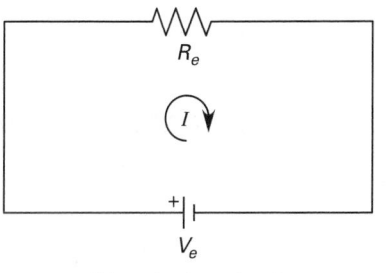

(b) equivalent circuit

11. SIMPLE PARALLEL CIRCUITS

Figure 83.3 illustrates a simple parallel DC circuit with one voltage source and its equivalent circuit.

- The voltage drop is the same across all legs.

$$
\begin{aligned}
V &= V_{R_1} = V_{R_2} = V_{R_3} \\
&= I_1 R_1 = I_2 R_2 = I_3 R_3
\end{aligned}
\qquad 83.21
$$

- The reciprocal of the equivalent resistance is the sum of the reciprocals of the individual resistances.

$$
\frac{1}{R_e} = \frac{1}{R_1} + \frac{1}{R_2} + \frac{1}{R_3} \qquad 83.22
$$

$$
G_e = G_1 + G_2 + G_3 \qquad 83.23
$$

- The sum of all of the leg currents is equal to the total current. This fact is an extension of *Kirchhoff's current law*. The current flowing out of a connection (*node*) is equal to the current flowing into it.

$$
\begin{aligned}
I &= I_1 + I_2 + I_3 \\
&= \frac{V}{R_1} + \frac{V}{R_2} + \frac{V}{R_3} = V(G_1 + G_2 + G_3)
\end{aligned}
\qquad 83.24
$$

Figure 83.3 *Simple Parallel DC Circuit and Its Equivalent*

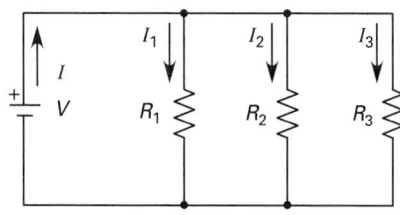

(a) original parallel circuit

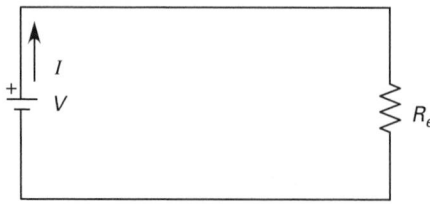

(b) equivalent circuit

12. VOLTAGE AND CURRENT DIVIDERS

Figure 83.4(a) illustrates a *voltage divider circuit*. The voltage across resistor 2 is

$$
V_2 = V\left(\frac{R_2}{R_1 + R_2}\right) \qquad 83.25
$$

Figure 83.4(b) illustrates a *current divider circuit*. The current through resistor 2 is

$$
I_2 = I\left(\frac{R_1}{R_1 + R_2}\right) \qquad 83.26
$$

Figure 83.4 *Divider Circuits*

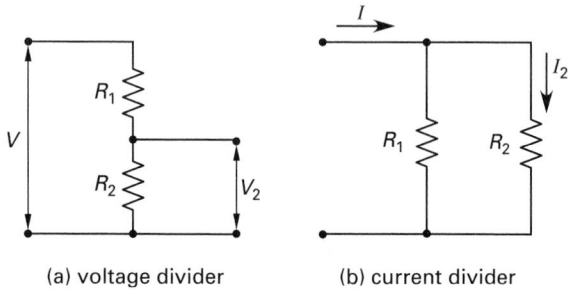

(a) voltage divider   (b) current divider

13. AC VOLTAGE SOURCES

The term *alternating current* (AC) almost always means that the current is produced from the application of a voltage with sinusoidal waveform.[5] Sinusoidal variables can be specified without loss of generality as either sines or cosines. If a sine waveform is used, Eq. 83.27 gives the instantaneous voltage as a function of time. V_m is the *maximum value*, also known as the *amplitude*, of the sinusoid. If $v(t)$ is not zero at $t = 0$, a *phase angle*, θ, must be included.

$$
v(t) = V_m \sin(\omega t + \theta) \quad \text{[trigonometric form]} \qquad 83.27
$$

Figure 83.5 *Sinusoidal Waveform*

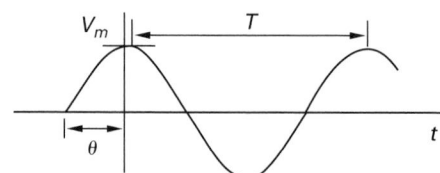

Figure 83.5 illustrates the *period*, T, of the waveform. The *frequency*, f (also known as *linear frequency*), of the sinusoid is the reciprocal of the period and is expressed in hertz (Hz).[6] *Angular frequency*, ω, in radians per second (rad/s), can also be specified.

$$
f = \frac{1}{T} = \frac{\omega}{2\pi} \qquad 83.28
$$

$$
\omega = 2\pi f = \frac{2\pi}{T} \qquad 83.29
$$

[5] Other alternating waveforms commonly encountered in commercial applications are the square, triangular, and sawtooth waveforms.

[6] In the United States, the standard frequency is 60 Hz. In Japan, the British Isles and Commonwealth countries, continental Europe, and some Mediterranean, Near Eastern, African, and South American countries, the standard is 50 Hz.

14. MAXIMUM, EFFECTIVE, AND AVERAGE VALUES

The *maximum value* (see Fig. 83.5), V_m, of a sinusoidal voltage is usually not specified in commercial and residential power systems. The *effective value*, also known as the *root-mean-square (rms) value*, is usually specified when referring to single- and three-phase voltages.[7] A DC current equal in magnitude to the effective value of a sinusoidal AC current produces the same heating effect as the sinusoid. The scale reading of a typical AC current meter is proportional to the effective current.

$$V_{\text{eff}} = \frac{V_m}{\sqrt{2}} \approx 0.707 V_m \qquad 83.30$$

The *average value* of a symmetrical sinusoidal waveform is zero. However, the average value of a rectified sinusoid (or the average value of a sinusoid taken over half of the cycle) is $V_{\text{ave}} = 2V_m/\pi$. A DC current equal to the average value of a *rectified AC current* has the same electrolytic action (e.g., capacitor charging, plating, and ion formation) as the rectified sinusoid.[8]

15. IMPEDANCE

Simple alternating current circuits can be composed of three different types of passive circuit components—resistors, inductors, and capacitors. Each type of component affects both the magnitude of the current flowing as well as the phase angle (Sec. 23) of the current. For both individual components and combinations of components, these two effects are quantified by the *impedance*, $\mathbf{Z}$. Impedance is a complex quantity with a magnitude (in ohms) and an associated *impedance angle*, ϕ. It is usually written in *phasor (polar) form* (e.g., $\mathbf{Z} \equiv Z\angle\phi$).

Multiple impedances in a circuit combine in the same manner as resistances: Impedances in series add; reciprocals of impedances in parallel add. However, the addition must use complex (i.e., vector) arithmetic.

16. REACTANCE

Impedance, like any complex quantity, can also be written in rectangular form. In this case, impedance is written as the complex sum of the resistive (R) and reactive (X) components, both having units of ohms. The resistive and reactive components combine trigonometrically in the impedance triangle. The reactive component, X, is known as the *reactance*.

$$\mathbf{Z} \equiv R \pm jX \qquad 83.31$$

---

[7]The standard U.S. 110 V household voltage (also commonly referred to as 115 V, 117 V, and 120 V) is an effective value. This is sometimes referred to as the *nominal system voltage* or just *system voltage*. Other standard voltages used around the world, expressed as effective values, are 208 V, 220 V, 230 V, 240 V, 480 V, 550 V, 575 V, and 600 V.

[8]A *rectified waveform* has had all of its negative values converted to positive values of equal absolute value.

$$R = Z\cos\phi \quad \text{[resistive part]} \qquad 83.32$$
$$X = Z\sin\phi \quad \text{[reactive part]} \qquad 83.33$$

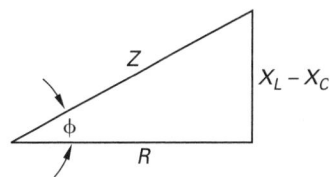

Figure 83.6 *Impedance Triangle of a Complex Circuit*

17. ADMITTANCE

The reciprocal of impedance is the complex quantity *admittance*, $\mathbf{Y}$. Admittance can be used to analyze parallel circuits, since admittances of parallel circuit elements add together.

$$\mathbf{Y} = \frac{1}{\mathbf{Z}} \equiv \frac{1}{Z}\angle-\phi \qquad 83.34$$

The reciprocal of the resistive part of the impedance is the *conductance*, G, with units of siemens, S. The reciprocal of the reactive part of impedance is the *susceptance*, B.

$$G = \frac{1}{R} \qquad 83.35$$

$$B = \frac{1}{X} \qquad 83.36$$

18. RESISTORS IN AC CIRCUITS

An ideal resistor, with resistance R, has no inductance or capacitance. The magnitude of the impedance is equal to the resistance, R, in ohms. The impedance angle is zero. Therefore, current and voltage are in phase in a purely resistive circuit.

$$\mathbf{Z}_R \equiv R\angle 0° \equiv R + j0 \qquad 83.37$$

19. CAPACITORS

A *capacitor* stores electrical charge. The charge on a capacitor is proportional to its *capacitance*, C (in farads, F), and voltage.[9]

$$Q = CV \qquad 83.38$$

An ideal capacitor has no resistance or inductance. The magnitude of the impedance is the *capacitive reactance*, X_C, in ohms. The impedance angle is $-\pi/2$ ($-90°$).

---

[9]Since a farad is a very large unit of capacitance, most capacitors are measured in microfarads, μF.

Therefore, current leads the voltage by 90° in a purely capacitive circuit.

$$\mathbf{Z}_C \equiv X_C \angle -90° \equiv 0 - jX_C \qquad \textit{83.39}$$

$$X_C = \frac{1}{\omega C} = \frac{1}{2\pi f C} \qquad \textit{83.40}$$

20. INDUCTORS

An ideal *inductor*, with an *inductance L* (in henries, H), has no resistance or capacitance. The magnitude of the impedance is the *inductive reactance*, X_L, in ohms. The impedance angle is $\pi/2$ (90°). Therefore, current lags the voltage by 90° in a purely inductive circuit.

$$\mathbf{Z}_L \equiv X_L \angle 90° \equiv 0 + jX_L \qquad \textit{83.41}$$

$$X_L = \omega L = 2\pi f L \qquad \textit{83.42}$$

21. TRANSFORMERS

Transformers are used to change voltages, isolate circuits, and match impedances. Transformers usually consist of two coils of wire wound on magnetically permeable cores. One coil, designated as the *primary coil*, serves as the input; the other coil, the *secondary coil*, is the output. The primary current produces a magnetic flux in the core; the magnetic flux, in turn, induces a voltage in the secondary coil. In an *ideal transformer* (*loss-less transformer* or *100% efficient transformer*), the coils have no electrical resistance, and all magnetic flux lines pass through both coils.

The ratio of the numbers of primary to secondary coil windings is the *turns ratio (ratio of transformation)*, a. If the turns ratio is greater than 1.0, the transformer decreases voltage and is a *step-down transformer*. If the turns ratio is less than 1.0, the transformer increases voltage and is a *step-up transformer*.

$$a = \frac{N_p}{N_s} \qquad \textit{83.43}$$

In an ideal transformer, the power transferred from the primary side equals the power received by the secondary side.

$$I_p V_p = I_s V_s \qquad \textit{83.44}$$

$$a = \frac{V_p}{V_s} = \frac{I_s}{I_p} = \sqrt{\frac{Z_p}{Z_s}} \qquad \textit{83.45}$$

22. OHM'S LAW FOR AC CIRCUITS

Ohm's law (Sec. 6) can be written in phasor (polar) form. Voltage and current can be represented by their maximum, effective (rms), or average values. However, both must be represented in the same manner.

$$\mathbf{V} = \mathbf{IZ} \qquad \textit{83.46}$$

$$V = IZ \quad \text{[magnitudes only]} \qquad \textit{83.47}$$

$$\phi_V = \phi_I + \phi_Z \quad \text{[angles only]} \qquad \textit{83.48}$$

23. PHASE ANGLE

The current and current phase angle of a circuit are determined by using Ohm's law (Sec. 6) in phasor form. The current *phase angle*, ϕ_I, is the angular difference between when the current and voltage waveforms peak.

$$\mathbf{I} = \frac{\mathbf{V}}{\mathbf{Z}} \qquad \textit{83.49}$$

$$I = \frac{V}{Z} \quad \text{[magnitudes only]} \qquad \textit{83.50}$$

$$\phi_I = \phi_V - \phi_Z \quad \text{[angles only]} \qquad \textit{83.51}$$

Example 83.2

An inventor's black box is connected across standard household voltage (110 V rms). The current drawn is 1.7 A with a lagging phase angle of 14° with respect to the voltage. What is the impedance of the black box?

Solution

From Eq. 83.49,

$$\mathbf{Z} = \frac{\mathbf{V}}{\mathbf{I}} = \frac{110 \text{ V} \angle 0°}{1.7 \text{ A} \angle 14°}$$

$$= 64.71 \ \Omega \angle -14°$$

24. POWER IN AC CIRCUITS

In a purely resistive circuit, all of the current drawn contributes to dissipated energy. The flow of current causes resistors to increase in temperature, and heat is transferred to the environment.

In a typical AC circuit containing inductors and capacitors as well as resistors, some of the current drawn does not cause heating.[10] Rather, the current charges capacitors and creates magnetic fields in inductors. Since the voltage alternates in polarity, capacitors alternately charge and discharge. Thus, energy is repeatedly drawn and returned by capacitors. Similarly, energy is repeatedly drawn and returned by inductors as their magnetic fields form and collapse.

Current through resistors in AC circuits causes heating, just as in DC circuits. The power dissipated is represented by the *real power* vector, **P**. Current through

---

[10]The notable exception is a *resonant circuit* in which the inductive and capacitive reactances are equal. In that case, the circuit is purely resistive in nature.

capacitors and inductors contributes to reactive power, represented by the *reactive power* vector, **Q**. Reactive power does not contribute to heating. For convenience, both real and reactive power are considered to be complex (vector) quantities, with magnitudes and associated angles. Real and reactive power combine as vectors into the *complex power* vector, **S**, as shown in Fig. 83.7. The angle, ϕ, is known as the overall *impedance angle*.

Figure 83.7 *Complex Power Triangle (lagging)*

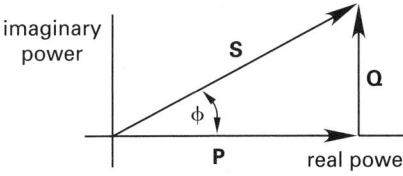

The magnitude of real power is known as the *average power*, P, and is measured in watts. The magnitude of the reactive power vector is known as *reactive power*, Q, and is measured in VARs (volt-amps-reactive). The magnitude of the complex power vector is the *apparent power*, S, measured in VAs (volt-amps). The apparent power is easily calculated from measurements of the line current and line voltage.

$$S = I_{\text{eff}} V_{\text{eff}} = \tfrac{1}{2} I_m V_m \qquad \textit{83.52}$$

25. POWER FACTOR

The complex power triangle shown in Fig. 83.7 is congruent to the impedance triangle (Fig. 83.6), and the *power angle*, ϕ, is identical to the overall impedance angle.

$$S^2 = P^2 + Q^2 \qquad \textit{83.53}$$
$$P = S \cos \phi \qquad \textit{83.54}$$
$$Q = S \sin \phi \qquad \textit{83.55}$$
$$\phi = \arctan \frac{Q}{P} \qquad \textit{83.56}$$

The *power factor*, pf (also occasionally referred to as the *phase factor*), for a sinusoidal voltage is $\cos\phi$. By convention, the power factor is usually given in percent value rather than by its equivalent decimal value. For a purely resistive circuit, pf = 100%; for a purely reactive circuit, pf = 0.

$$\text{pf} = \frac{P}{S} \qquad \textit{83.57}$$

Since the cosine is positive for both positive and negative angles, the terms "leading" and "lagging" are used to describe the nature of the circuit. In a circuit with a *leading power factor* (i.e., in a *leading circuit*), the load is primarily capacitive in nature. In a circuit with a *lagging power factor* (i.e., in a *lagging circuit*), the load is primarily inductive in nature.

26. COST OF ELECTRICAL ENERGY

Except for large industrial users, electrical meters at service locations usually measure and record real power only. Electrical utilities charge on the basis of the total energy used. Energy usage, commonly referred to as the *usage* or *demand*, is measured in kilowatt-hours, abbreviated kWh or kW-hr.

$$\text{cost} = (\text{cost per kW-hr}) \times E_{\text{kW-hr}} \qquad \textit{83.58}$$

The cost per kW-hr may not be a single value but may be tiered so that cost varies with cumulative usage. The lowest rate is the *baseline rate*.[11] To encourage conservation, the incremental cost of energy increases with increases in monthly usage.[12] To encourage cutbacks during the day, the cost may also increase during periods of peak demand.[13] The increase in cost for usage during peak demand may be accomplished by varying the rate, additively, or by use of a *peak demand multiplier*. There may also be different rates for summer and winter usage.

Although only real power is dissipated, reactive power contributes to total current. (Reactive power results from the current drawn in supplying the magnetization energy in motors and charges on capacitors.) Therefore, the distribution system (wires, transformers, etc.) must be sized to carry the total current, not just the current supplying the heating effect. When real power alone is measured at the service location, the power factor is routinely monitored and its effect is built into the charge per kW-hr. This has the equivalent effect of charging the user for apparent power usage.

Example 83.3

A small office normally uses 700 kW-hr per month of electrical energy. The company adds a 1.5 kW heater (to be used at the rate of 1000 kW-hr/month) and a 5 hp motor with a mechanical efficiency of 90% (to be used at a rate of 240 hr/mo). What is the incremental cost of adding the heater and motor? The tiered rate structure is

electrical usage (kW-hr)	rate ($/kW-hr)
less than 350	0.1255
350 to 999	0.1427
1000 to 3999	0.1693

---

[11] There might also be a *lifeline rate* for low-income individuals with very low usage.
[12] There are different rate structures for different categories of users. While increased use within certain categories of users (e.g., residential) results in a higher cost per kW-hr, larger users in another category may pay substantially less per kW-hr due to their volume "buying power."
[13] The day may be divided into *peak periods*, *partial-peak periods*, and *off-peak periods*.

Solution

Motors are rated by their real power output, which is less than their real power demand. The incremental electrical usage per month is

$$1000 \text{ kWh} + \frac{(5 \text{ hp}) \left(0.7457 \frac{\text{kW}}{\text{hp}}\right) (240 \text{ hr})}{0.90} = 1994 \text{ kWh}$$

The cumulative monthly electrical usage is

$$700 \text{ kW-hr} + 1994 \text{ kWh} = 2694 \text{ kWh}$$

The company was originally in the second tier. The new usage will be billed partially at the second and third tier rates.

The incremental cost is

$$(999 \text{ kWh} - 700 \text{ kWh}) \left(0.1427 \frac{\$}{\text{kW-hr}}\right)$$
$$+ (2694 \text{ kWh} - 999 \text{ kWh}) \left(0.1693 \frac{\$}{\text{kW-hr}}\right)$$
$$= \$329.63$$

27. POWER FACTOR CORRECTION

Inasmuch as apparent power is paid for but only real power is used to drive motors or provide light and heating, it may be possible to reduce electrical utility charges by reducing the power angle without changing the real power. This strategy, known as *power factor correction*, is routinely accomplished by changing the circuit reactance in order to reduce the reactive power. The change in reactive power needed to change the power angle from ϕ_1 to ϕ_2 is

$$\Delta Q = P(\tan \phi_1 - \tan \phi_2) \qquad \textbf{83.59}$$

When a circuit is capacitive (i.e., leading), induction motors (Sec. 39) can be connected across the line to improve the power factor. In the more common situation, when a circuit is inductive (i.e., lagging), capacitors or synchronous capacitors (Sec. 41) can be added across the line. The size (in farads) of capacitor required is

$$C = \frac{\Delta Q}{\pi f V_m^2} \quad [V_m \text{ maximum}] \qquad \textbf{83.60}$$

$$C = \frac{\Delta Q}{2\pi f V_{\text{eff}}^2} \quad [V_{\text{eff}} \text{ effective}] \qquad \textbf{83.61}$$

Capacitors for power factor correction are generally rated in kVA. Equation 83.59 can be used to find that rating.

Example 83.4

A 60 Hz, 5 hp induction motor draws 53 A (rms) at 117 V (rms) with a 78.5% electrical-to-mechanical energy conversion efficiency. What capacitance should be connected across the line to increase the power factor to 92%?

Solution

The apparent power is found from the observed voltage and the current. Use Eq. 83.52.

$$S = IV = \frac{(53 \text{ A})(117 \text{ V})}{1000 \frac{\text{VA}}{\text{kVA}}}$$
$$= 6.201 \text{ kVA}$$

The real power drawn from the line is calculated from the real work done by the motor. Use Eq. 83.72.

$$P_{\text{electrical}} = \frac{P_{\text{out}}}{\eta} = \frac{(5 \text{ hp}) (0.7457 \text{ kW/hp})}{0.785}$$
$$= 4.750 \text{ kW}$$

The reactive power and power angle are calculated from the real and apparent powers.

$$Q_1 = \sqrt{S^2 - P^2} = \sqrt{(6.201 \text{ kVA})^2 - (4.750 \text{ kW})^2}$$
$$= 3.986 \text{ kVAR}$$

From Eq. 83.53,

$$\phi_1 = \arccos \frac{4.750 \text{ kW}}{6.201 \text{ kVA}} = 40.00°$$

The desired power factor angle is

$$\phi_2 = \arccos 0.92 = 23.07°$$

The reactive power after the capacitor is installed is given by Eq. 83.56.

$$Q_2 = P \tan \phi_2 = (4.750 \text{ kW})(\tan 23.07°)$$
$$= 2.023 \text{ kVAR}$$

The required capacitance is found from Eq. 83.61.

$$C = \frac{\Delta Q}{2\pi f V_{\text{eff}}^2}$$
$$= \frac{(3.986 \text{ kVAR} - 2.023 \text{ kVAR}) \left(1000 \frac{\text{VAR}}{\text{kVAR}}\right)}{(2\pi)(60 \text{ Hz})(117 \text{ V})^2}$$
$$= 3.8 \times 10^{-4} \text{ F} \quad (380 \text{ } \mu\text{F})$$

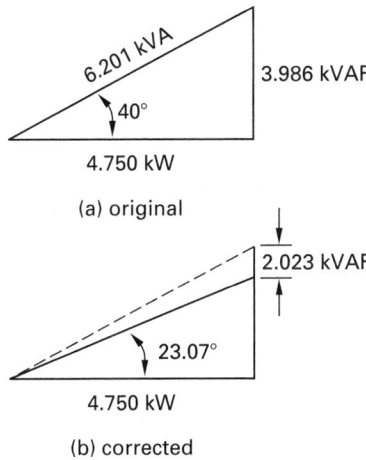

(a) original

(b) corrected

28. THREE-PHASE ELECTRICITY

Smaller electric loads, such as household loads, are served by single-phase power. The power company delivers a sinusoidal voltage of fixed frequency and amplitude connected between two wires—a phase wire and a neutral wire. Large electric loads, large buildings, and industrial plants are served by three-phase power. Three voltage signals are connected between three phase wires and a single neutral wire. The phases have equal frequency and amplitude, but they are out of phase by 120° (electrical) with each other. Such *three-phase systems* use smaller conductors to distribute electricity.[14] Thus, for the same delivered power, three-phase distribution systems have lower losses and are more efficient.

Three-phase motors provide a more uniform torque than do single-phase motors whose torque production pulsates.[15] Three-phase induction motors require no additional starting windings or associated switches. When rectified, three-phase voltage has a smoother waveform.

Figure 83.8 *Three-Phase Voltage*

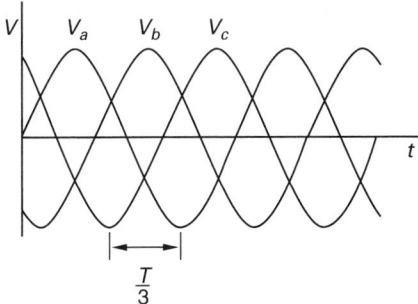

29. THREE-PHASE LOADS

Three impedances are required to fully load a three-phase voltage source. In three-phase motors and

---

[14]The uppercase Greek letter phi is often used as an abbreviation for the word "phase." For example, "3Φ" would be interpreted as "three-phase."
[15]Single-phase motors require auxiliary windings for starting, since one phase alone cannot get the magnetic field rotating.

---

other devices, these impedances are three separate motor windings. Similarly, three-phase transformers have three separate sets of primary and secondary windings.

The impedances in a three-phase system are said to be *balanced loads* when they are identical in magnitude and angle. The voltages, line currents, and real, apparent, and reactive powers are all identical in a balanced system. Also, the power factor is the same for each phase. Therefore, only one phase of a balanced system needs to be analyzed (i.e., a *one-line analysis*).

30. LINE AND PHASE VALUES

The *line current*, I_l, is the current carried by the distribution lines (wires). The *phase current*, I_p, is the current flowing through each of the three separate loads (i.e., the phase) in the motor or device. Line and phase currents are both vector quantities.

Depending on how the motor or device is internally wired, the line and phase currents may or may not be the same. Figure 83.9 illustrates delta- and wye-connected loads. For balanced *wye-connected loads*, the line and phase currents are the same. For balanced *delta-connected loads*, they are not.

$$I_p = I_l \quad \text{[wye]} \qquad 83.62$$

$$I_p = \frac{I_l}{\sqrt{3}} \quad \text{[delta]} \qquad 83.63$$

Similarly, the *line voltage*, V_l (same as *line-to-line voltage*, commonly referred to as the *terminal voltage*), and *phase voltage*, V_p, may not be the same. With balanced delta-connected loads, the full line voltage appears across each phase. With balanced wye-connected loads, the line voltage appears across two loads.

$$V_p = \frac{V_l}{\sqrt{3}} \quad \text{[wye]} \qquad 83.64$$

$$V_p = V_l \quad \text{[delta]} \qquad 83.65$$

31. INPUT POWER THREE-PHASE SYSTEMS

Each impedance in a balanced system dissipates the same real *phase power*, P_p. The power dissipated in a balanced three-phase system is three times the phase power and is calculated in the same manner for both delta- and wye-connected loads.

$$P_t = 3P_p = 3V_p I_p \cos\phi$$
$$= \sqrt{3}V_l I_l \cos\phi \qquad 83.66$$

The real power component is sometimes referred to as "power in kW." Apparent power is sometimes referred to as "kVA value" or "power in kVA."

$$S_t = 3S_p = 3V_p I_p$$
$$= \sqrt{3}V_l I_l \qquad 83.67$$

Figure 83.9 *Wye- and Delta-Connected Loads*

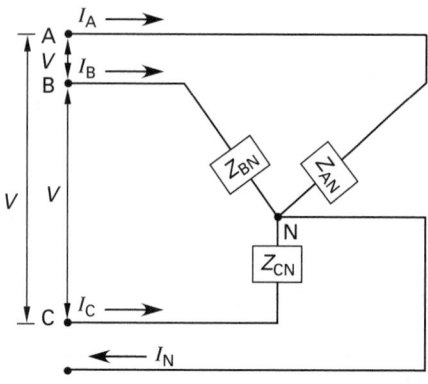

(a) wye-connected loads

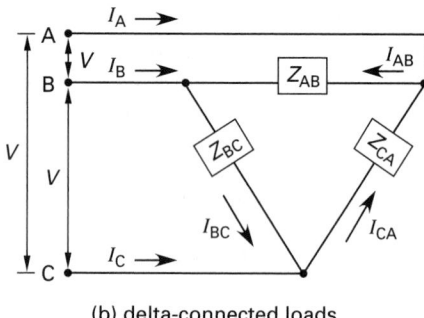

(b) delta-connected loads

32. ROTATING MACHINES

Rotating machines are categorized as AC and DC machines. Both categories include machines that use electrical power (i.e., motors) and those that generate electrical power (alternators and generators).[16] Machines can be constructed in either single-phase or poly-phase configurations, although single-phase machines may be inferior in terms of economics and efficiency. (See Sec. 28.)

Large AC motors are almost always three-phase. However, since the phases are balanced, it is necessary to analyze one phase only of the motor. Torque and power are divided evenly among the three phases.

33. REGULATION

Rotating machines (motors and alternators), as well as power supplies, are characterized by changes in voltage and speed under load. *Voltage regulation* is defined as

$$\text{VR} = \frac{\text{no-load voltage} - \text{full-load voltage}}{\text{full-load voltage}} \times 100\%$$

83.68

Speed regulation is defined as

$$\text{SR} = \frac{\text{no-load speed} - \text{full-load speed}}{\text{full-load speed}} \times 100\% \quad \textit{83.69}$$

---

[16]An *alternator* produces AC potential. A *generator* produces DC potential.

34. TORQUE AND POWER

For rotating machines, torque and power are basic operational parameters. It takes mechanical power to turn an alternator or generator. A motor converts electrical power into mechanical power. In SI units, power is given in watts (W) and kilowatts (kW). One horsepower (hp) is equivalent to 0.7457 kilowatts. The relationship between torque and power is

$$T_{\text{ft-lbf}} = (5252)\left(\frac{P_{\text{horsepower}}}{n_{\text{rpm}}}\right) \qquad \textit{83.70}$$

$$T_{\text{N·m}} = (9549)\left(\frac{P_{\text{kW}}}{n_{\text{rpm}}}\right) \qquad \textit{83.71}$$

There are many important torque parameters for motors. The *starting torque* (also known as *static torque, break-away torque,* and *locked-rotor torque*) is the turning effort exerted by the motor when starting from rest. *Pull-up torque (acceleration torque)* is the minimum torque developed during the period of acceleration from rest to full speed. *Pull-in torque* (as developed in synchronous motors) is the maximum torque that brings the motor back to synchronous speed (Sec. 39). *Nominal pull-in torque* is the torque that is developed at 95% of synchronous speed.

The *full-load torque (steady-state torque)* occurs at the rated speed and horsepower. Full-load torque is supplied to the load on a continuous basis. The full-load torque establishes the temperature increase that the motor must be able to withstand without deterioration. The *rated torque* is developed at rated speed and rated horsepower. The maximum torque that the motor can develop at its synchronous speed is the *pull-out torque*. *Breakdown torque* is the maximum torque that the motor can develop without stalling (i.e., without coming rapidly to a complete stop).

Figure 83.10 *Induction Motor Torque-Speed Characteristic (typical of design B frames)*

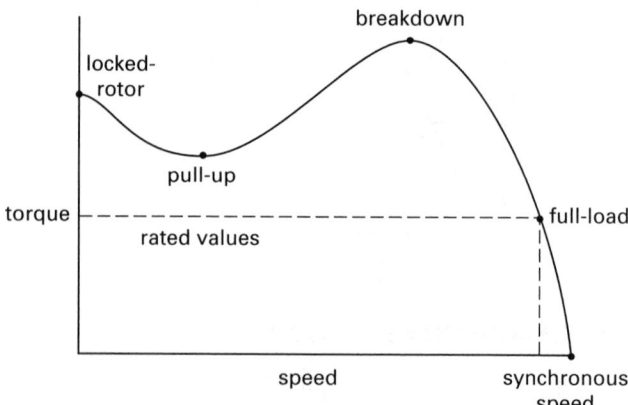

Motors are rated according to their output power (*rated power* or *brake power*). Thus, a 5 hp motor will normally deliver a full five horsepower when running at its rated speed. While the rated power is not affected by the motor's energy conversion efficiency, η, the electrical power input to the motor is.

$$P_{\text{electrical}} = \frac{P_{\text{rated}}}{\eta} \qquad 83.72$$

Table 83.1 lists standard motor sizes by rated horsepower.[17] The rated horsepower should be greater than the calculated brake power requirements. Since the rated power is the power actually produced, motors are not selected on the basis of their efficiency or electrical power input. The smaller motors listed in Table 83.1 are generally single-phase motors. The larger motors listed are three-phase motors.

Table 83.1 Typical Standard Motor Sizes (horsepower)^a

$\frac{1}{8}$	$\frac{1}{6}$	$\frac{1}{4}$	$\frac{1}{3}$	$\frac{1}{2}$	$\frac{3}{4}$			
1	$1\frac{1}{2}$	2	3	5	$7\frac{1}{2}$			
10	15	20	25	30	40	50	60	75
100	125	150	200	250				

(Multiply hp by 0.7457 to obtain kW.)
^{a} $1/8$ hp and $1/6$ hp motors are less common.

35. NEMA MOTOR CLASSIFICATIONS

Motors are classified by their NEMA design type—A, B, C, D, and F.[18,19] These designs are collectively referred to as *NEMA frame motors*. All NEMA motors of a given frame size are interchangeable as to bolt holes, shaft diameter, height, length, and various other dimensions.

- Frame design A: three-phase, squirrel-cage motor with high locked-rotor (starting) current but also higher breakdown torques; capable of handling intermittent overloads without stalling; 1 to 3% slip at full load.

- Frame design B: three-phase, squirrel-cage motor capable of withstanding full-voltage starting; locked-rotor and breakdown torques suitable for most applications; 1 to 3% slip at full load; often designated for "normal" usage; the most common design.[20]

- Frame design C: three-phase, squirrel-cage motor capable of withstanding full-voltage starting; high locked-rotor torque for special applications (e.g., conveyors and compressors); 1 to 3% slip at full load.

- Frame design D: three-phase, squirrel-cage motor capable of withstanding full-voltage starting;

develops 275% locked-rotor torque for special applications; high breakdown torque; depending on design, 5 to 13% slip at full load; commonly known as a *high-slip motor*; used for cranes, hoists, oil well pump jacks, and valve actuators.

- Frame design F: three-phase, squirrel-cage motor with low starting current and low starting torque; used to meet applicable regulations regarding starting current; limited availability.

36. NAMEPLATE VALUES

A *nameplate* is permanently affixed to each motor's housing or frame. This nameplate is embossed or engraved with the *rated values* of the motor (*nameplate values* or *full-load values*). Nameplate information may include some or all of the following: voltage,[21] frequency,[22] number of phases, rated power (output power), running speed, duty cycle (Sec. 38), locked-rotor and breakdown torques, starting current, current drawn at rated load (in kVA/hp), ambient temperature, temperature rise at rated load, temperature rise at service factor (Sec. 37), insulation rating, power factor, efficiency, frame size, and enclosure type.

It is important to recognize that the rated power of the motor is the actual output power (i.e., power delivered to the load), not the input power. Only the electrical power input is affected by the motor efficiency. (See Eq. 83.72.)

Nameplates are also provided on transformers. Transformer nameplate information includes the two winding voltages (either of which can be the primary winding), frequency, and kVA rating. Apparent power in kVA, not real power, is used in the rating because heating is proportional to the square of the supply current. As with motors, continuous operation at the rated values will not result in excessive heat build-up.

37. SERVICE FACTOR

The horsepower and torque ratings listed on the nameplate of an AC motor can be maintained on a continuous basis without overheating. However, motors can be operated at slightly higher loads without exceeding a safe temperature rise.[23] The ratio of the safe to rated loads (horsepower or torque) is the *service factor*, sf, usually expressed as a decimal.

$$\text{sf} = \frac{\text{maximum safe load}}{\text{nameplate load}} \qquad 83.73$$

---

[17]For economics, standard motor sizes should be specified.
[18]NEMA stands for the National Electrical Manufacturers Association.
[19]There is no type E.
[20]Design B is estimated to be used in 90% of all applications.

[21]Standard NEMA nameplate voltages (effective) are 200 V, 230 V, 460 V, and 575 V. NEMA motors are capable of operating in a range of only ±10% of their rated voltages. Thus, 230 V motors should not be used on 208 V systems.
[22]While some 60 Hz motors (notably those intended for 230 V operation) can be used at 50 Hz, most others (e.g., those intended for 200 V operation) are generally not suitable for use at 50 Hz.
[23]Higher temperatures have a deteriorating effect on the winding insulation. A general rule of thumb is that a motor loses two or three hours of useful life for each hour run at the service factor load.

Service factors vary from 1.15 to 1.4, with the lower values being applicable to the larger, more efficient motors. Typical values of service factor are 1.4 (up to $^1/_8$ hp motors), 1.35 ($^1/_6$ to $^1/_3$ hp motors), 1.25 ($^1/_2$ to 1 hp motors), and 1.15 (1 or $1^1/_2$ to 200 hp).

When running above the rated load, the motor speed, temperature, power factor, full-load current, and efficiency will differ from the nameplate values. However, the locked-rotor and breakdown torques will remain the same, as will the starting current.

Active current is proportional to torque, and hence, is proportional to the horsepower developed (Eq. 83.75). The active current (line or phase) drawn is also proportional to the service factor. The current drawn per phase is given by Eq. 83.75.[24]

$$I_{\text{active}} = I_l(\text{pf}) \qquad 83.74$$

$$I_{\text{actual},p} = \frac{(\text{sf})P_{\text{rated},p}}{V_p \eta (\text{pf})}$$

$$= \frac{P_{\text{actual},p}}{V_p \eta (\text{pf})} \quad [\text{per phase}; P \text{ in watts}] \qquad 83.75$$

Equation 83.75 can also be used when a motor is developing less than its rated power. In this case, the service factor can be considered as the fraction of the rated power being developed.

Example 83.5

What is the approximate phase current drawn by a three-phase, 75 hp, 230 V (rms) motor running at 88% power factor and its rated load?

Solution

No information is given about the efficiency, which is taken as 100%. The service factor is 1.00 because the motor is running at its rated load. From Eq. 83.75, the approximate phase current is

$$I_p = \frac{(\text{sf})P_p}{V_p \eta (\text{pf})}$$

$$= \frac{(1.00)\left(\dfrac{75 \text{ hp}}{3 \text{ phases}}\right)\left(0.7457 \dfrac{\text{kW}}{\text{hp}}\right)\left(1000 \dfrac{\text{W}}{\text{kW}}\right)}{(230 \text{ V})(1.00)(0.88)}$$

$$= 92.1 \text{ A} \quad [\text{per phase}]$$

38. DUTY CYCLE

Motors are categorized according to their *duty cycle: continuous-duty* (24 hr/day); *short-time duty* (15 to 30 min); and *special-duty* (application specific).

39. INDUCTION MOTORS

The three-phase induction motor is by far the most frequently used motor in industry. In an induction motor, the magnetic field rotates at the synchronous speed. The *synchronous speed* can be calculated from the number of poles and frequency. The frequency, f, is either 60 Hz (in the United States) or 50 Hz (in Europe and other locations). The number of poles, p, must be even.[25] The most common motors have 2, 4, and 6 poles.

$$n_{\text{synchronous}} = \frac{120f}{p} \quad [\text{rpm}] \qquad 83.76$$

Due to friction and other factors, rotors (and hence, the motor shafts) in induction motors run slightly slower than their synchronous speeds. The percentage difference is known as the *slip*, s. Slip is seldom greater than 10%, and it is usually much less than that. 4% is a typical value.

$$s = \frac{n_{\text{synchronous}} - n_{\text{actual}}}{n_{\text{synchronous}}} \qquad 83.77$$

The rotor's actual speed is[26]

$$n_{\text{actual}} = (1 - s)n_{\text{synchronous}} \qquad 83.78$$

Induction motors are usually specified in terms of the *kVA ratings*. The kVA rating is not the same as the motor power in kilowatts, although one can be calculated from the other if the motor's power factor is known. The power factor generally varies from 0.8 to 0.9 depending on the motor size.

$$\text{kVA rating} = \frac{P_{\text{kW}}}{\text{pf}} \qquad 83.79$$

$$P_{\text{kW}} = 0.7457 P_{\text{mechanical,hp}} \qquad 83.80$$

Induction motors can differ in the manner in which their rotors are constructed. A *wound rotor* is similar to an armature winding in a dynamo. Wound rotors have high-torque and soft-starting capabilities. There are no wire windings at all in a *squirrel-cage* rotor. Most motors use squirrel-cage rotors. Typical torque-speed characteristics of a design B induction motor are shown in Fig. 83.10.

Example 83.6

A pump is driven by a three-phase induction motor running at its rated values. The motor's nameplate lists the

---

[24]It is important to recognize the difference between the rated (i.e., nameplate) power and the actual power developed. The actual power should not be combined with the service factor, since the actual power developed is the product of the rated power and the service factor.

[25]There are various forms of Eq. 83.76. As written, the speed is given in rpm, and the number of poles, p, is twice the number of *pole pairs*. When the synchronous speed is specified as f/p, it is understood that the speed is in revolutions per second (rps) and p is the number of pole *pairs*.

[26]Some motors (i.e., *integral gear motors*) are manufactured with integral speed reducers. Common standard output speeds are 37, 45, 56, 68, 84, 100, 125, 155, 180, 230, 280, 350, 420, 520, and 640 rpm. While the integral gear motor is more compact, lower in initial cost, and easier to install than a separate motor with belt drive, coupling, and guard, the separate motor and reducer combination may nevertheless be preferred for its flexibility, especially in replacing and maintaining the motor.

following rated values: 50 hp, 440 V, 92% lagging power factor, 90% efficiency, 60 Hz, and 4 poles. The motor's windings are delta-connected. The pump efficiency is 80%. When running under the pump load, the slip is 4%. What are the (a) total torque developed, (b) torque developed per phase, and (c) line current?

Solution

(a) The synchronous speed is given by Eq. 83.76.

$$n_{\text{synchronous}} = \frac{120f}{p} = \frac{(120)(60\text{ Hz})}{4}$$
$$= 1800\text{ rpm}$$

From Eq. 83.78, the rotor speed is

$$n_{\text{actual}} = (1-s)n_{\text{synchronous}} = (1-0.04)(1800\text{ rpm})$$
$$= 1728\text{ rpm}$$

Since the motor is running at its rated values, the motor delivers 50 hp to the pump. The pump's efficiency is irrelevant. From Eq. 83.70, the total torque developed by all three phases is

$$T_t = (5252)\left(\frac{P_{\text{horsepower}}}{n_{\text{rpm}}}\right) = \frac{\left(5252\ \frac{\text{ft-lbf}}{\text{hp-min}}\right)(50\text{ hp})}{1728\text{ rpm}}$$
$$= 152.0\text{ ft-lbf} \quad [\text{total}]$$

(b) The torque developed per phase is one third of the total torque developed.

$$T_p = \frac{T_t}{3} = \frac{152.0\text{ ft-lbf}}{3\text{ phases}}$$
$$= 50.67\text{ ft-lbf} \quad [\text{per phase}]$$

(c) The total electrical input power is given by Eq. 83.72.

$$P_{\text{electrical}} = \frac{P_{\text{rated}}}{\eta} = \frac{(50\text{ hp})\left(0.7457\ \frac{\text{kW}}{\text{hp}}\right)}{0.90}$$
$$= 41.43\text{ kW} \quad [\text{total}]$$

Since the power factor is less than 1.0, more current is being drawn than is being converted into useful work. From Eq. 83.79, the apparent power in kVA per phase is

$$S_{\text{kVA}} = \frac{P_{\text{kW}}}{\text{pf}} = \frac{41.43\text{ kW}}{(3)(0.92)}$$
$$= 15.01\text{ kVA} \quad [\text{per phase}]$$

The phase current is given by Eq. 83.52.

$$I = \frac{S}{V} = \frac{(15.01\text{ kVA})\left(1000\ \frac{\text{VA}}{\text{kVA}}\right)}{440\text{ V}}$$
$$= 34.11\text{ A}$$

Since the motor's windings are delta-connected across the three lines, the line current is

$$I_l = \sqrt{3}I_p = (\sqrt{3})(34.11\text{ A})$$
$$= 59.08\text{ A}$$

40. TYPICAL INDUCTION MOTOR PERFORMANCE

The following rules of thumb can be used for initial estimates of induction motor performance.

- At 1800 rpm, a motor will develop a torque of 3 ft-lbf/hp.
- At 1200 rpm, a motor will develop a torque of 4.5 ft-lbf/hp.
- At 550 V, a three-phase motor will draw 1 A/hp.
- At 440 V, a three-phase motor will draw 1.25 A/hp.
- At 220 V, a three-phase motor will draw 2.5 A/hp.
- At 220 V, a single-phase motor will draw 5 A/hp.
- At 110 V, a single-phase motor will draw 10 A/hp.

41. SYNCHRONOUS MOTORS

Synchronous motors are essentially dynamo alternators operating in reverse. The stator field frequency is fixed, so regardless of load, the motor runs only at a single speed—the synchronous speed given by Eq. 83.76. Stalling occurs when the motor's counter-torque is exceeded. For some equipment that must be driven at constant speed, such as large air or gas compressors, the additional complexity of synchronous motors is justified.

Power factor can be adjusted manually by varying the field current. With *normal excitation* field current, the power factor is 1.0. With *over-excitation*, the power factor is leading, and the field current is greater than normal. With *under-excitation*, the power factor is lagging, and the field current is less than normal.

Since a synchronous motor can be adjusted to draw leading current, it can be used for power factor correction. A synchronous motor used purely for power factor correction is referred to as a *synchronous capacitor* or *synchronous condenser*. A power factor of 80% is often specified or used with synchronous capacitors.

42. DC MACHINES

DC motors and generators can be wired in one of three ways: series, shunt, and compound. Operational characteristics are listed in Table 83.2. Equations 83.70 and 83.71 can be used to calculate torque and power.

Table 83.2 Operational Characteristics of DC and AC Machines

	motors			generators		
	shunt	series	compound	shunt	series	compound
equivalent circuit						
line voltage, V	V	V	V	V	V	V
line current, I_l	I_l	$I_l = I_a$	I_l	$I_l = I_a - I_f$	$I_l = I_a$	$I_l = I_a - I_f$
field current, I_f	$I_f = \dfrac{V}{R_f}$	$I_f = I_a$	$I_f = \dfrac{V}{R_{f1}}$	$I_f = \dfrac{V}{R_f}$	$I_f = I_a$	$I_f = \dfrac{V}{R_{f1}}$
armature current, I_a	$I_a = I_l - I_f$	$I_a = I_l$	$I_a = I_l - I_f$	$I_a = I_l + I_f$	$I_a = I_l$	$I_a = I_l + I_f$
armature circuit loss, V_a	$V_a = I_a R_a$	$V_a = I_a(R_a + R_f)$	$V_a = I_a(R_a + R_{f2})$	$V_a = I_a R_a$	$V_a = I_a(R_a + R_f)$	$V_a = I_a(R_a + R_{f2})$
counter emf, E_S	$E_S = V - V_a$ $= V - I_a R_a$	$E_S = V - V_a$ $= V - I_a(R_a + R_f)$	$E_S = V - I_a$ $\times (R_a + R_{f2})$	$E_S = V + V_a$ $= V + I_a R_a$	$E_S = V + V_a$ $= V + I_a(R_a + R_f)$	$E_S = V + V_a$ $= V + I_a(R_a + R_{f2})$
power in kW or hp [DC machines]	$P = VI_l$	$P = VI_l$	$P = VI_l$	$\mathrm{hp} = \dfrac{2\pi nT}{33,000}$	$\mathrm{hp} = \dfrac{2\pi nT}{33,000}$	$\mathrm{hp} = \dfrac{2\pi nT}{33,000}$
power in kVA [AC machines]	$P = VI_l \cos\phi$	$P = VI_l \cos\phi$	$P = VI_l \cos\phi$	$\mathrm{hp} = \dfrac{2\pi nT}{33,000}$	$\mathrm{hp} = \dfrac{2\pi nT}{33,000}$	$\mathrm{hp} = \dfrac{2\pi nT}{33,000}$
power out in hp or kW [DC machines]	$\mathrm{hp} = \dfrac{2\pi nT}{33,000}$	$\mathrm{hp} = \dfrac{2\pi nT}{33,000}$	$\mathrm{hp} = \dfrac{2\pi nT}{33,000}$	$P = VI_l$	$P = VI_l$	$P = VI_l$
power out in kVA [AC machines]	$\mathrm{hp} = \dfrac{2\pi nT}{33,000}$	$\mathrm{hp} = \dfrac{2\pi nT}{33,000}$	$\mathrm{hp} = \dfrac{2\pi nT}{33,000}$	$P = VI_l \cos\phi$	$P = VI_l \cos\phi$	$P = VI_l \cos\phi$

43. CHOICE OF MOTOR TYPES

Squirrel-cage induction motors are commonly chosen because of their simple construction, low maintenance, and excellent efficiencies.[27] A wound-rotor induction motor should be used only if it is necessary to achieve a low starting kVA, controllable kVA, controllable torque, or variable speed.[28]

While induction motors are commonly used, synchronous motors are suitable for many applications normally handled by a NEMA design B squirrel-cage motor. They have adjustable power factors and higher efficiency. Their initial cost may also be less.

Selecting a motor type is greatly dependent on the power, torque, and speed requirements of the rotating load. Table 83.3, Fig. 83.11, and Apps. 83.A and 83.B can be used as starting points in the selection process.

Table 83.3 *Recommended Motor Voltage and Power Ranges*

voltage	horsepower
direct current	
115	0 to 30 (max)
230	0 to 200 (max)
550 or 600	$\frac{1}{2}$ and upward
alternating current, one-phase	
110, 115, or 120	0 to $1\frac{1}{2}$
220, 230, or 240	0 to 10
440 or 550	5 to 10[a]
alternating current, two- and three-phase	
110, 115, or 120	0 to 15
208, 220, 230, or 240	0 to 200
440 or 550	0 to 500
2200 or 2300	40 and upward
4000	75 and upward
6600	400 and upward

(Multiply hp by 0.7457 to obtain kW.)
[a]not recommended.

44. LOSSES IN ROTATING MACHINES

Losses in rotating machines are typically divided into the following categories: armature copper losses, field copper losses, mechanical losses (including friction and windage), core losses (including hysteresis losses, eddy current losses, and brush resistance losses), and stray losses.

Copper losses (also known as I^2R *losses*) are real power losses due to wire and winding resistance.

$$P_{\text{Cu}} = \Sigma I^2 R \qquad \text{83.81}$$

---

[27]The larger the motor and the higher the speed, the higher the efficiency. Large 3600 rpm induction motors have excellent performance.
[28]A constant-speed motor with a slip coupling could also be used.

Core losses (also known as *iron losses*) are constant losses that are independent of the load and, for that reason, are also known as *open-circuit* and *no-load losses*.

Figure 83.11 *Motor Rating According to Speed[a]*
(general guidelines)

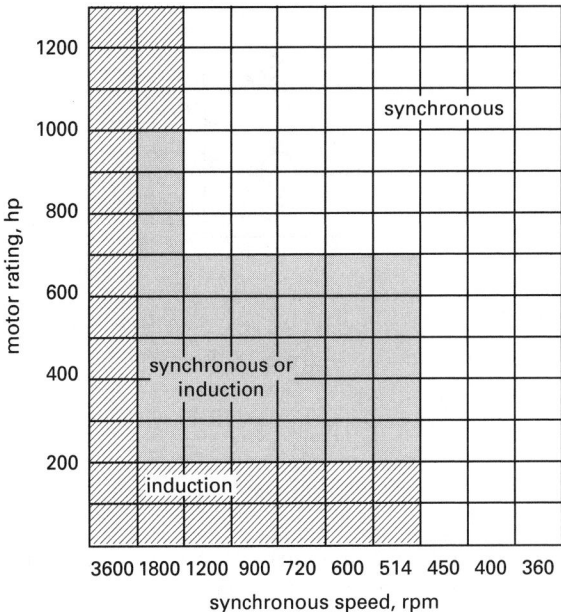

[a]Adapted from *Mechanical Engineering*, Design Manual NAVFAC DM-3, Department of the Navy, © 1972.

Mechanical losses (also known as *rotational losses*) include brush and bearing friction and *windage* (air friction). (Windage is a no-load loss but is not an electrical core loss.) Mechanical losses are determined by measuring the power input at the rated speed and with no load.

Stray losses are due to non-uniform current distribution in the conductors. Stray losses are approximately 1% for DC machines and zero for AC machines.

45. EFFICIENCY OF ROTATING MACHINES

Only real power is used to compute the efficiency of a rotating machine. This efficiency is sometimes referred to as *overall efficiency* and *commercial efficiency*.

$$\begin{aligned} \eta &= \frac{\text{output power}}{\text{input power}} \\ &= \frac{\text{output power}}{\text{output power} + \text{power losses}} \\ &= \frac{\text{input power} - \text{power losses}}{\text{input power}} \qquad \text{83.82} \end{aligned}$$

Example 83.7

A DC shunt motor draws 40 A at 112 V when fully loaded. When running without a load at the same speed, it draws only 3 A at 106 V. The field resistance is

100 Ω, and the armature resistance is 0.125 Ω. (a) What is the efficiency of the motor? (b) What power (in hp) does the motor deliver at full load?

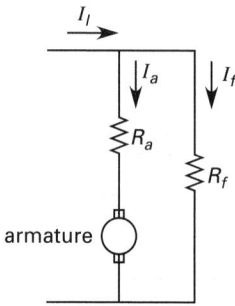

Solution

(a) The field current is given by Eq. 83.9.

$$I_f = \frac{V}{R_f} = \frac{112 \text{ V}}{100 \text{ } \Omega}$$
$$= 1.12 \text{ A}$$

The total full-load line current is known to be 40 A. The full-load armature current is

$$I_a = I_l - I_f = 40 \text{ A} - 1.12 \text{ A}$$
$$= 38.88 \text{ A}$$

The field copper loss is given by Eq. 83.11.

$$P_{\text{Cu},f} = I_f^2 R_f = (1.12 \text{ A})^2 (100 \text{ } \Omega)$$
$$= 125.4 \text{ W}$$

The armature copper loss is given by Eq. 83.11.

$$P_{\text{Cu},a} = I_a^2 R_a = (38.88 \text{ A})^2 (0.125 \text{ } \Omega)$$
$$= 189.0 \text{ W}$$

The total copper loss is

$$P_{\text{Cu},t} = P_{\text{Cu},f} + P_{\text{Cu},a} = 125.4 \text{ W} + 189.0 \text{ W}$$
$$= 314.4 \text{ W}$$

Stray power is determined from the no-load conditions. At no load, the field current is given by Eq. 83.9.

$$I_f = \frac{V}{R_f} = \frac{106 \text{ V}}{100 \text{ } \Omega}$$
$$= 1.06 \text{ A}$$

At no load, the armature current is

$$I_a = I_l - I_f = 3 \text{ A} - 1.06 \text{ A}$$
$$= 1.94 \text{ A}$$

The stray power loss is

$$\begin{aligned}
P_{\text{stray}} &= V_l I_a - I_a^2 R_a \\
&= (106 \text{ V})(1.94 \text{ A}) - (1.94 \text{ A})^2 (0.125 \text{ } \Omega) \\
&= 205.2 \text{ W}
\end{aligned}$$

The stray power loss is assumed to be independent of the load. The total losses at full load are

$$\begin{aligned}
P_{\text{loss}} &= P_{\text{Cu},t} + P_{\text{stray}} = 314.4 \text{ W} + 205.2 \text{ W} \\
&= 519.6 \text{ W}
\end{aligned}$$

The power input to the motor when fully loaded is

$$P_{\text{input}} = I_l V_l = (40 \text{ A})(112 \text{ V}) = 4480 \text{ W}$$

The efficiency is

$$\begin{aligned}
\eta &= \frac{\text{input power} - \text{power losses}}{\text{input power}} \\
&= \frac{4480 \text{ W} - 519.6 \text{ W}}{4480 \text{ W}} = 0.884 \quad (88.4\%)
\end{aligned}$$

(b) The real power (in hp) delivered at full load is

$$\begin{aligned}
P_{\text{real}} &= \text{input power} - \text{power losses} \\
&= \frac{4480 \text{ W} - 519.6 \text{ W}}{\left(1000 \frac{\text{W}}{\text{kW}}\right)\left(0.7457 \frac{\text{kW}}{\text{hp}}\right)} \\
&= 5.3 \text{ hp}
\end{aligned}$$

46. HIGH-EFFICIENCY MOTORS AND DRIVES

A premium, energy-efficient motor will have approximately 50% of the losses of a conventional motor. Due to the relatively high overall efficiency enjoyed by all motors, however, this translates into only a 5% increase in overall efficiency.

High-efficiency motors are often combined with *variable-frequency drives* (VFDs) to achieve continuous-variable speed control.[29] VFDs can substantially reduce the power drawn by the process. For example, with a motor-driven pump or fan, the motor can be slowed down instead of closing a valve or damper when the flow requirements decrease. Since the required motor horsepower varies with the cube root of the speed, for processes that do not always run at full-flow (fluid

---

[29] Prior to VFDs, there were two common ways to change the speed of an induction motor: (a) increasing the slip, and (b) changing the number of pole-pairs. Increasing the slip was accomplished by under-magnetizing the motor so that it received less input voltage than it was built for. Dropping or adding the number of active poles resulted in the speed changing by a factor of 2 (or $^1/_2$).

pumping, metering, flow control, etc.), the energy savings can be substantial.

A VFD uses an electronic controller to produce a variable-frequency signal that is not an ideal sine wave. This results in additional heating (e.g., an increase of 20 to 40%) from copper and core losses in the motor. In pumping applications, though, the process load drops off faster than additional heat is produced. Thus, the process power savings dominate.

Most low- and medium-power motors implement VFD with DC voltage intermediate circuits, a technique known as *voltage-source inversion*. Voltage-source inversion is further subdivided into *pulse-width modulation* (PWM) and *pulse-amplitude modulation* (PAM).

Under VFD control, the motor torque is approximately proportional to the applied voltage and drawn current but inversely proportional to the applied frequency.

$$T \propto \frac{VI}{f} \qquad \textit{83.83}$$

84 Instrumentation and Measurements[1]

Nomenclature

A	area	in^2	m^2
b	base length	in	m
BC	bridge constant	–	–
c	distance from neutral axis	in	m
C	concentration	various	various
d	diameter	in	m
E	modulus of elasticity	psi	Pa
F	force	lbf	N
G	shear modulus	psi	Pa
GF	gage factor	–	–
h	height	in	m
I	current	A	A
I	moment of inertia	in^4	m^4
J	polar moment of inertia	in^4	m^4
k	constant	various	various
k	deflection constant	lbf/in	N/m
K	factor	–	–
L	shaft length	in	m
M	moment	in-lbf	N·m
n	number	–	–
n	rotational speed	rpm	rpm
p	pressure	psi	Pa
P	permeability	various	various
P	power	hp	kW
Q	statical moment	in^3	m^3
r	radius	in	m
R	resistance	Ω	Ω
t	thickness	in	m
T	temperature	1/°R	K
T	torque	in-lbf	N·m
V	voltage	V	V
VR	voltage ratio	–	–
y	deflection	in	m

Symbols

α	temperature coefficient	1/°R	1/K
β	constant	°R	K
β	temperature coefficient	1/°R^2	1/K^2
γ	shear strain	–	–
ϵ	strain	–	–
η	efficiency		
θ	angle of twist	deg	deg
ν	Poisson's ratio	–	–
ρ	resistivity	Ω-in	Ω·cm
σ	stress	psi	Pa
τ	shear stress	psi	Pa
ϕ	angle of twist	rad	rad

Subscripts

b	battery
g	gage
o	original
r	ratio
ref	reference
t	transverse or total
T	at temperate T
x	in x-direction
y	in y-direction

---

[1]Measurement of fluid pressure is covered in Ch. 15; measurement of fluid flow is covered in Ch. 17.

1. ACCURACY

A measurement is said to be *accurate* if it is substantially unaffected by (i.e., is insensitive to) all variation outside of the measurer's control.

For example, suppose a rifle is aimed at a point on a distant target and several shots are fired. The target point represents the "true value" of a measurement—the value that should be obtained. The impact points represent the measured values—what is obtained. The distance from the centroid of the points of impact to the target point is a measure of the alignment accuracy between the barrel and the sights. This difference between the true and measured values is known as the measurement *bias*.

2. PRECISION

Precision is not synonymous with *accuracy*. Precision is concerned with the repeatability of the measured results. If a measurement is repeated with identical results, the experiment is said to be precise. The average distance of each impact from the centroid of the impact group is a measure of precision. Thus, it is possible to take highly precise measurements and still have a large bias.

Most measurement techniques (e.g., taking multiple measurements and refining the measurement methods or procedures) that are intended to improve accuracy actually increase the precision.

Sometimes the term *reliability* is used with regard to the precision of a measurement. A *reliable measurement* is the same as a *precise estimate*.

3. STABILITY

Stability and *insensitivity* are synonymous terms. (Conversely, *instability* and *sensitivity* are synonymous.) A stable measurement is insensitive to minor changes in the measurement process.

Example 84.1

At 65°F (18°C), the centroid of an impact group on a rifle target is 2.1 in (5.3 cm) from the sight-in point. At 80°F (27°C), the distance is 2.3 in (5.8 cm). What is the sensitivity to temperature?

SI Solution

$$
\begin{aligned}
\text{sensitivity to} \atop \text{temperature} &= \frac{\Delta \text{measurement}}{\Delta \text{temperature}} \\
&= \frac{5.8 \text{ cm} - 5.3 \text{ cm}}{27°C - 18°C} \\
&= 0.0556 \text{ cm/°C}
\end{aligned}
$$

Customary U.S. Solution

$$
\begin{aligned}
\text{sensitivity to} \atop \text{temperature} &= \frac{\Delta \text{measurement}}{\Delta \text{temperature}} \\
&= \frac{2.3 \text{ in} - 2.1 \text{ in}}{80°F - 65°F} \\
&= 0.0133 \text{ in/°F}
\end{aligned}
$$

4. CALIBRATION

Calibration is used to determine or verify the scale of the measurement device. In order to calibrate a measurement device, one or more known values of the quantity to be measured (temperature, force, torque, etc.) are applied to the device and the behavior of the device is noted. (If the measurement device is linear, it may be adequate to use just a single calibration value. This is known as *single-point calibration*.)

Once a measurement device has been calibrated, the calibration signal should be reapplied as often as necessary to prove the reliability of the measurements. In some electronic measurement equipment, the calibration signal is applied continuously.

5. ERROR TYPES

Measurement errors can be categorized as *systematic (fixed) errors*, *random (accidental) errors*, *illegitimate errors*, and *chaotic errors*.

Systematic errors, such as improper calibration, use of the wrong scale, and incorrect (though consistent) technique, are essentially constant or similar in nature over time. *Loading error* is a systematic error and occurs when the act of measuring alters the true value.[2] Some *human errors*, if present in each repetition of the measurement, are also systematic. Systematic errors can be reduced or eliminated by refinement of the experimental method.

Random errors are caused by random and irregular influences generally outside the control of the measurer. Such errors are introduced by fluctuations in the environment, changes in the experimental method, and variations in materials and equipment operation. Since the occurrence of these errors is irregular, their effects can be reduced or eliminated by multiple repetitions of the experiment.

There is no reason to expect or tolerate *illegitimate errors* (e.g., errors in computations and other blunders). These are essentially mistakes that can be avoided through proper care and attention.

Chaotic errors, such as resonance, vibration, or experimental "noise," essentially mask or entirely invalidate the experimental results. Unlike the random errors previously mentioned, chaotic disturbances are sufficiently large to reduce the experimental results to meaninglessness.[3] Chaotic errors must be eliminated.

---

[2]For example, inserting an air probe into a duct will change the flow pattern and velocity around the probe.
[3]Much has been written in recent years about *chaos theory*. This theory holds that, for many processes, the ending state is dependent on imperceptible differences in the starting state. Future weather conditions and the landing orientation of a finely balanced spinning top are often used as examples of states that are greatly affected by their starting conditions.

6. ERROR MAGNITUDES

If a single measurement is taken of some quantity whose true value is known, the *error* is simply the difference between the true and measured values. However, the true value is never known in an experiment, and measurements are usually taken several times, not just once. Therefore, many conventions exist for estimating the unknown error.[4]

When most experimental quantities are measured, the measurements tend to cluster around some "average value." The measurements will be distributed according to some distribution, such as linear, normal, Poisson, and so on. The measurements can be graphed in a *histogram* and the distribution inferred. Usually the data will be normally distributed.[5]

Certain error terms used with normally distributed data have been standardized. These are listed in Table 84.1.

Table 84.1 Normal Distribution Error Terms

term	number of standard deviations	percent certainty	approximate odds of being incorrect
probable error	0.6745	50	1 in 2
mean deviation	0.6745	50	1 in 2
standard deviation	1.000	68.3	1 in 3
one-sigma error	1.000	68.3	1 in 3
90% error	1.6449	90	1 in 10
two-sigma error	2.000	95	1 in 20
three-sigma error	3.000	99.7	1 in 370
maximum error[a]	3.29	99.9+	1 in 1000

[a]The true maximum error is theoretically infinite.

7. POTENTIOMETERS

A *potentiometer (potentiometer transducer, variable resistor)* is a resistor with a sliding third contact. It converts linear or rotary motion into a variable resistance (voltage).[6] It consists of a resistance oriented in a linear or angular manner and a variable-position contact point known as the *tap*. A voltage is applied across the entire resistance, causing current to flow through the resistance. The voltage at the tap will vary with tap position.

---

[4]This may be a good time to review the material in Ch. 11.
[5]The results of all numerical experiments are not automatically normally distributed. The throw of a die (one of two dice) is linearly distributed. Emissive power of a heated radiator is skewed with respect to wavelength. However, the means of sets of experimental data generally will be normally distributed, even if the raw measurements are not.
[6]There is a voltage-balancing device that shares the name *potentiometer (potentiometer circuit)*. An unknown voltage source can be measured by adjusting a calibrated voltage until a null reading is obtained on a voltage meter. The applications are sufficiently different that no confusion occurs when the "pot is adjusted."

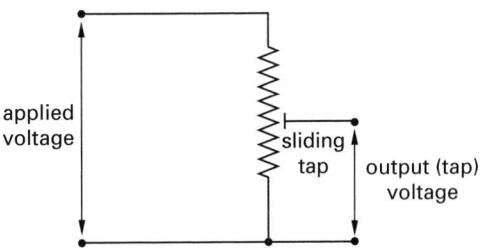

Figure 84.1 Potentiometer

8. TRANSDUCERS

Physical quantities are often measured with transducers (*detector-transducers*). A *transducer* converts the measured quantity to a second quantity that is measured. For example, a Bourdon tube pressure gage converts pressure to angular displacement; a strain gage converts stress to resistance change. Transducers are primarily mechanical in nature (e.g., pitot tube, spring devices, Bourdon tube pressure gage) or electrical in nature (e.g., thermocouple, strain gage, moving-core transformer).

9. SENSORS

While the term "transducer" is commonly used for devices that respond to mechanical input (force, pressure, torque, etc.), the term *sensor* is commonly applied to devices that respond to varying chemical conditions.[7] For example, an electrochemical sensor might respond to a specific gas, compound, or ion (known as a *target substance* or *species*). Two types of electrochemical sensors are in use today: potentiometric and amperometric.

Potentiometric sensors generate a measurable voltage at their terminals. In electrochemical sensors taking advantage of half-cell reactions at electrodes, the generated voltage is proportional to the absolute temperature, T, and is inversely proportional to the number of electrons, n, taking part in the chemical reaction at the half-cell. In Eq. 84.1, p_1 is the partial pressure of the target substance at the measurement electrode; p_2 is the partial pressure of the target substance at the reference electrode.

$$V \propto \left(\frac{T_{\text{absolute}}}{n} \right) \ln \frac{p_1}{p_2} \qquad 84.1$$

Amperometric sensors (also known as *voltammetric sensors*) generate a measurable current at their terminals. In the conventional electrochemical sensors known as *diffusion-controlled cells*, a high-conductivity acid or alkaline liquid electrolyte is used with a gas-permeable membrane that transmits ions from the outside to the inside of the sensor. A reference voltage is applied to two terminals within the electrolyte, and the current generated at a (third) sensing electrode is measured.

---

[7]The categorization is common but not universal. The terms "transducer," "sensor," and "pickup" are often used loosely.

The maximum current generated is known as the *limiting current*. Current is proportional to the concentration (C) of the target substance, the permeability (P), the exposed sensor (membrane) area (A), and the number of electrons transferred per molecule detected (n). The current is inversely proportional to the membrane thickness (t).

$$I \propto \frac{nPCA}{t} \qquad 84.2$$

10. VARIABLE-INDUCTANCE TRANSDUCERS

Inductive transducers contain a wire coil and a moving permeable *core*.[8] As the core moves, the flux linkage through the coil changes. The change in inductance affects the overall impedance of the detector circuit.

The *differential transformer* or *linear variable differential transformer* (LVDT) is an important type of *variable-inductance transducer*. It converts linear motion into a change in voltage. The transformer is supplied with a low AC voltage. When the core is centered between the two secondary windings, the LVDT is said to be in its *null position*.

Movement of the core changes the magnetic flux linkage between the primary and secondary windings. Over a reasonable displacement range, the output voltage is proportional to the displacement of the core from the null position, hence the description "linear." The voltage changes phase (by 180°) as the core passes through the null position.

Figure 84.2 *Linear Variable Differential Transformer Schematic and Performance Characteristic*

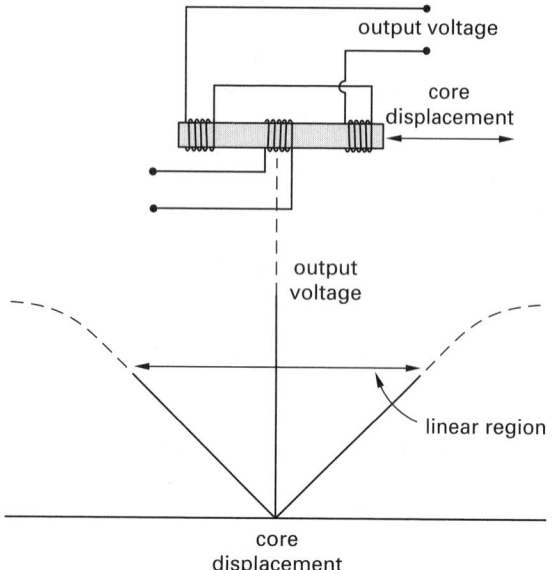

---

[8]The term *core* is used even if the cross sections of the coil and core are not circular.

Sensitivity of a LVDT is measured in mV/in (mV/cm) of core movement. The sensitivity and output voltage depend on the frequency of the applied voltage (i.e., the *carrier frequency*) and are directly proportional to the magnitude of the applied voltage.

11. VARIABLE-RELUCTANCE TRANSDUCERS

A *variable-reluctance transducer (pickup)* is essentially a permanent magnet and a coil in the vicinity of the process being monitored.[9] There are no moving parts in this type of transducer. However, some of the magnet's magnetic flux passes through the surroundings, and the presence or absence of the process changes the coil voltage. Two typical applications of variable-reluctance pickups are measuring liquid levels and determining the rotational speed of a gear.

12. VARIABLE-CAPACITANCE TRANSDUCERS

In *variable-capacitance transducers*, the capacitance of a device can be modified by changing the plate separation, plate area, or dielectric constant of the medium separating the plates.

13. OTHER ELECTRICAL TRANSDUCERS

The *piezoelectric effect* is the name given to the generation of an electrical voltage when placed under stress.[10] *Piezoelectric transducers* generate a small voltage when stressed (strained). Since voltage is developed during the application of changing strain but not while strain is constant, piezoelectric transducers are limited to dynamic applications. Piezoelectric transducers may suffer from low voltage output, instability, and limited ranges in operating temperature and humidity.

The *photoelectric effect* is the generation of an electrical voltage when a material is exposed to light.[11] Devices using this effect are known as *photocells, photovoltaic cells, photosensors*, or *light-sensitive detectors*, depending on the applications. The sensitivity need not be to light in the visible spectrum. Photoelectric detectors can be made that respond to infrared and ultraviolet radiation. The magnitude of the voltage (or of the

---

[9]*Reluctance* depends on the area, length, and permeability of the medium through which the magnetic flux passes.

[10]Quartz, table sugar, potassium sodium tartarate (Rochelle salt), and barium titanate are examples of piezoelectric materials. Quartz is commonly used to provide a stable frequency in electronic oscillators. Barium titanate is used in some ultrasonic cleaners and sonar-like equipment.

[11]While the *photogenerative (photovoltaic)* definition is the most common definition, the term "photoelectric" can also be used with *photoconductive devices* (those whose resistance changes with light) and *photoemissive devices* (those that emit light when a voltage is applied).

current in an attached circuit) depends on the amount of illumination. If the cell is reverse-biased by an external battery, its operation is similar to a constant-current source.[12]

Figure 84.3 *Photovoltaic Device Characteristic Curves*

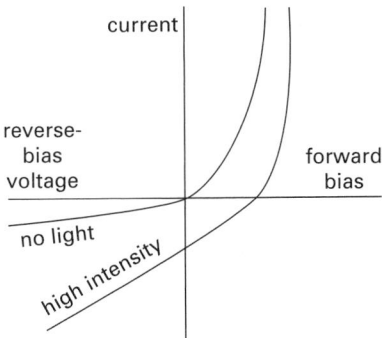

and β.[14] The variation of resistance with temperature is nonlinear, though β is small and is often insignificant over short temperature ranges. Therefore, a linear relationship is often assumed and only α is used. In Eq. 84.3, $R_{\rm ref}$ is the resistance at the reference temperature, $T_{\rm ref}$, usually 100 Ω at 32°F (0°C).

$$R_T \approx R_{\rm ref}(1 + \alpha \Delta T + \beta \Delta T^2) \qquad 84.3$$

$$\Delta T = T - T_{\rm ref} \qquad 84.4$$

In commercial RTDs, α is referred to by the literal term *alpha-value*. There are two applicable alpha values for platinum, depending on the purity. Commercial platinum RTDs produced in the United States generally have alpha values of 0.00392 1/°C, while RTDs produced in Europe and other countries generally have alpha values of 0.00385 1/°C.

14. PHOTOSENSITIVE CONDUCTORS

Cadmium sulfide and *cadmium selenide* are two compounds that decrease in resistance when exposed to light. Cadmium sulfide is most sensitive to light in the 5000Å to 6000Å (0.5 to 0.6 μm) range, while cadmium selenide shows peak sensitivities in the 7000Å to 8000Å range (0.7 to 0.8 μm). These compounds are used in *photosensitive conductors*. Due to hysteresis, photosensitive conductors do not react instantaneously to changes in intensity.[13] High-speed operation requires high light intensities and careful design.

15. RESISTANCE TEMPERATURE DETECTORS

Resistance temperature detectors (RTDs), also known as *resistance thermometers*, make use of changes in their resistance to determine changes in temperature. A fine wire is wrapped around a form and protected with glass or a ceramic coating. Nickel and copper are commonly used for industrial RTDs. Platinum is used when precision resistance thermometry is required. RTDs are connected through resistance bridges (Sec. 19) to compensate for lead resistance.

Resistance in most conductors increases with temperature. The resistance at a given temperature can be calculated from the *coefficients of thermal resistance*, α

16. THERMISTORS

Thermistors are temperature-sensitive semiconductors constructed from oxides of manganese, nickel, and cobalt and from sulfides of iron, aluminum, and copper. Thermistor materials are encapsulated in glass or ceramic materials to prevent penetration of moisture. Unlike RTDs, the resistance of thermistors decreases as the temperature increases.

Thermistor temperature-resistance characteristics are exponential. Depending on the brand, material, and construction, β typically varies between 3400 K and 3900 K.

$$R = R_o e^k \qquad 84.5$$

$$k = \beta \left(\frac{1}{T} - \frac{1}{T_o} \right) \qquad [T \text{ in K}] \qquad 84.6$$

Thermistors can be connected to measurement circuits with copper wire and soldered connections. Compensation of lead wire effects is not required because resistance of thermistors is very large, far greater than the resistance of the leads. Since the negative temperature characteristic makes it difficult to design customized detection circuits, some thermistor and instrumentation standardization has occurred. The most common thermistors have resistances of 2252 Ω at 77°F (25°C), and most instrumentation is compatible with them. Other standardized resistances are 5000 Ω and 10,000 Ω at 25°C.

Thermistors typically are less precise and more unpredictable than metallic resistors. Since resistance varies exponentially, most thermistors are suitable for use only up to approximately 550°F (290°C).

---

[12] A semiconductor device is *reverse-biased* when a negative battery terminal is connected to a *p*-type semiconductor material in the device, or when a positive battery terminal is connected to an *n*-type semiconductor material.

[13] *Hysteresis* is the tendency for the transducer to continue to respond (i.e., indicate) when the load is removed. Alternatively, hysteresis is the difference in transducer outputs when a specific load is approached from above and from below. Hysteresis is usually expressed in percent of the full-load reading during any single calibration cycle.

[14] Higher-order terms (third, fourth, etc.) are used when extreme accuracy is required.

Table 84.2 *Typical Resistivities and Coefficients of Thermal Resistance[a]*

conductor	resistivity[b] (Ω·cm)	$\alpha^{c,d}$ (1/°C)
alumel[e]	28.1×10^{-6}	0.0024 @ 212°F (100°C)
aluminum	2.82×10^{-6}	0.0039 @ 68°F (20°C)
		0.0040 @ 70°F (21°C)
brass	7×10^{-6}	0.002 @ 68°F (20°C)
constantan[f,g]	49×10^{-6}	0.00001 @ 68°F (20°C)
chromel[h]	—	—
copper, annealed	1.724×10^{-6}	0.0043 @ 32°F (0°C)
		0.0039 @ 70°F (21°C)
		0.0037 @ 100°F (38°C)
		0.0031 @ 200°F (93°C)
gold	2.44×10^{-6}	0.0034 @ 68°F (20°C)
iron (99.98% pure)	10×10^{-6}	0.005 @ 68°F (20°C)
isoelastic[i]	112×10^{-6}	0.00047
lead	22×10^{-6}	0.0039
magnesium	4.6×10^{-6}	0.004 @ 68°F (20°C)
manganin[j]	44×10^{-6}	0.0000 @ 68°F (20°C)
monel[k]	42×10^{-6}	0.002 @ 68°F (20°C)
nichrome[l]	100×10^{-6}	0.0004 @ 68°F (20°C)
nickel	7.8×10^{-6}	0.006 @ 68°F (20°C)
platinum	10×10^{-6}	0.0039 @ 32°F (0°C)
		0.0036 @ 70°F (21°C)
platinum-iridium[m]	24×10^{-6}	0.0013
platinum-rhodium[n]	18×10^{-6}	0.0017 @ 212°F (100°C)
silver	1.59×10^{-6}	0.004 @ 68°F (20°C)
tin	11.5×10^{-6}	0.0042 @ 68°F (20°C)
tungsten (drawn)	5.8×10^{-6}	0.0045 @ 70°F (21°C)

[a]Compiled from various sources. Data is not to be taken too literally, as values depend on composition and cold working.
[b]At 20°C (68°F)
[c]Values vary with temperature. Common values given when no temperature is specified.
[d]Multiply 1/°C by 5/9 to obtain 1/°F. Multiply ppm/°F by 1.8×10^{-6} to obtain 1/°C.
[e]Trade name for 94% Ni, 2.5% Mn, 2% Al, 1% Si, 0.5% Fe (TM of Hoskins Manufacturing Co.)
[f]60% Cu, 40% Ni, also known by trade names *Advance, Eureka,* and *Ideal.*
[g]Constantan is also the name given to the composition 55% Cu and 45% Ni, an alloy with slightly different properties.
[h]Trade name for 90% Ni, 10% Cr (TM of Hoskins Manufacturing Co.)
[i]36% Ni, 8% Cr, 0.5% Mo, remainder Fe
[j]9 to 18% Mn, 11 to 4% Ni, remainder Cu
[k]33% Cu, 67% Ni
[l]75% Ni, 12% Fe, 11% Cr, 2% Mn
[m]95% Pt, 5% Ir
[n]90% platinum, 10% rhodium

17. THERMOCOUPLES

A *thermocouple* consists of two wires of dissimilar metals joined at both ends.[15] One set of ends, typically called a *junction*, is kept at a known *reference*

[15]The joint may be made by simply twisting the ends together. However, to achieve a higher mechanical strength and a better electrical connection, the ends should be soldered, brazed, or welded.

temperature while the other junction is exposed to the unknown temperature.[16] In a laboratory, the reference junction is often maintained at the *ice point*, 32°F (0°C), in an ice/water bath for convenience in later analysis. In commercial applications, the reference temperature can be any value, with appropriate compensation being made.

Figure 84.4 *Thermocouple*

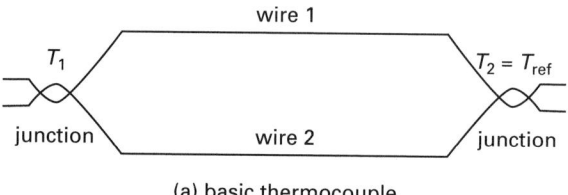

(a) basic thermocouple

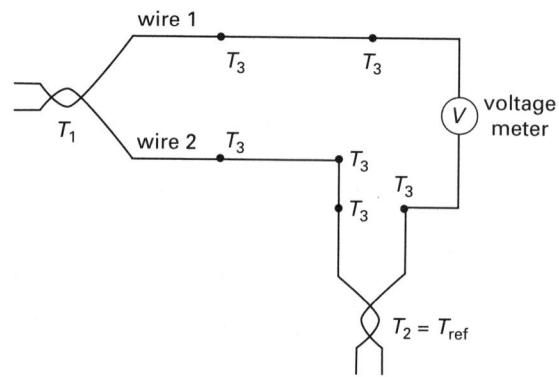

(b) thermocouple in measurement circuit

Thermocouple materials, standard ANSI designations, and approximate useful temperature ranges are given in Table 84.3.[17] The "usable" temperature range can be much larger than the useful range. The most significant factors limiting the useful temperature range, sometimes referred to as the *error limits range*, are linearity, the rate at which the material will erode due to oxidation at higher temperatures, irreversible magnetic effects above magnetic critical points, and longer stabilization periods at higher temperatures.

A voltage is generated when the temperatures of the two junctions are different. This phenomenon is known as the *Seebeck effect*.[18] Referring to the polarities of

[16]The *ice point* is the temperature at which liquid water and ice are in equilibrium. Other standardized temperature references are: *oxygen point,* −297.346°F (90.19K); *steam point,* 212.0°F (373.16K); *sulfur point,* 832.28°F (717.76K); *silver point,* 1761.4°F (1233.96K); and *gold point,* 1945.4°F (1336.16K).
[17]It is not uncommon to list the two thermocouple materials with "vs." (as in *versus*). For example, a copper/constantan thermocouple might be designated as "copper vs. constantan."
[18]The inverse of the Seebeck effect, that current flowing through a junction of dissimilar metals will cause either heating or cooling, is the *Peltier effect*, though the term is generally used in regard to cooling applications. An extension of the Peltier effect, known as the *Thompson effect*, is that heat will be carried along the conductor. Both the Peltier and Thompson effects occur simultaneously with the Seebeck effect. However, the Peltier and Thompson effects are so minuscule that they can be disregarded.

Table 84.3 *Typical Temperature Ranges*
of Thermocouple Materials[a]

materials	ANSI designation	useful range, °F (°C)
copper-constantan	T	−300 to 700 (−180 to 370)
chromel-constantan	E	32 to 1600 (0 to 870)
iron-constantan[b]	J	32 to 1400 (0 to 760)
chromel-alumel	K	32 to 2300 (0 to 1260)
platinum-10% rhodium	S	32 to 2700 (0 to 1480)
platinum-13% rhodium	R	32 to 2700 (0 to 1480)
Pt-6% Rh-Pt-30% Rh	B	1600 to 3100 (870 to 1700)
tungsten-Tu-25% rhenium	–	to 4200[c] (2320)
Tu-5% rhenium-Tu-26% rhenium	–	to 4200[c] (2320)
Tu-3% rhenium-Tu-25% rhenium	–	to 4200[c] (2320)
iridium-rhodium	–	to 3500[c] (1930)
nichrome-constantan	–	to 1600[c] (870)
nichrome-alumel	–	to 2200[c] (1200)

[a]Actual values will depend on wire gauge, atmosphere (oxidizing or reducing), use (continuous or intermittent), and manufacturer.
[b]Nonoxidizing atmospheres only.
[c]Approximate usable temperature range. Error limit range is less.

the voltage generated, one metal is known as the *positive element* while the other is the *negative element*. The generated voltage is small, and thermocouples are calibrated in $\mu V/°F$ or $\mu V/°C$. An amplifier may be required to provide usable signal levels, although thermocouples can be connected in series (a *thermopile*) to increase the value.[19] The accuracy (referred to as the *calibration*) of thermocouples is approximately $1/2$ to $3/4$%, though manufacturers produce thermocouples with various guaranteed accuracies.

The voltage generated by a thermocouple is given by Eq. 84.7. Since the *thermoelectric constant*, k_T, varies with temperature, thermocouple problems are solved with published tables of total generated voltage versus temperature. (See App. 84.A.)

$$V = k_T(T - T_{\text{ref}}) \qquad 84.7$$

Generation of thermocouple voltage in a measurement circuit is governed by three laws. The *law of homogeneous circuits* states that the temperature distribution along one or both of the thermocouple leads is irrelevant. Only the junction temperatures contribute to the generated voltage.

The *law of intermediate metals* states that an intermediate length of wire placed within one leg or at the junction of the thermocouple circuit will not affect the voltage generated as long as the two new junctions are

[19]There is no special name for a combination of thermocouples connected in parallel.

at the same temperature. This law permits the use of a measuring device, soldered connections, and extension leads.

The *law of intermediate temperatures* states that if a thermocouple generates voltage V_1 when its junctions are at T_1 and T_2, and it generates voltage V_2 when its junctions are at T_2 and T_3, then it will generate voltage $V_1 + V_2$ when its junctions are at T_1 and T_3.

Example 84.2

A type-K (chromel-alumel) thermocouple produces a voltage of 10.79 mV. The "cold" junction is kept at 32°F (0°C) by an ice bath. What is the temperature of the hot junction?

Solution

Since the cold junction temperature corresponds to the reference temperature, the hot junction temperature is read directly from App. 84.A as 510°F.

Example 84.3

A type-K (chromel-alumel) thermocouple produces a voltage of 10.87 mV. The "cold" junction is at 70°F. What is the temperature of the hot junction?

Solution

Use the law of intermediate temperatures. From App. 84.A, the thermoelectric constant for 70°F is 0.84 mV. If the cold junction had been at 32°C, the generated voltage would have been higher. The corrected reading is

$$10.87 \text{ mV} + 0.84 \text{ mV} = 11.71 \text{ mV}$$

The temperature corresponding to this voltage is 550°F.

Example 84.4

A type-T (copper-constantan) thermocouple is connected directly to a voltage meter. The temperature of the meter's screw-terminals is measured by a nearby thermometer as 70°F. The thermocouple generates 5.262 mV. What is the temperature of its hot junction?

Solution

There are two connections at the meter. However, both connections are at the same temperature, so the meter can be considered to be a length of different wire, and the law of intermediate metals applies. Since the meter connections are not at 32°F, the law of intermediate temperatures applies. The corrected voltage is

$$5.262 \text{ mV} + 0.832 \text{ mV} = 6.094 \text{ mV}$$

From App. 84.A, 6.094 mV corresponds to 280°F.

18. STRAIN GAGES

A *bonded strain gage* is a metallic resistance device that is cemented to the surface of the unstressed member.[20] The gage consists of a metallic conductor (known as the *grid*) on a backing (known as the *substrate*).[21] The substrate and grid experience the same strain as the surface of the member. The resistance of the gage changes as the member is stressed due to changes in conductor cross section and intrinsic changes in resistivity with strain. Temperature effects must be compensated for by the circuitry or by using a second unstrained gage as part of the bridge measurement system. (See Sec. 19.)

When simultaneous strain measurements in two or more directions are needed, it is convenient to use a commercial *rosette strain gage*. A rosette consists of two or more *grids* properly oriented for application as a single unit.

Figure 84.5 *Strain Gage*

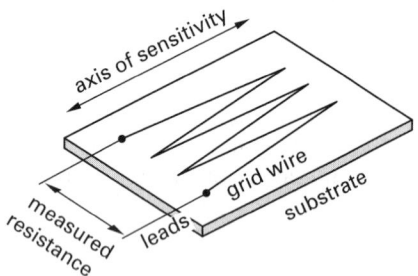

(a) folded-wire strain gage

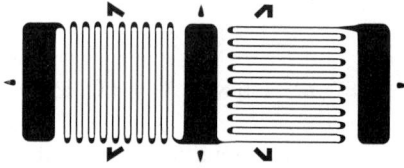

(b) commercial two-element rosette

The *gage factor (strain sensitivity factor)*, GF, is the ratio of the fractional change in resistance to the fractional change in length (strain) along the detecting axis of the gage. The gage factor is a function of the gage material. It can be calculated from the grid material's properties and configuration. The higher the gage factor, the greater the sensitivity of the gage. From a

---

[20] A *bonded strain gage* is constructed by bonding the conductor to the surface of the member. An *unbonded strain gage* is constructed by wrapping the conductor tightly around the member or between two points on the member.

Strain gages on rotating shafts are usually connected through *slip rings* to the measurement circuitry.

[21] The grids of strain gages were originally of the folded-wire variety. For example, nichrome wire with a total resistance under 1000 Ω was commonly used. Modern strain gages are generally of the foil type manufactured by printed circuit techniques. Semiconductor gages are also used when extreme sensitivity (i.e., gage factors in excess of 100) is required. However, semiconductor gages are extremely temperature-sensitive.

practical standpoint, however, the gage factor and gage resistance are provided by the gage manufacturer. Only the change in resistance is measured.

$$
\begin{aligned}
\mathrm{GF} &= 1 + 2\nu + \dfrac{\dfrac{\Delta\rho}{\rho_o}}{\dfrac{\Delta L}{L_o}} \\[2ex]
&= \dfrac{\dfrac{\Delta R_g}{R_g}}{\dfrac{\Delta L}{L_o}} \\[2ex]
&= \dfrac{\dfrac{\Delta R_g}{R_g}}{\epsilon}
\end{aligned}
$$

84.8

Table 84.4 *Approximate Gage Factors*[a]

material	GF
constantan	2.0
iron, soft	4.2
isoelastic	3.5
manganin	0.47
monel	1.9
nichrome	2.0
nickel	-12[b]
platinum	4.8
platinum-iridium	5.1

[a] Other properties of strain gage materials are listed in Table 84.2.
[b] Value depends on amount of preprocessing and cold working.

Constantan and isoelastic wires along with metal foil with gage factors of approximately 2 and initial resistances of less than 1000 Ω (typically 120 Ω, 350 Ω, 600 Ω, and 700 Ω) are commonly used. In practice, the gage factor and initial gage resistance, R_g, are specified by the manufacturer of the gage. Once the strain sensitivity factor is known, the strain, ϵ, can be determined from the change in resistance. Strain is often reported in units of $\mu\mathrm{in/in}$ ($\mu\mathrm{m/m}$) and is given the name *microstrain*.

$$
\epsilon = \frac{\Delta R_g}{(\mathrm{GF}) R_g}
$$

84.9

Theoretically, a strain gage should not respond to strain in its transverse direction. However, the turn-around end-loops are also made of strain-sensitive material, and the end-loop material contributes to a nonzero sensitivity to strain in the transverse direction. Equation 84.10 defines the *transverse sensitivity factor*, K_t, which is of academic interest in most problems. The transverse sensitivity factor is seldom greater than 2%.

$$
K_t = \frac{(\mathrm{GF})_{\mathrm{transverse}}}{(\mathrm{GF})_{\mathrm{longitudinal}}}
$$

84.10

Example 84.5

A strain gage with a nominal resistance of 120 Ω and gage factor of 2.0 is used to measure a strain of 1 μin/in. What is the change in resistance?

Solution

From Eq. 84.9,

$$\Delta R_g = (GF)R_g\epsilon$$
$$= (2.0)(120 \ \Omega)\left(1 \times 10^{-6} \ \frac{in}{in}\right) = 2.4 \times 10^{-4} \ \Omega$$

19. WHEATSTONE BRIDGES

The *Wheatstone bridge* shown in Fig. 84.6 is one type of *resistance bridge*.[22] The bridge can be used to determine the unknown resistance of a resistance transducer (e.g., thermistor or resistance-type strain gage), say R_1 in Fig. 84.6. The potentiometer is adjusted (i.e., the bridge is "balanced") until no current flows through the meter or until there is no voltage across the meter (hence the name *null indicator*).[23,24] When the bridge is balanced and no current flows through the meter leg, Eqs. 84.11 through 84.14 are applicable.

$$I_2 = I_4 \quad \text{[balanced]} \qquad 84.11$$
$$I_1 = I_3 \quad \text{[balanced]} \qquad 84.12$$
$$V_1 + V_3 = V_2 + V_4 \quad \text{[balanced]} \qquad 84.13$$
$$\frac{R_1}{R_2} = \frac{R_3}{R_4} \quad \text{[balanced]} \qquad 84.14$$

Figure 84.6 *Series-Balance Wheatstone Bridge*

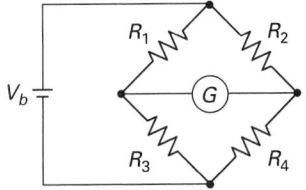

Since any one of the four resistances can be the unknown, up to three of the remaining resistances can be fixed or adjustable, and the battery and meter can be connected to either of two diagonal corners, it is sometimes confusing to apply Eq. 84.14 literally. However, the following bridge law statement can be used to help formulate the proper relationship: *When a series*

Wheatstone bridge is null-balanced, the ratio of resistance of any two adjacent arms equals the ratio of resistance of the remaining two arms, taken in the same sense. In this statement, "taken in the same sense" means that both ratios must be formed reading either left to right, right to left, top to bottom, or bottom to top.

20. STRAIN GAGE DETECTION CIRCUITS

The resistance of a strain gage can be measured by placing the gage in either a ballast circuit or bridge circuit. A *ballast circuit* consists of a voltage source (V_b) of less than 10 V (typical), a current-limiting ballast resistance (R_b), and the strain gage of known resistance (R_g) in series. This is essentially a voltage-divider circuit. (See Chap. 83.) The change in voltage (ΔV_g) across the strain gage is measured. The strain (ϵ) can be determined from Eq. 84.15.

$$\Delta V_g = \frac{(GF)\epsilon V_b R_b R_g}{(R_b + R_g)^2} \qquad 84.15$$

Figure 84.7 *Ballast Circuit*

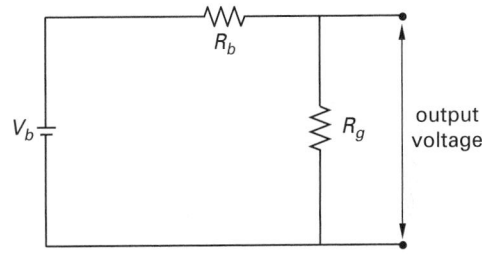

Ballast circuits do not provide temperature compensation, nor is their sensitivity adequate for measuring static strain. Ballast circuits, where used, are often limited to measurement of transient strains. A bridge detection circuit overcomes these limitations.

Figure 84.8 illustrates how a strain gage can be used with a resistance bridge. Gage 1 measures the strain, while *dummy gage* 2 provides temperature compensation.[25] The meter voltage is a function of the input (battery) voltage and the resistors. (As with bridge circuits, the input voltage is typically less than 10 V.) The variable resistance is used for balancing the bridge prior to the strain. When the bridge is balanced, V_{meter} is zero.

When the gage is strained, the bridge becomes unbalanced. Assuming the bridge is initially balanced, the

---

[22]Other types of resistance bridges are the *differential series balance bridge, shunt balance bridge,* and *differential shunt balance bridge.* These differ in the manner in which the adjustable resistor is incorporated into the circuit.

[23]This gives rise to the alternate names of *zero-indicating bridge* and *null-indicating bridge.*

[24]The unknown resistance can also be determined from the amount of voltage unbalance shown by the meter reading, in which case, the bridge is known as a *deflection bridge* rather than a null-indicating bridge. Deflection bridges are described in Sec. 20.

[25]This is a "quarter-bridge" or "1/4-bridge" configuration, as described in Sec. 22. The strain gage used for temperature compensation is not active.

Figure 84.8 Strain Gage in Resistance Bridge

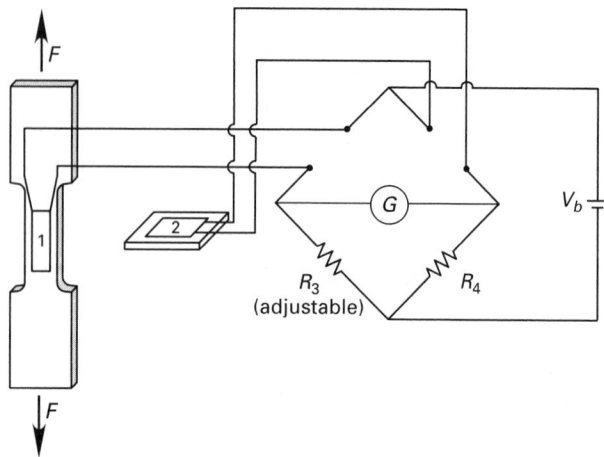

voltage at the meter (known as the *voltage deflection* from the null condition) will be[26]

$$V_{\text{meter}} = V_b \left(\frac{R_1}{R_1 + R_3} - \frac{R_2}{R_2 + R_4} \right) \quad [\tfrac{1}{4}\text{-bridge}] \quad 84.16$$

For a single strain gage in a resistance bridge and neglecting lead resistance, the voltage deflection is related to the strain by Eq. 84.17.

$$V_{\text{meter}} = \frac{(\text{GF})\epsilon V_b}{4 + 2(\text{GF})\epsilon}$$

$$\approx \tfrac{1}{4}(\text{GF})\epsilon V_b \quad [\tfrac{1}{4}\text{-bridge}] \quad 84.17$$

21. STRAIN GAGE IN UNBALANCED RESISTANCE BRIDGE

A resistance bridge does not need to be balanced prior to use as long as an accurate digital voltmeter is used in the detection circuit. The voltage ratio difference, ΔVR, is defined as the fractional change in the output voltage from the unstrained to the strained condition.

$$\Delta \text{VR} = \left(\frac{V_{\text{meter}}}{V_b} \right)_{\text{strained}} - \left(\frac{V_{\text{meter}}}{V_b} \right)_{\text{unstrained}} \quad 84.18$$

If the only resistance change between the strained and unstrained conditions is in the strain gage and lead resistance is disregarded, the fractional change in gage resistance for a single strain gage in a resistance bridge is

$$\frac{\Delta R_g}{R_g} = \frac{-4\Delta(\text{VR})}{1 + 2\Delta(\text{VR})} \quad [\tfrac{1}{4}\text{-bridge}] \quad 84.19$$

Since the fractional change in gage resistance also occurs in the definition of the gage factor (Eq. 84.9), the strain is

$$\epsilon = \frac{-4\Delta \text{VR}}{(\text{GF})(1 + 2\Delta \text{VR})} \quad [\tfrac{1}{4}\text{-bridge}] \quad 84.20$$

[26]Equation 84.16 applies to the unstrained condition as well. However, if the gage is unstrained and the bridge is balanced, the bracketed resistance term is zero.

22. BRIDGE CONSTANT

The voltage deflection can be doubled (or quadrupled) by using two (or four) strain gages in the bridge circuit. The larger voltage deflection is more easily detected, resulting in more accurate measurements.

Use of multiple strain gages is generally limited to configurations where symmetrical strain is available on the member. For example, a beam in bending experiences the same strain on the top and bottom faces. Therefore, if the temperature-compensation strain gage shown in Fig. 84.8 is bonded to the bottom of the beam, the resistance change would double.

The *bridge constant* (BC) is the ratio of the actual voltage deflection to the voltage deflection from a single gage. Depending on the number and orientation of the gages used, bridge constants of 1.0, 1.3, 2.0, 2.6, and 4.0 may be encountered (for materials with a Poisson's ratio of 0.3).

Figure 84.9 illustrates how (up to) four strain gages can be connected in a Wheatstone bridge circuit. The total strain indicated will be the algebraic sum of the four strains detected. For example, if all four strains are equal in magnitude, ϵ_1 and ϵ_4 are tensile (i.e., positive), and ϵ_2 and ϵ_3 are compressive (i.e., negative), then the bridge constant would be 4.

$$\epsilon_t = \epsilon_1 - \epsilon_2 - \epsilon_3 + \epsilon_4 \quad 84.21$$

Figure 84.9 Wheatstone Bridge Strain Gage Circuit[a]

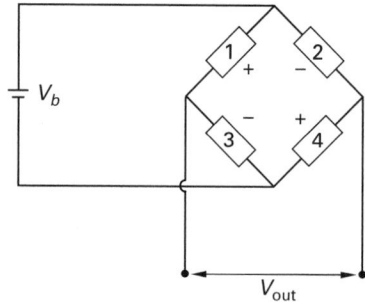

[a]The 45° orientations shown are figurative. Actual gage orientation can be in any direction.

23. STRESS MEASUREMENTS IN KNOWN DIRECTIONS

Strain gages are the most frequently used method of determining stress in a member. Stress can be calculated from strain, or the measurement circuitry can be calibrated to give the stress directly.

For stress in only one direction (i.e., the *uniaxial stress* case), such as a simple bar in tension, only one strain gage is required. The stress can be calculated from *Hooke's law.*

$$\sigma = E\epsilon \quad 84.22$$

When a surface, such as that of a pressure vessel, experiences simultaneous stresses in two directions (the *biaxial stress* case), the strain in one direction affects the strain in the other direction.[27] Therefore, two strain gages are needed, even if the stress in only one direction is needed. The strains actually measured by the gages are known as the *net strains*.

$$\epsilon_x = \frac{\sigma_x - \nu\sigma_y}{E} \qquad 84.23$$

$$\epsilon_y = \frac{\sigma_y - \nu\sigma_x}{E} \qquad 84.24$$

The stresses are determined by solving Eqs. 84.23 and 84.24 simultaneously.

$$\sigma_x = \frac{E(\epsilon_x + \nu\epsilon_y)}{1 - \nu^2} \qquad 84.25$$

$$\sigma_y = \frac{E(\epsilon_y + \nu\epsilon_x)}{1 - \nu^2} \qquad 84.26$$

Figure 84.9 shows how four strain gages can be interconnected in a bridge circuit. Figure 84.10 shows how (up to) four strain gages would be physically oriented on a test specimen to measure different types of stress. Not all four gages are needed in all cases. If four gages are used, the arrangement is said to be a *full bridge*. If only one or two gages are used, the terms *quarter-bridge* ($\frac{1}{4}$-bridge) and *half-bridge* ($\frac{1}{2}$-bridge), respectively, apply.

In the case of up to four gages applied to detect bending strain (Fig. 84.10(a)), the bridge constant (BC) can be 1.0 (one gage in position 1), 2.0 (two gages in positions 1 and 2), or 4.0 (all four gages). The relationships between the stress, strain, and applied force are

$$\sigma = E\epsilon = \frac{E\epsilon_t}{\text{BC}} \qquad 84.27$$

$$\sigma = \frac{Mc}{I} = \frac{Mh}{2I} \qquad 84.28$$

$$I = \frac{bh^3}{12} \quad \text{[rectangular section]} \qquad 84.29$$

For axial strain (Fig. 84.10(b)) and a material with a Poisson's ratio of 0.3, the bridge constant can be 1.0 (one gage in position 1), 1.3 (two gages in positions 1 and 2), 2.0 (two gages in positions 1 and 3), or 2.6 (all four gages).

$$\sigma = E\epsilon = \frac{E\epsilon_t}{\text{BC}} \qquad 84.30$$

$$\sigma = \frac{F}{A} \qquad 84.31$$

$$A = bh \quad \text{[rectangular section]} \qquad 84.32$$

---

[27]Thin-wall pressure vessel theory shows that the hoop stress is twice the longitudinal stress. However, the ratio of circumferential to longitudinal strains is closer to 4:1 than to 2:1.

Figure 84.10 *Orientation of Strain Gages*

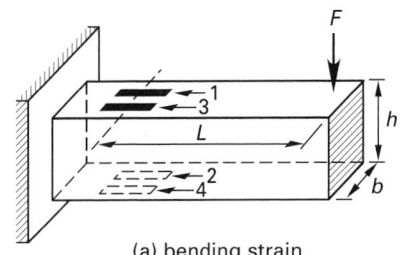

(a) bending strain

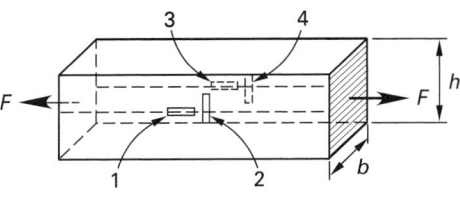

(b) axial strain

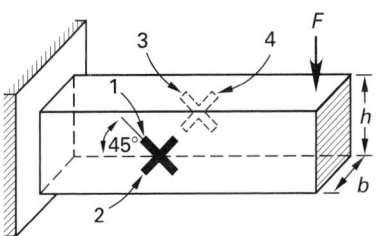

(c) shear strain

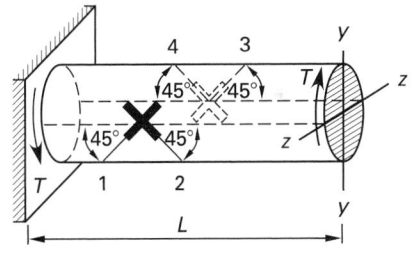

(d) torsional strain

For shear strain (Fig. 84.10(c)) and a material with a Poisson's ratio of 0.3, the bridge constant can be 2.0 (two gages in positions 1 and 2) or 4.0 (all four gages). The shear strain is twice the axial strain at 45°.

$$\tau = G\gamma = 2G\epsilon = \frac{2G\epsilon_t}{\text{BC}} \qquad 84.33$$

$$\tau = \frac{FQ}{bI} \qquad 84.34$$

$$\gamma = 2\epsilon \quad \text{[at 45°]} \qquad 84.35$$

$$Q_{\max} = \frac{bh^2}{8} \quad \text{[rectangular section]} \qquad 84.36$$

$$G = \frac{E}{2(1 + \nu)} \qquad 84.37$$

For torsional strain (Fig. 84.10(d)) and a material with a Poisson's ratio of 0.3, the bridge constant can be 2.0 (two gages in positions 1 and 2) or 4.0 (all four gages). The shear strain is twice the axial strain at 45°.

$$\tau = G\gamma = 2G\epsilon = \frac{2G\epsilon_t}{\text{BC}} \qquad 84.38$$

$$\tau = \frac{Tr}{J} = \frac{Td}{2J} \qquad 84.39$$

$$\gamma = 2\epsilon \quad \text{[at 45°]} \qquad 84.40$$

$$J = \frac{\pi d^4}{32} \quad \text{[solid circular]} \qquad 84.41$$

$$\phi = \frac{TL}{JG} \qquad 84.42$$

$$G = \frac{E}{2(1+\nu)} \qquad 84.43$$

24. STRESS MEASUREMENTS IN UNKNOWN DIRECTIONS

In order to calculate the maximum stresses (i.e., the principal stresses) on the surface shown in Sec. 23, the gages were oriented in the known directions of the principal stresses.

In most cases, however, the directions of the principal stresses are not known. Therefore, rosettes of at least three gages are used to obtain information in a third direction. Rosettes of three gages (*rectangular* and *equiangular (delta) rosettes*) are used for this purpose. *T-delta rosettes* include a fourth strain gage to refine and validate the results of the three primary gages. Table 84.5 can be used to calculate the principal stresses.

25. LOAD CELLS

Load cells are used to measure force. A load cell is a transducer that converts a tensile or compressive force into an electrical signal. Though the details of the load cell vary with the application, the basic elements are (a) a member that is strained by the force and (b) a strain detection system (e.g., strain gage). The force is calculated from the observed deflection, y. In Eq. 84.44, the spring constant, k, is known as the load cell's *deflection constant*.

$$F = ky \qquad 84.44$$

Because of their low cost and simple construction, *bending beam load cells* are the most common variety of load cell. Two strain gages, one on the top and the other mounted on the bottom of a cantilever bar, are used. *Shear beam load cells* (which detect force by measuring the shear stress) can be used where the shear does not vary considerably with location, as in the web of an

I-beam cross section.[28] The common S-shaped load cell constructed from a machined steel block can be instrumented as either a bending beam or shear beam load cell.

Load cell applications are categorized into classes, with class III (using a single load cell) being the most common. Commercial load cells meet standardized limits on errors due to temperature, nonlinearity, and hysteresis. The *temperature effect on output* (TEO) is typically stated in percentage change per 100°F (55.5°C) change in temperature.

Nonlinearity errors are reduced in proportion to the load cell's derating (i.e., using the load cell to measure forces less than its rated force). For example, a 2:1 derating will reduce the nonlinearity errors by 50%. Hysteresis is not normally reduced by derating.

The overall error of force measurement can be reduced by a factor of $1/\sqrt{n}$ (where n is the number of load cells that share the load equally) by using more than one load cell. Conversely, the applied force can vary by $\sqrt{n}$ times the known accuracy of a single load cell without decreasing the error.

26. DYNAMOMETERS

Torque from large motors and engines is measured by a *dynamometer*. *Absorption dynamometers* (e.g., the simple *friction brake, Prony brake, water brake,* and *fan brake*) dissipate energy as the torque is measured. Opposing torque in pumps and compressors must be supplied by a *driving dynamometer*, which has its own power input. *Transmission dynamometers* (e.g., *torque meters, torsion dynamometers*) use strain gages to sense torque. They do not absorb or provide energy.

Using a brake dynamometer involves measuring a force, a moment arm, and the angular speed of rotation. The familiar torque-power-speed relationships are used with absorption dynamometers.

$$T = Fr \qquad 84.45$$

$$P_{\text{ft-lbf/min}} = 2\pi T_{\text{ft-lbf}} n_{\text{rpm}} \qquad 84.46$$

$$P_{\text{kW}} = \frac{T_{\text{N·m}} n_{\text{rpm}}}{9549} \quad \text{[SI]} \qquad 84.47(a)$$

$$P_{\text{hp}} = \frac{2\pi F_{\text{lbf}} r_{\text{ft}} n_{\text{rpm}}}{33{,}000}$$

$$= \frac{2\pi T_{\text{ft-lbf}} n_{\text{rpm}}}{33{,}000} \quad \text{[U.S.]} \qquad 84.47(b)$$

---

[28]While shear in a rectangular beam varies parabolically with distance from the neutral axis, shear in the web of an I-beam is essentially constant at F/A. The flanges carry very little of the shear load.

Other advantages of the shear beam load cell include protection from the load and environment, high side load rejection, lower creep, faster RTZ (return to zero) after load removal, and higher tolerance of vibration, dynamic forces, and noise.

Table 84.5 *Stress-Strain Relationships for Strain Gage Rosettes*[a]

type of rosette	rectangular	equiangular (delta)	T-delta
principal strains, ϵ_p, ϵ_q	$\frac{1}{2}\left(\epsilon_a + \epsilon_c \pm\sqrt{2(\epsilon_a - \epsilon_b)^2 + 2(\epsilon_b - \epsilon_c)^2}\right)$	$\frac{1}{3}\left(\epsilon_a + \epsilon_b + \epsilon_c \pm\sqrt{\begin{array}{c}2(\epsilon_a - \epsilon_b)^2 + 2(\epsilon_b - \epsilon_c)^2 \\ + 2(\epsilon_c - \epsilon_a)^2\end{array}}\right)$	$\frac{1}{2}\left(\epsilon_a + \epsilon_d \pm\sqrt{(\epsilon_a - \epsilon_d)^2 + \left(\frac{4}{3}\right)(\epsilon_b - \epsilon_c)^2}\right)$
principal stresses, σ_1, σ_2	$\left(\frac{E}{2}\right)\left(\frac{\epsilon_a + \epsilon_c}{1 - \nu} \pm \frac{1}{1 + \nu}\right.$ $\left.\times\sqrt{2(\epsilon_a - \epsilon_b)^2 + 2(\epsilon_b - \epsilon_c)^2}\right)$	$\left(\frac{E}{3}\right)\left(\frac{\epsilon_a + \epsilon_b + \epsilon_c}{1 - \nu} \pm \frac{1}{1 + \nu}\right.$ $\left.\times\sqrt{\begin{array}{c}2(\epsilon_a - \epsilon_b)^2 + 2(\epsilon_b - \epsilon_c)^2 \\ + 2(\epsilon_c - \epsilon_a)^2\end{array}}\right)$	$\left(\frac{E}{2}\right)\left(\frac{\epsilon_a + \epsilon_d}{1 - \nu} \pm \frac{1}{1 + \nu}\right.$ $\left.\times\sqrt{(\epsilon_a - \epsilon_d)^2 + \left(\frac{4}{3}\right)(\epsilon_b - \epsilon_c)^2}\right)$
maximum shear, $\tau_{\max}$	$\left(\frac{E}{2(1 + \nu)}\right)$ $\times\sqrt{2(\epsilon_a - \epsilon_b)^2 + 2(\epsilon_b - \epsilon_c)^2}$	$\left(\frac{E}{3(1 + \nu)}\right)$ $\times\sqrt{\begin{array}{c}2(\epsilon_a - \epsilon_b)^2 + 2(\epsilon_b - \epsilon_c)^2 \\ + 2(\epsilon_c - \epsilon_a)^2\end{array}}$	$\left(\frac{E}{2(1 + \nu)}\right)$ $\times\sqrt{(\epsilon_a - \epsilon_d)^2 + \left(\frac{4}{3}\right)(\epsilon_b - \epsilon_c)^2}$
$\tan 2\theta$[b]	$\dfrac{2\epsilon_b - \epsilon_a - \epsilon_c}{\epsilon_a - \epsilon_c}$	$\dfrac{\sqrt{3}(\epsilon_c - \epsilon_b)}{2\epsilon_a - \epsilon_b - \epsilon_c}$	$\left(\dfrac{2}{\sqrt{3}}\right)\left(\dfrac{\epsilon_c - \epsilon_b}{\epsilon_a - \epsilon_d}\right)$
$0 < \theta < +90°$	$\epsilon_b > \dfrac{\epsilon_a + \epsilon_c}{2}$	$\epsilon_c > \epsilon_b$	$\epsilon_c > \epsilon_b$

[a]θ is measured in the counterclockwise direction from the a-axis of the rosette to the axis of the algebraically larger stress.
[b]θ is the angle from gage A axis to axis of maximum normal stress.

Some brakes and dynamometers are constructed with a "standard" brake arm whose length is 5.252 ft. In that case, the horsepower calculation conveniently reduces to

$$P_{\text{hp}} = \frac{F_{\text{lbf}} n_{\text{rpm}}}{1000} \quad \text{[``standard arm'' brake]} \qquad 84.48$$

If an absorption dynamometer uses a DC generator to dissipate energy, the generated voltage (V in volts) and line current (I in amps) are used to determine the power. Equations 84.47 and 84.48 are used to determine the torque.

$$P_{\text{hp}} = \frac{IV}{\eta \left(1000 \, \dfrac{\text{W}}{\text{kW}}\right)\left(0.7457 \, \dfrac{\text{W}}{\text{hp}}\right)} \quad \text{[absorption]}$$

$$\qquad 84.49$$

For a driving dynamometer using a DC motor,

$$P_{\text{hp}} = \frac{IV\eta}{\left(1000 \, \dfrac{\text{W}}{\text{kW}}\right)\left(0.7457 \, \dfrac{\text{W}}{\text{hp}}\right)} \quad \text{[driving]} \qquad 84.50$$

Torque can be measured directly by a *torque meter* mounted to the power shaft. Either the angle of twist (ϕ) or the shear strain (τ/G) are measured. The torque in a solid shaft of diameter d and length L is

$$T = \frac{JG\phi}{L} = \left(\frac{\pi}{32}\right) d^4 \left(\frac{G\phi}{L}\right)$$

$$= \left(\frac{\pi}{16}\right) d^3 \tau \quad \text{[solid round]} \qquad 84.51$$

27. INDICATOR DIAGRAMS

Indicator diagrams are plots of pressure versus volume and are encountered in the testing of reciprocating engines. In the past, indicator diagrams were actually drawn on an *indicator card* wrapped around a drum through a mechanical linkage of arms and springs. This method is mechanically complex and is not suitable for rotational speeds above 2000 rpm. Modern records of pressure and volume are produced by signals from electronic transducers recorded in real time by computers.

Analysis of indicator diagrams produced by mechanical devices requires knowing the spring constant, also known as the *spring scale*. The *mean effective pressure* (MEP) is calculated by dividing the area of the diagram by the width of the plot and then multiplying by the spring constant.

85 Project Management, Budgeting, and Scheduling

Nomenclature

D duration
EF   earliest finish
ES   earliest start
LF   latest finish
LS   latest start
t time
z standard normal variable

Symbols

μ mean
σ standard deviation

1. PROJECT MANAGEMENT

Project management is the coordination of the entire process of completing a job, from its inception to final move-in and post-occupancy follow-up. In many cases, project management is the responsibility of one person. Large projects can be managed with *partnering*. With this method, the various stakeholders of a project, such as the architect, owner, contractor, engineer, vendors, and others are brought into the decision making process. Partnering can produce much closer communication on a project and shared responsibilities. However, the day-to-day management of a project may be difficult with so many people involved. A clear line of communications and delegation of responsibility should be established and agreed to before the project begins.

Many project managers follow the procedures outlined in *A Guide to the Project Management Body of Knowledge* (PMBOK Guide), published by the Project Management Institute. The PMBOK Guide is an internationally recognized standard (IEEE Std 1490) that defines the fundamentals of project management as they apply to a wide range of projects, including construction, engineering, software, and many other industries. The PMBOK Guide is process-based, meaning it describes projects as being the outcome of multiple processes. Processes overlap and interact throughout the various phases of a project. Each process occurs within one of five process groups, which are related as shown in Fig. 85.1.

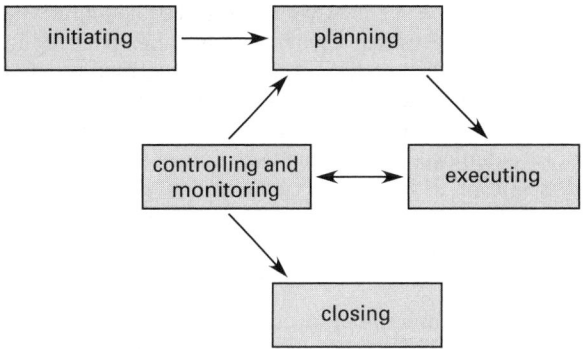

Figure 85.1 PMBOK Process Groups

The PMBOK Guide identifies nine project management knowledge areas that are typical of nearly all projects. The nine knowledge areas and their respective processes are summarized as follows.

1. Integration: develop project charter, project scope statement, and management plan; direct and manage execution; monitor and control work; integrate change control; close project

2. Scope: plan, define, create work breakdown structure (WBS), verify, and control scope

3. Time: define, sequence, estimate resources, and estimate duration of activities; develop and control schedule

4. Cost: estimate, budget, and control costs

5. Quality: plan; perform quality assurance and quality control

6. Human Resources: plan; acquire, develop, and manage project team

7. Communications: plan; distribute information; report on performance; manage stakeholders

8. Risk: identify risks; plan risk management and response; perform qualitative and quantitative risk analysis; monitor and control

9. Procurement: plan purchases, acquisitions, and contracts; request seller responses; select sellers; administer and close contracts

Each of the processes also falls into one of the five basic process groups, creating a matrix so that every process is related to one knowledge area and one process group. Additionally, processes are described in terms of inputs (documents, plans, designs, etc.), tools and techniques (mechanisms applied to inputs), and outputs (documents, products, etc.).

Establishing a budget (knowledge area 4) and scheduling design and construction (knowledge area 3) are two of the most important parts of project management because they influence many of the design decisions to follow and can determine whether a project is even feasible.

2. BUDGETING

Budgets may be set in several ways. For speculative or for-profit projects, the owner or developer works out a pro forma statement listing the expected income of the project and the expected costs to build it. An estimated selling price of the developed project or rent per square foot is calculated and balanced against all the various costs, one of which is the construction price. In order to make the project economically feasible, there will be a limit on the building costs. This becomes the budget within which the work must be completed.

Budgets for municipal and other public projects are often established through public funding or legislation. In these cases, the construction budget is often fixed without the architect's or engineer's involvement, and the project must be designed and built for the fixed amount. Unfortunately, when public officials estimate the cost to build a project, they sometimes neglect to include all aspects of development, such as professional fees, furnishings, and other items.

Budgets may also be based on the proposed project specifics. This is the most realistic and accurate way to establish a preliminary budget because it is based on a particular building type of a particular size on a particular site.

There are four basic variables in developing any construction budget: quantity, quality, available funds, and time. There is always a balance among these four variables, and changing one or more affects the others. For instance, if an owner needs a certain area (quantity), needs the project built at a certain time, and has a fixed amount of money to spend, then the quality of construction will have to be adjusted to meet the other constraints. If time, quality, and the total budget are fixed, then the area must be adjusted. For more information, see the "Time-Cost Trade-Off" section in this chapter. In some cases, *value engineering* can be performed during which individual systems and materials are reviewed to see if the same function can be accomplished in a less expensive way.

Fee Projections

A *fee projection* is one of the earliest and most important tasks that a project manager must complete. A fee projection takes the total fee the designer will receive for the project and allocates it to the schedule and staff members who will work on the project, after deducting amounts for profit, overhead, and other expenses that will not be used for professional time.

Ideally, fee projections should be developed from a careful projection of the scope of work, its associated costs (direct personnel expense, indirect expenses, and overhead), consultant fees, reimbursable expenses, and profit desired. These should be determined as a basis for setting the final fee agreement with the client. If this is done correctly, there should be enough money to complete the project within the allotted time.

There are many methods for estimating and allocating fees. Figure 85.2 shows a simple manual form that combines time scheduling with fee projections. In this example, the total working fee, that is, the fee available to pay people to do the job after subtracting for profit, consultants, and other expenses, is listed in the upper right corner of the chart. The various phases or work tasks needed to complete the job are listed in the left-hand column, and the time periods (most commonly in weeks) are listed across the top of the chart.

The project manager estimates the percentage of the total amount of work or fee that he or she thinks each phase will require. This estimate is based on experience and common rules of thumb the design or construction office may use. The percentages are placed in the third column on the right and multiplied by the total working fee to get the allotted fee for each phase (the figure in the second column on the right). This allotted fee is then divided among the number of time periods in the schedule and placed in the individual columns under each time period.

If phases or tasks overlap (as they do in Fig. 85.2), total the fees in each period and place this figure at the bottom of the chart. This dollar amount can then be divided by an average billing rate for the people working on the project to determine an approximate budgeted number of hours that the office can afford to spend on the project each week and still make a profit. Of course, if the number of weekly hours exceeds about 40, then more than one person will be needed to do the work.

By monitoring time sheets, the project manager can compare the actual hours (or fees) expended against the budgeted time (or fees) and take corrective action if actual time exceeds budgeted time.

Quality planning involves determining with the client what the expectations are concerning design, cost, and other aspects of the project. Quality does not simply mean high-cost finishes, but rather the requirements of the client based on his or her needs. These needs should be clearly defined in the programming phase of a project

Figure 85.2 *Fee Projection Chart*

Project: Mini-mall | Project No.: 9274 | Date: 10/14/2010
Completed by: JBL | Project Manager: JBL | Total Fee: $26,400

Phase or Task	Period / Date	1 (11/16–22)	2 (11/23)	3 (11/30)	4 (12/7)	5 (12/14)	6 (12/21)	7 (12/28)	8 (1/4)	9 (1/11)	% of total fee	fee allocation by phase or task	person-hrs. est.
SD-design		1320	1320								10	2640	
SD presentation			1320								5	1320	
DD—arch. work				1980	1980						15	3960	
DD—consultant coord.				530	790						5	1320	
DD—approvals					1320						5	1320	
CD—plans/elevs.						1056	1056	1056	1056	1056	20	5280	
CD—details								2640	2640		20	5280	
CD—consultant coord.						440		440	440		5	1320	
CD specs.									1320	1320	10	2640	
CD—material sel.						660	660				5	1320	
budgeted fees /period		1320	2640	2510	4090	2156	1716	4136	5456	2376	100%	$26,400	
person–weeks or hours		53 / 1.3	106 / 2.6	100 / 2.5	164 / 4	108 / 2.7	86 / 2.2	207 / 5	273 / 6.8	119 / 3			
staff assigned		JLK	JLK AST JBC	JLK AST EMW-(1/2)	JLK AST JBC EMW	JLK AST EMW	JLK AST	JLK AST EMW →	JLK SBS BFD	JLK AST EMW			
actual fees expended													

and written down and approved by the client before design work begins.

Cost Estimating

Estimators compile and analyze data on all of the factors that can influence costs, such as materials, labor, location, duration of the project, and special machinery requirements. The methods for estimating costs can differ greatly by industry. On a construction project, for example, the estimating process begins with the decision to submit a bid. After reviewing various preliminary drawings and specifications, the estimator visits the site of the proposed project. The estimator needs to gather information on access to the site; the availability of electricity, water, and other services; and surface topography and drainage. The estimator usually records this information in a signed report that is included in the final project estimate.

After the site visit, the estimator determines the quantity of materials and labor the firm will need to furnish. This process, called the quantity survey or "takeoff," involves completing standard estimating forms, filling in dimensions, numbers of units, and other information. Table 85.1 is a small part of a larger takeoff report illustrating the degree of detail needed to estimate the project cost. A cost estimator working for a general contractor, for example, uses a construction project's plans and specifications to estimate the materials dimensions and count the quantities of all items associated with the project.

Though the quantity takeoff process can be done manually using a printout, a red pen, and a clicker, it can also be done with a digitizer that enables the user to take measurements from paper bid documents, or with an integrated *takeoff viewer* program that interprets electronic bid documents. In any case, the objective is to generate a set of takeoff elements (counts, measurements, and other conditions that affect cost) that is used to establish cost estimates. Table 85.1 is an example of a typical quantity take-off report for lumber needed for a construction project.

Although subcontractors estimate their costs as part of their own bidding process, the general contractor's cost estimator often analyzes bids made by subcontractors. Also during the takeoff process, the estimator must make decisions concerning equipment needs, the sequence of operations, the size of the crew required, and physical constraints at the site. Allowances for wasted materials, inclement weather, shipping delays, and other factors that may increase costs also must be incorporated in the estimate. After completing the quantity surveys, the estimator prepares a cost summary for the entire project, including the costs of labor, equipment, materials, subcontracts, overhead, taxes, insurance, markup, and any other costs that may affect the project. The chief estimator then prepares the bid proposal for submission to the owner. Construction cost estimators also may be employed by the project's architect or owner to estimate costs or to track actual costs relative to bid specifications as the project develops.

Systems, Mgmt, and Professional

Table 85.1 *Partial Lumber and Hardware Take-Off Report*

size	description	usage	pieces	total length
foundation framing				
2×4	DF STD/BTR	stud	38	8
2×6	DF #2/BTR	stud	107	8
2×6	PTDF	mudsill	–	RL
2×6	DF #2/BTR	bracing	–	RL
2×4	DF STD/BTR	blocking	–	RL
$11\,^7/_8''$	TJI/250	floor joist	11	18
$11\,^7/_8''$	TJI/250	floor joist	12	10
2×4	DF STD/BTR	plate	–	RL
2×6	DF #2/BTR	plate	–	RL
4×4	PTDF	posts	7	4
4×6	PTDF	posts	23	4
6×6	PTDF	posts	1	4
2×12	DF #2/BTR	rim	–	RL
$^{15}/_{32}''$	CDX plywood	subfloor	12	4×8
pre-cut doors				
$3^1/_2'' \times 7^1/_4''$	LVL	header	1	$100''$
$3^1/_2'' \times 7^1/_4''$	LVL	header	1	$130''$
$3^1/_2'' \times 7^1/_4''$	LVL	header	10	$24''$
exterior sheathing and shearwall				
$^1/_2''$	CDX plywood	shear wall	100	4×8
$^1/_2''$	CDX plywood	exterior sheathing	89	4×8
roof sheathing				
$^1/_2''$	CDX plywood	roof sheathing	67	4×8
building A & B hardware				
Simpson HD64	6 pieces each			
Simpson HD22	23 pieces each			

Used with permission from Modern Estimating Services (MES), www.quantitytakeoffreports.com.

Estimators often specialize in large construction companies employing more than one estimator. For example, one may estimate only electrical work and another may concentrate on excavation, concrete, and forms.

Computers play an integral role in cost estimation because estimating often involves numerous mathematical calculations requiring access to various historical databases. For example, to undertake a parametric analysis (a process used to estimate costs per unit based on square footage or other specific requirements of a project), cost estimators use a computer database containing information on the costs and conditions of many other similar projects. Although computers cannot be used for the entire estimating process, they can relieve estimators of much of the drudgery associated with routine, repetitive, and time-consuming calculations.

Cost Influences

There are many variables that affect project cost. Construction cost is only one part of the total project development budget. Other factors include such things as site acquisition, site development, fees, and financing. Table 85.2 lists most of the items commonly found in a project budget and a typical range of values based on construction cost. Not all of these are part of every development, but they illustrate the things that must be considered.

Building cost is the money required to construct the building, including structure, exterior cladding, finishes, and electrical and mechanical systems. *Site development costs* are usually a separate item. They include such things as parking, drives, fences, landscaping, exterior lighting, and sprinkler systems. If the development is large and affects the surrounding area, a developer may be required to upgrade roads, extend utility lines, and do other major off-site work as a condition of getting approval from public agencies.

Movable equipment and furnishings include furniture, accessories, window coverings, and major equipment necessary to put the facility into operation. These are

Table 85.2 *Project Budget Line Items*

	line item		example
A	site acquisition		$1,100,000
B	building costs	area times cost per ft^2	(assume) $6,800,000
C	site development	10% to 20% of B	(15%) $1,020,000
D	total construction cost	B + C	$7,820,000
E	movable equipment	5% to 10% of B	(5%) $340,000
F	furnishings		$200,000
G	total construction and furnishings	D + E + F	$8,360,000
H	professional services	5% to 10% of D	(7%) $547,400
I	inspection and testing		$15,000
J	escalation estimate	2% to 20% of G per year	(10%) $836,000
K	contingency	5% to 10% of G	(8%) $668,800
L	financing costs		$250,000
M	moving expenses		(assume) $90,000
N	total project budget	G + H through M	$11,867,200

often listed as separate line items because the funding for them may come out of a separate budget and because they may be supplied under separate contracts.

Professional services are architectural and engineering fees as well as costs for such things as topographic surveys, soil tests, special consultants, appraisals and legal fees, and the like. Inspection and testing involve money required for special on-site, full-time inspection (if required), and testing of such things as concrete, steel, window walls, and roofing.

Because construction takes a great deal of time, a factor for inflation should be included. Generally, the present budget estimate is escalated to a time in the future at the expected midpoint of construction. Although it is impossible to predict the future, by using past cost indexes and inflation rates and applying an estimate to the expected condition of the construction, the architect can usually make an educated guess.

A *contingency cost* should also be added to account for unforeseen changes by the client and other conditions that add to the cost. For an early project budget, the percentage of the contingency should be higher than contingencies applied to later budgets, because there are more unknowns. Normally, from 5% to 10% should be included.

Financing includes not only the long-term interest paid on permanent financing but also the immediate costs of loan origination fees, construction loan interest, and other administrative costs. On long-term loans, the cost of financing can easily exceed all of the original building and development costs. In many cases, long-term interest, called debt service, is not included in the project budget because it is an ongoing cost to the owner, as are maintenance costs.

Finally, many clients include moving costs in the development budget. For large companies and other types of clients, the money required to physically relocate, including changing stationery, installing telephones, and the like, can be a substantial amount.

Methods of Budgeting

The costs described in the previous section and shown in Table 85.2 represent a type of budget done during programming or even prior to programming to test the feasibility of a project. The numbers are preliminary, often based on sketchy information. For example, the building cost may simply be an estimated cost per square foot multiplied by the number of gross square feet needed. The square footage cost may be derived from similar buildings in the area, from experience, or from commercially available cost books.

Budgeting, however, is an ongoing activity. At each stage of the design process, there should be a revised budget reflecting the decisions made to that time. As shown in the example, pre-design budgets are usually based only on area, but other units can also be used. For example, many companies have rules of thumb for making estimates based on cost per hospital bed, cost per student, cost per hotel room, or similar functional units.

After the pre-programming budget, the architect usually begins to concentrate on the building and site development costs. At this stage an average cost per square foot may still be used, or the building may be divided into several functional parts and different square footage prices may be assigned to each part. A school, for example, may be classified into classroom space, laboratory space, shop space, office space, and gymnasium space, each having a different cost per square foot. This type of division can be developed concurrently with the programming of the space requirements.

During schematic design, when more is known about the space requirements and general configuration of the building and site, *system budgeting* is based on major subsystems. Historical cost information on each type of subsystem can be applied to the design. At this point it is easier to see where the money is being used in the building. Design decisions can then be based on studies of alternative systems. A typical subsystem budget is shown in Table 85.3.

Table 85.3 *System Cost Budget of Office Buildings*

| | average cost | |
subsystem	($/ft^2)	(% of total)
foundations	3.96	5.2
floors on grade	3.08	4.0
superstructure	16.51	21.7
roofing	0.18	0.2
exterior walls	9.63	12.6
partitions	5.19	6.8
wall finishes	3.70	4.8
floor finishes	3.78	5.0
ceiling finishes	2.79	3.7
conveying systems	6.45	8.5
specialties	0.70	0.9
fixed equipment	2.74	3.6
HVAC	9.21	12.1
plumbing	3.61	4.6
electrical	4.68	6.1
	76.21	100.0

Values for low-, average-, and high-quality construction for different building types can be obtained from cost databases and published estimating manuals and applied to the structure being budgeted. The dollar amounts included in system cost budgets usually include markup for contractor's overhead and profit and other construction administrative costs.

During the later stages of schematic design and early stages of construction documents, more detailed estimates are made. The procedure most often used is the *parameter method*, which involves an expanded itemization of construction quantities and assignment of unit costs to these quantities. For example, instead of using one number for floor finishes, the cost is broken down into carpeting, vinyl tile, wood strip flooring, unfinished concrete, and so forth. Using an estimated cost per square foot, the cost of each type of flooring can be estimated based on the area. With *parametric budgeting*, it is possible to evaluate the cost implications of each building component and to make decisions concerning both quantity and quality in order to meet the original budget estimate. If floor finishes are over budget, the architect and the client can review the parameter estimate and decide, for example, that some wood flooring must be replaced with less expensive carpeting. Similar decisions can be made concerning any of the parameters in the budget.

Another way to compare and evaluate alternative construction components is with *matrix costing*. With this technique, a matrix is drawn showing, along one side, the various alternatives and, along the other side, the individual elements that combine to produce the total

cost of the alternatives. For example, in evaluating alternatives for workstations, all of the factors that would comprise the final cost could be compared. These factors might include the cost of custom-built versus pre-manufactured workstations, task lighting that could be planned with custom-built units versus higher-wattage ambient lighting, and so on.

Parameter line items are based on commonly used units that relate to the construction element under study. For instance, a gypsum board partition would have an assigned cost per square foot of complete partition of a particular construction type rather than separate costs for metal studs, gypsum board, screws, and finishing. There would be different costs for single-layer gypsum board partitions, 1-hour rated walls, 2-hour rated walls, and other partition types.

Overhead and Profit

Two additional components of construction cost are the contractor's overhead and profit. Overhead can be further divided into general overhead and project overhead. *General overhead* is the cost to run a contracting business, and involves office rent, secretarial help, heat, and other recurring costs. *Project overhead* is the money it takes to complete a particular job, not including labor, materials, or equipment. Temporary offices, project telephones, sanitary facilities, trash removal, insurance, permits, and temporary utilities are examples of project overhead. The total overhead costs, including both general and project expenses, can range from about 10% to 20% of the total costs for labor, materials, and equipment.

Profit is the last item a contractor adds onto an estimate and is listed as a percentage of the total of labor, materials, equipment, and overhead. This is one of the most highly variable parts of a budget. Profit depends on the type of project, its size, the amount of risk involved, how much money the contractor wants to make, the general market conditions, and, of course, whether or not the job is being bid.

During extremely difficult economic conditions, a contractor may cut the profit margin to almost nothing simply to get the job and keep his or her workforce employed. If the contract is being negotiated with only one contractor, the profit percentage will be much higher. In most cases, however, profit will range from 5% to 20% of the total cost of the job. Overall, overhead and profit can total about 15% to 40% of construction cost.

Cost Information

One of the most difficult aspects of developing project budgets is obtaining current, reliable prices for the kinds of construction units being used. There is no shortage of commercially produced cost books that are published yearly. These books list costs in different ways; some are very detailed, giving the cost for labor and materials for individual construction items, while others list parameter costs and subsystem costs. The detailed price listings are of little use to architects because they are too specific and make comparison of alternate systems difficult.

There are also computerized cost estimating services that only require the architect or engineer to provide general information about the project, location, size, major materials, and so forth. The computer service then applies its current price database to the information and produces a cost budget. Many architects and engineers also work closely with general contractors to develop a realistic budget.

Commercially available cost information, however, is the average of many past construction projects from around the country. Local variations and particular conditions may affect the value of their use on a specific project.

Two conditions that must be accounted for in developing any project budget are geographical location and inflation. These variables can be adjusted by using cost indexes that are published in a variety of sources, including the major architectural and construction trade magazines. Using a base year as index 1000, for example, for selected cities around the country, new indexes are developed each year that reflect the increase in costs (both material and labor) that year.

The indexes can be used to apply costs from one part of the country to another and to escalate past costs to the expected midpoint of construction of the project being budgeted.

Example 85.1

The cost index in your city is 1257 and the cost index for another city in which you are designing a building is 1308. If the expected construction cost is $1,250,000 based on prices for your city, what will be the expected cost in the other region?

Solution

$$\frac{\text{cost A}}{\text{index A}} = \frac{\text{cost B}}{\text{index B}}$$

$$\begin{aligned}
\text{cost in} \atop \text{other region} &= (\text{index in other region}) \\
&\quad \times \left(\frac{\text{your city cost}}{\text{your city index}}\right) \\
&= (1308)\left(\frac{\$1,250,000}{1257}\right) \\
&= \$1,300,716
\end{aligned}$$

Life-Cycle Cost Analysis

Life-cycle cost analysis (LCC) is a method for determining the total cost of a building or building component or system over a specific period of time. It takes into account the initial cost of the element or system under consideration as well as the cost of financing, operation, maintenance, and disposal. Any residual value of the components is subtracted from the other costs. The costs are estimated over a length of time called the *study period*. The duration of the study period varies with the needs of the client and the useful life of the material or system. For example, investors in a building may be interested in comparing various alternate materials over the expected investment time frame, while a city government may be interested in a longer time frame representing the expected life of the building. All future costs are discounted back to a common time, usually the base date, to account for the time value of money. The *discount rate* is used to convert future costs to their equivalent present values.

Using life-cycle cost analysis allows two or more alternatives to be evaluated and their total costs to be compared. This is especially useful when evaluating energy conservation measures where one design alternative may have a higher initial cost than another, but a lower overall cost because of energy savings. Some of the specific costs involved in an LCC of a building element include the following.

- initial costs, which include the cost of acquiring and installing the element

- operational costs for electricity, water, and other utilities

- maintenance costs for the element over the length of the study period, including any repair costs

- replacement costs, if any, during the length of the study period

- finance costs required during the length of the study period

- taxes, if any, for initial costs and operating costs

The residual value is the remaining value of the element at the end of the study period based on resale value, salvage value, value in place, or scrap value. All of the costs listed are estimated, discounted to their present value, and added together. Any residual value is discounted to its present value and then subtracted from the total to get the final life-cycle cost of the element.

A life-cycle cost analysis is not the same as a *life-cycle assessment* (LCA). An LCA analyzes the environmental impact of a product or building system over the entire life of the product or system.

3. SCHEDULING

There are two major elements affecting of a project schedule: design time and construction time. The architect has control over design and the production of contract documents, and the contractor has control over construction. The entire project should be scheduled for the best course of action to meet the client's goals. For example, if the client must move by a certain date and normal design and construction sequences make this impossible, the engineer, architect, or contractor may recommend a fast-track schedule or some other approach to meet the deadline.

Design Sequencing

The design process normally consists of several clearly defined phases, each of which must be substantially finished and approved by the client before the next one can begin. These are commonly referred to and accepted in the profession and are referred to in the American Institute of Architects' (AIA) Owner-Architect Agreement as well as in other documents.

Following programming, the first phase is *schematic design*. During this phase, the general layout of the project is developed along with preliminary alternate studies for materials and building systems. Once the direction of the project documented in schematic design drawings is reviewed and approved by the client, the *design development phase* starts. Here, the decisions made during the previous phase are refined and developed in more detail. Preliminary or outline specifications are written, and a more detailed cost budget is made.

Construction documents are produced next. These include the final working drawings as well as the full project manual and any bidding and contract documents required. These are used for the *bidding* or *negotiation phase*, which includes obtaining bids from several contractors and analyzing them or negotiating a contract with one contractor.

The time required for these phases is highly variable and depends on the following factors.

- The size and complexity of the project. Obviously, a 500,000 ft^2 (46 450 m^2) hospital will take much longer to design than a 30,000 ft^2 (2787 m^2) office building.

- The number of people working on the project. Although adding more people to the job can shorten the schedule, there is a point of diminishing returns. Having too many people only creates a management and coordination problem, and for some phases only a few people are needed, even for very large jobs.

- The abilities and design methodology of the project team. Younger, less-experienced designers will usually need more time to do the same amount of work than would a more senior staff.

- The type of client and the decision-making and approval processes of the client. Large corporations or public agencies are likely to have a multilayered decision-making and approval process. Getting necessary information or approval on one phase from a large client may take weeks or even months, while a small, single-authority client might make the same decision in a matter of days.

The construction schedule may be established by the contractor or construction manager, or it may be estimated by the architect during the programming phase so that the client has some idea of the total time required from project conception to move-in.

Many variables can affect construction time. Most can be controlled in one way or another, but others, like weather, are independent of anyone's control. Beyond the obvious variables of size and complexity, the following is a partial list of some of the more common variables.

- the management ability of the contractor to organize his or her own forces as well as those of the subcontractors

- material delivery times

- the quality and completeness of the architect's drawings and specification

- the weather

- labor availability and labor disputes

- new construction or remodeling (remodeling generally takes more time and coordination for equal areas than new buildings take)

- site conditions (construction sites or those with subsurface problems usually take more time to build on)

- the architect (some professionals are more diligent than others in performing their duties during construction)

- lender approvals

- agency and governmental approvals

Construction Sequencing

Construction sequencing involves creating and following a work schedule that balances the timing and sequencing of land disturbance activities, such as earthwork, and the installation of *erosion and sedimentation control* (ESC) measures. The objective of construction sequencing is to reduce on-site erosion and off-site sedimentation that might affect the water quality of nearby water bodies.

The project manager should confirm that the general construction schedule and the construction sequencing schedule are compatible. Key construction activities and associated ESC measures are listed in Table 85.4.

Time-Cost Trade-Off

A project's completion time and its cost are intricately related. Though some costs are not directly related to the time a project takes, many costs are. This is the essence of the time-cost trade-off: The cost increases as the project time is decreased and vice versa. A project manager's roles include understanding the time-cost relationship, optimizing a project's pace for minimal cost, and predicting the impact of a schedule change on project cost.

Table 85.4 Construction Activities and ESC Measures

construction activity	ESC measures
designate site access	Stabilize exposed areas with gravel and/or temporary vegetation. Immediately apply stabilization to areas exposed throughout site development.
protect runoff outlets and conveyance systems	Install principle sediment traps, fences, and basins prior to grading; stabilize stream banks and install storm drains, channels, etc.
land clearing	Mark trees and buffer areas for preservation.
site grading	Install additional ESC measures as needed during grading.
site stabilization	Install temporary and permanent seeding, mulching, sodding, riprap, etc.
building construction and utilities installation	Install additional ESC measures as needed during construction.
landscaping and final site stabilization	This is the last construction phase. Remove all temporary control measures; install topsoil, trees and shrubs, permanent seeding, mulching, sodding, riprap, etc.; stabilize all open areas, including borrow and spoil areas.

The costs associated with a project can be classified as direct costs or indirect costs. The project cost is the sum of the direct and indirect costs.

Direct costs, also known as variable costs, operating costs, prime costs, and on costs, are costs that vary directly with the level of output (e.g., labor, fuel, power, and the cost of raw material). Generally, direct costs increase as a project's completion time is decreased, since more resources need to be allocated to increase the pace.

Indirect costs, also known as fixed costs, are costs that are not directly related to a particular function or product. Indirect costs include taxes, administration, personnel, and security costs. Such costs tend to be relatively steady over the life of the project and decrease as the project duration decreases.

The time required to complete a project is determined by the critical path, so to compress (or "crash") a project schedule (accelerate the project activities in order to complete the project sooner), a project manager must focus on critical path activities.

A procedure for determining the optimal project time, or time-cost-trade-off, is to determine the normal completion time and direct cost for each critical path activity and compare it to its respective "crash time" and direct cost. The crash time is the shortest time in which an activity can be completed. If a new critical path emerges, consider this in subsequent time reductions. In this way, one can step through the critical path activities and calculate the total direct project cost versus the project time. (To minimize the cost, those activities that are not on the critical path can be extended without increasing the project completion time.) The indirect, direct, and total project costs can then be calculated for different project durations. The optimal duration is the one with the lowest cost. This model assumes that the normal cost for an activity is lower than the crash cost, the time and cost are linearly related, and the resources needed to shorten an activity are available. If these assumptions are not true, then the model would need to be adapted. Other cost considerations include incentive payments, marketing initiatives, and the like.

Fast Tracking

Besides efficient scheduling, construction time can be compressed with *fast-track scheduling*. This method overlaps the design and construction phases of a project. Ordering of long-lead materials and equipment can occur, and work on the site and foundations can begin before all the details of the building are completely worked out. With fast-track scheduling, separate contracts are established so that each major system can be bid and awarded by itself to avoid delaying other construction.

Although the fast-track method requires close coordination between the architect, contractor, subcontractors, owner, and others, it makes it possible to construct a high-quality building in 10% to 30% less time than with a conventional construction contract.

Scheduling Methods

Several methods are used to schedule both design and construction. The most common and easiest is the *bar chart* or *Gantt chart*, such as Fig. 85.3. The various activities of the schedule are listed along the vertical axis. Each activity is given a starting and finishing date, and overlaps are indicated by drawing the bars for each activity so that they overlap. Bar charts are simple to make and understand and are suitable for small to midsize projects. However, they cannot show all the sequences and dependencies of one activity on another.

Critical path techniques are used to graphically represent the multiple relationships between stages in a complicated project. The graphical network shows the *precedence relationships* between the various activities. The graphical network can be used to control and monitor the progress, cost, and resources of a project. A critical path technique will also identify the most critical activities in the project.

Critical path techniques use *directed graphs* to represent a project. These graphs are made up of *arcs* (arrows) and *nodes* (junctions). The placement of the arcs

Systems, Mgmt, and Professional

Figure 85.3 *Gantt Chart*

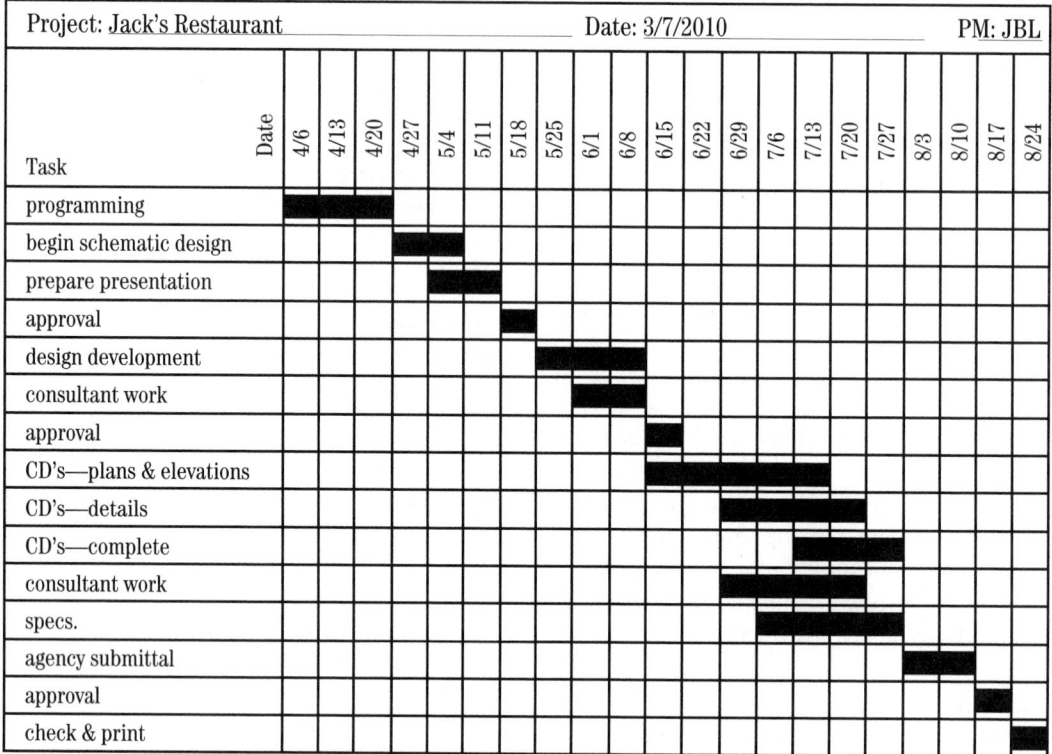

and nodes completely specifies the precedences of the project. Durations and precedences are usually given in a *precedence table (matrix)*.

Activity-on-Node Networks: The Critical Path Method

One technique is known as the *critical path method*, CPM. This deterministic method is applicable when all activity durations are known in advance. CPM is usually represented as an *activity-on-node* model since arcs are used to specify precedence and the nodes actually represent the activities. Events are not represented on the graph, other than as the heads and tails of the arcs. Two *dummy nodes* taking zero time may be used to specify the start and finish of the project.

A CPM chart graphically depicts all the tasks required to complete a project, the sequence in which they must occur, their duration, the earliest or latest possible starting time, and the earliest or latest possible finishing time. It also defines the sequence of critical tasks known as the *critical path*: those tasks that must be started and finished exactly on time if the total schedule is to be met. The critical path is the longest path through the network.

A CPM chart for a simple design project is shown in Fig. 85.4. Each solid arrow in the chart represents an activity with a beginning and end point (represented by the numbered circles). No activity can begin until all activities leading into a circle have been completed.

The dashed arrows indicate dependency relationships that are not activities themselves, and thus they have no duration. These arrows are called *dummies* and are used to give each activity a unique beginning and ending number and to allow establishment of dependency relationships without tying in nondependent activities.

The heavier arrows in the illustration show the critical path, or the sequence of events that must happen as scheduled if the deadline is to be met. The numbers under the activities give the duration of each activity in days. Delaying the starting time of any of the activities in the critical path or increasing their duration will delay the whole project. The noncritical activities can begin or finish earlier or later (within limits) without affecting the final completion date. This variable time is called the *float* of each activity. *Float* or *slack time* is the minimum time that an activity can be delayed without causing the project to fall behind schedule. Float is always zero or minimum along the critical path.

4. SOLVING A CPM PROBLEM

The solution to a critical path method problem reveals the earliest and latest times that an activity can be started and finished. It also identifies the *critical path* and generates the *slack times* for each activity.

The following procedure may be used to solve a CPM problem. To facilitate the solution, each node should be replaced by a square that has been quartered. The compartments have the meanings indicated by the key.

Figure 85.4 *CPM Schedule*

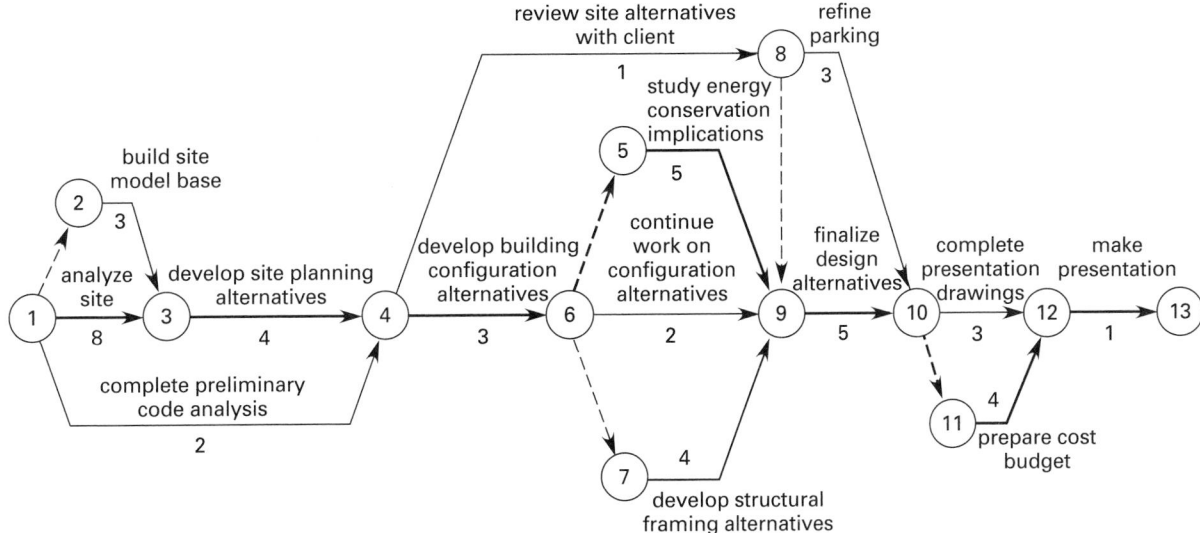

critical time path: 30 days

(Numbers in circles are beginning and ending points; numbers between circles indicate days.)

key

ES	**EF**
LS	**LF**

ES: Earliest Start

EF: Earliest Finish

LS: Latest Start

LF: Latest Finish

step 1: Place the project start time or date in the **ES** and **EF** positions of the start activity. The start time is zero for relative calculations.

step 2: Consider any unmarked activity, all of whose predecessors have been marked in the **EF** and **ES** positions. (Go to step 4 if there are none.) Mark in its **ES** position the largest number marked in the **EF** position of those predecessors.

step 3: Add the activity time to the **ES** time and write this in the **EF** box. Go to step 2.

step 4: Place the value of the latest finish date in the **LS** and **LF** boxes of the finish mode.

step 5: Consider unmarked predecessors whose successors have all been marked. Their **LF** is the smallest **LS** of the successors. Go to step 7 if there are no unmarked predecessors.

step 6: The **LS** for the new node is **LF** minus its activity time. Go to step 5.

step 7: The slack time for each node is **LS**−**ES** or **LF**−**EF**.

step 8: The critical path encompasses nodes for which the slack time equals **LS**−**ES** from the start node. There may be more than one critical path.

Example 85.2

Using the precedence table given, construct the precedence matrix and draw an activity-on-node network.

activity	duration (days)	predecessors
A, start	0	–
B	7	A
C	6	A
D	3	B
E	9	B,C
F	1	D,E
G	4	C
H, finish	0	F,G

Solution

The precedence matrix is

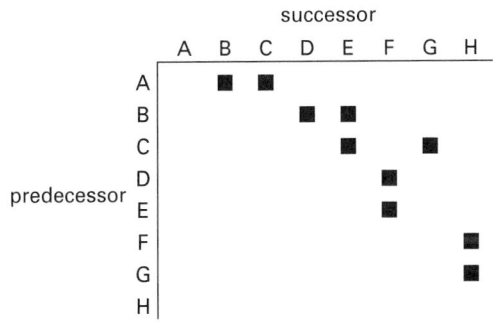

The activity-on-node network is

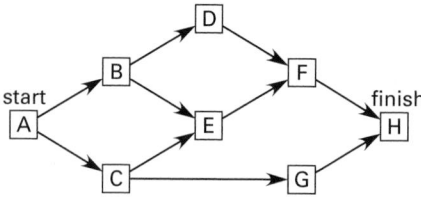

Example 85.3

Complete the network for the previous example and find the critical path. Assume the desired completion date is in 19 days.

Solution

The critical path is shown with darker lines.

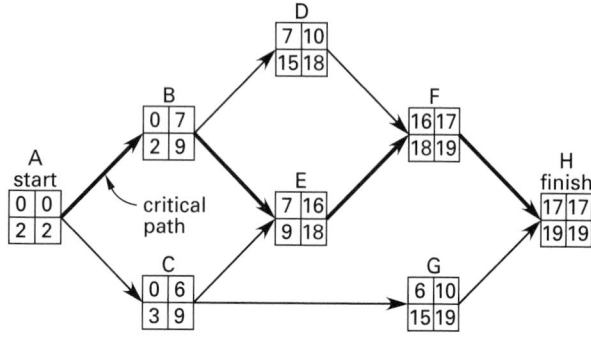

5. ACTIVITY-ON-BRANCH NETWORKS

In an *activity-on-branch* network, the arcs represent the activities, which are labeled with letters of the alphabet, and the nodes represent events, which are numbered. The activity durations may appear in parentheses near the activity letter.

The activity-on-branch method is complicated by the frequent requirement for *dummy activities* and nodes to maintain precedence. Consider the following part of a precedence table.

activity	predecessors
L	–
M	–
N	L,M
P	M

Note that activity P depends on the completion of only M. Figure 85.5(a) is an activity-on-branch representation of this precedence. However, N depends on the completion of both L and M. It would be incorrect to draw the network as Fig. 85.5(b) since the activity N appears twice. To represent the project, the dummy activity X must be used, as shown in Fig. 85.5(c).

Figure 85.5 *Activity-on-Branch Networks*

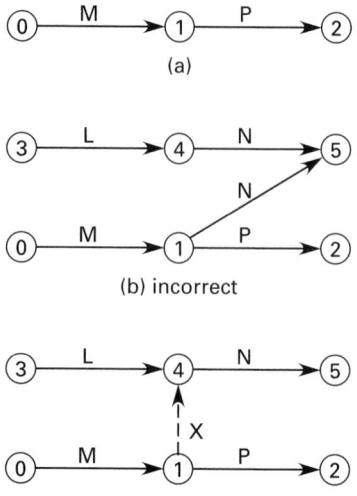

If two activities have the same starting and ending events, a dummy node is required to give one activity a uniquely identifiable completion event. This is illustrated in Fig. 85.6(b).

Figure 85.6 *Use of a Dummy Node*

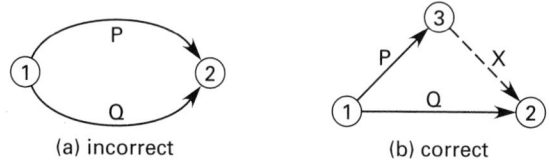

The solution method for an activity-on-branch problem is essentially the same as for the activity-on-node problem, requiring forward and reverse passes to determine earliest and latest dates.

Example 85.4

Represent the project in Ex. 85.2 as an activity-on-branch network.

Solution

event	event description
0	start project
1	finish B, start D
2	finish C, start G
3	finish B and C, start E
4	finish D and E, start F
5	finish F and G

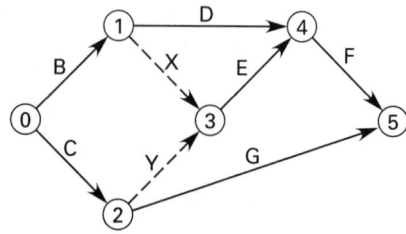

6. STOCHASTIC CRITICAL PATH MODELS

Stochastic models differ from deterministic models only in the way in which the activity durations are found. Whereas durations are known explicitly for the deterministic model, the time for a stochastic activity is distributed as a random variable.

This stochastic nature complicates the problem greatly since the actual distribution is often unknown. Such problems are solved as a deterministic model using the mean of an assumed duration distribution as the activity duration.

The most common stochastic critical path model is PERT, which stands for *program evaluation and review technique.* In PERT, all duration variables are assumed to come from a *beta distribution*, with mean and standard deviation given by Eqs. 85.1 and 85.2, respectively.

$$\mu = \tfrac{1}{6}(t_{minimum} + 4t_{most \; likely} + t_{maximum}) \qquad 85.1$$
$$\sigma = \tfrac{1}{6}(t_{maximum} - t_{minimum}) \qquad 85.2$$

The project *completion time* for large projects is assumed to be normally distributed with mean equal to the critical path length and with overall variance equal to the sum of the variances along the critical path.

The probability that a project duration will exceed some length (D) can be found from Eq. 85.3. z is the standard normal variable.

$$p\{duration > D\} = p\{t > z\} \qquad 85.3$$

$$z = \frac{D - \mu_{critical \; path}}{\sigma_{critical \; path}} \qquad 85.4$$

Example 85.5

The mean times and variances for activities along a PERT critical path are given. What is the probability that the project's completion time will be (a) less than 14 days, (b) more than 14 days, (c) more than 23 days, and (d) between 14 and 23 days?

activity mean time (days)	activity standard deviation (days)
9	1.3
4	0.5
7	2.6

Solution

The most likely completion time is the sum of the mean activity times.

$$\mu_{critical \; path} = 9 \text{ days} + 4 \text{ days} + 7 \text{ days}$$
$$= 20 \text{ days}$$

The variance of the project's completion times is the sum of the variances along the critical path. Variance, σ^2, is the square of the standard deviation, σ.

$$(\sigma_{critical \; path})^2 = (1.3 \text{ days})^2 + (0.5 \text{ days})^2$$
$$+ (2.6 \text{ days})^2$$
$$= 8.7 \text{ days}^2$$
$$\sigma_{critical \; path} = \sqrt{8.7 \text{ days}^2}$$
$$= 2.95 \text{ days} \quad [\text{use 3 days}]$$

The standard normal variable corresponding to 14 days is given by Eq. 85.4.

$$z = \frac{D - \mu_{critical \; path}}{\sigma_{critical \; path}}$$
$$= \frac{14 \text{ days} - 20 \text{ days}}{3 \text{ days}} = -2.0$$

The area under the standard normal curve is 0.4772 for $0.0 < z < -2.0$. Since the normal curve is symmetrical, the negative sign is irrelevant in determining the area.

The standard normal variable corresponding to 23 days is given by Eq. 85.4.

$$z = \frac{D - \mu_{critical \; path}}{\sigma_{critical \; path}}$$
$$= \frac{23 \text{ days} - 20 \text{ days}}{3 \text{ days}} = 1.0$$

The area under the standard normal curve is 0.3413 for $0.0 < z < 1.0$.

(a) $p\{duration < 14\} = p\{z < -2.0\} = 0.5 - 0.4772$
$$= 0.0228 \; (2.28\%)$$

(b) $p\{duration > 14\} = p\{z > -2.0\} = 0.4772 + 0.5$
$$= 0.9772 \; (97.72\%)$$

(c) $p\{duration > 23\} = p\{z > 1.0\} = 0.5 - 0.3413$
$$= 0.1587 \; (15.87\%)$$

(d) $p\{14 < duration < 23\} = p\{-2.0 < z < 1.0\}$
$$= 0.4772 + 0.3413$$
$$= 0.8185 \; (81.85\%)$$

7. MONITORING

Monitoring is keeping track of the progress of the job to see if the planned aspects of time, fee, and quality are being accomplished. The original fee projections can be monitored by comparing weekly time sheets with the original estimate. This can be done manually or with project management software. A manual method is shown in Fig. 85.7, which uses the same example project estimated in Fig. 85.1.

Figure 85.7 *Project Monitoring Chart*

Project: Mini-mall		time												
Phase/People/Departments		1	2	3	4	5	6	7	8	9	10	11	12	total
schematic design	budgeted	1320	2640											
schematic design	actual	2000	2900											
design development	budgeted			2510	4090									
design development	actual			3200										
construction docs.	budgeted					2156	1716	4136	5456	2376				
construction docs.	actual													
	budgeted													
	actual													
	budgeted													
	actual													
	budgeted													
	actual													
	budgeted													
	actual													
total (cumulative)	budgeted	1320	3960	6470	10,560	12,716	14,432	18,568	24,024	26,400				
total (cumulative)	actual	2000	4900	8100										

(left label for the blank rows: *fee dollars*)

At beginning of job, plot budgeted total dollars (or hours) on graph. Plot actual expended dollars (or hours) as job progresses. Also plot estimated percentage complete as job progresses.

Budgeted - - - -
Actual ———

(% completed axis: 20, 40, 60, 80, 100)

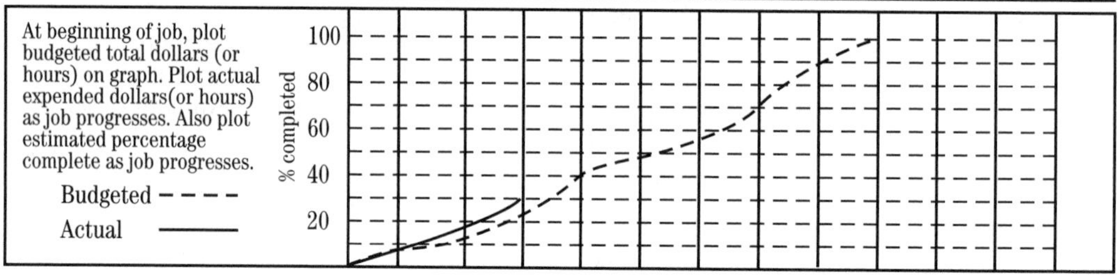

In Fig. 85.7, the budgeted weekly costs are placed in the table under the appropriate time-period column and phase-of-work row. The actual costs expended are written next to them. At the bottom of the chart, a cumulative graph is plotted that shows the actual money expended against the budgeted fees. The cumulative ratio of percentage completion to cost can also be plotted.

Monitoring quality is more difficult. At regular times during a project, the project manager, designers, and office principals should review the progress of the job to determine if the original project goals are being met and if the job is being produced according to the client's and design firm's expectations. The work in progress can also be reviewed to see whether it is technically correct and if all the contractual obligations are being met.

8. COORDINATING

During the project, the project manager must constantly coordinate the various people involved: the architect's staff, the consultants, the client, the building code officials, firm management, and, of course, the construction contractors. This may be done on a weekly, or even daily, basis to make sure the schedule is being maintained and the necessary work is getting done.

The coordination can be done by using checklists, holding weekly project meetings to discuss issues and assign work, and exchanging drawings or project files among the consultants.

9. DOCUMENTATION

Everything that is done on a project must be documented in writing. This documentation provides a record in case legal problems develop and serves as a project history to use for future jobs. Documentation is also a vital part of communication. An email or written memo is more accurate, communicates more clearly, and is more difficult to forget than a simple phone call, for example.

Most design firms have standard forms or project management software for documents such as transmittals, job observation reports, time sheets, and the like. Such software makes it easy to record the necessary information. In addition, all meetings should be documented with meeting notes. Phone call logs (listing date, time, participants, and discussion topics), emails, personal daily logs, and formal communications like letters and memos should also be generated and preserved to serve as documentation.

Two types of documents, *change orders* due to unexpected conditions or changes to the plans after bidding, and *as-built construction documents* to record what was actually installed (as opposed to what was shown in the original construction documents) are particularly important.

86 Engineering Economic Analysis

Nomenclature

A	annual amount	\$
B	present worth of all benefits	\$
BV_j	book value at end of the jth year	\$
C	cost or present worth of all costs	\$
d	declining balance depreciation rate	decimal
D	demand	various
D	depreciation	\$
DR	present worth of after-tax depreciation recovery	\$
e	constant inflation rate	decimal
E_0	initial amount of an exponentially growing cash flow	\$
EAA	equivalent annual amount	\$
$EUAC$	equivalent uniform annual cost	\$
f	federal income tax rate	decimal
F	forecasted quantity	various
F	future worth	\$
g	exponential growth rate	decimal
G	uniform gradient amount	\$
i	effective rate per period (usually per year)	decimal per unit time
i'	effective interest rate corrected for inflation	decimal
k	number of compounding periods per year	–
m	an integer	–
n	number of compounding periods or life of asset	–
P	present worth	\$

r	nominal rate per year (rate per annum)	decimal per unit time
ROI	return on investment	$
ROR	rate of return	decimal per unit time
s	state income tax rate	decimal
S_n	expected salvage value in year n	$
t	composite tax rate	decimal
t	time	years (typical)
T	a quantity equal to $\frac{1}{2}n(n+1)$	–
TC	tax credit	$
z	a quantity equal to $\dfrac{1+i}{1-d}$	decimal

Symbols

α	smoothing coefficient for forecasts	–
ϕ	effective rate per period (r/k)	decimal
$\mathcal{E}$	expected value	various

Subscripts

0	initial
j	at time j
n	at time n
t	at time t

1. IRRELEVANT CHARACTERISTICS

In its simplest form, an *engineering economic analysis* is a study of the desirability of making an investment.[1] The decision-making principles in this chapter can be applied by individuals as well as by companies. The nature of the spending opportunity or industry is not important. Farming equipment, personal investments, and multimillion dollar factory improvements can all be evaluated using the same principles.

Similarly, the applicable principles are largely insensitive to the monetary units. Although *dollars* are used in this chapter, it is equally convenient to use pounds, yen, or euros.

Finally, this chapter may give the impression that investment alternatives must be evaluated on a year-by-year basis. Actually, the *effective period* can be defined as a day, month, century, or any other convenient period of time.

2. MULTIPLICITY OF SOLUTION METHODS

Most economic conclusions can be reached in more than one manner. There are usually several different analyses that will eventually result in identical answers.[2] Other than the pursuit of elegant solutions in a timely manner,

there is no reason to favor one procedural method over another.[3]

3. PRECISION AND SIGNIFICANT DIGITS

The full potential of electronic calculators will never be realized in engineering economic analyses. Considering that calculations are based on estimates of far-future cash flows and that unrealistic assumptions (no inflation, identical cost structures of replacement assets, etc.) are routinely made, it makes little sense to carry cents along in calculations.

The calculations in this chapter have been designed to illustrate and review the principles presented. Because of this, greater precision than is normally necessary in everyday problems may be used. Though used, such precision is not warranted.

Unless there is some compelling reason to strive for greater precision, the following rules are presented for use in reporting final answers to engineering economic analysis problems.

- Omit fractional parts of the dollar (i.e., cents).

- Report and record a number to a maximum of four significant digits unless the first digit of that number is 1, in which case, a maximum of five significant digits should be written. For example,

$49	not	$49.43
$93,450	not	$93,453
$1,289,700	not	$1,289,673

4. NONQUANTIFIABLE FACTORS

An engineering economic analysis is a quantitative analysis. Some factors cannot be introduced as numbers into the calculations. Such factors are known as *nonquantitative factors, judgment factors*, and *irreducible factors*. Typical nonquantifiable factors are

- preferences
- political ramifications
- urgency
- goodwill
- prestige
- utility
- corporate strategy
- environmental effects
- health and safety rules
- reliability
- political risks

---

[1]This subject is also known as *engineering economics* and *engineering economy*. There is very little, if any, true economics in this subject.

[2]Because of round-off errors, particularly when factors are taken from tables, these different calculations will produce slightly different numerical results (e.g., $49.49 versus $49.50). However, this type of divergence is well known and accepted in engineering economic analysis.

[3]This does not imply that approximate methods, simplifications, and rules of thumb are acceptable.

Since these factors are not included in the calculations, the policy is to disregard the issues entirely. Of course, the factors should be discussed in a final report. The factors are particularly useful in breaking ties between competing alternatives that are economically equivalent.

5. YEAR-END CONVENTION

Except in short-term transactions, it is simpler to assume that all receipts and disbursements (cash flows) take place at the end of the year in which they occur.[4] This is known as the *year-end convention*. The exceptions to the year-end convention are initial project cost (purchase cost), trade-in allowance, and other cash flows that are associated with the inception of the project at $t = 0$.

On the surface, such a convention appears grossly inappropriate since repair expenses, interest payments, corporate taxes, and so on seldom coincide with the end of a year. However, the convention greatly simplifies engineering economic analysis problems, and it is justifiable on the basis that the increased precision associated with a more rigorous analysis is not warranted (due to the numerous other simplifying assumptions and estimates initially made in the problem).

There are various established procedures, known as *rules* or *conventions*, imposed by the Internal Revenue Service on U.S. taxpayers. An example is the *half-year rule*, which permits only half of the first-year depreciation to be taken in the first year of an asset's life when certain methods of depreciation are used. These rules are subject to constantly changing legislation and are not covered in this book. The implementation of such rules is outside the scope of engineering practice and is best left to accounting professionals.

6. CASH FLOW DIAGRAMS

Although they are not always necessary in simple problems (and they are often unwieldy in very complex problems), *cash flow diagrams* can be drawn to help visualize and simplify problems having diverse receipts and disbursements.

The following conventions are used to standardize cash flow diagrams.

- The horizontal (time) axis is marked off in equal increments, one per period, up to the duration (or *horizon*) of the project.

- Two or more transfers in the same period are placed end-to-end, and these may be combined.

- Expenses incurred before $t = 0$ are called *sunk costs*. Sunk costs are not relevant to the problem unless they have tax consequences in an after-tax analysis.

- *Receipts* are represented by arrows directed upward. *Disbursements* are represented by arrows directed downward. The arrow length is proportional to the magnitude of the cash flow.

Example 86.1

A mechanical device will cost $20,000 when purchased. Maintenance will cost $1000 each year. The device will generate revenues of $5000 each year for five years, after which the salvage value is expected to be $7000. Draw and simplify the cash flow diagram.

Solution

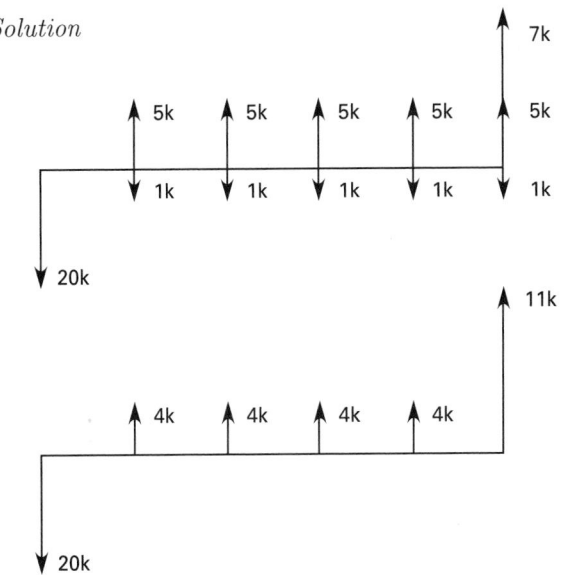

7. TYPES OF CASH FLOWS

To evaluate a real-world project, it is necessary to present the project's cash flows in terms of standard cash flows that can be handled by engineering economic analysis techniques. The standard cash flows are single payment cash flow, uniform series cash flow, gradient series cash flow, and the infrequently encountered exponential gradient series cash flow.

A *single payment cash flow* can occur at the beginning of the time line (designated as $t = 0$), at the end of the time line (designated as $t = n$), or at any time in between.

The *uniform series cash flow* consists of a series of equal transactions starting at $t = 1$ and ending at $t = n$. The symbol A is typically given to the magnitude of each individual cash flow.[5]

---

[4]A *short-term transaction* typically has a lifetime of five years or less and has payments or compounding that are more frequent than once per year.

[5]Notice that the cash flows do not begin at $t = 0$. This is an important concept with all of the series cash flows. This convention has been established to accommodate the timing of annual maintenance (and similar) cash flows for which the year-end convention is applicable.

The *gradient series cash flow* starts with a cash flow (typically given the symbol G) at $t = 2$ and increases by G each year until $t = n$, at which time the final cash flow is $(n-1)G$.

An *exponential gradient series cash flow* is based on a phantom value (typically given the symbol E_0) at $t = 0$ and grows or decays exponentially according to the following relationship.[6]

$$\text{amount at time } t = E_t = E_0(1+g)^t$$
$$[t = 1, 2, 3, \ldots, n] \quad \textbf{86.1}$$

In Eq. 86.1, g is the *exponential growth rate*, which can be either positive or negative. Exponential gradient cash flows are rarely seen in economic justification projects assigned to engineers.[7]

Figure 86.1 *Standard Cash Flows*

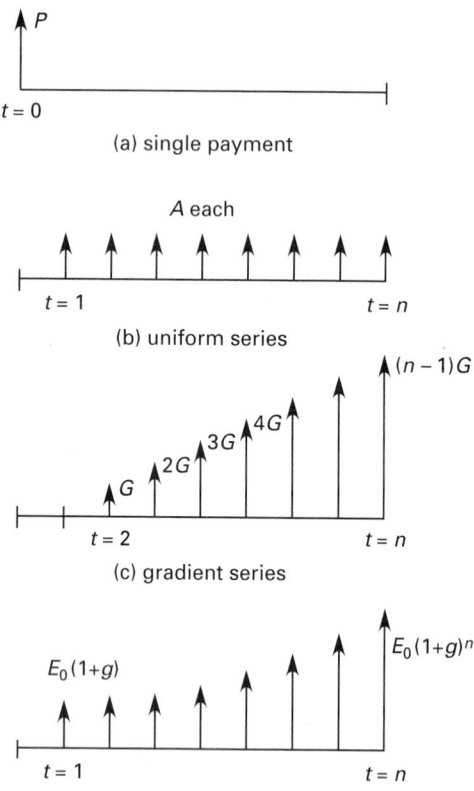

(a) single payment

(b) uniform series

(c) gradient series

(d) exponential gradient series

8. TYPICAL PROBLEM TYPES

There is a wide variety of problem types that, collectively, are considered to be engineering economic analysis problems.

---

[6]Notice the convention for an exponential cash flow series: The first cash flow E_0 is at $t = 1$, as in the uniform annual series. However, the first cash flow is $E_0(1 + g)$. The cash flow of E_0 at $t = 0$ is absent (i.e., is a *phantom cash flow*).

[7]For one of the few discussions on exponential cash flow, see *Capital Budgeting*, Robert V. Oakford, The Ronald Press Company, New York, 1970.

By far, the majority of engineering economic analysis problems are *alternative comparisons*. In these problems, two or more mutually exclusive investments compete for limited funds. A variation of this is a *replacement/retirement analysis*, which is repeated each year to determine if an existing asset should be replaced. Finding the percentage return on an investment is a *rate of return problem*, one of the alternative comparison solution methods.

Investigating interest and principal amounts in loan payments is a *loan repayment problem*. An *economic life analysis* will determine when an asset should be retired. In addition, there are miscellaneous problems involving economic order quantity, learning curves, break-even points, product costs, and so on.

9. IMPLICIT ASSUMPTIONS

Several assumptions are implicitly made when solving engineering economic analysis problems. Some of these assumptions are made with the knowledge that they are or will be poor approximations of what really will happen. The assumptions are made, regardless, for the benefit of obtaining a solution.

The most common assumptions are the following.

- The year-end convention is applicable.

- There is no inflation now, nor will there be any during the lifetime of the project.

- Unless otherwise specifically called for, a before-tax analysis is needed.

- The effective interest rate used in the problem will be constant during the lifetime of the project.

- Nonquantifiable factors can be disregarded.

- Funds invested in a project are available and are not urgently needed elsewhere.

- Excess funds continue to earn interest at the effective rate used in the analysis.

This last assumption, like most of the assumptions listed, is almost never specifically mentioned in the body of a solution. However, it is a key assumption when comparing two alternatives that have different initial costs.

For example, suppose two investments, one costing $10,000 and the other costing $8000, are to be compared at 10%. It is obvious that $10,000 in funds is available, otherwise the costlier investment would not be under consideration. If the smaller investment is chosen, what is done with the remaining $2000? The last assumption yields the answer: the $2000 is "put to work" in some investment earning (in this case) 10%.

10. EQUIVALENCE

Industrial decision makers using engineering economic analysis are concerned with the magnitude and timing of a project's cash flow as well as with the total profitability of that project. In this situation, a method is required to compare projects involving receipts and disbursements occurring at different times.

By way of illustration, consider $100 placed in a bank account that pays 5% effective annual interest at the end of each year. After the first year, the account will have grown to $105. After the second year, the account will have grown to $110.25.

Assume that you will have no need for money during the next two years, and any money received will immediately go into your 5% bank account. Then, which of the following options would be more desirable?

option a: $100 now

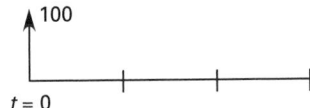

option b: $105 to be delivered in one year.

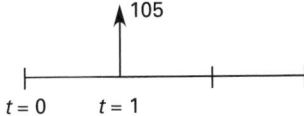

option c: $110.25 to be delivered in two years

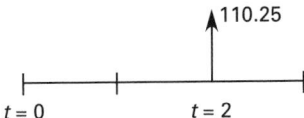

As illustrated, none of the options is superior under the assumptions given. If the first option is chosen, you will immediately place $100 into a 5% account, and in two years the account will have grown to $110.25. In fact, the account will contain $110.25 at the end of two years regardless of the option chosen. Therefore, these alternatives are said to be *equivalent*.

Equivalence may or may not be the case, depending on the interest rate. Thus, an alternative that is acceptable to one decision maker may be unacceptable to another. The interest rate that is used in actual calculations is known as the *effective interest rate*.[8] If compounding is

once a year, it is known as the *effective annual interest rate*. However, effective quarterly, monthly, daily, and so on, interest rates are also used.

The fact that $100 today grows to $105 in one year (at 5% annual interest) is an example of what is known as the *time value of money* principle. This principle simply articulates what is obvious: Funds placed in a secure investment will increase to an equivalent future amount. The procedure for determining the present investment from the equivalent future amount is known as *discounting*.

11. SINGLE-PAYMENT EQUIVALENCE

The equivalence of any present amount, P, at $t = 0$, to any future amount, F, at $t = n$, is called the *future worth* and can be calculated from Eq. 86.2.

$$F = P(1 + i)^n \qquad 86.2$$

The factor $(1 + i)^n$ is known as the single payment *compound amount factor* and has been tabulated in App. 86.B for various combinations of i and n.

Similarly, the equivalence of any future amount to any present amount is called the *present worth* and can be calculated from Eq. 86.3.

$$P = F(1 + i)^{-n} = \frac{F}{(1 + i)^n} \qquad 86.3$$

The factor $(1 + i)^{-n}$ is known as the *single payment present worth factor*.[9]

The interest rate used in Eqs. 86.2 and 86.3 must be the effective rate per period. Also, the basis of the rate (annually, monthly, etc.) must agree with the type of period used to count n. Thus, it would be incorrect to use an effective annual interest rate if n was the number of compounding periods in months.

Example 86.2

How much should you put into a 10% (effective annual rate) savings account in order to have $10,000 in five years?

Solution

This problem could also be stated: What is the equivalent present worth of $10,000 five years from now if money is worth 10% per year?

$$P = F(1 + i)^{-n} = (\$10,000)(1 + 0.10)^{-5}$$
$$= \$6209$$

The factor 0.6209 would usually be obtained from the tables.

---

[8]The adjective *effective* distinguishes this interest rate from other interest rates (e.g., nominal interest rates) that are not meant to be used directly in calculating equivalent amounts.

[9]The *present worth* is also called the *present value* and *net present value*. These terms are used interchangeably and no significance should be attached to the terms *value, worth,* and *net*.

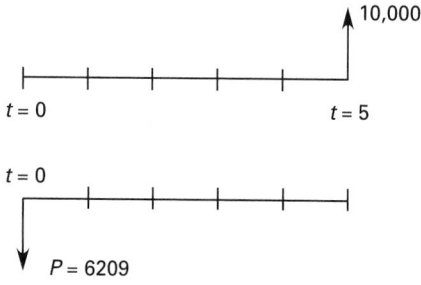

12. STANDARD CASH FLOW FACTORS AND SYMBOLS

Equations 86.2 and 86.3 may give the impression that solving engineering economic analysis problems involves a lot of calculator use, and, in particular, a lot of exponentiation. Such calculations may be necessary from time to time, but most problems are simplified by the use of tabulated values of the factors.

Rather than actually writing the formula for the compound amount factor (which converts a present amount to a future amount), it is common convention to substitute the standard functional notation of $(F/P, i\%, n)$. Thus, the future value in n periods of a present amount would be symbolically written as

$$F = P(F/P, i\%, n) \qquad 86.4$$

Similarly, the present worth factor has a functional notation of $(P/F, i\%, n)$. The present worth of a future amount n periods hence would be symbolically written as

$$P = F(P/F, i\%, n) \qquad 86.5$$

Values of these *cash flow (discounting) factors* are tabulated in App. 86.B. There is often initial confusion about whether the (F/P) or (P/F) column should be used in a particular problem. There are several ways of remembering what the functional notations mean.

One method of remembering which factor should be used is to think of the factors as conditional probabilities. The conditional probability of event **A** given that event **B** has occurred is written as $p\{\mathbf{A}|\mathbf{B}\}$, where the given event comes after the vertical bar. In the standard notational form of discounting factors, the given amount is similarly placed after the slash. What you want comes before the slash. (F/P) would be a factor to find F given P.

Another method of remembering the notation is to interpret the factors algebraically. Thus, the (F/P) factor could be thought of as the fraction F/P. Algebraically, Eq. 86.4 would be

$$F = P(F/P) \qquad 86.6$$

This algebraic approach is actually more than an interpretation. The numerical values of the discounting factors are consistent with this algebraic manipulation.

Thus, the (F/A) factor could be calculated as $(F/P) \times (P/A)$. This consistent relationship can be used to calculate other factors that might be occasionally needed, such as (F/G) or (G/P). For instance, the annual cash flow that would be equivalent to a uniform gradient may be found from

$$A = G(P/G, i\%, n)(A/P, i\%, n) \qquad 86.7$$

Formulas for the compounding and discounting factors are contained in Table 86.1. Normally, it will not be necessary to calculate factors from the formulas. Appendix 86.B is adequate for solving most problems.

Example 86.3

What factor will convert a gradient cash flow ending at $t = 8$ to a future value at $t = 8$? (That is, what is the $(F/G, i\%, 8)$ factor?) The effective annual interest rate is 10%.

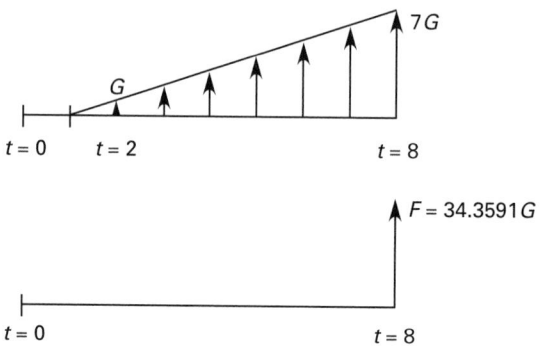

Solution

> *method 1:* From Table 86.1, the $(F/G, 10\%, 8)$ factor is
>
> $$(F/G, 10\%, 8)$$
> $$= \frac{(1+i)^n - 1}{i^2} - \frac{n}{i}$$
> $$= \frac{(1+0.10)^8 - 1}{(0.10)^2} - \frac{8}{0.10}$$
> $$= 34.3589$$

> *method 2:* The tabulated values of (P/G) and (F/P) in App. 86.B can be used to calculate the factor.
>
> $$(F/G, 10\%, 8)$$
> $$= (P/G, 10\%, 8)(F/P, 10\%, 8)$$
> $$= (16.0287)(2.1436) = 34.3591$$

> The (F/G) factor could also have been calculated as the product of the (A/G) and (F/A) factors.

Table 86.1 *Discount Factors for Discrete Compounding*

factor name	converts	symbol	formula
single payment compound amount	P to F	$(F/P, i\%, n)$	$(1+i)^n$
single payment present worth	F to P	$(P/F, i\%, n)$	$(1+i)^{-n}$
uniform series sinking fund	F to A	$(A/F, i\%, n)$	$\dfrac{i}{(1+i)^n - 1}$
capital recovery	P to A	$(A/P, i\%, n)$	$\dfrac{i(1+i)^n}{(1+i)^n - 1}$
uniform series compound amount	A to F	$(F/A, i\%, n)$	$\dfrac{(1+i)^n - 1}{i}$
uniform series present worth	A to P	$(P/A, i\%, n)$	$\dfrac{(1+i)^n - 1}{i(1+i)^n}$
uniform gradient present worth	G to P	$(P/G, i\%, n)$	$\dfrac{(1+i)^n - 1}{i^2(1+i)^n} - \dfrac{n}{i(1+i)^n}$
uniform gradient future worth	G to F	$(F/G, i\%, n)$	$\dfrac{(1+i)^n - 1}{i^2} - \dfrac{n}{i}$
uniform gradient uniform series	G to A	$(A/G, i\%, n)$	$\dfrac{1}{i} - \dfrac{n}{(1+i)^n - 1}$

13. CALCULATING UNIFORM SERIES EQUIVALENCE

A cash flow that repeats each year for n years without change in amount is known as an *annual amount* and is given the symbol A. As an example, a piece of equipment may require annual maintenance, and the maintenance cost will be an annual amount. Although the equivalent value for each of the n annual amounts could be calculated and then summed, it is more expedient to use one of the uniform series factors. For example, it is possible to convert from an annual amount to a future amount by use of the (F/A) factor.

$$F = A(F/A, i\%, n) \qquad 86.8$$

A *sinking fund* is a fund or account into which annual deposits of A are made in order to accumulate F at $t = n$ in the future. Since the annual deposit is calculated as $A = F(A/F, i\%, n)$, the (A/F) factor is known as the *sinking fund factor*. An *annuity* is a series of equal payments (A) made over a period of time.[10] Usually, it

---

[10]An annuity may also consist of a lump sum payment made at some future time. However, this interpretation is not considered in this chapter.

is necessary to "buy into" an investment (a bond, and insurance policy, etc.) in order to ensure the annuity. In the simplest case of an annuity that starts at the end of the first year and continues for n years, the purchase price (P) is

$$P = A(P/A, i\%, n) \qquad 86.9$$

The present worth of an *infinite (perpetual) series* of annual amounts is known as a *capitalized cost*. There is no $(P/A, i\%, \infty)$ factor in the tables, but the capitalized cost can be calculated simply as

$$P = \frac{A}{i} \qquad \text{[i in decimal form]} \qquad 86.10$$

Alternatives with different lives will generally be compared by way of *equivalent uniform annual cost* (EUAC). An EUAC is the annual amount that is equivalent to all of the cash flows in the alternative. The EUAC differs in sign from all of the other cash flows. Costs and expenses expressed as EUACs, which would normally be considered negative, are actually positive. The term *cost* in the designation EUAC serves to make clear the meaning of a positive number.

Example 86.4

Maintenance costs for a machine are $250 each year. What is the present worth of these maintenance costs over a 12-year period if the interest rate is 8%?

Solution

$$P = A(P/A, 8\%, 12) = (-\$250)(7.5361)$$
$$= -\$1884$$

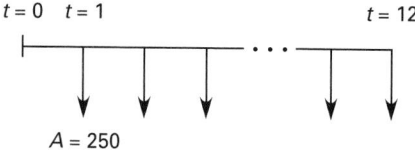

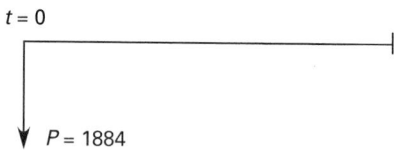

14. FINDING PAST VALUES

From time to time, it will be necessary to determine an amount in the past equivalent to some current (or future) amount. For example, you might have to calculate the original investment made 15 years ago given a current annuity payment.

Such problems are solved by placing the $t = 0$ point at the time of the original investment, and then calculating the past amount as a P value. For example, the original investment, P, can be extracted from the annuity, A, by using the standard cash flow factors.

$$P = A(P/A, i\%, n) \qquad 86.11$$

The choice of $t = 0$ is flexible. As a general rule, the $t = 0$ point should be selected for convenience in solving a problem.

Example 86.5

You are currently paying $250 per month to lease your office phone equipment. You have three years (36 months) left on the five-year (60-month) lease. What would have been an equivalent purchase price two years ago? The effective interest rate per month is 1%.

Solution

The solution of this example is not affected by the fact that investigation is being performed in the middle of the horizon. This is a simple calculation of present worth.

$$P = A(P/A, 1\%, 60)$$
$$= (-\$250)(44.9550) = -\$11,239$$

15. TIMES TO DOUBLE AND TRIPLE AN INVESTMENT

If an investment doubles in value (in n compounding periods and with $i\%$ effective interest), the ratio of current value to past investment will be 2.

$$F/P = (1 + i)^n = 2 \qquad 86.12$$

Similarly, the ratio of current value to past investment will be 3 if an investment triples in value. This can be written as

$$F/P = (1 + i)^n = 3 \qquad 86.13$$

It is a simple matter to extract the number of periods, n, from Eqs. 86.12 and 86.13 to determine the *doubling time* and *tripling time*, respectively. For example, the doubling time is

$$n = \frac{\log 2}{\log(1 + i)} \qquad 86.14$$

When a quick estimate of the doubling time is needed, the *rule of 72* can be used. The doubling time is approximately $72/i$.

The tripling time is

$$n = \frac{\log 3}{\log(1 + i)} \qquad 86.15$$

Equations 86.14 and 86.15 form the basis of Table 86.2.

Table 86.2 *Doubling and Tripling Times for Various Interest Rates*

interest rate ($i\%$)	doubling time (periods)	tripling time (periods)
1	69.7	110.4
2	35.0	55.5
3	23.4	37.2
4	17.7	28.0
5	14.2	22.5
6	11.9	18.9
7	10.2	16.2
8	9.01	14.3
9	8.04	12.7
10	7.27	11.5
11	6.64	10.5
12	6.12	9.69
13	5.67	8.99
14	5.29	8.38
15	4.96	7.86
16	4.67	7.40
17	4.41	7.00
18	4.19	6.64
19	3.98	6.32
20	3.80	6.03

16. VARIED AND NONSTANDARD CASH FLOWS

Gradient Cash Flow

A common situation involves a uniformly increasing cash flow. If the cash flow has the proper form, its present worth can be determined by using the *uniform gradient factor*, $(P/G, i\%, n)$. The uniform gradient factor finds the present worth of a uniformly increasing cash flow that starts in year two (not in year one).

There are three common difficulties associated with the form of the uniform gradient. The first difficulty is that the initial cash flow occurs at $t = 2$. This convention recognizes that annual costs, if they increase uniformly, begin with some value at $t = 1$ (due to the year-end convention) but do not begin to increase until $t = 2$. The tabulated values of (P/G) have been calculated to find the present worth of only the increasing part of the annual expense. The present worth of the base expense incurred at $t = 1$ must be found separately with the (P/A) factor.

The second difficulty is that, even though the $(P/G, i\%, n)$ factor is used, there are only $n - 1$ actual cash flows. It is clear that n must be interpreted as the *period number* in which the last gradient cash flow occurs, not the number of gradient cash flows.

Finally, the sign convention used with gradient cash flows may seem confusing. If an expense increases each year (as in Ex. 86.6), the gradient will be negative, since it is an expense. If a revenue increases each year, the gradient will be positive. In most cases, the sign of the gradient depends on whether the cash flow is an expense or a revenue.[11]

Example 86.6

Maintenance on an old machine is $100 this year but is expected to increase by $25 each year thereafter. What is the present worth of five years of the costs of maintenance? Use an interest rate of 10%.

Solution

In this problem, the cash flow must be broken down into parts. (Notice that the five-year gradient factor is used even though there are only four nonzero gradient cash flows.)

$$P = A(P/A, 10\%, 5) + G(P/G, 10\%, 5)$$
$$= (-\$100)(3.7908) - (\$25)(6.8618)$$
$$= -\$551$$

---

[11]This is not a universal rule. It is possible to have a uniformly decreasing revenue as in Fig. 86.2(c). In this case, the gradient would be negative.

Figure 86.2 *Positive and Negative Gradient Cash Flows*

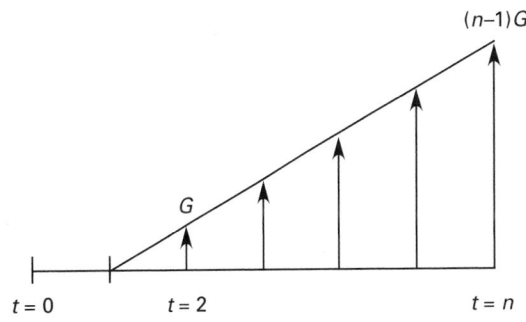

(a) positive gradient cash flow

$$P = G(P/G)$$

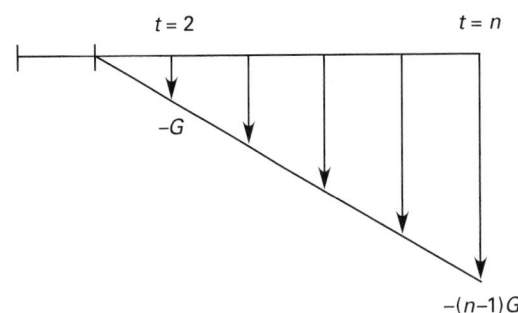

(b) negative gradient cash flow

$$P = -G(P/G)$$

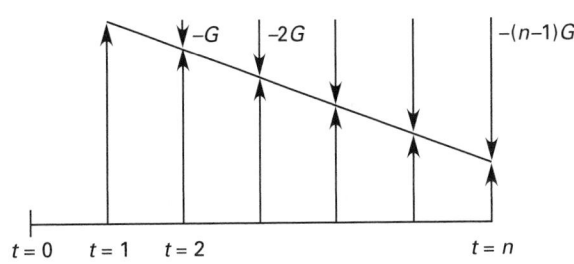

(c) decreasing revenue incorporating a negative gradient

$$P = A(P/A) - G(P/G)$$

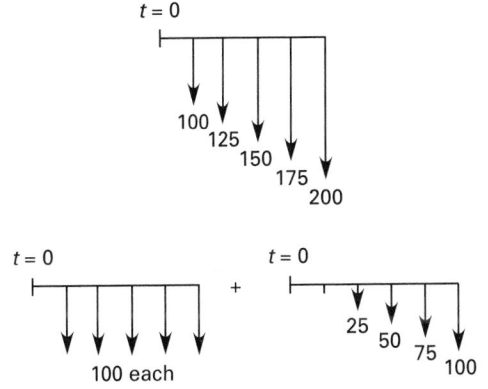

Systems, Mgmt, and Professional

Stepped Cash Flows

Stepped cash flows are easily handled by the technique of *superposition of cash flows*. This technique is illustrated by Ex. 86.7.

Example 86.7

An investment costing $1000 returns $100 for the first five years and $200 for the following five years. How would the present worth of this investment be calculated?

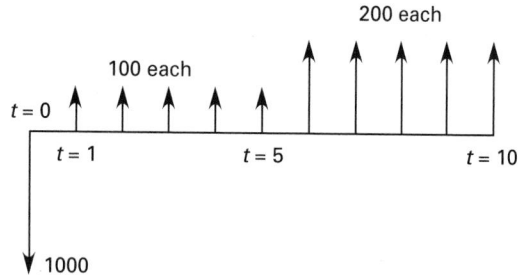

Solution

Using the principle of superposition, the revenue cash flow can be thought of as $200 each year from $t = 1$ to $t = 10$, with a negative revenue of $100 from $t = 1$ to $t = 5$. Superimposed, these two cash flows make up the actual performance cash flow.

$$P = -\$1000 + (\$200)(P/A, i\%, 10) - (\$100)(P/A, i\%, 5)$$

Missing and Extra Parts of Standard Cash Flows

A missing or extra part of a standard cash flow can also be handled by superposition. For example, suppose an annual expense is incurred each year for ten years, except in the ninth year. (The cash flow is illustrated in Figure 86.3.) The present worth could be calculated as a subtractive process.

$$P = A(P/A, i\%, 10) - A(P/F, i\%, 9) \qquad \textbf{86.16}$$

Alternatively, the present worth could be calculated as an additive process.

$$P = A(P/A, i\%, 8) + A(P/F, i\%, 10) \qquad \textbf{86.17}$$

Figure 86.3 *Cash Flow with a Missing Part*

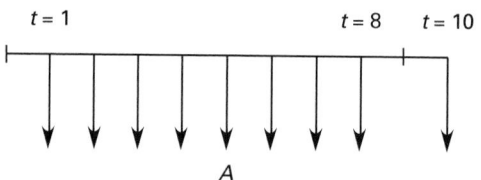

Delayed and Premature Cash Flows

There are cases when a cash flow matches a standard cash flow exactly, except that the cash flow is delayed or starts sooner than it should. Often, such cash flows can be handled with superposition. At other times, it may be more convenient to shift the time axis. This shift is known as the *projection method*. Example 86.8 illustrates the projection method.

Example 86.8

An expense of $75 is incurred starting at $t = 3$ and continues until $t = 9$. There are no expenses or receipts until $t = 3$. Use the projection method to determine the present worth of this stream of expenses.

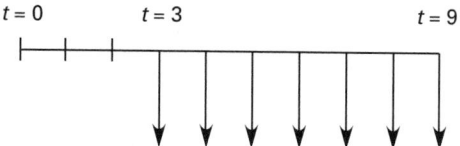

Solution

First, determine a cash flow at $t = 2$ that is equivalent to the entire expense stream. If $t = 0$ was where $t = 2$ actually is, the present worth of the expense stream would be

$$P' = (-\$75)(P/A, i\%, 7)$$

P' is a cash flow at $t = 2$. It is now simple to find the present worth (at $t = 0$) of this future amount.

$$P = P'(P/F, i\%, 2) = (-\$75)(P/A, i\%, 7)(P/F, i\%, 2)$$

Cash Flows at Beginnings of Years: The Christmas Club Problem

This type of problem is characterized by a stream of equal payments (or expenses) starting at $t = 0$ and ending at $t = n - 1$. It differs from the standard annual cash flow in the existence of a cash flow at $t = 0$ and the absence of a cash flow at $t = n$. This problem gets its name from the service provided by some savings institutions whereby money is automatically deposited each week or month (starting immediately, when the savings plan is opened) in order to accumulate money to purchase Christmas presents at the end of the year.

Figure 86.4 *Cash Flows at Beginnings of Years*

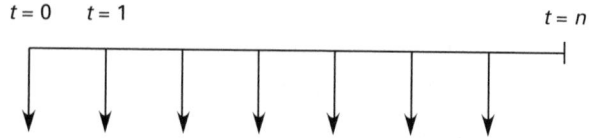

It may seem that the present worth of the savings stream can be determined by directly applying the (P/A) factor. However, this is not the case, since the Christmas

Club cash flow and the standard annual cash flow differ. The Christmas Club problem is easily handled by superposition, as illustrated by Ex. 86.9.

Example 86.9

How much can you expect to accumulate by $t = 10$ for a child's college education if you deposit $300 at the beginning of each year for a total of ten payments?

Solution

Notice that the first payment is made at $t = 0$ and that there is no payment at $t = 10$. The future worth of the first payment is calculated with the (F/P) factor. The absence of the payment at $t = 10$ is handled by superposition. Notice that this "correction" is not multiplied by a factor.

$$F = (\$300)(F/P, i\%, 10) + (\$300)(F/A, i\%, 10) - \$300$$
$$= (\$300)(F/A, i\%, 11) - \$300$$

17. THE MEANING OF PRESENT WORTH AND i

If $100 is invested in a 5% bank account (using annual compounding), you can remove $105 one year from now; if this investment is made, you will receive a *return on investment* (ROI) of $5. The cash flow diagram and the present worth of the two transactions are

$$P = -\$100 + (\$105)(P/F, 5\%, 1)$$
$$= -\$100 + (\$105)(0.9524) = 0$$

Figure 86.5 *Cash Flow Diagram*

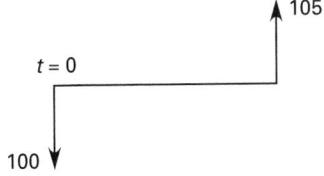

Notice that the present worth is zero even though you will receive a $5 return on your investment.

However, if you are offered $120 for the use of $100 over a one-year period, the cash flow diagram and present worth (at 5%) would be

$$P = -\$100 + (\$120)(P/F, 5\%, 1)$$
$$= -\$100 + (\$120)(0.9524) = \$14.29$$

Figure 86.6 *Cash Flow Diagram*

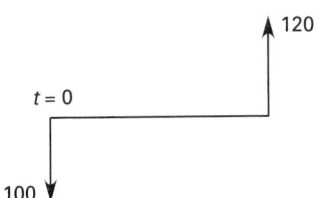

Therefore, the present worth of an alternative is seen to be equal to the equivalent value at $t = 0$ of the increase in return above that which you would be able to earn in an investment offering $i\%$ per period. In the previous case, $14.29 is the present worth of ($20 − $5), the difference in the two ROIs.

The present worth is also the amount that you would have to be given to dissuade you from making an investment, since placing the initial investment amount along with the present worth into a bank account earning $i\%$ will yield the same eventual return on investment. Relating this to the previous paragraphs, you could be dissuaded from investing $100 in an alternative that would return $120 in one year by a $t = 0$ payment of $14.29. Clearly, ($100 + $14.29) invested at $t = 0$ will also yield $120 in one year at 5%.

Income-producing alternatives with negative present worths are undesirable, and alternatives with positive present worths are desirable because they increase the average earning power of invested capital. (In some cases, such as municipal and public works projects, the present worths of all alternatives are negative, in which case, the least negative alternative is best.)

The selection of the interest rate is difficult in engineering economics problems. Usually, it is taken as the average rate of return that an individual or business organization has realized in past investments. Alternatively, the interest rate may be associated with a particular level of risk. Usually, i for individuals is the interest rate that can be earned in relatively *risk-free investments*.

18. SIMPLE AND COMPOUND INTEREST

If $100 is invested at 5%, it will grow to $105 in one year. During the second year, 5% interest continues to be accrued, but on $105, not on $100. This is the principle of *compound interest*: The interest accrues interest.[12]

If only the original principal accrues interest, the interest is said to be *simple interest*. Simple interest is rarely encountered in engineering economic analyses, but the concept may be incorporated into short-term transactions.

---

[12]This assumes, of course, that the interest remains in the account. If the interest is removed and spent, only the remaining funds accumulate interest.

19. EXTRACTING THE INTEREST RATE: RATE OF RETURN

An intuitive definition of the *rate of return* (ROR) is the effective annual interest rate at which an investment accrues income. That is, the rate of return of a project is the interest rate that would yield identical profits if all money were invested at that rate. Although this definition is correct, it does not provide a method of determining the rate of return.

It was previously seen that the present worth of a $100 investment invested at 5% is zero when $i = 5\%$ is used to determine equivalence. Thus, a working definition of rate of return would be the effective annual interest rate that makes the present worth of the investment zero. Alternatively, rate of return could be defined as the effective annual interest rate that will discount all cash flows to a total present worth equal to the required initial investment.

It is tempting, but impractical, to determine a rate of return analytically. It is simply too difficult to extract the interest rate from the equivalence equation. For example, consider a $100 investment that pays back $75 at the end of each of the first two years. The present worth equivalence equation (set equal to zero in order to determine the rate of return) is

$$P = 0 = -\$100 + (\$75)(1+i)^{-1} + (\$75)(1+i)^{-2} \quad \textbf{\textit{86.18}}$$

Solving Eq. 86.18 requires finding the roots of a quadratic equation. In general, for an investment or project spanning n years, the roots of an nth-order polynomial would have to be found. It should be obvious that an analytical solution would be essentially impossible for more complex cash flows. (The rate of return in this example is 31.87%.)

If the rate of return is needed, it can be found from a trial-and-error solution. To find the rate of return of an investment, proceed as follows.

step 1: Set up the problem as if to calculate the present worth.

step 2: Arbitrarily select a reasonable value for i. Calculate the present worth.

step 3: Choose another value of i (not too close to the original value), and again solve for the present worth.

step 4: Interpolate or extrapolate the value of i that gives a zero present worth.

step 5: For increased accuracy, repeat steps 2 and 3 with two more values that straddle the value found in step 4.

A common, although incorrect, method of calculating the rate of return involves dividing the annual receipts or returns by the initial investment. (See Sec. 54.) However, this technique ignores such items as salvage, depreciation, taxes, and the time value of money. This technique also is inadequate when the annual returns vary.

It is possible that more than one interest rate will satisfy the zero present worth criteria. This confusing situation occurs whenever there is more than one change in sign in the investment's cash flow.[13] Table 86.3 indicates the numbers of possible interest rates as a function of the number of sign reversals in the investment's cash flow.

Table 86.3 *Multiplicity of Rates of Return*

number of sign reversals	number of distinct rates of return
0	0
1	0 or 1
2	0, 1, or 2
3	0, 1, 2, or 3
4	0, 1, 2, 3, or 4
m	$0, 1, 2, 3, \ldots m-1, m$

Difficulties associated with interpreting the meaning of multiple rates of return can be handled with the concepts of external investment and external rate of return. An *external investment* is an investment that is distinct from the investment being evaluated (which becomes known as the internal investment). The *external rate of return*, which is the rate of return earned by the external investment, does not need to be the same as the rate earned by the internal investment.

Generally, the multiple rates of return indicate that the analysis must proceed as though money will be invested outside of the project. The mechanics of how this is done are not covered here.

Example 86.10

What is the rate of return on invested capital if $1000 is invested now with $500 being returned in year 4 and $1000 being returned in year 8?

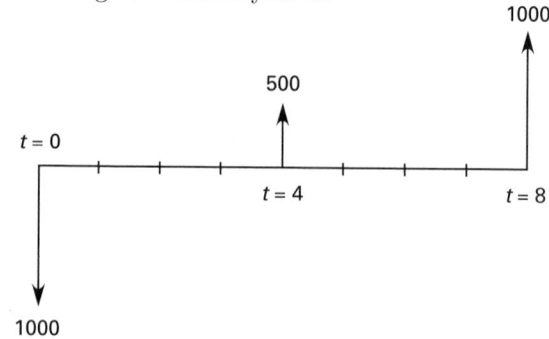

---

[13]There will always be at least one change of sign in the cash flow of a legitimate investment. (This excludes municipal and other tax-supported functions.) At $t = 0$, an investment is made (a negative cash flow). Hopefully, the investment will begin to return money (a positive cash flow) at $t = 1$ or shortly thereafter. Although it is possible to conceive of an investment in which all of the cash flows were negative, such an investment would probably be classified as a *hobby*.

Solution

First, set up the problem as a present worth calculation. Try $i = 5\%$.

$$P = -\$1000 + (\$500)(P/F, 5\%, 4) + (\$1000)$$
$$\times (P/F, 5\%, 8)$$
$$= -\$1000 + (\$500)(0.8227) + (\$1000)(0.6768)$$
$$= \$88$$

Next, try a larger value of i to reduce the present worth. If $i = 10\%$,

$$P = -\$1000 + (\$500)(P/F, 10\%, 4) + (\$1000)$$
$$\times (P/F, 10\%, 8)$$
$$= -\$1000 + (\$500)(0.6830) + (\$1000)(0.4665)$$
$$= -\$192$$

Using simple interpolation, the rate of return is

$$\text{ROR} = 5\% + \left(\frac{\$88}{\$88 + \$192} \right)(10\% - 5\%)$$
$$= 6.6\%$$

A second iteration between 6% and 7% yields 6.39%.

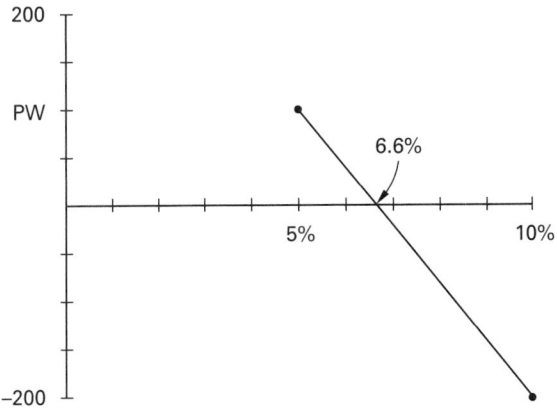

Example 86.11

An existing biomedical company is developing a new drug. A venture capital firm gives the company $25 million initially and $55 million more at the end of the first year. The drug patent will be sold at the end of year 5 to the highest bidder, and the biomedical company will receive $80 million. (The venture capital firm will receive everything in excess of $80 million.) The firm invests unused money in short-term commercial paper earning 10% effective interest per year through its bank. In the meantime, the biomedical company incurs development expenses of $50 million annually for the first three years. The drug is to be evaluated by a government agency and there will be neither expenses nor revenues during the fourth year. What is the biomedical company's rate of return on this investment?

Solution

Normally, the rate of return is determined by setting up a present worth problem and varying the interest rate until the present worth is zero. Writing the cash flows, though, shows that there are two reversals of sign: one at $t = 2$ (positive to negative) and the other at $t = 5$ (negative to positive). Therefore, there could be two interest rates that produce a zero present worth. (In fact, there actually are two interest rates: 10.7% and 41.4%.)

time	cash flow (millions)
0	+25
1	+55 − 50 = +5
2	−50
3	−50
4	0
5	+80

However, this problem can be reduced to one with only one sign reversal in the cash flow series. The initial $25 million is invested in commercial paper (an *external investment* having nothing to do with the drug development process) during the first year at 10%. The accumulation of interest and principal after one year is

$$(25)(1 + 0.10) = 27.5$$

This 27.5 is combined with the 5 (the money remaining after all expenses are paid at $t = 1$) and invested externally, again at 10%. The accumulation of interest and principal after one year (i.e., at $t = 2$) is

$$(27.5 + 5)(1 + 0.10) = 35.75$$

This 35.75 is combined with the development cost paid at $t = 2$.

The cash flow for the development project (the internal investment) is

time	cash flow (millions)
0	0
1	0
2	35.75 − 50 = −14.25
3	−50
4	0
5	+80

Now, there is only one sign reversal in the cash flow series. The *internal rate of return* on this development project is found by the traditional method to be 10.3%. Notice that this is different from the rate the company can earn from investing externally in commercial paper.

20. RATE OF RETURN VERSUS RETURN ON INVESTMENT

Rate of return (ROR) is an effective annual interest rate, typically stated in percent per year. *Return on investment* (ROI) is a dollar amount. Thus, *rate of return* and *return on investment* are not synonymous.

Return on investment can be calculated in two different ways. The accounting method is to subtract the total of all investment costs from the total of all net profits (i.e., revenues less expenses). The time value of money is not considered.

In engineering economic analysis, the return on investment is calculated from equivalent values. Specifically, the present worth (at $t = 0$) of all investment costs is subtracted from the future worth (at $t = n$) of all net profits.

When there are only two cash flows, a single investment amount and a single payback, the two definitions of return on investment yield the same numerical value. When there are more than two cash flows, the returns on investment will be different depending on which definition is used.

21. MINIMUM ATTRACTIVE RATE OF RETURN

A company may not know what effective interest rate, i, to use in engineering economic analysis. In such a case, the company can establish a minimum level of economic performance that it would like to realize on all investments. This criterion is known as the *minimum attractive rate of return* (MARR). Unlike the effective interest rate, i, the minimum attractive rate of return is not used in numerical calculations.[14] It is used only in comparisons with the rate of return.

Once a rate of return for an investment is known, it can be compared to the minimum attractive rate of return. To be a viable alternative, the rate of return must be greater than the minimum attractive rate of return.

The advantage of using comparisons to the minimum attractive rate of return is that an effective interest rate, i, never needs to be known. The minimum attractive rate of return becomes the correct interest rate for use in present worth and equivalent uniform annual cost calculations.

22. TYPICAL ALTERNATIVE-COMPARISON PROBLEM FORMATS

With the exception of some investment and rate of return problems, the typical problem involving engineering economics will have the following characteristics.

- An interest rate will be given.
- Two or more alternatives will be competing for funding.
- Each alternative will have its own cash flows.
- It is necessary to select the best alternative.

23. DURATIONS OF INVESTMENTS

Because they are handled differently, short-term investments and short-lived assets need to be distinguished from investments and assets that constitute an infinitely lived project. Short-term investments are easily identified: a drill press that is needed for three years or a temporary factory building that is being constructed to last five years.

Investments with perpetual cash flows are also (usually) easily identified: maintenance on a large flood control dam and revenues from a long-span toll bridge. Furthermore, some items with finite lives can expect renewal on a repeated basis.[15] For example, a major freeway with a pavement life of 20 years is unlikely to be abandoned; it will be resurfaced or replaced every 20 years.

Actually, if an investment's finite lifespan is long enough, it can be considered an infinite investment because money 50 or more years from now has little impact on current decisions. The $(P/F, 10\%, 50)$ factor, for example, is 0.0085. Thus, one dollar at $t = 50$ has an equivalent present worth of less than one penny. Since these far-future cash flows are eclipsed by present cash flows, long-term investments can be considered finite or infinite without significant impact on the calculations.

24. CHOICE OF ALTERNATIVES: COMPARING ONE ALTERNATIVE WITH ANOTHER ALTERNATIVE

Several methods exist for selecting a superior alternative from among a group of proposals. Each method has its own merits and applications.

Present Worth Method

When two or more alternatives are capable of performing the same functions, the superior alternative will have the largest present worth. The *present worth method* is restricted to evaluating alternatives that are mutually exclusive and that have the same lives. This method is suitable for ranking the desirability of alternatives.

Example 86.12

Investment A costs $10,000 today and pays back $11,500 two years from now. Investment B costs $8000 today and pays back $4500 each year for two years. If an interest rate of 5% is used, which alternative is superior?

---

[14]Not everyone adheres to this rule. Some people use "minimum attractive rate of return" and "effective interest rate" interchangeably.

[15]The term *renewal* can be interpreted to mean replacement or repair.

Solution

$$P(A) = -\$10,000 + (\$11,500)(P/F, 5\%, 2)$$
$$= -\$10,000 + (\$11,500)(0.9070)$$
$$= \$431$$
$$P(B) = -\$8000 + (\$4500)(P/A, 5\%, 2)$$
$$= -\$8000 + (\$4500)(1.8594)$$
$$= \$367$$

Alternative A is superior and should be chosen.

Capitalized Cost Method

The present worth of a project with an infinite life is known as the *capitalized cost* or *life cycle cost*. Capitalized cost is the amount of money at $t = 0$ needed to perpetually support the project on the earned interest only. Capitalized cost is a positive number when expenses exceed income.

In comparing two alternatives, each of which is infinitely lived, the superior alternative will have the lowest capitalized cost.

Normally, it would be difficult to work with an infinite stream of cash flows since most economics tables do not list factors for periods in excess of 100 years. However, the (A/P) discounting factor approaches the interest rate as n becomes large. Since the (P/A) and (A/P) factors are reciprocals of each other, it is possible to divide an infinite series of equal cash flows by the interest rate in order to calculate the present worth of the infinite series. This is the basis of Eq. 86.19.

$$\text{capitalized cost} = \text{initial cost} + \frac{\text{annual costs}}{i} \qquad \textit{86.19}$$

Equation 86.19 can be used when the annual costs are equal in every year. If the operating and maintenance costs occur irregularly instead of annually, or if the costs vary from year to year, it will be necessary to somehow determine a cash flow of equal annual amounts (EAA) that is equivalent to the stream of original costs.

The equal annual amount may be calculated in the usual manner by first finding the present worth of all the actual costs and then multiplying the present worth by the interest rate (the (A/P) factor for an infinite series). However, it is not even necessary to convert the present worth to an equal annual amount since Eq. 86.20 will convert the equal amount back to the present worth.

$$\text{capitalized cost} = \text{initial cost} + \frac{\text{EAA}}{i}$$
$$= \text{initial cost} + \frac{\text{present worth}}{\text{of all expenses}} \qquad \textit{86.20}$$

Example 86.13

What is the capitalized cost of a public works project that will cost \$25,000,000 now and will require \$2,000,000 in maintenance annually? The effective annual interest rate is 12%.

Solution

Worked in millions of dollars, from Eq. 86.19, the capitalized cost is

$$\text{capitalized cost} = 25 + (2)(P/A, 12\%, \infty)$$
$$= 25 + \frac{2}{0.12} = 41.67$$

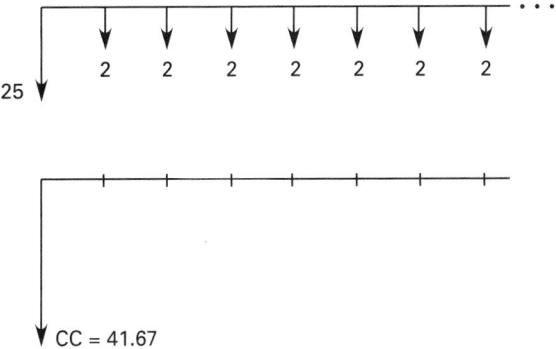

Annual Cost Method

Alternatives that accomplish the same purpose but that have unequal lives must be compared by the *annual cost method*.[16] The annual cost method assumes that each alternative will be replaced by an identical twin at the end of its useful life (infinite renewal). This method, which may also be used to rank alternatives according to their desirability, is also called the *annual return method* or *capital recovery method*.

Restrictions are that the alternatives must be mutually exclusive and repeatedly renewed up to the duration of the longest-lived alternative. The calculated annual cost is known as the *equivalent uniform annual cost* (EUAC) or just *equivalent annual cost*. Cost is a positive number when expenses exceed income.

Example 86.14

Which of the following alternatives is superior over a 30-year period if the interest rate is 7%?

	alternative A	alternative B
type	brick	wood
life	30 years	10 years
initial cost	\$1800	\$450
maintenance	\$5/year	\$20/year

[16]Of course, the annual cost method can be used to determine the superiority of assets with identical lives as well.

Solution

$$EUAC(A) = (\$1800)(A/P, 7\%, 30) + \$5$$
$$= (\$1800)(0.0806) + \$5$$
$$= \$150$$
$$EUAC(B) = (\$450)(A/P, 7\%, 10) + \$20$$
$$= (\$450)(0.1424) + \$20$$
$$= \$84$$

Alternative B is superior since its annual cost of operation is the lowest. It is assumed that three wood facilities, each with a life of 10 years and a cost of $450, will be built to span the 30-year period.

25. CHOICE OF ALTERNATIVES: COMPARING AN ALTERNATIVE WITH A STANDARD

With specific economic performance criteria, it is possible to qualify an investment as acceptable or unacceptable without having to compare it with another investment. Two such performance criteria are the benefit-cost ratio and the minimum attractive rate of return.

Benefit-Cost Ratio Method

The *benefit-cost ratio* method is often used in municipal project evaluations where benefits and costs accrue to different segments of the community. With this method, the present worth of all benefits (irrespective of the beneficiaries) is divided by the present worth of all costs. The project is considered acceptable if the ratio equals or exceeds 1.0, that is, if $B/C \geq 1.0$.

When the benefit-cost ratio method is used, disbursements by the initiators or sponsors are *costs*. Disbursements by the users of the project are known as *disbenefits*. It is often difficult to determine whether a cash flow is a cost or a disbenefit (whether to place it in the numerator or denominator of the benefit-cost ratio calculation).

Regardless of where the cash flow is placed, an acceptable project will always have a benefit-cost ratio greater than or equal to 1.0, although the actual numerical result will depend on the placement. For this reason, the benefit-cost ratio method should not be used to rank competing projects.

The benefit-cost ratio method of comparing alternatives is used extensively in transportation engineering where the ratio is often (but not necessarily) written in terms of annual benefits and annual costs instead of present worths. Another characteristic of highway benefit-cost ratios is that the route (road, highway, etc.) is usually already in place and that various alternative upgrades are being considered. There will be existing benefits

and costs associated with the current route. Therefore, the *change* (usually an increase) in benefits and costs is used to calculate the benefit-cost ratio.[17]

$$B/C = \frac{\Delta^{\text{user}}_{\text{benefits}}}{\Delta^{\text{investment}}_{\text{cost}} + \Delta\text{maintenance} - \Delta^{\text{residual}}_{\text{value}}} \quad 86.21$$

Notice that the change in *residual value (terminal value)* appears in the denominator as a negative item. An increase in the residual value would decrease the denominator.

Example 86.15

By building a bridge over a ravine, a state department of transportation can shorten the time it takes to drive through a mountainous area. Estimates of costs and benefits (due to decreased travel time, fewer accidents, reduced gas usage, etc.) have been prepared. Should the bridge be built? Use the benefit-cost ratio method of comparison.

	millions
initial cost	40
capitalized cost of perpetual annual maintenance	12
capitalized value of annual user benefits	49
residual value	0

Solution

If Eq. 86.21 is used, the benefit-cost ratio is

$$B/C = \frac{49}{40 + 12 + 0} = 0.942$$

Since the benefit-cost ratio is less than 1.00, the bridge should not be built.

If the maintenance costs are placed in the numerator (per Ftn. 17), the benefit-cost ratio value will be different but the conclusion will not change.

$$B/C_{\text{alternate method}} = \frac{49 - 12}{40} = 0.925$$

Rate of Return Method

The minimum attractive rate of return (MARR) has already been introduced as a standard of performance against which an investment's actual *rate of return*

---

[17]This discussion of highway benefit-cost ratios is not meant to imply that everyone agrees with Eq. 86.21. In *Economic Analysis for Highways* (International Textbook Company, Scranton, PA, 1969), author Robley Winfrey takes a strong stand on one aspect of the benefits versus disbenefits issue: highway maintenance. According to Winfrey, regular highway maintenance costs should be placed in the numerator as a subtraction from the user benefits. Some have called this mandate the *Winfrey method*.

(ROR) is compared. If the rate of return is equal to or exceeds the minimum attractive rate of return, the investment is qualified. This is the basis for the *rate of return method* of alternative selection.

Finding the rate of return can be a long, iterative process. Usually, the actual numerical value of rate of return is not needed; it is sufficient to know whether or not the rate of return exceeds the minimum attractive rate of return. This *comparative analysis* can be accomplished without calculating the rate of return simply by finding the present worth of the investment using the minimum attractive rate of return as the effective interest rate (i.e., $i = \text{MARR}$). If the present worth is zero or positive, the investment is qualified. If the present worth is negative, the rate of return is less than the minimum attractive rate of return.

26. RANKING MUTUALLY EXCLUSIVE MULTIPLE PROJECTS

Ranking of multiple investment alternatives is required when there is sufficient funding for more than one investment. Since the best investments should be selected first, it is necessary to be able to place all investments into an ordered list.

Ranking is relatively easy if the present worths, future worths, capitalized costs, or equivalent uniform annual costs have been calculated for all the investments. The highest ranked investment will be the one with the largest present or future worth, or the smallest capitalized or annual cost. Present worth, future worth, capitalized cost, and equivalent uniform annual cost can all be used to rank multiple investment alternatives.

However, neither rates of return nor benefit-cost ratios should be used to rank multiple investment alternatives. Specifically, if two alternatives both have rates of return exceeding the minimum acceptable rate of return, it is not sufficient to select the alternative with the highest rate of return.

An *incremental analysis*, also known as a *rate of return on added investment study*, should be performed if rate of return is used to select between investments. An incremental analysis starts by ranking the alternatives in order of increasing initial investment. Then, the cash flows for the investment with the lower initial cost are subtracted from the cash flows for the higher-priced alternative on a year-by-year basis. This produces, in effect, a third alternative representing the costs and benefits of the added investment. The added expense of the higher priced investment is not warranted unless the rate of return of this third alternative exceeds the minimum attractive rate of return as well. The choice criterion is to select the alternative with the higher initial investment if the incremental rate of return exceeds the minimum attractive rate of return.

An incremental analysis is also required if ranking is to be done by the benefit-cost ratio method. The incremental analysis is accomplished by calculating the ratio of differences in benefits to differences in costs for each possible pair of alternatives. If the ratio exceeds 1.0, alternative 2 is superior to alternative 1. Otherwise, alternative 1 is superior.[18]

$$\frac{B_2 - B_1}{C_2 - C_1} \geq 1 \quad [\text{alternative 2 superior}] \qquad 86.22$$

27. ALTERNATIVES WITH DIFFERENT LIVES

Comparison of two alternatives is relatively simple when both alternatives have the same life. For example, a problem might be stated: "Which would you rather have: car A with a life of three years, or car B with a life of five years?"

However, care must be taken to understand what is going on when the two alternatives have different lives. If car A has a life of three years and car B has a life of five years, what happens at $t = 3$ if the five-year car is chosen? If a car is needed for five years, what happens at $t = 3$ if the three-year car is chosen?

In this type of situation, it is necessary to distinguish between the length of the need (the *analysis horizon*) and the lives of the alternatives or assets intended to meet that need. The lives do not have to be the same as the horizon.

Finite Horizon with Incomplete Asset Lives

If an asset with a five-year life is chosen for a three-year need, the disposition of the asset at $t = 3$ must be known in order to evaluate the alternative. If the asset is sold at $t = 3$, the salvage value is entered into the analysis (at $t = 3$) and the alternative is evaluated as a three-year investment. The fact that the asset is sold when it has some useful life remaining does not affect the analysis horizon.

Similarly, if a three-year asset is chosen for a five-year need, something about how the need is satisfied during the last two years must be known. Perhaps a rental asset will be used. Or, perhaps the function will be "farmed out" to an outside firm. In any case, the costs of satisfying the need during the last two years enter the analysis, and the alternative is evaluated as a five-year investment.

If both alternatives are "converted" to the same life, any of the alternative selection criteria (present worth method, annual cost method, etc.) can be used to determine which alternative is superior.

---

[18]It goes without saying that the benefit-cost ratios for all investment alternatives by themselves must also be equal to or greater than 1.0.

Finite Horizon with Integer Multiple Asset Lives

It is common to have a long-term horizon (need) that must be met with short-lived assets. In special instances, the horizon will be an integer number of asset lives. For example, a company may be making a 12-year transportation plan and may be evaluating two cars: one with a three-year life, and another with a four-year life.

In this example, four of the first car or three of the second car are needed to reach the end of the 12-year horizon.

If the horizon is an integer number of asset lives, any of the alternative selection criteria can be used to determine which is superior. If the present worth method is used, all alternatives must be evaluated over the entire horizon. (In this example, the present worth of 12 years of car purchases and use must be determined for both alternatives.)

If the equivalent uniform annual cost method is used, it may be possible to base the calculation of annual cost on one lifespan of each alternative only. It may not be necessary to incorporate all of the cash flows into the analysis. (In the running example, the annual cost over three years would be determined for the first car; the annual cost over four years would be determined for the second car.) This simplification is justified if the subsequent asset replacements (renewals) have the same cost and cash flow structure as the original asset. This assumption is typically made implicitly when the annual cost method of comparison is used.

Infinite Horizon

If the need horizon is infinite, it is not necessary to impose the restriction that asset lives of alternatives be integer multiples of the horizon. The superior alternative will be replaced (renewed) whenever it is necessary to do so, forever.

Infinite horizon problems are almost always solved with either the annual cost or capitalized cost method. It is common to (implicitly) assume that the cost and cash flow structure of the asset replacements (renewals) are the same as the original asset.

28. OPPORTUNITY COSTS

An *opportunity cost* is an imaginary cost representing what will not be received if a particular strategy is rejected. It is what you will lose if you do or do not do something. As an example, consider a growing company with an existing operational computer system. If the company trades in its existing computer as part of an upgrade plan, it will receive a *trade-in allowance*. (In other problems, a *salvage value* may be involved.)

If one of the alternatives being evaluated is not to upgrade the computer system at all, the trade-in allowance (or, salvage value in other problems) will not be realized. The amount of the trade-in allowance is an opportunity cost that must be included in the problem analysis.

Similarly, if one of the alternatives being evaluated is to wait one year before upgrading the computer, the *difference in trade-in allowances* is an opportunity cost that must be included in the problem analysis.

29. REPLACEMENT STUDIES

An investigation into the retirement of an existing process or piece of equipment is known as a *replacement study*. Replacement studies are similar in most respects to other alternative comparison problems: An interest rate is given, two alternatives exist, and one of the previously mentioned methods of comparing alternatives is used to choose the superior alternative. Usually, the annual cost method is used on a year-by-year basis.

In replacement studies, the existing process or piece of equipment is known as the *defender*. The new process or piece of equipment being considered for purchase is known as the *challenger*.

30. TREATMENT OF SALVAGE VALUE IN REPLACEMENT STUDIES

Since most defenders still have some market value when they are retired, the problem of what to do with the salvage arises. It seems logical to use the salvage value of the defender to reduce the initial purchase cost of the challenger. This is consistent with what would actually happen if the defender were to be retired.

By convention, however, the defender's salvage value is subtracted from the defender's present value. This does not seem logical, but it is done to keep all costs and benefits related to the defender with the defender. In this case, the salvage value is treated as an opportunity cost that would be incurred if the defender is not retired.

If the defender and the challenger have the same lives and a present worth study is used to choose the superior alternative, the placement of the salvage value will have no effect on the net difference between present worths for the challenger and defender. Although the values of the two present worths will be different depending on the placement, the difference in present worths will be the same.

If the defender and the challenger have different lives, an annual cost comparison must be made. Since the salvage value would be "spread over" a different number of years depending on its placement, it is important to abide by the conventions listed in this section.

There are a number of ways to handle salvage value in retirement studies. The best way is to calculate the cost of keeping the defender one more year. In addition to the usual operating and maintenance costs, that cost includes an opportunity interest cost incurred by not selling the defender, and also a drop in the salvage value if the defender is kept for one additional year. Specifically,

$$
\begin{aligned}
\text{EUAC (defender)} =\ & \text{next year's maintenance costs} \\
& + i(\text{current salvage value}) \\
& + \text{current salvage} \\
& - \text{next year's salvage} \qquad 86.23
\end{aligned}
$$

It is important in retirement studies not to double count the salvage value. That is, it would be incorrect to add the salvage value to the defender and at the same time subtract it from the challenger.

Equation 86.23 contains the difference in salvage value between two consecutive years. This calculation shows that the defender/challenger decision must be made on a year-by-year basis. One application of Eq. 86.23 will not usually answer the question of whether the defender should remain in service indefinitely. The calculation must be repeatedly made as long as there is a drop in salvage value from one year to the next.

31. ECONOMIC LIFE: RETIREMENT AT MINIMUM COST

As an asset grows older, its operating and maintenance costs typically increase. Eventually, the cost to keep the asset in operation becomes prohibitive, and the asset is retired or replaced. However, it is not always obvious when an asset should be retired or replaced.

As the asset's maintenance cost is increasing each year, the amortized cost of its initial purchase is decreasing. It is the sum of these two costs that should be evaluated to determine the point at which the asset should be retired or replaced. Since an asset's initial purchase price is likely to be high, the amortized cost will be the controlling factor in those years when the maintenance costs are low. Therefore, the EUAC of the asset will decrease in the initial part of its life.

However, as the asset grows older, the change in its amortized cost decreases while maintenance cost increases. Eventually, the sum of the two costs reaches a minimum and then starts to increase. The age of the asset at the minimum cost point is known as the *economic life* of the asset. The economic life generally is less than the length of need and the technological lifetime of the asset.

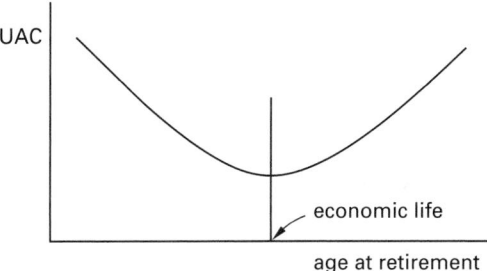

Figure 86.7 *EUAC versus Age at Retirement*

The determination of an asset's economic life is illustrated by Ex. 86.16.

Example 86.16

Buses in a municipal transit system have the characteristics listed. When should the city replace its buses if money can be borrowed at 8%?

initial cost of bus: $120,000

year	maintenance cost	salvage value
1	35,000	60,000
2	38,000	55,000
3	43,000	45,000
4	50,000	25,000
5	65,000	15,000

Solution

The annual maintenance is different each year. Each maintenance cost must be spread over the life of the bus. This is done by first finding the present worth and then amortizing the maintenance costs. If a bus is kept for one year and then sold, the annual cost will be

$$
\begin{aligned}
\text{EUAC}(1) =\ & (\$120,000)(A/P, 8\%, 1) \\
& + (\$35,000)(A/F, 8\%, 1) \\
& - (\$60,000)(A/F, 8\%, 1) \\
=\ & (\$120,000)(1.0800) + (\$35,000)(1.000) \\
& - (\$60,000)(1.000) \\
=\ & \$104,600
\end{aligned}
$$

If a bus is kept for two years and then sold, the annual cost will be

$$
\begin{aligned}
\text{EUAC}(2) =\ & \big(\$120,000 + (\$35,000)(P/F, 8\%, 1)\big) \\
& \times (A/P, 8\%, 2) \\
& + (\$38,000 - \$55,000)(A/F, 8\%, 2) \\
=\ & \big(\$120,000 + (\$35,000)(0.9259)\big)(0.5608) \\
& + (\$38,000 - \$55,000)(0.4808) \\
=\ & \$77,296
\end{aligned}
$$

If a bus is kept for three years and then sold, the annual cost will be

$$
\begin{aligned}
\text{EUAC(3)} = & \big(\$120{,}000 + (35{,}000)(P/F, 8\%, 1) \\
& + (\$38{,}000)(P/F, 8\%, 2)\big)(A/P, 8\%, 3) \\
& + (\$43{,}000 - \$45{,}000)(A/F, 8\%, 3) \\
= & \big(\$120{,}000 + (\$35{,}000)(0.9259) \\
& + (\$38{,}000)(0.8573)\big)(0.3880) \\
& - (\$2000)(0.3080) \\
= & \ \$71{,}158
\end{aligned}
$$

This process is continued until the annual cost begins to increase. In this example, EUAC(4) is $71,700. Therefore, the buses should be retired after three years.

32. LIFE-CYCLE COST

The *life-cycle cost* of an alternative is the equivalent value (at $t = 0$) of the alternative's cash flow over the alternative's lifespan. Since the present worth is evaluated using an effective interest rate of i (which would be the interest rate used for all engineering economic analyses), the life-cycle cost is the same as the alternative's present worth. If the alternative has an infinite horizon, the life-cycle cost and capitalized cost will be identical.

33. CAPITALIZED ASSETS VERSUS EXPENSES

High expenses reduce profit, which in turn reduces income tax. It seems logical to label each and every expenditure, even an asset purchase, as an expense. As an alternative to this *expensing the asset*, it may be decided to capitalize the asset. *Capitalizing the asset* means that the cost of the asset is divided into equal or unequal parts, and only one of these parts is taken as an expense each year. Expensing is clearly the more desirable alternative, since the after-tax profit is increased early in the asset's life.

There are long-standing accounting conventions as to what can be expensed and what must be capitalized.[19] Some companies capitalize everything—regardless of cost—with expected lifetimes greater than one year. Most companies, however, expense items whose purchase costs are below a cut-off value. A cut-off value in the range of $250 to $500, depending on the size of the company, is chosen as the maximum purchase cost of an expensed asset. Assets costing more than this are capitalized.

It is not necessary for a large corporation to keep track of every lamp, desk, and chair for which the purchase price is greater than the cut-off value. Such assets, all

of which have the same lives and have been purchased in the same year, can be placed into groups or *asset classes*. A group cost, equal to the sum total of the purchase costs of all items in the group, is capitalized as though the group was an identifiable and distinct asset itself.

34. PURPOSE OF DEPRECIATION

Depreciation is an *artificial expense* that spreads the purchase price of an asset or other property over a number of years.[20] Depreciating an asset is an example of capitalization, as previously defined. The inclusion of depreciation in engineering economic analysis problems will increase the after-tax present worth (profitability) of an asset. The larger the depreciation, the greater will be the profitability. Therefore, individuals and companies eligible to utilize depreciation want to maximize and accelerate the depreciation available to them.

Although the entire property purchase price is eventually recognized as an expense, the net recovery from the expense stream never equals the original cost of the asset. That is, depreciation cannot realistically be thought of as a fund (an annuity or sinking fund) that accumulates capital to purchase a replacement at the end of the asset's life. The primary reason for this is that the depreciation expense is reduced significantly by the impact of income taxes, as will be seen in later sections.

35. DEPRECIATION BASIS OF AN ASSET

The *depreciation basis* of an asset is the part of the asset's purchase price that is spread over the *depreciation period*, also known as the *service life*.[21] Usually, the depreciation basis and the purchase price are not the same.

A common depreciation basis is the difference between the purchase price and the expected salvage value at the end of the depreciation period. That is,

$$
\text{depreciation basis} = C - S_n \qquad \textit{86.24}
$$

There are several methods of calculating the year-by-year depreciation of an asset. Equation 86.24 is not universally compatible with all depreciation methods.

---

[19] For example, purchased vehicles must be capitalized; payments for leased vehicles can be expensed. Repainting a building with paint that will last five years is an expense, but the replacement cost of a leaking roof must be capitalized.

[20] The IRS tax regulations allow depreciation on almost all forms of *business property* except land. The following types of property are distinguished: *real* (e.g., buildings used for business), *residential* (e.g., buildings used as rental property), and *personal* (e.g., equipment used for business). Personal property does *not* include items for personal use (such as a personal residence), despite its name. *Tangible personal property* is distinguished from *intangible property* (goodwill, copyrights, patents, trademarks, franchises, agreements not to compete, etc.).

[21] The *depreciation period* is selected to be as short as possible within recognized limits. This depreciation will not normally coincide with the *economic life* or *useful life* of an asset. For example, a car may be capitalized over a depreciation period of three years. It may become uneconomical to maintain and use at the end of an economic life of nine years. However, the car may be capable of operation over a useful life of 25 years.

Some methods do not consider the salvage value. This is known as an *unadjusted basis*. When the depreciation method is known, the depreciation basis can be rigorously defined.[22]

36. DEPRECIATION METHODS

Generally, tax regulations do not allow the cost of an asset to be treated as a deductible expense in the year of purchase. Rather, portions of the depreciation basis must be allocated to each of the n years of the asset's depreciation period. The amount that is allocated each year is called the *depreciation*.

Various methods exist for calculating an asset's depreciation each year.[23] Although the depreciation calculations may be considered independently (for the purpose of determining book value or as an academic exercise), it is important to recognize that depreciation has no effect on engineering economic analyses unless income taxes are also considered.

Straight Line Method

With the *straight line method*, depreciation is the same each year. The depreciation basis $(C - S_n)$ is allocated uniformly to all of the n years in the depreciation period. Each year, the depreciation will be

$$D = \frac{C - S_n}{n} \qquad 86.25$$

Constant Percentage Method

The *constant percentage method*[24] is similar to the straight line method in that the depreciation is the same each year. If the fraction of the basis used as depreciation is $1/n$, there is no difference between the constant percentage and straight line methods. The two methods differ only in what information is available. (With the straight line method, the life is known. With the constant percentage method, the depreciation fraction is known.)

Each year, the depreciation will be

$$D = (\text{depreciation fraction})(\text{depreciation basis})$$
$$= (\text{depreciation fraction})(C - S_n) \qquad 86.26$$

Sum-of-the-Years' Digits Method

In *sum-of-the-years' digits* (SOYD) depreciation, the digits from 1 to n inclusive are summed. The total, T, can also be calculated from

$$T = \tfrac{1}{2}n(n + 1) \qquad 86.27$$

The depreciation in year j can be found from Eq. 86.28. Notice that the depreciation in year j, D_j, decreases by a constant amount each year.

$$D_j = \frac{(C - S_n)(n - j + 1)}{T} \qquad 86.28$$

Double Declining Balance Method[25]

Double declining balance[26] (DDB) depreciation is independent of salvage value. Furthermore, the book value never stops decreasing, although the depreciation decreases in magnitude. Usually, any book value in excess of the salvage value is written off in the last year of the asset's depreciation period. Unlike any of the other depreciation methods, double declining balance depends on accumulated depreciation.

$$D_{\text{first year}} = \frac{2C}{n} \qquad 86.29$$

$$D_j = \frac{2\left(C - \displaystyle\sum_{m=1}^{j-1} D_m\right)}{n} \qquad 86.30$$

Calculating the depreciation in the middle of an asset's life appears particularly difficult with double declining balance, since all previous years' depreciation amounts seem to be required. It appears that the depreciation in the sixth year, for example, cannot be calculated unless the values of depreciation for the first five years are calculated. However, this is not true.

Depreciation in the middle of an asset's life can be found from the following equations. (d is known as the *depreciation rate*.)

$$d = \frac{2}{n} \qquad 86.31$$

$$D_j = dC(1 - d)^{j-1} \qquad 86.32$$

Statutory Depreciation Systems

In the United States, property placed into service in 1981 and thereafter must use the *Accelerated Cost Recovery System* (ACRS), and after 1986, the *Modified*

---

[22]For example, with the Accelerated Cost Recovery System (ACRS) the *depreciation basis* is the total purchase cost, regardless of the expected salvage value. With declining balance methods, the depreciation basis is the purchase cost less any previously taken depreciation.

[23]This discussion gives the impression that any form of depreciation may be chosen regardless of the nature and circumstances of the purchase. In reality, the IRS tax regulations place restrictions on the higher-rate (accelerated) methods, such as declining balance and sum-of-the-years' digits methods. Furthermore, the *Economic Recovery Act of 1981* and the *Tax Reform Act of 1986* substantially changed the laws relating to personal and corporate income taxes.

[24]The *constant percentage method* should not be confused with the declining balance method, which used to be known as the *fixed percentage on diminishing balance method*.

[25]In the past, the *declining balance method* has also been known as the *fixed percentage of book value* and *fixed percentage on diminishing balance method*.

[26]Double declining balance depreciation is a particular form of *declining balance depreciation*, as defined by the IRS tax regulations. Declining balance depreciation includes 125% declining balance and 150% declining balance depreciations that can be calculated by substituting 1.25 and 1.50, respectively, for the 2 in Eq. 86.29.

Accelerated Cost Recovery System (MACRS) or other statutory method. Other methods (straight line, declining balance, etc.) cannot be used except in special cases.

Property placed into service in 1980 or before must continue to be depreciated according to the method originally chosen (e.g., straight line, declining balance, or sum-of-the-years' digits). ACRS and MACRS cannot be used.

Under ACRS and MACRS, the cost recovery amount in the jth year of an asset's cost recovery period is calculated by multiplying the initial cost by a factor.

$$D_j = C \times \text{factor} \qquad 86.33$$

The initial cost used is not reduced by the asset's salvage value for ACRS and MACRS calculations. The factor used depends on the asset's cost recovery period. Such factors are subject to continuing legislation changes. Current tax publications should be consulted before using this method.

Table 86.4 *Representative MACRS Depreciation Factors[a]*

depreciation rate for recovery period (n)

year (j)	3 years	5 years	7 years	10 years
1	33.33%	20.00%	14.29%	10.00%
2	44.45%	32.00%	24.49%	18.00%
3	14.81%	19.20%	17.49%	14.40%
4	7.41%	11.52%	12.49%	11.52%
5		11.52%	8.93%	9.22%
6		5.76%	8.92%	7.37%
7			8.93%	6.55%
8			4.46%	6.55%
9				6.56%
10				6.55%
11				3.28%

[a]Values are for the "half-year" convention. This table gives typical values only. Since these factors are subject to continuing revision, they should not be used without consulting an accounting professional.

Production or Service Output Method

If an asset has been purchased for a specific task and that task is associated with a specific lifetime amount of output or production, the depreciation may be calculated by the fraction of total production produced during the year. The depreciation is not expected to be the same each year.

$$D_j = (C - S_n) \left(\frac{\text{actual output in year } j}{\text{estimated lifetime output}} \right) \qquad 86.34$$

Sinking Fund Method

The *sinking fund method* is seldom used in industry because the initial depreciation is low. The formula for sinking fund depreciation (which increases each year) is

$$D_j = (C - S_n)(A/F, i\%, n)(F/P, i\%, j - 1) \qquad 86.35$$

Disfavored Methods

Three other depreciation methods are mentioned here, not because they are currently accepted or in widespread use, but because they are still occasionally encountered in the literature.[27]

The *sinking fund plus interest on first cost* depreciation method, like the following two methods, is an attempt to include the *opportunity interest cost* on the purchase price with the depreciation. That is, the purchasing company not only incurs an annual expense due to the drop in book value, but it also loses the interest on the purchase price. The formula for this method is

$$D = (C - S_n)(A/F, i\%, n) + Ci \qquad 86.36$$

The *straight line plus interest on first cost* method is similar. Its formula is

$$D = \left(\frac{1}{n} \right)(C - S_n) + Ci \qquad 86.37$$

The *straight line plus average interest method* assumes that the opportunity interest cost should be based on the book value only, not on the full purchase price. Since the book value changes each year, an average value is used. The depreciation formula is

$$D = \left(\frac{C - S_n}{n} \right) \left(1 + \frac{i(n + 1)}{2} \right) + iS_n \qquad 86.38$$

Example 86.17

An asset is purchased for \$9000. Its estimated economic life is ten years, after which it will be sold for \$200. Find the depreciation in the first three years using straight line, double declining balance, and sum-of-the-years' digits depreciation methods.

---

[27]These three depreciation methods should not be used in the usual manner (e.g., in conjunction with the income tax rate). These methods are attempts to calculate a more accurate annual cost of an alternative. Sometimes they give misleading answers. Their use cannot be recommended. They are included in this chapter only for the sake of completeness.

Solution

SL: $\quad D = \dfrac{\$9000 - \$200}{10} \quad = \$880$ each year

DDB: $\quad D_1 = \dfrac{(2)(\$9000)}{10} \quad = \$1800$ in year 1

$\quad D_2 = \dfrac{(2)(\$9000 - \$1800)}{10} = \$1440$ in year 2

$\quad D_3 = \dfrac{(2)(\$9000 - \$3240)}{10} = \$1152$ in year 3

SOYD: $\quad T = \left(\frac{1}{2}\right)(10)(11) = 55$

$\quad D_1 = \left(\frac{10}{55}\right)(\$9000 - \$200) = \$1600$ in year 1

$\quad D_2 = \left(\frac{9}{55}\right)(\$8800) \quad\quad = \$1440$ in year 2

$\quad D_3 = \left(\frac{8}{55}\right)(\$8800) \quad\quad = \$1280$ in year 3

37. ACCELERATED DEPRECIATION METHODS

An *accelerated depreciation method* is one that calculates a depreciation amount greater than a straight line amount. Double declining balance and sum-of-the-years' digits methods are accelerated methods. The ACRS and MACRS methods are explicitly accelerated methods. Straight line and sinking fund methods are not accelerated methods.

Use of an accelerated depreciation method may result in unexpected tax consequences when the depreciated asset or property is disposed of. Professional tax advice should be obtained in this area.

38. BOOK VALUE

The difference between original purchase price and accumulated depreciation is known as *book value*.[28] At the end of each year, the book value (which is initially equal to the purchase price) is reduced by the depreciation in that year.

It is important to properly synchronize depreciation calculations. It is difficult to answer the question, "What is the book value in the fifth year?" unless the timing of the book value change is mutually agreed upon. It is better to be specific about an inquiry by identifying when the book value change occurs. For example, the following question is unambiguous: "What is the book

value at the end of year 5, after subtracting depreciation in the fifth year?" or "What is the book value after five years?"

Unfortunately, this type of care is seldom taken in book value inquiries, and it is up to the respondent to exercise reasonable care in distinguishing between beginning-of-year book value and end-of-year book value. To be consistent, the book value equations in this chapter have been written in such a way that the year subscript (j) has the same meaning in book value and depreciation calculations. That is, BV_5 means the book value at the end of the fifth year, after five years of depreciation, including D_5, has been subtracted from the original purchase price.

There can be a great difference between the book value of an asset and the *market value* of that asset. There is no legal requirement for the two values to coincide, and no intent for book value to be a reasonable measure of market value.[29] Therefore, it is apparent that book value is merely an accounting convention with little practical use. Even when a depreciated asset is disposed of, the book value is used to determine the consequences of disposal, not the price the asset should bring at sale.

The calculation of book value is relatively easy, even for the case of the declining balance depreciation method.

For the straight line depreciation method, the book value at the end of the jth year, after the jth depreciation deduction has been made, is

$$ BV_j = C - \frac{j(C - S_n)}{n} = C - jD \qquad 86.39 $$

For the sum-of-the-years' digits method, the book value is

$$ BV_j = (C - S_n)\left(1 - \frac{j(2n + 1 - j)}{n(n + 1)}\right) + S_n \qquad 86.40 $$

For the declining balance method, including double declining balance (see Ftn. 26), the book value is

$$ BV_j = C(1 - d)^j \qquad 86.41 $$

For the sinking fund method, the book value is calculated directly as

$$ BV_j = C - (C - S_n)(A/F, i\%, n)(F/A, i\%, j) \qquad 86.42 $$

---

[28]The balance sheet of a corporation usually has two asset accounts: the *equipment account* and the *accumulated depreciation account*. There is no book value account on this financial statement, other than the implicit value obtained from subtracting the accumulated depreciation account from the equipment account. The book values of various assets, as well as their original purchase cost, date of purchase, salvage value, and so on, and accumulated depreciation appear on detail sheets or other peripheral records for each asset.

[29]Common examples of assets with great divergences of book and market values are buildings (rental houses, apartment complexes, factories, etc.) and company luxury automobiles (Porsches, Mercedes, etc.) during periods of inflation. Book values decrease, but actual values increase.

Of course, the book value at the end of year j can always be calculated for any method by successive subtractions (i.e., subtraction of the accumulated depreciation), as Eq. 86.43 illustrates.

$$BV_j = C - \sum_{m=1}^{j} D_m \qquad 86.43$$

Figure 86.8 illustrates the book value of a hypothetical asset depreciated using several depreciation methods. Notice that the double declining balance method initially produces the fastest write-off, while the sinking fund method produces the slowest write-off. Note also that the book value does not automatically equal the salvage value at the end of an asset's depreciation period with the double declining balance method.[30]

Figure 86.8 *Book Value with Different Depreciation Methods*

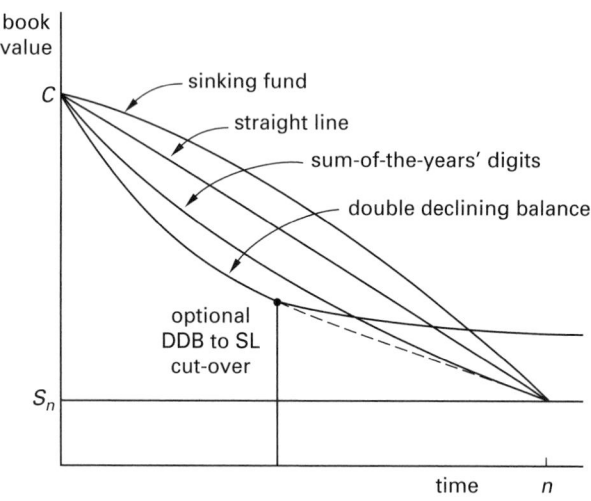

Example 86.18

For the asset described in Ex. 86.17, calculate the book value at the end of the first three years if sum-of-the-years' digits depreciation is used. The book value at the beginning of year 1 is $9000.

Solution

From Eq. 86.43,

$$BV_1 = \$9000 - \$1600 = \$7400$$
$$BV_2 = \$7400 - \$1440 = \$5960$$
$$BV_3 = \$5960 - \$1280 = \$4680$$

---

[30]This means that the straight line method of depreciation may result in a lower book value at some point in the depreciation period than if double declining balance is used. A *cut-over* from double declining balance to straight line may be permitted in certain cases. Finding the *cut-over point*, however, is usually done by comparing book values determined by both methods. The analytical method is complicated.

39. AMORTIZATION

Amortization and depreciation are similar in that they both divide up the cost basis or value of an asset. In fact, in certain cases, the term "amortization" may be used in place of the term "depreciation." However, depreciation is a specific form of amortization.

Amortization spreads the cost basis or value of an asset over some base. The base can be time, units of production, number of customers, and so on. The asset can be tangible (e.g., a delivery truck or building) or intangible (e.g., goodwill or a patent).

If the asset is tangible, if the base is time, and if the length of time is consistent with accounting standards and taxation guidelines, then the term "depreciation" is appropriate. However, if the asset is intangible, if the base is some other variable, or if some length of time other than the customary period is used, then the term "amortization" is more appropriate.[31]

Example 86.19

A company purchases complete and exclusive patent rights to an invention for $1,200,000. It is estimated that once commercially produced, the invention will have a specific but limited market of 1200 units. For the purpose of allocating the patent right cost to production cost, what is the amortization rate in dollars per unit?

Solution

The patent should be amortized at the rate of

$$\frac{\$1,200,000}{1200 \text{ units}} = \$1000 \text{ per unit}$$

40. DEPLETION

Depletion is another artificial deductible operating expense, designed to compensate mining organizations for decreasing mineral reserves. Since original and remaining quantities of minerals are seldom known accurately, the *depletion allowance* is calculated as a fixed percentage of the organization's gross income. These percentages are usually in the 10 to 20% range and apply to such mineral deposits as oil, natural gas, coal, uranium, and most metal ores.

41. BASIC INCOME TAX CONSIDERATIONS

The issue of income taxes is often overlooked in academic engineering economic analysis exercises. Such a position is justifiable when an organization (e.g., a nonprofit school, a church, or the government) pays no

---

[31]From time to time, the U.S. Congress has allowed certain types of facilities (e.g., emergency, grain storage, and pollution control) to be written off more rapidly than would otherwise be permitted in order to encourage investment in such facilities. The term "amortization" has been used with such write-off periods.

income taxes. However, if an individual or organization is subject to income taxes, the income taxes must be included in an economic analysis of investment alternatives.

Assume that an organization pays a fraction f of its profits to the federal government as income taxes. If the organization also pays a fraction s of its profits as state income taxes and if state taxes paid are recognized by the federal government as tax-deductible expenses, then the composite tax rate is

$$t = s + f - sf \qquad 86.44$$

The basic principles used to incorporate taxation into engineering economic analyses are the following.

- Initial purchase expenditures are unaffected by income taxes.
- Salvage revenues are unaffected by income taxes.
- Deductible expenses, such as operating costs, maintenance costs, and interest payments, are reduced by the fraction t (e.g., multiplied by the quantity $1 - t$).
- Revenues are reduced by the fraction t (e.g., multiplied by the quantity $1 - t$).
- Since tax regulations allow the depreciation in any year to be handled as if it were an actual operating expense, and since operating expenses are deductible from the income base prior to taxation, the after-tax profits will be increased. If D is the depreciation, the net result to the after-tax cash flow will be the addition of tD. Depreciation is multiplied by t and added to the appropriate year's cash flow, increasing that year's present worth.

For simplicity, most engineering economics practice problems involving income taxes specify a single income tax rate. In practice, however, federal and most state tax rates depend on the income level. Each range of incomes and its associated tax rate are known as *income bracket* and *tax bracket*, respectively. For example, the state income tax rate might be 4% for incomes up to and including $30,000, and 5% for incomes above $30,000. The income tax for a taxpaying entity with an income of $50,000 would have to be calculated in two parts.

$$\text{tax} = (0.04)(\$30,000) + (0.05)(\$50,000 - \$30,000)$$
$$= \$2200$$

Income taxes and depreciation have no bearing on municipal or governmental projects since municipalities, states, and the U.S. government pay no taxes.

Example 86.20

A corporation that pays 53% of its profit in income taxes invests $10,000 in an asset that will produce a $3000 annual revenue for eight years. If the annual expenses are $700, salvage after eight years is $500, and 9% interest is used, what is the after-tax present worth? Disregard depreciation.

Solution

$$P = -\$10,000 + (\$3000)(P/A, 9\%, 8)(1 - 0.53)$$
$$- (\$700)(P/A, 9\%, 8)(1 - 0.53)$$
$$+ (\$500)(P/F, 9\%, 8)$$
$$= -\$10,000 + (\$3000)(5.5348)(0.47)$$
$$- (\$700)(5.5348)(0.47) + (\$500)(0.5019)$$
$$= -\$3766$$

42. TAXATION AT THE TIMES OF ASSET PURCHASE AND SALE

There are numerous rules and conventions that governmental tax codes and the accounting profession impose on organizations. Engineers are not normally expected to be aware of most of the rules and conventions, but occasionally it may be necessary to incorporate their effects into an engineering economic analysis.

Tax Credit

A *tax credit* (also known as an *investment tax credit* or *investment credit*) is a one-time credit against income taxes.[32] Therefore, it is added to the after-tax present worth as a last step in an engineering economic analysis. Such tax credits may be allowed by the government from time to time for equipment purchases, employment of various classes of workers, rehabilitation of historic landmarks, and so on.

A tax credit is usually calculated as a fraction of the initial purchase price or cost of an asset or activity.

$$\text{TC} = \text{fraction} \times \text{initial cost} \qquad 86.45$$

When the tax credit is applicable, the fraction used is subject to legislation. A professional tax expert or accountant should be consulted prior to applying the tax credit concept to engineering economic analysis problems.

Since the investment tax credit reduces the buyer's tax liability, a tax credit should be included only in after-tax engineering economic analyses. The credit is assumed to be received at the end of the year.

Gain or Loss on the Sale of a Depreciated Asset

If an asset that has been depreciated over a number of prior years is sold for more than its current book value, the difference between the book value and selling price is taxable income in the year of the sale. Alternatively, if the asset is sold for less than its current book value, the difference between the selling price and book value is an expense in the year of the sale.

---

[32]Strictly, *tax credit* is the more general term, and applies to a credit for doing anything creditable. An *investment tax credit* requires an investment in something (usually real property or equipment).

Example 86.21

One year, a company makes a \$5000 investment in a historic building. The investment is not depreciable, but it does qualify for a one-time 20% tax credit. In that same year, revenue is \$45,000 and expenses (exclusive of the \$5000 investment) are \$25,000. The company pays a total of 53% in income taxes. What is the after-tax present worth of this year's activities if the company's interest rate for investment is 10%?

Solution

The tax credit is

$$TC = (0.20)(\$5000) = \$1000$$

This tax credit is assumed to be received at the end of the year. The after-tax present worth is

$$
\begin{aligned}
P &= -\$5000 + (\$45,000 - \$25,000)(1 - 0.53) \\
&\quad \times (P/F, 10\%, 1) + (\$1000)(P/F, 10\%, 1) \\
&= -\$5000 + (\$20,000)(0.47)(0.9091) \\
&\quad + (\$1000)(0.9091) \\
&= \$4455
\end{aligned}
$$

43. DEPRECIATION RECOVERY

The economic effect of depreciation is to reduce the income tax in year j by tD_j. The present worth of the asset is also affected: The present worth is increased by $tD_j(P/F, i\%, j)$. The after-tax present worth of all depreciation effects over the depreciation period of the asset is called the *depreciation recovery* (DR).[33]

$$DR = t \sum_{j=1}^{n} D_j(P/F, i\%, j) \qquad 86.46$$

Straight line depreciation recovery from an asset is easily calculated, since the depreciation is the same each year. Assuming the asset has a constant depreciation of D and depreciation period of n years, the depreciation recovery is

$$DR = tD(P/A, i\%, n) \qquad 86.47$$

$$D = \frac{C - S_n}{n} \qquad 86.48$$

Sum-of-the-years' digits depreciation recovery is also relatively easily calculated, since the depreciation decreases uniformly each year.

$$DR = \left(\frac{t(C - S_n)}{T}\right)\left(n(P/A, i\%, n) - (P/G, i\%, n)\right) \qquad 86.49$$

---

[33]Since the depreciation benefit is reduced by taxation, depreciation cannot be thought of as an annuity to fund a replacement asset.

Finding *declining balance depreciation recovery* is more involved. There are three difficulties. The first (the apparent need to calculate all previous depreciations in order to determine the subsequent depreciation) has already been addressed by Eq. 86.32.

The second difficulty is that there is no way to ensure (that is, to force) the book value to be S_n at $t = n$. Therefore, it is common to write off the remaining book value (down to S_n) at $t = n$ in one lump sum. This assumes $\text{BV}_n \geq S_n$.

The third difficulty is that of finding the present worth of an *exponentially decreasing cash flow*. Although the proof is omitted here, such exponential cash flows can be handled with the *exponential gradient factor*, (P/EG).[34]

$$(P/EG, z - 1, n) = \frac{z^n - 1}{z^n(z - 1)} \qquad 86.50$$

$$z = \frac{1 + i}{1 - d} \qquad 86.51$$

Then, as long as $\text{BV}_n > S_n$, the declining balance depreciation recovery is

$$DR = tC\left(\frac{d}{1 - d}\right)(P/EG, z - 1, n) \qquad 86.52$$

Example 86.22

For the asset described in Ex. 86.17, calculate the after-tax depreciation recovery with straight line and sum-of-the-years' digits depreciation methods. Use 6% interest with 48% income taxes.

Solution

SL:
$$
\begin{aligned}
DR &= (0.48)(\$880)(P/A, 6\%, 10) \\
&= (0.48)(\$880)(7.3601) \\
&= \$3109
\end{aligned}
$$

SOYD: The depreciation series can be thought of as a constant \$1600 term with a negative \$160 gradient.

$$
\begin{aligned}
DR &= (0.48)(\$1600)(P/A, 6\%, 10) \\
&\quad - (0.48)(\$160)(P/G, 6\%, 10) \\
&= (0.48)(\$1600)(7.3601) \\
&\quad - (0.48)(\$160)(29.6023) \\
&= \$3379
\end{aligned}
$$

---

[34]The (P/A) columns in App. 86.B can be used for (P/EG) as long as the interest rate is assumed to be $z - 1$.

Table 86.5 *Depreciation Calculation Summary*

method	depreciation basis	depreciation in year $j (D_j)$	book value after jth depreciation (BV_j)	after-tax depreciation recovery (DR)	supplementary formulas
straight line (SL)	$C - S_n$	$\dfrac{C - S_n}{n}$ (constant)	$C - jD$	$tD(P/A, i\%, n)$	
constant percentage	$C - S_n$	$\begin{array}{c}\text{fraction} \times (C - S_n)\\ \text{(constant)}\end{array}$	$C - jD$	$tD(P/A, i\%, n)$	
sum-of-the-years' digits (SOYD)	$C - S_n$	$\dfrac{(C - S_n)}{T} \times (n - j + 1)$	$(C - S_n) \times \left(1 - \dfrac{j(2n + 1 - j)}{n(n+1)}\right) + S_n$	$\dfrac{t(C - S_n)}{T} \times (n(P/A, i\%, n) - (P/G, i\%, n))$	$T = \frac{1}{2}n(n + 1)$
double declining balance (DDB)	C	$dC(1 - d)^{j-1}$	$C(1 - d)^j$	$tC\left(\dfrac{d}{1 - d}\right) \times (P/EG, z - 1, n)$	$d = \dfrac{2}{n}; \ z = \dfrac{1 + i}{1 - d}$ $(P/EG, z - 1, n) = \dfrac{z^n - 1}{z^n(z - 1)}$
sinking fund (SF)	$C - S_n$	$\begin{array}{l}(C - S_n)\\ \times (A/F, i\%, n)\\ \times (F/P, i\%, j - 1)\end{array}$	$\begin{array}{l}C - (C - S_n)\\ \times (A/F, i\%, n)\\ \times (F/A, i\%, j)\end{array}$	$\dfrac{t(C - S_n)(A/F, i\%, n)}{1 + i}$	
accelerated cost recovery system (ACRS/ MACRS)	C	$C \times \text{factor}$	$C - \displaystyle\sum_{m=1}^{j} D_m$	$t\displaystyle\sum_{j=1}^{n} D_j(P/F, i\%, j)$	
production or service output	$C - S_n$	$\begin{array}{l}(C - S_n)\\ \times \left(\dfrac{\text{actual output in year } j}{\text{lifetime output}}\right)\end{array}$	$C - \displaystyle\sum_{m=1}^{j} D_m$	$t\displaystyle\sum_{j=1}^{n} D_j(P/F, i\%, j)$	

Notice that the ten-year (P/G) factor is used even though there are only nine years in which the gradient reduces the initial $1600 amount.

Example 86.23

What is the after-tax present worth of the asset described in Ex. 86.20 if straight line, sum-of-the-years' digits, and double declining balance depreciation methods are used?

Solution

Using SL, the depreciation recovery is

$$\text{DR} = (0.53)\left(\frac{\$10,000 - \$500}{8}\right)(P/A, 9\%, 8)$$

$$= (0.53)\left(\frac{\$9500}{8}\right)(5.5348)$$

$$= \$3483$$

Using SOYD, the depreciation recovery is calculated as follows.

$$T = \left(\tfrac{1}{2}\right)(8)(9) = 36$$

$$\text{depreciation base} = \$10,000 - \$500 = \$9500$$

$$D_1 = \left(\tfrac{8}{36}\right)(\$9500) = \$2111$$

$$G = \text{gradient} = \left(\tfrac{1}{36}\right)(\$9500)$$

$$= \$264$$

$$\text{DR} = (0.53)\big((\$2111)(P/A, 9\%, 8)$$

$$- (\$264)(P/G, 9\%, 8)\big)$$

$$= (0.53)\big((\$2111)(5.5348)$$

$$- (\$264)(16.8877)\big)$$

$$= \$3830$$

Using DDB, the depreciation recovery is calculated as follows.[35]

$$d = \tfrac{2}{8} = 0.25$$

$$z = \frac{1 + 0.09}{1 - 0.25} = 1.4533$$

$$(P/EG, z - 1, n) = \frac{(1.4533)^8 - 1}{(1.4533)^8 (0.4533)} = 2.095$$

From Eq. 86.52,

$$DR = (0.53)\left(\frac{(0.25)(\$10,000)}{0.75}\right)(2.095)$$

$$= \$3701$$

The after-tax present worth, neglecting depreciation, was previously found to be −$3766.

The after-tax present worths, including depreciation recovery, are

SL: $P = -\$3766 + \$3483 = -\$283$
SOYD: $P = -\$3766 + \$3830 = \quad\$64$
DDB: $P = -\$3766 + \$3701 = \quad-\$65$

44. OTHER INTEREST RATES

The *effective interest rate per period*, i (also called *yield* by banks), is the only interest rate that should be used in equivalence equations. The interest rates at the top of the factor tables in App. 86.B are implicitly all effective interest rates. Usually, the period will be one year, hence the name *effective annual interest rate*. However, there are other interest rates in use as well.

The term *nominal interest rate*, r (*rate per annum*), is encountered when compounding is more than once per year. The nominal rate does not include the effect of compounding and is not the same as the effective rate. And, since the effective interest rate can be calculated from the nominal rate only if the number of compounding periods per year is known, nominal rates cannot be compared unless the method of compounding is specified. The only practical use for a nominal rate per year is for calculating the effective rate per period.

45. RATE AND PERIOD CHANGES

If there are k compounding periods during the year (two for semiannual compounding, four for quarterly compounding, twelve for monthly compounding, etc.) and the nominal rate is r, the *effective rate per compounding period* is

$$\phi = \frac{r}{k} \qquad\qquad \textit{86.53}$$

[35]This method should start by checking that the book value at the end of the depreciation period is greater than the salvage value. In this example, such is the case. However, the step is not shown.

The effective annual rate, i, can be calculated from the effective rate per period, ϕ, by using Eq. 86.54.

$$i = (1 + \phi)^k - 1$$

$$= \left(1 + \frac{r}{k}\right)^k - 1 \qquad \textit{86.54}$$

Sometimes, only the effective rate per period (e.g., per month) is known. However, that will be a simple problem since compounding for n periods at an effective rate per period is not affected by the definition or length of the period.

The following rules may be used to determine which interest rate is given in a problem.

- Unless specifically qualified in the problem, the interest rate given is an annual rate.

- If the compounding is annual, the rate given is the effective rate. If compounding is other than annual, the rate given is the nominal rate.

The effective annual interest rate determined on a *daily compounding basis* will not be significantly different than if *continuous compounding* is assumed.[36] In the case of continuous (or daily) compounding, the discounting factors can be calculated directly from the nominal interest rate and number of years, without having to find the effective interest rate per period. Table 86.6 can be used to determine the discount factors for continuous compounding.

Table 86.6 Discount Factors for Continuous Compounding
(n is the number of years)

symbol	formula
$(F/P, r\%, n)$	e^{rn}
$(P/F, r\%, n)$	e^{-rn}
$(A/F, r\%, n)$	$\dfrac{e^r - 1}{e^{rn} - 1}$
$(F/A, r\%, n)$	$\dfrac{e^{rn} - 1}{e^r - 1}$
$(A/P, r\%, n)$	$\dfrac{e^r - 1}{1 - e^{-rn}}$
$(P/A, r\%, n)$	$\dfrac{1 - e^{-rn}}{e^r - 1}$

Example 86.24

A savings and loan offers a nominal rate of $5\tfrac{1}{4}\%$ compounded daily over 365 days in a year. What is the effective annual rate?

[36]The number of *banking days in a year* (250, 360, etc.) must be specifically known.

Solution

method 1: Use Eq. 86.54.

$$r = 0.0525, \ k = 365$$

$$i = \left(1 + \frac{0.0525}{365}\right)^{365} - 1 = 0.0539$$

method 2: Assume daily compounding is the same as continuous compounding.

$$i = (F/P, r\%, 1) - 1$$
$$= e^{0.0525} - 1 = 0.0539$$

Example 86.25

A real estate investment trust pays $7,000,000 for a 100-unit apartment complex. The trust expects to sell the complex in ten years for $15,000,000. In the meantime, it expects to receive an average rent of $900 per month from each apartment. Operating expenses are expected to be $200 per month per occupied apartment. A 95% occupancy rate is predicted. In similar investments, the trust has realized a 15% effective annual return on its investment. Compare the expected present worth of this investment when calculated assuming (a) annual compounding (i.e., the year-end convention), and (b) monthly compounding. Disregard taxes, depreciation, and all other factors.

Solution

(a) The net annual income will be

$$(0.95)(100 \text{ units})\left(\$900 \ \frac{\$}{\text{unit-month}} \right.$$
$$\left. - \$200 \ \frac{\$}{\text{unit-month}}\right)\left(12 \ \frac{\text{months}}{\text{year}}\right)$$
$$= \$798,000/\text{year}$$

The present worth of ten years of operation is

$$P = -\$7,000,000 + (\$798,000)(P/A, 15\%, 10)$$
$$+ (\$15,000,000)(P/F, 15\%, 10)$$
$$= -\$7,000,000 + (\$798,000)(5.0188)$$
$$+ (\$15,000,000)(0.2472)$$
$$= \$713,000$$

(b) The net monthly income is

$$(0.95)(100 \text{ units})\left(\$900 \ \frac{\$}{\text{unit-month}} \right.$$
$$\left. - \$200 \ \frac{\$}{\text{unit-month}}\right) = \$66,500/\text{month}$$

Equation 86.54 is used to calculate the effective monthly rate, ϕ, from the effective annual rate, $i = 15\%$, and the number of compounding periods per year, $k = 12$.

$$\phi = (1 + i)^{1/k} - 1$$
$$= (1 + 0.15)^{1/12} - 1 = 0.011715 \ (1.1715\%)$$

The number of compounding periods in ten years is

$$n = (10 \text{ years})\left(12 \ \frac{\text{months}}{\text{year}}\right) = 120 \text{ months}$$

The present worth of 120 months of operation is

$$P = -\$7,000,000 + (\$66,500)(P/A, 1.1715\%, 120)$$
$$+ (\$15,000,000)(P/F, 1.1715\%, 120)$$

Since table values for 1.1715% discounting factors are not available, the factors are calculated from Table 86.1.

$$(P/A, 1.1715\%, 120) = \frac{(1+i)^n - 1}{i(1+i)^n}$$
$$= \frac{(1+0.011715)^{120} - 1}{(0.011715)(1+0.011715)^{120}}$$
$$= 64.261$$
$$(P/F, 1.1715\%, 120) = (1+i)^{-n} = (1+0.011715)^{-120}$$
$$= 0.2472$$

The present worth over 120 monthly compounding periods is

$$P = -\$7,000,000 + (\$66,500)(64.261)$$
$$+ (\$15,000,000)(0.2472)$$
$$= \$981,357$$

46. BONDS

A *bond* is a method of long-term financing commonly used by governments, states, municipalities, and very large corporations.[37] The bond represents a contract to pay the bondholder specific amounts of money at specific times. The holder purchases the bond in exchange for specific payments of interest and principal. Typical municipal bonds call for quarterly or semiannual interest payments and a payment of the *face value of the bond* on the *date of maturity* (end of the bond period).[38] Due to the practice of discounting in the bond market, a bond's face value and its purchase price generally will not coincide.

---

[37] In the past, 30-year bonds were typical. Shorter term 10-year, 15-year, 20-year, and 25-year bonds are also commonly issued.
[38] A *fully amortized bond* pays back interest and principal throughout the life of the bond. There is no balloon payment.

In the past, a bondholder had to submit a coupon or ticket in order to receive an interim interest payment. This has given rise to the term *coupon rate*, which is the nominal annual interest rate on which the interest payments are made. Coupon books are seldom used with modern bonds, but the term survives. The coupon rate determines the magnitude of the semiannual (or otherwise) interest payments during the life of the bond. The bondholder's own effective interest rate should be used for economic decisions about the bond.

Actual *bond yield* is the bondholder's actual rate of return of the bond, considering the purchase price, interest payments, and face value payment (or, value realized if the bond is sold before it matures). By convention, bond yield is calculated as a nominal rate (rate per annum), not an effective rate per year. The bond yield should be determined by finding the effective rate of return per payment period (e.g., per semiannual interest payment) as a conventional rate of return problem. Then, the nominal rate can be found by multiplying the effective rate per period by the number of payments per year, as in Eq. 86.54.

Example 86.26

What is the maximum amount an investor should pay for a 25-year bond with a $20,000 face value and 8% coupon rate (interest only paid semiannually)? The bond will be kept to maturity. The investor's effective annual interest rate for economic decisions is 10%.

Solution

For this problem, take the compounding period to be six months. Then, there are 50 compounding periods. Since 8% is a nominal rate, the effective bond rate per period is calculated from Eq. 86.53 as

$$\phi_{bond} = \frac{r}{k} = \frac{8\%}{2} = 4\%$$

The bond payment received semiannually is

$$(0.04)(\$20,000) = \$800$$

10% is the investor's effective rate per year, so Eq. 86.54 is again used to calculate the effective analysis rate per period.

$$0.10 = (1 + \phi)^2 - 1$$
$$\phi = 0.04881 \quad (4.88\%)$$

The maximum amount that the investor should be willing to pay is the present worth of the investment.

$$P = (\$800)(P/A, 4.88\%, 50)$$
$$+ (\$20,000)(P/F, 4.88\%, 50)$$

Table 86.1 can be used to calculate the following factors.

$$(P/A, 4.88\%, 50) = \frac{(1 + 0.0488)^{50} - 1}{(0.0488)(1.0488)^{50}} = 18.600$$

$$(P/F, 4.88\%, 50) = \frac{1}{(1 + 0.0488)^{50}} = 0.09233$$

Then, the present worth is

$$P = (\$800)(18.600) + (\$20,000)(0.09233)$$
$$= \$16,727$$

47. PROBABILISTIC PROBLEMS

If an alternative's cash flows are specified by an implicit or explicit probability distribution rather than being known exactly, the problem is *probabilistic*.

Probabilistic problems typically possess the following characteristics.

- There is a chance of loss that must be minimized (or, rarely, a chance of gain that must be maximized) by selection of one of the alternatives.

- There are multiple alternatives. Each alternative offers a different degree of protection from the loss. Usually, the alternatives with the greatest protection will be the most expensive.

- The magnitude of loss or gain is independent of the alternative selected.

Probabilistic problems are typically solved using annual costs and expected values. An *expected value* is similar to an *average value* since it is calculated as the mean of the given probability distribution. If cost 1 has a probability of occurrence, p_1, cost 2 has a probability of occurrence, p_2, and so on, the expected value is

$$\mathcal{E}\{cost\} = p_1(cost\ 1) + p_2(cost\ 2) + \cdots \qquad \textbf{86.55}$$

Example 86.27

Flood damage in any year is given according to the following table. What is the present worth of flood damage for a ten-year period? Use 6% as the effective annual interest rate.

damage	probability
0	0.75
$10,000	0.20
$20,000	0.04
$30,000	0.01

Solution

The expected value of flood damage in any given year is

$$\mathcal{E}\{damage\} = (0)(0.75) + (\$10,000)(0.20)$$
$$+ (\$20,000)(0.04) + (\$30,000)(0.01)$$
$$= \$3100$$

The present worth of ten years of expected flood damage is

$$\text{present worth} = (\$3100)(P/A, 6\%, 10)$$
$$= (\$3100)(7.3601)$$
$$= \$22,816$$

Example 86.28

A dam is being considered on a river that periodically overflows and causes \$600,000 damage. (The damage is essentially the same each time the river causes flooding.) The project horizon is 40 years. A 10% interest rate is being used.

Three different designs are available, each with different costs and storage capacities.

design alternative	cost	maximum capacity
A	\$500,000	1 unit
B	\$625,000	1.5 units
C	\$900,000	2.0 units

The national weather service has provided a statistical analysis of annual rainfall runoff from the watershed draining into the river.

units annual rainfall	probability
0	0.10
0.1 to 0.5	0.60
0.6 to 1.0	0.15
1.1 to 1.5	0.10
1.6 to 2.0	0.04
2.1 or more	0.01

Which design alternative would you choose assuming the dam is essentially empty at the start of each rainfall season?

Solution

The sum of the construction cost and the expected damage should be minimized. If alternative A is chosen, it will have a capacity of 1 unit. Its capacity will be exceeded (causing \$600,000 damage) when the annual rainfall exceeds 1 unit. Therefore, the expected value of the annual cost of alternative A is

$$\mathcal{E}\{\text{EUAC(A)}\}$$
$$= (\$500,000)(A/P, 10\%, 40)$$
$$\quad + (\$600,000)(0.10 + 0.04 + 0.01)$$
$$= (\$500,000)(0.1023) + (\$600,000)(0.15)$$
$$= \$141,150$$

Similarly,

$$\mathcal{E}\{\text{EUAC(B)}\} = (\$625,000)(A/P, 10\%, 40)$$
$$\quad + (\$600,000)(0.04 + 0.01)$$
$$= (\$625,000)(0.1023) + (\$600,000)(0.05)$$
$$= \$93,938$$

$$\mathcal{E}\{\text{EUAC(C)}\} = (\$900,000)(A/P, 10\%, 40)$$
$$\quad + (\$600,000)(0.01)$$
$$= (\$900,000)(0.1023) + (\$600,000)(0.01)$$
$$= \$98,070$$

Alternative B should be chosen.

48. FIXED AND VARIABLE COSTS

The distinction between fixed and variable costs depends on how these costs vary when an independent variable changes. For example, factory or machine production is frequently the independent variable. However, it could just as easily be vehicle miles driven, hours of operation, or quantity (mass, volume, etc.).

If a cost is a function of the independent variable, the cost is said to be a *variable cost*. The change in cost per unit variable change (i.e., what is usually called the *slope*) is known as the *incremental cost*. Material and labor costs are examples of variable costs. They increase in proportion to the number of product units manufactured.

If a cost is not a function of the independent variable, the cost is said to be a *fixed cost*. Rent and lease payments are typical fixed costs. These costs will be incurred regardless of production levels.

Some costs have both fixed and variable components, as Fig. 86.9 illustrates. The fixed portion can be determined by calculating the cost at zero production.

Figure 86.9 Fixed and Variable Costs

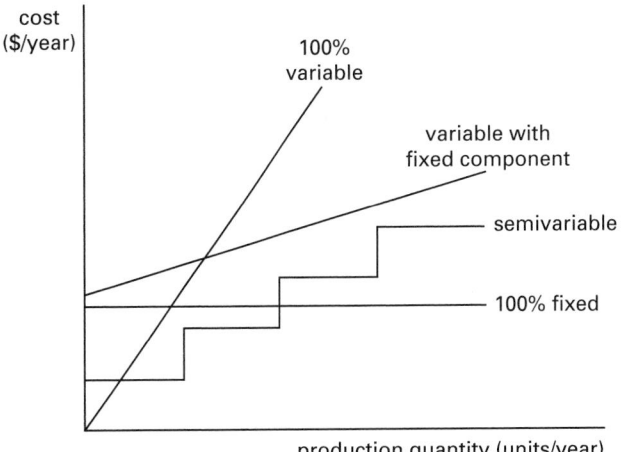

An additional category of cost is the *semivariable cost*. This type of cost increases step-wise. Semivariable cost structures are typical of situations where *excess capacity* exists. For example, supervisory cost is a step-wise function of the number of production shifts. Also, labor cost for truck drivers is a step-wise function of weight (volume) transported. As long as a truck has room left (i.e., excess capacity), no additional driver is needed. As soon as the truck is filled, labor cost will increase.

Table 86.7 *Summary of Fixed and Variable Costs*

> *fixed costs*
>> rent
>> property taxes
>> interest on loans
>> insurance
>> janitorial service expense
>> tooling expense
>> setup, cleanup, and tear-down expenses
>> depreciation expense
>> marketing and selling costs
>> cost of utilities
>> general burden and overhead expense
>
> *variable costs*
>> direct material costs
>> direct labor costs
>> cost of miscellaneous supplies
>> payroll benefit costs
>> income taxes
>> supervision costs

49. ACCOUNTING COSTS AND EXPENSE TERMS

The accounting profession has developed special terms for certain groups of costs. When annual costs are incurred due to the functioning of a piece of equipment, they are known as *operating and maintenance* (O&M) *costs*. The annual costs associated with operating a business (other than the costs directly attributable to production) are known as *general, selling, and administrative* (GS&A) *expenses*.

Direct labor costs are costs incurred in the factory, such as assembly, machining, and painting labor costs. *Direct material costs* are the costs of all materials that go into production.[39] Typically, both direct labor and direct material costs are given on a per-unit or per-item basis. The sum of the direct labor and direct material costs is known as the *prime cost*.

There are certain additional expenses incurred in the factory, such as the costs of factory supervision, stock-picking, quality control, factory utilities, and miscellaneous supplies (cleaning fluids, assembly lubricants, routing tags, etc.) that are not incorporated into the

final product. Such costs are known as *indirect manufacturing expenses* (IME) or *indirect material and labor costs*.[40] The sum of the per-unit indirect manufacturing expense and prime cost is known as the *factory cost*.

Research and development (R&D) *costs* and *administrative expenses* are added to the factory cost to give the *manufacturing cost* of the product.

Additional costs are incurred in marketing the product. Such costs are known as *selling expenses* or *marketing expenses*. The sum of the selling expenses and manufacturing cost is the *total cost* of the product.

Figure 86.10 illustrates these terms.[41]

Figure 86.10 *Costs and Expenses Combined*

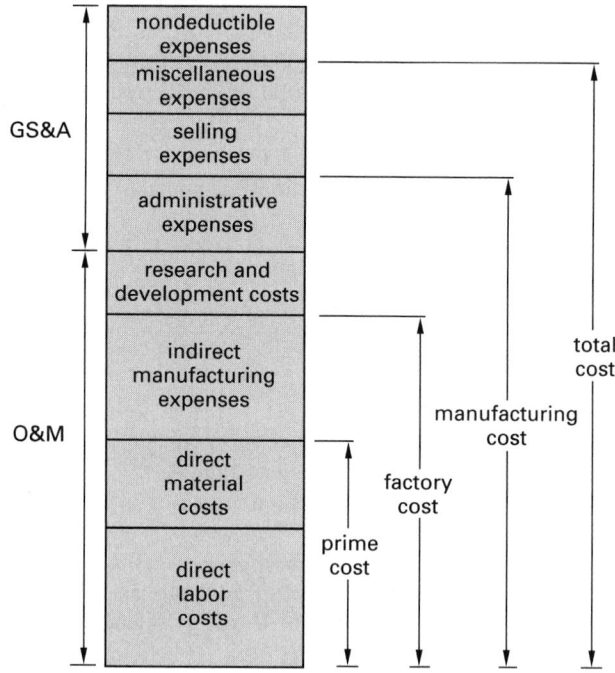

The distinctions among the various forms of cost (particularly with overhead costs) are not standardized. Each company must develop a classification system to deal with the various cost factors in a consistent manner. There are also other terms in use (e.g., *raw materials, operating supplies, general plant overhead*), but these terms must be interpreted within the framework of each company's classification system. Table 86.8 is typical of such classification systems.

---

[39]There may be problems with pricing the material when it is purchased from an outside vendor and the stock on hand derives from several shipments purchased at different prices.

[40]The *indirect material* and *labor costs* usually exclude costs incurred in the office area.

[41]Notice that *total cost* does not include income taxes.

Table 86.8 *Typical Classification of Expenses*

direct labor expenses
    machining and forming
    assembly
    finishing
    inspection
    testing
direct material expenses
    items purchased from other vendors
    manufactured assemblies
factory overhead expenses (indirect manufacturing expenses)
    supervision
    benefits
        pension
        medical insurance
        vacations
    wages overhead
        unemployment compensation taxes
        social security taxes
        disability taxes
    stock-picking
    quality control and inspection
    expediting
    rework
    maintenance
    miscellaneous supplies
        routing tags
        assembly lubricants
        cleaning fluids
        wiping cloths
        janitorial supplies
    packaging (materials and labor)
    factory utilities
    laboratory
    depreciation on factory equipment
research and development expenses
    engineering (labor)
    patents
    testing
    prototypes (material and labor)
    drafting
    O&M of R&D facility
administrative expenses
    corporate officers
    accounting
    secretarial/clerical/reception
    security (protection)
    medical (nurse)
    employment (personnel)
    reproduction
    data processing
    production control
    depreciation on nonfactory equipment
    office supplies
    office utilities
    O&M of offices
selling expenses
    marketing (labor)
    advertising
    transportation (if not paid by customer)
    outside sales force (labor and expenses)
    demonstration units
    commissions
    technical service and support
    order processing
    branch office expenses
miscellaneous expenses
    insurance
    property taxes
    interest on loans
nondeductible expenses
    federal income taxes
    fines and penalties

50. ACCOUNTING PRINCIPLES

Basic Bookkeeping

An accounting or *bookkeeping system* is used to record historical financial transactions. The resultant records are used for product costing, satisfaction of statutory requirements, reporting of profit for income tax purposes, and general company management.

Bookkeeping consists of two main steps: recording the transactions, followed by categorization of the transactions.[42] The transactions (receipts and disbursements) are recorded in a *journal (book of original entry)* to complete the first step. Such a journal is organized in a simple chronological and sequential manner.[43] The transactions are then categorized (into interest income, advertising expense, etc.) and posted (i.e., entered or written) into the appropriate *ledger account*.

The ledger accounts together constitute the *general ledger* or *ledger*. All ledger accounts can be classified into one of three types: *asset accounts, liability accounts,* and *owners' equity accounts*. Strictly speaking, income and expense accounts, kept in a separate journal, are included within the classification of owners' equity accounts.

Together, the journal and ledger are known simply as "the books" of the company.

Balancing the Books

In a business environment, *balancing the books* means more than reconciling the checkbook and bank statements. All accounting entries must be posted in such a way as to maintain the equality of the *basic accounting equation*,

$$\text{assets} = \text{liability} + \text{owners' equity} \qquad 86.56$$

In a *double-entry bookkeeping system*, the equality is maintained within the ledger system by entering each transaction into two balancing ledger accounts. For example, paying a utility bill would decrease the cash account (an asset account) and decrease the utility expense account (a liability account) by the same amount.

Transactions are either *debits* or *credits*, depending on their sign. Increases in asset accounts are debits; decreases are credits. For liability and equity accounts, the opposite is true: Increases are credits, and decreases are debits.[44]

---

[42]These two steps are not to be confused with the *double-entry bookkeeping method*.
[43]The two-step process is more typical of a *manual bookkeeping system* than a computerized *general ledger system*. However, even most computerized systems produce reports in journal entry order, as well as account summaries.
[44]There is a difference in sign between asset and liability accounts. Thus, an increase in an expense account is actually a decrease. The accounting profession, apparently, is comfortable with the common confusion that exists between debits and credits.

Cash and Accrual Systems[45]

The simplest form of bookkeeping is based on the *cash system*. The only transactions that are entered into the journal are those that represent cash receipts and disbursements. In effect, a checkbook register or bank deposit book could serve as the journal.

During a given period (e.g., month or quarter), expense liabilities may be incurred even though the payments for those expenses have not been made. For example, an invoice (bill) may have been received but not paid. Under the *accrual system*, the obligation is posted into the appropriate expense account before it is paid.[46] Analogous to expenses, under the accrual system, income will be claimed before payment is received. Specifically, a sales transaction can be recorded as income when the customer's order is received, when the outgoing invoice is generated, or when the merchandise is shipped.

Financial Statements

Each period, two types of corporate financial statements are typically generated: the *balance sheet* and *profit and loss (P&L) statements*.[47] The profit and loss statement, also known as a *statement of income and retained earnings*, is a summary of sources of *income* or *revenue* (interest, sales, fees charged, etc.) and *expenses* (utilities, advertising, repairs, etc.) for the period. The expenses are subtracted from the revenues to give a *net income* (generally, before taxes).[48] Figure 86.11 illustrates a simple profit and loss statement.

The *balance sheet* presents the *basic accounting equation* in tabular form. The balance sheet lists the major categories of assets and outstanding liabilities. The difference between asset values and liabilities is the *equity*, as defined in Eq. 86.56. This equity represents what would be left over after satisfying all debts by liquidating the company.

---

[45]There is also a distinction made between cash flows that are known and those that are expected. It is a *standard accounting principle* to record losses in full, at the time they are recognized, even before their occurrence. In the construction industry, for example, losses are recognized in full and projected to the end of a project as soon as they are foreseeable. Profits, on the other hand, are recognized only as they are realized (typically, as a percentage of project completion). The difference between cash and accrual systems is a matter of *bookkeeping*. The difference between loss and profit recognition is a matter of *accounting convention*. Engineers seldom need to be concerned with the accounting tradition.
[46]The expense for an item or service might be accrued even *before* the invoice is received. It might be recorded when the purchase order for the item or service is generated, or when the item or service is received.
[47]Other types of financial statements (*statements of changes in financial position, cost of sales statements*, inventory and asset reports, etc.) also will be generated, depending on the needs of the company.
[48]Financial statements also can be prepared with percentages (of total assets and net revenue) instead of dollars, in which case they are known as *common size financial statements*.

Figure 86.11 *Simplified Profit and Loss Statement*

revenue		
interest	2000	
sales	237,000	
returns	(23,000)	
net revenue		216,000
expenses		
salaries	149,000	
utilities	6000	
advertising	28,000	
insurance	4000	
supplies	1000	
net expenses		188,000

period net income	28,000
beginning retained earnings	63,000
net year-to-date earnings	91,000

There are several terms that appear regularly on balance sheets.

- *current assets:* cash and other assets that can be converted quickly into cash, such as accounts receivable, notes receivable, and merchandise (inventory). Also known as *liquid assets*.
- *fixed assets:* relatively permanent assets used in the operation of the business and relatively difficult to convert into cash. Examples are land, buildings, and equipment. Also known as *non-liquid assets*.
- *current liabilities:* liabilities due within a short period of time (e.g., within one year) and typically paid out of current assets. Examples are accounts payable, notes payable, and other accrued liabilities.
- *long-term liabilities:* obligations that are not totally payable within a short period of time (e.g., within one year).

Figure 86.12 is a simplified balance sheet.

Analysis of Financial Statements

Financial statements are evaluated by management, lenders, stockholders, potential investors, and many other groups for the purpose of determining the *health of the company*. The health can be measured in terms of *liquidity* (ability to convert assets to cash quickly), *solvency* (ability to meet debts as they become due), and *relative risk* (of which one measure is *leverage*—the portion of total capital contributed by owners).

The analysis of financial statements involves several common ratios, usually expressed as percentages. The following are some frequently encountered ratios.

- *current ratio:* an index of short-term paying ability.

$$\text{current ratio} = \frac{\text{current assets}}{\text{current liabilities}}$$

Figure 86.12 *Simplified Balance Sheet*

ASSETS

current assets

cash	14,000
accounts receivable	36,000
notes receivable	20,000
inventory	89,000
prepaid expenses	3000
total current assets	162,000

*plant, property,
and equipment*

land and buildings	217,000
motor vehicles	31,000
equipment	94,000
accumulated depreciation	(52,000)
total fixed assets	290,000
total assets	452,000

LIABILITIES AND OWNERS' EQUITY

current liabilities

accounts payable	66,000
accrued income taxes	17,000
accrued expenses	8000
total current liabilities	91,000

long-term debt

notes payable	117,000
mortgage	23,000
total long-term debt	140,000

*owners' and stockholders'
equity*

stock	130,000
retained earnings	91,000
total owners' equity	221,000

total liabilities and owners' equity 452,000

- *quick (or acid-test) ratio:* a more stringent measure of short-term debt-paying ability. The *quick assets* are defined to be current assets minus inventories and prepaid expenses.

$$\text{quick ratio} = \frac{\text{quick assets}}{\text{current liabilities}}$$

- *receivable turnover:* a measure of the average speed with which accounts receivable are collected.

$$\text{receivable turnover} = \frac{\text{net credit sales}}{\text{average net receivables}}$$

- *average age of receivables:* number of days, on the average, in which receivables are collected.

$$\text{average age of receivables} = \frac{365}{\text{receivable turnover}}$$

- *inventory turnover:* a measure of the speed with which inventory is sold, on the average.

$$\text{inventory turnover} = \frac{\text{cost of goods sold}}{\text{average cost of inventory on hand}}$$

- *days supply of inventory on hand:* number of days, on the average, that the current inventory would last.

$$\text{days supply of inventory on hand} = \frac{365}{\text{inventory turnover}}$$

- *book value per share of common stock:* number of dollars represented by the balance sheet owners' equity for each share of common stock outstanding.

$$\text{book value per share of common stock} = \frac{\text{common shareholders' equity}}{\text{number of outstanding shares}}$$

- *gross margin:* gross profit as a percentage of sales. (Gross profit is sales less cost of goods sold.)

$$\text{gross margin} = \frac{\text{gross profit}}{\text{net sales}}$$

- *profit margin ratio:* percentage of each dollar of sales that is net income.

$$\text{profit margin} = \frac{\text{net income before taxes}}{\text{net sales}}$$

- *return on investment ratio:* shows the percent return on owners' investment.

$$\text{return on investment} = \frac{\text{net income}}{\text{owners' equity}}$$

- *price-earnings ratio:* indication of relationship between earnings and market price per share of common stock, useful in comparisons between alternative investments.

$$\text{price-earnings} = \frac{\text{market price per share}}{\text{earnings per share}}$$

51. COST ACCOUNTING

Cost accounting is the system that determines the cost of manufactured products. Cost accounting is called *job cost accounting* if costs are accumulated by part number or contract. It is called *process cost accounting* if costs are accumulated by departments or manufacturing processes.

Cost accounting is dependent on historical and recorded data. The unit product cost is determined from actual expenses and numbers of units produced. Allowances (i.e., budgets) for future costs are based on these historical figures. Any deviation from historical figures is called a *variance*. Where adequate records are available, variances can be divided into *labor variance* and *material variance*.

When determining a unit product cost, the direct material and direct labor costs are generally clear-cut and easily determined. Furthermore, these costs are 100% variable costs. However, the indirect cost per unit of product is not as easily determined. Indirect costs (*burden*, *overhead*, etc.) can be fixed or semivariable costs. The amount of indirect cost allocated to a unit will depend on the unknown future overhead expense as well as the unknown future production (*vehicle size*).

A typical method of allocating indirect costs to a product is as follows.

step 1: Estimate the total expected indirect (and overhead) costs for the upcoming year.

step 2: Determine the most appropriate vehicle (basis) for allocating the overhead to production. Usually, this vehicle is either the number of units expected to be produced or the number of direct hours expected to be worked in the upcoming year.

step 3: Estimate the quantity or size of the overhead vehicle.

step 4: Divide expected overhead costs by the expected overhead vehicle to obtain the unit overhead.

step 5: Regardless of the true size of the overhead vehicle during the upcoming year, one unit of overhead cost is allocated per unit of overhead vehicle.

Once the prime cost has been determined and the indirect cost calculated based on projections, the two are combined into a *standard factory cost* or *standard cost*, which remains in effect until the next budgeting period (usually a year).

During the subsequent manufacturing year, the standard cost of a product is not generally changed merely because it is found that an error in projected indirect costs or production quantity (vehicle size) has been made. The allocation of indirect costs to a product is assumed to be independent of errors in forecasts. Rather, the difference between the expected and actual expenses, known as the *burden (overhead) variance*, experienced during the year is posted to one or more *variance accounts*.

Burden (overhead) variance is caused by errors in forecasting both the actual indirect expense for the upcoming year and the overhead vehicle size. In the former case, the variance is called *burden budget variance*; in the latter, it is called *burden capacity variance*.

Example 86.29

A company expects to produce 8000 items in the coming year. The current material cost is $4.54 each. Sixteen minutes of direct labor are required per unit. Workers are paid $7.50 per hour. 2133 direct labor hours are forecasted for the product. Miscellaneous overhead costs are estimated at $45,000.

Find the per-unit (a) expected direct material cost, (b) direct labor cost, (c) prime cost, (d) burden as a function of production and direct labor, and (e) total cost.

Solution

(a) The direct material cost was given as $4.54.

(b) The direct labor cost is

$$\left(\frac{16 \text{ min}}{60 \frac{\text{min}}{\text{hr}}}\right)(\$7.50) = \$2.00$$

(c) The prime cost is

$$\$4.54 + \$2.00 = \$6.54$$

(d) If the burden vehicle is production, the burden rate is $45,000/8000 = $5.63 per item, making the total cost

$$\$4.54 + \$2.00 + \$5.63 = \$12.17$$

(e) If the burden vehicle is direct labor hours, the burden rate is $45,000/2133 = $21.10 per hour, making the total cost

$$\$4.54 + \$2.00 + \left(\frac{16 \text{ min}}{60 \frac{\text{min}}{\text{hr}}}\right)(\$21.10) = \$12.17$$

Example 86.30

The actual performance of the company in Ex. 86.29 is given by the following figures.

$$\text{actual production: } 7560$$
$$\text{actual overhead costs: } \$47,000$$

What are the burden budget variance and the burden capacity variance?

Solution

The burden capacity variance is

$$\$45,000 - (7560)(\$5.63) = \$2437$$

The burden budget variance is

$$\$47,000 - \$45,000 = \$2000$$

The overall burden variance is

$$\$47{,}000 - (7560)(\$5.63) = \$4437$$

The sum of the burden capacity and burden budget variances should equal the overall burden variance.

$$\$2437 + \$2000 = \$4437$$

52. COST OF GOODS SOLD

Cost of goods sold (COGS) is an accounting term that represents an inventory account adjustment.[49] Cost of goods sold is the difference between the starting and ending inventory valuations. That is,

$$\text{COGS} = \text{starting inventory valuation}$$
$$- \text{ending inventory valuation} \quad \textit{86.57}$$

Cost of goods sold is subtracted from *gross profit* to determine the *net profit* of a company. Despite the fact that cost of goods sold can be a significant element in the profit equation, the inventory adjustment may not be made each accounting period (e.g., each month) due to the difficulty in obtaining an accurate inventory valuation.

With a *perpetual inventory system*, a company automatically maintains up-to-date inventory records, either through an efficient stocking and stock-releasing system or through a *point of sale* (POS) *system* integrated with the inventory records. If a company only counts its inventory (i.e., takes a *physical inventory*) at regular intervals (e.g., once a year), it is said to be operating on a *periodic inventory system*.

Inventory accounting is a source of many difficulties. The inventory value is calculated by multiplying the quantity on hand by the standard cost. In the case of completed items actually assembled or manufactured at the company, this standard cost usually is the manufacturing cost, although factory cost also can be used. In the case of purchased items, the standard cost will be the cost per item charged by the supplying vendor. In some cases, delivery and transportation costs will be included in this standard cost.

It is not unusual for the elements in an item's inventory to come from more than one vendor, or from one vendor in more than one order. Inventory valuation is more difficult if the price paid is different for these different purchases. There are four methods of determining the cost of elements in inventory. Any of these methods can be used (if applicable), but the method must be used consistently from year to year. The four methods are as follows.

[49]The cost of goods sold inventory adjustment is posted to the *COGS expense account.*

- *specific identification method:* Each element can be uniquely associated with a cost. Inventory elements with serial numbers fit into this costing scheme. Stock, production, and sales records must include the serial number.

- *average cost method:* The standard cost of an item is the average of (recent or all) purchase costs for that item.

- *first-in, first-out* (FIFO) *method:* This method keeps track of how many of each item were purchased each time and the number remaining out of each purchase, as well as the price paid at each purchase. The inventory system assumes that the oldest elements are issued first.[50] Inventory value is a weighted average dependent on the number of elements from each purchase remaining. Items issued no longer contribute to the inventory value.

- *last-in, first-out* (LIFO) *method:* This method keeps track of how many of each item were purchased each time and the number remaining out of each purchase, as well as the price paid at each purchase.[51] The inventory value is a weighted average dependent on the number of elements from each purchase remaining. Items issued no longer contribute to the inventory value.

53. BREAK-EVEN ANALYSIS

Special Nomenclature

f	fixed cost that does not vary with production
a	*incremental cost* to produce one additional item (also called *marginal cost* or *differential cost*)
Q	quantity sold
p	*incremental value* (price)
R	total revenue
C	total cost

Break-even analysis is a method of determining when the value of one alternative becomes equal to the value of another. A common application is that of determining when costs exactly equal revenue. If the manufactured quantity is less than the break-even quantity, a loss is incurred. If the manufactured quantity is greater than the break-even quantity, a profit is made.

Assuming no change in the inventory, the *break-even point* can be found by setting costs equal to revenue ($C = R$).

$$C = f + aQ \quad \textit{86.58}$$
$$R = pQ \quad \textit{86.59}$$
$$Q^* = \frac{f}{p - a} \quad \textit{86.60}$$

[50]If all elements in an item's inventory are identical, and if all shipments of that item are agglomerated, there will be no way to guarantee that the oldest element in inventory is issued first. But, unless *spoilage* is a problem, it really does not matter.
[51]See Ftn. 50.

Figure 86.13 *Break-Even Quantity*

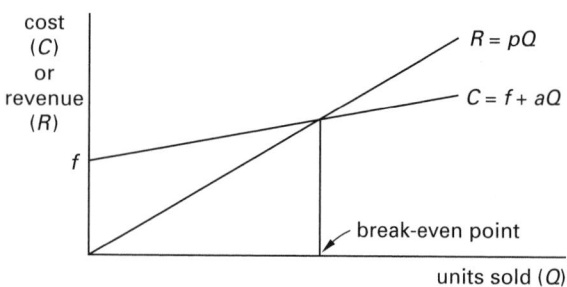

An alternative form of the break-even problem is to find the number of units per period for which two alternatives have the same total costs. Fixed costs are to be spread over a period longer than one year using the equivalent uniform annual cost (EUAC) concept. One of the alternatives will have a lower cost if production is less than the break-even point. The other will have a lower cost for production greater than the break-even point.

Example 86.31

Two plans are available for a company to obtain automobiles for its sales representatives. How many miles must the cars be driven each year for the two plans to have the same costs? Use an interest rate of 10%. (Use the year-end convention for all costs.)

plan A: Lease the cars and pay $0.15 per mile.

plan B: Purchase the cars for $5000. Each car has an economic life of three years, after which it can be sold for $1200. Gas and oil cost $0.04 per mile. Insurance is $500 per year.

Solution

Let x be the number of miles driven per year. Then, the EUAC for both alternatives is

$$\text{EUAC(A)} = 0.15x$$
$$\begin{aligned}\text{EUAC(B)} &= 0.04x + \$500 + (\$5000)(A/P, 10\%, 3) \\ &\quad - (\$1200)(A/F, 10\%, 3) \\ &= 0.04x + \$500 + (\$5000)(0.4021) \\ &\quad - (\$1200)(0.3021) \\ &= 0.04x + 2148\end{aligned}$$

Setting EUAC(A) and EUAC(B) equal and solving for x yields 19,527 miles per year as the break-even point.

54. PAY-BACK PERIOD

The *pay-back period* is defined as the length of time, usually in years, for the cumulative net annual profit to equal the initial investment. It is tempting to introduce equivalence into pay-back period calculations, but by convention, this is generally not done.[52]

$$\text{pay-back period} = \frac{\text{initial investment}}{\text{net annual profit}} \qquad 86.61$$

Example 86.32

A ski resort installs two new ski lifts at a total cost of $1,800,000. The resort expects the annual gross revenue to increase by $500,000 while it incurs an annual expense of $50,000 for lift operation and maintenance. What is the pay-back period?

Solution

From Eq. 86.61,

$$\text{pay-back period} = \frac{\$1,800,000}{\$500,000 - \$50,000} = 4 \text{ years}$$

55. MANAGEMENT GOALS

Depending on many factors (market position, age of the company, age of the industry, perceived marketing and sales windows, etc.), a company may select one of many production and marketing strategic goals. Three such strategic goals are

- maximization of product demand

- minimization of cost

- maximization of profit

Such goals require knowledge of how the dependent variable (e.g., demand quantity or quantity sold) varies as a function of the independent variable (e.g., price). Unfortunately, these three goals are not usually satisfied simultaneously. For example, minimization of product cost may require a large production run to realize economies of scale, while the actual demand is too small to take advantage of such economies of scale.

If sufficient data are available to plot the independent and dependent variables, it may be possible to optimize the dependent variable graphically. Of course, if the relationship between independent and dependent variables is known algebraically, the dependent variable can be optimized by taking derivatives or by use of other numerical methods.

---

[52]Equivalence (i.e., interest and compounding) generally is not considered when calculating the "pay-back period." However, if it is desirable to include equivalence, then the term *pay-back period* should not be used. Other terms, such as *cost recovery period* or *life of an equivalent investment*, should be used. Unfortunately, this convention is not always followed in practice.

Figure 86.14 *Graphs of Management Goal Functions*

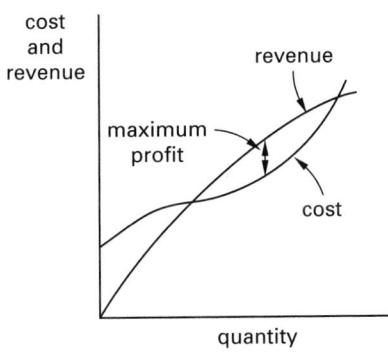

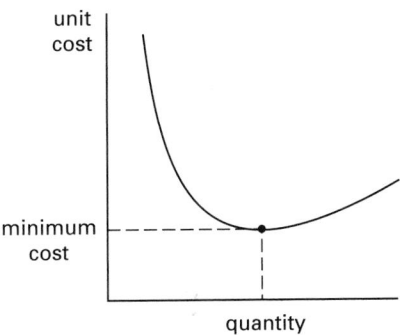

56. INFLATION

It is important to perform economic studies in terms of *constant value dollars*. One method of converting all cash flows to constant value dollars is to divide the flows by some annual *economic indicator* or price index.

If indicators are not available, cash flows can be adjusted by assuming that inflation is constant at a decimal rate (e) per year. Then, all cash flows can be converted to $t = 0$ dollars by dividing by $(1 + e)^n$, where n is the year of the cash flow.

An alternative is to replace the effective annual interest rate (i) with a value corrected for inflation. This corrected value (i') is

$$i' = i + e + ie \qquad \qquad \textbf{86.62}$$

This method has the advantage of simplifying the calculations. However, precalculated factors are not available for the non-integer values of i'. Therefore, Table 86.1 must be used to calculate the factors.

Example 86.33

What is the uninflated present worth of a $2000 future value in two years if the average inflation rate is 6% and i is 10%?

Solution

$$P = \frac{\$2000}{(1 + 0.10)^2(1 + 0.06)^2} = \$1471$$

Example 86.34

Repeat Ex. 86.33 using i'.

Solution

$$i' = 0.10 + 0.06 + (0.10)(0.06) = 0.166$$

$$P = \frac{\$2000}{(1 + 0.166)^2} = \$1471$$

57. CONSUMER LOANS

Special Nomenclature

BAL_j	balance after the jth payment
LV	principal total value loaned (cost minus down payment)
j	payment or period number
N	total number of payments to pay off the loan
PI_j	jth interest payment
PP_j	jth principal payment
PT_j	jth total payment
ϕ	effective rate per period (r/k)

Many different arrangements can be made between a borrower and a lender. With the advent of creative financing concepts, it often seems that there are as many variations of loans as there are loans made. Nevertheless, there are several traditional types of transactions. Real estate or investment texts, or a financial consultant, should be consulted for more complex problems.

Simple Interest

Interest due does not compound with a *simple interest loan*. The interest due is merely proportional to the length of time that the principal is outstanding. Because of this, simple interest loans are seldom made for long periods (e.g., more than one year). (For loans less than one year, it is commonly assumed that a year consists of 12 months of 30 days each.)

Example 86.35

A $12,000 simple interest loan is taken out at 16% per annum interest rate. The loan matures in two years with no intermediate payments. How much will be due at the end of the second year?

Solution

The interest each year is

$$\mathrm{PI} = (0.16)(\$12,000) = \$1920$$

The total amount due in two years is

$$\mathrm{PT} = \$12,000 + (2)(\$1920) = \$15,840$$

Example 86.36

$4000 is borrowed for 75 days at 16% per annum simple interest. How much will be due at the end of 75 days?

Solution

$$\text{amount due} = \$4000 + (0.16)\left(\frac{75 \text{ days}}{360 \frac{\text{days}}{\text{bank yr}}}\right)(\$4000)$$

$$= \$4133$$

Loans with Constant Amount Paid Toward Principal

With this loan type, the payment is not the same each period. The amount paid toward the principal is constant, but the interest varies from period to period. The equations that govern this type of loan are

$$\text{BAL}_j = \text{LV} - (j)(\text{PP}) \qquad 86.63$$

$$\text{PI}_j = \phi(\text{BAL})_{j-1} \qquad 86.64$$

$$\text{PT}_j = \text{PP} + \text{PI}_j \qquad 86.65$$

$$\text{PP} = \frac{\text{LV}}{N} \qquad 86.66$$

$$N = \frac{\text{LV}}{\text{PP}} \qquad 86.67$$

$$\text{LV} = (\text{PP} + \text{PI}_1)(P/A, \phi, N)$$
$$\qquad - \text{PI}_N(P/G, \phi, N) \qquad 86.68$$

$$1 = \left(\frac{1}{N} + \phi\right)(P/A, \phi, N)$$
$$\qquad - \left(\frac{\phi}{N}\right)(P/G, \phi, N) \qquad 86.69$$

Figure 86.15 *Loan with Constant Amount Paid Toward Principal*

Example 86.37

A $12,000 six-year loan is taken from a bank that charges 15% effective annual interest. Payments toward the principal are uniform, and repayments are made at

the end of each year. Tabulate the interest, total payments, and the balance remaining after each payment is made.

Solution

The amount of each principal payment is

$$\text{PP} = \frac{\$12,000}{6} = \$2000$$

At the end of the first year (before the first payment is made), the principal balance is $12,000 (i.e., $BAL_0 = \$12,000$). From Eq. 86.64, the interest payment is

$$\text{PI}_1 = (0.15)(\$12,000) = \$1800$$

The total first payment is

$$\text{PT}_1 = \text{PP} + \text{PI} = \$2000 + \$1800$$
$$= \$3800$$

The following table is similarly constructed.

j	BAL_j	PP_j	PI_j	PT_j
	(in dollars)			
0	12,000	–	–	–
1	10,000	2000	1800	3800
2	8000	2000	1500	3500
3	6000	2000	1200	3200
4	4000	2000	900	2900
5	2000	2000	600	2600
6	0	2000	300	2300

Direct Reduction Loans

This is the typical "interest paid on unpaid balance" loan. The amount of the periodic payment is constant, but the amounts paid toward the principal and interest both vary.

$$\text{BAL}_{j-1} = \text{PT}\left(\frac{1 - (1 + \phi)^{j-1-N}}{\phi}\right) \qquad 86.70$$

$$\text{PI}_j = \phi(\text{BAL})_{j-1} \qquad 86.71$$

$$\text{PP}_j = \text{PT} - \text{PI}_j \qquad 86.72$$

$$\text{BAL}_j = \text{BAL}_{j-1} - \text{PP}_j \qquad 86.73$$

$$N = \frac{-\ln\left(1 - \frac{\phi(\text{LV})}{\text{PT}}\right)}{\ln(1 + \phi)} \qquad 86.74$$

Equation 86.74 calculates the number of payments necessary to pay off a loan. This equation can be solved with effort for the total periodic payment (PT) or the initial value of the loan (LV). It is easier, however, to use the $(A/P, i\%, n)$ factor to find the payment and loan value.

$$PT = LV(A/P, \phi\%, N) \qquad 86.75$$

If the loan is repaid in yearly installments, then i is the effective annual rate. If the loan is paid off monthly, then i should be replaced by the effective rate per month (ϕ from Eq. 86.54). For monthly payments, N is the number of months in the loan period.

Figure 86.16 *Direct Reduction Loan*

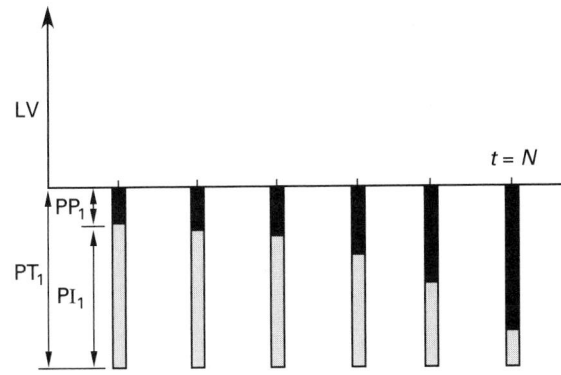

Example 86.38

A \$45,000 loan is financed at 9.25% per annum. The monthly payment is \$385. What are the amounts paid toward interest and principal in the 14th period? What is the remaining principal balance after the 14th payment has been made?

Solution

The effective rate per month is

$$\phi = \frac{r}{k} = \frac{0.0925}{12}$$
$$= 0.0077083\ldots \quad [\text{say } 0.007708]$$

$$N = \frac{-\ln\left(1 - \dfrac{(0.007708)(45,000)}{385}\right)}{\ln(1 + 0.007708)} = 301$$

$$BAL_{13} = (\$385)\left(\frac{1 - (1 + 0.007708)^{14-1-301}}{0.007708}\right)$$
$$= \$44,476.39$$

$$PI_{14} = (0.007708)(\$44,476.39) = \$342.82$$

$$PP_{14} = \$385 - \$342.82 = \$42.18$$

$$BAL_{14} = \$44,476.39 - \$42.18 = \$44,434.21$$

Direct Reduction Loans with Balloon Payments

This type of loan has a constant periodic payment, but the duration of the loan is insufficient to completely pay back the principal (i.e, the loan is not fully amortized).

Therefore, all remaining unpaid principal must be paid back in a lump sum when the loan matures. This large payment is known as a *balloon payment*.[53]

Equations 86.70 through 86.74 also can be used with this type of loan. The remaining balance after the last payment is the balloon payment. This balloon payment must be repaid along with the last regular payment calculated.

Figure 86.17 *Direct Reduction Loan with Balloon Payment*

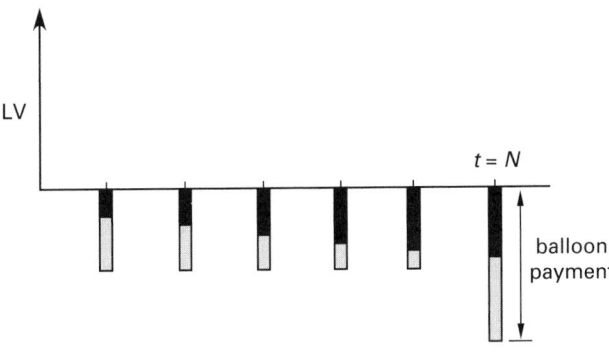

58. FORECASTING

There are many types of forecasting models, although most are variations of the basic types.[54] All models produce a *forecast* (F_{t+1}) of some quantity (*demand* is used in this section) in the next period based on actual measurements (D_j) in current and prior periods. All of the models also try to provide *smoothing* (or *damping*) of extreme data points.

Forecasts by Moving Averages

The method of *moving average forecasting* weights all previous demand data points equally and provides some smoothing of extreme data points. The amount of smoothing increases as the number of data points, n, increases.

$$F_{t+1} = \frac{1}{n} \sum_{m=t+1-n}^{t} D_m \qquad 86.76$$

Forecasts by Exponentially Weighted Averages

With *exponentially weighted forecasts*, the more current (most recent) data points receive more weight. This method uses a *weighting factor* (α), also known as a *smoothing coefficient*, which typically varies between

---

[53]The term *balloon payment* may include the final interest payment as well. Generally, the problem statement will indicate whether the balloon payment is inclusive or exclusive of the regular payment made at the end of the loan period.

[54]For example, forecasting models that take into consideration steady (linear), cyclical, annual, and seasonal trends are typically variations of the exponentially weighted model. A truly different forecasting tool, however, is *Monte Carlo simulation*.

0.01 and 0.30. An initial forecast is needed to start the method. Forecasts immediately following are sensitive to the accuracy of this first forecast. It is common to choose $F_0 = D_1$ to get started.

$$F_{t+1} = \alpha D_t + (1 - \alpha) F_t \qquad 86.77$$

59. LEARNING CURVES

Special Nomenclature

R	decimal learning curve rate (2^{-b})
T_1	time or cost for the first item
T_n	time or cost for the nth item
n	total number of items produced
b	learning curve constant

The more products that are made, the more efficient the operation becomes due to experience gained. Therefore, direct labor costs decrease.[55] Usually, a *learning curve* is specified by the decrease in cost each time the cumulative quantity produced doubles. If there is a 20% decrease per doubling, the curve is said to be an 80% learning curve (i.e., the *learning curve rate*, R, is 80%).

Then, the time to produce the nth item is

$$T_n = T_1 n^{-b} \qquad 86.78$$

The total time to produce units from quantity n_1 to n_2 inclusive is approximately given by Eq. 86.79. T_1 is a constant, the time for item 1, and does not correspond to n unless $n_1 = 1$.

$$\int_{n_1}^{n_2} T_n \, dn \approx \left(\frac{T_1}{1-b} \right) \left(\left(n_2 + \tfrac{1}{2} \right)^{1-b} - \left(n_1 - \tfrac{1}{2} \right)^{1-b} \right)$$
$$86.79$$

The *average time per unit* over the production from n_1 to n_2 is the above total time from Eq. 86.79 divided by the quantity produced, $(n_2 - n_1 + 1)$.

$$T_{\text{ave}} = \frac{\displaystyle\int_{n_1}^{n_2} T_n \, dn}{n_2 - n_1 + 1} \qquad 86.80$$

Table 86.9 lists representative values of the *learning curve constant* (b). For learning curve rates not listed in the table, Eq. 86.81 can be used to find b.

$$b = \frac{-\log_{10} R}{\log_{10}(2)} = \frac{-\log_{10} R}{0.301} \qquad 86.81$$

---

[55]Remember that learning curve reductions apply only to direct labor costs. They are not applied to indirect labor or direct material costs.

Table 86.9 *Learning Curve Constants*

learning curve rate (R)	b
0.70 (70%)	0.515
0.75 (75%)	0.415
0.80 (80%)	0.322
0.85 (85%)	0.234
0.90 (90%)	0.152
0.95 (95%)	0.074

Example 86.39

A 70% learning curve is used with an item whose first production time is 1.47 hr. (a) How long will it take to produce the 11th item? (b) How long will it take to produce the 11th through 27th items?

Solution

(a) From Eq. 86.78,

$$T_{11} = (1.47 \text{ hr})(11)^{-0.515} = 0.428 \text{ hr}$$

(b) The time to produce the 11th item through 27th item is given by Eq. 86.79.

$$T \approx \left(\frac{1.47 \text{ hr}}{1 - 0.515} \right) \left((27.5)^{1-0.515} - (10.5)^{1-0.515} \right)$$
$$= 5.643 \text{ hr}$$

60. ECONOMIC ORDER QUANTITY

Special Nomenclature

a	constant depletion rate (items/unit time)
h	inventory storage cost (\$/item-unit time)
H	total inventory storage cost between orders (\$)
K	fixed cost of placing an order (\$)
Q	order quantity (original quantity on hand)

The *economic order quantity* (EOQ) is the order quantity that minimizes the inventory costs per unit time. Although there are many different EOQ models, the simplest is based on the following assumptions.

- Reordering is instantaneous. The time between order placement and receipt is zero.

- Shortages are not allowed.

- Demand for the inventory item is deterministic (i.e., is not a random variable).

- Demand is constant with respect to time.

- An order is placed when the inventory is zero.

Figure 86.18 *Inventory with Instantaneous Reorder*

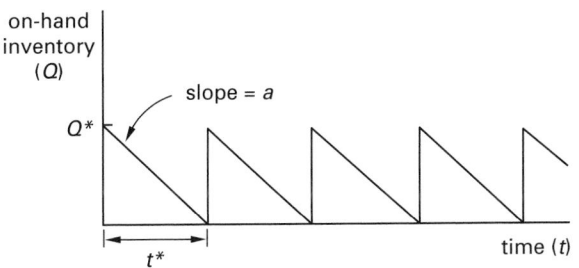

Figure 86.19 *Types of Sensitivity*

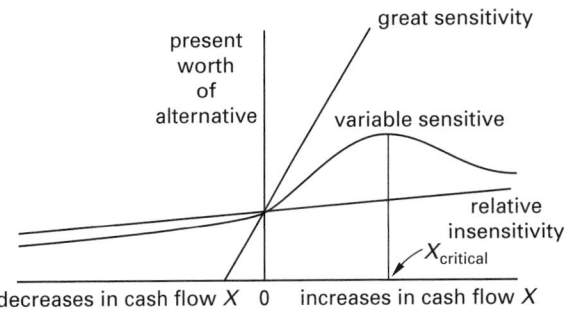

If the original quantity on hand is Q, the stock will be depleted at

$$t^* = \frac{Q}{a} \qquad 86.82$$

The total inventory storage cost between t_0 and t^* is

$$H = \tfrac{1}{2}Qht^* = \frac{Q^2 h}{2a} \qquad 86.83$$

The total inventory and ordering cost per unit time is

$$C_t = \frac{aK}{Q} + \frac{hQ}{2} \qquad 86.84$$

C_t can be minimized with respect to Q. The economic order quantity and time between orders are

$$Q^* = \sqrt{\frac{2aK}{h}} \qquad 86.85$$

$$t^* = \frac{Q^*}{a} \qquad 86.86$$

61. SENSITIVITY ANALYSIS

Data analysis and forecasts in economic studies require estimates of costs that will occur in the future. There are always uncertainties about these costs. However, these uncertainties are insufficient reason not to make the best possible estimates of the costs. Nevertheless, a decision between alternatives often can be made more confidently if it is known whether or not the conclusion is sensitive to moderate changes in data forecasts. Sensitivity analysis provides this extra dimension to an economic analysis.

The sensitivity of a decision is determined by inserting a range of estimates for critical cash flows and other parameters. If radical changes can be made to a cash flow without changing the decision, the decision is said to be *insensitive* to uncertainties regarding that cash flow. However, if a small change in the estimate of a cash flow will alter the decision, that decision is said to be very *sensitive* to changes in the estimate. If the decision is sensitive only for a limited range of cash flow values, the term *variable sensitivity* is used. Figure 86.19 illustrates these terms.

An established semantic tradition distinguishes between risk analysis and uncertainty analysis. *Risk analysis* addresses variables that have a known or estimated probability distribution. In this regard, statistics and probability theory can be used to determine the probability of a cash flow varying between given limits. On the other hand, *uncertainty analysis* is concerned with situations in which there is not enough information to determine the probability or frequency distribution for the variables involved.

As a first step, sensitivity analysis should be applied one at a time to the dominant factors. Dominant cost factors are those that have the most significant impact on the present value of the alternative.[56] If warranted, additional investigation can be used to determine the sensitivity to several cash flows varying simultaneously. Significant judgment is needed, however, to successfully determine the proper combinations of cash flows to vary. It is common to plot the dependency of the present value on the cash flow being varied in a two-dimensional graph. Simple linear interpolation is used (within reason) to determine the critical value of the cash flow being varied.

62. VALUE ENGINEERING

The *value* of an investment is defined as the ratio of its return (performance or utility) to its cost (effort or investment). The basic object of *value engineering* (VE, also referred to as *value analysis*) is to obtain the maximum per-unit value.[57]

Value engineering concepts often are used to reduce the cost of mass-produced manufactured products. This is done by eliminating unnecessary, redundant, or superfluous features, by redesigning the product for a less expensive manufacturing method, and by including features for easier assembly without sacrificing utility and

---

[56]In particular, engineering economic analysis problems are sensitive to the choice of effective interest rate (i) and to accuracy in cash flows at or near the beginning of the horizon. The problems will be less sensitive to accuracy in far-future cash flows, such as salvage value and subsequent generation replacement costs.

[57]Value analysis, the methodology that has become today's value engineering, was developed in the early 1950s by Lawrence D. Miles, an analyst at General Electric.

function.[58] However, the concepts are equally applicable to one-time investments, such as buildings, chemical processing plants, and space vehicles. In particular, value engineering has become an important element in all federally funded work.[59]

Typical examples of large-scale value engineering work are using stock-sized bearings and motors (instead of custom manufactured units), replacing rectangular concrete columns with round columns (which are easier to form), and substituting custom buildings with prefabricated structures.

Value engineering is usually a team effort. And, while the original designers may be on the team, usually outside consultants are utilized. The cost of value engineering is usually returned many times over by reduced construction and life-cycle costs.

---

[58]Some people say that value engineering is "the act of going over the plans and taking out everything that is interesting."
[59]U.S. Government Office of Management and Budget Circular A-131 outlines value engineering for federally funded construction projects.

87 Engineering Law[1]

1. FORMS OF COMPANY OWNERSHIP

There are three basic forms of company ownership in the United States: (a) sole proprietorship, (b) partnership, and (c) corporation.[2] Each of these forms of ownership has advantages and disadvantages.

2. SOLE PROPRIETORSHIPS

A *sole proprietorship (single proprietorship)* is the easiest form of ownership to establish. Other than the necessary licenses (which apply to all forms of ownership), no legal formalities are required to start business operations. A sole proprietor (the owner) has virtually total control of the business and makes all supervisory and management decisions.

Legally, there is no distinction between the sole proprietor and the sole proprietorship (the business). This is the greatest disadvantage of this form of business. The

---

[1]The author is not giving legal advice in this chapter, nor is this chapter intended to be a substitute for professional advice. Law is not always black and white. For every rule there are exceptions. For every legal principle, there are variations. For every type of injury, there are numerous legal precedents. This chapter covers the superficial basics of a small subset of U.S. law affecting engineers.
[2]The discussion of forms of company ownership in Sections 2, 3, and 4 applies equally to service-oriented companies (e.g., consulting engineering firms) and product-oriented companies.

owner is solely responsible for the operation of the business, even if the owner hires others for assistance. The owner assumes personal, legal, and financial liability for all acts and debts of the company. If the company debts remain unpaid, or in the event there is a legal judgment against the company, the owner's personal assets (home, car, savings, etc.) can be seized or attached.

Another disadvantage of the sole proprietorship is the lack of significant organizational structure. In times of business crisis or trouble, there may be no one to share the responsibility or to help make decisions. When the owner is sick or dies, there may be no way to continue the business.

There is also no distinction between the incomes of the business and the owner. Therefore, the business income is taxed at the owner's income tax rate. Depending on the owner's financial position, the success of the business, and the tax structure, this can be an advantage or a disadvantage.[3]

3. PARTNERSHIPS

A *partnership* (also known as a *general partnership*) is ownership by two or more persons known as *general partners*. Legally, this form is very similar to a sole proprietorship, and the two forms of business have many of the same advantages and disadvantages. For example, with the exception of an optional *partnership agreement*, there are a minimum of formalities to setting up business. The partners make all business and management decisions themselves according to an agreed-upon process. The business income is split among the partners and taxed at the partners' individual tax rates.[4] Continuity of the business is still a problem since most partnerships are automatically dissolved upon the withdrawal or death of one of the partners.[5]

One advantage of a partnership over a sole proprietorship is the increase in available funding. Not only do more partners bring in more start-up capital, but the resource pool may make business credit easier to obtain. Also, the partners bring a diversity of skills and talents.

---

[3]To use a simplistic example, if the corporate tax rates are higher than the individual tax rates, it would be *financially* better to be a sole proprietor because the company income would be taxed at a lower rate.
[4]The percentage split is specified in the partnership agreement.
[5]Some or all of the remaining partners may want to form a new partnership, but this is not always possible.

Unless the partnership agreement states otherwise, each partner can individually obligate (i.e., *bind*) the partnership without the consent of the other partners. Similarly, each partner has personal responsibility and liability for the acts and debts of the partnership company, just as sole proprietors do. In fact, each partner assumes the *sole* responsibility, not just a proportionate share. If one or more partners are unable to pay, the remaining partners shoulder the entire debt. The possibility of one partner having to pay for the actions of another partner must be considered when choosing this form of business ownership.

A *limited partnership* differs from a general partnership in that one or more of the partners is silent. The *limited partners* make a financial contribution to the business and receive a share of the profit but do not participate in the management and cannot bind the partnership. While *general partners* have unlimited personal liabilities, limited partners are generally liable only to the extent of their investment.[6] A written partnership agreement is required, and the agreement must be filed with the proper authorities.

4. CORPORATIONS

A corporation is a legal entity (i.e., a legal person) distinct from the founders and owners. The separation of ownership and management makes the corporation a fundamentally different kind of business form than a sole proprietorship or partnership, with very different advantages and disadvantages.

A corporation becomes legally distinct from its founders upon formation and proper registration. Ownership of the corporation is through shares of stock, which are distributed to the founders and investors according to some agreed-upon investment and distribution rule. Thus, the founders and investors become the stockholders (i.e., owners) of the corporation. A *closely held (private) corporation* is one in which all stock is owned by a family or small group of coinvestors. A *public corporation* is one whose stock is available for the public-at-large to purchase.

There is no mandatory connection between ownership and management functions. The decision-making power is vested in the executive officers and a *board of directors* that governs by majority vote. The stockholders elect the board of directors which, in turn, hires the executive officers, management, and other employees. Employees of the corporation may or may not be stockholders.

Disadvantages (at least for a person or persons who could form a partnership or sole proprietorship instead) include the higher corporate tax rate, difficulty in forming (some states require a minimum number of persons on the board of directors), and additional legal and accounting paperwork.

However, since a corporation is distinctly separate from its founders and investors, those individuals are not liable for the acts and debts of the corporation. Debts are paid from the corporate assets. Income to the corporation is not taxable income to the owners. (Only the salaries, if any, paid to the employees by the corporation are taxable to the employees.) Even if the corporation were to go bankrupt, the assets of the owners would not ordinarily be subject to seizure or attachment.

A corporation offers the best guarantee of continuity of operation in the event of the death, incapacity, or retirement of the founders since, as a legal entity, it is distinct from the founders and owners.

5. AGENCY

In some contracts, decision-making authority and right of action are transferred from one party (the owner, or *principal*) who would normally have that authority to another person (the *agent*). For example, in construction contracts, the engineer is ordinarily the agent of the owner. Agents are limited in what they can do by the scope of the agency agreement. Within that scope, however, an agent acts on behalf of the principal, and the principal is liable for the acts of the agent and is bound by contracts made in the principal's name by the agent.

Agents are required to execute their work with care, skill, and diligence. Specifically, agents have *fiduciary responsibility* toward their principal, meaning that agent must be honest and loyal. Agents will be liable for damages resulting from a lack of diligence, loyalty, and/or honesty. If the agents misrepresented their skills when obtaining the agency, they can be liable for breach of contract or fraud.

6. GENERAL CONTRACTS

A *contract* is a legally binding agreement or promise to exchange goods or services.[7] A written contract is merely a documentation of the agreement. Some agreements must be in writing, but most agreements for engineering services can be verbal, particularly if the parties to the agreement know each other well.[8] Written contract documents do not need to contain intimidating legal language, but all agreements must satisfy three basic requirements to be enforceable (binding).

---

[6]That is, if the partnership fails or is liquidated to pay debts, the limited partners lose no more than their initial investments.

[7]Not all agreements are legally binding (i.e., enforceable). Two parties may agree on something, but unless the agreement meets all of the requirements and conditions of a contract, the parties cannot hold each other to the agreement.

[8]All states have a *statute of frauds* that, among other things, specifies what types of contracts must be in writing to be enforceable. These include contracts for the sale of land, contracts requiring more than one year for performance, contracts for the sale of goods over $500 in value, contracts to satisfy the debts of another, and marriage contracts. Contracts to provide engineering services do not fall under the statute of frauds.

- There must be a clear, specific, and definite *offer* with no room for ambiguity or misunderstanding.

- There must be some form of conditional future *consideration* (i.e., payment).[9]

- There must be an *acceptance* of the offer.

There are other conditions that the agreement must meet to be enforceable. These conditions are not normally part of the explicit agreement but represent the conditions under which the agreement was made.

- The agreement must be *voluntary* for all parties.

- All parties must have *legal capacity* (i.e., be mentally competent, of legal age, and uninfluenced by drugs).

- The purpose of the agreement must be *legal.*

For small projects, a simple *letter of agreement* on one party's stationery may suffice. For larger, complex projects, a more formal document may be required. Some clients prefer to use a *purchase order*, which can function as a contract if all basic requirements are met.

Regardless of the format of the written document— letter of agreement, purchase order, or standard form— a contract should include the following features.[10]

- introduction, preamble, or preface indicating the purpose of the contract

- name, address, and business forms of both contracting parties

- signature date of the agreement

- effective date of the agreement (if different from the signature date)

- duties and obligations of both parties

- deadlines and required service dates

- fee amount

- fee schedule and payment terms

- agreement expiration date

- standard boilerplate clauses

- signatures of parties or their agents

- declaration of authority of the signatories to bind the contracting parties

- supporting documents

7. STANDARD BOILERPLATE CLAUSES

It is common for full-length contract documents to include important *boilerplate clauses*. These clauses have specific wordings that should not normally be changed, hence the name "boilerplate." Some of the most common boilerplate clauses are paraphrased here.

- Delays and inadequate performance due to war, strikes, and acts of God and nature are forgiven (*force majeure*).

- The contract document is the complete agreement, superseding all previous verbal and written agreements.

- The contract can be modified or canceled only in writing.

- Parts of the contract that are determined to be void or unenforceable shall not affect the enforceability of the remainder of the contract (*severability*). Alternatively, parts of the contract that are determined to be void or unenforceable shall be rewritten to accomplish their intended purpose without affecting the remainder of the contract.

- None (or one, or both) of the parties can (or cannot) assign its (or their) rights and responsibilities under the contract (*assignment*).

- All notices provided for in the agreement must be in writing and sent to the address in the agreement.

- Time is of the essence.[11]

- The subject headings of the agreement paragraphs are for convenience only and do not control the meaning of the paragraphs.

- The laws of the state in which the contract is signed must be used to interpret and govern the contract.

- Disagreements shall be arbitrated according to the rules of the American Arbitration Association.

- Any lawsuits related to the contract must be filed in the county and state in which the contract is signed.

- Obligations under the agreement are unique, and in the event of a breach, the defaulting party waives the defense that the loss can be adequately compensated by monetary damages (*specific performance*).

- In the event of a lawsuit, the prevailing party is entitled to an award of reasonable attorneys' and court fees.[12]

- Consequential damages are not recoverable in a lawsuit.

---

[9]Actions taken or payments made prior to the agreement are irrelevant. Also, it does not matter to the courts whether the exchange is based on equal value or not.

[10]*Construction contracts* are unique unto themselves. Items that might also be included as part of the *contract documents* are the agreement form, the general conditions, drawings, specifications, and addenda.

[11]Without this clause in writing, damages for delay cannot be claimed.

[12]Without this clause in writing, attorneys' fees and court costs are rarely recoverable.

8. SUBCONTRACTS

When a party to a contract engages a third party to perform the work in the original contract, the contract with the third party is known as a *subcontract*. Whether or not responsibilities can be subcontracted under the original contract depends on the content of the *assignment clause* in the original contract.

9. PARTIES TO A CONSTRUCTION CONTRACT

A specific set of terms has developed for referring to parties in consulting and construction contracts. The *owner* of a construction project is the person, partnership, or corporation that actually owns the land, assumes the financial risk, and ends up with the completed project. The *developer* contracts with the architect and/or engineer for the design and with the contractors for the construction of the project. In some cases, the owner and developer are the same, in which case the term *owner-developer* can be used.

The *architect* designs the project according to established codes and guidelines but leaves most stress and capacity calculations to the *engineer*.[13] Depending on the construction contract, the engineer may work for the architect, or vice versa, or both may work for the developer.

Once there are approved plans, the developer hires *contractors* to do the construction. Usually, the entire construction project is awarded to a *general contractor*. Due to the nature of the construction industry, separate *subcontracts* are used for different tasks (electrical, plumbing, mechanical, framing, fire sprinkler installation, finishing, etc.). The general contractor who hires all of these different *subcontractors* is known as the *prime contractor* (or *prime*). (The subcontractors can also work directly for the owner-developer, although this is less common.) The prime contractor is responsible for all of the acts of the subcontractors and is liable for any damage suffered by the owner-developer due to those acts.

Construction is managed by an agent of the owner-developer known as the *construction manager*, who may be the engineer, the architect, or someone else.

10. STANDARD CONTRACTS FOR DESIGN PROFESSIONALS

Several of the design engineering societies have produced standard agreement forms and other standard documents for design professionals.[14] Among other standard forms, notices, and agreements, the following standard contracts are available.[15]

- standard contract between engineer and client
- standard contract between engineer and architect
- standard contract between engineer and contractor
- standard contract between owner and construction manager

The major advantage of the standard contracts is that the meanings of the clauses are well established, not only among the design professionals and their clients but also in the courts. The clauses in these contracts have already been litigated many times. Where a clause has been found to be unclear or ambiguous, it has been rewritten to accomplish its intended purpose.

11. CONSULTING FEE STRUCTURE

Compensation for consulting engineering services can incorporate one or more of the following concepts.

- *lump-sum fee*: This is a predetermined fee agreed upon by client and engineer. This payment can be used for small projects where the scope of work is clearly defined.
- *cost plus fixed fee*: All costs (labor, material, travel, etc.) incurred by the engineer are paid by the client. The client also pays a predetermined fee as profit. This method has an advantage when the scope of services cannot be determined accurately in advance. Detailed records must be kept by the engineer in order to allocate costs among different clients.
- *per diem fee*: The engineer is paid a specific sum for each day spent on the job. Usually, certain direct expenses (e.g., travel and reproduction) are billed in addition to the per diem rate.
- *salary plus*: The client pays for the employees on an engineer's payroll (the salary) plus an additional percentage to cover indirect overhead and profit plus certain direct expenses.
- *retainer*: This is a minimum amount paid by the client, usually in total and in advance, for a normal amount of work expected during an agreed-upon period. None of the retainer is returned, regardless of how little work the engineer performs. The engineer can be paid for additional work

---

[13]On simple small projects, such as wood-framed residential units, the design may be developed by a *building designer*. The legal capacities of building designers vary from state to state.

[14]There are two main sources of standard forms. The American Consulting Engineers' Council (ACEC), National Society of Professional Engineers (NSPE), and American Society of Civil Engineers (ASCE) have produced one set. Working independently, the American Institute of Architects (AIA) and the Associated General Contractors of America (AGC) have produced another.
[15]The Construction Specifications Institute (CSI) has produced standard specifications for materials.

beyond what is normal, however. Some direct costs, such as travel and reproduction expenses, may be billed directly to the client.

- *percentage of construction cost*: This method, which is widely used in construction design contracts, pays the architect and/or the engineer a percentage of the final total cost of the project. Cost of land, financing, and legal fees are generally not included in the construction cost, and other costs (plan revisions, project management labor, value engineering, etc.) are billed separately.

12. DISCHARGE OF A CONTRACT

A contract is normally discharged when all parties have satisfied their obligations. However, a contract can also be terminated for the following reasons.

- mutual agreement of all parties to the contract

- impossibility of performance (e.g., death of a party to the contract)

- illegality of the contract

- material breach by one or more parties to the contract

- fraud on the part of one or more parties

- failure (i.e., loss or destruction) of consideration (e.g., the burning of a building one party expected to own or occupy upon satisfaction of the obligations)

Some contracts may be dissolved by actions of the court (e.g., bankruptcy), passage of new laws and public acts, or a declaration of war.

Extreme difficulty (including economic hardship) in satisfying the contract does not discharge it, even if it becomes more costly or less profitable than originally anticipated.

13. TORTS

A *tort* is a civil wrong committed by one person causing damage to another person or person's property, emotional well-being, or reputation.[16] It is a breach of the rights of an individual to be secure in person or property. In order to correct the wrong, a civil lawsuit (*tort action* or *civil complaint*) is brought by the alleged injured party (the *plaintiff*) against the *defendant*. To be a valid *tort action* (i.e., lawsuit), there must have been

injury (i.e., damage). Generally, there will be no contract between the two parties, so the tort action cannot claim a breach of contract.[17]

Tort law is concerned with compensation for the injury, not punishment. Therefore, tort awards usually consist of general, compensatory, and special damages and rarely include punitive and exemplary damages. (See Sec. 17 for definitions of these damages.)

14. BREACH OF CONTRACT, NEGLIGENCE, MISREPRESENTATION, AND FRAUD

A *breach of contract* occurs when one of the parties fails to satisfy all of its obligations under a contract. The breach can be *willful* (as in a contractor walking off a construction job) or *unintentional* (as in providing less than adequate quality work or materials). A *material breach* is defined as nonperformance that results in the injured party receiving something substantially less than or different from what the contract intended.

Normally, the only redress that an *injured party* has through the courts in the event of a breach of contract is to force the breaching party to provide *specific performance*—that is, to satisfy all remaining contract provisions and to pay for any damage caused. Normally, *punitive damages* (to punish the breaching party) are unavailable.

Negligence is an action, willful or unwillful, taken without proper care or consideration for safety, resulting in damages to property or injury to persons. "Proper care" is a subjective term, but in general it is the diligence that would be exercised by a reasonably prudent person.[18] Damages sustained by a negligent act are recoverable in a tort action. (See Sec. 13.) If the plaintiff was partially at fault (as in the case of *comparative negligence*), the defendant will be liable only for the portion of the damage caused by the defendant.

Punitive damages are available, however, if the breaching party was fraudulent in obtaining the contract. In addition, the injured party has the right to void (nullify) the contract entirely. A *fraudulent act* is basically a special case of misrepresentation (i.e., an intentionally false statement known to be false at the time it is made). Misrepresentation that does not result in a contract is a tort. When a contract is involved, misrepresentation can be a breach of that contract (i.e., *fraud*).

---

[16]The difference between a *civil tort (lawsuit)* and a *criminal lawsuit* is the alleged injured party. A *crime* is a wrong against society. A criminal lawsuit is brought by the state against a defendant.

[17]It is possible for an injury to be both a breach of contract and a tort. Suppose an owner has an agreement with a contractor to construct a building, and the contract requires the contractor to comply with all state and federal safety regulations. If the owner is subsequently injured on a stairway because there was no guardrail, the injury could be recoverable both as a tort and as a breach of contract. If a third party unrelated to the contract was injured, however, that party could recover only through a tort action.

[18]Negligence of a design professional (e.g., an engineer or architect) is the absence of a *standard of care* (i.e., customary and normal care and attention) that would have been provided by other engineers. It is highly subjective.

Unfortunately, it is extremely difficult to prove *compensatory fraud* (i.e., fraud for which damages are available). Proving fraud requires showing *beyond a reasonable doubt* (a) a reckless or intentional misstatement of a material fact (b) meant to deceive, (c) resulting in misleading the innocent party to contract (d) to the innocent party's detriment.

For example, if an engineer claims to have experience in designing steel buildings but actually has none, the court might consider the misrepresentation a fraudulent action. If, however, the engineer has some experience, but an insufficient amount to do an adequate job, the engineer probably will not be considered to have acted fraudulently.

15. STRICT LIABILITY IN TORT

Strict liability in tort means that the injured party wins if the injury can be proven. It is not necessary to prove negligence, breach of explicit or implicit warranty, or the existence of a contract (*privity of contract*). Strict liability in tort is most commonly encountered in product liability cases. A defect in a product, regardless of how the defect got there, is sufficient to create strict liability in tort.

Case law surrounding defective products has developed and refined the following requirements for winning a strict liability in tort case. The following points must be proved.

- The product was defective in manufacture, design, labeling, and so on.
- The product was defective when used.
- The defect rendered the product unreasonably dangerous.
- The defect caused the injury.
- The specific use of the product that caused the damage was reasonably foreseeable.

16. MANUFACTURING AND DESIGN LIABILITY

Case law makes a distinction between *design professionals* (architects, structural engineers, building designers, etc.) and manufacturers of consumer products. Design professionals are generally consultants whose primary product is a design service sold to sophisticated clients. Consumer product manufacturers produce a specific product line sold through wholesalers and retailers to the unsophisticated public.

The law treats design professionals favorably. Such professionals are expected to meet a *standard of care* and skill that can be measured by comparison with the conduct of other professionals. However, professionals are not expected to be infallible. In the absence of a contract provision to the contrary, design professionals are not held to be guarantors of their work in the strict sense of legal liability. Damages incurred due to design errors are recoverable through tort actions, but proving a breach of contract requires showing negligence (i.e., not meeting the standard of care).

On the other hand, the law is much stricter with consumer product manufacturers, and perfection is (essentially) expected of them. They are held to the standard of strict liability in tort without regard to negligence. A manufacturer is held liable for all phases of the design and manufacturing of a product being marketed to the public.[19]

Prior to 1916, the court's position toward product defects was exemplified by the expression *caveat emptor* ("let the buyer beware").[20] Subsequent court rulings have clarified that "... a manufacturer is strictly liable in tort when an article [it] places on the market, knowing that it will be used without inspection, proves to have a defect that causes injury to a human being."[21]

Although all defectively designed products can be traced back to a design engineer or team, only the manufacturing company is usually held liable for injury caused by the product. This is more a matter of economics than justice. The company has liability insurance; the product design engineer (who is merely an employee of the company) probably does not. Unless the product design or manufacturing process is intentionally defective, or unless the defect is known in advance and covered up, the product design engineer will rarely be punished by the courts.[22]

17. DAMAGES

An injured party can sue for *damages* as well as for specific performance. Damages are the award made by the court for losses incurred by the injured party.

- *General* or *compensatory damages* are awarded to make up for the injury that was sustained.
- *Special damages* are awarded for the direct financial loss due to the breach of contract.

---

[19]The reason for this is that the public is not considered to be as sophisticated as a client who contracts with a design professional for building plans.

[20]1916, *McPherson vs. Buick*. McPherson bought a Buick from a car dealer. The car had a defective wooden steering wheel, and there was evidence that reasonable inspection would have uncovered the defect. The steering wheel injured McPherson, who then sued Buick. Buick defended itself under the ancient *prerequisite of privity* (i.e., the requirement of a face-to-face contractual relationship in order for liability to exist), since the dealer, not Buick, had sold the car to McPherson, and no contract between Buick and McPherson existed. The judge disagreed, thus establishing the concept of *third party liability* (i.e., manufacturers are responsible to consumers even though consumers do not buy directly from manufacturers).

[21]1963, *Greenman vs. Yuba Power Products*. Greenman purchased and was injured by an electric power tool.

[22]Of course, the engineer can expect to be discharged from the company. However, for strategic reasons, this discharge probably will not occur until after the company loses the case.

- *Nominal damages* are awarded when responsibility has been established but the injury is so slight as to be inconsequential.

- *Liquidated damages* are amounts that are specified in the contract document itself for nonperformance.

- *Punitive* or *exemplary damages* are awarded, usually in tort and fraud cases, to punish and make an example of the defendant (i.e., to deter others from doing the same thing).

- *Consequential damages* provide compensation for indirect losses that are incurred by the injured party but are not directly related to the contract.

18. INSURANCE

Most design firms and many independent design professionals carry *errors and omissions insurance* to protect them from claims due to their mistakes. Such policies are costly, and for that reason, some professionals choose to "go bare."[23] Policies protect against inadvertent mistakes only, not against willful, knowing, or conscious efforts to defraud or deceive.

---

[23]Going bare appears foolish at first glance, but there is a perverted logic behind the strategy. One-person consulting firms (and perhaps, firms that are not profitable) are "judgment-proof." Without insurance or other assets, these firms would be unable to pay any large judgments against them. When damage victims (and their lawyers) find this out in advance, they know that judgments will be uncollectable. So, often the lawsuit never makes its way to trial.

88 Engineering Ethics

1. CREEDS, CODES, CANONS, STATUTES, AND RULES

It is generally conceded that an individual acting on his or her own cannot be counted on to always act in a proper and moral manner. Creeds, statutes, rules, and codes all attempt to complete the guidance needed for an engineer to do "...the correct thing."

A *creed* is a statement or oath, often religious in nature, taken or assented to by an individual in ceremonies. For example, the *Engineers' Creed* adopted by the National Society of Professional Engineers is[1]

> I pledge...
>
> ... to give the utmost of performance;
>
> ... to participate in none but honest enterprise;
>
> ... to live and work according to the laws of man and the highest standards of professional conduct;
>
> ... to place service before profit, the honor and standing of the profession before personal advantage, and the public welfare above all other considerations.
>
> In humility and with need for Divine Guidance, I make this pledge.

A *code* is a system of nonstatutory, nonmandatory canons of personal conduct. A *canon* is a fundamental belief that usually encompasses several rules. For example, the code of ethics of the American Society of Civil Engineers (ASCE) contains the following seven canons.

[1]The *Faith of an Engineer* adopted by the Accreditation Board for Engineering and Technology (ABET) is a similar but more detailed creed.

1. Engineers shall hold paramount the safety, health, and welfare of the public in the performance of their professional duties.

2. Engineers shall perform services only in areas of their competence.

3. Engineers shall issue public statements only in an objective and truthful manner.

4. Engineers shall act in professional matters for each employer or client as faithful agents or trustees and shall avoid conflicts of interest.

5. Engineers shall build their professional reputation on the merit of their service and shall not compete unfairly with others.

6. Engineers shall act in such a manner as to uphold and enhance the honor, integrity, and dignity of the engineering profession.

7. Engineers shall continue their professional development throughout their careers and shall provide opportunities for the professional development of those engineers under their supervision.

A *rule* is a guide (principle, standard, or norm) for conduct and action in a certain situation. A *statutory rule* is enacted by the legislative branch of state or federal government and carries the weight of law. Some U.S. engineering registration boards have statutory *rules of professional conduct*.

2. PURPOSE OF A CODE OF ETHICS

Many different sets of *codes of ethics* (*canons of ethics, rules of professional conduct*, etc.) have been produced by various engineering societies, registration boards, and other organizations.[2] The purpose of these ethical guidelines is to guide the conduct and decision making of engineers. Most codes are primarily educational. Nevertheless, from time to time they have been used

[2]All of the major engineering technical and professional societies in the United States (ASCE, IEEE, ASME, AIChE, NSPE, etc.) and throughout the world have adopted codes of ethics. Most U.S. societies have endorsed the *Code of Ethics of Engineers* developed by the Accreditation Board for Engineering and Technology (ABET), formerly the Engineers' Council for Professional Development (ECPD). The National Council of Examiners for Engineering and Surveying (NCEES) has developed its *Model Rules of Professional Conduct* as a guide for state registration boards in developing guidelines for the professional engineers in those states.

by the societies and regulatory agencies as the basis for disciplinary actions.

Fundamental to ethical codes is the requirement that engineers render faithful, honest, professional service. In providing such service, engineers must represent the interests of their employers or clients and, at the same time, protect public health, safety, and welfare.

There is an important distinction between what is legal and what is ethical. Many legal actions can be violations of codes of ethical or professional behavior.[3] For example, an engineer's contract with a client may give the engineer the right to assign the engineer's responsibilities, but doing so without informing the client would be unethical.

Ethical guidelines can be categorized on the basis of who is affected by the engineer's actions—the client, vendors and suppliers, other engineers, or the public at large.[4]

3. ETHICAL PRIORITIES

There are frequently conflicting demands on engineers. While it is impossible to use a single decision-making process to solve every ethical dilemma, it is clear that ethical considerations will force engineers to subjugate their own self-interests. Specifically, the ethics of engineers dealing with others need to be considered in the following order from highest to lowest priority.

- society and the public
- the law
- the engineering profession
- the engineer's client
- the engineer's firm
- other involved engineers
- the engineer personally

4. DEALING WITH CLIENTS AND EMPLOYERS

The most common ethical guidelines affecting engineers' interactions with their employer (the *client*) can be summarized as follows.[5]

- Engineers should not accept assignments for which they do not have the skill, knowledge, or time.

---

[3]Whether the guidelines emphasize ethical behavior or professional conduct is a matter of wording. The intention is the same: to provide guidelines that transcend the requirements of the law.
[4]Some authorities also include ethical guidelines for dealing with the employees of an engineer. However, these guidelines are no different for an engineering employer than they are for a supermarket, automobile assembly line, or airline employer. Ethics is not a unique issue when it comes to employees.
[5]These general guidelines contain references to contractors, plans, specifications, and contract documents. This language is common, though not unique, to the situation of an engineer supplying design services to an owner-developer or architect. However, most of the ethical guidelines are general enough to apply to engineers in industry as well.

- Engineers must recognize their own limitations. They should use associates and other experts when the design requirements exceed their abilities.

- The client's interests must be protected. The extent of this protection exceeds normal business relationships and transcends the legal requirements of the engineer-client contract.

- Engineers must not be bound by what the client wants in instances where such desires would be unsuccessful, dishonest, unethical, unhealthy, or unsafe.

- Confidential client information remains the property of the client and must be kept confidential.

- Engineers must avoid conflicts of interest and should inform the client of any business connections or interests that might influence their judgment. Engineers should also avoid the *appearance* of a conflict of interest when such an appearance would be detrimental to the profession, their client, or themselves.

- The engineers' sole source of income for a particular project should be the fee paid by their client. Engineers should not accept compensation in any form from more than one party for the same services.

- If the client rejects the engineer's recommendations, the engineer should fully explain the consequences to the client.

- Engineers must freely and openly admit to the client any errors made.

All courts of law have required an engineer to perform in a manner consistent with normal professional standards. This is not the same as saying an engineer's work must be error-free. If an engineer completes a design, has the design and calculations checked by another competent engineer, and an error is subsequently shown to have been made, the engineer may be held responsible, but will probably not be considered negligent.

5. DEALING WITH SUPPLIERS

Engineers routinely deal with manufacturers, contractors, and vendors (*suppliers*). In this regard, engineers have great responsibility and influence. Such a relationship requires that engineers deal justly with both clients and suppliers.

An engineer will often have an interest in maintaining good relationships with suppliers since this often leads to future work. Nevertheless, relationships with suppliers must remain highly ethical. Suppliers should not be encouraged to feel that they have any special favors coming to them because of a long-standing relationship with the engineer.

The ethical responsibilities relating to suppliers are listed as follows.

- The engineer must not accept or solicit gifts or other valuable considerations from a supplier during, prior to, or after any job. An engineer should not accept discounts, allowances, commissions, or any other indirect compensation from suppliers, contractors, or other engineers in connection with any work or recommendations.

- The engineer must enforce the plans and specifications (i.e., the *contract documents*) but must also interpret the contract documents fairly.

- Plans and specifications developed by the engineer on behalf of the client must be complete, definite, and specific.

- Suppliers should not be required to spend time or furnish materials that are not called for in the plans and contract documents.

- The engineer should not unduly delay the performance of suppliers.

6. DEALING WITH OTHER ENGINEERS

Engineers should try to protect the engineering profession as a whole, to strengthen it, and to enhance its public stature. The following ethical guidelines apply.

- An engineer should not attempt to maliciously injure the professional reputation, business practice, or employment position of another engineer. However, if there is proof that another engineer has acted unethically or illegally, the engineer should advise the proper authority.

- An engineer should not review someone else's work while the other engineer is still employed unless the other engineer is made aware of the review.

- An engineer should not try to replace another engineer once the other engineer has received employment.

- An engineer should not use the advantages of a salaried position to compete unfairly (i.e., moonlight) with other engineers who have to charge more for the same consulting services.

- Subject to legal and proprietary restraints, an engineer should freely report, publish, and distribute information that would be useful to other engineers.

7. DEALING WITH (AND AFFECTING) THE PUBLIC

In regard to the social consequences of engineering, the relationship between an engineer and the public is essentially straightforward. Responsibilities to the public demand that the engineer place service to humankind

above personal gain. Furthermore, proper ethical behavior requires that an engineer avoid association with projects that are contrary to public health and welfare or that are of questionable legal character.

- Engineers must consider the safety, health, and welfare of the public in all work performed.

- Engineers must uphold the honor and dignity of their profession by refraining from self-laudatory advertising, by explaining (when required) their work to the public, and by expressing opinions only in areas of knowledge.

- When engineers issue a public statement, they must clearly indicate if the statement is being made on anyone's behalf (i.e., if anyone is benefitting from their position).

- Engineers must keep their skills at a state-of-the-art level.

- Engineers should develop public knowledge and appreciation of the engineering profession and its achievements.

- Engineers must notify the proper authorities when decisions adversely affecting public safety and welfare are made.[6]

8. COMPETITIVE BIDDING

The ethical guidelines for dealing with other engineers presented here and in more detailed codes of ethics no longer include a prohibition on *competitive bidding*. Until 1971, most codes of ethics for engineers considered competitive bidding detrimental to public welfare, since cost cutting normally results in a lower quality design.

However, in a 1971 case against the National Society of Professional Engineers that went all the way to the U.S. Supreme Court, the prohibition against competitive bidding was determined to be a violation of the Sherman Antitrust Act (i.e., it was an unreasonable restraint of trade).

The opinion of the Supreme Court does not *require* competitive bidding—it merely forbids a prohibition against competitive bidding in NSPE's code of ethics. The following points must be considered.

- Engineers and design firms may individually continue to refuse to bid competitively on engineering services.

- Clients are not required to seek competitive bids for design services.

- Federal, state, and local statutes governing the procedures for procuring engineering design services, even those statutes that prohibit competitive bidding, are not affected.

---

[6] This practice has come to be known as *whistle-blowing*.

- Any prohibitions against competitive bidding in individual state engineering registration laws remain unaffected.

- Engineers and their societies may actively and aggressively lobby for legislation that would prohibit competitive bidding for design services by public agencies.

89 Engineering Licensing in the United States

1. WHAT LICENSING IS

Engineering licensing (also known as *engineering registration*) in the United States is an examination process by which a state's *board of engineering licensing* (typically referred to as the "engineers' board" or "board of registration") determines and certifies that an engineer has achieved a minimum level of competence.[1] This process is intended to protect the public by preventing unqualified individuals from offering engineering services.

Most engineers in the United States do not need to be licensed.[2] In particular, most engineers who work for companies that design and manufacture products are exempt from the licensing requirement. This is known as the *industrial exemption*, something that is built into the laws of most states.[3]

Nevertheless, there are many good reasons for wanting to become a licensed engineer. For example, you cannot offer consulting engineering services in any state unless you are licensed in that state. Even within a product-oriented corporation, you may find that employment, advancement, and managerial positions are limited to licensed engineers.

Once you have met the licensing requirements, you will be allowed to use the titles *Professional Engineer* (PE), *Registered Engineer* (RE), or *Consulting Engineer* (CE) as permitted by your state.

---

[1]Licensing of engineers is not unique to the United States. However, the practice of requiring a degreed engineer to take an examination is not common in other countries. Licensing in many countries requires a degree and may also require experience, references, and demonstrated knowledge of ethics and law, but no technical examination.

[2]Less than one-third of the degreed engineers in the United States are licensed.

[3]Only one or two states have abolished the industrial exemption. There has always been a lot of "talk" among engineers about abolishing it, but there has been little success in actually trying to do so. One of the reasons is that manufacturers' lobbies are very strong.

Although the licensing process is similar in each of the 50 states, each has its own licensing law. Unless you offer consulting engineering services in more than one state, however, you will not need to be licensed in the other states.

2. THE U.S. LICENSING PROCEDURE

The licensing procedure is similar in all states. You will take two eight-hour written examinations. The full process requires you to complete two applications, one for each of the two examinations. The first examination is the *Fundamentals of Engineering* (FE) *examination*, formerly known (and still commonly referred to) as the *Engineer-In-Training* (E-I-T) *examination*.[4] This examination covers basic subjects from all of the mathematics, physics, chemistry, and engineering courses you took during your university years.

The second examination is the *Professional Engineering* (PE) *examination*, also known as the *Principles and Practices* (P&P) *examination*. This examination covers only the subjects in your engineering discipline (e.g., civil, mechanical, electrical, and others).

The actual details of licensing qualifications, experience requirements, minimum education levels, fees, oral interviews, and examination schedules vary from state to state. Contact your state's licensing board for more information.

3. NATIONAL COUNCIL OF EXAMINERS FOR ENGINEERING AND SURVEYING

The *National Council of Examiners for Engineering and Surveying* (NCEES) in Clemson, South Carolina, writes, prints, distributes, and scores the national FE and PE examination.[5] The individual states purchase the examinations from NCEES and administer them. NCEES does not distribute applications to take the examinations, administer the examinations or appeals, or notify examinees of the results. These tasks are all performed by the individual states.

---

[4]The terms *engineering intern* (EI) and *intern engineer* (IE) have also been used in the past to designate the status of an engineer who has passed the first exam. These uses are rarer but may still be encountered in some states.

[5]National Council of Examiners for Engineering and Surveying, P.O. Box 1686, Clemson, SC 29633, (803) 654-6824.

4. RECIPROCITY AMONG STATES

With minor exceptions, having a license from one state will not permit you to practice engineering in another state. You must have a professional engineering license from each state in which you work. For most engineers, this is not a problem, but for some it is. Luckily, it is not too difficult to get a license from every state you work in once you have a license from one of them.

All states use the NCEES examinations. If you take and pass the FE or PE examination in one state, your certificate or license will be honored by all of the other states. Upon proper application, payment of fees, and proof of your license, you will be issued a license by the new state. Although there may be other special requirements imposed by a state, it will not be necessary to retake the FE or PE examinations.[6] The issuance of an engineering license based on another state's licensing is known as *reciprocity* or *comity*.

5. UNIFORM EXAMINATIONS

Although each state has its own licensing law and is, theoretically, free to administer its own exams, none does so for the major disciplines. All states have chosen to use the NCEES exams. Each state administers the exams on the same days. The exams from all the states are sent to NCEES and are graded by the same graders. Each state adopts the cut-off passing scores recommended by NCEES. These practices have led to the term *uniform examination*.

6. APPLYING FOR THE EXAMINATION

While the exam administration process is essentially standardized among the states, there is a lot of variation in the application process. Each state has its own application forms, charges different fees, and has different age, education, and experience requirements. Therefore, you will have to request an application from the state in which you want to become licensed in order to find out what is required. It is generally sufficient for you to phone for this application; a written request is unnecessary. You'll find contact information (websites, telephone numbers, email addresses, etc.) for all U.S. state and territorial boards of registration at **www.ppi2pass.com**. Click on the State Boards link. Some states have a fee for the application.

As with any other important document, it is a good idea to keep a copy of your examination application and send the original application by certified mail, requesting a receipt of delivery. Keep your proof of mailing and delivery receipt with your copy of the application.

All states make special accommodations for persons who are physically challenged or who have other special needs. Be sure to communicate your need to the state board well in advance of the examination day.

7. EXAMINATION DATES

The national FE and PE examinations are administered twice a year (usually in mid-April and late October), on the same weekends in all states. Check **www.ppi2pass.com** for a current exam schedule. Click on the Exam FAQs link.

8. FE EXAMINATION FORMAT

The FE examination consists of two four-hour sessions, separated by a one-hour lunch period. There are 120 multiple-choice problems in the morning and 60 multiple-choice problems in the afternoon. All problems have four answer choices, from which you are to choose the best single answer. Afternoon problems are slightly more difficult and have double weight.

Questions from all undergraduate technical courses appear in the examination, including mathematics, chemistry, physics, and engineering.[7]

The FE exam is essentially all in SI units. There are a few non-SI problems and some opportunities to choose between identical SI and non-SI problems.

NCEES provides its own reference handbook for use in the examination. You may bring your own calculator and pencils, but you are not allowed to use any of your own books or notes.

9. PE EXAMINATION FORMAT

The NCEES PE examination consists of two four-hour sessions, separated by a one-hour lunch period. It covers subjects in your major field of study only (e.g., civil engineering). There are several variations in PE exam format, but the most common type has 40 questions each session.

All questions are *objectively scored*. This means that each question is graded by computer. There is no penalty for guessing or wrong answers, but there is also no partial credit for correct methods and assumptions.

Unlike the FE exam where SI units are used extensively, virtually all of the questions on the PE exam are in customary U.S. units. There are a few exceptions (e.g., environmental topics), but most problems use pounds, feet, seconds, gallons, and British thermal units.

The PE examination is open book. While some states have a few restrictions, generally you may bring in the calculators and books of your choice.

[6]For example, California requires all civil engineering applicants to pass special examinations in seismic design and surveying in addition to their regular eight-hour PE exams. Licensed engineers from other states only have to pass these two special exams. They do not need to retake the PE exam.

[7]The format of the FE exam, other valuable information, and tips on passing are given in greater detail in *FE Review Manual*, (Lindeburg), and *Engineer-In-Training Reference Manual* (Lindeburg), both published by Professional Publications.

Topic IX: Support Material

Appendices

Glossary

APPENDIX 1.A
Conversion Factors

multiply	by	to obtain	multiply	by	to obtain
acres	0.4047	hectares	foot-pounds	1.285×10^{-3}	Btu
	43,560.0	square feet		5.051×10^{-7}	horsepower-hours
	1.5625×10^{-3}	square miles		3.766×10^{-7}	kilowatt-hours
ampere-hours	3600.0	coulombs	foot-pound/sec	4.6272	Btu/hr
angstrom units	3.937×10^{-9}	inches		1.818×10^{-3}	horsepower
	1×10^{-4}	microns		1.356×10^{-3}	kilowatts
astronomical units	1.496×10^{8}	kilometers	furlongs	660.0	feet
atmospheres	76.0	centimeters of		0.125	miles (statute)
		mercury	gallons	0.1337	cubic feet
atomic mass unit	9.3146×10^{8}	electron-volts		3.785	liters
	1.492×10^{-10}	joules	gallons H_2O	8.3453	pounds H_2O
	1.66×10^{-27}	kilograms	gallons/min	8.0208	cubic feet/hr
BeV (also GeV)	1×10^{9}	electron-volts		0.002228	cubic feet/sec
Btu	3.93×10^{-4}	horsepower-hours	GeV (also BeV)	1×10^{9}	electron-volts
	778.3	foot-pounds	grams	1×10^{-3}	kilograms
	2.93×10^{-4}	kilowatt hours		3.527×10^{-2}	ounces (avoirdupois)
	1.0×10^{-5}	therms		3.215×10^{-2}	ounces (troy)
Btu/hr	0.2161	foot-pounds/sec		2.205×10^{-3}	pounds
	3.929×10^{-4}	horsepower	hectares	2.471	acres
	0.293	watts		1.076×10^{5}	square feet
bushels	2150.4	cubic inches	horsepower	2545.0	Btu/hr
calories, gram (mean)	3.9683×10^{-3}	Btu (mean)		42.44	Btu/min
centares	1.0	square meters		550	foot-pounds/sec
centimeters	1×10^{-5}	kilometers		0.7457	kilowatts
	1×10^{-2}	meters		745.7	watts
	10.0	millimeters	horsepower-hours	2545.0	Btu
	3.281×10^{-2}	feet		1.976×10^{-6}	foot-pounds
	0.3937	inches		0.7457	kilowatt-hours
chains	792.0	inches	hours	4.167×10^{-2}	days
coulombs	1.036×10^{-5}	faradays		5.952×10^{-3}	weeks
cubic centimeters	0.06102	cubic inches	inches	2.540	centimeters
	2.113×10^{-3}	pints (U.S. liquid)		1.578×10^{-5}	miles
cubic feet	0.02832	cubic meters	inches, H_2O	5.199	pounds force/ft^2
	7.4805	gallons		0.0361	psi
cubic feet/min	62.43	pounds H_2O/min		0.0735	inches, mercury
cubic feet/sec	448.831	gallons/min	inches, mercury	70.7	pounds force/ft^2
	0.64632	millions of gallons		0.491	pounds force/in^2
		per day		13.60	inches, H_2O
cubits	18.0	inches	joules	6.705×10^{9}	atomic mass units
days	86,400.0	seconds		9.480×10^{-4}	Btu
degrees (angle)	1.745×10^{-2}	radians		1×10^{7}	ergs
degrees/sec	0.1667	revolutions/min		6.242×10^{18}	electron-volts
dynes	1×10^{-5}	newtons		1.113×10^{-17}	kilograms
electron-volts	1.074×10^{-9}	atomic mass units	kilograms	6.025×10^{26}	atomic mass units
	1×10^{-9}	BeV (also GeV)		5.610×10^{35}	electron-volts
	1.602×10^{-19}	joules		8.987×10^{16}	joules
	1.783×10^{-36}	kilograms		2.205	pounds
	1×10^{-6}	MeV	kilometers	3281.0	feet
faradays/sec	96,500	amperes (absolute)		1000.0	meters
fathoms	6.0	feet		0.6214	miles
feet	30.48	centimeters	kilometers/hr	0.5396	knots
	0.3048	meters	kilowatts	3412.9	Btu/hr
	1.645×10^{-4}	miles (nautical)		737.6	foot-pounds/sec
	1.894×10^{-4}	miles (statute)		1.341	horsepower
feet/min	0.5080	centimeters/sec	kilowatt-hours	3413.0	Btu
feet/sec	0.592	knots			*(continued)*
	0.6818	miles/hr			

APPENDIX 1.A *(continued)*
Conversion Factors

multiply	by	to obtain
knots	6076.0	feet/hr
	1.0	nautical miles/hr
	1.151	statute miles/hr
light years	5.9×10^{12}	miles
links (surveyor)	7.92	inches
liters	1000.0	cubic centimeters
	61.02	cubic inches
	0.2642	gallons (U.S. liquid)
	1000.0	milliliters
	2.113	pints
MeV	1×10^6	electron-volts
meters	100.0	centimeters
	3.281	feet
	1×10^{-3}	kilometers
	5.396×10^{-4}	miles (nautical)
	6.214×10^{-4}	miles (statute)
	1000.0	millimeters
microns	1×10^{-6}	meters
miles (nautical)	6076	feet
	1.853	kilometers
	1.1516	miles (statute)
miles (statute)	5280.0	feet
	1.609	kilometers
	0.8684	miles (nautical)
miles/hr	88.0	feet/min
milligrams/liter	1.0	parts/million
milliliters	1×10^{-3}	liters
millimeters	3.937×10^{-2}	inches
newtons	1×10^5	dynes
ohms (international)	1.0005	ohms (absolute)
ounces	28.349527	grams
	6.25×10^{-2}	pounds
ounces (troy)	1.09714	ounces (avoirdupois)
parsecs	3.086×10^{13}	kilometers
	1.9×10^{13}	miles
pascal-sec	1000	centipoise
	10	poise
	0.02089	pound force-sec/ft^2
	0.6720	pound mass/ft-sec
	0.02089	slug/ft-sec
pints (liquid)	473.2	cubic centimeters
	28.87	cubic inches
	0.125	gallons
	0.5	quarts (liquid)

multiply	by	to obtain
poise	0.002089	pound-sec/ft^2
pounds	0.4536	kilograms
	16.0	ounces
	14.5833	ounces (troy)
	1.21528	pounds (troy)
pounds/ft^2	0.006944	pounds/in^2
pounds/in^2	2.308	feet, H_2O
	27.7	inches, H_2O
	2.037	inches, mercury
	144	pounds/ft^2
quarts (dry)	67.20	cubic inches
quarts (liquid)	57.75	cubic inches
	0.25	gallons
	0.9463	liters
radians	57.30	degrees
	3438.0	minutes
revolutions	360.0	degrees
revolutions/min	6.0	degrees/sec
rods	16.5	feet
	5.029	meters
rods (surveyor)	5.5	yards
seconds	1.667×10^{-2}	minutes
square meters/sec	1×10^6	centistokes
	10.76	square feet/sec
	1×10^4	stokes
slugs	32.174	pounds mass
stokes	0.0010764	square feet/sec
tons (long)	1016.0	kilograms
	2240.0	pounds
	1.120	tons (short)
tons (short)	907.1848	kilograms
	2000.0	pounds
	0.89287	tons (long)
volts (absolute)	3.336×10^{-3}	statvolts
watts	3.4129	Btu/hr
	1.341×10^{-3}	horsepower
yards	0.9144	meters
	4.934×10^{-4}	miles (nautical)
	5.682×10^{-4}	miles (statute)

APPENDIX 1.B
Common SI Unit Conversion Factors

multiply	by	to obtain
AREA		
circular mil	506.7	square micrometer
square foot	0.0929	square meter
square kilometer	0.3861	square mile
square meter	10.764	square foot
	1.196	square yard
square micrometer	0.001974	circular mil
square mile	2.590	square kilometer
square yard	0.8361	square meter
ENERGY		
Btu (international)	1.0551	kilojoule
erg	0.1	microjoule
foot-pound	1.3558	joule
horsepower-hour	2.6485	megajoule
joule	0.7376	foot-pound
	0.10197	meter-kilogram force
kilogram-calorie (international)	4.1868	kilojoule
kilojoule	0.9478	Btu
	0.2388	kilogram-calorie
kilowatt-hour	3.6	megajoule
megajoule	0.3725	horsepower-hour
	0.2778	kilowatt-hour
	0.009478	therm
meter-kilogram force	9.8067	joule
microjoule	10.0	erg
therm	105.506	megajoule
FORCE		
dyne	10.0	micronewton
kilogram force	9.8067	newton
kip	4448.2	newton
micronewton	0.1	dyne
newton	0.10197	kilogram force
	0.0002248	kip
	3.597	ounce force
	0.2248	pound force
ounce force	0.2780	newton
pound force	4.4482	newton
HEAT		
Btu/ft^2-hr	3.1546	$watt/m^2$
Btu/hr-ft^2-$°F$	5.6783	$watt/m^2 \cdot °C$
Btu/ft^3	0.0373	$megajoule/m^3$
Btu/ft^3-$°F$	0.06707	$megajoule/m^3 \cdot °C$
Btu/hr	0.2931	watt
Btu/lbm	2326	joule/kg
Btu/lbm-$°F$	4186.8	$joule/kg \cdot °C$
Btu-in/ft^2-hr-$°F$	0.1442	$watt/m \cdot °C$
joule/kg	0.000430	Btu/lbm
$joule/kg \cdot °C$	0.0002388	Btu/lbm-$°F$
$megajoule/m^3$	26.839	Btu/ft^3
$megajoule/m^3 \cdot °C$	14.911	Btu/ft^3-$°F$
watt	3.4121	Btu/hr
$watt/m \cdot °C$	6.933	Btu-in/ft^2-hr-$°F$
$watt/m^2$	0.3170	Btu/ft^2-hr
$watt/m^2 \cdot °C$	0.1761	Btu/hr-ft^2-$°F$

(*continued*)

APPENDIX 1.B *(continued)*
Common SI Unit Conversion Factors

multiply	by	to obtain
LENGTH		
angstrom	0.1	nanometer
foot	0.3048	meter
inch	25.4	millimeter
kilometer	0.6214	mile
	0.540	mile (nautical)
meter	3.2808	foot
	1.0936	yard
micrometer	1.0	micron
micron	1.0	micrometer
mil	0.0254	millimeter
mile	1.6093	kilometer
mile (nautical)	1.852	kilometer
millimeter	0.0394	inch
	39.370	mil
nanometer	10.0	angstrom
yard	0.9144	meter
MASS (weight)		
grain	64.799	milligram
gram	0.0353	ounce (avoirdupois)
	0.03215	ounce (troy)
kilogram	2.2046	pound mass
	0.068522	slug
	0.0009842	ton (long—2240 lbm)
	0.001102	ton (short—2000 lbm)
milligram	0.0154	grain
ounce (avoirdupois)	28.350	gram
ounce (troy)	31.1035	gram
pound mass	0.4536	kilogram
slug	14.5939	kilogram
ton (long—2240 lbm)	1016.047	kilogram
ton (short—2000 lbm)	907.185	kilogram
PRESSURE		
bar	100.0	kilopascal
inch, H_2O ($20°C$)	0.2486	kilopascal
inch, Hg ($20°C$)	3.3741	kilopascal
kilogram force/cm^2	98.067	kilopascal
kilopascal	0.01	bar
	4.0219	inch, H_2O ($20°C$)
	0.2964	inch, Hg ($20°C$)
	0.0102	kilogram force/cm^2
	7.528	millimeter Hg ($20°C$)
	0.1450	pound force/in^2
	0.009869	standard atmosphere (760 torr)
	7.5006	torr
millimeter Hg ($20°C$)	0.13332	kilopascal
pound force/in^2	6.8948	kilopascal
standard atmosphere (760 torr)	101.325	kilopascal
torr	0.13332	kilopascal

(continued)

APPENDIX 1.B *(continued)*
Common SI Unit Conversion Factors

multiply	by	to obtain
POWER		
Btu (international)/hr	0.2931	watt
foot-pound/sec	1.3558	watt
horsepower	0.7457	kilowatt
kilowatt	1.341	horsepower
	0.2843	ton of refrigeration
meter·kilogram force/sec	9.8067	watt
ton of refrigeration	3.517	kilowatt
watt	3.4122	Btu (international)/hr
	0.7376	foot-pound/sec
	0.10197	meter·kilogram force/sec
TEMPERATURE		
Celsius	$\frac{9}{5}°C + 32$	Fahrenheit
Fahrenheit	$\frac{5}{9}(°F - 32)$	Celsius
Kelvin	$\frac{9}{5}$	Rankine
Rankine	$\frac{5}{9}$	Kelvin
TORQUE		
gram force·centimeter	0.098067	millinewton·meter
kilogram force·meter	9.8067	newton·meter
millinewton·meter	10.197	gram force·centimeter
newton·meter	0.10197	kilogram force·meter
	0.7376	foot·pound
	8.8495	inch·pound
foot-pound	1.3558	newton·meter
inch-pound	0.1130	newton·meter
VELOCITY		
feet/sec	0.3048	meters/sec
kilometers/hr	0.6214	miles/hr
meters/sec	3.2808	feet/sec
	2.2369	miles/hr
miles/hr	1.60934	kilometers/hr
	0.44704	meters/sec
VISCOSITY		
centipoise	1.0	millipascal·second
centistoke	1.0	millimeter^2/sec
millipascal·second	1.0	centipoise
millimeter^2/sec	1.0	centistoke

(continued)

APPENDIX 1.B *(continued)*
Common SI Unit Conversion Factors

multiply	by	to obtain
VOLUME (capacity)		
cubic centimeter	0.06102	cubic inch
cubic foot	28.3168	liter
cubic inch	16.3871	cubic centimeter
cubic meter	1.308	cubic yard
cubic yard	0.7646	cubic meter
gallon (U.S.)	3.785	liter
liter	0.2642	gallon (U.S.)
	2.113	pint (U.S. fluid)
	1.0567	quart (U.S. fluid)
	0.03531	cubic foot
milliliter	0.0338	ounce (U.S. fluid)
ounce (U.S. fluid)	29.574	milliliter
pint (U.S. fluid)	0.4732	liter
quart (U.S. fluid)	0.9464	liter
VOLUME FLOW (gas–air)		
cubic meter/sec	2119	cubic foot/min
liter/sec	2.119	cubic foot/min
microliter/sec	0.000127	cubic foot/hr
milliliter/sec	0.002119	cubic foot/min
	0.127133	cubic foot/hr
cubic foot/min	0.0004719	cubic meter/sec
	0.4719	liter/sec
	471.947	milliliter/sec
cubic foot/hr	7866	microliter/sec
	7.8658	milliliter/sec
VOLUME FLOW (liquid)		
gallon/hr (U.S.)	0.001052	liter/sec
gallon/min (U.S.)	0.06309	liter/sec
liter/sec	951.02	gallon/hr (U.S.)
	15.850	gallon/min (U.S.)

APPENDIX 7.A
Mensuration of Two-Dimensional Areas

Nomenclature
A total surface area
b base
c chord length
d distance
h height
L length
p perimeter
r radius
s side (edge) length, arc length
θ vertex angle, in radians
ϕ central angle, in radians

Triangle

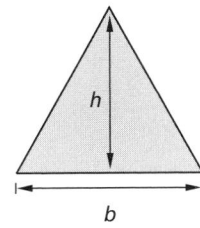 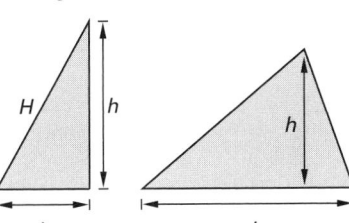

equilateral   right   oblique

$A = \frac{1}{2}bh = \frac{\sqrt{3}}{4}b^2$ $A = \frac{1}{2}bh$ $A = \frac{1}{2}bh$

$h = \frac{\sqrt{3}}{2}b$ $H^2 = b^2 + h^2$

Circle

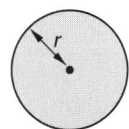

$p = 2\pi r$

$A = \pi r^2 = \frac{p^2}{4\pi}$

Circular Segment

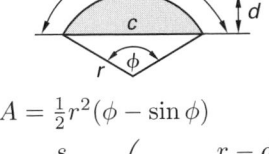

$A = \frac{1}{2}r^2(\phi - \sin\phi)$

$\phi = \frac{s}{r} = 2\left(\arccos\frac{r-d}{r}\right)$

$c = 2r\sin\left(\frac{\phi}{2}\right)$

Circular Sector

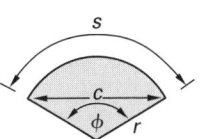

$A = \frac{1}{2}\phi r^2 = \frac{1}{2}sr$

$\phi = \frac{s}{r}$

$s = r\phi$

$c = 2r\sin\left(\frac{\phi}{2}\right)$

Parabola

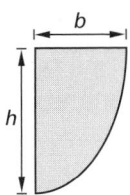

$A = \frac{2}{3}bh$

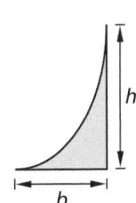

$A = \frac{1}{3}bh$

Ellipse

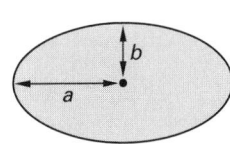

$A = \pi ab$

$p \approx 2\pi\sqrt{\frac{1}{2}(a^2 + b^2)}$ $\begin{bmatrix} \text{Euler's} \\ \text{upper bound} \end{bmatrix}$

(continued)

APPENDIX 7.A *(continued)*
Mensuration of Two-Dimensional Areas

Trapezoid

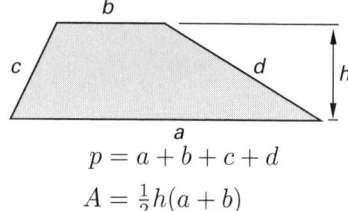

$$p = a + b + c + d$$

$$A = \tfrac{1}{2}h(a + b)$$

The trapezoid is isosceles if $c = d$.

Parallelogram

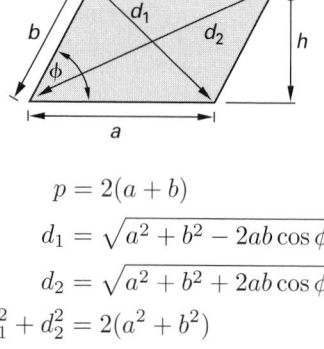

$$p = 2(a + b)$$

$$d_1 = \sqrt{a^2 + b^2 - 2ab\cos\phi}$$

$$d_2 = \sqrt{a^2 + b^2 + 2ab\cos\phi}$$

$$d_1^2 + d_2^2 = 2(a^2 + b^2)$$

$$A = ah = ab\sin\phi$$

If $a = b$, the parallelogram is a rhombus.

Regular Polygon (*n* equal sides)

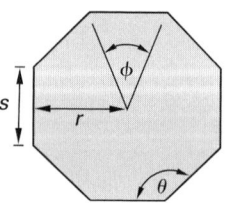

$$\phi = \frac{2\pi}{n}$$

$$\theta = \frac{\pi(n - 2)}{n} = \pi - \phi$$

$$p = ns$$

$$s = 2r\tan\left(\frac{\theta}{2}\right)$$

$$A = \tfrac{1}{2}nsr$$

sides	name	area (A) when diameter of inscribed circle = 1	area (A) when side = 1	radius (r) of circumscribed circle when side = 1	length (L) of side when radius (r) of circumscribed circle = 1	length (L) of side when perpendicular to center = 1	perpendicular (p) to center when side = 1
3	triangle	1.299	0.433	0.577	1.732	3.464	0.289
4	square	1.000	1.000	0.707	1.414	2.000	0.500
5	pentagon	0.908	1.720	0.851	1.176	1.453	0.688
6	hexagon	0.866	2.598	1.000	1.000	1.155	0.866
7	heptagon	0.843	3.634	1.152	0.868	0.963	1.038
8	octagon	0.828	4.828	1.307	0.765	0.828	1.207
9	nonagon	0.819	6.182	1.462	0.684	0.728	1.374
10	decagon	0.812	7.694	1.618	0.618	0.650	1.539
11	undecagon	0.807	9.366	1.775	0.563	0.587	1.703
12	dodecagon	0.804	11.196	1.932	0.518	0.536	1.866

regular polygons

APPENDIX 7.B
Mensuration of Three-Dimensional Volumes

Nomenclature
A area
b base
h height
r radius
R radius
s side (edge) length
V volume

Sphere

$$V = \frac{4\pi r^3}{3}$$
$$A = 4\pi r^2$$

Right Circular Cone

$$V = \frac{\pi r^2 h}{3}$$
$$A = \pi r \sqrt{r^2 + h^2}$$

(does not include base area)

Right Circular Cylinder

$$V = \pi r^2 h$$
$$A = 2\pi r h$$

(does not include end area)

Spherical Segment (Spherical Cap)

Surface area of a spherical segment of radius r cut out by an angle θ_0 rotated from the center about a radius, r, is

$$A = 2\pi r^2 \left(1 - \cos\theta_0\right)$$
$$\omega = \frac{A}{r^2} = 2\pi \left(1 - \cos\theta_0\right)$$

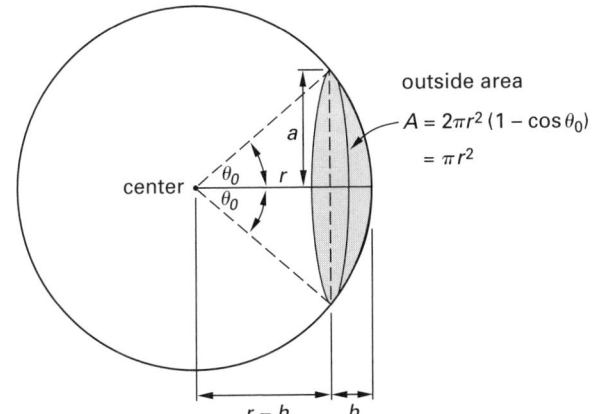

outside area

$A = 2\pi r^2 \left(1 - \cos\theta_0\right)$

$= \pi r^2$

center

$$V_{\text{cap}} = \frac{\pi}{6}h(3a^2 + h^2)$$
$$= \frac{\pi}{3}h^2(3r - h)$$
$$a = \sqrt{h(2r - h)}$$

Paraboloid of Revolution

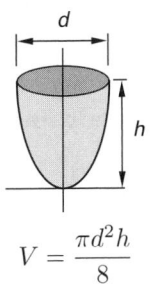

$$V = \frac{\pi d^2 h}{8}$$

Torus

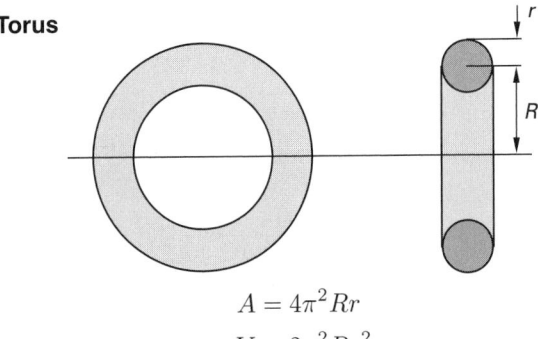

$$A = 4\pi^2 R r$$
$$V = 2\pi^2 R r^2$$

Regular Polyhedra (identical faces)

name	number of faces	form of faces	total surface area	volume
tetrahedron	4	equilateral triangle	$1.7321s^2$	$0.1179s^3$
cube	6	square	$6.0000s^2$	$1.0000s^3$
octahedron	8	equilateral triangle	$3.4641s^2$	$0.4714s^3$
dodecahedron	12	regular pentagon	$20.6457s^2$	$7.6631s^3$
isosahedron	20	equilateral triangle	$8.6603s^2$	$2.1817s^3$

The radius of a sphere inscribed within a regular polyhedron is

$$r = \frac{3V_{\text{polyhedron}}}{A_{\text{polyhedron}}}$$

APPENDIX 9.A
Abbreviated Table of Indefinite Integrals
(In each case, add a constant of integration.
All angles are measured in radians.)

General Formulas

1. $\int dx = x$
2. $\int c\,dx = c \int dx$
3. $\int (dx + dy) = \int dx + \int dy$
4. $\int u\,dv = uv - \int v\,du$ (integration by parts)

Algebraic Forms

5. $\int x^n\,dx = \dfrac{x^{n+1}}{n+1} \qquad [n \neq -1]$

6. $\int x^{-1}dx = \int \dfrac{dx}{x} = \ln|x|$

7. $\int (ax+b)^n\,dx = \dfrac{(ax+b)^{n+1}}{a(n+1)} \qquad [n \neq -1]$

8. $\int \dfrac{dx}{ax+b} = \dfrac{1}{a}\ln(ax+b)$

9. $\int \dfrac{x\,dx}{ax+b} = \dfrac{1}{a^2}\left[ax+b - b\ln(ax+b)\right]$

10. $\int \dfrac{x\,dx}{(ax+b)^2} = \dfrac{1}{a^2}\left[\dfrac{b}{ax+b} + \ln(ax+b)\right]$

11. $\int \dfrac{dx}{x(ax+b)} = \dfrac{1}{b}\ln\left(\dfrac{x}{ax+b}\right)$

12. $\int \dfrac{dx}{x(ax+b)^2} = \dfrac{1}{b(ax+b)} + \dfrac{1}{b^2}\ln\left(\dfrac{x}{ax+b}\right)$

13. $\int \dfrac{dx}{x^2+a^2} = \dfrac{1}{a}\tan^{-1}\left(\dfrac{x}{a}\right)$

14. $\int \dfrac{dx}{a^2-x^2} = \dfrac{1}{a}\tanh^{-1}\left(\dfrac{x}{a}\right)$

15. $\int \dfrac{x\,dx}{ax^2+b} = \dfrac{1}{2a}\ln(ax^2+b)$

16. $\int \dfrac{dx}{x(ax^n+b)} = \dfrac{1}{bn}\ln\left(\dfrac{x^n}{ax^n+b}\right)$

17. $\int \dfrac{dx}{ax^2+bx+c} = \dfrac{1}{\sqrt{b^2-4ac}}\ln\left(\dfrac{2ax+b-\sqrt{b^2-4ac}}{2ax+b+\sqrt{b^2-4ac}}\right) \quad [b^2 > 4ac]$

18. $\int \dfrac{dx}{ax^2+bx+c} = \dfrac{2}{\sqrt{4ac-b^2}}\tan^{-1}\left(\dfrac{2ax+b}{\sqrt{4ac-b^2}}\right) \quad [b^2 < 4ac]$

19. $\int \sqrt{a^2-x^2}\,dx = \dfrac{x}{2}\sqrt{a^2-x^2} + \dfrac{a^2}{2}\sin^{-1}\left(\dfrac{x}{a}\right)$

20. $\int x\sqrt{a^2-x^2}\,dx = -\tfrac{1}{3}(a^2-x^2)^{3/2}$

21. $\int \dfrac{dx}{\sqrt{a^2-x^2}} = \sin^{-1}\left(\dfrac{x}{a}\right)$

22. $\int \dfrac{x\,dx}{\sqrt{a^2-x^2}} = -\sqrt{a^2-x^2}$

APPENDIX 10.A
Laplace Transforms

$f(t)$	$\mathcal{L}(f(t))$	$f(t)$	$\mathcal{L}(f(t))$
$\delta(t)$ [unit impulse at $t = 0$]	1	e^{at}	$\dfrac{1}{s-a}$
$\delta(t-c)$ [unit impulse at $t = c$]	e^{-cs}	$e^{at}\sin bt$	$\dfrac{b}{(s-a)^2 + b^2}$
1 or u_0 [unit step at $t = 0$]	$\dfrac{1}{s}$	$e^{at}\cos bt$	$\dfrac{s-a}{(s-a)^2 + b^2}$
u_c [unit step at $t = c$]	$\dfrac{e^{-cs}}{s}$	$e^{at}t^n$ (n is a positive integer)	$\dfrac{n!}{(s-a)^{n+1}}$
t [unit ramp at $t = 0$]	$\dfrac{1}{s^2}$	$1 - e^{-at}$	$\dfrac{a}{s(s+a)}$
$\dfrac{t^{n-1}}{(n-1)!}$	$\dfrac{1}{s^n}$	$e^{-at} + at - 1$	$\dfrac{a^2}{s^2(s+a)}$
$\sin at$	$\dfrac{a}{s^2 + a^2}$	$\dfrac{e^{-at} - e^{-bt}}{b-a}$	$\dfrac{1}{(s+a)(s+b)}$
$at - \sin at$	$\dfrac{a^3}{s^2(s^2 + a^2)}$	$\dfrac{(c-a)e^{-at} - (c-b)e^{-bt}}{b-a}$	$\dfrac{s+c}{(s+a)(s+b)}$
$\sinh at$	$\dfrac{a}{s^2 - a^2}$	$\dfrac{1}{ab} + \dfrac{be^{-at} - ae^{-bt}}{ab(a-b)}$	$\dfrac{1}{s(s+a)(s+b)}$
$t\sin at$	$\dfrac{2as}{(s^2 + a^2)^2}$	$t\sinh at$	$\dfrac{2as}{(s^2 - a^2)^2}$
$\cos at$	$\dfrac{s}{s^2 + a^2}$	$t\cosh at$	$\dfrac{s^2 + a^2}{(s^2 - a^2)^2}$
$1 - \cos at$	$\dfrac{a^2}{s(s^2 + a^2)}$	rectangular pulse, magnitude M, duration a	$\left(\dfrac{M}{s}\right)(1 - e^{-as})$
$\cosh at$	$\dfrac{s}{s^2 - a^2}$	triangular pulse, magnitude M, duration $2a$	$\left(\dfrac{M}{as^2}\right)(1 - e^{-as})^2$
$t\cos at$	$\dfrac{s^2 - a^2}{(s^2 + a^2)^2}$	sawtooth pulse, magnitude M, duration a	$\left(\dfrac{M}{as^2}\right)[1 - (as+1)e^{-as}]$
t^n (n is a positive integer)	$\dfrac{n!}{s^{n+1}}$	sinusoidal pulse, magnitude M, duration π/a	$\left(\dfrac{Ma}{s^2 + a^2}\right)(1 + e^{-\pi s/a})$

APPENDIX 11.A
Areas Under the Standard Normal Curve
(0 to z)

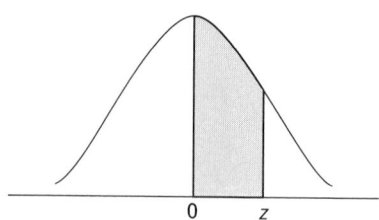

z	0	1	2	3	4	5	6	7	8	9
0.0	0.0000	0.0040	0.0080	0.0120	0.0160	0.0199	0.0239	0.0279	0.0319	0.0359
0.1	0.0398	0.0438	0.0478	0.0517	0.0557	0.0596	0.0636	0.0675	0.0714	0.0754
0.2	0.0793	0.0832	0.0871	0.0910	0.0948	0.0987	0.1026	0.1064	0.1103	0.1141
0.3	0.1179	0.1217	0.1255	0.1293	0.1331	0.1368	0.1406	0.1443	0.1480	0.1517
0.4	0.1554	0.1591	0.1628	0.1664	0.1700	0.1736	0.1772	0.1808	0.1844	0.1879
0.5	0.1915	0.1950	0.1985	0.2019	0.2054	0.2088	0.2123	0.2157	0.2190	0.2224
0.6	0.2258	0.2291	0.2324	0.2357	0.2389	0.2422	0.2454	0.2486	0.2518	0.2549
0.7	0.2580	0.2612	0.2642	0.2673	0.2704	0.2734	0.2764	0.2794	0.2823	0.2852
0.8	0.2881	0.2910	0.2939	0.2967	0.2996	0.3023	0.3051	0.3078	0.3106	0.3133
0.9	0.3159	0.3186	0.3212	0.3238	0.3264	0.3289	0.3315	0.3340	0.3365	0.3389
1.0	0.3413	0.3438	0.3461	0.3485	0.3508	0.3531	0.3554	0.3577	0.3599	0.3621
1.1	0.3643	0.3665	0.3686	0.3708	0.3729	0.3749	0.3770	0.3790	0.3810	0.3830
1.2	0.3849	0.3869	0.3888	0.3907	0.3925	0.3944	0.3962	0.3980	0.3997	0.4015
1.3	0.4032	0.4049	0.4066	0.4082	0.4099	0.4115	0.4131	0.4147	0.4162	0.4177
1.4	0.4192	0.4207	0.4222	0.4236	0.4251	0.4265	0.4279	0.4292	0.4306	0.4319
1.5	0.4332	0.4345	0.4357	0.4370	0.4382	0.4394	0.4406	0.4418	0.4429	0.4441
1.6	0.4452	0.4463	0.4474	0.4484	0.4495	0.4505	0.4515	0.4525	0.4535	0.4545
1.7	0.4554	0.4564	0.4573	0.4582	0.4591	0.4599	0.4608	0.4616	0.4625	0.4633
1.8	0.4641	0.4649	0.4656	0.4664	0.4671	0.4678	0.4686	0.4693	0.4699	0.4706
1.9	0.4713	0.4719	0.4726	0.4732	0.4738	0.4744	0.4750	0.4756	0.4761	0.4767
2.0	0.4772	0.4778	0.4783	0.4788	0.4793	0.4798	0.4803	0.4808	0.4812	0.4817
2.1	0.4821	0.4826	0.4830	0.4834	0.4838	0.4842	0.4846	0.4850	0.4854	0.4857
2.2	0.4861	0.4864	0.4868	0.4871	0.4875	0.4878	0.4881	0.4884	0.4887	0.4890
2.3	0.4893	0.4896	0.4898	0.4901	0.4904	0.4906	0.4909	0.4911	0.4913	0.4916
2.4	0.4918	0.4920	0.4922	0.4925	0.4927	0.4929	0.4931	0.4932	0.4934	0.4936
2.5	0.4938	0.4940	0.4941	0.4943	0.4945	0.4946	0.4948	0.4949	0.4951	0.4952
2.6	0.4953	0.4955	0.4956	0.4957	0.4959	0.4960	0.4961	0.4962	0.4963	0.4964
2.7	0.4965	0.4966	0.4967	0.4968	0.4969	0.4970	0.4971	0.4972	0.4973	0.4974
2.8	0.4974	0.4975	0.4976	0.4977	0.4977	0.4978	0.4979	0.4979	0.4980	0.4981
2.9	0.4981	0.4982	0.4982	0.4983	0.4984	0.4984	0.4985	0.4985	0.4986	0.4986
3.0	0.4987	0.4987	0.4987	0.4988	0.4988	0.4989	0.4989	0.4989	0.4990	0.4990
3.1	0.4990	0.4991	0.4991	0.4991	0.4992	0.4992	0.4992	0.4992	0.4993	0.4993
3.2	0.4993	0.4993	0.4994	0.4994	0.4994	0.4994	0.4994	0.4995	0.4995	0.4995
3.3	0.4995	0.4995	0.4996	0.4996	0.4996	0.4996	0.4996	0.4996	0.4996	0.4997
3.4	0.4997	0.4997	0.4997	0.4997	0.4997	0.4997	0.4997	0.4997	0.4997	0.4998
3.5	0.4998	0.4998	0.4998	0.4998	0.4998	0.4998	0.4998	0.4998	0.4998	0.4998
3.6	0.4998	0.4998	0.4999	0.4999	0.4999	0.4999	0.4999	0.4999	0.4999	0.4999
3.7	0.4999	0.4999	0.4999	0.4999	0.4999	0.4999	0.4999	0.4999	0.4999	0.4999
3.8	0.4999	0.4999	0.4999	0.4999	0.4999	0.4999	0.4999	0.4999	0.4999	0.4999
3.9	0.5000	0.5000	0.5000	0.5000	0.5000	0.5000	0.5000	0.5000	0.5000	0.5000

APPENDIX 14.A
Properties of Water at Atmospheric Pressure
(English Units)

temperature (°F)	density (lbm/ft^3)	absolute viscosity (lbf-sec/ft^2)	kinematic viscosity (ft^2/sec)	surface tension (lbf/ft)	vapor pressure head^{a,b,c} (ft)	bulk modulus (lbf/in^2)
32	62.42	3.746×10^{-5}	1.931×10^{-5}	0.518×10^{-2}	0.20	293×10^3
40	62.43	3.229×10^{-5}	1.664×10^{-5}	0.514×10^{-2}	0.28	294×10^3
50	62.41	2.735×10^{-5}	1.410×10^{-5}	0.509×10^{-2}	0.41	305×10^3
60	62.37	2.359×10^{-5}	1.217×10^{-5}	0.504×10^{-2}	0.59	311×10^3
70	62.30	2.050×10^{-5}	1.059×10^{-5}	0.500×10^{-2}	0.84	320×10^3
80	62.22	1.799×10^{-5}	0.930×10^{-5}	0.492×10^{-2}	1.17	322×10^3
90	62.11	1.595×10^{-5}	0.826×10^{-5}	0.486×10^{-2}	1.62	323×10^3
100	62.00	1.424×10^{-5}	0.739×10^{-5}	0.480×10^{-2}	2.21	327×10^3
110	61.86	1.284×10^{-5}	0.667×10^{-5}	0.473×10^{-2}	2.97	331×10^3
120	61.71	1.168×10^{-5}	0.609×10^{-5}	0.465×10^{-2}	3.96	333×10^3
130	61.55	1.069×10^{-5}	0.558×10^{-5}	0.460×10^{-2}	5.21	334×10^3
140	61.38	0.981×10^{-5}	0.514×10^{-5}	0.454×10^{-2}	6.78	330×10^3
150	61.20	0.905×10^{-5}	0.476×10^{-5}	0.447×10^{-2}	8.76	328×10^3
160	61.00	0.838×10^{-5}	0.442×10^{-5}	0.441×10^{-2}	11.21	326×10^3
170	60.80	0.780×10^{-5}	0.413×10^{-5}	0.433×10^{-2}	14.20	322×10^3
180	60.58	0.726×10^{-5}	0.385×10^{-5}	0.426×10^{-2}	17.87	313×10^3
190	60.36	0.678×10^{-5}	0.362×10^{-5}	0.419×10^{-2}	22.29	313×10^3
200	60.12	0.637×10^{-5}	0.341×10^{-5}	0.412×10^{-2}	27.61	308×10^3
212	59.83	0.593×10^{-5}	0.319×10^{-5}	0.404×10^{-2}	35.38	300×10^3

^{a}Based on actual densities, not on standard "cold, clear water."
^{b}Can also be calculated from steam tables as $(p_{saturation})\,(144\text{ in}^2/\text{ft}^2)\,(v)\,(g/g_c)$.
^{c}Multiply the vapor pressure head by the density and divide by 144 in^2/ft^2 to obtain psi.

APPENDIX 14.B
Properties of Water at Atmospheric Pressure
(SI Units)

temperature (°C)	density (kg/m^3)	absolute viscosity (Pa·s)	kinematic viscosity (m^2/s)	vapor pressure (kPa)	bulk modulus (kPa)
0	999.87	1.7921×10^{-3}	1.792×10^{-6}	0.611	204×10^4
4	1000.00	1.5674×10^{-3}	1.567×10^{-6}	0.813	206×10^4
10	999.73	1.3077×10^{-3}	1.371×10^{-6}	1.228	211×10^4
20	998.23	1.0050×10^{-3}	1.007×10^{-6}	2.338	220×10^4
25	997.08	0.8937×10^{-3}	8.963×10^{-7}	3.168	221×10^4
30	995.68	0.8007×10^{-3}	8.042×10^{-7}	4.242	223×10^4
40	992.25	0.6560×10^{-3}	6.611×10^{-7}	7.375	227×10^4
50	988.07	0.5494×10^{-3}	5.560×10^{-7}	12.333	230×10^4
60	983.24	0.4688×10^{-3}	4.768×10^{-7}	19.92	228×10^4
70	977.81	0.4061×10^{-3}	4.153×10^{-7}	31.16	225×10^4
80	971.83	0.3565×10^{-3}	3.668×10^{-7}	47.34	221×10^4
90	965.34	0.3165×10^{-3}	3.279×10^{-7}	70.10	216×10^4
100	958.38	0.2838×10^{-3}	2.961×10^{-7}	101.325	207×10^4

Compiled from various sources.

APPENDIX 14.C
Viscosity of Water in Other Units
(English Units)

temperature (°F)	absolute viscosity (cP)	kinematic viscosity	
		(cSt)	(SSU)
32	1.79	1.79	33.0
50	1.31	1.31	31.6
60	1.12	1.12	31.2
70	0.98	0.98	30.9
80	0.86	0.86	30.6
85	0.81	0.81	30.4
100	0.68	0.69	30.2
120	0.56	0.57	30.0
140	0.47	0.48	29.7
160	0.40	0.41	29.6
180	0.35	0.36	29.5
212	0.28	0.29	29.3

Reprinted with permission from *The Hydraulic Handbook*, © 1988, by Fairbanks Morse Pump, Kansas City, Kansas.

APPENDIX 14.D
Properties of Air at Atmospheric Pressure
(English Units)

temperature (°F)	density (lbm/ft^3)	absolute viscosity (lbf-sec/ft^2)	kinematic viscosity (ft^2/sec)
0	0.0862	3.28×10^{-7}	12.6×10^{-5}
20	0.0827	3.50×10^{-7}	13.6×10^{-5}
40	0.0794	3.62×10^{-7}	14.6×10^{-5}
60	0.0763	3.74×10^{-7}	15.8×10^{-5}
68	0.0752	3.78×10^{-7}	16.0×10^{-5}
80	0.0735	3.85×10^{-7}	16.9×10^{-5}
100	0.0709	3.96×10^{-7}	18.0×10^{-5}
120	0.0684	4.07×10^{-7}	18.9×10^{-5}
250	0.0559	4.74×10^{-7}	27.3×10^{-5}

APPENDIX 14.E
Properties of Air at Atmospheric Pressure
(SI Units)

temperature (°C)	density (kg/m^3)	absolute viscosity (Pa·s)	kinematic viscosity (m^2/s)
0	1.293	1.709×10^{-5}	1.322×10^{-5}
20	1.20	1.80×10^{-5}	1.51×10^{-5}
50	1.093	1.951×10^{-5}	1.785×10^{-5}
100	0.946	2.175×10^{-5}	2.474×10^{-5}
150	0.834	2.385×10^{-5}	3.077×10^{-5}
200	0.746	2.582×10^{-5}	3.724×10^{-5}
250	0.675	2.770×10^{-5}	4.416×10^{-5}
300	0.616	2.946×10^{-5}	5.145×10^{-5}
350	0.567	3.113×10^{-5}	5.907×10^{-5}
400	0.525	3.277×10^{-5}	6.721×10^{-5}
450	0.488	3.433×10^{-5}	7.750×10^{-5}
500	0.457	3.583×10^{-5}	8.436×10^{-5}

APPENDIX 16.A
Area, Wetted Perimeter, and Hydraulic Radius
of Partially Filled Circular Pipes

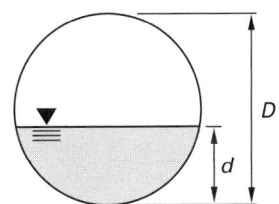

$\frac{d}{D}$	$\frac{\text{area}}{D^2}$	$\frac{\text{wetted perimeter}}{D}$	$\frac{r_h}{D}$	$\frac{d}{D}$	$\frac{\text{area}}{D^2}$	$\frac{\text{wetted perimeter}}{D}$	$\frac{r_h}{D}$
0.01	0.0013	0.2003	0.0066	0.51	0.4027	1.5908	0.2531
0.02	0.0037	0.2838	0.0132	0.52	0.4127	1.6108	0.2561
0.03	0.0069	0.3482	0.0197	0.53	0.4227	1.6308	0.2591
0.04	0.0105	0.4027	0.0262	0.54	0.4327	1.6509	0.2620
0.05	0.0147	0.4510	0.0326	0.55	0.4426	1.6710	0.2649
0.06	0.0192	0.4949	0.0389	0.56	0.4526	1.6911	0.2676
0.07	0.0242	0.5355	0.0451	0.57	0.4625	1.7113	0.2703
0.08	0.0294	0.5735	0.0513	0.58	0.4723	1.7315	0.2728
0.09	0.0350	0.6094	0.0574	0.59	0.4822	1.7518	0.2753
0.10	0.0409	0.6435	0.0635	0.60	0.4920	1.7722	0.2776
0.11	0.0470	0.6761	0.0695	0.61	0.5018	1.7926	0.2797
0.12	0.0534	0.7075	0.0754	0.62	0.5115	1.8132	0.2818
0.13	0.0600	0.7377	0.0813	0.63	0.5212	1.8338	0.2839
0.14	0.0688	0.7670	0.0871	0.64	0.5308	1.8546	0.2860
0.15	0.0739	0.7954	0.0929	0.65	0.5404	1.8755	0.2881
0.16	0.0811	0.8230	0.0986	0.66	0.5499	1.8965	0.2899
0.17	0.0885	0.8500	0.1042	0.67	0.5594	1.9177	0.2917
0.18	0.0961	0.8763	0.1097	0.68	0.5687	1.9391	0.2935
0.19	0.1039	0.9020	0.1152	0.69	0.5780	1.9606	0.2950
0.20	0.1118	0.9273	0.1206	0.70	0.5872	1.9823	0.2962
0.21	0.1199	0.9521	0.1259	0.71	0.5964	2.0042	0.2973
0.22	0.1281	0.9764	0.1312	0.72	0.6054	2.0264	0.2984
0.23	0.1365	1.0003	0.1364	0.73	0.6143	2.0488	0.2995
0.24	0.1449	1.0239	0.1416	0.74	0.6231	2.0714	0.3006
0.25	0.1535	1.0472	0.1466	0.75	0.6318	2.0944	0.3017
0.26	0.1623	1.0701	0.1516	0.76	0.6404	2.1176	0.3025
0.27	0.1711	1.0928	0.1566	0.77	0.6489	2.1412	0.3032
0.28	0.1800	1.1152	0.1614	0.78	0.6573	2.1652	0.3037
0.29	0.1890	1.1373	0.1662	0.79	0.6655	2.1895	0.3040
0.30	0.1982	1.1593	0.1709	0.80	0.6736	2.2143	0.3042
0.31	0.2074	1.1810	0.1755	0.81	0.6815	2.2395	0.3044
0.32	0.2167	1.2025	0.1801	0.82	0.6893	2.2653	0.3043
0.33	0.2260	1.2239	0.1848	0.83	0.6969	2.2916	0.3041
0.34	0.2355	1.2451	0.1891	0.84	0.7043	2.3186	0.3038
0.35	0.2450	1.2661	0.1935	0.85	0.7115	2.3462	0.3033
0.36	0.2546	1.2870	0.1978	0.86	0.7186	2.3746	0.3026
0.37	0.2642	1.3078	0.2020	0.87	0.7254	2.4038	0.3017
0.38	0.2739	1.3284	0.2061	0.88	0.7320	2.4341	0.3008
0.39	0.2836	1.3490	0.2102	0.89	0.7384	2.4655	0.2995
0.40	0.2934	1.3694	0.2142	0.90	0.7445	2.4981	0.2980
0.41	0.3032	1.3898	0.2181	0.91	0.7504	2.5322	0.2963
0.42	0.3130	1.4101	0.2220	0.92	0.7560	2.5681	0.2944
0.43	0.3229	1.4303	0.2257	0.93	0.7612	2.6061	0.2922
0.44	0.3328	1.4505	0.2294	0.94	0.7662	2.6467	0.2896
0.45	0.3428	1.4706	0.2331	0.95	0.7707	2.6906	0.2864
0.46	0.3527	1.4907	0.2366	0.96	0.7749	2.7389	0.2830
0.47	0.3627	1.5108	0.2400	0.97	0.7785	2.7934	0.2787
0.48	0.3727	1.5308	0.2434	0.98	0.7816	2.8578	0.2735
0.49	0.3827	1.5508	0.2467	0.99	0.7841	2.9412	0.2665
0.50	0.3927	1.5708	0.2500	1.00	0.7854	3.1416	0.2500

APPENDIX 16.B
Dimensions of Welded and Seamless Steel Pipe[a,b]
(selected sizes)[c]
(English units)

nominal diameter (in)	schedule	outside diameter (in)	wall thickness (in)	internal diameter (in)	internal area (in^2)	internal diameter (ft)	internal area (ft^2)
$\frac{1}{8}$	40 (S)	0.405	0.068	0.269	0.0568	0.0224	0.00039
	80 (X)		0.095	0.215	0.0363	0.0179	0.00025
$\frac{1}{4}$	40 (S)	0.540	0.088	0.364	0.1041	0.0303	0.00072
	80 (X)		0.119	0.302	0.0716	0.0252	0.00050
$\frac{3}{8}$	40 (S)	0.675	0.091	0.493	0.1909	0.0411	0.00133
	80 (X)		0.126	0.423	0.1405	0.0353	0.00098
$\frac{1}{2}$	40 (S)	0.840	0.109	0.622	0.3039	0.0518	0.00211
	80 (X)		0.147	0.546	0.2341	0.0455	0.00163
	160		0.187	0.466	0.1706	0.0388	0.00118
	(XX)		0.294	0.252	0.0499	0.0210	0.00035
$\frac{3}{4}$	40 (S)	1.050	0.113	0.824	0.5333	0.0687	0.00370
	80 (X)		0.154	0.742	0.4324	0.0618	0.00300
	160		0.218	0.614	0.2961	0.0512	0.00206
	(XX)		0.308	0.434	0.1479	0.0362	0.00103
1	40 (S)	1.315	0.133	1.049	0.8643	0.0874	0.00600
	80 (X)		0.179	0.957	0.7193	0.0798	0.00500
	160		0.250	0.815	0.5217	0.0679	0.00362
	(XX)		0.358	0.599	0.2818	0.0499	0.00196
$1\frac{1}{4}$	40 (S)	1.660	0.140	1.380	1.496	0.1150	0.01039
	80 (X)		0.191	1.278	1.283	0.1065	0.00890
	160		0.250	1.160	1.057	0.0967	0.00734
	(XX)		0.382	0.896	0.6305	0.0747	0.00438
$1\frac{1}{2}$	40 (S)	1.900	0.145	1.610	2.036	0.1342	0.01414
	80 (X)		0.200	1.500	1.767	0.1250	0.01227
	160		0.281	1.338	1.406	0.1115	0.00976
	(XX)		0.400	1.100	0.9503	0.0917	0.00660
2	40 (S)	2.375	0.154	2.067	3.356	0.1723	0.02330
	80 (X)		0.218	1.939	2.953	0.1616	0.02051
	160		0.343	1.689	2.240	0.1408	0.01556
	(XX)		0.436	1.503	1.774	0.1253	0.01232
$2\frac{1}{2}$	40 (S)	2.875	0.203	2.469	4.788	0.2058	0.03325
	80 (X)		0.276	2.323	4.238	0.1936	0.02943
	160		0.375	2.125	3.547	0.1771	0.02463
	(XX)		0.552	1.771	2.464	0.1476	0.01711
3	40 (S)	3.500	0.216	3.068	7.393	0.2557	0.05134
	80 (X)		0.300	2.900	6.605	0.2417	0.04587
	160		0.437	2.626	5.416	0.2188	0.03761
	(XX)		0.600	2.300	4.155	0.1917	0.02885
$3\frac{1}{2}$	40 (S)	4.000	0.226	3.548	9.887	0.2957	0.06866
	80 (X)		0.318	3.364	8.888	0.2803	0.06172
4	40 (S)	4.500	0.237	4.026	12.73	0.3355	0.08841
	80 (X)		0.337	3.826	11.50	0.3188	0.07984
	120		0.437	3.626	10.33	0.3022	0.07171
	160		0.531	3.438	9.283	0.2865	0.06447
	(XX)		0.674	3.152	7.803	0.2627	0.05419
5	40 (S)	5.563	0.258	5.047	20.01	0.4206	0.1389
	80 (X)		0.375	4.813	18.19	0.4011	0.1263
	120		0.500	4.563	16.35	0.3803	0.1136
	160		0.625	4.313	14.61	0.3594	0.1015
	(XX)		0.750	4.063	12.97	0.3386	0.09004

(continued)

APPENDIX 16.B *(continued)*
Dimensions of Welded and Seamless Steel Pipe^{a,b}
(selected sizes)^c
(English units)

nominal diameter (in)	schedule	outside diameter (in)	wall thickness (in)	internal diameter (in)	internal area (in^2)	internal diameter (ft)	internal area (ft^2)
6	40 (S)	6.625	0.280	6.065	28.89	0.5054	0.2006
	80 (X)		0.432	5.761	26.07	0.4801	0.1810
	120		0.562	5.501	23.77	0.4584	0.1650
	160		0.718	5.189	21.15	0.4324	0.1469
	(XX)		0.864	4.897	18.83	0.4081	0.1308
8	20	8.625	0.250	8.125	51.85	0.6771	0.3601
	30		0.277	8.071	51.16	0.6726	0.3553
	40 (S)		0.322	7.981	50.03	0.6651	0.3474
	60		0.406	7.813	47.94	0.6511	0.3329
	80 (X)		0.500	7.625	45.66	0.6354	0.3171
	100		0.593	7.439	43.46	0.6199	0.3018
	120		0.718	7.189	40.59	0.5990	0.2819
	140		0.812	7.001	38.50	0.5834	0.2673
	(XX)		0.875	6.875	37.12	0.5729	0.2578
	160		0.906	6.813	36.46	0.5678	0.2532
10	20	10.75	0.250	10.250	82.52	0.85417	0.5730
	30		0.307	10.136	80.69	0.84467	0.5604
	40 (S)		0.365	10.020	78.85	0.83500	0.5476
	60 (X)		0.500	9.750	74.66	0.8125	0.5185
	80		0.593	9.564	71.84	0.7970	0.4989
	100		0.718	9.314	68.13	0.7762	0.4732
	120		0.843	9.064	64.53	0.7553	0.4481
	140 (XX)		1.000	8.750	60.13	0.7292	0.4176
	160		1.125	8.500	56.75	0.7083	0.3941
12	20	12.75	0.250	12.250	117.86	1.0208	0.8185
	30		0.330	12.090	114.80	1.0075	0.7972
	(S)		0.375	12.000	113.10	1.0000	0.7854
	40		0.406	11.938	111.93	0.99483	0.7773
	(X)		0.500	11.750	108.43	0.97917	0.7530
	60		0.562	11.626	106.16	0.96883	0.7372
	80		0.687	11.376	101.64	0.94800	0.7058
	100		0.843	11.064	96.14	0.92200	0.6677
	120 (XX)		1.000	10.750	90.76	0.89583	0.6303
	140		1.125	10.500	86.59	0.87500	0.6013
	160		1.312	10.126	80.53	0.84383	0.5592
14 O.D.	10	14.00	0.250	13.500	143.14	1.1250	0.9940
	20		0.312	13.376	140.52	1.1147	0.9758
	30 (S)		0.375	13.250	137.89	1.1042	0.9575
	40		0.437	13.126	135.32	1.0938	0.9397
	(X)		0.500	13.000	132.67	1.0833	0.9213
	60		0.593	12.814	128.96	1.0679	0.8956
	80		0.750	12.500	122.72	1.0417	0.8522
	100		0.937	12.126	115.48	1.0105	0.8020
	120		1.093	11.814	109.62	0.98450	0.7612
	140		1.250	11.500	103.87	0.95833	0.7213
	160		1.406	11.188	98.31	0.93233	0.6827
16 O.D.	10	16.00	0.250	15.500	188.69	1.2917	1.3104
	20		0.312	15.376	185.69	1.2813	1.2895
	30 (S)		0.375	15.250	182.65	1.2708	1.2684
	40 (X)		0.500	15.000	176.72	1.2500	1.2272
	60		0.656	14.688	169.44	1.2240	1.1767
	80		0.843	14.314	160.92	1.1928	1.1175
	100		1.031	13.938	152.58	1.1615	1.0596
	120		1.218	13.564	144.50	1.1303	1.0035
	140		1.437	13.126	135.32	1.0938	0.9397
	160		1.593	12.814	128.96	1.0678	0.8956

(continued)

APPENDIX 16.B *(continued)*
Dimensions of Welded and Seamless Steel Pipe[a,b]
(selected sizes)[c]
(English units)

nominal diameter (in)	schedule	outside diameter (in)	wall thickness (in)	internal diameter (in)	internal area (in²)	internal diameter (ft)	internal area (ft²)
18 O.D.	10	18.00	0.250	17.500	240.53	1.4583	1.6703
	20		0.312	17.376	237.13	1.4480	1.6467
	(S)		0.375	17.250	233.71	1.4375	1.6230
	30		0.437	17.126	230.36	1.4272	1.5997
	(X)		0.500	17.000	226.98	1.4167	1.5762
	40		0.562	16.876	223.68	1.4063	1.5533
	60		0.750	16.500	213.83	1.3750	1.4849
	80		0.937	16.126	204.24	1.3438	1.4183
	100		1.156	15.688	193.30	1.3073	1.3423
	120		1.375	15.250	182.65	1.2708	1.2684
	140		1.562	14.876	173.81	1.2397	1.2070
	160		1.781	14.438	163.72	1.2032	1.1370
20 O.D.	10	20.00	0.250	19.500	298.65	1.6250	2.0739
	20 (S)		0.375	19.250	291.04	1.6042	2.0211
	30 (X)		0.500	19.000	283.53	1.5833	1.9689
	40		0.593	18.814	278.00	1.5678	1.9306
	60		0.812	18.376	265.21	1.5313	1.8417
	80		1.031	17.938	252.72	1.4948	1.7550
	100		1.281	17.438	238.83	1.4532	1.6585
	120		1.500	17.000	226.98	1.4167	1.5762
	140		1.750	16.500	213.83	1.3750	1.4849
	160		1.968	16.064	202.67	1.3387	1.4075
24 O.D.	10	24.00	0.250	23.500	433.74	1.9583	3.0121
	20 (S)		0.375	23.250	424.56	1.9375	2.9483
	(X)		0.500	23.000	415.48	1.9167	2.8852
	30		0.562	22.876	411.01	1.9063	2.8542
	40		0.687	22.626	402.07	1.8855	2.7922
	60		0.968	22.060	382.20	1.8383	2.6542
	80		1.218	21.564	365.21	1.7970	2.5362
	100		1.531	20.938	344.32	1.7448	2.3911
	120		1.812	20.376	326.92	1.6980	2.2645
	140		2.062	19.876	310.28	1.6563	2.1547
	160		2.343	19.310	292.87	1.6092	2.0337
30 O.D.	10	30.00	0.312	29.376	677.76	2.4480	4.7067
	(S)		0.375	29.250	671.62	2.4375	4.6640
	20 (X)		0.500	29.000	660.52	2.4167	4.5869
	30		0.625	28.750	649.18	2.3958	4.5082

(Multiply in by 25.4 to obtain mm.)
(Multiply in² by 645 to obtain mm².)
[a]Designations per ANSI B36.10.
[b]The "S" wall thickness was formerly designated as "standard weight." Standard weight and schedule-40 are the same for all diameters through 10 in. For diameters between 12 in and 24 in, standard weight pipe has a wall thickness of 0.375 in. The "X" wall thickness was formerly designated as "extra strong." Extra strong weight and schedule-80 are the same for all diameters through 8 in. For diameters between 10 in and 24 in, extra strong weight pipe has a wall thickness of 0.50 in. The "XX" wall thickness was formerly designed as "double extra strong." Double extra strong weight pipe does not have a corresponding schedule number.
[c]Pipe sizes and weights in most common usage listed. Other weights and sizes exist.

APPENDIX 16.C
Dimensions of Welded and Seamless Steel Pipe
Schedules 40 and 80
(SI units)

nominal pipe size	schedule number	wall thickness	inside diameter	inside cross-sectional area
in		mm	mm	m^2
$\frac{1}{8}$	40	1.73	6.83	0.3664×10^{-4}
	80	2.41	5.46	0.2341×10^{-4}
$\frac{1}{4}$	40	2.24	9.25	0.6720×10^{-4}
	80	3.02	7.67	0.4620×10^{-4}
$\frac{3}{8}$	40	2.31	12.52	1.231×10^{-4}
	80	3.20	10.74	0.9059×10^{-4}
$\frac{1}{2}$	40	2.77	15.80	1.961×10^{-4}
	80	3.73	13.87	1.511×10^{-4}
$\frac{3}{4}$	40	2.87	20.93	3.441×10^{-4}
	80	3.91	18.85	2.791×10^{-4}
1	40	3.38	26.64	5.574×10^{-4}
	80	4.45	24.31	4.641×10^{-4}
$1\frac{1}{4}$	40	3.56	35.05	9.648×10^{-4}
	80	4.85	32.46	8.275×10^{-4}
$1\frac{1}{2}$	40	3.68	40.89	13.13×10^{-4}
	80	5.08	38.10	11.40×10^{-4}
2	40	3.91	52.50	21.65×10^{-4}
	80	5.54	49.25	19.05×10^{-4}
$2\frac{1}{2}$	40	5.16	62.71	30.89×10^{-4}
	80	7.01	59.00	27.30×10^{-4}
3	40	5.49	77.92	47.69×10^{-4}
	80	7.62	73.66	42.61×10^{-4}
$3\frac{1}{2}$	40	5.74	90.12	63.79×10^{-4}
	80	8.08	85.45	57.35×10^{-4}
4	40	6.02	102.3	82.19×10^{-4}
	80	8.56	97.18	74.17×10^{-4}
5	40	6.55	128.2	129.1×10^{-4}
	80	9.53	122.3	117.5×10^{-4}
6	40	7.11	154.1	186.5×10^{-4}
	80	10.97	146.3	168.1×10^{-4}
8	40	8.18	202.7	322.7×10^{-4}
	80	12.70	193.7	294.7×10^{-4}
10	40	9.27	254.5	508.6×10^{-4}
	80	15.06	242.9	463.4×10^{-4}
12	40	10.31	303.2	721.9×10^{-4}
	80	17.45	289.0	655.6×10^{-4}
14	40	11.10	333.4	872.8×10^{-4}
	80	19.05	317.5	791.5×10^{-4}
16	40	12.70	381.0	1140×10^{-4}
	80	21.41	363.6	1038×10^{-4}
18	40	14.05	428.7	1443×10^{-4}
	80	23.80	409.6	1317×10^{-4}
20	40	15.06	477.9	1793×10^{-4}
	80	26.19	455.6	1630×10^{-4}
24	40	17.45	574.7	2593×10^{-4}
	80	30.94	547.7	2356×10^{-4}

APPENDIX 16.D
Dimensions of Small Diameter PVC Pipe[a,b]
(English units)

nominal size (in)	schedule	wall thickness[c] (in)	O.D. (in)	I.D. (in)
$\frac{1}{4}$	40	0.088	0.540	0.364
	80	0.119	0.540	0.302
$\frac{1}{2}$	40	0.109	0.840	0.622
	80	0.147	0.840	0.546
$\frac{3}{4}$	40	0.113	1.050	0.824
	80	0.154	1.050	0.742
1	40	0.133	1.315	1.049
	80	0.179	1.315	0.957
$1\frac{1}{4}$	40	0.140	1.660	1.380
	80	0.191	1.660	1.278
$1\frac{1}{2}$	40	0.145	1.900	1.610
	80	0.200	1.900	1.500
2	40	0.154	2.375	2.067
	80	0.218	2.375	1.939
3	40	0.216	3.500	3.068
	80	0.300	3.500	2.900
4	40	0.237	4.500	4.026
	80	0.337	4.500	3.826

(Multiply in by 25.4 to obtain mm.)
(Multiply in^2 by 645 to obtain mm^2.)
[a]Abstracted from ASTM Specification D1785-85.
[b]Two strengths of PVC are in use. A maximum (bursting) stress of 6400 psig (44 MPa) applies to PVC types 1120, 1220, and 4116. A maximum (bursting) stress of 5000 psig (35 MPa) applies to PVC types 2112, 2116, and 2120.
[c]Minimum wall thickness, with tolerances of −0%, +10%.

APPENDIX 16.E
Dimensions of Large Diameter PVC Sewer and Water Pipe
(English units)

nominal size (in)	designation	minimum wall thickness (in)	O.D. (in)
ASTM 2729			
3		0.070	3.250
4		0.075	4.215
6		0.100	6.275
ASTM D3034			
4	regular	0.120	4.215
	HW	0.162	4.215
6	regular	0.180	6.275
	HW	0.241	6.275
8	regular	0.240	8.400
	HW	0.323	8.400
10	regular	0.300	10.50
	HW	0.404	10.50
12	regular	0.360	12.50
	HW	0.481	12.50
15	regular	0.437	15.30
	HW	0.588	15.30
18	regular		18.701
	HW		18.701
21	regular		22.047
	HW		22.047
24	regular		24.803
	HW		24.803
27	regular		27.95
	HW		27.95
ASTM F679			
18	regular	0.536	18.701
	HW	0.719	18.801
ASTM AWWA C900			
4	CL150	0.267	4.800
6	CL150	0.383	6.900
8	CL150	0.503	9.060
10	CL150	0.617	11.100
12	CL150	0.733	13.200

(Multiply in by 25.4 to obtain mm.)

APPENDIX 16.F
Dimensions and Weights of Concrete Sewer Pipe
(English units)

3000 psi			3500 psi			4000 psi		
internal diameter (in)	minimum shell thickness (in)	mass per foot[a] (lbm/ft)	internal diameter (in)	minimum shell thickness (in)	mass per foot[a] (lbm/ft)	internal diameter (in)	minimum shell thickness (in)	mass per foot[a] (lbm/ft)
12	2	93	12	$1\frac{3}{4}$	79	12		
15	$2\frac{1}{4}$	127	15	2	111	15		
18	$2\frac{1}{2}$	168	18			18	2	131
21	$2\frac{3}{4}$	214	21			21	$2\frac{1}{4}$	171
24	3	264	24	$2\frac{5}{8}$	229	24	$2\frac{1}{2}$	217
27	3	295	27	$2\frac{3}{4}$	268	27	$2\frac{5}{8}$	255
30	$3\frac{1}{2}$	384	30	3	324	30	$2\frac{3}{4}$	295
33	$3\frac{3}{4}$	451	33	$3\frac{1}{4}$	396	33	$2\frac{3}{4}$	322
36	4	524	36	$3\frac{3}{8}$	435	36	3	383
42	$4\frac{1}{2}$	686	42	$3\frac{3}{4}$	561	42	$3\frac{3}{8}$	500
48	5	867	48	$4\frac{1}{4}$	727	48	$3\frac{3}{4}$	635
54	$5\frac{1}{2}$	1068	54	$4\frac{5}{8}$	887	54	$4\frac{1}{4}$	810
60	6	1295	60	5	1064	60	$4\frac{1}{2}$	950
66	$6\frac{1}{2}$	1542	66	$5\frac{3}{8}$	1256	66	$4\frac{3}{4}$	1100
72	7	1811	72	$5\frac{3}{4}$	1463	72	5	1260
78	$7\frac{1}{2}$	2100						
84	8	2409						
90	8	2565						
96	$8\frac{1}{2}$	2906						
108	9	3446						

(Multiply in by 25.4 to obtain mm.)
(Multiply psi by 6.89 to obtain kPa.)
(Multiply lbm/ft by 1.488 to obtain kg/m.)
[a]Masses given are for reinforced tongue-and-groove pipe. Reinforced bell-and-spigot is heavier. Based on 150 lbm/ft^3 concrete.

APPENDIX 16.G
Dimensions of Cast Iron Pipe
(English units)

(all dimensions in inches)

nominal diameter	class A 100 ft head 43 psig		class B 200 ft head 86 psig		class C 300 ft head 130 psig		class D 400 ft head 173 psig	
	outside diameter	inside diameter	outside diameter	inside diameter	outside diameter	inside diameter	outside diameter	inside diameter
3	3.80	3.02	3.96	3.12	3.96	3.06	3.96	3.00
4	4.80	3.96	5.00	4.10	5.00	4.04	5.00	3.96
6	6.90	6.02	7.10	6.14	7.10	6.08	7.10	6.00
8	9.05	8.13	9.05	8.03	9.30	8.18	9.30	8.10
10	11.10	10.10	11.10	9.96	11.40	10.16	11.40	10.04
12	13.20	12.12	13.20	11.96	13.50	12.14	13.50	12.00
14	15.30	14.16	15.30	13.98	15.65	14.17	15.65	14.01
16	17.40	16.20	17.40	16.00	17.80	16.20	17.80	16.02
18	19.50	18.22	19.50	18.00	19.92	18.18	19.92	18.00
20	21.60	20.26	21.60	20.00	22.06	20.22	22.06	20.00
24	25.80	24.28	25.80	24.02	26.32	24.22	26.32	24.00
30	31.74	29.98	32.00	29.94	32.40	30.00	32.74	30.00
36	37.96	35.98	38.30	36.00	38.70	39.98	39.16	36.00
42	44.20	42.00	44.50	41.94	45.10	42.02	45.58	42.02
48	50.50	47.98	50.80	47.96	51.40	47.98	51.98	48.06
54	56.66	53.96	57.10	54.00	57.80	54.00	58.40	53.94
60	62.80	60.02	63.40	60.06	64.20	60.20	64.82	60.06
72	75.34	72.10	76.00	72.10	76.88	72.10		
84	87.54	84.10	88.54	84.10				

nominal diameter	class E 500 ft head 217 psig		class F 600 ft head 260 psig		class G 700 ft head 304 psig		class H 800 ft head 347 psig	
	outside diameter	inside diameter	outside diameter	inside diameter	outside diameter	inside diameter	outside diameter	inside diameter
6	7.22	6.06	7.22	6.00	7.38	6.08	7.38	6.00
8	9.42	8.10	9.42	8.00	9.60	8.10	9.60	8.00
10	11.60	10.12	11.60	10.00	11.84	10.12	11.84	10.00
12	13.78	12.14	13.78	12.00	14.08	12.14	14.08	12.00
14	15.98	14.18	15.98	14.00	16.32	14.18	16.32	14.00
16	18.16	16.20	18.16	16.00	18.54	16.18	18.54	16.00
18	20.34	18.20	20.34	18.00	20.78	18.22	20.78	18.00
20	22.54	20.24	22.54	20.00	23.02	20.24	23.02	20.00
24	26.90	24.28	26.90	24.00	27.76	24.26	27.76	24.00
30	33.10	30.00	33.46	30.00				
36	39.60	36.00	40.04	36.00				

(Multiply in by 25.4 to obtain mm.)

APPENDIX 16.H
Standard ANSI Piping Symbols

	flanged	screwed	bell and spigot	welded	soldered
joint					
elbow—90°					
elbow—45°					
elbow—turned up					
elbow—turned down					
elbow—long radius					
reducing elbow					
tee					
tee—outlet up					
tee—outlet down					
side outlet tee—outlet up					
cross					
reducer—concentric					
reducer—eccentric					
lateral					
gate valve					
globe valve					
check valve					
stop cock					
safety valve					
expansion joint					
union					
sleeve					
bushing					

APPENDIX 17.A
Specific Roughness and Hazen-Williams Constants
for Various Water Pipe Materials[b]

type of pipe or surface	ϵ (ft)		C		
	range	design	range	clean	design[a]
steel					
welded and seamless	0.0001–0.0003	0.0002	150–80	140	100
interior riveted, no projecting rivets				139	100
projecting girth rivets				130	100
projecting girth and horizontal rivets				115	100
vitrified, spiral-riveted, flow with lap				110	100
vitrified, spiral-riveted, flow against lap				100	90
corrugated				60	60
mineral					
concrete	0.001–0.01	0.004	152–85	120	100
cement-asbestos			160–140	150	140
vitrified clays					110
brick sewer					100
iron					
cast, plain	0.0004–0.002	0.0008	150–80	130	100
cast, tar (asphalt) coated	0.0002–0.0006	0.0004	145–50	130	100
cast, cement lined	0.000008	0.000008		150	140
cast, bituminous lined	0.000008	0.000008	160–130	148	140
cast, centrifugally spun	0.00001	0.00001			
galvanized, plain	0.0002–0.0008	0.0005			
wrought, plain	0.0001–0.0003	0.0002	150–80	130	100
miscellaneous					
copper and brass	0.000005	0.000005	150–120	140	130
wood stave	0.0006–0.003	0.002	145–110	120	110
transite	0.000008	0.000008			
lead, tin, glass		0.000005	150–120	140	130
plastic (PVC and ABS)		0.000005	150–120	150–140	130
fiberglass	0.000017	0.000017	160–150	150	150

(Multiply ft by 0.3 to obtain m.)

[a] The following guidelines are provided for selecting Hazen-Williams coefficients for cast-iron pipes of different ages. Values for welded steel pipe are similar to those of cast-iron pipe 5 years older. New pipe, all sizes: $C = 130$. 5 year-old pipe: $C = 120$ (d < 24 in); $C = 115$ (d $\geq$ 24 in). 10 year-old pipe: $C = 105$ (d = 4 in); $C = 110$ (d = 12 in); $C = 85$ (d $\geq$ 30 in). 40 year-old pipe: $C = 65$ (d = 4 in); $C = 80$ (d = 16 in).

[b] C values for sludge are 20 to 40% less than the corresponding water pipe values.

APPENDIX 17.B
Darcy Friction Factors
(turbulent flow)

relative roughness (ϵ/D)

Reynolds no.	0.00000	0.000001	0.0000015	0.00001	0.00002	0.00004	0.00005	0.00006	0.00008
2×10^3	0.0495	0.0495	0.0495	0.0495	0.0495	0.0495	0.0495	0.0495	0.0495
2.5×10^3	0.0461	0.0461	0.0461	0.0461	0.0461	0.0461	0.0461	0.0461	0.0461
3×10^3	0.0435	0.0435	0.0435	0.0435	0.0435	0.0436	0.0436	0.0436	0.0436
4×10^3	0.0399	0.0399	0.0399	0.0399	0.0399	0.0399	0.0400	0.0400	0.0400
5×10^3	0.0374	0.0374	0.0374	0.0374	0.0374	0.0374	0.0374	0.0375	0.0375
6×10^3	0.0355	0.0355	0.0355	0.0355	0.0355	0.0356	0.0356	0.0356	0.0356
7×10^3	0.0340	0.0340	0.0340	0.0340	0.0340	0.0341	0.0341	0.0341	0.0341
8×10^3	0.0328	0.0328	0.0328	0.0328	0.0328	0.0328	0.0329	0.0329	0.0329
9×10^3	0.0318	0.0318	0.0318	0.0318	0.0318	0.0318	0.0318	0.0319	0.0319
1×10^4	0.0309	0.0309	0.0309	0.0309	0.0309	0.0309	0.0310	0.0310	0.0310
1.5×10^4	0.0278	0.0278	0.0278	0.0278	0.0278	0.0279	0.0279	0.0279	0.0280
2×10^4	0.0259	0.0259	0.0259	0.0259	0.0259	0.0260	0.0260	0.0260	0.0261
2.5×10^4	0.0245	0.0245	0.0245	0.0245	0.0246	0.0246	0.0246	0.0247	0.0247
3×10^4	0.0235	0.0235	0.0235	0.0235	0.0235	0.0236	0.0236	0.0236	0.0237
4×10^4	0.0220	0.0220	0.0220	0.0220	0.0220	0.0221	0.0221	0.0222	0.0222
5×10^4	0.0209	0.0209	0.0209	0.0209	0.0210	0.0210	0.0211	0.0211	0.0212
6×10^4	0.0201	0.0201	0.0201	0.0201	0.0201	0.0202	0.0203	0.0203	0.0204
7×10^4	0.0194	0.0194	0.0194	0.0194	0.0195	0.0196	0.0196	0.0197	0.0197
8×10^4	0.0189	0.0189	0.0189	0.0189	0.0190	0.0190	0.0191	0.0191	0.0192
9×10^4	0.0184	0.0184	0.0184	0.0184	0.0185	0.0186	0.0186	0.0187	0.0188
1×10^5	0.0180	0.0180	0.0180	0.0180	0.0181	0.0182	0.0183	0.0183	0.0184
1.5×10^5	0.0166	0.0166	0.0166	0.0166	0.0167	0.0168	0.0169	0.0170	0.0171
2×10^5	0.0156	0.0156	0.0156	0.0157	0.0158	0.0160	0.0160	0.0161	0.0163
2.5×10^5	0.0150	0.0150	0.0150	0.0151	0.0152	0.0153	0.0154	0.0155	0.0157
3×10^5	0.0145	0.0145	0.0145	0.0146	0.0147	0.0149	0.0150	0.0151	0.0153
4×10^5	0.0137	0.0137	0.0137	0.0138	0.0140	0.0142	0.0143	0.0144	0.0146
5×10^5	0.0132	0.0132	0.0132	0.0133	0.0134	0.0137	0.0138	0.0140	0.0142
6×10^5	0.0127	0.0128	0.0128	0.0129	0.0131	0.0133	0.0135	0.0136	0.0139
7×10^5	0.0124	0.0124	0.0124	0.0126	0.0127	0.0131	0.0132	0.0134	0.0136
8×10^5	0.0121	0.0121	0.0121	0.0123	0.0125	0.0128	0.0130	0.0131	0.0134
9×10^5	0.0119	0.0119	0.0119	0.0121	0.0123	0.0126	0.0128	0.0130	0.0133
1×10^6	0.0116	0.0117	0.0117	0.0119	0.0121	0.0125	0.0126	0.0128	0.0131
1.5×10^6	0.0109	0.0109	0.0109	0.0112	0.0114	0.0119	0.0121	0.0123	0.0127
2×10^6	0.0104	0.0104	0.0104	0.0107	0.0110	0.0116	0.0118	0.0120	0.0124
2.5×10^6	0.0100	0.0100	0.0101	0.0104	0.0108	0.0113	0.0116	0.0118	0.0123
3×10^6	0.0097	0.0098	0.0098	0.0102	0.0105	0.0112	0.0115	0.0117	0.0122
4×10^6	0.0093	0.0094	0.0094	0.0098	0.0103	0.0110	0.0113	0.0115	0.0120
5×10^6	0.0090	0.0091	0.0091	0.0096	0.0101	0.0108	0.0111	0.0114	0.0119
6×10^6	0.0087	0.0088	0.0089	0.0094	0.0099	0.0107	0.0110	0.0113	0.0118
7×10^6	0.0085	0.0086	0.0087	0.0093	0.0098	0.0106	0.0110	0.0113	0.0118
8×10^6	0.0084	0.0085	0.0085	0.0092	0.0097	0.0106	0.0109	0.0112	0.0118
9×10^6	0.0082	0.0083	0.0084	0.0091	0.0097	0.0105	0.0109	0.0112	0.0117
1×10^7	0.0081	0.0082	0.0083	0.0090	0.0096	0.0105	0.0109	0.0112	0.0117
1.5×10^7	0.0076	0.0078	0.0079	0.0087	0.0094	0.0104	0.0108	0.0111	0.0116
2×10^7	0.0073	0.0075	0.0076	0.0086	0.0093	0.0103	0.0107	0.0110	0.0116
2.5×10^7	0.0071	0.0073	0.0074	0.0085	0.0093	0.0103	0.0107	0.0110	0.0116
3×10^7	0.0069	0.0072	0.0073	0.0084	0.0092	0.0103	0.0107	0.0110	0.0116
4×10^7	0.0067	0.0070	0.0071	0.0084	0.0092	0.0102	0.0106	0.0110	0.0115
5×10^7	0.0065	0.0068	0.0070	0.0083	0.0092	0.0102	0.0106	0.0110	0.0115

(continued)

APPENDIX 17.B *(continued)*
Darcy Friction Factors
(turbulent flow)

relative roughness (ϵ/D)

Reynolds no.	0.0001	0.00015	0.00020	0.00025	0.00030	0.00035	0.0004	0.0006	0.0008
2×10^3	0.0495	0.0496	0.0496	0.0496	0.0497	0.0497	0.0498	0.0499	0.0501
2.5×10^3	0.0461	0.0462	0.0462	0.0463	0.0463	0.0463	0.0464	0.0466	0.0467
3×10^3	0.0436	0.0437	0.0437	0.0437	0.0438	0.0438	0.0439	0.0441	0.0442
4×10^3	0.0400	0.0401	0.0401	0.0402	0.0402	0.0403	0.0403	0.0405	0.0407
5×10^3	0.0375	0.0376	0.0376	0.0377	0.0377	0.0378	0.0378	0.0381	0.0383
6×10^3	0.0356	0.0357	0.0357	0.0358	0.0359	0.0359	0.0360	0.0362	0.0365
7×10^3	0.0341	0.0342	0.0343	0.0343	0.0344	0.0345	0.0345	0.0348	0.0350
8×10^3	0.0329	0.0330	0.0331	0.0331	0.0332	0.0333	0.0333	0.0336	0.0339
9×10^3	0.0319	0.0320	0.0321	0.0321	0.0322	0.0323	0.0323	0.0326	0.0329
1×10^4	0.0310	0.0311	0.0312	0.0313	0.0313	0.0314	0.0315	0.0318	0.0321
1.5×10^4	0.0280	0.0281	0.0282	0.0283	0.0284	0.0285	0.0285	0.0289	0.0293
2×10^4	0.0261	0.0262	0.0263	0.0264	0.0265	0.0266	0.0267	0.0272	0.0276
2.5×10^4	0.0248	0.0249	0.0250	0.0251	0.0252	0.0254	0.0255	0.0259	0.0264
3×10^4	0.0238	0.0239	0.0240	0.0241	0.0243	0.0244	0.0245	0.0250	0.0255
4×10^4	0.0223	0.0224	0.0226	0.0227	0.0229	0.0230	0.0232	0.0237	0.0243
5×10^4	0.0212	0.0214	0.0216	0.0218	0.0219	0.0221	0.0223	0.0229	0.0235
6×10^4	0.0205	0.0207	0.0208	0.0210	0.0212	0.0214	0.0216	0.0222	0.0229
7×10^4	0.0198	0.0200	0.0202	0.0204	0.0206	0.0208	0.0210	0.0217	0.0224
8×10^4	0.0193	0.0195	0.0198	0.0200	0.0202	0.0204	0.0206	0.0213	0.0220
9×10^4	0.0189	0.0191	0.0194	0.0196	0.0198	0.0200	0.0202	0.0210	0.0217
1×10^5	0.0185	0.0188	0.0190	0.0192	0.0195	0.0197	0.0199	0.0207	0.0215
1.5×10^5	0.0172	0.0175	0.0178	0.0181	0.0184	0.0186	0.0189	0.0198	0.0207
2×10^5	0.0164	0.0168	0.0171	0.0174	0.0177	0.0180	0.0183	0.0193	0.0202
2.5×10^5	0.0158	0.0162	0.0166	0.0170	0.0173	0.0176	0.0179	0.0190	0.0199
3×10^5	0.0154	0.0159	0.0163	0.0166	0.0170	0.0173	0.0176	0.0188	0.0197
4×10^5	0.0148	0.0153	0.0158	0.0162	0.0166	0.0169	0.0172	0.0184	0.0195
5×10^5	0.0144	0.0150	0.0154	0.0159	0.0163	0.0167	0.0170	0.0183	0.0193
6×10^5	0.0141	0.0147	0.0152	0.0157	0.0161	0.0165	0.0168	0.0181	0.0192
7×10^5	0.0139	0.0145	0.0150	0.0155	0.0159	0.0163	0.0167	0.0180	0.0191
8×10^5	0.0137	0.0143	0.0149	0.0154	0.0158	0.0162	0.0166	0.0180	0.0191
9×10^5	0.0136	0.0142	0.0148	0.0153	0.0157	0.0162	0.0165	0.0179	0.0190
1×10^6	0.0134	0.0141	0.0147	0.0152	0.0157	0.0161	0.0165	0.0178	0.0190
1.5×10^6	0.0130	0.0138	0.0144	0.0149	0.0154	0.0159	0.0163	0.0177	0.0189
2×10^6	0.0128	0.0136	0.0142	0.0148	0.0153	0.0158	0.0162	0.0176	0.0188
2.5×10^6	0.0127	0.0135	0.0141	0.0147	0.0152	0.0157	0.0161	0.0176	0.0188
3×10^6	0.0126	0.0134	0.0141	0.0147	0.0152	0.0157	0.0161	0.0176	0.0187
4×10^6	0.0124	0.0133	0.0140	0.0146	0.0151	0.0156	0.0161	0.0175	0.0187
5×10^6	0.0123	0.0132	0.0139	0.0146	0.0151	0.0156	0.0160	0.0175	0.0187
6×10^6	0.0123	0.0132	0.0139	0.0145	0.0151	0.0156	0.0160	0.0175	0.0187
7×10^6	0.0122	0.0132	0.0139	0.0145	0.0151	0.0155	0.0160	0.0175	0.0187
8×10^6	0.0122	0.0131	0.0139	0.0145	0.0150	0.0155	0.0160	0.0175	0.0187
9×10^6	0.0122	0.0131	0.0139	0.0145	0.0150	0.0155	0.0160	0.0175	0.0187
1×10^7	0.0122	0.0131	0.0138	0.0145	0.0150	0.0155	0.0160	0.0175	0.0186
1.5×10^7	0.0121	0.0131	0.0138	0.0144	0.0150	0.0155	0.0159	0.0174	0.0186
2×10^7	0.0121	0.0130	0.0138	0.0144	0.0150	0.0155	0.0159	0.0174	0.0186
2.5×10^7	0.0121	0.0130	0.0138	0.0144	0.0150	0.0155	0.0159	0.0174	0.0186
3×10^7	0.0120	0.0130	0.0138	0.0144	0.0150	0.0155	0.0159	0.0174	0.0186
4×10^7	0.0120	0.0130	0.0138	0.0144	0.0150	0.0155	0.0159	0.0174	0.0186
5×10^7	0.0120	0.0130	0.0138	0.0144	0.0150	0.0155	0.0159	0.0174	0.0186

(continued)

APPENDIX 17.B *(continued)*
Darcy Friction Factors
(turbulent flow)

relative roughness (ϵ/D)

Reynolds no.	0.001	0.0015	0.002	0.0025	0.003	0.0035	0.004	0.006	0.008
2×10^3	0.0502	0.0506	0.0510	0.0513	0.0517	0.0521	0.0525	0.0539	0.0554
2.5×10^3	0.0469	0.0473	0.0477	0.0481	0.0485	0.0489	0.0493	0.0509	0.0524
3×10^3	0.0444	0.0449	0.0453	0.0457	0.0462	0.0466	0.0470	0.0487	0.0503
4×10^3	0.0409	0.0414	0.0419	0.0424	0.0429	0.0433	0.0438	0.0456	0.0474
5×10^3	0.0385	0.0390	0.0396	0.0401	0.0406	0.0411	0.0416	0.0436	0.0455
6×10^3	0.0367	0.0373	0.0378	0.0384	0.0390	0.0395	0.0400	0.0421	0.0441
7×10^3	0.0353	0.0359	0.0365	0.0371	0.0377	0.0383	0.0388	0.0410	0.0430
8×10^3	0.0341	0.0348	0.0354	0.0361	0.0367	0.0373	0.0379	0.0401	0.0422
9×10^3	0.0332	0.0339	0.0345	0.0352	0.0358	0.0365	0.0371	0.0394	0.0416
1×10^4	0.0324	0.0331	0.0338	0.0345	0.0351	0.0358	0.0364	0.0388	0.0410
1.5×10^4	0.0296	0.0305	0.0313	0.0320	0.0328	0.0335	0.0342	0.0369	0.0393
2×10^4	0.0279	0.0289	0.0298	0.0306	0.0315	0.0323	0.0330	0.0358	0.0384
2.5×10^4	0.0268	0.0278	0.0288	0.0297	0.0306	0.0314	0.0322	0.0352	0.0378
3×10^4	0.0260	0.0271	0.0281	0.0291	0.0300	0.0308	0.0317	0.0347	0.0374
4×10^4	0.0248	0.0260	0.0271	0.0282	0.0291	0.0301	0.0309	0.0341	0.0369
5×10^4	0.0240	0.0253	0.0265	0.0276	0.0286	0.0296	0.0305	0.0337	0.0365
6×10^4	0.0235	0.0248	0.0261	0.0272	0.0283	0.0292	0.0302	0.0335	0.0363
7×10^4	0.0230	0.0245	0.0257	0.0269	0.0280	0.0290	0.0299	0.0333	0.0362
8×10^4	0.0227	0.0242	0.0255	0.0267	0.0278	0.0288	0.0298	0.0331	0.0361
9×10^4	0.0224	0.0239	0.0253	0.0265	0.0276	0.0286	0.0296	0.0330	0.0360
1×10^5	0.0222	0.0237	0.0251	0.0263	0.0275	0.0285	0.0295	0.0329	0.0359
1.5×10^5	0.0214	0.0231	0.0246	0.0259	0.0271	0.0281	0.0292	0.0327	0.0357
2×10^5	0.0210	0.0228	0.0243	0.0256	0.0268	0.0279	0.0290	0.0325	0.0355
2.5×10^5	0.0208	0.0226	0.0241	0.0255	0.0267	0.0278	0.0289	0.0325	0.0355
3×10^5	0.0206	0.0225	0.0240	0.0254	0.0266	0.0277	0.0288	0.0324	0.0354
4×10^5	0.0204	0.0223	0.0239	0.0253	0.0265	0.0276	0.0287	0.0323	0.0354
5×10^5	0.0202	0.0222	0.0238	0.0252	0.0264	0.0276	0.0286	0.0323	0.0353
6×10^5	0.0201	0.0221	0.0237	0.0251	0.0264	0.0275	0.0286	0.0323	0.0353
7×10^5	0.0201	0.0221	0.0237	0.0251	0.0264	0.0275	0.0286	0.0322	0.0353
8×10^5	0.0200	0.0220	0.0237	0.0251	0.0263	0.0275	0.0286	0.0322	0.0353
9×10^5	0.0200	0.0220	0.0236	0.0251	0.0263	0.0275	0.0285	0.0322	0.0353
1×10^6	0.0199	0.0220	0.0236	0.0250	0.0263	0.0275	0.0285	0.0322	0.0353
1.5×10^6	0.0198	0.0219	0.0235	0.0250	0.0263	0.0274	0.0285	0.0322	0.0352
2×10^6	0.0198	0.0218	0.0235	0.0250	0.0262	0.0274	0.0285	0.0322	0.0352
2.5×10^6	0.0198	0.0218	0.0235	0.0249	0.0262	0.0274	0.0285	0.0322	0.0352
3×10^6	0.0197	0.0218	0.0235	0.0249	0.0262	0.0274	0.0285	0.0321	0.0352
4×10^6	0.0197	0.0218	0.0235	0.0249	0.0262	0.0274	0.0284	0.0321	0.0352
5×10^6	0.0197	0.0218	0.0235	0.0249	0.0262	0.0274	0.0284	0.0321	0.0352
6×10^6	0.0197	0.0218	0.0235	0.0249	0.0262	0.0274	0.0284	0.0321	0.0352
7×10^6	0.0197	0.0218	0.0234	0.0249	0.0262	0.0274	0.0284	0.0321	0.0352
8×10^6	0.0197	0.0218	0.0234	0.0249	0.0262	0.0274	0.0284	0.0321	0.0352
9×10^6	0.0197	0.0218	0.0234	0.0249	0.0262	0.0274	0.0284	0.0321	0.0352
1×10^7	0.0197	0.0218	0.0234	0.0249	0.0262	0.0273	0.0284	0.0321	0.0352
1.5×10^7	0.0197	0.0217	0.0234	0.0249	0.0262	0.0273	0.0284	0.0321	0.0352
2×10^7	0.0197	0.0217	0.0234	0.0249	0.0262	0.0273	0.0284	0.0321	0.0352
2.5×10^7	0.0196	0.0217	0.0234	0.0249	0.0262	0.0273	0.0284	0.0321	0.0352
3×10^7	0.0196	0.0217	0.0234	0.0249	0.0262	0.0273	0.0284	0.0321	0.0352
4×10^7	0.0196	0.0217	0.0234	0.0249	0.0262	0.0273	0.0284	0.0321	0.0352
5×10^7	0.0196	0.0217	0.0234	0.0249	0.0262	0.0273	0.0284	0.0321	0.0352

(continued)

APPENDIX 17.B *(continued)*
Darcy Friction Factors
(turbulent flow)

relative roughness (ϵ/D)

Reynolds no.	0.01	0.015	0.02	0.025	0.03	0.035	0.04	0.045	0.05
2×10^3	0.0568	0.0602	0.0635	0.0668	0.0699	0.0730	0.0760	0.0790	0.0819
2.5×10^3	0.0539	0.0576	0.0610	0.0644	0.0677	0.0709	0.0740	0.0770	0.0800
3×10^3	0.0519	0.0557	0.0593	0.0628	0.0661	0.0694	0.0725	0.0756	0.0787
4×10^3	0.0491	0.0531	0.0570	0.0606	0.0641	0.0674	0.0707	0.0739	0.0770
5×10^3	0.0473	0.0515	0.0555	0.0592	0.0628	0.0662	0.0696	0.0728	0.0759
6×10^3	0.0460	0.0504	0.0544	0.0583	0.0619	0.0654	0.0688	0.0721	0.0752
7×10^3	0.0450	0.0495	0.0537	0.0576	0.0613	0.0648	0.0682	0.0715	0.0747
8×10^3	0.0442	0.0489	0.0531	0.0571	0.0608	0.0644	0.0678	0.0711	0.0743
9×10^3	0.0436	0.0484	0.0526	0.0566	0.0604	0.0640	0.0675	0.0708	0.0740
1×10^4	0.0431	0.0479	0.0523	0.0563	0.0601	0.0637	0.0672	0.0705	0.0738
1.5×10^4	0.0415	0.0466	0.0511	0.0553	0.0592	0.0628	0.0664	0.0698	0.0731
2×10^4	0.0407	0.0459	0.0505	0.0547	0.0587	0.0624	0.0660	0.0694	0.0727
2.5×10^4	0.0402	0.0455	0.0502	0.0544	0.0584	0.0621	0.0657	0.0691	0.0725
3×10^4	0.0398	0.0452	0.0499	0.0542	0.0582	0.0619	0.0655	0.0690	0.0723
4×10^4	0.0394	0.0448	0.0496	0.0539	0.0579	0.0617	0.0653	0.0688	0.0721
5×10^4	0.0391	0.0446	0.0494	0.0538	0.0578	0.0616	0.0652	0.0687	0.0720
6×10^4	0.0389	0.0445	0.0493	0.0536	0.0577	0.0615	0.0651	0.0686	0.0719
7×10^4	0.0388	0.0443	0.0492	0.0536	0.0576	0.0614	0.0650	0.0685	0.0719
8×10^4	0.0387	0.0443	0.0491	0.0535	0.0576	0.0614	0.0650	0.0685	0.0718
9×10^4	0.0386	0.0442	0.0491	0.0535	0.0575	0.0613	0.0650	0.0684	0.0718
1×10^5	0.0385	0.0442	0.0490	0.0534	0.0575	0.0613	0.0649	0.0684	0.0718
1.5×10^5	0.0383	0.0440	0.0489	0.0533	0.0574	0.0612	0.0648	0.0683	0.0717
2×10^5	0.0382	0.0439	0.0488	0.0532	0.0573	0.0612	0.0648	0.0683	0.0717
2.5×10^5	0.0381	0.0439	0.0488	0.0532	0.0573	0.0611	0.0648	0.0683	0.0716
3×10^5	0.0381	0.0438	0.0488	0.0532	0.0573	0.0611	0.0648	0.0683	0.0716
4×10^5	0.0381	0.0438	0.0487	0.0532	0.0573	0.0611	0.0647	0.0682	0.0716
5×10^5	0.0380	0.0438	0.0487	0.0531	0.0572	0.0611	0.0647	0.0682	0.0716
6×10^5	0.0380	0.0438	0.0487	0.0531	0.0572	0.0611	0.0647	0.0682	0.0716
7×10^5	0.0380	0.0438	0.0487	0.0531	0.0572	0.0611	0.0647	0.0682	0.0716
8×10^5	0.0380	0.0437	0.0487	0.0531	0.0572	0.0611	0.0647	0.0682	0.0716
9×10^5	0.0380	0.0437	0.0487	0.0531	0.0572	0.0610	0.0647	0.0682	0.0716
1×10^6	0.0380	0.0437	0.0487	0.0531	0.0572	0.0610	0.0647	0.0682	0.0716
1.5×10^6	0.0379	0.0437	0.0487	0.0531	0.0572	0.0610	0.0647	0.0682	0.0716
2×10^6	0.0379	0.0437	0.0487	0.0531	0.0572	0.0610	0.0647	0.0682	0.0716
2.5×10^6	0.0379	0.0437	0.0487	0.0531	0.0572	0.0610	0.0647	0.0682	0.0716
3×10^6	0.0379	0.0437	0.0487	0.0531	0.0572	0.0610	0.0647	0.0682	0.0716
4×10^6	0.0379	0.0437	0.0486	0.0531	0.0572	0.0610	0.0647	0.0682	0.0716
5×10^6	0.0379	0.0437	0.0486	0.0531	0.0572	0.0610	0.0647	0.0682	0.0716
6×10^6	0.0379	0.0437	0.0486	0.0531	0.0572	0.0610	0.0647	0.0682	0.0716
7×10^6	0.0379	0.0437	0.0486	0.0531	0.0572	0.0610	0.0647	0.0682	0.0716
8×10^6	0.0379	0.0437	0.0486	0.0531	0.0572	0.0610	0.0647	0.0682	0.0716
9×10^6	0.0379	0.0437	0.0486	0.0531	0.0572	0.0610	0.0647	0.0682	0.0716
1×10^7	0.0379	0.0437	0.0486	0.0531	0.0572	0.0610	0.0647	0.0682	0.0716
1.5×10^7	0.0379	0.0437	0.0486	0.0531	0.0572	0.0610	0.0647	0.0682	0.0716
2×10^7	0.0379	0.0437	0.0486	0.0531	0.0572	0.0610	0.0647	0.0682	0.0716
2.5×10^7	0.0379	0.0437	0.0486	0.0531	0.0572	0.0610	0.0647	0.0682	0.0716
3×10^7	0.0379	0.0437	0.0486	0.0531	0.0572	0.0610	0.0647	0.0682	0.0716
4×10^7	0.0379	0.0437	0.0486	0.0531	0.0572	0.0610	0.0647	0.0682	0.0716
5×10^7	0.0379	0.0437	0.0486	0.0531	0.0572	0.0610	0.0647	0.0682	0.0716

APPENDIX 17.C
Flow of Water Through Schedule-40 Steel Pipe

pressure drop per 1000 ft of schedule-40 steel pipe, in lbf/in²

discharge (gal/min)	velocity (ft/sec)	pressure drop	velocity (ft/sec)	pressure drop	velocity (ft/sec)	pressure drop	velocity (ft/sec)	pressure drop	velocity (ft/sec)	pressure drop	velocity (ft/sec)	pressure drop	velocity (ft/sec)	pressure drop	velocity (ft/sec)	pressure drop	velocity (ft/sec)	pressure drop
1	1 in: 0.37	0.49																
2	0.74	1.70	1¼ in: 0.43	0.45														
3	1.12	3.53	0.64	0.94	1½ in: 0.47	0.44												
4	1.49	5.94	0.86	1.55	0.63	0.74												
5	1.86	9.02	1.07	2.36	0.79	1.12	2 in											
6	2.24	12.25	1.28	3.30	0.95	1.53	0.57	0.46										
8	2.98	21.1	1.72	5.52	1.26	2.63	0.76	0.75	2½ in									
10	3.72	30.8	2.14	8.34	1.57	3.86	0.96	1.14	0.67	0.48								
15	5.60	64.6	3.21	17.6	2.36	8.13	1.43	2.33	1.00	0.99	3 in							
20	7.44	110.5	4.29	29.1	3.15	13.5	1.91	3.86	1.34	1.64	0.87	0.59	3½ in					
25			5.36	43.7	3.94	20.2	2.39	5.81	1.68	2.48	1.08	0.67	0.81	0.42				
30			6.43	62.9	4.72	29.1	2.87	8.04	2.01	3.43	1.30	1.21	0.97	0.60	4 in			
35			7.51	82.5	5.51	38.2	3.35	10.95	2.35	4.49	1.52	1.58	1.14	.079	0.88	0.42		
40					6.30	47.8	3.82	13.7	2.68	5.88	1.74	2.06	1.30	1.00	1.01	0.53		
45					7.08	60.6	4.30	17.4	3.00	7.14	1.95	2.51	1.46	1.21	1.13	0.67		
50					7.87	74.7	4.78	20.6	3.35	8.82	2.17	3.10	1.62	1.44	1.26	0.80		
60							5.74	29.6	4.02	12.2	2.60	4.29	1.95	2.07	1.51	1.10	5 in	
70							6.69	38.6	4.69	15.3	3.04	5.84	2.27	2.71	1.76	1.50	1.12	0.48
80							7.65	50.3	5.37	21.7	3.48	7.62	2.59	3.53	2.01	1.87	1.28	0.63
90	6 in						8.60	63.6	6.04	26.1	3.91	9.22	2.92	4.46	2.26	2.37	1.44	0.80
100	1.11	0.39					9.56	75.1	6.71	32.3	4.34	11.4	3.24	5.27	2.52	2.81	1.60	0.95
125	1.39	0.56							8.38	48.2	5.42	17.1	4.05	7.86	3.15	4.38	2.00	1.48
150	1.67	0.78							10.06	60.4	6.51	23.5	4.86	11.3	3.78	6.02	2.41	2.04
175	1.94	1.06							11.73	90.0	7.59	32.0	5.67	14.7	4.41	8.20	2.81	2.78
200	2.22	1.32	8 in								8.68	39.7	6.48	19.2	5.04	10.2	3.21	3.46
225	2.50	1.66	1.44	0.44							9.77	50.2	7.29	23.1	5.67	12.9	3.61	4.37
250	2.78	2.05	1.60	0.55							10.85	61.9	8.10	28.5	6.30	15.9	4.01	5.14
275	3.06	2.36	1.76	0.63							11.94	75.0	8.91	34.4	6.93	18.3	4.41	6.22
300	3.33	2.80	1.92	0.75							13.02	84.7	9.72	40.9	7.56	21.8	4.81	7.41
325	3.61	3.29	2.08	0.88									10.53	45.5	8.18	25.5	5.21	8.25
350	3.89	3.62	2.24	0.97									11.35	52.7	8.82	29.7	5.61	9.57
375	4.16	4.16	2.40	1.11									12.17	60.7	9.45	32.3	6.01	11.0
400	4.44	4.72	2.56	1.27									12.97	68.9	10.08	36.7	6.41	12.5
425	4.72	5.34	2.72	1.43									13.78	77.8	10.70	41.5	6.82	14.1
450	5.00	5.96	2.88	1.60	10 in								14.59	87.3	11.33	46.5	7.22	15.0
475	5.27	6.66	3.04	1.69	1.93	0.30									11.96	51.7	7.62	16.7
500	5.55	7.39	3.20	1.87	2.04	0.63									12.59	57.3	8.02	18.5
550	6.11	8.94	3.53	2.26	2.24	0.70									13.84	69.3	8.82	22.4
600	6.66	10.6	3.85	2.70	2.44	0.86									15.10	82.5	9.62	26.7
650	7.21	11.8	4.17	3.16	2.65	1.01	12 in										10.42	31.3
700	7.77	13.7	4.49	3.69	2.85	1.18	2.01	0.48									11.22	36.3
750	8.32	15.7	4.81	4.21	3.05	1.35	2.15	0.55									12.02	41.6
800	8.88	17.8	5.13	4.79	3.26	1.54	2.29	0.62	14 in								12.82	44.7
850	9.44	20.2	5.45	5.11	3.46	1.74	2.44	0.70	2.02	0.43							13.62	50.5
900	10.00	22.6	5.77	5.73	3.66	1.94	2.58	0.79	2.14	0.48							14.42	56.6
950	10.55	23.7	6.09	6.38	3.87	2.23	2.72	0.88	2.25	0.53							15.22	63.1
1000	11.10	26.3	6.41	7.08	4.07	2.40	2.87	0.98	2.38	0.59							16.02	70.0
1100	12.22	31.8	7.05	8.56	4.48	2.74	3.16	1.18	2.61	0.68	16 in						17.63	84.6
1200	13.32	37.8	7.69	10.2	4.88	3.27	3.45	1.40	2.85	0.81	2.18	0.40						
1300	14.43	44.4	8.33	11.3	5.29	3.86	3.73	1.56	3.09	0.95	2.36	0.47						
1400	15.54	51.5	8.97	13.0	5.70	4.44	4.02	1.80	3.32	1.10	2.54	0.54						
1500	16.65	55.5	9.62	15.0	6.10	5.11	4.30	2.07	3.55	1.19	2.73	0.62						
1600	17.76	63.1	10.26	17.0	6.51	5.46	4.59	2.36	3.80	1.35	2.91	0.71	18 in					
1800	19.98	79.8	11.54	21.6	7.32	6.91	5.16	2.98	4.27	1.71	3.27	0.85	2.58	0.48				
2000	22.20	98.5	12.83	25.0	8.13	8.54	5.73	3.47	4.74	2.11	3.63	1.05	2.88	0.56				
2500			16.03	39.0	10.18	12.5	7.17	5.41	5.92	3.09	4.54	1.63	3.59	0.88	20 in			
3000			19.24	52.4	12.21	18.0	8.60	7.31	7.12	4.45	5.45	2.21	4.31	1.27	3.45	0.73		
3500			22.43	71.4	14.25	22.9	10.03	9.95	8.32	6.18	6.35	3.00	5.03	1.52	4.03	0.94	24 in	
4000			25.65	93.3	16.28	29.9	11.48	13.0	9.49	7.92	7.25	3.92	5.74	2.12	4.61	1.22	3.19	0.51
4500					18.31	37.8	12.90	15.4	10.67	9.36	8.17	4.97	6.47	2.50	5.19	1.55	3.59	0.60
5000					20.35	46.7	14.34	18.9	11.84	11.6	9.08	5.72	7.17	3.08	5.76	1.78	3.99	0.74
6000					24.42	67.2	17.21	27.3	14.32	15.4	10.88	8.24	8.62	4.45	6.92	2.57	4.80	1.00
7000					28.50	85.1	20.08	37.2	16.60	21.0	12.69	12.2	10.04	6.06	8.06	3.50	5.68	1.36
8000							22.95	45.1	18.98	27.4	14.52	13.6	11.48	7.34	9.23	4.57	6.38	1.80
9000							25.80	57.0	21.35	34.7	16.32	17.2	12.92	9.20	10.37	5.36	7.19	2.25
10,000							28.63	70.4	23.75	42.9	18.16	21.2	14.37	11.5	11.53	6.63	7.96	2.78
12,000							34.38	93.6	28.50	61.8	21.80	30.9	17.23	16.5	13.83	9.54	9.57	3.71
14,000									33.20	84.0	25.42	41.6	20.10	20.7	16.14	12.0	11.18	5.05
16,000											29.05	54.4	22.96	27.1	18.43	15.7	12.77	6.60

(Multiply gal/min by 0.0631 to obtain L/s.)
(Multiply in by 25.4 to obtain mm.)
(Multiply ft/sec by 0.3 to obtain m/s.)
(Multiply lbf/in²-1000 ft by 2.3 to obtain kPa/100 m.)

Reprinted with permission from *Design of Fluid Systems Hook-Ups*, copyright © 1997, by Spirax Sarco®.

APPENDIX 17.D
Equivalent Length of Straight Pipe for Various (Generic) Fittings
(in feet, turbulent flow only, for any fluid)

fittings			¼	⅜	½	¾	1	1¼	1½	2	2½	3	4	5	6	8	10	12	14	16	18	20	24
reg-ular 90° ell	screwed	steel	2.3	3.1	3.6	4.4	5.2	6.6	7.4	8.5	9.3	11.0	13.0										
		cast iron										9.0	11.0										
	flanged	steel			0.92	1.2	1.6	2.1	2.4	3.1	3.6	4.4	5.9	7.3	8.9	12.0	14.0	17.0	18.0	21.0	23.0	25.0	30.0
		cast iron										3.6	4.8		7.2	9.8	12.0	15.0	17.0	19.0	22.0	24.0	28.0
long radius 90° ell	screwed	steel	1.5	2.0	2.2	2.3	2.7	3.2	3.4	3.6	3.6	4.0	4.6										
		cast iron										3.3	3.7										
	flanged	steel			1.1	1.3	1.6	2.0	2.3	2.7	2.9	3.4	4.2	5.0	5.7	7.0	8.0	9.0	9.4	10.0	11.0	12.0	14.0
		cast iron										2.8	3.4		4.7	5.7	6.8	7.8	8.6	9.6	11.0	11.0	13.0
reg-ular 45° ell	screwed	steel	0.34	0.52	0.71	0.92	1.3	1.7	2.1	2.7	3.2	4.0	5.5										
		cast iron										3.3	4.5										
	flanged	steel			0.45	0.59	0.81	1.1	1.3	1.7	2.0	2.6	3.5	4.5	5.6	7.7	9.0	11.0	13.0	15.0	16.0	18.0	22.0
		cast iron										2.1	2.9		4.5	6.3	8.1	9.7	12.0	13.0	15.0	17.0	20.0
tee-line flow	screwed	steel	0.79	1.2	1.7	2.4	3.2	4.6	5.6	7.7	9.3	12.0	17.0										
		cast iron										9.9	14.0										
	flanged	steel			0.69	0.82	1.0	1.3	1.5	1.8	1.9	2.2	2.8	3.3	3.8	4.7	5.2	6.0	6.4	7.2	7.6	8.2	9.6
		cast iron										1.9	2.2		3.1	3.9	4.6	5.2	5.9	6.5	7.2	7.7	8.8
tee-branch flow	screwed	steel	2.4	3.5	4.2	5.3	6.6	8.7	9.9	12.0	13.0	17.0	21.0										
		cast iron										14.0	17.0										
	flanged	steel			2.0	2.6	3.3	4.4	5.2	6.6	7.5	9.4	12.0	15.0	18.0	24.0	30.0	34.0	37.0	43.0	47.0	52.0	62.0
		cast iron										7.7	10.0		15.0	20.0	25.0	30.0	35.0	39.0	44.0	49.0	57.0
180° return bend — regular	screwed	steel	2.3	3.1	3.6	4.4	5.2	6.6	7.4	8.5	9.3	11.0	13.0										
		cast iron										9.0	11.0										
	flanged	steel			0.92	1.2	1.6	2.1	2.4	3.1	3.6	4.4	5.9	7.3	8.9	12.0	14.0	17.0	18.0	21.0	23.0	25.0	30.0
		cast iron										3.6	4.8		7.2	9.8	12.0	15.0	17.0	19.0	22.0	24.0	28.0
180° return bend — long radius	flanged	steel			1.1	1.3	1.6	2.0	2.3	2.7	2.9	3.4	4.2	5.0	5.7	7.0	8.0	9.0	9.4	10.0	11.0	12.0	14.0
		cast iron										2.8	3.4		4.7	5.7	6.8	7.8	8.6	9.6	11.0	11.0	13.0
globe valve	screwed	steel	21.0	22.0	22.0	24.0	29.0	37.0	42.0	54.0	62.0	79.0	110.0										
		cast iron										65.0	86.0										
	flanged	steel			38.0	40.0	45.0	54.0	59.0	70.0	77.0	94.0	120.0	150.0	190.0	260.0	310.0	390.0					
		cast iron										77.0	99.0		150.0	210.0	270.0	330.0					
gate valve	screwed	steel	0.32	0.45	0.56	0.67	0.84	1.1	1.2	1.5	1.7	1.9	2.5										
		cast iron										1.6	2.0										
	flanged	steel								2.6	2.7	2.8	2.9	3.1	3.2	3.2	3.2	3.2	3.2	3.2	3.2	3.2	3.2
		cast iron										2.3	2.4		2.6	2.7	2.8	2.9	2.9	3.0	3.0	3.0	3.0
angle valve	screwed	steel	12.8	15.0	15.0	15.0	17.0	18.0	18.0	18.0	18.0	18.0	18.0										
		cast iron										15.0	15.0										
	flanged	steel			15.0	15.0	17.0	18.0	18.0	21.0	22.0	28.0	38.0	50.0	63.0	90.0	120.0	140.0	160.0	190.0	210.0	240.0	300.0
		cast iron										23.0	31.0		52.0	74.0	98.0	120.0	150.0	170.0	200.0	230.0	280.0
swing check valve	screwed	steel	7.2	7.3	8.0	8.8	11.0	13.0	15.0	19.0	22.0	27.0	38.0										
		cast iron										22.0	31.0										
	flanged	steel			3.8	5.3	7.2	10.0	12.0	17.0	21.0	27.0	38.0	50.0	63.0	90.0	120.0	140.0					
		cast iron										22.0	31.0		52.0	74.0	98.0	120.0					
coup-ling or union	screwed	steel	0.14	0.18	0.21	0.24	0.29	0.36	0.39	0.45	0.47	0.53	0.65										
		cast iron										0.44	0.52										
inlet — bell mouth inlet		steel	0.04	0.07	0.10	0.13	0.18	0.26	0.31	0.43	0.52	0.67	0.95	1.3	1.6	2.3	2.9	3.5	4.0	4.7	5.3	6.1	7.6
		cast iron										0.55	0.77		1.3	1.9	2.4	3.0	3.6	4.3	5.0	5.7	7.0
square mouth inlet		steel	0.44	0.68	0.96	1.3	1.8	2.6	3.1	4.3	5.2	6.7	9.5	13.0	16.0	23.0	29.0	35.0	40.0	47.0	53.0	61.0	76.0
		cast iron										5.5	7.7		13.0	19.0	24.0	30.0	36.0	43.0	50.0	57.0	70.0
re-entrant pipe		steel	0.88	1.4	1.9	2.6	3.6	5.1	6.2	8.5	10.0	13.0	19.0	25.0	32.0	45.0	58.0	70.0	80.0	95.0	110.0	120.0	150.0
		cast iron										11.0	15.0		26.0	37.0	49.0	61.0	73.0	86.0	100.0	110.0	140.0

(Multiply in by 25.4 to obtain mm. Multiply ft by 0.3 to obtain m.)

APPENDIX 17.E
Hazen-Williams Nomograph
$(C = 100)$

Quantity (i.e., flow rate) and velocity are proportional to the C-value. For values of C other than 100, the quantity and velocity must be converted according to $\dot{V}_{\text{actual}} = \dot{V}_{\text{chart}} C_{\text{actual}}/100$. When quantity is the unknown, use the chart with known values of diameter, slope, or velocity to find $\dot{V}_{\text{chart}}$, and then convert to $\dot{V}_{\text{actual}}$. When velocity is the unknown, use the chart with the known values of diameter, slope, or quantity to find V_{chart}, then convert to V_{actual}. If $\dot{V}_{\text{actual}}$ is known, it must be converted to $\dot{V}_{\text{chart}}$ before this nomograph can be used. In that case, the diameter, loss, and quantity are as read from this chart.

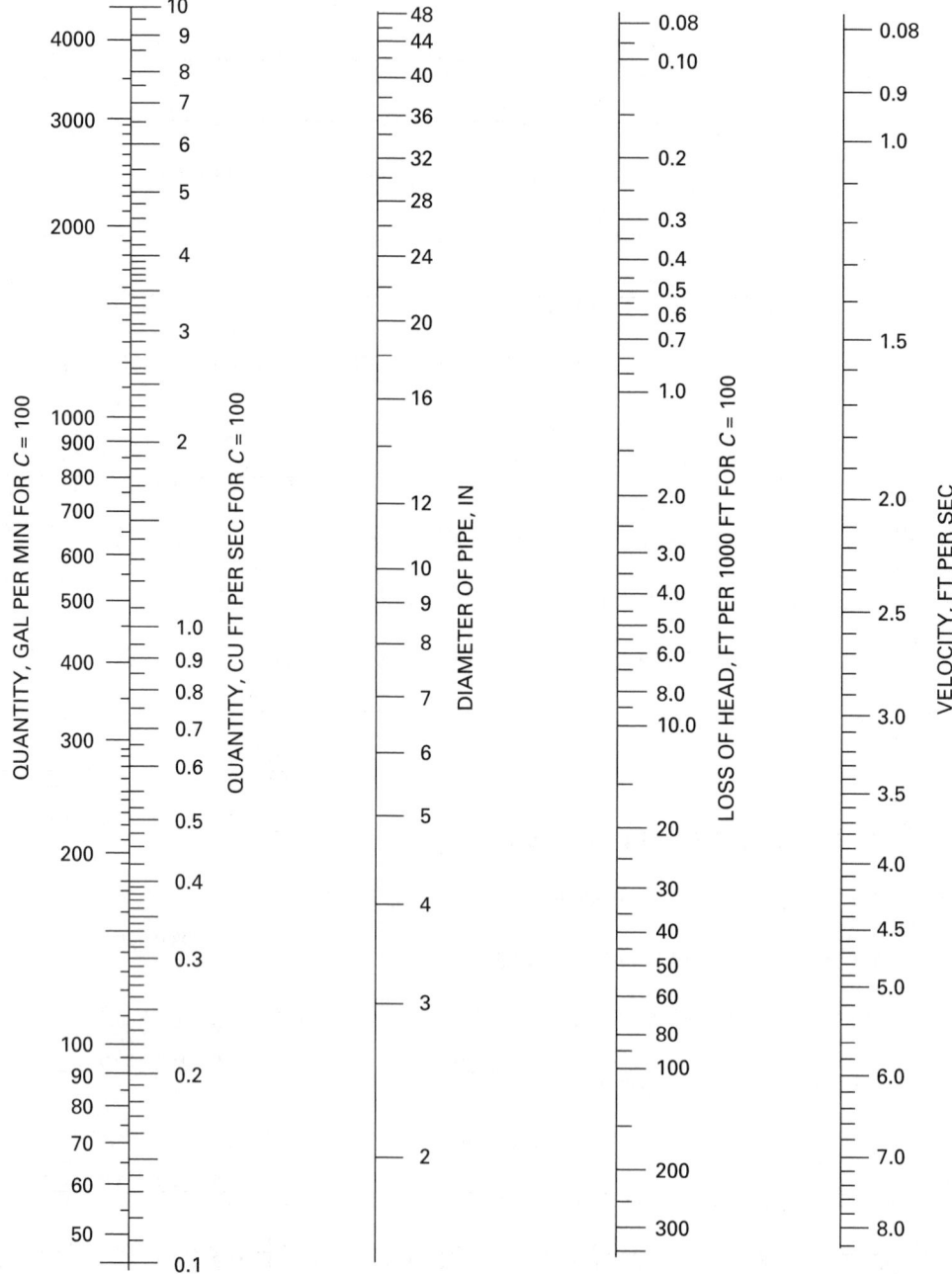

(Multiply gal/min by 0.0631 to obtain L/s.)
(Multiply ft³/sec by 28.3 to obtain L/s.)
(Multiply in by 25.4 to obtain mm.)
(Multiply ft/1000 ft by 0.1 to obtain m/100 m.)
(Multiply ft/sec by 0.3 to obtain m/s.)

APPENDIX 18.A
International Standard Atmosphere

customary U.S. units

atmospheric layer	altitude (ft)	temperature (°R)	pressure (psia)
troposphere	0	518.7	14.696
	1000	515.1	14.175
	2000	511.6	13.664
	3000	508.0	13.168
	4000	504.4	12.692
	5000	500.9	12.225
	6000	497.3	11.778
	7000	493.7	11.341
	8000	490.2	10.914
	9000	486.6	10.501
	10,000	483.0	10.108
	11,000	479.5	9.720
	12,000	475.9	9.347
	13,000	472.3	8.983
	14,000	468.8	8.630
	15,000	465.2	8.291
	16,000	461.6	7.962
	17,000	458.1	7.642
	18,000	454.5	7.338
	19,000	450.9	7.038
	20,000	447.4	6.753
	21,000	443.8	6.473
	22,000	440.2	6.203
	23,000	436.7	5.943
	24,000	433.1	5.693
	25,000	429.5	5.451
	26,000	426.0	5.216
	27,000	422.4	4.990
	28,000	418.8	4.774
	29,000	415.3	4.563
	30,000	411.7	4.362
	31,000	408.1	4.165
	32,000	404.6	3.978
	33,000	401.0	3.797
	34,000	397.5	3.625
tropopause	35,000	393.9	3.458
	36,000	392.7	3.296
	37,000	392.7	3.143
	38,000	392.7	2.996
	39,000	392.7	2.854
	40,000	392.7	2.721
	41,000	392.7	2.593
	42,000	392.7	2.475
	43,000	392.7	2.358
	44,000	392.7	2.250
	45,000	392.7	2.141
	46,000	392.7	2.043
	47,000	392.7	1.950
	48,000	392.7	1.857
	49,000	392.7	1.768
	50,000	392.7	1.690
	51,000	392.7	1.611
	52,000	392.7	1.532
	53,000	392.7	1.464
	54,000	392.7	1.395
	55,000	392.7	1.331
	56,000	392.7	1.267
	57,000	392.7	1.208
	58,000	392.7	1.154
	59,000	392.7	1.100
	60,000	392.7	1.046
	61,000	392.7	0.997
	62,000	392.7	0.953
	63,000	392.7	0.909
	64,000	392.7	0.864
stratosphere (to approximately 160,000 ft)	65,000	392.7	0.825

SI units

atmospheric layer	altitude (m)	temperature (K)	pressure (bar)
troposphere	0	288.15	1.01325
	500	284.9	0.9546
	1000	281.7	0.8988
	1500	278.4	0.8456
	2000	275.2	0.7950
	2500	271.9	0.7469
	3000	268.7	0.7012
	3500	265.4	0.6578
	4000	262.2	0.6166
	4500	258.9	0.5775
	5000	255.7	0.5405
	5500	252.4	0.5054
	6000	249.2	0.4722
	6500	245.9	0.4408
	7000	242.7	0.4111
	7500	239.5	0.3830
	8000	236.2	0.3565
	8500	233.0	0.3315
	9000	229.7	0.3080
	9500	226.5	0.2858
	10 000	223.3	0.2650
tropopause	10 500	220.0	0.2454
	11 000	216.8	0.2270
	11 500	216.7	0.2098
	12 000	216.7	0.1940
	12 500	216.7	0.1793
	13 000	216.7	0.1658
	13 500	216.7	0.1533
	14 000	216.7	0.1417
	14 500	216.7	0.1310
	15 000	216.7	0.1211
	15 500	216.7	0.1120
	16 000	216.7	0.1035
	16 500	216.7	0.09572
	17 000	216.7	0.08850
	17 500	216.7	0.08182
	18 000	216.7	0.07565
	18 500	216.7	0.06995
	19 000	216.7	0.06467
	19 500	216.7	0.05980
	20 000	216.7	0.05529
stratosphere (to approximately 50 000 m)	22 000	218.6	0.04047
	24 000	220.6	0.02972
	26 000	222.5	0.02188
	28 000	224.5	0.01616
	30 000	226.5	0.01197
	32 000	228.5	0.00889

APPENDIX 18.B
Properties of Saturated Steam by Temperature

temp. (°F)	absolute pressure (psia)	specific volume (ft³/lbm)		internal energy (Btu/lbm)		enthalpy (Btu/lbm)			entropy (Btu/lbm-°R)	
		sat. liquid (v_f)	sat. vapor (v_g)	sat. liquid (u_f)	sat. vapor (u_g)	sat. liquid (h_f)	evap. (h_{fg})	sat. vapor (h_g)	sat. liquid (s_f)	sat. vapor (s_g)
32	0.0886	0.01602	3302	−0.02	1021.0	−0.02	1075.2	1075.2	−0.00004	2.1868
35	0.1000	0.01602	2945	3.00	1022.0	3.00	1074.0	1077.0	0.00609	2.1760
40	0.1217	0.01602	2443	8.03	1024.0	8.03	1071.0	1079.0	0.01620	2.1590
45	0.1476	0.01602	2036	13.05	1025.0	13.05	1068.0	1081.0	0.02620	2.1420
50	0.1781	0.01602	1703	18.07	1026.9	18.07	1065.0	1083.1	0.03609	2.1257
52	0.1918	0.01603	1587	20.07	1027.6	20.07	1063.8	1083.9	0.04001	2.1192
54	0.2065	0.01603	1481	22.07	1028.2	22.07	1062.7	1084.8	0.04392	2.1128
56	0.2221	0.01603	1382	24.07	1028.9	24.08	1061.6	1085.7	0.04781	2.1065
58	0.2387	0.01603	1291	26.08	1029.6	26.08	1060.5	1086.6	0.05168	2.1002
60	0.2564	0.01604	1206	28.08	1030.2	28.08	1059.3	1087.4	0.05554	2.0940
62	0.2752	0.01604	1128	30.08	1030.9	30.08	1058.2	1088.3	0.05938	2.0879
64	0.2953	0.01604	1055	32.08	1031.5	32.08	1057.1	1089.2	0.06321	2.0818
66	0.3166	0.01604	987.7	34.08	1032.2	34.08	1055.9	1090.0	0.06702	2.0758
68	0.3393	0.01605	925.2	36.08	1032.8	36.08	1054.8	1090.9	0.07081	2.0698
70	0.3634	0.01605	867.1	38.07	1033.5	38.08	1053.7	1091.8	0.07459	2.0639
72	0.3889	0.01606	813.2	40.07	1034.1	40.07	1052.5	1092.6	0.07836	2.0581
74	0.4160	0.01606	763.0	42.07	1034.8	42.07	1051.4	1093.5	0.08211	2.0523
76	0.4448	0.01606	716.3	44.07	1035.4	44.07	1050.3	1094.4	0.08585	2.0466
78	0.4752	0.01607	672.9	46.07	1036.1	46.07	1049.1	1095.2	0.08957	2.0409
80	0.5075	0.01607	632.4	48.06	1036.7	48.07	1048.0	1096.1	0.09328	2.0353
82	0.5416	0.01608	594.7	50.06	1037.4	50.06	1046.9	1097.0	0.09697	2.0297
84	0.5778	0.01608	559.5	52.06	1038.0	52.06	1045.7	1097.8	0.1007	2.0242
86	0.6160	0.01609	526.7	54.05	1038.7	54.06	1044.6	1098.7	0.1043	2.0187
88	0.6564	0.01610	496.0	56.05	1039.3	56.05	1043.4	1099.5	0.1080	2.0133
90	0.6990	0.01610	467.4	58.05	1039.9	58.05	1042.4	1100.4	0.1116	2.0079
92	0.7441	0.01611	440.7	60.04	1040.6	60.05	1041.3	1101.3	0.1152	2.0026
94	0.7917	0.01611	415.6	62.04	1041.2	62.04	1040.1	1102.1	0.1189	1.9974
96	0.8418	0.01612	392.3	64.04	1041.9	64.04	1039.0	1103.0	0.1225	1.9922
98	0.8947	0.01613	370.3	66.03	1042.5	66.04	1037.8	1103.8	0.1260	1.9870
100	0.9505	0.01613	349.8	68.03	1043.2	68.03	1036.7	1104.7	0.1296	1.9819
110	1.277	0.01617	265.0	78.01	1046.4	78.02	1031.0	1109.0	0.1473	1.9570
120	1.695	0.01621	203.0	88.00	1049.6	88.00	1025.2	1113.2	0.1647	1.9333
130	2.226	0.01625	157.1	97.99	1052.7	97.99	1019.4	1117.4	0.1818	1.9106
140	2.893	0.01629	122.8	107.98	1055.8	107.99	1013.6	1121.6	0.1986	1.8888
150	3.723	0.01634	96.9	117.98	1059.0	117.99	1007.7	1125.7	0.2151	1.8680
160	4.747	0.01639	77.2	127.98	1062.0	128.00	1001.8	1129.8	0.2314	1.8481
170	6.000	0.01645	62.0	138.00	1065.1	138.01	995.9	1133.9	0.2474	1.8290
180	7.520	0.01651	50.2	148.02	1068.1	148.04	989.9	1137.9	0.2632	1.8106
190	9.350	0.01657	40.9	158.05	1071.0	158.08	983.7	1141.8	0.2788	1.7930
200	11.538	0.01663	33.6	168.09	1074.0	168.13	977.6	1145.7	0.2941	1.7760
210	14.14	0.01670	27.79	178.2	1076.8	178.2	971.3	1149.5	0.3092	1.7597
220	17.20	0.01677	23.13	188.2	1079.6	188.3	965.0	1153.3	0.3242	1.7440
230	20.80	0.01685	19.37	198.3	1082.4	198.4	958.5	1156.9	0.3389	1.7288
240	24.99	0.01692	16.31	208.4	1085.1	208.5	952.0	1160.5	0.3534	1.7141
250	29.84	0.01700	13.82	218.5	1087.7	218.6	945.4	1164.0	0.3678	1.7000
260	35.45	0.01708	11.76	228.7	1090.3	228.8	938.6	1167.4	0.3820	1.6863
270	41.88	0.01717	10.06	238.9	1092.8	239.0	931.7	1170.7	0.3960	1.6730
280	49.22	0.01726	8.64	249.0	1095.2	249.2	924.7	1173.9	0.4099	1.6601
290	57.57	0.01735	7.46	259.3	1097.5	259.5	917.6	1177.0	0.4236	1.6476

(continued)

APPENDIX 18.B *(continued)*
Properties of Saturated Steam by Temperature

temp. (°F)	absolute pressure (psia)	specific volume (ft³/lbm) sat. liquid (v_f)	sat. vapor (v_g)	internal energy (Btu/lbm) sat. liquid (u_f)	sat. vapor (u_g)	enthalpy (Btu/lbm) sat. liquid (h_f)	evap. (h_{fg})	sat. vapor (h_g)	entropy (Btu/lbm-°R) sat. liquid (s_f)	sat. vapor (s_g)
300	67.03	0.01745	6.466	269.5	1099.8	269.7	910.3	1180.0	0.4372	1.6354
310	77.69	0.01755	5.626	279.8	1101.9	280.1	902.8	1182.8	0.4507	1.6236
320	89.67	0.01765	4.914	290.1	1103.9	290.4	895.1	1185.5	0.4640	1.6120
330	103.07	0.01776	4.308	300.5	1105.9	300.8	887.2	1188.0	0.4772	1.6007
340	118.02	0.01787	3.788	310.9	1107.7	311.2	879.3	1190.5	0.4903	1.5897
350	134.63	0.01799	3.343	321.3	1109.4	321.7	871.0	1192.7	0.5032	1.5789
360	153.03	0.01811	2.958	331.8	1111.0	332.3	862.5	1194.8	0.5161	1.5684
370	173.36	0.01823	2.625	342.3	1112.5	342.9	853.8	1196.7	0.5289	1.5580
380	195.74	0.01836	2.336	352.9	1113.9	353.5	845.0	1198.5	0.5415	1.5478
390	220.3	0.01850	2.084	363.5	1115.1	364.3	835.8	1200.1	0.5541	1.5378
400	247.3	0.01864	1.864	374.2	1116.2	375.1	826.4	1201.4	0.5667	1.5279
410	276.7	0.01879	1.671	385.0	1117.1	385.9	816.7	1202.6	0.5791	1.5182
420	308.8	0.01894	1.501	395.8	1117.8	396.9	806.8	1203.6	0.5915	1.5085
430	343.6	0.01910	1.351	406.7	1118.4	407.9	796.4	1204.3	0.6038	1.4990
440	381.5	0.01926	1.218	417.6	1118.9	419.0	785.8	1204.8	0.6161	1.4895
450	422.5	0.01944	1.0999	428.7	1119.1	430.2	774.9	1205.1	0.6283	1.4802
460	466.8	0.01962	0.9951	439.8	1119.2	441.5	763.6	1205.1	0.6405	1.4708
470	514.5	0.01981	0.9015	451.1	1119.0	452.9	752.0	1204.9	0.6527	1.4615
480	566.0	0.02001	0.8179	462.4	1118.7	464.5	739.8	1204.3	0.6648	1.4522
490	621.2	0.02022	0.7429	473.8	1118.1	476.1	727.4	1203.5	0.6769	1.4428
500	680.6	0.02044	0.6756	485.4	1117.2	487.9	714.4	1202.3	0.6891	1.4335
520	812.1	0.02092	0.5601	508.9	1114.8	512.0	686.9	1198.9	0.7134	1.4146
540	962.2	0.02146	0.4655	533.0	1111.1	536.8	657.2	1194.0	0.7378	1.3952
560	1132.7	0.02208	0.3874	557.8	1106.0	562.4	624.8	1187.2	0.7625	1.3753
580	1325.5	0.02279	0.3223	583.5	1099.2	589.1	589.3	1178.3	0.7876	1.3543
600	1542.5	0.02363	0.2674	610.3	1090.3	617.0	549.6	1166.6	0.8133	1.3320
620	1786.2	0.02465	0.2206	638.5	1078.6	646.7	504.8	1151.5	0.8401	1.3077
640	2059.2	0.02593	0.1802	668.9	1063.2	678.7	453.1	1131.8	0.8683	1.2803
660	2365	0.02766	0.1444	702.4	1042.2	714.5	390.8	1105.3	0.8991	1.2482
680	2707	0.03036	0.1113	742.2	1011.1	757.4	309.4	1066.8	0.9355	1.2070
700	3093	0.03665	0.0748	801.5	948.2	822.5	168.5	991.0	0.9900	1.1353
705.1028	3200.11	0.049746	0.049746	866.6	866.6	896.1	0	896.1	1.0530	1.0530
32.018	0.08871	0.01602	3300	0.0	1021.0	0.000263	1075	1075.2	0	2.1868

APPENDIX 19.A
Manning's Roughness Coefficient[a,b]
(design use)

channel material	n
plastic (PVC and ABS)	0.009
clean, uncoated cast iron	0.013–0.015
clean, coated cast iron	0.012–0.014
dirty, tuberculated cast iron	0.015–0.035
riveted steel	0.015–0.017
lock-bar and welded steel pipe	0.012–0.013
galvanized iron	0.015–0.017
brass and glass	0.009–0.013
wood stave	
small diameter	0.011–0.012
large diameter	0.012–0.013
concrete	
average value used	0.013
typical commercial, ball and spigot	
rubber gasketed end connections	
– full (pressurized and wet)	0.010
– partially full	0.0085
with rough joints	0.016–0.017
dry mix, rough forms	0.015–0.016
wet mix, steel forms	0.012–0.014
very smooth, finished	0.011–0.012
vitrified sewer	0.013–0.015
common-clay drainage tile	0.012–0.014
asbestos	0.011
planed timber (flume)	0.012 (0.010–0.014)
canvas	0.012
unplaned timber (flume)	0.013 (0.011–0.015)
brick	0.016
rubble masonry	0.017
smooth earth	0.018
firm gravel	0.023
corrugated metal pipe (CMP)	0.024 (see App. 17.F)
natural channels, good condition	0.025
rip rap	0.035
natural channels with stones and weeds	0.035
very poor natural channels	0.060

[a]Compiled from various sources.
[b]Values outside these ranges have been observed, but these values are typical.

APPENDIX 19.B
Manning Equation Nomograph

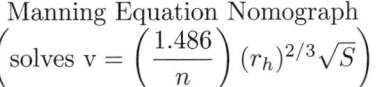

$$\left(\text{solves } v = \left(\frac{1.486}{n}\right)(r_h)^{2/3}\sqrt{S}\right)$$

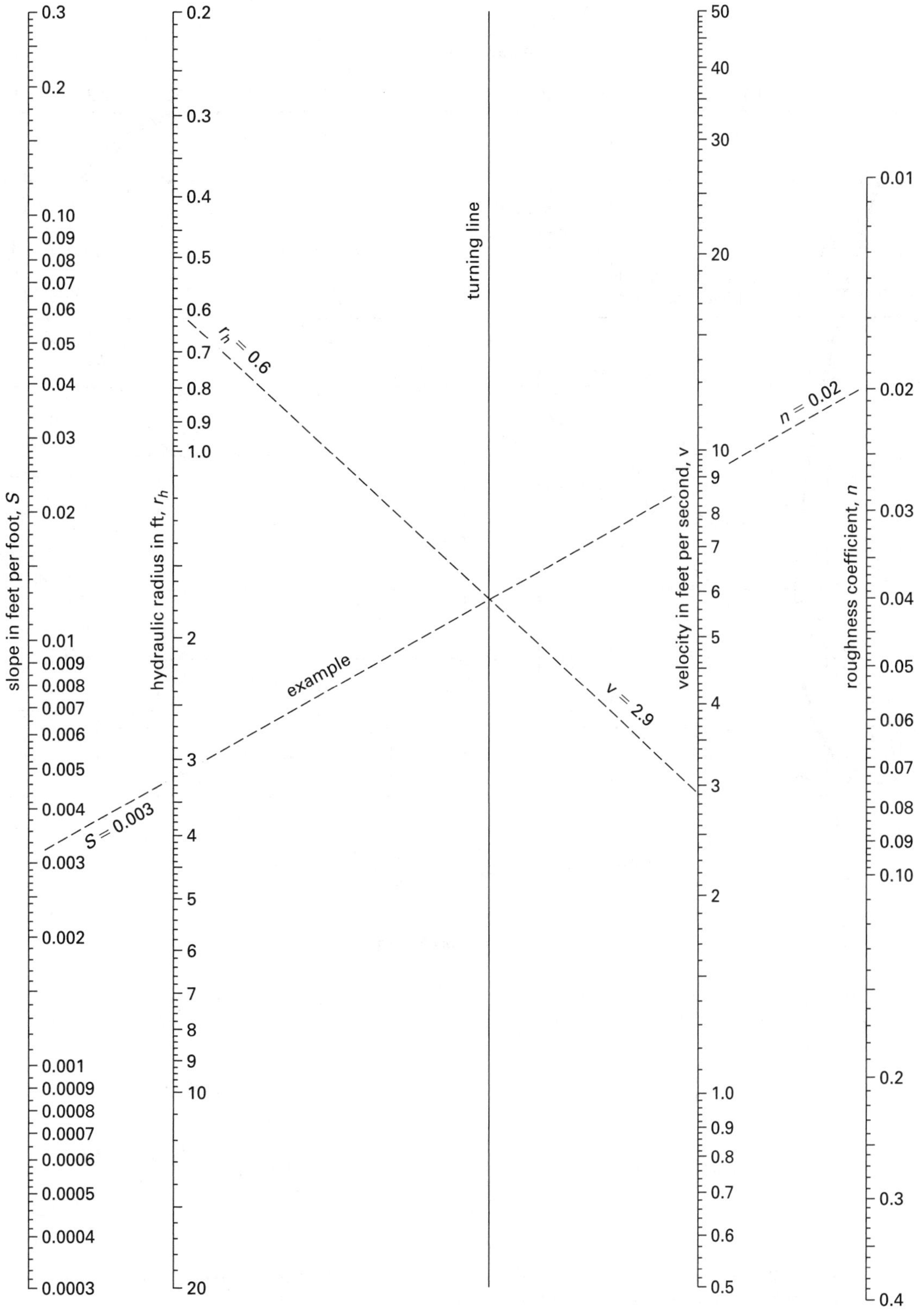

slope in feet per foot, S

- 0.3
- 0.2
- 0.10
- 0.09
- 0.08
- 0.07
- 0.06
- 0.05
- 0.04
- 0.03
- 0.02
- 0.01
- 0.009
- 0.008
- 0.007
- 0.006
- 0.005
- 0.004
- 0.003
- 0.002
- 0.001
- 0.0009
- 0.0008
- 0.0007
- 0.0006
- 0.0005
- 0.0004
- 0.0003

S = 0.003

hydraulic radius in ft, r_h

- 0.2
- 0.3
- 0.4
- 0.5
- 0.6
- 0.7
- 0.8
- 0.9
- 1.0
- 2
- 3
- 4
- 5
- 6
- 7
- 8
- 9
- 10
- 20

$r_h = 0.6$

example

turning line

velocity in feet per second, v

- 50
- 40
- 30
- 20
- 10
- 9
- 8
- 7
- 6
- 5
- 4
- 3
- 2
- 1.0
- 0.9
- 0.8
- 0.7
- 0.6
- 0.5

v = 2.9

roughness coefficient, n

- 0.01
- 0.02
- 0.03
- 0.04
- 0.05
- 0.06
- 0.07
- 0.08
- 0.09
- 0.10
- 0.2
- 0.3
- 0.4

n = 0.02

APPENDIX 19.C
Circular Channel Ratios[a,b]

Experiments have shown that n varies slightly with depth. This figure gives velocity and flow rate ratios for varying n (solid line) and constant n (broken line) assumptions.

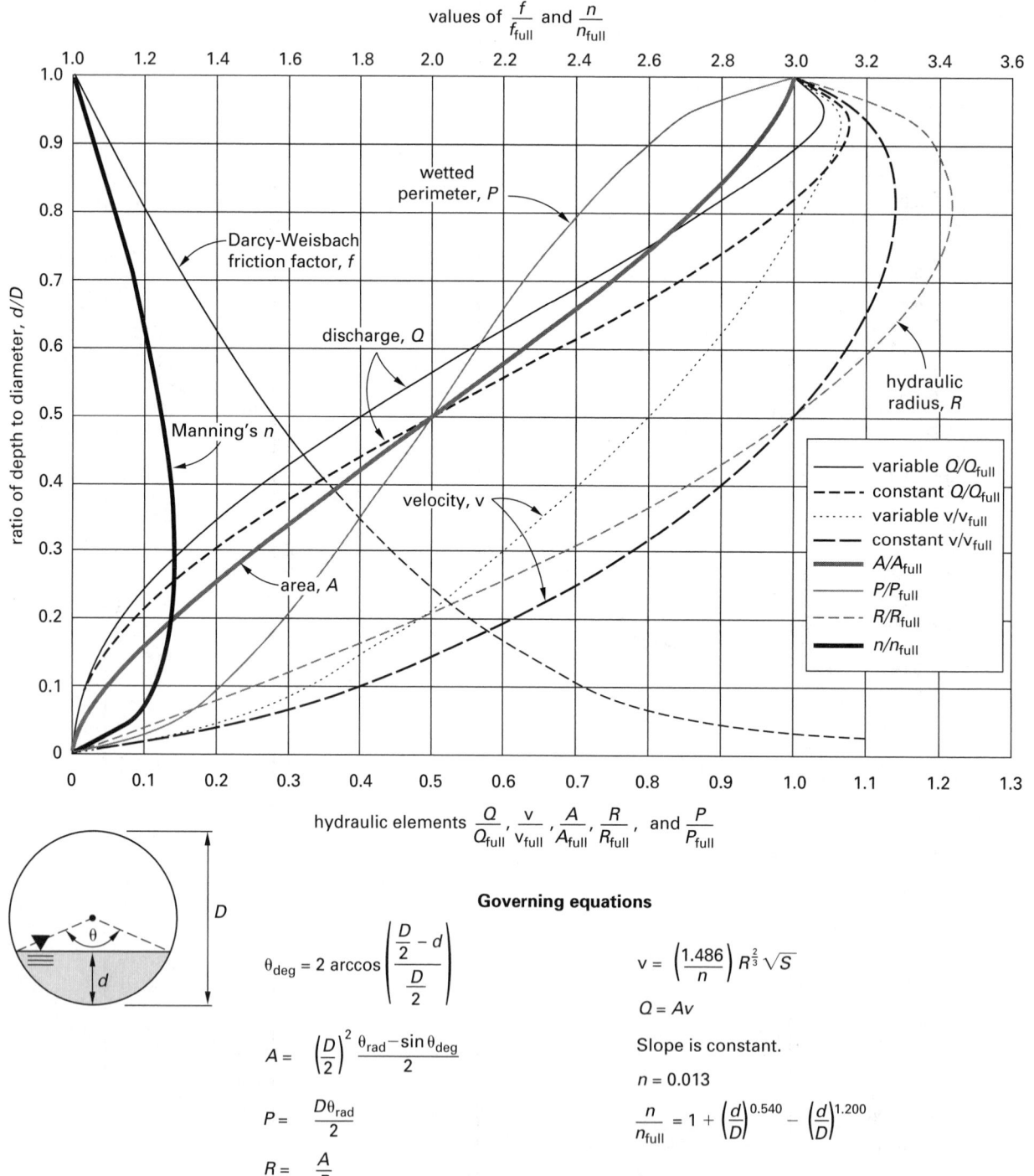

[a] Adapted from *Design and Construction of Sanitary and Storm Sewers*, p. 87, ASCE, 1969, as originally presented in "Design of Sewers to Facilitate Flow," Camp, T. R., *Sewage Works Journal*, 18, 3 (1946).
[b] For $n = 0.013$

APPENDIX 19.D
Critical Depths in Circular Channels

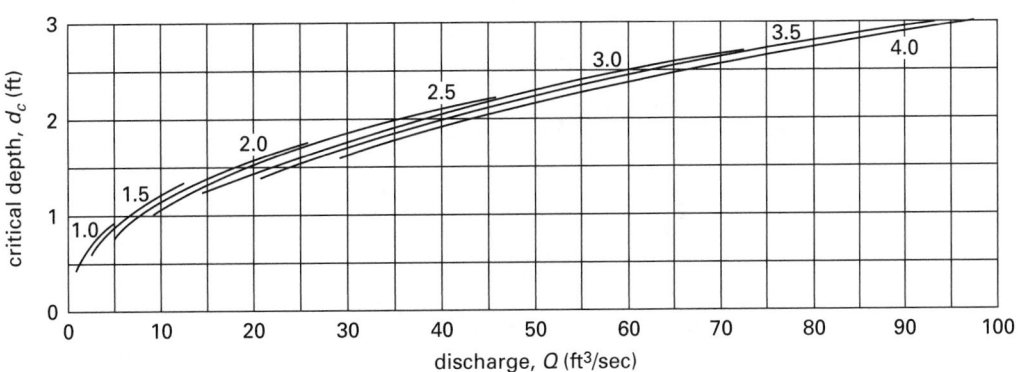

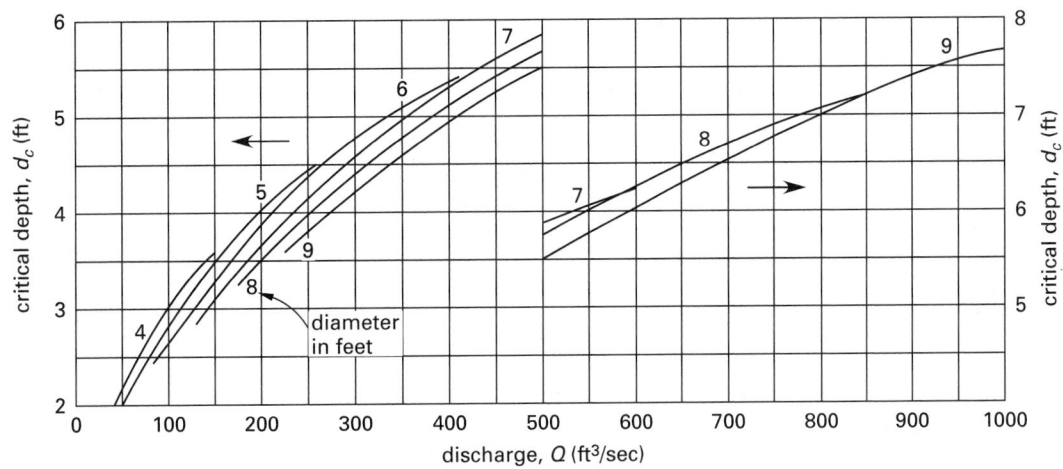

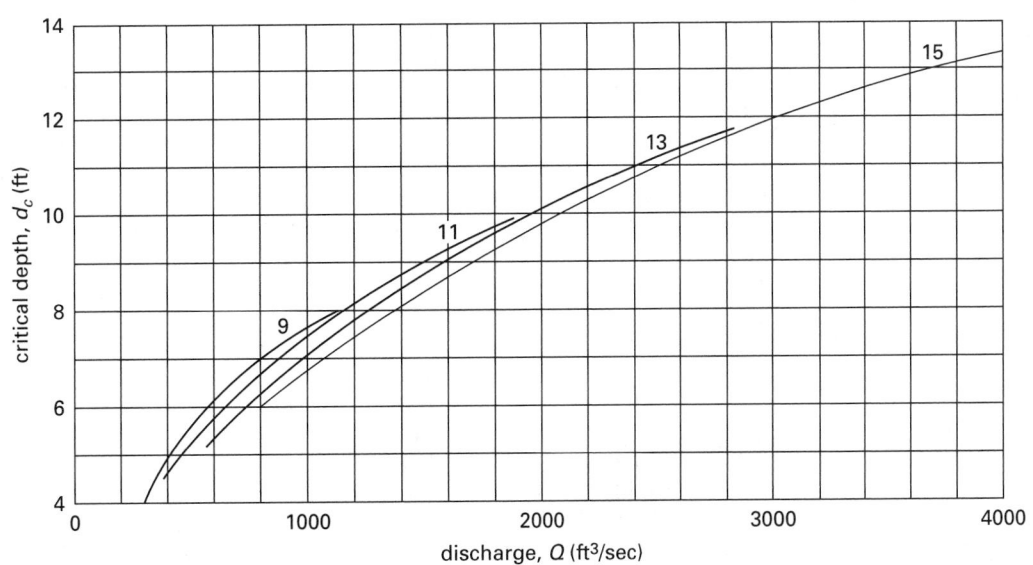

APPENDIX 19.E
Conveyance Factor, K
Symmetrical Rectangular,[a] Trapezoidal, and V-Notch[b] Open Channels
(use for determining Q or b when d is known)

[Customary U.S. Units[c,d]]

$$K \text{ in } Q = K\left(\frac{1}{n}\right)d^{8/3}\sqrt{S}$$

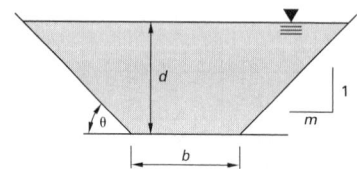

					m and θ					
	0.0	0.25	0.5	0.75	1.0	1.5	2.0	2.5	3.0	4.0
$x=d/b$	90°	76.0°	63.4°	53.1°	45.0°	33.7°	26.6°	21.8°	18.4°	14.0°
0.01	146.7	147.2	147.6	148.0	148.3	148.8	149.2	149.5	149.9	150.5
0.02	72.4	72.9	73.4	73.7	74.0	74.5	74.9	75.3	75.6	76.3
0.03	47.6	48.2	48.6	49.0	49.8	49.8	50.2	50.6	50.9	51.6
0.04	35.3	35.8	36.3	36.6	36.9	37.4	37.8	38.2	38.6	39.3
0.05	27.9	28.4	28.9	29.2	29.5	30.0	30.5	30.9	31.2	32.0
0.06	23.0	23.5	23.9	24.3	24.6	25.1	25.5	26.0	26.3	27.1
0.07	19.5	20.0	20.4	20.8	21.1	21.6	22.0	22.4	22.8	23.6
0.08	16.8	17.3	17.8	18.1	18.4	18.9	19.4	19.8	20.2	21.0
0.09	14.8	15.3	15.7	16.1	16.4	16.9	17.4	17.8	18.2	19.0
0.10	13.2	13.7	14.1	14.4	14.8	15.3	15.7	16.2	16.6	17.4
0.11	11.83	12.33	12.76	13.11	13.42	13.9	14.4	14.9	15.3	16.1
0.12	10.73	11.23	11.65	12.00	12.31	12.8	13.3	13.8	14.2	15.0
0.13	9.80	10.29	10.71	11.06	11.37	11.9	12.4	12.8	13.3	14.1
0.14	9.00	9.49	9.91	10.26	10.57	11.1	11.6	12.0	12.5	13.4
0.15	8.32	8.80	9.22	9.67	9.88	10.4	10.9	11.4	11.8	12.7
0.16	7.72	8.20	8.61	8.96	9.27	9.81	10.29	10.75	11.2	12.1
0.17	7.19	7.67	8.08	8.43	8.74	9.28	9.77	10.23	10.68	11.6
0.18	6.73	7.20	7.61	7.96	8.27	8.81	9.30	9.76	10.21	11.1
0.19	6.31	6.78	7.19	7.54	7.85	8.39	8.88	9.34	9.80	10.7
0.20	5.94	6.40	6.81	7.16	7.47	8.01	8.50	8.97	9.43	10.3
0.22	5.30	5.76	6.16	6.51	6.82	7.36	7.86	8.33	8.79	9.70
0.24	4.77	5.22	5.62	5.96	6.27	6.82	7.32	7.79	8.26	9.18
0.26	4.32	4.77	5.16	5.51	5.82	6.37	6.87	7.35	7.81	8.74
0.28	3.95	4.38	4.77	5.12	5.48	5.98	6.48	6.96	7.43	8.36
0.30	3.62	4.05	4.44	4.78	5.09	5.64	6.15	6.63	7.10	8.04
0.32	3.34	3.77	4.15	4.49	4.80	5.35	5.86	6.34	6.82	7.75
0.34	3.09	3.51	3.89	4.23	4.54	5.10	5.60	6.09	6.56	7.50
0.36	2.88	3.29	3.67	4.01	4.31	4.87	5.38	5.86	6.34	7.28
0.38	2.68	3.09	3.47	3.81	4.11	4.67	5.17	5.66	6.14	7.09
0.40	2.51	2.92	3.29	3.62	3.93	4.48	4.99	5.48	5.96	6.91
0.42	2.36	2.76	3.13	3.46	3.77	4.32	4.83	5.32	5.80	6.75
0.44	2.22	2.61	2.98	3.31	3.62	4.17	4.68	5.17	5.66	6.60
0.46	2.09	2.48	2.85	3.18	3.48	4.04	4.55	5.04	5.52	6.47
0.48	1.98	2.36	2.72	3.06	3.36	3.91	4.43	4.92	5.40	6.35
0.50	1.87	2.26	2.61	2.94	3.25	3.80	4.31	4.81	5.29	6.24
0.55	1.65	2.02	2.37	2.70	3.00	3.55	4.07	4.56	5.05	6.00

(continued)

APPENDIX 19.E *(continued)*
Conveyance Factor, K
Symmetrical Rectangular,[a] Trapezoidal, and V-Notch[b] Open Channels
(use for determining Q or b when d is known)

[Customary U.S. Units[c,d]]

$$K \text{ in } Q = K \left(\frac{1}{n}\right) d^{8/3}\sqrt{S}$$

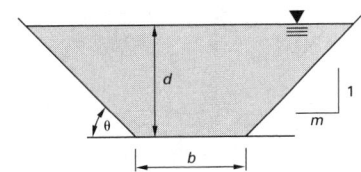

	m and θ									
	0.0	0.25	0.5	0.75	1.0	1.5	2.0	2.5	3.0	4.0
$x = d/b$	90°	76.0°	63.4°	53.1°	45.0°	33.7°	26.6°	21.8°	18.4°	14.0°
0.60	1.46	1.83	2.17	2.50	2.80	3.35	3.86	4.36	4.84	5.80
0.70	1.18	1.53	1.87	2.19	2.48	3.03	3.55	4.04	4.53	5.49
0.80	0.982	1.31	1.64	1.95	2.25	2.80	3.31	3.81	4.30	5.26
0.90	0.831	1.15	1.47	1.78	2.07	2.62	3.13	3.63	4.12	5.08
1.00	0.714	1.02	1.33	1.64	1.93	2.47	2.99	3.48	3.97	4.93
1.20	0.548	0.836	1.14	1.43	1.72	2.26	2.77	3.27	3.76	4.72
1.40	0.436	0.708	0.998	1.29	1.57	2.11	2.62	3.12	3.60	4.57
1.60	0.357	0.616	0.897	1.18	1.46	2.00	2.51	3.00	3.49	4.45
1.80	0.298	0.546	0.820	1.10	1.38	1.91	2.42	2.91	3.40	4.36
2.00	0.254	0.491	0.760	1.04	1.31	1.84	2.35	2.84	3.33	4.29
2.25	0.212	0.439	0.700	0.973	1.24	1.77	2.28	2.77	3.26	4.22
∞	0.00	0.091	0.274	0.499	0.743	1.24	1.74	2.23	2.71	3.67

[a]For rectangular channels, use the 0.0 (90°, vertical sides) column.
[b]For V-notch triangular channels, use the $d/b = \infty$ row.
[c]Q = flow rate, ft^3/sec; d = depth of flow, ft; b = bottom width of channel, ft; S = geometric slope, ft/ft; n = Manning's roughness constant.
[d]For SI units (e.g., Q in m^3/s and d and b in m), divide each table value by 1.486.

APPENDIX 19.F
Conveyance Factor, K'
Symmetrical Rectangular,[a] Trapezoidal Open Channels
(use for determining Q or d when b is known)

[Customary U.S. Units[b,c]]

$$K' \text{ in } Q = K'\left(\frac{1}{n}\right)b^{8/3}\sqrt{S}$$

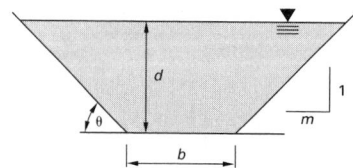

| | \multicolumn{10}{c}{m and θ} | | | | | | | | | |
$x = d/b$	0.0 90°	0.25 76.0°	0.5 63.4°	0.75 53.1°	1.0 45.0°	1.5 33.7°	2.0 26.6°	2.5 21.8°	3.0 18.4°	4.0 14.0°
0.01	0.00068	0.00068	0.00069	0.00069	0.00069	0.00069	0.00069	0.00069	0.00070	0.00070
0.02	0.00213	0.00215	0.00216	0.00217	0.00218	0.00220	0.00221	0.00222	0.00223	0.00225
0.03	0.00414	0.00419	0.00423	0.00426	0.00428	0.00433	0.00436	0.00439	0.00443	0.00449
0.04	0.00660	0.00670	0.00679	0.00685	0.00691	0.00700	0.00708	0.00716	0.00723	0.00736
0.05	0.00946	0.00964	0.00979	0.00991	0.01002	0.01019	0.01033	0.01047	0.01060	0.01086
0.06	0.0127	0.0130	0.0132	0.0134	0.0136	0.0138	0.0141	0.0148	0.0145	0.0150
0.07	0.0162	0.0166	0.0170	0.0173	0.0175	0.0180	0.0183	0.0187	0.0190	0.0197
0.08	0.0200	0.0206	0.0211	0.0215	0.0219	0.0225	0.0231	0.0236	0.0240	0.0250
0.09	0.0241	0.0249	0.0256	0.0262	0.0267	0.0275	0.0282	0.0289	0.0296	0.0310
0.10	0.0284	0.0294	0.0304	0.0311	0.0318	0.0329	0.0339	0.0348	0.0358	0.0376
0.11	0.0329	0.0343	0.0354	0.0364	0.0373	0.0387	0.0400	0.0413	0.0424	0.0448
0.12	0.0376	0.0393	0.0408	0.0420	0.0431	0.0450	0.0466	0.0482	0.0497	0.0527
0.13	0.0425	0.0446	0.0464	0.0480	0.0493	0.0516	0.0537	0.0556	0.0575	0.0613
0.14	0.0476	0.0502	0.0524	0.0542	0.0559	0.0587	0.0612	0.0636	0.0659	0.0706
0.15	0.0528	0.0559	0.0585	0.0608	0.0627	0.0662	0.0692	0.0721	0.0749	0.0805
0.16	0.0582	0.0619	0.0650	0.0676	0.0700	0.0740	0.0777	0.0811	0.0845	0.0912
0.17	0.0638	0.0680	0.0716	0.0748	0.0775	0.0823	0.0866	0.0907	0.0947	0.1026
0.18	0.0695	0.0744	0.0786	0.0822	0.0854	0.0910	0.0960	0.1008	0.1055	0.1148
0.19	0.0753	0.0809	0.0857	0.0899	0.0936	0.1001	0.1059	0.1115	0.1169	0.1277
0.20	0.0812	0.0876	0.0931	0.0979	0.1021	0.1096	0.1163	0.1227	0.1290	0.1414
0.22	0.0934	0.1015	0.109	0.115	0.120	0.130	0.139	0.147	0.155	0.171
0.24	0.1061	0.1161	0.125	0.133	0.140	0.152	0.163	0.173	0.184	0.204
0.26	0.119	0.131	0.142	0.152	0.160	0.175	0.189	0.202	0.215	0.241
0.28	0.132	0.147	0.160	0.172	0.182	0.201	0.217	0.234	0.249	0.281
0.30	0.146	0.163	0.179	0.193	0.205	0.228	0.248	0.267	0.287	0.324
0.32	0.160	0.180	0.199	0.215	0.230	0.256	0.281	0.304	0.327	0.371
0.34	0.174	0.198	0.219	0.238	0.256	0.287	0.316	0.343	0.370	0.423
0.36	0.189	0.216	0.241	0.263	0.283	0.319	0.353	0.385	0.416	0.478
0.38	0.203	0.234	0.263	0.288	0.312	0.353	0.392	0.429	0.465	0.537
0.40	0.218	0.253	0.286	0.315	0.341	0.389	0.434	0.476	0.518	0.600
0.42	0.233	0.273	0.309	0.342	0.373	0.427	0.478	0.526	0.574	0.668
0.44	0.248	0.293	0.334	0.371	0.405	0.467	0.525	0.580	0.633	0.740
0.46	0.264	0.313	0.359	0.401	0.439	0.509	0.574	0.636	0.696	0.816
0.48	0.279	0.334	0.385	0.432	0.474	0.553	0.625	0.695	0.763	0.897
0.50	0.295	0.355	0.412	0.463	0.511	0.598	0.679	0.757	0.833	0.983
0.55	0.335	0.410	0.482	0.548	0.609	0.722	0.826	0.926	1.025	1.22

(continued)

APPENDIX 19.F *(continued)*
Conveyance Factor, K'
Symmetrical Rectangular,[a] Trapezoidal Open Channels
(use for determining Q or d when b is known)

[Customary U.S. Units[b,c]]

$$K' \text{ in } Q = K'\left(\frac{1}{n}\right) b^{8/3}\sqrt{S}$$

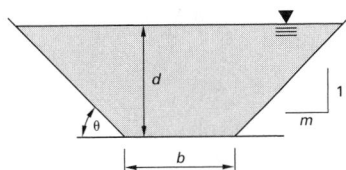

					m and θ					
	0.0	0.25	0.5	0.75	1.0	1.5	2.0	2.5	3.0	4.0
$x = d/b$	90°	76.0°	63.4°	53.1°	45.0°	33.7°	26.6°	21.8°	18.4°	14.0°
0.60	0.375	0.468	0.557	0.640	0.717	0.858	0.990	1.117	1.24	1.49
0.70	0.457	0.592	0.722	0.844	0.959	1.17	1.37	1.56	1.75	2.12
0.80	0.542	0.725	0.906	1.078	1.24	1.54	1.83	2.10	2.37	2.90
0.90	0.628	0.869	1.11	1.34	1.56	1.98	2.36	2.74	3.11	3.83
1.00	0.714	1.022	1.33	1.64	1.93	2.47	2.99	3.48	3.97	4.93
1.20	0.891	1.36	1.85	2.33	2.79	3.67	4.51	5.32	6.11	7.67
1.40	1.07	1.74	2.45	3.16	3.85	5.17	6.42	7.64	8.84	11.2
1.60	1.25	2.16	3.14	4.14	5.12	6.99	8.78	10.52	12.2	15.6
1.80	1.43	2.62	3.93	5.28	6.60	9.15	11.6	14.0	16.3	20.9
2.00	1.61	3.12	4.82	6.58	8.32	11.7	14.9	18.1	21.2	27.3
2.25	1.84	3.81	6.09	8.46	10.8	15.4	19.8	24.1	28.4	36.7

[a] For rectangular channels, use the 0.0 (90°, vertical sides) column.
[b] Q = flow rate, ft³/sec; d = depth of flow, ft; b = bottom width of channel, ft; S = geometric slope, ft/ft; n = Manning's roughness constant.
[c] For SI units (e.g., Q in m³/s, and d and b in m), divide each table value by 1.486.

APPENDIX 20.A
Rational Method Runoff C-Coefficients

categorized by surface

forested	0.059–0.2
asphalt	0.7–0.95
brick	0.7–0.85
concrete	0.8–0.95
shingle roof	0.75–0.95
lawns, well-drained (sandy soil)	
up to 2% slope	0.05–0.1
2% to 7% slope	0.10–0.15
over 7% slope	0.15–0.2
lawns, poor drainage (clay soil)	
up to 2% slope	0.13–0.17
2% to 7% slope	0.18–0.22
over 7% slope	0.25–0.35
driveways, walkways	0.75–0.85

categorized by use

farmland	0.05–0.3
pasture	0.05–0.3
unimproved	0.1–0.3
parks	0.1–0.25
cemeteries	0.1–0.25
railroad yards	0.2–0.35
playgrounds (except asphalt or concrete)	0.2–0.35
business districts	
neighborhood	0.5–0.7
city (downtown)	0.7–0.95
residential	
single family	0.3–0.5
multiplexes, detached	0.4–0.6
multiplexes, attached	0.6–0.75
suburban	0.25–0.4
apartments, condominiums	0.5–0.7
industrial	
light	0.5–0.8
heavy	0.6–0.9

APPENDIX 20.B
Random Numbers[a]

78466	83326	96589	88727	72655	49682	82338	28583	01522	11248
78722	47603	03477	29528	63956	01255	29840	32370	18032	82051
06401	87397	72898	32441	88861	71803	55626	77847	29925	76106
04754	14489	39420	94211	58042	43184	60977	74801	05931	73822
97118	06774	87743	60156	38037	16201	35137	54513	68023	34380
71923	49313	59713	95710	05975	64982	79253	93876	33707	84956
78870	77328	09637	67080	49168	75290	50175	34312	82593	76606
61208	17172	33187	92523	69895	28284	77956	45877	08044	58292
05033	24214	74232	33769	06304	54676	70026	41957	40112	66451
95983	13391	30369	51035	17042	11729	88647	70541	36026	23113
19946	55448	75049	24541	43007	11975	31797	05373	45893	25665
03580	67206	09635	84610	62611	86724	77411	99415	58901	86160
56823	49819	20283	22272	00114	92007	24369	00543	05417	92251
87633	31761	99865	31488	49947	06060	32083	47944	00449	06550
95152	10133	52693	22480	50336	49502	06296	76414	18358	05313
05639	24175	79438	92151	57602	03590	25465	54780	79098	73594
65927	55525	67270	22907	55097	63177	34119	94216	84861	10457
59005	29000	38395	80367	34112	41866	30170	84658	84441	03926
06626	42682	91522	45955	23263	09764	26824	82936	16813	13878
11306	02732	34189	04228	58541	72573	89071	58066	67159	29633
45143	56545	94617	42752	31209	14380	81477	36952	44934	97435
97612	87175	22613	84175	96413	83336	12408	89318	41713	90669
97035	62442	06940	45719	39918	60274	54353	54497	29789	82928
62498	00257	19179	06313	07900	46733	21413	63627	48734	92174
80306	19257	18690	54653	07263	19894	89909	76415	57246	02621
84114	84884	50129	68942	93264	72344	98794	16791	83861	32007
58437	88807	92141	88677	02864	02052	62843	21692	21373	29408
15702	53457	54258	47485	23399	71692	56806	70801	41548	94809
59966	41287	87001	26462	94000	28457	09469	80416	05897	87970
43641	05920	81346	02507	25349	93370	02064	62719	45740	62080
25501	50113	44600	87433	00683	79107	22315	42162	25516	98434
98294	08491	25251	26737	00071	45090	68628	64390	42684	94956
52582	89985	37863	60788	27412	47502	71577	13542	31077	13353
26510	83622	12546	00489	89304	15550	09482	07504	64588	92562
24755	71543	31667	83624	27085	65905	32386	30775	19689	41437
38399	88796	58856	18220	51056	04976	54062	49109	95563	48244
18889	87814	52232	58244	95206	05947	26622	01381	28744	38374
51774	89694	02654	63161	54622	31113	51160	29015	64730	07750
88375	37710	61619	69820	13131	90406	45206	06386	06398	68652
10416	70345	93307	87360	53452	61179	46845	91521	32430	74795

[a]To use, enter the table randomly and arbitrarily select any direction (i.e., up, down, to the right, left, or diagonally).

APPENDIX 22.A
Atomic Numbers and Weights of the Elements
(referred to Carbon-12)

name	symbol	atomic number	atomic weight	name	symbol	atomic number	atomic weight
actinium	Ac	89	–	mercury	Hg	80	200.59
aluminum	Al	13	26.9815	molybdenum	Mo	42	95.94
americium	Am	95	–	neodymium	Nd	60	144.24
antimony	Sb	51	121.75	neon	Ne	10	20.183
argon	Ar	18	39.948	neptunium	Np	93	237.048
arsenic	As	33	74.9216	nickel	Ni	28	58.71
astatine	At	85	–	niobium	Nb	41	92.906
barium	Ba	56	137.34	nitrogen	N	7	14.0067
berkelium	Bk	97	–	nobelium	No	102	–
beryllium	Be	4	9.0122	osmium	Os	76	190.2
bismuth	Bi	83	208.980	oxygen	O	8	15.9994
boron	B	5	10.811	palladium	Pd	46	106.4
bromine	Br	35	79.904	phosphorus	P	15	30.9738
cadmium	Cd	48	112.40	platinum	Pt	78	195.09
calcium	Ca	20	40.08	plutonium	Pu	94	–
californium	Cf	98	–	polonium	Po	84	–
carbon	C	6	12.01115	potassium	K	19	39.102
cerium	Ce	58	140.12	praseodymium	Pr	59	140.907
cesium	Cs	55	132.905	promethium	Pm	61	–
chlorine	Cl	17	35.453	protactinium	Pa	91	231.036
chromium	Cr	24	51.996	radium	Ra	88	–
cobalt	Co	27	58.9332	radon	Rn	86	226.025
copper	Cu	29	63.546	rhenium	Re	75	186.2
curium	Cm	96	–	rhodium	Rh	45	102.905
dysprosium	Dy	66	162.50	rubidium	Rb	37	85.47
einsteinium	Es	99	–	ruthenium	Ru	44	101.07
erbium	Er	68	167.26	samarium	Sm	62	150.35
europium	Eu	63	151.96	scandium	Sc	21	44.956
fermium	Fm	100	–	selenium	Se	34	78.96
fluorine	F	9	18.9984	silicon	Si	14	28.086
francium	Fr	87	–	silver	Ag	47	107.868
gadolinium	Gd	64	157.25	sodium	Na	11	22.9898
gallium	Ga	31	69.72	strontium	Sr	38	87.62
germanium	Ge	32	72.59	sulfur	S	16	32.064
gold	Au	79	196.967	tantalum	Ta	73	180.948
hafnium	Hf	72	178.49	technetium	Tc	43	–
helium	He	2	4.0026	tellurium	Te	52	127.60
holmium	Ho	67	164.930	terbium	Tb	65	158.924
hydrogen	H	1	1.00797	thallium	Tl	81	204.37
indium	In	49	114.82	thorium	Th	90	232.038
iodine	I	53	126.9044	thulium	Tm	69	168.934
iridium	Ir	77	192.2	tin	Sn	50	118.69
iron	Fe	26	55.847	titanium	Ti	22	47.90
krypton	Kr	36	83.80	tungsten	W	74	183.85
lanthanum	La	57	138.91	uranium	U	92	238.03
lead	Pb	82	207.19	vanadium	V	23	50.942
lithium	Li	3	6.939	xenon	Xe	54	131.30
lutetium	Lu	71	174.97	ytterbium	Yb	70	173.04
magnesium	Mg	12	24.312	yttrium	Y	39	88.905
manganese	Mn	25	54.9380	zinc	Zn	30	65.37
mendelevium	Md	101	–	zirconium	Zr	40	91.22

APPENDIX 22.B
Periodic Table of the Elements
(referred to Carbon-12)

The Periodic Table of Elements (Long Form)

The number of electrons in filled shells is shown in the column at the extreme left; the remaining electrons for each element are shown immediately below the symbol for each element. Atomic numbers are enclosed in brackets. Atomic weights (rounded, based on carbon-12) are shown above the symbols. Atomic weight values in parentheses are those of the isotopes of longest half-life for certain radioactive elements whose atomic weights cannot be precisely quoted without knowledge of origin of the element.

metals — transition metals — nonmetals

periods	I A	II A	III B	IV B	V B	VI B	VII B	VIII	VIII	VIII	I B	II B	III A	IV A	V A	VI A	VII A	0
1 (0)	1.0079 H[1] 1																	4.0026 He[2] 2
2 (2)	6.939 Li[3] 1	9.0122 Be[4] 2											10.81 B[5] 3	12.0115 C[6] 4	14.0067 N[7] 5	15.9994 O[8] 6	18.994 F[9] 7	20.183 Ne[10] 8
3 (2,8)	22.9898 Na[11] 1	24.312 Mg[12] 2											26.9815 Al[13] 3	28.086 Si[14] 4	30.9738 P[15] 5	32.064 S[16] 6	35.453 Cl[17] 7	39.948 Ar[18] 8
4 (2,8)	39.098 K[19] 8,1	40.08 Ca[20] 8,2	44.956 Sc[21] 9,2	47.90 Ti[22] 10,2	50.942 V[23] 11,2	51.996 Cr[24] 13,1	54.938 Mn[25] 13,2	55.847 Fe[26] 14,2	58.933 Co[27] 15,2	58.71 Ni[28] 16,2	63.546 Cu[29] 18,1	65.38 Zn[30] 18,2	69.72 Ga[31] 18,3	72.59 Ge[32] 18,4	74.922 As[33] 18,5	78.96 Se[34] 18,6	79.904 Br[35] 18,7	83.80 Kr[36] 18,8
5 (2,8,18)	85.47 Rb[37] 8,1	87.62 Sr[38] 8,2	88.905 Y[39] 9,2	91.22 Zr[40] 10,2	92.906 Nb[41] 12,1	95.94 Mo[42] 13,1	(98) Tc[43] 14,1	101.07 Ru[44] 15,1	102.905 Rh[45] 16,1	106.4 Pd[46] 18	107.868 Ag[47] 18,1	112.40 Cd[48] 18,2	114.82 In[49] 18,3	118.69 Sn[50] 18,4	121.75 Sb[51] 18,5	127.60 Te[52] 18,6	126.904 I[53] 18,7	131.30 Xe[54] 18,8
6 (2,8,18)	132.905 Cs[55] 18,8,1	137.34 Ba[56] 18,8,2	* (57–71)	178.49 Hf[72] 32,10,2	180.948 Ta[73] 32,11,2	183.85 W[74] 32,12,2	186.2 Re[75] 32,13,2	190.2 Os[76] 32,14,2	192.2 Ir[77] 32,15,2	195.09 Pt[78] 32,17,1	196.967 Au[79] 32,18,1	200.59 Hg[80] 32,18,2	204.37 Tl[81] 32,18,3	207.19 Pb[82] 32,18,4	208.980 Bi[83] 32,18,5	(210) Po[84] 32,18,6	(210) At[85] 32,18,7	(222) Rn[86] 32,18,8
7 (2,8,18,32)	(223) Fr[87] 18,8,1	226.025 Ra[88] 18,8,2	† (89–103)	Rf[104] 32,10,2	Ha[105] 32,11,2	[106] 32,12,2	[107]	[108]										

*lanthanide series

138.91 La[57] 18,9,2	140.12 Ce[58] 20,8,2	140.907 Pr[59] 21,8,2	144.24 Nd[60] 22,8,2	(147) Pm[61] 23,8,2	150.35 Sm[62] 24,8,2	151.96 Eu[63] 25,8,2	157.25 Gd[64] 25,9,2	158.924 Tb[65] 27,8,2	162.50 Dy[66] 28,8,2	164.930 Ho[67] 29,8,2	167.26 Er[68] 30,8,2	168.934 Tm[69] 31,8,2	173.04 Yb[70] 32,8,2	174.97 Lu[71] 32,9,2

†actinide series

(227) Ac[89] 18,9,2	232.038 Th[90] 18,10,2	231.036 Pa[91] 20,9,2	238.03 U[92] 21,9,2	237.048 Np[93] 23,8,2	(242) Pu[94] 24,8,2	(243) Am[95] 25,8,2	(247) Cm[96] 25,9,2	(247) Bk[97] 26,9,2	(249) Cf[98] 28,8,2	(254) Es[99] 29,8,2	(254) Fm[100] 30,8,2	(256) Md[101] 31,8,2	(254) No[102] 32,8,2	(257) Lr[103] 32,9,2

APPENDIX 22.C
Water Chemistry $CaCO_3$ Equivalents

cations	formula	ionic weight	equivalent weight	factor
aluminum	Al^{+3}	27.0	9.0	5.56
ammonium	NH_4^+	18.0	18.0	2.78
calcium	Ca^{+2}	40.1	20.0	2.50
cupric copper	Cu^{+2}	63.6	31.8	1.57
cuprous copper	Cu^{+3}	63.6	21.2	2.36
ferric iron	Fe^{+3}	55.8	18.6	2.69
ferrous iron	Fe^{+2}	55.8	27.9	1.79
hydrogen	H^+	1.0	1.0	50.00
manganese	Mn^{+2}	54.9	27.5	1.82
magnesium	Mg^{+2}	24.3	12.2	4.10
potassium	K^+	39.1	39.1	1.28
sodium	Na^+	23.0	23.0	2.18

anions	formula	ionic weight	equivalent weight	factor
bicarbonate	HCO_3^-	61.0	61.0	0.82
carbonate	CO_3^{-2}	60.0	30.0	1.67
chloride	Cl^-	35.5	35.5	1.41
fluoride	F^-	19.0	19.0	2.66
hydroxide	OH^-	17.0	17.0	2.94
nitrate	NO_3^-	62.0	62.0	0.81
phosphate (tribasic)	PO_4^{-3}	95.0	31.7	1.58
phosphate (dibasic)	HPO_4^{-2}	96.0	48.0	1.04
phosphate (monobasic)	$H_2PO_4^-$	97.0	97.0	0.52
sulfate	SO_4^{-2}	96.1	48.0	1.04
sulfite	SO_3^{-2}	80.1	40.0	1.25

compounds	formula	molecular weight	equivalent weight	factor
aluminum hydroxide	$Al(OH)_3$	78.0	26.0	1.92
aluminum sulfate	$Al_2(SO_4)_3$	342.1	57.0	0.88
aluminum sulfate	$Al_2(SO_4)_3 \cdot 18H_2O$	666.1	111.0	0.45
alumina	Al_2O_3	102.0	17.0	2.94
sodium aluminate	$Na_2Al_2O_4$	164.0	27.3	1.83
calcium bicarbonate	$Ca(HCO_3)_2$	162.1	81.1	0.62
calcium carbonate	$CaCO_3$	100.1	50.1	1.00
calcium chloride	$CaCl_2$	111.0	55.5	0.90
calcium hydroxide (pure)	$Ca(OH)_2$	74.1	37.1	1.35
calcium hydroxide (90%)	$Ca(OH)_2$	—	41.1	1.22
calcium oxide (lime)	CaO	56.1	28.0	1.79
calcium sulfate (anhydrous)	$CaSO_4$	136.2	68.1	0.74
calcium sulfate (gypsum)	$CaSO_4 \cdot 2H_2O$	172.2	86.1	0.58
calcium phosphate	$Ca_3(PO_4)_2$	310.3	51.7	0.97
disodium phosphate	$Na_2HPO_4 \cdot 12H_2O$	358.2	119.4	0.42
disodium phosphate (anhydrous)	Na_2HPO_4	142.0	47.3	1.06

(continued)

APPENDIX 22.C (*continued*)
Water Chemistry CaCO$_3$ Equivalents

compounds	formula	molecular weight	equivalent weight	factor
ferric oxide	Fe$_2$O$_3$	159.6	26.6	1.88
iron oxide (magnetic)	Fe$_3$O$_4$	321.4	–	–
ferrous sulfate (copperas)	FeSO$_4$·7H$_2$O	278.0	139.0	0.36
magnesium oxide	MgO	40.3	20.2	2.48
magnesium bicarbonate	Mg(HCO$_3$)$_2$	146.3	73.2	0.68
magnesium carbonate	MgCO$_3$	84.3	42.2	1.19
magnesium chloride	MgCl$_2$	95.2	47.6	1.05
magnesium hydroxide	Mg(OH)$_2$	58.3	29.2	1.71
magnesium phosphate	Mg$_3$(PO$_4$)$_2$	263.0	43.8	1.14
magnesium sulfate	MgSO$_4$	120.4	60.2	0.83
monosodium phosphate	NaH$_2$PO$_4$·H$_2$O	138.1	46.0	1.09
monosodium phosphate (anhydrous)	NaH$_2$PO$_4$	120.1	40.0	1.25
metaphosphate	NaPO$_3$	102.0	34.0	1.47
silica	SiO$_2$	60.1	30.0	1.67
sodium bicarbonate	NaHCO$_3$	84.0	84.0	0.60
sodium carbonate	Na$_2$CO$_3$	106.0	53.0	0.94
sodium chloride	NaCl	58.5	58.5	0.85
sodium hydroxide	NaOH	40.0	40.0	1.25
sodium nitrate	NaNO$_3$	85.0	85.0	0.59
sodium sulfate	Na$_2$SO$_4$	142.0	71.0	0.70
sodium sulfite	Na$_2$SO$_3$	126.1	63.0	0.79
tetrasodium EDTA	(CH$_2$)$_2$N$_2$(CH$_2$COONa)$_4$	380.2	95.1	0.53
trisodium phosphate	Na$_3$PO$_4$ · 12H$_2$O	380.2	126.7	0.40
trisodium phosphate (anhydrous)	Na$_3$PO$_4$	164.0	54.7	0.91
trisodium NTA	(CH$_2$)$_3$N(COONa)$_3$	257.1	85.7	0.58

gases				
ammonia	NH$_3$	17	17	2.94
carbon dioxide	CO$_2$	44	22	2.27
hydrogen	H$_2$	2	1	50.00
hydrogen sulfide	H$_2$S	34	17	2.94
oxygen	O$_2$	32	8	6.25

acids				
carbonic	H$_2$CO$_3$	62.0	31.0	1.61
hydrochloric	HCl	36.5	36.5	1.37
phosphoric	H$_3$PO$_4$	98.0	32.7	1.53
sulfuric	H$_2$SO$_4$	98.1	49.1	1.02

(Multiply the concentration (in mg/L) of the substance by the corresponding factors to obtain the equivalent concentration in mg/L as CaCO$_3$. For example, 70 mg/L of Mg^{++} as substance would be (70 mg/L)(4.1) = 287 mg/L as CaCO$_3$.)

APPENDIX 22.D
Saturation Values of Dissolved Oxygen in Water[a]

temp (°C)	chloride concentration in water (mg/L)			difference per 100 mg chloride	vapor pressure (mm/Hg)
	0	5000	10,000		
	dissolved oxygen (mg/L)				
0	14.62	13.79	12.97	0.017	5
1	14.23	13.41	12.61	0.016	5
2	13.84	13.05	12.28	0.015	5
3	13.48	12.72	11.98	0.015	6
4	13.13	12.41	11.69	0.014	6
5	12.80	12.09	11.39	0.014	7
6	12.48	11.79	11.12	0.014	7
7	12.17	11.51	10.85	0.013	8
8	11.87	11.24	10.61	0.013	8
9	11.59	10.97	10.36	0.012	9
10	11.33	10.73	10.13	0.012	9
11	11.08	10.49	9.92	0.011	10
12	10.83	10.28	9.72	0.011	11
13	10.60	10.05	9.52	0.011	11
14	10.37	9.85	9.32	0.010	12
15	10.15	9.65	9.14	0.010	13
16	9.95	9.46	8.96	0.010	14
17	9.74	9.26	8.78	0.010	15
18	9.54	9.07	8.62	0.009	16
19	9.35	8.89	8.45	0.009	17
20	9.17	8.73	8.30	0.009	18
21	8.99	8.57	8.14	0.009	19
22	8.83	8.42	7.99	0.008	20
23	8.68	8.27	7.85	0.008	21
24	8.53	8.12	7.71	0.008	22
25	8.38	7.96	7.56	0.008	24
26	8.22	7.81	7.42	0.008	25
27	8.07	7.67	7.28	0.008	27
28	7.92	7.53	7.14	0.008	28
29	7.77	7.39	7.00	0.008	30
30	7.63	7.25	6.86	0.008	32

[a] For saturation at barometric pressures other than 760 mm (29.92 in), C'_s is related to the corresponding tabulated values, C_s, by the following equation.

$$C'_s = C_s \left(\frac{P - p}{760 - p} \right)$$

C'_s = solubility at barometric pressure P and given temperature, mg/L
C_s = saturation at given temperature from table, mg/L
P = barometric pressure, mm
p = pressure of saturated water vapor at temperature of the water selected from table, mm

APPENDIX 22.E
Names and Formulas of Important Chemicals

common name	chemical name	chemical formula
acetone	acetone	$(CH_3)_2CO$
acetylene	acetylene	C_2H_2
ammonia	ammonia	NH_3
ammonium	ammonium hydroxide	NH_4OH
aniline	aniline	$C_6H_5NH_2$
bauxite	hydrated aluminum oxide	$Al_2O_3 \cdot 2H_2O$
bleach	calcium hypochlorite	$CaCl(OCl)$
borax	sodium tetraborate	$Na_2B_4O_7 \cdot 10H_2O$
carbide	calcium carbide	CaC_2
carbolic acid	phenol	C_6H_5OH
carbon dioxide	carbon dioxide	CO_2
carborundum	silicon carbide	SiC
caustic potash	potassium hydroxide	KOH
caustic soda/lye	sodium hydroxide	$NaOH$
chalk	calcium carbonate	$CaCO_3$
cinnabar	mercuric sulfide	HgS
ether	diethyl ether	$(C_2H_5)_2O$
formic acid	methanoic acid	$HCOOH$
Glauber's salt	decahydrated sodium sulfate	$Na_2SO_4 \cdot 10H_2O$
glycerine	glycerine	$C_3H_5(OH)_3$
grain alcohol	ethanol	C_2H_5OH
graphite	crystalline carbon	C
gypsum	calcium sulfate	$CaSO_4 \cdot 2H_2O$
halite	sodium chloride	$NaCl$
iron chloride	ferrous chloride	$FeCl_2 \cdot 4H_2O$
laughing gas	nitrous oxide	N_2O
limestone	calcium carbonate	$CaCO_3$
magnesia	magnesium oxide	MgO
marsh gas	methane	CH_4
muriate of potash	potassium chloride	KCl
muriatic acid	hydrochloric acid	HCl
niter	sodium nitrate	$NaNO_3$
niter cake	sodium bisulfate	$NaHSO_4$
oleum	fuming sulfuric acid	SO_3 in H_2SO_4
potash	potassium carbonate	K_2CO_3
prussic acid	hydrogen cyanide	HCN
pyrites	ferrous sulfide	FeS
pyrolusite	manganese dioxide	MnO_2
quicklime	calcium oxide	CaO
sal soda	decahydrated sodium carbonate	$NaCO_3 \cdot 10H_2O$
salammoniac	ammonium chloride	NH_4Cl
sand or silica	silicon dioxide	SiO_2
salt cake	sodium sulfate (crude)	Na_2SO_4
slaked lime	calcium hydroxide	$Ca(OH)_2$
soda ash	sodium carbonate	Na_2CO_3
soot	amorphous carbon	C
stannous chloride	stannous chloride	$SnCl_2 \cdot 2H_2O$
superphosphate	monohydrated primary calcium phosphate	$Ca(H_2PO_4)_2 \cdot H_2O$
table salt	sodium chloride	$NaCl$
table sugar	sucrose	$C_{12}H_{22}O_{11}$
trilene	trichloroethylene	C_2HCl_3
urea	urea	$CO(NH_2)_2$
vinegar (acetic acid)	ethanoic acid	CH_3COOH
washing soda	decahydrated sodium carbonate	$Na_2CO_3 \cdot 10H_2O$
wood alcohol	methanol	CH_3OH
zinc blende	zinc sulfide	ZnS

APPENDIX 22.F
Approximate Solubility Product Constants at 25°C

substance	formula	K_{sp}
aluminum hydroxide	$Al(OH)_3$	1.3×10^{-33}
aluminum phosphate	$AlPO_4$	6.3×10^{-19}
barium carbonate	$BaCO_3$	5.1×10^{-9}
barium chromate	$BaCrO_4$	1.2×10^{-10}
barium fluoride	BaF_2	1.0×10^{-6}
barium hydroxide	$Ba(OH)_2$	5×10^{-3}
barium sulfate	$BaSO_4$	1.1×10^{-10}
barium sulfite	$BaSO_3$	8×10^{-7}
barium thiosulfate	BaS_2O_3	1.6×10^{-6}
bismuthyl chloride	$BiOCl$	1.8×10^{-31}
bismuthyl hydroxide	$BiOOH$	4×10^{-10}
cadmium carbonate	$CdCO_3$	5.2×10^{-12}
cadmium hydroxide	$Cd(OH)_2$	2.5×10^{-14}
cadmium oxalate	CdC_2O_4	1.5×10^{-8}
cadmium sulfide*	CdS	8×10^{-28}
calcium carbonate	$CaCO_3$	2.8×10^{-9}
calcium chromate	$CaCrO_4$	7.1×10^{-4}
calcium fluoride	CaF_2	5.3×10^{-9}
calcium hydrogen phosphate	$CaHPO_4$	1×10^{-7}
calcium hydroxide	$Ca(OH)_2$	5.5×10^{-6}
calcium oxalate	CaC_2O_4	2.7×10^{-9}
calcium phosphate	$Ca_3(PO_4)_2$	2.0×10^{-29}
calcium sulfate	$CaSO_4$	9.1×10^{-6}
calcium sulfite	$CaSO_3$	6.8×10^{-8}
chromium (II) hydroxide	$Cr(OH)_2$	2×10^{-16}
chromium (III) hydroxide	$Cr(OH)_3$	6.3×10^{-31}
cobalt (II) carbonate	$CoCO_3$	1.4×10^{-13}
cobalt (II) hydroxide	$Co(OH)_2$	1.6×10^{-15}
cobalt (III) hydroxide	$Co(OH)_3$	1.6×10^{-44}
cobalt (II) sulfide*	CoS	4×10^{-21}
copper (I) chloride	$CuCl$	1.2×10^{-6}
copper (I) cyanide	$CuCN$	3.2×10^{-20}
copper (I) iodide	CuI	1.1×10^{-12}
copper (II) arsenate	$Cu_3(AsO_4)_2$	7.6×10^{-36}
copper (II) carbonate	$CuCO_3$	1.4×10^{-10}
copper (II) chromate	$CuCrO_4$	3.6×10^{-6}
copper (II) ferrocyanide	$Cu[Fe(CN)_6]$	1.3×10^{-16}
copper (II) hydroxide	$Cu(OH)_2$	2.2×10^{-20}
copper (II) sulfide*	CuS	6×10^{-37}
iron (II) carbonate	$FeCO_3$	3.2×10^{-11}
iron (II) hydroxide	$Fe(OH)_2$	8.0×10^{-16}
iron (II) sulfide*	FeS	6×10^{-19}
iron (III) arsenate	$FeAsO_4$	5.7×10^{-21}
iron (III) ferrocyanide	$Fe_4[Fe(CN)_6]_3$	3.3×10^{-41}
iron (III) hydroxide	$Fe(OH)_3$	4×10^{-38}
iron (III) phosphate	$FePO_4$	1.3×10^{-22}

(continued)

APPENDIX 22.F (*continued*)
Approximate Solubility Product Constants at 25°C

substance	formula	K_{sp}
lead (II) arsenate	$Pb_3(AsO_4)_2$	4×10^{-36}
lead (II) azide	$Pb(N_3)_2$	2.5×10^{-9}
lead (II) bromide	$PbBr_2$	4.0×10^{-5}
lead (II) carbonate	$PbCO_3$	7.4×10^{-14}
lead (II) chloride	$PbCl_2$	1.6×10^{-5}
lead (II) chromate	$PbCrO_4$	2.8×10^{-13}
lead (II) fluoride	PbF_2	2.7×10^{-8}
lead (II) hydroxide	$Pb(OH)_2$	1.2×10^{-15}
lead (II) iodide	PbI_2	7.1×10^{-9}
lead (II) sulfate	$PbSO_4$	1.6×10^{-8}
lead (II) sulfide*	PbS	3×10^{-28}
lithium carbonate	Li_2CO_3	2.5×10^{-2}
lithium fluoride	LiF	3.8×10^{-3}
lithium phosphate	Li_3PO_4	3.2×10^{-9}
magnesium ammonium phosphate	$MgNH_4PO_4$	2.5×10^{-13}
magnesium arsenate	$Mg_3(AsO_4)_2$	2×10^{-20}
magnesium carbonate	$MgCO_3$	3.5×10^{-8}
magnesium fluoride	MgF_2	3.7×10^{-8}
magnesium hydroxide	$Mg(OH)_2$	1.8×10^{-11}
magnesium oxalate	MgC_2O_4	8.5×10^{-5}
magnesium phosphate	$Mg_3(PO_4)_2$	1×10^{-25}
manganese (II) carbonate	$MnCO_3$	1.8×10^{-11}
manganese (II) hydroxide	$Mn(OH)_2$	1.9×10^{-13}
manganese (II) sulfide*	MnS	3×10^{-14}
mercury (I) bromide	Hg_2Br_2	5.6×10^{-23}
mercury (I) chloride	Hg_2Cl_2	1.3×10^{-18}
mercury (I) iodide	Hg_2I_2	4.5×10^{-29}
mercury (II) sulfide*	HgS	2×10^{-53}
nickel (II) carbonate	$NiCO_3$	6.6×10^{-9}
nickel (II) hydroxide	$Ni(OH)_2$	2.0×10^{-15}
nickel (II) sulfide*	NiS	3×10^{-19}
scandium fluoride	ScF_3	4.2×10^{-18}
scandium hydroxide	$Sc(OH)_3$	8.0×10^{-31}
silver acetate	$AgC_2H_3O_2$	2.0×10^{-3}
silver arsenate	Ag_3AsO_4	1.0×10^{-22}
silver azide	AgN_3	2.8×10^{-9}
silver bromide	$AgBr$	5.0×10^{-13}
silver chloride	$AgCl$	1.8×10^{-10}
silver chromate	Ag_2CrO_4	1.1×10^{-12}
silver cyanide	$AgCN$	1.2×10^{-16}
silver iodate	$AgIO_3$	3.0×10^{-8}
silver iodide	AgI	8.5×10^{-17}
silver nitrite	$AgNO_2$	6.0×10^{-4}
silver sulfate	Ag_2SO_4	1.4×10^{-5}
silver sulfide*	Ag_2S	6×10^{-51}
silver sulfite	Ag_2SO_3	1.5×10^{-14}
silver thiocyanate	$AgSCN$	1.0×10^{-12}
strontium carbonate	$SrCO_3$	1.1×10^{-10}
strontium chromate	$SrCrO_4$	2.2×10^{-5}

(*continued*)

APPENDIX 22.F (*continued*)
Approximate Solubility Product Constants at 25°C

substance	formula	K_{sp}
strontium fluoride	SrF_2	2.5×10^{-9}
strontium sulfate	$SrSO_4$	3.2×10^{-7}
thallium (I) bromide	$TlBr$	3.4×10^{-6}
thallium (I) chloride	$TlCl$	1.7×10^{-4}
thallium (I) iodide	TlI	6.5×10^{-8}
thallium (III) hydroxide	$Tl(OH)_3$	6.3×10^{-46}
tin (II) hydroxide	$Sn(OH)_2$	1.4×10^{-28}
tin (II) sulfide*	SnS	1×10^{-26}
zinc carbonate	$ZnCO_3$	1.4×10^{-11}
zinc hydroxide	$Zn(OH)_2$	1.2×10^{-17}
zinc oxalate	ZnC_2O_4	2.7×10^{-8}
zinc phosphate	$Zn_3(PO_4)_2$	9.0×10^{-33}
zinc sulfide*	ZnS	2×10^{-25}

*Sulfide equilibrium of the type:
$$MS(s) + H_2O(l) \rightleftharpoons M^{2+}(aq) + HS^-(aq) + OH^-(aq)$$

APPENDIX 24.A
Heats of Combustion for Common Compounds[a]

substance	formula	molecular weight	specific volume (ft^3/lbm)	heat of combustion			
				Btu/ft^3		Btu/lbm	
				gross (high)	net (low)	gross (high)	net (low)
carbon	C	12.01				14,093	14,093
carbon dioxide	CO_2	44.01	8.548				
carbon monoxide	CO	28.01	13.506	322	322	4347	4347
hydrogen	H_2	2.016	187.723	325	275	60,958	51,623
nitrogen	N_2	28.016	13.443				
oxygen	O_2	32.000	11.819				
paraffin series (alkanes)							
methane	CH_4	16.041	23.565	1013	913	23,879	21,520
ethane	C_2H_6	30.067	12.455	1792	1641	22,320	20,432
propane	C_3H_8	44.092	8.365	2590	2385	21,661	19,944
n-butane	C_4H_{10}	58.118	6.321	3370	3113	21,308	19,680
isobutane	C_4H_{10}	58.118	6.321	3363	3105	21,257	19,629
n-pentane	C_5H_{12}	72.144	5.252	4016	3709	21,091	19,517
isopentane	C_5H_{12}	72.144	5.252	4008	3716	21,052	19,478
neopentane	C_5H_{12}	72.144	5.252	3993	3693	20,970	19,396
n-hexane	C_6H_{14}	86.169	4.398	4762	4412	20,940	19,403
olefin series (alkenes and alkynes)							
ethylene	C_2H_4	28.051	13.412	1614	1513	21,644	20,295
propylene	C_3H_6	42.077	9.007	2336	2186	21,041	19,691
n-butene	C_4H_8	56.102	6.756	3084	2885	20,840	19,496
isobutene	C_4H_8	56.102	6.756	3068	2869	20,730	19,382
n-pentene	C_5H_{10}	70.128	5.400	3836	3586	20,712	19,363
aromatic series							
benzene	C_6H_6	78.107	4.852	3751	3601	18,210	17,480
toluene	C_7H_8	92.132	4.113	4484	4284	18,440	17,620
xylene	C_8H_{10}	106.158	3.567	5230	4980	18,650	17,760
miscellaneous gases							
acetylene	C_2H_2	26.036	14.344	1499	1448	21,500	20,776
air		28.9	13.063				
ammonia	NH_3	17.031	21.914	441	365	9668	8001
digester gas[b]	–	25.8	18.3	658	593	15,521	13,988
ethyl alcohol	C_2H_5OH	46.067	8.221	1600	1451	13,161	11,929
hydrogen sulfide	H_2S	34.076	10.979	647	596	7100	6545
iso-octane	C_8H_{18}	114.2	0.0232[c]		98.9[c]	20,590	19,160
methyl alcohol	CH_3OH	32.041	11.820	868	768	10,259	9078
naphthalene	$C_{10}H_8$	128.162	2.955	5854	5654	17,298	16,708
sulfur	S	32.06				3983	3983
sulfur dioxide	SO_2	64.06	5.770				
water vapor	H_2O	18.016	21.017				

(Multiply Btu/lbm by 2.326 to obtain kJ/kg.)
(Multiply Btu/ft^3 by 37.25 to obtain kJ/m^3.)
[a]Gas volumes listed are at 60°F (16°C) and 1 atm.
[b]Digester gas from wastewater treatment plants is approximately 65% methane and 35% carbon dioxide by volume. For different compositions, use composite properties of these two gases.
[c]liquid form

APPENDIX 24.B
Approximate Properties of Selected Gases

| gas | symbol | temp °F | MW | customary U.S. units | | | SI units | | | k |
				R $\frac{\text{ft-lbf}}{\text{lbm-°R}}$	c_p $\frac{\text{Btu}}{\text{lbm-°R}}$	c_v $\frac{\text{Btu}}{\text{lbm-°R}}$	R $\frac{\text{J}}{\text{kg·K}}$	c_p $\frac{\text{J}}{\text{kg·K}}$	c_v $\frac{\text{J}}{\text{kg·K}}$	
acetylene	C_2H_2	68	26.038	59.35	0.350	0.274	319.32	1465	1146	1.279
air		100	28.967	53.35	0.240	0.171	287.03	1005	718	1.400
ammonia	NH_3	68	17.032	90.73	0.523	0.406	488.16	2190	1702	1.287
argon	Ar	68	39.944	38.69	0.124	0.074	208.15	519	311	1.669
n-butane	C_4H_{10}	68	58.124	26.59	0.395	0.361	143.04	1654	1511	1.095
carbon dioxide	CO_2	100	44.011	35.11	0.207	0.162	188.92	867	678	1.279
carbon monoxide	CO	100	28.011	55.17	0.249	0.178	296.82	1043	746	1.398
chlorine	Cl_2	100	70.910	21.79	0.115	0.087	117.25	481	364	1.322
ethane	C_2H_6	68	30.070	51.39	0.386	0.320	276.50	1616	1340	1.206
ethylene	C_2H_4	68	28.054	55.08	0.400	0.329	296.37	1675	1378	1.215
Freon (R-12)[a]	CCl_2F_2	200	120.925	12.78	0.159	0.143	68.76	666	597	1.115
helium	He	100	4.003	386.04	1.240	0.744	2077.03	5192	3115	1.667
hydrogen	H_2	100	2.016	766.53	3.420	2.435	4124.18	14 319	10 195	1.405
hydrogen sulfide	H_2S	68	34.082	45.34	0.243	0.185	243.95	1017	773	1.315
krypton	Kr		83.800	18.44	0.059	0.035	99.22	247	148	1.671
methane	CH_4	68	16.043	96.32	0.593	0.469	518.25	2483	1965	1.264
neon	Ne	68	20.183	76.57	0.248	0.150	411.94	1038	626	1.658
nitrogen	N_2	100	28.016	55.16	0.249	0.178	296.77	1043	746	1.398
nitric oxide	NO	68	30.008	51.50	0.231	0.165	277.07	967	690	1.402
nitrous oxide	NO_2	68	44.01	35.11	0.221	0.176	188.92	925	736	1.257
octane vapor	C_8H_{18}		114.232	13.53	0.407	0.390	72.78	1704	1631	1.045
oxygen	O_2	100	32.000	48.29	0.220	0.158	259.82	921	661	1.393
propane	C_3H_8	68	44.097	35.04	0.393	0.348	188.55	1645	1457	1.129
sulfur dioxide	SO_2	100	64.066	24.12	0.149	0.118	129.78	624	494	1.263
water vapor[a]	H_2O	212	18.016	85.78	0.445	0.335	461.50	1863	1402	1.329
xenon	Xe		131.300	11.77	0.038	0.023	63.32	159	96	1.661

(Multiply Btu/lbm-°F by 4186.8 to obtain J/kg·K.)
(Multiply ft-lbf/lbm-°R by 5.3803 to obtain J/kg·K.)
[a]Values for steam and Freon are approximate and should be used only for low pressures and high temperatures.

APPENDIX 25.A
National Primary Drinking Water Regulations
Code of Federal Regulations (CFR), Title 40, Ch. I, Part 141, October 2003

microorganisms	MCLG^a (mg/L)^b	MCL or TT^a (mg/L)^b	potential health effects from ingestion of water	sources of contaminant in drinking water
Cryptosporidium	0	TT^c	gastrointestinal illness (e.g., diarrhea, vomiting, cramps)	human and animal fecal waste
Giardia lamblia	0	TT^c	gastrointestinal illness (e.g., diarrhea, vomiting, cramps)	human and animal fecal waste
heterotrophic plate count	n/a	TT^c	HPC has no health effects; it is an analytic method used to measure the variety of bacteria that are common in water. The lower the concentration of bacteria in drinking water, the better maintained the water is.	HPC measures a range of bacteria that are naturally present in the environment.
Legionella	0	TT^c	Legionnaire's disease, a type of pneumonia	found naturally in water; multiplies in heating systems
total coliforms (including fecal coliform and *E. coli*)	0	5.0%^{d}	Not a health threat in itself; it is used to indicate whether other potentially harmful bacteria may be present.^{e}	Coliforms are naturally present in the environment as well as in feces; fecal coliforms and *E. coli* only come from human and animal fecal waste.
turbidity	n/a	TT^c	Turbidity is a measure of the cloudiness of water. It is used to indicate water quality and filtration effectiveness (e.g., whether disease-causing organisms are present). Higher turbidity levels are often associated with higher levels of disease-causing microorganisms such as viruses, parasites, and some bacteria. These organisms can cause symptoms such as nausea, cramps, diarrhea, and associated headaches.	soil runoff
viruses (enteric)	0	TT^c	gastrointestinal illness (e.g., diarrhea, vomiting, cramps)	human and animal fecal waste

disinfection products	MCLG^a (mg/L)^b	MCL or TT^a (mg/L)^b	potential health effects from ingestion of water	sources of contaminant in drinking water
bromate	0	0.010	increased risk of cancer	byproduct of drinking-water disinfection
chlorite	0.8	1.0	anemia in infants and young children; nervous system effects	byproduct of drinking-water disinfection
haloacetic acids (HAA5)	n/a^f	0.060	increased risk of cancer	byproduct of drinking-water disinfection
total trihalomethanes (TTHMs)	n/a^f	0.080	liver, kidney, or central nervous system problems; increased risk of cancer	byproduct of drinking-water disinfection

(continued)

APPENDIX 25.A *(continued)*
National Primary Drinking Water Regulations
Code of Federal Regulations (CFR), Title 40, Ch. I, Part 141, October 2003

disinfectants	MRDLG[a] $(mg/L)^b$	MRDL[a] $(mg/L)^b$	potential health effects from ingestion of water	sources of contaminant in drinking water
chloramines (as Cl_2)	4[a]	4.0[a]	eye/nose irritation, stomach discomfort, anemia	water additive used to control microbes
chlorine (as Cl_2)	4[a]	4.0[a]	eye/nose irritation, stomach discomfort	water additive used to control microbes
chlorine dioxide (as ClO_2)	0.8[a]	0.8[a]	anemia in infants and young children, nervous system effects	water additive used to control microbes

inorganic chemicals	MCLG[a] $(mg/L)^b$	MCL or TT[a] $(mg/L)^b$	potential health effects from ingestion of water	sources of contaminant in drinking water
antimony	0.006	0.006	increase in blood cholesterol; decrease in blood sugar	discharge from petroleum refineries; fire retardants; ceramics; electronics; solder
arsenic	0[g]	0.010 as of January 23, 2006	skin damage or problems with cirulatory systems; may increase cancer risk	erosion of natural deposits; runoff from orchards; runoff from glass and electronics production wastes
asbestos (fiber > 10 micrometers)	7 million fibers per liter	7 MFL	increased risk of developing benign intestinal polyps	decay of asbestos cement in water mains; erosion of natural deposits
barium	2	2	increase in blood pressure	discharge of drilling wastes; discharge from metal refineries; erosion of natural deposits
beryllium	0.004	0.004	intestinal lesions	discharge from metal refineries and coal-burning factories; discharge from electrical, aerospace, and defense industries
cadmium	0.005	0.005	kidney damage	corrosion of galvanized pipes; erosion of natural deposits; discharge from metal refineries; runoff from waste batteries and paints
chromium (total)	0.1	0.1	allergic dermatitis	discharge from steel and pulp mills; erosion of natural deposits
copper	1.3	TT[h], action level=1.3	short-term exposure: gastrointestinal distress long-term exposure: liver or kidney damage People with Wilson's disease should consult their personal doctor if the amount of copper in their water exceeds the action level.	corrosion of household plumbing systems; erosion of natural deposits
cyanide (as free cyanide)	0.2	0.2	nerve damage or thyroid problems	discharge from steel/metal factories; discharge from plastic and fertilizer factories
fluoride	4.0	4.0	bone disease (pain and tenderness of the bones); children may get mottled teeth	water additive that promotes strong teeth; erosion of natural deposits; discharge from fertilizer and aluminum factories

(continued)

APPENDIX 25.A *(continued)*
National Primary Drinking Water Regulations
Code of Federal Regulations (CFR), Title 40, Ch. I, Part 141, October 2003

inorganic chemicals	MCLG[a] (mg/L)[b]	MCL or TT[a] (mg/L)[b]	potential health effects from ingestion of water	sources of contaminant in drinking water
lead	0	TT[h], action level=0.015	infants and children: delays in physical or mental development; children could show slight deficits in attention span and learning disabilities adults: kidney problems, high blood pressure	corrosion of houshold plumbing systems; erosion of natural deposits
mercury (inorganic)	0.002	0.002	kidney damage	erosion of natural deposits; discharge from refineries and factories; runoff from landfills and croplands
nitrate (measured as nitrogen)	10	10	Infants below the age of six months who drink water containing nitrate in excess of the MCL could become seriously ill and, if untreated, may die. Symptoms include shortness of breath and blue baby syndrome.	runoff from fertilizer use; leaching from septic tanks/sewage; erosion of natural deposits
nitrite (measured as nitrogen)	1	1	Infants below the age of six months who drink water containing nitrite in excess of the MCL could become seriously ill and, if untreated, may die. Symptoms include shortness of breath and blue baby syndrome.	runoff from fertilizer use; leaching from septic tanks/sewage; erosion of natural deposits
selenium	0.05	0.05	hair and fingernail loss; numbness in fingers or toes; circulatory problems	discharge from petroleum refineries; erosion of natural deposits; discharge from mines
thalium	0.0005	0.002	hair loss; changes in blood; kidney, intestine, or liver problems	leaching from ore-processing sites; discharge from electronics, glass, and drug factories

organic chemicals	MCLG[a] (mg/L)[b]	MCL or TT[a] (mg/L)[b]	potential health effects from ingestion of water	sources of contaminant in drinking water
acrylamide	0	TT[i]	nervous system or blood problems; increased risk of cancer	added to water during sewage/ wastewater treatment
alachlor	0	0.002	eye, liver, kidney, or spleen problems; anemia; increased risk of cancer	runoff from herbicide used on row crops
atrazine	0.003	0.003	cardiovascular system or reproductive problems	runoff from herbicide used on row crops
benzene	0	0.005	anemia; decrease in blood platelets; increased risk of cancer	discharge from factories; leaching from gas storage tanks and landfills
benzo(a)pyrene (PAHs)	0	0.0002	reproductive difficulties; increased risk of cancer	leaching from linings of water storage tanks and distribution lines
carbofuran	0.04	0.04	problems with blood, nervous system, or reproductive system	leaching of soil fumigant used on rice and alfalfa
carbon tetrachloride	0	0.005	liver problems; increased risk of cancer	discharge from chemical plants and other industrial activities

(continued)

APPENDIX 25.A *(continued)*
National Primary Drinking Water Regulations
Code of Federal Regulations (CFR), Title 40, Ch. I, Part 141, October 2003

organic chemicals	MCLG[a] (mg/L)[b]	MCL or TT[a] (mg/L)[b]	potential health effects from ingestion of water	sources of contaminant in drinking water
chlordane	0	0.002	liver or nervous system problems; increased risk of cancer	residue of banned termiticide
chlorobenzene	0.1	0.1	liver or kidney problems	discharge from chemical and agricultural chemical factories
2,4-D	0.07	0.07	kidney, liver, or adrenal gland problems	runoff from herbicide used on row crops
dalapon	0.2	0.2	minor kidney changes	runoff from herbicide used on rights of way
1,2-dibromo-3-chloropropane (DBCP)	0	0.0002	reproductive difficulties; increased risk of cancer	runoff/leaching from soil fumigant used on soybeans, cotton, pineapples, and orchards
o-dichloro-benzene	0.6	0.6	liver, kidney, or circulatory system problems	discharge from industrial chemical factories
p-dichloro-benzene	0.075	0.075	anemia; liver, kidney, or spleen damage; changes in blood	discharge from industrial chemical factories
1,2-dichloro-ethane	0	0.005	increased risk of cancer	discharge from industrial chemical factories
1,1-dichloro-ethylene	0.007	0.007	liver problems	discharge from industrial chemical factories
cis-1,2-dichloro-ethylene	0.07	0.07	liver problems	discharge from industrial chemical factories
$trans$-1,2-dichloro-ethylene	0.1	0.1	liver problems	discharge from industrial chemical factories
dichloromethane	0	0.005	liver problems; increased risk of cancer	discharge from industrial chemical factories
1,2-dichloro-propane	0	0.005	increased risk of cancer	discharge from industrial chemical factories
di(2-ethylhexyl) adipate	0.4	0.4	general toxic effects or reproductive difficulties	discharge from chemical factories
di(2-ethylhexyl) phthalate	0	0.006	reproductive difficulties; liver problems; increased risk of cancer	discharge from rubber and chemical factories
dinoseb	0.007	0.007	reproductive difficulties	runoff from herbicide used on soybeans and vegetables
dioxin (2,3,7,8-TCDD)	0	0.00000003	reproductive difficulties; increased risk of cancer	emissions from waste incineration and other combustion; discharge from chemical factories
diquat	0.02	0.02	cataracts	runoff from herbicide use
endothall	0.1	0.1	stomach and intestinal problems	runoff from herbicide use
endrin	0.002	0.002	liver problems	residue of banned insecticide
epichlorohydrin	0	TT[i]	increased cancer risk; over a long period of time, stomach problems	discharge from industrial chemical factories; an impurity of some water treatment chemicals
ethylbenzene	0.7	0.7	liver or kidney problems	discharge from petroleum refineries
ethylene dibromide	0	0.00005	problems with liver, stomach, reproductive system, or kidneys; increased risk of cancer	discharge from petroleum refineries

(continued)

APPENDIX 25.A *(continued)*
National Primary Drinking Water Regulations
Code of Federal Regulations (CFR), Title 40, Ch. I, Part 141, October 2003

organic chemicals	MCLG[a] (mg/L)[b]	MCL or TT[a] (mg/L)[b]	potential health effects from ingestion of water	sources of contaminant in drinking water
glyphosate	0.7	0.7	kidney problems; reproductive difficulties	runoff from herbicide use
heptachlor	0	0.0004	liver damage; increased risk of cancer	residue of banned termiticide
heptachlor epoxide	0	0.0002	liver damage; increased risk of cancer	breakdown of heptachlor
hexachlorobenzene	0	0.001	liver or kidney problems; reproductive difficulties; increased risk of cancer	discharge from metal refineries and agricultural chemical factories
hexachlorocyclopentadiene	0.05	0.05	kidney or stomach problems	discharge from chemical factories
lindane	0.0002	0.0002	liver or kidney problems	runoff/leaching from insecticide used on cattle, lumber, and gardens
methoxychlor	0.04	0.04	reproductive difficulties	runoff/leaching from insecticide used on fruits, vegetables, alfalfa, and livestock
oxamyl (vydate)	0.2	0.2	slight nervous system effects	runoff/leaching from insecticide used on apples, potatoes, and tomatoes
polychlorinated biphenyls (PCBs)	0	0.0005	skin changes; thymus gland problems; immune deficiencies; reproductive or nervous system difficulties; increased risk of cancer	runoff from landfills; discharge of waste chemicals
pentachlorophenol	0	0.001	liver or kidney problems; increased cancer risk	discharge from wood preserving factories
picloram	0.5	0.5	liver problems	herbicide runoff
simazine	0.004	0.004	problems with blood	herbicide runoff
styrene	0.1	0.1	liver, kidney, or circulatory system problems	discharge from rubber and plastic factories; leaching from landfills
tetrachloroethylene	0	0.005	liver problems; increased risk of cancer	discharge from factories and dry cleaners
toluene	1	1	nervous system, kidney, or liver problems	discharge from petroleum factories
toxaphene	0	0.003	kidney, liver, or thyroid problems; increased risk of cancer	runoff/leaching from insecticide used on cotton and cattle
2,4,5-TP (silvex)	0.05	0.05	liver problems	residue of banned herbicide
1,2,4-trichlorobenzene	0.07	0.07	changes in adrenal glands	discharge from textile finishing factories
1,1,1-trichloroethane	0.20	0.2	liver, nervous system, or circulatory problems	discharge from metal degreasing sites and other factories
1,1,2-trichloroethane	0.003	0.005	liver, kidney, or immune system problems	discharge from industrial chemical factories
trichloroethylene	0	0.005	liver problems; increased risk of cancer	discharge from metal degreasing sites and other factories
vinyl chloride	0	0.002	increased risk of cancer	leaching from PVC pipes; discharge from plastic factories
xylenes (total)	10	10	nervous system damage	discharge from petroleum factories; discharge from chemical factories

(continued)

APPENDIX 25.A *(continued)*
National Primary Drinking Water Regulations
Code of Federal Regulations (CFR), Title 40, Ch. I, Part 141, October 2003

radionuclides	MCLG[a] $(mg/L)^b$	MCL or TT[a] $(mg/L)^b$	potential health effects from ingestion of water	sources of contaminant in drinking water
alpha particles	none[g]	15 pCi/L	increased risk of cancer	erosion of natural deposits of certain minerals that are radioactive and may emit a form of radiation known as alpha radiation
beta particles emitters	none[g]	4 mrem/yr	increased risk of cancer	decay of natural and artificial deposits of certain minerals that are radioactive and may emit forms of radiation known as photons and beta radiation
radium 226 and radium 228 (combined)	none[g]	5 pCi/L	increased risk of cancer	erosion of natural deposits
uranium	0	30 µg/L as of December 8, 2003	increased risk of cancer; kidney toxicity	erosion of natural deposits

[a]Definitions:

Maximum Contaminant Level (MCL) - The highest level of a contaminant that is allowed in drinking water. MCLs are set as close to MCLGs as feasible using the best available treatment technology and taking cost into consideration. MCLs are enforceable standards.

Maximum Contaminant Level Goal (MCLG) - The level of a contaminant in drinking water below which there is no known or expected risk to health. MCLGs allow for a margin of safety and are non-enforceable public health goals.

Maximum Residual Disinfectant Level (MRDL) - The highest level of a disinfectant allowed in drinking water. There is convincing evidence that addition of a disinfectant is necessary for control of microbial contaminants.

Maximum Residual Disinfectant Level Goal (MRDLG) - The level of a drinking water disinfectant below which there is no known or expected risk to health. MRDLGs do not reflect the benefits of the use of disinfectants to control microbial contaminants.

Treatment Technique - A required process intended to reduce the level of a contaminant in drinking water.

[b]Units are in milligrams per liter (mg/L) unless otherwise noted. Milligrams per liter are equivalent to parts per million.

[c]The EPA's surface water treatment rules require systems using surface water or ground water under the direct influence of surface water to (1) disinfect their water, and (2) filter their water or meet criteria for avoiding filtration so that the following contaminants are controlled at the following levels.

- cryptosporidium (as of January 1, 2002, for systems serving >10,000 and January 14, 2005, for systems serving <10,000): 99% removal
- *Giardia lamblia*: 99.9% removal/inactivation
- Viruses: 99.99% removal/inactivation
- *Legionella*: No limit, but the EPA believes that if Giardia and viruses are removed/inactivated, Legionella will also be controlled.
- Turbidity: At no time can turbidity (cloudiness of water) go above 5 nephelolometric turbidity units (NTU); systems that filter must ensure that the turbidity go no higher than 1 NTU (0.5 NTU for conventional or direct filtration) in at least 95% of the daily samples in any month. As of January 1, 2002, turbidity may never exceed 1 NTU, and must not exceed 0.3 NTU in 95% of daily samples in any month.
- heterotrophic plate count (HPC): No more than 500 bacterial colonies per milliliter.
- Long Term 1 Enhanced Surface Water Treatment (as of January 14, 2005): Surface water systems or ground water under direct influence (GWUDI) systems serving fewer than 10,000 people must comply with the applicable Long Term 1 Enhanced Surface Water Treatment Rule provisions (e.g., turbidity standards, individual filter monitoring, cryptosporidium removal requirements, updated watershed control requirements for unfiltered systems).
- Filter Backwash Recycling: The Filter Backwash Recycling Rule requires systems that recycle to return specific recycle flows through all processes of the systems' existing conventional or direct filtration system or at an alternate location approved by the state.

[d]More than 5.0% of samples are total coliform-positive in a month. (For water systems that collect fewer than 40 routine samples per month, no more than one sample can be total coliform-positive per month.) Every sample that has total coliform must be analyzed for either fecal coliforms or *E. coli*: If two consecutive samples are TC-positive, and one is also positive for *E. coli* or fecal coliforms, the system has an acute MCL violation.

(continued)

APPENDIX 25.A *(continued)*
National Primary Drinking Water Regulations
Code of Federal Regulations (CFR), Title 40, Ch. I, Part 141, October 2003

[e]Fecal coliform and *E. coli* are bacteria whose presence indicates that the water may be contaminated with human or animal wastes. Disease-causing microbes (pathogens) in these wastes can cause diarrhea, cramps, nausea, headaches, or other symptoms. These pathogens may pose a special health risk for infants, young children, and people with severely compromised immune systems.

[f]Although there is no collective MCLG for this contaminant group, there are individual MCLGs for some of the individual contaminants.
- Haloacetic acids: dichloroacetic acid (0); trichloroacetic acid (0.3 mg/L). Monochloroacetic acid, bromoacetic acid, and dibromoacetic acid are regulated with this group but have no MCLGs.
- Trihalomethanes: bromodichloromethane (0); bromoform (0); dibromochloromethane (0.06 mg/L). Chloroform is regulated with this group but has no MCLG.

[g]MCLGs were not established before the 1986 Amendments to the Safe Drinking Water Act. Therefore, there is no MCLG for this contaminant.

[h]Lead and copper are regulated by a treatment technique that requires systems to control the corrosiveness of their water. If more than 10% of tap water samples exceed the action level, water systems must take additional steps. For copper, the action level is 1.3 mg/L, and for lead it is 0.015 mg/L.

[i]Each water system agency must certify, in writing, to the state (using third party or manufacturers' certification) that, when acrylamide and epichlorohydrin are used in drinking water systems, the combination (or product) of dose and monomer level does not exceed the levels specified, as follows.
- acrylamide = 0.05% dosed at 1 mg/L (or equivalent)
- epichlorohydrin = 0.01% dosed at 20 mg/L (or equivalent)

APPENDIX 26.A
Properties of Chemicals Used in Water Treatment

chemical name	formula	use	molecular weight	equivalent weight
activated carbon	C	taste and odor control	12.0	–
aluminum hydroxide	$Al(OH)_3$	(hypothetical combination)	78.0	26.0
aluminum sulfate (filter alum)	$Al_2(SO_4)_3 \cdot 14.3H_2O$	coagulation	600	100
ammonia	NH_3	chloramine disinfection	17.0	–
ammonium fluosilicate	$(NH_4)_2SiF_6$	fluoridation	178	89.0
ammonium sulfate	$(NH_4)_2SO_4$	coagulation	132	66.1
calcium bicarbonate	$Ca(HCO_3)_2$	(hypothetical combination)	162	81.0
calcium carbonate	$CaCO_3$	corrosion control	100	50.0
calcium fluoride	CaF_2	fluoridation	78.1	39.0
calcium hydroxide	$Ca(OH)_2$	softening	74.1	37.0
calcium hypochlorite	$Ca(ClO)_2 \cdot 2H_2O$	disinfection	179	–
calcium oxide (lime)	CaO	softening	56.1	28.0
carbon dioxide	CO_2	recarbonation	44.0	22.0
chlorine	Cl_2	disinfection	71.0	–
chlorine dioxide	ClO_2	taste and odor control	67.0	–
copper sulfate	$CuSO_4$	algae control	160	79.8
ferric chloride	$FeCl_3$	coagulation	162	54.1
ferric hydroxide	$Fe(OH)_3$	(hypothetical combination)	107	35.6
ferric sulfate	$Fe_2(SO_4)_3$	coagulation	400	66.7
ferrous sulfate (copperas)	$FeSO_4 \cdot 7H_2O$	coagulation	278	139
fluosilicic acid	H_2SiF_6	fluoridation	144	72.0
hydrochloric acid	HCl	pH adjustment	36.5	36.5
magnesium hydroxide	$Mg(OH)_2$	defluoridation	58.3	29.2
oxygen	O_2	aeration	32.0	16.0
ozone	O_3	disinfection	48.0	16.0
potassium permanganate	$KMnO_4$	oxidation	158	158
sodium aluminate	$NaAlO_2$	coagulation	82.0	82.0
sodium bicarbonate (baking soda)	$NaHCO_3$	alkalinity adjustment	84.0	84.0
sodium carbonate (soda ash)	Na_2CO_3	softening	106	53.0
sodium chloride (common salt)	$NaCl$	ion-exchange regeneration	58.4	58.4
sodium diphosphate	Na_2HPO_4	corrosion control	142	71.0
sodium fluoride	NaF	fluoridation	42.0	42.0
sodium fluosilicate	Na_2SiF_6	fluoridation	188	99.0
sodium hexametaphosphate	$(NaPO_3)_n$	corrosion control	–	–
sodium hydroxide	$NaOH$	pH adjustment	40.0	40.0
sodium hypochlorite	$NaClO$	disinfection	74.0	–
sodium phosphate	NaH_2PO_4	corrosion control	120	120
sodium silicate	Na_2OSiO_2	corrosion control	184	92.0
sodium thiosulfate	$Na_2S_2O_3$	dechlorination	158	79.0
sodium tripolyphosphate	$Na_5P_3O_{10}$	corrosion control	368	–
sulfur dioxide	SO_2	dechlorination	64.1	–
sulfuric acid	H_2SO_4	pH adjustment	98.1	49.0
trisodium phosphate	Na_3PO_4	corrosion control	118	–
water	H_2O	–	18.0	–

APPENDIX 29.A
Selected *Ten States' Standards*[a]

[11.243] **Hydraulic Load:** Use 100 gpcd (0.38 m^3/day) for new systems in undeveloped areas unless other information is available.

[42.3] **Pumps:** At least two pumps are required. Both pumps should have the same capacity if only two pumps are used. This capacity must exceed the total design flow. If three or more pumps are used, the capacities may vary, but capacity (peak hourly flow) pumping must be possible with one pump out of service.

[42.7] **Pump Well Ventilation:** Provide 30 complete air changes per hour for both wet and dry wells using intermittent ventilation. For continuous ventilation, the requirement is reduced to 12 (for wet wells) and 6 (for dry wells) air changes per hour. In general, ventilation air should be forced in, as opposed to air being extracted and replaced by infiltration.

[61.12] **Racks and Bar Screens:** All racks and screens shall have openings less than 1.75 in (45 mm) wide. The smallest opening for manually cleaned screens is 1 in (2.54 cm). The smallest opening for mechanically cleaned screens may be smaller. Flow velocity must be 1.25 to 3.0 ft/sec (38 to 91 cm/s).

[62.2–62.3] **Grinders and Shredders:** Comminutors are required if there is no screening. Gravel traps or grit-removal equipment should precede comminutors.

[63.3–63.4] **Grit Chambers:** Grit chambers are required when combined storm and sanitary sewers are used. A minimum of two grit chambers in parallel should be used, with a provision for bypassing. For channel-type grit chambers, the optimum velocity is 1 ft/sec (30 cm/s) throughout. The detention time is dependent on the particle sizes to be removed.

[71–72] **Settling Tanks:** Multiple units are desirable, and multiple units must be provided if the average flow exceeds 100,000 gal/day (379 m^3/day). For primary settling, the sidewater depth should be 10 ft (3.0 m) or greater. For tanks not receiving activated sludge, the design average overflow rate is 1000 gal/day-ft^2 (41 m^3/day·m^2); the maximum peak settling rate is 1500–3000 gal/day-ft^2 (60–120 m^3/day·m^2). The maximum weir loading is 30,000 gal/day-ft (375 m^3/day·m). The basin size shall also be calculated based on the average design flow rate and a maximum settling rate of 1000 gal/day-ft^2. The larger of the two sizes shall be used. If the flow rate is less than 1 MGD (3785 m^3/day), the maximum weir loading is reduced to 20,000 gal/day-ft (250 m^3/day·m).

For settling tanks following trickling filters and rotating biological contactors, the maximum settling rate is 1200 gal/day-ft^2 (49 m^3/day·m^2).

For settling tanks following activated sludge processes, the maximum hydraulic loadings are: 1200 gal/day-ft^2 (49 m^3/day·m^2) for conventional, step, complete mix, and contact units; 1000 gal/day-ft^2 (41 m^3/day·m^2) for extended aeration units; and 800 gal/day-ft^2 (33 m^3/day·m^2) for separate two-stage nitrification units.

[84] **Anaerobic Digesters:** Multiple units are required. For completely mixed digesters, maximum loading of volatile solids is 80 lbm/day-1000 ft^3 (1.3 kg/day·m^3) of digester volume. For moderately mixed digesters, the limit is 40 lbm/day-1000 ft^3 (0.65 kg/day·m^3). Minimum sidewater depth is 20 ft (6.1 m).

[88.22] **Sludge Drying Beds:** For design purposes, the maximum sludge depth is 8 in (200 mm).

[91.3] **Trickling Filters:** All media should have a minimum depth of 6 ft (1.8 m). Rock media depth should not exceed 10 ft (3 m). Manufactured media depth should not exceed the manufacturer's recommendations. The rock media should be 1 to 4.5 in (2.5 to 11.4 cm) in size. Freeboard of 4 ft (1.2 ft) or more is required for manufactured media. The drain should slope at 1% or more, and the average drain velocity should be 2 ft/sec (0.6 m/s).

[92.3] **Activated Sludge Processes:** The maximum BOD loading shall be 40 lbm/day-1000 ft^3 (0.64 kg/day·m^3) for conventional, step, and complete-mix units; 50 lbm/day-1000 ft^3 (0.8 kg/day·m^3) for contact stabilization units; and 15 lbm/day-1000 ft^3 (0.24 kg/day·m^3) for extended aeration units and oxidation ditch designs.

Aeration tank depths should be 10 to 30 ft (3 to 9 m). At least two aeration tanks shall be used. The dissolved oxygen content should not be allowed to drop below 2 mg/L at any time. The aeration rate should be 1500 ft^3 of oxygen per pound of BOD$_5$ (94 m^3/kg). For extended aeration, the rate should be 2050 ft^3 of oxygen per pound of BOD$_5$ (128 m^3/kg).

(continued)

APPENDIX 29.A (*continued*)
Selected *Ten States' Standards*[a]

[93] **Wastewater Treatment Ponds:** (Applicable to controlled-discharge, flow-through (facultative), and aerated pond systems.) Pond bottoms must be at least 4 ft (1.2 m) above the highest water table elevation. Pond primary cells are designed based on average BOD_5 loads of 15 to 35 lbm/ac-day (17 to 40 kg/ha-day) at the mean operating depth. Detention time for controlled-discharge ponds shall be at least 180 days between a depth of 2 ft (0.61 m) and the maximum operating depth. Detention time for flow-through (facultative) ponds should be 90 to 120 days, modified for cold weather. Detention time for aerated ponds may be estimated from the percentage, E, of BOD_5 to be removed: $t_{\text{days}} = E/2.3k_1 (100\% - E)$, where k_1 is the base-10 reaction coefficient, assumed to be 0.12/day at 68°F (20°C) and 0.06/day at 34°F (1°C). Although two cells can be used in very small systems, a minimum of three cells is normally required. Maximum cell size is 40 ac (16 ha). Ponds may be round, square, or rectangular (with length no greater than three times the width). Islands and sharp corners are not permitted. Dikes shall be at least 8 ft (2.4 m) wide at the top. Inner and outer dike slopes cannot be steeper than 1:3 ($V{:}H$). Inner slopes cannot be flatter than 1:4 ($V{:}H$). At least 3 ft (0.91 m) freeboard is required unless the pond system is very small, in which case 2 ft (0.6 m) is acceptable. Pond depth shall never be less than 2 ft (0.6 m). Depth shall be: (a) for controlled discharged and flow-through (facultative) ponds, a maximum of 6 ft (1.8 m) for primary cells; greater for subsequent cells with mixing and aeration as required; and (b) for aerated ponds, a design depth of 10 to 15 ft (3 to 4.5 m)

[93.3] **Facultative Treatment Ponds:** BOD loading for lagoons following primary treatment is 15 to 35 lbm BOD/day-ac (17 to 40 kg/day·ha) for both controlled-discharge and flow-through stabilization ponds.

[102.4] **Chlorination:** Minimum contact time is 15 min at peak flow.

[a]Numbers in square brackets refer to sections in *Recommended Standards for Sewage Works*, 1997 ed., Great Lakes-Upper Mississippi River Board of State Sanitary Engineers, published by Health Education Service, NY, from which these guidelines were extracted. Refer to the original document for complete standards.

APPENDIX 35.A
USCS Soil Boring, Well, and Geotextile Symbols

USCS soil boring

- well-graded gravels, gravel-sand mixtures, little or no fines
- poorly graded gravels, gravel-sand mixtures, little or no fines
- silty gravels, gravel-sand silt mixtures
- clayey gravels, gravel-sand-clay mixtures
- well-graded sands, gravelly sands, little or no fines
- poorly graded sands, gravelly sands, little or no fines
- silty sands, sand-silt mixtures
- clayey sands, sand-clay mixtures
- inorganic silts and very fine sands, rock flour, silty or clayey fine sands
- inorganic clays of low to medium plasticity, gravelly clays, sandy clays, silty
- organic silts and organic silty clays of low plasticity
- inorganic silts, micaceous or diatomaceous fine sandy or silty soils, elastic
- inorganic clays of high plasticity, fat clays
- organic clays of medium to high plasticity, organic silts
- peat, humus, swamp, and other highly organic soils

Well symbols
Pipes and screens
- none (NONE)
- pipe
- double-walled pipe
- sealed pipe
- fine screen
- coarse screen
- screen 1
- screen 2

Top fittings
- none (NONE)
- cap
- flush-mount cap
- above-ground cap
- connector
- reducer
- pipe break
- packer

Bottom fittings
- none (NONE)
- cap
- cone
- screw-on cap
- connector
- enlarger
- pipe break
- packer

Packing and backfill
- none (NONE)
- bentonite
- clay
- silt
- cement
- sand
- sand and gravel
- gravel

Sample symbols
- split spoon
- auger
- core
- grab
- shelby tube
- excavation
- undisturbed
- no recovery

(continued)

APPENDIX 35.A *(continued)*
USCS Soil Boring, Well, and Generic Geotextile Symbols

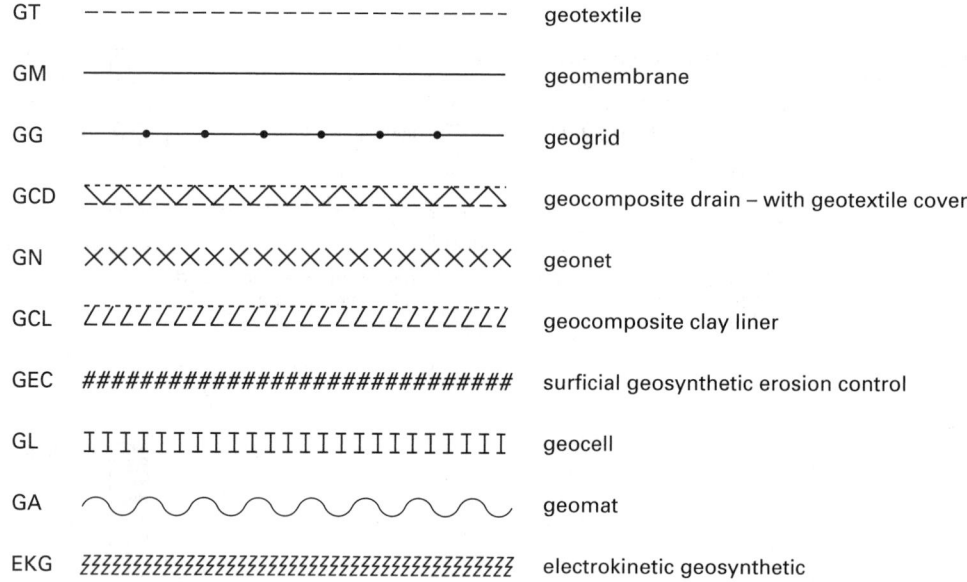

GT	– – – – – – – – – – – – – – – – – – –	geotextile
GM	————————————	geomembrane
GG	•—•—•—•—•—•—•—	geogrid
GCD	∨∨∨∨∨∨∨∨∨∨∨∨∨∨	geocomposite drain – with geotextile cover
GN	××××××××××××××××××	geonet
GCL	ℤℤℤℤℤℤℤℤℤℤℤℤℤℤℤℤℤℤℤ	geocomposite clay liner
GEC	############################	surficial geosynthetic erosion control
GL	IIIIIIIIIIIIIIIIIIIIIIIII	geocell
GA	∿∿∿∿∿∿∿∿∿	geomat
EKG	ƵƵƵƵƵƵƵƵƵƵƵƵƵƵƵƵƵƵƵƵƵƵƵƵƵƵƵƵ	electrokinetic geosynthetic

The following function symbols may be used where a description of the role of the geosynthetic material is needed.

S: separation geotextile

Ⓢ
– – – – – – – – – –

R: reinforcement geotextile

Ⓡ
– – – – – – – – – –

B: barrier (fluid)

D: drainage (fluid)

E: surficial erosion control

F: filtration

P: protection

R: reinforcement

S: separation

APPENDIX 37.A
Active Components for Retaining Walls
(Straight Slope Backfill)
(walls not over 20 ft high)
(lbf/ft^2 per lineal foot)

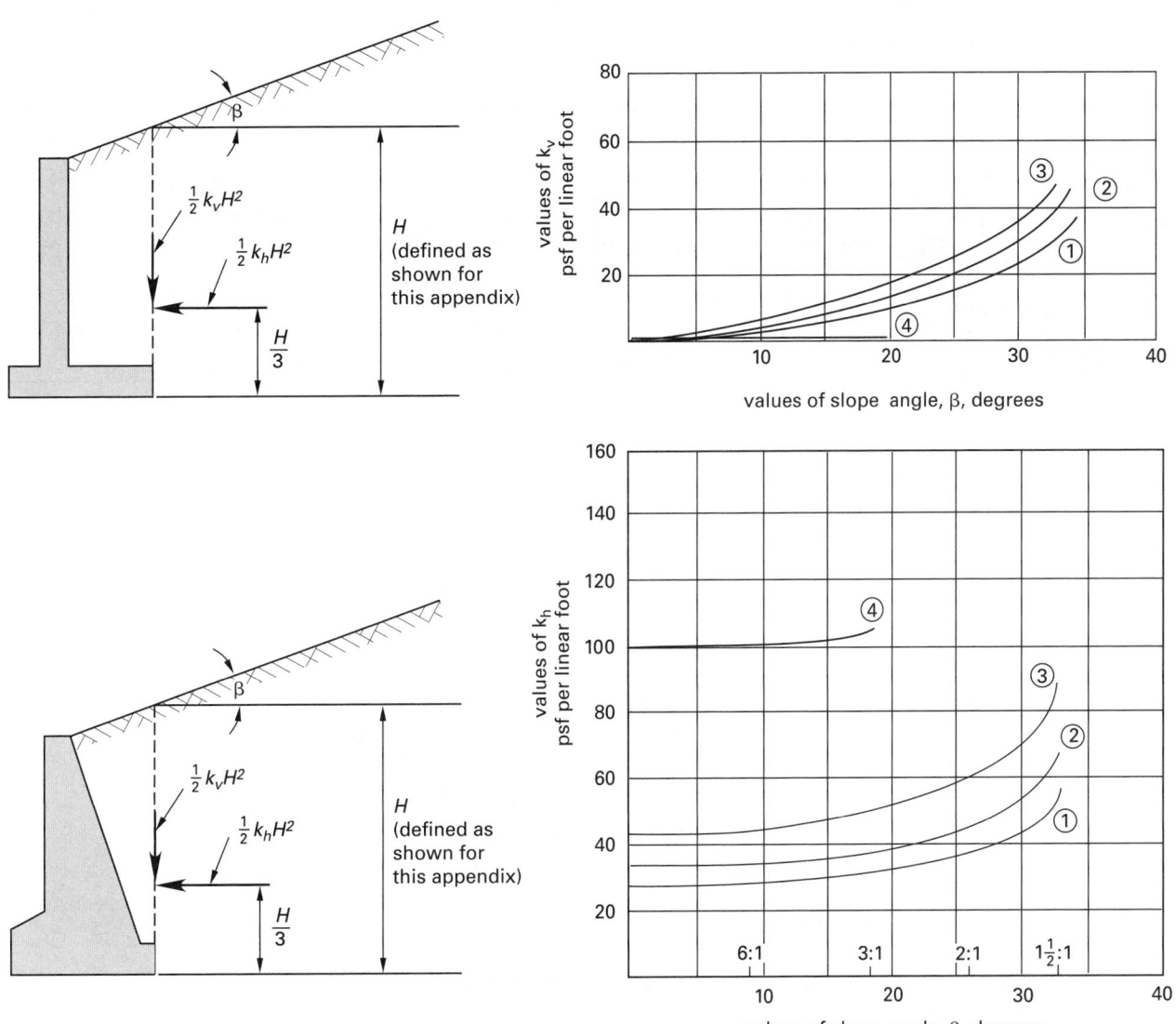

Circled numbers indicate the following soil types.

1. Clean sand and gravel: GW, GP, SW, SP.
2. Dirty sand and gravel of restricted permeability: GM, GM-GP, SM, SM-SP.
3. Stiff residual silts and clays, silty fine sands, and clayey sands and gravels:
   CL, ML, CH, MH, SM, SC, GC.
4. Soft or very soft clay, organic silt, or silty clay.

Source: *Foundations and Earth Structures*, NAVFAC DM-7.2 (1982), p. 7.2-86, Fig. 16

APPENDIX 37.B
Active Components for Retaining Walls
(Broken Slope Backfill)
(lbf/ft^2 per lineal foot)

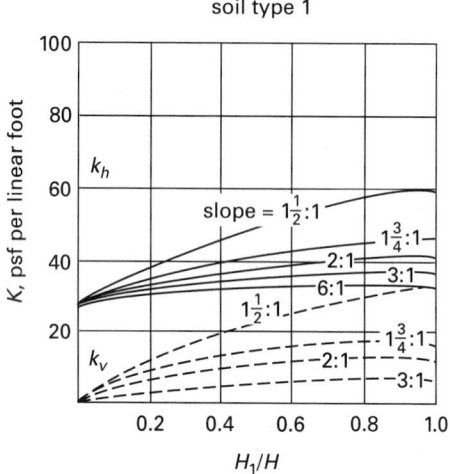

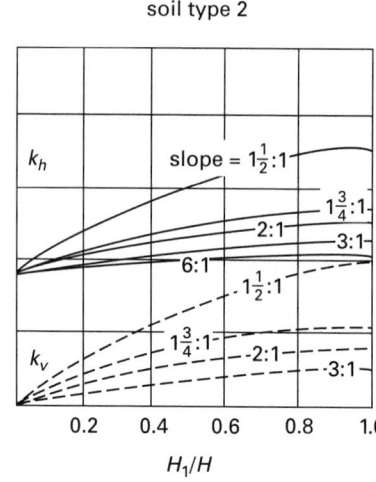

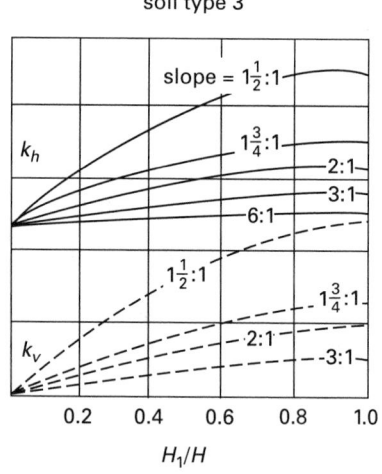

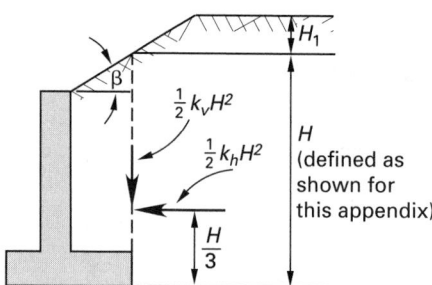

Soil types:

1. Clean sand and gravel: GW, GP, SW, SP.
2. Dirty sand and gravel of restricted permeability: GM, GM-GP, SM, SM-SP.
3. Stiff residual silts and clays, silty fine sands, and clayey sands and gravels: CL, ML, CH, MH, SM, SC, GC.

Source: *Foundations and Earth Structures*, NAVFAC DM-7.2 (1982), p. 7.2-87, Fig. 17

APPENDIX 40.A
Boussinesq Stress Contour Chart
(Infinitely Long and Square Footings)

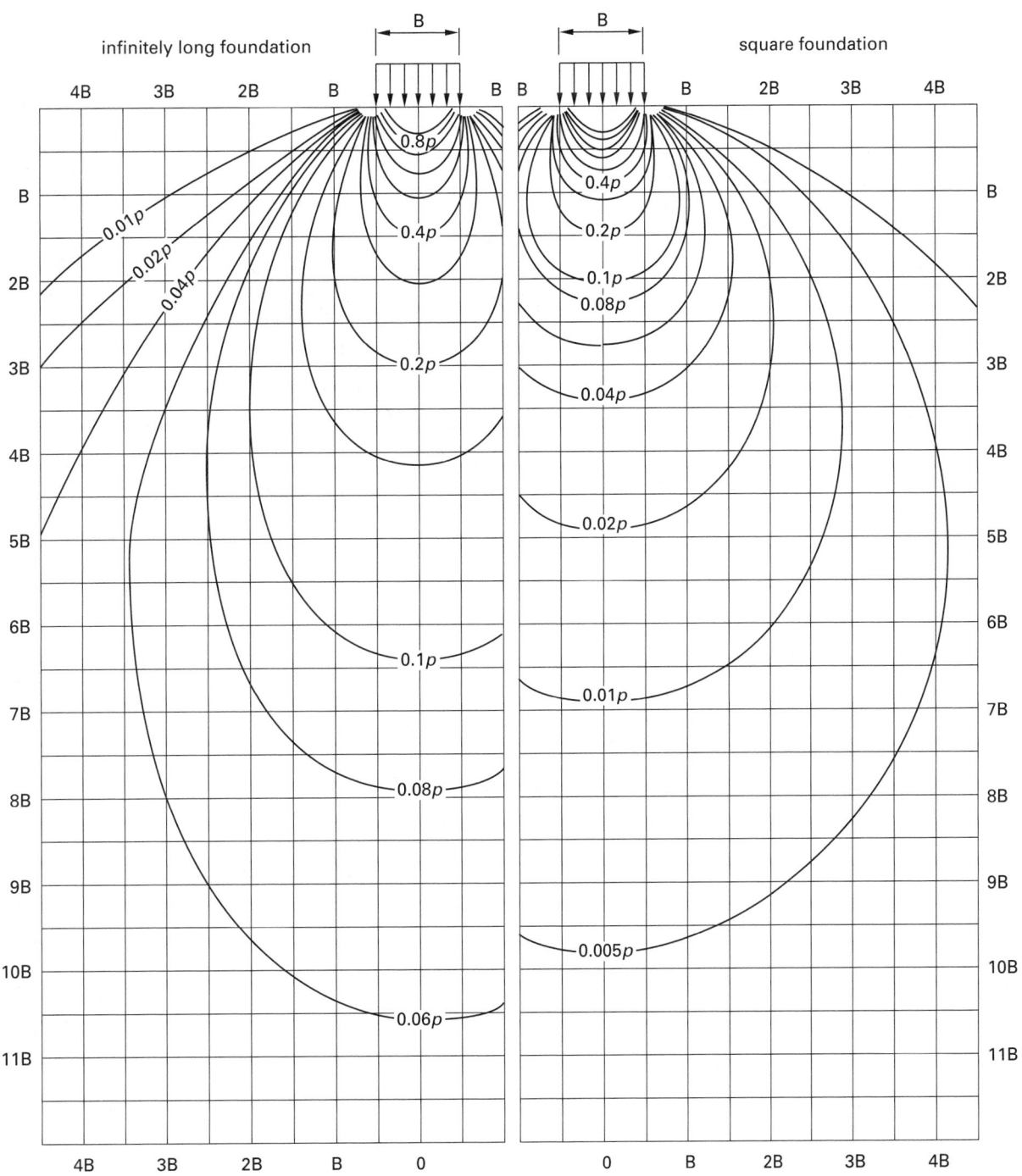

APPENDIX 40.B
Boussinesq Stress Contour Chart
(Uniformly Loaded Circular Footings)

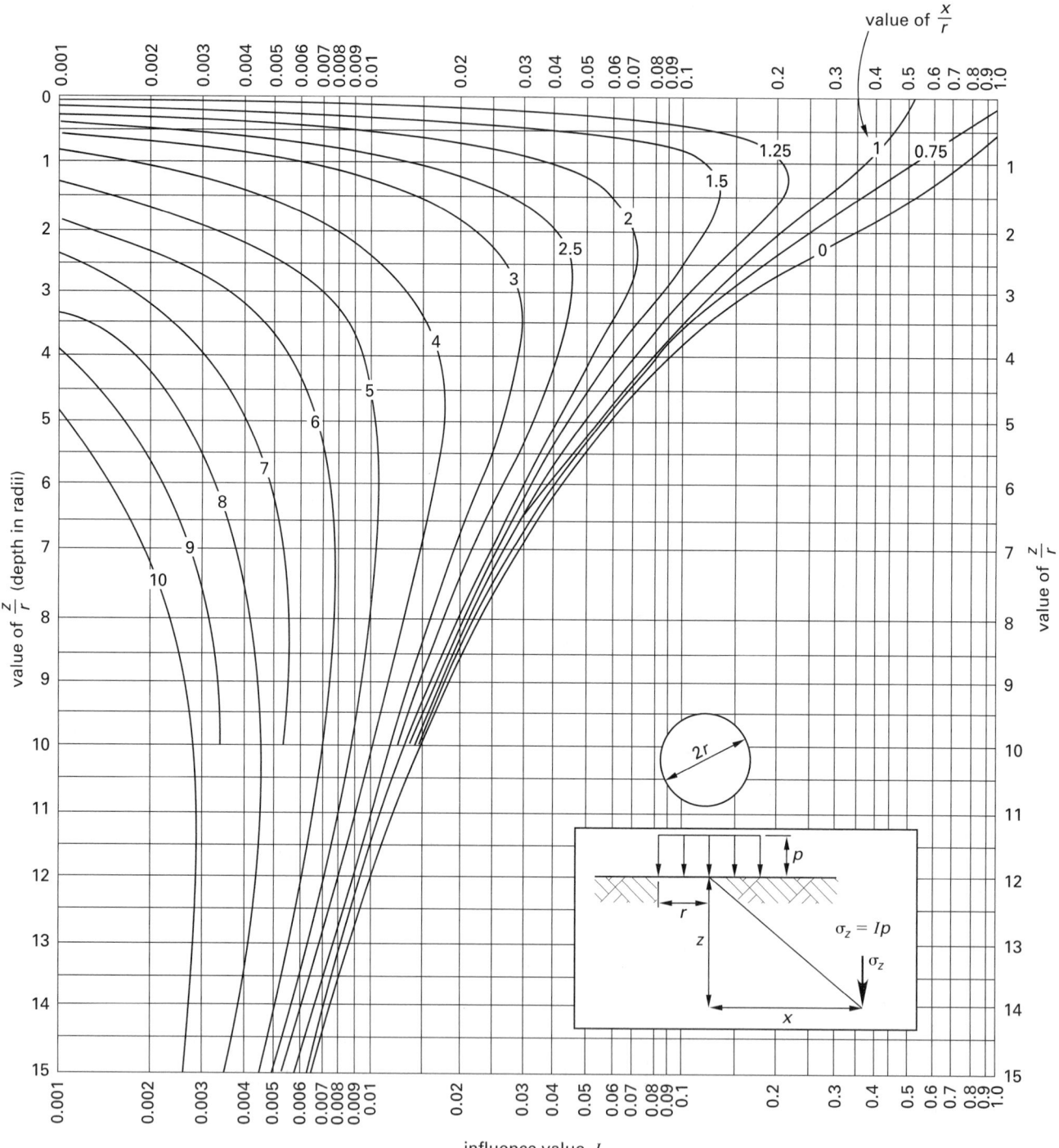

influence value, I

APPENDIX 42.A
Centroids and Area Moments of Inertia
for Basic Shapes

shape		centroidal location		area, A	area moment of inertia (rectangular and polar), I, J	radius of gyration, r
		x_c	y_c			
rectangle		$\dfrac{b}{2}$	$\dfrac{h}{2}$	bh	$I_x = \dfrac{bh^3}{3}$ $I_{c,x} = \dfrac{bh^3}{12}$ $J_c = \left(\dfrac{1}{12}\right)bh(b^2 + h^2)*$	$r_x = \dfrac{h}{\sqrt{3}}$ $r_{c,x} = \dfrac{h}{2\sqrt{3}}$
triangular area			$\dfrac{h}{3}$	$\dfrac{bh}{2}$	$I_x = \dfrac{bh^3}{12}$ $I_{c,x} = \dfrac{bh^3}{36}$	$r_x = \dfrac{h}{\sqrt{6}}$ $r_{c,x} = \dfrac{h}{3\sqrt{2}}$
trapezoid			$h\left(\dfrac{b+2t}{3b+3t}\right)$	$\dfrac{(b+t)h}{2}$	$I_x = \dfrac{(b+3t)h^3}{12}$ $I_{c,x} = \dfrac{(b^2+4bt+t^2)h^3}{36(b+t)}$	$r_x = \left(\dfrac{h}{\sqrt{6}}\right)\sqrt{\dfrac{b+3t}{b+t}}$ $r_{c,x} = \dfrac{h\sqrt{2(b^2+4bt+t^2)}}{6(b+t)}$
circle		0	0	πr^2	$I_x = I_y = \dfrac{\pi r^4}{4}$ $J_c = \dfrac{\pi r^4}{2}$	$r_x = \dfrac{r}{2}$
quarter-circular area		$\dfrac{4r}{3\pi}$	$\dfrac{4r}{3\pi}$	$\dfrac{\pi r^2}{4}$	$I_x = I_y = \dfrac{\pi r^4}{16}$ $J_o = \dfrac{\pi r^4}{8}$	
semicircular area		0	$\dfrac{4r}{3\pi}$	$\dfrac{\pi r^2}{2}$	$I_x = I_y = \dfrac{\pi r^4}{8}$ $I_{c,x} = 0.1098r^4$ $J_o = \dfrac{\pi r^4}{4}$ $J_c = 0.5025r^4$	$r_x = \dfrac{r}{2}$ $r_{c,x} = 0.264r$
quarter-elliptical area		$\dfrac{4a}{3\pi}$	$\dfrac{4b}{3\pi}$	$\dfrac{\pi ab}{4}$	$I_x = \dfrac{\pi ab^3}{8}$ $I_y = \dfrac{\pi a^3 b}{8}$ $J_o = \dfrac{\pi ab(a^2 + b^2)}{8}$	
semielliptical area		0	$\dfrac{4b}{3\pi}$	$\dfrac{\pi ab}{2}$		
semiparabolic area		$\dfrac{3a}{8}$	$\dfrac{3h}{5}$	$\dfrac{2ah}{3}$		
parabolic area		0	$\dfrac{3h}{5}$	$\dfrac{4ah}{3}$	$I_x = \dfrac{4ah^3}{7}$ $I_y = \dfrac{4ha^3}{15}$ $I_{c,x} = \dfrac{16ah^3}{175}$	$r_x = h\sqrt{\dfrac{3}{7}}$ $r_y = \dfrac{a}{\sqrt{5}}$
parabolic spandrel		$\dfrac{3a}{4}$	$\dfrac{3h}{10}$	$\dfrac{ah}{3}$	$I_x = \dfrac{ah^3}{21}$ $I_y = \dfrac{3ha^3}{15}$	
general spandrel		$\left(\dfrac{n+1}{n+2}\right)a$	$\left(\dfrac{n+1}{4n+2}\right)h$	$\dfrac{ah}{n+1}$	*Theoretical definition based on $J = I_x + I_y$. However, in torsion, not all parts of the shape are effective. Effective values will be lower.	
circular sector [α in radians]		$\dfrac{2r\sin\alpha}{3\alpha}$	0	αr^2	$J = C\left(\dfrac{b^2 + h^2}{b^3 h^3}\right)$	

b/h	C
1	3.56
2	3.50
4	3.34
8	3.21

APPENDIX 43.A
Typical Properties of Structural Steel,
Aluminum, and Magnesium
(all values in ksi)

structural steel

designation	application	S_u	S_y	approximate S_e
A36	shapes	58–80	36	29–40
	plates	58–80	36	29–40
A53	pipe	60	35	30
A242	shapes	70	50	35
	plates to $\frac{3}{4}$ in	70	50	35
A440	shapes	70	50	35
	plates to $\frac{3}{4}$ in	70	50	35
A441	shapes	70	50	35
	plates to $\frac{3}{4}$ in	70	50	35
A500	tubes	45	33	22
A501	tubes	58	36	29
A514	plates to $\frac{3}{4}$ in	115–135	100	55
A529	shapes	60–85	42	30–42
	plates to $\frac{1}{2}$ in	60–85	42	30–42
A570	sheet/strip	55	40	27
A572	shapes	65	50	30
	plates	60	42	30
A588	shapes	70	50	35
	plates to 4 in	70	50	35
A606	hot-rolled sheet	70	50	35
	cold-rolled sheet	65	45	32
A607	sheet	60	45	30
A618	shapes	70	50	35
	tubes	70	50	35
A913	shapes	65	50	30
A992	shapes	65	50	–

structural aluminum

designation	application	S_u	S_y	approximate S_e (@ 10^8 cyc.)
2014-T6	shapes/bars	63	55	19
6061-T6	all	42	35	14.5

structural magnesium (ksi)

designation	application	S_u	S_y	approximate S_e (@ 10^7 cyc.)
AZ31	shapes	38	29	19
AZ61	shapes	45	33	19
AZ80	shapes	55	40	–

(Multiply ksi by 6.895 to obtain MPa.)

APPENDIX 43.B
Typical Mechanical Properties of Representative Metals
(room temperature)

The following mechanical properties are not guaranteed since they are averages for various sizes, product forms, and methods of manufacture. Thus, this data is not for design use, but is intended only as a basis for comparing alloys and tempers.

material designation, composition, typical use, and source if applicable	condition, heat treatment	S_{ut} (ksi)	S_{yt} (ksi)
IRON BASED			
Armco ingot iron, for fresh and saltwater piping	normalized	44	24
AISI 1020, plain carbon steel, for general machine parts and screws and carburized parts	hot rolled cold worked	65 78	43 66
AISI 1030, plain carbon steel, for gears, shafts, levers, seamless tubing, and carburized parts	cold drawn	87	74
AISI 1040, plain carbon steel, for high-strength parts, shafts, gears, studs, connecting rods, axles, and crane hooks	hot rolled cold worked hardened	91 100 113	58 88 86
AISI 1095, plain carbon steel, for handtools, music wire springs, leaf springs, knives, saws, and agricultural tools such as plows and disks	annealed hot rolled hardened	100 142 180	53 84 118
AISI 1330, manganese steel, for axles and drive shafts	annealed cold drawn hardened	97 113 122	83 93 100
AISI 4130, chromium-molybdenum steel, for high-strength aircraft structures	annealed hardened	81 161	52 137
AISI 4340, nickel-chromium-molybdenum steel, for large-scale, heavy-duty, high-strength structures	annealed as rolled hardened	119 192 220	99 147 200
AISI 2315, nickel steel, for carburized parts	as rolled cold drawn	85 95	56 75
AISI 2330, nickel steel	as rolled cold drawn annealed normalized	98 110 80 95	65 90 50 61
AISI 3115, nickel-chromium steel for carburized parts	cold drawn as rolled annealed	95 75 71	70 60 62
STAINLESS STEELS			
AISI 302, stainless steel, most widely used, same as 18-8	annealed cold drawn	90 105	35 60
AISI 303, austenitic stainless steel, good machineability	annealed cold worked	90 110	35 75
AISI 304, austenitic stainless steel, good machineability and weldability	annealed cold worked	85 110	30 75
AISI 309, stainless steel, good weldability, high strength at high temperatures, used in furnaces and ovens	annealed cold drawn	90 110	35 65
AISI 316, stainless steel, excellent corrosion resistance	annealed cold drawn	85 105	35 60
AISI 410, magnetic, martensitic, can be quenched and tempered to give varying strength	annealed cold drawn oil quenched and drawn	60 180 110	32 150 91
AISI 430, magnetic, ferritic, used for auto and architectural trim and for equipment in food and chemical industries	annealed cold drawn	60 100	35
AISI 502, magnetic, ferritic, low cost, widely used in oil refineries	annealed	60	25

(continued)

APPENDIX 43.B *(continued)*
Typical Mechanical Properties of Representative Metals
(room temperature)

material designation, composition, typical use, and source if applicable	condition, heat treatment	S_{ut} (ksi)	S_{yt} (ksi)
ALUMINUM BASED			
2011, for screw machine parts, excellent machineability, but not weldable, and corrosion sensitive	T3	55	43
	T8	59	45
2014, for aircraft structures, weldable	T3	63	40
	T4, T451	61	37
	T6, T651	68	60
2017, for screw machine parts	T4, T451	62	40
2018, for engine cylinders, heads, and pistons	T61	61	46
2024, for truck wheels, screw machine parts, and aircraft structures	T3	65	45
	T4, T351	64	42
	T361	72	57
2025, for forgings	T6	58	37
2117, for rivets	T4	43	24
2219, high-temperature applications (up to 600°F), excellent weldability and machineabilty	T31, T351	52	36
	T37	57	46
	T42	52	27
3003, for pressure vessels and storage tanks, poor machine-ability but good weldability, excellent corrosion resistance	0	16	6
	H12	19	18
	H14	22	21
	H16	26	25
3004, same characteristics as 3003	0	26	10
	H32	31	25
	H34	35	29
	H36	38	33
4032, pistons	T6	55	46
5083, unfired pressure vessels, cryogenics, towers, and drilling rigs	0	42	21
	H116, H117, H321	46	33
5154, saltwater services, welded structures, and storage tanks	0	35	17
	H32	39	30
	H34	42	33
5454, same characteristics as 5154	0	36	17
	H32	40	30
	H34	44	35
5456, same characteristics as 5154	0	45	23
	H111	47	33
	H321, H116, H117	51	37
6061, corrosion resistant and good weldability, used in railroad cars	T4	33	19
	T6	42	37
7178, Alclad, corrosion-resistant	0	33	15
	T6	88	78

CAST IRON (note redefinition of columns)		S_{ut} (ksi)	S_{us} (ksi)	S_{uc} (ksi)
gray cast iron	class 20	30	32.5	30
	class 25	25	34	100
	class 30	30	41	110
	class 35	35	49	125
	class 40	40	52	135
	class 50	50	64	160
	class 60	60	60	150

(continued)

APPENDIX 43.B *(continued)*
Typical Mechanical Properties of Representative Metals
(room temperature)

material designation, composition, typical use, and source if applicable	condition, heat treatment	S_{ut} (ksi)	S_{yt} (ksi)
COPPER BASED			
copper, commercial purity	annealed (furnace cool from 400°C)	32	10
	cold drawn	45	40
cartridge brass: 70% Cu, 30% Zn	cold rolled (annealed 400°C, furnace cool)	76	63
copper-beryllium (1.9% Be, 0.25% Co)	annealed, wqf 1450°F	70	
	cold rolled	200	
	hardened after annealing	200	150
phosphor-bronze, for springs	wire, 0.025 in and under	145	
	0.025 in to 0.0625 in	135	
	0.125 in to 0.250 in	125	
monel metal	cold-drawn bars, annealed	70	30
red brass	sheet and strip half-hard	51	
	hard	63	
	spring	78	
yellow brass	sheet and strip half-hard	55	
	hard	68	
	spring	86	
NICKEL BASED			
pure nickel, magnetic, high corrosion resistance	annealed (ht 1400°F, acrt)	46	8.5
	annealed at 2050°F	125	75
Inconel X, type 550, excellent high temperature properties	annealed and age hardened	175	110
	annealed (wqf 1600°F)	100	45
K-monel, excellent high temperature properties and corrosion resistance	age hardened spring stock	185	160
Invar, 36% Ni, 64% Fe, low coefficient of expansion (0.9×10^{-6} %/°C)	annealed (wqf 800°C)	71	40
REFRACTORY METALS (properties at room temperature)			
molybdenum	as rolled	100	75
tantalum	annealed at 1050°C in vacuum	60	45
	as rolled	110	100
titanium, commercial purity	annealed at 1200°F	95	80
titanium, 6% Al, 4% V	annealed at 1400°F, acrt	135	130
	heat treated (wqf 1750°F, ht 1000°F, acrt)	170	150
titanium, 4% Al, 4% Mn OR 5% Al, 2.75% Cr, 1.25% Fe OR 5% Al, 1.5% Fe, 1.4% Cr, 1.2% Mo	wqf 1450°F, ht 900°F, acrt	185	170
tungsten, commercial purity	hard wire	600	540

MAGNESIUM		S_{ut} (ksi)	S_{yt} (ksi)	S_{us} (ksi)
AZ92, for sand and permanent-mold casting	as cast	24	14	
	solution treated	39	14	
	aged	39	21	
AZ91, for die casting	as cast	33	21	
AZ31X (sheet)	annealed	35	20	
	hard	40	31	
AZ80X, for structural shapes	extruded	48	32	
	extruded and aged	52	37	
ZK60A, for structural shapes	extruded	49	38	
	extruded and aged	51	42	
AZ31B (sheet and plate), for structural shapes in use below 300°F	temper 0	32	15	17
	temper 1124	34	18	18
	temper 1126	35	21	18
	temper F	32	16	17

Abbreviations:
  wqf: water-quench from
  acrt: air-cooled to room temperature
  ht: heated to

(Multiply ksi by 6.895 to obtain MPa.)

APPENDIX 44.A
Elastic Beam Deflection Equations
(w is the load per unit length.)
(y is positive downward.)

Case 1: Cantilever with End Load

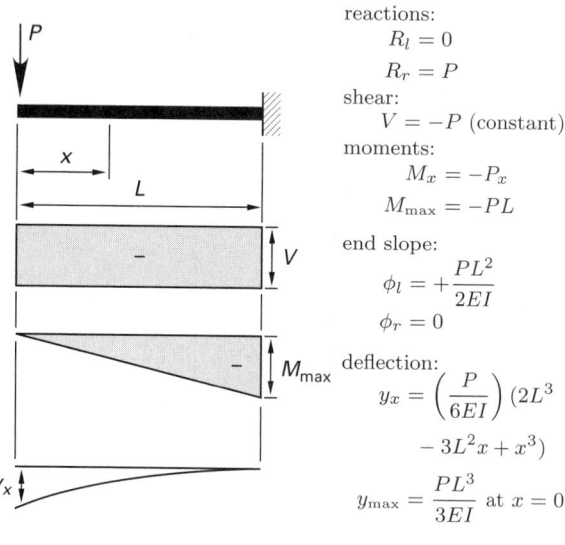

reactions:
$$R_l = 0$$
$$R_r = P$$

shear:
$$V = -P \text{ (constant)}$$

moments:
$$M_x = -P_x$$
$$M_{max} = -PL$$

end slope:
$$\phi_l = +\frac{PL^2}{2EI}$$
$$\phi_r = 0$$

deflection:
$$y_x = \left(\frac{P}{6EI}\right)(2L^3 - 3L^2x + x^3)$$
$$y_{max} = \frac{PL^3}{3EI} \text{ at } x = 0$$

Case 2: Cantilever with Uniform Load

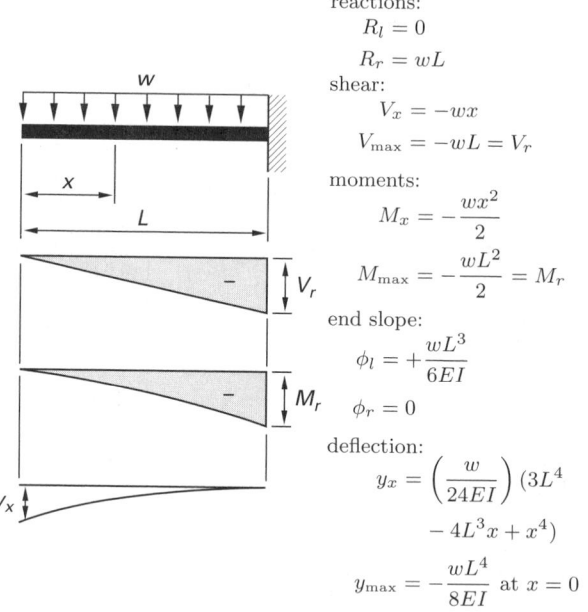

reactions:
$$R_l = 0$$
$$R_r = wL$$

shear:
$$V_x = -wx$$
$$V_{max} = -wL = V_r$$

moments:
$$M_x = -\frac{wx^2}{2}$$
$$M_{max} = -\frac{wL^2}{2} = M_r$$

end slope:
$$\phi_l = +\frac{wL^3}{6EI}$$
$$\phi_r = 0$$

deflection:
$$y_x = \left(\frac{w}{24EI}\right)(3L^4 - 4L^3x + x^4)$$
$$y_{max} = -\frac{wL^4}{8EI} \text{ at } x = 0$$

Case 3: Cantilever with Triangular Load

reactions:
$$R_l = 0$$
$$R_r = \frac{wL}{2}$$

shear:
$$V_x = -\frac{wx^2}{2L}$$
$$V_{max} = \frac{wL}{2} \text{ at } x = L$$

moments:
$$M_x = -\frac{wx^3}{6L}$$
$$M_{max} = \frac{wL^2}{6} \text{ at } x = L$$

end slope:
$$\phi_l = +\frac{wL^3}{24EI}$$
$$\phi_r = 0$$

deflection:
$$y_x = \left(\frac{w}{120EIL}\right)(4L^5 - 5L^4x + x^5)$$
$$y_{max} = \frac{wL^4}{30EI} \text{ at } x = 0$$

Case 4: Propped Cantilever with Uniform Load

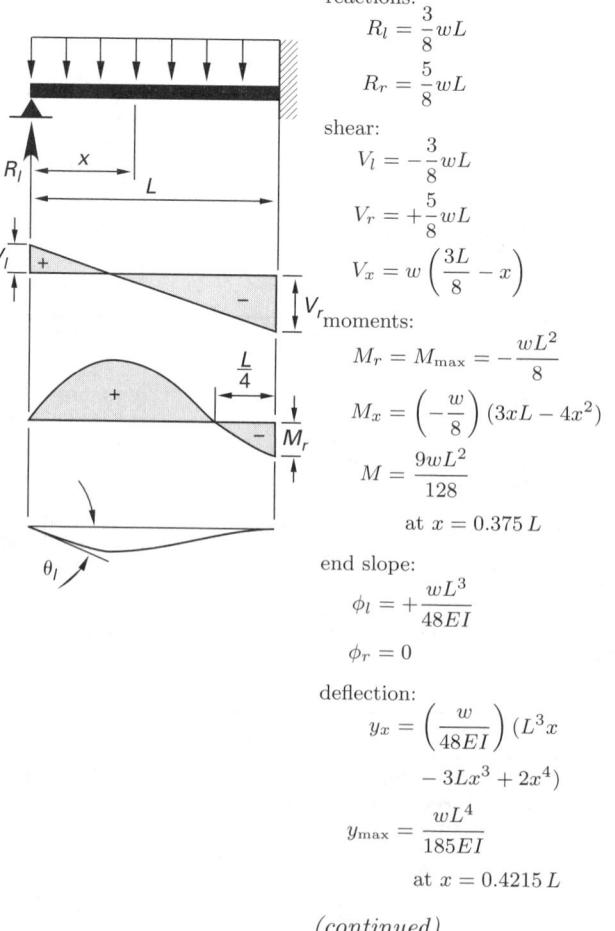

reactions:
$$R_l = \frac{3}{8}wL$$
$$R_r = \frac{5}{8}wL$$

shear:
$$V_l = -\frac{3}{8}wL$$
$$V_r = +\frac{5}{8}wL$$
$$V_x = w\left(\frac{3L}{8} - x\right)$$

moments:
$$M_r = M_{max} = -\frac{wL^2}{8}$$
$$M_x = \left(-\frac{w}{8}\right)(3xL - 4x^2)$$
$$M = \frac{9wL^2}{128}$$
$$\text{at } x = 0.375\,L$$

end slope:
$$\phi_l = +\frac{wL^3}{48EI}$$
$$\phi_r = 0$$

deflection:
$$y_x = \left(\frac{w}{48EI}\right)(L^3x - 3Lx^3 + 2x^4)$$
$$y_{max} = \frac{wL^4}{185EI}$$
$$\text{at } x = 0.4215\,L$$

(continued)

APPENDIX 44.A *(continued)*
Elastic Beam Deflection Equations
(w is the load per unit length.)
(y is positive downward.)

Case 5: Cantilever with End Moment

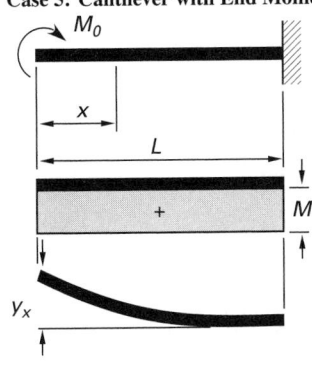

reactions:
$$R_l = 0$$
$$R_r = 0$$
shear:
$$V = 0$$
moments:
$$M = M_0 = M_{max}$$
end slope:
$$\phi_l = -\frac{M_0 L}{EI}$$
$$\phi_r = 0$$
deflection:
$$y_x = \left(-\frac{M_0}{2EI}\right)$$
$$\times (L^2 - 2xL + x^2)$$
$$y_{max} = -\frac{M_0 L^2}{2EI} \text{ at } x = 0$$

Case 6: Simple Beam with Center Load

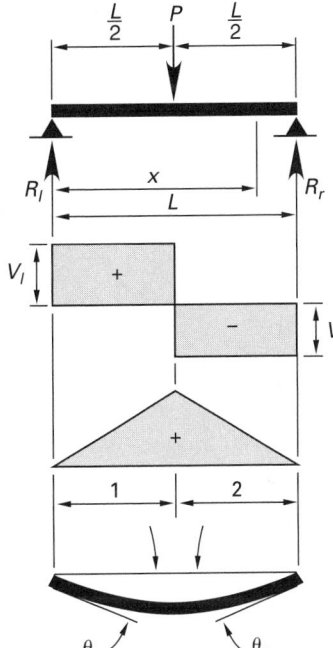

reactions:
$$R_l = R_r = \frac{P}{2}$$
shear:
$$V_l = \frac{P}{2}$$
$$V_r = -\frac{P}{2}$$
moments:
$$M_{x1} = \frac{Px}{2}$$
$$M_{x2} = \left(\frac{P}{2}\right)(L - x)$$
$$M_{max} = \frac{PL}{4}$$
end slope:
$$\phi_l = -\frac{PL^2}{16EI}$$
$$\phi_r = +\frac{PL^2}{16EI}$$
deflection:
$$y_{x1} \ (x < \tfrac{L}{2})$$
$$= \left(\frac{P}{48EI}\right)(3xL^2 - 4x^3)$$
$$y_{max} = \frac{PL^3}{48EI} \text{ at } x = \frac{L}{2}$$

Case 7: Simple Beam with Intermediate Load

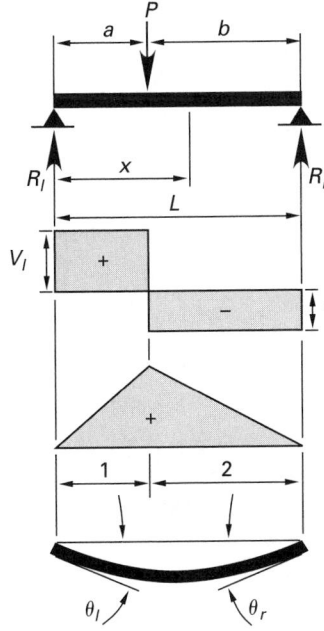

reactions:
$$R_l = \frac{Pb}{L}$$
$$R_r = \frac{Pa}{L}$$
shear:
$$V_l = +\frac{Pb}{L}$$
$$V_r = -\frac{Pa}{L}$$
moments:
$$M_{x1} = \frac{Pbx}{L}$$
$$M_{x2} = \frac{Pa(L - x)}{L}$$
$$M_{max} = \frac{Pab}{L} \text{ at } x = a$$
end slope:
$$\phi_l = -\frac{Pab\left(1 + \frac{b}{L}\right)}{6EI}$$
$$\phi_r = \frac{Pab\left(1 + \frac{a}{L}\right)}{6EI}$$
deflection:
$$y_{x1} \ (x < a)$$
$$= \left(\frac{Pb}{6EIL}\right)(L^2 x - b^2 x - x^3)$$
$$y_{x2} \ (x > a)$$
$$= \left(\frac{Pb}{6EIL}\right)\left[\left(\frac{L}{b}\right)(x - a)^3\right.$$
$$\left. + (L^2 - b^2)x - x^3\right]$$
$$y = \frac{Pa^2 b^2}{3EIL} \text{ at } x = a$$
$$y_{max} = \left(\frac{0.06415\,Pb}{EIL}\right)(L^2 - b^2)^{3/2}$$
$$\text{at } x = \sqrt{\frac{a(L + b)}{3}}$$

Case 8: Simple Beam with Two Loads

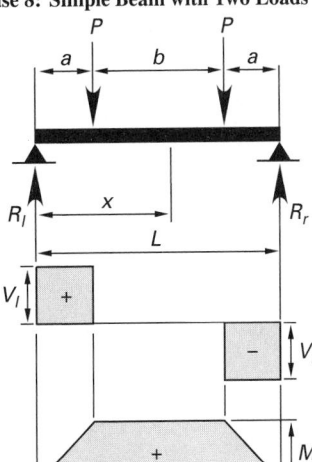

reactions:
$$R_l = R_r = P$$
shear:
$$V_l = +P$$
$$V_r = -P$$
moments:
$$M_{x1} = Px$$
$$M_{x2} = Pa$$
$$M_{x3} = P(L - x)$$
end slope:
$$\phi_l = -\frac{Pa(a + b)}{2EI}$$
$$\phi_r = +\frac{Pa(a + b)}{2EI}$$
deflection:
$$y_{x1} \ (x < a)$$
$$= \left(\frac{P}{6EI}\right)(3Lax - 3a^2 x - x^3)$$
$$y_{x2} \ (a < x < a + b)$$
$$= \left(\frac{P}{6EI}\right)(3Lax - 3ax^2 - a^3)$$
$$y_{max} = \left(\frac{P}{24EI}\right)(3L^2 a - 4a^3)$$
$$\text{at } x = \frac{L}{2}$$

(continued)

APPENDIX 44.A *(continued)*
Elastic Beam Deflection Equations
(*w* is the load per unit length.)
(*y* is positive downward.)

Case 9: Simple Beam with Uniform Load

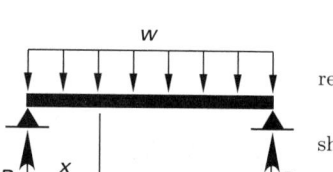

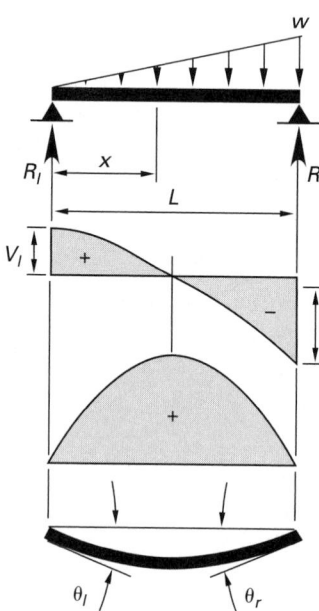

reactions:
$$R_l = R_r = \frac{wL}{2}$$
shear:
$$V_l = +\frac{wL}{2}$$
$$V_r = -\frac{wL}{2}$$
moments:
$$M = \left(\frac{w}{2}\right)(Lx - x^2)$$
$$M_{max} = \frac{wL^2}{8}$$
end slope:
$$\phi_l = \frac{-wL^3}{24EI}$$
$$\phi_r = +\frac{wL^3}{24EI}$$
deflection:
$$y_x = \left(\frac{w}{24EI}\right)(L^3 x$$
$$- 2Lx^3 + x^4)$$
$$y_{max} = \frac{5wL^4}{384EI} \text{ at } x = \frac{L}{2}$$

Case 10: Simple Beam with Triangular Load
(*w* is the maximum loading per unit length at the right end, not the total load, $W = \frac{1}{2}Lw$.)

reactions:
$$R_l = \frac{wL}{6}$$
$$R_r = \frac{wL}{3}$$
shear:
$$V_l = +\frac{wL}{6}$$
$$V_r = -\frac{wL}{3}$$
$$V_x = \left(\frac{wL}{6}\right)\left(1 - 3\left(\frac{x}{L}\right)^2\right)$$
moments:
$$M_x = \left(\frac{w}{6}\right)\left(Lx - \frac{x^3}{L}\right)$$
$$M_{max} = \frac{wL^2}{9\sqrt{3}} = 0.0642wL^2$$
$$\text{at } x = 0.577L$$
end slope:
$$\phi_l = \frac{-7wL^3}{360EI}$$
$$\phi_r = +\frac{wL^3}{45EI}$$
deflection:
$$y_x = \left(\frac{w}{360EI}\right)\left(7L^3 x\right.$$
$$\left. -10Lx^3 + \frac{3x^5}{L}\right)$$
$$y_{max} = (0.00652)\left(\frac{wL^4}{EI}\right)$$
$$\text{at } x = 0.519L$$

(continued)

APPENDIX 44.A *(continued)*
Elastic Beam Deflection Equations
(*w* is the load per unit length.)
(*y* is positive downward.)

Case 11: Simple Beam with Overhung Load

reactions:

$$R_l = \left(\frac{P}{b}\right)(b+a)$$

$$R_r = \frac{-Pa}{b}$$

shear:

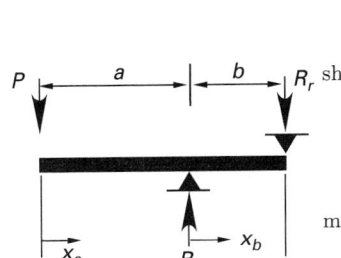

$$V_l = -P$$

$$V_r = \frac{Pa}{b}$$

moments:

$$M_a = Px_a$$

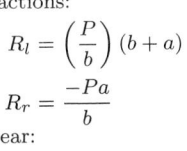

$$M_b = \left(\frac{Pa}{b}\right)(b - x_b)$$

$$M_{\max} = Fa \text{ at } x_a = a$$

deflections:

$$y_a = \left(\frac{F}{3EI}\right)$$

$$\times \left((a^2 + ab)(a - x_a)\right.$$

$$\left. + \left(\frac{x_a}{2}\right)(x_a^2 - a^2)\right)$$

$$y_b = \left(\frac{Fax_b}{6EI}\right)\left(3x_b - \left(\frac{x_b^2}{b}\right) - 2b\right)$$

$$y_{\text{tip}} = \left(\frac{Fa^2}{3EI}\right)(a + b) \quad [\text{max down}]$$

$$y_{\max} = (0.06415)\left(\frac{Fab^2}{EI}\right) \text{ at } x_b$$

$$= 0.4226b \quad [\text{max up}]$$

**Case 12: Simple Beam with Uniform Load
Distributed over Half of Beam**

reactions:

$$R_l = \frac{3wL}{8}$$

$$R_r = \frac{wL}{8}$$

shear:

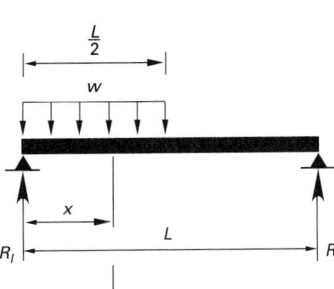
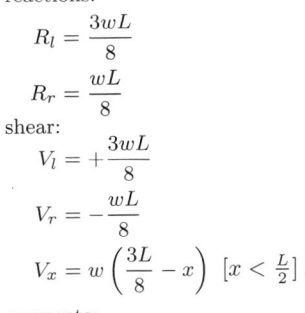

$$V_l = +\frac{3wL}{8}$$

$$V_r = -\frac{wL}{8}$$

$$V_x = w\left(\frac{3L}{8} - x\right) \quad [x < \tfrac{L}{2}]$$

moments:

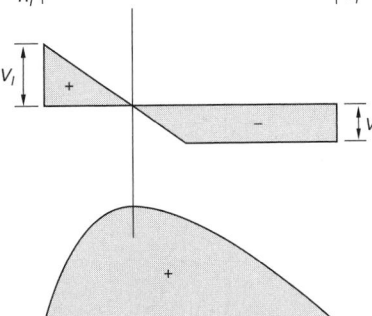

$$M_x = \left(\frac{w}{8}\right)\left(3Lx - 4x^2\right) \quad [x < \tfrac{L}{2}]$$

$$M_x = \left(\frac{wL^2}{8}\right)\left(1 - \frac{x}{L}\right) \quad [x > \tfrac{L}{2}]$$

$$M_{\max} = \frac{9wL^2}{128} \quad [x = \tfrac{3L}{8}]$$

deflections:

$$y_x = \left(\frac{wx}{384EI}\right)\left(9L^3 - 24Lx^2 + 16x^3\right)$$

$$[x < \tfrac{L}{2}]$$

$$y_x = \left(\frac{wL(L-x)}{384EI}\right)\left(16xL - 8x^2 - L^2\right)$$

$$[x > \tfrac{L}{2}]$$

APPENDIX 45.A
Properties of Weld Groups
(treated as lines)

weld configuration	centroid location	section modulus $S = I_{c,x}/\overline{y}$	polar moment of inertia $J = I_{c,x} + I_{c,y}$
	$\overline{y} = \dfrac{d}{2}$	$\dfrac{d^2}{6}$	$\dfrac{d^3}{12}$
	$\overline{y} = \dfrac{d}{2}$	$\dfrac{d^2}{3}$	$\dfrac{d(3b^2 + d^2)}{6}$
	$\overline{y} = \dfrac{d}{2}$	bd	$\dfrac{b(3d^2 + b^2)}{6}$
	$\overline{y} = \dfrac{d^2}{2(b+d)}$ $\overline{x} = \dfrac{b^2}{2(b+d)}$	$\dfrac{4bd + d^2}{6}$	$\dfrac{(b+d)^4 - 6b^2d^2}{12(b+d)}$
	$\overline{x} = \dfrac{b^2}{2b+d}$	$bd + \dfrac{d^2}{6}$	$\dfrac{8b^3 + 6bd^2 + d^3}{12} - \dfrac{b^4}{2b+d}$
	$\overline{y} = \dfrac{d^2}{b+2d}$	$\dfrac{2bd + d^2}{3}$	$\dfrac{b^3 + 6b^2d + 8d^3}{12} - \dfrac{d^4}{2d+b}$
	$\overline{y} = \dfrac{d}{2}$	$bd + \dfrac{d^2}{3}$	$\dfrac{(b+d)^3}{6}$
	$\overline{y} = \dfrac{d^2}{b+2d}$	$\dfrac{2bd + d^2}{3}$	$\dfrac{b^3 + 8d^3}{12} - \dfrac{d^4}{b+2d}$
	$\overline{y} = \dfrac{d}{2}$	$bd + \dfrac{d^2}{3}$	$\dfrac{b^3 + 3bd^2 + d^3}{6}$
	$\overline{y} = r$	πr^2	$2\pi r^3$

APPENDIX 47.A
Elastic Fixed-End Moments

1.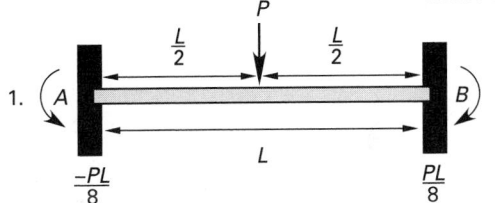

$$\frac{-PL}{8} \qquad \frac{PL}{8}$$

8.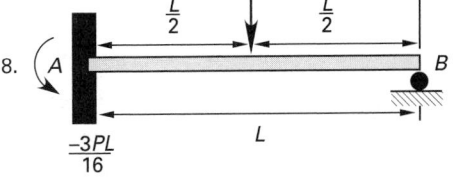

$$\frac{-3PL}{16}$$

2.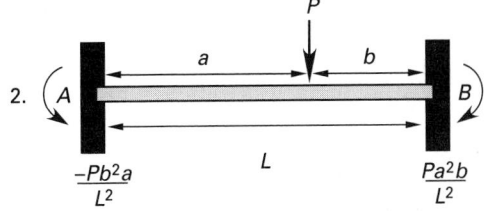

$$\frac{-Pb^2a}{L^2} \qquad \frac{Pa^2b}{L^2}$$

9.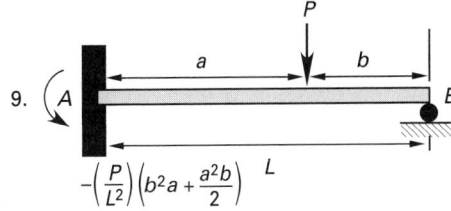

$$-\left(\frac{P}{L^2}\right)\left(b^2a + \frac{a^2b}{2}\right)$$

3.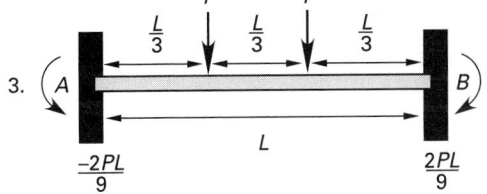

$$\frac{-2PL}{9} \qquad \frac{2PL}{9}$$

10.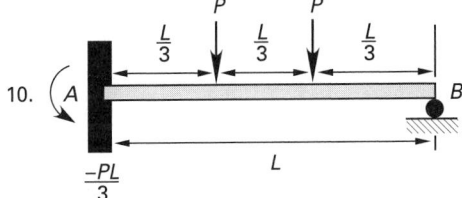

$$\frac{-PL}{3}$$

4.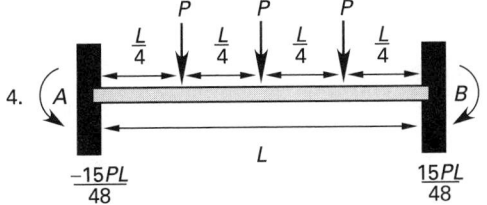

$$\frac{-15PL}{48} \qquad \frac{15PL}{48}$$

11.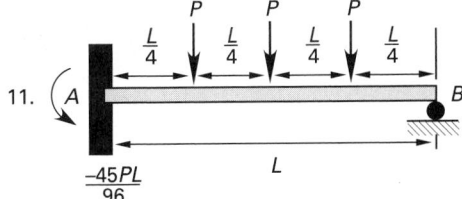

$$\frac{-45PL}{96}$$

5.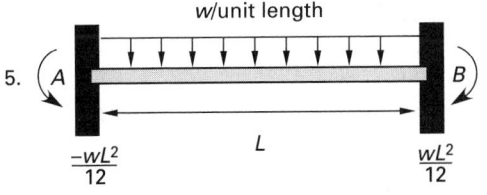

$$\frac{-wL^2}{12} \qquad \frac{wL^2}{12}$$

12.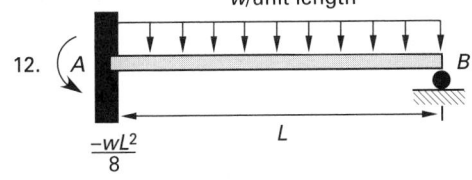

$$\frac{-wL^2}{8}$$

6.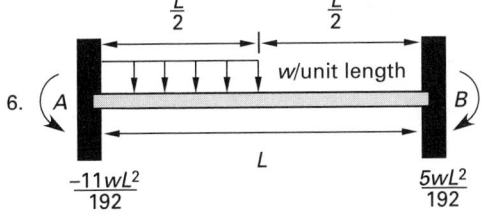

$$\frac{-11wL^2}{192} \qquad \frac{5wL^2}{192}$$

13.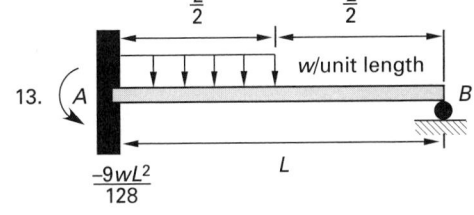

$$\frac{-9wL^2}{128}$$

7.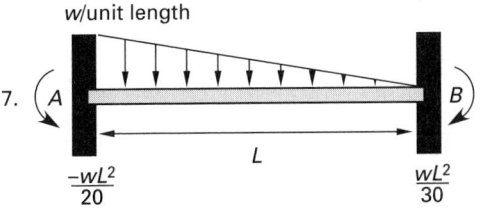

$$\frac{-wL^2}{20} \qquad \frac{wL^2}{30}$$

14.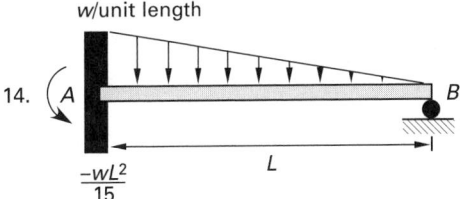

$$\frac{-wL^2}{15}$$

(continued)

APPENDIX 47.A *(continued)*
Elastic Fixed-End Moments

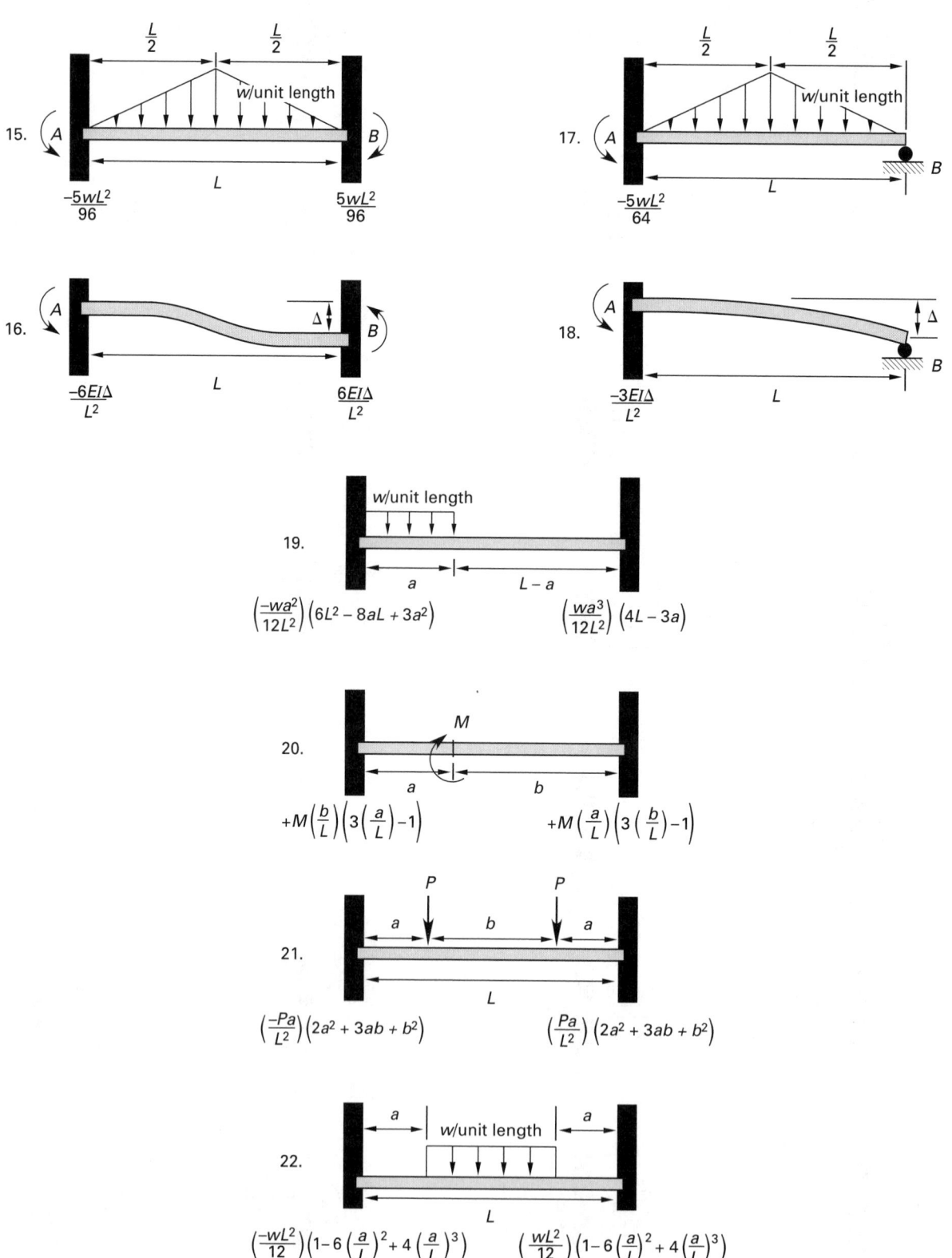

15.

$$\frac{-5wL^2}{96} \qquad \frac{5wL^2}{96}$$

16.

$$\frac{-6EI\Delta}{L^2} \qquad \frac{6EI\Delta}{L^2}$$

17.

$$\frac{-5wL^2}{64}$$

18.

$$\frac{-3EI\Delta}{L^2}$$

19.

$$\left(\frac{-wa^2}{12L^2}\right)\left(6L^2 - 8aL + 3a^2\right) \qquad \left(\frac{wa^3}{12L^2}\right)\left(4L - 3a\right)$$

20.

$$+M\left(\frac{b}{L}\right)\left(3\left(\frac{a}{L}\right)-1\right) \qquad +M\left(\frac{a}{L}\right)\left(3\left(\frac{b}{L}\right)-1\right)$$

21.

$$\left(\frac{-Pa}{L^2}\right)\left(2a^2 + 3ab + b^2\right) \qquad \left(\frac{Pa}{L^2}\right)\left(2a^2 + 3ab + b^2\right)$$

22.

$$\left(\frac{-wL^2}{12}\right)\left(1 - 6\left(\frac{a}{L}\right)^2 + 4\left(\frac{a}{L}\right)^3\right) \qquad \left(\frac{wL^2}{12}\right)\left(1 - 6\left(\frac{a}{L}\right)^2 + 4\left(\frac{a}{L}\right)^3\right)$$

APPENDIX 47.B
Indeterminate Beam Formulas

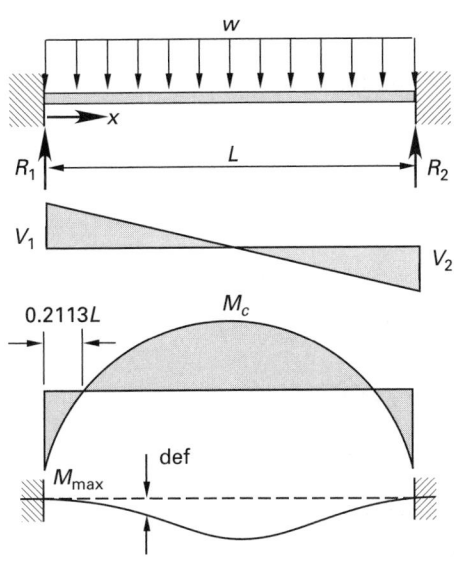

Uniformly distributed load: w in lbf/unit length

Total load: $W = wL$

Reactions: $R_1 = R_2 = \dfrac{W}{2}$

Shear forces: $V_1 = +\dfrac{W}{2}$

$$V_2 = -\dfrac{W}{2}$$

Maximum (negative) bending moment:

$$M_{\max} = -\frac{wL^2}{12} = -\frac{WL}{12}, \text{ at end}$$

Maximum (positive) bending moment:

$$M_c = \frac{wL^2}{24} = \frac{WL}{24}, \text{ at center}$$

Maximum deflection: $\dfrac{wL^4}{384\,EI} = \dfrac{WL^3}{384\,EI}$, at center

$$\text{def} = \left(\frac{wx^2}{24\,EI}\right)(L-x)^2, \ 0 \le x \le L$$

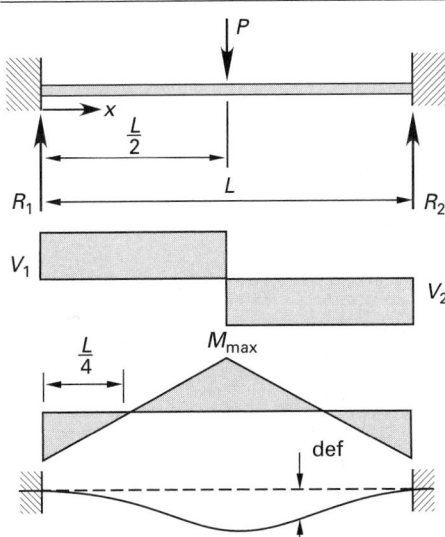

Concentrated load, P, at center

Reactions: $R_1 = R_2 = \dfrac{P}{2}$

Shear forces: $V_1 = +\dfrac{P}{2}$; $V_2 = -\dfrac{P}{2}$

Maximum bending moment:

$$M_{\max} = \frac{PL}{8}, \text{ at center}$$

$$M_{\max} = -\frac{PL}{8}, \text{ at ends}$$

Maximum deflection: $\dfrac{PL^3}{192\,EI}$, at center

$$\text{def} = \left(\frac{Px^2}{48\,EI}\right)(3L - 4x), \ 0 \le x \le \frac{L}{2}$$

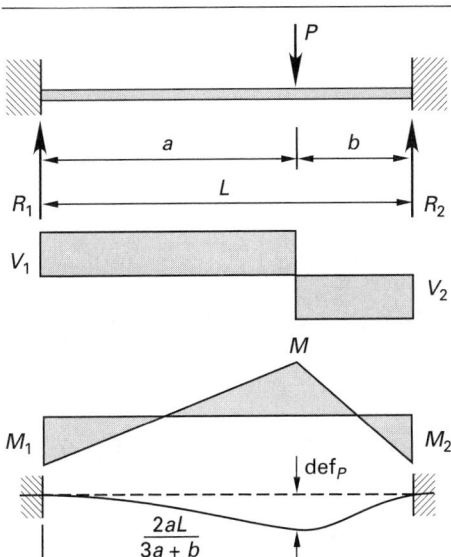

Concentrated load, P, at any point

Reactions: $R_1 = \left(\dfrac{Pb^2}{L^3}\right)(3a + b)$

$$R_2 = \left(\frac{Pa^2}{L^3}\right)(3b + a)$$

Shear forces: $V_1 = R_1$; $V_2 = -R_2$

Bending moments:

$$M_1 = -\frac{Pab^2}{L^2}, \text{ max. when } a < b$$

$$M_2 = -\frac{Pa^2b}{L^2}, \text{ max. when } a > b$$

$$M_P = +\frac{2Pa^2b^2}{L^3}, \text{ at point of load}$$

def_P: $\dfrac{Pa^3b^3}{3\,EIL^3}$, at point of load

max. def. $= \dfrac{2Pa^3b^2}{3\,EI(3a+b)^2}$, at $x = \dfrac{2aL}{3a+b}$, for $a > b$

(continued)

APPENDIX 47.B *(continued)*
Indeterminate Beam Formulas

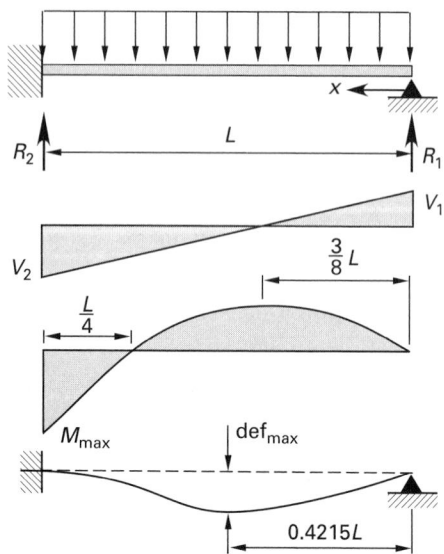

Uniformly distributed load: w in lbf/unit length
Total load: $W = wL$

Reactions: $R_1 = \dfrac{3wL}{8}$, $R_2 = \dfrac{5wL}{8}$

Shear forces: $V_1 = +R_1$; $V_2 = -R_2$

Bending moments:

Max. negative moment: $-\dfrac{wL^2}{8}$, at left end

Max. positive moment: $\dfrac{9}{128}wL^2$, $x = \dfrac{3}{8}L$

$M = \dfrac{3wLx}{8} - \dfrac{wx^2}{2}$, $0 \le x \le L$

Maximum deflection: $\dfrac{wL^4}{185\,EI}$, $x = 0.4215L$

$\text{def} = \left(\dfrac{wx}{48\,EI}\right)(L^3 - 3Lx^2 + 2x^3)$, $0 \le x \le L$

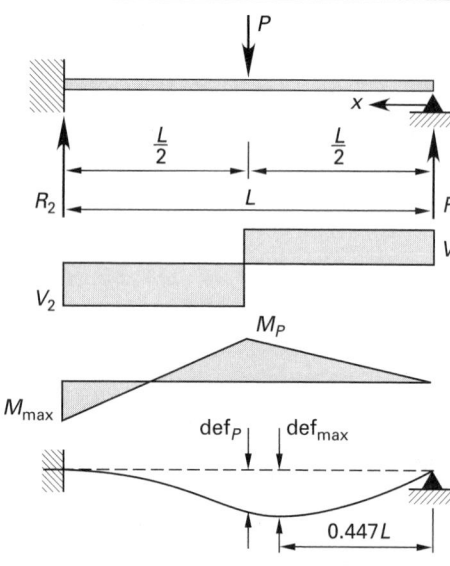

Concentrated load, P, at center

Reactions: $R_1 = \dfrac{5}{16}P$; $R_2 = \dfrac{11}{16}P$

Shear forces: $V_1 = R_1$; $V_2 = -R_2$

Bending moments:

Max. negative moment: $-\dfrac{3PL}{16}$, at fixed end

Max. positive moment: $\dfrac{5PL}{32}$, at center

Maximum deflection: $(0.009317)\left(\dfrac{PL^3}{EI}\right)$, at $x = 0.447L$

Deflection at center under load: $\dfrac{7PL^3}{768\,EI}$

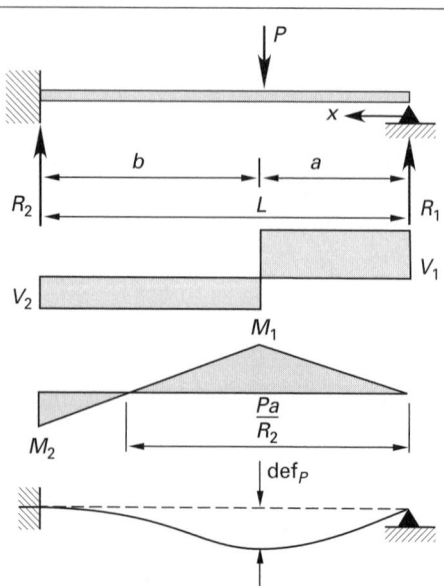

Concentrated load, P, at any point

Reactions: $R_1 = \left(\dfrac{Pb^2}{2L^3}\right)(a + 2L)$, $R_2 = \left(\dfrac{Pa}{2L^3}\right)(3L^2 - a^2)$

Shear forces: $V_1 = R_1$; $V_2 = -R_2$

Bending moments:

Max. negative moment: $M_2 = \left(-\dfrac{Pab}{2L^2}\right)(a + L)$, at fixed end

Max. positive moment: $M_1 = \left(\dfrac{Pab^2}{2L^3}\right)(a + 2L)$, at load

Deflections: $\text{def}_P = \left(\dfrac{Pa^2b^3}{12\,EIL^3}\right)(3L + a)$, at load

$\text{def}_{max} = \dfrac{Pa(L^2 - a^2)^3}{3\,EI(3L^2 - a^2)^2}$, at $x = \dfrac{L^2 + a^2}{3L^2 - a^2}L$, when $a < 0.414L$

$\text{def}_{max} = \dfrac{Pab^2}{6\,EI}\sqrt{\dfrac{a}{2L + a}}$, at $x = L\sqrt{\dfrac{2}{2L + a}}$, when $a > 0.414L$

(continued)

APPENDIX 47.B *(continued)*
Indeterminate Beam Formulas

Continuous beam of two equal spans—equal concentrated loads, P, at center of each span

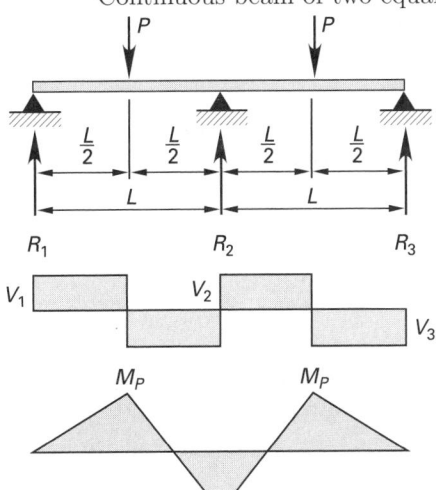

Reactions: $R_1 = R_3 = \dfrac{5}{16}P$

$R_2 = 1.375P$

Shear forces: $V_1 = -V_3 = \dfrac{5}{16}P$

$V_2 = \pm\dfrac{11}{16}P$

Bending moments:

$M_{\max} = -\dfrac{6}{32}PL$, at R_2

$M_P = \dfrac{5}{32}PL$, at point of load

Continuous beam of two equal spans—concentrated loads, P, at third points of each span

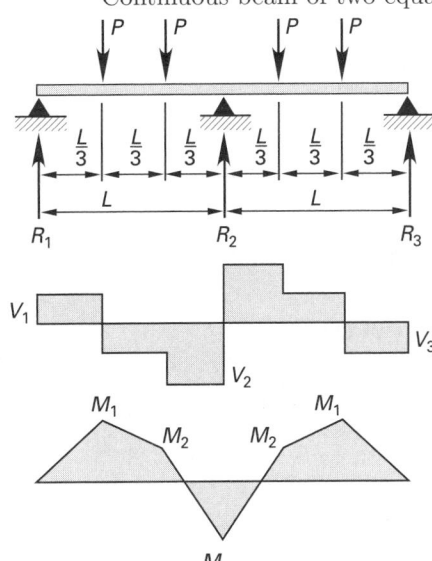

Reactions: $R_1 = R_3 = \dfrac{2}{3}P$

$R_2 = \dfrac{8}{3}P$

Shear forces: $V_1 = -V_3 = \dfrac{2}{3}P$

$V_2 = \pm\dfrac{4}{3}P$

Bending moments:

$M_{\max} = -\dfrac{1}{3}PL$, at R_2

$M_1 = \dfrac{2}{9}PL$

$M_2 = \dfrac{1}{9}PL$

Continuous beam of two equal spans—uniformly distributed load of w in lbf/unit length

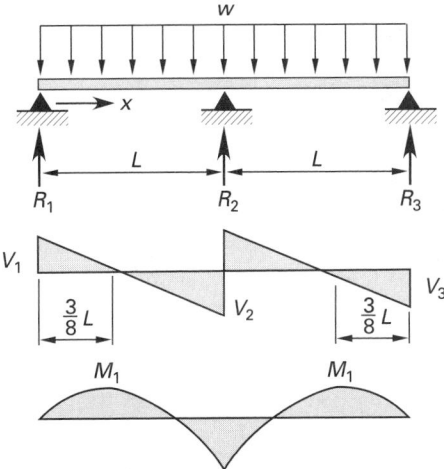

Reactions: $R_1 = R_3 = \dfrac{3}{8}wL$

$R_2 = 1.25wL$

Shear forces: $V_1 = -V_3 = \dfrac{3}{8}wL$

$V_2 = \pm\dfrac{5}{8}wL$

Bending moments:

$M_{\max} = -\dfrac{1}{8}wL^2$

$M_1 = \dfrac{9}{128}wL^2$

Maximum deflection: $0.00541\left(\dfrac{wL^4}{EI}\right)$, at $x = 0.4215L$

$\text{def} = \left(\dfrac{w}{48\,EI}\right)(L^3x - 3Lx^3 + 2x^4),\ 0 \le x \le L$

(continued)

APPENDIX 47.B *(continued)*
Indeterminate Beam Formulas

Continuous beam of two equal spans—uniformly distributed load of w lbf/unit length on one span

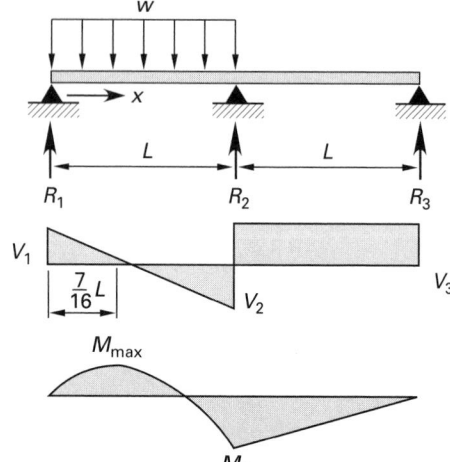

Reactions: $R_1 = \dfrac{7}{16}wL$, $R_2 = \dfrac{5}{8}wL$, $R_3 = -\dfrac{1}{16}wL$

Shear forces: $V_1 = \dfrac{7}{16}wL$, $V_2 = -\dfrac{9}{16}wL$, $V_3 = \dfrac{1}{16}wL$

Bending moments:

$$M_{\max} = \frac{49}{512}wL^2, \text{ at } x = \frac{7}{16}L$$

$$M_R = -\frac{1}{16}wL^2, \text{ at } R_2$$

$$M = \left(\frac{wx}{16}\right)(7L - 8x), \ 0 \leq x \leq L$$

Continuous beam of two equal spans—concentrated load, P, at center of one span

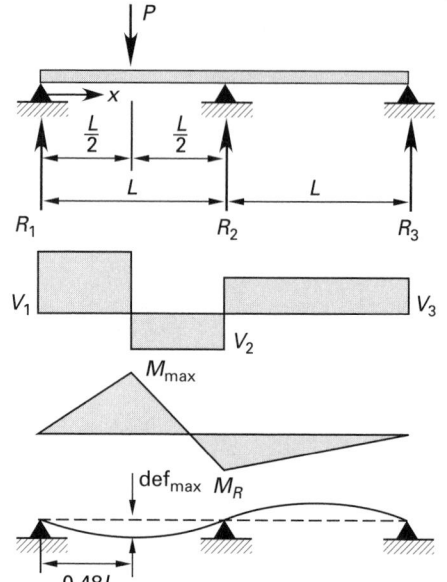

Reactions: $R_1 = \dfrac{13}{32}P$, $R_2 = \dfrac{11}{16}P$, $R_3 = -\dfrac{3}{32}P$

Shear forces: $V_1 = \dfrac{13}{32}P$, $V_2 = -\dfrac{19}{32}P$, $V_3 = \dfrac{3}{32}P$

Bending moments:

$$M_{\max} = \frac{13}{64}PL, \text{ at point of load}$$

$$M_R = -\frac{3}{32}PL, \text{ at support of load}$$

Maximum deflection: $\dfrac{0.96\,PL^3}{64\,EI}$, at $x = 0.48L$

Continuous beam of two equal spans—concentrated load, P, at any point on one span

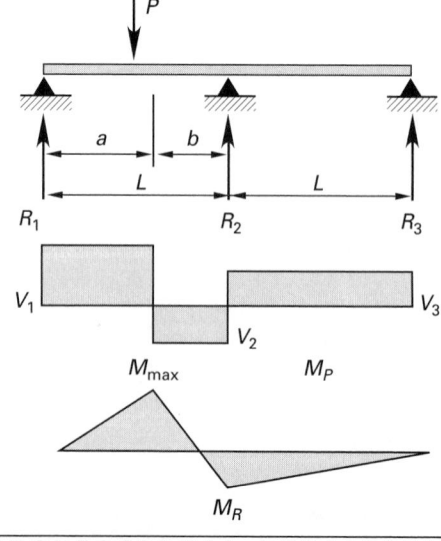

Reactions: $R_1 = \left(\dfrac{Pb}{4L^3}\right)\left(4L^2 - a(L+a)\right)$

$$R_2 = \left(\frac{Pa}{2L^3}\right)\left(2L^2 + b(L+a)\right)$$

$$R_3 = \left(-\frac{Pab}{4L^3}\right)(L+a)$$

Shear forces: $V_1 = \left(\dfrac{Pb}{4L^3}\right)\left(4L^2 - a(L+a)\right)$

$$V_2 = \left(-\frac{Pa}{4L^3}\right)\left(4L^2 + b(L+a)\right)$$

$$V_3 = \left(\frac{Pab}{4L^3}\right)(L+a)$$

Bending moments:

$$M_{max} = \left(\frac{Pab}{4L^3}\right)\left(4L^2 - a(L+a)\right)$$

$$M_R = \left(-\frac{Pab}{4L^2}\right)(L+a) \hspace{2em} \text{(continued)}$$

APPENDIX 47.B *(continued)*
Indeterminate Beam Formulas

Continuous beam of three equal spans—concentrated load, P, at center of each span

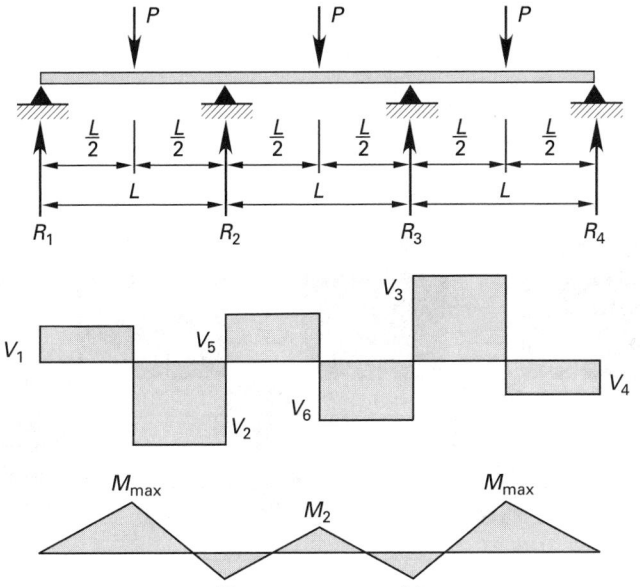

Reactions:
$$R_1 = R_4 = \frac{7}{20}P$$
$$R_2 = R_3 = \frac{23}{20}P$$

Shear forces:
$$V_1 = -V_4 = \frac{7}{20}P$$
$$V_3 = -V_2 = \frac{13}{20}P$$
$$V_5 = -V_6 = \frac{P}{2}$$

Bending moments:
$$M_{\max} = \frac{7}{40}PL$$
$$M_1 = -\frac{3}{20}PL$$
$$M_2 = \frac{1}{10}PL$$

Continuous beam of three equal spans—concentrated loads, P, at third points of each span

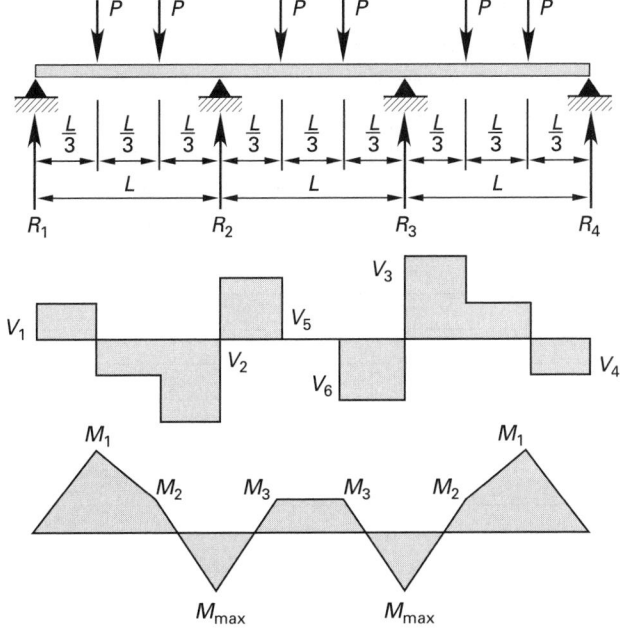

Reactions:
$$R_1 = R_4 = \frac{11}{15}P$$
$$R_2 = R_3 = \frac{34}{15}P$$

Shear forces:
$$V_1 = -V_4 = \frac{11}{15}P$$
$$V_3 = -V_2 = \frac{19}{15}P$$
$$V_5 = -V_6 = P$$

Bending moments:
$$M_{\max} = -\frac{12}{45}PL$$
$$M_1 = \frac{11}{45}PL$$
$$M_2 = \frac{7}{45}PL$$
$$M_3 = \frac{3}{45}PL$$

APPENDIX 47.C
Moment Distribution Worksheet

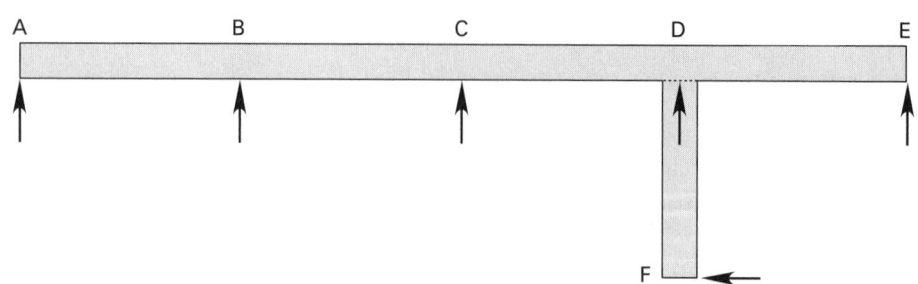

	AB		BA	BC		CB	CD		DC	DE		ED	DF		FD
L															
EI															
R															
F															
K															
D															
C															

FEM

BAL
COM

BAL
COM

BAL
COM

BAL
COM

BAL
COM

BAL
COM

total

Nomenclature

BAL	balance amount		*F*	fixity factor
C	carry-over factor		FEM	fixed-end moment
COM	carry-over moment		*K*	stiffness factor
D	distribution factor		*L*	length
EI	product of modulus of elasticity		*R*	rigidity factor
	and moment of inertia			

APPENDIX 48.A
ASTM Standards for Reinforcing Bars

customary U.S. bar no.	soft metric bar no.	nominal diameter (in)	nominal diameter (mm)	nominal area (in²)	nominal weight (lbf/ft)
3	10	0.375	9.5	0.11	0.376
4	13	0.500	12.7	0.20	0.668
5	16	0.625	15.9	0.31	1.043
6	19	0.750	19.1	0.44	1.502
7	22	0.875	22.2	0.60	2.044
8	25	1.000	25.4	0.79	2.670
9	29	1.128	28.7	1.00	3.400
10	32	1.270	32.3	1.27	4.303
11	36	1.410	35.8	1.56	5.313
14	43	1.693	43.0	2.25	7.650
18	57	2.257	57.3	4.00	13.600

(a) Customary US[a] and Soft Metric Units[b]

hard metric bar number	diameter (mm)	area (cm²)	mass (kg/m)
10M	11.3	1.0	0.784
15M	16.0	2.0	1.568
20M	19.5	3.0	2.352
25M	25.2	5.0	3.920
30M	29.9	7.0	5.488
35M	35.7	10.0	7.840
45M	43.7	15.0	11.760
55M	56.4	25.0	19.600

(b) (Obsolete Canadian) Hard SI Bar Dimensions

(Multiply in by 25.4 to obtain mm.)
(Multiply in² by 6.45 to obtain cm².)
(Multiply lbm/ft by 1.49 to obtain kg/m.)
[a] as adopted by ASTM (A-615 and A-706) and ACI
[b] as adopted by ASTM (A-615M and A-706M) and AASHTO

APPENDIX 48.B
ASTM Standards for Wire Reinforcement[a]

W&D size		nominal diameter (in)	nominal area (in²)	nominal weight (lbf/ft)	area, in²/ft of width for various spacings						
					center-to-center spacing, in						
plain	deformed				2	3	4	6	8	10	12
W31	D31	0.628	0.310	1.054	1.86	1.24	0.93	0.62	0.465	0.372	0.31
W30	D30	0.618	0.300	1.020	1.80	1.20	0.90	0.60	0.45	0.366	0.30
W28	D28	0.597	0.280	0.952	1.68	1.12	0.84	0.56	0.42	0.336	0.28
W26	D26	0.575	0.260	0.934	1.56	1.04	0.78	0.52	0.39	0.312	0.26
W24	D24	0.553	0.240	0.816	1.44	0.96	0.72	0.48	0.36	0.288	0.24
W22	D22	0.529	0.220	0.748	1.32	0.88	0.66	0.44	0.33	0.264	0.22
W20	D20	0.504	0.200	0.680	1.20	0.80	0.60	0.40	0.30	0.24	0.20
W18	D18	0.478	0.180	0.612	1.08	0.72	0.54	0.36	0.27	0.216	0.18
W16	D16	0.451	0.160	0.544	0.96	0.64	0.48	0.32	0.24	0.192	0.16
W14	D14	0.422	0.140	0.476	0.84	0.56	0.42	0.28	0.21	0.168	0.14
W12	D12	0.390	0.120	0.408	0.72	0.48	0.36	0.24	0.18	0.144	0.12
W11	D11	0.374	0.110	0.374	0.66	0.44	0.33	0.22	0.165	0.132	0.11
W10.5		0.366	0.105	0.357	0.63	0.42	0.315	0.21	0.157	0.126	0.105
W10	D10	0.356	0.100	0.340	0.60	0.40	0.30	0.20	0.15	0.12	0.10
W9.5		0.348	0.095	0.323	0.57	0.38	0.285	0.19	0.142	0.114	0.095
W9	D9	0.338	0.090	0.306	0.54	0.36	0.27	0.18	0.135	0.108	0.09
W8.5		0.329	0.085	0.289	0.51	0.34	0.255	0.17	0.127	0.102	0.085
W8	D8	0.319	0.080	0.272	0.48	0.32	0.24	0.16	0.12	0.096	0.08
W7.5		0.309	0.075	0.255	0.45	0.30	0.225	0.15	0.112	0.09	0.075
W7	D7	0.298	0.070	0.238	0.42	0.28	0.21	0.14	0.105	0.084	0.07
W6.5		0.288	0.065	0.221	0.39	0.26	0.195	0.13	0.097	0.078	0.065
W6	D6	0.276	0.060	0.204	0.36	0.24	0.18	0.12	0.09	0.072	0.06
W5.5		0.264	0.055	0.187	0.33	0.22	0.165	0.11	0.082	0.066	0.055
W5	D5	0.252	0.050	0.170	0.30	0.20	0.15	0.10	0.075	0.06	0.05
W4.5		0.240	0.045	0.153	0.27	0.18	0.135	0.09	0.067	0.054	0.045
W4	D4	0.225	0.040	0.136	0.24	0.16	0.12	0.08	0.06	0.048	0.04
W3.5		0.211	0.035	0.119	0.21	0.14	0.105	0.07	0.052	0.042	0.035
W3		0.195	0.030	0.102	0.18	0.12	0.09	0.06	0.045	0.036	0.03
W2.9		0.192	0.029	0.098	0.174	0.116	0.087	0.058	0.043	0.035	0.029
W2.5		0.178	0.025	0.085	0.15	0.10	0.075	0.05	0.037	0.03	0.025
W2		0.159	0.020	0.068	0.12	0.08	0.06	0.04	0.03	0.024	0.02
W1.4		0.135	0.014	0.049	0.084	0.056	0.042	0.028	0.021	0.017	0.014
W1.7	(9 gage)	0.148	0.017								
W2.1	(8 gage)	0.162	0.020								
W1.1	(11 gage)	0.121	0.011								
W2.8	(³/₁₆ wire)	0.187	0.027								
W4.9	(¹/₄ wire)	0.250	0.049								

(Multiply in by 25.4 to obtain mm.)
(Multiply in² by 6.45 to obtain cm².)
(Multiply lbm/ft by 1.49 to obtain kg/m.)
(Multiply in²/ft by 21.2 to obtain cm²/m.)
[a] as adopted by ACI

APPENDIX 52.A
Reinforced Concrete Interaction Diagram
(round, 4 ksi concrete, 60 ksi steel, $\gamma = 0.60$)

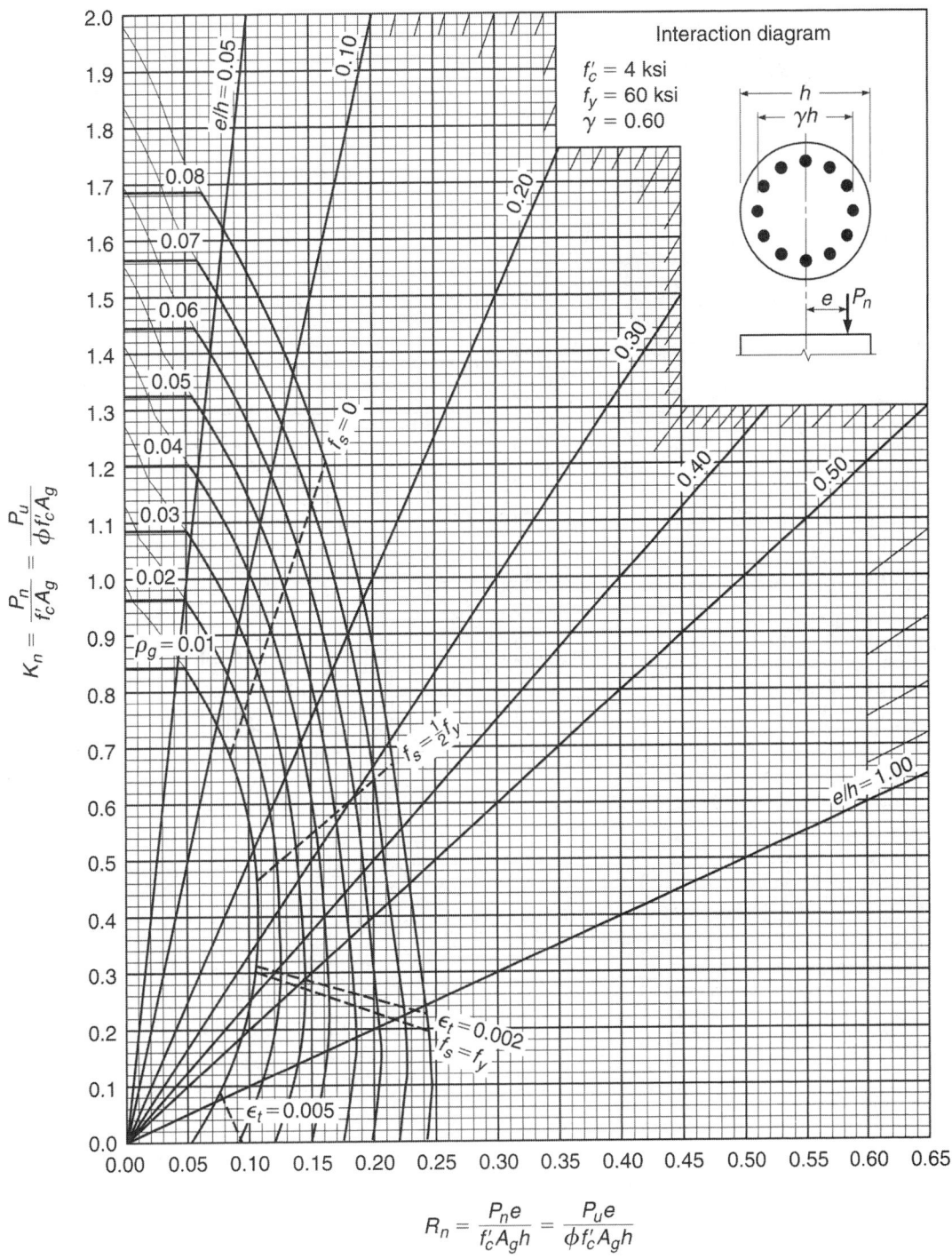

Interaction diagram

$f'_c = 4$ ksi
$f_y = 60$ ksi
$\gamma = 0.60$

$$K_n = \frac{P_n}{f'_c A_g} = \frac{P_u}{\phi f'_c A_g}$$

$$R_n = \frac{P_n e}{f'_c A_g h} = \frac{P_u e}{\phi f'_c A_g h}$$

(continued)

APPENDIX 52.A *(continued)*
Reinforced Concrete Interaction Diagram
(round, 4 ksi concrete, 60 ksi steel, $\gamma = 0.70$)

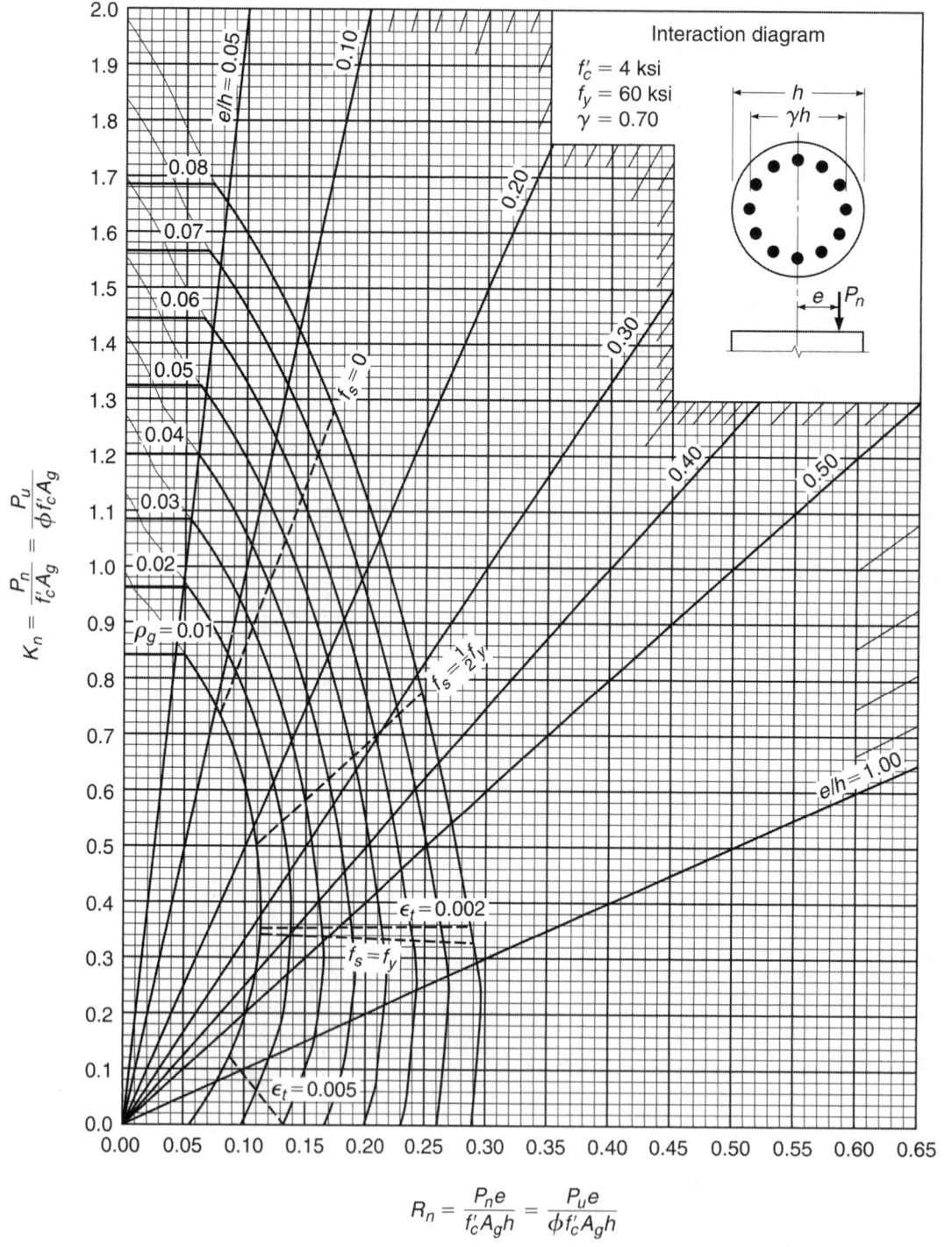

$$R_n = \frac{P_n e}{f_c' A_g h} = \frac{P_u e}{\phi f_c' A_g h}$$

Reprinted with permission from Arthur H. Nilson, David Darwin, and Charles W. Dolan, *Design of Concrete Structures*, 13th ed., copyright © 2004, by the McGraw-Hill Companies.

(continued)

APPENDIX 52.A *(continued)*
Reinforced Concrete Interaction Diagram
(round, 4 ksi concrete, 60 ksi steel, $\gamma = 0.75$)

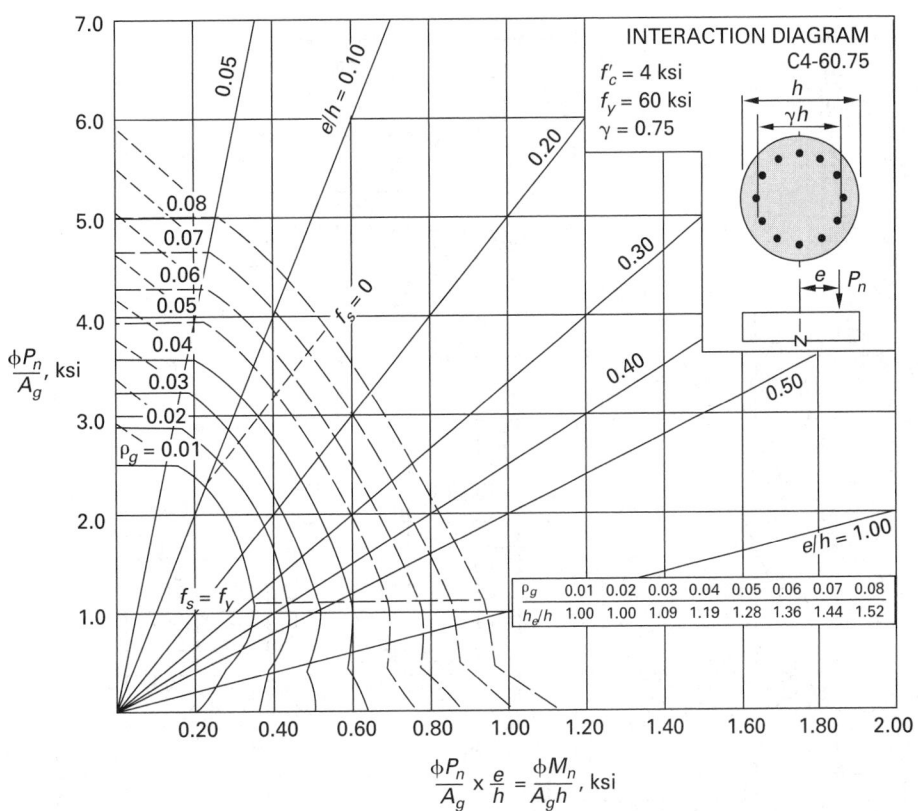

Reprinted with permission of the American Concrete Institute from *ACI Publication SP-17A: Design Handbook*, Volume 2, Columns.

(continued)

APPENDIX 52.A *(continued)*
Reinforced Concrete Interaction Diagram
(round, 4 ksi concrete, 60 ksi steel, $\gamma = 0.80$)

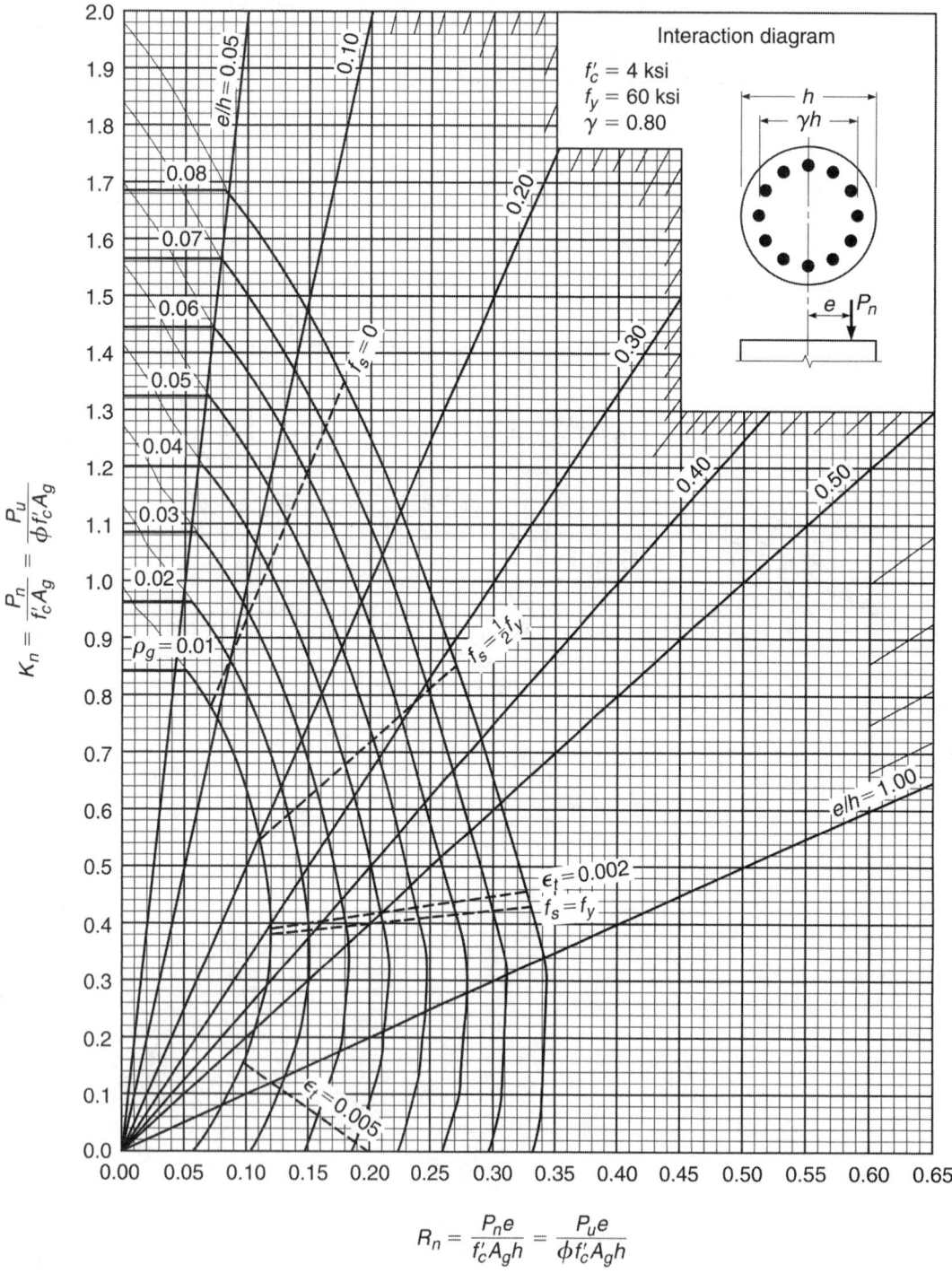

$$R_n = \frac{P_n e}{f'_c A_g h} = \frac{P_u e}{\phi f'_c A_g h}$$

Reprinted with permission from Arthur H. Nilson, David Darwin, and Charles W. Dolan, *Design of Concrete Structures*, 13th ed., copyright © 2004, by the McGraw-Hill Companies.

(continued)

APPENDIX 52.A *(continued)*
Reinforced Concrete Interaction Diagram
(round, 4 ksi concrete, 60 ksi steel, $\gamma = 0.90$)

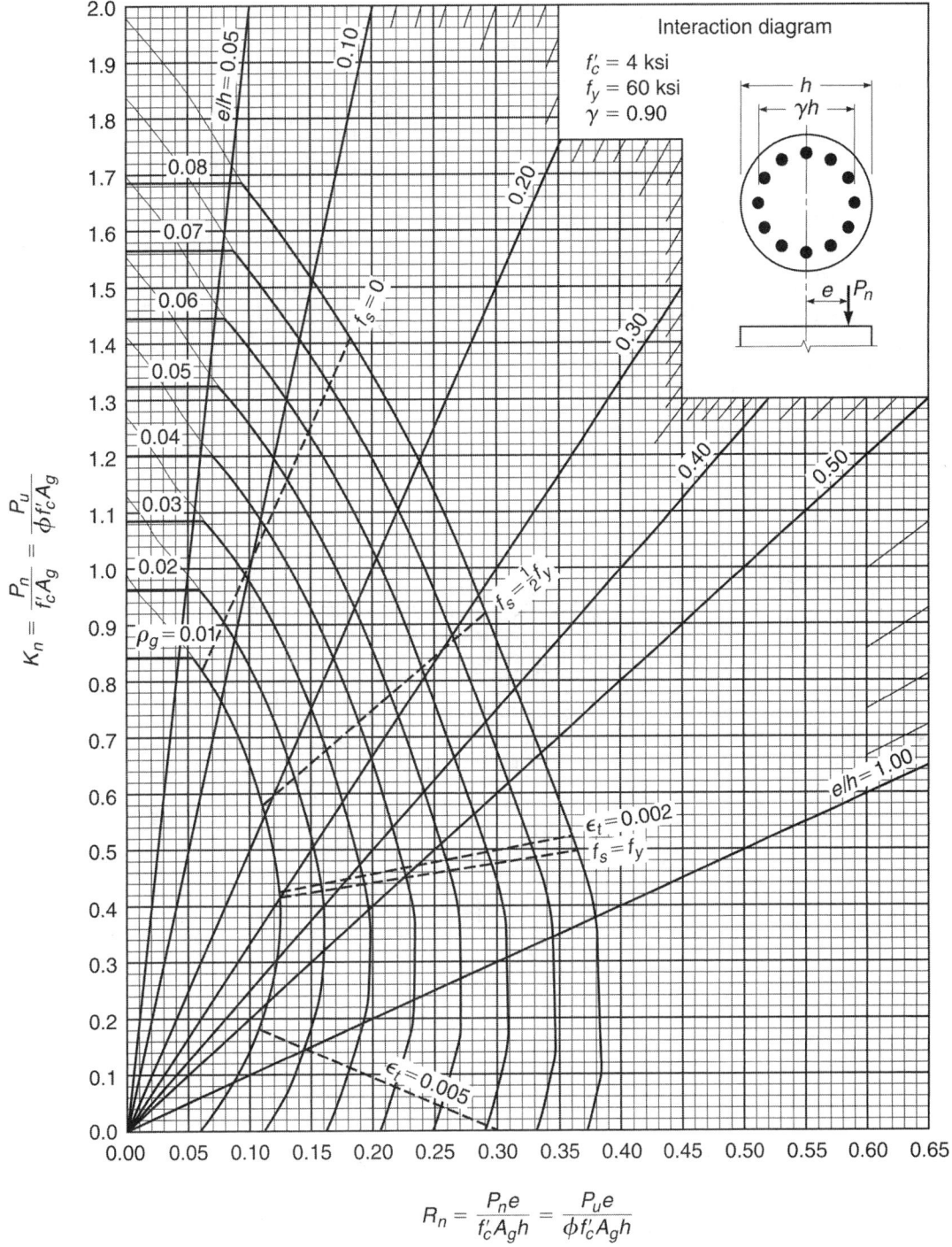

$$R_n = \frac{P_n e}{f'_c A_g h} = \frac{P_u e}{\phi f'_c A_g h}$$

(continued)

APPENDIX 52.A *(continued)*
Reinforced Concrete Interaction Diagram
(round, 4 ksi concrete, 60 ksi steel, $\gamma = 0.90$)

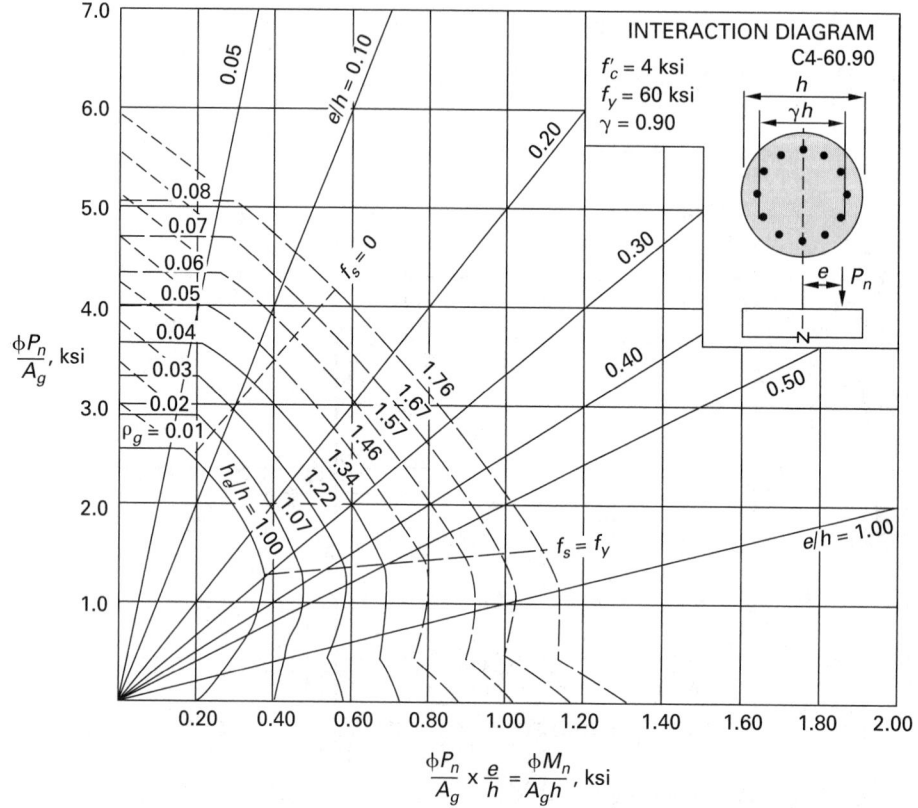

Reprinted with permission of the American Concrete Institute from *ACI Publication SP-17A: Design Handbook*, Volume 2, Columns.

(continued)

APPENDIX 52.A *(continued)*
Reinforced Concrete Interaction Diagram
(square, 4 ksi concrete, 60 ksi steel, $\gamma = 0.60$)

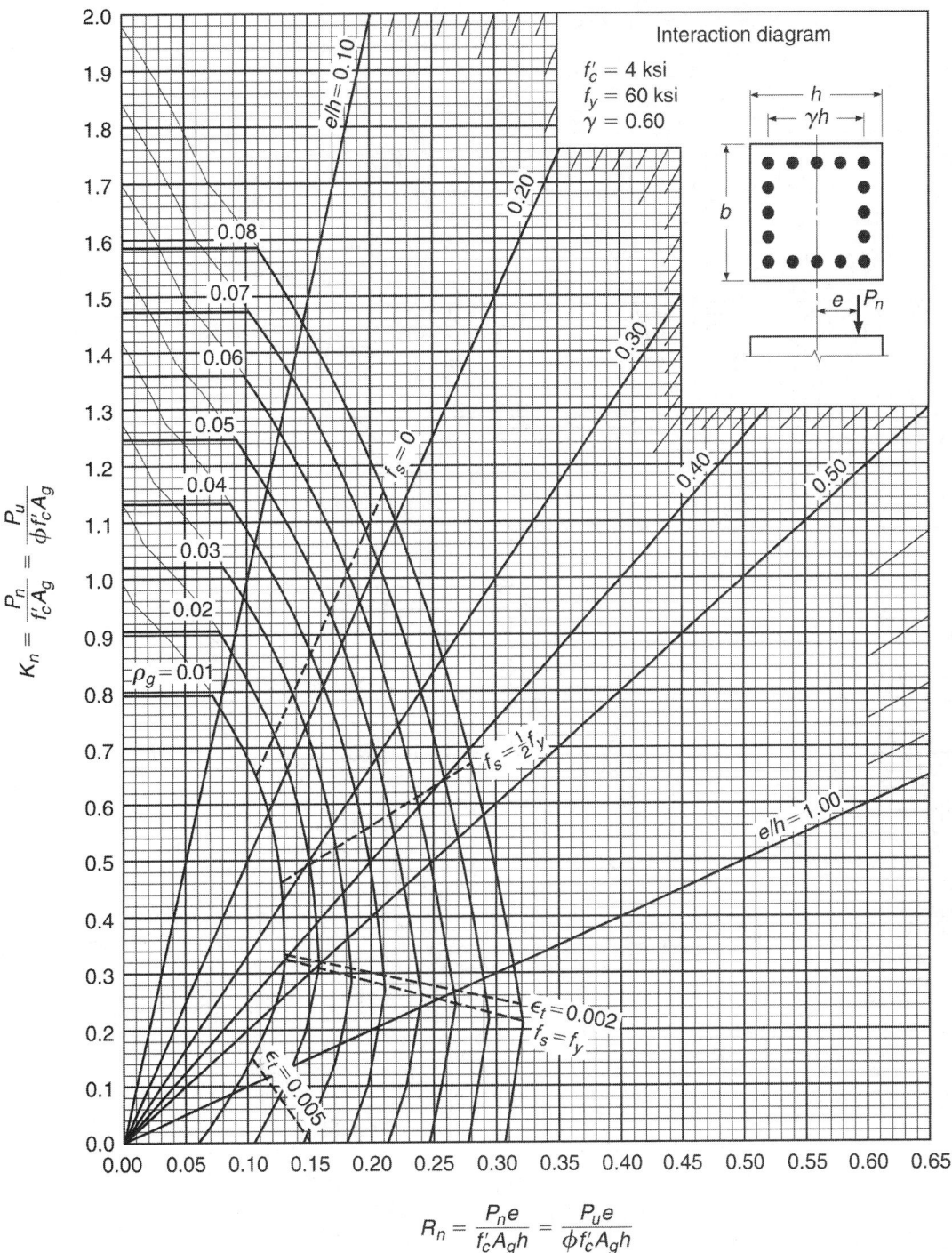

$$K_n = \frac{P_n}{f'_c A_g} = \frac{P_u}{\phi f'_c A_g}$$

$$R_n = \frac{P_n e}{f'_c A_g h} = \frac{P_u e}{\phi f'_c A_g h}$$

(continued)

APPENDIX 52.A *(continued)*
Reinforced Concrete Interaction Diagram
(square, 4 ksi concrete, 60 ksi steel, $\gamma = 0.70$)

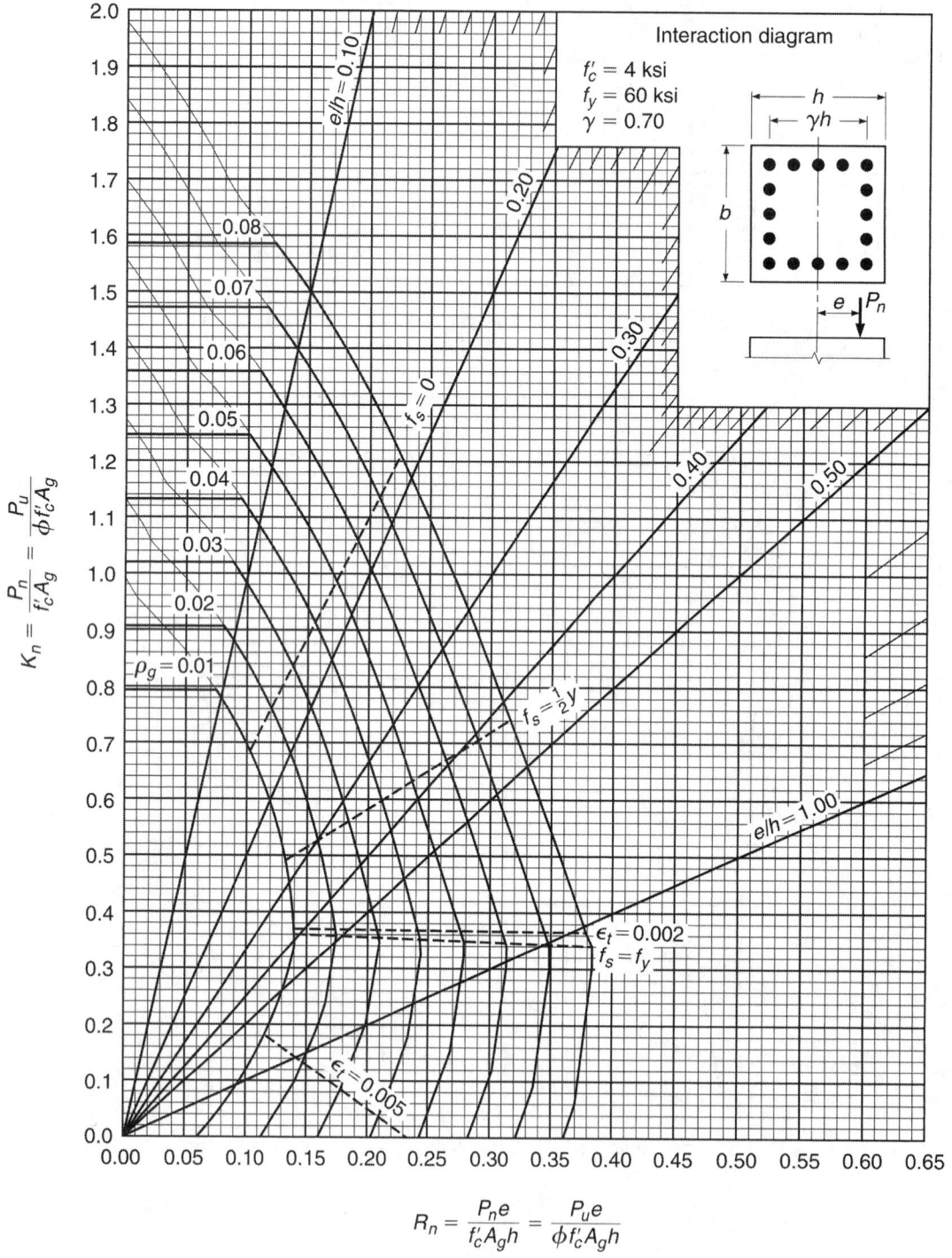

Reprinted with permission from Arthur H. Nilson, David Darwin, and Charles W. Dolan, *Design of Concrete Structures*, 13th ed., copyright © 2004, by the McGraw-Hill Companies.

(continued)

APPENDIX 52.A *(continued)*
Reinforced Concrete Interaction Diagram
(square, 4 ksi concrete, 60 ksi steel, $\gamma = 0.75$)

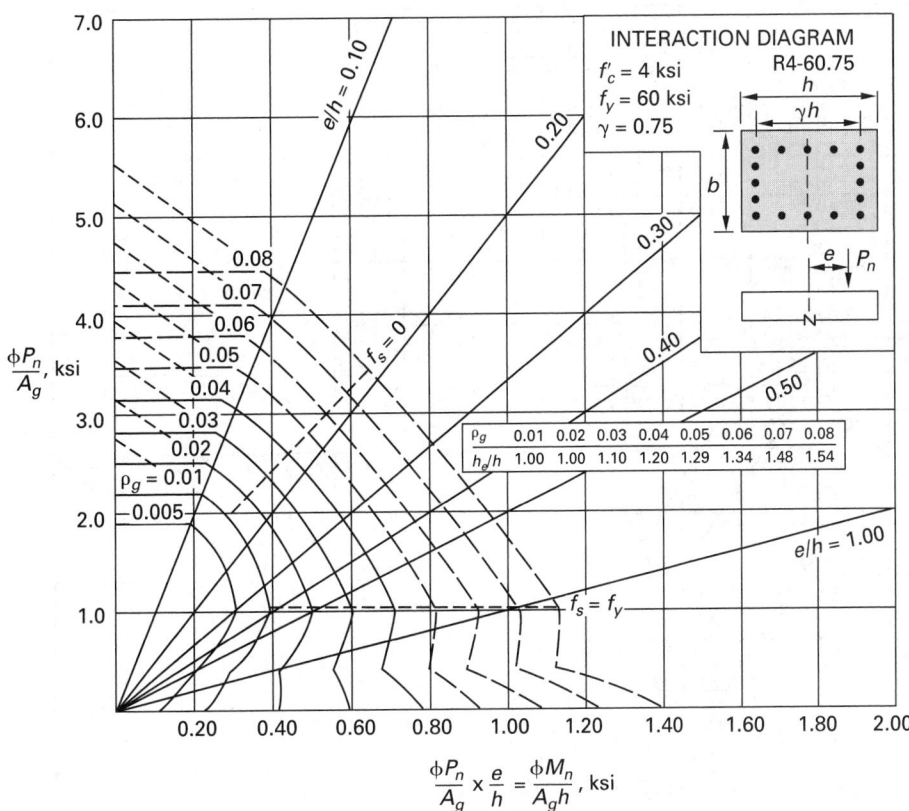

Reprinted with permission of the American Concrete Institute from *ACI Publication SP-17A: Design Handbook*, Volume 2, Columns.

(continued)

APPENDIX 52.A *(continued)*
Reinforced Concrete Interaction Diagram
(square, 4 ksi concrete, 60 ksi steel, $\gamma = 0.80$)

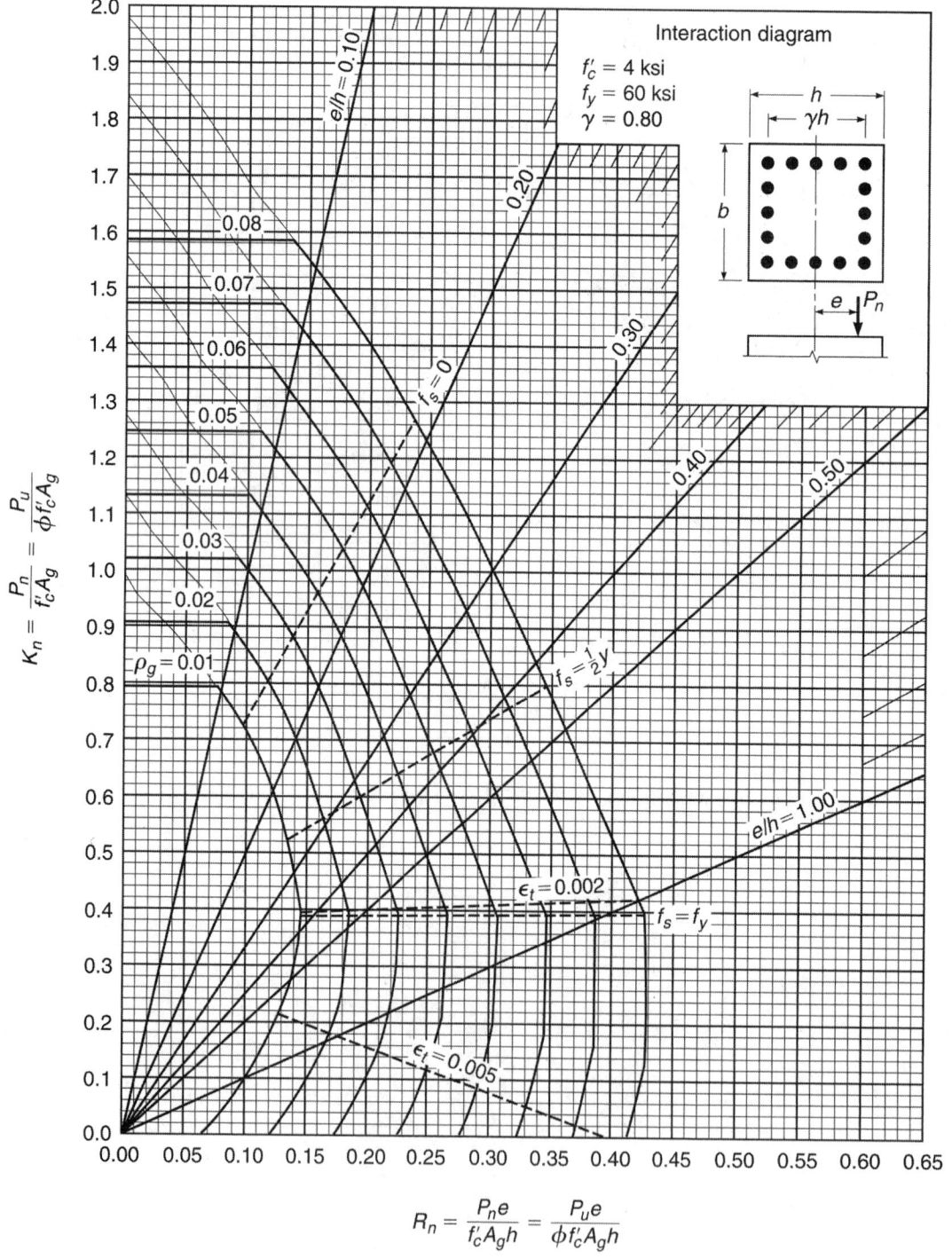

$$R_n = \frac{P_n e}{f'_c A_g h} = \frac{P_u e}{\phi f'_c A_g h}$$

Reprinted with permission from Arthur H. Nilson, David Darwin, and Charles W. Dolan, *Design of Concrete Structures*, 13th ed., copyright © 2004, by the McGraw-Hill Companies.

(continued)

APPENDIX 52.A *(continued)*
Reinforced Concrete Interaction Diagram
(square, 4 ksi concrete, 60 ksi steel, $\gamma = 0.90$)

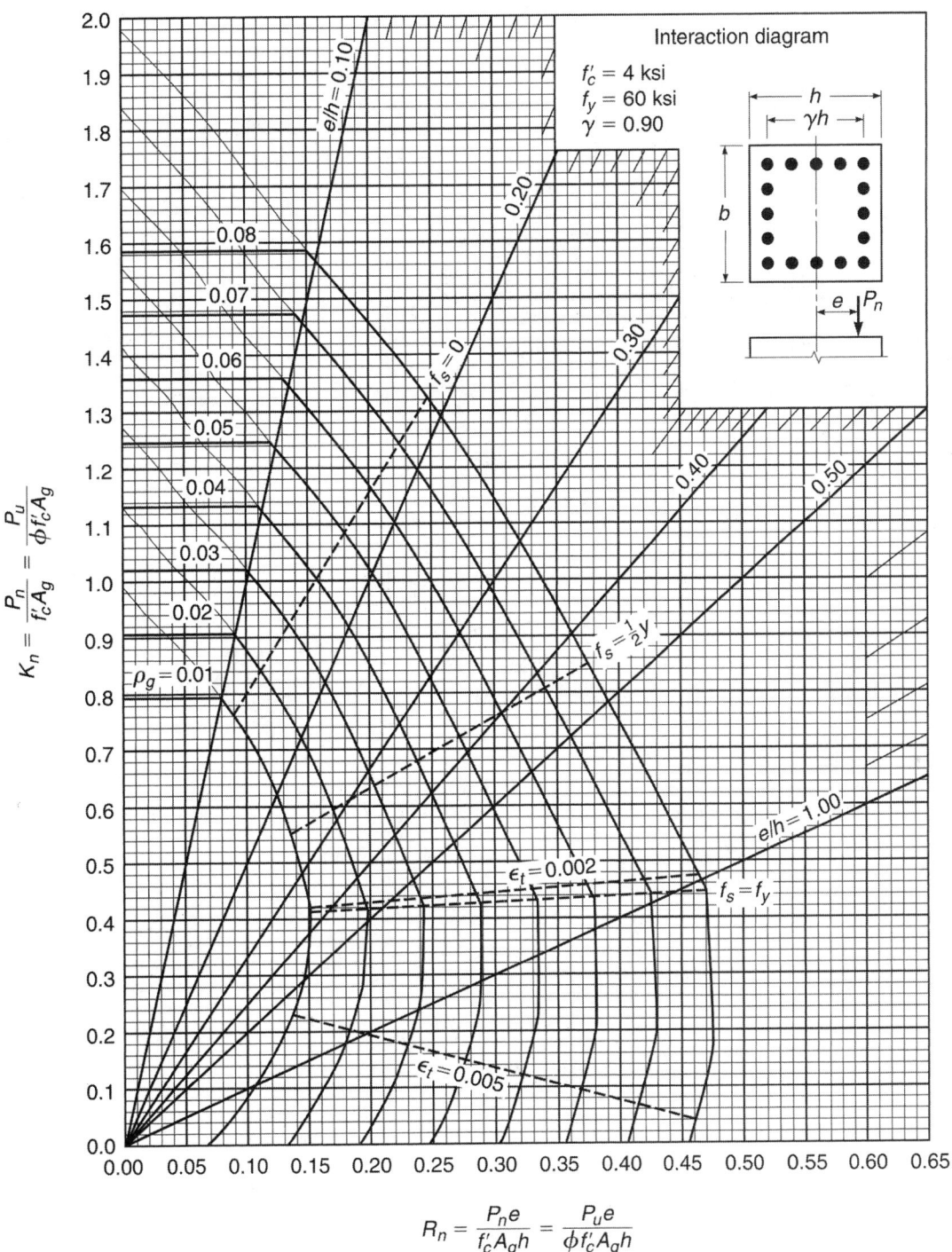

Interaction diagram

$f'_c = 4$ ksi
$f_y = 60$ ksi
$\gamma = 0.90$

$$K_n = \frac{P_n}{f'_c A_g} = \frac{P_u}{\phi f'_c A_g}$$

$$R_n = \frac{P_n e}{f'_c A_g h} = \frac{P_u e}{\phi f'_c A_g h}$$

(continued)

APPENDIX 52.A *(continued)*
Reinforced Concrete Interaction Diagram
(square, 4 ksi concrete, 60 ksi steel, $\gamma = 0.90$)

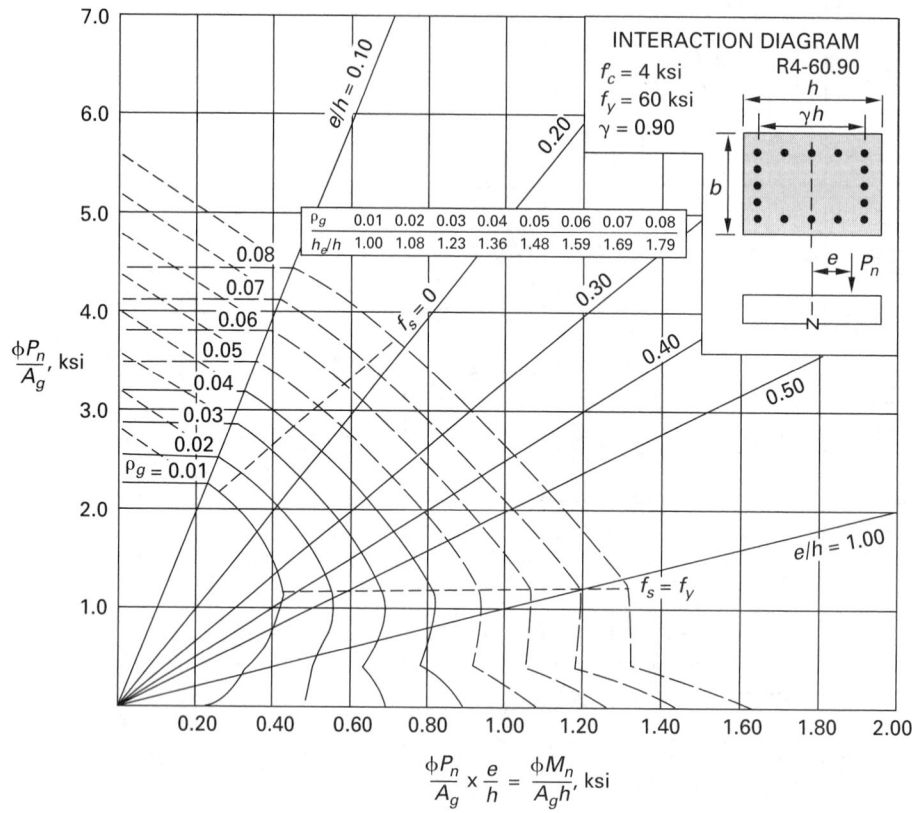

$$\frac{\phi P_n}{A_g} \times \frac{e}{h} = \frac{\phi M_n}{A_g h}, \text{ ksi}$$

Reprinted with permission of the American Concrete Institute from *ACI Publication SP-17A: Design Handbook*, Volume 2, Columns.

(continued)

APPENDIX 52.A *(continued)*
Reinforced Concrete Interaction Diagram
(uniplane, 4 ksi concrete, 60 ksi steel, $\gamma = 0.60$)

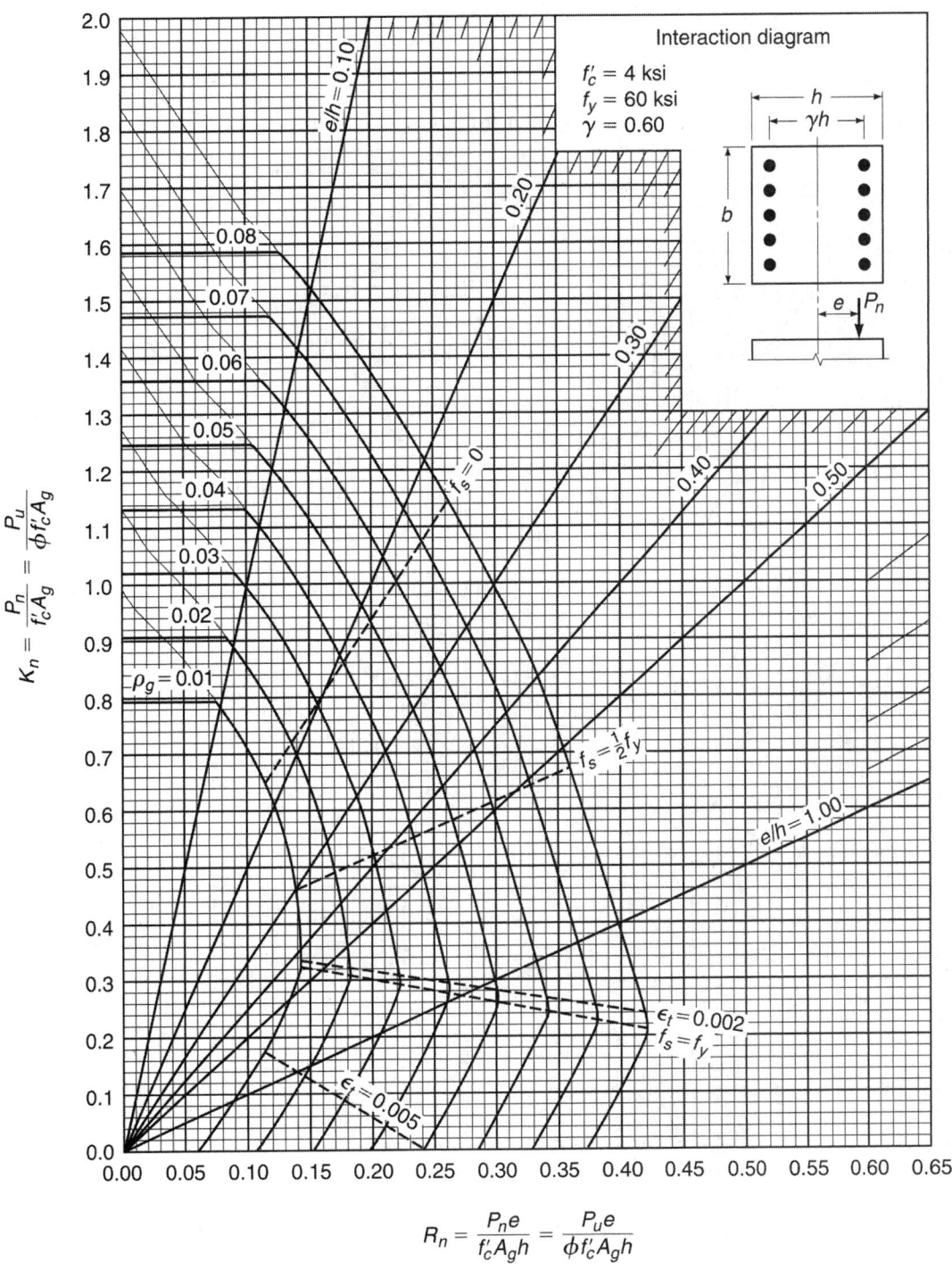

(continued)

APPENDIX 52.A *(continued)*
Reinforced Concrete Interaction Diagram
(uniplane, 4 ksi concrete, 60 ksi steel, $\gamma = 0.70$)

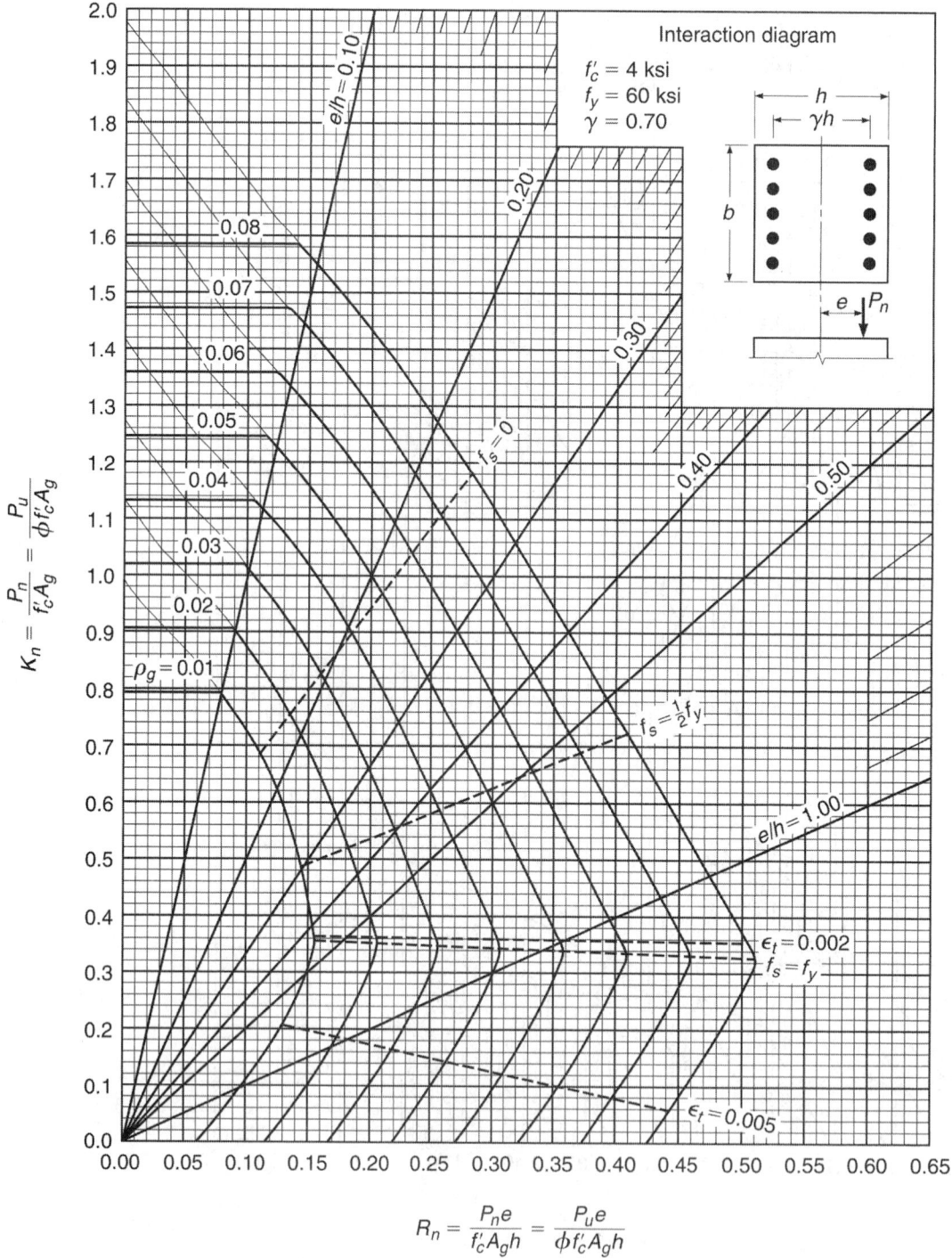

Reprinted with permission from Arthur H. Nilson, David Darwin, and Charles W. Dolan, *Design of Concrete Structures*, 13th ed., copyright © 2004, by the McGraw-Hill Companies.

(continued)

APPENDIX 52.A *(continued)*
Reinforced Concrete Interaction Diagram
(uniplane, 4 ksi concrete, 60 ksi steel, $\gamma = 0.75$)

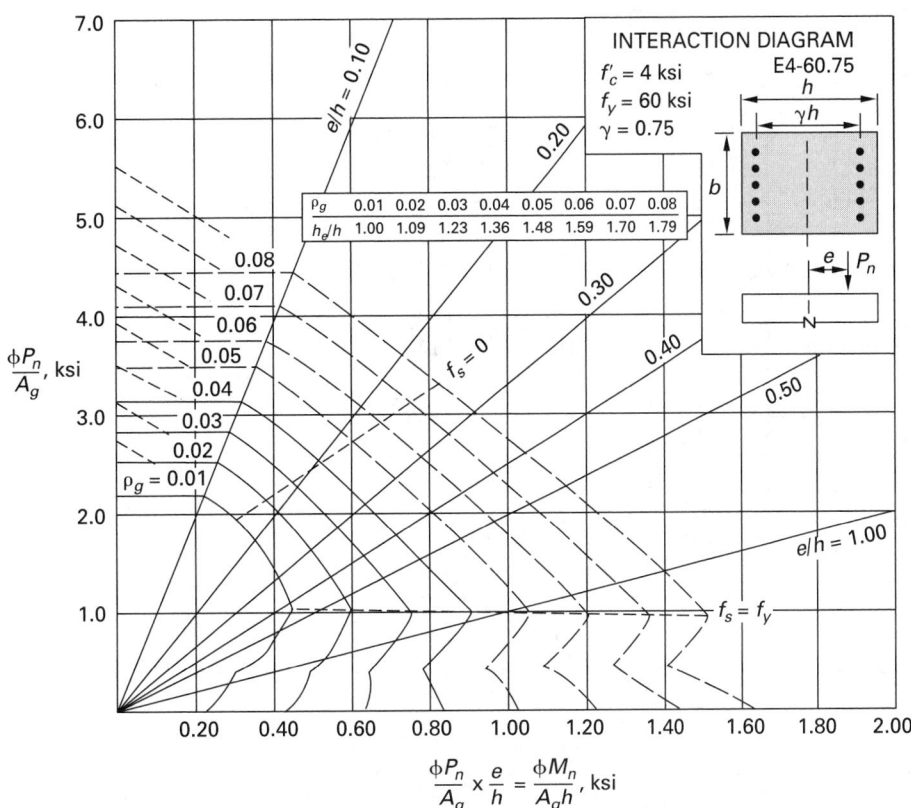

Reprinted with permission of the American Concrete Institute from *ACI Publication SP-17A: Design Handbook*, Volume 2, Columns.

(continued)

APPENDIX 52.A *(continued)*
Reinforced Concrete Interaction Diagram
(uniplane, 4 ksi concrete, 60 ksi steel, $\gamma = 0.80$)

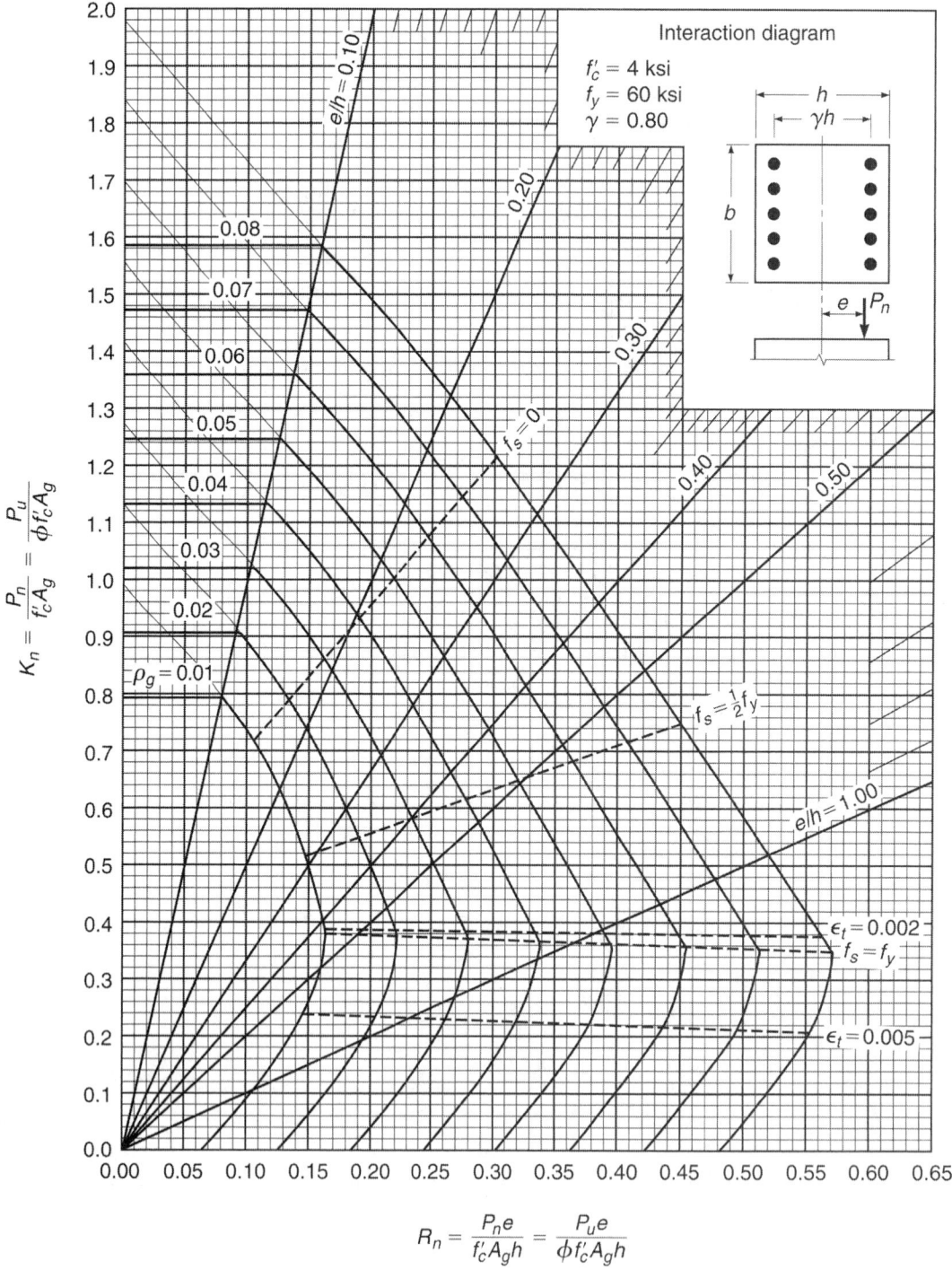

$$R_n = \frac{P_n e}{f_c' A_g h} = \frac{P_u e}{\phi f_c' A_g h}$$

Reprinted with permission from Arthur H. Nilson, David Darwin, and Charles W. Dolan, *Design of Concrete Structures*, 13th ed., copyright © 2004, by the McGraw-Hill Companies.

(continued)

APPENDIX 52.A *(continued)*
Reinforced Concrete Interaction Diagram
(uniplane, 4 ksi concrete, 60 ksi steel, $\gamma = 0.90$)

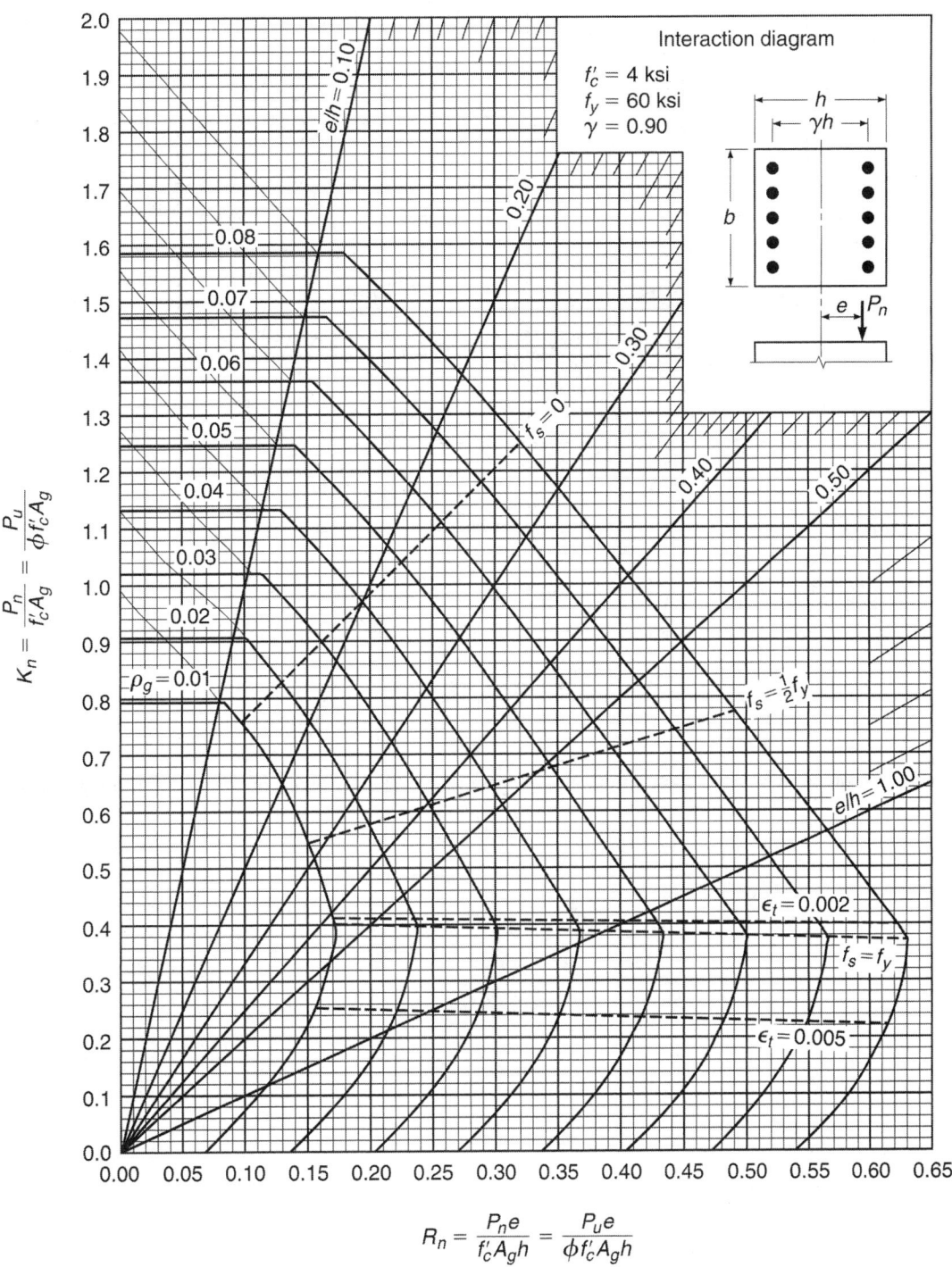

Reprinted with permission from Arthur H. Nilson, David Darwin, and Charles W. Dolan, *Design of Concrete Structures*, 13th ed., copyright © 2004, by the McGraw-Hill Companies.

(continued)

APPENDIX 52.A *(continued)*
Reinforced Concrete Interaction Diagram
(uniplane, 4 ksi concrete, 60 ksi steel, $\gamma = 0.90$)

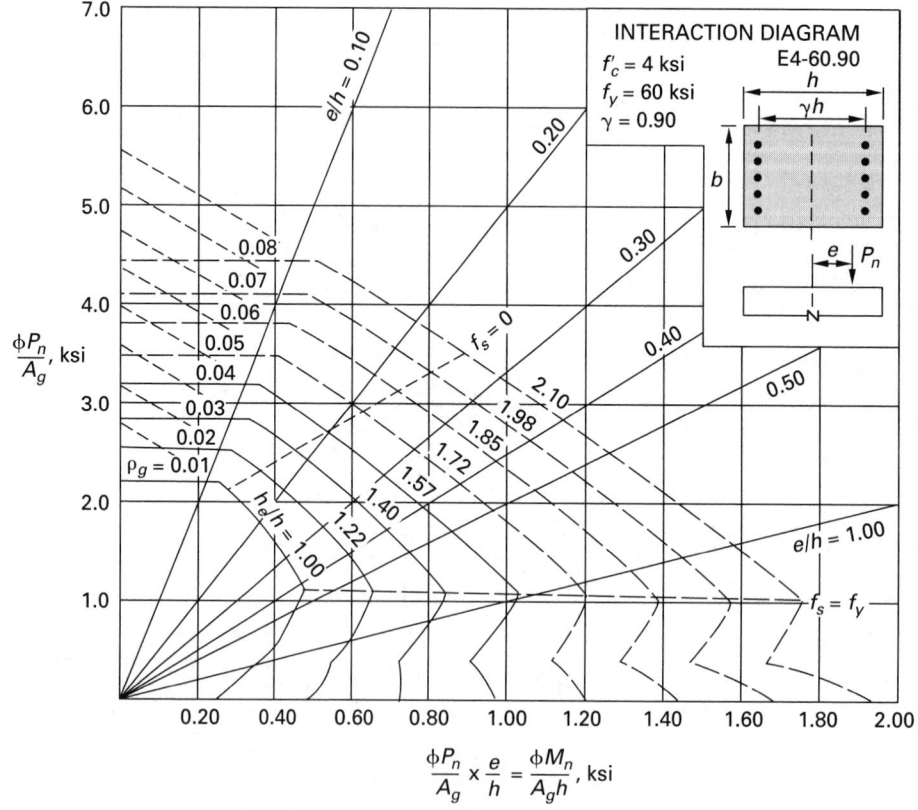

$$\frac{\phi P_n}{A_g} \times \frac{e}{h} = \frac{\phi M_n}{A_g h}, \text{ ksi}$$

Reprinted with permission of the American Concrete Institute from *ACI Publication SP-17A: Design Handbook*, Volume 2, Columns.

APPENDIX 58.A
Common Structural Steels

ASTM designation	type of steel	forms	recommended uses	minimum yield stress F_y, ksi[a]	specified minimum tensile strength F_u, ksi[b]
A36	carbon	shapes, bars, and plates	bolted or welded buildings and bridges and other structural uses	36, but 32 if thickness > 8 in	58–80
A529	carbon	shapes and plates to $\frac{1}{2}$ in	similar to A36	42–50	60–100
A572	high-strength low-alloy columbium-vanadium	shapes, plates, and bars to 6 in	bolted or welded construction. Not for welded bridges for F_y grades 55 and above	42–65	60–80
A242	atmospheric corrosion-resistant high-strength low-alloy	shapes, plates, and bars to 5 in	bolted or welded construction; welding technique very important	42–50	63–70
A588	atmospheric corrosion-resistant high-strength low-alloy	plates and bars to 4 in	bolted construction	42–50	63–70
A852	quenched and tempered alloy	plates only to 4 in	welded or bolted construction, primarily for welded bridges and buildings. Welding technique of fundamental importance	70	90–110
A514	quenched and tempered low-alloy	plates only $2\frac{1}{2}$ to 6 in	welded structures with great attention technique, discouraged if ductility important	90–100	100–130
A992	high-strength low-alloy manganese-silicon-vanadium	shapes	wide flange shapes	42–65 (design 50)	60–80 (design 65)

(Multiply in by 25.4 to obtain mm.)
(Multiply ksi by 6.9 to obtain MPa.)
[a] F_y values vary with thickness and group.
[b] F_u values vary by grade and type.

Adapted from *Structural Steel Design*, fourth edition, by McCormac, © 1995. Reprinted with permission of Prentice-Hall, Inc., Upper Saddle River, NJ.

APPENDIX 58.B
Properties of Structural Steel at High Temperatures

ASTM designation	temperature (°F)	yield strength 0.2% offset (ksi)	ultimate tensile strength (ksi)
A36	80	36.0	64.0
	300	30.2	64.0
	500	27.8	63.8
	700	25.4	57.0
	900	21.5	44.0
	1100	16.3	25.2
	1300	7.7	9.0
A242	80	54.1	81.3
	200	50.8	76.2
	400	47.6	76.4
	600	41.1	81.3
	800	39.9	76.4
	1000	35.2	52.8
	1200	20.6	27.6
A588	80	58.6	78.5
	200	57.3	79.5
	400	50.4	74.8
	600	42.5	77.7
	800	37.6	70.7
	1000	32.6	46.4
	1200	17.9	23.3

(Multiply in by 25.4 to obtain mm.)
(Multiply ksi by 6.9 to obtain MPa.)

APPENDIX 59.A
Values of C_b for Simply Supported Beams
× designates a point of lateral bracing.

load	lateral bracing along span	C_b
P (load at midpoint)	none load at midpoint	1.32
	at load point	1.67 1.67
P P	none loads at third points	1.14
	at load points loads symmetrically placed	1.67 1.00 1.67
P P P	none loads at quarter points	1.14
	at load points loads at quarter points	1.67 1.11 1.11 1.67
w	none	1.14
	at midpoint	1.30 1.30
	at third points	1.45 1.01 1.45
	at quarter points	1.52 1.06 1.06 1.52
	at fifth points	1.56 1.12 1.00 1.12 1.56

Note: Per *AISC Specification* Sec. F, lateral bracing must be provided at points of support.

APPENDIX 68.A
Section Properties of Masonry Horizontal Cross Sections
(loading parallel to wall face)

units	grouted cores[a]	mortar bedding	A_n		I_x		S_x		r	
			in²/ft	(10³ mm²/m)	in⁴/ft	(10⁶ mm⁴/m)	in³/ft	(10⁶ mm³/m)	in	(mm)
4 in single wythe walls[b]										
hollow	none	face shell	18.0	(38.1)	39.4	(53.8)	21.0	(1.13)	1.35	(34.3)
hollow	none	full	21.6	(45.7)	39.4	(53.8)	21.7	(1.17)	1.35	(34.3)
solid	none	full	43.5	(92.1)	47.4	(64.7)	26.3	(1.41)	1.04	(26.5)
6 in single wythe walls										
hollow	none	face shell	24.0	(50.8)	139.3	(190)	46.3	(2.49)	2.08	(52.9)
hollow	none	full	32.2	(68.1)	139.3	(190)	49.5	(2.66)	2.08	(52.9)
solid	none	full	67.5	(143)	176.9	(242)	63.3	(3.40)	1.62	(41.1)
hollow	8 in OC	full	67.5	(143)	176.9	(242)	63.3	(3.40)	1.62	(41.1)
hollow	16 in OC	face shell	46.6	(98.6)	158.1	(216)	55.1	(2.96)	1.79	(45.5)
hollow	24 in OC	face shell	39.1	(82.7)	151.8	(207)	52.2	(2.81)	1.87	(47.4)
hollow	32 in OC	face shell	35.3	(74.7)	148.7	(203)	50.7	(2.73)	1.91	(48.5)
hollow	40 in OC	face shell	33.0	(69.9)	146.8	(200)	49.9	(2.68)	1.94	(49.3)
hollow	48 in OC	face shell	31.5	(66.7)	145.5	(199)	49.3	(2.65)	1.96	(49.8)
hollow	56 in OC	face shell	30.5	(64.5)	144.6	(198)	48.9	(2.63)	1.98	(50.2)
hollow	64 in OC	face shell	29.6	(62.8)	144.0	(197)	48.5	(2.61)	1.99	(50.5)
hollow	72 in OC	face shell	29.0	(61.4)	143.5	(196)	48.3	(2.60)	2.00	(50.7)
8 in single wythe walls										
hollow	none	face shell	30.0	(63.5)	334.0	(456)	81.0	(4.35)	2.84	(72.0)
hollow	none	full	41.5	(87.9)	334.0	(456)	87.6	(4.71)	2.84	(72.0)
solid	none	full	91.5	(194)	440.2	(601)	116.3	(6.25)	2.19	(55.7)
hollow	8 in OC	full	91.5	(194)	440.2	(601)	116.3	(6.25)	2.19	(55.7)
hollow	16 in OC	face shell	62.0	(131)	387.1	(529)	99.3	(5.34)	2.43	(61.6)
hollow	24 in OC	face shell	51.3	(109)	369.4	(504)	93.2	(5.01)	2.53	(64.3)
hollow	32 in OC	face shell	46.0	(97.3)	360.5	(492)	90.1	(4.85)	2.59	(65.8)
hollow	40 in OC	face shell	42.8	(90.6)	355.2	(485)	88.3	(4.75)	2.63	(66.9)
hollow	48 in OC	face shell	40.7	(86.0)	351.7	(480)	87.1	(4.68)	2.66	(67.6)
hollow	56 in OC	face shell	39.1	(82.8)	349.1	(477)	86.2	(4.64)	2.68	(68.2)
hollow	64 in OC	face shell	38.0	(80.4)	347.2	(474)	85.6	(4.60)	2.70	(68.6)
hollow	72 in OC	face shell	37.1	(78.5)	345.8	(472)	85.0	(4.57)	2.71	(69.0)

[a]OC = on center.

[b]Values in these tables are based on minimum face shell and web thicknesses defined in ASTM C 90. Manufactured units generally exceed these dimensions, making 4 inch concrete masonry units difficult or impossible to grout.

Reprinted with permission of the National Concrete Masonry Association from *NCMA TEK Section Properties of Concrete Masonry Walls*, TEK 14-1, Structural, Table 1, copyright © 1993.

APPENDIX 68.B
Section Properties of Masonry Vertical Cross Sections
(loading perpendicular to wall face)

units	grouted cores[a]	mortar bedding	A_n		I_y		S_y	
			in²/ft	(10³ mm²/m)	in⁴/ft	(10⁶ mm⁴/m)	in³/ft	(10⁶ mm³/m)
4 in single wythe walls[b]								
hollow	none	face shell	8.0	(38.1)	38.0	(51.9)	21.0	(1.13)
solid	none	full	43.5	(92.1)	47.6	(65.0)	26.3	(1.41)
6 in single wythe walls								
hollow	none	face shell	24.0	(50.8)	130.3	(178)	46.3	(2.49)
solid	none	full	67.5	(143)	178.0	(243)	63.3	(3.40)
hollow	8 in OC	full	67.5	(143)	178.0	(243)	63.3	(3.40)
hollow	16 in OC	face shell	44.7	(94.7)	154.2	(211)	54.8	(2.95)
hollow	24 in OC	face shell	37.8	(80.1)	146.2	(200)	52.0	(2.80)
hollow	32 in OC	face shell	34.4	(72.7)	142.3	(194)	50.6	(2.72)
hollow	40 in OC	face shell	32.3	(68.4)	139.9	(191)	49.7	(2.67)
hollow	48 in OC	face shell	30.9	(65.4)	138.3	(189)	49.2	(2.64)
hollow	56 in OC	face shell	29.9	(63.3)	137.1	(187)	48.8	(2.62)
hollow	64 in OC	face shell	29.2	(61.8)	136.3	(186)	48.5	(2.61)
hollow	72 in OC	face shell	28.6	(60.6)	135.6	(185)	48.2	(2.59)
8 in single wythe walls								
hollow	none	face shell	30.0	(63.5)	308.7	(422)	81.0	(4.35)
solid	none	full	91.5	(194)	443.3	(605)	116.3	(6.25)
hollow	8 in OC	full	91.5	(194)	443.3	(605)	116.3	(6.25)
hollow	16 in OC	face shell	59.3	(126)	376.0	(513)	98.6	(5.30)
hollow	24 in OC	face shell	49.5	(105)	353.6	(483)	92.7	(4.99)
hollow	32 in OC	face shell	44.7	(94.5)	342.4	(468)	89.8	(4.83)
hollow	40 in OC	face shell	41.7	(88.3)	335.6	(458)	88.0	(4.73)
hollow	48 in OC	face shell	39.8	(84.2)	331.1	(452)	86.9	(4.67)
hollow	56 in OC	face shell	38.4	(81.2)	327.9	(448)	86.0	(4.62)
hollow	64 in OC	face shell	37.3	(79.0)	325.5	(445)	85.4	(4.59)
hollow	72 in OC	face shell	36.5	(77.3)	323.7	(442)	84.9	(4.56)
10 in single wythe walls								
hollow	none	face shell	33.0	(69.9)	566.7	(774)	117.8	(6.33)
solid	none	full	115.5	(244)	891.7	(1220)	185.3	(9.96)
hollow	8 in OC	full	115.5	(244)	891.7	(1220)	185.3	(9.96)
hollow	16 in OC	face shell	72.3	(153)	729.2	(996)	151.5	(8.15)
hollow	24 in OC	face shell	59.2	(125)	675.0	(922)	140.3	(7.54)
hollow	32 in OC	face shell	52.7	(111)	648.0	(885)	134.6	(7.24)
hollow	40 in OC	face shell	48.7	(103)	631.7	(863)	131.3	(7.06)
hollow	48 in OC	face shell	46.1	(97.6)	620.9	(848)	129.0	(6.94)
hollow	56 in OC	face shell	44.2	(93.6)	613.1	(837)	127.4	(6.85)
hollow	64 in OC	face shell	42.8	(90.7)	607.3	(829)	126.2	(6.78)
hollow	72 in OC	face shell	41.7	(88.3)	602.8	(823)	125.3	(6.73)

(continued)

APPENDIX 68.B *(continued)*
Section Properties of Vertical Cross Sections

units	grouted cores[a]	mortar bedding	A_n		I_y		S_y	
			in^2/ft	(10^3 mm^2/m)	in^4/ft	(10^6 mm^4/m)	in^3/ft	(10^6 mm^3/m)
12 in single wythe walls[b]								
hollow	none	face shell	36.0	(76.2)	929.4	(1270)	159.9	(8.60)
solid	none	full	139.5	(295)	1571.0	(2150)	270.3	(14.5)
hollow	8 in OC	full	139.5	(295)	1571.0	(2150)	270.3	(14.5)
hollow	16 in OC	face shell	85.3	(181)	1250.2	(1710)	215.1	(11.6)
hollow	24 in OC	face shell	68.9	(146)	1143.3	(1560)	196.7	(10.6)
hollow	32 in OC	face shell	60.7	(128)	1089.8	(1490)	187.5	(10.1)
hollow	40 in OC	face shell	55.7	(118)	1057.7	(1440)	182.0	(9.78)
hollow	48 in OC	face shell	52.4	(111)	1036.3	(1420)	178.3	(9.59)
hollow	56 in OC	face shell	50.1	(106)	1021.1	(1390)	175.7	(9.44)
hollow	64 in OC	face shell	48.3	(102)	1009.6	(1380)	173.7	(9.34)
hollow	72 in OC	face shell	47.0	(99.4)	1000.7	(1370)	172.2	(9.26)

[a]OC = on center.
[b]Values in these tables are based on minimum face shell and web thicknesses defined in ASTM C 90 (ref. 5), as shown in Table 3. Manufactured units generally exceed these dimensions, making 4 inch concrete masonry units difficult or impossible to grout.

Reprinted with permission of the National Concrete Masonry Association from *NCMA TEK Section Properties of Concrete Masonry Walls*, TEK 14-1, Structural, Table 2, copyright © 1993.

APPENDIX 68.C
Ungrouted Wall Section Properties[a]

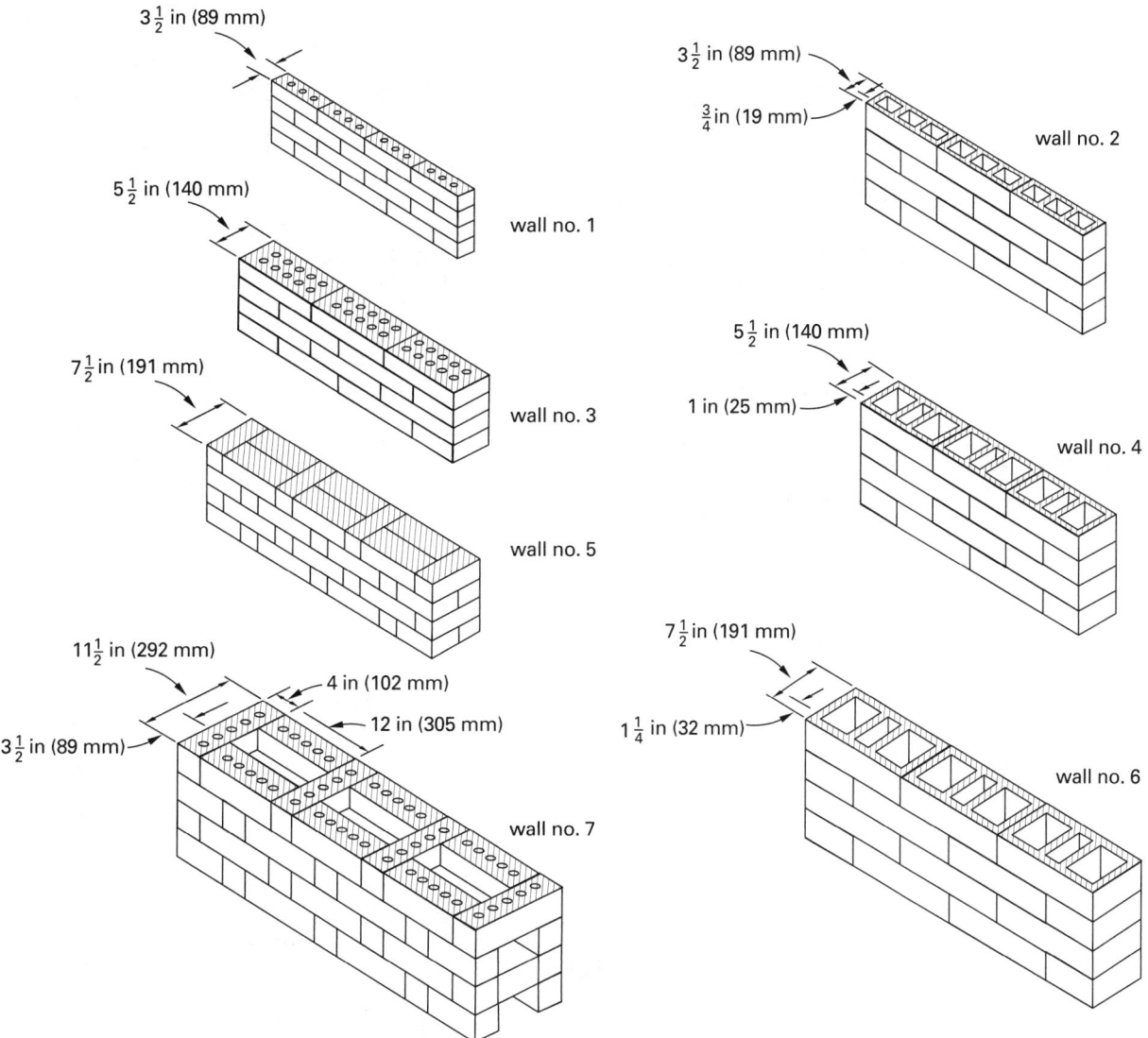

wall no.	A_n in²	(mm²)	I in⁴	(×10⁻⁵ m⁴)	S in³	(×10⁻⁵ m³)	r in	(mm)	Q in³	(×10⁻⁵ m³)
1	42	(27 100)	42.9	(1.79)	24.5	(40.1)	1.01	(25.7)	18.4	(30.2)
2[b]	18	(11 600)	34.9	(1.45)	19.9	(32.6)	1.39	(35.3)	12.4	(20.3)
3	66	(42 600)	166.4	(6.93)	60.5	(99.1)	1.59	(40.4)	45.4	(74.4)
4[b]	24	(15 500)	123.5	(5.14)	44.9	(73.6)	2.27	(57.7)	27.0	(44.2)
5	90	(58 100)	421.9	(17.6)	112.5	(184)	2.17	(55.1)	84.4	(138)
6[b]	30	(19 400)	296.9	(12.4)	79.2	(130)	3.15	(80.0)	46.9	(76.9)
7	100	(64 500)	1456.3	(60.6)	253.3	(415)	3.82	(97.0)	176.9	(290)

[a]per foot (305 mm) of wall
[b]Section properties are based on minimum solid face shell thickness and face shell bedding.

Reprinted with permission of the Brick Industry Association from *Technical Notes on Brick Construction*, Table 4, copyright © 1996.

APPENDIX 68.D
Grouted Wall Section Properties[a]

wall no.	cell grout spacing		A_n		I		r	
	in	(mm)	in²	(m²)	in⁴	(×10⁻⁵ m⁴)	in	(mm)
1	8	(200)	28.5	(18 400)	38.4	(1.60)	1.16	(29.5)
	12	(300)	25.0	(16 100)	37.2	(1.55)	1.22	(31.0)
	16	(410)	23.3	(15 000)	36.6	(1.52)	1.26	(32.0)
	24	(610)	21.5	(13 900)	36.0	(1.50)	1.30	(33.0)
	32	(810)	20.6	(13 300)	35.8	(1.49)	1.32	(33.5)
	48	(1200)	19.8	(12 800)	35.5	(1.48)	1.34	(34.0)
2	12	(300)	41.5	(26 800)	141	(5.87)	1.85	(47.0)
	24	(610)	32.8	(21 200)	132	(5.49)	2.01	(51.1)
	36	(910)	29.8	(19 200)	129	(5.37)	2.08	(52.8)
	48	(1200)	28.4	(18 300)	128	(5.33)	2.12	(53.8)
3	16	(410)	56.3	(36 300)	352	(14.7)	2.50	(63.5)
	24	(610)	47.5	(30 600)	333	(13.9)	2.65	(67.3)
	32	(810)	43.1	(27 800)	324	(13.5)	2.74	(69.6)
	48	(1200)	38.8	(25 000)	315	(13.1)	2.85	(72.4)
4	fully grouted		120	(77 400)	1000	(41.6)	2.89	(73.4)
5	12	(300)	120	(77 400)	1390	(57.9)	3.41	(86.6)
	24	(610)	111	(71 600)	1240	(51.6)	3.34	(84.8)
	36	(910)	108	(69 700)	1180	(49.1)	3.31	(84.1)
	48	(1200)	106	(68 400)	1150	(47.9)	3.29	(83.6)

[a]per foot (305 mm) of wall. Section properties are based on minimum solid face shell thickness and face shell bedding of hollow unit masonry.
[b]nominal dimensions

Reprinted with permission of the Brick Industry Association from *Technical Notes on Brick Construction*, Table 5, copyright © 1996.

APPENDIX 69.A
Column Interaction Diagram
(compression controls, $g = 0.4$)

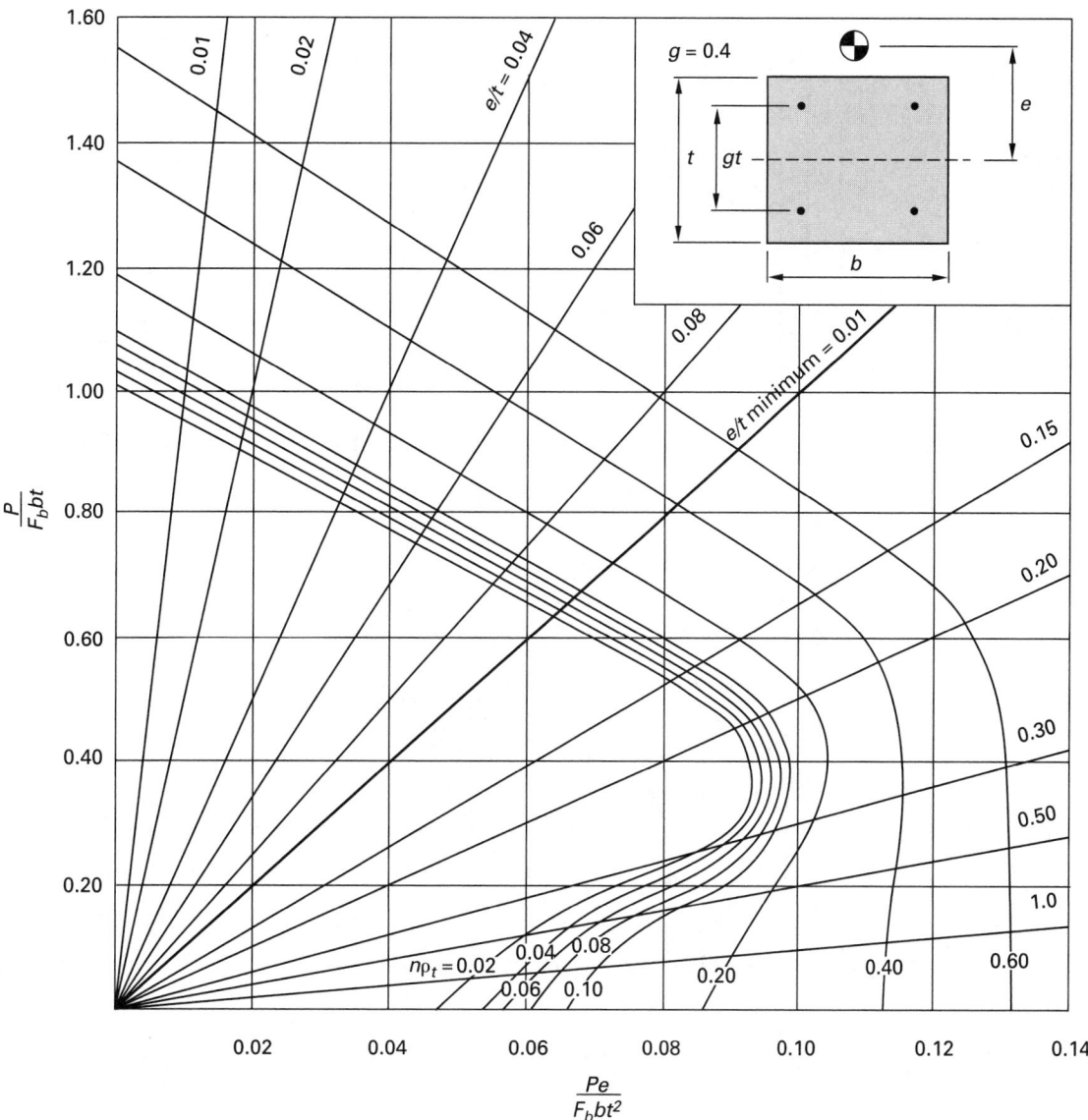

Reprinted with permission of the American Concrete Institute from *Masonry Designer's Guide*, copyright © 1993.

APPENDIX 69.B
Column Interaction Diagram
(compression controls, $g = 0.6$)

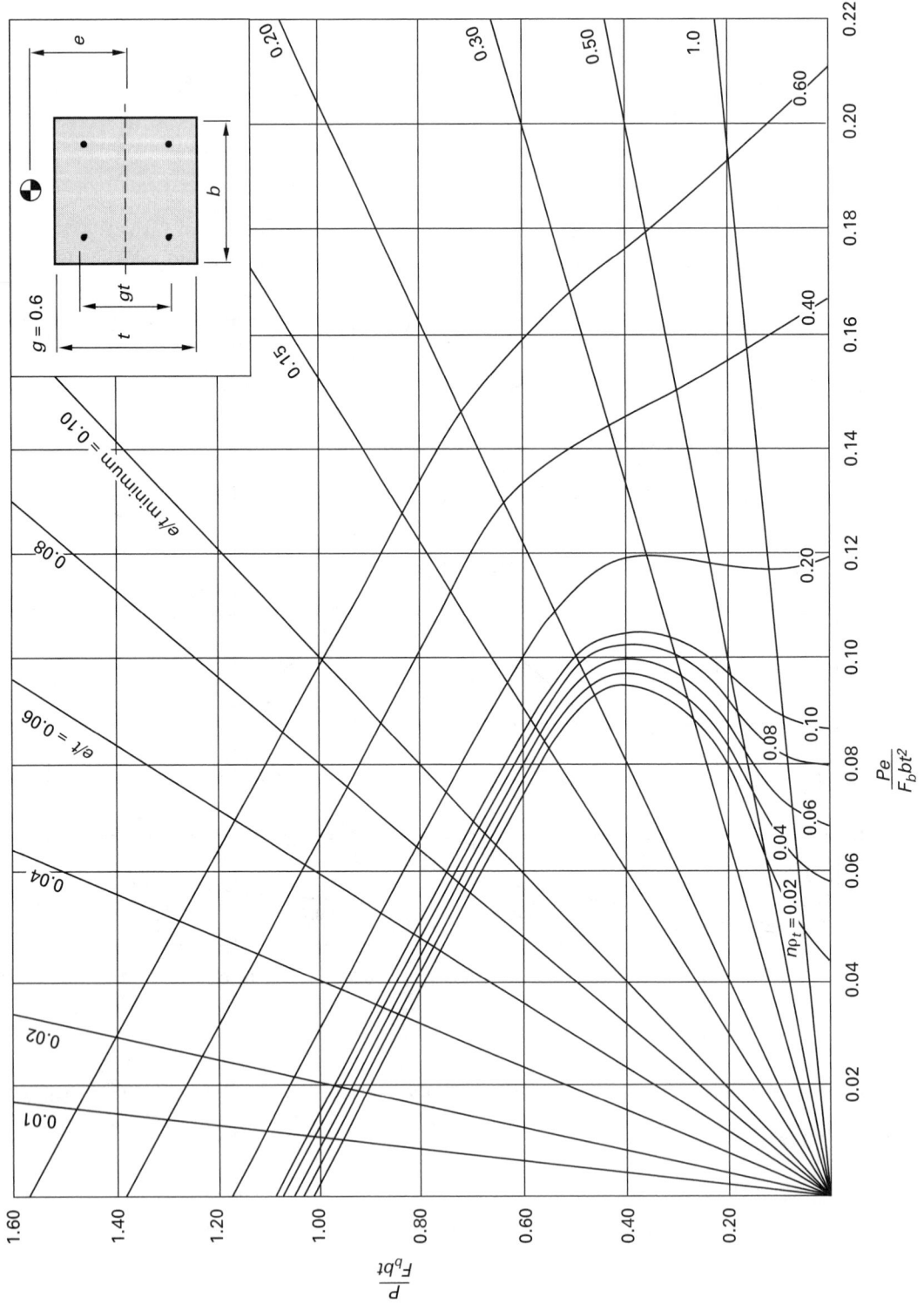

APPENDIX 69.C
Column Interaction Diagram
(compression controls, $g = 0.8$)

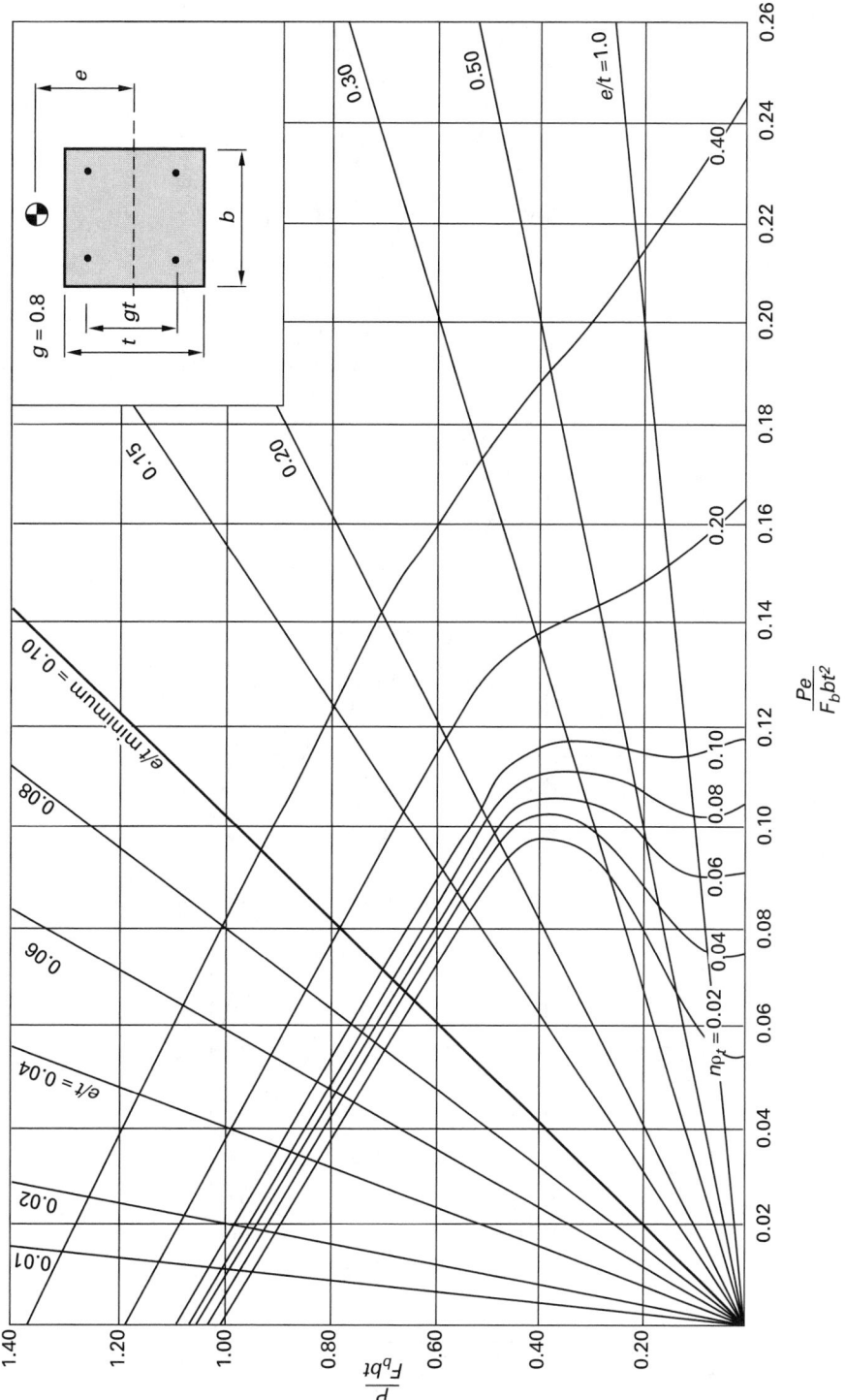

Reprinted with permission of the American Concrete Institute from *Masonry Designer's Guide*, copyright © 1993.

APPENDIX 69.D
Column Interaction Diagram
(tension controls, $g = 0.4$)

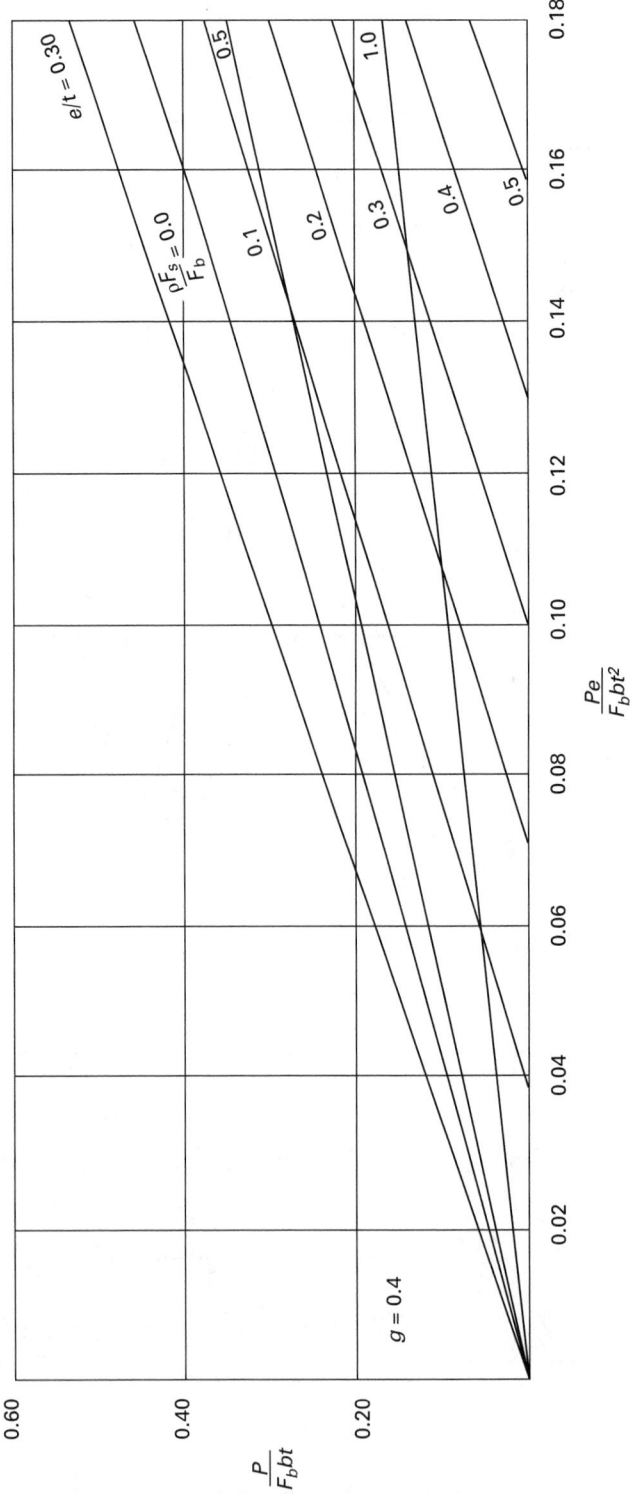

Reprinted with permission of the American Concrete Institute from *Masonry Designer's Guide*, copyright © 1993.

APPENDIX 69.E
Column Interaction Diagram
(tension controls, $g = 0.6$)

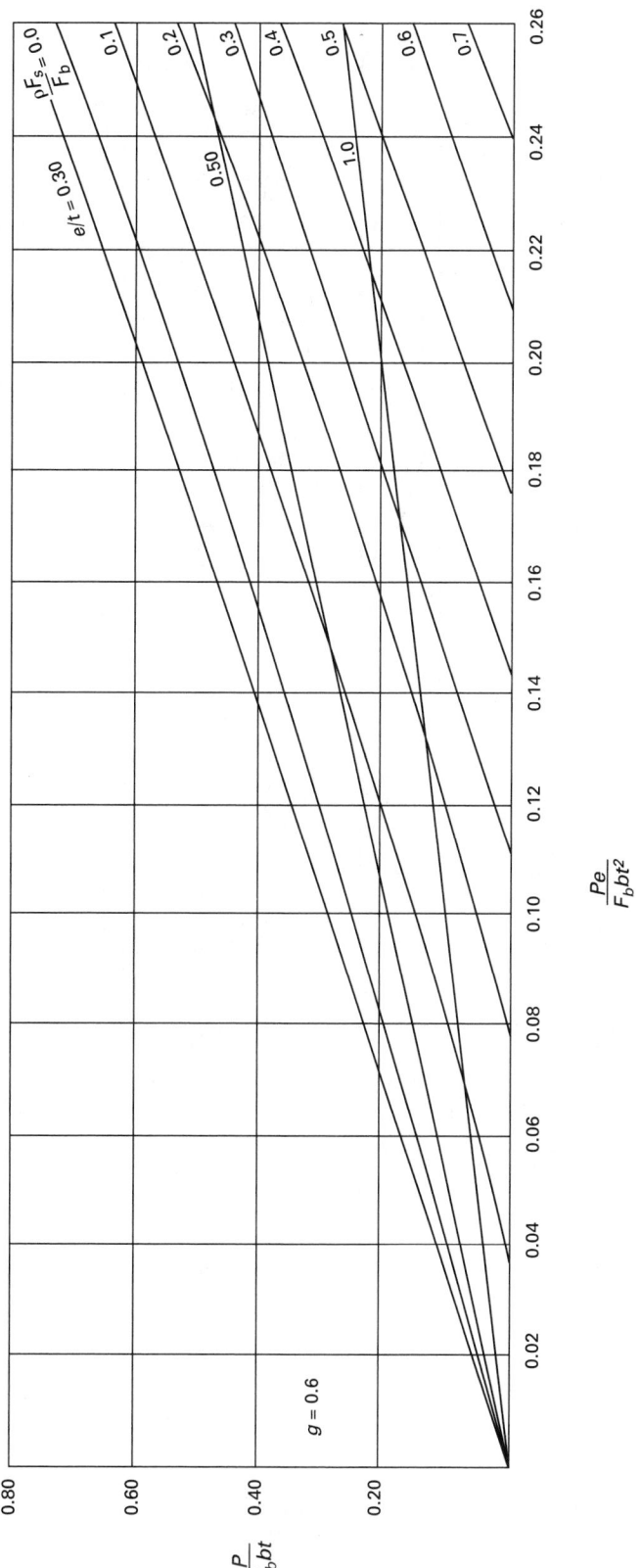

APPENDIX 69.F
Column Interaction Diagram
(tension controls, $g = 0.8$)

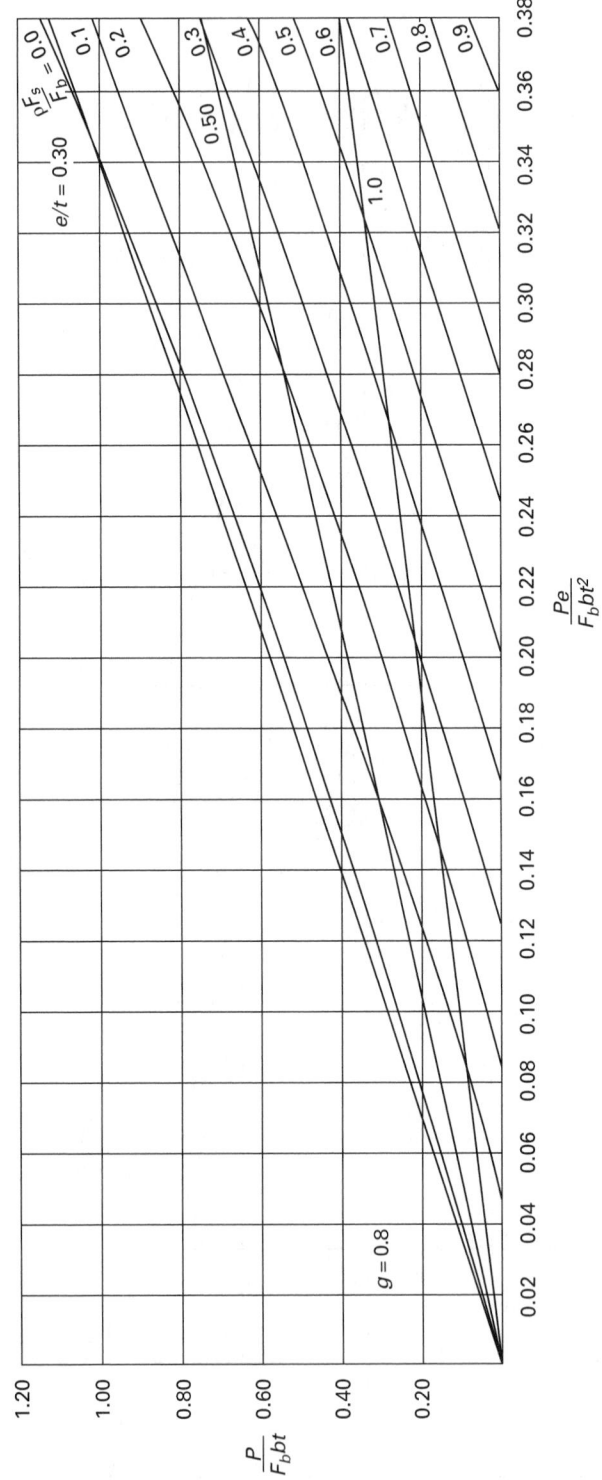

Reprinted with permission of the American Concrete Institute from *Masonry Designer's Guide*, copyright © 1993.

APPENDIX 70.A
Mass Moments of Inertia
(centroids at points labeled C)

slender rod		$I_y = I_z = \dfrac{mL^2}{12}$ $I'_y = I'_z = \dfrac{mL^2}{3}$
solid circular cylinder, radius r		$I_x = \dfrac{mr^2}{2}$ $I_y = I_z = \dfrac{m(3r^2 + L^2)}{12}$
hollow circular cylinder, inner radius r_i, outer radius r_o		$I_x = \dfrac{m\left(r_o^2 + r_i^2\right)}{2}$ $= \left(\dfrac{\pi\rho L}{2}\right)\left(r_o^4 - r_i^4\right)$ $I_y = I_z = \dfrac{\pi\rho L}{12}\left(3(r_2^4 - r_1^4)\right.$ $\left. + L^2(r_2^2 - r_1^2)\right)$
thin disk, radius r		$I_x = \dfrac{mr^2}{2}$ $I_y = I_z = \dfrac{mr^2}{4}$
solid circular cone, base radius r		$I_x = \dfrac{3mr^2}{10}$ $I_y = I_z = \left(\dfrac{3m}{5}\right)\left(\dfrac{r^2}{4} + h^2\right)$
thin rectangular plate		$I_x = \dfrac{m(b^2 + c^2)}{12}$ $I_y = \dfrac{mc^2}{12}$ $I_z = \dfrac{mb^2}{12}$
rectangular parallelepiped		$I_x = \dfrac{m(b^2 + c^2)}{12}$ $I_y = \dfrac{m(c^2 + a^2)}{12}$ $I_z = \dfrac{m(a^2 + b^2)}{12}$ $I_{x'} = \dfrac{m(4b^2 + c^2)}{12}$
sphere, radius r		$I_x = I_y = I_z = \dfrac{2mr^2}{5}$

APPENDIX 75.A
Axle Load Equivalency Factors for Flexible Pavements
(single axles and p_t of 2.5)

axle load (kips)	pavement structural number (SN)					
	1	2	3	4	5	6
2	0.0004	0.0004	0.0003	0.0002	0.0002	0.0002
4	0.003	0.004	0.004	0.003	0.002	0.002
6	0.011	0.017	0.017	0.013	0.010	0.009
8	0.032	0.047	0.051	0.041	0.034	0.031
10	0.078	0.102	0.118	0.102	0.088	0.080
12	0.168	0.198	0.229	0.213	0.189	0.176
14	0.328	0.358	0.399	0.388	0.360	0.342
16	0.591	0.613	0.646	0.645	0.623	0.606
18	1.00	1.00	1.00	1.00	1.00	1.00
20	1.61	1.57	1.49	1.47	1.51	1.55
22	2.48	2.38	2.17	2.09	2.18	2.30
24	3.69	3.49	3.09	2.89	3.03	3.27
26	5.33	4.99	4.31	3.91	4.09	4.48
28	7.49	6.98	5.90	5.21	5.39	5.98
30	10.3	9.5	7.9	6.8	7.0	7.8
32	13.9	12.8	10.5	8.8	8.9	10.0
34	18.4	16.9	13.7	11.3	11.2	12.5
36	24.0	22.0	17.7	14.4	13.9	15.5
38	30.9	28.3	22.6	18.1	17.2	19.0
40	39.3	35.9	28.5	22.5	21.1	23.0
42	49.3	45.0	35.6	27.8	25.6	27.7
44	61.3	55.9	44.0	34.0	31.0	33.1
46	75.5	68.8	54.0	41.4	37.2	39.3
48	92.2	83.9	65.7	50.1	44.5	46.5
50	112	102	79	60	53	55

From *Guide for Design of Pavement and Structures*, Table D.4, copyright © 1993 by the American Association of State Highway and Transportation Officials, Washington, D.C. Used by permission.

APPENDIX 75.B
Axle Load Equivalency Factors for Flexible Pavements
(tandem axles and p_t of 2.5)

axle load (kips)	pavement structural number (SN)					
	1	2	3	4	5	6
2	0.0001	0.0001	0.0001	0.0000	0.0000	0.0000
4	0.0005	0.0005	0.0004	0.0003	0.0003	0.0002
6	0.002	0.002	0.002	0.001	0.001	0.001
8	0.004	0.006	0.005	0.004	0.003	0.003
10	0.008	0.013	0.011	0.009	0.007	0.006
12	0.015	0.024	0.023	0.018	0.014	0.013
14	0.026	0.041	0.042	0.033	0.027	0.024
16	0.044	0.065	0.070	0.057	0.047	0.043
18	0.070	0.097	0.109	0.092	0.077	0.070
20	0.107	0.141	0.162	0.141	0.121	0.110
22	0.160	0.198	0.229	0.207	0.180	0.166
24	0.231	0.273	0.315	0.292	0.260	0.242
26	0.327	0.370	0.420	0.401	0.364	0.342
28	0.451	0.493	0.548	0.534	0.495	0.470
30	0.611	0.648	0.703	0.695	0.658	0.633
32	0.813	0.843	0.889	0.887	0.857	0.834
34	1.06	1.08	1.11	1.11	1.09	1.08
36	1.38	1.38	1.38	1.38	1.38	1.38
38	1.75	1.73	1.69	1.68	1.70	1.73
40	2.21	2.16	2.06	2.03	2.08	2.14
42	2.76	2.67	2.49	2.43	2.51	2.61
44	3.41	3.27	2.99	2.88	3.00	3.16
46	4.18	3.98	3.58	3.40	3.55	3.79
48	5.08	4.80	4.25	3.98	4.17	4.49
50	6.12	5.76	5.03	4.64	4.86	5.28
52	7.33	6.87	5.93	5.38	5.63	6.17
54	8.72	8.14	6.95	6.22	6.47	7.15
56	10.3	9.6	8.1	7.2	7.4	8.2
58	12.1	11.3	9.4	8.2	8.4	9.4
60	14.2	13.1	10.9	9.4	9.6	10.7
62	16.5	15.3	12.6	10.7	10.8	12.1
64	19.1	17.6	14.5	12.2	12.2	13.7
66	22.1	20.3	16.6	13.8	13.7	15.4
68	25.3	23.3	18.9	15.6	15.4	17.2
70	29.0	26.6	21.5	17.6	17.2	19.2
72	33.0	30.3	24.4	19.8	19.2	21.3
74	37.5	34.4	27.6	22.2	21.3	23.6
76	42.5	38.9	31.1	24.8	23.7	26.1
78	48.0	43.9	35.0	27.8	26.2	28.8
80	54.0	49.4	39.2	30.9	29.0	31.7
82	60.6	55.4	43.9	34.4	32.0	34.8
84	67.8	61.9	49.0	38.2	35.3	38.1
86	75.7	69.1	54.5	42.3	38.8	41.7
88	84.3	76.9	60.6	46.8	42.6	45.6
90	93.7	85.4	67.1	51.7	46.8	49.7

APPENDIX 75.C
Axle Load Equivalency Factors for Flexible Pavements
(triple axles and p_t of 2.5)

axle load (kips)	pavement structural number (SN)					
	1	2	3	4	5	6
2	0.0000	0.0000	0.0000	0.0000	0.0000	0.0000
4	0.0002	0.0002	0.0002	0.0001	0.0001	0.0001
6	0.0006	0.0007	0.0005	0.0004	0.0003	0.0003
8	0.001	0.002	0.001	0.001	0.001	0.001
10	0.003	0.004	0.003	0.002	0.002	0.002
12	0.005	0.007	0.006	0.004	0.003	0.003
14	0.008	0.012	0.010	0.008	0.006	0.006
16	0.012	0.019	0.018	0.013	0.011	0.010
18	0.018	0.029	0.028	0.021	0.017	0.016
20	0.027	0.042	0.042	0.032	0.027	0.024
22	0.038	0.058	0.060	0.048	0.040	0.036
24	0.053	0.078	0.084	0.068	0.057	0.051
26	0.072	0.103	0.114	0.095	0.080	0.072
28	0.098	0.133	0.151	0.128	0.109	0.099
30	0.129	0.169	0.195	0.170	0.145	0.133
32	0.169	0.213	0.247	0.220	0.191	0.175
34	0.219	0.266	0.308	0.281	0.246	0.228
36	0.279	0.329	0.379	0.352	0.313	0.292
38	0.352	0.403	0.461	0.436	0.393	0.368
40	0.439	0.491	0.554	0.533	0.487	0.459
42	0.543	0.594	0.661	0.644	0.597	0.567
44	0.666	0.714	0.781	0.769	0.723	0.692
46	0.811	0.854	0.918	0.911	0.868	0.838
48	0.979	1.015	1.072	1.069	1.033	1.005
50	1.17	1.20	1.24	1.25	1.22	1.20
52	1.40	1.41	1.44	1.44	1.43	1.41
54	1.66	1.66	1.66	1.66	1.66	1.66
56	1.95	1.93	1.90	1.90	1.91	1.93
58	2.29	2.25	2.17	2.16	2.20	2.24
60	2.67	2.60	2.48	2.44	2.51	2.58
62	3.09	3.00	2.82	2.76	2.85	2.95
64	3.57	3.44	3.19	3.10	3.22	3.36
66	4.11	3.94	3.61	3.47	3.62	3.81
68	4.71	4.49	4.06	3.88	4.05	4.30
70	5.38	5.11	4.57	4.32	4.52	4.84
72	6.12	5.79	5.13	4.80	5.03	5.41
74	6.93	6.54	5.74	5.32	5.57	6.04
76	7.84	7.37	6.41	5.88	6.15	6.71
78	8.83	8.28	7.14	6.49	6.78	7.43
80	9.92	9.28	7.95	7.15	7.45	8.21
82	11.1	10.4	8.8	7.9	8.2	9.0
84	12.4	11.6	9.8	8.6	8.9	9.9
86	13.8	12.9	10.8	9.5	9.8	10.9
88	15.4	14.3	11.9	10.4	10.6	11.9
90	17.1	15.8	13.2	11.3	11.6	12.9

From *Guide for Design of Pavement and Structures*, Table D.6, copyright © 1993 by the American Association of State Highway and Transportation Officials, Washington, D.C. Used by permission.

APPENDIX 76.A
Axle Load Equivalency Factors for Rigid Pavements
(Single Axles and p_t of 2.5)

axle load (kips)	slab thickness, D (in)								
	6	7	8	9	10	11	12	13	14
2	0.0002	0.0002	0.0002	0.0002	0.0002	0.0002	0.0002	0.0002	0.0002
4	0.003	0.002	0.002	0.002	0.002	0.002	0.002	0.002	0.002
6	0.012	0.011	0.010	0.010	0.010	0.010	0.010	0.010	0.010
8	0.039	0.035	0.033	0.032	0.032	0.032	0.032	0.032	0.032
10	0.097	0.089	0.084	0.082	0.081	0.080	0.080	0.080	0.080
12	0.203	0.189	0.181	0.176	0.175	0.174	0.174	0.173	0.173
14	0.376	0.360	0.347	0.341	0.338	0.337	0.336	0.336	0.336
16	0.634	0.623	0.610	0.604	0.601	0.599	0.599	0.599	0.598
18	1.00	1.00	1.00	1.00	1.00	1.00	1.00	1.00	1.00
20	1.51	1.52	1.55	1.57	1.58	1.58	1.59	1.59	1.59
22	2.21	2.20	2.28	2.34	2.38	2.40	2.41	2.41	2.41
24	3.16	3.10	3.22	3.36	3.45	3.50	3.53	3.54	3.55
26	4.41	4.26	4.42	4.67	4.85	4.95	5.01	5.04	5.05
28	6.05	5.76	5.92	6.29	6.61	6.81	6.92	6.98	7.01
30	8.16	7.67	7.79	8.28	8.79	9.14	9.35	9.46	9.52
32	10.8	10.1	10.1	10.7	11.4	12.0	12.3	12.6	12.7
34	14.1	13.0	12.9	13.6	14.6	15.4	16.0	16.4	16.5
36	18.2	16.7	16.4	17.1	18.3	19.5	20.4	21.0	21.3
38	23.1	21.1	20.6	21.3	22.7	24.3	25.6	26.4	27.0
40	29.1	26.5	25.7	26.3	27.9	29.9	31.6	32.9	33.7
42	36.2	32.9	31.7	32.2	34.0	36.3	38.7	40.4	41.6
44	44.6	40.4	38.8	39.2	41.0	43.8	46.7	49.1	50.8
46	54.5	49.3	47.1	47.3	49.2	52.3	55.9	59.0	61.4
48	66.1	59.7	56.9	56.8	58.7	62.1	66.3	70.3	73.4
50	79.4	71.7	68.2	67.8	69.6	73.3	78.1	83.0	87.1

From *Guide for Design of Pavement and Structures*, Table D.13, copyright © 1993 by the American Association of State Highway and Transportation Officials, Washington, D.C. Used by permission.

APPENDIX 76.B
Axle Load Equivalency Factors for Rigid Pavements
(Double Axles and p_t of 2.5)

axle load (kips)	slab thickness, D (in)								
	6	7	8	9	10	11	12	13	14
2	0.0001	0.0001	0.0001	0.0001	0.0001	0.0001	0.0001	0.0001	0.0001
4	0.0006	0.0006	0.0005	0.0005	0.0005	0.0005	0.0005	0.0005	0.0005
6	0.002	0.002	0.002	0.002	0.002	0.002	0.002	0.002	0.002
8	0.007	0.006	0.006	0.005	0.005	0.005	0.005	0.005	0.005
10	0.015	0.014	0.013	0.013	0.012	0.012	0.012	0.012	0.012
12	0.031	0.028	0.026	0.026	0.025	0.025	0.025	0.025	0.025
14	0.057	0.052	0.049	0.048	0.047	0.047	0.047	0.047	0.047
16	0.097	0.089	0.084	0.082	0.081	0.081	0.080	0.080	0.080
18	0.155	0.143	0.136	0.133	0.132	0.131	0.131	0.131	0.131
20	0.234	0.220	0.211	0.206	0.204	0.203	0.203	0.203	0.203
22	0.340	0.325	0.313	0.308	0.305	0.304	0.303	0.303	0.303
24	0.475	0.462	0.450	0.444	0.441	0.440	0.439	0.439	0.439
26	0.644	0.637	0.627	0.622	0.620	0.619	0.618	0.618	0.618
28	0.855	0.854	0.852	0.850	0.850	0.850	0.849	0.849	0.849
30	1.11	1.12	1.13	1.14	1.14	1.14	1.14	1.14	1.14
32	1.43	1.44	1.47	1.49	1.50	1.51	1.51	1.51	1.51
34	1.82	1.82	1.87	1.92	1.95	1.96	1.97	1.97	1.97
36	2.29	2.27	2.35	2.43	2.48	2.51	2.52	2.52	2.53
38	2.85	2.80	2.91	3.03	3.12	3.16	3.18	3.20	3.20
40	3.52	3.42	3.55	3.74	3.87	3.94	3.98	4.00	4.01
42	4.32	4.16	4.30	4.55	4.74	4.86	4.91	4.95	4.96
44	5.26	5.01	5.16	5.48	5.75	5.92	6.01	6.06	6.09
46	6.36	6.01	6.14	6.53	6.90	7.14	7.28	7.36	7.40
48	7.64	7.16	7.27	7.73	8.21	8.55	8.75	8.86	8.92
50	9.11	8.50	8.55	9.07	9.68	10.14	10.42	10.58	10.66
52	10.8	10.0	10.0	10.6	11.3	11.9	12.3	12.5	12.7
54	12.8	11.8	11.7	12.3	13.2	13.9	14.5	14.8	14.9
56	15.0	13.8	13.6	14.2	15.2	16.2	16.8	17.3	17.5
58	17.5	16.0	15.7	16.3	17.5	18.6	19.5	20.1	20.4
60	20.3	18.5	18.1	18.7	20.0	21.4	22.5	23.2	23.6
62	23.5	21.4	20.8	21.4	22.8	24.4	25.7	26.7	27.3
64	27.0	24.6	23.8	24.4	25.8	27.7	29.3	30.5	31.3
66	31.0	28.1	27.1	27.6	29.2	31.3	33.2	34.7	35.7
68	35.4	32.1	30.9	31.3	32.9	35.2	37.5	39.3	40.5
70	40.3	36.5	35.0	35.3	37.0	39.5	42.1	44.3	45.9
72	45.7	41.4	39.6	39.8	41.5	44.2	47.2	49.8	51.7
74	51.7	46.7	44.6	44.7	46.4	49.3	52.7	55.7	58.0
76	58.3	52.6	50.2	50.1	51.8	54.9	58.6	62.1	64.8
78	65.5	59.1	56.3	56.1	57.7	60.9	65.0	69.0	72.3
80	73.4	66.2	62.9	62.5	64.2	67.5	71.9	76.4	80.2
82	82.0	73.9	70.2	69.6	71.2	74.7	79.4	84.4	88.8
84	91.4	82.4	78.1	77.3	78.9	82.4	87.4	93.0	98.1
86	102	92	87	86	87	91	96	102	108
88	113	102	96	95	96	100	105	112	119
90	125	112	106	105	106	110	115	123	130

From *Guide for Design of Pavement and Structures*, Table D.14, copyright © 1993 by the American Association of State Highway and Transportation Officials, Washington, D.C. Used by permission.

APPENDIX 76.C
Axle Load Equivalency Factors for Rigid Pavements
(Triple Axles and p_t of 2.5)

axle load (kips)	slab thickness, D (in)								
	6	7	8	9	10	11	12	13	14
2	0.0001	0.0001	0.0001	0.0001	0.0001	0.0001	0.0001	0.0001	0.0001
4	0.0003	0.0003	0.0003	0.0003	0.0003	0.0003	0.0003	0.0003	0.0003
6	0.001	0.001	0.001	0.001	0.001	0.001	0.001	0.001	0.001
8	0.003	0.002	0.002	0.002	0.002	0.002	0.002	0.002	0.002
10	0.006	0.005	0.005	0.005	0.005	0.005	0.005	0.005	0.005
12	0.011	0.010	0.010	0.009	0.009	0.009	0.009	0.009	0.009
14	0.020	0.018	0.017	0.017	0.016	0.016	0.016	0.016	0.016
16	0.033	0.030	0.029	0.028	0.027	0.027	0.027	0.027	0.027
18	0.053	0.048	0.045	0.044	0.044	0.043	0.043	0.043	0.043
20	0.080	0.073	0.069	0.067	0.066	0.066	0.066	0.066	0.066
22	0.116	0.107	0.101	0.099	0.098	0.097	0.097	0.097	0.097
24	0.163	0.151	0.144	0.141	0.139	0.139	0.138	0.138	0.138
26	0.222	0.209	0.200	0.195	0.194	0.193	0.192	0.192	0.192
28	0.295	0.281	0.271	0.265	0.263	0.262	0.262	0.262	0.262
30	0.384	0.371	0.359	0.354	0.351	0.350	0.349	0.349	0.349
32	0.490	0.480	0.468	0.463	0.460	0.459	0.458	0.458	0.458
34	0.616	0.609	0.601	0.596	0.594	0.593	0.592	0.592	0.592
36	0.765	0.762	0.759	0.757	0.756	0.755	0.755	0.755	0.755
38	0.939	0.941	0.946	0.948	0.950	0.951	0.951	0.951	0.951
40	1.14	1.15	1.16	1.17	1.18	1.18	1.18	1.18	1.18
42	1.38	1.38	1.41	1.44	1.45	1.46	1.46	1.46	1.46
44	1.65	1.65	1.70	1.74	1.77	1.78	1.78	1.78	1.79
46	1.97	1.96	2.03	2.09	2.13	2.15	2.16	2.16	2.16
48	2.34	2.31	2.40	2.49	2.55	2.58	2.59	2.60	2.60
50	2.76	2.71	2.81	2.94	3.02	3.07	3.09	3.10	3.11
52	3.24	3.15	3.27	3.44	3.56	3.62	3.66	3.68	3.68
54	3.79	3.66	3.79	4.00	4.16	4.26	4.30	4.33	4.34
56	4.41	4.23	4.37	4.63	4.84	4.97	5.03	5.07	5.09
58	5.12	4.87	5.00	5.32	5.59	5.76	5.85	5.90	5.93
60	5.91	5.59	5.71	6.08	6.42	6.64	6.77	6.84	6.87
62	6.80	6.39	6.50	6.91	7.33	7.62	7.79	7.88	7.93
64	7.79	7.29	7.37	7.82	8.33	8.70	8.92	9.04	9.11
66	8.90	8.28	8.33	8.83	9.42	9.88	10.17	10.33	10.42
68	10.1	9.4	9.4	9.9	10.6	11.2	11.5	11.7	11.9
70	11.5	10.6	10.6	11.1	11.9	12.6	13.0	13.3	13.5
72	13.0	12.0	11.8	12.4	13.3	14.1	14.7	15.0	15.2
74	14.6	13.5	13.2	13.8	14.8	15.8	16.5	16.9	17.1
76	16.5	15.1	14.8	15.4	16.5	17.6	18.4	18.9	19.2
78	18.5	16.9	16.5	17.1	18.2	19.5	20.5	21.1	21.5
80	20.6	18.8	18.3	18.9	20.2	21.6	22.7	23.5	24.0
82	23.0	21.0	20.3	20.9	22.2	23.8	25.2	26.1	26.7
84	25.6	23.3	22.5	23.1	24.5	26.2	27.8	28.9	29.6
86	28.4	25.8	24.9	25.4	26.9	28.8	30.5	31.9	32.8
88	31.5	28.6	27.5	27.9	29.4	31.5	33.5	35.1	36.1
90	34.8	31.5	30.3	30.7	32.2	34.4	36.7	38.5	39.8

From *Guide for Design of Pavement and Structures*, Table D.15, copyright © 1993 by the American Association of State Highway and Transportation Officials, Washington, D.C. Used by permission.

APPENDIX 77.A
Oblique Triangle Equations

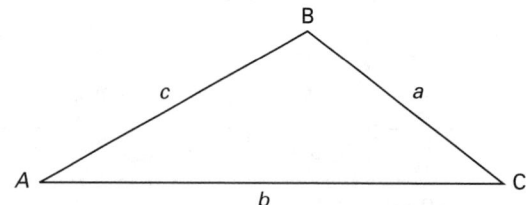

given	equation
A, B, a	$C = 180° - (A + B)$
	$b = \left(\dfrac{a}{\sin A}\right)\sin B$
	$c = \left(\dfrac{a}{\sin A}\right)\sin(A + B) = \left(\dfrac{a}{\sin A}\right)\sin C$
	$\text{area} = \frac{1}{2}ab\sin C = \dfrac{a^2 \sin B \sin C}{2\sin A}$
A, a, b	$\sin B = \left(\dfrac{\sin A}{a}\right)b$
	$C = 180° - (A + B)$
	$c = \left(\dfrac{a}{\sin A}\right)\sin C$
	$\text{area} = \frac{1}{2}ab\sin C$
C, a, b	$c = \sqrt{a^2 + b^2 - 2ab\cos C}$
	$\frac{1}{2}(A + B) = 90° - \frac{1}{2}C$
	$\tan\frac{1}{2}(A - B) = \left(\dfrac{a - b}{a + b}\right)\tan\frac{1}{2}(A + B)$
	$A = \frac{1}{2}(A + B) + \frac{1}{2}(A - B)$
	$B = \frac{1}{2}(A + B) - \frac{1}{2}(A - B)$
	$c = (a + b)\left(\dfrac{\cos\frac{1}{2}(A + B)}{\cos\frac{1}{2}(A - B)}\right) = (a - b)\left(\dfrac{\sin\frac{1}{2}(A + B)}{\sin\frac{1}{2}(A - B)}\right)$
	$\text{area} = \frac{1}{2}ab\sin C$
a, b, c	$\text{Let } s = \dfrac{a + b + c}{2}$
	$\sin\frac{1}{2}A = \sqrt{\dfrac{(s - b)(s - c)}{bc}}$
	$\cos\frac{1}{2}A = \sqrt{\dfrac{s(s - a)}{bc}}$
	$\tan\frac{1}{2}A = \sqrt{\dfrac{(s - b)(s - c)}{s(s - a)}}$
	$\sin A = 2\sqrt{\dfrac{s(s - a)(s - b)(s - c)}{bc}}$
	$\cos A = \dfrac{b^2 + c^2 - a^2}{2bc}$
	$\text{area} = \sqrt{s(s - a)(s - b)(s - c)}$

APPENDIX 83.A
Polyphase Motor Classifications and Characteristics

speed regulations	speed control	starting torque	breakdown torque	application
general-purpose squirrel cage (NEMA design B)				
Drops about 3% for large to 5% for small sizes.	None, except multi-speed types, designed for two to four fixed speeds.	100% for large; 275% for 1 hp 4 pole unit.	200% of full load.	Constant-speed service where starting is not excessive. Fans, blowers, rotary compressors, and centrifugal pumps.
high-torque squirrel cage (NEMA design C)				
Drops about 3% for large to 6% for small sizes.	None, except multi-speed types, designed for two and four fixed speeds.	250% of full load for high-speed to 200% for low-speed designs.	200% of full load.	Constant-speed where fairly high starting torque is required infrequently with starting current about 550% of full load. Reciprocating pumps and compressors, crushers, etc.
high-slip squirrel cage (NEMA design D)				
Drops about 10 to 15% from no load to full load.	None, except multi-speed types, designed for two to four fixed speeds.	225 to 300% full load, depending on speed with rotor resistance.	200%. Will usually not stall until loaded to max torque, which occurs at standstill.	Constant-speed and high starting torque, if starting is not too frequent, and for high-peak loads with or without flywheels. Punch presses, shears, elevators, etc.
low-torque squirrel cage (NEMA design F)				
Drops about 3% for large to 5% for small sizes.	None, except multi-speed types, designed for two to four fixed speeds.	50% of full load for high-speed to 90% for low-speed designs.	135 to 170% of full load.	Constant-speed service where starting duty is light. Fans, blowers, centrifugal pumps and similar loads.
wound rotor				
With rotor rings short circuited, drops about 3% for large to 5% for small sizes.	Speed can be reduced to 50% by rotor resistance. Speed varies inversely as load.	Up to 300% depending on external resistance in rotor circuit and how distributed.	300% when rotor slip rings are short circuited.	Where high starting torque with low starting current or where limited speed control is required. Fans, centrifugal and plunger pumps, compressors, conveyors, hoists, cranes, etc.
synchronous				
Constant.	None, except special motors designed for two fixed speeds.	40% for slow to 160% for medium-speed 80% pf. Specials develop higher.	Unity-pf motors 170%, 80%-pf motors 225%. Specials, up to 300%	For constant-speed service, direct connection to slow-speed machines and where pf correction is required.

Adapted from *Mechanical Engineering*, Design Manual NAVFAC DM-3, Department of the Navy, copyright © 1972.

APPENDIX 83.B
DC and Single-Phase Motor Classifications and Characteristics

speed regulations	speed control	starting torque	breakdown torque	application
series				
Varies inversely as load. Races on light loads and full voltage.	Zero to maximum depending on control and load.	High. Varies as square of voltage. Limited by commutation, heating, and line capacity.	High. Limited by commutation, heating, and line capacity.	Where high starting torque is required and speed can be regulated. Traction, bridges, hoists, gates, car dumpers, car retarders.
shunt				
Drops 3 to 5% from no load to full load.	Any desired range depending on design, type of system.	Good. With constant field, varies directly as voltage applied to armature.	High. Limited by commutation, heating, and line capacity.	Where constant or adjustable speed is required and starting conditions are not severe. Fans, blowers, centrifugal pumps, conveyors, wood- and metal-working machines, elevators.
compound				
Drops 7 to 20% from no load to full load depending on amount of compounding.	Any desired range, depending on design, type of control.	Higher than for shunt, depending on amount of compounding.	High. Limited by commutation, heating and line capacity.	Where high starting torque and fairly constant speed is required. Plunger pumps, punch presses, shears, bending rolls, geared elevators, conveyors, hoists.
split-phase				
Drops about 10% from no load to full load.	None.	75% for large to 175% for small sizes.	150% for large to 200% for small sizes.	Constant-speed service where starting is easy. Small fans, centrifugal pumps and light-running machines, where polyphase is not available.
capacitors				
Drops about 5% for large to 10% for small sizes.	None.	150 to 350% of full load depending on design and size.	150% for large to 200% for small sizes.	Constant-speed service for any starting duty and quiet operation, where polyphase current cannot be used.
commutator type				
Drops about 5% for large to 10% for small sizes.	Repulsion induction, none. Brush-shifting types, four to one at full load.	250% for large to 350% for small sizes.	150% for large to 250%.	Constant-speed service for any starting duty where speed control is required and polyphase current cannot be used.

Adapted from *Mechanical Engineering*, Design Manual NAVFAC DM-3, Department of the Navy, copyright © 1972.

APPENDIX 84.A
Thermoelectric Constants for Thermocouples
(mV, reference 32°F (0°C))

(a) chromel-alumel (type K)

°F	0	10	20	30	40	50	60	70	80	90
					millivolts					
−300	−5.51	−5.60								
−200	−4.29	−4.44	−4.58	−4.71	−4.84	−4.96	−5.08	−5.20	−5.30	−5.41
−100	−2.65	−2.83	−3.01	−3.19	−3.36	−3.52	−3.69	−3.84	−4.00	−4.15
−0	−0.68	−0.89	−1.10	−1.30	−1.50	−1.70	−1.90	−2.09	−2.28	−2.47
+0	−0.68	−0.49	−0.26	−0.04	0.18	0.40	0.62	0.84	1.06	1.29
100	1.52	1.74	1.97	2.20	2.43	2.66	2.89	3.12	3.36	3.59
200	3.82	4.05	4.28	4.51	4.74	4.97	5.20	5.42	5.65	5.87
300	6.09	6.31	6.53	6.76	6.98	7.20	7.42	7.64	7.87	8.09
400	8.31	8.54	8.76	8.98	9.21	9.43	9.66	9.88	10.11	10.34
500	10.57	10.79	11.02	11.25	11.48	11.71	11.94	12.17	12.40	12.63
600	12.86	13.09	13.32	13.55	13.78	14.02	14.25	14.48	14.71	14.95
700	15.18	15.41	15.65	15.88	16.12	16.35	16.59	16.82	17.06	17.29
800	17.53	17.76	18.00	18.23	18.47	18.70	18.94	19.18	19.41	19.65
900	19.89	20.13	20.36	20.60	20.84	21.07	21.31	21.54	21.78	22.02
1000	22.26	22.49	22.73	22.97	23.20	23.44	23.68	23.91	24.15	24.39
1100	24.63	24.86	25.10	25.34	25.57	25.81	26.05	26.28	26.52	26.75
1200	26.98	27.22	27.45	27.69	27.92	28.15	28.39	28.62	28.86	29.09
1300	29.32	29.56	29.79	30.02	30.25	30.49	30.72	30.95	31.18	31.42
1400	31.65	31.88	32.11	32.34	32.57	32.80	33.02	33.25	33.48	33.71
1500	33.93	34.16	34.39	34.62	34.84	35.07	35.29	35.52	35.75	35.97
1600	36.19	36.42	36.64	36.87	37.09	37.31	37.54	37.76	37.98	38.20
1700	38.43	38.65	38.87	39.09	39.31	39.53	39.75	39.96	40.18	40.40
1800	40.62	40.84	41.05	41.27	41.49	41.70	41.92	42.14	42.35	42.57
1900	42.78	42.99	43.21	43.42	43.63	43.85	44.06	44.27	44.49	44.70
2000	44.91	45.12	45.33	45.54	45.75	45.96	46.17	46.38	46.58	46.79

(b) iron-constantan (type J)

°F	0	10	20	30	40	50	60	70	80	90
					millivolts					
−300	−7.52	−7.66								
−200	−5.76	−5.96	−6.16	−6.35	−6.53	−6.71	−6.89	−7.06	−7.22	−7.38
−100	−3.49	−3.73	−3.97	−4.21	−4.44	−4.68	−4.90	−5.12	−5.34	−5.55
−0	−0.89	−1.16	−1.43	−1.70	−1.96	−2.22	−2.48	−2.74	−2.99	−3.24
+0	−0.89	−0.61	−0.34	−0.06	0.22	0.50	0.79	1.07	1.36	1.65
100	1.94	2.23	2.52	2.82	3.11	3.41	3.71	4.01	4.31	4.61
200	4.91	5.21	5.51	5.81	6.11	6.42	6.72	7.03	7.33	7.64
300	7.94	8.25	8.56	8.87	9.17	9.48	9.79	10.10	10.41	10.72
400	11.03	11.34	11.65	11.96	12.26	12.57	12.88	13.19	13.50	13.81
500	14.12	14.42	14.73	15.04	15.34	15.65	15.96	16.26	16.57	16.88
600	17.18	17.49	17.80	18.11	18.41	18.72	19.03	19.34	19.64	19.95
700	20.26	20.56	20.87	21.18	21.48	21.79	22.10	22.40	22.71	23.01
800	23.32	23.63	23.93	24.24	24.55	24.85	25.16	25.47	25.78	26.09
900	26.40	26.70	27.02	27.33	27.64	27.95	28.26	28.58	28.89	29.21
1000	29.52	29.84	30.16	30.48	30.80	31.12	31.44	31.76	32.08	32.40
1100	32.72	33.05	33.37	33.70	34.03	34.36	34.68	35.01	35.35	35.68
1200	36.01	36.35	36.69	37.02	37.36	37.71	38.05	38.39	38.74	39.08
1300	39.43	39.78	40.13	40.48	40.83	41.19	41.54	41.90	42.25	42.61

(c) copper-constantan (type T)

°F	0	10	20	30	40	50	60	70	80	90
					millivolts					
−300	−5.284	−5.379								
−200	−4.111	−4.246	−4.377	−4.504	−4.627	−4.747	−4.863	−4.974	−5.081	−5.185
−100	−2.559	−2.730	−2.897	−3.062	−3.223	−3.380	−3.533	−3.684	−3.829	−3.972
−0	−0.670	−0.872	−1.072	−1.270	−1.463	−1.654	−1.842	−2.026	−2.207	−2.385
+0	−0.670	−0.463	−0.254	−0.042	0.171	0.389	0.609	0.832	1.057	1.286
100	1.517	1.751	1.987	2.226	2.467	2.711	2.958	3.207	3.458	3.712
200	3.967	4.225	4.486	4.749	5.014	5.280	5.550	5.821	6.094	6.370
300	6.647	6.926	7.208	7.491	7.776	8.064	8.352	8.642	8.935	9.229
400	9.525	9.823	10.123	10.423	10.726	11.030	11.336	11.643	11.953	12.263
500	12.575	12.888	13.203	13.520	13.838	14.157	14.477	14.799	15.122	15.447

Based on National Bureau of Standards Circular No. 561, April 1955.

APPENDIX 86.A
Standard Cash Flow Factors and Formulas

multiply	by	to obtain

$$P = F(1+i)^{-n}$$
$$(P/F, i\%, n)$$

F P

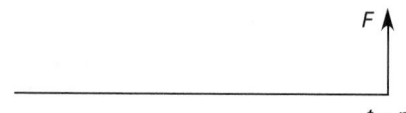

$$F = P(1+i)^n$$
$$(F/P, i\%, n)$$

P F

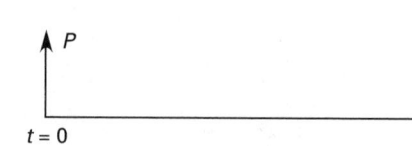

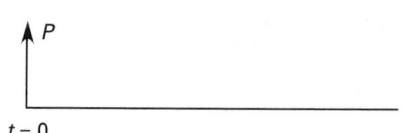

$$P = A\left(\frac{(1+i)^n - 1}{i(1+i)^n}\right)$$
$$(P/A, i\%, n)$$

A P

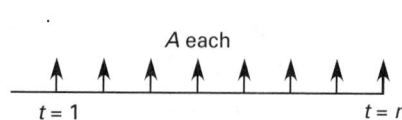

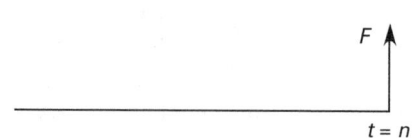

$$A = P\left(\frac{i(1+i)^n}{(1+i)^n - 1}\right)$$
$$(A/P, i\%, n)$$

P A

$$F = A\left(\frac{(1+i)^n - 1}{i}\right)$$
$$(F/A, i\%, n)$$

A F

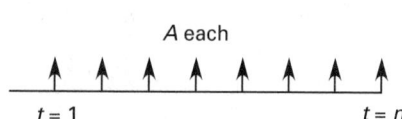

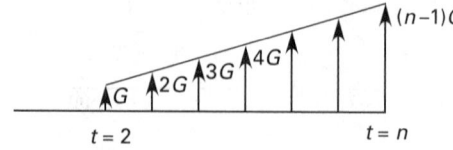

$$A = F\left(\frac{i}{(1+i)^n - 1}\right)$$
$$(A/F, i\%, n)$$

F A

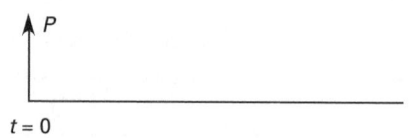

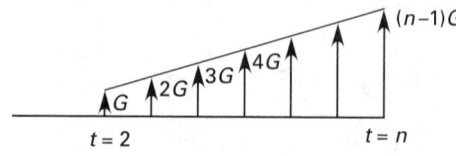

$$P = G\left(\frac{(1+i)^n - 1}{i^2(1+i)^n} - \frac{n}{i(1+i)^n}\right)$$
$$(P/G, i\%, n)$$

G P

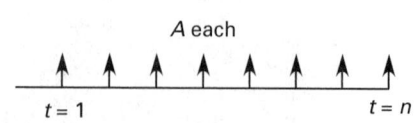

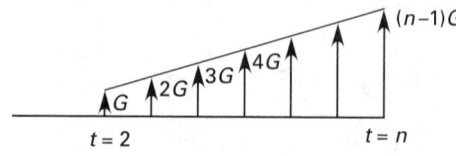

$$A = G\left(\frac{1}{i} - \frac{n}{(1+i)^n - 1}\right)$$
$$(A/G, i\%, n)$$

G A

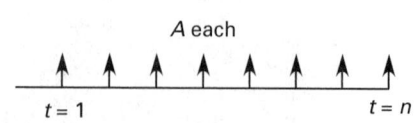

Support
Material

APPENDIX 86.B
Factor Tables

$I = 0.50\%$

n	P/F	P/A	P/G	F/P	F/A	A/P	A/F	A/G	n
1	0.9950	0.9950	0.0000	1.0050	1.0000	1.0050	1.0000	0.0000	1
2	0.9901	0.9851	0.9901	1.0100	2.0050	0.5038	0.4988	0.4988	2
3	0.9851	2.9702	2.9604	1.0151	3.0150	0.3367	0.3317	0.9967	3
4	0.9802	3.9505	5.9011	1.0202	4.0301	0.2531	0.2481	1.4938	4
5	0.9754	4.9259	9.8026	1.0253	5.0503	0.2030	0.1980	1.9900	5
6	0.9705	5.8964	14.6552	1.0304	6.0755	0.1696	0.1646	2.4855	6
7	0.9657	6.8621	20.4493	1.0355	7.1059	0.1457	0.1407	2.9801	7
8	0.9609	7.8230	27.1755	1.0407	8.1414	0.1278	0.1228	3.4738	8
9	0.9561	8.7791	34.8244	1.0459	9.1821	0.1139	0.1089	3.9668	9
10	0.9513	9.7304	43.3865	1.0511	10.2280	0.1028	0.0978	4.4589	10
11	0.9466	10.6770	52.8526	1.0564	11.2792	0.0937	0.0887	4.9501	11
12	0.9419	11.6189	63.2136	1.0617	12.3356	0.0861	0.0811	5.4406	12
13	0.9372	12.5562	74.4602	1.0670	13.3972	0.0796	0.0746	5.9302	13
14	0.9326	13.4887	86.5835	1.0723	14.4642	0.0741	0.0691	6.4190	14
15	0.9279	14.4166	99.5743	1.0777	15.5365	0.0694	0.0644	6.9069	15
16	0.9233	15.3399	113.4238	1.0831	16.6142	0.0652	0.0602	7.3940	16
17	0.9187	16.2586	128.1231	1.0885	17.6973	0.0615	0.0565	7.8803	17
18	0.9141	17.1728	143.6634	1.0939	18.7858	0.0582	0.0532	8.3658	18
19	0.9096	18.0824	160.0360	1.0994	19.8797	0.0553	0.0503	8.8504	19
20	0.9051	18.9874	177.2322	1.1049	20.9791	0.0527	0.0477	9.3342	20
21	0.9006	19.8880	195.2434	1.1104	22.0840	0.0503	0.0453	9.8172	21
22	0.8961	20.7841	214.0611	1.1160	23.1944	0.0481	0.0431	10.2993	22
23	0.8916	21.6757	233.6768	1.1216	24.3104	0.0461	0.0411	10.7806	23
24	0.8872	22.5629	254.0820	1.1272	25.4320	0.0443	0.0393	11.2611	24
25	0.8828	23.4456	275.2686	1.1328	26.5591	0.0427	0.0377	11.7407	25
26	0.8784	24.3240	297.2281	1.1385	27.6919	0.0411	0.0361	12.2195	26
27	0.8740	25.1980	319.9523	1.1442	28.8304	0.0397	0.0347	12.6975	27
28	0.8697	26.0677	343.4332	1.1499	29.9745	0.0384	0.0334	13.1747	28
29	0.8653	26.9330	367.6625	1.1556	31.1244	0.0371	0.0321	13.6510	29
30	0.8610	27.7941	392.6324	1.1614	32.2800	0.0360	0.0310	14.1265	30
31	0.8567	28.6508	418.3348	1.1672	33.4414	0.0349	0.0299	14.6012	31
32	0.8525	29.5033	444.7618	1.1730	34.6086	0.0339	0.0289	15.0750	32
33	0.8482	30.3515	471.9055	1.1789	35.7817	0.0329	0.0279	15.5480	33
34	0.8440	31.1955	499.7583	1.1848	36.9606	0.0321	0.0271	16.0202	34
35	0.8398	32.0354	528.3123	1.1907	38.1454	0.0312	0.0262	16.4915	35
36	0.8356	32.8710	557.5598	1.1967	39.3361	0.0304	0.0254	16.9621	36
37	0.8315	33.7025	587.4934	1.2027	40.5328	0.0297	0.0247	17.4317	37
38	0.8274	34.5299	618.1054	1.2087	41.7354	0.0290	0.0240	17.9006	38
39	0.8232	35.3531	649.3883	1.2147	42.9441	0.0283	0.0233	18.3686	39
40	0.8191	36.1722	681.3347	1.2208	44.1588	0.0276	0.0226	18.8359	40
41	0.8151	36.9873	713.9372	1.2269	45.3796	0.0270	0.0220	19.3022	41
42	0.8110	37.7983	747.1886	1.2330	46.6065	0.0265	0.0215	19.7678	42
43	0.8070	38.6053	781.0815	1.2392	47.8396	0.0259	0.0209	20.2325	43
44	0.8030	39.4082	815.6087	1.2454	49.0788	0.0254	0.0204	20.6964	44
45	0.7990	40.2072	850.7631	1.2516	50.3242	0.0249	0.0199	21.1595	45
46	0.7950	41.0022	886.5376	1.2579	51.5758	0.0244	0.0194	21.6217	46
47	0.7910	41.7932	922.9252	1.2642	52.8337	0.0239	0.0189	22.0831	47
48	0.7871	42.5803	959.9188	1.2705	54.0978	0.0235	0.0185	22.5437	48
49	0.7832	43.3635	997.5116	1.2768	55.3683	0.0231	0.0181	23.0035	49
50	0.7793	44.1428	1035.6966	1.2832	56.6452	0.0227	0.0177	23.4624	50
51	0.7754	44.9182	1074.4670	1.2896	57.9284	0.0223	0.0173	23.9205	51
52	0.7716	45.6897	1113.8162	1.2961	59.2180	0.0219	0.0169	24.3778	52
53	0.7677	46.4575	1153.7372	1.3026	60.5141	0.0215	0.0165	24.8343	53
54	0.7639	47.2214	1194.2236	1.3091	61.8167	0.0212	0.0162	25.2899	54
55	0.7601	47.9814	1235.2686	1.3156	63.1258	0.0208	0.0158	25.7447	55
60	0.7414	51.7256	1448.6458	1.3489	69.7700	0.0193	0.0143	28.0064	60
65	0.7231	55.3775	1675.0272	1.3829	76.5821	0.0181	0.0131	30.2475	65
70	0.7053	58.9394	1913.6427	1.4178	83.5661	0.0170	0.0120	32.4680	70
75	0.6879	62.4136	2163.7525	1.4536	90.7265	0.0160	0.0110	34.6679	75
80	0.6710	65.8023	2424.6455	1.4903	98.0677	0.0152	0.0102	36.8474	80
85	0.6545	69.1075	2695.6389	1.5280	105.5943	0.0145	0.0095	39.0065	85
90	0.6383	72.3313	2976.0769	1.5666	113.3109	0.0138	0.0088	41.1451	90
95	0.6226	75.4757	3265.3298	1.6061	121.2224	0.0132	0.0082	43.2633	95
100	0.6073	78.5426	3562.7934	1.6467	129.3337	0.0127	0.0077	45.3613	100

(continued)

APPENDIX 86.B *(continued)*
Factor Tables

$I = 0.75\%$

n	P/F	P/A	P/G	F/P	F/A	A/P	A/F	A/G	n
1	0.9926	0.9926	0.0000	1.0075	1.0000	1.0075	1.0000	0.0000	1
2	0.9852	1.9777	0.9852	1.0151	2.0075	0.5056	0.4981	0.4981	2
3	0.9778	2.9556	2.9408	1.0227	3.0226	0.3383	0.3308	0.9950	3
4	0.9706	3.9261	5.8525	1.0303	4.0452	0.2547	0.2472	1.4907	4
5	0.9633	4.8894	9.7058	1.0381	5.0756	0.2045	0.1970	1.9851	5
6	0.9562	5.8456	14.4866	1.0459	6.1136	0.1711	0.1636	2.4782	6
7	0.9490	6.7946	20.1808	1.0537	7.1595	0.1472	0.1397	2.9701	7
8	0.9420	7.7366	26.7747	1.0616	8.2132	0.1293	0.1218	3.4608	8
9	0.9350	8.6716	34.2544	1.0696	9.2748	0.1153	0.1078	3.9502	9
10	0.9280	9.5996	42.6064	1.0776	10.3443	0.1042	0.0967	4.4384	10
11	0.9211	10.5207	51.8174	1.0857	11.4219	0.0951	0.0876	4.9253	11
12	0.9142	11.4349	61.8740	1.0938	12.5076	0.0875	0.0800	5.4110	12
13	0.9074	12.3423	72.7632	1.1020	13.6014	0.0810	0.0735	5.8954	13
14	0.9007	13.2430	84.4720	1.1103	14.7034	0.0755	0.0680	6.3786	14
15	0.8940	14.1370	96.9876	1.1186	15.8137	0.0707	0.0632	6.8606	15
16	0.8873	15.0243	110.2973	1.1270	16.9323	0.0666	0.0591	7.3413	16
17	0.8807	15.9050	124.3887	1.1354	18.0593	0.0629	0.0554	7.8207	17
18	0.8742	16.7792	139.2494	1.1440	19.1947	0.0596	0.0521	8.2989	18
19	0.8676	17.6468	154.8671	1.1525	20.3387	0.0567	0.0492	8.7759	19
20	0.8612	18.5080	171.2297	1.1612	21.4912	0.0540	0.0465	9.2516	20
21	0.8548	19.3628	188.3253	1.1699	22.6524	0.0516	0.0441	9.7261	21
22	0.8484	20.2112	206.1420	1.1787	23.8223	0.0495	0.0420	10.1994	22
23	0.8421	21.0533	224.6682	1.1875	25.0010	0.0475	0.0400	10.6714	23
24	0.8358	21.8891	243.8923	1.1964	26.1885	0.0457	0.0382	11.1422	24
25	0.8296	22.7188	263.8029	1.2054	27.3849	0.0440	0.0365	11.6117	25
26	0.8234	23.5422	284.3888	1.2144	28.5903	0.0425	0.0350	12.0800	26
27	0.8173	24.3595	305.6387	1.2235	29.8047	0.0411	0.0336	12.5470	27
28	0.8112	25.1707	327.5416	1.2327	31.0282	0.0397	0.0322	13.0128	28
29	0.8052	25.9759	350.0867	1.2420	32.2609	0.0385	0.0310	13.4774	29
30	0.7992	26.7751	373.2631	1.2513	33.5029	0.0373	0.0298	13.9407	30
31	0.7932	27.5683	397.0602	1.2607	34.7542	0.0363	0.0288	14.4028	31
32	0.7873	28.3557	421.4675	1.2701	36.0148	0.0353	0.0278	14.8636	32
33	0.7815	29.1371	446.4746	1.2796	37.2849	0.0343	0.0268	15.3232	33
34	0.7757	29.9128	472.0712	1.2892	38.5646	0.0334	0.0259	15.7816	34
35	0.7699	30.6827	498.2471	1.2989	39.8538	0.0326	0.0251	16.2387	35
36	0.7641	31.4468	524.9924	1.3086	41.1527	0.0318	0.0243	16.6946	36
37	0.7585	32.2053	552.2969	1.3185	42.4614	0.0311	0.0236	17.1493	37
38	0.7528	32.9581	580.1511	1.3283	43.7798	0.0303	0.0228	17.6027	38
39	0.7472	33.7053	608.5451	1.3383	45.1082	0.0297	0.0222	18.0549	39
40	0.7416	34.4469	637.4693	1.3483	46.4465	0.0290	0.0215	18.5058	40
41	0.7361	35.1831	666.9144	1.3585	47.7948	0.0284	0.0209	18.9556	41
42	0.7306	35.9137	696.8709	1.3686	49.1533	0.0278	0.0203	19.4040	42
43	0.7252	36.6389	727.3297	1.3789	50.5219	0.0273	0.0198	19.8513	43
44	0.7198	37.3587	758.2815	1.3893	51.9009	0.0268	0.0193	20.2973	44
45	0.7145	38.0732	789.7173	1.3997	53.2901	0.0263	0.0188	20.7421	45
46	0.7091	38.7823	821.6283	1.4102	54.6898	0.0258	0.0183	21.1856	46
47	0.7039	39.4862	854.0056	1.4207	56.1000	0.0253	0.0178	21.6280	47
48	0.6986	40.1848	886.8404	1.4314	57.5207	0.0249	0.0174	22.0691	48
49	0.6934	40.8782	920.1243	1.4421	58.9521	0.0245	0.0170	22.5089	49
50	0.6883	41.5664	953.8486	1.4530	60.3943	0.0241	0.0166	22.9476	50
51	0.6831	42.2496	988.0050	1.4639	61.8472	0.0237	0.0162	23.3850	51
52	0.6780	42.9276	1022.5852	1.4748	63.3111	0.0233	0.0158	23.8211	52
53	0.6730	43.6006	1057.5810	1.4859	64.7859	0.0229	0.0154	24.2561	53
54	0.6680	44.2686	1092.9842	1.4970	66.2718	0.0226	0.0151	24.6898	54
55	0.6630	44.9316	1128.7869	1.5083	67.7688	0.0223	0.0148	25.1223	55
60	0.6387	48.1734	1313.5189	1.5657	75.4241	0.0208	0.0133	27.2665	60
65	0.6153	51.2963	1507.0910	1.6253	83.3709	0.0195	0.0120	29.3801	65
70	0.5927	54.3046	1708.6065	1.6872	91.6201	0.0184	0.0109	31.4634	70
75	0.5710	57.2027	1917.2225	1.7514	100.1833	0.0175	0.0100	33.5163	75
80	0.5500	59.9944	2132.1472	1.8180	109.0725	0.0167	0.0092	35.5391	80
85	0.5299	62.6838	2352.6375	1.8873	118.3001	0.0160	0.0085	37.5318	85
90	0.5104	65.2746	2577.9961	1.9591	127.8790	0.0153	0.0078	39.4946	90
95	0.4917	67.7704	2807.5694	2.0337	137.8225	0.0148	0.0073	41.4277	95
100	0.4737	70.1746	3040.7453	2.1111	148.1445	0.0143	0.0068	43.3311	100

(continued)

APPENDIX 86.B *(continued)*
Factor Tables

$I = 1.00\%$

n	P/F	P/A	P/G	F/P	F/A	A/P	A/F	A/G	n
1	0.9901	0.9901	0.0000	1.0100	1.0000	1.0100	1.0000	0.0000	1
2	0.9803	1.9704	0.9803	1.0201	2.0100	0.5075	0.4975	0.4975	2
3	0.9706	2.9410	2.9215	1.0303	3.0301	0.3400	0.3300	0.9934	3
4	0.9610	3.9020	5.8044	1.0406	4.0604	0.2563	0.2463	1.4876	4
5	0.9515	4.8534	9.6103	1.0510	5.1010	0.2060	0.1960	1.9801	5
6	0.9420	5.7955	14.3205	1.0615	6.1520	0.1725	0.1625	2.4710	6
7	0.9327	6.7282	19.9168	1.0721	7.2135	0.1486	0.1386	2.9602	7
8	0.9235	7.6517	26.3812	1.0829	8.2857	0.1307	0.1207	3.4478	8
9	0.9143	8.5660	33.6959	1.0937	9.3685	0.1167	0.1067	3.9337	9
10	0.9053	9.4713	41.8435	1.1046	10.4622	0.1056	0.0956	4.4179	10
11	0.8963	10.3676	50.8067	1.1157	11.5668	0.0965	0.0865	4.9005	11
12	0.8874	11.2551	60.5687	1.1268	12.6825	0.0888	0.0788	5.3815	12
13	0.8787	12.1337	71.1126	1.1381	13.8093	0.0824	0.0724	5.8607	13
14	0.8700	13.0037	82.4221	1.1495	14.9474	0.0769	0.0669	6.3384	14
15	0.8613	13.8651	94.4810	1.1610	16.0969	0.0721	0.0621	6.8143	15
16	0.8528	14.7179	107.2734	1.1726	17.2579	0.0679	0.0579	7.2886	16
17	0.8444	15.5623	120.7834	1.1843	18.4304	0.0643	0.0543	7.7613	17
18	0.8360	16.3983	134.9957	1.1961	19.6147	0.0610	0.0510	8.2323	18
19	0.8277	17.2260	149.8950	1.2081	20.8109	0.0581	0.0481	8.7017	19
20	0.8195	18.0456	165.4664	1.2202	22.0190	0.0554	0.0454	9.1694	20
21	0.8114	18.8570	181.6950	1.2324	23.2392	0.0530	0.0430	9.6354	21
22	0.8034	19.6604	198.5663	1.2447	24.4716	0.0509	0.0409	10.0998	22
23	0.7954	20.4558	216.0660	1.2572	25.7163	0.0489	0.0389	10.5626	23
24	0.7876	21.2434	234.1800	1.2697	26.9735	0.0471	0.0371	11.0237	24
25	0.7798	22.0232	252.8945	1.2824	28.2432	0.0454	0.0354	11.4831	25
26	0.7720	22.7952	272.1957	1.2953	29.5256	0.0439	0.0339	11.9409	26
27	0.7644	23.5596	292.0702	1.3082	30.8209	0.0424	0.0324	12.3971	27
28	0.7568	24.3164	312.5047	1.3213	32.1291	0.0411	0.0311	12.8516	28
29	0.7493	25.0658	333.4863	1.3345	33.4504	0.0399	0.0299	13.3044	29
30	0.7419	25.8077	355.0021	1.3478	34.7849	0.0387	0.0287	13.7557	30
31	0.7346	26.5423	377.0394	1.3613	36.1327	0.0377	0.0277	14.2052	31
32	0.7273	27.2696	399.5858	1.3749	37.4941	0.0367	0.0267	14.6532	32
33	0.7201	27.9897	422.6291	1.3887	38.8690	0.0357	0.0257	15.0995	33
34	0.7130	28.7027	446.1572	1.4026	40.2577	0.0348	0.0248	15.5441	34
35	0.7059	29.4086	470.1583	1.4166	41.6603	0.0340	0.0240	15.9871	35
36	0.6989	30.1075	494.6207	1.4308	43.0769	0.0332	0.0232	16.4285	36
37	0.6920	30.7995	519.5329	1.4451	44.5076	0.0325	0.0225	16.8682	37
38	0.6852	31.4847	544.8835	1.4595	45.9527	0.0318	0.0218	17.3063	38
39	0.6784	32.1630	570.6616	1.4741	47.4123	0.0311	0.0211	17.7428	39
40	0.6717	32.8347	596.8561	1.4889	48.8864	0.0305	0.0205	18.1776	40
41	0.6650	33.4997	623.4562	1.5038	50.3752	0.0299	0.0199	18.6108	41
42	0.6584	34.1581	650.4514	1.5188	51.8790	0.0293	0.0193	19.0424	42
43	0.6519	34.8100	677.8312	1.5340	53.3978	0.0287	0.0187	19.4723	43
44	0.6454	35.4555	705.5853	1.5493	54.9318	0.0282	0.0182	19.9006	44
45	0.6391	36.0945	733.7037	1.5648	56.4811	0.0277	0.0177	20.3273	45
46	0.6327	36.7272	762.1765	1.5805	58.0459	0.0272	0.0172	20.7524	46
47	0.6265	37.3537	790.9938	1.5963	59.6263	0.0268	0.0168	21.1758	47
48	0.6203	37.9740	820.1460	1.6122	61.2226	0.0263	0.0163	21.5976	48
49	0.6141	38.5881	849.6237	1.6283	62.8348	0.0259	0.0159	22.0178	49
50	0.6080	39.1961	879.4176	1.6446	64.4632	0.0255	0.0155	22.4363	50
51	0.6020	39.7981	909.5186	1.6611	66.1078	0.0251	0.0151	22.8533	51
52	0.5961	40.3942	939.9175	1.6777	67.7689	0.0248	0.0148	23.2686	52
53	0.5902	40.9844	970.6057	1.6945	69.4466	0.0244	0.0144	23.6823	53
54	0.5843	41.5687	1001.5743	1.7114	71.1410	0.0241	0.0141	24.0945	54
55	0.5785	42.1472	1032.8148	1.7285	72.8525	0.0237	0.0137	24.5049	55
60	0.5504	44.9550	1192.8061	1.8167	81.6697	0.0222	0.0122	26.5333	60
65	0.5237	47.6266	1358.3903	1.9094	90.9366	0.0210	0.0110	28.5217	65
70	0.4983	50.1685	1528.6474	2.0068	100.6763	0.0199	0.0099	30.4703	70
75	0.4741	52.5871	1702.7340	2.1091	110.9128	0.0190	0.0090	32.3793	75
80	0.4511	54.8882	1879.8771	2.2167	121.6715	0.0182	0.0082	34.2492	80
85	0.4292	57.0777	2059.3701	2.3298	132.9790	0.0175	0.0075	36.0801	85
90	0.4084	59.1609	2240.5675	2.4486	144.8633	0.0169	0.0069	37.8724	90
95	0.3886	61.1430	2422.8811	2.5735	157.3538	0.0164	0.0064	39.6265	95
100	0.3697	63.0289	2605.7758	2.7048	170.4814	0.0159	0.0059	41.3426	100

(continued)

APPENDIX 86.B (continued)
Factor Tables
$$I = 1.50\%$$

n	P/F	P/A	P/G	F/P	F/A	A/P	A/F	A/G	n
1	0.9852	0.9852	0.0000	1.0150	1.0000	1.0150	1.0000	0.0000	1
2	0.9707	1.9559	0.9707	1.0302	2.0150	0.5113	0.4963	0.4963	2
3	0.9563	2.9122	2.8833	1.0457	3.0452	0.3434	0.3284	0.9901	3
4	0.9422	3.8544	5.7098	1.0614	4.0909	0.2594	0.2444	1.4814	4
5	0.9283	4.7826	9.4229	1.0773	5.1523	0.2091	0.1941	1.9702	5
6	0.9145	5.6972	13.9956	1.0934	6.2296	0.1755	0.1605	2.4566	6
7	0.9010	6.5982	19.4018	1.1098	7.3230	0.1516	0.1366	2.9405	7
8	0.8877	7.4859	25.6157	1.1265	8.4328	0.1336	0.1186	3.4219	8
9	0.8746	8.3605	32.6125	1.1434	9.5593	0.1196	0.1046	3.9008	9
10	0.8617	9.2222	40.3675	1.1605	10.7027	0.1084	0.0934	4.3772	10
11	0.8489	10.0711	48.8568	1.1779	11.8633	0.0993	0.0843	4.8512	11
12	0.8364	10.9075	58.0571	1.1956	13.0412	0.0917	0.0767	5.3227	12
13	0.8240	11.7315	67.9454	1.2136	14.2368	0.0852	0.0702	5.7917	13
14	0.8118	12.5434	78.4994	1.2318	15.4504	0.0797	0.0647	6.2582	14
15	0.7999	13.3432	89.6974	1.2502	16.6821	0.0749	0.0599	6.7223	15
16	0.7880	14.1313	101.5178	1.2690	17.9324	0.0708	0.0558	7.1839	16
17	0.7764	14.9076	113.9400	1.2880	19.2014	0.0671	0.0521	7.6431	17
18	0.7649	15.6726	126.9435	1.3073	20.4894	0.0638	0.0488	8.0997	18
19	0.7536	16.4262	140.5084	1.3270	21.7967	0.0609	0.0459	8.5539	19
20	0.7425	17.1686	154.6154	1.3469	23.1237	0.0582	0.0432	9.0057	20
21	0.7315	17.9001	169.2453	1.3671	24.4705	0.0559	0.0409	9.4550	21
22	0.7207	18.6208	184.3798	1.3876	25.8376	0.0537	0.0387	9.9018	22
23	0.7100	19.3309	200.0006	1.4084	27.2251	0.0517	0.0367	10.3462	23
24	0.6995	20.0304	216.0901	1.4295	28.6335	0.0499	0.0349	10.7881	24
25	0.6892	20.7196	232.6310	1.4509	30.0630	0.0483	0.0333	11.2276	25
26	0.6790	21.3986	249.6065	1.4727	31.5140	0.0467	0.0317	11.6646	26
27	0.6690	22.0676	267.0002	1.4948	32.9867	0.0453	0.0303	12.0992	27
28	0.6591	22.7267	284.7958	1.5172	34.4815	0.0440	0.0290	12.5313	28
29	0.6494	23.3761	302.9779	1.5400	35.9987	0.0428	0.0278	12.9610	29
30	0.6398	24.0158	321.5310	1.5631	37.5387	0.0416	0.0266	13.3883	30
31	0.6303	24.6461	340.4402	1.5865	39.1018	0.0406	0.0256	13.8131	31
32	0.6210	25.2671	359.6910	1.6103	40.6883	0.0396	0.0246	14.2355	32
33	0.6118	25.8790	379.2691	1.6345	42.2986	0.0386	0.0236	14.6555	33
34	0.6028	26.4817	399.1607	1.6590	43.9331	0.0378	0.0228	15.0731	34
35	0.5939	27.0756	419.3521	1.6839	45.5921	0.0369	0.0219	15.4882	35
36	0.5851	27.6607	439.8303	1.7091	47.2760	0.0362	0.0212	15.9009	36
37	0.5764	28.2371	460.5822	1.7348	48.9851	0.0354	0.0204	16.3112	37
38	0.5679	28.8051	481.5954	1.7608	50.7199	0.0347	0.0197	16.7191	38
39	0.5595	29.3646	502.8576	1.7872	52.4807	0.0341	0.0191	17.1246	39
40	0.5513	29.9158	524.3568	1.8140	54.2679	0.0334	0.0184	17.5277	40
41	0.5431	30.4590	546.0814	1.8412	56.0819	0.0328	0.0178	17.9284	41
42	0.5351	30.9941	568.0201	1.8688	57.9231	0.0323	0.0173	18.3267	42
43	0.5272	31.5212	590.1617	1.8969	59.7920	0.0317	0.0167	18.7227	43
44	0.5194	32.0406	612.4955	1.9253	61.6889	0.0312	0.0162	19.1162	44
45	0.5117	32.5523	635.0110	1.9542	63.6142	0.0307	0.0157	19.5074	45
46	0.5042	33.0565	657.6979	1.9835	65.5684	0.0303	0.0153	19.8962	46
47	0.4967	33.5532	680.5462	2.0133	67.5519	0.0298	0.0148	20.2826	47
48	0.4894	34.0426	703.5462	2.0435	69.5652	0.0294	0.0144	20.6667	48
49	0.4821	34.5247	726.6884	2.0741	71.6087	0.0290	0.0140	21.0484	49
50	0.4750	34.9997	749.9636	2.1052	73.6828	0.0286	0.0136	21.4277	50
51	0.4680	35.4677	773.3629	2.1368	75.7881	0.0282	0.0132	21.8047	51
52	0.4611	35.9287	796.8774	2.1689	77.9249	0.0278	0.0128	22.1794	52
53	0.4543	36.3830	820.4986	2.2014	80.0938	0.0275	0.0125	22.5517	53
54	0.4475	36.8305	844.2184	2.2344	82.2952	0.0272	0.0122	22.9217	54
55	0.4409	37.2715	868.0285	2.2679	84.5296	0.0268	0.0118	23.2894	55
60	0.4093	39.3803	988.1674	2.4432	96.2147	0.0254	0.0104	25.0930	60
65	0.3799	41.3378	1109.4752	2.6320	108.8028	0.0242	0.0092	26.8393	65
70	0.3527	43.1549	1231.1658	2.8355	122.3638	0.0232	0.0082	28.5290	70
75	0.3274	44.8416	1352.5600	3.0546	136.9728	0.0223	0.0073	30.1631	75
80	0.3039	46.4073	1473.0741	3.2907	152.7109	0.0215	0.0065	31.7423	80
85	0.2821	47.8607	1592.2095	3.5450	169.6652	0.0209	0.0059	33.2676	85
90	0.2619	49.2099	1709.5439	3.8189	187.9299	0.0203	0.0053	34.7399	90
95	0.2431	50.4622	1824.7224	4.1141	207.6061	0.0198	0.0048	36.1602	95
100	0.2256	51.6247	1937.4506	4.4320	228.8030	0.0194	0.0044	37.5295	100

(continued)

APPENDIX 86.B *(continued)*
Factor Tables

$I = 2.00\%$

n	P/F	P/A	P/G	F/P	F/A	A/P	A/F	A/G	n
1	0.9804	0.9804	0.0000	1.0200	1.0000	1.0200	1.0000	0.0000	1
2	0.9612	1.9416	0.9612	1.0404	2.0200	0.5150	0.4950	0.4950	2
3	0.9423	2.8839	2.8458	1.0612	3.0604	0.3468	0.3268	0.9868	3
4	0.9238	3.8077	5.6173	1.0824	4.1216	0.2626	0.2426	1.4752	4
5	0.9057	4.7135	9.2403	1.1041	5.2040	0.2122	0.1922	1.9604	5
6	0.8880	5.6014	13.6801	1.1262	6.3081	0.1785	0.1585	2.4423	6
7	0.8706	6.4720	18.9035	1.1487	7.4343	0.1545	0.1345	2.9208	7
8	0.8535	7.3255	24.8779	1.1717	8.5830	0.1365	0.1165	3.3961	8
9	0.8368	8.1622	31.5720	1.1951	9.7546	0.1225	0.1025	3.8681	9
10	0.8203	8.9826	38.9551	1.2190	10.9497	0.1113	0.0913	4.3367	10
11	0.8043	9.7868	46.9977	1.2434	12.1687	0.1022	0.0822	4.8021	11
12	0.7885	10.5753	55.6712	1.2682	13.4121	0.0946	0.0746	5.2642	12
13	0.7730	11.3484	64.9475	1.2936	14.6803	0.0881	0.0681	5.7231	13
14	0.7579	12.1062	74.7999	1.3195	15.9739	0.0826	0.0626	6.1786	14
15	0.7430	12.8493	85.2021	1.3459	17.2934	0.0778	0.0578	6.6309	15
16	0.7284	13.5777	96.1288	1.3728	18.6393	0.0737	0.0537	7.0799	16
17	0.7142	14.2919	107.5554	1.4002	20.0121	0.0700	0.0500	7.5256	17
18	0.7002	14.9920	119.4581	1.4282	21.4123	0.0667	0.0467	7.9681	18
19	0.6864	15.6785	131.8139	1.4568	22.8406	0.0638	0.0438	8.4073	19
20	0.6730	16.3514	144.6003	1.4859	24.2974	0.0612	0.0412	8.8433	20
21	0.6598	17.0112	157.7959	1.5157	25.7833	0.0588	0.0388	9.2760	21
22	0.6468	17.6580	171.3795	1.5460	27.2990	0.0566	0.0366	9.7055	22
23	0.6342	18.2922	185.3309	1.5769	28.8450	0.0547	0.0347	10.1317	23
24	0.6217	18.9139	199.6305	1.6084	30.4219	0.0529	0.0329	10.5547	24
25	0.6095	19.5235	214.2592	1.6406	32.0303	0.0512	0.0312	10.9745	25
26	0.5976	20.1210	229.1987	1.6734	33.6709	0.0497	0.0297	11.3910	26
27	0.5859	20.7069	244.4311	1.7069	35.3443	0.0483	0.0283	11.8043	27
28	0.5744	21.2813	259.9392	1.7410	37.0512	0.0470	0.0270	12.2145	28
29	0.5631	21.8444	275.7064	1.7758	38.7922	0.0458	0.0258	12.6214	29
30	0.5521	22.3965	291.7164	1.8114	40.5681	0.0446	0.0246	13.0251	30
31	0.5412	22.9377	307.9538	1.8476	42.3794	0.0436	0.0236	13.4257	31
32	0.5306	23.4683	324.4035	1.8845	44.2270	0.0426	0.0226	13.8230	32
33	0.5202	23.9886	341.0508	1.9222	46.1116	0.0417	0.0217	14.2172	33
34	0.5100	24.4986	357.8817	1.9607	48.0338	0.0408	0.0208	14.6083	34
35	0.5000	24.9986	374.8826	1.9999	49.9945	0.0400	0.0200	14.9961	35
36	0.4902	25.4888	392.0405	2.0399	51.9944	0.0392	0.0192	15.3809	36
37	0.4806	25.9695	409.3424	2.0807	54.0343	0.0385	0.0185	15.7625	37
38	0.4712	26.4406	426.7764	2.1223	56.1149	0.0378	0.0178	16.1409	38
39	0.4619	26.9026	444.3304	2.1647	58.2372	0.0372	0.0172	16.5163	39
40	0.4529	27.3555	461.9931	2.2080	60.4020	0.0366	0.0166	16.8885	40
41	0.4440	27.7995	479.7535	2.2522	62.6100	0.0360	0.0160	17.2576	41
42	0.4353	28.2348	497.6010	2.2972	64.8622	0.0354	0.0154	17.6237	42
43	0.4268	28.6616	515.5253	2.3432	67.1595	0.0349	0.0149	17.9866	43
44	0.4184	29.0800	533.5165	2.3901	69.5027	0.0344	0.0144	18.3465	44
45	0.4102	29.4902	551.5652	2.4379	71.8927	0.0339	0.0139	18.7034	45
46	0.4022	29.8923	569.6621	2.4866	74.3306	0.0335	0.0135	19.0571	46
47	0.3943	30.2866	587.7985	2.5363	76.8172	0.0330	0.0130	19.4079	47
48	0.3865	30.6731	605.9657	2.5871	79.3535	0.0326	0.0126	19.7556	48
49	0.3790	31.0521	624.1557	2.6388	81.9406	0.0322	0.0122	20.1003	49
50	0.3715	31.4236	642.3606	2.6916	84.5794	0.0318	0.0118	20.4420	50
51	0.3642	31.7878	660.5727	2.7454	87.2710	0.0315	0.0115	20.7807	51
52	0.3571	32.1449	678.7849	2.8003	90.0164	0.0311	0.0111	21.1164	52
53	0.3501	32.4950	696.9900	2.8563	92.8167	0.0308	0.0108	21.4491	53
54	0.3432	32.8383	715.1815	2.9135	95.6731	0.0305	0.0105	21.7789	54
55	0.3365	33.1748	733.3527	2.9717	98.5865	0.0301	0.0101	22.1057	55
60	0.3048	34.7609	823.6975	3.2810	114.0515	0.0288	0.0088	23.6961	60
65	0.2761	36.1975	912.7085	3.6225	131.1262	0.0276	0.0076	25.2147	65
70	0.2500	37.4986	999.8343	3.9996	149.9779	0.0267	0.0067	26.6632	70
75	0.2265	38.6771	1084.6393	4.4158	170.7918	0.0259	0.0059	28.0434	75
80	0.2051	39.7445	1166.7868	4.8754	193.7720	0.0252	0.0052	29.3572	80
85	0.1858	40.7113	1246.0241	5.3829	219.1439	0.0246	0.0046	30.6064	85
90	0.1683	41.5869	1322.1701	5.9431	247.1567	0.0240	0.0040	31.7929	90
95	0.1524	42.3800	1395.1033	6.5617	278.0850	0.0236	0.0036	32.9189	95
100	0.1380	43.0984	1464.7527	7.2446	312.2323	0.0232	0.0032	33.9863	100

(continued)

APPENDIX 86.B *(continued)*
Factor Tables

$I = 3.00\%$

n	P/F	P/A	P/G	F/P	F/A	A/P	A/F	A/G	n
1	0.9709	0.9709	0.0000	1.0300	1.0000	1.0300	1.0000	0.0000	1
2	0.9426	1.9135	0.9426	1.0609	2.0300	0.5226	0.4926	0.4926	2
3	0.9151	2.8286	2.7729	1.0927	3.0909	0.3535	0.3235	0.9803	3
4	0.8885	3.7171	5.4383	1.1255	4.1836	0.2690	0.2390	1.4631	4
5	0.8626	4.5797	8.8888	1.1593	5.3091	0.2184	0.1884	1.9409	5
6	0.8375	5.4172	13.0762	1.1941	6.4684	0.1846	0.1546	2.4138	6
7	0.8131	6.2303	17.9547	1.2299	7.6625	0.1605	0.1305	2.8819	7
8	0.7894	7.0197	23.4806	1.2668	8.8923	0.1425	0.1125	3.3450	8
9	0.7664	7.7861	29.6119	1.3048	10.1591	0.1284	0.0984	3.8032	9
10	0.7441	8.5302	36.3088	1.3439	11.4639	0.1172	0.0872	4.2565	10
11	0.7224	9.2526	43.5330	1.3842	12.8078	0.1081	0.0781	4.7049	11
12	0.7014	9.9540	51.2482	1.4258	14.1920	0.1005	0.0705	5.1485	12
13	0.6810	10.6350	59.4196	1.4685	15.6178	0.0940	0.0640	5.5872	13
14	0.6611	11.2961	68.0141	1.5126	17.0863	0.0885	0.0585	6.0210	14
15	0.6419	11.9379	77.0002	1.5580	18.5989	0.0838	0.0538	6.4500	15
16	0.6232	12.5611	86.3477	1.6047	20.1569	0.0796	0.0496	6.8742	16
17	0.6050	13.1661	96.0280	1.6528	21.7616	0.0760	0.0460	7.2936	17
18	0.5874	13.7535	106.0137	1.7024	23.4144	0.0727	0.0427	7.7081	18
19	0.5703	14.3238	116.2788	1.7535	25.1169	0.0698	0.0398	8.1179	19
20	0.5537	14.8775	126.7987	1.8061	26.8704	0.0672	0.0372	8.5229	20
21	0.5375	15.4150	137.5496	1.8603	28.6765	0.0649	0.0349	8.9231	21
22	0.5219	15.9369	148.5094	1.9161	30.5368	0.0627	0.0327	9.3186	22
23	0.5067	16.4436	159.6566	1.9736	32.4529	0.0608	0.0308	9.7093	23
24	0.4919	16.9355	170.9711	2.0328	34.4265	0.0590	0.0290	10.0954	24
25	0.4776	17.4131	182.4336	2.0938	36.4593	0.0574	0.0274	10.4768	25
26	0.4637	17.8768	194.0260	2.1566	38.5530	0.0559	0.0259	10.8535	26
27	0.4502	18.3270	205.7309	2.2213	40.7096	0.0546	0.0246	11.2255	27
28	0.4371	18.7641	217.5320	2.2879	42.9309	0.0533	0.0233	11.5930	28
29	0.4243	19.1885	229.4137	2.3566	45.2189	0.0521	0.0221	11.9558	29
30	0.4120	19.6004	241.3613	2.4273	47.5754	0.0510	0.0210	12.3141	30
31	0.4000	20.0004	253.3609	2.5001	50.0027	0.0500	0.0200	12.6678	31
32	0.3883	20.3888	265.3993	2.5751	52.5028	0.0490	0.0190	13.0169	32
33	0.3770	20.7658	277.4642	2.6523	55.0778	0.0482	0.0182	13.3616	33
34	0.3660	21.1318	289.5437	2.7319	57.7302	0.0473	0.0173	13.7018	34
35	0.3554	21.4872	301.6267	2.8139	60.4621	0.0465	0.0165	14.0375	35
36	0.3450	21.8323	313.7028	2.8983	63.2759	0.0458	0.0158	14.3688	36
37	0.3350	22.1672	325.7622	2.9852	66.1742	0.0451	0.0151	14.6957	37
38	0.3252	22.4925	337.7956	3.0748	69.1594	0.0445	0.0145	15.0182	38
39	0.3158	22.8082	349.7942	3.1670	72.2342	0.0438	0.0138	15.3363	39
40	0.3066	23.1148	361.7499	3.2620	75.4013	0.0433	0.0133	15.6502	40
41	0.2976	23.4124	373.6551	3.3599	78.6633	0.0427	0.0127	15.9597	41
42	0.2890	23.7014	385.5024	3.4607	82.0232	0.0422	0.0122	16.2650	42
43	0.2805	23.9819	397.2852	3.5645	85.4839	0.0417	0.0117	16.5660	43
44	0.2724	24.2543	408.9972	3.6715	89.0484	0.0412	0.0112	16.8629	44
45	0.2644	24.5187	420.6325	3.7816	92.7199	0.0408	0.0108	17.1556	45
46	0.2567	24.7754	432.1856	3.8950	96.5015	0.0404	0.0104	17.4441	46
47	0.2493	25.0247	443.6515	4.0119	100.3965	0.0400	0.0100	17.7285	47
48	0.2420	25.2667	455.0255	4.1323	104.4084	0.0396	0.0096	18.0089	48
49	0.2350	25.5017	466.3031	4.2562	108.5406	0.0392	0.0092	18.2852	49
50	0.2281	25.7298	477.4803	4.3839	112.7969	0.0389	0.0089	18.5575	50
51	0.2215	25.9512	488.5535	4.5154	117.1808	0.0385	0.0085	18.8258	51
52	0.2150	26.1662	499.5191	4.6509	121.6962	0.0382	0.0082	19.0902	52
53	0.2088	26.3750	510.3742	4.7904	126.3471	0.0379	0.0079	19.3507	53
54	0.2027	26.5777	521.1157	4.9341	131.1375	0.0376	0.0076	19.6073	54
55	0.1968	26.7744	531.7411	5.0821	136.0716	0.0373	0.0073	19.8600	55
60	0.1697	27.6756	583.0526	5.8916	163.0534	0.0361	0.0061	21.0674	60
65	0.1464	28.4529	631.2010	6.8300	194.3328	0.0351	0.0051	22.1841	65
70	0.1263	29.1234	676.0869	7.9178	230.5941	0.0343	0.0043	23.2145	70
75	0.1089	29.7018	717.6978	9.1789	272.6309	0.0337	0.0037	24.1634	75
80	0.0940	30.2008	756.0865	10.6409	321.3630	0.0331	0.0031	25.0353	80
85	0.0811	30.6312	791.3529	12.3357	377.8570	0.0326	0.0026	25.8349	85
90	0.0699	31.0024	823.6302	14.3005	443.3489	0.0323	0.0023	26.5667	90
95	0.0603	31.3227	853.0742	16.5782	519.2720	0.0319	0.0019	27.2351	95
100	0.0520	31.5989	879.8540	19.2186	607.2877	0.0316	0.0016	27.8444	100

(continued)

APPENDIX 86.B *(continued)*
Factor Tables

$I = 4.00\%$

n	P/F	P/A	P/G	F/P	F/A	A/P	A/F	A/G	n
1	0.9615	0.9615	0.0000	1.0400	1.0000	1.0400	1.0000	0.0000	1
2	0.9246	1.8861	0.9246	1.0816	2.0400	0.5302	0.4902	0.4902	2
3	0.8890	2.7751	2.7025	1.1249	3.1216	0.3603	0.3203	0.9739	3
4	0.8548	3.6299	5.2670	1.1699	4.2465	0.2755	0.2355	1.4510	4
5	0.8219	4.4518	8.5547	1.2167	5.4163	0.2246	0.1846	1.9216	5
6	0.7903	5.2421	12.5062	1.2653	6.6330	0.1908	0.1508	2.3857	6
7	0.7599	6.0021	17.0657	1.3159	7.8983	0.1666	0.1266	2.8433	7
8	0.7307	6.7327	22.1806	1.3686	9.2142	0.1485	0.1085	3.2944	8
9	0.7026	7.4353	27.8013	1.4233	10.5828	0.1345	0.0945	3.7391	9
10	0.6756	8.1109	33.8814	1.4802	12.0061	0.1233	0.0833	4.1773	10
11	0.6496	8.7605	40.3772	1.5395	13.4864	0.1141	0.0741	4.6090	11
12	0.6246	9.3851	47.2477	1.6010	15.0258	0.1066	0.0666	5.0343	12
13	0.6006	9.9856	54.4546	1.6651	16.6268	0.1001	0.0601	5.4533	13
14	0.5775	10.5631	61.9618	1.7317	18.2919	0.0947	0.0547	5.8659	14
15	0.5553	11.1184	69.7355	1.8009	20.0236	0.0899	0.0499	6.2721	15
16	0.5339	11.6523	77.7441	1.8730	21.8245	0.0858	0.0458	6.6720	16
17	0.5134	12.1657	85.9581	1.9479	23.6975	0.0822	0.0422	7.0656	17
18	0.4936	12.6593	94.3498	2.0258	25.6454	0.0790	0.0390	7.4530	18
19	0.4746	13.1339	102.8933	2.1068	27.6712	0.0761	0.0361	7.8342	19
20	0.4564	13.5903	111.5647	2.1911	29.7781	0.0736	0.0336	8.2091	20
21	0.4388	14.0292	120.3414	2.2788	31.9692	0.0713	0.0313	8.5779	21
22	0.4220	14.4511	129.2024	2.3699	34.2480	0.0692	0.0292	8.9407	22
23	0.4057	14.8568	138.1284	2.4647	36.6179	0.0673	0.0273	9.2973	23
24	0.3901	15.2470	147.1012	2.5633	39.0826	0.0656	0.0256	9.6479	24
25	0.3751	15.6221	156.1040	2.6658	41.6459	0.0640	0.0240	9.9925	25
26	0.3607	15.9828	165.1212	2.7725	44.3117	0.0626	0.0226	10.3312	26
27	0.3468	16.3296	174.1385	2.8834	47.0842	0.0612	0.0212	10.6640	27
28	0.3335	16.6631	183.1424	2.9987	49.9676	0.0600	0.0200	10.9909	28
29	0.3207	16.9837	192.1206	3.1187	52.9663	0.0589	0.0189	11.3120	29
30	0.3083	17.2920	201.0618	3.2434	56.0849	0.0578	0.0178	11.6274	30
31	0.2965	17.5885	209.9556	3.3731	59.3283	0.0569	0.0169	11.9371	31
32	0.2851	17.8736	218.7924	3.5081	62.7015	0.0559	0.0159	12.2411	32
33	0.2741	18.1476	227.5634	3.6484	66.2095	0.0551	0.0151	12.5396	33
34	0.2636	18.4112	236.2607	3.7943	69.8579	0.0543	0.0143	12.8324	34
35	0.2534	18.6646	244.8768	3.9461	73.6522	0.0536	0.0136	13.1198	35
36	0.2437	18.9083	253.4052	4.1039	77.5983	0.0529	0.0129	13.4018	36
37	0.2343	19.1426	261.8399	4.2681	81.7022	0.0522	0.0122	13.6784	37
38	0.2253	19.3679	270.1754	4.4388	85.9703	0.0516	0.0116	13.9497	38
39	0.2166	19.5845	278.4070	4.6164	90.4091	0.0511	0.0111	14.2157	39
40	0.2083	19.7928	286.5303	4.8010	95.0255	0.0505	0.0105	14.4765	40
41	0.2003	19.9931	294.5414	4.9931	99.8265	0.0500	0.0100	14.7322	41
42	0.1926	20.1856	302.4370	5.1928	104.8196	0.0495	0.0095	14.9828	42
43	0.1852	20.3708	310.2141	5.4005	110.0124	0.0491	0.0091	15.2284	43
44	0.1780	20.5488	317.8700	5.6165	115.4129	0.0487	0.0087	15.4690	44
45	0.1712	20.7200	325.4028	5.8412	121.0294	0.0483	0.0083	15.7047	45
46	0.1646	20.8847	332.8104	6.0748	126.8706	0.0479	0.0079	15.9356	46
47	0.1583	21.0429	340.0914	6.3178	132.9454	0.0475	0.0075	16.1618	47
48	0.1522	21.1951	347.2446	6.5705	139.2632	0.0472	0.0072	16.3832	48
49	0.1463	21.3415	354.2689	6.8333	145.8337	0.0469	0.0069	16.6000	49
50	0.1407	21.4822	361.1638	7.1067	152.6671	0.0466	0.0066	16.8122	50
51	0.1353	21.6175	367.9289	7.3910	159.7738	0.0463	0.0063	17.0200	51
52	0.1301	21.7476	374.5638	7.6866	167.1647	0.0460	0.0060	17.2232	52
53	0.1251	21.8727	381.0686	7.9941	174.8513	0.0457	0.0057	17.4221	53
54	0.1203	21.9930	387.4436	8.3138	182.8454	0.0455	0.0055	17.6167	54
55	0.1157	22.1086	393.6890	8.6464	191.1592	0.0452	0.0052	17.8070	55
60	0.0951	22.6235	422.9966	10.5196	237.9907	0.0442	0.0042	18.6972	60
65	0.0781	23.0467	449.2014	12.7987	294.9684	0.0434	0.0034	19.4909	65
70	0.0642	23.3945	472.4789	15.5716	364.2905	0.0427	0.0027	20.1961	70
75	0.0528	23.6804	493.0408	18.9453	448.6314	0.0422	0.0022	20.8206	75
80	0.0434	23.9154	511.1161	23.0498	551.2450	0.0418	0.0018	21.3718	80
85	0.0357	24.1085	526.9384	28.0436	676.0901	0.0415	0.0015	21.8569	85
90	0.0293	24.2673	540.7369	34.1193	827.9833	0.0412	0.0012	22.2826	90
95	0.0241	24.3978	552.7307	41.5114	1012.7846	0.0410	0.0010	22.6550	95
100	0.0198	24.5050	563.1249	50.5049	1237.6237	0.0408	0.0008	22.9800	100

(continued)

APPENDIX 86.B (continued)
Factor Tables
$$I = 5.00\%$$

n	P/F	P/A	P/G	F/P	F/A	A/P	A/F	A/G	n
1	0.9524	0.9524	0.0000	1.0500	1.0000	1.0500	1.0000	0.0000	1
2	0.9070	1.8594	0.9070	1.1025	2.0500	0.5378	0.4878	0.4878	2
3	0.8638	2.7232	2.6347	1.1576	3.1525	0.3672	0.3172	0.9675	3
4	0.8227	3.5460	5.1028	1.2155	4.3101	0.2820	0.2320	1.4391	4
5	0.7835	4.3295	8.2369	1.2763	5.5256	0.2310	0.1810	1.9025	5
6	0.7462	5.0757	11.9680	1.3401	6.8019	0.1970	0.1470	2.3579	6
7	0.7107	5.7864	16.2321	1.4071	8.1420	0.1728	0.1228	2.8052	7
8	0.6768	6.4632	20.9700	1.4775	9.5491	0.1547	0.1047	3.2445	8
9	0.6446	7.1078	26.1268	1.5513	11.0266	0.1407	0.0907	3.6758	9
10	0.6139	7.7217	31.6520	1.6289	12.5779	0.1295	0.0795	4.0991	10
11	0.5847	8.3064	37.4988	1.7103	14.2068	0.1204	0.0704	4.5144	11
12	0.5568	8.8633	43.6241	1.7959	15.9171	0.1128	0.0628	4.9219	12
13	0.5303	9.3936	49.9879	1.8856	17.7130	0.1065	0.0565	5.3215	13
14	0.5051	9.8986	56.5538	1.9799	19.5986	0.1010	0.0510	5.7133	14
15	0.4810	10.3797	63.2880	2.0789	21.5786	0.0963	0.0463	6.0973	15
16	0.4581	10.8378	70.1597	2.1829	23.6575	0.0923	0.0423	6.4736	16
17	0.4363	11.2741	77.1405	2.2920	25.8404	0.0887	0.0387	6.8423	17
18	0.4155	11.6896	84.2043	2.4066	28.1324	0.0855	0.0355	7.2034	18
19	0.3957	12.0853	91.3275	2.5270	30.5390	0.0827	0.0327	7.5569	19
20	0.3769	12.4622	98.4884	2.6533	33.0660	0.0802	0.0302	7.9030	20
21	0.3589	12.8212	105.6673	2.7860	35.7193	0.0780	0.0280	8.2416	21
22	0.3418	13.1630	112.8461	2.9253	38.5052	0.0760	0.0260	8.5730	22
23	0.3256	13.4886	120.0087	3.0715	41.4305	0.0741	0.0241	8.8971	23
24	0.3101	13.7986	127.1402	3.2251	44.5020	0.0725	0.0225	9.2140	24
25	0.2953	14.0939	134.2275	3.3864	47.7271	0.0710	0.0210	9.5238	25
26	0.2812	14.3752	141.2585	3.5557	51.1135	0.0696	0.0196	9.8266	26
27	0.2678	14.6430	148.2226	3.7335	54.6691	0.0683	0.0183	10.1224	27
28	0.2551	14.8981	155.1101	3.9201	58.4026	0.0671	0.0171	10.4114	28
29	0.2429	15.1411	161.9126	4.1161	62.3227	0.0660	0.0160	10.6936	29
30	0.2314	15.3725	168.6226	4.3219	66.4388	0.0651	0.0151	10.9691	30
31	0.2204	15.5928	175.2333	4.5380	70.7608	0.0641	0.0141	11.2381	31
32	0.2099	15.8027	181.7392	4.7649	75.2988	0.0633	0.0133	11.5005	32
33	0.1999	16.0025	188.1351	5.0032	80.0638	0.0625	0.0125	11.7566	33
34	0.1904	16.1929	194.4168	5.2533	85.0670	0.0618	0.0118	12.0063	34
35	0.1813	16.3742	200.5807	5.5160	90.3203	0.0611	0.0111	12.2498	35
36	0.1727	16.5469	206.6237	5.7918	95.8363	0.0604	0.0104	12.4872	36
37	0.1644	16.7113	212.5434	6.0814	101.6281	0.0598	0.0098	12.7186	37
38	0.1566	16.8679	218.3378	6.3855	107.7095	0.0593	0.0093	12.9440	38
39	0.1491	17.0170	224.0054	6.7048	114.0950	0.0588	0.0088	13.1636	39
40	0.1420	17.1591	229.5452	7.0400	120.7998	0.0583	0.0083	13.3775	40
41	0.1353	17.2944	234.9564	7.3920	127.8398	0.0578	0.0078	13.5857	41
42	0.1288	17.4232	240.2389	7.7616	135.2318	0.0574	0.0074	13.7884	42
43	0.1227	17.5459	245.3925	8.1497	142.9933	0.0570	0.0070	13.9857	43
44	0.1169	17.6628	250.4175	8.5572	151.1430	0.0566	0.0066	14.1777	44
45	0.1113	17.7741	255.3145	8.9850	159.7002	0.0563	0.0063	14.3644	45
46	0.1060	17.8801	260.0844	9.4343	168.6852	0.0559	0.0059	14.5461	46
47	0.1009	17.9810	264.7281	9.9060	178.1194	0.0556	0.0056	14.7226	47
48	0.0961	18.0772	269.2467	10.4013	188.0254	0.0553	0.0053	14.8943	48
49	0.0916	18.1687	273.6418	10.9213	198.4267	0.0550	0.0050	15.0611	49
50	0.0872	18.2559	277.9148	11.4674	209.3480	0.0548	0.0048	15.2233	50
51	0.0831	18.3390	282.0673	12.0408	220.8154	0.0545	0.0045	15.3808	51
52	0.0791	18.4181	286.1013	12.6428	232.8562	0.0543	0.0043	15.5337	52
53	0.0753	18.4934	290.0184	13.2749	245.4990	0.0541	0.0041	15.6823	53
54	0.0717	18.5651	293.8208	13.9387	258.7739	0.0539	0.0039	15.8265	54
55	0.0683	18.6335	297.5104	14.6356	272.7126	0.0537	0.0037	15.9664	55
60	0.0535	18.9293	314.3432	18.6792	353.5837	0.0528	0.0028	16.6062	60
65	0.0419	19.1611	328.6910	23.8399	456.7980	0.0522	0.0022	17.1541	65
70	0.0329	19.3427	340.8409	30.4264	588.5285	0.0517	0.0017	17.6212	70
75	0.0258	19.4850	351.0721	38.8327	756.6537	0.0513	0.0013	18.0176	75
80	0.0202	19.5965	359.6460	49.5614	971.2288	0.0510	0.0010	18.3526	80
85	0.0158	19.6838	366.8007	63.2544	1245.0871	0.0508	0.0008	18.6346	85
90	0.0124	19.7523	372.7488	80.7304	1597.6073	0.0506	0.0006	18.8712	90
95	0.0097	19.8059	377.6774	103.0347	2040.6935	0.0505	0.0005	19.0689	95
100	0.0076	19.8479	381.7492	131.5013	2610.0252	0.0504	0.0004	19.2337	100

(continued)

APPENDIX 86.B *(continued)*
Factor Tables

$I = 6.00\%$

n	P/F	P/A	P/G	F/P	F/A	A/P	A/F	A/G	n
1	0.9434	0.9434	0.0000	1.0600	1.0000	1.0600	1.0000	0.0000	1
2	0.8900	1.8334	0.8900	1.1236	2.0600	0.5454	0.4854	0.4854	2
3	0.8396	2.6730	2.5692	1.1910	3.1836	0.3741	0.3141	0.9612	3
4	0.7921	3.4651	4.9455	1.2625	4.3746	0.2886	0.2286	1.4272	4
5	0.7473	4.2124	7.9345	1.3382	5.6371	0.2374	0.1774	1.8836	5
6	0.7050	4.9173	11.4594	1.4185	6.9753	0.2034	0.1434	2.3304	6
7	0.6651	5.5824	15.4497	1.5036	8.3938	0.1791	0.1191	2.7676	7
8	0.6274	6.2098	19.8416	1.5938	9.8975	0.1610	0.1010	3.1952	8
9	0.5919	6.8017	24.5768	1.6895	11.4913	0.1470	0.0870	3.6133	9
10	0.5584	7.3601	29.6023	1.7908	13.1808	0.1359	0.0759	4.0220	10
11	0.5268	7.8869	34.8702	1.8983	14.9716	0.1268	0.0668	4.4213	11
12	0.4970	8.3838	40.3369	2.0122	16.8699	0.1193	0.0593	4.8113	12
13	0.4688	8.8527	45.9629	2.1329	18.8821	0.1130	0.0530	5.1920	13
14	0.4423	9.2950	51.7128	2.2609	21.0151	0.1076	0.0476	5.5635	14
15	0.4173	9.7122	57.5546	2.3966	23.2760	0.1030	0.0430	5.9260	15
16	0.3936	10.1059	63.4592	2.5404	25.6725	0.0990	0.0390	6.2794	16
17	0.3714	10.4773	69.4011	2.6928	28.2129	0.0954	0.0354	6.6240	17
18	0.3503	10.8276	75.3569	2.8543	30.9057	0.0924	0.0324	6.9597	18
19	0.3305	11.1581	81.3062	3.0256	33.7600	0.0896	0.0296	7.2867	19
20	0.3118	11.4699	87.2304	3.2071	36.7856	0.0872	0.0272	7.6051	20
21	0.2942	11.7641	93.1136	3.3996	39.9927	0.0850	0.0250	7.9151	21
22	0.2775	12.0416	98.9412	3.6035	43.3923	0.0830	0.0230	8.2166	22
23	0.2618	12.3034	104.7007	3.8197	46.9958	0.0813	0.0213	8.5099	23
24	0.2470	12.5504	110.3812	4.0489	50.8156	0.0797	0.0197	8.7951	24
25	0.2330	12.7834	115.9732	4.2919	54.8645	0.0782	0.0182	9.0722	25
26	0.2198	13.0032	121.4684	4.5494	59.1564	0.0769	0.0169	9.3414	26
27	0.2074	13.2105	126.8600	4.8223	63.7058	0.0757	0.0157	9.6029	27
28	0.1956	13.4062	132.1420	5.1117	68.5281	0.0746	0.0146	9.8568	28
29	0.1846	13.5907	137.3096	5.4184	73.6398	0.0736	0.0136	10.1032	29
30	0.1741	13.7648	142.3588	5.7435	79.0582	0.0726	0.0126	10.3422	30
31	0.1643	13.9291	147.2864	6.0881	84.8017	0.0718	0.0118	10.5740	31
32	0.1550	14.0840	152.0901	6.4534	90.8898	0.0710	0.0110	10.7988	32
33	0.1462	14.2302	156.7681	6.8406	97.3432	0.0703	0.0103	11.0166	33
34	0.1379	14.3681	161.3192	7.2510	104.1838	0.0696	0.0096	11.2276	34
35	0.1301	14.4982	165.7427	7.6861	111.4348	0.0690	0.0090	11.4319	35
36	0.1227	14.6210	170.0387	8.1473	119.1209	0.0684	0.0084	11.6298	36
37	0.1158	14.7368	174.2072	8.6361	127.2681	0.0679	0.0079	11.8213	37
38	0.1092	14.8460	178.2490	9.1543	135.9042	0.0674	0.0074	12.0065	38
39	0.1031	14.9491	182.1652	9.7035	145.0585	0.0669	0.0069	12.1857	39
40	0.0972	15.0463	185.9568	10.2857	154.7620	0.0665	0.0065	12.3590	40
41	0.0917	15.1380	189.6256	10.9029	165.0477	0.0661	0.0061	12.5264	41
42	0.0865	15.2245	193.1732	11.5570	175.9505	0.0657	0.0057	12.6883	42
43	0.0816	15.3062	196.6017	12.2505	187.5076	0.0653	0.0053	12.8446	43
44	0.0770	15.3832	199.9130	12.9855	199.7580	0.0650	0.0050	12.9956	44
45	0.0727	15.4558	203.1096	13.7646	212.7435	0.0647	0.0047	13.1413	45
46	0.0685	15.5244	206.1938	14.5905	226.5081	0.0644	0.0044	13.2819	46
47	0.0647	15.5890	209.1681	15.4659	241.0986	0.0641	0.0041	13.4177	47
48	0.0610	15.6500	212.0351	16.3939	256.5645	0.0639	0.0039	13.5485	48
49	0.0575	15.7076	214.7972	17.3775	272.9584	0.0637	0.0037	13.6748	49
50	0.0543	15.7619	217.4574	18.4202	290.3359	0.0634	0.0034	13.7964	50
51	0.0512	15.8131	220.0181	19.5254	308.7561	0.0632	0.0032	13.9137	51
52	0.0483	15.8614	222.4823	20.6969	328.2814	0.0630	0.0030	14.0267	52
53	0.0456	15.9070	224.8525	21.9387	348.9783	0.0629	0.0029	14.1355	53
54	0.0430	15.9500	227.1316	23.2550	370.9170	0.0627	0.0027	14.2402	54
55	0.0406	15.9905	229.3222	24.6503	394.1720	0.0625	0.0025	14.3411	55
60	0.0303	16.1614	239.0428	32.9877	533.1282	0.0619	0.0019	14.7909	60
65	0.0227	16.2891	246.9450	44.1450	719.0829	0.0614	0.0014	15.1601	65
70	0.0169	16.3845	253.3271	59.0759	967.9322	0.0610	0.0010	15.4613	70
75	0.0126	16.4558	258.4527	79.0569	1300.9487	0.0608	0.0008	15.7058	75
80	0.0095	16.5091	262.5493	105.7960	1746.5999	0.0606	0.0006	15.9033	80
85	0.0071	16.5489	265.8096	141.5789	2342.9817	0.0604	0.0004	16.0620	85
90	0.0053	16.5787	268.3946	189.4645	3141.0752	0.0603	0.0003	16.1891	90
95	0.0039	16.6009	270.4375	253.5463	4209.1042	0.0602	0.0002	16.2905	95
100	0.0029	16.6175	272.0471	339.3021	5638.3681	0.0602	0.0002	16.3711	100

(continued)

APPENDIX 86.B (continued)
Factor Tables

$I = 7.00\%$

n	P/F	P/A	P/G	F/P	F/A	A/P	A/F	A/G	n
1	0.9346	0.9346	0.0000	1.0700	1.0000	1.0700	1.0000	0.0000	1
2	0.8734	1.8080	0.8734	1.1449	2.0700	0.5531	0.4831	0.4831	2
3	0.8163	2.6243	2.5060	1.2250	3.2149	0.3811	0.3111	0.9549	3
4	0.7629	3.3872	4.7947	1.3108	4.4399	0.2952	0.2252	1.4155	4
5	0.7130	4.1002	7.6467	1.4026	5.7507	0.2439	0.1739	1.8650	5
6	0.6663	4.7665	10.9784	1.5007	7.1533	0.2098	0.1398	2.3032	6
7	0.6227	5.3893	14.7149	1.6058	8.6540	0.1856	0.1156	2.7304	7
8	0.5820	5.9713	18.7889	1.7182	10.2598	0.1675	0.0975	3.1465	8
9	0.5439	6.5152	23.1404	1.8385	11.9780	0.1535	0.0835	3.5517	9
10	0.5083	7.0236	27.7156	1.9672	13.8164	0.1424	0.0724	3.9461	10
11	0.4751	7.4987	32.4665	2.1049	15.7836	0.1334	0.0634	4.3296	11
12	0.4440	7.9427	37.3506	2.2522	17.8885	0.1259	0.0559	4.7025	12
13	0.4150	8.3577	42.3302	2.4098	20.1406	0.1197	0.0497	5.0648	13
14	0.3878	8.7455	47.3718	2.5785	22.5505	0.1143	0.0443	5.4167	14
15	0.3624	9.1079	52.4461	2.7590	25.1290	0.1098	0.0398	5.7583	15
16	0.3387	9.4466	57.5271	2.9522	27.8881	0.1059	0.0359	6.0897	16
17	0.3166	9.7632	62.5923	3.1588	30.8402	0.1024	0.0324	6.4110	17
18	0.2959	10.0591	67.6219	3.3799	33.9990	0.0994	0.0294	6.7225	18
19	0.2765	10.3356	72.5991	3.6165	37.3790	0.0968	0.0268	7.0242	19
20	0.2584	10.5940	77.5091	3.8697	40.9955	0.0944	0.0244	7.3163	20
21	0.2415	10.8355	82.3393	4.1406	44.8652	0.0923	0.0223	7.5990	21
22	0.2257	11.0612	87.0793	4.4304	49.0057	0.0904	0.0204	7.8725	22
23	0.2109	11.2722	91.7201	4.7405	53.4361	0.0887	0.0187	8.1369	23
24	0.1971	11.4693	96.2545	5.0724	58.1767	0.0872	0.0172	8.3923	24
25	0.1842	11.6536	100.6765	5.4274	63.2490	0.0858	0.0158	8.6391	25
26	0.1722	11.8258	104.9814	5.8074	68.6765	0.0846	0.0146	8.8773	26
27	0.1609	11.9867	109.1656	6.2139	74.4838	0.0834	0.0134	9.1072	27
28	0.1504	12.1371	113.2264	6.6488	80.6977	0.0824	0.0124	9.3289	28
29	0.1406	12.2777	117.1622	7.1143	87.3465	0.0814	0.0114	9.5427	29
30	0.1314	12.4090	120.9718	7.6123	94.4608	0.0806	0.0106	9.7487	30
31	0.1228	12.5318	124.6550	8.1451	102.0730	0.0798	0.0098	9.9471	31
32	0.1147	12.6466	128.2120	8.7153	110.2182	0.0791	0.0091	10.1381	32
33	0.1072	12.7538	131.6435	9.3253	118.9334	0.0784	0.0084	10.3219	33
34	0.1002	12.8540	134.9507	9.9781	128.2588	0.0778	0.0078	10.4987	34
35	0.0937	12.9477	138.1353	10.6766	138.2369	0.0772	0.0072	10.6687	35
36	0.0875	13.0352	141.1990	11.4239	148.9135	0.0767	0.0067	10.8321	36
37	0.0818	13.1170	144.1441	12.2236	160.3374	0.0762	0.0062	10.9891	37
38	0.0765	13.1935	146.9730	13.0793	172.5610	0.0758	0.0058	11.1398	38
39	0.0715	13.2649	149.6883	13.9948	185.6403	0.0754	0.0054	11.2845	39
40	0.0668	13.3317	152.2928	14.9745	199.6351	0.0750	0.0050	11.4233	40
41	0.0624	13.3941	154.7892	16.0227	214.6096	0.0747	0.0047	11.5565	41
42	0.0583	13.4524	157.1807	17.1443	230.6322	0.0743	0.0043	11.6842	42
43	0.0545	13.5070	159.4702	18.3444	247.7765	0.0740	0.0040	11.8065	43
44	0.0509	13.5579	161.6609	19.6285	266.1209	0.0738	0.0038	11.9237	44
45	0.0476	13.6055	163.7559	21.0025	285.7493	0.0735	0.0035	12.0360	45
46	0.0445	13.6500	165.7584	22.4726	306.7518	0.0733	0.0033	12.1435	46
47	0.0416	13.6916	167.6714	24.0457	329.2244	0.0730	0.0030	12.2463	47
48	0.0389	13.7305	169.4981	25.7289	353.2701	0.0728	0.0028	12.3447	48
49	0.0363	13.7668	171.2417	27.5299	378.9990	0.0726	0.0026	12.4387	49
50	0.0339	13.8007	172.9051	29.4570	406.5289	0.0725	0.0025	12.5287	50
51	0.0317	13.8325	174.4915	31.5190	435.9860	0.0723	0.0023	12.6146	51
52	0.0297	13.8621	176.0037	33.7253	467.5050	0.0721	0.0021	12.6967	52
53	0.0277	13.8898	177.4447	36.0861	501.2303	0.0720	0.0020	12.7751	53
54	0.0259	13.9157	178.8173	38.6122	537.3164	0.0719	0.0019	12.8500	54
55	0.0242	13.9399	180.1243	41.3150	575.9286	0.0717	0.0017	12.9215	55
60	0.0173	14.0392	185.7677	57.9464	813.5204	0.0712	0.0012	13.2321	60
65	0.0123	14.1099	190.1452	81.2729	1146.7552	0.0709	0.0009	13.4760	65
70	0.0088	14.1604	193.5185	113.9894	1614.1342	0.0706	0.0006	13.6662	70
75	0.0063	14.1964	196.1035	159.8760	2269.6574	0.0704	0.0004	13.8136	75
80	0.0045	14.2220	198.0748	224.2344	3189.0627	0.0703	0.0003	13.9273	80
85	0.0032	14.2403	199.5717	314.5003	4478.5761	0.0702	0.0002	14.0146	85
90	0.0023	14.2533	200.7042	441.1030	6287.1854	0.0702	0.0002	14.0812	90
95	0.0016	14.2626	201.5581	618.6697	8823.8535	0.0701	0.0001	14.1319	95
100	0.0012	14.2693	202.2001	867.7163	12381.6618	0.0701	0.0001	14.1703	100

(continued)

APPENDIX 86.B *(continued)*
Factor Tables
$$I = 8.00\%$$

n	P/F	P/A	P/G	F/P	F/A	A/P	A/F	A/G	n
1	0.9259	0.9259	0.0000	1.0800	1.0000	1.0800	1.0000	0.0000	1
2	0.8573	1.7833	0.8573	1.1664	2.0800	0.5608	0.4808	0.4808	2
3	0.7938	2.5771	2.4450	1.2597	3.2464	0.3880	0.3080	0.9487	3
4	0.7350	3.3121	4.6501	1.3605	4.5061	0.3019	0.2219	1.4040	4
5	0.6806	3.9927	7.3724	1.4693	5.8666	0.2505	0.1705	1.8465	5
6	0.6302	4.6229	10.5233	1.5869	7.3359	0.2163	0.1363	2.2763	6
7	0.5835	5.2064	14.0242	1.7138	8.9228	0.1921	0.1121	2.6937	7
8	0.5403	5.7466	17.8061	1.8509	10.6366	0.1740	0.0940	3.0985	8
9	0.5002	6.2469	21.8081	1.9990	12.4876	0.1601	0.0801	3.4910	9
10	0.4632	6.7101	25.9768	2.1589	14.4866	0.1490	0.0690	3.8713	10
11	0.4289	7.1390	30.2657	2.3316	16.6455	0.1401	0.0601	4.2395	11
12	0.3971	7.5361	34.6339	2.5182	18.9771	0.1327	0.0527	4.5957	12
13	0.3677	7.9038	39.0463	2.7196	21.4953	0.1265	0.0465	4.9402	13
14	0.3405	8.2442	43.4723	2.9372	24.2149	0.1213	0.0413	5.2731	14
15	0.3152	8.5595	47.8857	3.1722	27.1521	0.1168	0.0368	5.5945	15
16	0.2919	8.8514	52.2640	3.4259	30.3243	0.1130	0.0330	5.9046	16
17	0.2703	9.1216	56.5883	3.7000	33.7502	0.1096	0.0296	6.2037	17
18	0.2502	9.3719	60.8426	3.9960	37.4502	0.1067	0.0267	6.4920	18
19	0.2317	9.6036	65.0134	4.3157	41.4463	0.1041	0.0241	6.7697	19
20	0.2145	9.8181	69.0898	4.6610	45.7620	0.1019	0.0219	7.0369	20
21	0.1987	10.0168	73.0629	5.0338	50.4229	0.0998	0.0198	7.2940	21
22	0.1839	10.2007	76.9257	5.4365	55.4568	0.0980	0.0180	7.5412	22
23	0.1703	10.3711	80.6726	5.8715	60.8933	0.0964	0.0164	7.7786	23
24	0.1577	10.5288	84.2997	6.3412	66.7648	0.0950	0.0150	8.0066	24
25	0.1460	10.6748	87.8041	6.8485	73.1059	0.0937	0.0137	8.2254	25
26	0.1352	10.8100	91.1842	7.3964	79.9544	0.0925	0.0125	8.4352	26
27	0.1252	10.9352	94.4390	7.9881	87.3508	0.0914	0.0114	8.6363	27
28	0.1159	11.0511	97.5687	8.6271	95.3388	0.0905	0.0105	8.8289	28
29	0.1073	11.1584	100.5738	9.3173	103.9659	0.0896	0.0096	9.0133	29
30	0.0994	11.2578	103.4558	10.0627	113.2832	0.0888	0.0088	9.1897	30
31	0.0920	11.3498	106.2163	10.8677	123.3459	0.0881	0.0081	9.3584	31
32	0.0852	11.4350	108.8575	11.7371	134.2135	0.0875	0.0075	9.5197	32
33	0.0789	11.5139	111.3819	12.6760	145.9506	0.0869	0.0069	9.6737	33
34	0.0730	11.5869	113.7924	13.6901	158.6267	0.0863	0.0063	9.8208	34
35	0.0676	11.6546	116.0920	14.7853	172.3168	0.0858	0.0058	9.9611	35
36	0.0626	11.7172	118.2839	15.9682	187.1021	0.0853	0.0053	10.0949	36
37	0.0580	11.7752	120.3713	17.2456	203.0703	0.0849	0.0049	10.2225	37
38	0.0537	11.8289	122.3579	18.6253	220.3159	0.0845	0.0045	10.3440	38
39	0.0497	11.8786	124.2470	20.1153	238.9412	0.0842	0.0042	10.4597	39
40	0.0460	11.9246	126.0422	21.7245	259.0565	0.0839	0.0039	10.5699	40
41	0.0426	11.9672	127.7470	23.4625	280.7810	0.0836	0.0036	10.6747	41
42	0.0395	12.0067	129.3651	25.3395	304.2435	0.0833	0.0033	10.7744	42
43	0.0365	12.0432	130.8998	27.3666	329.5830	0.0830	0.0030	10.8692	43
44	0.0338	12.0771	132.3547	29.5560	356.9496	0.0828	0.0028	10.9592	44
45	0.0313	12.1084	133.7331	31.9204	386.5056	0.0826	0.0026	11.0447	45
46	0.0290	12.1374	135.0384	34.4741	418.4261	0.0824	0.0024	11.1258	46
47	0.0269	12.1643	136.2739	37.2320	452.9002	0.0822	0.0022	11.2028	47
48	0.0249	12.1891	137.4428	40.2106	490.1322	0.0820	0.0020	11.2758	48
49	0.0230	12.2122	138.5480	43.4274	530.3427	0.0819	0.0019	11.3451	49
50	0.0213	12.2335	139.5928	46.9016	573.7702	0.0817	0.0017	11.4107	50
51	0.0197	12.2532	140.5799	50.6537	620.6718	0.0816	0.0016	11.4729	51
52	0.0183	12.2715	141.5121	54.7060	671.3255	0.0815	0.0015	11.5318	52
53	0.0169	12.2884	142.3923	59.0825	726.0316	0.0814	0.0014	11.5875	53
54	0.0157	12.3041	143.2229	63.8091	785.1141	0.0813	0.0013	11.6403	54
55	0.0145	12.3186	144.0065	68.9139	848.9232	0.0812	0.0012	11.6902	55
60	0.0099	12.3766	147.3000	101.2571	1253.2133	0.0808	0.0008	11.9015	60
65	0.0067	12.4160	149.7387	148.7798	1847.2481	0.0805	0.0005	12.0602	65
70	0.0046	12.4428	151.5326	218.6064	2720.0801	0.0804	0.0004	12.1783	70
75	0.0031	12.4611	152.8448	321.2045	4002.5566	0.0802	0.0002	12.2658	75
80	0.0021	12.4735	153.8001	471.9548	5886.9354	0.0802	0.0002	12.3301	80
85	0.0014	12.4820	154.4565	693.4565	8655.7061	0.0801	0.0001	12.3772	85
90	0.0010	12.4877	154.9925	1018.9151	12723.9386	0.0801	0.0001	12.4116	90
95	0.0007	12.4917	155.3524	1497.1205	18701.5069	0.0801	0.0001	12.4365	95
100	0.0005	12.4943	155.6107	2199.7613	27484.5157	0.0800	0.0000	12.4545	100

(continued)

APPENDIX 86.B *(continued)*
Factor Tables

$I = 9.00\%$

n	P/F	P/A	P/G	F/P	F/A	A/P	A/F	A/G	n
1	0.9174	0.9174	0.0000	1.0900	1.0000	1.0900	1.0000	0.0000	1
2	0.8417	1.7591	0.8417	1.1881	2.0900	0.5685	0.4785	0.4785	2
3	0.7722	2.5313	2.3860	1.2950	3.2781	0.3951	0.3051	0.9426	3
4	0.7084	3.2397	4.5113	1.4116	4.5731	0.3087	0.2187	1.3925	4
5	0.6499	3.8897	7.1110	1.5386	5.9847	0.2571	0.1671	1.8282	5
6	0.5963	4.4859	10.0924	1.6771	7.5233	0.2229	0.1329	2.2498	6
7	0.5470	5.0330	13.3746	1.8280	9.2004	0.1987	0.1087	2.6574	7
8	0.5019	5.5348	16.8877	1.9926	11.0285	0.1807	0.0907	3.0512	8
9	0.4604	5.9952	20.5711	2.1719	13.0210	0.1668	0.0768	3.4312	9
10	0.4224	6.4177	24.3728	2.3674	15.1929	0.1558	0.0658	3.7978	10
11	0.3875	6.8052	28.2481	2.5804	17.5603	0.1469	0.0569	4.1510	11
12	0.3555	7.1607	32.1590	2.8127	20.1407	0.1397	0.0497	4.4910	12
13	0.3262	7.4869	36.0731	3.0658	22.9534	0.1336	0.0436	4.8182	13
14	0.2992	7.7862	39.9633	3.3417	26.0192	0.1284	0.0384	5.1326	14
15	0.2745	8.0607	43.8069	3.6425	29.3609	0.1241	0.0341	5.4346	15
16	0.2519	8.3126	47.5849	3.9703	33.0034	0.1203	0.0303	5.7245	16
17	0.2311	8.5436	51.2821	4.3276	36.9737	0.1170	0.0270	6.0024	17
18	0.2120	8.7556	54.8860	4.7171	41.3013	0.1142	0.0242	6.2687	18
19	0.1945	8.9501	58.3868	5.1417	46.0185	0.1117	0.0217	6.5236	19
20	0.1784	9.1285	61.7770	5.6044	51.1601	0.1095	0.0195	6.7674	20
21	0.1637	9.2922	65.0509	6.1088	56.7645	0.1076	0.0176	7.0006	21
22	0.1502	9.4424	68.2048	6.6586	62.8733	0.1059	0.0159	7.2232	22
23	0.1378	9.5802	71.2359	7.2579	69.5319	0.1044	0.0144	7.4357	23
24	0.1264	9.7066	74.1433	7.9111	76.7898	0.1030	0.0130	7.6384	24
25	0.1160	9.8226	76.9265	8.6231	84.7009	0.1018	0.0118	7.8316	25
26	0.1064	9.9290	79.5863	9.3992	93.3240	0.1007	0.0107	8.0156	26
27	0.0976	10.0266	82.1241	10.2451	102.7231	0.0997	0.0097	8.1906	27
28	0.0895	10.1161	84.5419	11.1671	112.9682	0.0989	0.0089	8.3571	28
29	0.0822	10.1983	86.8422	12.1722	124.1354	0.0981	0.0081	8.5154	29
30	0.0754	10.2737	89.0280	13.2677	136.3076	0.0973	0.0073	8.6657	30
31	0.0691	10.3428	91.1024	14.4618	149.5752	0.0967	0.0067	8.8083	31
32	0.0634	10.4062	93.0690	15.7633	164.0370	0.0961	0.0061	8.9436	32
33	0.0582	10.4644	94.9314	17.1820	179.8003	0.0956	0.0056	9.0718	33
34	0.0534	10.5178	96.6935	18.7284	196.9823	0.0951	0.0051	9.1933	34
35	0.0490	10.5668	98.3590	20.4140	215.7108	0.0946	0.0046	9.3083	35
36	0.0449	10.6118	99.9319	22.2512	236.1247	0.0942	0.0042	9.4171	36
37	0.0412	10.6530	101.4162	24.2538	258.3759	0.0939	0.0039	9.5200	37
38	0.0378	10.6908	102.8158	26.4367	282.6298	0.0935	0.0035	9.6172	38
39	0.0347	10.7255	104.1345	28.8160	309.0665	0.0932	0.0032	9.7090	39
40	0.0318	10.7574	105.3762	31.4094	337.8824	0.0930	0.0030	9.7957	40
41	0.0292	10.7866	106.5445	34.2363	369.2919	0.0927	0.0027	9.8775	41
42	0.0268	10.8134	107.6432	37.3175	403.5281	0.0925	0.0025	9.9546	42
43	0.0246	10.8380	108.6758	40.6761	440.8457	0.0923	0.0023	10.0273	43
44	0.0226	10.8605	109.6456	44.3370	481.5218	0.0921	0.0021	10.0958	44
45	0.0207	10.8812	110.5561	48.3273	525.8587	0.0919	0.0019	10.1603	45
46	0.0190	10.9002	111.4103	52.6767	574.1860	0.0917	0.0017	10.2210	46
47	0.0174	10.9176	112.2115	57.4176	626.8628	0.0916	0.0016	10.2780	47
48	0.0160	10.9336	112.9625	62.5852	684.2804	0.0915	0.0015	10.3317	48
49	0.0147	10.9482	113.6661	68.2179	746.8656	0.0913	0.0013	10.3821	49
50	0.0134	10.9617	114.3251	74.3575	815.0836	0.0912	0.0012	10.4295	50
51	0.0123	10.9740	114.9420	81.0497	889.4411	0.0911	0.0011	10.4740	51
52	0.0113	10.9853	115.5193	88.3442	970.4908	0.0910	0.0010	10.5158	52
53	0.0104	10.9957	116.0593	96.2951	1058.8349	0.0909	0.0009	10.5549	53
54	0.0095	11.0053	116.5642	104.9617	1155.1301	0.0909	0.0009	10.5917	54
55	0.0087	11.0140	117.0362	114.4083	1260.0918	0.0908	0.0008	10.6261	55
60	0.0057	11.0480	118.9683	176.0313	1944.7921	0.0905	0.0005	10.7683	60
65	0.0037	11.0701	120.3344	270.8460	2998.2885	0.0903	0.0003	10.8702	65
70	0.0024	11.0844	121.2942	416.7301	4619.2232	0.0902	0.0002	10.9427	70
75	0.0016	11.0938	121.9646	641.1909	7113.2321	0.0901	0.0001	10.9940	75
80	0.0010	11.0998	122.4306	986.5517	10950.5741	0.0901	0.0001	11.0299	80
85	0.0007	11.1038	122.7533	1517.9320	16854.8003	0.0901	0.0001	11.0551	85
90	0.0004	11.1064	122.9758	2335.5266	25939.1842	0.0900	0.0000	11.0726	90
95	0.0003	11.1080	123.1287	3593.4971	39916.6350	0.0900	0.0000	11.0847	95
100	0.0002	11.1091	123.2335	5529.0408	61422.6755	0.0900	0.0000	11.0930	100

(continued)

APPENDIX 86.B *(continued)*
Factor Tables

$I = 10.00\%$

n	P/F	P/A	P/G	F/P	F/A	A/P	A/F	A/G	n
1	0.9091	0.9091	0.0000	1.1000	1.0000	1.1000	1.0000	0.0000	1
2	0.8264	1.7355	0.8264	1.2100	2.1000	0.5762	0.4762	0.4762	2
3	0.7513	2.4869	2.3291	1.3310	3.3100	0.4021	0.3021	0.9366	3
4	0.6830	3.1699	4.3781	1.4641	4.6410	0.3155	0.2155	1.3812	4
5	0.6209	3.7908	6.8618	1.6105	6.1051	0.2638	0.1638	1.8101	5
6	0.5645	4.3553	9.6842	1.7716	7.7156	0.2296	0.1296	2.2236	6
7	0.5132	4.8684	12.7631	1.9487	9.4872	0.2054	0.1054	2.6216	7
8	0.4665	5.3349	16.0287	2.1436	11.4359	0.1874	0.0874	3.0045	8
9	0.4241	5.7590	19.4215	2.3579	13.5795	0.1736	0.0736	3.3724	9
10	0.3855	6.1446	22.8913	2.5937	15.9374	0.1627	0.0627	3.7255	10
11	0.3505	6.4951	26.3963	2.8531	18.5312	0.1540	0.0540	4.0641	11
12	0.3186	6.8137	29.9012	3.1384	21.3843	0.1468	0.0468	4.3884	12
13	0.2897	7.1034	33.3772	3.4523	24.5227	0.1408	0.0408	4.6988	13
14	0.2633	7.3667	36.8005	3.7975	27.9750	0.1357	0.0357	4.9955	14
15	0.2394	7.6061	40.1520	4.1772	31.7725	0.1315	0.0315	5.2789	15
16	0.2176	7.8237	43.4164	4.5950	35.9497	0.1278	0.0278	5.5493	16
17	0.1978	8.0216	46.5819	5.0545	40.5447	0.1247	0.0247	5.8071	17
18	0.1799	8.2014	49.6395	5.5599	45.5992	0.1219	0.0219	6.0526	18
19	0.1635	8.3649	52.5827	6.1159	51.1591	0.1195	0.0195	6.2861	19
20	0.1486	8.5136	55.4069	6.7275	57.2750	0.1175	0.0175	6.5081	20
21	0.1351	8.6487	58.1095	7.4002	64.0025	0.1156	0.0156	6.7189	21
22	0.1228	8.7715	60.6893	8.1403	71.4027	0.1140	0.0140	6.9189	22
23	0.1117	8.8832	63.1462	8.9543	79.5430	0.1126	0.0126	7.1085	23
24	0.1015	8.9847	65.4813	9.8497	88.4973	0.1113	0.0113	7.2881	24
25	0.0923	9.0770	67.6964	10.8347	98.3471	0.1102	0.0102	7.4580	25
26	0.0839	9.1609	69.7940	11.9182	109.1818	0.1092	0.0092	7.6186	26
27	0.0763	9.2372	71.7773	13.1100	121.0999	0.1083	0.0083	7.7704	27
28	0.0693	9.3066	73.6495	14.4210	134.2099	0.1075	0.0075	7.9137	28
29	0.0630	9.3696	75.4146	15.8631	148.6309	0.1067	0.0067	8.0489	29
30	0.0573	9.4269	77.0766	17.4494	164.4940	0.1061	0.0061	8.1762	30
31	0.0521	9.4790	78.6395	19.1943	181.9434	0.1055	0.0055	8.2962	31
32	0.0474	9.5264	80.1078	21.1138	201.1378	0.1050	0.0050	8.4091	32
33	0.0431	9.5694	81.4856	23.2252	222.2515	0.1045	0.0045	8.5152	33
34	0.0391	9.6086	82.7773	25.5477	245.4767	0.1041	0.0041	8.6149	34
35	0.0356	9.6442	83.9872	28.1024	271.0244	0.1037	0.0037	8.7086	35
36	0.0323	9.6765	85.1194	30.9127	299.1268	0.1033	0.0033	8.7965	36
37	0.0294	9.7059	86.1781	34.0039	330.0395	0.1030	0.0030	8.8789	37
38	0.0267	9.7327	87.1673	37.4043	364.0434	0.1027	0.0027	8.9562	38
39	0.0243	9.7570	88.0908	41.1448	401.4478	0.0125	0.0025	9.0285	39
40	0.0221	9.7791	88.9525	45.2593	442.5926	0.1023	0.0023	9.0962	40
41	0.0201	9.7991	89.7560	49.7852	487.8518	0.1020	0.0020	9.1596	41
42	0.0183	9.8174	90.5047	54.7637	537.6370	0.1019	0.0019	9.2188	42
43	0.0166	9.8340	91.2019	60.2401	592.4007	0.1017	0.0017	9.2741	43
44	0.0151	9.8491	91.8508	66.2641	652.6408	0.1015	0.0015	9.3258	44
45	0.0137	9.8628	92.4544	72.8905	718.9048	0.1014	0.0014	9.3740	45
46	0.0125	9.8753	93.0157	80.1795	791.7953	0.1013	0.0013	9.4190	46
47	0.0113	9.8866	93.5372	88.1975	871.9749	0.1011	0.0011	9.4610	47
48	0.0103	9.8969	94.0217	97.0172	960.1723	0.1010	0.0010	9.5001	48
49	0.0094	9.9063	94.4715	106.7190	1057.1896	0.1009	0.0009	9.5365	49
50	0.0085	9.9148	94.8889	117.3909	1163.9085	0.1009	0.0009	9.5704	50
51	0.0077	9.9226	95.2761	129.1299	1281.2994	0.1008	0.0008	9.6020	51
52	0.0070	9.9296	95.6351	142.0429	1410.4293	0.1007	0.0007	9.6313	52
53	0.0064	9.9360	95.9679	156.2472	1552.4723	0.1006	0.0006	9.6586	53
54	0.0058	9.9418	96.2763	171.8719	1708.7195	0.1006	0.0006	9.6840	54
55	0.0053	9.9471	96.5619	189.0591	1880.5914	0.1005	0.0005	9.7075	55
60	0.0033	9.9672	97.7010	304.4816	3034.8164	0.1003	0.0003	9.8023	60
65	0.0020	9.9796	98.4705	490.3707	4893.7073	0.1002	0.0002	9.8672	65
70	0.0013	9.9873	98.9870	789.7470	7887.4696	0.1001	0.0001	9.9113	70
75	0.0008	9.9921	99.3317	1271.8954	12708.9537	0.1001	0.0001	9.9410	75
80	0.0005	9.9951	99.5606	2048.4002	20474.0021	0.1000	0.0000	9.9609	80
85	0.0003	9.9970	99.7120	3298.9690	32979.6903	0.1000	0.0000	9.9742	85
90	0.0002	9.9981	99.8118	5313.0226	53120.2261	0.1000	0.0000	9.9831	90
95	0.0001	9.9988	99.8773	8556.6760	85556.7605	0.1000	0.0000	9.9889	95
100	0.0001	9.9993	99.9202	13780.6123	137796.1234	0.1000	0.0000	9.9927	100

(continued)

APPENDIX 86.B *(continued)*
Factor Tables

$I = 12.00\%$

n	P/F	P/A	P/G	F/P	F/A	A/P	A/F	A/G	n
1	0.8929	0.8929	0.0000	1.1200	1.0000	1.1200	1.0000	0.0000	1
2	0.7972	1.6901	0.7972	1.2544	2.1200	0.5917	0.4717	0.4717	2
3	0.7118	2.4018	2.2208	1.4049	3.3744	0.4163	0.2963	0.9246	3
4	0.6355	3.0373	4.1273	1.5735	4.7793	0.3292	0.2092	1.3589	4
5	0.5674	3.6048	6.3970	1.7623	6.3528	0.2774	0.1574	1.7746	5
6	0.5066	4.1114	8.9302	1.9738	8.1152	0.2432	0.1232	2.1720	6
7	0.4523	4.5638	11.6443	2.2107	10.0890	0.2191	0.0991	2.5515	7
8	0.4039	4.9676	14.4714	2.4760	12.2997	0.2013	0.0813	2.9131	8
9	0.3606	5.3282	17.3563	2.7731	14.7757	0.1877	0.0677	3.2574	9
10	0.3220	5.6502	20.2541	3.1058	17.5487	0.1770	0.0570	3.5847	10
11	0.2875	5.9377	23.1288	3.4785	20.6546	0.1684	0.0484	3.8953	11
12	0.2567	6.1944	25.9523	3.8960	24.1331	0.1614	0.0414	4.1897	12
13	0.2292	6.4235	28.7024	4.3635	28.0291	0.1557	0.0357	4.4683	13
14	0.2046	6.6282	31.3624	4.8871	32.3926	0.1509	0.0309	4.7317	14
15	0.1827	6.8109	33.9202	5.4736	37.2797	0.1468	0.0268	4.9803	15
16	0.1631	6.9740	36.3670	6.1304	42.7533	0.1434	0.0234	5.2147	16
17	0.1456	7.1196	38.6973	6.8660	48.8837	0.1405	0.0205	5.4353	17
18	0.1300	7.2497	40.9080	7.6900	55.7497	0.1379	0.0179	5.6427	18
19	0.1161	7.3658	42.9979	8.6128	63.4397	0.1358	0.0158	6.8375	19
20	0.1037	7.4694	44.9676	9.6463	72.0524	0.1339	0.0139	6.0202	20
21	0.0926	7.5620	46.8188	10.8038	81.6987	0.1322	0.0122	6.1913	21
22	0.0826	7.6446	48.5543	12.1003	92.5026	0.1308	0.0108	6.3514	22
23	0.0738	7.7184	50.1776	13.5523	104.6029	0.1296	0.0096	6.5010	23
24	0.0659	7.7843	51.6929	15.1786	118.1552	0.1285	0.0085	6.6406	24
25	0.0588	7.8431	53.1046	17.0001	133.3339	0.1275	0.0075	6.7708	25
26	0.0525	7.8957	54.4177	19.0401	150.3339	0.1267	0.0067	6.8921	26
27	0.0469	7.9426	55.6369	21.3249	169.3740	0.1259	0.0059	7.0049	27
28	0.0419	7.9844	56.7674	23.8839	190.6989	0.1252	0.0052	7.1098	28
29	0.0374	8.0218	57.8141	26.7499	214.5828	0.1247	0.0047	7.2071	29
30	0.0334	8.0552	58.7821	29.9599	241.3327	0.1241	0.0041	7.2974	30
31	0.0298	8.0850	59.6761	33.5551	271.2926	0.1237	0.0037	7.3811	31
32	0.0266	8.1116	60.5010	37.5817	304.8477	0.1233	0.0033	7.4586	32
33	0.0238	8.1354	61.2612	42.0915	342.4294	0.1229	0.0029	7.5302	33
34	0.0212	8.1566	61.9612	47.1425	384.5210	0.1226	0.0026	7.5965	34
35	0.0189	8.1755	62.6052	52.7996	431.6635	0.1223	0.0023	7.6577	35
36	0.0169	8.1924	63.1970	59.1356	484.4631	0.1221	0.0021	7.7141	36
37	0.0151	8.2075	63.7406	66.2318	543.5987	0.1218	0.0018	7.7661	37
38	0.0135	8.2210	64.2394	74.1797	609.8305	0.1216	0.0016	7.8141	38
39	0.0120	8.2330	64.6967	83.0812	684.0102	0.1215	0.0015	7.8582	39
40	0.0107	8.2438	65.1159	93.0510	767.0914	0.1213	0.0013	7.8988	40
41	0.0096	8.2534	65.4997	104.2171	860.1424	0.1212	0.0012	7.9361	41
42	0.0086	8.2619	65.8509	116.7231	964.3595	0.1210	0.0010	7.9704	42
43	0.0076	8.2696	66.1722	130.7299	1081.0826	0.1209	0.0009	8.0019	43
44	0.0068	8.2764	66.4659	146.4175	1211.8125	0.1208	0.0008	8.0308	44
45	0.0061	8.2825	66.7342	163.9876	1358.2300	0.1207	0.0007	8.0572	45
46	0.0054	8.2880	66.9792	183.6661	1522.2176	0.1207	0.0007	8.0815	46
47	0.0049	8.2928	67.2028	205.7061	1705.8838	0.1206	0.0006	8.1037	47
48	0.0043	8.2972	67.4068	230.3908	1911.5898	0.1205	0.0005	8.1241	48
49	0.0039	8.3010	67.5929	258.0377	2141.9806	0.1205	0.0005	8.1427	49
50	0.0035	8.3045	67.7624	289.0022	2400.0182	0.1204	0.0004	8.1597	50
51	0.0031	8.3076	67.9169	323.6825	2689.0204	0.1204	0.0004	8.1753	51
52	0.0028	8.3103	68.0576	362.5243	3012.7029	0.1203	0.0003	8.1895	52
53	0.0025	8.3128	68.1856	406.0273	3375.2272	0.1203	0.0003	8.2025	53
54	0.0022	8.3150	68.3022	454.7505	3781.2545	0.1203	0.0003	8.2143	54
55	0.0020	8.3170	68.4082	509.3206	4236.0050	0.1202	0.0002	8.2251	55
60	0.0011	8.3240	68.8100	897.5969	7471.6411	0.1201	0.0001	8.2664	60
65	0.0006	8.3281	69.0581	1581.8725	13173.9374	0.1201	0.0001	8.2922	65
70	0.0004	8.3303	69.2103	2787.7998	23223.3319	0.1200	0.0000	8.3082	70
75	0.0002	8.3316	69.3031	4913.0558	40933.7987	0.1200	0.0000	8.3181	75
80	0.0001	8.3324	69.3594	8658.4831	72145.6925	0.1200	0.0000	8.3241	80
85	0.0001	8.3328	69.3935	15259.2057	127151.7140	0.1200	0.0000	8.3278	85
90	0.0000	8.3330	69.4140	26891.9342	224091.1185	0.1200	0.0000	8.3300	90
95	0.0000	8.3332	69.4263	47392.7766	394931.4719	0.1200	0.0000	8.3313	95
100	0.0000	8.3332	69.4336	83522.2657	696010.5477	0.1200	0.0000	8.3321	100

(continued)

APPENDIX 86.B *(continued)*
Factor Tables
$$I = 15.00\%$$

n	P/F	P/A	P/G	F/P	F/A	A/P	A/F	A/G	n
1	0.8696	0.8696	0.0000	1.1500	1.0000	1.1500	1.0000	0.0000	1
2	0.7561	1.6257	0.7561	1.3225	2.1500	0.6151	0.4651	0.4651	2
3	0.6575	2.2832	2.0712	1.5209	3.4725	0.4380	0.2880	0.9071	3
4	0.5718	2.8550	3.7864	1.7490	4.9934	0.3503	0.2003	1.3263	4
5	0.4972	3.3522	5.7751	2.0114	6.7424	0.2983	0.1483	1.7228	5
6	0.4323	3.7845	7.9368	2.3131	8.7537	0.2642	0.1142	2.0972	6
7	0.3759	4.1604	10.1924	2.6600	11.0668	0.2404	0.0904	2.4498	7
8	0.3269	4.4873	12.4807	3.0590	13.7268	0.2229	0.0729	2.7813	8
9	0.2843	4.7716	14.7548	3.5179	16.7858	0.2096	0.0596	3.0922	9
10	0.2472	5.0188	16.9795	4.0456	20.3037	0.1993	0.0493	3.3832	10
11	0.2149	5.2337	19.1289	4.6524	24.3493	0.1911	0.0411	3.6549	11
12	0.1869	5.4206	21.1849	5.3503	29.0017	0.1845	0.0345	3.9082	12
13	0.1625	5.5831	23.1352	6.1528	34.3519	0.1791	0.0291	4.1438	13
14	0.1413	5.7245	24.9725	7.0757	40.5047	0.1747	0.0247	4.3624	14
15	0.1229	5.8474	26.9630	8.1371	47.5804	0.1710	0.0210	4.5650	15
16	0.1069	5.9542	28.2960	9.3576	55.7175	0.1679	0.0179	4.7522	16
17	0.0929	6.0472	29.7828	10.7613	65.0751	0.1654	0.0154	4.9251	17
18	0.0808	6.1280	31.1565	12.3755	75.8364	0.1632	0.0132	5.0843	18
19	0.0703	6.1982	32.4213	14.2318	88.2118	0.1613	0.0113	5.2307	19
20	0.0611	6.2593	33.5822	16.3665	102.4436	0.1598	0.0098	5.3651	20
21	0.0531	6.3125	34.6448	18.8215	118.8101	0.1584	0.0084	5.4883	21
22	0.0462	6.3587	35.6150	21.6447	137.6316	0.1573	0.0073	5.6010	22
23	0.0402	6.3988	36.4988	24.8915	159.2764	0.1563	0.0063	5.7040	23
24	0.0349	6.4338	37.3023	28.6252	184.1678	0.1554	0.0054	5.7979	24
25	0.0304	6.4641	38.0314	32.9190	212.7930	0.1547	0.0047	5.8834	25
26	0.0264	6.4906	38.6918	37.8568	245.7120	0.1541	0.0041	5.9612	26
27	0.0230	6.5135	39.2890	43.5353	283.5688	0.1535	0.0035	6.0319	27
28	0.0200	6.5335	39.8283	50.0656	327.1041	0.1531	0.0031	6.0960	28
29	0.0174	6.5509	40.3146	57.5755	377.1697	0.1527	0.0027	6.1541	29
30	0.0151	6.5660	40.7526	66.2118	434.7451	0.1523	0.0023	6.2066	30
31	0.0131	6.5791	41.1466	76.1435	500.9569	0.1520	0.0020	6.2541	31
32	0.0114	6.5905	41.5006	87.5651	577.1005	0.1517	0.0017	6.2970	32
33	0.0099	6.6005	41.8184	100.6998	664.6655	0.1515	0.0015	6.3357	33
34	0.0086	6.6091	42.1033	115.8048	765.3654	0.1513	0.0013	6.3705	34
35	0.0075	6.6166	42.3586	133.1755	881.1702	0.1511	0.0011	6.4019	35
36	0.0065	6.6231	42.5872	153.1519	1014.3457	0.1510	0.0010	6.4301	36
37	0.0057	6.6288	42.7916	176.1246	1167.4975	0.1509	0.0009	6.4554	37
38	0.0049	6.6338	42.9743	202.5433	1343.6222	0.1507	0.0007	6.4781	38
39	0.0043	6.6380	43.1374	232.9248	1546.1655	0.1506	0.0006	6.4985	39
40	0.0037	6.6418	43.2830	267.8635	1779.0903	0.1506	0.0006	6.5168	40
41	0.0032	6.6450	43.4128	308.0431	2046.9539	0.1505	0.0005	6.5331	41
42	0.0028	6.6478	43.5286	354.2495	2354.9969	0.1504	0.0004	6.5478	42
43	0.0025	6.6503	43.6317	407.3870	2709.2465	0.1504	0.0004	6.5609	43
44	0.0021	6.6524	43.7235	468.4950	3116.6334	0.1503	0.0003	6.5725	44
45	0.0019	6.6543	43.8051	538.7693	3585.1285	0.1503	0.0003	6.5830	45
46	0.0016	6.6559	43.8778	619.5847	4123.8977	0.1502	0.0002	6.5923	46
47	0.0014	6.6573	43.9423	712.5224	4743.4824	0.1502	0.0002	6.6006	47
48	0.0012	6.6585	43.9997	819.4007	5456.0047	0.1502	0.0002	6.6080	48
49	0.0011	6.6596	44.0506	942.3108	6275.4055	0.1502	0.0002	6.6146	49
50	0.0009	6.6605	44.0958	1083.6574	7217.7163	0.1501	0.0001	6.6205	50
51	0.0008	6.6613	44.1360	1246.2061	8301.3737	0.1501	0.0001	6.6257	51
52	0.0007	6.6620	44.1715	1433.1370	9547.5798	0.1501	0.0001	6.6304	52
53	0.0006	6.6626	44.2031	1648.1075	10980.7167	0.1501	0.0001	6.6345	53
54	0.0005	6.6631	44.2311	1895.3236	12628.8243	0.1501	0.0001	6.6382	54
55	0.0005	6.6636	44.2558	2179.6222	14524.1479	0.1501	0.0001	6.6414	55
60	0.0002	6.6651	44.3431	4383.9987	29219.9916	0.1500	0.0000	6.6530	60
65	0.0001	6.6659	44.3903	8817.7874	58778.5826	0.1500	0.0000	6.6593	65
70	0.0001	6.6663	44.4156	17735.7200	118231.4669	0.1500	0.0000	6.6627	70
75	0.0000	6.6665	44.4292	35672.8680	237812.4532	0.1500	0.0000	6.6646	75
80	0.0000	6.6666	44.4364	71750.8794	478332.5293	0.1500	0.0000	6.6656	80
85	0.0000	6.6666	44.4402	144316.6470	962104.3133	0.1500	0.0000	6.6661	85
90	0.0000	6.6666	44.4422	290272.3252	1935142.1680	0.1500	0.0000	6.6664	90
95	0.0000	6.6667	44.4433	583841.3276	3892268.8509	0.1500	0.0000	6.6665	95
100	0.0000	6.6667	44.4438	1174313.4507	7828749.6713	0.1500	0.0000	6.6666	100

(continued)

Support Material

APPENDIX 86.B *(continued)*
Factor Tables

$I = 20.00\%$

n	P/F	P/A	P/G	F/P	F/A	A/P	A/F	A/G	n
1	0.8333	0.8333	0.0000	1.2000	1.0000	1.2000	1.0000	0.0000	1
2	0.6944	1.5278	0.6944	1.4400	2.2000	0.6545	0.4545	0.4545	2
3	0.5787	2.1065	1.8519	1.7280	3.6400	0.4747	0.2747	0.8791	3
4	0.4823	2.5887	3.2986	2.0736	5.3680	0.3863	0.1863	1.2742	4
5	0.4019	2.9906	4.9061	2.4883	7.4416	0.3344	0.1344	1.6405	5
6	0.3349	3.3255	6.5806	2.9860	9.9299	0.3007	0.1007	1.9788	6
7	0.2791	3.6046	8.2551	3.5832	12.9159	0.2774	0.0774	2.2902	7
8	0.2326	3.8372	9.8831	4.2998	16.4991	0.2606	0.0606	2.5756	8
9	0.1938	4.0310	11.4335	5.1598	20.7989	0.2481	0.0481	2.8364	9
10	0.1615	4.1925	12.8871	6.1917	25.9587	0.2385	0.0385	3.0739	10
11	0.1346	4.3271	14.2330	7.4301	32.1504	0.2311	0.0311	3.2893	11
12	0.1122	4.4392	15.4667	8.9161	39.5805	0.2253	0.0253	3.4841	12
13	0.0935	4.5327	16.5883	10.6993	48.4966	0.2206	0.0206	3.6597	13
14	0.0779	4.6106	17.6008	12.8392	59.1959	0.2169	0.0169	3.8175	14
15	0.0649	4.6755	18.5095	15.4070	72.0351	0.2139	0.0139	3.9588	15
16	0.0541	4.7296	19.3208	18.4884	87.4421	0.2114	0.0114	4.0851	16
17	0.0451	4.7746	20.0419	22.1861	105.9306	0.2094	0.0094	4.1976	17
18	0.0376	4.8122	20.6805	26.6233	128.1167	0.2078	0.0078	4.2975	18
19	0.0313	4.8435	21.2439	31.9480	154.7400	0.2065	0.0065	4.3861	19
20	0.0261	4.8696	21.7395	38.3376	186.6880	0.2054	0.0054	4.4643	20
21	0.0217	4.8913	22.1742	46.0051	225.0256	0.2044	0.0044	4.5334	21
22	0.0181	4.9094	22.5546	55.2061	271.0307	0.2037	0.0037	4.5941	22
23	0.0151	4.9245	22.8867	66.2474	326.2369	0.2031	0.0031	4.6475	23
24	0.0126	4.9371	23.1760	79.4968	392.4842	0.2025	0.0025	4.6943	24
25	0.0105	4.9476	23.4276	95.3962	471.9811	0.2021	0.0021	4.7352	25
26	0.0087	4.9563	23.6460	114.4755	567.3773	0.2018	0.0018	4.7709	26
27	0.0073	4.9636	23.8353	137.3706	681.8528	0.2015	0.0015	4.8020	27
28	0.0061	4.9697	23.9991	164.8447	819.2233	0.2012	0.0012	4.8291	28
29	0.0051	4.9747	24.1406	197.8136	984.0680	0.2010	0.0010	4.8527	29
30	0.0042	4.9789	24.2628	237.3763	1181.8816	0.2008	0.0008	4.8731	30
31	0.0035	4.9824	24.3681	284.8516	1419.2579	0.2007	0.0007	4.8908	31
32	0.0029	4.9854	24.4588	341.8219	1704.1095	0.2006	0.0006	4.9061	32
33	0.0024	4.9878	24.5368	410.1863	2045.9314	0.2005	0.0005	4.9194	33
34	0.0020	4.9898	24.6038	492.2235	2456.1176	0.2004	0.0004	4.9308	34
35	0.0017	4.9915	24.6614	590.6682	2948.3411	0.2003	0.0003	4.9406	35
36	0.0014	4.9929	24.7108	708.8019	3539.0094	0.2003	0.0003	4.9491	36
37	0.0012	4.9941	24.7531	850.5622	4247.8112	0.2002	0.0002	4.9564	37
38	0.0010	4.9951	24.7894	1020.6747	5098.3735	0.2002	0.0002	4.9627	38
39	0.0008	4.9959	24.8204	1224.8096	6119.0482	0.2002	0.0002	4.9681	39
40	0.0007	4.9966	24.8469	1469.7716	7343.8578	0.2001	0.0001	4.9728	40
41	0.0006	4.9972	24.8696	1763.7259	8813.6294	0.2001	0.0001	4.9767	41
42	0.0005	4.9976	24.8890	2116.4711	10577.3553	0.2001	0.0001	4.9801	42
43	0.0004	4.9980	24.9055	2539.7653	12693.8263	0.2001	0.0001	4.9831	43
44	0.0003	4.9984	24.9196	3047.7183	15233.5916	0.2001	0.0001	4.9856	44
45	0.0003	4.9986	24.9316	3657.2620	18281.3099	0.2001	0.0001	4.9877	45
46	0.0002	4.9989	24.9419	4388.7144	21938.5719	0.2000	0.0000	4.9895	46
47	0.0002	4.9991	24.9506	5266.4573	26327.2863	0.2000	0.0000	4.9911	47
48	0.0002	4.9992	24.9581	6319.7487	31593.7436	0.2000	0.0000	4.9924	48
49	0.0001	4.9993	24.9644	7583.6985	37913.4923	0.2000	0.0000	4.9935	49
50	0.0001	4.9995	24.9698	9100.4382	45497.1908	0.2000	0.0000	4.9945	50
51	0.0001	4.9995	24.9744	10920.5258	54597.6289	0.2000	0.0000	4.9953	51
52	0.0001	4.9996	24.9783	13104.6309	65518.1547	0.2000	0.0000	4.9960	52
53	0.0001	4.9997	24.9816	15725.5571	78622.7856	0.2000	0.0000	4.9966	53
54	0.0001	4.9997	24.9844	18870.6685	94348.3427	0.2000	0.0000	4.9971	54
55	0.0000	4.9998	24.9868	22644.8023	113219.0113	0.2000	0.0000	4.9976	55
60	0.0000	4.9999	24.9942	56347.5144	281732.5718	0.2000	0.0000	4.9989	60
65	0.0000	5.0000	24.9975	140210.6469	701048.2346	0.2000	0.0000	4.9995	65
70	0.0000	5.0000	24.9989	348888.9569	1744439.7847	0.2000	0.0000	4.9998	70
75	0.0000	5.0000	24.9995	868147.3693	4340731.8466	0.2000	0.0000	4.9999	75

(continued)

APPENDIX 86.B *(continued)*
Factor Tables

$I = 25.00\%$

n	P/F	P/A	P/G	F/P	F/A	A/P	A/F	A/G	n
1	0.8000	0.8000	0.0000	1.2500	1.0000	1.2500	1.0000	0.0000	1
2	0.6400	1.4400	0.6400	1.5625	2.2500	0.6944	0.0444	0.4444	2
3	0.5120	1.9520	1.6640	1.9531	3.8125	0.5123	0.2623	0.8525	3
4	0.4096	2.3616	2.8928	2.4414	5.7656	0.4234	0.1734	1.2249	4
5	0.3277	2.6893	4.2035	3.0518	8.2070	0.3718	0.1218	1.5631	5
6	0.2621	2.9514	5.5142	3.8147	11.2588	0.3383	0.0888	1.8683	6
7	0.2097	3.1611	6.7725	4.7684	15.0735	0.3163	0.0663	2.1424	7
8	0.1678	3.3289	7.9469	5.9605	19.8419	0.3004	0.0504	2.3872	8
9	0.1342	3.4631	9.0207	7.4506	25.8023	0.2888	0.0388	2.6048	9
10	0.1074	3.5705	9.9870	9.3132	33.2529	0.2801	0.0301	2.7971	10
11	0.0859	3.6564	10.8460	11.6415	42.5661	0.2735	0.0235	2.9663	11
12	0.0687	3.7251	11.6020	14.5519	54.2077	0.2684	0.0184	3.1145	12
13	0.0550	3.7801	12.2617	18.1899	68.7596	0.2645	0.0145	3.2437	13
14	0.0440	3.8241	12.8334	22.7374	86.9495	0.2615	0.0115	3.3559	14
15	0.0352	3.8593	13.3260	28.4217	109.6868	0.2591	0.0091	3.4530	15
16	0.0281	3.8874	13.7482	35.5271	138.1085	0.2572	0.0072	3.5366	16
17	0.0225	3.9099	14.1085	44.4089	173.6357	0.2558	0.0058	3.6084	17
18	0.0180	3.9279	14.4147	55.5112	218.0446	0.2546	0.0046	3.6698	18
19	0.0144	3.9424	14.6741	69.3889	273.5558	0.2537	0.0037	3.7222	19
20	0.0115	3.9539	14.8932	86.7362	342.9447	0.2529	0.0029	3.7667	20
21	0.0092	3.9631	15.0777	108.4202	429.6809	0.2523	0.0023	3.8045	21
22	0.0074	3.9705	15.2326	135.5253	538.1011	0.2519	0.0019	3.8365	22
23	0.0059	3.9764	15.3625	169.4066	673.6264	0.2515	0.0015	3.8634	23
24	0.0047	3.9811	15.4711	211.7582	843.0329	0.2512	0.0012	3.8861	24
25	0.0038	3.9849	15.5618	264.6978	1054.7912	0.2509	0.0009	3.9052	25
26	0.0030	3.9879	15.6373	330.8722	1319.4890	0.2508	0.0008	3.9212	26
27	0.0024	3.9903	15.7002	413.5903	1650.3612	0.2506	0.0006	3.9346	27
28	0.0019	3.9923	15.7524	516.9879	2063.9515	0.2505	0.0005	3.9457	28
29	0.0015	3.9938	15.7957	646.2349	2580.9394	0.2504	0.0004	3.9551	29
30	0.0012	3.9950	15.8316	807.7936	3227.1743	0.2503	0.0003	3.9628	30
31	0.0010	3.9960	15.8614	1009.7420	4034.9678	0.2502	0.0002	3.9693	31
32	0.0008	3.9968	15.8859	1262.1774	5044.7098	0.2502	0.0002	3.9746	32
33	0.0006	3.9975	15.9062	1577.7218	6306.8872	0.2502	0.0002	3.9791	33
34	0.0005	3.9980	15.9229	1972.1523	7884.6091	0.2501	0.0001	3.9828	34
35	0.0004	3.9984	15.9367	2465.1903	9856.7613	0.2501	0.0001	3.9858	35
36	0.0003	3.9987	15.9481	3081.4879	12321.9516	0.2501	0.0001	3.9883	36
37	0.0003	3.9990	15.9574	3851.8599	15403.4396	0.2501	0.0001	3.9904	37
38	0.0002	3.9992	15.9651	4814.8249	19255.2994	0.2501	0.0001	3.9921	38
39	0.0002	3.9993	15.9714	6018.5311	24070.1243	0.2500	0.0000	3.9935	39
40	0.0001	3.9995	15.9766	7523.1638	30088.6554	0.2500	0.0000	3.9947	40
41	0.0001	3.9996	15.9809	9403.9548	37611.8192	0.2500	0.0000	3.9956	41
42	0.0001	3.9997	15.9843	11754.9435	47015.7740	0.2500	0.0000	3.9964	42
43	0.0001	3.9997	15.9872	14693.6794	58770.7175	0.2500	0.0000	3.9971	43
44	0.0001	3.9998	15.9895	18367.0992	73464.3969	0.2500	0.0000	3.9976	44
45	0.0000	3.9998	15.9915	22958.8740	91831.4962	0.2500	0.0000	3.9980	45
46	0.0000	3.9999	15.9930	28698.5925	114790.3702	0.2500	0.0000	3.9984	46
47	0.0000	3.9999	15.9943	35873.2407	143488.9627	0.2500	0.0000	3.9987	47
48	0.0000	3.9999	15.9954	44841.5509	179362.2034	0.2500	0.0000	3.9989	48
49	0.0000	3.9999	15.9962	56051.9386	224203.7543	0.2500	0.0000	3.9991	49
50	0.0000	3.9999	15.9969	70064.9232	280255.6929	0.2500	0.0000	3.9993	50
51	0.0000	4.0000	15.9975	87581.1540	350320.6161	0.2500	0.0000	3.9994	51
52	0.0000	4.0000	15.9980	109476.4425	437901.7701	0.2500	0.0000	3.9995	52
53	0.0000	4.0000	15.9983	136845.5532	547378.2126	0.2500	0.0000	3.9996	53
54	0.0000	4.0000	15.9986	171056.9414	684223.7658	0.2500	0.0000	3.9997	54
55	0.0000	4.0000	15.9989	213821.1768	855280.7072	0.2500	0.0000	3.9997	55
60	0.0000	4.0000	15.9996	652530.4468	2610117.7872	0.2500	0.0000	3.9999	60

(continued)

APPENDIX 86.B *(continued)*
Factor Tables
$$I = 30.00\%$$

n	P/F	P/A	P/G	F/P	F/A	A/P	A/F	A/G	n
1	0.7692	0.7692	0.0000	1.3000	1.0000	1.3000	1.0000	0.000	1
2	0.5917	1.3609	0.5917	1.6900	2.3000	0.7348	0.4348	0.434	2
3	0.4552	1.8161	1.5020	2.1970	3.9900	0.5506	0.2506	0.827	3
4	0.3501	2.1662	2.5524	2.8561	6.1870	0.4616	0.1616	1.178	4
5	0.2693	2.4356	3.6297	3.7129	9.0431	0.4106	0.1106	1.490	5
6	0.2072	2.6427	4.6656	4.8268	12.7560	0.3784	0.0784	1.765	6
7	0.1594	2.8021	5.6218	6.2749	17.5828	0.3569	0.0569	2.006	7
8	0.1226	2.9247	6.4800	8.1573	23.8577	0.3419	0.0419	2.215	8
9	0.0943	3.0190	7.2343	10.6045	32.0150	0.3312	0.0312	2.396	9
10	0.0725	3.0915	7.8872	13.7858	42.6195	0.3235	0.0235	2.551	10
11	0.0558	3.1473	8.4452	17.9216	56.4053	0.3177	0.0177	2.683	11
12	0.0429	3.1903	8.9173	23.2981	74.3270	0.3135	0.0135	2.795	12
13	0.0330	3.2233	9.3135	30.2875	97.6250	0.3102	0.0102	2.889	13
14	0.0254	3.2487	9.6437	39.3738	127.9125	0.3078	0.0078	2.968	14
15	0.0195	3.2682	9.9172	51.1859	167.2863	0.3060	0.0060	3.034	15
16	0.0150	3.2832	10.1426	66.5417	218.4722	0.3046	0.0046	3.089	16
17	0.0116	3.2948	10.3276	86.5042	285.0139	0.3035	0.0035	3.134	17
18	0.0089	3.3037	10.4788	112.4554	371.5180	0.3027	0.0027	3.171	18
19	0.0068	3.3105	10.6019	146.1920	483.9734	0.3021	0.0021	3.202	19
20	0.0053	3.3158	10.7019	190.0496	630.1655	0.3016	0.0016	3.227	20
21	0.0040	3.3198	10.7828	247.0645	820.2151	0.3012	0.0012	3.248	21
22	0.0031	3.3230	10.8482	321.1839	1067.2796	0.3009	0.0009	3.264	22
23	0.0024	3.3254	10.9009	417.5391	1388.4635	0.3007	0.0007	3.278	23
24	0.0018	3.3272	10.9433	542.8008	1806.0026	0.3006	0.0006	3.289	24
25	0.0014	3.3286	10.9773	705.6410	2348.8033	0.3004	0.0004	3.297	25
26	0.0011	3.3297	11.0045	917.3333	3054.4443	0.3003	0.0003	3.305	26
27	0.0008	3.3305	11.0263	1192.5333	3971.7776	0.3003	0.0003	3.310	27
28	0.0006	3.3312	11.0437	1550.2933	5164.3109	0.3002	0.0002	3.315	28
29	0.0005	3.3317	11.0576	2015.3813	6714.6042	0.3001	0.0001	3.318	29
30	0.0004	3.3321	11.0687	2619.9956	8729.9855	0.3001	0.0001	3.321	30
31	0.0003	3.3324	11.0775	3405.9943	11349.9811	0.3001	0.0001	3.324	31
32	0.0002	3.3326	11.0845	4427.7926	14755.9755	0.3001	0.0001	3.326	32
33	0.0002	3.3328	11.0901	5756.1304	19183.7681	0.3001	0.0001	3.327	33
34	0.0001	3.3329	11.0945	7482.9696	24939.8985	0.3000	0.0000	3.328	34
35	0.0001	3.3330	11.0980	9727.8604	32422.8681	0.3000	0.0000	3.329	35
36	0.0001	3.3331	11.1007	12646.2186	42150.7285	0.3000	0.0000	3.330	36
37	0.0001	3.3331	11.1029	16440.0841	54796.9471	0.3000	0.0000	3.331	37
38	0.0000	3.3332	11.1047	21372.1094	71237.0312	0.3000	0.0000	3.331	38
39	0.0000	3.3332	11.1060	27783.7422	92609.1405	0.3000	0.0000	3.331	39
40	0.0000	3.3332	11.1071	36118.8648	120392.8827	0.3000	0.0000	3.332	40
41	0.0000	3.3333	11.1080	46954.5243	156511.7475	0.3000	0.0000	3.332	41
42	0.0000	3.3333	11.1086	61040.8815	203466.2718	0.3000	0.0000	3.332	42
43	0.0000	3.3333	11.1092	79353.1460	264507.1533	0.3000	0.0000	3.332	43
44	0.0000	3.3333	11.1096	103159.0898	343860.2993	0.3000	0.0000	3.332	44
45	0.0000	3.3333	11.1099	134106.8167	447019.3890	0.3000	0.0000	3.333	45
46	0.0000	3.3333	11.1102	174338.8617	581126.2058	0.3000	0.0000	3.333	46
47	0.0000	3.3333	11.1104	226640.5202	755465.0675	0.3000	0.0000	3.333	47
48	0.0000	3.3333	11.1105	294632.6763	982105.5877	0.3000	0.0000	3.333	48
49	0.0000	3.3333	11.1107	383022.4792	1276738.2640	0.3000	0.0000	3.333	49
50	0.0000	3.3333	11.1108	497929.2230	1659760.7433	0.3000	0.0000	3.333	50

(continued)

APPENDIX 86.B *(continued)*
Factor Tables

$I = 40.00\%$

n	P/F	P/A	P/G	F/P	F/A	A/P	A/F	A/G	n
1	0.7143	0.7143	0.0000	1.4000	1.0000	1.4000	1.0000	0.000	1
2	0.5102	1.2245	0.5102	1.9600	2.4000	0.8167	0.4167	0.416	2
3	0.3644	1.5889	1.2391	2.7440	4.3600	0.6294	0.2294	0.779	3
4	0.2603	1.8492	2.0200	3.8416	7.1040	0.5408	0.1408	1.092	4
5	0.1859	2.0352	2.7637	5.3782	10.9456	0.4914	0.0914	1.358	5
6	0.1328	2.1680	3.4278	7.5295	16.3238	0.4613	0.0613	1.581	6
7	0.0949	2.2628	3.9970	10.5414	23.8534	0.4419	0.0419	1.766	7
8	0.0678	2.3306	4.4713	14.7579	34.3947	0.4291	0.0291	1.918	8
9	0.0484	2.3790	4.8585	20.6610	49.1526	0.4203	0.0203	2.042	9
10	0.0346	2.4136	5.1696	28.9255	69.8137	0.4143	0.0143	2.141	10
11	0.0247	2.4383	5.4166	40.4957	98.7391	0.4101	0.0101	2.221	11
12	0.0176	2.4559	5.6106	56.6939	139.2348	0.4072	0.0072	2.284	12
13	0.0126	2.4685	5.7618	79.3715	195.9287	0.4051	0.0051	2.334	13
14	0.0090	2.4775	5.8788	111.1201	275.3002	0.4036	0.0036	2.372	14
15	0.0064	2.4839	5.9688	155.5681	386.4202	0.4026	0.0026	2.403	15
16	0.0046	2.4885	6.0376	217.7953	541.9883	0.4018	0.0018	2.426	16
17	0.0033	2.4918	6.0901	304.9135	759.7837	0.4013	0.0013	2.444	17
18	0.0023	2.4941	6.1299	426.8789	1064.6971	0.4009	0.0009	2.457	18
19	0.0017	2.4958	6.1601	597.6304	1491.5760	0.4007	0.0007	2.468	19
20	0.0012	2.4970	6.1828	836.6826	2089.2064	0.4005	0.0005	2.476	20
21	0.0009	2.4979	6.1998	1171.3556	2925.8889	0.4003	0.0003	2.482	21
22	0.0006	2.4985	6.2127	1639.8978	4097.2445	0.4002	0.0002	2.486	22
23	0.0004	2.4989	6.2222	2295.8569	5737.1423	0.4002	0.0002	2.490	23
24	0.0003	2.4992	6.2294	3214.1997	8032.9993	0.4001	0.0001	2.492	24
25	0.0002	2.4994	6.2347	4499.8796	11247.1990	0.4001	0.0001	2.494	25
26	0.0002	2.4996	6.2387	6299.8314	15747.0785	0.4001	0.0001	2.495	26
27	0.0001	2.4997	6.2416	8819.7640	22046.9099	0.4000	0.0000	2.496	27
28	0.0001	2.4998	6.2438	12347.6696	30866.6739	0.4000	0.0000	2.497	28
29	0.0001	2.4999	6.2454	17286.7374	43214.3435	0.4000	0.0000	2.498	29
30	0.0000	2.4999	6.2466	24201.4324	60501.0809	0.4000	0.0000	2.498	30
31	0.0000	2.4999	6.2475	33882.0053	84702.5132	0.4000	0.0000	2.499	31
32	0.0000	2.4999	6.2482	47434.8074	118584.5185	0.4000	0.0000	2.499	32
33	0.0000	2.5000	6.2487	66408.7304	166019.3260	0.4000	0.0000	2.499	33
34	0.0000	2.5000	6.2490	92972.2225	232428.0563	0.4000	0.0000	2.499	34
35	0.0000	2.5000	6.2493	130161.1116	325400.2789	0.4000	0.0000	2.499	35
36	0.0000	2.5000	6.2495	182225.5562	455561.3904	0.4000	0.0000	2.499	36
37	0.0000	2.5000	6.2496	255115.7786	637786.9466	0.4000	0.0000	2.499	37
38	0.0000	2.5000	6.2497	357162.0901	892902.7252	0.4000	0.0000	2.499	38
39	0.0000	2.5000	6.2498	500026.9261	1250064.8153	0.4000	0.0000	2.499	39
40	0.0000	2.5000	6.2498	700037.6966	1750091.7415	0.4000	0.0000	2.499	40
41	0.0000	2.5000	6.2499	980052.7752	2450129.4381	0.4000	0.0000	2.500	41
42	0.0000	2.5000	6.2499	1372073.8853	3430182.2133	0.4000	0.0000	2.500	42
43	0.0000	2.5000	6.2499	1920903.4394	4802256.0986	0.4000	0.0000	2.500	43
44	0.0000	2.5000	6.2500	2689264.8152	6723159.5381	0.4000	0.0000	2.500	44
45	0.0000	2.5000	6.2500	3764970.7413	9412424.3533	0.4000	0.0000	2.500	45

Glossary

A

AASHTO: American Association of State and Highway Transportation Officials.

Abandonment: The reversion of title to the owner of the underlying fee where an easement for highway purposes is no longer needed.

Absorbed dose: The energy deposited by radiation as it passes through a material.

Absorption: The process by which a liquid is drawn into and tends to fill permeable pores in a porous body. Also, the increase in weight of a porous solid body resulting from the penetration of liquid into its permeable pores.

Abutment: A retaining wall that also supports a vertical load.

Accelerated flow: A form of varied flow in which the velocity is increasing and the depth is decreasing.

Acid: Any compound that dissociates in water into H^+ ions. (The combination of H^+ and water, H_3O^+, is known as the hydronium ion.) Acids conduct electricity in aqueous solutions, have a sour taste, turn blue litmus paper red, have a pH between 0 and 7, and neutralize bases, forming salts and water.

Activated sludge: Solids from aerated settling tanks that are rich in bacteria.

Active pressure: Pressure causing a wall to move away from the soil.

Adenosine triphosphate (ATP): The macromolecule that functions as an energy carrier in cells. The energy is stored in a high-energy bond between the second and third phosphates.

Adiabatic expansion: The process that occurs when an air mass rises and expands without exchanging heat with its surroundings.

Adjudication: A court proceeding to determine rights to the use of water on a particular stream or aquifer.

Admixture: Material added to a concrete mixture to increase its workability, strength, or imperviousness, or to lower its freezing point.

Adsorbed water: Water held near the surface of a material by electrochemical forces.

Adsorption: The physical adherence or bonding of ions and molecules onto the surface of another molecule. The preferred process used in cleanup of environmental spills.

Adsorption edge: The pH range where solute adsorption changes sharply.

Advection: The transport of solutes along stream lines at the average linear seepage flow velocity.

Aerated lagoon: A holding basin into which air is mechanically introduced to speed up aerobic decomposition.

Aeration: Mixing water with air, either by spraying water or diffusing air through water.

Aerobe: A microorganism whose growth requires free oxygen.

Aerobic: Requiring oxygen. Descriptive of a bacterial class that functions in the presence of free dissolved oxygen.

Aggregate: Inert material that is mixed with portland cement and water to produce concrete.

Aggregate, coarse: Aggregate retained on a no. 4 sieve.

Aggregate, fine: Aggregate passing the no. 4 sieve and retained on the no. 200 sieve.

Aggregate, lightweight: Aggregate having a dry density of 70 lbm/ft^3 (32 kg/m^3) or less.

Agonic line: A line with no magnetic declination.

Air change (Air flush): A complete replacement of the air in a room or other closed space.

Algae: Simple photosynthetic plants having neither roots, stems, nor leaves.

Alidade: A tachometric instrument consisting of a telescope similar to a transit, an upright post that supports the standards of the horizontal axis of the telescope, and a straightedge whose edges are essentially in the same direction as the line of sight. An alidade is used in the field in conjunction with a plane table.

Alkalinity: A measure of the capacity of a water to neutralize acid without significant pH change. It is usually associated with the presence of hydroxyl, carbonate, and/or bicarbonate radicals in the water.

Alternate depths: For a particular channel geometry and discharge, two depths at which water flows with the same specific energy in uniform flow. One depth corresponds to subcritical flow; the other corresponds to supercritical flow.

Alluvial deposit: A material deposited within the alluvium.

Alluvium: Sand, silt, clay, gravel, etc., deposited by running water.

Amictic: Experiencing no overturns or mixing. Typical of polar lakes.

Amphoteric behavior: Ability of an aqueous complex or solid material to have a negative, neutral, or positive charge.

Anabolism: The phase of metabolism involving the formation of organic compounds; usually an energy-utilizing process.

Anabranch: The intertwining channels of a braided stream.

Anaerobe: A microorganism that grows only or best in the absence of free oxygen.

Anaerobic: Not requiring oxygen. Descriptive of a bacterial class that functions in the absence of free dissolved oxygen.

Anion: Negative ion that migrates to the positive electrode (anode) in an electrolytic solution.

Anticlinal spring: A portion of an exposed aquifer (usually on a slope) between two impervious layers.

Apparent specific gravity of asphalt mixture: A ratio of the unit weight of an asphalt mixture (excluding voids permeable to water) to the unit weight of water.

Appurtenance: That which belongs with or is designed to complement something else. For example, a manhole is a sewer appurtenance.

Apron: An underwater "floor" constructed along the channel bottom to prevent scour. Aprons are almost always extensions of spillways and culverts.

Aquiclude: A saturated geologic formation with insufficient porosity to support any significant removal rate or contribute to the overall groundwater regime. In groundwater analysis, an aquiclude is considered to confine an aquifer at its boundaries.

Aquifer: Rock or sediment in a formation, group of formations, or part of a formation that is saturated and sufficiently permeable to transmit economic quantities of water to wells and springs.

Aquifuge: An underground geological formation that has absolutely no porosity or interconnected openings through which water can enter or be removed.

Aquitard: A saturated geologic unit that is permeable enough to contribute to the regional groundwater flow regime, but not permeable enough to supply a water well or other economic use.

Arterial highway: A general term denoting a highway primarily for through traffic, usually on a continuous route.

Artesian formation: An aquifer in which the piezometric height is greater than the aquifer height. In an artesian formation, the aquifer is confined and the water is under hydrostatic pressure.

Artesian spring: Water from an artesian formation that flows to the ground surface naturally, under hydrostatic pressure, due to a crack or other opening in the formation's confining layer.

Asphalt emulsion: A mixture of asphalt cement with water. Asphalt emulsions are produced by adding a small amount of emulsifying soap to asphalt and water. The asphalt sets when the water evaporates.

Atomic mass unit (AMU): One AMU is one twelfth the atomic weight of carbon.

Atomic number: The number of protons in the nucleus of an atom.

Atomic weight: Approximately, the sum of the numbers of protons and neutrons in the nucleus of an atom.

Autotroph: An organism than can synthesize all of its organic components from inorganic sources.

Auxiliary lane: The portion of a roadway adjoining the traveled way for truck climbing, speed change, or other purposes supplementary to through traffic movement.

Avogadro's law: A gram-mole of any substance contains 6.022×10^{23} molecules.

Azimuth: The horizontal angle measured from the plane of the meridian to the vertical plane containing the line. The azimuth gives the direction of the line with respect to the meridian and is usually measured in a clockwise direction with respect to either the north or south meridian.

B

Backward pass: The steps in a critical path analysis in which the latest start times are determined, usually after the earliest finish times have been determined in the forward pass.

Backwater: Water upstream from a dam or other obstruction that is deeper than it would normally be without the obstruction.

Backwater curve: A plot of depth versus location along the channel containing backwater.

Base: (a) A layer of selected, processed, or treated aggregate material of planned thickness and quality placed immediately below the pavement and above the subbase or subgrade soil. (b) Any compound that dissociates in water into OH$^-$ ions. Bases conduct electricity in aqueous solutions, have a bitter taste, turn red litmus paper blue, have a pH between 7 and 14, and neutralize acids, forming salts and water.

Base course: The bottom portion of a pavement where the top and bottom portions are not the same composition.

Base flow: Component of stream discharge that comes from groundwater flow. Water infiltrates and moves through the ground very slowly; up to 2 years may elapse between precipitation and discharge.

Batter pile: A pile inclined from the vertical.

Bed: A layer of rock in the earth. Also the bottom of a body of water such as a river, lake, or sea.

Bell: An enlarged section at the base of a pile or pier used as an anchor.

Belt highway: An arterial highway carrying traffic partially or entirely around an urban area.

Bent: A supporting structure (usually of a bridge) consisting of a beam or girder transverse to the supported roadway and that is supported, in turn, by columns at each end, making an inverted "U" shape.

Benthic zone: The bottom zone of a lake, where oxygen levels are low.

Benthos: Organisms (typically anaerobic) occupying the benthic zone.

Bentonite: A volcanic clay that exhibits extremely large volume changes with moisture content changes.

Berm: A shelf, ledge, or pile.

Bifurcation ratio: The average number of streams feeding into the next side (order) waterway. The range is usually 2 to 4.

Binary fission: An asexual reproductive process in which one cell splits into two independent daughter cells.

Bioaccumulation: The process by which chemical substances are ingested and retained by organisms, either from the environment directly or through consumption of food containing the chemicals.

Bioaccumulation factor: *See* Bioconcentration factor.

Bioactivation process: A process using sedimentation, trickling filter, and secondary sedimentation before adding activated sludge. Aeration and final sedimentation are the follow-up processes.

Bioassay: The determination of kinds, quantities, or concentrations, and in some cases, the locations of material in the body, whether by direct measurement (in vivo counting) or by analysis and evaluation of materials excreted or removed (in vitro) from the body.

Bioavailability: A measure of what fraction, how much, or the rate that a substance (ingested, breathed in,

dermal contact) is actually absorbed biologically. Bioavailability measurements are typically based on absorption into the blood or liver tissue.

Biochemical oxygen demand (BOD): The quantity of oxygen needed by microorganisms in a body of water to decompose the organic matter present. An index of water pollution.

Bioconcentration: The increase in concentration of a chemical in an organism resulting from tissue absorption (bioaccumulation) levels exceeding the rate of metabolism and excretion (biomagnification).

Bioconcentration factor: The ratio of chemical concentration in an organism to chemical concentration in the surrounding environment.

Biodegradation: The use of microorganisms to degrade contaminants.

Biogas: A mixture of approximately 55% methane and 45% carbon dioxide that results from the digestion of animal dung.

Biomagnification: The cumulative increase in the concentration of a persistent substance in successively higher levels of the food chain.

Biomagnification factor: *See* Bioconcentration factor.

Biomass: Renewable organic plant and animal material such as plant residue, sawdust, tree trimmings, rice straw, poultry litter and other animal wastes, some industrial wastes, and the paper component of municipal solid waste that can be converted to energy.

Biosorption process: A process that mixes raw sewage and sludge that have been pre-aerated in a separate tank.

Biosphere: The part of the world in which life exists.

Biota: All of the species of plants and animals indigenous to an area.

Bleeding: A form of segregation in which some of the water in the mix tends to rise to the surface of freshly placed concrete.

Blind drainage: Geographically large (with respect to the drainage basin) depressions that store water during a storm and therefore stop it from contributing to surface runoff.

Bloom: A phenomenon whereby excessive nutrients within a body of water results in an explosion of plant life, resulting in a depletion of oxygen and fish kill. Usually caused by urban runoff containing fertilizers.

Bluff: A high and steep bank or cliff.

Body burden: The total amount of a particular chemical in the body.

Braided stream: A wide, shallow stream with many anabranches.

Branch sewer: A sewer off the main sewer.

Breaking chain: A technique used when the slope is too steep to permit bringing the full length of the chain or tape to a horizontal position. When breaking chain, the distance is measured in partial tape lengths.

Breakpoint chlorination: Application of chlorine that results in a minimum of chloramine residuals. No significant free chlorine residual is produced unless the breakpoint is reached.

Bulking: *See* Sludge bulking.

Bulk specific gravity of asphalt mixture: Ratio of the unit weight of an asphalt mixture (including permeable and impermeable voids) to the unit weight of water.

Butte: A hill with steep sides that usually stands away from other hills.

C

Caisson: An air- and watertight chamber used as a foundation and/or used to work or excavate below the water level.

Capillary water: Water just above the water table that is drawn up out of an aquifer due to capillary action of the soil.

Carbonaceous demand: Oxygen demand due to biological activity in a water sample, exclusive of nitrogenous demand.

Carbonate hardness: Hardness associated with the presence of bicarbonate radicals in the water.

Carcinogen: A cancer-causing agent.

Carrier (biological): An individual harboring a disease agent without apparent symptoms.

Cascade impactor: *See* Impactor.

Cased hole: An excavation whose sides are lined or sheeted.

Catabolism: The chemical reactions by which food materials or nutrients are converted into simpler substances for the production of energy and cell materials.

Catena: A group of soils of similar origin occurring in the same general locale but that differ slightly in properties.

Cation: Positive ion that migrates to the negative electrode (cathode) in an electrolytic solution.

Cell wall: The cell structure exterior to the cell membrane of plants, algae, bacteria, and fungi. It give cells form and shape.

Cement-treated base: A base layer constructed with good-quality, well-graded aggregate mixed with up to 6% cement.

CFR: The Code of (U.S.) Federal Regulations, a compilation of all federal documents that have general applicability and legal effect, as published by the Office of the Federal Register.

Channelization: The separation or regulation of conflicting traffic movements into definite paths of travel by use of pavement markings, raised islands, or other means.

Chat: Small pieces of crushed rock and gravel. May be used for paving roads and roofs.

Check: A short section of built-up channel placed in a canal or irrigation ditch and provided with gates or flashboards to control flow or raise upstream level for diversion.

Chelate: A chemical compound into which a metallic ion (usually divalent) is tightly bound.

Chelation: (a) The process in which a compound or organic material attracts, combines with, and removes a metallic ion. (b) The process of removing metallic contaminants by having them combine with special added substances.

Chemical precipitation: Settling out of suspended solids caused by adding coagulating chemicals.

Chemocline: A steep chemical (saline) gradient separating layers in meromictic lakes.

Chloramine: Compounds of chlorine and ammonia (e.g., NH_2Cl, $NHCl_2$, or NCl_3).

Chlorine demand: The difference between applied chlorine and the chlorine residual. Chlorine demand is chlorine that has been reduced in chemical reactions and is no longer available for disinfection.

Class: A major taxonomic subdivision of a phylum. Each class is composed of one or more related orders.

Clean-out: A pipe through which snakes can be pushed to unplug a sewer.

Clearance: Distance between successive vehicles as measured between the vehicles, back bumper to front bumper.

Coliform: Gram-negative, lactose-fermenting rods, including escherichieae coli and similar species that normally inhabit the colon (large intestine). Commonly included in the coliform are Enterobacteria aerogenes, Klebsiella species, and other related bacteria.

Colloid: A fine particle ranging in size from 1 to 500 millimicrons. Colloids cause turbidity because they do not easily settle out.

Combined residuals: Compounds of an additive (such as chlorine) that have combined with something else. Chloramines are examples of combined residuals.

Combined system: A system using a single sewer for disposal of domestic waste and storm water.

Comminutor: A device that cuts solid waste into small pieces.

Compaction: Densification of soil by mechanical means, involving the expulsion of excess air.

Compensation level: In a lake, the depth of the limnetic area where oxygen production from light and photosynthesis are exactly balanced by depletion.

Support
Material

Complete mixing: Mixing accomplished by mechanical means (stirring).

Compound: A homogeneous substance composed of two or more elements that can be decomposed by chemical means only.

Concrete: A mixture of portland cement, fine aggregate, coarse aggregate, and water, with or without admixtures.

Concrete, normal weight: Concrete having a hardened density of approximately 150 lbm/ft^3.

Concrete, plain: Concrete that is not reinforced with steel.

Concrete, structural lightweight: A concrete containing lightweight aggregate.

Condemnation: The process by which property is acquired for public purposes through legal proceedings under power of eminent domain.

Cone of depression: The shape of the water table around a well during and immediately after use. The cone's water surface level differs from the original water table by the well's drawdown.

Confined water: Artesian water overlaid with an impervious layer, usually under pressure.

Confirmed test: A follow-up test used if the presumptive test for coliforms is positive.

Conflagration: Total involvement or engulfment (as in a fire).

Conjugate depths: The depths on either side of a hydraulic jump.

Connate water: Pressurized water (usually, high in mineral content) trapped in the pore spaces of sedimentary rock at the time it was formed.

Consolidation: Densification of soil by mechanical means, involving expulsion of excess water.

Contraction: A decrease in the width or depth of flow caused by the geometry of a weir, orifice, or obstruction.

Control of access: The condition where the right of owners or occupants of abutting land or other persons to access in connection with a highway is fully or partially controlled by public authority.

Critical depth: The depth that minimizes the specific energy of flow.

Critical flow: Flow at the critical depth and velocity. Critical flow minimizes the specific energy and maximizes discharge.

Critical slope: The slope that produces critical flow.

Critical velocity: The velocity that minimizes specific energy. When water is moving at its critical velocity, a disturbance wave cannot move upstream since the wave moves at the critical velocity.

Cuesta: (*from the Spanish word for cliff*) A hill with a steep slope on one side and a gentle slope on the other.

Cunette: A small channel in the invert of a large combined sewer for dry weather flow.

Curing: The process and procedures used for promoting the hydration of cement. It consists of controlling the temperature and moisture from and into the concrete.

Cyclone impactor: *See* Impactor.

D

Dead load: An inert, inactive load, primarily due to the structure's own weight.

Decision sight distance: Sight distance allowing for additional decision time in cases of complex conditions.

Delta: A deposit of sand and other sediment, usually triangular in shape. Deltas form at the mouths of rivers where the water flows into the sea.

Deoxygenation: The act of removing dissolved oxygen from water.

Deposition: The laying down of sediment such as sand, soil, clay, or gravel by wind or water. It may later be compacted into hard rock and buried by other sediment.

Depression storage: Initial storage of rain in small surface puddles.

Depth-area-duration analysis: A study made to determine the maximum amounts of rain within a given time period over a given area.

Detrial mineral: Mineral grain resulting from the mechanical disintegration of a parent rock.

Dewatering: Removal of excess moisture from sludge waste.

Digestion: Conversion of sludge solids to gas.

Dilatancy: The tendency of a material to increase in volume when undergoing shear.

Dilution disposal: Relying on a large water volume (lake or stream) to dilute waste to an acceptable concentration.

Dimictic: Experiencing two overturns per year. Dimictic lakes are usually found in temperate climates.

Dimiper lake: The freely circulating surface water with a small but variable temperature gradient.

Dimple spring: A depression in the earth below the water table.

Distribution coefficient: *See* Partition coefficient.

Divided highway: A highway with separated roadbeds for traffic in opposing directions.

Domestic waste: Waste that originates from households.

Downpull: A force on a gate, typically less at lower depths than at upper depths due to increased velocity, when the gate is partially open.

Drainage density: The total length of streams in a watershed divided by the drainage area.

Drawdown: The lowering of the water table level of an unconfined aquifer (or of the potentiometric surface of a confined aquifer) by pumping of wells.

Drawdown curve: *See* Cone of depression.

Dredge line: *See* Mud line.

Dry weather flow: *See* Base flow.

Dystrophic: Receiving large amounts of organic matter from surrounding watersheds, particularly humic materials from wetlands that stain the water brown. Dystrophic lakes have low plankton productivity, except in highly productive littoral zones.

E

Easement: A right to use or control the property of another for designated purposes.

Effective specific gravity of an asphalt mixture: Ratio of the unit weight of an asphalt mixture (excluding voids permeable to asphalt) to the unit weight of water.

Effluent: That which flows out of a process.

Effluent stream: A stream that intersects the water table and receives groundwater. Effluent streams seldom go completely dry during rainless periods.

Element: A pure substance that cannot be decomposed by chemical means.

Elutriation: A counter-current sludge washing process used to remove dissolved salts.

Elutriator: A device that purifies, separates, or washes material passing through it.

Embankment: A raised structure constructed of natural soil from excavation or borrow sources.

Eminent Domain: The power to take private property for public use without the owner's consent upon payment of just compensation.

Emulsion: *See* Asphalt emulsion.

Encroachment: Use of the highway right-of-way for non-highway structures or other purposes.

Energy gradient: The slope of the specific energy line (i.e., the sum of the potential and velocity heads).

Enteric: Intestinal.

Enzyme: An organic (protein) catalyst that causes changes in other substances without undergoing any alteration itself.

Ephemeral stream: A stream that goes dry during rainless periods.

Epilimnion: (Gr. for "upper lake"); the freely circulating surface water with a small but variable temperature gradient.

Equivalent weight: The amount of substance (in grams) that supplies one mole of reacting units. It is calculated as the molecular weight divided by the change in oxidation number experienced in a chemical reaction. An alternative calculation is the atomic weight of an element divided by its valence or the molecular weight of a radical or compound divided by its valence.

Escarpment: A steep slope or cliff.

Escherichieae coli (E. coli): *See* Coliform.

Estuary: An area where fresh water meets salt water.

Eutrophic: Nutrient-rich; a eutrophic lake typically has a high surface area-to-volume ratio.

Eutrophication: The enrichment of water bodies by nutrients (e.g., phosphorus). Eutroficaction of a lake normally contributes to a slow evolution into a bog, marsh, and ultimately, dry land.

Evaporite: Sediment deposited when sea water evaporates. Gypsum, salt, and anhydrite are evaporites.

Evapotranspiration: Evaporation of water from a study area due to all sources including water, soil, snow, ice, vegetation, and transpiration.

F

Facultative: Able to live under different or changing conditions. Descriptive of a bacterial class that functions either in the presence or absence of free dissolved oxygen.

Fecal coliform: Coliform bacterium present in the intestinal tracts and feces of warm-blooded animals.

Fines: Silt- and/or clay-sized particles.

First-stage demand: *See* Carbonaceous demand.

Flexible pavement: A pavement having sufficiently low bending resistance to maintain intimate contact with the underlying structure, yet having the required stability furnished by aggregate interlock, internal friction, and cohesion to support traffic.

Float: The amount of time that an activity can be delayed without delaying any succeeding activities.

Floc: Agglomerated colloidal particles.

Flotation: Addition of chemicals and bubbled air to liquid waste in order to get solids to float to the top as scum.

Flowing well: A well that flows under hydrostatic pressure to the surface. Also called an Artesian well.

Flow regime: (a) Subcritical, critical, or supercritical. (b) Entrance control or exit control.

Flume: In general, any open channel for carrying water. More specifically, an open channel constructed above the earth's surface, usually supported on a trestle or on piers.

Force main: A sewer line that is pressured.

Forebay: A reservoir holding water for subsequent use after it has been discharged from a dam.

Forward pass: The steps in a critical path analysis in which the earliest finish times are determined, usually before the latest start times are determined in the backward pass.

Freeboard distance: The vertical distance between the water surface and the crest of a dam or top of a channel side. The distance the water surface can rise before it overflows.

Freehaul: Pertaining to hauling "for free" (i.e., without being able to bill an extra amount over the contract charge).

Free residuals: Ions or compounds not combined or reduced. The presence of free residuals signifies excess dosage.

Freeway: A divided arterial highway with full control of access.

Freeze (in piles): A large increase in the ultimate capacity (and required driving energy) of a pile after it has been driven some distance.

Friable: Easily crumbled.

Frontage road: A local street or road auxiliary to and located on the side of an arterial highway for service to abutting property and adjacent areas, and for control of access.

Frost susceptibility: Susceptible to having water continually drawn up from the water table by capillary action, forming ice crystals below the surface (but above the frost line).

Fulvic acid: The alkaline-soluble portion of organic material (i.e., humus) that remains in solution at low pH and is of lower molecular weight. A breakdown product of cellulose from vascular plants.

Fungi: Aerobic, multicellular, nonphotosynthetic heterotrophic, eucaryote protists that degrade dead organic matter, releasing carbon dioxide and nitrogen.

G

Gap: Corresponding time between successive vehicles (back bumper to front bumper) as they pass a point on a roadway.

Gap-graded: A soil with a discontinuous range of soil particle sizes; for example, containing large particles and small particles but no medium-sized particles.

Geobar: A polymeric material in the form of a bar.

Geocell: A three-dimensional, permeable, polymeric (synthetic or natural) honeycomb or web structure, made of alternating strips of geotextiles, geogrids, or geomembranes.

Geocomposite: A manufactured or assembled material using at least one geosynthetic product among the components.

Geofoam: A polymeric material that has been formed by the application of the polymer in semiliquid form through the use of a foaming agent. Results in a lightweight material with high void content.

Geographic Information System (GIS): A digital database containing geographic information.

Geogrid: A planar, polymeric (synthetic or natural) structure consisting of a regular open network of integrally connected tensile elements that may be linked or formed by extrusion, bonding, or interlacing (knitting or lacing).

Geomat: A three-dimensional, permeable, polymeric (synthetic or natural) structure made of bonded filaments, used for soil protection and to bind roots and small plants in erosion control applications.

Geomembrane: A planar, relatively impermeable, polymeric (synthetic or natural) sheet. May be bituminous, elastomeric, or plastomeric.

Geonet: A planar, polymeric structure consisting of a regular dense network whose constituent elements are linked by knots or extrusions and whose openings are much larger than the constituents.

Geopipe: A polymeric pipe.

Geospacer: A three-dimensional polymeric structure with large void spaces.

Geostrip: A polymeric material in the form of a strip, with a width less than approximately 200 mm.

Geosynthetic: A planar, polymeric (synthetic or natural) material.

Geotextile: A planar, permeable, polymeric (synthetic or natural) textile material that may be woven, nonwoven, or knitted.

Glacial till: Soil resulting from a receding glacier, consisting of mixed clay, sand, gravel, and boulders.

GMT: *See* UTC.

Gobar gas: *See* Biogas.

Gore: The area immediately beyond the divergence of two roadways bounded by the edges of those roadways.

Gradient: The energy (head) loss per unit distance. *See also* Slope.

Gravel: Granular material retained on a no. 4 sieve.

Gravitational water: Free water in transit downward through the vadose (unsaturated) zone.

Grillage: A footing or part of a footing consisting of horizontally laid timbers or steel beams.

Groundwater: Loosely, all water that is underground as opposed to on the surface of the ground. Usually refers to water in the saturated zone below the water table.

Gumbo: Silty soil that becomes soapy, sticky, or waxy when wet.

H

Hardpan: A shallow layer of earth material that has become relatively hard and impermeable, usually through the decomposition of minerals.

Hard water: Water containing dissolved salts of calcium and magnesium, typically associated with bicarbonates, sulfates, and chlorides.

Head: (Total hydraulic) the sum of the elevation head, pressure head, and velocity head at a given point in an aquifer.

Headwall: Entrance to a culvert or sluiceway.

Headway: The time between successive vehicles as they pass a common point.

Heat of hydration: The exothermic heat given off by concrete as it cures.

Horizon: A layer of soil with different color or composition than the layers above and below it.

HOV: High-occupancy vehicle (e.g., bus).

Humic acid: The alkaline-soluble portion of organic material (i.e., humus) that precipitates from solution at low pH and is of higher molecular weight. A breakdown product of cellulose from vascular plants.

Humus: A grayish-brown sludge consisting of relatively large particle biological debris, such as the material sloughed off from a trickling filter.

Hydration: The chemical reaction between water and cement.

Hydraulic depth: Ratio of area in flow to the width of the channel at the fluid surface.

Hydraulic jump: An abrupt increase in flow depth that occurs when the velocity changes from supercritical to subcritical.

Hydraulic radius: Ratio of area in flow to wetted perimeter.

Hydrogen ion: The hydrogen atom stripped of its one orbital electron (H^+). It associates with a water molecule to form the hydronium ion (H_3O^+).

Hydrological cycle: The cycle experienced by water in its travel from the ocean, through evaporation and precipitation, percolation, runoff, and return to the ocean.

Hydrometeor: Any form of water falling from the sky.

Hydronium ion: *See* Hydrogen ion.

Hydrophilic: Seeking or liking water.

Hydrophobic: Avoiding or disliking water.

Hygroscopic: Absorbing moisture from the air.

Hygroscopic water: Moisture tightly adhering in a thin film to soil grains that is not removed by gravity or capillary forces.

Hypolimnion: (Gr. for "lower lake"); the deep, cold layer in a lake, below the epiliminion and metalimnion, cut off from the air above.

I

Igneous rock: Rock that forms when molten rock (magma or lava) cools and hardens.

Impactor: An environmental device that removes and measures micron-sized dusts, particles, and aerosols from an air stream.

Impervious layer: A geologic layer through which no water can pass.

Independent float: The amount of time that an activity can be delayed without affecting the float on any preceding or succeeding activities.

Infiltration: (a) Groundwater that enters sewer pipes through cracks and joints. (b) The movement of water downward from the ground surface through the upper soil.

Influent: Flow entering a process.

Influent stream: A stream above the water table that contributes to groundwater recharge. Influent streams may go dry during the rainless season.

Initial loss: The sum of interception and depression loss, excluding blind drainage.

In situ: "In place"; without removal; in original location.

Interception: The process by which precipitation is captured on the surfaces of vegetation and other impervious surfaces and evaporates before it reaches the land surface.

Interflow: Infiltrated subsurface water that travels to a stream without percolating down to the water level.

Intrusion: An igneous rock formed from magma that pushed its way through other rock layers. Magma often moves through rock fractures, where it cools and hardens.

Inverse condemnation: The legal process that may be initiated by a property owner to compel the payment of fair compensation when the owner's property has been taken or damaged for a public purpose.

Inversion layer: An extremely stable layer in the atmosphere in which temperature increases with elevation and mobility of airborne particles is restricted.

Inverted siphon: A sewer line that drops below the hydraulic grade line.

In vitro: Removed or obtained from an organism.

In vivo: Within an organism.

Ion: An atom that has either lost or gained one or more electrons, becoming an electrically charged particle.

Isogonic line: A line representing the magnetic declination.

Isotopes: Atoms of the same atomic number but having different atomic weights due to a variation in the number of neutrons.

J

Jam density: The density at which vehicles or pedestrians come to a halt.

Juvenile water: Water formed chemically within the earth from magma that has not participated in the hydrologic cycle.

K

Kingdom: A major taxonomic category consisting of several phyla or divisions.

Krause process: Mixing raw sewage, activated sludge, and material from sludge digesters.

L

Lagging: Heavy planking used to construct walls in excavations and braced cuts.

Lamp holes: Sewer inspection holes large enough to lower a lamp into but too small for a person.

Lane occupancy (ratio): The ratio of a lane's occupied time to the total observation time. Typically measured by a lane detector.

Lapse rate, dry: The rate that the atmospheric temperature decreases with altitude for a dry, adiabatic air mass.

Lapse rate, wet: The rate that the atmospheric temperature decreases with altitude for a moist, adiabatic air mass. The exact rate is a function of the moisture content.

Lateral: A sewer line that branches off from another.

Lava: Hot, liquid rock above ground. Also called lava once it has cooled and hardened.

Limnetic: Open water; extending down to the compensation level. Limnetic lake areas are occupied by suspended organisms (plankton) and free swimming fish.

Limnology: The branch of hydrology that pertains to the study of lakes.

Lipids: A group of organic compounds composed of carbon and hydrogen (e.g., fats, phospholipids, waxes, and steroids) that are soluble in a nonpolar, organic liquid (e.g., ether or chloroform); a constituent of living cells.

Lipiphilic: Having an affinity for lipids.

Littoral: Shallow; heavily oxygenated. In littoral lake areas, light penetrates all the way through to the bottom, and the zone is usually occupied by a diversity of rooted plants and animals.

Live load: The weight of all nonpermanent objects in a structure, including people and furniture. Live load does not include seismic or wind loading.

Loess: A deposit of wind-blown silt.

Lysimeter: A container used to observe and measure percolation and mineral leaching losses due to water percolating through the soil in it.

M

Magma: Hot, liquid rock under the earth's surface.

Main: A large sewer at which all other branches terminate.

Malodorous: Offensive smelling.

Marl: An earthy substance containing 35 to 65% clay and 65 to 35% carbonate formed under marine or freshwater conditions.

Meander corner: A survey point set where boundaries intersect the bank of a navigable stream, wide river, or large lake.

Meandering stream: A stream with large curving changes of direction.

Median: The portion of a divided highway separating traffic traveling in opposite directions.

Median lane: A lane within the median to accommodate left-turning vehicles.

Meridian: A great circle of the earth passing through the poles.

Meromictic lake: A lake with a permanent hypolimnion layer that never mixes with the epilimnion. The hypolimnion layer is perennially stagnant and saline.

Mesa: A flat-topped hill with steep sides.

Mesophyllic bacteria: Bacteria growing between $10°C$ and $40°C$, with an optimum temperature of $37°C$. $40°C$ is, therefore, the upper limit for most wastewater processes.

Metabolism: All cellular chemical reactions by which energy is provided for vital processes and new cell substances are assimilated.

Metalimnion: A middle portion of a lake, between the epilimnion and hypolimnion, characterized by a steep and rapid decline in temperature (e.g., $1°C$ for each meter of depth).

Metamorphic rock: Rock that has changed from one form to another by heat or pressure.

Meteoric water: *See* Hydrometeor.

Methylmercury: A form of mercury that is readily absorbed through the gills of fish, resulting in large bioconcentration factors. Methymercury is passed on to organisms higher in the food chain.

Microorganism: A microscopic form of life.

Mixture: A heterogeneous physical combination of two or more substances, each retaining its identity and specific properties.

Mohlman index: *See* Sludge volume index.

Mole: A quantity of substance equal to its molecular weight in grams (gmole or gram-mole) or in pounds (pmole or pound-mole).

Molecular weight: The sum of the atomic weights of all atoms in a molecule.

Monomictic: Experiencing one overturn per year. Monomictic lakes are typically very large and/or deep.

Mud line: The lower surface of an excavation or braced cut.

N

Nephelometric turbidity unit: The unit of measurement for visual turbidity in water and other solutions.

Net rain: That portion of rain that contributes to surface runoff.

Nitrogen fixation: The formation of nitrogen compounds (NH_3, organic nitrogen) from free atmospheric nitrogen (N_2).

Nitrogenous demand: Oxygen demand from nitrogen consuming bacteria.

Node: An activity in a precedence (critical path) diagram.

Nonpathogenic: Not capable of causing disease.

Nonpoint source: A pollution source caused by sediment, nutrients, and organic and toxic substances originating from land-use activities, usually carried to lakes and streams by surface runoff from rain or snowmelt. Nonpoint pollutants include fertilizers, herbicides, insecticides, oil and grease, sediment, salt, and bacteria. Nonpoint sources do not generally require NPDES permits.

Nonstriping sight distance: Nonstriping sight distances are in between stopping and passing distances, and they exceed the minimum sight distances required for marking no-passing zones. They provide a practical distance to complete the passing maneuver in a reasonably safe manner, eliminating the need for a no-passing zone pavement marking.

Normal depth: The depth of uniform flow. This is a unique depth of flow for any combination of channel conditions. Normal depth can be determined from the Manning equation.

Normally consolidated soil: Soil that has never been consolidated by a greater stress than presently existing.

NPDES: National Pollutant Discharge Elimination System.

NTU: *See* Nephelometric turbidity unit.

O

Observation well: A nonpumping well used to observe the elevation of the water table or the potentiometric surface. An observation well is generally of larger diameter than a piezometer well and typically is screened or slotted throughout the thickness of the aquifer.

Odor number: *See* Threshold odor number.

Oligotrophic: Nutrient-poor. Oligotrophic lakes typically have low surface area-to-volume ratios and largely inorganic sediments and are surrounded by nutrient-poor soil.

Order: In taxonomy, a major subdivision of a class. Each order consists of one or more related families.

Orthotropic bridge deck: A bridge deck, usually steel plate covered with a wearing surface, reinforced in one direction with integral cast-in-place concrete ribs. Used to reduce the bridge deck mass.

Orthotropic material: A material with different strengths (stiffnesses) along different axes.

Osmosis: The flow of a solvent through a semipermeable membrane separating two solutions of different concentrations.

Outcrop: A natural exposure of a rock bed at the earth's surface.

Outfall: A pipe that discharges treated wastewater into a lake, stream, or ocean.

Overchute: A flume passing over a canal to carry floodwaters away without contaminating the canal water below. An elevated culvert.

Overhaul: Pertaining to billable hauling (i.e., with being able to bill an extra amount over the contract charge).

Overland flow: Water that travels over the ground surface to a stream.

Overturn: (a) The seasonal (fall) increase in epilimnion depth to include the entire lake depth, generally aided by unstable temperature/density gradients. (b) The seasonal (spring) mixing of lake layers, generally aided by wind.

Oxidation: The loss of electrons in a chemical reaction. Opposite of *reduction*.

Oxidation number: An electrical charge assigned by a set of prescribed rules, used in predicting the formation of compounds in chemical reactions.

P

Pan: A container used to measure surface evaporation rates.

Support Material

Parkway: An arterial highway for noncommercial traffic, with full or partial control of access, usually located within a park or a ribbon of park-like development.

Partial treatment: Primary treatment only.

Partition coefficient: The ratio of the contaminant concentration in the solid (unabsorbed) phase to the contaminant concentration in the liquid (absorbed) phase when the system is in equilibrium; typically represented as K_d.

Passing sight distance: The length of roadway ahead required to pass another vehicle without meeting an oncoming vehicle.

Passive pressure: A pressure acting to counteract active pressure.

Pathogenic: Capable of causing disease.

Pathway: An environmental route by which chemicals can reach receptors.

Pay as you throw: An administration scheme by which individuals are charged based on the volume of municipal waste discarded.

Pedology: The study of the formation, development, and classification of natural soils.

Penetration treatment: Application of light liquid asphalt to the roadbed material. Used primarily to reduce dust.

Perched spring: A localized saturated area that occurs above an impervious layer.

Percolation: The movement of water through the subsurface soil layers, usually continuing downward to the groundwater table.

Permanent hardness: Hardness that cannot be removed by heating.

Person-rem: A unit of the amount of total radiation received by a population. It is the product of the average radiation dose in rems times the number of people exposed in the population group.

pH: A measure of a solution's hydrogen ion concentration (acidity).

Phreatic zone: The layer the water table down to an impervious layer.

Phreatophytes: Plants that send their roots into or below the capillary fringe to access groundwater.

Phytoplankton: Small drifting plants.

Pier shaft: The part of a pier structure that is supported by the pier foundation.

Piezometer: A nonpumping well, generally of small diameter, that is used to measure the elevation of the water table or potentiometric surface. A piezometer generally has a short well screen through which water can enter.

Piezometer nest: A set of two or more piezometers set close to each other but screened to different depths.

Piezometric level: The level to which water will rise in a pipe due to its own pressure.

Pile bent: A supporting substructure of a bridge consisting of a beam or girder transverse to the roadway and that is supported, in turn, by a group of piles.

Pitot tube traverse: A volume or velocity measurement device that measures the impact energy of an air or liquid flow simultaneously at various locations in the flow area.

Planimeter: A device used to measure the area of a drawn shape.

Plant mix: A paving mixture that is not prepared at the paving site.

Plat: (a) A plan showing a section of land. (b) A small plot of land.

pOH: A measure of a solution's hydroxyl radical concentration (alkalinity).

Point source: A source of pollution that discharges into receiving waters from easily identifiable locations (e.g., a pipe or feedlot). Common point sources are factories and municipal sewage treatment plants. Point sources typically require NPDES permits.

Pollutant: Any solute or cause of change in physical properties that renders water, soil, or air unfit for a given use.

Polymictic: Experiencing numerous or continual overturns. Polymictic lakes, typically in the high mountains of equatorial regions, experience little seasonable temperature change.

Porosity: The ratio of pore volume to total rock, sediment, or formation volume.

Post-chlorination: Addition of chlorine after all other processes have been completed.

Potable: Suitable for human consumption.

Pre-chlorination: Addition of chlorine prior to sedimentation to help control odors and to aid in grease removal.

Presumptive test: A first-stage test in coliform fermentation. If positive, it is inconclusive without follow-up testing. If negative, it is conclusive.

Prime coat: The initial application of a low-viscosity liquid asphalt to an absorbent surface, preparatory to any subsequent treatment, for the purpose of hardening or toughening the surface and promoting adhesion between it and the superimposed constructed layer.

Probable maximum rainfall: The rainfall corresponding to some given probability (e.g., 1 in 100 years).

Protium: The stable ^{1}H isotope of hydrogen.

Protozoa: Single-celled aquatic animals that reproduce by binary fission. Several classes are known pathogens.

Putrefaction: Anaerobic decomposition of organic matter with accompanying foul odors.

Pycnometer: A closed flask with graduations.

Q

q-curve: A plot of depth of flow versus quantity flowing for a channel with a constant specific energy.

R

Rad: Abbreviation for "radiation absorbed dose." A unit of the amount of energy deposited in or absorbed by a material. A rad is equalto 62.5×10^6 MeV per gram of material.

Radical: A charged group of atoms that act together as a unit in chemical reactions.

Ranger: *See* Wale.

Rapid flow: Flow at less than critical depth, typically occurring on steep slopes.

Rating curve: A plot of quantity flowing versus depth for a natural watercourse.

Reach: A straight section of a channel, or a section that is uniform in shape, depth, slope, and flow quantity.

Redox reaction: A chemical reaction in which oxidation and reduction occur.

Reduction: The loss of oxygen or the gain of electrons in a chemical reaction. Opposite of *oxidation*.

Refractory: Dissolved organic materials that are biologically resistant and difficult to remove.

Regulator: A weir or device that diverts large volume flows into a special high-capacity sewer.

Rem: Abbreviation for "Roentgen equivalent mammal." A unit of the amount of energy absorbed by human tissue. It is the product of the absorbed dose (rad) times the quality factor.

Residual: A chemical that is left over after some of it has been combined or inactivated.

Resilient modulus: The modulus of elasticity of the soil.

Respiration: Any biochemical process in which energy is released. Respiration may be aerobic (in the presence of oxygen) or anaerobic (in the absence of oxygen).

Restraint: Any limitation (e.g., scarcity of resources, government regulation, or nonnegativity requirement) placed on a variable or combination of variables.

Resurfacing: A supplemental surface or replacement placed on an existing pavement to restore its riding qualities or increase its strength.

Retarded flow: A form of varied flow in which the velocity is decreasing and the depth is increasing.

Retrograde solubility: Solubility that decreases with increasing temperature. Typical of calcite (calcium carbonate, $CaCO_3$) and radon.

Right of access: The right of an abutting land owner for entrance to or exit from a public road.

Rigid pavement: A pavement structure having portland cement concrete as one course.

Rip rap: Pieces of broken stone used as lining to protect the sides of waterways from erosion.

Roadbed: That portion of a roadway extending from curb line to curb line or from shoulder line to shoulder line. Divided highways are considered to have two roadbeds.

Road mix: A low-quality asphalt surfacing produced from liquid asphalts and used when plant mixes are not available or economically feasible and where volume is low.

Roentgen: The amount of energy absorbed in air by the passage of gamma or x-rays. A roentgen is equal to 5.4×10^7 MeV per gram of air or 0.87 rad per gram of air and 0.96 rad per gram of tissue.

S

Safe yield: The maximum rate of water withdrawal that is economically, hydrologically, and ecologically feasible.

Sag pipe: *See* Inverted siphon.

Saline: Dominated by anionic carbonate, chloride, and sulfate ions.

Salt: An ionic compound formed by direct union of elements, reactions between acids and bases, reaction of acids and salts, and reactions between different salts.

Sand: Granular material passing through a no. 4 sieve but predominantly retained on a no. 200 sieve.

Sand trap: A section of channel constructed deeper than the rest of the channel to allow sediment to settle out.

Scour: Erosion typically occurring at the exit of an open channel or toe of a spillway.

Scrim: An open-weave, woven or nonwoven, textile product that is encapsulated in a polymer (e.g., polyester) material to provide strength and reinforcement to a watertight membrane.

Seal coat: An asphalt coating, with or without aggregate, applied to the surface of a pavement for the purpose of waterproofing and preserving the surface, altering the surface texture of the pavement, providing delineation, or providing resistance to traffic abrasion.

Second-stage demand: *See* Nitrogenous demand.

Sedimentary rock: Rocks formed from sediment, broken rocks, or organic matter. Sedimentary rocks are formed when wind or water deposits sediment into layers, which are pressed together by more layers of sediment above.

Seed: The activated sludge initially taken from a secondary settling tank and returned to an aeration tank to start the activated sludge process.

Seep: *See* Spring.

Seiche: External: an oscillation of the surface of a landlocked body of water; Internal: an alternating pattern in the directions of layers of lake water movement.

Sensitivity: The ratio of a soil's undisturbed strength to its disturbed strength.

Separate system: A system with separate sewers for domestic and storm wastewater.

Septic: Produced by putrefaction.

Settling basin: A large, shallow basin through which water passes at low velocity, where most of the suspended sediment settles out.

Sheeted pit: *See* Cased hole.

Shooting flow: *See* Rapid flow.

Sight distance: The length of roadway that a driver can see.

Sinkhole: A natural dip or hole in the ground formed when underground salt or other rocks are dissolved by water and the ground above collapses into the empty space.

Sinuosity: The stream length divided by the valley length.

Slickenside: A surface (plane) in stiff clay that is a potential slip plane.

Slope: The tangent of the angle made by the channel bottom. *See also* Gradient.

Sludge: The precipitated solid matter produced by water and sewage treatment.

Sludge bulking: Failure of suspended solids to completely settle out.

Sludge volume index (SVI): The volume of sludge that settles in 30 min out of an original volume of 1000 mL. May be used as a measure of sludge bulking potential.

Sol: A homogenous suspension or dispersion of colloidal matter in a fluid.

Soldier pile: An upright pile used to hold lagging.

Solution: A homogeneous mixture of solute and solvent.

Sorption: A generic term covering the processes of absorption and adsorption.

Space mean speed: One of the measures of average speed of a number of vehicles over a common (fixed) distance. Determined as the inverse of the average time per unit distance. Usually less than time mean speed.

Spacing: Distance between successive vehicles, measured front bumper to front bumper.

Specific activity: The activity per gram of a radioisotope.

Specific storage: *See* Specific yield.

Specific yield: (a) The ratio of water volume that will drain freely (under gravity) from a sample to the total volume. Specific yield is always less than porosity. (b) The amount of water released from or taken into storage per unit volume of a porous medium per unit change in head.

Split chlorination: Addition of chlorine prior to sedimentation and after final processing.

Spring: A place where water flows or ponds on the surface due to the intersection of an aquifer with the earth surface.

Stadia method: Obtaining horizontal distances and differences in elevation by indirect geometric methods.

Stage: Elevation of flow surface above a fixed datum.

Standing wave: A stationary wave caused by an obstruction in a water course. The wave cannot move (propagate) because the water is flowing at its critical speed.

Steady flow: Flow in which the flow quantity does not vary with time at any location along the channel.

Stilling basin: An excavated pool downstream from a spillway used to decrease tailwater depth and to produce an energy-dissipating hydraulic jump.

Stoichiometry: The study of how elements combine in fixed proportions to form compounds.

Stopping sight distance: The distance that allows a driver traveling at the maximum speed to stop before hitting an observed object.

Stratum: Layer.

Stream gaging: A method of determining the velocity in an open channel.

Stream order: An artificial categorization of stream genealogy. Small streams are first order. Second-order streams are fed by first-order streams, third-order streams are fed by second-order streams, and so on.

Stringer: *See* Wale.

Structural section: The planned layers of specific materials, normally consisting of subbase, base, and pavement, placed over the subbase soil.

Subbase: A layer of aggregate placed on the existing soil as a foundation for the base.

Subcritical flow: Flow with depth greater than the critical depth and velocity less than the critical velocity.

Subgrade: The portion of a roadbed surface that has been prepared as specified, upon which a subbase, base, base course, or pavement is to be constructed.

Submain: *See* Branch sewer.

Substrate: A substance acted upon by an organism, chemical, or enzyme. Sometimes used to mean organic material.

Subsurface runoff: *See* Interflow.

Superchlorination: Chlorination past the breakpoint.

Supercritical flow: Flow with depth less than the critical depth and velocity greater than the critical velocity.

Superelevation: Roadway banking on a horizontal curve for the purpose of allowing vehicles to maintain the traveled speed.

Supernatant: The clarified liquid rising to the top of a sludge layer.

Surcharge: An additional loading. (a) In geotechnical work, any force loading added to the in situ soil load. (b) In water resources, any additional pressurization of a fluid in a pipe.

Surcharged sewer: (a) A sewer that is flowing under pressure (e.g., as a force main). (b) A sewer that is supporting an additional loading (e.g., a truck parked above it).

Surface detention: *See* Surface retention.

Surface retention: The part of a storm that does not contribute to runoff. Retention is made up of depression storage, interception, and evaporation.

Surface runoff: Water flow over the surface that reaches a stream after a storm.

Surficial: Pertaining to the surface.

Swale: (a) A low-lying portion of land, below the general elevation of the surroundings. (b) A natural ditch or long, shallow depression through which accumulated water from adjacent watersheds drains to lower areas.

T

Tack coat: The initial application of asphalt material to an existing asphalt or concrete surface to provide a bond between the existing surface and the new material.

Tail race: An open waterway leading water out of a dam spillway and back to a natural channel.

Tailwater: The water into which a spillway or outfall discharges.

Taxonomy: The description, classification, and naming of organisms.

Temporary hardness: Hardness that can be removed by heating.

Theodolite: A survey instrument used to measure or lay off both horizontal and vertical angles.

Thermocline: The temperature gradient in the metalimnion.

Thermophilic bacteria: Bacteria that thrive in the 45°C to 75°C range. The optimum temperature is near 55°C.

Thixotropy: A property of a soil that regains its strength over time after being disturbed and weakened.

Threshold odor number: A measure of odor strength, typically the number of successive dilutions required to reduce an odorous liquid to undetectable (by humans) level.

Till: *See* Glacial till.

Time mean speed: One of the measures of average speed of a number of vehicles over a common (fixed) distance. Determined as the average vehicular speed over a distance. Usually greater than space mean speed.

Time of concentration: The time required for water to flow from the most distant point on a runoff area to the measurement or collection point.

TON: *See* Threshold odor number.

Topography: Physical features such as hills, valleys, and plains that shape the surface of the earth.

Total float: The amount of time that an activity in the critical path (e.g., project start) can be delayed without delaying the project completion date.

Township: A square parcel of land 6 mi on each side.

Toxin: A toxic or poisonous substance.

Tranquil flow: Flow at greater than the critical depth.

Transmissivity: The rate at which water moves through a unit width of an aquifer or confining bed under a unit hydraulic gradient. It is a function of properties of the liquid, the porous media, and the thickness of the porous media.

Transpiration: The process by which water vapor escapes from living plants (principally from the leaves) and enters the atmosphere.

Traveled way: The portion of the roadway for the movement of vehicles, exclusive of shoulders and auxiliary lanes.

Turbidity: (a) Cloudiness in water caused by suspended colloidal material. (b) A measure of the light-transmitting properties of water.

Turbidity unit: *See* Nephelometric turbidity unit.

Turnout: (a) A location alongside a traveled way where vehicles may stop off of the main road surface without impeding following vehicles. (b) A pipe placed through a canal embankment to carry water from the canal for other uses.

U

Uniform flow: Flow that has constant velocity along a streamline. For an open channel, uniform flow implies constant depth, cross-sectional area, and shape along its course.

Unit process: A process used to change the physical, chemical, or biological characteristics of water or wastewater.

Unit stream power: The product of velocity and slope, representing the rate of energy expenditure per unit mass of water.

Uplift: Elevation or raising of part of the earth's surface through forces within the earth.

UTC: Coordinated Universal Time, the international time standard; previously referred to as Greenwich Meridian Time (GMT).

V

Vadose water: All underground water above the water table, including soil water, gravitational water, and capillary water.

Vadose zone: A zone above the water table containing both saturated and empty soil pores.

Valence: The relative combining capacity of an atom or group of atoms compared to that of the standard hydrogen atom. Essentially equivalent to the oxidation number.

Varied flow: Flow with different depths along the water course.

Varve: A layer of different material in the soil; fine layers of alluvium sediment deposited in glacial lakes.

Vitrification: Encapsulation in or conversion to an extremely stable, insoluble, glasslike solid by melting (usually electrically) and cooling. Used to destroy or immobilize hazardous compounds in soils.

Volatile organic chemicals (VOCs): A class of toxic chemicals that easily evaporate or mix with the atmosphere and environment.

Volatile solid: Solid material in a water sample or in sludge that can be burned away or vaporized at high temperature.

Volatilization: The driving off or evaporation of a liquid in a solid or one or more phases in a liquid mixture.

W

Wah gas: *See* Biogas.

Wale: A horizontal brace used to hold timbers in place against the sides of an excavation or to transmit the braced loads to the lagging.

Wasteway: A canal or pipe that returns excess irrigation water to the main channel.

Water table: The piezometric surface of an aquifer, defined as the locus of points where the water pressure is equal to the atmospheric pressure.

Waving the rod: A survey technique used when reading a rod reading for elevation data. By waving the rod, the highest rod reading will indicate the point at which the rod is most vertical. The reading at that point is then used to determine the difference in elevation.

Wet well: A short-term storage tank from which liquid is pumped.

Wetted perimeter: The length of the channel cross section that has water contact. The air-water interface is not included in the wetted perimeter.

X

Xeriscape: Creative landscaping for water and energy efficiency and lower maintenance.

Xerophytes: Drought-resistant plants, typically with root systems well above the water table.

Z

Zone of aeration: *See* Vadose zone.

Zone of saturation: *See* Phreatic zone.

Zoogloea: The gelatinous film (i.e., "slime") of aerobic organisms that covers the exposed surfaces of a biological filter.

Zooplankton: Small, drifting animals capable of independent movement.

Index

Italicized page numbers represent sources of data in tables, figures, and appendices. "G" indicates a glossary entry.

Italicized page numbers represent sources of data in tables, figures, and appendices. "G" indicates a glossary entry.

Italicized page numbers represent sources of data in tables, figures, and appendices. "G" *indicates a glossary entry.*

under the standard normal curve, *A-12*
weaving, 73-21
Areal rain, 20-19
Argument, 3-8
Arithmetic
sequence, 3-11
series, 3-11
Arm, moment, 41-2
Aromatic, *23-3*
alcohol, *23-2*
liquid, 14-9 (ftn)
Array, 4-1
Arrival type, 73-13
Arterial, 73-3
highway, G
Artesian
formation, G
spring, G
well, 21-2
Artificial expense, 86-20
Aryl, *23-1*
halide, *23-3*
Asbestos, 32-4
-cement pipe, 16-10, 26-10, 28-4
substitutes, 32-4, *32-5*
Asbestosis, 32-4
As-built documents, 85-14
Ascaris lumbricoides, 27-2
ASCE (*see* American Society of Civil
Engineers)
ASD, 58-4
Aseptic meningitis, 27-2
As fired, 24-3
Ash, 24-3
air heater, 24-3
bottom, 24-3, 32-5
coal, 24-3
combined, 32-5
economizer, 24-3
fly, 24-3
pit, 24-3
refractory, 24-3
soda, 26-13
ASME long-radius nozzle, 17-30 (ftn)
Aspect ratio, 17-37
diaphragm, *68-12*
ellipse, 7-11
Aspergillus, 27-2
Asphalt
absorbed, 75-9
binder, 75-2
grade, Superpave, *75-3*
cement, 75-2
concrete, 75-1
mix, *75-4*
content
effective, 75-9
optimum, 75-11
cutback, 75-2
emulsion, 75-2, G
grading, 75-2
mix, 75-3
hot rubber, 75-24
mixer, 75-6
modifier, 75-5

pavement, 75-1
full-depth, *75-20, 75-21*
recycling, 75-22
polymer-modified, 75-5
rubber, 75-23
stone matrix, 75-23
synthetic, 75-24
Asphaltic coating, pipe, 16-11
ASR, 48-2
Assembly, door, 81-4
Assessment, life-cycle, 85-7
Asset
account, 86-33
current, 86-34
fixed, 86-34
Assignment, 87-3
Associative law
addition, 3-3
matrix, 4-4
multiplication, 3-3
sets, 11-2
Assumed
inflection point, 47-18
meridian, 77-11
Assumption, Whitney, 50-9
AST, 33-1
ASTM, 81-2, 81-3
C330, 48-9
E-119, 81-4
prestressing tendon, *56-5*
soil tests, 35-15, 35-16, *35-16*
standards
pipe, 16-11
steel reinforcing, *A-91, A-92*
Asymmetrical function, 7-4
Asymptote, 7-2
Asymptotic, 7-2
Atmosphere,
air, *30-8*
properties, *A-14*
earth, 15-12
pressure, *14-3*
standard, 14-3, *A-33*
Atmospheric
air, 24-8, 30-8
fluidized bed combustion, 24-5
head, 18-6 (ftn)
refraction, 77-9
Atom, 22-1
Atomic
mass unit, 22-1
number, 22-2, *A-46*, G
numbers and weights of the elements,
A-51
structure, 22-1
weight, 22-1, *A-46*, G
Atrazine, 32-11
At-rest
earth pressure coefficient, 37-5
soil pressure, 37-5
Attack, intergranular, 22-16
Attenuator, impact, 74-8
Atterberg limit test, 35-4, 35-21

Auger
boring, 40-11
hole method, 35-23
Augmented matrix, 4-1
Austenitic stainless steel, 22-16 (ftn)
Authority having jurisdiction, 81-2
Autochthonous material, 25-7
Autogenous
fuel, 31-11
waste, 34-8 (ftn)
Autoignition temperature, 24-8
Automatic traffic recorder, 73-5
Autotroph, 27-1, 27-3, G
Autotrophic bacteria, 28-9
Auxiliary
equation, 10-1
lane, G
view, 2-2
Available
combined residual, 28-13
compressive strength, 62-2
hydrogen, 24-3, 24-14
method, 32-13
moment table, 59-4
Average
annual daily
flow, 26-18, 73-4
traffic, 73-5
daily traffic, 73-4
end area method, 79-4
highway speed, 73-4
power, 83-7
precipitation, 20-2
pressure, 15-6
running speed, 73-4
spot speed, 73-4
travel speed, 73-4
value
by integration, 9-4
sinusoid, 83-5
velocity, 16-8
Avogadro's
hypothesis, 22-4
law, 22-4, G
number, 22-4
AWWA standards, pipe, 16-11
Axial
compression
large, 62-2
small, 62-2
flow
impeller, 18-4, 26-9
turbine, 18-24
loading, 44-11
member, 41-11
strain, 43-4
stress, 35-25
tensile stress, 60-1
Axis,
bending, 45-15
conjugate, 7-11
neutral, 44-10
oblique, 2-2
parabolic, 7-10
principal, 42-8, 59-14, 70-2

Italicized page numbers represent sources of data in tables, figures, and appendices. "G" indicates a glossary entry.

Italicized page numbers represent sources of data in tables, figures, and appendices. "G" indicates a glossary entry.

Italicized page numbers represent sources of data in tables, figures, and appendices. "G" indicates a glossary entry.

Index

Italicized page numbers represent sources of data in tables, figures, and appendices. "G" indicates a glossary entry.

Italicized page numbers represent sources of data in tables, figures, and appendices. "G" indicates a glossary entry.

Italicized page numbers represent sources of data in tables, figures, and appendices. "G" indicates a glossary entry.

Italicized page numbers represent sources of data in tables, figures, and appendices. "G" indicates a glossary entry.

Italicized page numbers represent sources of data in tables, figures, and appendices. "G" indicates a glossary entry.

Italicized page numbers represent sources of data in tables, figures, and appendices. "G" indicates a glossary entry.

Italicized page numbers represent sources of data in tables, figures, and appendices. "G" indicates a glossary entry.

Index

Italicized page numbers represent sources of data in tables, figures, and appendices. "G" *indicates a glossary entry.*

Italicized page numbers represent sources of data in tables, figures, and appendices. "G" indicates a glossary entry.

Italicized page numbers represent sources of data in tables, figures, and appendices. "G" indicates a glossary entry.

Italicized page numbers represent sources of data in tables, figures, and appendices. "G" indicates a glossary entry.

Italicized page numbers represent sources of data in tables, figures, and appendices. "G" indicates a glossary entry.

Italicized page numbers represent sources of data in tables, figures, and appendices. "G" indicates a glossary entry.

Index

Italicized page numbers represent sources of data in tables, figures, and appendices. "G" indicates a glossary entry.

Italicized page numbers represent sources of data in tables, figures, and appendices. "G" indicates a glossary entry.

Index

Italicized page numbers represent sources of data in tables, figures, and appendices. "G" indicates a glossary entry.

Italicized page numbers represent sources of data in tables, figures, and appendices. "G" indicates a glossary entry.

Italicized page numbers represent sources of data in tables, figures, and appendices. "G" indicates a glossary entry.

Italicized page numbers represent sources of data in tables, figures, and appendices. "G" indicates a glossary entry.

Italicized page numbers represent sources of data in tables, figures, and appendices. "G" indicates a glossary entry.

Italicized page numbers represent sources of data in tables, figures, and appendices. "G" indicates a glossary entry.

Italicized page numbers represent sources of data in tables, figures, and appendices. "G" indicates a glossary entry.

Italicized page numbers represent sources of data in tables, figures, and appendices. "G" indicates a glossary entry.

Italicized page numbers represent sources of data in tables, figures, and appendices. "G" indicates a glossary entry.

Italicized page numbers represent sources of data in tables, figures, and appendices. "G" indicates a glossary entry.

Italicized page numbers represent sources of data in tables, figures, and appendices. "G" indicates a glossary entry.

Italicized page numbers represent sources of data in tables, figures, and appendices. "G" indicates a glossary entry.

Italicized page numbers represent sources of data in tables, figures, and appendices. "G" indicates a glossary entry.

Index

Italicized page numbers represent sources of data in tables, figures, and appendices. "G" indicates a glossary entry.

real, 3-1
Reynolds, 16-7
  similarity, 17-43
smoke spot, 24-13, 32-13
stability, 39-2, 40-7
structural, 75-12
threshold odor, 26-13
type, 3-1
Weber, similarity, 17-45
Numbering system, 3-1
Numerical method, 3-3, 12-1
Nusselt number, *1-9*
Nut
  coal, 24-4
  factor, 45-12

O

Objectively scored problems, 89-3
Obligate aerobe, 27-3
Obligate anaerobe, 27-3
Oblique
  axis, 2-2
  impact, 72-16
  perspective, 2-3
  triangle, 6-5, *A-132*
  view, 2-2
Observation well, G
Observed yield, 30-8
Obsidian, 35-31
Obstruction meter, 17-25
Obtuse angle, 6-1
Occupancy, 85-6
  category, 81-6
  class, 81-6
  group, 81-6, *81-7*, *85-2*, 85-6
  mixed, 81-6
  types of, 85-6
Occupancy ratio, lane, G
Occupational Safety and Health Act, 82-1
Occurrence, frequency of, 20-5
Ocean dumping, sludge, 30-19
O'Connor and Dobbins formula, 28-7
Octane number, 24-6
Odd symmetry, 7-4, *9-7*
Odor, 32-10
  in water, 26-12
  number, G
    threshold, 26-13
  sewage, 29-3
Oedometer test, 35-23, 40-5
Offer, 87-3
Off-gas, 34-16
Official benchmark, 77-7
Offset, 74-15
  chord, method, 78-5
  parallel, 43-3
  stake, 80-1, 80-3
  station, 78-5
  tangent, 78-4
  tangent, spiral, 78-18
Ogee spillway, 19-13
Ohm's law, 83-3, 83-6
Oil
  Bunker C, 24-5
  distillate, 24-6

fuel, 24-5, *24-6*
  residual fuel, 24-6
  spill, 32-10
Olefin series, 24-2
Oligotrophic, G
Once-through cooling water, 32-6
On cost, 85-9
One
  percent flood, 20-6
  -tail limit, 11-12
  -third rule, 77-17
  -way
    draining, 40-5
    moment, 44-8 (ftn)
    shear, 44-7 (ftn), 55-2
    slab, 51-1
Opacity, smoke, 32-13
Open
  area, well screen, 21-4
  channel, 19-2
    flow, nomograph, *A-46*
    flow, types, *19-2*
    hydraulic parameters, *19-3*
  circuit, electrical, 83-2
  -cut method, 40-11
  -graded friction course, 75-3
  -graded mix, 75-3
  loop recovery, 34-3
  manometer, 15-3
  traverse, 77-13
Openings, sieve, 35-2, 35-2
Operating
  cost, 85-9
  expense, 86-32
  point, pump, 18-17
  speed, 73-4
Operational form, log, 3-5
Operations on complex numbers, 3-8
Opportunity cost, 86-18
Opposite side, 6-2
Optical
  density, 32-13
  holography, 43-12
  plummet, 77-8
Optimum
  density, 73-6
  moisture content, *35-19*
  speed, 73-6
  water content, 35-17
Orbit
  eccentricity, 72-19, *72-19*
  geostationary, 72-18
  law of, 72-18
Order, G
  differential equation, 10-1
  matrix, 4-1
  of accuracy, 77-4, *77-5*
  of the reaction, 22-10
  stream, G
Ordinance, zoning, 81-2
Ordinate, 7-2
Organic
  acid, *23-2*
  chemical, volatile, G
  chemistry, 23-1
  color, in water, 25-8

compound, 23-1
  family, 23-1
  synthesis route, *23-3*
  volatile, 28-14, 32-14, 34-3
family, *23-2*, *23-3*
functional group, 23-1, *23-1*, *23-3*
loading, 28-3, 29-10
matter, 35-5
nitrogen, 25-8, 28-13
polymer, 26-6
precursor, 25-11, 25-12
refractory, 28-6
sedimentary rock, 35-31
solid, dissolved, 29-3
sulfur, 24-3
trace, 29-3
Organism, indicator, 27-4
Organohalogen, 32-10
Organochlorine pesticide, 32-10
Organophosphate, 32-11
Organotroph, 27-1
Orientation, angular, 5-1
Orifice, 17-16
  coefficients, *17-17*
  large, 17-18
  meter, 17-29
  plate, 17-29
    flow coefficients, *17-30*
  pressure loss, 17-29 (ftn)
Orsat apparatus, 24-11
Orthogonal, 7-2
  vectors, 5-3
  view, 2-2
Orthographic
  oblique view, 2-3
  view, 2-2
Orthophosphate, 25-7
Orthotrophic, G
Orthotropic
  bridge deck, G
  material, G
OSHA (*see also* Occupational Safety and
    Health Act)
  noise limit, 82-3
  scaffold regulations, 82-3
Osmosis, G
  reverse, 26-18
Osmotic pressure, 14-9
Othmer correlation, 14-11
Outfall, G
Outlet control, 19-26
Overall
  coefficient of heat transfer, 30-16
  efficiency, pump, 18-9
  stability constant, 22-12
Overautogenous waste, 34-8 (ftn)
Overburden, 36-4
  correction, *36-7*
  stress, 35-14
Overchute, G
Overconsolidated
  curve, 40-4
  soil, 35-24, 40-4
Overconsolidation ratio, 35-24
Overdraft, 21-5
Overdrive gear, 74-2

Italicized page numbers represent sources of data in tables, figures, and appendices. "G" indicates a glossary entry.

Italicized page numbers represent sources of data in tables, figures, and appendices. "G" indicates a glossary entry.

rigid, 76-1, 76-4, G
  AASHTO nomograph, *76-6, 76-7*
  structural number, 75-18
Paving
  machine, 75-7
  temperature, 75-7
Pay as you throw, G
Payback period, 86-38
PCB, 32-10
PCC, 48-1
PDC (*see* Peak discharge compartment)
P-delta
  analysis, 47-16
  effect, 53-1, 62-1
PDO accident, 74-7
Pea coal screening, 24-4
Peak
  demand multiplier, 83-7
  -discharge compartment, 32-12
  discharge, NRCS method, 20-19
  flame temperature, 32-9
  hour
    factor, 73-5
    warrant, signalization, 73-14
  runoff, 20-11
    rational method, 20-13
  time to, 20-10
Peaking factor, 28-2
  wastewater, *28-2*
Peclet number, *1-9*
Pedestal, 52-1
Pedestrian
  circulation area, 73-15
  density, 73-17
  speed, 73-17
  unit flow rate, 73-17
  volume warrant, signalization, 73-15
Pedology, G
Peltier effect, 84-6 (ftn)
Pelton wheel, 18-21
Pendulum, ballistic, 72-3
Penetration
  grading, 75-2
  resistance, 35-17
  test, standard, 35-17, 36-7
  treatment, G
Penetrometer test, cone, 35-17
Penstock, 18-21
Per capita daily demand, 26-18
Percent
  mole, 14-6
  pore space, 35-7
  VMA, 75-9
Percentage yield, 22-7
Percentile, 11-10
  rank, 11-10
  speed, 73-4
Perception-reaction time, 74-5
Perched spring, G
Percolation, G
  field, 29-2
Per diem fee, 87-4
Perfect
  gases, *A-63*
  reaction, 22-6
Perforated cover plate, 61-2

Performance
  curve, pump, 18-16
  function, 11-8
  grading, 75-2
  incinerator, *34-10*
  number, 24-6
  period, 75-14
Perigee, 72-19
Perihelion, 72-19
Perimeter, wetted, 16-6, G
Period, 86-2 (*see also type*)
  aeration, 30-8
  chemical, 22-2
  detention, 26-5
  electrical, 83-4
  law of, 72-18
  of waveform, 9-6
Periodic
  chart, *A-47*
  table, 22-2, *A-47*
  time, 72-18
  waveform, 9-6
Permanent
  hardness, 25-4, G
  set, 43-2, 48-7
Permeability, 21-2, *21-2*
  coefficient of, 21-2
  intrinsic, 21-2
  soil, *35-22*
  specific, 21-2
  test, 35-22
Permeator, 35-22
Permissible exposure limit, 82-2
Permutation, 11-2
  circular, 5-5
Perpendicular
  axis theorem, 42-5
  line, 7-8
  line principle, 2-1
Perpetual inventory system, 86-37
Persistence, 27-6
  pesticide, 32-10
Personal
  property, 86-20 (ftn)
  protective equipment, 82-2
Person-rem, G
Perspective
  angular, 2-3
  oblique, 2-3
  parallel, 2-3
  view, 2-2, 2-3
PERT, 85-13
Pesticide, 32-10 (*see also* Volatile Organic
  Compounds)
  chlorinated, 32-10
PFR, 30-6
pH, 22-8, G
Phase, 83-9
  angle, 83-4, 83-6
  current, 83-9
  factor, 83-7
  signal, 73-11
Phases, project scheduling, 85-8
Phasor form, 3-8, 83-5
Phenol, *23-3*
Philadelphia rod, 77-9

Phosphated surface, 65-3
Phosphorus
  ion, in water, 25-7
  removal, in wastewater, 29-3, 29-12
Photoautotroph, 27-3
Photochemical
  reaction, 32-10
  smog, 32-13
Photoelectric effect, 84-4
Photogrammetric survey, 77-5
Photogrammetry, 77-17
Photosensitive conductor, 84-5
Phototroph, 27-1
Photovoltaic cell, 84-4
Phreatic zone, 21-1, G
Phreatophytes, G
Physical
  constants, *back endpaper*
  inventory, 86-37
  strain, 43-5
  stress, 43-5
Phytoplankton, G
Pick-up, 84-3 (ftn)
Pictorial drawing, 2-2 (ftn)
Pier, 38-6, 44-12, 45-2
  shaft, G
PIEV time, 74-5
Piezoelectric
  effect, 15-2, 84-4
  transducer, 84-4
Piezometer, G
  nest, G
  ring, 15-2 (ftn)
  tube, 15-2
Piezometric
  height, 21-2
  level, G
Pi-group, 1-8
Pilaster, 69-1
Pile, 38-1, 79-3 (*see also type*)
  batter, G
  bearing capacity, ultimate, 38-1
  capacity, tensile, 38-5
  composite, 38-1
  design capacity, 38-3
  friction, 38-1
  group, 38-5
    efficiency, 38-5
  grouted, 38-6
  machine-drilled, 38-6
  Meyerhof values, *38-2*
  micro-, 38-6
  mini-, 38-6
  needle, 38-6
  pin, 38-6
  point-bearing, 38-1
  resistance, 38-3
  root, 38-6
  safe bearing value, 38-2
  safe load, 38-2
  settlement, 38-5
  small-diameter, 38-6
  soldier, 39-1, G
  tension, 38-5
  volume, 79-3
Pile bent, G

Italicized page numbers represent sources of data in tables, figures, and appendices. "G" indicates a glossary entry.

Italicized page numbers represent sources of data in tables, figures, and appendices. "G" indicates a glossary entry.

Italicized page numbers represent sources of data in tables, figures, and appendices. "G" indicates a glossary entry.

Italicized page numbers represent sources of data in tables, figures, and appendices. "G" indicates a glossary entry.

Index

Italicized page numbers represent sources of data in tables, figures, and appendices. "G" indicates a glossary entry.

Italicized page numbers represent sources of data in tables, figures, and appendices. "G" *indicates a glossary entry.*

Index

Italicized page numbers represent sources of data in tables, figures, and appendices. "G" indicates a glossary entry.

Italicized page numbers represent sources of data in tables, figures, and appendices. "G" indicates a glossary entry.

Italicized page numbers represent sources of data in tables, figures, and appendices. "G" indicates a glossary entry.

Italicized page numbers represent sources of data in tables, figures, and appendices. "G" indicates a glossary entry.

Italicized page numbers represent sources of data in tables, figures, and appendices. "G" indicates a glossary entry.

overconsolidated, 40-4
parameter formulas, *35-9*, *35-10*
particles, classification, 35-2, *35-2*
permeability, *21-2*, *35-22*
pressure
  at-rest, 37-5
  backward, 37-2
  compressed, 37-2
  forward, 37-2
  tensioned, 37-2
  vertical, 37-3
resilient modulus, 75-15
resistance value, 75-15
strength, *35-25*
tests
  ASTM, *35-16*
  standard, 35-15
types, *79-2*
washing, 34-15
well-graded, 35-3
Sol, G
Soldier, 39-1
  beam, cofferdam, 39-6, 39-8
  pile, 39-1, G
Sole proprietorship, 87-1
Solid
  angle, 6-6
  bowl centrifuge, 30-17
  contact unit, 26-10
  density, 35-7
  volatile, G
  volume method, 49-2
  waste, 32-2
    municipal, 31-1
Solidity, 79-7
Solids, 28-6
  dissolved, 28-6
  fixed, 30-5
  in water, 25-8 (see also type)
  nonvolatile, 30-5
  refractory, 28-6
  residence time, 30-6
  suspended, 28-6, 29-3
  total, 28-6
  volatile, 28-6, 30-5
  washout, 30-12
Solubility, 31-8, 34-17
  product, 22-13
  retrograde, G
Soluble BOD, 30-6
Solute, 14-9, 22-7
Solution, 14-9, 22-7, G
  buffer, 22-9
  complementary, 10-4
  indicator, 25-2, *25-2*
  neutral, 22-9, 25-1
  particular, 10-4
  regeneration, 26-15
  saturated, 22-8
  trivial, 3-7 (ftn)
Solutions, properties of, 14-15
Solvency, 86-34
Solvent, 14-9, 22-7
  recovery, 34-3

Sonic
  boom, 14-14
  velocity, 14-14
Sorbent injection, 34-15
Sorption, G
Sound
  limit, 82-2
  speed of, 14-14
Source
  energy, 17-15
  nonpoint, G
  point, G
Space
  mean speed, 73-4, G
  mechanics, 72-18
  pedestrian, 73-17
  -time diagram, 73-16
Spacing, G
  bracing, 59-3
  gage, 60-2
  hole, 65-1
  molecule, 14-2
  pitch, 60-2
  stirrup, 50-18, 50-30
  vehicular, 73-6
Spalling, 75-5
Span, 17-37
  shear, 50-30
  -to-width ratio, floor diaphragm, *68-10*
  maximum, without calculations, 50-18
Spandrel, 59-2
Sparging, 34-15
Special
  damages, 87-6
  triangle, 6-2
Specific
  activity, G
  area, filter, 29-9
  capacity, 21-3
  collection area, 34-7
  discharge, 21-4
  energy, 13-1, 16-1, 16-9, 19-14
    diagram, 19-15
  feed characteristic, 31-11, 34-9
  force, 19-15
  gravity, 14-4, 35-7
    apparent, 75-8, G
    bulk, 75-8, 75-9, G
    effective, 75-8, G
    maximum 75-8
  heat, 13-4
    gas, *24-2*
    molar, 13-4
    ratio of, 15-13
  heats, *13-4*
  performance, 87-3, 87-5
  permeability, 21-2
  retention, 21-3
  roughness, 17-4, *17-4*, *A-25*
  speed, 18-12, *18-12*
    suction, 18-16
    turbine, 18-23
  storage, G
  substrate utilization, 30-7
  utilization, 30-7
  volume, 14-4, 14-5

weight
  concrete, 48-4
  equivalent, 37-6
  yield, 21-3, G
Specifications, 58-4
  DOT geotextile, *40-10*
  pipe, 16-11
Specimen, 43-1
Spectrum analyzer, 9-7
Speed, 71-3 (see also type)
  control, motor, 83-15
  degradation on grade, 74-7
  design, 73-3, *73-4*
  free-flow, 73-4, 73-6, 73-8, 73-10
  legal, 73-3
  limit, 73-4
  of
    efflux, 17-16
    reaction, 22-10
    sound, 14-14
  parameter, traffic, 73-3
  pedestrian, 73-17
  regulation, 83-10
  rule, 78-8
  runaway, 17-34
  space mean, G
  specific, 18-12, *18-12*
  specific, turbine, 18-23
  synchronous, 18-10, *18-10*, 83-12
  tip, 18-4
SPF, 20-6
Sphere, 7-12
  unit, 6-6
Spherical
  coordinate system, *7-3*, *7-4*
  defect, 6-6
  drop, 10-10
  excess, 6-6
  tank, 45-5
  triangle, 6-5
  trigonometry, 6-5
Sphericity, 17-40, *17-41*
Spheroid, 77-6
Spill, hazardous, 32-13
Spillway, 19-13
  coefficient, 19-13
  ogee, 19-13
Spiral
  angle, 78-18
  column, 52-2
  curve, 78-17
  curve, length, 78-18
  deflection angle, 78-18
  entrance, 78-17
  exit, 78-17
  tangent offset, 78-18
  wire, column, *52-2*
  wire, splice, 52-2
Spitzglass formula, 17-9
Splice, spiral wire, 52-2
Split
  beam tee construction, 65-8
  chlorination, G
  cylinder, 48-5
  mastic, 75-23

Italicized page numbers represent sources of data in tables, figures, and appendices. "G" indicates a glossary entry.

Index

Italicized page numbers represent sources of data in tables, figures, and appendices. "G" indicates a glossary entry.

Italicized page numbers represent sources of data in tables, figures, and appendices. "G" indicates a glossary entry.

Italicized page numbers represent sources of data in tables, figures, and appendices. "G" *indicates a glossary entry.*

Italicized page numbers represent sources of data in tables, figures, and appendices. "G" *indicates a glossary entry.*

Index

Italicized page numbers represent sources of data in tables, figures, and appendices. "G" indicates a glossary entry.

Italicized page numbers represent sources of data in tables, figures, and appendices. "G" *indicates a glossary entry.*

Index

Italicized page numbers represent sources of data in tables, figures, and appendices. "G" indicates a glossary entry.

Italicized page numbers represent sources of data in tables, figures, and appendices. "G" *indicates a glossary entry.*

Italicized page numbers represent sources of data in tables, figures, and appendices. "G" indicates a glossary entry.

Italicized page numbers represent sources of data in tables, figures, and appendices. "G" indicates a glossary entry.